MW01050571

NEUROBIOLOGY OF LANGUAGE

NEUROBIOLOGY OF LANGUAGE

Edited by

GREGORY HICKOK
Department of Cognitive Sciences, University of California, Irvine, CA, USA

STEVEN L. SMALL
Department of Neurology, University of California, Irvine, CA, USA

AMSTERDAM • BOSTON • HEIDELBERG • LONDON
NEW YORK • OXFORD • PARIS • SAN DIEGO
SAN FRANCISCO • SINGAPORE • SYDNEY • TOKYO
Academic Press is an imprint of Elsevier

Academic Press is an imprint of Elsevier
32 Jamestown Road, London NW1 7BY, UK
525 B Street, Suite 1800, San Diego, CA 92101-4495, USA
225 Wyman Street, Waltham, MA 02451, USA
The Boulevard, Langford Lane, Kidlington, Oxford OX5 1GB, UK

ISBN: 978-0-12-407794-2

British Library Cataloguing-in-Publication Data
A catalogue record for this book is available from the British Library

Library of Congress Cataloging-in-Publication Data
A catalog record for this book is available from the Library of Congress

For Information on all Academic Press publications
visit our website at http://store.elsevier.com/

Typeset by MPS Limited, Chennai, India
www.adi-mps.com

Publisher: Mica Haley
Acquisition Editor: Mica Haley
Editorial Project Manager: Kathy Padilla
Production Project Manager: Julia Haynes
Designer: Maria Inês Cruz

Dedication

We would like to dedicate Neurobiology of Language to all researchers and clinicians who spend their careers trying to understand human language, its neurobiological basis, and/or its neurological breakdown and rehabilitation. We would also like to thank our families for their support of our own careers that have aimed to understand all three of these aspects.

Contents

SECTION D

LARGE-SCALE MODELS

SECTION E

DEVELOPMENT, LEARNING, AND PLASTICITY

SECTION F

PERCEPTUAL ANALYSIS OF THE SPEECH SIGNAL

SECTION G
WORD PROCESSING

SECTION H
SENTENCE PROCESSING

SECTION J

SPEAKING

SECTION K

CONCEPTUAL SEMANTIC KNOWLEDGE

SECTION P
LANGUAGE TREATMENT

List of Contributors

Hermann Ackermann Department of General Neurology/Center for Neurology, Hertie Institute for Clinical Brain Research, University of Tübingen, Tübingen, Germany

Mauro Adenzato Center for Cognitive Science, Department of Psychology, University of Torino, Italy; Neuroscience Institute of Turin, Italy

Diana R. Alkire Center for the Study of Learning, Georgetown University, Washington, DC, USA

Luc H. Arnal Department of Neurosciences, Biotech Campus, University of Geneva, Geneva, Switzerland; Department of Psychology, New York University, New York, NY, USA

Cesar Ávila Dpt. Psicologia Bàsica, Clínica i Psicobiologia, Universitat Jaume I, Castelló de la Plana, Spain

Bruno G. Bara Center for Cognitive Science, Department of Psychology, University of Torino, Italy; Neuroscience Institute of Turin, Italy

Brian Barton Department of Cognitive Sciences, University of California, Irvine, Irvine, CA, USA; Center for Cognitive Neuroscience and Engineering, University of California, Irvine, Irvine, CA, USA

Shari R. Baum School of Communication Sciences and Disorders and Centre for Research on Brain, Language & Music, McGill University, Montréal, QC, Canada

Michael S. Beauchamp Department of Neurosurgery and Core for Advanced MRI, Baylor College of Medicine, Houston, TX, USA

Jeffrey R. Binder Department of Neurology, Medical College of Wisconsin, Milwaukee, WI, USA

Ferdinand Christoph Binkofski Section Clinical Cognition Sciences, Department of Neurology, University Hospital, RWTH Aachen University, Aachen, Germany; Institute of Neuroscience and Medicine (INM-1), Research Centre Jülich, Jülich, Germany

Shane Blau Department of Linguistics, Center for Mind and Brain, University of California, Davis, CA, USA

Sheila E. Blumstein Department of Cognitive Linguistic and Psychological Sciences, Brown University and the Brown Institute for Brain Sciences, Providence, RI, USA

Tobias Bormann Department of Neurology and Clinical Neuroscience, University Hospital Freiburg, Freiburg, Germany

Ina Bornkessel-Schlesewsky Cognitive Neuroscience Laboratory, School of Psychology, Social Work and Social Policy, University of South Australia, Adelaide, SA, Australia; Department of Germanic Linguistics, University of Marburg, Marburg, Germany

Francesca M. Branzi Center for Brain and Cognition (CBC), Universitat Pompeu Fabra, Barcelona, Spain

Bettina Brendel Department of General Neurology/Center for Neurology, Hertie Institute for Clinical Brain Research, University of Tübingen, Tübingen, Germany

Alyssa A. Brewer Department of Cognitive Sciences; Center for Cognitive Neuroscience and Engineering; Center for Hearing Research, University of California, Irvine, Irvine, CA, USA

Iris Broce Department of Psychology, Florida International University, Miami, FL, USA

Timothy T. Brown Department of Neurosciences, University of California, San Diego, School of Medicine, La Jolla, CA, USA

Bradley R. Buchsbaum Rotman Research Institute, Baycrest, University of Toronto, Toronto, ON, Canada

David Caplan Neuropsychology Laboratory, Department of Neurology, Massachusetts General Hospital, Boston, MA, USA

Svenja Caspers Institute of Neuroscience and Medicine (INM-1), Research Centre Jülich, Jülich, Germany

Tracy M. Centanni School of Behavioral and Brain Sciences, The University of Texas at Dallas, Richardson, TX, USA

Edward F. Chang Department of Neurological Surgery, University of California, San Francisco, CA, USA

Jennifer Chesters Department of Experimental Psychology, University of Oxford, Oxford, UK

Derya Çokal Institute for Brain and Mind, University of South Carolina, Columbia, SC, USA

Emily L. Connally Department of Experimental Psychology, University of Oxford, Oxford, UK

David P. Corina Department of Linguistics; Department of Psychology, Center for Mind and Brain, University of California, Davis, CA, USA

H. Branch Coslett Department of Neurology, Perelman School of Medicine at the University of Pennsylvania, Philadelphia, PA, USA

Albert Costa Center for Brain and Cognition (CBC), Universitat Pompeu Fabra, Barcelona, Spain; Institució

Catalana de Recerca i Estudis Avançats (ICREA), Barcelona, Spain

Steven C. Cramer Departments of Neurology, Anatomy and Neurobiology, and PM&R, Sue and Bill Gross Stem Cell Research Center, University of California, Irvine, CA, USA

Suzanne Curtin Speech Development Laboratory, Department of Psychology, University of Calgary, Calgary, AB, Canada

Matthew H. Davis Medical Research Council, Cognition and Brain Sciences Unit, Cambridge, UK

Gary S. Dell University of Illinois, Urbana-Champaign, Beckman Institute, University of Illinois, Urbana, IL, USA

Özlem Ece Demir Department of Communication Sciences and Disorders, Northwestern University, Evanston, IL, USA

Isabelle Deschamps Centre de Recherche de l'Institut Universitaire en Santé Mentale de Québec, Québec City, QC, Canada; Département de Réadaptation, Faculté de Médecine, Université Laval, Québec City, QC, Canada

Anthony Steven Dick Department of Psychology, Florida International University, Miami, FL, USA

Frederic Dick Birkbeck/UCL Centre for NeuroImaging (BUCNI), London, United Kingdom; Centre for Brain and Cognitive Development (CBCD), Department of Psychological Sciences, Birkbeck College, University of London, London, United Kingdom

Danielle S. Dickson Department of Psychology, University of Illinois, Urbana, IL, USA

Hugues Duffau Department of Neurosurgery, Hôpital Gui de Chauliac, Montpellier University Medical Center, Montpellier, France; Team "Plasticity of Central Nervous System, Stem Cells and Glial Tumors," INSERM U1051, Institute for Neuroscience of Montpellier, Montpellier University Medical Center, Montpellier, France

E. Susan Duncan Solodkin/Small Brain Circuits Laboratory, Department of Neurology, University of California, Irvine, Irvine, CA, USA

Guinevere F. Eden Center for the Study of Learning, Georgetown University, Washington, DC, USA

Crystal T. Engineer School of Behavioral and Brain Sciences, The University of Texas at Dallas, Richardson, TX, USA

Ivan Enrici Center for Cognitive Science, Department of Psychology, University of Torino, Italy; Neuroscience Institute of Turin, Italy; Department of Philosophy and Educational Sciences, University of Torino, Italy

Julia L. Evans School of Behavioral and Brain Sciences, The University of Texas at Dallas, Richardson, TX, USA; Center for Research in Language, University of California, San Diego, La Jolla, CA, USA

Tanya M. Evans Center for the Study of Learning, Georgetown University, Washington, DC, USA

Luciano Fadiga Department of Human Physiology, University of Ferrara, Ferrara, Italy; Italian Institute of Technology, Genoa, Italy

Kara D. Federmeier Department of Psychology, Program in Neuroscience, Beckman Institute for Advanced Science and Technology, University of Illinois, Urbana, IL, USA

Fernanda Ferreira Department of Psychology and Center for Mind and Brain, University of California, Davis, CA, USA

Evelyn C. Ferstl Institute for Informatics and Society, Centre of Cognitive Science, Albert-Ludwigs-University, Freiburg, Germany

Julie A. Fiez Department of Psychology, Department of Neuroscience, Department of Communication Science and Disorders, Center for Learning Research and Development and Center for the Neural Basis of Cognition, University of Pittsburgh, PA, USA

Simon E. Fisher Language and Genetics Department, Max Planck Institute for Psycholinguistics, Nijmegen, the Netherlands; Donders Institute for Brain, Cognition, and Behaviour, Radboud University, Nijmegen, the Netherlands

Carol A. Fowler Department of Psychology, University of Connecticut, Storrs, CT, USA

Julius Fridriksson The Aphasia Lab, Department of Communication Sciences and Disorders, The University of South Carolina, Columbia, SC, USA

Angela D. Friederici Department of Neuropsychology, Max Planck Institute for Human Cognitive and Brain Sciences, Leipzig, Germany

Jackson T. Gandour Department of Speech Language Hearing Sciences, Purdue University, West Lafayette, IN, USA

Fatemeh Geranmayeh Computational, Cognitive and Clinical Neuroimaging Laboratory (C3NL), Imperial College London, Hammersmith Hospital, London, UK

Morton Ann Gernsbacher Psychology, University of Wisconsin-Madison, Madison, WI, USA

Marta Ghio Institute for Experimental Psychology, Heinrich-Heine-University, Dusseldorf, Germany

Anne-Lise Giraud Department of Neurosciences, Biotech Campus, University of Geneva, Geneva, Switzerland

Susan Goldin-Meadow Department of Psychology, University of Chicago, Chicago, IL, USA

Maria Luisa Gorno-Tempini UCSF Memory and Aging Center, Sandler Neurosciences Center, University of California, San Francisco, CA, USA

Vincent L. Gracco Centre for Research on Brain, Language and Music; School of Communication Sciences and Disorders, McGill University, Montreal, QC, Canada; Haskins Laboratories, New Haven, CT, USA

Elizabeth J. Grace Special Education, National Louis University, Chicago, IL, USA

Deanna J. Greene Department of Psychiatry; Department of Radiology, Washington University School of Medicine in St. Louis, St. Louis, MO, USA

Frank H. Guenther Department of Speech, Language, and Hearing Sciences, Department of Biomedical Engineering, Boston University, Boston, MA, USA

Peter Hagoort Donders Institute for Brain, Cognition and Behaviour, Max Planck Institute for Psycholinguistics, Nijmegen, The Netherlands

Uri Hasson Center for Mind and Brain Sciences (CIMeC), University of Trento, Mattarello (TN), Italy

Olaf Hauk MRC Cognition and Brain Sciences Unit, Cambridge, UK

Shannon Heald Department of Psychology, The University of Chicago, Chicago, IL, USA

Arturo E. Hernandez Department of Psychology, University of Houston, Houston, TX, USA

Gregory Hickok Department of Cognitive Sciences, Center for Language Science, Center for Cognitive Neuroscience, University of California, Irvine, CA, USA

Argye E. Hillis Department of Physical Medicine and Rehabilitation; Department of Neurology; Department of Cognitive Science, Johns Hopkins University, School of Medicine, Baltimore, MD, USA

Lori L. Holt Department of Psychology and the Center for the Neural Basis of Cognition, Carnegie Mellon University, Pittsburgh, PA, USA

Norbert Hornstein Department of Linguistics, University of Maryland, College Park, MD, USA

John F. Houde Department of Otolaryngology—Head and Neck Surgery, University of California, San Francisco, CA, USA

William J. Idsardi Department of Linguistics; Neuroscience and Cognitive Science Program, University of Maryland, College Park, MD, USA

Cassandra L. Jacobs Beckman Institute, University of Illinois, Urbana, IL, USA

Ned Jenkinson Nuffield Department of Clinical Neuroscience, University of Oxford, John Radcliffe Hospital, Oxford, UK; School of Sport, Exercise and Rehabilitation Sciences, The University of Birmingham, Birmingham, UK

Ingrid S. Johnsrude Department of Psychology and Centre for Neuroscience Studies, Queen's University, Kingston, ON, Canada

Michael P. Kilgard School of Behavioral and Brain Sciences, The University of Texas at Dallas, Richardson, TX, USA

Tilo Kircher Department of Psychiatry and Psychotherapy, Philipps University Marburg, Marburg, Germany

Juliane Klann Section Clinical Cognition Sciences, Department of Neurology, University Hospital, RWTH Aachen University, Aachen, Germany

Serena Klos Department of Psychology, The University of Chicago, Chicago, IL, USA

Sonja A. Kotz School of Psychological Sciences, University of Manchester, Manchester, UK; Department of Neuropsychology, Max Planck Institute for Human Cognitive and Brain Sciences, Leipzig, Germany

Anthony J. Krafnick Center for the Study of Learning, Georgetown University, Washington, DC, USA

Ananthanarayan Krishnan Department of Speech Language Hearing Sciences, Purdue University, West Lafayette, IN, USA

Saloni Krishnan Birkbeck/UCL Centre for NeuroImaging (BUCNI), London, United Kingdom; Centre for Brain and Cognitive Development (CBCD), Department of Psychological Sciences, Birkbeck College, University of London, London, United Kingdom

Dorothee Kuemmerer Department of Neurology and Clinical Neuroscience, University Hospital Freiburg, Freiburg, Germany

Marta Kutas Department of Cognitive Science, Department of Neurosciences, Center for Research in Language, University of California, San Diego, CA, USA

Robert Leech Computational, Cognitive and Clinical Neuroimaging Laboratory (C3NL), Imperial College London, London, United Kingdom

Matthew K. Leonard Department of Neurological Surgery, University of California, San Francisco, CA, USA

Christina N. Lessov-Schlaggar Department of Psychiatry, Washington University School of Medicine in St. Louis, St. Louis, MO, USA

Susan C. Levine Department of Psychology, Department of Comparative Human Development, and Committee on Education, University of Chicago, Chicago, IL, USA

Daniel A. Llano Department of Molecular and Integrative Physiology, University of Illinois at Urbana-Champaign, Champaign, IL, USA

Andrew J. Lotto Speech, Language, & Hearing Sciences, University of Arizona, Tucson, AZ, USA

Alec Marantz Department of Linguistics, New York University, New York, NY, USA; Department of Psychology, New York University, New York, NY, USA; NYUAD Institute, New York University Abu Dhabi, Abu Dhabi, United Arab Emirates

Conor T. McLennan Department of Psychology, Cleveland State University, Cleveland, OH, USA

Lars Meyer Deparment of Neuropsychology, Max Planck Institute for Human Cognitive and Brain Sciences, Leipzig, Germany

Lee M. Miller Center for Mind & Brain, and Department of Neurobiology, Physiology, & Behavior, University of California, Davis, CA, USA

Bettina Mohr Department of Psychiatry, Campus Benjamin Franklin, Charité Universitätsmedizin, Berlin, Germany

Philip J. Monahan Centre for French and Linguistics, University of Toronto Scarborough, Toronto, ON, Canada; Department of Linguistics, University of Toronto, Toronto, ON, Canada

Emily M. Morson Psychology and Neuroscience, Indiana University, Bloomington, IN, USA

Mariachristina Musso Department of Neurology and Clinical Neuroscience, University Hospital Freiburg, Freiburg, Germany

Srikantan S. Nagarajan Department of Radiology and Biomedical Imaging, University of California, San Francisco, CA, USA

Arne Nagels Department of Psychiatry and Psychotherapy, Philipps University Marburg, Marburg, Germany

Hal X. Nguyen Mind Research Unit, Sue and Bill Gross Stem Cell Research Center, University of California, Irvine, CA, USA

Nazbanou Nozari Department of Neurology, Johns Hopkins University, Baltimore, MD, USA; Department of Cognitive Science, Johns Hopkins University, Baltimore, MD, USA

Howard Nusbaum Department of Psychology, The University of Chicago, Chicago, IL, USA

Olumide A. Olulade Center for the Study of Learning, Georgetown University, Washington, DC, USA

Karalyn Patterson Neurology Unit, Department of Clinical Neurosciences, University of Cambridge, Cambridge, UK; MRC Cognition and Brain Sciences Unit, Cambridge, UK

Silke Paulmann Department of Psychology, Centre for Brain Science, University of Essex, Colchester, UK

Michael Petrides Montreal Neurological Institute, McGill University, Montreal, Quebec, Canada

David B. Pisoni Department of Psychological and Brain Sciences, Indiana University, Bloomington, IN, USA

David Poeppel Department of Psychology, New York University, New York, NY, USA; Max-Planck-Institute for Empirical Aesthetics, Frankfurt, Germany

Peter Pressman UCSF Memory and Aging Center, Sandler Neurosciences Center, University of California, San Francisco, CA, USA

Friedemann Pulvermüller Brain Language Laboratory, Department of Philosophy and Humanities, Freie Universtät Berlin, Berlin, Germany; Berlin School of Mind and Brain, Humboldt-Universität zu Berlin, Berlin, Germany

Liina Pylkkänen Department of Linguistics, Department of Psychology, New York University, New York, NY, USA; NYUAD Institute, New York University Abu Dhabi, Abu Dhabi, United Arab Emirates

Anjali Raja Beharelle Laboratory for Social and Neural Systems Research, University of Zurich, Zurich, Switzerland; Department of Economics, University of Zürich, Zürich, Switzerland

Matthew A. Lambon Ralph Neuroscience and Aphasia Research Unit, School of Psychological Sciences, University of Manchester, Manchester, UK

Kathleen Rastle Department of Psychology, Royal Holloway, University of London, Egham, Surrey, UK

Josef P. Rauschecker Laboratory of Integrative Neuroscience and Cognition, Georgetown University Medical Center, NW, Washington, DC, USA; Institute for Advanced Study, TU München, München-Garching, Germany

Jessica D. Richardson Department of Communication Sciences and Disorders, The University of South Carolina, Columbia, SC, USA

Michel Rijntjes Department of Neurology and Clinical Neuroscience, University Hospital Freiburg, Freiburg, Germany

Giacomo Rizzolatti Department of Neuroscience, University of Parma, Parma, Italy; Brain Center for Motor and Social Cognition, Italian Institute of Technology, Parma, Italy

Jennifer M. Rodd Department of Experimental Psychology, University College London, London UK

Corianne Rogalsky Department of Speech and Hearing Science, Arizona State University, Tempe, AZ, USA

Stefano Rozzi Department of Neuroscience, University of Parma, Parma, Italy

Ayşe Pinar Saygin Department of Cognitive Science, University of California—San Diego, La Jolla, CA, USA

Bradley L. Schlaggar Department of Neurology; Department of Radiology; Department of Psychiatry; Department of Pediatrics; Department of Anatomy & Neurobiology, Washington University School of Medicine in St. Louis, St. Louis, MO, USA

Gottfried Schlaug Department of Neurology, Neuroimaging, and Stroke Recovery Laboratories, Beth Israel Deaconess Medical Center and Harvard Medical School, Boston, MA, USA

Matthias Schlesewsky Department of English and Linguistics, Johannes Gutenberg-University Mainz, Mainz, Germany

Myrna F. Schwartz Moss Rehabilitation Research Institute, Elkins Park, PA, USA

Michael Schwartze School of Psychological Sciences, University of Manchester, Manchester, UK

Sophie K. Scott Institute of Cognitive Neuroscience, University College London, London, UK

Steven L. Small Department of Neurology, University of California, Irvine, CA, USA

Kimberly Smith The Aphasia Lab, Department of Communication Sciences and Disorders, University of South Carolina, Columbia, SC, USA

Jon Sprouse Department of Linguistics, University of Connecticut, Storrs, CT, USA

Anja Staiger Clinical Neuropsychology Research Group, Clinic for Neuropsychology, City Hospital, Munich, Germany

Craig E.L. Stark Department of Neurobiology and Behavior, University of California, Irvine, CA, USA

Shauna M. Stark Department of Neurobiology and Behavior, University of California, Irvine, CA, USA

Adrian Staub Department of Psychological and Brain Sciences, University of Massachusetts, Amherst, MA, USA

Edward Taub Department of Psychology, University of Alabama at Birmingham, Birmingham, AL, USA

Marco Tettamanti Department of Nuclear Medicine and Division of Neuroscience, San Raffaele Scientific Institute, Milano, Italy

Sharon L. Thompson-Schill Department of Psychology, University of Pennsylvania, Philadelphia, PA, USA

Donna C. Tippett Department of Otolaryngology—Head and Neck Surgery; Department of Physical Medicine and Rehabilitation; Department of Neurology, Johns Hopkins University School of Medicine, Baltimore, MD, USA

Pascale Tremblay Centre de Recherche de l'Institut Universitaire en Santé Mentale de Québec, Québec City, QC, Canada; Département de Réadaptation, Faculté de Médecine, Université Laval, Québec City, QC, Canada

Peter Turkeltaub Department of Neurology, Georgetown University School of Medicine, Washington, DC, USA; MedStar National Rehabilitation Hospital, Washington, DC, USA

Michael T. Ullman Brain and Language Laboratory, Department of Neuroscience, Georgetown University, Washington, DC, USA

Kate E. Watkins Department of Experimental Psychology, University of Oxford, Oxford, UK

Cornelius Weiller Department of Neurology and Clinical Neuroscience, University Hospital Freiburg, Freiburg, Germany

Richard J.S. Wise Computational, Cognitive and Clinical Neuroimaging Laboratory (C3NL), Imperial College London, Hammersmith Hospital, London, UK

Jeffrey M. Zacks Department of Psychology, Washington University, Saint Louis, MO, USA

Wolfram Ziegler Clinical Neuropsychology Research Group, Clinic for Neuropsychology, City Hospital, Munich, Germany

Acknowledgement

The editors would like to thank the following individuals for their important participation in the editorial process by reviewing and helping edit the chapters of this volume. We are highly appreciative of their support for this major project.

Hermann Ackermann
Lisa Aziz
Helen Barbas
Lawrence Barsalou
Ferdinand Binkofski
Robert Blumenfeld
Sheila Blumstein
James Booth
Ina Bornkessel
Bradley Buchsbaum
Laurie Cutting
Gary Dell
Ece Demir
Anthony Dick
Joseph Duffy
Susan Duncan
Julie Fiez
Anne Foundas
Felipe Fregni
Julius Fridriksson
Sean Fulop
Ted Gibson
Matthew Goldrick
Bethanie Gouldthorp
Murray Grossman
Sara Guediche
Argye Hillis
Fumiko Hoeft
Lori Holt
Rich Ivry
Howard Kirschner
Brock Kirwan
Robert Kluender
Judith Kroll
Christina Leonard
James Lewis
Daniel Llano
Paola Marangolo
Alex Martin
Martin Meyer
Lee Miller
Howard Nusbaum
Jonas Obleser
Manuel Parea
Lisa Pearl
Jonathan Peelle
Daniela Perani
Amy Price
Anjali Raja
Alexander Rapp
Khaleel Razak
Kourosh Saberi
Matthias Scheslewesky
Gottfried Schlaug
Mohamed Seghier
Doris Trauner
Pascale Tremblay
Sarah Tune
Jonathan Venezia
Ryan Walsh
Kate Watkins
Daniel Weiss
Stephen Wilson
Patrick Wong
Jie Yang
Yang Zhang

SECTION A

INTRODUCTION

CHAPTER

1

The Neurobiology of Language

Steven L. Small[1] *and Gregory Hickok*[2]

[1]Department of Neurology, University of California, Irvine, CA, USA; [2]Department of Cognitive Sciences, Center for Language Science, Center for Cognitive Neuroscience, University of California, Irvine, CA, USA

1.1 HISTORY

For many centuries, the biological basis of human thought has been an important focus of attention in medicine, with particular interest in the brain basis of language sparked by the famous patients of Pierre Paul Broca in the mid 19th century (Broca, 1861a,c). The patient Louis Victor LeBorgne (Domanski, 2013) presented to the Hôpital Bicêtre in Paris with severe difficulty speaking, purportedly only uttering the syllable "tan," sometimes as a pair "tan, tan," and often accompanied by gestures (Domanski, 2013). The diagnosis was not clear until autopsy, when Broca found on gross inspection that some neurological process (he reported a resulting collection of serous fluid) had destroyed a portion of the left posterior inferior frontal gyrus (Broca, 1861c) (Figure 1.1). A subsequent patient, LeLong, had a similar paucity of speech output (five words were reported) with a lesion not dissimilar to that of LeBorgne (Broca, 1861a). Given the ongoing debates at the time about brain localization of language, including attribution of the "seat of language" to the frontal lobes (Auburtin, 1861; Bouillaud, 1825; Gall & Spurtzheim, 1809)—which led Broca to investigate this case in the first place—he presented this patient with "aphémie" (LeBorgne) to the Société d'Anthropologie de Paris in 1861 (Broca, 1861b). These brains remain preserved to this day, and brain imaging studies have confirmed the original findings and extended them to demonstrate lesions to deeper structures (Signoret, Castaigne, Lhermitte, Abelanet, & Lavorel, 1984) and to white matter underlying the original descriptions (Dronkers, Plaisant, Iba-Zizen, & Cabanis, 2007).

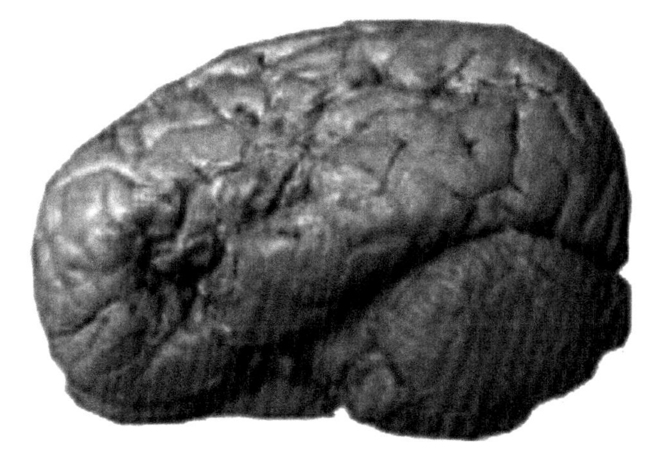

FIGURE 1.1 The exterior surface of the brain of LeBorgne ("tan").

1.2 LESION ANALYSIS

The era of brain localization for language blossomed after this, with the famous doctoral dissertation of Wernicke (1874), the diagram-making of Lichtheim (1885) (Figure 1.2) and Grashey (1885), the anatomy of Déjerine (1895), and of course many other contributors. In the past century, Norman Geschwind recapitulated and added to the language "center" models that preceded him and presented a reconceptualized "connectionist" view of the brain mechanisms of language (Geschwind, 1965, 1970). Whereas the 19th century investigators relied on simple views of behavior and postmortem brain pathology, those of the mid to late 20th century were able to take advantage of significant advances in both the study of behavior (information processing psychology and formal linguistic theory), allowing much more robust characterizations of language performance than had been possible previously (Caramazza & Berndt, 1978; Chomsky, 1965), and the technology of structural brain imaging with computed tomography and magnetic resonance imaging (MRI), permitting the elucidation of brain pathology *in vivo* (e.g., Cappa & Vignolo, 1983; Metter et al., 1984). These advances led to a blossoming of brain lesion analysis (neuropsychological) studies in the second half of the

Neurobiology of Language. DOI: http://dx.doi.org/10.1016/B978-0-12-407794-2.00001-8

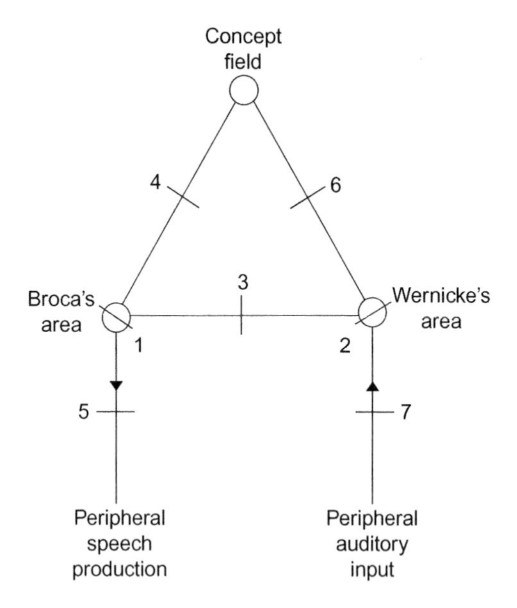

FIGURE 1.2 Lichtheim's model. Note the emphasis on brain anatomy.

20th century (for reviews, see Damasio & Damasio, 1989; Kertesz, 1983; Shallice, 1988a). Recent advances in image analysis (e.g., Ashburner & Friston, 2000; Bates et al., 2003) have improved the lesion analysis method, and it continues to be a valuable method for biological inquiries in language.

By their very nature, lesion analysis studies tend to relate single foci of brain injury to single psychological or linguistic phenomena. The goal of the enterprise is to "double dissociate" functions and brain regions, such that an underlying focal substrate of brain can be tied unequivocally to a single function (and not another) (e.g., Damasio & Tranel, 1993) or to a specific "locus" in a cognitive or linguistic model (e.g., Garrard & Hodges, 2000; Shallice, 1988b). Although not without its controversy (e.g., Plaut, 1995), this approach has been quite successful in giving insights into the neurobiology of language, that is, attributing functions to aspects of the brain (in this case, brain areas damaged by vascular lesions—or sometimes other types of lesions).

1.3 FROM NEUROPSYCHOLOGY TO COGNITIVE NEUROSCIENCE

The advent of high-resolution functional brain imaging in the past decade of the 20th century (e.g., Fox, Raichle, & Thach, 1985; Raichle, Martin, & Herscovitch, 1983), its initial applications to the study of language (Petersen, Fox, Posner, Mintun, & Raichle, 1988, 1989), and its widespread acceptance for the study of brain/cognition relations in the early part of this century have dramatically changed the conduct of studies of brain and behavior. Although the earliest functional anatomical studies were conducted with positron emission tomography, the most recent work uses functional magnetic resonance imaging (fMRI), a noninvasive approach that does not require intravenous administration of radioactive agents. These methods permit the investigation *in vivo* of brain regions that participate in the performance of any type of task that can be performed inside of an imaging machine. By contrast, task-dependent electroencephalography (EEG), more commonly known as "event-related potentials" (ERP), permits the characterization of temporal aspects of task performance. An important, if less commonly used, method for analyzing task-dependent brain function is magnetoencephalography (MEG), which can have finer spatial resolution than EEG and higher temporal resolution than fMRI, and thus can play a particularly important role in characterizing brain processing over time. Besides these methods of activating the brain, the method of transcranial magnetic stimulation (TMS) can be used to create "reversible lesions" in the brain. The majority of contemporary research in brain–behavior relations for language uses either the older method of lesion analysis or one of these newer methods of functional activation (fMRI, ERP, MEG) or ablation (TMS). Intracranial electrical recordings in humans undergoing elective brain surgery, including both surface (electrocorticography) and deep, are becoming more commonplace. A host of other less prevalent methods also play valuable roles.

Approximately 20 years ago, the burgeoning use of these brain measurement techniques to study psychological and linguistic processes led to the creation of a new field of cognitive neuroscience. During its existence, this field has evolved into an important discipline, with the majority of top cognitive programs (and some neuroscience programs) incorporating cognitive neuroscience as an important component of curriculum and, in some cases, degree-granting status. Importantly, the evolution of this discipline has focused more on using biological methods than on asking biological questions, for example, addressing linguistic or psychological questions by measuring brain responses constitutes a significant portion—if not the majority—of scientific studies in this area. In other words, the fraction of studies that develop and test "linking hypotheses" between neural and computational systems is smaller than one would hope for in a field that targets an understanding of the *relation* between mind and brain. Addressing biological issues is far less common in this field, and most practitioners of this discipline neither study biology nor concern themselves with biological questions. Students in this area are typically not required to study cellular and molecular neuroscience, and only a portion study systems neuroscience, neuroanatomy, neurophysiology, or neurology/neuropsychology. Of course, it is also true that from the biological side, students and researchers are not typically required to study linguistics or experimental psychology. Despite this high prevalence of nonbiological studies of cognition using brain measurement, it has become increasingly clear that the new methods of functional brain imaging, along with other methods of human and animal neuroscience, provide new avenues to investigate the actual biological substrate and computations for human language.

1.4 THE NEUROBIOLOGY OF LANGUAGE

Thus, a neurobiology of language is now possible. Until several years ago, this term was barely used, if at all, with most research focusing on neuropsychology, cognitive neuroscience, or vaguely "brain and language." We define "neurobiology of language" as the biological implementation and linking relations for representations and processes necessary and sufficient for production and understanding of speech and language in context. Biological disciplines that are highly relevant to the neurobiology of language include the anatomy and physiology of the human brain, the network connectivity of the brain, and the multiple roles of different brain areas (relevant cognitive areas are discussed later). We also note that the basic scientific findings in this area will necessarily lead to a physiological approach to therapy for speech and language disorders (Small, Buccino, & Solodkin, 2013).

Importantly, we define the neurobiology of language as a subfield of neuroscience, requiring substantial knowledge of psychology and linguistics, sharing in its primary assumptions, methods, and questions. By way of explanation, whereas psychology is the scientific study of the human mind and its functions, especially those affecting behavior in particular contexts, and linguistics is the scientific study of language and its structure, including the study of morphology, syntax, phonetics, and semantics, neurobiology is the study of the biology of the nervous system. Whereas cognitive neuroscience emphasizes experimental psychology and brain measurements, the neurobiology of cognition emphasizes cellular and molecular neuroscience, systems neuroscience, and other biological bases of language, such as genetics. We believe that both cognitive neuroscience and the neurobiology of cognition must address the *links* between cognitive and neuroscience data, although they come from different directions. Although the neurobiology of language is primarily a field involving human subjects—and has benefited from the novel and highly robust methods of brain measurement discussed—it must also adopt the central tenets of modern biology, including the evolutionary imperative that the human brain evolved from that of nonhuman primates, and thus the study of other species plays an important role in the investigations of the neurobiology of language (Bornkessel-Schlesewsky, Schlesewsky, Small, & Rauschecker, 2015).

A number of brain imaging methods are used for the neurobiology of cognition but do not play a significant role in cognitive neuroscience. An enormous amount of information about the neurobiology of language has been gleaned from high-resolution quantitative structural imaging of both gray matter and white matter. For the gray matter, it is possible to measure precisely and accurately whole brain and regional volumes, and for white matter, we can measure accurately anisotropic diffusion (and several other diffusivity characteristics). In both cases, the anatomical measures can be related to behavior or to functional imaging measures. At a network level, white matter pathways can be reconstructed from the diffusivity data and can be related to regional correlations of gray matter characteristics or functional brain activations. These relations between structural connectivity and functional connectivity are leading to a preliminary understanding of brain circuits for language processing.

It is also possible to make direct measurements of human neural cells when medically indicated, as in people with intractable epilepsy who are undergoing ablative brain surgery. Early work demonstrated that direct cortical stimulation could lead to alterations in language performance in humans *in vivo* (Ojemann, Ojemann, Lettich, & Berger, 1989; Penfield & Roberts, 1959). Although this technique remains viable (e.g., Duffau, 2008), recent work has used cortical recording (electrocorticography) to begin to elucidate cellular and circuit features of language representations at the level of small neuronal populations (e.g., Chang et al., 2010; Flinker et al., 2015). Many additional biological techniques, such as genetic mapping, single cell recording, pharmacological manipulation, molecular imaging, and others, are beginning to be viable for human neuroscience research.

1.5 SOME COMMON FALLACIES

From a philosophical point of view, several assumptions commonly made in contemporary research in brain–behavior relations related to language are problematic. We identify four such assumptions, which we call the (i) methodological fallacy; (ii) theoretical fallacy; (iii) uniqueness fallacy; and (iv) the mind/brain fallacy. These fallacies relate to research assumptions by investigators in the cognitive sciences that lead to inappropriate biological conclusions from otherwise well-conducted studies in cognition (not infrequently accompanying perfectly interesting and valid cognitive conclusions). We explain these in turn.

The *methodological fallacy* assumes that if a brain measurement technique is used in a study, that study is a study of the brain. Such an assumption essentially focuses on the method that is used, for example, if a research study uses EEG or MRI, then it is a neurobiological study. This is not the case. Among the most common uses of ERP is to distinguish one linguistic or psychological model from another, and fMRI is used very commonly to distinguish individuals with one condition (medical or psychological) from another. In neither case would an understanding of the biology be necessary to achieve the goals of the study. However, it is important to note that some EEG/MEG research does aim to understand the relation between neural network rhythms and cognitive processes, and much fMRI research also addresses neural computations. In summary, studies of this sort do not address linking hypotheses, although data from these studies could be used to inform linking relations.

The *theoretical fallacy* assumes that a theoretical model developed exclusively from assumptions about linguistic data or cognitive processing explains brain function. Such an assumption essentially ignores data from neuroscience while focusing on theoretical aspects of the legal utterances (linguistics) or processes (psychology) in one or more (or even "all") languages. Even if a theory accounts for all legal utterances in all languages but does not say how brain structures actually compute them, it is not a theory of the biology of language. Of course, if a biological theory ignores the *linkages* between the neurobiology and cognition, it also is not a theory of the neurobiology of language.

The *uniqueness fallacy* is perhaps the most important because it ignores evolution, which is one of the most important principles of modern biology. This fallacy typically arises in the form of arguments that human language is so unique and so special that it is not computationally intertwined with other systems. But we know that language systems of the brain are fundamentally and inherently intertwined with those that process motor function, sensation, emotion, affect, and other aspects of human experience, and that the human brain—including those parts of the brain that participate in language processing—had to evolve from the brains of species (i.e., nonhuman primates) that do not use language. The evolutionary principles of phylogeny and ontogeny apply to the human brain as with other organ systems, and thus language representations in the brain have necessarily co-opted neural structures with evolutionary history. Relevant examples include auditory object perception, motor sequence control and planning, action understanding, and social interaction, all of which are found in other species.

In summary, we are concerned with the notion that linguistic computations are so unique that they bear no evolutionary resemblance to other systems. However, we acknowledge that language uniqueness cannot be ignored. Language ability is certainly unique to our species, suggesting an evolutionary process in which language computations developed on a framework of other computational structures in the brain. It is thus critical that we take the biology seriously and search for both computational commonalities and differences between language and nonlanguage systems within our species and in our cousin species.

Finally, the most common of all the fallacies is the *mind/brain fallacy*, which argues that all studies of human psychology are *ipso facto* studies of the brain, because the brain is the biological organ that subserves thought. Of course, this is perfectly correct, to a point, and the Society for Neuroscience includes human behavior among its disciplinary mandates. However, it is completely possible—even commonplace—for studies of human behavior to completely ignore the constraints placed on the behavior by the physical aspects of the brain or the plausibility of the behavioral conclusions in the context of brain structure and function. Among the most common examples in language research are explanations of experimental data in the form of "conceptual" or "cognitive" or "mental" models that could not possibly be implemented in brain hardware.

We certainly have no objection to cognitive models and indeed value them, whether or not they make immediate contact with neural implementation. These models help us think about information flow (and/or dynamics) and computational descriptions of the process. As such, they can sometimes generate testable hypotheses related to the brain as well as provide prospective points of contact for descriptions that relate brain to behavior. If we tried to build a neurobiology of language without linguistics or psychology, as the classical neurologists were forced to do, we would not have been able to transcend the 19th century models (e.g., characterizations such as "auditory word images"). But an important point here (for the cognocentric among us) is that the constraints must work *both* ways. Knowledge of biology must constrain our models of cognitive and linguistic function, just as knowledge of cognition must constrain our biological interpretations. Both are obviously critical in the development of linking hypotheses.

1.6 HUMANS IN PARTICULAR

There is no question that the human brain is unique in its capacity to manage speech understanding and production, in addition to their evolutionarily younger derivatives, reading and writing. That the human brain can do this and other animal brains cannot is something that must be explained in a way that preserves the basic tenets of biology. Certainly, humans have vocal communication skills of very high complexity, and these depend on a large and complex variety of sensory and motor skills, memory capacities, and learning abilities, as well as the capability to represent and process intricate messages of many types. Although it is not clear exactly what changed between nonhuman primates and humans to effectuate this, the expansion of the cerebral cortex increased the human capacity for learning, remembering, and executing motor sequences, complex perceptions, and recursive processing (e.g., theory of mind, syntax) (Hauser, Chomsky, & Fitch, 2002; Yoshida, Dolan, & Friston, 2008). It remains to be elucidated just how the processes of language work in the human brain, and through this—perhaps with concomitant work in nonhuman primates to test their limits—we should find the answer to why humans have language and other species do not.

1.7 COGNITION AND THE NEUROBIOLOGY OF LANGUAGE

The neurobiology of cognition will only be elucidated by direct investigation of how the brain works, that is, the detailed implementation of understanding and producing speech in the hardware that is unique to the human brain. At the same time, such direct investigation must be informed by scientific evidence about the nature of cognition. For the neurobiology of language, this entails understanding the nature of languages and its processing, that is, investigations in linguistics and experimental psychology. The first step must involve spoken language, because that is the evolutionary basis on which the brain architecture for written language sits.

Understanding the nature of human languages is crucial, particularly research into the wide variety of different types of languages and how they express thoughts (e.g., MacWhinney & Bates, 1989). There is nothing in the basic brain structure of people born to parents who speak Basque, Farsi, or Navaho to distinguish them from speakers of English, Italian, or German. Yet, most research studies relating brain and language have been conducted in the latter three languages (and now, increasingly, Mandarin as well), and rarely in the other approximately 7,000 languages of the world (Lewis, Simons, & Fennig, 2009), some of which have properties that are quite different from those of the most commonly studied languages. Thus, it is important in characterizing the biological basis of language to know just what needs to be accounted for both in terms of variation of language forms and the computational mechanisms used to process them. The study of linguistics is therefore critical to a successful neurobiology of language.

Equally important is characterizing the nature of speech production and understanding in processing terms. Without detailed knowledge about what people can understand, in what contexts, and under what constraints, we cannot develop a solid neurobiology of understanding. The same is true for production, that is, we need to know what people can produce at a given time and place, particularly with memories, emotions, and goals. Such investigations are the purview of experimental psychology and, as such, research in such areas as psycholinguistics and the psychology of memory has fundamental importance to inform investigations in the biological aspects of language.

1.8 BRAIN DISEASE, TREATMENT, AND THE NEUROBIOLOGY OF LANGUAGE

There are many diseases and/or types of injury to the human brain that lead to language disturbances, and these are important to our enterprise for two reasons: (i) helping to understand the basic neurobiology and (ii) advancing knowledge useful for development of treatments for conditions that affect language. Both of these endeavors have considerable importance to the field.

By understanding the nature of disease processes, it may be possible to gain insight into the fundamental neurobiology of language itself. For example, the cellular and molecular biology of language has been given a first hearing with the distinct pathological markers that differentiate the types of primary progressive aphasia (Gorno-Tempini et al., 2011). The entire history of research in the biological basis of language has depended significantly on inferences from individuals with ischemic stroke (e.g., Mohr, 1976). Patients surviving herpes simplex encephalitis, a multifocal disease, have provided important fodder for inferences about semantic representations in the brain (Warrington & Shallice, 1984). Language problems arising in people with dementia, tumors, Parkinson's disease, schizophrenia, and many other diseases—combined with the nature of the pathology specific to that disease (particularly in the context of the pathology of each individual involved)—are highly informative with respect to the neurobiology of language. Additionally, people undergoing surgery for diseases such as epilepsy and brain tumors, even when they do not affect language, have generously allowed investigators to study their language intraoperatively.

Of course, a very important purpose in understanding the biology underlying language is to develop new therapies (Small et al., 2013; Small & Llano, 2009). Novel treatments including pharmacotherapy, electrical and magnetic stimulation, and even behavioral (speech-language therapy) interventions, have been developed based on rationales that come from a better understanding of brain anatomy and physiology.

1.9 SUMMARY

In summary, investigation into the neurobiology of language has a long history, beginning even before the seminal work of the 19th century neurologists, who associated brain lesions in certain focal areas (but not others) with language impairments. However, recently, this field has had a resurgence due to the advent of extraordinary new technologies that allow *in vivo* imaging of brain structure, structure/functional relations, and temporal processing features. Although lesion analysis remains quite important—and even more valuable in conjunction with advanced imaging—the current era is marked by a predominance of investigations using brain imaging technologies (broadly speaking). These methods can be used to advance our understanding of linguistics or psychology, thus contributing in a different way to the neurobiology of language, or to advance our knowledge of the anatomy, physiology, pharmacology, and even cellular and molecular biology of language directly. Such direct investigations of the brain in conjuction with investigations into the representations and computations behind language processes represent basic prerequisites in building the future of this discipline.

The two editors of this book—without a consensus on all issues—have come to the many common understandings of the field articulated here. We are in agreement that it is important to have "theoretically precise, computationally explicit, biologically grounded explanatory models of the human brain's ability to comprehend and produce speech and language" (Poeppel & Hickok, 2004). We are also in agreement that the basic evolutionary biology of the brain must play an important role in such a theory, for example, by understanding the relation between computational implementations of sensory and motor functions (among others) to linguistic computations (Skipper, Nusbaum, & Small, 2005). By linking biological and cognitive hypotheses into well-designed studies, it is now possible to build a true neurobiology of language.

The reviews in this book reflect a combination of articles on the neurobiology of language *per se* and on the supporting knowledge from the neurosciences and from the cognitive sciences (including cognitive neuroscience) that neurobiologists of language require to accurately characterize the neural structures and functions that implement language. Ultimately, we hope that this reference book will encourage and support the efforts of practicing scientists and their students to delve directly into this challenging field.

References

Ashburner, J., & Friston, K. J. (2000). Voxel-based morphometry—The methods. *NeuroImage, 11*, 805–821.

Auburtin, E. (1861). Reprise de la discussion sur la forme et le volume du cerveau. *Bulletins de la Société d'Anatomie Paris, 2*, 209–220.

Bates, E., Wilson, S. M., Saygin, A. P., Dick, F., Sereno, M. I., Knight, R. T., et al. (2003). Voxel-based lesion-symptom mapping. *Nature Neuroscience, 6*(5), 448–450.

Bornkessel-Schlesewsky, I., Schlesewsky, M., Small, S. L., & Rauschecker, J. P. (2015). Neurobiological roots of language in primate audition: Common computational properties. *Trends in Cognitive Science*. Available from: http://dx.doi.org/10.1016/j.tics.2014.12.008.

Bouillaud, J.-B. (1825). Recherches cliniques propres à démontrer que la perte de la parole correspond à la lésion des lobules antérieurs du cerveau, et à confirmer l'opinion de Gall, sur le siège du langage articulé. *Archives of General Medicine (Paris), 8*, 25–45.

Broca, P. P. (1861a). Nouvelle Observation d'Aphémie produite par une Lesion de la Partie Postérieure des Deuxième et Troisième Circonvolutions Frontales. *Bulletins de la Société d'Anatomie Paris, 6*, 398–407.

Broca, P. P. (1861b). Perte de la parole, ramollisement chronique et destruction partielle du lobe antérieur gauche du cerveau. *Bulletins de la Société d'Anthropologie, 2*, 235–238.

Broca, P. P. (1861c). Remarques sur le siège de la faculté du langage articulé, suivies d'une observation d'aphémie (perte de la parole). *Bulletins de la Société d'Anatomie (Paris), 6*(2e serie), 330–357.

Cappa, S. F., & Vignolo, L. A. (1983). CT scan studies of aphasia. *Human Neurobiology, 2*(3), 129–134.

Caramazza, A., & Berndt, R. S. (1978). Semantic and syntactic processes in aphasia: A review of the literature. *Psychological Bulletin, 85*(4), 898–918.

Chang, E. F., Rieger, J. W., Johnson, K., Berger, M. S., Barbaro, N. M., & Knight, R. T. (2010). Categorical speech representation in human superior temporal gyrus. *Nature Neuroscience, 13*(11), 1428–1432. Available from: http://dx.doi.org/10.1038/nn.2641.

Chomsky, N. (1965). *Aspects of the theory of syntax*. Cambridge, MA: The MIT Press.

Damasio, A. R., & Tranel, D. (1993). Nouns and verbs are retrieved with differently distributed neural systems. *Proceedings of the National Academy of Sciences of the United States of America, 90*(11), 4957–4960.

Damasio, H., & Damasio, A. R. (1989). *Lesion analysis in neuropsychology*. New York, NY: Oxford University Press.

Déjerine, J. (1895). *Anatomie des centres nerveux* (Vol. i–ii). Paris: Rueff et Cie.

Domanski, C. W. (2013). Mysterious "Monsieur Leborgne": The mystery of the famous patient in the history of neuropsychology is explained. *Journal of the History of the Neurosciences, 22*(1), 47–52. Available from: http://dx.doi.org/10.1080/0964704x.2012.667528.

Dronkers, N. F., Plaisant, O., Iba-Zizen, M. T., & Cabanis, E. A. (2007). Paul Broca's historic cases: High resolution MR imaging of the brains of Leborgne and Lelong. *Brain, 130*(Pt 5), 1432–1441.

Duffau, H. (2008). The anatomo-functional connectivity of language revisited. New insights provided by electrostimulation and tractography. *Neuropsychologia, 46*(4), 927–934. Available from: http://dx.doi.org/10.1016/j.neuropsychologia.2007.10.025.

Flinker, A., Korzeniewska, A., Shestyuk, A. Y., Franaszczuk, P. J., Dronkers, N. F., Knight, R. T., et al. (2015). Redefining the role of

Broca's area in speech. *Proceedings of the National Academy of Sciences of the United States of America, 112*(9), 2871–2875. Available from: http://dx.doi.org/10.1073/pnas.1414491112.

Fox, P. T., Raichle, M. E., & Thach, W. T. (1985). Functional mapping of the human cerebellum with positron emission tomography. *Proceedings of the National Academy of Sciences of the United States of America, 82*, 7462–7466.

Gall, F. J., & Spurtzheim, J. G. (1809). *Recherches sur le système nerveux en général et sur celui de cerveau en particulier*. Paris: F. Schoell.

Garrard, P., & Hodges, J. R. (2000). Semantic dementia: Clinical, radiological and pathological perspectives. *Journal of Neurology, 247*(6), 409–422.

Geschwind, N. (1965). Disconnection syndromes in animals and man. *Brain, 88*, 237–294. 585–644.

Geschwind, N. (1970). The organization of language and the brain. *Science, 170*(961), 940–944.

Gorno-Tempini, M. L., Hillis, A. E., Weintraub, S., Kertesz, A., Mendez, M., Cappa, S. F., et al. (2011). Classification of primary progressive aphasia and its variants. *Neurology, 76*(11), 1006–1014. Available from: http://dx.doi.org/10.1212/WNL.0b013e31821103e6.

Grashey (1885). Ueber aphasie und ihre Beziehungen zur Wahrnehmung. *Archiv für Psychiatrie und Nervenkrankheiten, 16*.

Hauser, M. D., Chomsky, N., & Fitch, W. T. (2002). The faculty of language: What is it, who has it, and how did it evolve? *Science, 298*(5598), 1569–1579. Available from: http://dx.doi.org/10.1126/science.298.5598.1569.

Kertesz, A. (Ed.), (1983). *Localization in neuropsychology* New York, NY: Academic Press.

Lewis, M. P., Simons, G. F., & Fennig, C. D. (2009). (16th ed.). *Ethnologue: Languages of the world*, (Vol. 9). Dallas, TX: SIL international.

Lichtheim, L. (1885). On aphasia. *Brain, 7*, 433–484.

MacWhinney, B., & Bates, E. (Eds.), (1989). *The crosslinguistic study of sentence processing* Cambridge, England: Cambridge University Press.

Metter, E. J., Riege, W. H., Hanson, W. R., Camras, L. R., Phelps, M. E., & Kuhl, D. E. (1984). Correlations of glucose metabolism and structural damage to language function in aphasia. *Brain and Language, 21*(2), 187–207.

Mohr, J. P. (1976). Broca's area and Broca's aphasia. In H. Whitaker, & H. Whitaker (Eds.), *Studies in neurolinguistics* (Vol. 1, pp. 201–233). New York, NY: Academic Press.

Ojemann, G., Ojemann, J., Lettich, E., & Berger, M. (1989). Cortical language localization in left, dominant hemisphere: An electrical stimulation mapping investigation in 117 patients. *Journal of Neurosurgery, 71*, 316–326.

Penfield, W., & Roberts, L. (1959). *Speech and brain mechanisms*. Princeton, NJ: Princeton University Press.

Petersen, S. E., Fox, P. T., Posner, M. I., Mintun, M. A., & Raichle, M. E. (1988). Positron emission tomographic studies of the cortical anatomy of single-word processing. *Nature, 331*, 585–589.

Petersen, S. E., Fox, P. T., Posner, M. I., Mintun, M. A., & Raichle, M. E. (1989). Positron emission tomographic studies of the processing of single words. *Journal of Cognitive Neuroscience, 1*, 153–170.

Plaut, D. C. (1995). Double dissociation without modularity: Evidence from connectionist neuropsychology. *Journal of Clinical and Experimental Neuropsychology, 17*(2), 291–321. Available from: http://dx.doi.org/10.1080/01688639508405124.

Poeppel, D., & Hickok, G. (2004). Towards a new functional anatomy of language. *Cognition, 92*(1–2), 1–12.

Raichle, M. E., Martin, W. R. W., & Herscovitch, P. (1983). Brain blood flow measured with intravenous $H_2{}^{15}O$ II: Implementation and validation. *Journal of Nuclear Medicine, 24*, 790–798.

Shallice, T. (1988a). *From neuropsychology to mental structure*. Cambridge, England: Cambridge University Press.

Shallice, T. (1988b). Specialisation within the semantic system. *Cognitive Neuropsychology, 5*(1), 133–142.

Signoret, J. L., Castaigne, P., Lhermitte, F., Abelanet, R., & Lavorel, P. (1984). Rediscovery of Leborgne's brain: Anatomical description with CT scan. *Brain and Language, 22*(2), 303–319.

Skipper, J. I., Nusbaum, H. C., & Small, S. L. (2005). Listening to talking faces: Motor cortical activation during speech perception. *Neuroimage, 25*(1), 76–89.

Small, S. L., Buccino, G., & Solodkin, A. (2013). Brain repair after stroke—A novel neurological model. *Nature Reviews Neurology, 9* (12), 698–707. Available from: http://dx.doi.org/10.1038/nrneurol.2013.222.

Small, S. L., & Llano, D. A. (2009). Biological approaches to aphasia treatment. *Current Neurology and Neuroscience Reports, 9*(6), 443–450.

Warrington, E. K., & Shallice, T. (1984). Category specific semantic impairments. *Brain, 107*(Pt 3), 829–854.

Wernicke, C. (1874). *Der aphasische symptomenkomplex*. Breslau, ON: Cohn & Weigert.

Yoshida, W., Dolan, R. J., & Friston, K. J. (2008). Game theory of mind. *PLoS Computers in Biology, 4*(12), e1000254. Available from: http://dx.doi.org/10.1371/journal.pcbi.1000254.

SECTION B

NEUROBIOLOGICAL FOUNDATIONS

CHAPTER

2

A Molecular Genetic Perspective on Speech and Language

Simon E. Fisher

Language and Genetics Department, Max Planck Institute for Psycholinguistics, Nijmegen, the Netherlands; Donders Institute for Brain, Cognition, and Behaviour, Radboud University, Nijmegen, the Netherlands

2.1 INTRODUCTION

For decades there has been speculation about the potential contribution of inherited factors to human capacities for speech and language. Arguments for a genetic basis have drawn from an array of diverse fields and approaches, marshaling threads of evidence taken from formal linguistics, child development, twin studies, biological anthropology, comparative psychology, and so on. In recent years, through advances in molecular biology techniques, it has become possible to move beyond these indirect sources and let the genome speak for itself (Graham & Fisher, 2013). In particular, by studying neurodevelopmental disorders that disproportionately disturb speech and language skills, researchers started to identify individual genes that may be involved in the relevant neurobiological pathways.

Rather than being seriously rooted in biology (Fisher, 2006), much of the prior debate on genetic foundations of spoken language has treated genes as abstract entities that can mysteriously yet directly determine linguistic functions. Accounts that depend on "genes for grammar" and other such magic bullets are simply untenable in light of all that is known about molecular and cellular processes and how these processes are able to impact development and function of brain circuitry. The human genome comprises approximately 20,000 different protein-coding genes. Each such gene is a string of G, C, T, and A nucleotides, the specific order of which is used by the cellular machinery to thread together a specific polypeptide sequence of amino acid residues taken from 20 different types of amino acids that are available as building blocks. (Linguists might enjoy the fact that this is a discrete combinatorial system with the potential to yield an infinite number of different amino acid strings.) The sequence of amino acids in a protein determines the way that it folds into a three-dimensional shape, and the protein's shape determines the function that it will have in cells and in the body. In this way, the different DNA sequences of different genes are able to specify a plethora of distinct cellular proteins—enzymes, structural molecules, receptors, signaling factors, transporters, and others. Some of these proteins play roles in the ways that cells of the nervous system proliferate, migrate (move to their final position), differentiate, and form connections with each other during development; some might be important neurotransmitters or other factors that help strengthen or weaken synapses during learning. Overall, intricate webs of genes and proteins acting through complicated sequences of developmental events and via continual interactions with the environment lead to assembly of complex networks of functioning neural circuits, and it is the latter providing the behavioral and cognitive outputs of the system that we call the human brain.

Based on this, we should never expect simple direct connections between DNA and language, but this does not mean that we cannot track down genes that are most relevant to our human capacities. To the contrary, by pinpointing crucial genes (e.g., those for which mutations lead to language impairments) it is possible to gain entirely novel entry points into the critical neural pathways and use those to work toward mechanistic accounts that are properly built on biologically plausible foundations. In what follows, the promise and challenges of the approach are illustrated by focusing on *FOXP2*, a gene that is at the heart of this new paradigm.

Neurobiology of Language. DOI: http://dx.doi.org/10.1016/B978-0-12-407794-2.00002-X

2.2 THE DISCOVERY OF *FOXP2*

The starting point for the *FOXP2* story was the identification of an unusual family in which multiple close relatives suffered from similar disruptions of speech and language skills. This family, dubbed the KE family, spanned three generations and included 15 affected members as well as a similar number of unaffected relatives. Because the disorder was present in each successive generation and affected approximately half of the family members, it attracted the attention of geneticists who recognized that the pattern was consistent with dominant monogenic inheritance (Hurst, Baraitser, Auger, Graham, & Norell, 1990). In other words, it raised the remarkable possibility that this family's speech and language problems might be explained by a mutation affecting one copy of a single gene. Before any DNA investigations had even begun, commentators already began to speculate excitedly about the discovery of a "language gene" (see Fisher (2006) for a detailed account). At the same time, the affected members of the KE family became the focus of intensive neuropsychological studies to gain more insights into their profile of impairments.

According to these investigations, the most prominent aspect of the disorder is a problem mastering the coordinated movement sequences that underlie fluent speech (Vargha-Khadem et al., 1998). The affected people make articulation errors that are inconsistent (they can differ from one utterance to the next) and that become worse as the length and complexity of the utterance increases (Watkins, Dronkers, & Vargha-Khadem, 2002). These are characteristic features of a syndrome known as developmental verbal dyspraxia (DVD) or childhood apraxia of speech (CAS). The difficulties can be robustly captured by tests in which the participant is asked to repeat a series of pronounceable nonsense words of differing length and complexity (Liegeois, Morgan, Connelly, & Vargha-Khadem, 2011). CAS is generally thought of as a disorder of speech learning and production underpinned by neural deficits in the motor planning of sequences of orofacial movements. Intriguingly, the impairments in the affected KE family members are not confined to speech; they extend to the written domain, disturbing a wide range of linguistic skills, both expressive and receptive. To give some examples, affected relatives perform significantly worse than their unaffected siblings on lexical decision tasks, spoken and written tests of verbal fluency, nonsense word spelling, and processing of sentence- and word-level syntax (Watkins, Dronkers, et al., 2002). Given that these skills have developed in the context of a severe restriction in expressive skills, it is possible that such impairments are secondary consequences rather than primary deficits. In general, many members of this family (regardless of CAS diagnosis) have a lower than average nonverbal IQ, which complicates discussions over the selectivity of the phenotype. Nevertheless, because nonverbal cognitive difficulties do not appear to cosegregate with the inherited disorder, it is argued that this is primarily a disturbance of speech and language rather than some form of general intellectual disability (Watkins, Dronkers, et al., 2002). These issues are discussed in more depth elsewhere (Fisher, Lai, & Monaco, 2003).

Screening of different parts of the genome revealed that the KE family disorder was strongly linked to genetic markers on one particular section of chromosome 7 (Fisher, Vargha-Khadem, Watkins, Monaco, & Pembrey, 1998). These markers were passed on from the grandmother to all other affected family members, but not to any unaffected relatives; that is, there was perfect cosegregation with the disorder. The molecular mapping data thus provided experimental confirmation that the speech and language problems of the family had a genetic origin and localized the responsible gene to a particular region of chromosome 7, which was given the name *SPCH1* (Fisher et al., 1998). After intensive analyses of this *SPCH1* interval (Lai et al., 2000), and aided by clues from another case (discussed later), the researchers eventually pinpointed a causative mutation in a novel gene given the name *FOXP2* (Lai, Fisher, Hurst, Vargha-Khadem, & Monaco, 2001).

FOXP2 encodes a transcription factor—a regulatory protein that is able to modulate the activities of other genes (Vernes et al., 2006). The protein does so by directly binding to the DNA of these target genes and affecting how efficiently they are transcribed into messenger RNA molecules (the templates that are used for building proteins). *FOXP2* belongs to one particular group of transcription factors defined by the presence of a special type of DNA-binding motif referred to as a forkhead-box (or FOX) domain (Benayoun, Caburet, & Veitia, 2011). All the affected people in the KE family carry the same single nucleotide change in *FOXP2*, a G-to-A transition in a crucial part of the gene (Lai et al., 2001). This missense mutation leads to alteration of the amino acid sequence of the encoded protein. The mutated protein carries a histidine (H) residue, instead of arginine (R), at a key point of the forkhead domain, that prevents it from binding to the usual target sequences and severely disrupts its function as a transcription factor (Vernes et al., 2006). (Because this amino acid substitution is at the 553rd residue from the start of the protein, it is denoted by the symbol R553H.) The mutation is in a heterozygous state in the affected KE family members, meaning that one gene copy is intact and functioning normally but the other is dysfunctional.

Thus, it was hypothesized that two functioning copies of *FOXP2* are necessary for development of proficient speech and language skills (Lai et al., 2001).

2.3 *FOXP2* MUTATIONS IN SPEECH AND LANGUAGE DISORDERS

Clearly, studies of the KE pedigree were pivotal in enabling the first identification of a gene contributing to speech and language functions. This family represents the most well-characterized example in the literature with respect to both the neuropsychological profile of the associated disorder and the functional impacts of the etiological mutation (Fisher, 2007). However, contrary to the usual story popularized in media reports and many scientific commentaries, the KE family is not the sole documented case of *FOXP2* mutation. Over the years, distinct etiological disruptions of this gene have been reported in several different families and cases, ranging from point mutations (change of a single nucleotide of DNA) to gross rearrangements of chromosome 7 that disturb the entire *FOXP2* locus (Newbury & Monaco, 2010). In fact, the original *FOXP2* paper included not only the KE family mutation but also an independent case of similar speech/language problems with a chromosome 7 rearrangement disturbing the locus (case CS, which is described later) (Lai et al., 2001), something that is often overlooked by commentators.

The predominant isoform of the FOXP2 protein is 715 amino acids long and encoded by 2,145 nucleotides of DNA (split between several different coding exons); a disruptive mutation could potentially occur anywhere within this coding sequence. For rare dominant causal variants with large effect size, such as the R553H mutation found in the KE family, it is likely that the sequence change will be "private," meaning that it is exclusive to just one family or case. Thus, when screening *FOXP2* in new cohorts of people with speech and language problems, it is necessary to thoroughly search for any variants across the entire known coding sequence rather than simply testing for presence/absence of a previously reported mutation. When such screening efforts have been performed in cohorts of people diagnosed with speech disorders, a number of novel *FOXP2* point mutations have been uncovered. For example, Laffin and colleagues (2012) sequenced *FOXP2* in 24 probands with a strict diagnosis of CAS and found that one case carried a heterozygous missense mutation yielding an amino acid substitution (asparagine-to-histidine at position 597, i.e., N597H) just beyond the end of the FOX domain. In a previous screening study of 49 children with clinical reports of CAS, MacDermot and colleagues (2005) identified another type of causal variant, a nonsense mutation that inserts a stop codon halfway through the gene (arginine-to-stop at position 328, i.e., R328X) that is predicted to yield a severely truncated FOXP2 protein. This variant was in the heterozygous state, like the other etiological *FOXP2* mutations. It was found in three family members, the proband, his sister who also had a CAS diagnosis, and his mother who had a history of speech problems. A small number of other potential mutations of interest were identified by the MacDermot study, including a Q17L substitution (glutamine-to-leucine at position 17, near the start of the protein), but in those cases the causal significance was unclear because they did not cosegregate with the disorder in affected siblings (MacDermot et al., 2005; Vernes et al., 2006). Most recently, an individual with CAS was identified carrying an intragenic deletion of two nucleotides in the *FOXP2* locus predicted to yield an abnormal truncated protein (Turner et al., 2013). Specifically, the loss of the two nucleotides yields a shift in the reading frame of the coding sequence at position 415 of the protein; after this point, five novel amino acids are incorporated immediately followed by a premature stop codon. Just as for the R328X mutation, the resulting mutant protein completely lacks the FOX domain.

So far, two types of gross chromosomal rearrangements have been reported to affect *FOXP2*: translocations (Feuk et al., 2006; Kosho et al., 2008; Lai et al., 2001; Shriberg et al., 2006) and deletions (Feuk et al., 2006; Lennon et al., 2007; Palka et al., 2012; Rice et al., 2012; Zeesman et al., 2006; Zilina et al., 2012). In the translocation cases, part of chromosome 7 is exchanged with part of another chromosome; because the chromosome 7 breakpoint in these cases lies directly within (or close to) the *FOXP2* locus, this is expected to interfere with the activity of the disrupted copy (Feuk et al., 2006; Kosho et al., 2008; Lai et al., 2001; Shriberg et al., 2006). The first example of a *FOXP2* translocation was found in a child known as CS, as reported in the same paper that uncovered the KE family mutation (Lai et al., 2001). Most reported *FOXP2* translocations are *de novo*—the rearrangement is present in the case but not found in parents or siblings. However, Shriberg and colleagues identified a family (TB) in which a mother and daughter both carried the same translocation directly disrupting *FOXP2* and reported that the associated speech problems (Shriberg et al., 2006), language impairments, and cognitive profiles (Tomblin et al., 2009) were notably consistent with those previously observed for people carrying the missense mutation in the KE family (Watkins, Dronkers, et al., 2002).

In the reported large-scale deletion cases, one copy of *FOXP2* is completely lost from the genome, often together with other flanking genes (Feuk et al., 2006;

Lennon et al., 2007; Palka et al., 2012; Rice et al., 2012; Zeesman et al., 2006; Zilina et al., 2012). Investigations of the phenotypes observed in these cases again support the idea that damage to one copy of *FOXP2* is sufficient to derail speech and language development, although the larger deletions that encompass multiple other genes are often noted to include additional problems. As with the translocations, although most cases are *de novo*, there is at least one report of an inherited rearrangement: a mother and a son carrying the same deletion of *FOXP2* (as well as neighboring genes *MDFIC* and *PPP1R3A*) and both diagnosed with CAS (Rice et al., 2012). Interestingly, there are no reports of any human with disruption of both copies of *FOXP2*, presumably because a total absence of the gene would be lethal (Fisher & Scharff, 2009).

2.4 FUNCTIONS OF *FOXP2*: THE VIEW FROM THE BENCH

The identification of a particular gene underlying a trait is often portrayed as the endpoint of a scientific study. In reality, this kind of discovery is more akin to a new beginning because it opens up entirely novel avenues for investigating the basis of the trait from the perspective of the gene in question. Thus, the identification of *FOXP2* may have been something of a paradigm shift for the language sciences because it facilitated a series of innovative molecular investigations into the neurobiological pathways and evolutionary history of spoken language using this gene as a unique entry point (Fisher & Scharff, 2009). Such work has called on a diverse array of experimental strategies and model systems, ranging from neuronal cells investigated at a laboratory bench, to genetic manipulations in animals, to studies of humans (Graham & Fisher, 2013).

Laboratory experiments using genetically modified human cells are important for establishing whether putative etiological mutations impact gene function (Deriziotis & Fisher, 2013). As noted, several different point mutations of *FOXP2* have been found in people with CAS; some cosegregate with disorders in a family, like the R553H substitution and the R328X truncation, whereas others are found in just a single proband, such as the Q17L (MacDermot et al., 2005) and N597H substitutions (Laffin et al., 2012). Vernes and colleagues (2006) studied the functional significance of R553H, R328X, and Q17L variants by expressing the mutated proteins in cultured human cell lines, assessing properties such as protein stability, intracellular localization (normal FOXP2 protein is located in the nucleus of the cell), DNA-binding capacity, and ability to repress target genes. R553H and R328X showed obvious disruptions in most or all of these assays, strongly supporting their causal roles, whereas Q17L did not show any functional differences from the normal protein in this system, so its etiological relevance remains uncertain (Vernes et al., 2006). At the time of writing this book, no functional analyses of the N597H substitution had yet been reported.

Crucially, even though they involve rather basic model systems (as compared with neural circuits or living brains), cell-based analyses can go well beyond simply validating disruptive effects of mutations. By applying state-of-the-art genomic and proteomic techniques, researchers can use cellular models to gain new insights into neurogenetic mechanisms, which can have direct relevance to human biology (Deriziotis & Fisher, 2013). The *FOXP2* literature provides particularly apt illustrations of this principle in action. Because *FOXP2* encodes a transcription factor working to regulate the expression of other genes, it can be thought of as a hub in a network of molecules, a number of which might also be related to speech and language development. Thus, over the years, several studies have used cellular models to screen parts, or all, of the genome, searching for target genes regulated by *FOXP2* (Konopka et al., 2009; Vernes et al., 2007, 2008).

In 2008, a study of human neuron-like cells grown in the laboratory found that the FOXP2 protein binds directly to a regulatory sequence within a gene called *contactin-associated protein-like-2*, or *CNTNAP2* (Vernes et al., 2008). The researchers went on to show that when they artificially increased expression of *FOXP2* in cultured cells, this caused a significant reduction in *CNTNAP2* mRNA levels, a finding that was further supported by analyses of developing cortical tissue from human fetuses, in which there was an inverse correlation between expression levels of the two genes. To test for connections between *CNTNAP2* and language development, the team assessed sets of common DNA variations (single-nucleotide polymorphisms [SNPs]) from different parts of the gene in a cohort of 184 families with typical forms of specific language impairment (SLI) previously collected by the UK SLI consortium. They identified a cluster of SNPs in one section of the gene (around exons 13–15) that showed association with measures of performance on language tasks, most notably the nonsense word repetition test; children who carried a particular set of risk variants scored significantly lower than others (Vernes et al., 2008). Intriguingly, in a prior study screening *CNTNAP2* in children with autism, the same risk variants had been associated with delayed language, as indexed by "age at first word" (Alarcon et al., 2008). Because the Vernes et al. (2008) study explicitly excluded any children diagnosed with autism, the convergent findings suggest that the *CNTNAP2* risk variants might be implicated in language-related problems across distinct clinical boundaries of neurodevelopmental disorders.

In a later study, the same variants were shown to be consistently associated with assessments of early language acquisition (at 2 years of age) in 1,149 children from the general population, suggesting that the effects extend beyond disorder into normal variation (Whitehouse, Bishop, Ang, Pennell, & Fisher, 2011).

CNTNAP2 is a member of the neurexin superfamily that encodes a transmembrane protein that has been implicated in multiple fundamental processes in the developing and mature nervous system (Rodenas-Cuadrado, Ho, & Vernes, 2013). It helps to cluster potassium channels at nodes of Ranvier in myelinated axons, and it has also been linked to mechanisms of neuronal migration, dendritic arborization, and spine formation during development (Anderson et al., 2012). Diverse *CNTNAP2* variants (rare mutations and common polymorphisms) have been associated with a range of neurodevelopmental disorders, including not only SLI and autism but also epilepsy, schizophrenia, Tourette syndrome, and intellectual disability (Rodenas-Cuadrado et al., 2013).

After the identification of the *CNTNAP2* connection, additional functional reports have further demonstrated the value of tracing *FOXP2* networks for understanding language-related disorders. *FOXP2* has been shown to regulate *uPAR* and *SRPX2*, genes potentially implicated in a form of rolandic epilepsy that also involves speech apraxia (Roll et al., 2010). (However, see Lesca et al. (2013) for evidence that casts doubt on the role of uPAR/SRPX2 in this disorder, instead implicating a different gene, *GRIN2A*). Intriguingly, *SRPX2* regulation by *FOXP2* is thought to be an important mediator of synaptogenesis (Sia, Clem, & Huganir, 2013). Other *FOXP2* targets of particular clinical relevance include the receptor tyrosine kinase *MET*, proposed as a candidate for autism (Mukamel et al., 2011), and *DISC1*, a gene that was originally implicated in schizophrenia (Walker et al., 2012).

It is not only the downstream targets of *FOXP2* that may be informative for making links to human phenotypes. Transcription factors never act alone; they work together with other interacting proteins to regulate their targets. *FOXP1* is the most similar gene in the genome to *FOXP2*. In some cells in the central nervous system, these two genes are coexpressed (Teramitsu, Kudo, London, Geschwind, & White, 2004), and the resulting proteins have the capacity to directly interact with each other, acting together to regulate targets in a coordinated manner (Li, Weidenfeld, & Morrisey, 2004). Rare causative mutations of *FOXP1* have been implicated in a small number of cases of autism and/or intellectual disability, accompanied by notably severe speech and language problems (Bacon & Rappold, 2012). Moreover, it has been shown that *FOXP1* actively represses the *CNTNAP2* gene, and an autism screening study that sequenced all human protein-coding genes identified an affected child who carried disruptive mutations in both *FOXP1* and *CNTNAP2*, "hits" in two different parts of the same functional pathway (O'Roak et al., 2011). Efforts are underway to identify and characterize all the other key protein interactors in this pathway (Deriziotis & Fisher, 2013).

2.5 INSIGHTS FROM ANIMAL MODELS

The human capacity for acquiring complex spoken language appears to be unique in the natural world (Fisher & Marcus, 2006). At first glance this may seem to preclude any chance of biologically meaningful genetic studies in animal models. However, the majority of human genes did not appear spontaneously in our species (Varki & Altheide, 2005). So, after human studies have identified a gene implicated in speech and language, an obvious next step is to examine the broader evolutionary history of the gene and assess whether its function(s) in nonspeaking species can be informative for understanding its contributions to human brain development (Fisher & Marcus, 2006).

FOXP2 has a particularly deep evolutionary history, with versions of the gene described in many different vertebrate species, including monkeys (Takahashi et al., 2008), ferrets (Iwai et al., 2013), mice (Ferland, Cherry, Preware, Morrisey, & Walsh, 2003; Lai, Gerrelli, Monaco, Fisher, & Copp, 2003), rats (Takahashi, Liu, Hirokawa, & Takahashi, 2003), bats (Li, Wang, Rossiter, Jones, & Zhang, 2007), birds (Haesler et al., 2004; Teramitsu et al., 2004), reptiles (Haesler et al., 2004), and fish (Bonkowsky et al., 2008). Researchers have investigated neural expression patterns for most of these species, determining where and when the gene is transcribed and/or translated in developing and mature brain tissue. These studies found striking similarities in distantly related vertebrates, with concordant expression in neuronal subpopulations of cortex, thalamus, basal ganglia, and cerebellum. Thus, it seems likely that activities of *FOXP2* in the human brain are built on evolutionarily ancient functions in the vertebrate central nervous system (Fisher & Marcus, 2006).

As is apparent from the previous paragraph, there have been a large number of studies characterizing the corresponding versions of this gene found in different species. A proper discussion of the many findings from this research area is beyond the scope of this chapter. The interested reader is referred to recent reviews (French & Fisher, 2014; Wohlgemuth, Adam, & Scharff, 2014). Here, a sample of the work is provided, focusing on two of the most extensively studied model systems: mice and (briefly) birds (Fisher & Scharff, 2009). Much progress has already been made in uncovering relevant neural mechanisms via work with these two complementary models, and there is promise of more insights as the field develops.

The laboratory mouse is widely used in the field of neurogenetics, in large part due to the availability of a comprehensive toolkit for genetic manipulations (French & Fisher, 2014). Mice carry their own version of the *FOXP2* gene, which has the symbol *Foxp2*. The most recent common ancestor of humans and mice lived more than 75 million years ago but, despite this lengthy time since divergence, the sequence of the human FOXP2 protein differs very little from that of its mouse counterpart (Enard et al., 2002). In a sequence of more than 700 amino acid residues, there is one small change in the length of a stretch of glutamines and three sites where one amino acid is substituted for another. In contrast to the substitutions that cause disorder, these evolutionary substitutions occur outside known domains of the protein and are predicted to have only subtle effects on function (see "*FOXP2* in Human Evolution" section for further commentary). In addition to very high conservation of protein sequence, the neural expression patterns are remarkably consistent; for example, in both humans and mice the gene is particularly highly expressed in deep layers in the cortex, medium spiny neurons in the striatum, and Purkinje cells in the cerebellum (Ferland et al., 2003; Lai et al., 2003).

Researchers have generated several different mouse models for studying Foxp2 functions, including animals in which the gene is completely knocked out (Shu et al., 2005) and others that carry known etiological mutations that cause speech problems in humans (Groszer et al., 2008). If both copies of *Foxp2* are damaged (e.g., when mutations are in the homozygous state), then the mice cannot survive; they live for only 3 or 4 weeks after birth, during which time they develop at a substantially slower rate than normal siblings, show significant delays in maturation of the cerebellum, and have severe general problems with their motor system (Groszer et al., 2008; Shu et al., 2005). Thus, a total absence of functional Foxp2 protein is lethal, which is consistent with the lack of any reports of humans carrying homozygous mutations in the gene. The cause of death in homozygous animals is unknown but may relate to one of the various non-neural sites in the body where Foxp2 is expressed; for example, it is switched on in subtypes of cells in the lungs and cardiovascular system (Li et al., 2004). As a brief aside, transcription factors and other regulatory molecules are typically expressed in a range of tissues and cell types in different organs of the body. They exert distinct effects at different sites, depending on the sets of cofactors that they interact with, which is another example where biology takes advantage of the power of combinatorial systems and is a reminder of why specific "language genes" are unlikely to exist (Fisher, 2006).

Despite the associated lethality, investigations of mice that completely lack functional *Foxp2* have revealed some fundamental roles of the gene in early development and patterning of the central nervous system (French & Fisher, 2014). The results from such studies are helping to inform hypotheses about the contributions of the human gene to development and patterning of neural circuits in our species. For example, one report used *Foxp2* mouse models to uncover networks of direct and indirect target genes during embryonic brain development (Vernes et al., 2011). The researchers found that there was an overrepresentation of genes implicated in biological processes like neurite outgrowth and axon guidance, consistent with prior findings from human cells (Spiteri et al., 2007; Vernes et al., 2007). They went on to validate this putative functional role in striatal precursor cells taken from the mouse embryos, finding that an absence of functional Foxp2 led to reduced branching and shorter neurites in these cells (Vernes et al., 2011). Other studies of embryonic mouse cortex using different techniques (genetic manipulations in utero) have confirmed roles for *Foxp2* in neurite outgrowth (Clovis, Enard, Marinaro, Huttner, & De Pietri Tonelli, 2012) and also suggest potential functional impacts on other developmental processes such as neurogenesis (Tsui, Vessey, Tomita, Kaplan, & Miller, 2013) and neuronal migration (Clovis et al., 2012).

In stark contrast to the severe consequences of damage to both copies of *Foxp2*, mice that carry disruptions in the heterozygous state (i.e., only one copy is mutated or knocked out) live long healthy lives, usually without any obvious adverse outcome (Groszer et al., 2008). Such findings are concordant with descriptions of humans with heterozygous *FOXP2* mutations, who typically do not have associated medical problems or gross general developmental impairments (Laffin et al., 2012; Lai et al., 2001; Lennon et al., 2007; MacDermot et al., 2005; Rice et al., 2012; Shriberg et al., 2006; Turner et al., 2013).

Several studies of cognition, behavior, and electrophysiology in the heterozygous mouse models have built on prior observations that corticobasal ganglia and corticocerebellar circuits are key conserved sites of expression. Groszer et al. (2008) investigated heterozygous mice carrying the same mutation as the KE family and reported delays in learning to run on accelerating rotarods and voluntary running wheel systems against a background of normal motor behaviors. In slices taken from the brains of these mice, they also observed altered synaptic plasticity in corticostriatal and corticocerebellar circuits, most notably a lack of long-term depression for glutamatergic synapses on medium spiny neurons of the striatum (Groszer et al., 2008). A follow-up study used *in vivo* electrophysiology to record directly from medium spiny neurons in live behaving mice while the animals learned to run on accelerating rotarods (French et al., 2012). In mice that were heterozygous for the KE family mutation, compared with normal littermates, these neurons had significantly elevated basal firing rates as well as striking abnormalities in both their modulation and their temporal coordination during motor skill learning. The discovery of disturbed striatal plasticity during learning of a complex motor task in mice is intriguing because neuroimaging studies of humans with the same mutation have independently suggested striatal dysfunction as a potential core feature of their disorder (Liegeois et al., 2003, 2011; Vargha-Khadem et al., 1998; Watkins, Dronkers, et al., 2002). Another behavioral study of this mouse model demonstrated reduced performance on a learning task in which the animals had to associate auditory signals with motor outputs (Kurt, Fisher, & Ehret, 2012). This last investigation also compared the learning dynamics with those of another mouse line that carried a different etiological mutation of *Foxp2*, reporting that the degree of impairment seemed to be affected by the type of mutation (Kurt et al., 2012).

It is interesting to note that these mouse studies uncovered effects on auditory-motor associations and motor skill learning that are not confined to the orofacial system. It remains unresolved whether such effects might be detectable in humans with *FOXP2* dysfunction (Peter et al., 2011), or if this instead points to a refinement of gene function in the human lineage. Studies of impacts of rodent *Foxp2* on vocal behaviors have yielded somewhat conflicting data (Fisher & Scharff, 2009; French & Fisher, 2014). What has been consistently established is that mouse pups that totally lack functional *Foxp2* have greatly *reduced* vocal output (Gaub, Groszer, Fisher, & Ehret, 2010; Groszer et al., 2008; Shu et al., 2005). Normally, when a young mouse pup is isolated from its mother and/or the nest, it produces ultrasonic calls that elicit its retrieval. When *Foxp2* is completely missing, pups produce few (if any) isolation calls; however, they do emit ultrasonic calls with complex properties when put in situations of greater stress. Although some researchers interpret these findings as evidence of specific roles of *Foxp2* in pup vocalization (Shu et al., 2005), others have pointed out that pups that lack this gene have very severe general motor problems and global developmental delay, making it impossible to draw conclusions about selective effects (Gaub et al., 2010; Groszer et al., 2008). For heterozygous mouse pups, which carry one damaged and one normal copy of *Foxp2*, there is debate regarding whether there are differences in amounts of vocalization (Groszer et al., 2008; Shu et al., 2005), and in-depth studies of properties of the vocalizations that are produced failed to find significant differences in normal littermates (Gaub et al., 2010). A study of rats reported that amounts of Foxp2 protein are higher in brains of male pups, and that this correlates with production of a higher number of isolation calls as compared with female pups (Bowers, Perez-Pouchoulen, Edwards, & McCarthy, 2013). The researchers went on to assess sex differences of FOXP2 protein levels in Brodmann Area 44 of the human brain by using postmortem tissue from a small number of 3- to 5-year-old children (five boys and five girls). One caveat is that although the male versus female samples were age-matched, they differed greatly in ethnic background, introducing a major confound. Bowers and colleagues (2013) observed higher amounts of FOXP2 protein in the human females and interpreted this as evidence that elevated protein levels "are associated with the more communicative sex." Given the very small number of data points (particularly from humans) and the fact that there are never going to be simplistic mappings from genes and proteins to communication skills (Fisher, 2006), this wide-reaching conclusion may be premature (French & Fisher, 2014).

At the time of writing this chapter, reports of impacts of rodent *Foxp2* on vocalization skills have focused exclusively on pup calls without describing, for example, effects on the ultrasonic "songs" of adolescent males (Fisher & Scharff, 2009). Nevertheless, although rodent vocalizations can provide a useful readout for studying the bases of social behaviors, it is thought that mice have very restricted abilities for using auditory experience to shape their vocal output (Hammerschmidt et al., 2012). Auditory-guided vocal learning is an important skill that underlies our abilities for acquiring speech, and mice are unlikely to provide an appropriate animal model for investigating this particular trait. Luckily, by looking further afield in the animal kingdom, it has been possible to find alternative model systems. Perhaps the most informative of these has been the zebra finch, a songbird that has provided entry points into both the neurobiology and neurogenetics of vocal learning.

A young male zebra finch learns its song during a critical developmental period by matching it to a template that it hears from an adult tutor (Bolhuis, Okanoya, & Scharff, 2010). Zebra finches have their own version of *FOXP2*, known as *FoxP2*. Intriguingly, expression levels of FoxP2 in a key site of the songbird brain are correlated with changes in vocal plasticity (Haesler et al., 2004; Teramitsu, Poopatanapong, Torrisi, & White, 2010; Thompson et al., 2013). This key site is Area X, a striatal nucleus that is an essential part of a neural circuit known to mediate vocal learning. The zebra finch studies have gone beyond simply observing correlations by adopting cutting-edge molecular genetic tools to selectively reduce ("knock down") levels of FoxP2 expression in the living songbird brain. In a landmark paper, Haesler and colleagues (2007) reported that such FoxP2 knockdown in Area X (but not surrounding areas) during the developmental period of song acquisition led to incomplete and inaccurate imitation of tutor song. Further studies of knockdown birds indicate that *FoxP2* loss yields reduced density for dendritic spines of spiny neurons (Schulz, Haesler, Scharff, & Rochefort, 2010) and interferes with dopamine modulation of activity propagation in a corticostriatal pathway involved in song variability (Murugan, Harward, Scharff, & Mooney, 2013). As with the mouse models, neural plasticity in striatal circuitry emerges as a common theme associated with this gene (Murugan et al., 2013; Schulz et al., 2010). There is insufficient space available in this chapter to give a full account of all the relevant songbird studies; for further information on this burgeoning area of work, the interested reader is referred to reviews by Bolhuis et al. (2010), Scharff and Petri (2011), and Wohlgemuth et al. (2014).

2.6 *FOXP2* IN HUMAN EVOLUTION

As shown, *FOXP2* has a deep evolutionary history with conserved functions in neural plasticity of a subset of vertebrate brain circuits. However, given its known links to human speech and language, it is reasonable to ask whether the gene has changed in any interesting ways during the evolution of our species (Fisher & Marcus, 2006). Of the three amino acid substitutions that distinguish the human and mouse version of the protein, two occurred on the human lineage after splitting from the chimpanzee at some point within the past 5 or 6 million years (Enard et al., 2002). These evolutionary changes occur outside the known domains of the FOXP2 protein and are predicted to have only subtle (if any) effects on function. Moreover, they are entirely distinct from the known mutations that have been implicated in human speech and language disorder. Nevertheless, several investigations support the idea that the differences between the human and chimpanzee proteins have some functional significance. For example, one investigation compared human neuron-like cells expressing each version of the protein and reported quantitative differences in the regulation of some of the downstream targets of this transcription factor (Konopka et al., 2009). To study their effects on a living brain, Enard and colleagues (2009) inserted the human-specific amino acids into the mouse *Foxp2* locus. Remarkably, they observed effects on neurite outgrowth and plasticity that were in the opposite direction to those seen for loss-of-function mutations of this gene (Enard et al., 2009; Groszer et al., 2008; Vernes et al., 2011), and that seemed to be specific for corticobasal ganglia circuitry (Reimers-Kipping, Hevers, Paabo, & Enard, 2011).

The timing of these amino acid substitutions has been a matter of some debate. Initial studies estimated that they had arisen within the past 200,000 years, concordant with evidence that the *FOXP2* locus had been subject to Darwinian selection during the origin of modern humans (Enard et al., 2002). However, subsequent work has indicated that the supposedly human-specific substitutions are also found in Neanderthal samples, indicating an earlier origin and predating the human–Neanderthal split several hundred thousand years ago (Krause et al., 2007). Further investigations revealed noncoding changes (i.e., those that do not affect amino acid sequences) that occurred on the human lineage after the split from Neanderthals, and that might have affected the way that the expression of *FOXP2* is regulated. It is possible that these later changes may impact on functions of the gene and could explain the evidence of relatively recent Darwinian selection at the locus (Ptak et al., 2009). Thus, human *FOXP2* may have been subject to multiple selective events during human evolution, which might have involved modifications of its functions in neural circuitry. Obviously, we cannot go back in time to formally test whether such modifications were really relevant for the emergence of language. Regardless, based on its deeper evolutionary history, it is unlikely that *FOXP2* was a sole trigger for appearance of this complex suite of skills, but instead it represents one piece of a complex puzzle involving other factors (Fisher & Ridley, 2013).

2.7 CONCLUSIONS

FOXP2 is not the only gene to have been implicated in speech and language, although it provides perhaps the clearest links to the underlying biology. For example, studies of common forms of SLI have suggested several other candidates, like *CNTNAP2* (introduced above), *ATP2C2*, and *CMIP* (Newbury & Monaco, 2010; Newbury et al., 2009). As genomic technologies continue to advance at an astonishing rate, we can expect more and more of the critical molecules to be uncovered (Deriziotis & Fisher, 2013). This chapter has sought to provide an illustration of the new vistas that open up after the identification of a language-related gene, emphasizing that such a discovery is just a starting point for functional investigations. One emerging approach in the language sciences that has not yet achieved its full potential is that of neuroimaging genetics, that is, testing for associations between genetic variants and variability in structure and/or function of language-related circuits of the human brain. Investigations of structural and functional consequences of rare mutations associated with disorder have been informative (Liegeois et al., 2003, 2011; Vargha-Khadem et al., 1998; Watkins, Vargha-Khadem et al., 2002). However, the effects of common variations in genes of interest have been more difficult to decipher (Hoogman et al., 2014), and reports have generally been underpowered for detecting biologically meaningful effects (Bishop, 2013; Graham & Fisher, 2013). It is likely that large-scale sophisticated studies involving genetic information, structural and function neuroimaging data, and performance on cognitive measures will yield exciting new insights. Overall, these developments in bridging genes, neurons, circuits, brains, and cognition are bringing us closer to understanding the basis of our most mysterious human capacities.

References

Alarcon, M., Abrahams, B. S., Stone, J. L., Duvall, J. A., Perederiy, J. V., Bomar, J. M., et al. (2008). Linkage, association, and gene-expression analyses identify CNTNAP2 as an autism-susceptibility gene. *American Journal of Human Genetics, 82*(1), 150–159. Available from: http://dx.doi.org/10.1016/j.ajhg.2007.09.005.

Anderson, G. R., Galfin, T., Xu, W., Aoto, J., Malenka, R. C., & Sudhof, T. C. (2012). Candidate autism gene screen identifies critical role for cell-adhesion molecule CASPR2 in dendritic arborization and spine development. *Proceedings of the National Academy of Sciences of the United States of America, 109*(44), 18120–18125. Available from: http://dx.doi.org/10.1073/pnas.1216398109.

Bacon, C., & Rappold, G. A. (2012). The distinct and overlapping phenotypic spectra of FOXP1 and FOXP2 in cognitive disorders. *Human Genetics, 131*(11), 1687–1698. Available from: http://dx.doi.org/10.1007/s00439-012-1193-z.

Benayoun, B. A., Caburet, S., & Veitia, R. A. (2011). Forkhead transcription factors: Key players in health and disease. *Trends in Genetics: TIG, 27*(6), 224–232. Available from: http://dx.doi.org/10.1016/j.tig.2011.03.003.

Bishop, D. V. (2013). Cerebral asymmetry and language development: Cause, correlate, or consequence? *Science, 340*(6138), 1230531. Available from: http://dx.doi.org/10.1126/science.1230531.

Bolhuis, J. J., Okanoya, K., & Scharff, C. (2010). Twitter evolution: Converging mechanisms in birdsong and human speech. *Nature Reviews Neuroscience, 11*(11), 747–759. Available from: http://dx.doi.org/10.1038/nrn2931.

Bonkowsky, J. L., Wang, X., Fujimoto, E., Lee, J. E., Chien, C. B., & Dorsky, R. I. (2008). Domain-specific regulation of foxP2 CNS expression by lef1. *BMC Developmental Biology, 8*, 103. Available from: http://dx.doi.org/10.1186/1471-213X-8-103.

Bowers, J. M., Perez-Pouchoulen, M., Edwards, N. S., & McCarthy, M. M. (2013). Foxp2 mediates sex differences in ultrasonic vocalization by rat pups and directs order of maternal retrieval. *The Journal of Neuroscience: The Official Journal of the Society for Neuroscience, 33*(8), 3276–3283. Available from: http://dx.doi.org/10.1523/JNEUROSCI.0425-12.2013.

Clovis, Y. M., Enard, W., Marinaro, F., Huttner, W. B., & De Pietri Tonelli, D. (2012). Convergent repression of Foxp2 3′UTR by miR-9 and miR-132 in embryonic mouse neocortex: Implications for radial migration of neurons. *Development, 139*(18), 3332–3342. Available from: http://dx.doi.org/10.1242/dev.078063.

Deriziotis, P., & Fisher, S. E. (2013). Neurogenomics of speech and language disorders: The road ahead. *Genome Biology, 14*(4), 204. Available from: http://dx.doi.org/10.1186/gb-2013-14-4-204.

Enard, W., Gehre, S., Hammerschmidt, K., Holter, S. M., Blass, T., Somel, M., et al. (2009). A humanized version of Foxp2 affects cortico-basal ganglia circuits in mice. *Cell, 137*(5), 961–971. Available from: http://dx.doi.org/10.1016/j.cell.2009.03.041.

Enard, W., Przeworski, M., Fisher, S. E., Lai, C. S., Wiebe, V., Kitano, T., et al. (2002). Molecular evolution of FOXP2, a gene involved in speech and language. *Nature, 418*(6900), 869–872. Available from: http://dx.doi.org/10.1038/nature01025.

Ferland, R. J., Cherry, T. J., Preware, P. O., Morrisey, E. E., & Walsh, C. A. (2003). Characterization of Foxp2 and Foxp1 mRNA and protein in the developing and mature brain. *The Journal of Comparative Neurology, 460*(2), 266–279. Available from: http://dx.doi.org/10.1002/cne.10654.

Feuk, L., Kalervo, A., Lipsanen-Nyman, M., Skaug, J., Nakabayashi, K., Finucane, B., et al. (2006). Absence of a paternally inherited FOXP2 gene in developmental verbal dyspraxia. *American Journal of Human Genetics, 79*(5), 965–972. Available from: http://dx.doi.org/10.1086/508902.

Fisher, S. E. (2006). Tangled webs: Tracing the connections between genes and cognition. *Cognition, 101*(2), 270–297. Available from: http://dx.doi.org/10.1016/j.cognition.2006.04.004.

Fisher, S. E. (2007). Molecular windows into speech and language disorders. *Folia Phoniatrica et Logopaedica: Official Organ of the International Association of Logopedics and Phoniatrics, 59*(3), 130–140. Available from: http://dx.doi.org/10.1159/000101771.

Fisher, S. E., & Marcus, G. F. (2006). The eloquent ape: Genes, brains and the evolution of language. *Nature Reviews Genetics, 7*(1), 9–20. Available from: http://dx.doi.org/10.1038/nrg1747.

Fisher, S. E., & Ridley, M. (2013). Evolution. Culture, genes, and the human revolution. *Science, 340*(6135), 929–930. Available from: http://dx.doi.org/10.1126/science.1236171.

Fisher, S. E., & Scharff, C. (2009). FOXP2 as a molecular window into speech and language. *Trends in Genetics: TIG, 25*(4), 166–177. Available from: http://dx.doi.org/10.1016/j.tig.2009.03.002.

Fisher, S. E., Lai, C. S., & Monaco, A. P. (2003). Deciphering the genetic basis of speech and language disorders. *Annual Review of Neuroscience, 26*, 57–80. Available from: http://dx.doi.org/10.1146/annurev.neuro.26.041002.131144.

Fisher, S. E., Vargha-Khadem, F., Watkins, K. E., Monaco, A. P., & Pembrey, M. E. (1998). Localisation of a gene implicated in a severe speech and language disorder. *Nature Genetics, 18*(2), 168–170. Available from: http://dx.doi.org/10.1038/ng0298-168.

French, C. A., & Fisher, S. E. (2014). What can mice tell us about Foxp2 function? *Current Opinion in Neurobiology, 28C*, 72–79. Available from: http://dx.doi.org/10.1016/j.conb.2014.07.003.

French, C. A., Jin, X., Campbell, T. G., Gerfen, E., Groszer, M., Fisher, S. E., et al. (2012). An aetiological Foxp2 mutation causes aberrant striatal activity and alters plasticity during skill learning. *Molecular Psychiatry, 17*(11), 1077–1085. Available from: http://dx.doi.org/10.1038/mp.2011.105.

Gaub, S., Groszer, M., Fisher, S. E., & Ehret, G. (2010). The structure of innate vocalizations in Foxp2-deficient mouse pups. *Genes, Brain, and Behavior, 9*(4), 390–401. Available from: http://dx.doi.org/10.1111/j.1601-183X.2010.00570.x.

Graham, S. A., & Fisher, S. E. (2013). Decoding the genetics of speech and language. *Current Opinion in Neurobiology, 23*(1), 43–51. Available from: http://dx.doi.org/10.1016/j.conb.2012.11.006.

Groszer, M., Keays, D. A., Deacon, R. M., de Bono, J. P., Prasad-Mulcare, S., Gaub, S., et al. (2008). Impaired synaptic plasticity and motor learning in mice with a point mutation implicated in human speech deficits. *Current Biology: CB, 18*(5), 354–362. Available from: http://dx.doi.org/10.1016/j.cub.2008.01.060.

Haesler, S., Rochefort, C., Georgi, B., Licznerski, P., Osten, P., & Scharff, C. (2007). Incomplete and inaccurate vocal imitation after knockdown of FoxP2 in songbird basal ganglia nucleus Area X. *PLoS Biology, 5*(12), e321. Available from: http://dx.doi.org/10.1371/journal.pbio.0050321.

Haesler, S., Wada, K., Nshdejan, A., Morrisey, E. E., Lints, T., Jarvis, E. D., et al. (2004). FoxP2 expression in avian vocal learners and non-learners. *The Journal of Neuroscience: The Official Journal of the Society for Neuroscience, 24*(13), 3164–3175. Available from: http://dx.doi.org/10.1523/JNEUROSCI.4369-03.2004.

Hammerschmidt, K., Reisinger, E., Westekemper, K., Ehrenreich, L., Strenzke, N., & Fischer, J. (2012). Mice do not require auditory input for the normal development of their ultrasonic vocalizations. *BMC Neuroscience*, *13*, 40. Available from: http://dx.doi.org/10.1186/1471-2202-13-40.

Hoogman, M., Guadalupe, T., Zwiers, M. P., Klarenbeek, P., Francks, C., & Fisher, S. E. (2014). Assessing the effects of common variation in the FOXP2 gene on human brain structure. *Frontiers in Human Neuroscience*, *8*, 473. Available from: http://dx.doi.org/10.3389/fnhum.2014.00473.

Hurst, J. A., Baraitser, M., Auger, E., Graham, F., & Norell, S. (1990). An extended family with a dominantly inherited speech disorder. *Developmental Medicine and Child Neurology*, *32*(4), 352–355.

Iwai, L., Ohashi, Y., van der List, D., Usrey, W. M., Miyashita, Y., & Kawasaki, H. (2013). FoxP2 is a parvocellular-specific transcription factor in the visual thalamus of monkeys and ferrets. *Cerebral Cortex*, *23*(9), 2204–2212. Available from: http://dx.doi.org/10.1093/cercor/bhs207.

Konopka, G., Bomar, J. M., Winden, K., Coppola, G., Jonsson, Z. O., Gao, F., et al. (2009). Human-specific transcriptional regulation of CNS development genes by FOXP2. *Nature*, *462*(7270), 213–217. Available from: http://dx.doi.org/10.1038/nature08549.

Kosho, T., Sakazume, S., Kawame, H., Wakui, K., Wada, T., Okoshi, Y., et al. (2008). De-novo balanced translocation between 7q31 and 10p14 in a girl with central precocious puberty, moderate mental retardation, and severe speech impairment. *Clinical Dysmorphology*, *17*(1), 31–34. Available from: http://dx.doi.org/10.1097/MCD.0b013e3282f17688.

Krause, J., Lalueza-Fox, C., Orlando, L., Enard, W., Green, R. E., Burbano, H. A., et al. (2007). The derived FOXP2 variant of modern humans was shared with Neanderthals. *Current Biology: CB*, *17*(21), 1908–1912. Available from: http://dx.doi.org/10.1016/j.cub.2007.10.008.

Kurt, S., Fisher, S. E., & Ehret, G. (2012). Foxp2 mutations impair auditory-motor association learning. *PloS One*, *7*(3), e33130. Available from: http://dx.doi.org/10.1371/journal.pone.0033130.

Laffin, J. J., Raca, G., Jackson, C. A., Strand, E. A., Jakielski, K. J., & Shriberg, L. D. (2012). Novel candidate genes and regions for childhood apraxia of speech identified by array comparative genomic hybridization. *Genetics in Medicine: Official Journal of the American College of Medical Genetics*, *14*(11), 928–936. Available from: http://dx.doi.org/10.1038/gim.2012.72.

Lai, C. S., Fisher, S. E., Hurst, J. A., Levy, E. R., Hodgson, S., Fox, M., et al. (2000). The SPCH1 region on human 7q31: Genomic characterization of the critical interval and localization of translocations associated with speech and language disorder. *American Journal of Human Genetics*, *67*(2), 357–368. Available from: http://dx.doi.org/10.1086/303011.

Lai, C. S., Fisher, S. E., Hurst, J. A., Vargha-Khadem, F., & Monaco, A. P. (2001). A forkhead-domain gene is mutated in a severe speech and language disorder. *Nature*, *413*(6855), 519–523. Available from: http://dx.doi.org/10.1038/35097076.

Lai, C. S., Gerrelli, D., Monaco, A. P., Fisher, S. E., & Copp, A. J. (2003). FOXP2 expression during brain development coincides with adult sites of pathology in a severe speech and language disorder. *Brain: A Journal of Neurology*, *126*(Pt 11), 2455–2462. Available from: http://dx.doi.org/10.1093/brain/awg247.

Lennon, P. A., Cooper, M. L., Peiffer, D. A., Gunderson, K. L., Patel, A., Peters, S., et al. (2007). Deletion of 7q31.1 supports involvement of FOXP2 in language impairment: Clinical report and review. *American Journal of Medical Genetics Part A*, *143A*(8), 791–798. Available from: http://dx.doi.org/10.1002/ajmg.a.31632.

Lesca, G., Rudolf, G., Bruneau, N., Lozovaya, N., Labalme, A., Boutry-Kryza, N., et al. (2013). GRIN2A mutations in acquired epileptic aphasia and related childhood focal epilepsies and encephalopathies with speech and language dysfunction. *Nature Genetics*, *45*(9), 1061–1066. Available from: http://dx.doi.org/10.1038/ng.2726.

Li, G., Wang, J., Rossiter, S. J., Jones, G., & Zhang, S. (2007). Accelerated FoxP2 evolution in echolocating bats. *PloS One*, *2*(9), e900. Available from: http://dx.doi.org/10.1371/journal.pone.0000900.

Li, S., Weidenfeld, J., & Morrisey, E. E. (2004). Transcriptional and DNA binding activity of the Foxp1/2/4 family is modulated by heterotypic and homotypic protein interactions. *Molecular and Cellular Biology*, *24*(2), 809–822.

Liegeois, F., Baldeweg, T., Connelly, A., Gadian, D. G., Mishkin, M., & Vargha-Khadem, F. (2003). Language fMRI abnormalities associated with FOXP2 gene mutation. *Nature Neuroscience*, *6*(11), 1230–1237. Available from: http://dx.doi.org/10.1038/nn1138.

Liegeois, F., Morgan, A. T., Connelly, A., & Vargha-Khadem, F. (2011). Endophenotypes of FOXP2: Dysfunction within the human articulatory network. *European Journal of Paediatric Neurology: EJPN: Official Journal of the European Paediatric Neurology Society*, *15*(4), 283–288. Available from: http://dx.doi.org/10.1016/j.ejpn.2011.04.006.

MacDermot, K. D., Bonora, E., Sykes, N., Coupe, A. M., Lai, C. S., Vernes, S. C., et al. (2005). Identification of FOXP2 truncation as a novel cause of developmental speech and language deficits. *American Journal of Human Genetics*, *76*(6), 1074–1080. Available from: http://dx.doi.org/10.1086/430841.

Mukamel, Z., Konopka, G., Wexler, E., Osborn, G. E., Dong, H., Bergman, M. Y., et al. (2011). Regulation of MET by FOXP2, genes implicated in higher cognitive dysfunction and autism risk. *The Journal of Neuroscience: The Official Journal of the Society for Neuroscience*, *31*(32), 11437–11442. Available from: http://dx.doi.org/10.1523/JNEUROSCI.0181-11.2011.

Murugan, M., Harward, S., Scharff, C., & Mooney, R. (2013). Diminished FoxP2 levels affect dopaminergic modulation of corticostriatal signaling important to song variability. *Neuron*, *80*(6), 1464–1476. Available from: http://dx.doi.org/10.1016/j.neuron.2013.09.021.

Newbury, D. F., & Monaco, A. P. (2010). Genetic advances in the study of speech and language disorders. *Neuron*, *68*(2), 309–320. Available from: http://dx.doi.org/10.1016/j.neuron.2010.10.001.

Newbury, D. F., Winchester, L., Addis, L., Paracchini, S., Buckingham, L. L., Clark, A., et al. (2009). CMIP and ATP2C2 modulate phonological short-term memory in language impairment. *American Journal of Human Genetics*, *85*(2), 264–272. Available from: http://dx.doi.org/10.1016/j.ajhg.2009.07.004.

O'Roak, B. J., Deriziotis, P., Lee, C., Vives, L., Schwartz, J. J., Girirajan, S., et al. (2011). Exome sequencing in sporadic autism spectrum disorders identifies severe de novo mutations. *Nature Genetics*, *43*(6), 585–589. Available from: http://dx.doi.org/10.1038/ng.835.

Palka, C., Alfonsi, M., Mohn, A., Cerbo, R., Guanciali Franchi, P., Fantasia, D., et al. (2012). Mosaic 7q31 deletion involving FOXP2 gene associated with language impairment. *Pediatrics*, *129*(1), e183–e188. Available from: http://dx.doi.org/10.1542/peds.2010-2094.

Peter, B., Raskind, W. H., Matsushita, M., Lisowski, M., Vu, T., Berninger, V. W., et al. (2011). Replication of CNTNAP2 association with nonword repetition and support for FOXP2 association with timed reading and motor activities in a dyslexia family sample. *Journal of Neurodevelopmental Disorders*, *3*(1), 39–49. Available from: http://dx.doi.org/10.1007/s11689-010-9065-0.

Ptak, S. E., Enard, W., Wiebe, V., Hellmann, I., Krause, J., Lachmann, M., et al. (2009). Linkage disequilibrium extends across putative selected sites in FOXP2. *Molecular Biology and Evolution*, *26*(10), 2181–2184. Available from: http://dx.doi.org/10.1093/molbev/msp143.

Reimers-Kipping, S., Hevers, W., Paabo, S., & Enard, W. (2011). Humanized Foxp2 specifically affects cortico-basal ganglia circuits. *Neuroscience, 175*, 75–84. Available from: http://dx.doi.org/10.1016/j.neuroscience.2010.11.042.

Rice, G. M., Raca, G., Jakielski, K. J., Laffin, J. J., Iyama-Kurtycz, C. M., Hartley, S. L., et al. (2012). Phenotype of FOXP2 haploinsufficiency in a mother and son. *American Journal of Medical Genetics Part A, 158A*(1), 174–181. Available from: http://dx.doi.org/10.1002/ajmg.a.34354.

Rodenas-Cuadrado, P., Ho, J., & Vernes, S. C. (2013). Shining a light on CNTNAP2: Complex functions to complex disorders. *European Journal of Human Genetics: EJHG, 22*(2), 171–178. Available from: http://dx.doi.org/10.1038/ejhg.2013.100.

Roll, P., Vernes, S. C., Bruneau, N., Cillario, J., Ponsole-Lenfant, M., Massacrier, A., et al. (2010). Molecular networks implicated in speech-related disorders: FOXP2 regulates the SRPX2/uPAR complex. *Human Molecular Genetics, 19*(24), 4848–4860. Available from: http://dx.doi.org/10.1093/hmg/ddq415.

Scharff, C., & Petri, J. (2011). Evo-devo, deep homology and FoxP2: Implications for the evolution of speech and language. *Philosophical Transactions of the Royal Society of London Series B, Biological Sciences, 366*(1574), 2124–2140. Available from: http://dx.doi.org/10.1098/rstb.2011.0001.

Schulz, S. B., Haesler, S., Scharff, C., & Rochefort, C. (2010). Knockdown of FoxP2 alters spine density in Area X of the zebra finch. *Genes, Brain, and Behavior, 9*(7), 732–740. Available from: http://dx.doi.org/10.1111/j.1601-183X.2010.00607.x.

Shriberg, L. D., Ballard, K. J., Tomblin, J. B., Duffy, J. R., Odell, K. H., & Williams, C. A. (2006). Speech, prosody, and voice characteristics of a mother and daughter with a 7;13 translocation affecting FOXP2. *Journal of Speech, Language, and Hearing Research: JSLHR, 49*(3), 500–525. Available from: http://dx.doi.org/10.1044/1092-4388(2006/038).

Shu, W., Cho, J. Y., Jiang, Y., Zhang, M., Weisz, D., Elder, G. A., et al. (2005). Altered ultrasonic vocalization in mice with a disruption in the Foxp2 gene. *Proceedings of the National Academy of Sciences of the United States of America, 102*(27), 9643–9648. Available from: http://dx.doi.org/10.1073/pnas.0503739102.

Sia, G. M., Clem, R. L., & Huganir, R. L. (2013). The human language-associated gene SRPX2 regulates synapse formation and vocalization in mice. *Science, 342*(6161), 987–991. Available from: http://dx.doi.org/10.1126/science.1245079.

Spiteri, E., Konopka, G., Coppola, G., Bomar, J., Oldham, M., Ou, J., et al. (2007). Identification of the transcriptional targets of FOXP2, a gene linked to speech and language, in developing human brain. *American Journal of Human Genetics, 81*(6), 1144–1157. Available from: http://dx.doi.org/10.1086/522237.

Takahashi, K., Liu, F. C., Hirokawa, K., & Takahashi, H. (2003). Expression of Foxp2, a gene involved in speech and language, in the developing and adult striatum. *Journal of Neuroscience Research, 73*(1), 61–72. Available from: http://dx.doi.org/10.1002/jnr.10638.

Takahashi, K., Liu, F. C., Oishi, T., Mori, T., Higo, N., Hayashi, M., et al. (2008). Expression of FOXP2 in the developing monkey forebrain: Comparison with the expression of the genes FOXP1, PBX3, and MEIS2. *The Journal of Comparative Neurology, 509*(2), 180–189. Available from: http://dx.doi.org/10.1002/cne.21740.

Teramitsu, I., Kudo, L. C., London, S. E., Geschwind, D. H., & White, S. A. (2004). Parallel FoxP1 and FoxP2 expression in songbird and human brain predicts functional interaction. *The Journal of Neuroscience: The Official Journal of the Society for Neuroscience, 24*(13), 3152–3163. Available from: http://dx.doi.org/10.1523/JNEUROSCI.5589-03.2004.

Teramitsu, I., Poopatanapong, A., Torrisi, S., & White, S. A. (2010). Striatal FoxP2 is actively regulated during songbird sensorimotor learning. *PloS One, 5*(1), e8548. Available from: http://dx.doi.org/10.1371/journal.pone.0008548.

Thompson, C. K., Schwabe, F., Schoof, A., Mendoza, E., Gampe, J., Rochefort, C., et al. (2013). Young and intense: FoxP2 immunoreactivity in Area X varies with age, song stereotypy, and singing in male zebra finches. *Frontiers in Neural Circuits, 7*, 24. Available from: http://dx.doi.org/10.3389/fncir.2013.00024.

Tomblin, J. B., O'Brien, M., Shriberg, L. D., Williams, C., Murray, J., Patil, S., et al. (2009). Language features in a mother and daughter of a chromosome 7;13 translocation involving FOXP2. *Journal of Speech, Language, and Hearing Research: JSLHR, 52*(5), 1157–1174. Available from: http://dx.doi.org/10.1044/1092-4388(2009/07-0162).

Tsui, D., Vessey, J. P., Tomita, H., Kaplan, D. R., & Miller, F. D. (2013). FoxP2 regulates neurogenesis during embryonic cortical development. *The Journal of Neuroscience: The Official Journal of the Society for Neuroscience, 33*(1), 244–258. Available from: http://dx.doi.org/10.1523/JNEUROSCI.1665-12.2013.

Turner, S. J., Hildebrand, M. S., Block, S., Damiano, J., Fahey, M., Reilly, S., et al. (2013). Small intragenic deletion in FOXP2 associated with childhood apraxia of speech and dysarthria. *American Journal of Medical Genetics Part A, 161*(9), 2321–2326. Available from: http://dx.doi.org/10.1002/ajmg.a.36055.

Vargha-Khadem, F., Watkins, K. E., Price, C. J., Ashburner, J., Alcock, K. J., Connelly, A., et al. (1998). Neural basis of an inherited speech and language disorder. *Proceedings of the National Academy of Sciences of the United States of America, 95*(21), 12695–12700.

Varki, A., & Altheide, T. K. (2005). Comparing the human and chimpanzee genomes: Searching for needles in a haystack. *Genome Research, 15*(12), 1746–1758. Available from: http://dx.doi.org/10.1101/gr.3737405.

Vernes, S. C., Newbury, D. F., Abrahams, B. S., Winchester, L., Nicod, J., Groszer, M., et al. (2008). A functional genetic link between distinct developmental language disorders. *The New England Journal of Medicine, 359*(22), 2337–2345. Available from: http://dx.doi.org/10.1056/NEJMoa0802828.

Vernes, S. C., Nicod, J., Elahi, F. M., Coventry, J. A., Kenny, N., Coupe, A. M., et al. (2006). Functional genetic analysis of mutations implicated in a human speech and language disorder. *Human Molecular Genetics, 15*(21), 3154–3167. Available from: http://dx.doi.org/10.1093/hmg/ddl392.

Vernes, S. C., Oliver, P. L., Spiteri, E., Lockstone, H. E., Puliyadi, R., Taylor, J. M., et al. (2011). Foxp2 regulates gene networks implicated in neurite outgrowth in the developing brain. *PLoS Genetics, 7*(7), e1002145. Available from: http://dx.doi.org/10.1371/journal.pgen.1002145.

Vernes, S. C., Spiteri, E., Nicod, J., Groszer, M., Taylor, J. M., Davies, K. E., et al. (2007). High-throughput analysis of promoter occupancy reveals direct neural targets of FOXP2, a gene mutated in speech and language disorders. *American Journal of Human Genetics, 81*(6), 1232–1250. Available from: http://dx.doi.org/10.1086/522238.

Walker, R. M., Hill, A. E., Newman, A. C., Hamilton, G., Torrance, H. S., Anderson, S. M., et al. (2012). The DISC1 promoter: Characterization and regulation by FOXP2. *Human Molecular Genetics, 21*(13), 2862–2872. Available from: http://dx.doi.org/10.1093/hmg/dds111.

Watkins, K. E., Dronkers, N. F., & Vargha-Khadem, F. (2002). Behavioural analysis of an inherited speech and language disorder: Comparison with acquired aphasia. *Brain: A Journal of neurology, 125*(Pt 3), 452–464.

Watkins, K. E., Vargha-Khadem, F., Ashburner, J., Passingham, R. E., Connelly, A., Friston, K. J., et al. (2002). MRI analysis of an inherited speech and language disorder: Structural brain abnormalities. *Brain: A Journal of Neurology, 125*(Pt 3), 465–478.

Whitehouse, A. J., Bishop, D. V., Ang, Q. W., Pennell, C. E., & Fisher, S. E. (2011). CNTNAP2 variants affect early language development in the general population. *Genes, Brain, and Behavior, 10*(4), 451–456. Available from: http://dx.doi.org/10.1111/j.1601-183X.2011.00684.x.

Wohlgemuth, S., Adam, I., & Scharff, C. (2014). FoxP2 in songbirds. *Current Opinion in Neurobiology, 28C*, 86–93. Available from: http://dx.doi.org/10.1016/j.conb.2014.06.009.

Zeesman, S., Nowaczyk, M. J., Teshima, I., Roberts, W., Cardy, J. O., Brian, J., et al. (2006). Speech and language impairment and oromotor dyspraxia due to deletion of 7q31 that involves FOXP2. *American Journal of Medical Genetics Part A, 140*(5), 509–514. Available from: http://dx.doi.org/10.1002/ajmg.a.31110.

Zilina, O., Reimand, T., Zjablovskaja, P., Mannik, K., Mannamaa, M., Traat, A., et al. (2012). Maternally and paternally inherited deletion of 7q31 involving the FOXP2 gene in two families. *American Journal of Medical Genetics Part A, 158A*(1), 254–256. Available from: http://dx.doi.org/10.1002/ajmg.a.34378.

CHAPTER

3

The Ventrolateral Frontal Region

Michael Petrides

Montreal Neurological Institute, McGill University, Montreal, Quebec, Canada

The lateral frontal cortex is a heterogeneous region comprising several distinct areas that differ both in terms of their cellular architecture (cytoarchitecture) and their connections with other cortical and subcortical areas (Figure 3.1). The posterior part of the frontal lobe includes several motor and premotor areas that lie mostly on the precentral gyrus (region in white in Figure 3.1). The compact layer of small neurons (layer IV) clearly separating the pyramids of layer III from those of layer V in primary sensory cortex and other isocortical areas is difficult to discern in the motor/premotor areas because these small neurons are intermixed with larger pyramidal neurons, leading to the tradition of referring to these areas as "agranular." However, these so-called agranular isocortical motor areas do have small interneurons and must not be confused with the truly agranular phylogenetically older areas of the limbic region of the brain (García-Cabezas & Barbas, 2014). Anterior to these motor areas are several cytoarchitectonic areas that exhibit a compact and distinct layer IV, and this region is often referred to as the "granular" frontal cortex or the "prefrontal" cortex. These prefrontal cortical areas have been shown to participate in several higher-order control processes that regulate attention to the environment, working memory, various aspects of controlled memory retrieval, and behavioral adjustment to changes in the environment (see reviews in Stuss & Knight, 2013). The part of the lateral frontal cortex that extends anterior to the precentral gyrus is traditionally divided into three gyri: the superior, middle, and inferior frontal gyri. The caudal part of the prefrontal cortex on the superior and middle frontal gyri is occupied by subdivisions of area 8, a cortical region regulating attentional processes, which is succeeded anteriorly by the mid-dorsolateral prefrontal region (area 46 and the related areas 9/46) that plays a major role in certain aspects of working memory, such as the tracking of self-generated and externally generated events in working memory (Petrides, 1996, 2013). Although we might expect the specific roles of these areas in attentional control and working memory to be reflected in language processing, the dorsolateral prefrontal areas are not core language areas in the sense that neither fundamental language comprehension nor production is impaired.

The part of the frontal lobe that is most relevant to language processing is the inferior frontal gyrus and the adjacent ventral part of the precentral gyrus, namely the ventrolateral frontal region (Figure 3.2). The motor representation of the orofacial part of the body is found on the ventral part of the precentral gyrus (Penfield & Boldrey, 1937; Penfield & Rasmussen, 1950). The ventral part of the primary motor cortex (Brodmann area 4) that represents the orofacial musculature is largely hidden in the anterior bank of the central sulcus and, therefore, most of the cortex on the crown of the ventral precentral gyrus is occupied by premotor cortex (i.e., area 6) (Brodmann, 1909). More recent studies have identified two premotor cortical areas on the ventral precentral gyrus, a caudal one, area 6VC (ventrocaudal part of area 6; also known as area F4), and a rostral area, area 6VR (ventrorostral part of area 6; also known as area F5) (see Petrides, 2014, for details). The terms "area F1" (corresponding to area 4), "area F4" (corresponding to area 6VC), and "area F5" (corresponding to area 6VR) were proposed for comparable areas in the macaque monkey by Matelli, Luppino, and Rizzolatti (1985). Immediately anterior to the orofacial part of the precentral gyrus lies the posterior part of the inferior frontal gyrus, namely the pars opercularis, which is occupied by cortical area 44 and is succeeded anteriorly by the pars triangularis (area 45) (Figures 3.1 and 3.2).

The posterior part of the inferior frontal gyrus in the language-dominant hemisphere is traditionally considered to be the classical Broca's region, namely the

Neurobiology of Language. DOI: http://dx.doi.org/10.1016/B978-0-12-407794-2.00003-1

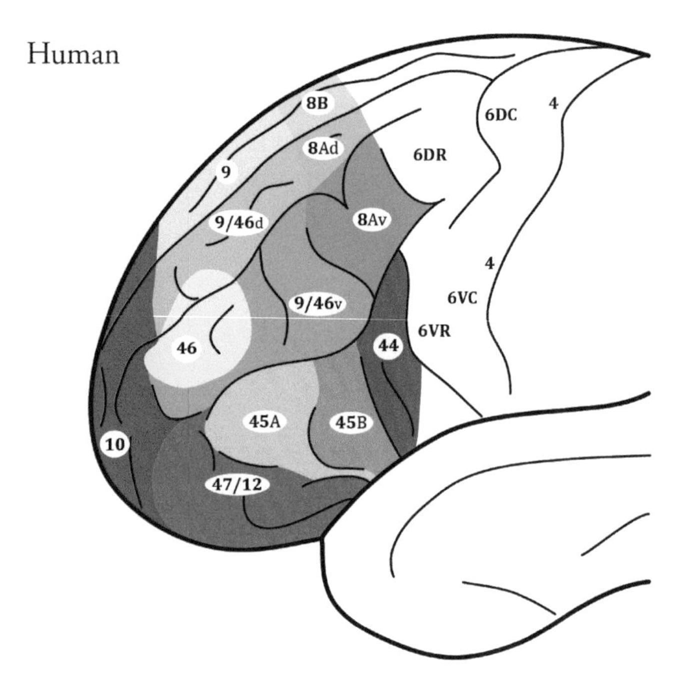

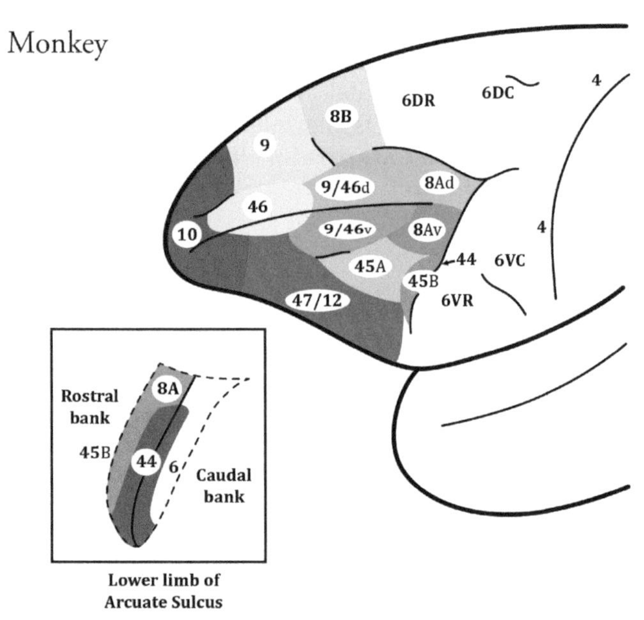

FIGURE 3.1 Cytoarchitectonic map of the lateral surface of the human and the macaque monkey frontal lobe by Petrides and Pandya (1994). The white region on the precentral gyrus is the primary motor cortex (area 4) and the various subdivisions of the premotor region (area 6). The inset shows the location of area 44 in the macaque monkey in the fundus of the inferior limb (ramus) of the arcuate sulcus.

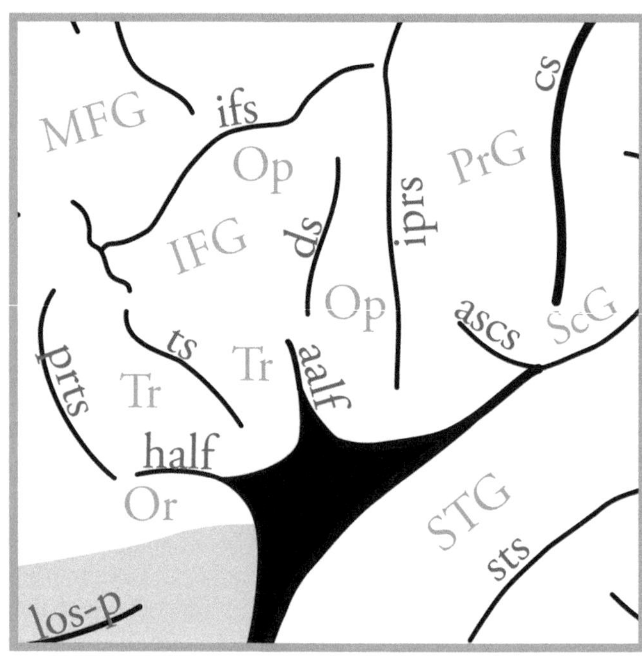

FIGURE 3.2 The sulcal and gyral morphology of the ventrolateral frontal region in the human brain. The shaded region represents the orbitofrontal cortex that is continuous with the pars orbitalis of the inferior frontal gyrus. Abbreviations: aalf, anterior ascending ramus of the lateral fissure (ascending sulcus, vertical sulcus); ascs, anterior subcentral sulcus; cs, central sulcus; ds, diagonal sulcus; half, horizontal anterior ramus of the lateral fissure (horizontal sulcus); IFG, inferior frontal gyrus; ifs, inferior frontal sulcus; iprs, inferior precentral sulcus; los-p, posterior ramus of the lateral orbital sulcus; MFG, middle frontal gyrus; Op, pars opercularis of the inferior frontal gyrus; Or, pars orbitalis of the inferior frontal gyrus; PrG, precentral gyrus; prts, pretriangular sulcus; ScG, subcentral gyrus; STG, superior temporal gyrus; sts, superior temporal sulcus; Tr, pars triangularis of the inferior frontal gyrus; ts, triangular sulcus (incisura capitis).

frontal cortical region that plays a critical role in certain aspects of language production (Friederici, 2011; Geschwind, 1970; Grodzinsky, 2000). Several attempts have been made to specify more precisely the critical zone for language within the inferior frontal gyrus on the basis of clinical–anatomical correlation studies, but these efforts had only limited success because lesions in human subjects are rarely restricted to specific subdivisions of the inferior frontal region (Mohr, 1976; Mohr et al., 1978). The syndrome of Broca's aphasia, which is characterized by severe impairment in language production (including impaired syntactic processing), is the result of massive damage to the territory of the upper division of the middle cerebral artery and involves not only the cortical structures in the posterior part of the inferior frontal gyrus (i.e., areas 44 and 45) but also the adjacent frontoparietal opercular region and the anterior parts of the insula (Ackermann & Riecker, 2004; Baldo, Wilkins, Ogar, Willock, & Dronkers, 2011; Dronkers, 1996; Mohr, 1976). The best evidence thus far linking specific parts of the inferior frontal gyrus to language production has been obtained from electrical stimulation of the cerebral cortex under local anesthesia during brain surgery. In this approach that is motivated by the need to spare cortex critical for language during brain surgery, the critical region for speech is considered to be the part of the cortex from which dysphasic speech arrest can be evoked by the application of electrical stimulation (Duffau, Moritz-Gasser, & Mandonnet, 2014; Ojemann, 1992; Ojemann, Ojemann, Lettich, & Berger, 1989; Penfield & Roberts, 1959; Rasmussen & Milner, 1975). Dysphasic speech arrest occurs most reliably from stimulation of the pars opercularis (area 44) (Rasmussen & Milner, 1975), although speech arrest can also be evoked from stimulation of the posterior part of area 45. Stimulation of the ventral precentral region, where the orofacial musculature is represented, also interferes with speech, primarily in the form of dysarthria and evoked vocalization responses caused by disruption of normal activity in the motor circuits necessary for speech articulation (Penfield & Roberts, 1959; Rasmussen & Milner, 1975).

The studies of Penfield and colleagues established another important region for speech on the posterior part of the dorsomedial surface of the frontal lobe, the supplementary motor area. Vocalization, as well as interference with speech, can result from stimulation of the supplementary motor area (Penfield & Welch, 1951). Several studies have shown that lesions of the dorsomedial frontal region, which include the supplementary motor area but are not restricted to it, lead to significant reduction in speech (Chapados & Petrides, 2013; Goldberg, 1985; Krainik et al., 2003; Nachev, Kennard, & Husain, 2008; Rostomily, Berger, Ojemann, & Lettich, 1991). Furthermore, three somatotopically organized motor areas just ventral and anterior to the supplementary motor region, originally shown in the monkey brain (Dum & Strick, 1993), have also been recently demonstrated in the human brain (Amiez & Petrides, 2014), and there is some evidence that the cingulate motor region may also play a role in speech (Paus, Petrides, Evans, & Meyer, 1993).

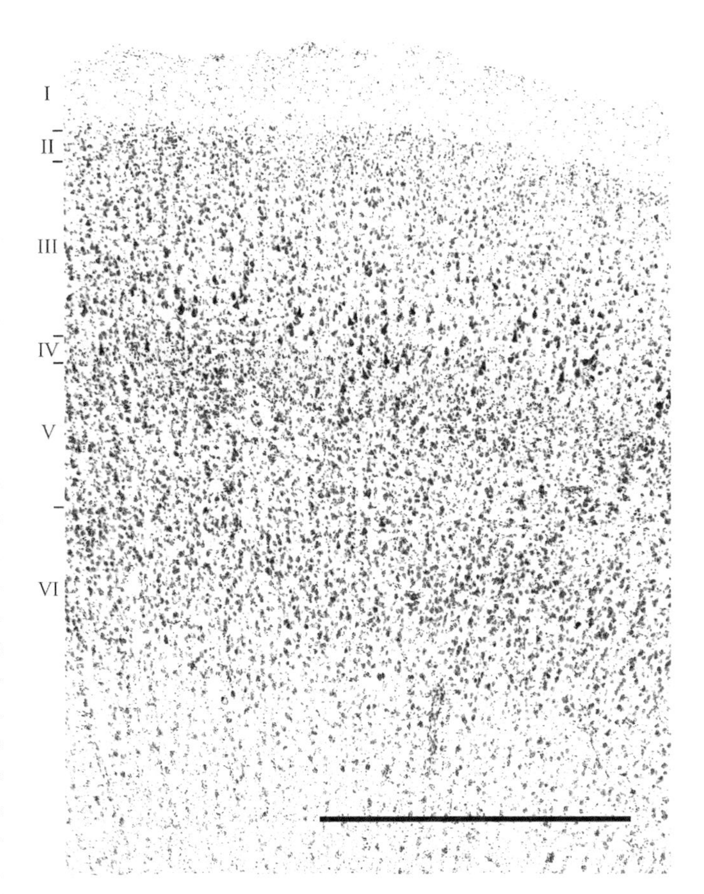

FIGURE 3.3 Photomicrograph of area 44 in the human brain. Note the interrupted layer IV, highlighted with yellow. The Roman numerals I–VI mark the six layers of the cortex. Calibration bar equals 1 mm. *From Petrides (2014) with permission from the publisher.*

3.1 CYTOARCHITECTONIC AREAS OF THE VENTROLATERAL PREFRONTAL CORTEX

Anterior to the ventral premotor region lies a cortical area in which irregular patches of small neurons appear between the pyramidal neurons of layers III and V (Figure 3.3). This area that occupies the most caudal subdivision of the inferior frontal gyrus, the pars opercularis, is Brodmann area 44 (area FCBm of Economo & Koskinas, 1925) (Figure 3.1). Area 44 is succeeded anteriorly by prefrontal area 45, which lies on the pars triangularis of the inferior frontal gyrus. In area 45, the small neurons of layer IV create a compact layer and, therefore, the pyramidal neurons of layers III and V are clearly separated (compare Figures 3.3 and 3.4) (Amunts et al., 1999; Petrides & Pandya, 1994, 2002). Area 45 is further characterized by clusters of unusually large and deeply stained pyramidal neurons in the deep part of layer III, a characteristic that unambiguously differentiates area 45 from the surrounding prefrontal areas. This unusual characteristic of area 45 led Economo and Koskinas (1925) to refer to it as area FDΓ; the Greek letter Γ refers to the clusters of giant-like neurons in layer III (Figure 3.4). In conclusion, starting from the ventral part of the central sulcus where the orofacial part of the primary motor cortical area 4 is represented, and proceeding in an anterior direction, there are two premotor areas, 6VC and 6VR, that are succeeded by the transitional area 44 and, further anterior, by the prefrontal cortical area 45 (Figure 3.1).

Anterior and ventral to area 45 lies area 47/12, which occupies the pars orbitalis of the inferior frontal gyrus (Figure 3.1). Although area 47/12 has not been traditionally considered as a core language area, recent functional neuroimaging studies have suggested that it may play a major role in the controlled access to stored conceptual representations (Badre & Wagner, 2007) and semantic unification (Zhu et al., 2012). Thus, we provide a brief discussion of its identification and cytoarchitecture here because the architectonic description of

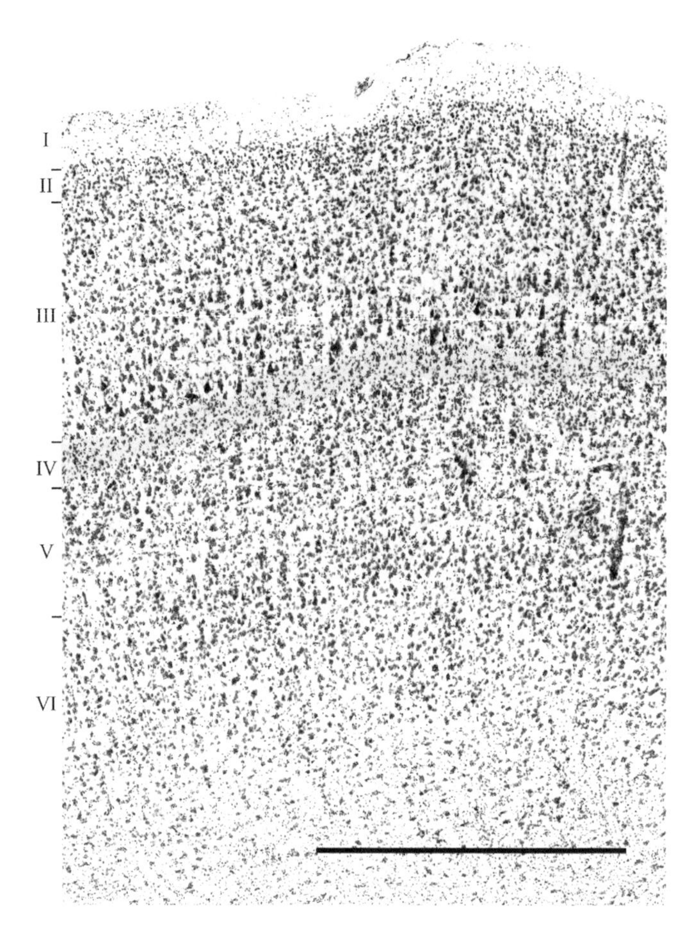

FIGURE 3.4 Photomicrograph of area 45 of the human brain. Note the continuous compact layer IV, highlighted with yellow, which clearly separates layer III pyramidal neurons from those of layer V. Calibration bar equals 1 mm. *From Petrides (2014) with permission from the publisher.*

this cortical region has generated considerable confusion. Brodmann (1909) used the term "area 47" to refer to a region ventral to area 45 that extends through most of the caudal orbital frontal region. Brodmann (1909) commented on the heterogeneity of this region and pointed out that he did not delineate its various parts. The region included in area 47 by Brodmann ranges from a ventrolateral granular part that lies just below area 45 and exhibits a distinct layer IV which separates clearly the pyramidal neurons of layer III from those of layer V, followed by a dysgranular part in which the separation between layers III and V is interrupted and, finally, an agranular part close to the medial end of the posterior orbitofrontal region. The ventrolateral part of Brodmann area 47 corresponds to the caudal part of the area that Petrides and Pandya (1994, 2002) labeled as area 47/12 (Figure 3.1). Area 47/12 corresponds to a ventrolateral prefrontal area in the macaque monkey brain that had previously been referred to as area 12 by Walker (1940). The dysgranular part of area 47 of Brodmann, which lies on the orbital frontal cortex, was named area 13 by Petrides and Pandya (1994, 2002) to be consistent with the label of the homologous region in the macaque monkey and should not be confused with area 47/12. Area 47/12 is a prefrontal area with a compact layer IV. It differs from area 45 in that its layer III lacks the clusters of unusually large neurons found in area 45. Thus, the distinction between area 47/12 and area 45 can be made easily, permitting a reliable placing of the border between these two areas. Nonetheless, both are related granular frontal areas and constitute the mid-ventrolateral prefrontal system for active controlled memory retrieval (Petrides, 1996, 2002).

3.2 PARIETAL AND TEMPORAL CORTICO-CORTICAL CONNECTION PATTERNS OF THE LANGUAGE PRODUCTION AREAS IN THE VENTROLATERAL FRONTAL REGION

We know from classical gross dissection studies of the white matter in human cadaver brains that various bundles (fasciculi) of cortico-cortical axons connect the inferior parietal lobule, the lateral temporal cortex, and the occipito-temporal junction region with the ventrolateral frontal cortical areas (Klingler, 1935; Ludwig & Klingler, 1956). However, neither the classical gross dissection of these fasciculi in cadaver brains nor their modern reconstructions with diffusion MRI methods in vivo can demonstrate the precise origin of these axons within specific cytoarchitectonic cortical areas or their precise termination within particular cortical areas (Campbell & Pike, 2014; Petrides, 2014; Thomas et al., 2014). The relatively recent demonstration of the homologues of the ventrolateral frontal areas 44 and 45 in the macaque monkey (Petrides, Cadoret, & Mackey, 2005; Petrides & Pandya, 1994, 2002) has permitted exploration of their precise cortico-cortical connections using invasive anterograde and retrograde tract tracing methods that are the gold standard for establishing precise connections in the brain (Frey, Mackey, & Petrides, 2014; Petrides & Pandya, 2002, 2009). More recent studies with resting state functional MRI have permitted the testing of hypotheses, based on the macaque monkey studies, regarding the connectivity patterns of the various ventrolateral frontal cortical areas in the human brain. Such studies have provided evidence that the cortico-cortical connectivity patterns of the ventrolateral frontal areas in the human brain are comparable with those established in the macaque monkey (Kelly et al., 2010; Margulies & Petrides, 2013).

As a rule, all adjacent cortical areas are interconnected. Thus, it is the distant connections of the

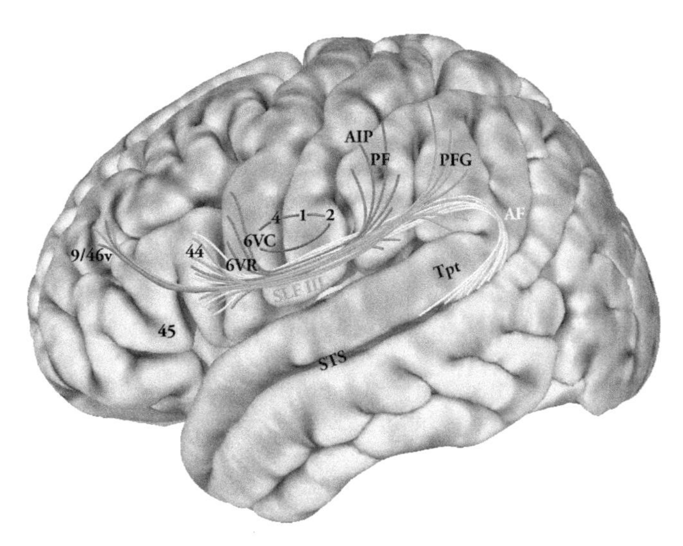

FIGURE 3.5 Lateral view of the left cerebral hemisphere of the human brain to illustrate the local peri-central association pathway (in red) linking the primary motor cortical area 4 and the caudal area 6 (i.e., 6VC) with somatosensory cortex on the postcentral gyrus. The cortex of the supramarginal gyrus (areas PF and PFG) and the adjacent anterior intraparietal sulcus (AIP) connect with the rostral premotor cortex (area 6VR) and area 44 via the third branch of the superior longitudinal fasciculus (SLF III) shown in green. Area 44 is also connected with the posterior part of the superior temporal region via the arcuate fasciculus (AF) shown in yellow. Note that adjacent cortical areas are connected with each other and areas 4 and 6VC also have local connections with adjacent frontoparietal opercular areas and insula, but these local connections are not shown in order not to crowd the diagram. *From Petrides (2014) with permission from the publisher.*

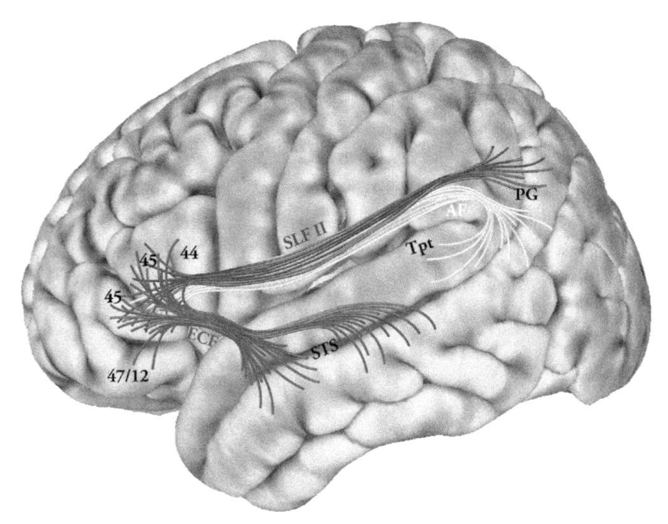

FIGURE 3.6 Lateral view of the left cerebral hemisphere of the human brain to illustrate parietal and temporal connections of prefrontal area 45. The cortex of the angular gyrus (area PG) connects with the cortex of the pars triangularis (area 45) via the second branch of the superior longitudinal fasciculus (SLF II). Area 45 is also connected with the posterior part of the superior temporal region via the arcuate fasciculus (AF) shown in yellow and the lateral part of the temporal lobe via the extreme capsule fasciculus (ECF). *From Petrides (2014) with permission from the publisher.*

various ventrolateral frontal areas with specific parietal, temporal, and occipital areas that require clarification. Primary motor cortical area 4 and the adjacent area 6VC are strongly connected with somatosensory cortical areas on the postcentral gyrus (Figure 3.5), therefore forming a local peri-central association system for orofacial sensorimotor control (Pons & Kaas, 1986; Vogt & Pandya, 1978). By contrast, area 6VR and area 44 are connected with the most anterior part of the inferior parietal lobule via the third branch of the superior longitudinal fasciculus (SLF III) (Figure 3.5). This is a distinct part of the parieto-frontal association fiber system, first identified in the macaque monkey (Petrides & Pandya, 1984) and later confirmed with diffusion MRI in the human brain (Frey, Campbell, Pike, & Petrides, 2008; Makris et al., 2005). In the human brain, the rostral part of the inferior parietal lobule forms the morphological entity known as the supramarginal gyrus that is occupied by area PF, anteriorly, and area PFG, posteriorly (Petrides, 2014). The rostral part of the ventral premotor cortex (area 6VR; also known as F5) has been shown to be most strongly linked with area PF while area 44 with area PFG in the monkey brain (Frey et al., 2014; Petrides & Pandya, 2009) and in the human brain (Margulies & Petrides, 2013; Petrides, 2014) (Figure 3.5). Areas 6VR and 44 are also linked with the cortex in the anterior part of the intraparietal sulcus (area AIP).

Prefrontal area 45 has a strikingly different cortico-cortical connection profile. Area 45 is linked with the cortex in the caudal part of the inferior parietal lobule, area PG, and the nearby cortex in the intraparietal sulcus (area LIP) via the second division of the superior longitudinal fasciculus (SLF II) in the monkey brain (Frey et al., 2014; Petrides & Pandya, 2009) and in the human brain (Margulies & Petrides, 2013; Petrides, 2014) (Figure 3.6). Importantly, area 45 is additionally linked with auditory association cortical areas in the mid-section of the superior temporal gyrus and adjacent multisensory cortex in the superior temporal sulcus and adjacent middle temporal gyrus via the extreme capsule fasciculus, a bundle of axons first discovered in the macaque monkey (Petrides & Pandya, 1988, 2009) and later confirmed to exist also in the human brain (Frey et al., 2008). Area 44 receives relatively minor connections from the mid-temporal auditory and multimodal temporal cortex via the extreme capsule fasciculus (Figure 3.6).

The accumulating recent evidence regarding the precise cortico-cortical connectivity of areas 44 and 45, which in the language-dominant hemisphere of the human brain are known as Broca's region, has revealed a much richer connectivity profile than what had traditionally been presented in standard textbooks of the neural basis of language. Under the influence of Geschwind (1970), the arcuate fasciculus, a bundle of axons arching around the end of the lateral fissure, was considered to be the main language pathway linking the temporal language comprehension region with the ventrolateral frontal region critical for speech production (Catani & Mesulam, 2008; Petrides, 2014). Diffusion MRI can easily reconstruct an arching set of fibers around the end of the lateral fissure, but it is not possible to establish the precise origin and termination of these fibers. Examination of the arcuate fasciculus in the macaque monkey brain has revealed a number of interesting details. First, the fibers originating from the crown of the most posterior part of the superior temporal gyrus do not terminate in area 44, but rather in the dorsal premotor region and adjacent dorsal prefrontal area 8Ad (Petrides & Pandya, 1988). Later, we learned that the neurons whose axons project to the ventrolateral frontal area 44 lie in the adjacent superior temporal sulcus (Frey et al., 2014; Petrides, 2014).

3.3 FUNCTIONAL IMPLICATIONS

What clues can the anatomical studies provide about the functional organization of the ventrolateral frontal region? First, it is clear that there is a distinct peri-central sensorimotor circuit formed by local association fibers that link primary motor area 4 and the adjacent caudal premotor area 6VC with the somatosensory cortex in the postcentral gyrus (Figure 3.5) and somatomotor areas in the nearby frontoparietal operculum; there are also longer connections from the motor region to the insula and the supplementary motor cortex. It is reasonable to assume that this local sensorimotor circuit underlies the most detailed aspects of orofacial articulation. Thus, lesions relatively limited to the peri-central sensorimotor cortex (including the frontoparietal operculum and adjacent insula) might be expected to result in articulatory deficits, and there is considerable evidence for such impairments in the literature (Ackermann & Riecker, 2004; Baldo et al., 2011; Dronkers, 1996; Graff-Radford et al., 2014).

The local peri-central sensorimotor circuit is succeeded anteriorly by the longer association fiber system of the third branch of the superior longitudinal fasciculus (SLF III) that links the anterior premotor cortical area 6VR and the transitional area 44 with the supramarginal cortex and the cortex in the anterior part of the intraparietal sulcus (Figure 3.5). The anterior part of the inferior parietal lobule (i.e., the supramarginal gyrus) is a multimodal cortical region with a focus on an integrated representation of the body. In the monkey, neurons in the anterior inferior parietal lobule exhibit complex body-centered responses (Hyvarinen & Shelepin, 1979; Leinonen, Hyvarinen, Nyman, & Linnankoski, 1979; Robinson & Burton, 1980; Taira, Mine, Georgopoulos, Murata, & Tanaka, 1990). It should be noted that, in this older macaque monkey literature, areas PF and PFG were referred to as area 7b. In the left hemisphere of the human brain, this parieto-frontal circuit formed by the third branch of the superior longitudinal fasciculus (SLF III) and perhaps the arcuate fasciculus has been shown to be involved in phonological processing (Saur et al., 2008). For instance, a recent functional neuroimaging study has provided strong evidence that the supramarginal gyrus in the left hemisphere is involved in phonological processing, but there was no evidence of such processing in the posterior inferior parietal lobule, such as in area PG in the left angular gyrus (Church, Balota, Petersen, & Schlaggar, 2011). Thus, we can assume that a phylogenetically old primate parieto-ventrolateral premotor circuit controlling orofacial action has been adapted in the language-dominant hemisphere of the human brain for phonological processing.

The parietal connectivity of prefrontal area 45 is different from that of area 44 (compare Figures 3.5 and 3.6). Area 45 is linked via the second branch of the superior longitudinal fasciculus (SLF II) with the caudal part of the inferior parietal region, which in the human brain constitutes the angular gyrus (area PG) and the adjacent intraparietal sulcal cortex (Petrides & Pandya, 1984, 2009). In the human brain, there is evidence for involvement of the left angular gyrus in semantic processing (Binder, Desai, Graves, & Conant, 2009) and reading (Segal & Petrides, 2013). Furthermore, area 45 has massive connections with the auditory association region of the superior temporal gyrus and the adjacent multisensory cortex in the superior temporal sulcus and adjacent middle temporal gyrus via the extreme capsule fasciculus (Frey et al., 2008; Petrides, 2014; Petrides & Pandya, 2009). The extreme capsule fasciculus, first discovered in the macaque monkey (Petrides & Pandya, 1988) and later shown to exist also in the human brain (Frey et al., 2008), has become the focus of intense interest in recent years as possibly the ventral language system underlying semantic processing (Saur et al., 2008). Area 45 is a rather special prefrontal area. In addition to its strong connections with auditory and multisensory lateral temporal cortex related to the semantic system and its connections with the angular gyrus, it has widespread and strong connections with virtually all the other prefrontal areas. This unique prefrontal cortical area has direct access to the mid-dorsolateral prefrontal monitoring system (areas 46 and 9/46), the posterior dorsolateral attentional systems (areas 8Av, 8Ad, and 8B), and orbitofrontal emotional and motivational control systems, while at the same time having access to auditory and semantic temporal and parietal information processing systems. Thus, it is in a critical position to mediate between widespread prefrontal control systems and the processing of language information at the highest levels. This unique connectional profile led us to refer to this area as the great prefrontal integrator (Petrides, 2014).

What might be the special role of ventrolateral prefrontal area 45 and the related nearby area 47/12? There is now considerable evidence that this ventrolateral prefrontal region is critical for the active controlled retrieval of information that lies in posterior parietal cortex and the lateral temporal cortex (Petrides, 1996, 2002). The role of area 45 in the left hemisphere of the human brain can be viewed as the adapting of its general role in the active controlled retrieval of information to the retrieval of verbal information in the language-dominant hemisphere. If this argument were to be true, one might expect the left ventrolateral prefrontal region (area 45) to be critical for the retrieval of verbal information acquired in specific contexts (Petrides, Alivisatos, & Evans, 1995), the retrieval of words that belong to particular categories (Amunts et al., 2004; Poldrack et al., 1999), the retrieval of synonyms within the same language or corresponding words across languages (Klein, Milner, Zatorre, Meyer, & Evans, 1995), and, more generally, processing in the semantic system (Saur et al., 2008).

In conclusion, in the ventrolateral frontal region of the primate brain there are three distinct distributed neural circuits. First, a circuit focused on the prefrontal granular cortical area 45 (and the adjacent prefrontal area 47/12) that, via its links with the angular gyrus of the inferior parietal lobule and the lateral temporal region (auditory, multisensory, and high-level visual processing), underlies the active controlled retrieval of semantic information (Badre & Wagner, 2007; Petrides, 1996, 2002). Second, a high-level premotor circuit focused on the transitional area 44 and the adjacent rostral premotor area 6VR, via links to the supramarginal gyrus and the caudal superior temporal gyrus, appears to underlie phonological processing in the language-dominant hemisphere of the human brain. Finally, the caudal premotor area 6VC and the primary motor cortex (area 4) together with their local somatic sensory connections underlie fine aspects of orofacial articulation. How can we link these findings to language production? We argue that, regardless of whether a verbal propositional utterance is self-generated or driven by an external question, it must be initiated by the selective controlled retrieval of information that is relevant to the question that the subject must address. This type of controlled retrieval depends on ventrolateral prefrontal cortical areas 45 and 47/12. These areas are linked, in turn, with transitional area 44, which by virtue of both its location and connections, may act as the intermediary between pure cognitive retrieval and motor articulation, such as the go-between truly prefrontal ventrolateral cortex (areas 45 and 47/12) and purely motor/premotor cortex (areas 6VC and 4). This transitional area, through its linkage with the multisensory but somatomotor focused supramarginal gyrus and the posterior superior temporal region, appears to form a critical phonological circuit. Finally, the sensorimotor peri-central circuit controls the details of articulation in speech. Thus, we can construe three levels of hierarchical processing leading from cognitive selection of what must be articulated (prefrontal areas 45 and 47/12) to high-level planning of motor/syntactical aspects of linguistic action (area 44 and rostral area 6VR) to the details of movement articulation (areas 4 and 6VC and their somatomotor connections).

3.4 NON-VENTROLATERAL PREFRONTAL AREAS AND THEIR POSSIBLE ROLE IN LANGUAGE

What might be the role of dorsolateral prefrontal areas lying on the middle and superior frontal gyrus in language? Damage to these areas yields severe impairments in certain aspects of working memory and in attention control, but there are no obvious impairments in language production or comprehension. However, one might expect the specific contribution of these areas to cognitive processing to be reflected in verbal information processing, particularly in the left hemisphere, because it is obvious that language processing involves virtually all aspects of higher cortical processing. For instance, it has been established that the mid-dorsolateral prefrontal cortical region (areas 46 and 9/46) is critical for the monitoring of information in working memory (Petrides, 2013). Damage to the mid-dorsolateral prefrontal cortex in the language-dominant left hemisphere yields an impairment in the monitoring of verbal information in working memory and neuroimaging studies show increased activity in the mid-dorsolateral prefrontal region during verbal working memory tasks when the monitoring requirements are taxed (Petrides, Alivisatos, Meyer, & Evans, 1993). Of special interest is a particular component of the mid-dorsolateral prefrontal region, area 9/46v, which is strongly connected with areas 44 and 45 and also with the supramarginal gyrus (Figure 3.5). This area is an essential component of the mid-dorsolateral prefrontal region for the monitoring of verbal information.

Acknowledgments

I thank Jennifer Novek for help with the preparation of figures. This work was supported by grants from the Canadian Institutes of Health Research and the Natural Sciences and Engineering Research Council of Canada.

References

Ackermann, H., & Riecker, A. (2004). The contribution of the insula to motor aspects of speech production: A review and a hypothesis. *Brain and Language*, *89*, 320–328.

Amiez, C., & Petrides, M. (2014). Neuroimaging evidence of the anatomo-functional organization of the human cingulate motor areas. *Cerebral Cortex*, *24*, 563–578.

Amunts, K., Schleicher, A., Burgel, U., Mohlberg, H., Uylings, H. B. M., & Zilles, K. (1999). Broca's region revisited: Cytoarchitecture and intersubject variability. *Journal of Comparative Neurology*, *412*, 319–341.

Amunts, K., Weiss, P. H., Mohlberg, H., Pieperhoff, P., Eickhoff, S., Gurd, J. M., et al. (2004). Analysis of neural mechanisms underlying verbal fluency in cytoarchitectonically defined stereotaxic space—The roles of Brodmann areas 44 and 45. *NeuroImage*, *22*, 42–56.

Badre, D., & Wagner, A. D. (2007). Left ventrolateral prefrontal cortex and the cognitive control of memory. *Neuropsychologia*, *45*, 2883–2901.

Baldo, J. V., Wilkins, D. P., Ogar, J., Willock, S., & Dronkers, N. F. (2011). Role of the precentral gyrus of the insula in complex articulation. *Cortex*, *47*, 800–807.

Binder, J. R., Desai, R. H., Graves, W. W., & Conant, L. L. (2009). Where is the semantic system? A critical review and meta-analysis of 120 functional neuroimaging studies. *Cerebral Cortex*, *19*, 2767–2796.

Brodmann, K. (1909). *Vergleichende Lokalisationslehre der Grosshirnrinde in ihren Prinzipien dargestellt auf Grund des Zellenbauses*. Leipzig: Barth.

Campbell, J. S. W., & Pike, G. B. (2014). Potential and limitations of diffusion MRI tractography for the study of language. *Brain and Language*, *131*, 65–73.

Catani, M., & Mesulam, M. (2008). The arcuate fasciculus and the disconnection theme in language and aphasia: History and current state. *Cortex*, *44*, 953–961.

Chapados, C., & Petrides, M. (2013). Impairment only on the fluency subtest of the frontal assessment battery after prefrontal lesions. *Brain*, *136*, 2966–2978.

Church, J. A., Balota, D. A., Petersen, S. E., & Schlaggar, B. L. (2011). Manipulation of length and lexicality localizes the functional neuroanatomy of phonological processing in adult readers. *Journal of Cognitive Neuroscience*, *23*, 1475–1493.

Dronkers, N. F. (1996). A new brain region for coordinating speech articulation. *Nature*, *384*, 159–161.

Duffau, H., Moritz-Gasser, S., & Mandonnet, E. (2014). A re-examination of neural basis of language processing: Proposal of a dynamic hodotopical model from data provided by brain stimulation mapping during picture naming. *Brain and Language*, *131*, 1–10.

Dum, R. P., & Strick, P. L. (1993). Cingulate motor areas. In B. A. Vogt, & M. Gabriel (Eds.), *Neurobiology of cingulate cortex and limbic thalamus. A comprehensive handbook* (pp. 415–441). New York, NY: Springer Science and Business Media.

Economo, C., & Koskinas, G. N. (1925). *Die Cytoarchitektonik der Hirnrinde deserwachsenen Menschen*. Wien und Berlin: Springer.

Frey, S., Campbell, J. S., Pike, G. B., & Petrides, M. (2008). Dissociating the human language pathways with high angular resolution diffusion fiber tractography. *Journal of Neuroscience*, *28*, 11435–11444.

Frey, S., Mackey, S., & Petrides, M. (2014). Cortico-cortical connections of areas 44 and 45B in the macaque monkey. *Brain and Language*, *131*, 36–55.

Friederici, A. (2011). The brain basis of language processing: From structure to function. *Physiological Reviews*, *91*, 1357–1392.

García-Cabezas, M. A., & Barbas, H. (2014). Area 4 has layer IV in adult primates. *European Journal of Neuroscience*, *39*, 1824–1834.

Geschwind, N. (1970). The organization of language and the brain. *Science*, *170*, 940–944.

Goldberg, G. (1985). Supplementary motor area structure and function: Review and hypotheses. *Behavioral Brain Sciences*, *8*, 567–588.

Graff-Radford, J., Jones, D. T., Strand, E. A., Rabinstein, A. A., Duffy, J. R., & Josephs, K. A. (2014). The neuroanatomy of pure apraxia of speech in stroke. *Brain and Language*, *129*, 43–46.

Grodzinsky, Y. (2000). The neurology of syntax: Language use without Broca's area. *Behavioral and Brain Sciences*, *23*, 1–71.

Hyvarinen, J., & Shelepin, Y. (1979). Distribution of visual and somatic functions in the parietal associative areas 7 of the monkey. *Brain Research*, *169*, 561–564.

Kelly, C., Uddin, L. Q., Shehzad, Z., Margulies, D. S., Castellanos, F. X., Milham, M. P., et al. (2010). Broca's region: Linking human brain functional connectivity data and non-human primate tracing anatomy studies. *European Journal of Neuroscience*, *32*, 383–398.

Klein, D., Milner, B., Zatorre, R. J., Meyer, E., & Evans, A. C. (1995). The neural substrates underlying word generation: A bilingual functional-imaging study. *Proceedings of the National Academy of Sciences of the United States of America*, *92*, 2899–2903.

Klingler, J. (1935). Erleichterung der makroskopischen Präparation des Gehirns durch den Gefrierprozeß. *Schweizer Archiv für Neurologie und Psychiatrie*, *36*, 247–256.

Krainik, A., Lehéricy, S., Duffau, H., Capelle, L., Chainay, H., Cornu, P., et al. (2003). Post-operative speech disorder after medial frontal surgery: Role of the supplementary motor area. *Neurology*, *60*, 587–594.

Leinonen, L., Hyvarinen, J., Nyman, G., & Linnankoski, I. (1979). Functional properties of neurons in lateral part of associative area 7 in awake monkeys. *Experimental Brain Research*, *34*, 299–320.

Ludwig, E., & Klingler, J. (1956). *Atlas Cerebri Humani*. Basel: Karger.

Makris, N., Kennedy, D. N., McInerney, S., Sorensen, A. G., Wang, R., Caviness, V. S., Jr., et al. (2005). Segmentation of subcomponents within the superior longitudinal fascicle in humans: A quantitative, in vivo, DT-MRI study. *Cerebral Cortex*, *15*, 854–869.

Margulies, D. S., & Petrides, M. (2013). Distinct parietal and temporal connectivity profiles of ventrolateral frontal areas involved in language production. *Journal of Neuroscience*, *33*, 16846–16852.

Matelli, M., Luppino, G., & Rizzolatti, G. (1985). Patterns of cytochrome oxidase activity in the frontal agranular cortex of the macaque monkey. *Behavioural Brain Research*, *18*, 125–136.

Mohr, J. P. (1976). Broca's area and Broca's aphasia. In H. Whitaker, & H. A. Whitaker (Eds.), *Studies in neurolinguistics* (Vol. 1, pp. 201–233). New York, NY: Academic Press.

Mohr, J. P., Pessin, M. S., Finkelstein, S., Funkenstein, H. H., Duncan, G. W., & Davis, K. R. (1978). Broca aphasia: Pathologic and clinical. *Neurology*, *28*, 311–324.

Nachev, P., Kennard, C., & Husain, M. (2008). Functional role of the supplementary and pre-supplementary motor areas. *Nature Reviews Neuroscience*, *9*, 856–869.

Ojemann, G. (1992). Localization of language in frontal cortex. *Advances in Neurology*, *57*, 361–368.

Ojemann, G., Ojemann, J., Lettich, E., & Berger, M. (1989). Cortical language localization in left, dominant hemisphere. An electrical stimulation mapping investigation in 117 patients. *Journal of Neurosurgery*, *71*, 316–326.

Paus, T., Petrides, M., Evans, A. C., & Meyer, E. (1993). Role of the human anterior cingulate cortex in the control of oculomotor, manual and speech responses: A positron emission tomography study. *Journal of Neurophysiology*, *70*, 453–469.

Penfield, W., & Boldrey, E. (1937). Somatic motor and sensory representation in cerebral cortex of man as studied by electrical stimulation. *Brain*, *60*, 389–443.

Penfield, W., & Rasmussen, T. (1950). *The cerebral cortex of man. A clinical study of localization of function*. New York, NY: Macmillan.

Penfield, W., & Roberts, L. (1959). *Speech and brain mechanisms*. Princeton, NJ: Princeton University Press.

Penfield, W., & Welch, K. (1951). The supplementary motor area of the cerebral cortex. A clinical and experimental study. *Archives of Neurology and Psychiatry*, *66*, 289–317.

Petrides, M. (1996). Specialized systems for the processing of mnemonic information within the primate frontal cortex. *Philosophical Transactions of the Royal Society, London B*, *351*, 1455–1462.

Petrides, M. (2002). The mid-ventrolateral prefrontal cortex and active mnemonic retrieval. *Neurobiology of Learning and Memory*, *78*, 528–538.

Petrides, M. (2013). The mid-dorsolateral prefronto-parietal network and the epoptic process. In D. T. Stuss, & R. T. Knight (Eds.), *Principles of frontal lobe function* (2nd ed., pp. 79–89). New York, NY: Oxford University Press.

Petrides, M. (2014). *Neuroanatomy of language regions of the human brain*. New York, NY: Academic Press.

Petrides, M., Alivisatos, B., & Evans, A. C. (1995). Functional activation of the human ventrolateral frontal cortex during mnemonic retrieval of verbal information. *Proceedings of the National Academy of Sciences of the United States of America*, *92*, 5803–5807.

Petrides, M., Alivisatos, B., Meyer, E., & Evans, A. C. (1993). Functional activation of the human frontal cortex during the performance of verbal working memory tasks. *Proceedings of the National Academy of Sciences of the United States of America*, *90*, 878–882.

Petrides, M., Cadoret, G., & Mackey, S. (2005). Orofacial somatomotor responses in the macaque monkey homologue of Broca's area. *Nature*, *435*, 1235–1238.

Petrides, M., & Pandya, D. N. (1984). Projections to the frontal cortex from the posterior parietal region in the rhesus monkey. *Journal of Comparative Neurology*, *228*, 105–116.

Petrides, M., & Pandya, D. N. (1988). Association fiber pathways to the frontal cortex from the superior temporal region in the rhesus monkey. *Journal of Comparative Neurology*, *273*, 52–66.

Petrides, M., & Pandya, D. N. (1994). Comparative architectonic analysis of the human and the macaque frontal cortex. In F. Boller, & J. Grafman (Eds.), *Handbook of neuropsychology* (Vol. 9, pp. 17–58). Amsterdam: Elsevier.

Petrides, M., & Pandya, D. N. (2002). Comparative architectonic analysis of the human and the macaque ventrolateral prefrontal cortex and corticocortical connection patterns in the monkey. *European Journal of Neuroscience*, *16*, 291–310.

Petrides, M., & Pandya, D. N. (2009). Distinct parietal and temporal pathways to the homologues of Broca's area in the monkey. *Public Library of Science Biology*, *7*, e1000170.

Poldrack, R. A., Wagner, A. D., Prull, M. W., Desmond, J. E., Glover, G. H., & Gabrieli, J. D. E. (1999). Functional specialization for semantic and phonological processing in the left inferior prefrontal cortex. *NeuroImage*, *10*, 15–35.

Pons, T. P., & Kaas, J. H. (1986). Corticocortical connections of area 2 of somatosensory cortex in macaque monkeys: A correlative anatomical and electrophysiological study. *Journal of Comparative Neurology*, *248*, 313–335.

Rasmussen, T., & Milner, B. (1975). Clinical and surgical studies of the cerebral speech areas in man. In K. J. Zulch, O. Creutzfeldt, & G. C. Galbraith (Eds.), *Cerebral localization* (pp. 238–257). New York, NY: Springer-Verlag.

Robinson, C. J., & Burton, H. (1980). Organization of somatosensory receptive fields in cortical areas 7b, retroinsula, postauditory, granular insula of M. fascicularis. *Journal of Comparative Neurology*, *192*, 69–92.

Rostomily, R. C., Berger, M. S., Ojemann, G. A., & Lettich, E. (1991). Postoperative deficits and functional recovery following removal of tumors involving the dominant hemisphere supplementary motor area. *Journal of Neurosurgery*, *75*, 62–68.

Saur, D., Kreher, B. W., Schnell, S., Kummerer, D., Kellmeyer, P., Vry, M. S., et al. (2008). Ventral and dorsal pathways for language. *Proceedings of the National Academy of Sciences of the United States of America*, *105*, 18035–18040.

Segal, E., & Petrides, M. (2013). Functional activation during reading in relation to the sulci of the angular gyrus region. *European Journal of Neuroscience*, *38*, 2793–2801.

Stuss, D. T., & Knight, R. T. (Eds.), (2013). *Principles of frontal lobe function* (2nd ed.). New York, NY: Oxford University Press.

Taira, M., Mine, S., Georgopoulos, A. P., Murata, A., & Tanaka, Y. (1990). Parietal cortex neurons of the monkey related to the visual guidance of hand movement. *Experimental Brain Research*, *83*, 29–36.

Thomas, C., Ye, F. Q., Irfanoglu, M. O., Modi, P., Saleem, K. S., Leopold, D. A., et al. (2014). Anatomical accuracy of brain connections derived from diffusion MRI tractography is inherently limited. *Proceedings of the National Academy of Sciences of the United States of America*, *111*, 16574–16579.

Vogt, B. A., & Pandya, D. N. (1978). Cortico-cortical connections of somatic sensory cortex (areas 3, 1 and 2) in the rhesus monkey. *Journal of Comparative Neurology*, *177*, 179–191.

Walker, A. E. (1940). A cytoarchitectural study of the prefrontal area of the macaque monkey. *Journal of Comparative Neurology*, *73*, 59–86.

Zhu, Z., Hagoort, P., Zhang, J. X., Feng, G., Chen, H.-C., Bastiaansen, M. C. M., et al. (2012). The anterior left inferior frontal gyrus contributes to semantic unification. *NeuroImage*, *60*, 2230–2237.

CHAPTER

4

On the Neuroanatomy and Functional Role of the Inferior Parietal Lobule and Intraparietal Sulcus

Ferdinand Christoph Binkofski[1,2], *Juliane Klann*[1] *and Svenja Caspers*[2]

[1]Section Clinical Cognition Sciences, Department of Neurology, University Hospital, RWTH Aachen University, Aachen, Germany; [2]Institute of Neuroscience and Medicine (INM-1), Research Centre Jülich, Jülich, Germany

4.1 GROSS ANATOMY OF THE IPL AND IPS

The inferior parietal lobule (IPL) covers the ventral aspect of the posterior parietal cortex (Figure 4.1, top right). The rostral and dorsal borders are well-defined, being the postcentral sulcus and the intraparietal sulcus (IPS), respectively. The ventral rim is provided by the Sylvian fissure in its rostral aspect, but there is no such clear definition for the caudal aspect, where the IPL directly merges into the posterior parts of the superior and middle temporal gyri. The caudal border of the IPL is equally loosely defined because there is no prominent demarcation from the occipital lobe. The IPL is composed of two gyri, the supramarginal gyrus rostrally and the angular gyrus caudally. These are separated by the primary intermediate sulcus of Jenssen, which is highly variable across brains because it is only present in 24% of the right hemisphere and 80% of the left hemisphere (Ono, Kubik, & Abernathey, 1990). The supramarginal gyrus is situated dorsal to the caudal aspect of the Sylvian fissure. The angular gyrus typically bends around the angular sulcus, which is a continuation of the dorsal-posterior segment of the superior temporal sulcus.

The IPS is the most prominent sulcus within the posterior parietal cortex. It separates the superior from the IPL, and thus runs mainly in the rostrocaudal direction. In most cases, the IPS is directly connected to the postcentral sulcus (64% of the right hemisphere and 72% of the left hemisphere; Ono et al., 1990) or starts immediately adjacent to it (36% of the right hemisphere and 28% of the left hemisphere; Ono et al., 1990). It usually ends in or close to the transverse occipital sulcus and the lateral aspect of the parieto-occipital sulcus. The IPS typically exists as one continuous sulcus (28% of the right hemisphere and 72% of the left hemisphere; Ono et al., 1990) or as a sulcus with two segments (68% of the right hemisphere and 28% of the left hemisphere; Ono et al., 1990).

4.2 MODERN PARCELLATION OF THE IPL AND IPS

4.2.1 Human

Microstructural parcellations of the IPL can already be found in most of the classical brain maps of the early 20th century. They are based on cytoarchitectonic as well as myeloarchitectonic criteria. Campbell (1905) described one common type of parietal architecture that encompassed the IPL as well as the IPS and the superior parietal lobule (SPL). The first distinction within the IPL was mentioned by Smith (1907), who found not only a division into rostral and caudal IPL (areas pariB and pariA) according to the two main gyri but also a further subdivision within rostral IPL, covering the lateral aspect of the parietal operculum (area pariC). Two years later, Brodmann (1909) provided the most famous parcellation of the IPL, identifying the well-known bipartition into a rostral and a caudal part (BA 40 and BA 39). Using myeloarchitectonic criteria, Vogt (1911) and

Neurobiology of Language. DOI: http://dx.doi.org/10.1016/B978-0-12-407794-2.00004-3

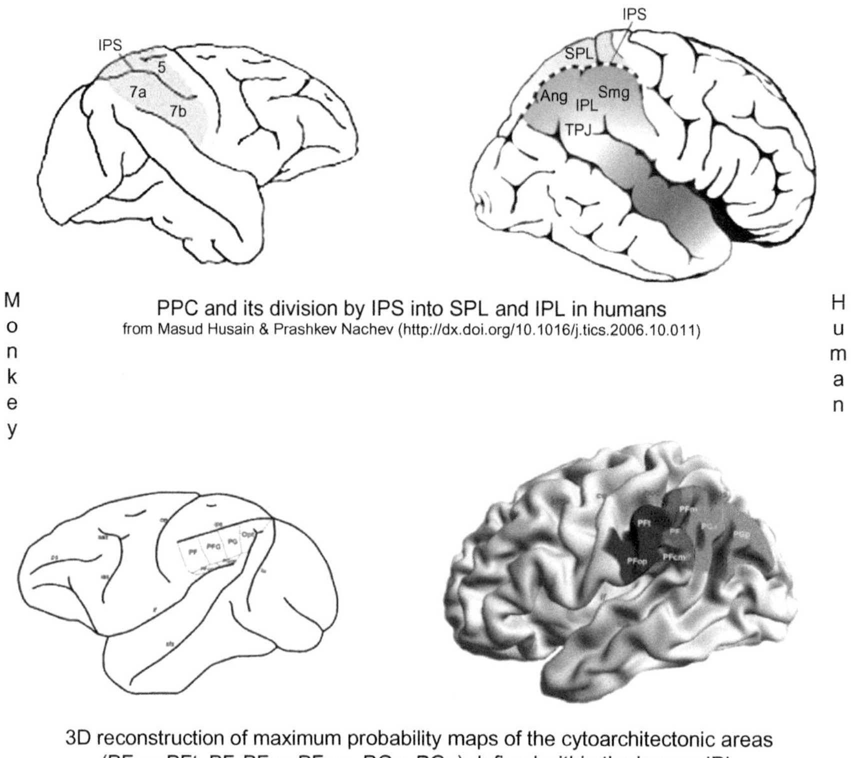

FIGURE 4.1 Gross anatomy (top) and parcellation (bottom) of IPL and IPS in humans (right) and monkeys (left). *From Caspers et al. (2006, 2008, 2011) and Husain & Nachev (2006) with permission from the publisher.*

Vogt and Vogt (1919) found four major areas within the IPL (areas 74, 88, 89, and 90) with several subareas with modulations of the overall myeloarchitectonic pattern. In 1920, Flechsig proposed a rostrocaudal tripartition of the IPL (areas 26, 37, and 42). von Economo and Koskinas (1925) again adopted the idea of Brodmann of having two major cytoarchitectonic patterns within the IPL (rostral area parietal F (PF) and caudal area parietal G (PG)). Additionally, they identified four local variations of the PF pattern, that is, an opercular area PFop, caudal to area PFcm (columnata magnocellularis), a rostral area PFt (tenuicorticalis), and a caudal variation at the border to PG, area PFm (magnocellularis).

Later, these pioneering maps were largely replicated. Gerhardt (1940) and Batsch (1956) mainly adopted the parcellation scheme of Vogt and Vogt (1919), amending it by additional subdivisions of the areas. Similar results were obtained by Hopf and Vitzthum (1957) and Hopf (1969) who suggested a similar complex myeloarchitectonic organization within the IPL. Sarkissov, Filimonoff, and Preobrashenskaya (1949) again built on Brodmann's bipartition and described additional subfields within these areas (40op, 40p, 40i, and 39p).

In 1984, Eidelberg and Galaburda described a rostrocaudal sequence of areas within the IPL (areas PF, parietal F-G (PFG), PG, and occipito-parietal G (OPG)) that largely resembled that found in macaques at approximately the same time (Pandya & Seltzer, 1982). Additionally, they described an asymmetry in the caudal part of the IPL: the volume of area PG was larger in the left hemisphere than in the right hemisphere, but only in those brains in which the planum temporale was also larger in the left as compared with the right hemisphere. They assumed this asymmetry to be relevant for the left-lateralized language network because it resembled known asymmetries toward the left in language-relevant areas.

Recently, a novel parcellation scheme of the IPL into seven cytoarchitectonically distinct areas was proposed (Caspers et al., 2006, 2008) using an observer-independent, quantitative, statistically testable mapping approach in a sample of 10 postmortem brains (Figure 4.1, bottom right). There are five areas in a rostrocaudal sequence (areas PFt, PF, PFm, PGa, and PGp) and two ventral areas in the caudal aspect of the parietal operculum (areas PFop and PFcm). All five areas on the lateral surface of the IPL abut on areas of the IPS, whereas the ventral areas directly border the areas of the secondary somatosensory cortex (i.e., areas OP1 and OP4; Eickhoff, Amunts, Mohlberg, & Zilles, 2006; Eickhoff, Schleicher, Zilles, & Amunts, 2006). The most rostro-dorsal area PFt additionally borders area 2 of the primary somatosensory cortex. It is characterized by well-separated layers: a cell-dense layer II, mid to large pyramidal cells in lower layer III, and layer V is separated into an upper part and lower part. Area PFop is located immediately ventral to it. The layers are less well-separated, layer IV is sparsely developed, and layer V cannot be subdivided further. Caudal to these areas, areas PF and PFcm abut in the dorsal and ventral aspects, respectively. Area PF is overall very cell-dense; neurons of neighboring layers, particularly those of layers III/IV, IV/V, and V/VI, are heavily intermingled with each other, and there is a pronounced superficial-to-deep increase in pyramidal cell size in layer III. Area PFcm is characterized by a pronounced columnar arrangement of the cells across layers III to VI and large pyramidal cells in lower layer III. Area PFm is located at the transition between the supramarginal and angular gyrus, which is also reflected at the cytoarchitectonic level, having characteristics of both. Layer IV is better developed, there are large pyramidal cells in lower layer III, and layer V is again separable into an upper and lower part. The angular gyrus is covered by two areas: PGa and PGp. Area PGa is characterized by a shift of layer IV to a more superficial position, causing the supragranular layers to be smaller, a thin layer II, a sudden change in pyramidal cell size from upper to lower layer III, and a well-developed layer IV. Area PGp appears more homogeneous, with an overall decrease in cell size, layers II and III are barely separable from each other, and layer IV stands out very clearly against lower layer III and upper layer V.

These seven IPL areas differ not only in cytoarchitectonic characteristics but also in their receptor distribution patterns, reflecting their molecular architecture (Caspers et al., 2013). The concentrations of the glutamatergic α-amino-3-hydroxy-5-methyl-4-isoxazolepropionic acid (AMPA), the GABAergic $GABA_A$, and the cholinergic nicotinic receptors increase from rostral to caudal areas, whereas the concentrations of the serotoninergic $5\text{-}HT_{1A}$ and $5\text{-}HT_2$ receptors decrease. Investigating the similarities in receptor distributions between the IPL areas revealed a tripartition into a rostral (areas PFt, PFop, and PFcm), middle (areas PF and PFm), and caudal part (areas PGa and PGp; for more details, see Figure 8 in Caspers et al., 2013). All parts show a receptor distribution pattern similar to that of Broca's region, whereas middle IPL showed additional similarity with the caudolateral aspect of the SPL, and caudal IPL was also similar to extrastriate visual areas.

Considerably less is known about subdivisions within the IPS. Except for a quite detailed parcellation into 15 cytoarchitectonic areas by Gerhardt (1940), no other "classical" brain map provided information about the architecture within the IPS. Recently, the anterior part of the IPS has been subdivided into three cytoarchitectonically distinct areas: areas hIP1 and hIP2 at the bottom and the lateral wall (adjacent to the IPL) of the sulcus, respectively (Choi et al., 2006), and area hIP3 on the medial wall (adjacent to the SPL; Scheperjans, Hermann, et al., 2008; Scheperjans, Eickhoff, et al., 2008). hIP1 is characterized by larger pyramidal cells in lower layer III, which are clearly separated from thin layer IV by a cell-sparse thin stripe, as well as by homogeneous layer V with small pyramidal cells spread over the whole extent of the layer. hIP2 contains smaller pyramidal cells within all layers; layer II appears more cell-dense as compared with hIP1, and layers V and VI form one continuous infragranular band of cells. In hIP3, layers II and III are clearly separable from each other, large pyramidal cells dominate the architecture within lower layer III, and layer V is comparably cell-sparse. Based on functional studies, additional areas similar to those described in monkeys are expected for the posterior IPS (Grefkes & Fink, 2005; Seitz & Binkofski, 2003). At the time of writing, this needs to be elucidated in future cytoarchitectonical studies.

4.2.2 Monkey

Because quite a lot of evidence regarding the functions and the structural–functional relations of the IPL and IPS have been obtained from studies of macaque monkeys, an overview of the respective microstructural parcellations is provided here as a basis for comparison with the human data.

Brodmann (1905) identified one large area within the IPL of the macaque, BA 7. This area has been subdivided into rostral area 7b and caudal area 7a (Vogt & Vogt, 1919; Figure 4.1, top left). Later, a rostrocaudal sequence of areas within the IPL, namely areas PF, PFG, PG, and Opt, and the adjacent caudal part of the parietal operculum, areas PFop and PGop, were described (Gregoriou, Borra, Matelli, & Luppino, 2006; Pandya & Seltzer, 1982; Figure 4.1, bottom left).

Within the IPS, several areas in the lateral and medial wall have been identified. In the anterior part, there are the anterior, ventral, and medial intraparietal areas (AIP, VIP, MIP), whereas the lateral, posterior, and caudal intraparietal areas (LIP, PIP, CIP) are located in the posterior part (Lewis & Van Essen, 2000; Rizzolatti & Luppino, 2001; Ungerleider & Desimone, 1986). Further subdivisions of these areas have been partially suggested using immunohistochemical staining, that is, a medial and lateral aspect of VIP (VIPm and VIPl) or a ventral and dorsal part of LIP (LIPv and LIPd) (Lewis & Van Essen, 2000).

4.3 CONNECTIVITY OF THE IPL AND IPS

4.3.1 Human

The IPL is anatomically connected mainly via association fibers to other cortical areas. Two major fiber bundles dominate the architecture of the white matter underlying the IPL: the arcuate and the superior longitudinal fasciculus (branch III). The arcuate fasciculus was found to be lateralized to the left and to be separable into three major parts: one continuous part, running between Broca's region and posterior temporal cortex, and a parallel path (separated into two parts), that connects Broca's region with rostral and middle IPL and the latter with posterior temporal cortex (Catani, Jones, & Ffytche, 2005; Parker et al., 2005; Powell et al., 2006). Therefore, this pathway was often referred to as the main connection structure within the language system of the brain. The superior longitudinal fasciculus (branch III) in principle provides comparable, but more widespread, connections between the IPL and the frontal cortex (Rushworth, Behrens, & Johansen-Berg, 2006). However, this pathway seems to be lateralized to the right, which is assumed to be the structural connectional architectural framework for the visuospatial network (Thiebaut de Schotten et al., 2011). Regarding this discussion of lateralization, it needs to be stressed that the arcuate fasciculus has repeatedly been reported as being a part of the superior longitudinal fasciculus (Martino et al., 2013).

Via these fiber bundles, the IPL areas entertain distinct connections with other cortical and subcortical areas. Rostral IPL areas are mainly connected to inferior frontal, premotor, primary and secondary somatosensory, rostral superior parietal, and posterior temporal areas. Moving to middle and caudal IPL areas, the connection patterns shift to more lateral and medial prefrontal, caudal SPL, lateral occipital, posterior cingulate, and anterior middle and inferior temporal areas (Caspers et al., 2011). Caudal IPL is also strongly connected to the parahippocampal gyrus via the inferior longitudinal fascicle (Rushworth et al., 2006), the hippocampus, and the basal ganglia (Uddin et al., 2010).

Anatomical connectivity across the whole brain has been used to parcellate the IPL into areas within distinct connection patterns. This revealed a sequence of five areas within the left IPL and the right IPL, which largely corresponded to the cytoarchitectonic areas PFop, PFt, PFm, PGa, and PGp (Mars et al., 2011; Wang et al., 2012). With a similar approach, Ruschel et al. (2013) identified a tripartition of the IPL, which is in line with the subdivision of the IPL based on similarities in receptor distribution patterns. Using whole-brain functional connectivity patterns as a basis for the parcellation, a subdivision into seven functionally distinct clusters was identified within the IPL, which could be grouped into rostral, middle, and caudal groups of clusters (Zhang & Li, 2014). This corresponds well to the parcellations obtained by the postmortem cytoarchitectonic and receptorarchitectonic studies (Caspers et al., 2006, 2013).

The functional connectivity patterns of the IPL as obtained in the resting state were found to be similar to the structural ones. Rostral IPL is functionally connected to inferior frontal and adjacent operculum and supplementary motor and posterior temporal cortex, whereas caudal IPL is connected with lateral and medial prefrontal, posterior cingulate, anterior middle, and inferior temporal cortex, and parahippocampal gyrus (Mars et al., 2011; Yeo et al., 2011).

Within the IPS, major differences in connectivity patterns can be found between its anterior and posterior parts. The anterior IPS has predominant fiber tracts to prefrontal regions, whereas the posterior IPS is mainly connected to the posterior superior temporal gyrus and retinotopically defined visual areas of the occipital cortex (Bray, Arnold, Iaria, & MacQueen, 2013; Greenberg et al., 2012). Specifically, areas hIP1 and hIP2 of anterior IPS were found to be mainly connected with the ventral premotor cortex and the middle frontal gyrus, whereas area hIP3 was already more connected with extrastriate areas, thus resembling the major connections of the posterior IPS. hIP1 was additionally connected with the insular cortex (Uddin et al., 2010). All parts of the IPS have comparable connections with the striatum and the thalamus.

4.3.2 Monkey

Using invasive tracer techniques, the connectivity patterns of the IPL and IPS areas in the monkey have been studied extensively. This section provides an overview of the major results to allow for comparisons with the indirect measure of structural connectivity obtained in humans via diffusion imaging.

The rostral part of macaque IPL (area 7b) is reversely connected with the insular and retroinsular cortex. It projects to secondary somatosensory and superior parietal cortex, anterior intraparietal areas AIP and VIP, as well as premotor area F5. The rostral-most area PF is additionally connected to premotor area F4, whereas area PFG is additionally connected with prefrontal area 46v, orbitofrontal cortex, and posterior superior temporal cortex. The caudal part of the IPL, area 7a, is strongly connected to IPS areas VIP, MIP, LIP, and PIP, parieto-occipital areas, superior, middle and medial temporal areas, as well as prefrontal area 46 and the frontal eye field. Subdivision PG is mainly connected with intraparietal area MIP, insular and retroinsular cortex, and middle temporal cortex, whereas caudal-most area Opt is more strongly connected with intraparietal area LIP, medial superior parietal and parieto-occipital cortex, as well as premotor area F7 (Andersen, Asanuma, Essick, & Siegel, 1990; Cavada & Goldman-Rakic, 1989a, 1989b; Gregoriou et al., 2006; Mesulam, van Hoesen, Pandya, & Geschwind, 1977; Neal, Pearson, & Powell, 1990a, 1990b; Rozzi et al., 2006; Seltzer & Pandya, 1984).

The IPS areas are densely interconnected among each other and with the surrounding areas of the IPL and SPL. Beyond that, the anterior IPS areas AIP, VIP, and PEip are additionally strongly connected with primary and secondary somatosensory cortex, different areas of the premotor and part of the prefrontal cortex, as well as the posterior temporal cortex. The posterior IPS areas MIP, LIP, and PIP are predominantly connected to anterior and medial temporal cortex as well as striate and extrastriate visual areas (Andersen et al., 1990; Barbas, 1988; Borra et al., 2008; Felleman & Van Essen, 1991; Matelli, Camarda, Glickstein, & Rizzolatti, 1988; Tanné-Gariepy, Rouiller, & Boussaoud, 2002; Ungerleider & Desimone, 1986)

4.4 ANATOMICAL DIFFERENCES BETWEEN HUMANS AND MONKEYS

Comparing the microstructural gray matter parcellations between monkeys and humans gives the impression of a largely similar organization of the IPL between these species. Furthermore, there is converging evidence based on structural and functional connectivity of the IPL areas that several networks are largely preserved across species. But some networks might have evolved more from monkeys to humans than others. In particular, the connections between IPL and rostral aspects of prefrontal cortex, which could be found in humans but have not been reported in monkeys, give rise to the idea that this network evolved in accordance with the pronounced development of the frontal lobe. Comparing the anatomical connectivity of human and monkey, IPL points to potential homology between human areas PFt and PF with monkey area PF, human area PFm with monkey area PFG, human area PGa and monkey area Opt, and human area PGp (Caspers et al., 2011). This anatomical connectivity and topological homology only provide hints to answers to the question of interspecies homologies. Studies on the role of monkey areas PF and PFG and human areas PFt and PF in action processing (Bonini et al., 2010; Caspers, Zilles, Laird, & Eickhoff, 2010; Rozzi, Ferrari, Bonini, Rizzolatti, & Fogassi, 2008) and tool use (Orban & Rizzolatti, 2012; Peeters et al., 2009) provide additional hints with regard to a functional homology.

It needs to be stressed, though, that a direct comparison between the species based on the functional role of these areas remains difficult, because physiological properties of cells are not available, functional experiments are not always comparable, and lesions to the IPL do not necessarily cause the same symptoms in monkeys and humans (Caminiti et al., 2010).

4.5 FUNCTIONS AND FUNCTIONAL CONNECTIVITY OF THE IPL AND IPS

In humans, the parietal cortex is involved in several functions, including sensory-motor control, motor skills, object and tool use, spatial reception, and language, speech, verbal working memory, and number processing. Similar to the comparable anatomical parcellation, there is some overlap with the functional role of homologue regions in the monkey brain.

4.5.1 Language and Speech

Besides classical perisylvian regions, language and speech functions are also related to left supramarginal (rostral IPL—PFop, PFt, and PFcm) and angular gyri (caudal IPL—PGa and PGp), which are, among others, anatomically interconnected with frontal and temporal parts of the perisylvian language network (Caspers et al., 2011; Seghier, 2013). The cytoarchitectonical parcellation of the IPL into seven different areas as demonstrated motivates the idea of parallel functional subdivisions. Individual language studies have documented the parietal cortex to be involved in diverse functions and processes as reaching from the selection of articulatory gestures (Tremblay & Gracco, 2010) to verbal integration within complex comprehensive contexts like sentences (Lerner, Honey, Silbert, & Hasson, 2011).

Convergent with results in visual perception, recent imaging and electrophysiological studies resulted in growing evidence for two major pathways for auditory perception in humans as well as in the monkey; connecting temporal auditory and frontal regions seems to be supported by two major pathways. Data are in favor of a "ventral" pathway, which may be dedicated to the projection from the primary auditory cortex to prefrontal regions along the superior temporal gyrus via the uncinate fasciculus, and a "dorsal" stream connecting the same regions but involving the IPL. The anatomical pendant of this dual route model of auditory language input was displayed for the first time by using diffusion Magnet Resonance Imaging (MRI) tractography by Parker et al. (2005). More specifically, they found white matter connections along the dorsal stream connecting classical speech areas and the IPL in the left hemisphere, which hints at a crucial role of the IPL in speech processing. Notably, this dorsal stream is anatomically represented by the arcuate fasciculus, which interconnects BA 44 and BA 45 with IPL (Parker et al., 2005; Saur et al., 2008; Weiller, Bormann, Saur, Musso, & Rijntjes, 2011), and with posterior superior and middle temporal cortex. This is in line with the predictions of the DIVA (Directions Into Velocities of Articulators) model of speech production (Guenther, 2006), which maintains a feedback process for motor speech planning relying on Broca's area and involving a somatosensory error map that lies within the rostral part of the IPL.

Apart from that, there is robust proof for the angular gyrus (PGa and PGp as caudal parts of the IPL; Seghier, 2013) being involved in reading, writing, naming, and verbal repetition (Dronkers, Wilkins, Van Valin, Redfern, & Jaeger, 2004; Henseler, Regenbrecht, & Obrig, 2014; Zukic, Mrkonjic, Sinanovic, Vidovic, & Kojic, 2012). Lesion studies demonstrate converging evidence for dysgraphia/agraphia, dyslexia/agraphia, and anomia resulting from damage to left angular gyrus. Dysgraphia as part of Gerstmann's syndrome is additionally associated with lesions of the supramarginal gyrus (rostral part of the IPL), which furthermore gives rise to relatively isolated repetition problems and conduction aphasia (Damasio & Damasio, 1980; Fridriksson et al., 2010; Zukic et al., 2012). The cardinal symptom of conduction aphasia is a proportionally heavy deficit in verbal repetition, implying a central functional role of the rostral IPL in verbal working memory. In recent literature, the repetition deficit is especially associated with phonological problems in short-term memory (Baldo, Klostermann, & Dronkers, 2008; Dell, Martin, & Schwartz, 2007), which is in line with imaging data revealing that the left supramarginal gyrus fosters phonological short-term memory functions (Ravizza, Delgado, Chein, Becker, & Fiez, 2004). In contrast, lesion data as well as meta-analyses of language studies have demonstrated that the angular gyrus (caudal part of IPL) supports the processing of semantic content (Binder, Desai, Graves, & Conant, 2009; Borovsky, Saygin, Bates, & Dronkers, 2007; De Leon et al., 2007; Price, 2000; Vigneau et al., 2006). This might be associated with the grounding of language functions in somatosensory experiences. According to the "action–perception theory" (Pulvermueller, 1999), it is not a coincidence that both linguistic semantics and action perception rely on action related networks, including the angular gyrus (see Rizzolatti & Rozzi chapter in this book). Confirmative evidence comes from investigations involving lesion data as well as fMRI, uncovering the frontoparietal junction underlying action knowledge just as well as the comprehension of action words (Higuchi, Chaminade, Imamizu, & Kawatoa, 2009; Perani et al., 1999; Pulvermueller, Shtyrov, & Illmoniemi, 2005).

Another speech- and language-related function subserved by the IPL is linked to sequential processing. Increased neural activity in the caudal IPL (corresponding to PGa and PGp; BA 39) and in the anterior speech regions were found with increased syllable sequence complexity (Bohland & Guenther, 2006). Moreover, Moser, Baker, Sanchez, Rorden, and Fridriksson (2009) proposed that the IPL, especially its rostral parts (BA 40), may play a role in processing the temporal order of speech syllables. The rostral parts of the IPL may even be important for processing more complex constructions like sentences. An fMRI study that investigated single sentence production (e.g., "The child throws the ball") displayed activations of the supramarginal gyrus (BA 40) together with frontal regions (left and right inferior frontal gyri, left superior frontal and precentral gyrus, right medial frontal gyrus and right insula) and left SPL in sentence production as compared with word reading (Haller, Radue, Erb, Grodd, & Kircher, 2005).

Few functional imaging studies were dedicated to the exploration of discourse. Some of these studies investigated relations between expressive and receptive skills in spoken language (Awad, Warren, Scott, Turkheimer, & Wise, 2007), sign language communication (Braun, Guillemin, Hosey, & Varga, 2001), and reading (Brownsett & Wise, 2010). They underpin the important role of the SPL, IPL, and, more specifically, the left angular gyrus (caudal IPL) in the production of language, regardless of the output modality (spoken, signed, written language). Regarding the aspect of expressive versus receptive language skills, this result was generally supported by a recent fMRI study that, using the independent component analysis (ICA) of resting state data, demonstrated the participation of the left IPL in spoken discourse rather than nonspeech-related repetitive tongue movements (Geranmayeh et al., 2012). This was manifest in the identification of a left-lateralized frontal–temporal–parietal component involving rostral and caudal IPL (angular and supramarginal gyri) in connection with perisylvian classical language regions, including Broca's area (Geranmayeh et al., 2012). Using voxel-based lesion–symptom mapping (VLSM) on 50 aphasic patients, Borovsky et al. (2007) analyzed lesion correlates of impaired expressive discourse behavior. The authors found damage of frontal and/or temporal regions combined with a strong focus on the left angular gyrus (caudal IPL), especially when language production was impaired in terms of semantic context. As in most lesion studies of aphasia, nearly all lesions in this study were large and extended to subcortical white matter tracts (Friederici, 2009).

In summary, it can be assumed that the IPL area and especially its rostral and caudal parts are involved in many different speech and language processes on different linguistic levels up to discourse, as well as independently from input-and/or output modality. Therein, the rostral part seems to be more related to temporal ordering of sound and articulation (speech functions), which is in line with the anatomical connection (arcuate fasciculus; dorsal route of auditory dual-stream model) of PFop, PFt, and PFcm to BA 44, an area of the language network functionally more dedicated to modality-specific oral speech and articulation processes (Horwitz et al., 2003). In contrast, the caudal parts seem to be involved in modality-independent conceptual and semantic processes, including the connection between language and action networks that is linked to the idea of language being grounded in action and sensual experiences. This is supported by the anatomical alliance (arcuate fasciculus; dorsal route of auditory dual-stream model) of PGa and PGp with BA 45 (Caspers et al., 2011), a language area associated with modality-independent language processes (Horwitz et al., 2003). Regarding the caudal parts (PGa and PGp), this is very much in line with the results of a recent review in which the angular gyrus was interpreted, because of its specific involvement in many different functional processes including and beyond speech and language, as a "cross-modal hub where converging multisensory information is combined and integrated to comprehend and give sense to events, manipulate mental representations, solve familiar problems, and reorient attention to relevant information" (Seghier, 2013, p. 43).

4.5.2 Motor Functions and Interaction with Objects

Areas PFop, PFt, and PF in the rostral part of the IPL are strongly associated with motor and action-related functions, from simple motor behavior and tactile reception to motor control of spatially/timely complex movements and tool use. The last two functional processes are also related to the middle IPL consisting of PF and PFm.

Although left rostral IPL is dedicated to tactile perception, that is, its damage impairs the tactile recognition of an object (astereognosis), right rostral and middle IPL are more concerned with spatial coding and spatial attention (Mesulam, 1999). This holds even for motor functions and motor learning. These have been shown to generally recruit a frontoparietal network complemented by cerebellar and subcortical regions (Seitz, 2001). The rostral IPL seems to be specifically associated with learning and execution of complex sequential motor movements, thereby serving the processing of complexity in terms of time and space. Damage to the supramarginal gyrus, for example, may lead to deficits in the control of complex motor movements (Goldenberg, Hermsdörfer, Glindeman, Rorden, & Karnath, 2007; Goldenberg & Karnath, 2006). In line with the lesion data, functional brain imaging studies revealed that the IPL and especially its middle and rostral parts play a specific role in the control of spatial characteristics of complex motor behavior or attention (Halsband & Lange, 2006; Rauch et al., 1995).

Derived from anatomical and functional data in monkeys, Rizzolatti and Matelli (2003) further subdivided the dorsal stream of visual perception, formerly defined by Mishkin and colleagues (Mishkin & Ungerleider, 1982; Mishkin, Ungerleider, & Macko, 1983), into two separate pathways: the dorso-dorsal stream and the ventrodorsal stream. The dorso-dorsal stream seems to involve the IPS and SPL, whereas the ventrodorsal stream mainly targets the IPL. In humans, recent evidence supports a functional segregation that mirrors anatomically identified routes along the ventrodorsal stream (Binkofski & Buxbaum, 2013; Pisella, Binkofski, Lasek, Toni, & Rossetti, 2006). In contrast to the dorso-dorsal stream, the ventrodorsal stream appears to underlie processing of sensorimotor information based on longer-term object use representations. Damage within the ventrodorsal stream leads to impaired pantomimic and/or real object use, functions that require knowledge about the skilled handling of objects and therefore signify more overtly "cognitive" aspects of action representation. Deficits in object-related actions are a hallmark of limb apraxia (LA). Because optic ataxia (OA) is typically related to damage within the dorso-dorsal stream, one can derive that online motor performance should be spared in LA. A number of studies show reaching and grasping actions within normal range in LA as far as vision of the limb and target are spared, but they usually decrease when they have to be executed "off line" (e.g., when subjects are blindfolded prior to movement performance) (Buxbaum, Johnson, & Bartlett-Williams, 2005; Haaland, Harrington, & Knight, 1999; Jax, Buxbaum, & Moll, 2006; Laimgruber, Goldenberg, & Hermsdörfer, 2005). This and other observations (Dawson, Buxbaum, & Duff, 2010) suggest underlying deficits in anticipatory planning that entails over-reliance on online movement correction in this cohort.

A specific example of object use in which the ventrodorsal stream and, more precisely, area PFt in the rostral part of the IPL (Peeters et al., 2009) play a major role is tool use. Although behavioral studies on more naturalistic multistep tasks involving several tools/objects (e.g., preparing coffee, fixing a cassette recorder) make it evident that the right hemisphere is important for such complex functions, there is an unequivocal observation of only patients with left brain damage suffering from problems with single familiar tools or tool/object pairs (Hartmann, Goldenberg, Daumüller, & Hermsdörfer, 2005; Schwartz et al., 1998). Specifically, it has been reported that left lesions only lead to errors in matching objects to pantomime actions (mainly caudal IPL; Buxbaum, Kyle, & Menon, 2005; Kalénine, Buxbaum, & Coslett, 2010; Vaina, Goodglass, & Daltroy, 1995; Varney, 1978; Vignolo, 1990), pantomiming an action stimulated by an object stimulus (mainly rostral IPL; Barbieri & De Renzi, 1988; Goldenberg, Hartmann, & Schlott, 2003; Goodglass & Kaplan, 1963), or matching objects subserving the same purpose (Caspers et al., 2010; De Renzi, Scotti, & Spinnler, 1969; Rumiati, Zanini, Vorano, & Shallice, 2001; Vignolo, 1990). The same holds for the ability to infer possible functions from structure to apply novel tools linked to their complementary objects by transparent mechanical relationships (Goldenberg & Hagmann, 1998; Heilman, Maher, Greenwald, & Rothi, 1997) or to discover alternative uses of familiar tools (e.g., a coin for turning a screw; Heilman et al., 1997; Roy & Square, 1985). In the same line of evidence, Randerath, Goldenberg, Spijkers, Li, and Hermsdörfer (2010) state, by observation, that inappropriate nonfunctional grasping was committed extremely seldom in a huge group consisting of 42 left hemisphere stroke patients. The authors assumed the preserved dorso-dorsal route to underlie this effect. Furthermore, lesion studies revealed that the left rostral IPL is crucial for inferring the function of an object from its structure correctly (Barbieri & De Renzi, 1988). The anterior portion of IPS in the IPL together with the ventral premotor cortex in the left hemisphere appear to be tool-responsive regions in the ventrodorsal pathway (Binkofski, Buccino, Posse, et al., 1999; Binkofski, Buccino, Stephan, et al., 1999; Boronat et al., 2005; Chao & Martin, 2000; Handy, Grafton, Shroff, Ketay, & Gazzaniga, 2003; Johnson-Frey, 2004; Kellenbach, Brett, & Patterson, 2003). Even in the monkey, the homologue of this region, AIP, is a part of a functional circuit with the ventral premotor cortex related to the coordination of fine-grain hand movements (Jeannerod, Arbib, Rizzolatti, & Sakata, 1995).

In summary, research has identified two distinctive action systems hosted in the IPS, SPL, and IPL: a bilateral system associated with the dorso-dorsal stream (IPS and SPL), which is specialized for online actions directed at currently visible stimuli on the basis of their structure (size, shape, and orientation), and a left ventrodorsal stream (mainly IPL), which is dedicated to capabilities of skilled object-related actions (Buxbaum, 2001; Buxbaum & Kalénine, 2010; Fridman et al., 2006; Glover, Rosenbaum, Graham, & Dixon, 2004; Johnson-Frey, 2004; Pisella et al., 2006; Randerath et al., 2010; Vingerhoets, Acke, Vandemaele, & Achten, 2009). Integrating the structural and functional knowledge of these different action systems remains to be elucidated at the time of writing. It might be assumed that the cytoarchitectonic subdivisions reflect different processing modules, which are highly connected to each other and help to map visual input onto the motor system.

4.5.3 Spatial Functions

In addition to speech, language, motor, and action-related skills, the parietal cortex also subserves spatial functions, including general spatial perception (mainly PGa and PGp—caudal IPL), estimation of directions, processing of extrapersonal and peripersonal space, as well as the localization of objects. In contrast to language functions that are predominantly assigned to the left hemisphere, spatial functions seem to focus the right parietal cortex. Nevertheless, both hemispheres are involved in both functional systems. For example, the visuospatial component that is also evident in reading and in calculation (as long as the spatial imagery of a magnitude is required) is associated with left caudal IPL and IPS (Seghier, 2013; Simon, Mangin, Cohen, Le Bihan, & Dehaene, 2002).

In humans as in monkeys, these parietal regions, and particularly the IPS, play a role in visuomotor and visuospatial coordination and in constructing spatial characteristics during visuospatial perception (e.g., localizing objects and one's own body in a perceived multidimensional surrounding) (Grefkes & Fink, 2005). Parallel to its anatomical parcellation, functions are segregated into processing of extrapersonal space in area 7a and processing of peripersonal space in area 7b. Area 7 is connected to frontal area BA 8, which is known to be central in eye movement control (frontal eye field). This interconnection might be a prerequisite of goal-directed (hand) movements because it guarantees the further administration of visually noticed impulses as being localized in relation to one's own body and in the spatial surrounding.

The most common disorder associated with damage to the right caudal IPL is spatial hemineglect (Bisiach, Luzzatti, & Perani, 1979; Colombo, De Renzi, & Faglioni, 1976; Driver, Baylis, & Rafal, 1992; Gainotti, Messerli, & Tissot, 1972). Patients with hemineglect typically ignore the visual items in their contralesional side of space even though vision itself is spared. This deficit manifests itself in shaving only the right side of the face and sparing a meal on the left side of a plate while eating just the food placed on the right side. Besides these perceptive deficits, patients also show expressive problems, for instance, drawing the right side of their face when asked to make a self-portrait or positioning all numbers of a watch in the right half of a given circle when asked to draw a clock. In line with these symptoms, the most popular screening test to detect hemineglect consists of a line cancellation task (Albert's Neglect Test [ANT]; Albert, 1973). There are many hypotheses to explain the underlying mechanisms and causes of hemineglect. The most widespread assumption is an underlying attention deficit (Mesulam, 1999, 1981). In contrast, several theories proposed a deficit in the mental representation of sensory stimuli (Bisiach, Capitani, Luzzatti, & Perani, 1981; Rizzolatti & Berti, 1990) or a disorder in terms of transformation of (multimodal) sensory input into an egocentric coordinates (Karnath, 1994, 1997; Vallar, 1997).

The correlation of lesion site (right caudal IPL) and syndrome (hemineglect) was challenged by a lesion study that showed a significant lesion-deficit association between neglect and right temporal lesions (Karnath, Ferber, & Himmelbach, 2001). Because spatial attention is processed by distributed cortical networks, this finding can be interpreted as proof for a temporal node playing a crucial role within the whole right fronto-temporo-parietal attention network (Corbetta & Shulman, 2011, 2002). This assumption is confirmed by connectivity analyses showing interconnections between these parts of the cortex. According to fMRI studies, the IPL herein is associated with nonspatial functions related to control of spatial attention rather than spatial attention itself (e.g., arousal, reorienting, detection of behaviorally important novel input) (Corbetta & Shulman, 2011).

4.6 SUMMARY

Although there is clear evidence for a cytoarchitectonic parcellation of the IPL into seven subareas, the investigation of potentially equally distributed functions is only at its beginning. So far, the IPL is considered a multifunctional area involved in several different processes and functional networks. To this end, it already has been demonstrated that the IPL's cytoarchtectonic parcellation is reflected in terms of functional fragmentaion to some degree. The rostral parts of the IPL (areas PFop, PFt, and PFcm; supramarginal gyrus) have been related with processes of speech and articulation, phonological short-term memory, simple motor behavior, tactile reception, and general higher somatosensory functions. The middle part of the IPL as represented by PF and PFm seems to be dedicated to motor attention, motor control, and motor planning, especially in the left hemisphere. Last but not least, the caudal part comprising PGa and PGp, which together represent the angular gyrus, was reported to be involved in processes related to spatial attention, spatial perception, spatial memory, mathematical cognition, and modality-independent language functions, as well as conceptual and discourse-related processes and visuomotor integration. In line with this high degree of variable functions within the caudal IPL, it also seems to play a more general role as a cross-modal hub that guarantees comprehension, manipulation, and orientation based on processes of multisensory combination and integration. In contrast to the IPL, the functional and cytoarchitectonical distribution as well as interconnectivity of the IPS in humans still are less clear and need further investigation. The IPS seems to constitute the border between the dorso-dorsal and ventro-dorsal stream within a two-stream dorsal action system. Herein, the dorso-dorsal stream along the IPS and SPL seems to be specialized for online actions directed at currently visible stimuli on the basis of their structure (size, shape, and orientation), and the ventrodorsal stream along the IPL seems to enable skilled object-related actions. The dual-stream model of visuomotor coordination has to be distinguished from the dual-stream model of auditory language processing, which connects the two parts of Broca's area (BA 44 and BA 45) with superior temporal gyrus (STG) and/or IPL via a dorsal and ventral stream.

Based on the fact that the three parts of the IPL are somehow present in the monkey, future work will also have to explore evolutionary aspects of the relationship between cytoarchitectonic and functional parcellation. So far, it is already evident that the human IPL cytoarchtectonically parallels that of the nonhuman primate in respect to PF (human: PFt and PF), PFG (human: PFm), PG (human: PGa), and Opt (human: PGp). Furthermore, hints toward a similar functional role in action processing and tool use are apparent for monkey areas PF and PFG and human areas PFt and PF. Another important aspect that needs further investigation concerns the correlation between each part of the IPL and its function within different cortical networks using the evolving knowledge over the past years about the relation between structure and function of the IPL in both species.

References

Albert, M. L. (1973). A simple test of visual neglect. *Neurology, 1*(23), 658–664.

Andersen, R. A., Asanuma, C., Essick, G., & Siegel, R. M. (1990). Corticocortical connections of anatomically and physiologically defined subdivisions within the inferior parietal lobule. *The Journal of Comparative Neurology, 296*, 65–113.

Awad, M., Warren, J. E., Scott, S. K., Turkheimer, F. E., & Wise, R. J. S. (2007). A common system for the comprehension and production of speech. *Journal of Neuroscience, 27*, 11455–11464.

Baldo, J. V., Klostermann, E. C., & Dronkers, N. F. (2008). It's either a cook or a baker: Patients with conduction aphasia get the gist but lose the trace. *Brain and Language, 105*, 134–140.

Barbas, H. (1988). Anatomic organization of basoventral and mediodorsal visual recipient prefrontal regions in the rhesus monkey. *The Journal of Comparative Neurology, 276*(3), 313–342.

Barbieri, C., & De Renzi, E. (1988). The executive and ideational components of apraxia. *Cortex, 24*(4), 535–543.

Batsch, E. G. (1956). Die myeloarchitektonische Untergliederung des Isocortex parietalis beim Menschen. *Journal fur Hirnforschung, 2*, 225–258.

Binder, J. R., Desai, R. H., Graves, W. W., & Conant, L. L. (2009). Where is the semantic system? A critical review and meta-analysis of 120 functional neuroimaging studies. *Cerebral Cortex, 19*, 2767–2796.

Binkofski, F., Buccino, G., Stephan, K. M., Rizzolatti, G., Seitz, R. J., & Freund, H.-J. (1999). Parieto-premotor network for object manipulation: Evidence from neuroimaging. *Experimental Brain Research, 128*, 210–213.

Binkofski, F. C., Buccino, G., Posse, S., Seitz, R. J., Rizzolatti, G., & Freund, H.-J. (1999). A fronto-parietal circuit for object manipulation in man: Evidence from an fMRI study. *The European Journal of Neuroscience, 11*, 3276–3286.

Binkofski, F. C., & Buxbaum, L. (2013). Two action systems in the human brain. *Brain and Language, 127*(2), 222–229.

Bisiach, E., Capitani, E., Luzzatti, C., & Perani, D. (1981). Brain and conscious representation of outside reality. *Neuropsychologia, 19*, 543–551.

Bisiach, E., Luzzatti, C., & Perani, D. (1979). Unilateral neglect, representational schema and consciousness. *Brain, 102*, 609–661.

Bohland, J. W., & Guenther, F. H. (2006). An fMRI investigation of syllable sequence production. *NeuroImage, 32*(2), 821–841.

Bonini, L., Rozzi, S., Serventi, F. U., Simone, L., Ferrari, P. F., & Fogassi, L. (2010). Ventral premotor and inferior parietal cortices make distinct contribution to action organization and intention understanding. *Cerebral Cortex, 20*(6), 1372–1385.

Boronat, C., Buxbaum, L. J., Coslett, H. B., Tang, K., Saffran, E. M., Kimberg, D. Y., et al. (2005). Distinctions between function and manipulation knowledge of objects: Evidence from functional magnetic resonance imaging. *Cognitive Brain Research, 23*(2–3), 361–373.

Borovsky, A., Saygin, A. P., Bates, E., & Dronkers, N. (2007). Lesion correlates of conversational speech production deficits. *Neuropsychologia, 45*, 2525–2533.

Borra, E., Belmahli, A., Calzavara, R., Gerbella, M., Murata, A., Rozzi, S., et al. (2008). Cortical connections of macaque anterior intraparietal (AIP) area. *Cerebral Cortex, 18*, 1094–1111.

Braun, A. R., Guillemin, A., Hosey, L., & Varga, M. (2001). The neural organization of discourse: An H2 15O-PET study of narrative production in English and American sign language. *Brain, 124*, 2028–2044.

Bray, S., Arnold, A. E., Iaria, G., & MacQueen, G. (2013). Structural connectivity of visuotopic intraparietal sulcus. *NeuroImage, 82*, 137–145.

Brodmann, K. (1905). Beiträge zur histologischen Lokalisation der Grosshirnrinde. III. Die Rindenfelder der niederen Affen. *Journal fuer Psychologie und Neurologie, 4*, 177–226.

Brodmann, K. (1909). *Vergleichende Lokalisationslehre der Grosshirnrinde*. Leipzig: Barth.

Brownsett, S. L. E., & Wise, R. J. S. (2010). The contribution of the parietal lobes to speaking and writing. *Cerebral Cortex, 20*, 517–523.

Buxbaum, L. J. (2001). Ideomotor apraxia: A call to action. *Neurocase, 7*, 445–458.

Buxbaum, L. J., Johnson, S. H., & Bartlett-Williams, M. (2005). Deficient internal models for planning hand-object interactions in ideomotor apraxia. *Neuropsychologia, 43*(6), 917–929.

Buxbaum, L. J., & Kalénine, S. (2010). Action knowledge, visuomotor activation, and embodiment in the two action systems. *Annals of the New York Academy of Sciences, 1191*, 201–218.

Buxbaum, L. J., Kyle, K., & Menon, R. (2005). On beyond mirror neurons: Internal representations subserving imitation and recognition of skilled object-related actions in humans. *Cognitive Brain Research, 25*(1), 226–239.

Caminiti, R., Chafee, M. V., Battaglia-Meyer, A., Averbeck, B. B., Crowe, D. A., & Georgopoulos, A. P. (2010). Understanding the parietal lobe syndrome from a neurophysiological and evolutionary perspective. *The European Journal of Neuroscience, 31*, 2320–2340.

Campbell, A. W. (1905). *Histological studies on the localization of cerebral function*. Cambridge: University Press.

Caspers, S., Eickhoff, S. B., Geyer, S., Scheperjans, F., Mohlberg, H., Zilles, K., et al. (2008). The human inferior parietal lobule in stereotaxic space. *Brain Structure & Function, 212*, 481–495.

Caspers, S., Eickhoff, S. B., Rick, T., von Kapri, A., Kuhlen, T., Huang, R., et al. (2011). Probabilistic fibre tract analysis of cytoarchitectonically defined human inferior parietal lobule areas reveals similarities to macaques. *NeuroImage, 58*, 362–380.

Caspers, S., Geyer, S., Schleicher, A., Mohlberg, H., Amunts, K., & Zilles, K. (2006). The human inferior parietal cortex: Cytoarchitectonic parcellation and interindividual variability. *NeuroImage, 33*, 430–448.

Caspers, S., Schleicher, A., Bacha-Trams, M., Palomero-Gallagher, N., Amunts, K., & Zilles, K. (2013). Organization of the human inferior parietal lobule based on receptor architectonics. *Cerebral Cortex, 23*, 615–628.

Caspers, S., Zilles, K., Laird, A. R., & Eickhoff, S. B. (2010). ALE meta-analysis of action observation and imitation in the human brain. *NeuroImage, 50*, 1148–1167.

Catani, M., Jones, D. K., & Ffytche, D. H. (2005). Perisylvian language networks of the human brain. *Annals of Neurology, 57*, 8–16.

Cavada, C., & Goldman-Rakic, P. S. (1989a). Posterior parietal cortex in rhesus monkey: I. Parcellation of areas based on distinctive limbic and sensory corticocortical connections. *The Journal of Comparative Neurology, 287*, 393–421.

Cavada, C., & Goldman-Rakic, P. S. (1989b). Posterior parietal cortex in rhesus monkey: II. Evidence of segregated corticocortical networks linking sensory and limbic areas with the frontal lobe. *The Journal of Comparative Neurology, 287*, 422–445.

Chao, L. L., & Martin, A. (2000). Representation of manipulable manmade objects inthe dorsal stream. *NeuroImage, 12*, 478–484.

Choi, H. J., Zilles, K., Mohlberg, H., Schleicher, A., Fink, G. R., Armstrong, E., et al. (2006). Cytoarchitectonic identification and probabilistic mapping of two distinct areas within the anterior ventral bank of the human intraparietal sulcus. *The Journal of Comparative Neurology, 495*, 53–69.

Colombo, A., De Renzi, E., & Faglioni, P. (1976). The occurrence of visual neglect in patients with unilateral cerebral disease. *Cortex, 12*, 221–231.

Corbetta, M., & Shulman, G. L. (2002). Control of goal-directed and stimulus-driven attention in the brain. *Nature Reviews Neuroscience, 3*, 201–215.

Corbetta, M., & Shulman, G. L. (2011). Spatial neglect and attention networks. *Annual Review of Neuroscience, 34*, 569–599.

Damasio, H., & Damasio, A. (1980). The anatomical basis of conduction aphasia. *Brain, 103*(2), 337–350.

Dawson, A. M., Buxbaum, L. J., & Duff, S. (2010). The impact of left hemisphere stroke on force control with familiar and novel objects: Neuroanatomic substrates and relationship to apraxia. *Brain Research, 1317*, 124–136.

De Leon, M. P., Benatti, P., Di Gregorio, C., Losi, L., Pedroni, M., Ponti, G., et al. (2007). Genotype-phenotype correlations in individuals with a founder mutation in the MLH1 gene and hereditary non-polyposis colorectal cancer. *Scandinavian Journal of Gastroenterology, 42*(6), 746–753.

Dell, G. S., Martin, N., & Schwartz, M. F. (2007). A case-series test of the interactive two-step model of lexical access: Predicting word repetition from picture naming. *Journal of Memory and Language, 56*(4), 490–520.

De Renzi, E., Scotti, G., & Spinnler, H. (1969). Perceptual and associative disorders of visual recognition. *Neurology, 19*, 634–642.

Driver, J., Baylis, G. C., & Rafal, R. D. (1992). Preserved figure-ground segregation and symmetry perception in visual neglect. *Nature, 360*, 73–75.

Dronkers, N. F., Wilkins, D. P., Van Valin, R. D., Jr., Redfern, B. B., & Jaeger, J. J. (2004). Lesion analysis of the brain areas involved in language comprehension. *Cognition, 92*, 145–177.

Eickhoff, S. B., Amunts, K., Mohlberg, H., & Zilles, K. (2006). The human parietal operculum. II. Stereotaxic maps and correlation with functional imaging results. *Cerebral Cortex, 16*, 268–279.

Eickhoff, S. B., Schleicher, A., Zilles, K., & Amunts, K. (2006). The human parietal operculum. I. Cytoarchitectonic mapping of subdivisions. *Cerebral Cortex, 16*, 254–267.

Eidelberg, D., & Galaburda, A. M. (1984). Inferior parietal lobule—Divergent architectonic asymmetries in the human brain. *Archives of Neurology, 41*, 843–852.

Felleman, D. J., & Van Essen, D. C. (1991). Distributed hierarchical processing in the primate cerebral cortex. *Cerebral Cortex, 1*, 1–47.

Flechsig, P. (1920). *Anatomie des menschlichen Gehirns und Rückenmarks of myelogenetischer Grundlage*. Leipzig: Thieme.

Fridman, E. A., Immisch, I., Hanakawa, T., Bohlhalter, S., Waldvogel, D., Kansaku, K., et al. (2006). The role of the dorsal stream for gesture production. *NeuroImage, 29*, 417–428.

Fridriksson, J., Kjartansson, O., Morgan, P. S., Hjaltason, H., Magnusdottir, S., Bonilha, L., et al. (2010). Impaired speech repetition and left parietal lobe damage. *Journal of Neuroscience, 30*(33), 11057–11061.

Friederici, A. (2009). Pathways to language: Fiber tracts in the human brain. *Trends in Cognitive Neuroscience, 13*(4), 175–181.

Gainotti, G., Messerli, P., & Tissot, R. (1972). Qualitative analysis of unilateral spatial neglect in relation to laterality of cerebral lesions. *Journal of Neurology, Neurosurgery, and Psychiatry, 35*, 545–550.

Geranmayeh, F., Brownsett, S. L., Leech, R., Beckmann, C. F., Woodhead, Z., & Wise, R. J. (2012). The contribution of the inferior parietal cortex to spoken language production. *Brain and Language, 121*(1), 47–57.

Gerhardt, E. (1940). Die Cytoarchitektonik des Isocortex parietalis beim Menschen. *Journal fuer Psychologie und Neurologie, 49*, 367–419.

Glover, S., Rosenbaum, D. A., Graham, J., & Dixon, P. (2004). Grasping the meaning of words. *Experimental Brain Research, 154*(1), 103–108.

Goldenberg, G., & Hagmann, S. (1998). Tool use and mechanical problem solving in apraxia. *Neuropsychologia, 36*, 581–589.

Goldenberg, G., Hartmann, K., & Schlott, I. (2003). Defective pantomime of object use in left brain damage: Apraxia or asymbolia? *Neuropsychologia, 41*, 1565–1573.

Goldenberg, G., Hermsdörfer, J., Glindeman, R., Rorden, C., & Karnath, H. O. (2007). Pantomime of tool use depends on integrity of left inferior frontal cortex. *Cerebral Cortex, 17*(12), 2769–2776.

Goldenberg, G., & Karnath, H. O. (2006). The neural basis of imitation is body part specific. *The Journal of Neuroscience, 26*(23), 6282–6287.

Goodglass, H., & Kaplan, E. (1963). Disturbance of gesture and pantomime in aphasia. *Behavioral and Brain Sciences, 86*, 703–720.

Greenberg, A. S., Verstynen, T., Chiu, Y. C., Yantis, S., Schneider, W., & Behrmann, M. (2012). Visuotopic cortical connectivity underlying attention revealed with white-matter tractography. *The Journal of Neuroscience, 32*(8), 2773–2782.

Grefkes, C., & Fink, G. R. (2005). The functional organization of the intraparietal sulcus in humans and monkeys. *Journal of Anatomy, 207*(1), 3–17.

Gregoriou, G. G., Borra, E., Matelli, M., & Luppino, G. (2006). Architectonic organization of the inferior parietal convexity of the macaque monkey. *The Journal of Comparative Neurology, 496*(3), 422–451.

Guenther, F. H. (2006). Cortical interactions underlying the production of speech sounds. *Journal of Communication Disorders, 39*(5), 350–365.

Haaland, K. Y., Harrington, D. L., & Knight, R. T. (1999). Spatial deficits in ideomotor limb apraxia: A kinematic analysis of aiming movements. *Behavioral & Brain Sciences, 122*, 1169–1182.

Haller, S., Radue, E. W., Erb, M., Grodd, W., & Kircher, T. (2005). Overt sentence production in event-related fMRI. *Neuropsychologia, 43*, 807–814.

Halsband, U., & Lange, R. K. (2006). Motor learning in man: A review of functional and clinical studies. *Journal of Physiology—Paris, 99*, 414–424.

Handy, T. C., Grafton, S. T., Shroff, N. M., Ketay, S., & Gazzaniga, M. S. (2003). Graspable objects grab attention when the potential for action is recognised. *Nature Neuroscience, 6*(421–427), 974.

Hartmann, K., Goldenberg, G., Daumüller, M., & Hermsdörfer, J. (2005). It takes the whole brain to make a cup of coffee: The neuropsychology of naturalistic actions involving technical devices. *Neuropsychologia, 43*, 625–637.

Heilman, K. M., Maher, L. M., Greenwald, M. L., & Rothi, L. J. G. (1997). Conceptual apraxia from lateralized lesions. *Neurology, 49*, 457–464.

Henseler, I., Regenbrecht, F., & Obrig, H. (2014). Lesion correlates of patholinguistic profiles in chronic aphasia: Comparisons of syndrome-,modality- and symptom-level assessment. *Brain, 137*, 918–930.

Higuchi, S., Chaminade, T., Imamizu, H., & Kawatoa, M. (2009). Shared neural correlates for language and tool use in Broca's area. *Neuroreport, 20*(15), 1376–1381.

Hopf, A. (1969). Photometric studies on the myeloarchitecture of the human parietal lobe. I. Parietal region. *Journal fur Hirnforschung, 11*, 253–265.

Hopf, A., & Vitzthum, H. G. (1957). Über die Verteilung myeloarchitektonischer Merkmale in der Scheitellappenrinde beim Menschen. *Journal fur Hirnforschung, 3*, 79–104.

Horwitz, B., Amunts, K., Bhattacharyya, R., Patkin, D., Jeffries, K., Zilles, K., et al. (2003). Activation of Broca's area during the production of spoken and signed language: A combined cytoarchitectonic mapping and PET analysis. *Neusropsychologia, 41*(14), 1868–1876.

Husain, M., & Nachev, P. (2006). Space and the parietal cortex. *Trends in Cognitive Sciences, 11*(1), 30–36.

Jax, S., Buxbaum, L. J., & Moll, A. (2006). Deficits in movement planning and intrinsic coordinate control in ideomotor apraxia. *Journal of Cognitive Neuroscience, 18*(12), 2063–2076.

Jeannerod, M., Arbib, M. A., Rizzolatti, G., & Sakata, H. (1995). Grasping objects: the cortical mechanisms of visuomotor transformation. *Trends in Neurosciences, 18*(7), 314–320.

Johnson-Frey, S. H. (2004). The neural bases of complex tool use in humans. *Trends in Cognitive Sciences*, *8*, 71–78.

Kalénine, S., Buxbaum, L. J., & Coslett, H. B. (2010). Critical brain regions for action recognition: Lesion-symptom mapping in left hemisphere stroke. *Behavioral and Brain Sciences*, *133*(11), 3269–3280.

Karnath, H.-O. (1994). Disturbed coordinate transformation in the neural representation of space as the crucial mechanism leading to neglect. *Neuropsychological Rehabilitation*, *4*, 147–150.

Karnath, H.-O. (1997). Spatial orientation and the representation of space with parietal lobe lesions. *Philosophical Transactions of the Royal Society of London Series B, Biological Sciences*, *352*, 1411–14119.

Karnath, H. O., Ferber, S., & Himmelbach, M. (2001). Spatial awareness is a function of the temporal not the posterior parietal lobe. *Nature*, *411*, 950–953.

Kellenbach, M. L., Brett, M., & Patterson, K. (2003). Actions speak louder than functions: The importance of manipulability and action in tool representation. *Journal of Cognitive Neuroscience*, *15*, 30–46.

Laimgruber, K., Goldenberg, G., & Hermsdörfer, J. (2005). Manual and hemispheric asymmetries in the execution of actual and pantomimed prehension. *Neuropsychologia*, *43*, 682–692.

Lerner, Y., Honey, C. J., Silbert, L. J., & Hasson, U. (2011). Topographic mapping of a hierarchy of temporal receptive windows using a narrated story. *Journal of Neuroscience*, *31*(8), 2906–2915.

Lewis, J. W., & Van Essen, D. C. (2000). Mapping of architectonic subdivisions in the macaque monkey, with emphasis on parieto-occipital cortex. *The Journal of Comparative Neurology*, *428*(1), 79–111.

Mars, R. B., Jbabdi, S., Sallet, J., O'Reilly, J. X., Croxson, P. L., Olivier, E., et al. (2011). Diffusion-weighted imaging tractography-based parcellation of the human parietal cortex and comparison with human and macaque resting-state functional connectivity. *Journal of Neuroscience*, *31*, 4087–4100.

Martino, J., De Witt Hamer, P. C., Berger, M. S., Lawton, M. T., Arnold, C. M., de Lucas, E. M., et al. (2013). Analysis of the subcomponents and cortical terminations of the perisylvian superior longitudinal fasciculus: A fiber dissection and DTI tractography study. *Brain Structure and Function*, *218*, 105–121.

Matelli, M., Camarda, R., Glickstein, M., & Rizzolatti, G. (1988). Functional organization of inferior area 6 in the macaque monkey: II. Area F5 and the control of distal movements. *Experimental Brain Research*, *71*, 491–507.

Mesulam, M. M. (1981). A cortical network for directed attention and unilateral neglect. *Annals of Neurology*, *10*(4), 309–325.

Mesulam, M. M. (1999). Spatial attention and neglect: Parietal, frontal and cingulate contributions to the mental representation and attentional targeting of salient extrapersonal events. *Philosophical Transactions of the Royal Society of London Series B*, *354*(1387), 1325–1346.

Mesulam, M. M., van Hoesen, G. W., Pandya, D. N., & Geschwind, N. (1977). Limbic and sensory connections of the inferior parietal lobule (area PG) in the rhesus monkey: A study with a new method for horseradish peroxidase histochemistry. *Brain Research*, *136*, 393–414.

Mishkin, M., & Ungerleider, L. G. (1982). Contribution of striate inputs to the visuospatial functions of parieto-preoccipital cortex in monkeys. *Behavioural Brain Research*, *6*(1), 57–77.

Mishkin, M., Ungerleider, L. G., & Macko, K. A. (1983). Object vision and spatial vision: Two cortical pathways. *Trends in Neurosciences*, *6*, 414–417.

Moser, D., Baker, J. M., Sanchez, C. E., Rorden, C., & Fridriksson, J. (2009). Temporal order processing of syllables in the left parietal lobe. *The Journal of Neuroscience*, *29*(40), 12568–12573.

Neal, J. W., Pearson, R. C. A., & Powell, T. P. S. (1990a). The ipsilateral cortico-cortical connections of area 7b, PF, in the parietal and temporal lobes of the monkey. *Brain Research*, *524*, 119–132.

Neal, J. W., Pearson, R. C. A., & Powell, T. P. S. (1990b). The connections of area PG, 7a, with cortex in the parietal, occipital, and temporal lobes of the monkey. *Brain Research*, *532*, 249–264.

Ono, M., Kubik, S., & Abernathey, C. D. (1990). *Atlas of the cerebral sulci*. Stuttgart/New York, NY: Thieme.

Orban, G. A., & Rizzolatti, G. (2012). An area specifically devoted to tool use in human left inferior parietal lobule. *The Behavioral and Brain Sciences*, *35*(4), 234.

Pandya, D. N., & Seltzer, B. (1982). Intrinsic connections and architectonics of posterior parietal cortex in the rhesus monkey. *The Journal of Comparative Neurology*, *204*, 196–210.

Parker, G. J. M., Luzzi, S., Alexander, D. C., Wheeler-Kingshott, C. A. M., Ciccarelli, O., & Ralph, M. A. L. (2005). Lateralization of ventral and dorsal auditory-language pathways in the human brain. *NeuroImage*, *24*, 656–666.

Peeters, R. R., Simone, L., Nelissen, K., Fabbri-Destro, M., Vanduffel, W., Rizzolatti, G., et al. (2009). The representation of tool use in humans and monkeys: Common and uniquely human features. *The Journal of Neuroscience*, *29*(37), 11523–11539.

Perani, I., Cappa, S., Schnur, T., Tettamanti, M., Collina, S., Rosa, M. M., et al. (1999). The neural correlates of verb and noun processing. A PET study. *Brain*, *122*(12), 2337–2344.

Pisella, L., Binkofski, F. C., Lasek, K., Toni, I., & Rossetti, Y. (2006). No double-dissociation between optic ataxia and visual agnosia: Multiple sub-streams for multiple visuo-manual integrations. *Neuropsychologia*, *44*(13), 2734–2748.

Powell, H. W., Parker, G. J., Alexander, D. C., Symms, M. R., Boulby, P. A., Wheeler-Kingshott, C. A., et al. (2006). Hemispheric asymmetries in language-related pathways: A combined functional MRI and tractography study. *NeuroImage*, *32*, 388–399.

Price, C. (2000). The anatomy of language: Contributions from functional neuroimaging. *Journal of Anatomy*, *197*, 335–359.

Pulvermueller, F. (1999). Words in the brain's language. *Behavioral and Brain Sciences*, *22*, 253–336.

Pulvermueller, F., Shtyrov, Y., & Ilmoniemi, R. (2005). Brain signatures of meaning access in action word recognition. *Journal of Cognitive Neuroscience*, *17*(6), 884–892.

Randerath, J., Goldenberg, G., Spijkers, W., Li, Y., & Hermsdörfer, J. (2010). Different left brain regions are essential for grasping a tool compared with its subsequent use. *NeuroImage*, *53*(1), 171–180.

Rauch, S. L., Savage, C. R., Brown, H. D., Curran, T., Alpert, N. M., Kendrick, A., et al. (1995). A PET investigation of implicit and explicit sequence learning. *Human Brain Mapping*, *3*, 271–286.

Ravizza, S. M., Delgado, M. R., Chein, J. M., Becker, J. T., & Fiez, J. A. (2004). Functional dissociations within the inferior parietal cortex in verbal working memory. *NeuroImage*, *22*, 562–573.

Rizzolatti, G., & Berti, A. (1990). Neglect as a neural representation deficit. *Revue Neurologique*, *146*, 626–634.

Rizzolatti, G., & Luppino, G. (2001). The cortical motor system. *Neuron*, *31*(6), 889–901.

Rizzolatti, G., & Matelli, M. (2003). Two different streams form the dorsal visual system: Anatomy and functions. *Experimental Brain Research*, *153*, 146–157.

Roy, E. A., & Square, P. A. (1985). Common considerations in the study of limb, verbal and oral apraxia. In E. A. Roy (Ed.), *Neuropsychological studies of apraxia and related disorders* (pp. 111–162). Amsterdam: North-Holland.

Rozzi, S., Calzavara, R., Belmalih, A., Borra, E., Gregoriou, G. G., Matelli, M., et al. (2006). Cortical connections of the inferior parietal cortical convexity of the macaque monkey. *Cerebral Cortex*, *16*(10), 1389–1417.

Rozzi, S., Ferrari, P. F., Bonini, L., Rizzolatti, G., & Fogassi, L. (2008). Functional organization of inferior parietal lobule convexity in the macaque monkey: Electrophysiological characterization of motor, sensory and mirror responses and their correlation with cytoarchitectonic areas. *The European Journal of Neuroscience, 28*(8), 1569–1588.

Rumiati, R. I., Zanini, S., Vorano, L., & Shallice, T. (2001). A form of ideatonal apraxia as a selective deficit of contention scheduling. *Cognitive Neuropsychology, 18*, 617–642.

Ruschel, M., Knösche, T. R., Friederici, A. D., Turner, R., Geyer, S., & Anwander, A. (2013). Connectivity architecture and subdivisions of the human inferior parietal cortex revealed by diffusion MRI. *Cerebral Cortex*epub, ahead of print. doi:10.1093/cercor/bht098.

Rushworth, M. F. S., Behrens, T. E. J., & Johansen-Berg, H. (2006). Connection patterns distinguish 3 regions of human parietal cortex. *Cerebral Cortex, 16*, 1418–1430.

Sarkissov, S. A., Filimonoff, I. N., & Preobrashenskaya, N. S. (1949). Cytoarchitecture of the human cortex cerebri *(Russian)*. Moscow: Medgiz.

Saur, D., Kreher, B. W., Schnell, S., Kuemmerer, D., Kellermeyer, P., Vrya, M.-S., et al. (2008). Ventral and dorsal pathways for language. *Proceedings of the National Academy of Sciences of the United States of America, 105*(46), 18035–18040.

Scheperjans, F., Eickhoff, S. B., Hömke, L., Mohlberg, H., Hermann, K., Amunts, K., et al. (2008). Probabilistic maps, morphometry, and variability of cytoarchitectonic areas in the human superior parietal cortex. *Cerebral Cortex, 18*(9), 2141–2157.

Scheperjans, F., Hermann, K., Eickhoff, S. B., Amunts, K., Schleicher, A., & Zilles, K. (2008). Observer-independent cytoarchitectonic mapping of the human superior parietal cortex. *Cerebral Cortex, 18*(4), 846–867.

Schwartz, M. F., Buxbau, L. J., Montgomery, M. W., Fitzpatrick-DeSalme, E., Hart, T., Ferraro, M., et al. (1998). Naturalistic action production following right hemisphere stroke. *Neuropsychologia, 37*(1), 51–66.

Seghier, M. L. (2013). The angular gyrus: Multiple functions and multiple subdivisions. *The Neuroscientist, 19*(1), 43–61.

Seitz, R. J. (2001). Motorisches Lernen: Untersuchungen mit der funktionellen Bildgebung Motor learning: Functional neuroimaging studies. *Deutsche ZS für Sportmedizin, 52*(12), 343–349.

Seitz, R. J., & Binkofski, F. C. (2003). Modular organization of parietal lobe functions as revealed by functional activation studies. *Advances in Neurology, 93*, 281–292.

Seltzer, B., & Pandya, D. N. (1984). Further observations on parieto-temporal connections in the rhesus monkey. *Experimental Brain Research, 55*, 301–312.

Simon, O., Mangin, J. F., Cohen, L., Le Bihan, D., & Dehaene, S. (2002). Topographical layout of hand, eye, calculation, and language-related areas in the human parietal lobe. *Neuron, 33*(3), 475–487.

Smith, G. E. (1907). A new topographical survey of the human cerebral cortex, being an account of the distribution of the anatomically distinct cortical areas and their relationship to the cerebral sulci. *Journal of Anatomy, 41*, 237–254.

Tanné-Gariepy, J., Rouiller, E. M., & Boussaoud, D. (2002). Parietal inputs to dorsal versus ventral premotor areas in the macaque monkey: Evidence for largely segregated visuomotor pathways. *Experimental Brain Research, 145*, 91–103.

Thiebaut de Schotten, M., Dell'Acqua, F., Forkel, S. J., Simmons, A., Vergani, F., Murphy, D. G. M., et al. (2011). A lateralized brain network for visuospatial attention. *Nature Neuroscience, 14*, 1245–1246.

Tremblay, P., & Gracco, V. L. (2010). On the selection of words and oral motor responses: Evidence of a response-independent fronto-parietal network. *Cortex, 46*(1), 15–28.

Uddin, L. Q., Supekar, K., Amin, H., Rykhlevskaia, E., Nguyen, D. A., Greicius, M. D., et al. (2010). Dissociable connectivity within human angular gyrus and intraparietal sulcus: Evidence from functional and structural connectivity. *Cerebral Cortex, 20*, 2636–2646.

Ungerleider, L. G., & Desimone, R. (1986). Cortical connections of visual area MT in the macaque. *The Journal of Comparative Neurology, 248*, 190–222.

Vaina, L. M., Goodglass, H., & Daltroy, L. (1995). Inference of object use from pantomimed actions by aphasics and patients with right hemisphere lesions. *Synthese, 104*, 43–57.

Vallar, G. (1997). Spatial frames of reference and somatosensory processing: A neuropsychological perspective. *Philosophical Transactions of the Royal Society, London, B, 352*, 1401–1409.

Varney, N. R. (1978). Linguistic correlates of pantomime recognition in aphasic patients. *Journal of Neurology Neurosurgery and Psychiatry, 41*, 564–568.

Vigneau, M., Beaucousin, V., Herve, P. Y., Duffau, H., Crivello, F., Houde, O., et al. (2006). Meta-analyzing left hemisphere language areas: Phonology, semantics, and sentence processing. *NeuroImage, 30*, 1414–1432.

Vignolo, L. A. (1990). Non-verbal conceptual impairment in aphasia. In F. Boller, & J. Grafman (Eds.), *Handbook of clinical neuropsychology* (pp. 185–206). Amsterdam: Elsevier.

Vingerhoets, G., Acke, F., Vandemaele, P., & Achten, E. (2009). Tool responsive regions in the posterior parietal cortex: Effect of differences in motor goal and target object during imagined transitive movements. *NeuroImage, 47*, 1832–1843.

Vogt, C., & Vogt, O. (1919). Allgemeinere Ergebnisse unserer Hirnforschung. *Journal fuer Psychologie und Neurologie, 25*, 279–461.

Vogt, O. (1911). Die Myeloarchitektonik des Isocortex parietalis. *Journal fuer Psychologie und Neurologie, 18*, 107–118.

von Economo, K., & Koskinas, G. (1925). *Die Cytoarchitektonik des Hirnrinde des erwachsenen Menschen*. Wien: Springer.

Wang, J., Fan, L., Zhang, Y., Liu, Y., Jiang, D., Zhang, Y., et al. (2012). Tractography-based parcellation of the human left inferior parietal lobule. *NeuroImage, 63*, 641–652.

Weiller, C., Bormann, T., Saur, D., Musso, M., & Rijntjes, M. (2011). How the ventral pathway got lost: And what its recovery might mean. *Brain and Language, 118*, 29–39.

Yeo, B. T., Krienen, F. M., Sepulcre, J., Sabuncu, M. R., Lashkari, D., Hollinshead, M., et al. (2011). The organization of the human cerebral cortex estimated by intrinsic functional connectivity. *Journal of Neurophysiology, 106*, 1125–1165.

Zhang, S., & Li, C. S. (2014). Functional clustering of the human inferior parietal lobule by whole-brain connectivity mapping of resting-state functional magnetic resonance imaging signals. *Brain Connect, 4*(1), 53–69.

Zukic, S., Mrkonjic, Z., Sinanovic, O., Vidovic, M., & Kojic, B. (2012). Gerstmann's syndrome in acute stroke patients. *Acta Informatica Medica, 20*(4), 242–243.

CHAPTER

5

Human Auditory Cortex

Brian Barton[1,2] and Alyssa A. Brewer[1,2,3]

[1]Department of Cognitive Sciences, University of California, Irvine, Irvine, CA, USA; [2]Center for Cognitive Neuroscience and Engineering, University of California, Irvine, Irvine, CA, USA; [3]Center for Hearing Research, University of California, Irvine, Irvine, CA, USA

5.1 INTRODUCTION

One of the fundamental discoveries of neuroscience is that sensory regions of cortex are formed of many functionally specialized areas that are organized into hierarchical networks (Kaas & Hackett, 2000; Krubitzer, 2007; Schreiner & Winer, 2007; Van Essen, Felleman, DeYoe, Olavarria, & Knierim, 1990). The simplest features are processed in low-level areas that then pass that information up the hierarchy to perform increasingly complex computations. A general feature of these systems is that the topography of the sensing organ represents the most fundamental stimulus information, which is preserved through much or all of the hierarchy (Brewer & Barton, 2012). It has been suggested that this preservation of topographical organization allows for efficient connectivity between neurons that represent nearby portions of sensory space, likely necessary for processes such as lateral inhibition and gain control (Chklovskii & Koulakov, 2004; Mitchison, 1991; Moradi & Heeger, 2009; Shapley, Hawken, & Xing, 2007). From a researcher's perspective, this topographic preservation allows us to use one set of stimuli to localize a number of sensory areas rather than designing specialized stimuli for each sensory area (Brewer & Barton, 2012).

Many details remain to be elucidated in each cortical network, and it is the goal of this chapter to provide an overview of the current understanding of the cortical organization of the low-level sound processing areas of the human auditory system. It is not within the scope of this chapter to discuss the organization of the entire human auditory system; we only touch on the subcortical areas and higher-order processing areas. However, we note that low-level cortical organization has much to teach us about both subcortical and higher-order processing areas.

Researchers have long been aware that low-level auditory processing occurs on and near Heschl's gyrus (HG) of the human temporal lobe in the lateral (Sylvian) fissure, but many questions have remained unanswered. How many distinct auditory areas exist in this region? What is the proper method to functionally localize these areas? Which computations are performed in which areas?

5.2 CORTICAL FIELD MAPS

The accurate delineation of cortical areas is important to review here. Cortical areas have traditionally been identified in visual cortex, the most studied of the sensory cortices, based on a combination of the following measurements: (i) cytoarchitecture; (ii) connectivity patterns; (iii) cortical field topography; and (iv) functional characteristics. This definition has led to many controversies because these measurements have conflicted at times. Thus, in vision and in audition, investigations have primarily been limited to the measurement of cortical field maps (arising from topographical measurements), which is the principal measurement of cortical areas in the *in vivo* human brain currently available.

The presence of a cortical field map is established according to several criteria. First, by definition, each cortical field map contains a single representation for each point in the sensory domain (DeYoe et al., 1996; Press, Brewer, Dougherty, Wade, & Wandell, 2001; Sereno et al., 1995; Wandell, Dumoulin, & Brewer, 2007). For this to be valid, orthogonal gradients of fundamental sensory dimensions must comprise each field map (Brewer & Barton, 2012; Wandell et al., 2007). In addition, each field map should represent a substantial

Neurobiology of Language. DOI: http://dx.doi.org/10.1016/B978-0-12-407794-2.00005-5

portion of sensory space, although cortical magnification of specific subsets of sensory space and measurement limitations may reduce the measurable portion. Second, each representation of the sensory domain must be organized as an orderly gradient that is generally contiguous. Third, the general features of the gradient representations comprising the field maps should be consistent across individuals. It is important to note, however, that even well-accepted cortical field maps in visual cortex can vary dramatically in size and anatomical location (Brewer, Liu, Wade, & Wandell, 2005; Dougherty et al., 2003). Even so, the topographical pattern of adjacency among specific cortical field maps should be preserved across individuals.

An important step forward in understanding the organization of human auditory cortex was the recent discovery of 11 auditory field maps (AFMs) on HG (Barton, Venezia, Saberi, Hickok, & Brewer, 2012). These measurements provide an important framework for the organization of individual AFMs in humans as well as the organization of the AFMs with respect to one another. Whereas this information still leaves much to be discovered, it is a fundamental rethinking of the organization of auditory cortex that has implications that resonate throughout the auditory processing hierarchy.

In particular, this rethinking has important implications for speech perception, which occurs at the upper end of the auditory processing hierarchy (Hickok & Poeppel, 2007). To date, research of the neural basis of speech perception overwhelmingly emphasizes the identification of relatively high-level auditory systems that are specialized for coding speech categories (e.g., phonemes). Such work often compares the cortical response to speech with various acoustic controls to factor out low-level acoustic processes. While valuable, this approach ignores the fact that the input to these higher-level systems is derived from an acoustic signal that is already highly processed. Our increased understanding of the inputs to speech perception systems is critical to understanding what kind of categorical information is ultimately extracted from the speech stream and, from a computational perspective, how that information is extracted (Poeppel, Emmorey, Hickok, & Pylkkanen, 2012).

5.3 TONOTOPY: THE FIRST DIMENSION OF AFMs

Beginning in the ear, the auditory system takes a complex sound wave and breaks it down into individual component frequencies, analogous to a Fourier analysis (Spoendlin, 1979). A topographic gradient of low-to-high frequencies, or tones, is referred to as tonotopy (or less commonly, cochleotopy). This basic auditory information and tonotopic organization are preserved through multiple subcortical areas and in low-level auditory cortex (for review, see Ress & Chandrasekaran, 2013; Saenz & Langers, 2014).

Each small band of frequency channels is thus processed largely independently of the others in its own computational pipeline and can be thought of as a common topographic reference frame between individual areas. Tonotopy is thus one aspect of the fundamental auditory reference frame. Each auditory area performs one or more computations across the entire reference frame, such as sound onset, offset, duration, intensity, localization, and others. The degree to which tonotopy is preserved remains unclear above the lower levels of the hierarchy (Barton et al., 2012; Humphries, Liebenthal, & Binder, 2010; Kaas & Hackett, 1998). However, it would be surprising if this information were only preserved to a certain point of the system and then abolished (Brewer & Barton, 2012). It is much more efficient to put that information to use, even if it is only a part of the information necessary to perform a high-order computation.

Although this understanding of one aspect of the auditory reference frame is important, it is incomplete for the purposes of delineating individual AFMs (Barton et al., 2012; Brewer & Barton, 2012; Wandell, Brewer, & Dougherty, 2005; Wandell et al., 2007). To identify tonotopic gradients using a single set of stimuli that activate most or all of the maps, researchers present an array of pure tones (or tone complexes or narrowband noise; Barton et al., 2012; Da Costa et al., 2011; Formisano et al., 2003; Humphries et al., 2010; Sweet, Dorph-Petersen, & Lewis, 2005; Talavage et al., 2004; Upadhyay et al., 2007; Woods et al., 2010). However, tonotopy is one-dimensional, whereas the cortical surface, on which we would like to draw the boundaries between auditory areas, is two-dimensional. Thus, any tonotopic gradient with a given width can be divided into any number of arbitrary individual areas with a complete low-to-high tonotopic gradient (Figure 5.1). As such, we can be certain that auditory areas exist where tonotopic gradients exist, but their number and characteristics require additional measurement, as described. To accurately define an AFM, measurements of a second dimension of the auditory reference frame that is orthogonal to tonotopy are needed (Barton et al., 2012; Brewer & Barton, 2012). Auditory researchers have lacked confirmation of a second dimension until recently; as such, they have relied heavily on complimentary methods that are invasive and typically performed using animal models.

5.4 CORTICAL ORGANIZATION OF THE MONKEY AUDITORY SYSTEM

In monkeys, a standard model of the cortical auditory system has been developed through a

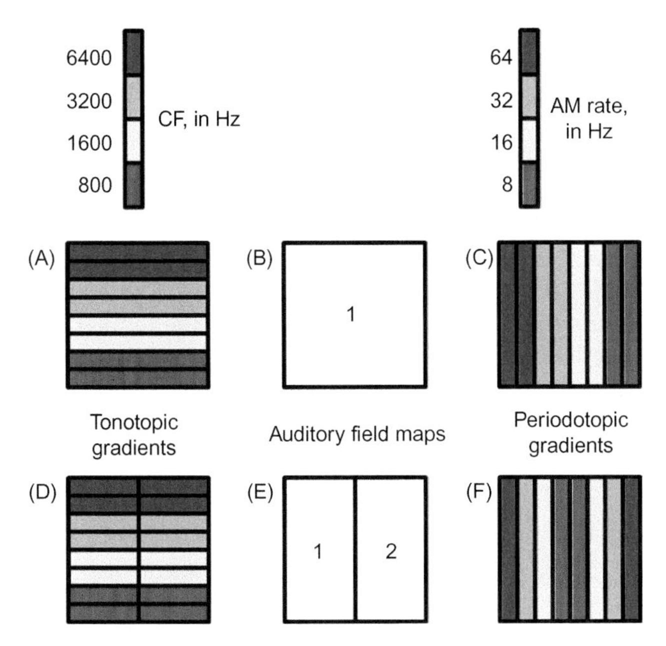

FIGURE 5.1 Orthogonal tonotopic and periodotopic gradients. A cartoon representation of a flattened section of the cortical surface to demonstrate the delineation of AFMs. Inset color legends indicate preferred frequency (tonotopy) or AM rate (periodotopy). The top row is an example showing one AFM existing in the area (B), whereas the bottom row is an example showing two existing in the area (E). (A) and (D) Examples of a tonotopic gradient in the area. Note that without the additional information contained in the periodotopic gradients, it is impossible to know whether there are one, two, or more AFMs in the area. With the information contained in (C), one could conclude that there are orthogonal tonotopic and periodotopic gradients; therefore, one AFM exists in that location (B). Similarly, a combination of the tonotopic gradients in (D) with the periodotopic gradients in (F) would yield two AFMs in the area (E).

convergence of evidence drawn from postmortem cytoarchitectural measurements and tracer studies of anatomical connectivity (de la Mothe, Blumell, Kajikawa, & Hackett, 2006; Fullerton & Pandya, 2007; Hackett, Preuss, & Kaas, 2001; Kaas & Hackett, 1998, 2000), *in vivo* neurophysiological recordings from penetrating electrodes (Kusmierek & Rauschecker, 2009; Merzenich & Brugge, 1973; Morel, Garraghty, & Kaas, 1993; Rauschecker & Tian, 2004; Tian & Rauschecker, 2004), and fMRI measurements (Petkov, Kayser, Augath, & Logothetis, 2006; Petkov, Kayser, Augath, & Logothetis, 2009; Tanji et al., 2010). The model consists of a core comprising three auditory areas (A1, R, and, with less certainty, RT) surrounded by a belt of eight auditory areas (CM, RM, MM, RTM, RTL, RL, AL, and CL). The axis of orientation in the monkey model is indicated by the naming scheme, whereby areas are named by location with respect to A1. The caudal portion of the core is area A1, whereas the more rostral portion contains areas rostral (R) and rostral temporal (RT). The anatomical naming scheme has been adopted for the other maps as well, with four medial and four lateral maps in the belt encircling the core. Several additional areas have been proposed to comprise the parabelt region abutting the belt region (Kaas & Hackett, 2000; Sweet et al., 2005; Tanji et al., 2010), but these areas are not discussed further here because they are not yet relevant for comparison with the current understanding of human AFMs.

The concept of a "core" can generally be ascribed to studies of cytoarchitectural staining and thalamic inputs, which locate a primary-like region of initial auditory processing along monkey superior temporal gyrus (STG; for review see Kaas & Hackett, 2000). Within the definition of the core, three areas, A1, R, and RT, have further been differentiated on the basis of three tonotopic gradients, with one complete gradient per area. These three gradients are oriented in a "high-to-low-to-high" pattern, with high tones represented in the broadly caudal aspect of A1 and low tones in the broadly rostral aspect that is mirrored in R. RT then mirrors R, which creates two abutting tonotopic gradients that reverse from one gradient to the next at the shared boundary between the areas (Figure 5.3). These gradients were measured using electrode penetrations and recordings from relatively small numbers of neurons and later confirmed by measurements of tonotopy using fMRI in monkey (de la Mothe et al., 2006; Kaas & Hackett, 1998, 2000; Kusmierek & Rauschecker, 2009; Merzenich & Brugge, 1973; Morel et al., 1993; Petkov et al., 2006, 2009; Rauschecker & Tian, 2004; Tian & Rauschecker, 2004). In the belt, some areas have been proposed to contain more coarsely organized tonotopic gradients; however, the tonotopy in other belt areas have primarily been measured using fMRI, and thus assume the boundaries between AFMs in the core and belt without directly measuring them with two orthogonal gradients, leaving open the exact organization of each area (Petkov et al., 2006, 2009; Tanji et al., 2010; Tian & Rauschecker, 2004). Furthermore, it is always possible that there are differences among the species of primate studied (Hackett et al., 2001; Kaas & Hackett, 1998, 2000).

5.5 CORTICAL ORGANIZATION OF THE HUMAN AUDITORY SYSTEM

Naturally, the model of low-level auditory cortical organization in the human is largely a transposition of the monkey model. However, the macaque monkey brains typically studied are much smaller than human brains and diverged from human development during evolution more than 25 million years ago (Kumar & Hedges, 1998). As such, the homology of cortical areas among species is not immediately apparent, nor should we assume that the monkey model is a correct representation of the human cortical organization. Rather, the monkey model should be viewed as an especially useful but approximate model of human cortex, where differences are expected.

Similar, yet not identical, cytoarchitectural features of core and belt in macaque STG indicate that HG should be the location of the auditory core and belt in humans. Similar cytoarchitectural techniques that identified monkey core reveal in humans a homologous auditory core region on HG, surrounded by regions similar to monkey belt (Dick et al., 2012; Fullerton & Pandya, 2007; Galaburda & Sanides, 1980; Rivier & Clarke, 1997; Sweet et al., 2005). These results suggest that human and monkey STG are not perfectly homologous, with perhaps a portion of monkey STG evolving into human HG. The same data suggest that the human analogue to CM is located on the medial wall of the lateral fissure, near the tip of HG. This is important, because it anchors the expected orientation of the maps from a strictly rostral–caudal axis for A1 to R to RT in monkeys to a medial-lateral axis in humans.

The majority of measurements of low-level auditory cortex are measures of tonotopy using sets of pure tones, tone bursts, tone complexes, and narrowband noise (Da Costa et al., 2011; Formisano et al., 2003; Humphries et al., 2010; Saenz & Langers, 2014; Sweet et al., 2005; Talavage et al., 2004; Upadhyay et al., 2007; Woods et al., 2010). The data across these studies are actually quite similar, but the models put forth based on the interpretation of the data vary widely. In all cases, there is a central low-frequency representation centered on HG with increasing frequencies represented in surrounding bands that form an approximately circular shape. In some cases, portions of the circle are weak, such that the higher-frequency bands resemble a horseshoe shape nearly encircling the low frequencies (Humphries et al., 2010). It is likely that higher-order auditory areas also contain tonotopic gradients, because it is unlikely that the tonotopic information has simply been discarded at this level of processing. Such a persistence of topographic organization into higher-order sensory processing regions has now been measured in the visual system (for reviews, see Brewer & Barton, 2012; Wandell et al., 2007).

Sometimes different naming schemes for these tonotopic measurements of auditory areas have been adopted, but the majority of studies place A1 on the medial or posterior aspect of HG (Da Costa et al., 2011; Formisano et al., 2003; Humphries et al., 2010; Sweet et al., 2005; Talavage et al., 2004; Upadhyay et al., 2007; Woods et al., 2010). Then, R is placed variably on the lateral or anterior aspect of HG, depending on where the strongest "high-to-low-to-high" tonotopic reversal pattern can be identified. Sometimes the "high-to-low-to-high" pattern is a relatively straight path, but in other cases it is bent. Typically, researchers have guessed at boundaries of cortical areas based on the tonotopic gradient and the potentially homologous monkey model, resulting in either a medial–lateral (consistent with the cytoarchitecture) or anterior–posterior axis of orientation (Da Costa et al., 2011; Formisano et al., 2003; Humphries et al., 2010; Sweet et al., 2005; Talavage et al., 2004; Upadhyay et al., 2007; Woods et al., 2010). In sum, there has been little agreement on how to interpret the similar datasets; there are simply too many ways to interpret sets of one-dimensional gradients when trying to divide them on the two-dimensional cortical surface (for review of similar issues in visual cortex see Brewer & Barton, 2012; Wandell et al., 2007).

5.6 PERIODOTOPY: THE SECOND DIMENSION OF AFMs

Converging evidence suggests that the human homologue of monkey auditory core is located on HG, but the fact that tonotopy is one-dimensional makes it difficult to use tonotopic gradients alone to identify AFMs. An orthogonal gradient is necessary, but for that another fundamental type of information that is a component of the auditory reference frame must be identified. Humans can differentiate sounds based on their pitch, temporal information, loudness, and timber. Of these, timber is very likely a process of recognizing combinations of the other three characteristics when one differentiates a tone played by, for example, a flute or an oboe; thus, it is thought that this is unlikely to be part of the fundamental auditory reference frame, but more likely processed along specific computational processing pathways in auditory cortex (Menon et al., 2002; Rauschecker & Scott, 2009; Zatorre, Belin, & Penhune, 2002). Intensity is a good candidate; it is used in sound localization (Middlebrooks & Green, 1991) (i.e., comparing intensity of the same sound detected by each ear) and sound motion localization (McBeath & Neuhoff, 2002) (e.g., the Doppler effect). However, intensity is very similar to brightness in the visual domain, which is not one of the fundamental gradients used to identify visual field maps (VFMs), and it is encoded in the cochlea and potentially in cortex as increases in firing rates of tonotopically tuned neurons rather than in a gradient organization (Shapley et al., 2007; Spoendlin, 1979; Tanji et al., 2010; Wandell et al., 2007).

Recent human psychophysical studies indicate that there are separable filter banks (neurons with receptive fields or tunings) for not only frequency spectra (as expected given tonotopy) but also temporal information (Dau, Kollmeier, & Kohlrausch, 1997; Ewert & Dau,

2000; Hsieh & Saberi, 2010). Furthermore, gradients based on temporal information have recently been discovered in cat primary auditory cortex and the macaque midbrain (Baumann et al., 2011; Langner, Dinse, & Godde, 2009). In addition, such gradients, known as periodotopic gradients, were measured in both cases to be in the same location as, but orthogonal to, tonotopic gradients. Periodotopy refers to the topographic organization of neurons that respond differentially to sounds of different temporal envelope modulation rates.

Inspired by these studies, Barton et al. (2012) recently presented amplitude-modulated (AM) broadband noise to human subjects to measure the cortical periodotopic responses in humans using fMRI. With these stimuli, temporal duration refers to the length of time from peak-to-peak of the AM noise. This stimulus is designed to activate neurons with tuning to sounds that last for a particular duration; in other words, the stimuli differentiate temporal tuning. These stimuli likely drive neurons that respond to the onset and offset of sounds with different amounts of lag time before they can be reactivated, as well as neurons that respond to sounds that exist for a certain duration. Like the monkey and cat studies, Barton et al. (2012) also presented tonotopic mapping stimuli (narrowband noise with varying AM rates) to the same subjects and investigated the responses in low-level auditory cortex.

Three primary findings resulted from the work of Barton et al. (2012). First, temporal information is the second fundamental type of sound information of the human auditory reference frame, complimenting spectral (frequency) information. Second, tonotopy and periodotopy are represented orthogonally to one another in human cortex, allowing for the localization of individual AFMs. By identifying both tonotopic and periodotopic gradients in the same locations and measuring that these gradients are orthogonal to one another, Barton et al. (2012) were able to localize 11 independent AFMs that largely resemble the 11 AFMs of the monkey model (Figure 5.3). Taking into account many characteristics of their data, as well as the correspondence to the monkey model and the underlying human cytoarchitecture, they named each of the AFMs based on those of the monkey model (hA1, hR, hML, hAL, hMM, hRM, hCM, hCL, hRT, hRTM, and hRTL). Because the monkey areas were named based on orientation and because the human AFMs are oriented medial–lateral rather than caudal–rostral, the human AFMs are only the abbreviated letters, not the full title used for monkeys (e.g., hRM stands for human RM, not human rostral medial), and "h" has been appended to mean "human." Third, these individual AFMs are organized into at least one "clover leaf" cluster (see Section 5.8) (Figures 5.2 and 5.3).

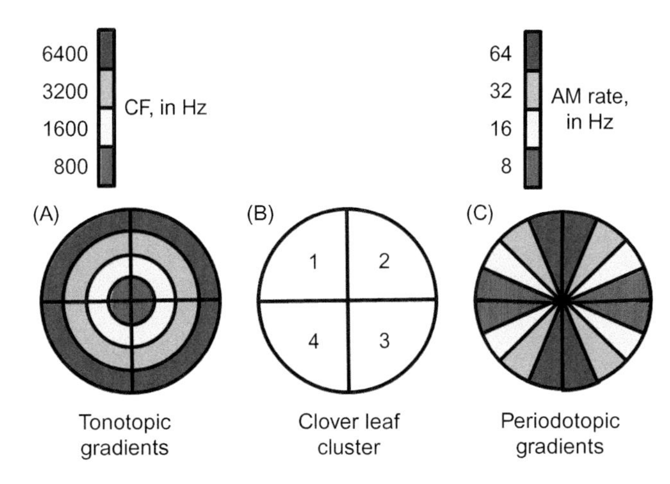

FIGURE 5.2 "Clover leaf" cluster organization. A cartoon example of a "clover leaf" cluster of AFMs on a flattened section of the cortical surface. Inset color legends indicate preferred frequency (tonotopy) or AM rate (periodotopy). This example contains four AFMs, indicated in (B). (A) Each AFM has a complete tonotopic gradient organized such that low-frequency bands are represented in the center of the cluster, with increasing frequencies represented in more peripheral bands of the cluster. (C) Each AFM also has a complete periodotopic gradient organized such that all AM rate bands span from the center to the periphery of the cluster like spokes on a wheel. Each AFM thus has orthogonal tonotopic and periodotopic gradients.

5.7 SIMILARITIES TO AFM ORGANIZATION IN THE HUMAN VISUAL SYSTEM

Visual information is pooled in a series of steps in the retina before being passed through the optic nerve to the thalamus. From there, this information is passed through optic radiations to primary visual cortex into an area known as V1. From there, visual information branches out to visual areas that perform various computations in a generally hierarchical manner from low-level simple visual feature processing to high-level complex feature analysis (for review see Dacey, 2000; Van Essen et al., 1990).

Several characteristics are shared between these areas. At each area, one or more computations are performed for locations of varying size across the entire visual scene (Brewer et al., 2005; Van Essen et al., 1990; Wandell, 1999; Wandell et al., 2005, 2007). These computations are performed by neurons with receptive fields of a portion of visual space, typically in a mutually inhibitory center-surround organization, such that a neuron will receive inputs from other related neurons (Burkhalter & Bernardo, 1989; Carandini & Heeger, 1994; Carandini, Horton, & Sincich, 2007). A very efficient way to accomplish this task is to keep

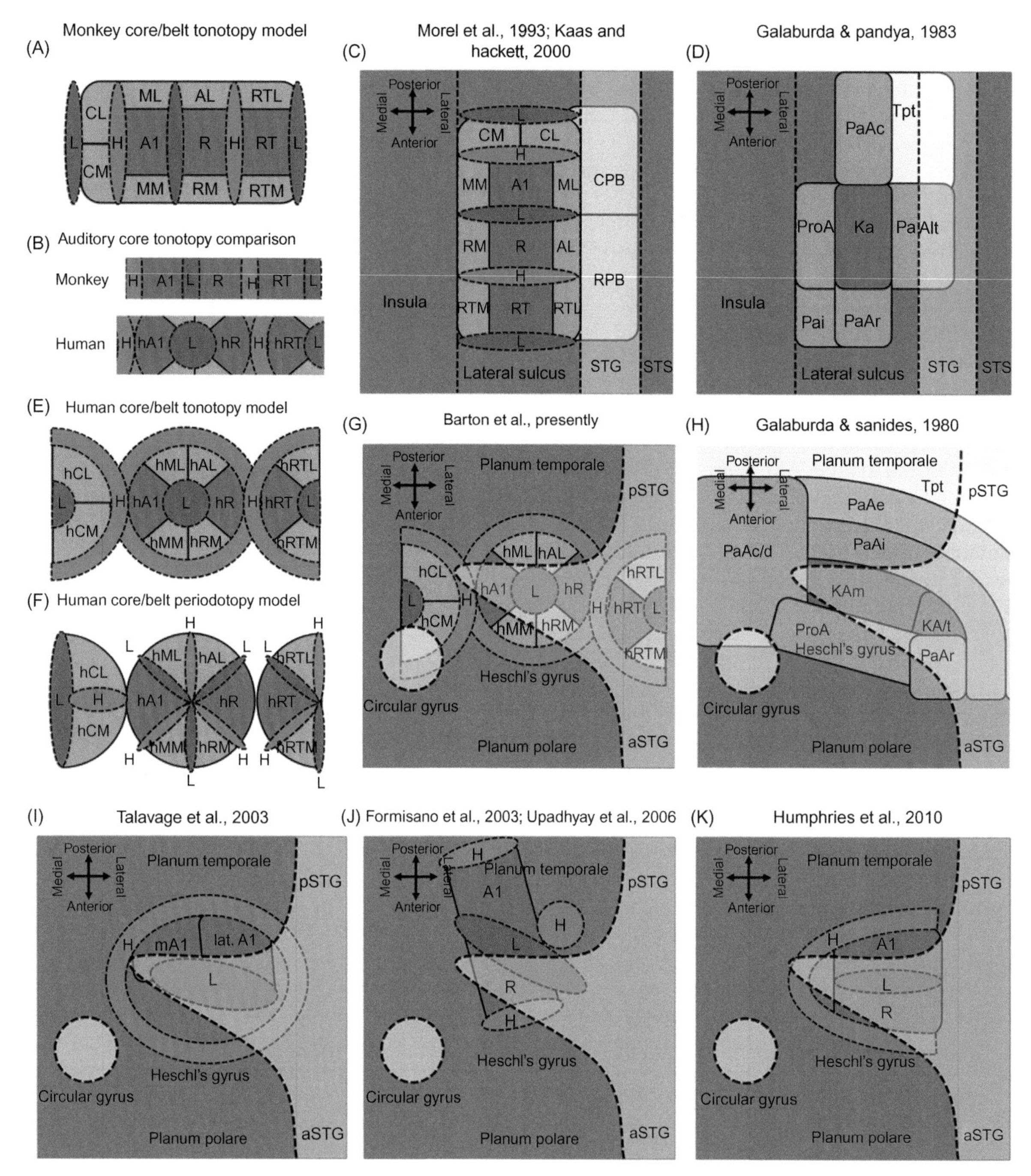

FIGURE 5.3 Comparison of present "clover leaf" cluster organization with human and monkey tonotopic and anatomical data. Cartoon models depict representations of the original models cited above each figure, modified for consistency here. "L" stands for "low" and "H" stands for "high," referring to low (red regions) or high (blue regions) model tonotopic or periodotopic responses. Dark gray indicates sulci or the plane of the lateral fissure, whereas light gray indicates gyri. Purple regions represent auditory core. Orange regions indicate auditory belt. Green regions indicate auditory parabelt. Yellow regions indicate temporal planum temporal (Tpt). All figures are oriented along the same global axes (see direction legends). (A) Monkey core/belt tonotopy model. (B) Auditory core tonotopy comparison. Top: monkey. Bottom: human. (C) Tonotopic model of monkey auditory core, belt, and parabelt. (D) Cytoarchitectonic model of monkey auditory cortex. (E) Our human core/belt tonotopy model. (F) Our human core/belt periodotopy model. (G) Our tonotopic model of human belt and core. (H) Cytoarchitectonic model of human auditory cortex. (I–K) Recent human fMRI tonotopy models.

neurons that analyze nearby locations in space close to one another (Chklovskii & Koulakov, 2004; Mitchison, 1991; Moradi & Heeger, 2009; Shapley et al., 2007). As a result, visual areas are organized retinotopically; that is, nearby points of visual space are represented by neurons in nearby locations in cortex after the organization in the retina (Brewer et al., 2005; Wandell, 1999; Wandell et al., 2005, 2007). Because visual space is basically two-dimensional, we can chart two gradients: eccentricity (i.e., distance from a central fixation point) and polar angle (i.e., angular distance around a central fixation point). A VFM can be defined where these two gradients are measured in the same location of cortex, each representing most or all of one hemifield of visual space and positioned orthogonally to one another.

Retinotopy allows for efficient connections within a VFM, but what about between VFMs? One organizational pattern that would maintain efficient connections between nearby points in space across maps would be to have the representations of a visuospatial gradient reverse from one visual area to the next at the visual area boundaries, such that two VFMs abut at a merged representation of, for example, the upper vertical meridian of the visual field, as seen in the adjoining AFMs elsewhere (Barton et al., 2012; Figure 5.2). To be efficient for two gradient dimensions, an efficient pattern is to have approximately circular clusters of maps. Then, reversals of one dimension occur between maps within the cluster (e.g., polar angle) and reversals for the other occur between maps in different clusters (e.g., eccentricity). The result is concentric rings of increasingly eccentric iso-eccentricity representations expanding from the center of a cluster and lines of iso-polar angles extending from the center to the periphery of the cluster like spokes on a wheel. Such clusters of VFMs have been discovered in human and monkey visual cortex and have been termed "clover leaf" clusters (Brewer & Barton, 2011; Kolster, Peeters, & Orban, 2010; Wandell et al., 2005, 2007). For a recent detailed discussion of VFM organization and the implications of "clover leaf" clusters, see Brewer and Barton (2012).

5.8 "CLOVER LEAF" CLUSTERS ACROSS SENSES

Revealing a similar macrostructural organization to the visual system, the first "clover leaf" cluster was discovered in the human auditory system on HG (Barton et al., 2012). The same benefit of efficient connectivity is achieved in the same way, by representing nearby portions of auditory acoustic dimensions (i.e., tonotopy and periodotopy) in nearby portions of cortex. Specifically, the first auditory "clover leaf" cluster consists of six AFMs: hA1, hAL, hML, hR, hRM, and hMM. Concentric circles of increasing iso-tone bands expand from the low-frequency representation centered on HG, with iso-period bands extending from the center to the periphery of the cluster like spokes on a wheel. Abutting the cluster where HG meets the STG, there exists a tonotopic reversal into three additional AFMs: hRT, hRTL, and hRTM. Additionally, two more AFMs abut the HG cluster medially: hCM and hCL. Although hRT, hRTL, hRTM, hCM, and hCL have not been shown to be organized into complete clusters of AFMs yet, we suspect that they are as well; more research is required to determine whether that is the case.

It is important to consider the monkey cortical model now that the first human auditory "clover leaf" cluster has been discovered. Because tonotopy and cytoarchitecture indicate that the early auditory cortical organization is very similar across these primate species, it is very likely that the monkey areas are organized into "clover leaf" clusters as well. Additionally, "clover leaf" clusters in the visual domain have been found to exist in macaque in homologous areas to clusters in humans (Kolster et al., 2009, 2010), suggesting that the same will be true in the auditory domain. It is impossible to know for certain until orthogonal periodotopic gradients are identified in the same monkey cortex as tonotopic gradients (Barton et al., 2012).

One interesting feature of "clover leaf" clusters is that there must be an even number of maps in a cluster, likely to minimize connection length by always having gradient reversals at the adjoining boundaries between maps. With an odd number of maps in a cluster, there would need to be at least one discrete jump between periodotopic gradient representations, which would reduce the connection efficiency. So far, "clover leaf" clusters have only been observed with even numbers of maps in a cluster (Brewer & Barton, 2012; Kolster et al., 2009, 2010; Wandell et al., 2005, 2007). Another interesting feature is that there need not be the same number of maps in each cluster; so far, two, four, and six maps have been observed in "clover leaf" clusters. Finally, the fact that these clusters of sensory field maps have been observed with such similar characteristics across two senses strongly implies that similar organizational schemes are common for sensory systems in the brain in general.

5.9 CONCLUSION

The cortical organization of the human auditory system so far has been incompletely measured. Although we know some of the features and locations of low-level processing in cortex, until recently we did not have the tools to localize individual AFMs. The key insight to the second dimension of AFMs, periodotopy, was recently discovered. Because tonotopic and periodotopic gradients are represented orthogonally to one another along the cortical surface, it is possible to accurately differentiate the locations of individual cortical AFMs rather than to attempt to estimate map boundaries based only on tonotopy. Additionally, a new organizational scheme of AFMs has been revealed: the "clover leaf" cluster. With these new insights into AFM organization, researchers can now better localize and identify specific regions of auditory cortex across subjects and more accurately investigate which computations each of the AFMs subserves. Human visual and auditory cortex interestingly share a common organizational scheme, with each sensory system compartmentalized into cortical field maps that are themselves arranged on a larger scale into "clover leaf" clusters. Such similarity may be common across many sensory systems, which may aid in the future identification of cortical field maps in the representation of other senses.

Naturally, many important questions remain unanswered. The AFMs identified so far are unlikely to be the last; novel human cortical VFMs continue to be discovered after two decades of research, and we expect that additional AFMs will be measured in human auditory cortex outside of HG. Because complex language is a uniquely human trait, animal models offer little guidance as to which AFMs to expect to perform relevant computations. However, evidence from monkey perception of monkey vocalizations (Petkov et al., 2008) suggests distinct cortical regions as strong candidates for investigation of specific AFMs subserving human speech perception. Armed with greater knowledge of the inputs to and organization of the initial tiers of the auditory processing hierarchy, researchers will finally be in a position to rigorously investigate the nature of higher-order speech perception computations (Hickok & Poeppel, 2007).

References

Barton, B., Venezia, J. H., Saberi, K., Hickok, G., & Brewer, A. A. (2012). Orthogonal acoustic dimensions define auditory field maps in human cortex. [Research Support, N.I.H., Extramural Research Support, Non-U.S. Gov't]. *Proceedings of the National Academy of Sciences of the United States of America*, *109*(50), 20738–20743. Available from: http://dx.doi.org/10.1073/pnas.1213381109.

Baumann, S., Griffiths, T. D., Sun, L., Petkov, C. I., Thiele, A., & Rees, A. (2011). Orthogonal representation of sound dimensions in the primate midbrain. *Nature Neuroscience*, *14*(4), 423–425. Available from: http://dx.doi.org/10.1038/nn.2771.

Brewer, A.A., & Barton, B. (2011) 'Clover Leaf' Clusters in Human Visual Cortex. Toulouse, France: European Conference on Visual Perception. Perception 40 (ECVP Abstract Supplement), 48.

Brewer, A. A., & Barton, B. (2012). Visual field map organization in human visual cortex. In S. Molotchnikoff (Ed.), *Visual Cortex*. New York: InTech.

Brewer, A. A., Liu, J., Wade, A. R., & Wandell, B. A. (2005). Visual field maps and stimulus selectivity in human ventral occipital cortex. *Nature Neuroscience*, *8*(8), 1102–1109. Available from: http://dx.doi.org/10.1038/nn1507.

Burkhalter, A., & Bernardo, K. L. (1989). Organization of corticocortical connections in human visual cortex. *Proceedings of the National Academy of Sciences of the United States of America*, *86*(3), 1071–1075.

Carandini, M., & Heeger, D. J. (1994). Summation and division by neurons in primate visual cortex. [Research Support, Non-U.S. Gov't Research Support, U.S. Gov't, Non-P.H.S. Research Support, U.S. Gov't, P.H.S.]. *Science*, *264*(5163), 1333–1336.

Carandini, M., Horton, J. C., & Sincich, L. C. (2007). Thalamic filtering of retinal spike trains by postsynaptic summation. *Journal of Vision*, *7*(14), 20, 1–11. Available from: http://dx.doi.org/10.1167/7.14.20/7/14/20/ [pii].

Chklovskii, D. B., & Koulakov, A. A. (2004). Maps in the brain: What can we learn from them? *Annual Review of Neuroscience*, *27*, 369–392. Available from: http://dx.doi.org/10.1146/annurev.neuro.27.070203.144226.

Dacey, D. M. (2000). Parallel pathways for spectral coding in primate retina. [Research Support, U.S. Gov't, P.H.S. Review]. *Annual Review of Neuroscience*, *23*, 743–775. Available from: http://dx.doi.org/10.1146/annurev.neuro.23.1.743.

Da Costa, S., van der Zwaag, W., Marques, J. P., Frackowiak, R. S., Clarke, S., & Saenz, M. (2011). Human primary auditory cortex follows the shape of Heschl's gyrus. *The Journal of Neuroscience: The Official Journal of the Society for Neuroscience*, *31*(40), 14067–14075. Available from: http://dx.doi.org/10.1523/JNEUROSCI.2000-11.2011.

Dau, T., Kollmeier, B., & Kohlrausch, A. (1997). Modeling auditory processing of amplitude modulation. II. Spectral and temporal integration. *The Journal of the Acoustical Society of America*, *102* (5 Pt 1), 2906–2919.

de la Mothe, L. A., Blumell, S., Kajikawa, Y., & Hackett, T. A. (2006). Cortical connections of the auditory cortex in marmoset monkeys: Core and medial belt regions. *The Journal of Comparative Neurology*, *496*(1), 27–71. Available from: http://dx.doi.org/10.1002/cne.20923.

DeYoe, E. A., Carman, G. J., Bandettini, P., Glickman, S., Wieser, J., Cox, R., et al. (1996). Mapping striate and extrastriate visual areas in human cerebral cortex. [Research Support, U.S. Gov't, P.H.S.]. *Proceedings of the National Academy of Sciences of the United States of America*, *93*(6), 2382–2386.

Dick, F., Tierney, A. T., Lutti, A., Josephs, O., Sereno, M. I., & Weiskopf, N. (2012). *In vivo* functional and myeloarchitectonic mapping of human primary auditory areas. [Research Support, N.I.H., Extramural Research Support, Non-U.S. Gov't]. *The Journal of Neuroscience: The Official Journal of the Society for Neuroscience*, *32*(46), 16095–16105. Available from: http://dx.doi.org/10.1523/JNEUROSCI.1712-12.2012.

Dougherty, R. F., Koch, V. M., Brewer, A. A., Fischer, B., Modersitzki, J., & Wandell, B. A. (2003). Visual field representations and locations of visual areas V1/2/3 in human visual cortex. *Journal of Vision*, *3*(10), 586–598 . Available from: http://dx.doi.org/10.1167/3.10.1/3/10/1/ [pii].

Ewert, S. D., & Dau, T. (2000). Characterizing frequency selectivity for envelope fluctuations. *The Journal of the Acoustical Society of America*, *108*(3 Pt 1), 1181–1196.

Formisano, E., Kim, D. S., Di Salle, F., van de Moortele, P. F., Ugurbil, K., & Goebel, R. (2003). Mirror-symmetric tonotopic maps in human primary auditory cortex. *Neuron*, *40*(4), 859–869. Available from: http://dx.doi.org/10.1016/s0896-6273(03)00669-X [pii].

Fullerton, B. C., & Pandya, D. N. (2007). Architectonic analysis of the auditory-related areas of the superior temporal region in human brain. *The Journal of Comparative Neurology*, *504*(5), 470–498. Available from: http://dx.doi.org/10.1002/cne.21432.

Galaburda, A., & Sanides, F. (1980). Cytoarchitectonic organization of the human auditory cortex. *The Journal of Comparative Neurology*, *190*(3), 597–610. Available from: http://dx.doi.org/10.1002/cne.901900312.

Hackett, T. A., Preuss, T. M., & Kaas, J. H. (2001). Architectonic identification of the core region in auditory cortex of macaques, chimpanzees, and humans. *The Journal of Comparative Neurology*,

441(3), 197–222. Available from: http://dx.doi.org/10.1002/cne.1407.

Hickok, G., & Poeppel, D. (2007). The cortical organization of speech processing. [Research Support, N.I.H., Extramural Review]. *Nature Reviews Neuroscience*, *8*(5), 393–402. Available from: http://dx.doi.org/10.1038/nrn2113.

Hsieh, I. H., & Saberi, K. (2010). Detection of sinusoidal amplitude modulation in logarithmic frequency sweeps across wide regions of the spectrum. *Hearing Research*, *262*(1–2), 9–18. Available from: http://dx.doi.org/10.1016/j.heares.2010.02.002.

Humphries, C., Liebenthal, E., & Binder, J. R. (2010). Tonotopic organization of human auditory cortex. *Neuroimage*, *50*(3), 1202–1211. Available from: http://dx.doi.org/10.1016/j.neuroimage.2010.01.046.

Kaas, J. H., & Hackett, T. A. (1998). Subdivisions of auditory cortex and levels of processing in primates. *Audiology and Neuro-Otology*, *3*(2–3), 73–85. Available from: http://dx.doi.org/10.1159/000013783 [pii].

Kaas, J. H., & Hackett, T. A. (2000). Subdivisions of auditory cortex and processing streams in primates. *Proceedings of the National Academy of Sciences of the United States of America*, *97*(22), 11793–11799. Available from: http://dx.doi.org/10.1073/pnas.97.22.1179397/22/11793 [pii].

Kolster, H., Mandeville, J. B., Arsenault, J. T., Ekstrom, L. B., Wald, L. L., & Vanduffel, W. (2009). Visual field map clusters in macaque extrastriate visual cortex. *The Journal of Neuroscience: The Official Journal of the Society for Neuroscience*, *29*(21), 7031–7039. Available from: http://dx.doi.org/10.1523/JNEUROSCI.0518-09.2009.

Kolster, H., Peeters, R., & Orban, G. A. (2010). The retinotopic organization of the human middle temporal area MT/V5 and its cortical neighbors. *The Journal of Neuroscience: The Official Journal of the Society for Neuroscience*, *30*(29), 9801–9820. Available from: http://dx.doi.org/10.1523/JNEUROSCI.2069-10.2010.

Krubitzer, L. (2007). The magnificent compromise: Cortical field evolution in mammals. *Neuron*, *56*(2), 201–208. Available from: http://dx.doi.org/10.1016/j.neuron.2007.10.002.

Kumar, S., & Hedges, S. B. (1998). A molecular timescale for vertebrate evolution. [Research Support, U.S. Gov't, Non-P.H.S. Research Support, U.S. Gov't, P.H.S.]. *Nature*, *392*(6679), 917–920. Available from: http://dx.doi.org/10.1038/31927.

Kusmierek, P., & Rauschecker, J. P. (2009). Functional specialization of medial auditory belt cortex in the alert rhesus monkey. [Research Support, N.I.H., Extramural]. *Journal of Neurophysiology*, *102*(3), 1606–1622. Available from: http://dx.doi.org/10.1152/jn.00167.2009.

Langner, G., Dinse, H. R., & Godde, B. (2009). A map of periodicity orthogonal to frequency representation in the cat auditory cortex. *Frontiers in Integrative Neuroscience*, *3*, 27. Available from: http://dx.doi.org/10.3389/neuro.07.027.2009.

McBeath, M. K., & Neuhoff, J. G. (2002). The Doppler effect is not what you think it is: Dramatic pitch change due to dynamic intensity change. [Research Support, Non-U.S. Gov't Research Support, U.S. Gov't, Non-P.H.S.]. *Psychonomic Bulletin & Review*, *9*(2), 306–313.

Menon, V., Levitin, D. J., Smith, B. K., Lembke, A., Krasnow, B. D., Glazer, D., et al. (2002). Neural correlates of timbre change in harmonic sounds. [Research Support, Non-U.S. Gov't Research Support, U.S. Gov't, P.H.S.]. *Neuroimage*, *17*(4), 1742–1754.

Merzenich, M. M., & Brugge, J. F. (1973). Representation of the cochlear partition of the superior temporal plane of the macaque monkey. *Brain Research*, *50*(2), 275–296.

Middlebrooks, J. C., & Green, D. M. (1991). Sound localization by human listeners. [Research Support, Non-U.S. Gov't Research Support, U.S. Gov't, Non-P.H.S. Research Support, U.S. Gov't, P.H.S. Review]. *Annual Review of Psychology*, *42*, 135–159. Available from: http://dx.doi.org/10.1146/annurev.ps.42.020191.001031.

Mitchison, G. (1991). Neuronal branching patterns and the economy of cortical wiring. *Proceedings Biological Sciences/The Royal Society*, *245*(1313), 151–158. Available from: http://dx.doi.org/10.1098/rspb.1991.0102.

Moradi, F., & Heeger, D. J. (2009). Inter-ocular contrast normalization in human visual cortex. *Journal of Vision*, *9*(3), 13 11-22. Available from: http://dx.doi.org/10.1167/9.3.13/9/3/13/ [pii].

Morel, A., Garraghty, P. E., & Kaas, J. H. (1993). Tonotopic organization, architectonic fields, and connections of auditory cortex in macaque monkeys. *The Journal of Comparative Neurology*, *335*(3), 437–459. Available from: http://dx.doi.org/10.1002/cne.903350312.

Petkov, C. I., Kayser, C., Augath, M., & Logothetis, N. K. (2006). Functional imaging reveals numerous fields in the monkey auditory cortex. *PLoS Biology*, *4*(7), e215. Available from: http://dx.doi.org/10.1371/journal.pbio.0040215.

Petkov, C. I., Kayser, C., Augath, M., & Logothetis, N. K. (2009). Optimizing the imaging of the monkey auditory cortex: Sparse vs. continuous fMRI. *Magnetic Resonance Imaging*, *27*(8), 1065–1073. Available from: http://dx.doi.org/10.1016/j.mri.2009.01.018.

Petkov, C. I., Kayser, C., Steudel, T., Whittingstall, K., Augath, M., & Logothetis, N. K. (2008). A voice region in the monkey brain. [Research Support, Non-U.S. Gov't]. *Nature Neuroscience*, *11*(3), 367–374. Available from: http://dx.doi.org/10.1038/nn2043.

Poeppel, D., Emmorey, K., Hickok, G., & Pylkkanen, L. (2012). Towards a new neurobiology of language. [Research Support, N. I.H., Extramural Research Support, U.S. Gov't, Non-P.H.S. Review]. *The Journal of Neuroscience: The Official Journal of the Society for Neuroscience*, *32*(41), 14125–14131. Available from: http://dx.doi.org/10.1523/JNEUROSCI.3244-12.2012.

Press, W. A., Brewer, A. A., Dougherty, R. F., Wade, A. R., & Wandell, B. A. (2001). Visual areas and spatial summation in human visual cortex. *Vision Research*, *41*(10-11), 1321–1332. Available from: http://dx.doi.org/10.1016/S0042-6989(01)00074-8.

Rauschecker, J. P., & Scott, S. K. (2009). Maps and streams in the auditory cortex: Nonhuman primates illuminate human speech processing. [Research Support, N.I.H., Extramural Research Support, Non-U.S. Gov't Research Support, U.S. Gov't, Non-P.H. S. Review]. *Nature Neuroscience*, *12*(6), 718–724. Available from: http://dx.doi.org/10.1038/nn.2331.

Rauschecker, J. P., & Tian, B. (2004). Processing of band-passed noise in the lateral auditory belt cortex of the rhesus monkey. [Research Support, U.S. Gov't, P.H.S.]. *Journal of Neurophysiology*, *91*(6), 2578–2589. Available from: http://dx.doi.org/10.1152/jn.00834.2003.

Ress, D., & Chandrasekaran, B. (2013). Tonotopic organization in the depth of human inferior colliculus. *Frontiers in Integrative Neuroscience*, *7*, 586. Available from: http://dx.doi.org/10.3389/fnhum.2013.00586.

Rivier, F., & Clarke, S. (1997). Cytochrome oxidase, acetylcholinesterase, and NADPH-diaphorase staining in human supratemporal and insular cortex: Evidence for multiple auditory areas. *Neuroimage*, *6*(4), 288–304. Available from: http://dx.doi.org/10.1006/nimg.1997.0304.

Saenz, M., & Langers, D. R. (2014). Tonotopic mapping of human auditory cortex. [Review]. *Hearing Research*, *307*, 42–52. Available from: http://dx.doi.org/10.1016/j.heares.2013.07.016.

Schreiner, C. E., & Winer, J. A. (2007). Auditory cortex mapmaking: Principles, projections, and plasticity. [Research Support, N.I.H., Extramural Review]. *Neuron*, *56*(2), 356–365. Available from: http://dx.doi.org/10.1016/j.neuron.2007.10.013.

Sereno, M. I., Dale, A. M., Reppas, J. B., Kwong, K. K., Belliveau, J. W., Brady, T. J., et al. (1995). Borders of multiple visual areas in humans revealed by functional magnetic resonance imaging. [Research Support, Non-U.S. Gov't Research Support, U.S. Gov't, P.H.S.]. *Science*, *268*(5212), 889–893.

Shapley, R., Hawken, M., & Xing, D. (2007). The dynamics of visual responses in the primary visual cortex. *Progress in Brain Research*,

165, 21–32. Available from: http://dx.doi.org/10.1016/S0079-6123(06)65003-6.

Spoendlin, H. (1979). Sensory neural organization of the cochlea. [Review]. *The Journal of Laryngology and Otology*, *93*(9), 853–877.

Sweet, R. A., Dorph-Petersen, K. A., & Lewis, D. A. (2005). Mapping auditory core, lateral belt, and parabelt cortices in the human superior temporal gyrus. *The Journal of Comparative Neurology*, *491*(3), 270–289. Available from: http://dx.doi.org/10.1002/cne.20702.

Talavage, T. M., Sereno, M. I., Melcher, J. R., Ledden, P. J., Rosen, B. R., & Dale, A. M. (2004). Tonotopic organization in human auditory cortex revealed by progressions of frequency sensitivity. *Journal of Neurophysiology*, *91*(3), 1282–1296 . Available from: http://dx.doi.org/10.1152/jn.01125.200201125.2002 [pii].

Tanji, K., Leopold, D. A., Ye, F. Q., Zhu, C., Malloy, M., Saunders, R. C., et al. (2010). Effect of sound intensity on tonotopic fMRI maps in the unanesthetized monkey. *Neuroimage*, *49*(1), 150–157. Available from: http://dx.doi.org/10.1016/j.neuroimage.2009.07.029.

Tian, B., & Rauschecker, J. P. (2004). Processing of frequency-modulated sounds in the lateral auditory belt cortex of the rhesus monkey. [Research Support, U.S. Gov't, P.H.S.]. *Journal of Neurophysiology*, *92*(5), 2993–3013. Available from: http://dx.doi.org/10.1152/jn.00472.2003.

Upadhyay, J., Ducros, M., Knaus, T. A., Lindgren, K. A., Silver, A., Tager-Flusberg, H., et al. (2007). Function and connectivity in human primary auditory cortex: A combined fMRI and DTI study at 3 Tesla. *Cerebral Cortex*, *17*(10), 2420–2432. Available from: http://dx.doi.org/10.1093/cercor/bhl150.

Van Essen, D. C., Felleman, D. J., DeYoe, E. A., Olavarria, J., & Knierim, J. (1990). Modular and hierarchical organization of extrastriate visual cortex in the macaque monkey. *Cold Spring Harbor Symposia on Quantitative Biology*, *55*, 679–696.

Wandell, B. A. (1999). Computational neuroimaging of human visual cortex. *Annual Review of Neuroscience*, *22*, 145–173. Available from: http://dx.doi.org/10.1146/annurev.neuro.22.1.145.

Wandell, B. A., Brewer, A. A., & Dougherty, R. F. (2005). Visual field map clusters in human cortex. *Philosophical Transactions of the Royal Society of London Series B, Biological Sciences*, *360*(1456), 693–707. Available from: http://dx.doi.org/10.1098/rstb.2005.1628.

Wandell, B. A., Dumoulin, S. O., & Brewer, A. A. (2007). Visual field maps in human cortex. *Neuron*, *56*(2), 366–383. Available from: http://dx.doi.org/10.1016/j.neuron.2007.10.012.

Woods, D. L., Herron, T. J., Cate, A. D., Yund, E. W., Stecker, G. C., Rinne, T., et al. (2010). Functional properties of human auditory cortical fields. *Frontiers in Systems Neuroscience*, *4*, 155. Available from: http://dx.doi.org/10.3389/fnsys.2010.00155.

Zatorre, R. J., Belin, P., & Penhune, V. B. (2002). Structure and function of auditory cortex: Music and speech. *Trends in Cognitive Sciences*, *6*(1), 37–46.

CHAPTER

6

Motor Cortex and Mirror System in Monkeys and Humans

Giacomo Rizzolatti[1,2] and Stefano Rozzi[1]

[1]Department of Neuroscience, University of Parma, Parma, Italy; [2]Brain Center for Motor and Social Cognition, Italian Institute of Technology, Parma, Italy

6.1 INTRODUCTION

There are many ways in which individuals communicate. Undoubtedly, at the core of human communication is speech. However, humans also communicate using gestures, body postures, and facial expressions. Whereas speech is uniquely human, gestural communication is also present in other species of primates. There is still controversy regarding the relation between these two types of communication, yet it is very likely that they have a common evolutionary root.

Starting from these premises, we first review the anatomy and physiology of the motor system in the monkey, with emphasis on those areas that are involved in communicative functions. We then compare the anatomical and functional properties of these areas with those of humans. A particular emphasis is given to the mirror mechanism and its role in communication.

An important point to stress is that communication is not necessarily intentional. Many messages that we receive from conspecifics as well as from individuals of other species are not intentional. Of these two types of communications, the nonintentional one is the most primitive. At the end of this chapter, we argue that intentional communication is an evolutionarily development of nonintentional communication.

6.2 ANATOMY OF THE MONKEY MOTOR CORTEX

6.2.1 The Agranular Frontal Cortex

According to the classical cortical map of Brodmann (1909), the motor cortex is formed by two cytoarchitectonic areas: area 4 and area 6. Although Brodmann considered area 6 as a single entity, various authors subdivided it into various sub-areas (Vogt & Vogt, 1919; Von Bonin & Bailey, 1947). In recent years, the combination of the cytoarchitectonic techniques with the neurochemical ones has proven to be useful for a more objective assessment of areal borders (see Belmalih et al., 2007; Geyer et al., 2000).

This multi-architectonic approach led to the parcellation of the agranular frontal cortex shown in Figure 6.1 (Belmalih et al., 2009; Matelli, Luppino, & Rizzolatti, 1985, 1991). Over the years, this map has been validated by evidence showing that each of its subdivisions has its own characterizing functional and connectional features, thus fulfilling all the criteria generally accepted for the definition of a cortical area (Van Essen, 1985).

The most caudal agranular frontal area is the primary motor cortex. This area, often referred to as area M1, corresponds to area F1 in the classification of Figure 6.1. The motor areas located rostral to area F1 can be grouped into two major classes: the caudal premotor areas *F2*, *F3*, *F4*, *F5p*, and *F5c* and the rostral premotor areas *F5a*, *F6*, and *F7* (Rizzolatti & Luppino, 2001).

This subdivision, mostly based on their cortical connectivity, fits the organization of the corticospinal and corticobulbar projections of the premotor areas. The caudal premotor areas send direct projection to both the spinal chord and the brain stem (Dum & Strick, 1991; He, Dum, & Strick, 1993, 1995; Keizer & Kuypers, 1989). In contrast, the rostral ones do not project to the spinal cord; their descending fibers terminate in the brain stem (Keizer & Kuypers, 1989). It is interesting to note that while caudal premotor areas display somatotopically organized connections with each other and with the primary motor area F1, rostral

Neurobiology of Language. DOI: http://dx.doi.org/10.1016/B978-0-12-407794-2.00006-7

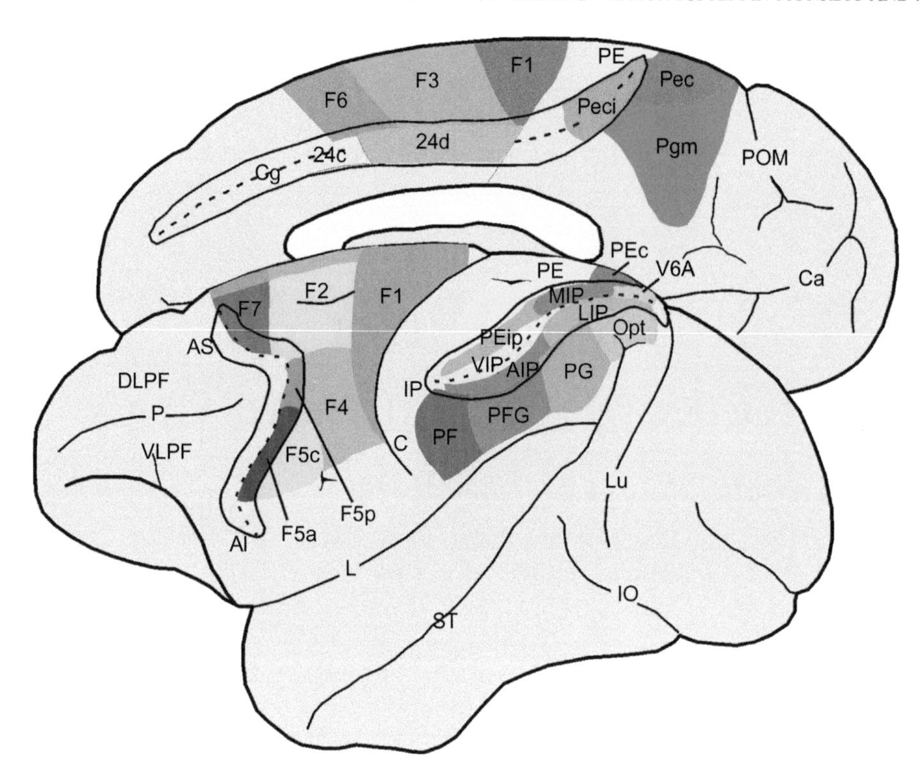

FIGURE 6.1 Lateral and mesial view of the monkey brain showing a modern detailed anatomical and functional parcellation of the agranular frontal cortex and the posterior parietal cortex. Intraparietal, arcuate, and cingulated sulci are shown unfolded. DLPF: dorsolateral prefrontal cortex; AI: inferior arcuate sulcus; AS: superior arcuate sulcus; C: central sulcus; Ca: calcarine fissure; Cg: cingulated sulcus; DLPF: dorsolateral prefrontal cortex; IO: inferior occipital sulcus; IP: intraparietal sulcus; L: lateral fissure; Lu: lunate sulcus; P: principal sulcus; PO: parieto-occipital sulcus; ST: superior temporal sulcus; VLPF: ventrolateral prefrontal cortex. *Modified from Matelli et al. (1991).*

premotor areas are not connected with F1 and, especially in the case of F6, have widespread connections with other motor areas.

6.2.2 Cortical Connections of the Motor Areas

Cortical afferents to the frontal motor areas originate from three main regions: the parietal cortex, the prefrontal cortex, and the agranular cingulate cortex (Rizzolatti & Luppino, 2001).

The strong, reciprocal connections with the parietal cortex are the major source of input to the F1 and the caudal premotor areas. As the agranular frontal cortex, the posterior parietal cortex is formed by a mosaic of areas (Figure 6.1), each of which deals with specific aspects of sensory information and with the control of specific effectors. Both the inferior parietal lobule (IPL) and the superior parietal lobule (SPL) receive somatosensory and visual inputs, originating from the "dorsal visual stream" (Colby, 1998; Rizzolatti & Matelli, 2003). In general, areas of IPL and the posterior areas of the SPL process either visual only or somatosensory and visual information, whereas the rostral areas of the SPL deal mostly with somatosensory information (Caminiti, Ferraina, & Johnson, 1996; Rizzolatti, Fogassi, & Gallese, 1997).

Examining the parieto-frontal organization in more detail, it turns out that each motor area is reciprocally connected to a specific set of parietal areas. Thus, within the general framework of parieto-frontal connections it is possible to identify a series of largely segregated circuits formed by parietal and motor areas linked by predominant connections. Functional evidence indicates that parietal and frontal areas forming each circuit share common functional properties. Therefore, the functional correlate of this anatomical organization is that each of these circuits is specifically involved in transforming sensory stimuli into motor terms (Rizzolatti et al., 1997; Rizzolatti, Luppino, & Matelli, 1998). Thus, these parieto-frontal circuits, and not the frontal motor areas in isolation, should be considered the functional units of the cortical motor system (Rizzolatti et al., 1998).

Prefrontal projections to the motor cortex are primarily directed to rostral premotor areas (Borra, Gerbella, Rozzi, & Luppino, 2011; Gerbella, Belmalih, Borra, Rozzi, & Luppino, 2010; Gerbella, Borra, Tonelli, Rozzi, & Luppino, 2013; Lu, Preston, & Strick, 1994; Luppino, Matelli, Camarda, & Rizzolatti, 1993). Prefrontal input to area F7 originates only from the dorsal part of the lateral prefrontal cortex (DLPF), that to F6 originates from both the dorsal and the ventral part of the lateral prefrontal cortex (VLPF), and that to F5a originates only from the VLPF. In addition, F6 also receives strong afferents from the rostral cingulate cortex (area 24c). Cingulate connections are considerably weaker for F7 and F5a. However, F5a is densely connected with rostral opercular frontal areas.

Altogether, these data indicate that the caudal and rostral premotor sectors play a different functional role in motor control. Caudal premotor areas are involved in transformations of sensory information into potential motor acts. Rostral premotor areas appear to play a hierarchically higher role. These areas convey information related to high-order action organization, working memory, and motivation from the prefrontal, cingulate, and opercular frontal cortex to the caudal premotor areas. This information is used by the caudal premotor areas for determining which motor acts will be executed and when, according to external and internal contingencies.

6.2.3 Area F5: Anatomical Subdivisions

The area crucially involved, besides motor control, in communication is area F5. In this and the next sections, we describe its anatomical and functional properties, with particular emphasis on its role in action understanding.

Area F5 forms the rostral part of the ventral premotor cortex. Recent evidence showed that F5 is subdivided into three sectors (Belmalih et al., 2009): F5c, F5p, and F5a. F5c (convexity) extends on the convexity of the postarcuate cortex adjacent to the inferior arcuate sulcus; F5p (posterior) and F5a (anterior) lie within the postarcuate bank, at different antero-posterior levels.

F5p and F5c sectors encode hand and hand and face/mouth motor acts. They host motor and visuomotor neurons, including mirror neurons. F5p is tightly connected with the hand field of the primary motor area and is a source of projections to the brain stem and to the cervical spinal cord (Borra, Belmalih, Gerbella, Rozzi, & Luppino, 2010; Dum & Strick, 1991). These connections enable F5p to generate object-oriented hand motor acts (Prabhu et al., 2009; Shimazu, Maier, Cerri, Kirkwood, & Lemon, 2004).

F5a sector, although basically part of the agranular frontal cortex, displays "transitional" architectonic features between those of PMv and those of the granular frontal cortex (Belmalih et al., 2009). Recent data (Borra et al., 2011; Gerbella et al., 2010; Gerbella, Belmalih, Borra, Rozzi, & Luppino, 2011; Gerbella et al., 2013) showed that F5a is the PMv subdivision most strongly connected to ventral prefrontal areas 12 and ventral 46. Furthermore, its connections with rostral opercular frontal sectors are much stronger than those of the other F5 subdivisions. These data suggest that F5a is a privileged site of integration of sensory motor signals originating from the parietal cortex and higher-order information originating from prefrontal and rostral frontal opercular areas.

Finally, an area located on the fundus of the inferior arcuate sulcus is considered by some authors as the homolog of human area 44 (Petrides & Pandya, 1994). However, the fundal location of this area renders it very difficult to assess its cytoarchitectonic structure and to study its connectivity. For this reason, naming this fundal region as monkey area 44 could be accepted with some caution.

6.2.4 Motor Properties of Area F5: The Vocabulary of Motor Acts

Single neuron studies revealed that most F5 neurons code specific motor acts, rather than individual movements (Rizzolatti et al., 1988). Using the effective motor act as the classification criterion, F5 neurons were subdivided into various classes. Among them, the most represented are "grasping," "holding," and "manipulating" neurons. Neurons discharging for a specific motor act typically do not discharge during the execution of similar movements aimed at a different goal. For example, a neuron that discharges during finger movements for grasping an object does not discharge during similar movements aimed at scratching. In contrast, F5 neurons typically discharge when the same goal is achieved by using different effectors (e.g., the right hand, the left hand, or the mouth), thus requiring completely different movements (Figure 6.2A).

On these bases, it has been proposed that F5 contains a "vocabulary" of motor acts (Rizzolatti et al., 1988). This motor vocabulary comprises "words," each of which is represented by a population of F5 neurons. Some of them encode the general goal of a motor act, others encode how a specific goal-directed motor act must be executed, and others specify the temporal aspects of the motor act to be executed (Jeannerod, Arbib, Rizzolatti, & Sakata, 1995). A further demonstration that F5 neurons encode motor acts has been recently provided by a study in which the same motor goal (i.e., taking possession of food) was achieved by means of opposite movements (Umiltà et al., 2008). Monkeys grasped objects using "normal pliers" (i.e., pliers that require hand closure to take possession of the object) and "reverse pliers" (i.e., pliers that instead require hand-opening to achieve the same goal) (Figure 6.2B). In both cases the neural discharge correlated with food grasping, regardless of whether it was achieved by closing the hand or by opening it.

6.2.5 Canonical Neurons and the Visuomotor Transformation for Grasping

Beside purely motor neurons, area F5 also contains visuomotor neurons (Rizzolatti et al., 1988). Among them, two main categories were described: canonical and mirror neurons. Canonical neurons are mostly located in area F5p. Their visual discharge is triggered by the presentation of 3D objects (Murata et al., 1997; Raos, Umiltá, Murata, Fogassi, & Gallese, 2006). The majority of canonical neurons respond selectively to objects of a

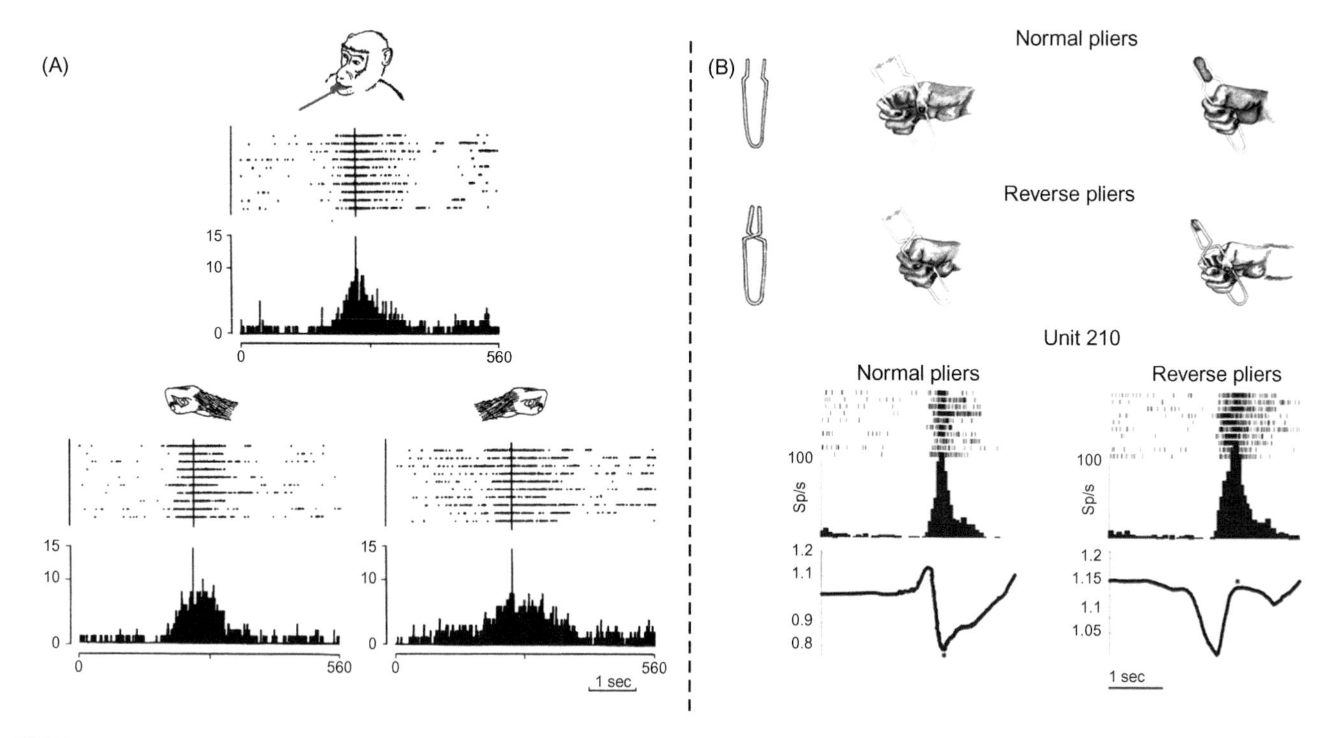

FIGURE 6.2 Goal coding in area F5. (A) Discharge of an F5 neuron active during grasping with the mouth, the right hand, and the left hand. Abscissae: time; ordinates: spikes per bin; bin width: 20 ms. (B) Example of an F5 neuron discharging during grasping with normal and reverse pliers (*modified from Rizzolatti et al., 1988*). (Upper) Pliers and hand movements necessary for grasping with the two types of pliers. (Lower) Rasters and histograms of the neurons' discharge during grasping with pliers. The alignments are with the end of the grasping closure phase (asterisks). The traces below each histogram indicate the hand position, recorded with a potentiometer, expressed as function of the distance between the pliers handles. When the trace goes down, the hand closes; when it goes up, it opens. The values on the vertical axes indicate the voltage change measured with the potentiometer. Other conventions as in Figure 6.2A. *Modified from Umiltà et al. (2008).*

certain size, shape, and orientation. Typically, the visual specificity is congruent with the motor one (e.g., motor selectivity: precision grip/visual selectivity of small objects; motor selectivity: whole hand prehension/visual selectivity of large objects). The visual responses of canonical neurons cannot simply be explained in terms of motor preparation because they are also present when no response toward the object is required (Murata et al., 1997). The most accepted interpretation of their discharge is that object presentation activates a representation of the observed stimulus in motor terms. In other words, when an object is seen, the discharge of canonical neurons codes a potential motor act congruent with the properties of the presented object, independent of whether the act will be executed. In other words, when an object is seen, the discharge of specific canonical neurons codes a *potential* motor act congruent with the properties of the presented object.

Area F5p is strongly connected with the intraparietal area AIP (Borra et al., 2008; Luppino, Murata, Govoni, & Matelli, 1999), where neurons with similar functional properties have been described (Murata, Gallese, Luppino, Kaseda, & Sakata, 2000; Sakata, Taira, Murata, & Mine, 1995; Taira, Mine, Georgopoulos, Murata, & Sakata, 1990). How do the visuomotor transformations for grasping occur? There are various models that attempt to explain the role that AIP and F5 play in this process (Fagg & Arbib, 1998; Jeannerod et al., 1995; Rizzolatti & Luppino, 2001; Taira et al., 1990). A common idea underlying these models is that when an object is observed, AIP neurons extract specific aspects of its physical properties and provide F5 with the description of the *possible ways* in which the object could be grasped (*affordances* as defined by Gibson, 1979). On the basis of the intention of the individual and the context, the prefrontal lobe activates AIP visuomotor neurons and neurons in F5 coding the appropriate grip. The information relative to the chosen grip is then sent from F5 to F1, where the different movements necessary to grasp the object are selected and the final command for its execution is generated.

As already mentioned, F5 canonical neurons code potential motor acts that eventually may become actual movements. What is the mechanism that allows the transformation of these potential motor acts into actual movements? Mesial area F6 (pre-SMA) is anatomically connected with F5 (Luppino et al., 1993). In turn, this area receives inputs from prefrontal and cingulate cortex, likely providing it with contextual and motivational information. Thus, area F6 may be a key area to determine, according to the external contingencies, whether the potential motor act will be actually executed.

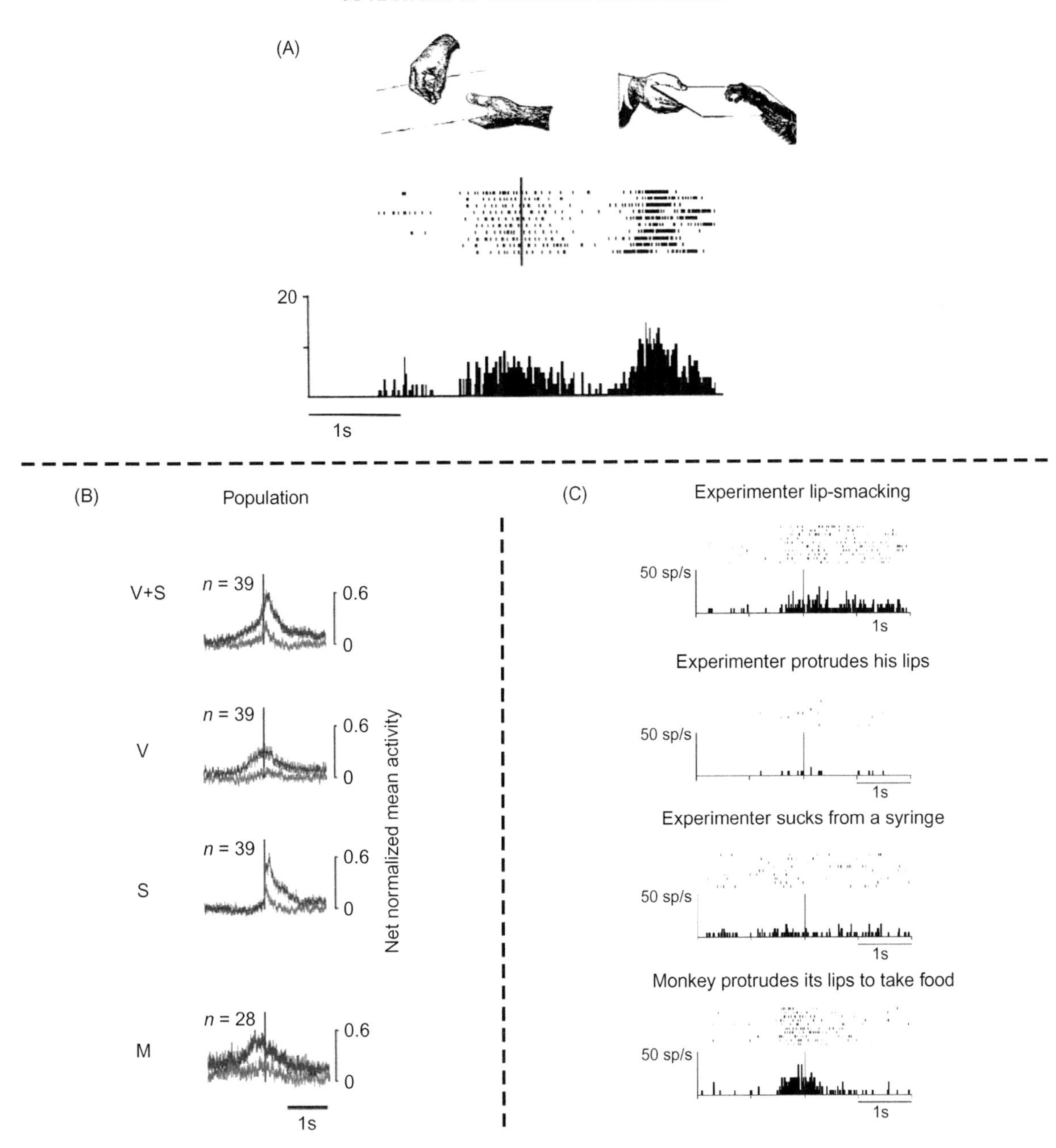

FIGURE 6.3 Examples of F5 mirror neurons. (A) Mirror neuron responding during observation and execution of a hand-grasping motor act. Conventions as in Figure 6.2 (*modified from di Pellegrino et al., 1992*). (B) Responses of the population of tested neurons selectively activated by the vision-and-sound, vision-only, sound-only, and motor conditions. Responses to the most effective stimuli are shown in blue, and responses to the poorly effective stimuli are in red. Vertical lines indicate the auditory response onset in the population. *Y* axes are in normalized units for the population (*modified from Kohler et al., 2002*). (C) Mirror neuron responding during the observation of an intransitive mouth action (lip-smacking). During observation of intransitive actions, the rasters and histograms alignment is at the moment in which the action is fully expressed; other conventions as in Figure 6.2 (*modified from Ferrari et al., 2003*).

6.2.6 Mirror Neurons and Action Understanding

A second class of visuomotor neurons present in area F5 is that of mirror neurons. Mirror neurons are more frequently located in F5c. They discharge both when the monkey *performs* a goal-directed motor act and when it observes the same, or a similar, motor act performed by another individual (Figure 6.3A; Di Pellegrino, Fadiga, Fogassi, Gallese, & Rizzolatti, 1992; Gallese, Fadiga, Fogassi, & Rizzolatti, 1996; Rizzolatti, Fadiga, Gallese, & Fogassi, 1996). Unlike canonical neurons, they do not respond to object presentation. Furthermore, they do not discharge or discharge less during the observation of biological movements devoid of a goal, including mimicked actions.

The observed hand motor acts more effective in eliciting mirror neurons discharge are grasping, manipulating, and holding. The visual response of many mirror neurons is invariant with respect to visual aspects of the observed action. However, other mirror neurons show specificity for the direction of the hand movement or the space sector (left or right; close or far) in which the observed motor act is presented or the hand (left or right) used by the observed agent (Caggiano, Fogassi, Rizzolatti, & Thier, 2009; Gallese et al., 1996). As far as the location in depth is concerned, it was found that half of mirror neurons sensitive to this parameter discharged more strongly when the motor act was performed within the monkey peripersonal space, whereas the other half responded better when the same motor act was performed in the extrapersonal space. Interestingly, when the peripersonal space (defined as the space within which grasping is possible) of the monkey was reduced by introducing a transparent barrier, a set of extrapersonal neurons started to discharge within the old peripersonal space transformed into a space no longer reachable (Caggiano et al., 2009). These data indicate that this specific set of mirror neurons code the actions of others in relation to interaction possibilities.

It has been proposed that the property shown by mirror neurons of encoding the visual description of a goal-directed act in motor terms allows the observer to understand *what* another individual is doing, because the observation of a motor act performed by others determines an automatic retrieval of a potential motor act from the "vocabulary" of the observer.

The hypothesis that mirror neurons have an important role in understanding the motor acts of others has been supported by various studies. In one of them it was shown that grasping mirror neurons discharge not only when the monkey observes a grasping motor act (effective visual stimulus) but also when it sees the agent's hand moving toward the target hidden by an opaque screen (Umiltà et al., 2001). The discharge was absent when the monkey knew that there was no object behind the screen. This finding suggests that mirror neurons use prior information to retrieve the motor representation of the observed motor act.

In another study, sensory information concerning the motor act was presented to the monkey in an acoustic and/or a visual format (Kohler et al., 2002). It was found that a subset of mirror neurons, called "audio-visual mirror neurons," discharged not only during execution and observation of a motor act (e.g., breaking a peanut) but also by listening to the typical sound produced by that act. This indicates that a motor act is understood regardless of how the information reaches the mirror neurons (Figure 6.3B).

Besides mirror neurons encoding hand motor acts, mouth mirror neurons have also been described. These neurons are mostly found in the lateral part of area F5. The majority of them respond to the observation and execution of ingestive motor acts such as biting, sucking, and licking (Ferrari, Gallese, Rizzolatti, & Fogassi, 2003). They do not respond to object presentation or to mouth-mimed motor acts. A smaller but significant population of mouth mirror neurons responds specifically to the observation of mouth communicative gestures, such as smacking of the lips or tongue protrusion (Figure 6.3C). Mouth mirror neurons of this sub-category do not respond, or respond very weakly, to the observation of ingestive motor acts. These mirror neurons are of particular interest because they could constitute an evolutionary ancient communicative system evolving from neurons coding ingestive motor acts (Fogassi & Ferrari, 2012; Rizzolatti & Arbib, 1998).

In the experiment just described, it was very difficult to elicit mouth communicative gestures and even more difficult to elicit communicative calls. In a subsequent experiment, however, the monkey turned out to be able, after long periods of training, to emit voluntarily "coo-calls" (Coudé et al., 2011). Single neuron recording demonstrated that in the ventral part of area F5, there are neurons whose discharge correlates with the voluntary call emission (Coudé et al., 2011). This finding is of great interest because it suggests that this F5 sector might be the precursor of the region that controls the voluntary emission of voice in humans.

6.2.7 Mirror Neurons in the Parietal and the Primary Motor Cortex

6.2.7.1 Primary Motor Cortex

The issue of whether the output of premotor and motor cortex contains neurons endowed with mirror properties has been recently addressed. Kraskov, Dancause, Quallo, Shepherd, and Lemon (2009) investigated the activity of corticospinal neurons (PTNs) located in area F5 and in area F1. They found that approximately half of F5 corticospinal neurons responded to action observation. Interestingly, approximately 25% of these PTNs showed a suppression of their discharge when the monkey observed the experimenter grasping an object. The authors suggested that the suppression of the PTNs during grasping observation may play a role in inhibiting the movement of the observer triggered by the observed action.

The same paradigm was applied to F1 PTNs (Kraskov et al., 2009). As in F5, approximately half of these neurons were modulated during action observation. The majority increased their discharge during action observation ("facilitation-type" mirror neurons), whereas some reduced or stopped their firing ("suppression-type" mirror neurons). A comparison of the properties of PTNs F1 and F5 PTNs mirror neurons showed that the visual response in F1 was much weaker than in F5. Taken together, these data indicate that the understanding of motor goals is not only a function of F5 mirror neurons but also a function of the activation of a complex motor pattern that involves corticospinal tract neurons, including those originating in F1.

6.2.7.2 Parietal Cortex

Mirror neurons were also recorded from IPL and, in particular, from area PFG (Fogassi et al., 2005; Gallese, Fadiga, Fogassi, & Rizzolatti, 2002; Rozzi, Ferrari, Bonini, Rizzolatti, & Fogassi, 2008). There is evidence, however, that mirror neurons are also present in AIP (personal data). PFG grasping neurons have been specifically studied to elucidate whether their discharge was modulated by the overarching action intention (Fogassi et al., 2005). For this purpose their activity was recorded while the monkey executed a motor task in which the same motor act (grasping) was embedded into two different actions (grasping to eat and grasping to place). The neurons were then tested with the monkey observing the same task, performed by an experimenter.

The results showed that a high percentage of parietal neurons discharge with different intensity during grasping execution, depending on overarching goal of the actions. On the basis of these findings, it was proposed (Fogassi et al., 2005) that parietal neurons form prewired chains in which a neuron coding a given motor act is facilitated by the neuron coding that previously executed. Any time an agent has the intention (overarching goal) to perform an action, a specific neuronal chain is activated. This model accounts for the fluidity with which the different motor acts of an action are executed one after another (Jeannerod, Paulignan, & Weiss, 1998; Rosenbaum, Cohen, Jax, Weiss, & van der Wel, 2007).

In the visual task, as in the motor task, it was found that most mirror neurons discharged differently during grasping, depending on overarching goal of the actions. Because in this case grasping was performed by the observed agent, it was suggested that the neuronal selectivity for the action goal during grasping observation activated the chain of motor neurons corresponding to a specific intention. Similar results were also obtained in area F5, where the same paradigm was applied (Bonini et al., 2010).

6.3 THE HUMAN MOTOR CORTEX

The organization of human agranular frontal cortex presents strong similarities with that of the monkey (Von Bonin & Bailey, 1947). First, as in the monkey, the human motor cortex is formed by a mosaic of areas. Area M1, the primary motor cortex, is located caudally. It is mostly buried inside the central sulcus. Second, as in the monkey, there are two areas located on the mesial cortical surface: the supplementary motor area (SMA/F3) and the presupplementary motor area (pre-SMA/F6; see Zilles et al., 1996). Third, the motor cortex of the lateral convexity consists of two main regions: the dorsal and the ventral "premotor" region. They are frequently referred as PMd and PMv. Unlike in the monkey, where the approximate border between these two regions is easy to recognize because it is marked by the spur of the arcuate sulcus (Figure 6.1), their border is difficult to identify in humans. On the basis of embryological considerations, Rizzolatti and Arbib (1998) suggested that the human superior frontal sulcus plus the superior precentral sulcus should correspond to the monkey's superior limb of the arcuate sulcus plus the arcuate spur. Thus, according to them, the border between PMd and PMv should lie approximately at the Z level 50 of Talairach coordinates. Recent diffusion tractography data confirmed this location (Mars et al., 2011; Schubotz, Anwander, Knösche, von Cramon, & Tittgemeyer, 2010; Tomassini et al., 2007).

Functionally, human M1 is somatotopically organized with the leg field located medially, the mouth field laterally, and the forelimb field in between. This map does not extend rostrally into area 6, as was found in macroelectrode surface stimulation studies (Penfield & Welch, 1951). The various areas forming area 6 are independent functional entities involved, as in the monkey, in different aspects of sensory motor transformation and in movement control.

Broca's area (area 44 and area 45) is located rostral to PMv. On the basis of receptor-architectonic data (Amunts et al., 2010; Amunts & Zilles, 2012), these two areas are clustered together with opercular and ventral premotor areas. Most interestingly, recent data obtained using meta-analytic connectivity-based parcellation (Clos, Amunts, Laird, Fox, & Eickhoff, 2013) revealed five separate functional "clusters" in area 44. The two posterior clusters are primarily related to action processes, including phonology and speech, whereas the three anterior clusters are primarily associated with language and cognition. Cluster 4, located in the posterior ventral part of area 44, is associated with the observation of hand action and action imagery.

As stated, although a clear anatomical and functional homology is recognizable between the agranular frontal areas of monkeys and humans, such a homology is difficult to draw for Broca's area. A homology between this area and the tiny cytoarchitectonic area 44 of monkeys is possible, but not very likely. More interesting for the homology issue is that the data indicating that this region codes not only speech but also communicative actions (Clos et al., 2013). From this perspective, it is plausible that some functions located in area F5 in monkeys "moved" to the evolutionary new Broca's region. This point is elaborated in the following sections after presentation of studies of the mirror mechanism in humans.

6.3.1 The Mirror Mechanism in Humans

The existence of the mirror mechanism in humans has been demonstrated by a large number of neurophysiological (EEG, MEG, and TMS) and neuroimaging (PET and fMRI) studies (Rizzolatti & Craighero, 2004). These studies have shown that the mirror network for hand actions includes two main regions: the dorsal part of the IPL comprising the cortex located inside the intraparietal sulcus and the ventral premotor cortex plus the caudal part of the inferior frontal gyrus (area 44). Additional areas, such as the dorsal premotor cortex and the SPL, were also found to be active during action observation (Di Dio et al., 2013; Filimon, Nelson, Hagler, & Sereno, 2007). These last areas are activated by the observation of reaching movements.

Do the mirror neurons of human premotor and parietal cortex encode motor acts, such as movement with a specific goal, as in the monkey? There is clear evidence from fMRI studies that this is the case. Gazzola, Rizzolatti, Wicker, and Keysers (2007) presented volunteers with video-clips depicting either a human or a robot arm grasping objects. The results showed that the parieto-frontal mirror circuit was activated by both these types of stimuli. This observation was extended by Peeters et al. (2009). They investigated the cortical activations in response to the observation of motor acts performed by a human hand, a robot hand, and a variety of tools in both humans and monkeys. Regardless of the type of effector used, the hand grasping circuit was activated in humans as well as in monkeys. In humans, the observation of tool motor acts also activated a rostral sector of the left anterior supramarginal gyrus. Such activation was absent in monkeys even when they observed actions made with the tools they learned to use.

A series of experiments addressed the issue of the somatotopic organization of the areas endowed with the mirror mechanism (Buccino, Binkofski, & Fink, 2001; Sakreida, Schubotz, Wolfensteller, & von Cramon, 2005; Saygin, Wilson, Hagler, Bates, & Sereno, 2004; Shmuelof & Zohary, 2005; Ulloa & Pineda, 2007; Wheaton, Thompson, Syngeniotis, Abbott, & Puce, 2004). These studies showed that the observed motor acts are encoded in the precentral gyrus according to an approximate somatotopic organization similar to that of the classical motor physiology. A somatotopic organization was also found to be present in the IPL along and within the intraparietal sulcus. The mouth is located rostrally, the hand is in an intermediate position, and the leg is located caudally (Buccino et al., 2001). A recent study by Jastorff, Begliomini, Fabbri-destro, Rizzolatti, and Orban (2010) tried to better define the general principles underlying the somatotopic organization in the parietal and frontal cortex. Four motor acts (grasping, dragging, dropping, and pushing) performed with the mouth, hand, and foot were presented to volunteers. The results confirmed the data from previous authors concerning the premotor cortex. As for the parietal lobe, they showed that different sectors of IPL were activated by the observation of motor acts with the same behavioral valence, independent of the observed effector. More specifically, there was a subdivision between the localization of self-directed (grasping and dragging) and outward-directed motor acts (dropping and pushing). Therefore, it appears that while in the premotor cortex, motor acts executed with the same effector tend to cluster together, but in the parietal cortex the encoding is biased by the action valence.

A few studies showed that in humans the mirror network is involved in intention understanding (Iacoboni et al., 2005; Ortigue, Sinigaglia, Rizzolatti, & Grafton, 2010). In an fMRI experiment, Iacoboni et al. (2005) tested volunteers during three conditions: (i) "context;" (ii) "action;" and (iii) "intention." During the context condition, individuals were presented with a scene showing either a "ready breakfast" or a "finished breakfast;" in the action condition, they saw pictures of a hand grasping a mug, without context; and in the intention condition, the individuals saw the same hand grasping the mug within one of the two contexts. The context represented the clue that allowed the participants to understand the agent's intention. The comparison between conditions showed that intention understanding determined the strongest increase in the activity of the mirror system and, in particular, of its frontal node.

6.3.2 Imitation

It is important to stress at the outset of this section that the human mirror system encodes not only the observed motor acts but also the individual movements that form it. Evidence in favor of the encoding of the observed movements came from transcranic magnetic stimulation (TMS) experiments. Fadiga, Fogassi, Pavesi, and Rizzolatti (1995) asked volunteers to observe grasping acts or meaningless arm gestures. The results showed that the observation of meaningless gestures, and not just of motor acts, produced an increase in the motor-evoked potentials (MEPs). This increase was present in the same muscles that are used to produce the observed movements. These findings were subsequently confirmed and extended by other TMS experiments (Alaerts, Heremans, Swinnen, & Wenderoth, 2009; Cattaneo, Caruana, Jezzini, & Rizzolatti, 2009; Gangitano, Mottaghy, & Pascual-Leone, 2001; Strafella & Paus, 2000).

The existence of the mirror mechanism for simple movements is a fundamental prerequisite of human capacity to imitate others. The term "imitation" includes various phenomena. Two are of interest here. The first is the capacity to *replicate* immediately the observed movements (Prinz, 1990). The second is the capacity to learn a new motor behavior by observing and then by repeating the *same* movements that the teacher executed (Byrne, 2003). In both cases, imitation requires the capacity to translate an observed movement into a motor copy of it.

Evidence that the mirror neuron mechanism is involved in imitation was provided by Iacoboni et al. (1999). Subjects were tested in two main conditions: "observation" and "observation/execution." In the first, subjects saw a moving finger or a cross depicted on a stationary finger or a cross presented on an empty background. The subjects had to observe the stimuli. In the "observation/execution" condition, the same stimuli were presented, but with the instruction to lift the right finger as quickly as possible when they occurred. The crucial contrast was between the trials of the observation/execution condition in which the subjects made the movement in response to the observed movement ("imitation") and the trials in which the response was triggered by the cross ("nonimitative behavior"). The main result was that the activation of the caudal part of the inferior frontal gyrus was significantly stronger during "imitation" than during the "nonimitative behavior."

The mechanism involved in imitation learning is much more complex. Two fMRI studies addressed this issue. In the first, naive participants had to imitate guitar chords made by a teacher (Buccino, Vogt, Ritzl, Fink, & Zilles, 2004). Cortical activations were studied in three epochs: chord observation, chord execution, and the pause between the two. The results showed that during new motor pattern formation (pause epoch) there was, besides activation of the mirror circuit, an activation of area 46 and of the anterior mesial cortex. The importance of area 46 in imitation learning was subsequently confirmed in a study performed with expert players (Vogt et al., 2007).

Following Byrne (2003), a two-step model of imitation learning was proposed: (i) segmentation of the observed action into its individual movements and their transformation, thanks to the mirror mechanism, into corresponding potential movements; and (ii) the organization of these potential movements into specific temporal and spatial patterns replicating that shown by the teacher. Parsing and sensory motor transformation are performed by the mirror mechanism, whereas the recombination of these elements into a new pattern could be a function of the prefrontal lobe and, in particular, of area 46 (Buccino et al., 2004; Vogt et al., 2007).

6.4 MOTOR SYSTEM AND COMMUNICATION

Some years ago Rizzolatti and Arbib (1998) proposed that the mirror mechanism could represent the mechanism through which language evolved. This proposal was based on the consideration that mirror neurons create a direct link between the sender of a message and its receiver. Thus, thanks to the mirror mechanism, observing and doing become manifestations of a single communicative faculty.

The proposal that speech evolved from gestural communication was by no means new (Armstrong, Stokoe, & Wilcox, 1995; Corballis, 2002, 2003). Its novelty existed in the fact that it identified a neurophysiological mechanism that could create a nonarbitrary link between the communicating individuals. In contrast to theories claiming a gestural origin of communication, other theories maintain that human speech derived from animal calls. Although intuitively appealing, these theories are, however, difficult to accept for various reasons. First, the anatomical basis for speech and animal calls are markedly different. Animal calls are mediated by the cingulate cortex and by diencephalic and brain stem structures (Jürgens, 2002). In contrast, the core circuit underlying human speech is formed by perisylvian areas. Second, human speech, unlike animal calls, is not necessarily related to emotional behavior. Third, speech is mostly a person-to-person communication, whereas animal calls are, typically, directed to all possible listeners. Finally, speech is endowed with combinatorial properties that are absent in animal communication.

Let us examine the assets and the weaknesses of the mirror neuron-based hypothesis of language evolution. The main asset is that mirror neurons solve two fundamental communication problems: parity and direct comprehension of the action. Parity requires the following: what counts for the sender of the message also counts for the receiver. Direct comprehension means that there is no need for an agreement between individuals to understand each other. The comprehension is inherent to neural organization of the individual.

The major weakness of the mirror mechanism-based theory, at least in its original formulation, was that the properties of the mirror neurons known when it was formulated indicate a close system limited to object-directed actions. As discussed, there is now evidence that humans also possess mirror neurons encoding observed intransitive movements and not just observed motor acts. Thus, the human mirror system is an open system endowed with communicative properties.

How did it occur that the monkey closed mirror system became an open system able to describe actions and objects without directly referring to them? How did this evolutionary leap occur? A possible way to make a hypothesis regarding this point is by examining child development. A splendid example comes from studies by Vygotsky (1934). He described that when objects are located close to a child, the child grasps them; when they are located far from the child, the child extends his/her arm and hand toward them. Thus, an object-related action becomes, progressively, an intransitive communicative gesture. An evolutionary ancient goal-directed motor act becomes "pointing," a fundamental gesture for communicating.

Another fundamental step in the evolution of speech should be the appearance of a link between gestures and sounds. Is there evidence that such a link exists? A series of recent studies provided evidence in this sense. By using TMS, it was shown that the excitability of the motor field of the right hand increases during reading and during spontaneous speech (Meister et al., 2003; Seyal, Mull, Bhullar, Ahmad, & Gage, 1999; Tokimura, Tokimura, Oliviero, Asakura, & Rothwell, 1996). Control experiments showed that this increase in excitability is not related to word articulation.

The presence of a link between speech and hand gesture was also demonstrated by Gentilucci, Benuzzi, Gangitano, and Grimaldi (2001) using a different approach. In an initial series of experiments, they asked participants to grasp two objects of different size with their mouth and, simultaneously, to open their right hand. The results showed that the maximal finger aperture and time to maximal finger aperture increased when the mouth grasped the large object. A subsequent experiment was crucial. In this experiment they asked the participants to pronounce a syllable, such as GU or GA, during observation and subsequent hand grasping of objects of different size. The syllables were written on the objects. The results showed that both sound production and mouth opening were affected by the size of the grasped object (Gentilucci et al., 2001).

Taken together, these results show that buccal and oro-laryngeal synergies necessary for syllable emission are linked to manual gestures. These findings obviously do not solve the problem of how meaning becomes associated with gestures. The discussion on how this occurred is centered on the possible relations between the sound of a word and its meaning. On the one side, some authors postulate a *natural* origin of the words, that is, that there are intrinsic links between sounds and what they represent; however, other authors regard the faculty of speech as the result of *cultural* factors that led to an agreement on the word meaning among speakers of the same group. There is, at the moment, no convincing evidence to disentangle this dichotomy, although the hypothesis of a natural origin of language seems to be more satisfactory intellectually.

6.5 CONCLUSION

Traditionally, the caudal, motor part of the frontal lobe was considered as a region not involved in cognitive functions. The discovery of the mirror mechanism radically changed this view. There is now rich evidence that a series of cognitive functions such as space perception and action recognition not only require the activity of the motor system but also are deeply embedded in motor organization. In particular, action understanding relies on potential motor acts that originally evolved for motor behavior and subsequently became the substrate for understanding others (Figure 6.4).

Similar considerations are valid for language. There is clear evidence that the motor system is involved in perception of phonemes. By using TMS, it was shown that during speech listening there is an increase of MEPs recorded from the listeners' tongue muscles when the presented words strongly involve, when pronounced, tongue movements (Figure 6.5; Fadiga, Craighero, Buccino, & Rizzolatti, 2002). Although these data do not provide crucial evidence for the motor theory of speech perception (Liberman & Mattingly, 1985), they certainly indicate that the motor system is involved in phoneme perception.

As far as other aspects of language are concerned, it has been proposed that the semantic representations are grounded in those brain circuits that are also responsible for action organization and perception (Pulvermüller, 1999). This account is referred to as "action perception theory." It purports that, besides the core perisylvian language circuits, there are sensory motor circuits that become active during linguistic communication. This activation would lead to the understanding of the semantics of actions. In favor of this hypothesis are neuroimaging data that found specific activations of parieto-frontal circuits when subjects listen to action words (Aziz-Zadeh, Wilson, Rizzolatti, & Iacoboni, 2006; Hauk, Johnsrude, & Pulvermüller, 2004; Tettamanti et al., 2005). Moreover, TMS data and lesion studies show contributions of motor circuits to the comprehension of phonemes and word semantics (Pulvermüller & Fadiga, 2010). Altogether, these data show that language comprehension relies, at least in part, on parieto-frontal motor systems and

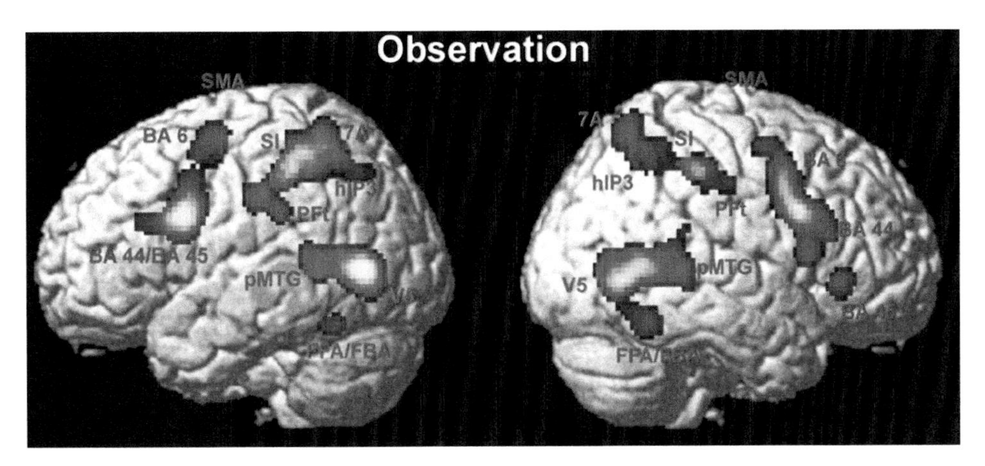

FIGURE 6.4 Cortical areas active during action observation in humans. Three main regions are active: a portion of the superior temporal sulcus; the IPL, including the intraparietal sulcus; and a small part of SPL, the premotor cortex, mainly in its ventral part, and the posterior part of the inferior frontal gyrus. The activated regions depicted on the left and right hemispheres result from a meta-analysis performed in 87 studies. *Modified from Caspers, Zilles, Laird, & Eickhoff (2010).*

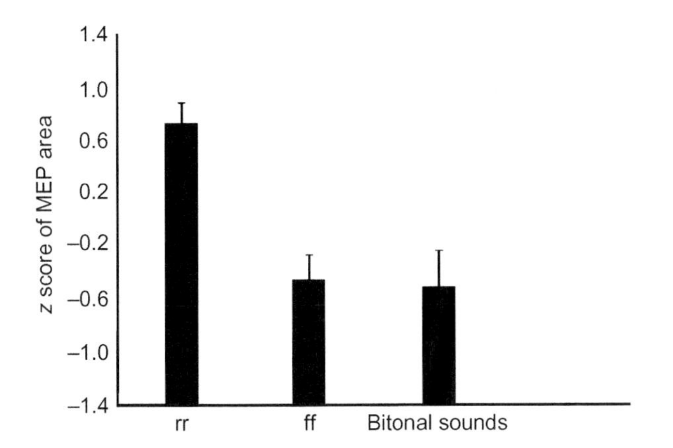

FIGURE 6.5 Activation of tongue muscles during word listening. Average value (±SEM) of intersubject normalized MEP total areas recorded from tongue muscles of subjects listening to words containing a double lingua-palatal fricative consonant (rr), a double labiodental fricative consonant (ff), or bitonal sounds. *Modified from Fadiga et al. (2002).*

indicate that action and language-perception circuits are interdependent. Finally, at present there is no neurophysiological evidence that allows one to identify the neural bases of syntactic aspects of language.

Acknowledgment

This study was supported by the European Grant "Cogsystems" and by Italian Institute of Technology (IIT).

References

Alaerts, K., Heremans, E., Swinnen, S. P., & Wenderoth, N. (2009). How are observed actions mapped to the observer's motor system? Influence of posture and perspective. *Neuropsychologia, 47*(2), 415–422.

Amunts, K., Lenzen, M., Friederici, A. D., Schleicher, A., Morosan, P., Palomero-Gallagher, N., et al. (2010). Broca's region: Novel organizational principles and multiple receptor mapping. *PLoS Biology, 8*(9).

Amunts, K., & Zilles, K. (2012). Architecture and organizational principles of Broca's region. *Trends in Cognitive Sciences, 16*(8), 418–426.

Armstrong, D. F., Stokoe, W. C., & Wilcox, S. E. (1995). *Gesture and the nature of language*. New York, NY: Cambridge University Press.

Aziz-Zadeh, L., Wilson, S. M., Rizzolatti, G., & Iacoboni, M. (2006). Congruent embodied representations for visually presented actions and linguistic phrases describing actions. *Current Biology, 16*(18), 1818–1823.

Belmalih, A., Borra, E., Contini, M., Gerbella, M., Rozzi, S., & Luppino, G. (2007). A multiarchitectonic approach for the definition of functionally distinct areas and domains in the monkey frontal lobe. *Journal of Anatomy, 211*(2), 199–211.

Belmalih, A., Borra, E., Contini, M., Gerbella, M., Rozzi, S., & Luppino, G. (2009). Multimodal architectonic subdivision of the rostral part (area F5) of the macaque ventral premotor cortex. *Journal of Comparative Neurology, 512*(2), 183–217.

Bonini, L., Rozzi, S., Serventi, F. U., Simone, L., Ferrari, P. F., & Fogassi, L. (2010). Ventral premotor and inferior parietal cortices make distinct contribution to action organization and intention understanding. *Cerebral Cortex, 20*(6), 1372–1385.

Borra, E., Belmalih, A., Calzavara, R., Gerbella, M., Murata, A., Rozzi, S., et al. (2008). Cortical connections of the macaque anterior intraparietal (AIP) area. *Cerebral Cortex, 18*(5), 1094–1111.

Borra, E., Belmalih, A., Gerbella, M., Rozzi, S., & Luppino, G. (2010). Projections of the hand field of the macaque ventral premotor area F5 to the brainstem and spinal cord. *Journal of Comparative Neurology, 518*(13), 2570–2591.

Borra, E., Gerbella, M., Rozzi, S., & Luppino, G. (2011). Anatomical evidence for the involvement of the macaque ventrolateral prefrontal area 12r in controlling goal-directed actions. *Journal of Neuroscience, 31*(34), 12351–12363.

Brodmann, K. (1909). *Vergleichende lokalisationslehre der großhirnrinde*. Leipzig: Barth.

Buccino, G., Binkofski, F., & Fink, G. (2001). Action observation activates premotor and parietal areas in a somatotopic manner: An fMRI study. *European Journal of Neuroscience, 13*, 400–404.

Buccino, G., Vogt, S., Ritzl, A., Fink, G., & Zilles, K. (2004). Neural circuits underlying imitation learning of hand actions: An event-related fMRI study. *Neuron, 42*, 323–334.

Byrne, R. W. (2003). Imitation as behaviour parsing. *Philosophical Transactions of the Royal Society of London Series B, Biological Sciences, 358*(1431), 529–536.

Caggiano, V., Fogassi, L., Rizzolatti, G., & Thier, P. (2009). Mirror neurons differentially encode the peripersonal and extrapersonal space of monkeys. *Science, 324*(5925), 403–406.

Caminiti, R., Ferraina, S., & Johnson, P. B. (1996). The sources of visual information to the primate frontal lobe: A novel role for the superior parietal lobule. *Cerebral Cortex, 6*(3), 319–328.

Caspers, S., Zilles, K., Laird, A. R., & Eickhoff, S. B. (2010). ALE meta-analysis of action observation and imitation in the human brain. *NeuroImage, 50*(3), 1148–1167.

Cattaneo, L., Caruana, F., Jezzini, A., & Rizzolatti, G. (2009). Representation of goal and movements without overt motor behavior in the human motor cortex: A transcranial magnetic stimulation study. *The Journal of Neuroscience, 29*(36), 11134–11138.

Clos, M., Amunts, K., Laird, A. R., Fox, P. T., & Eickhoff, S. B. (2013). Tackling the multifunctional nature of Broca's region meta-analytically: Co-activation-based parcellation of area 44. *NeuroImage, 83C*, 174–188.

Colby, C. L. (1998). Action-oriented spatial reference frames in cortex. *Neuron, 20*(1), 15–24.

Corballis, M. C. (2002). *From hand to mouth: The origins of language*. Princeton, NJ: Princeton University Press.

Corballis, M. C. (2003). From mouth to hand: Gesture, speech, and the evolution of right-handedness. *The Behavioral and Brain Sciences, 26*(2), 199–208 (discussion 208–260).

Coudé, G., Ferrari, P. F., Rodà, F., Maranesi, M., Borelli, E., Veroni, V., et al. (2011). Neurons controlling voluntary vocalization in the macaque ventral premotor cortex. *PloS One, 6*(11), e26822.

Di Dio, C., Di Cesare, G., Higuchi, S., Roberts, N., Vogt, S., & Rizzolatti, G. (2013). The neural correlates of velocity processing during the observation of a biological effector in the parietal and premotor cortex. *NeuroImage, 64*, 425–436.

Di Pellegrino, G., Fadiga, L., Fogassi, L., Gallese, V., & Rizzolatti, G. (1992). Understanding motor events: A neurophysiological study. *Experimental Brain Research, 91*(1), 176–180.

Dum, R. P., & Strick, P. L. (1991). The origin of corticospinal projections from the premotor areas in the frontal lobe. *Journal of Neuroscience, 11*(3), 667–689.

Fadiga, L., Craighero, L., Buccino, G., & Rizzolatti, G. (2002). Speech listening specifically modulates the excitability of tongue muscles: A TMS study. *European Journal of Neuroscience, 15*, 399–402.

Fadiga, L., Fogassi, L., Pavesi, G., & Rizzolatti, G. (1995). Motor facilitation during action observation: A magnetic stimulation study. *Journal of Neurophysiology, 73*(6), 2608–2611.

Fagg, A. H., & Arbib, M. A. (1998). Modeling parietal-premotor interactions in primate control of grasping. *Neural Networks, 11*(7–8), 1277–1303.

Ferrari, P. F., Gallese, V., Rizzolatti, G., & Fogassi, L. (2003). Mirror neurons responding to the observation of ingestive and communicative mouth actions in the monkey ventral premotor cortex. *European Journal of Neuroscience, 17*(8), 1703–1714.

Filimon, F., Nelson, J. D., Hagler, D. J., & Sereno, M. I. (2007). Human cortical representations for reaching: Mirror neurons for execution, observation, and imagery. *NeuroImage, 37*(4), 1315–1328.

Fogassi, L., & Ferrari, P. (2012). Cortical motor organization, mirror neurons, and embodied language: An evolutionary perspective. *Biolinguistic*, 308–337.

Fogassi, L., Ferrari, P. F., Gesierich, B., Rozzi, S., Chersi, F., & Rizzolatti, G. (2005). Parietal lobe: From action organization to intention understanding. *Science, 308*(5722), 662–667.

Gallese, V., Fadiga, L., Fogassi, L., & Rizzolatti, G. (1996). Action recognition in the premotor cortex. *Brain, 119*, 593–609.

Gallese, V., Fadiga, L., Fogassi, L., & Rizzolatti, G. (2002). Common mechanism in perception and action. In H. Prinz, & B. Hommel (Eds.), *Attention and performance XIX (p. attention and performance XIX)*. Oxford: Oxford University Press.

Gangitano, M., Mottaghy, F. M., & Pascual-Leone, A. (2001). Phase-specific modulation of cortical motor output during movement observation. *Neuroreport, 12*(7), 1489–1492.

Gazzola, V., Rizzolatti, G., Wicker, B., & Keysers, C. (2007). The anthropomorphic brain: The mirror neuron system responds to human and robotic actions. *NeuroImage, 35*(4), 1674–1684.

Gentilucci, M., Benuzzi, F., Gangitano, M., & Grimaldi, S. (2001). Grasp with hand and mouth: A kinematic study on healthy subjects. *Journal of Neurophysiology, 86*(4), 1685–1699.

Gerbella, M., Belmalih, A., Borra, E., Rozzi, S., & Luppino, G. (2010). Cortical connections of the macaque caudal ventrolateral prefrontal areas 45A and 45B. *Cerebral Cortex, 20*(1), 141–168.

Gerbella, M., Belmalih, A., Borra, E., Rozzi, S., & Luppino, G. (2011). Cortical connections of the anterior (F5a) subdivision of the macaque ventral premotor area F5. *Brain Structure & Function, 216*(1), 43–65.

Gerbella, M., Borra, E., Tonelli, S., Rozzi, S., & Luppino, G. (2013). Connectional heterogeneity of the ventral part of the macaque area 46. *Cerebral Cortex, 23*(4), 967–987.

Geyer, S., Matelli, M., Luppino, G., & Zilles, K. (2000). Functional neuroanatomy of the primate isocortical motor system. *Anatomy and Embryology, 202*(6), 443–474.

Gibson, J. J. (1979). *The ecological approach to visual perception*. Boston: Houghton Mifflin.

Hauk, O., Johnsrude, I., & Pulvermüller, F. (2004). Somatotopic representation of action words in human motor and premotor cortex. *Neuron, 41*(2), 301–307.

He, S. Q., Dum, R. P., & Strick, P. L. (1993). Topographic organization of corticospinal projections from the frontal lobe: Motor areas on the lateral surface of the hemisphere. *Journal of Neuroscience, 13*(3), 952–980.

He, S. Q., Dum, R. P., & Strick, P. L. (1995). Topographic organization of corticospinal projections from the frontal lobe: Motor areas on the medial surface of the hemisphere. *Journal of Neuroscience, 15*, 3284–3306.

Iacoboni, M., Molnar-Szakacs, I., Gallese, V., Buccino, G., Mazziotta, J. C., & Rizzolatti, G. (2005). Grasping the intentions of others with one's own mirror neuron system. *PLoS Biology, 3*(3), e79.

Iacoboni, M., Woods, R. P., Brass, M., Bekkering, H., Mazziotta, J. C., & Rizzolatti, G. (1999). Cortical mechanisms of human imitation. *Science, 286*(5449), 2526–2528.

Jastorff, J., Begliomini, C., Fabbri-destro, M., Rizzolatti, G., & Orban, G. A. (2010). Coding observed motor acts: Different organizational principles in the parietal and premotor cortex of humans. *Journal of Neurophysiology, 104*, 128–140.

Jeannerod, M., Arbib, M. A., Rizzolatti, G., & Sakata, H. (1995). Grasping objects: The cortical mechanisms of visuomotor transformation. *Trends in Neurosciences, 18*(7), 314–320.

Jeannerod, M., Paulignan, Y., & Weiss, P. (1998). Grasping an object: One movement, several components. *Novartis Foundation Symposium, 218*, 5–16 (discussion 16–20).

Jürgens, U. (2002). Neural pathways underlying vocal control. *Neuroscience and Biobehavioral Reviews, 26*(2), 235–258.

Keizer, K., & Kuypers, H. G. (1989). Distribution of corticospinal neurons with collaterals to the lower brain stem reticular formation in monkey (*Macaca fascicularis*). *Experimental Brain Research, 74*(2), 311–318.

Kohler, E., Keysers, C., Umiltà, M. A., Fogassi, L., Gallese, V., & Rizzolatti, G. (2002). Hearing sounds, understanding actions: Action representation in mirror neurons. *Science, 297*(5582), 846–848.

Kraskov, A., Dancause, N., Quallo, M. M., Shepherd, S., & Lemon, R. N. (2009). Corticospinal neurons in macaque ventral premotor cortex with mirror properties: A potential mechanism for action suppression? *Neuron, 64*(6), 922–930.

Liberman, A. M., & Mattingly, I. G. (1985). The motor theory of speech perception revised. *Cognition, 21*(1), 1–36.

Lu, M. T., Preston, J. B., & Strick, P. L. (1994). Interconnections between the prefrontal cortex and the premotor areas in the frontal lobe. *Journal of Comparative Neurology, 341*(3), 375–392.

Luppino, G., Matelli, M., Camarda, R., & Rizzolatti, G. (1993). Corticocortical connections of area F3 (SMA-proper) and area F6 (pre-SMA) in the macaque monkey. *Journal of Comparative Neurology, 338*(1), 114–140.

Luppino, G., Murata, A., Govoni, P., & Matelli, M. (1999). Largely segregated parietofrontal connections linking rostral intraparietal cortex (areas AIP and VIP) and the ventral premotor cortex (areas F5 and F4). *Experimental Brain Research, 128*(1–2), 181–187.

Mars, R. B., Jbabdi, S., Sallet, J., O'Reilly, J. X., Croxson, P. L., Olivier, E., et al. (2011). Diffusion-weighted imaging tractography-based parcellation of the human parietal cortex and comparison with human and macaque resting-state functional connectivity. *The Journal of Neuroscience, 31*(11), 4087–4100.

Matelli, M., Luppino, G., & Rizzolatti, G. (1985). Patterns of cytochrome oxidase activity in the frontal agranular cortex of the macaque monkey. *Behavioural Brain Research, 18*(2), 125–136.

Matelli, M., Luppino, G., & Rizzolatti, G. (1991). Architecture of superior and mesial area 6 and the adjacent cingulate cortex in the macaque monkey. *Journal of Comparative Neurology, 311*(4), 445–462.

Meister, I. G., Boroojerdi, B., Foltys, H., Sparing, R., Huber, W., & Töpper, R. (2003). Motor cortex hand area and speech: Implications for the development of language. *Neuropsychologia, 41*(4), 401–406.

Murata, A., Fadiga, L., Fogassi, L., Gallese, V., Raos, V., Rizzolatti, G., et al. (1997). Object representation in the ventral premotor cortex (area F5) of the monkey. *Journal of Neurophysiology, 78*(4), 2226–2230.

Murata, A., Gallese, V., Luppino, G., Kaseda, M., & Sakata, H. (2000). Selectivity for the shape, size, and orientation of objects for grasping in neurons of monkey parietal area AIP. *Journal of Neurophysiology, 83*(5), 2580–2601.

Ortigue, S., Sinigaglia, C., Rizzolatti, G., & Grafton, S. T. (2010). Understanding actions of others: The electrodynamics of the left and right hemispheres. A high-density EEG neuroimaging study. *PloS One, 5*(8), e12160.

Peeters, R., Simone, L., Nelissen, K., Fabbri-Destro, M., Vanduffel, W., Rizzolatti, G., et al. (2009). The representation of tool use in humans and monkeys: Common and uniquely human features. *Journal of Neuroscience, 29*(37), 11523–11539.

Penfield, W., & Welch, K. (1951). The supplementary motor area of the cerebral cortex: A clinical and experimental study. *Archives of Neurology and Psychiatry, 66*(3), 289–317.

Petrides, M., & Pandya, D. N. (1994). Comparative architectonic analysis of the human and the macaque frontal cortex. In F. Boller, & J. Grafman (Eds.), *Handbook of neuropsychology* (pp. 17–58). Amsterdam: Elsevier.

Prabhu, G., Shimazu, H., Cerri, G., Brochier, T., Spinks, R. L., Maier, M. A., et al. (2009). Modulation of primary motor cortex outputs from ventral premotor cortex during visually guided grasp in the macaque monkey. *Journal of Physiology, 587*(Pt 5), 1057–1069.

Prinz, W. (1990). In O. Neumann, & W. Prinz (Eds.), *Relationships between perception and action*. Berlin, Heidelberg: Springer.

Pulvermüller, F. (1999). Words in the brain's language. *The Behavioral and Brain Sciences, 22*(2), 253–279 (discussion 280–336).

Pulvermüller, F., & Fadiga, L. (2010). Active perception: Sensorimotor circuits as a cortical basis for language. *Nature Reviews Neuroscience, 11*(5), 351–360.

Raos, V., Umiltá, M.-A., Murata, A., Fogassi, L., & Gallese, V. (2006). Functional properties of grasping-related neurons in the ventral premotor area F5 of the macaque monkey. *Journal of Neurophysiology, 95*(2), 709–729.

Rizzolatti, G., & Arbib, M. (1998). Language within our grasp. *Trends in Neurosciences, 2236*(1988), 1667–1669.

Rizzolatti, G., Camarda, R., Fogassi, L., Gentilucci, M., Luppino, G., & Matelli, M. (1988). Functional organization of inferior area 6 in the macaque monkey. II. Area F5 and the control of distal movements. *Experimental Brain Research, 71*(3), 491–507.

Rizzolatti, G., & Craighero, L. (2004). The mirror-neuron system. *Annual Review of Neuroscience, 27*, 169–192.

Rizzolatti, G., Fadiga, L., Gallese, V., & Fogassi, L. (1996). Premotor cortex and the recognition of motor actions. *Brain Research Cognitive Brain Research, 3*(2), 131–141.

Rizzolatti, G., Fogassi, L., & Gallese, V. (1997). Parietal cortex: From sight to action. *Current Opinion in Neurobiology, 7*(4), 562–567.

Rizzolatti, G., & Luppino, G. (2001). The cortical motor system. *Neuron, 31*, 889–901.

Rizzolatti, G., Luppino, G., & Matelli, M. (1998). The organization of the cortical motor system: New concepts. *Electroencephalography and Clinical Neurophysiology, 106*(4), 283–296.

Rizzolatti, G., & Matelli, M. (2003). Two different streams form the dorsal visual system: Anatomy and functions. *Experimental Brain Research, 153*(2), 146–157.

Rosenbaum, D. A, Cohen, R. G., Jax, S. A, Weiss, D. J., & van der Wel, R. (2007). The problem of serial order in behavior: Lashley's legacy. *Human Movement Science, 26*(4), 525–554.

Rozzi, S., Ferrari, P. F., Bonini, L., Rizzolatti, G., & Fogassi, L. (2008). Functional organization of inferior parietal lobule convexity in the macaque monkey: Electrophysiological characterization of motor, sensory and mirror responses and their correlation with cytoarchitectonic areas. *European Journal of Neuroscience, 28*(8), 1569–1588.

Sakata, H., Taira, M., Murata, A., & Mine, S. (1995). Neural mechanisms of visual guidance of hand action in the parietal cortex of the monkey. *Cerebral Cortex, 5*(5), 429–438.

Sakreida, K., Schubotz, R. I., Wolfensteller, U., & von Cramon, D. Y. (2005). Motion class dependency in observers' motor areas revealed by functional magnetic resonance imaging. *The Journal of Neuroscience, 25*(6), 1335–1342.

Saygin, A. P., Wilson, S. M., Hagler, D. J., Bates, E., & Sereno, M. I. (2004). Point-light biological motion perception activates human premotor cortex. *Journal of Neuroscience, 24*(27), 6181–6188.

Schubotz, R. I., Anwander, A., Knösche, T. R., von Cramon, D. Y., & Tittgemeyer, M. (2010). Anatomical and functional parcellation of the human lateral premotor cortex. *NeuroImage, 50*(2), 396–408.

Seyal, M., Mull, B., Bhullar, N., Ahmad, T., & Gage, B. (1999). Anticipation and execution of a simple reading task enhance corticospinal excitability. *Clinical Neurophysiology, 110*(3), 424–429.

Shimazu, H., Maier, M. A., Cerri, G., Kirkwood, P. A., & Lemon, R. N. (2004). Macaque ventral premotor cortex exerts powerful facilitation of motor cortex outputs to upper limb motoneurons. *Journal of Neuroscience, 24*(5), 1200–1211.

Shmuelof, L., & Zohary, E. (2005). Dissociation between ventral and dorsal fMRI activation during object and action recognition. *Neuron, 47*(3), 457–470.

Strafella, A. P., & Paus, T. (2000). Modulation of cortical excitability during action observation: A transcranial magnetic stimulation study. *Neuroreport, 11*(10), 2289–2292.

Taira, M., Mine, S., Georgopoulos, A. P., Murata, A., & Sakata, H. (1990). Parietal cortex neurons of the monkey related to the visual guidance of hand movement. *Experimental Brain Research, 83*(1), 29–36.

Tettamanti, M., Buccino, G., Saccuman, M. C., Gallese, V., Danna, M., Scifo, P., et al. (2005). Listening to action-related sentences activates fronto-parietal motor circuits. *Journal of Cognitive Neuroscience, 17*(2), 273–281.

Tokimura, H., Tokimura, Y., Oliviero, A., Asakura, T., & Rothwell, J. C. (1996). Speech-induced changes in corticospinal excitability. *Annals of Neurology, 40*(4), 628–634.

Tomassini, V., Jbabdi, S., Klein, J. C., Behrens, T. E. J., Pozzilli, C., Matthews, P. M., et al. (2007). Diffusion-weighted imaging tractography-based parcellation of the human lateral premotor cortex identifies dorsal and ventral subregions with anatomical and functional specializations. *Journal of Neuroscience, 27*(38), 10259–10269.

Ulloa, E. R., & Pineda, J. A. (2007). Recognition of point-light biological motion: Mu rhythms and mirror neuron activity. *Behavioural Brain Research, 183*(2), 188–194.

Umiltà, M., Escola, L., Intskirveli, I., Grammont, F., Rochat, M., Caruana, F., et al. (2008). When pliers become fingers in the monkey motor system. *Proceedings of the National Academy of Sciences of the United States of America, 105*(6), 2209–2213.

Umiltà, M. A., Kohler, E., Gallese, V., Fogassi, L., Fadiga, L., Keysers, C., et al. (2001). I know what you are doing. A neurophysiological study. *Neuron, 31*(1), 155–165.

Van Essen, D. (1985). Functional organization of primate visual cortex. In E. G. Jones, & A. Peters (Eds.), *Cerebral cortex* (pp. 259–329).

Vogt, C., & Vogt, O. (1919). Allgemeinere ergebnisse unserer hirnforschung. *Journal für Psychologie und Neurologie, 25*, 279–461.

Vogt, S., Buccino, G., Wohlschläger, A. M., Canessa, N., Shah, N. J., Zilles, K., et al. (2007). Prefrontal involvement in imitation learning of hand actions: Effects of practice and expertise. *NeuroImage, 37*(4), 1371–1383.

Von Bonin, G., & Bailey, P. (1947). *The neocortex of macaca mulatta.* Urbana, IL: University of Illinois Press.

Vygotsky, L. S. (1986). *Thought and language*. Cambridge, MA: MIT Press.

Wheaton, K. J., Thompson, J. C., Syngeniotis, A., Abbott, D. F., & Puce, A. (2004). Viewing the motion of human body parts activates different regions of premotor, temporal, and parietal cortex. *NeuroImage, 22*(1), 277–288.

Zilles, K., Schlaug, G., Geyer, S., Luppino, G., Matelli, M., Qü, M., et al. (1996). Anatomy and transmitter receptors of the supplementary motor areas in the human and nonhuman primate brain. In H. O. Lüders (Ed.), *Supplementary sensorimotor area.* Philadelphia: Lippincott-Raven.

CHAPTER

7

Cerebellar Contributions to Speech and Language

Hermann Ackermann and Bettina Brendel

Department of General Neurology/Center for Neurology, Hertie Institute for Clinical Brain Research, University of Tübingen, Tübingen, Germany

7.1 INTRODUCTION

Tracing back to clinical observations in the late 19th and early 20th century, cerebellar disorders have been assumed to yield a range of distinct motor abnormalities, including a syndrome of speech motor deficits called ataxic dysarthria, but to spare perceptual and cognitive aspects of verbal communication. During the past two decades, however, a variety of higher-order abnormalities of spoken language (e.g., more or less exclusive agrammatism, amnesic, or transcortical motor aphasia) have been noted—although rather sporadically—in patients with cerebellar vascular lesions and, more frequently, during the course of recovery from a syndrome of transient mutism after resection of posterior fossa tumors in children. In addition, the identification of distinct speech sound categories appears to critically depend on the integrity of cerebellar structures. Besides motor execution, recent functional imaging data point to a contribution of the cerebellum, in concert with dorsolateral and medial frontal areas, to prearticulatory processes of verbal communication such as the sequencing or parsing of an "inner speech" code. Apart from their well-established contribution to vocal tract innervation, these cerebello-frontal interactions might support a variety of other perceptual, linguistic, and executive tasks under specific conditions (Desmond & Fiez, 1998; Ivry, 1996; Thach, 1998; for a recent review see Marien et al., 2014).

7.2 MACROSCOPIC AND MICROSCOPIC ANATOMY OF THE HUMAN CEREBELLUM

Transverse "trenches" of a varying depth subdivide the cerebellum ("small or lesser brain") into lobes, lobules, and folia (Nieuwenhuys, Voogd, & van Huijzen, 2008). As the most salient sulci, the posterolateral fissure separates the "small brain" into the corpus cerebelli and the flocculonodulus, and the primary fissure splits the former component into the anterior and posterior lobe (Figure 7.1A). Furthermore, two longitudinal sulci demarcate a medial vermis from the two lateral hemispheres. Unfolded on a two-dimensional surface, these subsections would translate into three rather narrow bands of tissue extending over a length of more than 100 cm each. In their actual convoluted state, the rostral and caudal ends of these strips meet each other near the dorsal surface of the fourth ventricle. An outer mantle of cortex overlies white matter encompassing two pairs of interconnected nuclei at either side: the medio-caudal group comprises the fastigial and globose, and the latero-rostral pair consists of the emboliform and dentate nucleus. In subhuman species, the globose and emboliform gray matter masses are often combined into the interpositus nucleus. Altogether, three "cables" connect the cerebellum with the spinal cord, the brain stem, and the thalamus:

1. The inferior peduncle encompasses several afferent bundles (altogether approximately 0.5 million fibers), especially the spinocerebellar and olivocerebellar tracts (restiform body), as well as direct efferent connections between flocculonodulus and vestibular nuclei (juxtarestiform body).
2. The approximately 20 million fibers of the medial peduncle (brachium pontis) arise from the pontine nuclei of the brain stem.
3. The superior peduncle (brachium conjunctivum) represents the major efferent pathway of the cerebellum, originating in the deep nuclei of the white matter and targeting predominantly the

Neurobiology of Language. DOI: http://dx.doi.org/10.1016/B978-0-12-407794-2.00007-9

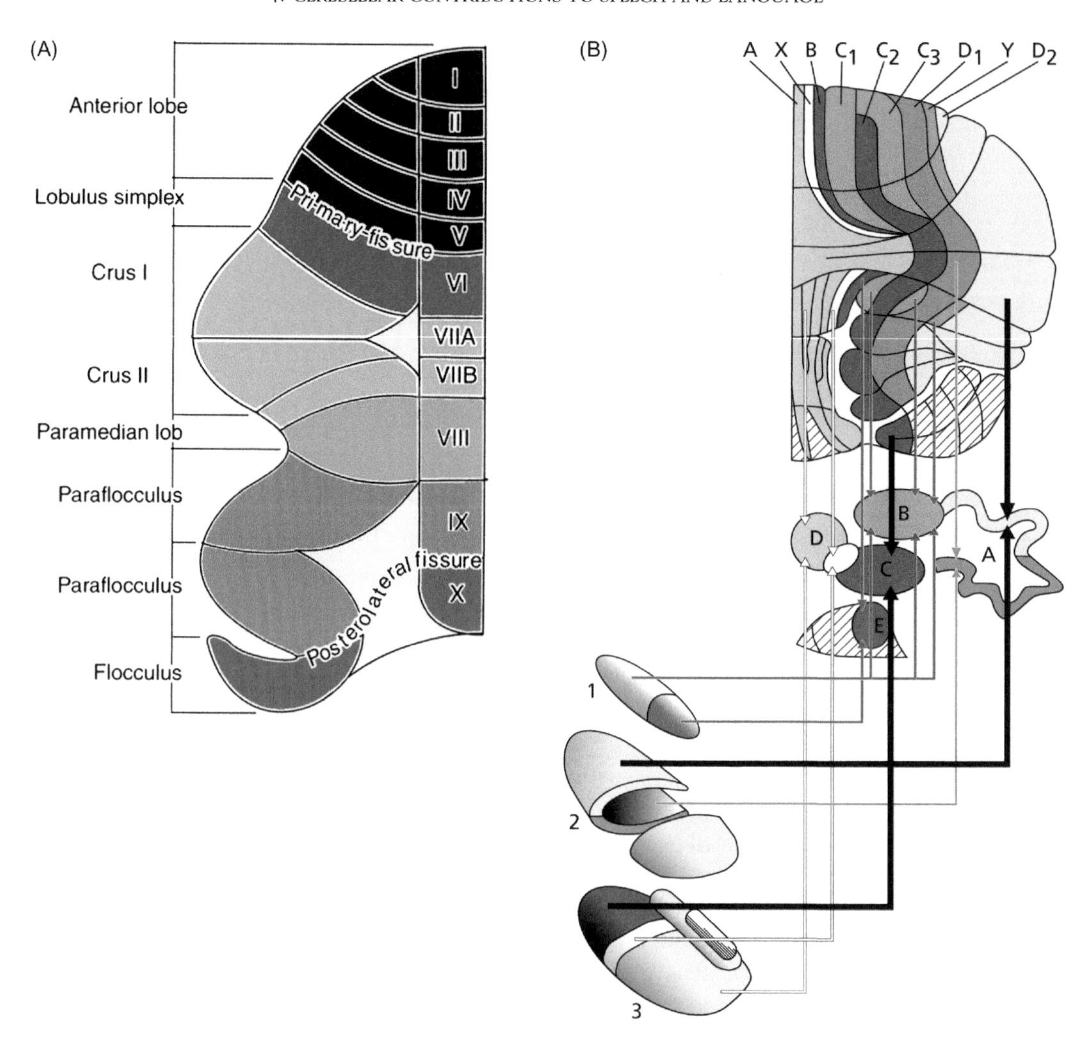

FIGURE 7.1 (A) The major macroscopic subdivisions of the cerebellum (flattened surface of the left hemisphere) concomitant with the widely used nomenclature introduced by Larsell: lobules are labeled consecutively by roman numerals (projected on the left vermis). The ansiform lobule is divided into Crus I and Crus II, separated by the intercrural fissure. (B) The cerebellar cortex is arranged into a multitude of longitudinal zones (A–D2): the Purkinje cells of a given zone project exclusively to the same component of the target nuclei (A = dentate, B = emboliform, C = globose, D = fastigial cerebellar nucleus, E = vestibular nuclei); furthermore, the Purkinje cells of each zone and their respective targets receive climbing fiber input from a particular subcomponent of the inferior olive each (1 = dorsal accessory olive, 2 = principal olive, 3 = medial accessory olive); for example, the longitudinal zones C2 and D2 exclusively project to (a subcomponent of) the dentate nucleus or the globose nucleus, respectively, which in turn are targeted by distinct subcomponents of the olivary complex (heavy black lines). *Redrawn from Nieuwenhuys et al. (2008) with permission from the publisher.*

nucleus ruber (red nucleus) as well as various thalamic nuclei.

Afferent and efferent tracts are related to each other by a ratio of approximately 40:1 fibers, forming a massive convergence of information processing within the "small brain."

Pontine nuclei and several subcomponents of the thalamus represent the relay stations of a cascading cerebrocerebellar circuitry: the afferent (feedforward) limb encompasses sequential corticopontine and pontocerebellar (mossy fibers) projections, and the efferent (feedback) loop encompasses cerebellothalamic and thalamocortical pathways (Schmahmann & Pandya, 1997). The traditional model of cerebrocerebellar interactions assumed a broad range of frontal and parietal regions to project via pontine nuclei to the contralateral cerebellar cortex, whereas the respective efferent connections arising in the deep nuclei predominantly focus—via thalamic structures—on contralateral primary motor and premotor areas (Rothwell, 1994). However, more recent transneuronal tract-tracing studies in nonhuman primates found that distinct subcomponents of the deep white matter nuclei project to different areas of the cerebral cortex, including primary motor, premotor, and prefrontal areas, which, in turn, represent a major source of input to the respective sites of origin, giving rise to a series of parallel closed loops (Strick, Dum, & Fiez, 2009). More advanced magnetic resonance imaging (MRI) technologies are now beginning to provide the opportunity to visualize the structure of cerebellar circuits in the intact human brain, for example, the connections between the cerebellar cortex and the thalamus (Granziera et al., 2009).

The cortex of the cerebellum segregates into three stacked layers: a superficial cell-poor molecular sheet, an adjacent thin tier composed of the perikarya of the large Purkinje cells, and a deep zone of densely packed small granule cells (Figure 7.2). The axons of the Purkinje cells represent the sole "output channel" of the cerebellar cortex, exerting a strong inhibitory impact on their target structures (i.e., the deep nuclei of the cerebellar white matter and the vestibular nuclei of the brain stem). Within some limits, the arrangement of these neural elements and their connections reiterate the longitudino-transversal organization of the cerebellar surface. The flattened dendritic trees of the Purkinje cells are confined to a plane perpendicular to the axons of the granule cells (parallel fibers) "running" in the direction of the transverse fissures. Each Purkinje cell may bear tens of thousands of synaptic connections with the parallel fibers traversing its dendritic tree. However, most of these contacts appear to be "closed" at any given moment due to long-term depression effects (Nieuwenhuys et al., 2008). These processes may "sculpt the cerebellar cortical network, according to previous experience" and, thus, provide a basis for cerebellum-dependent optimization of movement sequences. Two major afferent systems access the cerebellar cortex: (i) restiform body and middle peduncle conduct mossy fibers, arising from the spinal cord as well as the brain stem and targeting the granule cells; (ii) in addition, the restiform body contains climbing fibers originating in the inferior olive and terminating in the contralateral middle layer (Figure 7.2). Both fiber types, which also give off collateral branches to the deep cerebellar nuclei, have an excitatory effect on Purkinje cells. In anesthetized animal preparations, the cerebellar cortex displays "two complete somatotopic representations of the body," one in the anterior lobe and one in the posterior lobe (Rothwell, 1994). Perioperative electrical surface stimulation revealed in our species a motor map bound to the posterior cerebellar cortex (other areas were not investigated; Mottolese et al., 2013). Beyond these findings, functional imaging studies suggest that the "small brain" houses up to four distinct somatomotor homunculi (Buckner, Krienen, Castellanos, Diaz, & Yeo, 2011).

The rather uniform architectural design of the "small brain" does not allow for a cytoarchitectonic parcellation similar to the cerebral cortex. Nevertheless, specific input/output arrangements as well as the distribution of marker molecules give rise to a functional compartmentalization of the cerebellum in terms of parallel

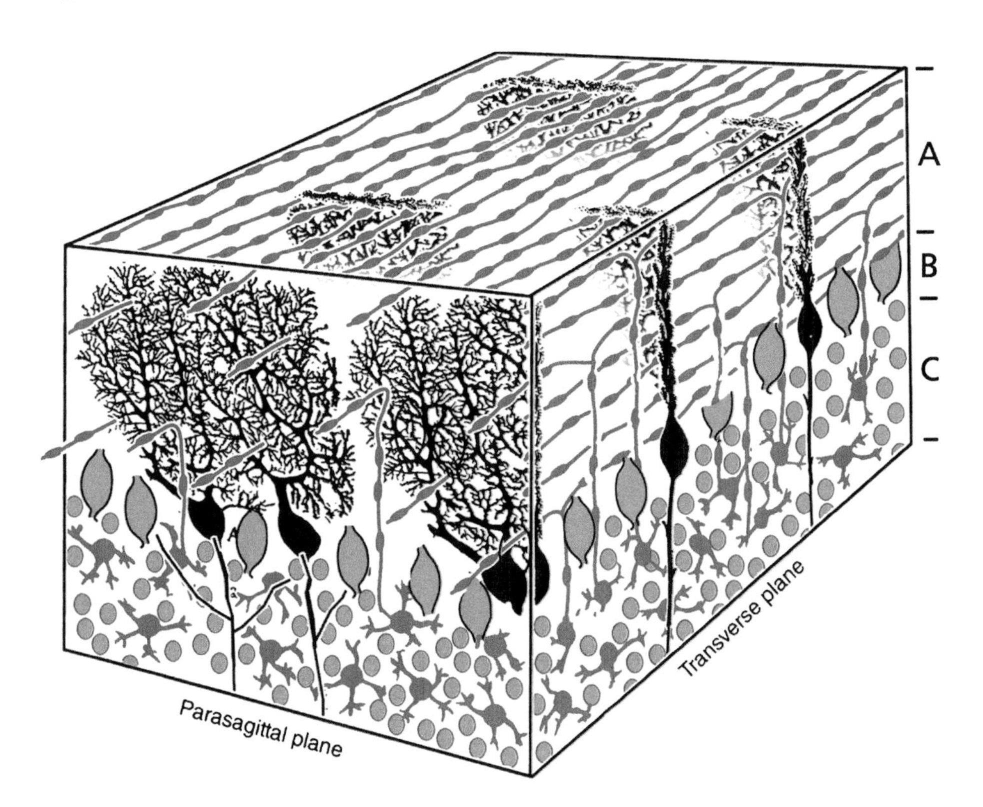

FIGURE 7.2 Microscopic organization of the three-layer cerebellar cortex (A = molecular layer, B = Purkinje cell layer, C = granular layer). The flat dendritic trees of the Purkinje cells (large black bodies) are spread out in a parasagittal plane perpendicular to the parallel fibers, running along the transverse plane. *Modified from Nieuwenhuys et al. (2008) with permission from the publisher.*

longitudinal stripes that intersect the transverse fissures at a right angle (Figure 7.1B). More recent formulations assume that each half of the vermis contains two or three and that each hemisphere contains five or more such elongated but rather narrow cortical zones (Nieuwenhuys et al., 2008). In addition, gene expression patterns provide the basis for a further compartmentalization of the cerebellum in at least four transverse bands, more or less, independent of lobulation (Oberdick & Sillitoe, 2011). In contrast, the traditional separation into vestibulocerebellar, spinocerebellar, and neocerebellar components does not correspond to strict boundaries of the projections of the respective afferent sources (Coffman, Dum, & Strick, 2011). Given a rather uniform internal circuitry, the cerebellar cortex has been assumed to support rather uniform computational processes throughout its entire extent (Ramnani, 2006). Against this background, compartmentalization of the "small brain" could depend upon the functional architecture of the interconnected extracerebellar areas.

7.3 COMPARATIVE ANATOMIC PERSPECTIVES ON SIZE AND COMPOSITION OF THE CEREBELLUM

The overall volume of the cerebellum varies considerably across vertebrate classes—even after scaling to body or hindbrain size (Butler & Hodos, 2005). First, mammals and birds are endowed with a (relatively) larger cerebellum than reptiles and amphibians. Second, electroreceptive fishes such as mormyrids show an exceptionally huge "small brain," by far exceeding its dimensions in nonelectric relatives and in all four-limbed vertebrates. Furthermore, the various macroscopic components of the cerebellum may show a differential enlargement across species within the same taxon—allegedly related to specific behavioral traits such as beak control in woodpeckers (Sultan & Glickstein, 2007) or acoustic communication in marine mammals (Oelschläger, Ridgway, & Knauth, 2010). Given these rather complicated comparative anatomic relationships, the differentiation between an archicerebellum, paleocerebellum, and neocerebellum, based on the suggestion that the major macroscopic constituents of the "small brain" can be assigned to successive stages of evolution, must be considered with some precautions. As a side note, the neural networks supporting vocal behavior in birds, including song learning, do not seem to encroach on the cerebellum (Ackermann, 2008), whereas motor coordination and skill acquisition in our species, including the domain of speech production, clearly depend on the integrity of the "small brain."

Brain volume underwent a significant increase during the course of human evolution—allegedly concomitant with a disproportionate expansion of the prefrontal cortex relative to motor areas (see Teffer & Semendeferi, 2012 for a critical review). Furthermore, structural MRI points at a conjoint relative enlargement of the cerebellar lobules connected with prefrontal areas in our species (Balsters et al., 2010). Such morphological correlations have been assumed to indicate a contribution of the cerebellum to cognitive functions such as language. In light of these findings, one model claims that "the cerebellum couples the motor function of articulating speech to the mental function that selects the language to be spoken, thus helping to produce fluent human speech" (Leiner, 2010).

7.4 CEREBELLAR SUPPORT OF MOTOR CONTROL IN HUMANS: UPPER LIMB MOVEMENTS

The systematic experimental analysis of cerebellar functions traces back to Luigi Rolando, who conducted ablation studies in a variety of species around the turn of the 18th to 19th century (Schmahmann, 2010). He found that damage to the "small brain" compromised motor activities of the homolateral body side but spared the (in more recent parlance) sensory, autonomic, and cognitive domains. Holmes (1917, 1939) provided the classic description of cerebellar movement abnormalities in humans. Besides reduced muscle tone (hypotonia) and strength (asthenia)—observed at least in the acute stage after cerebellar damage with significant improvement during further follow-up—"the most prominent disturbances of function that result from destructive lesions of the cerebellum are distinct disorders of voluntary movement," usually referred to as "ataxia" or "incoordination." More specifically, Holmes noted a "delay in starting motion," slowed and irregular execution of a movement, an excessive range or premature arrest of goal-directed motor activities (hypermetria/hypometria), and increased oscillations (tremor), especially during the final approach of an object. Furthermore, cerebellar injuries may give rise to vertigo, "spontaneous deviations of the limbs," unsteady postural equilibrium and gait ataxia, compromised ocular movements (especially, nystagmus), reflex abnormalities, a loss of associated movements (e.g., of the ipsilateral arm during walking), and "disturbances of speech." For the most part, these early analyses of the motor deficits of cerebellar disorders have stood the test of time and found their way into generations of subsequent textbooks of clinical neurology (Adams & Victor, 1989; Adams, Victor, & Ropper, 1997).

More recent kinematic and electromyographic (EMG) studies provide evidence for at least two pathomechanisms of cerebellar ataxia in case of single-joint rapid arm movements: deficient timing of agonist/antagonist EMG bursts and compromised implementation of phasic muscle forces (Berardelli et al., 1996). In line with the latter suggestion, functional imaging studies revealed separate subregions of the cerebellum to scale with distinct force-related parameters each (amplitude versus rate of change) during isometric pinch-grip tasks (Spraker et al., 2012). Rather than a failure to generate sufficient force levels *per se*, anticipatory adjustment of and compensation for the arising torques during movement execution appears to be disrupted (see Nowak, 2012 for a brief recent review). An alternative model considers the distorted triphasic innervation patterns of ballistic gestures, giving rise to hypermetria of movements (Rothwell, 1994), as well as the disruption of multijoint sequences in cerebellar ataxia (Ito, 2012, p. 156) to reflect motor timing deficits. Indeed, considerable evidence indicates that the "small brain" engages in the "precise representation of temporal information" (Ivry & Fiez, 2000, p. 1005; see Ivry, 2012 for a recent review).

A variety of clinical and functional imaging data suggest that motor skill acquisition in our species critically depends on the "small brain." Early computational models of the 1960s considered the neuronal circuitry of the cerebellum a potential substrate of its motor learning capacities (see Ito, 2012 for references). More recent formulations associate these capabilities with internal forward or inverse representations of body–environment interactions (Passot, Luque, & Arleo, 2013). However, the cerebellar subcomponents specifically engaged in motor skill acquisition appear to vary with the experimental paradigm considered (sensorimotor adaptation versus motor sequencing) and the stage of the training process (e.g., early versus late; Hardwick, Rottschy, Miall, & Eickhoff, 2013; see Orban et al., 2011). At the molecular level, durable experience-dependent changes in synaptic transmission efficacy of Purkinje cells through convergent activation of parallel and climbing fibers have been demonstrated (Ito, 2012). Thus, evidence based on different methodological approaches corroborates the notion of a cerebellar contribution to motor learning mechanisms.

7.5 CONTRIBUTIONS OF THE CEREBELLUM TO SPEECH MOTOR CONTROL

7.5.1 The Profile of Ataxic Dysarthria: Auditory-Perceptual and Instrumental Studies

Holmes (1917) provided the first more detailed account of the motor speech deficits after cerebellar damage, more specifically, unilateral and bilateral gunshot injuries. His early observations are in line with the still authoritative auditory-perceptual investigation of the Mayo Clinic based on patients with different etiological variants of a syndrome of ataxia (Darley, Aronson, & Brown, 1975). Imprecise consonants and distorted vowels, irregular articulatory breakdown, "excess and equal stress" (i.e., scanning speech rhythm), reduced speaking rate, as well as harsh voice quality emerged as the most salient features. Assuming the same underlying pathomechanisms as in other motor subsystems, the label "ataxic dysarthria" was assigned to the profile of speech abnormalities (see Ackermann, Mathiak, & Riecker, 2007; Duffy, 2005 for recent reviews). However, the extent to which extracerebellar pathology contributed to the observed deficits—given the rather limited diagnostic opportunities available at that time—remains unsettled. More recent studies interested in the specific contribution of the "small brain" to speech motor control tried to recruit patients with a disease process restricted to the cerebellum—as documented by neuroradiological techniques—or subjects with a molecular-genetic diagnosis of a distinct hereditary degenerative disease entity such as Friedreich's ataxia (FRDA; Brendel et al., 2013) or the various subtypes of a spinocerebellar ataxia (SCA; Sidtis, Ahn, Gomez, & Sidtis, 2011; Schalling & Hartelius, 2013). Comparison of various syndromes of a hereditary ataxia (matched for disease severity and disease duration) revealed both commonalities and differences in the respective profiles of speech motor deficits (Brendel et al., 2015). Voice quality and stability, articulatory preciseness, and respiratory support turned out to be the most affected auditory-perceptual dimensions of speech production in patients with FRDA, SCA3, or SCA6. Such comparative analyses may provide the opportunity to further delineate the differential contribution of the various cerebellar subcomponents to speech motor control. At least in initial disease stages, FRDA predominantly affects the afferent fiber tracts that project to the cerebellum, whereas SCA6 pathology appears mostly restricted to the cortical layer of the "small brain" and SCA3 seems, by contrast, to encroach upon the deep cerebellar nuclei (Schalling & Hartelius, 2013). As compared with the other two ataxia variants, SCA6 patients showed a more pronounced reduction in speaking rate and more salient abnormalities of the prosodic modulation of verbal utterances (Brendel et al., 2015). Most noteworthy, several investigations reported the dysarthric signs of degenerative cerebellar diseases (FRDA, SCA subtypes) to be limited to rather mild deficits (e.g., in terms of intelligibility) even after an extended disease duration of more than two decades (Brendel et al., 2013; Folker et al., 2010; Sidtis et al., 2011). Although patient selection bias cannot be ruled out in these cases, cerebellar disorders appear to have a less devastating impact on motor aspects of spoken language than does damage to left-hemisphere ventrolateral frontal cortex (apraxia of speech) or to the corticobulbar tracts (spastic dysarthria), which in its extreme gives rise to a complete breakdown of articulation and/or phonation (anarthria/aphonia; Ackermann & Ziegler, 2010). A similar dissociation between upper motor neuron disorders and cerebellar dysfunctions at the behavioral level has been observed during tasks involving highly overlearned finger force control (Brandauer et al., 2012).

Acoustic analyses of verbal utterances of patients with ataxic dysarthria concentrated mainly on measurements of durational parameters, such as voice onset time (VOT), syllable length, and vowel duration (see Ackermann et al., 2007 for a review). In line with auditory-perceptual data, an increased duration of syllables, vowels, and smaller segments—giving rise to a reduced speaking or articulation rate—has been repeatedly observed in ataxic patients as compared with healthy control speakers (Brendel et al., 2013; Schalling & Hartelius, 2013). Most noteworthy, the decline of speech rate during syllable repetition tasks appears to approach a plateau at approximately 2.5 to 3 Hz (Ackermann, 2008). In some instances of a cerebellar disorder, acoustic analyses were able to document voice tremor of a similar frequency or enhanced pitch fluctuations during sustained vowel productions (Ackermann et al., 2007; Boutsen, Duffy, Dimassi, & Christman, 2011). "Stabilization at a fixed posture or state" represents one of the fundamental control functions of the cerebellum (Massaquoi, 2012). In light of this notion, the respective cerebellar vocal dysfunctions may result from compromised maintenance of laryngeal sound production.

Because of an inherent nonlinear relationship between vocal tract configuration and the emitted speech signal, acoustic data often do not allow for unambiguous characterization of the underlying vocal tract movements. Among other things, kinematic analyses of lip and jaw excursions found a reduced mass-normalized stiffness (i.e., a decreased ratio of maximum velocity and displacement) in cerebellar patients as compared with control speakers (Ackermann et al., 2007). These findings have been assumed to indicate an impaired ability to generate adequate muscular forces under time-critical conditions, similar to upper limb movements. As an alternative, increased movement durations in cerebellar patients could point to larger articulatory distances during speech production rather than reduced peak velocity (Folker et al., 2011).

7.5.2 The Syndrome of Cerebellar Mutism: Impaired Initiation of Speech Production?

Speechlessness (mutism) after resection of a posterior fossa tumor or a stereotactic lesion of the dentate nucleus predominantly emerges in children at an incidence of up to 30%, but it may sporadically occur in adults as well (Küper & Timmann, 2013). Rarely, cerebellar mutism has also been observed in patients with traumatic injury to or viral inflammation of the "small brain." As a rule, disrupted speech production develops several hours or days after surgery and frequently is associated with a variety of further behavioral and/or cognitive-mental abnormalities (posterior fossa syndrome). The initial—and always transient—mutism may evolve into speech motor deficits ("syndrome of cerebellar mutism and subsequent dysarthria"), into a syndrome of agrammatism concomitant with impaired executive functions, and/or into emotional/motivational abnormalities (Ackermann et al., 2007). In contrast, auditory speech comprehension as well as repetition capabilities are well-preserved. A recent review points to "crossed cerebello-cerebral diaschisis" as a potential pathomechanism of the initial mutism of the posterior fossa syndrome (i.e., "functional depression" of cerebral areas reciprocally interconnected with the damaged cerebellar region), translating into reduced blood flow (hypoperfusion), decreased oxygen consumption, and hypometabolism of the affected supratentorial structures (Küper & Timmann, 2013, see Botez, Léveillé, Lambert, & Botez, 1991 for further details on this concept). Bilateral lesions of anterior aspects of the medial wall of the frontal lobe may give rise to speechlessness (akinetic mutism), and damage to the left-hemisphere supplementary motor area has been found to compromise speech initiation mechanisms (Ackermann & Ziegler, 2010). Thus, transient cerebellar mutism conceivably reflects bilateral functional depression of mesiofrontal structures, whereas involvement of dorsolateral and/or orbital parts of the frontal lobe may give rise to the associated additional cognitive and/or behavioral abnormalities. In contrast, a persisting ataxic dysarthria that occurs subsequent to cerebellar mutism could reflect local cerebellar injury due to surgical intervention, because mesiofrontal lesions do not give rise to articulatory/phonatory deficits. Therefore, the so-called cerebellar mutism does not necessarily arise from cerebellar dysfunctions proper. Such an interpretation would also be more consistent with the suggestion that damage to the "small brain" yields rather mild speech motor deficits.

7.5.3 Functional Imaging Data Related to Speech Production

A meta-analysis of the "localization" of speech motor deficits in infarctions of the "small brain" found this constellation to be associated with damage to superior aspects of the cerebellar cortex—more specifically, the area of blood supply of the superior cerebellar artery (Urban, 2013). In line with these clinical findings, functional imaging studies focusing on motor aspects of spoken language in healthy subjects reported that speech production tasks elicit robust bilateral hemodynamic responses centered around lobule VI (Figure 7.3). Three pieces of evidence support the notion of a crucial role of lobule VI in speech production:

1. A bilateral somatomotor map of the face has been documented at this level.
2. The execution of nonspeech movements of the tongue and lips yields bilateral signal changes within the same area.
3. Resting-state functional connectivity MRI was able to demonstrate tight functional coupling between the topographic motor representations of the frontal cortex and cerebellar lobule VI (Buckner et al., 2011; Grodd, Hülsmann, & Ackermann, 2005; O'Reilly, Beckmann, Tomassini, Ramnani, & Johansen-Berg, 2010).

Taken together, these findings suggest that the superior corpus cerebelli must be considered an essential component of the brain network of (motor) speech production engaged, presumably, in the online control of articulatory movement sequences, regardless of stimulus type.

Signal changes in lobule VI lateralized toward the right side have been documented even during inner (covert, subvocal) speech tasks (see Ackermann & Riecker, 2010 for references). Psycholinguistic models (e.g., Levelt, 1989) often equate covert utterances with overt speech production—minus motor execution. As a consequence, inner speech must be expected to rely on a prearticulatory but otherwise fully specified verbal code. From this perspective, speech motor control separates into at least two stages: (i) the generation of a prearticulatory code (i.e., preparatory/planning processes) and (ii) its subsequent execution.

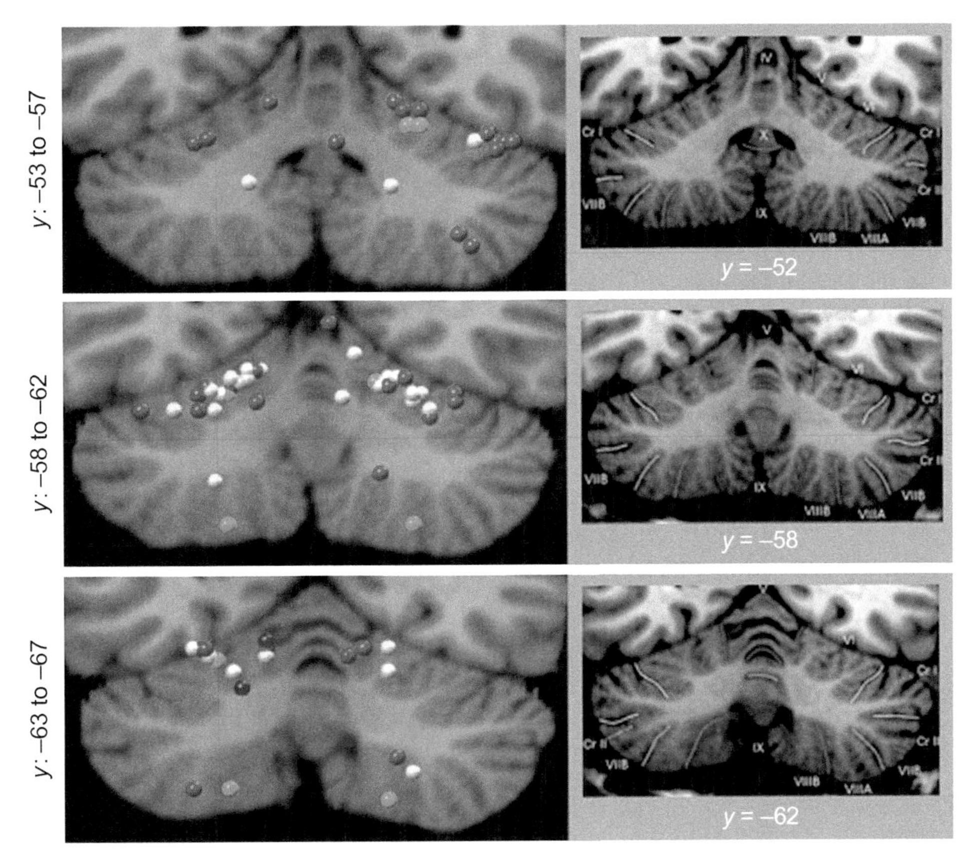

FIGURE 7.3 Spatial distribution of the signal maxima (Montreal Neurological Institute [MNI] space) of a series of published functional imaging studies addressing various aspects of speech and nonspeech motor control (Brendel et al., in preparation): pseudowords (red), repetitive syllables (green), single vowels (blue), lexical items (yellow), as well as nonspeech oral movements of tongue, lips, and jaw (white). Signal peaks represent main effects (experimental condition compared with a baseline, i.e., rest condition; coordinates are superimposed on the Collins brain template) and extend in the anterior–posterior direction (*y*-dimension) from −53 to −67 (divided into three segments). The majority of the peaks fall in this range, although a few signal maxima have either a more anterior or posterior location, respectively; the anatomical landmarks (right column) are based on Schmahmann, Doyon, Toga, Petrides, and Evans (2000).

In light of these findings, hemodynamic activation bound to inner speech may be related to a prearticulatory level of spoken language—although the computational processes involved are still a topic of debate (e.g., Geva et al., 2011). Some preliminary evidence for cerebellar involvement in these higher-level processes derives from the observation of functional coupling between superior aspects of the right cerebellar hemisphere and left premotor cortex, insula, and Broca's area (Buckner et al., 2011; Riecker et al., 2005). Importantly, these latter supratentorial areas of the language-dominant hemisphere are assumed to be crucially engaged in speech motor planning (Ackermann & Ziegler, 2010).

In contrast to the clinical data on the cerebellar topography of speech motor control, functional imaging yielded a more complicated picture because additional activation of caudal-inferior parts (lobule VIIIA) emerged across speech and nonspeech vocal tract tasks—although in a more inconsistent and variable way. Whereas this area also incorporates a somatomotor map of the vocal tract and displays strong resting-state functional connectivity with supratentorial sensorimotor cortex, the functional role of lobule VIIIA within the framework of speech production is still a matter of speculation. Given that inner speech is assumed to represent an important rehearsal strategy to temporarily maintain information in the phonological buffer of the verbal working memory circuit, a particularly puzzling aspect of these functional neuroimaging findings is that signal changes in lobule VIIIA have been repeatedly observed during verbal working memory tasks (Chen & Desmond, 2005; Durisko & Fiez, 2010), but not during subvocal speech. Hypotheses about the contributions of the inferior cerebellum to speech production range from a strictly "basic" motor/sensorimotor function (Stoodley & Schmahmann, 2009) to an involvement in "higher-level" linguistic processes, such as those supporting "fast-loading" mechanisms of phonological representations (Bohland, Bullock, & Guenther, 2010).

7.6 ENGAGEMENT OF THE CEREBELLUM IN NONMOTOR FUNCTIONS

7.6.1 The Cerebellar Cognitive Affective Syndrome

A classical tenet of clinical neurology tracing back to the early 19th century and passed down via generations of textbooks until approximately two decades ago generally restricted cerebellar symptomatology to the motor domain (compare Adams & Victor, 1989 with Adams et al., 1997, p. 92). However, the lateral hemispheres of the "small brain" showed an overproportional increase of size during the course of human evolution, and a large portion of corticopontine projections arises within the prefrontal cortex. Considering, among other things, such (comparative) neuroanatomic data, the notion of an engagement of the "small brain" in mental skills attracted more and more attention from the 1970s and 1980s onward (Leiner, 2010; Schmahmann, 2010). Whereas the morphological findings referred to could just reflect enhanced capabilities for visuomotor control of manual tasks, systematic neuropsychological studies of larger groups of patients with pathology restricted to the "small brain" revealed—beyond the preceding (rather anecdotal) clinical observations—altered nonmotor functions such as compromised executive operations (verbal working memory, set-shifting, etc.), impaired visual-spatial capacities, disrupted memory as well as language deficits (Schmahmann & Sherman, 1998). Besides mental-cognitive disorders, personality changes in terms of blunted affect and inappropriate behavior could be observed as well. This "cerebellar cognitive affective syndrome" has been reported to include deviations of speech prosody (i.e., a high-pitched voice of a "whining, childish, and hypophonic quality") emerging, especially in bilateral or generalized disease processes (Schmahmann & Sherman, 1998, p. 564). Rather than a cognitive-linguistic deficit, these abnormalities might reflect disruption of laryngeal motor control mechanisms. In accordance with these clinical data, the available imaging studies point to a functional compartmentalization of the cerebellar cortex in that motor, cognitive, and "limbic" tasks elicit activation patterns of a distinct topographical distribution each (Stoodley & Schmahmann, 2009). As an interesting side note, nonmotor abnormalities appear to be more prominent in acute rather than chronic pathologies.

7.6.2 Lexical-Semantic and Syntactic Disorders of Spoken Language

In addition to executive/visual-spatial dysfunctions and behavioral abnormalities, lexical-semantic deficits in terms of compromised picture naming, verb-for-noun generation, word stem completion, and verbal fluency have been documented in a variety of cerebellar disorders such as ischemic infarctions, cerebellitis, or degenerative diseases (Fiez, Petersen, Cheney, & Raichle, 1992; Schmahmann & Sherman, 1998; Stoodley & Schmahmann, 2009). Furthermore, functional imaging studies and transcranial magnetic stimulation techniques (implementation of a virtual transient lesion) point to a predominant engagement of posterolateral aspects of the right cerebellar hemisphere in these tasks (De Smet, Paquier, Verhoeven, & Marien, 2013; Stoodley & Schmahmann, 2009). Rather than compromised access to the mental vocabulary, the observed lexical-semantic disorders have been assumed to reflect impaired cognitive search strategies and error monitoring (Stoodley & Schmahmann, 2009)—operations that are conceivably related to interactions of the cerebellar hemispheres with those parts of the prefrontal cortex supporting executive functions.

Silveri et al. (1994) reported for the first time signs of agrammatic speech concomitant with slightly reduced verbal fluency in a patient who had a right-hemisphere cerebellar ischemic lesion. Verbal utterances were characterized by the omission of free-standing grammatic morphemes and the use of infinitives in place of inflected verb forms. Otherwise, language examination was entirely unremarkable (see Zettin et al., 1997 for a further example). Whereas structural neuroimaging failed to detect any supratentorial lesions in the two aforementioned case studies, measurements of hemodynamic functions revealed marked hypoperfusion of the entire dominant hemisphere (crossed cerebello-cerebral diaschisis). During further follow-up, an increase of left-hemisphere cerebral blood flow paralleled the improvement of language deficits. A subsequent group study noted—mostly mild—agrammatism in 6 out of a total of 20 patients with a cerebellar disorder (Schmahmann & Sherman, 1998). Apart from a single exception (midline tumor removal), all of them had a diagnosis of bilateral or right-hemisphere infarction, whereas subjects with cerebellitis or cortical atrophy did not display any abnormalities of the syntax of spoken language at bedside testing. Beyond exclusive agrammatism, right-hemispheric cerebellar infarctions may give rise to a broader profile of speech-language abnormalities resembling transcortical motor or amnesic aphasia (for a recent review, see De Smet et al., 2013). Similar to cerebellar agrammatism, these more extensive constellations of disrupted verbal communication may also be associated with crossed cerebello-cerebral diaschisis effects. For example, left-frontoparietal perfusion was again found to increase in parallel with the improvement of speech-language pathology during follow-up.

Variants of an acquired dyslexia/dysgraphia have—although rarely—been observed in cerebellar stroke patients as well (De Smet et al., 2013). Most presumably, these constellations are also related to the pathomechanism of crossed cerebello-cerebral diaschisis, "the most plausible explanation posited to date" (Murdoch & Barwood, 2014) for higher-level language deficits, at least in cerebellar lesions of a vascular origin. As a further disorder of written language capabilities, developmental dyslexia is often associated with signs of a cerebellar dysfunction (up to approximately 80% of cases; Nicolson, Fawcett, & Dean, 2001). Therefore, this syndrome has been assumed to reflect a more general impairment of cerebellum-dependent automatic performance of learned skills including, but not restricted to, the acquisition of written language capabilities ("cerebellar deficit hypothesis" of developmental dyslexia; Nicolson et al., 2001; for a recent brief review see Nicolson & Fawcett, 2014).

7.6.3 Contributions of the Cerebellum to Nonmotor Functions: Speech and Nonspeech Auditory Perception

Although Holmes (1917) did not find unambiguous evidence for perceptual deficits in patients with acute gunshot injuries of the cerebellum, more recent investigations reported that disorders of the "small brain" may compromise, for example, visual motion discrimination (Bastian, 2011) or the assessment of spatio-temporal relationships. In line with preceding observations of a disrupted evaluation of the length of time intervals bound by tones (see Ivry & Fiez, 2000 for a review), a series of studies was able to further document impairments in perceptual encoding of temporal aspects of speech sounds in cerebellar patients (Mathiak, Hertrich, Grodd, & Ackermann, 2002). Both the English and the German language comprise pairs of syllables or lexical items exclusively differing in a single durational parameter of the acoustic signal. For example, the English word "rapid" is characterized by a short period of silence (closure or occlusion time [CLT]) signaling the intraword stop consonant /p/. Variation of CLT from 10 to 110 ms gives rise to a phoneme-boundary effect: long intraword pauses yield the percept "rapid," whereas short variants lead to the recognition of "rabbit," with a rather abrupt transition from one response type to the other in-between. Under these conditions, word recognition solely depends on the processing of a durational acoustic cue. Using a German analogue of the "rabbit/rapid" paradigm ("Boden" [bodn] = short CLT, Engl. "floor" versus "Boten" [botn] = long occlusion time, "messengers"), patients with diffuse cerebellar atrophy did not show any significant phoneme-boundary effect (Ackermann, Gräber, Hertrich, & Daum, 1997). Apart from CLT, the difference in sound structure between the lexical items "Boten" and "Boden" can be signaled by VOT of the wordmedial stop consonants /t/ and /d/ ([bo:then] versus [bo:den], with the /t/-sound of the first item being characterized by a short aspiration noise). An fMRI study asked subjects to discriminate the same lexical items (i.e., "Boten" versus "Boden") either by analysis of a durational parameter such as CLT (experimental condition) or a noise segment such as VOT (control condition). Subtraction of the hemodynamic responses to the "noise stimuli" from the activation pattern obtained during application of the CLT items yielded a circumscript activation focus within the right cerebellar hemisphere. Most noteworthy, a subsequent fMRI study found the discrimination of nonspeech time intervals bound by tones to elicit hemodynamic activation at the level of the right cerebellar hemisphere concomitant with a left-sided prefrontal cluster (Mathiak, Hertrich, Grodd, & Ackermann, 2004). Thus, a right-cerebellar/left-prefrontal loop appears to be engaged in the decoding of durational parameters of speech and nonspeech acoustic signals.

7.7 CONCLUSION

(i) It is well established that the cerebellum engages in movement preparation/execution as well as motor skill acquisition, although the underlying mechanisms remain to be further elucidated. These capacities also extend, at least partially, to the control of up to approximately 100 vocal tract muscles engaged in speech production. Disorders of the "small brain" may give rise to the syndrome of ataxic dysarthria, characterized by (among other things) compromised stability of sound production, slowed execution of single articulatory gestures, especially under enhanced temporal constraints, and disrupted coordination/sequencing of orofacial and laryngeal activities. These abnormalities accord quite well with the pathophysiological deficits observed in upper limb ataxia. Most noteworthy, reduced maximum speaking rate appears to approach a plateau at approximately 2.5–3 Hz in cerebellar patients. Therefore, the processing capabilites of the cerebellum seem to provide a necessary prerequisite to push the verbal sequencing capacities beyond this level and, furthermore, to modulate the rhythmic structure of verbal utterances (see Ackermann, 2008 for more details). (ii) From a phylogenetic perspective, inner speech mechanisms—based on a prearticulatory verbal code—may have "emerged from overt speech and motor systems as an evolutionary adaptive way to boost cognitive processes that rely on working memory, such as language acquisition" (Marvel & Desmond, 2010, p. 8). Conceivably, the computational power of the cerebellum also subserves the sequential organization of internal verbal codes. Therefore, cerebellar disorders may compromise cognitive capacities associated with "inner speech" such as the subvocal rehearsal component of verbal working memory or may impede the linguistic scaffolding of executive functions. Conceivably, a prearticulatory verbal code might also support—under specific conditions—the perceptual resolution of temporal aspects of speech sounds. (iii) As a further pathomechanism of nonmotor higher-order language disorders, the tight reciprocal interactions between prefrontal cortex and lateral cerebellar hemispheres during a variety of mental tasks may elicit cerebello-cerebral diaschisis effects, especially in acute vascular disorders, giving rise to aphasia-like or mutism-like syndromes.

References

Ackermann, H. (2008). Cerebellar contributions to speech production and speech perception: Psycholinguistic and neurobiological perspectives. *Trends in Neurosciences*, *31*, 265–272.

Ackermann, H., Gräber, S., Hertrich, I., & Daum, I. (1997). Categorical speech perception in cerebellar disorders. *Brain and Language*, *60*, 323–331.

Ackermann, H., Mathiak, K., & Riecker, A. (2007). The contribution of the cerebellum to speech production and speech perception: Clinical and functional imaging data. *Cerebellum*, *6*, 202–213.

Ackermann, H., & Riecker, A. (2010). Cerebral control of motor aspects of speech production: Neurophysiological and functional imaging data. In B. Maassen, & P. van Lieshout (Eds.), *Speech motor control: New developments in basic and applied research* (pp. 117–134). Oxford: Oxford University Press.

Ackermann, H., & Ziegler, W. (2010). Brain mechanisms underlying speech motor control. In W. J. Hardcastle, J. Laver, & F. E. Gibbon (Eds.), *The handbook of phonetic sciences* (2nd ed., pp. 202–250). Malden, MA: Wiley-Blackwell.

Adams, R. D., & Victor, M. (1989). *Principles of neurology* (4th ed.). New York, NY: McGraw-Hill.

Adams, R. D., Victor, M., & Ropper, A. H. (1997). *Principles of neurology* (6th ed.). New York, NY: McGraw-Hill.

Balsters, J. H., Cussans, E., Diedrichsen, J., Phillips, K. A., Preuss, T. M., Rilling, J. K., et al. (2010). Evolution of the cerebellar cortex: The selective expansion of prefrontal-projecting cerebellar lobules. *NeuroImage*, *49*, 2045–2052.

Bastian, A. J. (2011). Moving, sensing and learning with cerebellar damage. *Current Opinion in Neurobiology*, *21*, 596–601.

Berardelli, A., Hallett, M., Rothwell, J. C., Agostino, R., Manfredi, M., Thompson, P. D., et al. (1996). Single-joint rapid arm movements in normal subjects and in patients with motor disorders. *Brain*, *119*, 661–674.

Bohland, J. W., Bullock, D., & Guenther, F. H. (2010). Neural representations and mechanisms for the performance of simple speech sequences. *Journal of Cognitive Neuroscience*, *22*, 1504–1529.

Botez, M. I., Léveillé, J., Lambert, R., & Botez, T. (1991). Single photon emission computed tomography (SPECT) in cerebellar disease: Cerebello-cerebral diaschisis. *European Neurology*, *31*, 405–412.

Boutsen, F., Duffy, J. R., Dimassi, H., & Christman, S. S. (2011). Long-term phonatory instability in ataxic dysarthria. *Folia Phoniatrica et Logopaedica*, *63*, 213–220.

Brandauer, B., Hermsdörfer, J., Geißendörfer, T., Schoch, B., Gizewski, E. R., & Timmann, D. (2012). Impaired and preserved aspects of independent finger control in patients with cerebellar damage. *Journal of Neurophysiology*, *107*, 1080–1093.

Brendel, B., Ackermann, H., Berg, D., Lindig, T., Schölderle, T., Schöls, L., et al. (2013). Friedreich ataxia; Dysarthria profile and clinical data. *Cerebellum*, *12*, 475–484.

Brendel, B., Synofzik, M., Ackermann, H., Lindig, T., Schölderle, T., Schöls, L., et al. (2015). Dysarthria in hereditary ataxias: Commonalities and differences [Epub ahead of print 30 Sept].

Buckner, R. L., Krienen, F. M., Castellanos, A., Diaz, J. C., & Yeo, B. T. T. (2011). The organization of the human cerebellum estimated by intrinsic functional connectivity. *Journal of Neurophysiology*, *106*, 2322–2345.

Butler, A. B., & Hodos, W. (2005). *Comparative vertebrate neuroanatomy: Evolution and adaptation* (2nd ed.). Hoboken, NJ: Wiley.

Chen, S. H. A., & Desmond, J. E. (2005). Cerebrocerebellar networks during articulatory rehearsal and verbal working memory tasks. *NeuroImage*, *24*, 332–338.

Coffman, K. A., Dum, R. P., & Strick, P. L. (2011). Cerebellar vermis is a target of projections from the motor areas in the cerebral cortex. *Proceedings of the National Academy of Sciences of the United States of America*, *108*, 16068–16073.

Darley, F. L., Aronson, A. E., & Brown, J. R. (1975). *Motor speech disorders*. Philadelphia, PA: WB Saunders.

De Smet, H. J., Paquier, P., Verhoeven, J., & Marien, P. (2013). The cerebellum: Its role in language and related cognitive and affective functions. *Brain and Language*, *127*, 334–342.

Desmond, J. E., & Fiez, J. A. (1998). Neuroimaging studies of the cerebellum: Language, learning and memory. *Trends in Cognitive Sciences*, *2*, 355–362.

Duffy, J. R. (2005). *Motor speech disorders: Substrates, differential diagnosis, and management* (2nd ed.). St. Louis, MO: Elsevier Mosby.

Durisko, C., & Fiez, J. A. (2010). Functional activation in the cerebellum during working memory and simple speech tasks. *Cortex*, *46*, 896–906.

Fiez, J. A., Petersen, S. E., Cheney, M. K., & Raichle, M. E. (1992). Impaired non-motor learning and error detection associated with cerebellar damage: A single case study. *Brain*, *115*, 155–178.

Folker, J., Murdoch, B., Cahill, L., Delatycki, M., Corben, L., & Vogel, A. (2010). Dysarthria in Friedreich's ataxia: A perceptual analysis. *Folia Phoniatrica et Logopaedica*, *62*, 97–103.

Folker, J. E., Murdoch, B. E., Cahill, L. M., Delatycki, M. B., Corben, L. A., & Vogel, A. P. (2011). Kinematic analysis of lingual movements during consonant productions in dysarthric speakers with Friedreich's ataxia: A case-by-case analysis. *Clinical Linguistics and Phonetics*, *25*, 66–79.

Geva, S., Jones, P. S., Crinion, J. T., Price, C. J., Baron, J.-C., & Warburton, E. A. (2011). The neural correlates of inner speech defined by voxel-based lesion–symptom mapping. *Brain*, *134*, 3071–3082.

Granziera, C., Schmahmann, J. D., Hadjikhani, N., Meyer, H., Meuli, R., Wedeen, V., et al. (2009). Diffusion spectrum imaging shows the structural basis of functional cerebellar circuits in the human cerebellum *in vivo*. *PLoS ONE*, *4*, e5101.

Grodd, W., Hülsmann, E., & Ackermann, H. (2005). Functional MRI localizing in the cerebellum. *Neurosurgery Clinics of North America*, *16*, 77–99.

Hardwick, R. M., Rottschy, C., Miall, R. C., & Eickhoff, S. B. (2013). A quantitative meta-analysis and review of motor learning in the human brain. *NeuroImage*, *67*, 283–297.

Holmes, G. (1917). The symptoms of acute cerebellar injuries due to gunshot injuries. *Brain*, *40*, 461–535.

Holmes, G. (1939). The cerebellum of man. *Brain*, *62*, 1–30.

Ito, M. (2012). *The cerebellum: Brain for an implicit self*. Upper Saddle River, NJ: FT Press.

Ivry, R. (1996). Cerebellar timing systems. In J. D. Schmahmann (Ed.), *The cerebellum and cognition* (Vol. 41, pp. 555–573). San Diego, CA: Academic Press (Int Rev Neurobiol).

Ivry, R. (2012). The cerebellum and timing. *Cerebellum*, *11*, 469–471 (In Consensus paper: Roles of the cerebellum in motor control—The diversity of ideas on cerebellar involvement in movement. *Cerebellum*, 11, 457-487).

Ivry, R. B., & Fiez, J. A. (2000). Cerebellar contributions to cognition and imagery. In M. S. Gazzaniga (Ed.), *The new cognitive neurosciences* (2nd ed., pp. 999–1011). Cambridge, MA: MIT Press.

Küper, M., & Timmann, D. (2013). Cerebellar mutism. *Brain and Language*, *127*, 327–333.

Leiner, H. C. (2010). Solving the mystery of the human cerebellum. *Neuropsychology Review*, *20*, 229–235.

Levelt, W. J. M. (1989). *Speaking. From intention to articulation*. Cambridge, MA: MIT Press.

Marien, P., Ackermann, H., Adamaszek, M., Barwood, C. H. S., Beaton, A., Desmond, J., et al. (2014). Consensus paper: Language and the cerebellum—An ongoing enigma. *Cerebellum*, *13*, 386–410.

Marvel, C. L., & Desmond, J. E. (2010). Functional topography of the cerebellum in verbal working memory. *Neuropsychology Review*, *20*, 271–279.

Massaquoi, S. G. (2012). Physiology of clinical dysfunction of the cerebellum. *Handbook of Clinical Neurology*, *103*, 37–62.

Mathiak, K., Hertrich, I., Grodd, W., & Ackermann, H. (2002). Cerebellum and speech perception: A functional magnetic resonance imaging study. *Journal of Cognitive Neuroscience*, *14*, 902–912.

Mathiak, K., Hertrich, I., Grodd, W., & Ackermann, H. (2004). Discrimination of temporal information at the cerebellum: Functional magnetic resonance imaging of nonverbal auditory memory. *NeuroImage*, *21*, 154–162.

Mottolese, C., Richard, N., Harquel, S., Szathmari, A., Sirigu, A., & Desmurget, M. (2013). Mapping motor representations in the human cerebellum. *Brain*, *136*, 330–342.

Murdoch, B. E., & Barwood, C. H. S. (2014). Cerebellar-induced aphasia. *Cerebellum*, *13*, 395–396 (In Marien et al. 2014).

Nicolson, R. I., & Fawcett, A. J. (2014). Reading and the cerebellum. *Cerebellum*, *13*, 398–399 (In Marien et al. 2014).

Nicolson, R. I., Fawcett, A. J., & Dean, P. (2001). Developmental dyslexia; The cerebellar deficit hypothesis. *Trends in Neurosciences*, *24*, 508–511.

Nieuwenhuys, R., Voogd, J., & van Huijzen, C. (2008). *The human central nervous system* (4th ed.). Berlin: Springer.

Nowak (2012). The cerebellum and timing. *Cerebellum*, *11*, 469–471 (In Consensus paper: Roles of the cerebellum in motor control—The diversity of ideas on cerebellar involvement in movement. *Cerebellum*, 11, 457-487).

Oberdick, J., & Sillitoe, R. V. (2011). Cerebellar zones: History, development, and function. *Cerebellum*, *10*, 301–306.

Oelschläger, H. H., Ridgway, S. H., & Knauth, M. (2010). Cetacean brain evolution: Dwarf sperm whale (Kogia sima) and common dolphin (Delphinus delphis)—An investigation with high-resolution 3D MRI. *Brain, Behavior and Evolution*, *75*, 33–62.

Orban, P., Peigneux, P., Lungo, O., Debas, K., Barakat, M., Bellec, P., et al. (2011). Functional neuroanatomy associated with the expression of distinct movement kinematics in motor sequence learning. *Neuroscience*, *179*, 94–103.

O'Reilly, J. X., Beckmann, C. F., Tomassini, V., Ramnani, N., & Johansen-Berg, H. (2010). Distinct and overlapping functional zones in the cerebellum defined by resting state functional connectivity. *Cerebral Cortex*, *20*, 953–965.

Passot, J. B., Luque, N. R., & Arleo, A. (2013). Coupling internal cerebellar models enhances online adaptation and supports offline consolidation in sensorimotor tasks. *Frontiers in Computational Neuroscience*, *7*, 95.

Ramnani, N. (2006). The primate cortico-cerebellar system: Anatomy and function. *Nature Reviews Neuroscience*, *7*, 511–522.

Riecker, A., Mathiak, K., Wildgruber, D., Erb, M., Grodd, W., & Ackermann, H. (2005). fMRI reveals two distinct cerebral networks subserving speech motor control. *Neurology*, *64*, 700–706.

Rothwell, J. (1994). *Control of human voluntary movement* (2nd ed.). London: Chapman & Hall.

Schalling, E., & Hartelius, L. (2013). Speech in spinocerebellar ataxia. *Brain and Language*.

Schmahmann, J. D. (2010). The role of the cerebellum in cognition and emotion: Personal reflections since 1982 on the dysmetria of thought hypothesis, and its historical evolution from theory to therapy. *Neuropsychology Review*, *20*, 236–260.

Schmahmann, J. D., Doyon, J., Toga, A. W., Petrides, M., & Evans, A. C. (2000). *MRI atlas of the human cerebellum*. San Diego, CA: Academic Press.

Schmahmann, J. D., & Pandya, D. N. (1997). The cerebrocerebellar system. *International Review of Neurobiology*, *41*, 31–60.

Schmahmann, J. D., & Sherman, J. C. (1998). The cerebellar cognitive affective syndrome. *Brain*, *121*, 561–579.

Sidtis, J. J., Ahn, J. S., Gomez, C., & Sidtis, D. (2011). Speech characteristics associated with three genotypes of ataxia. *Journal of Communication Disorders*, *44*, 478–492.

Silveri, M. C., Leggio, M. G., & Molinari, M. (1994). The cerebellum contributes to linguistic production: A case of agrammatic speech following a right cerebellar lesion. *Neurology*, *44*, 2047–2050.

Spraker, M. B., Corcos, D. M., Kurani, A. S., Prodoehl, J., Swinnen, S. P., & Vaillancourt, D. E. (2012). Specific cerebellar regions are related to force amplitude and rate of force development. *NeuroImage*, *59*, 1647–1656.

Stoodley, C. J., & Schmahmann, J. D. (2009). Functional topography in the human cerebellum: A meta-analysis of neuroimaging studies. *NeuroImage*, *44*, 489–501.

Strick, P. L., Dum, R. P., & Fiez, J. A. (2009). Cerebellum and nonmotor function. *Annual Review of Neuroscience*, *32*, 413–434.

Sultan, F., & Glickstein, M. (2007). The cerebellum: Comparative and animal studies. *Cerebellum*, *6*, 168–176.

Teffer, K., & Semendeferi, K. (2012). Human prefrontal cortex: Evolution, development, and pathology. *Progress in Brain Research*, *195*, 191–218.

Thach, W. T. (1998). What is the role of the cerebellum in motor learning and cognition? *Trends in Cognitive Sciences*, *2*, 331–337.

Urban, P. P. (2013). Speech motor deficits in cerebellar infarctions. *Brain and Language*, *127*, 323–326.

Zettin, M., Cappa, S. F., D'Amico, A., Rago, R., Perino, C., Perani, D., et al. (1997). Agrammatic speech production after a right cerebellar haemorrhage. *Neurocase*, *3*, 375–380.

CHAPTER

8

The Anatomy of the Basal Ganglia

Kate E. Watkins[1] *and Ned Jenkinson*[2,3]

[1]Department of Experimental Psychology, University of Oxford, Oxford, UK; [2]Nuffield Department of Clinical Neuroscience, University of Oxford, John Radcliffe Hospital, Oxford, UK; [3]School of Sport, Exercise and Rehabilitation Sciences, The University of Birmingham, Birmingham, UK

8.1 INTRODUCTION

The basal ganglia are a collection of highly interconnected subcortical nuclei in the brain. Classically, the basal ganglia comprise the striatum (caudate nucleus and putamen), the globus pallidus, subthalamic nucleus (STN), and substantia nigra. The principal input to the basal ganglia arises in the cerebral cortex and its principal output targets are frontal areas of the cortex. As such, the basal ganglia appear to form a series of parallel functionally segregated loops originating in partially discrete cortical areas with which they form a circuit. Historically, the basal ganglia have been associated with the control of movement, and it is the motor function of the basal ganglia that has been most extensively studied. The closed cortico-striatal-thalamo-cortical loops serve a role in selecting between competing possible actions such that appropriate behaviors are chosen and less appropriate ones are suppressed. By considering the broader implications of this principal function of the basal ganglia, we can see that these nuclei, and pathways through them, can contribute to a number of processes, including planning, decision-making, action selection, learning, sequencing, and the initiation and timing of movement. Because diseases of the basal ganglia impair gross motor control of the limbs and manifest in symptoms such as involuntary movement, akinesia, rigidity, and tremor, the likely role of the basal ganglia in speech motor control has been largely overlooked. Given that basal ganglia circuits encompass expansive cortical territories, including those known to contribute to linguistic processes such as lexical selection, cognitive control over competing languages in multilingual speakers, and learning of grammar rules, it seems likely that the basal ganglia themselves contribute to these processes.

8.2 HISTORICAL BACKGROUND

The basal ganglia play an undisputed role in the control of movement. This is largely known because of the obvious motor symptoms of two major diseases of the basal ganglia, Parkinson's disease (PD) and Huntington's disease (HD). These diseases and their symptoms were first described in the nineteenth century, well before the advent of modern neurology. Although, at the time, the involvement of the basal ganglia could not have been known, subsequent study of these diseases has defined our investigations of the functions of the basal ganglia. As such, the basal ganglia have been exclusively associated with motor control for most of the history of neuroscience. In the past few decades, however, it has become increasingly clear that the structures of the basal ganglia are crucial to cognitive functions such as learning and memory. Regardless, comparatively little is known about the contribution of the basal ganglia to behaviors such as speech and language. This is somewhat surprising given that neurologists contemporary with Paul Broca, such as Pierre Marie, observed language impairment in association with lesions in these nuclei (Marie, 1906). Even Broca himself observed that the lesion in his famous patient M. Leborgne extended to the striatum, which MRI scans of the preserved specimen recently confirmed (Dronkers, Plaisant, Iba-Zizen, & Cabanis, 2007).

Here, we review the anatomy of the basal ganglia and current models of their function. We then consider these functions in the context of the human brain's unique abilities to communicate using speech and language. Critical questions concern whether the human basal ganglia has speech-specific and language-specific circuitry, or if the speech and language impairments associated with basal ganglia dysfunction reflect more general processes.

Neurobiology of Language. DOI: http://dx.doi.org/10.1016/B978-0-12-407794-2.00008-0

8.3 OVERVIEW OF BASAL GANGLIA ANATOMY

Classically, the basal ganglia is considered to be a group of subcortical nuclei located in each hemisphere, including the caudate nucleus, putamen, globus pallidus, substantia nigra, and STN.

The caudate nucleus and putamen are two spatially distinct gray matter masses that comprise the dorsal striatum. In the human brain, the caudate nucleus is a C-shaped structure that lies lateral to the lateral ventricle and medial to the putamen (Figure 8.1). It has a large head at the rostral end, which extends caudally to a narrower body and slender tail that curves around into the anterior temporal lobe. The putamen is located lateral and posterior to the head of the caudate nucleus and medial to the insula cortex from which it is separated by the external capsule, claustrum, and extreme capsule. The two portions of the dorsal striatum become separated during development when fibers to and from the cortex become numerous enough to form a contiguous internal capsule that divides the caudate nucleus from the putamen. However, the structures always remain connected dorsally by small bridges of tissue interdigitated by white matter projections traveling in the internal capsule. The stripy appearance created by this arrangement of the bundles of projection neurons gives the structure its Latin name the *corpus striatum* (striated body). In lower mammals, such as the rat, the caudate nucleus and putamen are not divided but instead form a single complex with fibers passing through in the form of numerous fascicles. Regardless of the anatomical relationship of these two cell masses, their cellular anatomy is identical; together, they comprise the largest subcortical cell mass in the mammalian brain. The majority of cells in the striatum—approximately 75% of the total—are medium-sized neurons with extensive dendritic branches. The branches are packed with dendritic spines, giving the cells a characteristic appearance and their name, medium spiny neurons (MSNs). All MSNs are GABAergic; therefore, they inhibit the targets to which they project. In addition to these projection neurons, there are several groups of intrinsic interneurons, all of which have smooth aspiny cell bodies. Of these groups, most are inhibitory and there is one excitatory group with noticeable large cell bodies that express acetylcholine.

The globus pallidus (pale globe) is a triangular nucleus immediately medial to the putamen and separated from it by a thin white-matter sheet (Figure 8.1). The globus pallidus contains many large, sparsely distributed fusiform (spindle-shaped) cells. Because of the low cellular density and the large number of myelinated axons that course through the nucleus, the globus pallidus is pale in appearance in fresh specimens compared with the putamen or caudate nucleus, hence its name. The globus pallidus is split by the medial medullary lamina into the internal and external segments, also known as the medial and lateral segments, respectively. Both segments of the globus pallidus

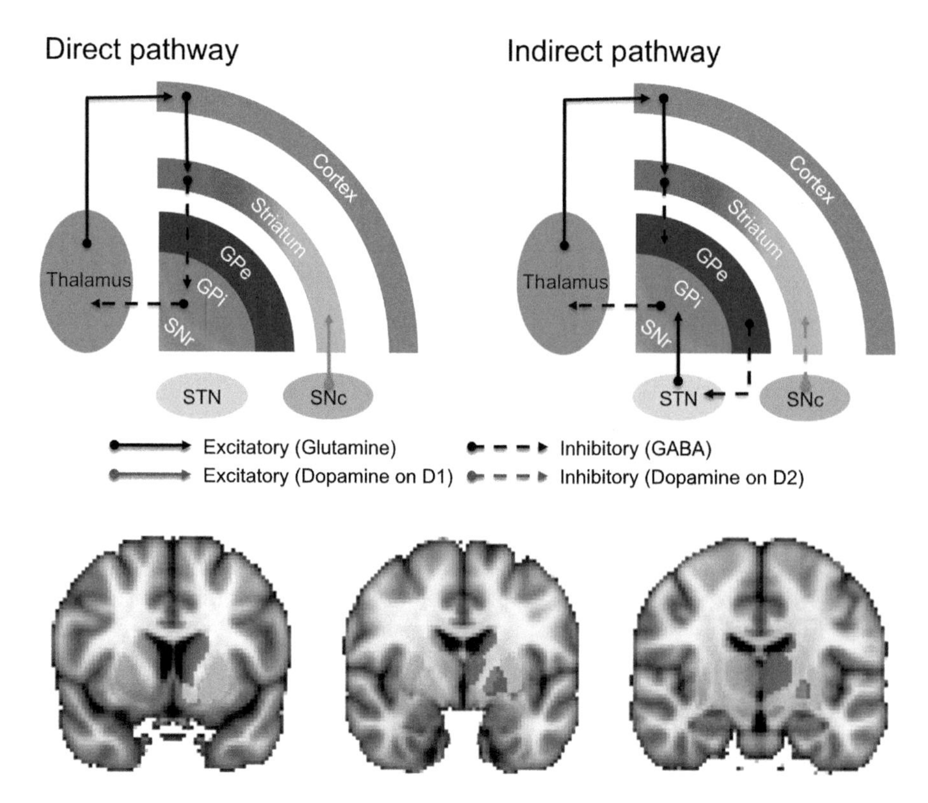

FIGURE 8.1 Basal ganglia nuclei and connections. Top row: schematic representations of the direct and indirect pathways through the basal ganglia. Bottom row: Coronal sections through the T1-weighted MNI152 average brain are shown from left to right at 8 mm in front of the vertical plane through the anterior commissure and at 4 and 14 mm behind it. Colored areas on the right side of the images correspond to the structures labeled in the schematics above. Red, caudate nucleus; green, putamen; yellow, nucleus accumbens; pink, thalamus; dark blue, globus pallidus, external segment (GPe); light blue, globus pallidus, internal segment (GPi); dark orange, substantia nigra (pars reticulata, SNr; pars compacta, SNc); light orange, subthalamic nucleus (STN). Black arrows indicate projections within the cortico-striatal-thalamo-cortical loops. Gray arrows indicate the dopamine innervation of the striatum from the SNc.

receive afferent input from MSNs of the striatum. Efferent projections from both segments of the globus pallidus are GABAergic (inhibitory), but the internal segment (GPi) projects to the thalamus, whereas the external segment (GPe) projects to the STN.

The STN is a small nucleus located in the midbrain ventral to the zona incerta and dorsal to the internal capsule/cerebral peduncle junction (Figure 8.1). The STN consists of large triangular and polygonal cells with dendritic trees that form ellipsoid domains within the structure. The STN receives a large inhibitory GABAergic input from the GPe and an excitatory glutaminergic input from the cortex. Neurons within the STN are glutaminergic and send excitatory projections to the GPi and pars reticularis of the substantia nigra.

The substantia nigra is the largest cell mass in the mesencephalon and located ventral to the STN (Figure 8.1). The nucleus was initially split cytoarchitectonically (on the basis of cell density and appearance) into two subdivisions: the cell-rich, dorsal pars compacta (SNc) and the less cell-dense, ventral pars reticularis (SNr). More modern cytochemical methods have confirmed these divisions, demonstrating that the pars compacta comprises large darkly pigmented cells that synthesize dopamine (and also give the structure its name due to its dark appearance in fresh specimens). The pars reticularis, however, consists of smaller GABAergic cells. The cyto-architecture and chemo-architecture of the SNr are strikingly similar to that of the GPi, and both receive excitatory glutaminergic inputs from the STN.

Here, we have defined the structures of the basal ganglia. Next, we outline the major connections between these structures.

8.3.1 Inputs to the Basal Ganglia

The majority of extrinsic input to the basal ganglia originates from excitatory glutaminergic neurons in layer 5 of the cortex. There are two points of input to the basal ganglia from the cortex. The first, and largest, is the corticostriatal input. The whole of the neocortex sends projections that terminate topographically throughout the whole striatum (caudate nucleus and putamen). The organization of these projections is such that restricted areas of cerebral cortex project to longitudinal territories that run the length of the striatum but are limited in their medio-lateral aspect so that the cortical input to the striatum is arranged in a number of strips or bands orientated along the rostrocaudal axis (Selemon & Goldman-Rakic, 1985). The excitatory glutaminergic input from the cortex terminates directly onto the output neurons of the striatum (i.e., the MSN).

The second cortical input to the basal ganglia is an input from the cortex to the STN. These cortical afferents arise mainly (though not exclusively) in the primary, premotor, and supplementary motor cortices (Nambu, Tokuno, & Takada, 2002). It should be noted that the STN was not considered a classical input nucleus of the basal ganglia. Instead, this input has been viewed as a "shortcut" through the basal ganglia.

8.3.2 Outputs from the Basal Ganglia

The two major output nuclei of the basal ganglia are the GPi and the SNr. As previously stated, these two nuclei share a similar cellular anatomy and provide the major output of the basal ganglia in the form of GABAergic projections that terminate in the thalamus (Haber & McFarland, 2001). Pallido-thalamic projections initially arise in two separate fiber bundles, the lenticular fasicularis and the ansa fasicularis. These fascicles merge to form the thalamic fasiculus whose fibers cross the internal capsule and terminate principally in the ventral anterior (VA) nucleus and oral subdivision of the ventral lateral (VL) nucleus of the thalamus, which in turn project back to the cortex (Nauta & Mehler, 1966). As well as projection to these principal thalamic nuclei, collateral fibers are given off to the centromedian nucleus. The centromedian nucleus is part of the intralaminar thalamic nuclei, which completes an internal loop by sending projections back into the striatum. Nigrothalamic fibers also terminate in VA as well as the paralaminar portion of the mediodorsal thalamus (Carpenter, Nakano, & Kim, 1976). VA and VL project to the primary, premotor, and supplementary motor cortices, whereas the mediodorsal nucleus projects to the prefrontal cortex, as well as the frontal eye fields. These thalamocortical projections are excitatory.

Therefore, although the entire cortex sends input topographically to the striatum, the frontal cortex uniquely receives the output from the basal ganglia via the thalamus. In the next section, we see that although the inputs and outputs of the basal ganglia appear the same, there are actually two segregated pathways through the basal ganglia that have opposing effects on behavior.

8.3.3 Pathways Through the Basal Ganglia

The segregated pathways through the basal ganglia start at a cellular level in the striatum, where MSN can be grouped according to the type of dopamine receptors and peptides that they express: D1-type and substance P or D2-type and enkephalin (Gerfen et al., 1990). The two groups of cells project to different targets: the SNr and the GPi in the case of D1 MSN and GPe in the case of D2 MSN. The specificity of this segregation is very high (~95%; Bertran-Gonzalez, Herve, Girault, & Valjent, 2010) and is the start of two distinct pathways that are known as the direct and indirect pathway (Figure 8.1). The direct pathway is called such because this pathway runs directly through the basal ganglia from the input (striatum) directly to the output nuclei (GPi/SNr). The indirect pathway connects the input to the output via the GPe and the STN. In terms of basal ganglia output, these two pathways have opposite effects. In the direct pathway, input from the cortex increases firing in the D1 MSN in the striatum that project directly to the output nuclei GPi/SNr. Because both sets of projection neurons in these nuclei are GABAergic, the net effect is decreased inhibition (i.e., disinhibition) of the thalamus, releasing it to excite the cortex. Therefore, activity in the direct pathway increases activity in the cortical targets and, as such, acts as a "go" signal and can provide positive feedback. In the case of the indirect pathway, input from the cortex increases firing in the D2 MSN projection from the striatum to GPe, which is inhibitory and therefore reduces the inhibition of GPe neurons on the STN. This increases the excitatory influence of the STN on the inhibitory output nuclei (GPi/SNr), thereby increasing the inhibition of the thalamic relay to the cortex. Therefore, the net effect of activity in the indirect pathway is to decrease activity in the cortex, producing the opposite action to that of the direct pathway (i.e., it acts as a "stop" or "no-go" signal) and can provide negative feedback.

The final pathway through the basal ganglia starts at the input from the cortex to the STN. Due to the shortness of the path and the fact that this pathway avoids the classical input to the basal ganglia, it is called the hyperdirect pathway. As the excitatory input from the cortex increases STN firing, which in turn excites the inhibitory output of the GPi/SNr, the hyperdirect pathway is thought to provide rapid inhibition of basal ganglia output (Nambu et al., 2002).

It is thought that normal function of the basal ganglia is produced by a balanced combination of these pathways, and as such it has been suggested that diseases of the basal ganglia are the result of imbalance between these pathways (Mink, 1996). For example, the paucity of movement in PD is associated with an overactivation of the indirect pathway and underactivity in the direct pathway, whereas in Huntington's chorea it is associated with the opposite pattern of abnormal activation in these pathways (Wichmann & DeLong, 1996). Experimental findings have challenged this view of basal ganglia function, however. It is proposed that the activity in these two pathways is more coordinated than first thought (Cui et al., 2013), and that there is considerable interaction both structurally and functionally between them (see Calabresi, Picconi, Tozzi, Ghiglieri, & Di Filippo, 2014 for further discussion).

The anatomy up to this point has described the basal ganglia in the classical context. However, the last three decades have seen an expansion of the concept of the basal ganglia to include a ventral striatal complex. The ventral striatum—as opposed to dorsal striatum—is centered on the nucleus accumbens. The cellular and histochemical make-up of the nucleus accumbens is similar to that of the caudate nucleus, with the substantia innominate, or ventral pallidum, analogous to the GP. These nodes of the ventral striatum form a loop with the cortex in a similar manner as those described in the classical description of the basal ganglia, or dorsal striatum. The ventral striatum primarily receives input from mesocortex and allocortex (inputs that are shared with the dorsal striatum). Specifically, parts of the "limbic" system, including the hippocampal formation, amygdala, and orbitofrontal and temporal cortex, send inputs to the ventral striatum, which is thought to play a crucial role in gating behavior due to emotional or motivational stimuli or both (Haber, Lynd, Klein, & Groenewegen, 1990). Another crucial similarity of the two striatal systems (dorsal and ventral) that comprise the basal ganglia is the powerful modulatory role of dopamine on both.

8.3.4 Dopamine in the Basal Ganglia

The basal ganglia receive dopaminergic input from two dopaminergic nuclei in the mesencephalon. The pars compacta subdivision of the substantia nigra provides a massive dopaminergic input to the entire striatum. The ventral striatum receives input from the ventral tegmental area, with a lesser input from the substantia nigra. Dopamine has been ascribed many functions in the brain since it was first described as an independent neurotransmitter in 1957 (Carlsson, Lindqvist, & Magnusson, 1957), and the substrate of most of those functions is the basal ganglia. A loss of cells in the SNc that leads to a decrease in the ambient levels of dopamine—or dopaminergic tone—in the brain is known to cause PD. A huge body of work has described the importance of dopamine for normal motor function. However, a growing understanding of the role of temporal phasic release of dopamine in the dorsal and ventral striatum and beyond are beginning to unravel the role of dopamine in other behaviors, such as associative learning, response association, decision-making, and working memory. In addition, the finding that the input to the direct and indirect pathway is segregated by dopamine receptor type suggests a key role for dopamine in the delicate balancing of input to these pathways.

The normal action of dopamine is to selectively promote activity in the direct pathway. Activation of D1 and D2 receptors produces opposite effects; D1 neurons are excited by dopamine, whereas D2 neurons are inhibited (Gerfen, Keefe, & Gauda, 1995). The D1 and D2 receptors are predominantly expressed on the neurons comprising the direct and indirect pathways, respectively. Therefore, dopamine release selects for activity in the direct (D1) pathway and inhibits the competing indirect (D2) pathway (Bamford et al., 2004).

8.3.5 Functional Circuits Through the Basal Ganglia

As noted, the cortical input to the striatum arises from the cortical mantle, and the output from the thalamus projects back to the cortex. Several segregated loops are proposed to run in parallel through the basal ganglia (Alexander, DeLong, & Strick, 1986). Their purported function is based on the cortical areas from which the basal ganglia input arises and also the part of the striatum receiving this input. For example, a motor loop via the putamen to the supplementary motor area receives input from premotor, primary motor, and somatosensory cortex. Another loop is formed by input from the dorsolateral prefrontal cortex to the head of the caudate nucleus and output back to this area from the basal ganglia via the thalamus. The function of this loop is thought to be "associative" or "executive," contributing to frontal-lobe cognitive processes such as decision-making, working memory, and attention, for example. The ventral striatum receives its main input from medial and orbitofrontal cortex and amygdala and, in turn, projects back to these areas via the medial dorsal nucleus of the thalamus. The function of this latter circuit is often described as "limbic," indicating a role in emotional processing, motivational states, and reward-based learning. Studies in nonhuman primates describe further functional loops through the basal ganglia, for example, one involved in oculomotor control. It is probable that the human brain contains specialist loops through the basal ganglia for auditory-motor control and vocal learning, homologous to those described in songbirds (Bolhuis, Okanoya, & Scharff, 2010; Nottebohm, 2005).

Although the functional loops through the basal ganglia are described as parallel, they are not strictly segregated; information flowing through these loops is convergent, allowing integration of information across areas. For example, the MSNs in the putamen receive convergent input from primary motor and primary somatosensory representations of the same body part (Flaherty & Graybiel, 1993). One cortical area may innervate multiple targets in the striatum, however, and these divergent projections reconverge on pallidal structures downstream (Flaherty & Graybiel, 1994). Furthermore, there may be cross-talk between the parallel loops mediated by recurrent striato-nigral-striatal circuits through the three major divisions of the striatum (Haber, 2003).

8.3.6 Disorders of the Basal Ganglia

Much of our understanding of the motor function of the basal ganglia has been gleaned from studies of two neurodegenerative diseases, namely PD and HD. In PD, the dopamine-containing neurons in the SNc degenerate, leading to the loss of dopamine innervation to the striatum. This loss typically starts posteriorly, affecting the putamen and resulting in motor characteristics that are symptomatic of the disease. As more anterior regions of the striatum become affected during the disease progression, the effects on cognitive and emotional processes are observed. Motor symptoms in PD are characterized by impairment in initiating movement (akinesia), slower and smaller movements (bradykinesia), resting tremor, muscle rigidity, and postural instability. The lack of facial movements results in a loss of expression and characteristic "mask." In HD, movements are uncontrollable and described as choreiform ("dancing"); there are also emotional and cognitive disturbances that are attributed to degeneration of cortical areas. The disease is caused by an expanded CAG triplet repeat in the Huntington gene. The MSNs in the indirect pathway degenerate in HD. At the simplest level, excessive inhibition of movement in PD appears to result from overactivity in the indirect pathway; the dopamine depletion results in reduced activity of the direct pathway and a corollary reduction in the inhibition of the indirect pathway, with the net result being increased inhibition of the thalamus. Conversely, the choreiform movements associated with HD would result from overactivity in the direct pathway due to the selective loss of the MSN in the indirect pathway that normally inhibit the GPi/SNr output via GABAergic projection to the STN. As noted, this direct/indirect pathway model of basal ganglia function has been challenged, in particular, with respect to its explanation of pathological conditions such as PD and HD. Even so, recent optogenetic studies support the notion that this model can help explain the pathophysiological basis for the cardinal signs of PD. For example, using optogenetics, selective excitation of the MSNs in the indirect pathway produced Parkinsonian symptoms in a rodent model that were reduced by activation of direct pathway MSNs (Kravitz et al., 2010).

In addition to causing disorders of movement control, such as PD and HD, pathology within the basal ganglia has been linked to a range of other disorders, including neuropsychiatric ones such as obsessive–compulsive disorder, Tourette's syndrome, and addiction. Common to these disorders is impairment in the control of behavior in general, not just in the control of movement. One explanation for this is found in the role of dopamine in linking rewards to actions, thereby providing motivation or "will." Dopamine modulation is critical to signaling reward, lack of expected reward, and predicting rewards during learning (Schultz, 1997). A loss of dopamine might lead to reduced motivation or apathy. Similarly, too much dopamine might lead to excessive or risky behaviors, such as pathological checking or gambling.

8.3.7 Learning and Memory and the Basal Ganglia

The role of the basal ganglia in learning and memory is largely thought to be nondeclarative (procedural or implicit), leading to refinement of motor skills through practice and the acquisition of behavioral routines or habits (Graybiel, 2008; Wickens, Horvitz, Costa, & Killcross, 2007). Behaviors can be learned and habits can be acquired over long time periods (days or years) and, once established, they are performed almost automatically, without awareness. Such habits are typically sequences of motor or cognitive behaviors evoked in response to a specific stimulus or context (Yin & Knowlton, 2006). In pathological states, these habits or routines become motor stereotypes or repetitive behaviors and thoughts that can feature in diseases such as HD, schizophrenia, obsessive–compulsive disorder, and others. The basal ganglia also contribute to another form of nondeclarative memory, namely, the development of stimulus-response associations underlying conditioning or reinforcement learning (Schultz, Dayan, & Montague, 1997).

8.3.8 Summary

The basal ganglia contribute to the control of movement and selection of actions through the balance of activity in the direct and indirect pathways that serve to release the thalamic output to the cortex or further inhibit it. In this way, the basal ganglia gains control over the smooth execution of movements. Therefore, diseases affecting the structure or function of the basal ganglia result primarily in impairment in movement control. The role of the basal ganglia in promotion of desired motor behaviors and suppression of undesired ones can be generalized to the cognitive domain. Similarly, their role in learning is not restricted to the motor domain. For the remainder of this chapter, we briefly outline how these basal ganglia functions might contribute to the control of speech and language.

8.4 THE ROLE OF THE BASAL GANGLIA IN SPEECH MOTOR CONTROL

As noted, the most commonly studied function of the basal ganglia is their role in the control of movement. Speech production requires precise control of movement of a large number of muscles, from the diaphragm and intercostal muscles involved in breath control to those involved in very rapid, tiny movements of the tongue to alter the shape of the vocal tract and passage of air through it. This complex process requires coordination of activity in multiple brain regions to plan, sequence, time, execute, and monitor these movements. The coordination of sequences of articulatory movements that comprise the elements of speech—phonemes and syllables—is normally achieved smoothly and with little effort. It seems likely that this control, like limb-movement control, is achieved by basal ganglia regulation of thalamocortical outputs to prefrontal cortical areas. These cortical areas include ventrolateral prefrontal cortex, ventral and dorsal premotor cortex, presupplementary and supplementary motor areas, as well as primary sensorimotor cortex.

Patients with damage to basal ganglia nuclei are commonly reported to have disturbances affecting speech production, such as apraxia of speech and dysarthria (Pickett, Kuniholm, Protopapas, Friedman, & Lieberman, 1998). Hypophonia, reduced speech output and initiation, and poor articulatory and phonatory control, in general, appear consequential to lesions of the striatum (particularly the putamen) and pallidum. In accord with these lesion data, speech production in PD appears to be similar to the hypokinesia described for limb movements in these patients; pitch variation and loudness are reduced so that speech is monotonous and low (Ho, Iansek, Marigliani, Bradshaw, & Gates, 1999). In contrast, patients with HD can produce erratic speech with increased variation and loss of control of pitch and loudness (Hartelius, Carlstedt, Ytterberg, Lillvik, & Laakso, 2003).

Sequencing the individual elements of speech utterances relies on timing cues. These internal cues are thought to signal the end of one submovement and trigger the initiation of the next submovement in a sequence via phasic changes in basal ganglia output to the supplementary motor area (SMA) (Brotchie, Iansek, & Horne, 1991). When the relative timing of speech elements is altered in speech production, the listener often perceives a change in accent. The meaning of an individual word or its emphasis can also be altered by timing changes and the timing of words within an utterance affects prosody, which, in turn, changes meaning. One example of this is seen in foreign accent syndrome, which can occur after lesions to the basal ganglia, particularly to the left putamen (Gurd, Bessell, Bladon, & Bamford, 1988). In this disorder, previously fluent speakers appear to produce their native language with a foreign accent mostly likely caused by disturbed timing, prosody, and articulation that, combined, give the impression of nonnative speech production (Blumstein, Alexander, Ryalls, Katz, & Dworetzky, 1987).

Strikingly, cortico-striatal-thalamo-cortical circuit abnormalities are a key feature of a genetic disorder of speech and language demonstrated by affected members of a large multigenerational family who have a mutation in the gene FOXP2 (see Chapter 2). Family members carrying the mutation have developmental verbal dyspraxia. They are impaired at repeating sequences of syllables that are either familiar (words) or unfamiliar (nonwords), and they are impaired at imitating sequences of nonverbal orofacial movements, but not of limb movements (Watkins, Dronkers, & Vargha-Khadem, 2002). Brain imaging revealed structural and functional abnormalities of the dorsal striatum in individuals carrying the mutation (Watkins, Vargha-Khadem, et al., 2002).

FOXP2 is expressed in MSN in the striatum (among many other cortical and subcortical areas). Mice with one copy of *Foxp2* knocked out show abnormal synaptic plasticity at the level of the MSNs and are impaired at learning to run on an accelerating rotating rod (Groszer et al., 2008). Songbirds express *FoxP2* in striatal spiny neurons. One portion of the avian striatum, Area X, is specialized for vocal learning and shows increased *FoxP2* expression during vocal learning. When levels of FoxP2 are reduced in Area X, juvenile zebra finches show impaired learning of their tutor's song (Haesler et al., 2007).

Basal ganglia dysfunction is suspected to cause another developmental speech disorder, namely stuttering (Alm, 2004); neurogenic stuttering acquired after brain injury is also frequently associated with basal ganglia lesions. The main difficulty in stuttering is with the initiation of speech segments and producing smooth transitions between them. Dopamine blockers, such as haloperidol, can improve speech fluency, and early imaging studies showed abnormal levels of dopamine metabolism in people who stutter. More recent imaging studies confirm functional abnormalities in the basal ganglia and their cortical targets (see Chapter 79). People who stutter often experience periods of fluency and can achieve fluency through practice or via external cues and altered feedback. Interestingly, patients with PD also benefit from external cues to initiate and perform sequences of limb movements fluently (Glickstein & Stein, 1991).

The motor circuitry involving the striatum is critical for the control of movement sequences more generally and not specifically for speech, yet neither the affected members of the KE family nor people who stutter exhibit impairments in the control of other movements. It could be that speech-specific corticostriatal circuits are affected in these disorders, or that fluent speech production requires relatively more rapid and complex coordination of movements than the control of other effectors. It could also be the case that speech is special because it requires integration with auditory feedback, and that this is critical for accurate timing of movement. The inputs to the basal ganglia allow convergence of related motor and sensory representations as described.

8.5 THE ROLE OF THE BASAL GANGLIA IN LANGUAGE

The role of the basal ganglia in language is less clearly established relative to their role in speech motor control. Language disturbance is commonly reported in association with damage to nuclei in the left hemisphere (Fabbro, Clarici, & Bava, 1996). Aside from the speech fluency deficits associated with lesions to the putamen and pallidum described, problems with lexico-semantics are typically associated with damage to the head of the caudate nucleus. However, lesions that cause language impairment tend to be extensive, with damage often extending to the white matter tracts adjacent to the striatum, including the internal, external, and extreme capsules. Such damage would interrupt communication to and from the thalamus or between temporal and frontal language areas, which could explain the subsequent language impairment rather than the deficits being the result of specific lesions to the basal ganglia (Nadeau & Crosson, 1997). Furthermore, the language impairments described in patients with PD, such as comprehension of syntactically complex sentences, could be explained by more general processing deficits in attention, working memory, or general slowing (Grossman et al., 2003). In sum, a consensus opinion has emerged that selective pathology of the basal ganglia nuclei does not cause aphasic symptoms like those seen after damage to cortical language areas. Rather, it may result in more subtle disturbances that contribute to complex language functions.

One example of a complex language function to which the basal ganglia likely contribute is in the implicit learning and application of morphosyntactic rules. Language learning is clearly analogous to other forms of learned behaviors that become habits. Language acquisition typically occurs over many years. The rules that govern the construction of sequences of morphemes and words are learned implicitly. These rule-governed sequences are produced accurately, effortlessly, and without awareness, even when operating on novel combinations of language elements. According to one model, the basal ganglia contribute to the use of rules such as the one for producing the regular past tense in English (add -ed) (see Chapter 76). Patients with PD have difficulty applying this rule to novel verbs (e.g., "Everyday, he *plags* his lawn. Yesterday, he....?" [*plagged*]), whereas patients with HD overregularize (e.g., "Everyday, he prunes his roses. Yesterday, he....?" [*pruneded*]) (Ullman et al., 1997). Both groups of patients are unimpaired at production of irregular past tense word forms because these are learned as individual lexical items and do not require rules to produce them. However, another study found that similar groups of patients with PD and HD or with basal ganglia vascular lesions were unimpaired on priming tasks that tested these past-tense relationships (Longworth, Keenan, Barker, Marslen-Wilson, & Tyler, 2005). In that study, some patients had problems inhibiting semantically appropriate alternatives to novel word forms, which is consistent with a role for the basal ganglia in inhibiting competing alternatives during selection.

8.6 SEGREGATED FUNCTIONAL LOOPS FOR SPEECH AND LANGUAGE

Previously, we described two loops through the basal ganglia labeled as "motor" and "executive." The putamen was the primary recipient of inputs from premotor, primary motor, and somatosensory cortex in the "motor" loop, whereas the caudate nucleus received inputs from prefrontal cortex as part of the "executive" loop. Findings from functional imaging studies of bilinguals and lesion and brain stimulation studies of patients support a similar division for motor and cognitive control of speech and language. For example, the caudate nucleus is robustly activated when bilinguals switch between languages and need to exert control over language production (Crinion et al., 2006). Language switching requires selection of the desired output (the correct language) and suppression of the undesired one (the competing alternative), which could be achieved by activation of the direct and indirect pathways through the basal ganglia, respectively. In contrast, the left putamen contributes to articulatory processes when speaking a second language, but only if the speaker is not highly proficient in the second language (Abutalebi et al., 2013). Further evidence for a differential role of the caudate nucleus and putamen in language functions comes from a functional mapping study of the dominant striatum in patients undergoing tumor removal. Electrical stimulation of the head of the caudate nucleus elicited perseveration of previously named pictures, consistent with a role for this nucleus in selection of linguistic items. However, stimulation of the putamen produced anarthria with no accompanying interruption of hand movements or facial muscular contraction. This is consistent with the role of the putamen in motor coordination for speech articulation (Gil Robles, Gatignol, Capelle, Mitchell, & Duffau, 2005).

8.7 SUMMARY

The role of the basal ganglia in motor function in general is evident in the contribution they make to speech motor control. Even so, the impairments noted in speech due to pathology of the basal ganglia are not as severe as those seen for limb-movement control in the same patients. More severe speech and language disruptions are seen in developmental disorders with known or suspected basal ganglia pathology, which points to a more critical role played by these circuits in language acquisition. The significance of the basal ganglia in language acquisition is also supported by the literature on second language learning. However, a definite role for the basal ganglia in more cognitive aspects of language processing has not yet been established, because specific linguistic impairments could be explained by more general cognitive control mechanisms that are known to be impaired by basal ganglia pathology. In this brief review of basal ganglia contributions to speech and language processing, we focused primarily on the production side. Nonetheless, speech perception and language comprehension (in particular the comprehension of syntax) rely on accurate perception of timing and the ability to predict cues in auditory sequences of speech and language (Kotz, Schwartze, & Schmidt-Kassow, 2009). The basal ganglia circuitry is a strong candidate substrate not only for the production of speech but also for the processes involved in abstracting statistical information from perceptual sequences, a function that might be critical for learning rules and producing apparently rule-governed behavior, such as syntax.

References

Abutalebi, J., Della Rosa, P. A., Gonzaga, A. K., Keim, R., Costa, A., & Perani, D. (2013). The role of the left putamen in multilingual language production. *Brain and Language*, *125*(3), 307–315.

Alexander, G. E., DeLong, M. R., & Strick, P. L. (1986). Parallel organization of functionally segregated circuits linking basal ganglia and cortex. *Annual Review of Neuroscience, 9*, 357–381.

Alm, P. A. (2004). Stuttering and the basal ganglia circuits: A critical review of possible relations. *Journal of Communication Disorders, 37*(4), 325–369.

Bamford, N. S., Robinson, S., Palmiter, R. D., Joyce, J. A., Moore, C., & Meshul, C. K. (2004). Dopamine modulates release from corticostriatal terminals. *The Journal of Neuroscience, 24*(43), 9541–9552.

Bertran-Gonzalez, J., Herve, D., Girault, J. A., & Valjent, E. (2010). What is the Degree of Segregation between Striatonigral and Striatopallidal Projections? *Frontiers in Neuroanatomy, 4*. pii: 136.

Blumstein, S. E., Alexander, M. P., Ryalls, J. H., Katz, W., & Dworetzky, B. (1987). On the nature of the foreign accent syndrome: A case study. *Brain and Language, 31*(2), 215–244.

Bolhuis, J. J., Okanoya, K., & Scharff, C. (2010). Twitter evolution: Converging mechanisms in birdsong and human speech. *Nature Reviews Neuroscience, 11*(11), 747–759.

Brotchie, P., Iansek, R., & Horne, M. K. (1991). Motor function of the monkey globus pallidus. 2. Cognitive aspects of movement and phasic neuronal activity. *Brain, 114*(Pt 4), 1685–1702.

Calabresi, P., Picconi, B., Tozzi, A., Ghiglieri, V., & Di Filippo, M. (2014). Direct and indirect pathway of basal ganglia: A critical reappraisal. *Nature Neuroscience, 17*(8), 1022–1030.

Carlsson, A., Lindqvist, M., & Magnusson, T. (1957). 3,4-Dihydroxyphenylalanine and 5-hydroxytryptophan as reserpine antagonists. *Nature, 180*(4596), 1200.

Carpenter, M. B., Nakano, K., & Kim, R. (1976). Nigrothalamic projections in the monkey demonstrated by autoradiographic technics. *The Journal of Comparative Neurology, 165*(4), 401–415.

Crinion, J., Turner, R., Grogan, A., Hanakawa, T., Noppeney, U., Devlin, J. T., et al. (2006). Language control in the bilingual brain. *Science, 312*(5779), 1537–1540.

Cui, G., Jun, S. B., Jin, X., Pham, M. D., Vogel, S. S., Lovinger, D. M., et al. (2013). Concurrent activation of striatal direct and indirect pathways during action initiation. *Nature, 494*, 238–242.

Dronkers, N. F., Plaisant, O., Iba-Zizen, M. T., & Cabanis, E. A. (2007). Paul Broca's historic cases: High resolution MR imaging of the brains of Leborgne and Lelong. *Brain, 130*(Pt 5), 1432–1441.

Fabbro, F., Clarici, A., & Bava, A. (1996). Effects of left basal ganglia lesions on language production. *Perceptual and Motor Skills, 82*(3 Pt 2), 1291–1298.

Flaherty, A. W., & Graybiel, A. M. (1993). Two input systems for body representations in the primate striatal matrix: Experimental evidence in the squirrel monkey. *The Journal of Neuroscience, 13*(3), 1120–1137.

Flaherty, A. W., & Graybiel, A. M. (1994). Input–output organization of the sensorimotor striatum in the squirrel monkey. *The Journal of Neuroscience, 14*(2), 599–610.

Gerfen, C. R., Engber, T. M., Mahan, L. C., Susel, Z., Chase, T. N., Monsma, F. J., Jr., et al. (1990). D1 and D2 dopamine receptor-regulated gene expression of striatonigral and striatopallidal neurons. *Science, 250*(4986), 1429–1432.

Gerfen, C. R., Keefe, K. A., & Gauda, E. B. (1995). D1 and D2 dopamine receptor function in the striatum: Coactivation of D1- and D2-dopamine receptors on separate populations of neurons results in potentiated immediate early gene response in D1-containing neurons. *The Journal of Neuroscience, 15*(12), 8167–8176.

Gil Robles, S., Gatignol, P., Capelle, L., Mitchell, M. C., & Duffau, H. (2005). The role of dominant striatum in language: A study using intraoperative electrical stimulations. *Journal of Neurology, Neurosurgery, and Psychiatry, 76*(7), 940–946.

Glickstein, M., & Stein, J. (1991). Paradoxical movement in Parkinson's disease. *Trends in Neurosciences, 14*(11), 480–482.

Graybiel, A. M. (2008). Habits, rituals, and the evaluative brain. *Annual Review of Neuroscience, 31*, 359–387.

Grossman, M., Cooke, A., DeVita, C., Lee, C., Alsop, D., Detre, J., et al. (2003). Grammatical and resource components of sentence processing in Parkinson's disease: An fMRI study. *Neurology, 60*(5), 775–781.

Groszer, M., Keays, D. A., Deacon, R. M., de Bono, J. P., Prasad-Mulcare, S., Gaub, S., et al. (2008). Impaired synaptic plasticity and motor learning in mice with a point mutation implicated in human speech deficits. *Current Biology, 18*(5), 354–362.

Gurd, J. M., Bessell, N. J., Bladon, R. A., & Bamford, J. M. (1988). A case of foreign accent syndrome, with follow-up clinical, neuropsychological and phonetic descriptions. *Neuropsychologia, 26*(2), 237–251.

Haber, S. N. (2003). The primate basal ganglia: Parallel and integrative networks. *Journal of Chemical Neuroanatomy, 26*(4), 317–330.

Haber, S. N., Lynd, E., Klein, C., & Groenewegen, H. J. (1990). Topographic organization of the ventral striatal efferent projections in the rhesus monkey: An anterograde tracing study. *The Journal of Comparative Neurology, 293*(2), 282–298.

Haber, S. N., & McFarland, N. R. (2001). The place of the thalamus in frontal cortical-basal ganglia circuits. *Neuroscientist, 7*(4), 315–324.

Haesler, S., Rochefort, C., Georgi, B., Licznerski, P., Osten, P., & Scharff, C. (2007). Incomplete and inaccurate vocal imitation after knockdown of FoxP2 in songbird basal ganglia nucleus Area X. *PLoS Biology, 5*(12), e321.

Hartelius, L., Carlstedt, A., Ytterberg, M., Lillvik, M., & Laakso, K. (2003). Speech disorders in mild and moderate Huntington disease: Results of dysarthria assessments of 19 individuals. *Journal of Medical Speech-Language Pathology, 11*(14), 1–14.

Ho, A. K., Iansek, R., Marigliani, C., Bradshaw, J. L., & Gates, S. (1999). Speech impairment in a large sample of patients with Parkinson's disease. *Behavioural Neurology, 11*(3), 131–137.

Kotz, S. A., Schwartze, M., & Schmidt-Kassow, M. (2009). Non-motor basal ganglia functions: A review and proposal for a model of sensory predictability in auditory language perception. *Cortex, 45*(8), 982–990.

Kravitz, A. V., Freeze, B. S., Parker, P. R., Kay, K., Thwin, M. T., Deisseroth, K., et al. (2010). Regulation of parkinsonian motor behaviours by optogenetic control of basal ganglia circuitry. *Nature, 466*, 622–626.

Longworth, C. E., Keenan, S. E., Barker, R. A., Marslen-Wilson, W. D., & Tyler, L. K. (2005). The basal ganglia and rule-governed language use: Evidence from vascular and degenerative conditions. *Brain, 128*(Pt 3), 584–596.

Marie, P. (1906). Revision de la question de l'aphasie: La troisieme circonvolution frontale gauche ne joue aucun role special dans la fonction du langage. *Semaine Medicale, 26*, 241–247.

Mink, J. W. (1996). The basal ganglia: Focused selection and inhibition of competing motor programs. *Progress in Neurobiology, 50*(4), 381–425.

Nadeau, S. E., & Crosson, B. (1997). Subcortical aphasia. *Brain and Language, 58*(3), 355–402.

Nambu, A., Tokuno, H., & Takada, M. (2002). Functional significance of the cortico-subthalamo-pallidal "hyperdirect" pathway. *Neuroscience Research, 43*(2), 111–117.

Nauta, W. J., & Mehler, W. R. (1966). Projections of the lentiform nucleus in the monkey. *Brain Research, 1*(1), 3–42.

Nottebohm, F. (2005). The neural basis of birdsong. *PLoS Biology, 3*(5), e164.

Pickett, E. R., Kuniholm, E., Protopapas, A., Friedman, J., & Lieberman, P. (1998). Selective speech motor, syntax and cognitive deficits associated with bilateral damage to the putamen and the head of the caudate nucleus: A case study. *Neuropsychologia, 36*(2), 173–188.

Schultz, W. (1997). Dopamine neurons and their role in reward mechanisms. *Current Opinion in Neurobiology, 7*(2), 191–197.

Schultz, W., Dayan, P., & Montague, P. R. (1997). A neural substrate of prediction and reward. *Science, 275*(5306), 1593–1599.

Selemon, L. D., & Goldman-Rakic, P. S. (1985). Longitudinal topography and interdigitation of corticostriatal projections in the rhesus monkey. *The Journal of Neuroscience, 5*(3), 776–794.

Ullman, M. T., Corkin, S., Coppola, M., Hickok, G., Growdon, J. H., Koroshetz, W. J., et al. (1997). A neural dissociation within language: Evidence that the mental dictionary is part of declarative memory, and that grammatical rules are processed by the procedural system. *Journal of Cognitive Neuroscience, 9*(2), 266–276.

Watkins, K. E., Dronkers, N. F., & Vargha-Khadem, F. (2002). Behavioural analysis of an inherited speech and language disorder: Comparison with acquired aphasia. *Brain, 125*(Pt 3), 452–464.

Watkins, K. E., Vargha-Khadem, F., Ashburner, J., Passingham, R. E., Connelly, A., Friston, K. J., et al. (2002). MRI analysis of an inherited speech and language disorder: Structural brain abnormalities. *Brain, 125*(Pt 3), 465–478.

Wichmann, T., & DeLong, M. R. (1996). Functional and pathophysiological models of the basal ganglia. *Current Opinion in Neurobiology, 6*(6), 751–758.

Wickens, J. R., Horvitz, J. C., Costa, R. M., & Killcross, S. (2007). Dopaminergic mechanisms in actions and habits. *The Journal of Neuroscience, 27*(31), 8181–8183.

Yin, H. H., & Knowlton, B. J. (2006). The role of the basal ganglia in habit formation. *Nature Reviews Neuroscience, 7*(6), 464–476.

CHAPTER

9

The Thalamus and Language

Daniel A. Llano

Department of Molecular and Integrative Physiology, University of Illinois at Urbana-Champaign, Champaign, IL, USA

Is there a role for the thalamus in language? This question has vexed investigators for decades. As early as 1959, Penfield and Roberts postulated that certain areas of the thalamus may integrate information across distant cortical areas (Figure 9.1). This was based on their awareness of clinical findings that patients with left thalamic lesions may have aphasia, and because of their own findings that anterior and posterior language areas appeared to be highly functionally integrated, even when intervening areas of peri-Sylvian cortex were lesioned, suggesting a subcortical integration site (Penfield & Roberts, 1959). Over the ensuing decades, a great deal has been learned about the functional organization of the thalamus as well as the clinical impact of small lesions within the thalamus. Based on this body of work, there have been recent attempts to integrate the clinical literature (reviewed in De Witte et al., 2011) as well as the imaging literature (reviewed in Llano, 2013) with modern theories about thalamic function into a coherent view of the potential role for the thalamus in language function (Crosson, 2013; Hebb & Ojemann, 2013; Klostermann, Krugel, & Ehlen, 2013). This chapter summarizes these contributions and offers suggestions for future approaches to link human and animal studies to better understand the role of the thalamus in language.

9.1 OVERVIEW OF THALAMIC ORGANIZATION

The thalamus comprises at least a dozen subnuclei that project in an approximately topographic fashion to all areas of the cerebral cortex (Table 9.1). Therefore, a reasonable starting point in the study of the potential role for the thalamus in language is to ask which thalamic nuclei project to the frontal, temporal, and parietal cortical regions that are typically reported as playing a role in language. However, this question is complicated by the fact that connectivity data are mostly available for nonhuman primates and are relatively scarce in humans. Thus, we are left to examine thalamic inputs to cortical regions that bear some structural and functional similarities to human language areas, such as the macaque ventral premotor cortex and the superior temporal gyrus, including the supratemporal plane (Gil-da-Costa et al., 2006). A restricted retrograde tracer injection placed into the caudal portion of the macaque ventral premotor cortex, a putative homologue of Broca's area, produces substantial retrograde labeling in no fewer than 10 thalamic nuclei: ventrolateral nucleus, ventral anterior nucleus, ventral medial nucleus, centrolateral nucleus, centré-median nucleus, medial dorsal nucleus, area X, lateral posterior nucleus, medial pulvinar, and ventral posterior nucleus (Morel, Liu, Wannier, Jeanmonod, & Rouiller, 2005), thus revealing the difficulty in assigning thalamic nuclei to cortical regions in a one-to-one fashion. More recently, a study by Bruce Crosson's group using high-resolution diffusion-weighted imaging tractography in humans identified the ventral anterior thalamus and pulvinar as sites of connectivity with Broca's area in humans (Ford et al., 2012, 2013). Injections of retrograde tracers into the macaque caudal superior temporal gyrus, a putative homologue of Wernicke's area, retrogradely labeled neurons in multiple thalamic nuclei: the medial pulvinar, lateral posterior nucleus, suprageniculate-limitans nucleus, and the medial division of the medial geniculate body (Hackett, Stepniewska, & Kaas, 1998). In each of the tracer studies, a convergence of projections of neurons from both principal and intralaminar or paralaminar nuclei to a single cortical area is observed. This type of convergent organization is not restricted to language cortex homologues, because other cortical areas in the macaque receive inputs from multiple thalamic nuclei (Schmahmann & Pandya, 1990). These findings

Neurobiology of Language. DOI: http://dx.doi.org/10.1016/B978-0-12-407794-2.00009-2

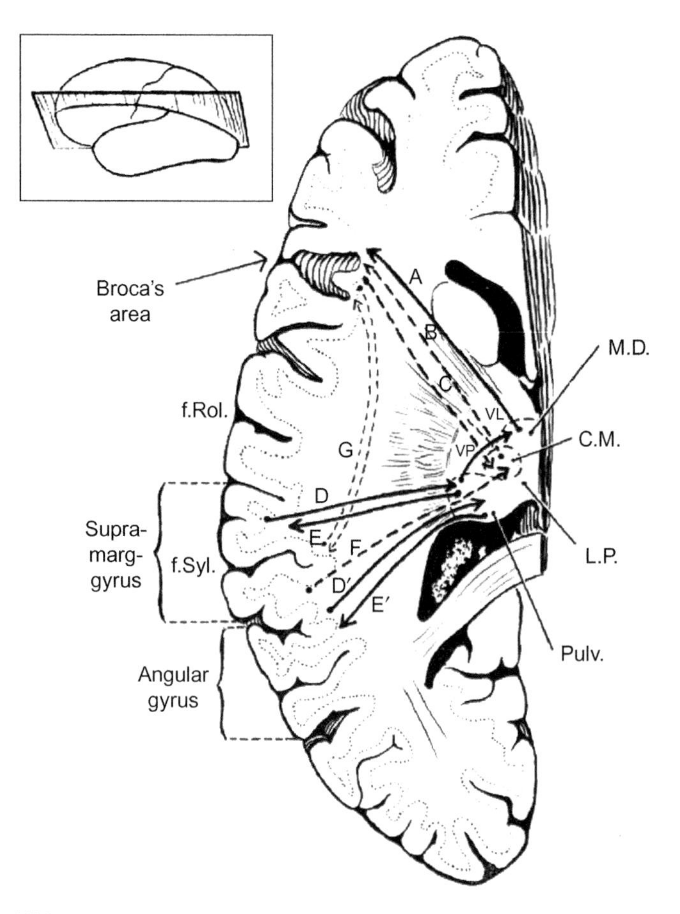

FIGURE 9.1 Penfield and Roberts' model of thalamic involvement in language. Penfield postulated that there were multiple thalamic regions that were reciprocal relationships with language-related cortical structures. Solid lines represent connections that (at the time) were established via monkey tract-tracing and electrophysiological recordings. Dotted lines represent connections only established via electrophysiological recordings. *From Penfield & Roberts (1959) with permission from Princeton University Press.*

suggest that any viable model of the thalamic contribution to cognitive function should account for the multiplicity of thalamic inputs to individual cortical regions.

9.2 DEFINING THE ROLE OF THE THALAMUS IN LANGUAGE

There are substantial numbers of case reports and case series documenting an impairment of language function after thalamic damage. However, the anatomical precision of the lesions and the quality of behavioral assessments (both language and other cognitive functions) performed in these studies are extremely variable. To restrict this review of the literature to studies in which reasonable structure–function relationships can be established, this chapter focuses on patients with ischemic stroke or who have had stereotactic neurosurgery and does not include literature in the meta-analysis from patients with intracranial hemorrhage and with thalamic tumors, given the potential for remote CNS damage and the development of compensatory mechanisms in these clinical scenarios, respectively. In addition, the studies that have performed more than a cursory language and cognitive assessment will be included in this analysis and, as a benchmark, only studies that permit classification into the Wernicke–Geschwind model (i.e., studies that have assessed naming, speech production, speech comprehension, and repetition) have been included.

Given these restrictions, the author has compiled a list of 36 case reports and series from the English-language literature and these are displayed in Table 9.2. Several generalizations can be made from these data. First, although nearly all lesions causing language dysfunction

TABLE 9.1 Major Groups of Thalamic Nuclei, Their Putative Domain of Function and Major Projection Targets

Thalamic nucleus	Putative domain of function	Major projection target
Anterior nuclear group	Memory	Cingulate cortex
Mediodorsal nucleus	Executive function	Prefrontal cortex
Intralaminar + midline nuclei (including centré-median and parafascicular group)	Attention, arousal	Frontal and parietal cortices
Pulvinar + lateral posterior	Higher order sensation	Frontal and parietal cortices
Ventral anterior and ventrolateral nuclei	Motor	Primary and supplementary motor cortex
Lateral geniculate nucleus	Vision	Visual cortex
Ventral posterior nucleus	Somatosensation	Somatosensory cortex
Medial geniculate body	Audition	Auditory cortex

The nomenclature proposed by Hirai and Jones (1989) will be used here rather than the nomenclature commonly seen in the older literature (Hassler, 1959).

TABLE 9.2 Case Reports of Thalamic Aphasia

Reference	Location	Naming	Repetition	Production	Comprehension	Other findings
ISCHEMIC INFARCTS						
Cohen, Gelfer, and Sweet (1980)	Left anterior thalamus	Impaired	Preserved	?	Impaired	Decreased spontaneous speech
Archer, Ilinsky, Goldfader, and Smith (1981)	Left ventrolateral or ventral anterior nucleus	Impaired	Impaired	Preserved	Impaired	Perseveration Fluctuating performance Improvement to mild anomia at 2 months
Case 1 McFarling, Rothi, and Heilman (1982)	Left lateral thalamus	Impaired	Preserved	Impaired	Impaired	Perseveration Poor spontaneous speech Fluctuating performance Grammar preserved No paraphasic errors No followup data
Case 2 McFarling et al. (1982)	Left lateral thalamus	Preserved	Preserved	Impaired	Preserved	Poor verbal fluency on FAS test Grammar preserved No paraphasic errors Etiology not known, but no hemorrhage Authors note remarkable improvement while in hospital
Lhermitte (1984)	Infarct of whole left thalamus	Preserved	Impaired	Impaired	Preserved	Semantic paraphasic errors Poor short-term verbal memory
Case 1 Graff-Radford, Eslinger, Damasio, and Yamada, (1984)	Left anterior thalamus	Impaired	Preserved	Preserved	Impaired	Perseveration Paraphasic errors On SSEP: Delay in first three negative waves after P14
Case 2 Graff-Radford et al. (1984)	Left thalamus	Impaired	Preserved	Preserved	Impaired	Perseveration Semantic paraphasic errors Significant improvement after 2 months, but continued word-finding-difficulty and difficulties with comprehension after 4 years
Case 1 Bogousslavsky, Regli et al. (1986)	Left anterior thalamus—tuberothalamic artery distribution	Impaired	Preserved	Preserved	Impaired	Naming deficit worse than comprehension deficit Decrease in spontaneous speech Semantic paraphasic errors Grammar preserved Dyscalculia
Case 2 Bogousslavsky, Regli et al. (1986)	Left anterior thalamus—tuberothalamic artery distribution	Impaired	Preserved	Impaired	Impaired	Many semantic and phonological paraphasic errors Perseveration

(Continued)

TABLE 9.2 (Continued)

Reference	Location	Naming	Repetition	Production	Comprehension	Other findings
Case 3 Bogousslavsky, Regli et al. (1986)	Left anterior thalamus—tuberothalamic artery distribution	Impaired	Preserved	Preserved	Impaired	
Case 18 Bruyn (1989)	Left anterior and ventral anterior nucleus	Impaired	Preserved	?	Preserved	Perseveration No paraphasic errors No neologisms
Case 20 Bruyn (1989)	Dorsal medial (side not given)	Impaired	Preserved	?	Preserved	No paraphasic errors No perseveration Neologisms
Demeurisse et al. (1979)	Left thalamus	Impaired	?	Preserved	Preserved	Perseveration Inattentive Hypophonic Paraphasic errors Remained aphasic for >1 year
Puel et al. (1992)	Case 1 Left anterior thalamic nucleus and ventral anterior nucleus	Impaired	Preserved	?	Impaired	Many semantic paraphasic errors No phonemic paraphasic errors Perseveration Deficits persisted for at least 1 year Hypoperfusion in left caudate, left temporal cortex, right temporal cortex on SPECT
Case 7 Neau and Bogousslavsky (1996)	Left pulvinar—posterior choroidal territory	Impaired	Preserved	Impaired	Preserved	
Case 8 Neau and Bogousslavsky (1996)	Left pulvinar—posterior choroidal territory	Impaired	Preserved	Impaired	Preserved	
Case BD Raymer, Moberg, Crosson, Nadeau, and Rothi (1997)	Left anterior thalamus—tuberothalamic artery distribution	Impaired	Preserved	Preserved	Impaired	Lethargy Dysarthria Verbal paraphasias No perseveration General improvement at 5 months, but persistent dysnoma (WAB score = 92.2, BNT = 48/60)
Case WT Raymer et al. (1997)	Left thalamus	Impaired	Preserved	Preserved	Preserved	Poor spontaneous speech Paraphasic errors Neologisms Grammar preserved

(Continued)

TABLE 9.2 (Continued)

Reference	Location	Naming	Repetition	Production	Comprehension	Other findings
Ebert, Vinz, Görtler, Wallesch, and Herrmann (1999)	Right nucleus lateropolaris, mamillothalamic tract, nucleus ventrooralis externus, ventrooralis internus (patient was left handed)	Impaired	?	Impaired	Impaired	Semantic and phonological paraphasic errors Marked fluctuations in language performance Non-aphasic at 16 months Patient had pre-existing left subcortical white matter lesion
Case 3 Karussis, Leker, and Abramsky (2000)	Left tuberothalamic artery territory	Impaired	Preserved	Impaired	Preserved	Paraphasic errors
Case 4 Karussis et al. (2000)	Left tuberothalamic artery territory	Impaired	Impaired	Impaired	Preserved	Poor grammar No improvement after 1 year
Case 7 Karussis et al. (2000)	Left inferior lateral artery territory	Impaired	Impaired	Impaired	Preserved	
Case 8 Karussis et al. (2000)	Left anterior choridal artery territory	Impaired	Impaired	Impaired	Preserved	Poor grammar Improved at followup (followup interval not given)
Weisman, Hisama, Waxman, and Blumenfeld (2003)	Left ventrolateral thalamus lacune	Preserved	Impaired	Preserved	Preserved	No paraphasic errors
Case OGy Szirmai, Vastagh, Szombathelyi, and Kamondi (2002)	Left LP, CM, VP	Impaired	Preserved	Impaired	Preserved	Apathy Poor memory Disorientation Hypoperfusion in left temporal cortex, left insular cortex and right cerebellum on SPECT
Radanovic, Azambuja, Mansur, Porto, and Scaff (2003)	Left medial thalamus	Impaired	Impaired	?	Impaired	Perseveration Poor performance on Trail-Making-Test A and B Semantic paraphasia Evaluated 6 months post-insult
Radanovic et al. (2003)	Right posterior-lateral thalamus (patient was right-handed)	Impaired	Impaired	?	Impaired	Poor performance on Trail-Making-Test A and B Evaluated 9 years post-insult
Segal, Williams, Kraut, and Hart (2003)	Left anterior thalamus	Impaired	Preserved	Impaired	Impaired	Poor animal fluency Impaired semantic recall
Levin, Ben-Hur, Biran, and Wertman (2005)	Left anterior thalamus—tuberothalamic artery distribution	Impaired	Preserved	Preserved	Preserved	Category-specific dysnomia Lethargic Poor verbal memory Executive dysfunction Perseveration Symptoms persisted for at least 1 month.

(Continued)

TABLE 9.2 (Continued)

Reference	Location	Naming	Repetition	Production	Comprehension	Other findings
Margolin et al. (2008)	Left anterior and paramedian thalamus	Impaired	Preserved	Impaired	Impaired	Hypophonia Skew deviation Phonemic paraphasic errors General improvement over first 8 days, but naming still poor (BNT = 8/60) No language deficits at 1 year
De Witte et al. (2011)	Right midline—lateral thalamus (patient right-handed)	Impaired	Preserved	Preserved	Impaired	Semantic paraphasic errors Neologisms Persistent dysnomia at 18 months Perfusion SPECT at 1 year showed severe hypoperfusion in right temporo-parietal, anterior frontal cortex and right thalamus
Hoffmann (2012)	Left lateral posterior thalamus	Preserved	Preserved	Impaired	Preserved	Semantic paraphasias Impaired oral word and sentence reading Impaired picture word matching and writing
SURGICAL LESIONS						
Bell (1968)	10 cases of VL thalamotomy	Impaired	Preserved	Impaired	Preserved	Perseveration seen 5 patients better within 3 weeks 4 patients had gradual improvement with some persistent deficit 1 with persistent deficit after 4 years
Samra et al. (1969)	8 patients with ventrolateral thalamotomy (7 on left, 1 on right)	?	Impaired in 2/8	Impaired in 8/8	Impaired in 4/8	All cases pathologically confirmed to be within ventrolateral nucleus
Darley, Brown, and Swenson (1975)	Left ventrolateral thalamotomy	Preserved	?	Impaired	Impaired	
Vilkki and Laitinen (1976)	Left pulvinar or ventrolateral thalamotomy	Impaired	?	?	Impaired	Recovery at 3 months

are in the dominant thalamus, the specific thalamic nucleus involved varies considerably (see Figure 9.2 for a summary of lesion locations). This anatomic heterogeneity is further addressed later. Clinically, there is an almost universal deficit in naming, often severe, and, importantly, most naming errors in these studies take the form of semantic substitutions, rather than omissions. Second, across most studies, there is a relative preservation of repetition, often including the repetition of complex utterances. Beyond these features, the findings are more variable. Not all studies describe a clear deficit of language production or comprehension, but of those that do, most describe relatively mild deficits, more commonly in comprehension than in production, with general congruence of verbal and orthographic deficits. Therefore, applying a Wernicke–Geschwind scheme to these data (with all appropriate provisos regarding the Wernicke–Geschwind taxonomy), one would describe most of these patients as having a transcortical sensory aphasia, a transcortical motor aphasia, or, most commonly, an anomic aphasia.

Another notable feature in this population is the high degree of individual performance variability, often moment to moment, which seems to track with the level of arousal of the patient (Archer et al., 1981; Ebert et al., 1999; McFarling et al., 1982), seen in both ischemic and

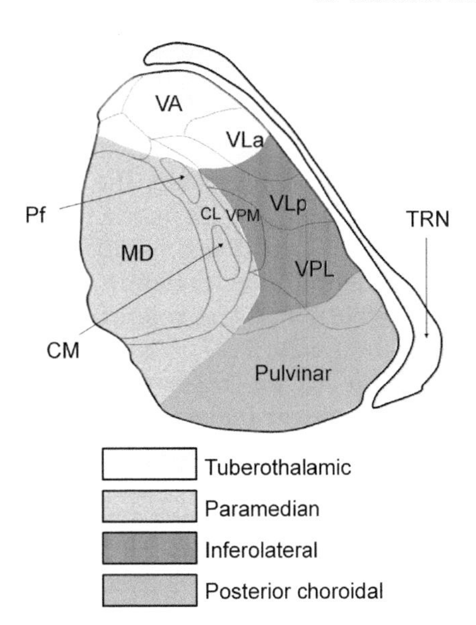

FIGURE 9.2 Thalamic blood supply. CL = centrolateral division, CM = center median, MD = medial dorsal, Pf = parafascicular, TRN = thalamic reticular nucleus, VA = ventral anterior, VLa = ventrolateral anterior region, VLp = ventrolateral posterior region, VPL = ventral posterior lateral, VPM = ventral posterior medial. *Redrawn and modified from Carrera & Bogousslavsky (2006).*

hemorrhagic stroke, and detailed in an early study of a thalamic hemorrhage patient (Mohr, Watters, & Duncan, 1975). In addition, many authors have noted a high frequency of semantic paraphasic errors (Demeurisse et al., 1979; Ebert et al., 1999; Karussis et al., 2000; Radanovic & Scaff, 2003; Raymer, Moberg, Crosson, Nadeau, & Rothi, 1997) and perseverations (Bell, 1968; Bogousslavsky, Regli, & Assal, 1986; Bruyn, 1989; Demeurisse et al., 1979; Graff-Radford et al., 1984; Levin et al., 2005; McFarling et al., 1982; Puel et al., 1992). Finally, a commonly reported feature of these patients is a relatively rapid recovery of language deficits. When recovery has been described, most patients recover to a significant degree within 6 months of the ictus (Archer et al., 1981; Graff-Radford et al., 1984; McFarling et al., 1982; Raymer et al., 1997; Vilkki & Laitinen, 1976), although several patients with persistent aphasic deficits after focal thalamic lesions have been described as well (Bell, 1968; Demeurisse et al., 1979; Graff-Radford et al., 1984; Karussis et al., 2000; Puel et al., 1992; Radanovic & Scaff, 2003). Although not included in Table 9.2 for the reasons described previously, it is worth mentioning here that a large number of patients with thalamic hemorrhage have been described (etiologically more common than isolated thalamic infarction) with a virtually identical symptom complex (Karussis et al., 2000; Mohr et al., 1975; Radanovic & Scaff, 2003; Wallesch, 1997).

Although the Wernicke–Geschwind formulation provides a bird's eye view of thalamic aphasia as primarily a disorder of naming and mild comprehension deficits with spared repetition, it does not provide sufficient detail about language and cognitive behavior to assist in the generation of neural models to explain the role of the thalamus in language. Several authors have provided more detailed accounts of the language deficits in patients with thalamic lesions, and these data are briefly reviewed here. Note that several patients with thalamic hemorrhage have been included in this analysis because hemorrhage sizes were small and the quality of the language data is high. Raymer et al. (1997) and Crosson (1999) described the language deficits of two thalamic stroke patients, one with a stroke damaging the ventrolateral, ventral anterior, centré-median, and thalamic reticular nuclei and another with damage to the ventral anterior, ventrolateral, mediodorsal, centré-median, parafascicular, and thalamic reticular nuclei. Both subjects had significant difficulty with naming across modalities but had no difficulty with writing to dictation or reading aloud. Furthermore, the majority of the naming errors (84% and 63% in the two subjects) were semantic. These suggested to Crosson that the core deficit was neither purely lexical nor purely semantic. Instead, the deficit was one of *retrieval of lexical items from semantic input* (Crosson, 1999).

Another study of a patient with Parkinson's disease undergoing left thalamotomy of the ventrolateral nucleus and studied before and 3 months after thalamotomy showed no change in performance on the Boston Naming Test but demonstrated decreases in category fluency and deficits in higher-level tests of language function, such as combining lexical units into sentences, as well as significant difficulty remembering word pairs. If a semantic strategy were used to attempt to remember word pairs, then both the word pair deficit and the deficit in sentence creation could potentially be explained by a deficit in the mapping of semantic to lexical information (Whelan, Murdoch, Theodoros, Silburn, & Hall, 2002). Further breakdown in lexical-semantic processing was noted by Segal, Williams, Kraut, and Hart (2003) in a case of left tuberothalamic artery stroke. Their patient had poor naming, halting speech, impaired comprehension, and impaired performance in category fluency, with preserved reading and repetition of words and sentences. Other deficits were also present, including mild impairments on property judgment, category judgment, and semantic association, and severe disruption of semantic recall (i.e., the patient was given two words or a word plus picture and asked to generate the concept of an item that is associated with both). The patient reported that he could only generate a semantic item 64% of the time, but more importantly he was not able to provide the name of the semantic item for any of the pairs. These data point to a deficit in the mapping of semantic information to lexical units and may also suggest a deficit in the generation of new conceptual representations.

In addition to these lexical-semantic difficulties, patients with thalamic lesions have marked variability in language performance and often have significant perseveration. Two particularly dramatic cases were described by Mohr et al. (1975) in the setting of left thalamic hemorrhage. In the acute stage, both patients ranged in performance from essentially normal language to marked logorrheic jargonaphasia with perseverative components. An excerpt from the logorrheic state:

Examiner	Patient
How are things going?	*I thinks, a, going fine, and ... I can say, s'quit alright, si sings say, rou, rup...*
What happened?	*Well, ... See up chup a lupdup. Cheche den, etc., and, a, she is quite please at that.*
What did you say?	*She said, that was fudal wedel wedl And she also, when she comes in, eetosaid You have it every time it nevel a dedal wedel.*

Seven days later, the patient had generally improved but continued to have periods of jargonaphasia with more overt perseveration:

Examiner	Patient
How are you today?	*Just fine.*
Anything wrong?	*Yes, I feeling fine.*
How old are you?	*Ahh, 59.*
Where are you?	*59.*
No, where are you?	*Where, 59.*

Mohr commented that increases in the level of arousal were associated with improved language performance. Another example was provided by Luria (1977), who described a patient with a lesion secondary to aneurysmal hemorrhage into the left thalamus. This patient was fluent, with normal grammar and prosody, but had severe loss of spontaneous speech, repetition, and naming, and could understand only fragments of speech. Many of his errors were perseverative in nature and his performance fluctuated significantly. Luria noted that the patient's speech was marked by "extraneous associations, influences of the immediate situation, and fragments from former traces" which Luria termed "pathological inertia." The patient appeared to do better when structure was imposed on the range of semantic information that needed to be culled for a task (i.e., his performance during picture description was better than during spontaneous speech). A similar improvement was noted during discussion of a restricted range of familiar versus unfamiliar topics in a patient with hemorrhage into the lateral nucleus of the thalamus and the anterior superior pulvinar (Crosson et al., 1986). Although it is possible that the fluctuations described may be related to secondary effects of the hemorrhage such as edema or nonconvulsive seizures, similar findings have been described (although without similar vivid descriptions) in both ischemic thalamic stroke and thalamotomy patients.

Given the heterogeneity of lesion location in these patients, the marked performance variability, the lack of a uniform aphasic syndrome, the absence of language deficits in many patients with injury to the dominant thalamus, and the generally rapid recovery of language function after thalamic damage, several authors have questioned the role of the thalamus in language at all (Luria, 1977; Van Buren, 1975). For example, Cappa et al. reported five consecutive patients with dominant thalamic lesions who did not have aphasia (Cappa, 1986), and Wallesch et al. found few differences in language performance between patients with small left-side versus right-side thalamic strokes (Wallesch et al., 1983). Given these conflicting clinical data, it seems likely that subjects with focal thalamic damage without language deficits must fall into one of three categories: (i) patients without damage to structures and/or connections that play a role in language performance; (ii) patients who recovered rapidly and thus escaped detection of any deficits because they were not present at the time of testing; or (iii) patients with actual deficits that were not detected for other reasons, such as inadequate examination.

The relative consistency in naming deficits despite heterogeneity in thalamic lesion locus is not dissimilar to what has been described for the cerebral cortex. For example, Ojemann (1991) and Penfield & Roberts (1959) described dysnomia in response to cortical stimulation across virtually the whole peri-Sylvian cortex, without clear location specificity to the patterns of errors observed. These data point to two features of naming. First, it requires many different component processes to achieve, from early vision to visual object recognition to modality-dependent (and/or modality-independent) conceptual access to lexical access to phonological assembly to phonemic mapping and motor output, functions that are subserved by networks extending throughout the whole peri-Sylvian region. Second, many regions throughout the peri-Sylvian cortex and associated thalamic nuclei probably participate in a distributed network that encodes word meanings and conceptual information, such that a lesion in any part of the network causes a breakdown in lexical-semantic processing. This similarity in the clinical presentation of focal thalamic and cortical lesions extends to recovery in language function after focal thalamic lesions. Rather than being evidence for the absence of a role for the thalamus in language, we note that both thalamic and focal cortical lesions demonstrate recovery (Mohr et al., 1978; Penfield & Roberts, 1959) and point to a high level of redundancy and plasticity in thalamocortical networks that are important for language.

Another criticism of "thalamic aphasia" is one that has been articulated by Luria and others, stating that thalamic language deficits are not bona fide *language* deficits at all in that there is no deficit in the formulation or understanding of grammatical structure, but rather they reflect instability of cortical language traces. As described, it may be that the absence of grammatical formulation errors in patients with dominant thalamic injury serves as an important constraint on any model of thalamic function for language. Given the well-known role of the thalamus in arousal function, it may be that a key role for the thalamus is to sustain cortical representations of language during production or comprehension of language, which is explored in Section 9.4.1.

9.3 A THALAMIC "LOCUS" FOR LANGUAGE?

At a first approximation, one might think it could be possible to resolve some of the heterogeneity in clinical presentation from careful analysis of lesion location or vascular distributions of thalamic infarcts. Unfortunately, this approach has proven to be quite challenging. For example, there is not a single thalamic blood vessel that has been exclusively associated with thalamic aphasia. The thalamus receives its blood supply from the tuberothalamic, paramedian, inferolateral, and posterior choroidal arteries, which are derived from the posterior circulation and all have been implicated in thalamic aphasia (Bogousslavsky, Miklossy, Deruaz, Regli, & Assal, 1986; Bogousslavsky, Regli, et al., 1986; Karussis et al., 2000; Levin et al., 2005; McFarling et al., 1982; Neau & Bogousslavsky, 1996; Perren, Clarke, & Bogousslavsky, 2005; Radanovic et al., 2003; Raymer et al., 1997). Analysis of the distributions of these infarcts is further complicated by the fact that certain regions of the thalamus, such as the pulvinar, receive a dual vascular supply (Morandi et al., 1996; Takahashi et al., 1994), and because patients with pulvinar lesions tend to also have associated (often devastating) midbrain or brainstem lesions, making any analysis of language difficult. This may create the impression that the pulvinar is not involved in language. However, given the widespread projections of the pulvinar to regions of the cortex that are involved with language (Romanski, Giguere, Bates, & Goldman-Rakic, 1997) and findings using other clinical sources (hemorrhage, deep brain stimulation, tracing degenerating axons), all point to a potential role for the pulvinar in language (Crosson et al., 1986; Ojemann, Fedio, & van Buren, 1968; Van Buren & Borke, 1969).

Another approach is to examine lesion location and perform a traditional lesion-deficit analysis. One starting hypothesis may be that anterior thalamic infarctions would be more likely to cause deficits in language motor production, and posterior thalamic infarctions would be more likely to cause deficits in language comprehension. Such an analysis has been performed, and the anterior/posterior distinction has not solved the problem. For example, several investigators have described infarction of the tuberothalamic artery, which serves the anterior thalamic nuclei, ventral pole of the dorsal medial nucleus, ventral anterior nucleus, and intralaminar and rostral ventral lateral nucleus (Figure 9.2). In these patients, the clinical syndromes span the full range of language behaviors, without a clear predominance of "sensory" or "motor" manifestations. Of the patients described in Table 9.2, seven can be described as having primarily a "sensory" aphasia, and one might predict that the lesions in these cases predominantly affect portions of the thalamus known to project to superior temporal, inferior parietal, or adjacent areas of the cortex (i.e., "Wernicke's area"). However, five out of seven of these patients had lesions in the territory of the tuberothalamic artery, which supplies anterior thalamic structures (ventral anterior nucleus, ventral lateral nucleus, mediodorsal nucleus, anterior thalamic nuclei, and the anterior intralaminar nuclei), which are generally not known to project to areas within or near Wernicke's area.

The author has examined the locations and extents of lesions in the case reports and series (when figures were available from the original publications) and has superimposed this information onto generic maps of the thalamus (taken from Morel, 2007); these maps are shown in Figure 9.3. As shown, the distribution of lesion sites demonstrates a propensity for sites to be located in the anterior nuclei (possibly for the reasons stated), but importantly demonstrates that lesions producing aphasia can be found in nearly all regions of the thalamus.

It may be possible to avoid some of the biases introduced by analysis of the vascular distributions of thalamic ischemic infarctions and the nonspecific effects of hemorrhage by analyzing data from patients undergoing therapeutic thalamotomy or deep brain stimulation. Here, for therapeutic reasons, most of the lesion/stimulation sites have been in the ventral lateral nucleus, intralaminar nuclei, and pulvinar. Much of the classic work in this field was performed by Ojemann and his colleagues. They found that stimulation in adjacent areas of the left ventral lateral nucleus and pulvinar caused dysnomia, and that the dysnomia was not related to motor speech difficulties. Most of the errors were substitution errors, which stands in contrast to errors made in a similar paradigm using stimulation of subcortical white matter beneath parietal cortex, which mostly causes omission errors (Ojemann et al., 1968; Ojemann & Ward, 1971).

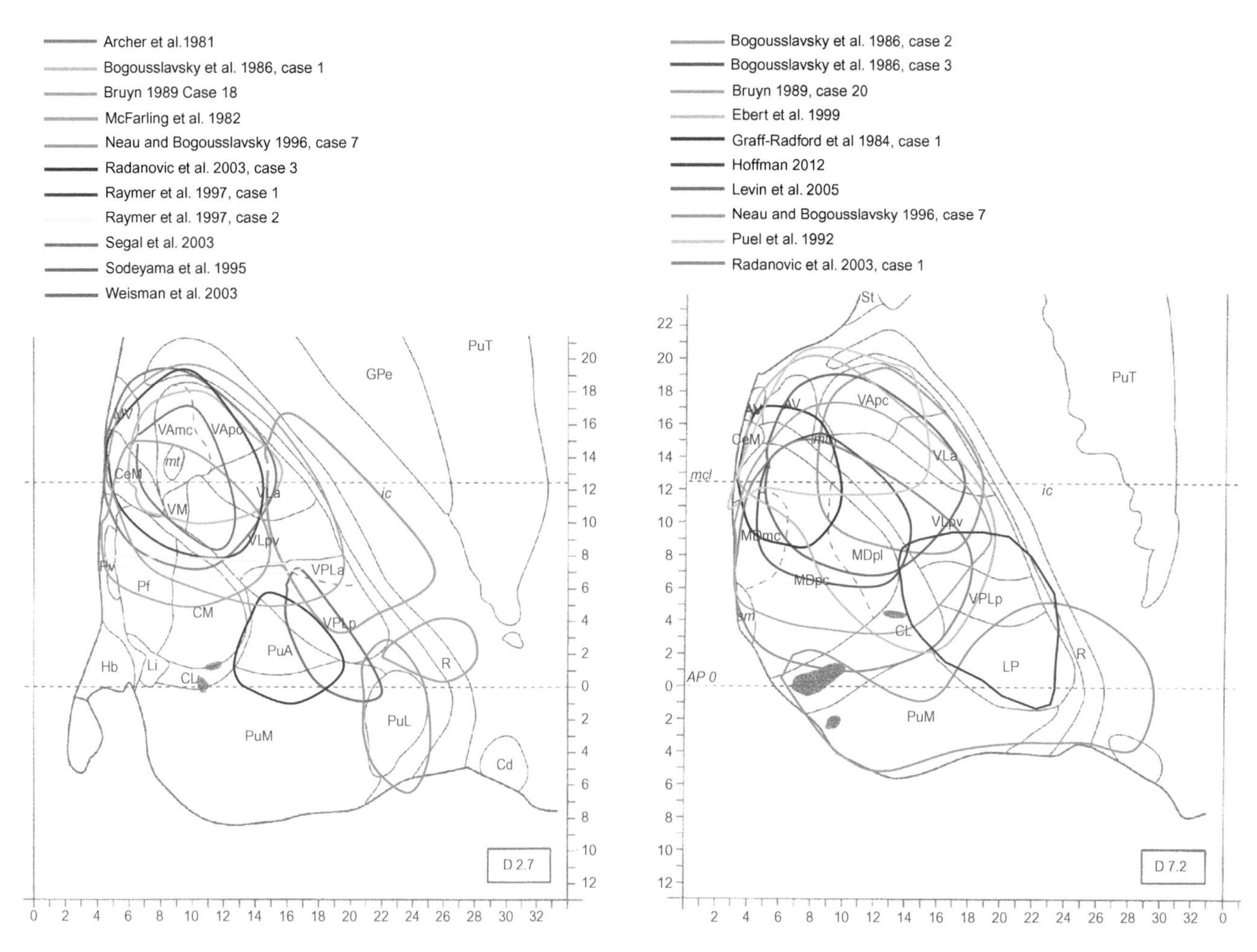

FIGURE 9.3 Perimeters of lesions taken from original publications referenced in Table 9.2 (when good-quality images were available), overlaid on a generic axial map of the thalamus, taken from (Morel, 2007). Two different superior-inferior levels are shown (left 2.3 mm, right 7.2 mm dorsal to posterior commissure). Background images of the thalamus obtained with permission from Taylor and Francis Group LLC Books.

The notion that the pulvinar plays a role in language, which was not obvious from the lesion data (Figure 9.3), was supported by work by Fedio and Van Buren, who also performed a direct brain stimulation study of picture naming. They predominantly found substitution errors during stimulation of the left pulvinar, but no errors during stimulation of areas anterior or inferior to the left pulvinar (or the right pulvinar). These findings contrast with those of Vilkki and Laitinen (1976), who compared language performance of patients undergoing thalamotomy of the ventral lateral nucleus with those undergoing pulvinotomy. They found decreases in word fluency and token test performance in those undergoing left ventrolateral thalamotomy, but only nonsignificant trends toward worsening token test performance and naming for patients undergoing pulvinotomy. Finally, more recent work involving patients undergoing deep brain stimulation of the centré-median nucleus of the intralaminar group of nuclei has induced enhancement across a number of cognitive domains (language, memory, and reasoning). Many language tasks showed improvement, including object naming, recall, antonym recall, word fluency, sentence comprehension (Token test), and verbal expression (Bhatnagar, 2005). These studies, coupled with the previously described neurosurgical literature, interpreted conservatively, implicate the ventrolateral nucleus, the pulvinar, and possibly the centré-median nucleus in engaging in language.

A third approach to determining the anatomical site within the thalamus for language performance is to attempt to trace specific areas of the cortex known to play a role in language (e.g., inferior frontal gyrus, posterior superior temporal gyrus) to specific thalamic nuclei. There are very few data using this approach in humans, but the data generally point to the same nuclei that emerged from the thalamic lesion literature. For example, limited human autopsy studies examining patterns of degeneration after cortical aphasic stroke involving posterior regions important for language (e.g., superior temporal gyrus, supramarginal gyrus, angular gyrus) produced degeneration in the central lateral nucleus (an intralaminar nucleus), lateral posterior (part of the pulvinar complex), and pulvinar proper. For example, a patient with stroke suffering from a "motor aphasia with perseverations" involving cortical damage to the postcentral gyrus, superior insula, and parietal operculum, yielded thalamic degeneration in ventral posterior medial nucleus, medial geniculate body, ventral anterior nucleus, and the centrolateral nucleus (Van Buren & Borke, 1969). Most recently, as mentioned, Ford and colleagues identified the ventral anterior nucleus as demonstrating strong connectivity with Broca's area (Ford et al., 2013), suggesting that this region is involved in the motor and/or grammatical aspects of speech. Overall, these data again implicate the pulvinar, intralaminar nuclei, the ventrolateral nucleus, and the ventral anterior in language function, with some regional anatomical specificity such that ventral anterior nucleus may have direct projections to Broca's area and more posterior language areas having connectivity with pulvinar and possibly the intralaminar nuclei. These analyses are very preliminary and, as tractography techniques advance, we are likely to learn much more about thalamic correlates to physiologically identified cortical regions involved with language.

Given what appears to be a general conservation of the topology of the cortical structures involved in language superimposed on the thalamus, it is peculiar that the functional distinctions between the anterior and posterior language systems, evident in the cortex, are not evident in the thalamus. In other words, although naming tends to be distributed widely throughout brain language centers, other more specific types of language deficits appear to have a more segregated anatomical distribution in the cortex but do not appear to do so in the thalamus. Some light on this issue has been shed by work examining cortical perfusion in the setting of acute thalamic stroke. Puel et al. (1992) demonstrated cortical hypoperfusion via SPECT imaging in the peri-Sylvian, posterior temporal, and temporo-occipital regions of the left hemisphere, regions not innervated by the anterior group of nuclei and occupying a larger area than would be predicted by a lesion of one or more of these nuclei. These results were confirmed and extended by Shim et al. (2008), who demonstrated decreased regional cerebral blood flow in several peri-Sylvian cortical regions, including the left inferior frontal gyrus, supramarginal gyrus, and superior temporal gyrus after infarction of left anterior thalamic structures manifested by dysnomia and poor semantic and phonemic fluency (Figure 9.4). Similar decreases across broad regions of cortex were also seen by Lanna et al. (2012). These data point to an apparent lack of a 1:1 structural–functional relationship of the thalamic nuclei and their cortical targets with respect to language. This difficulty was noted by Crosson (1999), who suggested that the common deficits produced by disparate lesions point to the possibility of damage to different portions of a common neural system, rather than separable neural systems with separate roles each. This distinction will become important as different models of thalamic involvement in language are discussed.

9.4 IMAGING OF THE THALAMUS IN LANGUAGE TASKS

A complementary approach to the lesion-deficit analysis presented is to measure thalamic activation during language tasks using functional imaging. The author has systematically reviewed the relevant fMRI and PET literature previously (Llano, 2013) and summarizes the findings here. Despite a large number of functional imaging studies having been performed over the past 20–30 years measuring brain activity during language tasks, very few have described the patterns of activation seen in the thalamus (or any other subcortical structures). There are several reasons for this. First, thalamic nuclei are relatively small (a few millimeters in diameter, typically) relative to the smoothing windows for most studies, which range from 5 to 20 mm, full width at half maximum. Second, movement artifact due to respiration and heart rate is more prominent in subcortical rather than cortical structures (Guimaraes et al., 1998). Third, it is likely that neurometabolic coupling is different in cortical regions versus thalamus (Llano, Theyel, Mallik, Sherman, & Issa, 2009), such that canonical hemodynamic response functions, optimized for cortex, may not be optimal for thalamus. There may also be fundamental differences in redundancies and integrative properties between thalamus and cortex, yielding different metabolic requirements during the activations of these structures. Finally, and possibly most importantly, most investigators do not place regions of interest in the thalamus and/or do not optimize their spatial resolution on subcortical structures, leading to a lack of detection of activation in thalamus during a language task.

Despite these limitations, the author previously identified 50 studies that demonstrated thalamic activation while normal subjects performed a language task (Llano, 2013). Two of the most common tasks leading to thalamic activation were word or sentence generation tasks and naming. Less commonly, activation was seen during lexical decision, reading, and working memory tasks. Activation was generally seen throughout the thalamus, but with a predominance of activation sites in the left thalamus, and was associated with activations in frontal and temporal cortical regions typically associated

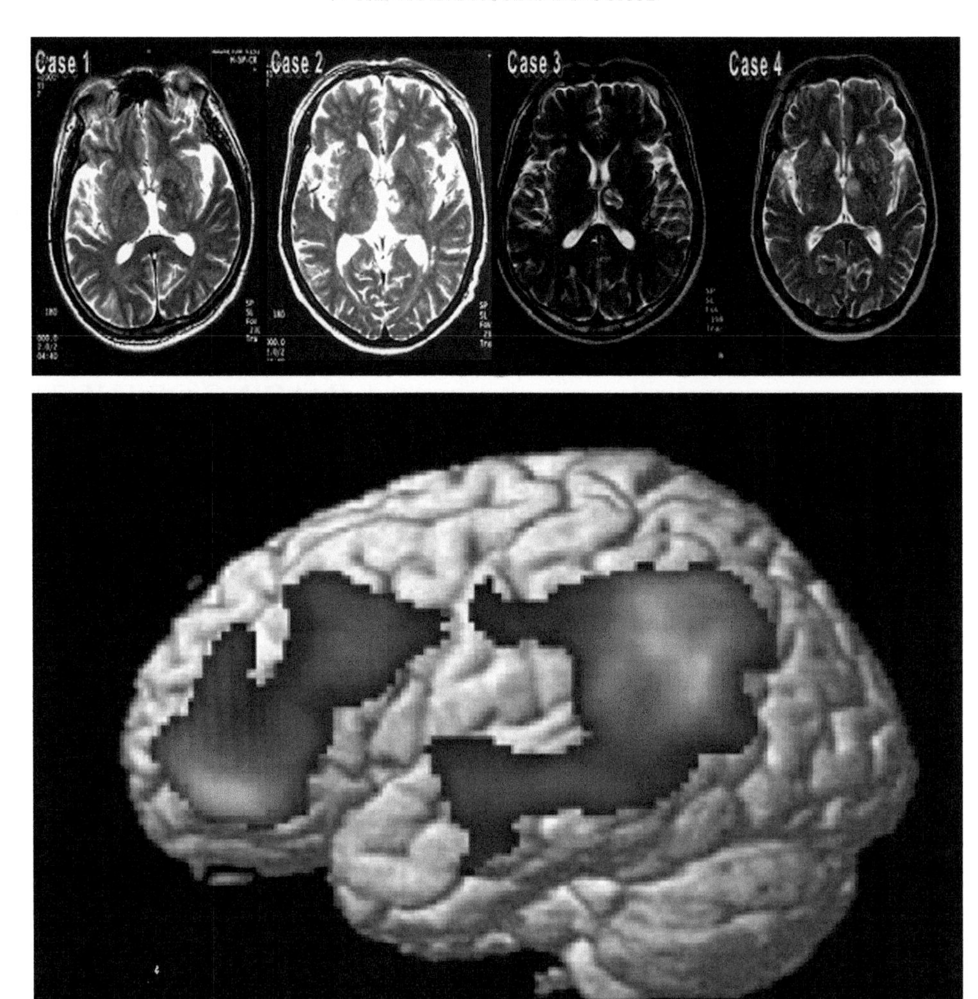

FIGURE 9.4 (Top) T2-weighted images of four patients with isolated infarcts to the left tuberothalamic artery. (Bottom) Average regional cerebral blood flow map, averaged across all four patients, demonstrating hypoperfusion across multiple anterior and posterior language-related cortical areas. *Image taken from Shim et al. (2008) with permission from the publishers.*

with language tasks (Figure 9.5). These analyses indicate that the thalamus may be involved in processes that involve manipulations of lexical information, particularly when based on semantic cues.

9.4.1 Models of the Role of the Thalamus in Language

The data presented suggest that the thalamus plays an important role in language, although its specific role or roles are not yet known. Several models of thalamic function in language are considered here. Any model of thalamic function in language should be able to reproduce the key findings described in the clinical literature on thalamic aphasia and should be constrained by the known anatomical and physiological features of thalamic circuitry. As such, it is proposed that a reasonable model should be able to reproduce the following elements:

1. The primary cognitive deficits in thalamic aphasia involve the following three features:
   a. The deficit is one of lexical selection based on semantic information
   b. The deficits fluctuate with level of arousal
   c. The deficits do not involve repetition
2. These deficits can be produced via lesions in multiple different loci in the dominant thalamus, with the primary nuclei involved being the ventrolateral, centré-median, and pulvinar.
3. The model should be constrained by the known anatomical and physiological properties of the thalamus.

Several investigators have developed descriptive models of thalamic mechanisms that underlie language

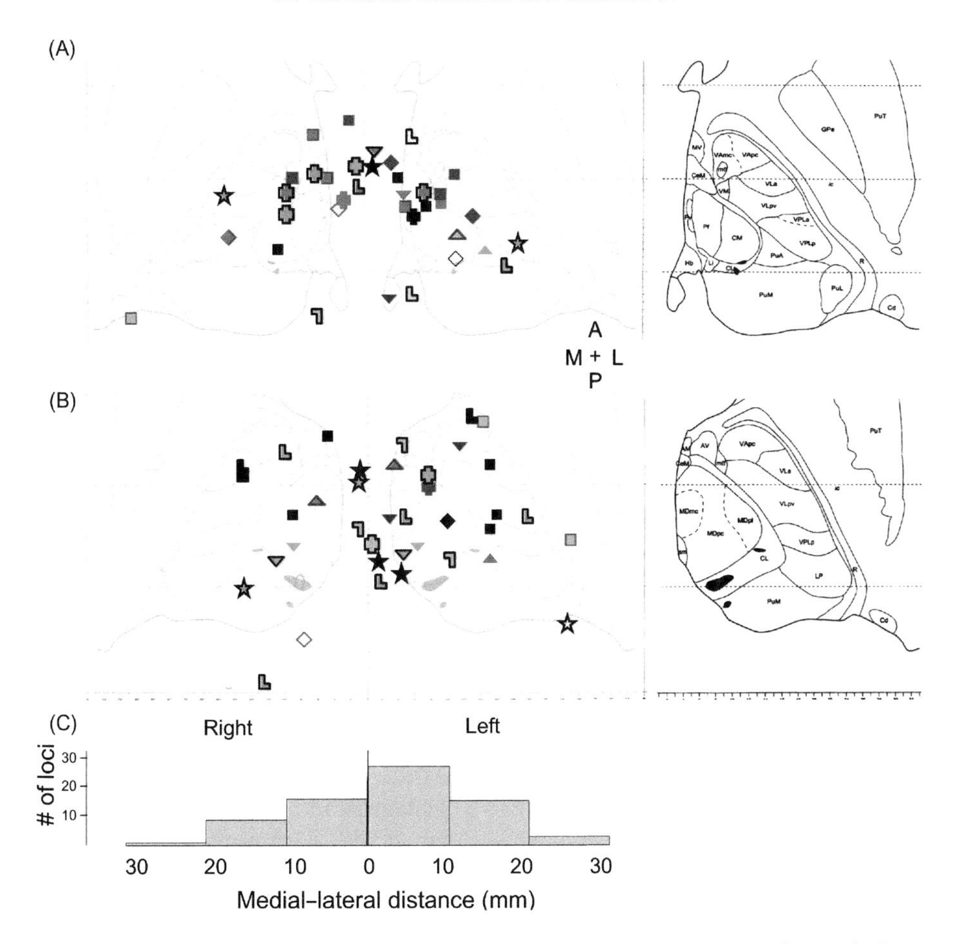

FIGURE 9.5 Compilation of the center of activation foci taken from Talairach and Tournoux coordinates (either given in original publications or transformed from MNI coordinates). Two axial sections are shown. (A) Taken from 2.7 mm superior–inferior (base image reproduced from Morel, 2007). (B) Taken from 8.1 mm superior–inferior. Each study is represented by a different symbol. (C) Histogram of the number of activation loci within each 10 mm of midline in the medio-lateral dimension. Different symbols represent activations from different studies. A = anterior, P = posterior, M = medial, L = lateral, AM = anteromedial nucleus, AV = anteroventral nucleus, Cd = caudate nucleus, CeM = central medial nucleus, GPe = globus pallidus externa, Hb = habenular nucleus, ic = internal capsule, Li = limitans nucleus, LP = lateral posterior nucleus, MDmc = mediodorsal nucleus, magnocellular division, MDpc = mediodorsal nucleus, parvocellular division, MDpl = mediodorsal nucleus, paralamellar division, mtt = mammillothalamic tract, Pf = parafascicular nucleus, PuA = anterior pulvinar, PuL = lateral pulvinar, PuM = medial pulvinar, PuT = putamen, Pv = paraventricular, R = thalamic reticular nucleus, sm = stria medullaris, VAmc = ventral anterior nucleus magnocellular, VApc = ventral anterior nucleus parvocellular, VLa = ventrolateral anterior, VLpd = ventrolateral posterior nucleus dorsal division, VLpv = ventrolateral posterior nucleus ventral division, VM = ventromedial, VPLa = ventral posterior nucleus anterior division, VPLp = ventral posterior nucleus posterior division. *Anatomical image reproduced with permission from Informa Healthcare. Figure reproduced from Llano (2013). Please see Llano (2013) for details.*

processing. Most of these models have ascribed an attentional role to the thalamus (Crosson, 1999; Johnson & Ojemann, 2000; McFarling et al., 1982; Riklan & Cooper, 1975). Two of the more well-developed and most referenced are described here after a brief description of relevant thalamic anatomy and physiology.

9.5 THALAMIC CIRCUITRY AND PHYSIOLOGY

One challenge in understanding the role of the thalamus in language is the heterogeneity of organization within this structure. There have been many attempts to bring order to the thalamus by classifying thalamic nuclei and thalamocortical neurons. The main points of distinction between thalamic neurons and thalamic nuclei have been the types of inputs received by a thalamic neuron, and therefore the direction of information flow through the thalamus, and the layer of termination in the cortex of thalamocortical neurons and, therefore, the types of effects that a thalamic neuron can have on the cortex. For example, it is possible to distinguish between principal thalamic nuclei, such as the lateral geniculate nucleus for vision, which sends specific sensory information to middle layers of the occipital cortex, compared with the intralaminar nuclei, whose cortical projections show a more widespread distribution and tend to terminate in upper layers and have been implicated in the control of arousal (Purpura & Schiff, 1997; Van der Werf, Witter, & Groenewegen, 2002) (Figure 9.6A). Although this distinction has heuristic value, it is muddied by the presence of layer 1-projecting neurons in principal nuclei (Abramson & Chalupa, 1985; Rockland, Andresen, Cowie, & Robinson, 1999). An additional parcellation scheme is based on the types of inputs received by thalamic neurons. Thalamic neurons receive inputs from many sources: primary sensory afferents (e.g., the retina for the lateral geniculate nucleus), brainstem monoaminergic and cholinergic nuclei, the thalamic reticular nucleus, and the cerebral cortex. It has been proposed that these inputs can be divided into "drivers," which are inputs with morphological and physiological specializations characterized by very high synaptic efficacy, and "modulators," which are physiologically better suited to regulate the tuning functions of thalamic neurons (Sherman & Guillery, 1998). Based on these distinctions, Sherman and Guillery divided the thalamus into "first-order" nuclei, such as the lateral geniculate nucleus, which receive their drivers primarily from the sensory periphery and "higher-order" nuclei, such as the pulvinar, which receive their drivers from the cerebral cortex (Figure 9.6B). The relevance here for language is that higher-order nuclei may serve as an alternative route for the transmission of long-range cortical communication via a cortico-thalamo-cortical route (Theyel, Llano, & Sherman, 2010).

The other important and relevant features of the thalamus concern the physiological properties of thalamic neurons and the main readout of thalamocortical neurons: the thalamocortical synapse. Most thalamocortical neurons contain T-type calcium channels, which de-inactivate with protracted (100 s of milliseconds) hyperpolarization and, then, with a low threshold, produce a large depolarization leading a burst of spikes (Figure 9.6C). Therefore, thalamocortical neurons may exist in one of two response modes: *tonic* or *burst* (Jones, 2007; Sherman, 2001). In addition, thalamocortical synapses can show profound *short-term depression* (Boudreau & Ferster, 2005; Castro-Alamancos & Oldford, 2002; Chung, Li, & Nelson, 2002; Gil, Connors, & Amitai, 1999), which limits the types of patterns of thalamic discharges that are likely to elicit cortical responses. Therefore, high-frequency tonic patterns of spiking activity, often assumed to be the "relay" mode of thalamic neurons, may be nearly completely suppressed at the level of the cortex, and other patterns of thalamic activity, such as bursts and pauses, may, in fact, be more suited to drive postsynaptic cortical activation.

The presence of bursting in thalamocortical neurons and synaptic depression in thalamocortical synapses opens the door to coding schemes beyond the simple gated relay model often referenced in the literature. Of particular interest is the micro-organization of circuits that engage GABAergic neurons in the thalamic reticular nucleus, which is closely associated with the thalamus and implicated in thalamic models of language (Crosson, 2013). The thalamic reticular nucleus receives input from thalamus and cortex, as well as the basal forebrain and amygdala (Asanuma & Porter, 1990; Zikopoulos & Barbas, 2012), and sends GABAergic projections to thalamocortical neurons. By hyperpolarizing thalamocortical cells, the thalamic reticular nucleus may modify cortical activation by: (i) diminishing spiking in thalamocortical neurons; (ii) introducing a pause in the thalamic spike train, resulting in relief of tonic depression at thalamocortical synapses; (iii) synchronizing populations of thalamic cells; and/or (iv) inducing low-threshold bursting behavior, which has been shown to be a highly effective mechanism to drive postsynaptic activation

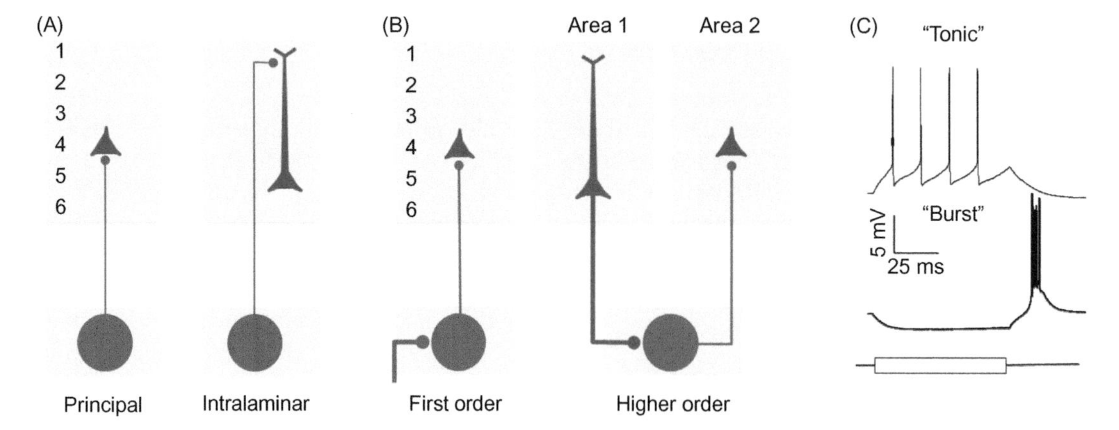

FIGURE 9.6 Models of thalamic organization. (A) Principal nuclei primarily project to middle cortical layers while intralaminar nuclei project to layer 1. (B) First-order nuclei receive their driving input from the sensory periphery, whereas higher-order thalamic nuclei receive their input from layer 5 of cortex, and relay that input to other areas of cortex. (C) An example of a neuron in tonic mode (top) and burst mode (bottom). *Data from Llano laboratory.*

(Denning & Reinagel, 2005; Lisman, 1997; Mukherjee & Kaplan, 1995; Person & Perkel, 2005; Reinagel, Godwin, Sherman, & Koch, 1999; Smith, Cox, Sherman, & Rinzel, 2000; Swadlow & Gusev, 2001). Therefore, the influence of the thalamic reticular nucleus may be an increase or decrease of cortical activation, depending on the ongoing activity in thalamic cells and the relative timing of inputs from the thalamic reticular nucleus and other structures. These data indicate that simple linear models of thalamic function are unlikely to capture the range of influences that the thalamus may have on cortical function.

9.6 MODELS OF THALAMUS AND LANGUAGE

Two influential models of thalamic function and language are considered here in more detail: that of Ojemann (Johnson & Ojemann, 2000, updated in Hebb & Ojemann, 2013) and Crosson (1999) (updated in Crosson, 2013). George Ojemann and colleagues have attributed a "specific alerting response" to the thalamus, generated by ventrolateral thalamus and pulvinar, which gates the entry of language information to the peri-Sylvian regions. The specific alerting response is also required for the retrieval of specific lexical information, which is embedded into peri-Sylvian circuits. This descriptive model is based on four types of evidence driven by Ojemann's body of work using deep brain electrical stimulation in human subjects. First, as described, naming deficits occur during electrical stimulation of ventrolateral thalamus and pulvinar (Ojemann et al., 1968; Ojemann & Ward, 1971). These errors are typically misnaming errors rather than omissions or speech arrest. Second, stimulation of the ventrolateral nucleus during the presentation of verbal material enhanced the ability to recall this material, but stimulation during recall of previously learned material impaired this process (Ojemann, Blick, & Ward, 1971). Third, stimulation of the left ventrolateral nucleus during the presentation of nonverbal information impaired the retention of this information. This suggests that the enhancement effects seen with left ventrolateral nucleus stimulation are restricted to verbal information. Fourth, silent naming induced desynchronization in the electrocorticogram across peri-Sylvian structures known to be important for naming (Ojemann, Fried, & Lettich, 1989). In this context, it is worth noting that Slotnick et al. noted a decrease in low-frequency power during a semantic recall task across multiple cortical sites that was correlated with a decrease in low-frequency power in the thalamus, bilaterally (Slotnick, Moo, Kraut, Lesser, & Hart, 2002). The four points described led Ojemann to speculate that specific left thalamic areas (most likely ventrolateral nucleus and pulvinar) selectively engage language cortex to support the retrieval of lexical information (Johnson & Ojemann, 2000). The most recent update of the model (Hebb & Ojemann, 2013) offers speculation about regional specialization and circuit-based mechanisms such that the ventral tier nuclei (such as the ventrolateral nucleus) and intralaminar nuclei modulate the specific alerting response either by recruiting striatal networks or via intralaminar thalamocortical or thalamoreticular projections. They also speculate that the pulvinar may synchronize oscillations of related cortical regions representing different features of an object to be named, a hypothesis for which there is recent support from the literature involving monkeys (Saalmann, Pinsk, Wang, Li, & Kastner, 2012).

Bruce Crosson and colleagues have proposed several potential models of the role of the thalamus in language (Crosson, 2013), and these have been derived primarily by the detailed study of patients with thalamic damage and aphasia. In the most recent instantiation, Crosson and colleagues hypothesized that the language deficits observed in their patients were caused by a failure to activate cortical circuits needed to retrieve lexical information from semantic representations. Further, they proposed that the failure was due to a lesion in the fronto-inferior thalamic peduncle-nucleus reticularis-centré-median thalamus system. This is based on the idea that the frontal cortex can gate the transmission of information through the thalamus to other areas of cortex (Brunia, 1993; Yingling & Skinner, 1975). This model focuses on the intralaminar nuclei (centré-median nucleus) rather than the ventrolateral nucleus and pulvinar, based on several arguments. They claim that there is a lack of anatomical specificity in electrical stimulation studies of ventrolateral nucleus and pulvinar due to spread of current to adjacent nuclei or fibers of passage, and that infarctions may have encroached on neighboring structures (such as the inferior thalamic peduncle, centré-median, or their connections). In addition, they note that their patients with strokes in apparently disparate areas within the thalamus have in common the disruption of the inferior thalamic peduncle-nucleus reticularis-centré-median thalamus system. They further argue that the intralaminar system, by virtue of its unique anatomical and physiological properties, is more suited to serve an attentional language function (Nadeau & Crosson, 1997).

This attentional model is derived from the cardinal findings that their patients with thalamic aphasia had particular difficulties with retrieval of lexical information from semantic information. They have argued that this process is more dependent on attention than is pure lexical processing or pure semantic processing. This postulate is supported by data from Crosson's group, which showed that incorporation of a lexical-semantic task (semantic associate generation) was far more distracting than a lexical-lexical task (word reading) in a modified Posner paradigm (Crosson, 1999). A more specific form of attention, such as selective engagement, may also be utilized during these tasks. Specifically, they have proposed that the role of this type of selective engagement of cortex is to enhance differences in the activation level of semantically related lexical items. Failure of selective engagement increases the likelihood that inappropriate lexical items will be selected.

Most recently, Crosson has updated this model by attempting to unify the multiple different processing streams that exist in thalamocortical and thalamocortical networks. He proposed four different circuits responsible for different aspects of language function and used the process of naming a hammer to illustrate these mechanisms. First, for example, to generate the name "hammer," Broca's area exerts control over disparate areas of cortex responsible for the semantic representation of a hammer (e.g., fusiform gyrus for the visual form of the hammer and inferior parietal lobule for motor sequences). This control occurs via layer 6 corticothalamic projections to the thalamic reticular nucleus, which, in turn, controls activation of pulvino-cortical circuits, potentially by inducing gamma oscillations (Figure 9.7A). This is a modification of the specific alerting response mentioned, and the suggestion that pulvinar may conjoin disparate areas of cortex using oscillatory assemblies is supported by some recent experimental data (Saalmann et al., 2012). It should also be mentioned that this type of frontal cortex-thalamic organization has long been speculated to be part of a mechanism of frontal cortical control over widespread cortical networks (Brunia, 1993; Skinner & Yingling, 1977) and is consistent with recent observations of the existence of prefrontal cortical circuits that target portions of the thalamic reticular nucleus that influence parts of the thalamus that do not necessarily send a return projection to the prefrontal cortex (Zikopoulos & Barbas, 2006).

Once appropriate areas of the cortex are activated via the alerting response described, a cortico-thalamocortical route may transfer semantic feature information from these areas to the posterior parietal cortex (Figure 9.7B). There is some support for the existence of cortico-thalamocortical transmission in animal models (Llano & Sherman, 2008; Theyel et al., 2010), although all of the current work to date has focused on sensory processing within a single modality. A third mechanism involves the activation of modulator-type corticothalamic projections, which have been shown to augment and potentially sharpen thalamic activations within sensory systems (Alitto & Usrey, 2003; Sillito, Cudeiro, & Jones, 2006), and may facilitate word retrieval by activating particular regions of the thalamus (Figure 9.7C). Finally, word selection, analogous to movement selection, may occur via pathways that involve presupplementary motor areas that project to basal ganglia circuits that project to the ventral anterior nucleus (Figure 9.7D). This may diminish the chance of errors in lexical selection. For further discussion of the potential role of the basal ganglia in language, please see Chapter 8 in this volume on the basal ganglia.

This multi-level model of thalamic function may help to explain: (i) the number of thalamic nuclei implicated in thalamic aphasia; (ii) the widespread cortical metabolic impact of restricted thalamic lesions (Figure 9.4); and (iii) the convergence of multiple thalamic nuclei to individual cortical areas (Hackett et al., 1998; Morel et al., 2005), such that different thalamic subnuclei may play distinct roles in cortical processing.

9.7 SUMMARY AND CONCLUSIONS

The clinical literature demonstrates that lesions of the dominant thalamus can cause dysnomia that is frequently coupled with perseverations and fluctuations in performance related to arousal. The dysnomia is likely driven by a deficit in lexical selection based on semantic information. Anatomically, the clinical and the imaging literature have not consistently pointed to a particular responsible thalamic nucleus, although certain nuclei, such as the ventrolateral nucleus, the centré-median nucleus, and medial pulvinar are frequently implicated. Review of previous models of large-scale thalamocortical organization suggests that multiple mechanisms may underlie thalamic involvement in language. Specifically, the thalamus may play a key role in arousal of specific cortical networks necessary for language function and may use the ubiquitous layer 6 corticothalamic projections to refine these representations. Further, cortico-thalamocortical communication may be required to route information represented in different regions of the cortex to the posterior parietal cortex. Finally, lexical selection may take place via pallidothalamic circuits. This type of multifunction model of the thalamus helps to reconcile some of the disparate clinical and imaging data, which point to lesions or activations in virtually all thalamic nuclei in aphasic patients and imaging studies, respectively.

Many questions remain, however, and many of these have to do with fundamental questions regarding the physiologic organization of thalamic functions. For example, core features of thalamic physiology, such as the role of transitions between tonic and burst modes, which will have substantially different effects on their postsynaptic targets, are not specified in these models. Further, the role of convergent thalamic inputs, particularly those to layer 1 and layer 4, speculated to play a key role in the ability of the thalamus to activate cortex (Llinas, Leznik, & Urbano, 2002; Purpura & Schiff, 1997)

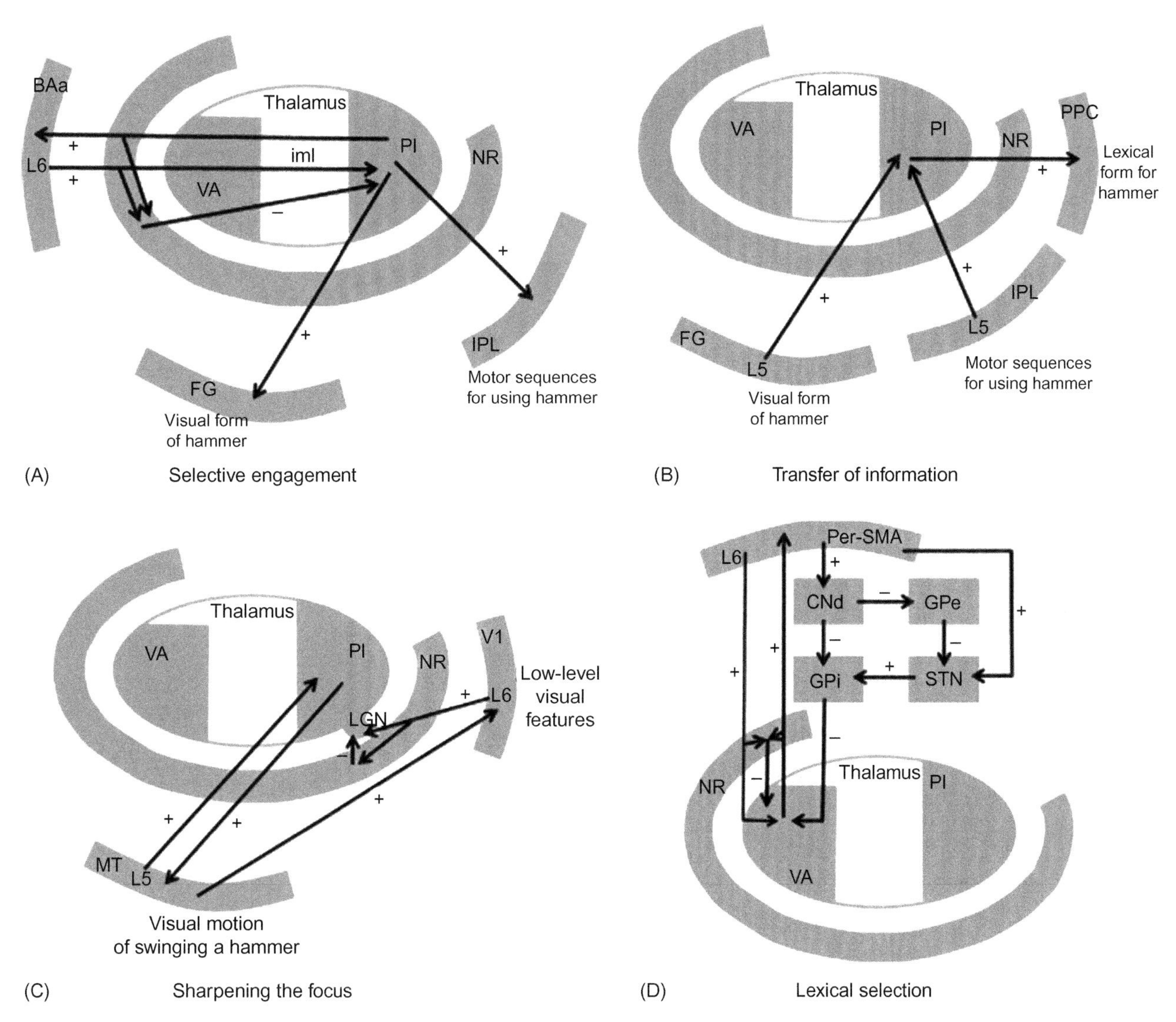

FIGURE 9.7 Four mechanisms proposed by Crosson (2013) to explain thalamic involvement in language. (A) A selective engagement model whereby corticothalamic projections from Broca's area to the thalamic reticular nucleus and/or pulvinar initiate more widespread activation of cortical areas involved in the representation of a particular object to be named. (B) A transfer of information model whereby information from cortical areas representing features of the object to be named is routed to the inferior parietal lobe via cortico-thalamocortical routes. (C) Layer 6 corticothalamic neurons sharpen the focus of activation in the thalamus in its primary sensory representation (in this case, the visual cortex). (D) Lexical selection is performed using basal ganglia circuits involving the direct, indirect, and hyper-direct loops from pre-supplementary motor cortical areas via the basal ganglia and the ventral anterior nucleus. Sharpening may also be possible here via layer 6 projections to the thalamic reticular nucleus. *From Crosson (2013) with permission from the publishers. BAa = anterior Broca's area, GPi = internal globus pallidus, GPe = external globus pallidus, CNd = dorsal caudate nucleus, FG = fusiform gyrus, iml = internal medullary lamina, IPL = inferior parietal lobule, L6 = cortical layer 6, LGN = lateral geniculate nucleus, NR nucleus reticularis, Pl = pulvinar, PPC = posterior perisylvian cortex, STN = subthalamic nucleus, V1 = primary visual cortex, VA = ventral anterior nucleus.*

have not explicitly been tied to language function, although such convergence models could support the "specific alerting system," as conceived of by Ojemann and colleagues (Hebb & Ojemann, 2013; Johnson & Ojemann, 2000). Finally, the thalamic reticular nucleus, whose function remains quite mysterious, given its relative inaccessibility to experimentation, has been speculated to be involved in language (Crosson, 2013; Nadeau & Crosson, 1997), but basic questions about this structure remain unanswered. For example, previous models postulate that the thalamic reticular nucleus can impact the flow of information between thalamic nuclei. Although this possibility is intriguing and some data do support this (Crabtree & Isaac, 2002; Kimura, Imbe, Donishi, & Tamai, 2007; Pinault & Deschênes, 1998), to date there is a very poor understanding of how such "open-loop" networks are organized, or even if they cause inhibition or paradoxical excitation (via low-threshold bursting) in their thalamic targets.

Therefore, to make progress in this area, it will be necessary to develop a better basic understanding of the role of the thalamus in long-range cortical processing and, specifically, how the individual circuit elements found in the thalamus alter patterns of activation found in the cortex. In addition, more work will be needed using human subjects to better delineate the thalamic nuclei involved in task-based activations and in structural imaging studies to more precisely define the connectivity of thalamic nuclei with functionally defined regions of cortex. Finally, longitudinal functional and structural imaging studies of patients recovering from thalamic lesions will help us to understand the degree to which alternate or redundant pathways may contribute to cortical language processing. Therefore, the basic and the clinical literatures have served complementary roles in further developing our understanding of the thalamus and language to this point, and both will continue to shape our understanding of this critical yet poorly understood structure.

Acknowledgments

The author thanks Joe Beatty, Bruce Crosson, Anthony Dick, Uri Hasson, Iraklis Petrof, and Steve Small for their contributions to this and earlier versions of this manuscript.

References

Abramson, B. P., & Chalupa, L. M. (1985). The laminar distribution of cortical connections with the tecto- and cortico-recipient zones in the cat's lateral posterior nucleus. *Neuroscience*, *15*(1), 81–95.

Alitto, H. J., & Usrey, W. M. (2003). Corticothalamic feedback and sensory processing. *Current Opinion in Neurobiology*, *13*(4), 440–445.

Archer, C., Ilinsky, I., Goldfader, P., & Smith, K. (1981). Case report. Aphasia in thalamic stroke: CT stereotactic localization. *Journal of Computer Assisted Tomography*, *5*(3), 427–432.

Asanuma, C., & Porter, L. L. (1990). Light and electron microscopic evidence for a GABAergic projection from the caudal basal forebrain to the thalamic reticular nucleus in rats. *The Journal of Comparative Neurology*, *302*(1), 159–172. Available from: http://dx.doi.org/10.1002/cne.903020112.

Bell, D. S. (1968). Speech functions of the thalamus inferred from the effects of thalamotomy. *Brain*, *91*(4), 619–638. Available from: http://dx.doi.org/10.1093/brain/91.4.619.

Bhatnagar, S., & Mandybur, G. (2005). Effects of intralaminar thalamic stimulation on language functions. *Brain and Language*, *92*, 1–11.

Bogousslavsky, J., Miklossy, J., Deruaz, J., Regli, F., & Assal, G. (1986). Unilateral left paramedian infarction of thalamus and midbrain: A clinico-pathological study. *Journal of Neurology, Neurosurgery and Psychiatry*, *49*(6), 686–694.

Bogousslavsky, J., Regli, F., & Assal, G. (1986). The syndrome of unilateral tuberothalamic artery territory infarction. *Stroke*, *17*(3), 434–441.

Boudreau, C. E., & Ferster, D. (2005). Short-term depression in thalamocortical synapses of cat primary visual cortex. *The Journal of Neuroscience*, *25*(31), 7179–7190.

Brunia, C. (1993). Waiting in readiness: Gating in attention and motor preparation. *Psychophysiology*, *30*(4), 327–339.

Bruyn, R. (1989). Thalamic aphasia. A conceptional critique. *Journal of Neurology*, *236*, 21–25.

Cappa, S. (1986). Aphasia does not always follow left thalamic hemorrhage: A study of five negative cases. *Cortex*, *22*(4), 639–647.

Carrera, E., & Bogousslavsky, J. (2006). The thalamus and behavior effects of anatomically distinct strokes. *Neurology*, *66*(12), 1817–1823.

Castro-Alamancos, M. A., & Oldford, E. (2002). Cortical sensory suppression during arousal is due to the activity-dependent depression of thalamocortical synapses. *The Journal of Physiology*, *541*(1), 319–331.

Chung, S., Li, X., & Nelson, S. B. (2002). Short-term depression at thalamocortical synapses contributes to rapid adaptation of cortical sensory responses *in vivo*. *Neuron*, *34*(3), 437–446.

Cohen, J., Gelfer, C., & Sweet, R. (1980). Thalamic infarction producing aphasia. *Mount Sinai Journal of Medicine*, *47*(4), 398–404.

Crabtree, J. W., & Isaac, J. T. R. (2002). New intrathalamic pathways allowing modality-related and cross-modality switching in the dorsal thalamus. *The Journal of Neuroscience*, *22*(19), 8754–8761.

Crosson, B. (1999). Subcortical mechanisms in language: Lexical-semantic mechanisms and the thalamus. *Brain and Cognition*, *40*(2), 414–438.

Crosson, B. (2013). Thalamic mechanisms in language: A reconsideration based on recent findings and concepts. *Brain and Language*, *126*(1), 73–88.

Crosson, B., Parker, J., Kim, A., Warren, R., Kepes, J., & Tully, R. (1986). A case of thalamic aphasia with postmortem verification. *Brain and Language*, *29*, 301–314.

Darley, F., Brown, J., & Swenson, W. (1975). Language changes after neurosurgery for Parkinsonism. *Brain and Language*, *2*(1), 65–69.

Demeurisse, G., Derouck, M., Coekaerts, M., Deltenre, P., Van Nechel, C., Demol, O., et al. (1979). Study of two cases of aphasia by infarction of the left thalamus, without cortical lesion. *Acta Neurologica Belgica*, *79*, 450–459.

Denning, K. S., & Reinagel, P. (2005). Visual control of burst priming in the anesthetized lateral geniculate nucleus. *The Journal of Neuroscience*, *25*(14), 3531–3538. Available from: http://dx.doi.org/10.1523/jneurosci.4417-04.2005.

De Witte, L., Brouns, R., Kavadias, D., Engelborghs, S., De Deyn, P. P., & Mariën, P. (2011). Cognitive, affective and behavioural disturbances following vascular thalamic lesions: A review. *Cortex: A Journal Devoted to the Study of the Nervous System and Behavior*, *47*(3), 273–319.

Ebert, A., Vinz, B., Görtler, M., Wallesch, C., & Herrmann, M. (1999). Is there a syndrome of tuberothalamic artery infarction? A case report and critical review. *Journal of Clinical and Experimenal Neuropsychology*, *21*(3), 397–411.

Ford, A., Triplett, W., Sudhyadhom, S., Gullett, J., McGregor, K., FitzGerald, D., et al. (2012). Thalamus and its connections with Broca's area: A diffusion-MRI tractography study. *Society for Neuroscience Abstracts*, #526.09.

Ford, A. A., Triplett, W., Sudhyadhom, A., Gullett, J., McGregor, K., FitzGerald, D. B., et al. (2013). Broca's area and its striatal and thalamic connections: A diffusion-MRI tractography study. *Frontiers in Neuroanatomy*, *7*, 1–12.

Gil, Z., Connors, B. W., & Amitai, Y. (1999). Efficacy of thalamocortical and intracortical synaptic connections: Quanta, innervation, and reliability. *Neuron*, *23*(2), 385–397.

Gil-da-Costa, R., Martin, A., Lopes, M. A., Munoz, M., Fritz, J. B., & Braun, A. R. (2006). Species-specific calls activate homologs of Broca's and Wernicke's areas in the macaque. *Nature Neuroscience*, *9*(8), 1064–1070.

Graff-Radford, N., Eslinger, P., Damasio, A., & Yamada, T. (1984). Nonhemorrhagic infarction of the thalamus: Behavioral, anatomic, and physiologic correlates. *Neurology*, *34*, 14–23.

Guimaraes, A., Melcher, J., Talavage, T., Baker, J., Ledden, P., Rosen, B., et al. (1998). Imaging subcortical auditory activity in humans. *Human Brain Mapping*, *6*(1), 33–41.

Hackett, T., Stepniewska, I., & Kaas, J. (1998). Thalamocortical connections of the parabelt auditory cortex in macaque monkeys. *The Journal of Comparative Neurology*, *400*(2), 271–286.

Hassler, R. (1959). Introduction to stereotaxis with an Atlas of the human brain. *Anatomy of the Thalamus*, 230–290.

Hebb, A. O., & Ojemann, G. A. (2013). The thalamus and language revisited. *Brain and Language*, *126*(1), 99–108.

Hirai, T., & Jones, E. (1989). A new parcellation of the human thalamus on the basis of histochemical staining. *Brain Research Reviews*, *14*(1), 1–34.

Hoffmann, M. (2012). Thalamic semantic paralexia. *Neurology International*, *4*(1), 24–25.

Johnson, M. D., & Ojemann, G. A. (2000). The role of the human thalamus in language and memory: Evidence from electrophysiological studies. *Brain and Cognition*, *42*(2), 218–230.

Jones, E. G. (2007). *The thalamus*. Cambridge, UK: Cambridge University Press.

Karussis, D., Leker, R. R., & Abramsky, O. (2000). Cognitive dysfunction following thalamic stroke: A study of 16 cases and review of the literature. *Journal of the Neurological Sciences*, *172*(1), 25–29.

Kimura, A., Imbe, H., Donishi, T., & Tamai, Y. (2007). Axonal projections of single auditory neurons in the thalamic reticular nucleus: Implications for tonotopy-related gating function and cross-modal modulation. *European Journal of Neuroscience*, *26*(12), 3524–3535. Available from: http://dx.doi.org/10.1111/j.1460-9568.2007.05925.x.

Klostermann, F., Krugel, L. K., & Ehlen, F. (2013). Functional roles of the thalamus for language capacities. *Frontiers in Systems Neuroscience*, *7*, 1–8.

Lanna, M. E. d. O., Alves, C. E. O., Sudo, F. K., Alves, G., Valente, L., Moreira, D. M., et al. (2012). Cognitive disconnective syndrome by single strategic strokes in vascular dementia. *Journal of the Neurological Sciences*, *322*, 176–183.

Levin, N., Ben-Hur, T., Biran, I., & Wertman, E. (2005). Category specific dysnomia after thalamic infarction: A case–control study. *Neuropsychologia*, *43*, 1385–1390.

Lhermitte, F. (1984). Language disorders and their relationship to thalamic lesions. *Advances in Neurology*, *42*, 99–113.

Lisman, J. E. (1997). Bursts as a unit of neural information: Making unreliable synapses reliable. *Trends in Neurosciences*, *20*(1), 38–43.

Llano, D. A. (2013). Functional imaging of the thalamus in language. *Brain and Language*, *126*(1), 62–72. Available from: http://dx.doi.org/10.1016/j.bandl.2012.06.004.

Llano, D. A., & Sherman, S. M. (2008). Evidence for nonreciprocal organization of the mouse auditory thalamocortical-corticothalamic projection systems. *The Journal of Comparative Neurology*, *507*(2), 1209–1227.

Llano, D. A., Theyel, B. B., Mallik, A. K., Sherman, S. M., & Issa, N. P. (2009). Rapid and sensitive mapping of long-range connections in vitro using flavoprotein autofluorescence imaging combined with laser photostimulation. *Journal of Neurophysiology*, *101*(6), 3325–3340. Available from: http://dx.doi.org/10.1152/jn.91291.2008.

Llinas, R. R., Leznik, E., & Urbano, F. J. (2002). Temporal binding via cortical coincidence detection of specific and nonspecific thalamocortical inputs: A voltage-dependent dye-imaging study in mouse brain slices. *Proceedings of the National Academy of Sciences*, *99*(1), 449–454. Available from: http://dx.doi.org/10.1073/pnas.012604899.

Luria, A. R. (1977). On quasi-aphasic speech disturbances in lesions of the deep structures of the brain. *Brain and Language*, *4*(3), 432–459.

Margolin, E., Hanifan, D., Berger, M. K., Ahmad, O. R., Trobe, J. D., & Gebarski, S. S. (2008). Skew deviation as the initial manifestation of left paramedian thalamic infarction. *Journal of Neuro-Ophthalmology*, *28*(4), 283–286. Available from: http://dx.doi.org/10.1097/WNO.0b013e318183cb79.

McFarling, D., Rothi, L., & Heilman, K. (1982). Transcortical aphasia from ischaemic infarcts of the thalamus: A report of two cases. *Journal of Neurology, Neurosurgery and Psychiatry*, *45*, 107–112.

Mohr, J., Pessin, M., Finkelstein, S., Funkenstein, H., Duncan, G., & Davis, K. (1978). Broca aphasia: Pathologic and clinical. *Neurology*, *28*(4), 311–324.

Mohr, J., Watters, W., & Duncan, G. (1975). Thalamic hemorrhage and aphasia. *Brain and Language*, *2*, 3–17.

Morandi, X., Brassier, G., Darnault, P., Mercier, P., Scarabin, J., & Duval, J. (1996). Microsurgical anatomy of the anterior choroidal artery. *Surgical and Radiologic Anatomy*, *18*, 275–280.

Morel, A. (2007). *Stereotactic Atlas of the human thalamus and basal ganglia*. Informa Health Care.

Morel, A., Liu, J., Wannier, T., Jeanmonod, D., & Rouiller, E. (2005). Divergence and convergence of thalamocortical projections to premotor and supplementary motor cortex: A multiple tracing study in the macaque monkey. *European Journal of Neuroscience*, *21*(4), 1007–1029.

Mukherjee, P., & Kaplan, E. (1995). Dynamics of neurons in the cat lateral geniculate nucleus: *In vivo* electrophysiology and computational modeling. *Journal of Neurophysiology*, *74*(3), 1222–1243.

Nadeau, S., & Crosson, B. (1997). Subcortical aphasia. *Brain and Language*, *58*(3), 355–402.

Neau, J. P., & Bogousslavsky, J. (1996). The syndrome of posterior choroidal artery territory infarction. *Annals of Neurology*, *39*, 779–788.

Ojemann, G. (1991). Cortical organization of language. *Journal of Neuroscience*, *11*(8), 2281–2287.

Ojemann, G. A., Blick, K. I., & Ward, A. A., Jr. (1971). Improvement and disturbance of short-term verbal memory with human ventrolateral thalamic stimulation. *Brain*, *94*(2), 225–240. Available from: http://dx.doi.org/10.1093/brain/94.2.225.

Ojemann, G. A., Fedio, P., & van Buren, J. M. (1968). Anomia from pulvinar and subcortical parietal stimulation. *Brain*, *91*(1), 99–116. Available from: http://dx.doi.org/10.1093/brain/91.1.99.

Ojemann, G. A., Fried, I., & Lettich, E. (1989). Electrocorticographic (ECoG) correlates of language. I. Desynchronization in temporal language cortex during object naming. *Electroencephalography and Clinical Neurophysiology*, *73*, 453–463.

Ojemann, G. A., & Ward, A. A. J. (1971). Speech represenation in the ventrolateral thalamus. *Brain*, *94*(4), 669–680. Available from: http://dx.doi.org/10.1093/brain/94.4.669.

Penfield, W., & Roberts, L. (1959). *Speech and brain-mechanisms*. Princeton University Press.

Perren, F., Clarke, S., & Bogousslavsky, J. (2005). The syndrome of combined polar and paramedian thalamic infarction. *Archives of Neurology*, *62*(8), 1212–1216. Available from: http://dx.doi.org/10.1001/archneur.62.8.1212.

Person, A. L., & Perkel, D. J. (2005). Unitary IPSPs drive precise thalamic spiking in a circuit required for learning. *Neuron*, *46*(1), 129–140. Available from: http://dx.doi.org/10.1016/j.neuron.2004.12.057.

Pinault, D., & Deschênes, M. (1998). Anatomical evidence for a mechanism of lateral inhibition in the rat thalamus. *European Journal of Neuroscience*, *10*(11), 3462–3469.

Puel, M., Demonet, J. F., Cardebat, D., Berry, I., Celsis, P., Marc-Vergnes, J. P, et al. (1992). Three topographical types of thalamic aphasia: A neurolingusitic, MRI, and SPECT study. In G. Valla, S. F. Cappa, & C. W. Wallesch (Eds.), *Neuropsychological disorders associated with subcortical lesions*. Oxford University Press.

Purpura, K. P., & Schiff, N. D. (1997). The thalamic intralaminar nuclei: A role in visual awareness. *Neuroscientist*, *3*, 8–15.

Radanovic, M., Azambuja, M., Mansur, L., Porto, C., & Scaff, M. (2003). Thalamus and language: Interface with attention, memory and executive functions. *Arquivos de neuro-psiquiatria*, *61*(1), 34–42.

Radanovic, M., & Scaff, M. (2003). Speech and language disturbances due to subcortical lesions. *Brain and Language*, *84*(3), 337–352.

Raymer, A., Moberg, P., Crosson, B., Nadeau, S., & Rothi, L. (1997). Lexical-semantic deficits in two patients with dominant thalamic infarction. *Neuropsychologia*, *35*(2), 211–219.

Reinagel, P., Godwin, D., Sherman, S. M., & Koch, C. (1999). Encoding of visual information by LGN bursts. *Journal of Neurophysiology*, *81*(5), 2558–2569.

Riklan, M., & Cooper, I. S. (1975). Psychometric studies of verbal functions following thalamic lesions in humans. *Brain and Language*, *2*, 45–64.

Rockland, K. S., Andresen, J., Cowie, R. J., & Robinson, D. L. (1999). Single axon analysis of pulvinocortical connections to several visual areas in the Macaque. *The Journal of Comparative Neurology*, *406*(2), 221–250.

Romanski, L. M., Giguere, M., Bates, J. F., & Goldman-Rakic, P. S. (1997). Topographic organization of medial pulvinar connections with the prefrontal cortex in the rhesus monkey. *The Journal of Comparative Neurology*, *379*(3), 313–332.

Saalmann, Y. B., Pinsk, M. A., Wang, L., Li, X., & Kastner, S. (2012). The pulvinar regulates information transmission between cortical areas based on attention demands. *Science*, *337*(6095), 753–756. Available from: http://dx.doi.org/10.1126/science.1223082.

Samra, K., Riklan, M., Levita, E., Zimmerman, J., Waltz, J. M., Bergmann, L., et al. (1969). Language and speech correlates of anatomically verified lesions in thalamic surgery for parkinsonism. *Journal of Speech, Language and Hearing Research*, *12*(3), 510.

Schmahmann, J., & Pandya, D. (1990). Anatomical investigation of projections from thalamus to posterior parietal cortex in the rhesus monkey: A WGA-HRP and fluorescent tracer study. *Journal of Comparative Neurology*, *295*(2), 299–326.

Segal, J. B., Williams, R., Kraut, M. A., & Hart, J., Jr. (2003). Semantic memory deficit with a left thalamic infarct. *Neurology*, *61*(2), 252–254.

Sherman, S. M. (2001). Tonic and burst firing: Dual modes of thalamocortical relay. *Trends in Neurosciences*, *24*(2), 122–126.

Sherman, S. M., & Guillery, R. (1998). On the actions that one nerve cell can have on another: Distinguishing "drivers" from "modulators". *Proceedings of the National Academy of Sciences*, *95*(12), 7121–7126.

Shim, Y., Kim, J., Shon, Y., Chung, Y., Ahn, K., & Yang, D. (2008). A serial study of regional cerebral blood flow deficits in patients with left anterior thalamic infarction: Anatomical and neuropsychological correlates. *Journal of the Neurological Sciences*, *266*(1), 84–91.

Sillito, A. M., Cudeiro, J., & Jones, H. E. (2006). Always returning: Feedback and sensory processing in visual cortex and thalamus. *Trends in Neurosciences*, *29*(6), 307–316.

Skinner, J. E., & Yingling, C. D. (1977). Reconsideration of the cerebral mechanisms underlying selective attention and slow potential shifts. Attention, voluntary contraction and event-related cerebral potentials. *Progress in Clinical Neurophysiology*, *1*, 30–69.

Slotnick, S. D., Moo, L. R., Kraut, M. A., Lesser, R. P., & Hart, J. (2002). Interactions between thalamic and cortical rhythms during semantic memory recall in human. *Proceedings of the National Academy of Sciences of the United States of America*, *99*(9), 6440–6443. Available from: http://dx.doi.org/10.1073/pnas.092514899.

Smith, G. D., Cox, C. L., Sherman, S. M., & Rinzel, J. (2000). Fourier analysis of sinusoidally driven thalamocortical relay neurons and a minimal integrate-and-fire-or-burst model. *Journal of Neurophysiology*, *83*(1), 588–610.

Swadlow, H., & Gusev, A. (2001). The impact of "bursting" thalamic impulses at a neocortical synapse. *Nature Neuroscience*, *4*(4), 402–408.

Szirmai, I., Vastagh, I., Szombathelyi, É., & Kamondi, A. (2002). Strategic infarcts of the thalamus in vascular dementia. *Journal of the Neurological Sciences*, *203–204*, 91–97.

Takahashi, S., Suzuki, M., Matsumoto, K., Ishii, K., Higano, S., Fukasawa, H., et al. (1994). Extent and location of cerebral infarcts on multiplanar MR images: Correlation with distribution of perforating arteries on cerebral angiograms and on cadaveric microangiograms. *American Journal of Roentgenology*, *163*(5), 1215–1222.

Theyel, B. B., Llano, D. A., & Sherman, S. M. (2010). The corticothalamocortical circuit drives higher-order cortex in the mouse. *Nature Neuroscience*, *13*(1), 84–88. Available from: http://dx.doi.org/10.1038/nn.2449. http://www.nature.com/neuro/journal/v13/n1/suppinfo/nn.2449_S1.html.

Van Buren, J. (1975). The question of thalamic participation in speech mechanisms. *Brain and Language*, *2*, 31–44.

Van Buren, J. M., & Borke, R. C. (1969). Alterations in speech and the pulvinar. A serial section sudy of cerebrothalamic relationships in cases of acquired speech disorders. *Brain*, *92*(2), 255–284. Available from: http://dx.doi.org/10.1093/brain/92.2.255.

Van der Werf, Y. D., Witter, M. P., & Groenewegen, H. J. (2002). The intralaminar and midline nuclei of the thalamus. Anatomical and functional evidence for participation in processes of arousal and awareness. *Brain Research Reviews*, *39*(2–3), 107–140.

Vilkki, J., & Laitinen, L. (1976). Effects of pulvinotomy and ventrolateral thalamotomy on some cognitive functions. *Neuropsychologia*, *14*(1), 67–78.

Wallesch, C. (1997). Sympomatology of subcortical aphasia. *Journal of Neurolinguistics*, *10*(4), 267–275.

Wallesch, C.-W., Kornhuber, H. H., Brunner, R. J., Kunz, T., Hollerbach, B., & Suger, G. (1983). Lesions of the basal ganglia, thalamus, and deep white matter: Differential effects on language functions. *Brain and Language*, *20*(2), 286–304.

Weisman, D., Hisama, F. M., Waxman, S. G., & Blumenfeld, H. (2003). Going deep to cut the link: Cortical disconnection syndrome caused by a thalamic lesion. *Neurology*, *60*(11), 1865–1866.

Whelan, B., Murdoch, B., Theodoros, D., Silburn, P., & Hall, B. (2002). A role for the dominant thalamus in language? A linguistic comparison of two cases subsequent to unilateral thalamotomy procedures in the dominant and non-dominant hemispheres. *Aphasiology*, *16*(12), 1213–1226.

Yingling, C., & Skinner, J. (1975). Regulation of unit activity in nucleus reticularis thalami by the mesencephalic reticular formation and the frontal granular cortex. *Electroencephalography and Clinical Neurophysiology*, *39*(6), 635–642.

Zikopoulos, B., & Barbas, H. (2006). Prefrontal projections to the thalamic reticular nucleus form a unique circuit for attentional mechanisms. *The Journal of Neuroscience*, *26*(28), 7348–7361. Available from: http://dx.doi.org/10.1523/jneurosci.5511-05.2006.

Zikopoulos, B., & Barbas, H. (2012). Pathways for emotions and attention converge on the thalamic reticular nucleus in primates. *The Journal of Neuroscience*, *32*(15), 5338–5350. Available from: http://dx.doi.org/10.1523/jneurosci.4793-11.2012.

CHAPTER

10

The Insular Cortex

Jessica D. Richardson and Julius Fridriksson

Department of Communication Sciences and Disorders, The University of South Carolina, Columbia, SC, USA

10.1 GROSS ANATOMY

Rudimentary sketches of the insular cortex appeared as early as the 1500s (Shelley & Trimble, 2004). The first notable description of the insula was published in 1796 by Johann-Christian Reil (Binder, Schaller, & Clusmann, 2007), who later named it "die Insel" in 1809 (Shelley & Trimble, 2004). Illustrations appeared in the first edition of *Gray's Anatomy* (1858), with the earliest detailed descriptions published by Korbinian Brodmann (1909), Binder et al. (2007), and Kurth et al. (2010). The insular cortex, also referred to as the "island (or isle) of Reil," "Reil's island," "insula Reili," and the "central lobe," is a discrete lobe in the cerebral cortex covered by considerable vasculature and the frontal, temporal, and parietal opercula (Afif & Mertens, 2010; Stephani, Vaca, Maciunas, Koubeissi, & Luders, 2011; Ture, Yasargil, Al-Mefty, & Yasargil, 1999). Popular illustrations of this hidden cortical island reveal it by locating the Sylvian fissure and pulling away the opercula to view the structure underneath (Figures 10.1 and 10.2).

The insula is divided into anterior and posterior portions by a central insular sulcus (CIS). The CIS is generally continuous with the central cerebral sulcus (Rolandic fissure), sharing an axis of development and orientation with the overlying cortex (Afif & Mertens, 2010; Varnavas & Grand, 1999). The insula is generally described as triangular in shape, with three peri-insular sulci (anterior/inferior/superior). Recently, a morphometric study performed using fresh human cadaver brain specimens revealed that, in all hemispheres dissected (10 left, 10 right), there were four peri-insular sulci (anterior/inferior/posterior/superior) separating a *trapezoidal* insula from surrounding structures (Afif & Mertens, 2010). There are eight sulci facing the underside of the operculum. Anterior to the CIS (moving outward from center) are the pre-CIS, anterior IS (AIS), and the anterior peri-IS (ApIS). Posterior to the CIS are the post-CIS and the posterior peri-IS (PpIS). The remaining two sulci are the inferior and superior boundaries to the insula (inferior and superior peri-IS [IpIS, SpIS]). There are six insular gyri, four in the anterior insula and two in the posterior insula. Located anterior to the CIS are the short posterior insular gyrus (sPIG), short middle IG (sMIG), and short anterior IG (sAIG). There is a smaller transverse insular gyrus (TIG) below the apex of the insula, readily viewed in a majority of brain specimens (Ture et al., 1999). In approximately half of the brain specimens, there is an accessory gyrus on the ventral anterior insular cortex (it is underdeveloped or absent in the other half; Ture et al., 1999). Located posterior to the CIS are the long anterior insular gyrus (lAIG) and the long posterior IG (lPIG) (Figure 10.3).

10.2 CYTOARCHITECTURE

Brodmann (1909) first identified two distinct areas within the insula—an anterior agranular area and a posterior granular area (Kurth et al., 2010). Since then, cytoarchitectural parcellations ranging from 3 to 30+ subdivisions have been proposed (for review, see Kurth et al., 2010; Nieuwenhuys, 2012). Presently, the most popular scheme for humans involves three divisions following a posterosuperior to anteroinferior gradient of decreasing granularity—posterior insula-granular (Ig), intermediate insula-dysgranular (Idg), and anterior insula-agranular (Iag) (Figure 10.3). Several additional subdivisions have been independently proposed with some convincing and/or

Neurobiology of Language. DOI: http://dx.doi.org/10.1016/B978-0-12-407794-2.00010-9

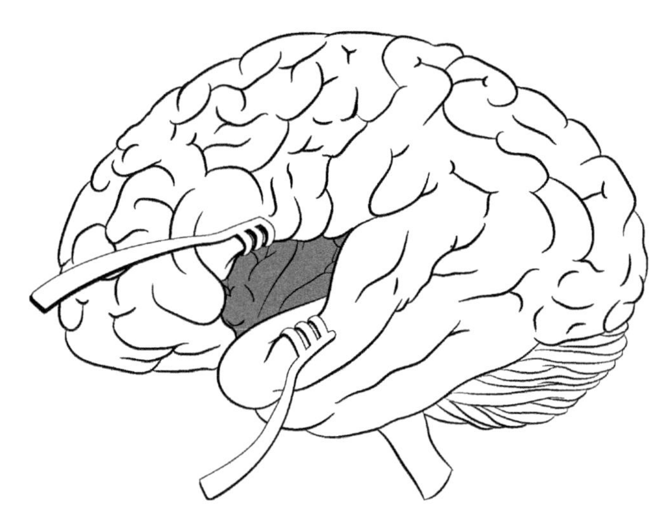

FIGURE 10.1 The hidden insula is revealed by locating the Sylvian fissure and separating the overlying opercula. Artist: C. Vincent Collier.

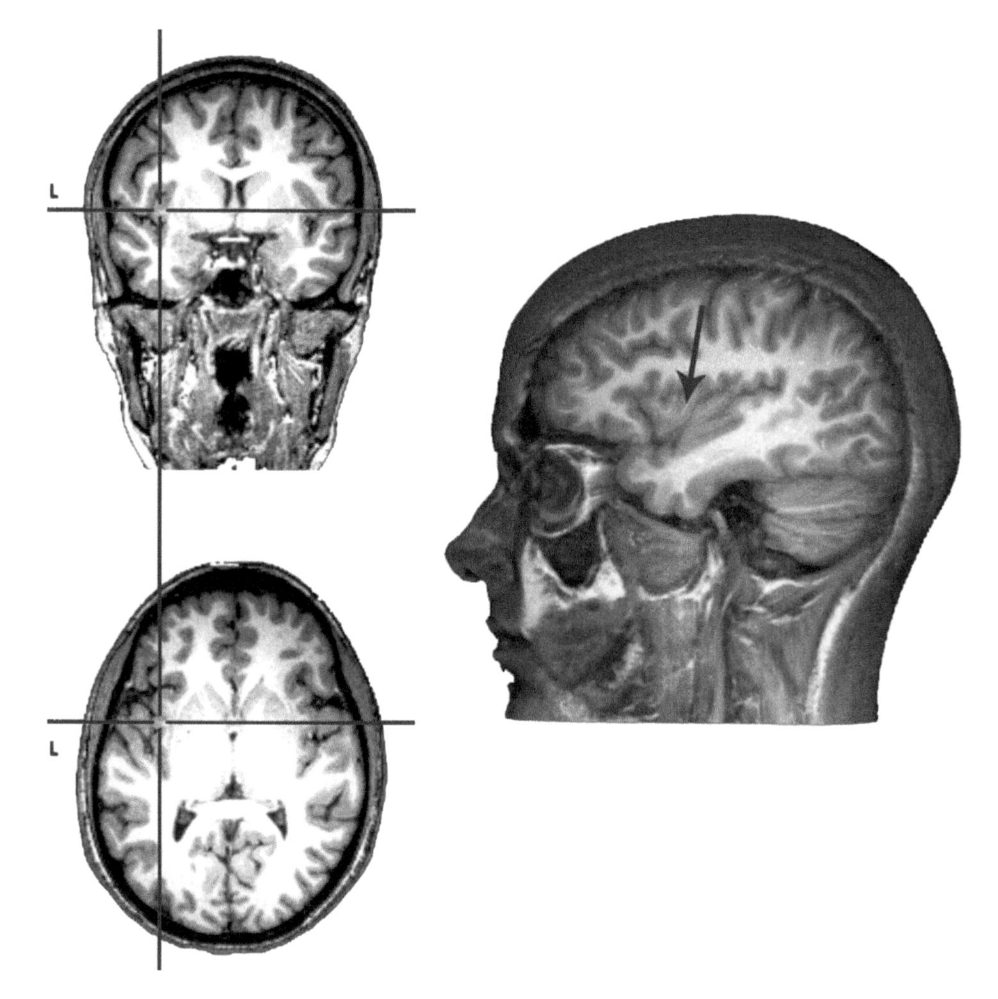

FIGURE 10.2 High-resolution T1 MRI with left hemisphere insula featured in coronal (upper left), axial (lower left), and sagittal (right) planes.

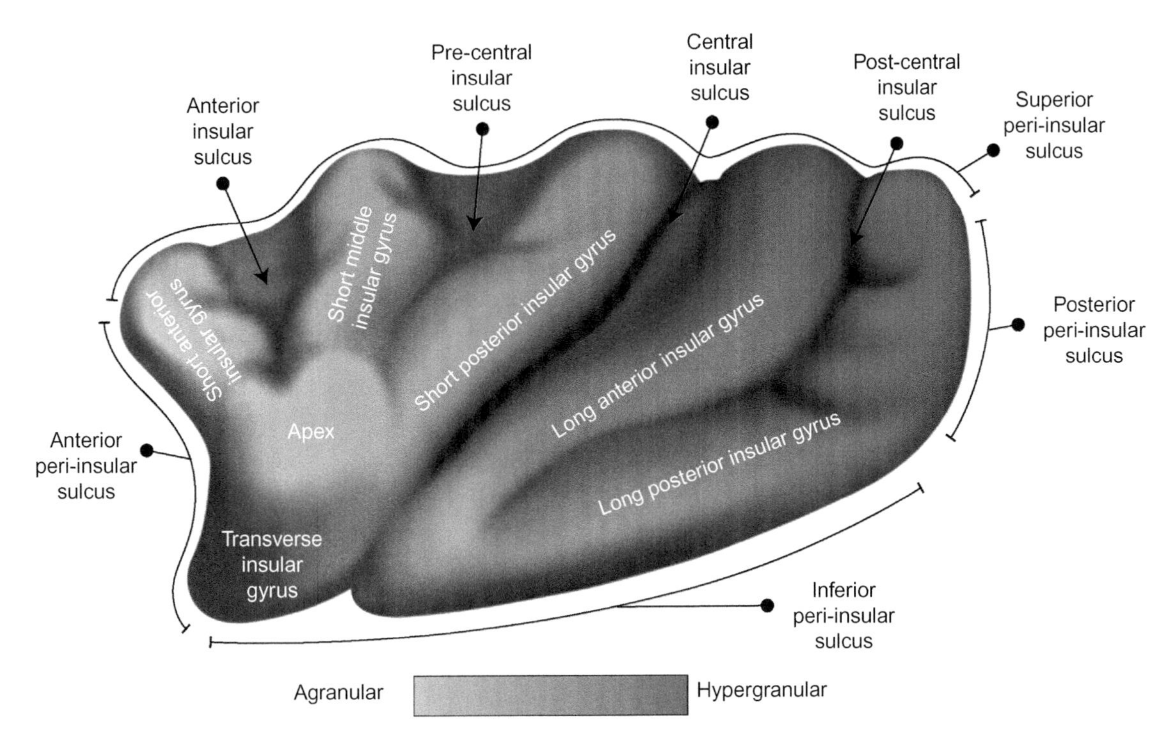

FIGURE 10.3 The left hemisphere insula is shown with sulci and gyri labeled. The color gradient represents the posterosuperior to anteroinferior gradient of decreasing granularity. Artist: C. Vincent Collier.

convergent results (despite differences in boundaries and terminology):

- Two Ig areas (posterior), with the most extreme posterosuperior section described as "hypergranular"
  - G and Ig (Morel, Gallay, Baechler, Wyss, & Gallay, 2013)
  - Alternatively, Ig1 and Ig2 (Kurth et al., 2010)
- Three Idg areas (intermediate), with progressively decreasing granule cell layer representation
  - Id3, Id2, and Id1, from greatest to least layer II/IV representation (Kurth et al., 2010; Morel et al., 2013)
- Two Iag areas (anterior), with no granule cell layer representation and the presence of von Economo neurons (VENs; Box 10.1)
  - Ia2 and Ia1 (Morel et al., 2013)
  - Or, agranular anterior zone (AA) and insular limbic cortex (ILC) (Nieuwenhuys, 2012)
  - Or, "frontoinsular cortex" (referring to the section of anterior insula heavily populated with VENs) and agranular insular cortex (Allman et al., 2011; Bauernfeind et al., 2013; Butti, Santos, Uppal, & Hof, 2013; Cauda et al., 2013).

10.3 VASCULATURE

The insula receives its blood supply from the middle cerebral artery (MCA), which is generally divided into four segments (Gibo, Carver, Rhoton, Lenkey, & Mitchell, 1981; Tanrovier, Rhoton, Kawashima, Ulm, & Yasuda, 2004; Tatu, Moulin, Bogousslavsky, & Duvernoy, 1998; Ture, Yasargil, Al-Mefty, & Yasargil, 2000):

M1: sphenoidal (the main trunk, before bi- or trifurcation; Kahilogullari, Ugur, Comert, Tekdemir, & Kanpolat, 2012; Ture et al., 2000)
M2: insular
M3: opercular
M4: cortical (which may be further divided into separate "parasylvian" and "terminal" segments; Ture et al., 2000).

The insula is perfused primarily by M2, with some additional contribution from M1 and M3 (Tanrovier et al., 2004; Ture et al., 2000). The accessory MCA, which originates from the anterior cerebral artery, has also been observed to serve the insula (Tanrovier et al., 2004). The insula is drained by deep (deep middle cerebral vein) and superficial (superficial Sylvian vein) venous drainage systems (Tanrovier et al., 2004). Although deep venous drainage predominates for the insula as a whole, there are cortical areas within the insula primarily drained by superficial drainage systems (e.g., sAIG, sMIG, insula apex) and many areas draining into both deep and superficial systems.

Isolated insular infarction is rare. Most often, the insula is just one of many regions damaged when MCA is compromised. However, if infarction extends

BOX 10.1

VON ECONOMO NEURONS (VENs)

In the early 1900s, Ramon y Cajal first described large spindle cells in the human cortex that seemed to only appear in the cingulate and insular cortices (Butti et al., 2013; Seeley et al., 2012). Soon thereafter, Constantin von Economo (and George Koskinas) provided the first detailed illustration and description of these neurons in 1925 (Butti et al., 2013; Sak & Grzybowski, 2013; Seeley et al., 2012), referring to them as rod cells (Stabzellen) and corkscrew cells (Korkzieherzellen) (Seeley et al., 2012). In humans, VENs appear in the cingulate and insular cortices and are also present in the frontal cortex and hippocampal formation (Butti et al., 2013; Fajardo et al., 2008; Nimchinsky et al., 1999). VENs have been reported in nonhuman primates, whales, elephants, hippopotamus, manatees, zebras, walrus, and dolphins (Bauernfeind et al., 2013; Butti & Hof, 2010, Butti et al., 2013; Sak & Grzybowski, 2013) and are thought to be related to social cognition (Allman et al., 2011; Bauernfeind et al., 2013) and vocalizations (Butti & Hof, 2010). Interestingly, although they are present in other species, they only appear in cluster formations in humans and great apes (Bauernfeind et al., 2013). VENs are thought to be specialized for rapid and long-distance signal transmission (Butti et al., 2013; Fajardo et al., 2008; Menon & Uddin, 2010; Sridharan, Levitin, & Menon, 2008) and are present in agranular and dysgranular areas, specifically layers V and III (Butti & Hof, 2010; Butti et al., 2013; Morel et al., 2013).

Another large cell appears alongside VENs in the anterior insula. These cells are similarly large and long, but they have two large dendrites pointing away from the cell soma and have thus been described as fork cells (Gabelzellen) (Seeley et al., 2012). These cells are less studied but are of interest because they only appear in the anterior insula.

to the insula after MCA damage, then it is thought to be a special risk factor for increased tissue damage (Ay, Arsava, Koroshetz, & Sorensen, 2008; Kamalian et al., 2013). Ay et al. (2008) revealed that when MCA infarcts lead to insular damage, they resulted in significantly greater volume of infarct progression (ischemic-to-infarcted tissue conversion) than MCA infarcts without insular damage, even when controlling for volume of ischemic tissue and site of occlusion. Further, when more than 25% of the insular cortex is infarcted at admission, it is a stronger predictor of infarct progression than stroke scale score (NIHSS), infarct volume at admission as judged by diffusion weighted imaging, or other imaging-based scores (Kamalian et al., 2013). The reason for insular damage as a risk factor is unclear—whether it is due to its numerous connections, whether it initiates other harmful events (Ay et al., 2008), or whether greater insula damage leads to further stroke complications (e.g., vasoconstriction, immunodepression, and sympathetic hyperactivation; Kamalian et al., 2013; Walter et al., 2013) requires further scrutiny.

10.4 CONNECTIVITY

The anterior insula is considered to be a paralimbic region (Eslinger, 2011; Mesulam, 2003; Nieuwenhuys, 2012; Stephani et al., 2011). Briefly, there are four stages of cortical differentiation (Mesulam, 2003):

1. Corticoid—Basic differentiation into "cortex-like structures;" inconsistent organization/patterning
2. Allocortex—Moderate differentiation of bands of neurons; includes paleocortex and archicortex
   *Corticoid plus Allocortex form the limbic region
3. Paralimbic (or periallocortex)—Greater differentiation; transition zone from limbic regions to neocortex; divided into olfactocentric and hippocampocentric groupings, with the insula as part of the olfactocentric group (Eslinger, 2011)
4. Neocortex—Most complex differentiation

As a paralimbic structure, the anterior insula has extensive interconnections with other paralimbic structures along a cortical differentiation continuum (i.e., the more allocortical-like insula is connected to more allocortical-like portions of other paralimbic structures, and the more neocortical-like insula is connected to the more neocortical-like portions of other paralimbic structures) (Eslinger, 2011). Additionally, the insula is well connected with primitive limbic regions and more advanced neocortex (Augustine, 1996; Eslinger, 2011; Nieuwunhuys, 2012).

Since Augustine's (1985, 1996) seminal reviews on primate (human and nonhuman) insular circuitry, the topic has been extensively studied with varied approaches. Because of the different techniques,

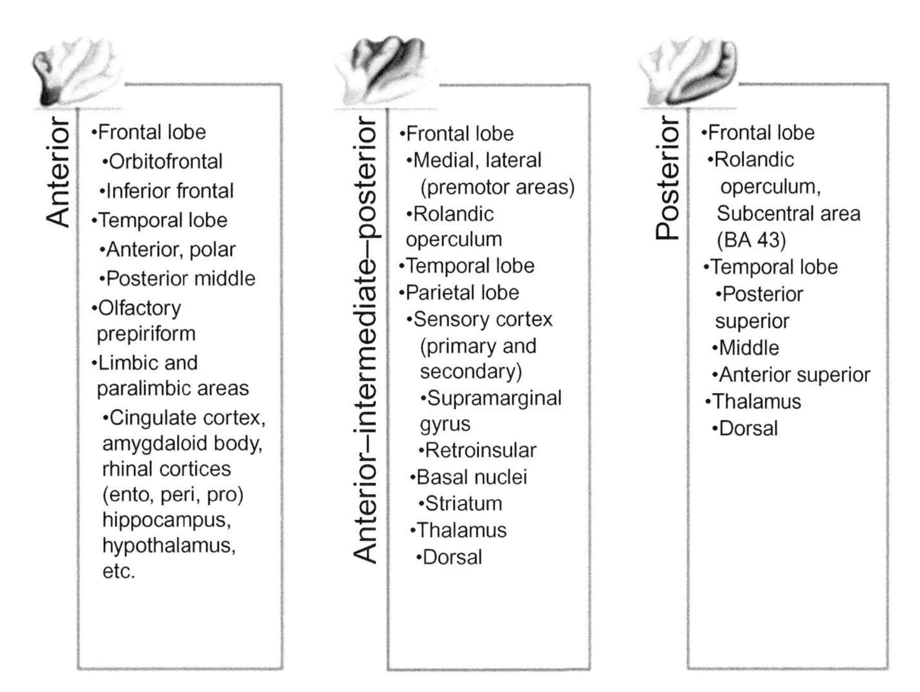

FIGURE 10.4 Basic illustration of insular structural circuitry based on reviews and investigations of primate (human and nonhuman) insula (Augustine, 1985, 1996; Catani et al., 2012; Cloutman et al., 2012; Mesulam & Mufson, 1982, 1985; Nieuwenhuys, 2012). The shaded areas represent approximate estimates of the insular territory associated with the specified connections: anterior insula (left), anterior–intermediate–posterior (center), and posterior (right). Artist: C. Vincent Collier.

regions of interest, methodologies, and terminology utilized, it is difficult to generate a coherent picture of insular circuitry. For example, translating the circuitry information from the mixed human and nonhuman primate literature should be performed with caution, given that recent tractography work has revealed some inconsistencies between the two populations, specifically regarding connections between the insula and the anterior cingulate cortex (Cerliani et al., 2012; Cloutman, Binney, Drakesmith, Parker, & Lambon Ralph, 2012). Additionally, nomenclature may be similar, but definitions often differ; not all anterior, intermediate, or posterior insular divisions are the same. Divisions may be determined by granularity, sulcal boundaries, functional connectivity, or other structural or functional features. Even if authors use a similar technique but different thresholds for determining boundaries, location and size of divisions will vary. The expanse of these divisions is also known to differ between humans and nonhuman primates, with positive allometric relationships (i.e., hyperallometric) between insula volume and total brain volume observed when comparing across primate brain volumes. This means that as total brain volume increases in primates, insula volume increases to a slightly greater degree (Bauernfeind et al., 2013). Interestingly, the Iag exhibits greater hyperallometry than Idg and Ig (Bauernfeind et al., 2013).

In Figure 10.4, we merge connectivity findings in an attempt to provide a basic illustration of structural circuitry. No information about direction of information flow is provided here, because previous reports often differ and/or may stem from nonhuman primate or mixed human/nonhuman primate reports. Further, the boundaries and number of divisions (e.g., anterior/posterior versus anterior/intermediate/posterior) vary. We have included:

- Aforementioned (reciprocal) connections with paralimbic and limbic structures (to and from anterior insula)
- Neocortical structures
  - Frontal, temporal, parietal lobes
- Basal nuclei
  - Striatum
- Olfactory prepiriform cortex
- Dorsal thalamus

Cloutman et al. (2012) recently utilized *in vivo* probabilistic tractography in humans to investigate insular circuitry. Relying on gross anatomical gyral landmarks for seed positioning, they hypothesized that the pathways carrying information between neocortical and insula structures identified in their research might include the following fasciculi: arcuate, inferior fronto-occipital, middle longitudinal, and uncinate. They also hypothesized that the extreme capsule (EC) was a path of information flow, but some consider the EC not as a separate tract but rather as a thoroughfare for more distinct tracts.

In addition to these connections, animal and human research has revealed abundant and reciprocal intra-insular connections (not included in Figure 10.4). In nonhuman primates, the informational flow has been described as traveling from anterior to posterior (i.e., abundant anterior to posterior, little posterior to anterior; Flynn, 1999; Mesulam & Mufson, 1982). In humans, Cloutman et al. (2012) revealed strong interconnectivity within anterior and posterior divisions, and an intermediate division with heavy connections to both anterior and posterior divisions. In addition, there were significant connections between the anterior apex and the posterior pole. Almashaikhi et al. (2014) performed an electrophysiological study complementary to Cloutman et al. (2012) due to their use of gyral landmarks for electrode placement (placed electrodes in all gyri used as seed regions in Cloutman et al., excluding only apical and polar regions). Interpreting their electrophysiological findings within the anterior–intermediate–posterior framework used by Cloutman et al., they demonstrated predominantly unidirectional flow of information from sAIG to more posterior regions (with only stimulation of sMIG resulting in sAIG response) and reciprocal connections between intermediate and posterior regions. The agreement between these tractography and electrophysiology results bolsters confidence in the findings. Commissural fibers, via anterior commissure and corpus callosum, have received little attention (Flynn, 1999), and the recent study by Almashaikhi et al. produced surprising results because no electrodes stimulated resulted in evoked potentials in the contralateral hemisphere. (Almashaikhi et al. urge caution when interpreting the predominantly unilateral information flow from sAIG as well as the contralateral findings because of small sample sizes specifically for these regions.)

Knowledge of these unique structural connections is helpful in the quest for determining insular function. The insula seems uniquely situated and connected to serve important integrative functions. Indeed, "paralimbic and related limbic system structures may be particularly important for developmentally establishing and mediating links between cognition, visceral states, and emotion" (Eslinger, 2011; see also Mesulam, 2000; Mesulam & Mufson, 1985).

10.5 INSULAR CORTEX AND BEHAVIOR

Research of the function of the insula using a variety of techniques has revealed functional divisions that are generally consistent with structural connectivity data. Although the number of divisions identified differs, there is some agreement between different research groups and methods that point toward three functional divisions—posterior, dorsal anterior, and ventral anterior (Chang, Yarkoni, Khaw, & Sanfey, 2013; Deen, Pitskel, & Pelphrey, 2011; Kelly et al., 2012; Kurth et al., 2010). *Resting state fMRI* and *metanalytic* research conducted by Chang et al. (2013) revealed the following functional relationships: posterior insula is coactivated with *supplementary motor area (frontal), somatosensory cortex (parietal), posterior temporal lobes, hippocampus, and rostral ACC*; dorsal anterior insula is coactivated with *dorsal ACC, dorsolateral prefrontal cortex (DLPFC), dorsal striatum, and temporoparietal junction (TPJ)*; and ventral anterior insula is coactivated with *amygdala, ventral tegmental area (VTA)*, superior temporal sulcus (STS), *temporal poles*, postero*lateral orbitofrontal cortex (pLOFC), medial prefrontal cortex (MPFC)*, and *ventral striatum*. Posterior (and central) insula has been implicated in sensory, motor, interoceptive, and language behaviors (Chang et al., 2013; Kelly et al., 2012; Kurth et al., 2010). Dorsal anterior insula has been associated with goal-directed cognitive tasks, including speech and language, and memory (Chang et al., 2013; Kelly et al., 2012; Kurth et al., 2010). Finally, ventral anterior insula is implicated in social/emotional, autonomic, and chemosensory behaviors (Chang et al., 2013; Kelly et al., 2012; Kurth et al., 2010).

Discussion of every suggested role of the insula is beyond the scope of this chapter and textbook because the insula has been tied to nearly every human behavior as well as numerous disorders (Box 10.2). It has been proposed that because of the insula's unique positioning at the junction of structures that mediate lower level and higher level functions, the insula performs an important integrative role, specifically regarding the salience of stimuli and events (Craig, 2009, 2010, 2011; Menon & Uddin, 2010; Seeley, et al., 2007). Craig (2009, 2010) has proposed that: (i) the feelings we experience (e.g., anger, trust, surprise, and joy) are a result of the neural encoding of "the unified representation of all salient conditions;" (ii) these feelings are updated continually, with each sampling of feelings at a point in time constituting a "global emotional moment;" (iii) these "global emotional moments" beget a sense of awareness, or sentience, which is used to evaluate and predict energy utilization; (iv) *all* feelings are correlated with anterior insula activation; and (v) therefore, the processing that takes place within the anterior insula is the foundation for sentience.

Relatedly, recent cortical network research has identified two prominent networks in brains during tasks and at rest—the central executive network (CEN) and the default mode network (DMN) (Greicius, Krasnow,

BOX 10.2

INSULAR FUNCTION

The insular cortex in humans is unique in that it includes allocortical, paralimbic, and isocortical sections. Because of its unique architecture and positioning within the central nervous system, and because of the numerous connections with limbic, paralimbic, and isocortical association cortices, it is no surprise that the insula has been credited with many and varied functions and disorders.

Visceral

- Sensory (esophageal/gastric/abdominal)
- Motor (cardiovascular function, respiration, vomiting)

Sensory

- Audition, vestibular sense
- Gustation, olfaction
- Somatosensation (tactile, pain, temperature, itch)

Motor

- Articulation (speech)
- Ocular movements
- Swallowing
- Voluntary respiratory control

Integration

- Awareness, bodily awareness, consciousness, emotion processing and recognition, interoception, mood, motivation
- Time awareness
- Music (listening, singing)
- Sensorimotor
- Self-recognition

Higher-Order/Complex/Cognitive

- Attention, decision-making, error awareness, working memory
- Speech and language

Disorders

- Addiction, craving
- Agnosia, neglect
- Alzheimer's disease, frontotemporal dementia, primary progressive aphasia
- Anxiety, depression .
- AOS, speech fluency
- Dysphagia
- Eating disorders (anorexia nervosa, bulimia nervosa)
- Schizophrenia
- Seizures

Plasticity/Recovery

Ackermann and Riecker (2010), Augustine (1996), Bates et al. (2003), Couto et al. (2013), Craig (2009, 2010, 2011), Dronkers (1996), Garavan (2010), Ibanez, Gleichgerrcht, and Manes (2010), Kim, Ku, Lee, Lee, and Jung (2012), Klein, Ullsperger, and Danielmeier (2013), Kosillo and Smith (2010), Lee and Reeve (2013), Manoliu et al. (2013), Nagai, Kishi, and Kato (2007), Nieuwenhuys (2012), Palaniyappan and Liddle (2012), Rossi et al. (2013), Sliz and Hayley (2012).

Reiss, & Menon, 2003; Menon & Uddin, 2010). Consistently demonstrated is a pattern of increased CEN activation and decreased DMN activation at onset of tasks requiring cognitive effort, with decreased DMN activation correlated with "accurate behavioral performance" (Kelly et al., 2008; Menon & Uddin, 2010). Menon and Uddin (2010) have expanded this model by identifying with rs-fMRI research a salience network (SN) that includes as a key component the anterior insula, along with dorsal ACC, amygdala, VTA, and thalamus (Menon & Uddin, 2010; Seeley et al., 2007). The anterior insula seems to be crucial for integrating internal state with external state/stimuli, determining the salience of external state/stimuli, and communicating with higher-order structures that are capable of initiating goal-directed behavior (Menon & Uddin, 2010). In other words, the anterior insula seems to be the structure that switches between CEN and DMN modes. This is supported by fine-grained analysis of the timing of network responses, which shows the anterior insular activation occurring before the increase in either CEN or DMN activation changes. Such a role is consistent with the structural and functional connections we have so far reviewed in this chapter. Further, the existence of the VENs in CEN, DMN, and SN structures would provide the necessary rapid signaling required for this crucial and time-sensitive network communication because VENs are thought to be specialized for quickly transmitting information over long distances (see Box 10.1).

10.6 ASSOCIATION WITH SPEECH–LANGUAGE BEHAVIOR

The insula has been associated with speech–language production since the beginning of the search for the "seat" of this complex behavior. One of the earliest appearances of the insula in association with (or actually in dissociation from) speech–language behavior was in Broca's (1861) report of the famous case of Leborgne ("Tan-Tan," Mr. Tan, Patient Tan). When describing changes over time, Broca divided the behavioral deficits into the first clinical period and the second clinical period. The first period involved more restricted deficits (i.e., speech production) and the second clinical period involved the appearance of additional motor and cognitive deficits. Broca hypothesized that the speech–language deficits demonstrated by Leborgne were attributed to damage (softening) of the left inferior frontal gyrus and that the additional deficits occurred only when the softening progressed to other structures (Broca, 1861; Kann, 1950).

Broca's early descriptions of Leborgne's speech referred to reduced "faculty to coordinate the movements which belong to the articulate language, or simpler, it is the faculty of articulate language" (Broca, 1861). This description is more consistent with our current understanding of apraxia of speech (AOS), a motor speech impairment, rather than aphasia (Berker, Berker, & Smith, 1986; Lazar & Mohr, 2011; Richardson, Fillmore, Rorden, Lapointe, & Fridriksson, 2012). Another patient of Broca's, Lelong, demonstrated expressive deficits (could only utter five words, some mispronounced) and similar damage to the left inferior frontal lobe, but with sparing of the insula (Dronkers, Plaisant, Iba-Zizen, & Cabanis, 2007; Pearce, 2009). His early judgment was then that the insula was not necessarily a contributor to speech articulation.

Many and conflicting reports of insular involvement in speech–language disorders ensued. Jules Parrot (1863) reported a case in which the entire insular cortex and inferior frontal cortex of the *right* hemisphere were damaged but speech and cognition were preserved, lending support to language lateralization theories (Bateman, 1890; Klingbeil, 1939). M. Flaubert (1866) reported a case of extensive damage, including "flattening" of the insula in both hemispheres, with no observable motor or speech–language deficits (Bateman, 1890; Klingbeil, 1939), supporting the notion that the insula was not necessary for motor or speech–language functions. However, William Sanders (1866) and D.C. Finkelnburg (1870) theorized, based on separate case observations of insular lesions, that damage to the insula, rather than inferior frontal convolutions (i.e., Broca's area), was responsible for speech impairment or "palsy" (Bateman, 1890; Duffy & Liles, 1979; Elder, 1897; Klingbeil, 1939).

In the late 1860s to early 1900s, there was much discussion and debate regarding the relationship between insular damage and conduction aphasia, a type of fluent aphasia in which individuals produce output containing phonemic paraphasias, have little to no auditory comprehension deficits, and demonstrate marked difficulty with speech repetition (Goodglass, 1992). Theodor Meynert, Karl Wernicke, and Ludwig Lichtheim all separately hypothesized that damage to the insular region was related to conduction aphasia (Ackermann & Riecker, 2010; Goldstein, 1948; Henderson, 1992), with Lichtheim (1885) even proposing use of the term "insular aphasia" (Henderson, 1992). However, a close examination of their discourse reveals that they all eventually attributed conduction aphasia to damage to the white matter fibers coursing near or through the insula, particularly the arcuate fasciculus fibers that branch off and travel through the EC (Ackermann & Riecker, 2010; Goldstein, 1948; Henderson, 1992; Palumbo, Alexander, & Naeser, 1992). Between the early and late 1900s, the insula was consistently linked to reports of conduction aphasia, although damage was rarely restricted to the insula (but see Goldstein, 1948). Over time, discussions of conduction aphasia included less of an emphasis on the insula and more of an emphasis on the arcuate fasciculus, EC, inferior parietal lobule, Sylvian–parietal–temporal, and others (Damasio & Damasio, 1980; Geschwind, 1965; Goodglass, 1992; Hickok, 2009; Palumbo et al., 1992; Saur et al., 2008). Schuell et al. (1964) summarized this ever-evolving debate—"Speech functions have been ascribed to [the insula], although conclusive evidence for this is lacking" (p. 69).

Notions about the involvement of the insula in speech processing changed considerably with the publication of a seminal paper by Dronkers (1996). In this study involving stroke patients, a strong relationship was revealed between damage to the anterior portion of the left insula and AOS. Specifically, each of the 25 patients who presented with AOS also had damage to the anterior insula, whereas 19 patients who were not diagnosed with AOS had cortical damage that did not involve the same portion of the anterior insula. Thus, this study presented a strong dissociation between patients with or without AOS whose stroke either did or did not involve the left anterior insula, respectively. Based on these findings, Dronkers concluded that the left anterior insula, and not Broca's area, was a crucial region for speech articulation, a claim that challenged the long-held view that speech articulation was associated with Broca's area. Although this study was particularly remarkable for its unexpected findings, it was also important in that it was one of the first to use a lesion overlay method to compare and contrast patients with different behavioral profiles. It is important to note, however, that several years before, Naeser et al. (1989) had utilized statistical methods to compare lesion locations among patients with different levels of speech fluency and found that damage to the subcallosal fasciculus (deep to Broca's area, with projections to cingulate gyrus, supplementary motor area, and striatum) and periventricular white matter (deep to sensorimotor cortical areas for mouth) was particularly detrimental to speech fluency in stroke patients with left hemisphere damage. Nevertheless, Dronkers' study probably had greater effects on the field based on its implications of the insula, instead of Broca's area, in speech articulation. Had Dronkers relied on the statistical methods outlined by Naeser and colleagues, it is unlikely that their conclusions would have changed based on the reported "double dissociation," where 25 out of 25 patients with AOS had insular damage and 19 out of 19 patients without AOS did not have insular damage.

In a follow-up study that included some of the same patients from the studies of Dronkers (1996), Baldo, Wilkins, Ogar, Willock, and Dronkers (2011) found that complex speech articulation (i.e., speech sound clusters, high articulatory travel, greater number of syllables) is particularly vulnerable in patients who have damage to a subregion of the anterior insula, the superior portion of the short posterior insular gyrus (superior sPIG), referred to as the superior precentral gyrus of the insula (SPGI). Similar to the study by Dronkers (1996), a clear association between Broca's area damage and speech articulation was absent. It is important to note that this study relied on relatively few patients ($N = 33$) and that their lesion overlay map yielded limited statistical power in regions outside the anterior insula. Therefore, it is perhaps not surprising that this study implicated the anterior insula because it had limited potential to highlight other regions that might potentially contribute to speech articulation. Because the insula receives its blood supply from the MCA, maximum lesion overlap in the anterior insula is fairly typical among lesion studies and, as a result, attributing specific behavioral processes to the insula should be performed with caution unless it can be demonstrated that ample statistical power is achieved in other cortical regions.

Although the studies by Dronkers (1996) and Baldo et al. (2011) provided concordant evidence regarding the role of the anterior insula, specifically the SPGI, in speech articulation, others have provided contradictory evidence. In a study that included 80 left MCA stroke patients (40 with insular damage, 40 without insular damage), Hillis et al. (2004) found limited involvement of the anterior insula in AOS. In contrast, damage to Broca's area was found to be a far stronger predictor of AOS. The Hillis et al. study only included acute patients and damage to Broca's area, and the anterior insula was evaluated by raters who were blinded to clinical examination (i.e., the presence of AOS).

To provide a clearer picture of the role of the left insula in speech articulation, Richardson et al. (2012) attempted to replicate the seminal study by Dronkers (1996). Two notable improvements were included by Richardson et al. First, brain damage among the patients examined by Dronkers was determined based on MRI or CT scans. The spatial resolution of early CT scans was typically limited, making it difficult to ascertain lesion boundaries. Patients ($N = 50$) included in the Richardson et al. study all underwent high-resolution MRI scans collected using a 3T MRI scanner, making lesion detection relatively easier compared with lower resolution modalities. Second, Richardson et al. utilized voxel-wise lesion symptom mapping (VLSM; Bates et al., 2003), the current standard for determining lesion–behavior relationships and a method that was not available to Dronkers in the 1990s. VLSM relies on binary maps of cortical damage that typically is demarcated by a trained expert and compares patients' behavioral scores on voxel-by-voxel basis in standardized brain space. The lesion overlay comparisons among patients with and without AOS in the Richardson et al. study were relatively similar to Dronkers, with one major exception: patients with AOS had the greatest lesion overlap in the SPGI, whereas the greatest overlap in damage was found in the left middle temporal lobe among patients without AOS. (Also, unlike Dronkers' patients without AOS, several of the AOS-negative patients examined by Richardson et al. had cortical damage that involved the anterior insula, including the SPGI.) However, the VLSM analysis used by Richardson et al. yielded very different results compared with those of Dronkers. Instead of implicating the anterior insula, their findings showed a strong relationship between AOS and damage to the posterior portion of Broca's area, the *pars opercularis*, as well as areas of the left precentral gyrus. In addition to relating frank cortical damage to AOS, Richardson and colleagues examined cerebral perfusion in relation to AOS. Both structural damage and hypoperfusion in Broca's area were strongly associated with AOS, a finding that corroborates findings by Hillis et al (2004) but strongly contradicts Dronkers and the later study by Baldo et al. (2011).

One final lesion study is of particular relevance to the role of the insula in speech production. A unique study by Graff-Radford et al. (2014) included patients who had pure AOS as a result of stroke. Although AOS is fairly common after left frontal stroke, the majority of AOS patients also have aphasia, making pure AOS a very rare syndrome. Among seven patients, five of whom had pure AOS and two who had AOS plus only very minor language impairment, the greatest lesion overlap was found in the left premotor cortex with relative or complete sparing of the left insula. Although this lesion study included a small patient sample, its finding of nearly complete sparing of the insula among patients with pure AOS further suggests that the insula does not play a specific role in speech articulation.

So far, we have focused mostly on lesion studies that examined speech deficits in stroke patients. Hillis et al. (2004), Richardson et al. (2012), and Graff-Radford et al. (2014) provide strong evidence against a central role of the insula in speech production. Evidence from dementia patients tells a similar story. A recently identified syndrome, primary progressive apraxia of speech (PPAOS), is a form of dementia where the primary behavioral deterioration manifests in impaired speech articulation with relative sparing of language and other cognitive functions (Josephs et al., 2012). Although PPAOS is a very rare syndrome, a few studies have shown that the primary areas of deterioration include gray matter in the premotor and motor cortex with relative sparing of other regions such as the insula. Decreased metabolism is observed in the same cortical regions and in the underlying white matter areas. Along with the aforementioned stroke studies, evidence from dementia patients further suggests that the insula is probably not the cortical seat of speech articulation.

Whereas several lesion studies have yielded compelling evidence against the insula as playing an important role in speech, numerous functional imaging studies have revealed insula activation during speech processing (for a review, see Ackermann & Riecker, 2010). As is typical among functional studies, it is difficult to determine if insula activation is crucial for speech or if such activation represents other processes involved in task completion but not specific to speech. To address this issue, Fedorenko, Fillmore, Smith, and Fridriksson (in press) used fMRI to examine speech-related activity in the anterior insula, including the SPGI. This research differed from previous studies (where insula activation was typically revealed in a whole-brain group analysis) by functionally defining speech-related regions of interest (ROI) within the insula on a subject-by-subject basis and then comparing activation across different functional tasks. In addition to the insula, functional activation was also examined in the left inferior frontal gyrus (LIFG), an area commonly implicated in speech production. Twenty control participants completed several tasks during fMRI scanning: speech production (easy versus difficult articulation), vowel production (/a/, /i/, /u/), breath patterns (e.g., breathe in slowly, breathe out fast), and nonspeech oral movements (e.g., move tongue to touch the upper and lower lips). In addition, all participants completed a visual–spatial working memory task to verify that functional activation reflected process-specific activation and not effort related to completing the tasks. The study revealed very limited activation in the SPGI or other areas of the insula during speech production, vowel production, breathing, or working memory tasks. As importantly, the SPGI and other insula regions did not respond preferentially to more difficult speech articulation. In contrast, the LIFG showed significant modulation during speech articulation and enhanced activation during the difficult (compared with easy) speech articulation condition. Although the LIFG was active during other tasks, it showed by far the greatest activation during speech articulation. Somewhat surprisingly, the SPGI seemed to be selectively recruited during the nonspeech oral movement task. Fedorenko and colleagues concluded that although the left SPGI or other subregions of the insula are somewhat engaged during speech tasks, it is probably not specific to speech articulation. Further study is needed to better understand insula involvement during oral movements that are not speech-specific.

It seems less likely that the insula is essential for speech articulation and more likely that the insula is important for other behaviors that accompany or facilitate speech articulation. For example, it could be that the insula is primarily involved in the social aspects of speech (or other communication modalities). Alternatively, the insula could be more involved in the emotional tone of speech production, which would require modulation of respiratory, phonatory, and articulatory systems to alter rhythm and prosody appropriately. Given the wealth of knowledge about insula involvement in interoception and salience networks, and given the connections discussed in this chapter (Catani et al., 2012), it seems most likely that the insula is involved in: (i) monitoring internal states (visceral and emotional, Catani et al., 2012) and perhaps articulatory and auditory states (Hickok & Poeppel, 2004; Tourville & Guenther, 2011); (ii) monitoring external state/stimuli; and (iii) communicating with other structures to assist with modulation of speech. Whatever the role in speech, language, or communication, the only thing that can be said with certainty is that there is much more to be learned about the role of the insula in human behavior, and that research in the area of communication, both typical and disordered, will continue to inform this knowledge base. As reviewed in this chapter, there are many qualities of the insula unique to humans, and it is an exciting time to explore the potential relationships between this structure and behaviors, such as speech and language, that are also unique to humans.

References

Ackermann, H., & Riecker, A. (2010). The contribution(s) of the insula to speech production: A review of the clinical and functional imaging literature. *Brain Structure and Function, 214*, 419–433.

Afif, A., & Mertens, P. (2010). Description of sulcal organization of the insular cortex. *Surgical and Radiologic Anatomy, 32*, 491–498.

Allman, J. M., Tetreault, N. A., Hakeem, A. Y., Manaye, K. F., Semendeferi, K., Erwin, J. M., et al. (2011). The von Economo neurons in fronto-insular and anterior cingulate cortex. *Annals of the New York Academy of Sciences, 1225*, 59–71.

Almashaikhi, T., Rheims, S., Ostrowsky-Coste, K., Montavont, A., Jung, J., De Bellescize, J., et al. (2014). Intrainsular functional connectivity in human. *Human Brain Mapping, 35*(6), 2779–2788.

Augustine, J. R. (1985). The insular lobe in primates including humans. *Neurological Research, 7*(1), 2–10.

Augustine, J. R. (1996). Circuitry and functional aspects of the insular lobe in primates including humans. *Brain Research Reviews, 22*, 229–244.

Ay, H., Arsava, M., Koroshetz, W. J., & Sorensen, A. G. (2008). Middle cerebral artery infarcts encompassing the insula are more prone to growth. *Stroke, 39*, 373–378.

Baldo, J. V., Wilkins, D. P., Ogar, J., Willock, S., & Dronkers, N. F. (2011). Role of the precentral gyrus of the insula in complex articulation. *Cortex, 47*(7), 800–807.

Bateman, F. (1890). *On aphasia or loss of speech and the localization of the faculty of articulate language* (2nd ed.). London, England: J.&A. Churchill.

Bates, E., Wilson, S. M., Saygin, A. P., Dick, F., Sereno, M. I., Knight, R. T., et al. (2003). Voxel-based lesion-symptom mapping. *Nature Neuroscience, 6*(5), 448–450.

Bauernfeind, A. L., de Sousa, A. A., Avasthi, T., Dobson, S. D., Raghanti, M. A., Lewandowski, A. H., et al. (2013). A volumetric comparison of the insular cortex and its subregions in primates. *Journal of Human Evolution, 64*, 263–279.

Berker, E. A., Berker, A. H., & Smith, A. (1986). Localization of speech in the third left frontal convolution. *Archives of Neurology, 43*, 1065–1072.

Binder, D. K., Schaller, K., & Clusmann, H. (2007). The seminal contributions of Johann–Christian Reil to anatomy, physiology, and psychiatry. *Neurosurgery, 61*, 1091–1096.

Broca, P. (1861). Remarques sur le siége de la faculté du langage articulé: Suivies d'une observation d'aphemie. *The Bulletin of the Society of Anatomy (Paris), 6*, 330–357.

Brodmann, K. (1909). *Vergleichende Lokalisationslehre der Großhirnrinde: in ihren Prinzipien dargestellt auf Grund des Zellenbaues*. Leipzig: Barth.

Butti, C., & Hof, P. R. (2010). The insular cortex: A comparative perspective. *Brain Structure and Function, 214*, 477–493.

Butti, C., Santos, M., Uppal, N., & Hof, P. R. (2013). Von Economo neurons: Clinical and evolutionary perspectives. *Cortex, 49*, 312–326.

Catani, M., Dell'Acqua, F., Vergani, F., Malik, F., Hodge, H., Roy, P., et al. (2012). Short frontal lobe connections of the human brain. *Cortex, 48*, 273–291.

Cauda, F., Torta, D. M. E., Sacco, K., D'Agata, F., Geda, E., Duca, S., et al. (2013). Functional anatomy of cortical areas characterized by Von Economo neurons. *Brain Structure and Function, 218*, 1–20.

Cerliani, L., Thomas, R. M., Jbabdi, S., Siero, J. C. W., Nanettie, L., Crippa, A., et al. (2012). Probabilistic tractography recovers a rostrocaudal trajectory of connectivity variability in the human insular cortex. *Human Brain Mapping, 33*, 2005–2034.

Chang, L. J., Yarkoni, T., Khaw, M. W., & Sanfey, A. G. (2013). Decoding the role of the insula in human cognition: Functional parcellation and large-scale reverse inference. *Cerebral Cortex, 23*, 739–749.

Cloutman, L. L., Binney, R. J., Drakesmith, M., Parker, G. J. M., & Lambon Ralph, M. A. (2012). The variation of function across the human insula mirrors its patterns of structural connectivity: Evidence from *in vivo* probabilistic tractography. *NeuroImage, 59*, 3514–3521.

Couto, B., Manes, F., Mantanes, P., Matallana, D., Reyes, P., Velasquez, M., et al. (2013). Structural neuroimaging of social cognition in progressive non-fluent aphasia and behavioral variant frontotemporal dementia. *Frontiers in Human Neuroscience, 7*, 467.

Craig, A. D. (2009). How do you feel—now? The anterior insula and human awareness. *Nature Reviews Neuroscience, 10*, 59–70.

Craig, A. D. (2010). The sentient self. *Brain Structure and Function, 214* (5–6), 563–577.

Craig, A. D. (2011). Significance of the insula for the evolution of human awareness of feelings from the body. *Annals of the New York Academy of Sciences, 1225*, 72–82.

Damasio, H., & Damasio, A. R. (1980). The anatomical basis of conduction aphasia. *Brain, 103*, 337–350.

Deen, B., Pitskel, N. B., & Pelphrey, K. A. (2011). Three systems of insular functional connectivity identified with cluster analysis. *Cerebral Cortex, 21*(7), 1498–1506.

Dronkers, N. F. (1996). A new brain region for coordinating speech articulation. *Nature, 384*(6605), 159–161.

Dronkers, N. F., Plaisant, O., Iba-Zizen, M. T., & Cabanis, E. A. (2007). Paul Broca's historic cases: High resolution MR imaging of the brains of Leborgne and Lelong. *Brain, 130*(5), 1432–1441.

Duffy, R. J., & Liles, B. Z. (1979). A translation of Finkelnburg's (1870) lecture on aphasia as "asymbolia" with commentary. *Journal of Speech and Hearing Disorders, 44*(2), 156–168.

Elder, W. (1897). *Aphasia and the cerebral speech mechanism*. London, England: H.K. Lewis.

Eslinger, P. J. (2011). Functional neuroanatomy of the limbic system. In A. S. Davis (Ed.), *The handbook of pediatric neuropsychology* (pp. 137–146). New York, NY: Springer.

Fajardo, C., Escobar, M. I., Buritica, E., Arteaga, G., Umbarila, J., Casnova, M. F., et al. (2008). Von Economo neurons are present in the dorsolateral (dysgranular) prefrontal cortex of humans. *Neuroscience Letters, 435*, 215–218.

Fedorenko, E., Fillmore, P., Smith, K., & Fridriksson, J. (in press). The superior precentral gyrus of the insula (SPGI) does not appear to be functionally specialized for articulation or complex articulation. *Journal of Neurophysiology*.

Flynn, F. G. (1999). Anatomy of the insula: Functional and clinical correlates. *Aphasiology, 13*(1), 55–78.

Garavan, H. (2010). Insula and drug cravings. *Brain Structure and Function, 214*(5–6), 593–601.

Geschwind, N. (1965). Disconnection syndromes in animals and man. *Brain, 88*, 237–294.

Gibo, H., Carver, C. C., Rhoton, A. L., Jr., Lenkey, C., & Mitchell, R. J. (1981). Microsurgical anatomy of the middle cerebral artery. *Journal of Neurosurgery, 54*(2), 151–169.

Goldstein, K. (1948). *Language and language disturbances: Aphasic symptom complexes and their significance for medicine and theory of language*. New York, NY: Grune & Stratton.

Goodglass, H. (1992). Diagnosis of conduction aphasia. In S. E. Kohn (Ed.), *Conduction aphasia* (pp. 39–50). Hillsdale, NJ: Lawrence Erlbaum.

Graff-Radford, J., Jones, D.T., Strand, E.A., Rabinstein, A.A., Duffy, J. R., & Josephs, K.A. (2014). The neuroanatomy of pure apraxia of speech in stroke. *Brain and Language*, *129*, 43–46.

Gray, H. (1858). *Anatomy: Descriptive and surgical*. London, England: John W. Parker and Son.

Greicius, M. D., Krasnow, B., Reiss, A. L., & Menon, V. (2003). Functional connectivity in the resting brain: A network analysis of the default mode hypothesis. *Proceedings of the National Academy of Sciences of the United States of America*, *100*(1), 253–258.

Henderson, V. W. (1992). Early concepts of conduction aphasia. In S. E. Kohn (Ed.), *Conduction aphasia* (pp. 23–38). Hillsdale, NJ: Lawrence Erlbaum.

Hickok, G. (2009). The functional neuroanatomy of language. *Physics of Life Reviews*, *6*(3), 121–143.

Hickok, G., & Poeppel, D. (2004). Dorsal and ventral streams: A framework for understanding aspects of the functional anatomy of language. *Cognition*, *91*(1–2), 67–99.

Hillis, A. E., Work, M., Barker, P. B., Jacobs, M. A., Breese, E. L., & Maurer, K. (2004). Re-examining the brain regions crucial for orchestrating speech articulation. *Brain*, *127*, 1479–1487.

Ibanez, A., Gleichgerrcht, E., & Manes, F. (2010). Clinical effects of insular damage in humans. *Brain Structure and Function*, *214*, 397–410.

Josephs, K. A., Duffy, J. R., Strand, E. A., Machulda, M. M., Senjem, M. L., Master, A. V., et al. (2012). Characterizing a neurodegenerative syndrome: Primary progressive apraxia of speech. *Brain*, *135* (5), 1522–1536.

Kahilogullari, G., Ugur, H. C., Comert, A., Tekdemir, I., & Kanpolat, Y. (2012). The branching pattern of the middle cerebral artery: Is the intermediate trunk real or not? An anatomical study correlating with simple angiography. *Journal of Neurosurgery*, *116*(5), 1024–1034.

Kamalian, S., Kemmling, A., Borgie, R. C., Morais, L. T., Payabvash, S., Franceschi, A. M., et al. (2013). Admission insular infarction >25% is the strongest predictor of large mismatch loss in proximal middle cerebral artery stroke. *Stroke*, *44*, 3084–3089.

Kann, J. (1950). A translation of Broca's original article on the location of the speech center. *Journal of Speech and Hearing Disorders*, *15*, 16–20.

Kelly, A. M. C., Uddin, L. Q., Biswal, B. B., Castellanos, F. X., & Milham, M. P. (2008). Competition between functional brain networks mediates behavioral variability. *NeuroImage*, *39*, 527–537.

Kelly, C., Toro, R., Di Martino, A., Cox, C. L., Bellec, P., Castellanos, F. X., et al. (2012). A convergent functional architecture of the insula emerges across imaging modalities. *NeuroImage*, *61*(4), 1129–1142.

Kim, K. R., Ku, J., Lee, J. H., Lee, H., & Jung, Y. C. (2012). Functional and effective connectivity of anterior insula in anorexia nervosa and bulimia nervosa. *Neuroscience Letters*, *521*(2), 152–157.

Klein, T. A., Ullsperger, M., & Danielmeier, C. (2013). Error awareness and the insula: Links to neurological and psychiatric diseases. *Frontiers in Human Neuroscience*, *7*, 14.

Klingbeil, G. M. (1939). The historical background of the modern speech clinic. Part two—Aphasia. *Journal of Speech Disorders*, *4*, 267–284.

Kosillo, P., & Smith, A. T. (2010). The role of the human anterior insular cortex in time processing. *Brain Structure and Function*, *214* (5–6), 623–628.

Kurth, F., Eickhoff, S. B., Schleicher, A., Hoemke, L., Zilles, K., & Amunts, K. (2010). Cytoarchitecture and probabilistic maps of the human posterior insular cortex. *Cerebral Cortex*, *20*, 1448–1461.

Lazar, R. M., & Mohr, J. P. (2011). Revisiting the contributions of Paul Broca to the study of aphasia. *Neuropsychology Review*, *21*(3), 236–239.

Lee, W., & Reeve, J. (2013). Self-determined, but not non-self-determined, motivation predicts activations in the anterior insular cortex: An fMRI study of personal agency. *Social, Cognitive, and Affective Neuroscience*, *8*(5), 538–545.

Manoliu, A., Riedl, V., Doll, A., Bauml, J. G., Muhlau, M., Schwerthoffer, D., et al. (2013). Insular dysfunction reflects altered between-network connectivity and severity of negative symptoms in schizophrenia during psychotic remission. *Frontiers in Human Neuroscience*, *7*, 216.

Menon, V., & Uddin, L. Q. (2010). Saliency, switching, attention, and control: A network model of insula function. *Brain Structure and Function*, *214*(5–6), 655–667.

Mesulam, M. M. (Ed.), (2000). *Principles of behavioral and cognitive neurology* (2nd ed.). New York, NY: Oxford University Press.

Mesulam, M. M. (2003). Some anatomic principles related to behavioral neurology and neuropsychology. In T. E. Feinberg, & M. J. Farah (Eds.), *Behavioral neurology and neuropsychology* (2nd ed., pp. 45–56). New York, NY: McGraw-Hill Professional.

Mesulam, M. M., & Mufson, E. G. (1982). Insula of the old world monkey. III. Efferent cortical output and comments on function. *Journal of Comparative Neurology*, *212*(1), 38–52.

Mesulam, M. M., & Mufson, E. G. (1985). The insula of Reil in man and monkey. Architectonics, connectivity and function. In A. Peters, & E. G. Jones (Eds.), *Cerebral cortex* (Vol. 4, pp. 179–226). New York, NY: Plenum.

Morel, A., Gallay, M. N., Baechler, A., Wyss, M., & Gallay, D. S. (2013). The human insula: Architectonic organization and postmortem MRI registration. *Neuroscience*, *236*, 117–135.

Naeser, M. A., Palumbo, C. L., Helm-Estabroks, N., Stiassny-Eder, D., & Albert, M. L. (1989). Severe nonfluency in aphasia: Role of the medial subcallosal fasciculus and other white matter pathways in recovery of spontaneous speech. *Brain*, *112*, 1–38.

Nagai, M., Kishi, K., & Kato, S. (2007). Insular cortex and neuropsychiatric disorders: A review of recent literature. *European Psychiatry*, *22*(6), 387–394.

Nieuwenhuys, R. (2012). The myeloarchitectonic studies on the human cerebral cortex of the Vogt-Vogt school, and their significance for the interpretation of functional neuroimaging data. *Brain Structure and Function*, *218*, 303–352.

Nimchinsky, E. A., Gilissen, E., Allman, J. M., Perl, D. P., Erwin, J. M., & Hof, P. R. (1999). A neuronal morphologic type unique to humans and great apes. *Proceedings of the National Academy of Sciences of the United States of America*, *96*(9), 5266–5273.

Palaniyappan, L., & Liddle, P. F. (2012). Does the salience network play a cardinal role in psychosis? An emerging hypothesis of insular dysfunction. *Journal of Psychiatry and Neuroscience*, *37*(1), 17–27.

Palumbo, C. L., Alexander, M. P., & Naeser, M. A. (1992). CT scan lesion sites associated with conduction aphasia. In S. E. Kohn (Ed.), *Conduction aphasia* (pp. 51–76). Hillsdale, NJ: Lawrence Erlbaum.

Pearce, J. M. (2009). Broca's aphasiacs. *European Neurology*, *61*(3), 183–189.

Richardson, J. D., Fillmore, P., Rorden, C., Lapointe, L. L., & Fridriksson, J. (2012). Re-establishing Broca's initial findings. *Brain and Language*, *123*(2), 125–130.

Rossi, S., Lubin, A., Simon, G., Lanoe, C., Poirel, N., Cachia, A., et al. (2013). Structural brain correlates of executive engagement in working memory: Children's inter-individual differences are reflected in the anterior insular cortex. *Neuropsychologia*, *51*(7), 1145–1150.

Sak, J., & Grzybowski, A. (2013). Brain and aviation: On the 80th anniversary of Constantin von Economo's (1876–1931) death. *Neurological Sciences*, *34*(3), 387–391.

Saur, D., Kreher, B. W., Schnell, S., Kummerer, D., Kellmeyer, P., Vry, M.-S., et al. (2008). Ventral and dorsal pathways for language. *Proceedings of the National Academy of Sciences of the United States of America, 105*(46), 18035–18040.

Schuell, H., Jenkins, J. J., & Jimenez-Pabon, E. (1964). *Aphasia in adults: Diagnosis, prognosis, and treatment*. New York, NY: Harper & Row.

Seeley, W. W., Menon, V., Schatzberg, A. F., Keller, J., Glover, G. H., Kenna, H., et al. (2007). Dissociable intrinsic connectivity networks for salience processing and executive control. *Journal of Neuroscience, 27*(9), 2349–2356.

Seeley, W. W., Merkle, F. T., Gaus, S. E., Craig, A. D., Allman, J. M., & Hof, P. R. (2012). Distinctive neurons of the anterior cingulated and frontoinsular cortex: A historical perspective. *Cerebral Cortex, 22*, 245–250.

Shelley, B. P., & Trimble, M. R. (2004). The insular lobe of Reil—Its anatomico-functional, behavioural and neuropsychiatric attributes in humans: A review. *World Journal of Biological Psychiatry, 5*, 176–200.

Sliz, D., & Hayley, S. (2012). Major depressive disorder and alterations in insular cortical activity: A review of current functional magnetic imaging research. *Frontiers in Human Neuroscience, 6*, 323.

Sridharan, D., Levitin, D. J., & Menon, V. (2008). A critical role for the right fronto-insular cortex in switching between central-executive and default-mode networks. *Proceedings of the National Academy of Sciences of the United States of America, 105*(34), 12569–12574.

Stephani, C., Vaca, G. F.-B., Maciunas, R., Koubeissi, M., & Luders, H. O. (2011). Functional neuroanatomy of the insular lobe. *Brain Structure and Function, 216*, 137–149.

Tanrovier, N., Rhoton, A. L., Jr., Kawashima, M., Ulm, A. J., & Yasuda, A. (2004). Microsurgical anatomy of the insula and the Sylvian fissure. *Journal of Neurosurgery, 100*(5), 891–922.

Tatu, L., Moulin, T., Bogousslavsky, J., & Duvernoy, H. (1998). Arterial territories of the human brain: Cerebral hemispheres. *Neurology, 50*(6), 1699–1708.

Tourville, J. A., & Guenther, F. H. (2011). The DIVA model: A neural theory of speech acquisition and production. *Language and Cognitive Processes, 26*(7), 952–981.

Ture, U., Yasargil, D. C. H., Al-Mefty, O., & Yasargil, M. G. (1999). Topographic anatomy of the insular region. *Journal of Neurosurgery, 90*, 720–733.

Ture, U., Yasargil, M. G., Al-Mefty, O., & Yasargil, D. C. H. (2000). Arteries of the insula. *Journal of Neurosurgery, 92*, 676–687.

Varnavas, G. G., & Grand, W. (1999). The insular cortex: Morphological and vascular anatomic characteristics. *Neurosurgery, 44*, 127–136.

Walter, U., Kolbaske, S., Patejdl, R., Steinhagen, V., Abu-Mugheisib, M., Grossman, A., et al. (2013). Insular stroke is associated with acute sympathetic hyperactivation and immunodepression. *European Journal of Neurology, 20*, 153–159.

CHAPTER

11

White Matter Pathways in the Human

Hugues Duffau[1,2]

[1]Department of Neurosurgery, Hôpital Gui de Chauliac, Montpellier University Medical Center, Montpellier, France; [2]Team "Plasticity of Central Nervous System, Stem Cells and Glial Tumors," INSERM U1051, Institute for Neuroscience of Montpellier, Montpellier University Medical Center, Montpellier, France

11.1 INTRODUCTION

For many decades, the neural basis underlying cognitive functions was conceived in a localizationist and fixed framework. According to this view, the brain was thought to be organized in highly specialized critical zones (e.g., Broca's and Wernicke's areas) for which any injury induces severe and irrevocable functional impairment, and in "noneloquent" regions (that is, areas that are not generating neurological deficits when they are damaged). This dogma of a static functional organization of the brain had numerous implications for neuroscience, notably resulting in the elaboration of simplistic and rigid models of cognition. In this setting, the subcortical connectivity has received less attention, especially in humans. Although many studies on the structural anatomy of white matter pathways have been performed in monkeys because of the possibility of using anterograde and retrograde tracers (Schmahmann & Pandya, 1990), in the 20th century only a few works of anatomic dissection on cadavers were reported in humans (Déjerine, 1895; Klingler, 1935; Ludwig & Klinger, 1956).

Because of the development of connectomics (i.e., the map of neural connections), an alternative hodotopical account was proposed in which brain functions are subserved by the dynamic interactions of large-scale distributed and parallel subnetworks (Catani, 2007; de Benedictis & Duffau, 2011; Duffau, 2013). The recent development of diffusion tensor imaging (DTI), allowing a tractography of white matter tracts *in vivo*, provided new insights into the cerebral connectivity (Catani, Jones, & ffytche, 2005). However, it is noteworthy that every recent review article about DTI has noted its shortcomings and underlined the need for rigorous anatomical validation (Hubbard & Parker, 2009; Jbabdi & Johansen-Berg, 2011; Le Bihan, Poupon, Amadon, & Lethimonnier, 2006). In this context, Klingler's work has received renewed interest because it is applicable for correlating the DTI results to actual anatomic evidence provided by postmortem dissection of white matter bundles (Fernandez-Miranda et al., 2008, 2012). However, the classical fiber dissection methodology consists of extensive removal of surrounding brain tissue and hampers analysis of cortical terminations. Recently, a modification of Klinger's technique was proposed in which removal of brain tissue was kept to a minimum to preserve the cortex and relationships within the brain until the end of the dissection (Martino et al., 2011). Using this cortex-sparing fiber dissection, the trajectory and the orientation of white matter tracts as well as their cortical terminations can now be identified reliably and reproducibly, including in humans (Martino et al., 2013; Sarubbo, De Benedictis, Maldonado, Basso, & Duffau, 2013). A validation of this new technique in a macaque would nonetheless be useful.

Beyond structural considerations, a better understanding of the functional role of subcortical pathways was made possible thanks to advances in intraoperative mapping achieved in patients who underwent awake surgery for brain lesions. Direct electrical stimulation (DES) of the brain offers a unique opportunity to investigate functional anatomy. It has become common clinical practice to awaken patients to assess the functional role of restricted cortical and subcortical regions, and to avoid neurological impairments. Patients perform cognitive tasks while DES temporarily inactivates discrete brain areas: if the patients produce wrong response, then the stimulated site is preserved. DES interacts locally with a small cortical or axonal site, but also nonlocally, as the focal perturbation disrupts the

Neurobiology of Language. DOI: http://dx.doi.org/10.1016/B978-0-12-407794-2.00011-0

whole (sub)network sustaining a given function. Therefore, DES induces a transient virtual lesion by inhibiting a subcircuit during a few seconds. By gathering all cortical and axonal sites where the same type of errors were observed when stimulated, one can build-up the subnetwork of the disrupted subfunction. DES identifies with a great accuracy (approximately 5 mm) and reproducibility, *in vivo* in humans, the structures—not only cortex but also white matter tracts—crucial for cognitive functions (Duffau, 2011). Combining transient disturbances elicited by DES with the anatomical data provided by preoperative and postoperative MRI enables performance of real-time anatomo-functional correlations both at cortical and subcortical levels, supporting a network organization of the brain and leading to the reappraisal of cognitive models—notably regarding language (Duffau, Moritz-Gasser, & Mandonnet, 2013).

Here, the goal is to review the recent literature on the structure and function of white matter pathways in humans. In addition, the limiting role of subcortical tracts in cerebral plasticity is discussed.

11.2 PROJECTION PATHWAYS

11.2.1 Motor Corticospinal Tract and Somatosensory Thalamocortical Pathways

Recent anatomical–functional studies have transformed our understanding of cerebral motor control away from a hierarchical structure and toward parallel and interconnected specialized circuits. First, anatomic dissection, DTI, and DES in humans have confirmed that the corticospinal tracts come from the primary motor cortex run with a somatotopical organization within the corona radiata (with, from lateral to medial, the pyramidal tracts of the face, upper limb, and lower limb) and then within the posterior limb of the internal capsule (with, from anterior to posterior, the pyramidal tracts of the face, upper limb, and lower limb) before reaching the brainstem and the spinal cord (Duffau, Capelle, Denvil, Sichez, et al., 2003; Maldonado, Mandonnet, & Duffau, 2012). Furthermore, the existence of an additional "modulatory motor network" was recently evidenced, eliciting movement arrest or acceleration when stimulated in awake patients, with no loss of consciousness. The subcortical stimulation sites were distributed veil-like, anterior to the primary motor fibers, suggesting descending pathways originating from premotor areas known for negative motor response characteristics. Further stimulation sites in the anterior arm of the internal capsule indicate a large-scale motor control circuit (Schucht, Moritz-Gasser, Herbet, Raabe, & Duffau, 2012). More recently, the first evidence of bilateral negative motor responses elicited by unilateral subcortical DES has been reported. Such findings support the existence of a bilateral cortico-subcortical network connecting the premotor cortices, basal ganglia, and spinal cord involved in the control of bimanual coordination (Rech, Herbet, Moritz-Gasser, & Duffau, 2013).

Posterior thalamocortical somatosensory pathways and their somatotopy have also been investigated by anatomic studies and using DES, which generate dysesthesias or tingling in awake patients (Duffau, Capelle, Denvil, Sichez, et al., 2003). Of note, stimulation of the white matter under the retrocentral gyrus may also induce disturbances in movement control (Almairac, Herbet, Moritz-Gasser, & Duffau, 2014), possibly due to transient inhibition of U fibers within the rolandic region (Catani et al., 2012). Regarding a frontoparietal sensory-motor network connection through thalamic nuclei, few studies in nonhuman primates demonstrate connections between these two structures (Gharbawie, Stepniewska, Burish, & Kaas, 2010). Anatomical tracer studies have shown that the posterior parietal cortex receives input from motor nuclei in the thalamus and shares overlapping thalamocortical connections with the frontal cortex (Gharbawie et al., 2010). In human, on the basis of DES findings, a fronto-thalamo-parietal network cannot be excluded (Almairac et al., 2014; Schucht et al., 2012). Moreover, in patients who experienced interference with movement during subcortical DES, fibers that induced inhibition or acceleration were located immediately posterior to thalamocortical somatosensory pathways. Therefore, a thalamo-parietal connection distinct from somatosensory pathways remains a possibility (Almairac et al., 2014; Schucht et al., 2012).

11.2.2 Optic Radiations

The optic radiations arise from the lateral geniculate body in three bundles. The anterior bundle curves anterolaterally above the temporal horn (Meyer's loop), usually reaching beyond the anterior limit of the temporal horn, and then loops backward along the inferolateral wall of the atrium. The middle bundle courses laterally around and turns posteriorly along the lateral wall of the atrium and the occipital horn. The posterior bundle courses directly backward, also along the lateral wall of the atrium and occipital horn.

Recently, visual pathways have also been mapped in awake patients. Interestingly, their stimulation generates a "shadow" (negative effect) or phosphenes (positive effect) in the controlateral visual field, possibly associated with metamorphopsia (i.e., visual illusion) or visual hemiagnosia. In all cases, these various and complex phenomena lead to a transitory visual deficit in the contralateral hemifield (Gras-Combes, Moritz-Gasser, Herbet, & Duffau, 2012).

11.3 LONG-DISTANCE ASSOCIATION PATHWAYS

Since the seminal work by Underleider and Haxby (1994), the process of visual information has been divided in a dorsal stream dedicated to the analysis of the spatial position ("where") and in a ventral stream specialized in object identification ("what"). By analogy, a dual-stream model for auditory language processing was suggested, with a dorsal stream involved in mapping sound to articulation and a ventral stream involved in mapping sound to meaning (Hickok & Poeppel, 2004). New insights into the connectivity subserving this model have recently been provided in humans (Dick, Bernal, & Tremblay, 2013).

11.3.1 The Dorsal Superior Longitudinal Fascicle/Arcuate Fascicle Complex

11.3.1.1 Anatomy

Recent fiber dissection and DTI tractography studies in humans have investigated the anatomical connectivity of the superior longitudinal fascicle (SLF)/arcuate fascicle (AF) complex (Catani et al., 2005; Martino et al., 2013). The different components of the perisylvian SLF were isolated and the fibers were followed until their cortical terminations. Three segments of the perisylvian SLF were identified: (i) anterior segment of the lateral SLF, connecting the supramarginal gyrus and superior temporal gyrus (in the region just posterior to the Heschl's gyrus) with the ventral portion of the precentral gyrus (ventral premotor cortex); (ii) posterior segment of the lateral SLF, connecting the posterior portion of the middle temporal gyrus with the angular gyrus; and (iii) long segment of the AF, deeply located, stemming from the caudal part of the temporal lobe, mainly the inferior and middle temporal gyri, that arches around the insula and advances forward to end within the frontal lobe, essentially within the precentral gyrus and posterior portion of the inferior and middle frontal gyri. Based on these original results, and challenging the traditional view, it was suggested that the fibers from the posterior part of the superior temporal gyrus are part of the anterior portion of the perisylvian SLF and not of the AF (Martino et al., 2013).

11.3.1.2 Structural–Functional Correlations

In awake patients performing a picture-naming task, cortically, phonemic paraphasia can be elicited by DES of the inferior parietal lobule and inferior frontal gyrus in the dominant hemisphere (Maldonado, Moritz-Gasser, & Duffau, 2011). Axonally speaking, phonemic paraphasias were elicited when stimulating the AF (Duffau et al., 2002; Martino et al., 2013). Geschwind (1970) previously postulated that lesions of this tract would produce conduction aphasia, including phonemic paraphasia, and this supports the role of the subpart of the dorsal stream mediated by the AF in phonological processing. Interestingly, the posterior cortical origin of the AF within the posterior part of the inferior temporal gyrus corresponds to the visual object form area (Martino et al., 2013). This region represents a functional hub involved in semantic and phonological processing dedicated to visual material (Vigneau et al., 2006). Thus, phonological processing subserved by the AF is performed in parallel to the semantic processes implemented by the ventral route. In addition to this direct dorsal route, the indirect dorsal stream of the lateral SLF is implied in articulation and phonological working memory, as demonstrated by DES. Cortical areas eliciting articulatory disorders are located in the ventral premotor cortex, supramarginal gyrus, and posterior part of the superior temporal gyrus (Duffau, Capelle, Denvil, Gatignol, et al., 2003). Axonally, stimulation of the white matter under the frontoparietal operculum and supramarginal gyrus, laterally and ventrally to the AF, induces anarthria as well (van Geemen, Herbet, Moritz-Gasser, & Duffau, 2014). This bundle corresponds to part III of the SLF according to Makris et al. (2005). This lateral operculo-opercular component of the SLF constitutes the articulatory loop by connecting the supramarginal gyrus/posterior portion of the superior temporal gyrus (which receives feedback information from somatosensory and auditory areas) with the frontal operculum (which receives afferences bringing the phonological/phonetic information to be translated into articulatory motor programs and efferences toward the primary motor area) (Duffau, Capelle, Denvil, Gatignol, et al., 2003).

Using the same paradigm, DES also demonstrated that syntactic processing was subserved by delocalized cortical regions (left inferior frontal gyrus and posterior middle temporal gyrus) connected by a subpart of the left SLF. Interestingly, this subcircuit is interacting but independent of the subnetwork involved in naming, as demonstrated by a double dissociation between syntactic (especially grammatical gender) and naming processing during DES. These findings support a parallel rather than serial theory, calling into question the principle of "lemma" (Vidorreta, Garcia, Moritz-Gasser, & Duffau, 2011).

The left SLF seems also to underpin word repetition (Moritz-Gasser & Duffau, 2013).

In the same vein, DES showed that spatial cognition was subserved by several cortical areas, including the right supramarginal gyrus, connected together by a subpart of the SLF (Thiebaut de Schotten et al., 2005). In such a "connectionist" view of brain organization, interactions between different systems have also been described. DES evidenced the existence of an executive system (including prefrontal cortex, anterior cingulate, and caudate nucleus) involved in the cognitive control of a more dedicated subcircuit for language switching—itself constituted by a wide cortico-subcortical network comprising postero-temporal areas, supramarginal and angular gyri, inferior frontal gyrus, and a subpart of the SLF (Moritz-Gasser & Duffau, 2009). In addition, it seems that the frontal aslant tract, which connects the presupplementary motor area and anterior cingulate with the inferior frontal gyrus (Catani et al., 2013; Thiebaut de Schotten, Dell'Acqua, Valabregue, & Catani, 2012; Vergani et al., 2014), might play a role in language control, especially with regard to planning of speech articulation (Kinoshita et al., 2014). In the same vein, a cortico-subcortical loop involving the deep gray nuclei, especially the caudate nucleus was also demonstrated as participating in the control of language (selection/inhibition) because DES of the head of the caudate nucleus in the left hemisphere generated perseverations with a high level of reliability (Gil Robles, Gatignol, Capelle, Mitchell, & Duffau, 2005). This cortico-striatal loop could be anatomically supported by the frontostriatal tract (Kinoshita et al., 2014).

11.3.2 The Ventral Stream

11.3.2.1 Anatomy

The ventral route connects the occipital and posterior temporal areas with the frontal lobe. This ventral stream is referred to by some authors as "extreme capsule fiber system" with reference to connectivity studies in the primate (Makris et al., 2009; Saur et al., 2008). It nonetheless seems more adapted to speak about fascicles rather than the "extreme capsule," because the latter only considers a discrete anatomical structure and the former considers actual neural pathways with their cortical termination in a hodotopical view. If one takes account of the sole subcortical region without any considerations regarding the cortical epicenters connected by the white matter tracts, then it does not allow the understanding of the whole eloquent network.

Using both anatomic dissection and DTI, it has recently been demonstrated that the ventral stream is supported by direct and indirect pathways. The direct pathway is represented by the inferior fronto-occipital fascicle (IFOF). This IFOF has never been described in animals, explaining the debate about its role (Schmahmann & Pandya, 2006). In humans, the IFOF is a ventral associative bundle that connects the occipital lobe, parietal lobe, and the postero-temporal cortex with the frontal lobe. Recent anatomic studies (Martino, Brogna, Gil Robles, Vergani, & Duffau, 2010) combined with DTI have investigated the main course of the IFOF (Sarubbo et al., 2013). From the posterior cortex, it runs within the sagittal stratum in the superior and lateral part of the atrium; it reaches the roof of the sphenoidal horn in the temporal lobe; it joins the ventral part of the external/extreme capsule and runs under the insula at the posterior two-thirds of the temporal stem; and then it joins the frontal lobe (Martino et al., 2010). Two layers of the IFOF have been described (Sarubbo et al., 2013). The superficial and dorsal layers connect the posterior portion of the superior and middle occipital gyri, the superior parietal lobule, and the posterior part of the superior temporal gyrus to the inferior frontal gyrus (pars triangularis and opercularis). The deep and ventral subcomponent connects the posterior portion of the inferior occipital gyrus, the posterior temporal-basal area including the Fusa (fusiform area at the occipito-temporal junction), and the posterior part of the middle temporal gyrus to the frontal lobe—orbitofrontal cortex, middle frontal gyrus, and dorsolateral prefrontal cortex.

In parallel, the ventral stream is subserved by an indirect pathway comprising the anterior part of the inferior longitudinal fascicle (ILF) (running below the IFOF) that links the posterior occipitotemporal region (Fusa) and the temporal pole (TP) and is then relayed by the uncinate fasciculus (UF) that connects the TP to the basifrontal areas by running within the anterior third of the temporal stem (in front of the IFOF) (Mandonnet, Nouet, Gatignol, Capelle, & Duffau, 2007). Of note, the posterior part of the ILF links the occipital lobe to the posterior occipitotemporal junction (visual object form area) (Mandonnet, Gatignol, & Duffau, 2009). This means that this indirect route connects the occipital/Fusa to the orbitofrontal cortex, which is partially overlapped with the IFOF. Finally, although previously observed in monkey, another pathway has recently been described in humans: the middle longitudinal fascicle (MdLF) (Makris et al., 2009). It connects the angular gyrus with the superior temporal gyrus up to the TP and courses under the superior temporal sulcus, lateral and superior to the IFOF (Maldonado et al., 2013; Menjot et al., 2013).

11.3.2.2 Structural–Functional Correlations

In the awake patient during picture naming, DES of the IFOF, at least in the left dominant hemisphere, elicited semantic paraphasias either associative (e.g.,/key/ for /padlock/) or coordinate (e.g., /tiger/ for /lion/) in more than 85% of cases. It did not matter what portion of the IFOF was stimulated (parieto-occipital junction, temporal, subinsular, or frontal part) (Duffau et al., 2005). These language disorders were mainly generated by stimulating the superficial layer of the IFOF. Interestingly, semantic paraphasias were never observed during stimulation of the dorsal route (SLF) (Maldonado et al., 2011). Of note, IFOF stimulation may also generate verbal perseveration (Khan, Herbet, Moritz-Gasser, & Duffau, 2013).

DES of the IFOF also induced nonverbal comprehension disturbances in more than 90% of cases during nonverbal semantic association test (e.g., Pyramid and Palm Trees Test). The patients were no longer able to make a semantic choice during DES, with some of them still able to join a short verbal description of their feelings, like "I don't know at all," "what do I have to do?," and "I don't understand anything" (Moritz-Gasser, Herbet, & Duffau, 2013). These comprehension disorders were mainly generated by stimulating the deep layer of the IFOF and elicited a double dissociation: semantic paraphasia with normal nonverbal semantic choice during DES of superficial IFOF and *vice versa* during DES of deep IFOF. Thus, it was suggested that there exists a superficial component involved in verbal semantics and a deep component involved in amodal semantic processing.

These data are in agreement with the cortical terminations of the IFOF (prefrontal, temporal-basal, and parietal areas) that correspond with the cortical network involved in semantic control (Whitney, Kirk, O'Sullivan, Lambon-Ralph, & Jefferies, 2011). Consequently, an original anatomo-functional model of semantic processing has recently been proposed in which the crucial pathway is represented by the IFOF. In this model, visual information is processed at the level of the occipital and temporal-basal associative cortices, and auditory information is processed at the level of the temporal and parietal associative cortices. They are transmitted directly on an amodal shape to the prefrontal areas, which exert top-down control over this amodal information to achieve successful semantic processing in a given context. DES of this fascicle generates a disruption of these rapid direct connections. The transient semantic disorganization observed when stimulating the IFOF would therefore be caused by dis-synchronization within this large-scale network, simultaneously interrupting the bottom-up transmission and the top-down control mechanisms (Moritz-Gasser et al., 2013). Thus, IFOF might play a crucial role not only in verbal and nonverbal semantic processing but also in the awareness of amodal semantic knowledge, namely noetic consciousness. From a phylogenic perspective, because recent studies in the primate failed to identify this tract, one could suggest that the IFOF is the proper human fascicle. This multi-function fascicle allows humans to produce and understand language, to manipulate concepts, and to comprehend the world (i.e., metalinguistics, conceptualization, and awareness of knowledge), and it contributes to making the human what he/she is, with his/her infinite wealth of mind.

The functional role of the indirect ventral pathway is still debated. This indirect route connects areas involved in semantic processing such as Fusa and lateral frontal cortex (Vigneau et al., 2006). Moreover, the major cortical relay between the ILF and UF is the TP, which is a "hub" (i.e., a functional epicenter allowing a plurimodal integration of the multiple data coming from the unimodal systems [subserved by ILF, UF, and MdLF]), explaining its role in semantics and its implication in semantic dementia when (bilaterally) damaged (Holland & Lambon-Ralph, 2010). However, except for the posterior part of the ILF for which injury generates alexia (Chan-Seng, Moritz-Gasser, & Duffau, 2014; Mandonnet et al., 2009), the indirect pathway can be functionally compensated when (unilaterally) damaged. DES of both the anterior ILF and UF never elicited any naming or nonverbal semantic disorders (Duffau, Gatignol, Moritz-Gasser, & Mandonnet, 2009; Mandonnet et al., 2007). This was also confirmed by functional recovery after anterior temporal lobectomy in tumor and in epilepsy surgery (Duffau, Thiebaut de Schotten, & Mandonnet, 2008). Even if very mild and selective deficit may persist, as with proper name retrieval after resection of the UF (Papagno, Miracapillo, et al., 2011), this is a good illustration of the concept of "subcortical plasticity" in which a subnetwork (IFOF, direct pathway) is able to bypass another subnetwork (indirect pathway) and functionally compensate for it (Duffau, 2009). Similarly, DES of MdLF and resection of its anterior part failed to induce any functional disorders (De Witt Hamer, Moritz-Gasser, Gatignol, & Duffau, 2011), demonstrating that this fascicle converging to the TP can also be compensated.

11.4 IMPLICATION OF A HODOTOPICAL VIEW OF BRAIN ORGANIZATION IN HUMANS: RETHINKING THE CONNECTIVITY OF LANGUAGE AND ITS RELATIONSHIPS WITH COGNITION

These original data based on the better knowledge of functional neuroanatomy of the subcortical white matter pathway result in the elaboration of new models of cognition. For example, the connectivity underpinning picture naming based on multiple direct and indirect cortico-subcortical interacting subnetworks involved in

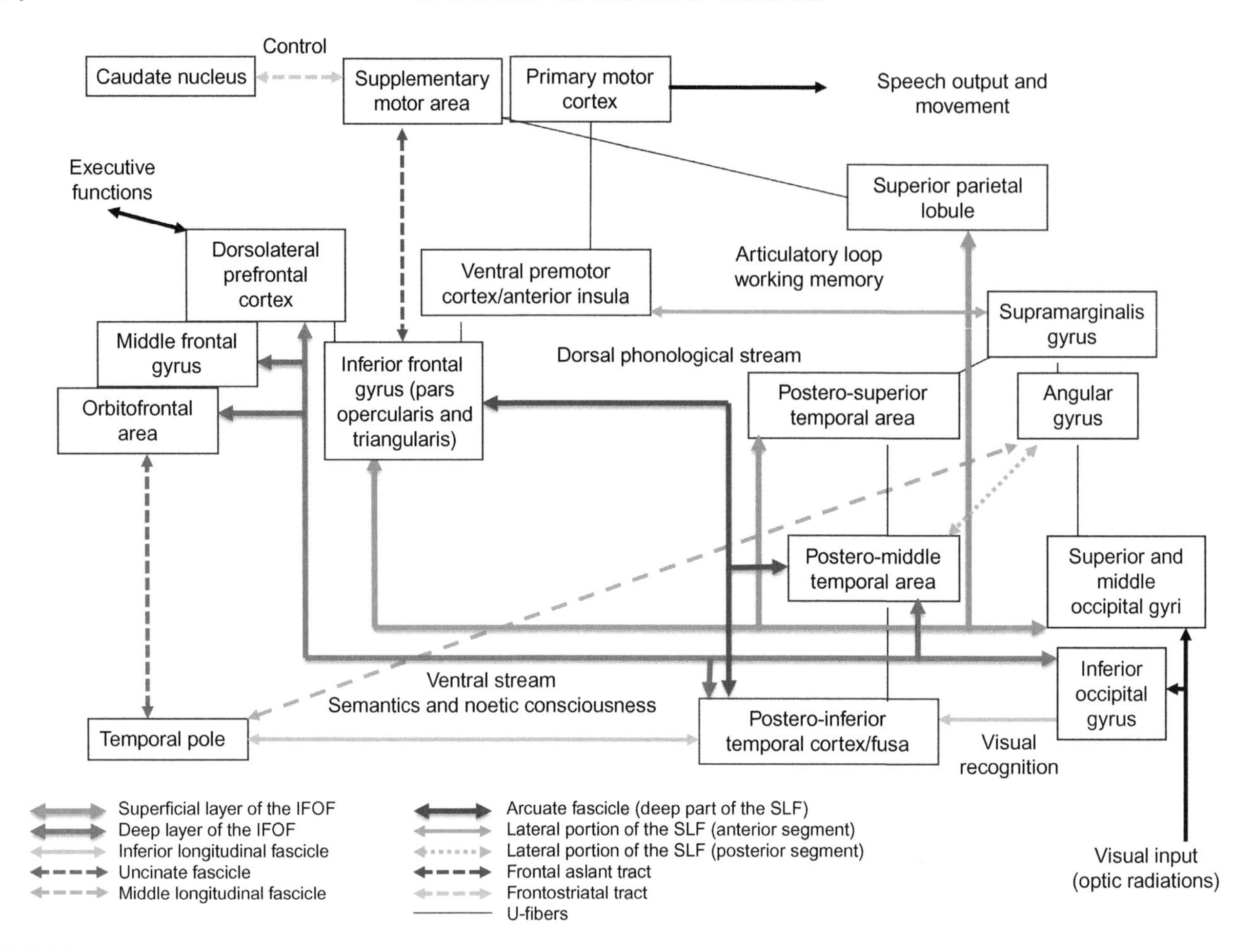

FIGURE 11.1 Proposal of a new model of connectivity underlying language processing and its relationships with executive functions (as judgment, decision-making, or problem-solving), with incorporation of anatomic subcortical constraints and elaborated on the basis of structural–functional correlations provided by dissection, DTI, and intraoperative DES. *Modified from Duffau et al. (2013) with permission from the publisher.*

semantic, phonological, and articulatory processes has recently been reevaluated. This model offers several advantages in comparison with previous ones: it explains double dissociations during axonal DES (e.g., semantic versus phonemic paraphasias); it takes into account the subcortical anatomic constraints; and it explains the possible recovery of aphasia after a lesion within the "classical" language areas (Figure 11.1).

Beyond language, by using subcortical DES and performing anatomo-functional correlations, it has recently been demonstrated that low-level and high-level mentalizing accuracy were correlated with the degree of (dis) connection in the AF and the cingulum, respectively (Herbet, Lafargue, Bonnetblanc, et al., 2014). These findings, which constitute the first experimental data on the structural connectivity of the mentalizing network, suggest the existence of a dual-stream hodotopical model and could lead to a better understanding of disorders that affect social cognition. Finally, intraoperative DES of the posterior cingulate connectivity may also elicit transient behavioral unresponsiveness with loss of external connectedness, supporting the role of functional integrity of this subcortical network for maintaining consciousness of external environment (Herbet, Lafargue, de Champfleur, et al., 2014).

11.5 THE LIMITING ROLE OF AXONAL CONNECTIVITY IN BRAIN PLASTICITY

Although a huge plastic potential has been demonstrated in humans at the cortical level, subcortical plasticity is low, implying that axonal connectivity should be preserved to allow postlesional compensation (Papagno, Gallucci, et al., 2011). Lessons from stroke studies have taught us that damage of the white matter pathways generates a more severe neurological outcome than lesions of the cortex. Recently, we proposed the elaboration of a probabilistic postsurgical residue atlas computed using a series of patients who underwent incomplete resection for a glioma on the basis of intraoperative DES (Ius, Angelini, de Schotten, Mandonnet, & Duffau, 2011). We gathered 58 postoperative MRI results of patients who underwent operation for WHO grade II glioma under direct electrical cortico-subcortical stimulation with a tumor resection performed according to functional boundaries. Postoperative images were registered on the Montreal Neurological Institute (MNI) template to construct an atlas of functional resectability for which each voxel represents the probability of observing residual nonresectable glioma, that is, noncompensable area (Ius et al., 2011). The anatomo-functional correlations obtained by combining the intrasurgical functional data with postoperative anatomical MRI findings provided new insights into the potentials and limitations of cerebral plasticity. This probabilistic atlas highlighted the crucial role of the axonal pathways in the reorganization of the brain after a lesion. It provided a general framework to establish anatomo-functional correlations by computing for each brain voxel its probability to be left—due to its functional role—on the postoperative MRI. Its overlap with the cortical MNI template and a DTI atlas offered a unique tool to analyze the potentialities and the limitations of interindividual variability and plasticity for cortical areas and axonal pathways. We observed low probability of residual tumors on the cortical surface, whereas most of the regions with high probability of residual tumor were located in the deep white matter. Thus, projection and association axonal pathways seem to play a critical role in the proper functioning of the brain. In other words, the functions subserved by long-range axonal pathways seem to be less subject to interindividual variability and reorganization than cortical sites (Duffau, 2009; Ius et al., 2011). The reproducibility of these results may suggest the existence of a "minimal common brain" necessary for the basic cognitive functions, even if likely insufficient for more complex functions such as multiprocessing.

11.6 CONCLUSION

Neuroscientists should improve their knowledge of white matter circuitry. Understanding axonal connectivity is crucial to optimize the models of neurocognition (movement, language, behavior, and even consciousness), which should integrate the anatomic constraint represented by the subcortical pathways. In addition, according to the principle of connectomics, cerebral plasticity seems to be possible only on the condition that the white matter fibers are preserved to allow spatial communication and temporal synchronization among large interconnected networks.

References

Almairac, F., Herbet, G., Moritz-Gasser, S., & Duffau, H. (2014). Parietal network underlying movement control: Disturbances during subcortical electrostimulation. *Neurosurgical Review, 37*, 513–516.

Catani, M. (2007). From hodology to function. *Brain, 130*, 602–605.

Catani, M., Dell'acqua, F., Vergani, F., Malik, F., Hodge, H., Roy, P., et al. (2012). Short frontal lobe connections of the human brain. *Cortex, 48*, 273–291.

Catani, M., Jones, D. K., & ffytche, D. H. (2005). Perisylvian language networks of the human brain. *Annals of Neurology, 57*, 8–16.

Catani, M., Mesulam, M. M., Jakobsen, E., Malik, F., Martersteck, A., Wieneke, C., et al. (2013). A novel frontal pathway underlies verbal fluency in primary progressive aphasia. *Brain, 136*, 2619–2628.

Chan-Seng, E., Moritz-Gasser, S., & Duffau, H. (2014). Awake mapping for low-grade gliomas involving the left sagittal stratum: Anatomofunctional and surgical considerations. *Journal of Neurosurgery, 120*, 1069–1077.

de Benedictis, A., & Duffau, H. (2011). Brain hodotopy: From esoteric concept to practical surgical applications. *Neurosurgery, 68*, 1709–1723.

Déjerine, J. (1895). *Anatomie des centres nerveux. Tome 1*. Paris: Rueff et Cie.

De Witt Hamer, P., Moritz-Gasser, S., Gatignol, P., & Duffau, H. (2011). Is the human left middle longitudinal fascicle essential for language? A brain electrostimulation study. *Human Brain Mapping, 32*, 962–973.

Dick, A. S., Bernal, B., & Tremblay, P. (2013). The language connectome: New pathways, new concepts. *Neuroscientist, 20*, 453–467.

Duffau, H. (2009). Does post-lesional subcortical plasticity exist in the human brain? *Neuroscience Research, 65*, 131–135.

Duffau, H. (2011). *Brain mapping: From neural basis of cognition to surgical applications*. New York, NY: Springer Wien.

Duffau, H. (2013). The huge plastic potential of adult brain and the role of connectomics: New insights provided by serial mappings in glioma surgery. *Cortex, 58*, 325–337.

Duffau, H., Capelle, L., Denvil, D., Gatignol, P., Sichez, N., Lopes, M., et al. (2003). The role of dominant premotor cortex in language: A study using intraoperative functional mapping in awake patients. *NeuroImage, 20*, 1903–1914.

Duffau, H., Capelle, L., Denvil, D., Sichez, N., Gatignol, P., Taillandier, L., et al. (2003). Usefulness of intraoperative electrical subcortical mapping during surgery for low-grade gliomas located within eloquent brain regions: Functional results in a consecutive series of 103 patients. *Journal of Neurosurgery, 98*, 764–778.

Duffau, H., Capelle, L., Sichez, N., Denvil, D., Bitar, A., Sichez, J. P., et al. (2002). Intraoperative mapping of the subcortical language pathways using direct stimulations. An anatomo-functional study. *Brain, 125*, 199–214.

Duffau, H., Gatignol, P., Mandonnet, E., Peruzzi, P., Tzourio-Mazoyer, N., & Capelle, L. (2005). New insights into the anatomo-functional connectivity of the semantic system: A study using cortico-subcortical electrostimulations. *Brain, 128*, 797–810.

Duffau, H., Gatignol, P., Moritz-Gasser, S., & Mandonnet, E. (2009). Is the left uncinate fasciculus essential for language? A cerebral stimulation study. *Journal of Neurology, 256*, 382–389.

Duffau, H., Moritz-Gasser, S., & Mandonnet, E. (2013). A re-examination of neural basis of language processing: Proposal of a dynamic hodotopical model from data provided by brain stimulation mapping during picture naming. *Brain and Language, 131*, 1–10.

Duffau, H., Thiebaut de Schotten, M., & Mandonnet, E. (2008). White matter functional connectivity as an additional landmark for dominant temporal lobectomy. *Journal of Neurology, Neurosurgery, and Psychiatry, 79*, 492–495.

Fernandez-Miranda, J. C., Pathak, S., Engh, J., Jarbo, K., Verstynen, T., Yeh, F. C., et al. (2012). High-definition fiber tractography of the human brain: Neuroanatomical validation and neurosurgical applications. *Neurosurgery, 71*, 430–453.

Fernandez-Miranda, J. C., Rhoton, A. L., Jr., Alvarez-Linera, J., Kakizawa, Y., Choi, C., & de Oliveira, E. P. (2008). Three-dimensional microsurgical and tractographic anatomy of the white matter of the human brain. *Neurosurgery, 62*, 989–1026.

Geschwind, N. (1970). The organization of language and the brain. *Science, 170*, 940–944.

Gharbawie, O. A., Stepniewska, I., Burish, M. J., & Kaas, J. H. (2010). Thalamocortical connections of functional zones in posterior parietal cortex and frontal cortex motor regions in new world monkeys. *Cerebral Cortex, 20*, 2391–2410.

Gil Robles, S., Gatignol, P., Capelle, L., Mitchell, M. C., & Duffau, H. (2005). The role of dominant striatum in language: A study using intraoperative electrical stimulations. *Journal of Neurology, Neurosurgery, and Psychiatry, 76*, 940–946.

Gras-Combes, G., Moritz-Gasser, S., Herbet, G., & Duffau, H. (2012). Intraoperative subcortical electrical mapping of optic radiations in awake surgery for glioma involving visual pathways. *Journal of Neurosurgery, 117*, 466–473.

Herbet, G., Lafargue, G., Bonnetblanc, F., Moritz-Gasser, S., Menjot de Champfleur, N., & Duffau, H. (2014). Inferring a dual-stream model of mentalizing from associative white matter fibres disconnection. *Brain, 137*, 944–959.

Herbet, G., Lafargue, G., de Champfleur, N. M., Moritz-Gasser, S., le Bars, E., Bonnetblanc, F., et al. (2014). Disrupting posterior cingulate connectivity disconnects consciousness from the external environment. *Neuropsychologia, 56*, 239–244.

Hickok, G., & Poeppel, D. (2004). Dorsal and ventral streams: A framework for understanding aspects of the functional anatomy of language. *Cognition, 92*, 67–99.

Holland, R., & Lambon-Ralph, M. A. (2010). The anterior temporal lobe semantic hub is a part of the language neural network: Selective disruption of irregular past tense verb by rTMS. *Cerebral Cortex, 20*, 2771–2775.

Hubbard, P. L., & Parker, G. J. M. (2009). Validation of tractography. In H. Johansen-Berg, & T. Behrens (Eds.), *Diffusion MRI* (pp. 353–376). Burlington: Academic Press.

Ius, T., Angelini, E., de Schotten, M. T., Mandonnet, E., & Duffau, H. (2011). Evidence for potentials and limitations of brain plasticity using an atlas of functional respectability of WHO grade II gliomas: towards a "minimal common brain". *NeuroImage, 56*, 992–1000.

Jbabdi, S., & Johansen-Berg, H. (2011). Tractography: Where do we go from here? *Brain Connect, 1*, 169–183.

Khan, O. H., Herbet, G., Moritz-Gasser, S., & Duffau, H. (2013). The role of left inferior fronto-occipital fascicle in verbal perseveration: A brain electrostimulation mapping study. *Brain Topography, 27*, 403–411.

Kinoshita, M., Menjot de Champfleur, N., Deverdun, J., Moritz-Gasser, S., Herbet, G., & Duffau, H. (2014). Role of fronto-striatal tract and frontal aslant tract in movement and speech: An axonal mapping study. *Brain Structure and Function*, August 3 [Epub ahead of print].

Klingler, J. (1935). Erleichterung der makroskopischen praeparation des gehirns durch den gefrierprozess. *Schweizer Archiv fur Neurologie und Psychiatrie, 36*, 247–256.

Le Bihan, D., Poupon, C., Amadon, A., & Lethimonnier, F. (2006). Artifacts and pitfalls in diffusion MRI. *Journal of Magnetic Resonance Imaging, 24*, 478–488.

Ludwig, E., & Klinger, J. (1956). *Atlas cerebri humani*. Boston, Toronto: Little, Brown.

Makris, N., Kennedy, D. N., McInerney, S., Sorensen, A. G., Wang, R., Caviness, V. S., Jr., et al. (2005). Segmentation of subcomponents within the superior longitudinal fascicle in humans: A quantitative, *in vivo*, DT-MRI study. *Cerebral Cortex, 15*, 854–869.

Makris, N., Papadimitriou, G. M., Kaiser, J. R., Sorg, S., Kennedy, D. N., & Pandya, D. N. (2009). Delineation of the middle longitudinal fascicle in humans: A quantitative, *in vivo*, DT-MRI study. *Cerebral Cortex, 19*, 777–785.

Maldonado, I. L., de Champfleur, N. M., Velut, S., Destrieux, C., Zemmoura, I., & Duffau, H. (2013). Evidence of a middle longitudinal fasciculus in the human brain from *in vitro* dissection. *Journal of Anatomy, 223*, 38–45.

Maldonado, I. L., Mandonnet, E., & Duffau, H. (2012). Dorsal fronto-parietal connections of the human brain: A fiber dissection study of their composition and anatomical relationships. *The Anatomical Record (Hoboken), 295*, 187–195.

Maldonado, I. L., Moritz-Gasser, S., & Duffau, H. (2011). Does the left superior longitudinal fascicle subserve language semantics? A brain electrostimulation study. *Brain Structure and Function, 216*, 263–264.

Mandonnet, E., Gatignol, P., & Duffau, H. (2009). Evidence for an occipito-temporal tract underlying visual recognition in picture naming. *Clinical Neurology and Neurosurgery, 111*, 601–605.

Mandonnet, E., Nouet, A., Gatignol, P., Capelle, L., & Duffau, H. (2007). Does the left inferior longitudinal fasciculus play a role in language? A brain stimulation study. *Brain, 130*, 623–629.

Martino, J., Brogna, C., Gil Robles, S., Vergani, F., & Duffau, H. (2010). Anatomic dissection of the inferior fronto-occipital fasciculus revisited in the lights of brain stimulation data. *Cortex, 46*, 691–699.

Martino, J., De Witt Hamer, P. C., Berger, M. S., Lawton, M. T., Arnold, C. M., de Lucas, E. M., et al. (2013). Analysis of the subcomponents and cortical terminations of the perisylvian superior longitudinal fasciculus: A fiber dissection and DTI tractography study. *Brain Structure and Function, 218*, 105–121.

Martino, J., De Witt Hamer, P. C., Vergani, F., Brogna, C., de Lucas, E. M., Vázquez-Barquero, A., et al. (2011). Cortex-sparing fiber dissection: An improved method for the study of white matter anatomy in the human brain. *Journal of Anatomy, 219*, 531–541.

Menjot de Champfleur, N., Maldonado, I. L., Moritz-Gasser, S., Machi, P., Le Bars, E., Bonafé, A., et al. (2013). Middle longitudinal fasciculus delineation within language pathways: A diffusion tensor imaging study in human. *European Journal of Radiology, 82*, 151–157.

Moritz-Gasser, S., & Duffau, H. (2009). Cognitive processes and neural basis of language switching: Proposal of a new model. *Neuroreport, 20*, 1577–1580.

Moritz-Gasser, S., & Duffau, H. (2013). The anatomo-functional connectivity of word repetition: Insights provided by awake brain tumor surgery. *Frontiers in Human Neuroscience, 7*, 405.

Moritz-Gasser, S., Herbet, G., & Duffau, H. (2013). Mapping the connectivity underlying multimodal (verbal and non-verbal) semantic processing: A brain electrostimulation study. *Neuropsychologia, 51*, 1814–1822.

Papagno, C., Gallucci, M., Casarotti, A., Castellano, A., Falini, A., Fava, E., et al. (2011). Connectivity constraints on cortical reorganization of neural circuits involved in object naming. *NeuroImage, 55*, 1306–1313.

Papagno, C., Miracapillo, C., Casarotti, A., Romero Lauro, L. J., Castellano, A., Falini, A., et al. (2011). What is the role of the uncinate fasciculus? Surgical removal and proper name retrieval. *Brain, 134*, 405–414.

Rech, F., Herbet, G., Moritz-Gasser, S., & Duffau, H. (2013). Disruption of bimanual movement by unilateral subcortical electrostimulation. *Human Brain Mapping, 35*, 3439–3445.

Sarubbo, S., De Benedictis, A., Maldonado, I. L., Basso, G., & Duffau, H. (2013). Frontal terminations for the inferior fronto-occipital fascicle: Anatomical dissection, DTI study and functional considerations on a multi-component bundle. *Brain Structure and Function, 218*, 21–37.

Saur, D., Kreher, B. W., Schnell, S., Kümmerer, D., Kellmeyer, P., Vry, M. S., et al. (2008). Ventral and dorsal pathways for language. *Proceedings of the National Academy of Sciences of the United States of America, 105*, 18035–18040.

Schmahmann, J. D., & Pandya, D. N. (1990). Anatomical investigation of projections from thalamus to posterior parietal cortex in the rhesus monkey: A WGA-HRP and fluorescent tracer study. *The Journal of Comparative Neurology, 295*, 299–326.

Schmahmann, J. D., & Pandya, D. N. (2006). *Fiber pathways of the brain*. New York, NY: Oxford University Press.

Schucht, P., Moritz-Gasser, S., Herbet, G., Raabe, A., & Duffau, H. (2012). Subcortical electrostimulation to identify network subserving motor control. *Human Brain Mapping, 34*, 3023–3030.

Thiebaut de Schotten, M., Dell'Acqua, F., Valabregue, R., & Catani, M. (2012). Monkey to human comparative anatomy of the frontal lobe association tracts. *Cortex, 48*, 82–96.

Thiebaut de Schotten, M., Urbanski, M., Duffau, H., Volle, E., Levy, R., Dubois, B., et al. (2005). Direct evidence for a parietal-frontal pathway subserving spatial awareness in humans. *Science, 309*, 2226–2228.

Underleider, L. G., & Haxby, J. V. (1994). "What" and "where" in the human brain. *Current Opinion in Neurobiology, 4*, 157–165.

van Geemen, K., Herbet, G., Moritz-Gasser, S., & Duffau, H. (2014). Limited plastic potential of the left ventral premotor cortex in speech articulation: Evidence from intraoperative awake mapping in glioma patients. *Human Brain Mapping, 35*, 1587–1596.

Vergani, F., Lacerda, L., Martino, J., Attems, J., Morris, C., Mitchell, P., et al. (2014). White matter connections of the supplementary motor area in humans. *Journal of Neurology, Neurosurgery, and Psychiatry, 85*, 1377–1385.

Vidorreta, J. G., Garcia, R., Moritz-Gasser, S., & Duffau, H. (2011). Double dissociation between syntactic gender and picture naming processing: A brain stimulation mapping study. *Human Brain Mapping, 32*, 331–340.

Vigneau, M., Beaucousin, V., Herve, P. Y., Duffau, H., Crivello, F., Houdé, O., et al. (2006). Meta-analyzing left hemisphere language areas: Phonology, semantics, and sentence processing. *NeuroImage, 30*, 1414–1432.

Whitney, C., Kirk, M., O'Sullivan, J., Lambon Ralph, M. A., & Jefferies, E. (2011). The neural organization of semantic control: TMS evidence for a distributed network in left inferior frontal and posterior middle temporal gyrus. *Cerebral Cortex, 21*, 1066–1075.

S E C T I O N C

BEHAVIORAL FOUNDATIONS

CHAPTER

12

Phonology

William J. Idsardi[1,2] *and Philip J. Monahan*[3,4]

[1]Department of Linguistics, University of Maryland, College Park, MD, USA; [2]Neuroscience and Cognitive Science Program, University of Maryland, College Park, MD, USA; [3]Centre for French and Linguistics, University of Toronto Scarborough, Toronto, ON, Canada; [4]Department of Linguistics, University of Toronto, Toronto, ON, Canada

12.1 INTRODUCTION

Phonology is typically defined as "the study of speech sounds of a language or languages, and the laws governing them,"[1] particularly the laws governing the composition and combination of speech sounds in language. This definition reflects a segmental bias in the historical development of the field and we can offer a more general definition: the study of the knowledge and representations of the sound system of human languages. From a neurobiological or cognitive neuroscience perspective, one can consider phonology as the study of the mental model for human speech. In this brief review, we restrict ourselves to spoken language, although analogous concerns hold for signed language (Brentari, 2011). Moreover, we limit the discussion to what we consider the most important aspects of phonology. These include: (i) the mappings between three systems of representation: action, perception, and long-term memory; (ii) the fundamental components of speech sounds (i.e., distinctive features); (iii) the laws of combinations of speech sounds, both adjacent and long-distance; and (iv) the chunking of speech sounds into larger units, especially syllables.

To begin, consider the word-form "glark." Given this string of letters, native speakers of English will have an idea of how to pronounce it and what it would sound like if another person said it. They would have little idea, if any, of what it means.[2] The meaning of a word is arbitrary given its form, and it could mean something else entirely. Consequently, we can have very specific knowledge about a word's *form* from a single presentation and can recognize and repeat such word-forms without much effort, all without knowing its meaning. Phonology studies the regularities of form (i.e., "rules without meaning") (Staal, 1990) and the laws of combination for speech sounds and their sub-parts.

Any account needs to address the fact that speech is produced by one anatomical system (the mouth) and perceived with another (the auditory system). Our ability to repeat new word-forms, such as "glark," is evidence that people effortlessly map between these two systems. Moreover, new word-forms can be stored in both short-term and long-term memory. As a result, phonology must confront the conversion of representations (i.e., data structures) between three broad neural systems: memory, action, and perception (the MAP loop; Poeppel & Idsardi, 2011). Each system has further sub-systems that we ignore here. The basic proposal is that this is done through the use of phonological primitives (features), which are temporally organized (chunked, grouped, coordinated) on at least two fundamental time scales: the feature or segment and the syllable (Poeppel, 2003).

12.2 SPEECH SOUNDS AND THE MAP LOOP

The alphabet is an incredible human invention, but its ubiquity overly influences our ideas regarding the

[1]Longman Dictionary of Contemporary English.

[2]Urban Dictionary (http://www.urbandictionary.com/) states that it means "to slowly grasp the meaning of a word or concept, based on the situation in which it is used" (i.e., almost grokking a concept).

Neurobiology of Language. DOI: http://dx.doi.org/10.1016/B978-0-12-407794-2.00012-2

basic units of speech. This continues to this day and is evident in the influence of the International Phonetic Alphabet (IPA; http://www.langsci.ucl.ac.uk/ipa/) for transcribing speech. Not all writing systems are alphabetic, however. Some languages choose orthographic units larger than single sounds (moras, syllables) and a few, such as Bell's Visible Speech (Bell, 1867) and the Korean orthographic system Hangul (Kim-Renaud, 1997), decompose sounds into their component articulations, all of which constitute important, interconnected representations for speech.

12.2.1 Action or Articulation of Speech

The musculature of the mouth has historically been somewhat more accessible to investigation than audition or memory, and linguistic phonetics has often displayed a bias toward classifying speech sounds in terms of the actions needed to produce them (i.e., the articulation of the speech sounds by the mouth). For example, the standard IPA charts for consonants and vowels (Figure 12.1) are organized by how speech sounds are articulated. The columns in Figure 12.1A arrange consonants with respect to *where* they are articulated in the mouth (note: right to left corresponds to anterior to posterior position within the oral cavity), and the rows correspond to how they are articulated (i.e., their manner of articulation). The horizontal dimension in Figure 12.1B represents the relative frontness-backness of the tongue, and the vertical dimension represents the aperture of the mouth during production. These are the standard methods for organizing consonant and vowel inventories in languages.

Within the oral cavity, there are several controllable structures used to produce speech sounds. These include the larynx, the velum, the tongue (which is further divided into three relatively independently moveable sections: the tongue blade, the tongue dorsum, and the tongue root), and the lips (see Figure 12.2 reproduced from Bell, 1867; for more detail see Zemlin, 1998).

Each of these structures has some degrees of freedom of movement, which we describe in terms of their deflection from a neutral posture for speaking. The position for the mid-central vowel schwa, [ə], is considered to be the neutral posture of the speech articulators. In most structures, two opposite directions of movement are possible, yielding three stable regions of articulation, that is, the tongue dorsum can be put into a high, mid (neutral), or low position. In the neutral posture the velum is closed, but it can be opened to allow air to flow through the nose, and such speech sounds are classified as nasal (as in English "m" [m]). The lips can deflect from the neutral posture by being rounded (as in English "oo" [u]) or drawn back (as in English "ee" [i]). The tongue tip can be curled concavely or convexly either along its length (yielding retroflex and laminal sounds, respectively) or across its width (yielding grooved and lateral sounds, respectively). The tongue dorsum (as mentioned) can be moved vertically (high or low) and horizontally (front or back), and the tongue root can be moved horizontally (advanced or retracted).

The larynx (Figure 12.3) is particularly complex and can be moved along three different dimensions: modifying its vertical position (raised or lowered), modifying its tilt (rotated forward to slacken the vocal folds or rotated backwards to stiffen them), and changing the degree of separation of the vocal folds (adducted or abducted). Furthermore, the lips and the tongue blade and dorsum can close off the mouth to different degrees (termed the "manner" of production): completely closed (stops), nearly closed with turbulent airflow (fricatives), or substantially open (approximants). Taken together, these articulatory maneuvers describe how to make various speech sounds. For example, an English [s], as in "sea", is an abducted (voiceless, high glottal airflow) grooved fricative. Furthermore, as described in Section 12.3, the antagonistic relationships between articulator movements serve as the basis for the featural distinctions (whether

(A)

Place of articulation

Manner of articulation	Bilabial		Labio-dental		Interdental		Alveolar		Alveo-palatal		Palatal		Velar		Glottal	
Stop	p	b					t	d					k	g		
Fricative			f	v	θ	ð	s	z	ʃ	ʒ						h
Affricate									tʃ	dʒ						
Nasal		m						n						ŋ		
Lateral								l								
Retroflex								ɹ								
Glide		w										j				

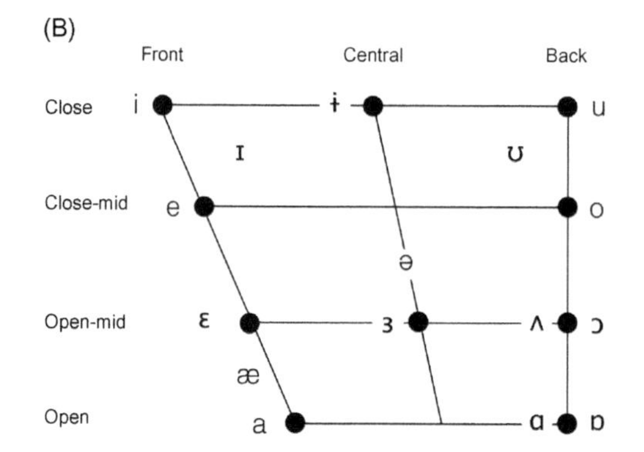

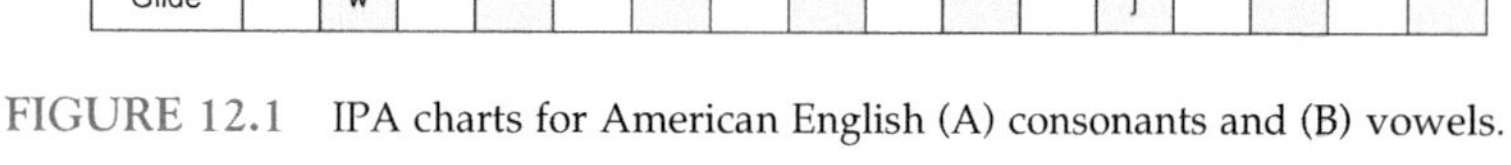

FIGURE 12.1 IPA charts for American English (A) consonants and (B) vowels.

monovalent, equipollent, or binary; see Fant, 1973; Trubetzkoy, 1969) that have proven so powerful in understanding not only the composition of speech sounds but also the phonology of human language.

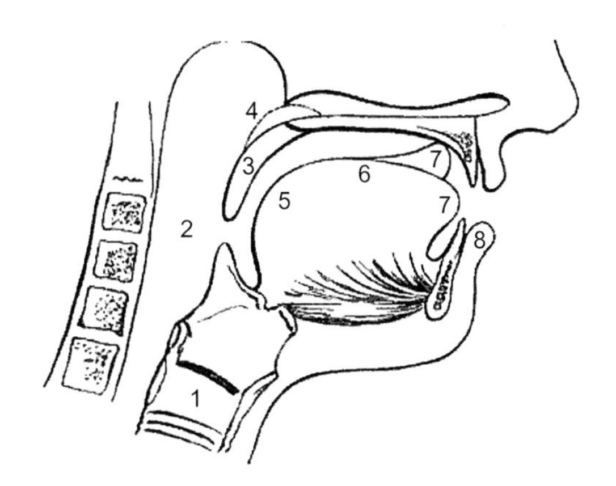

FIGURE 12.2 Speech articulators. Note that terminology in 1–8 are all still in current usage (Zemlin, 1998:251) and that there are many synonymous terms in common use. In the current chapter, articulator 5 is known as the tongue dorsum, 6 is the tongue blade or corona, and 7 is the tongue tip or apex. *From Bell (1867) with permission from the publishers.*

12.2.2 Perception or Audition of Speech

A great deal of the literature regarding speech perception deals with how "special" speech is (Liberman, 1996) or is not. Often, this is cast as a debate between the motor theory of speech perception (Liberman & Mattingly, 1985) and speech as an area of expertise within general auditory perception (Carbonnell & Lotto, 2014). The motor theory of speech perception posits speech-specific mechanisms that recover the intended articulatory gestures that produced the physical auditory stimulus. General auditory perception models, however, posit that the primary representational modality of speech perception is auditory and the mechanisms used during speech perception are the same as those responsible for nonspeech auditory perception. This dichotomy, in some ways, parallels debates about face perception (Rhodes, Calder, Johnson, & Haxby, 2011). Since the development of the sound spectrograph (Potter, Kopp, & Green, 1947) and the Haskins pattern playback machine (Cooper, Liberman, & Borst, 1951), it has been known that it is technologically feasible to analyze and accurately reproduce speech with time–frequency–amplitude analysis techniques, as in the spectrogram in Figure 12.4, where time is on the horizontal axis, frequency is on the vertical axis, and amplitude is illustrated in the relative darkness of the pixels.

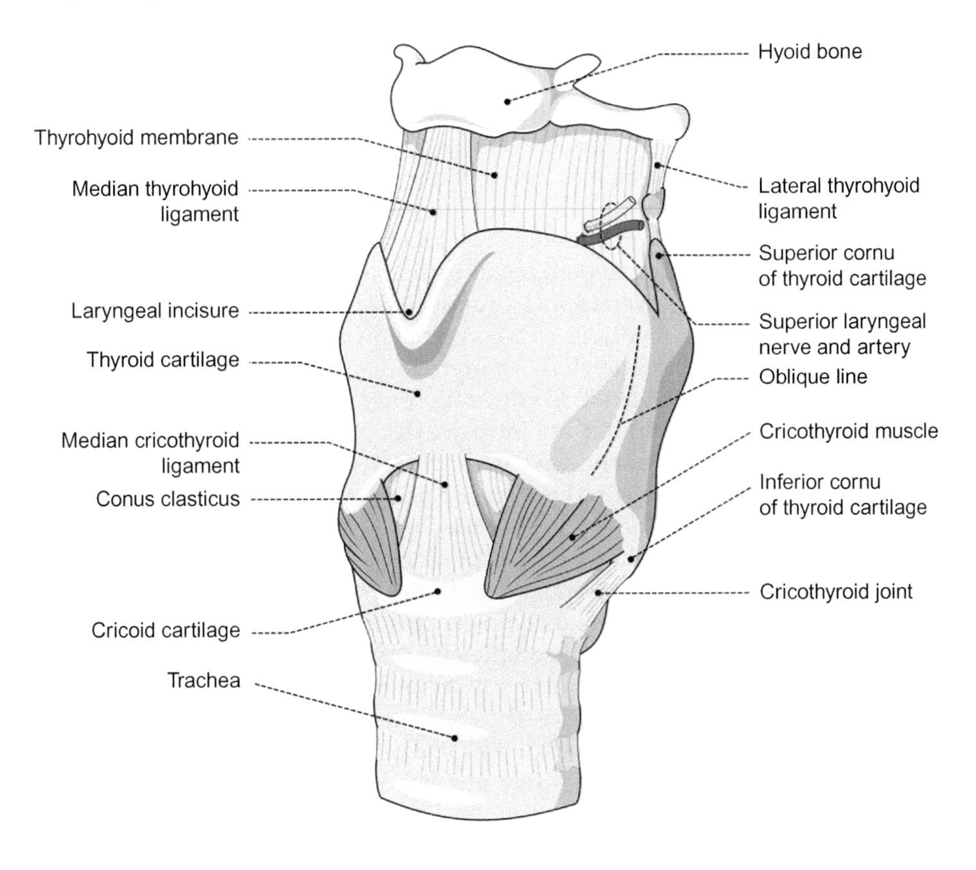

FIGURE 12.3 The external view of the larynx. *From Olek Remesz, http://commons.wikimedia.org/wiki/File:Larynx_external_en.svg, with permission from the publisher.*

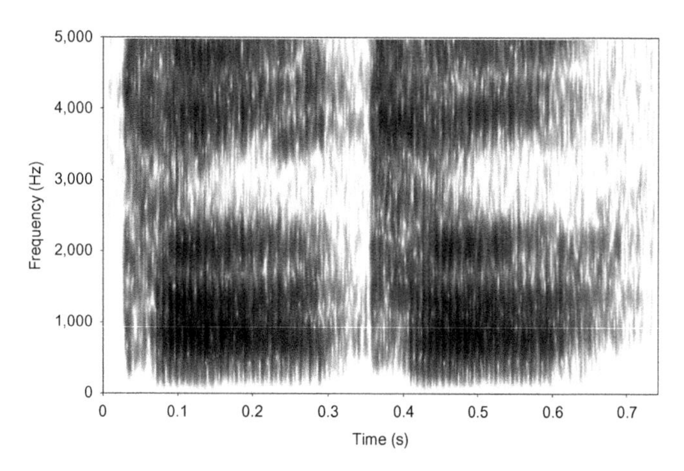

FIGURE 12.4 Sound spectrogram of a male voice saying "tata."

Linking such technologies with neuroscience, Aertsen and Johannesma (1981) demonstrated that auditory neurons display particular spectro-temporal receptive fields (STRFs) to auditory stimuli, akin to a set of building blocks for spectrograms. So, this sets a strong goal for phonology: find lawful relationships between the available articulator movements and their acoustic/auditory consequences (as in Stevens, 1998), especially in terms of STRFs. We return to this question in the discussion of features. More recently, Mesgarani, David, Fritz, and Shamma (2014) have shown that STRFs derived from measuring responses in ferret primary auditory cortex can be used to "clean" speech in various kinds of noise, including reverberation, demonstrating that the auditory system enhances the neural representation of speech events against background noise.

12.2.3 Memory or the Long-Term Storage of Speech

It is a remarkable fact that humans can retain detailed knowledge about a great number of words, arbitrarily pairing forms with meanings across tens of thousands of cases. This, again, is reminiscent of our memory abilities for faces. More remarkably, the landscape of the form-meaning relation is not at all smooth, because small physical changes in form between two words (what linguists call "minimal pairs") can have profound differences in meaning. If the form-meaning relation were smooth, then there would be many more pairs like "ram" and "lamb," which differ only in their first sound and share a great deal in meaning ("a male sheep" and "a young sheep," respectively, definitions from Merriam Webster's Collegiate Dictionary 11th edition). Instead, most cases are like "ramp" and "lamp," which differ in the same small sound attribute ("r" versus "l"), but share nothing discernible in meaning (see Hinton, Nichols, & Ohala, 1994 for cases of phonaesthesia).

The fundamental question of long-term memory representations in speech is one of abstraction: attending to and storing critical differences between forms while ignoring irrelevant differences. This is homomorphic to a fundamental problem in vision. A primary problem for visual object recognition is to account for "discrimination among diagnostic object features and generalization across nondiagnostic features" (Hoffman & Logothetis, 2009). The traditional linguistic solution to this problem has been to posit a single long-term memory representation (called the underlying representation [UR], notated with / /) and a set of transformations that yield the observed pronunciation variants (called surface representations [SRs], notated with [ ]). Determining URs for word-forms and the nature of the transformations responsible for SRs (given a particular UR) has been the core goal of generative phonology for the past 50 years (Chomsky & Halle, 1968) and survives as the dominant question in modern generative models of phonology, such as Optimality Theory (OT), which posit ranked violable constraints instead of transformations (Prince & Smolensky, 2004; see Idsardi, 2006 for an argument that OT may not be computationally tractable). This is the analog of "view invariant" models of visual object recognition (Booth & Rolls, 1998). The relation between various URs and SRs can become quite complicated, but it boils down to two kinds of basic cases: derived homophones and derived differences.

Consider the two words "intense" and "intents," which can both be pronounced so that their pronunciations end in [nts]. This is a case of derived homophones. Yoo and Blankenship (2003) show that in an experimental setting such pairs have substantially overlapping pronunciations, and also that a statistical analysis of a corpus shows that the [t] in "intense" is somewhat shorter on average. However, another related word-form, "intensive," is much less likely to exhibit an intrusive [t], with the "ns" usually being pronounced [ns]. Ohala (1997) provides a compelling misproduction explanation for this effect due to the difficulty in precisely coordinating the closing of the velum during production in /ns/ with the simultaneous transition from a complete closure in [n] to a narrow incomplete closure in [s]. If these two changes are not completely synchronized, then for a short period there can be a complete closure in the mouth and a closed velum, resulting in the presence of an intrusive [t]. As Ohala (1997: 85) notes: "the emergent stop is purely epiphenomenal and, indeed, such brief unintended stops are often observed in spontaneous speech." Now the listener faces the following problem: is that [t] a critical aspect of the word-form or is it an irrelevant detail? Presumably relying on statistics and the pronunciations of related word-forms, such as "intensive" and "intent," learners generally do arrive at different long-term memory representations, one with a /t/ ("intents") and one without ("intense"), as the usual English spellings reflect. But, Ohala argues, the listener cannot always reliably reconstruct the speaker's intent, and this is one cause of historical language change.

The other situation, that of derived differences, is already also illustrated by part of the previous example in that "intensive" is less likely to have an intrusive [t] in its pronunciation than "intense" is, so the pronunciations of the "intense" portion diverge. However, much more dramatic examples can be found in other English words. Consider the word "atom" in Standard American English, for instance. On its own, it can be pronounced as a homophone with "Adam" (the /t/ and /d/ both pronounced as a flap, notated as [ɾ]). When the suffix "-ic" is added to "atom" to create "atomic," its pronunciation contains a portion homophonous with "Tom," where the /t/ now has a pronunciation canonically associated with word-initial position, [tʰ]. Adding "-ic" to "Adam," however, gives "Adamic" with a portion pronounced as "damn." So, ideally we would like to know how to recognize /t/ from the speech signal. That is, what invariant aspects are obtained across [ɾ] and [tʰ] (and any other pronunciations of /t/)? Such speech sound sized memorized units are known as phonemes, or as archi-phonemes (Trubetzkoy, 1969) when their pronunciation ranges over an even wider variety of pronunciations, as in the final sound of the negative prefix "in-," which is pronounced [l] in "illegal," [ɹ] in "irregular," [m] in "impossible," and [ŋ] in "incomplete."

The goal, then, is to discover representations that abstract away from irrelevant changes in pronunciation but include diagnostic differences and differentiate between word-forms when they do have distinct pronunciations, such as "atomic" and "Adamic." These two words differ in [tʰ] and [d], motivating a /t/ versus /d/ difference in long-term memory. Note that they also differ in their middle vowels, /ɑ/ versus /æ/, a difference also missing in the corresponding portion of "atom" and "Adam."

Most importantly, memory serves as the mediating representation between audition and articulation (Gow, 2012). That is, the repetition of an auditorily presented word-form must be first mapped from a perceptual representation to a memory representation, and only then mapped to an articulatory representation. There is no short-cut directly from audition to articulation.

12.3 FEATURES OR THE INTERNAL COMPOSITION OF SOUNDS

Let us now return to the "glark" example, but this time consider its cursive written form in English, as in Figure 12.5.

The motions necessary to produce cursive "glark" can be accomplished with a finger (as in finger painting), a handheld stylus or pencil, a computer mouse (as was used to produce Figure 12.5), the tip of the nose, the elbow, or other body parts or instruments. This wide variety of available "articulators" to produce a cursive "glark" is a simple demonstration of the notion of motor equivalence (Wing, 2000). Motor equivalence is one kind of many-to-many mapping between causes and effects, and often is a symptom of an ill-posed inverse problem (of reasoning backward from effects to their causes). In contrast, speech exhibits only limited motor equivalence. It is not generally possible to substitute other coordinated sound-producing gestures (such as finger snapping) even when their acoustic effect might be appropriate (for example, as substitutes for the click sounds found in southern African languages, some of which are described as sounding like "sharply snapped fingers;" see http://en.wikipedia.org/wiki/Click_consonant). One stereotypical (and racist) example of a strong form of motor equivalence is the portrayal of Native American war chants with a "woo woo" sound produced by repetitively covering the mouth with a hand (and infants often explore such activities, e.g., https://www.youtube.com/watch?v=tcWmMPNUVb4). In this case, the acoustic effect of closing and opening an acoustic tube at one end can be accomplished either by closing and opening the lips or by placing and removing the palm of the hand from the lips; however, even though infants explore such gestures, no language uses the hand to make the bilabial approximant [w].

However, it is true that some conditions (ventriloquism, speaking with an object held by the teeth) show that some limited, approximate compensation is possible, but other simple "experiments" (such as the children's taunt of saying "I was born on a pirate ship" while holding your tongue with your fingers) show that some speech features can really only be performed by a single articulator, and no adequate compensation is possible. Speakers robustly compensate during production when their vocal tract is obstructed (Guenther, 2006), and vowels exhibit more

glark

FIGURE 12.5 The word "glark" as drawn with a mouse.

compensation possibilities than consonants generally (see Ladefoged & Maddieson, 1996: 300ff on tongue root position, for example), and English /r/ is notorious for its variety of articulations (Zhou, Espy-Wilson, Tiede, & Boyce, 2007). Although we have no solution to offer for the problem of /r/, we simply note that the range of different articulations for /r/ includes flaps, trills, and uvulars, which have a similarly wide range of acoustic manifestations, so this goes well beyond what is meant by motor equivalence (where there are multiple ways to achieve the same acoustic consequence). Whatever the reason for the extreme variability in /r/ realizations, it cannot be just motor equivalence.

Thus, given the limited amount of motor equivalence, we may have a reasonable inverse problem; therefore, we may also have the possibility of finding lawful relationships between articulatory actions and their acoustic/auditory consequences following the lead of Halle (1983), Jakobson, Fant, and Halle (1952), and especially Stevens (1998). The general neurobiological conceit here is that pre-existing auditory feature detectors for ecologically important events for mammals were paired with actions that the mouth could perform and thereby packaged as discrete units (features) for storage in long-term memory. Returning once again to the traditional articulatory descriptions of speech sounds, we describe them in terms of manner, place of articulation (POA), and laryngeal posture. Importantly, the cues for POA and laryngeal posture are dependent on the manner of articulation, so the cues for manner serve as "landmarks" for the analysis of the speech signal (Juneja & Espy-Wilson, 2008; Stevens, 2002) and constrain the subsequent search for POA and laryngeal attributes. We provide here only a very oversimplified catalog of these relationships.

One result that has persisted throughout theoretical innovation within phonology is the notion that speech sounds are composed of smaller units of representations (i.e., distinctive features; Jakobson et al., 1952). These features have traditionally been binary in nature (i.e., either a positive or negative value) and relate to a speech sound's articulation. For example, a consonant is either produced with the lips (i.e., [+labial]) or not (i.e., [-labial]), or a vowel is either produced with an advanced tongue root (i.e, [+ATR]) or not (i.e., [-ATR]). These binary distinctions incorporate the apparent antagonistic relationship between articulator positions discussed in Section 12.2.1. The power of distinctive features in generative phonology has been their ability to explain why certain cross-linguistic phonological patterns are observed and why others are not. Consider the case of word-final consonant devoicing, which occurs in German and Dutch among other languages. When the sounds /b, d, v, z, g/ occur in the final position of word, they are pronounced as their devoiced counterparts [p, t, f, s, k] (e.g., bewei[z]en "to prove" is produced as bewei[s] "proof"). The sounds that undergo this process form a natural class (i.e., voiced obstruents) that can be represented as [+ voiced, + obstruent]. The expectation, then, is that all consonants that are [+ voiced, + obstruent] that exist in German or Dutch are to participate in this process. Thus, a very straightforward transformation can be formulated to account for this process: /+voiced, +obstruent/ → [-voiced]/___ # (in word final position). What we do not observe in natural language is for phonological rules to target non-natural classes of sounds, such as /i, g, θ, j, a/. These sounds do not form a natural class and, consequently, are not predicted to undergo systematic alternations like what we observed in German word-final obstruent devoicing. Thus, the utility of distinctive features in explaining observed cross-linguistic phonological patterns is obvious. That the features themselves are cast in articulatory terms or binary in nature have been points of substantial debate. We would argue that neither is necessary to maintain their explanatory power. As a point of fact, distinctive features were initially described in auditory/acoustic terms (Jakobson et al., 1952; see Clements and Hume, 1995 for a discussion of different proposals for features valences).

There are four broad classes for manner of articulation: stops (plosives), fricatives, nasals, and approximants. Stops have a complete closure in the mouth and are characterized acoustically by a period of very low energy or silence often followed by a burst, similar to percussive environmental events with discontinuities such as twigs snapping or rocks colliding. Fricatives are made with a narrow channel causing turbulent airflow and are characterized acoustically by sustained aperiodic noise (which may be overlaid on a periodic signal) as with wind noise or rushing water. Nasals are made with a closed oral cavity and an open velum, and they are characterized acoustically by having a single, strong single frequency resonance, a hum such as some insects produce (but see Pruthi & Espy-Wilson 2004 for a comprehensive discussion of other important attributes for nasals). Finally, approximants are made with a relatively open vocal tract and are characterized by having a rich resonance structure of multiple formants; these sounds are characteristic of animal vocal tracts in particular and would have served as a useful kind of animal detector within the mammalian auditory system. Neurobiologically, this four-way division into "major class" features also seems to be well-reflected in a coherent cortical map. Mesgarani, Cheung, Johnson, and Chang (2014) measured the responses in implanted electrical cortical grids (ECOG) placed along the superior temporal gyrus (STG) in presurgical epilepsy patients and found remarkable correlations between articulatory characteristics (consistent with phonological classes) and single electrode sites. In the supplemental materials for their article, we can see that electrode e1 responds to stops, e2 responds to fricatives, e5 responds to nasals, and e3 and e4 respond to approximants with dorsal and coronal POA, respectively.

POA is primarily characterized by the most active (most displaced) articulator, as described: the lips (labial), tongue blade (coronal), tongue body (dorsal), tongue root (pharyngeal), or the larynx alone (laryngeal). As already noted, the acoustic correlates of POA are heavily dependent on the manner of articulation. In the case of approximants, POA is signaled by the relative frequencies of the first three formants (Monahan & Idsardi, 2010), in fricatives by the dispersion and center of the frication noise (often termed center of gravity), and in stops and fricatives by the formant transitions with neighboring approximants. POA also appears to be topologically organized in cortex. ECOG recordings (Bouchard, Mesgarani, Johnson, & Chang, 2013) find a two-dimensional dorso-ventral by anterior–posterior POA map in STG, and cluster analysis of the electrode responses recapitulates the POA categories labial, coronal, and dorsal. Most importantly, the POA categories cut across the various manner classifications (stop, fricative, nasal, and approximant). Taking the ECOG findings together, this suggests a complex spatially entwined set of multidimensional maps for manner and POA, broadly consistent with other speech sound featural maps found using magnetoencephalography (Scharinger, Idsardi, & Poe, 2012).

Phonetic feature information is encoded in the brain in several different ways. Along with the recent cortical topographic findings, it is also known that the latency of the evoked magnetoencephalographic M100 response tracks vowel height (Roberts, Flagg, & Gage, 2004), manner of articulation of consonants (Gage, Poeppel, Roberts, & Hickok, 1998), and POA of consonants (Gage, Roberts, & Hickok, 2002), and that the electrophysiological mismatch negativity (MMN) response is sensitive to native language phonetic and phonological category representations (Kazanina, Phillips, & Idsardi, 2006; Näätänen et al., 1997; Sharma & Dorman, 2000). So, although maps are often a satisfying answer to the question of neural coding, the brain seems to be using all the methods at its disposal in coding speech.

Now that we have features at our disposal, we can redefine our intuitive notion of segments as "feature bundles" (Chomsky & Halle, 1968: 64), overlapping features that are phonologically coordinated during a period of time.

12.4 LOCAL SOUND COMBINATIONS AND CHUNKING

Speech sounds can be combined together, but not all combinations are possible. Given the set of sounds {a, g, k, l, r}, some combinations are possible English words (e.g., the now-familiar "glark" and others like "gralk"), but other combinations are not licit, *"rlgka." One important constraint on sound sequences in word-forms is (approximately) that they must form a legal sequence of syllables in the language, and within syllables there are a limited set of pre-vowel and post-vowel consonant sequences. Moreover, these local sequence constraints often depend on the manner features of the segments. For example, a stop-approximant sequence is an acceptable pre-vowel sequence in English (as in "blue"), but approximant-stop is not (*"lbue"). Berent (2013) summarizes a number of investigations into this preference, which is exhibited even when listeners lack language experience to both kinds of consonant sequences ("bl" and "lb"). Building on this work, Berent et al. (2014) show modulation of the fMRI BOLD response across such contrasts, with anterior portions of BA 45 showing less activation to syllables beginning with "lb" and posterior portions showing more activation relative to that for "bl." Additionally, listeners are sensitive to vowel–consonant sequence restrictions, exhibiting negative deflections in event-related potentials in response to illicit sequences (Steinberg, Truckenbrodt, & Jacobsen, 2011). In short, listeners are clearly sensitive to licit and illicit local sequences within syllables, and these can be detected with various different methods.

But what is the motivation for another layer of organization that groups segments together? Ghitza (2012) and Ghitza, Giraud, and Poeppel (2012) suggest that the dual time-scale organization is related to endogenous brain rhythms in the theta (syllable) and gamma (segment) bands. The idea here is that syllables can be identified from gross, easily tracked properties of the speech signal, namely its envelope. Once syllables have been chunked, the signal can be further analyzed to yield segment and feature information, guided by the syllable parsing. This proposal is similar to the landmarks proposal reviewed in that an initial *coarse* coding of the signal is performed and then elaborated into finer distinctions as necessary. Taken together, these proposals suggest another possible indexation method for the mental lexicon, similar to hash tables. If the signal is coarsely processed into manner classes (Plosive, Fricative, Nasal, Approximant) and syllable chunks are identified, then listeners can retrieve the lexical items matching that coarse form (e.g., both "nest" [nɛst] and "mashed" [mæʃt] would fall into the

<NAFP> group). We could then do further segment and feature discovery guided by the manner class, the syllabic position, local sequence constraints, and the pool of retrieved candidates, which would significantly be a reduction in the search space of a couple orders of magnitude compared with the entire lexicon.

So, then do we need segments at all, or could we do with just organizing features into syllables without segmental organization? This has been proposed at least for perception (Hickok, 2014). The strongest argument for segments comes from resyllabification effects. Russian provides a particularly clear example. Consider the name "Ivan." Alone, this name is pronounced with two syllables, indicated with parentheses: [(i)(van)]. Russian nouns are inflected for case with suffixes, and so the form for "to Ivan" adds a prefix "k-" meaning "to," and the appropriate case suffix "-u," so that we have (abstractly) "k-Ivan-u." But Russian strongly prefers that every consonant–vowel sequence be grouped together into a syllable and, consequently, the syllabification in the pronounced form is [(kɨ)(va)(nu)] (the vowel change from [i] to [ɨ] is symptomatic of the resyllabification). Notice that now none of the original syllables survive into the derived form, but the segment sequence [...ivan...] does (with a slight vowel change). If we store forms as syllables and features without a segment level of representation, then we would have to build large tables of syllable correspondences such as (kɨ)⇔(i), (va)⇔(van) to recognize "Ivan" in "k-Ivan-u." But such syllable correspondences are straightforwardly captured if segments are available to us. Moreover, the extent of resyllabification is usually limited to a single segment, so, for example, Russian [(gart)] "printer type metal nom. sg." resyllabifies to [(gar)(ta)] in the plural, even though syllables can begin with /rt/, [(rtut')] "mercury nom. sg.". In languages with small syllable inventories and restricted syllable types (e.g., Hawaiian, Japanese), eschewing segmental representations might suffice; however, once we consider languages with complex syllable structures (e.g., Russian, Polish), frequent resyllabification (e.g., Korean), and complex morphology (e.g., Navajo), it becomes more difficult to maintain segment-less representations during perception. As such, resyllabification remains the greatest conceptual challenge to understanding the appropriate data structure for the organization of the mental lexicon.

12.5 NONLOCAL SOUND COMBINATIONS

The local sound sequence restrictions discussed are widely known and form the conceptual basis for n-gram models in natural language processing. Less well-known are the nonlocal, action-at-a-distance phonological effects such as vowel and consonant harmony and disharmony. For example, in languages with vowel harmony, the set of vowels of the language are divided into two classes and an individual word-form will canonically draw all of its vowels from only one of the two sets. To illustrate vowel harmony, consider the following paradigm from Turkish (Clements & Sezer, 1982: 216) providing the nominative (nom) and genitive (gen) forms for the singular (sg) and plural (pl) versions of representative nouns.

	nom.sg	**gen.sg**	**nom.pl**	**gen.pl**
"rope"	ip	ip-in	ip-ler	ip-ler-in
"girl"	kɨz	kɨz-ɨn	kɨz-lar	kɨz-lar-ɨn
"face"	yüz	yüz-ün	yüz-ler	yüz-ler-in
"stamp"	pul	pul-un	pul-lar	pul-lar-ɨn
"hand"	el	el-in	el-ler	el-ler-in
"stalk"	sap	sap-ɨn	sap-lar	sap-lar-ɨn
"village"	köy	köy-ün	köy-ler	köy-ler-in
"end"	son	son-un	son-lar	son-lar-ɨn

The suffix in the genitive singular forms alternates between [in]/[ɨn] and [ün]/[un]. The suffix is produced as [in]/[ɨn] when the root vowel is [-round] (/i, ɨ, e, a/) and [un]/[ün] when the root vowel is [+round] (/u, ü, o, ö/). Moreover, the suffix is produced with a front vowel [in]/[ün] when the root vowel is [-back] (/i, ü, e, ö/) and with a back vowel [ɨn]/[un] when the root vowel is [+back] (/ɨ, u, a, o/). In short, two dimensions of the Turkish vowel space (i.e., backness and roundedness) participate in the harmony process (and the suffixes then need only contain vowel height information, a classic case of archi-phonemes). Similar analyses account for the nominative plural and genitive plural paradigms. Vowel harmony illustrates "action-at-a-distance" because these patterns hold despite the presence of intervening consonants. That is, these are nonlocal phonological dependencies, unique from local assimilation patterns we find in many languages (in English, /a/ is pronounced as a nasalized /ã/ before nasal consonants, e.g., [bæ̃n] "ban"). Consonant harmony is similar to vowel harmony, except that the harmony process is between consonants and not vowels. Disharmony refers to processes that cause two sounds (at a distance) to be less similar.

A remnant of a process of consonant disharmony in Latin survives statistically in English in the choice of the adjective forming suffixes "-al" and "-ar," which tend to alternate with a preceding "l" or "r" in the

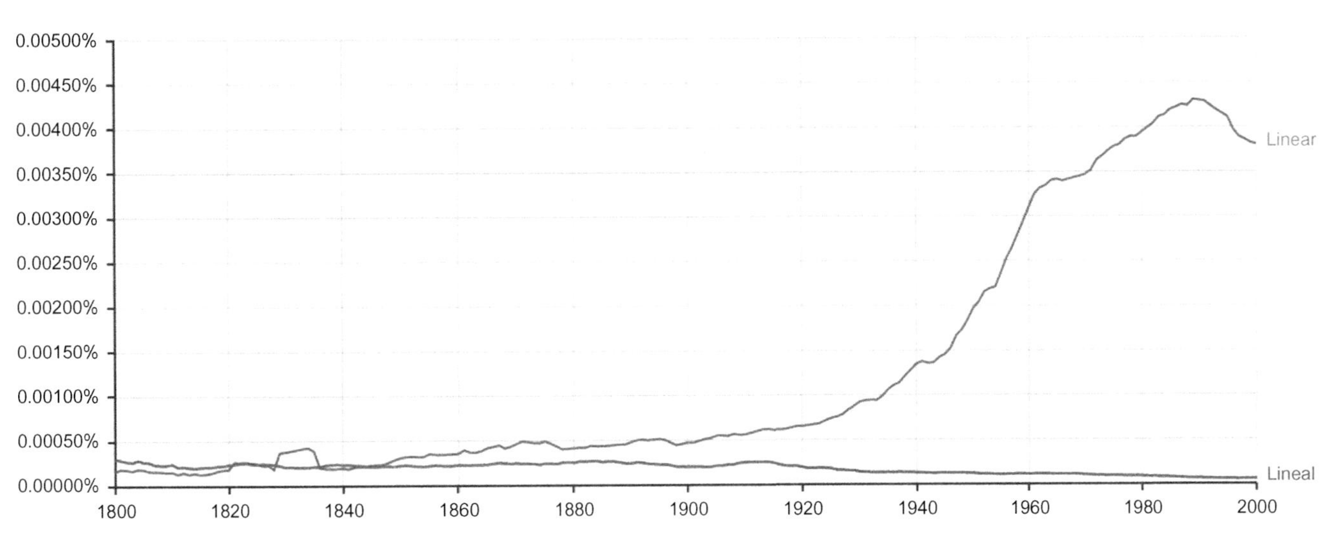

FIGURE 12.6 Google n-gram frequencies for "linear" (blue) and "lineal" (red).

word so that "circle" derives "circul-ar" but "flower" derives "flor-al," and the higher frequency of "line-ar" as compared with "line-al" (see Figure 12.6), even though the "l" in "linear" is three syllables apart from the "r."

Heinz and Idsardi (2011, 2013) argue that such effects cannot be reduced to iterated local effects of coarticulation across the intervening sounds and constitute a separate type of phonological generalization with distinct computational properties (perhaps also motivating larger grouping structures such as feet and phonological phrases that we are ignoring here). So far, there are few studies examining brain responses to vowel and consonant harmony (see Scharinger, Poe, and Idsardi, 2011 for a preliminary study of Turkish vowel harmony and see Monahan, 2013 for Basque sibilant harmony). Even though such action-at-a-distance effects may seem exotic, and even though the experimental materials are more difficult to construct (because they must eventually test the effects across various distances), it is important to examine the neuropsychological properties of these phonological laws to determine how they differ from the local sequence laws (for instance, whether the discontiguous sequence effects decrease with increasing distance).

12.6 SUMMARY

In this chapter, we have provided neurobiological motivations for what we feel are the core concepts of phonology: features, segments, syllables, abstraction, and laws of combination (both local and long-distance). Although we obviously do not yet understand how speech is mentally represented for action, perception, and memory, we feel confident that features, segments, syllables, abstraction, and laws of form will be crucial in explicating the mental representations and computations used in listening and speaking. We have deliberately not attempted to provide a comprehensive review of the neuropsychological findings in speech relevant for phonology. For two reviews in that vein, see Idsardi and Poeppel (2012) and Monahan, Lau, and Idsardi (2013).

References

Aertsen, A. M. H. J., & Johannesma, P. I. M. (1981). Spectro-temporal receptive field: A functional characteristic of auditory neurons. *Biological Cybernetics*, *42*, 133–143.

Bell, A. M. (1867). *Visible speech: The science of universal alphabetics*. London: Simkin, Marshall & Co.

Berent, I. (2013). *The phonological mind*. Cambridge: Cambridge University Press.

Berent, I., Pan, H., Zhao, X., Epstein, J., Bennett, M. L., Deshpande, V., et al. (2014). Language universals engage Broca's area. *PLoS One*, *17*, e95155.

Booth, M. C. A., & Rolls, E. T. (1998). View-invariant representations of familiar objects by neurons in the inferior temporal visual cortex. *Cerebral Cortex*, *8*, 510–523.

Bouchard, K. E., Mesgarani, N., Johnson, K., & Chang, E. F. (2013). Functional organization of human sensorimotor cortex for speech articulation. *Nature*, *495*, 327–332.

Brentari, D. (2011). Sign language phonology. In J. Goldsmith, J. Riggle, & A. Yu (Eds.), *Handbook of phonological theory* (2nd ed.) (pp. 692–721). Oxford: Blackwells.

Carbonnell, K. M., & Lotto, A. J. (2014). Speech is not special... again. *Frontiers in Psychology*, *5*, 427.

Chomsky, N., & Halle, M. (1968). *The sound pattern of English*. Cambridge, MA: MIT Press.

Clements, G. N., & Hume, E. V. (1995). The internal organization of speech sounds. In J. A. Goldsmith (Ed.), *The handbook of phonological theory* (pp. 245–306). Cambridge, MA: Blackwell Publishing.

Clements, G. N., & Sezer, E. (1982). Vowel and consonant disharmony in Turkish. In H. van der Hulst, & N. Smith (Eds.), *The structure of phonological representations, II* (pp. 213–254). Dordrecht: Foris.

Cooper, F. S., Liberman, A. M., & Borst, J. M. (1951). The interconversion of audible and visible patterns as a basis for research in the perception of speech. *Proceedings of the National Academy of Sciences, 37*, 318–325.

Fant, G. (1973). *Speech sounds and features*. Cambridge, MA: MIT Press.

Gage, N., Poeppel, D., Roberts, T. P. L., & Hickok, G. (1998). Auditory evoked M100 reflects onset acoustics of speech sounds. *Brain Research, 814*, 236–239.

Gage, N., Roberts, T. P. L., & Hickok, G. (2002). Hemispheric asymmetries in auditory evoked neuromagnetic fields in response to place of articulation contrasts. *Cognitive Brain Research, 14*, 303–306.

Ghitza, O. (2012). On the role of theta-driven syllabic parsing in decoding speech: Intelligibility of speech with a manipulated modulation spectrum. *Frontiers in Psychology, 3*, 238.

Ghitza, O., Giraud, A.-L., & Poeppel, D. (2012). Neuronal oscillations and speech perception: Critical-band temporal envelopes are the essence. *Frontiers in Human Neuroscience, 6*, 340.

Gow, D. (2012). The cortical organization of lexical knowledge: A dual lexicon model of spoken language processing. *Brain and Language, 121*, 273–288.

Guenther, F. H. (2006). Cortical interactions underlying the production of speech sounds. *Journal of Communication Disorders, 39*, 350–365.

Halle, M. (1983). On distinctive features and their articulatory implementation. *Natural Language and Linguistic Theory, 1*, 91–105.

Heinz, J., & Idsardi, W. J. (2011). Sentence and word complexity. *Science, 333*, 295–297.

Heinz, J., & Idsardi, W. J. (2013). What complexity domains reveal about domains in language. *Topics in Cognitive Science, 5*, 111–131.

Hickok, G. (2014). The architecture of speech production and the role of the phoneme in speech processing. *Language, Cognition and Neuroscience, 29*, 2–20.

Hinton, L., Nichols, J., & Ohala, J. J. (Eds.), (1994). *Sound symbolism*. Cambridge: Cambridge University Press.

Hoffman, K. L., & Logothetis, N. K. (2009). Cortical mechanisms of sensory learning and object recognition. *Philosophical Transactions of the Royal Society B, 364*, 321–329.

Idsardi, W. J. (2006). A simply proof that optimality theory is not computationally tractable. *Linguistic Inquiry, 37*, 271–275.

Idsardi, W. J., & Poeppel, D. (2012). Neurophysiological techniques in laboratory phonology. In A. Cohn, C. Fougeron, & M. Huffman (Eds.), *The Oxford handbook of laboratory phonology* (pp. 593–605). Oxford: Oxford University Press.

Jakobson, R., Fant, G., & Halle, M. (1952). *Preliminaries to speech analysis*. Cambridge, MA: MIT Press.

Juneja, A., & Espy-Wilson, C. (2008). A probabilistic framework for landmark detection based on phonetic features for automatic speech recognition. *Journal of the Acoustical Society of America, 117*, 1154–1168.

Kazanina, N., Phillips, C., & Idsardi, W. J. (2006). The influence of meaning on the perception of speech sounds. *Proceedings of the National Academy of Sciences, 103*, 11381–11386.

Kim-Renaud, Y.-K. (1997). *The Korean alphabet: Its history and structure*. Honolulu: University of Hawaii Press.

Ladefoged, P., & Maddieson, I. (1996). *The sounds of the world's languages*. Hoboken, NJ: Wiley.

Liberman, A. M. (1996). *Speech: A special code*. Cambridge, MA: MIT Press.

Liberman, A. M., & Mattingly, I. G. (1985). The motor theory of speech perception revised. *Cognition, 21*, 1–36.

Mesgarani, N., Cheung, C., Johnson, K., & Chang, E. F. (2014). Phonetic feature encoding in human superior temporal gyrus. *Science, 343*, 1006–1010.

Mesgarani, N., David, S. V., Fritz, J. B., & Shamma, S. A. (2014). Mechanisms of noise robust representation of speech in primary auditory cortex. *Proceedings of the National Academy of Sciences, 111*, 6792–6797.

Monahan, P. J. (2013). Using long-distance harmony to probe prediction in speech perception: ERP evidence from Basque. Presented at the 5th Annual Meeting of the Society for the Neurobiology of Language, San Diego, CA, November 6–8, 2013.

Monahan, P. J., & Idsardi, W. J. (2010). Auditory sensitivity to formant ratios: Toward an account of vowel normalization. *Language and Cognitive Processes, 25*, 808–839.

Monahan, P. J., Lau, E. F., & Idsardi, W. J. (2013). Computational primitives in phonology and their neural correlates. In C. Boeckx, & K. K. Grohmann (Eds.), *The Cambridge handbook of biolinguistics* (pp. 233–256). Cambridge: Cambridge University Press.

Näätänen, R., Lehtokoski, A., Lennes, M., Cheour, M., Huotilainen, M., Iivonen, A., et al. (1997). Language-specific phoneme representations revealed by electric and magnetic brain responses. *Nature, 385*, 432–434.

Ohala, J. J. (1997). Emergent stops. In *Proceedings of the 4th Seoul international conference on linguistics* (pp. 84–91). Seoul: Linguistic Society of Korea.

Poeppel, D. (2003). The analysis of speech in different temporal integration windows: Cerebral lateralization as 'asymmetric sampling in time'. *Speech Communication, 41*, 245–255.

Poeppel, D., & Idsardi, W. J. (2011). Recognizing words from speech: The perception-action-memory loop. In G. Gaskell, & P. Zwitserlood (Eds.), *Lexical representation: A multidisciplinary approach* (pp. 171–196). Berlin: de Gruyter.

Potter, R. K., Kopp, G. A., & Green, H. C. (1947). *Visible speech*. New York: van Nostrand.

Prince, A., & Smolensky, P. (2004). *Optimality theory: Constraint interaction in generative grammar*. Malden, MA: Blackwell Publishing.

Pruthi, T., & Espy-Wilson, C. (2004). Acoustic parameters for automatic detection of nasal manner. *Speech Communication, 43*, 225–239.

Rhodes, G., Calder, A., Johnson, M., & Haxby, J. V. (2011). *Oxford handbook of face perception*. Oxford: Oxford University Press.

Roberts, T. P. L., Flagg, E. J., & Gage, N. M. (2004). Vowel categorization induces departure of M100 latency from acoustic prediction. *Neuroreport, 15*, 1679–1682.

Scharinger, M., Idsardi, W. J., & Poe, S. (2012). A comprehensive three-dimensional cortical map of vowel space. *Journal of Cognitive Neuroscience, 12*, 3972–3982.

Scharinger, M., Poe, S., & Idsardi, W. J. (2011). Neuromagnetic reflections of harmony and constraint violations in Turkish. *Laboratory Phonology, 2*, 99–123.

Sharma, A., & Dorman, M. F. (2000). Neurophysiologic correlates of cross-language phonetic perception. *Journal of the Acoustical Society of America, 107*, 2697–2703.

Staal, F. (1990). *Rules without meaning: Ritual, mantras and the human science*. New York: Peter Lang.

Steinberg, J., Truckenbrodt, H., & Jacobsen, T. (2011). Phonotactic constraint violations in German grammar are detected automatically in auditory speech processing: A human event-related potentials study. *Psychophysiology, 48*, 1208–1216.

Stevens, K. N. (1998). *Acoustic Phonetics*. Cambridge, MA: MIT Press.

Stevens, K. N. (2002). Toward a model for lexical access based on acoustic landmarks and distinctive features. *Journal of the Acoustical Society of America, 111*, 1872–1891.

Trubetzkoy, N. S. (1969). *Principles of phonology*. Berkeley: University of California Press.

Wing, A. M. (2000). Motor control: Mechanisms of motor equivalence in handwriting. *Current Biology*, *10*, R245–R248.

Yoo, I. H., & Blankenship, B. (2003). Duration of epenthetic [t] in polysyllabic American English words. *Journal of the International Phonetic Association*, *33*, 153–164.

Zemlin, W. R. (1998). *Speech and hearing science: Anatomy and physiology* (4th ed.). Boston: Allyn and Bacon.

Zhou, X., Espy-Wilson, C. Y., Tiede, M., & Boyce, S. (2007). An articulatory and acoustic study of "retroflex" and "bunched" American English rhotic sound based on MRI. In *Proceedings of INTERSPEECH'07* (pp. 54–57).

CHAPTER

13

Morphology

Alec Marantz

Department of Linguistics, New York University, New York, NY, USA;
Department of Psychology, New York University, New York, NY, USA;
NYUAD Institute, New York University Abu Dhabi, Abu Dhabi, United Arab Emirates

13.1 INTRODUCTION

Within linguistics, morphology is the subdiscipline devoted to the study of the distribution and form of "morphemes," taken to be the minimal combinatorial unit languages use to build words and phrases. For example, it is a fact about English morphology that information about whether a sentence is in the past tense occurs at the end of verbs. This fact reduces to a generalization about the distribution of the tense morpheme in English, which is a fact about "morphotactics" (the distribution and ordering of morphemes) in morphology. It is also a fact about English morphology that the ("regular") past-tense morpheme is pronounced /t/ after a class of voiceless consonants (*walked, tipped, kissed*) and /d/ after a class of voiced consonants and after vowels (*gagged, ribbed, fizzed, played*). This fact is a fact about "allomorphy" (alternations in the pronunciation of morphemes). Traditionally, then, morphology concerns itself with morphotactics and allomorphy.

Although the division or decomposition of words and phrases into smaller units seems relatively intuitive, linguistic morphologists have repeatedly questioned basic assumptions about morphemes. With one view, instead of dealing with the distribution and pronunciation of small pieces of language, morphology is about the form of words, where, for example, *kick, kicks, kicking*, and *kicked*, are all forms of the same verb *kick* (Matthews, 1965) but are not composed of a sequence of morphemes. With this view, languages are claimed to make a strict distinction between words and phrases, with only the latter having an internal structure of organized pieces. From this morphemeless perspective, *kicked* is a form of the stem *kick*, not the combination of *kick* + PAST TENSE, where PAST TENSE is realized as /t/. Other morphologists also endorse a strict division between words and phrases but still analyze words as consisting of morphemes; with this view, the internal arrangement of morphemes within words falls under a different set of principles than the arrangement of words into sentences. However, in the morphological theory most closely associated with generative grammar in this century, distributed morphology, there is no strict word/phrase distinction (Matushansky & Marantz, 2013). The internal arrangement of morphemes both within words and within phrases and sentences is explained by a single *syntactic* theory, and morphology provides an account of the way in which these morphemes are realized phonologically (in sound), whether inside words or independently arranged in phrases.

This chapter explains aspects of the theory of morphology with a view of the way that morphology has been explored in neurolinguistics. An important conclusion of the chapter is that although the types of investigation of morphology currently found in neurolinguistics might seem to rely on motivated linguistic distinctions, such as that between derivational and inflectional morphology, linguistic theory itself does not support such distinctions in the manner required to motivate neurolinguistic experiments. Although research in the neurobiology of language does at least sometimes adopt the vocabulary of linguistic morphology in investigating the neural bases of morphology, the attention given to linguistic analysis is often superficial, with the consequence that experimental results are difficult to interpret with respect to central questions of language processing in the brain. There is hope that recent advances in psycholinguistics and in computational linguistics may help bridge the gap between linguistic theory and the theory of neurolinguistic processing, such that linguistics' deep understanding of the

Neurobiology of Language. DOI: http://dx.doi.org/10.1016/B978-0-12-407794-2.00013-4

nature of language can inform neurolinguistics and, in turn, neurolinguistic findings can help shape linguistic theory.

Although some controversies within linguistics over the correct analysis of morphological phenomena are explained in this chapter, in general I adopt the assumptions and results of distributed morphology. As explained in Marantz (2013b), distributed morphology is relatively conservative from a historical perspective, preserving the insights of mainstream linguistics from the 20th century. In experimental work, one can attempt to explicitly test differing predictions made by competing representational theories of language, and so an experimentalist could choose to pit predictions of distributed morphology against available alternatives. However, experimental work related to morphology must make some theoretical commitments; it is not possible to be agnostic over issues such as whether words decompose into morphemes.

13.2 WHY MORPHOLOGY?

There seems to be an obvious need for a theory of syntax that would explain the constraints on how smaller linguistic units combine into and distribute across words and phrases, or for a theory of phonology that would explain the way that the pronunciation of units is conditioned by their environments in sentences, or for a theory of semantics that would explain the way that meanings of smaller units combine to form meanings of larger units. Syntax, phonology, and semantics together represent the essential structure of grammar: an engine of combination (syntax) and two interpretive "components" that translate the combinations into sound (phonology) and meaning (semantics). However, the role of morphology in language presents more of a puzzle. If we think of morphology as exemplified by the stuff we add to English words—things like the past-tense ending or a prefix like *re-* in *repaint*—the question arises: *why is there morphology* in addition to syntax, phonology, and semantics? For the investigation of the nature of language, we can ask why languages appear to add stuff to words and why that stuff takes the particular shapes we observe cross-linguistically. For linguistics, we can ask whether an account of this stuff requires a special (sub-)theory of "morphology" in addition to syntax, phonology, and semantics—a theory of an independent morphological component of grammar—or whether the theory of morphemes can be reduced to these other theories (with the properties of morphemes distributed across the syntactic, phonological, and semantic components, as in distributed morphology). This chapter explains why contemporary morphologists claim that morphology is not a special component of grammar and that the interaction of syntax, phonology, and semantics produces the morphological phenomena we observe and allows for the variation in the expression of morphemes that is exhibited cross-linguistically.

The *why morphology?* question can be usefully divided into three issues. The first issue concerns the reason why certain information is sometimes indicated by attaching sounds to a word while the same information could be carried by an independent word. Why should the past tense in English be indicated by a suffix on *walk* in a statement, *He* ***walked***, but by an auxiliary verb *did* in a yes/no question, ***Did*** *he walk?* Why should we say *repaint* when the two words *paint again* can be used with the same meaning? Why do languages ever use prefixes or suffixes, given that independent words can serve the same function? If every language chose independent words for these functions, there is a sense in which there would be no morphology.

A second, related question is why such diversity exists in the way that morphemes are realized across the world's languages. Languages can signal information like past tense by copying part of a verb stem (reduplication), by tucking phonological material inside of a verb stem (infixing), and by other means, as well as by concatenating a stem and a tense prefix or suffix or by using a phonologically independent word. Do these different modes of signaling information correspond to deep grammatical differences between languages?

A third question involves particular types of morphemes such as agreement and case affixes. For a prefix like *re-*, it should be clear why English might want to use the morpheme for the expression of a meaning also expressible by the independent word *again* (*John* ***repainted*** *the house* = *John painted the house* ***again***): the meaning contrast between *repaint* and *paint* hinges on the presence of the prefix on the first verb. However, the necessity of agreement morphology on verbs (*He runs every day, They run every day*) and case marking on nouns and pronouns (***He*** *saw* ***him***, where the subject is "nominative" and the object "accusative") is less obvious, given that many languages lack such markings. Even in languages like English that show some limited agreement and case marking, any help that such morphology might provide in disambiguating word strings is minimal. For example, modal verbs like *may* in English show no agreement (*he may, they may*), whereas auxiliary verbs like *be* do (*he is, they are*), but the lack of agreement on *may* does not cause comprehension difficulties.

For each of these questions about the necessity of morphology for language, the answer that comes from the theory of morphology is this: do not be misled to generalize a language-particular choice along a spectrum of possibilities to a universal about the nature of language. That is, from a general perspective of the structure of grammar, the variations in morphologies cross-linguistically can be seen as superficial variations on strong universal themes.

Taking "morpheme" to be defined as the smallest unit combined by the syntax of a language, morphology makes a strong distinction between the roots of the so-called lexical categories—nouns, verbs, and adjectives—and all other morphemes. This distinction between roots and other morphemes underlies an often cited but less technical distinction between content words and function words and morphemes. The root morphemes, like *cat*, describe properties of entities, states, and events. Their meanings are strongly connected to our a-linguistic cultural and conceptual knowledge, and the set of root morphemes varies considerably across languages and people. However, nonroot morphemes are sets of grammatical features that operate in a uniform way across languages and are central to the grammatical system. A past-tense morpheme would consist of features available to the grammar of every language and would likely be shared across speakers of the language. Root morphemes need to combine with the lexical category morphemes that create nouns, verbs, and adjectives to be used in phrases and sentences; each root morpheme + category-determining morpheme complex will, in general, anchor a phonological word, part of the "open class" vocabulary of a language (cases of compounding like *blackbird* and similar phenomena in other languages allow for multiple roots in a single phonological word). Nonroot morphemes, so-called functional morphemes that form the "closed class" of items in a language, find their phonological realization either by joining a root in a phonological word anchored by the root or by forming a phonological word of their own, perhaps with other functional morphemes. In general, then, the answer to the first question about the existence of morphology—why some languages express certain types of information piled up in a single word while other languages might express the same information on separate words—follows from the generalization that language makes a cut between root morphemes and functional morphemes. The general principles about the sound realization of root morphemes demand that they anchor independent phonological words cross-linguistically. The general principles about the sound realization of functional morphemes allow languages plenty of room for variation, giving the appearance that some languages have more morphology—more of the functional morphemes appear as affixes on root-based phonological words—whereas others have less—more of the functional morphemes appear as or as part of root-less phonological words (i.e., as function words). In addition, the phonological realization of a functional morpheme is not forced by any general, universal principle; such morphemes often are silent. Thus, the inventory of phonologically realized morphemes also differs across languages, in addition to the differences in how pronounced morphemes are phonologically realized.

The distinction between root and functional morphemes and the distinction between morphemes as combinable constituents and the phonological (sound) forms of these morphemes also provide the answer to the second issue regarding the variety of alternatives cross-linguistically for the sound expression of various morphemes. The organization of morphemes within a word or phrase is determined by the syntax and can be displayed in a hierarchical tree structure. Linguists have developed a general account of the way that individual morphemes are realized that covers not only roots like *cat* that have the phonological form of a syllable but also suffixes like past tense /d/ that are a single consonant, and the roots of Semitic languages like Arabic, which might consist of just three consonants without vowels or syllable structure. The same generalized account that assigns each morpheme a bit of phonological substance can describe cases of reduplication, where the added phonological material is in a sense borrowed from the stem, and even truncation, where it looks like the addition of a morpheme results in a shortening of the stem. That is, the general picture has each morpheme determining a bit of phonological content (perhaps phonologically null), with the phonological forms of the morphemes combining according to the phonological principles of the language. The exoticness of infixing and reduplication—or the expression of a morpheme as a tone on the vowel of a stem or via gemination of a stem consonant—is relative to the phonological structure of English; from a cross-linguistic standpoint there is no reason to treat these morphemes' expression as more unusual than affixation.

Finally, we may address the issue of why certain types of syntactically dependent morphology should exist, particularly agreement and case marking. The overt realization of agreement between subject and verb (or verb and object, etc.) and the overt realization of case marking on constituents of noun phrases (nouns, adjectives, determiners, numerals) are certainly not necessary for the processing of language; many languages do fine with minimal expression of such morphology. But the general characteristic of such morphology is the reflection of grammatical relations among constituents of a sentence that are present and necessary with or without the morphology. Subject–verb agreement, for example, reflects the computation of a relation between (features of) a subject and (a constituent of) a verb that underlies the grammatical analysis of a sentence. The general picture of the syntax of language, then, is one of the recursive combination of morphemes into hierarchical constituent structures PLUS the computation of certain relations between morphemes—like the subject–verb (tense) relation—that are not completely reducible to constituency (like "sister" relations in a tree) or linear relations (like "next to"). Such grammatical relations (with traditional names like "subject" and "object") involve the transfer or checking of features on functional morphemes. These features are phonologically realized as case and agreement morphology in some languages, but the syntactic operations and computations that underlie this morphology are present in every language.

13.3 WHAT MAKES MORPHOLOGY, MORPHOLOGY

In answering the question of why language includes morphology, we have sketched a picture of a structure of grammar that denies a distinction between the organization of morphemes within words and the organization of words within sentences; that is, the contemporary linguistic perspective on morphology makes no principled distinction between syntax and morphology. Nevertheless, the sound system of a language organizes the pronunciation of sentences into units of different sizes, with smaller units nested inside larger units, that is, "phonological words" combined into "phonological phrases" (Hall, 1999). These phonological units are the locus of certain phonological processes and generalizations. For example, in many European languages, including German and Russian, voiced consonants like /b/ or /d/ are pronounced voiceless—/p/, /t/—at the end of a phonological word. In English, some of the units we write as separate words do not contain enough phonological material to be pronounced as phonological words; for example, *the* and *a* pronounced as they usually are before a consonant-initial noun must be joined with the noun in a phonological word. In general, the phonological constraints on these phonological constituents require a certain mismatch between the syntactic structure and the phonological structure, as is well-known from such examples as *the queen of England's hat*, which is pronounced with the possessive morpheme joined with *England*, although the hat belongs to the monarch rather than the country. The grammar of a language must describe how the syntactic arrangement of hierarchically organized morphemes is realized as the phonological organization of phonological words and phrases, and which morphemes will come together into single phonological words, as *'s* and *England* come together in our example. What we call the morphology of a language includes an account of how the grammar of the language packages some morphemes into phonological words, and how these morphemes are pronounced.

Contemporary generative grammar, as described within the "Minimalist Program" associated with Noam Chomsky (Chomsky, 1995), describes the essence of linguistic structure as involving the recursive "merger" of morphemes. Two morphemes are merged into a constituent, which might undergo further merger with another morpheme or with some previously constructed complex of morphemes. Repeated operation of merger yields familiar hierarchical constituent structures. For morphemes interpreted as semantic functions or operators, the hierarchical structure determines their scope. For example, if morphemes interpreted as *make, want,* and *go* are merged into a structure like *[make[want[go]]]*, then the interpretation would involve causing the desire to leave, while a structure that swaps the structural positions of *want* and *make*—*[want[make[go]]]*—would involve the desire to cause leaving. If we imagine that these hierarchical structures built by repeated merge operations hold no implications for the linear order of the two constituents joined by each merge, the usual pronunciation of these syntactic structures must involve a decision, for each merger, regarding the order of each merged pair. Because the decision to order a complex constituent X before a complex constituent Y causes each subconstituent of X to be ordered before each subconstituent of Y, the linearization of a hierarchical syntactic structure will result in morpheme order reflecting the hierarchical structure. Within words, this correspondence between hierarchical structure and linear order of morphemes has come to be called the "Mirror Principle" (Baker, 1985), which is not really a "principle" but an observation about this correspondence.

The interface between the syntax of a language that determines the hierarchical structures of morphemes and the phonology of a language that determines the pronunciation of these structures must decide which morphemes to group together into phonological words and how to pronounce these morphemes. The environment in which a morpheme appears will determine how the morpheme is pronounced to a greater or lesser extent, depending on properties of the language and the particular morpheme itself. Several factors may condition the form of a morpheme. For example, as morphemes are linearized in the phonological interpretation of a hierarchical syntactic structure, a morpheme might show a different form (a different allomorph) depending on the phonological shape or the actual identity of a neighboring morpheme. The indefinite article *a/an* in English is pronounced *a* before a consonant-initial word and *an* before a vowel-initial word—allomorphy conditioned by the phonology of a neighboring morpheme. The plural suffix is pronounced *-en* after *ox* but as silence after *sheep*—allomorphy conditioned by the identity of the morpheme to which it attaches, not its phonological shape. Similarly, an allomorph of a morpheme might be conditioned by general phonological properties of the language and of the morpheme, or the allomorph might be unconnected to phonological properties of the language. The pronunciation of the past-tense morpheme as /t/ after voiceless consonants in English but as /d/ after voiced consonants follows from general phonological facts about English. However, the choice of *an* before vowel-initial stems in the English indefinite article is not determined by general phonological properties of English; English does not generally insert an /n/ between vowels to avoid a hiatus.

From this short sketch, the reader can already project some of the investigations that occupy morphologists. For example, questions arise regarding whether there are constraints on how close two morphemes must be to influence each other's pronunciation, and whether closeness should be measured in terms of the hierarchical structure of morphemes or in terms of a linear string of their pronunciations. What kinds of information associated with morphemes could serve as triggers for particular phonological alternations on morphemes, and do the constraints on types of contextual information depend on the relative location of the morphemes within a hierarchical tree? Morphologists have discovered that these interactions between morphemes are very local and are very dependent on the hierarchical syntactic structures in which the morphemes appear (Embick, 2010). The locality of so-called contextual allomorphy (the pronunciation of a morpheme triggered by its linguistic context) seems convenient for language processing and would seem to constrain accounts of how language might be acquired.

13.4 TYPES OF MORPHEMES, TYPES OF MORPHOLOGIES, TYPES OF MORPHOLOGICAL THEORIES

Given this general picture of morphology as the exploration of principles governing the organization of morphemes into words and their pronunciation in context, we can turn to certain contrasts between sets of morphemes and between theories of morphemes that hold potential importance for the study of language in the brain. A commonly invoked division between morphemes divides the "inflectional" from the "derivational." On some definitions, inflectional morphology creates different forms of an individual word, while derivation creates new words from words. For example, English tense would be inflectional, because the past-tense form of a verb is arguably a form of the verb, rather than a word with its own distribution and meaning, whereas the suffix *-able* is derivational, creating an adjective from an input verb, as in *knowable* from *know*. Although there is no doubt that tense morphemes and category-changing morphemes like *-able* differ in many ways, one must ask two general questions about the distinction.

First, does language make a binary distinction between two classes of morphemes such that it is coherent and important that we can identify a given morpheme as either inflectional or derivational, with certain characteristics deducible from the identification (e.g., whether it changes grammatical category or whether it may be realized as reduplication)? Second, given a division into inflection and derivation, do any linguistic generalizations or principles rely on the feature of being inflectional or derivational, or do properties that characterize one or the other class of morphemes follow from specific features of the morphemes themselves, not the inflection versus derivation label? That is, does the theory of linguistics call on the features "inflectional" and/or "derivational"? The emerging answer to both questions from contemporary morphology is "no."

Recall that the syntax of a language places various constraints and requirements on different sets of morphemes. For example, a main clause in English requires a tense morpheme, and a present-tense verb in English must agree with a third-person singular subject, usually with the *-s* suffix (*He walks*). However, a prefix like *re-* in English behaves like an optional modifier—like the independent word *again*. A suffix like *-able* parallels to some degree the adjective *able*: *This game is winnable* parallels *Someone is able to win this game*. The question arises whether morphemes split into large classes such that properties of morphemes follow from their membership in these classes. A recurrent proposal has been to divide morphemes into "inflection" versus "derivation," where inflection would include "grammatical" morphemes like tense and agreement that are relevant to the syntax, and derivational morphemes would include those that derive words of one category (noun, verb, adjective) from words of a different category. However, although linguists have discovered some features of morphemes that determine their behavior, no characterization of the inflection versus derivation split has proved relevant within morphological theory.

For example, within distributed morphology it has been observed that morphemes that attach directly to roots, such as the morphemes that create nouns, verbs, and adjectives from roots, share properties governing the conditioning of allomorphy on roots that are not shared by morphemes that attach to constituents that already contain these category-determining morphemes (Embick & Marantz, 2008; Marantz, 2013a). Subclasses of functional morphemes, then, share properties that a theory of morphology should explain.

However, there is no particular evidence that, for example, tense and number morphemes form a coherent class with case and agreement morphemes (the putative class of inflection) as opposed to category-changing morphemes (the putative class of derivation) that might motivate a broad distinction between inflection and derivation.

The search for contrastive properties for inflection versus derivation was motivated by the observation that some morphemes, such as tense, were required by a syntactic environment and selective regarding their host words (e.g., tense is required in main clauses in English and attaches to verbs), whereas other words, while perhaps selective regarding their hosts (*-able* attaches generally to verbs), were not required by the broader sentential context in any important sense (although one might construct a syntactic environment that requires an adjective, this environment would not require an adjective made with the *-able* suffix). Historically, linguists have explored two main hypotheses about properties that might follow from the inflection versus derivation distinction, characterized in this way by the morphemes' sensitivity to their syntactic environment. First, it has been claimed that inflection is paradigmatic while derivation is not. Second, and relatedly, it has been claimed that inflection and derivation involve different mechanisms for the phonological realization of the information carried by the morphemes.

The notion of a paradigm should be familiar for readers learning a classical language or a highly "inflected" language like Finnish or Russian. For verbs, the various combinations of tense, aspect, and mood features that modify verbs, along with the features of agreement (usually subject agreement) that can be signaled on the verb construct a multidimensional grid of feature values such that each cell of the grid is filled by a form of the verb whose paradigm is being displayed. In English, the paradigm for the present tense of the verb *to be* would have six cells, one for each of the combinations of person and number for the subject of the verb: (*I*) *am*; (*you*) *are*; (*he*) *is*; (*we*) *are*; (*you* PLURAL) *are*; and (*they*) *are*. Two important features of paradigms are that for each verb, noun, or adjective associated with an inflectional paradigm, there is expected to be a form for each cell in the paradigm (although one form may fill multiple cells, as *are* does for *to be* in English) and, in general, only a single form fills each cell. The latter property is behind the notion of "blocking": an irregular form that fills a cell specific to a particular stem or class of stems "blocks" the creation of a regular, predicted form for that stem to fill that paradigm cell. For example, the irregular *is* blocks the creation of regular *be* for the third-person singular cell of the paradigm for the verb *to be*.

Although the possible role of paradigms in a speaker's knowledge of language is still a somewhat controversial subject in linguistics (Bachrach & Nevins, 2008), properties of paradigms do not motivate a distinction between inflection and derivation, for two main reasons. First, to the extent that a derivational relation is made available by the general properties of language, derived forms can be displayed paradigmatically such that a form is predicted for each cell for each noun, verb, or adjective stem and such that there is a blocking relation between an irregular form (specific to a stem or set of stems) and a regular form. For example, the function served by the English *-er* suffix, attaching to verbs to create a noun referring to a person that habitually does what the verb describes, is generally available cross-linguistically, and every verb with the appropriate meaning is predicted to allow an *-er* formation. A definition of inflection that related to the formation of paradigms would include category-changing *-er* as inflection. Blocking effects in canonical category-changing derivational morphology are less easy to illustrate with simple English examples for the reasons described in Embick and Marantz (2008); however, such effects are definitely found. For example, the productive phonological realization of the morpheme creating nouns from adjectives with the meaning "the quality or property of being *adjective*" is the suffix *-ness*. The suffix *-ity* is a phonological realization of the same morpheme, but with a more restricted environment. In particular, when a stem ends in the affix *-able*, *-ity* is the preferred pronunciation, and *-ness* is "blocked": we get *transferable* and *transferability*, but not *transferableness*. The function of the morpheme *-ness*/*-ity*, then, creates a type of paradigm for adjectives, as well as a blocking relation between forms.

In addition to the fact that canonical derivational morphemes can be paradigmatic and exhibit blocking, a crucial fact that undermines any categorical distinction between inflection and derivation is that apparently inflected forms are often used in languages as category-changing derivation. In English, for example, the present participles of verbs, as *-ing* forms, are often used to create nouns, as in gerunds (*the running of the race*), and passive participles of verbs are often used to create adjectives (*the closed door*, *a typed note*).

In addition to the possible correlation between paradigms and inflectional morphology, linguists have explored the possibility that the phonological expression of morphemes differs fundamentally between derivation and inflection. For example, in Anderson's A-Morphous Morphology theory (Anderson, 1992), inflection involves phonological alterations to a stem, whereas derivation involves the concatenation of the stem with morphemes with independent phonological form (standard affixation, for example). However, a major outcome of research in the 1970s and 1980s was the demonstration that derivation and inflection cannot be distinguished phonologically. Apparent "processes" altering the forms of stems such as reduplication and truncation are not limited to canonical inflection, and in general it is not possible to predict anything about the phonology of a morpheme based on the inflection versus derivation distinction.

This conclusion about phonological realization is more general: the organization of grammar computes the phonological realization of morphemes and combinations of morphemes without universal constraints based on their syntactic or semantic features. Individual languages may impose constraints on the phonological forms of classes of morphemes; for example, the roots of verbs in Arabic are sequences of two to four consonants, with further constraints observed regarding possible sequences of identical consonants. But the general principles of language do not impose such constraints on roots, nor do they restrict reduplication to any subset of functional morphemes. As already mentioned, these principles do impose locality constraints on what might serve as context for choice of phonological realizations, and other properties of syntactic structures may determine aspects of phonological realization due to the very architecture of the grammar. For example, if phonological realization is restricted to syntactic units of a particular size, as it is in theories in which linguistic computation is "cyclic," then information carried by morphemes outside these cyclic realizational ("spell-out") domains cannot influence the pronunciation of morphemes inside these domains. But grammatical principles governing the phonological realization of morphemes do not seem to refer directly to classes of morphemes.

The uniformity of phonological realization across types of morphemes extends to the distinction between morphemes like case and agreement that reflect grammatical relations between elements in a syntactic structure and all other morphemes. The principles of morphology that govern how root or functional morphemes are pronounced do not singleout case and agreement morphemes, although features of such morphemes do require special mechanisms in the syntax above and beyond the merger of morphemes into hierarchical structure. The fact that case and agreement cannot be identified solely on the basis of their phonology holds crucial clues regarding the organization of grammar and the nature of language acquisition. Although syntactic structure feeds semantic and phonological interpretation, its operation is autonomous and opaque in the phonological forms that language learners encounter.

In addition to investigating the significance of certain pretheoretical distinctions between classes of morphemes, like that between inflection and derivation, linguists have asked whether apparent typological differences between classes of languages, organized by their morphological systems, are theoretically meaningful. For example, languages are sometimes divided among the isolating and synthetic languages, with synthetic languages further divided into the fusional and the agglutinative. Isolating languages, to some degree, seem to lack morphology altogether. Within the morphological framework used in this chapter, these languages would be said to realize each root morpheme and each functional morpheme in an independent phonological word, with no affixation. Synthetic languages place multiple morphemes within a single phonological word, including multiple functional morphemes with a root. Latin is often cited as an example of a synthetic language, because, for example, a single verb might include information about tense (e.g., past), aspect (e.g., perfect), voice (e.g., active), person of the subject (e.g., first person), and number of the subject (e.g., plural): *portavimus* ("we carried (PERFECT)"). Among synthetic languages, fusional languages seem to use a single phonological piece, such as a single suffix, to express multiple grammatical features simultaneously (as English *-s* expresses both present-tense and third-person singular subject agreement), whereas agglutinative languages might string together long sequences of functional morphemes, each of which expresses an independent feature. This is exemplified in Turkish *çekoslovakyalılaŞtıramayacaklarımızdanmıydınız*, discussed in Lieber (2010) from Inkelas and Orgun (1998: 368):

1. çekoslovakya- lı- laŞ- tır- ama-
Czechoslovakia- from- become- CAUSE- unable-

yacak- lar- ımız- dan- mı- ydı- nız
FUT- PL- 1PL- ABL- INTERR- PAST- 2PL

"Were you one of those whom we are not going to be able to turn into Czechoslovakians?"

Even in describing these typological differences we have implied that languages do not fall into pure categories along the dimensions implied by their descriptions (English can illustrate isolation, fusion, and agglutination, for example). That is, languages are more or less isolating, and more or less fusional. Moreover, there do not appear to be any linguistic principles or generalizations that follow from the classification. That is, although being mostly isolating might be statistically correlated with other typological properties at a descriptive level of analysis, there is no direct relationship between the expression of functional morphemes as independent words and any syntactic or morphological properties of a language. The isolating/synthetic continuum, then, is descriptive rather than essential.

A final distinction between possible approaches to morphological theory will serve to illustrate another crucial property of the phonological realization of morphemes, that of the "default" realization. Traditional generative approaches to morphology have been "lexical" in the sense that they proposed that the "lexical" or storage form of morphemes included both the sets of features they displayed to the syntax and the phonological form they carried to the sound realization of sentences. In contrast, this chapter has been describing a "realizational" theory of morphology in which the syntax combines "formless" (phonology-free) morphemes and the phonological form is part of the phonological interpretation of a morpheme, constrained by the syntactic features the morpheme contains but separated from these features. For example, a past-tense morpheme would be identified as "past" and "tense" in the syntax, but its pronunciation would be determined after syntactic combination, during phonological interpretation.

A crucial difference between the lexical and realizational theories is their approach to "syncretism": a situation in which a single phonological form is used across a variety of sets of features. For example, the form *walk* in English is syncretic as a present-tense verb, being identical across first-person and second-person subjects as well as third-person plural subjects, but contrasting with *walks* for third-person singular subjects. With a realizational approach, a present-tense morpheme is realized as *-s* when it carries third-person singular features but is realized as null elsewhere (as a default). For the null suffix to work as a default, the grammar must setup a competition between *-s* and null for the realization of a present-tense feature with subject agreement features that are already present in the structure being realized. The *-s* realization wins the competition if the agreement features are third-person singular; otherwise, the null realization is used. With a lexical approach, phonologically identified morphemes must carry syntactic features into the syntactic derivation. With this approach, defaults that clearly behave as "elsewhere" cases in opposition to more featurally specified morphemes cannot exist because the notion of "elsewhere" implies the pre-existence of a structure with syntactic features, yet on the lexical approach it is the morphemes themselves that provide the features for the syntax. Lexical approaches would postulate five homophonous *walk*s for the various non-third-person singular sets of agreement features and attempt to provide other ways of explaining the apparent redundancy here.

The notion of default in phonological realization extends to "suppletion," where the same set of features is realized by different phonological forms in different environments (a type of contextual allomorphy, as described). For example, *-en* is the suppletive allomorph of the plural morpheme in English used in the environment of the stem *ox*; the default plural allomorph is *-s*. The *-en*/*-s* alternation is not an example of syncretism because the same feature—plural—is realized by both allomorphs. Such suppletion, however, is an example of the general asymmetry between a more specified phonological realization (*-en* for plural in the environment of *ox*) and a default realization (*-s* elsewhere) that covers syncretism in a realizational account. In general, the cross-linguistic properties of asymmetric syncretism (with less specific phonological realizations acting as defaults with respect to more specific realizations) and of (asymmetric) suppletion have supported realizational theories over lexical theories.

13.5 THE VIEW FROM ABOVE

This chapter describes a theory of grammar in which morphemes are the minimal units of syntactic combination. Within such a theory, morphemes are subject to a recursive merge operation that builds hierarchical structures of constituents. In addition, certain syntactic relations between constituents are computed, leading to the features that are realized as case and agreement morphology. Languages differ in their vocabularies of morphemes, particularly with respect to the root morphemes that anchor the major syntactic categories of nouns, verbs, and adjectives. Differences in the vocabularies of functional morphemes across languages directly influence typological differences in syntax, as described by syntacticians concerned with the "parameters" of variation between languages.

The linguistic subfield of morphology concerns itself with a number of topics surrounding morphemes and their realization. For example, what is the substantive inventory of functional morphemes from which individual languages must choose for their vocabularies? How do features of morphemes interact in the syntax, particularly with respect to the computation of case and agreement? The bulk of research specific to morphology concerns the phonological realization of morphemes, both the manner in which sets of morphemes get packaged into phonological words and the computation of allomorphy—the choice of phonological realizations for a morpheme and the ways in which phonological realization might be influenced by context and by competition among phonological forms (as for syncretism, suppletion, and blocking, described previously). The theory of morphology is therefore about the choice of functional morphemes, the way that language-specific choices in morphemes and features interact with the syntax, the syntactic principles that distribute and constrain features like case and agreement on morphemes, the principles by which morphemes receive their phonological and semantic interpretation, and the way that the phonology of a language packages the phonological material of morphemes into words and phrases.

Morphology presents an account of a speaker's knowledge of language inconsistent with the idea that words could be studied in isolation from larger syntactic constituents. The word itself—to the extent that it corresponds to anything real in a person's grammar—is a phonological unit, not a unit of syntactic combination, and even as a phonological unit its properties are dependent on other phonological words within its phonological phrase. That is, linguistics provides no basis for the notion of a mental lexicon consisting only of stored words with gestalt properties as wholes. The recognition or production of a word—to the extent that the word is being recognized or produced as belonging to a speaker's language—necessarily involves a sequence of grammatical computations, including syntactic merger of morphemes and their phonological realization and packaging.

13.6 WORDS AND RULES: THE MODERN CONSENSUS ON DECOMPOSITION

Psycholinguistic and neurolinguistic work since the 1990s has taken as its starting point the perspective of Pinker's "words and rules" framework (Pinker, 1999). From the viewpoint of linguistics, Pinker's theory was based on a fundamental analytic mistake, because it postulated a grammatical difference between regular and irregular inflectional morphology. For Pinker, it was a syntactic fact that an irregular past-tense form like *taught* was a single computational unit, represented as a nonbranching tree structure, as compared with a regular form like *walked*, which would consist of a separate stem and past-tense morpheme in hierarchical syntactic structure. In fact, from any consideration of syntax and morphotactics, irregular and regular forms behave identically—the difference is entirely within the realm of the realization of morphemes phonologically (allomorphy). This linguistic error of the words and rules framework was paired with an interesting but empirically falsified separation between memory for morphemes and experience with "rules," such as the combination of morphemes. For the retrieval of morphemes, Pinker hypothesized frequency effects, but he proposed that no frequency effects would be observed from the operation of (regular) rules.[1] Frequency modulates behavior across both retrieval of the memorized forms of morphemes and the composition of these forms. This observation does not undermine the essential distinction between the atoms of linguistic composition—the morphemes—and combinatory operations that produce and modify structures of morphemes, but it does put pressure on any distinction between words and phrases.

From the viewpoint of linguistic morphology, experiments using single word processing aimed at uncovering properties of a mental or neural lexicon are choosing a somewhat arbitrary unit for their stimuli. Phonological words (from open-class categories) consist at least of a root and a morpheme carrying syntactic category information, as well as the various functional morphemes from the root's syntactic environment that are required to join the root in the same word. A verb in English, for example, as a token of the language, would consist of at least three morphemes: the root, a morpheme that carries the syntactic category "verb," and a syntactically required tense morpheme. Single word experiments, then, might be seen as "small syntax" experiments; such experiments are not necessarily misguided, but they should not be presented as somehow in opposition to experiments using sequences of words. Even a simple experimental paradigm like confrontational picture naming requires the participants to produce minimal units of linguistic articulation; these would be phonological words, which are the output of a computation involving the syntactic combination of morphemes and the phonological interpretation of the resulting structure.

As explained in this chapter, morphology is not an independent "component" of grammar for linguistics, which recognizes a generative syntactic component and two "interfaces." One interface is concerned with the realization of syntactic structures in sound (or sign or orthography), and one is concerned with the semantic interpretation of those structures. Morphologists study particular aspects of the syntactic component and the interfaces that center on the minimal units of syntactic composition, but their special interests do not pick out a subsystem of grammar with linguistically significant autonomy. Therefore, neurobiological research aimed at denying the presence of "morphology" in the brain (Devlin, Jamison, Matthews, & Gonnerman, 2004) is not targeting the claim that morphology is neurologically isolatable—because the claim would be incoherent from a linguistic perspective—but rather the claim that the connection between sound

[1]For some discussion of the behavioral evidence apparently supporting a binary distinction between irregular and regular inflection, from the point of view of morpheme-based linguistics, see Albright and Hayes (2003), Embick and Marantz (2005), Fruchter, Stockall, and Marantz (2013), and the references cited therein.

and meaning involves an autonomous computation of a syntactic structure.

Computational linguists that seriously consider both the linguistic and the experimental evidence also come to the conclusion that there is no principled distinction between the structure and realization of morphemes within (phonological) words and the structure and realization of combinations of words and phrases. However, such linguists may also deny the existence of morphemes, claiming that the appearance of structured units in the mapping between sound and meaning is an emergent property of systems learning such mappings, not an a priori feature of the learner's language acquisition system, as in Baayen, Milin, Đurđević, Hendrix, and Marelli (2011). That is, for a generative linguist, the language learner's task is to learn the grammar—the morphemes, constraints on their combination, rules for their phonological realization, and semantic interpretation—to account for observations about the correlation of sound and meaning. Regarding the opposing view of Baayen et al. (2011), learners would be acquiring unmediated sound/meaning correspondences, and morphemes would reflect general sound/meaning correspondences that converge, for example, on contiguous sequences of sounds.

There are linguists who might question the analysis of words as hierarchical organizations of morphemes. In this chapter, we have seen some reasons why the consensus within generative grammar strongly supports the decomposition of words into such syntactic structures. Despite these disagreements over morphological decomposition within words and the possibility of an a-morphous morphology, no productive research program in linguistics has pursued the idea that sentences are a-syntactic. All competing accounts of the well-formedness of sentences and the connections between sound and meaning at the sentential level assume a syntactic analysis that involves structures of morphemes—both the elements and the relations between the elements—that are relatively abstract with respect to their interpretations. For example, all major theories of syntax assume a set of syntactic categories—like noun, verb, and adjective—that although associated with distributional categories and connected to meanings can be reduced to neither. The absence of a motivated analytic dividing line between words and phrases pushes the linguists' conclusions into the interior of words: the connections between sound and meaning within and between words involve the computation of syntactic structures of morphemes.

Because the most striking claims of morphology involve the decomposition of words into syntactic structures of morphemes, much of the neurolinguistic research in the era of brain imaging and brain monitoring of healthy intact human brains has, along with corresponding psycholinguistic work, centered on demonstrating that speakers decompose words into morphemes in visual and auditory word recognition (see Ettinger, Linzen, & Marantz, 2014; Fruchter et al., 2013; Lewis, Solomyak, & Marantz, 2011; Rastle, Davis, & New, 2004; Solomyak & Marantz, 2010; Zweig & Pylkkänen, 2009 and the references cited therein). These studies provide striking, although from a linguist's perspective inevitable, support for "full decomposition" models of recognition in which readers and listeners recognize morphologically complex forms via recognition and combination of their parts.

As Marcus Taft (2004) has pointed out, the apparent incompatibility of full decomposition models with the observation that the surface frequency of a complex word is the primary predictor of reaction time to the word in lexical decision experiments is tied to the unsupported claim that the frequency of regular computations and their results do not affect behavior. If the frequency of combination of a stem with a regular past-tense ending affected the speed with which this combination could be computed in the future, then surface frequency effects for regular past-tense forms could be attributed to the stage of processing in a full decomposition model in which the morphemes that result from decomposition of a word are recomposed for evaluation as a whole. When brains are monitored with MEG, for example, early correlations of brain activity with properties of morphologically complex words show sensitivity to properties of the constituent morphemes and their relations to other morphemes, not to the surface frequency of the forms, although the latter correlates well with button-pressing in responses (Solomyak & Marantz, 2010). The late timing of surface frequency effects supports Taft's interpretation: that they reflect the frequency of the computation of combination of morphemes, not the frequency of the static whole forms themselves.

Once we understand the necessity of composing morphemes via a syntactic derivation to create words, experiments using words provide a testing ground for general theories of language processing. Neurolinguistic work of the next decades should uncover how the brain accesses representations of morphemes, combines these representations into hierarchical syntactic structures, and realizes these structures in form and meaning. Theories of morphology in linguistics make explicit the types of knowledge that must be manipulated in these computations, as well as specifics about the computations and their constraints and detailed phenomena that any account of language processing must explain. Linguistic morphology, then, should be a key element in the neurobiology of language enterprise.

Acknowledgments

This work was supported in part by grant G1001 from the NYUAD Institute, New York University Abu Dhabi.

I thank Tal Linzen and Greg Hickok for comments regarding an earlier draft of this chapter, and Phoebe Gaston for editorial assistance.

References

Albright, A., & Hayes, B. (2003). Rules vs. analogy in English past tenses: A computational/experimental study. *Cognition, 90*(2), 119–161.

Anderson, S. R. (1992). *A-morphous morphology*. Cambridge: Cambridge University Press.

Baayen, R. H., Milin, P., Đurđević, D. F., Hendrix, P., & Marelli, M. (2011). An amorphous model for morphological processing in visual comprehension based on naive discriminative learning. *Psychological Review, 118*(3), 438.

Baker, M. (1985). The mirror principle and morphosyntactic explanation. *Linguistic Inquiry, 16*(3), 373–415.

Bobaljik, J. (2008). Paradigms (optimal and otherwise): A case for skepticism. In A. Bachrach, & A. Nevins (Eds.), *Inflectional identity* (pp. 29–54). Oxford: OUP.

Chomsky, N. (1995). *The minimalist program*. Cambridge, MA: MIT press.

Devlin, J. T., Jamison, H. L., Matthews, P. M., & Gonnerman, L. M. (2004). Morphology and the internal structure of words. *Proceedings of the National Academy of Sciences of the United States of America, 101*(41), 14984–14988.

Embick, D. (2010). *Localism versus globalism in morphology and phonology*. Cambridge, MA: MIT Press.

Embick, D., & Marantz, A. (2005). Cognitive neuroscience and the English past tense: Comments on the paper by Ullman et al. *Brain and Language, 93*(2), 243–247.

Embick, D., & Marantz, A. (2008). Architecture and blocking. *Linguistic Inquiry, 39*(1), 1–53.

Ettinger, A., Linzen, T., & Marantz, A. (2014). The role of morphology in phoneme prediction: Evidence from MEG. *Brain and Language, 129*, 14–23.

Fruchter, J., Stockall, L., & Marantz, A. (2013). MEG masked priming evidence for form-based decomposition of irregular verbs. *Frontiers in Human Neuroscience, 7*.

Hall, T. A. (1999). The phonological word: A review. In T. A. Hall, & U. Kleinhenz (Eds.), *Studies on the phonological word* (pp. 1–22). Philadelphia, PA: John Benjamins.

Inkelas, S., & Orhan Orgun, C. (1998). Level (non)ordering in recursive morphology: Evidence from Turkish. In S. Lapointe, D. Brentari, & P. Farrell (Eds.), *Morphology and its relation to phonology and syntax* (pp. 360–410). Stanford, CA: CSLI.

Lewis, G., Solomyak, O., & Marantz, A. (2011). The neural basis of obligatory decomposition of suffixed words. *Brain and Language, 118*(3), 118–127.

Lieber, R. (2010). *Introducing morphology*. Cambridge: Cambridge University Press.

Marantz, A. (2013a). Locality domains for contextual allomorphy across the interfaces. In O. Matushansky, & A. Marantz (Eds.), *Distributed morphology today* (pp. 95–115). Cambridge, MA: MIT Press.

Marantz, A. (2013b). No escape from morphemes in morphological processing. *Language and Cognitive Processes, 28*(7), 905–916.

Matthews, P. H. (1965). The inflectional component of a word-and-paradigm grammar. *Journal of Linguistics, 1*(2), 139–171.

Matushansky, O., & Marantz, A. (2013). *Distributed morphology today*. Cambridge, MA: MIT Press.

Pinker, S. (1999). *Words and rules: The ingredients of language*. New York, NY: Basic Books.

Rastle, K., Davis, M. H., & New, B. (2004). The broth in my brother's brothel: Morpho-orthographic segmentation in visual word recognition. *Psychonomic Bulletin & Review, 11*(6), 1090–1098.

Solomyak, O., & Marantz, A. (2010). Evidence for early morphological decomposition in visual word recognition. *Journal of Cognitive Neuroscience, 22*(9), 2042–2057.

Taft, M. (2004). Morphological decomposition and the reverse base frequency effect. *Quarterly Journal of Experimental Psychology Section A, 57*(4), 745–765.

Zweig, E., & Pylkkänen, L. (2009). A visual M170 effect of morphological complexity. *Language and Cognitive Processes, 24*(3), 412–439.

CHAPTER

14

Syntax and the Cognitive Neuroscience of Syntactic Structure Building

Jon Sprouse[1] *and Norbert Hornstein*[2]

[1]Department of Linguistics, University of Connecticut, Storrs, CT, USA; [2]Department of Linguistics, University of Maryland, College Park, MD, USA

14.1 INTRODUCTION

One goal of cognitive neuroscience, if not *the* goal of cognitive neuroscience, is to uncover how neural systems can give rise to the computations that underlie human cognition. Assuming, as most do, that the relevant biological description can be found at the level of neurons, then another way of stating this is that cognitive neuroscience is (at least) the search for the neuronal computations that underlie human cognition (Carandini, 2012; Carandini & Heeger, 2012). To the extent that this is an accurate formulation of the goal(s) of the field, any research program in cognitive neuroscience will have three components: (i) a cognitive theory that specifies the potential computations that underlie cognition; (ii) a neuroscientific theory that specifies how neurons (or populations of neurons) perform different types of computations; and (iii) a linking theory that maps between the cognitive theory and the neuroscientific theory (Marantz, 2005; Poeppel, 2012; Poeppel & Embick, 2005). We take all of this to be relatively uncontroversial; however, we mention it explicitly because we believe that modern syntactic theories, under a certain conception, are well-positioned to provide the first component (a theory of computations) for a cognitive neuroscientific theory of syntactic structure building. Our goal in this chapter is to make a case for this belief. We hope to demonstrate that the potential for a productive cross-fertilization exists between theoretical syntacticians and neuroscientists, and we suggest that developments in syntactic theory over the past two decades make this an optimal time to engage seriously in this collaboration.

For ease of exposition, we call our view that the theory of syntax can be, and should be, viewed as a theory of syntactic structure-building computations *the computational view of syntax*. This view is simply that the syntactic operations that have been proposed in syntactic theory (e.g., merge in Minimalism, substitution in Tree-Adjoining Grammar [TAG]) are a plausible cognitive theory of the structure-building computations that neurons must perform to process language. Therefore, a plausible research program for cognitive neuroscience would be to search for a theory of: (i) how (populations of) neurons could perform these computations and (ii) which (populations of) neurons are performing these computations during any given language processing event. As syntacticians, this strikes us as the natural evolution of the goals of the cognitive revolution of the 1950s in general, and of the goals of generative linguistics in particular. However, we are also aware that this is not how many would describe current syntactic theory. Therefore, we attempt to make our case in a series of steps. In Section 14.2, we provide a brief history of the field of syntax. The goal of this section is to contextualize modern syntactic theories such that it becomes clear that modern theories are not simply lists of grammatical rules (although older theories were), but instead theories of cognitive computations. In Section 14.3, we present two concrete examples of potential structure-building computations (from two distinct contemporary syntactic theories) to illustrate the computational view of syntax. In Section 14.4, we lay out several of the properties of modern syntactic theories that we believe make them well-suited for the computational view of syntax. We believe that these properties will be easily recognizable to all cognitive neuroscientists as the properties of a theory of cognitive computations. In Section 14.5, we discuss the large-scale collaboration between syntacticians, psycholinguists, and neuroscientists that will be necessary to construct a cognitive

Neurobiology of Language. DOI: http://dx.doi.org/10.1016/B978-0-12-407794-2.00014-6

neuroscience of syntactic structure building. In Section 14.6, we discuss some of the challenges that this collaboration might face. Section 14.7 concludes.

Before making our case for the computational view of syntax, a small clarification about the scope of this chapter is in order. We have explicitly chosen to focus on the issue of *why* syntactic theories will be useful for a cognitive neuroscience of language, and not *how* syntactic theorizing is conducted today. In other words, this chapter is intended to lay out arguments in favor of a large-scale collaboration between syntacticians and neuroscientists and is not intended to be a review chapter on syntax. We assume that if our arguments are successful, then syntacticians within these collaborations can carry the burden of doing the syntax. That being said, for readers interested in reviews of topics in contemporary syntax, we can recommend the review chapters in the recently published *Cambridge Handbook of Generative Syntax* (den Dikken, 2013), which contains 26 excellent review chapters covering everything from the history and goals of syntactic theory, to overviews of several major contemporary theories, to reviews of specific phenomena in syntax.

14.2 A BRIEF HISTORY OF SYNTACTIC THEORY

Syntactic theory starts from two critical observations. The first is that there is no upper bound on the number of possible phrases/sentences within any given language (i.e., languages are, for all practical purposes, "infinite"). This implies that successful language learning is not only the memorization of a set of expressions (otherwise infinity would be impossible) but also the acquisition of a grammar, which is just a finite specification of a recursive set of combinatory rules. The second observation is that any child can acquire any language (e.g., a child born to US citizens living in Kenya will successfully learn Swahili if exposed to Swahili speakers during childhood). Given that the first observation suggests that languages should be viewed as grammars, the second observation translates as any child can acquire any grammar. These two observations lead to the two driving questions for the field of syntax:

1. What are the properties of the grammars of all of the world's languages?
2. What are the mental mechanisms that allow humans to learn human languages?

The goal of Generative Syntax (GS) over the past 60 years has been to explore the properties of human grammars (question 1) in such a way to make it possible to explore the mental mechanisms that are required for successful language acquisition (question 2). As with any specialized science, the pursuit of these dual driving questions has led to the development of specific research programs and technical terminology, both of which have at times been opaque to other cognitive scientists working outside of syntax. Our goal in this section is to provide a brief history of the way the field has pursued these driving questions (to contextualize the modern syntactic theories discussed in Section 2.2) and to clarify some of the major points of miscommunication that have historically arisen between syntacticians and other cognitive scientists.

GS began by describing specific rules found in particular languages (and so contained in the grammars of these languages). This is not surprising; if one is interested in the kinds of rules natural language grammars contain, then a good way to begin is by looking for particular examples of such rules. Thus, in the earliest period of GS, syntacticians built mini-grammars describing how various constructions in particular languages were built (e.g., relative clauses in Chamorro, questions in English, topic constructions in German, reflexivization in French, etc.) and how they interacted with one another to generate a reasonably robust "fragment" of the language.

With models of such grammars in hand, the next step was to factor out the common properties of these language particular grammars and organize them into rule types (e.g., movement rules, phrase structure rules, construal rules). This more abstract categorization allowed for the radical simplification of the language that particular rules investigated in the prior period, with constructions reducing to congeries of simpler operations (although analogies are dangerous, this seems similar to the way other sciences often discover that seemingly distinct phenomena are in fact related, such as the unification of planetary motion, projectile motion, and tidal motion as instances of gravitational attraction in physics). By the mid 1980s there were several reasonably well-articulated candidate theories of syntax (e.g., Government and Binding Theory, Lexical-Functional Grammar, Tree-Adjoining Grammar), each specifying various rule types and their properties and each illuminating commonalities across constructions and across languages.

The simplification of grammatical rule types also led to progress on the second driving question. By reducing syntactic theories to only a few rule types, syntacticians could reduce the number of learning mechanisms required to learn human grammars (here we use the term "learning mechanisms" as a cover term for all of the components of learning theories: biases to attend to certain input, specifications of hypothesis spaces, algorithms for searching hypothesis spaces, etc.). With fewer learning mechanisms in the theory, syntacticians were able to investigate (and debate) the nature of the learning mechanisms themselves. Although there are a number of dimensions along which learning mechanisms might vary, syntactic theory has often focused on two in particular. The first is specificity: the learning mechanisms either can be domain-general, meaning that they are shared by several (or all) cognitive domains, or can be domain-specific, meaning that they are specific to language learning. The second dimension is nativity: the learning mechanisms either can be innate, meaning that they arise due to the genetic makeup of the organism, or can be derived, meaning that they are constructed from the combination of experience and other innate mechanisms. This leads to a 2 × 2 grid that can be used to classify any postulated learning mechanism (see also Pearl & Sprouse, 2013):

		Specificity	
		Domain-specific	Domain-general
Nativity	Innate	Universal Grammar	e.g., statistical learning
	Derived	e.g., learning to read	e.g., n-grams

What is particularly interesting about this grid is that it helps to clarify some of the miscommunications that have often arisen between syntacticians and other cognitive scientists surrounding terms like "innate," "domain-specific," and, worst of all, "Universal Grammar (UG)." This grid highlights the fact that the classification of any given learning mechanism is an empirical one. In other words, given a rule type X and a learning mechanism Y that could give rise to X, which cell does Y occupy in the grid? It may be the case that one or more of the cells are never used. Second, this grid highlights the fact that a complete specification of all of the rule types underlying human grammars and all of the learning mechanisms deployed to learn human grammars could involve any combination of the four types of learning mechanisms. As cognitive scientists, syntacticians are interested in all of the mechanisms underlying human syntax, not just the ones that get all of the attention in debates. Finally, this grid clarifies exactly what syntacticians mean when they use the term "Universal Grammar." Universal Grammar is just a special term for potential learning mechanisms that are simultaneously domain-specific and innate. Despite this rhetorical flourish, we hope it is clear that syntacticians view UG mechanisms (if they exist at all) as only a subset of the learning mechanisms that give rise to human language.[1]

The progress made in the 1980s regarding simplifying the rule types in human grammars also laid the foundation for the current research program within modern GS: to distill the computational commonalities found among the various *kinds* of rules (i.e., the computational features common to movement rules, phrase building rules, and construal rules). Here, again, the dimension of domain-generality and domain-specificity plays a role in theoretical discussions, but this time at the level of cognitive computation rather than at the level of learning mechanisms. As syntacticians have made progress distilling the computational properties of grammatical rules, they have found that some of the suggested computations appear similar to computations in other domains of cognition (e.g., the binding, or concatenation, of two mental representations), whereas others still retain some amount of domain-specificity (see Section 14.3 for a concrete example). Current GS work is pursuing this program in full force: attempting to identify the basic computations and determine which are specific to the syntax and which are shared with other cognitive domains.

Note the odyssey described: the field of syntax moved from the study of very specific descriptions of particular rules in particular languages to very general descriptions of the properties of linguistic computations and their relationship, and finally to cognitive computation more generally. This shift in the "grain" of linguistic analysis (in the sense of Poeppel & Embick, 2005) has had two important effects. First, it has reduced the special "linguistic" character of syntactic computations, making them more similar to the cognitive computations we find in other domains. Second, it has encouraged investigation of

[1]As a quick side note on Universal Grammar, the reason that UG receives so much attention, both within the syntax literature and across cognitive science, is that the other three types of learning mechanisms are generally uncontentious. For example, it is widely assumed that learning cannot occur in a blank slate (i.e., every learning system needs some built in biases if there is to be any generalization beyond the input); therefore, at least one learning mechanism must be innate. Nearly every postulated neural architecture (both symbolic and subsymbolic) assumes some form of statistical learning, which is presumably a learning mechanism (or set of mechanisms) that is domain-general and innate. The domain-general/derived cell is likely filled with the more complex statistical learning mechanisms required by different domains of cognition, such as the ability to track the probabilities of different sized sequences (n-grams). Similarly, the domain-specific/derived cell could potentially contain the learning mechanisms tailored to specific areas of higher-order cognition, such as reading (or maybe even language itself), but built from cognitive mechanisms available more broadly. It is the final cell, domain-specific/innate, that is the most contentious (and therefore, to some, the most interesting). In syntax, we call learning mechanisms that potentially fall into this cell Universal Grammar to highlight their significance. Currently, as we note here, a very hot area of syntactic investigation aims to reduce these domain-specific innate mechanisms to a minimum without losing explanations for the linguistic phenomena and generalizations that syntacticians have discovered over the past 60 years of syntactic research.

how syntactic computations might be used in real-time tasks such are parsing, production, and learning. Both these effects have had the consequence of bringing syntactic theory much closer to the empirical interests of others working in cognitive neuroscience.

Unfortunately, this shift in syntactic theory and its implications for cognitive neuroscience has not always been widely appreciated. Although the field of syntax was a central player in the cognitive revolution of the 1950s, in the intervening decades, syntax and the other domains of cognitive science have drifted apart. Some of this drift is the inevitable consequence of scientific specialization, and some of it reflects the internal logic of the different research programs (i.e., that the rule-based theories of the past were a necessary step in the evolution of syntactic theories). However, some of the drift reflects the view that syntactic theory has little to contribute to other domains of language research (including cognitive neuroscience). We worry that part of this problem may be that syntacticians have done a less-than-adequate job of conveying the general computational character of modern syntactic theories. In the absence of such discussions, it would not be surprising to learn that some cognitive neuroscientists still view syntax in terms of the phrase structure rules and transformations that typified syntactic theory in the 1950s and 1960s (and in varying forms up through the 1980s), rather than the more cognitively general computations common in current practice.[2] In the following sections, we provide two examples of how contemporary syntax might fruitfully make contact with cognitive neuroscience.

14.3 TWO CONCRETE EXAMPLES OF SYNTACTIC STRUCTURE-BUILDING COMPUTATIONS

Although early formulations of syntactic theories postulated complex rules that applied to entire constructions (often permuting, adding, or deleting multiple words in different positions in the constructions), as noted in Section 2.1, there has been a steady evolution toward theories that postulate a small number of structure-building operations that can be applied mechanistically (or derivationally) to construct more elaborate syntactic structures in a piecewise fashion. With very few exceptions, the primitives of contemporary syntactic theories are units and the computations that apply to those units. The following are two concrete examples[3]:

The syntactic theory known as *Minimalism* (or the Minimalist Program) postulates a single structure-building computation called `merge`, which takes two units and combines them to form a third. The units in Minimalism are lexical and sublexical items (something akin to the notion of word or morpheme, although the details can vary by analysis). `Merge` applies to these units directly, and also applies recursively to the output of previous instances of `merge`. In this way, `merge` can be used to iteratively construct complex syntactic structures from a basic inventory of lexical atoms. Of course, `merge` cannot freely concatenate any two units together. This means that restrictions on `merge` must be built into the lexical items themselves (only certain lexical items are compatible with each other), and in the case of merging units with the output of previous merges, this means that the outputs of `merge` must also contain restrictive properties. This is accomplished through a labeling computation, let us call it `label`, that applies a label to the output of `merge`, which can then be used to determine what that output can be merged with in the future.

The goal of syntactic theory is to capture the major properties of human syntactic structures with the proposed units and computations. For concreteness, we illustrate how `merge` and `label` succeed in capturing two such properties. The first is the distinction between local dependencies and nonlocal dependencies. A local dependency is simply the relationship between two adjacent items in a sentence. Local dependencies are

[2]The rule and transformation view of syntax has other problems as well. This conception of syntax is considered problematic for the computational view of syntax, because there are well-known empirical results from the 1950s and 1960s that appear to demonstrate that rule-based syntactic theories of that sort are poor models for real-time sentence processing (or, more specifically, poor predictors of complexity effects in language processing, as captured by the Derivational Theory of Complexity; for reviews see Fodor, Bever, & Garrett, 1974; but see Phillips, 1996 for a useful reevaluation of these claims). This problem is compounded by the fact that syntactic theories are at best only theories of syntactic structure building, with little to nothing to say about other components that are necessary for a complete theory of sentence processing, such as ambiguity resolution, memory/resource allocation, semantic structure building, and discourse structure building. Therefore, if one views syntactic theory as a rule-based theory, then it might appear to be a poor theory of only one small corner of language processing. Even as syntacticians, we understand why other cognitive scientists might find this version of syntactic theory difficult to engage with.

[3]There are, of course, a number of other syntactic theories that propose different types of computations (and different types of units). For example, Head-Driven Phrase Structure Grammar (HPSG) proposes a computation similar to merge, but without the possibility of internal merge (nonlocal dependencies involve a special slash unit instead). Construction grammar proposes a tree-unification computation similar to substitution in TAG, but operating over much larger units (entire constructions) and with the possibility of multiple unification points in a single construction. We assume that a full-fledged research program on the computational view of syntax would investigate all of these possible theories.

captured by `merge` by concatenating two distinct elements together. A nonlocal dependency is a relationship between two elements that are not adjacent in a sentence, such as the word *what* and *buy* in the question *What did John buy?* Nonlocal dependencies can be modeled by `merge` by concatenating a phrase with an element that is already properly contained within that phrase. Syntacticians call the former instantiation `external merge`, because the two elements are external to each other, and call the latter instantiation `internal merge`, because one element is properly contained within the other (Chomsky, 2004). The second property is the distinction between structures that contain verbs and their arguments (e.g., *eat bananas*) and structures that contain modifiers (e.g., *eat quickly*). The former, which we can call nonadjunction structures, are built from a combination of `merge` and `label`; the latter, which we can call adjunction structures, are built from `merge` alone (no `label`) (Hornstein, 2009). In this way, the two primitive computations `merge` and `label` can be used to construct syntactic structures capable of modeling the variety of structures one finds within natural language.

The syntactic theory known as *Tree-Adjoining Grammar* postulates two structure-building computations called `substitution` and `adjunction`. The units in TAG are small chunks of syntactic structure, or trees (hence the name of the theory). The `substitution` computation allows two elementary trees to be concatenated into locally dependent, nonadjunction structures. The `adjunction` computation, as the name implies, allows two trees to be concatenated into locally dependent, adjunction structures. TAG captures nonlocal dependencies that are only a single clause in length with a single elementary tree (so, *What did John buy?* is a single tree without any application of `substitution` or `adjunction`). For dependencies that are more than one clause in length, the `adjunction` computation is applied to a special type of tree called an auxiliary tree to extend the dependency length. In this way, the two primitive computations `substitution` and `adjunction` can be used to construct syntactic structures from elementary and auxiliary trees, and they give rise to the important distinctions of human syntax (for accessible introductions to TAG, see Frank, 2002, 2013) (Box 14.1).

Although both theories capture the same set of phenomena in human syntax, and although both theories postulate structure-building computations, they do so using different computations, different units, and different combinations of computations for each phenomenon. For nonadjunction structures that involve only local dependencies, Minimalism uses `external merge` and `label`, whereas TAG uses `substitution` with two elementary trees. For adjunction structures, Minimalism uses `external merge` alone, whereas TAG uses `adjunction` with two elementary trees. For nonlocal dependencies, Minimalism uses `internal merge` and `label`, whereas TAG uses `adjunction` with one elementary tree and one auxiliary tree. The similarities between these two syntactic theories (i.e., both use two basic computations to capture a wide range of characteristics of human syntax) suggest that both are

BOX 14.1

STRUCTURE-BUILDING COMPUTATIONS IN MINIMALISM AND TAG

The structure-building computation in Minimalism is called `merge`. It takes two syntactic objects and concatenates them into a third object. When the two syntactic objects are distinct, it is called `external merge`. When one of the objects is contained within the other, it is called `internal merge`:

`External merge`: [eat] + [bananas] = [[eat] [bananas]]

`Internal merge`: [did John buy what] + [what] = [[what] [did John buy what]]

The `label` computation determines the properties of the new syntactic object constructed by merge by applying a label based on the properties of one of the merged objects (the head). `Label` is mandatory for the merge of argument relationships (e.g., verbs and their arguments), but it appears to be optional for the merge of adjuncts (e.g., verbs and modifiers):

`Merge with Label`: [$_{V}$ eat] + [$_{NP}$ bananas] = [$_{VP}$ [$_{V}$ eat] [$_{NP}$ bananas]]

`Merge without Label`: [$_{VP}$ [$_{V}$ run]] + [$_{AdvP}$ quickly] = [[$_{VP}$ [$_{V}$ run]] [$_{AdvP}$ quickly]]

TAG proposes two structure-building operations. `Substitution` combines two elementary trees to form argument relationships, whereas `adjunction` combines elementary trees and adjunct trees to form adjunction structures:

`Substitution`: [$_{DP}$ John] + [$_{TP}$ [$_{DP}$] [$_{VP}$ eats bananas]] = [$_{TP}$ [$_{DP}$ John] [$_{VP}$ eats bananas]]

`Adjunction`: [$_{TP}$ [$_{DP}$ John] [$_{VP}$ [$_{V}$ runs]]] + [$_{VP}$ [$_{VP}$] quickly] = [$_{TP}$ [$_{DP}$ John] [$_{VP}$ [$_{VP}$ [$_{V}$ runs]] quickly]]

tapping into deeper truths about human structure-building computations. However, the subtle differences in the character of the proposed computations suggest that one might be able to derive competing predictions from each theory about the presence or absence of computations in different constructions. This combination of abstract similarities and subtle differences strikes us as a potentially fruitful starting point for a search for neuronal structure-building computations.

14.4 ADDITIONAL PROPERTIES OF SYNTACTIC THEORIES THAT ONE WOULD EXPECT FROM A THEORY OF COGNITIVE COMPUTATIONS

In addition to focusing on structure-building computations, there are a number of additional properties of contemporary syntactic theories that make them ideal candidates for the computational view of syntax. Here, we review three.

First, contemporary syntactic theories attempt to minimize the number of computations while maximizing the number of phenomena captured by the theory. This is a general desideratum of scientific theories in general (it is sometimes called unification, or reductionism, or just Occam's razor), and syntax, as a science, has adopted it as well. In fact, the name Minimalism was chosen to reflect the fact that years of investigations using earlier theories had yielded enough information about the properties of language as a cognitive system that it was finally possible to fruitfully incorporate unification/reduction/Occam's razor as a core principle of the research program. Other syntactic theories have been less blunt about this in their naming conventions, but the principles are obvious in the shape of the theories. Commitment to Occam has led to syntactic theories based on simple computations with wide applicability across the thousands of syntactic constructions in human languages. One nice side benefit of the ratio of computations to constructions is that it may make the search for neurophysiological correlates of these computations more fruitful, especially given concerns about spurious correlations in high-dimensional neurophysiological data.

Second, syntactic theories attempt to minimize the number of domain-specific computations and maximize the number of domain-general computations (to the extent possible given the overall minimization of the number of computations). This is an important, and often overlooked, point within syntax. The `merge` computation in Minimalism and the `substitution` computation in TAG are both plausibly domain-general computations similar to the binding computations that occur in multiple cognitive domains (vision, hearing, etc.), albeit operating over language-specific representations. The formulation of these plausibly domain-general computations stems directly from the premium that syntactic theories now place on unification/reductionism. In contrast, the `label` computation and the `adjunction` computation are potentially domain-specific, because there are no obvious correlates in other cognitive domains, although that could just be a consequence of our current state of knowledge. The question of whether plausibly domain-specific computations like `label` and `adjunction` can be learned or must be innate is an open area of research in language acquisition.

Finally, syntactic theories have mapped a sizable portion of the potential hypothesis space of syntactic structure-building computations. As we have mentioned, with few exceptions, every contemporary syntactic theory has the potential to serve as a theory of cognitive structure-building computations. Although the sheer number of competing theories may seem daunting from outside of syntax, from inside of syntax we believe this is a necessary step in the research. We need to explore every possible combination of unit-size and computation type that captures the empirical facts of human languages (and to be clear, not every combination does) to provide neuroscientists with a list of possible cognitive computations. To be sure, there is more work to be done on this front. And it goes without saying that syntacticians actively debate the empirical coverage of the different theories, and also how well each theory can achieve empirical coverage without inelegant stipulations. But from the perspective of cognitive neuroscience, the value is in the hypothesis space—each theory represents a different hypothesis about the types of fundamental structure-building computations (and the distribution of those functions across different sentences in any given language).[4]

[4]Inside of the field of syntax there is a recurring debate about whether different syntactic theories (e.g., Minimalism and TAG) are in some sense notational variants of one another. There are various mathematical proofs demonstrating that many theories are identical in terms of weak generative capacity (i.e., the ability to create certain strings of symbols and not others; e.g., Joshi, Vijay-Shanker & Weir 1991; Michaelis, 1998; Stabler, 1997). However, it is an open question whether these theories are equivalent in other terms, such as strong generative capacity (the types of structures that they can generate) or empirical adequacy for human languages. It is interesting to note that inside of syntax this debate is often couched in terms of theoretical "elegance" (i.e., how elegantly one theory captures a specific phenomenon relative to another theory). However, the research program suggested here would make such debates purely empirical: the "correct" syntactic theory would be the one that specifies the correct distribution of syntactic computations (and therefore their neuronal instantiations) across all of the constructions of a given language.

14.5 THE COLLABORATION NECESSARY TO ENGAGE IN THIS PROGRAM

The research program that the computational view of syntax suggests will require close collaboration between different types of researchers. The first step is for syntacticians to identify the structure-building computations that are deployed at each point in constructions from human syntax. From these analyses, syntacticians could identify two types of interesting cases. The first interesting case would be constructions that predict the same type of structure-building computation at the same location in all theories (e.g., Minimalism predicts merge at the same location in the construction as TAG predicts substitution). These areas of convergence may be fruitful places to begin the search for neuronal computations. A second interesting case would be constructions that require diverging computations across theories (e.g., Minimalism predicts merge but TAG predicts adjunction). If these analyses could be identified across a large number of constructions, then it should be possible to construct a type of comparison/subtractive logic that could uncover neuronal correlates of these computations. It seems to us that phenomena that vary along the major dimensions of human syntax, such as nonadjunction versus adjunction structures, or local versus nonlocal dependencies, will be most likely to lead to these types of convergences and divergences. But over the long-term, every phenomenon of syntax should be investigated (to the extent possible given some of the challenges discussed in Section 14.4).

The second step is for syntacticians and theoretical neuroscientists to figure out how neurons deploy the structure-building computations that underlie the phenomenon in each theory. In practice, this step might require several substeps. For example, the typical form of syntactic theories is "bottom-up": the most deeply embedded constituents are constructed first, followed by the next most deeply embedded, and so on. This is largely the reverse order from sentence comprehension and production. Given that the empirical studies required by later steps will be based on comprehension (and perhaps production), it may be necessary to convert the "bottom-up" computations of syntactic theories into the "left-to-right" or "top-down" computations of parsing theories. There exist several computational models for how to relate bottom-up grammars with left–right parsers. This step will most likely involve collaboration among mathematical linguists (to rigorously formalize the syntactic computations (Collins & Stabler, 2011; Stabler, 1997)), mathematical psycholinguists to convert those computations into parsing computations (e.g., Berwick & Weinberg, 1984; Marcus, 1980; Stabler, 2011, 2013; and for issues beyond structure-building: Hale, 2003; Kobele, Gerth, & Hale, 2013), and neuroscientists to identify candidate neurocomputational systems. Although this sounds straightforward, it is likely that the space of possible computations will expand at each step, from syntactic computations to mathematically formalized computations, from formalized computations to parsing computations, and from parsing computations to neuronal computations. It is quite possible that this step will result in hypothesis spaces for the possible neuronal computations for each syntactic theory relevant to each phenomenon.

Once the structure-building computations have been translated into potential neuronal computations (or hypothesis spaces of potential neuronal computations), the final step is to look for evidence of those computations in neural systems. Again, although we state this as a single step in principle, we assume that it will be a multifaceted process in practice, drawing on neuroscientists of all types: electrophysiologists (EEG/MEG), neuroimagers (fMRI), and even neurosurgeons (ECoG). As syntacticians, this step is the furthest beyond our area of expertise, but we could imagine a process like the following. First, (extracranial) electrophysiological work (either EEG or MEG) could be used to identify the gross neuronal signatures in either the amplitude domain (ERP/ERF) or frequency domain (oscillations) that occur at the critical regions in the sentences of interest. Depending on the similarities and differences predicted by the different syntactic theories, and the different classes of neuronal populations that follow from the formalization of those theories in the previous step, the neuronal signatures (ERP/ERFs or oscillations) may be useful in eliminating competing computations from consideration. Recently, there have been exciting examples of work of this type in both syntax and semantics research. For example, Pylkkanen and colleagues have been searching for neurological correlates of fundamental semantic combinatory processes in the time–amplitude domain using MEG, with results pointing to increased activity in left anterior temporal lobe (LATL) and ventromedial prefrontal cortex (vmPFC) (e.g., Bemis & Pylkkänen, 2011; Brennan & Pylkkänen, 2008; and many others). As another example, Bastiaansen and colleagues have been searching for neurological correlates of both syntactic and semantic combinatory processes in the time–frequency domain using EEG, with results pointing to the gamma frequency band (>30 Hz) for semantic processes and the lower beta frequency band (13–18 Hz) for syntactic processes (e.g., Bastiaansen, Magyari, & Hagoort, 2010; Bastiaansen, Van Berkum, & Hagoort, 2002; for a review see Bastiaansen, Mazaheri, & Jensen, 2012).

Once electrophysiological correlates have been identified, localization studies, either with MEG (if the orientation of the generators is appropriate) or concurrent EEG and fMRI, could be used to identify cortical areas associated with the neuronal activity of interest. There is a large and ever-growing literature on localization in language processing, and we are sure the other chapters in this volume provide more enlightening reviews of that literature. However, we would like to point to Pallier, Devauchelle, and Dehaene (2011) as an example of localization work that shares the same spirit as the program advocated here. Pallier et al. searched for brain areas that respond to the size of the syntactic constituent being processed (from 1 to 12 words), in essence using the number of syntactic computations deployed as a measure of complexity, and finding activity in a number of regions, including left inferior frontal gyrus (LIFG), left anterior superior temporal sulcus (LaPSTS), and left posterior superior temporal sulcus (LpSTS). Finally, when suitable location and electrophysiological hypotheses are established, intracranial recordings (ECoG) could be used to identify the single unit information necessary to begin to identify the specific neuronal computation and observe its implementation.

We admit that the brief sketch of the collaboration suggested by the computational view is based on our incomplete understanding of the various fields that would be part of the collaboration. We also admit that the space of possible neuronal computations is likely much larger than the space of extant structure-building operations, making the search for the actual neuronal computations that much more difficult. But it seems to us that the size of the hypothesis space is irrelevant to the question of how to move the fields of syntax and neuroscience forward (and together). This is either the right hypothesis space to be searching or it is not. It seems to us that multiple domains of cognition are converging on both the need for identifying neuronal computations and the plausibility of conducting such a search in the 21st century (Carandini, 2012; Poeppel, Emmorey, Hickok, & Pylkkänen, 2012). We believe that the wider field of syntax is ready to join the search that researchers such as Bastiaansen, Dehaene, Pallier, Pylkkanen, and colleagues have begun.

14.6 CHALLENGES TO THIS RESEARCH PROGRAM

Beyond the obvious challenge of engaging in the interdisciplinary work presented in Section 14.5, there are numerous smaller challenges that need to be addressed for the collaboration to be successful. In this section we discuss five, in some cases in an attempt to dispel the challenge and in others simply to raise the issue for future work.

One obvious challenge is the concern from some cognitive scientists that syntactic theories are not built on solid empirical foundations. This concern has been expressed since the earliest days of syntactic theorizing (Hill, 1961), and recently with several high-profile publications (Gibson & Fedorenko, 2010, 2013). This concern is driven by the idea that the typical data collection methods are too informal to provide reliable data; therefore, the theories built on that data are themselves unreliable. The persistence of this concern speaks to a fundamental failure on the part of syntacticians to make the argument either that the data type that they are collecting (acceptability judgments) are robust enough that the informality of the collection methods have no impact or that there are unreported safeguards in the informal methods to prevent the kind of unreliability that they are concerned about (see Marantz, 2005; Phillips, 2009 for discussions of these issues). Sprouse and Almeida (2012) and Sprouse, Schütze, and Almeida (2013) have begun to address this concern directly by exhaustively retesting all of the phenomena in a popular Minimalist textbook using formal methods, and by retesting a large random sample of phenomena from a popular syntax journal using formal methods. These retests have replicated 98% and 95% of the phenomena, respectively, suggesting that the informal methods used in syntax have the reliability that syntacticians claim. Given recent concerns about replicability inside of some areas of psychology, it is heartening to see that large-scale replications inside of syntax yield potential error rates at or below the conventional type I error rate of 5%.

Despite the substantial evidence that the acceptability judgments that form the basis of syntactic theory are reliable, one could imagine potential collaborators being concerned that a theory built on offline data (like acceptability judgments) would be irrelevant for a theory built on real-time language processing data (like the electrophysiological data required by the research program proposed here). We agree that this could be a reasonable concern *a priori*. However, there is also a growing body of research in the sentence processing literature demonstrating that real-time sentence processing behavior respects grammatical conditions on well-formedness. For example, several studies have shown that complex constraints on the formation of nonlocal dependencies (called island constraints in the syntax literature) are respected by the parsing mechanism that form these dependencies in real time (Stowe, 1986; Traxler & Pickering, 1996). In addition, several studies have demonstrated that these same processing mechanisms respect the sophisticated *exceptions* to these constraints postulated by syntactic theories

(Phillips, 2006; Wagers & Phillips, 2009). Similarly, several studies have demonstrated that complex constraints on the dependencies that give pronouns their referents (called binding constraints in the syntax literature) are also respected by real-time referential processing mechanisms (Kazanina, Lau, Lieberman, Yoshida, & Phillips, 2007; Sturt, 2003; Van Gompel & Liversedge, 2003). Several recent studies also show these effects to be the result of grammatical constraints and not the consequences of nongrammatical processing mechanisms (Dillon & Hornstein, 2013; Kush, Omaki, & Hornstein, 2013; Sprouse, Wagers, & Phillips, 2012; Yoshida, Kazanina, Pablos, & Sturt, 2013). In sum, there is a growing body of convincing evidence that syntactic theories capture structure-building properties that are relevant for real-time sentence processing, despite having initially been empirically based on offline data.

A third potential challenge for the computational view of syntax is that not every syntactician agrees that syntactic theories should serve as a theory of cognitive structure-building computations. The potential for a logical distinction between theories of syntax and theories of cognitive structure-building is clearest in examples of nonmentalistic, or Platonic, linguistic theories, which seek to study the mathematical properties of language without making any claims about how those properties are instantiated in a brain. Even within GS, which is mentalistic, it is not uncommon to hear theories of syntax described as theories of knowledge (or competence) and not theories of use (or performance). The computational view of syntax goes beyond simple knowledge description. The computational view sees syntactic theories as making substantive claims about how syntactic structure building is instantiated in the human brain. It may be the case that there is a one-to-many relationship between syntactic theories and neuronal structure-building computations, but the relationship is there (see Lewis & Phillips, 2014 for a deeper discussion of this challenge).

A final challenge to the computational view of syntax is the problem of isolating structure-building computations from other sentence processing computations in real-time processing data. Real-time language processing data are going to contain signals from both structure-building computations and all of the nonstructure-building computations that syntactic theory abstracts away from (parsing strategies, resource allocation, task specific strategies in the sense of Rogalksy & Hickok, 2011, etc). This means that the actual construction of neurophysiological experiments discussed in Section 14.5 will require quite a bit of ingenuity to isolate the structure-building computations, especially given the high-dimensionality of neural data, and the likelihood of spurious correlations. And even assuming that *logically* isolating a computation of interest is possible in the experimental stimuli, *physically* isolating a neuronal computation in human neural systems is probably orders of magnitude more difficult. To our knowledge, there are no existing neuronal computations that can be used as a guide (a Rosetta stone of sorts) to mark the beginning or end of a computation being physically performed. We assume that as more and more computations are investigated, combining them in novel ways will eventually allow the physical boundaries of computations to be mapped, but this is currently a promissory note. In summary, the narrow focus of syntactic theories on structure-building computations is in some ways positive, because it provides a hypothesis space for a problem that is potentially tractable, but it is also negative, because the computations left out of that hypothesis space may be either confounds or necessary additions to solve the physical localization problem.

14.7 CONCLUSION

We believe that modern syntactic theory is well-suited to serve as a cognitive theory of syntactic structure-building computations, and that the time is right for a large-scale collaboration between syntacticians, mathematical linguists and psycholinguists, and theoretical and experimental neuroscientists to identify the neuronal instantiations of those computations. Such a research program will be a collaborative project of unprecedented scope and will face numerous theoretical and technological challenges, but in the histories of cognitive science, linguistics, and neuroscience, there has never been a better time to try.

References

Bastiaansen, M. C. M., Magyari, L., & Hagoort, P. (2010). Tactic unification operations are reflected in oscillatory dynamics during online sentence comprehension. *Journal of Cognitive Neuroscience*, *22*, 1333–1347.

Bastiaansen, M. C. M., Mazaheri, A., & Jensen, O. (2012). Beyond ERPs: Oscillatory neuronal dynamics. In S. J. Luck, & E. S. Kappenman (Eds.), *The Oxford handbook of event-related potential components* (pp. 31–50). New York, NY: Oxford University Press.

Bastiaansen, M. C. M., Van Berkum, J. J. A., & Hagoort, P. (2002). Syntactic processing modulates the θ rhythm of the human EEG. *NeuroImage*, *17*, 1479–1492.

Bemis, D. K., & Pylkkänen, L. (2011). Simple composition: An MEG investigation into the comprehension of minimal linguistic phrases. *Journal of Neuroscience*, *31*(8), 2801–2814.

Berwick, R. C., & Weinberg, A. S. (1984). *The grammatical basis of linguistic performance*. Cambridge, MA: MIT Press.

Brennan, J., & Pylkkänen, L. (2008). Processing events: Behavioral and neuromagnetic correlates of aspectual coercion. *Brain and Language*, *106*, 132–143.

Carandini, M. (2012). From circuits to behavior: A bridge too far? *Nature Neuroscience, 15*(4), 507–509.

Carandini, M., & Heeger, D. J. (2012). Normalization as a canonical neural computation. *Nature Reviews Neuroscience, 13*, 51–62.

Chomsky, N. (2004). Beyond explanatory adequacy. In A. Belletti (Ed.), *The cartography of syntactic structure vol 3: Structures and beyond* (pp. 104–131). Oxford: Oxford University.

Collins, C., & Stabler E. (2011). A formalization of minimalist syntax. <http://ling.auf.net/lingbuzz/001691>.

den Dikken, M. (Ed.), (2013). *The Cambridge handbook of generative syntax*. Cambridge, UK: Cambridge University Press.

Dillon, B., & Hornstein, N. (2013). On the structural nature of island constraints. In J. Sprouse, & N. Hornstein (Eds.), *Experimental syntax and island effects* (pp. 208–222). Cambridge, UK: Cambridge University Press.

Fodor, J., Bever, T., & Garrett, M. (1974). *The psychology of language*. New York, NY: McGraw Hill.

Frank, R. (2002). *Phrase structure composition and syntactic dependencies*. Cambridge, MA: MIT Press.

Frank, R. (2013). Tree adjoining grammar. In M. den Dikken (Ed.), *The Cambridge handbook of generative syntax* (pp. 226–261). Cambridge, UK: Cambridge University Press.

Gibson, E., & Fedorenko, E. (2010). Weak quantitative standards in linguistics research. *Trends in Cognitive Sciences, 14*, 233–234.

Gibson, E., & Fedorenko, E. (2013). The need for quantitative methods in syntax and semantics research. *Language and Cognitive Processes, 28*, 88–124.

Hale, J. (2003). *Grammar, uncertainty, and sentence processing*. Baltimore, MD: Johns Hopkins University.

Hill, A. A. (1961). Grammaticality. *Word, 17*, 1–10.

Hornstein, N. (2009). *A theory of syntax: Minimal operations and universal grammar*. Cambridge, UK: Cambridge University Press.

Joshi, A. K., Vijay-Shanker, K., & Weir, D. J. (1991). The convergence of mildly context-sensitive grammar formalisms. In P. Sells, S. Sheiber, & T. Wasow (Eds.), *Foundational issues in natural language processing* (pp. 31–81). Cambridge, MA: MIT Press.

Kazanina, N., Lau, E. F., Lieberman, M., Yoshida, M., & Phillips, C. (2007). The effect of syntactic constraints on the processing of backwards anaphora. *Journal of Memory and Language, 56*, 384–409.

Kobele, G. M., Gerth, S., & Hale, J. T. (2013). Memory resource allocation in top-down minimalist parsing. In G. Morrill, & M.-J. Nederhof (Eds.), FG 2012/2013, volume 8036 of lecture notes in computer science (pp. 32–51). Heidelberg: Springer.

Kush, D., Omaki, A., & Hornstein, N. (2013). Microvariation in islands? In J. Sprouse, & N. Hornstein (Eds.), *Experimental syntax and island effects* (pp. 239–264). Cambridge University Press.

Lewis, S., & Phillips, C. (2014). Aligning grammatical theories and language processing models. *Journal of Psycholinguistic Research*. doi:10.1007/s10936-014-9329-z

Marantz, A. (2005). Generative linguistics within the cognitive neuroscience of language. *The Linguistic Review, 22*, 429–445.

Marcus, M. (1980). *A theory of syntactic recognition for natural language*. Cambridge. MA: MIT Press.

Michaelis, J. (1998). Derivational minimalism is mildly context sensitive. In M. Moortgat (Ed.), *Logical aspects of computational linguistics, lecture notes in artificial intelligence volume 1* (pp. 179–198). Heidelberg: Springer.

Pallier, C., Devauchelle, A.-D., & Dehaene, S. (2011). Cortical representation of the constituent structure of sentences. *Proceedings of the National Academy of Sciences, 108*(6), 2522–2527.

Pearl, L., & Sprouse, J. (2013). Syntactic islands and learning biases: Combining experimental syntax and computational modeling to investigate the language acquisition problem. *Language Acquisition, 20*, 23–68.

Phillips, C. (1996). *Order and structure*. Cambridge, MA: Massachusetts Institute of Technology.

Phillips, C. (2006). The real-time status of island phenomena. *Language, 82*, 795–823.

Phillips, C. (2009). Should we impeach armchair linguists? In S. Iwasaki, H. Hoji, P. Clancy, & S.-O. Sohn (Eds.), Proceedings from Japanese/Korean linguistics 17. Stanford, CA: CSLI Publications.

Poeppel, D. (2012). The maps problem and the mapping problem: Two challenges for a cognitive neuroscience of speech and language. *Cognitive Neuropsychology, 29*, 34–55.

Poeppel, D., & Embick, D. (2005). The relation between linguistics and neuroscience. In A. Cutler (Ed.), *Twenty-first century psycholinguistics: Four cornerstones* (pp. 103–118). Mahwah, NJ: Lawrence Erlbaum Associates.

Poeppel, D., Emmorey, K., Hickok, G., & Pylkkänen, L. (2012). Towards a new neurobiology of language. *Journal of Neuroscience, 32*(41), 14125–14131.

Rogalksy, C., & Hickok, G. (2011). The role of Broca's area in sentence comprehension. *Journal of Cognitive Neuroscience, 23*, 1664–1680.

Sprouse, J., & Almeida, D. (2012). Assessing the reliability of textbook data in syntax: Adger's Core Syntax. *Journal of Linguistics, 48*, 609–652.

Sprouse, J., Schütze, C. T., & Almeida, D. (2013). A comparison of informal and formal acceptability judgments using a random sample from Linguistic Inquiry 2001–2010. *Lingua, 134*, 219–248.

Sprouse, J., Wagers, M., & Phillips, C. (2012). A test of the relation between working memory capacity and syntactic island effects. *Language, 88*(1), 82–123.

Stabler, E. (1997). Derivational minimalism. In C. Retoré (Ed.), *Logical aspects of computational linguistics* (pp. 68–95). New York, NY: Springer.

Stabler, E. (2011). Top-down recognizers for MCFGs and MGs. In: *Proceedings of the second workshop on cognitive modeling and computational linguistics (CMCL '11)* (pp. 39–48). Stroudsburg, PA: Association for Computational Linguistics.

Stabler, E. (2013). Two models of minimalist, incremental syntactic analysis. *Topics in Cognitive Science, 5*(3), 611–633.

Stowe, L. A. (1986). Evidence for on-line gap-location. *Language and Cognitive Processes, 1*, 227–245.

Sturt, P. (2003). The time-course of the application of binding constraints in reference resolution. *Journal of Memory and Language, 48*, 542–562.

Traxler, M. J., & Pickering, M. J. (1996). Plausibility and the processing of unbounded dependencies: An eye-tracking study. *Journal of Memory and Language, 35*, 454–475.

Van Gompel, R. P. G., & Liversedge, S. P. (2003). The influence of morphological information on cataphoric pronoun assignment. *Journal of Experimental Psychology: Learning, Memory, and Cognition, 29*, 128–139.

Wagers, M., & Phillips, C. (2009). Multiple dependencies and the role of the grammar in real-time comprehension. *Journal of Linguistics, 45*, 395–433.

Yoshida, M., Kazanina, N., Pablos, L., & Sturt, P. (2014). On the origin of islands. *Language and Cognitive Processes, 29*, 761–770.

CHAPTER

15

Speech Perception as a Perceptuo-Motor Skill

Carol A. Fowler

Department of Psychology, University of Connecticut, Storrs, CT, USA

15.1 INTRODUCTION

Among theories of phonetic perception, there are "general auditory approaches" (Diehl, Lotto, & Holt, 2004) that contrast with "gesture theories" (Fowler, 1986; Liberman & Mattingly, 1985). These approaches are contrasted in many publications (Diehl et al., 2004; Fowler & Iskarous, 2013). The present chapter focuses on gesture theories, in which the integrality of speech production and perception is central to the accounts.

The best known of the gesture approaches to speech perception, the motor theory (Liberman, Cooper, Shankweiler, & Studdert-Kennedy, 1967; Liberman & Mattingly, 1985), claims that speech perceivers perceive linguistically significant ("phonetic") gestures of the vocal tract as immediate perceptual objects (rather than auditory transforms of the acoustic speech signal). Another claim is that perceiving speech necessarily involves speech motor system recruitment. A final claim is that, in these respects, speech perception is special, particularly in relation to other auditorily perceived events in which auditory transforms of acoustic signals are perceptual objects. The present chapter suggests that the first two claims are accurate (except possibly the *necessity* of motor recruitment). However, the third claim is not. Perception of distal events (gestures, for speech) is generally what perceptual systems achieve (Gibson, 1966), and motor system recruitment is widespread in perception and cognition.

An alternative gesture theory, direct realism (Fowler, 1986, 1996), agrees that listeners to speech perceive the distal events of speaking, phonetic gestures, but disagrees that, in regard to perceiving distal events, rather than proximal (e.g., acoustic) stimulation, speech perception is special (see Carello, Wagman, & Turvey, 2005 for a review of "ecological acoustics"; see also Rosenblum, 2008). In direct realism, recruitment of the motor system is not expected for speech perception, because information for gestures is available in the acoustic speech signal. Required or not, however, evidence shows that recruitment is widespread.

In both gesture theories, speech perception is a perceptuo-motor skill. Understanding why it is and why that does not make it special require embedding its study in a larger context of investigations of the ecology of perceiving and acting. Therefore, the context for the literature reviewed in this chapter is not that of processing in the brain. Rather, it is about language users in their world, in which perceiving and acting are inextricably intertwined. Presumably, the brains of language users, like the rest of language users, will be adapted to such a world. Therefore, findings of speech motor system activation in the brain during ordinary speech perception perhaps should be unsurprising. For an alternative perspective, see Lotto and Holt (Chapter 16).

15.1.1 Perception and Action are Inextricably Integrated

In the econiche, animals' activities are necessarily perceptually guided. For example, locomotion usually involves visual guidance so that walkers can move along a route toward a goal location while avoiding collision with obstacles (Warren, 2006). Sometimes, however, for example, crossing a street at a curve so that oncoming traffic is not visible (or crossing anywhere for those texting while walking) may involve detecting approaching vehicles by listening. It also involves maintaining postural balance with the help of vestibular systems and proprioceptive detection of the forces exerted by the walker on the support surface and on the walker by the support surface. Walking, like other activities, is a multimodal perceptuo-motor activity.

Neurobiology of Language. DOI: http://dx.doi.org/10.1016/B978-0-12-407794-2.00015-8

Sometimes animals' goals are more exploratory than performatory. For example, human sightseers may walk around (as it were) to enable their perceptual systems to intercept new sights or sounds or feels or smells or tastes. In that case, complementarily to performatory actions whereby perception serves action, action systems serve primarily perceptual aims. Either way, acting and perceiving both are perceptuo-motor skills.

Aside from being perceptuo-motor in nature, however, animals' actions and perceptions share something else that is crucial to their lives, namely the econiche. The world in which they act and the world in which they obtain perceptual information is the *same world*. Because survival depends on felicitous acting, it also depends on perceptual systems that accurately expose properties of the econiche relevant to their actions (the "affordances" of the econiche; Gibson, 1979). In short, perceptual systems as well as action systems have to be adapted to the animals' "way of life" (Gibson, 1994). There has to be a strong likelihood of a relation of parity between properties of the econiche implied by an animal's actions and properties of the econiche perceived in support of those actions.

Most analogous to language use in the world are coordinative social activities (e.g., moving a piece of furniture together, paddling a canoe, playing a duet). The foregoing characterizations are true of these activities as well, but now parity has an additional cross-person dimension. Participants have to perceive social affordances (Marsh, Richardson, Baron, & Schmidt, 2006), and their actions generally should be true to them. In addition, co-participants' perceptions should be shared and co-participants should coordinate their actions in relation to those shared social affordances. Speech perception and production in the econiche are coordinative social activities.

15.1.2 Parity in Speech

Because speaking and listening are social activities, much of this discussion applies to them. But they are different from some nonlinguistic, nonsocial actions in a notable way. At one level of description, an aim of speaking is to cause patterning in the air. That is how speaking can get the attention of a listener and how it can get a chance to have its intended impact on him or her. This is different from the activity of locomoting, for example, which typically is done to get somewhere, not to cause patterning in light or air. However, it is not so different in that respect from performances meant to be seen or heard, such as ballet or competitive diving or musical performances.

Regarding perceiving and acting generally, Liberman and Mattingly (1989) and Liberman and Whalen (2000) have remarked that parity is central to speaking and perceiving speech. For Liberman and Whalen, parity in language use has three essential aspects. One relates to the observation that the same language or languages are involved in a language user's dual roles of talking and of listening to the speech of others. There must be a relation of parity (sometimes called a "common code;" Schütz-Bosbach & Prinz, 2007) between language forms produced and perceived by the same person. That is how perception of one's own speech can guide its production (e.g., Houde & Jordan, 1998).

A second component of parity in language relates to between-person language use. For language use to serve its communicative role in a between-person exchange, there has to be a relation of sufficient parity[1] between forms uttered by a talker and forms intercepted by listeners. For many theorists (Pierrehumbert, 1990), the "common code" within a speaker–hearer and between them is "mental" and not physical. However, for speech perception not to be special, perceptual objects *have* to be physical. Only physical things can causally structure informational media such as light and air and therefore can have effects that *can* be perceived. Language forms as physical events do not prevent their being psychological in nature as well (Ryle, 1949). In the account of Liberman and Mattingly (1985) and others (Browman & Goldstein, 1986; Goldstein & Fowler, 2003), the smallest language forms are phonetic gestures of the vocal tract. They are physical actions that have linguistic and, hence, psychological significance.

The third component of parity for Liberman and Whalen (2000) is that brain systems for production and perception of language forms must have co-evolved, because each is specialized for the unique problems to which coarticulation gives rise, and neither specialized system would be useful without the other. This component is not addressed further here beyond commenting that coarticulation and its effects are not special to speech.

[1]The hedge "sufficient" is meant to forestall misunderstanding (Remez & Pardo, 2006). Talkers and listeners do not have to share their dialect, and listeners do not have to detect every phone, even every word, produced by talkers for language to "work" in public use. However, sharing of forms has to be sufficient for the talker's message to get across in events involving talking. Relatedly, at a slower time scale, in a language community, language forms and structures serve as conventions (Millikan, 2003) that *are* conventional just because they are reproduced (with the same hedge: reproduced with sufficient fidelity to count as being reproduced) across community members. The capacity to reproduce perceived forms implies perception–production parity.

The foregoing discussion is meant to underscore that, in the econiche, life, including linguistic life, is perceptuo-motor in nature. Nothing discussed here logically requires that *mechanisms* involved in action must be incorporated in perceptual systems or vice versa. However, it would be surprising if they were not, and in both the linguistic and nonlinguistic domains, they appear to be. A brief review is provided within the domain of speech and then, outside of it showing that evidence for the perceptuo-motor nature of perception and cognition is quite general.

15.2 RESEARCH FINDINGS

15.2.1 Speech

Liberman and Mattingly (1985) identify their motor theory of speech perception as motor in two respects. First, in the theory, listeners perceive speech gestures. Second, to achieve perception of gestures, they recruit their own speech motor system.

The first claim does not, in fact, make the theory a motor theory. No one would identify a theory of visual perception as a motor theory if it made the (uncontroversial) claim that, when a person walks by in a perceiver's line of sight, the perceiver sees someone walking. Perceiving a motor event when a motor event occurs in the econiche does not require a theoretical account deserving the descriptor "motor." The first claim of motor theorists should imply that there is nothing special about speech perception in regard to perceptual objects (Rosenblum, 2008). Listeners perceive speech gestures as they hear people knocking on doors and see walkers walking, because perceivers intercept structure in media, such as structured air and light that inform about what they need to perceive: the objects and events that compose the econiche.

Although the claim that speech listeners perceive phonetic gestures is controversial in the field of speech perception, within the context of the foregoing discussion of perception in the econiche, phonetic gestures are expected perceptual objects. Compatibly, there is converging evidence for gesture perception. A sampling is offered here.

Speech perception is a perceptuo-motor skill with respect to its objects of perception. In articles published in 1952 and 1954 (Liberman, Delattre, & Cooper, 1952; Liberman, Delattre, Cooper, & Gerstman, 1954), Liberman and colleagues reported their first findings that led them to develop a motor theory. Suggestions, such as that of Lotto, Hickok and Holt (2009), that the claims of the motor theory were motivated by a theoretical issue, the problem of coarticulation, are mistaken. The theory was motivated by research findings, the first of which were published in 1952. The claim of motor system involvement was *rationalized* in terms of coarticulation later. The first (Liberman et al., 1952) was a finding that the *same* acoustic pattern (a stop burst centered at 1440 Hz) that had to have been produced by *different* consonantal gestures coarticulated with different vowels was heard as *different* consonants (the "/pi/-/ka/-/pu/" phenomenon). The second (Liberman et al., 1954) was a finding that *different* acoustic patterns (second formant transitions) that were produced by the *same* (alveolar) consonant constriction gesture of the vocal tract coarticulated with different vowel gestures were heard as the *same* consonants (the "/di/-du/" phenomenon). In both cases, when listeners' perceptual tracking of acoustic "cues" could be dissociated from their tracking of gesture production, listeners were found to track gestures.[2]

These findings imply that listeners somehow "parse" (Fowler & Smith, 1986) the acoustic signal along gestural lines. That is, to perceive the same consonant, /d/, from different formant transitions for the syllables /di/ and /du/, they must extract acoustic information that supports perceptual separation of temporally overlapping consonantal and vocalic gestures. Many distinct findings support that parsing occurs (Fowler & Brown, 1997; Fowler & Smith, 1986; Pardo & Fowler, 1997; Silverman, 1986, 1987). Perceivers do the same kind of parsing in the visual domain (e.g., in perception of walking and other biological motions; Johanssen, 1973; Runeson & Frykholm, 1981).

[2]Intuitively, these findings can be explained as pattern learning instead of gesture perception. A pattern learner can learn to classify the very different acoustic patterns for /di/ and /du/ into the same /d/ category while learning to classify the same stop bursts in /pi/ and /ka/ differently. Learning these classifications requires a systematic basis for learning, of course, and the basis must be the articulatory sameness of the consonantal gestures in /di/ and /du/ and differentness in /pi/ and /ka/. By this account, pattern learners acquire the acoustic-articulatory links when they hear the acoustic consequences of their own speech. No presumption that listeners perceive articulation is required. This, in fact, was the earliest motor theory (Liberman, 1957), with the proposed learning underlying "acquired similarity" of the /d/ s in /di/ and /du/ and "acquired distinctiveness" of the stop bursts of /pi/ and /ka/. The account fails, however, as argued by early opponents of the motor theory, because individuals exist who perceive speech successfully without being able to produce what they perceive (MacNeilage, Rootes, & Chase, 1967). Information about articulation has to come from information in the acoustic signals from others' speech as well as one's own when that is possible.

One such line of investigation is that regarding compensation for coarticulation. This is a finding that listeners make speech judgments that reflect sensitivity to acoustic consequences of coarticulatory gestural overlap. The research line has a long and controversial history. In a seminal finding by Mann (1980), members of a /da/-/ga/ continuum differing only in the third formant (F3) onset frequency of the initial consonant (high for /da/, low for /ga/) were identified differently after a precursor /al/ (high ending F3) than /ar/ (low ending F3) syllable. Listeners identified more continuum syllables as "ga" after /al/ than /ar/. As Mann (1980) explained, this can be interpreted as perceptual compensation for the acoustic consequences of gestural overlap, that is, the coarticulatory fronting/backing pulls that /l/ and /r/ would exert, respectively, on a following /da/ or /ga/ in natural speech production.

However, Mann (1980) also remarked that her findings can be interpreted in another way. They can be seen as evidence for spectral contrast rather than for listeners' perceptual parsing of coarticulatory gestural overlap. In the contrast account, frequencies in a context segment render the perceptual system temporarily insensitive to frequencies in neighboring speech. For example, a high F3 transition in /al/ makes frequencies in a following syllable that is ambiguous between /da/ (high F3) and /ga/ (low F3) sound lower and therefore more /ga/-like. A preceding /ar/, with a low F3, has the opposite contrastive effect, leading the syllable to sound more /da/-like. This mimics true perceptual parsing of the coarticulatory effects of /l/ and /r/ on /d/ and /g/.

In support of this view, investigators have reported that nonspeech contexts can yield compensation-like perceptual judgments (Kingston et al., 2014; Lotto & Kluender, 1998). For example, in research by Lotto and Kluender (1998), high- and low-frequency tones replaced /al/ and /ar/ syllables and had qualitatively the same effect on /da/-/ga/ judgments as the context syllables had.

However, other findings oppose that account and favor an interpretation that listeners, in fact, track gestural overlap in speech. For example, compensation is achieved perceptually when contrast is ruled out because compensation is cross-modal (Mitterer, 2006). In this study, context syllables were distinguished only visually (in audiovisual presentations), whereas the continuum syllables were distinguished only acoustically. Because a visible speech gesture cannot be the source of spectral contrast on a following acoustic syllable, contrast is not a viable account of the perceptual compensation that occurred in this study. Contrast is also ruled out when the gestural overlap for which listeners compensate has simultaneous, rather than successive, acoustic consequences (Silverman, 1987), because contrast affects perceptual sensitivity of *neighbors* of the source of contrast, not of the source itself. Finally, it is ruled out when gestural overlap is both cross-modal and simultaneous (Fowler, 2006). Moreover, in the only two (difficult-to-find) instances in which predictions of gestural parsing and contrast accounts have been dissociated in speech stimuli (Johnson, 2011; Viswanathan, Magnuson, & Fowler, 2010), results supported gestural parsing, not contrast.

As noted, findings most supportive of the contrast account show that nonspeech contexts trigger compensation-like responses in speech. (Lotto & Kluender, 1998). However, Viswanathan, Magnuson, and Fowler (2013) distinguished a contrast from a masking account of nonspeech effects experimentally and found that the nonspeech contexts used in these studies induced energetic masking rather than contrast. Masking cannot explain the speech effects, however.

In a different line of investigation, perceivers are shown to integrate cross-modal information about speech gestures. A striking and seminal finding by McGurk and MacDonald (1976; MacDonald and McGurk, 1978) showed that appropriately selected pairings of acoustic consonant–vowel (CV) syllables and synchronous dubbed, visible mouthings of different CVs led listeners to report hearing a syllable that integrates information across the modalities. For example, acoustic /ma/ dubbed onto mouthed /da/ leads listeners to hear /na/, an integration of the visually specified place of articulation of the consonant with its acoustically specified nasal and voicing properties. A gestural account of the finding is that listeners integrate information about gestures that are specified cross-modally. That is, they experience an event of talking and integrate cross-modal information about that event. Although there can be other accounts of the finding that invoke past experience associating the sights and sounds of talking (Diehl & Kluender, 1989; Stephens & Holt, 2010), these accounts are challenged by findings of cross-modal integration, among others. For example, Gick and Derrick (2009) showed that puffs of air against the neck of listeners transformed their reports of acoustic /ba/ to /pa/ and of /da/ to /ta/. The puff of air is evidence of the aspiration or breathiness in production of voiceless (/p/, /t/) stops. The gestural account is that puffs of air, acoustic signals, reflected light, and more (Fowler & Dekle, 1991) have impacts on perceivers in ways that specify their ecological source.

This is a subset of research that provides converging evidence for gesture perception in speech. This work is consistent with the framework in which perceivers must perceive the econiche as it is specified multimodally (Stoffregen and Bardy, 2001) for perception to support action and for action to support life.

However, nothing in this review indicates that perception of speech gestures reflects recruitment of the motor system as the motor theory of Liberman and colleagues proposes (Liberman et al., 1967; Liberman & Mattingly, 1985; see Scott, McGettigan, & Eisner, 2009). Listeners perceive speech gestures because acoustic signals, having been lawfully and distinctively structured by speech gestures, specify the gestures (Fowler, 1986, 1996)[3] just as reflected light that has been lawfully and distinctively structured as the act of walking specifies walking. Even so, there is considerable evidence that the motor system *is* active in effective ways during phonetic perception. The following review is meant to show only that there are effective perceptuomotor links in speech perception. It does not show that motor involvement is *required* to extract phonetic (gestural) primitives from speech signals.

Speech perception is a perceptuo-motor skill in respect to mechanisms that support it. The following review is restricted to behavioral evidence and is only illustrative. Moreover, summaries of evidence of brain activation patterns that support speech perception and evidence complementary to that provided here showing that perceptual information changes speech production (Houde & Jordan, 1998; Tremblay, Shiller, & Ostry, 2003) are omitted.

A direct connection between the speech perception and action systems is reported by Yuen, Davis, Brysbaert, and Rastle (2010). These investigators collected electropalatalographic data as talkers produced syllables starting with /k/ (produced with a constriction gesture of the tongue dorsum against the velum) or /s/ (produced with a narrow constriction between the tongue tip and the alveolar region of the palate). While articulating either of these syllables, the talkers heard either congruent syllables or /t/-initial syllables (/t/, like /s/, is an alveolar consonant; however, because it is a stop, not a fricative like /s/, there is more alveolar contact for /t/). The remarkable finding was that the heard syllable left traces in the production of /k/- and /s/-initial syllables in the form of increased alveolo-palatal contact by the tongue when the distractor was /t/-initial. The effect was absent when distractor syllables were presented in print form. This finding is important because acoustically presented syllables were distractors to which participants did not explicitly respond. The explicit task involved talking, not listening; even so, listening had an impact on articulation, presumably because, in this experimental set-up, listening automatically and unintentionally involves motor system activation.

The findings are consistent with those of Fadiga and colleagues (Fadiga, Craighero, Buccino, & Rizzolatti, 2002). They found that transcranial magnetic stimulation (TMS) of the tongue region of the speech motor system of the brain leads to more activation of tongue muscles when words or nonwords being perceived include lingual as compared with labial intervocalic consonants. That is, motor activation that occurred during perception of speech was specific to the gestural properties of the words or nonwords to which the listener was exposed.

Compatibly, D'Ausilio and colleagues (D'Ausilio et al., 2009) used TMS either to the tongue or to the lip region of the motor system of the brain as listeners identified stop-initial syllables in noise. Response times were faster and accuracy was higher to identify lingual consonants (/d/, /t/) than labial consonants (/b/, /p/) when TMS was to the lingual region. The pattern reversed when stimulation was to the labial region (see Hickok [2009, 2014] for an alternative interpretation).

Finally, Ostry and colleagues (Ito, Tiede, & Ostry, 2009; Nasir & Ostry, 2009) have shown that changes in the way that talkers produce a vowel also lead to changes in the way that they perceive it. A striking finding in that regard is reported by Nasir and Ostry (2009). In their study, talkers produced monosyllables including the vowel /æ/. Talkers' jaws were perturbed in the direction of protrusion as they produced the monosyllables. This perturbation did not have measurable or audible acoustic consequences, but, despite that, most participants compensated for it. That is, their jaw trajectories after compensation were closer to the preperturbation path than it was before compensation. Before and after the perturbation experience, participants classified vowels perceptually along a *head* to *had* acoustic continuum (where /æ/ is the vowel in *had*). They showed a boundary shift after compensation identifying fewer vowels as /æ/ than they did before compensation to the jaw perturbations. No shifts in identification were found for control participants who followed the same protocol except that perturbations were never applied during the perturbation phase of the experiment. In addition, among participants in the experimental group, the size of the compensation to perturbation was correlated with the size of the perceptual shift.

[3]Claims that acoustic signals necessarily lack the required specificity for gestural recovery (Diehl, Lotto, & Holt, 2004, p. 172: "[T]he inverse problem appears to be intractable") are overstated. Most approaches to the inversion problem make simplifying assumptions about the vocal tract that are false and consequential (Iskarous, 2010; Iskarous, Fowler, & Whalen, 2010). Moreover, characteristically approaches are designed to recover individual vocal tract *states*, not gestures, and so cannot take advantage of constraints provided by the tract's history, an analog of the retinal image fallacy in visual perception (Gibson, 1966).

Therefore, clearly, the perceptual shift was tied to the adaptive learning. As talkers changed the way they produced a vowel (in response to perturbations that had no measurable or audible acoustic consequences), they also changed how they extracted acoustic information about that vowel in perception.

As noted, there have been proposals that motor system activation during speech perception may only occur under special circumstances, such as when the signal is noisy or only when special kinds of tasks are being performed (Osnes, Hugdahl, & Specht, 2011). However, the foregoing review suggests that motor activation occurs whether or not the signal is noisy or distorted and in a variety of tasks. Stimuli are sometimes meaningless syllables and sometimes meaningful words; they are sometimes presented in the clear and sometimes in noise. Moreover, the review that follows should lead skeptics to question whether motor activation during speech perception should be limited only to special circumstances. The review shows that motor system recruitment is widespread in other domains of perception and elsewhere in the realm of language. Its pervasiveness likely reflects humans' adaptation to the fundamentally perceptuo-motor nature of life in the econiche.

15.2.2 Nonspeech

15.2.2.1 Nonlanguage

As noted, evidence that perceivers perceive a motor event when one occurs in their vicinity is not in itself evidence for a motor theory. Even so, there *is* evidence for motor activation during visual perception of walking in research by Takahashi, Kamibayashi, Nakajima, Akai, and Nakazawa (2008). They applied TMS to the motor systems of observers' brains to potentiate muscles of their legs as the observers watched actors either walking or standing on a treadmill. Activity in the observers' leg muscles was measured. Findings were analogous to those of Fadiga et al. (2002) for speech perception. Greater muscle activity occurred in muscles of the leg as observers watched walking as contrasted with standing. As it does during speech perception, muscle activation occurs that is specific to the event being perceived during perception of biological motion.

Motor system activation consequent on observing action is not restricted to visual observation. Activation also occurs as listeners hear sounds (in a study by Caetano, Jousmäki, & Hari, 2007, the sound of a drum membrane being tapped with the finger) that is comparable in some, but not all respects, with activation that occurs when listeners produce the same sounds themselves or see someone else producing them with or without sound.

A finding that is conceptually analogous to that of D'Ausilio et al. (2009) but in the visual domain has been reported as well. D'Ausilio et al. (2009) showed that potentiation of lip or tongue muscles facilitated perception of labial or lingual consonants, respectively, in noise. Blaesi and Wilson (2010) had observers classify facial expressions that had been morphed along a continuum between a smile and a frown. In half of the trials, the observers clenched a pen lengthwise in their mouth to enforce an expression similar to a smile without (directly) evoking the associated emotion. Findings were that "smile" judgments increased in those trials as compared with trials in which the observers' facial expression was not manipulated.

Findings similar to those reported by Blaesi and Wilson (2010) abound in the literature. Barsalou (2008) provides a summary of some of them in the domain of "embodied" or "grounded" cognition.

The studies reviewed so far reveal motor activation in perception. Research by Goldin-Meadow and colleagues (see Goldin-Meadow & Beilock, 2010, for a review) uncovered a role for motor recruitment in problem-solving and thought processes more generally. In their review they show that the manual gesturing that accompanies language use not only reflects thought but also can guide thought. Research with children acquiring Piagetian conservation (Ping & Goldin-Meadow, 2008) showed that children learned more from instruction involving both gestures and speech than from instruction involving speech only and that the advantage accrued whether the containers of liquid used in the conservation problems were present or absent (so that gestures were not points to the critical properties of the objects).

In a study of adults, Beilock and Goldin-Meadow (2010) presented participants with variants of the tower of Hanoi problem.[4] Participants solved the problem and then were videotaped describing how they had solved it. After that, they solved a variant of the same problem. An important finding was that gestures produced during the description phase that were appropriate to the initial solution of the problem but inappropriate for the solution of the variant were associated with poorer performance on the variant.

[4]The tower of Hanoi tasks present solvers with four disks of different sizes and three pegs on which they may be placed. At the beginning, the four disks are stacked on the left-most peg in order of size, with the smallest disk on top. The solver's task is to shift the disks one at a time so that, eventually, they are stacked in the same size order on the last peg. A constraint is that a larger disk cannot sit above a smaller one.

15.2.2.2 *Language, Not Speech*

This review shows that motor activation occurs and is effective in perception and cognition outside of language. It is not special to speech perception. One conclusion from this is that motor activation is pervasive in perception and cognition. A second is that the view of Liberman and colleagues (Liberman et al., 1967) that motor recruitment in speech perception solves a problem that is special to speech is not particularly suggested by findings of motor activation in that domain. An explanation that is more likely to be valid for the occurrence of motor activation in speech perception will be one that is shared with explanations for motor activation elsewhere in perception and cognition.

Motor activation within the domain of language is not special to speech perception either. It occurs in word recognition and in language understanding more generally. The generality of motor activation to larger chunks of language than consonants and vowels reflects the fact that language use is a perceptuo-motor activity. It is fundamentally a between-person activity in the world; as such, it is inherently and pervasively perceptuo-motor in its nature (Fowler, 2013). Some examples of findings are presented.

Regarding word recognition, Pulvermüller and Fadiga (2010) reviewed evidence that words with action-related meanings (*grasp* or *kick*) activate the associated part of the motor system (Hauk, Johnsrude, & Pulvermüller, 2004) and do so with a sufficiently short latency that the activation is likely integral to word understanding, not consequent of it (Pulvermüller, Shtyrov, & Ilmoniemi, 2005). Compatibly, a TMS study (Pulvermüller, Hauk, Nikulin, & Ilmoniemi, 2005) showed that stimulation of the arm–hand motor region of the left hemispheres of right-handed participants facilitated lexical decisions (made as lip-movement responses) to words with arm–hand-related meanings compared with words with leg-related meanings, whereas stimulation of leg motor regions had a complementary effect on lexical decision times.

de Zubicaray, Arciuli, and McMahon (2013) challenged these findings, in part, by showing that localizations of motor activations in response to linguistic stimuli in the literature are questionable. They also show that some findings of motor activation to words (and nonwords) reflect sensitivity, not to the words' content but rather to their orthographic and phonological properties that statistically distinguish words by syntactic class. They show that nonwords having the statistical properties of verbs activate the motor system despite being mostly meaningless.

Despite these findings, there is clear behavioral evidence for motor activation specific to actions implied by sentence meanings (see Taylor & Zwaan, 2009, for a review). These findings are not subject to concerns about where in the brain motor activation occurs, because they show specificity in the motor actions themselves that are primed by linguistic meanings. Moreover, most are not subject to reinterpretation in terms of the orthographic or phonological properties of the stimuli. For example, Glenberg and Kaschak (2002) presented listeners with sentences such as *Andy delivered the pizza to you* or *You delivered the pizza to Andy* (two sentences with identical orthographic and phonological properties but describing different actions). Participants made a speeded response whether each sentence made sense. For one participant group, the *yes* response was a motion toward the body from a home button, whereas the *no* response was a motion in the opposite direction from the same home button. In a second group of participants, the mapping was opposite. Findings were that latencies to respond *yes* to sentences like the first sentence were faster for participants whose responses were toward the body, the same direction as the pizza's motion. Latencies to respond *yes* to sentences like the second sentence were faster for participants whose responses were away from the body. Compatibly, Zwaan and Taylor (2006) presented participants with sentences visually in sequential groups of one to three words (separated here by slashes):

To quench/his/thirst/the/marathon/runner/eagerly/opened/the/water bottle.

Participants turned a knob to see each new word or word sequence. In one block of trials, they turned the knob counterclockwise; in another block, they turned it in the opposite direction. Half of the critical sentences described a clockwise motion; half (as in the example) described a counterclockwise motion. Findings were that reading times for the critical region of the sentence (*opened* in the example) were faster when the direction of the knob turn matched the rotation direction implied by the sentence.

15.3 CONCLUSION

Set in the context of the many recent research findings showing motor system activation and effective involvement in perception, cognition, and language generally, the previously highly controversial claim of Liberman's motor theory (Liberman et al., 1967; Liberman & Mattingly, 1985) that there is motor system recruitment in speech perception appears quite plausible, even mundane. Even so, the associated claim of motor theorists that speech motor system recruitment evolved to solve a perceptual problem that is special to speech recedes in plausibility. Whether acoustic speech signals present an especially difficult obstacle to perception because of coarticulation (and, most likely it does not; Fowler, 1986, 1996; Fowler & Iskarous, 2013), motor recruitment occurs pervasively in instances in which this obstacle, if it is one, is absent.

The present review suggests that motor recruitment occurs generally in perception and cognition, including in language perception and comprehension. This is likely because life in the econiche is pervasively perceptuo-motor in nature, and animals, including humans, are adapted to that kind of life. Perception generally incorporates exploratory activity as an essential part and performatory actions are perceptually guided. Moreover, for activities of either sort to be felicitous requires both acting and perceiving to be true to the nature of the econiche. Actions have to be appropriate to the affordances of the econiche, and perception has to reveal the affordances. The econiche has to be shared (there must be a relation of parity) between perceiving and acting.

This kind of action–perception parity is required for interpersonal action in which participants in joint activities have to coordinate. Participants in joint actions have to perceive accurately their own participation in the action and their partner's; complementarily, their actions have to be true to the joint situation and their perception of it. This is no less true for language use than it is for activities such as dancing, paddling a canoe, or playing a duet (Clark, 1996). Perceivers of linguistic utterances produced by others have, in general, to perceive accurately what has been said by themselves and by interlocutors. There has to be a relation of (sufficient) parity between utterances produced and perceived on the parts of all participants in a linguistic interchange. In this case, the shared part of the econiche is the utterance composed at the level of language forms of appropriately sequenced linguistic actions of the vocal tract.

Where does this leave Liberman's motor theory? Although Liberman was by no means the first motor theorist, he should be recognized as among the earliest theorists to recognize the perceptuo-motor link in the domain of speech (Galantucci, Fowler, & Turvey, 2006). A task for motor theorists of speech perception now is to bring the theory into alignment with developments outside the domain of speech such as those reviewed here.

References

Barsalou, L. (2008). Grounded cognition. *Annual Review of Psychology, 59*, 617–645.

Beilock, S., & Goldin-Meadow, S. (2010). Gesture changes thought by grounding it in action. *Psychological Science, 21*, 1605–1610.

Blaesi, S., & Wilson, M. (2010). The mirror reflects both ways: Action influences perception of others. *Brain and Cognition, 72*, 306–309.

Browman, C., & Goldstein, L. (1986). Towards an articulatory phonology. *Phonology Yearbook, 3*, 219–252.

Caetano, G., Jousmäki, V., & Hari, R. (2007). Actor's and observer's primary motor cortices stabilize similarly after seen or heard motor actions. *Proceedings of the National Academy of Sciences of the United States of America, 104*, 9058–9062.

Carello, C., Wagman, J. B., & Turvey, M. T. (2005). Acoustic specification of object properties. In J. Anderson, & B. Anderson (Eds.), *Moving image theory: Ecological considerations* (pp. 79–104). Carbondale, IL: Southern Illinois Press.

Clark, H. (1996). *Using language*. Cambridge: Cambridge University Press.

D'Ausilio, A., Pulvermüller, F., Selmas, P., Bufalari, I., Begliomini, C., & Fadiga, L. (2009). The motor somatotopy of speech perception. *Current Biology, 19*, 381–385.

de Zubicaray, G., Arciuli, J., & McMahon, K. (2013). Putting an "end" to the motor cortex representations of action words. *Journal of Cognitive Neuroscience, 25*, 1957–1974. Available from: http://dx.doi.org/10.1162/jocn_a_00437.

Diehl, R., & Kluender, K. (1989). On the objects of speech perception. *Ecological Psychology, 1*, 121–144.

Diehl, R. L., Lotto, A. J., & Holt, L. L. (2004). Speech perception. *Annual Review of Psychology, 55*, 149–179.

Fadiga, L., Craighero, L., Buccino, G., & Rizzolatti, G. (2002). Speech listening specifically modulates the excitability of tongue muscles: A TMS study. *European Journal of Neuroscience, 15*, 399–402.

Fowler, C. (1986). An event approach to the study of speech perception from a direct-realist perspective. *Journal of Phonetics, 14*, 3–28.

Fowler, C. A. (1996). Listeners do hear sounds, not tongues. *Journal of the Acoustical Society of America, 99*, 1730–1741.

Fowler, C. A. (2006). Compensation for coarticulation reflects gesture perception, not spectral contrast. *Perception and Psychophysics, 68*, 161–177.

Fowler, C. A. (2013). An ecological alternative to a "sad response": Public language use transcends the boundaries of the skin. *Behavioral and Brain Sciences, 36*, 356–357.

Fowler, C. A., & Brown, J. (1997). Intrinsic f0 differences in spoken and sung vowels and their perception by listeners. *Perception and Psychophysics, 59*, 729–738.

Fowler, C. A., & Dekle, D. J. (1991). Listening with eye and hand: Crossmodal contributions to speech perception. *Journal of Experimental Psychology. Human Perception and Performance, 17*, 816–828.

Fowler, C. A., & Iskarous, K. (2013). Speech perception and production. In A. F. Healey, R. W. Proctor, & I. B. Weiner (Eds.), *Handbook of psychology, Vol. 4: Experimental psychology* (2nd ed., pp. 236–263). Hoboken, NJ: John Wiley & Sons Inc.

Fowler, C. A., & Smith, M. (1986). Speech perception as "vector analysis": An approach to the problems of segmentation and invariance. In J. Perkell, & D. Klatt (Eds.), *Invariance and variability of speech processes* (pp. 123–136). Hillsdale, NJ: Lawrence Erlbaum Associates.

Galantucci, B., Fowler, C. A., & Turvey, M. T. (2006). The motor theory of speech perception reviewed. *Psychonomic Bulletin & Review, 13*, 361.

Gibson, E. J. (1994). Has psychology a future? *Psychological Science, 5*, 69–76.

Gibson, J. J. (1966). *The senses considered as perceptual systems*. Boston, MA: Houghton Mifflin.

Gibson, J. J. (1979). *The ecological approach to visual perception*. Boston, MA: Houghton Mifflin.

Gick, B., & Derrick, D. (2009). Aero-tactile integration in speech perception. *Nature, 462*, 502–504.

Glenberg, A. M., & Kaschak, M. P. (2002). Grounding language in action. *Psychonomic Bulletin and Review, 9*, 558–565.

Goldin-Meadow, S., & Beilock, S. (2010). Action's influence on thought: The case of gesture. *Perspectives on Psychological Science, 5*, 664–674.

Goldstein, L., & Fowler, C. A. (2003). Articulatory phonology: A phonology for public language use. In N. Schiller, & A. Meyer (Eds.), *Phonetics and phonology in language comprehension and production: Differences and similarities* (pp. 159–207). Berlin: Mouton de Gruyter.

Hauk, O., Johnsrude, I., & Pulvermüller, F. (2004). Somatotopic representation of action words in human motor and remotor cortex. *Neuron, 41*, 301–307.

Hickok, G. (2009). Speech perception does not rely on motor cortex. Available from: <http://www.cell.com/current-biology/comments/S0960-9822(09)00556-9>.

Hickok, G. (2014). *The myth of mirror neurons: The real neuroscience of communication and cognition*. New York, NY: Norton.

Houde, J. F., & Jordan, M. I. (1998). Sensorimotor adaptation in speech production. *Science, 227*, 1213–1216.

Iskarous, K. (2010). Vowel constrictions are recoverable from formants. *Journal of Phonetics, 78*, 375–387.

Iskarous, K., Fowler, C. A., & Whalen, D. H. (2010). Locus equations are an acoustic signature of articulatory synergy. *Journal of the Acoustical Society of America, 128*, 2021–2032.

Ito, T., Tiede, M., & Ostry, D. J. (2009). Somatosensory function in speech perception. *Proceedings of the National Academy of Sciences of the United States of America, 106*, 1245–1248.

Johanssen, G. (1973). Visual perception of biological motion and a model for its analysis. *Perception and Psychophysics, 14*, 201–211.

Johnson, K. (2011). Retroflex versus bunched [r] in compensation for coarticulation. *UC Berkeley Phonology Lab Annual Report, 2011*, 114–127.

Kingston, J., Kawahara, S., Chambless, D., Key, M., Mash, D., & Watsky, S. (2014). Context effects as auditory contrast. *Attention, Perception, and Psychophysics, 76*, 1437–1464. Available from: http://dx.doi.org/10.3758/s13414-013-0593-z.

Liberman, A. M. (1957). Some results of research on speech perception. *Journal of the Acoustical Society of America, 29*, 117–123.

Liberman, A., Cooper, F. S., Shankweiler, D., & Studdert-Kennedy, M. (1967). Perception of the speech code. *Psychological Review, 74*, 431–461.

Liberman, A., Delattre, P., & Cooper, F. S. (1952). The role of selected stimulus variables in the perception of the unvoiced-stop consonants. *American Journal of Psychology, 65*, 497–516.

Liberman, A. M., Delattre, P., Cooper, F. S., & Gerstman, L. (1954). The role of consonant-vowel transitions in the perception of the stop and nasal consonants. *Psychological Monographs: General and Applied, 68*, 1–13.

Liberman, A. M., & Mattingly, I. (1985). The motor theory revised. *Cognition, 21*, 1–36.

Liberman, A. M., & Mattingly, I. (1989). A specialization for speech perception. *Science, 243*, 489–494.

Liberman, A. M., & Whalen, D. H. (2000). On the relation of speech to language. *Trends in Cognitive Sciences, 4*, 187–196.

Lotto, A., Hickok, G., & Holt, L. L. (2009). Reflections on mirror neurons and speech perception. *Trends in Cognitive Sciences, 13*, 110–114.

Lotto, A., & Kluender, K. (1998). General contrast effects in speech perception: Effect of preceding liquid on stop consonant identification. *Perception and Psychophysics, 60*, 602–619.

MacDonald, J., & McGurk, H. (1978). Visual influences on speech perception. *Perception and Psychophysics, 24*, 253–257.

MacNeilage, P. F., Rootes, T. A., & Chase, T. P. (1967). Speech production and perception in a patient with severe impairment of somesthetic perception and motor control. *Journal of Speech and Hearing Research, 10*, 449–467.

Mann, V. A. (1980). Influence of preceding liquid on stop-consonant perception. *Perception & Psychophysics, 28*, 407–412.

Marsh, K., Richardson, M., Baron, R., & Schmidt, R. (2006). Contrasting approach to perceiving and acting with others. *Ecological Psychology, 18*, 1–36.

McGurk, H., & MacDonald, J. (1976). Hearing lips and seeing voices. *Nature, 264*, 746–748.

Millikan, R. G. (2003). In defense of public language. In L. M. Antony, & N. Hornstein (Eds.), *Chomsky and his critics* (pp. 215–237). Malden, MA: Blackwell Publishing, Ltd.

Mitterer, H. (2006). On the causes of compensation for coarticulation: Evidence for phonological mediation. *Perception and Psychophysics, 68*, 1227–1240.

Nasir, S., & Ostry, D. J. (2009). Auditory plasticity and speech motor learning. *Proceedings of the National Academy of Sciences of the United States of America, 106*, 20470–20475.

Osnes, B., Hugdahl, K., & Specht, K. (2011). Effective connectivity demonstrates involvement of premotor cortex during speech perception. *Neuroimage, 54*, 2437–2445.

Pardo, J., & Fowler, C. A. (1997). Perceiving the causes of coarticulatory acoustic variation: Consonant voicing and vowel pitch. *Perception and Psychophysics, 59*, 1141–1152.

Pierrehumbert, J. (1990). Phonological and phonetic representations. *Journal of Phonetics, 18*, 375–394.

Ping, R., & Goldin-Meadow, S. (2008). Hands in the air: Using ungrounded iconic gestures to teach children conservation of quantity. *Developmental Psychology, 44*, 1277–1287.

Pulvermüller, F., & Fadiga, L. (2010). Active perception: Sensorimotor circuits as a cortical basis for language. *Nature Reviews Neuroscience, 11*, 351–360.

Pulvermüller, F., Hauk, O., Nikulin, V. V., & Ilmoniemi, R. J. (2005). Functional links between motor and language systems. *European Journal of Neuroscience, 21*, 793–797.

Pulvermüller, F., Shtyrov, U., & Ilmoniemi, R. J. (2005). Brain signatures of meaning access in action word recognition. *Journal of Cognitive Neuroscience, 17*, 884–892.

Remez, R. E., & Pardo, J. S. (2006). The perception of speech. In M. Traxler, & M. A. Gernsbacher (Eds.), *The handbook of psycholinguistics* (pp. 201–248). New York, NY: Academic Press.

Rosenblum, L. (2008). Primacy of multimodal speech perception. In D. B. Pisoni, & R. E. Remez (Eds.), *Handbook of speech perception* (pp. 51–78). Malden, MA: Blackwell Publishing.

Runeson, S., & Frykholm, G. (1981). Visual perception of lifted weight. *Journal of Experimental Psychology. Human Perception and Performance, 7*, 733–774-730.

Ryle, G. (1949). *The concept of mind*. New York, NY: Barnes and Noble.

Schütz-Bosbach, S., & Prinz, W. (2007). Perceptual resonance: Action-induced modulation of perception. *Trends in Cognitive Sciences, 11*, 349–355.

Scott, S. K., McGettigan, C., & Eisner, F. (2009). A little more conversation, a little less action—candidate roles for the motor cortex in speech perception. *Nature Reviews Neuroscience, 10*, 295–302.

Silverman, K. (1986). FO cues depend on intonation: The case of the rise after voiced stops. *Phonetica, 43*, 76–92.

Silverman, K. (1987). The structure and processing of fundamental frequency contours. Unpublished Ph.D. dissertation, Cambridge University.

Stephens, J. D. W., & Holt, L. (2010). Learning to use an artificial visual cue in speech identification. *Journal of the Acoustical Society of America, 128*, 2138–2149.

Stoffregen, T. A., & Bardy, B. G. (2001). On specification and the senses. *Behavioral and Brain Sciences*, *24*, 195–261.

Takahashi, M., Kamibayashi, K., Nakajima, T., Akai, J., & Nakazawa, K. (2008). Changes in corticospinal excitability during observation of walking. *Neuroreport*, *19*, 727–731.

Taylor, L. J., & Zwaan, R. A. (2009). Action in cognition: The case of language. *Language and Cognition*, *1*, 45–58.

Tremblay, S., Shiller, D. M., & Ostry, D. J. (2003). Somatosensory basis of speech production. *Nature*, *423*, 866–869.

Viswanathan, N., Magnuson, J. S., & Fowler, C. A. (2010). Compensation for coarticulation: Disentangling auditory and gestural theories of perception of coarticulatory effects in speech. *Journal of Experimental Psychology. Human Perception and Performance*, *35*, 1005–1015.

Viswanathan, N., Magnuson, J. S., & Fowler, C. A. (2013). Similar response patterns do not imply identical origins: An energetic masking account of nonspeech effects in compensation for coarticulation. *Journal of Experimental Psychology. Human Perception and Performance*, *39*, 1181–1192.

Warren, W. H. (2006). The dynamics of perception and action. *Psychological Review*, *113*, 358–389.

Yuen, I., Davis, M. H., Brysbaert, M., & Rastle, K. (2010). Activation of articulatory information in speech perception. *Proceedings of the National Academy of Sciences of the United States of America*, *107*, 592–597.

Zwaan, R. A., & Taylor, L. J. (2006). Seeing, acting, understanding; Motor resonance in language understanding. *Journal of Experimental Psychology. General*, *135*, 1–11.

CHAPTER

16

Speech Perception
The View from the Auditory System

Andrew J. Lotto[1] *and Lori L. Holt*[2]

[1]Speech, Language, & Hearing Sciences, University of Arizona, Tucson, AZ, USA; [2]Department of Psychology and the Center for the Neural Basis of Cognition, Carnegie Mellon University, Pittsburgh, PA, USA

16.1 INTRODUCTION

For much of the past 50 years, the main theoretical debate in the scientific study of speech perception has focused on whether the processing of speech sounds relies on neural mechanisms that are specific to speech and language or whether general perceptual/cognitive processes can account for all of the relevant phenomena. Starting with the first presentations of the Motor Theory of Speech Perception by Alvin Liberman and colleagues (Liberman, Cooper, Harris, MacNeilage, & Studdert-Kennedy, 1964; Liberman, Cooper, Shankweiler, & Studdert-Kennedy, 1967; Studdert-Kennedy, Liberman, Harris, & Cooper, 1970) and the critical reply from Harlan Lane (1965), many scientists defended "all-or-none" positions on the necessity of specialized speech processes, and much research was dedicated to demonstrations of phenomena that were purported to require general or speech-specific mechanisms (see Diehl, Lotto, & Holt, 2004 for a review of the theoretical commitments behind these positions). Whereas the "speech-is-special" debate continues to be relevant (Fowler, 2008; Lotto, Hickok, & Holt, 2009; Massaro & Chen, 2008; Trout, 2001), the focus of the field has moved toward more subtle distinctions concerning the relative roles of perceptual, cognitive, motor, and linguistic systems in speech perception and how each of these systems interacts in the processing of speech sounds. The result has been an opportunity to develop more plausible and complete models of speech perception/production (Guenther & Vladusich, 2012; Hickok, Houde, & Rong, 2011).

In line with this shift in focus, in this chapter we concentrate not on whether the general auditory system is sufficient for speech perception but rather on the ways that human speech communication appears to be constrained and structured on the basis of the operating characteristics of the auditory system. The basic premise is simple, with a long tradition in the scientific study of speech perception: the form of speech (at the level of phonetics and higher) takes advantage of what the auditory system does well, resulting in a robust and efficient communication system. We review here three aspects of auditory perception—discriminability, context interactions, and effects of experience—and discuss how the structure of speech appears to respect these general characteristics of the auditory system.

It should be noted that we include in our conception of the "auditory system" processes and constructs that are often considered to be "cognition," such as memory, learning, categorization, and attention (Holt & Lotto, 2010). This is in contrast to previous characterizations of "Auditorist" positions in speech perception that appeared to constrain explanations of speech phenomena to peculiarities of auditory encoding at the periphery. Most researchers who have advocated for general auditory accounts of speech perception actually propose explanations within a larger general auditory cognitive science framework (Holt & Lotto, 2008; Kluender & Kiefte, 2006). Recent findings in auditory neuroscience provide support for moving beyond simple dichotomies of perception versus cognition or top-down versus bottom-up or peripheral versus central. There have been demonstrations that manipulation of attention may affect the earliest stages of auditory encoding in the cochlea (Froehlich, Collet, Chanal, & Morgon, 1990; Garinis, Glattke, & Cone, 2011; Giard, Collet, Bouchet, & Pernier, 1994; Maison, Micheyl, & Collet, 2001) and experience with music and language

Neurobiology of Language. DOI: http://dx.doi.org/10.1016/B978-0-12-407794-2.00016-X

changes the neural representation of sound in the brain stem (Song, Skoe, Wong, & Kraus, 2008; Wong, Skoe, Russo, Dees, & Kraus, 2007). In line with these findings, we treat attention, categorization, and learning as intrinsic aspects of auditory processing.

16.2 EFFECTS OF AUDITORY DISTINCTIVENESS ON THE FORM OF SPEECH

At the most basic level, the characteristics of the auditory system must constrain the form of speech because the information-carrying aspects of the signal must be encoded by the system and must be able to be discriminated by listeners. Given the remarkable ability of normal-hearing listeners to discriminate spectral-temporal changes in simple sounds such as tones and noises, the resolution of the auditory system does not appear to provide much of a constraint on the possible sounds used for speech communication. The smallest discriminable frequency change for a tone of 1,000 Hz is just over 1 Hz (Wier, Jesteadt, & Green, 1977), and an increment in intensity of 1 dB for that tone will likely be detected by the listener (Jesteadt, Wier, & Green, 1977). However, it is a mistake to make direct inferences from discriminability of simple acoustic stimuli to the perception of complex sounds, such as speech. Speech perception is not a simple detection or discrimination task; it is more similar to a pattern recognition task in which the information is carried through changes in relative patterns across a complex multidimensional space. These patterns must be robustly encoded and perceptually discriminable for efficient speech communication.

To the extent that some patterns are more readily discriminable by the auditory system, they will presumably be more effective as vehicles for communication. Liljencrants and Lindblom (1972) demonstrated that one could predict the vowel inventories of languages relatively well by maximizing intervowel distances within a psychophysically scaled vowel space defined by the first two formant frequencies (in Mel scaling). For example, /i/, /a/, and /u/ are correctly predicted to be the most common set of vowels for a three-vowel language system based on the presumption that they would be most auditorily discriminable given that their formant patterns are maximally distinct in the vowel space. Vowel inventory predictions become even more accurate as one more precisely models the auditory representation of each vowel (Diehl, Lindblom, & Creeger, 2003; Lindblom, 1986). These demonstrations are in agreement with proposals that languages tend to use sounds that maximize auditory distinctiveness in balance with the value of reducing articulatory effort, such as Stevens' (1972, 1989) Quantal Theory, Lindblom's (1991) H&H Theory (which we return to below), and Ohala's (1993) models of sound change in historical linguistics.

The proposal that auditory distinctiveness is important for effective speech communication was pushed even further by the Auditory Enhancement Theory from Diehl and colleagues (Diehl & Kluender, 1987, 1989; Diehl, Kluender, Walsh, & Parker, 1991). According to Auditory Enhancement, speakers tend to combine articulations that result in acoustic changes that mutually enhance distinctiveness of the resulting sounds for the listener. For example, in English the voicing contrast between /b/ and /p/ when spoken between two vowels, such as *rabid* versus *rapid*, is signaled in part by the duration of a silent interval that corresponds to the lip closure duration, which is shorter for /b/. However, speakers also tend to lengthen the duration of the preceding vowel when producing a /b/. Kluender, Diehl, and Wright (1988) demonstrated that preceding a silent gap with a long-duration sound results in the perception of a shorter silent gap, even for nonspeech sounds; this can be considered a kind of durational contrast. Thus, when talkers co-vary short lip closure durations with longer preceding vowels and vice versa, they produce a clearer auditory distinction between /b/ and /p/. This is just one of numerous examples appearing to indicate that the need for auditory distinctiveness drives the phonetic structure of languages (Diehl, Kluender, & Walsh, 1990; Kingston & Diehl, 1995).

In addition to providing constraints on the global structure of spoken languages, there is good evidence that the *individual* behavior of speakers is influenced by the local needs of listeners for auditory distinctiveness. According to Lindblom's (1991) H(yper) & H(ypo) Theory of speech communication, speakers vary their productions from hyperarticulation to hypoarticulation depending on the contextual needs of the listener. Spoken utterances that are redundant with other sources of information or with prior knowledge may be spoken with reduced effort, resulting in reduced auditory distinctiveness. However, novel information or words that are likely to be misperceived by a listener are produced with greater clarity or hyperarticulation. In accordance with this theory, there have been many demonstrations that speakers modulate productions when speaking to listeners who may have perceptual challenges, such as hearing-impaired listeners or non-native language learners (Bradlow & Bent, 2002; Picheny, Durlach, & Braida, 1985, 1986).

Despite the continued success of the theories described, it remains a challenge to derive a valid metric of "auditory distinctiveness" for complex time-varying signals like speech (and equally difficult to quantify "articulatory effort"). The classic psychophysical measures of frequency, intensity, and temporal resolution are simply not sufficient. The pioneering work of David Green regarding auditory profile analysis in which listeners discriminate amplitude pattern changes across a multitonal complex (Green, 1988; Green, Mason, & Kidd, 1984) was a step in the right direction because it could conceivably be applied to measuring the ability to discriminate steady-state vowel acoustics. However, vowel acoustics in real speech are much more complex and it is not clear that these measures scale up to predict intelligibility of speech at even the level of words. The future prospects of understanding how the operating characteristics of the auditory system constrain the acoustic elements used in speech communication are brighter given more recent approaches to psychoacoustic research that investigate the roles of context, attention, learning, and memory in general auditory processing (Kidd, Richards, Streeter, Mason, & Huang, 2011; Krishnan, Leech, Aydelott, & Dick, 2013; Ortiz & Wright, 2010; Snyder & Weintraub, 2013).

16.3 EFFECTS OF AUDITORY INTERACTION ON THE FORM OF SPEECH

The patterns of acoustic change that convey information in speech are notoriously complex. Speech sounds like /d/ and /g/ are not conveyed by a necessary or sufficient acoustic cue and there is no canonical acoustic template that definitively signals a linguistic message. Furthermore, variability is the norm. The detailed acoustic signature of a particular phoneme, syllable, or word varies a great deal across different contexts, utterances, and talkers. The inherent multidimensionality of the acoustic signatures that convey speech sounds and the variability along these dimensions presents a challenge for understanding how listeners readily map the continuous signal to discrete linguistic representations. This has been the central issue of speech perception research. Although some researchers have suggested that acoustic variability may serve useful functions in speech communication (Elman & McClelland, 1986; Liberman, 1996), the prevailing approach has been to explore how listeners accommodate or compensate for the messy physical acoustic signal to align it with native-language linguistic knowledge.

Although this framing of speech perception has dominated empirical research and theory, the focus on acoustic variability may lead us to pursue answers to the wrong questions. Like all perceptual systems, the auditory system transforms sensory input; it is not a linear system. It is possible that the nature of auditory perceptual transformations is such that the challenge of *acoustic* variability is mitigated when analyzed through the lens of *auditory* perception. Some of the more daunting mysteries about the ability of humans to accommodate acoustic variability in speech may arise from a lack of understanding of how the auditory system encodes complex sounds, generally.

Coarticulation is a case in point. As we talk, the mouth, jaw, and other articulators move very quickly, but not instantaneously, from target to target. Consequently, at any point in time the movement of the articulators is a function of the articulatory demands of previous and subsequent phonetic sequences as well as the "current" intended production. As a direct result, the acoustic signature of a speech sound is context-dependent. When /al/ precedes /ga/, for example, the tongue must quickly move from anterior to posterior occlusions to form the consonants. The effect of coarticulation is to draw /ga/ to a more anterior position (toward /al/). This context-sensitive shift in production impacts the resultant acoustic realization, making it more "da"-like because the place of tongue occlusion slides forward in the mouth toward the articulation typical of "da." Likewise, when /da/ is spoken after the more posteriorly articulated /ar/, the opposite pattern occurs; the acoustics of /da/ become more "ga"-like. This means that, due to coarticulation, the acoustic signature of the second syllables in "alga" and "arda" can be highly similar (Mann, 1980).

Viewed from the perspective of acoustic variability, this issue seems intractable. If the second consonant of "alga" and "arda" is signaled by highly similar acoustics, then how is it that we hear the distinct syllables "ga" and "da"? The answer lies in the incredible context dependence of speech perception; perception appears to compensate for coarticulation. This can be demonstrated by preceding a perceptually ambiguous syllable between /ga/ and /da/ with /al/ or /ar/. Whereas the acoustics of /ga/ produced after /al/ are more "da"-like, a preceding /al/ shifts perception of the ambiguous sound toward "ga." Similarly, /ar/ shifts perception of the same ambiguous sound toward "da." This pattern opposes the coarticulatory effects in speech production. In this example and many replications with other tasks and stimuli, coarticulation assimilates speech acoustics, but perception "compensates" in the opposing direction (Mann, 1980; Mann & Repp, 1980).

The traditional interpretation of these findings highlights that theoretical approaches have tended to discount what the auditory system can contribute to the challenges of speech perception. The flexibility of speech perception to make use of so many acoustic dimensions to signal a particular speech sound and the dependence of this mapping on context has suggested to many that it is infeasible for these effects to arise from auditory processing. This challenge is part of what led to the proposal that motor representations might be better suited to serve as the basis of speech communication. But, by virtue of being sound, acoustic speech necessarily interfaces with early auditory perceptual operations. As noted, these operations are not linear; they do not simply convey raw acoustic input, they transform it. Thus, although acoustics are readily observable and provide a straightforward means of estimating input to the linguistic system, this representation is not equivalent to the *auditory* information available to the linguistic system. What might be gained by considering *auditory*—rather than *acoustic*—information?

Lotto and Kluender (1998) approached this question by examining whether perceptual compensation for coarticulation like that described for "alga" and "arda" really requires information about speech articulation, or whether the context sounds need to be speech at all. They did this by creating nonspeech sounds that had some of the acoustic energy that distinguishes /al/ from /ar/. These nonspeech signals do not carry information about articulation, talker identity, or any other speech-specific details. The result was two nonspeech tone sweeps, one with energy like /al/ and the other with energy mimicking /ar/. Lotto and Kluender found that when these nonspeech acoustic signals preceded /ga/ and /da/ sounds, the tone sweeps had the same influence as the /al/ and /ar/ sounds they modeled. So-called perceptual compensation for coarticulation is observed even for nonspeech contexts that convey no information about speech articulation.

This finding has been directly replicated (Fowler, 2006; Lotto, Sullivan, & Holt, 2003) and extended to other stimulus contexts (Coady, Kluender, & Rhode, 2003; Fowler, Brown, & Mann, 2000; Holt, 1999; Holt & Lotto, 2002) many times. Across these replications, the pattern of results reveals that a basic characteristic of auditory perception is to exaggerate contrast. Preceded by a high-frequency sound (whether speech or nonspeech), subsequent sounds are perceived to be lower-frequency. This is also true in the temporal domain; preceded by longer sounds or sounds presented at a slower rate, subsequent sounds are heard as shorter (Diehl & Walsh, 1989; Wade & Holt, 2005a, 2005b). Further emphasizing the generality of these effects, Japanese quail exhibit the pattern of speech context dependence that had been thought to be indicative of perceptual compensation for coarticulation (Lotto, Kluender, & Holt, 1997).

This example underscores the fact that *acoustic* and *auditory* are not one and the same. Whereas there is considerable variability in speech acoustics, some of this variability is accommodated by auditory perceptual processing. In this way, the form of speech can have coarticulation and still be an effective communication signal because the operating characteristics of the auditory system include exaggeration of spectral and temporal contrast. Lotto et al. (1997) argue that the symmetry of assimilated speech production and contrastive perception is not serendipitous, but rather is a consequence of organisms having evolved within natural environments in which sound sources are physically constrained in the sounds they can produce. Because of mass and inertia, natural sound sources tend to be assimilative, like speech articulators. Perceptual systems, audition included, tend to emphasize signs of change, perhaps because in comparison with physical systems' relative sluggishness rapid change is ecologically significant information. Having evolved like other perceptual systems to respect regularities of the natural environment, auditory processing transforms coarticulated acoustic signals to exaggerate contrast and, thus, eliminates some of the apparent challenges of coarticulation. We can communicate efficiently as our relatively sluggish articulators perform acrobatics across tens of milliseconds to produce speech, in part because our auditory system evolved to use acoustic signals from natural sound sources that face the same physical constraints.

These results also highlight the importance of considering higher-level auditory processing in constraining models of speech perception. Subsequent research has shown that the auditory system exhibits spectral and temporal contrast for more complex sound input (Holt, 2005, 2006a, 2006b; Laing, Liu, Lotto, & Holt, 2012). These studies indicate that the auditory system tracks the long-term average spectra (or rate; Wade & Holt, 2005a) of sounds, and that subsequent perception is relative to, and contrastive with, these distributional characteristics of preceding acoustic signals (Watkins, 1991; Watkins & Makin, 1994). These effects, described graphically in Figure 16.1, cannot be explained by low-level peripheral auditory processing; effects persist over more than a second of silence or intervening sound (Holt, 2005) and require the system to track distributional regularity across acoustic events (Holt, 2006a). These findings are significant for understanding talker and rate normalization, which refer to the challenges introduced to speech perception by acoustic variability arising from different speakers and different rates of speech. What is important is that the preceding context of sounds possess acoustic energy in the spectral (Laing et al., 2012) or temporal (Wade & Holt, 2005a, 2005b) region distinguishing the target phonemes, and not that the context carries articulatory or speech-specific information. Here, too, some of the challenges apparent from speech acoustics may be resolved in the transformation from acoustic to auditory.

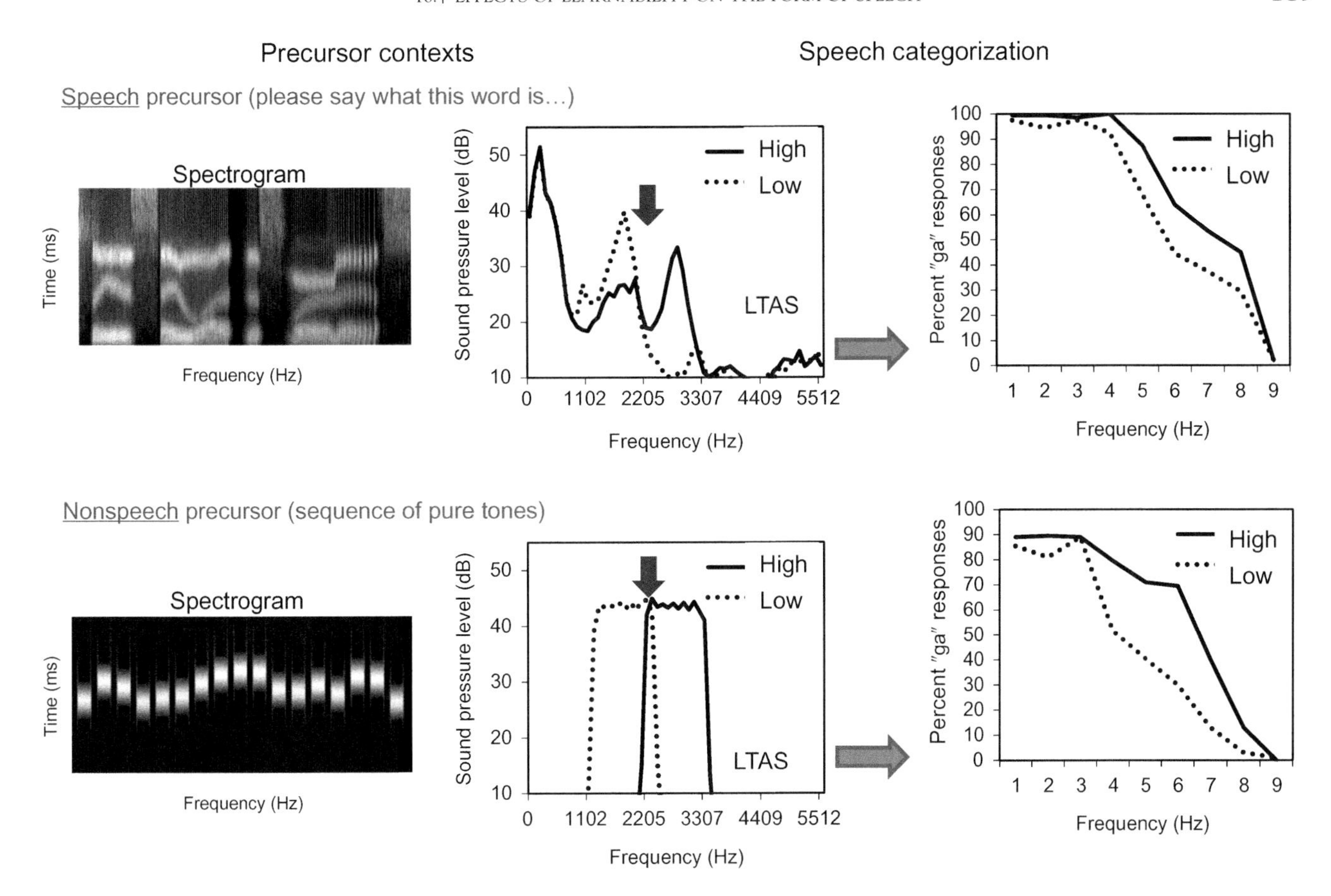

FIGURE 16.1 Precursor contexts and their effect on adult /ga/-/da/ categorization. Manipulation of the Long-Term Average Spectrum (LTAS) of both speech (top) and nonspeech (bottom) has a strong, contrastive influence on speech categorization. *From Laing, Liu, Lotto, and Holt (2012) with permission from the publishers.*

16.4 EFFECTS OF LEARNABILITY ON THE FORM OF SPEECH

Auditory representations are influenced greatly by both short-term and long-term experience. Categorical perception, the classic textbook example among speech perception phenomena, exemplifies this. When native-language speech varying gradually in its acoustics is presented to listeners, the patterns of identification change abruptly, not gradually, from one phoneme (or syllable or word) to another. Likewise, there is a corresponding discontinuity in discrimination such that pairs of speech sounds are more discriminable if they lie on opposite sides of the sharp identification boundary than if they lie on the same side of the identification curve's slope, even when they are matched in acoustic difference. Said another way, acoustically distinct speech sounds identified with the same label are difficult to discriminate, whereas those with different labels are readily discriminated. Despite the renown of categorical perception for speech, it is now understood that it is not specific to speech (Beale & Keil, 1995; Bimler & Kirkland, 2001; Krumhansl, 1991; Livingston, Andrews, & Harnad, 1998; Mirman, Holt, & McClelland, 2004), and that even speech is not entirely "categorical" (Eimas, 1963; Harnad, 1990; Liberman, Harris, Hoffman, & Griffith, 1957; Pisoni, 1973). Infants (Kuhl, 1991; McMurray & Aslin, 2005) and adults (Kluender, Lotto, Holt, & Bloedel, 1998; McMurray, Aslin, Tanenhaus, Spivey, & Subik, 2008) remain sensitive to within-category acoustic variation. Speech categories exhibit graded internal structure such that instances of a speech sound are treated as relatively better or worse exemplars of the category (Iverson & Kuhl, 1995; Iverson et al., 2003; Johnson, Flemming, & Wright, 1993; Miller & Volaitis, 1989).

We have argued that it may be more productive to consider speech perception as *categorization*, as opposed to *categorical* (Holt & Lotto, 2010). This may seem like a small difference in designation, but it has important consequences. Considering speech perception as an example of general auditory categorization provides a means of understanding how the system comes to exhibit relative perceptual constancy in the face of acoustic variability and does so in a native-language–specific manner. The reason for this is that although there is a great deal of variability in speech acoustics, there also exist underlying regularities in the distributions of experienced native-language speech sounds. This is the computational challenge of categorization; discriminably different exemplars come to be treated as functionally equivalent. A system that can generalize across variability to discover underlying patterns and distributional regularities—a system that can *categorize*—may cope with the acoustic variability inherent in speech without need for invariance. Seeking invariance in the acoustic signatures of speech becomes less essential if we take a broader view that extends beyond pattern matching to consider active auditory processing that involves higher-order and multimodal perception, categorization, attention, and learning.

From this perspective, learning about how listeners acquire auditory categories can constrain behavioral and neurobiological models of speech perception. Whereas the acquisition of first and second language phonetic systems provides an opportunity to observe the development of complex auditory categories, our ability to model these categorization processes is limited because it is extremely difficult to control or even accurately measure a listener's history of experience with speech sounds. However, we are beginning to develop insights into auditory categorization from experiments using novel artificial nonspeech sound categories that inform our understanding about how speech perception and acquisition are constrained by general perceptual learning mechanisms (Desai, Liebenthal, Waldron, & Binder, 2008; Guenther, Husain, Cohen, & Shinn-Cunningham, 1999; Holt & Lotto, 2006; Holt, Lotto, & Diehl, 2004; Ley et al., 2012; Liebenthal et al., 2010).

One example of what this approach can reveal about how auditory learning constrains speech relates to a classic early example of the "lack of invariance" in speech acoustics. If one examines the formant frequencies corresponding most closely with /d/ as it precedes different vowels, then it is impossible to define a single acoustic dimension that uniquely distinguishes the sound as a /d/; the acoustics are greatly influenced by the following vowel (Liberman, Delattre, Cooper, & Gerstman, 1954). This kind of demonstration fueled theoretical commitments that speech perception is accomplished via the speech motor system in the hopes that this would provide a more invariant mapping than acoustics (Liberman et al., 1967). Viewed from the perspective of acoustics, perceptual constancy for /d/ seemed an intractable problem for auditory processing.

Wade and Holt (2005a, 2005b) modeled this perceptual challenge with acoustically complex nonspeech sound exemplars that formed categories signaled only by higher-order acoustic structure and not by any invariant acoustic cue (see Figure 16.2 for a representation of the stimulus set). Naïve participants experienced these sounds in the context of a videogame in which learning sound categories facilitated advancement in the game but was never explicitly required or rewarded. Within just a half-hour of game play, participants categorized the sounds and generalized their category learning to novel exemplars. This learning led to an exaggeration of between-category discriminability (of the sort traditionally attributed to categorical perception) as measured with electroencephalography (EEG; Liu & Holt, 2011). The seemingly intractable lack of acoustic invariance is, in fact, readily learnable even in an incidental task.

This is proof that the auditory system readily uses multimodal environmental information (modeled in the videogame as sound-object links, as in natural environments) to facilitate discovery of the distributional regularities that define the relations between category exemplars while generalizing across acoustic variability within categories. More than this, however, the approach can reveal details of auditory processing that constrain behavioral and neurobiological models of speech perception. Using the same nonspeech categories and training paradigm, Leech, Holt, Devlin, and Dick (2009) discovered that the extent to which participants learn to categorize nonspeech sounds is strongly correlated with the pretraining to post-training recruitment of left posterior temporal sulcus (pSTS) during presentation of the nonspeech sound category exemplars. This is unexpected because left pSTS has been described as selective for specific acoustic and informational properties of speech signals (Price, Thierry, & Griffiths, 2005). In recent work, Lim, Holt, and Fiez (2013) have found that left pSTS is recruited online in the videogame category training task in a manner that correlates with behavioral measures of learning. These results also demonstrate that recruitment of left pSTS by the nonspeech sound categories cannot be attributed to their superficial acoustic signal similarity to speech or to mere exposure. When highly similar nonspeech sounds are sampled such that category membership is random instead of structured, left pSTS activation is not related to behavioral performance.

As in the examples from the preceding sections, this series of studies demonstrates that there is danger in presuming that speech is fundamentally different from other sounds in either its acoustic structure or in the basic perceptual processes it requires. The selectivity of left pSTS for speech should not be understood to be selectivity for intrinsic properties of acoustic speech signals, such as the articulatory information that speech may carry. Instead, this region seems to meet the computational demands presented by learning to treat structured distributions of acoustically variable sounds as functionally equivalent.

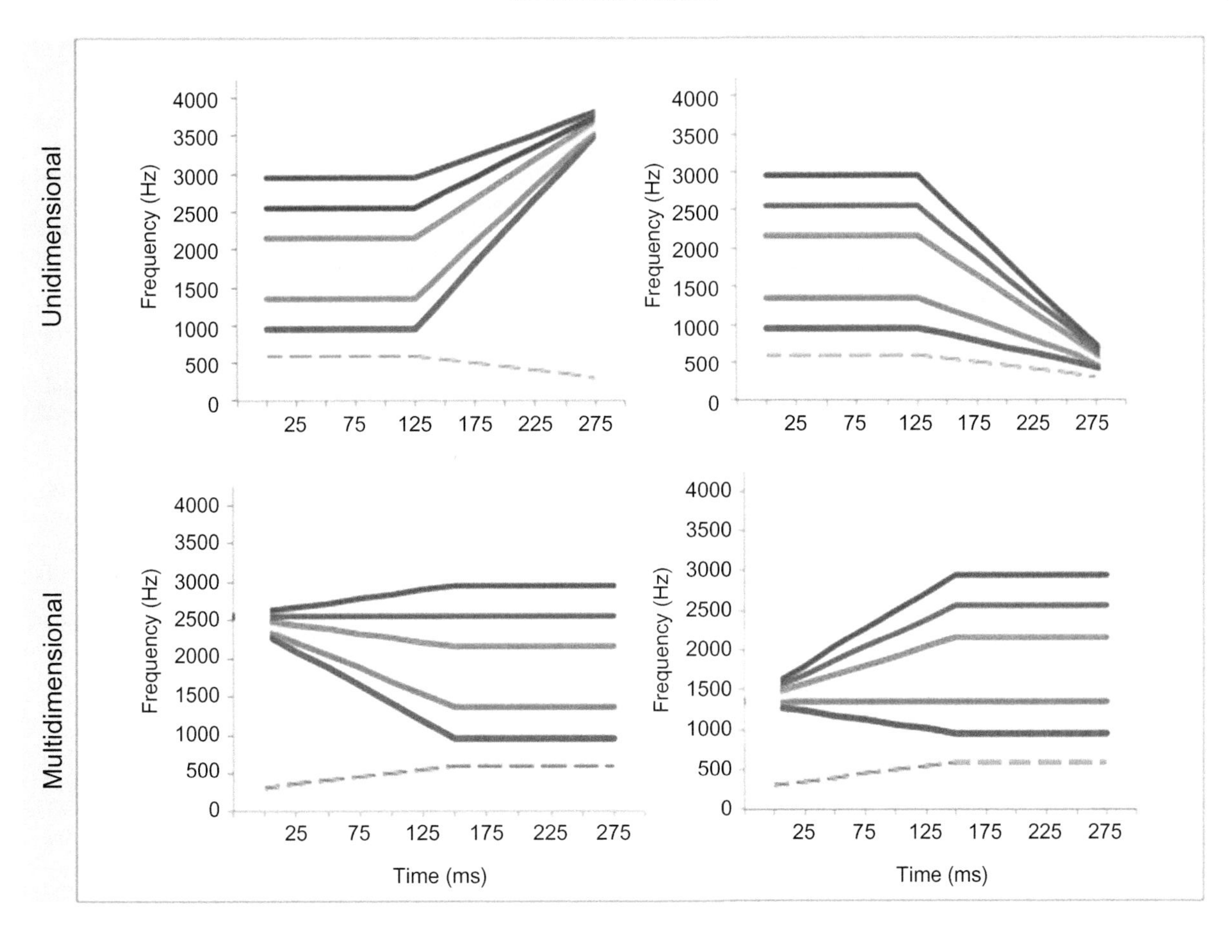

FIGURE 16.2 Schematic spectrograms showing the artificial nonspeech auditory categories across time and frequency. The dashed gray lines show the lower-frequency spectral peak, P1. The colored lines show the higher-frequency spectral peak, P2. The six exemplars of each category are composed of P1 and one of the colored P2 components pictured. Note that unidimensional categories are characterized by an offset glide that increases (top left) or decreases (top right) in frequency across all exemplars. No such unidimensional cue differentiates the multidimensional categories. *From Wade and Holt (2005a, 2005b) with permission from the publishers.*

Likewise, caution is warranted in presuming that the transformation from acoustic to auditory involves only a static mapping to stable, unchanging linguistic representations. The recruitment of putatively speech-selective left pSTS was driven by category learning in less than an hour (Lim et al., 2013). Thus, the behavioral relevance of the artificial, novel auditory categories drove reorganization of their transformations from acoustic to auditory. The examples we present here illustrate the facile manner by which auditory categories can be acquired. On an even shorter time scale, there is considerable evidence that the mapping of speech acoustics to linguistic representation is "tuned" by multiple information sources in an adaptive manner such as may be required to adapt to foreign accented speech or to speech in adverse, noisy environments (Kraljic, Brennan, & Samuel, 2008; Mehler et al., 1993; Vitela, Carbonell, & Lotto, 2012). The active, flexible nature of auditory processing puts learning in the spotlight and positions questions of speech perception in greater contact with other neurobiological approaches to understanding perception, cognition, and language.

16.5 MOVING FORWARD

The preceding sections provide a few brief examples of how general auditory processing may influence the perception of speech sounds as well as the structure of phonetic systems. These examples demonstrate that, at the very least, human speech communication appears to take advantage of the things that the auditory system does well—phonetic inventories tend to include sounds whose differences are well-encoded in the auditory system. The acoustic effects of coarticulation are just the types of interactions that the auditory system can accommodate, and the multidimensional structure of speech sounds form just the kinds of categories that are easily learned by the auditory system. Whether there are additional specialized processes required for speech perception, it is likely that the auditory system constrains the way we talk to and perceive each other to a greater extent than has been acknowledged.

One of the beneficial outcomes of the fact that the auditory system plays a strong role in speech perception is that there is the opportunity for synergy between research of speech and of general auditory processing. Speech perception phenomena shine a light on auditory processes that have remained unilluminated by research of simpler acoustic stimuli. The theories regarding the auditory distinctiveness of speech sounds have inspired the search for better models of auditory encoding of complex stimuli and better functions for computing distinctiveness (Lotto et al., 2003). The existence of perceptual compensation for coarticulation and talker normalization provide evidence for spectral and temporal interactions in general auditory processing that are not evident when presenting stimuli in isolation (Holt, 2006a, 2006b; Holt & Lotto, 2002; Watkins & Makin, 1994). The complexity of speech categories along with the ease with which humans learn them is the starting point for most of the current work on auditory categorization (Goudbeek, Smits, Swingly, & Cutler, 2005; Lotto, 2000; Maddox, Molis, & Diehl, 2002; Smits, Sereno, & Jongman, 2006; Wade & Holt, 2005a, 2005b).

The vitality of auditory and speech cognitive neuroscience depends on continuing this trend of using speech and auditory phenomena to mutually inform and inspire each field.

References

Beale, J. M., & Keil, F. C. (1995). Categorical effects in the perception of faces. *Cognition*, *57*(3), 217–239.

Bimler, D., & Kirkland, J. (2001). Categorical perception of facial expressions of emotion: Evidence from multidimensional scaling. *Cognition and Emotion*, *15*(5), 633–658.

Bradlow, A., & Bent, T. (2002). The clear speech effect for non-native listeners. *Journal of the Acoustical Society of America*, *112*(1), 272–284.

Coady, J. A., Kluender, K. R., & Rhode, W. S. (2003). Effects of contrast between onsets of speech and other complex spectra. *Journal of the Acoustical Society of America*, *114*, 2225.

Desai, R., Liebenthal, E., Waldron, E., & Binder, J. R. (2008). Left posterior temporal regions are sensitive to auditory categorization. *Journal of Cognitive Neuroscience*, *20*(7), 1174–1188.

Diehl, R., Lotto, A. J., & Holt, L. L. (2004). Speech perception. *Annual Review of Psychology*, *55*, 149–179.

Diehl, R. L., & Kluender, K. R. (1987). On the categorization of speech sounds. In S. Harnad (Ed.), *Categorical perception: The groundwork of cognition* (pp. 226–253). London: Cambridge University Press.

Diehl, R. L., & Kluender, K. R. (1989). On the objects of speech perception. *Ecological Psychology*, *1*(2), 121–144.

Diehl, R. L., Kluender, K. R., & Walsh, M. A. (1990). Some auditory bases of speech perception and production. *Advances in Speech, Hearing and Language Processing*, *1*, 243–268.

Diehl, R. L., Kluender, K. R., Walsh, M. A., & Parker, E. M. (1991). Auditory enhancement in speech perception and phonology. In R. Hoffman, & D. Palermo (Eds.), *Cognition and the symbolic process: Analytical and ecological perspectives* (pp. 59–75). Hillsdale, NJ: Lawrence Erlbaum Associates, Inc.

Diehl, R. L., Lindblom, B., & Creeger, C. P. (2003). *Increasing realism of auditory representations yields further insights into vowel phonetics, Proceedings of the fifteenth international congress of phonetic sciences* (Vol. 2, pp. 1381–1384). Adelaide: Causal Publications.

Diehl, R. L., & Walsh, M. A. (1989). An auditory basis for the stimulus-length effect in the perception of stops and glides. *Journal of the Acoustical Society of America*, *85*(5), 2154–2164.

Eimas, P. D. (1963). The relationship between identification and discrimination along speech and non-speech continua. *Language and Speech*, *6*(4), 206–217.

Elman, J. L., & McClelland, J. L. (1986). Exploiting lawful variability in the speech wave. In J. S. Perkell, & D. H. Klatt (Eds.), *Invariance and variability of speech processes* (pp. 360–385). Hillsdale, NJ: Lawrence Erlbaum Associates, Inc.

Fowler, C. A. (2006). Compensation for coarticulation reflects gesture perception, not spectral contrast. *Perception and Psychophysics*, *68* (2), 161–177.

Fowler, C. A. (2008). The FLMP STMPed. *Psychonomic Bulletin and Review*, *15*(2), 458–462.

Fowler, C. A., Brown, J. M., & Mann, V. A. (2000). Contrast effects do not underlie effects of perceding liquids on stop-consonant identification by humans. *Journal of Experimental Psychology: Human Perception and Performance*, *26*(3), 877–888.

Froehlich, P., Collet, L., Chanal, J.-M., & Morgon, A. (1990). Variability of the influence of a visual task on the active micromechanical properties of the cochlea. *Brain Research*, *508*(2), 286–288.

Garinis, A. C., Glattke, T., & Cone, B. K. (2011). The MOC reflex during active listening to speech. *Journal of Speech, Language, and Hearing Research*, *54*(5), 1464–1476.

Giard, M.-H., Collet, L., Bouchet, P., & Pernier, J. (1994). Auditory selective attention in the human cochlea. *Brain Research*, *633*(1), 353–356.

Goudbeek, M., Smits, R., Cutler, A., & Swingley, D. (2005). Acquiring auditory and phonetic categories. In H. Cohen, & C. Lefebvre (Eds.), *Handbook of categorization in cognitive science* (pp. 497–513). Amsterdam: Elsevier.

Green, D. M. (1988). *Profile analysis: Auditory intensity discrimination*. New York: Oxford University Press.

Green, D. M., Mason, C. R., & Kidd, G. (1984). Profile analysis: Critical bands and duration. *The Journal of the Acoustical Society of America*, *75*(4), 1163–1167.

Guenther, F. H., Husain, F. T., Cohen, M. A., & Shinn-Cunningham, B. G. (1999). Effects of categorization and discrimination training on auditory perceptual space. *Journal of the Acoustical Society of America*, *106*(5), 2900–2912.

Guenther, F. H., & Vladusich, T. (2012). A neural theory of speech acquisition and production. *Journal of Neurolinguistics*, *25*(5), 408–422.

Harnad, S. R. (1990). *Categorical perception: The groundwork of cognition*. New York: Cambridge University Press.

Hickok, G. S., Houde, J., & Rong, F. (2011). Sensorimotor integration in speech processing: Computational basis and neural organization. *Neuron*, *69*(3), 407–422.

Holt, L., & Lotto, A. (2006). Cue weighting in auditory categorization: Implications for first and second language acquisition. *The Journal of the Acoustical Society of America*, *119*, 3059.

Holt, L., & Lotto, A. (2010). Speech perception as categorization. *Attention, Perception, and Psychophysics*, *72*(5), 1218–1227.

Holt, L., Lotto, A., & Diehl, R. (2004). Auditory discontinuities interact with categorization: Implications for speech perception. *The Journal of the Acoustical Society of America*, *116*, 1763.

Holt, L., & Lotto, A. J. (2002). Behavioral examinations of the level of auditory processing of speech context effects. *Hearing Research*, *167*, 156–169.

Holt, L. L. (1999). *Auditory constraints on speech perception: An examination of spectral contrast*. Ph.D. thesis, University of Wisconsin-Madison.
Holt, L. L. (2005). Temporally nonadjacent nonlinguistic sounds affect speech categorization. *Psychological Science*, *16*(4), 305–312.
Holt, L. L. (2006a). The mean matters: Effects of statistically defined nonspeech spectral distributions on speech categorization. *Journal of the Acoustical Society of America*, *120*, 2801–2817.
Holt, L. L. (2006b). Speech categorization in context: Joint effects of nonspeech and speech precursors. *Journal of the Acoustical Society of America*, *119*(6), 4016–4026.
Holt, L. L., & Lotto, A. J. (2008). Speech perception within an auditory cognitive neuroscience framework. *Current Directions in Psychological Science*, *17*(1), 42–46.
Iverson, P., Kuhl, P., Akahane-Yamada, R., Diesch, E., Tohkura, Y., Kettermann, A., et al. (2003). A perceptual interference account of acquisition difficulties for non-native phonemes. *Cognition*, *87*(1), B47–B57.
Iverson, P., & Kuhl, P. K. (1995). Mapping the perceptual magnet effect for speech using signal detection theory and multidimensional scaling. *Journal of the Acoustical Society of America*, *97*(1), 553–562.
Jesteadt, W., Wier, C. C., & Green, D. M. (1977). Intensity discrimination as a function of frequency and sensation level. *Journal of the Acoustical Society of America*, *61*(1), 169–177.
Johnson, K., Flemming, E., & Wright, R. (1993). The hyperspace effect: Phonetic targets are hyperarticulated. *Language*, *69*(3), 505–528.
Kidd, G. J., Richards, V. M., Streeter, T., Mason, C. R., & Huang, R. (2011). Contextual effects in the identification of nonspeech auditory patterns. *Journal of the Acoustical Society of America*, *130*(6), 3926–3938.
Kingston, J., & Diehl, R. L. (1995). Intermediate properties in the perception of distinctive feature values. *Papers in Laboratory Phonology*, *4*, 7–27.
Kluender, K. R., Diehl, R. L., & Wright, B. A. (1988). Vowel length differences before voiced and voiceless consonants: An auditory explanation. *Journal of Phonetics*, *16*, 153–169.
Kluender, K. R., & Kiefte, M. (2006). Speech perception within a biologically realistic information-theoretic framework. In M. A. Gernsbacher, & M. Traxler (Eds.), *Handbook of psycholinguistics* (2nd ed., pp. 153–199). London: Elsevier.
Kluender, K. R., Lotto, A. J., Holt, L. L., & Bloedel, S. L. (1998). Role of experience for language-specific functional mappings of vowel sounds. *Journal of the Acoustical Society of America*, *104*(6), 3568–3582.
Kraljic, T., Brennan, S. E., & Samuel, A. G. (2008). Accommodating variation: Dialects, idiolects, and speech processing. *Cognition*, *107*(2), 54–81.
Krishnan, S., Leech, R., Aydelott, J., & Dick, F. (2013). School-age children's environmental object identification in natural auditory scenes: Effects of masking and contextual congruence. *Hearing Research*, *300*, 46–55.
Krumhansl, C. L. (1991). Music psychology: Tonal structures in perception and memory. *Annual Review of Psychology*, *42*(1), 277–303.
Kuhl, P. K. (1991). Human adults and human infants show a "perceptual magnet effect" for the prototypes of speech categories, monkeys do not. *Perception and Psychophysics*, *50*(2), 93–107.
Laing, E. J. C., Liu, R., Lotto, A. J., & Holt, L. L. (2012). Tuned with a tune: Talker normalization via general auditory processes. *Frontiers in Psychology*, *3*, 203–227.
Lane, H. (1965). The motor theory of speech perception: A critical review. *Psychological Review*, *72*(4), 275–309.
Leech, R., Holt, L. L., Devlin, J. T., & Dick, F. (2009). Expertise with artificial nonspeech sounds recruits speech-sensitive cortical regions. *The Journal of Neuroscience*, *29*(16), 5234–5239.
Ley, A., Vroomen, J., Hausfeld, L., Valente, G., de Weerd, P., & Formisano, E. (2012). Learning of new sound categories shapes neural response patterns in human auditory cortex. *The Journal of Neuroscience*, *32*(38), 13273–13280.
Liberman, A., Harris, K., Hoffman, H., & Griffith, B. (1957). The discrimination of speech sounds within and across phoneme boundaries. *Journal of Experimental Psychology*, *54*, 358–368.
Liberman, A. M. (1996). *Speech: A special code*. Cambridge: MIT Press.
Liberman, A. M., Cooper, F. S., Harris, K. S., MacNeilage, P. F., & Studdert-Kennedy, M. (1964). *Some observations on a model for speech perception. Proceedings of the AFCRL symposium on models for the perception of speech and visual form*. Cambridge: MIT Press.
Liberman, A. M., Cooper, F. S., Shankweiler, D. P., & Studdert-Kennedy, M. (1967). Perception of the speech code. *Psychological Review*, *74*(6), 431–461.
Liberman, A. M., Delattre, P. C., Cooper, F. S., & Gerstman, L. J. (1954). The role of consonant-vowel transitions in the perception of the stop and nasal consonants. *Psychological Monographs: General and Applied*, *68*(8), 1–13.
Liebenthal, E., Desai, R., Ellingson, M. M., Ramachandran, B., Desai, A., & Binder, J. R. (2010). Specialization along the left superior temporal sulcus for auditory categorization. *Cerebral Cortex*, *20*(12), 2958–2970.
Liljencrants, J., & Lindblom, B. (1972). Numerical simulation of vowel quality systems: The role of perceptual contrast. *Language*, *48*(4), 839–862.
Lim, S., Holt, L.L., & Fiez, J.A. (2013). Context-dependent modulation of striatal systems during incidental auditory category learning. *Poster presentation at the 43rd Annual Conference of the Society for Neuroscience*. San Diego, CA.
Lindblom, B. (1986). Phonetic universals in vowel systems. In J. Ohala, & J. Jaeger (Eds.), *Experimental phonology* (pp. 13–44). Orlando, FL: Academic Press.
Lindblom, B. (1991). The status of phonetic gestures. In I. G. Mattingly, & M. Studdert-Kennedy (Eds.), *Modularity and the motor theory of speech perception* (pp. 7–24). Hillsdale, NJ: Lawrence Erlbaum Associates.
Liu, R., & Holt, L. L. (2011). Neural changes associated with nonspeech auditory category learning parallel those of speech category acquisition. *Journal of Cognitive Neuroscience*, *23*(3), 683–698.
Livingston, K. R., Andrews, J. K., & Harnad, S. (1998). Categorical perception effects induced by category learning. *Journal of Experimental Psychology: Learning, Memory, and Cognition*, *24*(3), 732–753.
Lotto, A. J. (2000). Language acquisition as complex category formation. *Phonetica*, *57*, 189–196.
Lotto, A. J., Hickok, G. S., & Holt, L. L. (2009). Reflections on mirror neurons and speech perception. *Trends in Cognitive Sciences*, *13*(3), 110–114.
Lotto, A. J., & Kluender, K. R. (1998). General contrast effects in speech perception: Effect of preceding liquid on stop consonant identification. *Perception and Psychophysics*, *60*(4), 602–619.
Lotto, A. J., Kluender, K. R., & Holt, L. L. (1997). Perceptual compensation for coarticulation by Japanese quail (*Coturnix coturnix japonica*). *Journal of the Acoustical Society of America*, *102*, 1134–1140.
Lotto, A. J., Sullivan, S. C., & Holt, L. L. (2003). Central locus for nonspeech context effects on phonetic identification. *Journal of the Acoustical Society of America*, *113*(1), 53–56.
Maddox, W. T., Molis, M. R., & Diehl, R. L. (2002). Generalizing a neuropsychological model of visual categorization to auditory categorization of vowels. *Perception and Psychophysics*, *64*(4), 584–597.
Maison, S., Micheyl, C., & Collet, L. (2001). Influence of focused auditory attention on cochlear activity in humans. *Psychophysiology*, *38* (1), 35–40.

Mann, V., & Repp, B. (1980). Influence of vocalic context on perception of the [s]-[S] distinction. *Attention, Perception, and Psychophysics*, *28*(3), 213–228.

Mann, V. A. (1980). Influence of preceding liquid on stop-consonant perception. *Perception and Psychophysics*, *28*(5), 407–412.

Massaro, D. W., & Chen, T. H. (2008). The motor theory of speech perception revisited. *Psychonomic Bulletin and Review*, *15*(2), 453–457.

McMurray, B., & Aslin, R. N. (2005). Infants are sensitive to within-category variation in speech perception. *Cognition*, *95*(2), B15–B26.

McMurray, B., Aslin, R. N., Tanenhaus, M. K., Spivey, M. J., & Subik, D. (2008). Gradient sensitivity to within-category variation in words and syllables. *Journal of Experimental Psychology: Human Perception and Performance*, *34*(6), 1609–1631.

Mehler, J., Sebastian, N., Altmann, G., Dupoux, E., Christophe, A., & Pallier, C. (1993). Understanding compressed sentences: The role of rhythm and meaning. *Annals of the New York Academy of Sciences*, *682*(1), 272–282.

Miller, J. L., & Volaitis, L. E. (1989). Effect of speaking rate on the perceptual structure of a phonetic category. *Perception and Psychophysics*, *46*(6), 505–512.

Mirman, D., Holt, L., & McClelland, J. (2004). Categorization and discrimination of nonspeech sounds: Differences between steady-state and rapidly-changing acoustic cues. *Journal of the Acoustical Society of America*, *116*(2), 1198–1207.

Ohala, J. J. (1993). Sound change as nature's speech perception experiment. *Speech Communication*, *13*(1–2), 155–161.

Ortiz, J. A., & Wright, B. A. (2010). Differential rates of consolidation of conceptual and stimulus learning following training on an auditory skill. *Experimental Brain Research*, *201*(3), 441–451.

Picheny, M., Durlach, N., & Braida, L. (1985). Speaking clearly for the hard of hearing I: Intelligibility differences between clear and conversational speech. *Journal of Speech, Language, and Hearing Research*, *28*, 96–103.

Picheny, M. A., Durlach, N. I., & Braida, L. (1986). Speaking clearly for the hard of hearing II: Acoustic characteristics of clear and conversational speech. *Journal of Speech, Language, and Hearing Research*, *29*, 434–446.

Pisoni, D. B. (1973). Auditory and phonetic memory codes in the discrimination of consonants and vowels. *Perception and Psychophysics*, *13*, 253–260.

Price, C., Thierry, G., & Griffiths, T. (2005). Speech-specific auditory processing: Where is it? *Trends in Cognitive Sciences*, *9*(6), 271–276.

Smits, R., Sereno, J., & Jongman, A. (2006). Categorization of sounds. *Journal of Experimental Psychology: Human Perception and Performance*, *32*(3), 733–754.

Snyder, J. S., & Weintraub, D. M. (2013). Loss and persistence of implicit memory for sound: Evidence from auditory stream segregation context effects. *Attention, Perception, and Psychophysics*, *75*, 1056–1074.

Song, J. H., Skoe, E., Wong, P., & Kraus, N. (2008). Plasticity in the adult human auditory brainstem following short-term linguistic training. *Journal of Cognitive Neuroscience*, *20*(10), 1892–1902.

Stevens, K. N. (1972). The quantal nature of speech: Evidence from articulatory-acoustic data. In E. E. David, & P. B. Denes (Eds.), *Human communication: A unified view* (pp. 51–66). New York, NY: McGraw-Hill.

Stevens, K. N. (1989). On the quantal nature of speech. *Journal of Phonetics*, *17*, 3–45.

Studdert-Kennedy, M., Liberman, A. M., Harris, K. S., & Cooper, F. S. (1970). Motor theory of speech perception: A reply to Lane's critical review. *Psychological Review*, *77*(3), 234–249.

Trout, J. D. (2001). The biological basis of speech: What to infer from talking to the animals. *Psychological Review*, *108*(3), 523–549.

Vitela, A. D., Carbonell, K. M., & Lotto, A. J. (2012). Predicting the effects of carrier phrases in speech perception. *Poster presentation at the 53rd meeting of the Psychonomics Society*. Minneapolis, MN.

Wade, T., & Holt, L. (2005a). Incidental categorization of spectrally complex non-invariant auditory stimuli in a computer game task. *Journal of the Acoustical Society of America*, *118*(4), 2618–2633.

Wade, T., & Holt, L. L. (2005b). Perceptual effects of preceding nonspeech rate on temporal properties of speech categories. *Perception and Psychophysics*, *67*(6), 939–950.

Watkins, A. J. (1991). Central, auditory mechanisms of perceptual compensation for spectral-envelope distortion. *Journal of the Acoustical Society of America*, *90*(6), 2942–2955.

Watkins, A. J., & Makin, S. J. (1994). Perceptual compensation for speaker differences and for spectral-envelope distortion. *Journal of the Acoustical Society of America*, *96*(3), 1263–1282.

Wier, C. C., Jesteadt, W., & Green, D. M. (1977). Frequency discrimination as a function of frequency and sensation level. *Journal of the Acoustical Society of America*, *61*(1), 178–184.

Wong, P., Skoe, E., Russo, N. M., Dees, T., & Kraus, N. (2007). Musical experience shapes human brainstem encoding of linguistic pitch patterns. *Nature Neuroscience*, *10*(4), 420–422.

CHAPTER

17

Understanding Speech in the Context of Variability

Shannon Heald, Serena Klos and Howard Nusbaum

Department of Psychology, The University of Chicago, Chicago, IL, USA

In listening to spoken language, speech perception subjectively seems to be a simple, direct pattern matching system (cf. Fodor, 1983) because of the immediacy with which we understand what is said. This subjective simplicity is misleading given the difficulty of developing speech recognition devices with human-level performance over the range of conditions under which we have little difficulty understanding speech. A typical approach to understanding the mechanisms of human speech perception is to treat the system as if composed of two parts, a simple acoustic–linguistic pattern matcher and some kind of noise reducing filter. However, we argue that this simple form of speech recognition, which is based largely on subjective impression, misconstrues the nature of noise–robust processing in speech. Rather than construe noise robustness as a separate system, we argue that it is intrinsic to the definition of the human speech perception system and suggestive of the processes that are necessary to explain how we understand spoken language. We argue here that understanding the mechanisms of speech perception depends on adaptive processing that can respond to contextual variability that can take a wide range of noise and distortion. Further, what counts as signal and what counts as noise (or contextual variability) may well depend on the listener's goals at the moment of listening.

17.1 SPEECH AND SPEAKERS

In perceiving speech, we typically listen to understand what someone is saying (the content of their message), as well as to understand something about who is saying it (speaker identity). Of course the message (word by word) changes much faster than who is delivering the message. But this means that much of speech understanding takes place in the context of a particular speaker, and if the speaker changes, the context changes. Although what is being said changes more frequently in a conversation, there can also be changes between speakers, and such changes are important for the listener to recognize. A shift between talkers can pose a perceptual challenge to a listener due to an increase in the variability of how acoustic patterns map onto phonetic categories. This perceptual challenge is often referred to as the problem of talker variability. For different talkers, a given specific acoustic pattern may correspond to different phonemes perceptually, whereas conversely, across talkers, a given phoneme may be represented in speech by different acoustic patterns (Dorman, Studdert-Kennedy, & Raphael, 1977; Liberman, Cooper, Shankweiler, & Studdert-Kennedy, 1967; Peterson & Barney, 1952). For this reason, as well as others, changes in talker are important because they may mark significant changes in how acoustic patterns map onto phonetic categories (cf. Nusbaum & Magnuson, 1997).

Additionally, recognizing a change in speaker may be important because a listener's attitudes and behavior toward a speaker are often informed by what a listener knows about a speaker (e.g., Thackerar & Giles, 1981). For example, indirect requests are understood in the context of a speaker's status (Holtgraves, 1994). More directly relevant to speech perception, however, a listener's belief about the speaker's social group can alter the perceived intelligibility of the speech (Rubin, 1992). Additionally, dialect (Niedzielski, 1999) and gender (Johnson, Strand, & D'Imperio, 1999) expectations can meaningfully alter phoneme perception, highlighting that social knowledge about a speaker can affect relatively "low-level" perceptual processing of a

Neurobiology of Language. DOI: http://dx.doi.org/10.1016/B978-0-12-407794-2.00017-1

speaker's message, much in the same way that knowledge of vocal tract characteristics can (Ladefoged & Broadbent, 1957; although see Huang & Holt, 2012; Laing, Lotto, & Holt, 2012 for an auditory explanation of the mechanism that could underlie this). To understand speech, it is important to know something about the speaker's speech. Of course, speaker recognition is important in its own right, over and above the way that it informs message understanding, although speaker identification for its own sake is much less frequent as a typical listener goal.

In general, there have been two broad views regarding how talker information is used during perception. One account, often called "talker normalization" (Nearey, 1989; Nusbaum & Magnuson, 1997), suggests that listeners use talker information to calibrate or frame the interpretation of a given message to overcome the considerable uncertainty (e.g., acoustic variability, reference resolution) that arises from talker differences. An alternative view suggests that talker information is not used as a context to frame message understanding at all. Rather the alternative view is that speaker recognition is something listeners need to do for social information or just to identify the speaker as a known individual (see Goldinger, 1998). However, this then suggests that there are two general, independent processes for spoken language understanding—word recognition (or phoneme or syllable) and speaker identification.

The kind of knowledge and information that is used during talker normalization is different from the knowledge used to account for phoneme recognition. To carry out talker normalization, it is necessary to derive information about the talker's vocal characteristics. For example, in Gerstman's (1968) model, the point vowels are used to scale the location of the F1-F2 space to infer all the other vowels produced by a given talker. Because the point vowels represent the extremes of a talker's vowel space, they can be used to characterize the talker's vocal tract extremes and therefore bound the recognition space. Similarly, Syrdal and Gopal's (1986) model scales F1 and F2 using the talker's fundamental frequency and F3 because these are considered to be more characteristic of the talker's vocal characteristics than vowel quality (e.g., Fant, 1973; Peterson & Barney, 1952). Thus, talker normalization models use information about the talker rather than information about the specific message or phonetic context, as in models of phoneme perception such as Trace (McClelland & Elman, 1986), Motor Theory (Liberman, Cooper, Harris, & MacNeilage, 1962; Liberman, Cooper, Harris, MacNeilage, & Studdert-Kennedy, 1967; Liberman & Mattingly, 1985), analysis-by-synthesis (Stevens & Halle, 1967), or the Fuzzy Logical Model of Perception (Massaro, 1987; Massaro & Oden, 1980).

Traditionally, speech perception has been described as classifying linguistic units (e.g., phonemes, words) from a mixture of detailed acoustic information that contains both phonetically relevant and irrelevant information. In other words, there is an assumed (typical or idealized) pattern that corresponds to linguistic information combined with noise or distortion of that pattern as a consequence of the process of speaking, including aspects of speech production that are specific to the speaker. Given that the acoustic information about a talker (as opposed to the message) might be viewed as noise in relation to the canonical linguistic units on which speech perception relies, it is sometimes assumed that talker information is lost or stripped away during this process of message recognition (e.g., Blandon, Henton, & Pickering, 1984; Disner, 1980; Green, Kuhl, Meltzoff, & Stevens, 1991). Although it is possible that talker information, even in a talker normalization theory, is preserved in parallel representational structures for other listening goals[1] (e.g., Hasson, Skipper, Nusbaum, & Small, 2007), the concern about losing talker-specific information and recognition of the need for this information for other perceptual goals prompted the alternative view. In this view all auditory information in an utterance is putatively represented in a more veridical fashion that maintains talker-specific auditory information along with phonetically relevant auditory information (e.g., Goldinger, 1998) as well as any environmental noise or distortion. In the details of such theories, there is separate coding that represents both talker-specific auditory information such as fundamental frequency and acoustic-phonetic information.[2] However, because this is an auditory-trace model, there are no specific provisions for the representation or processing of other aspects of talker information such as knowledge about the social group of the talker, the dialect of the talker, or the gender of the talker that might come from glottal waveform or other source information. Further, the echoic encoding account does not explain how talker-specific information that is not acoustic (e.g., visual talker information) can affect speech processing or how other kinds of auditory information (e.g., noise or competing talkers) is filtered out.

[1]There are many examples of parallel representations for other sensory systems: there are multiple somatosensory maps (e.g., Kaas, 2004), visual maps (e.g., Bartels & Zeki, 1998), and auditory maps (e.g., Hackett & Kaas, 2004) in the brain.

[2]As specified, there is no process to determine what is talker-specific information (e.g., glottal waveshape or idiolectal cue combinations) versus phonetically relevant information, nor is there a way of separating noise, although the model suggests that there are separate codes for talker-specific and phonetic-specific information (Goldinger, 1998).

Moreover, and perhaps more importantly, this view privileges speaker differences as a problem for speech perception. But variation in speaking rate within a single talker can have the same kind of effect (Francis & Nusbaum, 1996) such that one acoustic pattern corresponds to different phonemes and one phoneme can be produced with different acoustic patterns (Miller, O'Rourke, & Volaitis, 1997; Miller & Volaitis, 1989). For example, at slow rates of speech, the acoustic pattern of /b/ is similar to /w/ produced at a fast speaking rate, meaning that one acoustic pattern can correspond to either a /b/ or /w/, depending on the rate of speech. Conversely, any phoneme can be produced at a fast or slow speaking rate, resulting in different acoustic patterns. Of course, changes in phonetic and lexical context can restructure the relationship between acoustic patterns and phonetic categories to produce the lack of invariance problem (Liberman, Cooper, Shankweiler, et al., 1967; Nusbaum & Magnuson, 1997). Thus, talker differences are simply one example of the different kinds of variability that can affect the pattern properties of speech.

Noise can also come from environmental sources such as conversations or machinery, and distortion can be introduced by transmission (e.g., cell phone) or room acoustics. These are all modifications of the speech signal that are extrinsic to the utterance that was emitted from the lips of the person talking. However, some signal modifications (that can affect pattern structure) of the speech signal arise within the speaker during speech production, such as varying speaking rate or voice amplitude or fundamental frequency changes. This distinction between extrinsic and intrinsic modifications of speech assumes that there is an idealized acoustic pattern corresponding to a linguistic message and these kinds of signal modifications impair the ability of the listener to recover that putative idealized form from within the noise and distortions. In addition, there is an assumption that the idealized forms of linguistic messages are distinctive in sufficiently differentiating among messages such that once the noise is eliminated, the similarity of a cleaned-up acoustic pattern to known representations of linguistic messages can be determined. While these assumptions underlie almost all theories of speech perception, the problems entailed by these assumptions and how they shape our understanding of the neurobiology of language are seldom examined explicitly.

17.2 THE LACK OF INVARIANCE PROBLEM

In speech perception, the rubric of a lack of invariance between acoustic patterns in speech and the linguistic interpretation of those patterns is a core challenge to theories of speech perception. Many simple recognition systems assume that given some pattern as input, the features or structure of the input pattern can be compared mathematically to a set of stored representations of patterns and the distance between the input and each can be computed in some Minkowski metric (e.g., city block or Euclidean space). This distance can then serve as the basis for the decision criterion for selecting the recognized interpretation. In other words, when stimulus patterns are sufficiently different, recognition is simply a comparison process between the input pattern and the stored representations with recognition determined by the stored representation that is most similar to the input. However, this kind of approach has traditionally failed for speech recognition.

The problem of lack of invariance in the relationship between the acoustic patterns of speech and the linguistic interpretation of those patterns is a fundamental problem. Although the many-to-many mapping between acoustic patterns of speech and perceptual interpretations is a longstanding, well-known issue (e.g., Liberman, Cooper, Shankweiler, et al., 1967), there are two aspects of this problem—many-to-one versus one-to-many mappings—that are not distinguished, but may be important to understanding the neural architecture of the language processing system. The core computational problem associated with the many-to-many mapping problem only truly emerges when a particular pattern has many different interpretations or can be classified in many different ways. Nusbaum and Magnuson (1997) argued that a many-to-one mapping can be understood with a simple deterministic class of mechanisms, whereas a one-to-many mapping can only be solved by nondeterministic mechanisms. In essence, a deterministic system establishes one-to-one mappings between inputs and outputs and thus can be computed by passive mechanisms such as feature detectors. To achieve many-to-one simply requires a set of one-to-one detectors for different input signals. In other words, a many-to-one mapping (e.g., rising formant transitions signaling a labial stop and diffuse consonant release spectrum signaling a labial stop) can be instantiated as a collection of one-to-one mappings. However, in the case of a one-to-many mapping (e.g., a formant pattern that could signal either the vowel in BIT or BET) there is ambiguity about the interpretation of the input without additional information. One solution is that additional context or information could eliminate some alternative interpretations, such as talker information (Nusbaum & Magnuson, 1997). However, this leaves the problem of determining the nature of the constraining context and how it is processed, which together are arguably contingent on the nature of the ambiguity itself. This suggests that there is no automatic or passive means of identifying and using the constraining information. Thus, an active mechanism that tests hypotheses about interpretations and tentatively identifies sources of constraining information (see Nusbaum & Schwab, 1986) is needed.

In spoken language understanding, active cognitive processing is vital to achieve flexibility and generativity (Nusbaum & Magnuson, 1997; Nusbaum & Schwab, 1986). Active cognitive processing is contrasted with passive processing in terms of the control mechanisms that organize the nature and sequence of cognitive operations (Nusbaum & Schwab, 1986). A passive process is one in which inputs map directly to outputs with no hypothesis testing or information-contingent operations as in the simple distance-based recognition system already described. Automatized cognitive systems (see Shiffrin & Schneider, 1977) behave as though passive, in that stimuli are mandatorily mapped onto responses without flexibility or any demand on cognitive resources. However, it is important to note that cognitive automatization does not have strong implications for the nature of the mediating control system such that different mechanisms have been proposed to account for the appearance of automatic processing (e.g., Logan, 1988). By comparison, active cognitive systems have a control structure that permits "information contingent processing" or the ability to change the sequence or nature of operations in the context of new information or uncertainty. In principle, active systems can generate hypotheses to be tested as new information arrives or is derived (Nusbaum & Schwab, 1986) and thus provide substantial cognitive flexibility to respond to novel situations and demands. Understanding how and why such active cognitive processes are involved in speech perception is fundamental to the development of a theory of speech perception. However, what is important for understanding the neurobiology of speech perception is the notion that an active control system could, in principle, implicate brain regions that are outside the traditional perisylvian language processing regions. This assumes that the active control of speech perception requires changes in attention to pattern information, as well as the recruitment of brain regions involved in long-term memory for nonlinguistic knowledge, and working memory systems to maintain alternative linguistic interpretations.

When there are multiple alternative interpretations for a particular acoustic pattern, the information needed to constrain the selection depends on the source of variability that produced the nondeterminism, and this could arise due to variation in speaking rate, or talker, or linguistic context, or other signal modifications. Whether the system uses articulatory or linguistic information or other contextual knowledge as a constraint, the perceptual system needs to flexibly use context as a guide in determining the relevant properties needed for recognition (Nusbaum & Schwab, 1986). The process of eliminating or weighing potential interpretations may involve working memory and changes in attention as alternative interpretations are considered, as well as adapting to new sources of lawful variability in context (Elman & McClelland, 1986). Similar mechanisms may be implicated at higher levels of linguistic processing in spoken language comprehension, although the neural implementation of such mechanisms might well differ depending on whether the nondeterminism occurred at the level of the speech signal, the lexical level, or the sentential level.

The involvement of cognitive mechanisms (e.g., working memory, attention) in speech perception remains controversial. In particular, one such mechanism, adaptability or plasticity in processing, has long been a point of theoretical contention. Although much of the controversy about learning in language processing has focused on syntax, there is also some disagreement about the plasticity of speech processing. At the center of this debate is how the long-term memory structures that guide speech processing are modified to allow for this plasticity while at the same time maintaining and protecting previously learned information from being expunged. This is especially important because often newly acquired information may represent irrelevant information to the system in a long-term sense (Born & Wilhelm, 2012; Carpenter & Grossberg, 1988).

17.3 ADAPTIVE PROCESSING AND PERCEPTUAL LEARNING

To overcome this problem, researchers have proposed various theories, and although there is no consensus, a hallmark characteristic of these accounts is that learning occurs in two stages. In the first stage, the memory system is able to use fast learning and temporary storage to achieve adaptability and in a subsequent stage, during an offline period such as sleep, this information is consolidated into long-term memory structures if the information is germane (Ashby, Ennis, & Spiering, 2007; Marr, 1971; McClelland, McNaughton, & O'Reilly, 1995). However, this kind of mechanism does not figure into speech recognition theories despite its arguable importance. Traditionally theories for speech recognition focus less on the formation of category representations and the need for plasticity during recognition than on the stability and structure of the categories (e.g., phonemes) to be recognized. Theories of speech perception often avoid the plasticity–stability trade-off problem by proposing that the basic categories of speech are established early in life, tuned by exposure, and subsequently only operate as a passive detection system (e.g., Abbs & Sussman, 1971; Fodor, 1983; McClelland & Elman, 1986). However, even these kinds of theories, suggest that early exposure to a phonological system has important effects on speech processing.

Research has established that adult listeners can learn a variety of new phonetic contrasts from outside their native language (Best, McRoberts, & Sithole, 1988; Lively, Logan, & Pisoni, 1993; Logan, Lively, & Pisoni, 1991; Pisoni, Aslin, Perey, & Hennessy, 1982; Yamada & Tohkura, 1992). For example, Francis and Nusbaum (2002) demonstrated that listeners are able to learn to direct perceptual attention to acoustic cues that were not previously used to form phonetic distinctions in their native language. This change in perceptual processing can be described as a shift in attention (Nusbaum & Schwab, 1986), although other descriptions are used as well. Auditory receptive fields may be tuned (e.g., Cruikshank & Weinberger, 1996; Wehr & Zador, 2003; Weinberger, 1998; Znamenskiy & Zador, 2013) or reshaped as a function of appropriate feedback (cf. Moran & Desimone, 1985) or context (Asari & Zador, 2009). This is consistent with theories of category learning (e.g., Schyns, Goldstone, & Thibaut, 1998) in which category structures are related to corresponding sensory patterns (Francis, Kaganovich, & Driscoll-Huber, 2008; Francis, Nusbaum, & Fenn, 2007). This learning could also be described as cue weighting as observed in the development of phonetic categories (e.g., McMurray & Jongman, 2011; Nittrouer & Lowenstein, 2007; Nittrouer & Miller, 1997). Yamada and Tohkura (1992) describe native Japanese listeners as typically directing attention to acoustic properties of /r/-/l/ stimuli that are not the dimensions used by English speakers, and as such are not able to discriminate between these categories. This is because Japanese and English listeners distribute attention in the acoustic pattern space for /r/ and /l/ differently as determined by the phonological function of this space in their respective languages. Perceptual learning of these categories by Japanese listeners suggests a shift of attention to the English phonetically relevant cues. Recently, McMurray and Jongman (2011) proposed the C-Cure model of phoneme classification in which the relative importance of cues varies with context. Although the model does not specify a neural mechanism by which such plasticity is implemented, there are a number of possibilities. This kind of approach to learning or modifying phonetic categories provides a mechanism that can support adaptation to contextual variability such as talker variability or other kinds of speech distortion.

Given that learning specific phonetic contrasts depends on changes in perceptual attention, learning the phonetic properties of a particular talker (distinct from more typical talkers) is relevant to the kinds of variability that reflect talker differences or might be produced by variation in speaking rate. For this reason, research on the way listeners learn to recognize low-quality synthetic speech produced, by rule, should be informative about the processes that underlie adaptation to talker variability and variation in speaking rate (Schwab, Nusbaum, & Pisoni, 1985). Synthetic speech learning has been demonstrated to generalize beyond the training exemplars to novel spoken words and contexts (Greenspan, Nusbaum, & Pisoni, 1988). Thus, listeners can learn the acoustic-phonetics that are idiosyncratic to a particular talker (Dorman et al., 1977). Furthermore, this kind of perceptual learning of the acoustic phonetics of a particular talker results in changes in attention to the speech signal (Francis, Baldwin, & Nusbaum, 2000). Moreover, this shift in attention between acoustic cues produces a restructuring of the perceptual space for that talker's speech (Francis & Nusbaum, 2002). Perceptual learning of a novel talker's speech results in a reduction in the cognitive load on working memory for recognizing the speech (Francis & Nusbaum, 2009). In other words, systematic experience listening to a novel talker allows a listener to learn the acoustic-phonetic mapping for a given talker. This learning increases the intelligibility of the talker's speech, which results from shifting attention to phonetically more relevant cues, which in turn lowers the working memory demands of speech perception. This is a hallmark of an active processing system—information about perceptual classification can be used to direct attention to improve performance and reduce the demands on working memory. However, although this demonstrates the operation of active cognitive processes during speech perception, it is more about learning the categories of speech or modifying those categories to be specific to improving recognition of that talker's speech.

Taken together, this work demonstrates that listeners are able to detect variance with known acoustic-phonetic patterns and to shift attention to appropriate cues, given feedback, in order to reduce uncertainty in interpretation and provides a computational solution to the one-to-many lack of invariance problem. Given a set of possible interpretations of a particular acoustic pattern, listeners may shift attention to the cues that discriminate among the alternative interpretations. The question, then, is whether such a mechanism seems to operate in circumstances of contextual (e.g., speaker or speaking rate) variability. If it is the case that sources of variability impose a nondeterministic computational structure on the problem of perception, then it must follow that we can reject all theories that have an inherently passive control structure. In other words, it is our contention that phonetic constancy must be achieved by an active computational system (e.g., Nusbaum & Morin, 1992; Nusbaum & Schwab, 1986).

17.4 EMPIRICAL EVIDENCE FOR ACTIVE PROCESSING IN TALKER NORMALIZATION

Active control systems use a feedback loop structure to systematically modify computation to converge on a single, stable interpretation (MacKay, 1951, 1956). By comparison, passive control structures represent invariant computational mappings between inputs and outputs. In consideration of this distinction, there are two general patterns of behavioral performance that can be taken as empirical evidence for the operation of an active control system (see Nusbaum & Schwab, 1986, for a discussion). First, evidence of load sensitivity in processing should provide an argument for active processing. There are several ways to justify this claim. For example, automatized processing in perception occurs when there is an invariant mapping between targets and responses whereas controlled—load-sensitive—processing occurs when there is uncertainty regarding the identity of targets and distractors over trials or when there is no simple single feature difference to distinguish targets and distractors (e.g., Shiffrin & Schneider, 1977; Treisman & Gelade, 1980). In other words, when there are multiple possible interpretations of a stimulus pattern, processing shows load sensitivity, which may be manifest as an increase in processing time, a decrease in recognition accuracy, or an interaction with an independent manipulation of cognitive load (Navon & Gopher, 1979) such as digit preload (e.g., Baddeley, 1986; Logan, 1979).

Second, the appearance of processing flexibility as demonstrated by the effects of listener expectations, context effects, learning, or other forms of online strategic processing should indicate active processing. Although an active process need not demonstrate this kind of flexibility, a passive process by virtue of its invariant computational mapping certainly cannot. This means, for example, that evidence for the effects of higher-order linguistic knowledge on a lower-level perceptual task such as lexical influence on phonetic recognition (e.g., Samuel, 1986) should implicate an active control system in processing.

There is definitely a great deal of evidence arguing that speech perception is load-sensitive under conditions of talker variability. For example, the accuracy of word recognition in noise and word recall is reduced when there is talker variability (speech produced by several talkers) compared with a condition in which a single talker produced the speech (Creelman, 1957; Martin, Mullennix, Pisoni, & Summers, 1989; Mullennix, Pisoni, & Martin, 1989). Talker variability also slows recognition time for vowels, consonants, and spoken words in a number of different experiments using a range of different paradigms (Mullennix & Pisoni, 1990; Nusbaum & Morin, 1992; Summerfield & Haggard, 1975). This provides some basic evidence that perception of speech is sensitive to talker variability, but it does not really indicate why this occurs.

Our view is that the evidence regarding the load sensitivity of the human listener when there is talker variability provides strong evidence that speech perception is performed by an active process. Furthermore, evidence of the flexibility of human listeners in processing speech, given talker variability, provides additional support. For example, we have found that listeners shift attention to different acoustic cues when there is a single talker and when there is talker variability (Nusbaum & Morin, 1992). In one condition, subjects monitored a sequence of spoken vowels for a specified target vowel, and the vowels were produced by one talker. In a second condition, a mix of different talkers produced the vowels. Both of these conditions were given with four different sets of vowels that were produced by LPC resynthesis of natural vowels used in our other experiments (Nusbaum & Morin, 1992). One set consisted of intact, four-formant voiced vowels. A second set consisted of the same vowels with voicing turned off to produce whispered counterparts. A third set was produced by filtering all information. A fourth set combined whispering with filtering to eliminate F0 and formant information above F2.

If listeners recognize vowels using a mechanism similar to the one described by Syrdal and Gopal (1986), then fundamental frequency and F3 information (although see Johnson, 1989, 1990a) should be necessary to recognition under all circumstances because their view is this information provides a talker-independent specification of vowel identity. This predicts that in both the single-talker and mixed-talker conditions, the intact voiced vowels should be recognized most accurately with whispering or filtering reducing performance somewhat and the combination reducing performance the most, because these modifications eliminate critical information for vowel recognition. Our results showed that in the single-talker condition, recognition performance was uniformly high across all four sets of stimuli. In the mixed-talker condition, however, accuracy dropped systematically as a function of the modifications of the stimuli with the voiced, intact vowels recognized most accurately and the whispered, filtered vowels recognized least accurately (Nusbaum & Morin, 1992). If vowel recognition were performed by a passive, talker-independent mechanism (e.g., Syrdal & Gopal, 1986), then the same pattern of results should have been obtained in both the single-talker and mixed-talker conditions. The results we obtained suggest that listeners only direct attention to F0 and F3 when there is talker variability (cf. Johnson, 1989, 1990a). This kind of strategic flexibility in recognition is strong evidence of an active mechanism. Furthermore, it suggests that the reason for the increase in cognitive load given talker variability may be because the listener must distribute attention over more cues in the signal than when there is a single talker. Wong, Nusbaum, and Small (2004) demonstrated that talker variability increased brain activity consistent with an increase in cognitive load and mobilization of attention in superior parietal cortex.

Listener expectations affect talker normalization processes as well. In a previous study, we found that not all talker differences increase recognition time in a mixed-talker condition (Nusbaum & Morin, 1992; also see Johnson, 1990a). When the vowel spaces of talkers are sufficiently similar and their fundamental frequencies are similar, there may be no difference in recognizing targets when speech from these talkers is presented in separate blocks or in the same block of trials. Magnuson and Nusbaum (2007) performed a study designed to investigate more specifically under what conditions talker variability increases recognition time. In this study, two sets of monosyllabic words were synthesized with two different mean F0s differing by 10 Hz. In one condition, a small passage was played to subjects in which two synthetic talkers, differing in F0 by 10 Hz, have a short dialogue. In a second condition, another group of subjects heard a passage in which one synthetic talker used a 10-Hz pitch increment to accent certain words. Both groups then listened to exactly the same set of single-pitch and mixed-pitch recognition trials using the monosyllabic stimuli. The subjects who listened to the dialogue between two talkers showed longer recognition times when there was a mix of the two different F0s in a trial compared to trials that consisted of words produced at a single F0. By comparison, subjects who expected that the 10-Hz pitch difference was not a talker difference showed no difference in recognition times or accuracy between the single-pitch and mixed-pitch trials. This demonstrates two things. First, the effect of increased recognition time in trials with a mix of F0s cannot be attributed to a simple contrast effect (see Johnson, 1990b) because both groups received exactly the same stimuli. Instead, the increased recognition times in the mixed-pitch trials seem to reflect processing specific to the attribution of the pitch difference to a talker difference and not something about the pitches themselves. Second, and perhaps more important for the present argument, the listeners' expectations affected whether they showed any processing sensitivity to pitch variability. This kind of processing flexibility cannot be accounted for by a simple passive computational system and argues strongly for an active perceptual mechanism (Nusbaum & Schwab, 1986).

17.5 TOWARD AN ACTIVE THEORY OF CONTEXTUAL NORMALIZATION

First and foremost, our view is that contextual normalization—using contextual variability as a perceptual frame for phoneme recognition—is carried out by the normal process of speech perception. In other words, talker or rate normalization is not carried out by a separate module or computational system, but it is a consequence of the basic computational structure of the normal operations of speech perception. This stands in sharp contrast to most previous approaches to normalization that emphasized the problem of computing talker vocal tract limits and scaling vowel spaces or base rates of speaking. It may be more productive to treat the processing of lawful variation as a single perceptual problem and focus on the commonalties rather than separating these problems based on the specific sources of information and knowledge needed to support normalization and recognition.

Second, the effects of talker or rate variability on perceptual processing directly reflect the computational operations needed to achieve phonetic constancy. Increased recognition times and interactions of varying cognitive load with recognition reflect the increased processing demands on capacity that are incurred by talker variability. Contextual variability increases the number of possible alternative interpretations of the signal, thereby increasing the processing demands on the listener. As a corollary of our first point, we predict that the same kinds of processing demands will be observed whenever there is any nondeterministic relationship between acoustic cues and linguistic categories during perceptual processing. Furthermore, even though there may be some relationship between the information used in talker identification and talker normalization, we claim that the perceptual effects of talker variability are not a consequence of talker identification processes competing with speech understanding. This is likely true of any aspect of contextual variability, although listeners are not typically called on to explicitly identify speaking rate or other forms of contextual variability.

Third, to achieve phonetic constancy, given a nondeterministic relationship between cues and categories, different sources of information and knowledge beyond the immediate acoustic pattern to be recognized must be brought to bear on the recognition problem. For example, if the F1 and F2 extracted from an utterance could have been intended as either of two different vowels given talker variability, information about the vocal tract that produced the vowels (e.g., from F0 and F3) will be used to provide the context for interpretation. Whenever there is a one-to-many mapping between a particular acoustic pattern and linguistic categories, listeners will have to use information outside the specific pattern to resolve the uncertainty. This information could come from other parts of the signal, previous utterances, linguistic knowledge, or subsequent parts of the utterance.

To realize the kind of computational flexibility required for this approach, it is important to reconceptualize the basic process of speech perception. The standard view of speech perception is that phoneme recognition or auditory word recognition is a process of comparing auditory patterns extracted from an utterance with stored mental representations of pattern information associated with linguistic categories. Our view is that speech perception, as an active process, is basically a cognitive process as described by Neisser (1967) and is more akin to hypothesis testing than pattern matching (cf. Nusbaum & Schwab, 1986). Nusbaum and Henly (1992) have argued that linguistic categories need to be represented by structures that are much more flexible than have been previously proposed. They claimed that a particular linguistic category such as the phoneme /b/ might be better represented by a theory of what a /b/ is. This view is an extension of the argument of Murphy and Medin (1985) regarding more consciously processed, higher-order categories. From this perspective, a theory is a set of statements that provide an explanation that accounts for membership in a category. Rather than view a theory as a set of explicit verbal statements, our view is that a theory representation of a linguistic category is an abstract, general specification regarding the identity and function of that linguistic category. Although this could be couched as a set of features, it is more reasonable to think of a theory as something that would generate a set of features given particular contextual constraints.

Recognizing a particular phoneme or word is a process of generating a set of candidate hypotheses regarding the classification of the pattern structure of an utterance. Conjectures about possible categories that could account for a section of utterance are proposed based on the prior context, listener expectations, and information in the signal. Given a set of alternative classifications for a stretch of signal information, the perceptual system may then carry out tests that are intended to diagnose the specific differences among the alternative classifications. Cognitive load increases as a function of the number of alternatives to be considered and the number of diagnostic tests that must be carried out.

By this view, phonetic constancy is the result of a process of testing hypotheses that have been tailored to distinguish between alternative linguistic interpretations of an utterance. An active control system mediates this process of hypothesis formation and testing. An abstract representation of linguistic categories in terms of theories provides the flexibility to apply diverse forms of evidence to this classification process allowing the perceptual system to resolve the nondeterministic structure produced by talker variability. These components taken together form a complex inferential system that has much in common with conceptual classification (Murphy & Medin, 1985) and other cognitive processes.

17.6 NEUROBIOLOGICAL THEORIES OF SPEECH PERCEPTION

There are a number of recently proposed theories of speech perception that have been framed in terms of brain regions identified as active during speech processing. To the extent that such theories are described as purely bottom-up recognition systems, wherein auditory coding leads to phonetic coding and then word recognition, it is difficult to reconcile that kind of architecture with evidence suggesting active processing in speech perception. Thus, neurobiological theories that are candidates for explaining recognition given contextual variability need to incorporate feedback or possibly feedforward information, although the difference in such models could well be testable in speech perception experiments. Moreover, such models seldom explicitly address the problem of talker or rate normalization, leaving the issues of the lack of invariance out of the domain of explanation. However, recent models have explicitly divided the neural processing of speech into dorsal and ventral streams following the neurobiology of visual processing models (Ungerleider & Mishkin, 1982). The difference among the models is typically in the functions attributed to these streams and their relationship. In considering visual perception, Bar (2003) explicitly proposed that the visual dorsal stream, typically conceived of as functional for object location or use (Milner & Goodale, 1995), may also serve as a fast pathway for coarse object classification, projecting through prefrontal cortex to ultimately connect with the ventral stream for object recognition. This proposal of interacting dorsal and ventral streams, interacting with prefrontal mechanisms for working memory, attention control, memory encoding, and goal and value maintenance is quite different from some of the neurobiological models of speech perception because it explicitly incorporates both feedback and feedforward active processing using neural networks that are not typically viewed as "perceptual." By contrast, neurobiological models of speech perception typically stay close to the perisylvian language areas, even when taking into account task effects in speech processing, albeit not explicitly considering active processing involving more general cognitive systems.

For example, Hickok and Poeppel (2007) have proposed a neurobiological model that explicitly separates ventral and dorsal speech processing streams identifying these largely with speech object recognition (ventral) and speech perception-production (dorsal). There is a somewhat unusual distinction made in this theory between speech perception and speech recognition as processes that double dissociate both functionally and cortically. Hickok and Poeppel define speech perception (as dissociated from speech recognition) as any sublexical task that involves the discrimination or categorization of auditory input. It is an active process that requires both working memory and executive control, but it does not necessarily lead to the lexical-sentential understanding of the speech signal. One could posit that such a network could play an important role in resolving the lack of invariance problem. Auditory–phonological representations that are "ambiguous" (in the ventral stream) mappings with more than one linguistic interpretation could, in principle, be resolved using the dorsal projections into auditory working memory and adaptive processing to shift attention between cues. However, this is not how Hickok and Poeppel describe the dorsal stream, which seems more functionally focused on word learning and metalinguistic task performance in speech perception experiments but is not identified as having any role in recognizing spoken language, although the connections are present within the model for this possibility. By contrast, utterances are recognized and understood by the process of speech recognition, which takes place solely within the ventral stream, transforming acoustic signals into mental lexicon representations.

This dual-stream neural network reflects a passive approach to speech recognition. An active process model of speech recognition would suggest that contextual influences processed in the dorsal stream may contribute to this comparison of multiple acoustic cues. However, because the cortical model proposed by Hickok and Poeppel does not explain how the ventral and dorsal streams interact, nor does it make clear how additional conceptual networks cited in the illustration of the model may influence these two pathways, the role of contextually based attentional changes cannot be explained by this model. Elsewhere, Hickok (2012) has argued that the speech motor system (within the dorsal stream) does not play a causal role in speech perception, despite evidence demonstrating that activity within the putative dorsal stream is affecting speech perception (Davis & Johnsrude, 2007; Skipper, van Wassenhove, Nusbaum, & Small, 2007).

By contrast, other models (e.g., Davis & Johnsrude, 2007; Friederici, 2012; Rauschecker & Scott, 2009) propose a more direct interaction between different pathways and brain regions. Rauschecker and Scott (2009) argue for a forward mapping ventral stream and an inverse mapping dorsal stream, which provides more explicitly for active processing. In the forward mapping pathway, the speech signal is decoded into linguistic categories in the inferior frontal cortex (IFC), which are then translated into articulatory/motor movements in premotor cortex (PMC). These articulatory representations of the speech signal are then sent to the inferior parietal lobe (IPL) as an efference copy. The inverse mapping stream essentially follows the same pathway in the reverse direction. Attentional and intentional demands originating in the IPL moderate the context-dependent motor plans that are activated in PMC and prefrontal cortex. These predictive motor plans are then compared with the sensory input processed by the IFC. Rauschecker and Scott posit that these two processing streams are active simultaneously, as the ventral stream solves the lack of invariance problem of the speech signal while the dorsal stream engages in domain-general linguistic processing beyond the maintenance of the phonological–articulatory loop.

Friederici (2012) argues for a more complex system in which four pathways are involved in speech processing. In many respects, this approach adds complexity to the dual-pathway model to account for sentence-level effects and explicit top-down processing as well as cognitive control mechanisms from prefrontal cortex. In doing so, this model goes beyond the more traditional perisylvian networks, but this is largely to accommodate the demands of syntactic complexity and sentence processing rather than the fundamental problems of lack of invariance in speech. This contrasts with Davis and Johnsrude (2007), who focus more on the role of active processing in basic speech perception. They argue that in the ventral pathway, multiple lexical interpretations of the speech input at various levels of representation must be activated in the IFC so that they may be compared with an echoic record of the incoming acoustic signal in the temporal cortex. This constant maintenance of the speech signal at multiple levels of representation allows top-down projections from the IFC to retune the perception of the acoustic signal at lower levels of the auditory pathway. Similarly, somato-motor representations of the acoustic input in the dorsal stream are projected both to the

IFC and downstream to areas of the temporal cortex to further influence perception of the signal at both upper and lower levels of the pathway. In this way, Davis and Johnsrude have created a neural model that definitively depicts speech perception as an active process.

Despite this improvement from the interpretation of speech recognition as a passive process, and the increased specificity and breadth of brain region interactions, perceptual learning of speech is not sufficiently explained. These models also fail to go beyond the cortico-cortical connections of speech processing. The auditory pathway begins well before the acoustic signal even reaches the primary auditory cortex and it has been well-established that there are more descending projections to these components of the peripheral nervous system than ascending projections (Huffman & Henson, 1990). Such evidence would suggest that a complete model of the neural substrates of speech perception should include the interaction between the cortical structures of the network and lower-level areas of the nervous system such as the thalamus, the auditory brainstem, and even the cochlea. To fully understand how speech perception adapts to the many variant cues in the speech signal, all components of the auditory pathway must be included in the neural model.

17.7 SUBCORTICAL STRUCTURES AND ADAPTIVE PROCESSING

The restriction of neural models of speech perception to cortical systems is in sharp contrast to the more cognitive–neurobiological models, such as the Ashby and Maddox (2005) model in which thalamus and striatum play important roles in fast mapping of category representations before slower sensorimotor cortical learning occurs or the complementary learning systems model (McClelland et al., 1995) in which fast learning occurs in the hippocampus and slower learning occurs in neocortical circuits. Acoustic input travels from the cochlea through the cochlear nucleus, the superior olive, the inferior colliculus, and the medial geniculate nucleus in the thalamus before reaching the primary auditory cortex. The connections between these structures contain twice as many descending projections from cortex as ascending projections to cortex. Evidence from animal models suggests that cortical structures utilize the corticofugal system to engage in egocentric selection, whereby they improve their own input from the brainstem through feedback and lateral inhibition (Suga, Gao, Zhang, Ma, & Olsen, 2002). Such processes allow for rapid readjustment of subcortical processing and long-term adjustments in cortex to facilitate associative learning. In humans, higher-level cognitive functions clearly have an effect on subcortical structures as low as the cochlea, because selective attention has been shown to enhance the spectral peaks of evoked otoacoustic emissions (Giard, Collet, Bouchet, & Pernier, 1994; Maison, Micheyl, & Collet, 2001) and discrimination training has been directly related to enhanced suppression of click-evoked otoacoustic emissions (de Boer & Thornton, 2008). Despite the growing evidence that the corticofugal system plays an important role in audition, researchers of speech perception continue to overlook this network when delineating the neurobiological underpinnings of speech. If speech perception utilizes the same basic categorization processes as auditory perception in general, then the subcortical structures that play a large role in audition must be included in these neural networks.

The influence of top-down cortical processes on subcortical structures is most apparent in the auditory brainstem. Electrophysiological recordings of the auditory brainstem response have demonstrated that the frequency following response (FFR), a sustained response phase-locked to the fundamental frequency of a periodic stimulus and/or the envelope of the stimulus (Krishnan, 2007), reflects changes in higher-level cognitive processes such as attention and learning. Galbraith and Arroyo (1993) determined that the FFR is modulated by selective attention to dichotic tones, with the attended tone eliciting larger peak amplitudes in the FFR than the ignored tone. The FFR is also affected by the reallocation of attentional resources to another modality. When listeners are presented with auditory and visual stimuli simultaneously but instructed to only attend to the visual stimulus, the signal-to-noise ratio of the FFR to the auditory stimulus decreases compared with when attention is directed to the auditory stimulus (Galbraith, Olfman, & Huffman, 2003). Although in both of these examples the auditory input consisted of tones rather than speech, they clearly establish an attentional influence on the activity of the auditory brainstem, which may modify the signal that ultimately reaches the auditory cortex.

When the FFR is examined in response to speech stimuli, top-down influences of the linguistic categorization processes that occur at the level of the cortex can also be seen. The FFR to synthetic English vowels contains prominent spectral peaks at the first formant harmonics of the signal and smaller peaks at the harmonics between formants (Krishnan, 2002). The enhanced peaks found in the FFR at the first formant suggest that some form of categorization is already occurring at the level of the auditory brainstem. The representation of the input is modified to strengthen the important cues so that they are more prominent than the rest of the signal, indicating that the translation of signal to lexical representations that occurs in the cortico-cortical connections of the speech network may influence the way in which the auditory brainstem represents the signal as it transfers it to higher points along the auditory pathway. Krishnan, Xu, Gandour, and Cariani (2005) compared the FFRs of Mandarin speakers and English speakers with four lexical tones used in Mandarin. They determined that Mandarin speakers had stronger pitch representations and smoother pitch tracking in their FFRs than did English speakers. They also had stronger representations of the second harmonic (F2) for all four tones. Based on these results, the researchers concluded that language experience may induce changes in the subcortical transfer of auditory input to enhance the representation of relevant linguistic features that are transmitted in the signal. The interaction between experience and brainstem activity is not exclusive to language experience. Musically trained individuals show earlier and larger FFRs and better phase-locking to the fundamental frequency in response to music stimuli as well as speech stimuli from their native language (Musacchia, Sama, Skoe, & Kraus, 2007; Wong, Skoe, Russo, Dees, & Kraus, 2007).

These studies demonstrate the influence of perceptual experience and training on the responses of the auditory brainstem, a relatively low-level neural structure. However, such effects are currently outside the domain of neurobiological theories of speech perception. These data do, however, demonstrate descending and experiential effects on the processing of speech and other acoustic stimuli and can be taken to reflect an active processing system. Speech input varies in many ways, both between talkers and within a single talker. The top-down control of subcortical structures allows the system to adapt to these changes in the signal by enhancing the most relevant spectral cues in the auditory input before it even reaches the cortical speech recognition networks.

17.8 CONCLUSION

With the increase in neuroimaging methods and studies of speech perception, there are a number of new theories that are grounded in brain regions attempting to explain speech perception. In some respects these theories can be viewed as modifications of the longstanding theory of speech perception proposed in the 1800s by Wernicke focusing mostly on perisylvian brain regions and incorporating aspects of neural architecture from the dorsal-ventral distinction in vision. While these theories may differ in the degree to which there are feedback connections among regions, incorporation of brain regions outside the "traditional" Wernicke language areas, and specificity of phenomena accounted for, there are two aspects of speech processing that are unaccounted for. First, the basic problem of lack of invariance in mapping acoustic patterns onto perceptual interpretations is not directly addressed. This problem has hindered the development of computer speech recognition systems, and yet neurobiological models of speech perception do not seem to recognize the need for such explanations. Moreover, this is a general problem of language understanding, not just speech perception. A similar many-to-many mapping can also be found between patterns at the syllabic, lexical, prosodic, and sentential level in speech and the interpretations of those patterns as linguistic messages. This is due to the fact that across linguistic contexts, speaker differences (idiolect, dialect, etc.), and other contextual variations, there are no patterns (acoustic, phonetic, syllabic, prosodic, lexical, etc.) in speech that have an invariant relationship to the interpretation of those patterns. For this reason, it could be beneficial to consider how these phenomena of acoustic perception, phonetic perception, syllabic perception, prosodic perception, lexical perception, and others are related computationally to one another and understand the computational similarities among the mechanisms that may subserve them (Marr, 2010).

Second, the plasticity of human speech perception and language processing, which is likely tied closely to the solution to the lack of invariance problem, is also not taken seriously. Such adaptive processing is at the core of human speech understanding rather than some kind of added-on system and is necessary to explain how listeners cope with talker and rate variability, as well as environmental noise and distortion. Explaining the kinds of neural mechanisms that mediate this kind of adaptive processing is key to explaining human speech perception as well as explaining language understanding more generally. Such explanations are unlikely to reside in exclusively cortical systems and need to take into account the fact that speech perception is carried out within the auditory pathway that has descending projections all the way down to the cochlea. Simply recognizing the need to develop a broader view of speech perception that incorporates brain regions outside the traditional perisylvian network (e.g., prefrontal attention-working memory regions, striatum, thalamus) and the need to explain speech perception in the context of an intrinsically active auditory system that has descending innervation to the cochlea is an advance over theories derived from 1800s neurology. However, now there is a need to develop theories that are explicit and testable based on this broader view that treats the perceptual processing of contextual variability in speech as central to understanding speech recognition rather than as a separable system.

Acknowledgments

Preparation of this manuscript was supported in part by an ONR grant DoD/ONR N00014-12-1-0850, and in part by the Division of Social Sciences at the University of Chicago.

References

Abbs, J. H., & Sussman, H. M. (1971). Neurophysiological feature detectors and speech perception: A discussion of theoretical implications. *Journal of Speech and Hearing Research*, *14*, 23–36.

Asari, H., & Zador, A. M. (2009). Long-lasting context dependence constrains neural encoding models in rodent auditory cortex. *Journal of Neurophysiology*, *102*, 2638–2656.

Ashby, F. G., Ennis, J. M., & Spiering, B. J. (2007). A neurobiological theory of automaticity in perceptual categorization. *Psychological Review*, *114*, 632–656.

Ashby, F. G., & Maddox, W. T. (2005). Human category learning. *Annual Review of Psychology*, *56*, 149–178.

Baddeley, A. D. (1986). *Working memory*. Oxford: Oxford Science Publications.

Bar, M. (2003). A cortical mechanism for triggering top-down facilitation in visual object recognition. *Journal of Cognitive Neuroscience*, *15*, 600–609.

Bartels, A., & Zeki, S. (1998). The theory of multistage integration in the visual brain. *Philosophical Transactions of the Royal Society of London B: Biological Sciences*, *265*, 2327–2332.

Best, C. T., McRoberts, G. W., & Sithole, N. M. (1988). Examination of perceptual reorganization for nonnative speech contrasts: Zulu click discrimination by English-speaking adults and infants. *Journal of Experimental Psychology: Human Perception and Performance*, *14*, 345.

Blandon, R. A. W., Henton, C. G., & Pickering, J. B. (1984). Towards an auditory theory of speaker normalization. *Language and Communication*, *4*, 59–69.

Born, J., & Wilhelm, I. (2012). System consolidation of memory during sleep. *Psychological Research*, *76*(2), 192–203.

Carpenter, G. A., & Grossberg, S. (1988). The ART of adaptive pattern recognition by a self-organizing neural network. *Computer*, *21*, 77–88.

Creelman, C. D. (1957). Case of the unknown talker. *Journal of the Acoustical Society of America*, *29*, 655.

Cruikshank, S. J., & Weinberger, N. M. (1996). Receptive-field plasticity in the adult auditory cortex induced by Hebbian covariance. *Journal of Neuroscience*, *16*, 861–875.

Davis, M. H., & Johnsrude, I. S. (2007). Hearing speech sounds: Top-down influences on the interface between audition and speech perception. *Hearing Research*, *229*, 132–147.

de Boer, J., & Thornton, A. R. D. (2008). Neural correlates of perceptual learning in the auditory brainstem: Efferent activity predicts and reflects improvement at a speech-in-noise discrimination task. *The Journal of Neuroscience*, *28*, 4929–4937.

Disner, S. F. (1980). Evaluation of vowel normalization procedures. *Journal of the Acoustical Society of America*, *67*, 253–261.

Dorman, M. F., Studdert-Kennedy, M., & Raphael, L. J. (1977). Stop consonant recognition: Release bursts and formant transitions as functionally equivalent, context-dependent cues. *Perception & Psychophysics*, *22*, 109–122.

Elman, J., & McClelland, J. (1986). Exploiting lawful variability in the speech wave. In J. S. Perkell, & D. H. Klatt (Eds.), *Invariance and variability in speech processes* (pp. 360–380). Hillsdale, NJ: Lawrence Erlbaum Associates.

Fant, G. (1973). *Speech sounds and features*. Cambridge: MIT Press.

Fodor, J. A. (1983). *The modularity of mind: An essay on faculty psychology*. Cambridge, MA: MIT press.

Francis, A., & Nusbaum, H. C. (2009). Effects of intelligibility on working memory demand for speech perception. *Attention, Perception, & Psychophysics*, *71*, 1360–1374.

Francis, A. L., Baldwin, K., & Nusbaum, H. C. (2000). Learning to listen: The effects of training on attention to acoustic cues. *Perception & Psychophysics*, *62*, 1668–1680.

Francis, A. L., Kaganovich, N., & Driscoll-Huber, C. J. (2008). Cue-specific effects of categorization training on the relative weighting of acoustic cues to consonant voicing in English. *Journal of the Acoustical Society of America*, *124*, 1234–1251.

Francis, A. L., & Nusbaum, H. C. (1996). Paying attention to speaking rate. *Proceedings of the International Conference on Spoken Language Processing*, Philadelphia.

Francis, A. L., & Nusbaum, H. C. (2002). Selective attention and the acquisition of new phonetic categories. *Journal of Experimental Psychology: Human Perception and Performance*, *28*, 349–366.

Francis, A. L., Nusbaum, H. C., & Fenn, K. (2007). Effects of training on the acoustic phonetic representation of synthetic speech. *Journal of Speech, Language and Hearing Research*, *50*, 1445–1465.

Friederici, A. D. (2012). The cortical language circuit: From auditory perception to sentence comprehension. *Trends in Cognitive Sciences*, *16*, 262–268.

Galbraith, G. C., & Arroyo, C. (1993). Selective attention and brainstem frequency-following responses. *Biological Psychology*, *37*, 3–22.

Galbraith, G. C., Olfman, D. M., & Huffman, T. D. (2003). Selective attention affects human brain stem frequency-following response. *Neuroreport*, *14*(15), 735–738.

Gerstman, L. J. (1968). Classification of self-normalized vowels. *IEEE Transactions on Audio Electroacoustics*, AU-16, 78–80.

Giard, M., Collet, L., Bouchet, P., & Pernier, J. (1994). Auditory selective attention in the human cochlea. *Brain Research*, *633*, 353–356.

Goldinger, S. D. (1998). Echoes of echoes? An episodic theory of lexical access. *Psychological Review*, *105*, 251–279.

Green, K. P., Kuhl, P. K., Meltzoff, A. N., & Stevens, E. B. (1991). Integrating speech information across talkers, gender, and sensory modalities: Female faces and male voices in the McGurk effect. *Perception & Psychophysics*, *50*, 524–536.

Greenspan, S. L., Nusbaum, H. C., & Pisoni, D. B. (1988). Perceptual learning of synthetic speech produced by rule. *Journal of Experimental Psychology: Learning, Memory, and Cognition*, *14*, 421–433.

Hackett, T. A., & Kaas, J. H. (2004). Auditory cortex in primates: Functional subdivisions and processing streams. In M. S. Gazzaniga (Ed.), *The cognitive neurosciences* (3rd ed.). Cambridge: MIT Press.

Hasson, U., Skipper, J. I., Nusbaum, H. C., & Small, S. L. (2007). Abstract coding of audiovisual speech: Beyond sensory representation. *Neuron*, *56*, 1116–1126.

Hickok, G. (2012). The cortical organization of speech processing: Feedback control and predictive coding the context of a dual-stream model. *Journal of Communication Disorders*, *45*, 393–402.

Hickok, G., & Poeppel, D. (2007). The cortical organization of speech processing. *Nature Reviews Neuroscience*, *8*, 393–402.

Holtgraves, T. M. (1994). Communication in context: The effects of speaker status on the comprehension of indirect requests. *Journal of Experimental Psychology: Learning, Memory and Cognition*, *20*, 1205–1218.

Huang, J., & Holt, L. L. (2012). Listening for the norm: Adaptive coding in speech categorization. *Frontiers in Perception Science*, *3*, 10. Available from: http://dx.doi.org/10.3389/fpsyg.2012.00010, PMC3078024.

Huffman, R. F., & Henson, O. W. (1990). The descending auditory pathway and acousticomotor systems: Connections with the inferior colliculus. *Brain Research Reviews*, *15*(3), 295–323.

Johnson, K. (1989). Higher formant normalization results from integration of F2 and F3. *Perception & Psychophysics*, *46*, 174–180.

Johnson, K. (1990a). The role of perceived speaker identity in F0 normalization of vowels. *Journal of the Acoustical Society of America, 88*, 642–654.

Johnson, K. (1990b). Contrast and normalization in vowel perception. *Journal of Phonetics, 18*, 229–254.

Johnson, K., Strand, E. A., & D'Imperio, M. (1999). Auditory-visual integration of talker gender in vowel perception. *Journal of Phonetics, 27*, 359–384.

Kaas, J. H. (2004). Somatosensory system. In G. Paxinos, & J. K. Mai (Eds.), *The human nervous system* (2nd ed., pp. 1059–1092). New York, NY: Elsevier Academic Press.

Krishnan, A. (2002). Human frequency-following responses: Representation of steady-state synthetic vowels. *Hearing Research, 166*, 192–201.

Krishnan, A. (2007). Frequency-following response. In R. F. Burkard, M. Don, & J. J. Eggermont (Eds.), *Auditory evoked potentials: Basic Principles and clinical application* (pp. 313–333). Baltimore, MD: Lippincott, Williams & Wilkins.

Krishnan, A., Xu, Y., Gandour, J. T., & Cariani, P. A. (2005). Encoding of pitch in the human brainstem is sensitive to language experience. *Cognitive Brain Research, 25*, 161–168.

Ladefoged, P., & Broadbent, D. E. (1957). Information conveyed by vowels. *Journal of the Acoustical Society of America, 29*, 98–104.

Laing, E. J. C., Lotto, A. J., & Holt, L. L. (2012). Tuned with a tune: Talker normalization via general auditory processes. *Frontiers in Psychology, 3*, 203. Available from: http://dx.doi.org/10.3389/fpsyg.2012.00203, PMC3381219.

Liberman, A.M., Cooper, F.S., Harris, K.S., & MacNeilage, P.F. (1962). A motor theory of speech perception. *Proceedings of the speech communication seminar* (Vol. 2). Stockholm: Royal Institute of Technology.

Liberman, A. M., Cooper, F. S., Harris, K. S., MacNeilage, P. F., & Studdert-Kennedy, M. (1967). Some observations on a model for speech perception. In W. Wathen-Dunn (Ed.), *Models for the perception of speech and visual form* (pp. 68–87). Cambridge: MIT Press.

Liberman, A. M., Cooper, F. S., Shankweiler, D. P., & Studdert-Kennedy, M. (1967). Perception of the speech code. *Psychological Review, 74*, 431–461.

Liberman, A. M., & Mattingly, I. G. (1985). The motor theory of speech perception revised. *Cognition, 21*, 1–36.

Lively, S. E., Logan, J. S., & Pisoni, D. B. (1993). Training Japanese listeners to identify English /r/ and /l/. II: The role of phonetic environment and talker variability in learning new perceptual categories. *The Journal of the Acoustical Society of America, 94*, 1242–1255.

Logan, G. D. (1979). On the use of a concurrent memory load to measure attention and automaticity. *Journal of Experimental Psychology: Human Perception and Performance, 5*, 189–207.

Logan, G. D. (1988). Toward an instance theory of automatization. *Psychological Review, 95*, 492–527.

Logan, J. S., Lively, S. E., & Pisoni, D. B. (1991). Training Japanese listeners to identify English /r/ and /l/: A first report. *The Journal of the Acoustical Society of America, 89*, 874–886.

MacKay, D. M. (1951). Mindlike behavior in artefacts. *British Journal for the Philosophy of Science, 2*, 105–121.

MacKay, D. M. (1956). The epistemological problem for automata. In C. E. Shannon, & J. McCarthy (Eds.), *Automata studies*. Princeton, NJ: Princeton University Press.

Magnuson, J. S., & Nusbaum, H. C. (2007). Acoustic differences, listener expectations, and the perceptual accommodation of talker variability. *Journal of Experimental Psychology: Human Perception and Performance, 33*, 391–409.

Maison, S., Micheyl, C., & Collet, L. (2001). Influence of focused auditory attention on cochlear activity in humans. *Psychophysiology, 38*, 35–40.

Marr, D. (1971). Simple memory: A theory for archicortex. *Philosophical Transactions of the Royal Society B: Biological Sciences, 262*, 23–81.

Marr, D. (2010). *Vision. A computational investigation into the human representation and processing of visual information*. Cambridge, MA: The MIT Press.

Martin, C. S., Mullennix, J. W., Pisoni, D. B., & Summers, W. V. (1989). Effects of talker variability on recall of spoken word lists. *Journal of Experimental Psychology Learning, Memory and Cognition, 15*, 676–684.

Massaro, D. W. (1987). *Speech perception by ear and eye: A paradigm for psychological inquiry*. Hillsdale, NJ: LEA.

Massaro, D. W., & Oden, G. C. (1980). Speech perception: A framework for research and theory. In N. J. Lass (Ed.), *Speech and language: Advances in basic research and practice* (Vol. 3, pp. 129–165). New York, NY: Academic Press.

McClelland, J. L., & Elman, J. L. (1986). The TRACE model of speech perception. *Cognitive Psychology, 18*, 1–86.

McClelland, J. L., McNaughton, B. L., & O'Reilly, R. C. (1995). Why there are complementary learning systems in the hippocampus and neocortex: Insights from the successes and failures of connectionist models of learning and memory. *Psychological Review, 102*, 419–457.

McMurray, B., & Jongman, A. (2011). What information is necessary for speech categorization? Harnessing variability in the speech signal by integrating cues computed relative to expectations. *Psychological Review, 118*, 219–246.

Miller, J. L., O'Rourke, T. B., & Volaitis, L. E. (1997). Internal structure of phonetic categories: Effects of speaking rate. *Phonetica, 54*, 121–137.

Miller, J. L., & Volaitis, L. E. (1989). Effect of speaking rate on the perceptual structure of a phonetic category. *Perception & Psychophysics, 46*, 505–512.

Milner, A. D., & Goodale, M. A. (1995). *The visual brain in action*. Oxford: Oxford University Press.

Moran, J., & Desimone, R. (1985). Selective attention gates visual processing in the extrastriate cortex. *Science, 229*, 782–784.

Mullennix, J. W., & Pisoni, D. B. (1990). Stimulus variability and processing dependencies in speech perception. *Perception & Psychophysics, 47*, 379–380.

Mullennix, J. W., Pisoni, D. B., & Martin, C. S. (1989). Some effects of talker variability on spoken word recognition. *Journal of the Acoustical Society of America, 85*, 365–378.

Murphy, G. L., & Medin, D. L. (1985). The role of theories in conceptual coherence. *Psychological Review, 92*, 289–316.

Musacchia, G., Sama, M., Skoe, E., & Kraus, N. (2007). Musicians have enhanced subcortical auditory and audiovisual processing of speech and music. *Proceedings of the National Academy of Sciences, 104*, 15894–15898.

Navon, D., & Gopher, D. (1979). On the economy of the human-processing system. *Psychological Review, 86*, 214–255.

Nearey, T. M. (1989). Static, dynamic, and relational properties in vowel perception. *Journal of the Acoustical Society of America, 85*, 2088–2113.

Neisser, U. (1967). *Cognitive psychology*. New York, NY: Appleton-Century-Crofts.

Niedzielski, N. (1999). The effects of social information on the perception of sociolinguistic variables. *Journal of Language and Social Psychology, 18*, 62–85.

Nittrouer, S., & Lowenstein, J. H. (2007). Children's weighting strategies for word-final stop voicing are not explained by auditory capacities. *Journal of Speech Language and Hearing Research, 50*, 58–73.

Nittrouer, S., & Miller, M. E. (1997). Predicting developmental shifts in perceptual weighting schemes. *Journal of the Acoustical Society of America, 101*, 2253–2266.

Nusbaum, H. C., & Henly, A. S. (1992). Listening to speech through an adaptive window of analysis. In B. Schouten (Ed.), *The processing of speech: From the auditory periphery to word recognition* (pp. 339–348). Berlin: Mouton-De Gruyter.

Nusbaum, H. C., & Magnuson, J. (1997). Talker normalization: Phonetic constancy as a cognitive process. In K. Johnson, & J. W. Mullennix (Eds.), *Talker variability in speech processing* (pp. 109–132). San Diego, CA: Academic Press.

Nusbaum, H. C., & Morin, T. M. (1992). Paying attention to differences among talkers. In Y. Tohkura, Y. Sagisaka, & E. Vatikiotis-Bateson (Eds.), *Speech perception, production, and linguistic structure* (pp. 113–134). Tokyo: OHM Publishing Company.

Nusbaum, H. C., & Schwab, E. C. (1986). The role of attention and active processing in speech perception. In E. C. Schwab, & H. C. Nusbaum (Eds.), *Pattern recognition by humans and machines: Vol. 1. Speech perception* (pp. 113–157). San Diego, CA: Academic Press.

Peterson, G., & Barney, H. (1952). Control methods used in a study of the vowels. *Journal of the Acoustical Society of America, 24*, 175–184.

Pisoni, D. B., Aslin, R. N., Perey, A. J., & Hennessy, B. L. (1982). Some effects of laboratory training on identification and discrimination of voicing contrasts in stop consonants. *Journal of Experimental Psychology: Human perception and performance, 8*, 297–314.

Rauschecker, J. P., & Scott, S. K. (2009). Maps and streams in the auditory cortex: Nonhuman primates illuminate human speech processing. *Nature Neuroscience, 12*, 718–724.

Rubin, D. L. (1992). Nonlanguage factors affecting undergraduate's judgments of nonnative English-speaking teaching assistants. *Research in Higher Education, 33*, 511–531.

Samuel, A. G. (1986). The role of the lexicon in speech perception. In E. C. Schwab, & H. C. Nusbaum (Eds.), *Pattern recognition by humans and machines: Vol. 1. Speech perception* (pp. 89–112). San Diego, CA: Academic Press.

Schwab, E. C., Nusbaum, H. C., & Pisoni, D. B. (1985). Effects of training on the perception of synthetic speech. *Human Factors, 27*, 395–408.

Schyns, P. G., Goldstone, R. L., & Thibaut, J. P. (1998). The development of features in object concepts. *Behavioral and Brain Sciences, 21*(01), 1–17.

Shiffrin, R. M., & Schneider, W. (1977). Controlled and automatic human information processing: II. Perceptual learning, automatic attending and a general theory. *Psychological Review, 84*, 127–190.

Skipper, J. I., van Wassenhove, V., Nusbaum, H. C., & Small, S. L. (2007). Hearing lips and seeing voices: How cortical areas supporting speech production mediate audiovisual speech perception. *Cerebral Cortex, 17*(10), 2387–2399.

Stevens, K. N., & Halle, M. (1967). Remarks on analysis by synthesis and distinctive features. In W. Walthen-Dunn (Ed.), *Models for the perception of speech and visual form* (pp. 88–102). Cambridge: MIT Press.

Suga, N., Gao, E., Zhang, Y., Ma, X., & Olsen, J. F. (2002). The corticofugal system for hearing: Recent progress. *Proceedings of the National Academy of Sciences, 97*, 11807–11814.

Summerfield, Q., & Haggard, M. (1975). Vocal tract normalization as demonstrated by reaction times. In G. Fant, & M. Tatham (Eds.), *Auditory analysis and perception of speech* (pp. 115–141). London: Academic Press.

Syrdal, A. K., & Gopal, H. S. (1986). A perceptual model of vowel recognition based on the auditory representation of American English vowels. *Journal of the Acoustic Society of America, 79*, 1086–1100.

Thackerar, J. N., & Giles, H. (1981). They are—so they spoke: Noncontent speech stereotypes. *Language and Communication, 1*, 255–261.

Treisman, A. M., & Gelade, G. (1980). A feature integration theory of attention. *Cognitive Psychology, 12*, 97–136.

Ungerleider, L. G., & Mishkin, M. (1982). Two cortical visual systems. In D. J. Ingle, M. A. Goodale, & R. J. W. Mansfield (Eds.), *Analysis of visual behavior* (pp. 549–586). Cambridge, MA: MIT Press.

Wehr, M., & Zador, A. (2003). Balanced inhibition underlies tuning and sharpens spike timing in auditory cortex. *Nature, 426*, 442–446.

Weinberger, N. M. (1998). Tuning the brain by learning and by stimulation of the nucleus basalis. *Trends in Cognitive Sciences, 2*, 271–273.

Wong, P. C. M., Nusbaum, H. C., & Small, S. (2004). Neural bases of talker normalization. *Journal of Cognitive Neuroscience., 16*, 1173–1184.

Wong, P. C. M., Skoe, E., Russo, N. M., Dees, T., & Kraus, N. (2007). Musical experience shapes human brainstem encoding of linguistic pitch patterns. *Nature Neuroscience, 10*(4), 420–422.

Yamada, R. A., & Tohkura, Y. (1992). The effects of experimental variables on the perception of American English /r/ and /l/ by Japanese listeners. *Perception & Psychophysics, 52*, 376–392.

Znamenskiy, P., & Zador, A. (2013). Corticostriatal neurons in auditory cortex drive decisions during auditory discrimination. *Nature, 497*, 482–486.

CHAPTER

18

Successful Speaking: Cognitive Mechanisms of Adaptation in Language Production

Gary S. Dell and Cassandra L. Jacobs

Beckman Institute, University of Illinois, Urbana, IL, USA

The language production system works. If a person is older than the age of 4, has no major brain pathology, and has been exposed to linguistic input that accords with their perceptual and motor abilities, then they will have developed a production system that transmits what they want to say. It works when the goal is only to say "hi," and when the speaker attempts to communicate a complicated novel thought that takes several sentences to convey.

Successful linguistic communication is achieved by a division of labor between the speaker and the listener (Ferreira, 2008). Both the production and comprehension systems have to do their job. The speaker has to say something apt and understandable, and the listener must do the rest, which can include compensating for any of the speaker's errors or other infelicities.

In this chapter, we focus on how the production system keeps up its end so that the listener is not overly burdened. Our central claim is that the production system benefits from a number of what we call *speaker tuning mechanisms*. Speaker tuning mechanisms are properties of the system that adapt it to current circumstances and to circumstances that are generally more likely. These include *implicit learning* mechanisms that create long-term adaptive changes in the production system, and a variety of short-term adaptive devices, including *error monitoring*, *availability-based retrieval*, *information-density sensitivity*, and, finally, *audience design*. Although we characterize these mechanisms in cognitive rather than neural terms, we include some pointers to relevant neurobiological data and mechanisms. In the following, we describe the production system generally and then focus on the long-term and then short-term speaker tuning mechanisms.

18.1 LANGUAGE PRODUCTION

The production system turns thoughts into sequences of words, which can be spoken aloud, inwardly spoken, or written down. Traditionally (Levelt, 1989), the production process consists of determining the semantic content of one's utterance (*conceptualization*), translating that content into linguistic form (*formulation*), and *articulation*, as illustrated in Figure 18.1. Here, we focus on the second of these stages, which describes how intended meaning, sometimes called the *message*, is turned into an ordered set of words that are specified for their phonological content. That is, the formulation stage describes how CHASE (CAT1, RAT1, past) becomes /ðə.kæt′.čest′.ðə.ræt′/. (The "1" in "CAT1" represents a particular definite CAT). Much of the psycholinguistic and neuroscience research on formulation has concerned three subprocesses: (i) *lexical access*, the retrieval of appropriate words; (ii) *grammatical encoding*, the specification of the order and grammatical forms of those words; and (iii) *phonological encoding*, determining the pronunciation of the sequence of words. These are discussed in turn.

18.1.1 Lexical Access

Most experimental, clinical, and theoretical research on production has concerned lexical access and focuses on the production of single-word utterances. When given a picture that has been identified as the concept CAT, how does the speaker retrieve the word "cat"? Lexical access has been characterized as a two-step process (Garrett, 1975). First, the concept is mapped

Neurobiology of Language. DOI: http://dx.doi.org/10.1016/B978-0-12-407794-2.00018-3

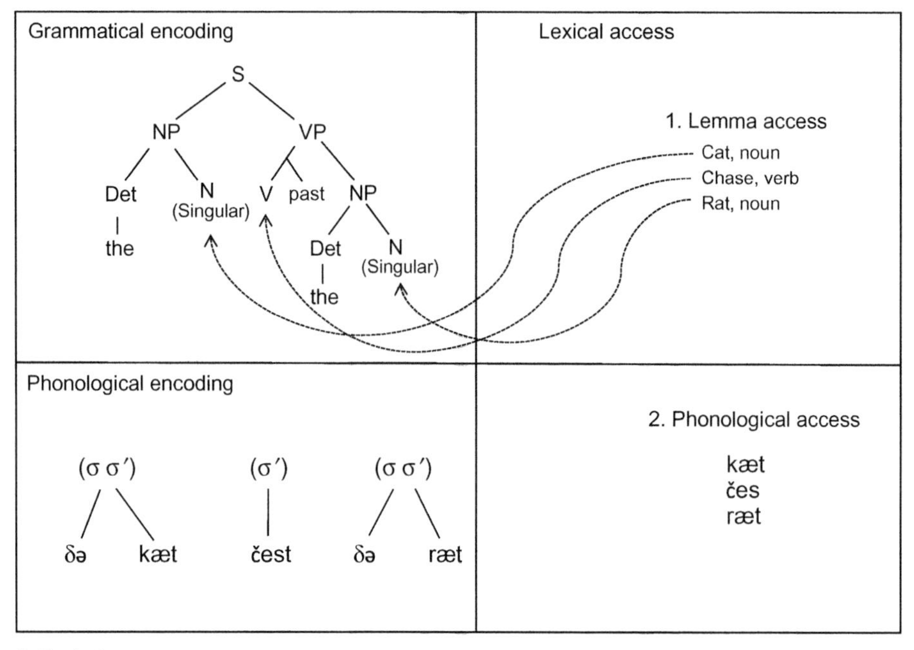

FIGURE 18.1 Components of the language production system.

onto an abstract lexical representation, variously called the *lemma*, the *L-level* representation, or simply the *word node*. This abstraction identifies the grammatical properties of the word such as its syntactic category (e.g., noun) and other grammatically relevant features (e.g., number, grammatical gender). Importantly, this level does not specify anything about pronunciation. That comes in the second step, where the word's phonological form, most often viewed as a sequence of phonemes, is retrieved. Intuitive support for the two-step notion comes from speech errors (Fromkin, 1971). Slips can profitably be divided into those that might have arisen during the first step (e.g., semantic errors such as "dog" for "cat") and those that could have happened in the second step (e.g., "cap" for "cat"). Furthermore, the tip-of-the tongue state ("I know that word! It's on the tip of my tongue") can be characterized as getting stuck between the steps.

Much of the research on lexical access has concerned just how separate the two steps are. For example, the *modular discrete-step* view states that the first step must be completed before the second step can begin (Levelt, Roelofs, & Meyer, 1999). Alternatively, one could allow for *cascading*, which blurs the boundaries between the steps by allowing for phonological properties of potential word candidates to be retrieved before a single lexical item has been settled on in the first step. Or, one could allow for *interaction*, which blurs the steps even further by allowing for relevant representations at each step to influence one another through the interactive spread of activation (see Dell, Nozari, & Oppenheim, 2014 for a recent review of the evidence for interaction between the steps, and see Dell, Schwartz, Nozari, Faseyitan, & Coslett, 2013 and Ueno, Saito, Rogers, & Lambon Ralph, 2011 for proposals regarding the neural correlates of the steps).

18.1.2 Grammatical Encoding

Although most production research concerns single-word utterances, the hallmark of production is the ability to construct multiword utterances, particularly those that the speaker has never said or even heard before. For example, William Blake famously used the phrase, "fearful symmetry" to characterize the tiger. And it is not just the poets who are linguistically inventive. Since Chomsky (1959) emphasized the creativity of language, it is a psycholinguistic cliché that most of what speakers say is novel. Regardless of whether this claim is strictly true, there is no doubt that theories must explain the production of novel utterances. The usual explanation is that the production system uses syntactic-sequential abstractions that specify how word categories can combine to express structured messages. For Blake's phrase, the relevant abstractions would dictate that, in English, adjectives (*fearful*) precede nouns (*symmetry*). Production models of grammatical encoding (Bock, 1982; Chang, Dell, & Bock, 2006; Kempen & Hoenkamp, 1987) differ considerably, but all recognize the distinction between categorically specified abstractions and lexical items. Typically, the abstractions are characterized as *frames*, that is, structures that specify the sequence and phrasal membership of syntactically categorized word-sized slots (Dell, 1986; Garrett, 1975). So, there might be a noun-phrase frame with slots for a singular indefinite determiner, an adjective, and a singular count noun. And this frame may occupy a larger slot in a clausal frame, and so on. Because of the separation between words and their slots, the system has the means to encode new phrases (e.g., "a poetic tiger") by putting known words into known frames in new combinations. Evidence for such a system comes from dissociations in aphasia between individuals with lexical retrieval deficits and those with deficits in syntactic-sequential processes (e.g., see Gordon & Dell, 2003, for review), from functional imaging data that identify different brain areas for word-retrieval and word-combination mechanisms (e.g., Hagoort, 2013), and from structural priming studies, which are reviewed later.

18.1.3 Phonological Encoding

The retrieval of the phonological form of a word results in a sequence of phonological segments: k æ t. The segments are then put together with the segments of surrounding words, and the resulting sequence must be characterized in terms of its syllables and how those syllables are stressed (Levelt et al., 1999). These processes must respect the phonological properties of the language being spoken, including how segments combine to make syllables (*phonotactic* knowledge), how syllables are organized into higher-level prosodic structures, and how timing, pitch, and intensity vary as a function of those structures. Ultimately, this organized phonological structure guides the articulatory process. The phonological encoding process has been studied by assessing the response time to produce words and syllables (e.g., Cholin, Dell, & Levelt, 2011; Meyer, 1991), by examining phonological speech errors (Warker & Dell, 2006), by measuring the articulatory and acoustic details of utterances (e.g., Goldrick & Blumstein, 2006; Goldstein, Pouplier, Chen, Saltzman, & Byrd, 2007; Lam & Watson, 2010), and more recently by event-related brain potentials and other imaging techniques (e.g., Qu, Damian, & Kazanina, 2012).

18.2 LONG-TERM SPEAKER TUNING: IMPLICIT LEARNING

The production system does its job because it has learned to do so, and the basis for that learning is experience in comprehending and speaking (Chang et al., 2006). Learning, however, is not just something that children do. The typical adult speaker says approximately 16,000 words per day (Mehl, Vazire, Ramirez-Esparza, Slatcher, & Pennebaker, 2007) and hears and reads many more. This experience adapts the production system so that it is able to make effective choices in the particular circumstances that it finds itself. We refer to this continual process of adaptation as *implicit learning*. We claim that this adaptation is a kind of learning because the changes induced are not short-lived, and that the learning is implicit because it is an automatic consequence of linguistic experience that occurs without any intention to learn or awareness of what has been learned. In the remainder of this section, we review implicit-learning research in each of the three production subprocesses mentioned previously. For lexical access, we consider mechanisms of lexical repetition priming and frequency effects, and the possibility of phrasal frequency effects. For grammatical encoding, we discuss the hypothesis that structural priming in production is a form of implicit learning. And, for phonological encoding, we review studies that find implicit learning of novel phonotactic patterns.

18.2.1 Implicit Learning of Words and Phrases

The production system adapts to make the words that it is most likely to use easier to retrieve and articulate. In particular, words that we have said recently are easier to say than words that we have not said recently (*repetition priming*; e.g., Mitchell & Brown, 1988). We are also, in general, faster and more accurate at producing words that we have more experience saying, that is, frequent words (Caramazza, Costa, Miozzo, & Bi, 2001; Jescheniak & Levelt, 1994). Both repetition priming and frequency effects are thought to arise from and can be explained by implicit learning, which optimizes the production system for situations that are more likely to happen.

One manifestation of implicit learning in word production is *cumulative semantic interference* (e.g., Howard, Nickels, Coltheart, & Cole-Virtue, 2006). When we have to name a picture of the same thing twice (e.g., *crow*), we benefit from repetition priming. But, if instead of repeating the picture's name we next have to name something that is similar in meaning, but not the same word (e.g., *finch*), then we produce this word more slowly and have a greater chance of error (e.g., Schnur, Schwartz, Brecher, & Hodgson, 2006). This negative effect is semantic interference. Oppenheim, Dell, and Schwartz (2010) investigated the "dark" (semantic interference) and "light" (repetition priming) sides of word production using a computational model, aptly called the "dark-side" model. In the model, each experience with a word tunes the production system by prioritizing words that are recently used and, importantly, deprioritizing their competitors, that is, semantically similar words. This tuning consists of the strengthening of connections to words when they are used, but weakening of connections to these words' competitors. As a result, when a word is repeated it becomes relatively more active in the lexical network, effectively by leeching activation from similar words. In this way, repeating the word *crow* becomes easier, whereas naming different, but semantically similar, words in a sequence (e.g., *crow, finch, gull*) becomes increasingly difficult. This effect shows that the production system is adaptive, because words that are used and will likely be used again become easier to say, whereas words that could potentially interfere with those words are rendered less accessible and, hence, less disruptive.

We described how lexical access in production involves two steps, retrieval of the abstract lexical item and then retrieval of the item's phonological form. We also noted that semantic errors such as "dog" for "cat" can occur at the first step, but phonological errors such as "cap" or "dat" for "cat" occur during the second step. We also said that recently spoken or high-frequency words (e.g., "cat" as opposed to "feline") are less vulnerable to error because an implicit learning process enhances their retrieval. But does the greater ease associated with common or repeated words apply to both steps or just one of them? Jescheniak and Levelt (1994) proposed that frequency effects in word retrieval are felt largely during the second step. Others (e.g., Knobel, Finkbeiner, & Caramazza, 2008) claim that both steps benefit when the target word is frequent, because implicit learning should have an effect throughout the retrieval process. Kittredge, Dell, Verkuilen, and Schwartz (2008) addressed this question by looking at how target-word frequency affects semantic and phonological errors during picture naming. They presented aphasic participants with pictures to name that varied, among other factors, in their word frequency. They found, as expected, that the odds of saying the right word increased with the frequency of the target, demonstrating that common words are "protected" by their frequency. This protective power was found to prevent both semantic and phonological errors, suggesting that both steps of lexical retrieval benefit from frequency and, more generally that the production system keeps track of likely events at all levels.

The production system also seems to keep track of and adapts to the degree to which words combine. Janssen and Barber (2012) explored this by looking at whether the frequency of the combination of two words (e.g., *red car* or *red hammer*) predicted how easily that phrase was generated. In particular, they presented participants with pictures of colored objects and had them name the object and its color with an appropriate phrase. They found that frequent phrases had faster naming latencies than would be predicted just by the frequency of the first or second word. This suggests that the production system tunes itself to probable events beyond the word level by keeping track of word combinations as well.

18.2.2 Structural Priming

One of the classic findings in psycholinguistics is structural priming, also known as syntactic priming, or structural repetition. Structural priming is the tendency for speakers to reuse recently experienced structures. Bock (1986a) gave experimental participants pictures that can be described with either of the two kinds of dative structures (double objects, "The woman handed the boy the paint brush," versus prepositional datives, "The woman handed the paint brush to the boy"). Participants described these pictures after saying an unrelated prime sentence that used either a double-object or prepositional dative structure. Priming was seen in the tendency for speakers to use the structure of the prime when describing the picture. Similar effects were seen for other structural alternations such as active transitive sentences ("Lightning is striking the church") versus passives ("The church is struck by lightning"). The important aspect of this priming is that it appears to be the persistence of an abstract syntactically characterized structure (e.g., the frame: *Noun_phrase Auxillary_verb Main_verb Prepositional_phrase* for a full passive), and not the lexical content of the utterance, its meaning, or its intonational properties (Bock & Loebell, 1990). As such, structural priming provides evidence for a production process that uses structural abstractions during grammatical encoding.

Bock and Griffin (2000) claimed that structural priming is not just a temporary change to the system, but instead it is a form of implicit learning, akin to the connection weight changes that characterize learning in connectionist models. They provided evidence for this claim by showing that the effect of a prime persists undiminished over at least 10 unrelated sentences (several minutes). If priming were due to temporary activation of a structure, then the prime's influence would rapidly decay. The evidence that the learning is implicit is that it occurs in brain-damaged speakers who have no explicit memory of the prime sentence (Ferreira, Bock, Wilson, & Cohen, 2008).

Chang et al. (2006) created a computational model that reflected the idea that structural priming is implicit learning. They trained a connectionist model to simulate a child experiencing sentences one word at a time. The model was also given a representation of the intended meaning of some of the sentences that it experienced, with this meaning presumably having been inferred by the child from context. The model learned by "listening" to each sentence and trying to predict each word. When the actual next word was heard, the model then compared its prediction to that word, thus generating a prediction error signal. This error signal was the impetus for the model to change its connection weights so that its future predictions were more accurate. By using prediction error, the model learned the linguistic patterns in the language (e.g., syntactic structures) and how those patterns mapped onto meaning (e.g., Elman, 1993). After this learning, the model was able to produce because "prediction is production" (Dell & Chang, 2014); generating the next word from a representation of previous words and intended meaning is, computationally, a production process. When given a representation of intended meaning, the model's sequence of word predictions constituted the production of a sentence. The key aspect of this model, for our purposes, is that it accounted for structural priming through learning. Even after the model attained "adult" status, it continued to learn. When a prime sentence was experienced, the model's connection weights were changed ever so slightly to favor the subsequent production of sentences with the same structure. Experiencing, for example, a double-object dative inclined the model to produce that structure later. Because the priming was based on weight change, it is a form of learning, thus accounting for Bock and Griffin's finding that structural priming is undiminished over time. Also, the evidence that the implicit learning that characterizes structural priming is based on prediction error comes from demonstrations that less common, and hence more surprising, prime structures lead to more priming than common ones (e.g., Jaeger & Snider, 2013).

18.2.3 Phonotactic Learning

Young children implicitly learn the phonotactic patterns of their language through experience. Such patterns include knowledge about where certain consonants can occur in the syllables in their language; for example, in English, /h/ only occurs at the beginning of a syllable (the *onset*) and /ng/ occurs only at the end (the *coda*). Because of their phonotactic knowledge, English speakers can readily produce the phonotactically legal nonword "heng," but not the illegal "ngeh." Evidence that the production system actively uses this knowledge comes from the *phonotactic regularity effect* on speech errors: slips tend to be phonotactically legal. One might mistakenly produce "nun" as "nung," a phonotactically legal nonword, but not as "ngun" (Wells, 1951).

Warker and Dell (2006) and Dell, Reed, Adams, and Meyer (2000) created an experimental analogue to the phonotactic regularity effect in which participants recited four-syllable tongue twisters such as "hes feng kem neg" at a fast pace. Unbeknownst to the participants, the syllables followed artificial phonotactic patterns that were present only in the experimental materials. For example, a participant's syllables might follow the pattern: *During the experiment, /f/ can only be a syllable onset and /s/ can only be a syllable coda* (as in the example four-syllable sequence above). Participants would recite several hundred of these sequences in each of four experimental sessions on consecutive days. Because of the fast speech rate, slips were reasonably common. Most often, these involved movements of consonants from one place to another, such as "hes feng kem neg" being spoken as "**f**es feng kem neg," in which /f/ moved to the first syllable. The crucial feature of the study was whether these slips respected the phonotactics of the experienced syllables. As expected, slips of /h/ and /ng/ respected English phonotactics; /h/ is always moving to an onset position and /ng/ is always moving to a coda position. The crucial finding, though, was that slips of the artificially restricted consonants (/f/ and /s/ in our example) also respected the local phonotactics of the experiment. Notice in the example that /f/ slips to an onset position, that is, the slip is "legal" with regard to the experimental phonotactic patterns. And this was not just a small statistical tendency; 98% of the slips of experimentally restricted consonants were "legal" in this respect, whereas consonants that were not experimentally restricted the way that /f/ and /s/ were often slipped from onset to coda or vice versa (Dell et al., 2000).

Finding that slips respected the experimental distributions of consonants suggests that participants implicitly learned these distributions, and this learning affected their slips. But is this effect truly one of learning, as opposed to some very temporary priming of preexisting knowledge (e.g., priming of a rule that /f/ can be an onset in English)? Evidence that true learning is occurring comes from exposing participants to more complex "second-order" constraints such as: *if the vowel is /ae/, then /f/ must be an onset and /s/ must be a coda, but if the vowel is /I/, then /s/ must be an onset and*

/f/ must be a coda. Warker and Dell (2006) found that participants' slips did *not* follow this vowel-dependent second-order constraint on the first day of a 4-day experiment. On the second and subsequent days, though, the slips did obey the constraint (e.g., more than 90% of slips were legal). This suggests that the effect requires *consolidation*, a period of time (possibly involving sleep; Warker, 2013) in which the results of the experience are registered in a relatively permanent way in the brain. After consolidation, the effects appear to remain at least for 1 week (Warker, 2013). Because the effect requires consolidation and is persistent in time, it appears to be a form of learning. Thus, phonotactic-like knowledge and its expression in speech production errors can be tuned by an implicit learning process.

18.3 SHORT-TERM SPEAKER TUNING

Implicit learning is not the only mechanism that allows the production system to fluently generate appropriate, grammatically correct utterances that listeners can easily interpret. There are several adaptive phenomena in production that involve immediate processing, rather than long-term learning. These short-term tuning mechanisms include error monitoring, availability-based production choices, sensitivity to information density, and audience design.

18.3.1 Error Monitoring

Speakers help their listeners by avoiding making speech errors or, when an error occurs, by attempting to correct it. Catching slips before they happen or fixing them after they do requires that speakers do error monitoring. Studies of monitoring suggest that we detect at least half of our overt slips after they happen, and that we detect and block potential errors before they can occur (Baars, Motley, & MacKay, 1975; Levelt, 1983). Evidence that errors can be detected before they are spoken comes from the existence of very rapid detections of overt errors. Levelt (1983) gave the example of "v—horizontal." The speaker started to say "vertical," but quickly stopped and replaced it with the correct "horizontal." The fact that speech was stopped right away (within 100 msec of the onset of the erroneous /v/) demonstrates that the error was almost certainly detected before articulation began. How is this possible? There are two theories of error detection. One is that speakers detect errors by comprehending their own speech and noting if there is a mismatch with what they intended (Hartsuiker & Kolk, 2001; Levelt, 1983). This view—the *perceptual loop theory*—allows for the comprehension of internal speech before it is produced to explain the fact that errors can be detected before articulation. The alternative is that error detection occurs within the production system itself. An example of this is the *conflict detection theory* of Nozari, Dell, and Schwartz (2011), which proposes that the production system can assess the extent to which its decisions are conflicted and assumes, when conflict is high, that an error is likely. For example, suppose that during word access, the word CAT was selected during the first lexical-access step, but DOG was also nearly as activated as CAT. That can be taken as a sign that there was a possible error during that step. Similarly, if a particular speech sound, for example, /d/, is selected while another, /k/, is almost as active, again that can be a signal that there may have been a mis-selection, this time during the second access step. Nozari et al. used a computational model to demonstrate that the association between high conflict and error likelihood is a strong one, but also that the association no longer holds when the production system is functioning very poorly. Thus, for some aphasic individuals, conflict would not be an effective predictor of error and such individuals would be expected to have trouble detecting their own errors.

To test the conflict detection theory of monitoring and the competing perceptual loop theory, Nozari et al. (2011) examined how successful aphasic individuals were at detecting their own errors in a picture-naming task. The perceptual loop account predicts that good error detection should be associated with good comprehension because detection is performed by the comprehension system in that theory. In contrast, the conflict detection theory expects good detection to be associated with production rather than comprehension skill. The results supported the conflict detection account. The aphasic patients with higher rates of error detection had relatively good production skills, and comprehension ability was unrelated to error detection rate. Furthermore, Nozari and colleagues showed that patients who were relatively better at the first step of lexical access, but poor at the second step, could detect their first-step errors (e.g., semantic errors) but not their second-step errors (phonological errors). The complement was true as well—doing better on the second step in production implied better detection of second-step errors in particular. These results show that dissociations in production abilities for the lexical-access steps are mirrored in differential abilities to detect errors at the two steps, exactly as expected by the conflict detection theory.

Do the results of Nozari et al. (2011) mean that we do not detect errors by comprehending our own speech? No. These results only point to another possible mechanism for error detection, particularly a mechanism that can detect errors before they happen. It seems likely that many overtly spoken slips are detected simply by hearing them, as proposed in the perceptual loop theory. In support of this claim, Lackner and Tuller (1979) found that using noise to mask a speaker's speech diminished the speaker's ability to detect their overt phonological errors, demonstrating that perception of the auditory signal plays a role in detection. It is therefore likely that multiple mechanisms contribute to the monitoring process. For example, speakers appear to guard their speech against slips that create taboo words (Motley, Camden, & Baars, 1982; see also Severens, Kühn, Hartsuiker, & Brass, 2012 for an fMRI study of frontal brain regions involved in taboo-word monitoring).

18.3.2 Availability-Based Production

The adage, "think before you speak" advises speakers to fully plan their utterances before saying them. The fact that the adage exists suggests that speakers do not routinely do this. Instead, language production involves some degree of *incrementality* (Kempen & Hoenkamp, 1987): utterances are often constructed and spoken in a piecemeal fashion, with the result that one might start talking before having planned the entire sentence. Because production can be incremental, the retrievability of the various parts of the utterance can influence its structure. For example, when attempting to produce the message illustrated in Figure 18.1, suppose that we are able to retrieve "rat," but have not yet retrieved "cat." We can start the utterance as "The rat..." and then, because English allows for a passive structure, can continue with "was chased by the cat" as we eventually retrieve the other words. Thus, the production system may opportunistically take advantage of the words that are retrieved first and may start with those words. This is the essence of availability-based production. What is retrieved first tends to be said first. More generally, what is available tends to be spoken as soon as it can. Although this strategy occasionally results in false starts, it makes for an efficient production system (Bock, 1982).

Bock (1986b) provided support for availability-based production by asking speakers to describe pictures such as one in which lightning is striking a church. This can be described with either an active ("Lightning is striking a church") or a passive ("The church is struck by lightning") structure. Earlier in this chapter, we showed how this structural choice can be influenced by structural priming. It turns out that this choice is also sensitive to the relative availability of the words "lightning" and "church." Bock found that participants who had recently experienced the word "thunder," which presumably makes "lightning" more available, were more likely to describe the picture with the active form, making the primed word come out earlier. Similarly, priming "church" made the passive more likely.

One can also see availability-based production at work in choices about optional words. A sentence such as "The coach knew **that** you missed practice" can be produced with no "that," without changing the meaning of the sentence. So, what determines whether you include the "that"? One possibility is that speakers engage in *audience design*: when faced with a production choice, they choose what will make the sentence easier for their listener to understand. Notice that if the "that" is missing, then the sentence has a temporary ambiguity when "you" is heard. The "you" can be either the direct object of "knew" or the subject of a new embedded clause. Including the "that" removes the ambiguity. Ferreira (2008) and Ferreira and Dell (2000) suggested an alternative explanation for when "that" is present in these sentences. It has to do with the availability of the material after "that." If "you" has already been retrieved and is ready to go at the point in the sentence after "The coach knew...," then the speaker is more likely to omit "that." But if the speaker is not quite ready with "you," then including "that" is a convenient way to pause and buy time. As described in the subsequent section on information-density sensitivity, there is evidence that speakers do attempt to stretch time out at certain points in a sentence, and this can be thought of as an example of this. Here the issue is whether speakers produce "that" to disambiguate the utterance for their listeners, or because their production systems naturally produce whatever is available.[1] If the "you" is immediately available after "The coach knew," then the sentence can grammatically continue without the "that." Ferreira and Dell tested these ideas by comparing the production of four kinds of sentences:

I knew (that) ***I*** *missed practice.* (embedded pronoun is repeated and unambiguously nominative)

You knew (that) ***you*** *missed practice.* (embedded pronoun is repeated and ambiguous)

I knew (that) ***you*** *missed practice.* (embedded pronoun is not repeated and ambiguous)

You knew (that) ***I*** *missed practice.* (embedded pronoun is not repeated and unambiguously nominative)

The sentences were presented and then recalled in situations in which the participants could not remember whether there had been a "that" in the sentence

[1]Of course, the grammar does not always allow you to eliminate "that" as a complementizer; "the girl that saw the boy is here" must have it in Standard American English.

and, hence, they tended to use their natural inclinations about whether to include "that." The key variable was the percentage of recalled sentences with "that." The hypothesis that speakers include "that" to help their listeners predicts that the two sentences with the ambiguous embedded "you" will include more instances of "that" than the unambiguous conditions that have the clearly nominative pronoun, "I," as the embedded subject. The availability hypothesis predicts that because of repetition priming, the embedded pronoun (**I** or **you**) will be more available if it had just been said as the subject of the main clause. Because their embedded pronouns should be quite available, the two conditions with repeated pronouns are expected to have fewer instances of "that." Across several experiments, there was no tendency for more "that"s in the ambiguous sentences, but repeating the pronoun caused the percentage of "that"s to decrease by approximately 9%. The results clearly supported the availability hypothesis, providing another demonstration that the production system's decisions are opportunistically guided by what is easily retrieved. In the next section, we approach the question of the production of optional words like "that" from another angle.

18.3.3 Information-Density Sensitivity

One way that people may alter production in the short-term is by monitoring for and adjusting the probabilistic characteristics of what they are about to say. Taking a cue from information theory (Shannon, 1948), it has been proposed that speakers control the *rate of information* conveyed in their utterances so that there are as few as possible points in which the rate is extremely high or extremely low. Recall that, on a formal level, words or structures that are *less* likely contain *more* information and that, in the reverse case, *redundant* or predictable items are associated with less information. The idea is that keeping the information rate constant at a level that listeners can handle maximizes the effective transmission to the listener. Too fast a rate leads to loss of transmission, and too slow a rate wastes time. The hypothesized information constancy in production is termed the *smooth signal redundancy hypothesis* or, alternatively, *uniform information density* (UID). This tendency can be assumed to apply at all levels of language production, including lexical choice (Mahowald, Fedorenko, Piantadosi, & Gibson, 2013), syntactic structure (Jaeger, 2006, 2010), and phonetic and phonological output (Aylett & Turk, 2004).

Lexical, syntactic, phonological, and pragmatic predictability and given-ness, as constrained by the discourse or experiment, strongly influence the durations of individual words as would be expected from the UID. For example, the word *nine* in "A stitch in time saves *nine*" is shorter than in the phrase "I'd like *nine*" (Lieberman, 1963) because the *nine* in the first example is highly predicted by the previous words. Speakers moderate duration and other prosodic cues in response to these linguistic factors, as has been demonstrated experimentally and in the wild, and this effect is robust even when a large number of other factors are taken into account (Jurafsky, Bell, Gregory, & Raymond, 2001).

Aylett and Turk (2004) examined the relationship between reduction and redundancy, or the contribution of statistical predictability to the short-term manifestation of phonetic output. They modeled the durations of syllables as a function of the degree to which they were predicted by the preceding information and the predictability of the word itself in discourse as well as its frequency of occurrence. They found evidence that individuals regulate the distribution of information in the signal (modulating various prosodic cues like duration, volume, and pitch) in that these cues represent a tradeoff between predictability and acoustic prominence. So, when a word is less predictable, it will carry more information in the sense that it is unexpected, but speakers take this into account by providing additional cues as to the identity of an upcoming word or syllable, such as articulating the word more loudly.

Jaeger (2006, 2010) identified analogous behavior in syntactic flexibility as a function of information density. Using evidence from the optional *that* structure that we introduced in the previous section, he demonstrated that the choice of whether to include a "that" provides the language production system with a means of redistributing information so that information density is more uniform across the utterance. For example, in sentences such as "My boss confirmed/thinks (that) we were absolutely crazy," speakers were more likely to include "that" when the presence of a complement clause (e.g., "...we were absolutely crazy") is unexpected given the main verb (e.g., "confirmed"). This is because *confirm* is a verb that most often takes a noun as its argument (e.g., "...confirmed the result"), and so the presence of a complement clause is less probable and therefore more surprising. In contrast, the verb *think* often takes a complement clause and, because that is more expected, "that" was less likely to be included in the utterance. In general, including a "that" when the complement clause is unexpected makes upcoming linguistic material less surprising for the listener, because "that" very commonly signals for a complement clause. The resulting overall structure is much more even in its syntactic surprisal. In this way, the speaker's choices about optional *that* are sensitive to the goal of minimizing peaks and valleys in the information conveyed during the incremental production of the sentence.

These choices presumably translate into less effort for the listener. It is also possible that producing sentences with more UID directly aids the fluency of the production process, because what may be highly surprising to a listener may be relatively more difficult for speakers to create.

When we say that speakers "monitor" and "adjust" information density, this implies active online control. But control of information rate is not necessarily the result of an active short-term adaptation. Instead, the mechanisms that achieve good information rates may be learned as speakers gain experience about what production choices lead to effective comprehension (Jaeger & Ferreira, 2013). For example, speakers may consistently include "that" in their complement clauses introduced by main verbs such as "confirm" because they have learned that failure to do so leads to misunderstanding. With this view, there is no active control of information density; only the retention of successful speaking habits.

18.3.4 Audience Design

The language production system adapts to one's partner, not only by avoiding high information rates but also by considering the partner's specific needs and abilities; a speaker uses syntax, words, and phonology that the partner will likely understand. As we have outlined throughout this chapter, the production system can adapt to internal moment-by-moment demands, and it can change itself in the long-term as a function of experience. In this final section, we consider how the production system goes beyond what is easy for *it* to do, and instead considers what might best help the other person understand. This consideration is known as *audience design*. We discuss two examples of such design. First, we consider the way individuals use words that result in more effective communication on a cooperative task via a process called *entrainment* (Brennan & Hanna, 2009; Clark & Wilkes-Gibbs, 1986), and then how talkers can change their own pronunciation of words to facilitate understanding in *phonetic convergence* (Pardo, 2006).

Entrainment, or the convergence on a single term between two talkers in a conversation, is a necessary part of communication. It is estimated that approximately 50% of discourse entities are mentioned multiple times in a conversation or text (Recasens, de Marneffe, & Potts, 2013). Given this degree of repetition, it would be useful if speakers could agree on labels for those entities. If one party to a conversation referred to a particular plant as a "bush" and the other called it a "tree," then confusion is likely. Agreement on terms through entrainment removes the confusion. In experimental settings, entrainment has been examined by looking at how participants in a cooperative task describe an object and how that description changes as the participants continue to interact. In Clark and Wilkes-Gibbs (1986), participants had to cooperate to sort a set of abstract visual shapes (made up of "tangrams") often resembling people or animals. Over the course of several turns, both partners came to use similar, eventually convergent terms or short phrases to describe the items. Early on, a speaker might recognize that the other person does not understand their initial description (e.g., "The next one is the rabbit." "huh?"), requiring that the speaker elaborate (e.g., "That's asleep, you know, it looks like it's got ears and a head pointing down."). As the experiment continues, the objects' labels become increasingly shorter and the listener's errors in interpreting what is said become rare, suggesting that talkers have optimized label length and form for communicative efficiency. The description of such a figure can go from the very complex on the first exchange ("looks like a person who's ice skating, except they're sticking two arms out in front") to shorter, multiphrasal ("the person ice skating, with two arms") to finally a single noun phrase ("the ice skater"). Thus, the entrainment process adapts the production systems of both participants in such a manner that communication success, rather than production ease, is the goal.

Phonetic convergence is a phenomenon where individuals adopt the phonetic and phonological representations of the other talker during a conversation. A person may adopt features of another's accent, such as the famous US southern "pin-pen" merger or a northern cities vowel shift (e.g., "dawg" becomes "dahg"), or even more subtle features such as differences in voice-onset time. Pardo (2006) demonstrated such convergence experimentally. Participants completed a map task where one partner's map (the receiver's) needs to be drawn to look like the other's (the giver's). Like the tangram task used by Clark and Wilkes-Gibbs, the map task requires cooperative communication. There were many places on the map with standard names provided to the talkers (e.g., *abandoned monastery*, *wheat field*, etc.). Phonetic convergence of the speech of the talker pairs was assessed by naïve participants who were asked to judge the degree of similarity of the pairs' pronunciations for these place names as they did the task. Not only did all conversation partners show some degree of convergence, but also these effects arose after very little interaction time—many partners showed convergent phonetics as early as before the halfway point in their dialogue, with convergence persisting into the second half as well. This convergence demonstrates that individuals engage in audience design by adopting the phonetic features of their conversation partner during a cooperative task. Because it is presumably easier for each speaker to use his or her own accent, phonetic convergence counts as another example in which the adaptation suits the goal of communication, rather than the immediate ease of the production systems of the individual speakers.

18.4 CONCLUSION

Language production is, in one sense, difficult. The speaker has to decide on something worth saying, choose words (out of a vocabulary of 40,000), appropriate syntax, morphology, and prosody, and ultimately has to articulate at the rate of two to three words per second. In another sense, production is easy. We think it takes little effort. Particularly when we are talking about familiar topics, we can at the same time walk, drive, or even play the piano (Becic et al., 2010). The seeming paradox that something so difficult is yet so easy is resolved when we consider the mechanisms presented in this chapter. The production system is continually being tuned by the extraor dinary amount of experience we have. We say 16,000 words per day and hear and read a lot more. The implicit learning that results from this input effectively trains the system and tunes it well to its current circumstances. But implicit learning is not the whole story. The production system also makes use of a variety of moment-by-moment mechanisms to compensate for and prevent errors, to promote fluency, and to make the job of the listener easier.

Acknowledgments

Preparation of this chapter was supported by NIH DC000191 and by an NSF fellowship to Cassandra Jacobs.

References

Aylett, M., & Turk, A. (2004). The smooth signal redundancy hypothesis: A functional explanation for relationships between redundancy, prosodic prominence, and duration in spontaneous speech. *Language and Speech*, *47*, 31–56.

Baars, B. J., Motley, M. T., & MacKay, D. G. (1975). Output editing for lexical status from artificially elicited slips of the tongue. *Journal of Verbal Learning and Verbal Behavior*, *14*, 382–391.

Becic, E., Dell, G. S., Bock, K., Garnsey, S. M., Kubose, T., & Kramer, A. F. (2010). Driving impairs talking. *Psychonomic Bulletin & Review*, *17*, 15–21.

Bock, J. K. (1982). Towards a cognitive psychology of syntax: Information processing contributions to sentence formulation. *Psychological Review*, *89*, 1–47.

Bock, K. (1986a). Syntactic persistence in language production. *Cognitive Psychology*, *18*, 355–387.

Bock, K. (1986b). Meaning, sound, and syntax: Lexical priming in sentence production. *Journal of Experimental Psychology: Learning, Memory, and Cognition*, *12*, 575–586.

Bock, K., & Griffin, Z. M. (2000). The persistence of structural priming: Transient activation or implicit learning? *Journal of Experimental Psychology: General*, *129*, 177–192.

Bock, K., & Loebell, H. (1990). Framing sentences. *Cognition*, *35*, 1–39.

Brennan, S. E., & Hanna, J. E. (2009). Partner-specific adaptation in dialog. *Topics in Cognitive Science*, *1*, 274–291.

Caramazza, A., Costa, A., Miozzo, M., & Bi, Y. (2001). The specific-word frequency effect: Implications for the representation of homophones in speech production. *Journal of Experimental Psychology: Learning, Memory, and Cognition*, *27*, 1430–1450.

Chang, F., Dell, G. S., & Bock, K. (2006). Becoming syntactic. *Psychological Review*, *113*, 234–272.

Cholin, J., Dell, G. S., & Levelt, W. J. M. (2011). Planning and articulation in incremental word production: Syllable-frequency effects in English. *Journal of Experimental Psychology: Learning, Memory, and Cognition*, *37*, 109–122.

Chomsky, N. (1959). A review of Skinner's Verbal Behavior. *Language*, *35*, 26–58.

Clark, H. H., & Wilkes-Gibbs, D. (1986). Referring as a collaborative process. *Cognition*, *22*, 1–39.

Dell, G. S. (1986). A spreading activation theory of retrieval in sentence production. *Psychological Review*, *93*, 283–321.

Dell, G. S., & Chang, F. (2014). The P-chain: Relating sentence production and its disorders to comprehension and acquisition. *Philosophical Transactions of the Royal Society B*, *369*, 20120394.

Dell, G. S., Nozari, N., & Oppenheim, G. M. (2014). Word production: Behavioral and computational considerations. In M. Goldrick, V. S. Ferreira, & M. Miozzo (Eds.), *The Oxford handbook of language production*. Oxford, UK: Oxford University Press.

Dell, G. S., Reed, K. D., Adams, D. R., & Meyer, A. S. (2000). Speech errors, phonotactic constraints, and implicit learning. A study of the role of experience in language production. *Journal of Experimental Psychology: Learning, Memory, and Cognition*, *26*, 1355–1367.

Dell, G. S., Schwartz, M. F., Nozari, N., Faseyitan, O., & Coslett, H. B. (2013). Voxel-based lesion-parameter mapping: Identifying the neural correlates of a computational model of word production. *Cognition*, *128*, 380–396.

Elman, J. L. (1993). Learning and development in neural networks: The importance of starting small. *Cognition*, *48*, 71–99.

Ferreira, V. S. (2008). Ambiguity, availability, and a division of labor for communicative success. *Psychology of Learning and Motivation*, *49*, 209–246.

Ferreira, V. S., Bock, K., Wilson, M. P., & Cohen, N. J. (2008). Memory for syntax despite amnesia. *Psychological Science*, *19*, 940–946.

Ferreira, V. S., & Dell, G. S. (2000). The effect of ambiguity and lexical availability on syntactic and lexical production. *Cognitive Psychology*, *40*, 296–340.

Fromkin, V. A. (1971). The non-anomalous nature of anomalous utterances. *Language*, *47*, 27–52.

Garrett, M. F. (1975). The analysis of sentence production. *Psychology of Learning and Motivation*, *9*, 133–177.

Goldrick, M., & Blumstein, S. (2006). Cascading activation from phonological planning to articulatory processes: Evidence from tongue twisters. *Language and Cognitive Processes*, *21*, 649–683.

Goldstein, L., Pouplier, M., Chen, L., Saltzman, E., & Byrd, D. (2007). Dynamic action units slip in speech production errors. *Cognition*, *103*, 386–412.

Gordon, J. K., & Dell, G. S. (2003). Learning to divide the labor: An account of deficits in light and heavy verb production. *Cognitive Science*, *27*, 1–40.

Hagoort, P. (2013). MUC (Memory, Unification, Control) and beyond. *Frontiers in Psychology*, *4*, 416.

Hartsuiker, R. J., & Kolk, H. H. J. (2001). Error monitoring in speech production: A computational test of the perceptual loop theory. *Cognitive Psychology, 42*, 113–157.

Howard, D., Nickels, L., Coltheart, M., & Cole-Virtue, J. (2006). Cumulative semantic inhibition in picture naming: Experimental and computational studies. *Cognition, 100*, 464–482.

Jaeger, T. F. (2006). *Redundancy and syntactic reduction in spontaneous speech*. Ph.D. dissertation. Stanford University.

Jaeger, T. F. (2010). Redundancy and reduction: Speakers manage syntactic information density. *Cognitive Psychology, 61*, 23–62.

Jaeger, T. F., & Ferreira, V. S. (2013). Seeking predictions from a predictive framework. *Behavioral and Brain Sciences, 36*, 359–360.

Jaeger, T. F., & Snider, N. E. (2013). Alignment as a consequence of expectation adaptation: Syntactic priming is affected by the prime's prediction error given both prior and recent experience. *Cognition, 127*, 57–83.

Janssen, N., & Barber, H. A. (2012). Phrase frequency effects in language production. *PLoS ONE, 7*, e33202.

Jescheniak, J. D., & Levelt, W. J. M. (1994). Word frequency effects in speech production: Retrieval of syntactic information and of phonological form. *Journal of Experimental Psychology: Learning, Memory, and Cognition, 20*, 824–843.

Jurafsky, D., Bell, A., Gregory, M., & Raymond, W. D. (2001). Probabilistic relations between words: Evidence from reduction in lexical production. *Typological Studies in Language, 45*, 229–254.

Kempen, G., & Hoenkamp, E. (1987). An incremental procedural grammar for sentence formulation. *Cognitive Science, 11*, 201–258.

Kittredge, A. K., Dell, G. S., Verkuilen, J., & Schwartz, M. F. (2008). Where is the effect of frequency in word production? Insights from aphasic picture-naming errors. *Cognitive Neuropsychology, 25*, 463–492.

Knobel, M., Finkbeiner, M., & Caramazza, A. (2008). The many places of frequency: Evidence for a novel locus of the frequency effect in word production. *Cognitive Neuropsychology, 25*, 256–286.

Lackner, J. R., & Tuller, B. H. (1979). Role of efference monitoring in the detection of self-produced speech errors. In W. E. Cooper, & E. C. T. Walker (Eds.), *Sentence processing: Psycholinguistic studies presented to Merrill Garrett*. Hillsdale, NJ: Erlbaum.

Lam, T. Q., & Watson, D. G. (2010). Repetition is easy: Why repeated referents have reduced prominence. *Memory & Cognition, 38*, 1137–1146.

Levelt, W. J. M. (1983). Monitoring and self-repair in speech. *Cognition, 14*, 41–104.

Levelt, W. J. M. (1989). *Speaking: From intention to articulation*. Cambridge, MA: MIT Press.

Levelt, W. J. M., Roelofs, A., & Meyer, A. S. (1999). A theory of lexical access in speech production. *Behavioral and Brain Sciences, 22*, 1–38.

Lieberman, P. (1963). Some effects of semantic and grammatical context on the production and perception of speech. *Language and Speech, 6*, 172–187.

Mahowald, K., Fedorenko, E., Piantadosi, S. T., & Gibson, E. (2013). Info/information theory: Speakers choose shorter words in predictive contexts. *Cognition, 126*, 313–318.

Mehl, M. R., Vazire, S., Ramirez-Esparza, N., Slatcher, R. B., & Pennebaker, J. W. (2007). Are women really more talkative than men? *Science, 317*, 82.

Meyer, A. S. (1991). The time course of phonological encoding in language production: Phonological encoding inside a syllable. *Journal of Memory and Language, 30*, 69–89.

Mitchell, D. B., & Brown, A. S. (1988). Persistent repetition priming in picture naming and its dissociation from recognition memory. *Journal of Experimental Psychology: Learning, Memory, and Cognition, 14*, 213–222.

Motley, M. T., Camden, C. T., & Baars, B. J. (1982). Covert formulation and editing of anomalies in speech production: Evidence from experimentally elicited slips of the tongue. *Journal of Verbal Learning and Verbal Behavior, 21*, 578–594.

Nozari, N., Dell, G. S., & Schwartz, M. F. (2011). Is comprehension necessary for error detection? A conflict-based account of monitoring in speech production. *Cognitive Psychology, 63*, 1–33.

Oppenheim, G. M., Dell, G. S., & Schwartz, M. F. (2010). The dark side of incremental learning: A model of cumulative semantic interference during lexical access in speech production. *Cognition, 114*, 227–252.

Pardo, J. S. (2006). On phonetic convergence during conversational interaction. *The Journal of the Acoustical Society of America, 119*, 2382–2393.

Qu, Q., Damian, M. F., & Kazanina, N. (2012). Sound-sized segments are significant for Mandarin speakers. *Proceedings of the National Academy of Sciences, 109*, 14265–14270.

Recasens, M., de Marneffe, M. C., & Potts, C. (2013). The life and death of discourse entities: Identifying singleton mentions. In *Proceedings of NAACL-HLT*, (pp. 627–633). Atlanta, GA.

Schnur, T. T., Schwartz, M. F., Brecher, A., & Hodgson, C. (2006). Semantic inference during blocked-cyclic naming: Evidence from aphasia. *Journal of Memory and Language, 54*, 199–227.

Severens, E., Kühn, S., Hartsuiker, R. J., & Brass, M. (2012). Functional mechanisms involved in the internal inhibition of taboo words. *Social, Cognitive, and Affective Neuroscience, 7*, 431–435.

Shannon, C. E. (1948). A mathematical theory of communication. *Bell System Technical Journal, 27*, 623–656.

Ueno, T., Saito, S., Rogers, T. T., & Lambon Ralph, M. A. (2011). Lichtheim 2: synthesizing aphasia and the neural basis of language in a neurocomputational model of the dual dorsal-ventral language pathways. *Neuron, 72*, 385–396.

Warker, J. A. (2013). Investigating the retention and time course for phonotactic constraint learning from production experience. *Journal of Experimental Psychology: Learning, Memory, and Cognition, 39*, 96–109.

Warker, J. A., & Dell, G. S. (2006). Speech errors reflect newly learned phonotactic constraints. *Journal of Experimental Psychology: Learning, Memory, and Cognition, 32*, 387–398.

Wells, R. (1951). Predicting slips of the tongue. *Yale Scientific Magazine, 3*, 9–30.

CHAPTER

19

Speech Motor Control from a Modern Control Theory Perspective

John F. Houde[1] *and Srikantan S. Nagarajan*[1,2]

[1]Department of Otolaryngology—Head and Neck Surgery, University of California, San Francisco, CA, USA;
[2]Department of Radiology and Biomedical Imaging, University of California, San Francisco, CA, USA

19.1 INTRODUCTION

Speech motor control is unique among motor behaviors in that it is a crucial part of the language system. It is the final neural processing step in speaking, where intended messages drive articulator movements that create sounds conveying those messages to a listener (Levelt, 1989). Many questions arise concerning this neural process we call speech motor control. What is its neural substrate? Is it qualitatively different from other motor control processes? Recently, research into other areas of motor control has benefited from a vigorous interplay between people who study the psychophysics and neurophysiology of motor control and engineers that develop mathematical approaches to the abstract problem of control. One of the key results of these collaborations has been the application of state feedback control (SFC) theory to modeling the role of the higher central nervous system (i.e., cortex, the cerebellum, thalamus, and basal ganglia—hereafter referred to as "the CNS") in motor control (Arbib, 1981; Guigon, Baraduc, & Desmurget, 2008b; Shadmehr & Krakauer, 2008; Todorov, 2004; Todorov & Jordan, 2002). SFC postulates that the CNS controls motor output by estimating the current state of the thing (e.g., arm) being controlled and by generating controls based on this estimated state. SFC has successfully predicted a great range of the phenomena seen in nonspeech motor control, but as yet it has not received attention in the speech motor control community. Here, we review some of the key characteristics of how sensory feedback appears to be used during speaking and what this says about the role of the CNS in the speech motor control process. Along the way, we discuss prior efforts to model this role, but ultimately we argue that such models can be seen as approximating characteristics best modeled by SFC. We conclude by presenting an SFC model of the role of the CNS in speech motor control and discuss its neural plausibility.

19.2 THE ROLE OF THE CNS IN PROCESSING SENSORY FEEDBACK DURING SPEAKING

It is not controversial that the CNS plays a role in speech motor output: cortex appears to be a main source of motor commands in speaking. In humans, the speech-relevant areas of motor cortex (M1) make direct connections with the motor neurons of the lips, tongue, and other speech articulators (Jürgens, 1982, 2002; Ludlow, 2004). Damage to these M1 areas causes mutism and dysarthria (Duffy, 2005; Jürgens, 2002). However, it is much less clear what the role of the CNS is in processing the sensory feedback from speaking. Sensory feedback, especially auditory feedback, is critically important for children learning to speak (Borden, Harris, & Raphael, 1994; Levitt, Stromberg, Smith, & Gold, 1980; Oller & Eilers, 1988; Osberger & McGarr, 1982; Ross & Giolas, 1978; Smith, 1975). However, once learned, the control of speech has the characteristics of being both responsive to, yet not completely dependent on, sensory feedback. In the absence of sensory feedback, speaking is only selectively disrupted. Somatosensory nerve block impacts only certain aspects of speech (e.g., lip rounding, fricative constrictions) and, even for these, the impact is not sufficient to prevent intelligible speech (Scott & Ringel, 1971). In postlingually deafened speakers, the control of pitch and loudness degrades rapidly after hearing loss, yet their speech will remain intelligible for decades

Neurobiology of Language. DOI: http://dx.doi.org/10.1016/B978-0-12-407794-2.00019-5

(Cowie & Douglas-Cowie, 1992; Lane et al., 1997). Normal speakers also produce intelligible speech with their hearing temporarily blocked by loud masking noise (Lane & Tranel, 1971; Lombard, 1911).

But this does not mean speaking is largely a feedforward control process that is unaffected by feedback. Delaying auditory feedback (DAF) by approximately a syllable's production time (100–200 ms) is very effective at disrupting speech (Fairbanks, 1954; Lee, 1950; Yates, 1963). Masking noise feedback causes increases in speech loudness (Lane & Tranel, 1971; Lombard, 1911), whereas amplifying feedback causes compensatory decreases in speech loudness (Chang-Yit, Pick, Herbert, & Siegel, 1975). Speakers compensate for mechanical perturbations of their articulators (Abbs & Gracco, 1984; Nasir & Ostry, 2006; Saltzman, Lofqvist, Kay, Kinsella-Shaw, & Rubin, 1998; Shaiman & Gracco, 2002; Tremblay, Shiller, & Ostry, 2003), and compensatory changes in speech production are seen when auditory feedback is altered in its pitch (Burnett, Freedland, Larson, & Hain, 1998; Elman, 1981; Hain et al., 2000; Jones & Munhall, 2000a; Larson, Altman, Liu, & Hain, 2008), loudness (Bauer, Mittal, Larson, & Hain, 2006; Heinks-Maldonado & Houde, 2005), formant frequencies (Houde & Jordan, 1998, 2002; Purcell & Munhall, 2006), or, in the case of fricative production, when the center of spectral energy is shifted (Shiller, Sato, Gracco, & Baum, 2007).

Taken together, such phenomena reveal a complex role for feedback in the control of speaking—a role not easily modeled as simple feedback control. Beyond this, however, there are also more basic difficulties with modeling the control of speech as being based on sensory feedback. In biological systems, sensory feedback is noisy due to environment noise and the stochastic firing properties of neurons (Kandel, Schwartz, & Jessell, 2000). Furthermore, when considering the role of the CNS in particular, an even more significant problem is that sensory feedback is delayed. There are several obvious reasons why sensory feedback to the CNS is delayed (e.g., by axon transmission times and synaptic delays; Kandel et al., 2000), but a less obvious reason involves the time needed to process raw sensory feedback into features useful in controlling speech. For example, in the auditory domain, there are several key features of the acoustic speech waveform that are important for discriminating between speech utterances. For some of these features, like pitch, spectral envelope, and formant frequencies, signal processing theory dictates that the accuracy in which the features are estimated from the speech waveform depends on the duration of the time window used to calculate them (Parsons, 1987). In practice, this means such features are estimated from the acoustic waveform using sliding time windows with lengths of approximately 30–100 ms in duration. Such integration-window-based feature estimation methods are slow to respond to changes in the speech waveform, and thus they will effectively introduce additional delays in the detection of such changes. Consistent with this theoretical account, studies show that response latencies of auditory areas to changes in higher-level auditory features can range from 30 ms to more than 100 ms (Cheung, Nagarajan, Schreiner, Bedenbaugh, & Wong, 2005; Godey, Atencio, Bonham, Schreiner, & Cheung, 2005; Heil, 2003). A particularly relevant example is the long (~100 ms) response latency of neurons in a recently discovered area of pitch-sensitive neurons in auditory cortex (Bendor & Wang, 2005). As a result, while auditory responses can be seen within 10–15 ms of a sound at the ear (Heil & Irvine, 1996; Lakatos et al., 2005), there are important reasons to suppose that the features needed for controlling speech are not available to the CNS until a significant time (~30–100 ms) after they are peripherally present. This is a problem for feedback control models, because direct feedback control based on delayed feedback is inherently unstable, particularly for fast movements (Franklin, Powell, & Emami-Naeini, 1991).

19.3 THE CNS AS A FEEDFORWARD SOURCE OF SPEECH MOTOR COMMANDS

Given these problems with controlling speech via sensory feedback control, it is not surprising that, in some models of speech motor control, the role of the CNS has been relegated to being a pure feedforward source, outputting desired trajectories for the lower motor system to follow (Ostry, Flanagan, Feldman, & Munhall, 1991, 1992; Payan & Perrier, 1997; Perrier, Ostry, & Laboissiere, 1996; Sanguineti, Laboissiere, & Ostry, 1998; Sanguineti, Laboissiere, & Payan, 1997). In these models, it is the lower motor system (e.g., brainstem and spinal cord) that implements feedback control and responds to feedback perturbations. The inspiration for these models comes from consideration of biomechanics and neurophysiology. A muscle has mechanical spring-like properties that naturally resist perturbations (Hill, 1925; Zajac, 1989), and these spring-like properties are further enhanced by somatosensory feedback to the motor neurons in the brainstem and spinal cord that control the muscle (e.g., for the jaw: Pearce, Miles, Thompson, & Nordstrom, 2003; see also the stretch reflex: Hulliger, 1984; Matthews, 1931; Merton, 1951). This local feedback control of the muscle makes it look, on first approximation, like a spring with an adjustable rest-length that can be set by control descending from the higher levels of the CNS (Asatryan & Feldman, 1965). The muscles affecting an articulator's position (e.g., the muscles controlling the position of the tongue tip) always come in opposing pairs—agonists and antagonists—whose contractions have opposite effects on articulator position. Thus, for any given set of muscle activations, an articulator will always come to rest at an *equilibrium point* where the muscle forces are balanced. In response to perturbations from its current equilibrium point, the articulator will naturally generate forces that return it to the equilibrium point, without any higher-level intervention. This characteristic was the inspiration for models of motor control based on equilibrium point control (EPC) (Bizzi, Accornero, Chapple, & Hogan, 1982; Feldman, 1986; Polit & Bizzi, 1979). EPC models postulate that to control an articulator's movement, the higher-level CNS need only provide the lower motor system with a sequence of desired equilibrium point to specify the trajectory of that articulator. The lower motor system handles responses to perturbations.

In speech, EPC models can explain the phenomenon of "undershoot," or "carryover," *coarticulation* (Lindblom, 1963). This can be seen when a speaker produces a vowel in a consonant-vowel-consonant (CVC) context: as the duration of the vowel segment is made shorter, the formants of the vowel do not reach (i.e., they undershoot) their normal steady-state values. This undershoot is easily explained by supposing that successive equilibrium points are generated faster than they can be achieved. In the case of a rapidly produced CVC syllable, undershoot of vowel formants would happen if, while it was still moving toward the equilibrium point for the vowel, the tongue was retargeted to the equilibrium point of the following consonant.

There are, however, several problems with the EPC account of the lower motor system being solely responsible for feedback control. First, although both somatosensory (Jürgens, 2002; Kandel et al., 2000) and auditory (Burnett et al., 1998; Jürgens, 2002) pathways make subcortical connections with descending motor pathways, the latencies of responses to somatosensory and auditory feedback perturbations (approximately 50–150 ms) are longer than would be expected for subcortical feedback loops (Abbs & Gracco, 1983). Instead, such response delays appear sufficiently long enough for neural signals to go to and come from cortex (Kandel et al., 2000). By themselves, such timing estimates do not prove involvement of cortex, but a study by Ito and Gomi using transcranial magnetic stimulation (TMS) gives further evidence (Ito, Kimura, & Gomi, 2005). The authors examined the facilitatory effect of applying a subthreshold TMS pulse to mouth motor cortex on two oral reflexes: the compensatory response by the upper lip to a jaw-lowering perturbation during the production of /ph/ (a soft version of /f/ in Japanese made only with the lips) and a response to upper lip stimulation know to be subcortically mediated called the perioral reflex. The TMS pulse was applied approximately 10 ms before the time of the reflex response (i.e., at the time motor cortex would be activated if it governed the response). The authors found motor TMS only facilitated the response to jaw perturbation during /ph/, implicating cortex involvement specifically in only the task-dependent perturbation response during speaking.

Perhaps a larger problem with ascribing feedback control to only subcortical levels is that responses to sensory feedback perturbations in speaking often look task-specific. For example, perturbation of the upper lip will induce compensatory movement of the lower lip, but only in the production of bilabials. The upper lip is not involved in the production of /f/ and perturbation of the upper lip before /f/ in /afa/ induces no lower lip response. However, the upper lip is involved in the production of /p/ and, here, perturbation of the upper lip before /p/ in /apa/ does induce compensatory movement of the lower lip (Shaiman & Gracco, 2002). Task-dependence is also seen in responses to auditory feedback. The production of vowels in stressed syllables appears to be more sensitive to immediate auditory feedback than vowels in unstressed syllables (Kalveram & Jancke, 1989; Natke, Grosser, & Kalveram, 2001; Natke & Kalveram, 2001), responses to pitch perturbations are modulated by how fast the subject is changing pitch (Larson, Burnett, Kiran, & Hain, 2000), and responses to loudness perturbations appear to be modulated by syllable emphasis (Liu, Zhang, Xu, & Larson, 2007). Such task-dependent perturbation responses cannot be simply explained with pure feedback control by setting stiffness levels (i.e., muscle impedance) for individual articulators (e.g., upper lip or lower lip) and instead suggest that, depending on the task (i.e., the particular speech target being produced), the higher-level CNS uses sensory feedback to couple the behavior of different articulators in ways that accomplish a higher-level goal (e.g., closing of the lip opening) (Bernstein, 1967; Kelso, Tuller, Vatikiotis-Bateson, & Fowler, 1984; Saltzman & Munhall, 1989).

There is also evidence that the CNS is sensitive to the dynamics of the articulators. In controlling fast movements, the CNS behaves as if it does anticipate that the articulators will have dynamical responses to its motor commands. For example, arm movement studies have shown that fast movements are characterized by a "three-phase" muscle activation sequence whereby an initial burst of activation of the agonist muscle accelerates the articulator quickly toward its target, followed by, at approximately mid-movement, a "breaking" burst of antagonist muscle activation that decelerates the articulator, causing it to come to rest near the target (followed, in turn, by a weaker agonist burst to further correct the articulator's position) (Hallett, Shahani, & Young, 1975; Shadmehr & Wise, 2005; Wachholder & Altenburger, 1926). Such activation patterns appear to take advantage of the momentum of the arm. When equilibrium points are determined for such muscle activations, they appear to follow complex trajectories, initially racing far ahead of the target position before finally converging back to it (Gomi & Kawato, 1996). Yet, in such cases, the actual trajectory of the arm is always a smooth path to the target that greatly differs from the complex equilibrium point trajectory. This mismatch suggests that even if the CNS were outputting "desired" articulatory trajectories to the lower motor system, it does so by taking into account dynamical responses to these trajectory requests, such that a fast smooth motion is achieved.

This ability of the CNS to take articulator dynamics into account can also be seen in speech production. A series of experiments has shown that speakers will learn to compensate for perturbations of jaw protrusion that are dependent on jaw velocity (Nasir & Ostry, 2008, 2009; Tremblay, Houle, & Ostry, 2008; Tremblay et al., 2003). In learning to compensate for such altered articulator dynamics, speakers show that they are formulating articulator movement commands that anticipate and cancel out the effects of those altered dynamics. Thus, the ability to anticipate articulator dynamics is not only a theoretically desirable property of a model of speech motor control but also a property required to account for real experimental results.

Taken together, these several lines of evidence suggest that, rather than simply instructing the lower motor system on what its goals are, the CNS instead likely plays an active role in responding to sensory information about deviations from task goals.

19.4 CURRENT MODELS OF THE ROLE OF THE CNS IN SPEECH MOTOR CONTROL

Current models of speech motor control can trace their lineage back to Fairbanks' early model. With the advent of cybernetic theory (Wiener, 1948) and the discovery of the effects of DAF soon after (Lee, 1950), it was natural that at conference in 1953, Fairbanks would propose a model of speech motor control based in large part on the principles of feedback control (Fairbanks, 1954). A key element of Fairbanks' model was a "comparator" that subtracted sensory feedback (including auditory feedback) from a target "input signal," creating an "error signal" that was used in the control of the vocal tract articulators. However, given the aforementioned phenomena concerning auditory feedback, it is not surprising that current models of speech motor control are significantly more complicated than simple feedback control models. Even Fairbanks appeared to hedge on proposing a model completely based on feedback control. In his model, the "error signal" output of his feedback control subsystem does not drive the vocal tract directly. Instead, it is first combined with the "input signal" (the output of a feedforward control subsystem) by a "mixer" element to create the "effective driving signal" that directly controls the vocal tract. This combination of feedback and feedforward control subsystems is similar in design to that of the current Directions into Velocities of Articulators (DIVA) model of speech motor control (Guenther, 1995; Guenther, Ghosh, & Tourville, 2006; Guenther, Hampson, & Johnson, 1998; Guenther & Vladusich, 2012), although the feedforward control subsystem in DIVA is implemented as an internal feedback loop, which we describe further.

Feedback control models can be considered the most extreme implementation of the efference copy hypothesis, where the motor-derived prediction functions as the target output, and comparison with this target/prediction results in a prediction error that directly drives the motor control output. In current speech motor control models, the efference-copy/feedback prediction process is still used to create a correction, but that correction does not directly generate output controls. Instead, it is a contributing factor in the generation of output controls. These models retain the concept of feedback control but put the feedback loop inside the CNS, where processing delays are minimal, with actual sensory feedback forming a slower, possibly delayed and intermittent, external loop that updates the internal feedback loop (Guenther & Vladusich, 2012; Hickok, Houde, & Rong, 2011; Houde & Nagarajan, 2011; Price, Crinion, & Macsweeney, 2011; Tian & Poeppel, 2010). It turn out that such models can be described as variations in the general theory of SFC, developed in the domain of modern control engineering theory, which is based on the concept of a dynamic state.

19.5 THE CONCEPT OF DYNAMICAL STATE

A key feature of these current models of speech motor control is their outer sensory processing loops are based on comparing incoming feedback with a prediction of that feedback. To make such sensory predictions, the CNS would ideally base them not on what the current articulatory target was, but instead on the actual articulatory commands currently being sent to the articulators (i.e., true efference copy of the descending motor commands output to the motor units of the articulators). However, without a model of how these motor commands affect articulator dynamics, accurate feedback predictions cannot be made because it is only through their effects on the dynamics of the articulators that motor commands affect articulator positions and velocities, and thus acoustic output and somatosensory feedback from the vocal tract.

But how can we model the effects of motor commands on articulator dynamics? To say that vocal tract articulators have "dynamics" is another way of saying that how they will move in the future and how they will react to applied controls are dependent on their immediate history (e.g., the direction they were last moving in). The past can only affect the future via the present; in engineering terms, the description of the present sufficient to predict how a system's past affects its future is called the dynamical *state* of the system. It is this concept of dynamical state that is basis for engineering models of systems and how they respond to applied controls.

Based on these ideas, Figure 19.1 illustrates how the problem of controlling speaking can be phrased in terms of the control of vocal tract state. This discrete time description represents a snapshot of the speech motor control process at time t, where the controls $\mathbf{u}_{t-1}$ formulated at the previous timestep $t-1$ have now been applied to the muscles of the vocal tract, changing its dynamic state to $\mathbf{x}_t$, which in turn results in the vocal tract outputting $\mathbf{y}_t$. In this process, $\mathbf{x}_t$ represents an instantaneous dynamical description of the vocal tract (e.g., positions and velocities of various parts of the tongue, lips, or jaw) sufficient to predict its future behavior and $\mathbf{vtdyn}(\mathbf{u}_{t-1}, \mathbf{x}_{t-1})$ expresses the physical processes (e.g., inertia) that dictate what next state $\mathbf{x}_t$ will result from controls $\mathbf{u}_{t-1}$ being applied to prior state $\mathbf{x}_{t-1}$. The next state $\mathbf{x}_t$ is also partly determined by random disturbances $\mathbf{w}_{t-1}$ (called state noise). A key part of this formulation is that $\mathbf{x}_t$ is not directly observable from sensory feedback. Instead, output function $\mathbf{vtout}(\mathbf{x}_t)$ represents all the physical and biophysical processes causing $\mathbf{x}_t$ to generate sensory consequences $\mathbf{y}_t$. $\mathbf{y}_t$ is also corrupted by noise $\mathbf{v}_t$ and delayed by $\boxed{z^{-N}}$, where $\mathbf{N}$ is a vector of time delays representing the time taken to neurally transmit each element of $\mathbf{y}_t$ to the higher CNS and process it into a control-useable form (e.g., into pitch, formant frequencies, tongue height). Furthermore, certain elements of $\mathbf{y}_t$ can be intermittently unavailable, as when auditory feedback is blocked by noise. Therefore, from this description, the control of vocal tract state can be summarized as follows: how can the higher CNS correctly formulate the next controls $\mathbf{u}_t$ to be applied to the vocal tract given access only to previously applied controls $\mathbf{u}_{t-1}$ and noisy, delayed, and possibly intermittent feedback $\mathbf{y}_{t-N}$?

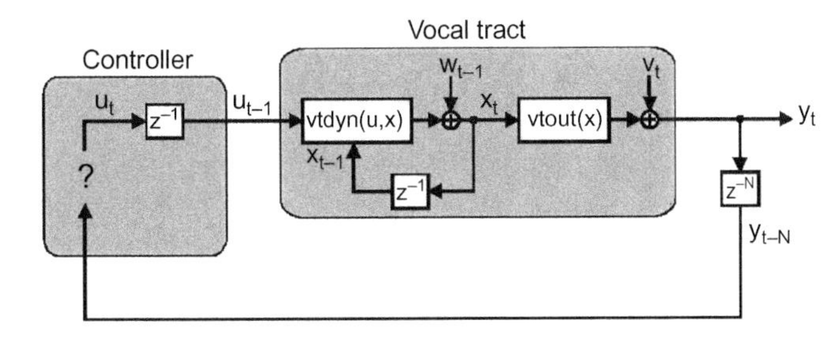

FIGURE 19.1 The control problem in speech motor control. The figure shows a snapshot at time t, when the vocal tract has produced output $\mathbf{y}_t$ in response to the previously applied control $\mathbf{u}_{t-1}$.

19.6 A MODEL OF SPEECH MOTOR CONTROL BASED ON STATE FEEDBACK

An approach to this problem is based on the following idealization shown in Figure 19.2: If the state $\mathbf{x}_t$ of the vocal tract is available to the CNS via immediate feedback, then the CNS could control vocal tract state directly via feedback control. For this reason, this control approach is referred to as SFC. However, as discussed, because $\mathbf{x}_t$ is not directly observable from any type of sensory feedback, and because the sensory feedback that comes to the higher CNS is both noisy and delayed, the scheme as shown is unrealizable. As a result, a fundamental principle of SFC is that control must instead be based on a running internal *estimate* of the state $\mathbf{x}_t$ (Jacobs, 1993). The first step toward getting this estimate is another idealization. Suppose, as shown in Figure 19.3, the higher CNS had an internal model of the vocal tract, $\widehat{\textbf{vocal tract}}$, which had accurate forward models of the dynamics $\widehat{\mathbf{vtdyn}}(\mathbf{u}_{t-1}, \hat{\mathbf{x}}_{t-1})$ and output function $\widehat{\mathbf{vtout}}(\hat{\mathbf{x}}_t)$ (i.e., its acoustics, auditory and somatosensory transformations) of the actual vocal tract. Such an internal model could mimic the response of the real vocal tract to applied controls and provide an estimate $\hat{\mathbf{x}}_t$ of the actual vocal tract state. In this situation, the controller could permanently ignore the feedback $\mathbf{y}_{t-N}$ of the actual vocal tract and perform ideal SFC $\mathbf{U}_t(\hat{\mathbf{x}})$ based only on $\hat{\mathbf{x}}_t$. The controls $\mathbf{u}_t$ thus generated would correctly control both $\widehat{\textbf{vocal tract}}$ and the actual vocal tract.

But this situation is still idealized. The vocal tract state $\mathbf{x}_t$ is subject to disturbances $\mathbf{w}_{t-1}$, and the forward

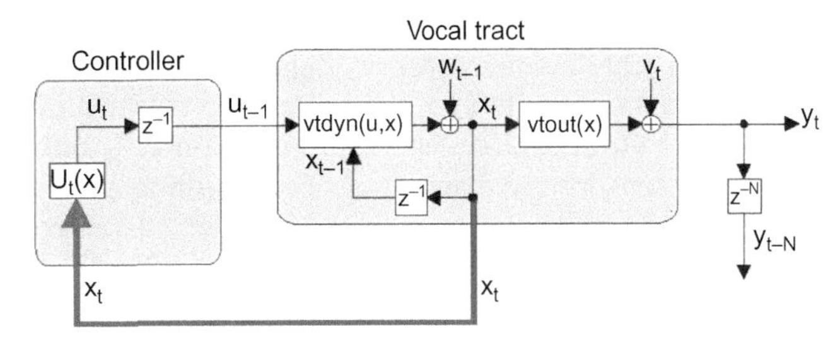

FIGURE 19.2 Ideal SFC. If the controller in the CNS had access to the full internal state $\mathbf{x}_t$ of the vocal tract system (red path), then it could ignore feedback $\mathbf{y}_{t-N}$ and formulate an SFC law $\mathbf{U}_t(\mathbf{x}_t)$ that would optimally guide the vocal tract articulators to produce the desired speech output $\mathbf{y}_t$. However, as discussed in the text, the internal vocal tract state $\mathbf{x}_t$ is, by definition, not directly available.

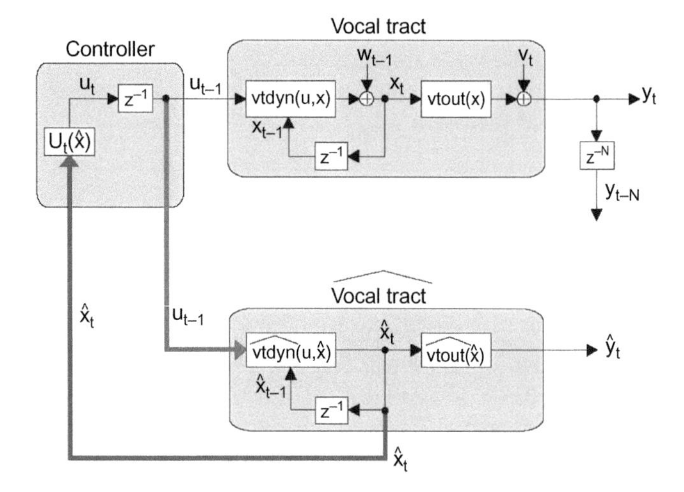

FIGURE 19.3 A more realizable model of SFC based on an estimate $\hat{x}_t$ of the true internal vocal tract state x_t. If the CNS had an internal model of the vocal tract $\widehat{\textbf{vocal tract}}$, comprising dynamics model $\widehat{\textbf{vtdyn}}(\mathbf{u}_{t-1}, \hat{x}_{t-1})$ and sensory feedback model $\widehat{\textbf{vtout}}(\hat{x}_t)$, then it could send efference copy (green path) of vocal tract controls $\mathbf{u}_{t-1}$ to the internal model, whose state $\hat{x}_t$ is accessible and could be used in place of x_t in the controller's feedback control law $\mathbf{U}_t(\hat{x})$ (red path). However, this scheme only works if $\hat{x}_t$ always closely tracks x_t, which is not a realistic assumption.

models $\widehat{\textbf{vtdyn}}(\mathbf{u}_{t-1}, \hat{\mathbf{x}}_{t-1})$ and $\widehat{\textbf{vtout}}(\hat{\mathbf{x}}_t)$ could never be assumed to be perfectly accurate. Furthermore, $\widehat{\textbf{vocal tract}}$ could not be assumed to start out in the same state as the actual vocal tract. Thus, without corrective help, $\hat{\mathbf{x}}_t$ will not, in general, track $\mathbf{x}_t$. Unfortunately, only noisy and delayed sensory feedback $\mathbf{y}_{t-N}$ is available to the controller, and $\mathbf{y}_{t-N}$ is not tightly correlated with the current vocal tract state $\mathbf{x}_t$. Nevertheless, because $\mathbf{y}_{t-N}$ is not completely uncorrelated with $\mathbf{x}_t$, it carries some information about $\mathbf{x}_t$ that can be used to correct $\hat{\mathbf{x}}_t$. Figure 19.4 shows how this can be done by augmenting the idealization shown in Figure 19.3 to include the following prediction/correction process. First, in the prediction (green) direction, efference copy of the previous vocal tract control $\mathbf{u}_{t-1}$ is input to forward dynamics model $\widehat{\textbf{vtdyn}}(\mathbf{u}_{t-1}, \hat{\mathbf{x}}_{t-1})$ to generate a prediction $\hat{\mathbf{x}}_{t|t-1}$ of the next vocal tract state. $\hat{\mathbf{x}}_{t|t-1}$ is then delayed by $\boxed{z^{-\hat{N}}}$ to match the actual sensory delays. The resulting delayed state estimate $\hat{\mathbf{x}}_{(t|t-1)-\hat{N}}$ is input to forward output model $\widehat{\textbf{vtout}}(\hat{\mathbf{x}}_t)$ to generate a prediction $\hat{\mathbf{y}}_{t-\hat{N}}$ of the expected sensory feedback $\mathbf{y}_{t-N}$. The resulting sensory feedback prediction error $\tilde{\mathbf{y}}_{t-\hat{N}} = \mathbf{y}_{t-N} - \hat{\mathbf{y}}_{t-\hat{N}}$ is a measure of how well $\hat{\mathbf{x}}_t$ is currently tracking $\mathbf{x}_t$ (note, for example, if $\hat{\mathbf{x}}_t$ was perfectly tracking $\mathbf{x}_t$, then $\tilde{\mathbf{y}}_{t-\hat{N}}$ would be approximately zero). Next, in the correction (red) direction, feedback prediction error $\tilde{\mathbf{y}}_{t-\hat{N}}$ is converted into state estimate correction $\hat{\mathbf{e}}_t$ by the function $\mathbf{K}_t(\tilde{\mathbf{y}})$. Finally, $\hat{\mathbf{e}}_t$ is added to the original next state prediction $\hat{\mathbf{x}}_{t|t-1}$ to derive the corrected state estimate $\hat{\mathbf{x}}_t$. By this process, therefore, an accurate estimate of the true vocal tract state $\mathbf{x}_t$ can be derived in a feasible way and used by the SFC law $\mathbf{U}_t(\hat{\mathbf{x}}_t)$ to determine the next controls $\mathbf{u}_t$ output to the vocal tract.

As Figure 19.4 indicates, the combination of $\widehat{\textbf{vocal tract}}$ plus this feedback-based correction process is called an *observer* (Jacobs, 1993; Stengel, 1994; Tin & Poon, 2005; Wolpert, 1997), which in this case, because it includes allowances for feedback delays, is also a variant of a *Smith Predictor* (Mehta & Schaal, 2002; Miall, Weir, Wolpert, & Stein, 1993; Smith, 1959). Within the observer, $\mathbf{K}_t(\tilde{\mathbf{y}})$ converts changes in feedback to changes in state. When it is optimally determined, $\mathbf{K}_t(\tilde{\mathbf{y}})$ is a feedback gain proportional to how correlated the feedback prediction error $\tilde{\mathbf{y}}_{t-\hat{N}}$ is with the state prediction error $(\mathbf{x}_t - \hat{\mathbf{x}}_{t|t-1})$. Thus, if $\tilde{\mathbf{y}}_{t-\hat{N}}$ is highly uncorrelated with $(\mathbf{x}_t - \hat{\mathbf{x}}_{t|t-1})$, as happens with large feedback delays or feedback being blocked, $\mathbf{K}_t(\tilde{\mathbf{y}})$ largely attenuates the influence of feedback prediction errors on correcting the current state estimate. When $\mathbf{K}_t(\tilde{\mathbf{y}})$ is so optimally determined, it is referred to as the *Kalman gain function* and the observer is referred to as a *Kalman filter* (Jacobs, 1993; Kalman, 1960; Stengel, 1994; Todorov, 2006). We also refer to $\mathbf{K}_t(\tilde{\mathbf{y}})$ as the *Kalman gain function* because we assume the speech motor control system would seek an optimal value for this function.

Therefore, SFC is the combination of a control law acting on a state estimate provided by an observer. This is a relatively new way to model speech motor control, but SFC models are well-known in other areas of motor control research. Interest in SFC models of motor control has a long history that can trace its roots all the way back to Nikolai Bernstein, who suggested that the CNS would need to take into account the current state of the body (both the nervous system and articulatory biomechanics) to know the sensory outcomes of motor commands it issued (Bernstein, 1967; Whiting, 1984). Since then, the problem of motor control has been formulated in state-space terms like those discussed (Arbib, 1981), and observer-based SFC models of reaching motor control have been advanced to explain how people optimize their movements (Guigon et al., 2008b; Shadmehr & Krakauer, 2008; Todorov, 2004; Todorov & Jordan, 2002).

19.7 SFC MODELS MOTOR ACTIONS AS AN OPTIMAL CONTROL PROCESS

Experiments show that people appear to move optimally, not just on average, but in each movement, making optimal responses to perturbations of their movements that take advantage of task constraints (Guigon, Baraduc, & Desmurget, 2008a; Guigon et al., 2008b; Izawa, Rane, Donchin, & Shadmehr, 2008; Kording, Tenenbaum, & Shadmehr, 2007; Li, Todorov, & Pan, 2004; Liu & Todorov, 2007; Shadmehr & Krakauer, 2008;

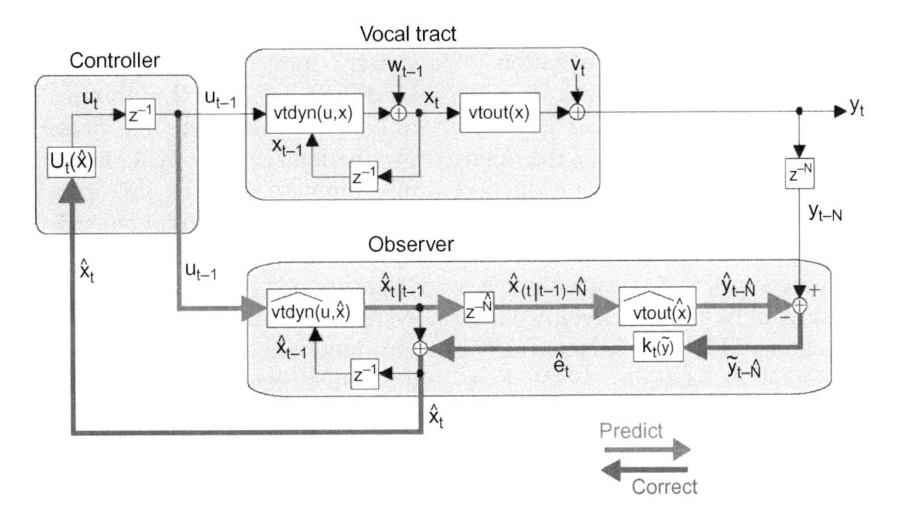

FIGURE 19.4 SFC model of speech motor control. The model is similar to that depicted in Figure 19.4 (i.e., the forward models $\widehat{\mathbf{vtdyn}}(\mathbf{u}_{t-1}, \hat{\mathbf{x}}_{t-1})$ and $\widehat{\mathbf{vtout}}(\hat{\mathbf{x}}_t)$ constitute the internal model of the vocal tract $\widehat{\textbf{vocal tract}}$ shown in Figure 19.4), but here sensory feedback $\mathbf{y}_{t-N}$ is used to keep the state estimate $\hat{\mathbf{x}}_t$ tracking the true vocal tract state $\mathbf{x}_t$. This is accomplished with a prediction/correction process in which, in the prediction (green) direction, efference copy of vocal motor commands $\mathbf{u}_{t-1}$ are passed through dynamics model $\widehat{\mathbf{vtdyn}}(\mathbf{u}_{t-1}, \hat{\mathbf{x}}_{t-1})$ to generate the next state prediction $\hat{\mathbf{x}}_{t|t-1}$, which is delayed by $\boxed{\mathbf{z}^{-\hat{N}}}$. $\boxed{\mathbf{z}^{-\hat{N}}}$ outputs the next state prediction $\hat{\mathbf{x}}_{(t|t-1)-\hat{N}}$ from $\hat{\mathbf{N}}$ seconds ago to match the sensory transduction delay of $\mathbf{N}$ seconds. $\hat{\mathbf{x}}_{(t|t-1)-\hat{N}}$ is passed through sensory feedback model $\widehat{\mathbf{vtout}}(\hat{\mathbf{x}}_t)$ to generate feedback prediction $\hat{\mathbf{y}}_{t-\hat{N}}$. Then, in the correction (red) direction, incoming sensory feedback $\mathbf{y}_{t-N}$ is compared with prediction $\hat{\mathbf{y}}_{t-\hat{N}}$, resulting in sensory feedback prediction error $\tilde{\mathbf{y}}_{t-\hat{N}}$. $\tilde{\mathbf{y}}_{t-\hat{N}}$ is converted by Kalman gain function $\mathbf{K}_t(\tilde{\mathbf{y}})$ into state correction $\hat{\mathbf{e}}_t$, which is added to $\hat{\mathbf{x}}_{t|t-1}$ to make corrected state estimate $\hat{\mathbf{x}}_t$. Finally, as in Figure 19.4, $\hat{\mathbf{x}}_t$ is used by SFC law $\mathbf{U}_t(\hat{\mathbf{x}}_t)$ in the controller to generate the controls u_t that will be applied at the next timestep to the vocal tract.

Todorov, 2007; Todorov & Jordan, 2002). Furthermore, people quickly reoptimize their movements as task requirements change. They flexibly discover and adaptively adjust control of different aspects of their movements (e.g., contact force, final velocity) to take advantage of any aspect of the task that lets them reduce control effort (e.g., reaching to a target they must stop in front of versus reaching to a target when they can use impact with the target to slow their reach; Liu & Todorov, 2007).

19.8 SPEAKING BEHAVES LIKE AN OPTIMAL CONTROL PROCESS

Like learning to reach, the process of learning to speak could be described as an *optimization* process, with the speaker attempting to learn articulatory controls that strike a balance between the idiosyncrasies of his/her own vocal tract and the sounds demanded by his/her language. The reason speakers can, in general, find such an optimal balance is that the speaking task, if defined only as "be understood by the listener," is underspecified with respect to the available articulatory degrees of freedom. This is especially true because of the many-to-one nature of the articulatory–acoustic relationship. For example, during the initial closure portion of /b/, the lips are closed and phonation has not begun; the position of the tongue at this time is therefore acoustically irrelevant and thus unconstrained by the task.

Many studies have shown that speakers appear to systematically take advantage of this underspecification in their articulation of speech. This is manifest in the trading relations seen in the production of /u/, with tongue position being made similar to the surrounding phonetic context and lip rounding accommodating the resulting tongue position (Perkell, Matthies, Svirsky, & Jordan, 1993). This same context-dependent choice of tongue position is also seen in the production of /r/, with the bunched articulation used in velar contexts (e.g., /grg/) and the retroflex articulation used in alveolar contexts (e.g., /drd/) (Espy-Wilson & Boyce, 1994; Guenther et al., 1998; Guenther et al., 1999; Zhou, 2008). These effects of phonetic context on the articulation of a speech sound are broadly referred to as coarticulation—a term introduced in the discussion of the undershoot as a phenomenon that the equilibrium point hypothesis can explain. However, coarticulation is often more complicated than this simple undershoot. In their running speech, speakers appear to anticipate the future need of currently noncritical articulators by moving them in advance to their ultimately needed positions: in the production of /ba/, the tongue is already moved to the position for /a/ during the production of /b/ (Farnetani & Recasens, 1999). How critical an articulator is for a given speech target is also often language-dependent; for example, in English, nasalization of vowels is not a critical perceptual distinction, leaving speakers free in the production of /am/ to begin in advance the nasalization needed for /m/ (i.e., the opening of the velo-pharyngeal port) during the production of /a/. However, in French, where nasalization of vowels is critical distinction, this advance nasalization of /m/ is not seen (Clumeck, 1976). Coarticulation has also been shown to vary widely across different native speakers of the same language (Kühnert & Nolan, 1999; Lubker & Gay, 1982; Perkell & Matthies, 1992). Even within the same speaker, the instruction to speak "clearly" reduces undershoot coarticulation (Moon & Lindblom, 1994), showing that speaking style controls some types of coarticulation (but not all types; see Matthies, Perrier, Perkell, & Zandipour, 2001).

These observations suggest that coarticulation is partly a learned phenomenon, but what exactly is learned is a matter of debate that distinguishes theories of coarticulation. One theory explains coarticulation purely in terms of "phonological" rules (i.e., rules based on the assumption that speech sounds are stored in memory as groups of discreet valued features) (Chomsky & Halle, 1968). For example, the representation of the /a/ includes the binary tongue features [+high] and [+back], whereas these features are left unspecified in /b/. In the look-ahead model of coarticulation, there is a feature spreading process that considers the full utterance to be spoken and that fills all unspecified features with any values they take on in the future (Henke, 1966). Thus, in the case of /ba/, the unspecified [high] and [back] features would be set to [+high] and [+back], based on looking ahead to the features specified for /a/. Unfortunately, not all coarticulation phenomena can be accounted for as an all-or-nothing feature spreading to unspecified features (remember the example of partial undershoot coarticulation described), and rule-based theories have been expanded to include learning of continuously variable, target-specific coarticulation resistance values (Bladon & Al-Bamerni, 1976). Another attempt to explain the continuous nature of coarticulation dismisses the idea of discreet feature targets in speech, instead postulating speakers learn whole trajectories (timecourses) for different features (e.g., lip opening or tongue-tip height) called gestures; coarticulation then naturally results when these gestures overlap in time (Browman & Goldstein, 1986; Fowler & Saltzman, 1993). Unfortunately, not all coarticulation can be modeled as the linear overlap of such feature time courses, necessitating once again the supposition that speakers learn a resistance to coarticulation (in this case, called blending strength) for different speech targets.

An alternative to explaining coarticulation with such explicit rules is to model it as resulting from an optimization process, like that postulated for non-speech movements. This was first suggested by Lindblom in 1983 (Lindblom, 1983), and later more fully elaborated in his "H&H" theory of speech production (Lindblom, 1990). In it, Lindblom explains phenomena observed in speech production as variations between "hyperspeech" (speech determined by demanding constraints on acoustic output) and "hypospeech" (speech more determined by production system [e.g., minimal effort] constraints). Acoustic output demands are determined by two things: (i) acoustic distinctiveness (i.e., how confusable a given speech sound is with nearest neighbors in "acoustic space") and (ii) how easily the listener can predict the next sound to be produced based on any number of sources of information the listener has available (e.g., semantic, linguistic, or phonetic contextual constraints). These factors can be approximately summarized by the more general constraint, mentioned previously, that what the speaker says should be understood by the listener. In this way, the complexities of coarticulation are explained as a "tug-of-war" between acoustic output and production system constraints, with coarticulated speech resulting when acoustic distinctiveness constraints are sufficiently lax that production system constraints can determine a minimal effort articulation. The acoustic output constraint accounts for the language-dependent nature of coarticulation, whereas production system constraint accounts for variations in coarticulation across speakers, because what counts as "minimal effort" depends on an individual speaker's vocal tract geometry and musculature. A variant of this idea that makes the listener-oriented acoustic distinctiveness constraints more explicit is Keating's window theory of coarticulation (Keating, 1990). The theory postulates that speech targets are not single-valued, but instead are specified as windows—permissible ranges for different speech features, where these permissible feature ranges are learned by a speaker from listening to other speakers of the language. Coarticulation happens because articulatory trajectory planning in speaking is a process of satisfying a language constraint that features bounds for speech targets must be respected and a minimal effort constraint.

Historically, the problem with explaining the control of speaking as resulting from an optimization process (i.e., that speaking is an *optimal control* process) is that, by itself, this is only a descriptive theory; qualitatively, speakers do behave as if they are working to minimize different movement cost terms, but this does not explain how this minimization is accomplished. This incompleteness of the theory makes it difficult to use it as a model of the role of the CNS in speaking. As a result, for example, in early versions of the DIVA model, one of the best current attempts at providing a mechanistic account of speech motor control, the optimization process was implemented by adopting a specific version of Keating's window theory; the phonetic feature ranges on speech targets provide an explicit rule for specifying how targets are achieved (i.e., the articulatory trajectory must pass within the target's feature ranges), and the minimal effort constraint is replaced with an explicit rule that says that minimal distance trajectories in phonetic feature space will be followed between successive phonetic targets (Guenther, 1995).

Without resorting to such explicit rules, however, the more general question about speech motor control remains unanswered: just how could the CNS choose the next articulatory control to be output in ongoing speech, such that, ultimately, some overall movement cost constraints are satisfied? How does a distal goal, whose achievement is only known after a word is produced, guide the selection of the next articulatory control at a given point in the ongoing production of the word? Optimal control theory provides a solution: the algorithms at the heart of the theory provide mechanisms for translating an overall movement goal into moment-by-moment controls (Stengel, 1994). In this framework, overall movement goals are expressed as cost functions to minimize (Todorov, 2004, 2006; Todorov & Jordan, 2002; Scott, 2004). These cost functions are a composite of terms reflecting the competing constraints governing movements; for example, one term (accuracy) could express the constraint that the listener should understand what was said (e.g., it could be obtained by an evaluation of the probability that the listener confuses what was said with a different utterance during the state corresponding to movement finished). Another term (effort) could express the fact that actions incur a metabolic cost depending on the forces the articulator muscles are commanded to generate (e.g., cost as the sum of all force magnitudes over the whole movement). The total cost is then a weighting of these cost terms, where the weighting is determined by the current task (e.g., high weighting of accuracy and low weighting of effort for clear speech, and low weighting of accuracy and high weighting of effort for casual speech).

To find the control law that minimizes a given cost function, the concept of state is crucial. Intuitively, if you knew exactly how the system being controlled would respond to your commands (i.e., if you knew its state), you could choose one that minimized the need for future corrective commands and thus minimize control efforts. This intuitive idea is the principle behind the algorithms that determine optimal controls. Perhaps the most understandable versions of these algorithms are those based on dynamic programming and reinforcement learning, in which each state has a cost-to-go, which is the movement cost incurred if only optimal control actions are taken after that state. Control actions also have a cost (as discussed) and, in one optimal control algorithm (dynamic programming), the optimal next control action for the current state is chosen to minimize the cost of that action plus the cost-to-go of the state it leads to (Bellman, 1957; Bertsekas, 2000). That minimal cost also becomes the new cost-to-go for the current state. Thus, over repeated utterance variations, costs-to-go for later states are propagated backward to earlier states, with the cost-to-go for the end state defined as the probability that the listener will misunderstand what was said. And, in this process of back-propagating costs-to-go, an optimal next control action is chosen for every state visited. In this way, the complete control law mapping states to output controls is eventually learned.

This process for learning a control law is of particular interest because neurophysiological processes appear to be in the CNS that mimics it. The basal ganglia (BG) in particular appear to represent several of the needed quantities, which is significant because the BG are thought to be involved in action selection. Dopaminergic neurons have been implicated in both reward prediction (Hollerman & Schultz, 1998; Schultz, Dayan, & Montague, 1997) and the detection of novel sensory outcomes of actions (Redgrave & Gurney, 2006). Interestingly, these neurons also display a back-propagation of their responses: they initially respond vigorously to delivery of a reward but soon habituate and instead respond to earlier sensory cues that predict the reward (Schultz, 1998). If it can be considered rewarding to improve the outcome of a movement (i.e., minimize its cost), then the behavior of dopaminergic neurons suggests the possibility of a back-propagation of state cost-to-go values would be a way for the CNS to learn control laws (Daw & Doya, 2006; Doya, 2000). Several studies have shown that neurons in the BG striatum appear to represent and retain expected reward returns for different possible future actions (Samejima, Ueda, Doya, & Kimura, 2005; Wang, Miura, & Uchida, 2013)—characteristics that are well-suited for representing costs-to-go of different states. In addition, other studies have shown that GPi output appears to represent movement effort (Desmurget, Grafton, Vindras, Grea, & Turner, 2003; Grafton, Vindras, Grea, & Turner, 2004; Turner, Desmurget, Grethe, Crutcher, & Grafton, 2003). And although many of the detailed studies of the role of BG in action learning have been done at a relatively high level of discreet action choice, the BG are also likely involved in the learning of new "lower level" sensorimotor skills. The subthalamic nucleus (STN) has been shown to react more strongly to more successful movement outcomes in the production of simple movements (Brown et al., 2006). Taken together, there is ample evidence to implicate the BG in the learning of state-based control laws, and studies have modeled is role as a "critic" that uses cost-to-go values to learn movement control laws (Barto, 1995; Berthier, Rosenstein, & Barto, 2005).

19.9 SFC EXPLAINS THE TASK-SPECIFIC ROLE OF THE CNS IN SPEECH FEEDBACK PROCESSING

Besides providing a mechanistic explanation for how optimal control laws could be learned, the SFC framework also provides an explanation for how and why the CNS would process sensory feedback during ongoing speaking. This is because, in its most general form, the process of estimating the current state of the system being controlled relies on more than just tracking the sequence of controls sent to the system. Crucially, it also relies on sensory feedback to correct errors in the state estimate, as is described here. This full state estimation process not only serves as a model of the role of CNS in feedback processing but also explains how task-specific responses to feedback perturbations would occur: In SFC, such perturbations cause corrections to the current state estimate, and the corrected state, if it has been visited before, has a task-specific optimal control law associated with it. If the state has not been visited before, then the process updating the cost-to-go for that state will begin the process of learning a task-optimal control response in that state.

In this way, a state estimation process that includes sensory feedback explains how task-specific responses to feedback perturbations could be learned without recourse to assuming speech is perceived in terms of certain specialized features. This is important because experiments that test whether speakers use task-relevant feature representations often have mixed results, with some speakers behaving as if they use a certain representation and others behaving as if they do not. To return to an earlier example, it is often reported that speakers exhibit trading relations in the production of /u/, with variations in tongue height being compensated by covariations in lip extension (and vice versa) such that an acoustic representation of their /u/ production (i.e., its formant pattern) is preserved. And when such covariation has been looked for in experiments, it is observed in many speakers (Perkell et al., 1993), consistent with their use of an acoustic representation to constrain their /u/ productions. Critically, however, it is not observed in other speakers. Related experiments have looked for acoustic /u/ representations by examining how speakers produce /u/ when required to hold tubes at their lips. These tubes function as artificial perturbations of a speaker's lip extension, requiring compensatory adjustment of other articulators like the tongue to maintain the original /u/ formant pattern. Some speakers compensated for the lip tubes, suggesting they worked to maintain an acoustic /u/ representation, whereas others did not (Savariaux, Perrier, & Orliaguet, 1995). Because these other speakers nevertheless had normal speech and hearing, what explains their results? It could be that unknown experimental factors confounded the results for these speakers, but another distinct possibility is that there are many viable solutions to the speech motor control problem, and that different speakers learn different solutions. Thus, it may be that the control solution based on an acoustic representation of /u/ is only discovered and used by some speakers. An optimal SFC model of speaking can explain such variability across speakers because it specifies only the way that speakers learn task-dependent perturbation responses rather than the specific representations of the task that different speakers learn.

19.10 IS SFC NEURALLY PLAUSIBLE?

For speech, the SFC model suggests not only that auditory processing is used by the CNS for comprehension during listening, but also that the CNS uses auditory information in a distinctly different way during speech production; it is compared with a prediction derived from efference copy of motor output, with the resulting prediction error used to keep an internal model tracking the state of the vocal tract. There are a number of lines of evidence supporting the neural plausibility of this second, production-specific mode of sensory processing. First, even in other primates, there appear to be at least two distinct pathways, or streams, of auditory processing. The concept of multiple sensory processing streams in both the auditory (Deutsch & Roll, 1976; Evans & Nelson, 1973; Poljak, 1926) and visual (Held, 1968; Ingle, 1967) systems of the brain has been around for decades, but the idea gained most attention when it was advanced as an organizational principle of the cortical regions involved in visual information processing. A dorsal "where" stream leading to parietal cortex that was concerned with object location and a ventral "what" stream leading to the temporal pole was concerned with object recognition were hypothesized (Mishkin, Ungerleider, & Macko, 1983; Ungerleider & Mishkin, 1982). Subsequently, studies of the auditory system found a match to this visual system organization (Romanski et al., 1999). Neurons responding to auditory source location were found in a dorsal pathway leading up to parietal cortex, and neurons responding to auditory source type were found in a ventral pathway leading down toward the temporal pole (Rauschecker & Tian, 2000). More recent evidence, however, has refined the view of the dorsal stream's task to be one of sensorimotor integration. The dorsal visual stream was found to be closely linked with non-speech motor control systems (e.g., reaching, head, and eye movement control) (Andersen, 1997; Rizzolatti, Fogassi, & Gallese, 1997) while in humans, the dorsal auditory stream was found to be closely linked with the vocal motor control system. In particular, a variety of studies have implicated the posterior superior temporal gyrus (STG) (Zheng, Munhall, & Johnsrude, 2010) and the superior parietal temporal area (Spt) (Buchsbaum, Hickok, & Humphries, 2001; Hickok, Buchsbaum, Humphries, & Muftuler, 2003) as serving auditory feedback processing specifically related to speech production. Consistent with this, studies of stroke victims have shown a double dissociation between ability to perform discreet production-related perceptual judgments and ability to understand continuous speech that depends on lesion location (dorsal and ventral stream lesions, respectively) (Baker, Blumstein, & Goodglass, 1981; Miceli, Gainotti, Caltagirone, & Masullo, 1980). This has led to refined looped and "dual stream" models of speech processing (Hickok et al., 2011; Hickok & Poeppel, 2007; Rauschecker & Scott, 2009), with a ventral stream serving speech comprehension and a dorsal stream serving feedback processing related to speaking. This two-stream model is in fact a close match with that originally proposed by Wernicke more than 100 years earlier (Wernicke, 1874/1977).

When the production-oriented auditory processing of the dorsal stream is disrupted, a number of speech sensorimotor disorders appear to result (Hickok et al., 2011). Conduction aphasia is a neurological condition resulting from stroke in which production and comprehension of speech are preserved but the ability to repeat speech sound sequences just heard is impaired (Geschwind, 1965). Conduction aphasia appears to result from damage to area Spt in the dorsal auditory processing stream (Buchsbaum et al., 2011). Consistent with this, the impairment is particularly apparent in the task of repeating nonsense speech sounds, because when the sound sequences do not form meaningful words, the intact speech comprehension system (the ventral stream) cannot aid in remembering what was heard. More speculatively, stuttering may also result from impairments in auditory feedback processing in the dorsal stream. It is well-known that altering auditory feedback (e.g., altering pitch (Howell, El-Yaniv, & Powell, 1987), masking feedback with noise (Maraist & Hutton, 1957), and DAF (Soderberg, 1968)) can make many persons who stutter speak fluently. Evidence for dorsal stream involvement in these fluency enhancements comes from a study relating DAF-induced fluency to structural MRIs of the brains of persons who stutter (Foundas et al., 2004). The planum temporale (PT) is an area of temporal cortex encompassing dorsal stream areas like Spt, and the study found that right PT was aberrantly larger than left PT in those stutterers whose fluency was enhanced by DAF. Several other anatomical studies have also implicated dorsal stream dysfunction in stuttering, including studies showing impaired white matter connectivity in this region (Cykowski, Fox, Ingham, Ingham, & Robin, 2010) as well as aberrant gyrification patterns (Foundas, Bollich, Corey, Hurley, & Heilman, 2001).

19.11 SFC ACCOUNTS FOR EFFERENCE COPY PHENOMENA

There are a number of studies that have found evidence that production-specific feedback processing involves comparison of incoming feedback with a feedback prediction derived from motor efference copy. Nonspeech evidence for this is seen when a robot creates delay between the tickle action subjects produce and when they feel it on their own hand (Blakemore, Wolpert, & Frith, 1998, 1999, 2000). With increasing delay, subjects report a more ticklish sensation, as expected if the delay created mismatch between a sensory prediction derived from the tickle action and the actual somatosensory feedback. By using different neuroimaging techniques, an analogous effect can be seen in speech production; the response of a subject's auditory cortices to his/her own self-produced speech is significantly smaller than their response to similar but externally produced speech (e.g., tape playback of the subject's previous self-productions). This effect, which we call speaking-induced suppression (SIS), has been seen using positron emission tomography (PET) (Hirano et al., 1996; Hirano, Kojima et al., 1997; Hirano, Naito et al., 1997), electroencephalography (EEG) (Ford et al., 2001; Ford & Mathalon, 2004), and magnetoencephalography (MEG) (Curio, Neuloh, Numminen, Jousmaki, & Hari, 2000; Heinks-Maldonado, Nagarajan, & Houde, 2006; Houde, Nagarajan, Sekihara, & Merzenich, 2002; Numminen & Curio, 1999; Numminen, Salmelin, & Hari, 1999; Ventura, Nagarajan, & Houde, 2009). An analog of the SIS effect has also been seen in nonhuman primates (Eliades & Wang, 2003, 2005, 2008). Our own MEG experiments have shown that the SIS effect is only minimally explained by a general suppression of auditory cortex during speaking and that this suppression is not happening in the more peripheral parts of the CNS (Houde et al., 2002). We have also shown that the observed suppression goes away if the subject's feedback is altered to mismatch his/her expectations (Heinks-Maldonado et al., 2006; Houde et al., 2002), as is consistent with some of the PET study findings. Finally, if SIS depends on a precise match between feedback and prediction, then precise time alignment of prediction with feedback would be critical for complex rapidly changing productions (e.g., rapidly speaking "ah-ah-ah"), and less critical for slow or static productions (e.g., speaking "ah"). Assuming a given level of time alignment inaccuracy, the prediction/feedback match should therefore be better (and SIS stronger) for slower, less dynamic productions, which is what we found in a recent study (Ventura et al., 2009).

By itself, evidence of feedback being compared with a prediction derived from efference copy implies the existence of predictive forward models within the CNS, but another line of evidence for forward models comes from sensorimotor adaptation experiments (Ghahramani, Wolpert, & Jordan, 1996; Wolpert & Ghahramani, 2000; Wolpert, Ghahramani, & Jordan, 1995). Such experiments have been conducted with speech production, where subjects are shown to alter and then retain compensatory production changes in response to extended exposure to artificially altered audio feedback (Houde & Jordan, 1997, 1998, 2002; Jones & Munhall, 2000a, 2000b, 2002, 2003, 2005; Jones, Munhall, & Vatikiotis-Bateson, 1998; Purcell & Munhall, 2006; Shiller, Sato, Gracco, & Baum, 2009; Villacorta, Perkell, & Guenther, 2007) or altered somatosensory feedback (Nasir & Ostry, 2006, 2008, 2009; Tremblay et al., 2003; Tremblay et al., 2008). For example, in the original speech sensorimotor adaptation experiment, subjects produced the vowel /ɛ/ (as in "head"), first hearing normal audio feedback and then hearing their formants shifted toward /i/ (as in "heed"). Over repeated productions while hearing the altered feedback, subjects gradually shifted their productions of /ɛ/ in the opposite direction (i.e., they shifted their produced formants toward /ɑ/, as in "hot"). This had the effect of making the altered feedback sound more like /ɛ/ again. These changes in the production of /ɛ/ were retained even when feedback was subsequently blocked by noise (Houde & Jordan, 1997, 1998, 2002). The retained production changes are consistent with the existence of a forward model making feedback predictions that are modified by experience. In addition to providing evidence for forward models, such adaptation experiments also allow investigation of the organization of forward models in the speech production system. By examining how compensation trained in the production of one phonetic task (e.g., the production of /eh/) generalizes to another untrained phonetic task (e.g., the production of /ah/), such experiments can determine if there are shared representations like forward models used in the control of both tasks. Some of these experiments have found generalization of adaptation across speech tasks (Houde & Jordan, 1997, 1998; Jones & Munhall, 2005), but other experiments have not found such generalization (Pile, Dajani, Purcell, & Munhall, 2007; Tremblay et al., 2008), suggesting that, in many cases, forward models used in the control of different speech tasks are perhaps not shared across tasks.

19.12 NEURAL SUBSTRATE OF THE SFC MODEL

Based partly on the discussion here, Figure 19.5 suggests a putative neural substrate for the SFC model. Basic neuroanatomical facts dictate the neural substrates on both ends of the SFC prediction/correction processing loop. On one end of the loop, motor cortex (M1) is the likely area where the feedback control law $\mathbf{U}_t(\hat{\mathbf{x}}_t)$ generates neuromuscular controls applied to the vocal tract. Motor cortex is the main source of motor fibers of the pyramidal tract, which synapse directly with motor neurons in the brainstem and spinal cord and enable fine motor movements (Kandel et al., 2000). As mentioned, damage to the vocal tract areas of motor cortex often results in mutism (Duffy, 2005; Jürgens, 2002). On the other end of the loop, auditory and somatosensory information first reaches the higher CNS in the primary auditory (A1) and somatosensory (S1) cortices, respectively (Kandel et al., 2000). Based on our SIS studies, we hypothesize this end of the loop is where the operation comparing the feedback prediction with incoming feedback occurs. Between these endpoints, the model also predicts the need for an additional area that mediates the prediction (green) and correction (red) processes running between motor and the sensory cortices. The premotor cortices are ideally placed for such an intermediary role: premotor cortex is both bidirectionally well-connected to motor cortex (Kandel et al., 2000) and, via the arcuate and longitudinal fasiculi (Glasser & Rilling, 2008; Schmahmann et al., 2007; Upadhyay, Hallock, Ducros, Kim, & Ronen, 2008), bidirectionally connected to the higher order somatosensory (S2/inferior parietal lobe [IPL]) and auditory (Spt/PT) cortices, respectively. In this way, the key parts of the SFC model are a good fit for a known network of sensorimotor areas that are, in turn, well-placed to receive task-dependent, modulatory connections (blue dashed arrows in Figure 19.5) from other frontal areas.

What evidence is there for premotor cortex playing such an intermediary role in speech production? First, reciprocal connections with sensory areas suggest the possibility that premotor cortex could also be active during passive listening to speech, and this appears to be the case. Wilson et al. found the superior ventral premotor area (svPMC), bilaterally, was activated by both listening to and speaking meaningless syllables, but not by listening to nonspeech sounds (Wilson, Saygin, Sereno, & Iacoboni, 2004). In a follow-up study, Wilson et al. found that this area, bilaterally, showed greater activation when subjects heard non-native speech sounds than they did when they heard native sounds. In this same study, auditory areas were also activated more for speech sounds rated least producible, and that svPMC

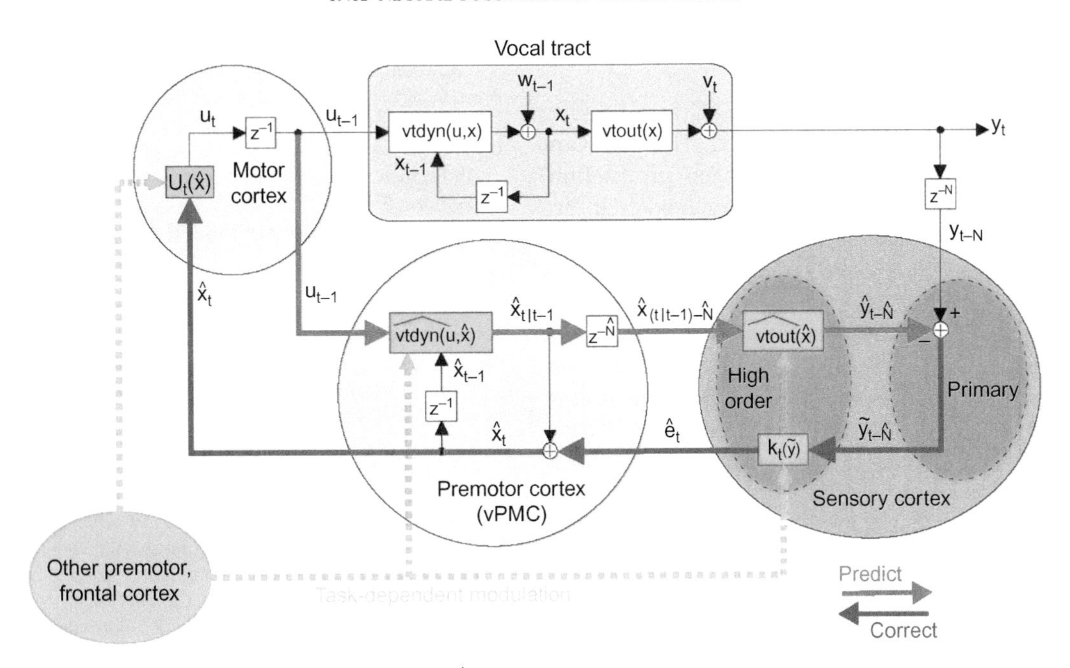

FIGURE 19.5 SFC model of speech motor control with putative neural substrate. The figure depicts the same operations as those shown in Figure 19.5, but with suggested cortical locations of the operations (motor areas are in yellow, sensory areas are in pink). The current model is largely doubtful regarding hemispheric specialization for these operations. Also, for diagrammatic simplicity, the operations in the auditory and somatosensory cortices are depicted in the single area marked "sensory cortex," with the understanding that it represents analogous operations occurring in both of these sensory cortices, that is, the delayed state estimate $\hat{x}_{(t|t-1)-\hat{N}}$ is sent to both high-order somatosensory and auditory cortex, each with separate feedback prediction modules $\widehat{\mathbf{vtout}}(\hat{x}_t)$ for predicting auditory feedback in high-order auditory cortex and $\widehat{\mathbf{vtout}}(\hat{x}_t)$ for predicting somatosensory feedback in high-order somatosensory cortex. The feedback prediction errors $\tilde{y}_{t-\hat{N}}$ generated in auditory and somatosensory cortex are converted into separate state corrections $\hat{e}_t$ based on auditory and somatosensory feedback by auditory and somatosensory Kalman gain functions $\mathbf{K}_t(\tilde{y})$ in the high-order auditory and somatosensory cortices, respectively. The auditory-based and somatosensory-based state corrections are then added to $\hat{x}_{t|t-1}$ in premotor cortex to make the next state estimate $\hat{x}_t$. Finally, the key operations depicted in blue are all postulated to be modulated by the current speech task goals (e.g., what speech sound is currently meant to be produced) that are expressed in other areas of frontal cortex.

was functionally connected to these auditory areas during listening (Wilson & Iacoboni, 2006). This activation of premotor cortex when speech is heard has also been seen in other functional imaging studies (Skipper, Nusbaum, & Small, 2005) and studies based on TMS (Watkins & Paus, 2004).

Second, altering sensory feedback during speech production should create feedback prediction errors in sensory cortices, increasing activations in these areas, and the resulting state estimate corrections should be passed back to premotor cortex, increasing its activation as well. A study that tested this prediction was performed by Tourville et al., who used fMRI to examine how cortical activations changed when subjects spoke with their auditory feedback altered (Tourville, Reilly, & Guenther, 2008). In the study, subjects spoke simple CVC words with the frequency of first formant occasionally altered in their audio feedback of some of their productions. When they looked for areas more active in altered feedback versus nonaltered trials, Tourville et al. found auditory areas (pSTG, including Spt in both hemispheres), and they also found areas in the right frontal cortex: a motor area (vMC), a premotor area (vPMC), and an area (IFt) in the inferior frontal gyrus, pars triangularis (Broca's) region. When they looked at the functional connectivity of these right frontal areas, they found that the presence of the altered feedback significantly increased the functional connectivity only of the left and right auditory areas, as well as the functional connectivity of these auditory areas with vPMC and IFt. The result suggests that the auditory feedback correction information from higher auditory areas has a bigger effect on premotor/pars triangularis regions than motor cortex regions, which is consistent with our SFC model if we expand the neural substrate of our state estimation process beyond premotor cortex to also include Broca's area. The results of Tourville et al. are partly confirmed by another fMRI study. Toyomura et al. had subjects continuously phonate a vowel and, in some trials, the pitch of the subjects' audio feedback was briefly perturbed higher or lower by two semitones (Toyomura et al., 2007). In examining the contrast between perturbed and unperturbed trials, Toyomura et al. found premotor activation in the left hemisphere and a number of activations in the right hemisphere, including auditory cortex (STG) and frontal area BA9, which is near the IFt activation found by Tourville et al.

19.13 CONCLUSION

In this review, the applicability of SFC to modeling speech motor control has been explored. The phenomena related to the role of CNS in speech production, especially its role in processing sensory feedback, are complex and suggest that speech motor control is not an example of pure feedback control or feedforward control. The task-specificity of responses to feedback perturbations in speech further argues that feedback control is not only a function of the lower motor system but also one in which CNS plays an active role in the online processing of sensory feedback during speaking. Current models of this role are described as variations of concept of SFC from engineering control theory. Thus, SFC is put forth as an appropriate and neurally plausible model of how the CNS processes feedback and controls the vocal tract.

References

Abbs, J. H., & Gracco, V. L. (1983). Sensorimotor actions in the control of multi-movement speech gestures. *Trends in Neurosciences, 6*, 391.

Abbs, J. H., & Gracco, V. L. (1984). Control of complex motor gestures: Orofacial muscle responses to load perturbations of lip during speech. *Journal of neurophysiology, 51*(4), 705–723.

Andersen, R. A. (1997). Multimodal integration for the representation of space in the posterior parietal cortex. *Philosophical Transactions of the Royal Society of London Series B-Biological Sciences, 352*(1360), 1421–1428.

Arbib, M. A. (1981). Perceptual structures and distributed motor control. In J. M. Brookhart, V. B. Mountcastle, & V. B. Brooks (Eds.), *Handbook of physiology, Section 1: The nervous system, volume 2: Motor control, part 2* (pp. 1449–1480). Bethesda, MD: American Phsyiological Society.

Asatryan, D. G., & Feldman, A. G. (1965). Biophysics of complex systems and mathematical models. Functional tuning of nervous system with control of movement or maintenance of a steady posture. I. Mechanographic analysis of the work of the joint on execution of a postural task. *Biophysics, 10*, 925–935.

Baker, E., Blumstein, S. E., & Goodglass, H. (1981). Interaction between phonological and semantic factors in auditory comprehension. *Neuropsychologia, 19*(1), 1–15.

Barto, A. G. (1995). Adaptive critics and the basal ganglia. In J. C. Houk, J. Davis, & D. Beiser (Eds.), *Models of information processing in the basal ganglia* (pp. 215–232). Cambridge, MA: MIT Press.

Bauer, J. J., Mittal, J., Larson, C. R., & Hain, T. C. (2006). Vocal responses to unanticipated perturbations in voice loudness feedback: An automatic mechanism for stabilizing voice amplitude. *The Journal of the Acoustical Society of America, 119*(4), 2363–2371.

Bellman, R. (1957). *Dynamic programming*. Princeton, NJ: Princeton University Press.

Bendor, D., & Wang, X. (2005). The neuronal representation of pitch in primate auditory cortex. *Nature, 436*(7054), 1161–1165.

Bernstein, N. A. (1967). *The co-ordination and regulation of movements*. Oxford: Pergamon Press.

Berthier, N. E., Rosenstein, M. T., & Barto, A. G. (2005). Approximate optimal control as a model for motor learning. *Psychological Review, 112*(2), 329–346.

Bertsekas, D. P. (2000). *Dynamic programming and optimal control* (2nd ed.). Belmont, MA: Athena Scientific.

Bizzi, E., Accornero, N., Chapple, W., & Hogan, N. (1982). Arm trajectory formation in monkeys. *Experimental Brain Research, 46*(1), 139–143.

Bladon, R. A. W., & Al-Bamerni, A. (1976). Coarticulation resistance in English /l/. *Journal of Phonetics, 4*, 137–150.

Blakemore, S. J., Wolpert, D. M., & Frith, C. D. (1998). Central cancellation of self-produced tickle sensation. *Nature Neuroscience, 1*(7), 635–640.

Blakemore, S. J., Wolpert, D. M., & Frith, C. D. (1999). The cerebellum contributes to somatosensory cortical activity during self-produced tactile stimulation. *Neuroimage, 10*(4), 448–459.

Blakemore, S. J., Wolpert, D. M., & Frith, C. D. (2000). Why can't you tickle yourself? *Neuroreport, 11*(11), R11–R16.

Borden, G. J., Harris, K. S., & Raphael, L. J. (1994). *Speech science primer: Physiology, acoustics, and perception of speech* (3rd ed.). Baltimore, MD: Williams & Wilkins.

Browman, C. P., & Goldstein, L. M. (1986). Towards an articulatory phonology. *Phonology Yearbook, 3*, 219–252.

Brown, P., Chen, C. C., Wang, S., Kuhn, A. A., Doyle, L., Yarrow, K., et al. (2006). Involvement of human basal ganglia in offline feedback control of voluntary movement. *Current Biology, 16*(21), 2129–2134.

Buchsbaum, B. R., Baldo, J., Okada, K., Berman, K. F., Dronkers, N., D'Esposito, M., et al. (2011). Conduction aphasia, sensory-motor integration, and phonological short-term memory—An aggregate analysis of lesion and fMRI data. *Brain and Language, 119*(3), 119–128.

Buchsbaum, B. R., Hickok, G., & Humphries, C. (2001). Role of left posterior superior temporal gyrus in phonological processing for speech perception and production. *Cognitive Science, 25*(5), 663–678.

Burnett, T. A., Freedland, M. B., Larson, C. R., & Hain, T. C. (1998). Voice F0 responses to manipulations in pitch feedback. *Journal of the Acoustical Society of America, 103*(6), 3153–3161.

Chang—Yit, R., Pick, J., Herbert, L., & Siegel, G. M. (1975). Reliability of sidetone amplification effect in vocal intensity. *Journal of Communication Disorders, 8*(4), 317–324.

Cheung, S. W., Nagarajan, S. S., Schreiner, C. E., Bedenbaugh, P. H., & Wong, A. (2005). Plasticity in primary auditory cortex of monkeys with altered vocal production. *Journal of Neuroscience, 25*(10), 2490–2503.

Chomsky, N., & Halle, M. (1968). *The sound pattern of English*. New York, NY: Harper & Row.

Clumeck, H. (1976). Patterns of soft palate movement in six languages. *Journal of Phonetics, 4*, 337–351.

Cowie, R., & Douglas-Cowie, E. (1992). *Postlingually acquired deafness: Speech deterioration and the wider consequences*. Hawthorne, NY: Mouton de Gruyter.

Curio, G., Neuloh, G., Numminen, J., Jousmaki, V., & Hari, R. (2000). Speaking modifies voice-evoked activity in the human auditory cortex. *Human Brain Mapping, 9*(4), 183–191.

Cykowski, M. D., Fox, P. T., Ingham, R. J., Ingham, J. C., & Robin, D. A. (2010). A study of the reproducibility and etiology of diffusion anisotropy differences in developmental stuttering: A potential role for impaired myelination. *Neuroimage, 52*(4), 1495–1504.

Daw, N. D., & Doya, K. (2006). The computational neurobiology of learning and reward. *Current Opinion in Neurobiology, 16*(2), 199–204.

Desmurget, M., Grafton, S. T., Vindras, P., Grea, H., & Turner, R. S. (2003). Basal ganglia network mediates the control of movement amplitude. *Experimental Brain Research, 153*(2), 197–209.

Desmurget, M., Grafton, S. T., Vindras, P., Grea, H., & Turner, R. S. (2004). The basal ganglia network mediates the planning of movement amplitude. *European Journal of Neuroscience, 19*(10), 2871–2880.

Deutsch, D., & Roll, P. L. (1976). Separate "what" and "where" decision mechanisms in processing a dichotic tonal sequence. *Journal of Experimental Psychology. Human Perception and Performance, 2*(1), 23–29.

Doya, K. (2000). Complementary roles of basal ganglia and cerebellum in learning and motor control. *Current Opinion in Neurobiology, 10* (6), 732–739.

Duffy, J. R. (2005). *Motor speech disorders: Substrates, differential diagnosis, and management* (2nd ed.). Saint Louis, MO: Elsevier Mosby.

Eliades, S. J., & Wang, X. (2003). Sensory-motor interaction in the primate auditory cortex during self-initiated vocalizations. *Journal of Neurophysiology, 89*(4), 2194–2207.

Eliades, S. J., & Wang, X. (2005). Dynamics of auditory-vocal interaction in monkey auditory cortex. *Cerebral Cortex, 15*(10), 1510–1523.

Eliades, S. J., & Wang, X. (2008). Neural substrates of vocalization feedback monitoring in primate auditory cortex. *Nature, 453* (7198), 1102–1106.

Elman, J. L. (1981). Effects of frequency-shifted feedback on the pitch of vocal productions. *The Journal of the Acoustical Society of America, 70*(1), 45–50.

Espy-Wilson, C., & Boyce, S. (1994). Acoustic differences between "bunched" and "retroflex" variants of American English /r/. *The Journal of the Acoustical Society of America, 95*(5), 2823.

Evans, E. F., & Nelson, P. G. (1973). On the functional relationship between the dorsal and ventral divisions of the cochlear nucleus of the cat. *Experimental Brain Research, 17*(4), 428–442.

Fairbanks, G. (1954). Systematic research in experimental phonetics: 1. A theory of the speech mechanism as a servosystem. *Journal of Speech and Hearing Disorders, 19*(2), 133–139.

Farnetani, E., & Recasens, D. (1999). Coarticulation models in recent speech production theories. In W. Hardcastle, & N. Hewlett (Eds.), *Coarticulation: Theory, data, and techniques* (pp. 31–65). Cambrigde, UK: Cambridge University Press.

Feldman, A. G. (1986). Once more on the equilibrium-point hypothesis (lambda model) for motor control. *Journal of Motor Behavior, 18* (1), 17–54.

Ford, J. M., & Mathalon, D. H. (2004). Electrophysiological evidence of corollary discharge dysfunction in schizophrenia during talking and thinking. *Journal of Psychiatric Research, 38*(1), 37–46.

Ford, J. M., Mathalon, D. H., Heinks, T., Kalba, S., Faustman, W. O., & Roth, W. T. (2001). Neurophysiological evidence of corollary discharge dysfunction in schizophrenia. *American Journal of Psychiatry, 158*(12), 2069–2071.

Foundas, A. L., Bollich, A. M., Corey, D. M., Hurley, M., & Heilman, K. M. (2001). Anomalous anatomy of speech-language areas in adults with persistent developmental stuttering. *Neurology, 57*(2), 207–215.

Foundas, A. L., Bollich, A. M., Feldman, J., Corey, D. M., Hurley, M., Lemen, L. C., et al. (2004). Aberrant auditory processing and atypical planum temporale in developmental stuttering. *Neurology, 63* (9), 1640–1646.

Fowler, C. A., & Saltzman, E. (1993). Coordination and coarticulation in speech production. *Language and Speech, 36*(Pt 2–3), 171–195.

Franklin, G. F., Powell, J. D., & Emami-Naeini, A. (1991). *Feedback control of dynamic systems* (2nd ed.). Reading, MA: Addison-Wesley.

Geschwind, N. (1965). Disconnexion syndromes in animals and man, Part II. *Brain, 88*(3), 585–644.

Ghahramani, Z., Wolpert, D. M., & Jordan, M. I. (1996). Generalization to local remappings of the visuomotor coordinate transformation. *The Journal of Neuroscience, 16*(21), 7085–7096.

Glasser, M. F., & Rilling, J. K. (2008). DTI tractography of the human brain's language pathways. *Cerebral Cortex, 18*(11), 2471–2482.

Godey, B., Atencio, C. A., Bonham, B. H., Schreiner, C. E., & Cheung, S. W. (2005). Functional organization of squirrel monkey primary auditory cortex: Responses to frequency-modulation sweeps. *Journal of Neurophysiology, 94*(2), 1299–1311.

Gomi, H., & Kawato, M. (1996). Equilibrium-point control hypothesis examined by measured arm stiffness during multijoint movement. *Science, 272*(5258), 117–120.

Guenther, F. H. (1995). Speech sound acquisition, coarticulation, and rate effects in a neural network model of speech production. *Psychological Review, 102*(3), 594–621.

Guenther, F. H., Espy-Wilson, C. Y., Boyce, S. E., Matthies, M. L., Zandipour, M., & Perkell, J. S. (1999). Articulatory tradeoffs reduce acoustic variability during American English /r/ production. *The Journal of the Acoustical Society of America, 105*(5), 2854–2865.

Guenther, F. H., Ghosh, S. S., & Tourville, J. A. (2006). Neural modeling and imaging of the cortical interactions underlying syllable production. *Brain and Language, 96*(3), 280–301.

Guenther, F. H., Hampson, M., & Johnson, D. (1998). A theoretical investigation of reference frames for the planning of speech movements. *Psychological Review, 105*(4), 611–633.

Guenther, F. H., & Vladusich, T. (2012). A neural theory of speech acquisition and production. *Journal of Neurolinguistics, 25*(5), 408–422.

Guigon, E., Baraduc, P., & Desmurget, M. (2008a). Computational motor control: Feedback and accuracy. *The European Journal of Neuroscience, 27*(4), 1003–1016.

Guigon, E., Baraduc, P., & Desmurget, M. (2008b). Optimality, stochasticity, and variability in motor behavior. *Journal of Computational Neuroscience, 24*(1), 57–68.

Hain, T. C., Burnett, T. A., Kiran, S., Larson, C. R., Singh, S., & Kenney, M. K. (2000). Instructing subjects to make a voluntary response reveals the presence of two components to the audio-vocal reflex. *Experimental Brain Research, 130*(2), 133–141.

Hallett, M., Shahani, B. T., & Young, R. R. (1975). EMG analysis of stereotyped voluntary movements in man. *Journal of Neurology, Neurosurgery and Psychiatry, 38*(12), 1154–1162.

Heil, P. (2003). Coding of temporal onset envelope in the auditory system. *Speech Communication, 41*(1), 123–134. Available from: http://dx.doi.org/10.1016/S0167-6393(02)00099-7.

Heil, P., & Irvine, D. R. (1996). On determinants of first-spike latency in auditory cortex. *Neuroreport, 7*(18), 3073–3076.

Heinks-Maldonado, T. H., & Houde, J. F. (2005). Compensatory responses to brief perturbations of speech amplitude. *Acoustics Research Letters Online, 6*(3), 131–137.

Heinks-Maldonado, T. H., Nagarajan, S. S., & Houde, J. F. (2006). Magnetoencephalographic evidence for a precise forward model in speech production. *Neuroreport, 17*(13), 1375–1379.

Held, R. (1968). Dissociation of visual functions by deprivation and rearrangement. *Psychologische Forschung, 31*(4), 338–348.

Henke, W.L. (1966). *Dynamic articulatory model of speech production using computer simulation.* Unpublished Ph.D., Cambridge, MA: MIT.

Hickok, G., Buchsbaum, B., Humphries, C., & Muftuler, T. (2003). Auditory-motor interaction revealed by fMRI: Speech, music, and working memory in area Spt. *Journal of Cognitive Neuroscience, 15* (5), 673–682.

Hickok, G., Houde, J. F., & Rong, F. (2011). Sensorimotor integration in speech processing: Computational basis and neural organization. *Neuron, 69*(3), 407–422.

Hickok, G., & Poeppel, D. (2007). The cortical organization of speech processing. *Nature Reviews Neuroscience, 8*(5), 393–402.

Hill, A. V. (1925). Length of muscle, and the heat and tension developed in an isometric contraction. *The Journal of Physiology, 60*(4), 237–263.

Hirano, S., Kojima, H., Naito, Y., Honjo, I., Kamoto, Y., Okazawa, H., et al. (1996). Cortical speech processing mechanisms while vocalizing visually presented languages. *Neuroreport, 8*(1), 363–367.

Hirano, S., Kojima, H., Naito, Y., Honjo, I., Kamoto, Y., Okazawa, H., et al. (1997). Cortical processing mechanism for vocalization with auditory verbal feedback. *Neuroreport, 8*(9–10), 2379–2382.

Hirano, S., Naito, Y., Okazawa, H., Kojima, H., Honjo, I., Ishizu, K., et al. (1997). Cortical activation by monaural speech sound stimulation demonstrated by positron emission tomography. *Experimental Brain Research, 113*(1), 75–80.

Hollerman, J. R., & Schultz, W. (1998). Dopamine neurons report an error in the temporal prediction of reward during learning. *Nature Neuroscience*, *1*(4), 304–309.

Houde, J. F., & Jordan, M. I. (1997). Adaptation in speech motor control. In M. I. Jordan, M. J. Kearns, & S. A. Solla (Eds.), *Advances in neural information processing systems* (Vol. 10, pp. 38–44). Cambridge, MA: MIT Press.

Houde, J. F., & Jordan, M. I. (1998). Sensorimotor adaptation in speech production. *Science*, *279*(5354), 1213–1216.

Houde, J. F., & Jordan, M. I. (2002). Sensorimotor adaptation of speech I: Compensation and adaptation. *Journal of Speech, Language, and Hearing Research*, *45*(2), 295–310.

Houde, J. F., & Nagarajan, S. S. (2011). Speech production as state feedback control. *Frontiers in Human Neuroscience*, *5*, 82.

Houde, J. F., Nagarajan, S. S., Sekihara, K., & Merzenich, M. M. (2002). Modulation of the auditory cortex during speech: An MEG study. *Journal of Cognitive Neuroscience*, *14*(8), 1125–1138.

Howell, P., El-Yaniv, N., & Powell, D. J. (1987). Factors affecting fluency in stutterers when speaking under altered auditory feedback. In H. F. Peters, & W. Hulstijn (Eds.), *Speech motor dynamics in stuttering* (pp. 361–369). New York, NY: Springer Press.

Hulliger, M. (1984). The mammalian muscle spindle and its central control. *Reviews of Physiology Biochemistry & Pharmacology*, *101*, 1–110.

Ingle, D. (1967). Two visual mechanisms underlying the behavior of fish. *Psychologische Forschung*, *31*(1), 44–51.

Ito, T., Kimura, T., & Gomi, H. (2005). The motor cortex is involved in reflexive compensatory adjustment of speech articulation. *Neuroreport*, *16*(16), 1791–1794.

Izawa, J., Rane, T., Donchin, O., & Shadmehr, R. (2008). Motor adaptation as a process of reoptimization. *Journal of Neuroscience*, *28* (11), 2883–2891.

Jacobs, O. L. R. (1993). *Introduction to control theory* (2nd ed.). Oxford, UK: Oxford University Press.

Jones, J. A., & Munhall, K. G. (2000a). Perceptual calibration of F0 production: Evidence from feedback perturbation. *Journal of the Acoustical Society of America*, *108*(3 Pt 1), 1246–1251.

Jones, J.A., & Munhall, K.G. (2000b). Perceptual contributions to fundamental frequency production. Paper presented at the 5th Seminar on Speech Production: Models and Data, Kloster Seeon, Germany.

Jones, J. A., & Munhall, K. G. (2002). The role of auditory feedback during phonation: Studies of mandarin tone production. *Journal of Phonetics*, *30*(3), 303–320.

Jones, J. A., & Munhall, K. G. (2003). Learning to produce speech with an altered vocal tract: The role of auditory feedback. *Journal of the Acoustical Society of America*, *113*(1), 532–543.

Jones, J. A., & Munhall, K. G. (2005). Remapping auditory-motor representations in voice production. *Current Biology*, *15*(19), 1768–1772.

Jones, J.A., Munhall, K.G., & Vatikiotis-Bateson, E. (1998). Adaptation to altered feedback in speech. Paper presented at The 136th Meeting of the Acoustical Society of America, Norfolk, VA.

Jürgens, U. (1982). Afferents to the cortical larynx area in the monkey. *Brain Research*, *239*(2), 377–389.

Jürgens, U. (2002). Neural pathways underlying vocal control. *Neuroscience and Biobehavioral Reviews*, *26*(2), 235–258.

Kalman, R. E. (1960). A new approach to linear filtering and prediction problems. *Transactions of the ASME-Journal of Basic Engineering*, *82* (Series. D), 35–45.

Kalveram, K. T., & Jancke, L. (1989). Vowel duration and voice onset time for stressed and nonstressed syllables in stutterers under delayed auditory feedback condition. *Folia Phoniatrica*, *41*(1), 30–42.

Kandel, E. R., Schwartz, J. H., & Jessell, T. M. (2000). *Principles of neural science* (4th ed.). New York, NY: McGraw-Hill.

Keating, P. (1990). The window model of coarticulation: Articulatory evidence. In J. Kingston, & M. Beckman (Eds.), *Papers in laboratory phonology I: Between the grammar and physics of speech* (pp. 451–470). Cambridge: Cambridge University Press.

Kelso, J. A. S., Tuller, B., Vatikiotis-Bateson, E., & Fowler, C. A. (1984). Functionally specific articulatory cooperation following jaw perturbations during speech: Evidence for coordinative structures. *Journal of Experimental Psychology: Human Perception and Performance*, *10*(6), 812–832.

Kording, K. P., Tenenbaum, J. B., & Shadmehr, R. (2007). The dynamics of memory as a consequence of optimal adaptation to a changing body. *Nature Neuroscience*, *10*(6), 779–786.

Kühnert, B., & Nolan, F. (1999). The origin of coarticulation. In W. J. Hardcastle, & N. Hewlett (Eds.), *Coarticulation: Theory, data and techniques* (pp. 7–30). Cambridge, UK: Cambridge University Press.

Lakatos, P., Pincze, Z., Fu, K. M., Javitt, D. C., Karmos, G., & Schroeder, C. E. (2005). Timing of pure tone and noise-evoked responses in macaque auditory cortex. *Neuroreport*, *16*(9), 933–937.

Lane, H., & Tranel, B. (1971). The lombard sign and the role of hearing in speech. *Journal of Speech and Hearing Research*, *14*(4), 677–709.

Lane, H., Wozniak, J., Matthies, M., Svirsky, M., Perkell, J., O'Connell, M., et al. (1997). Changes in sound pressure and fundamental frequency contours following changes in hearing status. *The Journal of the Acoustical Society of America*, *101*(4), 2244–2252.

Larson, C. R., Altman, K. W., Liu, H. J., & Hain, T. C. (2008). Interactions between auditory and somatosensory feedback for voice F0 control. *Experimental Brain Research*, *187*(4), 613–621.

Larson, C. R., Burnett, T. A., Kiran, S., & Hain, T. C. (2000). Effects of pitch-shift velocity on voice F-0 responses. *Journal Of The Acoustical Society Of America*, *107*(1), 559–564.

Lee, B. S. (1950). Some effects of side-tone delay. *Journal of the Acoustical Society of America*, *22*, 639–640.

Levelt, W. J. M. (1989). *Speaking: From intention to articulation*. Cambridge, MA: The MIT Press.

Levitt, H., Stromberg, H., Smith, C., & Gold, T. (1980). The structure of segmental errors in the speech of deaf children. *Journal of Communication Disorders*, *13*(6), 419–441.

Li, W., Todorov, E., & Pan, X. (2004). Hierarchical optimal control of redundant biomechanical systems. *Conference Proceedings IEEE Engineering in Medicine and Biology Society*, *6*, 4618–4621.

Lindblom, B. (1963). Spectrographic study of vowel reduction. *The Journal of the Acoustical Society of America*, *35*(11), 1773–1781.

Lindblom, B. (1983). Economy of speech gestures. In P. F. MacNeilage (Ed.), *The production of speech* (pp. 217–245). New York, NY: Springer-Verlag.

Lindblom, B. (1990). Explaining phonetic variation: A sketch of the H&H theory. In W. J. Hardcastle, & A. Marchal (Eds.), *Speech production and speech modelling* (Vol. 55, pp. 403–439). Dordrecht, Netherlands: Kluwer Academic Publishers.

Liu, D., & Todorov, E. (2007). Evidence for the flexible sensorimotor strategies predicted by optimal feedback control. *Journal of Neuroscience*, *27*(35), 9354–9368.

Liu, H., Zhang, Q., Xu, Y., & Larson, C. R. (2007). Compensatory responses to loudness-shifted voice feedback during production of mandarin speech. *The Journal of the Acoustical Society of America*, *122*(4), 2405–2412.

Lombard, E. (1911). Le signe de l'elevation de la voix. *Annales des Maladies de l'oreille, du Larynx, du Nez et du Pharynx*, *37*, 101–119.

Lubker, J., & Gay, T. (1982). Anticipatory labial coarticulation: Experimental, biological, and linguistic variables. *The Journal of the Acoustical Society of America*, *71*(2), 437–448.

Ludlow, C. L. (2004). Recent advances in laryngeal sensorimotor control for voice, speech and swallowing. *Current Opinion in Otolaryngology & Head and Neck Surgery*, *12*(3), 160–165.

Maraist, J. A., & Hutton, C. (1957). Effects of auditory masking upon the speech of stutterers. *The Journal of Speech and Hearing Disorders, 22*(3), 385–389.

Matthews, B. H. (1931). The response of a single end organ. *Journal of Physiology, 71*(1), 64–110.

Matthies, M., Perrier, P., Perkell, J. S., & Zandipour, M. (2001). Variation in anticipatory coarticulation with changes in clarity and rate. *Journal of Speech, Language, and Hearing Research, 44*(2), 340–353.

Mehta, B., & Schaal, S. (2002). Forward models in visuomotor control. *Journal of Neurophysiology, 88*(2), 942–953.

Merton, P. A. (1951). The silent period in a muscle of the human hand. *Journal of Physiology, 114*(1–2), 183–198.

Miall, R. C., Weir, D. J., Wolpert, D. M., & Stein, J. F. (1993). Is the cerebellum a smith predictor? *Journal of Motor Behavior, 25*(3), 203–216.

Miceli, G., Gainotti, G., Caltagirone, C., & Masullo, C. (1980). Some aspects of phonological impairment in aphasia. *Brain and Language, 11*(1), 159–169.

Mishkin, M., Ungerleider, L. G., & Macko, K. A. (1983). Object vision and spatial vision: Two cortical pathways. *Trends in Neurosciences, 6*, 414–417.

Moon, S. J., & Lindblom, B. (1994). Interaction between duration, context, and speaking style in English stressed vowels. *Journal of the Acoustical Society of America, 96*(1), 40–55.

Nasir, S. M., & Ostry, D. J. (2006). Somatosensory precision in speech production. *Current Biology, 16*(19), 1918–1923.

Nasir, S. M., & Ostry, D. J. (2008). Speech motor learning in profoundly deaf adults. *Nature Neuroscience, 11*(10), 1217–1222.

Nasir, S. M., & Ostry, D. J. (2009). Auditory plasticity and speech motor learning. *Proceedings of the National Academy of Sciences, 106* (48), 20470–20475.

Natke, U., Grosser, J., & Kalveram, K. T. (2001). Fluency, fundamental frequency, and speech rate under frequency-shifted auditory feedback in stuttering and nonstuttering persons. *Journal of Fluency Disorders, 26*(3), 227–241.

Natke, U., & Kalveram, K. T. (2001). Effects of frequency-shifted auditory feedback on fundamental frequency of long stressed and unstressed syllables. *Journal of Speech, Language, and Hearing Research, 44*(3), 577–584.

Numminen, J., & Curio, G. (1999). Differential effects of overt, covert and replayed speech on vowel- evoked responses of the human auditory cortex. *Neuroscience Letters, 272*(1), 29 32.

Numminen, J., Salmelin, R., & Hari, R. (1999). Subject's own speech reduces reactivity of the human auditory cortex. *Neuroscience Letters, 265*(2), 119–122.

Oller, D. K., & Eilers, R. E. (1988). The role of audition in infant babbling. *Child Development, 59*(2), 441–449.

Osberger, M. J., & McGarr, N. S. (1982). Speech production characteristics of the hearing-impaired. In N. J. Lass (Ed.), *Speech and language: Advances in basic research and practice* (pp. 221–284). New York, NY: Academic Press.

Ostry, D. J., Flanagan, J. R., Feldman, A. G., & Munhall, K. G. (1991). Human jaw motion control in mastication and speech. In J. Requin, & G. E. Stelmach (Eds.), *Tutorials in motor neuroscience. NATO ASI series; Series D: Behavioral and social sciences* (Vol. 62, pp. 535–543). New York, NY: Kluwer Academic/Plenum Publishers.

Ostry, D. J., Flanagan, J. R., Feldman, A. G., & Munhall, K. G. (1992). Human jaw movement kinematics and control. In G. E. Stelmach, & J. Requin (Eds.), *Tutorials in motor behavior, 2. Advances in psychology* (Vol. 87, pp. 647–660). Oxford, England: North-Holland.

Parsons, T. W. (1987). *Voice and speech processing*. New York, NY: McGraw-Hill Book Company.

Payan, Y., & Perrier, P. (1997). Synthesis of V-V sequences with a 2D biomechanical tongue model controlled by the equilibrium point hypothesis. *Speech Communication, 22*(2–3), 185–205.

Pearce, S. L., Miles, T. S., Thompson, P. D., & Nordstrom, M. A. (2003). Is the long-latency stretch reflex in human masseter transcortical? *Experimental Brain Research, 150*(4), 465–472.

Perkell, J. S., & Matthies, M. L. (1992). Temporal measures of anticipatory labial coarticulation for the vowel /u/: Within- and cross-subject variability. *The Journal of the Acoustical Society of America, 91*(5), 2911–2925.

Perkell, J. S., Matthies, M. L., Svirsky, M. A., & Jordan, M. I. (1993). Trading relations between tongue-body raising and lip rounding in production of the vowel /u/: a pilot "motor equivalence" study. *Journal of the Acoustical Society of America, 93*(5), 2948–2961.

Perrier, P., Ostry, D. J., & Laboissiere, R. (1996). The equilibrium point hypothesis and its application to speech motor control. *Journal of Speech and Hearing Research, 39*(2), 365–378.

Pile, E.J.S., Dajani, H.R., Purcell, D.W., & Munhall, K.G. (2007, August 6–10). Talking under conditions of altered auditory feedback: Does adaptation of one vowel generalize to other vowels? Paper presented at the International Congress of Phonetic Sciences, Saarland University, Saarbrücken, Germany.

Polit, A., & Bizzi, E. (1979). Characteristics of motor programs underlying arm movements in monkeys. *Journal of Neurophysiology, 42* (1 Pt 1), 183–194.

Poljak, S. (1926). The connections of the acoustic nerve. *Journal of Anatomy, 60*(4), 465–469.

Price, C. J., Crinion, J. T., & Macsweeney, M. (2011). A generative model of speech production in Broca's and Wernicke's areas. *Frontiers in Psychology, 2*, 237.

Purcell, D. W., & Munhall, K. G. (2006). Adaptive control of vowel formant frequency: Evidence from real-time formant manipulation. *Journal of the Acoustical Society of America, 120*(2), 966–977.

Rauschecker, J. P., & Scott, S. K. (2009). Maps and streams in the auditory cortex: Nonhuman primates illuminate human speech processing. *Nature Neuroscience, 12*(6), 718–724.

Rauschecker, J. P., & Tian, B. (2000). Mechanisms and streams for processing of "what" and "where" in auditory cortex. *Proceedings of the National Academy of Sciences of the United States of America, 97* (22), 11800–11806.

Redgrave, P., & Gurney, K. (2006). The short-latency dopamine signal: A role in discovering novel actions? *Nature Reviews Neuroscience, 7* (12), 967–975.

Rizzolatti, G., Fogassi, L., & Gallese, V. (1997). Parietal cortex: From sight to action. *Current Opinion in Neurobiology, 7*(4), 562–567.

Romanski, L. M., Tian, B., Fritz, J., Mishkin, M., Goldman-Rakic, P. S., & Rauschecker, J. P. (1999). Dual streams of auditory afferents target multiple domains in the primate prefrontal cortex. *Nature Neuroscience, 2*(12), 1131–1136.

Ross, M., & Giolas, T. G. (1978). *Auditory management of hearing-impaired children: Principles and prerequisites for intervention*. Baltimore, MD: University Park Press.

Saltzman, E. L., Lofqvist, A., Kay, B., Kinsella-Shaw, J., & Rubin, P. (1998). Dynamics of intergestural timing: A perturbation study of lip-larynx coordination. *Experimental Brain Research, 123*(4), 412–424.

Saltzman, E. L., & Munhall, K. G. (1989). A dynamical approach to gestural patterning in speech production. *Ecological Psychology, 1* (4), 333–382.

Samejima, K., Ueda, Y., Doya, K., & Kimura, M. (2005). Representation of action-specific reward values in the striatum. *Science, 310*(5752), 1337–1340.

Sanguineti, V., Laboissiere, R., & Ostry, D. J. (1998). A dynamic biomechanical model for neural control of speech production. *The Journal of the Acoustical Society of America, 103*(3), 1615–1627.

Sanguineti, V., Laboissiere, R., & Payan, Y. (1997). A control model of human tongue movements in speech. *Biological Cybernetics, 77*(1), 11–22.

Savariaux, C., Perrier, P., & Orliaguet, J.-P. (1995). Compensation strategies for the perturbation of the rounded vowel [u] using a lip tube: A study of the control space in speech production. *The Journal of the Acoustical Society of America*, *98*(5), 2428.

Schmahmann, J. D., Pandya, D. N., Wang, R., Dai, G., D'Arceuil, H. E., de Crespigny, A. J., et al. (2007). Association fibre pathways of the brain: Parallel observations from diffusion spectrum imaging and autoradiography. *Brain*, *130*(Pt 3), 630–653.

Schultz, W. (1998). Predictive reward signal of dopamine neurons. *Journal of Neurophysiology*, *80*(1), 1–27.

Schultz, W., Dayan, P., & Montague, P. R. (1997). A neural substrate of prediction and reward. *Science*, *275*(5306), 1593–1599.

Scott, C. M., & Ringel, R. L. (1971). Articulation without oral sensory control. *Journal of Speech and Hearing Research*, *14*(4), 804–818.

Scott, S. H. (2004). Optimal feedback control and the neural basis of volitional motor control. *Nature Reviews Neuroscience*, *5*(7), 534–546.

Shadmehr, R., & Krakauer, J. W. (2008). A computational neuroanatomy for motor control. *Experimental Brain Research*, *185*(3), 359–381.

Shadmehr, R., & Wise, S. P. (2005). *The computational neurobiology of reaching and pointing: A foundation for motor learning*. Cambridge, MA: MIT Press.

Shaiman, S., & Gracco, V. L. (2002). Task-specific sensorimotor interactions in speech production. *Experimental Brain Research*, *146*(4), 411–418.

Shiller, D.M., Sato, M., Gracco, V.L., & Baum, S.R. (2007). Motor and sensory adaptation following auditory perturbation of /s/ production. Paper presented at the 154th Meeting of the Acoustical Society of America, New Orleans, LA.

Shiller, D. M., Sato, M., Gracco, V. L., & Baum, S. R. (2009). Perceptual recalibration of speech sounds following speech motor learning. *Journal of the Acoustical Society of America*, *125*(2), 1103–1113.

Skipper, J. I., Nusbaum, H. C., & Small, S. L. (2005). Listening to talking faces: Motor cortical activation during speech perception. *Neuroimage*, *25*(1), 76–89.

Smith, C. R. (1975). Residual hearing and speech production in deaf children. *Journal of Speech and Hearing Research*, *18*(4), 795–811.

Smith, O. J. M. (1959). A controller to overcome deadtime. *ISA Journal*, *6*, 28–33.

Soderberg, G. A. (1968). Delayed auditory feedback and stuttering. *Journal of Speech and Hearing Disorders*, *33*(3), 260–267.

Stengel, R. F. (1994). *Optimal control and estimation*. Mineola, NY: Dover Publications, Inc.

Tian, X., & Poeppel, D. (2010). Mental imagery of speech and movement implicates the dynamics of internal forward models. *Frontiers in Psychology*, *1*, 166.

Tin, C., & Poon, C.-S. (2005). Internal models in sensorimotor integration: Perspectives from adaptive control theory. *Journal of Neural Engineering*, *2*(3), S147–S163.

Todorov, E. (2004). Optimality principles in sensorimotor control. *Nature Neuroscience*, *7*(9), 907–915.

Todorov, E. (2006). Optimal control theory. In K. Doya, S. Ishii, A. Pouget, & R. P. N. Rao (Eds.), *Bayesian brain: Probabilistic approaches to neural coding* (pp. 269–298). Cambridge, MA: MIT Press.

Todorov, E. (2007). *Mixed muscle-movement representations emerge from optimization of stochastic sensorimotor transformations*. Unpublished manuscript, UCSD, La Jolla, CA.

Todorov, E., & Jordan, M. I. (2002). Optimal feedback control as a theory of motor coordination. *Nature Neuroscience*, *5*(11), 1226–1235.

Tourville, J. A., Reilly, K. J., & Guenther, F. H. (2008). Neural mechanisms underlying auditory feedback control of speech. *Neuroimage*, *39*(3), 1429–1443.

Toyomura, A., Koyama, S., Miyamaoto, T., Terao, A., Omori, T., Murohashi, H., et al. (2007). Neural correlates of auditory feedback control in human. *Neuroscience*, *146*(2), 499–503.

Tremblay, S., Houle, G., & Ostry, D. J. (2008). Specificity of speech motor learning. *Journal of Neuroscience*, *28*(10), 2426–2434.

Tremblay, S., Shiller, D. M., & Ostry, D. J. (2003). Somatosensory basis of speech production. *Nature*, *423*(6942), 866–869.

Turner, R. S., Desmurget, M., Grethe, J., Crutcher, M. D., & Grafton, S. T. (2003). Motor subcircuits mediating the control of movement extent and speed. *Journal of Neurophysiology*, *90*(6), 3958–3966.

Ungerleider, L. G., & Mishkin, M. (1982). Two cortical visual systems. In D. J. Ingle, M. A. Goodale, & R. J. W. Mansfield (Eds.), *Analysis of visual behavior* (pp. 549–586). Cambridge, MA: MIT Press.

Upadhyay, J., Hallock, K., Ducros, M., Kim, D.-S., & Ronen, I. (2008). Diffusion tensor spectroscopy and imaging of the arcuate fasciculus. *Neuroimage*, *39*(1), 1–9.

Ventura, M. I., Nagarajan, S. S., & Houde, J. F. (2009). Speech target modulates speaking induced suppression in auditory cortex. *BMC Neuroscience*, *10*, 58.

Villacorta, V. M., Perkell, J. S., & Guenther, F. H. (2007). Sensorimotor adaptation to feedback perturbations of vowel acoustics and its relation to perception. *Journal of The Acoustical Society of America*, *122*(4), 2306–2319.

Wachholder, K., & Altenburger, H. (1926). Beiträge zur physiologie der willkürlichen bewegung X. Mitteilung. Einzelbewegungen. *Pflugers Archiv fur die gesamte Physiologie des Menschen und der Tiere*, *214*(1), 642–661.

Wang, A. Y., Miura, K., & Uchida, N. (2013). The dorsomedial striatum encodes net expected return, critical for energizing performance vigor. *Nature Neuroscience*, *16*(5), 639–647.

Watkins, K., & Paus, T. (2004). Modulation of motor excitability during speech perception: The role of Broca's area. *Journal of Cognitive Neuroscience*, *16*(6), 978–987.

Wernicke, C. (1874). Der aphasische symptomencomplex: Eine psychologische studie auf anatomischer basis. In G. H. Eggert (Ed.), *Wernicke's works on aphasia: A sourcebook and review* (pp. 91–145). The Hague: Mouton.

Whiting, H. T. A. (Ed.), (1984). *Human motor actions: Bernstein reassessed*. Amsterdam, NL: North-Holland.

Wiener, N. (1948). *Cybernetics: Control and communication in the animal and the machine*. New York, NY: John Wiley & sons, Inc.

Wilson, S. M., & Iacoboni, M. (2006). Neural responses to non-native phonemes varying in producibility: Evidence for the sensorimotor nature of speech perception. *Neuroimage*, *33*(1), 316–325.

Wilson, S. M., Saygin, A. P., Sereno, M. I., & Iacoboni, M. (2004). Listening to speech activates motor areas involved in speech production. *Nature Neuroscience*, *7*(7), 701–702. Available from: http://dx.doi.org/10.1038/nn1263.

Wolpert, D. M. (1997). Computational approaches to motor control. *Trends in Cognitive Sciences*, *1*(6), 209.

Wolpert, D. M., & Ghahramani, Z. (2000). Computational principles of movement neuroscience. *Nature Neuroscience*, *3*(Suppl.), 1212–1217.

Wolpert, D. M., Ghahramani, Z., & Jordan, M. I. (1995). An internal model for sensorimotor integration. *Science*, *269*(5232), 1880–1882.

Yates, A. J. (1963). Delayed auditory feedback. *Psychological Bulletin*, *60*(3), 213–232.

Zajac, F. E. (1989). Muscle and tendon: Properties, models, scaling, and application to biomechanics and motor control. *Critical Reviews in Biomedical Engineering*, *17*(4), 359–411.

Zheng, Z. Z., Munhall, K. G., & Johnsrude, I. S. (2010). Functional overlap between regions involved in speech perception and in monitoring one's own voice during speech production. *Journal of Cognitive Neuroscience*, *22*(8), 1770–1781.

Zhou, X., Espy-Wilson, C. Y., Boyce, S., Tiede, M., Holland, C., & Choe, A. (2008). A magnetic resonance imaging-based articulatory and acoustic study of "retroflex" and "bunched" American English /r/. *The Journal of the Acoustical Society of America*, *123*(6), 4466–4481. Available from: http://dx.doi.org/10.1121/1.2902168.

CHAPTER

20

Spoken Word Recognition: Historical Roots, Current Theoretical Issues, and Some New Directions

David B. Pisoni[1] *and Conor T. McLennan*[2]

[1]Department of Psychological and Brain Sciences, Indiana University, Bloomington, IN, USA; [2]Department of Psychology, Cleveland State University, Cleveland, OH, USA

20.1 INTRODUCTION

This is an exciting time to be working in the field of human *spoken word recognition* (SWR). Many long-standing assumptions about spoken language processing are being reevaluated in light of new experimental and computational methods, empirical findings, and theoretical developments (Gaskell, 2007a, 2007b; Hickok & Poeppel, 2007; McQueen, 2007; Pisoni & Levi, 2007). Moreover, recent findings on SWR also have direct applications to issues related to hearing impairment in deaf children and adults, non-native speakers of English, bilinguals, and older adults.

The fundamental problems in the field of SWR, such as invariance and variability, neural coding, representational specificity, and perceptual constancy in the face of diverse sensory input, are similar to the perceptual problems studied in other areas of cognitive psychology and neuroscience. Although these well-known theoretical problems have occupied speech research since the early 1950s, until recently research and theory on spoken language processing have been intellectually isolated from mainstream developments in neurobiology and cognitive science (Arlinger, Lunner, Lyxell, & Pichora-Fuller, 2009; Dahan & Magnuson, 2006; Hickok & Poeppel, 2007; Magnuson, Mirman, & Harris, 2012; Magnuson, Mirman, & Myers, 2013; Rönnberg et al., 2013). The isolation of speech communication evolved because speech scientists and communication engineers relied heavily on linguistically motivated theoretical assumptions about the core properties of speech and the computational processes underlying spoken language processing (see Chomsky & Miller, 1963). These assumptions embodied the *conventional segmental linguistic view*, which assumes that speech signals consist of a linear sequence of abstract, idealized, context-free segments ordered temporally in time much like the discrete letters of the alphabet or bricks on the wall (Halle, 1985; Hockett, 1955; Licklider, 1952; Peterson, 1952). The assumption that the continuously varying speech signal can be represented as a sequence of discrete units has played a central role in all theoretical accounts of spoken language research (Lindgren, 1965a, 1965b).

The present chapter is organized into four sections. First, we briefly review the historical roots of the field of SWR. Second, we discuss the principle theoretical issues and contemporary models of SWR. Third, we contrast the conventional segmental view of speech and SWR with an alternative proposal that moves beyond abstract linguistic representations. Finally, we briefly consider how basic research in SWR has led to several new research directions and additional challenges.

20.2 HISTORICAL ROOTS AND PRECURSORS TO SWR

This is a chapter about human SWR. However, before jumping right into our discussion of SWR, some brief historical background is necessary to place the current research and theory into a broader historical context. By discussing some older theoretical issues and empirical findings from research on speech and

Neurobiology of Language. DOI: http://dx.doi.org/10.1016/B978-0-12-407794-2.00020-1

hearing, we are able to illustrate important changes in the way researchers think about representational and processing issues in SWR today. There are, of course, also some connections and parallels with studies of visual word recognition (Carreiras, Armstrong, Perea, & Frost, 2014) and neuroimaging methods of language processing (Price, 2012; Sharp, Scott, & Wise, 2004), which are discussed in other chapters of this book.

The field of speech and hearing science has a well-documented history dating back to the end of the 19th century when researchers began to use electrically recorded audio signals to assess hearing loss supplementing traditional clinical measures that relied on simple acoustical signals (Flanagan, 1965; Fletcher, 1929; Miller, 1951; Wilson & McArdle, 2005). Except for the seminal, but otherwise obscure, early findings reported by Bagley (1900) on SWR using a novel experimental methodology involving mispronunciation detection (see Cole, 1973; Cole & Rudnicky, 1983), most of what we currently know about the basic acoustical and perceptual foundations of speech and hearing comes from pioneering research performed at Bell Telephone Laboratories in the 1920s (Flanagan, 1965; Fletcher, 1929, 1953; Fletcher & Galt, 1950). This extensive body of research established the minimal necessary and sufficient acoustical conditions for effective and highly reliable speech transmission and reception over conventional telephone circuits and provided an enormous body of empirical data and acoustic measurements on the foundations of hearing and speech communication under limited telephone bandwidth conditions (Allen, 1994, 2005; Fletcher, 1953).

20.2.1 Speech Intelligibility

Most of the quantitative experimental methods developed for assessing speech intelligibility routinely used today can be traced directly back to these early empirical studies (Fletcher, 1929, 1953; Hirsh, 1947; Konkle & Rintelmann, 1983; Wilson & McArdle, 2005). The primary focus of this research was on speech intelligibility; there was little interest in describing the human listener's perceptual and cognitive abilities to recognize spoken words (Allen, 1994, 2005). Further applied research on speech communication in noise was performed at the Psycho-Acoustic Laboratory at Harvard University during World War II (WW II) (see Hudgins, Hawkins, Karlin, & Stevens, 1947; Licklider & Miller, 1951; Rosenzweig & Stone, 1948 for reviews). Although these two applied research programs provided much of the core knowledge about hearing and speech communication, almost all of these investigations were focused on practical telephone and military-related communications issues. Little effort was devoted to broader theoretical and conceptual issues in SWR. One exception was a brief theoretical work by Licklider (1952) on *process models* of human speech perception. Surprisingly, several ideas proposed by Licklider are still relevant to current theoretical issues today. After WW II, speech and hearing scientists and acoustical engineers turned their attention to the human listener. Several research programs were initiated to understand how SWR is performed so efficiently under highly impoverished conditions (Cooper, Delattre, Liberman, Borst, & Gerstman, 1952). Efforts were also begun to develop methods for speech synthesis-by-rule that could be used in reading machines for blind veterans returning from the war (Allen, Hunnicutt, & Klatt, 1987; Cooper et al., 1952; Klatt, 1987).

20.2.2 Source-Filter Theory and Speech Cues

The development of the source-filter theory of speech acoustics at MIT (Stevens, Kasowski, & Fant, 1953) and the well-known *pattern playback* studies of speech cues using highly simplified hand-painted spectrographic patterns of synthetic speech at Haskins Laboratories (Cooper et al., 1952) provided the foundations of modern speech science and acoustic-phonetics. The modeling work at MIT focused on the acoustic nature of the speech signal (Stevens, 1998); research at Haskins investigated listeners' perceptual skills in making efficient use of minimal acoustic-phonetic cues as central components in the *speech chain* (Denes & Pinson, 1963; Liberman, 1996; Liberman, Cooper, Shankweiler, & Studdert-Kennedy, 1967; Moore, 2007a, 2007b).

These early studies were directly responsible for uncovering many core theoretical problems in spoken language processing, especially problems related to the articulatory dynamics of speech production and the context-dependent nature of the acoustic cues to speech perception that have remained among the major theoretical issues in the field (see Klatt, 1979; Liberman, 1996; Pisoni, 1978; Studdert-Kennedy, 1974; and see chapters in Gaskell, 2007a; Pisoni & Remez, 2005 for additional reviews and discussion).

20.3 PRINCIPLE THEORETICAL ISSUES IN SWR

In this section, we discuss the principle theoretical issues in SWR and then briefly review several contemporary models of SWR (for more detailed reviews, see Jusczyk & Luce, 2002; Magnuson et al., 2012, 2013; Marslen-Wilson, 1989; Pisoni & Levi, 2007). The fundamental problem in SWR is to understand how listeners recover the talker's intended message from the complex time-varying speech waveform. This problem is typically broken-down into a series of more manageable subquestions (Pisoni, 1978; Studdert-Kennedy, 1974). What stages of perceptual analysis intervene between the presentation of the speech signal and recognition of the talker's intended linguistic message? What types of processing operations occur at each stage of analysis? What are the primary processing units of speech? What is the nature of the neural and cognitive representations of spoken words? Finally, what specific perceptual, neurocognitive, and linguistic processing operations are used in SWR, and how are they coordinated into an integrated system?

Although many of these questions have remained basically the same since the early 1950s, the answers have changed, reflecting new theoretical and methodological developments and novel ways of thinking about sensory and neurocognitive processes used to recognize spoken words (Hickok & Poeppel, 2007; Luce & Pisoni, 1998; Moore, 2007a, 2007b; Scott & Johnsrude, 2003). As discussed, applied research on hearing and speech performed at Bell Labs and Harvard was primarily concerned with assessing the adequacy of telephone communication equipment and investigating factors that affected speech intelligibility in noise. Other research focused on methods to improve speech intelligibility in military combat conditions (Black, 1946). Although these applied research programs were created to address practical real-world problems, many theoretically important empirical findings were uncovered. The significance of these discoveries for theories of human SWR were discussed only briefly in numerous research reports from Harvard in the early 1940s (Abrams et al., 1944; Karlin, Abrams, Sanford, & Curtis, 1944; Wiener & Miller, 1946). Many of these empirical observations played substantial roles in theoretical accounts and models of perceptual and cognitive processes (Lindgren, 1965a, 1965b, 1967). Next, we briefly consider three theoretically significant findings originally uncovered by speech scientists at Harvard: (i) word frequency effects; (ii) word length effects; and (iii) sentence context effects. These findings, among others, later played central roles in theoretical discussions of SWR (Broadbent, 1967; Morton, 1979), have shaped the direction of the field, and need to be accounted for in any model of SWR.

20.3.1 Word Frequency, Word Length, and Sentence Context Effects

Numerous investigations have reported that high-frequency words presented in noise are identified more accurately than low-frequency words (Howes, 1954, 1957; Savin, 1963). At the time the word frequency effect was first discovered in the late 1940s, researchers believed that word frequency was equivalent to *experienced frequency* and that word counts of printed text of English (Francis & Kucera, 1964; Thorndike-Lorge, 1944) could serve as a good proxy for word frequency. Experienced frequency reflects how often a listener encountered a specific word form. Significant theoretical work was performed by Broadbent (1967) and many others to understand the basis of frequency effects. The word frequency effect is one of the distinctive hallmarks of SWR and has played a central role in research and theory development for many years (Forster, 1976; Morton, 1979; Oldfield, 1966).

Computational analyses and theoretical work performed by Landauer and Streeter (1973) suggested that although *experienced frequency* may play a role in word recognition processes, frequency effects may also reflect more subtle underlying differences in the structural properties of high- and low-frequency words, and that *experienced frequency* may simply be a byproduct of the statistical regularities of the sound patterns of words in the language. In an unpublished seminal study using phonotactically legal nonwords, Eukel (1980) demonstrated frequency effects for novel sound patterns. These results suggested that phonotactics—the frequency and patterning of sound segments and syllables within words—may be responsible for the robust perceptual differences observed between high- and low-frequency words in English (Pisoni, Nusbaum, Luce, & Slowiaczek, 1985; Vitevitch & Luce, 1999).

Computational and behavioral studies performed by Luce and Pisoni (1998) revealed that the sound similarity neighborhoods of high- and low-frequency words differed significantly, and that spoken words are recognized *relationally* in the context of other phonetically similar words in the mental lexicon. Spoken words are not recognized left-to-right segment-by-segment in serial order as traditionally assumed by the conventional linguistic view of speech perception. Instead, spoken words are recognized by processes involving activation and competition among word form candidates or *lexical neighbors* of spoken words (Pisoni & Luce, 1986).

Results from one of the earliest studies on word length effects are illustrated in Figure 20.1. The data from this study, described by Wiener and Miller (1946), demonstrate effects of word length on SWR scores. In marked contrast to word length effects observed in visual perception and memory, which consistently show that longer words are more difficult to recognize and recall, word length effects in SWR show precisely the opposite result—longer words are easier to perceive and recognize (Rosenzweig & Postman, 1957; Savin, 1963). Weiner and Miller's results also demonstrate that the longer a spoken word, the less often it will be confused with phonetically similar sounding words (Savin, 1963).

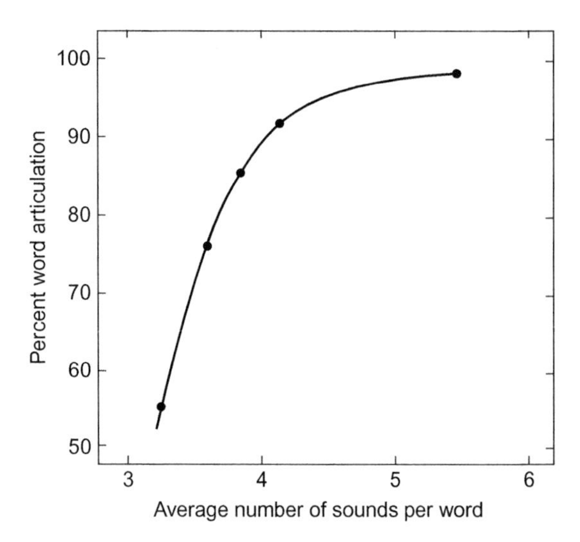

FIGURE 20.1 The word length effect in SWR illustrating the improvement in speech articulation as the average number of speech sounds per word is increased. *Adapted from Wiener and Miller (1946) with permission from the publisher.*

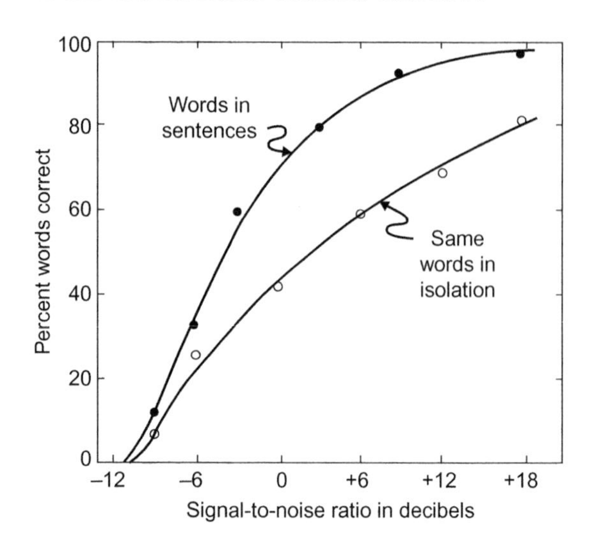

FIGURE 20.2 The effects of sentence context on the intelligibility of spoken words in noise as a function of signal-to-noise ratio in decibels. The filled circles show percentage of words correct in meaningful English sentences. The open circles show percentage of words correct for the same words presented in isolation. *Adapted from Miller et al. (1951) with permission from the publisher.*

Word frequency and word length effects suggest that spoken words are recognized *relationally* in the context of other words in lexical memory and are not processed in a left-to-right fashion segment-by-segment as many theorists had assumed. In *open-set* speech intelligibility tests in which no response alternatives are given to the listener, spoken words are recognized in relation to other perceptually similar words the listener knows that serve as potential lexical candidates for responses (Pollack, Rubenstein, & Decker, 1959, 1960). When listeners receive compromised sensory information, they make use of *sophisticated guessing* strategies, generating lexical candidates and strongly biased responses in a systematic manner reflecting the sound similarity relations among words in lexical memory (Broadbent, 1967; Morton, 1969; Savin, 1963; Treisman, 1978a, 1978b).

When words occur in meaningful sentences, listeners also make use of additional knowledge and linguistic constraints that are unavailable when the same words are presented in isolation (Marks & Miller, 1964; Miller, Heise, & Lichten, 1951; Miller & Isard, 1963). SWR is an active and highly automatized process that takes place very rapidly with little conscious awareness of the underlying sensory, cognitive, and linguistic processes. Many different sources of knowledge are used to recognize words depending on the context, test materials, and specific task demands (Jenkins, 1979). One of the most important and powerful sources of contextual constraint comes from sentences. Figure 20.2 shows the pioneering speech intelligibility results obtained by Miller et al. (1951). In one condition, Miller et al. presented sentences containing key words mixed in noise. In a second condition, the same words were presented in isolation. The results in Figure 20.2 illustrate that spoken words are much more difficult to recognize in isolation than in meaningful sentences.

These results establish that sentence context constrains word recognition and demonstrate the contributions of the listener's prior linguistic knowledge to SWR (Miller, 1962; Miller & Selfridge, 1950). When words are encoded in sentences, multiple sources of information automatically become available and their associated brain circuits are recruited to support the recognition process (Scott, Blank, Rosen, & Wise, 2000).

20.3.2 Contemporary Approaches to SWR

One of the most important changes in research on spoken language processing over the past 40 years has been a dramatic shift from a focus on the perception of individual speech sounds in isolated nonsense syllables to the study of the underlying cognitive and linguistic processes involved in SWR. For many speech scientists, the domain of speech perception was narrowly confined to the study of the perception of speech features and phonetic segments in highly controlled experimental contexts using simplified synthesized nonsense syllables (Liberman, 1996). The widespread view at the time was that speech perception was a necessary prerequisite for SWR. In most discussions in the early 1950s, the linguistic construct of the phoneme was considered to be the central elementary building block of speech (Peterson, 1952). The theoretical assumption was that if we can understand the processes used to recognize individual phonemes in nonsense syllables, then this knowledge could be scaled up to SWR. Such a narrow reductionist research strategy is not surprising; researchers in all areas of science typically work on tractable problems that can be studied with existing paradigms and experimental methodologies. Despite the voluminous literature on isolated phoneme perception in the 1950s, 1960s, and 1970s, until the mid-1970s very little was known about how listeners use acoustic-phonetic information in the speech signal to support SWR (Marslen-Wilson, 1975).

Several reasons can be identified for the shift in research efforts from the study of phonetic segments to SWR. First, the number of speech scientists and psycholinguists increased. Second, the cost of performing speech research decreased significantly with the widespread availability of low-cost digital computers and high-powered sophisticated digital signal processing techniques. Third, with more interest in the field of speech perception and more powerful research tools, younger investigators were able to turn their attention and creative efforts to a wider range of challenges related to how acoustic-phonetic information in the speech signal makes contact with stored lexical representations in memory. Importantly, new studies on SWR also used novel experimental paradigms and techniques that required listeners to actively use phonological, lexical, syntactic, and semantic knowledge in assigning a meaningful interpretation to the input. The shift in emphasis from perception of phonetic segments to SWR was also motivated by the belief that by investigating SWR, new insights would be obtained about the role of context, lack of acoustic-phonetic invariance, linearity, and the interaction of multiple sources of knowledge in spoken language processing (Miller, 1962).

20.3.3 Theoretical Accounts of SWR

Early theories of SWR were based on models and research findings in visual word recognition. Three basic families of models have been proposed to account for mapping of speech waveforms onto lexical representations. One approach, represented by the Autonomous Search Model developed by Forster (1976, 1989), is based on the assumption that words are accessed using a *frequency-ordered* search process. In this model, the initial search is performed based on frequency, with high-frequency words searched before low-frequency words. Search theories are no longer considered viable models of SWR and are not considered any further in this chapter.

The second family of models assumes that words are recognized through processes of activation and competition. Early pure activation models like Morton's Logogen Theory assumed that words are recognized based on sensory evidence in the input signal (Morton, 1969). Passive sensing devices called logogens were associated with individual words in the lexicon. These *word detectors* collected information from the input. Once a Logogen reached a threshold, it became activated. To account for frequency effects, common high-frequency words had lower thresholds than rare low-frequency words. There were a number of problems with the Logogen model. It failed to specify precisely the perceptual units used to map acoustic phonetic input onto logogens or how different sources of linguistic information are combined together to alter the activation levels of individual logogens. Finally, the Logogen model was also unable to account for lexical neighborhood effects and the effects of lexical competition among phonetically similar words because the logogens for individual words are activated independently and have no input from other phonetically similar words in memory.

The third family of models combined assumptions from both search and activation models. One example of a hybrid model of SWR is Klatt's Lexical Access From Spectra (LAFS) model (Klatt, 1979), which relies extensively on real-speech input in the form of power spectra that change over time, unlike other models of SWR that rely on preprocessed coded speech signals as input. Klatt argued that earlier models failed to acknowledge the important role of fine phonetic detail because they uniformly assumed the existence of an intermediate abstract level of representation that eliminated potentially useful acoustic information from the speech signal (Klatt, 1986). Based on a detailed analysis of the design architecture of the HARPY speech recognition system (Lowerre & Reddy, 1980), Klatt suggested that intermediate representations may not be optimal for human or machine SWR because they are always potentially error-prone, especially in noise (Klatt, 1977). Instead, Klatt suggested that spoken words could be recognized directly from an analysis of the input power spectrum using a large network of diphones combined with a "backward beam search" technique like the one originally incorporated in HARPY that eliminated weak lexical candidates from further processing (Klatt, 1979). LAFS is the only model of SWR that attempted to deal with fine phonetic variation in speech, which in recent years has come to occupy the attention of many speech and hearing scientists as well as computer engineers who are interested in designing psychologically plausible models of SWR that are robust under challenging conditions (Moore, 2005, 2007b).

20.3.4 Activation and Competition

Almost all current models of SWR assume two fundamental processes: activation and competition

(Gaskell & Marslen-Wilson, 2002; Luce & Pisoni, 1998; McClelland & Elman, 1986; Norris, 1994). Although there is widespread agreement that acoustic-phonetic input activates a set of lexical candidates that are subsequently selected as a response, the precise details of activation and competition remain a matter of continuing debate (Magnuson et al., 2012). As we discuss similarities and differences between activation-competition models of SWR, it is important to emphasize that all of these SWR models deal with somewhat different theoretical issues and empirical findings, making it difficult to draw direct comparisons among specific models. This is a problem that needs to be addressed in the future because all of the current models target different problems and often focus on specific issues (i.e., the role of top-down feedback, competition dynamics, or representational specificity; see Magnuson et al., 2013 for further discussion). Moreover, none of the current models of SWR deal satisfactorily, if at all, with the indexical channel of information encoded in the speech signal, especially vocal source information about the speaker's voice quality, vocal tract transfer function, and environmental context conditions.

Logogen, TRACE, Shortlist, PARSYN, and the Distributed Cohort Model (DCM) all assume that form-based lexical and sublexical representations can be activated at any point in the speech signal, referred to as *radical activation* (Luce & McLennan, 2005). Radical activation differs from the earlier proposal of *constrained activation* in the original Cohort Theory in which the initial activation of a set of lexical candidates was strictly limited to word initial onsets (Marslen-Wilson & Welsh, 1978). The original Cohort Theory was based on the hypothesis that the beginnings of words played a special role in activating a set of word initial lexical candidates or *cohorts*. As more sensory information is acquired, words that became inconsistent with the input signal were dropped from the cohort until only one word remained. Although the first version of the Cohort Theory was quite influential and generated many novel studies that shaped the field of SWR (see papers in Altman, 1990), the model has been significantly revised and updated in the DCM described below.

In current models of SWR, a central defining feature is the assumption of competition among multiple activated lexical candidates. Lexical competition is one of the major areas of current research and theory on SWR (see Hannagan, Magnuson, & Grainger, 2013; Scharenborg & Boves, 2010). Although there is now considerable evidence for competition in SWR, debate continues over the precise cognitive and neural mechanisms underlying lexical competition dynamics (see Magnuson et al., 2013). In TRACE, Shortlist, and PARSYN, competition involves lateral inhibition among lexical representations. Lateral inhibition refers to competition within the same *level* (e.g., words competing with words or segments competing with segments). In contrast, in the DCM, lateral inhibition among local units is replaced by an active process that results from the *blending* of multiple representations distributed across processing levels (Gaskell & Marslen-Wilson, 2002).

20.3.5 TRACE

TRACE is a highly influential interactive-activation localist connectionist model of SWR (McClelland & Elman, 1986). Localist models of processing assume the existence of discrete stand-alone representations or processing units that have meaning and can be interpreted directly, whereas distributed models make use of patterns of activation across a collection of representations that are dependent on each other and cannot be interpreted alone by looking at individual units in the network. TRACE contains three types of processing units corresponding to features, phonemes, and words (Elman & McClelland, 1986; McClelland & Elman, 1986). Connection weights raise or lower activation levels of the nodes at each level depending on the input and activity of the system. Although TRACE has had considerable influence in the field, the model has several weaknesses and relies extensively on a psychologically and neurally implausible processing architecture. One weakness is that TRACE is only concerned with the recognition of isolated spoken words and has little to say about the recognition of spoken words in connected fluent speech. Furthermore, nodes and connections in TRACE are reduplicated to deal with the temporal dynamics of SWR. Recently, Hannagan et al. (2013) developed a new model called TISK that combines time-specific representations with higher-level representations based on string kernels. TISK reduces the number of units and connections by several orders of magnitude relative to TRACE.

20.3.6 Shortlist, Merge, and Shortlist B

Shortlist is another localist connectionist model of SWR (Norris, 1994). First, a short list of lexical candidates is activated consisting of word forms that match the speech signal. In the second stage, a subset of hypothesized lexical items enters a smaller network of word units. Lexical units then compete with one another for recognition via lateral inhibition. Shortlist also attempts to account for segmentation of words in fluent speech via lexical competition. Shortlist simulates the temporal dynamics of SWR without having to rely on the unrealistic processing architecture of

TRACE. Shortlist is also an autonomous model of SWR. Unlike TRACE, Shortlist does not allow for any top-down lexical influences to affect the initial activation of its phoneme nodes. In Merge, an extension of Shortlist, the flow of information between phoneme and word levels is unidirectional and strictly bottom-up (Norris, McQueen, & Cutler, 2000) A revised version of the Shortlist model, Shortlist B, retains many key assumptions of the original model but differs radically in two ways (Norris & McQueen, 2008). First, Shortlist B is based on Bayesian principles. Second, input to Shortlist B is a sequence of phoneme probabilities obtained from human listeners, rather than discrete phonemes. Two other closely related models, SpeM and Fine-Tracker, have been developed recently to accept real speech waveforms (Weber & Scharenborg, 2012).

20.3.7 NAM and PARSYN

One of the most successful models of SWR is the Neighborhood Activation Model (NAM) developed by Luce and Pisoni (1998). NAM was designed to confront the acoustic-phonetic invariance problem in speech perception. NAM assumes that listeners recognize spoken words *relationally* in the context of other phonetically similar words rather than by strictly bottom-up processing of a sequence of abstract phonetic segments. NAM uses a simple similarity metric for estimating phonological distances of spoken words based on the one-phoneme deletion, addition, or substitution rule developed by Greenberg and Jenkins (1964). This computational method of assessing lexical similarity provides an efficient and powerful way of quantifying the *relations* between spoken words.

The approach embodied in NAM avoids the long-standing intractable problem of trying to recognize individual context-free abstract idealized sound segments (e.g., phonemes) from bottom-up linguistic analysis of invariant acoustic-phonetic properties in the speech waveform. The search for unique acoustic-phonetic invariants for phonemes is no longer a necessary prerequisite when the primary recognition problem is viewed as lexical discrimination and selection among similar sounding words in memory rather than context-independent identification of phonetic segments. Perceptual constancy and lexical abstraction emerge naturally in NAM and are an automatic byproduct of processing interactions between initial sensory information in the signal and lexical knowledge the listener has about possible words and phonological contrasts (also, see Grossberg, 2003; Grossberg & Myers, 2000).

PARSYN, another localist model of SWR, is a connectionist instantiation of the design principles originally incorporated in NAM. The model contains three levels of interconnected processing units: (i) input allophones; (ii) pattern allophones; and (iii) words (Luce, Goldinger, & Vitevitch, 2000). Lateral connections between nodes are mutually inhibitory. PARSYN was originally developed to account for lexical competition and probabilistic phonotactics in SWR studies motivated by predictions based on NAM (Vitevitch & Luce, 1999). Unlike TRACE and Shortlist, however, PARSYN assumes the existence of an intermediate allophonic level of representation in the network that encodes fine content-dependent phonetic details (see Wickelgren, 1969).

20.3.8 Distributed Cohort Model

In the DCM, unlike the original Cohort Theory, activation corresponding to a specific word is distributed over a large set of simple processing units (Gaskell & Marslen-Wilson, 2002). In contrast to the three localist SWR models discussed, the DCM assumes distributed representations in which featural input is *projected* onto simple semantic and phonological units. Because the DCM is a distributed model of SWR, there are no intermediate or sublexical units of representation. Moreover, in contrast to lateral inhibition used by TRACE, Shortlist, and PARSYN, lexical competition in DCM is expressed as a *blending* of multiple lexical items that are consistent with the input.

The differences that exist among the four models of SWR reviewed here are relatively modest. Some unresolved issues need to be explored further, such as segmentation of words in sentences and connected fluent speech, the contributions of sentence and discourse context, and the interaction of other sources of knowledge used in spoken language understanding. These issues are known to affect SWR in normal-hearing listeners and in clinical populations with hearing loss, language delay, and cognitive aging. Because the current group of SWR models are all very similar, it is doubtful that these issues will prove to be critically important in deciding which model provides the most realistic and valid account of SWR.

In many ways, it appears that the field of SWR research has reached a modeling plateau. Although there remain numerous unresolved empirical issues, it is also clear that substantial progress has already been made in dealing with many long-standing foundational issues surrounding SWR and how information in the speech signal is mapped onto lexical representations. Given recent findings documenting the increasingly important role of fine phonetic variation and representational specificity on SWR, especially in hearing-impaired populations, it is likely that several of the unconventional design features of the LAFS architecture may find their way into revised versions of these four basic models of SWR in the near future (see recent models developed by Moore, 2007a, 2007b; Rönnberg et al., 2013).

20.4 SWR AND THE MENTAL LEXICON

One solution to long-standing problems associated with the lack of acoustic-phonetic invariance, segmentation, and the context-conditioned nature of speech has been to reframe the perceptual invariance issue by proposing that the primary function of speech perception is the recognition of spoken words rather than the recognition and identification of phonetic segments (Luce & Pisoni, 1998). The proposal to recast research on speech perception and SWR as a lexical selection problem has had a significant influence in the field because it drew attention away from traditional studies of *speech cues* in isolated nonsense syllables to somewhat broader theoretical and empirical issues related to SWR, the mental lexicon and the organization of spoken words in lexical memory (Marslen-Wilson, 1975; Marslen-Wilson & Welsh, 1978). It also emphasized the central role of SWR in language comprehension and production, topics that had been ignored by previous work focused exclusively on phonetic perception of speech sounds.

20.4.1 The Conventional View

The conventional view of speech assumes a bottom-up sensory-based approach to speech perception and SWR in which segments are first recognized from elementary cues and distinctive features in the speech signal are then parsed into words (see Lindgren, 1965a, 1965b for reviews of early theories). Historically, within the conventional *segmental linguistic* approach, variability in speech was treated as an undesirable source of *noise* that needed to be reduced in order to reveal the *hidden* idealized linguistic message (Chomsky & Miller, 1963; Halle, 1985; Miller & Chomsky, 1963). Many factors known to produce acoustic-phonetic variability were deliberately eliminated or systematically controlled for by speech scientists in experimental protocols used to study speech. As a consequence, very little basic research was specifically devoted to understanding how variability in the speech signal and listening environment is encoded, processed, and stored, especially in noise and under adverse listening conditions.

20.4.2 Linearity, Invariance, and Segmentation

Several aspects of the conventional view of speech are difficult to reconcile with the continuous nature of the acoustic waveform produced by a speaker. Importantly, the acoustic consequences of coarticulation, as well as other sources of contextually conditioned variability, result in the failure of the acoustic signal to meet two formal conditions: linearity and invariance. This failure gives rise to a third problem: the absence of explicit segmentation of the acoustic speech signal into discrete units (Chomsky & Miller, 1963). The linearity condition requires that each segment corresponds to a stretch of sound in the utterance. The linearity condition is not met in speech because extensive coarticulation and other contextual effects *smear* acoustic features for adjacent segments. The smearing or *parallel transmission* of acoustic features results in stretches of the speech waveform in which acoustic features of more than one segment are present simultaneously (Liberman, 1996). Acoustic-phonetic invariance means that every segment must have a specific set of defining acoustic attributes in all contexts. Because of coarticulatory effects in speech production, the acoustic properties of a particular speech sound vary as a function of the phonetic environment it is embedded in. Acoustic-phonetic invariance is absent due to within-speaker variation as well as when we look across different speakers (Peterson & Barney, 1952). The absence of acoustic-phonetic invariance is inconsistent with the theoretical assumption that speech can be represented as an idealized context-free linear sequence of discrete linguistic segments. A large body of research over the past 60 years demonstrates clearly that the speech signal cannot be reliably segmented into discrete acoustically defined units; in fluent connected speech, especially casual or reduced speech, it is impossible to identify where one word ends and another begins using acoustic criteria alone. Precisely how the continuous speech signal is mapped onto discrete segmental representations by the listener still remains one of the most important and challenging problems in speech research today and suggests the existence of additional representations that encode and process the graded continuous properties of the speech signal (McMurray & Jongman, 2011).

20.4.3 An Alternative Proposal

Theoretical developments in neurobiology, cognitive science, and brain modeling along with the availability of powerful new computational tools and large digital speech databases have led researchers to reconceptualize the major theoretical problems in speech perception and how they should be approached in light of what we now know about neural development and brain function (Sporns, 1998, 2003; Sporns, Tononi, & Edelman, 2000). In particular, several new *exemplar-based* approaches to SWR have emerged in recent years from independent developments in the field of human categorization (Kruschke, 1992; Nosofsky, 1986), phonetics and laboratory phonology (Johnson, 2002, 2006), and frequency-based phonology in linguistics (Bybee, 2001; Pierrehumbert, 2001). These novel approaches to SWR offer new insights into many traditional problems related to variability and the lack of acoustic-phonetic invariance in speech perception (Moore, 2007a, 2007b; Pisoni, 1997).

Research findings over the past 25 years provide converging support for an approach to SWR that is compatible with a large and growing body of literature in cognitive science dealing with *episodic* models of categorization and *multiple-trace* models of human memory (Erickson & Kruschke, 1998; Hintzman, 1986, 1988; Kruschke, 1992; Nosofsky, 1986; Shiffrin & Steyvers, 1997). This theoretical approach emphasizes the importance of temporal context and the encoding of specific instances in memory. Such accounts of SWR assume that highly detailed stimulus information in the speech signal and listening environment is encoded, processed, and stored by the listener and subsequently becomes an inseparable component of rich and highly detailed lexical representations of spoken words (Port, 2010a, 2010b).

A critical assumption of this approach to speech perception and SWR is that variability in speech is useful and *highly informative* to the listener, rather than being a source of noise that degrades the underlying idealized abstract linguistic representations (Elman & McClelland, 1986). Exemplar-based views assume that listeners encode and store highly detailed *records* of episodic experiences, rather than prototypes or abstractions (Kruschke, 1992; Nosofsky, 1986). According to these accounts, abstraction and categorization occur but they are assumed to *emerge* from computational processes that take place at retrieval not at encoding. Thus, the fine-grained continuous acoustic-phonetic and indexical details of speech and episodic contextual information are not discarded as a consequence of early sensory processing and perceptual encoding of spoken words into lexical representations.

20.4.4 Indexical Properties of Speech

Numerous studies on the perception of talker variability in SWR have shown that *indexical* information in the speech (Abercrombie, 1967), including details about vocal sound source and optical information about the speaker's face, as well as highly detailed contextual information about speaking rate, speaking mode, and other detailed episodic properties, such as the acoustic properties of the listening environment, are encoded into lexical representations (Brandewie & Zahorik, 2010; Lachs, McMichael, & Pisoni, 2003). These additional sources of information are assumed to become an integral part of the long-term representations that a listener stores about the sound patterns of spoken words in his or her language and the talkers he or she has been exposed to (Pisoni, 1997; Port, 2010a, 2010b; Remez, Fellows, & Nagel, 2007; Remez, Rubin, Pisoni, & Carrell, 1981). Viewed from this approach, speech variability is an important theoretical problem to study and understand because it has been ignored by speech scientists over the years despite its central role in all aspects of speech communication, cognition, learning, and memory (Jacoby & Brooks, 1984; Klatt, 1986, 1989; Stevens, 1996). Until recently, very little was known about the contribution of the indexical properties of speech to speech perception and the role these complementary attributes play in SWR (see Van Lancker & Kreiman, 1987).

An example of the parallel encoding of linguistic and indexical information in speech is displayed in Figure 20.3. The absolute frequencies of the vowel formants shown by peaks in the spectrum provide cues to speaker identification (A), whereas the relative differences among the formants specify information

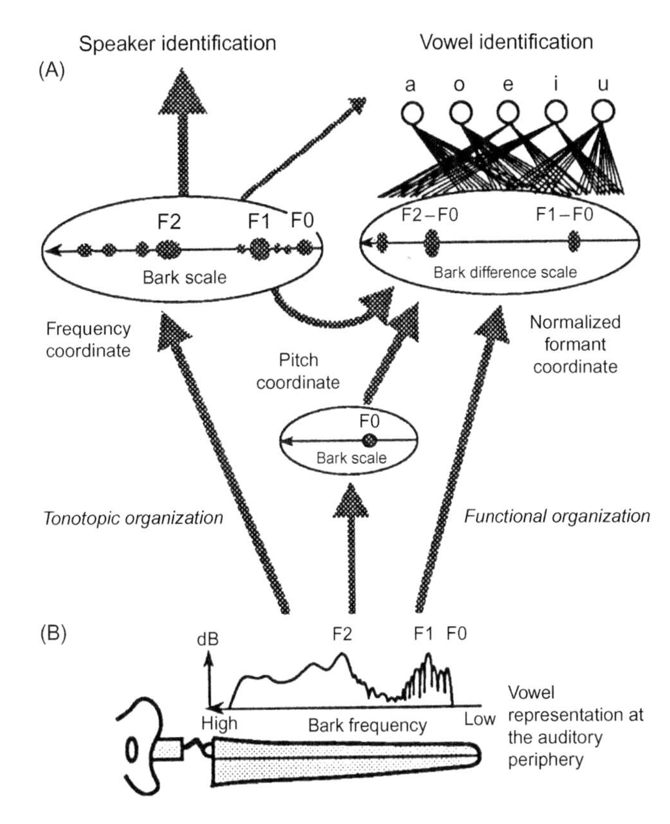

FIGURE 20.3 (A) Vowel projections at the auditory periphery reveal that information for speaker identification and (B) perception of vowel quality is carried simultaneously and in parallel by the same acoustic signal. The tonotopic organization of the absolute frequencies using a bark scale provides reliable cues to speaker identification, whereas the relations among the formant (F1, F2, and F3) patterns in terms of difference from F0 in barks provide reliable cues to vowel identification. *Adapted from Hirahara and Kato (1992) with permission from the publisher.*

used for vowel identification (B). Both channels are carried simultaneously by the same acoustic signal and both sources of information are encoded by the peripheral and central auditory mechanisms used to process speech signals.

Despite recent evidence in favor of episodic approaches to speech perception and SWR, we believe a hybrid account of SWR that also incorporates abstract representations espoused in the conventional view is a promising direction for future research (see proposals by Moore, 2007a, 2007b; Rönnberg et al., 2013). Substantial evidence has been reported for both abstract and episodic coding of speech by human listeners. The challenge for the future is to identify the conditions under which these types of representations are used in SWR. For example, there is mounting evidence in support of the time course hypothesis of SWR (Luce & Lyons, 1998; Luce & McLennan, 2005) that posits that during the process of recognizing spoken words, abstract linguistic information is typically processed prior to vocal sound source information and other types of indexical and episodic information in speech (Krestar & McLennan, 2013; Mattys & Liss, 2008; McLennan & Luce, 2005; Vitevitch & Donoso, 2011). Time course data on SWR are also consistent with hemispheric differences demonstrating that abstract information is processed more efficiently in the left hemisphere and more specific episodic details of speech, including indexical information, are processed more efficiently in the right hemisphere. There are data in support of such hemispheric differences in the visual domain as well, both for visual words (Marsolek, 2004) and nonlinguistic information (Burgund & Marsolek, 2000), and in the auditory domain for spoken words (González & McLennan, 2007), talker identification (González, Cervera-Crespo, & McLennan, 2012), and nonlinguistic environmental sounds (González & McLennan, 2009). Given that hemispheric differences have been observed in the visual and auditory domains—and for both linguistic and nonlinguistic stimuli—these results suggest that having both abstract and more detailed specific representations may be a general property of cognition and the human information processing system. Finally, controlled attention and other neurocognitive and linguistic factors also affect listeners' processing of abstract and indexical information in speech (Maibauer, Markis, Newell, & McLennan, 2014; McLennan, 2006; Theodore & Blumstein, 2011).

20.5 SOME NEW DIRECTIONS AND FUTURE CHALLENGES

After more than 50 years of research on human speech perception, many of the foundational assumptions of the conventional segmental linguistic view of speech have turned out to be misguided, given our current understanding of how the peripheral and central auditory system work and how the brain and nervous system function to recognize spoken language (Davis & Johnsrude, 2003; Hickok & Poeppel, 2007; Scott et al., 2000; Scott & Johnsrude, 2003). Long-standing assumptions about SWR are being critically revaluated (Luce & McLennan, 2005). Deeper theoretical insights have also emerged in recent years, encouraging further empirical research on normal-hearing populations, as well as clinical populations with hearing loss (Niparko et al., 2009; Pichora-Fuller, 1995) and language delays (Beckage, Smith, & Hills, 2011).

We have suggested that speech should no longer be viewed as just a linear sequence of idealized abstract context-free linguistic segments (see Port, 2010a, 2010b). It is also becoming clear that any principled theoretically motivated account of SWR will also have to be compatible with what we currently know about human information processing, including episodic memory and learning (Pisoni, 2000). SWR does not take place in a vacuum isolated from the rest of cognition; core computational processes used in SWR are inseparable from basic memory, learning, and cognitive control processes that reflect the operation of many separate neurobiological components working together as a functionally integrated system (Belin, Zattore, Lafaille, Ahad, & Pike, 2000). As Nauta said more than 50 years ago, "no part of the brain functions on its own but only through the other parts of the brain with which it is connected" (Nauta, 1964, p. 125). In other words, it takes a whole brain to recognize words and understand spoken language—the ear is connected to the brain and the brain is connected to the ear.

Research on how early sensory information in speech is *mapped* onto lexical representations of spoken words and how these detailed representations are accessed in SWR tasks has numerous clinical implications for listeners with hearing loss, especially hearing-impaired children and adults who have received cochlear implants (Kirk, 2000; Kirk & Choi, 2009; Kronenberger, Pisoni, Henning, & Colson, 2013; Niparko et al., 2009; Pisoni, 2005).

In addition to clinical implications related to hearing impairment, basic research in SWR has led to other new directions. Research on the recognition of foreign-accented speech contributes to our understanding of lexical activation and selection processes (e.g., Chan & Vitevitch, 2015) and provides new insights into the circumstances under which variability in indexical information affects listeners' ability to recognize words by non-native speakers (McLennan & González, 2012).

SWR in bilinguals represents another promising area of research. Studies with bilinguals make important contributions to our understanding of basic representational and processing issues in SWR (Vitevitch, 2012).

For example, a recent study with bilinguals demonstrates that it is easier to learn to recognize the voices of previously unfamiliar talkers in a language learned early in life (Bregman & Creel, 2014). Studies of bilingual SWR can also be used as a tool to investigate differences in cognitive control processes between monolinguals and bilinguals (Kroll & Bialystok, 2013; however, see de Bruin, Treccani, & Della Sala, 2015).

Finally, scientists have extended SWR studies to investigations of the aging lexicon and the decline in language processes (e.g., Ben-David et al., 2011; Meister et al., 2013; Sommers, 2005; Yonan & Sommers, 2000). Many aspects of language processing are less likely to show age-related declines—and some show improvements—as a consequence of normal aging compared with many other perceptual and cognitive domains (Taler, Aaron, Steinmetz, & Pisoni, 2010).

20.6 SUMMARY AND CONCLUSIONS

Listeners bring an enormous amount of prior knowledge to every spoken language task they are asked to perform in the research laboratory, clinic, or daily life. In this chapter, we have argued that it is important to keep these broad observations in mind in understanding how spoken words are recognized so efficiently and how listeners manage to reliably recover the talker's intended linguistic message from highly degraded sensory inputs under challenging conditions. The findings reviewed in this chapter suggest that SWR processes are highly robust because listeners are able to make use of multiple sources of information encoded in the speech signal—the traditional linguistic pathway that encodes acoustic-phonetic information specifying the talker's intended message, the indexical pathway that encodes and carries detailed episodic contextual attributes specifying the vocal sound source such as the talker's gender, regional dialect, and mental and physical states, as well as other downstream sources of linguistic knowledge that support word prediction strategies, sentence parsing, and linguistic interpretation. Variability in the speech signal was once considered an undesirable source of noise and signal degradation that needed to be eliminated or normalized away to recover the idealized abstract segmental content of the talker's intended linguistic message. We now realize that this long-standing conventional view of speech perception and SWR is fundamentally incorrect and that variability in speech is highly informative and an extremely valuable source of contextual information that listeners encode, process, and routinely make use of in recognizing spoken words, especially in noise and under other adverse listening conditions.

We suggested that the field of SWR appears to have reached a plateau in terms of major theoretical or modeling advancements. However, there have been a number of important new empirical and methodological contributions. Among these include time course findings, due in large part to innovative techniques such as eye-tracking using the visual world paradigm (Allopenna, Magnuson, & Tanenhaus, 1998) and, more recently, mouse-tracking (Spivey, Grosjean, & Knoblich, 2005). Moving forward into the future, these new methodologies should contribute to improved theories and more precise models of SWR. We also expect to see increases in the number of studies investigating the recognition of casually spoken phonetically reduced words (e.g., Ernestus, Baayen, & Schreuder, 2002) and novel approaches to modeling SWR, including the integration of current computational models with existing hybrid frameworks that incorporate the roles of both abstract and indexical information in SWR (Luce et al., 2000). Finally, there is a rapidly growing body of research on lexical organization and lexical connectivity of words using theory and methodology developed in the field of complex networks and graph theory that has provided additional new insights into spoken language processing and holds promise for dealing with more global aspects of SWR in typical and atypical populations (Altieri, Gruenenfelder, & Pisoni, 2010; Beckage et al., 2011; Kenett, Wechsler-Kashi, Kenett, Schwartz, & Ben-Jacob, 2013; Vitevitch, 2008; Vitevitch, Chan, & Goldstein, 2014).

Acknowledgments

Preparation of this chapter was supported, in part, by research grants R01 DC-000111 and R01 DC-009581 to Indiana University from NIH NIDCD. We are grateful to Luis Hernandez for his dedication, help, and continued assistance and advice over many years in maintaining our research laboratory and contributing to the creation of a highly productive and stimulating research environment at Indiana University in Bloomington. We also thank Terren Green for her help and assistance in the preparation of the present manuscript. Finally, D.B.P. expresses his thanks and deepest appreciation to Professor Kenneth N. Stevens, his postdoctoral mentor in the Speech Communications Group at the Research Laboratory of Electronics, MIT, who passed away last year. As everyone in our field knows, Ken Stevens was one of the early pioneers and major architects in the field of speech communications research, speech acoustics, acoustics-phonetics, and speech perception, and he provided wonderful guidance and advice to many graduate students and postdocs during his more than 50 years as a member of the faculty at MIT. Ken—this one's for you! We'll miss you.

References

Abercrombie, D. (1967). *Elements of general phonetics*. Edinburgh: Edinburgh University.

Abrams, M. H., Goffard, S. J., Kryter, K. D., Miller, G. A., Miller, J., & Sanford, F. H. (1944). *Speech in noise: A study of the factors determining its intelligibility. OSRD Report 4023*. Cambridge, MA: Research on Sound Control. Psycho-Acoustic Laboratory, Harvard University.

Allen, J., Hunnicutt, M. S., & Klatt, D. (1987). *From text to speech: The MITalk system*. Cambridge, UK: Cambridge University Press.

Allen, J. B. (1994). How do humans process and recognize speech? *IEEE Transactions on Speech Audio*, *2*(4), 567–577.

Allen, J. B. (2005). *Articulation and intelligibility*. San Rafael, CA: Morgan & Claypool Publishers.

Allopenna, P. D., Magnuson, J. S., & Tanenhaus, M. K. (1998). Tracking the time course of spoken word recognition using eye movements: Evidence for continuous mapping models. *Journal of Memory and Language*, *38*, 419–439.

Altieri, N., Gruenenfelder, T., & Pisoni, D. B. (2010). Clustering coefficients of lexical neighborhoods. *The Mental Lexicon*, *5*(1), 1–21.

Altman, G. T. M. (Ed.), (1990). *Cognitive models of speech processing: Psycholinguistic and computational perspectives*. Cambridge, MA: MIT Press.

Arlinger, S., Lunner, T., Lyxell, B., & Pichora-Fuller, M. K. (2009). The emergence of cognitive hearing science. *Scandinavian Journal of Psychology*, *50*, 371–384.

Bagley, W. C. (1900). The apperception of the spoken sentence: A study in the psychology of language. *American Journal of Psychology*, *12*(1), 80–130.

Beckage, N., Smith, L., & Hills, T. (2011). Small worlds and semantic network growth in typical and late talkers. *PLoS One*, *6*(5), e19348.

Belin, P., Zattore, R. J., Lafaille, P., Ahad, P., & Pike, B. (2000). Voice selective areas in human auditory cortex. *Nature*, *403*, 309–312.

Ben-David, B. M., Chambers, C. G., Daneman, M., Pichora-Fuller, M. K., Reingold, E. M., & Schneider, B. A. (2011). Effects of aging and noise on real-time spoken word recognition: Evidence from eye movements. *Journal of Speech, Language, and Hearing Research*, *54*, 243–262.

Black, J. W (1946). Effects of voice communication training. *Speech Monographs*, *13*, 64–68.

Brandewie, E., & Zahorik, P. (2010). Prior listening in rooms improves speech intelligibility. *Journal of the Acoustical Society of America*, *128*(1), 291–299.

Bregman, M. R., & Creel, S. C. (2014). Gradient language dominance affects talker learning. *Cognition*, *130*, 85–95.

Broadbent, D. E. (1967). Word-frequency effect and response bias. *Psychological Review*, *74*(1), 1–15.

Burgund, E. D., & Marsolek, C. J. (2000). Viewpoint-invariant and viewpoint-dependent object recognition in dissociable neural subsystems. *Psychonomic Bulletin & Review*, *7*, 480–489.

Bybee, J. (2001). *Phonology and language use*. Cambridge, UK: Cambridge University Press.

Carreiras, M., Armstrong, B. C., Perea, M., & Frost, R. (2014). The what, when, where, and how of visual word recognition. *Trends in Cognitive Sciences*, *18*, 90–98.

Chan, K. Y., & Vitevitch, M. S. (2015). The influence of neighborhood density on the recognition of Spanish-accented words. *Journal of Experimental Psychology: Human Perception and Performance*, *41*(1), 65–85.

Chomsky, N., & Miller, G. A. (1963). Introduction to the formal analysis of natural languages. In R. D. Luce, R. R. Bush, & E. Galanter (Eds.), *Handbook of mathematical psychology* (pp. 269–321). New York, NY: Wiley.

Cole, R. A. (1973). Listening to mispronunciations: A measure of what we hear during speech. *Perception & Psychophysics*, *13*, 153–156.

Cole, R. A., & Rudnicky, A. I. (1983). What's new in speech perception? The research and ideas of William Chandler Bagley, 1874–1946. *Psychological Review*, *90*, 94–101.

Cooper, F. S., Delattre, P. C., Liberman, A. M., Borst, J. M., & Gerstman, L. J. (1952). Some experiments on the perception of synthetic speech sounds. *Journal of the Acoustical Society of America*, *24*, 597–606.

Dahan, D., & Magnuson, J. S. (2006). Spoken-word recognition. In M. J. Traxler, & M. A. Gernsbacher (Eds.), *Handbook of psycholinguistics* (pp. 249–283). Amsterdam: Academic Press.

Davis, M. H., & Johnsrude, I. S. (2003). Hierarchical processing in spoken language comprehension. *Journal of Neuroscience*, *23*, 3423–3431.

de Bruin, A., Treccani, B., & Della Sala, S. (2015). Cognitive advantage in bilingualism: An example of publication bias? *Psychological Science*, *26*(1), 99–107.

Denes, P. B., & Pinson, E. N. (1963). *The speech chain: The physics and biology of spoken language*. New York, NY: Bell Telephone Laboratories.

Elman, J. L., & McClelland, J. L. (1986). Exploiting lawful variability in the speech waveform. In J. S. Perkell, & D. H. Klatt (Eds.), *Invariance and variability in speech processing* (pp. 360–385). Hillsdale, NJ: Erlbaum.

Erickson, M. A., & Kruschke, J. K. (1998). Rules and exemplars in category learning. *Journal of Experimental Psychology: General*, *127*, 107–140.

Ernestus, M., Baayen, H., & Schreuder, R. (2002). The recognition of reduced word forms. *Brain and Language*, *81*, 162–173.

Eukel, B. (1980). Phonotactic basis for word effects: Implications for lexical distance metrics. *Journal for the Acoustical Society of America*, *68*(S1), S33.

Flanagan, J. L. (1965). *Speech analysis synthesis and perception*. Heidelberg: Springer-Verlag.

Fletcher, H. (1929). *Speech and hearing* (1st ed.). New York, NY: D. Van Nostrand Company, Inc.

Fletcher, H. (1953). *Speech and hearing in communication*. Huntington, NY: Krieger.

Fletcher, H., & Galt, R. H. (1950). The perception of speech and its relation to telephony. *Journal of the Acoustical Society of America*, *22* (2), 89–151.

Forster, K. I. (1976). Accessing the mental lexicon. In R. J. Wales, & E. Walker (Eds.), *New approaches to language mechanisms*. Amsterdam: North Holland.

Forster, K. I. (1989). Basic issues in lexical processing. In W. Marslen-Wilson (Ed.), *Lexical representation and process* (pp. 75–107). Cambridge, MA: MIT Press.

Francis, W. N., & Kucera, H. (1964). *A standard corpus of present-day edited American English*. Providence, RI: Department of Linguistics, Brown University.

Gaskell, M. G. (Ed.), (2007a). *The Oxford handbook of psycholinguistics* New York, NY: Oxford University Press.

Gaskell, M. G. (2007b). Statistical and connectionist models of speech perception and word recognition. In M. G. Gaskell (Ed.), *The Oxford handbook of psycholinguistics* (pp. 55–69). New York, NY: Oxford University Press.

Gaskell, M. G., & Marslen-Wilson, W. M. (2002). Representation and competition in the perception of spoken words. *Cognitive Psychology*, *45*, 220–266.

González, J., Cervera-Crespo, T., & McLennan, C. T. (2012). Hemispheric differences in specificity effects in talker identification. *Attention, Perception, & Psychophysics*, *72*, 2265–2273.

González, J., & McLennan, C. T. (2007). Hemispheric differences in indexical specificity effects in spoken word recognition. *Journal of Experimental Psychology: Human Perception and Performance*, *33*, 410–424.

González, J., & McLennan, C. T. (2009). Hemispheric differences in the recognition of environmental sounds. *Psychological Science*, *20*, 887–894.

Greenberg, J. H., & Jenkins, J. J. (1964). Studies in the psychological correlates of the sound system of American English. *Word*, *20*, 157–177.

Grossberg, S. (2003). Resonant neural dynamics of speech perception. *Journal of Phonetics*, *31*, 423–445.

Grossberg, S., & Myers, C. W. (2000). The resonant dynamics of speech perception: Interword integration and duration-dependent backward effects. *Psychological Review, 107*, 735–767.

Halle, M. (1985). Speculations about the representation of words in memory. In V. A. Fromkin (Ed.), *Phonetic linguistics* (pp. 101–104). New York, NY: Academic Press.

Hannagan, T., Magnuson, J. S., & Grainger, J. (2013). Spoken word recognition without a trace. *Frontiers in Psychology, 4*, 563.

Hickok, G., & Poeppel, D. (2007). The cortical organization of speech processing. *Nature Reviews Neuroscience, 8*, 393–402.

Hintzman, D. L. (1986). "Schema abstraction" in a multiple-trace memory model. *Psychological Review, 93*, 411–428.

Hintzman, D. L. (1988). Judgments of frequency and recognition memory in a multiple-trace memory model. *Psychological Review, 95*, 528–551.

Hirahara, T., & Kato, H (1992). The effect of F0 on vowel identification. In Y. Tohkura, E. Vatikiotis-Bateson, & Y Sagisaka (Eds.), *Speech perception, production and linguistic structure* (pp. 89–112). Tokyo: Ohmsha Publishing.

Hirsh, I. J. (1947). Clinical application of two Harvard auditory tests. *Journal of Speech and Hearing Disorders, 12*, 151–158.

Hockett, C. F. (1955). *A manual of phonology.* Baltimore, MD: Waverly Press.

Howes, D. (1954). On the interpretation of word frequency as a variable affecting speed of recognition. *Journal of Experimental Psychology, 48*(2), 106–112.

Howes, D. (1957). On the relation between the intelligibility and frequency of occurrence of English words. *The Journal of the Acoustical Society of America, 29*(2), 296–305.

Hudgins, C. V., Hawkins, J. E., Karlin, J. E., & Stevens, S. S. (1947). The development of recorded auditory tests for measuring hearing loss for speech. *The Laryngoscope, 57*(1), 57–89.

Jacoby, L. L., & Brooks, L. R. (1984). Nonanalytic cognition: Memory, perception, and concept learning. In G. Bower (Ed.), *The psychology of learning and motivation* (pp. 1–47). New York, NY: Academic Press.

Jenkins, J. J. (1979). Four points to remember: A tetrahedral model of memory experiments. In L. S. Cermak, & F. I. M. Craik (Eds.), *Levels of processing in human memory* (pp. 429–446). Hillsdale, NJ: Erlbaum Associates.

Johnson, K. (2002). *Acoustic and auditory phonetics* (2nd ed., (1st edition, 1997)).). Oxford: Blackwell.

Johnson, K. (2006). Resonance in an exemplar-based lexicon: The emergence of social identity and phonology. *Journal of Phonetics, 34*, 485–499.

Jusczyk, P. W., & Luce, P. A. (2002). Speech perception and spoken word recognition: Past and present. *Ear & Hearing, 23*, 2–40.

Karlin, J. E., Abrams, M. H., Sanford, F. H., & Curtis, J. F. (1944). *Auditory tests of the ability to hear speech in noise. OSRD Report 3516.* Cambridge, MA: Research on Sound Control. Psycho-Acoustic Laboratory, Harvard University.

Kenett, Y. N., Wechsler-Kashi, D., Kenett, D. Y., Schwartz, R. G, & Ben-Jacob, E. (2013). Semantic organization in children with cochlear implants: Computational analysis of verbal fluency. *Frontiers in Psychology, 4*(543), 1–11.

Kirk, K. I. (2000). *Communication skills in early-implanted children.* Washington, DC: American Speech-Language-Hearing Association.

Kirk, K. I., & Choi, S. (2009). Clinical investigations of cochlear implant performance. In J. K. Niparko (Ed.), *Cochlear implants: Principles & practices* (2nd ed., pp. 191–222). Philadelphia, PA: Lippincott Williams & Wilkins.

Klatt, D. H. (1977). Review of the ARPA speech understanding project. *Journal of the Acoustical Society of America, 62*, 1345–1366.

Klatt, D. H. (1979). Speech perception: A model of acoustic-phonetic analysis and lexical access. *Journal of Phonetics, 7*, 279–312.

Klatt, D. H. (1986). The problem of variability in speech recognition and in models of speech perception. In J. S. Perkell, & D. H. Klatt (Eds.), *Invariance and variability in speech processing* (pp. 300–319). Hillsdale, NJ: Erlbaum.

Klatt, D. H. (1987). Review of text-to-speech conversion for English. *Journal of the Acoustical Society of America, 82*(3), 737–793.

Klatt, D. H. (1989). Review of selected models of speech perception. In W. Marslen-Wilson (Ed.), *Lexical representation and process* (pp. 169–226). Cambridge, MA: MIT Press.

Konkle, D. F., & Rintelmann, W. F. (1983). *Principles of speech audiometry perspectives in audiology series.* Baltimore, MD: University Park Press.

Krestar, M. L., & McLennan, C. T. (2013). Examining the effects of emotional tone of voice on spoken word recognition. *The Quarterly Journal of Experimental Psychology, 66*, 1793–1802.

Kroll, J. F., & Bialystok, E. (2013). Understanding the consequences of bilingualism for language processing and cognition. *Journal of Cognitive Psychology, 25.* Available from: http://dx.doi.org/10.1080/20445911.2013.799170.

Kronenberger, W. G., Pisoni, D. B., Henning, S. C., & Colson, B. G. (2013). Executive functioning skills in long-term users of cochlear implants: A case control study. *Journal of Pediatric Psychology, 38*(8), 902–914.

Kruschke, J. K. (1992). Alcove: An exemplar-based connectionist model of category learning. *Psychological Review, 99*, 22–44.

Lachs, L., McMichael, K., & Pisoni, D. B. (2003). Speech perception and implicit memory: Evidence for detailed episodic encoding. In J. Bowers, & C. Marsolek (Eds.), *Rethinking implicit memory* (pp. 215–235). Oxford: Oxford University Press.

Landauer, T. K., & Streeter, L. A. (1973). Structural differences between common and rare words: Failure of equivalence assumptions for theories of word recognition. *Journal of Verbal Learning and Verbal Behavior, 7*, 291–295.

Liberman, A. M. (1996). *Speech: A special code.* Cambridge, MA: MIT Press.

Liberman, A. M., Cooper, F. S., Shankweiler, D. P., & Studdert-Kennedy, M. (1967). Perception of the speech code. *Psychological Review, 74*(6), 431–461.

Licklider, J. C. R. (1952). On the process of speech perception. *Journal of the Acoustical Society of America, 24*, 590–594.

Licklider, J. C. R., & Miller, G. A. (1951). The perception of speech. In S. S. Stevens (Ed.), *Handbook of experimental psychology* (pp. 1040–1074). Oxford, England: Wileyxi, 1436 pp.

Lindgren, N. (1965a). Machine recognition of human language part I-Automatic speech recognition. *IEEE Spectrum, 2*(3), 114–136.

Lindgren, N. (1965b). Machine recognition of human language part II-Theoretical models of speech perception and language. *IEEE Spectrum, 2*(4), 45–59.

Lindgren, N. (1967). Speech—Man's natural communication. *IEEE Spectrum, 4*, 75–86.

Lowerre, B., & Reddy, R. (1980). The Harpy speech understanding system. In W. A. Lea (Ed.), *Trends in speech recognition.* Englewood Cliffs, NJ: Prentice-Hall.

Luce, P. A., Goldinger, S. D., & Vitevitch, M. S. (2000). It's good ... but is it ART? *Behavioral and Brain Sciences, 23*, 336.

Luce, P. A., & Lyons, E. A. (1998). Specificity of memory representations for spoken words. *Memory & Cognition, 36*, 708–715.

Luce, P. A., & McLennan, C. T. (2005). Spoken word recognition: The challenge of variation. In D. B. Pisoni, & R. E. Remez (Eds.), *Handbook of speech perception* (pp. 591–609). Malden, MA: Blackwell.

Luce, P. A., & Pisoni, D. B. (1998). Recognizing spoken words: The neighborhood activation model. *Ear & Hearing, 19*, 1–36.

Magnuson, J. S., Mirman, D., & Harris, H. D. (2012). Computational models of spoken word recognition. In M. Spivey, K. McRae, & M. Joanisse (Eds.), *The Cambridge handbook of psycholinguistics.* Cambridge, UK: Cambridge University Press.

Magnuson, J. S., Mirman, D., & Myers, E. (2013). Spoken word recognition. In D. Reisberg (Ed.), *The Oxford handbook of cognitive psychology* (pp. 412–441). New York, NY: Oxford University Press.

Maibauer, A. M., Markis, T. A., Newell, J., & McLennan, C. T. (2014). Famous talker effects in spoken word recognition. *Attention, Perception, & Psychophysics, 76*, 11–18.

Marks, L. E., & Miller, G. A. (1964). The role of semantic and syntactic constraints in the memorization of English sentences. *Journal of Verbal Learning and Verbal Behavior, 3*, 1–5.

Marslen-Wilson, W. D. (1975). Sentence perception as an interactive parallel process. *Science, 189*(4198), 226–228.

Marslen-Wilson, W. D. (1989). *Lexical representation and process.* Cambridge, MA: MIT Press.

Marslen-Wilson, W. D., & Welsh, A. (1978). Processing interactions and lexical access during word recognition in continuous speech. *Cognitive Psychology, 10*, 29–63.

Marsolek, C. J. (2004). Abstractionist versus exemplar-based theories of visual word priming: A subsystems resolution. *Quarterly Journal of Experimental Psychology: Section A, 57*, 1233–1259.

Mattys, S. L., & Liss, J. M. (2008). On building models of spoken-word recognition: Where there is as much to learn from natural "oddities" as artificial normality. *Perception & Psychophysics, 70*, 1235–1242.

McClelland, J. L., & Elman, J. L. (1986). The TRACE model of speech perception. *Cognitive Psychology, 18*, 1–86.

McLennan, C. T. (2006). The time course of variability effects in the perception of spoken language: Changes across the lifespan. *Language and Speech, 49*, 113–125.

McLennan, C. T., & González, J. (2012). Examining talker effects in the perception of native- and foreign-accented speech. *Attention, Perception, & Psychophysics, 74*, 824–830.

McLennan, C. T., & Luce, P. A. (2005). Examining the time course of indexical specificity effects in spoken word recognition. *Journal of Experimental Psychology: Learning, Memory, and Cognition, 31*, 306–321.

McMurray, B., & Jongman, A. (2011). What information is necessary for speech categorization? Harnessing variability in the speech signal by integrating cues computed relative to expectations. *Psychological Review, 118*(2), 219–246.

McQueen, J. M. (2007). Eight questions about spoken-word recognition. In M. G. Gaskell (Ed.), *The Oxford handbook of psycholinguistics* (pp. 37–53). Oxford: Oxford University Press.

Meister, H., Schreitmüller, S., Grugel, L., Ortmann, M., Beutner, D., Walger, M., et al. (2013). Cognitive resources related to speech recognition with a competing talker in younger and older listeners. *Neuroscience, 232*, 74–82.

Miller, G., & Isard, S. (1963). Some perceptual consequences of linguistic rules. *Journal of Verbal Learning and Verbal Behavior, 2*, 217–228.

Miller, G. A. (1951). *Language and communication.* New York, NY: McGraw-Hill.

Miller, G. A. (1962). Decision units in the perception of speech. *Information Theory, IRE Transactions, 8*(2), 81–83.

Miller, G. A., & Chomsky, N. (1963). Finitary models of language users. In R. D. Luce, R. R. Bush, & E. Galanter (Eds.), *Handbook of mathematical psychology* (Vol. 2, pp. 419–491). New York, NY: Wiley.

Miller, G. A., Heise, G. A., & Lichten, W. (1951). The intelligibility of speech as a function of the context of the test material. *Journal of Experimental Psychology, 41*, 329–335.

Miller, G. A., & Selfridge, J. A. (1950). Verbal context and the recall of meaningful material. *American Journal of Psychology, 63*, 176–185.

Moore, R. (2007a). Spoken language processing: Piecing together the puzzle. *Speech Communication, 49*, 418–443.

Moore, R.K. (2005). Towards a unified theory of spoken language processing. *Proceeding 4th IEEE international conference on cognitive informatics.* Irvine, USA, 8–10.

Moore, R. K. (2007b). Presence: A human-inspired architecture for speech-based human-machine interaction. *IEEE Transactions on Computers, 56*(9), 1176–1188.

Morton, J. (1969). Interaction of information in word recognition. *Psychological Review, 76*, 165–178.

Morton, J. (1979). Word recognition. In J. Morton, & J. D. Marshall (Eds.), *Psycholinguistics 2: Structures and processes* (pp. 107–156). Cambridge, MA: MIT Press.

Nauta, W. J. H. (1964). Discussion of 'retardation and faciliatation in learning by stimulation of frontal cortex in monkeys'. In J. M Warren, & K. Akert (Eds.), *The frontal granular cortex and behavior* (p. 125). New York, NY: McGraw-Hill.

Niparko, J. K., Kirk, K. I., McConkey-Robbins, A., Mellon, N. K., Tucci, D. L., & Wilson, B. S. (2009). *Cochlear implants: Principles & practices* (2nd ed.). Philadelphia, PA: Lippincott Williams & Wilkins.

Norris, D. (1994). Shortlist: A connectionist model of continuous speech recognition. *Cognition, 52*, 189–234.

Norris, D., & McQueen, J. M. (2008). Shortlist B: A Bayesian model of continuous speech recognition. *Psychological Review, 115*(2), 357–395.

Norris, D. S., McQueen, J. M., & Cutler, A. (2000). Merging information in speech recognition: Feedback is never necessary. *Behavioral and Brain Sciences, 23*(3), 299–325.

Nosofsky, R. M. (1986). Attention, similarity, and the identification-categorization relationship. *Journal of Experimental Psychology: General, 115*, 39–57.

Oldfield, R. C. (1966). Things, words, and the brain. *The Quarterly Journal of Experimental Psychology, 18*(4), 340–353.

Peterson, G. E. (1952). The information-bearing elements of speech. *Acoustical Society of America Journal, 24*(6), 629–637.

Peterson, G. E., & Barney, H. L. (1952). Control methods used in the study of a vowel. *Journal of the Acoustical Society of America, 24*(2), 175–184.

Pichora-Fuller, M. K. (1995). How young and old adults listen to and remember speech in noise. *Journal of the Acoustical Society of America, 97*(1), 593–608.

Pierrehumbert, J. B. (2001). Exemplar dynamics: Word frequency, lenition and contrast. In J. Bybee, & P. Hopper (Eds.), *Frequency and the emergence of linguistic structure* (pp. 137–158). Amsterdam: John Benjamins.

Pisoni, D. B. (1978). Speech perception. In W. K. Estes (Ed.), *Handbook of learning and cognitive processes* (Vol. 6, pp. 167–233). Hillsdale, NJ: Erlbaum Associates.

Pisoni, D. B. (1997). Some thoughts on "normalization" in speech perception. In K. Johnson, & J. W. Mullennix (Eds.), *Talker variability in speech processing* (pp. 9–32). San Diego, CA: Academic Press.

Pisoni, D. B. (2000). Cognitive factors and cochlear implants: Some thoughts on perception, learning, and memory in speech perception. *Ear & Hearing, 21*, 70–78.

Pisoni, D. B. (2005). Speech perception in deaf children with cochlear implants. In D. B. Pisoni, & R. E. Remez (Eds.), *Handbook of speech perception* (pp. 494–523). Oxford: Blackwell Publishers.

Pisoni, D. B., & Levi, S. V. (2007). Representations and representational specificity in speech perception and spoken word recognition. In M. G. Gaskell (Ed.), *The Oxford handbook of psycholinguistics* (pp. 3–18). Oxford: Oxford University Press.

Pisoni, D. B., & Luce, P. A. (1986). Speech perception: Research, theory and the principal issues. In E. C. Schwab, & H. C. Nusbaum (Eds.), *Pattern recognition by humans and machines: Speech perception* (Vol. 1, pp. 1–50). New York, NY: Academic Press.

Pisoni, D. B., Nusbaum, H. C., Luce, P. A., & Slowiaczek, L. M. (1985). Speech perception, word recognition and the structure of the lexicon. *Speech Communication, 4*, 75–95.

Pisoni, D. B., & Remez, R. E. (2005). *The handbook of speech perception.* Oxford: Blackwell.

Pollack, I., Rubenstein, H., & Decker, L. (1959). Intelligibility of known and unknown message sets. *Journal of the Acoustical Society of America, 31*, 273–279.

Pollack, I., Rubenstein, H., & Decker, L. (1960). Analysis of incorrect responses to an unknown message set. *Journal of the Acoustical Society of America, 32*, 454–457.

Port, R. F. (2010a). Language is a social institution: Why phonemes and words do not have explicit psychological form. *Ecological Psychology, 22*, 304–326.

Port, R. F. (2010b). Rich memory and distributed phonology. *Language Sciences, 32*(1), 43–55.

Price, C. J. (2012). A review and synthesis of the first 20 years of PET and fMRI studies of heard speech, spoken language and reading. *NeuroImage, 62*, 816–847.

Remez, R. E., Fellows, J. M., & Nagel, D. S. (2007). On the perception of similarity among talkers. *Journal of the Acoustical Society of America, 122*, 3688–3696.

Remez, R. E., Rubin, P. E., Pisoni, D. B., & Carrell, T. D. (1981). Speech perception without traditional speech cues. *Science, 212* (4497), 947–950.

Rönnberg, J., Lunner, T., Zekveld, A., Sörqvist, P., Danielsson, H., Lyxell, B., et al. (2013). The ease of language understanding (ELU) model: Theoretical, empirical, and clinical advances. *Frontiers in Systems Neuroscience, 7*(article 31). Available from: http://dx.doi.org/10.3389/fnsys.2013.00031.

Rosenzweig, M. R., & Postman, L. (1957). Intelligibility as a function of frequency of usage. *Journal of Experimental Psychology, 54*(6), 412–422.

Rosenzweig, M. R., & Stone, G. (1948). Wartime research in psychoacoustics. *Review of Educational Research Special Edition: Psychological Research in the Armed Forces, 18*(6), 642–654.

Savin, H. B. (1963). Word-frequency effect and errors in the perception of speech. *Journal of the Acoustical Society of America, 35*, 200–206.

Scharenborg, O., & Boves, L. (2010). Computational modelling of spoken word recognition processes. *Pragmatics & Cognition, 18*(1), 136–164.

Scott, S. K., & Johnsrude, I. S. (2003). The neuroanatomical and functional organization of speech perception. *Trends in Neuroscience, 26*, 100–107.

Scott, S. K. C. C., Blank, C. C., Rosen, S., & Wise, R. J. S. (2000). Identification of a pathway for intelligible speech in the left temporal lobe. *Brain, 123*(12), 2400–2406.

Sharp, D. J., Scott, S. K., & Wise, R. J. S. (2004). Retrieving meaning after temporal lobe infarction: The role of the basal language area. *Annals of Neurology, 56*, 836–846.

Shiffrin, R. M., & Steyvers, M. (1997). A model for recognition memory: REM—retrieving effectively from memory. *Psychonomic Bulletin & Review, 4*, 145–166.

Sommers, M. S. (2005). Age-related changes in spoken word recognition. In D. B. Pisoni, & R. E. Remez (Eds.), *Handbook of speech perception* (pp. 469–493). Malden, MA: Blackwell.

Spivey, M. J., Grosjean, M., & Knoblich, G. (2005). Continuous attraction toward phonological competitors. *PNAS, 102*, 10393–10398.

Sporns, O. (1998). Biological variability and brain function. In J. Cornwell (Ed.), *Consciousness and human identity* (pp. 38–56). Oxford: Oxford University Press.

Sporns, O. (2003). Network analysis, complexity, and brain function. *Complexity, 8*, 56–60.

Sporns, O., Tononi, G., & Edelman, G. M. (2000). Connectivity and complexity: The relationship between neuroanatomy and brain dynamics. *Neural Networks, 13*, 909–922.

Stevens, K. N. (1996). Understanding variability in speech: A requisite for advances in speech synthesis and recognition. *Journal of the Acoustical Society of America, 100*, 2634.

Stevens, K. N. (1998). *Acoustic phonetics*. Cambridge, MA: MIT Press.

Stevens, K. N., Kasowski, S., & Fant, G. (1953). An electrical analog of the vocal tract. *Journal of the Acoustical Society of America, 25*, 734–742.

Studdert-Kennedy, M. (1974). The perception of speech. In T. A. Sebeok (Ed.), *Current trends in linguistics* (Vol. 12, pp. 2349–2385). The Hague: Mouton.

Taler, V., Aaron, G. P., Steinmetz, L. G., & Pisoni, D. B. (2010). Lexical neighborhood density effects on spoken word recognition and production in healthy aging. *Journal of Gerontology: Psychological Sciences, 65*, 551–560.

Theodore, R. M., & Blumstein, S. E. (2011). Attention modulates the time-course of talker-specificity effects in lexical retrieval. *Poster presented at the 162nd meeting of the Acoustical Society of America*. San Diego, CA.

Thorndike, E. L., & Lorge, I. (1944). *The teacher's word book of 30,000 words*. New York: Teachers College Bureau of Publications, Columbia University.

Treisman, M. (1978a). Space or lexicon? *Journal of Verbal Learning and Verbal Behavior, 17*, 37–59.

Treisman, M. (1978b). A theory of the identification of complex stimuli with an application to word recognition. *Psychological Review, 85*, 525–570.

Van Lancker, D., & Kreiman, J. (1987). Voice discrimination and recognition are separate abilities. *Neuropsychologia, 25*, 829–834.

Vitevitch, M. S. (2008). What can graph theory tell us about word learning and lexical retrieval? *Journal of Speech Language Hearing Research, 51*, 408–422.

Vitevitch, M. S. (2012). What do foreign neighbors say about the mental lexicon? *Bilingualism: Language and Cognition, 15*, 167–172.

Vitevitch, M. S., Chan, K. Y., & Goldstein, R. (2014). Insights into failed lexical retrieval from network science. *Cognitive Psychology, 68*, 1–32.

Vitevitch, M. S., & Donoso, A. (2011). Processing of indexical information requires time: Evidence from change deafness. *The Quarterly Journal of Experimental Psychology, 64*, 1484–1493.

Vitevitch, M. S., & Luce, P. A. (1999). Probabilistic phonotactics and neighborhood activation in spoken word recognition. *Journal of Memory & Language, 40*, 374–408.

Weber, A., & Scharenborg, O. (2012). Models of spoken-word recognition. WIREs. *Cognitive Science, 3*, 387–401. Available from: http://dx.doi.org/10.1002/wcs.1178.

Wickelgren, W. A. (1969). Context-sensitive coding, associative memory, and serial order in (speech) behavior. *Psychological Review., 76*(1), 1–15.

Wiener, F.M., & Miller, G.A. (1946). Some characteristics of human speech. In C. E. Waring (Ed.), *Transmission and reception of sounds under combat conditions* (Vol. 3, pp. 58–68). Summary Technical Report of NDRC Division 17. Washington, DC: NDRC.

Wilson, R. H., & McArdle, R. (2005). Speech signals used to evaluate functional status of the auditory system. *Journal of Rehabilitation Research & Development, 42*(4 Suppl. 2), 79–94.

Yonan, C. A., & Sommers, M. S. (2000). The effects of talker familiarity on spoken word identification in younger and older listeners. *Psychology & Aging, 15*, 88–99.

CHAPTER

21

Visual Word Recognition

Kathleen Rastle

Department of Psychology, Royal Holloway, University of London, Egham, Surrey, UK

Reading is one of the most remarkable of our language abilities. Skilled readers are able to recognize printed words and compute their associated sounds and meanings with astonishing speed and a great deal of accuracy. Yet, unlike our inborn capacity for spoken language, reading is not a universal part of the human experience. Reading is a cultural invention and a learned skill, acquired only through years of instruction and practice. Understanding the functional mechanisms that underpin reading and learning to read has been a question of interest since the beginnings of psychology as a scientific discipline (Cattell, 1886; Huey, 1908) and remains a central aim of modern psycholinguistics. This chapter considers one aspect of the reading process—visual word recognition—which is the process whereby we identify a printed letter string as a unique word and compute its meaning. I focus on the understanding of this process that we have gained through the analysis of *behavior* and draw particular attention to those aspects of this process that have been the object of recent debate.

21.1 THE ARCHITECTURE OF VISUAL WORD RECOGNITION

Although the earliest theories of visual word recognition claimed that words were recognized as wholes on the basis of their shapes (Cattell, 1886), there is a strong consensus among modern theories that words are recognized in a hierarchical manner on the basis of their constituents, as in the interactive-activation model (McClelland & Rumelhart, 1981; Rumelhart & McClelland, 1982) shown in Figure 21.1 and its subsequent variants (Coltheart, Rastle, Perry, Langdon, & Ziegler, 2001; Grainger & Jacobs, 1996; Perry, Ziegler, & Zorzi, 2007).

Information from the printed stimulus maps onto stored representations about the visual features that make up letters (e.g., horizontal bar), and information from this level of representation then maps onto stored representations of letters. Some theories assert that letter information goes on to activate higher-level subword representations at increasing levels of abstraction, including orthographic rimes (e.g., the -and in "band"; Taft, 1992), morphemes (Rastle, Davis, & New, 2004), and syllables (Carreiras & Perea, 2002), before activating stored representations of the spellings of known whole words in an orthographic lexicon. Representations in the orthographic lexicon can then activate information about their respective sounds and/or meanings. The major theories of visual word recognition posit that word recognition is achieved when a unique representation in the orthographic lexicon reaches a critical level of activation (Coltheart et al., 2001; Grainger & Jacobs, 1996; Perry et al., 2007).

In recent years, a different class of theory based on distributed-connectionist principles has made a substantial impact on our understanding of processes involved in mapping orthography to phonology (Plaut, McClelland, Seidenberg, & Patterson, 1996) and mapping orthography to meaning (Harm & Seidenberg, 2004). This chapter highlights some of the most important insights that these models have offered to our understanding of reading. However, although these models have been very effective in helping us to understand the acquisition of quasi-regular mappings (as in spelling-to-sound relationships in English), they have been less successful in describing performance in the most frequently used visual word recognition tasks. They offer no coherent account of the most elementary of these tasks—deciding whether a letter string is a known word (i.e., visual lexical decision). Therefore, this chapter assumes a theoretical perspective based on the interactive-activation model and its

Neurobiology of Language. DOI: http://dx.doi.org/10.1016/B978-0-12-407794-2.00021-3

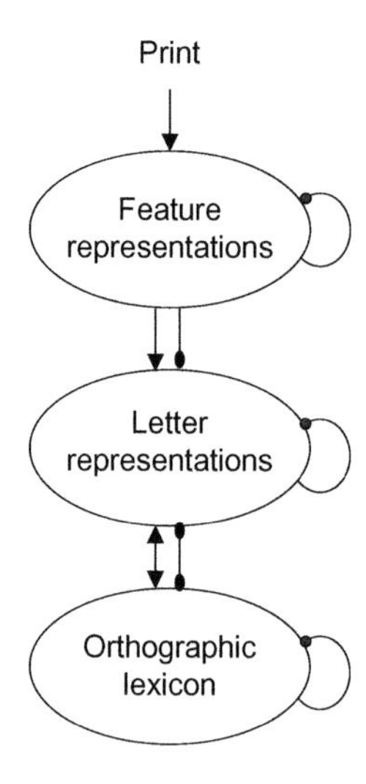

FIGURE 21.1 The interactive-activation model of visual word recognition (McClelland & Rumelhart, 1981; Rumelhart & McClelland, 1982).

subsequent variants but directs the reader to further discussion of this issue in relation to distributed-connectionist models (Coltheart, 2004; Rastle & Coltheart, 2006).

21.2 ORTHOGRAPHIC REPRESENTATION

21.2.1 Letters and Letter Position

There is widespread agreement that stored representations of letters are *abstract letter identities*, meaning that they are activated independently of font, size, case, color, or retinal location (Bowers, 2000). This abstraction is a key part of skilled reading because it permits rapid recognition of words presented in unfamiliar surface contexts (e.g., handwriting, typeface). In addition to encoding information about abstract letter *identities*, letter representations must also encode information about letter position. Otherwise, the visual word recognition system would not be able to distinguish words like SALT and SLAT, which share the same letters but in different positions. The classical solution to this problem implemented in the interactive-activation model (McClelland & Rumelhart, 1981; Rumelhart & McClelland, 1982) and its subsequent variants (Coltheart et al., 2001; Grainger & Jacobs, 1996; Perry et al., 2007) involves slot-based coding. In this scheme, there are slots for each position in a letter string, and a full set of letters within each of those positions. Thus, SALT is coded as $S_1A_2L_3T_4$ and is therefore easy to distinguish from SLAT ($S_1L_2A_3T_4$), because these stimuli overlap only in the initial and final letters (S_1T_4).

However, recent research has demonstrated convincingly that information about letter position is not represented through this type of slot-based coding. The general problem with slot-based coding is that words such as SLAT and SALT are judged by skilled readers to be perceptually very similar, despite the fact that their slot-based codes overlap by only 50% (S_1T_4). In fact, the slot-based codes for SLAT and SALT overlap to the same degree as do those for SPIT and SALT (S_1T_4), which are judged to be much less similar. The following e-mail message, which was circulated globally some years ago, demonstrates this principle very well:

> Aoccdrnig to rseearch at Cmabrigde Uinervtisy, it deosn't mttaer in waht oredr the ltteers in a wrod are, the olny iprmoetnt tihng is taht the frist and lsat ltteer be at the rghit pclae.

Indeed, the reason that we can read this passage so easily is that words with transposed letters are perceived as being very similar to their base words. This issue has been studied experimentally using the masked form priming technique (Forster & Davis, 1984) in which a target stimulus presented for recognition is preceded by a consciously imperceptible prime stimulus. For example, Schoonbaert and Grainger (2004) demonstrated that recognition of a target stimulus like SERVICE is faster when it is preceded by a masked transposed-letter prime like *sevrice* than when it is preceded by a masked substitution prime stimulus like *sedlice*. This result is important because according to slot-based coding, transposed-letter prime *sevrice* and substitution prime *sedlice* have equivalent perceptual overlap with target SERVICE, and thus should speed target recognition to the same degree. Similar results are observed when transpositions are nonadjacent; for example, the recognition of CASINO is speeded by the prior masked presentation of the prime *caniso* relative to the prime *caviro* (Perea & Lupker, 2003). Finally, these kinds of results extend to even more extreme modifications; for example, the recognition of SANDWICH is speeded by the prior masked presentation of the prime *snawdcih* relative to the prime *skuvgpah* (Guerrera & Forster, 2008).

These and other findings highlight an intriguing problem in visual word recognition. Readers are clearly able to distinguish anagram stimuli such as SNAWDCIH and SANDWICH, so letter representations must be coded for position. However, classical theories of how this is achieved (McClelland & Rumelhart, 1981) are clearly inadequate. The evidence now seems to suggest that orthographic representations must code position in a *relative* rather than *absolute* manner, and probably with some degree of uncertainty or sloppiness, as in the spatial coding scheme used by the SOLAR model (Davis, 2010). Further, while these kinds of effects motivating position uncertainty have been reported across a variety of alphabetic languages, it is also important to observe that they are not universal. Primes with transposed letters do not facilitate recognition of their base words (relative to substitution primes) in Hebrew, for example (Velan & Frost, 2009). The reasons for this are not yet well-understood, but it seems likely that the greater the density of the orthographic space, the greater the pressure to develop very precise orthographic representations (Frost, 2012). Further research is necessary to determine the exact nature of orthographic coding and why position uncertainty appears to vary as a function of the nature of the writing system.

21.2.2 Frequency, Cumulative Frequency, and Age of Acquisition

There is a broad consensus that an individual's previous experience with a word is the most powerful determinant of how rapidly that word is identified. But what is meant by "an individual's previous experience"? The most common proxy for this is word frequency—the number of times a particular word occurs in some large corpus of text (Baayen, Piepenbrock, & van Rijn, 1993; New, Brysbaert, Veronis, & Pallier, 2007). Effects of word frequency have been reported in lexical decision (Balota, Cortese, Sergent-Marshall, Spieler, & Yap, 2004; Forster & Chambers, 1973) along with every other speeded task thought to reflect access to orthographic representations, including perceptual identification (Broadbent, 1967), reading aloud (Balota & Chumbley, 1984), and eye-fixation times in sentence reading (Schilling, Rayner, & Chumbley, 1998). Provided frequency estimates are derived from a suitably large corpus of text (approximately half of the frequency effect occurs for words between 0 and 1 occurrences per million; van Heuven, Mandera, Keuleers, & Brysbaert, 2014), word frequency estimates can explain more than 40% of the variance in lexical decision time (Brysbaert & New, 2009). In light of these data, there is wide agreement that one's experience with words is somehow encoded in the orthographic representations of those words and influences the ease with which they can be identified. One long-standing theory is that orthographic representations for high-frequency words have higher resting levels of activation than those for lower-frequency words, making them easier to reach a critical recognition threshold (McClelland & Rumelhart, 1981).

Recently, an interesting debate has emerged over whether the age at which a word is acquired might also be an important aspect of lexical experience, with words acquired earlier processed more easily in visual word recognition tasks. Although several studies claim to have observed independent effects of frequency and age-of-acquisition on visual word recognition when these factors are manipulated orthogonally (Gerhand & Barry, 1999; Morrison & Ellis, 1995), these claims have been very difficult to assess for a number of reasons. For one, the age-of-acquisition metrics used in these studies are typically *subjective estimates* given by adults of the age at which they acquired particular words. It is not unlikely that the frequency with which a word occurs influences those subjective estimates provided by adults. Further, word frequency and age-of-acquisition are very tightly correlated (i.e., high-frequency words are typically the ones acquired earliest; $r = -0.68$; Carroll & White, 1973), making it extremely difficult to design experiments that examine independent effects of these variables, particularly given that there are multiple corpora from which to draw both of these metrics. Finally, it has been proposed that word frequency and age-of-acquisition are just two dimensions of a single variable—cumulative frequency (i.e., the frequency with which a word occurs over an individual's lifetime; e.g., Zevin & Seidenberg, 2002).

Although it is now fairly well-accepted that cumulative frequency provides a better description of our experience with words than printed word frequency (Brysbaert & Ghyselinck, 2006), the more difficult question is whether the age-of-acquisition effects observed on visual word recognition can be accounted for by cumulative frequency. It now seems that the answer is "no." Recent empirical work has demonstrated that the impact of age-of-acquisition on a number of word processing tasks is greater than would be predicted by cumulative frequency (Ghyselinck, Lewis, & Brysbaert, 2004). Further, work using connectionist models has shown that age-of-acquisition effects may be a fundamental property of models that learn incrementally over time (Monaghan & Ellis, 2010). This computational work has also suggested that age-of-acquisition effects may be more prevalent when input-to-output mappings are less systematic—the reason being that the solution space for early-acquired items will be less helpful for later-acquired items when the mapping is more arbitrary (Monaghan & Ellis, 2010). This observation is consistent with the empirical literature that finds particularly robust effects of age-of-acquisition in tasks that require semantic involvement such as object naming (Ghyselinck, Lewis et al., 2004), translation judgment (Izura & Ellis, 2004), and living versus nonliving decisions (Ghyselinck, Custers, & Brysbaert, 2004).

21.2.3 Morphology

The majority of words in English, and in virtually all of the world's languages, are built by combining and recombining a finite set of morphemes. These combinatorial processes in English start with a small number of stem morphemes (e.g., *trust*) and pair them with other stem morphemes to form compound words (e.g., *trustworthy*), or with derivational (e.g., *trusty, distrust*) or inflectional (e.g., *trusted, trusts*) affixes to form the much larger proportion of the words that we use. Despite the fact that words with just a single morpheme (e.g., *trust*) are in the extreme minority, the major computational models of visual word recognition (Coltheart et al., 2001; Grainger & Jacobs, 1996; Perry et al., 2007) have focused on those. Even when affixed words are included in these models, they are treated in exactly the same way as are nonaffixed words. This treatment is likely to be inadequate, because there is now substantial evidence that words comprising more than one morpheme are recognized in terms of their morphemic constituents.

One of the main sources of evidence for this claim comes from studies that investigate whether the frequency of the stem in a morphologically complex word (e.g., the *trust* in *distrust*) plays any role in the time taken to recognize that morphologically complex word. The answer is virtually unequivocal that it does. For example, Taft and Ardasinski (2006) demonstrated that visual lexical decisions for prefixed words with high-frequency stems (e.g., *rediscover*) were significantly faster than those for prefixed words with low-frequency stems (e.g., *refuel*), despite the fact that these two sets of words were matched on whole-word frequency (e.g., Ford, Davis, & Marslen-Wilson, 2010). This result appears to indicate that participants access the stems of these words during the recognition process, or alternatively that the representation for morphologically complex words like *rediscover* is somehow strengthened during acquisition by experience with their constituent stems (in this case, *discover*). Interestingly, the findings of Taft and Ardasinski (2006) held even in cases in which the nonword fillers for the lexical decision task comprised a prefix attached to a real-word stem (e.g., *relaugh*), which, if anything, should have biased participants against segmenting the stimuli into their morphemic constituents.

The other major source of evidence for the claim that printed words are recognized in terms of their morphemic constituents comes from masked priming data. Multiple studies have now demonstrated that the recognition of a stem target (e.g., DARK) is speeded by the prior masked presentation of a morphologically related prime (e.g., *darkness*). The locus of this facilitation appears to reside in orthographic representations, because the recognition of stem targets is speeded to the same degree by the prior presentation of morphologically simple masked primes that have the *appearance* of morphological complexity (e.g., the prime *corner* speeds recognition of CORN). Critically, this facilitation cannot be ascribed to overlap in letter representations between primes and targets, because masked primes that share letters but no apparent morphological relationship with targets (e.g., brothel-BROTH; -el is not a possible suffix in English) yield no facilitation (Rastle et al., 2004; see Rastle & Davis, 2008 for review and relevant neural evidence). These data again indicate that readers activate representations of the stems of morphologically structured words during the recognition process.

21.3 PROCESSING DYNAMICS AND MECHANISMS OF SELECTION

The discussion thus far has described a hierarchical theory of visual word recognition that involves multiple layers of orthographic representation. Representations of the features are used to make letters map onto representations that code abstract letter identity and letter position. There is evidence that these representations then activate representations of sublexical units (e.g., morphemes) before activating representations of the spellings of known words in an orthographic lexicon. This section considers how information flows through this architecture, and how amidst activation of multiple candidates a unique word representation reaches a recognition threshold. This discussion is based largely on principles of the interactive-activation model (McClelland & Rumelhart, 1981; Rumelhart & McClelland, 1982), which many still consider to be the cornerstone of our understanding of visual word recognition.

21.3.1 Interactive Processing

One of the key principles of the interactive-activation model and its subsequent variants is interactive, or bidirectional, processing. In the hierarchical model described, there are assumed to be connections between adjacent levels of representation, which are both excitatory and inhibitory. Information flows in a bidirectional manner across these connections (e.g., letter representations activate word representations, and word representations activate letter representations), and this is what allows the model to explain how higher-level knowledge can influence processing at a lower level.

Two empirical effects were particularly important in identifying the role of bidirectional processing in visual word recognition—the word superiority effect (Reicher, 1969; Wheeler, 1970) and the pseudoword superiority effect (Carr, Davidson, & Hawkins, 1978; McClelland & Johnston, 1977). In the Reicher–Wheeler word superiority experiments, a letter string was flashed very briefly and then replaced by a pattern mask. Participants were then asked to decide which of two letters, positioned below or above the previous target letter, were in the original target stimulus. The key manipulation was whether the original target was a word (e.g., WORD) or nonword (e.g., OWRK). Results revealed that letter identification was far superior when the original flashed target was a word. These data suggest that letter representations receive top-down support through bidirectional connections from whole-word representations activated on presentation of the target stimulus. Intriguingly, a similar letter identification benefit is observed when the stimulus is a pronounceable pseudoword (e.g., TARK) as opposed to a nonpronounceable string (e.g., ATRK) (Carr et al., 1978; McClelland & Johnston, 1977). Even though pseudowords like TARK are not represented in the orthographic lexicon, this result indicates that they may activate whole-word representations for similar words (e.g., DARK, TALK, PARK), which then feed activation back to the letter level, thus explaining the letter identification benefit observed. Overall, these effects support the notion of interactive processing, because they suggest that a decision based on activation at the letter level is influenced by higher-level information from the orthographic lexicon. More recent research has revealed top-down influences of semantic and phonological variables on visual lexical decision, which can only be explained through interactive processing in the reading system. These semantic and phonological effects are discussed in Section 21.4.

21.3.2 Competition as a Mechanism for Selection

The explanation of the pseudoword superiority effect suggests that printed letter strings activate multiple candidates at the word level. Thus, a letter string like WORD may activate the whole-word representation for WORD, along with whole-word representations for WORK, WARD, CORD, LORD, and others. In the interactive-activation model, the activation of multiple candidates is achieved through *cascaded processing* (McClelland, 1979). Representations at every level excite and inhibit representations at adjacent levels continuously, without having to reach some threshold (as was the case in the "logogen" model; Morton, 1969). However, the situation in which multiple candidates are activated from a single printed stimulus raises the question of how the recognition system selects a unique representation corresponding to the target. The interactive-activation model solves this problem through competition. In addition to connections between levels of representation, the interactive-activation model posits intralevel inhibitory connections. These lateral inhibitory connections between whole-word representations allow the most active unit (typically that of the target) to drive down activation of its competitors. Of course, representations for any competing alternative candidates will also be exerting inhibition, which will serve to drive down activation of other competitors as well as the representation of the target, making it more difficult for the target to reach a recognition threshold.

21.3.2.1 Neighborhood Effects

One way in which this prediction has been tested is by looking at the impact of lexical similarity on visual word recognition. If a letter string is similar to many words (and thus activates multiple candidates), then it should be more difficult to recognize than a letter string that is similar to few words (and thus does not activate multiple candidates). Before describing the literature around this prediction, it is important to consider what is meant by the phrase "similar to many words." This is a key point: what counts as similar depends entirely on the nature of the scheme adopted for coding letter position (see Section 21.2.1). Two stimuli that have large orthographic overlap according to one scheme for coding letter position may have much less overlap according to another scheme. This principle is nicely illustrated by considering the example words BLAND and LAND. In the slot-based coding scheme described on Section 21.2.1, these stimuli share no overlap whatsoever. However, in a letter coding scheme based on relative position, such as the spatial coding scheme of the SOLAR model (Davis, 2010), these stimuli share substantial overlap. Thus, research on the consequences of lexical similarity for visual word recognition is impeded by the lack of consensus around the nature of orthographic input coding.

Until recently, much of the work regarding the impact of lexical similarity on visual word recognition has been based on a metric known as Coltheart's N (Coltheart, Davelaar, Jonasson, & Besner, 1977). N is defined as the number of words of the same length that can be created by changing one letter of a stimulus, such that a word like CAKE has a very large neighborhood (e.g., BAKE, LAKE, CARE, CAVE) and a word like TUFT has no neighbors. Coltheart et al. (1977) reported that participants rejected high-N nonwords (e.g., PAKE) more slowly than low-N nonwords (e.g., PLUB) in a lexical decision task, an effect replicated several times (Forster & Shen, 1996; McCann, Besner, & Davelaar, 1988). High-N nonwords are thought to be more difficult to reject in lexical decision because they activate many units at the word level, and this activation makes it difficult to decide that the stimulus is not a word. The situation is more complicated for word targets, however. In contrast to the predictions of competitive models, Coltheart et al. (1977) reported no effect of N on lexical decisions to word targets. Andrews (1989) then went on to report that words with many neighbors are responded to more *quickly* than words with few neighbors. The same year, however, Grainger, O'Regan, Jacobs, and Segui (1989) reported that words with at least one higher-frequency neighbor are recognized more slowly than words with no higher-frequency neighbors. This latter result makes the Andrews (1989) findings particularly perplexing given that words with many neighbors will almost certainly have at least one higher-frequency neighbor.

Although some investigators have continued to report facilitatory effects of N on recognition latency (Balota et al., 2004; Forster & Shen, 1996), most reports are in line with the prediction from competitive models (i.e., inhibitory effects of N) (Carreiras, Perea, & Grainger, 1997; Grainger & Jacobs, 1996; Perea & Pollatsek, 1998). Grainger and Jacobs (1996) put forward one of the most compelling explanations for these divergent effects, arguing that the inhibitory pattern is the "true" pattern and that facilitatory effects might be the result of strategic processes involved in making lexical decisions. Specifically, they argued that participants in the lexical decision task may be able to make a fast "YES" response if the *total* activation in the orthographic lexicon is high. The idea is that a large neighborhood is likely to lead to high total activation (i.e., because of the large number of word units activated), and hence it is through this fast guess mechanism that facilitatory effects of neighborhood are deemed to arise. This explanation has received support from studies showing that the direction of the neighborhood size effect can be influenced by instructions that stress speed or accuracy (De Moor, Verguts, & Brysbaert, 2005; Grainger & Jacobs, 1996). When participants need to be very accurate, inhibitory neighborhood effects are observed, presumably because their decisions are based on the activation of a single orthographic unit (which will be influenced by lateral inhibition). Conversely, when participants need to be very fast, facilitatory effects are observed, presumably because participants' decisions can be based on the fast guess mechanism that does not require access to an individual orthographic unit.

21.3.2.2 Masked Form Priming Effects

Masked form priming effects have been another powerful source of evidence in support of models that use competition as the mechanism for selection. In the interactive-activation model, priming is conceptualized as a balance between facilitation and inhibition. Primes activate visually similar targets, thus producing savings in the time taken for those targets to reach a recognition threshold. However, primes can also activate orthographic units for whole words, which compete with targets for recognition. Davis (2003) therefore argued that the lexical status of a prime should be an important factor in determining the magnitude of form priming effects. Nonword primes (e.g., azle-AXLE) should yield facilitation, because primes activate units for their corresponding targets without also activating any units for competing words. Conversely, word primes (e.g., able-AXLE) should yield inhibition, because although the prime will still activate the target, it will activate the orthographic unit for itself much more strongly, which will compete with the target for recognition. Results strongly favor competitive models. Nonword masked primes always facilitate recognition of visually similar targets (e.g., azle-AXLE), whereas word masked primes typically inhibit or yield no effect on the recognition of visually similar targets (e.g., able-AXLE; see Davis & Lupker, 2006 for review). The only exception is when word primes appear morphologically related to targets (e.g., darker-DARK); in these cases, primes clearly facilitate rather than inhibit recognition of their targets (Rastle et al., 2004; see Section 21.2).

21.4 VISUAL WORD RECOGNITION AND THE READING SYSTEM

This chapter has put forward an understanding of visual word recognition based on a hierarchical analysis of visual features, letters, subword units (e.g., morphemes), and, ultimately, orthographic representations of whole words. Although visual word recognition is typically regarded in modern theories as based on the analysis of orthography, this system is embedded in a larger reading system that comprises processes to compute the sounds and meanings associated with known spellings. Further, although visual word recognition *remains possible* in the face of severe semantic and/or phonological impairment due to brain damage (Coltheart, 2004; Coltheart & Coltheart, 1997), it is undisputed that semantic and phonological information *can contribute* to visual word recognition.

One computational model that may help us to understand semantic and phonological influences on visual word recognition is the DRC model (Coltheart et al., 2001) shown in Figure 21.2.

This model postulates three processing pathways: (i) one pathway in which a printed letter string is translated to sound in the absence of lexical information; (ii) one pathway in which the phonological form of a word is retrieved directly after its activation in the orthographic lexicon; and (iii) one pathway in which

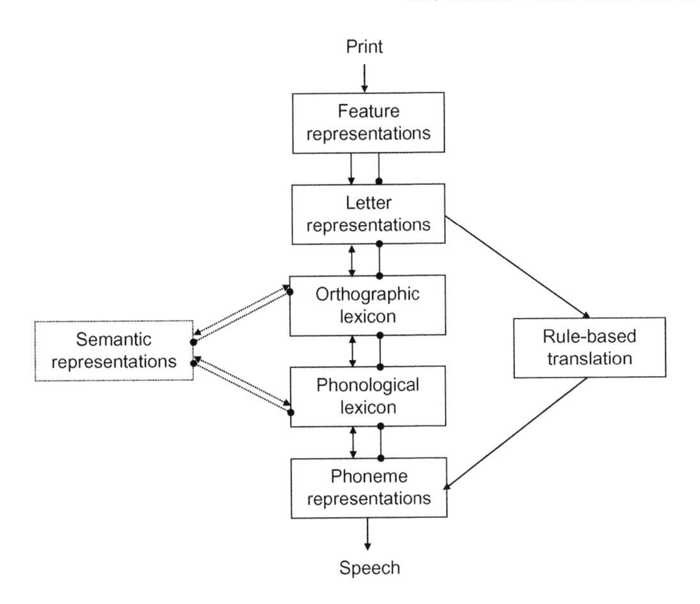

FIGURE 21.2 The DRC model (Coltheart et al., 2001).

the phonological form of a word is retrieved via its meaning representation. The architecture of this model maps fairly well to our understanding of the neural underpinnings of reading (see Taylor, Rastle, & Davis, 2013 for review).

The activation of orthographic whole-word units in the DRC model is based on the interactive-activation model (McClelland & Rumelhart, 1981; Rumelhart & McClelland, 1982). Because information about the printed stimulus flows through all three pathways in cascade, it is entirely possible in this model that information about the semantic and phonological characteristics of a letter string will be activated before any unit in the orthographic lexicon reaches a critical recognition threshold. Further, and critically, there are bidirectional connections between semantic, phonological, and orthographic bodies of knowledge, which make it possible for semantic and phonological information to impact on the activation of units in the orthographic lexicon.

21.4.1 Phonological Influences on Visual Word Recognition

It has been apparent for more than 40 years (actually, going all the way back to Huey, 1908) that the sounds associated with a printed letter string can influence its recognition. Huey (1908) described reading as involving auditory imagery or a "voice in the head," and empirical effects reported during the cognitive renaissance (Rubenstein, Lewis, & Rubenstein, 1971) led theorists of reading to consider that phonological representations may be not only *involved* in visual word recognition but also a *requirement* of it (Frost, 1998; Lukatela & Turvey, 1994). Although this *strong phonological theory* of visual word recognition (Frost, 1998) has fallen out of favor in more recent years, the evidence is unequivocal that sound-based representations are computed as a matter of routine during reading (see Rastle & Brysbaert, 2006 for review).

The use of homophones and pseudohomophones (i.e., nonwords that sound like words; e.g., BRANE) has proven especially useful in delineating the role of phonological representations in visual word recognition. Rubenstein et al. (1971) observed that YES responses in lexical decision were slower for homophones than for nonhomophones (e.g., recognition of MAID slower than PAID), and that NO responses were slower for pseudohomophones than for nonpseudohomophones (e.g., KOAT slower than FOAT). Both of these effects have been replicated and are well-accepted (e.g., homophone effect: Pexman, Lupker, & Hino, 2002; pseudohomophone effect: Ziegler, Jacobs, & Kluppel, 2001). Both of these effects are easy to understand in the light of bidirectional connections in the reading system. The homophone effect arises because after presentation of the stimulus MAID, activation of the phonological unit corresponding to MAID goes on to activate the competitor MADE in the orthographic lexicon, thus slowing recognition of MAID. The pseudohomophone effect arises because the stimulus KOAT will be translated nonlexically to a phonological representation that will activate a unit in the phonological lexicon. This phonological unit will then send activation back to the orthographic unit for COAT, making it difficult to classify the stimulus as a nonword.

Pseudohomophones have also been used extensively in the context of masked form priming to elucidate the role of phonology in visual word recognition. There are now a number of studies showing that the recognition of a target stimulus (e.g., COAT) is facilitated by the prior presentation of a masked pseudohomophone prime (e.g., KOAT) relative to an orthographic control prime (e.g., POAT; Ferrand & Grainger, 1992). This masked phonological priming effect arises in lexical decision, reading aloud, perceptual identification, and eye-movement paradigms, although a meta-analysis conducted by Rastle and Brysbaert (2006) revealed that the effect is small. These effects suggest not only that phonological representations can play a role in visual word processing but also that phonology is activated remarkably quickly in the recognition process. It was these demonstrations of "fast phonology" that led to excitement around the strong phonological theory of reading (Frost, 1998), although modern theorizing suggests that these effects can be explained within *weak phonological theories* like the DRC model (Figure 21.2), in which visual word recognition is characterized by an orthographic analysis that *can be influenced by* phonological representations (Rastle & Brysbaert, 2006).

21.4.2 Semantic Influences on Visual Word Recognition

The DRC model (Coltheart et al., 2001) asserts that skilled readers can recognize printed words in the absence of semantic information, and this claim is backed by evidence that brain-damaged patients with severe semantic impairments can nevertheless recognize printed words accurately (Coltheart, 2004). However, as in the case of phonological information, there is a broad consensus that semantic information can influence visual word recognition. Multiple studies now suggest that words that have particularly rich semantic representations are recognized more quickly than words with more impoverished semantic representations. Visual word recognition is speeded by high imageability (e.g., Balota et al., 2004), high semantic neighborhood density (Locker, Simpson, & Yates, 2003), a large number of meanings (Hino & Lupker, 1996) and related meanings (Azuma & Van Orden, 1997), and a large number of related senses (Rodd, Gaskell, & Marslen-Wilson, 2002). Here, again, bidirectional connections provide the mechanism for explaining these kinds of effects. Printed words activate their semantic representations, and activation at this level feeds back to support orthographic representations for word targets.

Priming studies also reveal semantic influences on visual word recognition. In a seminal study, Meyer and Schvaneveldt (1971) observed that lexical decisions to words (e.g., DOCTOR) were speeded by the prior presentation of semantically related primes (e.g., NURSE) relative to unrelated primes (e.g., BREAD). This finding has been replicated numerous times and has motivated a literature all of its own (e.g., see Hutchison, 2003; Lucas, 2000 for reviews). Semantic priming is usually conceptualized in terms of spreading activation between localist units (Collins & Loftus, 1975), overlap in distributed featural representations (McRae, de Sa, & Seidenberg, 1997), or Euclidean distance between high-dimensional vectors derived from lexical co-occurrence matrices (Landauer & Dumais, 1997). Bidirectional connections between semantic and orthographic levels of representation also play a role in explaining the semantic priming effect. If semantic information about the target is activated after presentation of the prime (e.g., information about *doctor* activated after presentation of *nurse*), then this activation can feed back to the orthographic unit for the target (e.g., DOCTOR), thus speeding recognition time.

21.5 CONCLUSION

Humans are born to speak, but they have to learn to read. This fact is part of what makes visual word recognition in the literate adult such an astonishing ability. Readers are faced with considerable variability in the forms of the symbols presented to them, and the density of the orthographic space (particularly in writing systems such as Hebrew) renders words highly confusable. Further, the reading system must develop in such a way that it is closely linked to phonological and semantic bodies of knowledge, and there is substantial evidence that this stored knowledge is activated very soon after presentation of a printed letter string. Research over the past 40 years on the functional mechanisms that underpin visual word processing has been a great success story. This research provides a sound basis for which to discover how the brain supports the mind in regard to this remarkable human achievement.

Acknowledgment

This work was supported by research grants from the ESRC (RES-000-62-2268, ES/L002264/1) and Leverhulme Trust (RPG-2013-024).

References

Andrews, S. (1989). Frequency and neighborhood effects on lexical access: Activation or search? *Journal of Experimental Psychology: Learning, Memory, and Cognition, 15*(5), 802–814.

Azuma, T., & Van Orden, G. (1997). Why safe is better than fast: The relatedness of a word's meaning affects lexical decision times. *Journal of Memory and Language, 36*(4), 484–504.

Baayen, R. H., Piepenbrock, R., & van Rijn, H. (1993). *The CELEX lexical database (CD-ROM). Linguistic Data Consortium.* Philadelphia, PA: University of Pennsylvania.

Balota, D. A., & Chumbley, J. I. (1984). Are lexical decisions a good measure of lexical access? The role of word frequency in the neglected decision stage. *Journal of Experimental Psychology: Human Perception and Performance, 10*(3), 340–357.

Balota, D. A., Cortese, M. J., Sergent-Marshall, S. D., Spieler, D. H., & Yap, M. J. (2004). Visual word recognition of single-syllable words. *Journal of Experimental Psychology: General, 133*(2), 283–316.

Bowers, J. S. (2000). In defense of abstractionist theories of repetition priming and word identification. *Psychonomic Bulletin & Review, 7* (1), 83–99.

Broadbent, D. E. (1967). Word frequency effects and response bias. *Psychological Review, 74*(1), 1–15.

Brysbaert, M., & Ghyselinck, M. (2006). The effect of age-of-acquisition: Partly frequency-related, partly frequency-independent. *Visual Cognition, 13*(7–8), 992–1011.

Brysbaert, M., & New, B. (2009). Moving beyond Kucera and Francis: A critical evaluation of current word frequency norms and the introduction of a new and improved word frequency measure for American English. *Behavior Research Methods, 41*(4), 977–990.

Carr, T. H., Davidson, B. J., & Hawkins, H. L. (1978). Perceptual flexibility in word recognition: Strategies affect orthographic computation but not lexical access. *Journal of Experimental Psychology: Human Perception and Performance, 4*(4), 674–690.

Carreiras, M., & Perea, M. (2002). Masked priming effects with syllabic neighbors in the lexical decision task. *Journal of Experimental Psychology: Human Perception and Performance, 28*(5), 1228–1242.

Carreiras, M., Perea, M., & Grainger, J. (1997). Effects of the orthographic neighborhood in visual word recognition: Cross-task comparisons. *Journal of Experimental Psychology: Learning, Memory, and Cognition, 23*(4), 857–871.

Carroll, J. B., & White, M. N. (1973). Word frequency and age of acquisition as determiners of picture naming latencies. *Quarterly Journal of Experimental Psychology, 24*(1), 85–95.

Cattell, J. (1886). The time it takes to see and name objects. *Mind, 11*, 63–65.

Collins, A. M., & Loftus, E. F. (1975). A spreading-activation theory of semantic processing. *Psychological Review, 82*(6), 407–428.

Coltheart, M. (2004). Are there lexicons? *Quarterly Journal of Experimental Psychology, 57A*(7), 1153–1171.

Coltheart, M., & Coltheart, V. (1997). Reading comprehension is not exclusively reliant upon phonological representation. *Cognitive Neuropsychology, 14*(1), 167–175.

Coltheart, M., Davelaar, E., Jonasson, J. T., & Besner, D. (1977). Access to the internal lexicon. In S. Dornic (Ed.), *Attention and Performance VI* (pp. 535–555). Hillsdale, NJ: Erlbaum.

Coltheart, M., Rastle, K., Perry, C., Langdon, R., & Ziegler, J. (2001). DRC: A dual route cascaded model of visual word recognition and reading aloud. *Psychological Review, 108*(1), 204–256.

Davis, C. J. (2003). Factors underlying masked priming effects in competitive network models of visual word recognition. In S. Kinoshita, & S. J. Lupker (Eds.), *Masked Priming: The State of the Art* (pp. 121–170). Hove, UK: Psychology Press.

Davis, C. J. (2010). The spatial coding model of visual word identification. *Psychological Review, 117*(3), 713–758.

Davis, C. J., & Lupker, S. J. (2006). Masked inhibitory priming in English: Evidence for lexical inhibition. *Journal of Experimental Psychology: Human Perception and Performance, 32*(3), 668–687.

De Moor, W., Verguts, T., & Brysbaert, M. (2005). Testing the "multiple" in the multiple read-out model of visual word recognition. *Journal of Experimental Psychology: Learning, Memory, and Cognition, 31*(6), 1502–1508.

Ferrand, L., & Grainger, J. (1992). Phonology and orthography in visual word recognition: Evidence from masked non-word priming. *Quarterly Journal of Experimental Psychology, 45A*(3), 353–372.

Ford, M. A., Davis, M. H., & Marslen-Wilson, W. D. (2010). Derivational morphology and base morpheme frequency. *Journal of Memory and Language, 63*(1), 117–130.

Forster, K. I., & Chambers, S. (1973). Lexical access and naming time. *Journal of Verbal Learning and Verbal Behavior, 12*(6), 627–635.

Forster, K. I., & Davis, C. (1984). Repetition priming and frequency attenuation in lexical access. *Journal of Experimental Psychology: Learning, Memory, and Cognition, 10*(4), 680–689.

Forster, K. I., & Shen, D. (1996). No enemies in the neighborhood: Absence of inhibitory neighborhood effects in lexical decision and semantic categorization. *Journal of Experimental Psychology: Learning, Memory, & Cognition, 22*(3), 696–713.

Frost, R. (1998). Toward a strong phonological theory of visual word recognition: True issues and false trails. *Psychological Bulletin, 123* (1), 71–99.

Frost, R. (2012). Towards a universal model of reading. *Behavioral and Brain Sciences, 35*(5), 263–279.

Gerhand, S., & Barry, C. (1999). Age of acquisition, word frequency, and the role of phonology in the lexical decision task. *Memory and Cognition, 27*(4), 592–602.

Ghyselinck, M., Custers, R., & Brysbaert, M. (2004). The effect of age of acquisition in visual word processing: Further evidence for the semantic hypothesis. *Journal of Experimental Psychology: Learning, Memory, and Cognition, 30*(2), 550–554.

Ghyselinck, M., Lewis, M. B., & Brysbaert, M. (2004). Age of acquisition and the cumulative-frequency hypothesis: A review of the literature and a new multi-task investigation. *Acta Psychologica, 115*(1), 43–67.

Grainger, J., & Jacobs, A. M. (1996). Orthographic processing in visual word recognition: A multiple read-out model. *Psychological Review, 103*(3), 518–565.

Grainger, J., O'Regan, J. K., Jacobs, A. M., & Segui, J. (1989). On the role of competing word units in visual word recognition: The neighborhood frequency effect. *Perception and Psychophysics, 51*(1), 49–56.

Guerrera, C., & Forster, K. I. (2008). Masked form priming with extreme transpositions. *Language and Cognitive Processes, 23*(1), 117–142.

Harm, M., & Seidenberg, M. S. (2004). Computing the meanings of words in reading: Cooperative division of labor between visual and phonological processes. *Psychological Review, 111*(3), 662–720.

Hino, Y., & Lupker, S. J. (1996). Effects of polysemy in lexical decision and naming: An alternative to lexical access accounts. *Journal of Experimental Psychology: Human Perception and Performance, 22* (6), 1331–1356.

Huey, E. B. (1908). *The Psychology and Pedagogy of Reading. Repr. 1968.* Cambridge, MA: MIT Press.

Hutchison, K. A. (2003). Is semantic priming due to association strength or feature overlap? A micro-analytic review. *Psychonomic Bulletin and Review, 10*(4), 785–813.

Izura, C., & Ellis, A. W. (2004). Age of acquisition effects in translation judgement tasks. *Journal of Memory and Language, 50*(2), 165–181.

Landauer, T. K., & Dumais, S. T. (1997). A solution to Plato's problem: The Latent Semantic Analysis theory of the acquisition, induction, and representation of knowledge. *Psychological Review, 104*(2), 211–240.

Locker, L., Simpson, G. B., & Yates, M. (2003). Semantic neighborhood effects on the recognition of ambiguous words. *Memory and Cognition, 31*(4), 505–515.

Lucas, M. (2000). Semantic priming without association: A meta-analytic review. *Psychonomic Bulletin and Review, 7*(4), 618–630.

Lukatela, G., & Turvey, M. T. (1994). Visual lexical access is initially phonological: 1. Evidence from associative priming by words, homophones, and pseudohomophones. *Journal of Experimental Psychology: General, 123*(2), 107–128.

McCann, R. S., Besner, D., & Davelaar, E. (1988). Word recognition and identification. Do word-frequency effects reflect lexical access? *Journal of Experimental Psychology: Human Perception and Performance, 14*(4), 693–706.

McClelland, J. L. (1979). On the time relations of mental processes: A framework for analyzing processes in cascade. *Psychological Review, 86*(4), 287–330.

McClelland, J. L., & Johnston, J. C. (1977). The role of familiar units in perception of words and nonwords. *Perception and Psychophysics, 22*(3), 249–261.

McClelland, J. L., & Rumelhart, D. E. (1981). An interactive activation model of context effects in letter perception: Part 1. An account of basic findings. *Psychological Review, 88*(5), 375–407.

McRae, K., de Sa, V. R., & Seidenberg, M. S. (1997). On the nature and scope of featural representations of word meaning. *Journal of Experimental Psychology: General, 126*(2), 99–130.

Meyer, D. E., & Schvaneveldt, R. W. (1971). Facilitation in recognizing pairs of words: Evidence of a dependence between retrieval operations. *Journal of Experimental Psychology, 90*(2), 227–234.

Monaghan, P., & Ellis, A. W. (2010). Modeling reading development: Cumulative, incremental learning in a computational model of word naming. *Journal of Memory and Language, 63*(4), 506–525.

Morrison, C. M., & Ellis, A. W. (1995). Roles of word frequency and age of acquisition in word naming and lexical decision. *Journal of Experimental Psychology: Learning, Memory, and Cognition, 21*(1), 116–133.

Morton, J. (1969). Interaction of information in word recognition. *Psychological Review*, *76*(2), 165–178.

New, B., Brysbaert, M., Veronis, J., & Pallier, C. (2007). The use of film subtitles to estimate word frequencies. *Applied Psycholinguistics*, *28* (4), 661–677.

Perea, M., & Lupker, S. J. (2003). Does jugde activate COURT? Transposed-letter similarity effects in masked associative priming. *Memory and Cognition*, *31*(6), 829–841.

Perea, M., & Pollatsek, A. (1998). The effects of neighborhood frequency in reading and lexical decision. *Journal of Experimental Psychology: Human Perception and Performance*, *24*(3), 767–779.

Perry, C., Ziegler, J. C., & Zorzi, M. (2007). Nested incremental modeling in the development of computational theories: The CDP+ model of reading aloud. *Psychological Review*, *114*(2), 273–315.

Pexman, P. M., Lupker, S. J., & Hino, Y. (2002). The impact of feedback semantics in visual word recognition: Number of features effects in lexical decision and naming tasks. *Psychonomic Bulletin and Review*, *9*(3), 542–549.

Plaut, D. C., McClelland, J. L., Seidenberg, M. S., & Patterson, K. (1996). Understanding normal and impaired word reading: Computational principles in quasi-regular domains. *Psychological Review*, *103*(1), 56–115.

Rastle, K., & Brysbaert, M. (2006). Masked phonological priming effects in English: Are they real? Do they matter? *Cognitive Psychology*, *53*, 97–145.

Rastle, K., & Coltheart, M. (2006). Is there serial processing in the reading system; and are there local representations? In S. Andrews (Ed.), *From inkmarks to ideas: Current issues in lexical processing*. Hove: Psychology Press.

Rastle, K., & Davis, M. H. (2008). Morphological decomposition based on the analysis of orthography. *Language and Cognitive Processes*, *23*(7–8), 942–971.

Rastle, K., Davis, M., & New, B. (2004). The broth in my brother's brothel: Morpho-orthographic segmentation in visual word recognition. *Psychonomic Bulletin and Review*, *11*(6), 1090–1098.

Reicher, G. M. (1969). Perceptual recognition as a function of meaningfulness of stimulus material. *Journal of Experimental Psychology*, *81*(2), 274–280.

Rodd, J., Gaskell, G., & Marslen-Wilson, W. (2002). Making sense of semantic ambiguity: Semantic competition in lexical access. *Journal of Memory and Language*, *46*(2), 245–266.

Rubenstein, H., Lewis, S. S., & Rubenstein, M. A. (1971). Evidence for phonemic recoding in visual word recognition. *Journal of Verbal Learning and Verbal Behavior*, *10*(6), 645–657.

Rumelhart, D. E., & McClelland, J. L. (1982). An interactive activation model of context effects in letter perception: Part 2. The contextual enhancement effect and some tests and extensions of the model. *Psychological Review*, *89*(1), 60–94.

Schilling, H. E. H., Rayner, K., & Chumbley, J. I. (1998). Comparing naming, lexical decision, and eye fixation times: Word frequency effects and individual differences. *Memory and Cognition*, *26*(6), 1270–1281.

Schoonbaert, S., & Grainger, J. (2004). Letter position coding in printed word perception: Effects of repeated and transposed letters. *Language and Cognitive Processes*, *19*(3), 333–367.

Taft, M. (1992). The body of the BOSS: Sub-syllabic units in the lexical processing of polysyllabic words. *Journal of Experimental Psychology: Human Perception and Performance*, *18* (4), 1004–1014.

Taft, M., & Ardasinski (2006). Obligatory decomposition in reading prefixed words. *The Mental Lexicon*, *1*(2), 183–199.

Taylor, J., Rastle, K., & Davis, M. H. (2013). Can cognitive models explain brain activation during word and pseudoword reading? A meta-analysis of 36 neuroimaging studies. *Psychological Bulletin*, *139*(4), 766–791.

van Heuven, W. J. B., Mandera, P., Keuleers, E., & Brysbaert, M. (2014). SUBTLEX-UK: A new and improve word frequency database for British English. *The Quarterly Journal of Experimental Psychology*, *67*(6), 1176–1190.

Velan, H., & Frost, R. (2009). Letter-transposition effects are not universal: The impact of transposing letters in Hebrew. *Journal of Memory and Language*, *61*(3), 285–302.

Wheeler, D. D. (1970). Processes in visual word recognition. *Cognitive Psychology*, *1*(1), 59–85.

Zevin, J. D., & Seidenberg, M. S. (2002). Age of acquisition effects in word reading and other tasks. *Journal of Memory and Language*, *47* (1), 1–29.

Ziegler, J. C., Jacobs, A. M., & Kluppel, D. (2001). Pseudohomophone effects in lexical decisions: Still a challenge for current word recognition models. *Journal of Experimental Psychology: Human Perception and Performance*, *27*(3), 547–559.

CHAPTER

22

Sentence Processing

Fernanda Ferreira[1] *and Derya Çokal*[2]

[1]Department of Psychology and Center for Mind and Brain, University of California, Davis, CA, USA; [2]Institute for Brain and Mind, University of South Carolina, Columbia, SC, USA

The existence of a field called "sentence processing" attests to the implicit agreement among most psycholinguists that the sentence is a fundamental unit of language. In addition, by convention, the term "processing" in this context tends to refer to comprehension rather than production, and thus the topic of this chapter is people's interpretations of sentences. Our goal is to provide an overview of the findings, theories, and debates that are discussed in more detail in the chapters in this volume comprising the section on "Sentence Processing" (Chapters 47–52). The relevant issues include syntactic and semantic processing, the time-course of interpretation, and the role of other cognitive systems such as working memory in forming sentence interpretations. In this chapter, we begin by examining the sources of information that are used during sentence processing. We then review the major theoretical controversies and debates in the field: the incremental nature of interpretation, serial versus parallel processing, and the extent of interaction among information sources during online processing. Then, we go over the major models of sentence processing, including syntax-based models, constraint-based models, the good-enough approach, and the very recent rational analysis approaches. We end with a few conclusions and speculations concerning future research directions.

22.1 SOURCES OF INFORMATION FOR SENTENCE PROCESSING

Since the 1980s, when psycholinguistics experienced a renaissance (Clifton, 1981) and returned to the question of how to relate formal and psychological approaches to language, the field of sentence processing has been associated with a commitment to the idea that syntactic information is critical to successful language comprehension. Not all theorists agree on the nature of those syntactic representations or the relative importance of information sources that are nonsyntactic, but almost all assume that structure-building operations are essential for successful comprehension (Fodor, Bever, & Garrett, 1974; Frazier & Rayner, 1990). One key component is phrase-structure parsing, which refers to the process of identifying constituents and grouping them into a hierarchical structure. For example, in a sentence such as *While Mary bathed the baby played in the crib*, the parser must create a structural analysis that postulates the existence of a subordinate and a main clause; moreover, the main verb of the subordinate clause must be analyzed as intransitive and reflexive, and the subject of the main clause must be identified as *the baby*. With this analysis, the correct meaning can be derived, which is that Mary is bathing herself, and the baby is the agent of playing.

As the same example makes clear, one of the challenges to the parser is syntactic ambiguity. At various points in a sentence, a sequence of words can be given more than one grammatical analysis. In the example, the phrase *the baby* appears to be the object of bathed, but in fact it turns out to be the subject of *played*. The result is a so-called "garden-path." The parser first builds an incorrect analysis, and reanalysis processes are triggered on receipt of a constituent that cannot be incorporated into the existing structure. Because the parser obeys the rules of the grammar, including the rule mandating overt subjects, the parse will fail at *played*, and the sentence processing system must locate the alternative analysis in which *the baby* is a subject. How this happens is another point of divergence between competing sentence processing models, as is discussed in Section 22.2.

An additional complication regarding the syntactic analysis of a sentence is that the grammar allows

Neurobiology of Language. DOI: http://dx.doi.org/10.1016/B978-0-12-407794-2.00022-5

constituents to be moved from their canonical positions. One classic example is the passive, in which the theme of an action is also the sentential subject, contrary to the general preference to align agency and subjecthood (Fillmore, 1968; Grimshaw, 1990; Jackendoff, 1990). Another type of moved constituent is wh-phrases; in English, as in many other languages, wh-phrases must be moved from their canonical position to a position at the beginning of the clause, leaving behind a trace or "gap." For example, in *Which man did the dog bite?*, the phrase *which man* receives its thematic role from *bite*. The job of the parser is to find the gap and relate it to the wh-phrase so that the sentence can receive a correct interpretation. This task is made difficult by two challenges. First, the gap is a phonetically null element in the string, and therefore the parser must identify the gap based on the application of a range of linguistic constraints. The second challenge concerns ambiguity. Because many verbs have multiple argument structures, the parser may end up postulating a gap incorrectly. The result is so-called decoy gaps, as illustrated in *Who will the zombie eat with?* The parser initially assumes that *who* was moved from the direct object position after *eat*, and then must reanalyze that structure when *with* is encountered.

Studies investigating the processing of filler-gap dependencies have found evidence for a filled-gap effect, which is closely related to decoy gaps. Consider the example *Which patient did the doctor expect the nurse to call?* Most comprehenders will assume that *which patient* is the object of *expect*, but the noun phrase (NP) *the nurse* occupies that position, which means that the parser must look further along for the correct gap (located after *call*). The existence of filled-gap effects has led researchers to postulate two parsing preferences for creating filler-gap dependencies. One is that the parser adopts an active or early filler strategy (Frazier, Clifton, & Randall, 1983; Frazier & Flores D'Arcais, 1989), according to which a gap is postulated at the first syntactically permissible location. The second is that the parser makes use of verb argument structure information to guide the postulation of gaps. If a verb has a strong intransitive bias, then the parser is less likely to postulate a gap after it; if the verb is strongly transitive, then a postverbal gap will be more compelling.

As we have been discussing the importance of syntactic information for parsing, we have had numerous occasions to refer to lexical information as well. This is because lexical information is the fundamental bottom-up information source for sentence processing. In lexicalist theories, syntactic information is attached to specific words so that when a word is retrieved, its associated structural possibilities become available as well (Joshi & Schabes, 1997; MacDonald, Pearlmutter, & Seidenberg, 1994). For example, retrieval of the verb *bathe* would bring up not only information associated with the syntactic category and meaning of that word but also the word's syntactic dependents in the form of what are known as argument structures. An optionally transitive verb like *bathe* would have at least two argument structures, one specifying an agent and a patient and the other specifying an agent and an obligatory reflexive null element. Nonlexicalist theories also assume a major role for this type of information; however, in contrast with lexicalist theories, argument structures are used not to generate a parse, but rather to filter or reinforce a particular analysis and to facilitate recovery from a garden-path.

Another type of lexical information that can be critical for parsing relates to semantic features such as number and animacy. Number information can affect how an ambiguous phrase is attached during online processing; for example, in a sentence such as *While John and Mary kissed the baby slept*, the verb *kissed* is interpreted as intransitive because the plural subject triggers a reciprocal reading of *kissed*. A singular subject does not license this reciprocal interpretation (Ferreira & McClure, 1997; Patson & Ferreira, 2009). Similarly, animacy can help the parser avoid a garden-path, or help it recover more easily (Ferreira & Clifton, 1986; Trueswell, Tanenhaus, & Kello, 1993). Specifically, if a subject is inanimate, then it is unlikely to be an agent, and that analysis, in turn, might lead the parser to adopt a less frequent passive or reduced relative parse (e.g., *The evidence examined by the lawyer*). These examples also show how word properties such as number and animacy interact with lexical argument structures, because those features can lead the parser to select one argument structure (e.g., a reciprocal one for a verb such as *kiss*) over another.

Next, let us consider the question of how prosodic information might influence sentence processing. The starting point for most studies published on this topic is that syntactic and prosodic structures are related and, in particular, major syntactic boundaries such as those separating clauses are usually marked by phrase-final lengthening and changes in pitch (Ferreira, 1993). Some clause-internal phrasal boundaries are also marked, although much less reliably (Allbritton, McKoon, & Ratcliff, 1996); for example, in the sentence *John hit the thief with the bat*, the higher attachment of *with the bat*, which supports the instrument interpretation, is sometimes (but not always) associated with lengthening of *thief*. The logic of the research enterprise is to see whether prosodic "cues" can signal syntactic structure and help the parser to avoid going down a garden-path. One of the earliest studies to consider this question was conducted by Beach (1991), who demonstrated that meta-linguistic judgments about sentence structure are influenced by the availability of durational and pitch information linked to the final structures of the sentences. A few decades later, more sensitive online techniques including recording of event-related potentials (ERPs) and eye-tracking have yielded a wealth of information about the comprehension of spoken sentences, and one of the ideas on which there is now a general consensus is that prosody does influence the earliest stages of parsing (Nakamura, Arai, & Mazuka, 2012).

Another potentially influential source of information for sentence processing is context, both discourse and visual. An early analysis of the role of discourse context is known as Referential Theory (Crain & Steedman, 1985). It has been observed that many of the sentence forms identified as syntactically dispreferred by the two-stage model are also presuppositionally more complex. For example, the sentence *John hit the thief with the bat* allows for two interpretations: the *with*-phrase may be interpreted as an instrument or a modifier; the latter interpretation requires a more complex structure (on some theories of syntax). The "confound" here is that the more complex structure also involves modification, whereas the simpler analysis does not. Moreover, a modified phrase such as *the thief with the bat* presupposes the existence of more than one thief, and thus the difficulty of the more complex structure might not be due to its syntax, but rather to the lack of a context to motivate the modified phrase. Crain and Steedman predicted that sentences processed in presuppositionally appropriate contexts would be easy to process, a prediction that Ferreira and Clifton (1986) examined using eye movement monitoring in reading. Their data were consistent with the idea that context did not affect initial parsing decisions: Supportive contexts led to shorter global reading times and more accurate question-answering behavior, but early measures of processing revealed that processing times were longer for structurally complex sentences compared with their structurally simpler counterparts.

The potential role of visual context became a topic of intense interest in the 1990s with the emergence of the Visual World Paradigm (VWP) for studying sentence processing. The idea behind the paradigm is simple. From reading studies, it was known that fixations are closely tied to attention and processing (Rayner, 1977). The VWP extends this logic to spoken language processing by pairing spoken utterances with simple displays containing mentioned and unmentioned objects. The "linking hypothesis" (Tanenhaus, Magnuson, Dahan, & Chambers, 2000) is that as a word is heard, its representation in memory becomes activated, which triggers eye movements toward the named object as well as objects semantically and even phonologically associated with it (Huettig & McQueen, 2007). The widespread adoption of the VWP occurred in part because the idea of multimodal processing was also catching on, with many cognitive scientists wanting to understand the way different cognitive systems might work together—in this case, the auditory language processing system and the visuo-attention system associated with object recognition (Henderson & Ferreira, 2004; Jackendoff, 1996). There was also a growing interest in auditory language processing generally, and in the investigation of how prosodic information might be used during comprehension, as discussed previously. By now, hundreds of studies have been reported making use of it in one way or another (Ferreira, Foucart, & Engelhardt, 2013; Huettig, Olivers, & Hartsuiker, 2011; Huettig, Rommers, & Meyer, 2011).

The reports that triggered the widespread use of the VWP are those by Spivey, Tanenhaus, Eberhard, and Sedivy (2002) and Tanenhaus, Spivey-Knowlton, Eberhard, and Sedivy (1995). This study adapted the ideas of Crain and Steedman (1985) concerning presuppositional support to the domain of visual contexts and spoken sentences that could be evaluated against them. To illustrate, consider the sentence *Put the apple on the towel in the box*. At the point at which the listener hears *on the towel*, two interpretations are possible: either *on the towel* is the location to which the apple should be moved or it is a modifier of *apple*. The phrase *into the box* forces the latter interpretation because it is unambiguously a location. Referential Theory specifies that speakers should provide modifiers only when modification is necessary to establish reference. It follows that if two apples are present in the visual world and one of them is supposed to be moved, then right from the earliest stages of processing, the phrase *on the towel* will be taken to be a modifier, because the modifier allows a unique apple to be identified. The listener faced with this visual world containing two referents should therefore immediately interpret the phrase as a modifier and avoid being garden-pathed (Farmer, Cargill, Hindy, Dale, & Spivey, 2007; Novick, Thompson-Schill, & Trueswell, 2008; Spivey et al., 2002; Tanenhaus et al., 1995). Recently, however, the interpretation of these findings has been challenged. Ferreira et al. (2013) conducted three experiments manipulating properties of the utterances and the visual worlds. They concluded that listeners engage in a fairly atypical mode of processing in VWP experiments with simple visual worlds and utterances that are highly similar to each other in all experimental trials. Rather than processing utterances normally, they instead form a skeleton, underspecified representation of what they are likely to hear based on the content of the display, and then they evaluate that prediction against the utterance itself. These issues concerning the use of the VWP require additional investigation.

In summary, a range of sources of information is used for successful sentence processing. Lexical and syntactic constraints are central for defining the structural alternatives considered by the language processing system, and information associated with the prosody of the sentence as well as the discourse and visual context in which the sentence occurs helps to reinforce some interpretations and flesh out the full meaning of the sentence. In the following section, we consider some of the theoretical controversies concerning the architecture of the language system and the way these sources of information are coordinated. This discussion sets the stage for our discussion of theoretical models of sentence processing.

22.2 THEORETICAL CONTROVERSIES

In this section, we consider four issues that help distinguish among competing models of sentence processing: (i) incremental interpretation; (ii) serial versus parallel processing; (iii) interactivity versus modularity; and (iv) sources of complexity in comprehension, including those that arise due to working memory constraints.

Incremental interpretation refers to whether the sentence processing system builds the meaning of a sentence word-by-word, as the input unfolds, or whether the system either falls behind or gets ahead of the input. Falling behind the input would indicate delays in interpretation; getting ahead would indicate anticipation or prediction. Essentially all current models of processing assume that interpretations are built incrementally and, in particular, that there are no delays in incorporating new words into the ongoing representation of sentence meaning. In addition, there is some evidence that comprehenders engage in prediction (Levy, 2008; Rayner, Li, Juhasz, & Yan, 2005; Van Berkum, Brown, Zwitserlood, Kooijman, & Hagoort, 2005). The classic demonstration of prediction comes from Altmann and Kamide (1999), who used the VWP and semantically constrained sentences such as *The boy will eat the cake*. They observed that listeners made anticipatory eye movements to a depicted cake prior to hearing the word *cake*, indicating that they predicted that continuation. In the structural domain, Staub and Clifton (2006) found that when readers processed a clause beginning with the word *either*, they predicted an upcoming *or*-clause based on the syntactic constraint that the latter must follow the former. These and other studies have been taken as evidence that the sentence processing system is not just an incremental, but actually a predictive, anticipating structure and even specific lexical content.

At the same time, there is some evidence that additional processing takes place at major syntactic boundaries. So-called end of sentence wrap-up refers to the finding that reading times at the ends of clauses and sentences are longer than in other sentential positions (Aaronson & Scarborough, 1976; Just & Carpenter, 1980; Rayner, Kambe, & Duffy, 2000; Rayner, Sereno, Morris, Schmauder, & Clifton, 1989). Wrap-up effects indicate that some elements of meaning are computed over a more global domain. In addition, clause boundaries might be the locations where the comprehension system evaluates the entire structure to ensure that all relevant constraints are satisfied, for example, to check that a verb has all its obligatory arguments. Evidence for underspecified representations also suggests some tendency on the part of the processing system to delay interpretations (for an excellent summary, see Frisson, 2009). Words with multiple senses (e.g., *book* as an object versus its content) seem to be processed by initially activating an underspecified meaning, and then by filling out the semantics once contextually disambiguating information becomes available. Some syntactic ambiguities may also be handled in a similar manner, for example, comprehenders leave open the interpretation of ambiguous relative clauses (*the servant of the actress who was on the balcony*), making a specific attachment decision only once it is necessary to do so (Swets, Desmet, Clifton, & Ferreira, 2008). Pronouns are also often not assigned specific antecedents (McKoon, Greene, & Ratcliff, 1993).

The second theoretical issue in which theories of sentence processing differ is serial versus parallel processing, which typically refers to assumptions about whether the system considers only one interpretation at a time or multiple interpretations. For example, consider *The defendant examined by the lawyer turned out to be unreliable* (Ferreira & Clifton, 1986). The sequence *the defendant examined* could mean that the defendant examined something or that the defendant is the thing being examined (the ultimately correct analysis). The issue is whether only one of these interpretations is built and evaluated at any one time, or whether all the interpretations are simultaneously activated and assessed. In the serial view, first the system considers one analysis—in most theories, the one that assumes that the defendant is the agent of examining, given that this analysis is syntactically simpler and more frequent—and then reanalyzes it if a revision signal is encountered. The sentence processing system then goes into "reanalysis mode," attempting to adjust the syntactic structure that has been built to create a grammatical analysis (Ferreira & Henderson, 1991; Fodor & Ferreira, 1998; Fodor & Inoue, 1994). Ease of reanalysis depends on the extent to which the sentence processing system can find lexical and grammatical information that motivates an alternative structure.

The parallel view assumes that the sentence processing system activates all grammatically licensed analyses simultaneously. Considering our example, both the incorrect and the ultimately correct interpretations of *the defendant examined* would be available in parallel, initially weighted by their frequency. The agent analysis of *defendant* is more frequent; therefore, at first, it will be stronger than the ultimately correct analysis. But when the word *by* is encountered, the sentence processing system must shift to the other activated interpretation. Ease of reanalysis depends on the relative activation levels of the two interpretations. If the ultimately correct interpretation is infrequent, then it will be difficult to retrieve and reanalysis might even fail. If the right interpretation has some strength based on the extent to which it conforms to a wide range of linguistic and nonlinguistic constraints, then reanalysis will be easier, and so will overall comprehension of the sentence.

A careful reader might have noticed subtle differences in the terminology used in our discussion of serial versus parallel processing. For the former, interpretations are typically described as being "built," whereas for the latter they are often referred to as being "activated" or "retrieved." These different terms reflect fundamentally different ideas about how interpretations are stored in memory and accessed during sentence processing. The serial view tends to assume that syntactic rules are stored in memory and then used online to create a structural representation bit by bit. Reanalysis processes are a matter of editing the structure. The parallel view tends to assume that structures are stored in chunks, typically corresponding to an argument-taking word such as a verb and its arguments. Online processing involves not so much building a structure as much as activating one. These issues are raised again when we consider models of sentence processing.

The third issue in which theories of processing differ is interactivity versus modularity. Almost since the earliest days of psycholinguistics, debate has centered around the issue of whether the system considers only linguistic (and possibly even only syntactic) information when parsing a sentence versus a system that considers all potentially relevant sources of information. Modular models assume sentence structures are assigned to words at least initially without any consideration of whether the structure will map to a sentence interpretation that makes sense given prior knowledge or given the contents of the immediate linguistic, visual, or social context. For example, the sentence processing system would be garden-pathed not only by *the defendant examined by the lawyer* but also by *the evidence examined by the lawyer*, even though *evidence* is inanimate and therefore cannot engage in an act of examination. In contrast, interactive models assume the immediate use of all relevant constraints. At this stage, there is widespread belief in the field that the preponderance of evidence supports interactive models, although it is possible to argue that this conclusion goes somewhat beyond the evidence (Ferreira & Nye, in progress).

Thus far we have mainly focused on structural ambiguity, which is certainly one source of difficulty or complexity in processing. But structure-building processes independent of ambiguity resolution are also a potential source of complexity, as first argued by Gibson (1991). One source of complexity is structural frequency. All things being equal, a structure encountered more frequently will be easier to comprehend than one that is rare (Ferreira, 2003; Gibson, 1991, 1998; MacDonald et al., 1994). The demands that structures place on working memory are an additional source of processing difficulty for both ambiguous and unambiguous structures (Chomsky & Miller, 1963; Gibson, 1991, 1998, 2000; Lewis & Vasishth, 2005; Yngve, 1960). For example, nested structures (*The reporter who the senator who John met attacked disliked the editor*) are harder to process than right-branching structures (*John met the senator who attached the reporter who disliked the editor*), a generalization that holds across typologically different languages (e.g., English, which is a subject verb object [SVO] language, and Japanese, which is subject object verb [SOV]). This contrast between nested and right-branching structures can be explained by appealing to the greater demands the former structures place on working memory. More specifically, two kinds of demands increase processing complexity for unambiguous as well as ambiguous structures: (i) storage costs and (ii) distance-based integration costs. Storage costs are incurred when incomplete materials must be held in working memory, for example, a verb that needs its arguments (Chen, Gibson, & Wolf, 2005; Gibson, 1998; Nakatani & Gibson, 2010). Distance-based costs are those that arise from attempts to integrate a word into the structure already built and seem to be proportional to the difficulty of reactivating an earlier word, for example, an argument that must be linked back to its verb (Gibson, 1998, 2000; Gordon, Hendrick, & Johnson, 2001, 2004; Grodner & Gibson, 2005; Lewis & Vasishth, 2005; Lewis, Vasishth, & Van Dyke, 2006). Distance-based costs also account for the well-known preference for subject-extracted over object-extracted relative clauses (Grodner & Gibson, 2005) and arise in part due to similarity-based interference in working memory (Gordon et al., 2001).

22.3 CLASSES OF MODELS OF SENTENCE PROCESSING

We begin with the so-called two-stage model or garden-path model, first developed by Lyn Frazier (Ferreira & Clifton, 1986; Frazier & Fodor, 1978; Rayner, Carlson, & Frazier, 1983). The model assumes that a single parse is constructed for any sentence based on the operation of Minimal Attachment, which constrains the parser to construct no potentially unnecessary syntactic nodes, and Late Closure, which causes the parser to attach new linguistic input to the current constituent. In addition, the model assumes that the only information that the parser has access to when building a syntactic structure is its database of phrase-structure rules; therefore, the parser cannot consult information associated with lexical items. For example, in the sequence *Mary knew Bill*, the noun phrase *Bill* would be assigned the role of direct object because that analysis is simpler than the alternative subject-of-complement-clause analysis, and the information that *know* takes sentence complements more frequently than direct objects could not be used to inform the initial parse.

The two-stage model has evolved over the past three decades to take into account changes in linguistic theory and significant findings in psycholinguistics. One important addition is the notion of "Construal" (Frazier & Clifton, 1997; Frisson & Pickering, 2001), which allows some constituents to be merely associated with a specific thematic domain in a sentence rather than definitively attached to the structure. Evidence for Construal comes from the finding that readers process sentences with ambiguous relative clauses more quickly than those that have a unique attachment (e.g., *the servants of the actress who was on the balcony*), unless the sentence is followed by a question that forces the reader to provide a specific interpretation; in that case, readers take longer to read the ambiguous versions, presumably because they are trying to choose between the attachment options. Another important revision of the two-stage model is that, now, prosody plays an essential role in determining how parsing proceeds from the earliest stages of processing (Millotte, Wales, & Christophe, 2007; Nakamura et al., 2012; Price, Ostendorf, Shattuck-Hufnagel, & Fong, 1991). Pitch and durational information associated with different kinds of prosodic and intonational phrasing are used to constrain the parser's syntactic analyses and assist in the construction of semantic meanings such as focus and presupposition. Nonetheless, the essential features of the two-stage model remain. The model assumes that (i) information is used incrementally to build an interpretation; (ii) different possible interpretations are built and evaluated serially, rather than in parallel; and (iii) only certain kinds of information can be used during the initial stages of sentence processing, particularly information stated in the syntactic and prosodic vocabulary of the sentence processing module.

The two-stage model was soon challenged by researchers in sentence processing who were strongly influenced by the connectionist architectures popular in the 1980s and 1990s (Rumelhart & McClelland, 1985, 1986; Seidenberg & McClelland, 1989). These architectures contrast with the assumptions of the two-stage model in two defining ways. First, in connectionist systems, alternative possibilities are activated and evaluated in parallel; second, any relevant source of information can be used to modulate the activation levels and allow one possible analysis to win at the expense of the others (MacDonald et al., 1994). Applying these ideas to sentence processing, the connectionist alternative assumed the following principles. First, rather than analyses being built with the help of grammatical rules, a great deal of the burden of syntactic representation is put into the lexicon. Adapting ideas that were then timely in linguistic theory (Pesetsky, 1995), lexical representations were assumed to activate not only words and word meanings but also syntactic frames. In this view, syntactic rules are redundant because almost all the necessary information is already stated in the lexicon. Thus, with syntactic structures being stored rather than built, it is easy to imagine an architecture in which all possible analyses are considered in parallel, weighted by their frequency of use. Lexical, contextual, and pragmatic constraints can be used to further modulate the activation levels. With this approach the sentence processing system is incremental, but different possible interpretations are activated in parallel. In addition, any potential source of information can be used at any stage of sentence processing, making the system interactive rather than modular.

Other classes of models emphasize the role of complexity in sentence processing (Gibson, 1991, 1998). The significance of these models is two-fold. First, they highlight sources of information that lead to processing costs for both ambiguous and unambiguous structures and thus capture well-known findings such as the preference for subject- over object-relative clauses (Gibson, 1991, 1998, 2000; Grodner & Gibson, 2005; Lewis & Vasishth, 2005; Yngve, 1960). Second, these models are specifically designed to take into account the role of working memory in sentence processing, with an emphasis on how costs associated with maintaining and integrating items in working memory affect complexity and, therefore, processing difficulty. These models make important predictions about phenomena in sentence processing that are of particular interest to researchers attempting to uncover the neural mechanisms that underlie sentence comprehension, including structures such as passives and relative clauses.

In the past 15 years or so, a new class of models has emerged with roots in all the approaches that have been described thus far. There are many variants with important distinctions among them, but what they share is the idea that comprehenders sometimes end up with an interpretation that differs from the actual input received—the interpretation is simpler (Construal), somewhat distorted (Late Assignment of Syntax Theory (LAST); Good-Enough Processing), or outright inconsistent (Noisy Channel Approaches) with the sentence's true content. Let us begin with the models that assume representations that reduce the input in some way. One implementation is to allow representations to be underspecified (Sanford & Sturt, 2002). Consider Construal. As mentioned, this model assumes that syntactic structures are not always fully connected and adjunct phrases in particular (e.g., relative clauses, modifying prepositional phrases) may instead simply get associated with a certain processing domain, "floating" until disambiguating information arrives. The parser thus remains uncommitted (Pickering, McElree, Frisson, Chen, & Traxler, 2006) concerning the attachment of the relative clause and the interpretation that would follow from any particular attachment (Frisson & Pickering, 2001; Sanford & Graesser, 2006; Sturt, Sanford, Stewart, & Dawydiak, 2004). Other studies support the idea of underspecified representations for global syntactic structures (Tyler & Warren, 1987), semantic information (Frazier & Rayner, 1990), and coercion structures (Pickering et al., 2006).

More radical variants of shallow processing models allow the comprehension system to generate an interpretation that is even more discrepant from the input. Researchers in the field of text processing have shown that readers are sometimes remarkably insensitive to contradictions in text (Otero & Kintsch, 1992), and they also often fail to update their interpretations when later information undermines a fact stated previously (Albrecht & O'Brien, 1993). These ideas from text processing were exported to the sentence processing literature in a series of experiments showing that people do not seem to fully recover from garden-paths (Christianson, Hollingworth, Halliwell, & Ferreira, 2001). Participants read sentences such as *While the woman bathed the baby played in the crib* and then answered a question such as *Did the woman bathe the baby?* The surprising finding was that most people answered "yes," even though the meaning of the reflexive verb *bathe* requires that the object be interpreted as coreferential with the subject (see also Slattery, Sturt, Christianson, Yoshida, & Ferreira, 2013). It appears that comprehenders are not entirely up to the task of syntactic reanalysis and sometimes fail to revise either all pieces of the syntactic structure or all elements of the semantic consequences of the initial, incorrect parse. And the more semantically compelling the original misinterpretation, the more likely people are to want to retain it.

Townsend and Bever's (2001) model implements an architecture similar to what has been suggested for decision-making (Gigerenzer, 2004; Kahneman, 2003), where researchers sometimes distinguish between so-called System 1 and System 2 (or Type 1 and Type 2) reasoning. System 1 is fast, automatic, and operates via the application of simple heuristics. System 2, however, is slow and attention-demanding, and it consults a wide range of beliefs—essentially anything the organism knows and has stored in memory. In Townsend and Bever's (2001) LAST, sentences are essentially processed twice. First, heuristics are accessed that yield a quick meaning, and then syntactic computations are performed on the same word string to yield a fully connected, syntactic analysis. The second process ensures that the meaning that is obtained for a sentence is consistent with its actual form. Townsend and Bever also assume that the first stage is nonmodular and that the second is modular; this is to account for the use of semantics in the first stage and the use of essentially only syntactic constraints in the second.

Two models similar in spirit to LAST but that assume a modular architecture for the first stage have been proposed by Ferreira (2003) and by Garrett (2000). The Ferreira model assumes that the first stage consults just two heuristics—a version of the noun verb noun (NVN) strategy in which people assume an agent–patient mapping of semantic roles to syntactic positions and an animacy heuristic, in which animate entities are biased toward subjecthood. The 2003 Ferreira model explains comprehenders' tendencies to misinterpret passive sentences, particularly when they express an implausible event with reversible semantic roles, as in *the dog was bitten by the man*. The application of heuristics in the first stage yields the dog-bit-man interpretation; a proper syntactic parse will deliver the opposite, correct interpretation, but the model assumes that it is fragile and susceptible to interference. Garrett (2000) offers a more explicitly analysis-by-synthesis model that incorporates the production system to generate what are generally thought of as top-down effects. A first pass, bottom-up process uses syntactic information to create a simple parse that, in turn, allows for a rudimentary interpretation. Then, the language production system takes over and uses that representation to generate the detailed syntactic structure that would support the initial parse and interpretation.

Finally, a family of models has recently been proposed that assume people engage in rational behavior over a noisy communication channel. The channel is noisy because listeners sometimes mishear or misread due to processing error or environmental contamination, and because speakers sometimes make mistakes when they communicate. Thus, a rational comprehender whose goal is to recover the intention behind the utterance will normalize the input according to Bayesian priors. A body of evidence from research using ERPs helped to motivate these ideas (Kim & Osterhout, 2005; Van Herten, Kolk, & Chwilla, 2005). In these experiments, it is reported that subjects who encounter a sentence such as *The fox that hunted the poachers stalked through the woods* experience a P600 rather than an N400 on encountering the semantically anomalous word, even though an N400 might be expected given that it is presumed to reflect problems related to meaning. There is still not a great deal of consensus regarding what triggers P600s, but an idea that has been gaining traction is that it reflects a need to engage in some type of structural reanalysis or revision. The conclusion, then, is that when a person encounters a sentence that seems to say that the fox hunted the poachers, that person "fixes" it so it makes sense, resulting in a P600. Other models have taken this idea and developed it further (Gibson, Bergen, & Piantadosi, 2013; Levy, 2011; Levy, Bicknell, Slattery, & Rayner, 2009). These models are generally interactive, because the information that is accessed to establish the priors can range from biases related to structural forms and all the way to beliefs concerning speaker characteristics (Van Berkum, Van den Brink, Tesink, Kos, & Hagoort, 2008). However, these noisy channel models have not yet been rigorously tested using a methodology that allows early processes to be distinguished from later ones. For example, it remains possible that comprehenders create a simple parse in a manner compatible with modularity and then consult information outside the module to revise that interpretation, right down to actually normalizing the input. Models designed to explain the comprehension of sentences containing self-repairs and other disfluencies (*turn left uh right at the light*) assume mechanisms that allow input to be deleted so that the speaker's intended meaning can be recovered (Ferreira, Lau, & Bailey, 2004).

22.4 CONCLUSION

The field of sentence processing has changed significantly since the 1980s. Current models emphasize more detailed, context-specific information such as speaker, and there is a great deal of interest in mechanisms that allow the input to be rationally evaluated and corrected. Future work will continue to make use of behavioral techniques as well as methods from neuroscience to expand our understanding of these topics. The critical next stage is to determine how the processes assumed in models of sentence processing are actually implemented in the human brain. Our view is that the field is well-positioned for this challenge given the sophistication of extant sentence processing models.

References

Aaronson, D., & Scarborough, H. S. (1976). Performance theories for sentence coding: Some quantitative evidence. *Journal of Experimental Psychology: Human Perception and Performance*, *2*(1), 56–70.

Albrecht, J. E., & O'Brien, E. J. (1993). Updating a mental model: Maintaining both local and global coherence. *Journal of Experimental Psychology: Learning, Memory, and Cognition*, *19*(5), 1061–1070.

Allbritton, D. W., McKoon, G., & Ratcliff, R. (1996). Reliability of prosodic cues for resolving syntactic ambiguity. *Journal of Experimental Psychology: Learning, Memory, and Cognition*, *22*(3), 714–735.

Altmann, G. T. M., & Kamide, Y. (1999). Incremental interpretation at verbs: Restricting the domain of subsequent reference. *Cognition*, *73*, 247–264.

Beach, C. M. (1991). The interpretation of prosodic patterns at points of syntactic structure ambiguity: Evidence for cue trading relations. *Journal of Memory and Language*, *30*(6), 644–663.

Chen, E., Gibson, E., & Wolf, F. (2005). Online syntactic storage costs in sentence comprehension. *Journal of Memory and Language*, *52*, 144–169.

Chomsky, N., & Miller, G. A. (1963). Introduction to the formal analysis of natural languages. In R. D. Luce, R. R. Bush, & E. Galanter (Eds.), *Handbook of mathematical psychology* (pp. 269–321). New York, NY: Wiley.

Christianson, K., Hollingworth, A., Halliwell, J. F., & Ferreira, F. (2001). Thematic roles assigned along the garden path linger. *Cognitive Psychology*, *42*(4), 368–407.

Clifton, C., Jr. (1981). Psycholinguistic renaissance? *Contemporary Psychology*, *26*, 919–921.

Crain, S., & Steedman, M. (1985). On not being led up the garden path: The use of context by the psychological parser. In D. Dowty, L. Karttunen, & A. Zwicky (Eds.), *Natural language parsing: Psychological, computational, and theoretical perspectives* (pp. 320–358). Cambridge, UK: Cambridge University Press.

Farmer, T. A., Cargill, S. A., Hindy, N. C., Dale, R., & Spivey, M. J. (2007). Tracking the continuity of language comprehension: Computer mouse trajectories suggest parallel syntactic processing. *Cognitive Science*, *31*(5), 889–909.

Ferreira, F. (1993). The creation of prosody during sentence production. *Psychological Review*, *100*, 233–253.

Ferreira, F. (2003). The misinterpretation of noncanonical sentences. *Cognitive Psychology*, *47*(2), 14–203.

Ferreira, F., & Clifton, C. (1986). The independence of syntactic processing. *Journal of Memory and Language*, *25*(3), 348–368.

Ferreira, F., Foucart, A., & Engelhardt, P. E. (2013). Language processing in the visual world: Effects of preview, visual complexity, and prediction. *Journal of Memory and Language*, *69*(3), 165–182.

Ferreira, F., & Henderson, J. M. (1991). Recovery from misanalyses of garden-pathsentences. *Journal of Memory and Language, 30*, 725–745.

Ferreira, F., Lau, E. F., & Bailey, K. G. (2004). Disfluencies, language comprehension, and tree adjoining grammars. *Cognitive Science, 28*(5), 721–749.

Ferreira, F., & McClure, K. (1997). Parsing of garden-path sentences with reciprocal verbs. *Language and Cognitive Processes, 12*, 273–306.

Ferreira, F., & Nye, J. (in press). The modularity of sentence processing reconsidered. In R. G. De Almeida & L. Gleitman (Eds.), *Minds on language and thought*. Oxford, UK: Oxford University Press.

Fillmore, C. J. (1968). The creation of prosody during sentence production. *Psychological Review, 100*, 233–253.

Fodor, J. A., Bever, T. G., & Garrett, M. (1974). *The psychology of language: An introduction to psycholinguistics and generative grammar*. New York, NY: McGraw-Hill.

Fodor, J. D., & Ferreira, F. (1998). *Reanalysis in sentence processing*. Dordrecht, The Netherlands: Kluwer.

Fodor, J. D., & Inoue, A. (1994). The diagnosis and cure of garden paths. *Journal of Psycholinguistic Research, 23*, 407–434.

Frazier, L., & Clifton, C., Jr. (1997). Construal: Overview, motivation, and some new evidence. *Journal of Psycholinguistic Research, 26*(3), 277–295.

Frazier, L., Clifton, C., & Randall, J. (1983). Filling gaps: Decision principles and structure in sentence comprehension. *Cognition, 13*, 187–222.

Frazier, L., & Flores D'Arcais, G. B. (1989). Filler driven parsing: A study of filling gap in Dutch. *Journal of Memory and Language, 28*, 331–444.

Frazier, L., & Fodor, J. D. (1978). The sausage machine: A new two-stage parsing model. *Cognition, 6*(4), 291–325.

Frazier, L., & Rayner, K. (1990). Taking on semantic commitments: Processing multiple meanings vs. multiple senses. *Journal of Memory and Language, 29*(2), 181–200.

Frisson, S. (2009). Semantic underspecification in language processing. *Language and Linguistics Compass, 3*(1), 111–127.

Frisson, S., & Pickering, M. J. (2001). Obtaining a figurative interpretation of a word: Support for underspecification. *Metaphor and Symbol, 16*(3–4), 149–171.

Garrett, M. (2000). Remarks on the architecture of language processing systems. In Y. Grodzinsky, & L. Shapiro (Eds.), *Language and the brain: Representation and processing* (pp. 31–69). San Diego, CA: Academic Press.

Gibson, E. (1991). *A computational theory of human linguistic processing: Memory limitations and processing breakdown*. Pittsburgh, PA: Carnegie Mellon University.

Gibson, E. (1998). Linguistic complexity: Locality of syntactic dependencies. *Cognition, 68*(1), 1–76.

Gibson, E. (2000). The dependency locality theory: A distance-based theory of linguistic complexity. In Y. Miyashita, A. Marantz, & W. O'Neil (Eds.), *Image, language, brain* (pp. 95–126). Cambridge, MA: MIT Press.

Gibson, E., Bergen, L., & Piantadosi, S. T. (2013). Rational integration of noisy evidence and prior semantic expectations in sentence interpretation. *Proceedings of the National Academy of Sciences of the United States of America, 110*(20), 8051–8056.

Gigerenzer, G. (2004). Fast and frugal heuristics: The tools of bounded rationality. In D. J. Koehler, & N. Harvey (Eds.), *Handbook of judgment and decision making* (pp. 62–88). Oxford, UK: Blackwell.

Gordon, P. C., Hendrick, R., & Johnson, M. (2001). Memory interference during language processing. *Journal of Experimental Psychology: Learning, Memory, and Cognition, 27*, 1411–1423.

Gordon, P. C., Hendrick, R., & Johnson, M. (2004). Effects of noun phrase type on sentence complexity. *Journal of Memory and Language, 51*(1), 97–114.

Grimshaw, J. (1990). *Argument structure*. Cambridge, MA: MIT Press.

Grodner, D., & Gibson, E. (2005). Consequences of the serial nature of linguistic input. *Cognitive Science, 29*, 261–291.

Henderson, J. M., & Ferreira, F. (2004). Scene perception for psycholinguists. In J. M. Henderson, & F. Ferreira (Eds.), *The interface of language, vision, and action: Eye movements and the visual world* (pp. 1–58). New York, NY: Psychology Press.

Huettig, F., & McQueen, J. M. (2007). The tug of war between phonological, semantic and shape information in language-mediated visual search. *Journal of Memory and Language, 57*(4), 460–482.

Huettig, F., Olivers, C. N., & Hartsuiker, R. J. (2011). Looking, language, and memory: Bridging research from the visual world and visual search paradigms. *Acta Psychologica, 137*(2), 138–150.

Huettig, F., Rommers, J., & Meyer, A. S. (2011). Using the visual world paradigm to study language processing: A review and critical evaluation. *Acta Psychologica, 137*(2), 151–171.

Jackendoff, R. (1996). The architecture of the linguistic-spatial interface. In P. Bloom, M. A. Peterson, L. Nadel, & M. F. Garrett (Eds.), *Language and space. Language, speech, and communication* (pp. 1–30). Cambridge, MA: MIT Press.

Jackendoff, R. S. (1990). *Semantic structures*. Cambridge, MA: MIT Press.

Joshi, A. K., & Schabes, Y. (1997). Tree-adjoining grammars. In G. Rozenberg, & A. Salomaa (Eds.), *Handbook of formal languages* (pp. 69–123). Berlin: Springer.

Just, M. A., & Carpenter, P. A. (1980). A theory of reading: From eye fixations to comprehension. *Psychological Review, 87*(4), 329–354.

Kahneman, D. (2003). Maps of bounded rationality: Psychology for behavioral economics. *The American Economic Review, 93*(5), 1449–1475.

Kim, A., & Osterhout, L. (2005). The independence of combinatory semantic processing: Evidence from event-related potentials. *Journal of Memory and Language, 52*(2), 205–225.

Levy, R. (2008). Expectation-based syntactic comprehension. *Cognition, 106*(3), 1126–1177.

Levy, R. (2011). Probabilistic linguistic expectations, uncertain input, and implications. *Studies of Psychology and Behavior, 9*(1), 52–63.

Levy, R., Bicknell, K., Slattery, T., & Rayner, K. (2009). Eye movement evidence that readers maintain and act on uncertainty about past linguistic input. *Proceedings of the National Academy of Sciences of the United States of America, 106*(50), 21086–21090.

Lewis, R. L., & Vasishth, S. (2005). An activation-based model of sentence processing as skilled memory retrieval. *Cognitive Science, 29*, 375–419.

Lewis, R. L., Vasishth, S., & Van Dyke, J. A. (2006). Computational principles of working memory in sentence comprehension. *Trends in Cognitive Sciences, 10*, 44–54.

MacDonald, M. C., Pearlmutter, N. J., & Seidenberg, M. S. (1994). The lexical nature of syntactic ambiguity resolution. *Psychological Review, 101*(4), 676–703.

McKoon, G., Greene, S., & Ratcliff, R. (1993). Discourse models, pronoun resolution, and the implicit causality of verbs. *Journal of Experimental Psychology: Learning, Memory, and Cognition, 19*, 1–13.

Millotte, S., Wales, R., & Christophe, A. (2007). Phrasal prosody disambiguates syntax. *Language and Cognitive Processes, 22*(6), 898–909.

Nakamura, C., Arai, M., & Mazuka, R. (2012). Immediate use of prosody and context in predicting a syntactic structure. *Cognition, 125* (2), 317–325.

Nakatani, K., & Gibson, E. (2010). An on-line study of Japanese nesting complexity. *Cognitive Science, 34*, 94–112.

Novick, J. M., Thompson-Schill, S. L., & Trueswell, J. C. (2008). Putting lexical constraints in context into the visual-world paradigm. *Cognition, 107*(3), 850–903.

Otero, J., & Kintsch, W. (1992). Failures to detect contradictions in a text: What readers believe versus what they read. *Psychological Science*, *3*(4), 229–235.

Patson, N. D., & Ferreira, F. (2009). Conceptual plural information is used to guide early parsing decisions: Evidence from garden-path sentences with reciprocal verbs. *Journal of Memory and Language*, *60*, 464–486.

Pesetsky, D. (1995). *Zero Syntax. Experiencers and cascades.* Cambridge, MA: MIT Press.

Pickering, M. J., McElree, B., Frisson, S., Chen, L., & Traxler, M. J. (2006). Underspecification and aspectual coercion. *Discourse Processes*, *42*(2), 131–155.

Price, P. J., Ostendorf, M., Shattuck-Hufnagel, S., & Fong, C. (1991). The use of prosody in syntactic disambiguation. *The Journal of the Acoustical Society of America*, *90*, 2956–2970.

Rayner, K. (1977). Visual attention in reading: Eye movements reflect cognitive processes. *Memory and Cognition*, *5*(4), 443–448.

Rayner, K., Carlson, M., & Frazier, L. (1983). The interaction of syntax and semantics during sentence processing: Eye movements in the analysis of semantically biased sentences. *Journal of Verbal Learning and Verbal Behavior*, *22*(3), 358–374.

Rayner, K., Kambe, G, & Duffy, S. A. (2000). The effect of clause wrap-up on eye movements during reading. *The Quarterly Journal of Experimental Psychology*, *53*(4), 1061–1080.

Rayner, K., Li, X., Juhasz, B., & Yan, G. (2005). The effect of word predictability on the eye movements of Chinese readers. *Psychonomic Bulletin and Review*, *12*(6), 1089–1093.

Rayner, K, Sereno, S. C., Morris, R. K., Schmauder, A. R., & Clifton, C. (1989). Eye movements and on-line language comprehension processes. *Language and Cognitive Processes*, *4*, 21–49.

Rumelhart, D. E., & McClelland, J. L. (1985). Level's indeed! A response to Broadbent. *Journal of Experimental Psychology: General*, *114*(2), 193–197.

Rumelhart, D. E., & McClelland, J. L. (1986). PDP Models and general issues in cognitive science. In D. E. Rumelhart, J. L. McClelland, & the PDP Research Group (Eds.), *Parallel distributed processing* (pp. 111–146). Cambridge, MA: MIT Press.

Sanford, A. J., & Graesser, A. C. (2006). Shallow processing and underspecification. *Discourse Processes*, *42*(2), 99–108.

Sanford, A. J., & Sturt, P. (2002). Depth of processing in language comprehension: Not noticing the evidence. *Trends in Cognitive Sciences*, *6*(9), 382–386.

Seidenberg, M. S., & McClelland, J. L. (1989). A distributed, developmental model of word recognition and naming. *Psychological review*, *96*(4), 523.

Slattery, T. J., Sturt, P., Christianson, K., Yoshida, M., & Ferreira, F. (2013). Lingering misinterpretations of garden path sentences arise from competing syntactic representations. *Journal of Memory and Language*, *69*(2), 104–120.

Spivey, M. J., Tanenhaus, M. K., Eberhard, K. M., & Sedivy, J. C. (2002). Eye movements and spoken language comprehension: Effects of visual context on syntactic ambiguity resolution. *Cognitive Psychology*, *45*(4), 447–481.

Staub, A., & Clifton, C. (2006). Syntactic prediction in language comprehension: Evidence from either... or. *Journal of Experimental Psychology: Learning, Memory, and Cognition*, *32*, 425–436.

Sturt, P., Sanford, A. J., Stewart, A., & Dawydiak, E. (2004). Linguistic focus and good-enough representations: An application of the change-detection paradigm. *Psychonomic Bulletin and Review*, *11*(5), 882–888.

Swets, B., Desmet, T., Clifton, C., & Ferreira, F. (2008). Underspecification of syntactic ambiguities: Evidence from self-paced reading. *Memory and Cognition*, *36*(1), 201–216.

Tanenhaus, M. K., Magnuson, J. S., Dahan, D., & Chambers, C. (2000). Eye movements and lexical access in spoken-language comprehension: Evaluating a linking hypothesis between fixations and linguistic processing. *Journal of Psycholinguistic Research*, *29*(6), 557–580.

Tanenhaus, M. K., Spivey-Knowlton, M. J., Eberhard, K. M., & Sedivy, J. C. (1995). Integration of visual and linguistic information in spoken language comprehension. *Science*, *268*(5217), 1632–1634.

Townsend, D. J., & Bever, T. G. (2001). *Sentence comprehension: The integration of habits and rules (*Vol. *1950)*. Cambridge, MA: MIT Press.

Trueswell, J. C., Tanenhaus, M. K., & Kello, C. (1993). Verb-specific constraints in sentence processing: Separating effects of lexical preference from garden-paths. *Journal of Experimental Psychology: Learning, Memory and Cognition*, *19*(3), 528–553.

Tyler, L. K., & Warren, P. (1987). Local and global structure in spoken language comprehension. *Journal of Memory and Language*, *26* (6), 638–657.

Van Berkum, J., Van den Brink, D., Tesink, C., Kos, M., & Hagoort, P. (2008). The neural integration of speaker and message. *Journal of Cognitive Neuroscience*, *20*(4), 580–591.

Van Berkum, J. A., Brown, C. M., Zwitserlood, P., Kooijman, & Hagoort, P. (2005). Anticipating upcoming words in discourse: Evidence from erps and reading times. *Journal of Experimental Psychology: Learning, Memory and Cognition*, *31*(3), 443–467.

Van Herten, M., Kolk, H. H., & Chwilla, D. J. (2005). An ERP study of P600 effects elicited by semantic anomalies. *Cognitive Brain Research*, *22*(2), 241–255.

Yngve, V. H. (1960). A model and an hypothesis for language structure. *Proceedings of the American Philosophical Society*, *104*(5), 444–466.

CHAPTER

23

Gesture's Role in Learning and Processing Language

Özlem Ece Demir[1] *and Susan Goldin-Meadow*[2]

[1]Department of Communication Sciences and Disorders, Northwestern University, Evanston, IL, USA; [2]Department of Psychology, University of Chicago, Chicago, IL, USA

In all cultures and at all ages, speakers move their hands when they talk—they gesture. Even congenitally blind individuals who have never seen anyone gesture move their hands when they talk (Iverson & Goldin-Meadow, 1998), suggesting that gesturing is a robust part of speaking. Moreover, gesture and speech form an integrated system for expressing meaning. Gesture conveys the visual component of the meaning and uses imagistic and analog devices to do so; speech conveys the linguistic component and uses the linear-segmented, hierarchical devices characteristic of language (McNeill, 1992, 2008). Our goal in this chapter is to introduce neuroscientists interested in the neurobiology of language to gesture and the role it plays in language learning and language processing. We focus primarily on behavioral studies and include findings from neuroimaging studies only when relevant (we direct the reader interested in the neurobiology of gesture to Chapter 32 by Dick and Broce). We begin by briefly reviewing evidence showing that gesture can provide a unique window into the mind of a speaker—not only does gesture reflect a speaker's thoughts but also it can play a role in changing those thoughts. We then explore in detail the role gesture plays in how language is learned and how it is processed.

23.1 GESTURE NOT ONLY REFLECTS THOUGHT, IT CAN PLAY A ROLE IN CHANGING THOUGHT

Although gesture may seem like handwaving, it in fact conveys substantive information, often information that is not found in the speaker's words. For example, consider a child who is shown two rows of checkers. The child is first asked to verify that the two rows have the same number of checkers and is then asked whether the rows still have the same number after one row is spread out. The child says "no" and justifies his response by saying, "They are different because you moved them." But at the same time, the child produces the following gestures—he moves his finger between the first checker in row 1 and the first checker in row 2, then the second checker in rows 1 and 2, and so on. In his gestures, the child is demonstrating an understanding of one-to-one correspondence, a central concept underlying the conservation of number that does not appear in his speech (Church & Goldin-Meadow, 1986).

Two additional points are worth noting about gesture. The information conveyed uniquely in a speaker's gestures is often accessible *only* to gesture, that is, it is encapsulated knowledge not yet accessible to speech (Goldin-Meadow, Alibali, & Church, 1993). Speakers who produce gestures that convey information not found in their speech when explaining a task are ready to learn that task—when given instruction in the task, they are more likely to profit from that instruction than speakers whose gestures convey the same information as their speech, whether the speakers are children (Church & Goldin-Meadow, 1986; Perry, Church, & Goldin-Meadow, 1988; Pine, Lufkin, & Messer, 2004) or adults (Perry & Elder, 1997; Ping, Decatur, Larson, Zinchenko, & Goldin-Meadow, under review).

Gesture can thus reflect the state of a speaker's knowledge. But there is now good evidence that gesture can do more than display what speakers know and can play a role in changing what they know. Gesture can change thinking in at least two ways.

Neurobiology of Language. DOI: http://dx.doi.org/10.1016/B978-0-12-407794-2.00023-7

First, the gestures we see others produce can change our minds. Learners are more likely to profit from instruction when it is accompanied by gesture than when that same instruction is not accompanied by gesture (Perry, Berch, & Singleton, 1995; Valenzeno, Alibali, & Klatzky, 2003), even when the gestures are not directed at objects in the immediate environment (Ping & Goldin-Meadow, 2008). Gesture has been found to be particularly helpful in instruction when it conveys a correct strategy for solving a math problem that is different from the (also correct) strategy conveyed in the accompanying speech (Singer & Goldin-Meadow, 2005).

Second, the gestures that we ourselves produce can change our minds. To determine whether gesture can bring about change, we need to teach speakers to gesture in particular ways. If speakers can extract meaning from their gestures, then they should be sensitive to the particular movements in those gestures and change their minds accordingly. Alternatively, all that may matter is that speakers move their hands. If so, then they should change their minds regardless of which gestures they produce. To investigate these alternatives, Goldin-Meadow, Cook, and Mitchell (2009) manipulated gesturing during a math lesson. They found that children required to produce *correct* gestures learned more than children required to produce *partially correct* gestures, who in turn learned more than children required to produce *no* gestures. After the lesson, the children who had gestured were able to express in their own words the information that they had conveyed only in their gestures during the lesson (and that the teacher had not conveyed at all), that is, they had learned from their hands. These findings suggest that the gestures speakers produce can have an impact on what they learn.

Having found that gesture has cognitive significance in many contexts, we are now in a position to explore the role of gesture in learning and processing language.

23.2 ROLE OF GESTURE IN LANGUAGE LEARNING

Just like adults, children gesture as they speak. Children start communicating through gestures even before they are able to speak. In this section, we review the role gesture plays in developing vocabulary, syntax, and discourse skills in language comprehension and production.

23.2.1 Vocabulary

23.2.1.1 Vocabulary Comprehension

Children rely on gesture to help them comprehend words starting from approximately 8 to 12 months of age (Bates, 1976). Gesture affects children's language comprehension over both short and long periods of time. When seeing an experimenter label an object, infants look longer if the named object is not at the location indicated by the experimenter's gesture but instead is on the other side of the display, suggesting that infants expect concurrently occurring labels and deictic gestures to indicate the same referent (Gliga & Csibra, 2009). Not surprisingly, then, gesture can help children learn new object labels (Namy & Waxman, 1998). Infants are more likely to associate a label with an object if the label is accompanied by a pointing gesture to the object than if it is not (Woodward, 2004). When the newly learned object label needs to be retrieved, less scaffolding by pictures or gestures is needed if the labels were initially taught with accompanying gestures than if they were taught without gestures (Capone & McGregor, 2004). Parental gesture thus has the potential to facilitate children's language comprehension by providing nonverbal support. In naturalistic situations, parents frequently gesture when talking to their children, and most of these gestures reinforce the information conveyed in the accompanying speech (Iverson, Capirci, Longobardi, & Caselli, 1999). Moreover, when infants misunderstand their parents, their parents often provide additional gesture cues (e.g., through pointing) that repair the misunderstanding and enable the dyad to reach a consensus (Zukow-Goldring, 1996).

Looking over longer periods of time, early parent gesture use has been found to predict the size of children's comprehension vocabularies years later. Rowe and colleagues (Rowe, Özçalışkan, & Goldin-Meadow, 2008) found that the number of different meanings parents convey in their gestures to a child at age 14 months is a significant predictor of that child's vocabulary comprehension at age 42 months. However, early parent gesture did not have a direct effect on later child comprehension—early parent gesture was related to early child gesture, which, in turn, was related to later child comprehension. This study suggests that parent gesture might be indirectly related to children's vocabulary development through encouraging the child's own gestures. The relation between parent gesture and child vocabulary has been replicated by Pan and colleagues in a low-income sample and extended to language production—early parent pointing predicted their children's vocabulary production growth between 14 and 36 months of age (Pan, Rowe, Singer, & Snow, 2005). In an experimental manipulation, Goodwyn, Acredolo, and Brown (2000) trained a group of parents to use baby signs in addition to words when talking to their children. The children showed greater gains in vocabulary and used more gestures themselves than children of parents who were encouraged to use only words or who did not receive any training at all.

23.2.1.2 *Vocabulary Production*

In the earliest stages of language learning, infants produce few, if any, words. Their referential communication is primarily through gestures. Children start using gesture to communicate even before they say their first words, which are usually accompanied by meaningless vocalizations (Bates, 1976). For example, children point to places, people, or objects, they hold up an object to show it to others, or they extend their hands to request an object (Bates, 1976). Children in the United States are also often taught to communicate using "baby signs" (Acredolo, Goodwyn, & Gentieu, 2002).

Not much is known about whether producing a gesture helps children to actually produce a word (although this is an interesting question). But we do know that early gesture use is a strong predictor of later vocabulary production. Children's use of gesture for specific objects (e.g., pointing to a ball) predicts the appearance of verbal labels for these objects in their lexicon (e.g., producing the word *"ball"*) (Iverson & Goldin-Meadow, 2005). More remarkably, the number of meanings children convey through gesture at 14 months predicts not only the size of their comprehension vocabularies but also the size of their production vocabularies at 54 months (Rowe & Goldin-Meadow, 2009a). Moreover, early gestures can also be used to predict developmental trajectories of clinical populations. Sauer and colleagues showed that children with unilateral focal brain injury whose gesture production at 18 months is within the typical range (but whose speech production is below the range) will catch up to their peers and achieve production (and comprehension) spoken vocabularies later in development that are within the typical range. Importantly, children with brain injury whose gesture rate is below the typical range at 18 months continue to display delays in both their production and comprehension vocabularies (Sauer, Levine, & Goldin-Meadow, 2010).

23.2.2 Syntax

23.2.2.1 *Syntactic Comprehension*

Children rely on gesture to comprehend sentences from a very early age. Morford and Goldin-Meadow (1992) showed that comprehension of simple sentences, such as *"give the clock,"* is facilitated either by a pointing gesture at the clock or by a *give* gesture (hand extended, palm up) in 15- to 29-month-olds. In fact, children who were unable to produce two-word utterances were able to understand a "two-word idea" if one of those ideas was presented in gesture. For example, children responded appropriately to "give" plus a point at a clock significantly more often than they responded to the message when it was produced entirely in speech (i.e., "give the clock").

The role of gesture varies depending on the complexity of the spoken message—gesture is most beneficial when the listeners are young and the message is complex. For example, in a referential communication game, preschool children were given instructions to select a certain set of blocks from an array of blocks. Instructions were accompanied by a reinforcing gesture (saying *"up"* and producing an *up* gesture), a contradicting gesture (saying *"up"* and producing a *down* gesture), or no gesture. The spoken messages varied in complexity and that complexity influenced the impact the gestures had on comprehension. Reinforcing gestures facilitated comprehension only when the spoken message was complex. Interestingly, no gesture and contradicting gesture had a similar impact on comprehension and did not facilitate comprehension as much as reinforcing gestures (McNeil, Alibali, & Evans, 2000).

23.2.2.2 *Syntactic Production*

Gesture paves the way for children's ability to form sentences. Starting from approximately 10 months of age, children produce gestures along with single words. Children use three types of gesture–speech combinations during the one-word period. Gestures are used to *reinforce* the meanings conveyed in speech (e.g., pointing to a ball and saying *"ball"*), to *disambiguate* the meanings conveyed in speech (e.g., pointing to a ball and saying *"it"*), or to *add* to the meanings conveyed in speech (e.g., pointing to a ball and saying *"want"*).

Interestingly, using gesture–speech combinations to convey sentence-like ideas seems to set the stage for children's earliest sentences. The age at which children first produce combinations in which gesture conveys one idea and speech conveys another (e.g., point at bird + "nap" to describe a sleeping bird) predicts the age at which they first produce their two-word combinations (e.g., "bird sleep") (Iverson & Goldin-Meadow, 2005). Moreover, the number of these gesture–speech combinations that children produce at 18 months selectively predicts their syntactic skill, as measured by the Index of Productive Syntax (Scarborough, 1990) at 42 months (the measure does not predict vocabulary size at this age, Rowe & Goldin-Meadow, 2009b).

Gesture thus gives children a means to convey complex ideas before they are able to convey the same ideas entirely in speech. In fact, particular constructions produced across gesture and speech predict the emergence of the same constructions entirely in speech several months later. For example, saying *"bird"* in speech and producing a flying gesture (an argument + verb combination) precedes and predicts the onset of argument + verb combinations in speech (i.e., saying *"bird fly"*) (Özçalışkan & Goldin-Meadow, 2005). Once a construction is produced in speech, children do not seem to rely on gesture to further expand that construction.

For example, once children acquire the ability to produce a verb and one argument in a single utterance, they do not produce additional arguments first in gesture, that is, a two-argument + verb construction is just as likely to appear first in speech alone as in speech + gesture (Ozçalışkan, Goldin-Meadow, & Özçalışkan, 2009).

23.2.3 Discourse

23.2.3.1 Discourse Comprehension

With age, children face increasingly complex language tasks, such as understanding indirect requests or listening to stories. Gesture continues to support children's language comprehension in later stages of language learning. For example, 3- to 5-year-old children understand indirect requests better if those requests are accompanied by a gesture, for example, saying *"It's going to get loud in here"* is more easily understood as a request to close the door if the words are accompanied by a pointing gesture to the door than if they are not accompanied by gesture (Kelly, 2001).

Gesture can also have an impact on children's comprehension of longer stretches of discourse. In a recent study, Demir and colleagues (Demir, Fisher, Goldin-Meadow, & Levine, 2014) compared children's ability to retell a story presented to them under different conditions: a wordless cartoon; an audio recording of a storyteller (like listening to a story on radio); an audiovisual presentation of a storyteller who does not produce cospeech gestures (like listening to a storyteller holding a book while reading it); and an audiovisual presentation of a storyteller producing cospeech gestures while talking (like listening to an oral storyteller). Children told better-structured narratives in the gesture condition than in any of the other three conditions, consistent with findings that cospeech gesture can scaffold comprehension of complex language. The gestures were particularly beneficial to children who had difficulty telling a well-structured narrative, suggesting that gesture might be most helpful when language skill is low.

23.2.3.2 Discourse Production

With age, children start producing longer and more frequent utterances that need to be sensitive to more sophisticated discourse-pragmatic principles. As in earlier stages of language learning, gesture reveals that children have more understanding of discourse-pragmatics than they express in speech. For example, when a referent is new to the discourse, or is not perceptually available, proficient speakers know that the referents must be explicitly expressed and they use a noun to do so; conversely, if the referent is available in perceptual context or retrievable from discourse, then the referent can be expressed in a pronoun or omitted entirely. Although English-, Chinese-, and Turkish-speaking 4-year-olds do not have control of this discourse principle in speech, they do display an understanding of the principle in gesture—they produce more gestures when referring to referents that are new to the perceptual or discourse context, particularly when those referents are underspecified or ambiguous in speech (Demir, So, Ozyürek, & Goldin-Meadow, 2012; So, Demir, & Goldin-Meadow, 2010).

The gestures children produce in complex discourse early in their development have been found to predict their discourse skills in speech later in development. For example, age 5 to 6 years marks a transitional stage in narrative development. Children produce narratives on their own but rarely include the goal of the story or the perspective of the story characters in those narratives. However, some 5- to 6-year-olds use gesture to portray a character from a first person perspective (e.g., moving the arms back and forth to describe a character who is running from the character's point of view) as opposed to a third-person perspective (e.g., wiggling the fingers to describe the character from an observer's point of view). Whether children use character-viewpoint gestures in their narratives at age 5 predicts the structure of their spoken narratives 3 years later, controlling for early narrative structure in speech and language skill (Demir, Levine, & Goldin-Meadow, under review). Thus, gesture continues to be a harbinger of change in speech even in later stages of language development.

23.2.4 Does Gesture Play a Causal Role in Language Learning?

We have seen that the early gestures children produce reflect their cognitive potential for learning particular aspects of language. But early gesture could be doing more—it could be helping children realize their potential. Child gesture could have an impact on language learning in at least two ways.

First, gesture gives children an opportunity to practice producing particular meanings by hand at a time when those meanings are difficult to express by mouth. To accurately determine whether child gesture is playing a causal role in language learning, we need to manipulate the gestures children produce. LeBarton, Goldin-Meadow and Raudenbush, (2013) studied 15 toddlers (beginning at 17 months) in an 8-week at-home intervention study (6 weekly training sessions plus follow-up 2 weeks later) in which all children were exposed to object words, but only some were told to point at the named objects. Before each training session and at follow-up, children interacted naturally with their parents to establish a baseline against which changes in communication were measured. Children who were told to gesture increased the number of gesture meanings they conveyed not only when interacting with the experimenter during training but also when later interacting naturally with their parents. Critically, these experimentally induced increases in gesture led to larger spoken repertoires at follow-up, and thus suggest that gesturing can have an impact on language learning through the cognitive effect it has on the learner.

The second way in which child gesture could play a role in language learning is more indirect—child gesture could elicit timely speech from listeners. Supporting this view, Goldin-Meadow and colleagues (Goldin-Meadow, Goodrich, Sauer, & Iverson, 2007) found that when their children are in one-word stage, mothers translate the child's gestures into speech. For example, on seeing her child point to a ball and say *"kick,"* a mother might say, *"Do you want to kick the ball?,"* thus modeling for the child a two-word sentence that expresses the ideas the child conveyed in gesture + speech. Importantly, these maternal translations are reliable predictors of children's subsequent word and sentence learning, suggesting that gesturing can have an impact on language through the communicative effect it has on the learning environment.

23.3 ROLE OF GESTURE IN LANGUAGE PROCESSING

Children continue to use gesture long after they become proficient users of their native language(s). The tight relation between speech and gesture increases with development (Thompson & Massaro, 1986). In this section, we first describe the role gesture plays in how language is processed once language has been mastered, and then we describe the functions gesture serves for both listeners and speakers.

23.3.1 Gesturing is Involved in Language Processing at Every Level

23.3.1.1 Phonology

Gesture is linked to spoken language at every level of analysis. At the phonological level, producing gestures influences the voice spectra of the accompanying speech for deictic gestures (Chieffi, Secchi, & Gentilucci, 2009), emblem gestures (Barbieri, Buonocore, Dalla Volta, & Gentilucci, 2009; Bernardis & Gentilucci, 2006), and beat gestures (Krahmer & Swerts, 2007). When phonological production breaks down, as in stuttering or aphasia, gesture production stops as well (Mayberry & Jaques, 2000, McNeill, Pedelty, & Levy, 1990). There are phonological costs to producing gestures with speech—producing words and pointing gestures together leads to long initiation times for the accompanying speech, relative to producing speech alone (Feyereisen, 1997; Levelt, Richardson, & Laheij, 1985). Viewing gesture also affects voicing in listeners' vocal responses to audiovisual stimuli (Bernardis & Gentilucci, 2006).

23.3.1.2 Lexicon

At the lexical level, gesturing increases when the speaker is searching for a word (Morsella & Krauss, 2004). More generally, gestures reflect and compensate for gaps in a speaker's verbal lexicon. Gestures can package information in the same way that information is packaged in the lexicon of the speaker's language. For example, when speakers of English, Japanese, and Turkish are asked to describe a scene in which an animated figure swings on a rope, English speakers overwhelmingly use the verb "swing" along with an arced gesture (Kita & Özyürek, 2003). In contrast, speakers of Japanese and Turkish, languages that do not have single verbs that express an arced trajectory, use generic motion verbs along with the comparable gesture, that is, a straight gesture (Kita & Özyürek, 2003). But gesture can also compensate for gaps in the speaker's lexicon by conveying information that is not encoded in the accompanying speech. For example, complex shapes that are difficult to describe in speech can be conveyed in gesture (Emmorey & Casey, 2002).

23.3.1.3 Syntax

At the syntactic level, gestures are influenced by the structural properties of the accompanying speech. For example, English expresses manner and path within the same clause, whereas Turkish expresses the two in separate clauses. The gestures that accompany manner and path constructions in these two languages display a parallel structure—English speakers produce a single gesture combining manner and path (a rolling movement produced while moving the hand forward), whereas Turkish speakers produce two separate gestures (a rolling movement produced in place, followed by a moving forward movement) (Kita & Özyürek, 2003; Kita, Özyürek, Allen, Brown, Furman, & Ishizuka, 2007). A recent event-related potential (ERP) study illustrates how gesture can influence syntactic processing (Holle, Obermeier, Schmidt-Kassow, Friederici, Ward, & Gunter, 2012). Listeners were presented with two types of sentences, one with a less-preferred syntactic structure. Less-preferred syntactic structures commonly elicit P600 waves, which is usually associated with syntactic reanalysis. When the less-preferred syntactic structures were accompanied by rhythmic beat gestures, the P600 was eliminated, suggesting that gesture can reduce the processing cost associated with hearing a syntactically complex sentence. Supporting this finding, gesture has been found to play a greater role in language comprehension when the spoken message is syntactically complex than when it is simple (McNeil et al., 2000). Gesture production also reflects the amount of information encoded in a syntactic structure. Speakers gesture more when producing an unexpected (and, in this sense, more informative) syntactic structure than when producing an expected structure (Cook, Jaeger, & Tanenhaus, 2009).

23.3.1.4 Discourse

At the discourse level, speakers use recurring gestural features (e.g., the same hand shape or location) throughout a narrative when referring to a particular character, thus creating linkages across the narrative (McNeill, 2000). In terms of narrative comprehension, when listening to stories containing gestures, listeners activate regions that are responsive to semantic manipulations in speech (triangular and opercular portions of the left inferior frontal gyrus; left posterior middle temporal gyrus). However, the level of activation in these areas differs as a function of semantic relation between gesture and speech—stories in which gesture conveys information that differs from, but complements, the information conveyed in speech (e.g., a flying gesture produced along with a story about a "pet") activates the regions more than stories in which gesture conveys the same information as speech (e.g., a flying gesture produced along with a story about a "pet") (Dick, Mok, Beharelle, Goldin-Meadow, & Small, 2014).

23.3.2 Gesture Serves a Function for Both Listeners and Speakers

23.3.2.1 Impact of Gesture on Listeners

Speakers' gestures reveal their thoughts. Accordingly, one function that gesture could serve is to convey those thoughts to listeners. There is, in fact, considerable evidence that listeners can use gesture as a source of information about the speaker's thinking (e.g., Goldin-Meadow & Sandhofer, 1999; Graham & Argyle, 1975; McNeil et al., 2000). The ability of listeners to glean information from a speaker's gestures can be seen most clearly when the gestures convey information that cannot be found anywhere in the speaker's words (e.g., Cook & Tanenhaus, 2009). Gesture can even affect the information listeners glean from the accompanying speech. Listeners are quicker to identify a speaker's referent when speech is accompanied by gesture than when it is not (Silverman, Bennetto, Campana, & Tanenhaus, 2010). Moreover, listeners are *more* likely to glean the message from speech when that speech is accompanied by gesture conveying the same information than when the speech is accompanied by no gesture (Beattie & Shovelton, 1999, 2002; Graham & Argyle, 1975; McNeil et al., 2000; Thompson & Massaro, 1994). Conversely, listeners are *less* likely to glean the message from speech when the speech is accompanied by gesture conveying different information than when the speech is accompanied by no gesture (Goldin-Meadow & Sandhofer, 1999; Kelly & Church, 1998; McNeil et al., 2000). In addition, more incongruent gestures lead to greater processing difficulty than congruent gestures (Kelly, Özyürek, & Maris, 2010). The effect that gesture has on listeners' processing is thus linked to the meaning relation between gesture and speech. Moreover, listeners cannot ignore gesture even when given explicit instructions to do so (Kelly, Özyürek, & Maris, 2010; Langton, O'Malley, & Bruce, 1996), suggesting that the integration of gesture and speech is automatic.

23.3.2.2 Impact of Gesture Impact on Speakers

But gesture can also have an impact on the speakers themselves. Gestures have long been argued to help speakers "find" words, that is, to facilitate lexical access (Rauscher, Krauss, & Chen, 1996). Studies supporting this view show that gesture production increases when lexical retrieval is made difficult (Morsella & Krauss, 2004), and speakers are more successful in resolving tip of the tongue states when they are permitted to gesture than when they are prevented from gesturing (Frick-Horbury & Guttentag, 1998). Gestures have also been hypothesized to reduce demands on conceptualization, and speakers have been found to gesture more on problems that are conceptually difficult, even when lexical demands are equated (Alibali, Kita, & Young, 2000; Hostetter, Alibali, & Kita, 2007; Melinger & Kita, 2007). For example, adults were asked to describe complex geometric shapes under two different conditions. In the easy condition, the shapes that the adults were supposed to describe were outlined by dark lines; in the hard condition, the shapes were obscured by lines outlining alternative organizations. The adults produced more gestures in the hard condition than in the easy condition (Kita & Davies, 2009).

Although findings of this sort are consistent with the idea that gesturing reduces demands on conceptualization, to be certain that gesturing is playing a causal role in reducing demands (as opposed to merely reflecting those demands), we need to manipulate gesture and demonstrate that the manipulation reduces the demands on conceptualization. This type of manipulation has been done in some cases, and gesturing has been found to reduce demand on speakers' working memory (Goldin-Meadow, Nusbaum, Kelly, & Wagner, 2001; Wagner, Nusbaum, & Goldin-Meadow, 2004), to activate knowledge that speakers have but do not express

(Broaders, Cook, Mitchell, & Goldin-Meadow, 2007), and to build new knowledge (Goldin-Meadow, Cook, & Mitchell, 2009).

23.4 IMPLICATIONS FOR THE NEUROBIOLOGY OF LANGUAGE

Our review reveals that gesture can play a role in language learning and processing. Going forward, we suggest that the right question to ask is not *whether* gesture helps language processing, but rather *when* and *how* it does so to explore the mechanisms by which gesture exerts its influence on communication. In this regard, neural levels of analyses have the potential to provide insight into unanswered questions that behavioral analyses cannot. Behavioral studies reflect the combined influence of multiple processes in how gesture affects communication and cognition. By localizing brain networks underlying different cognitive functions and examining their contribution during various language tasks, neuroimaging studies may be able to help us tease apart the contribution of different cognitive processes.

There are a number of important neurobiologic questions that can be raised about gesture's role in communication and cognition. For example, does the neural basis for gesture–speech integration vary as a function of the content of speech or the linguistic skills of the listener? In this regard, it is important to point out that many neuroimaging studies examine the effects of gestures conveying information that contradicts the information conveyed in speech (e.g., Willems, Ozyürek, & Hagoort, 2007). Although these studies can offer important insights into what it possible, we note that contradictory gestures (i.e., gestures that convey information that contradicts, and therefore cannot under any conditions be integrated with, the information conveyed in speech) are *not* commonly observed in naturalistic, spontaneous conversation. We therefore encourage researchers to include in their studies gestures conveying information that is different from, but has the potential to be integrated with, the information conveyed in speech (as in Dick et al., 2014).

In sum, the gestures that we produce when we talk are not mindless handwaving. Gesture takes on significance simply because it can convey information about a speaker's thoughts that are not found in the speaker's words. Moreover, the information conveyed in gesture forecasts subsequent changes in a speaker's thinking and can even play a causal role in changing that thinking. Gesture thus offers a unique lens through which we can explore the mechanisms that underlie language learning and processing at the behavioral and the neurobiological levels.

References

Acredolo, L. P., Goodwyn, S., & Gentieu, P. (2002). *My first baby signs.* New York, NY: HarperFestival.

Alibali, M. W., Kita, S., & Young, A. J. (2000). Gesture and the process of speech production: We think, therefore we gesture. *Language and Cognitive Processes, 15*(6), 593–613.

Barbieri, F., Buonocore, A., Dalla Volta, R., & Gentilucci, M. (2009). How symbolic gestures and words interact with each other. *Brain and Language, 110*(1), 1–11.

Bates, E. (1976). *Langauge and context: The acquisition of pragmatics.* New York, NY: Academic Press.

Beattie, G., & Shovelton, H. (1999). Mapping the range of information contained in the iconic hand gestures that accompany spontaneous speech. *Journal of Language and Social Psychology, 18*(4), 438–462.

Beattie, G., & Shovelton, H. (2002). An experimental investigation of some properties of individual iconic gestures that mediate their communicative power. *British Journal of Psychology, 93*(2), 179–192.

Bernardis, P., & Gentilucci, M. (2006). Speech and gesture share the same communication system. *Neuropsychologia, 44*(2), 178–190.

Broaders, S. C., Cook, S. W., Mitchell, Z., & Goldin-Meadow, S. (2007). Making children gesture brings out implicit knowledge and leads to learning. *Journal of Experimental Psychology: General, 136*(4), 539.

Capone, N. C., & McGregor, K. K. (2004). Gesture development: A review for clinical and research practices. *Journal of Speech, Language, and Hearing Research, 47*(1), 173–186. Available from: http://dx.doi.org/doi:10.1044/1092-4388(2004/015).

Chieffi, S., Secchi, C., & Gentilucci, M. (2009). Deictic word and gesture production: Their interaction. *Behavioural Brain Research, 203* (2), 200–206.

Church, R. B., & Goldin-Meadow, S. (1986). The mismatch between gesture and speech as an index of transitional knowledge. *Cognition, 23*(1), 43–71.

Cook, S. W., Jaeger, T. F., & Tanenhaus, M. (2009). Producing less preferred structures: More gestures, less fluency. In *The 31st annual meeting of the cognitive science society (CogSci09)* (pp. 62–67).

Cook, S. W., & Tanenhaus, M. K. (2009). Embodied communication: Speakers' gestures affect listeners' actions. *Cognition, 113*(1), 98–104.

Demir, Ö. E., Fisher, J. A., Goldin-Meadow, S., & Levine, S. C. (2014). Narrative processing in typically developing children and children with early unilateral brain injury: Seeing gesture matters. *Developmental Psychology, 50*(3), 815.

Demir, O. E., So, W.-C., Ozyürek, A., & Goldin-Meadow, S. (2012). Turkish- and English-speaking children display sensitivity to perceptual context in the referring expressions they produce in speech and gesture. *Language and Cognitive Processes, 27*(6), 844–867. Available from: http://dx.doi.org/doi:10.1080/01690965.2011.589273.

Dick, A. S., Mok, E. H., Beharelle, A. R., Goldin-Meadow, S., & Small, S. L. (2014). Frontal and temporal contributions to understanding the iconic co-speech gestures that accompany speech. *Human Brain Mapping, 34*, 900–917. Available from: http://dx.doi.org/doi:10.1002/hbm.22222.

Emmorey, K., & Casey, S. (2002). *Gesture, thought, and spatial language* ((pp. 87–101)). *Spatial language.* Netherlands: Springer.

Frick-Horbury, D., & Guttentag, R. E. (1998). The effects of restricting hand gesture production on lexical retrieval and free recall. *The American Journal of Psychology*43–62.

Gliga, T., & Csibra, G. (2009). One-year-old infants appreciate the referential nature of deictic gestures and words. *Psychological Science, 20*(3), 347–353. Available from: http://dx.doi.org/doi:10.1111/j.1467-9280.2009.02295.x.

Goldin-Meadow, S., Alibali, M. W., & Church, R. B. (1993). Transitions in concept acquisition: Using the hand to read the mind. *Psychological Review*, *100*(2), 279.

Goldin-Meadow, S., Cook, S. W., & Mitchell, Z. A. (2009). Gesturing gives children new ideas about math. *Psychological Science*, *20*(3), 267–272.

Goldin-Meadow, S., Goodrich, W., Sauer, E., & Iverson, J. (2007). Young children use their hands to tell their mothers what to say. *Developmental Science*, *10*(6), 778–785. Available from: http://dx.doi.org/doi:10.1111/j.1467-7687.2007.00636.x.

Goldin-Meadow, S., Nusbaum, H., Kelly, S. D., & Wagner, S. (2001). Explaining math: Gesturing lightens the load. *Psychological Science*, *12*(6), 516–522.

Goldin-Meadow, S., & Sandhofer, C. M. (1999). Gestures convey substantive information about a child's thoughts to ordinary listeners. *Developmental Science*, *2*(1), 67–74.

Goodwyn, S. W., Acredolo, L. P., & Brown, C. A. (2000). Impact of symbolic gesturing on early language development. *Journal of Nonverbal Behavior*, *24*(2), 81–103.

Graham, J. A., & Argyle, M. (1975). A cross-cultural study of the communication of extra-verbal meaning by gestures (1). *International Journal of Psychology*, *10*(1), 57–67.

Holle, H., Obermeier, C., Schmidt-Kassow, M., Friederici, A. D., Ward, J., & Gunter, T. C. (2012). Gesture facilitates the syntactic analysis of speech. *Frontiers in Psychology*, *3*.

Hostetter, A. B., Alibali, M. W., & Kita, S. (2007). I see it in my hands' eye: Representational gestures reflect conceptual demands. *Language and Cognitive Processes*, *22*(3), 313–336.

Iverson, J. M., Capirci, O., Longobardi, E., & Caselli, M. C. (1999). Gesturing in mother–child interactions. *Cognitive Development*, 57–75.

Iverson, J. M., & Goldin-Meadow, S. (1998). Why people gesture when they speak. *Nature*, *396*(6708), 228.

Iverson, J. M., & Goldin-Meadow, S. (2005). Gesture paves the way for language development. *Psychological Science*, *16*(5), 367–371.

Kelly, S. D. (2001). Broadening the units of analysis in communication: Speech and nonverbal behaviours in pragmatic comprehension. *Journal of Child Language*, *28*(2), 325–349. Retrieved from: http://www.ncbi.nlm.nih.gov/pubmed/11449942.

Kelly, S. D., & Church, R. B. (1998). A comparison between children's and adults' ability to detect conceptual information conveyed through representational gestures. *Child Development*, *69*(1), 85–93.

Kelly, S. D., Özyürek, A., & Maris, E. (2010). Two sides of the same coin speech and gesture mutually interact to enhance comprehension. *Psychological Science*.

Kita, S., & Davies, T. S. (2009). Competing conceptual representations trigger co-speech representational gestures. *Language and Cognitive Processes*, *24*(5), 761–775.

Kita, S., & Özyürek, A. (2003). What does cross-linguistic variation in semantic coordination of speech and gesture reveal?: Evidence for an interface representation of spatial thinking and speaking. *Journal of Memory and Language*, *48*(1), 16–32.

Kita, S., Özyürek, A., Allen, S., Brown, A., Furman, R., & Ishizuka, T. (2007). Relations between syntactic encoding and co-speech gestures: Implications for a model of speech and gesture production. *Language and Cognitive Processes*, *22*(8), 1212–1236.

Krahmer, E., & Swerts, M. (2007). The effects of visual beats on prosodic prominence: Acoustic analyses, auditory perception and visual perception. *Journal of Memory and Language*, *57*(3), 396–414.

Langton, S. R., O'Malley, C., & Bruce, V. (1996). Actions speak no louder than words: Symmetrical cross-modal interference effects in the processing of verbal and gestural information. *Journal of Experimental Psychology: Human Perception and Performance*, *22*(6), 1357.

LeBarton, E. S., Goldin-Meadow, S., & Raudenbush, S. (2013). Experimentally-induced increases in early gesture lead to increases in spoken vocabulary. *Journal of Cognition and Development*.

Mayberry, R. I., & Jaques, J. (2000). 10 Gesture production during stuttered speech: Insights into the nature of gesture-speech integration. *Language and Gesture*, *2*, 199.

McNeill, D. (1992). *Hand and mind: What gestures reveal about thought*. Chicago, IL: University of Chicago Press.

McNeill, D. (Ed.), (2000). *Language and gesture* (Vol. 2).). Cambridge University Press.

McNeill, D. (2008). *Gesture and thought*. Chicago, IL: University of Chicago Press.

McNeill, D., Pedelty, L. L., & Levy, E. T. (1990). Speech and gesture. *Advances in Psychology*, *70*, 203–256.

McNeil, N. M., Alibali, M. W., & Evans, J. L. (2000). The role of gesture in children's comprehension of spoken language: Now they need it, now they don't. *Journal of Nonverbal Behavior*, *24*(2), 131–150.

Melinger, A., & Kita, S. (2007). Conceptualisation load triggers gesture production. *Language and Cognitive Processes*, *22*(4), 473–500.

Morford, M., & Goldin-Meadow, S. (1992). Comprehension and production of gesture in combination with speech in one-word speakers. *Journal of Child Language*, *19*(03), 559–580.

Morsella, E., & Krauss, R. M. (2004). The role of gestures in spatial working memory and speech. *The American Journal of Psychology*, *117*(3), 411–424. Retrieved from: http://www.ncbi.nlm.nih.gov/pubmed/15457809.

Namy, L. L., & Waxman, S. R. (1998). Words and gestures: Infants' interpretations of different forms of symbolic reference. *Child Development*, *69*(2), 295–308 Retrieved from: http://www.ncbi.nlm.nih.gov/pubmed/9586206.

Özçalışkan, S., & Goldin-Meadow, S. (2005). Gesture is at the cutting edge of early language development. *Cognition*, *96*(3), B101–B113. Available from: http://dx.doi.org/doi:10.1016/j.cognition.2005.01.001.

Ozçalışkan, S., Goldin-Meadow, S., & Özçalışkan, S. (2009). When gesture-speech combinations do and do not index linguistic change. *Language and Cognitive Processes*, *24*(2), 190. Available from: http://dx.doi.org/doi:10.1080/01690960801956911.When.

Pan, B. A., Rowe, M. L., Singer, J. D., & Snow, C. E. (2005). Maternal correlates of growth in toddler vocabulary production in low-income families. *Child Development*, *76*(4), 763–782. Available from: http://dx.doi.org/doi:10.1111/j.1467-8624.2005.00876.x.

Perry, M., Berch, D., & Singleton, J. (1995). Constructing shared understanding: The role of nonverbal input in learning contexts. *Journal of Contemporary Legal Issues*, *6*, 213.

Perry, M., Church, R. B., & Goldin-Meadow, S. (1988). Transitional knowledge in the acquisition of concepts. *Cognitive Development*, *3* (4), 359–400.

Perry, M., & Elder, A. D. (1997). Knowledge in transition: Adults' developing understanding of a principle of physical causality. *Cognitive Development*, *12*(1), 131–157.

Pine, K. J., Lufkin, N., & Messer, D. (2004). More gestures than answers: Children learning about balance. *Developmental Psychology*, *40*(6), 1059.

Ping, R., Decatur, M., Larson, S. W., Zinchenko, E., & Goldin-Meadow, S. (revision under review). Unpacking the gestures of chemistry learners: What the hands can tell us about correct and incorrect conceptions of stereochemistry.

Ping, R. M., & Goldin-Meadow, S. (2008). Hands in the air: using ungrounded iconic gestures to teach children conservation of quantity. *Developmental Psychology*, *44*(5), 1277.

Rauscher, F. H., Krauss, R. M., & Chen, Y. (1996). Gesture, speech, and lexical access: The role of lexical movements in speech production. *Psychological Science*, *7*(4), 226–231.

Rowe, M. L., & Goldin-Meadow, S. (2009a). Differences in early gesture explain SES disparities in child vocabulary size at school entry. *Science*, *323*, 951–953.

Rowe, M. L., & Goldin-Meadow, S. (2009b). Early gesture selectively predicts later language learning. *Developmental Science*, *323*(1), 182–187. Available from: http://dx.doi.org/doi:10.1111/j.1467-7687.2008.00764.x.

Rowe, M. L., Özçalışkan, S., & Goldin-Meadow, S. (2008). Learning words by hand: Gesture's role in predicting vocabulary development. *First Language*, *28*(2), 182–199. Available from: http://dx.doi.org/doi:10.1177/0142723707088310.Learning.

Sauer, E., Levine, S. C., & Goldin-Meadow, S. (2010). Early gesture predicts language delay in children with pre- or perinatal brain lesions. *Child Development*, *81*(2), 528–539. Available from: http://dx.doi.org/doi:10.1111/j.1467-8624.2009.01413.x.

Scarborough, H. S. (1990). Very early language deficits in dyslexic children. *Child Development*, *61*(6), 1728–1743.

Silverman, L. B., Bennetto, L., Campana, E., & Tanenhaus, M. K. (2010). Speech-and-gesture integration in high functioning autism. *Cognition*, *115*(3), 380–393.

Singer, M. A., & Goldin-Meadow, S. (2005). Children learn when their teacher's gestures and speech differ. *Psychological Science*, *16* (2), 85–89.

So, W. C., Demir, Ö. E., & Goldin-Meadow, S. (2010). When speech is ambiguous gesture steps in: Sensitivity to discourse-pragmatic principles in early childhood. *Applied Psycholinguistics*, *31*(1), 209–224. Available from: http://dx.doi.org/doi:10.1017/S0142716409990221.When.

Thompson, L. A., & Massaro, D. W. (1986). Evaluation and integration of speech and pointing gestures during referential understanding. *Journal of Experimental Child Psychology*, *42*(1), 144–168.

Thompson, L. A., & Massaro, D. W. (1994). Children' s integration of speech and pointing gestures in comprehension. *Journal of Experimental Child Psychology*, *57*(3), 327–354.

Valenzeno, L., Alibali, M. W., & Klatzky, R. (2003). Teachers' gestures facilitate students' learning: A lesson in symmetry. *Contemporary Educational Psychology*, *28*(2), 187–204.

Wagner, S. M., Nusbaum, H., & Goldin-Meadow, S. (2004). Probing the mental representation of gesture: Is handwaving spatial? *Journal of Memory and Language*, *50*(4), 395–407.

Willems, R. M., Ozyürek, A., & Hagoort, P. (2007). When language meets action: The neural integration of gesture and speech. *Cerebral Cortex*, *17*(10), 2322–2333. Available from: http://dx.doi.org/doi:10.1093/cercor/bhl141.

Woodward, A. L. (2004). Infants' use of action knowledge to get a grasp on words. *Weaving a Lexicon*149–172.

Zukow-Goldring, P. (1996). Sensitive caregiving fosters the comprehension of speech: When gestures speak louder than words. *Early Development and Parenting*, *5*(4), 195–211.

SECTION D

LARGE-SCALE MODELS

CHAPTER

24

Pathways and Streams in the Auditory Cortex

An Update on How Work in Nonhuman Primates has Contributed to Our Understanding of Human Speech Processing

Josef P. Rauschecker[1,2] *and Sophie K. Scott*[3]

[1]Laboratory of Integrative Neuroscience and Cognition, Georgetown University Medical Center, NW, Washington, DC, USA; [2]Institute for Advanced Study, TU München, München-Garching, Germany; [3]Institute of Cognitive Neuroscience, University College London, London, UK

24.1 HUMAN SPEECH PERCEPTION

Human speech is a sound of immensely spectrotemporal complexity, reflecting the intricacy of the articulatory movements that produce it. The aim of this chapter is to outline the contributions that the study of nonhuman primates has made to our understanding of the neural basis of human speech perception. However, this needs to be set in the context of insights about the neural basis of human speech processing that has arisen from the contribution by studies of patients with a receptive aphasia, often known as Wernicke's aphasia, which was first described by German neurologist Carl Wernicke.

24.2 WHERE IS "WERNICKE'S AREA"?

"Wernicke's area" is often defined as the posterior one-third of the superior temporal gyrus (STG), sometimes including the contiguous region of the inferior parietal lobule (IPL; in particular angular gyrus and supramarginal gyrus). However, the exact location of "Wernicke's area" has undergone various transformations since its initial discovery (Bogen & Bogen, 1976; Rauschecker & Scott, 2009). Wernicke himself (Wernicke, 1881) defined the speech-receptive region simply as the whole STG. Later neurologists, typically investigating single-case studies, shifted the emphasis to posterior STG and IPL, and "Wernicke's area" came to be identified with those posterior regions. Several highly popular and widely distributed textbooks in the early half of the 20th century disseminated this version. Norman Geschwind depicted the site of "Wernicke's area" as the "posterior language region" (in relation to the "anterior language region" of Broca's considered important for speech production). Although Broca's area was in the frontal lobe, both Broca's area and Wernicke's area were termed as "perisylvian" with regard to the Sylvian fissure, separating the temporal and frontal lobe. Both sides of the Sylvian fissure are often damaged together in stroke cases and are often recorded together as a single source in electroencephalogram (EEG) studies. The role of the STG as a whole, including its anterior portions, in speech perception was all but forgotten until the linguist Harry Whitaker (1971) and the neurosurgeon Wilder Penfield (Penfield and Roberts, 1959) separately noted the shift from Wernicke's original work.

Detailed neuropsychological studies involving quantitative lesion analysis have recently confirmed that the anterior STG makes a more significant contribution to speech perception than the posterior STG (Dronkers, Wilkins, Van Valin, Redfern, & Jaeger, 2004). At the phonological level, phonemic discrimination between real words and nonsense syllables was little affected after lesions of posterior STG, whereas this ability was severely impaired in aphasics with lesions of anterior STG (Blumstein, Baker, & Goodglass, 1977) but not in Broca aphasics, for which the anterior STG was spared. Generally, infarcts in bilateral anterior STG cause auditory agnosias, in which people fail to identify familiar sounds (Clarke, Bellmann, Meuli, Assal, & Steck, 2000; Vignolo, 1982), and the anterior temporal lobe is the location of atrophy in "semantic dementia" (Mummery et al., 2000; Patterson et al., 2006). Functional neuroimaging

Neurobiology of Language. DOI: http://dx.doi.org/10.1016/B978-0-12-407794-2.00024-9

studies also demonstrate anterior superior temporal (ST) regions to be activated more than posterior regions when subjects listen to speech (Obleser et al., 2006; Scott, Blank, Rosen, & Wise, 2000). Finally, in a meta-analysis by DeWitt and Rauschecker (2012) that was specific to syllables, words, and phrases, syllables were represented just lateral to Heschl's gyrus, words were antero-lateral (AL) to that, and short standard phrases were even more anteriorly in the ST.

24.3 DUAL PROCESSING STREAMS AND HIERARCHICAL ORGANIZATION IN THE AUDITORY CORTEX OF THE MONKEY

24.3.1 "What" and "Where" Pathways in Vision and Audition

The aim of the current chapter is to address the neural basis of human speech perception in light of neuroanatomical and neurophysiological evidence from nonhuman primate studies. A decade and a half ago it was suggested that auditory cortical processing pathways are organized in a dual fashion similar to those in the visual cortex (Figure 24.1) (Rauschecker, 1998; Rauschecker & Tian, 2000), with one major pathway projecting into the posterior parietal cortex (PPC) and another pathway into the anterior temporal cortex. As in the visual system (Mishkin, Ungerleider, & Macko, 1983), the posterior parietal pathway was considered to subserve spatial processing, and the temporal pathway was considered to participate in the decoding of complex patterns and identification of objects. Because the orientation of the projections in the auditory system differs from those in the visual system, these auditory pathways were referred to as the postero-dorsal and antero-ventral streams, respectively.

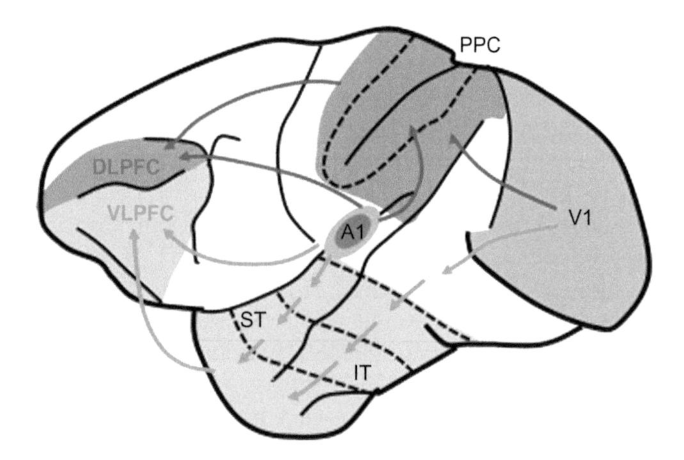

FIGURE 24.1 Dual-processing scheme for auditory "what" and "where" proposed for nonhuman primates on anatomical and physiological grounds (Rauschecker, 1998; Rauschecker & Tian, 2000) combined with the analogous scheme from the visual system (Mishkin et al., 1983). V1, primary visual cortex; A1, primary auditory cortex; IT, inferior temporal region; ST, superior temporal region; PPC, posterior parietal cortex; VLPFC, ventrolateral prefrontal cortex; DLPFC, dorsolateral prefrontal cortex.

This anterior/posterior projection scheme in auditory cortex (AC) has been supported by anatomical tracing studies (Hackett, Stepniewska, & Kaas, 1998; Kaas & Hackett, 2000), with long-range projections originating from the belt areas surrounding the primary-like core areas. Projections were found from anterior belt directly to ventrolateral prefrontal cortex (VLPFC), and from the caudal (posterior) belt to dorsolateral prefrontal cortex (DLPFC) (Romanski et al., 1999). This finding provided clear evidence for the existence of ventral and dorsal processing streams within AC, with strong anatomical and functional implications (Goldman-Rakic, 1996; Petrides, 2005).

Can comparisons be made between human and nonhuman primates? We have argued that they can (Rauschecker & Scott, 2009). Modern noninvasive neuroanatomical tract-tracing techniques in humans on the basis of high-resolution magnetic resonance imaging (MRI) and diffusion tensor imaging (DTI) have revealed that the projections from the core and belt areas of AC follow the same route as in the monkey: The main projection runs anteriorly from Heschl's gyrus to anterior STG, and then crosses the divide to the inferior frontal regions, including Broca's area (Brodmann's area 44/45), via the uncinate fascicle and extreme capsule (Frey, Campbell, Pike, & Petrides, 2008). The connection from posterior STG to Broca's area via the arcuate fascicle, assumed by Geschwind (1965) to be all important for connecting Wernicke's and Broca's areas, is much sparser, if it exists at all, as a direct connection. Most projections from posterior STG go (via the superior longitudinal fascicle) to the IPL and the frontal eye fields (BA 6), consistent with a role of the posterior STG in spatial functions.

24.3.2 Functional Dual Pathways

The anatomical finding of a dual pathway from auditory to prefrontal cortex (PFC) (Romanski et al., 1999) was paralleled by direct functional evidence for an auditory dual-processing scheme from single-unit studies in the lateral belt areas of macaques. Species-specific communication sounds were presented in varying spatial locations, and Tian, Reser, Durham, Kustov, and Rauschecker (2001) discovered that neurons in the antero-lateral belt (area AL) were more specific for the type of monkey call. Reversing the call abolished the response in AL neurons (B. Tian et al., unpublished), but not in the core or caudal belt (Recanzone, 2008). By contrast, neurons in the caudolateral belt (area CL) were more sensitive to spatial location than neurons in core or anterior belt. These data indicate that "what" and "where" processing dissociate in rhesus monkey AC.

The concept of dual-streams in auditory perceptual systems has found support from other studies (Recanzone & Sutter, 2008; Schreiner & Winer, 2007). Recanzone, Guard, Phan, and Su (2000) reported a correlation between spatial tuning of neurons and behavioral performance and found a tighter correlation of neuronal activity and sound localization in caudal belt than in primary AC, supporting a functional role for the posterior auditory stream in spatial processing of sound. Lewis and Van Essen (2000) described a direct auditory projection from the posterior ST (pST) region to area ventral intraparietal area (VIP) in the PPC of the monkey. Functional specialization was seen in single-unit studies as well as imaging studies in monkeys (Bendor & Wang, 2005; Poremba et al., 2003; Rauschecker & Tian, 2004; Rauschecker, Tian, & Hauser, 1995; Tian & Rauschecker, 2004).

Research using functional magnetic resonance imaging (fMRI) in nonhuman primates has identified tonotopic maps on the ST plane and gyrus (Petkov, Kayser, Augath, & Logothetis, 2006), and then a "voice region" in the anterior part of the right STG (Petkov et al., 2008), which projects further to the anterior superior temporal sulcus (STS) and VLPFC (Petkov et al., in press). An auditory voice region was also subsequently found in the AC of dogs (Andics, Gacsi, Farago, Kis, & Miklosi, 2014). Cortical cooling, which leads to reversible cortical inactivation, has been used in cat AC (Lomber & Malhotra, 2008) to show that cooling anterior areas caused a deterioration of auditory pattern discrimination, whereas cooling of posterior areas impaired spatial discrimination. These studies all suggest that an antero-ventral processing stream forms the substrate for the recognition of auditory objects, within which communication sounds may form an important category, whereas a postero-dorsal stream includes spatial perception as one of its functions.

Hierarchical processing is an organizational property of cortical perceptual systems, combining elements of serial as well as parallel processing (Bizley & Cohen, 2013). Thus, "lower" cortical areas with simpler receptive-field organization (Tian, Kusmierek, & Rauschecker, 2013), such as sensory core areas, project to "higher" areas with increasingly complex response properties, such as belt, parabelt, and PFC regions. These properties are generated by convergence and summation. Parallel processing principles in hierarchical organizations are evident if one considers that specialized cortical areas ("maps") with related functions (corresponding to submodalities or modules) are bundled into parallel processing "streams." As an alternative to hierarchical processing structures, neural networks have been proposed that are highly interconnected and dynamically modulated by different task demands or contexts. Feedback connections, although well-known from neuroanatomy, are often not sufficiently accounted for in hierarchical models.

24.3.3 "What" and "How" Pathways: The Perception–Action Cycle

In contrast to the "what/where" model in vision (Mishkin et al., 1983), Goodale and Milner (1992) proposed that two pathways differentially subserve behaviors related to perception and action. The ventral pathway's role in perception is largely consistent with a "what" pathway, whereas the dorsal pathway takes on a sensorimotor role involved in action ("how"), including spatial analysis. Fuster (2008) advocates a similar distinction with regard to PFC and unites the two pathways into a perception–action cycle. We argue here that the "what/where" and "perception/action" theories differ mainly as a matter of emphasis. The morphing of a "where" to a "how" pathway has largely to do with the addition of a temporal component, turning spatial into spatio-temporal information. Recent neurophysiological recordings in monkey caudal belt areas that have given more attention to the latency and temporal precision of responses to sounds have come up with surprising results. First, the latency of neurons in the caudo-medial belt area (CM) is significantly shorter than even in primary AC (Bendor & Wang, 2008; Kusmierek, Ortiz, & Rauschecker, 2012). Furthermore, fine-grained auditory information is better represented in the caudo-medial belt (Camalier, D'Angelo, Sterbing-D'Angelo, de la Mothe, & Hackett, 2012; Scott, Malone, & Semple, 2011). By contrast, spatially fine-grained information (i.e., narrow spatial tuning) is mostly a domain of caudo-lateral area CL, as would be expected from previous studies (Tian et al., 2001). Therefore, it is possible that substreams of the postero-dorsal stream exist early on and are later combined into a whole spatio-temporal main stream. Alternatively, the emphasis on different aspects of dorsal-stream processing may be maintained, or a split into substreams could occur later. For instance, Kravitz, Saleem, Baker, and Mishkin (2011) recently presented a model of the visual dorsal stream wherein three different functions (spatial working memory, visually guided actions, and spatial navigation) are performed by anatomically distinct components of the dorsal stream, all originating from parietal cortex.

24.4 DUAL PROCESSING STREAMS IN THE AUDITORY CORTEX OF HUMANS

The concepts of auditory streams of processing have proved to be extremely powerful as a framework for understanding functional imaging studies of speech perception (Scott, 2005; Scott & Johnsrude, 2003; Scott, McGettigan, & Eisner, 2009), and for understanding the aphasic stroke (Wise, 2003). Human studies also confirm the role of the postero-dorsal stream in the perception of auditory space and motion (see Rauschecker (2007) and Recanzone & Sutter (2008) for review). However, the question remains justified whether multiple (i.e., more than two) processing streams exist (Kaas & Hackett, 1999). In particular, the posterior STG and inferior parietal cortex have long been implicated in the processing of speech and language, so it would seem inappropriate to assign an exclusively spatial function to the postero-dorsal auditory stream. It is therefore essential to discuss how the *planum temporale (PT)*, the temporo-parietal junction (TPJ), and the inferior (IP) parietal cortex are involved in speech and language, and whether we can assign a common computational function to the postero-dorsal stream that encompasses both spatial and language functions.

24.4.1 Role of Antero-Ventral Auditory Pathway in Object Identification and Speech Perception

24.4.1.1 Hierarchical Organization

Binder et al. (2000), as part of a meta-analysis of imaging studies of speech processing, confirmed that an AL gradient can be established, along which the complexity of preferred stimuli increases from tones and noise bursts to words and sentences. As in non-human primates, frequency responses demonstrate tonotopy, whereas core regions responding to tones were surrounded by belt areas preferring bandpass noise bursts (Wessinger et al., 2001). Using high-field scanners, multiple tonotopic fields (Formisano et al., 2003) and multiple processing levels (core, belt, and parabelt) (Chevillet, Riesenhuber, & Rauschecker, 2011) can be identified in human AC. This notion was extended in a meta-analysis by DeWitt and Rauschecker (2012) specific to syllables, words, and phrases. In a clear hierarchy, syllables were represented just lateral to Heschl's gyrus, words AL to that, and short standard phrases even more anteriorly. This confirms both the functional organization in monkeys as well as the organization of speech in humans found earlier.

24.4.1.2 Auditory Object Identification

Hierarchical organization in the antero-ventral auditory pathway of humans is important in auditory pattern recognition and object identification. As in animal models, preferred features of lower-order neurons are combined to create selectivity for increasingly complex sounds (Patterson, Uppenkamp, Johnsrude, & Griffiths, 2002; Zatorre, Bouffard, & Belin, 2004), and regions can be seen that are specialized in different auditory object classes (Leaver & Rauschecker, 2010; Obleser et al., 2006). Although speech sounds are found more lateral and more prominently in the left hemisphere, musical instrument sounds were concentrated more anterior in the right hemisphere. Animal sounds are less consistently represented as a distinct category and may depend on expert training (Leaver & Rauschecker, 2010). Developments in how we conceive the structure of auditory objects (Kumar, Stephan, Warren, Friston, & Griffiths, 2007; Shamma, 2008) will help extend these kinds of investigations. Like their visual counterparts, auditory objects coexist based on a multitude of attributes, such as timbre, pitch, and loudness, that give each its distinctive perceptual identity (Shamma, 2008).

24.4.1.3 Speech and Voice Perception

Within speech perception, there is evidence that speech sounds are hierarchically encoded because anterior portions of ST cortex respond as a function of speech intelligibility, and not stimulus complexity alone (Narain et al., 2003; Scott et al., 2000; Scott, Rosen, Lang, & Wise, 2006). Similarly, Liebenthal, Binder, Spitzer, Possing, and Medler (2005) and Obleser, Wise, Alex Dresner, and Scott (2007) showed that the left middle and anterior STS is more responsive to consonant–vowel syllables than auditory baselines. Thus, regions within the "what" stream show the first clear responses to abstract, linguistic information in speech. These findings culminated in the postulate for an auditory word form area (AWFA) in the left anterior STS analogous to the visual word form area (VWFA) (DeWitt & Rauschecker, 2012, 2013), confirming earlier suggestions by Cohen, Russ, Gifford, Kiringoda, and MacLean (2004). Within the speech-specific regions of aST, subregions are beginning to emerge that are selective for particular speech-sound classes, such as vowels (Obleser et al., 2006; Obleser, Zimmermann, Van Meter, & Rauschecker, 2007), raising the possibility of phonetic maps having some anatomical implementation in anterior temporal lobe areas. However, speech-sound categories, such as consonants, for which fine differences in articulation play an important role, may be represented in dorsal-stream areas of (e.g., premotor cortex [PMC])

(Chevillet, Jiang, Rauschecker, & Riesenhuber, 2013) (see next section). This could lead to dissociations of speech processing in patients with corresponding lesions (Caramazza, Papagno, & Ruml, 2000).

Speaker recognition has also been shown to be represented in AL temporal lobe areas (Belin & Zatorre, 2003), sometimes extending into mid-temporal regions as well (Leaver & Rauschecker, 2010). These human voice regions may be homologous, according to crude topological criteria, to the areas in the monkey (Petkov et al., 2008) mentioned previously. The human "voice area" in the anterior auditory fields processes detailed spectral properties of talkers (Warren, Scott, Price, & Griffiths, 2006). Notably, speech perception and voice discrimination dissociate clinically, suggesting that the two are supported by different systems within the anterior and middle temporal lobes.

24.4.1.4 *Invariance, Categorization*

An important problem to be solved in speech perception is the problem of invariance against distortions in the scale of frequency (e.g., pitch changes) or time (e.g., compressions). Comparing spectrograms of clear speech and of noise-vocoded speech, the latter is quite coarse in its spectrotemporal representation (Shannon, Zeng, Kamath, Wygonski, & Ekelid, 1995), yet it is readily intelligible after a brief training session. Perceptual invariance is important in the perception of normal speech because the "same" phoneme can be acoustically very different (due to coarticulation) while still being identified as the same sound (Bailey & Summerfield, 1980). The sound /s/ is different at the start of "sue" than at the start of "see," but remains an /s/.

These examples of perceptual constancy appear relatively simple; however, they are computationally difficult to solve. The ability to deal with invariance problems is not unique to speech or audition, being a hallmark of all higher perceptual systems. The structural and functional organization of the anterior/ventral streams in both the visual and auditory systems could illustrate how the cerebral cortex solves this problem. For example, it has been suggested that visual categories are formed in the lateral PFC (Freedman, Riesenhuber, Poggio, & Miller, 2001), which receives input from higher-order object representations in the anterior temporal lobe (Goldman-Rakic, 1996). In the auditory domain, using species-specific communication sounds, Romanski, Averbeck, and Diltz (2005) found clusters of neurons in the macaque ventrolateral PFC encoding similar complex calls, and category-specific cells encoding single semantic categories have also been reported (Russ, Ackelson, Baker, & Cohen, 2008). In humans, rapid adaptation studies with functional MRI in the visual system have recently led to similar conclusions (Jiang et al., 2007). Potentially, the invariance problem in speech perception is solved in the inferior frontal cortex (IFC) or by interactions between IFC and aST.

24.4.1.5 *Hemispheric Asymmetry*

Speech perception and production are left-lateralized in the human brain (Dhanjal, Handunnetthi, Patel, & Wise, 2008; Scott et al., 2000; Wise, 2003), and there is considerable interest in the neural basis of this (Boemio, Fromm, Braun, & Poeppel, 2005). Hemispheric specialization is an important feature of the human brain, particularly in relation to speech, attention, and spatial processing, and linguistic phenomena have long been considered to be left-lateralized in most adult humans. Recent work in speech perception has argued that hemispheric asymmetries in human speech perception are driven by functional rather than acoustic properties (McGettigan & Scott, 2012; Scott & McGettigan, 2013): selective activation of left auditory fields is typically only seen for linguistic material rather than particular acoustic properties of sounds. In contrast, right auditory fields show selective responses to longer stimuli, to stimuli with pitch variation, as well as to stimuli with voice or talker cues. Further studies will elaborate on the extent animal models can develop our understanding of these asymmetries, although there is evidence that the right temporal lobe voice areas (Latinus, McAleer, Bestelmeyer, & Belin, 2013) have nonhuman primate homologues, which are also right-dominant (Perrodin, Kayser, Logothetis, & Petkov, 2011).

24.4.2 Role of Postero-Dorsal Auditory Pathway in Space Processing

Evidence for a role of the postero-dorsal stream in auditory spatial processing is just as strong in the human as in nonhuman primates. Classical stroke studies as well as modern neuroimaging have shown that spatial processing in the temporo-parietal cortex is often right lateralized in humans (i.e., contralateral to speech and language). For instance, spatial neglect is more frequent and severe after damage to the right hemisphere (see Rauschecker (2007) and Recanzone and Sutter (2008) for review).

The auditory "where" stream may originate in the brainstem, perhaps as early as the dorsal cochlear nucleus (Rauschecker, 1997; Yu & Young, 2000), and ultimately reaches PPC via areas in posterior ST (pST), which serve as a transformation stage between primary core areas and PPC. Studies of the macaque cited previously (Lewis & Van Essen, 2000) are paralleled by functional–anatomical studies in the human (Bremmer et al., 2001; Lewis, Beauchamp, & DeYoe, 2000).

A meta-analysis of 38 human imaging studies provided evidence that the majority of auditory spatial studies involved activation of the postero-dorsal pathway (pST, IPL, and superior frontal sulcus [SFS]), whereas only two activated antero-ventral areas (aST, IFC) (Arnott, Binns, Grady, & Alain, 2004; see also Rauschecker, 2007 for review). Likewise, studies of patients with lesions in the posterior auditory pathway have shown deficits in the processing of auditory space (Clarke et al., 2000; Griffiths et al., 1998).

Imaging studies in humans have also demonstrated specific activation in pST with auditory motion (Griffiths et al., 1998; Warren, Zielinski, Green, Rauschecker, & Griffiths, 2002) near the location of visual motion areas, the middle temporal area (MT) and the medial superior temporal area (MST). Imaging studies that have tested moving auditory stimuli in addition to stationary ones have reported that auditory motion leads to activation in areas of pST and PPC that are adjacent to each other, with motion being the more powerful stimulus (Krumbholz et al., 2005).

24.4.3 Role of Postero-Dorsal Auditory Pathway in Speech Processing

The pST region (or *PT*) in humans (and the dorsal stream emanating from it) has classically been assigned a role in speech perception (Geschwind, 1965). The apparent contradiction of this view with that of a spatial role for pST (as well as with the evidence for speech-sound decoding in aST), as mentioned, needs to be discussed.

One unifying view is that the PT is generally involved in the processing of spectro-temporally complex sounds (Obleser, Zimmermann, et al., 2007), which also includes the processing of music (Hyde, Peretz, & Zatorre, 2008). Following this view, it has been suggested that the PT operates as a "computational hub" (Griffiths & Warren, 2002). More specifically, one could argue, following some of the more recent monkey data, that the postero-dorsal stream contains fine details in both the spatial and temporal domains. These two representations could be combined in specific ways for either of its main functions of space and speech (as well as music).

The IPL, particularly the angular and supramarginal gyri (or Brodmann areas 39/40), has also been linked to linguistic functions (Caplan, Rochon, & Waters, 1992), such as the "phonological-articulatory loop" (Baddeley, Lewis, & Vallar, 1984). Functional imaging has confirmed this role, although the activity seen varies with the working memory task used (Buchsbaum & D'Esposito, 2008; Gelfand & Bookheimer, 2003). However the IPL does not seem to be driven by acoustic processing of speech; the angular gyrus (together with extensive prefrontal activation) is recruited when higher-order linguistic factors improve speech comprehension (Obleser, Wise, et al., 2007) rather than by acoustic influences on intelligibility.

Thus, the parietal cortex is associated with more domain-general, linguistic factors in speech comprehension, rather than acoustic-phonetic processing. Bornkessel-Schlesewsky and colleagues (2013, 2015) extend this view by postulating that the dorsal stream as a whole engages in the time-dependent combination of elements, subserving both syntactic structuring and a linkage to action. This unified spatio-temporal perspective has language rooted in mechanisms for spatial processing, which have been generalized to apply to other domains.

24.4.4 Multisensory Responses and Sensorimotor Integration

There is now neurophysiological evidence that auditory caudal belt areas are not solely responsive to auditory input but show multimodal responses (Fu et al., 2003; Kayser, Petkov, Augath, & Logothetis, 2007). This has been extended to show that both caudal medial and lateral belt fields receive input from somatosensory and multisensory cortex, as well as thalamic input. Thus, any spatial transformations conducted in the posterior-dorsal stream may be based on a multisensory reference frame (Andersen & Buneo, 2002; Colby & Goldberg, 1999).

These multisensory responses in caudal auditory areas have been extended into some functional specificity in humans. Several studies of silent articulation (Wise et al., 2001) and nonspeech auditory stimuli (Hickok, Buchsbaum, Humphries, & Muftuler, 2003) have identified activation in a posterior medial PT region within the postero-dorsal stream. Following this, the medial PT in humans (Warren, Wise, & Warren, 2005) has been associated with the representation of templates for "doable" articulations and sounds (not limited to speech sounds). This approach can be compared with the "affordance" model of Gibson (Gibson, 1977; Rizzolatti, Ferrari, Rozzi, & Fogassi, 2006), in which objects and events are described in terms of action possibilities. Such a sensorimotor role for the dorsal stream is consistent with the notion of an "action" stream in vision (Goodale & Milner, 1992). The concept can be extended to auditory-motor transformations in verbal working memory tasks (Hickok & Poeppel, 2000, 2007) that involve articulatory representations (Baddeley et al., 1984; Jacquemot & Scott, 2006). The postero-medial PT area has also been identified as a key node for the control of speech production (Dhanjal et al., 2008) because it shows a response to somatosensory input from articulators.

24.4.4.1 *Speech Perception-Production Links*

There is considerable neural convergence for the speech perception and production systems. For example, the postero-medial PT area described in the previous section is an auditory area that is important in the motor act of articulation. Conversely, activation by real or imagined speech sounds and music has been described within premotor areas important in overt production of speech (Wilson, Saygin, Sereno, & Iacoboni, 2004) and music (Chen, Penhune, & Zatorre, 2008; Leaver, Van Lare, Zielinski, Halpern, & Rauschecker, 2009). Within auditory areas, monkey studies have shown that auditory neurons are suppressed during the monkey's own vocalizations (Eliades & Wang, 2008; Müller-Preuss & Ploog, 1981). This finding is consistent with results from humans that indicate that auditory perception in (anterior) ST areas is suppressed by an audio-motor efference-copy signal during speech production (Houde, Nagarajan, Sekihara, & Merzenich, 2002; Numminen, Salmelin, & Hari, 1999) and even during silent lip-reading (Kauramäki et al., 2010). As a result, the response to one's own voice is always lower than that to someone else's.

At one level these findings may simply reflect the ways that sensory responses to actions caused by oneself are always differently processed to those caused by the actions of others (Blakemore, Wolpert, & Frith, 1998), and this may support mechanisms important in differentiating between one's own voice and the voices of others. However, in primate studies, auditory neurons that are suppressed during vocalizations are often more activated if the sound of the vocalizations is distorted (Eliades & Wang, 2008). This might indicate a specific role for these auditory responses in the comparison of feed-forward and feedback information from the motor and auditory system during speech production (Guenther, 2006). Distorting speech production in real time reveals enhanced activation in bilateral (posterior temporal) auditory fields to distorted feedback (Tourville, Reilly, & Guenther, 2008). New work using high-resolution DTI in humans has revealed that there are direct projections from the pars opercularis of Broca's area (BA44) to the IPL (Frey et al., 2008) in addition to the ones from the ventral premotor (vPM) cortex (Petrides & Pandya, 1984). With the known connections between parietal cortex and posterior auditory fields, this could form the basis for feed-forward connections between speech production areas and posterior temporal auditory areas (Figure 24.2).

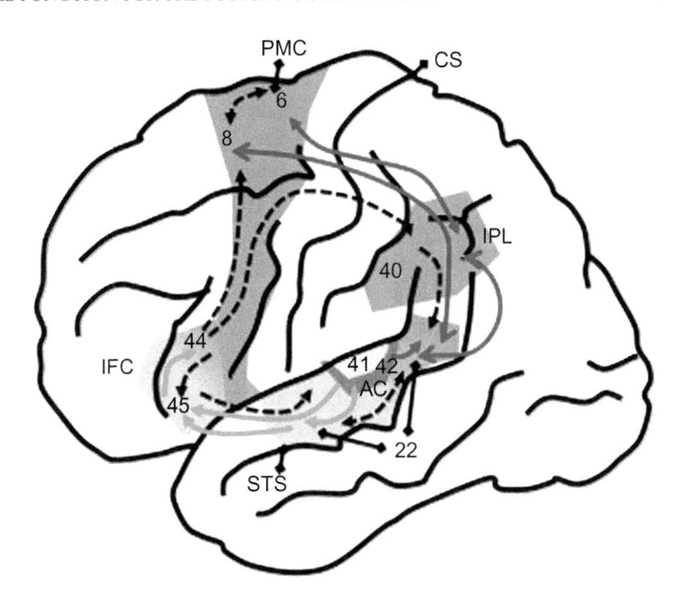

FIGURE 24.2 Dual auditory processing scheme of the human brain incorporating current understanding of the role of internal models in sensory systems (*modified from Rauschecker & Scott, 2009*). This model closes the loop between speech perception and production and proposes a common computational structure for space processing and speech control in the auditory dorsal stream. AC, auditory cortex; STS, superior temporal sulcus; IFC, inferior frontal cortex: PMC, premotor cortex; IPL, inferior parietal lobule; CS, central sulcus. Numbers correspond to Brodmann areas.

24.5 CONCLUSIONS: A COMMON COMPUTATIONAL FUNCTION FOR THE POSTERO-DORSAL STREAM?

The dual-stream processing model in audition (Rauschecker, 1998; Rauschecker & Tian, 2000; Rauschecker & Scott, 2009) has been a useful construct in hearing research, perceptual physiology, and, in particular, psycholinguistics, where it has spawned a number of further models (Hickok & Poeppel, 2000, 2007) that have tried to accommodate specific results from this field. Although we recognize the theoretical importance of dual-stream models in linguistics, we emphasize the importance of a firm grounding of such models in the anatomy and physiology of the AC, particularly as it has been worked out in nonhuman primates. Precursors of speech and language are bound to exist in nonhuman animals, and the consistency of linguistic models with such animal models is a *conditio-sine-qua-non*.

The role of a ventral stream in hierarchical processing of objects, as in the visual system, is now widely accepted, thanks to both individual and meta-analytic data. Specifically for speech, anterior regions of the ST respond to native speech sounds and intelligible speech, and these sounds are mapped along phonological parameter domains. By contrast, early posterior regions in and around the *PT* are involved in the processing of many different types of complex sounds. CL areas seem to concentrate more on spatial aspects of sounds, whereas caudo-medial areas process timing of sounds most efficiently. Later posterior regions participate in the processing of auditory motion-in-space but seem to integrate input from several modalities as well.

Thus, whereas evidence is strong for the role of the dorsal pathway (including posterior ST) in space processing, the dorsal pathway is in an excellent position to accommodate speech and language functions as well. Spatial transformations may be one example of fast adaptations used by "internal models" or "emulators," as first developed in motor control theory. Within these models, "forward models" (predictors) can be used to predict the consequences of actions, whereas "inverse models" (controllers) determine the motor commands required to produce a desired outcome (Wolpert, Doya, & Kawato, 2003). More recently, forward models have been used to describe the predictive nature of perception and imagery (Grush, 2004). The inferior parietal cortex (IPL) could provide an ideal interface where feed-forward signals from motor preparatory networks in the inferior frontal (IFC) and premotor (PMC) cortices are matched with feedback signals from sensory areas (Rizzolatti et al., 2006).

In speech perception and production, projections from articulatory networks in Broca's area and PMC to the IPL and pST interact with signals from AC (Figure 24.2). The feed-forward projection from BA44 (and ventral PMC) may provide an efference copy in the classical sense of von Holst and Mittelstaedt (1950), informing the sensory system of planned motor articulations that are about to happen. The activity arriving in the IPL and pST from frontal areas anticipates the sensory consequences of action. The feedback signal coming to the IPL from pST, however, could be considered an "afference copy" (Hershberger, 1976) with relatively short latencies and high temporal precision (Jääskeläinen, Ahveninen, Bonmassar, Dale, & Ilmoniemi, 2004)—a sparse but fast primal sketch of ongoing sensory events (Bar et al., 2006) that are compared with the predictive motor signal in the IPL at every instance.

"Internal model" structures in the brain are generally thought to enable smooth sequential motor behaviors from visuo-spatial reaching to articulation of speech. The goal of these models is to minimize the resulting error signal by adaptive mechanisms. At the same time, these motor behaviors also support aspects of motor control in perceptual processing, such as stabilization of the retinal image and disambiguation of phonological information, thus switching between forward and inverse modes. As Indefrey and Levelt (2004) point out, spoken language "constantly operates a dual system, [processing] and producing utterances. These systems not only alternate, but in many cases they partially or wholly operate in concert." What is more, both spatial processing and real-time speech processing make use of the same internal model structures.

In summary, our new model of the auditory cortical pathways builds on the previous model of dual-processing pathways for object identification and spatial analysis (Mishkin et al., 1983; Rauschecker & Tian, 2000), but it integrates the spatial (dorsal) pathway with findings from speech and music processing as well. The model is based on neuroanatomical data from nonhuman primates, operating under the assumption that mechanisms of speech and language in humans have built on structures available in other primates. The pivotal factor is the simultaneous processing of space and time, which combines dorsal-stream function into a homogeneous space–time continuum. Finally, our new model extends beyond language processing (Hickok & Poeppel, 2007), and applies in a very general sense to both vision and audition, in its relationship with prior models of perception and action (Fuster, 2008; Goodale & Milner, 1992) by using the notion of internal models as its common denominator.

Acknowledgments

We thank D. Klemm for help with graphic design. The work was supported by grants from the US National Institutes of Health (R01NS52494 and R01-DC03489) and the US National Science Foundation (BCS-0350041 and PIRE-OISE-0730255) to J.P.R., and by Wellcome Trust grant WT074414MA to S.K.S.

References

Andersen, R. A., & Buneo, C. A. (2002). Intentional maps in posterior parietal cortex. *Annual Review of Neuroscience*, *25*, 189–220.

Andics, A., Gacsi, M., Farago, T., Kis, A., & Miklosi, A. (2014). Voice-sensitive regions in the dog and human brain are revealed by comparative fMRI. *Current Biology*, *24*(5), 574–578.

Arnott, S. R., Binns, M. A., Grady, C. L., & Alain, C. (2004). Assessing the auditory dual-pathway model in humans. *NeuroImage*, *22*(1), 401–408.

Baddeley, A., Lewis, V., & Vallar, G. (1984). Exploring the articulatory loop. *The Quarterly Journal of Experimental Psychology. A, Human Experimental Psychology*, *36*, 233–252.

Bailey, P. J., & Summerfield, Q. (1980). Information in speech: Observations on the perception of [s]-stop clusters. *Journal of Experimental Psychology Human Perception and Performance*, *6*(3), 536–563.

Bar, M., Kassam, K. S., Ghuman, A. S., Boshyan, J., Schmid, A. M., Dale, A. M., et al. (2006). Top-down facilitation of visual recognition. *Proceedings of the National Academy of Sciences of the United States of America*, *103*(2), 449–454.

Belin, P., & Zatorre, R. J. (2003). Adaptation to speaker's voice in right anterior temporal lobe. *NeuroReport*, *14*(16), 2105–2109.

Bendor, D., & Wang, X. (2005). The neuronal representation of pitch in primate auditory cortex. *Nature*, *436*(7054), 1161–1165.

Bendor, D., & Wang, X. (2008). Neural response properties of primary, rostral, and rostrotemporal core fields in the auditory cortex of marmoset monkeys. *Journal of Neurophysiology*, *100*(2), 888–906.

Binder, J. R., Frost, J. A., Hammeke, T. A., Bellgowan, P. S., Springer, J. A., Kaufman, J. N., et al. (2000). Human temporal lobe activation by speech and nonspeech sounds. *Cerebral Cortex*, *10*(5), 512–528.

Bizley, J. K., & Cohen, Y. E. (2013). The what, where and how of auditory-object perception. *Nature Reviews Neuroscience*, *14*(10), 693–707.

Blakemore, S. J., Wolpert, D. M., & Frith, C. D. (1998). Central cancellation of self-produced tickle sensation. *Nature Neuroscience*, *1*(7), 635–640.

Blumstein, S. E., Baker, E., & Goodglass, H. (1977). Phonological factors in auditory comprehension in aphasia. *Neuropsychologia*, *15*(1), 19–30.

Boemio, A., Fromm, S., Braun, A., & Poeppel, D. (2005). Hierarchical and asymmetric temporal sensitivity in human auditory cortices. *Nature Neuroscience*, *8*(3), 389–395.

Bogen, J. E., & Bogen, G. M. (1976). Wernicke's region—where is it? *Annals of the New York Academy of Sciences*, *280*, 834–843.

Bornkessel-Schlesewsky, I., & Schlesewsky, M. (2013). Reconciling time, space and function: A new dorsal-ventral stream model of sentence comprehension. *Brain and Language*, *125*(1), 60–76.

Bornkessel-Schlesewsky, I., Schlesewsky, M., Small, S. L., & Rauschecker, J. P. (2015). Neurobiological roots of language in primate audition: Common computational properties. *Trends in Cognitive Science* (ePub ahead of print). Available from: http://dx.doi.org/10.1016/j.tics.2014.12.008.

Bremmer, F., Schlack, A., Shah, N. J., Zafiris, O., Kubischik, M., Hoffmann, K., et al. (2001). Polymodal motion processing in posterior parietal and premotor cortex: A human fMRI study strongly implies equivalencies between humans and monkeys. *Neuron*, *29*(1), 287–296.

Buchsbaum, B. R., & D'Esposito, M. (2008). The search for the phonological store: From loop to convolution. *Journal of Cognitive Neuroscience*, *20*(5), 762–778.

Camalier, C. R., D'Angelo, W. R., Sterbing-D'Angelo, S. J., de la Mothe, L. A., & Hackett, T. A. (2012). Neural latencies across auditory cortex of macaque support a dorsal stream supramodal timing advantage in primates. *Proceedings of the National Academy of Sciences of the United States of America*, *109*(44), 18168–18173.

Caplan, D., Rochon, E., & Waters, G. S. (1992). Articulatory and phonological determinants of word length effects in span tasks. *The Quarterly Journal of Experimental Psychology*, *45*(2), 177–192.

Caramazza, A., Papagno, C., & Ruml, W. (2000). The selective impairment of phonological processing in speech production. *Brain and Language*, *75*(3), 428–450.

Chen, J. L., Penhune, V. B., & Zatorre, R. J. (2008). Listening to musical rhythms recruits motor regions of the brain. *Cerebral Cortex*, *18* (12), 2844–2854.

Chevillet, M., Riesenhuber, M., & Rauschecker, J. P. (2011). Functional localization of the auditory "what" stream hierarchy. *Journal of Neuroscience*, *31*(25), 9345–9352.

Chevillet, M. A., Jiang, X., Rauschecker, J. P., & Riesenhuber, M. (2013). Automatic phoneme category selectivity in the dorsal auditory stream. *The Journal of Neuroscience: The Official Journal of the Society for Neuroscience*, *33*(12), 5208–5215.

Clarke, S., Bellmann, A., Meuli, R. A., Assal, G., & Steck, A. J. (2000). Auditory agnosia and auditory spatial deficits following left hemisphere lesions: Evidence for distinct processing pathways. *Neuropsychologia*, *38*, 797–807.

Cohen, Y. E., Russ, B. E., Gifford, G. W., 3rd, Kiringoda, R., & MacLean, K. A. (2004). Selectivity for the spatial and nonspatial attributes of auditory stimuli in the ventrolateral prefrontal cortex. *The Journal of Neuroscience: The Official Journal of the Society for Neuroscience*, *24*(50), 11307–11316.

Colby, C. L., & Goldberg, M. E. (1999). Space and attention in parietal cortex. *Annual Review of Neuroscience*, *22*, 319–349.

DeWitt, I., & Rauschecker, J. P. (2012). Phoneme and word recognition in the auditory ventral stream. *Proceedings of the National Academy of Sciences of the United States of America*, *109*(8), E505–514.

DeWitt, I., & Rauschecker, J. P. (2013). Wernicke's area revisited: Parallel streams and word processing. *Brain and Language*, *127*(2), 181–191.

Dhanjal, N. S., Handunnetthi, L., Patel, M. C., & Wise, R. J. (2008). Perceptual systems controlling speech production. *The Journal of Neuroscience: The Official Journal of the Society for Neuroscience*, *28* (40), 9969–9975.

Dronkers, N. F., Wilkins, D. P., Van Valin, R. D., Jr., Redfern, B. B., & Jaeger, J. J. (2004). Lesion analysis of the brain areas involved in language comprehension. *Cognition*, *92*(1–2), 145–177.

Eliades, S. J., & Wang, X. (2008). Neural substrates of vocalization feedback monitoring in primate auditory cortex. *Nature*, *453* (7198), 1102–1106.

Formisano, E., Kim, D. S., Di Salle, F., van de Moortele, P. F., Ugurbil, K., & Goebel, R. (2003). Mirror-symmetric tonotopic maps in human primary auditory cortex. *Neuron*, *40*(4), 859–869.

Freedman, D. J., Riesenhuber, M., Poggio, T., & Miller, E. K. (2001). Categorical representation of visual stimuli in the primate prefrontal cortex. *Science*, *291*(5502), 312–316.

Frey, S., Campbell, J. S., Pike, G. B., & Petrides, M. (2008). Dissociating the human language pathways with high angular resolution diffusion fiber tractography. *The Journal of Neuroscience: The Official Journal of the Society for Neuroscience*, *28*(45), 11435–11444.

Fu, K. G., Shah, A. S., Arnold, L., Garraghty, P. E., Smiley, J., Hackett, T. A., et al. (2003). Auditory cortical neurons respond to somatosensory stimulation. *The Journal of Neuroscience: The Official Journal of the Society for Neuroscience*, *23*, 7510–7515.

Fuster, J. (2008). *The prefrontal cortex*. Burlington, MA: Academic Press.

Gelfand, J. R., & Bookheimer, S. Y. (2003). Dissociating neural mechanisms of temporal sequencing and processing phonemes. *Neuron*, *38*(5), 831–842.

Geschwind, N. (1965). Disconnexion syndromes in animals and man. *Brain*, *88*(2), 237–294, 585-644.

Gibson, J. J. (1977). The theory of affordances. In R. Shaw, & J. Bransford (Eds.), *Perceiving, acting, and knowing: Toward an ecological psychology* (pp. 67–82). Hillsdale, NJ: Erlbaum.

Goldman-Rakic, P. S. (1996). The prefrontal landscape: Implications of functional architecture for understanding human mentation and the central executive. *Philosophical Transactions of the Royal Society of London. Series B, Biological Sciences*, *351*(1346), 1445–1453.

Goodale, M. A., & Milner, A. D. (1992). Separate visual pathways for perception and action. *Trends in Neurosciences*, *15*(1), 20–25.

Griffiths, T. D., Rees, G., Rees, A., Green, G. G., Witton, C., Rowe, D., et al. (1998). Right parietal cortex is involved in the perception of sound movement in humans. *Nature Neuroscience*, *1*(1), 74–79.

Griffiths, T. D., & Warren, J. D. (2002). The planum temporale as a computational hub. *Trends in Neurosciences*, *25*(7), 348–353.

Grush, R. (2004). The emulation theory of representation: Motor control, imagery, and perception. *Behavioral and Brain Sciences*, *27*(3), 377–396 (discussion 396–442).

Guenther, F. H. (2006). Cortical interactions underlying the production of speech sounds. *Journal of Communication Disorders, 39*(5), 350–365.

Hackett, T. A., Stepniewska, I., & Kaas, J. H. (1998). Subdivisions of auditory cortex and ipsilateral cortical connections of the parabelt auditory cortex in macaque monkeys. *The Journal of Comparative Neurology, 394*(4), 475–495.

Hershberger, W. (1976). Afference copy, the closed-loop analogue of von Holst's efference copy. *Cybernetics Forum, 8*, 97–102.

Hickok, G., Buchsbaum, B., Humphries, C., & Muftuler, T. (2003). Auditory-motor interaction revealed by fMRI: Speech, music, and working memory in area spt. *Journal of Cognitive Neuroscience, 15* (5), 673–682.

Hickok, G., & Poeppel, D. (2000). Towards a functional neuroanatomy of speech perception. *Trends in Cognitive Science, 4*(4), 131–138.

Hickok, G., & Poeppel, D. (2007). The cortical organization of speech processing. *Nature Reviews. Neuroscience, 8*(5), 393–402.

Houde, J. F., Nagarajan, S. S., Sekihara, K., & Merzenich, M. M. (2002). Modulation of the auditory cortex during speech: An MEG study. *Journal of Cognitive Neuroscience, 14*(8), 1125–1138.

Hyde, K. L., Peretz, I., & Zatorre, R. J. (2008). Evidence for the role of the right auditory cortex in fine pitch resolution. *Neuropsychologia, 46*(2), 632–639.

Indefrey, P., & Levelt, W. J. M. (2004). The spatial and temporal signatures of word production components. *Cognition, 92*, 101–144.

Jääskeläinen, I. P., Ahveninen, J., Bonmassar, G., Dale, A. M., Ilmoniemi, R. J. L. S., et al. (2004). Human posterior auditory cortex gates novel sounds to consciousness. *Proceedings of the National Academy of Sciences of the United States of America, 101*(17), 6809–6814.

Jacquemot, C., & Scott, S. K. (2006). What is the relationship between phonological short-term memory and speech processing? *Trends in Cognitive Science, 10*(11), 480–486.

Jiang, X., Bradley, E., Rini, R. A., Zeffiro, T., Vanmeter, J., & Riesenhuber, M. (2007). Categorization training results in shape- and category-selective human neural plasticity. *Neuron, 53*(6), 891–903.

Kaas, J. H., & Hackett, T. A. (1999). "What" and "where" processing in auditory cortex. *Nature Neuroscience, 2*(12), 1045–1047.

Kaas, J. H., & Hackett, T. A. (2000). Subdivisions of auditory cortex and processing streams in primates. *Proceedings of the National Academy of Sciences of the United States of America, 97*(22), 11793–11799.

Kauramäki, J., Jääskeläinen, I. P., Hari, R., Möttönen, R., Rauschecker, J. P., & Sams, M. (2010). Lipreading and covert speech production similarly modulate human auditory-cortex responses to pure tones. *The Journal of Neuroscience: The Official Journal of the Society for Neuroscience, 30*(4), 1314–1321.

Kayser, C., Petkov, C. I., Augath, M., & Logothetis, N. K. (2007). Functional imaging reveals visual modulation of specific fields in auditory cortex. *The Journal of Neuroscience: The Official Journal of the Society for Neuroscience, 27*(8), 1824–1835.

Kravitz, D. J., Saleem, K. S., Baker, C. I., & Mishkin, M. (2011). A new neural framework for visuospatial processing. *Nature Reviews Neuroscience, 12*(4), 217–230.

Krumbholz, K., Schonwiesner, M., von Cramon, D. Y., Rubsamen, R., Shah, N. J., Zilles, K., et al. (2005). Representation of interaural temporal information from left and right auditory space in the human planum temporale and inferior parietal lobe. *Cerebral Cortex, 15*(3), 317–324.

Kumar, S., Stephan, K. E., Warren, J. D., Friston, K. J., & Griffiths, T. D. (2007). Hierarchical processing of auditory objects in humans. *PLoS Computational Biology, 3*(6), e100.

Kusmierek, P., Ortiz, M., & Rauschecker, J. P. (2012). Sound-identity processing in early areas of the auditory ventral stream in the macaque. *Journal of Neurophysiology, 107*(4), 1123–1141.

Latinus, M., McAleer, P., Bestelmeyer, P. E., & Belin, P. (2013). Norm-based coding of voice identity in human auditory cortex. *Current Biology, 23*(12), 1075–1080.

Leaver, A., Van Lare, J. E., Zielinski, B. A., Halpern, A., & Rauschecker, J. P. (2009). Brain activation during anticipation of sound sequences. *The Journal of Neuroscience: The Official Journal of the Society for Neuroscience, 29*(8), 2477–2485.

Leaver, A. M., & Rauschecker, J. P. (2010). Cortical representation of natural complex sounds: Effects of acoustic features and auditory object category. *The Journal of Neuroscience: The Official Journal of the Society for Neuroscience, 30*(22), 7604–7612.

Lewis, J. W., Beauchamp, M. S., & DeYoe, E. A. (2000). A comparison of visual and auditory motion processing in human cerebral cortex. *Cerebral Cortex, 10*(9), 873–888.

Lewis, J. W., & Van Essen, D. C. (2000). Corticocortical connections of visual, sensorimotor, and multimodal processing areas in the parietal lobe of the macaque monkey. *The Journal of Comparative Neurology, 428*(1), 112–137.

Liebenthal, E., Binder, J. R., Spitzer, S. M., Possing, E. T., & Medler, D. A. (2005). Neural substrates of phonemic perception. *Cerebral Cortex, 15*(10), 1621–1631.

Lomber, S. G., & Malhotra, S. (2008). Double dissociation of "what" and "where" processing in auditory cortex. *Nature Neuroscience, 11*(5), 609–616.

McGettigan, C., & Scott, S. K. (2012). Cortical asymmetries in speech perception: What's wrong, what's right and what's left? *Trends in Cognitive Sciences, 16*(5), 269–276.

Mishkin, M., Ungerleider, L. G., & Macko, K. A. (1983). Object vision and spatial vision: Two cortical pathways. *Trends in Neurosciences, 6*(10), 414–417.

Müller-Preuss, P., & Ploog, D. (1981). Inhibition of auditory cortical neurons during phonation. *Brain Research, 215*(1–2), 61–76.

Mummery, C. J., Patterson, K., Price, C. J., Ashburner, J., Frackowiak, R. S., & Hodges, J. R. (2000). A voxel-based morphometry study of semantic dementia: Relationship between temporal lobe atrophy and semantic memory. *Annals of Neurology, 47*(1), 36–45.

Narain, C., Scott, S. K., Wise, R. J., Rosen, S., Leff, A., Iversen, S. D., et al. (2003). Defining a left-lateralized response specific to intelligible speech using fMRI. *Cerebral Cortex, 13*(12), 1362–1368.

Numminen, J., Salmelin, R., & Hari, R. (1999). Subject's own speech reduces reactivity of the human auditory cortex. *Neuroscience Letters, 265*(2), 119–122.

Obleser, J., Boecker, H., Drzezga, A., Haslinger, B., Hennenlotter, A., Roettinger, M., et al. (2006). Vowel sound extraction in anterior superior temporal cortex. *Human Brain Mapping, 27*(7), 562–571.

Obleser, J., Wise, R. J., Alex Dresner, M., & Scott, S. K. (2007). Functional integration across brain regions improves speech perception under adverse listening conditions. *The Journal of Neuroscience: The Official Journal of the Society for Neuroscience, 27* (9), 2283–2289.

Obleser, J., Zimmermann, J., Van Meter, J., & Rauschecker, J. P. (2007). Multiple stages of auditory speech perception reflected in event-related fMRI. *Cerebral Cortex, 17*(10), 2251–2257.

Patterson, K., Lambon Ralph, M. A., Jefferies, E., Woollams, A., Jones, R., Hodges, J. R., et al. (2006). "Presemantic" cognition in semantic dementia: Six deficits in search of an explanation. *Journal of Cognitive Neuroscience, 18*(2), 169–183.

Patterson, R. D., Uppenkamp, S., Johnsrude, I. S., & Griffiths, T. D. (2002). The processing of temporal pitch and melody information in auditory cortex. *Neuron, 36*(4), 767–776.

Penfield, W., & Roberts, L. (1959). *Speech and brain mechanisms*. Princeton, NJ: Princeton University Press.

Perrodin, C., Kayser, C., Logothetis, N. K., & Petkov, C. I. (2011). Voice cells in the primate temporal lobe. *Current Biology, 21*(16), 1408–1415.

Petkov, C. I., Kayser, C., Augath, M., & Logothetis, N. K. (2006). Functional imaging reveals numerous fields in the monkey auditory cortex. *PLoS Biology*, *4*(7), e215.

Petkov, C. I., Kayser, C., Steudel, T., Whittingstall, K., Augath, M., & Logothetis, N. K. (2008). A voice region in the monkey brain. *Nature Neuroscience*, *11*(3), 367–374.

Petkov, C. I., Kikuchi, Y., Milne, A. E., Mishkin, M., Rauschecker, J. P., & Logothetis, N. K. (2015). Different forms of effective connectivity in primate fronto-temporal pathways. *Nature Communications*, *6*, 1–12. Available from: http://dx.doi.org/10.1038/ncomms7000.

Petrides, M. (2005). Lateral prefrontal cortex: Architectonic and functional organization. *Philosophical Transactions of the Royal Society of London Series B, Biological Sciences*, *360*(1456), 781–795.

Petrides, M., & Pandya, D. N. (1984). Projections to the frontal cortex from the posterior parietal region in the rhesus monkey. *The Journal of Comparative Neurology*, *228*(1), 105–116.

Poremba, A., Saunders, R. C., Crane, A. M., Cook, M., Sokoloff, L., & Mishkin, M. (2003). Functional mapping of the primate auditory system. *Science*, *299*(5606), 568–572.

Rauschecker, J. P. (1997). Processing of complex sounds in the auditory cortex of cat, monkey and man. *Acta Oto-Laryngologica*, *532*, 34–38.

Rauschecker, J. P. (1998). Cortical processing of complex sounds. *Current Opinion in Neurobiology*, *8*, 516–521.

Rauschecker, J. P. (2007). Cortical processing of auditory space: Pathways and plasticity. In F. Mast, & L. Jäncke (Eds.), *Spatial processing in navigation, imagery and perception* (pp. 389–410). New York, NY: Springer-Verlag.

Rauschecker, J. P., & Scott, S. K. (2009). Maps and streams in the auditory cortex: Non-human primates illuminate human speech processing. *Nature Neuroscience*, *12*(6), 718–724.

Rauschecker, J. P., & Tian, B. (2000). Mechanisms and streams for processing of "what" and "where" in auditory cortex. *Proceedings of the National Academy of Sciences of the United States of America*, *97* (22), 11800–11806.

Rauschecker, J. P., & Tian, B. (2004). Processing of band-passed noise in the lateral auditory belt cortex of the rhesus monkey. *Journal of Neurophysiology*, *91*(6), 2578–2589.

Rauschecker, J. P., Tian, B., & Hauser, M. (1995). Processing of complex sounds in the macaque nonprimary auditory cortex. *Science*, *268*(5207), 111–114.

Recanzone, G. H. (2008). Representation of con-specific vocalizations in the core and belt areas of the auditory cortex in the alert macaque monkey. *The Journal of Neuroscience: The Official Journal of the Society for Neuroscience*, *28*(49), 13184–13193.

Recanzone, G. H., Guard, D. C., Phan, M. L., & Su, T. K. (2000). Correlation between the activity of single auditory cortical neurons and sound-localization behavior in the macaque monkey. *Journal of Neurophysiology*, *83*(5), 2723–2739.

Recanzone, G. H., & Sutter, M. L. (2008). The biological basis of audition. *Annual Review of Psychology*, *59*, 119–142.

Rizzolatti, G., Ferrari, P. F., Rozzi, S., & Fogassi, L. (2006). The inferior parietal lobule: Where action becomes perception. *Novartis Foundation Symposia*, *270*, 129–140 (discussion 140–145, 164–169).

Romanski, L. M., Averbeck, B. B., & Diltz, M. (2005). Neural representation of vocalizations in the primate ventrolateral prefrontal cortex. *Journal of Neurophysiology*, *93*(2), 734–747.

Romanski, L. M., Tian, B., Fritz, J., Mishkin, M., Goldman-Rakic, P. S., & Rauschecker, J. P. (1999). Dual streams of auditory afferents target multiple domains in the primate prefrontal cortex. *Nature Neuroscience*, *2*(12), 1131–1136.

Russ, B. E., Ackelson, A. L., Baker, A. E., & Cohen, Y. E. (2008). Coding of auditory-stimulus identity in the auditory non-spatial processing stream. *Journal of Neurophysiology*, *99*(1), 87–95.

Schreiner, C. E., & Winer, J. A. (2007). Auditory cortex mapmaking: Principles, projections, and plasticity. *Neuron*, *56*(2) 356–365.

Scott, B. H., Malone, B. J., & Semple, M. N. (2011). Transformation of temporal processing across auditory cortex of awake macaques. *Journal of Neurophysiology*, *105*(2), 712–730.

Scott, S. K. (2005). Auditory processing—speech, space and auditory objects. *Current Opinion in Neurobiology*, *15*(2), 197–201.

Scott, S. K., Blank, C. C., Rosen, S., & Wise, R. J. (2000). Identification of a pathway for intelligible speech in the left temporal lobe. *Brain*, *123*(Pt 12), 2400–2406.

Scott, S. K., & Johnsrude, I. S. (2003). The neuroanatomical and functional organization of speech perception. *Trends in Neurosciences*, *26*(2), 100–107.

Scott, S. K., & McGettigan, C. (2013). Do temporal processes underlie left hemisphere dominance in speech perception? *Brain and Language*, *127*(1), 36–45.

Scott, S. K., McGettigan, C., & Eisner, F. (2009). A little more conversation, a little less action—candidate roles for the motor cortex in speech perception. *Nature Reviews Neuroscience*, *10*(4), 295–302.

Scott, S. K., Rosen, S., Lang, H., & Wise, R. J. (2006). Neural correlates of intelligibility in speech investigated with noise vocoded speech—a positron emission tomography study. *The Journal of the Acoustical Society of America*, *120*(2), 1075–1083.

Shamma, S. (2008). On the emergence and awareness of auditory objects. *PLoS Biology*, *6*(6), e155.

Shannon, R. V., Zeng, F. G., Kamath, V., Wygonski, J., & Ekelid, M. (1995). Speech recognition with primarily temporal cues. *Science*, *270*, 303–304.

Tian, B., Kusmierek, P., & Rauschecker, J. P. (2013). Analogues of simple and complex cells in rhesus monkey auditory cortex. *Proceedings of the National Academy of Sciences of the United States of America*, *110*(19), 7892–7897.

Tian, B., & Rauschecker, J. P. (2004). Processing of frequency-modulated sounds in the lateral auditory belt cortex of the rhesus monkey. *Journal of Neurophysiology*, *92*(5), 2993–3013.

Tian, B., Reser, D., Durham, A., Kustov, A., & Rauschecker, J. P. (2001). Functional specialization in rhesus monkey auditory cortex. *Science*, *292*(5515), 290–293.

Tourville, J. A., Reilly, K. J., & Guenther, F. H. (2008). Neural mechanisms underlying auditory feedback control of speech. *NeuroImage*, *39*(3), 1429–1443.

Vignolo, L. A. (1982). Auditory agnosia. *Philosophical Transactions of the Royal Society of London. Series B, Biological Sciences*, *298*(1089), 49–57.

von Holst, E., & Mittelstaedt, H. (1950). Das reafferenzprinzip (wechselwirkungen zwischen zentralnervensystem und peripherie). *Die Naturwissenschaften*, *37*, 464–476.

Warren, J. D., Scott, S. K., Price, C. J., & Griffiths, T. D. (2006). Human brain mechanisms for the early analysis of voices. *NeuroImage*, *31*(3), 1389–1397.

Warren, J. D., Zielinski, B. A., Green, G. G. R., Rauschecker, J. P., & Griffiths, T. D. (2002). Analysis of sound source motion by the human brain. *Neuron*, *34*, 1–20.

Warren, J. E., Wise, R. J., & Warren, J. D. (2005). Sounds do-able: Auditory-motor transformations and the posterior temporal plane. *Trends in Neurosciences*, *28*(12), 636–643.

Wernicke, C. (1881). *Lehrbuch der Gehirnkrankheiten für Aerzte und Studirende*. Kassel, Berlin: Verlag Theodor Fischer.

Wessinger, C. M., VanMeter, J., Tian, B., Van Lare, J., Pekar, J., & Rauschecker, J. P. (2001). Hierarchical organization of the human auditory cortex revealed by functional magnetic resonance imaging. *Journal of Cognitive Neuroscience*, *13*(1), 1–7.

Whitaker, H. A. (1971). *On the representation of language in the human brain; Problems in the neurology of language and the linguistic analysis of aphasia*. Edmonton, AB: Linguistic Research.

Wilson, S. M., Saygin, A. P., Sereno, M. I., & Iacoboni, M. (2004). Listening to speech activates motor areas involved in speech production. *Nature Neuroscience*, *7*(7), 701–702.

Wise, R. J. (2003). Language systems in normal and aphasic human subjects: Functional imaging studies and inferences from animal studies. *British Medical Bulletin*, *65*, 95–119.

Wise, R. J., Scott, S. K., Blank, S. C., Mummery, C. J., Murphy, K., & Warburton, E. A. (2001). Separate neural subsystems within "Wernicke's area". *Brain*, *124*(Pt 1), 83–95.

Wolpert, D. M., Doya, K., & Kawato, M. (2003). A unifying computational framework for motor control and social interaction. *Philosophical Transactions of the Royal Society of London Series B, Biological Sciences*, *358*(1431), 593–602.

Yu, J. J., & Young, E. D. (2000). Linear and nonlinear pathways of spectral information transmission in the cochlear nucleus. *Proceedings of the National Academy of Sciences of the United States of America*, *97*(22), 11780–11786.

Zatorre, R. J., Bouffard, M., & Belin, P. (2004). Sensitivity to auditory object features in human temporal neocortex. *The Journal of Neuroscience: The Official Journal of the Society for Neuroscience*, *24* (14), 3637–3642.

CHAPTER

25

Neural Basis of Speech Perception

Gregory Hickok[1] *and David Poeppel*[2,3]

[1]Department of Cognitive Sciences, Center for Language Science, Center for Cognitive Neuroscience, University of California, Irvine, CA, USA; [2]Department of Psychology, New York University, New York, NY, USA; [3]Max-Planck-Institute for Empirical Aesthetics, Frankfurt, Germany

25.1 INTRODUCTION

The mind/brain must figure out at least two things when faced with the task of learning a natural language. One is how to transform the sound patterns of speech into a representation of the meaning of an utterance. The other is how to reproduce those sound patterns with the vocal tract (or in the case of signed languages with manual and facial gestures). Put differently, speech information must be processed along two different routes, an auditory-conceptual route and an auditory-motor route. These two processing streams involve partially segregated circuits in the brain and form the basis of the dual route model of speech processing (Hickok & Poeppel, 2000, 2004, 2007), which traces its routes to the classical model of Wernicke (Wernicke, 1874/1977) and parallels analogous proposals in the visual (Milner & Goodale, 1995) and somatosensory (Dijkerman & de Haan, 2007) systems. Thus, the division of labor proposed in dual route models, wherein one route is sensory-conceptual and the other is sensory-motor, appears to be a general organizational property of the cerebral cortex.

This chapter outlines the dual route model as a foundation for understanding the functional anatomy of speech and language processing.

25.2 THE DUAL ROUTE MODEL OF SPEECH PROCESSING

The dual route model (Figure 25.1) holds that a ventral stream, which involves structures in the superior and middle portions of the temporal lobe, is involved in processing speech signals for comprehension. A dorsal stream, which involves structures in the posterior planum temporale region (at the parietal-temporal junction) and the posterior frontal lobe, is involved in translating acoustic-based representations of speech signals into articulatory representations, essential for speech production. In contrast to the canonical view that speech processing is mainly left hemisphere–dependent, a wide range of evidence suggests that the ventral stream is bilaterally organized (although with important computational differences between the two hemispheres). The compelling extent to which neuroimaging data implicate both hemispheres has recently been reviewed (Price, 2012; Schirmer, Fox, & Grandjean, 2012; Turkeltaub & Coslett, 2010). The dorsal stream, however, is traditionally—and in the model outlined here—held to be strongly left-dominant.

25.2.1 Ventral Stream: Mapping from Sound to Meaning

25.2.1.1 Bilateral Organization and Parallel Computation

The ventral stream is bilaterally organized, although not computationally redundant, in the two hemispheres. This may not be obvious based on a cursory evaluation of the clinical data. After all, left hemisphere damage yields language deficits of a variety of sorts, including comprehension impairment, whereas, in most cases, right hemisphere damage has little effect on phonological, lexical, or sentence-level language abilities. A closer look tells a different story. In particular, research in the 1980s showed that auditory comprehension deficits in aphasia (caused by unilateral left hemisphere lesions) were not caused primarily by impairment in the ability to perceive speech sounds, as Wernicke and later Luria proposed (Luria, 1970; Wernicke, 1874/1969). For example, when Wernicke's aphasics are asked to match pictures to auditorily presented words, their overall performance is well above chance, and when they do

Neurobiology of Language. DOI: http://dx.doi.org/10.1016/B978-0-12-407794-2.00025-0

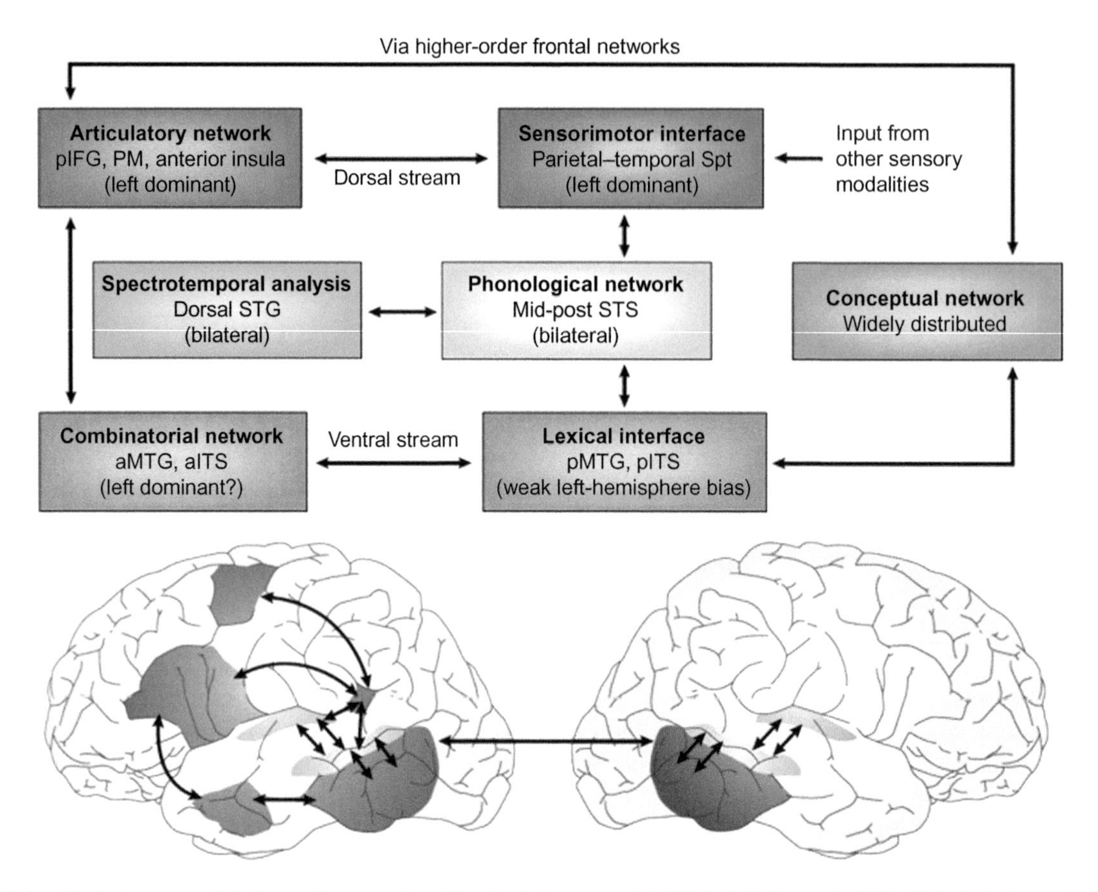

FIGURE 25.1 Dual stream model of speech processing. The dual stream model (Hickok & Poeppel, 2000; Hickok & Poeppel, 2004; Hickok & Poeppel, 2007) holds that early stages of speech processing occurs bilaterally in auditory regions on the dorsal STG (spectrotemporal analysis; green) and STS (phonological access/representation; yellow), and then diverges into two broad streams: a temporal lobe ventral stream supports speech comprehension (lexical access and combinatorial processes; pink), whereas a strongly left-dominant dorsal stream supports sensory-motor integration and involves structures at the parietal-temporal junction (Spt) and frontal lobe. The conceptual network (gray box) is assumed to be widely distributed throughout cortex. IFG, inferior frontal gyrus; ITS, inferior temporal sulcus; MTG, middle temporal gyrus; PM, premotor; Spt, Sylvian parietal-temporal; STG, superior temporal gyrus; STS, superior temporal sulcus. *From Hickok and Poeppel (2007) with permission from the publisher.*

make errors they tend to confuse the correct answer with *semantically* similar alternatives more often than with *phonemically* similar foils (Baker, Blumstein, & Goodglass, 1981; Miceli, Gainotti, Caltagirone, & Masullo, 1980; Rogalsky, Pitz, Hillis, & Hickok, 2008; Rogalsky, Love, Driscoll, Anderson, & Hickok, 2011). A similar pattern of performance has been observed after acute deactivation of the entire left hemisphere in Wada procedures (Figure 25.2) (Hickok et al., 2008). "Speech perception" deficits can be identified in left-injured patients, but only on metalinguistic tasks such as syllable discrimination that involve some level of conscious attention to phonemic structure and working memory; the involvement of the left hemisphere in these tasks likely follows from the relation between working memory and speech articulation (Hickok & Poeppel, 2000, 2004, 2007). In contrast to the (minimal) effects of unilateral lesions on the processing of phoneme level information during auditory comprehension, bilateral lesions involving the superior temporal lobe can have a devastating effect, as cases of word deafness attest (see Chapter 37) (Buchman, Garron, Trost-Cardamone, Wichter, & Schwartz, 1986; Poeppel, 2001).

Data from neuroimaging have been more controversial. One consistent and uncontroversial finding, however, is that, when contrasted with a resting baseline, listening to speech activates the STG bilaterally, including the dorsal STG and superior temporal sulcus (STS). But, when listening to connected, intelligible speech is contrasted against various acoustic baselines, some studies have reported left-dominant activation patterns (Narain et al., 2003; Scott, Blank, Rosen, & Wise, 2000), leading some authors to argue for a fully left-lateralized network for speech perception (Rauschecker & Scott, 2009; Scott et al., 2000). Other studies report bilateral activation even when acoustic controls are subtracted out of the activation pattern (Okada et al., 2010) (for a review see Hickok and Poeppel, 2007). The issue is still being actively debated within the functional imaging literature, although recent reviews and meta-analyses

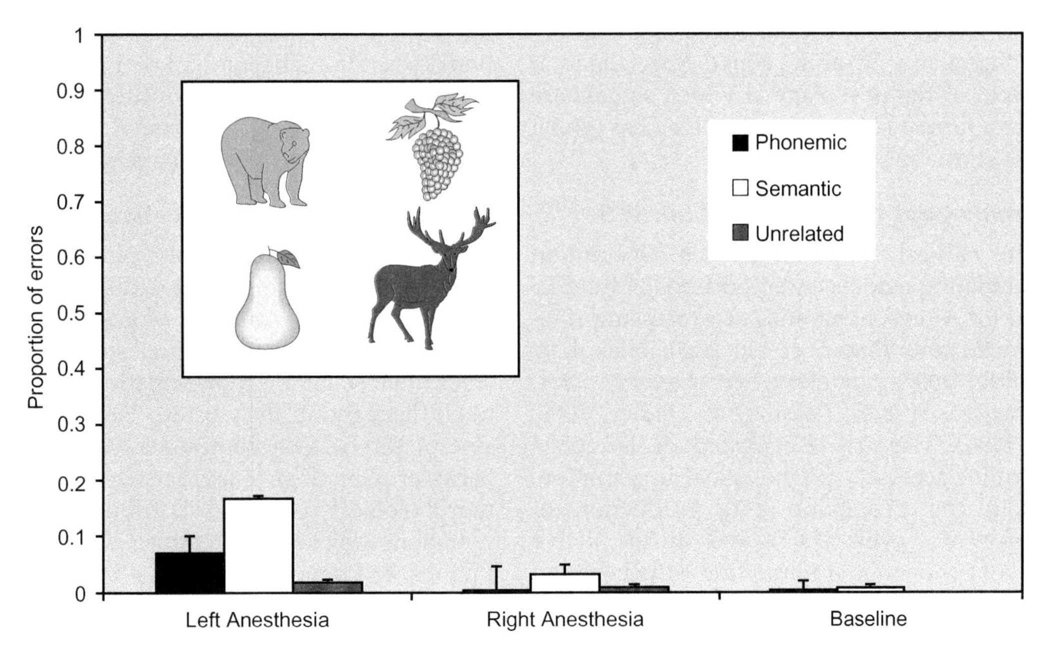

FIGURE 25.2 Auditory comprehension performance during Wada procedure. Data show mean error rate on a four-alternative forced choice auditory comprehension task with phonemic, semantic, and unrelated foils (inset shows sample stimulus card for the spoken target word, *bear*) in 20 patients undergoing clinically indicated Wada procedures. Error rate is shown as a function error type and amytal condition: left hemisphere injection, right hemisphere injection, and baseline. Note overall low error rate, even with left hemisphere injection, and the dominance of semantic mis-selections when errors occur. *Modified from Hickok et al. (2008).*

support the conjecture of bilateral STG/STS involvement (Price, 2012; Schirmer et al., 2012; Turkeltaub & Coslett, 2010).

25.2.1.2 Computational Asymmetries

The hypothesis that sublexical-level processes in speech recognition are bilaterally organized does not imply that the two hemispheres are computationally identical. In fact, there is strong evidence for hemispheric differences in the processing of acoustic/speech information (Abrams, Nicol, Zecker, & Kraus, 2008; Boemio, Fromm, Braun, & Poeppel, 2005; Giraud et al., 2007; Hickok & Poeppel, 2007; Zatorre, Belin, & Penhune, 2002). The basis of these differences is currently being debated. One view, arguing for a domain-general perspective for all sounds, is that the difference turns on the selectivity for temporal (left hemisphere) versus spectral (right hemisphere) resolution (Obleser, Eisner, & Kotz, 2008; Zatorre et al., 2002). That is, the left hemisphere may be particularly well-suited for resolving rapid acoustic change (such as a formant transition), whereas the right hemisphere may have an advantage in resolving spectral frequency information. A closely related proposal is that the two hemispheres differ in terms of their preferred "sampling rate," with left auditory cortical regions incorporating a bias for faster rate sampling (25–50 Hz) and the right hemisphere for slower rate sampling (4–8 Hz) (Poeppel, 2003). These two proposals are not incompatible because there is a relation between sampling rate and spectral versus temporal resolution; rapid sampling allows the system to detect changes that occur over short timescales but sacrifices spectral resolution, and vice versa (Zatorre et al., 2002).

Further research is needed to address these hypotheses. For the present purposes, the central point is that this asymmetry of function indicates that spoken word recognition involves parallel pathways—at least one in each hemisphere—in the mapping from sound to lexical meaning (Hickok & Poeppel, 2007), similar to well-accepted dual-route models of reading (phoneme-to-grapheme conversion and whole-word routes) (Coltheart, Curtis, Atkins, & Haller, 1993). Although the parallel pathway view differs from standard models of speech recognition (Luce & Pisoni, 1998; Marslen-Wilson, 1987; McClelland & Elman, 1986), wherein the processor proceeds from small to larger units in serial stages, it is consistent with the fact that speech contains redundant cues to phonemic information (e.g., in the speech envelope and fine spectral structure cues) and with behavioral evidence suggesting that the speech system can take advantage of these different cues (Remez, Rubin, Pisoni, & Carrell, 1981; Shannon, Zeng, Kamath, Wygonski, & Ekelid, 1995). It is worth bearing in mind that such computational asymmetries apply to all sounds that the auditory system analyzes.

They reflect properties of neuronal ensembles that are like filters acting on any incoming signal. Specialization is likely to occur at the next stage at which signals are translated into a format suitable for lexical access (given that words are stored in some format).

25.2.1.3 Phonological Processing and the STS

Beyond the earliest stages of speech recognition there is accumulating evidence that portions of the STS are important for representing and/or processing phonological information (Binder et al., 2000; Hickok & Poeppel, 2004, 2007; Indefrey & Levelt, 2004; Liebenthal, Binder, Spitzer, Possing, & Medler, 2005; Price et al., 1996). The STS is activated by language tasks that require access to phonological information, including both the perception and production of speech (Indefrey & Levelt, 2004), and during active maintenance of phonemic information (Buchsbaum, Hickok, & Humphries, 2001; Hickok, Buchsbaum, Humphries, & Muftuler, 2003). Portions of the STS seem to be relatively selective for acoustic signals that contain phonemic information when compared with complex nonspeech signals (yellow shaded portion of Figure 25.1) (Hickok & Poeppel, 2007; Liebenthal et al., 2005; Narain et al., 2003; Okada et al., 2010). STS activation can be modulated by the manipulation of psycholinguistic variables that tap phonological networks (Okada & Hickok, 2006), such as phonological neighborhood density (the number of words that sound similar to a target word), and this region shows neural adaptation effects to phonological level information (Vaden, Muftuler, & Hickok, 2009).

One currently unresolved question concerns the relative contribution of anterior versus posterior STS regions in phonological processing. Lesion evidence indicates that damage to posterior temporal lobe areas is most predictive of auditory comprehension deficits (Bates et al., 2003) and a majority of functional imaging studies targeting phonological processing in perception have identified regions in the posterior half of the STS (Hickok & Poeppel, 2007). Other studies, however, have reported *anterior* STS activation in perceptual speech tasks (Mazoyer et al., 1993; Narain et al., 2003; Scott et al., 2000; Spitsyna, Warren, Scott, Turkheimer, & Wise, 2006). These studies typically involved sentence-level stimuli, raising the possibility that anterior STS regions may be responding to some other aspect of the stimuli such as its syntactic or prosodic organization (Friederici, Meyer, & von Cramon, 2000; Humphries, Binder, Medler, & Liebenthal, 2006; Humphries, Love, Swinney, & Hickok, 2005; Humphries, Willard, Buchsbaum, & Hickok, 2001; Vandenberghe, Nobre, & Price, 2002). Recent electrophysiological work supports the hypothesis that left anterior temporal lobe (ATL) is critical to elementary structure building (Bemis & Pylkkanen, 2011), in line with the view that intelligibility tasks tap into additional operations beyond speech recognition. It will, in any case, be important in future work to understand the role of various portions of the STS in auditory speech perception and language processing.

25.2.1.4 Lexical-Semantic Access

During auditory comprehension, the goal of speech processing is to use phonological information to access words: conceptual-semantic representations that are critical to comprehension. The dual stream model holds that conceptual-semantic representations are widely distributed throughout the cortex. However, a more focal system serves as a computational interface that maps between phonological-level representations of words or morphological roots and distributed conceptual representations (Hickok & Poeppel, 2000, 2004, 2007; Lau, Phillips, & Poeppel, 2008). This interface is not the site for storage of conceptual information. Instead, it is hypothesized to store information regarding the *relation* (or correspondences) between phonological information and conceptual information. Most authors agree that the temporal lobe(s) play a critical role in this process, but again there is disagreement regarding the role of anterior versus posterior regions. The evidence for both of these viewpoints is briefly presented here.

Damage to posterior temporal lobe regions, particularly along the MTG, has long been associated with auditory comprehension deficits (Bates et al., 2003; Damasio, 1991; Dronkers, Redfern, & Knight, 2000), an effect confirmed in a large-scale study involving 101 patients (Bates et al., 2003). We infer that these deficits are primarily postphonemic in nature because phonemic deficits after unilateral lesions to these areas are mild (Hickok & Poeppel, 2004). Data from direct cortical stimulation studies corroborate the involvement of the MTG in auditory comprehension and also indicate the involvement of a much broader network involving most of the superior temporal lobe (including anterior portions) and the inferior frontal lobe (Miglioretti & Boatman, 2003). Functional imaging studies have also implicated posterior middle temporal regions in lexical-semantic processing (Binder et al., 1997; Rissman, Eliassen, & Blumstein, 2003; Rodd, Davis, & Johnsrude, 2005). These findings do not preclude the involvement of more anterior regions in lexical-semantic access, but they do make a strong case for significant involvement of posterior regions. Electrophysiological studies have successfully used paradigms building on the N400 response to study lexical-semantic processing. This response is very sensitive to a range of variables known to implicate lexical-level properties. A review of that literature (including source localization studies of the N400) also suggests that posterior MTG plays a key role, although embedded in a network of anterior temporal, parietal, and inferior frontal regions (Lau et al., 2008).

ATL regions have been implicated both in lexical-semantic and sentence-level processing (syntactic and/or semantic integration processes). Patients with semantic dementia, who have been used to argue for a lexical-semantic function (Scott et al., 2000; Spitsyna et al., 2006), have atrophy involving the ATL bilaterally along with deficits in lexical tasks such as naming, semantic association, and single-word comprehension (Gorno-Tempini et al., 2004). However, these deficits are not specific to the mapping between phonological and conceptual representations and appear to involve more general semantic integration (Patterson, Nestor, & Rogers, 2007). Further, because atrophy in semantic dementia involves a number of regions in addition to the lateral ATL, including bilateral inferior and medial temporal lobe, bilateral caudate nucleus, and right posterior thalamus, among others (Gorno-Tempini et al., 2004), linking the deficits specifically to the ATL is difficult.

Higher-level syntactic and compositional semantic processing might involve the ATL. Functional imaging studies have found portions of the ATL to be more active while subjects listen to or read sentences rather than unstructured lists of words or sounds (Friederici et al., 2000; Humphries et al., 2001; Humphries et al., 2005; Mazoyer et al., 1993; Vandenberghe et al., 2002). This structured versus unstructured effect is independent of the semantic content of the stimuli, although semantic manipulations can modulate the ATL response somewhat (Vandenberghe et al., 2002). Recent electrophysiological data (Brennan & Pylkkanen, 2012; Bemis & Pylkkanen, 2013) also implicate left ATL in elementary structure building. Damage to the ATL has also been linked to deficits in comprehending complex syntactic structures (Dronkers, Wilkins, Van Valin, Redfern, & Jaeger, 2004). However, data from semantic dementia are contradictory, because these patients are reported to have good sentence-level comprehension (Gorno-Tempini et al., 2004).

In summary, there is strong evidence that lexical-semantic access from auditory input involves the posterior lateral temporal lobe. In terms of syntactic and compositional semantic operations, neuroimaging evidence is converging on the ATL as an important component of the computational network (Humphries et al., 2005; Humphries et al., 2006; Vandenberghe et al., 2002); however, the neuropsychological evidence remains equivocal.

25.2.2 Dorsal Stream: Mapping from Sound to Action

The earliest proposals regarding the dorsal auditory stream argued that this system was involved in spatial hearing, a "where" function (Rauschecker, 1998), similar to the dorsal "where" stream proposal in the cortical visual system (Ungerleider & Mishkin, 1982). More recently, there has been some convergence on the idea that the dorsal stream supports auditory-motor integration (Hickok & Poeppel, 2000, 2004, 2007; Rauschecker, 2011; Rauschecker & Scott, 2009; Scott & Wise, 2004; Wise et al., 2001). Specifically, the idea is that the auditory dorsal stream supports an interface between auditory and motor representations of speech, a proposal similar to the claim that the dorsal visual stream has a sensory-motor integration function (Andersen, 1997; Milner & Goodale, 1995).

25.2.2.1 The Need for Auditory-Motor Integration

The idea of auditory-motor interaction in speech is not new. Wernicke's classic model of the neural circuitry of language incorporated a direct link between sensory and motor representations of words and argued explicitly that sensory systems participated in speech production (Wernicke, 1874/1969). More recently, research on motor control has revealed why this sensory-motor link is critical. Motor acts aim to hit sensory targets. In the visual-manual domain, we identify the location and shape of, say, a cup visually (the sensory target) and generate a motor command that allows us to move our limb toward that location and shape the hand to match the shape of the object. In the speech domain, the targets are not external objects but rather internal representations of the sound pattern (phonological form) of a word. We know that the targets are auditory in nature because manipulating a speaker's auditory feedback during speech production results in compensatory changes in motor speech acts (Houde & Jordan, 1998; Larson, Burnett, Bauer, Kiran, & Hain, 2001; Purcell & Munhall, 2006). For example, if a subject is asked to produce one vowel and the feedback that she hears is manipulated so that it sounds like another vowel, then the subject will change the vocal tract configuration so that the feedback sounds like the original vowel. In other words, talkers will readily modify their motor articulations to hit an auditory target, indicating that the goal of speech production is not a particular motor configuration but rather a speech *sound* (Guenther, Hampson, & Johnson, 1998). The role of auditory input is nowhere more apparent than in development, where the child must use acoustic information in the linguistic environment to shape vocal tract movements that must reproduce those sounds.

A great deal of progress has been made in mapping the neural organization of sensorimotor integration for speech. Early functional imaging studies identified an auditory-related area in the left planum temporale region as involved in speech *production* (Hickok et al., 2000;

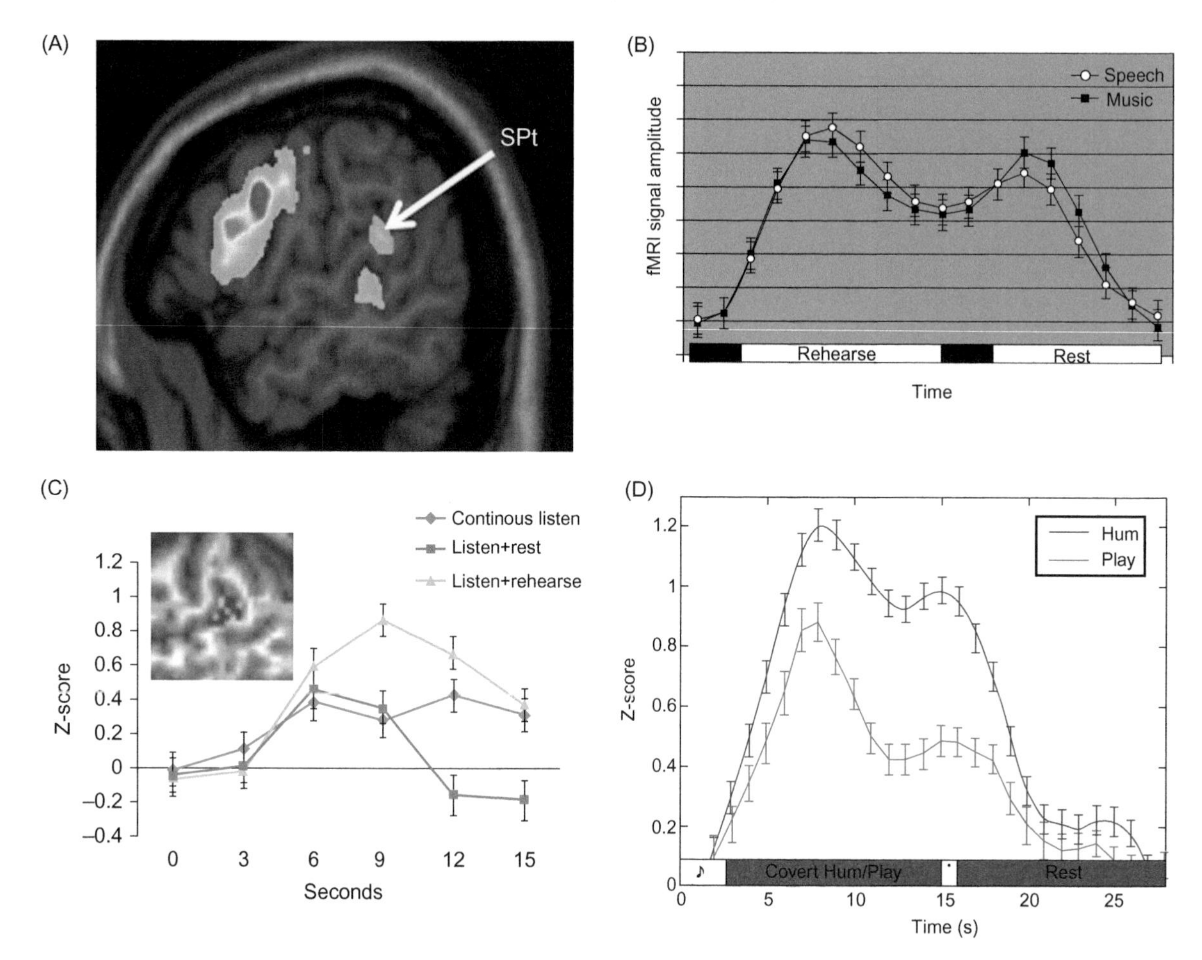

FIGURE 25.3 Location and functional properties of area Spt. (A) Activation map for covert speech articulation (rehearsal of a set of nonwords). (B) Activation timecourse (fMRI signal amplitude) in Spt during a sensorimotor task for speech and music. A trial is composed of 3 s of auditory stimulation, followed by 15 s of covert rehearsal/humming of the heard stimulus, followed by 3 s of auditory stimulation, followed by 15 s of rest. The two humps represent the sensory responses; the valley between the humps is the motor (covert rehearsal) response and the baseline values at the onset and offset of the trial represent resting activity levels. Note similar response to both speech and music. (C) Activation timecourse in Spt in three conditions: continuous speech (15 s, blue curve), listen + rest (3 s speech, 12 s rest, red curve), and listen + covert rehearse (3 s speech, 12 s rehearse, green curve). The pattern of activity within Spt (inset) was found to be different for listening to speech compared with rehearsing speech assessed at the end of the continuous listen versus listen + rehearse conditions despite the lack of a significant signal amplitude difference at that time point. (D) Activation timecourse in Spt in skilled pianists performing a sensorimotor task involving listening to novel melodies and then covertly humming them (blue curve) versus listening to novel melodies and imagining playing them on a keyboard (red curve). This indicates that Spt is relatively selective for vocal tract actions. *(A) From Hickok and Buchsbaum (2003) with permission from the publisher. (B) Adapted from Hickok et al. (2003). (C) Adapted from Hickok, Okada, and Serences (2009b). (D) From Hickok (2009) with permission from the publisher.*

Wise et al., 2001). Subsequent studies showed that this left-dominant region, dubbed Spt for its location in the Sylvian fissure at the parietal-temporal boundary (Figure 25.3A) (Hickok et al., 2003), exhibited a number of properties characteristic of sensorimotor integration areas such as those found in macaque parietal cortex (Andersen, 1997; Colby & Goldberg, 1999). Most fundamentally, Spt exhibits sensorimotor response properties, activating both during the passive perception of speech and during covert (subvocal) speech articulation (Buchsbaum et al., 2001; Buchsbaum, Olsen, Koch, & Berman, 2005; Hickok et al., 2003); further, the different subregional *patterns* of activity are apparent during the sensory and motor phases of the task (Hickok, Okada, & Serences, 2009a), likely reflecting the activation of different neuronal subpopulations (Dahl, Logothetis, & Kayser, 2009) (some sensory-weighted and others motor-weighted). Figure 25.3B–D shows examples of the sensory-motor response properties of Spt and the patchy organization of this region for sensory-weighted versus motor-weighted voxels (25.3C, inset).

Spt is not speech-specific; its sensorimotor responses are equally robust when the sensory stimulus consists of tonal melodies and (covert) humming is the motor task (see the two curves in Figure 25.3B) (Hickok et al., 2003).

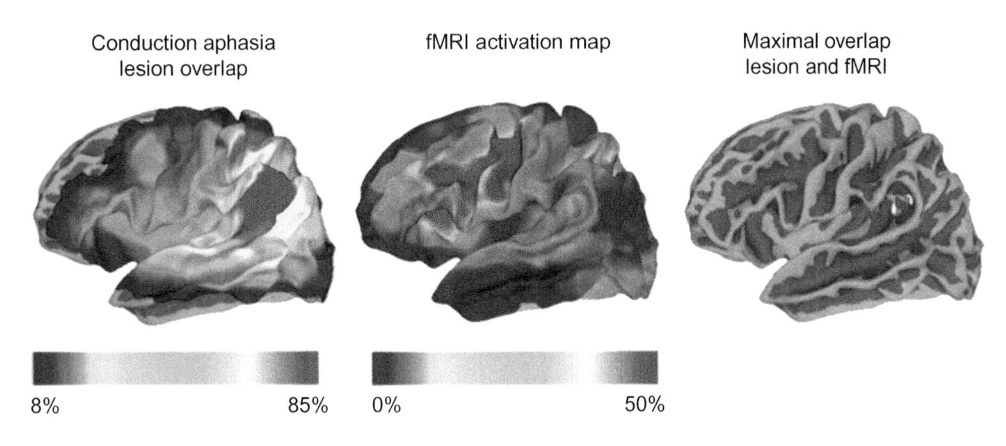

FIGURE 25.4 Relation between lesions associated with conduction aphasia and the cortical auditory-motor network. A comparison of conduction aphasia, an auditory-motor task (listening to and then repeating back speech) in fMRI, and their overlap. The uninflated surface in the left panel shows the regional distribution lesion overlap in patients with conduction aphasia (maximum is 12/14 or 85% overlap). Middle panel shows the auditory-motor network in the fMRI analysis. The right panel shows the area of maximal overlap between the lesion and fMRI surfaces (lesion >85% overlap and significant fMRI activity). *Modified from Buchsbaum et al. (2011).*

Activity in Spt is highly correlated with activity in the pars opercularis (Buchsbaum et al., 2001; Buchsbaum et al., 2005), which is the posterior sector of Broca's region. White matter tracts identified via diffusion tensor imaging suggest that Spt and the pars opercularis are densely connected anatomically (for review see Friederici, 2009, Rogalsky and Hickok, 2011). Finally, consistent with sensorimotor integration areas in the monkey parietal lobe (Andersen, 1997; Colby & Goldberg, 1999), Spt appears to be motor-effector–selective, responding more robustly when the motor task involves the vocal tract than when it involves the manual effectors (Figure 25.2D) (Pa & Hickok, 2008). More broadly, Spt is situated in the middle of a network of auditory (STS) and motor (pars opercularis, premotor cortex) regions (Buchsbaum et al., 2001; Buchsbaum et al., 2005; Hickok et al., 2003), perfectly positioned both functionally and anatomically to support sensorimotor integration for speech and related vocal tract functions.

Lesion evidence is consistent with the functional imaging data implicating Spt as part of a sensorimotor integration circuit. Damage to auditory-related regions in the left hemisphere often results in speech production deficits (Damasio, 1991, 1992), demonstrating that sensory systems participate in motor speech. More specifically, damage to the left temporal-parietal junction is associated with conduction aphasia, a syndrome that is characterized by good comprehension but frequent phonemic errors in speech production (Baldo, Klostermann, & Dronkers, 2008; Damasio & Damasio, 1980; Goodglass, 1992), and the lesion distribution overlaps with the location of functional area Spt (Figure 25.4) (Buchsbaum et al., 2011). Conduction aphasia has classically been considered to be a disconnection syndrome involving damage to the arcuate fasciculus. However, there is now good evidence that this syndrome results from cortical dysfunction (Anderson et al., 1999; Hickok et al., 2000). The production deficit is load-sensitive: errors are more likely on longer, lower-frequency words and verbatim repetition of strings of speech with little semantic constraint (Goodglass, 1992, 1993). In the context of this discussion, the effects of such lesions can be understood as an interruption of the system that serves at the interface between auditory target and the motor speech actions that can achieve them (Hickok & Poeppel, 2000, 2004, 2007).

Recent theoretical work has clarified the computational details underlying auditory-motor integration in the dorsal stream. Drawing on advances in understanding motor control generally, speech researchers have emphasized the role of internal forward models in speech motor control (Golfinopoulos, Tourville, & Guenther, 2010; Hickok, Houde, & Rong, 2011; Houde & Nagarajan, in press; Houde & Nagarajan, 2011). The basic idea is that to control action the nervous system makes forward predictions about the future state of the motor articulators and the sensory consequences of the predicted actions. The predictions are assumed to be generated by an internal model that receives copies of motor commands and integrates them with information about the current state of the system and past experience (learning) of the relation between particular motor commands and their sensory consequences. This internal model affords a mechanism for detecting and correcting motor errors (i.e., motor actions that fail to hit their sensory targets).

Several models have been proposed with similar basic assumptions but slightly different architectures (Golfinopoulos et al., 2010; Guenther et al., 1998; Hickok et al., 2011; Houde & Nagarajan, in press; also

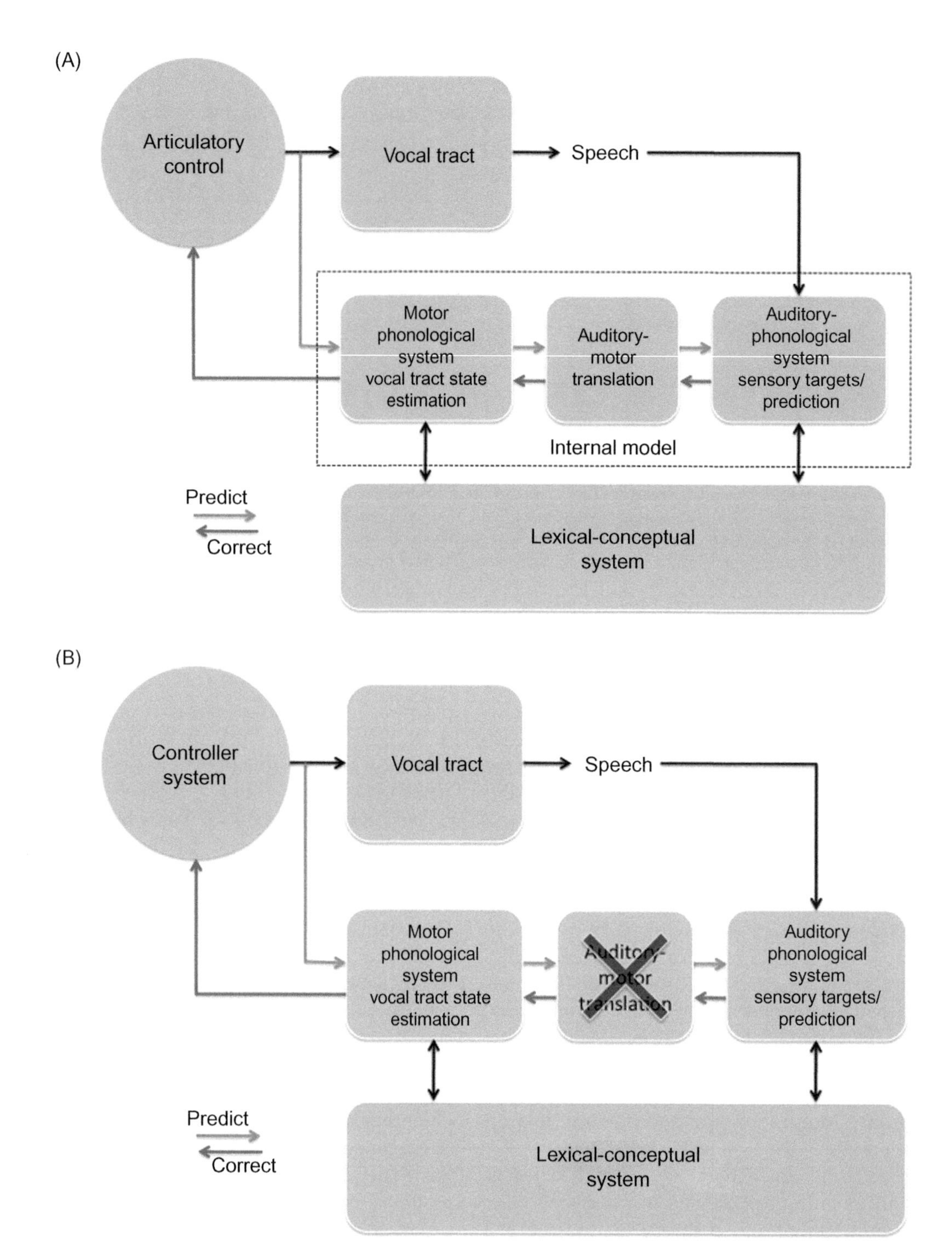

FIGURE 25.5 An integrated state feedback control (SFC) model of speech production. (A) Speech models derived from the feedback control, psycholinguistic, and neurolinguistic literatures are integrated into one framework, presented here. The architecture is fundamentally that of an SFC system with a controller, or set of controllers (Haruno, Wolpert, & Kawato, 2001), localized to primary motor cortex, which generates motor commands to the vocal tract and sends a corollary discharge to an internal model, which makes forward predictions about the dynamic state of the vocal tract and about the sensory consequences of those states. Deviations between predicted auditory states and the intended targets or actual sensory feedback generate an error signal that is used to correct and update the internal model of the vocal tract. The internal model of the vocal tract is instantiated as a "motor phonological system," which corresponds to the neurolinguistically elucidated *phonological output lexicon*, and is localized to premotor cortex. Auditory targets and forward predictions of sensory consequences are encoded in the same network, namely the "auditory phonological system," which corresponds to the neurolinguistically elucidated *phonological input lexicon* and is localized to the STG/STS. Motor and auditory phonological systems are linked via an auditory-motor translation system localized to area Spt. The system is activated via parallel inputs from the lexical-conceptual system to the motor and auditory phonological systems. (B) Proposed source of the deficit in conduction aphasia: damage to the auditory-motor translation system. Input from the lexical-conceptual system to motor and auditory phonological systems are unaffected, allowing for fluent output and accurate activation of sensory targets. However, internal forward sensory predictions are not possible, leading to an increase in error rate. Further, errors detected as a consequence of mismatches between sensory targets and actual sensory feedback cannot be used to correct motor commands. *From Hickok et al. (2011) and Hickok and Poeppel (2004) with permission from the publisher.*

see Chapter 58, this volume). One such model is shown in Figure 25.5 (Hickok et al., 2011). Input to the system comes from a lexical-conceptual network, as assumed by psycholinguistic models of speech production (Dell, Schwartz, Martin, Saffran, & Gagnon, 1997; Levelt, Roelofs, & Meyer, 1999). In between the input/output system is a phonological system that is split into two components, corresponding to sensory input and motor output subsystems and mediated by a sensorimotor translation system, which corresponds to area Spt (Buchsbaum et al., 2001; Hickok et al., 2003; Hickok et al., 2009a). Parallel inputs to sensory and motor systems are needed to explain neuropsychological observations (Jacquemot, Dupoux, & Bachoud-Levi, 2007), such as conduction aphasia, as shown here. Inputs to the auditory phonological network define the auditory targets of speech acts. As a motor speech unit (ensemble) begins to be activated, its predicted auditory consequences can be checked against the auditory target. If they match, then that unit will continue to be activated, resulting in an articulation that will hit the target. If there is a mismatch, then a correction signal can be generated to activate the correct motor unit.

This model provides a natural explanation of conduction aphasia. A lesion to Spt would disrupt the ability to generate forward predictions in auditory cortex and thereby the ability to perform internal feedback monitoring, making errors more frequent than in an unimpaired system (Figure 25.5B). However, this would not disrupt the activation of auditory targets via the lexical-semantic system, thus leaving the patient capable of detecting errors in their own speech, a characteristic of conduction aphasia. Once an error is detected, however, the correction signal will not be accurately translated to the internal model of the vocal tract due to disruption of Spt. The ability to detect but not accurately correct speech errors should result in repeated unsuccessful self-correction attempts, again a characteristic of conduction aphasia.

25.3 CLINICAL CORRELATES OF THE DUAL STREAM MODEL

The dual stream model, like the classical Wernicke-Lichtheim model, provides an account of the major clinical aphasia syndromes (Hickok & Poeppel, 2004). Within the dual stream model, Broca's aphasia and conduction aphasia are considered to be dorsal stream–related syndromes, whereas Wernicke's aphasia, word deafness, and transcortical sensory aphasia are considered ventral stream syndromes. We have already noted that conduction aphasia can be conceptualized as a disruption of auditory-motor integration resulting from damage to area Spt. Broca's aphasia can be viewed as a disruption to representations that code for speech-related actions at multiple levels from coding low-level phonetic features, to sequences of syllables, to sequences of words in structured sentences. Although Broca's area and Broca's aphasia are widely considered to be associated with deficits in *receptive* syntactic processing (Caramazza & Zurif, 1976; Grodzinsky, 2000), this issue is now being seriously questioned and remains debatable (Rogalsky et al., 2011).

Word deafness is the "lowest-level" ventral stream syndrome according to the dual stream model, affecting the processing of phonemic information during speech recognition. This differs from classic interpretations of word deafness as a disconnection syndrome (Geschwind, 1965). Due to the key role that auditory systems play in speech production, as discussed, we should expect that disruption to auditory speech systems, as in word deafness, will impact production as well. Although the canonical description of word deafness is a syndrome in which speech production is preserved, the majority of case descriptions that provide information on the speech output of word deaf patients report the presence of paraphasic errors (Buchman et al., 1986).

Wernicke's aphasia is explained in terms of damage to multiple ventral stream processing levels in the dual stream model. Given the rather extensive posterior lesions that are typically required to yield a chronic Wernicke's aphasia (Dronkers & Baldo, 2009), it is likely that this syndrome results from damage to auditory-motor area Spt, left hemisphere auditory areas, and posterior middle temporal lexical-semantic interface systems. Such damage can explain the symptom complex: relatively good phonological level speech recognition (due to the bilateral organization as described above), poor comprehension at the higher semantic level (due to damage to lexical-semantic interface systems), fluent speech (due to preserved motor speech systems), poor repetition (due to disruption of auditory-motor interface network), and paraphasic errors (due to disruption of auditory-motor interface network).

Transcortical sensory aphasia, which is similar to Wernicke's aphasia but with preserved repetition, is conceptualized as a functionally more focal deficit involving the lexical-semantic interface network but sparing the auditory-motor network. Damage to the lexical-semantic interface explains the poor comprehension, whereas sparing of the auditory-motor interface explains the preserved repetition.

25.4 SUMMARY

Dual stream models of cortical organization have proven useful in understanding both language and visual-related systems, and have been a recurrent theme in neural models stretching back more than a century (Wernicke, 1874/1977). Thus, the general concept underlying the model—that the brain must interface sensory information with two different systems, conceptual and motor—not only is intuitively appealing but also has a proven track record across domains. In the language domain, the dual stream model provides an explanation of classical language disorders (Hickok et al., 2011; Hickok & Poeppel, 2004) and provides a framework for integrating and unifying research across psycholinguistic, neurolinguistic, and neurophysiological traditions. Recent work has shown that still further integration with motor control models is possible (Hickok et al., 2011). All of this suggests that the dual stream framework is on the right track as a model of language organization and provides a rich context for guiding future research.

References

Abrams, D. A., Nicol, T., Zecker, S., & Kraus, N. (2008). Right-hemisphere auditory cortex is dominant for coding syllable patterns in speech. *Journal of Neuroscience, 28*(15), 3958–3965.

Andersen, R. (1997). Multimodal integration for the representation of space in the posterior parietal cortex. *Philosophical Transactions of the Royal Society of London B Biological Sciences, 352*, 1421–1428.

Anderson, J. M., Gilmore, R., Roper, S., Crosson, B., Bauer, R. M., Nadeau, S., et al. (1999). Conduction aphasia and the arcuate fasciculus: A reexamination of the Wernicke-Geschwind model. *Brain and Language, 70*, 1–12.

Baker, E., Blumstein, S. E., & Goodglass, H. (1981). Interaction between phonological and semantic factors in auditory comprehension. *Neuropsychologia, 19*, 1–15.

Baldo, J. V., Klostermann, E. C., & Dronkers, N. F. (2008). It's either a cook or a baker: Patients with conduction aphasia get the gist but lose the trace. *Brain and Language, 105*(2), 134–140.

Bates, E., Wilson, S. M., Saygin, A. P., Dick, F., Sereno, M. I., Knight, R. T., et al. (2003). Voxel-based lesion-symptom mapping. *Nature Neuroscience, 6*(5), 448–450.

Bemis, D. K., & Pylkkanen, L. (2011). Simple composition: A magnetoencephalography investigation into the comprehension of minimal linguistic phrases. *The Journal of Neuroscience, 31*(8), 2801–2814.

Bemis, D. K., & Pylkkanen, L. (2013). Basic linguistic composition recruits the left anterior temporal lobe and left angular gyrus during both listening and reading. *Cerebral Cortex, 23*(8), 1859–1873.

Binder, J. R., Frost, J. A., Hammeke, T. A., Bellgowan, P. S., Springer, J. A., Kaufman, J. N., et al. (2000). Human temporal lobe activation by speech and nonspeech sounds. *Cerebral Cortex, 10*, 512–528.

Binder, J. R., Frost, J. A., Hammeke, T. A., Cox, R. W., Rao, S. M., & Prieto, T. (1997). Human brain language areas identified by functional magnetic resonance imaging. *Journal of Neuroscience, 17*, 353–362.

Boemio, A., Fromm, S., Braun, A., & Poeppel, D. (2005). Hierarchical and asymmetric temporal sensitivity in human auditory cortices. *Nature Neuroscience, 8*(3), 389–395.

Brennan, J., & Pylkkanen, L. (2012). The time-course and spatial distribution of brain activity associated with sentence processing. *Neuroimage, 60*(2), 1139–1148.

Buchman, A. S., Garron, D. C., Trost-Cardamone, J. E., Wichter, M. D., & Schwartz, M. (1986). Word deafness: One hundred years later. *Journal of Neurology, Neurosurgury, and Psychiatry, 49*, 489–499.

Buchsbaum, B., Hickok, G., & Humphries, C. (2001). Role of left posterior superior temporal gyrus in phonological processing for speech perception and production. *Cognitive Science, 25*, 663–678.

Buchsbaum, B. R., Baldo, J., Okada, K., Berman, K. F., Dronkers, N., D'Esposito, M., et al. (2011). Conduction aphasia, sensory-motor integration, and phonological short-term memory - an aggregate analysis of lesion and fMRI data. *Brain and Language, 119*(3), 119–128.

Buchsbaum, B. R., Olsen, R. K., Koch, P., & Berman, K. F. (2005). Human dorsal and ventral auditory streams subserve rehearsal-based and echoic processes during verbal working memory. *Neuron, 48*(4), 687–697.

Caramazza, A., & Zurif, E. B. (1976). Dissociation of algorithmic and heuristic processes in sentence comprehension: Evidence from aphasia. *Brain and Language, 3*, 572–582.

Colby, C. L., & Goldberg, M. E. (1999). Space and attention in parietal cortex. *Annual Review of Neuroscience, 22*, 319–349.

Coltheart, M., Curtis, B., Atkins, P., & Haller, M. (1993). Models of reading aloud: Dual-route and parallel-distributed-processing approaches. *Psychological Review, 100*, 589–608.

Dahl, C. D., Logothetis, N. K., & Kayser, C. (2009). Spatial organization of multisensory responses in temporal association cortex. *The European Journal of Neuroscience, 29*(38), 11924–11932.

Damasio, A. R. (1992). Aphasia. *New England Journal of Medicine, 326*, 531–539.

Damasio, H. (1991). Neuroanatomical correlates of the aphasias. In M. Sarno (Ed.), *Acquired aphasia* (pp. 45–71). San Diego, CA: Academic Press.

Damasio, H., & Damasio, A. R. (1980). The anatomical basis of conduction aphasia. *Brain, 103*, 337–350.

Dell, G. S., Schwartz, M. F., Martin, N., Saffran, E. M., & Gagnon, D. A. (1997). Lexical access in aphasic and nonaphasic speakers. *Psychological Review, 104*, 801–838.

Dijkerman, H. C., & de Haan, E. H. (2007). Somatosensory processes subserving perception and action. *The Behavioral and Brain Sciences, 30*(2), 189–201 (discussion 201–239).

Dronkers, N., & Baldo, J. (2009). Language: Aphasia. In L. R. Squire (Ed.), *Encyclopedia of neuroscience* (Vol. 5, pp. 343–348). Oxford: Academic Press.

Dronkers, N. F., Redfern, B. B., & Knight, R. T. (2000). The neural architecture of language disorders. In M. S. Gazzaniga (Ed.), *The new cognitive neurosciences* (pp. 949–958). Cambridge, MA: MIT Press.

Dronkers, N. F., Wilkins, D. P., Van Valin, R. D., Jr., Redfern, B. B., & Jaeger, J. J. (2004). Lesion analysis of the brain areas involved in language comprehension. *Cognition, 92*(1–2), 145–177.

Friederici, A. D. (2009). Pathways to language: Fiber tracts in the human brain. *Trends in Cognitive Sciences, 13*(4), 175–181.

Friederici, A. D., Meyer, M., & von Cramon, D. Y. (2000). Auditory languge comprehension: An event-related fMRI study on the processing of syntactic and lexical information. *Brain and Language, 74*, 289–300.

Geschwind, N. (1965). Disconnexion syndromes in animals and man. *Brain, 88*, 237–294 585-644.

Giraud, A. L., Kleinschmidt, A., Poeppel, D., Lund, T. E., Frackowiak, R. S., & Laufs, H. (2007). Endogenous cortical rhythms determine cerebral specialization for speech perception and production. *Neuron, 56*(6), 1127–1134.

Golfinopoulos, E., Tourville, J. A., & Guenther, F. H. (2010). The integration of large-scale neural network modeling and functional brain imaging in speech motor control. *Neuroimage, 52*(3), 862–874.

Goodglass, H. (1992). Diagnosis of conduction aphasia. In S. E. Kohn (Ed.), *Conduction aphasia* (pp. 39–49). Hillsdale, N.J: Lawrence Erlbaum Associates.
Goodglass, H. (1993). *Understanding aphasia*. San Diego, CA: Academic Press.
Gorno-Tempini, M. L., Dronkers, N. F., Rankin, K. P., Ogar, J. M., Phengrasamy, L., Rosen, H. J., et al. (2004). Cognition and anatomy in three variants of primary progressive aphasia. *Annals of Neurology, 55*(3), 335–346.
Grodzinsky, Y. (2000). The neurology of syntax: Language use without Broca's area. *Behavioral and Brain Sciences, 23*, 1–21.
Guenther, F. H., Hampson, M., & Johnson, D. (1998). A theoretical investigation of reference frames for the planning of speech movements. *Psychological Review, 105*, 611–633.
Haruno, M., Wolpert, D. M., & Kawato, M. (2001). Mosaic model for sensorimotor learning and control. *Neural Computation, 13*(10), 2201–2220.
Hickok, G. (2009). The functional neuroanatomy of language. *Physics of Life Reviews, 6*, 121–143.
Hickok, G., & Buchsbaum, B. (2003). Temporal lobe speech perception systems are part of the verbal working memory circuit: Evidence from two recent fMRI studies. *The Behavioral and Brain Sciences, 26*, 740–741.
Hickok, G., Buchsbaum, B., Humphries, C., & Muftuler, T. (2003). Auditory-motor interaction revealed by fMRI: Speech, music, and working memory in area Spt. *Journal of Cognitive Neuroscience, 15*, 673–682.
Hickok, G., Erhard, P., Kassubek, J., Helms-Tillery, A. K., Naeve-Velguth, S., Strupp, J. P., et al. (2000). A functional magnetic resonance imaging study of the role of left posterior superior temporal gyrus in speech production: Implications for the explanation of conduction aphasia. *Neuroscience Letters, 287*, 156–160.
Hickok, G., Houde, J., & Rong, F. (2011). Sensorimotor integration in speech processing: Computational basis and neural organization. *Neuron, 69*(3), 407–422.
Hickok, G., Okada, K., Barr, W., Pa, J., Rogalsky, C., Donnelly, K., et al. (2008). Bilateral capacity for speech sound processing in auditory comprehension: Evidence from Wada procedures. *Brain and Language, 107*(3), 179–184.
Hickok, G., Okada, K., & Serences, J. T. (2009a). Area Spt in the human planum temporale supports sensory-motor integration for speech processing. *Journal of Neurophysiology, 101*(5), 2725–2732.
Hickok, G., Okada, K., & Serences, J. T. (2009b). Area Spt in the human planum temporale supports sensory-motor integration for speech processing. *Journal of Neurophysiology, 101*(5), 2725–2732.
Hickok, G., & Poeppel, D. (2000). Towards a functional neuroanatomy of speech perception. *Trends in Cognitive Sciences, 4*, 131–138.
Hickok, G., & Poeppel, D. (2004). Dorsal and ventral streams: A framework for understanding aspects of the functional anatomy of language. *Cognition, 92*, 67–99.
Hickok, G., & Poeppel, D. (2007). The cortical organization of speech processing. *Nature Reviews Neuroscience, 8*(5), 393–402.
Houde, J., & Nagarajan, S.S. (in press). Speech production as state feedback control. *Frontiers in Human Neuroscience*, 5. http://dx.doi.org/10.3389/fnhum.2011.00082.
Houde, J. F., & Jordan, M. I. (1998). Sensorimotor adaptation in speech production. *Science, 279*, 1213–1216.
Houde, J. F., & Nagarajan, S. S. (2011). Speech production as state feedback control. *Frontiers in Human Neuroscience, 5*.
Humphries, C., Binder, J. R., Medler, D. A., & Liebenthal, E. (2006). Syntactic and semantic modulation of neural activity during auditory sentence comprehension. *Journal of Cognitive Neuroscience, 18* (4), 665–679.
Humphries, C., Love, T., Swinney, D., & Hickok, G. (2005). Response of anterior temporal cortex to syntactic and prosodic manipulations during sentence processing. *Human Brain Mapping, 26*, 128–138.
Humphries, C., Willard, K., Buchsbaum, B., & Hickok, G. (2001). Role of anterior temporal cortex in auditory sentence comprehension: An fMRI study. *Neuroreport, 12*, 1749–1752.
Indefrey, P., & Levelt, W. J. (2004). The spatial and temporal signatures of word production components. *Cognition, 92*(1–2), 101–144.
Jacquemot, C., Dupoux, E., & Bachoud-Levi, A. C. (2007). Breaking the mirror: Asymmetrical disconnection between the phonological input and output codes. *Cognitive Neuropsychology, 24*(1), 3–22.
Larson, C. R., Burnett, T. A., Bauer, J. J., Kiran, S., & Hain, T. C. (2001). Comparison of voice F0 responses to pitch-shift onset and offset conditions. *Journal of the Acoustical Society of America, 110*(6), 2845–2848.
Lau, E. F., Phillips, C., & Poeppel, D. (2008). A cortical network for semantics: (de)constructing the N400. *Nature Reviews Neuroscience, 9*(12), 920–933.
Levelt, W. J. M., Roelofs, A., & Meyer, A. S. (1999). A theory of lexical access in speech production. *Behavioral & Brain Sciences, 22*(1), 1–75.
Liebenthal, E., Binder, J. R., Spitzer, S. M., Possing, E. T., & Medler, D. A. (2005). Neural substrates of phonemic perception. *Cerebral Cortex, 15*(10), 1621–1631.
Luce, P. A., & Pisoni, D. B. (1998). Recognizing spoken words: The neighborhood activation model. *Ear and Hearing, 19*, 1–36.
Luria, A. R. (1970). *Traumatic aphasia. The Hague*. Mouton.
Marslen-Wilson, W. D. (1987). Functional parallelism in spoken word-recognition. *Cognition, 25*, 71–102.
Mazoyer, B. M., Tzourio, N., Frak, V., Syrota, A., Murayama, N., Levrier, O., et al. (1993). The cortical representation of speech. *Journal of Cognitive Neuroscience, 5*, 467–479.
McClelland, J. L., & Elman, J. L. (1986). The TRACE model of speech perception. *Cognitive Psychology, 18*, 1–86.
Miceli, G., Gainotti, G., Caltagirone, C., & Masullo, C. (1980). Some aspects of phonological impairment in aphasia. *Brain and Language, 11*, 159–169.
Miglioretti, D. L., & Boatman, D. (2003). Modeling variability in cortical representations of human complex sound perception. *Experimental Brain Research, 153*(3), 382–387.
Milner, A. D., & Goodale, M. A. (1995). *The visual brain in action*. Oxford: Oxford University Press.
Narain, C., Scott, S. K., Wise, R. J., Rosen, S., Leff, A., Iversen, S. D., et al. (2003). Defining a left-lateralized response specific to intelligible speech using fMRI. *Cerebral Cortex, 13*(12), 1362–1368.
Obleser, J., Eisner, F., & Kotz, S. A. (2008). Bilateral speech comprehension reflects differential sensitivity to spectral and temporal features. *The Journal of Neuroscience, 28*(32), 8116–8123.
Okada, K., & Hickok, G. (2006). Identification of lexical-phonological networks in the superior temporal sulcus using fMRI. *Neuroreport, 17*, 1293–1296.
Okada, K., Rong, F., Venezia, J., Matchin, W., Hsieh, I. H., Saberi, K., et al. (2010). Hierarchical organization of human auditory cortex: Evidence from acoustic invariance in the response to intelligible speech. *Cerebral Cortex, 20*(10), 2486–2495.
Pa, J., & Hickok, G. (2008). A parietal-temporal sensory-motor integration area for the human vocal tract: Evidence from an fMRI study of skilled musicians. *Neuropsychologia, 46*, 362–368.
Patterson, K., Nestor, P. J., & Rogers, T. T. (2007). Where do you know what you know? The representation of semantic knowledge in the human brain. *Nature Reviews Neuroscience, 8*(12), 976–987.
Poeppel, D. (2001). Pure word deafness and the bilateral processing of the speech code. *Cognitive Science, 25*, 679–693.

Poeppel, D. (2003). The analysis of speech in different temporal integration windows: Cerebral lateralization as "asymmetric sampling in time". *Speech Communication, 41*, 245–255.

Price, C. J. (2012). A review and synthesis of the first 20 years of PET and fMRI studies of heard speech, spoken language and reading. *Neuroimage, 62*(2), 816–847.

Price, C. J., Wise, R. J. S., Warburton, E. A., Moore, C. J., Howard, D., Patterson, K., et al. (1996). Hearing and saying: The functional neuro-anatomy of auditory word processing. *Brain, 119*, 919–931.

Purcell, D. W., & Munhall, K. G. (2006). Compensation following real-time manipulation of formants in isolated vowels. *Journal of the Acoustical Society of America, 119*(4), 2288–2297.

Rauschecker, J. P. (1998). Cortical processing of complex sounds. *Current Opinion in Neurobiology, 8*(4), 516–521.

Rauschecker, J. P. (2011). An expanded role for the dorsal auditory pathway in sensorimotor control and integration. *Hearing Research, 271*(1–2), 16–25.

Rauschecker, J. P., & Scott, S. K. (2009). Maps and streams in the auditory cortex: Nonhuman primates illuminate human speech processing. *Nature Neuroscience, 12*(6), 718–724.

Remez, R. E., Rubin, P. E., Pisoni, D. B., & Carrell, T. D. (1981). Speech perception without traditional speech cues. *Science, 212*, 947–950.

Rissman, J., Eliassen, J. C., & Blumstein, S. E. (2003). An event-related FMRI investigation of implicit semantic priming. *Journal of Cognitive Neuroscience, 15*(8), 1160–1175.

Rodd, J. M., Davis, M. H., & Johnsrude, I. S. (2005). The neural mechanisms of speech comprehension: fMRI studies of semantic ambiguity. *Cerebral Cortex, 15*, 1261–1269.

Rogalsky, C., & Hickok, G. (2011). The role of Broca's area in sentence comprehension. *Journal of Cognitive Neuroscience, 23*(7), 1664–1680.

Rogalsky, C., Love, T., Driscoll, D., Anderson, S. W., & Hickok, G. (2011). Are mirror neurons the basis of speech perception? Evidence from five cases with damage to the purported human mirror system. *Neurocase, 17*(2), 178–187.

Rogalsky, C., Pitz, E., Hillis, A. E., & Hickok, G. (2008). Auditory word comprehension impairment in acute stroke: Relative contribution of phonemic versus semantic factors. *Brain and Language, 107*(2), 167–169.

Schirmer, A., Fox, P. M., & Grandjean, D. (2012). On the spatial organization of sound processing in the human temporal lobe: A meta-analysis. *Neuroimage, 63*(1), 137–147.

Scott, S. K., Blank, C. C., Rosen, S., & Wise, R. J. S. (2000). Identification of a pathway for intelligible speech in the left temporal lobe. *Brain, 123*, 2400–2406.

Scott, S. K., & Wise, R. J. (2004). The functional neuroanatomy of prelexical processing in speech perception. *Cognition, 92*(1–2), 13–45.

Shannon, R. V., Zeng, F.-G., Kamath, V., Wygonski, J., & Ekelid, M. (1995). Speech recognition with primarily temporal cues. *Science, 270*, 303–304.

Spitsyna, G., Warren, J. E., Scott, S. K., Turkheimer, F. E., & Wise, R. J. (2006). Converging language streams in the human temporal lobe. *Journal of Neuroscience, 26*(28), 7328–7336.

Turkeltaub, P. E., & Coslett, H. B. (2010). Localization of sublexical speech perception components. *Brain and Language, 114*(1), 1–15.

Ungerleider, L. G., & Mishkin, M. (1982). Two cortical visual systems. In D. J. Ingle, M. A. Goodale, & R. J. W. Mansfield (Eds.), *Analysis of visual behavior* (pp. 549–586). Cambridge, MA: MIT Press.

Vaden, K. I., Jr., Muftuler, L. T., & Hickok, G. (2009). Phonological repetition-suppression in bilateral superior temporal sulci. *Neuroimage, 23*(10), 2665–2674.

Vandenberghe, R., Nobre, A. C., & Price, C. J. (2002). The response of left temporal cortex to sentences. *Journal of Cognitive Neuroscience, 14*(4), 550–560.

Wernicke, C. (1874). The symptom complex of aphasia: A psychological study on an anatomical basis. In R. S. Cohen, & M. W. Wartofsky (Eds.), *Boston studies in the philosophy of science* (pp. 34–97). Dordrecht: D. Reidel Publishing Company.

Wernicke, C. (1874). Der aphasische symptomencomplex: Eine psychologische studie auf anatomischer basis. In G. H. Eggert (Ed.), *Wernicke's works on aphasia: A sourcebook and review* (pp. 91–145). The Hague: Mouton.

Wise, R. J. S., Scott, S. K., Blank, S. C., Mummery, C. J., Murphy, K., & Warburton, E. A. (2001). Separate neural sub-systems within Wernicke's area. *Brain, 124*, 83–95.

Zatorre, R. J., Belin, P., & Penhune, V. B. (2002). Structure and function of auditory cortex: Music and speech. *Trends in Cognitive Sciences, 6*, 37–46.

CHAPTER

26

Brain Language Mechanisms Built on Action and Perception

Friedemann Pulvermüller[1,2] *and Luciano Fadiga*[3,4]

[1]Brain Language Laboratory, Department of Philosophy and Humanities, Freie Universtät Berlin, Berlin, Germany; [2]Berlin School of Mind and Brain, Humboldt-Universität zu Berlin, Berlin, Germany; [3]Department of Human Physiology, University of Ferrara, Ferrara, Italy; [4]Italian Institute of Technology, Genoa, Italy

26.1 INTRODUCTION

Brain language theory can build on well-established wisdom and principles known from neuroscience research. A huge body of knowledge is available about the structure and function of the brain and, more specifically, the human cortex, which is probably the most important device for language. Fundamental neuroscience knowledge can be systematized and summarized by general neurobiological principles. For example, an important *neuroanatomical principle* is that of rich local but more selective and specific long-distance connectivity. Neighboring neurons in the same cubic millimeter of cortex have a relatively high probability of being connected with each other, neuron pools in adjacent areas are still likely to exhibit a direct next-neighbor between-area projection, but areas far apart are linked in a highly specific and selective fashion (Braitenberg & Schüz, 1998; Young, Scannell, Burns, & Blakemore, 1994). A general *physiological principle* is that of Hebbian learning, whereby neurons that fire together wire together and strengthen their mutual connections and, vice versa, neurons out of sync delink (Hebb, 1949; O'Reilly, 1998). Such activity-dependent synaptic plasticity results in a structural mapping of functional relationships. Correlated activity leads to stronger connections, given the necessary connections are available in the first place. Correlation learning is one of the main neurobiological mechanisms believed to underlie learning, language learning included. Neuroanatomical connectivity sets the stage for such learning and mapping of correlations, whereby cortical connectivity structure not only enables such learning but also limits it. Where there are no connections in the first place, no learning can occur.

Likewise, the available structural and functional knowledge about specific features of the human brain is very detailed. Figure 26.1 shows a recently discovered specific feature of human cortical anatomy, which is of special relevance to language: the dorsal fiber bundle connecting temporal and inferior-frontal cortex called the *arcuate fascicle*. This bundle is not available in monkeys, weakly developed in apes, and strong only in humans (Figure 26.1A–C). Because it is also strongly lateralized to the left hemisphere—the language-dominant hemisphere in most of us—it appears as a prime candidate for specifically human language mechanisms (Rilling, Glasser, Jbabdi, Andersson, & Preuss, 2011; Rilling et al., 2008).

Taking into account neuroscience principles and knowledge—including the correlation learning principle and detailed information about, in part, genetically determined cortical neuroanatomy—it becomes possible to deduce from them those mechanisms that underlie language processing and representation. The difference between this theory-driven neuroscience-grounded perspective and approaches that remain at the level of boxes and arrows and abstract labels for cortical areas (Hickok & Poeppel, 2007) is two-fold. First, there is the neuromechanistic aspect: mechanisms are being specified in terms of neuronal assemblies and connections, because the level of boxes is obviously not concrete enough and that of areas is too crude. Second, there is the *explanatory aspect*: explanations are offered regarding how the specified mechanisms come about. In our opinion, the progression from abstract descriptive to explanatory-neuromechanistic theories is essential in current brain language science.

Neurobiology of Language. DOI: http://dx.doi.org/10.1016/B978-0-12-407794-2.00026-2

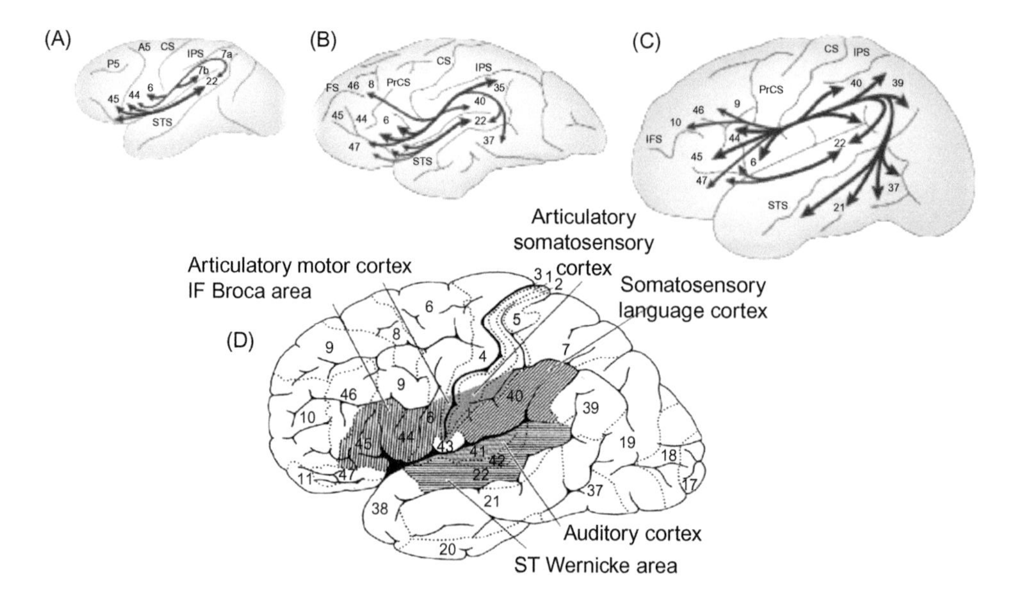

FIGURE 26.1 Major long-distance cortico-cortical connections in (A) macaques, (B) chimpanzees, and (C) humans (from Rilling et al., 2008 with permission from the publisher). Note the strong dorsal connection from inferior-frontal, precentral, and central sulcus (IFS, PrCS, SC) regions to areas around the inferior-parietal and superior-temporal sulcus (IPS, STS) of the arcuate fascicle, which is only present in humans. (D) Language core areas of the left cortical hemisphere surrounding the sylvian fissure (hatched). Brodmann area numbers are indicated. *Adapted from Braitenberg & Pulvermüller, 1992 with permission from the publisher.*

We briefly summarize an explanatory-neuromechanistic approach to language, progressing from the levels of speech sounds and signs to those of meaning and constructions, and on to communicative actions and action sequences. We highlight some issues of current debate in which predictions immanent to the explanatory-neuromechanistic approach diverge from those of most established brain language models, and then close with a summary and outlook.

26.2 PHONEMES

Phonemes, the smallest units of speech that distinguish between meaningful spoken words, are a characteristic ingredient of spoken human languages. The world's languages include approximately 100 overall, and each language has approximately 50 of them. In comparison, apes only have a few oral gestures (Call & Tomasello, 2007). Why is this so? The normal learning of phonemes requires that speech sounds be produced, which babies start doing at approximately 6 months and throughout the so-called *babbling phase* until the end of their first year of life (Locke, 1993). Babbling consists of productions of meaningless syllable sequences. Brain mechanisms implicated by such activity includes nerve cell activity in motor regions, where the articulatory gestures are initiated and controlled, and activity in the auditory and somatosensory cortex, because the infants also perceive the tactile self-stimulations produced by articulating along with the self-produced sounds. This means coactivation of nerve cells (or neurons) across a range of cortical areas close to the sylvian fissure (*perisylvian areas:* articulatory motor and somatosensory cortex, auditory cortex; see Figure 26.1D). If there were direct connections between the specific neurons involved in controlling articulatory movements and responding to sensory aspects of the sounds produced, then the neuroscience principle of correlation learning would imply that the distributed set of neurons strengthened these links (Fry, 1966). However, it seems that the frontotemporal connections available in the left hemisphere of humans do not strongly interlink these areas directly, but rather primarily connect areas *adjacent to* motor and auditory cortices—in inferior premotor and prefrontal areas and in superior and middle temporal cortex (Figure 26.1C; Braitenberg & Schüz, 1992). This means that the correlated neuronal activity in sensory and motor cortex can only be mapped indirectly, including neurons in "higher" zones adjacent to relevant primary areas (other hatched areas in Figure 26.1D). The connection structure of inferior-frontal and superior-temporal areas, sometimes called the *language areas of Broca and Wernicke*, is illustrated schematically in Figure 26.2A, where these larger areas are each further subdivided into premotor and prefrontal areas and auditory belt and parabelt areas, respectively. The corresponding connection diagram (Figure 26.2B) illustrates the available between-area connections, including long-distance connections through the arcuate fascicle, and points to the crucial role of the language areas as connection hubs

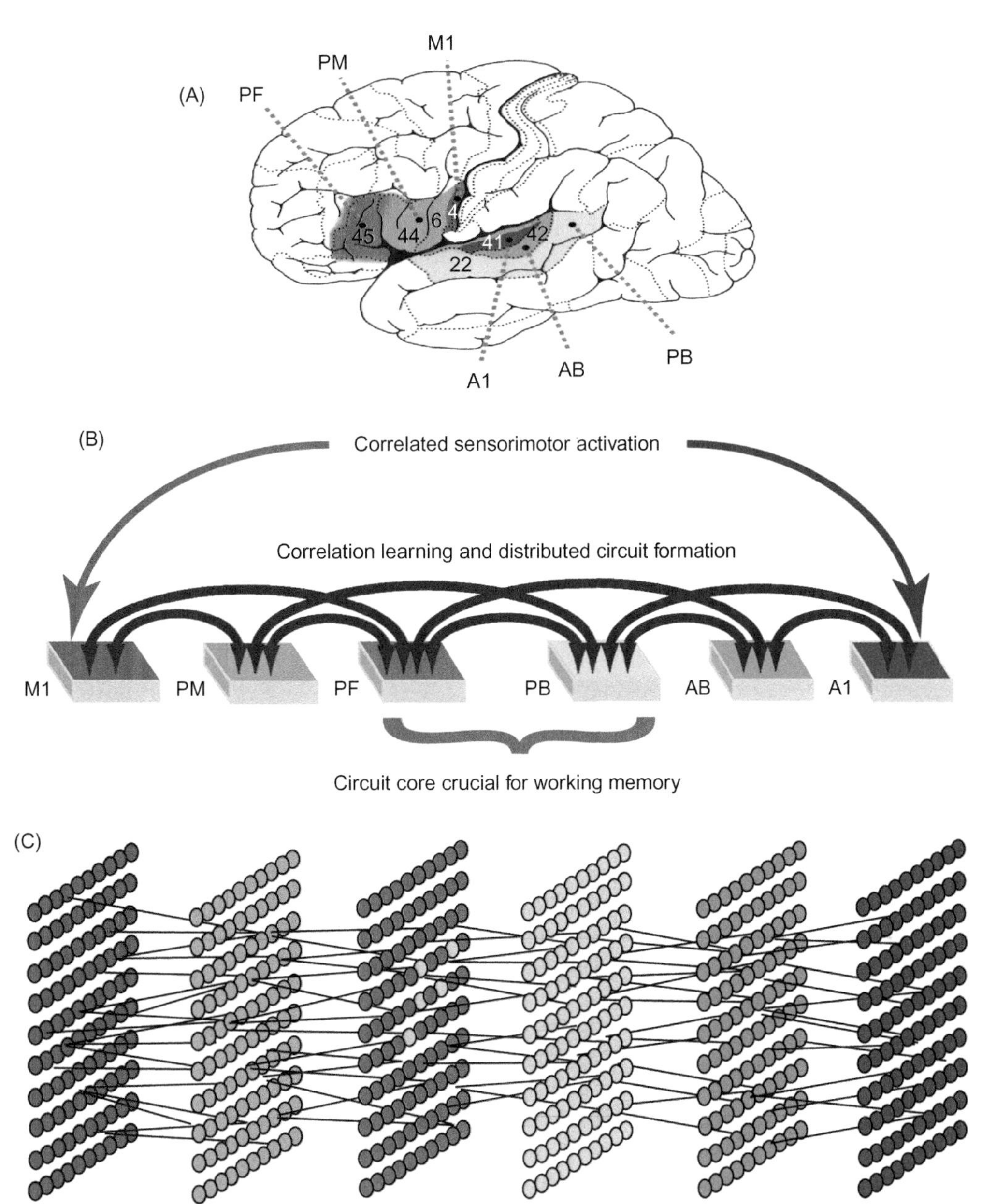

FIGURE 26.2 Formation of action perception circuits in the language network. (A) Neuroanatomical subdivision of inferior-frontal and superior-temporal cortex into six areas: M1, primary motor; PM, premotor; PF, prefrontal; A1, primary auditory; AB, auditory belt; and PB, auditory parabelt areas. (B) Schematic connection structure of the six areas highlighted. Correlated activation in M1 and A1 brought about by articulation leads to spreading activation in the network and distributed circuit formation for syllables and words. Their richer connectivity determines that PF and PB perisylvian hub areas develop circuit cores, where word form circuits link with each other in combinatorial learning. (C) Sketch of an *action perception circuit*, a distributed neuronal assembly driven by action perception correlation (indirect between-area connections are omitted; *adapted from Garagnani & Pulvermüller, 2013; Pulvermüller, 2013*).

within the language cortex (Garagnani & Pulvermüller, 2013; Garagnani, Wennekers, & Pulvermüller, 2008). Antonio Damasio called such hub areas *convergence zones* (Damasio, 1989).

Within this circuit structure, the arcuate fascicle may provide a prerequisite for building the more elaborate phoneme repertoire available to humans, but not to apes or monkeys. Together with the extreme capsule, which seems equally developed in apes and humans, the arcuate provides a powerful connection between a range of inferior-frontal (including prefrontal and premotor) and superior-temporal (including auditory belt and parabelt) areas (Rilling, 2014). The availability of a powerful data highway between articulation and auditory perception and the resultant more elaborate repertoire of articulatory gestures may have constituted a significant selection advantage for humans in their phylogenetic development.

26.3 SIGNS

Much more than the slim repertoire of approximately 50 phonemes is the huge vocabulary of tens to hundreds of thousands of words as a hallmark of language. Steven Pinker estimates the size of the vocabulary of a typical language to be approximately 40,000 words (Pinker, 1994), and some dictionaries (e.g., of Chinese) even list up to one million word forms. In contrast, the number of gestures in the repertoire of great apes appears to be limited to between 40 and approximately 100, verbal gestures included (Call & Tomasello, 2007; Genty, Breuer, Hobaiter, & Byrne, 2009; Hobaiter & Byrne, 2011). This corresponds to the small size of the phoneme repertoire of humans, who apparently manage to combine their phonemes sequentially to yield an extraordinarily rich *vocabulary* of signs stored in a *lexicon*. How can this most dramatic increase be explained? It seems straightforward that rich connectivity is required for storing such an immense vocabulary. Phonemes or, perhaps more likely, context-sensitive phoneme variants depending on adjacent phonemes within a syllable and word need to be combined with each other. When learning and articulating novel spoken word forms, correlated motor and auditory information needs to be mapped to yield knowledge about the articulatory-phonological and acoustic-phonological aspects of phonemes, syllables, and whole spoken words and constructions. For such mapping of numerous forms, a powerful connection pathway may be required. Here, once again, the arcuate fascicle appears essential.

The formation of action perception circuits for words can be studied at the theoretical level using neurocomputational models that imitate aspects of the relevant cortical structures along with neurofunctional principles (Garagnani et al., 2008; Westermann & Reck Miranda, 2004). These simulations bolster, illustrate, and further elaborate the proposed explanation of the linkage of inferior-frontal and superior-temporal neuron populations underpinning spoken word and phoneme processing (Figure 26.2). When action perception circuits form, the neurons that become incorporated into them—which may have been either sensory or motor neurons, or may not have had any role in neither production nor perception—acquire a role in both production and recognition of one or more spoken word(s) and are active during both types of processes. Giacomo Rizzolatti and his team called neurons responding during execution and perception *mirror neurons* (Rizzolatti & Craighero, 2004; Rizzolatti, Fadiga, Gallese, & Fogassi, 1996). The activation of inferior-frontal areas homologue to macaque premotor area F5, where mirror neurons are present, is involved in speech processing (Fadiga, Craighero, Buccino, & Rizzolatti, 2002; Pulvermüller et al., 2006; Watkins, Strafella, & Paus, 2003), and specific stimulation of motor representations of the articulators—for example, the tongue and lips—has a causal effect on the perceptual classification of speech sounds (D'Ausilio et al., 2009) and on the trajectories of articulatory movements (D'Ausilio et al., 2014). Thus, the perceptual language machinery of the human brain includes articulatory motor mechanisms as crucial elements connecting articulatory and acoustic aspects of phonemes and spoken word forms in the very same way as mirror neurons in monkeys' motor systems map actions, visual perception of that action, and sometimes even the sound of that action (Kohler et al., 2002).

If vocabularies are built by action perception circuit formation, then this should be physiologically manifest. Vocabulary learning involving word articulations and repetitions—as it frequently happens in first language acquisition—leads to the formation of neuronal circuits integrating action and perception information and distributed over the frontal and temporal areas of the perisylvian language cortex. Because the frontotemporal connections of the arcuate fascicle are especially rich in the left hemisphere, left-laterality due to strong left-frontal involvement is predicted. But note that speech can also be picked up passively, for example, when the learner hears a foreign language but does not engage in speaking and communicating (Saffran, Aslin, & Newport, 1996). Such a situation is also enforced in patients with articulatory motor deficits (Bishop, Brown, & Robson, 1990). For such purely perceptual learning, no involvement of frontal or motor circuits would be predicted. A recent study compared the brain correlates induced by passive perceptual and active articulatory learning of novel syllable combinations (Pulvermüller, Kiff, & Shtyrov, 2012). The results showed that perceptual novel word learning led to an activation enhancement in bilateral superior-temporal areas, whereas the brain correlates of articulatory learning were left-lateralized and especially prominent in inferior prefrontal and premotor cortex (Figure 26.3). Because passive presentation of novel spoken words elicited these different activation patterns depending on learning type, they provide support for the action perception model of vocabulary formation.

26.4 MEANING

Especially exciting, interesting, and rich, but equally controversial, in the neuroscience of language is the brain basis of meaning. Already the best definition of what meaning might be is under hot dispute; when it comes to the cortical meaning locus, opinions diverge dramatically, unlike in any other area of the cognitive and brain sciences. One fraction of respectable researchers believes that a unitary meaning center, which Karalyn Patterson

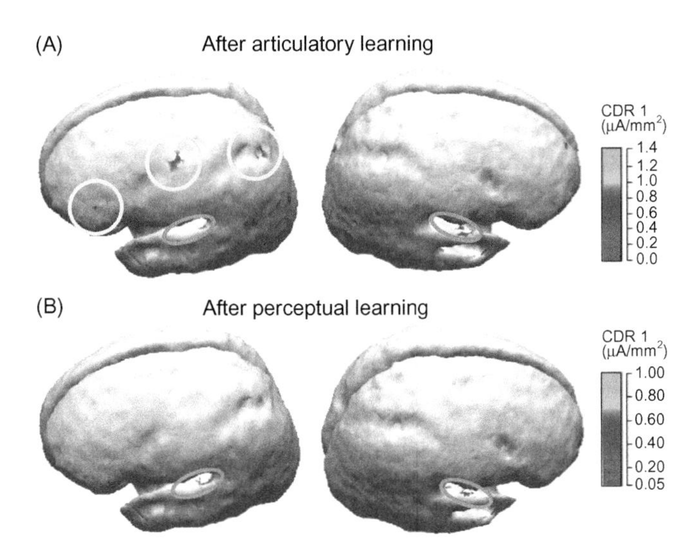

FIGURE 26.3 Enhanced cortical activity (sources of event-related potentials) elicited by novel spoken word forms that had been learned in an articulatory condition, by hearing and repeating the novel items (A), or in a perceptual learning paradigm by listening to repeated presentations of the same novel spoken word forms (B). Enhanced bilaterally symmetric superior-temporal activity to learned novel items (orange ovals, relative to before learning) is seen independently of the learning method. However, additional enhanced, fast, automatic, left-lateralized premotor, inferior-frontal, and inferior-parietal activity to spoken novel words (yellow ovals) is present after articulatory, but not after perceptual, learning (*adapted from Pulvermüller, Kiff, et al., 2012*). These results suggest that language laterality is driven by auditory-motor information linking, possibly by way of the left-lateralized arcuate fascicle.

and her colleagues once dubbed the *semantic hub*, carries the main burden of meaning processing (Patterson, Nestor, & Rogers, 2007). Such a hub might serve the role of binding together different features that comprise a concept or meaning and, in addition, might bind symbolic semantics to symbol forms, for example, to spoken or written word forms. Unfortunately, opinions widely diverge regarding where such a hub might be located. Some theories saw a special role of the right hemisphere here (Paivio, 1991), whereas many cognitive neuroscientists consider specific left-hemispheric areas, in inferior-frontal (Bookheimer, 2002), inferior-parietal (Binder & Desai, 2011), posterior-middle-temporal (Hickok & Poeppel, 2007), or anterior-temporal regions (Patterson et al., 2007) as the main semantic hub and center for interfacing knowledge about signs with knowledge about their meaning. Because each of these positions receives support from experimental evidence that the respective area becomes strongly active during semantic processing and/or that a lesion or functional change in the respective site affects the performance on general semantic tasks but none of them can be considered a unique hub, it appears best to speak about a *hub territory* in the periphery of and surrounding the perisylvian language cortex (Figure 26.4A; Pulvermüller, 2013). The hub territory includes frontal, parietal, and temporal association (convergence) areas, which are heavily connected to several modality-preferential (e.g., visual, auditory, motor) areas. Therefore, they serve as hubs in a neuroanatomical sense

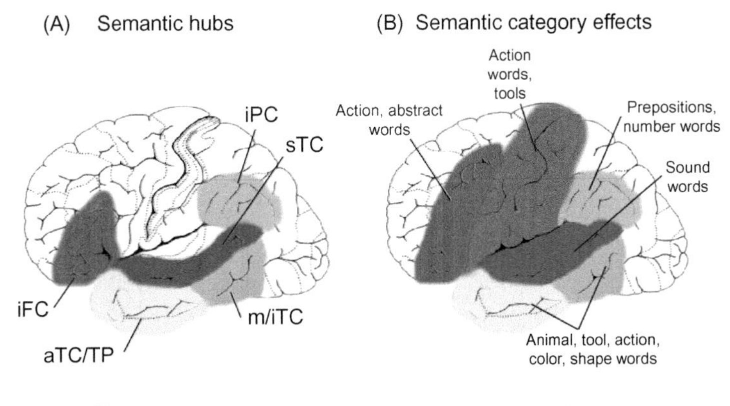

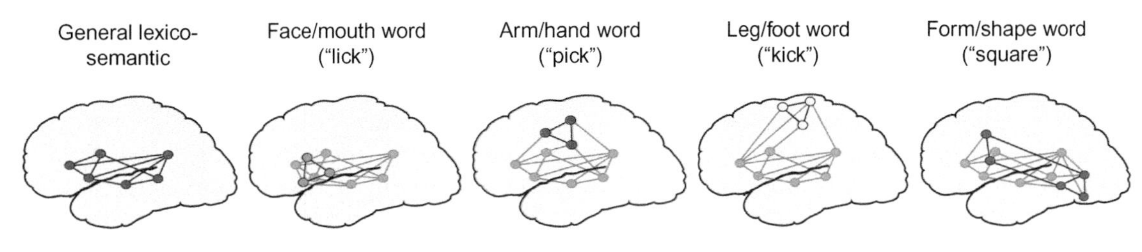

FIGURE 26.4 Cortical areas and circuits for semantics. (A) Proposed "semantic hubs" for general semantic processing. Abbreviations: iFC, inferior-frontal cortex; iPC, inferior-parietal cortex; sTC, superior-temporal cortex; m/iTC, middle/inferior-temporal cortex; aTC, anterior-temporal cortex; TP, temporal pole. (B) Cortical areas where semantic category-specific effects were reported in the literature. (C) Neuromechanistic model of action perception circuits for general lexicosemantic (leftmost graph) and category-specific referential-semantic information processing. Small circles represent local clusters of neurons and lines represent strong bidirectional connections between local clusters. *Adapted from Pulvermüller, 2013.*

(van den Heuvel & Sporns, 2013), which implies that they can also bridge between language areas and regions where modality-specific aspects of semantics are being processed.

Contrasting with the idea that there is one area—or entire widespread territory—that processes all meanings to the same degree, a different approach to meaning in the brain emphasizes the importance of *category-specific semantic mechanisms*. Interestingly, patients with focal cortical lesions frequently have deficits that primarily affect one or more semantic categories (e.g., animals or actions), and related neuroimaging studies show corresponding category-specific local brain activation. For example, the inferior-frontal cortex, including motor and premotor cortex, seems particularly important for processing action words and concepts (Pulvermüller, 2005), whereas part of the inferior-temporal lobe seems particularly relevant for animal words and concepts (Martin, 2007). Category-specific semantic deficits and brain activations are frequently present in and close to areas also carrying a role in action and perception, including fronto-central motor and premotor, superior-temporal auditory, inferior-temporal visual-object, and inferior-parietal visual-spatial processing areas, some of which are close to and overlap with the hub territory.

Why should specific brain parts contribute to meaning? Brain theory of meaning explains aspects of semantic mechanisms by neuroscience principles, including the correlation learning principle and specific features of between-area connectivity. When learning that a word such as "sun" relates to an object, the infant may experience such objects while the word is used at the same time. Therefore, neuronal populations for processing visual-object information and populations processing word forms coactivate and, according to the correlation learning rule, join together into a higher-order cell assembly linking the information about the word form with that of aspects of its meaning. Relevant areas where these semantic circuits are housed—or, more appropriately, *scattered*—include visual areas and visual-object processing areas in temporo-occipital and inferior-temporal cortex and language areas in perisylvian cortex (Farah & McClelland, 1991; McCarthy & Warrington, 1988; Pulvermüller, 1999). If words semantically relate to information from other modalities, for example, gustatory or auditory, then neuron population in the respective sensory areas plus interlinking connection hubs will come into play. For emotion words, motor circuits along with cortical and subcortical emotional-affective circuits of the limbic system seem relevant (Moseley, Carota, Hauk, Mohr, & Pulvermüller, 2012; Vigliocco et al., 2013). Words related to actions—for example "grasp"—may be heard when the learner performs related actions so that, in this case, the correlation learning principle predicts distribution of semantic circuits over motor and language areas, in addition to interlinking areas. Note that, from the point of view of this cell assembly model of meaning, both the category-specific nature of semantic processes in modality-specific areas as well as the involvement of adjacent association cortices (convergence zones, connection hubs) are explained based on neuronal correlation and neuroanatomical connection structure (Pulvermüller, 2013).

A hot debate in the cognitive neuroscience of language still persists about the relevance of sensorimotor areas for meaning processing. Hub-centered models predict that meaning binding and integration are functions of the hub(s), with sensorimotor areas not essential or necessary for meaning (Bedny & Caramazza, 2011). Critical tests are sometimes difficult, not least because, first, hub areas and areas of category-specific semantic processing are, in several cases, located side-by-side, even with substantial overlap (Figure 26.4A and B), although there are clear dissociations. Second, the areas where category-specific brain activation was found, in many cases, were not exactly *in* those areas where critical sensory information was processed, but rather *adjacent* to them (e.g., Martin, Haxby, Lalonde, Wiggs, & Ungerleider, 1995). Third, many studies of category specificity are subject to confounds—for example, when action verbs and object nouns were compared and features different from semantics (word frequency, lexical status, combinatorial features, etc.) may have influenced the patterns of activation or deficits observed (Bird, Lambon-Ralph, Patterson, & Hodges, 2000). Therefore, more recent studies addressed the role of sensory and motor areas in semantic processing.

Key results supporting semantic category specificity came from investigations of action words typically used to speak about actions performed with different body parts, for example, "lick" (mouth/face), "pick" (hand/arm), and "kick" (foot/leg). Passive perception (listening, reading) of these words, as well as action-related phrases and sentences, led to activation of motor areas normally involved in performing (and/or perceiving) actions with the respective body parts (Aziz-Zadeh, Wilson, Rizzolatti, & Iacoboni, 2006; Hauk, Johnsrude, & Pulvermüller, 2004; Pulvermüller, Cook, & Hauk, 2012; Tettamanti et al., 2005). Task contexts influence these motor activations, although a degree of motor activation sometimes seems present when subjects are actively distracted from processing linguistic input (Pulvermüller, Shtyrov, & Ilmoniemi, 2005). The activations appear as early as the earliest brain indicators of meaning processing observed so far (Moseley, Pulvermüller, & Shtyrov, 2013). Furthermore, functional changes in motor systems, which can be induced by magnetic or electrical stimulation or just behaviorally, led to differential causal influences on the processing of action word categories (Liuzzi et al., 2010; Pulvermüller, Hauk, Nikulin, & Ilmoniemi, 2005; Shebani & Pulvermüller, 2013; Willems, Labruna, D'Esposito, Ivry, & Casasanto, 2011). Lesions primarily affecting the motor system impair the processing of action meanings and action words (Bak, 2013; Kemmerer, Rudrauf, Manzel, & Tranel, 2012), although body part–specificity of functional degradation could so far be documented in healthy subjects (Shebani & Pulvermüller, 2013) but not in stroke patients (Arevalo, Baldo, & Dronkers, 2012). These results show that the motor system plays a causal and crucial role in semantic processing. Analogous data demonstrate the semantic relevance of sensory systems, including modality-specific areas for olfactory, gustatory, auditory, somatosensory, visual-object, and visual-spatial processing (Barrós-Loscertales et al., 2012; González et al., 2006; Kiefer, Sim, Herrnberger, Grothe, & Hoenig, 2008; Pulvermüller & Hauk, 2006; Simmons et al., 2007).

In essence, the seemingly diverging hub and category specificity positions can be integrated by viewing semantic processing as a function of neuronal circuits spread out over a range of areas whose formation and distribution are driven by neuronal correlation learning and neuroanatomical connectivity (Binder & Desai, 2011; Kiefer & Pulvermüller, 2012; Pulvermüller, 2013). Because semantic learning requires grounding and, thus, correlated information in the senses, category-specific semantic areas emerge in category-preferential areas, sometimes reaching out into sensorimotor cortex far away from core perisylvian language cortex. Because many of the connections bridging between modality-specific areas run through convergence zones in modality-unspecific association cortex, the hub territory in the periphery of and surrounding the perisylvian language cortex comes into play for all types of semantic processing.

26.5 COMBINATIONS AND CONSTRUCTIONS

Not all meaning is being picked up directly from perceptions and actions semantically related to signs. The meaning of most words is actually learned from context (Kintsch, 2008). After a stock of words has been semantically grounded in actions and perception knowledge, further semantic learning can be based on language contexts, where novel word forms can bind with semantic features of co-occurring familiar words. Note that in this case correlations are also essential, particularly the correlation structure of word forms co-occurring with other word forms in constructions and larger pieces of discourse. The brain mechanism for such "parasitic" semantic learning, or "symbolic theft" (Cangelosi, Greco, & Harnad, 2002), may be correlated activity of already established semantic circuits and newly forming circuits for novel word forms. If word meaning can be picked up indirectly from other signs, then this implies that semantic learning is not bound to sensorimotor constellations available in the world; potentially, new combinations of semantic feature combinations and, hence, concepts can arise.

Abstract words, such as "freedom" or "beauty," may be related to a range of variable contexts and equally variable constellations of actions, objects, and knowledge, so that it is more difficult to pick up their meaning from contextual or sensorimotor correlation. Because a sunset, face, flower, and piece of art may all be called "beautiful" based on sometimes radically different visual features and linguistic contexts in which they occur, the correlation learning principle does not predict massive strengthening of sensorimotor and linguistic circuit connections. Still, it is possible that, by way of convergence zones, weak links are being built between signs and the variable instantiations of their abstract meaning in sensorimotor systems (Pulvermüller, 2013). Neuroimaging evidence regarding the brain basis of abstract meaning is heterogeneous, but some studies indicate an involvement of association areas, for example, in dorsolateral prefrontal cortex (Binder, Westbury, McKiernan, Possing, & Medler, 2005).

Combinatorial semantic learning is particularly important when whole constructions acquire meanings unrelated to the semantics of their constituent words, for example, as in the case of idiomatic phrases such as "grasping ... ideas" or "kicking ... habits," where the idiom may denote cognitive processes unrelated to the motor acts of grasping or kicking (Gibbs & O'Brien, 1990). In this case, over and above the single words included, the whole phrase or construction may acquire a meaning of its own, which may be of an abstract nature. Again, some sensorimotor and combinatorial information may be relevant for such abstract idiomatic meaning, although the relevant contexts, both in terms of actions and perceptions and in terms of context words and constructions, are quite variable. As a consequence, the abstract meaning of idioms (as that of abstract words) may be mapped in convergence hubs, including dorsolateral prefrontal cortex and anterior-inferior temporal cortex. Neuroimaging studies of abstract idiom processing indeed confirm these hub area activations to idioms but not to matched literal constructions ("Berta kicked the habit" vs. "... the ball") (Boulenger, Hauk, & Pulvermüller, 2009; Lauro, Tettamanti, Cappa, & Papagno, 2008). Interestingly, motor systems activation reflecting the meaning of action words included in the constructions was seen for both literal and idiomatic phrases at a point in time when sentence meaning was processed (Boulenger et al., 2009; Desai, Conant, Binder, Park, & Seidenberg, 2013). This can be interpreted as an indication that, in idiomatic construction understanding, semantic information regarding the whole construction and the constituent words (here: action words) is being retrieved and, potentially, integrated. A study of the time course of semantic processing using magnetoencephalography showed that brain indexes of abstract idiom-related information in hub areas were manifest immediately and simultaneously with those of action word–related semantic information processing in sensorimotor systems (Boulenger, Shtyrov, & Pulvermüller, 2012). These neurophysiological results are consistent with the view that compositional and whole-construction–related semantic information are being processed simultaneously and interactively in idiom comprehension.

Some combinatorial aspects of language are captured by syntax, which describes abstract combinatorial rules for conjoining the members of large vocabulary groups (lexical categories) into phrases and increasingly complex sentences. In contrast to views dominating linguistics for many years, according to which most syntactic knowledge is genetically predetermined and learning only plays a modulatory role in language acquisition (Chomsky, 1980), neurocomputational investigations indicate that much combinatorial knowledge relevant to syntax can be picked up from the correlation structure of words in constructions (Chater & Manning, 2006; Elman et al., 1996). A further belief dominating linguistics for decades has been called into question by recent evidence, the claim that syntax and semantics belong to different cognitive systems that are informationally encapsulated toward each other. In contrast to this claim, neurocomputational simulation of combinatorial learning of constructions shows that syntactic mappings are colored by meaning so that, for example, nouns and verbs are being subcategorized into semantic subtypes (Elman, 1990). This conforms with postulates immanent to cognitive linguistics and the construction grammar framework, where syntax and semantics are viewed as intrinsically linked to each other (Goldberg, 2006). One further claim held by linguists has inspired much recent neurocomputational research, the claim that combinatorial mechanisms are discrete, that is, that they exhibit an all-or-none character rather than being probabilistic in nature. Whereas some researchers have strongly argued against this position (Elman et al., 1996), recent neurophysiological work and neurocomputational work seem to support the discreteness position (Pulvermüller, Cappelle, & Shtyrov, 2013; Pulvermüller & Knoblauch, 2009).

A close functional link between linguistic combinatorial knowledge and general action knowledge receives support from studies of patients with brain lesions usually causing grammar processing problems. These patients have been reported to exhibit difficulty in processing combinatorial information about general actions, further supporting the idea that language and action systems are tightly interwoven (Fazio et al., 2009). Although the grammar of sentences and that of complex human actions show structural similarities setting apart human combinatorial brain systems from those available to apes and monkeys (Pulvermüller & Fadiga, 2010), the discussion about precise analogies and differences between combinatorial systems of syntax and action is ongoing (Moro, 2014; Pulvermüller, 2014).

26.6 SPEECH ACTS AND SOCIAL-COMMUNICATIVE INTERACTION

Brain research on language so far has mostly dealt with linguistic structures, that is, words and sentences, without considering language use and communication as a social form of interaction. A word such as "apple" can be used as a tool to achieve different goals, for example, the goal to NAME an object in the context of a language exercise or, alternatively, the goal to REQUEST and obtain an apple. It is obvious that the same word or utterance can be a tool to reach different goals and that the utterance is therefore being linked with context-specific sets of assumptions and intentions (e.g., Alston, 1964 #11939; Fritz & Hundsnurscher, 1994 #396; Ehlich, 2007 #11562; Stalnaker, 2002 #12042) as follows: in the NAMING context with the expectation that the label can function as a tool to refer to the relevant object, but in the REQUEST context with a range of additional social-interactive expectations that the other party will hand over the object, that such an object is available to the other party, and that he or she is, in principle, willing to pass it. In close relationship to these different expectations, assumptions, intentions, and goals, the different speech acts in the present example NAMING and REQUEST are linked into different typical *action sequence schemas* (Figure 26.5A). When communicating successfully, interacting partners may have the same sequence schema representations active in their brains, together with the action perception circuits for the utterances with which the speech acts are performed. The context-dependent activation of different communicative sequence schemas by the same utterance has been shown to be manifest in brain activation. The action-heavy knowledge about the different sequences opened up by a REQUEST is manifest in activation of the motor and premotor cortex, whereas the emphasis of language-object links in NAMING draws on inferior-temporal areas. The different action sequence schemas characterizing specific social-communicative actions performed

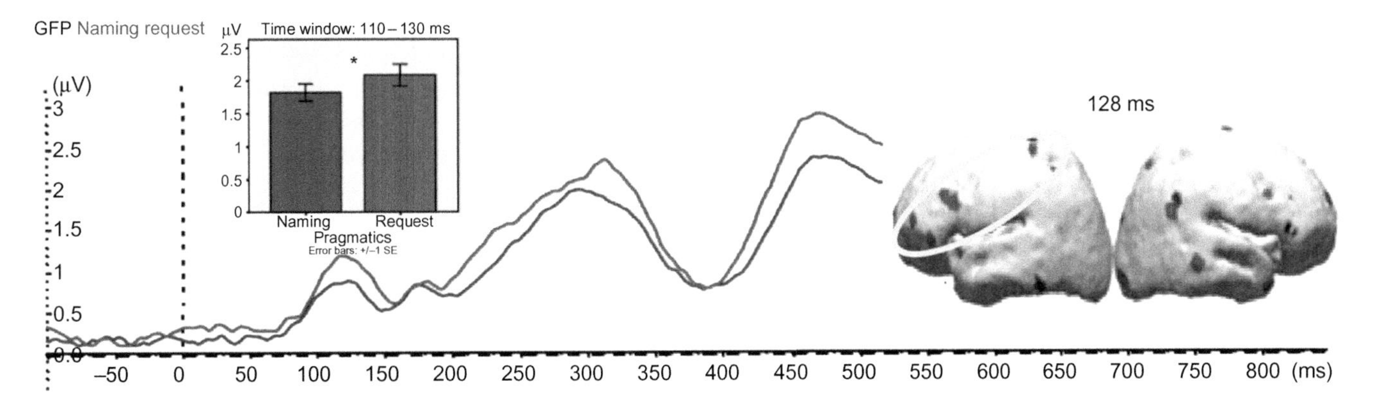

FIGURE 26.5 Brain basis of social-communicative speech acts. Results of an EEG experiment on understanding of NAMING and REQUEST actions performed by use of the same words. Compared with NAMING (in blue), stronger brain activation was seen in the REQUEST context (in red) starting already at 100–150 ms after critical (written) word onset. Plotted is global field power (GFP) calculated from multiple electrodes. Sources of this enhanced activity to REQUESTs (L1 minimum norm current estimates, see inset) were primarily localized in left fronto-central and right parieto-temporal cortex (*adapted from Egorova et al., 2013 with permission from the publisher*). These additional activations may reflect the processing of knowledge about interaction sequences, intentions, and action goals relevant to understanding REQUESTs.

with the same linguistic structures therefore have a brain correlate in speech act–specific cortical activations. These emerge rapidly (100–200 ms), thus demonstrating an almost instantaneous access to linguistic-pragmatic knowledge (Egorova, Pulvermüller, & Shtyrov, 2014; Egorova, Shtyrov, & Pulvermüller, 2013).

26.7 OUTLOOK: KEY ISSUES IN BRAIN LANGUAGE RESEARCH

The new field of the cognitive neuroscience of language is rapidly advancing toward a better understanding of the brain basis of a most remarkable human faculty. We comment on key issues in this progress, and future research will be particularly important.

26.7.1 Modeling Real Circuits

During the decade of the brain (2000–2010), a main effort addressed the role of different cortical areas (and other brain parts) in cognitive processing, language processing included. As a result, the box and arrow diagrams were, in many cases, supplied with cortical area labels (e.g., Hickok & Poeppel, 2007). However, a main insight from computational neuroscience is that neuronal processes can be shared by distributed neuronal assemblies so that it would be inaccurate to label one area with one function—because the function is, in fact, distributed over many areas—and each area could, in principle, carry a multitude of neurons belonging to different neuronal circuits with different functions. Therefore, it appears more appropriate to consider *distributed neuronal assemblies* (DNAs) instead of areas or whole systems as the carriers of functions, and thus to specify the *cortical topographies of functions*. The example of meaning not situated in one local hub but rather distributed over a set of areas that reflect relevant category-specific information has been discussed in detail. It seems an important step forward to replace the labeled box and arrow diagrams with brain topography maps of linguistic and cognitive functions and to make definite these topographic models in realistic neurocomputational simulations that mimic specific aspects of brain anatomy and physiology (Pulvermüller, Garagnani, & Wennekers, 2014).

26.7.2 Motor Involvement in Speech Perception and Comprehension

Some models consider primary and higher auditory cortices in superior-temporal cortex as the cortical locus of speech perception (Poeppel, Emmorey, Hickok, & Pylkkanen, 2012; Rauschecker & Scott, 2009). According to this perspective, frontal cortex, motor systems included, may play a role in predicting likely future perceptions in case of noisy environments or under high attentional demands. However, when speech is perceived passively, such models would not predict any inferior-frontal activation. Contrasting with this prediction, experimental neurophysiological studies showed that speech stimuli activate superior-temporal and inferior-frontal cortex activation in close temporal succession, even when subjects were actively distracted from language processing. Recent results suggest that motor system activation in speech processing and frontotemporal connectivity through the arcuate fascicle predict verbal short-term memory and word learning performance (Lopez-Barroso et al., 2013; Szenkovits, Peelle, Norris, & Davis, 2012), thus pointing to further features of their functional relevance.

The causal influence of motor and interior-frontal activation on speech perception is one more fact that makes it difficult to maintain the idea that, under normal conditions, temporal cortex processes speech without feedback from frontal areas. As mentioned, local stimulation of tongue and lip motor cortex biases perceptual classification of speech sounds (D'Ausilio et al., 2009; Möttönen, Dutton, & Watkins, 2013; Möttönen & Watkins, 2009). Although it has been argued that this behavioral change may be caused at the decision stage, leading to a decision bias rather than a perceptual change (Venezia, Saberi, Chubb, & Hickok, 2012), most recent evidence provides further support for a genuine causal effect of motor cortex activation on phoneme perception (Möttönen et al., 2013) and speech comprehension (Schomers et al., 2014). Data from patients still require further attention because some authors report phoneme classification deficits with inferior-frontal lesions (Caplan, Gow, & Makris, 1995), but others failed to replicate these (Rogalsky, Love, Driscoll, Anderson, & Hickok, 2011). A possible integration of these diverging results may draw on right inferior-frontal areas, which may contribute to different degrees, depending on the degree of laterality of the arcuate fascicle in specific individuals. Another most important issue is the involvement of inferior-frontal and motor systems in single word comprehension. Lesions in these key regions cause deficits in understanding single words under difficult perceptual conditions and, for clearly pronounced words, a significant processing delay has been documented (Mirman, Yee, Blumstein, & Magnuson, 2011; Moineau, Dronkers, & Bates, 2005). However, discussions are still ongoing about the status of these findings for brain language theory.

26.7.3 Localizing Semantics

As discussed in the semantics section, a broad area surrounding the sylvian fissure contributes to general semantic processing, and even more widely distributed networks are involved in category-specific semantics (Figure 26.4). Still, it may be asked whether the perisylvian cortex's *semantic rim* is indeed best conceived as a hub processing all semantics to the same degree and in a similar manner. Would these areas also show a degree of category specificity, as it has previously been shown for anterior-temporal, middle temporal, angular, and inferior-frontal cortex? Similar questions arise for areas contributing to category-specific semantics. Which are the specific facets of semantic knowledge that get lost if motor and other modality-preferential areas are lesioned? For which semantic categories would motor systems be important? Apart from genuine action words, a recent suggestion is that they are also crucial for abstract emotion words, idioms, and other abstract language (Boulenger et al., 2012; Moseley et al., 2012). What about abstract words with variable use? To which degree is abstract meaning processing shared or divided between action and connection hub systems of the human brain? A set of exciting questions may spark further brain research in the semantic domain, some of which also relate to context-dependent neuropragmatic function. Another important issue addresses the time course of semantic processing. Whereas the classic neurophysiological indexes of semantics, such as the N400 brain potential, emerge at 300 to 500 ms and later (Kutas & Federmeier, 2011), early brain reflections of meaning have been reported with the first 100 to 200 ms (Pulvermüller, Shtyrov, & Hauk, 2009). As for the localization in space, the localization of semantic subfunctions in time appears as an eminently important future research target.

26.7.4 Task Modulation and Attention

The interplay between language and attention mechanisms is one of the most delicate facets to be addressed in the neuroscience of language. This does not only imply that brain activation patterns change if linguistic items are being processed in different tasks. Note that the latter is quite trivial, because common psycholinguistic tasks (for example, attentive reading, lexical decision, and semantic categorization) differ so fundamentally (for example, in the need for overt responses, decisions, or comparison between stimuli) that their brain manifestations cannot be the same. But language-related brain activation is also under the influence of fine-grained task features that direct attention to different aspects of linguistic stimuli (Garagnani, Shtyrov, & Pulvermüller, 2009; Hoenig, Sim, Bochev, Herrnberger, & Kiefer, 2008). Early models had suggested that processing in specific "modules" can be switched on and off by the level of processing imposed by specific tasks. In this view, a semantic task may switch on semantic processing, but a phonological task may suppress it. Contrasting with this view, much evidence supports automatic semantic access, which seems very difficult to suppress (Stroop, 1935). Still, task demand may modulate the degree to which neuronal circuits processing word forms or meanings become active. One proposal is that a complex system of area-specific feedback regulation loops controls attention and that the gain in these regulation loops implements the level of attention to specific linguistic and cognitive aspects (Garagnani et al., 2008). Research about language attention interplay is still in its infancy. An avenue toward a better understanding may be opened up by conjoined performance of neurocomputational modeling and experimental neurophysiological research.

26.7.5 Neurocomputational Modeling and Temporal Dynamics

Using concrete neurobiologially grounded models—as illustrated in Figure 26.2, where specific connectivity features of perisylvian language cortex are implemented—it becomes possible to make predictions regarding brain physiology beyond the level previously possible. For example, the simulation of cell assemblies developing in perisylvian language cortex allows for testing their activation time course. When these distributed circuits become active, they follow a sequence of activation states that may be of interest for a better understanding of cognitive processing. The circuit is first stimulated from its sensory end. If this early stage of reception provides sufficient input activation, then it is followed by rapid spreading of activation throughout the circuit, a process called ignition. During ignition, most of the cell assembly neurons become active; however, because of biasing input, feedback regulation, and control mechanisms, the precise set of neurons partaking in the ignition may vary in a context-dependent and task-/attention-dependent manner. After ignition, activity declines because of inhibitory mechanisms set off by the strong excitation processes, but activation still survives and reverberates in those parts of the circuit that are most strongly connected. The different dynamic stages of the circuit activation process (stimulation, ignition, reverberation) can be related to different cognitive and linguistic processes (reception, word recognition, verbal working memory) and to different neurophysiological indexes (P50, N100/N160, N400 conjoined with high-frequency activity; Pulvermüller et al., 2014). Finding and refining the parallels between mechanistic-neurobiological, cognitive-linguistic, and neurophysiological mechanisms is one of the most exciting tasks for the future.

Acknowledgments

We wish to thank Malte Schomers, Rachel Moseley and Sabina Mollenhauer for discussion and help in preparing this ms. This work was supported by the Freie Universität Berlin, the Deutsche Forschungsgemeinschaft (Pu 97/15-1, Pu 97/16-1), and the Engineering and Physical Sciences Research Council and Behavioural and Brain Sciences Research Council (UK) (BABEL grant, EP/J004561/1). This work has been partially funded by Poeticon++ EU grant.

References

Alston, W. P. (1964). *Philosophy of language*. Englewood Cliffs, NJ: Prentice-Hall.

Arevalo, A. L., Baldo, J. V., & Dronkers, N. F. (2012). What do brain lesions tell us about theories of embodied semantics and the human mirror neuron system? *Cortex*, *48*(2), 242–254. Available from: http://dx.doi.org/10.1016/j.cortex.2010.06.001.

Aziz-Zadeh, L., Wilson, S. M., Rizzolatti, G., & Iacoboni, M. (2006). Congruent embodied representations for visually presented actions and linguistic phrases describing actions. *Current Biology*, *16*(18), 1818–1823.

Bak, T. H. (2013). The neuroscience of action semantics in neurodegenerative brain diseases. *Current Opinion in Neurology*, *28*(5), 271–277. Available from: http://dx.doi.org/10.1097/WCO.0000000000000039.

Barrós-Loscertales, A., González, J., Pulvermüller, F., Ventura-Campos, N., Bustamante, J. C., Costumero, V., et al. (2012). Reading salt activates gustatory brain regions: fMRI evidence for semantic grounding in a novel sensory modality. *Cerebral Cortex*, *22*(11), 2554–2563. Available from: http://dx.doi.org/10.1093/cercor/bhr324.

Bedny, M., & Caramazza, A. (2011). Perception, action, and word meanings in the human brain: The case from action verbs. *Annals of the New York Academy of Sciences*, *1224*, 81–95. Available from: http://dx.doi.org/10.1111/j.1749-6632.2011.06013.x.

Binder, J. R., & Desai, R. H. (2011). The neurobiology of semantic memory. *Trends in Cognitive Sciences*, *15*(11), 527–536. Available from: http://dx.doi.org/10.1016/j.tics.2011.10.001.

Binder, J. R., Westbury, C. F., McKiernan, K. A., Possing, E. T., & Medler, D. A. (2005). Distinct brain systems for processing concrete and abstract concepts. *Journal of Cognitive Neuroscience*, *17*(6), 905–917.

Bird, H., Lambon-Ralph, M. A., Patterson, K., & Hodges, J. R. (2000). The rise and fall of frequency and imageability: Noun and verb production in semantic dementia. *Brain and Language*, *73*(1), 17–49.

Bishop, D. V., Brown, B. B., & Robson, J. (1990). The relationship between phoneme discrimination, speech production, and language comprehension in cerebral-palsied individuals. *Journal of Speech and Hearing Research*, *33*(2), 210–219.

Bookheimer, S. (2002). Functional MRI of language: New approaches to understanding the cortical organization of semantic processing. *Annual Review of Neuroscience*, *25*, 151–188.

Boulenger, V., Hauk, O., & Pulvermüller, F. (2009). Grasping ideas with the motor system: Semantic somatotopy in idiom comprehension. *Cerebral Cortex*, *19*(8), 1905–1914.

Boulenger, V., Shtyrov, Y., & Pulvermüller, F. (2012). When do you grasp the idea? MEG evidence for instantaneous idiom understanding. *Neuroimage*, *59*(4), 3502–3513. Available from: http://dx.doi.org/10.1016/j.neuroimage.2011.11.011.

Braitenberg, V., & Pulvermüller, F. (1992). Entwurf einer neurologischen Theorie der Sprache. *Naturwissenschaften*, *79*, 103–117.

Braitenberg, V., & Schüz, A. (1992). Basic features of cortical connectivity and some considerations on language. In J. Wind, B. Chiarelli, B. H. Bichakjian, A. Nocentini, & A. Jonker (Eds.), *Language origin: A multidisciplinary approach* (pp. 89–102). Dordrecht: Kluwer.

Braitenberg, V., & Schüz, A. (1998). *Cortex: Statistics and geometry of neuronal connectivity* (2nd ed.). Berlin: Springer.

Call, J., & Tomasello, M. (2007). *The gestural communication of apes and monkeys*. New York, NY: Taylor & Francis Group/Lawrence Erlbaum Associates.

Cangelosi, A., Greco, A., & Harnad, S. (2002). Symbol grounding and the symbolic theft hypothesis. In A. Cangelosi, & D. Parisi (Eds.), *Simulating the evolution of language* (pp. 3–20). London: Springer.

Caplan, D., Gow, D., & Makris, N. (1995). Analysis of lesions by MRI in stroke patients with acoustic-phonetic processing deficits. *Neurology*, *45*(2), 293–298.

Chater, N., & Manning, C. D. (2006). Probabilistic models of language processing and acquisition. *Trends in Cognitive Sciences*, *10* (7), 335–344.

Chomsky, N. (1980). *Rules and representations*. New York, NY: Columbia University Press.

D'Ausilio, A., Maffongelli, L., Bartoli, E., Campanella, M., Ferrari, E., Berry, J., et al. (2014). Listening to speech recruits specific tongue motor synergies as revealed by transcranial magnetic stimulation and tissue-Doppler ultrasound imaging. *Philosophical Transactions of the Royal Society B: Biological Sciences*, *369*(1644), 20130418.

D'Ausilio, A., Pulvermüller, F., Salmas, P., Bufalari, I., Begliomini, C., & Fadiga, L. (2009). The motor somatotopy of speech perception. *Current Biology*, *19*(5), 381–385.

Damasio, A. R. (1989). Time-locked multiregional retroactivation: A systems-level proposal for the neural substrates of recall and recognition. *Cognition*, *33*, 25–62.

Desai, R. H., Conant, L. L., Binder, J. R., Park, H., & Seidenberg, M. S. (2013). A piece of the action: Modulation of sensory-motor regions by action idioms and metaphors. *Neuroimage*, *83*, 862–869. Available from: http://dx.doi.org/10.1016/j.neuroimage.2013.07.044.

Egorova, N., Pulvermüller, F., & Shtyrov, Y. (2014). Neural dynamics of speech act comprehension: An MEG study of Naming and Requesting. *Brain Topography*, *27*, 375–392. Available from: http://dx.doi.org/10.1007/s10548-013-0329-3.

Egorova, N., Shtyrov, Y., & Pulvermüller, F. (2013). Early and parallel processing of pragmatic and semantic information in speech acts: Neurophysiological evidence. *Frontiers in Human Neuroscience*, *7*(86), 1–13. Available from: http://dx.doi.org/10.3389/fnhum.2013.00086.

Ehlich, K. (2007). *Sprache und sprachliches Handeln / Language and linguistic action*. Berlin: Walter de Gruyter.

Elman, J. L. (1990). Finding structure in time. *Cognitive Science*, *14*, 179–211.

Elman, J. L., Bates, L., Johnson, M., Karmiloff-Smith, A., Parisi, D., & Plunkett, K. (1996). *Rethinking innateness. A connectionist perspective on development*. Cambridge, MA: MIT Press.

Fadiga, L., Craighero, L., Buccino, G., & Rizzolatti, G. (2002). Speech listening specifically modulates the excitability of tongue muscles: A TMS study. *European Journal of Neuroscience*, *15*(2), 399–402.

Farah, M. J., & McClelland, J. L. (1991). A computational model of semantic memory impairment: Modality specificity and emergent category specificity. *Journal of Experimental Psychology: General*, *120* (4), 339–357.

Fazio, P., Cantagallo, A., Craighero, L., D'Ausilio, A., Roy, A., Pozzo, T., et al. (2009). Encoding of human action in Broca's area. *Brain*, *132*(Pt 7), 1980–1988.

Fritz, G., & Hundsnurscher, F. (Eds.), (1994). *Handbook of dialogue analysis/Handbuch der Dialoganalyse* Niemeyer: Tübingen.

Fry, D. B. (1966). The development of the phonological system in the normal and deaf child. In F. Smith, & G. A. Miller (Eds.), *The genesis of language* (pp. 187–206). Cambridge, MA: MIT Press.

Garagnani, M., & Pulvermüller, F. (2013). Neuronal correlates of decisions to speak and act: Spontaneous emergence and dynamic topographies in a computational model of frontal and temporal areas. *Brain and Language*. Available from: http://dx.doi.org/10.1016/j.bandl.2013.02.001.

Garagnani, M., Shtyrov, Y., & Pulvermüller, F. (2009). Effects of attention on what is known and what is not: MEG evidence for functionally discrete memory circuits. *Frontiers in Human Neuroscience*, *3*(10). Available from: http://dx.doi.org/10.3389/neuro.3309.3010.2009.

Garagnani, M., Wennekers, T., & Pulvermüller, F. (2008). A neuroanatomically grounded Hebbian learning model of attention-language interactions in the human brain. *European Journal of Neuroscience*, *27*(2), 492–513.

Genty, E., Breuer, T., Hobaiter, C., & Byrne, R. W. (2009). Gestural communication of the gorilla (Gorilla gorilla): Repertoire, intentionality and possible origins. *Animal Cognition*, *12*(3), 527–546.

Gibbs, R. W., Jr., & O'Brien, J. E. (1990). Idioms and mental imagery: The metaphorical motivation for idiomatic meaning. *Cognition*, *36*(1), 35–68.

Goldberg, A. E. (2006). *Constructions at work: The nature of generalisation in language*. Oxford: Oxford University Press.

González, J., Barros-Loscertales, A., Pulvermüller, F., Meseguer, V., Sanjuán, A., Belloch, V., et al. (2006). Reading cinnamon activates olfactory brain regions. *Neuroimage*, *32*(2), 906–912.

Hauk, O., Johnsrude, I., & Pulvermüller, F. (2004). Somatotopic representation of action words in the motor and premotor cortex. *Neuron*, *41*, 301–307.

Hebb, D. O. (1949). *The organization of behavior. A neuropsychological theory*. New York, NY: John Wiley.

Hickok, G., & Poeppel, D. (2007). The cortical organization of speech processing. *Nature Reviews Neuroscience*, *8*(5), 393–402.

Hobaiter, C., & Byrne, R. W. (2011). The gestural repertoire of the wild chimpanzee. *Animal Cognition*, *14*(5), 745–767.

Hoenig, K., Sim, E. J., Bochev, V., Herrnberger, B., & Kiefer, M. (2008). Conceptual flexibility in the human brain: Dynamic recruitment of semantic maps from visual, motor, and motion-related areas. *Journal of Cognitive Neuroscience*, *20*(10), 1799–1814.

Kemmerer, D., Rudrauf, D., Manzel, K., & Tranel, D. (2012). Behavioural patterns and lesion sites associated with impaired processing of lexical and conceptual knowledge of action. *Cortex*, *48*(7), 826–848. Available from: http://dx.doi.org/10.1016/j.cortex.2010.11.001.

Kiefer, M., & Pulvermüller, F. (2012). Conceptual representations in mind and brain: Theoretical developments, current evidence and future directions. *Cortex*, *48*(7), 805–825. Available from: http://dx.doi.org/10.1016/j.cortex.2011.04.006.

Kiefer, M., Sim, E. J., Herrnberger, B., Grothe, J., & Hoenig, K. (2008). The sound of concepts: Four markers for a link between auditory and conceptual brain systems. *The Journal of Neuroscience*, *28*(47), 12224–12230.

Kintsch, W. (2008). Symbol systems and perceptual representations. In M. de Vega (Ed.), *Symbols and embodiment* (pp. 145–163). New York, NY: Oxford University Press.

Kohler, E., Keysers, C., Umilta, M. A., Fogassi, L., Gallese, V., & Rizzolatti, G. (2002). Hearing sounds, understanding actions: Action representation in mirror neurons. *Science*, *297*(5582), 846–848.

Kutas, M., & Federmeier, K. D. (2011). Thirty years and counting: Finding meaning in the N400 component of the event-related brain potential (ERP). *Annual Review of Psychology*, *62*, 621–647. Available from: http://dx.doi.org/10.1146/annurev.psych.093008.131123.

Lauro, L. J., Tettamanti, M., Cappa, S. F., & Papagno, C. (2008). Idiom comprehension: A prefrontal task? *Cerebral Cortex*, *18*(1), 162–170.

Liuzzi, G., Freundlieb, N., Ridder, V., Hoppe, J., Heise, K., Zimerman, M., et al. (2010). The involvement of the left motor cortex in learning of a novel action word lexicon. *Current Biology*, *20*(19), 1745–1751.

Locke, J. L. (1993). *The child's path to spoken language*. Cambridge, MA: Harvard University Press.

Lopez-Barroso, D., Catani, M., Ripolles, P., Dell'Acqua, F., Rodriguez-Fornells, A., & de Diego-Balaguer, R. (2013). Word learning is mediated by the left arcuate fasciculus. *Proceedings of the National Academy of Sciences of the United States of America*, *110*

(32), 13168–13173. Available from: http://dx.doi.org/10.1073/pnas.1301696110.

Martin, A. (2007). The representation of object concepts in the brain. *Annual Review of Psychology, 58*, 25–45.

Martin, A., Haxby, J. V., Lalonde, F. M., Wiggs, C. L., & Ungerleider, L. G. (1995). Discrete cortical regions associated with knowledge of color and knowledge of action. *Science, 270*, 102–105.

McCarthy, R. A., & Warrington, E. K. (1988). Evidence for modality-specific meaning systems in the brain. *Nature, 334* (6181), 428–430.

Mirman, D., Yee, E., Blumstein, S. E., & Magnuson, J. S. (2011). Theories of spoken word recognition deficits in aphasia: Evidence from eye-tracking and computational modeling. *Brain and Language, 117*(2), 53–68. Available from: http://dx.doi.org/10.1016/j.bandl.2011.01.004.

Moineau, S., Dronkers, N. F., & Bates, E. (2005). Exploring the processing continuum of single-word comprehension in aphasia. *Journal of Speech, Language, and Hearing Research, 48*(4), 884–896.

Moro, A. (2014). On the similarity between syntax and actions. *Trends in Cognitive Sciences, 18*(3), 109–110.

Moseley, R., Carota, F., Hauk, O., Mohr, B., & Pulvermüller, F. (2012). A role for the motor system in binding abstract emotional meaning. *Cerebral Cortex, 22*(7), 1634–1647. Available from: http://dx.doi.org/10.1093/cercor/bhr238.

Moseley, R. L., Pulvermüller, F., & Shtyrov, Y. (2013). Sensorimotor semantics on the spot: Brain activity dissociates between conceptual categories within 150 ms. *Scientific Reports, 3*, 1928.

Möttönen, R., Dutton, R., & Watkins, K. E. (2013). Auditory-motor processing of speech sounds. *Cerebral Cortex, 23*(5), 1190–1197. Available from: http://dx.doi.org/10.1093/cercor/bhs110.

Möttönen, R., & Watkins, K. E. (2009). Motor representations of articulators contribute to categorical perception of speech sounds. *The Journal of Neuroscience, 29*(31), 9819–9825.

O'Reilly, R. C. (1998). Six principles for biologically based computational models of cortical cognition. *Trends in Cognitive Sciences, 2* (11), 455–562.

Paivio, A. (1991). Dual coding theory: Retrospect and current status. *Canadian Journal of Psychology, 45*, 255–287.

Patterson, K., Nestor, P. J., & Rogers, T. T. (2007). Where do you know what you know? The representation of semantic knowledge in the human brain. *Nature Reviews Neuroscience, 8*(12), 976–987.

Pinker, S. (1994). *The language instinct. How the mind creates language.* New York, NY: Harper Collins Publishers.

Poeppel, D., Emmorey, K., Hickok, G., & Pylkkanen, L. (2012). Towards a new neurobiology of language. *The Journal of Neuroscience, 32*(41), 14125–14131. Available from: http://dx.doi.org/10.1523/JNEUROSCI.3244-12.2012.

Pulvermüller, F. (1999). Words in the brain's language. *Behavioral and Brain Sciences, 22*, 253–336.

Pulvermüller, F. (2005). Brain mechanisms linking language and action. *Nature Reviews Neuroscience, 6*(7), 576–582.

Pulvermüller, F. (2013). How neurons make meaning: Brain mechanisms for embodied and abstract-symbolic semantics. *Trends in Cognitive Sciences, 17*(9), 458–470. Available from: http://dx.doi.org/10.1016/j.tics.2013.06.004.

Pulvermüller, F. (2014). The syntax of action. *Trends in Cognitive Sciences, 8*(5), 219–220.

Pulvermüller, F., Cappelle, B., & Shtyrov, Y. (2013). Brain basis of meaning, words, constructions, and grammar. In T. Hoffmann, & G. Trousdale (Eds.), *Oxford handbook of construction grammar* (pp. 397–416). Oxford: Oxford University Press.

Pulvermüller, F., Cook, C., & Hauk, O. (2012). Inflection in action: Semantic motor system activation to noun- and verb-containing phrases is modulated by the presence of overt grammatical markers. *Neuroimage, 60*(2), 1367–1379. Available from: http://dx.doi.org/10.1016/j.neuroimage.2011.12.020.

Pulvermüller, F., & Fadiga, L. (2010). Active perception: Sensorimotor circuits as a cortical basis for language. *Nature Reviews Neuroscience, 11*(5), 351–360.

Pulvermüller, F., Garagnani, M., & Wennekers, T. (2014). Thinking in circuits: Towards neurobiological explanation in cognitive neuroscience. *Biological Cybernetics, 108*(5), 573–593.

Pulvermüller, F., & Hauk, O. (2006). Category-specific processing of color and form words in left fronto-temporal cortex. *Cerebral Cortex, 16*(8), 1193–1201.

Pulvermüller, F., Hauk, O., Nikulin, V. V., & Ilmoniemi, R. J. (2005). Functional links between motor and language systems. *European Journal of Neuroscience, 21*(3), 793–797.

Pulvermüller, F., Huss, M., Kherif, F., Moscoso del Prado Martin, F., Hauk, O., & Shtyrov, Y. (2006). Motor cortex maps articulatory features of speech sounds. *Proceedings of the National Academy of Sciences, USA, 103*(20), 7865–7870.

Pulvermüller, F., Kiff, J., & Shtyrov, Y. (2012). Can language-action links explain language laterality? An ERP study of perceptual and articulatory learning of novel pseudowords. *Cortex, 48*(7), 471–481. Available from: http://dx.doi.org/10.1016/j.cortex.2011.02.006.

Pulvermüller, F., & Knoblauch, A. (2009). Discrete combinatorial circuits emerging in neural networks: A mechanism for rules of grammar in the human brain? *Neural Networks, 22*(2), 161–172.

Pulvermüller, F., Shtyrov, Y., & Hauk, O. (2009). Understanding in an instant: Neurophysiological evidence for mechanistic language circuits in the brain. *Brain and Language, 110*(2), 81–94.

Pulvermüller, F., Shtyrov, Y., & Ilmoniemi, R. J. (2005). Brain signatures of meaning access in action word recognition. *Journal of Cognitive Neuroscience, 17*(6), 884–892.

Rauschecker, J. P., & Scott, S. K. (2009). Maps and streams in the auditory cortex: Nonhuman primates illuminate human speech processing. *Nature Neuroscience, 12*(6), 718–724.

Rilling, J. K. (2014). Comparative primate neuroimaging: Insights into human brain evolution. *Trends in Cognitive Sciences, 18*(1), 46–55.

Rilling, J. K., Glasser, M. F., Jbabdi, S., Andersson, J., & Preuss, T. M. (2011). Continuity, divergence, and the evolution of brain language pathways. *Frontiers in Evolutionary Neuroscience, 3*, 11. Available from: http://dx.doi.org/10.3389/fnevo.2011.00011.

Rilling, J. K., Glasser, M. F., Preuss, T. M., Ma, X., Zhao, T., Hu, X., et al. (2008). The evolution of the arcuate fasciculus revealed with comparative DTI. *Nature Neuroscience, 11*(4), 426–428.

Rizzolatti, G., & Craighero, L. (2004). The mirror-neuron system. *Annual Review in Neuroscience, 27*, 169–192.

Rizzolatti, G., Fadiga, L., Gallese, V., & Fogassi, L. (1996). Premotor cortex and the recognition of motor actions. *Cognitive Brain Research, 3*(2), 131–141.

Rogalsky, C., Love, T., Driscoll, D., Anderson, S. W., & Hickok, G. (2011). Are mirror neurons the basis of speech perception? Evidence from five cases with damage to the purported human mirror system. *Neurocase, 17*(2), 178–187. Available from: http://dx.doi.org/10.1080/13554794.2010.509318.

Saffran, J. R., Aslin, R. N., & Newport, E. L. (1996). Statistical learning by 8-month-old infants. *Science, 274*(5294), 1926–1928.

Schomers, M. R., Kirilina, E., Weigand, A., Bajbouj, M., & Pulvermüller, F. (2014). Causal influence of articulatory motor cortex on comprehending single spoken words: TMS evidence. *Cerebral Cortex.* Available from: http://dx.doi.org/10.1093/cercor/bhu274.

Shebani, Z., & Pulvermüller, F. (2013). Moving the hands and feet specifically impairs working memory for arm- and leg-related action words. *Cortex, 49*(1), 222–231. Available from: http://dx.doi.org/10.1016/j.cortex.2011.10.005.

Simmons, W. K., Ramjee, V., Beauchamp, M. S., McRae, K., Martin, A., & Barsalou, L. W. (2007). A common neural substrate for perceiving and knowing about color. *Neuropsychologia*, *45*(12), 2802–2810.

Stalnaker, R. C. (2002). Common ground. *Linguistics and Philosophy*, *25*(5), 701–721.

Stroop, J. R. (1935). Studies of interference in serial verbal reactions. *Journal of Experimental Psychology*, *18*, 643–662.

Szenkovits, G., Peelle, J. E., Norris, D., & Davis, M. H. (2012). Individual differences in premotor and motor recruitment during speech perception. *Neuropsychologia*, *50*(7), 1380–1392. Available from: http://dx.doi.org/10.1016/j.neuropsychologia.2012.02.023.

Tettamanti, M., Buccino, G., Saccuman, M. C., Gallese, V., Danna, M., Scifo, P., et al. (2005). Listening to action-related sentences activates fronto-parietal motor circuits. *Journal of Cognitive Neuroscience*, *17*(2), 273–281.

van den Heuvel, M. P., & Sporns, O. (2013). Network hubs in the human brain. *Trends in Cognitive Sciences*, *17*(12), 683–696. Available from: http://dx.doi.org/10.1016/j.tics.2013.09.012.

Venezia, J. H., Saberi, K., Chubb, C., & Hickok, G. (2012). Response bias modulates the speech motor system during syllable discrimination. *Frontiers in Psychology*, *3*, 157. Available from: http://dx.doi.org/10.3389/fpsyg.2012.00157.

Vigliocco, G., Kousta, S. T., Della Rosa, P. A., Vinson, D. P., Tettamanti, M., Devlin, J. T., et al. (2013). The neural representation of abstract words: The role of emotion. *Cerebral Cortex*. Available from: http://dx.doi.org/10.1093/cercor/bht025.

Watkins, K. E., Strafella, A. P., & Paus, T. (2003). Seeing and hearing speech excites the motor system involved in speech production. *Neuropsychologia*, *41*(8), 989–994.

Westermann, G., & Reck Miranda, E. (2004). A new model of sensorimotor coupling in the development of speech. *Brain and Language*, *89*(2), 393–400.

Willems, R. M., Labruna, L., D'Esposito, M., Ivry, R., & Casasanto, D. (2011). A functional role for the motor system in language understanding: Evidence from theta-burst transcranial magnetic stimulation. *Psychological Science*, *22*(7), 849–854. Available from: http://dx.doi.org/10.1177/0956797611412387.

Young, M. P., Scannell, J. W., Burns, G. A., & Blakemore, C. (1994). Analysis of connectivity: Neural systems in the cerebral cortex. *Reviews in the Neurosciences*, *5*(3), 227–250.

CHAPTER

27

The Dual Loop Model in Language

Cornelius Weiller, Tobias Bormann, Dorothee Kuemmerer, Mariachristina Musso and Michel Rijntjes

Department of Neurology and Clinical Neuroscience, University Hospital Freiburg, Freiburg, Germany

The idea of two parallel streams to process language seems almost ancient. For instance, Herder distinguished a more automatic production route from a close association of language with "self-consciousness" (Herder, 1772). The idea of conceptual representations independent of linguistic representations is commonly attributed to Lichtheim (1885), which may be due to his invention of the famous house diagram. Yet in his 1874 treatise Wernicke notes, "thinking and speaking are two independent processes, which even may inhibit each other" (Wernicke, 1874).

Based on the neuropsychological method, the distinction between phonological representations and nonlinguistic representations of conceptual knowledge has been incorporated in most modern (Dell, Martin, & Schwartz, 2007; Lambon Ralph, Graham, Ellis, & Hodges, 1998; Levelt, 1999; Morton, 1980) as well as historical (Freud, 1891; Goldstein, 1927; Kleist, 1905, see Weiller, Bormann, Saur, Musso, & Rijntjes, 2011) models of language processing. In all cognitive models of word processing, the mapping is achieved through different layers of representations, but all these models assume at least two parallel cognitive streams.

It was Carl Wernicke (at 26 years old) who, in his "physiological model on an anatomical basis," developed a central and very modern idea of how language is processed in the brain (Wernicke, 1874). Wernicke assumed only two centers in the brain, "the center for images of movement for sound production in the first frontal convolution" (i.e., the inferior frontal gyrus [IFG], called Broca's area) and "a sensory language center, containing the storage of sound images of speech in the first temporal convolution" (i.e., the superior temporal gyrus [STG], later called Wernicke's area), and all language functions would be derived from the interaction between both centers. This interaction is accomplished through two principal routes. The route for the "direct" interaction of Broca's and Wernicke's area would represent "sensorimotor mapping" and produce the so-called Wortbegriff (translated as "word-concept" or "word-form") (DeWitt & Rauschecker, 2013) as an intrinsic integration of sensory and motor aspects of the word, learned during development in early age through imitation and that is later used to produce speech. In parallel, there are concepts associated with a word distributed over the entire cortex and connected to Wernicke's and Broca's areas through the "indirect" route for understanding ("semantic route") and uttering thoughts. "Soon after we have learned to speak a word, we lose the intention only to reproduce sounds and plan to utter a meaning" (Wernicke, 1874). "We have to assume that (then) the majority of speech impulses reach the word concepts from the remaining cerebral cortex" (Wernicke, 1906). Here, we have the first description of a brain network with nodes and connections.

The ensuing confusion about the locations of association tracts for these two routes was partly due to the insistence by Wernicke that those tracts mediating the processing of the "Wortbegriff" (i.e., the "direct" route) should be located behind the insula, and conduction aphasia postulated as a consequence of an isolated damage of this direct route was not found in insular lesions (for more details see Weiller et al., 2011). Today, the function Wernicke ascribed to the direct route (e.g., sensorimotor mapping) is equivalent to what we today would ascribe to the dorsal route along the arcuate fascicle/superior longitudinal fascicle. Wernicke did not make any differentiation between dorsal and ventral tracts; however, he explicitly, clearly, and repeatedly mentioned the extreme capsule that is today seen as the major connection of the ventral stream. He referred to

Neurobiology of Language. DOI: http://dx.doi.org/10.1016/B978-0-12-407794-2.00027-4

the notion of his academic mentor Meynert, who was the first to ascribe a language function to a set of regions comprising the claustrum, the insula, as well as the ascending acoustic fibers in the external and extreme capsule (Meynert, 1866). Wernicke was aware of other association tracts (e.g., the arcuate fascicle [AF]) that he saw "not actually a special bundle, but a general system of association fibers, which must be considered in the anatomy of speech regions" (Wernicke, 1906). Later, under the influence of eminent researchers in Wernicke's time (von Monakow, Charcot, Dejerine; see Weiller et al., 2011), the arcuate fascicle was established as the main or sometimes only connecting tract between Wernicke's and Broca's areas. Geschwind's iconic diagram with an anterior "language center" in the inferior frontal region, commonly related to speech production, a posterior temporal region, related to speech comprehension connected through a single fiber system above the level of the ventricles, and the arcuate fascicle, along which language information is transferred, has been the blueprint that almost all textbooks refer to (Geschwind, 1972). Geschwind's preference for the arcuate fascicle (see Weiller et al., 2011) may be due to the importance he gave to the angular gyrus as multimodal integration area for language: "the function of Wernicke's area implies the existence of extensive connexions to the angular gyrus region" (Geschwind, 1965). Thus, the ventral pathway along the extreme capsule was out of sight until recently (Weiller et al., 2011).

Although the notion of a ventral pathway connecting temporal and frontal lobes for language processing got lost, the idea of a dual stream model emerged in the area of vision, assuming two parallel processes for perception of visual stimuli: "one ambient, determining space at large around the body, the other focal, which examines detail in small areas or space" (Trevarthen, 1968). This idea developed into the very well-known ventral "what" and dorsal "where/how" pathways (Kravitz, Saleem, Baker, & Mishkin, 2011; Milner & Goodale, 1995; Mishkin, Ungerleider, & Macko, 1983). The processing of a "where" and "what" of a visual stimulus take different courses, both starting in the primary visual cortex but with a subdivision in dorsal and ventral streams already present in extrastriate regions, extending toward the parietal and temporal lobe, respectively. Over the years, a subdivision of different processing streams was also proposed in the motor system (Rizzolatti & Matelli, 2003) or for visuospatial attention (Corbetta, Kincade, Lewis, Snyder, & Sapir, 2005). Studies of comparative anatomy revealing two processing streams for the acoustic modalities (Romanski et al., 1999) and a larger number of imaging studies in humans led to a "new" hypothesis of dual stream processing for auditory language system in the brain (Demonet, Thierry, & Cardebat, 2005; Hickok & Poeppel, 2004; Rauschecker, 1998; Scott, Blank, Rosen, & Wise, 2000; Wise, 2003): the dorsal stream is involved in mapping speech sound to articulation, whereas the ventral one would support speech to meaning correspondences. In these models, the frontal termination of the ventral stream has not exactly been determined and Broca's area is sometimes described only in the context of the dorsal stream. Wise (2003), favoring the anterior temporal lobe (ATL) as decisive for language comprehension, pointed to the uncinate fascicle (UF) as a ventral connection, as has been supported by DTI studies (Friederici, Bahlmann, Heim, Schubotz, & Anwander, 2006; Parker et al., 2005). Coming from the gold standard, tracing experiments in monkeys, Pandya's group suggested the extreme capsule as an important connecting pathway in language, and also in grammar, connecting the temporal lobe with what is seen as BA 45 in humans (Pandya & Yeterian, 1996; Petrides & Pandya, 1988, 2009; Petrides & Pandya, 2002, 2009; Schmahmann & Pandya, 2006).

This was the situation when we aimed to integrate functional connotations with tract morphology to test the anatomical basis of the "dual-stream model" as it was proposed by Hickok and Poeppel (2004) by combining fMRI with diffusion tensor imaging (DTI), assuming a distributed network comprising specialized brain areas (network nodes) and their interconnecting white matter fiber tracts (network connections) (Saur et al., 2008). The findings revealed that a sublexical, phonological speech task (pseudoword repetition) was sustained by a dorsal pathway connecting the superior temporal lobe and premotor cortices in the frontal lobe via superior longitudinal fascicles and the arcuate fascicle. In contrast, a higher-level language comprehension task was mediated by a ventral pathway connecting the middle temporal lobe and the ventrolateral prefrontal cortex via the extreme capsule. The finding of the extreme capsule as the place for connecting fibers within the ventral pathway was supported by similar publications on DTI trackings in humans (Croxson et al., 2005; Frey, Campbell, Pike, & Petrides, 2008; Friederici et al., 2006; Makris & Pandya, 2009; Parker et al., 2005).

Evidence from functional studies support the idea of the dual loop model; during presurgical assessment of epileptic patients, stereo-electroencephalographic recordings were performed and ERP signals were recorded during a phoneme task on a pair of pseudowords and a lexical-semantic task on adjective-nouns pairs. Phonological information processing predominantly occurred in the left supramarginal gyrus (SMG) (dorsal stream) and processing lexicosemantic information occurred in the anterior/middle temporal and fusiform gyrus (ventral stream) (Trebuchon, Demonet, Chauvel, & Liégeois-Chauvel, 2013). Stimulating the dorsal pathway with electrostimulation leads to phonological paraphasias (Leclercq et al., 2010), whereas electrostimulation of the ventral EmC leads to semantic paraphasias (Duffau et al., 2005).

27.1 PATIENTS

Results from patients with acute aphasia support the claim that language is processed within a dual pathway network. Repetition impairments were mainly associated with posterior temporoparietal regions and damage of the AF/SLF, whereas comprehension impairments were associated with lesions more ventrally in temporoprefrontal regions projecting on the ventral EmC (Kuemmerer et al., 2013). Although this study complemented and supported the previous DTI study, patients commonly present with syndromes (i.e., a typical collection of symptoms). Ueno, Saito, Rogers, and Lambon Ralph (2011) implemented the dual pathway network in a computational model and could virtually relate symptoms like repetition and comprehension impairments and also syndromes like Wernicke's and Broca's aphasia to the expected lesion sites. Allocation of symptom collections to syndromes could also be related to lesions in acute aphasic patients (Kuemmerer et al., 2012). Lesions associated with Broca's aphasia were in the IFG, Insula, and SMG, whereas lesions correlating with Wernicke's aphasia were in the STG, postcentral gyrus, and Insula. Semantic paraphasias and comprehension impairments in Wernicke's aphasia may be due to temporal lesions and damage to the ventral pathway, whereas phonemic paraphasias may be related to damage to the dorsal pathway. Thus, Wernicke aphasia affects both the ventral and the dorsal pathway.

Another aphasia syndrome, conduction aphasia (Damasio & Damasio, 1980; Poncet, Habib, & Robillard, 1987), is characterized by fluent, meaningful spontaneous speech with preserved auditory comprehension but frequent phonemic paraphasias and repetition impairments (Bernal & Ardila, 2009; Damasio & Damasio, 1980; Hickok, 2012; Weiller et al., 2011). The original conception of conduction aphasia as a disconnection syndrome (Wernicke, 1874) through damage to the AF/SLF (Geschwind, 1965) is supported by several recent studies (Duffau, Peggy Gatignol, Mandonnet, Capelle, & Taillandier, 2008; Yamada et al., 2007). However, many other contemporary studies favor the idea that lesions to cortical gray matter in the left inferior parietal and superior temporal lobe (Anderson et al., 1999; Bartha & Benke, 2003; Hickok, 2009, 2012; Quigg, Geldmacher, & Elias, 2006) or SMG (Axer, von Keyserlingk, Berks, & von Keyserlingk, 2001; Baldo & Dronkers, 2006; Baldo, Katseff, & Dronkers, 2012) lead to the syndrome of conduction aphasia. Tracts or cortical regions, both constellations, evidence the dorsal stream as responsible for conduction aphasia.

This consideration could also be taken into account when trying to understand symptoms in degenerative disease. The logopenic variant of primary progressive aphasia (PPA) has also similar symptoms as conduction aphasia and is characterized by intact word comprehension, but it has repetition and phonological impairments (Galantucci et al., 2011) and is associated with atrophy of temporoparietal regions. Impaired phonological processing can be explained by deficits in tasks that require phonological storage and the presence of phonological paraphasias (Gorno-Tempini et al., 2008). Results of studies using DTI, fractional anisotropy (FA), and volumetric studies (Galantucci et al., 2011; Gorno-Tempini et al., 2008; Hu et al., 2010) demonstrate the involvement of the dorsal pathway (phonological loop) and also gray matter atrophy in these patients, supporting the role of the dorsal stream in phonological processing.

Semantic dementia may also be seen in light of the dual loop model. This disorder is characterized by impaired comprehension and naming, but intact repetition (Jefferies, Patterson, Jones, & Lambon Ralph, 2009; Lambon Ralph et al., 1998; Mummery et al., 1999; Wilson et al., 2009). Atrophy in these patients is located in the ATL/temporal lobe/ventral semantic processing stream of the temporal lobe (Agosta et al., 2010; Galton et al., 2001; Hodges, Patterson, Oxbury, & Funnell, 1992). Within the computational model of Ueno et al. (2011), the specific symptom constellation in patients with semantic dementia could be elicited by virtual damage to the ATL and anterior STG, supporting the relevance of the ventral pathway in semantic dementia.

Taken together, damage within the dual pathway model has an effect on language impairments of patients with acute aphasia in stroke (Fridriksson et al., 2010; Hosomi et al., 2009; Kuemmerer et al., 2013), as well as on patients with chronic progressive language impairments, like semantic dementia or logopenic PPA (Schwindt et al., 2013). Within this model it is possible to relate different symptoms and syndromes to different lesion sites. However, most cognitive processes do naturally require an interaction of both pathways (Ueno et al., 2011; Weiller et al., 2011). Therefore, it is essential to further explore how different models of connectivity relate to language impairments to improve our understanding of the functional relevance of both pathways (Dick & Tremblay, 2012).

27.2 NEUROSPSYCHOLOGY

We have seen that patient studies have played an important role in the development of early models of language processing in the brain. The approach of Wernicke, Lichtheim, and others, however, eventually grew out of fashion, partially because predictions about the localization of cognitive functions proved to be wrong (Head, 1926). In a revival of this type of approach beginning in the 1970s, cognitive neuropsychology has been most influential in developing cognitive models of language processing (Shallice, 1988; Rapp, 2001). Early research in acquired dyslexia has led to the proposal of a dual-route model of reading based on dissociations among dyslexic individuals (Coltheart, Patterson, & Marshall, 1980). One route, the lexical-semantic route, involved mapping of the visual word onto a semantic representation and subsequent access of the phonological word form in the mental lexicon. The other route, the so-called sublexical route, involved the mapping of graphemes onto their respective sounds without access to the word's representation in the mental lexicon. Evidence for this dual-route model came from patients with phonological dyslexia who were unable to use their sublexical route: They are unable to read nonwords but are able to read regular and exception words. Please note that the "dual-route" in reading may not be identical with the proposed anatomical dual-route system along the AF/SLF and extreme capsule tracts.

The mirror impairment, surface dyslexia, can be observed in patients with a semantic impairment and is characterized by a predominant reliance on the sublexical reading route. Comparable dissociations and "syndromes" have also been reported for repetition, suggesting the involvement of at least two routes in this task. McCarthy and Warrington (1984) reported three individuals, two with conduction aphasia and one with transcortical motor aphasia. The authors argued that two routes were involved in repetition and speech production: a fast, automatic phonological route and a slow semantic route. McCarthy and Warrington (2001) reported a patient with a severe semantic impairment due to semantic dementia. The patient had a severe naming impairment and was also considerably impaired in word comprehension tasks. However, his digit span was preserved. A different syndrome is deep dysphasia, which consists of a severe repetition deficit. Patients with deep dysphasia are rare but present with an impressive, counterintuitive pattern of strengths and weaknesses. They usually have a severely limited digit span, not exceeding a single digit. They are unable to repeat nonwords, and they produce semantic errors in repetition tasks. One individual we worked with was asked to repeat the words "czar" and "raw" on two different occasions. She produced "castle" and "meat" instead. This syndrome suggests a severe impairment to the phonological loop and reliance on semantic information derived from the auditorily presented word. However, the semantic errors in repetition suggest the additional involvement of a mild semantic impairment in this syndrome besides the severe impairment in the nonlexical repetition route. A similar, yet milder, deficit is present in another syndrome, impaired phonological short-term memory (STM). This is sometimes also referred to as repetition conduction aphasia (Shallice & Warrington, 1977). These patients also suffer from reduced spans for digits and words and impaired repetition of long nonwords. Processing the meaning of stimuli presented for repetition is better preserved. In sentence repetition, STM patients appear to have the meaning of the sentence preserved, suggesting involvement of semantic processing in this task (Baldo, Klostermann, & Dronkers, 2008). It has been argued that deep dysphasia and STM deficit reflect the continuum of a deficit rather than two independent syndromes (Martin, Saffran, & Dell, 1996).

Cognitive neuropsychology has preferably used this type of dissociation among tasks to develop models of complex cognitive skills. More recently, however, associations of functional impairments have been used to develop models further. In a recent study, Dell, Schwartz, Nozari, Faseyitan, and Coslett (2013) have mapped the parameters of a dual-route model of naming and repetition onto neural regions. Their model is a dual-route interactive model, which has been implemented as a computer model. In this model, a representation of a phonological input is interactively linked to phonological output units, a lexical and a semantic layer. The connections between the input and the phonological output units represent the nonlexical, nonsemantic route. In contrast, during repetition of single words, lexical and semantic representations are activated in parallel, reflecting the contribution of a word's lexical status and its meaning. The model has been successfully "fitted" to large groups of aphasic individuals. With some exceptions, an aphasic person's pattern in naming and repetition can be simulated by changing parameters in the computer model (Dell et al., 2007). This means that the pattern of errors of an aphasic speaker can be simulated within that model, and each aphasic participant is represented by a specific combination of model parameters.

In their more recent study, Dell et al. (2013) mapped the individual patients' model parameters onto lesions. Artificial lesions to the lexical-semantic connections in the computer model were correlated with neural lesions in anterior temporal areas, inferior temporal areas, IFG, and the SMG. These regions had been identified as substrates of various aspects of processing of meaning in the brain. In contrast, artificial lesions to the phonological output representations in the computer model mapped onto the SMG, postcentral gyrus, precentral gyrus, and insula what the authors suggest is the "anterior part of the dorsal stream." The model's connections between auditory input and output phonemes mapped onto "the STG, the posterior third of the planum temporale and cortex at the juncture of the parietal and temporal lobes (area Spt) as well as the SMG and postcentral gyrus" (Dell et al., 2013).

27.3 FUNCTIONS OF THE DUAL LOOP MODEL

There is no reason to suppose that the organization of the acoustic language system is different from other modalities. Thus, a dual loop system has been postulated in various domains (see Rijntjes, Weiller, Bormann, & Musso, 2012 for an overview of various studies showing an involvement of a dorsal and a ventral pathway). Thus, the function ascribed to dorsal or ventral pathways in each modality may differ, but only in their modality-specific aspect (Rijntjes et al., 2012). Discussions have taken place regarding how the common aspects of processing in the dorsal and ventral streams in different modalities might be described (Hickok & Poeppel, 2007; Rauschecker & Scott, 2009; Rijntjes et al., 2012; Weiller et al., 2011; Weiller, Musso, Rijntjes, & Saur, 2009).

We postulated that a dual loop system, comprising long association tracts connecting prerolandic and postrolandic parts of the brain with different computational abilities, provides a scaffolding system around which various functions may have been developed (Rijntjes et al., 2012; Weiller et al., 2011). It may be the synergy between time-sensitive analysis of sequences comparing the (correct) serial alignment of segments with acquired (e.g., phonological) representations as a possible function of the dorsal stream and identification of an invariant set of auditory (and probably also visual) hierarchical structural relationships between the elements (e.g., words), that is, time-independent, along the ventral stream, which makes language possible (Weiller et al., 2011).

The dorsal stream is not limited to "where" or "how" functions; rather, it analyzes the sequence of elements in time and in space, allowing for a fast online integration between sensory event information and "internal models or emulators" (Rauschecker & Scott, 2009). Thus, the dorsal stream is responsible for the ordering of elements in a continuous string to assure a correct production of behavior (e.g., subsequent syllables, word order, movement trajectories, orienting to a visual stimulus, and updating the place value of an Arab digit), irrespective of whether this behavior has been executed before or has a meaning. Therefore, processing is "time-dependent." In language, correct phonology may be learned through imitation and repetition, and the dorsal stream may serve as an automated correction mechanism for correct production of speech. This function is typically attributed to what is called the dorso-dorsal stream (along SLF II) (Binkofski & Buxbaum, 2013; Hoeren et al., 2014; Vry et al., 2012; Vry et al., 2014), which connects with dorsal premotor cortex. Through exercise and experience, the sequences can be converted into blueprints (e.g., one's own signature; Rijntjes et al., 1999) that are stored in SMG and BA 44 in IFG ("canonical neurons") and may relate to context and semantics, thus accessing areas in the inferior parietal cortex (IPC) and posterior temporal lobe, constituting what is called the dorsal-ventral stream (SLF III and arcuate fascicle) (Binkofski & Buxbaum, 2013; Hoeren et al., 2014; Vry et al., 2014). Recall and execution of stored blueprints might be possible rapidly and without the dorso-dorsal stream, and thus may substitute, in part, the work of the dorso-dorsal stream. Adaptations in the blueprints (e.g., changing to a new swing of a golf-pro) may require temporary relearning through the dorso-dorsal stream.

Most scientists would agree that the ventral stream is involved in meaning. We postulated that, in more general terms, the ventral pathway would be involved in the identification of the structural (hierarchical) relations of elements that may not be adjacent, and thus "time-independent," independent of the sequence of occurrence (Weiller et al., 2011), as in grammar (Friederici et al., 2006; Rauschecker & Scott, 2009; Weiller et al., 2009), in tonal dependencies in music or in the categorical identification of the digits involved in difficult mental arithmetic (Klein, Korbinian, Glauche, Weiller, & Willmes, 2012) (Willmes, Korbinian, & Klein, 2014). Also in the visuospatial domain, relevance and meaning of stimuli require the ventral stream (Umarova et al., 2010) as well as imagery of movement (Vry et al., 2012) or pantomiming (Hoeren et al., 2014; Vry et al., 2014).

27.4 ANATOMY, HUBS, DIVISIONS

A clear distinction should be made between the terms "streams" and "tracts." Processing "streams" address the functions that are thought to be mediated by the ventral or dorsal pathways, which include tracts and the cortical regions connected by them. The term "tracts" should be reserved for the anatomical correlates (here, of long association tracts) connecting these cortical regions. Tracts themselves do not have functions; they only mediate between cortical areas, and the result of this interaction might lead to a constellation that we recognize as a "function" or, in the clinical setting, a symptom may be attributed to its disturbance. We know that inside one stream several tracts are known to connect different brain areas, and because we also know that certain types of functional processing take place in this stream, we may equate one cortico-cortical connection with one of these functions. But we have to be careful here. As stated elsewhere, "Interruption of the network has an impact on the remaining ("intact") parts of the network. In other words, the functions of Broca's or Wernicke's area with an intact arcuate fascicle may not be the same as after that fascicle's destruction." And "... the destruction of the interconnection (of the arcuate fascicle) may not result in a solitary repetition failure, but in a complete new phenomenological constellation, as the tract lesion affects the function of the regions it connects, and other regions in the remaining network, including in the other hemisphere, may become operational" (Rijntjes et al., 2012). However, this does not contradict the assumption that a particular connection in a network on its own is necessary for a specific function (e.g., repetition), as was shown recently (Kuemmerer et al., 2013).

Anatomically, the segregation between dorsal and ventral systems is alluded to by the orientation—in humans superior or inferior—of the involved cortical regions (e.g., in the motor system) (Rizzolatti & Matelli, 2003) or in the visuospatial attention system (Corbetta et al., 2005). Here, we take a different approach by defining dorsal and ventral systems by the course of the long association tracts connecting postrolandic and prerolandic brain regions, either above (i.e., "dorsal") or below (i.e., "ventral") the sylvian fissure. Note that this may give different results because cortical areas may have dorsal as well as ventral connections, and thus may serve as integration hubs.

It is common ground that dorsal and ventral streams use various anatomical tracts. The dorsal stream uses the superior longitudinal fascicles (SLF II, III) for parieto-frontal connections and the arcuate fascicle (AF) for temporo-frontal connections. Depending on the frontal endings, an anatomical correlate to the functional division originating from studies in the motor system [i.e., dorso-dorsal to premotor cortex and dorso-ventral connections to prefrontal cortex (Binkofski & Buxbaum, 2013)] starts to emerge (Hoeren et al., 2013, 2014). The ventral pathway comprises fibers running through the extreme capsule (Makris & Pandya, 2009) and the UF. Fibers in the ventral part of the external and extreme capsule have similar orientations and contribute to what in humans is called the IFOF (inferior frontal-occipital fasciculus) (Axer, Klingner, & Prescher, 2013; Fernández-Miranda, Rhoton, Kakizawa, Choi, & Alvarez-Linera, 2008; Gierhan, 2013). Thus, fibers from different cortical areas join and leave the IFOF, not only in the frontal or occipital lobes (as suggested by its name) but also in the temporal and parietal lobes (Umarova et al., 2010), which is not unlike entering and exiting a highway. The UF can be differentiated from the fibers of the extreme capsule and mainly connects medial and anterior temporal lobe regions with inferior frontal cortex. It is unclear whether there are direct ventral connections between inferior parietal lobule and prefrontal cortex, as suggested by human studies (Caspers et al., 2011; Catani, Howard, Pajevic, & Jones, 2002; Umarova et al., 2010; Vry et al., 2012) but without clear proof in primate studies (Schmahmann & Pandya, 2006). A connection between IPC and the temporal lobe was seen in primates studies; the middle longitudinal fascicle (mdlF) (Seltzer & Pandya, 1984) has been found in humans (Makris & Pandya, 2009; Saur et al., 2008; Wong, Chandrasekaran, Garibaldi, & Wong, 2011), but stimulation studies have not confirmed this connection so far (DeWitt Hamer, Morit-Gasser, Gatignol, & Duffau, 2011).

The exact anatomy of the fiber tracts and the cortical regions they connect, mainly based on tracing studies in nonhuman primates as well as a discussion on the pro and cons of human DTI-based fiber tracking in humans, are described elsewhere (Dick & Tremblay, 2012; Weiller et al., 2011) (Figure 27.1).

Conceptually, the dual loop model comprises at least three parallel layers. As the innermost layer, functionally close to the environmental end of the action–perception cycle, the dorsal streams use tracts to connect superior temporal lobe and IPC with premotor or posterior prefrontal cortex (BA 6,44, 45 B) (Bernal & Ardila, 2009; Parker et al., 2005; Saur et al., 2008, 2010). The next outer shell is provided by the ventral stream, which connects the middle and inferior temporal lobe with prefrontal cortices (BA 45A, 47) (Parker et al., 2005; Saur et al., 2008). Finally, prefrontal cortex (BA 47) and anterior temporal pole (ATL), functionally related to abstract and conceptual processing (Shallice & Cooper, 2013), are connected via the UF and may reflect the outermost layer. Also, in the parietal cortex, an onion bulb–like contribution to dorsal and ventral tracts has been suggested (Caspers et al., 2011) (Figures 27.2 and 27.3).

IFG is a region within the dual loop model that has direct ventral and dorsal connections with the two

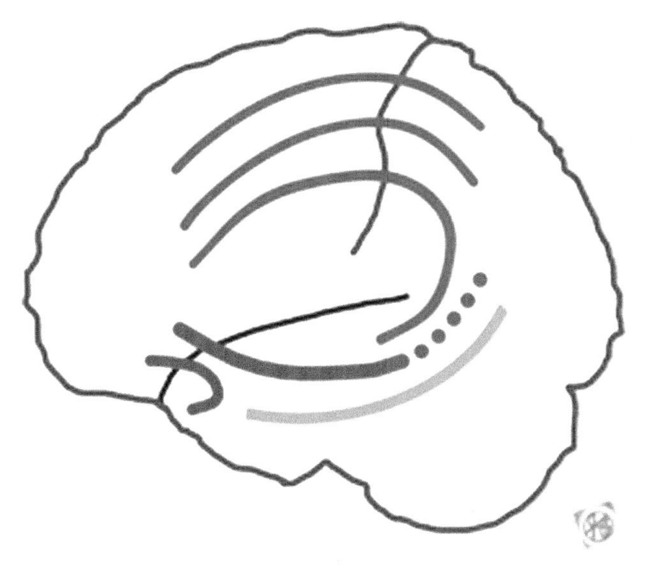

FIGURE 27.1 A schematic drawing of streams and tracts in the dual loop model. Blue denotes participation in the dorsal pathway and red denotes participation in the ventral pathway. There are at least three components within the dorsal system: the most superior tract relates to the "dorso-dorsal-stream" for "sensorimotor mapping" and may be mediated in humans by SLF II. The "dorso-ventral-stream" connects IPC with IFG along the SLF III. These two stream connotations are derived from the motor system. The AF connects the temporal lobe with Pmd and IFG, probably contributing to dorso-dorsal and dorso-ventral streams. The ventral stream has at least two components, the UF connecting the anterior part of the temporal lobe with the most inferior part of the IFG (BA 47). Fibers that may be related to the IFOF run through the extreme capsule and connect the middle and posterior parts of the temporal lobe with inferior anterior IFG (BA 45, 47). In this diagram we give the mdLF, which connects the temporal lobe with IPC a green color, thus attributing it neither to the dorsal nor to the ventral stream exclusively. Note that there are other opinions. Parietotemporal connections have been attributed to (an indirect part of) the arcuate fascicle (Catani et al., 2005), and direct connections between IPC and prefrontal cortex have been postulated. Clearly, this diagram is not comprehensive for the tracts involved in language processing (e.g., tracts connecting to SMA or to visual association cortex as Wernicke's perpendicular tract have been left out) but capitalizes on the dual loop system of long association tracts connecting prerolandic and postrolandic parts of the brain.

others lobes, temporal and parietal, which in turn might act in a kind of division of labor (Plaut et al., 1996) mediated by the mdLF, belonging to neither dorsal nor ventral streams, to master the extensive sensory processing (e.g., AG and pMTG may both play a role in "semantic control" (Noonan et al., 2013). IFG is also the only frontal area with connections to almost the entire rest of the prefrontal cortex, putting it in an ideal situation to integrate dorsal and ventral streams under frontal control. This was the reason we suggest that syntax, essentially to extract hierarchical relationships out of a sequence of elements, may rely on the integration of dorsal and ventral processing streams within Broca's area (Musso et al., 2003; Petrides & Pandya, 2002; Thompson-Schill, D'Esposito, Aguirre, & Farah, 1997; Weiller et al., 2009). There is a consensus that syntax, generally defined as a rule system capable of generating infinite sets of sequences, may not be processed by a single, monolithic cognitive computation, and thus may not be segregable in a single specific, isolated brain area, but rather it is processed in a widespread network. However, opinions are contrasting on how many and which pathways are involved in syntax processing, which tract is more relevant for language syntax, and what kind of functional processing may underlie syntactic-related pathways. There is an argument that the arcuate fascicle, especially the part connecting with BA 44, is essential or even specific for syntax (in the context of language) and decisive for child development and evolution of humans. We suggest that dorsal and ventral streams are both needed for syntax (Weiller et al., 2009). In a DTI study, Wilson et al. (2011) found reduced FA in patients with PPA in the AF as well as in the ventral tracts (UF and EmC). However, a correlation between severity of syntactic deficits and microstructural damage was related only with a decrease in FA in the arcuate within the AF/SLF tract, supporting the importance of the AF for syntax. These findings do not exclude the relevance of the ventral system for syntactic processing per se; instead, they show that interruption of AF fibers is critical in the breakdown of syntactic processing in this syndrome. Studies using healthy volunteers showed that syntax processing requires both ventral and dorsal tracts (Flöel, de Vries, Scholz, Breitenstein, & Johansen-Berg, 2009; Friederici et al., 2006; Lopez-Barroso et al., 2011). In the study by Friederici et al. (2006), processing of stimuli, generated by either a finite state or a phrase structure grammar (FSG versus PSG), was linked to the ventral route. Only stimuli generated by PSG were processed in the left pars opercularis of the IFG and thus belonged to the dorsal stream, which is in line, to some extent, with the Wilson et al. (2011) data. Thus, the studies of Wilson and Friederici appear to link the dorsal language route to syntactic complexity. Flöel et al. (2009) showed that FSG is also processed in the pars opercularis and the integrity of both ventral and dorsal tracts predicted high performance of grammar acquisition. In stroke patients, syntax was related to both the AF and the EmC systems (Griffiths, Marslen-Wilson, Stamatakis, & Tyler, 2012; Rolheiser, Stamatakis, & Tyler, 2011) and damage to the left ATL predicts impairment of complex syntax processing (Magnusdottir et al., 2013). Why should syntactic deficits in aphasia patients, but not in PPA patients, involve the ventral route? Etiology, disease course,

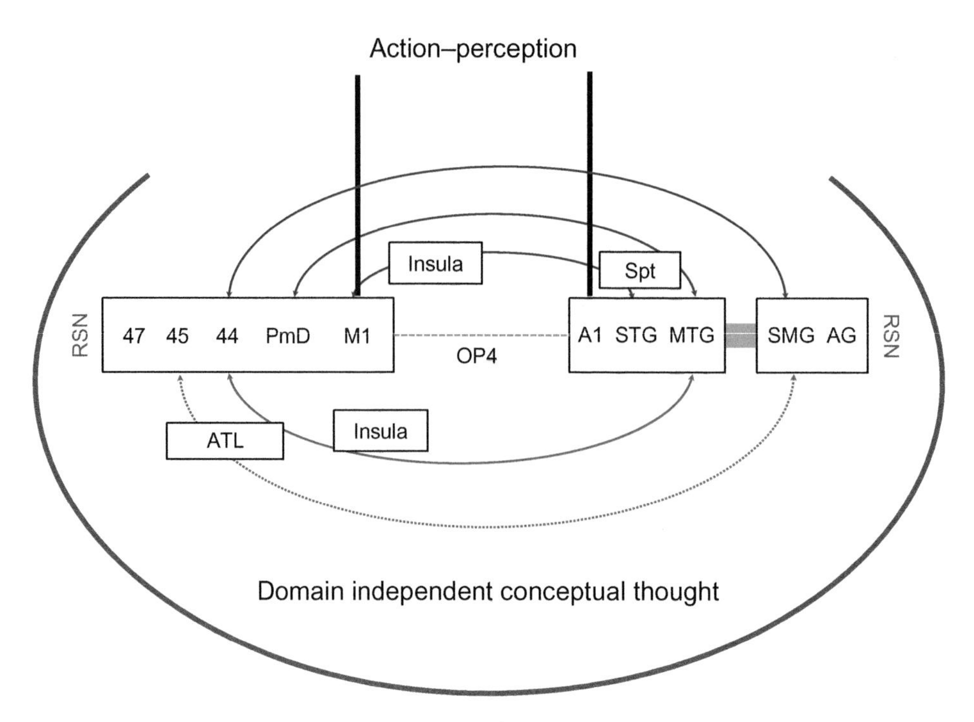

FIGURE 27.2 A diagram of connected regions of the dual loop model. Several dorsal (blue) tracts (see Figure 27.1) and ventral (red) tracts connect prerolandic (left) and postrolandic (right) brain regions. Most regions are connected by both dorsal and ventral tracts, resulting in a fully developed parallel dual loop system with reciprocal connections and equivalent pathways. Hierarchy is not determined by one specific pathway, although specific functions may primarily or crucially involve one or both, but rather by an extension of this system to regulatory, cytoarchitectonically more developed areas in prefrontal, temporal, and parietal neocortex in humans (Weiller et al., 2011). There are also regions that belong to either dorsal or ventral streams only (ATL). An onion-like structure is suggested by the different layers. For Op4, see Sepulcre (2013). RSN denotes resting-state network.

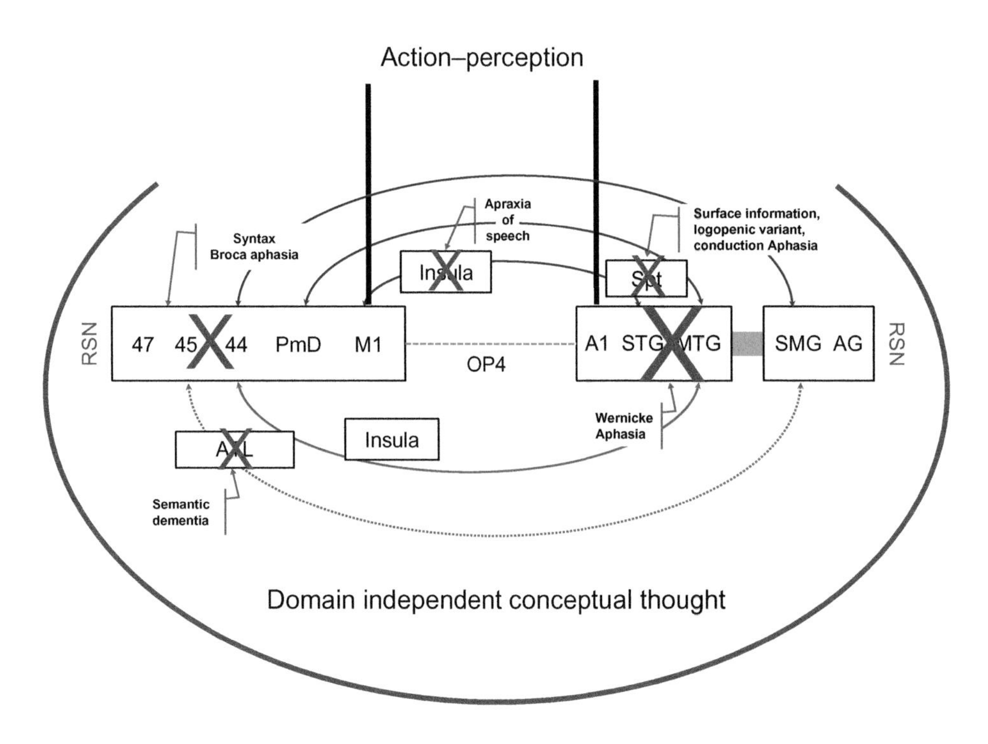

FIGURE 27.3 Hypothetical lesion sites of various forms of language disorders are put on the diagram in Figure 27.2, illustrating how symptoms or syndromes of patients might be fitted anatomically into the context of the dual loop model ("work in progress").

and functional reorganization of these two pathologies (aphasia and PPA) are different (Saur et al., 2006; Wilson et al., 2010). In PPA patients, progressive brain degeneration correlates with decreasing and, to some extent, abnormal brain plasticity (Wilson et al., 2010). Aphasia is due to an acute event that is followed by brain reorganization. Here, brain reorganization correlates with progressive recovery of language impairments in the acute, subacute, and chronic states (Weiller et al., 1995; Saur et al., 2006). There are also methodological differences between the studies. The extreme capsule mask used for extraction of FA values and used in the more recent DTI studies (Griffiths et al., 2012; Rolheiser et al., 2011) was centered on the left IFG (BA 45), whereas Wilson et al. (2011) used a mask starting in the insula. Different attributions to ventral or dorsal streams may follow these differences.

There are also claims of fundamentally different computational capacities of the various dorsal tracts, with (within the language domain) a specific role for syntax for the tract to IFG and a clear differentiation for language production for the other stream (Friederici et al., 2012) or a differentiation in one for sensorimotor integration and one for phonological processing (Catani, Jones, & ffytche, 2005). In the context of this chapter, similar differentiations have been made in other domains, such as with a dorsal-dorsal (for correct movement execution) and a dorsal-ventral (for the use of blueprints) stream in the motor system (Hoeren et al., 2013, 2014). We advocate that both dorsal streams have the capacity for time-dependent analysis (Rauschecker & Scott, 2009), which may be used for various functions. Also, the role of working memory or the phonological loop commonly attributed to the dorsal stream (to BA 44) (Paulesu, Frith, & Frackowiak, 1993) in mastering complex grammar should be taken into account. Syntactic relations, whether adjacent or long-distance, require strategies that not only are dependent on the temporal or spatial sequence of elements but also are optimized to test a limited number of possible combinations to convey meaningful relations. We attribute the latter features of syntax to the ventral route. One aspect, however, seems clear when looking at the current literature: syntax is served by both dorsal and ventral long association routes, which work in concert, albeit with different functional specializations. A clear division of the dorsal and ventral processing stream is artificial, resulting from experimental situations that do not reflect processing within the natural environment. For most functions, both streams would not be mutually exclusive but instead work in parallel (Cloutman, 2013; Makris & Pandya, 2009; Rauschecker & Scott, 2009; Rolheiser et al., 2011), constituting a loop (Weiller et al., 2011). Figure 27.3 potentially relates alterations of language to the dual loop model

27.5 DEVELOPMENT

A recent article shows that at birth, anterior and posterior language zones can be activated specifically but are not yet fully functionally connected (Perani et al., 2011), and that the interaction between the two regions becomes significantly synchronized at approximately 7 years of age (Friederici, Brauer, & Lohmann, 2011). This finding was considered in relation to the fact that, in contrast to the ventral pathway, the dorsal pathway has not yet fully matured (Brauer, Anwander, & Friederici, 2011; Brauer, Anwander, & Friederici, 2013) and that children up to the age of 7 are rather poor at comprehending syntactically complex sentences (Dittmar, Abbot-Smith, Lieven, & Tomasello, 2008; Dubois et al., 2008; Hahne, Eckstein, & Friederici, 2004).

Also, the full development of inner speech appears to occur around this age. According to Vygotsky, we do have conceptual awareness at birth, but no inner speech (Vygotzky, 1934). Young children approximately 4 years of age accompany their actions with speech in social situations. This speech is under the influence of continuous feedback from the environment and evolves from speech for others to overt speech for oneself, called "egocentric" or "private" speech that is still used in social situations, but its meaning is gradually directed at the child itself. During the process of a few years, until the age of approximately 7 years, egocentric speech is replaced by inner speech completely (Ehrich, 2006; Vygotzky, 1934).

It was suggested that it is only after the internalization of speech and after complete interaction of both pathways within the dual loop model that it is possible to *simultaneously* combine phonological and abstract thought proficiently (Figure 27.4). One assumption is that the development of inner speech is related to the mastering of complex grammar, which (until the age of 7) is something children fail in complex tasks because of metalinguistic demands. However, for both hypotheses the explanation could be that, for mastering these tasks, simultaneous analysis of both time-dependent and time-independent processing is required. Only then can we start using language as a tool to represent abstract concepts (Deutscher, 2005) and, as Jackendoff noted, use language as a "scaffolding that makes possible certain varieties of reasoning more complex than are available to nonlinguistic organisms"

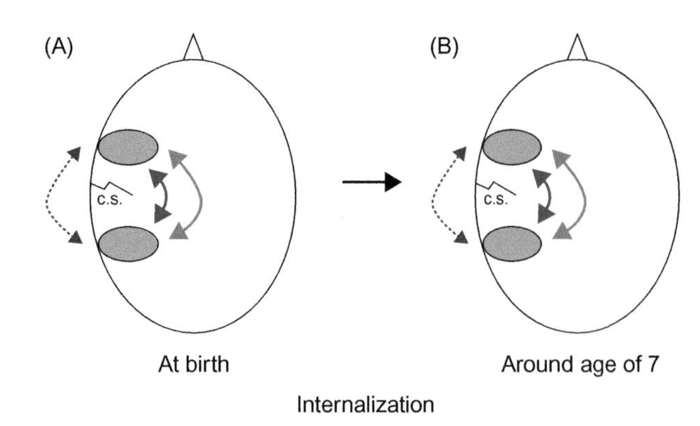

FIGURE 27.4 Heads seen from above with prerolandic and postrolandic areas around the central sulcus (c.s.). (A) At birth, dorsal anatomical connections (blue, dotted line) between prerolandic and postrolandic areas are present but immature. In the first years, influenced by the continuous percept and anticipation of the consequence of movement and speech via feedback (blue, continuous line), internal connections and representations synchronize and mature. (B) Over time, external behavior (blue, dotted line) is increasingly replaced by internal representations (blue, continuous line), whereas the interaction between dorsal (blue, continuous line) and ventral (red) pathways increases. For novel tasks, the external pathway can still be used (Rijntjes et al., 2012).

(Jackendoff, 1997), permitting self-description, reflection, self-questioning, and problem-solving (Barkley, 2001).

It is suggested that during this crucial age of 4–7 years in children, not only in language but also in all modalities, internalization takes place (Diaz & Berk, 1992; Vygotzky, 1978), "following the same general sequence of stages as the internalization of speech," and outer-directed behavior becomes turned on the self as a means to control one's own behavior (Barkley, 2001).

Of course, when encountering a new and difficult task, this process can be activated consciously, such as when reading a difficult text slowly and aloud to capture its meaning or when learning a new language and communicating with a teacher. In these cases, internalization will be a second step after externalization of the subject studied. It might be worthwhile to examine in patients with aphasia whether therapy should emphasize the externalization of language functions as a first step. Two new and promising therapies that were proven to be effective, CIAT (constraint-induced aphasia therapy) and PACE (promoting aphasics' communicative effectiveness), seem to have their focus exactly there.

References

Agosta, F., Henry, R. G., Migliaccio, R., Neuhaus, J., Miller, B. L., Dronkers, N. F., et al. (2010). Language networks in semantic dementia. *Brain*, *133*(1), 286–299.

Anderson, J. M., Gilmore, R., Roper, S., Crosson, B., Bauer, R. M., Nadeau, S., et al. (1999). Conduction aphasia and the arcuate fasciculus: A reexamination of the Wernicke–Geschwind model. *Brain and Language*, *70*, 1–12.

Axer, H., Klingner, C. M., & Prescher, A. (2013). Fiber anatomy of dorsal and ventral language streams. *Brain and Language*, *127*, 192–204.

Axer, H., von Keyserlingk, A. G., Berks, G., & von Keyserlingk, D. G. (2001). Supra- and infrasylvian conduction aphasia. *Brain and Language*, *76*, 317–331.

Baldo, J. V., & Dronkers, N. F. (2006). The role of inferior parietal and inferior frontal cortex in working memory. *Neuropsychology*, *20*(5), 529–538.

Baldo, J. V., Katseff, S., & Dronkers, N. F. (2012). Brain regions underlying repetition and auditory-verbal short-term memory deficits in aphasia: Evidence from voxel-based lesion symptom mapping. *Aphasiology*, *26*, 338–354.

Baldo, J. V., Klostermann, E. C., & Dronkers, N. F. (2008). It's either a cook or a baker: Patients with conduction aphasia get the gist but lose the trace. *Brain and Language*, *105*, 134–140.

Barkley, R. A. (2001). The executive functions and self-regulation: An evolutionary neuropsychological perspective. *Neuropsychology Review*, *11*, 1–29.

Bartha, L., & Benke, T. (2003). Acute conduction aphasia: An analysis of 20 cases. *Brain and Language*, *85*, 93–108.

Bernal, B., & Ardila, A. (2009). The role of the arcuate fasciculus in conduction aphasia. *Brain*, *132*, 2309–2316.

Binkofski, F., & Buxbaum, L. J. (2013). Two action systems in the human brain. *Brain and Language*, *127*, 222–229.

Brauer, J., Anwander, A., & Friederici, A. D. (2011). Neuroanatomical prerequisites for language functions in the maturing brain. *Cerebral Cortex*, *21*(2), 459–466.

Brauer, J., Anwander, A., & Friederici, A. D. (2013). Neuroanatomical prerequisites for language functions in the maturing brain. *Cerebral Cortex*, *21*, 459–466.

Caspers, S., Eickhoff, S. B., Rick, T., Von Kapri, A., Kuhlen, T., Huang, R., et al. (2011). Probabilistic fibre tract analysis of cytoarchitectonically defined human inferior parietal lobule areas reveals similarities to macaques. *Neuroimage*, *58*, 362–380.

Catani, M., Howard, R. J., Pajevic, S., & Jones, D. K. (2002). Virtual in vivo interactive dissection of white matter fasciculi in the human brain. *NeuroImage*, *17*, 77–94.

Catani, M., Jones, D. K., & ffytche, D. (2005). Perisylvian language networks of the human brain. *Annals of Neurology*, *57*, 8–16.

Cloutman, L. (2013). Interaction between dorsal and ventral processing streams: Where, when and how? *Brain and Language*, *127*, 251–263.

Coltheart, M., Patterson, K., & Marshall, J. C. (Eds.), (1980). *Deep dyslexia* London: Routledge & Kegan Paul.

Corbetta, M., Kincade, M. J., Lewis, C., Snyder, A. Z., & Sapir, A. (2005). Neural basis and recovery of spatial attention deficits in spatial neglect. *Nature Neuroscience*, *8*, 1603–1610.

Croxson, P. L., Johansen-Berg, H, Behrens, T. E., Robson, M. D., Pinsk, M. A., Gross, C. G., et al. (2005). Quantitavie investigation of connections of the prefrontal cortex in the huamn and macaque using probalistic diffusion tractography. *The Journal of Neuroscience*, *25*, 8854–8866.

Damasio, H., & Damasio, A. R. (1980). The anatomical basis of conduction aphasia. *Brain*, *103*, 337–350.

Dell, G. S., Martin, N., & Schwartz, M. F. (2007). A case-series test of the interactive two-step model of lexical access: Predicting word repetition from picture naming. *Journal of Memory and Language*, *56*, 490–520.

Dell, G. S., Schwartz, M. F., Nozari, N., Faseyitan, O., & Coslett, H. B. (2013). Voxel-based lesion-parameter mapping: Identifying the neural correlates of a computational model of word production. *Cognition*, *128*, 380–396.

Demonet, J. F., Thierry, G., & Cardebat, D. (2005). Renewal of the neurophysiology of language: Functional neuroimaging. *Physiological Reviews*, *85*(1), 49–95.

Deutscher, G. (2005). *The unfolding of language: The evolution of mankind's greatest invention*. London, UK: Random House.

DeWitt Hamer, P. C., Morit-Gasser, S., Gatignol, P., & Duffau, H. (2011). Is the human left middle longitudinal fascicle essential for language? A brain electrostimulation study. *Human Brain Mapping*, *32*, 962–973.

DeWitt, I., & Rauschecker, J. P. (2013). Wernicke's area revisited: Parallel streams and word processing. *Brain and Language*, *127*, 181–191.

Dick, A. S., & Tremblay, P. (2012). Beyond the arcuate fasciculus: Consensus and controversy in the connectional anatomy of language. *Brain*, *135*, 3529–3550.

Dittmar, M., Abbot-Smith, K., Lieven, E., & Tomasello, M. (2008). German children's comprehension of word order and casemarking in causative sentences. *Child Development*, *79*, 1152–1167.

Dubois, J., Dehaene-Lambertz, G., Perrin, M., Mangin, J. F., Cointepas, Y., Duchesnay, E., et al. (2008). Asynchrony of the early maturation of white matter bundles in healthy infants: Quantitative landmarks revealed noninvasively by diffusion tensor imaging. *Human Brain Mapping*, *29*, 14–27.

Duffau, H., Gatignol, P., Mandonnet, E., Peruzzi, P., Tzourio-Mazoyer, N., & Capelle, L. (2005). New insights into the anatomo-functional connectivity of the semantic system: A study using cortico-subcortical electrostimulations. *Brain*, *128*, 797–810.

Duffau, H., Peggy Gatignol, S. T., Mandonnet, E., Capelle, L., & Taillandier, L. (2008). Intraoperative subcortical stimulation mapping of language pathways in a consecutive series of 115 patients with Grade II glioma in the left dominant hemisphere. *Journal of Neurosurgery*, *109*(3), 461–471.

Ehrich, J. F. (2006). Vygotskian inner speech and the reading process. *Australian Journal of Educational & Developmental Psychology*, *6*, 12–25.

Fernández-Miranda, J. C., Rhoton, A. L., Jr., Kakizawa, Y., Choi, C., & Alvarez-Linera, J. (2008). The claustrum and its projection system in the human brain: A microsurgical and tractographic anatomical study. *Journal of Neurosurgery*, *108*(4), 764–774.

Flöel, A., de Vries, M. H., Scholz, J., Breitenstein, C., & Johansen-Berg, H. (2009). White matter integrity in the vicinity of Broca's area predicts grammar learning success. *Neuroimage*, *47*(4), 1974–1981.

Freud, S. (1891). *Zur Auffassung der Aphasien. Eine kritische Studie* (2nd ed.). Franz Deuticke: Leipzig und Wien.

Frey, S., Campbell, J. S. W., Pike, G. B., & Petrides, M. (2008). Dissociating the human language pathway with high angular resolution diffusion fiber tractography. *The Journal of Neuroscience: The Official Journal of the Society for Neuroscience*, *28*(5), 11435–11444.

Fridriksson, J., Kjartansson, O., Morgan, P. S., Hjaltason, H., Magnusdottir, S., Bonilha, L., et al. (2010). Impaired speech repetition and left parietal lobe damage. *The Journal of Neuroscience: The Official Journal of the Society for Neuroscience*, *30*, 11057–11061.

Friederici, A. D., Bahlmann, J., Heim, S., Schubotz, R. I., & Anwander, A. (2006). The brain differentiates human and nonhuman grammars: Functional localization and structural connectivity. *Proceedings of the National Academy of Sciences*, *103*(7), 2458–2463.

Friederici, A. D., Brauer, J., & Lohmann, G. (2011). Maturation of the language network: From inter- to intrahemispheric connectivities. *PLoS One*, *6*(6), e20726. Available from: http://dx.doi.org/10.1371/journal.pone.0020726.t001.

Galantucci, S., Tartaglia, M. C., Wilson, S. M., Henry, M. L., Filippi, M., Agosta, F., et al. (2011). White matter damage in primary progressive aphasias: A diffusion tensor tractography study. *Brain*, *134*(10), 3011–3029.

Galton, C. J., Patterson, K., Graham, K., Lambon-Ralph, M. A., Williams, G., Antoun, N., et al. (2001). Differing patterns of temporal atrophy in Alzheimer's disease and semantic dementia. *Neurology*, *57*, 216–225.

Geschwind, N. (1965). Disconnexion syndromes in animals and man. *Brain*, *88*(237–294), 585–644.

Geschwind, N. (1972). Language and the brain. *Scientific American*, *226*, 76–83.

Gierhan, S. M. E. (2013). Connections for auditory language in the brain. *Brain and Language*, *127*, 205–221.

Goldstein, K. (1927). *Die Lokalisation in der Großhirnrinde nach den Erfahrungen am kranken Menschen*. Berlin: Springer.

Gorno-Tempini, M. L., Brambati, S. M., Ginex, V., Ogar, J., Dronkers, N. F., Marcone, A., et al. (2008). The logopenic/phonological variant of primary progressive aphasia. *Neurology*, *71*(16), 1227–1234.

Griffiths, J. D., Marslen-Wilson, W. D., Stamatakis, E. A., & Tyler, L. K. (2012). Functional organization of the neural language system: Dorsal and ventral pathways are critical for syntax. *Cerebral Cortex*, *23*(1), 139–147. Available from: http://dx.doi.org/10.1093/cercor/bhr386.

Hahne, A., Eckstein, K., & Friederici, A. D. (2004). Brain signatures of syntactic and semantic processes during children's language development. *Journal of Cognitive Neuroscience*, *16*, 1302–1318.

Head, H. (1926). *Aphasia and kindred disorders of speech*. Cambridge: Cambridge University Press.

Herder, J.G. (1772). Abhandlung über den Ursprung der Sprache. Berlin Christian Friedrich Boß.

Hickok, G. (2009). The functional neuroanatomy of language. *Physical Life Review*, *6*, 121–143.

Hickok, G. (2012). Computational neuroanatomy of speech production. *Nature Reviews Neuroscience*, *13*, 135–145.

Hickok, G., & Poeppel, D. (2004). Dorsal and ventral streams: A framework for understanding aspects of the functional anatomy of language. *Cognition*, *92*, 67–99.

Hickok, G., & Poeppel, D. (2007). The cortical organization of speech processing. *Nature Reviews Neuroscience*, *8*, 393–402.

Hodges, J. R., Patterson, K., Oxbury, S., & Funnell, E. (1992). Semantic dementia: Progressive fluent aphasia with temporal-lobe atrophy. *Brain*, *115*, 1783–1806.

Hoeren, M., Kaller, C. P., Glauche, V., Vry, M. S., Rijntjes, M., Hamzei, F., et al. (2013). Action semantics and movement characteristics engage distinct processing streams during the observation of tool use. *Experimental Brain Research*, *229*(2), 243–260.

Hoeren, M., Kuemmerer, D., Bormann, T., Beume, L., Ludwig, L., Vry, M. S., et al. (2014). Neural bases of imitation and pantomime in acute stroke patients: Delineating dorsal and ventral streams for praxis. *Brain*, *137*, 2796–2810.

Hosomi, A., Nagakane, Y., Yamada, K., Kuriyama, N., Mizuno, T., Nishimura, T., et al. (2009). Assessment of arcuate fasciculus with diffusion-tensor tractography may predict the prognosis of aphasia in patients with left middle cerebral artery infarcts. *Neuroradiology*, *51*(9), 549–555.

Hu, W. T., McMillan, C., Libon, D., Leight, S., Forman, M., Lee, V. M., et al. (2010). Multimodal predictors for Alzheimer disease in nonfluent primary progressive aphasia. *Neurology*, *75*(7), 595–602.

Jackendoff, R. (1997). *The architecture of the language faculty*. Cambridge: The MIT Press.

Jefferies, E., Patterson, K., Jones, R. W., & Lambon Ralph, M. A. (2009). Comprehension of concrete and abstract words in semantic dementia. *Neuropsychology*, *23*(4), 492–499.

Klein, E., Korbinian, M., Glauche, V., Weiller, C., & Willmes, K. (2012). Processing pathways in mental arithmetic—evidence from probaalistic fiber tracking. *PLoS One*, *8*(1), e55455. Available from: http://dx.doi.org/10.1371/journal.pone.0055455.

Kleist, K. (1905). Über Leitungsaphasie. *Monatsschrift für Psychiatrie und Neurologie*, *17*, 503–532.

Kravitz, D. J., Saleem, K. S., Baker, C. I., & Mishkin, M. (2011). A new neural framework for visuospatial processing. *Nature reviews Neuroscience*, *12*, 217–230.

Kuemmerer, D., Bormann, T., Glauche, V., Mader, I., Rijntjes, M., Saur, D., et al. (2012). Acute aphasia syndromes in the context of the dual pathway model. *Neurobiology of language conference*, San Sebastian, 33.

Kuemmerer, D., Hartwigsen, G., Kellmeyer, P., Glauche, V., Mader, I., Kloppel, S., et al. (2013). Damage to ventral and dorsal language pathways in acute aphasia. *Brain*, *136*, 619–629.

Lambon Ralph, M. A., Graham, K. S., Ellis, A. W., & Hodges, J. R. (1998). Naming in semantic dementia—what matters? *Neuropsychologia*, *36*, 775–784.

Leclercq, D., Duffau, H., Delmaire, C., Capelle, L., Gatignol, P., Ducros, M., et al. (2010). Comparison of diffusion tensor imaging tractography of language tracts and intraoperative subcortical stimulations. *Journal of Neurosurgery*, *112*(3), 503–511.

Levelt, W. J. (1999). Models of word production. *Trends in Cognitive Sciences*, *3*(6), 223–232.

Lichtheim, L. (1885). On aphasia. *Brain*, *7*, 433–485.

Lopez-Barroso, D., de Diego-Balaguer, R., Cunillera, T., Camara, E., Münte, T. F., & Rodriguez-Fornells, A. (2011). Language learning under working memory constraints correlates with microstructural differences in the ventral language pathway. *Cerebral Cortex*, *21*, 2742–2750.

Magnusdottir, S., Fillmore, P., den Ouden, D. B., Hjaltason, H., Rorden, C., Kjartansson, O., et al. (2013). Damage to left anterior temporal cortex predicts impairment of complex syntax processing: A lesion-symptom mapping study. *Human Brain Mapping*, *34*, 2715–2723.

Makris, N., & Pandya, D. (2009). The extreme capsule in humans and rethinking of the language circuitry. *Brain Structure Function*, *213* (3), 343–358.

Martin, N., Saffran, E. M., & Dell, G. S. (1996). Recovery in deep dysphasia: Evidence for a relation between auditory verbal-STM capacity and lexical errors in repetition. *Brain and Language*, *52*, 83–113.

McCarthy, R., & Warrington, E. K. (1984). A two-route model of speech production: Evidence from aphasia. *Brain*, *107*, 463–485.

McCarthy, R., & Warrington, E. K. (2001). Repeating without semantics: Surface dysphasia? *Neurocase*, *7*, 77–87.

Meynert, T. (1866). Ein Fall von Sprachstörung, anatomisch begründet. In C. Braun, A. Duchek, & L. Schlager (Eds.), *XII. Band der Zeitschrift der K.u.K. Gesellschaft der Ärzte in Wien* (Vol. 22, pp. 152–189). Wien.

Milner, A. D., & Goodale, M. A. (1995). *The visual brain in action*. Oxford: Oxford University Press.

Mishkin, M., Ungerleider, L., & Macko, K. A. (1983). Object vision and spatial vision: Two visual pathways. *Trends in Neuroscience*, *6*, 414–417.

Morton, J. (1980). The logogen model and orthographic structure. In U. Frith (Ed.), *Cognitive processes in spelling* (pp. 117–133). London: Academic Press.

Mummery, C. J., Patterson, K., Wise, R. J., Vandenberghe, R., Price, C. J., & Hodges, J. R. (1999). Disrupted temporal lobe connections in semantic dementia. *Brain*, *122*(1), 61–73.

Musso, M., Moro, A., Glauche, V., Rijntjes, M., Reichenbach, J., Büchel, C., et al. (2003). Broca's area and the language instinct. *Nature Neuroscience*, *6*(7), 774–781.

Noonan, K. A., Jefferies, E., Visser, M., & Ralph, M. A. L. (2013). Going beyond inferior prefrontal involvement in semantic control: evidence for the additional contribution of dorsal angular gyrus and posterior middle temporal cortex. *Journal of cognitive neuroscience*, *25*(11), 1824–1850.

Pandya, D. N., & Yeterian, E. H. (1996). Comparison of prefrontal architecture and connections. *Philos Trans R Soc Lond B Biol Sci*, *351*(1346), 1423–1432.

Parker, G. J., Luzzi, S., Alexander, D. C., Wheeler-Kingshott, C. A., Ciccarelli, O., & Lambon Ralph, M. A. (2005). Lateralization of ventral and dorsal auditory-language pathways in the human brain. *Neuroimage*, *24*, 656–666.

Paulesu, E., Frith, C. D., & Frackowiak, R. S. J. (1993). The neural correlates of the verbal component of working memory. *Nature*, *362*, 342–344.

Petrides, M., & Pandya, D. (2002). Association pathways of the prefrontal cortex and functional observations. In D. T. Stuss, & R. T. Knight (Eds.), *Principles of frontal lobe function*. Oxford: Oxford University Press.

Petrides, M., & Pandya, D. (2009). Distinct parietal and temporal pathways to the homologues of Broca's area in the monkey. *PLoS Biology*, *7*(8), e1000170.

Petrides, M., & Pandya, D. N. (1988). Association fiber pathways to the frontal cortex from the superior temporal region in the rhesus monkey. *Journal of Comparative Neurology*, *273*(1), 52–66.

Plaut, D. C., McClelland, J. L., Seidenberg, M. S., & Patterson, K. (1996). Understanding normal and impaired word reading: Computational prinsiples in quasi-regular domains. *Psychological Review*, *103*, 56–115.

Poncet, M., Habib, M., & Robillard, A. (1987). Deep left parietal lobe syndrome: Conduction aphasia and other neurobehavioural disorders due to a small subcortical lesion. *Journal of Neurology, Neurosurgery, and Psychiatry*, *50*, 709–713.

Quigg, M., Geldmacher, D. S., & Elias, W. J. (2006). Conduction aphasia as a function of the dominant posterior perisylvian cortex. *Journal of Neurosurgery*, *104*, 845–848.

Rapp, B. (Ed.), (2001). *The handbook of cognitive neuropsychology: What deficits reveal about the human mind* Philadelphia, PA: Psychology Press.

Rauschecker, J. (1998). Cortical processing of complex sounds. *Current Opinion Neurobiology*, *8*, 516–521.

Rauschecker, J. P., & Scott, S. K. (2009). Maps and streams in the auditory cortex: Nonhuman primates illuminate human speech processing. *Nature Neuroscience*, *12*, 718–724.

Rijntjes, M., Dettmers, C., Buchel, C., Kiebel, S., Frackowiak, R. S., & Weiller, C. (1999). A blueprint for movement: Functional and anatomical representations in the human motor system. *The Journal of Neuroscience*, *19*, 8043–8048.

Rijntjes, M., Weiller, C., Bormann, T., & Musso, M. (2012). The dual loop model: Ist relation to language and other modalities. *Trends in Evolutionary Neuroscience*. Available from: http://dx.doi.org/10.3389/fnevo.2012.00009.

Rizzolatti, G., & Matelli, M. (2003). Two different streams form the dorsal visual system: Anatomy and functions. *Experimental Brain Research*, *153*, 146–157.

Rolheiser, T., Stamatakis, E. A., & Tyler, L. K. (2011). Dynamic processing in the human language system: Synergy between the arcuate fascicle and extreme capsule. *Journal of Neuroscience*, *31* (47), 16949–16957. Available from: http://dx.doi.org/10.1523/JNEUROSCI.2725-11.2011.

Romanski, L. M., Tian, B., Fritz, J., Mishkin, M., Goldman-Rakic, P. S., & Rauschecker, J. P. (1999). Dual streams of auditory afferents target multiple domains in the primate prefrontal cortex. *Nature Neuroscience*, *2*, 1131–1136.

Saur, D., Kreher, B. W., Schnell, S., Kuemmerer, D., Kellmeyer, P., Vry, M. S., et al. (2008). Ventral and dorsal pathways for language. *Proceedings of the National Academy of Sciences of the United States of America*, *105*, 18035–18040.

Saur, D., Schelter, B., Schnell, S., Kratochvil, D., Küpper, H., Kellmeyer, P., et al. (2010). Combining functional and anatomical connectivity reveals brain networks for auditory language comprehension. *NeuroImage*, *49*(4), 3187–3197.

Schmahmann, J. D., & Pandya, D. (2006). *Fiber pathways of the brain*. Oxford: Oxford University Press.

Schwindt, G. C., Graham, N. L., Rochon, E., Tang-Wai, D. F., Lobaugh, N. J., Chow, T. W., et al. (2013). Whole-brain white matter disruption in semantic and nonfluent variants of primary progressive aphasia. *Human Brain Mapping*, *34*(4), 973–984.

Scott, S. K., Blank, C. C., Rosen, S., & Wise, R. J. (2000). Identification of a pathway for intelligible speech in the left temporal lobe. *Brain*, *123*(Pt 12), 2400–2406.

Seltzer, B., & Pandya, D. N. (1984). Further observations on parieto-temporal connections in the rhesus monkey. *Experimental Brain Research*, *55*(2), 301–312.

Sepulcre, J. (2013). An OP4 functional stream in the language-related neuroarchitecture. *Cerebral Cortex*. Available from: http://dx.doi.org/10.1093/cercor/bht256.

Shallice, T. (1988). *From neuropsychology to mental structure*. Cambridge, UK: Cambridge University Press.

Shallice, T., & Cooper, R. P. (2013). Is there a semantic system for abstract words? *Frontiers in Human Neuroscience*. Available from: http://dx.doi.org/10.3389/fnhum.2013.00175.

Shallice, T., & Warrington, E. K. (1977). Auditory-verbal short-term memory impairment and conduction aphasia. *Brain and Language*, *4*, 479–491.

Thompson-Schill, S. L., D'Esposito, M., Aguirre, G. K., & Farah, M. J. (1997). Role of left inferior prefrontal cortex in retrieval of semantic knowledge: A reevaluation. *Proceedings of the National Academy of Sciences of the United States of America*, *94*, 14792–14797.

Trebuchon, A., Demonet, J. F., Chauvel, P., & Liégeois-Chauvel, C. (2013). Ventral and dorsal pathways of speech perception: An intracerebral ERP study. *Brain and Language*, *127*, 273–283.

Trevarthen, C. B. (1968). Two mechanisms of vision in primates. *Psychologische Forschung*, *31*, 299–337.

Ueno, T., Saito, S., Rogers, T. T., & Lambon Ralph, M. A. (2011). Lichtheim 2: Synthesizing aphasia and the neural basis of language in a neurocomputational model of the dual dorsal-ventral language pathways. *Neuron*, *72*, 385–396.

Umarova, R., Saur, D., Schnell, S., Kaller, C. P., Vry, M. S., Glauche, V., et al. (2010). Structural connectivity for visuospatial attention: Significance of ventral pathways. *Cerebral Cortex*, *20*(1), 121–129.

Vry, M. S., Saur, D., Rijntjes, M., Umarova, R., Kellmeyer, P., Schnell, S., et al. (2012). Ventral and dorsal fiber systems for imagined and executed movement. *Experimental Brain Research*, *219*, 203–216.

Vry, M. S., Tritschler, L. C., Hamzei, F., Rijntjes, J., Kaller, C., Hoeren, M., et al. (2014). The ventral pathway for pantomime of object use. *NeuroImage*, pii: S1053-8119(14)00904-5.

Vygotzky, L. S. (1934). *Thought and language*. Cambridge, MA: MIT Press.

Vygotzky, L. S. (1978). *Mind in society*. Cambridge, MA: Harvard University Press.

Weiller, C., Bormann, T., Saur, D., Musso, M., & Rijntjes, M. (2011). How the ventral pathway got lost: And what its recovery might mean. *Brain and Language*, *118*, 29–39.

Weiller, C., Musso, M., Rijntjes, M., & Saur, D. (2009). Please don't underestimate the ventral pathway in language. *Trends in Cognitive Science*, *13*(9), 369–370.

Wernicke, C. (1874). *Der aphasische Symptomenkomplex. Eine psychologische Studie auf anatomischer Basis*. Breslau: Cohn&Weigert.

Wernicke, C. (1906). Der aphasische Symptomenkomplex. In E. v. Leyden (Ed.), *Deutsche Klinik am Eingang des zwanzigsten Jahrhunderts in akademischen Vorlesungen* (Vol. VI, pp. 487–556). Berlin: Urban & Schwarzenberg.

Willmes, K., Korbinian, M., & Klein, E. (2014). Where numbers meet words: A common ventral network for semantic classification. *Scandinavian Journal of Psychology*. Available from: http://dx.doi.org/10.1111/sjop.12098.

Wilson, S. M., Brambati, S. M., Henry, R. G., Handwerker, D. A., Agosta, F., Miller, B. L., et al. (2009). The neural basis of surface dyslexia in semantic dementia. *Brain*, *132*(1), 71–86.

Wilson, S. M., Dronkers, N. F., Ogar, J. M., Jang, J., Growdon, M. E., Agosta, F., et al. (2010). Neural correlates of syntactic processing in the nonfluent variant of primary progressive aphasia. *Journal of Neuroscience*, *30*(50), 16845–16854.

Wilson, S. M., Galantucci, S., Tartaglia, M. C., Rising, K., Patterson, D. K., Henry, M. L., et al. (2011). Syntactic processing depends on dorsal language tracts. *Neuron*, *72*(2), 397–403.

Wise, R. J. (2003). Language systems in normal and aphasic human subjects: Functional imaging studies and inferences from animal studies. *British Medical Bulletin*, *65*, 95–119.

Wong, F. C. K., Chandrasekaran, B., Garibaldi, K., & Wong, P. C. M. (2011). White matter anisotropy in the ventral language pathway predicts sound-to-wort learning success. *The Journal of Neuroscience*, *31*(24), 8780–8785.

Yamada, K., Nagakane, Y., Mizuno, T., Hosomi, A., Nakagawa, M., & Nishimura, T. (2007). MR tractography depicting damage to the arcuate fasciculus in a patient with conduction aphasia. *Neurology*, *68*(10), 789.

CHAPTER

28

MUC (Memory, Unification, Control): A Model on the Neurobiology of Language Beyond Single Word Processing

Peter Hagoort

Donders Institute for Brain, Cognition and Behaviour, Max Planck Institute for Psycholinguistics, Nijmegen, The Netherlands

28.1 INTRODUCTION

Until not too long ago, the neurobiological model that has dominated our view on the neural architecture of language was the Wernicke-Lichtheim-Geschwind model. In this classical model, the human language faculty was situated in the left perisylvian cortex, with a division of labor between the frontal and temporal regions. Wernicke's area in left temporal cortex was assumed to subserve the comprehension of speech, whereas Broca's area in left inferior frontal cortex (LIFC) was claimed to subserve language production. The arcuate fasciculus connected these two areas. This model was based on single word processing. Since then, researchers interested in brain and language have realized that language is more than the concatenation of single words. Research focusing on sentence processing has found that lesions in Broca's region and adjacent cortex impair not only language production but also language comprehension (Caramazza & Zurif, 1976), whereas lesions in Wernicke's region not only affect language comprehension but also language production. More recent neuroimaging studies provided further evidence that central aspects of language production and comprehension are subserved by shared neural circuitry (Menenti, Gierhan, Segaert, & Hagoort, 2011; Segaert, Menenti, Weber, Petersson, & Hagoort, 2012). Since the advent of a whole toolkit of neuroimaging techniques, new models of the neural architecture of human language skills have been proposed. Here, I focus mainly on the Memory-Unification-Control (MUC) model as a model that tries to integrate knowledge about language processing beyond single words (Hagoort, 2005, 2013). After describing its three components, I discuss the evidence that has accumulated in support of the model.

28.2 MEMORY, UNIFICATION, AND CONTROL

The MUC model distinguishes three functional components of language processing: Memory, Unification, and Control. The Memory component refers to the linguistic knowledge that in the course of language acquisition gets consolidated in neocortical memory structures (see Davis & Gaskell, 2009, for the shift from medial temporal lobe to neocortical structures during consolidation). It is the only language-specific component of the model. The knowledge about the building blocks of language that is stored in memory (e.g., phonological, morphological, syntactic building blocks; jointly referred to as lexical items) is domain-specific and, hence, coded in a format that is different from, for example, color and visual object information.

However, language processing is more than memory retrieval and more than the simple concatenation of retrieved lexical items. The expressive power of human language derives from being able to combine elements from memory in novel ways. This process of deriving higher-level (i.e., sentence and beyond) meaning is referred to as Unification. Although as a result of the Chomskyan revolution in linguistics psycholinguistic studies of unification have mainly focused on

Neurobiology of Language. DOI: http://dx.doi.org/10.1016/B978-0-12-407794-2.00028-6

syntactic analysis, unification operations not only take place at the syntactic processing level but also are a hallmark of language across representational domains (Jackendoff, 2002). Thus, at the semantic and phonological levels, lexical elements are combined and integrated into larger structures. Hence, I distinguish between syntactic, semantic, and phonological unification (Hagoort, 2005).

Finally, the Control component relates language to joint action and social interaction, and it is invoked, for instance, when the contextually appropriate target language has to be selected, or for handling the joint action aspects of using language in conversational settings. Later, it is shown how languages have built-in linguistic devices that trigger the attentional control system into operation.

In the MUC model, the distribution of labor is as follows (Figure 28.1). Regions in the temporal cortex (in yellow) and the angular gyrus in parietal cortex subserve the knowledge representations that have been laid down in memory during acquisition. These regions store information, including phonological word forms, morphological information, word meanings, and the syntactic templates associated with noun, verbs, and adjectives (for details, see Hagoort, 2003, 2005, 2009). Dependent on knowledge type, different parts of temporal cortex are involved. Frontal regions (Broca's area and adjacent cortex; in blue) are crucial for unification operations. These operations generate larger structures from the building blocks that are retrieved from memory. Within LIFC (Unification Space), there seems to be a certain spatial distribution of recruitment dependent on the type of information that gets unified. Semantic unification recruits BA 47 and BA 45; syntactic unification has its focus in BA 45 and BA 44; phonological processes recruit BA 44 and ventral parts of BA 6 (Hagoort & Indefrey, 2014). In addition, executive control needs to be exerted such that the correct target language is selected, turn-taking in conversation is orchestrated, attention is given to the most relevant information in the input, and so forth. Control regions involve dorsolateral prefrontal cortex (in pink) and midline structure, including the anterior cingulate cortex and the parts of parietal cortex that are involved in attention (not shown in Figure 28.1).

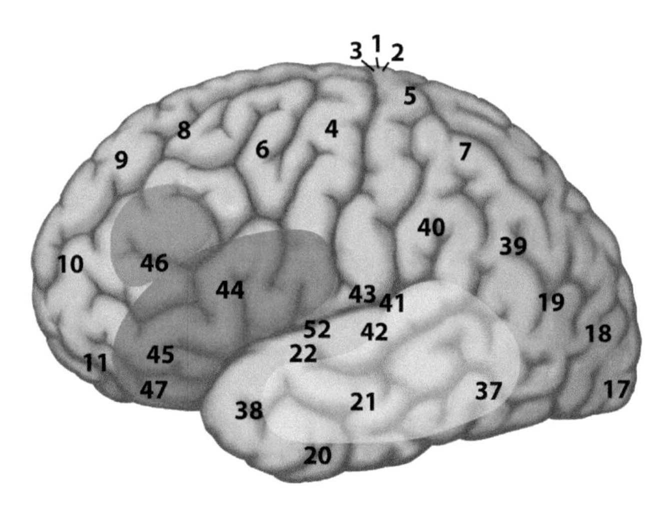

FIGURE 28.1 The MUC model of language. The figure displays a lateral view of the left hemisphere. The numbers indicate Brodmann areas. These are areas with differences in the cytoarchitectonics (i.e., composition of cell types). The memory areas are in the temporal cortex (in yellow) including the angular gyrus in parietal cortex. Unification requires the contribution of Broca's area (Brodmann areas 44 and 45) and adjacent cortex (Brodmann areas 47 and 6) in the frontal lobe. Control operations recruit another part of the frontal lobe (in pink) and the anterior cingulate cortex, as well as areas involved in attention (not shown in the figure).

The distribution of labor in the MUC model is not absolute. Language functions do not reside in single brain regions. Instead, language is subserved by dynamic networks of brain regions, including the ones outlined here. Ultimately, the mapping of a given language function onto the neural architecture of the brain is in terms of a network of brain areas instantiating that particular language function (McIntosh, 2008; Mesulam, 1998; Sporns, 2011). This is what Fedorenko and Thompson-Schill (2014) refer to as Networks of Interest. Typically, each node in such a network will participate dynamically in other functional networks as well. Although one can claim a certain contribution of a specific region (e.g., part of Broca's area), it is crucial to realize that such a contribution depends on the interaction with other regions that are part of the network. In short, "the mapping between neurons and cognition relies less on what individual nodes can do and more on the topology of their connectivity" (Sporns, 2011, p. 184). Therefore, before discussing the empirical evidence for the distribution of labor within the MUC framework, I discuss the connectivity profile of the language networks in the brain.

28.3 THE NETWORK TOPOLOGY OF THE LANGUAGE CORTEX

The classical model has given the arcuate fasciculus a central role in connecting the language-relevant parts of the brain. This was based on the idea that Broca's area and Wernicke's area were the two central nodes in the language network. The language network is much more extended than was assumed in the classical model and includes not only regions in the left hemisphere but also the right hemisphere areas. However, the evidence of additional activations in the right hemisphere and areas other than Broca's and Wernicke's does not take away the crucial role of left

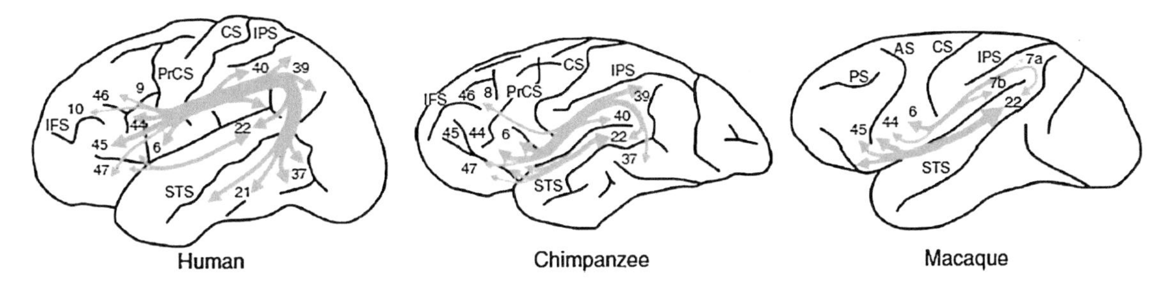

FIGURE 28.2 The arcuate fasciculus in a human, chimpanzee, and macaque in a schematic lateral view of the left hemisphere. *From Rilling et al. (2008), courtesy of Nature Publishing Group.*

perisylvian cortex. In a recent meta-analysis based on 128 neuroimaging studies, Vigneau et al. (2010) compared left and right hemisphere activations observed in relation to language processing. For phonological processing, lexico-semantic processing, and sentence or text processing, the number of activation peaks in the right hemisphere comprised less than one-third of the activation peaks in the left hemisphere. Moreover, in the majority of cases the right hemisphere activations were found in homotopic regions, suggesting a strong interhemispheric dependency. It is therefore justified to think that for the majority of the human population (e.g., with the exception of some portion of left-handers, cases of left hemispherectomy), the language-readiness of the human brain is strongly but not exclusively based on the organization of the left perisylvian cortex. This, however, does not deny the relevant contributions of the right hemisphere in, for instance, speech recognition (Hickok & Poeppel, 2007).

A recent technique for tracing fiber bundles in the living brain is diffusion tensor imaging (DTI). Using DTI, Rilling et al. (2008) tracked the arcuate fasciculus in humans, chimpanzees, and macaques. These authors found in humans a prominent temporal lobe projection of the arcuate fasciculus that is much smaller or absent in nonhuman primates (Figure 28.2). Moreover, connectivity with the middle temporal gyrus (MTG) was more widespread in the left than in the right hemisphere. This human specialization may be relevant for the evolution of language. Catani et al. (2007) found that the human arcuate fasciculus is strongly lateralized to the left, with quite some variation on the right. On the right, some people lack an arcuate fasciculus, in others it is smaller in size, and only in a minority of the population is this fiber bundle of equal size in both hemispheres. This pattern of lateralization was confirmed in a study involving 183 healthy right-handed volunteers in the age range between 5 and 30 years (Lebel & Beaulieu, 2009). The functionality of the arcuate fasciculus is not limited to single word processing. In a recent work, Wilson, Galantucci, Tartaglia, and Gorno-Tempini (2012) reported syntactic deficits in patients with primary

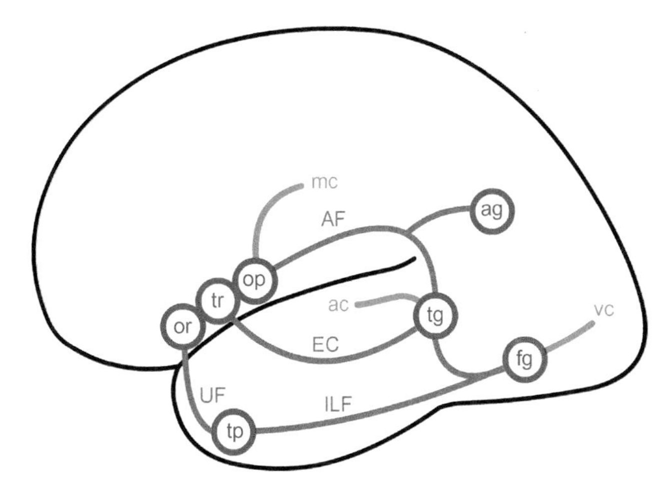

FIGURE 28.3 Simplified illustration of the anatomy and connectivity of the left hemisphere language network. Cortical areas are represented as red circles: pars orbitalis (or), pars triangularis (tr), and pars opercularis (op) of the LIFC, angular gyrus (ag), superior and middle temporal gyri (tg), fusiform gyrus (fg), and temporal pole (tp).

progressive aphasia after damage to the dorsal tracts but not after damage to the ventral tracts. This suggests that the dorsal tracts, including the arcuate fasciculus, are a key component in connecting frontal and temporal regions involved in syntactic processing. Again, exclusivity is difficult to establish. Part of these tracts might also subserve other aspects of language processing.

In addition to the arcuate fasciculus, other fiber bundles are important in connecting frontal with temporoparietal language regions (Figure 28.3). These include the superior longitudinal fasciculus (adjacent to the arcuate fasciculus) and the extreme capsule fasciculus, as well as the uncinate fasciculus, connecting Broca's area with superior and middle temporal cortex along a ventral path (Anwander, Tittgemeyer, von Cramon, Friederici, & Knosche, 2007; Friederici, 2009; Kelly et al., 2010). Figure 28.3 provides a schematic overview of the more extended connectivity profile of the left perisylvian cortex.

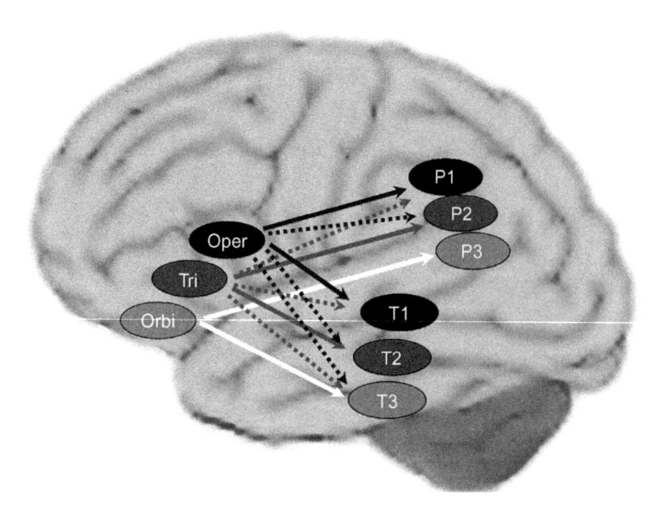

FIGURE 28.4 The topographical connectivity pattern between frontal and temporal/parietal cortex in the perisylvian language networks. Connections to the left pars opercularis (oper), pars triangularis (tri), and pars orbitalis (orbi) are shown in black, dark grey, and white arrows, respectively. The solid arrows represent the main (most significant) correlations and the dashed arrows represent the extending (overlapping) connections. Brain areas assumed to be mainly involved in phonological, syntactic, and semantic processing are shown in black, dark grey, and light grey circles, respectively. P1, supramarginal gyrus; P3, angular gyrus (AG); P2, the area between SMG and AG in the superior/inferior parietal lobule; T1, posterior superior temporal gyrus; T2, posterior MTG; P3, posterior inferior temporal gyrus.

Using resting state fMRI, Xiang, Fonteijn, Norris, and Hagoort (2010) found a clear topographically organized connectivity pattern in the left inferior frontal, parietal, and temporal regions (Figure 28.4). In the left—but not in the right—perisylvian cortex, functional connectivity patterns obeyed the tripartite nature of language processing (phonology, syntax, and semantics). These results support the assumption of the functional division for phonology, syntax, and semantics of the LIFC, including Broca's area, and revealed a topographical functional organization in the left perisylvian language network in which areas are most strongly connected according to information type (i.e., phonological, syntactic, and semantic). The dorsal pathways might be more relevant for phonological and syntactic processing, whereas the ventral pathways seem to be involved in connecting regions for semantic processing.

28.4 THE EMPIRICAL EVIDENCE FOR THE MUC MODEL

We have seen that there is a much more widespread connectivity profile in left perisylvian language cortex than was assumed in the classical model. The MUC model deviates from the classical model in the division of labor between Broca's area, Wernicke's area, and adjacent regions. However, the proposed distribution of labor is not absolute, but rather embedded and situated in the network skeleton of the language system's neural architecture.

What is the evidence for the relative division of labor proposed in the MUC model? Let us consider the syntactic network first. In comparison with phonological and semantic processing, which have compelling bilateral contributions, syntactic processing seems strongly lateralized to the left hemisphere perisylvian regions. Indirect support for a distinction between a memory component (i.e., the mental lexicon) and a unification component in syntactic processing comes from neuroimaging studies on syntactic processing. Two regions have been systematically reported in relation to syntactic processing (Hagoort & Indefrey, 2014): the left posterior superior/middle temporal gyrus (STG/MTG) and the LIFC. The left posterior temporal cortex is known to be involved in lexical processing (Hickok & Poeppel, 2004, 2007; Indefrey & Cutler, 2004; Lau, Stroud, Plesch, & Phillips, 2006). In connection to the MUC model, this part of the brain might be important for the retrieval of the syntactic frames that are stored in the lexicon. The idea of syntactic frames that specify the possible local syntactic environment of a given lexical item is in line with linguistic and computational approaches that assume syntactic knowledge to be lexically specified (Culicover & Jackendoff, 2006; Joshi & Schabes, 1997; Vosse & Kempen, 2000). The Unification Space, where individual frames are connected into a phrasal configuration for the whole utterance, might recruit the contribution of LIFC.

Direct empirical support for this distribution of labor between LIFC (Broca's area) and temporal cortex was found in a study of Snijders et al. (2009). These authors performed an fMRI study in which participants read sentences and word sequences containing word-category (noun-verb) ambiguous words (e.g., "watch") and the same materials with the unambiguous counterparts of the lexical-syntactic ambiguities. The ambiguous items were assumed to activate two independent syntactic frames, whereas the unambiguous counterparts result in the retrieval of only one syntactic frame. Solely based on a computational model of syntactic processing (Vosse & Kempen, 2000) and the hypothesized contribution of temporal and frontal cortex regions, it was predicted that the regions contributing to the syntactic unification process should show enhanced activation for sentences compared with words, and only within sentences should they display a larger signal for ambiguous than for unambiguous conditions. The posterior LIFC showed exactly this predicted pattern, confirming the hypothesis that

LIFC, particularly BA 44 and BA 45, contributes to syntactic unification. The left posterior MTG was activated more for ambiguous than unambiguous conditions, as predicted for regions subserving the retrieval of lexical-syntactic information from memory. It thus seems that the LIFC is crucial for syntactic processing in conjunction with the left posterior MTG, a finding supported by patient studies with lesions in these very same regions (Caplan & Waters, 1996; Rodd, Longe, Randall, & Tyler, 2010; Tyler et al., 2011). Presumably these regions are connected via the dorsal pathways.

In addition to syntactic unification, there is the need for semantic unification. One aspect of semantic unification is filling the slots in an abstract event schema. Semantic processing also recruits a left perisylvian network, albeit with a substantially weaker lateralization profile than syntactic processing. A series of fMRI studies aimed to identify the semantic processing network. These studies either compared sentences containing semantic/pragmatic anomalies with their correct counterparts (e.g., Friederici, Ruschemeyer, Hahne, & Fiebach, 2003; Hagoort, Hald, Bastiaansen, & Petersson, 2004; Kiehl, Laurens, & Liddle, 2002; Ruschemeyer, Zysset, & Friederici, 2006) or compared sentences with and without semantic ambiguities (Davis et al., 2007; Hoenig & Scheef, 2005; Rodd, Davis, & Johnsrude, 2005). In the latter case, there are multiple word meanings for a given lexical item that will induce competition and selection in relation to filling a particular slot in the event schema. As with syntactic unification, the availability of multiple candidates for a slot will therefore increase the unification load. In the case of the lexical-semantic ambiguities, there is no syntactic competition. Increased processing is therefore attributable to unification of meaning instead of syntax. The most consistent finding across studies on semantic unification is the activation of the LIFC, particularly BA 47 and BA 45 (Hagoort & Indefrey, 2014).

A further indication for the contribution of LIFC in semantic unification comes from a few studies investigating semantic unification of multimodal information with language. Using fMRI, Willems, Özyürek, and Hagoort (2007) assessed the neural integration of semantic information from spoken words and from co-speech gestures into a preceding sentence context. Spoken sentences were presented in which a critical word was accompanied by a co-speech gesture. Either the word or the gesture could be semantically incongruous with respect to the previous sentence context. Both an incongruous word as well as an incongruous gesture led to increased activation in LIFC (BA 45/47) as compared with congruous words and gestures (for a similar finding with pictures of objects, see Willems, Özyürek, & Hagoort, 2008). This supports the claim that LIFC is a key node in the semantic unification network, unifying semantic information from different modalities.

From these findings it seems that syntactic and semantic unification is realized in a dynamic interplay between LIFC as a multimodal unification site and also knowledge-specific regions. Again, it is important to stress that the interplay of these regions is crucial to realize the functional component of unification.

In other models, the anterior temporal lobe has been argued to be relevant for combinatorial operations (Hickok & Poeppel, 2007; Rogalsky et al., Chapter 47 of this volume). One possibility is that this is limited to conceptual combinations for which the mapping of grammatical roles (e.g., subject, object) onto thematic roles (e.g., agent, patient) is not required (Baron & Osherson, 2011). In the latter case, the contribution of Broca's region is presumably highly relevant.

28.5 A GENERAL ACCOUNT OF THE ROLE OF LIFC IN LANGUAGE PROCESSING

So far, we have seen that LIFC plays a central role in syntactic and semantic unification processes, albeit with different activation foci for these two types of unification. However, there is convincing evidence that LIFC also plays a role beneath the phrasal and sentence level. It is found to contribute to decomposition and unification at the word level. Words are not processed as unstructured, monolithic entities. Based on the morpho-phonological characteristics of a given word, a process of lexical decomposition takes place in which stems and affixes are separated. For spoken words, the trigger for decomposition can be something as simple as the inflectional rhyme pattern, which is a phonological pattern signaling the potential presence of an affix (Bozic, Tyler, Ives, Randall, & Marslen-Wilson, 2010). Decomposing lexical input appears to be a ubiquitous and mandatory perceptual strategy; that is, decompositional processes are triggered not only for words with obvious parts (e.g., work-ed) but also for semantically opaque words (e.g., bell-hop) and even nonwords with putative parts (e.g., blicket-s, blicket-ed). In a series of fMRI studies on the processing of inflectional morphology, Bozic et al. (2010) have found that LIFC, especially BA 45, subserves the process of morphological decomposition. Intracranial recordings in BA 45 from epileptic patients during presurgical preparation indicate that the same brain area is also involved in the generation of inflected forms during language production (Sahin, Pinker, Cash, Schomer, & Halgren, 2009; see also comments by Hagoort & Levelt, 2009).

The evidence for LIFC involvement at word- and sentence-level processing in both production and comprehension results in the question of how to account for its role more generally. This is still an open issue, but there is a possible answer. Notwithstanding the division of labor within LIFC, its overall contribution can be characterized in more general terms than hierarchical or even sentence-level processing. Instead, the LIFC is most likely involved in unification operations at the word and sentence level, in connection with temporal and parietal regions that are crucial for memory retrieval (Hagoort, 2005). Compositional and decompositional operations occur at multiple levels and at multiple time slices in the language processing system, but also outside the language system. Any time lexical and other building blocks enter into the process of utterance interpretation or construction, and any time the input string requires decomposition (presumably through analysis-by-synthesis) to contact the right lexical representations, LIFC is recruited.

This view is fully compatible with recent accounts in linguistics that view both morphology and syntax to involve the retrieval of pieces of stored structure with variables (Jackendoff, personal communication, 2014). Hence, no principled distinction is claimed between unification operations in syntax and morphology.

This account of LIFC's contribution is more general than is claimed in other models. For example, proposals have been made that LIFC (Broca's area) has a more specialized role in language processing, has more to do with linguistically motivated operations of syntactic movement (Grodzinsky & Santi, 2008), and is more involved in the processing of hierarchical structures (Friederici, Bahlmann, Heim, Schubotz, & Anwander, 2006). However, such proposals are difficult to reconcile with the LIFC contributions to morphological processes. Hence, the account specified here seems to have more empirical support.

28.6 THE DYNAMIC INTERPLAY BETWEEN MEMORY AND UNIFICATION

Although a connection is made between functional components of the cognitive architecture for language and specific brain regions, this is an idealization of the real neurophysiological dynamics of the perisylvian language network. Crucially, for language as for most other cognitive functions, the functional contribution of any area or region has to be characterized in the context of the network as a whole, where specialization of any given node is only relative and realized in a dynamic interaction with the other nodes in the network (Mesulam, 1990, 1998). How this can be viewed is specified in more detail for semantic unification by way of illustrating the principle of dynamic interaction.

In posterior and inferior temporal and parietal (angular gyrus) regions, neuronal populations are activated that represent lexical information associated with the incoming word, including its semantic features. From here, neural signals can follow two routes. The first exploits local connectivity within these posterior regions, resulting in a graded activation of neighboring neuronal populations, coding for related lexical-semantic information. Such local spread of activation contributes to setting up a lexical-semantic context in temporo-parietal cortex (Figure 28.5, green circle). The second route is based on long-distance connections to LIFC, through direct white matter fibers, resulting in the selective activation of populations of frontal cortex neurons. These will respond with a self-sustaining firing pattern (see Durstewitz, Seamans, & Sejnowski, 2000 for a review). Efferent signals in this case can only take the long-range route back. The most parsimonious account here is that frontal neurons will send efferent signals back to the same regions in temporo-parietal cortex from where afferent signals were received. This produces another spread of activation to neighboring temporo-parietal regions, which implies that connections representing a given semantic

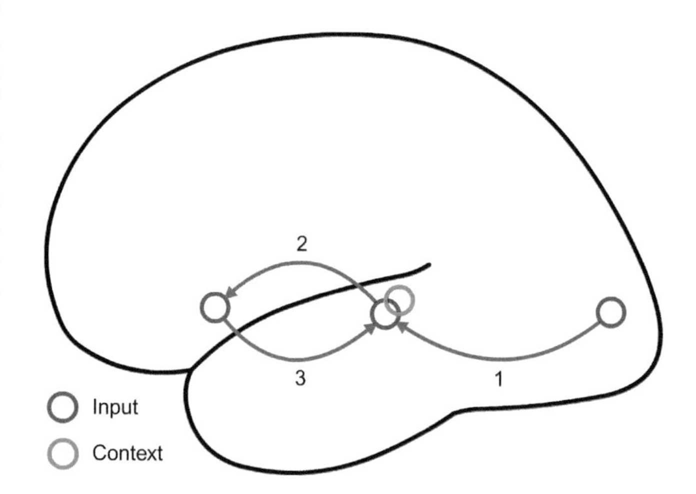

FIGURE 28.5 Processing cycle subserving semantic unification in the left hemisphere language network. Inputs are conveyed from sensory regions (here visual cortex) to the inferior, middle, and superior temporal gyri (1), where lexical information is activated. Signals are hence relayed to the inferior frontal gyrus (2), where neurons respond with a sustained firing pattern. Signals are then fed back into the same regions in temporal cortex from where they were received (3). A recurrent network is thus set-up, which allows information to be maintained online, a context (green circle) to be formed during subsequent processing cycles, and incoming words to be unified within the context. At each processing cycle a balance is achieved by letting input-driven activity find attractor states, that is, the maximum possible overlap with active populations in temporal cortex.

context will be strengthened. During each word processing cycle, the memory (temporo-parietal) and unification (inferior frontal) components interact by letting activation reverberate through the circuit in Figure 28.5. Achieving the necessary outcomes for language comprehension may be more or less demanding, depending on how close the relation is between input and context.

28.7 ATTENTIONAL CONTROL

The third component in the MUC model is referred to as Control. One form of control is attentional control. In classical models of sentence comprehension—of either the syntactic-structure-driven variety (Frazier, 1987) or in a constraint-based framework (Tanenhaus, Spivey-Knowlton, Eberhard, & Sedivy, 1995)—the implicit assumption is usually that a full phrasal configuration results and a complete interpretation of the input string is achieved. However, often the listener interprets the input on the basis of bits and pieces that are only partially analyzed. As a consequence, the listener might overhear semantic information (the Moses illusion; Erickson & Mattson, 1981) or syntactic information (the Chomsky illusion; Wang, Bastiaansen, Yang, & Hagoort, 2012). In the question "How many animals of each kind did Moses take on the ark?", people often answer "two," without noticing that it was Noah who was in command of the ark, not Moses. It was found that syntactic violations might not trigger a brain response if they are in a sentence constituent that provides no new information (Wang et al., 2012). Ferreira, Bailey, and Ferraro (2002) introduced the phrase "good-enough processing" to refer to the listeners' and readers' interpretation strategies. In a good-enough processing context, linguistic devices that highlight the most relevant parts of the input might help the listener/reader in allocating processing resources optimally. This aspect of linguistic meaning is known as "information structure" (Buring, 2007; Chafe, 1976; Halliday, 1967; Krifka, 2007). The information structure of an utterance essentially focuses the listener's attention on the crucial (new) information in it. In languages such as English and Dutch, prosody plays a crucial role in marking information structure. For instance, in question–answer pairs, the new or relevant information in the answer will typically be pitch accented. After a question like "What did Mary buy at the market?", the answer might be "Mary bought VEGETABLES" (accented word in capitals). In this case, the word "vegetables" is the focus constituent, which corresponds to the information provided for the Wh-element in the question. In a recent fMRI study (Kristensen, Wang, Petersson, & Hagoort, 2013), we tested the idea that pitch accent, which in Dutch is used to mark certain information as focus, recruits attentional networks in the service of more extended processing of the most relevant information. In our study, the attentional network was first localized in an auditory nonverbal attention task. This task activated, as expected, bilateral superior and inferior parietal cortex. In the language task, participants were listening to sentences with and sentences without semantic-pragmatic anomalies. In half of the cases these anomalies and their correct counterparts were marked as in focus by a pitch accent; in the other half of the cases they were not. The results showed an interaction in bilateral inferior parietal regions between prosody (pitch accent) and congruence; for incongruent sentences, but not for congruent ones, there was larger activation if the incongruent words carried a focus marker (i.e., the pitch accent).

Overall, the activation overlap in the attention networks between the localizer task and the sentence processing task indicated that marking of information structure modulated a domain-general attention network. Pitch accent signaled the saliency of the focused words and thereby recruited attentional resources for extended processing. This suggests that languages might have developed built-in linguistic devices (i.e., focus markers) that trigger the recruitment of attentional systems to safeguard against the possibility that in a good-enough processing system the most relevant information might go unnoticed. This provides one example of the interaction between a general demand/control system (Fedorenko, Duncan, & Kanwisher, 2012) and the core components of the language network.

28.8 BEYOND THE CLASSICAL MODEL

I have outlined the contours of a neurobiological model of language that is a substantial augmentation of the classical Wernicke-Lichtheim-Geschwind model, which was a model for single word processing mainly based on lesion and patient data.

Three major additions are worth highlighting. First, the connectivity of the language cortex in left perisylvian regions is much more extended than proposed in the classical model and is certainly not restricted to the arcuate fasciculus. Second, the distribution of labor between the core regions in left perisylvian cortex is not one in terms of production and comprehension. Shared circuitry has been established for core aspects of language production and comprehension. Both recruit temporal/parietal regions for retrieval of linguistic information that is laid down in memory during acquisition and LIFC for unification of building blocks into utterances or interpretations that are constructed online. Unification "enables words to cooperate to form new meanings" (Nowak, 2011, p. 179). Third, the operation of language in its full glory requires a much more extended network than what the classical model contained, which was mainly based on evidence from single word processing. The basic principle of brain organization for higher cognitive functions is that these are based on the interaction between a number of neuronal circuits and brain regions that support the different contributing functional components. These circuits are not necessarily specialized for language; nevertheless, they need to be recruited for the sake of successful language processing. One example is the general attentional networks that might be triggered into operation by specific linguistic devices to safeguard against missing out on the most relevant (new, focused) information in the language input. Another example, not further discussed here, is the Theory of Mind network that seems crucial for designing our utterances with knowledge of the listener in mind or, as a listener, to make the step from coded meaning to speaker meaning (Bašnáková, Weber, Petersson, van Berkum, & Hagoort, 2013; Hagoort & Levinson, 2015).

Acknowledgments

I am grateful to my colleagues in the NBL department and to Greg Hickok for their helpful comments. This chapter is an adapted version of an earlier discussion of the MUC model (Hagoort, 2013).

References

Anwander, A., Tittgemeyer, M., von Cramon, D. Y., Friederici, A. D., & Knosche, T. R. (2007). Connectivity-based parcellation of Broca's area. *Cerebral Cortex*, *17*, 816–825.

Baron, S. G., & Osherson, D. (2011). Evidence for conceptual combination in the left anterior temporal lobe. *Neuroimage*, *55*, 1847–1852.

Bašnáková, J., Weber, K., Petersson, K. M., van Berkum, J., & Hagoort, P. (2013). Beyond the language given: the neural correlates of inferring speaker meaning. *Cerebral Cortex*, *24*, 2572–2578. Available from: http://dx.doi.org/10.1093/cercor/bht112.

Bozic, M., Tyler, L. K., Ives, D. T., Randall, B., & Marslen-Wilson, W. D. (2010). Bihemispheric foundations for human speech comprehension. *Proceedings of the National Academy of Sciences of the United States of America*, *107*(40), 17439–17444.

Buring, D. (2007). Intonation, semantics and information structure. In G. Ramchand, & C. Reiss (Eds.), *The Oxford handbook of linguistic interfaces*. Oxford: Oxford University Press.

Caplan, D., & Waters, G. S. (1996). Syntactic processing in sentence comprehension under dual-task conditions in aphasic patients. *Language and Cognitive Processes*, *11*, 525–551.

Caramazza, A., & Zurif, E. B. (1976). Dissociation of algorithmic and heuristic processes in language comprehension: Evidence from aphasia. *Brain and Language*, *3*(4), 572–582.

Catani, M., Allin, M. P., Husain, M., Pugliese, L., Mesulam, M. M., Murray, R. M., et al. (2007). Symmetries in human brain language pathways correlate with verbal recall. *Proceedings of the National Academy of Sciences of the United States of America*, *104*(43), 17163–17168.

Chafe, W. L. (1976). Givenness, contrastiveness, definiteness, subjects, topics and point of view. In C. N. Li (Ed.), *Subject and topic*. New York, NY: Academic Press.

Culicover, P. W., & Jackendoff, R. (2006). The simpler syntax hypothesis. *Trends in Cognitive Sciences*, *10*(9), 413–418.

Davis, M. H., Coleman, M. R., Absalom, A. R., Rodd, J. M., Johnsrude, I. S., Matta, B. F., et al. (2007). Dissociating speech perception and comprehension at reduced levels of awareness. *Proceedings of the National Academy of Sciences of the United States of America*, *104*(41), 16032–16037.

Davis, M. H., & Gaskell, M. G. (2009). A complementary systems account of word learning: Neural and behavioural evidence. *Philosophical Transactions of the Royal Society of London. Series B*, *364*, 3773–3800.

Durstewitz, D., Seamans, J. K., & Sejnowski, T. J. (2000). Dopamine-mediated stabilization of delay-period activity in a network model of prefrontal cortex. *Journal of Neurophysiology*, *83*(3), 1733–1750.

Erickson, T. D., & Mattson, M. E. (1981). From words to meaning: A semantic illusion. *Journal of Verbal Learning and Verbal Behavior*, *20*(5), 540–551.

Fedorenko, E., Duncan, J., & Kanwisher, N. (2012). Language-selective and domain-general regions lie side by side within Broca's area. *Current Biology*, *22*(21), 2059–2062.

Fedorenko, E., & Thompson-Schill, S. L. (2014). Reworking the language network. *Trends in Cognitive Sciences*, *18*(3), 120–126.

Ferreira, F., Bailey, G. D. K., & Ferraro, V. (2002). Good-enough representations in language comprehension. *Current Directions in Psychological Science*, *11*(1), 11–15.

Frazier, L. (1987). Sentence processing: A tutorial review. In M. Coltheart (Ed.), *Attention and performance XII* (pp. 559–585). London, UK: Erlbaum.

Friederici, A. D. (2009). Pathways to language: Fiber tracts in the human brain. *Trends in Cognitive Sciences*, *13*, 175–181.

Friederici, A. D., Bahlmann, J., Heim, S., Schubotz, R. I., & Anwander, A. (2006). The brain differentiates human and non-human grammars: Functional localization and structural connectivity. *Proceedings of the National Academy of Sciences of the United States of America*, *103*(7), 2458–2463.

Friederici, A. D., Ruschemeyer, S. A., Hahne, A., & Fiebach, C. J. (2003). The role of left inferior frontal and superior temporal cortex in sentence comprehension: Localizing syntactic and semantic processes. *Cerebral Cortex*, *13*(2), 170–177.

Grodzinsky, Y., & Santi, A. (2008). The battle for Broca's region. *Trends in Cognitive Sciences*, *12*(12), 474–480.

Hagoort, P. (2003). How the brain solves the binding problem for language: A neurocomputational model of syntactic processing. *Neuroimage*, *20*, S18–S29.

Hagoort, P. (2005). On Broca, brain, and binding: A new framework. *Trends in Cognitive Sciences*, *9*, 416–423.

Hagoort, P. (2009). Reflections on the neurobiology of syntax. In D. Bickerton, & E. Szathmáry (Eds.), *Biological foundations and origin of syntax* (pp. 279–299). Cambridge, MA: MIT Press.

Hagoort, P. (2013). MUC (Memory, Unification, Control) and beyond. Frontiers in Psychology, 4(Article 416). Available from: http://dx.doi.org/10.3389/fpsyg.2013.00416.

Hagoort, P., Hald, L., Bastiaansen, M., & Petersson, K. M. (2004). Integration of word meaning and world knowledge in language comprehension. *Science*, *304*, 438–441.

Hagoort, P., & Indefrey, P. (2014). The neurobiology of language beyond single words. *Annual Review of Neuroscience*, *37*, 347–362.

Hagoort, P., & Levelt, W. J. M. (2009). The speaking brain. *Science*, *326*(5951), 372–373.

Hagoort, P., & Levinson, S. C. (2015). Neuropragmatics. In M. S. Gazzaniga (Ed.), *The cognitive neurosciences* (5th ed.). Cambridge, MA: MIT Press.

Halliday, M. A. K. (1967). Notes on transitivity and theme in English. Part 2. *Journal of Linguistics*, *3*, 177–274.

Hickok, G., & Poeppel, D. (2004). Dorsal and ventral streams: A framework for understanding aspects of the functional anatomy of language. *Cognition*, *92*, 67–99.

Hickok, G., & Poeppel, D. (2007). The cortical organization of speech processing. *Nature Reviews Neuroscience*, *8*(5), 393–402.

Hoenig, K., & Scheef, L. (2005). Mediotemporal contributions to semantic processing: fMRI evidence from ambiguity processing during semantic context verification. *Hippocampus*, *15*(5), 597–609.

Indefrey, P., & Cutler, A. (2004). Prelexical and lexical processing in listening. In M. S. Gazzaniga (Ed.), *The cognitive neurosciences III* (3rd ed., pp. 759–774). Cambridge, MA: MIT Press.

Jackendoff, R. (2002). *Foundations of language: Brain, meaning, grammar, evolution*. Oxford, UK: Oxford University Press.

Joshi, A. K., & Schabes, Y. (1997). Treeadjoining grammars. In A. Salomma, & G. Rosenberg (Eds.), *Handbook of formal languages and automata* (Vol. 3, pp. 69–124). Heidelberg: Springer-Verlag.

Kelly, C., Uddin, L. Q., Shehzad, Z., Margulies, D. S., Castellanos, F. X., Milham, M. P., et al. (2010). Broca's region: Linking human brain functional connectivity data and non-human primate tracing anatomy studies. *European Journal of Neuroscience*, *32*, 383–398.

Kiehl, K. A., Laurens, K. R., & Liddle, P. F. (2002). Reading anomalous sentences: An event-related fMRI study of semantic processing. *Neuroimage*, *17*(2), 842–850.

Krifka, M. (2007). Basic notions on information structure. In C. Féry, G. Fanselow, & M. Krifka (Eds.), *Working papers of the SFB632, interdisciplinary studies on information structure (ISIS) 6* (pp. 13–56). Potsdam: Universitätsverlag Potsdam.

Kristensen, L. B., Wang, L., Petersson, K. M., & Hagoort, P. (2013). The interface between language and attention: prosodic focus marking recruits a general attention network in spoken language comprehension. *Cerebral Cortex*, *23*(8), 1836–1848. Available from: http://dx.doi.org/10.1093/cercor/bhs164.

Lau, E., Stroud, C., Plesch, S., & Phillips, C. (2006). The role of structural prediction in rapid syntactic analysis. *Brain and Language*, *98*, 74–88.

Lebel, C., & Beaulieu, C. (2009). Lateralization of the arcuate fasciculus from childhood to adulthood and its relation to cognitive abilities in children. *Human Brain Mapping*, *30*(11), 3563–3573.

McIntosh, A. R. (2008). Large-scale network dynamics in neurocognitive function. In A. Fuchs, & V. K. Jirsa (Eds.), *Coordination: Neural, behavioral and social dynamics* (pp. 183–204). Springer: Berlin, Heidelberg.

Menenti, L., Gierhan, S. M. E., Segaert, K., & Hagoort, P. (2011). Shared language: Overlap and segregation of the neuronal infrastructure for speaking and listening revealed by functional MRI. *Psychological Science*, *22*, 1173–1182.

Mesulam, M.-M. (1990). Large-scale neurocognitive networks and distributed processing for attention, language, and memory. *Annals of Neurology*, *28*, 597–613.

Mesulam, M.-M. (1998). Form sensation to cognition. *Brain*, *121*, 1013–1052.

Nowak, M. (2011). *Super cooperators: Beyond the survival of the fittest. Why cooperation, not competition is the key to life*. Edinburgh-London: Canongate.

Rilling, J. K., Glasser, M. F., Preuss, T. M., Ma, X., Zhao, T., Hu, X., et al. (2008). The evolution of the arcuate fasciculus revealed with comparative DTI. *Nature Neuroscience*, *11*(4), 426–428.

Rodd, J. M., Davis, M. H., & Johnsrude, I. S. (2005). The neural mechanisms of speech comprehension: fMRI studies of semantic ambiguity. *Cerebral Cortex*, *15*(8), 1261–1269.

Rodd, J. M., Longe, O. A., Randall, B., & Tyler, L. K. (2010). Syntactic and semantic processing of spoken sentences: An fMRI study of ambiguity. *Paper presented at the 11th annual meeting Cognitive Neuroscience Society*. San Francisco.

Ruschemeyer, S. A., Zysset, S., & Friederici, A. D. (2006). Native and non-native reading of sentences: An fMRI experiment. *Neuroimage*, *31*(1), 354–365.

Sahin, N. T., Pinker, S., Cash, S. S., Schomer, D., & Halgren, E. (2009). Sequential processing of lexical, grammatical, and phonological information within Broca's area. *Science*, *326*(5951), 445–449.

Segaert, K., Menenti, L., Weber, K., Petersson, K. M., & Hagoort, P. (2012). Shared syntax in language production and language comprehension—an FMRI study. *Cerebral Cortex*, *22*(7), 1662–1670.

Snijders, T. M., Vosse, T., Kempen, G., van Berkum, J. J. A., Petersson, K. M., & Hagoort, P. (2009). Retrieval and unification of syntactic structure in sentence comprehension: An fMRI study using word-category ambiguity. *Cerebral Cortex*, *19*, 1493–1503.

Sporns, O. (2011). *Networks of the brain*. Cambridge, MA: MIT Press.

Tanenhaus, M. K., Spivey-Knowlton, M. J., Eberhard, K. M., & Sedivy, J. C. (1995). Integration of visual and linguistic information in spoken language comprehension. *Science*, *268*(5217), 1632–1634.

Tyler, L. K., Marslen-Wilson, W. D., Randall, B., Wright, P., Devereux, B. J., Zhuang, J., et al. (2011). Left inferior frontal cortex and syntax: Function, structure and behaviour in left-hemisphere damaged patients. *Brain*, *134*, 415–431.

Vigneau, M., Beaucousin, V., Herve, P. Y., Jobard, G., Petit, L., Crivello, F., et al. (2010). What is right-hemisphere contribution to phonological, lexico-semantic, and sentence processing? Insights from a meta-analysis. *Neuroimage*, *54*(1), 577–593.

Vosse, T., & Kempen, G. A. M. (2000). Syntactic structure assembly in human parsing: A computational model based on competitive inhibition and lexicalist grammar. *Cognition*, *75*, 105–143.

Wang, L., Bastiaansen, M. C. M., Yang, Y., & Hagoort, P. (2012). Information structure influences depth of syntactic processing: Event-related potential evidence for the Chomsky illusion. *PLoS One*, *7*(10), e47917.

Willems, R. M., Özyürek, A., & Hagoort, P. (2007). When language meets action: The neural integration of gesture and speech. *Cerebral Cortex*, *17*, 2322–2333.

Willems, R. M., Özyürek, A., & Hagoort, P. (2008). Seeing and hearing meaning: Event-related potential and functional magnetic resonance imaging evidence of word versus picture integration into a sentence context. *Journal of Cognitive Neuroscience*, *20*, 1235–1249.

Wilson, S. M., Galantucci, S., Tartaglia, M. C., & Gorno-Tempini, M. L. (2012). The neural basis of syntactic deficits in primary progressive aphasia. *Brain and Language*, *122*(3), 190–198.

Xiang, H., Fonteijn, H. M., Norris, D. G., & Hagoort, P. (2010). Topographical functional connectivity pattern in the perisylvian language networks. *Cerebral Cortex*, *20*, 549–560.

CHAPTER

29

The Neuroanatomical Pathway Model of Language: Syntactic and Semantic Networks

Angela D. Friederici

Department of Neuropsychology, Max Planck Institute for Human Cognitive and Brain Sciences, Leipzig, Germany

29.1 INTRODUCTION

Brain-related models of language have existed for more than 150 years, since Wernicke (1874) and Lichtheim (1884) published their models based on different aphasia syndromes. These models assumed separate centers for acoustic-sensory aspects supporting language comprehension and motor aspects supporting language production, as well as connections between them. It was assumed that if one of these centers or the connection between them was lesioned, then respective aphasic deficits would result, such as sensory aphasia, motor aphasia, or conduction aphasia. Pick (1909, 1913) was the first to incorporate syntactic and prosodic aspects in his model of language. However, the model remained vague with respect to the localization of these aspects, because it was again based on the analyses of language behavior in aphasic patients. In those days, the localization of the observed aphasia deficits had to wait until after the death of the patient, when the brain could be analyzed *ex vivo* (Broca, 1861, 1865; Wernicke, 1874).

Today, the existing brain imaging methodology allows us to localize different aspects of language in circumscribed brain regions and to specify their functional and structural connectivities *in vivo*. For the first time, this enables us to describe neural networks supporting semantic and syntactic processes in more detail and to formulate a functional neuroanatomical model of language.

Since the advent of functional magnetic resonance imaging (fMRI), a large number of studies have been conducted on word and sentence processing in different languages. These studies were able to localize different linguistic subfunctions such as phoneme discrimination, lexical retrieval, and syntactic phrase structure building in different brain regions. However, from these studies it also became clear that semantic and syntactic processes effective during sentence comprehension cannot be localized to a single brain area, but instead must be localized in definable neural networks.

This chapter describes the syntactic and sentence-level semantic networks taking into consideration fMRI data and, moreover, recent findings from diffusion-weighted MRI (dMRI) studies that allow the identification of the white matter fiber bundles connecting those brain regions that support particular language functions. Here, we focus on sentence-level processes and leave aside the studies on single word processing that are well-described in recent reviews (Binder, Desai, Graves, & Conant, 2009; Démonet, Thierry, & Cardebat, 2005; Price, 2000). Semantic processes at the word level may differ from semantic processes at the sentence level, where the most relevant issue is the relationship between words, not only between two nouns (*cat–dog*) but also between verbs and their argument nouns (*X chases Y*). Pick (1913) had already noted that the understanding of sentences requires more than just the process of word perception, which could be described as a reactive emergence of word meaning from memory. For the understanding of sentences, it holds that the meaning of a sentence is more than the sum of the meaning of the individual words.

29.2 FROM DORSAL AND VENTRAL STREAMS TO FIBER TRACTS

The functional MRI studies on syntactic and semantic aspects of sentence processing conducted over the past decade have consistently revealed an involvement

Neurobiology of Language. DOI: http://dx.doi.org/10.1016/B978-0-12-407794-2.00029-8

of the inferior frontal cortex and the temporal cortex. Meta-analyses indicate that both semantic and syntactic processes recruit partly different regions in the inferior frontal gyrus (IFG) and the superior and middle temporal gyrus (STG/MTG) (Chapter 48; Price, 2010; Vigneau et al., 2006). These brain regions are connected to each other via different fiber bundles (see Catani & de Schotten, 2008) that are referred to as the dorsal and the ventral pathway (for a review see Friederici, 2009a; Weiller, Musso, Rijntjes, & Saur, 2009).

Without basing their model on dMRI evidence of the fiber tracts connecting the language-related brain regions, Hickok and Poeppel (2004) discussed a functional "dorsal stream" as being responsible for sensory-to-motor mapping in speech processing and a functional "ventral stream" as supporting sound-to-meaning mapping. The method of dMRI that allows a neuroanatomical specification of fiber tracts within the human brain *in vivo* only emerged at that time (Behrens et al., 2003; Catani, Jones, & Ffytche, 2005). It is yet unclear what the relation is between the *functional* dorsal and ventral streams as defined by Hickok and Poeppel (2004) and the *anatomical* dorsal and ventral pathways as identified using dMRI. The early dMRI studies were able to identify the neuroanatomical connections between predefined brain regions such as Broca's area in the IFG and Wernicke's area in the prefrontal cortex (Catani et al., 2005). But because dMRI analyses only provide structural information about the fiber bundles, the possible functions of these can only be assigned quite indirectly by identifying the function of the fibers' termination regions (Friederici, 2009b).

A step toward a better description of the functions of fiber tracts was to use a combined fMRI–dMRI approach. In this approach, the specific function of particular brain regions is identified by an fMRI experiment. In a second step, these brain regions are used as seed regions from which the course of a fiber tract is calculated. The resulting fiber tract is interpreted to be relevant for the specific function processed in these brain regions. In principle, two methods of fiber tracking can be applied, probabilistic and deterministic fiber tracking. Probabilistic fiber tracking only takes one functionally defined region as the seed region and starting point of tractography, whereas deterministic fiber tracking takes two regions that are activated simultaneously by one function and calculates the fiber tract between the two.

A first study (Friederici, Bahlmann, Heim, Schubotz, & Anwander, 2006) using this combined fMRI–dMRI approach analyzed the dorsal and ventral fiber bundles connecting those prefrontal and temporal regions activated in response to a particular language function. The fMRI experiment applied an artificial grammar paradigm with rule-based syllable sequences after either adjacent or hierarchical nonadjacent dependency rules. Although the former activated the left frontal operculum, the latter additionally activated the posterior portion of Broca's area (BA 44). The dMRI data of probabilistic tracking from these two regions located in close vicinity revealed two distinct fiber tracts. First, a dorsal pathway connecting Broca's area to the posterior STG/MTG was found when seeding in Broca's area and, given this area's activation in the processing of nonadjacent hierarchical dependencies, this pathway was interpreted to support the processing of complex syntactic structures. Second, fiber tracking with a seed in the frontal operculum that was activated for the processing of adjacent dependencies revealed a ventral pathway to the temporal cortex; therefore, this pathway was viewed to support the processing of adjacent dependencies. Subsequently, a second study (Saur et al., 2008) using a combined fMRI–dMRI approach in language investigated the comprehension of simple sentences (*The pilot is flying the plane*) and the repetition of single words and pseudowords. The functional data show activation in the premotor cortex (PMC) and in the temporal cortex for repetition, whereas comprehension activated regions in the anterior frontal and the temporal cortex. Using a deterministic tracking approach, a dorsal and a ventral connection were identified. The authors interpret their data to show that the dorsal pathway supports sensory-to-motor mapping necessary for sensorimotor integration during speech processing and the ventral pathway sound-to-meaning mapping necessary for comprehension.

This apparent contradiction between the two studies in their functional interpretation of the dorsal and ventral pathway, however, can be explained on the basis of novel dMRI that allow analyses of those fiber tracts that connect the frontal cortex and temporal cortex dorsally and ventrally. These analyses indicate that the dorsal stream can be subdivided into two pathways with different termination points. A recent review (Friederici, 2011) proposed that this is also the case for the ventral stream. The present review discusses the relevant functional language processing studies and the combined fMRI–dMRI reports published more recently. Based on these findings, the Neuroanatomical Pathway Model of Language was formulated.

29.3 THE NEUROANATOMICAL PATHWAY MODEL OF LANGUAGE

The current model assumes four neuroanatomically distinguishable pathways connecting the language-relevant regions in the frontal cortex and the temporal cortex, two pathways that run dorsally and two pathways that run ventrally (see Figure 29.1).

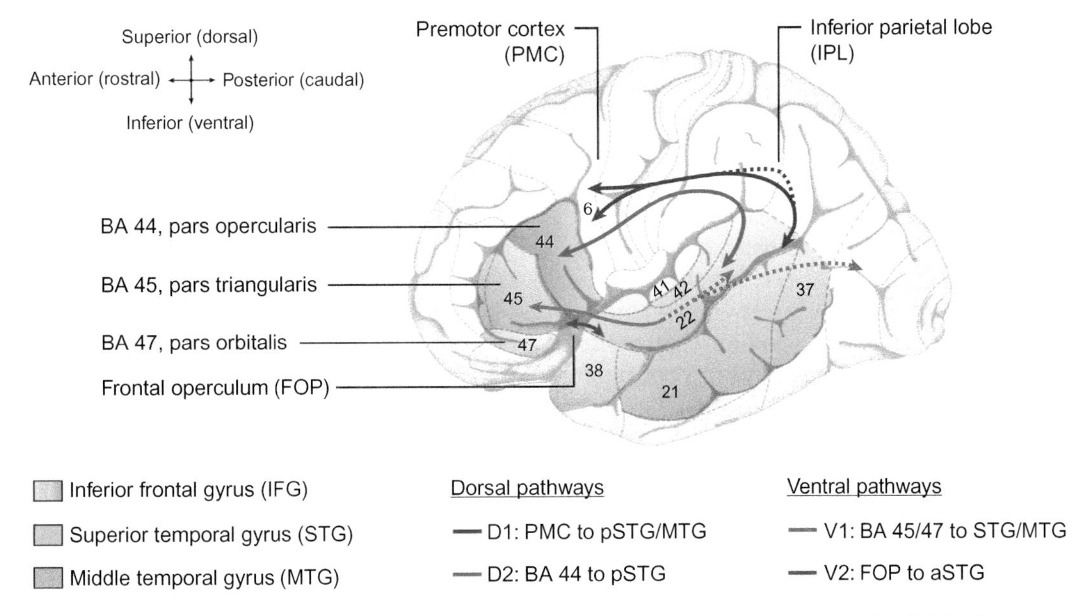

FIGURE 29.1 Language-relevant brain regions and schematic fiber tracts (displayed in the left hemisphere). Numbers represent cytoarchitectonically defined Brodmann areas (BAs). BA 44 and BA 45 together constitute Broca's area. BA 22 constitutes Wernicke's area. Different pathways are color-coded according to the color legend in the figure. Dorsal Pathway (D1) connects the PMC to the pSTG/MTG and involves the superior longitudinal fascile (SLF). Dorsal Pathway (D2) connects BA 44 to the pSTG and involves the arcuate fascile (AF). Ventral pathway (V1) connects BA 45/47 to the STG/MTG and involves the extreme capsule fiber system (ECFS)/longitudinal inferior-fronto-occipital fascile (IFOF). Ventral pathway (V2) connects the frontal operculum to the aSTG and involves the uncinate fascile (UF). PMC, premotor cortex; STG, superior temporal gyrus; pSTG, posterior superior temporal gyrus; aSTG, anterior superior temporal gyrus; MTG, middle temporal gyrus; pMTG, posterior middle temporal gyrus. *Adapted from Friederici and Gierhan (2013).*

29.3.1 Four Language-Related Pathways

Dorsally, one pathway (D1) connects the temporal cortex to the PMC and one pathway (D2) connects the temporal cortex to the posterior portion of Broca's area (BA 44). These two pathways can be distinguished neuroanatomically on the basis of their termination points, their relative location, and, moreover, they can be distinguished functionally. Neuroanatomically, the pathway from temporal cortex to PMC is located more dorsally and more laterally than the pathway to BA 44 (Perani et al., 2011). The pathway D1 that connects the STG and MTG with the PMC via the parietal cortex consists of fibers that are part of the SLF (Frey, Campbell, Pike, & Petrides, 2008; Saur et al., 2008). The SLF has been described as consisting of three different subparts (SLF I, II, III) (Makris et al., 2005). There is a fourth fiber bundle, the AF, that directly connects BA 44 in Broca's area with the posterior STG and MTG that constitute the pathway D2 (Frey et al., 2008; Friederici, Bahlmann, et al., 2006; Makris et al., 2005). Note that because this fascile runs closely parallel with the SLF, some researchers label this connection AF/SLF.

Ventrally, one pathway (V1) connects the anterior portion of Broca's area (BA 45) and BA 47 to the STG via a fiber bundle called the extreme fiber capsule system (EFCS) and is sometimes labeled IFOF. The other ventral pathway (V2) that connects the most ventral parts of the inferior frontal region including the frontal operculum to the temporal cortex via its anterior portion consists of the UF.

Only three of these four pathways are involved in sentence comprehension. The fourth pathway is the dorsal pathway (D1) that connects the temporal cortex to the PMC. In combined fMRI–dMRI studies, this pathway has clearly been shown to connect those brain regions that are involved in the repetition of speech (Gierhan, 2013; Saur et al., 2008). Thus, these findings are compatible with the model of Hickok and Poeppel (2004, 2007), which takes the dorsal pathway to support sensory-to-motor mapping. Furthermore, these authors proposed that this processing stream is most relevant during language acquisition (Hickok & Poeppel, 2007; but see Hickok, 2012; Hickok et al., 2011). Supporting evidence for this proposal, but at the same time evidence for a subdivision of the dorsal stream into two pathways, comes from a study comparing the fiber tracts in newborns and adults (Perani et al., 2011). Although the two dorsal pathways, the dorsal pathway (D1) connecting the temporal cortex to PMC and the dorsal pathway (D2) connecting the temporal cortex to posterior Broca's area, are present in adults, only the pathway (D1) to the PMC is present in newborns. Under the assumption that D1 supports sensory-to-motor mapping and that D2 supports the processing of complex hierarchical structures, it appears useful that D1 is established early in life because it allows the infant to enter the stage of babbling, during which the infant's output is tuned toward its perceived language environment.

The dorsal pathway D2 and the two ventral pathways are involved in language comprehension as they transmit information necessary for sentence processing and understanding. This information transfer during online sentence processing must take place within milliseconds. How the information transfer during language comprehension throughout the stages of auditory perception, word recognition, phrase structure building, and, finally, comprehension can be conceived has recently been outlined elsewhere when taking stimulus-driven bottom-up and context-driven top-down processes into account (Friederici, 2012). Here, we focus on the description of the neural basis of syntactic and semantic processes during sentence comprehension.

29.3.2 Syntactic and Semantic Networks

The present model is a weak syntax-first model that assumes that the processing system initially builds up a local phrase structure on the basis of the available word category information (Friederici, 2002; Friederici, 2011). Semantic and higher-order syntactic relations are only processed after that, unless the context is syntactically and semantically highly predictive. Thus, the model assumes two different stages of syntactic processing (Friederici, 2002) that are taken to be represented in two different syntactic networks.

29.3.2.1 Syntactic Networks

This first processing step of local structure building is functionally based on the grammatical knowledge of the target language. This concerns the basic knowledge about the structure of adjacent dependencies, such as local phrases, and there are only a few in each language, such as the determiner phrase and prepositional phrase. This knowledge must be acquired during language learning and its use becomes automatic as learning proceeds. In the adult brain, this process is highly automatic and it involves the frontal operculum and the anterior STG (Friederici, Rüschemeyer, Hahne, & Fiebach, 2003). If the process is less automatized, as in second language processing (Rüschemeyer et al., 2005) and during development (Brauer & Friederici, 2007), then BA 44 is also recruited. This is interesting because these two regions located adjacent to each other differ in their phylogeny, with the frontal operculum being phylogenetically older than BA 44 (Amunts & Zilles, 2012; Sanides, 1962). Thus, it appears that the more simple processes of processing adjacent dependencies are dealt with by a phylogenetically older cortex than is the more complex process of building structural hierarchies. The frontal operculum and the anterior STG are connected via the UF and comprise a network supporting local structure processing. The function of the respective regions could be defined as follows: in the adult brain, the anterior STG that receives its input from the auditory cortex represents templates of local phrases (determiner phrase, prepositional phrase), against which the incoming information is mapped. Thus, once the phrasal head (i.e., determiner, preposition) is encountered, the respective structure is made available in the anterior STG (Bornkessel & Schlesewsky, 2006). From here, the information is transferred via the UF to the frontal operculum, which in turn transmits this information to BA 44 for further processing. This ventral syntactic network is responsible for the most basic syntactic processes, that is, local syntactic computations.

A second syntactic network deals with more global computations. This refers to the processing of hierarchical dependencies as in syntactically complex sentences. The term complexity is used to cover different phenomena, including sentences with noncanonical word order (Friederici, Fiebach, Schlesewsky, Bornkessel, & von Cramon, 2006; Grewe et al., 2005; Meyer et al., 2012; Röder et al., 2002), sentences with varying degrees of embedding (Makuuchi et al., 2009), sentences with varying degrees of syntactically merged elements (Ohta et al., 2013), and the interplay of these sentence structures with working memory (see also Chapter 48). These studies indicate that across the different languages such as English, German, Hebrew, and Japanese, the factor of syntactic hierarchy operationalized as the reordering in noncanonical sentences or processing of embedded structures is localized in Broca's area, mostly in its posterior portion (BA 44). All these studies show that an increase in the level of hierarchy as defined in a syntactic tree leads to an increase in activation in BA 44.

A second region reported to be activated as a function of syntactic complexity and of verb-argument resolution is the posterior STG/STS (Ben-Shachar, Palti, & Grodzinsky, 2004; Friederici, Makuuchi, & Bahlmann, 2009; Kinno et al., 2008; Newman, Ikuta, & Burns, 2010; Santi & Grodzinsky, 2010). This region has also been activated when the semantic relation between a verb and its argument cannot be resolved (Friederici et al., 2003; Obleser & Kotz, 2010). Moreover, it was found that the factor of verb class and argument order interact in this region (Bornkessel, Zyssett, Friederici, von Cramon, & Schlesewsky, 2005). Thus, it appears that the posterior STG/STS is a region in which syntactic information and semantic verb-argument information are integrated (Grodzinsky & Friederici, 2006). Posterior Broca's area (BA 44) together with the posterior STG/STS constitute the second syntactic network, which is responsible for processing syntactically complex sentences. Within this dorsal syntactic network, BA 44 supports the build-up of hierarchical structures of nonadjacent elements, whereas the posterior STG/STS subserves the integration of semantic and syntactic information in complex sentences.

Thus, we have identified two syntactic networks, a ventral syntactic network and a dorsal syntactic network, each responsible for a different aspect of syntactic processing in the healthy adult brain. But the question arises whether there is further independent support for the view of two syntactic networks from either patient or developmental studies.

Unfortunately, patient studies do not allow us to distinguish between BA 44 and the frontal operculum, because both regions lie in the supply region of the middle artery. However, lesions in the IFG involving these two regions are reported to mostly result in syntactic processing deficits (for a review see Grodzinsky, 2000), although the diversity in group selection, design, and methodology leads to some diversity between the outcomes of different studies.

Recent fMRI and dMRI studies on patients revealed some interesting results concerning the dorsal and ventral syntactic network. A study by Griffiths, Marslen-Wilson, Stamatakis, and Tyler (2013) reported that patients with lesions in the left hemisphere involving either parts of the ventral network or the dorsal network showed some deficit in syntactic processing. Although this study does not systematically vary the complexity of the syntactic processes, the results generally support the idea that both networks are involved in syntactic processes. The study by Wilson et al. (2011) that investigated nonfluent progressive aphasics indicated that degeneration of the dorsal fiber tract connecting the temporal cortex (TC) and posterior Broca's area lead, in particular, to deficits in the processing of syntactically complex sentences. This finding is clearly in line with the current interpretation of D2.

Further support for the view that the pathway D2 subserves the processing of syntactically complex sentences stems from developmental studies on language that report that children, at an age when they are still deficient in processing noncanonical sentences, demonstrate a D2 that is not yet fully myelinized (Brauer, Anwander, & Friederici, 2011), and that the degree of myelination correlates with behavioral performance on processing noncanonical sentences (Skeide, 2012).

Concerning the ventral syntactic system, one must admit that reports on the relation between syntactic abilities and lesions in the temporal cortex are quite sparse because lesions in the temporal lobe are primarily related to semantic deficits. However, there is the interesting observation that only patients with temporal lesions that extend and include the anterior portion show syntactic comprehension deficits (Dronkers, Wilkins, Van Valin, Redfern, & Jaeger, 2004). Additionally there is the report of a correlational analysis of the white matter integrity in stroke patients and their behavioral performance indicating that syntax is processed both dorsally and ventrally (Rolheiser, Stamatakis, & Tyler, 2011).

Thus, quite a number of studies are in line with the view that there are two syntactic networks, a dorsal one and a ventral one. Moreover, few studies specifically demonstrate that the dorsal syntactic system is necessary to process hierarchically structured sentences (Brauer et al., 2011; Wilson et al., 2011).

29.3.2.2 Semantic Networks

The ventral stream has long been taken to support semantic processes (Hickok & Poeppel, 2004; Saur et al., 2008). As discussed, the ventral stream can be divided into two pathways, one involves the UF and the other involves the ECFS or IFOF. The functional allocation of these pathways is still undergoing debate. Many researchers see the UF as being involved in language processes (Catani & Mesulam, 2008; Duffau, 2008; Friederici, Bahlmann, et al., 2006; Parker et al., 2005); however, its particular function is a matter of discussion. The fMRI study by Friederici, Bahlmann, et al. (2006) suggests that the UF supports the processing of adjacent dependencies during sentence perception. In a parent contrast, the interoperative brain stimulation study with glioma patients by Duffau, Gatignol, Moritz-Gasser, and Mandonnet (2009) reported that language production interrupts when stimulating the ECFS, but not when stimulating the UF, which led the authors to deny the involvement of UF in language. However, the tasks used in this stimulation study were counting and picture naming; the former task taps the automatic production of number words and the latter task requires the retrieval of words, which would certainly rely more on a semantic network than on a syntactic network. Thus, the view that the UF supports basic syntactic processes in language is not to be disregarded on the basis of the findings from the interoperative stimulation study.

The ECFS has been reported to support semantic processes in many fMRI–dMRI and dMRI behavior studies (Saur et al., 2008; Turken & Dronkers, 2011; Wilson et al., 2010, 2011). This fiber system is also referred to as IFOF, because it runs from the ventral portion of the IFG along the temporal cortex to the occipital cortex. This way, those inferior frontal brain areas that are reported to be involved in semantic processes such as BA 45 and BA 47 (Bookheimer, 2002) are connected to the temporal cortex, which is known to support semantic processes including aspects of semantic memory (Patterson, Nestor, & Rogers, 2007).

BA 45/BA 47 as one of the regions in the semantic network is activated particularly when lexical semantic processes are under strategic control; that is when participants are required to perform some kind of semantic relatedness or plausibility judgement (Dapretto & Bookheimer, 1999; Fiez, 1997; Kuperberg et al., 2000; Newman et al., 2010; Thompson-Schill, D'Esposito, Aguirre, & Farah, 1997). The other region considered to be part of the semantic network is the anterior temporal lobe. Degeneration of this brain region leads to semantic deficits already at single-word level (Hodges, Patterson, Oxbury, & Funnell, 1992; Lambon Ralph & Patterson, 2008; Patterson et al., 2007).

Studies investigating sentence-level semantic processes are sparse. Different types of paradigms used to investigate sentence-level semantics revealed different findings. Studies that varied the semantic plausibility found activation in BA 45/47 in the IFG (Newman et al., 2010), whereas studies that investigated semantic predictability reported activation in the supramarginal and angular gyrus in the posterior temporo-parietal region (Obleser & Kotz, 2010).

At present, it is not entirely clear how to model these sentence-level semantic processes. What is clear, however, is that the anterior temporal lobe, the inferior frontal cortex, and the posterior temporo-parietal cortex are involved in sentence-level semantic processes, but that their interplay in the service of language understanding remains to be specified. The ventral pathway connecting the IFG and the STG/MTG plays a crucial part in semantic processes, but it is also conceived that the dorsal connection is involved whenever predictive processes are in dispute.

29.4 CONCLUSION

Sentence processing is based on three neuroanatomically defined networks: two syntactic networks and one semantic network. The semantic network involves the anterior temporal lobe, the anterior inferior frontal cortex, and the posterior temporo-parietal region. The former two regions are connected by a ventrally located pathway, the ECFS. Syntactic processes are based on two syntactic networks, a ventral and a dorsal network. The ventral network supporting the binding of adjacent dependencies, and thereby local syntactic building processes, involves the anterior STG and the frontal operculum connected via a ventrally located pathway (UF), whereas the dorsal network supporting the processing of syntactic hierarchies involves the posterior portion of Broca's area (BA 44) and the posterior STG connected via a dorsally located pathway (AF/SLF).

Acknowledgments

This work was supported by a grant from the European Research Council (ERC-2010-360 AdG 20100407 awarded to A.F.). The author thanks Elizabeth Kelly for English proofreading.

References

Amunts, K., & Zilles, K. (2012). Architecture and organizational principles of Broca's region. *Trends in Cognitive Sciences, 16*, 418–426.

Behrens, T. E. J, Johansen-Berg, H., Woolrich, M. W., Smith, S. M., Wheeler-Kingshott, C. A. M., Boulby, P. A., et al. (2003). Non-invasive mapping of connections between human thalamus and cortex using diffusion imaging. *Nature Neuroscience, 6*, 750–757.

Ben-Shachar, M., Palti, D., & Grodzinsky, Y. (2004). Neural correlates of syntactic movement: Converging evidence from two fMRI experiments. *NeuroImage, 21*, 1320–1336.

Binder, J. R., Desai, R. H., Graves, W. W., & Conant, L. L. (2009). Where is the semantic system? A critical review and meta-analysis of 120 functional neuroimaging studies. *Cerebral Cortex, 19*, 2767–2796.

Bookheimer, S. (2002). Functional MRI of language: New approaches to understanding the cortical organization of semantic processing. *Annual Review of Neuroscience, 25*, 151–188.

Bornkessel, I., & Schlesewsky, M. (2006). The extended argument dependency model: A neurocognitive approach to sentence comprehension across languages. *Psychologial Reviews, 113*, 787–821.

Bornkessel, I., Zyssett, S., Friederici, A. D., von Cramon, D. Y., & Schlesewsky, M. (2005). Who did what to whom? The neural basis of argument hierarchies during language comprehension. *NeuroImage, 26*, 221–233.

Brauer, J., Anwander, A., & Friederici, A. D. (2011). Neuroanatomical prerequisites for language functions in the maturing brain. *Cerebral Cortex, 21*, 459–466.

Brauer, J., & Friederici, A. D. (2007). Functional neural networks of semantic and syntactic processes in the developing brain. *Journal of Cognitive Neuroscience, 19*, 1609–1623.

Broca, P. (1861). Remarques sur le siége de la faculté du langage articulé, suivies d'une observation d'aphémie (parte de la parole). *Bulletins de la Société Anatomique de Paris, 6*, 330–357.

Broca, P. (1865). Sur le siege de la faculte du langage articule. *Bulletin de la Societe d'anthropologie, 6*, 337–393.

Catani, M., & de Schotten, M. T. (2008). A diffusion tensor imaging tractography atlas for virtual *in vivo* dissections. *Cortex, 44*, 1105–1132.

Catani, M., Jones, D. K., & Ffytche, D. H. (2005). Perisylvian language networks of the human brain. *Annals of Neurology, 57*, 8–16.

Catani, M., & Mesulam, M. (2008). The arcuate fasciculus and the disconnection theme in language and aphasia: History and current state. *Cortex, 44*, 953–961.

Dapretto, M., & Bookheimer, S. Y. (1999). Form and content: Dissociating syntax and semantics in sentence comprehension. *Neuron, 24*, 427–432.

Démonet, J. F., Thierry, G., & Cardebat, D. (2005). Renewal of the neurophysiology of language: Functional neuroimaging. *Physiologial Reviews, 85*, 49–95.

Dronkers, N. F., Wilkins, D. P., Van Valin, R. D., Redfern, B. B., & Jaeger, J. J. (2004). Lesion analysis of the brain areas involved in language comprehension. *Cognition, 92*, 145–177.

Duffau, H. (2008). The anatomo-functional connectivity of language revisited new insights provided by electrostimulation and tractography. *Neuropsychologia, 46*, 927–934.

Duffau, H., Gatignol, P., Moritz-Gasser, S., & Mandonnet, E. (2009). Is the left uncinate fasciculus essential for language? *Journal of Neurology, 256*, 382–389.

Fiez, J. A. (1997). Phonology, semantics, and the role of the left inferior prefrontal cortex. *Human Brain Mapping, 5*, 79–83.

Frey, S., Campbell, J. S. W., Pike, G. B., & Petrides, M. (2008). Dissociating the human language pathways with high angular resolution diffusion fiber tractography. *Journal of Neuroscience, 28*, 11435–11444.

Friederici, A. D. (2002). Towards a neural basis of auditory sentence processing. *Trends in Cognitive Sciences, 6*, 78–84.

Friederici, A. D. (2009a). Pathways to language: Fiber tracts in the human brain. *Trends in Cognitive Sciences, 13*, 175–181.

Friederici, A. D. (2009b). Allocating function to fiber tracts: Facing its indirectness. *Trends in Cognitive Sciences, 9*, 370–371.

Friederici, A. D. (2011). The brain basis of language processing: From structure to function. *Physiological Reviews, 91*, 1357–1392.

Friederici, A. D. (2012). The cortical language circuit: From auditory perception to sentence comprehension. *Trends in Cognitive Sciences, 16*, 262–268.

Friederici, A. D., Bahlmann, J., Heim, S., Schubotz, R. I., & Anwander, A. (2006). The brain differentiates human and non-human grammars: Functional localization and structural connectivity. *Proceedings of the National Academy of Sciences of the United States of America, 103*, 2458–2463.

Friederici, A. D., Fiebach, C. J., Schlesewsky, M., Bornkessel, I., & von Cramon, D. Y. (2006). Processing linguistic complexity and grammaticality in the left frontal cortex. *Cerebral Cortex, 16*, 1709–1717.

Friederici, A. D., & Gierhan, S. M. E. (2013). The language network. *Current Opinion in Neurobiology, 23*(2), 250–254.

Friederici, A. D., Makuuchi, M., & Bahlmann, J. (2009). The role of the posterior superior temporal cortex in sentence comprehension. *NeuroReport, 20*, 563–568.

Friederici, A. D., Rüschemeyer, S. A., Hahne, A., & Fiebach, C. J. (2003). The role of left inferior frontal and superior temporal cortex in sentence comprehension: Localizing syntactic and semantic processes. *Cerebral Cortex, 13*, 170–177.

Gierhan, S. M. E. (2013). Brain networks for language: Anatomy and functional roles of neural pathways supporting language comprehension and repetition (Doctoral dissertation). *MPI Series in Human Cognitive and Brain Sciences* (Vol. 144). Max Planck Institute for Human Cognitive and Brain Sciences, Leipzig.

Grewe, T., Bornkessel, I., Zysset, S., Wiese, R., von Cramon, D. Y., & Schlesewsky, M. (2005). The emergence of the unmarked: A new perspective on the language-specific function of Broca's brea. *Human Brain Mapping, 26*, 178–190.

Griffiths, J. D., Marslen-Wilson, W. D., Stamatakis, E. A., & Tyler, L. K. (2013). Functional organization of the neural language system: Dorsal and ventral pathways are critical for syntax. *Cerebral Cortex, 23*, 139–147.

Grodzinsky, Y. (2000). The neurology of syntax: Language use without Broca's area. *Behavioural Brain Sciences, 23*, 1–71.

Grodzinsky, Y., & Friederici, A. D. (2006). Neuroimaging of syntax and syntactic processing. *Current Opinion in Neurobiology, 16*, 240–246.

Hickok, G. (2012). The cortical organization of speech processing: Feedback control and predictive coding the context of a dual-stream model. *Journal of Communication Disorders, 45*, 393–402.

Hickok, G., Houde, J., & Rong, F. (2011). Sensorimotor integration in speech processing: Computational basis and neural organization. *Neuron, 69*, 407–422.

Hickok, G., & Poeppel, D. (2004). Dorsal and ventral streams: A framework for understanding aspects of the functional anatomy of language. *Cognition, 92*, 67–99.

Hickok, G., & Poeppel, D. (2007). The cortical organization of speech perception. *Nature Reviews Neuroscience, 8*, 393–402.

Hodges, J. R., Patterson, K., Oxbury, S., & Funnell, E. (1992). Semantic dementi. Progressive fluent aphasia with temporal lobe atrophy. *Brain, 115*, 1783–1806.

Kinno, R., Kawamura, M., Shioda, S., & Sakai, K. L. (2008). Neural correlates of noncanonical syntactic processing revealed by a picture-sentence matching task. *Human Brain Mapping, 29*, 1015–1027.

Kuperberg, G. R., McGuire, P. K., Bullmore, E. T., Brammer, M. J., Rabe-Hesketh, S., Wright, I. C., et al. (2000). Common and distinct neural substrates for pragmatic, semantic, and syntactic processing of spoken sentences: An fMRI study. *Journal of Cognitive Neuroscience, 12*, 321–341.

Lambon Ralph, M. A., & Patterson, K. (2008). Generalization and differentiation in semantic memory: Insights from semantic dementia. *Annals of the New York Academy of Sciences, 1124*, 61–76.

Lichtheim, L. (1884). Über Aphasie. *Deutsches Archiv für Klinische Medizin, 36*, 204–268 [reprinted and translated 1885. On aphasia. Brain, 7, 433-484.]

Makris, N., Kennedy, D. N., McInerney, S., Sorensen, A. G., Wang, R., Caviness, V. S., et al. (2005). Segmentation of subcomponents within the superior longitudinal fascicle in humans: A quantitative, *in vivo*, DT-MRI study. *Cerebral Cortex, 15*, 854–869.

Makuuchi, M., Bahlmann, J., Anwander, A., & Friederici, A. D. (2009). Segregating the core computational faculty of human language from working memory. *Proceedings of the National Academy of Sciences of the United States of America, 106*, 8362–8367.

Meyer, L., Obleser, J., Anwander, A., & Friederici, A. D. (2012). Linking ordering in Broca's area to storage in left temporo-parietal regions: The case of sentence processing. *NeuroImage, 62*, 1987–1998.

Newman, S. D., Ikuta, T., & Burns, T. (2010). The effect of semantic relatedness on syntactic analysis: an fMRI study. *Brain and Language, 113*, 51–58.

Obleser, J., & Kotz, S. A. (2010). Expectancy constraints in degraded speech modulate the language comprehension network. *Cerebral Cortex, 20*, 633–640.

Ohta, S., Fukui, N., & Sakai, K. L. (2013). Syntactic computation in the human brain: The degree of merger as a key factor. *PLoS ONE, 8*(2), e56230.

Parker, G. J., Luzzi, S., Alexander, D. C., Wheeler-Kingshott, C. A., Ciccarelli, O., & Lambon Ralph, M. A. (2005). Lateralization of ventral and dorsal auditory language pathways in the human brain. *NeuroImage, 24*, 656–666.

Patterson, K., Nestor, P. J., & Rogers, T. T. (2007). Where do you know what you know? The representation of semantic knowledge in the human brain. *Nature Reviews Neuroscience, 8*, 976–987.

Perani, D., Saccuman, M. C., Scifo, P., Anwander, A., Spada, D., Baldoli, C., et al. (2011). The neural language networks at birth. *Proceedings of the National Academy of Sciences of the United States of America, 108*, 16056–16061.

Pick, A. (1909). *Über das Sprachverständnis*. Leipzig: Barth.

Pick, A. (1913). *Die agrammatischen Sprachstörungen; Studien zur psychologischen Grundlegung der Aphasielehre*. Berlin: Springer.

Price, C. J. (2000). The anatomy of language: Contributions from functional neuroimaging. *Journal of Anatomy, 197*, 335–359.

Price, C. J. (2010). The anatomy of language: A review of 100 fMRI studies published in 2009. *Annals of the New York Academy of Sciences, 1191*, 62–88.

Röder, B., Stock, O., Neville, H. J., Bien, S., & Rösler, F. (2002). Brain activation modulated by the comprehension of normal and pseudo-word sentences of different processing demands a functional magnetic resonance imaging study. *NeuroImage, 15*, 1003–1014.

Rolheiser, T., Stamatakis, E. A., & Tyler, L. K. (2011). Dynamic processing in the human language system: Synergy between the arcuate fascicle and extreme capsule. *Journal of Neuroscience, 31*, 16949–16957.

Rüschemeyer, S.-A., Fiebach, C. J., Kempe, V., & Friederici, A. D. (2005). Processing lexical semantic and syntactic information in first and second language: fMRI evidence from German and Russian. *Human Brain Mapping, 25*, 266–286.

Sanides, F. (1962). The architecture of the human frontal lobe and the relation to its functional differentiation. *International Journal of Neurology, 5*, 247–261.

Santi, A., & Grodzinsky, Y. (2010). fMRI adaptation dissociates syntactic complexity dimensions. *NeuroImage, 51*, 1285–1293.

Saur, D., Kreher, B. W., Schnell, S., Kümmerer, D., Kellmeyer, P., Vry, M. S., et al. (2008). Ventral and dorsal pathways for language. *Proceedings of the National Academy of Sciences of the United States of America, 105*, 18035–18040.

Skeide, M.A. (2012). Syntax and semantics networks in the developing brain. (Doctoral dissertation). *MPI Series in Human Cognitive and Brain Sciences* (Vol. 143). Max Planck Institute for Human Cognitive and Brain Sciences, Leipzig.

Thompson-Schill, S. L., D'Esposito, M., Aguirre, G. K., & Farah, M. J. (1997). Role of left inferior prefrontal cortex in retrieval of semantic knowledge: A reevaluation. *Proceedings of the National Academy of Sciences of the United States of America, 94*, 14792–14797.

Turken, A. U., & Dronkers, N. F. (2011). The neural architecture of the language comprehension network: Converging evidence from lesion and connectivity analyses. *Frontiers in Systems Neuroscience, 5*, 1. Available from: http://dx.doi.org/10.3389/fnsys.2011.00001.

Vigneau, M., Beaucousin, V., Herve, P. Y., Duffau, H., Crivello, F., Houde, O., et al. (2006). Meta-analyzing left hemisphere language areas: Phonology, semantics, and sentence processing. *NeuroImage, 30*, 1414–1432.

Weiller, C., Musso, M., Rijntjes, M., & Saur, D. (2009). Please don't underestimate the ventral pathway in language. *Trends in Cognitive Sciences, 13*, 369–370.

Wernicke, C. (1874). *Der aphasische Symptomencomplex*. Berlin: Springer-Verlag.

Wilson, S. M., Dronkers, N. F., Ogar, J. M., Jang, J., Growdon, M. E., Agosta, F., et al. (2010). Neural correlates of syntactic processing in the nonfluent variant of primary progressive aphasia. *Journal of Neuroscience, 30*, 16845–16854.

Wilson, S. M., Galantucci, S., Tartaglia, M. C., Rising, K., Patterson, D. K., Henry, M. L., et al. (2011). Syntactic processing depends on dorsal language tracts. *Neuron, 72*, 397–403.

CHAPTER

30

The Argument Dependency Model

Ina Bornkessel-Schlesewsky[1,2] *and Matthias Schlesewsky*[3]

[1]Cognitive Neuroscience Laboratory, School of Psychology, Social Work and Social Policy, University of South Australia, Adelaide, SA, Australia; [2]Department of Germanic Linguistics, University of Marburg, Marburg, Germany; [3]Department of English and Linguistics, Johannes Gutenberg-University Mainz, Mainz, Germany

30.1 INTRODUCTION

The (extended) Argument Dependency Model (eADM) is somewhat of a newcomer to the scene of neurobiological models of language. It was originally formulated with the aim of accounting for cross-linguistic similarities and differences in the neurocognition of sentence processing (Bornkessel, 2002; Schlesewsky & Bornkessel, 2004) and has only recently aspired toward neurobiological grounding (Bornkessel-Schlesewsky & Schlesewsky, 2013b). In this way, the model's trajectory towards neurobiology has been somewhat unusual. However, this may be a strength rather than a weakness. In particular, the consideration of cross-linguistic diversity may provide some unique insights into the neurobiology of language. With approximately 7,000 living languages but only one human brain to process them, we argue that any biologically adequate model of language must be able to account for the striking differences in how individual languages are organized, and that an understanding of which patterns occur more frequently than others in the midst of this diversity can inform neurobiological models because these biases presumably reflect the organizational principles underlying the processing of language by the human brain. In this way, the current version of the eADM seeks to integrate neurobiological design principles with design principles gleaned from cross-linguistic considerations.

We first provide a very brief overview of the development of eADM and the changes that it has undergone (Section 30.2) before introducing the design principles—both language-based and neurobiological—underlying the current version of the model (Section 30.3). Section 30.4 then describes the current model architecture in more detail, followed by a discussion of the evidence supporting these architectural assumptions (Section 30.5). In Section 30.6, we show how the model's architectural assumptions translate into predictions for electrophysiology. The chapter concludes with a brief outlook (Section 30.7).

30.2 A BRIEF HISTORY OF THE DEVELOPMENT OF eADM

As already noted, the Argument Dependency Model (ADM) was originally formulated as a neurocognitive model of sentence processing across languages (Bornkessel, 2002; Schlesewsky & Bornkessel, 2004). The first model version was primarily based on electrophysiological data from German focusing specifically on the processing demands imposed by verb-final sentences. The model's name is derived from the proposal that interpretive dependencies ("who is acting on whom") can be set up between sentence participants ("arguments") even before the verb is encountered. In its original version, the ADM was an extension of Friederici's neurocognitive model of auditory language processing (Friederici, 1999, 2002) and therefore shared a number of assumptions with that model, perhaps most importantly the idea of different processing phases organized in a serial fashion.

On the basis of data from a range of languages, the ADM was expanded and modified to the extended Argument Dependency Model (eADM; Bornkessel & Schlesewsky, 2006; Bornkessel-Schlesewsky & Schlesewsky, 2008). As indicated by the modified name, the new model architecture incorporated some important changes, including the proposal of a cascaded rather than strictly serial organization. The eADM further posited that the processes underlying sentence comprehension are not

Neurobiology of Language. DOI: http://dx.doi.org/10.1016/B978-0-12-407794-2.00030-4

well-characterized by linguistic subdomains such as syntax and semantics. Finally, even more recent versions of the eADM have emphasized the importance of an "actor strategy" in sentence comprehension across languages, meaning that the language processing system endeavors to identify the participant primarily responsible for the state of affairs (the actor) as quickly and unambiguously as possible (Bornkessel-Schlesewsky & Schlesewsky, 2009, 2013a; for a computational formulation, see Alday, Schlesewsky, & Bornkessel-Schlesewsky, 2014).

All three assumptions—a cascaded architecture, no qualitative distinctions between syntactic and semantic cues, and the relevance of the actor strategy—play a central role in the current version of the eADM (Bornkessel-Schlesewsky & Schlesewsky, 2013b), which also introduced a number of neurobiological design principles. In the following sections, we first describe the language-based and neurobiological design principles underlying the current version of the eADM in more detail before introducing the latest version of the model architecture.

30.3 DESIGN PRINCIPLES

The design principles underlying the model are made explicit here. We begin by describing language-based design principles gleaned from cross-linguistic research (Section 30.3.1), and then discuss the relevant neurobiological design principles (Section 30.3.2).

30.3.1 Language-Based Design Principles

As described, one of the principal aims of the (e)ADM has always been to model cross-linguistic unity and diversity in the neurocognition of language. This goal is, in fact, inextricably intertwined with the goal of developing a neurobiologically plausible model of language processing, because any neurobiologically plausible model of language will need to ensure that linguistic diversity is adequately taken into account. Hence, the following design principles, all of which are motivated by cross-linguistic research, play an integral role in the current version of the eADM.

(L1) *Functional equivalence of syntactic and semantic cues*
The language processing literature often discusses the status of cues from different linguistic domains (e.g., syntax and semantics) in the comprehension architecture. However, since the seminal work conducted within the Competition Model (CM) framework from the late 1970s and early 1980s onward, systematic cross-linguistic comparisons have consistently provided evidence *against* a principled distinction between syntactic cues (e.g., word order, case marking, agreement) and semantic cues (e.g., animacy) in guiding comprehension (Bates, Devescovi, & Wulfeck, 2001; Bornkessel & Schlesewsky, 2006; MacWhinney & Bates, 1989; MacWhinney, Bates, & Kliegl, 1984). More recent neurocognitive studies even suggest that cues from different domains are associated with qualitatively similar neural correlates (Bornkessel-Schlesewsky & Schlesewsky, 2009). On the basis of these observations, the eADM posits that the distinction between syntax and semantics is *not* an organizational principle underlying the language processing architecture.

In accordance with principle (L1), the architecture of the eADM does not posit a principled distinction between syntax and semantics (as assumed by Friederici, 2002, 2009, 2012) or between stored lexical representations and combinatory operations (Hagoort, 2005, 2013). Rather, knowledge that would traditionally be characterized as semantic, syntactic, lexical, or combinatory is processed in both the dorsal and ventral auditory streams and the division of labor between the streams is based on other distinctions (see Sections 30.3.2 and 30.4).

A further linguistic distinction that is often assumed by models of language processing concerns the difference between word categories (e.g., nouns and verbs). Like the distinction between syntactic and semantic information sources, the eADM assumes that a principled noun–verb distinction does not play a crucial role in the language processing architecture (design principle L2).

(L2) *Transcategoriality*
A recent review of neurolinguistic studies on different word categories concluded that nouns and verbs are not differentially represented by the brain at the single word level, with apparent category differences arising primarily from the distinction between objects and actions (Vigliocco, Vinson, Druks, Barber, & Cappa, 2011; see also Moseley & Pulvermüller, 2014). Differences do begin to emerge in sentence context, however. This result accords well with observations from linguistic typology, which suggest that not all languages code category distinctions at the single word level ("transcategoriality" or "precategoriality," see Bisang, 2010) and that categories may differ between languages (Croft, 2001). In this way, findings from the neurocognition of language and linguistic typology converge in suggesting an "emergentist" view of categories, that is, a view in which word categories emerge from the interaction of form-based, meaning-based, and distributional properties.

Rather than assuming principled distinctions between purported linguistic modules such as syntax and semantics or categories such as nouns and verbs, the eADM posits that the combinatorial properties of higher-order language processing in the brain are based on a fundamental division of labor between two types of combinatory mechanisms, as described in (L3); see also Bornkessel-Schlesewsky, Schlesewsky, Small, & Rauschecker, in press.

(L3) *Sequence-based versus dependency-based combinatorics*
Although competing theories of grammar assume various different descriptive mechanisms for combining linguistic elements to form larger units, all existing theories (e.g., Bresnan, 2001; Chomsky, 2000; Goldberg, 2003; Van Valin, 2005)

assume two types of combinatory mechanisms in one form or another: (i) the combination of elements into sequences (i.e., combining elements A and B to form the sequence A-before-B); and (ii) the combination of elements to form dependencies, independent of sequential order. As an illustration, consider the phrase "the red boat" and its French counterpart "le bateau rouge" (literally, "the boat red"). In both cases, red describes a property of the boat (i.e., there is a dependency between boat and red), but the sequential order in which the two words are expressed differs between the two languages. While dependency formation correlates strongly with sequential information in some languages (e.g., English), most languages allow for a more flexible word order and, hence, a principled dissociation between sequence processing and dependency formation (e.g., German, Turkish, Hindi, Japanese, and Warlpiri, among many others).

Finally, it is important to note that the eADM rejects the notion that mechanisms particular to specific linguistic theories (e.g., merge, the notion of constructions) can be plausibly translated into neurobiologically valid constructs. It is well-known that competing theories of grammar are essentially equivalent in terms of their explanatory capacity, which can be demonstrated by expressing them in a single formalism (e.g., Stabler, 2010). In other words, competing theories overlap with regard to which sentences they classify as well-formed (part of a given language) or ill-formed (not part of a given language), but theories differ with respect to the descriptive computational mechanisms to which they attribute these differences in classification. Second, because no theory of grammar is currently constrained by knowledge regarding the structure/function of the human brain, it appears implausible to attempt to establish a neurobiological grounding for any of the computational operations specific to individual linguistic theories (for discussion, see Schlesewsky & Bornkessel-Schlesewsky, 2012). Hence, instead of drawing on mechanisms assumed in particular linguistic theories, the design principles L1–L3 are based primarily on the notion of cross-linguistic applicability.

30.3.2 Neurobiological Design Principles

The first neurobiological design principle underlying the eADM (NB1) concerns the existence of multiple streams of information processing. The assumption of a dorsal–ventral stream dissociation is shared with a wide range of other neurobiological models of speech and language processing (Friederici, 2012; Hagoort, 2013; Hickok & Poeppel, 2007; Rauschecker & Scott, 2009; Saur et al., 2008; Ueno, Saito, Rogers, & Lambon Ralph, 2011). Principle NB2 (hierarchical organization), while also not particularly controversial in neurobiology more generally, has not been explored in detail with respect to its potential consequences for the neurobiology of higher-order language processing. Finally, principles NB3 and NB4 are more controversial. NB3 posits a unified computational function for each stream (see Friederici, 2011, 2012, for a different perspective), whereas NB4 assumes that the computations performed by each stream in higher-order language processing are grounded in the computations performed by that stream in nonhuman primate audition.

(NB1) *Multiple streams of information processing*
The perception–action cycle is implemented by multiple streams of information transfer in the human brain. As in the visual domain (Goodale & Milner, 1992; Ungerleider & Mishkin, 1982), a dorsal and a ventral stream of audition can be distinguished (Rauschecker, 1998). The eADM posits that each stream can be associated with a distinguishable and internally unified function in information processing, irrespective of the fact that a stream comprises multiple fiber bundles. Furthermore, the computational properties of the dorsal and ventral streams are congruent with those of the dorsal and ventral streams in (human and nonhuman) primate audition (Bornkessel-Schlesewsky, Schlesewsky, Small, & Rauschecker, in press; Rauschecker, 2011; Rauschecker & Scott, 2009).

(NB2) *Hierarchical processing*
Following well-established findings in the visual (Felleman & Van Essen, 1991) and auditory (Rauschecker & Tian, 2000) domains (for an overarching synthesis concerning the relationship between perception and cognition, see Mesulam, 1998), the eADM assumes that the dorsal and ventral streams in language processing are organized in a hierarchical manner. Levels further downstream along a stream are sensitive to successively more intricate convergences of features (i.e., of features processed in levels further upstream). Hierarchical processing has been demonstrated convincingly for the ventral auditory stream using single unit recordings in monkeys (Rauschecker & Tian, 2000) and fMRI (DeWitt & Rauschecker, 2012). The eADM assumes that it also holds for the dorsal stream.

(NB3) *Computational functions of the dual streams*
In accordance with the classic assumption of a meaningful computational distinction between dorsal and ventral streams in both the visual and auditory domains (Goodale & Milner, 1992; Rauschecker, 1998; Ungerleider & Mishkin, 1982), the eADM assumes that the dorsal–ventral distinction in language amounts to a meaningful computational division of labor. Consequently, the dorsal and ventral streams have separable, but internally consistent, computational functions.

(NB4) *Computational grounding in primate audition*
The computations performed by the dorsal and ventral streams in higher-order language processing are grounded in the respective computational abilities of the two streams in primate audition.

Accordingly, the dorsal stream performs operations congruent with its role in auditory-motor mapping (such as the dense connectivity between auditory, parietal, and premotor regions in nonhuman primates, e.g., Lewis & Van Essen, 2000; Morecraft et al., 2012; Ward, Peden, & Sugar, 1946; and evidence for auditory-motor interactions in

monkeys based on these structures, e.g., Artchakov et al., 2012; Kohler et al., 2002) and as a "where" or "how" stream. Internal models, which serve to predict upcoming states of the world based on the current input and the properties of the model, have been proposed as a common denominator between the differing functions (Rauschecker, 2011).
The ventral stream, by contrast, recognizes auditory objects. Like visual objects, auditory objects are identified as a coherent *gestalt* via certain grouping cues (e.g., in audition, harmonicity, coherent changes over time, and a common temporal onset; Bizley & Cohen, 2013). Studies in nonhuman primates have demonstrated a sensitivity for successively more complex auditory objects along the course of the ventral stream, ranging from elementary auditory features such as frequency-modulated (FM) sweeps or bandpass noise bursts to species-specific vocalizations (Rauschecker, 2011; Rauschecker & Scott, 2009; Rauschecker & Tian, 2000). Auditory object formation is a form of categorization in which spectro-temporal properties are grouped into perceptual (Bizley & Cohen, 2013) and, at higher hierarchical levels, conceptual units. It thus provides the computational basis for an elementary mapping from spectro-temporal patterns to concepts.

The final neurobiological design assumption (NB5) concerns the role of frontal cortex within the language processing architecture. In contrast to several other neurobiological models of language (Friederici, 2012; Hagoort, 2013), the eADM does *not* assume that frontal regions—including IFG and vPMC—perform dedicated linguistic computations such as syntactic structuring or unification.

(NB5) *Control function of frontal cortex*
Frontal cortex, including vPMC and IFG, fulfill domain-general cognitive control functions in language processing (Bornkessel-Schlesewsky, Grewe, & Schlesewsky, 2012; January, Trueswell, & Thompson-Schill, 2009; Novick, Trueswell, & Thompson-Schill, 2005; Thompson-Schill, Bedny, & Goldberg, 2005; Thompson-Schill, D'Esposito, Aguirre, & Farah, 1997). Thus, they serve to select among multiple competing representations and provide a supervisory function that can shift the weighting attributed to the individual streams (Miller & Cohen, 2001). They are also crucial for the integration of information from the dorsal and ventral streams.

30.4 THE MODEL ARCHITECTURE

The basic model architecture of the eADM is shown in Figure 30.1. In the following subsections, we explain the assumptions underlying this architecture in more detail. We begin by focusing on the assumed computational division of labor between the dorsal and ventral streams (Section 30.4.1) before turning to hierarchical processing as a common organizational principle of both streams (Section 30.4.2). Sections 30.4.3 and 30.4.4 then discuss the assumed difference between the representation of actions in the two streams and cross-stream integration, respectively.

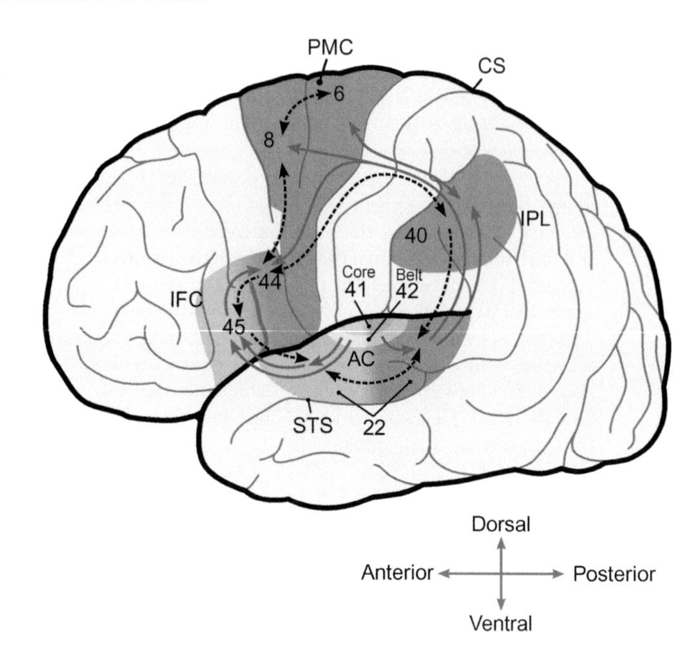

FIGURE 30.1 Schematic depiction of the eADM's neuroanatomical assumptions (*reproduced, with permission, from Bornkessel-Schlesewsky et al., in press*). Both the ventral (green) and dorsal (red) streams are assumed to emanate from auditory cortex (AC) and to perform information processing in a hierarchically organized manner. Abbreviations: AC, auditory cortex; CS, central sulcus; IFC, inferior frontal cortex; IPL, inferior parietal lobule; PMC, premotor cortex; STS, superior temporal sulcus. Numbers denote Brodmann area.

30.4.1 The Computational Division of Labor Between the Dorsal and Ventral Streams

Building on the design principles that were discussed in the preceding section, the current version of the eADM posits the following fundamental computational distinction between the postero-dorsal and antero-ventral streams (see L3 and NB3): the dorsal stream processes sequential information, whereas the ventral stream computes dependencies independent of sequential order. Both types of computations are combinatorial in that they serve to construct larger units via the combination of smaller units. However, they differ with respect to the informativity of temporal ordering: for the combination of two elements A and B to form a larger unit C, if the order in which A and B are encountered is relevant, then the computation is one of sequencing; if the order of A and B is irrelevant, then it is one of dependency formation. In other words, time-independent computations are commutative (like addition or multiplication in arithmetic), whereas time-dependent computations are not (like subtraction or division).

In accordance with NB4, the computational mechanisms posited here for the dual streams are grounded in their respective computational functions in primate audition. For the dorsal stream, it has been proposed that its basic functions as a where or how stream can be subsumed computationally under the construct of an internal model (Rauschecker, 2011; Rauschecker & Scott, 2009) that serves to predict the sensory consequences of actions (forward model) and to determine the necessary motor commands to produce a desired sensory outcome (inverse model). The eADM assumes that linguistic sequence processing within this stream can be understood using such internal model mechanisms. In particular, forward models provide the optimal means for predicting sequential information, as required at many different levels of language (see also Pickering & Garrod, 2007, 2013). The dorsal stream, which is generally viewed as the more "action-related" of the two auditory streams (Hickok & Poeppel, 2004, 2007; Saur et al., 2008; Ueno et al., 2011) and which shows considerable anatomical overlap with the "action observation network" (AON; Grafton, 2009), provides an ideal basis for this type of mechanism. In other words, because language and action share a number of constitutive characteristics (hierarchical organization, dynamic unfolding over time), neural architectures that are highly suited to implementing forward and inverse models of action should be able to serve a similar purpose with respect to the sequential properties of language.

In contrast to the sequence-based processing characteristics of the dorsal stream, the ventral stream processes feature structures (auditory objects) of increasing complexity. At the lower levels of the hierarchy (phoneme, syllable), we assume that these are represented as spectro-temporal patterns, similar to the monkey vocalizations discussed by Rauschecker and Tian (2000). On the basis of single unit recordings in macaques, these authors argued for a mechanism of spectral integration in which particular neurons show sensitivity to the combination of several spectral components comprised by a complex sound. They also proposed that similar mechanisms may be applicable in speech perception in humans, at least up to the level of the syllable. However, in accordance with the well-established idea that the ventral stream is somehow involved in mapping acoustic representations to semantic representations (e.g., Saur et al., 2008; Scott, Blank, Rosen, & Wise, 2000; Ueno et al., 2011; and, to a certain degree, Hickok & Poeppel, 2007), it appears clear that, at some level of the hierarchy, representations must begin to abstract from complex spectro-temporal acoustic patterns. The eADM posits that the level of the morpheme may be a promising candidate because there is general agreement in linguistics that morphemes are the smallest meaning-bearing units in language, and these are the smallest units that can plausibly be connected to an acoustically independent semantic representation.

The semantic representations assumed by the eADM are termed "actor event schemata" (AE schemata). They essentially reflect event scenarios comprising an event instigator or causer (the actor) and the action undertaken, as well as optional additional components such as the entity acted on, the time, and location of the action. The centrality of the actor as opposed to other participants in the event scenario is motivated by cross-linguistic considerations (Bornkessel-Schlesewsky & Schlesewsky, 2009, 2013b, 2014). The various components of a schema are combined via dependency formation, which we view as an extension of the combination sensitivity that is already attested in basic auditory objects in nonhuman primates. Thus, in accordance with the abstraction from spectro-temporal to conceptual auditory objects described, dependency formation is a mechanism that groups linguistic units into conceptual schemata as opposed to acoustic features into more complex acoustic representations. Like auditory object formation in the acoustic domain, dependency formation draws on grouping cues. These can include the likelihood of certain words occurring in a particular semantic relation (e.g., given "apple" and "eat," there is a high likelihood of the apple being interpreted as the thing eaten rather than the entity doing the eating) but also morphological cues such as case marking (such as the German noun phrase "den kleinen braunen Bären," "the.ACC small.ACC brown.ACC bear. ACC" in which overlapping accusative case inflections serve to make clear the dependency between the various elements of the noun phrase). Importantly, as noted, dependency formation is independent of sequential order as demonstrated, for example, by the fact that the component parts of a noun phrase can be separated from one another in many languages of the world. The following examples from Ukranian (Féry, Paslawska, & Fanselow, 2007) illustrate this property. Note the accusative feminine marker that serves as a grouping cue to identify the dependency between "interesting" and "book."

1. Dependency without sequential adjacency: split noun phrases in Ukrainian (Féry et al., 2007)
   a. Marija pročytala cikavu knyžku
   Mary has-read interesting-ACC.FEM book-ACC.FEM
   "Mary has read an interesting book"
   b. Cikavu Marija pročytala knyžku
   interesting-ACC.FEM Mary has-read book-ACC.FEM
   c. Knyžku Marija pročytala cikavu
   book-ACC.FEM Mary has-read interesting-ACC.FEM

In a more general sense, AE schemata can be compared with the notion of scripts (i.e., knowledge structures that represent information about events) (Schank, 1999; Schank & Abelson, 1977). Neuroanatomically, the utilization of scripts has been tied to the anterior temporal lobe (Frith & Frith, 2003; Funnell, Corballis, & Gazzaniga, 2001), which is in accordance with the eADM's notion of the antero-ventral stream. From a neurobiological perspective, increasing detail at successive levels of the hierarchy (from the activation of individual AE schemata to their combination) will lead to the activation of an increasing number of neuronal populations activated in "synchrony" in accordance with the assumption of distributed semantic representations (e.g., Martin, 2007; Patterson, Nestor, & Rogers, 2007, for reviews).

30.4.2 Hierarchical Organization as a Common Principle

Despite the qualitatively different mechanisms of information processing that they implement, the dorsal and ventral streams have in common that processing is organized in a hierarchical manner (see NB2). This means that information integration increases successively the further downstream a region is within a stream (i.e., the further removed from auditory cortex). For example, if the first level of the hierarchy processes features at the level of, for example, phonemes (e.g., /b/, /p/, /a/), then the second level of the hierarchy will be sensitive to combinations of these features (e.g., distinguishing syllables such as /ba/ and /pa/). The assumption that the hierarchy is "built up" successively from auditory cortex, hence implying an asymmetry in the directionality of information flow within the two streams, is based on the idea that the streams serve to link perception (primary sensory regions) with action (frontal cortex). It thus adopts the perspective that the purpose of the human brain is to plan and execute movements, thereby allowing us to interact with our environment:

> Movement is the only way we have of interacting with the world, whether foraging for food or attracting a waiter's attention. Direct information transmission between people, through speech, arm gestures or facial expressions, is mediated through the motor system which provides a common code for communication. From this viewpoint, the purpose of the human brain is to use sensory representations to determine future actions. ***Wolpert, Doya, and Kawato (2003, p. 593).***

Building on this view, we propose that the goal of language comprehension is to translate linguistic input (i.e., a stream of phonetic, graphematic, or gestural input) into a representation of the action or event being expressed to allow for a suitable behavioral response. Note that "behavioral" is used in very general terms here, referring both to overt action and to more internal operations such as memory updating or consolidation. Nevertheless, connectivity within the streams is of course inherently bidirectional.

The nature of the hierarchy within a stream is in accordance with the information processing function of that stream. In the dorsal stream, we assume that the hierarchy is defined via temporal integration windows of successively increasing size: phoneme sequences, prosodic chunks, category sequences, action sequences, and discourse sequences. The ventral stream, by contrast, comprises a hierarchy of successively more complex (i.e., feature-rich) auditory objects.

Combining the assumed division of labor between the computations performed by the two streams with hierarchical processing as a key organizational principle yields the process models in Figures 30.2 and 30.3 for the dorsal and ventral streams, respectively.

30.4.3 Representing Actions in the Dorsal and Ventral Streams

If, as suggested, the goal of (auditory) language comprehension is to translate an acoustic input into an action representation that can be used to guide future behavior, then this raises the question of how action representations differ between the dorsal and ventral streams. We

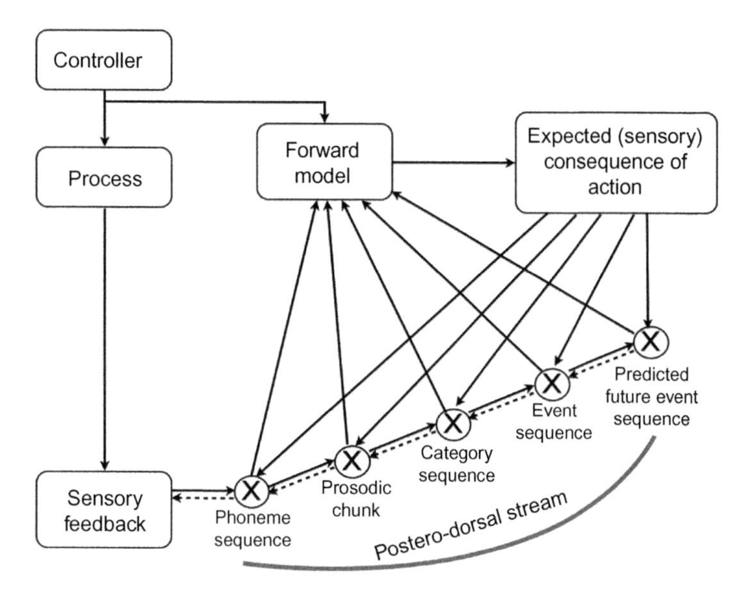

FIGURE 30.2 Functional organization of the postero-dorsal stream. Future input is predicted via a forward model and the expected consequences (sensory or more abstract) are compared with the actual input by a hierarchically organized series of comparators (designated by an X). Prediction errors arising in these comparisons are used to update the forward model.

propose that the crucial difference again lies in the dissociation between sequences and dependencies.

Let us consider the ventral stream first. Here, in accordance with the notion of AE schemata, we assume that action representations in the ventral stream are holistic in the sense that they encode the action and the participants involved in it, but not the precise manner in which the action unfolds in time.

The dorsal stream, by contrast, encodes an action in a sequence-based manner with the goal of predicting how the action/discourse event will continue to unfold over time (sequentially). We posit that, to do this, the dorsal stream draws on a set of action-relevant linguistic entities that are particularly potent in their predictive capacity (Figure 30.4). Importantly, we suggest that the dorsal stream's action representation is centered on these action-relevant entities and their role in the sequential unfolding of an action and that the stream does not represent the overall action itself. This is the role of the ventral stream, as described. This assumption is in line with the proposal that the (visual) dorsal stream only has a very limited memory capacity such that actions are possible when an object (i.e., an action-relevant entity) is currently present, but not when reliance on stored object information is required (e.g., when grasping an object in the dark) (Grafton, 2010; Milner, Dijkerman, McIntosh, Rossetti, & Pisella, 2003). Under such circumstances the interaction with the ventral stream becomes crucially important (Cohen, Cross, Tunik, Grafton, & Culham, 2009).

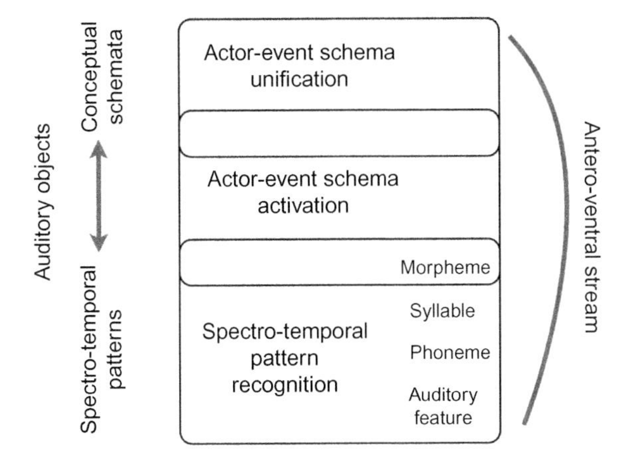

FIGURE 30.3 Functional organization of the antero-ventral stream that recognizes spectro-temporal patterns and maps these onto conceptual schemata (actor-event, AE schemata). AE schemata are combined to form more complex AE schemata via unification, a commutative combinatory operation. *Adapted, with permission, from Bornkessel-Schlesewsky et al. (in press).*

As an example of an action-relevant entity in language, consider the actor role. The actor refers to the participant that is primarily responsible for the successful execution/completion of the event or state of affairs described (Bornkessel & Schlesewsky, 2006). Cross-linguistic evidence suggests that actor identification constitutes a central goal of sentence understanding in languages of very different types (Bornkessel-Schlesewsky & Schlesewsky, 2009) and, accordingly, that different participants (arguments) compete for the actor role (Bornkessel-Schlesewsky & Schlesewsky, 2013a, 2014). Competition is resolved via a set of cues (also termed prominence features), some of which appear

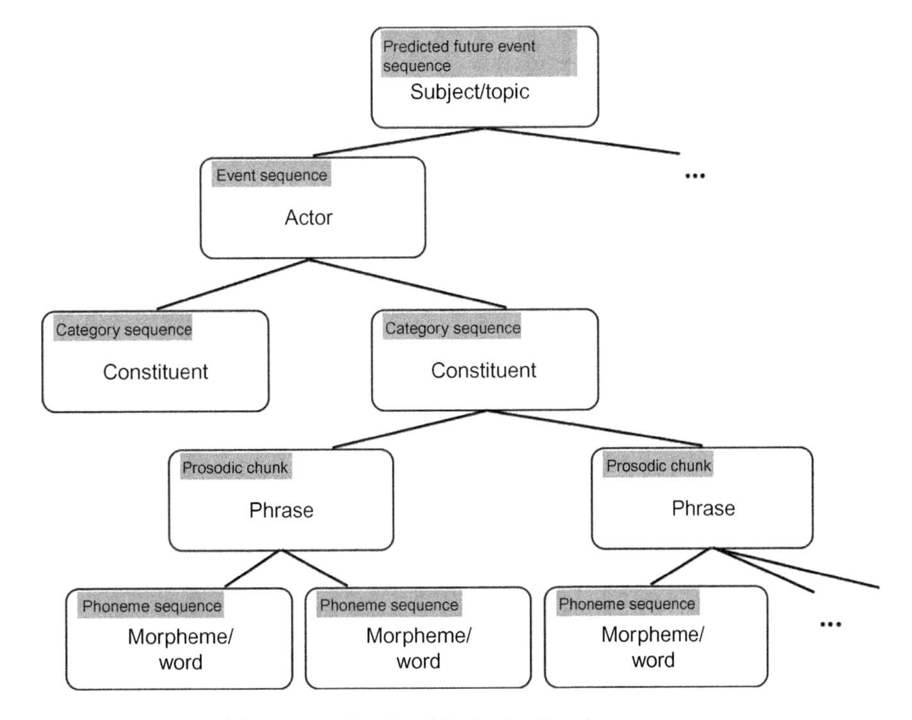

FIGURE 30.4 Hierarchical organization of the action-related entities in the dorsal stream.

to be universal (e.g., animacy, word order, first person), whereas others are language-specific (e.g., case marking, agreement). Individual cues are weighted differentially depending on the specific language being processed (Bates et al., 2001; Bates, McNew, MacWhinney, Devescovi, & Smith, 1982; Bornkessel & Schlesewsky, 2006; MacWhinney et al., 1984; MacWhinney & Bates, 1989). During language development, children learn which of the more abstract linguistic information sources (e.g., word order, case marking) correlate highly with actorhood in their native language—presumably based on more universal cues such as animacy or similarity to the first person (assuming that the self is used as a model for an optimal actor; Dahl, 2008; Haggard, 2008; Tomasello, 2003). Once the actor has been identified, it can be used to predict certain aspects of the event or state of affairs currently being described. For example, a human actor is likely to perform different types of actions—warranting different sets of linguistic descriptions—to an inanimate actor.

Predictions at a particular level of the hierarchy can also influence other levels (both higher and lower). For example, the anticipation of a particular action or set of possible actions depending on the actor can be used to constrain predictions with regard to the upcoming category sequence. Animate and particularly human actors are more likely than inanimate actors to perform goal-directed actions involving another person or object. Hence, in addition to a verb describing the action itself, the processing system may be more likely to anticipate an additional constituent describing an entity affected by the action (an undergoer), and the internal model will be modified to reflect this. At the same time, an actor participant is also very likely to be realized as the grammatical subject of a sentence. Following work in sentence production (see Myachykov, Thompson, Scheepers, & Garrod, 2011, for a review) and text analysis (Givón, 1983, 1994), the eADM assumes that a speaker codes an entity as the subject when he/she wishes to signal to the hearer that this entity will be particularly important in the upcoming discourse (Bornkessel-Schlesewsky & Schlesewsky, 2014). Hence, identifying the subject allows for predictions to be set up with regard to the next event to be described in the current discourse (i.e., a supra-sentential sequence prediction). Again, the internal model is dynamically modified in accordance with this prediction.

30.4.4 Integrating Information Between Streams

Finally, we turn to the interaction between the dorsal and ventral streams. Following Rauschecker and Scott (2009), the eADM posits that prefrontal cortex (specifically, the inferior frontal gyrus, IFG, and ventral premotor cortex [vPMC]) serves as a point of integration between the two streams. It also serves a control function whenever mediation between conflicting representations is required (Bornkessel-Schlesewsky et al., 2012; Novick et al., 2005; Thompson-Schill et al., 1997; Thompson-Schill et al., 2005). In this way, the model assumes that prefrontal regions—and particularly the IFG—do not perform any language-specific computations such as syntactic structure building, but rather fulfill a more general supervisory function (Miller & Cohen, 2001).

The integration of information from the two streams in frontal cortex further allows for information from one stream to induce a top-down modulation of the other. For example, an event representation in the form of an AE schema from the ventral stream can, when translated into a dynamic sequence-based representation (possibly in the interaction between IFG and vPMC), modify the forward model being used to generate predictions about upcoming input in the dorsal stream. This may lead to more concrete predictions about category sequences and phoneme sequences (i.e., specific words and morphemes) than would be possible on the basis of the dorsal stream alone due to the more specific information about actors and events represented in the ventral stream. Conversely, sequence-based information from the dorsal stream can play an important role in modulating AE schema unification in the ventral stream. Because AE schemata conceptualize both actors and events and can therefore represent both "nouny" and "verby" constituents depending on the context in which they occur (for more detailed discussion, Bornkessel-Schlesewsky & Schlesewsky, 2013b), current sequence information can be used to determine whether a given AE schema refers to the actor or the event in the overarching action being described (i.e., for the schema "run(ner)," whether we are currently processing an event involving a runner, someone who is running, or both).

Beyond an interaction of the two streams at their "termination" points in frontal cortex, a more dynamic transfer of information at multiple points along each stream appears very likely, possibly supported by subcortical connectivity. Functionally, this could help to consolidate representations across streams. However, the anatomy of these putative connections remains to be determined in detail and, accordingly, the functional consequences of a more direct inter-stream interaction remain an important topic for future research.

30.5 EVIDENCE FOR THE MODEL

In the following, we briefly summarize the existing evidence for the architectural claims described. Note that this discussion is not exhaustive for reasons of space.

Let us first consider the dorsal stream, for which the model makes several claims that are not shared by other approaches. The notion of hierarchical organization is supported by a recent fMRI study (Lerner, Honey, Silbert, & Hasson, 2011) on temporal receptive windows in the dorsal auditory stream. Lerner and colleagues' participants listened to stories that were presented either normally or scrambled (i.e., cut into segments and reassembled randomly) at several different levels or time scales: single words, sentences, or paragraphs. Activation changes were observed along the dorsal stream, with areas successively further downstream from auditory cortex only showing activation increases in response to coherent information at successively higher levels (i.e., larger and longer linguistic units). Hence, Lerner et al. (2011) argue for a hierarchical organization of temporal receptive windows, with window size increasing from "low-level sensory to high-level perceptual and cognitive areas" (p. 2906). This result provides strong converging support for the eADM's proposal of a hierarchically organized dorsal auditory stream.

A second key claim with regard to the dorsal stream concerns sequential processing, an aspect of language comprehension that has been examined in a number of neuroimaging studies that aimed at investigating syntactic processing. Evidence for sequence processing in the dorsal stream stems from an fMRI study on constituent processing in French (Pallier, Devauchelle, & Dehaene, 2011) that undertook a parametric manipulation of constituent size (1–12 words) using both real French words and pseudowords. Activation in the pSTS increased parametrically with constituent size, irrespective of the word/pseudoword distinction, whereas a comparable activation increase in the anterior STS and temporal pole was only observed for constituents consisting of real words. This finding is consistent with the view that the posterior temporal lobe is important for the processing of category sequences, whereas combinatory effects in the anterior temporal lobe can be modeled as AE schema unification. Although it is conceivable that pseudowords activate AE schemata on the basis of their similarity to existing words, this activation should be considerably less pronounced than that for real words, therefore yielding a larger schema unification effect for real word constituents. Category sequences, by contrast, can be constructed just as successfully from pseudowords as long as these preserve grammatical information.

Further evidence for the assumption that syntactic processing relies on the dorsal stream is provided by patient studies. Wilson et al. (2011) studied patients with primary progressive aphasia (PPA) and observed that deficits in processing noncanonically ordered sentences in English (e.g., passives, object relative clauses) correlated with damage to dorsal fiber tracts (the superior longitudinal fasciculus and arcuate fasciculus), but not ventral fiber tracts (the uncinate fasciculus and extreme capsule). An fMRI study using the same task that Wilson et al. (2010) used to assess patients' syntactic performance revealed activation in bilateral dorsal posterior IFG/IFS and anterior insulae, left mid-posterior STS and MTG, as well as left superior and medial (precuneus) parietal regions. Furthermore, in an additional study, Wilson et al. (2014) found that patients with semantic-type PPA and atrophy of the anterior temporal lobes showed statistically indistinguishable fMRI activation for noncanonical versus canonical sentences in comparison with healthy controls.

Coactivation of posterior superior temporal and posterior dorsal inferior frontal regions has also been reported in studies on word order variations (Bornkessel, Zysset, Friederici, von Cramon, & Schlesewsky, 2005; Röder, Stock, Neville, Bien, & Rösler, 2002) or for the contrast between locally (syntactically) ambiguous and unambiguous sentences (Snijders et al., 2009). Importantly, the eADM provides a principled explanation for the activation of both temporal and frontal regions, with temporal regions thought to perform the sequence processing necessary for recognizing syntactic structure and frontal regions providing the appropriate control mechanisms to supervise these processes. By contrast, models claiming that the left IFG is the core region for syntactic processing (Friederici, 2011) or linguistic combinatorics (Hagoort, 2013) do not straightforwardly account for the additional temporal activation in response to syntactic manipulations.

Turning now to the ventral stream, the claim that this stream processes auditory objects and provides a mapping from spectro-temporal patterns to conceptual schemata predicts that the anterior temporal lobe (aTL) should be sensitive to abstract semantic representations. There is evidence to suggest that auditory and visual processing converges in anterior temporal regions (e.g., Chan et al., 2011, for supramodal sensitivity to semantic categories in the aTL, as revealed by intracranial recordings) and that individual word meanings can be decoded from EEG and MEG recordings from anterior temporal cortex (Chan, Halgren, Marinkovic, & Cash, 2011). Dependency formation in the ventral stream is supported by combinatory effects in the aTL (i.e., increased activity in response to two-word sequences that yielded a meaningful modifier-head relation when combined, e.g., "red boat," in comparison with various controls) for both visual and auditory input (Bemis & Pylkkanen, 2013). The observation that semantic PPA patients with substantial atrophy of the aTL can process noncanonical word orders rather accurately and show similar dorsal stream activation to controls while doing so (Wilson et al., 2014) provides further converging evidence for the eADM's proposal that the combinatorics supported by the ventral stream are dependency-based (commutative) rather than sequence-based.

30.6 CONSEQUENCES FOR ELECTROPHYSIOLOGY

As noted in Section 30.1, the eADM was originally formulated on the basis of electrophysiological data. It therefore appears important to consider the consequences for electrophysiological measures of language processing arising from the architectural considerations outlined in the previous sections.

First, recall that the eADM posits no principled separation between syntax versus semantics, combinatorics versus lexicon, or linguistic rules versus representations. Hence, the model does not provide any basis for assuming a separation of ERP components based on such distinctions (e.g., the traditional assumption that N400 effects reflect semantic processing, whereas LAN and P600 effects reflect syntactic processing, e.g., Friederici, 2002; Hagoort, Brown, & Osterhout, 1999; Kutas, Van Petten, & Kluender, 2006; Osterhout & Nicol, 1999).

If linguistic subdomains are excluded, then what might be a plausible alternative for a functional classification of the ERP correlates of language processing? In this regard, let us first consider negative ERP deflections such as the left anterior negativity (LAN) and N400. Very broadly speaking, these are often regarded as reflecting mismatches with predicted information during language processing. From the perspective that forward models serve as a source of predictive processing within a dynamically unfolding linguistic input signal, it is apparent that prediction errors (i.e., mismatches between the predicted sensory consequences of an action and the sensory input that lead to an adjustment of the internal model) provide a unified mechanism for modeling prediction mismatches. The eADM thus proposes that endogenous language-related negativities can be subsumed under a unified mechanism: a prediction error during language processing. Rather than reflecting different linguistic domains, latency and topography of the negativity will vary depending on where in the underlying architecture the prediction error occurs (for a somewhat similar perspective, see Pulvermüller, Shtyrov, & Hauk, 2009). Note, however, that this does not mean that each component will be associated with a unique source. Rather, scalp ERP patterns likely reflect a mixing of a number of underlying sources and, hence, potentially, a range of prediction errors at multiple levels within the two streams.

With regard to positive ERP deflections, the eADM subscribes to the view that the P600 (and earlier language-related positivities) are instances of the domain-general P3 component (Coulson, King, & Kutas, 1998; Gunter, Stowe, & Mulder, 1997). A recent neurobiologically grounded model of the P3 proposes that this component reflects a systemic neuromodulatory response to motivationally significant events stemming from the release of norepinephrine (NE) from the Locus Coeruleus (LC; a brain stem nucleus) (Nieuwenhuis, Aston-Jones, & Cohen, 2005). The NE release serves to increase neural responsivity (gain) to these motivationally significant stimuli, thereby sharpening the dichotomy between the response to target versus nontarget inputs and facilitating an appropriate behavioral response ("noradrenergic potentiation of information processing"; Nieuwenhuis et al., 2005, p. 510). Note that, as in Section 30.4, the notion of behavior used here is very general, including both overt responses and internal state changes such as memory updating. From this perspective, language-related endogenous positivity effects reflect a response to motivationally significant stimuli, rather than indexing a particular type of structural processing (such as syntactic processing or reanalysis). This idea is in line with the finding that positivity effects occur for both syntactically and semantically deviant materials (for semantic violations, see Faustmann, Murdoch, Finnigan, & Copland, 2007; Gunter, Jackson, & Mulder, 1992; Kolk, Chwilla, van Herten, & Oor, 2003; Kuperberg, Sitnikova, Caplan, & Holcomb, 2003; Lotze, Tune, Schlesewsky, & Bornkessel-Schlesewsky, 2011; Roehm, Bornkessel-Schlesewsky, Rösler, & Schlesewsky, 2007; Sanford, Leuthold, Bohan, & Sanford, 2011) and that they are considerably less pronounced or even absent without a task (Batterink & Neville, 2013; Hasting & Kotz, 2008). Further converging evidence stems from the recent observation that, like P3 effects, P600 effects in language comprehension are reaction-time aligned—a necessary prediction of subsuming the P600 under the LC/NE account of the P3 (Sassenhagen, Schlesewsky, & Bornkessel-Schlesewsky, 2014). This account also leads to a number of testable predictions, such as that P600 effects should covary with other physiological indicators of LC/NE activity, including heart rate, pupil dilation, and skin conductance.

30.7 OUTLOOK

The model architecture described here is intended as a first step toward a neurobiologically and cross-linguistically plausible model of human language processing. As is the case with every model, it can only provide an approximation of reality and it will likely continue to undergo an evolution over the coming years. Development presupposes falsifying assumptions of the current model version and modifying them in accordance with the data collected. Thus, an emphasis has been placed on clear design principles, many of which may ultimately prove too simple in their current form. For example, it may be the case that, although both ventral and dorsal streams are organized hierarchically, the formulation of the hierarchy may need to be somewhat different in each case. To allow for the testing of such hypotheses, the current state of the model makes a range of testable and falsifiable predictions, some of which we have described in the present chapter.

Acknowledgments

Parts of the research reported here were supported by grants from the LOEWE program of the German state of Hesse (grant III L 4–518/70.004 to IBS) and the German Research Foundation (grants TRR 135/1 C05 to IBS and SCHL 544/6-1 to Ms). The authors thank Steve Small and Josef Rauschecker for many fruitful discussions related to this line of research, and Sarah Tune and Phillip Alday for helpful comments on a previous version of this manuscript.

References

Alday, P., Schlesewsky, M., & Bornkessel-Schlesewsky, I. (2014). Towards a computational model of actor-based language comprehension. *Neuroinformatics*, *12*, 143–179.

Artchakov, D., Ortiz, M., Kusmierek, P., Cui, D., VanMeter, I., Jääskeläinen, I., et al. (2012). Representation of sound sequences in the auditory dorsal stream after sensorimotor learning in the rhesus monkey. *Society for Neuroscience Abstracts* 368.04.

Bates, E., Devescovi, A., & Wulfeck, B. (2001). Psycholinguistics: A cross-language perspective. *Annual Review of Psychology*, *52*, 369–396.

Bates, E., McNew, S., MacWhinney, B., Devescovi, A., & Smith, S. (1982). Functional constraints on sentence processing: A cross-linguistic study. *Cognition*, *11*, 245–299.

Batterink, L., & Neville, H. J. (2013). The human brain processes syntax in the absence of conscious awareness. *The Journal of Neuroscience*, *33*(19), 8528–8533.

Bemis, D. K., & Pylkkanen, L. (2013). Basic linguistic composition recruits the left anterior temporal lobe and left angular gyrus during both listening and reading. *Cerebral Cortex*, *23*(8), 1859–1873.

Bisang, W. (2010). Word classes. In J. J. Song (Ed.), *The Oxford handbook of language typology*. Oxford: Oxford University Press.

Bizley, J. K., & Cohen, Y. E. (2013). The what, where and how of auditory-object perception. *Nature Reviews Neuroscience*, *14*(10), 693–707.

Bornkessel, I. (2002). *The Argument Dependency Model: A neurocognitive approach to incremental interpretation* (Vol. 28). Leipzig: MPI Series in Cognitive Neuroscience.

Bornkessel, I., & Schlesewsky, M. (2006). The Extended Argument Dependency Model: A neurocognitive approach to sentence comprehension across languages. *Psychological Review*, *113*, 787–821.

Bornkessel, I., Zysset, S., Friederici, A. D., von Cramon, D. Y., & Schlesewsky, M. (2005). Who did what to whom? The neural basis of argument hierarchies during language comprehension. *NeuroImage*, *26*(1), 221–233.

Bornkessel-Schlesewsky, I., Grewe, T., & Schlesewsky, M. (2012). Prominence vs. aboutness in sequencing: A functional distinction within the left inferior frontal gyrus. *Brain and Language*, *120*, 96–107.

Bornkessel-Schlesewsky, I., & Schlesewsky, M. (2008). An alternative perspective on "semantic P600" effects in language comprehension. *Brain Research Reviews*, *59*, 55–73.

Bornkessel-Schlesewsky, I., & Schlesewsky, M. (2009). The role of prominence information in the real time comprehension of transitive constructions: A cross-linguistic approach. *Language and Linguistics Compass*, *3*, 19–58.

Bornkessel-Schlesewsky, I., & Schlesewsky, M. (2013a). Neurotypology: Modelling cross-linguistic similarities and differences in the neurocognition of language comprehension. In M. Sanz, I. Laka, & M. K. Tanenhaus (Eds.), *Language down the garden path. The cognitive and biological basis for linguistic structures* (pp. 241–252). Oxford: Oxford University Press.

Bornkessel-Schlesewsky, I., & Schlesewsky, M. (2013b). Reconciling time, space and function: A new dorsal-ventral stream model of sentence comprehension. *Brain and Language*, *125*, 60–76.

Bornkessel-Schlesewsky, I., & Schlesewsky, M. (2014). Competition in argument interpretation: Evidence from the neurobiology of language. In B. MacWhinney, A. Malchukov, & E. Moravcsik (Eds.), *Competing motivations in grammar and usage* (pp. 107–126). Oxford: Oxford University Press.

Bornkessel-Schlesewsky, I., Schlesewsky, M., Small, S. L., & Rauschecker, J. P. Neurobiological roots of language in primate audition: Common computational properties. *Trends in Cognitive Sciences* (in press). Available from: http://dx.doi.org/10.1016/j.tics.2014.12.008.

Bresnan, J. (2001). *Lexical functional grammar*. Oxford: Blackwell.

Chan, A. M., Baker, J. M., Eskandar, E., Schomer, D., Ulbert, I., Marinkovic, K., et al. (2011). First-pass selectivity for semantic categories in human anteroventral temporal lobe. *The Journal of Neuroscience*, *31*(49), 18119–18129.

Chan, A. M., Halgren, E., Marinkovic, K., & Cash, S. S. (2011). Decoding word and category-specific spatiotemporal representations from MEG and EEG. *NeuroImage*, *54*(4), 3028–3039.

Chomsky, N. (2000). Minimalist inquiries: The framework. In R. Martin, D. Michaels, & J. Uriagereka (Eds.), *Step by step: Essays in minimalist syntax in honor of Howard Lasnik* (pp. 89–155). Cambridge, MA: MIT Press.

Cohen, N. R., Cross, E. S., Tunik, E., Grafton, S. T., & Culham, J. C. (2009). Ventral and dorsal stream contributions to the online control of immediate and delayed grasping: A TMS approach. *Neuropsychologia*, *47*(6), 1553–1562.

Coulson, S., King, J. W., & Kutas, M. (1998). Expect the unexpected: Event-related brain response to morphosyntactic violations. *Language and Cognitive Processes*, *13*, 21–58.

Croft, W. A. (2001). *Radical construction grammar: Syntactic theory in typological perspective*. Oxford: Oxford University Press.

Dahl, Ö. (2008). Animacy and egophoricity: Grammar, ontology and phylogeny. *Lingua*, *118*, 141–150.

DeWitt, I., & Rauschecker, J. P. (2012). Phoneme and word recognition in the auditory ventral stream. *Proceedings of the National Academy of Sciences of the United States of America*, *109*, E505–E514.

Faustmann, A., Murdoch, B. E., Finnigan, S. P., & Copland, D. A. (2007). Effects of advancing age on the processing of semantic anomalies in adults: Evidence from event-related brain potentials. *Experimental Aging Research*, *33*, 439–460.

Felleman, D. J., & Van Essen, D. C. (1991). Distributed hierarchical processing in the primate cerebral cortex. *Cerebral cortex*, *1*(1), 1–47.

Féry, C., Paslawska, A., & Fanselow, G. (2007). Nominal split constructions in Ukrainian. *Journal of Slavic Linguistics*, *15*, 3–48.

Friederici, A. D. (1999). The neurobiology of language comprehension. In A. D. Friederici (Ed.), *Language comprehension: A biological perspective* (pp. 263–301). Berlin/Heidelberg/New York: Springer.

Friederici, A. D. (2002). Towards a neural basis of auditory sentence processing. *Trends in Cognitive Sciences*, *6*(2), 78–84.

Friederici, A. D. (2009). Pathways to language: Fiber tracts in the human brain. *Trends in Cognitive Sciences*, *13*, 175–181.

Friederici, A. D. (2011). The brain basis of language processing: From structure to function. *Physiological Reviews*, *91*(4), 1357–1392.

Friederici, A. D. (2012). The cortical language circuit: From auditory perception to sentence comprehension. *Trends in Cognitive Sciences, 16*(5), 262–268.

Frith, U., & Frith, C. D. (2003). Development and neurophysiology of mentalizing. *Phil Transactions of the Royal Society London B, 358*, 459–473.

Funnell, M. G., Corballis, P. M., & Gazzaniga, M. S. (2001). Hemispheric processing asymmetries: Implications for memory. *Brain and Cognition, 46*, 135–139.

Givón, T. (1983). Topic continuity in discourse: An introduction. In T. Givón (Ed.), *Topic continuity in discourse: A quantitative cross-language study* (pp. 1–41). Amsterdam: John Benjamins.

Givón, T. (1994). The pragmatics of de-transitive voice: Functional and typological aspects of inversion. In T. Givón (Ed.), *Voice and inversion* (pp. 3–44). Amsterdam: John Benjamins.

Goldberg, A. E. (2003). Constructions: A new theoretical approach to language. *Trends in Cognitive Sciences, 7*, 219–224.

Goodale, M. A., & Milner, D. (1992). Separate visual pathways for perception and action. *Trends in Neurosciences, 15*, 20–25.

Grafton, S. T. (2009). Embodied cognition and the simulation of action to understand others. *The Year in Cognitive Neuroscience 2009: Annals of the New York Academy of Science, 1156*, 97–117.

Grafton, S. T. (2010). The cognitive neuroscience of prehension: Recent developments. *Experimental Brain Research, 204*, 475–491.

Gunter, T. C., Jackson, J. L., & Mulder, G. (1992). An electrophysiological study of semantic processes in young and middle-aged academics. *Psychophysiology, 29*, 38–54.

Gunter, T. C., Stowe, L. A., & Mulder, G. (1997). When syntax meets semantics. *Psychophysiology, 34*, 660–676.

Haggard, P. (2008). Human volition: Towards a neuroscience of will. *Nature Reviews Neuroscience, 9*, 934–946.

Hagoort, P. (2005). On Broca, brain, and binding: A new framework. *Trends in Cognitive Sciences, 9*, 416–423.

Hagoort, P. (2013). MUC (memory, unification, control) and beyond. *Frontiers in Psychology, 4*, 416.

Hagoort, P., Brown, C., & Osterhout, L. (1999). The neurocognition of syntactic processing. In C. Brown, & P. Hagoort (Eds.), *The neurocognition of language* (pp. 273–316). Oxford: Oxford University Press.

Hasting, A. S., & Kotz, S. A. (2008). Speeding up syntax: On the relative timing and automaticity of local phrase structure and morphosyntactic processing as reflected in event-related brain potentials. *Journal of Cognitive Neuroscience, 20*, 1207–1219.

Hickok, G., & Poeppel, D. (2004). Dorsal and ventral streams: A framework for understanding aspects of the functional neuroanatomy of language. *Cognition, 92*, 67–99.

Hickok, G., & Poeppel, D. (2007). The cortical organization of speech processing. *Nature Reviews Neuroscience, 8*(5), 393–402.

January, D., Trueswell, J. C., & Thompson-Schill, S. L. (2009). Co-localization of stroop and syntactic ambiguity resolution in Broca's area: Implications for the neural basis of sentence processing. *Journal of Cognitive Neuroscience, 21*, 2434–2444.

Kohler, E., Keysers, C., Umilta, M. A., Fogassi, L., Gallese, V., & Rizzolatti, G. (2002). Hearing sounds, understanding actions: Action representation in mirror neurons. *Science, 297*, 846–848.

Kolk, H. H. J., Chwilla, D. J., van Herten, M., & Oor, P. J. (2003). Structure and limited capacity in verbal working memory: A study with event-related potentials. *Brain and Language, 85*, 1–36.

Kuperberg, G. R., Sitnikova, T., Caplan, D., & Holcomb, P. (2003). Electrophysiological distinctions in processing conceptual relationships within simple sentences. *Cognitive Brain Research, 17*, 117–129.

Kutas, M., Van Petten, C., & Kluender, R. (2006). Psycholinguistics electrified II (1994–2005). In M. Traxler, & M. A. Gernsbacher (Eds.), *Handbook of psycholinguistics* (2nd ed., pp. 659–724). London: Elsevier.

Lerner, Y., Honey, C. J., Silbert, L. J., & Hasson, U. (2011). Topographic mapping of a hierarchy of temporal receptive windows using a narrated story. *The Journal of Neuroscience, 31*(8), 2906–2915.

Lewis, J. W., & Van Essen, D. C. (2000). Corticocortical connections of visual, sensorimotor, and multimodal processing areas in the parietal lobe of the macaque monkey. *Journal of Comparative Neurology, 428*, 112–137.

Lotze, N., Tune, S., Schlesewsky, M., & Bornkessel-Schlesewsky, I. (2011). Meaningful physical changes mediate lexical-semantic integration: Top-down and form-based bottom-up information sources interact in the N400. *Neuropsychologia, 49*, 3573–3582.

MacWhinney, B., & Bates, E. (Eds.), (1989). *The crosslinguistic study of sentence processing*. New York, NY: Cambridge University Press.

MacWhinney, B., Bates, E., & Kliegl, R. (1984). Cue validity and sentence interpretation in English, German and Italian. *Journal of Verbal Learning and Verbal Behavior, 23*, 127–150.

Martin, A. (2007). The representation of object concepts in the brain. *Annual Review of Psychology, 58*, 25–45.

Mesulam, M. M. (1998). From sensation to cognition. *Brain, 121*, 1013–1052.

Miller, E., & Cohen, J. D. (2001). An integrative theory of prefrontal cortex function. *Annual Review of Neuroscience, 24*, 167–202.

Milner, A. D., Dijkerman, H. C., McIntosh, R. D., Rossetti, Y., & Pisella, L. (2003). Delayed reaching and grasping in patients with optic ataxia. *Progress in Brain Research, 142*, 225–242.

Morecraft, R. J., Stillwell-Morecraft, K. S., Cipolloni, P. B., Ge, J., MacNeal, D. W., & Pandya, D. N. (2012). Cytoarchitecture and cortical connections of the anterior cingulate and adjacent somatomotor fields in the rhesus monkey. *Brain Research Bulletin, 87*, 457–497.

Moseley, R. L., & Pulvermüller, F. (2014). Nouns, verbs, objects, actions, and abstractions: Local fMRI activity indexes semantics, not lexical categories. *Brain and Language, 132*, 28–42.

Myachykov, A., Thompson, D., Scheepers, C., & Garrod, S. (2011). Visual attention and structural choice in sentence production across languages. *Language and Linguistics Compass, 5*, 95–107.

Nieuwenhuis, S., Aston-Jones, G., & Cohen, J. D. (2005). Decision making, the P3, and the locus coerulus-norepinephrine system. *Psychological Bulletin, 131*, 510–532.

Novick, J. M., Trueswell, J. C., & Thompson-Schill, S. L. (2005). Cognitive control and parsing: Reexamining the role of Broca's area in sentence comprehension. *Cognitive, Affective and Behavioral Neuroscience, 5*, 263–281.

Osterhout, L., & Nicol, J. (1999). On the distinctiveness, independence, and time course of the brain response to syntactic and semantic anomalies. *Language and Cognitive Processes, 14*, 283–317.

Pallier, C., Devauchelle, A. D., & Dehaene, S. (2011). Cortical representation of the constituent structure of sentences. *Proceedings of the National Academy of Sciences of the United States of America, 108*(6), 2522–2527.

Patterson, K., Nestor, P. J., & Rogers, T. T. (2007). Where do you know what you know? The representation of semantic knowledge in the human brain. *Nature Reviews Neuroscience, 8*, 976–988.

Pickering, M. J., & Garrod, S. (2007). Do people use language production to make predictions during comprehension? *Trends in Cognitive Sciences, 11*, 105–110.

Pickering, M. J., & Garrod, S. (2013). An integrated theory of language production and comprehension. *Behavioral and Brain Sciences, 36*(4), 329–347.

Pulvermüller, F., Shtyrov, Y., & Hauk, O. (2009). Understanding in an instant: Neurophysiological evidence for mechanistic language circuits in the brain. *Brain and Language, 110*, 81–94.

Rauschecker, J. P. (1998). Cortical processing of complex sounds. *Current Opinion in Neurobiology, 8*(4), 516–521.

Rauschecker, J. P. (2011). An expanded role for the dorsal auditory pathway in sensorimotor control and integration. *Hearing Research, 271*, 16–25.

Rauschecker, J. P., & Scott, S. K. (2009). Maps and streams in the auditory cortex: Nonhuman primates illuminate human speech processing. *Nature Neuroscience, 12*(6), 718–724.

Rauschecker, J. P., & Tian, B. (2000). Mechanisms and streams for processing of "what" and "where" in auditory cortex. *Proceedings of the National Academy of Sciences of the United States of America, 97*, 11800–11806.

Röder, B., Stock, O., Neville, H., Bien, S., & Rösler, F. (2002). Brain activation modulated by the comprehension of normal and pseudo-word sentences of different processing demands: A functional magnetic resonance imaging study. *NeuroImage, 15*, 1003–1014.

Roehm, D., Bornkessel-Schlesewsky, I., Rösler, F., & Schlesewsky, M. (2007). To predict or not to predict: Influences of task and strategy on the processing of semantic relations. *Journal of Cognitive Neuroscience, 19*, 1259–1274.

Sanford, A. J., Leuthold, H., Bohan, J., & Sanford, A. J. S. (2011). Anomalies at the borderline of awareness: An ERP study. *Journal of Cognitive Neuroscience, 93*, 514–523.

Sassenhagen, J., Schlesewsky, M., & Bornkessel-Schlesewsky, I. (2014). The P600-as-P3 hypothesis revisited: Single-trial analyses reveal that the late EEG positivity following linguistically deviant material is reaction time aligned. *Brain and Language, 137*, 29.

Saur, D., Kreher, B. W., Schnell, S., Kümmerer, D., Kellmeyer, P., Vry, M.-S., et al. (2008). Ventral and dorsal pathways for language. *Proceedings of the National Academy of Sciences of the United States of America, 105*, 18035–18040.

Schank, R. C. (1999). *Dynamic memory revisited*. Cambridge: Cambridge University Press.

Schank, R. C., & Abelson, R. P. (1977). *Scripts, plans goals and understanding: An inquiry into human knowledge structures*. Hillsdale, NJ: Erlbaum.

Schlesewsky, M., & Bornkessel, I. (2004). On incremental interpretation: Degrees of meaning accessed during sentence comprehension. *Lingua, 114*, 1213–1234.

Schlesewsky, M., & Bornkessel-Schlesewsky, I. (2012). Preface: The neurobiology of syntax. *Brain and Language, 120*, 79–82.

Scott, S., Blank, C., Rosen, S., & Wise, R. (2000). Identification of a pathway for intellgible speech in the left temporal lobe. *Brain, 123*, 2400–2406.

Snijders, T. M., Vosse, T., Kempen, G., van Berkum, J. J. A., Peterson, K. M., & Hagoort, P. (2009). Retrieval and unification of syntactic structure in sentence comprehension: An fMRI study using word category ambiguity. *Cerebral Cortex, 19*, 1493–1503.

Stabler, E. P. (2010). Computational perspectives. In C. Boeckx (Ed.), *Oxford handbook of linguistic minimalism* (pp. 616–641). Oxford: Oxford University Press.

Thompson-Schill, S. L., Bedny, M., & Goldberg, R. F. (2005). The frontal lobes and the regulation of mental activity. *Current Opinion in Neurobiology, 15*, 219–224.

Thompson-Schill, S. L., D'Esposito, M., Aguirre, G. K., & Farah, M. J. (1997). Role of left inferior prefrontal cortex in retrieval of semantic knowledge: A reevaluation. *Proceedings of the National Academy of Sciences of the United States of America, 94*, 14792–14797.

Tomasello, M. (2003). *Constructing a language: A usage-based theory of language acquisition*. Cambridge, MA: Harvard University Press.

Ueno, T., Saito, S., Rogers, T. T., & Lambon Ralph, M. A. (2011). Lichtheim 2: Synthesizing aphasia and the neural basis of language in a neurocomputational model of the dual dorsal-ventral language pathways. *Neuron, 72*(2), 385–396.

Ungerleider, L. G., & Mishkin, M. (1982). Two cortical visual streams. In D. J. Ingle, M. A. Goodale, & R. Mansfield (Eds.), *Analysis of visual behavior* (pp. 549–586). Cambridge, MA: MIT Press.

Van Valin, R. D., Jr. (2005). *Exploring the syntax-semantics interface*. Cambridge: Cambridge University Press.

Vigliocco, G., Vinson, D. P., Druks, J., Barber, H., & Cappa, S. F. (2011). Nouns and verbs in the brain: A review of behavioural, electrophysiological, neuropsychological and imaging studies. *Neuroscience and Biobehavioral Reviews, 35*, 407–426.

Ward, A. A., Jr., Peden, J. K., & Sugar, O. (1946). Cortico-cortical connections in the monkey with special reference to area 6. *Journal of Neurophysiology, 9*, 453–461.

Wilson, S. M., DeMarco, A. T., Henry, M. L., Gesierich, B., Babiak, M., Mandelli, M. L., et al. (2014). What role does the anterior temporal lobe play in sentence-level processing? Neural correlates of syntactic processing in semantic variant primary progressive aphasia. *Journal of Cognitive Neuroscience, 26*, 970–985.

Wilson, S. M., Dronkers, N. F., Ogar, J. M., Jang, J., Growdon, M. E., Agosta, F., et al. (2010). Neural correlates of syntactic processing in the nonfluent variant of primary progressive aphasia. *The Journal of Neuroscience, 30*(50), 16845–16854.

Wilson, S. M., Galantucci, S., Tartaglia, M. C., Rising, K., Patterson, D. K., Henry, M. L., et al. (2011). Syntactic processing depends on dorsal language tracts. *Neuron, 72*(2), 397–403.

Wolpert, D. M., Doya, K., & Kawato, M. (2003). A unifying computational framework for motor control and social interaction. *Philosophical Transactions of the Royal Society B, 358*(1431), 593–602.

SECTION E

DEVELOPMENT, LEARNING, AND PLASTICITY

CHAPTER

31

Language Development

Frederic Dick[1,2], Saloni Krishnan[1,2], Robert Leech[3] and Suzanne Curtin[4]

[1]Birkbeck/UCL Centre for NeuroImaging (BUCNI), London, United Kingdom; [2]Centre for Brain and Cognitive Development (CBCD), Department of Psychological Sciences, Birkbeck College, University of London, London, United Kingdom; [3]Computational, Cognitive and Clinical Neuroimaging Laboratory (C3NL), Imperial College London, London, United Kingdom; [4]Speech Development Laboratory, Department of Psychology, University of Calgary, Calgary, AB, Canada

Typically developing (TD) children will rapidly and comprehensively master at least one of the more than 6,000 languages that exist around the globe. The complexity of these language systems and the speed and apparent facility with which children master them have been the topic of philosophical and scientific speculation for millennia. In 397 AD, in reflecting on his own acquisition of language, St. Augustine wrote "... as I heard words repeatedly used in their proper places in various sentences, I gradually learnt to understand what objects they signified; and after I had trained my mouth to form these signs, I used them to express my own desires" (quoted in Wittgenstein, 1953/2001). St. Augustine's intuitions notwithstanding, more recent thinking and research on children's language acquisition suggest that the problem facing a child is much more intricate than simply remembering the association between a sound and an object and learning to reproduce the word's sound. The rich and multitiered nature of this problem—and the many and varied paths to its solution (Bates, Bretherton, & Snyder, 1988)—make the process of language acquisition a unique window into multiple low-level and high-level developmental processes.

Studies of language development have provided unparalleled views into broad neural and behavioral change in response to input and consolidation. In this chapter, we chart the multiple waves of change in language comprehension and production, beginning at birth with studies of speech perception, moving through babbling, phoneme, and word discrimination into the dawn of word comprehension and production, and the subsequent emergence of syntactic and pragmatic abilities. We also look at language's "fellow travelers," skills such as social cognition, gestural communication, and environmental sound recognition that appear to presage or accompany linguistic milestones. We also consider the neural bases underlying early (mostly electrophysiological studies) and later language development (predominantly functional magnetic resonance imaging). In particular, recent neuroimaging literature increasingly demonstrates the importance of experience and learning in the development of neural correlates of language development, as well as the absence of any straightforward and task-independent language-specific neural substrates. We emphasize the impressive degree of individual differences in language learning—something that is of prime importance when evaluating language development in atypical populations (as discussed in other chapters in this volume). We also highlight the importance of the structure and statistics of the input to multiple levels of language learning. Finally, we close with an overview of some of the important studies contributing to our understanding of the development of the neurobiology of language.

31.1 PRECURSORS TO LANGUAGE

The onset of language development is not signaled by the child's first word. Rather, even before birth infants are adapting to their language environment, mastering the necessary prelinguistic building blocks that support later language learning. Infants seem to prefer human speech over a number of other auditory signals, such as filtered speech (Spence & DeCasper, 1987), warbled tones (Samples & Franklin, 1978), white

Neurobiology of Language. DOI: http://dx.doi.org/10.1016/B978-0-12-407794-2.00031-6

noise (Butterfield & Siperstein, 1970; Colombo & Bundy, 1981), and sine-wave analogues of speech (Vouloumanos & Werker, 2004, 2007). Three-month-old infants prefer listening to speech compared with other naturally occurring sounds in their environment, even other human sounds such as laughter and coughing (Shultz & Vouloumanos, 2010). As we see in this section, during the first year of life children make huge strides in constructing the social, perceptual, and attentional tools that language needs to get off the ground.

Even before and soon after birth it is possible to see the effects of experience-dependent speech discrimination. For instance, the heartbeat of term fetuses tends to increase in response to hearing their mother's voice (Kisilevsky et al., 2003). Infants as young as 4 days old can use rhythm to discriminate between familiar and unfamiliar languages (Nazzi, Bertoncini, & Mehler, 1998; similar skills have also been noted in Tamarin monkeys; Ramus, Hauser, Miller, Morris, & Mehler, 2000). Newborn infants also demonstrate some evidence of being able to discriminate between some of the vowels and consonants from human natural languages (Eimas, Siqueland, Jusczyk, & Vigorito, 1971; see Kuhl, 2004 for a review). Under certain testing conditions, infants, like adults, classify sounds into different categories. That is to say, as some physical characteristic of a phonetic contrast varies along a continuum (such as voice onset time), we do not hear gradual variation in the sounds, but instead a sharp change from one sound to another. This finding has been further elaborated by research illustrating that 3- to 6-month-old infants show graded, within-category perception of voice onset time (VOT) under appropriate testing conditions (McMurray & Aslin, 2005; Miller & Eimas, 1996), suggesting that although phonetic discrimination may be the most easily revealed, gradient perception is also possible. Again, this ability to categorically perceive speech sounds is not specific to humans, with monkeys (Kuhl & Miller, 1975) and chinchillas (Kuhl & Padden, 1983) demonstrating similar phonetic discrimination as do infants. One suggestion is that human phonetic categories have evolved around more general characteristics of mammalian sensory systems (Dick, Saygin, Moineau, Aydelott, & Bates, 2004; Kuhl, 1986; Smith & Lewicki, 2006, but also see Tomasello, 1999, for evidence of more human-specific abilities that may play a part in language development).

Infants younger than 6 months appear to discriminate a wide range of speech sound contrasts, including those that do not occur in their native language (see Saffran, Werker, & Werner, 2006 for a review). However, not all speech sound contrasts are equally discriminable. Sometimes this is because of asymmetrical perception that results from the order in which speech sounds are presented, and sometimes this is the result of the salience of a contrast. Polka and Bohn (1996) found that an infant's perception of one vowel can shift their perception of a subsequent vowel. For example, if infants are first presented with one of point vowels (e.g., /i/) and then are presented with a more central vowel (e.g., /ɪ/), then the central vowel appears to be absorbed into the point vowel's auditory space, making it difficult for infants to discriminate between them (see Polka & Bohn, 2011 for a review). Kuhl (1991) found that the extreme productions of a vowel influence perception of a less extreme production of the same vowel. She argued that these more extreme tokens act as perceptual magnets attracting the less peripheral tokens into their categorical space. In the case of consonants, it has been shown that acoustic salience influences early speech sound discrimination. For example, Filipino and English infants 6 to 8 months old both have difficulty discriminating the alveolar /na/ versus the velar /ŋa/ nasal stop contrast in word initial position, even though the contrast exists in this position in Filipino and in syllable final positions in English. It is not until later in the development, approximately 10 months, that Filipino-learning infants discriminate this contrast, whereas English-learning infants at the same age still do not discriminate this contrast (Narayan, Werker, & Beddor, 2010), suggesting that perceptually difficult contrasts require more experience for infants to be able to discriminate between them.

Starting at approximately the half-year mark, infants' experience with their native language can be seen to influence their ability to distinguish between non-native speech sound contrasts, so that by approximately the first year infants form a strong bias towards *native language-specific phonetic perception*, beginning with vowels and subsequently extending to consonants (for a review see Werker & Desjardins, 1995). For instance, at 6 months an infant exposed to a Japanese-speaking environment can distinguish between the English /r/ and /l/ sounds, but by 12 months the same child discerns only a single phoneme, unlike an infant reared in an English speaking environment. Furthermore, children's abilities to make such phonetic classifications in their *native* language at 7 months *positively predict* language outcomes such as word production, mean length of utterance (MLU), and sentence complexity between 14 and 20 months, whereas ability regarding *non-native* phonetic contrasts is *inversely* related to later language measures (Kuhl, Conboy, Padden, Nelson, & Pruitt, 2005).

What underlies these developmental changes in infants' auditory discrimination? There is some evidence that in the first year of life, infants' phonetic discriminations in their native language can, in part, rely on the stochastic *distributional information* available in natural speech. Although the actual examples of any given phoneme that an infant hears vary considerably along many acoustic dimensions, some tend to conform to general statistical distributions that the infant may be able to use to identify the most informative boundaries for distinguishing phonemes (Jusczyk, Luce, & Charles-Luce, 1994; Kuhl, Williams, Lacerda, Stevens, & Lindblom, 1992; Maye, Werker, & Gerken, 2002; Saffran & Thiessen, 2003). However, in many (if not most) cases, such statistical information may be insufficient for perceiving such distinctions (Feldman, Griffiths, Goldwater, & Morgan, 2013; Swingley, 2009). Other evidence suggests the visual and lexical information play an important role in speech sound discrimination. In illustration, if a unimodal distribution of a continuum between /ba/ and /da/ is presented simultaneously with synchronous visual articulations of a canonical /ba/ and /da/, infants 6 months of age discriminate the contrast; however, if the articulations are incorrect, then they no longer show discrimination (Teinonen, Aslin, Alku, & Csibra, 2008). Using visual objects, Yeung and Werker (2009) found that while English-learning infants no longer discriminate non-native contrasts, such as the Hindi dental/retroflex stop contrast (Werker & Tees, 1984), they can discriminate this difficult contrast at 9 months of age if each individual speech sound is paired with a specific object. This suggests that infants are using information gleaned from visual context to help detect contrasts. In addition, lexical influences are likely supporting speech sound discrimination. Feldman, Myers, White, Griffiths, and Morgan (2013) found that 8-month-olds could distinguish similar vowels only after hearing those vowels consistently in distinct word environments. Modeling work supports the usefulness of lexical knowledge in determining sound categories. Martin and colleagues argue that learners have access to proto-lexicon consisting of high-frequency units (n-grams), which can be used for learning their native language phonemes (Martin, Peperkamp, & Dupoux, 2013). Together these findings suggest that a number of factors, including distributional, visual, and lexical, contribute to whether infants can discriminate speech sounds.

Although infants do lose the ability to make discriminations for phonetic contrasts in all the world's languages at approximately the first birthday, this is *not* a straightforward example of a "critical" or "sensitive" period in brain maturation. An elegant combined behavioral and fMRI study by Pallier and colleagues demonstrated that Korean-speaking children who were adopted into French-speaking families between the ages of 3 and 8—importantly, with no further exposure to Korean—did not differ from children born into French-speaking families when both were tested on phonetic contrasts in Korean and French as adults (Pallier et al., 2003).

In addition to learning to segment the speech stream into meaningful language, children are also faced with the task of producing meaningful speech themselves. Researchers interested in how sensorimotor factors might influence speech perception have examined the relationship between speech perception and production abilities. Using vocal motor schemes (VMS), which are defined as the consonants that are part of an infants' production inventory, DePaolis, Vihman, and Keren-Portnoy (2011) found that 10-month-olds who had multiple VMSs exhibited a novelty listening preference for passages whose nonwords contained consonants they did not consistently produce. Similar findings have been found for languages other than English, such as Italian (Majorano, Vihman, & DePaolis, 2013) and Welsh (DePaolis, Vihman, & Nakai, 2013). The link between perception and production has also been illustrated in studies exploring the patterns of neural activation in the temporal and frontal lobes of infants aged 6 and 12 months (Imada et al., 2006). The early *precursors of productive language* start with infants' preverbal vocalizations. From approximately 3 months, infants begin producing vowel sounds and appear to be able to imitate adult-modeled vowel sounds (Kuhl & Meltzoff, 1996). From 6 to 8 months, infants start *babbling*—making consonant vowel combinations (e.g., {ba}, {ata}). This early babbling is not obviously communicative, often occurring when the infant is on his/her own. As with phonetic perception, over the first year the sounds that an infant produces move from being "universal" (with respect to all of the world's languages) to increasingly resembling the sounds of the language(s) spoken around them. Infants produce native language–specific vowel and consonant sounds before they produce their first words, thus internalizing the acoustical or phonetic patterns of the language the child is exposed to (Boysson-Bardies, de, Halle, Sagart, & Durand, 1989; Boysson-Bardies & Vihman, 1991).

Although *social development* is not a direct precursor of word or syntax comprehension or production, it is entwined with language across early development. The early beneficial effects of social context on language learning are evident in infants' vocalizations and phonetic discriminations. For instance, Bloom, Russell, and Wassenberg (1987) showed that turn-taking can alter very young infants' vocalizations, and Goldstein, King, and West (2003) showed that parental feedback increases infant vocalizations. In addition, Kuhl, Tsao, and Liu (2003) found that North American infants learned non-native Mandarin phonemic contrasts in the presence of a Mandarin speaker but not from a video recording of the same information.

At approximately 3–6 months, social cognition in infants is perhaps most evident in *gaze following*, first directed at nearby targets (D'Entremont, Hains, & Muir, 1997) and expanding to further targets by approximately the first birthday (Corkum & Moore, 1995). As with other skills, this feature is not uniquely human and is present in at least several other nonhuman primate species (Tomasello, Carpenter, Call, Behne, & Moll, 2005). Starting at approximately 9 months, we also see a change from *diadic interactions* (i.e., the infant interacting with another object or another person) to *triadic interactions* (i.e., the infant and a caregiver jointly attending to each other and an object; Trevarthen & Hubley, 1978).

From the end of the first year and beyond, children become increasingly adept at understanding and directing other people's attention, using this information to make the task of language learning more tractable (Bates, 2004). In a longitudinal study of infants from 9 to 15 months, Carpenter, Nagell, and Tomasello (1998) showed a linked progression of gesture and *joint attention*, from infants initially sharing attention to subsequently following an adult's attention, to directing another's attention (for a detailed review see Tomasello et al., 2005). Slightly older infants use an adults' *communicative intent* to rapidly attach meaning to novel words (for a review see Tomasello, 2001). Given the prominence of joint attention and its relation to language development, it is not surprising that the quantity and type of joint attention between infant and caregiver predict children's early communicative abilities, with particular gains when the caregiver focuses on the object of the infant's attention (Carpenter et al., 1998; Tomasello & Todd, 1983).

31.2 FIRST WORDS

As the infant's native language discrimination improves, the child faces the simultaneous daunting task of using various cues in the speech input to *segment speech into words* and attach some meaning to them. Although adult listeners tend to perceive the speech stream as a series of discrete words presented one after another (at least in their native language), human speech actually affords no such luxury because there is generally no one-to-one mapping between pauses or silences and word boundaries. Despite this fact, by 7.5 months, infants can detect words they have been familiarized with from a stream of natural speech (Jusczyk & Aslin, 1995) and furthermore demonstrate longer-term memory for these words in speech (Jusczyk & Hohne, 1997). The speech stream contains a number of different clues regarding the location of word boundaries, such as *syllabic stress patterns* (Curtin, Mintz, & Christiansen, 2005; Jusczyk, Houston, & Newsome, 1999), *transitional probabilities* (i.e., the likelihood that one phonetic segment follows another; Saffran, Aslin, & Newport, 1996), and *familiar words* (such as *mommy*), which can help infants find and segment adjacent word forms (Bortfeld, Rathbun, Morgan, & Golinkoff, 2005). Infants use a combination of these different cues to segment speech into words, with the relative weighting of cues varying over development (Johnson & Jusczyk, 2001; Thiessen & Saffran, 2003).

The purpose of this segmentation is, of course, to identify the chunks of speech (words) to which meaning can be attached and/or extracted. Evidence has emerged suggesting that infants recognize highly common words as young as 6 months of age (Bergelson & Swingley, 2012; Tincoff & Jusczyk, 1999). By 12 months, in an associative learning task, English-learning infants will appropriately map a novel word to a novel object that has the phonological form of a typical noun, but will reject: (i) forms with illegal sound sequences such as "ptak" (MacKenzie, Curtin, & Graham, 2012a); (ii) forms that are more function-word like such as "iv" (MacKenzie, Curtin, & Graham, 2012b); or (iii) subminimal forms (i.e., communicative sounds such as "ooh" and single phonemes such as "l"; MacKenzie, Graham, & Curtin, 2011). These results show that the phonological knowledge acquired over the first year of life influences word-object mapping. Infants become increasingly rapid and skillful at forming these word-to-meaning associations. By approximately 2 years (and possibly earlier), infants demonstrate *"fast mapping"* (Carey, 1978), whereby word-to-meaning mappings are learned after a single exposure. As with previous examples, this skill is neither specific to language (Markson & Bloom, 1997) nor specific to humans (Kaminski, Call, & Fischer, 2004).

As any proud (yet weary) parent will attest, young children are not just consumers of language, but they also use it increasingly productively to communicate their needs, desires, and interests with others. Whereas *word comprehension* typically starts at approximately 9–10 months, *word production* typically follows several weeks later (Fenson et al., 1994). In general, the size of a toddler's early receptive vocabulary maintains a healthy numerical superiority over his or her productive vocabulary, although there is considerable individual variability in the extent of this relationship (again see Fenson et al., 1994). Infants' early productive vocabulary is mostly composed of nominal labels for objects or people, although they also produce non-nominal words (for instance, the relational label "up"). However, straightforward classifications of infants' language in terms of adult linguistic categories such as verbs or nouns are probably inaccurate. Because infant speech is driven primarily by the desire to communicate, Tomasello argues that many of the early one-word utterances are actually "holophrases"—expressing a holistic communicative function with a single label (Tomasello, 2006). For instance, an utterance such as "up" might serve as the infant's shorthand for an adult phrase such as "pick it up." Thus, the infant may be copying a part of the adult phrase as a way to express the communicative intent of the phrase as a whole.

31.3 INDIVIDUAL VARIABILITY, DEVELOPMENTAL TRAJECTORIES, AND THE VOCABULARY "BURST"

As noted in the introduction, to understand the mechanisms underlying both typical and atypical language development, it is vital to have an understanding of the trajectory of that development, both in the "average" child as well as in individual children. The MacArthur Bates Communicative Development Inventory (CDI) provides an excellent and carefully normed method of tracing an individual child's linguistic developmental trajectory from the tentative start of meaningful communication around the first birthday (for the "average" child) through the advent of complex sentence production and comprehension. An instrument based on parental report (and validated through laboratory observation; Fenson et al., 2000), the CDI is extremely useful for comparing typical and atypical populations, and for assessing individual variation within a given sample.

One interesting finding of the initial MacArthur Bates CDI norming studies (Fenson et al., 1994, 2000) is that there is *little overall difference between girls and boys* in terms of the trajectory of language development. Girls are, on average, 1 month ahead of boys, but this difference accounts for only 2% of the variation within and across age groups. Thus, these gender differences are relatively insignificant compared with the much greater variation between individuals. It is worth emphasizing how much *individual variation* there is between "typically" developing children—the idea of "typical" language development is something of a useful fiction. As an illustration of this, we have reprinted the cross-sectional growth curves from the Fenson et al. (1994) monograph showing the receptive and productive vocabulary size of children in the 10th, 25th, 50th, 75th, and 90th quartiles (Figures 31.1 and 31.2); we also show longitudinal growth curves for three different TD children for comparison (Figure 31.3).

The use of statistical averages to simplify complex and highly variable time series can sometimes mask more interesting phenomena. A case in point is the

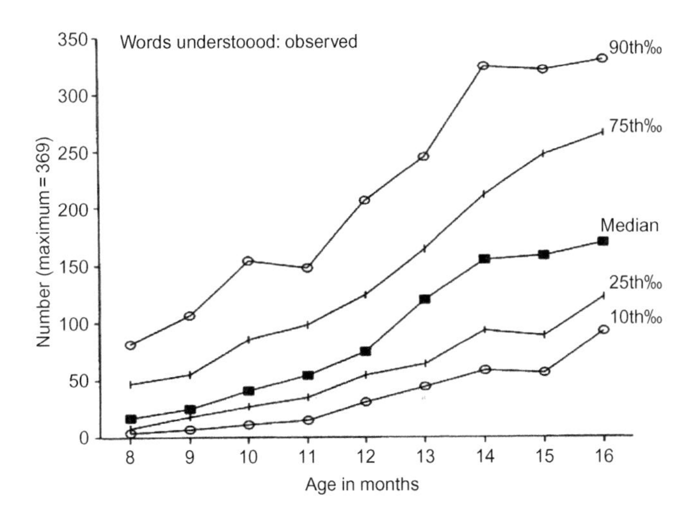

FIGURE 31.1 Number of phrases on the Infant form of the MacArthur Bates CDI reported to be understood by children at each month. Median values and percentile ranks are shown. *Reprinted from Fenson et al., 1994 with permission from the publisher.*

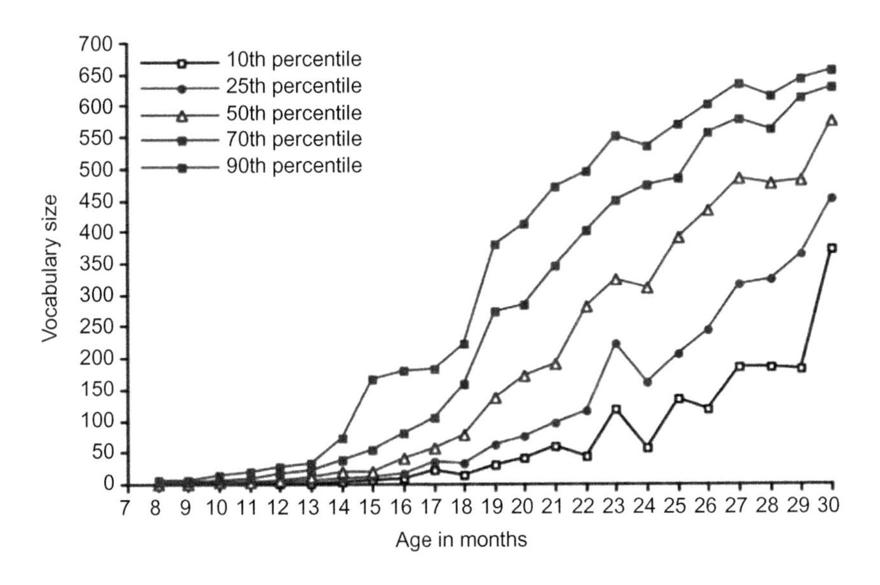

FIGURE 31.2 Number of words on the Toddler form of the MacArthur Bates CDI reported to be produced by children at each month. Median values and percentile ranks are shown. *Reprinted from Fenson et al., 1994 with permission from the publisher.*

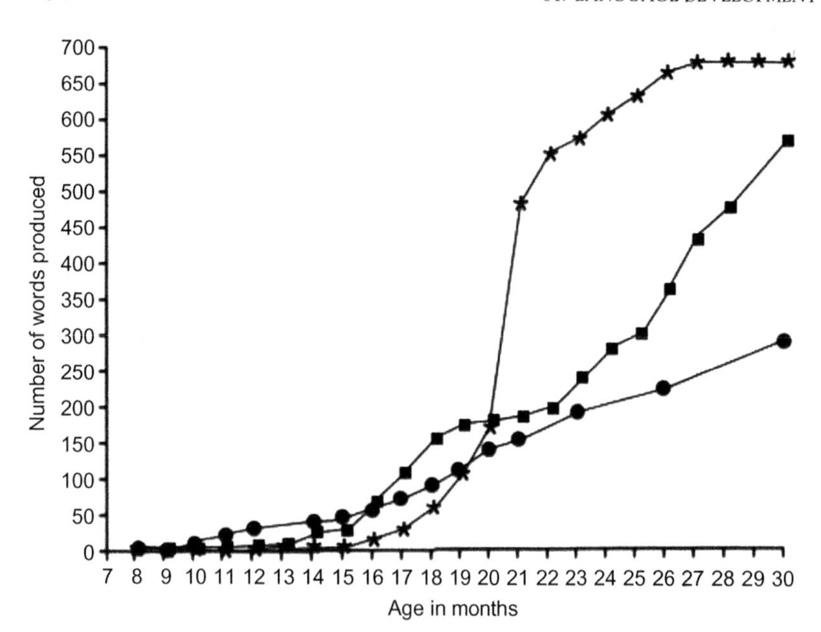

FIGURE 31.3 Longitudinal growth curves for the productive vocabulary of three different TD children, as measured by the MacArthur Bates CDI. *From Bates, personal communication.*

sudden acceleration at 16–20 months in a child's vocabulary, the so-called *"vocabulary burst"* that follows a period of very gradual increases in vocabulary size after the first few words. This vocabulary acceleration involves not only an increase in the total number of words a child produces but also changes in the content of the words, with a shift to a greater proportion of adjectives and verbs (Bates et al., 1988; Fenson et al., 1994; Hampson & Nelson, 1993; Nelson, 1981; Tomasello, 2006). This sudden change in vocabulary size has often been considered an indicator of the onset of a new cognitive ability, such as developing a "naming insight" (Dore, 1974). However, this "average" picture masks the wide gamut of individual developmental trajectories observed with the CDI for so-called normal children (i.e., children without obvious language problems). There is wide individual variation in productive vocabulary size at the point when the "average" child launches his/her vocabulary burst. At this time in chronological development (~16 months), children in the highest 10th percentile produce approximately 180 words, whereas those in the lowest 10th percentile produce fewer than 10 words. It is very important to note that despite this early variation, most of these children—including those in the 10th percentile who are slow getting language off the ground—will go on to have similar language outcomes as adults as their initially more able peers. There is also massive variation in the shape of the growth curves for different individuals' productive vocabulary with age. Some children show a recognizable burst, whereas others' vocabulary appears to grow at a much steadier pace, with still others advancing in a series of small successive bursts. A strong possibility is that the *relationship between vocabulary size and age is inherently nonlinear*, rather than the more frequently assumed form of one linear relationship giving way to another linear relationship with the onset of each new skill. The variability across individual's vocabulary growth curves can be captured most parsimoniously using nonlinear models (Bates & Carnevale, 1993; see also Elman et al., 1996, for a more general discussion of nonlinearity in development). Irrespective of the cause, recognizing the tremendous individual variability is paramount when assessing atypical populations and in understanding in what ways these children differ from TD infants.

31.4 EARLY LANGUAGE AND ITS RELATIONSHIP TO NONLINGUISTIC ABILITIES

Meaningful language production and comprehension develop in tandem with a raft of nonlinguistic cognitive and motor abilities. The close developmental relationship between *gesture and language* (see Bates & Dick, 2002, for a review) appears to begin from around 6 months, with a correlation between the onset of babbling and the onset of rhythmic hand-banging. Toward the end of the first year, first word comprehension tends to co-occur with the start of deictic gestures (e.g., pointing and showing gestures) and gestural routines (e.g., waving goodbye; Bates, Benigni, Bretherton, Camaioni, & Volterra, 1977). In a similar vein, the later onset of productive vocabulary also co-occurs with—or is slightly preceded by—early recognitory gestures, such as putting a phone to the ear or a brush to hair (Volterra, Bates, Benigni, Bretherton, & Camaioni, 1979).

The ability to consistently imitate both behavioral and vocal cues does not seem to develop until approximately 18 months (Jones, 2009). Interestingly, imitation in the form of repeating words emerges between 16 and 25 months and seems relatively stable despite language developing across this time (Bloom, Hood, & Lighbown, 1974). However, which words are imitated changes over time, with infants largely imitating words they do not know (i.e., would not produce spontaneously) and, then, once imitated enough, these words become part of spontaneous speech (Bloom et al., 1974). Motor skills are also related to some of the first forms of social communication, such as nonverbal requesting, initiating joint attention, and responding to joint attention. Impairment in any of these fundamental components could interfere with the foundations of interpersonal communication. Rhesus monkey infants who do not show imitative tongue protrusions are more likely to later develop stereotypes typical of those in human autism spectrum disorder (ASD) (Ferrari et al., 2009). Moreover, there is a negative correlation between the degree of tongue protrusion in infant imitation and language use in children with ASD (Gernsbacher, Sauer, Geye, Schweigert, & Goldsmith, 2008). Alcock and Krawczyk (2010) showed that oromotor control at 21 months is strongly associated with language production measures. The relationship between oromotor control and language production continues to be important through development. Using data from a large set of tasks tested on several samples of school-age children, Krishnan et al. (2013) found that oromotor praxis skills were the best predictor of nonword repetition skills, a key index of language proficiency.

Infants' word comprehension also shares a developmental trajectory similar to their understanding of familiar *environmental sounds* (i.e., meaningful, nonlinguistic sounds such as a cow mooing or a car starting). Cummings, Saygin, Bates, and Dick (2009) found that 15- to 25-month-old infants' accuracy in comprehending environmental sounds and spoken phrases was approximately equivalent (with a slight advantage for environmental sound recognition early in development). These results suggest that, in fact, speech does not appear to start out as being "privileged" as an acoustical transmitter of referential information.

During the course of development, the relationship between language, environmental sound comprehension, and gesture production changes as the infant gains more linguistic experience. Recognitory gesture is eclipsed by the exponential increase in a child's productive vocabulary as language "wins custody" over gesture as the prime means of expressive communication (Bates & Dick, 2002). Similarly, for environmental sound comprehension, infants with larger productive vocabularies show a significant accuracy advantage for comprehending spoken words over environmental sounds; this advantage of words over environmental sounds is not revealed when infants are grouped by chronological age. Nonetheless, throughout the lifespan there remains a close relationship between language and gesture (Saygin, Dick, & Bates, 2005) and between language and environmental sound (Cummings, Ceponiene, Dick, Saygin, & Townsend; 2008; Cummings et al., 2006; Dick et al., 2007; Leech & Saygin, 2011; Saygin, Dick, Wilson, Dronkers, & Bates, 2003). In sum, language is not an isolated ability "fenced off" from the rest of cognition. Instead, language appears to emerge from the interactions of many domain-general cognitive processes, including memory, attention, object recognition and categorization, social and emotional abilities, as well as the nonlinguistic motor and acoustical abilities (Bates, Thal, Clancy, & Finlay, 2002). Language might best be described as a "*new machine constructed entirely out of old parts*" (Bates & MacWhinney, 1989).

31.5 RELATIONSHIP BETWEEN EARLY DEVELOPMENT AND LATER LANGUAGE ABILITIES

Development over the first 2 years of life has implications for later language abilities. For example, developmental changes in oro-facial movement speeds in infants between 9 and 21 months have been correlated with developmental advances in language and cognitive skill (Nip, Green, & Marx, 2011). Six-month-old infants' ability to discriminate two acoustically distinct vowels (/u/ and /y/) is correlated with language abilities at 13–24 months of age (Tsao, Liu, & Kuhl, 2004). Five-month-old infants who prefer listening to their native language stress pattern have larger vocabularies at 12 months (Ference & Curtin, 2013), and the amount of time 12-month-olds spend listening to speech is related to their vocabulary size at 18 months (Vouloumanos & Curtin, 2014). Speech segmentation abilities at 12 months correlate with expressive vocabulary at 24 months and language skills at 4–6 years (Newman, Ratner, Jusczyk, Jusczyk, & Dow, 2006). Ability to map minimally differing words in a word-object associative learning task (e.g., bih, dih; Stager & Werker, 1997) at 17 and 20 months is related to children's performance on standardized tests of language comprehension and production 2.5 years later (Bernhardt, Kemp, & Werker, 2007).

31.6 THE RELATIONSHIP BETWEEN VOCABULARY AND GRAMMAR

The sudden acceleration in vocabulary growth is accompanied or followed by the first two-word

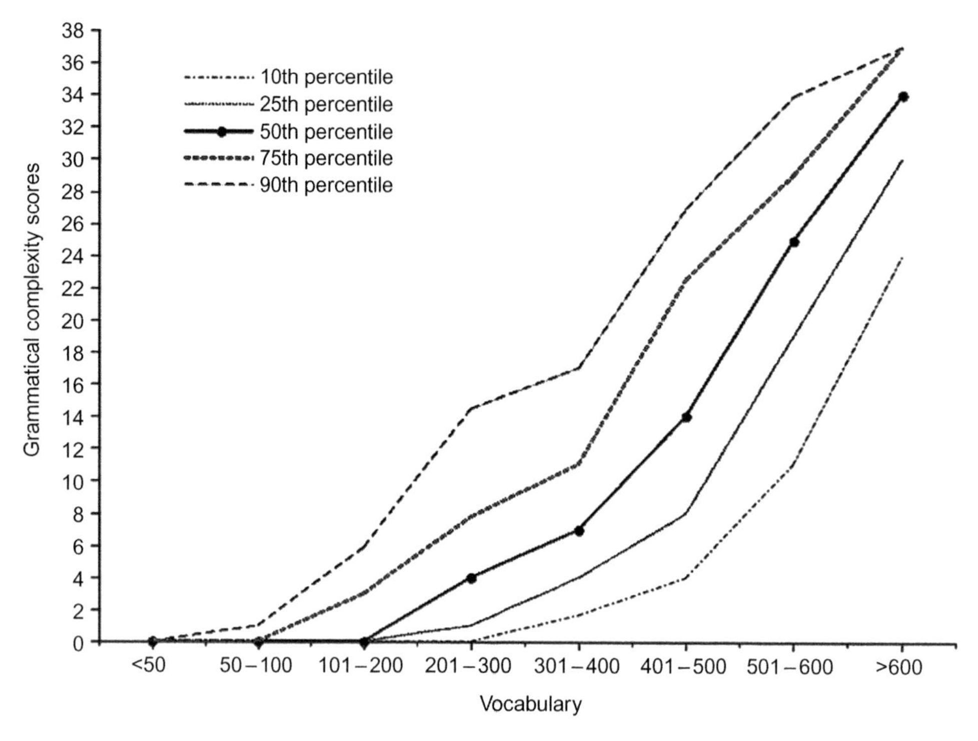

FIGURE 31.4 Relationship between grammar and vocabulary size: variation within each vocabulary level. *Reprinted from Bates & Goodman, 1997 with permission from the publisher.*

combinations at 18–20 months. Early word combinations mark the start of a second "burst" in the child's abilities, this time in the realm of productive grammatical complexity.[1] As with previous language milestones, this rapid increase in syntactic sophistication does not occur in a vacuum. Rather, toddlers' burgeoning syntactic abilities during the middle of the second year are closely yoked to their productive vocabularies—that is, *syntax is not independent from the lexicon* (Bates & Goodman, 1997).

Bates et al. (1988) found significant positive correlations between productive vocabulary size at 20–28 months and MLU[2] in the same period, with the strongest correlation between vocabulary at 20 months and MLU at 28 months—when the "average" child's complex grammatical language is changing most rapidly. It is important to note that this tight synchronous and diachronous relationship between lexical size and grammatical complexity is not driven by a latent variable, like "maturation." In this vein, Bates and Goodman (1997) used data from the large CDI norming study (Fenson et al., 1994) to demonstrate that total vocabulary size correlated with grammatical complexity equally as strongly as grammatical complexity correlates with itself—and that this relationship held true when chronological age was partialled out. In fact, calculated over the entire CDI sample, hierarchical stepwise regressions revealed that age uniquely accounted for only 0.8% of the variance for grammatical complexity, whereas vocabulary size accounted for 32.3% of unique grammatical variance. These results suggest a law-like relationship whereby *total vocabulary size, irrespective of age, predicts grammatical complexity.* Furthermore, there is very little variability between individuals around this relationship—including in all but one clinical population where this question has been investigated (Figure 31.4).

Remarkably, the lawful relationship between grammar and the lexicon also appears to hold over languages, despite the fact that languages differ tremendously in terms of the morphosyntactic cues that provide "clues" to meaning. For example, in English the most reliable grammatical cue to agency ("who is doing what to whom") is word order, whereas for

[1]Note that in terms of comprehension, infants younger than 12 months old can discriminate patterns analogous to simple grammars (Gomez & Gerken, 1999).

[2]MLU is a frequently used measure of grammatical complexity calculated by taking the average number of morphemes (the smallest units of meaning) per phrase. MLU is a somewhat problematic measure for comparing children's grammatical knowledge across ages and across languages (Bates et al., 1988). The CDI includes measures of grammatical complexity that have been normed across ages and languages, against benchmark laboratory studies, and so provide a more robust mechanism for investigating the grammar explosion seen in the third year.

some other languages (e.g., Italian), these sentential roles are often imparted through inflectional morphology. Prima facie, such languages show somewhat different grammatical developmental trajectories—for instance, in a highly inflected, regular, and transparent language like Turkish, the use of morphological particles is observed much earlier in development than in English (Slobin, 1985). However, this does not mean that the relation between the lexicon and syntax needs to be fundamentally different. The CDI has been used *cross-linguistically* to compare English and a language with rich inflectional morphology for tense, aspect, number, and gender—namely Italian. Despite the obvious differences in the languages, the same law-like predictive link between vocabulary size and grammatical complexity exists for Italian and English (Caselli, Casadio, & Bates, 1999), demonstrating the generality of this finding (Figure 31.2). The implication of this work is that grammar is not a completely separate process from word learning, nor does grammar simply require some word knowledge to get started. Instead, grammar and the lexicon are interwoven throughout early development.

31.7 THE NATURE OF CHILDREN'S EARLY GRAMMAR

One long-standing position in developmental psycholinguistics is that young children and adults fundamentally share the same syntactic "competence" (see Tomasello, 2000a for a detailed critical review). This *"continuity assumption"* is one offshoot of the theory that all human languages are built on a single innate universal grammar, with languages essentially differing only in the words they use (Pinker, 1989). An alternative developmental hypothesis postulates that children's early syntax is item-based. That is, young children initially produce grammatical language not through the utilization of general and abstract linguistic structures (e.g., "subject," "verb," "noun"), but rather through reproduction and very conservative and gradual tweaking of individual and specific linguistic "constructions" that they have learned from others' speech (Goldberg, 1995, 2006; Tomasello, 1992, 2000a, 2000b).

There is increasing evidence that at least some of *young children's grammar is item-based*. A number of observational studies of children's early language production (Pizutto & Caselli, 1992, 1994; Tomasello, 1992) have revealed that children's early production of verbs does not reveal a systematic pattern of usage. Instead, young children produce many verbs in only a single form with no transfer of structure from verb to verb. For instance, for the verb "cut," a child will only produce phrases of the form "cut_" (e.g., cut apple or cut bread). This phenomenon, termed the *"verb island"* hypothesis (Tomasello, 1992), has been reported cross-linguistically (Pizutto & Caselli, 1992, 1994; Rubino & Pine, 1998).

Experimental studies have also investigated the item-based nature of early syntax by considering how well children produce novel verb constructions. In a series of studies, Tomasello and colleagues (see Tomasello, 2000a, 2000b for reviews) investigated what linguistic forms a 2- to 3-year-old child produces when given a novel verb to use in a variety of linguistic situations. For instance, if a novel verb like "tam" is only modeled for the child in intransitive form, will a child produce the transitive form of the verb given an appropriate context? These studies repeatedly demonstrate that before the age of approximately 3 years, children will generally base their verb productions on the input that they have heard. In other words, children will not transfer the transitive structure to a novel verb that they encountered in the intransitive form, even when explicitly asked to do so. Akhtar (1999) demonstrated an even more extreme example of how the *structure of the input* determines children's linguistic productions. Exposing younger children (2- and 3-year-olds) to novel verbs in different word orders (i.e., subject verb object [SVO], SOV, and verb subject object [VSO]) led to framing the novel verb in ways that reflected the exposed verb order—even for noncanonical English word orders like SOV and VSO. In contrast, older children (at 4 years) generalized from their knowledge of SVO word order and used the novel verb only in a "canonical" English way, thus suggesting they were abstracting the verb away from its syntactic frame.

In general, these observational and experimental studies call into question the existence of abstract adult grammatical categories in children's early syntactic development. These novel verb production studies also indicate that *early language is highly sensitive to statistical patterns in the ambient language* (e.g., the frequency of a word order, such as SVO, is a key determinant of its production by a child). Gradually, as the typical child develops, by 3–4 years of age that child will be increasingly able to generalize to novel verbs using existing templates such as the transitive SVO structure. However, it is important to note that these word-order preferences are not immutable; even college-age adults are exquisitely sensitive to changes in the relative frequency of word orders.

31.8 LANGUAGE DEVELOPMENT IN OLDER CHILDREN

Children by the age of 3–4 years are increasingly proficient language users with large productive vocabularies, and they are able to fluently use and comprehend complex grammatical constructions (Bates & Goodman, 1997). However, this milestone does not mark the end of the development of language. Instead, *children's language abilities keep gradually improving into adolescence and beyond* (Nippold, 1998). One obvious area of improvement is vocabulary growth that continues throughout childhood, increasing by approximately 3,000 words per year (for a review see Graves, 1986). Similarly, auditory perception and speech perception continue to improve into adolescence (Ceponiene, Rinne, & Naatanen, 2002). Most surprisingly, perhaps, is that syntactic abilities also continue to develop into later life.

As with infants, school-age children remain sensitive to the frequency of syntactic constructions they hear. For instance, Huttenlocher, Vasilyeva, Cymerman, and Levine (2002) demonstrated that the proportion of complex syntactic constructions used by a child's primary school teacher predicts how well the child produces and understands difficult syntactic structures, over and above chronological age. Even approaching adolescence, children's syntactic comprehension can be demonstrated to vary from that of adults. Leech, Aydelott, Symons, Carnevale, and Dick (2007) explored children's (ages 5–17) and adults' (ages 18–51) comprehension of morphosyntactically diverse sentences under varying degrees of *attentional demands, auditory masking, and semantic interference*. The results indicated perceptual masking of the speech signal has an early and lasting impact on comprehension, particularly for more complex sentence structures, and that young children's syntactic comprehension is particularly vulnerable to disturbance. This study demonstrated not only that syntax follows an elongated developmental trajectory but also that other more general attentional and perceptual skills continue to play an important role in syntactic processing across the lifespan (see also Hayiou-Thomas, Bishop, & Plunkett, 2004).

31.9 NEURAL MEASURES OF LANGUAGE DEVELOPMENT

How is the child's brain reorganizing itself during this period of profound language development? A series of electrophysiological studies by Debra Mills and colleagues suggests that "*cerebral specialization for language emerges as a function of learning*, and to some extent depends on the rate of learning" (Sheehan & Mills, 2008). Mills and colleagues have shown that the relative lateralization of electrophysiological (EEG) components (P100, N200-400) during the first years of life is intimately related to language learning and expertise. In particular, the lateral distribution of the N200-400 component for known versus unknown words is related to the overall size of the infant's vocabulary in a particular language. Conboy and Mills (2006) showed that in 20-month-old bilingual toddlers classified as having high or low total vocabulary sizes, the N200-400 difference between known versus unknown words was lateralized only in children with higher vocabularies, and only in their dominant language. Conversely, this known-versus-unknown word N200-400 difference was bilaterally distributed in the nondominant language and in the toddlers with lower total vocabulary. A similar finding was reported in a rapid-word-learning experiment by Mills, Plunkett, Prat, and Schafer (2005). Thus, changes in the large-scale topography of neural responses to words were driven by infants' expertise with words in general, as well as by their knowledge of specific word exemplars.

More generally, in their review of the infant and child EEG language literature, Sheehan and Mills (2008) point out that the relative *lateralization of EEG components changes dynamically over the lifespan*. As an example, they cite the case of the P1 component evoked in response to auditory stimuli, which shows an early left-lateralization from 3 months to 3 years, a symmetrical distribution from 6 to 12 years, and a right-lateralized distribution from 13 years into adulthood. The existence of such complex developmental trajectories demonstrates that any putative "early lateralization" for speech or language stimuli must be understood in the context of changes over the lifespan—a particularly important point for interpreting functional and structural magnetic resonance imaging (MRI) studies of language development.

Functional and structural MRI studies of language development are often directed at questions of relative lateralization (and regionalization) of function. In this vein, Perani et al. (2010, 2011) scanned newborn Italian infants (~2 days old) as speech and music stimuli were played to them. Among other results, the authors found that newborn infants showed a substantially *right-lateralized* response in primary and secondary auditory regions for naturally produced speech *and* music, whereas *altered* speech and music showed a bilateral or even slightly left-lateralized profile of activation. There is some indication that this early lateralization profile for passive language listening may vary in the first months of life. An fMRI study of 3-month-old infants listening to meaningful *or* reversed speech while asleep or awake reported more left-lateralized activity in the superior temporal and angular gyri (Dehaene-Lambertz, Dehaene, & Hertz-Pannier, 2002). In an fMRI study with 7-month-old infants, Blasi et al. (2011) showed that nonlinguistic vocalizations evoked greater bilateral but right-lateralized superior temporal gyrus and sulcus activation relative to nonvocal environmental sounds. In terms of early underlying structural asymmetries related to language, O'Muircheartaigh et al. (2013) reported multiple age-contingent asymmetries in white matter myelination between ages 1 and 6 that were related to aspects of language proficiency as assessed by the Mullens Scales of Early Learning. In a study with early school-age children (ages 5–6), Brauer and Friederici (2007) showed that activation by passive sentence listening was somewhat less left-lateralized than for adults, but that children and adults were quite similar in their profile of perisylvian activation. Two early imaging studies of auditory language comprehension in young children also suggest that early language processing is predominantly bilateral, activating the inferior frontal gyrus (IFG) and the temporal cortices (Booth et al., 2000; Ulualp, Biswal, Yetkin, & Kidder, 1998).

Other developmental fMRI studies using language production have also focused on identifying patterns of language lateralization in individuals (Berl et al., 2014, Gaillard et al., 2004) over tasks (Bookheimer, Zeffiro, Blaxton, Gaillard, & Theodore, 2000; De Guibert et al., 2010; Lidzba, Schwilling, Grodd, Krägeloh-Mann, & Wilke, 2011) and over response modalities (Croft, Rankin, Liégeois, Banks, & Cross, 2013). By and large, developmental studies indicate that although most school-age children show left-lateralized responses, there are individual and task-dependent differences. Age-related changes in lateralization have also been described. For example, Szaflarski, Holland, Schmithorst, and Byars (2006) showed that neural activation for a covert verb generation task became increasingly left-lateralized between childhood and adolescence. In particular, age-related changes within so-called Broca's area have been a focus of many developmental language studies (reviewed in Berl et al., 2014). However, the conclusions that can be drawn appear to depend on the age range of the sample in question, as well as specific task demands. In general, studies that have used verbal fluency or categorization tasks tend to show age-related increases in activation over the left IFG (Holland et al., 2001), whereas studies using semantic association tasks tend to evoke differences over the right IFG (Booth et al., 2003; Chou et al., 2006). Yet other studies report that activation changes and increasing left-lateralization in the IFG are related to age (Berl et al., 2014) or performance (Bach et al., 2010; Blumenfeld, Booth, & Burman, 2006).

Much of what we know about the *neural development of language production* is based on a series of important developmental studies using well-characterized adult neuropsychological and/or fMRI language tasks. These include simple word repetition to an auditory cue (Church, Coalson, Lugar, Petersen, & Schlaggar, 2008), overt word reading (Church et al., 2008; Grande, Meffert, Huber, Amunts, & Heim, 2011; Heim et al., 2010; Schlaggar & McCandliss, 2007), word generation to a category (Gaillard et al., 2000, 2003) or more "metalinguistic" tasks such as verb, rhyme, or antonym generation to a read or heard cue word (Brown et al., 2005; Holland et al., 2001; Schapiro et al., 2004; Schlaggar et al., 2002; Szaflarski, Schmithorst, et al., 2006).

In a seminal study, Brown and colleagues (2005) studied a large group of children and adults performing overt word generation tasks (rhyme, verb, and opposite generation). By comparing adults and children with similar accuracy/reaction times on these tasks, they identified regions where age-related decreases (bilateral medial frontal, parietal, occipito-temporal, and cingulate cortex) and age-related increases in activity (left lateral and medial frontal regions) were observed. In contrast, by comparing adults and children whose performance differed, changes were noted over the right frontal cortex, medial parietal cortex, and posterior cingulate and occipital cortices bilaterally. These results were suggestive of increased activity in newly recruited regions such as frontal cortex over development and increased specialization of activity in earlier processing regions such as extrastriate cortex.

More recently, Krishnan, Leech, Mercure, Lloyd-Fox, and Dick (2014) used a picture naming fMRI paradigm with multiple levels of complexity to understand how school-age children (ages 7–12) and young adults responded to increasing processing demands. We found that neural organization for naming was largely similar in childhood and adulthood, where adults had greater activation in all naming conditions over inferior temporal gyri and superior temporal gyri/supramarginal gyri. However, naming complexity affected adults and children quite differently. Neural activation, especially over the dorsolateral prefrontal cortex but also in right posterior superior temporal sulcus (STS), showed complexity-dependent increases in adults but complexity-dependent *decreases* in children. These differences likely reflect adults' greater language repertoire, differential cognitive demands and strategies during word retrieval and production, as well as developmental changes in brain structure. It is clear that there must be considerable changes in the middle school and high school years in the neural organization for even fairly basic language production skills. In this vein, Ramsden et al. (2011) found that changes in individual subjects' verbal IQ between early and late adolescence were associated with changes in "gray matter density" in a left ventral somatomotor region where the same subjects also showed fMRI activation related to articulation.

Individual differences in language and grammatical ability have also been associated with activation for *comprehending* sentences. For instance, activation over the IFG for comprehending complex sentences correlates with individual differences in grammatical ability (Knoll, Obleser, Schipke, Friederici, & Brauer, 2012; Nuñez et al., 2011) and vocabulary knowledge (Yeatman, Ben-Shachar, Glover, & Feldman, 2010), rather than age.[3] In a combined functional and structural MRI study with school-age children, adolescents, and adults, Richardson, Thomas, Filippi, Harth, and Price (2010) found that individual differences in vocabulary size were positively correlated with activation for auditory sentence comprehension in the left posterior STS *and* with "gray matter density" in the same region. These results suggest that the process of learning language significantly sculpts the networks underlying language processing, even quite late in development.

31.10 CONCLUSION

Language development is inherently a process of change. Exploring the multiple and varied trajectories of language can provide us with insights into the development of more general cognitive processes. Studies of language development have been particularly useful in helping us to understand the emergence of specialization of function and the scale and flexibility of cognitive processes during learning. Novel approaches and technologies for capturing the linguistic environment that the developing child grows up in (Greenwood, Thiemeann-Bourque, Walker, Buzhardt, & Gilkerson, 2011)—and for capturing what the child is saying (Oller et al., 2010)—should allow for more fleshed out theories and models of how language development actually works. Correspondingly, new tools for understanding brain structure (Dick et al., 2012; Glasser & Van Essen, 2011; Sereno, Lutti, Weiskopf, & Dick, 2013), development (Dosenbach et al., 2010), representation (Huth, Nishimoto, Vu, & Gallant, 2012) and learning (Wiestler & Diedrichsen, 2013) should allow us to make much finer-grained predictions about when, where, and how language development changes the brain.

Acknowledgments

Thanks very much to Lori Holt and Teodora Gliga for suggestions on the chapter. Parts of this chapter also appeared in Dick, F., Leech, R., & Richardson, F. (2008). Language Development. Chapter in Child Neuropsychology: Concepts, Theory, & Practice. J. Reed & J. Warner Rogers, Eds. Blackwell.

References

Akhtar, N. (1999). Acquiring basic word order: Evidence for data-driven learning of syntactic structure. *Journal of Child Language, 26*, 261–278.

Alcock, K. J., & Krawczyk, K. (2010). Individual differences in language development: Relationship with motor skill at 21 months. *Developmental Science, 13*(5), 677–691. Available from: http://dx.doi.org/10.1111/j.1467-7687.2009.00924.x

Bach, S., Brandeis, D., Hofstetter, C., Martin, E., Richardson, U., & Brem, S. (2010). Early emergence of deviant frontal fMRI activity for phonological processes in poor beginning readers. *NeuroImage, 53*, 682–693.

Bates, E. (2004). Explaining and interpreting deficits in language development across clinical groups: Where do we go from here? *Brain and Language, 88*, 248–253.

Bates, E., Benigni, L., Bretherton, I., Camaioni, L., & Volterra, V. (1977). *The emergence of symbols: Cognition and communication in infancy*. New York, NY: Academic Press.

Bates, E., Bretherton, I., & Snyder, L. (1988). *From first words to grammar: Individual differences and dissociable mechanisms*. Cambridge: Cambridge University Press.

Bates, E., & Carnevale, G. F. (1993). New directions in language development. *Developmental Review, 13*, 436–470.

Bates, E., & Dick, F. (2002). Language, gesture, and the developing brain. *Developmental Psychobiology, Special Issue: Converging Method Approach to the Study of Developmental Science, 40*(3), 293–310.

Bates, E., & Goodman, J. (1997). On the inseparability of grammar and the lexicon: Evidence from acquisition, aphasia, and real-time processing. *Language and Cognitive Processes, 12*, 507–584.

Bates, E., & MacWhinney, B. (1989). Competition and connectionism. In B. MacWhinney, & E. Bates (Eds.), *The crosslinguistic study of sentence processing*. New York, NY: Cambridge University Press.

Bates, E., Thal, D., Clancy, B., & Finlay, B. (2002). Early language development and its neural correlates. In (2nd ed.). S. J. Segalowitz, & I. Rapin (Eds.), *Handbook of neuropsychology Part II* (Vol. 8). Amsterdam: Elsevier.

Bergelson, E., & Swingley, D. (2012). At 6 to 9 months, human infants know the meanings of many common nouns. *Proceedings of the National Academy of Sciences of the United States of America, 109*, 3253–3258.

Berl, M. M., Mayo, J., Parks, E. N., Rosenberger, L. R., VanMeter, J., Ratner, N. B., et al. (2014). Regional differences in the developmental trajectory of lateralization of the language network. *Human Brain Mapping, 35*(1), 270–284. Available from: http://dx.doi.org/10.1002/hbm.22179

Bernhardt, B. M., Kemp, N., & Werker, J. F. (2007). Early word-object associations and later language development. *First Language, 27*, 315–328.

Blasi, A., Mercure, E., Lloyd-Fox, S., Thomson, A., Brammer, M., Sauter, D., et al. (2011). Early specialization for voice and emotion processing in the infant brain. *Current Biology, 21*(14), 1220–1224. Available from: http://dx.doi.org/10.1016/j.cub.2011.06.009

Bloom, K., Russell, A., & Wassenberg, K. (1987). Turn taking affects the quality of infant vocalizations. *Journal of Child Language, 14*, 211–227.

Bloom, L., Hood, L., & Lighbown, P. (1974). Imitation in language development: If, when, and why. *Cognitive Psychology, 6*, 380–420.

Blumenfeld, H., Booth, J., & Burman, D. (2006). Differential prefrontal–temporal neural correlates of semantic processing in children. *Brain and Language, 99*, 226–235.

[3]It is worth noting that these studies have used fairly small sample sizes (22, 19, and 14, respectively) over different age ranges.

Bookheimer, S., Zeffiro, T., Blaxton, T., Gaillard, W., & Theodore, W. (2000). Activation of language cortex with automatic speech tasks. *Neurology, 55*, 1151–1157.

Booth, J. R., Burman, D. D., Meyer, J. R., Lei, Z., Choy, J., Gitelman, D. R., et al. (2003). Modality-specific and -independent developmental differences in the neural substrate for lexical processing. *Journal of Neurolinguistics, 16*(4–5), 383–405. Available from: http://dx.doi.org/10.1016/S0911-6044(03)00019-8.

Booth, J. R., MacWhinney, B., Thulborn, K. R., Sacco, K., Voyvodic, J. T., & Feldman, H. M. (2000). Developmental and lesion effects in brain activation during sentence comprehension and mental rotation. *Developmental Neuropsychology, 18*(2), 139–169.

Bortfeld, H., Rathbun, K., Morgan, J., & Golinkoff, R. (2005). Mommy and me. *Psychological Science, 16*, 298–304.

Boysson-Bardies, B., de, Halle, P., Sagart, L., & Durand, C. (1989). Across linguistic investigation of vowel formants in babbling. *Journal of Child Language, 16*, 1–17.

Boysson-Bardies, B. D., & Vihman, M. M. (1991). Adaptation to language: Evidence from babbling and first words in four languages. *Language, 67*, 297–319.

Brauer, J., & Friederici, A. D. (2007). Functional neural networks of semantic and syntactic processes in the developing brain. *Journal of Cognitive Neuroscience, 19*(10), 1609–1623. Available from: http://dx.doi.org/10.1162/jocn.2007.19.10.1609

Brown, T. T., Lugar, H. M., Coalson, R. S., Miezin, F. M., Petersen, S. E., & Schlaggar, B. L. (2005). Developmental changes in human cerebral functional organization for word generation. *Cerebral Cortex, 15*, 275–290.

Butterfield, E. C., & Siperstein, G. N. (1970). Influence of contingent auditory stimulation upon non-nutritional suckle. In J. F. Bosma (Ed.), *Third symposium on oral sensation and perception: The mouth of the infant* (pp. 313–334). Springfield, IL: Charles C. Thomas.

Carey, S. (1978). The child as word learner. In M. Halle, J. Bresnan, & G. Miller (Eds.), *Linguistic theory and psychological reality*. Cambridge, MA: MIT Press.

Carpenter, M., Nagell, K., & Tomasello, M. (1998). Social cognition, joint attention, and communicative competence from 9 to 15 months of age. *Monographs of the Society for Research in Child Development, 255*, 63.

Caselli, C., Casadio, P., & Bates, E. (1999). A comparison of the transition from first words to grammar in English and Italian. *Journal of Child Language, 26*, 69–111.

Ceponiene, R., Rinne, T., & Naatanen, R. (2002). Maturation of cortical sound processing as indexed by event-related potentials. *Clinical Neurophysiology, 113*, 870–882.

Chou, T., Booth, J., Burman, D., Bitan, T., Bigio, J., Lu, D., et al. (2006). Developmental changes in the neural correlates of semantic processing. *NeuroImage, 29*, 1141–1149.

Church, J. A., Coalson, R. S., Lugar, H. M., Petersen, S. E., & Schlaggar, B. L. (2008). A developmental fMRI study of reading and repetition reveals changes in phonological and visual mechanisms over age. *Cerebral Cortex, 18*(9), 2054–2065. Available from: http://dx.doi.org/10.1093/cercor/bhm228

Colombo, J. A., & Bundy, R. S. (1981). A method for the measurement of infant auditory selectivity. *Infant Behavior and Development, 4*(2), 219–223.

Conboy, B. T., & Mills, D. L. (2006). Two languages, one developing brain: Event-related potentials to words in bilingual toddlers. *Developmental Science, 9*, 1–12.

Corkum, V., & Moore, C. (1995). Development of joint visual attention in infants. In C. Moore, & P. J. Dunham (Eds.), *Joint attention: Its origins and role in development* (pp. 61–83). Hillsdale, NJ: Erlbaum.

Croft, L. J., Rankin, P., Liégeois, F., Banks, T., & Cross, J. (2013). To speak, or not to speak? The feasibility of imaging overt speech in children with epilepsy. *Epilepsy Research, 107*, 195–199.

Cummings, A., Ceponiene, R., Dick, F., Saygin, A. P., & Townsend, J. (2008). A developmental ERP study of verbal and non-verbal semantic processing. *Brain Research, 1208*, 137–149. Available from: http://dx.doi.org/10.1016/j.brainres.2008.02.015

Cummings, A., Ceponiene, R., Koyama, A., Saygin, A. P., Townsend, J., & Dick, F. (2006). Auditory semantic networks for words and natural sounds. *Brain Research, 1115*(1), 92–107. Available from: http://dx.doi.org/10.1016/j.brainres.2006.07.050

Cummings, A., Saygin, A. P., Bates, E., & Dick, F. (2009). Infants' recognition of meaningful verbal and nonverbal sounds. *Language Learning and Development: The Official Journal of the Society for Language Development, 5*(3), 172–190. Available from: http://dx.doi.org/10.1080/15475440902754086

Curtin, S., Mintz, T. H., & Christiansen, M. H. (2005). Stress changes the representational landscape: Evidence from word segmentation. *Cognition, 96*, 233–262.

De Guibert, C., Maumet, C., Ferré, J.-C., Jannin, P., Biraben, A., Allaire, C., et al. (2010). FMRI language mapping in children: A panel of language tasks using visual and auditory stimulation without reading or metalinguistic requirements. *NeuroImage, 51*(2), 897–909. Available from: http://dx.doi.org/10.1016/j.neuroimage.2010.02.054

Dehaene-Lambertz, G., Dehaene, S., & Hertz-Pannier, L. (2002). Functional neuroimaging of speech perception in infants. *Science, 298*, 2013–2015.

D'Entremont, B., Hains, S. M. J., & Muir, D. W. (1997). A demonstration of gaze following in 3-to 6-month-olds. *Infant Behavior and Development, 20*, 569–572.

DePaolis, R. A., Vihman, M. M., & Keren-Portnoy, T. (2011). Do production patterns influence the processing of speech in prelinguistic infants? *Infant Behavior & Development, 34*, 590–601.

DePaolis, R. A., Vihman, M. M., & Nakai, S. (2013). The influence of babbling patterns on the processing of speech. *Infant Behavior and Development, 36*, 642–649.

Dick, F., Saygin, A. P., Galati, G., Pitzalis, S., Bentrovato, S., D'Amico, S., et al. (2007). What is involved and what is necessary for complex linguistic and nonlinguistic auditory processing: Evidence from functional magnetic resonance imaging and lesion data. *Journal of Cognitive Neuroscience, 19*(5), 799–816. Available from: http://dx.doi.org/10.1162/jocn.2007.19.5.799

Dick, F., Saygin, A. P., Moineau, S., Aydelott, J., & Bates, E. (2004). Language in an embodied brain: The role of animal models. *Cortex, 40*, 226–227.

Dick, F., Tierney, A. T., Lutti, A., Josephs, O., Sereno, M. I., & Weiskopf, N. (2012). *In vivo* functional and myeloarchitectonic mapping of human primary auditory areas. *Journal of Neuroscience, 32*(46), 16095–16105. Available from: http://dx.doi.org/10.1523/JNEUROSCI.1712-12.2012

Dore, J. (1974). A pragmatic description of early language development. *Journal of Psycholinguistic Research, 4*, 423–430.

Dosenbach, N. U. F., Nardos, B., Cohen, A. L., Fair, D. A., Power, J. D., Church, J. A., et al. (2010). Prediction of individual brain maturity using fMRI. *Science, 329*(5997), 1358–1361. Available from: http://dx.doi.org/10.1126/science.1194144

Eimas, P. D., Siqueland, E. R., Jusczyk, P., & Vigorito, J. (1971). Speech perception in infants. *Science, 171*, 303–306.

Elman, J. L., Bates, E. A., Johnson, M. H., Karmiloff-Smith, A., Parisi, D., & Plunkett, K. (1996). *Rethinking innateness: A connectionist perspective on development*. Cambridge, MA: MIT Press.

Feldman, N. H., Griffiths, T. L., Goldwater, S., & Morgan, J. L. (2013). A role for the developing lexicon in phonetic category acquisition. *Psychological Review, 120*(4), 751–778.

Feldman, N. H., Myers, E. B., White, K. S., Griffiths, T. L., & Morgan, J. L. (2013). Word-level information influences phonetic learning in adults and infants. *Cognition, 127*, 427–438.

Fenson, L., Bates, E., Dale, P., Goodman, J., Reznick, J. S., & Thal, D. (2000). Reply: Measuring variability in early child language: Don't shoot the messenger. *Child Development, 71*(2), 323–328.

Fenson, L., Dale, P., Reznick, J., Bates, E., Thal, D., & Pethick, S. (1994). Variability in early communicative development. *Monographs of the Society for Research in Child Development, 59*.

Ference, J., & Curtin, S. (2013). Attention to lexical stress and early vocabulary growth in 5-month-olds at risk for autism spectrum disorder. *Journal of Experimental Child Psychology, 116*, 891–903.

Ferrari, P. F., Paukner, A., Ruggiero, A., Darcey, L., Unbehagen, S., & Suomi, S. J. (2009). *Child Development, 80*(4), 1057–1068.

Gaillard, W., Hertz-Pannier, L., Mott, S., Barnett, A., LeBihan, D., & Theodore, W. (2000). Functional anatomy of cognitive development: fMRI of verbal fluency in children and adults. *Neurology, 54*, 180–185.

Gaillard, W., Sachs, B., Whitnah, J., Ahmad, Z., Balsamo, L., Petrella, J., et al. (2003). Developmental aspects of language processing: fMRI of verbal fluency in children and adults. *Human Brain Mapping, 18*, 176–185.

Gaillard, W. D., Balsamo, L., Xu, B., McKinney, C., Papero, P. H., Weinstein, S., et al. (2004). FMRI language task panel improves determination of language dominance. *Neurology, 63*, 1403–1408.

Gernsbacher, M. A., Sauer, E. A., Geye, H. M., Schweigert, E. K., & Goldsmith, H. H. (2008). *Journal of Child Psychology and Psychiatry, 49*(1), 43–50.

Glasser, M. F., & Van Essen, D. C. (2011). Mapping human cortical areas *in vivo* based on myelin content as revealed by t1- and t2-weighted MRI. *Journal of Neuroscience, 31*(32), 11597–11616. Available from: http://dx.doi.org/10.1523/JNEUROSCI.2180-11.2011

Goldberg, A. (2006). *Constructions at work: The nature of generalization in language*. Oxford: Oxford University Press.

Goldberg, A. E. (1995). *Constructions: A construction grammar approach to argument structure*. Chicago: University of Chicago Press.

Goldstein, M. H., King, A. P., & West, M. J. (2003). Social interaction shapes babbling: Testing parallels between birdsong and speech. *Proceedings of the National Academy of Sciences of the United States of America, 100*, 8030–8035.

Gomez, R., & Gerken, L. (1999). Artificial grammar learn by 1-year-olds leads to specific and abstract knowledge. *Cognition, 70*, 109–135.

Grande, M., Meffert, E., Huber, W., Amunts, K., & Heim, S. (2011). Word frequency effects in the left IFG in dyslexic and normally reading children during picture naming and reading. *NeuroImage, 57*(3), 1212–1220.

Graves, M. F. (1986). Vocabulary learning and instruction. *Review of Research in Education, 13*, 49–89.

Greenwood, C., Thiemeann-Bourque, K., Walker, D., Buzhardt, J., & Gilkerson, J. (2011). Assessing children's home language environments using automatic speech recognition technology. *Communication Disorders Quarterly, 32*, 83–92. Available from: http://dx.doi.org/doi:10.1177/1525740110367826.

Hampson, J., & Nelson, K. (1993). The relation of maternal language to variation in rate and style of language acquisition. *Journal of Child Language, 20*(2), 313–342.

Hayiou-Thomas, M. E., Bishop, D. V., & Plunkett, K. (2004). Simulating SLIGeneral cognitive processing stressors can produce a specific linguistic profile. *Journal of Speech, Language, and Hearing Research, 47*(6), 1347–1362.

Heim, S., Grande, M., Meffert, E., Eickhoff, S. B., Schreiber, H., Kukolja, J., et al. (2010). Cognitive levels of performance account for hemispheric lateralisation effects in dyslexic and normally reading children. *NeuroImage, 53*(4), 1346–1358. Available from: http://dx.doi.org./10.1016/j.neuroimage.2010.07.009

Holland, S. K., Plante, E., Weber Byars, A., Strawsburg, R. H., Schmithorst, V. J., & Ball, W. S. (2001). Normal fMRI brain activation patterns in children performing a verb generation task. *NeuroImage, 14*(4), 837–843. Available from: http://dx.doi.org/10.1006/nimg.2001.0875

Huth, A. G., Nishimoto, S., Vu, A. T., & Gallant, J. L. (2012). A continuous semantic space describes the representation of thousands of object and action categories across the human brain. *Neuron, 76*(6), 1210–1224. Available from: http://dx.doi.org/10.1016/j.neuron.2012.10.014

Huttenlocher, J., Vasilyeva, M., Cymerman, E., & Levine, S. (2002). Language input and child syntax. *Cognitive Psychology, 45*, 337–374.

Imada, T., Zhang, Y., Cheour, M., Taulu, S., Ahonen, A., & Kuhl, P. K. (2006). Infant speech perception activates Broca's area: A developmental magnetoencephalography study. *Neuroreport, 17* (10), 957–962.

Johnson, E. K., & Jusczyk, P. W. (2001). Word segmentation by 8-month-olds: When speech cues count more than statistics. *Journal of Memory and Language, 44*, 548–567.

Jones, S. S. (2009). The development of imitation in infancy. *Philosophical Transactions of the Royal Society B: Biological Sciences, 364*, 2325–2335. Available from: http://dx.doi.org/doi:10.1098/rstb.2009.0045.

Jusczyk, P. W., & Aslin, R. N. (1995). Infants' detection of the sound patterns of words in fluent speech. *Cognitive Psychology, 29*, 1–23.

Jusczyk, P. W., & Hohne, E. A. (1997). Infants' memory for spoken words. *Science, 277*, 1984–1986.

Jusczyk, P. W., Houston, D. M., & Newsome, M. (1999). The beginnings of word-segmentation in English-learning infants. *Cognitive Psychology, 39*, 159–207.

Jusczyk, P. W., Luce, P. A., & Charles-Luce, J. (1994). Infants' sensitivity to phonotactic patterns in the native language. *Journal of Memory and Language, 33*, 630–645.

Kaminski, J., Call, J., & Fischer, J. (2004). Word learning in a domestic dog: Evidence for "fast mapping". *Science, 304*, 1682–1683.

Kisilevsky, B. S., Hains, S. M. J., Lee, K., Xie, X., Huang, H., Ye, H. H., et al. (2003). Effects of experience on fetal voice recognition. *Psychological Science, 14*(3), 220–224.

Knoll, L. J., Obleser, J., Schipke, C. S., Friederici, A. D., & Brauer, J. (2012). Left prefrontal cortex activation during sentence comprehension covaries with grammatical. Knowledge in children. *NeuroImage, 62*(1), 207–216. Available from: http://dx.doi.org/10.1016/j.neuroimage.2012.05.014

Krishnan, S., Alcock, K. J., Mercure, E., Leech, R., Barker, E., Karmiloff-Smith, A., et al. (2013). Articulating novel words: Children's oromotor skills predict non-word repetition abilities. *Journal of Speech, Language, and Hearing Research, 56*, 1800–1812.

Krishnan, S., Leech, R., Mercure, E., Lloyd-Fox, S., & Dick, F. (2014). Convergent and divergent fMRI responses in children and adults to increasing language production demands. *Cerebral Cortex*. Available from: http://dx.doi.org/10.1093/cercor/bhu120.

Kuhl, P., & Miller, J. (1975). Speech perception by the chinchilla: Voiced-voiceless distinction in alveolar plosive consonants. *Science, 190*, 69–72.

Kuhl, P. K. (1986). Theoretical contributions of tests on animals to the special-mechanisms debate in speech. *Experimental Biology, 45*, 233–265.

Kuhl, P. K. (1991). Human adults and human infants show a "perceptual magnet effect" for the prototypes of speech categories, monkeys do not. *Perception and Psychophysics, 50*, 93–107.

Kuhl, P. K. (2004). Early language acquisition: Cracking the speech code. *Nature Reviews Neuroscience, 5*, 831–843.

Kuhl, P. K., Conboy, B. T., Padden, D., Nelson, T., & Pruitt, J. C. (2005). Early speech perception and later language development: Implications for the "critical period." *Language Learning and Development, 1*, 237–264.

Kuhl, P. K., & Meltzoff, A. N. (1996). Infant vocalizations in response to speech: Vocal imitation and developmental change. *Journal of the Acoustical Society of America*, *100*, 2425–2438.

Kuhl, P. K., & Padden, D. M. (1983). Enhanced discriminability at the phonetic boundaries for the place feature in macaques. *Journal of the Acoustical Society of America*, *73*, 1003–1010.

Kuhl, P. K., Tsao, F. M., & Liu, H. M. (2003). Foreign-language experience in infancy: Effects of short-term exposure and social interaction on phonetic learning. *Proceedings of the National Academy of Sciences of the United States of America*, *100*, 9096–9101.

Kuhl, P. K., Williams, K. A., Lacerda, F., Stevens, K. N., & Lindblom, B. (1992). Language experience alters phonetic perception in infants by 6 months of age. *Science*, *255*, 606–608.

Leech, R., Aydelott, J., Symons, G., Carnevale, J., & Dick, F. (2007). The development of sentence interpretation: Effects of perceptual, attentional and semantic interference. *Developmental Science*, *10*(6), 794–813. Available from: http://dx.doi.org/10.1111/j.1467-7687.2007.00628.x

Leech, R., & Saygin, A. P. (2011). Distributed processing and cortical specialization for speech and environmental sounds in human temporal cortex. *Brain and Language*, *116*(2), 83–90. Available from: http://dx.doi.org/10.1016/j.bandl.2010.11.001

Lidzba, K., Schwilling, E., Grodd, W., Krägeloh-Mann, I., & Wilke, M. (2011). Language comprehension vs. language production: Age effects on fMRI activation. *Brain and Language*, *119*, 6–15.

MacKenzie, H., Curtin, S., & Graham, S. A. (2012a). Twelve-month-olds' phonotactic knowledge guides their word-object mappings. *Child Development*, *83*(4), 1129–1136.

MacKenzie, H., Curtin, S., & Graham, S. A. (2012b). Class matters: 12-month-olds' word-object associations privilege content over function words. *Developmental Science*, *15*(6), 753–761.

MacKenzie, H., Graham, S. A., & Curtin, S. (2011). Twelve-month-olds privilege words over other linguistic sounds in an associative learning task. *Developmental Science*, *14*(2), 249–255.

Majorano, M., Vihman, M. M., & DePaolis, R. A. (2013). The relationship between infants' production experience and their processing of speech. *Language Learning and Development*, *10*(2), 179–204.

Markson, L., & Bloom, P. (1997). Evidence against a dedicated system for word learning in children. *Nature*, *385*, 813–815.

Martin, A., Peperkamp, S., & Dupoux, E. (2013). Learning phonemes with a proto-lexicon. *Cognitive Science*, *37*, 103–124.

Maye, J., Werker, J. F., & Gerken, L. (2002). Infant sensitivity to distributional information can affect phonetic discrimination. *Cognition*, *82*, 101–111.

McMurray, B., & Aslin, R. N. (2005). Infants are sensitive to within-category variation in speech perception. *Cognition*, *95*(2), B15–B26.

Miller, J. L., & Eimas, P. D. (1996). Internal structure of voicing categories in early infancy. *Perception and Psychophysics*, *58*(8), 1157–1167.

Mills, D., Plunkett, K., Prat, C., & Schafer, G. (2005). Watching the infant brain learn words: Effects of language and experience. *Cognitive Development*, *20*, 19–31.

Narayan, C. R., Werker, J., & Beddor, P. (2010). The interaction between acoustic salience and language experience in developmental speech perception: Evidence from nasal place discrimination. *Developmental Science*, *13*(3), 407–420.

Nazzi, T., Bertoncini, J., & Mehler, J. (1998). Language discrimination by newborns: toward an understanding of the role of rhythm. *Journal of Experimental Psychology: Human perception and performance*, *24*(3), 756.

Nelson, K. (1981). Individual differences in language development: Implications for development and language. *Developmental Psychology*, *17*, 170–187.

Newman, R., Ratner, N. B., Jusczyk, A. M., Jusczyk, P. W., & Dow, K. A. (2006). Infants' early ability to segment the conversational speech signal predicts later language development: A retrospective analysis. *Developmental Psychology*, *42*, 643–655.

Nip, I. S., Green, J. R., & Marx, D. B. (2011). The co-emergence of cognition, language, and speech motor control in early development: A longitudinal correlation study. *Journal of Communication Disorders*, *44*, 149–160.

Nippold, M. A. (1998). *Later language development: The school-age and adolescent years*. Austin, TX: Pro-Ed.

Nuñez, S. C., Dapretto, M., Katzir, T., Starr, A., Bramen, J., Kan, E., et al. (2011). fMRI of syntactic processing in typically developing children: Structural correlates in the inferior frontal gyrus. *Developmental Cognitive Neuroscience*, *1*(3), 313–323.

Oller, D. K., Niyogi, P., Gray, S., Richards, J. A., Gilkerson, J., Xu, D., et al. (2010). Automated vocal analysis of naturalistic recordings from children with autism, language delay, and typical development. *Proceedings of the National Academy of Sciences of the United States of America*, *107*(30), 13354–13359. Available from: http://dx.doi.org/10.1073/pnas.1003882107

O'Muircheartaigh, J., Dean, D. C., Dirks, H., Waskiewicz, N., Lehman, K., Jerskey, B. A., et al. (2013). Interactions between white matter asymmetry and language during neurodevelopment. *Journal of Neuroscience*, *33*(41), 16170–16177. Available from: http://dx.doi.org/10.1523/JNEUROSCI.1463-13.2013

Pallier, C., Dehaene, S., Poline, J. B., LeBihan, D., Argenti, A. M., Dupoux, E., et al. (2003). Brain imaging of language plasticity in adopted adults: Can a second language replace the first? *Cerebral Cortex*, *13*, 155–161.

Perani, D., Saccuman, M. C., Scifo, P., Awander, A., Spada, D., Baldoli, C, et al. (2011). Neural language networks at birth. *Proceedings of the National Academy of Sciences*, *108*(38), 16056–16061.

Perani, D., Saccuman, M. C., Scifo, P., Spada, D., Andreolli, G., Rovelli, R., et al. (2010). Functional specializations for music processing in the human newborn brain. *Proceedings of the National Academy of Sciences of the United States of America*, *107*(10), 4758–4763. Available from: http://dx.doi.org/10.1073/pnas.0909074107

Pinker, S. (1989). *Learnability and cognition*. Cambridge, MA: MIT Press.

Pizutto, E., & Caselli, C. (1992). The acquisition of Italian morphology. *Journal of Child Language*, *19*, 491–557.

Pizutto, E., & Caselli, C. (1994). The acquisition of Italian verb morphology in a cross-linguistic perspective. In Y. Levy (Ed.), *Other children, other languages* (pp. 137–188). Hillsdale, NJ: Erlbaum.

Polka, L., & Bohn, O. S. (1996). A cross-language comparison of vowel perception in English-learning and German-learning infants. *Journal of the Acoustical Society of America*, *100*(1), 577–592.

Polka, L., & Bohn, O. S. (2011). Natural Referent Vowel (NRV) framework: An emerging view of early phonetic development. *Journal of Phonetics*, *39*(4), 467–478.

Ramsden, S., Richardson, F. M., Josse, G., Thomas, M. S. C., Ellis, C., Shakeshaft, C., et al. (2011). Verbal and non-verbal intelligence changes in the teenage brain. *Nature*, *479*(7371), 113–116.

Ramus, F., Hauser, M. D., Miller, C., Morris, D., & Mehler, J. (2000). Language discrimination by human newborns and by cotton-top tamarin monkeys. *Science*, *288*, 349–351.

Richardson, F. M., Thomas, M. S. C., Filippi, R., Harth, H., & Price, C. J. (2010). Contrasting effects of vocabulary knowledge on temporal and parietal brain structure across lifespan. *Journal of Cognitive Neuroscience*, *22*(5), 943–954. Available from: http://dx.doi.org/10.1162/jocn.2009.21238

Rubino, R., & Pine, J. (1998). Subject–verb agrement in Brazilian Portugese: What low error rates hide. *Journal of Child Language*, *25*, 35–60.

Saffran, J., Aslin, R., & Newport, E. (1996). Statistical learning by 8-month old infants. *Science*, *274*, 1926.

Saffran, J. R., & Thiessen, E. D. (2003). Pattern induction by infant language learners. *Developmental Psychology*, *39*, 484–494.

Saffran, J. R., Werker, J. F., & Werner, L. (2006). The infant's auditory world: Hearing, speech, and the beginnings of language. In W.

Damon, D. Kuhn, & R. Siegler (Eds.), *Handbook of child psychology volume 2: Cognition, perception, and language* (6th ed., pp. 58–108). New York, NY: John Wiley and Sons.

Samples, J. M., & Franklin, B. (1978). Behavioral responses in 7 to 9 month old infants to speech and non-speech stimuli. *Journal of Auditory Research*, *18*(2), 115–123.

Saygin, A. P., Dick, F., & Bates, E. (2005). An on-line task for contrasting auditory processing in the verbal and nonverbal domains and norms for younger and older adults. *Behavior Research Methods*, *37*(1), 99–110.

Saygin, A. P., Dick, F., Wilson, S. W., Dronkers, N. F., & Bates, E. (2003). Neural resources for processing language and environmental sounds evidence from aphasia. *Brain*, *126*(4), 928–945.

Schapiro, M., Schmithorst, V., Wilke, M., Byars, A., Strawsburg, R., & Holland, S. (2004). BOLD fMRI signal increases with age in selected brain regions in children. *Neuroreport*, *15*, 2575–2578.

Schlaggar, B. L., Brown, T. T., Lugar, H. M., Visscher, K. M., Miezin, F. M., & Petersen, S. E. (2002). *Science*, *296*, 1476–1479.

Schlaggar, B. L., & Mccandliss, B. D. (2007). Development of neural systems for reading. *Annual Review of Neuroscience*, *30*(1), 475–503. Available from: http://dx.doi.org/10.1146/annurev.neuro.28.061604.135645

Sereno, M. I., Lutti, A., Weiskopf, N., & Dick, F. (2013). Mapping the human cortical surface by combining quantitative T(1) with retinotopy. *Cerebral Cortex*, *23*(9), 2261–2268. Available from: http://dx.doi.org/10.1093/cercor/bhs213

Sheehan, E.A., & Mills, D.L. (2008). The effect of early word learning on brain development. *Early language development: Bridging brain and behaviour* (161–190). Amsterdam/Philadelphia: John Benjamins Publishing.

Shultz, S., & Vouloumanos, A. (2010). Three-month-olds prefer speech to other naturally occurring signals. *Language Learning and Development*, *6*(4), 241–257.

Slobin, D. (1985). Crosslinguistic evidence for the language-making capacity. In D. I. Slobin (Ed.), *The crosslinguistic study of language acquisition* (Vol. 1). Hillsdale, NJ: Lawrence Erlbaum Associates.

Smith, E. C., & Lewicki, M. S. (2006). Efficient auditory coding. *Nature*, *239*, 978–982.

Spence, M. J., & DeCasper, A. J. (1987). Prenatal experience with low-frequency maternal-voice sounds influence neonatal perception of maternal voice samples. *Infant Behavior and Development*, *10*(2), 133–142.

Stager, C. L., & Werker, J. F. (1997). Infants listen for more phonetic detail in speech perception than in word learning tasks. *Nature*, *388*, 381–382.

Swingley, D. (2009). Contributions of infant word learning to language development. *Philosophical Transactions of the Royal Society B*, *364*, 3617–3632.

Szaflarski, J. P., Holland, S. K., Schmithorst, V. J., & Byars, A. W. (2006). fMRI study of language lateralization in children and adults. *Human Brain Mapping*, *27*(3), 202–212. Available from: http://dx.doi.org/10.1002/hbm.20177

Szaflarski, J. P., Schmithorst, V. J., Altaye, M., Byars, A. W., Ret, J., Plante, E., et al. (2006). A longitudinal functional magnetic resonance imaging study of language development in children 5 to 11 years old. *Annals of Neurology*, *59*(5), 796–807. Available from: http://dx.doi.org/10.1002/ana.20817

Teinonen, T., Aslin, R. N., Alku, P., & Csibra, G. (2008). Visual speech contributes to phonetic learning in 6-month-old infants. *Cognition*, *108*(3), 850–855.

Thiessen, E. D., & Saffran, J. R. (2003). When cues collide: Use of stress and statistical cues to word boundaries by 7- to 9-month-old infants. *Developmental Psychology*, *39*(4), 706.

Tincoff, R., & Jusczyk, P. (1999). Some beginnings of word comprehension in 6-month-olds. At 6 to 9 months, human infants know the meanings of many common nouns. *Psychological Science*, *10*, 172–175.

Tomasello, M. (1992). *First verbs: A case study in early grammatical development*. Cambridge: Cambridge University Press.

Tomasello, M. (1999). *The cultural origins of human cognition*. Cambridge, MA: Harvard University Press.

Tomasello, M. (2000a). Do young children have adult syntactic competence? *Cognition*, *74*, 209–253.

Tomasello, M. (2000b). The item-based nature of children's early syntactic development. *Trends in Cognitive Sciences*, *4*, 156–163.

Tomasello, M. (2001). Perceiving intentions and learning words in the second year of life. In M. Bowerman, & S. C. Levinson (Eds.), *Language acquisition and conceptual development*. New York, NY: Cambridge University Press.

Tomasello, M. (2006). Acquiring linguistic constructions. In D. Kuhn, & R. Siegler (Eds.), *Handbook of Child Psychology*. New York, NY: Wiley.

Tomasello, M., Carpenter, M., Call, J., Behne, T., & Moll, H. (2005). Understanding and sharing intentions: The origins of cultural cognition. *Behavioral and Brain Sciences*, *28*, 675–691.

Tomasello, M., & Todd, J. (1983). Joint attention and lexical acquisition style. *First Language*, *4*, 197–212.

Trevarthen, C., & Hubley, P. (1978). Secondary intersubjectivity: Confidence, confiding and acts of meaning in the first year. In A. Lock (Ed.), *Action, gesture, and symbol: The emergence of language* (pp. 183–229). London: Academic.

Tsao, F. M., Liu, H. M., & Kuhl, P. K. (2004). Speech perception in infancy predicts language development in the second year of life: A longitudinal study. *Child Development*, *75*, 1067–1084.

Ulualp, S. O., Biswal, B. B., Yetkin, F. Z., & Kidder, T. M. (1998). Functional magnetic resonance imaging of auditory cortex in children. *Laryngoscope*, *108*, 1782–1786.

Volterra, V., Bates, E., Benigni, L., Bretherton, I., & Camaioni, L. (1979). First words in language and action: A qualitative look. In E. Bates, L. Benigni, I. Bretherton, L. Camaioni, & V. Volterra (Eds.), *The emergence of symbols: Cognition and communication in infancy* (pp. 141–222). New York, NY: Academic PressNeuroscience, 2, 661-670.

Vouloumanos, A., & Curtin, S. (2014). Foundational tuning: How infants' attention to speech predicts language development. *Cognitive Science*, *38*(8), 1675–1686. Available from: http://dx.doi.org/10.1111/cogs.12128.

Vouloumanos, A., & Werker, J. F. (2004). Tuned to the signal: The privileged status of speech for young infants. *Developmental Science*, *7*(3), 270–276.

Vouloumanos, A., & Werker, J. F. (2007). Listening to language at birth: Evidence for a bias for speech in neonates. *Developmental Science*, *10*(2), 159–164.

Werker, J. F., & Desjardins, R. N. (1995). Listening to speech in the 1st year of life: Experiential influences on phoneme perception. *Current Directions in Psychological Science*, *4*, 76–80.

Werker, J. F., & Tees, R. C. (1984). Cross-language speech perception: Evidence for perceptual reorganization during the first year of life. *Infant Behavior and Development*, *7*, 49–63.

Wiestler, T., & Diedrichsen, J. (2013). Skill learning strengthens cortical representations of motor sequences. *eLife*, *2*, e00801. Available from: http://dx.doi.org/10.7554/eLife.00801

Wittgenstein, L. (1953). *Philosophical investigations*. Oxford: Blackwell Publishing.

Yeatman, J. D., Ben-Shachar, M., Glover, G. H., & Feldman, H. M. (2010). Individual differences in auditory sentence comprehension in children: An exploratory event-related functional magnetic resonance imaging investigation. *Brain and Language*, *114*(2), 72–79. Available from: http://dx.doi.org/10.1016/j.bandl.2009.11.006

Yeung, H. H., & Werker, J. F. (2009). Learning words' sounds before learning how words sound: 9-month olds use distinct objects as cues to categorize speech information. *Cognition*, *113*, 234–243.

CHAPTER

32

The Neurobiology of Gesture and Its Development

Anthony Steven Dick and Iris Broce

Department of Psychology, Florida International University, Miami, FL, USA

The term gesture refers to the hand and arm movements that speakers routinely produce when they communicate, that occur naturally in face-to-face communication and that often accompany speech, and that play an important role in conveying a speaker's message (Goldin-Meadow, 2005; McNeill, 1992). The study of gesture has been important for understanding both thought and language, and the nature of the connection between them. The suggestion that symbolic gesture and spoken language are each pillars within a single semiotic system (Kelly, Ozyürek, & Maris, 2010; McNeill, 1992, 2005), that they develop together in children (Fenson et al., 1994), and that they share an overlapping neural substrate in the developing and adult brain (Bates & Dick, 2002) is supported by the research we review here. Moreover, the study of gesture and its development informs how we think about the neurobiology of language more generally.

32.1 EXPLORING GESTURE AND ITS DEVELOPMENT AT THE BEHAVIORAL LEVEL

To understand how the brain processes co-speech gesture, we must make distinctions among different types of gesture. There is a wealth of literature addressing these issues at the level of behavior, and this is expertly reviewed in a number of studies (Goldin-Meadow, 2005; Goldin-Meadow & Alibali, 2013; Kendon, 2004; McNeill, 1992, 2005). For our purposes, following the categorization outlined in McNeill (1992), we simply make some fundamental distinctions relevant to the neurobiology of gesture.

The first gesture types (at the "gesticulation" or "co-speech gesture" end of the continuum) are iconic and metaphoric gestures. Iconic gestures bear a formal relation to the content of speech, and they are most often understood only in the context of speech. For example, wiggling the fingers has a different meaning if it accompanies the sentence "She is quite the piano player" compared with "She tiptoed across the creaky floor." These gestures can be complementary and reinforce the information in speech, as indicated in the example, or supplementary to add new information—"She is quite the musician" can be accompanied by the finger-tapping movement to indicate that the person plays the piano. In the second example, fully understanding the speaker's intended message requires the listener to integrate information from separate auditory (speech) and visual (gesture) modalities into a unitary, coherent semantic representation. Metaphoric gestures can similarly be integrated with speech but convey abstract rather than concrete objects or events.

Other gesture types contribute less in the way of semantic information. Beat (rhythmic) gestures most often indicate significance of an accompanying word or phrase. Deictic gestures are pointing gestures and serve as indicators for objects and events in the world. At the other end of the continuum are pantomimes and emblems, which rely less on speech (or sometimes not at all) to convey meaning and have more codified, socially regulated meanings. Pantomimes are sequences of movements or "mimes" conveying a narrative (e.g., moving the hand in front of the mouth in a particular manner to convey "brushing teeth"). Emblems are even more

Neurobiology of Language. DOI: http://dx.doi.org/10.1016/B978-0-12-407794-2.00032-8

standardized and have formalized meanings understood by viewers outside the context of speech (e.g., "Thumbs-up" for "It's okay"). Signed language forms a third formalized category but is not be discussed in this chapter because it represents a highly codified, autonomous system akin to spoken language.

Children and adults produce gestures and are able to glean meaning from gestures when other people produce them (Fenson et al., 1994; Goldin-Meadow & Alibali, 2013; Hostetter, 2011). However, there are also significant behavioral changes in gesture comprehension and production across childhood (Botting, Riches, Gaynor, & Morgan, 2010; Mohan & Helmer, 1988; Ozçalişkan & Goldin-Meadow, 2011; Stefanini, Bello, Caselli, Iverson, & Volterra, 2009). In fact, gesture development extends at least into later childhood and coincides with the development of language at multiple levels from earliest spoken word production to narrative comprehension.

Pointing and referential gestures are the first to appear, accompany, and often precede the emergence of spoken language, and they predict later spoken language ability (Acredolo & Goodwyn, 1988; Capirici, Iverson, Pizzuto, & Volterra, 1996; Iverson, Capirci, & Caselli, 1994; Morford & Goldin-Meadow, 1992; Rowe & Goldin-Meadow, 2009). During the preschool period, the ability to comprehend and produce symbolic or "representational" gestures (Göksun, Hirsh-Pasek, & Golinkoff, 2010; Kidd & Holler, 2009; Kumin & Lazar, 1974; McNeil, Alibali, & Evans, 2000; Mohan & Helmer, 1988) and pantomime (Boyatzis & Watson, 1993; Dick, Overton, & Kovacs, 2005; O'Reilly 1995; Overton & Jackson, 1973) emerges, and children begin to be able to use gestures to learn concepts (McGregor, Rohlfing, Bean, & Marschner, 2009). As children acquire more complex language abilities, gesture develops as well. During early and later childhood, children increasingly produce gestures to accompany narrative-level language (Colletta, Pellenq, & Guidetti, 2010; Demir, So, Ozyürek, & Goldin-Meadow, 2012; Riseborough 1982; So, Demir, & Goldin-Meadow, 2010) or to indicate knowledge (e.g., spatial knowledge; Sauter, Uttal, Alman, Goldin-Meadow, & Levine, 2012). Further, the ability to comprehend and take advantage of gestural information that accompanies spoken words (Thompson & Massaro, 1994), instructions (Church, Ayman-Nolley, & Mahootian, 2004; Goldin-Meadow, Kim, & Singer, 1999; Perry, Berch, & Singleton, 1995; Ping & Goldin-Meadow, 2008; Singer & Goldin-Meadow, 2005; Valenzeno, Alibali, & Klatzky, 2003), and narrative explanations (Kelly & Church, 1997, 1998) improves. Age-related changes in gesture are thus evidenced across childhood at multiple levels of language processing.

32.2 GESTURE AND ITS DEVELOPMENT IN THE CONTEXT OF A BROADER NEUROBIOLOGY OF LANGUAGE

The study of the neurobiology of gesture has been influenced by and has contributed to a revision of the classical model of the neurobiology of language (Poeppel, Emmorey, Hickok, & Pylkkänen, 2012). With the "shedding" of the classical model, a dual-stream "dorsal-ventral" model has emerged (Hickok & Poeppel, 2007; Rauschecker & Scott, 2009). Within the dorsal stream, fronto-temporo-parietal regions are proposed to be involved in mapping auditory speech sounds to articulatory (motor) representations (Hickok & Poeppel, 2007), or in processing complex syntax (Friederici & Gierhan, 2013). The major fiber pathway proposed to connect these regions is the superior longitudinal fasciculus/arcuate fasciculus. In contrast, the ventral stream is proposed to be involved in mapping auditory speech sounds to meaning (Hickok & Poeppel, 2007), or in processing less complex syntax (Friederici & Gierhan, 2013). A number of fiber pathways have been proposed to anchor the ventral stream, including the uncinate fasciculus, extreme capsule, middle longitudinal fasciculus, inferior occipito-frontal fasciculus, and inferior longitudinal fasciculus (Dick & Tremblay, 2012 for review). Co-speech gesture may recruit both streams to process semantic, phonological, or syntactic information in the accompanying speech. At the present time, the dorsal-ventral architecture is a promising framework in which to investigate how the adult and developing brain process gesture.

32.3 THE NEUROBIOLOGY OF GESTURE: ELECTROPHYSIOLOGY

Electrophysiological studies have contributed significantly to understanding how the brain processes gesture, with most of these studies investigating the N400 component when gesture and speech are presented together. This ERP response is thought to index semantic integration during language comprehension (Kutas & Federmeier, 2011). These studies manipulate the semantic relation between gesture and speech (e.g., present gestures that are semantically congruent or incongruent with speech) and have shown that the N400 is affected by these manipulations (Holle & Gunter, 2007; Kelly, Kravitz, & Hopkins, 2004; Kelly, Ward, Creigh, & Bartolotti, 2007; Özyürek, Willems, Kita, & Hagoort, 2007; Wang & Chu, 2013; Wu & Coulson, 2005, 2007a, 2007b). However, the source of the N400 in language and in gesture has been difficult to determine, with some researchers emphasizing a left-lateralized temporo-parietal network (Friederici, Hahne, & von Cramon, 1998; Kwon et al., 2005; Simos, Basile, & Papanicolaou, 1997; Swaab, Brown, & Hagoort, 1997), and others suggesting a source in the inferior frontal gyrus (IFG) (Hagoort, Hald, Bastiaansen, & Petersson, 2004; Van Petten & Luka, 2006). Lau, Phillips, and Poeppel (2008) reviewed the literature on language without gesture and suggested that there is truth in both accounts. Specifically, they emphasized the contribution of the left IFG and left posterior middle temporal gyrus (MTGp) to processing semantic representations during language, with the left IFG proposed to be involved in semantic selection during retrieval, and the left MTGp proposed to be involved in lexical access/lexical storage. However, given the constraints on source localization, other functional imaging methods, such as magnetic resonance imaging (MRI), are necessary to characterize the brain regions that contribute to semantic processing during language comprehension.

32.4 THE NEUROBIOLOGY OF GESTURE: FUNCTIONAL IMAGING

32.4.1 Gesture Along the Ventral Stream

Within the broader dorsal-ventral model, the ventral stream is proposed to be involved in processing semantic information during language comprehension. Studies of gesture that have manipulated the semantic relation between co-speech gesture and speech have implicated the IFG (particularly the anterior *pars triangularis* but also more posterior *pars opercularis*; IFGTr and IFGOp), posterior superior temporal sulcus (STSp), and MTGp in gesture-speech integration at the semantic level (Dick, Goldin-Meadow, Hasson, Skipper, & Small, 2009; Dick, Goldin-Meadow, Solodkin, & Small, 2012; Dick, Mok, Beharelle, Goldin-Meadow, & Small, 2014; Green et al., 2009; Holle, Gunter, Rüschemeyer, Hennenlotter, & Iacoboni, 2008; Holle, Obleser, Rueschemeyer, & Gunter, 2010; Kircher et al., 2009; Skipper, Goldin-Meadow, Nusbaum, & Small, 2007, 2009; Straube, Green, Bromberger, & Kircher, 2011; Straube, Green, Jansen, Chatterjee, & Kircher, 2010; Straube, Green, Weis, Chatterjee, & Kircher, 2009; Straube, Green, Weis, & Kircher, 2012; Straube et al., 2013; Willems, Özyürek, & Hagoort, 2007, 2009; Wilson, Molnar-Szakacs, & Iacoboni, 2008).

These regions along the ventral stream, in particular left IFGTr and MTGp, are also implicated in semantic processing for receptive language that is not accompanied by gesture. Both regions are active when a linguistic task requires activating and selecting among multiple potential meanings at the word (Hoenig & Scheef, 2009) and sentence levels (Rodd, Davis, & Johnsrude, 2005; Zempleni, Renken, Hoeks, Hoogduin, & Stowe, 2007). The right hemisphere also makes contributions to language processing, particularly when demands on semantic selection are increased. For example, activity in the right IFGTr increases as a function of requirement for semantic selection among competing alternatives (Hein et al., 2007; Rodd et al., 2005), such as when the perceived spoken language requires considering multiple potential meanings (Lauro, Tettamanti, Cappa, & Papagno, 2008; Rodd et al., 2005; Stowe, Haverkort, & Zwarts, 2005; Zempleni et al., 2007).

Functional imaging studies in adults have also implicated these regions in processing the semantic relation between co-speech gestures and speech. For example, Willems et al. (2007) presented sentences accompanied by iconic gestures in which the gesture provided either congruent or incongruent information with speech (e.g., the verb "wrote" accompanied by a *hit* [incongruent] gesture or a *write* [congruent] gesture). Incongruent conditions elicited greater activity than the congruent conditions in the left inferior frontal gyrus (IFGTr). In a subsequent study, Willems et al. (2009) also showed that IFGTr responds more strongly to incongruent compared with congruent iconic gestures and pantomimes. Other studies report consistent results, suggesting the left IFG is recruited under conditions requiring additional semantic processing. For example, IFG responds more strongly to metaphoric gestures than to iconic gestures accompanying the same speech (Straube et al., 2011), and to iconic gestures that are unrelated to the accompanying speech (Green et al., 2009). Finally, Dick et al. (2014) found that the left IFGTr and IFGOp were more active when iconic gestures provided information supplementary compared with complementary to an accompanying narrative.

Sometimes the right IFG responds to manipulations of the semantic relation between gesture and speech (Dick et al., 2009; Straube et al., 2009, 2013). For example, Dick et al. (2009) presented spoken narratives accompanied by either meaningful iconic and metaphoric gestures or speech-unrelated self-adaptors (e.g., grooming movements such as pulling the shirt cuff). Although the left IFG was more active when hand movements accompanied speech compared to when it did not, this region did not respond differentially to meaningful iconic and metaphoric gestures compared with self-adaptors. In contrast, the right IFG responded more strongly when the hand movements were self-adaptors compared with gestures bearing a meaningful relation to speech. However, other research failed to find any response of either left or right IFG. Holle et al. (2008) reported no differential response in IFG for gestures that supported the dominant versus the subordinate meaning of a homonym, nor did it show particular sensitivity to whether the hand movement was an iconic gesture or a grooming movement.

The posterior temporal cortex—namely the STSp and MTGp—also contributes to processing gesture. Holle et al. (2008, 2010) have suggested that STSp is responsive to semantic information in gestures; for example, this region was shown to respond more strongly to speech accompanied by meaningful gestures than to speech accompanied by nonmeaningful self-adaptive movements (Holle et al., 2008). Other studies have also reported sensitivity of STSp to semantic information in gestures, but in some cases it appears the activity is predominantly in the adjacent MTGp and extends into the STSp (Straube et al., 2011; Willems et al., 2009). In two studies, Dick et al. (2009) have directly addressed this issue using regions of interest (ROI) analysis defined according to the individual anatomy of participants. Examining the entire STSp (2009) and the upper and lower STSp banks separately (2014), Dick et al. showed that the bilateral STSp responds more strongly to speech with gestures than to speech alone, but it is not sensitive to the meaning relation between gesture and accompanying speech. They suggested that these findings are consistent with the region's putative involvement in processing biologically relevant motion (Beauchamp, Lee, Haxby, & Martin, 2003; Grossman et al., 2000) or with the lower-level integration of auditory and visual information during speech perception and comprehension.

The adjacent MTGp, however, does seem to be sensitive to semantic manipulations of the relation between gesture and speech. Straube et al. (2011) found that the MTGp extending into STSp responds to metaphoric and iconic gestures that are integrated with speech. Willems et al. (2009) reported that bilateral MTGp extending into STSp responded more to speech accompanied by incongruent pantomimes than to the same speech accompanied by congruent pantomimes, but they did not find that the region was sensitive to iconic gestures that were incongruent versus congruent with speech. They suggested that the MTGp/STSp is involved in mapping the information conveyed in gesture and speech onto a common object representation in long-term memory. That is, the MTGp is not involved in the construction of a novel representation as a result of combining information in the visual and auditory input streams. However, Dick et al. (2014) did find that blood oxygenation level–dependent (BOLD) signal amplitude in the MTGp is modulated by the semantic relation between iconic gesture and speech. Like the left IFG, the left MTGp was more active for complementary compared with supplementary iconic gestures. The response of this region to gesture requires further investigation.

In summary, this brief review suggests that the IFG (especially the IFGTr) and MTGp have been the two regions most associated with semantic processing as part of the ventral stream in auditory language comprehension without gesture (Lau et al., 2008; Price, 2010; Vigneau et al., 2006 for review), and the same regions are implicated in gesture-speech integration. In Figure 32.1, we present a summary of these results focusing on activations in the inferior frontal and temporal lobes. Red and blue marked activation peaks represent a number of functional imaging studies that have manipulated semantic retrieval demands using similar paradigms used to assess N400 ERP responses during word, sentence, and narrative-level language comprehension (Bedny, McGill, & Thompson-Schill, 2008; Gennari, MacDonald, Postle, & Seidenberg, 2007; Giesbrecht, Camblin, & Swaab, 2004; Gold et al., 2006; Hoenig & Scheef, 2009; Kotz, Cappa, von Cramon, & Friederici, 2002; Matsumoto, Iidaka, Haneda, Okada, & Sadato, 2005; Rodd et al., 2005; Rossell, Price, & Nobre, 2003; Snijders et al., 2009; Wheatley, Weisberg, Beauchamp, & Martin, 2005; Whitney, Jefferies, & Kircher, 2011; Wible et al., 2006; Zempleni et al., 2007). For example, the typical study represented at the top of Figure 32.1 manipulated ambiguous versus unambiguous words or sentences, subordinate versus dominant concepts of homonyms, or high versus low demand for semantic retrieval.

Green marked activation peaks represent studies using similar manipulations for language produced with gesture. For example, the typical study represented manipulated whether gestures disambiguate the meaning of a homonym, whether they are semantically incongruent with or unrelated to the sentence context or a target word, or whether they supplemented versus complemented language in a narrative (Dick et al., 2009, 2012, 2014; Holle et al., 2008; Straube et al., 2011; Willems et al., 2007, 2009). The figure shows that the IFGTr and MTGp regions of the ventral language stream, particularly on the left but also on the right hemisphere, participate in processing semantic information from both gesture and speech and in the integration of the two modalities.

32.4.2 Gesture Along the Dorsal Stream

The notion that a perceiver's motor system is activated when watching another person performing actions—so-called observation-execution matching—is relatively uncontroversial. More controversial, though, is

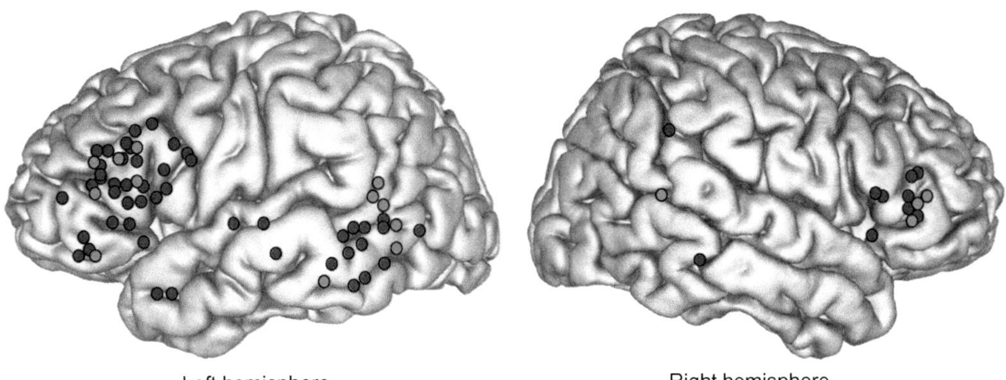

FIGURE 32.1 Temporal and inferior frontal activation peaks for semantic manipulations in speech and gesture. Red and blue marked activation peaks represent functional imaging studies that have manipulated semantic retrieval demands using similar paradigms used to assess N400 ERP responses during word-level and sentence-level language comprehension. Green marked activation peaks represent studies using similar manipulations for language produced with gesture. Only temporal and inferior frontal peaks are shown. Studies contributing to the figure are cited in the text.

the notion that understanding the meaning of actions is mediated, either in part or directly, through this motor system involvement (Kilner, 2011). Because gestures are observed actions, this controversy applies to gesture (Andric & Small, 2012; Willems & Hagoort, 2007).

Some behavioral evidence suggests that the motor system contributes to gesture understanding. For example, Ping, Goldin-Meadow, and Beilock (2013) found that moving the arms, but not the feet, interferes with gesture-speech integration, which implicates the motor system in understanding gesture at a semantic level. Behavioral literature suggesting that the motor system does *not* contribute to gesture understanding is more difficult to identify. The reason is that null findings in behavioral studies are difficult to publish (i.e., the "file-drawer problem"); consequently we were unable to find any that contributed to this side of the controversy. However, because null findings are published along with positive findings in neuroimaging, neuroimaging evidence provides one way to adjudicate this controversy.

Some studies have directly investigated this question (Holle et al., 2008; Skipper et al., 2007, 2009; Willems et al., 2007) with the expectation that premotor and inferior parietal regions along the dorsal stream, part of a putative "mirror neuron system" (Andric & Small, 2012; Kilner, 2011), would be sensitive to the semantic information contributed by gesture. For example, Skipper et al. (2009) reported that neural responses in premotor cortex and inferior parietal cortex are "tuned" to the co-occurrence of meaningful gestures with speech, but not with non-meaningful gestures with speech. Holle et al. (2008) also showed that the premotor cortex and inferior parietal cortex fired more strongly when gestures were present compared with grooming movements. Similarly, Willems et al. (2007) showed that the bilateral premotor cortex was more active when iconic gestures were incongruent compared with congruent with an accompanying verb. Finally, Josse, Joseph, Bertasi, and Giraud (2012) found that the relation between gesture and speech modulated the BOLD response in the premotor cortex. The BOLD response was suppressed after repetition of a word accompanied by a congruent gesture (e.g., "grasp" + grasping gesture), but not when the same action word was accompanied by an incongruent gesture (e.g., "grasp" + sprinkle).

However, other studies have not found that the premotor cortex and/or inferior parietal cortices show sensitivity to gesture meaning. For example, several investigations (Dick et al., 2009, 2012, 2014; Green et al., 2009; Straube et al., 2011; Willems et al., 2009) failed to report significant age-related change or modulation of activity in response to gesture meaning in these premotor or inferior parietal cortices. In addition, in some cases the premotor and/or inferior parietal activity identified during gesture is also identified when the same participants view speech without gestures. For example, the age-related changes in the response of the premotor cortex to gesture that

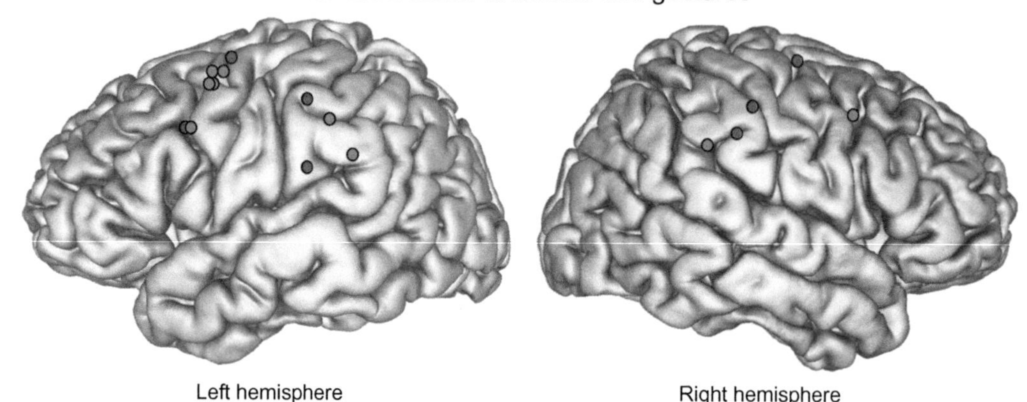

FIGURE 32.2 Premotor and inferior parietal activation peaks for studies of action observation and of gesture. Red marks show peaks identified by a meta-analysis of 125 studies of action observation (Molenberghs et al., 2012). Green marks show peaks of activation for studies in which premotor and inferior parietal regions indicated sensitivity to a semantic manipulation between gesture and speech. Studies contributing to the figure are cited in the text.

Wakefield, James, and James (2013) identified also occur for the speech-only condition without gestures, indicating that the response is not specific to processing gestures.

Other studies have suggested that the processing of motor and linguistic/semantic information in gesture can be dissociated in the brain. Thus, the evidence suggests that the motor system is recruited to process motor action information in gestures, but frontal and temporal regions contribute to processing linguistic/semantic aspects of gesture. Andric et al. (2013) investigated whether this is the case using fMRI while participants viewed two different kinds of hand actions—emblematic gestures (e.g., "thumbs-up!") and grasping movements—or speech without gesture conveying the same meaning as the emblem (e.g., "It's good!"). The results showed that when people observe emblems, regions of the brain involved in both observing grasping and listening to speech with the same meaning are active. Thus, lateral temporal (right MTGp and the anterior superior temporal gyrus) and inferior frontal (left IFGTr and IFGOp) were active in processing meaning, regardless of whether it was conveyed in gesture or speech (also see Xu, Gannon, Emmorey, Smith, & Braun, 2009). In contrast, premotor and inferior parietal regions were active in response to hand actions, regardless of whether the action was a grasping movement or an emblem. Thus, the motor system seems to contribute to processing gestures, but if it does contribute to processing meaning, then it does so in collaboration with brain regions also involved in processing linguistic aspects of gesture.

Studies of lesions provide further support for this notion. Mengotti et al. (2013) investigated the imitation of meaningful and meaningless gestures in people with left hemisphere stroke. They found that praxic performance and linguistic performance were associated when the gesture that needed to be imitated has meaning for the imitator, but they were dissociated when the gesture had no meaning. In other words, there is an overlap of brain regions underlying the linguistic abilities and the ability to imitate meaningful gestures (Andric et al., 2013). However, when the gestures are meaningless, they do not access the lexical semantic system; instead, they rely on visuo-motor processing. The authors suggested that this provides support for "two-pathway" models of processing gesture, and the recruitment of one or the other pathway may be determined by the specific kind of gesture and the context in which it occurs.

In summary, work continues to investigate whether and how the motor system contributes to understanding gesture at the semantic level. The answer to this question, as Mengotti et al. (2013) suggest, may depend on the specific kind of gesture, its relation to accompanying speech, and the context in which it is processed. In Figure 32.2, we summarize some of the relevant findings for premotor and inferior parietal regions. As Figure 32.2 shows, there is significant overlap for peaks of activity during observation of actions (as reviewed by Molenberghs, Cunnington, & Mattingley, 2012; red marks) and for manipulations of the semantic relationship between gesture and speech (green marks). Thus, premotor and parietal regions of the dorsal stream appear to participate in processing gesture semantics in some situations.

32.5 THE NEUROBIOLOGY OF GESTURE DEVELOPMENT

To our knowledge, only one study has explored gesture development using electrophysiological measures. In this study, Sheehan, Namy, and Mills (2007) presented word-picture and gesture-picture pairs to 18-month-old and 26-month-old children while they conducted EEG recordings. For both word-picture and gesture-picture pairs, the 18-month-old children showed a larger N400 response if the word or gesture mismatched the picture compared with when it matched the picture. However, 26-month-old children showed the significant effect only for word-picture pairs, although the trend was in the same direction. Together, the evidence suggests that gestures affect semantic processing, at least in 18-month-old children. Whether this effect changes as children enter toddlerhood needs further investigation.

Only two studies have used functional imaging methods to study gesture in the developing brain. In the first study, Dick et al. (2012) investigated adults and 8-year-old to 11-year-old children in two conditions. In the first condition, participants watched a storyteller tell a story while making meaningful gestures; in the other condition, she told the story while making nonmeaningful self-adaptive movements. The same regions implicated in processing gestures with speech in adults—STSp, IFGTr, and MTGp—showed age-related differences in response to gestures. In the STSp, compared with children, adults showed a greater BOLD response for both the meaningful gesture and nonmeaningful adaptor condition—this region did not differentiate the meaning of the gesture but did activate more strongly for adults in response to hand movements. The right IFGTr and left MTGp, however, did show sensitivity to the manipulation of gesture meaning. In both regions, adults showed greater activity for nonmeaningful compared with meaningful gestures, whereas children showed the opposite pattern of activity. This difference in the pattern of activation may index developmental changes in how meaning from gesture is activated and selected within an accompanying linguistic context. Further, as in adults, the results implicate brain regions associated with the ventral stream for semantic processing during language comprehension.

In a second study of the neurobiology of gesture development, Wakefield et al. (2013) studied the comprehension of co-speech gestures in 5-year-olds, 7-year-olds, 10-year-olds, and adults. All participants viewed a woman speaking a sentence with iconic content (e.g., the fingertips of right hand are placed together toward the mouth and spread apart as they are rotated away from mouth and outward, an action that accompanied the word "spoke"), speaking a sentence with iconic content and performing a corresponding iconic gesture, or performing a gesture in isolation. They showed that differences in activity in left MTGp between gesture and gesture and speech increased with age. Age-related differences were also detected in left precentral gyrus. This region was consistently recruited when adults viewed gestures, but it was not consistently recruited when children viewed gestures. The authors interpreted this finding to suggest that co-speech gesture production, which occurs with more frequency over the course of development and with which adults have more experience, contributes to the processing of co-speech gesture during perception. Notably, though, the same results were found for the speech-only condition without gesture; therefore, further research is needed to determine how motor regions contribute to gesture development.

In summary, although research on the neurobiology of gesture development is still emerging, there is the suggestion that the regions of both the ventral and dorsal streams involved in gesture processing in adults show age-related change. Further, there is the suggestion that the motor system contributes to the perception and understanding of gesture during development, but these findings need to be replicated in future studies.

32.6 CONCLUSION

The study of gesture and its development informs how we think about the neurobiology of language more broadly. Evidence from functional imaging in particular supports the dorsal-ventral model by showing that semantic processing of gestures influences activity in IFG and in MTGp regions of the ventral stream. Furthermore, in some situations the semantic contribution that gestures make influences activity in premotor and inferior parietal regions of the dorsal stream. This contributes to the emerging picture of how a broader language system processes communicative acts in multiple modalities.

References

Acredolo, L., & Goodwyn, S. (1988). Symbolic gesturing in normal infants. *Child Development*, *59*, 450–466.

Andric, M., & Small, S. L. (2012). Gesture's neural language. *Frontiers in Psychology*, *3*, 99. Available from: http://dx.doi.org/10.3389/fpsyg.2012.00099.

Andric, M., Solodkin, A., Buccino, G., Goldin-Meadow, S., Rizzolatti, G., & Small, S. L. (2013). Brain function overlaps when people observe emblems, speech, and grasping. *Neuropsychologia, 51*(8), 1619–1629. Available from: http://dx.doi.org/10.1016/j.neuropsychologia.2013.03.022.

Bates, E., & Dick, F. (2002). Language, gesture, and the developing brain. *Developmental Psychobiology, 40*, 293–310.

Beauchamp, M. S., Lee, K. E., Haxby, J. V., & Martin, A. (2003). FMRI responses to video and point-light displays of moving humans and manipulable objects. *Journal of Cognitive Neuroscience, 15*(7), 991–1001.

Bedny, M., McGill, M., & Thompson-Schill, S. L. (2008). Semantic adaptation and competition during word comprehension. *Cerebral Cortex, 18*(11), 2574–2585. Available from: http://dx.doi.org/10.1093/cercor/bhn018.

Botting, N., Riches, N., Gaynor, M., & Morgan, G. (2010). Gesture production and comprehension in children with specific language impairment. *British Journal of Developmental Psychology, 28*, 51–69.

Boyatzis, C. J., & Watson, M. W. (1993). Preschool children's symbolic representation objects through gestures. *Child Development, 64*, 729–735.

Capirici, O., Iverson, J. M., Pizzuto, E., & Volterra, V. (1996). Gestures and words during the transition to two-word speech. *Journal of Child Language, 23*, 645–673.

Church, R. B., Ayman-Nolley, S., & Mahootian, S. (2004). The role of gesture in bilingual education: Does gesture enhance learning? *International Journal of Bilingual Education and Bilingualism, 7*(4), 303–319.

Colletta, J.-M., Pellenq, C., & Guidetti, M. (2010). Age-related changes in co-speech gesture and narrative: Evidence from french children and adults. *Speech Communication, 52*(6), 565–576.

Demir, O. E., So, W.-C., Ozyürek, A., & Goldin-Meadow, S. (2012). Turkish- and English-speaking children display sensitivity to perceptual context in the referring expressions they produce in speech and gesture. *Language and Cognitive Processes, 27*(6), 844–867. Available from: http://dx.doi.org/10.1080/01690965.2011.589273.

Dick, A. S., Goldin-Meadow, S., Hasson, U., Skipper, J. I., & Small, S. L. (2009). Co-speech gestures influence neural activity in brain regions associated with processing semantic information. *Human Brain Mapping, 30*, 3509–3526.

Dick, A. S., Goldin-Meadow, S., Solodkin, A., & Small, S. L. (2012). Gesture in the developing brain. *Developmental Science, 15*, 165–180.

Dick, A. S., Mok, E., Raja Beharelle, A., Goldin-Meadow, S., & Small, S. L. (2014). Frontal and temporal contributions to understanding the iconic co-speech gestures that accompany speech. *Human Brain Mapping, 35*, 900–917.

Dick, A. S., Overton, W. F., & Kovacs, S. L. (2005). The development of symbolic coordination: Representation of imagined objects, executive function, and theory of mind. *Journal of Cognition and Development, 6*(1), 133–161.

Dick, A. S., & Tremblay, P. (2012). Beyond the arcuate fasciculus: Consensus and controversy in the connectional anatomy of language. *Brain: A Journal of Neurology, 135*, 3529–3550.

Fenson, L., Dale, P. S., Reznick, J. S., Bates, E., Thal, D. J., & Pethick, S. J. (1994). Variability in early communicative development. *Monographs of the Society for Research in Child Development, 59*. (5) Serial No. 242.

Friederici, A. D., & Gierhan, S. M. (2013). The language network. *Current Opinion in Neurobiology, 23*(2), 250–254.

Friederici, A. D., Hahne, A., & von Cramon, D. Y. (1998). First-pass versus second-pass parsing processes in a Wernicke's and a Broca's aphasic: Electrophysiological evidence for a double dissociation. *Brain and Language, 62*, 311–341.

Gennari, S. P., MacDonald, M. C., Postle, B. R., & Seidenberg, M. S. (2007). Context-dependent interpretation of words: Evidence for interactive neural processes. *NeuroImage, 35*(3), 1278–1286. Available from: http://dx.doi.org/10.1016/j.neuroimage.2007.01.015.

Giesbrecht, B., Camblin, C. C., & Swaab, T. Y. (2004). Separable effects of semantic priming and imageability on word processing in human cortex. *Cerebral Cortex, 14*(5), 521–529. Available from: http://dx.doi.org/10.1093/cercor/bhh014.

Göksun, T., Hirsh-Pasek, K., & Golinkoff, R. (2010). How do preschoolers express cause in gesture and speech? *Cognitive Development, 25*, 56–68.

Gold, B. T., Balota, D. A., Jones, S. J., Powell, D. K., Smith, C. D., & Andersen, A. H. (2006). Dissociation of automatic and strategic lexical-semantics: Functional magnetic resonance imaging evidence for differing roles of multiple frontotemporal regions. *The Journal of Neuroscience, 26*(24), 6523–6532. Available from: http://dx.doi.org/10.1523/JNEUROSCI.0808-06.2006.

Goldin-Meadow, S. (2005). *Hearing gesture: How our hands help us think*. Cambridge, MA: Belknap Press.

Goldin-Meadow, S., & Alibali, M. W. (2013). Gesture's role in speaking, learning, and creating language. *Annual Review of Psychology, 64*, 257–283. Available from: http://dx.doi.org/10.1146/annurev-psych-113011-143802.

Goldin-Meadow, S., Kim, S., & Singer, M. (1999). What the teacher's hands tell the student's mind about math. *Journal of Educational Psychology, 91*(4), 720–730.

Green, A., Straube, B., Weis, S., Jansen, A., Willmes, K., Konrad, K., et al. (2009). Neural integration of iconic and unrelated coverbal gestures: A functional MRI study. *Human Brain Mapping, 30*, 3309–3324.

Grossman, E., Donnelly, M., Price, R., Pickens, D., Morgan, V., Neighbor, G., et al. (2000). Brain areas involved in perception of biological motion. *Journal of Cognitive Neuroscience, 12*(5), 711–720.

Hagoort, P., Hald, L., Bastiaansen, M., & Petersson, K. M. (2004). Integration of word meaning and world knowledge in language comprehension. *Science, 304*(5669), 438–441.

Hein, G., Doehrmann, O., Müller, N. G., Kaiser, J., Muckli, L., & Naumer, M. J. (2007). Object familiarity and semantic congruency modulate responses in cortical audiovisual integration areas. *The Journal of Neuroscience, 27*(30), 7881–7887.

Hickok, G., & Poeppel, D. (2007). The cortical organization of speech processing. *Nature Reviews. Neuroscience, 8*(5), 393–402.

Hoenig, K., & Scheef, L. (2009). Neural correlates of semantic ambiguity processing during context verification. *NeuroImage, 45*(3), 1009–1019. Available from: http://dx.doi.org/10.1016/j.neuroimage.2008.12.044.

Holle, H., & Gunter, T. C. (2007). The role of iconic gestures in speech disambiguation: ERP evidence. *Journal of Cognitive Neuroscience, 19*(7), 1175–1192.

Holle, H., Gunter, T. C., Rüschemeyer, S. A., Hennenlotter, A., & Iacoboni, M. (2008). Neural correlates of the processing of co-speech gestures. *NeuroImage, 39*(4), 2010–2024.

Holle, H., Obleser, J., Rueschemeyer, S.-A., & Gunter, T. C. (2010). Integration of iconic gestures and speech in left superior temporal areas boosts speech comprehension under adverse listening conditions. *NeuroImage, 49*, 875–884.

Hostetter, A. B. (2011). When do gestures communicate? A meta-analysis. *Psychological Bulletin, 137*(2), 297–315. Available from: http://dx.doi.org/10.1037/a0022128.

Iverson, J. M., Capirci, O., & Caselli, M. C. (1994). From communication to language in two modalities. *Cognitive Development, 9*(1), 23–43.

Josse, G., Joseph, S., Bertasi, E., & Giraud, A.-L. (2012). The brain's dorsal route for speech represents word meaning: Evidence from gesture. *PLoS One, 7*(9), e46108. Available from: http://dx.doi.org/10.1371/journal.pone.0046108.

Kelly, S. D., & Church, R. B. (1997). Can children detect conceptual information conveyed through other children's nonverbal behavior's. *Cognition and Instruction, 15*(1), 107–134.

Kelly, S. D., & Church, R. B. (1998). A comparison between children's and adults' ability to detect conceptual information conveyed through representational gestures. *Child Development, 69*(1), 85–93.

Kelly, S. D., Kravitz, C., & Hopkins, M. (2004). Neural correlates of bimodal speech and gesture comprehension. *Brain and Language, 89*(1), 253–260.

Kelly, S. D., Ozyürek, A., & Maris, E. (2010). Two sides of the same coin: Speech and gesture mutually interact to enhance comprehension. *Psychological Science, 21*(2), 260–267. Available from: http://dx.doi.org/10.1177/0956797609357327.

Kelly, S. D., Ward, S., Creigh, P., & Bartolotti, J. (2007). An intentional stance modulates the integration of gesture and speech during comprehension. *Brain and Language, 101*(3), 222–233.

Kendon, A. (2004). *Gesture: Visible action as utterance*. Cambridge: Cambridge University Press.

Kidd, E., & Holler, J. (2009). Children's use of gesture to resolve lexical ambiguity. *Developmental Science, 12*(6), 903–913. Available from: http://dx.doi.org/10.1111/j.1467-7687.2009.00830.x.

Kilner, J. M. (2011). More than one pathway to action understanding. *Trends in Cognitive Sciences, 15*(8), 352–357.

Kircher, T., Straube, B., Leube, D., Weis, S., Sachs, O., Willmes, K., et al. (2009). Neural interaction of speech and gesture: Differential activations of metaphoric co-verbal gestures. *Neuropsychologia, 47* (1), 169–179.

Kotz, S. A., Cappa, S. F., von Cramon, D. Y., & Friederici, A. D. (2002). Modulation of the lexical-semantic network by auditory semantic priming: An event-related functional MRI study. *NeuroImage, 17*(4), 1761–1772.

Kumin, L., & Lazar, M. (1974). Gestural communication in preschool children. *Perceptual and Motor Skills, 38*, 708–710.

Kutas, M., & Federmeier, K. D. (2011). Thirty years and counting: Finding meaning in the N400 component of the event-related brain potential (ERP). *Annual Review of Psychology, 62*, 621–647. Available from: http://dx.doi.org/10.1146/annurev.psych.093008.131123.

Kwon, H., Kuriki, S., Kim, J. M., Lee, Y. H., Kim, K., & Nam, K. (2005). MEG study on neural activities associated with syntactic and semantic violations in spoken Korean sentences. *Neuroscience Research, 51*, 349–357.

Lau, E. F., Phillips, C., & Poeppel, D. (2008). A cortical network for semantics: (De)constructing the N400. *Nature Reviews. Neuroscience, 9*(12), 920–933.

Lauro, L. J. R., Tettamanti, M., Cappa, S. F., & Papagno, C. (2008). Idiom comprehension: A prefrontal task? *Cerebral Cortex, 18*(1), 162–170. Available from: http://dx.doi.org/10.1093/cercor/bhm042.

Matsumoto, A., Iidaka, T., Haneda, K., Okada, T., & Sadato, N. (2005). Linking semantic priming effect in functional MRI and event-related potentials. *NeuroImage, 24*(3), 624–634. Available from: http://dx.doi.org/10.1016/j.neuroimage.2004.09.008.

McGregor, K. K., Rohlfing, K. J., Bean, A., & Marschner, E. (2009). Gesture as a support for word learning: The case of under. *Journal of Child Language, 36*(4), 807–828. Available from: http://dx.doi.org/10.1017/S0305000908009173.

McNeil, N. M., Alibali, M. W., & Evans, J. L. (2000). The role of gesture in children's comprehension of spoken language: Now they need it, now they don't. *Journal of Nonverbal Behavior, 24*(2), 131–150.

McNeill, D. (1992). *Hand and mind: What gestures reveal about thought*. Chicago, IL: University of Chicago Press.

McNeill, D. (2005). *Gesture and thought*. Chicago, IL: University of Chicago Press.

Mengotti, P., Corradi-Dell'Acqua, C., Negri, G. A. L., Ukmar, M., Pesavento, V., & Rumiati, R. I. (2013). Selective imitation impairments differentially interact with language processing. *Brain: A Journal of Neurology, 136*(Pt. 8), 2602–2618. Available from: http://dx.doi.org/10.1093/brain/awt194.

Mohan, B., & Helmer, S. (1988). Context and second language development: Preschooler's comprehension of gestures. *Applied Linguistics, 9*(3), 275–292.

Molenberghs, P., Cunnington, R., & Mattingley, J. B. (2012). Brain regions with mirror properties: A meta-analysis of 125 human fMRI studies. *Neuroscience and Biobehavioral Reviews, 36*(1), 341–349. Available from: http://dx.doi.org/10.1016/j.neubiorev.2011.07.00.

Morford, M., & Goldin-Meadow, S. (1992). Comprehension and production of gesture in combination with speech in one-word speakers. *Journal of Child Language, 19*, 559–580.

O'Reilly, A. W. (1995). Using representations: Comprehension and production of actions with imagined objects. *Child Development, 66*(4), 999–1010.

Overton, W. F., & Jackson, J. P. (1973). The representation of imagined objects in action sequences: A developmental study. *Child Development, 44*, 309–314.

Ozçalişkan, S., & Goldin-Meadow, S. (2011). Is there an iconic gesture spurt at 26 months. In G. Stam, & M. Ishino (Eds.), *Integrating gestures: The interdisciplinary nature of gesture* (pp. 163–174). Amsterdam: John Benjamins.

Özyürek, A., Willems, R. M., Kita, S., & Hagoort, P. (2007). On-line integration of semantic information from speech and gesture: Insights from event-related brain potentials. *Journal of Cognitive Neuroscience, 19*(4), 605–616.

Perry, M., Berch, D., & Singleton, J. (1995). Constructing shared understanding: The role of nonverbal input in learning contexts. *Journal of Contemporary Legal Issues, 6*, 213–235.

Ping, R. M., & Goldin-Meadow, S. (2008). Hands in the air: Using ungrounded iconic gestures to teach children conservation of quantity. *Developmental Psychology, 44*(5), 1277–1287. Available from: http://dx.doi.org/10.1037/0012-1649.44.5.1277.

Ping, R. M., Goldin-Meadow, S., & Beilock, S. L. (2013). Understanding gesture: Is the listener's motor system involved? *Journal of Experimental Psychology. General*. Available from: http://dx.doi.org/10.1037/a0032246.

Poeppel, D., Emmorey, K., Hickok, G., & Pylkkänen, L. (2012). Towards a new neurobiology of language. *The Journal of Neuroscience: The Official Journal of the Society for Neuroscience, 32* (41), 14125–14131. Available from: http://dx.doi.org/10.1523/JNEUROSCI.3244-12.2012.

Price, C. J. (2010). The anatomy of language: A review of 100 fMRI studies published in 2009. *Annals of the New York Academy of Sciences, 1191*(1), 62–88.

Rauschecker, J. P., & Scott, S. K. (2009). Maps and streams in the auditory cortex: Nonhuman primates illuminate human speech processing. *Nature Neuroscience, 12*(6), 718–724. Available from: http://dx.doi.org/10.1038/nn.2331.

Riseborough, M. G. (1982). Meaning in movement: An investigation into the interrelationship of physiographic gestures and speech in seven-year-olds. *British Journal of Psychology, 73*, 497–503.

Rodd, J. M., Davis, M. H., & Johnsrude, I. S. (2005). The neural mechanisms of speech comprehension: FMRI studies of semantic ambiguity. *Cerebral Cortex, 15*(8), 1261–1269.

Rossell, S. L., Price, C. J., & Nobre, A. C. (2003). The anatomy and time course of semantic priming investigated by fMRI and ERPs. *Neuropsychologia, 41*(5), 550–564.

Rowe, M. L., & Goldin-Meadow, S. (2009). Early gesture selectively predicts later language learning. *Developmental Science, 12*(1), 182–187. Available from: http://dx.doi.org/10.1111/j.1467-7687.2008.00764.x.

Sauter, M., Uttal, D. H., Alman, A. S., Goldin-Meadow, S., & Levine, S. C. (2012). Learning what children know about space from looking at their hands: The added value of gesture in spatial communication. *Journal of Experimental Child Psychology, 111*(4), 587–606 http://dx.doi.org/10.1016/j.jecp.2011.11.009.

Sheehan, E. A., Namy, L. L., & Mills, D. L. (2007). Developmental changes in neural activity to familiar words and gestures. *Brain and Language, 101*, 246–259.

Simos, P. G., Basile, L. F. H., & Papanicolaou, A. C. (1997). Source localization of the N400 response in a sentence-reading paradigm using evoked magnetic fields and magnetic resonance imaging. *Brain Research, 762*, 29–39.

Singer, M., & Goldin-Meadow, S. (2005). Children learn when their teacher's gestures and speech differ. *Psychological Science, 16*(2), 85–89.

Skipper, J. I., Goldin-Meadow, S., Nusbaum, H. C., & Small, S. L. (2007). Speech-associated gestures, broca's area, and the human mirror system. *Brain and Language, 101*(3), 260–277.

Skipper, J. I., Goldin-Meadow, S., Nusbaum, H. C., & Small, S. L. (2009). Gestures orchestrate brain networks for language understanding. *Current Biology, 19*, 1–7.

Snijders, T. M., Vosse, T., Kempen, G., Van Berkum, J. J. A., Petersson, K. M., & Hagoort, P. (2009). Retrieval and unification of syntactic structure in sentence comprehension: An fMRI study using word-category ambiguity. *Cerebral Cortex, 19*, 1493–1503.

So, W. C., Demir, O. E., & Goldin-Meadow, S. (2010). When speech is ambiguous gesture steps in: Sensitivity to discourse-pragmatic principles in early childhood. *Applied Psycholinguistics, 31*(1), 209–224. Available from: http://dx.doi.org/10.1017/S0142716409990221.

Stefanini, S., Bello, A., Caselli, M. C., Iverson, J. M., & Volterra, V. (2009). Co-speech gestures in a naming task: Developmental data. *Language and Cognitive Processes, 24*(2), 168–189.

Stowe, L. A., Haverkort, M., & Zwarts, F. (2005). Rethinking the neurological basis of language. *Lingua, 115*(7), 997–1042.

Straube, B., Green, A., Bromberger, B., & Kircher, T. (2011). The differentiation of iconic and metaphoric gestures: Common and unique integration processes. *Human Brain Mapping, 32*(4), 520–533. Available from: http://dx.doi.org/10.1002/hbm.21041.

Straube, B., Green, A., Jansen, A., Chatterjee, A., & Kircher, T. (2010). Social cues, mentalizing and the neural processing of speech accompanied by gestures. *Neuropsychologia, 48*(2), 382–393. Available from: http://dx.doi.org/10.1016/j.neuropsychologia.2009.09.025.

Straube, B., Green, A., Weis, S., Chatterjee, A., & Kircher, T. (2009). Memory effects of speech and gesture binding: Cortical and hippocampal activation in relation to subsequent memory performance. *Journal of Cognitive Neuroscience, 21*(4), 821–836.

Straube, B., Green, A., Weis, S., & Kircher, T. (2012). A supramodal neural network for speech and gesture semantics: An fMRI study. *PLoS One, 7*(11), e51207. Available from: http://dx.doi.org/10.1371/journal.pone.0051207.

Straube, B., He, Y., Steines, M., Gebhardt, H., Kircher, T., Sammer, G., et al. (2013). Supramodal neural processing of abstract information conveyed by speech and gesture. *Frontiers in Behavioral Neuroscience, 7*, 120. Available from: http://dx.doi.org/10.3389/fnbeh.2013.00120.

Swaab, T. Y., Brown, C. M., & Hagoort, P. (1997). Spoken sentence comprehension in aphasia: Event-related potential evidence for a lexical integration deficit. *Journal of Cognitive Neuroscience, 9*(1), 39–66.

Thompson, L. A., & Massaro, D. W. (1994). Children's integration of speech and pointing gestures in comprehension. *Journal of Experimental Child Psychology, 57*(3), 327–354.

Valenzeno, L., Alibali, M. W., & Klatzky, R. (2003). Teachers' gestures facilitate students' learning: A lesson in symmetry. *Contemporary Educational Psychology, 28*(2), 187–204.

Van Petten, C., & Luka, B. J. (2006). Neural localization of semantic context effects in electromagnetic and hemodynamic studies. *Brain and Language, 97*, 279–293.

Vigneau, M., Beaucousin, V., Hervé, P. Y., Duffau, H., Crivello, F., Houdé, O., et al. (2006). Meta-analyzing left hemisphere language areas: Phonology, semantics, and sentence processing. *NeuroImage, 30*(4), 1414–1432. Available from: http://dx.doi.org/10.1016/j.neuroimage.2005.11.002.

Wakefield, E. M., James, T. W., & James, K. H. (2013). Neural correlates of gesture processing across human development. *Cognitive Neuropsychology, 30*(2), 58–76. Available from: http://dx.doi.org/10.1080/02643294.2013.794777.

Wang, L., & Chu, M. (2013). The role of beat gesture and pitch accent in semantic processing: An ERP study. *Neuropsychologia, 51*, 2847–2855. Available from: http://dx.doi.org/10.1016/j.neuropsychologia.2013.09.027.

Wheatley, T., Weisberg, J., Beauchamp, M. S., & Martin, A. (2005). Automatic priming of semantically related words reduces activity in the fusiform gyrus. *Journal of Cognitive Neuroscience, 17*(12), 1871–1885. Available from: http://dx.doi.org/10.1162/089892905775008689.

Whitney, C., Jefferies, E., & Kircher, T. (2011). Heterogeneity of the left temporal lobe in semantic representation and control: Priming multiple versus single meanings of ambiguous words. *Cerebral Cortex, 21*(4), 831–844. Available from: http://dx.doi.org/10.1093/cercor/bhq148.

Wible, C. G., Han, S. D., Spencer, M. H., Kubicki, M., Niznikiewicz, M. H., Jolesz, F. A., et al. (2006). Connectivity among semantic associates: An fMRI study of semantic priming. *Brain and Language, 97*(3), 294–305. Available from: http://dx.doi.org/10.1016/j.bandl.2005.11.006.

Willems, R. M., & Hagoort, P. (2007). Neural evidence for the interplay between language, gesture, and action: A review. *Brain and Language, 101*(3), 278–289.

Willems, R. M., Özyürek, A., & Hagoort, P. (2007). When language meets action: The neural integration of gesture and speech. *Cerebral Cortex, 17*(10), 2322.

Willems, R. M., Özyürek, A., & Hagoort, P. (2009). Differential roles for left inferior frontal and superior temporal cortex in multimodal integration of action and language. *NeuroImage, 47*, 1992–2004.

Wilson, S. M., Molnar-Szakacs, I., & Iacoboni, M. (2008). Beyond superior temporal cortex: Intersubject correlations in narrative speech comprehension. *Cerebral Cortex, 18*(1), 230–242. Available from: http://dx.doi.org/10.1093/cercor/bhm049.

Wu, Y. C., & Coulson, S. (2005). Meaningful gestures: Electrophysiological indices of iconic gesture comprehension. *Psychophysiology, 42*(6), 654–667.

Wu, Y. C., & Coulson, S. (2007a). How iconic gestures enhance communication: An ERP study. *Brain and Language, 101*(3), 234–245.

Wu, Y. C., & Coulson, S. (2007b). Iconic gestures prime related concepts: An ERP study. *Psychonomic Bulletin and Review, 14*(1), 57–63.

Xu, J., Gannon, P. J., Emmorey, K., Smith, J. F., & Braun, A. R. (2009). Symbolic gestures and spoken language are processed by a common neural system. *Proceedings of the National Academy of Sciences of the United States of America, 106*(49), 20664–20669.

Zempleni, M.-Z., Renken, R., Hoeks, J. C. J., Hoogduin, J. M., & Stowe, L. A. (2007). Semantic ambiguity processing in sentence context: Evidence from event-related fMRI. *NeuroImage, 34*(3), 1270–1279. Available from: http://dx.doi.org/10.1016/j.neuroimage.2006.09.048.

CHAPTER

33

Development of the Brain's Functional Network Architecture

Deanna J. Greene[1,2], *Christina N. Lessov-Schlaggar*[1] *and Bradley L. Schlaggar*[1,2,3,4,5]

[1]Department of Psychiatry, Washington University School of Medicine in St. Louis, St. Louis, MO, USA; [2]Department of Radiology, Washington University School of Medicine in St. Louis, St. Louis, MO, USA; [3]Department of Neurology, Washington University School of Medicine in St. Louis, St. Louis, MO, USA; [4]Department of Pediatrics, Washington University School of Medicine in St. Louis, St. Louis, MO, USA; [5]Department of Anatomy and Neurobiology, Washington University School of Medicine in St. Louis, St. Louis, MO, USA

33.1 WHAT IS A NETWORK AND HOW CAN WE STUDY BRAIN NETWORKS?

The human nervous system can be characterized at many levels of organization, from molecules to neurons to systems. The systems level is most amenable to studying the brain's network architecture using human neuroimaging techniques. Strictly speaking, a network is a set of well-defined items with well-defined pairwise relationships between those items. Network analysis, which is based on graph theory, allows for the quantification of those relationships. Graph theory is a branch of mathematics concerned with modeling the pairwise relationships (edges) between items (nodes). It allows for the formal measurement of many properties of networks, such as between-ness centrality, which is a measure of the frequency by which the shortest paths between all other nodes in the network pass through a given node, and modularity, which is a measure of clustering of nodes into subcommunities within a network (for a review, see Bullmore & Sporns, 2009). The ability to formalize and quantify network properties in this way makes graph theory highly attractive for measuring network properties in the brain. Hence, the number of studies in which graph theory tools are applied to brain imaging data is growing exponentially. Specifically at the systems level, recent investigations have utilized graph theory to measure and quantify relationships between brain regions, where nodes in the graph represent specified brain regions or voxels, and edges in the graph represent the pairwise relationships between the brain regions. The analytic method is not dependent on the type of data used; therefore, network organization can be characterized using structural and functional brain imaging measures. It is worth keeping in mind that applying graph theory to brain imaging data is subject to some methodological limitations relative to other types of network data (e.g., transportation networks and social networks). Because brain networks are represented as graphs of correlations (representing the relationships between brain regions), certain graph metrics may be biased; for example, the number of edges that a node has can be biased by network size. Reviews of the literature on structural brain networks and the development of structural networks can be found in Alexander-Bloch, Giedd, and Bullmore (2013) and Giedd and Rapoport (2010). Here, we focus our discussions on functionally defined brain systems, primarily emphasizing studies that implement resting state functional connectivity (RSFC) magnetic resonance imaging (MRI) (Figure 33.1).

RSFC is a technique that measures spontaneous low-frequency blood oxygen level–dependent (BOLD) activity while a subject is "at rest" (i.e., lying awake quietly, often with eyes open and foveating on a centrally positioned crosshair). Functional connectivity is defined as the temporal correlation (measured as a Pearson's *r*) in the high amplitude, low-frequency spontaneously generated BOLD signal between voxels

Neurobiology of Language. DOI: http://dx.doi.org/10.1016/B978-0-12-407794-2.00033-X

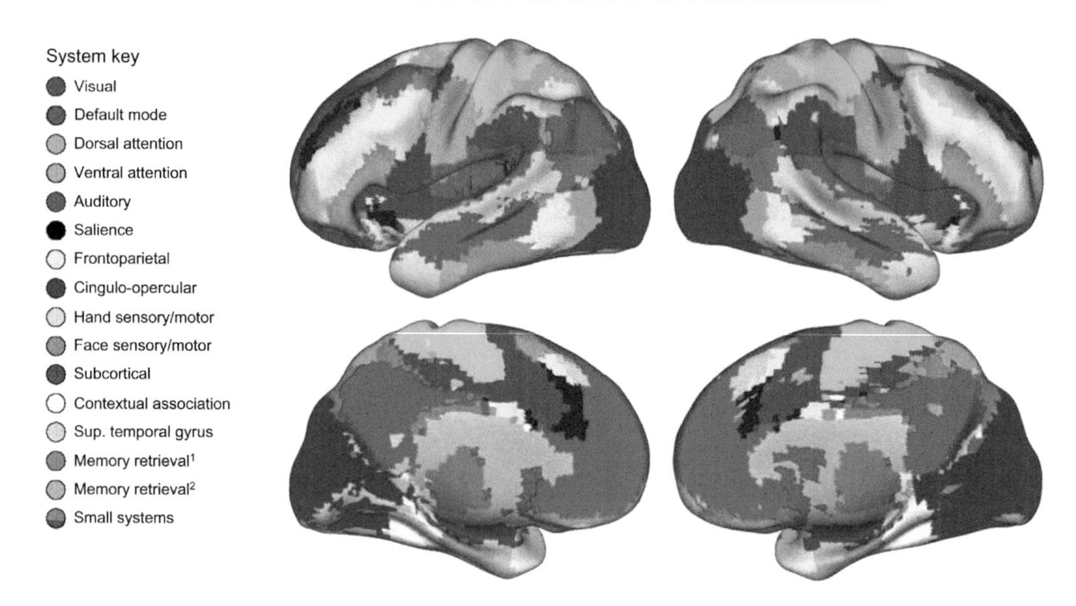

FIGURE 33.1 Functional brain network architecture derived using network analysis of RSFC data between every pair of voxels in the brain. *Adapted from Power et al. (2013) with permission from the publisher.*

(cubic "pixel" in a three-dimensional brain image) or brain regions (Fox & Raichle, 2007). Functional connectivity is spatially constrained by anatomy but does not necessarily reflect monosynaptic anatomical connections. For example, RSFC is high between homotopic nonfoveal V1 visual cortex regions despite the lack of direct callosal connections (Vincent et al., 2007). Thus, RSFC provides a measure of functional brain connectivity that may contain different information than strict anatomy. For example, it has been posited to reflect a "Hebbian-like" history of co-activation between brain regions (Dosenbach et al., 2007; Kelly et al., 2009; Lewis, Baldassarre, Committeri, Romani, & Corbetta, 2009; Wig, Schlaggar, & Petersen, 2011). Figure 33.1 shows results from the application of graph theory methods, specifically modularity, to RSFC correlations across every pair of voxels in the adult brain. Different functional systems (or networks) are indicated by different colors. The BOLD timecourses in distributed yet functionally related brain regions are correlated with each other; these region sets correspond to known functional systems, including sensorimotor (e.g., somato-motor; turquoise and orange colors in Figure 33.1), auditory (fuchsia in Figure 33.1), visual (blue in Figure 33.1), and higher-level control systems (e.g., frontoparietal [yellow in Figure 33.1] and dorsal attention [chartreuse in Figure 33.1]) (Biswal, Yetkin, Haughton, & Hyde, 1995; Dosenbach et al., 2007; Lowe, Mock, & Sorenson, 1998). RSFC has also enabled the discovery and interrogation of the default mode system (red in Figure 33.1) (Fox et al., 2005; Greicius, Krasnow, Reiss, & Menon, 2003; Raichle et al., 2001). Thus, the temporal correlations measured with RSFC provide a quantifiable metric suitable for network analysis.

In the functional MRI literature, the term "network" is often used to refer to a set of brain regions that co-activate or deactivate under certain task conditions, without consideration of the pairwise relationships between the regions. Designating networks in this sense is incomplete and can potentially misrepresent the true architecture of brain networks. Such misapprehension of network structure is common in the RSFC literature, with "resting state networks" or task activation networks being labeled without a sufficient examination of the pairwise relationships between brain regions. Some of the approaches commonly used to study these "networks," although valuable, stop short of true network analysis. One such approach is seed-based analysis, in which statistical maps are generated across the whole brain, representing the correlations between the BOLD timecourse of a "seed" region of interest and the BOLD timecourses of every other voxel in the brain. The resulting map displays regions (collections of voxels) with timecourses similar to the seed's, but without reflecting the relationships among all those regions. Similarly, independent component analysis approaches are commonly used to define resting state systems without accounting for the relationships between all the regions within a system or with other systems. Independent component analysis identifies voxels with shared temporal (and spatial) BOLD signal variance and separates them as orthogonal components. However, investigators most often stop at the point of identifying components and do not proceed to explore the relationships among regions within and between components. The statistical constraints of independent components analysis requiring zero correlation between components and not explicitly modeling the strength of relationships between nodes within each component could result in incomplete or distorted descriptions of functional networks (Wig et al., 2011). Thus, these approaches leave an incomplete depiction of the complex relationships within and between systems. Graph theory, by contrast, provides a framework with which to analyze networks in the true sense of the term. For RSFC data, the pairwise correlations comprise the edges in the graph. Thus, network properties can be quantified with respect to functional connectivity between regions (collection of voxels) or individual voxels.

33.2 ORGANIZATION OF THE BRAIN'S FUNCTIONAL NETWORK ARCHITECTURE

Investigating systems across the whole-brain is a sizeable task. Although many studies have examined particular targeted brain systems (e.g., frontoparietal and default mode), considering the whole brain's network architecture is key for obtaining a complete picture of brain organization. There have been a small number of studies aimed at characterizing whole-brain network organization using RSFC methods (Power et al., 2011; Yeo et al., 2011). These studies have effectively identified whole-brain network schemes that map reasonably well onto known functional systems. Utilizing graph theoretic approaches (Power et al., 2011), we have identified communities (modules, subnetworks, subgraphs) corresponding to sensorimotor systems (e.g., visual, auditory, and somato-motor), control systems (e.g., frontoparietal, involved in moment-to-moment control, and cingulo-opercular, involved in sustained task control) (Dosenbach, Fair, Cohen, Schlaggar, and Petersen (2008), attention systems (e.g., dorsal attention, involved in goal-directed attentional orienting, and ventral attention system, involved in orienting to relevant exogenous stimuli) Corbetta, Patel, and Shulman (2008), and others (e.g., salience and default mode) (Figure 33.1). Graph metrics allow for interrogation of network properties, revealing some intriguing findings that could only be gleaned by network analysis. For example, the default mode system has certain network properties that most closely resemble those of lower level processing systems in contrast to higher level control systems. Specifically, the default mode system has high local efficiency and low participation coefficients, such that it is relatively isolated from other systems but well-integrated within itself (Power, Schlaggar, Lessov-Schlaggar, & Petersen, 2013). Control systems, such as the cingulo-opercular, show the converse: low local efficiency and high participation coefficients, properties expected for systems whose putative role is to integrate information across systems (Power, Schlaggar, et al., 2013).

33.3 IS THERE A LANGUAGE NETWORK?

Descriptions of the brain's functional network architecture lead one to ask, what system(s) supports the uniquely human capacity for language? Some of the results from work using seed-based functional connectivity analysis have been interpreted as evidence for the presence of several language networks. One study used six left hemisphere regions that consistently activate in word reading tasks as seeds (inferior occipital gyrus, fusiform gyrus, superior temporal gyrus, temporoparietal junction, dorsal precentral gyrus, and inferior frontal gyrus) in RSFC connectivity analysis (Koyama et al., 2010). The whole-brain connectivity maps of each of these seeds differed substantially from each other, which was interpreted as evidence for the existence of six reading systems. Conjunction analysis aimed at identifying voxels that were significantly correlated with all of the seed regions showed that voxels in left posterior inferior frontal gyrus and left posterior middle temporal gyrus were common to five of the six correlation maps, suggesting that these regions are places of "functional interaction among the reading networks." Left inferior frontal and middle temporal gyrus regions were used as seeds in an investigation of functional connectivity of task-evoked activation data during acquisition of meaning of new nouns (Yeo et al., 2011). Based on the functional connectivity patterns of the two seeds across task conditions, the authors proposed the existence of two systems of meaning acquisition: one system that includes left inferior frontal gyrus, middle cingulate cortex/supplementary motor area, left inferior parietal lobule, thalamus, bilateral caudate, and bilateral middle frontal gyrus that subserves mapping meaning onto new words; and a second system that includes the left middle temporal gyrus, anterior and posterior cingulate cortex, and bilateral middle frontal gyrus that subserves semantic integration. Bilateral middle frontal cortex was functionally connected to both seed regions, implicating working memory processing as a shared domain between the two systems. However, no analysis was performed to examine the functional connectivity within each of the proposed system regions to test directly whether they belong to a separable network. Another study used task data but regressed out task-evoked activation to examine functional connectivity with left posterior superior temporal sulcus (Lohmann et al., 2010). They found this region to be correlated with seeds in Broca's area, a brain region commonly conceptualized as supporting language production, and with closely positioned left frontal operculum in data from language tasks but not in data from control tasks, possibly identifying language-specific temporal-frontal connectivity. It is important to reiterate that seed-based functional connectivity analysis does not constitute a true network analysis because, although it effectively identifies the seed's neighbors, it does not characterize the pairwise relationships between the regions that comprise the system of interest.

Our group has taken seed-based and network approaches to study reading-related brain regions. Seed-based RSFC analysis was used to examine the putative visual word form area in left occipitotemporal cortex and showed that it is functionally connected to regions comprising the dorsal attention system and not to other regions presumed to be involved in reading (Vogel et al., 2012). These results support a role for attention processing in reading and counter the idea that the putative visual word form area is functionally related to reading-related regions. We then asked whether a dedicated reading system could be identified and, if so, whether it changes over the course of development from age 7 years to young adulthood. We used RSFC data to perform a network analysis of 83 reading-related regions assembled from meta-analyses in adults and from developmental studies of reading. Results demonstrated that these regions segregated into previously reported systems, as in Figure 33.1, including visual, somato-motor, cognitive control, and default mode (Vogel et al., 2013). Regions most commonly identified in reading studies were part of communities in the left hemisphere and included the visual cortex (putative visual word form area), inferior frontal cortex, inferior temporal cortex, supramarginal gyrus, and angular gyrus, largely considered core regions of the default mode system. However, two left supramarginal gyrus regions, left inferior frontal cortex, and a few other regions (right medial superior parietal gyrus, left medial supplementary motor area, right temporal gyrus, and mesial thalamus) could not be definitively assigned to one of the known systems. Interestingly, left supramarginal gyrus and inferior frontal cortex have been singled out as comprising a language-related system in the task-based seed-based analyses discussed (Koyama et al., 2010; Lohmann et al., 2010; Yeo et al., 2011), suggesting some convergence in findings across different levels of analysis. We found no evidence for developmental effects of the network structure of reading-related regions (Vogel et al., 2013).

Most recently, a whole-brain RSFC study in a large dataset found that a language system could be reliably identified at the level of the individual brain (Hacker et al., 2013). This language system was composed of superior and middle temporal cortex and inferior frontal gyrus bilaterally, left supramarginal/angular gyrus, left dorsal premotor cortex, and right cerebellum. Interestingly, the language system emerged at much later iterations of the learning algorithm that was used to identify resting state networks than other systems such as the default mode network. The authors suggested that finer (more hierarchical) levels of analysis might be necessary to detect a language network. In our most recent efforts of mapping functional brain systems (Figure 33.1), bilateral inferior frontal gyrus, left dorsal premotor cortex areas along with relatively smaller areas in left inferior parietal lobule (including in the supramarginal and angular gyri), and left medial dorsal cortex were assigned to the ventral attention system (teal color in Figure 33.1). These areas closely overlap with areas considered to be part of the language and ventral attention systems in the study by Hacker et al. (2013). Bilateral superior temporal cortex (light pink in Figure 33.1) was not part of any system that we identified, raising the possibility that with higher order analysis, it could have integrated with areas of the ventral attention system into a language system.

It is also plausible and intriguing that language emerges from elements of other brain systems. Support for this idea comes from recent evidence for brain regions or "hubs" that sit at the intersection of multiple systems. Hubs are considered to be nodes with particularly important network properties, akin to high-traffic, densely connected airports. One property that has received considerable attention is the node "degree," which is defined as the number of edges (connections) on a node. The more edges a node has, the higher its degree and the more interconnected it is with other nodes. High-degree nodes have been identified within the default mode system, implying that the default mode system contains "hubs" of the human brain (Buckner et al., 2009). However, we have argued that degree-based nodes are confounded by the size of the system in which they belong in correlation-based systems (Power, Schlaggar, et al., 2013). Because the default mode system, with respect to the total volume of cerebral cortex devoted to it, is one of the largest systems in the brain, high-degree nodes within it may reflect membership in the largest system, rather than true hub-like qualities. True hubs in the brain would be expected to support and/or integrate multiple types of information and to cause widespread disruption if compromised. From this perspective, we have investigated regions in the brain that possess two properties related to but different from node degree. One property is referred to as "participation coefficient" and captures the number of regions in different systems to which a node is functionally connected; a second property is referred to as "articulation points," which captures voxels where multiple brain systems are represented. Convergence between these two properties identified candidate hub regions located in the anterior insula, dorsal medial prefrontal cortex, dorsal prefrontal cortex, lateral occipitotemporal cortex, and superior parietal cortex.

Intriguingly, some of the putative hub regions identified in Power, Schlaggar, et al. (2013) are positioned in locations of the cortex that overlap with, or are near, canonical language areas. In particular, hub-like regions identified in the posterior part of the superior temporal sulcus overlap with standard localization of Wernicke's area, a brain region commonly conceptualized as supporting language comprehension. Other hubs located in the anterior insula/frontal operculum are situated near the inferior frontal gyrus and the general region of Broca's area. Thus, locations in the brain with these hub-like features may support language function, suggesting the confluence of multiple brain systems for language. These hub-like regions may be susceptible to substantial disruption in brain function, including disruption in language domains, because of their location at the intersection of multiple brain systems. It has been shown that patients with focal lesions that included nodes located at the intersection of several networks showed significant decrease in modular organization of four tested networks, whereas patients with focal lesions that included nodes located within networks did not show decrease in network modularity (Gratton, Nomura, Pérez, & D'Esposito, 2012).

33.4 DEVELOPMENT OF BRAIN NETWORKS

Understanding how brain network organization develops from childhood to adulthood has been the focus of much study in the past decade. Initial efforts to investigate the development of brain systems resulted in some principles that were, unfortunately, confounded by an insidious movement artifact, resulting from submillimeter movements, that was not addressed by industry standards for dealing with movement. For a complete review of the literature on network development up until the discovery of the motion confound, see Power, Fair, Schlaggar, and Petersen (2010) and Power, Barnes, Snyder, Schlaggar, and Petersen (2012). Several groups, including ours, initially found two key results. First, within-network connectivity increased while between-network connectivity decreased over development, leading to more segregation of brain networks into adulthood (Fair et al., 2008, 2007; Supekar, Musen, & Menon, 2009). Second, development was associated with increased strength in long-range connections and decreased strength in short-range connections (Dosenbach et al., 2010; Fair et al., 2007; Supekar et al., 2009). Although these studies followed the best practices at the time for accounting for movement (e.g., frame-by-frame image realignment, excluding BOLD runs or subjects with average motion estimates exceeding set thresholds, matching groups on average motion estimates), they suffered from a previously unknown effect of motion on RSFC data. Head motion, even submillimeters in amplitude, that does not affect average motion estimates can cause spurious yet systematic changes in RSFC correlations (Power et al., 2012; Van Dijk, Sabuncu, & Buckner, 2012). Specifically, head motion increases the correlation strength between proximate brain regions and decreases the correlation strength between distant brain regions. Thus, if one compares RSFC correlations between two groups, a group with more micromovements will appear to have stronger short-range connections and weaker long-range connections, precisely mapping onto the "developmental" effect that we and others observed. Unfortunately, children move more than adults, patients move more than controls, and aging adults move more than younger adults, making it difficult to separate real group differences in RSFC correlations from motion artifacts.

The discovery of motion artifacts in RSFC correlation has prompted the development of data processing strategies to better account for motion (Power, Mitra, et al., 2014; Satterthwaite, Elliott, et al., 2013; Satterthwaite, Wolf, et al., 2013; Yan et al., 2013); however, there is no agreed-on standard. Still, these strategies have been shown to reduce significantly the spurious correlation between motion and BOLD signal intensity, thus mitigating some of the effects of motion-induced artifact. Studies that incorporate improved motion correction strategies support the first main finding of earlier studies: increased network segregation over the course of development defined by increased within-network connectivity and decreased between-network connectivity (Fair et al., 2012; Satterthwaite, Wolf, et al., 2013). Recent findings provide evidence that motion correction may even strengthen these effects (Satterthwaite, Wolf, et al., 2013). Earlier findings that a significant amount of the variance in brain maturation (55%) measured using whole-brain RSFC can be explained by chronological age (Dosenbach et al., 2010) are also supported using improved motion correction methods (Fair et al., 2012; Satterthwaite, Wolf, et al., 2013), with evidence that the variance in brain maturation attributable to age may actually increase after motion correction (Fair et al., 2012). Using seed-based analysis of four default mode system regions, one study showed that the increased positive correlations between these four regions across age were eliminated after movement artifact correction (Chai et al., 2014). At the same time, the amplitude of the negative correlations between default mode regions and attention-related regions increased over the course of development. Specifically, these correlations changed from positive in 8- to 12-year-olds to negative in 13- to 17-year-olds to strongly negative in 18- to 24-year-olds, and these effects were robust to motion correction. These results, suggesting age-related increases in between-system functional connectivity for negative correlations, are the first to our knowledge to examine the development of negative correlations.

The second developmental finding in earlier studies—increased connectivity of long-range connections and decreased connectivity of short-range connections, implying increased complexity of functional brain organization over development—seems to be almost entirely explained by the motion artifact (Power et al., 2012; Satterthwaite, Wolf, et al., 2013). There is still some evidence, however, for integration of distant anterior and posterior regions into the same system and for segregation of closely located regions into separate systems across development. In movement-corrected data, integration of regions in the frontoparietal and default mode systems was seen from childhood (ages 6–14 years) to young adulthood (ages 21–29 years), whereas segregation of medial and lateral visual system regions was seen across these age groups (Vogel et al., 2013). These results suggest that some changes do occur in the strength of the correlations in long-range and short-range connections, consistent with the original ideas (Fair et al., 2008, 2007; Supekar et al., 2009). The idea of increased complexity over the course of development is also supported by analyses of structural MRI data. A large study of 5- to 18-year-olds showed that gray matter intensity in seed regions in sensorimotor, salience, executive control, and default mode systems, as well as language-related regions, was correlated with age in a larger number of voxels across the brain (Zielinski, Gennatas, Zhou, & Seeley, 2010). Expansion of the seed-based structural connectivity was particularly dramatic between ages 12 and 14 years, with the exception of speech and executive control systems, where the largest expansion was seen in the oldest age group (16- to 18-year-olds). Of note, this study was a seed-based analysis that does not meet the definition of network analysis, as we have pointed out. In addition, with some exceptions (Rose et al., 2012), the impact of motion artifact on structural connectivity analysis has not been rigorously investigated at this time and will likely become an active area of research in the near future.

In contrast, motion correction of RSFC data is an active area of research and there is, at present, no standard approach to dealing with this issue. The details of the debate regarding optimizing data quality through removing movement artifact are beyond the scope of this chapter. The reader is directed to several recent publications on this topic (Chai et al., 2014; Power, Mitra, et al., 2014; Saad et al., 2012; Satterthwaite, Elliott, et al., 2013).

We have implemented an approach that includes, among other steps, censoring high movement volumes in RSFC data (Power et al., 2012; Power, Mitra, et al., 2014). One significant side effect of such an approach is substantial data loss (Yan et al., 2013), which could result in the inability to perform planned analyses. Data loss may be especially problematic for already collected developmental data that rely on *post hoc* statistical motion correction. In comparison with large datasets (Satterthwaite, Wolf, et al., 2013) or meta-analysis of data from multiple sites (Fair et al., 2012), smaller studies examining brain system development are at a relative disadvantage for showing robust effects. Efforts to minimize motion should be prioritized at the time of data collection, when study participants can be more rigorously secured and acquisition sequences can be optimized (Craddock et al., 2013). Collecting RSFC data for longer periods of time can also mitigate data loss by providing greater probability for sufficient data per subject to allow stringent movement correction.

33.5 IMPLICATIONS OF DEVELOPMENT OF BRAIN NETWORKS TO LANGUAGE-RELATED BRAIN REGIONS

We noted earlier that the seed-based whole-brain connectivity map of the putative visual word form area in adults parallels the dorsal attention system (Vogel et al., 2012). The same analysis in a sample of 7- to 9-year-old children showed a much reduced extent of putative visual word form area functional connectivity across the brain with positive correlations in visual cortex and negative correlations in superior frontal cortex (Vogel et al., 2012). Comparisons between children and adults matched on movement showed generally elevated connectivity in adults, although the matched sample was small ($n = 13$ adults, 6 children) and lacked statistical power. As mentioned, subsequent network analysis of 83 reading-related regions using movement-corrected data did not find a reading system in children and instead showed that reading-related regions largely segregated into known brain systems, including visual, sensorimotor, cognitive control, default mode, subcortical, and temporal. Although there was some modest evidence for integration (e.g., frontal and parietal regions of the frontoparietal cognitive control system) and segregation of regions (e.g., visual cortex into medial and lateral communities) across development, there was no evidence for regions that comprise a reading community in children (Vogel et al., 2013). To the extent that brain regions that serve as "hubs" for information processing include language-related regions, hub architecture so far appears to be largely stable across development from ages 10 to 20 years (Hwang, Hallquist, & Luna, 2013).

33.6 FUTURE DIRECTIONS

Investigating the development of the brain's functional network architecture is a highly active research area with rapidly evolving analytical methods. Some of the most important current challenges involve correcting resting state functional MRI data for submillimeter amplitude subject movements and applying network analysis methods that probe network architecture in the true sense of the word, considering all pairwise relationships between brain regions (collection of voxels) or individual voxels in a given analysis. Although the presence of a brain system uniquely devoted to language is still a subject of investigation, it is entirely possible that continued development of network analysis methods and increasing understanding of brain network architecture could uncover communities within the larger network structure, with language comprising one such community (Hacker et al., 2013). Considering the relatively recent evolutionary emergence of language for our species, it is also plausible that language and its development are the products of the interaction of multiple brain systems.

Acknowledgment

This effort was supported by a Tourette Syndrome Association fellowship (DJG), NIH K01DA027046 (CNLS), NIH R01HD057076 (BLS), and the Intellectual and Developmental Disabilities Research Center at Washington University NIH/NICHD P30HD062171 (BLS).

References

Alexander-Bloch, A., Giedd, J. N., & Bullmore, E. (2013). Imaging structural co-variance between human brain regions. *Nature Reviews Neuroscience, 14*(5), 322–336. Available from: http://dx.doi\org.10.1038/nrn3465.

Biswal, B., Yetkin, F. Z., Haughton, V. M., & Hyde, J. S. (1995). Functional connectivity in the motor cortex of resting human brain using echo-planar MRI. *Magnetic Resonance in Medicine, 34* (4), 537–541.

Buckner, R. L., Sepulcre, J., Talukdar, T., Krienen, F. M., Liu, H., Hedden, T., et al. (2009). Cortical hubs revealed by intrinsic functional connectivity: Mapping, assessment of stability, and relation to Alzheimer's disease. *The Journal of Neuroscience: The Official Journal of the Society for Neuroscience, 29*(6), 1860–1873.

Bullmore, E., & Sporns, O. (2009). Complex brain networks: Graph theoretical analysis of structural and functional systems. *Nature Reviews Neuroscience, 10*(3), 186–198.

Chai, X. J., Ofen, N., Gabrieli, J. D., & Whitfield-Gabrieli, S. (2014). Selective development of anticorrelated networks in the intrinsic functional organization of the human brain. *Journal of Cognitive Neuroscience, 26*(3), 501–513.

Corbetta, M., Patel, G., & Shulman, G. L. (2008). The reorienting system of the human brain: From environment to theory of mind. *Neuron, 58*(3), 306–324.

Craddock, R. C., Jbabdi, S., Yan, C. G., Vogelstein, J. T., Castellanos, F. X., Di Martino, A., et al. (2013). Imaging human connectomes at the macroscale. *Nature Methods, 10*(6), 524–539.

Dosenbach, N. U., Nardos, B., Cohen, A. L., Fair, D. A., Power, J. D., Church, J. A., et al. (2010). Prediction of individual brain maturity using fMRI. *Science, 329*(5997), 1358–1361.

Dosenbach, N. U. F., Fair, D. A., Cohen, A. L., Schlaggar, B. L., & Petersen, S. E. (2008). A dual-networks architecture of top-down control. *Trends in Cognitive Sciences, 12*(3), 99–105.

Dosenbach, N. U. F., Fair, D. A., Miezin, F. M., Cohen, A. L., Wenger, K. K., Dosenbach, R. A. T., et al. (2007). Distinct brain networks for adaptive and stable task control in humans. *Proceedings of the National Academy of Sciences of the United States of America, 104*(26), 11073–11078.

Fair, D. A., Cohen, A. L., Dosenbach, N. U., Church, J. A., Miezin, F. M., Barch, D. M., et al. (2008). The maturing architecture of the brain's default network. *Proceedings of the National Academy of Sciences of the United States of America, 105*(10), 4028–4032.

Fair, D. A., Dosenbach, N. U. F., Church, J. A., Cohen, A. L., Brahmbhatt, S., Miezin, F. M., et al. (2007). Development of distinct control networks through segregation and integration. *Proceedings of the National Academy of Sciences of the United States of America, 104*(33), 13507–13512.

Fair, D. A., Nigg, J. T., Iyer, S., Bathula, D., Mills, K. L., Dosenbach, N. U., et al. (2012). Distinct neural signatures detected for ADHD subtypes after controlling for micro-movements in resting state functional connectivity MRI data. *Frontiers in Systems Neuroscience, 6*, 80. Available from: http:\\dx.doi\org.10.3389/fnsys.2012.00080.

Fox, M. D., & Raichle, M. E. (2007). Spontaneous fluctuations in brain activity observed with functional magnetic resonance imaging. *Nature Reviews Neuroscience, 8*(9), 700–711.

Fox, M. D., Snyder, A. Z., Vincent, J. L., Corbetta, M., Van Essen, D. C., & Raichle, M. E. (2005). The human brain is intrinsically organized into dynamic, anticorrelated functional networks. *Proceedings of the National Academy of Sciences of the United States of America, 102*(27), 9673–9678.

Giedd, J. N., & Rapoport, J. L. (2010). Structural MRI of pediatric brain development: What have we learned and where are we going? *Neuron, 67*(5), 728–734.

Gratton, C., Nomura, E. M., Pérez, F., & D'Esposito, M. (2012). Focal brain lesions to critical locations cause widespread disruption of the modular organization of the brain. *Journal of Cognitive Neuroscience, 24*(6), 1275–1285.

Greicius, M. D., Krasnow, B., Reiss, A. L., & Menon, V. (2003). Functional connectivity in the resting brain: A network analysis of the default mode hypothesis. *Proceedings of the National Academy of Sciences of the United States of America, 100*(1), 253–258.

Hacker, C. D., Laumann, T. O., Szrama, N. P., Baldassarre, A., Snyder, A. Z., Leuthardt, E. C., et al. (2013). Resting state network estimation in individual subjects. *NeuroImage, 82*, 616–633.

Hwang, K., Hallquist, M. N., & Luna, B. (2013). The development of hub architecture in the human functional brain network. *Cerebral Cortex, 23*(10), 2380–2393.

Kelly, A. M. C., Di Martino, A., Uddin, L. Q., Shehzad, Z., Gee, D. G., Reiss, P. T., et al. (2009). Development of anterior cingulate functional connectivity from late childhood to early adulthood. *Cerebral Cortex, 19*(3), 640–657.

Koyama, M. S., Kelly, C., Shehzad, Z., Penesetti, D., Castellanos, F. X., & Milham, M. P. (2010). Reading networks at rest. *Cerebral Cortex, 20*(11), 2549–2559. Available from: http://dx.doi.org/10.1093/cercor/bhq005.

Lewis, C. M., Baldassarre, A., Committeri, G., Romani, G. L., & Corbetta, M. (2009). Learning sculpts the spontaneous activity of the resting human brain. *Proceedings of the National Academy of Sciences of the United States of America, 106*(41), 17558–17563.

Lohmann, G., Hoehl, S., Brauer, J., Danielmeier, C., Bornkessel-Schlesewsky, I., Bahlmann, J., et al. (2010). Setting the frame: The human brain activates a basic low-frequency network for language processing. *Cerebral Cortex, 20*(6), 1286–1292.

Lowe, M. J., Mock, B. J., & Sorenson, J. A. (1998). Functional connectivity in single and multislice echoplanar imaging using resting-state fluctuations. *NeuroImage, 7*(2), 119–132.

Power, J. D., Barnes, K. A., Snyder, A. Z., Schlaggar, B. L., & Petersen, S. E. (2012). Spurious but systematic correlations in functional connectivity MRI networks arise from subject motion. *NeuroImage, 59*(3), 2142–2154.

Power, J. D., Cohen, A. L., Nelson, S. M., Wig, G. S., Barnes, K. A., Church, J. A., et al. (2011). Functional network organization of the human brain. *Neuron, 72*(4), 665–678.

Power, J. D., Fair, D. A., Schlaggar, B. L., & Petersen, S. E. (2010). The development of human functional brain networks. *Neuron, 67*(5), 735–748.

Power, J. D., Mitra, A., Laumann, T. O., Snyder, A. Z., Schlaggar, B. L., & Petersen, S. E. (2014). Methods to detect, characterize, and remove motion artifact in resting state fMRI. *NeuroImage, 84*, 320–341.

Power, J. D., Schlaggar, B. L., Lessov-Schlaggar, C. N., & Petersen, S. E. (2013). Evidence for hubs in human functional brain networks. *Neuron, 79*(4), 798–813.

Raichle, M. E., MacLeod, A. M., Snyder, A. Z., Powers, W. J., Gusnard, D. A., & Shulman, G. L. (2001). A default mode of brain function. *Proceedings of the National Academy of Sciences of the United States of America, 98*(2), 676–682.

Rose, S., Pannek, K., Bell, C., Baumann, F., Hutchinson, N., Coulthard, A., et al. (2012). Direct evidence of intra- and interhemispheric corticomotor network degeneration in amyotrophic lateral sclerosis: An automated MRI structural connectivity study. *NeuroImage, 59*(3), 2661–2669.

Saad, Z. S., Gotts, S. J., Murphy, K., Chen, G., Jo, H. J., Martin, A., et al. (2012). Trouble at rest: How correlation patterns and group differences become distorted after global signal regression. *Brain Connect, 2*(1), 25–32.

Satterthwaite, T. D., Elliott, M. A., Gerraty, R. T., Ruparel, K., Loughead, J., Calkins, M. E., et al. (2013). An improved framework for confound regression and filtering for control of motion artifact in the preprocessing of resting-state functional connectivity data. *NeuroImage, 64*, 240–256.

Satterthwaite, T. D., Wolf, D. H., Ruparel, K., Erus, G., Elliott, M. A., Eickhoff, S. B., et al. (2013). Heterogeneous impact of motion on fundamental patterns of developmental changes in functional connectivity during youth. *NeuroImage, 83*, 45–57.

Supekar, K., Musen, M., & Menon, V. (2009). Development of large-scale functional brain networks in children. *PLoS Biology, 7*(7), e1000157. Available from: http:\\dx.doi.org\10.1371/journal.pbio.1000157.

Van Dijk, K. R., Sabuncu, M. R., & Buckner, R. L. (2012). The influence of head motion on intrinsic functional connectivity MRI. *NeuroImage, 59*(1), 431–438.

Vincent, J. L., Patel, G. H., Fox, M. D., Snyder, A. Z., Baker, J. T., Van Essen, D. C., et al. (2007). Intrinsic functional architecture in the anaesthetized monkey brain. *Nature, 447*(7140), 83–86.

Vogel, A. C., Miezin, F. M., Petersen, S. E., & Schlaggar, B. L. (2012). The putative visual word form area is functionally connected to the dorsal attention network. *Cerebral Cortex, 22*(3), 537–549. Available from: http://dx.doi.org/10.1093/cercor/bhr100. Epub 2011 Jun 20.

Vogel, A. C., Church, J. A., Power, J. D., Miezin, F. M., Petersen, S. E., & Schlaggar, B. L. (2013). Functional network architecture of reading-related regions across development. *Brain and Language, 125*(2), 231–243.

Wig, G. S., Schlaggar, B. L., & Petersen, S. E. (2011). Concepts and principles in the analysis of brain networks. *Annals of the New York Academy of Sciences, 1224*(1), 126–146.

Yan, C. G., Cheung, B., Kelly, C., Colcombe, S., Craddock, R. C., Di Martino, A., et al. (2013). A comprehensive assessment of regional variation in the impact of head micromovements on functional connectomics. *NeuroImage, 76*, 183–201.

Yeo, B. T. T., Krienen, F. M., Sepulcre, J., Sabuncu, M. R., Lashkari, D., Hollinshead, M., et al. (2011). The organization of the human cerebral cortex estimated by intrinsic functional connectivity. *Journal of Neurophysiology, 106*(3), 1125–1165.

Zielinski, B. A., Gennatas, E. D., Zhou, J., & Seeley, W. W. (2010). Network-level structural covariance in the developing brain. *Proceedings of the National Academy of Sciences of the United States of America, 107*(42), 18191–18196.

CHAPTER

34

Bilingual Development and Age of Acquisition

Arturo E. Hernandez

Department of Psychology, University of Houston, Houston, TX, USA

34.1 INTRODUCTION

Development by definition involves changes in age. The topic of maturational changes and effects due to critical periods in language has a long-standing tradition in the literature (Lenneberg, 1967; Pinker, 1994). This approach has also held true in the bilingual literature that has, in part, focused on the differences between adult and child language learners (Johnson & Newport, 1989). The traditional view in a purely maturational model would view language learning as being easiest early in life, with learning becoming progressively more difficult across the lifespan. Despite the evidence for such a view, there are findings that contradict a simple version of the critical period hypothesis. It has become increasingly clear that a classical view of development fails to capture some of the exceptions to this rule (Au, Knightly, Jun, & Oh, 2002; Oh, Au, & Jun, 2010; Oh, Jun, Knightly, & Au, 2003; Pallier et al., 2003; Ventureyra, Pallier, & Yoo, 2004). In this chapter, the nature of age effects in bilingual language development is considered. First, the ways in which age of acquisition (AoA) affects processing in monolinguals are described. This discussion involves the same effects within the bilingual population. Finally, the chapter concludes with a brief discussion of the theoretical implications of this review (for a longer discussion see Hernandez, 2013).

34.2 AGE OF ACQUISITION

In 2007, Hernandez and Li proposed that AoA is related to sensorimotor processing (Hernandez & Li, 2007). The clearest marker of this can be seen in brain development. Neurological changes that serve as markers of this development include neuronal proliferation, neuronal death, and dendritic pruning. Early in development, there is neuronal proliferation along with extensive dendritic connections between neurons. Over time neuronal death and dendritic pruning lead to a great reduction in the number of neurons and connections between them. At the same time, the density of the myelin sheath that helps speed up the electrical signal sent by the axons grows thicker (Campbell & Whitaker, 1986).

Interestingly, these changes during neural development do not occur at the same rate across all areas of the brain, at least not in humans. Unlike many other animal species, humans show a disjointed form of brain maturation. Neurons mature earlier at central points and then fan out to areas that are further and further away. In an attempt to consolidate all available data of the rate at which different brain regions mature, Best (1988) proposed that growth occurs in three different ways: right-to-left, primary-to-secondary-to-tertiary, and basal-to-cortical (from the middle of the brain out to the cortex).

More recent studies using a variety of techniques have confirmed the importance of these axes of brain development. In infants, sensory cortices are the earliest to develop in life. The development of sensory cortex is followed by development of association areas in the parietal lobe and motor cortex in the frontal lobe. The most anterior regions of the brain in the prefrontal cortex are known to develop the latest. Changes in brain structure continue into middle childhood, across adolescence and adulthood, and through older adulthood. These findings have been confirmed by studies using direct observation of brain tissue as well as those using indirect methods of measuring the brain such as

Neurobiology of Language. DOI: http://dx.doi.org/10.1016/B978-0-12-407794-2.00034-1

magnetic resonance imaging (MRI) and functional magnetic resonance imaging (fMRI).

Neuronal changes that occur across development were first investigated by observing the anatomy of neurons in the brains of a few individuals under a microscope (Huttenlocher, 1990, 1994; Huttenlocher & Dabholkar, 1997; Huttenlocher & de Courten, 1987; Huttenlocher, De Courten, Garey, & Van der Loos, 1982a, 1982b). These studies were resonant with the pattern discussed here. Differential development of areas under the microscope revealed changes in the production of synapses, in later synaptic pruning, and in myelination. Production of synapses is strongest in the occipital lobe (i.e., visual regions) of the cortex between 4 and 8 months of age (Huttenlocher & de Courten, 1987). In the frontal lobe, synapse production reaches its peak at 15 months of age (Huttenlocher & Dabholkar, 1997). The lag in the frontal lobes' overproduction of synapses is also observed in the reduction of synapses via pruning (Huttenlocher, 1994).

The advent of newer neuroimaging techniques has allowed researchers to look at brain development *in vivo*. Work using these newer techniques has found evidence that developmental changes in brain areas can be seen as involving both left–right gradient, anterior–posterior gradient, and primary-to-secondary-to-tertiary gradient. Recent neuroimaging studies of single word processing in children have shown changes in the magnitude of neural activity in adults and children centered in areas of the frontal lobe (Schlaggar et al., 2002), the area that has been found to reach adult levels of synaptic connectivity the latest. This is consistent with the view that children have less developed frontal lobes relative to adults.

In a groundbreaking study, Sowell et al. (2003) examined the gray matter density of high-resolution structural MRI scans in a group of individuals who ranged between 7 and 87 years of age. Results revealed a linear decrease in gray matter density across age. Regions in the frontal lobe showed decreases in gray matter density from ages 7 and beyond. The authors attribute some of this to an increase in myelination up until age 40, with a decrease in neural density ensuing into older adulthood. An interesting deviation from this general trend was observed in the left temporal regions (particularly the posterior portions of the middle and superior temporal gyrus), which revealed an increase in gray matter density until age 40. These studies converge with previous work that had found continued changes in left temporal cortices well into adulthood. The late maturation of the temporal lobes, particularly in the superior and posterior parts, is an example of primary-to-secondary-to-tertiary development. In this particular case, it would involve extensions of auditory processing into language processing.

Development affects both the patterns of neural specialization and brain anatomy. The notion of development has also played an interesting role in the bilingual literature as well as the monolingual word recognition literature. In the bilingual literature, it has been used to consider how second language processing differs from first language processing. The development of early sensorimotor effects and later developing frontal lobe functions can also be seen in the word recognition literature. First, we consider the literature with monolinguals before moving on to the bilingual literature in subsequent sections.

34.3 AoA IN A SINGLE LANGUAGE

Catriona Morrison and Andrew Ellis (Morrison & Ellis, 1995) asked participants to read or identify a visual stimulus as a word (i.e., shave) or not (i.e., mave) in a set of visually presented stimuli. They performed two separate manipulations. In one set of experiments, they chose a set of early-learned (fairy) and late-learned words (wharf) that were matched for lexical frequency. In a second set of experiments, they chose a set of high-frequency (market) and low-frequency (pigeon) words that were matched on AoA. By doing this, they could measure to what extent frequency of use or AoA influenced the speed with which a word was read or decisions were made. The results revealed effects of AoA but no effect of word frequency.

These findings were somewhat controversial at the time, because they called into question the results of previous research that had focused extensively on word frequency. These studies had identified word frequency as the primary determinant of speed and accuracy in word recognition experiments (Besner & Smith, 1992; Grainger, 1990; Ostergaard, 1998). Follow-up studies found that AoA and word frequency played a role in speed of picture naming (Meschyan & Hernandez, 2002). Thus, behavioral data, at least in some cases, indicate that word frequency may play a role independently of word AoA.

A second study, in collaboration with Christian Fiebach, Angela Friederici, and Sonja Kotz, sought to investigate the neural correlates associated with lexical AoA. The effect of frequency led to increased neural activity within the inferior frontal gyrus (IFG), a finding that was in line with previously published studies. What was most surprising was the effect of AoA (Fiebach, Friederici, Muller, von Cramon, & Hernandez, 2003). Words learned in early childhood led to activity in Heschl's gyrus and brain areas involved in the processing of speech sounds. Words learned in late childhood relied, to a greater extent, on brain areas in the lower portions of the inferior frontal lobe. These areas have been associated with the effortful access to a word's meaning (Badre, Poldrack, Pare-Blagoev, Insler, & Wagner, 2005; Bookheimer, 2002; Poldrack et al., 1999; Wagner, Maril, Bjork, & Schacter, 2001; Wagner, Pare-Blagoev, Clark, & Poldrack, 2001).

The divide between early-learned and late-learned words in the brain images bears an interesting resemblance to the way in which the brain develops. As discussed, areas in sensory cortex are the earliest to develop and those in the prefrontal cortex develop later. The results from the fMRI study investigating AoA at the word level were fascinating because they suggest that, even as adults, words are accessed in the way that they were learned. Adults were "sounding out" early-learned words in their heads. For late-learned words, their brains revealed that people were using meaning-based links to access their meaning.

34.4 THE RELATIONSHIP BETWEEN AoA AND SENSITIVE PERIODS

The rapid maturation of sensory and motor cortices has also been tied to the long-standing literature regarding the nature of critical periods. Evidence of AoA effects has been found in many nonlinguistic domains. For example, it is well-known that early deprivation or alteration of sensory input leads to impaired sensory perception in many species. This is known from the pioneering work of Hubel and Wiesel, two neurophysiologists who were interested in uncovering how the visual cortex develops. In a series of papers, they outlined how visual deprivation affected the neural structure of both the lateral geniculate nucleus of the thalamus and the primary visual cortex. This work found that young kittens had a sensitive period during which deprivation of visual input would lead to long-term visual impairments (Hubel & Wiesel, 1963; Wiesel & Hubel, 1963a, 1963b). Interruption of visual input in one eye, however, did not impair vision in adult cats. Since Hubel and Wiesel's groundbreaking findings, researchers in the field have found evidence of a sensitive period in cats, monkeys, rats, mice, ferrets, and humans (Banks, Aslin, & Letson, 1975; Fagiolini, Pizzorusso, Berardi, Domenici, & Maffei, 1994; Harwerth, Smith, Duncan, Crawford, & von Noorden, 1986; Huang et al., 1999; Issa, Trachtenberg, Chapman, Zahs, & Stryker, 1999; Olson & Freeman, 1980). Similar research also reveals critical period effects in the calibration of the auditory map by visual input (Brainard & Knudsen, 1998).

Sensory deprivation can also lead to problems in the motor system. For example, disruption of binocular experience can lead to problems that adversely affect smooth pursuit of moving objects and can lead to eyes drifting when viewing stationary targets (Norcia, 1996). Hence, problems in the sensory domain lead to abnormalities of motor function.

The effects of AoA have also been seen in the learning of birdsongs, which has been conceptualized as involving three phases, sensory, sensorimotor, and crystallized (Brainard & Doupe, 2002). During the sensory period, a bird listens to the song of a tutor to form a template. Lack of exposure to an adult song during this phase leads to irregular songs that contain some species-specific characteristics. In the second sensorimotor phase, the bird learns to match the song to the template. During this phase, songbirds fine-tune their songs through practice. Auditory feedback is crucial during this time. In the final crystallized phase, birds are mature and can produce the species-specific song but may no longer be able to learn other songs. In short, learning of birdsongs also points to a sensorimotor basis for this early form of learning followed by a longer period in which a bird is no longer able to produce different types of songs.

Finally, differences in AoA have also been observed in higher-level nonlinguistic functions. Work with musicians has found effects of AoA at both the behavioral and neural levels. There is evidence that absolute pitch can only be learned by speakers of nontonal languages before the age of 7 (Deutsch, Henthorn, Marvin, & Xu, 2006; Trainor, 2005). In addition, the ability to synchronize motor responses to a visually presented flashing square has been found to differ significantly between groups of professional musicians as a function of AoA, even when these groups are matched for years of musical experience, years of formal training, and hours of current practice (Watanabe, Savion-Lemieux, & Penhune, 2007). Work using fMRI has found that there is evidence that early musical training correlates with the size of the digit representation in motor regions of the cortex (Elbert, Pantev, Wienbruch, Rockstroh, & Taub, 1995). In a similar vein, Schlaug, Jancke, Huang, Staiger, and Steinmetz (1995) found the anterior corpus callosum (one of two tracts that connects the right and left hemispheres) to be larger in musicians than nonmusicians, and largest for those who learned to play before the age of 7. Hence, the effects of AoA on behavior as well as neural representations in the music domain can be viewed as reflecting sensorimotor processing.

34.5 AoA AND SECOND LANGUAGE LEARNING

So far, I have considered how reliance on sensorimotor processing changes as a function of AoA. Results from a number of studies have found that as brain maturation expands from primary to tertiary areas within sensorimotor cortex and extends into the frontal lobe, individuals begin to process information differently. The neurological changes that occur are also accompanied by changes in the concomitant behavior. This is captured in the way that researchers have conceptualized the learning of grammar as being based on the sounds of a particular language. It is also seen in the fact that sensitive periods during early development are observed mostly in basic sensory and motor functions. In the case of language, it involves multiple levels of processing that build on each other.

The fact that different aspects of language are differentially sensitive to age has also appeared in the second language learning literature. Second language learners often struggle learning to master the phonology and grammar of a new language. Furthermore, this effect changes as a function of age. The following section considers to what extent the effects of AoA on second language learning can also be conceptualized as involving sensorimotor processing.

34.6 PHONOLOGY IN A SECOND LANGUAGE

A number of studies have found that accent is related to the age of L2 acquisition. Asher and Garcia (1969) conducted a seminal study in this respect. A group of Cuban adults who had come to the United States between the ages of 6 and 19 were recorded as they spoke in English. A set of raters then decided how native-like their accents were. The results revealed a strong relationship between degree of nativeness and age of L2 learning. Since then, a number of studies have been conducted that show a clear linear relationship between AoA and second language accent (Piske, MacKay, & Flege, 2001).

If phonological development continues to occur across childhood and results in systems that are less and less able to adapt to a nonnative speech sound system, then we should expect that adults will also show these developmental effects. To investigate the link between language learning history and development of second language phonology, Susan Guion and colleagues have conducted a series of studies investigating the production of sounds. Guion's findings indicate that the independence of vowel categories in bilingual adults is dependent on when they learn their second language (Guion, 2003). In an early study, a group of Quichua–Spanish bilinguals were asked to pronounce a series of vowels in both languages. Quichua is a native dialect spoken in the area near Quito, Ecuador. Spanish has five vowels, a, e, i, o, and u. Quichua only has three vowels, which correspond to the Spanish a, i, and u. There are also differences in the ways in which these vowels are spread out. Figures of these vowels reveal a triangular shape, with each vowel in Quichua corresponding to each vertex. To accommodate two additional vowels, Spanish stretches the i and u further away, making space for an e and an o. This results in a compressed vowel space in Quichua relative to Spanish.

Not surprisingly, monolinguals of each language will speak their non-native language with an accent either because they are spreading the vowels out too far away or because they are constricting them too much. Quichuan accents will appear in Spanish because of the fact that the latter has more vowels. They will tend to produce a Spanish e and i or an o and u using a sound that lies between these two categories. Similarly, Spanish speakers will tend to slur Quichuan vowels by spreading them too far. In the case of Quicha and Spanish, accents appear either because of the fact that too many vowels are trying to be constrained into a space or because too few vowels are trying to be expanded into the same space.

Guion tested a group of Quichuan–Spanish bilinguals who had learned Spanish at various ages. The results revealed a strong relationship between the AoA of each language and a particular participant's ability to pronounce the vowels. Simultaneous bilinguals were able to pronounce vowels in both languages accurately. Bilinguals who learned a second language between the ages of 5 and 7 were close to native-like for both languages. As the age of second language increased, the ability to represent both sets of speech sounds diminished. The most extreme case of this was in the individuals who learned a second language between ages 15 and 25. In this group, there was a strong bias toward using the first language to guide pronunciation of vowels in the second language. That is, late learners tend to build their second language sound system around an already established first language system. Hence, the second language is more dependent on the first language in late learners.

Results from Guion's studies revealed that independence of the two systems is greatest in simultaneous bilinguals but much less so in sequential bilinguals. Specifically, those who learned a second language after 5 years of age exhibit differing amounts of dependence on the first language, depending on the age at which the L2 was learned. Childhood bilinguals show a mild dependence of L2 on L1, whereas the late bilinguals' L2 vowel categories are very dependent on L1. In short, the older an individual is at the age when beginning to learn a second language, the more the first language sound system is entrenched.

So far, evidence that exposure to an L2 is driven by AoA has been presented. Infant learners show incredible plasticity in the phonological system (Eimas, Siqueland, Jusczyk, & Vigorito, 1971; Kuhl et al., 2008; Werker & Curtin, 2005). They can learn to recognize sounds from another language even when that speech sound was lost during the first year (Kuhl, Tsao, & Liu, 2003). Children who learn a second language have softer accents than adult learners. One question that arises is how this dependence may play out in the recognition of phonemes by people with different second language AoAs.

Despite this early entrenchment, there is considerable evidence that later learners can recognize sounds in a second language (Flege). In one study, Flege, Munro, and MacKay (1995) tested a group of Italians who immigrated to Canada between the ages of 2 and 13 (early learners) or between the ages of 15 and 26 (late learners). The results revealed that early learners, in general, could distinguish English vowel sounds better than late learners. However, there was considerable variability. For example, early learners who used Italian often were found to differ from native speakers, whereas those who used it seldom were not. Furthermore, some of the late learners were able to distinguish these vowel sounds. Thus, early acquisition does not appear to be a guarantee that native-like L2 performance will be observed when recognizing speech sounds. Furthermore, it does appear that some late learners can learn to distinguish these sounds. The question is how late learners achieve speech recognition in their second language.

To uncover the mechanisms that differentiate early and late learning of a language, Archila-Suerte and colleagues asked a group of Spanish speakers who learned English at different points in childhood to rate whether two sound pairs were similar or different (Archila-Suerte, Zevin, Bunta, & Hernandez, 2012). The sounds were chosen such that some English sounds were similar to sounds in Spanish (SAF and SEF) and others were much less similar (SOF and SUF). Pairs of sounds were presented and people had to decide how similar the two sounds were using four different buttons that were marked as really similar, somewhat similar, somewhat different, or really different. Age of learning played a role when our participants had to decide that two items from the same category (SAF–SAF) were the same. Early learners, those who learned English before the age of 5, like monolinguals, showed four tight clusters for each sound. However, intermediate learners, who acquired English between ages 6 and 9, showed distinct but blurry clusters. Late learners, who acquired English after the age of 10, showed even blurrier clusters. Early learners, like monolinguals, were able to ignore slight variations in a sound to form a single sound category. Late learners used a very different strategy; they used a relative comparison strategy to compare different sounds. Hence, like Flege and MacKay noted, late learners can be quite good at distinguishing between different phonological categories. However, the processes that late learners use for L2 speech sound recognition differ from those seen in early learners and monolinguals.

34.7 AoA AND THE BILINGUAL BRAIN

One of the first studies looking at bilinguals using positron emission tomography (PET) was conducted by Daniela Perani and colleagues. In this study, a group of late Italian–English bilinguals were asked to listen to stories in Italian, in English, or in an unknown language. When listening to stories in Italian, the native language, participants revealed larger areas of activity that extended from the temporal lobes up into Broca's area. Many of the same areas were active when participants listened to the second language, English. However, the extent of this activation was reduced. Perani et al. suggest that later learned languages engage a smaller network than native languages when bilinguals listen to speech. Hence, the age at which an individual learns a particular language results in brain activity differences observed using PET.

In 1997, Kim and colleagues used fMRI to look at the nature of second language AoA in bilinguals *Nature* (Kim, Relkin, Lee, & Hirsch, 1997). While in the scanner, they were asked to say "in their head" what they had done the day before. They were given a cue that would tell them to think about something that happened at a particular time of day (morning, afternoon, or evening) and in a particular language (first or second). Researchers observed neural activity in both Broca's and Wernicke's areas to look at whether each language showed two distinct areas of activity or just one glob.

In Wernicke's area, both languages showed an overlapping area of activity in all the people tested whether they learned their second language early or late in life. Kim et al., however, found something different when comparing neural activity in Broca's area for early and late learners. In early bilinguals, brain activity for both languages was mostly overlapping. Brain activity for late bilinguals, however, showed two adjacent but clearly separate areas of activity. The findings from the study by Kim et al. have implications for results that we discussed previously. Specifically, the authors suggest that different areas of Broca's area may be a reflection of the specialization in producing language at an early age. Hence, early bilinguals may come to represent the motor programs for each language in overlapping areas because of the fact that they learn to speak both languages early in life. Late bilinguals learn one system first and then a second. The findings of Kim et al. support the view that once a first language is already firmly established, the second language builds around it in adjacent but clearly distinct areas.

The results from the Kim et al. study leave us with a few unanswered questions. The most important of these is, what aspect of the task leads to differences in neural activity across these two groups? The task used by Kim et al., with participants reciting to themselves what they did the day before, could involve a number of aspects of cognitive processing that go beyond simple motor planning. This task involves a memory component in that each participant had to recall activities of the previous day in a particular language. This task also involved self-talk in each language. Because we do not have a recording of exactly what each participant was saying, it is possible that participants may have produced a different amount or quality of speech in each language. Furthermore, it is unclear whether the retrieval of a memory in conjunction with having to turn it into a verbal code might have been responsible for the differences across groups. Finally, participants were producing sentences in each language. It is possible that differences in the way in which sentences were put together might have differed across groups and languages. Hence, we are not sure what aspect of language may have differed across languages. One possibility is that it was grammatical processing that differed, a point that we turn to next.

34.8 GRAMMATICAL PROCESSING AND AoA

The influence of AoA on grammatical processing has also been approached using neuorimaging methods. One of the earliest studies was performed by Weber-Fox and Neville (1996). In that study, a group of second language speakers who had learned English between the ages of 2 and 16 using electroencephalography (EEG) were recruited. Weber-Fox and Neville sought to test whether two EEG components might differ in learners who differed in their age of second language acquisition. The first component is a wave that peaks at approximately 200 msec after a stimulus has been presented. This early negative wave appears over left frontal sites and is hence termed the early left anterior negativity (Friederici, Hahne, & Mecklinger, 1996; Friederici & Meyer, 2004). It is commonly seen when listeners are presented with sentences that have phrase structure violations such as "The pizza was in the eaten." (Friederici et al., 1996; Friederici & Meyer, 2004). The second wave, a positivity that peaks at approximately 600 msec, has been found to occur in sentences that have grammatical violations and also in sentences that temporarily lure the reader into the wrong interpretation (Osterhout & Holcomb, 1992; Osterhout, Holcomb, & Swinney, 1994). For example, the sentence "The broker persuaded to sell the stock was tall" leads people astray because a noun is expected after persuaded (i.e., persuaded me, the audience, or the seller).

Weber-Fox and Neville asked participants to read sentences that had grammatical violations such as "The scientist criticized Max's **of** proof the theorem." These sentences were designed to go wrong at the bolded word. Not surprisingly, learning English later in childhood modulated the size of early and late component EEG waves relative to monolinguals. The early negativity that appears in sentences with blatant errors in the parts of speech used (i.e., The scientist criticized Max's **of** proof the theorem) was reduced in all second language learners, even when the age of L2 immersion occurred during the first 3 years of life. Furthermore, there was no indication of the brain signatures associated with reanalysis in individuals older than age 11. For those who learned the language between the ages of 1–10, however, this later component was present. The presence of a later component suggests that individuals who learn a second language during early or middle childhood perform reanalysis in a manner similar to that of monolinguals. In short, the early components of the brain waves were reduced in all second language learners. The late components were reduced in those who learned a second language after the age of 10.

In 2011, Eric Pakulak and Helen Neville followed-up this study by looking at the brain waves associated with grammatical errors in a group of late learners of English (who learned English between ages 10 and 12) who were matched to native English speakers in proficiency. Results revealed that native speakers showed both an early negativity and a late positivity for grammatical errors. Late learners, however, did not show such an early negativity. Furthermore, the late positivity showed a widespread pattern across the brain and extended for a longer time than that seen in native speakers. Similar results have also been observed by Manolo Carreiras, Horacio Barber, and colleagues (Dowens, Vergara, Barber, & Carreiras, 2010). These two studies confirm that late learners use other mechanisms to process grammar relative to native speakers.

Taken together, these three studies suggest that learning a second language, even during childhood, does not lead to a native-like pattern of electrical brain activity. The complete lack of an early negativity regardless of second language (L2) AoA suggests that automatic computation of grammatical processing may not occur in second language learners (for an alternative finding see Rossi, Gugler, Friederici, & Hahne, 2006).

Finally, one study that looked at artificial grammar learning showed that late learners are more likely to show an anterior negativity-P600 pattern of response after encountering a grammatical error when they have achieved relatively higher proficiency in that language (Morgan-Short, Steinhauer, Sanz, & Ullman, 2012). The use of a language that has the same phonotactics of the learners' language circumvents some of the difficulties associated with natural language acquisition. At the same time, phonological processing has been thought to play a role in early grammatical development (Demuth, Patrolia, Song, & Masapollo, 2012; Morgan & Demuth, 1996). Hence, the use of these artificial grammars may eliminate the varying source that distinguishes native and non-native speakers. It is important to note that learners of an artificial grammar did not reveal early negativity. The most common pattern across the majority of studies reveals that late learners do show later components of reanalysis necessary for controlled grammatical processing. Thus, detecting grammar errors in a second language relies to a greater extent on later components instead of early computation. Finally, there is some evidence that learning a second language later in life is associated with electrical activity in a larger brain network during reanalysis of grammatical errors relative to the network seen in native speakers.

In 2003, my colleague Isabell Wartenburger along with a group of other researchers, including Daniela Perani, wondered whether AoA might also influence the nature of brain activity using fMRI (Wartenburger et al., 2003). Building on Perani's previous work (Perani et al., 1998), Wartenburger compared groups of Italian–German bilinguals who had learned their second language either late or early in life. Participants were shown sentences that had errors of case, number, or gender in German as well as errors in number or gender in Italian. The results revealed increased activity for late bilinguals relative to early bilinguals in the prefrontal cortex in a region that was just above Broca's area. This superior portion of Broadmann area 44 is typically associated with the need to retrieve the speech sounds of a word (Poldrack et al., 1999). Hence, it appears that late learners appeared to access the speech sound of the article of a word to check whether it was correct. The early learners, however, showed no difference in the brain's blood metabolism when shown sentences with grammatical errors.

34.9 ISOLATING AoA

One issue that arises with studies that have looked at the brain bases of bilingualism is that AoA is often confounded with language proficiency. In the case of two studies reviewed earlier, the Perani et al. study involving language comprehension and the Kim et al. study involving production, there are potential confounds with language proficiency. Traditionally, researchers have used two complimentary approaches to isolate effects of proficiency. The first approach involves the use of a statistical method in which the effect is minimized by using regression. The second approach is to create two groups that differ in their second language AoA but are identical in proficiency. This was the approach used by Wartenburger and colleagues.

Both of these comparisons have some limitations. When comparing monolinguals and bilinguals, we are left with the possibility that any differences in the latter group have to do with speaking two languages and not with having learned one language later. That is, it is unclear if the difference between these groups might simply be due to the comparison between individuals who learned a second language with those who did not. In addition, comparing second language learners with different AoAs does not resolve the question of whether a first and second language differ due to AoA. An important complementary approach would be to compare performance in a second language with that of a first language while equating for language proficiency across groups in both languages. Although Pakulak and Neville did this with event-related potentials (ERPs) in 2011, there has been very little work exploring this approach using fMRI.

Work by Hernandez and colleagues had picked up on a different pattern of English dominance in a very different group of subjects. For many years, data had revealed that when testing Spanish native speakers who became bilingual in childhood but became dominant in their second language, English. This dominance in English was found in tasks such as picture naming or word reading (Hernandez, Bates, & Avila, 1996; Hernandez & Kohnert, 1999; Hernandez & Reyes, 2002). In addition, Hernandez and colleagues were able to find a group of late Spanish learners who were highly proficient. This allowed the comparison of two groups who showed very similar language proficiency profiles but differed in AoA.

34.10 AoA EFFECTS DURING GRAMMATICAL PROCESSING

The nature of grammatical processing in bilinguals had been a topic considered by Elizabeth Bates and colleagues during the 1980s and 1990s. Of particular interest was uncovering how different languages differed in their grammatical properties. For example, we could take a sentence like "The dog are kicking the cows" and ask native English speakers to choose the actor or subject of the sentence. The results reveal that even though there is a grammatical violation in subject–verb agreement, the majority of English speakers still choose the dog as the actor. Why is this? English is highly reliant on word order to indicate the subject of a sentence. There are some exceptions to this rule, but for the most part English speakers use the position of a noun in the sentence to determine its role.

Not all languages are built the same. Romance languages rely, to a much greater extent, on the verb to indicate the subject of a sentence. Other languages have particular case markings on a noun or its determiner to indicate a noun's role. In addition, morphology is particularly vulnerable even in languages that utilize it frequently for grammatical purposes (Bates, Friederici, & Wulfeck, 1987). The variability across languages leads to a complex mapping issue in which grammatical functions appear to be senseless to those outside of the language. Because of this variability and the complexity of mapping across languages, it is very difficult to show transfer from the grammar of one language to the other (MacWhinney, 2004).

One example that captures this complexity well is grammatical gender. For an English speaker, the notion that grammatical items are masculine or feminine is grounded purely in semantics. The tendency to break-up the world into masculine and feminine categories extends beyond nouns that refer to sex-based gender. Interestingly, this bias also appears in children who have a tendency to judge artificial categories (i.e., buildings, elevators, and cars) as masculine and natural categories (flowers, trees, and birds) as feminine (Sera, Berge, del Castillo Pintado, 1994). The conceptual basis of nongrammatical gender also appears across languages (Boroditsky, Scmidt, & Phillips, 2003; Konishi, 1993). Finally, semantic gender may also be marked explicitly on personal pronouns (he, she, and it) or on nouns such as waiter and waitress.

However, certain languages extend the semantic nature of gender further. In romance languages, for example, the gender of a noun can be masculine or feminine and must agree with an adjective that qualifies it. This can be seen in Spanish phrases such as *el carro rojo*, which translates as the_{masc} car_{masc} $\text{Red}_{\text{masc.}}$ People who speak nongender-marked languages such as English often wonder why grammatical gender even exists. Liz Bates argued that gender allowed a speaker to identify a noun without mentioning it (Bates, Devescovi, Hernandez, & Pizzamiglio, 1996; Bates, Devescovi, Pizzamiglio, D'Amico, & Hernandez, 1995). Spanish speakers may point to a pencil and a pen while exclaiming, *pasamela*, pass it to me. With enough proficiency, a Spanish speaker would pass the person a pen, *la pluma*, and not the pencil, *el lapiz*. In short, for a speaker of a gender-marked language, gender helps to reduce the search space of possible nouns.

The final dimension of interest is that grammatical gender, at least in Spanish, also has a regularity dimension to it. Nouns with regular gender carry an -o ending for masculine nouns (i.e., carro), and others carry an -a ending for feminine nouns (i.e., casa). However, other words that end in other letters (such as s, t, z, n, r, e) are much less regular. These words can be masculine or feminine. For example, la fuente (the fountain) is feminine, whereas el puente (the bridge) is masculine. The regularity of these mappings allows researchers to see whether the lack of direct connection between ending and gender mattered.

In a first study, a group of monolingual Spanish speakers judged whether a noun in Spanish was masculine or feminine while being scanned with fMRI (Hernandez et al., 2004). In every block of items, half of the items were masculine and half were feminine. In addition, we had blocks with regular gender signaled by an -a or -o ending and blocks with irregular gender markings that were signaled by words ending in s, z, n, r, t, z, or e.

Gender decisions for irregular nouns in Spanish led to increased neural activity in three distinct areas, the anterior insula, as well as a superior and inferior portions of Broca's area, relative to regular nouns. In addition, there was increased activity in the anterior cingulate gyrus, an area that is involved in tasks that require increased cognitive effort. The role of the anterior insula in articulation has been well-documented. Studies with patients who have difficulty with articulation, the physical movements associated with speech, have found that all of them have damage in the anterior insula, which lies just inside the surface of the cortex adjacent to Broca's area (Bates et al., 2003; Dronkers, 1996). Broca's area has also been associated with different aspects of speech. The lower portion has traditionally been associated with motor planning of speech (Broca, 1988; Graves, 1997). The upper portion has been found to be active in tasks that require people to access the speech sounds of a word. This area showed more activity when German native speakers were asked to produce the article (masculine, feminine, or neuter) when looking at the picture of a noun (Heim, Opitz, & Friederici, 2002).

From a psycholinguistic point of view, subjects were engaged in deeper phonological retrieval and more complex motor planning while using more cognitive effort. Irregular items are more difficult because they require monolingual Spanish speakers to create a small noun phrase when deciding whether a word is masculine or feminine. This involves accessing the speech sounds, planning a motor response, and preparing to articulate the word as revealed by the neural activity in the superior and inferior portions of Broca's area as well as the anterior insula. Finally, there was increased activity in the anterior cingulate gyrus that served as an indicator of the added cognitive effort required in this irregular condition. Hence, the brain signature associated with irregular items is substantially different than that observed for regular items. This is true even in speakers who have only been exposed to one language in their lives.

34.11 COMPARING FIRST AND SECOND LANGUAGES

Grammatical gender does not come easily to English speakers, although it can be learned (Morgan-Short, Sanz, Steinhauer, & Ullman, 2010). One interesting question is whether grammatical gender can also be learned by those who have uneven exposure to a language. To assess the influence of early exposure, Au and colleagues tested a group of overhearers who had been exposed to a language in childhood but never spoke it (Oh et al., 2010, 2003). This group showed sensitivity to speech sounds above those seen in individuals who had never been exposed to a language. Subsequent studies explored grammatical processing in this population by presenting two objects (a red cow or some red cows) and asking a group of overhearers to form a sentence with them (Au, Oh, Knightly, Jun, & Romo, 2008). Au made this task challenging because grammatical agreement in Spanish involves both number and gender. For example, the red cows would be *las vacas rojas.* Although the plural extends beyond the noun in Spanish, notice that it is marked by an -s, similar to the way it is done in English. However, overhearers also had to match the gender of the determiner, noun, and adjective. By looking at these two different ways to signal grammar, Au et al. could determine whether early exposure benefits grammatical functions that are not present in the dominant language. Interestingly, overhearers, like low-proficiency non-native speakers, were unable to correctly identify gender errors (la vaca rojo) but were able to identify errors in number (las vacas roja). These results show that early overhearing may help with the basic elements of a sound system in a particular language. However, hearing a language does not help with all aspects of a language. In this case, grammatical rules that are particular to a language are not learned when a child is exposed to that language in a passive way during early childhood.

Work in my laboratory sought to exploit the difference between English and Spanish to understand AoA effects better. Because gender does not exist in its grammatical form in English, the experience of non-native speakers would have differed substantially from that of native speakers. Specifically, native speakers would have early exposure to the grammatical rule, whereas native English speakers who learned Spanish in early adolescence would not. Hence, the fact that **grammatical** gender did not transfer easily from one language to the other allowed us to test the effects of AoA on grammar directly (MacWhinney, 2004).

In addition to the native English speakers who learned Spanish in adolescence, a group of early Spanish–English bilinguals was also recruited. This group had early exposure to Spanish at home but extensive exposure to English at school. Over time, education and immersion in the second language led to better language proficiency in English. This allowed us to match both early and late learners regarding English and Spanish proficiency. The greatest contributor to differences in brain activity would be the age at which Spanish was learned and not the particular proficiency in either language.

The results revealed increased activity for irregular items compared with regular items in the inferior portions of Broca's area. The surprising part was that both areas were adjacent to each other. The early Spanish learners showed activity in the area that was seen in monolinguals, one that is thought to be involved in motor planning. The late Spanish learners showed activity just below this. The activity was so extensive in the late learners that it spread into an adjacent portion that shows activity when people engage in retrieving the meaning of words.

These neural signatures clearly show that when a second language is learned, it has very different effects on people, even when ability in that language is matched across groups. For early learners, processing of gender involves more extensive motor planning much like monolingual Spanish speakers. Late learners, however, need to engage in a network that is involved in meaning retrieval. The problem is that this semantic extension is a less efficient way to represent grammatical gender. In short, English speakers appear to take advantage of existing language representations that they already have and then adapt them to learn Spanish during adolescence. The late learners tested in this study did not differ behaviorally from our early learners. Therefore, late learners can manage to retrieve gender correctly. However, their brains have to work harder and differently than that of native speakers.

34.12 AoA AND DEVELOPMENT

Earlier theoretical work in collaboration with Ping Li had suggested that early learning was associated with sensorimotor processing. There is clear evidence to support this view. Processing of sensorimotor information is more privileged early in life. Lack of exposure to certain types of sensory information can lead to lifelong deficiencies. In a similar vein, deprivation in the motor domain can also lead to lifelong deficiencies, but only when it occurs early in life. One implication of this fits nicely with emergentist views of language development (Elman, 1995; Elman, Bates, Johnson, & Karmiloff-Smith, 1996; MacWhinney, 1999, 2002a, 2002b, 2004). Specifically, these views suggest that development provides the tools with which to build from the bottom up. Children learn language differently than adults; hence, early acquisition can lead to a different pattern of bilingualism relative to late acquisition. The interesting part is that late acquisition does occur and can be quite successful. Adults have both the bottom-up route used by children that is based on sensorimotor processing and a top-down route that involves declarative forms of memory (Ullman, 2001, 2005) as well as cognitive control mechanisms (Abutalebi, 2008; Green & Abutalebi, 2008). Hence, we could define this as the difference between sensorimotor processing and controlled processing across development. It is mapping out how these two different types of processing interact across time that will continue to yield interesting results for language development and for the nature of bilingual language processing (for a longer discussion see Hernandez, 2013).

References

Abutalebi, J. (2008). Neural aspects of second language representation and language control. *Acta Psychologica, 128*(3), 466–478.

Archila-Suerte, P., Zevin, J., Bunta, F., & Hernandez, A. E. (2012). Age of acquisition and proficiency in a second language independently influence the perception of non-native speech. *Bilingualism: Language and Cognition, 15*(1), 190–201.

Asher, J. J., & Garcia, R. (1969). The optimal age to learn a foreign language. *Modern Language Journal, 53*(5), 334–341.

Au, T. K.-f., Knightly, L. M., Jun, S.-A., & Oh, J. S. (2002). Overhearing a language during childhood. *Psychological Science, 13*(3), 238–243.

Au, T. K.-f., Oh, J. S., Knightly, L. M., Jun, S.-A., & Romo, L. F. (2008). Salvaging a childhood language. *Journal of Memory and Language, 58*(4), 998–1011. Available from: http://dx.doi.org/10.1016/j.jml.2007.11.001.

Badre, D., Poldrack, R. A., Pare-Blagoev, E. J., Insler, R. Z., & Wagner, A. D. (2005). Dissociable controlled retrieval and generalized selection mechanisms in ventrolateral prefrontal cortex. *Neuron, 47*(6), 907–918.

Banks, M. S., Aslin, R. N., & Letson, R. D. (1975). Sensitive period for the development of human binocular vision. *Science, 190*, 675–677.

Bates, E., Devescovi, A., Hernandez, A. E., & Pizzamiglio, L. (1996). Gender priming in Italian. *Perception and Psychophysics, 58*(7), 992–1004.

Bates, E., Devescovi, A., Pizzamiglio, L., D'Amico, S., & Hernandez, A. E. (1995). Gender and lexical access in Italian. *Perception and Psychophysics, 57*(6), 847–862.

Bates, E., Friederici, A., & Wulfeck, B. (1987). Grammatical morphology in aphasia: Evidence from three languages. *Cortex, 23*, 545–574.

Bates, E., Wilson, S. M., Saygin, A. P., Dick, F., Sereno, M. I., Knight, R. T., et al. (2003). Voxel-based lesion-symptom mapping. *Nature Neuroscience, 6*(5), 448–450.

Besner, D., & Smith, M. (1992). Models of visual word recognition: When obscuring the stimulus yields a clearer view. *Journal of Experimental Psychology: Learning, Memory, and Cognition, 18*, 468–482.

Best, C. T. (1988). The emergence of cerebral asymmetries in early human development: A literature review and a neuroembryological model. In D. L. Molfese, & S. J. Segalowitz (Eds.), *Brain lateralization in children: Developmental implications* (pp. 5–34). New York, NY: Guilford.

Bookheimer, S. (2002). Functional MRI of language: New approaches to understanding the cortical organization of semantic processing. *Annual Review of Neuroscience, 25*, 151–188.

Boroditsky, L., Scmidt, L. A., & Phillips, W. (2003). Sex, syntax and semantics. In D. Gentner, & S. Goldin-Meadow (Eds.), *Language in mind: Advances in the study of language and thought* (pp. 61–79). Cambridge, MA: MIT Press.

Brainard, M., & Knudsen, E. (1998). Sensitive periods for visual calibration of the auditory space map in the barn own optic tectum. *Journal of Neuroscience, 18*, 3929–3942.

Brainard, M. S., & Doupe, A. J. (2002). What songbirds teach us about learning. *Nature, 417*(6886), 351–358.

Broca, P. (1988). On the speech center. In L. T. Benjamin (Ed.), *A history of psychology: Original sources and contemporary research*. New York, NY: McGraw-Hill.

Campbell, S., & Whitaker, H. A. (1986). Cortical maturation and developmental neurolinguistics. In J. E. Obrzut, & G. W. Hynd (Eds.), *Child neuropsychology: Vol. I. Theory and research* (pp. 55–72). New York, NY: Academic Press.

Demuth, K., Patrolia, M., Song, J. Y., & Masapollo, M. (2012). The development of articles in children's early Spanish: Prosodic interactions between lexical and grammatical form. *First Language, 32*(1–2), 17–37. Available from: http://dx.doi.org/10.1177/0142723710396796.

Deutsch, D., Henthorn, T., Marvin, E., & Xu, H.-S. (2006). Absolute pitch among American and Chinese conservatory students: Prevalence differences, and evidence for a speech-related critical period. *Journal of the Acoustical Society of America, 119*, 719–722.

Dowens, M. G., Vergara, M., Barber, H. A., & Carreiras, M. (2010). Morphosyntactic processing in late second-language learners. *Journal of Cognitive Neuroscience, 22*(8), 1870–1887.

Dronkers, N. F. (1996). A new brain region for coordinating speech articulation. *Nature, 384*(6605), 159–161.

Eimas, P. D., Siqueland, E. R., Jusczyk, P., & Vigorito, J. (1971). Speech perception in infants. *Science, 171*(3968), 303–306.

Elbert, T., Pantev, C., Wienbruch, C., Rockstroh, B., & Taub, E. (1995). Increased cortical representation of the fingers of the left hand in string players. *Science, 270*, 305–307.

Elman, J. L. (1995). Language as a dynamical system. In T. v. G. E. Robert, & F. Port (Eds.), *Mind as motion: Explorations in the dynamics of cognition* (pp. 195–225). Cambridge, MA: MIT Press.

Elman, J. L., Bates, E. A., Johnson, M. H., & Karmiloff-Smith, A. (1996). *Rethinking innateness: A connectionist perspective on development*. Cambridge, MA: MIT Press.

Fagiolini, M., Pizzorusso, T., Berardi, N., Domenici, L., & Maffei, L. (1994). Functional postnatal development of the rat primary visual cortex and the role of visual experience: Dark rearing and monocular deprivation. *Vision Research, 34*, 709–720.

Fiebach, C. J., Friederici, A. D., Muller, K., von Cramon, D. Y., & Hernandez, A. E. (2003). Distinct brain representations for early and late learned words. *NeuroImage, 19*(4), 1627–1637.

Flege, J. E., Munro, M. J., & MacKay, I. R. A. (1995). Effects of age of second-language learning on the production of English consonants. *Speech Communication, 16*, 1–26.

Friederici, A. D., Hahne, A., & Mecklinger, A. (1996). Temporal structure of syntactic parsing: Early and late event-related brain potential effects. *Journal of Experimental Psychology: Learning, Memory, and Cognition, 22*(5), 1219–1248.

Friederici, A. D., & Meyer, M. (2004). The brain knows the difference: Two types of grammatical violations. *Brain Research, 1000*(1–2), 72–77.

Grainger, J. (1990). Word frequency and neighborhood frequency effects in lexical decision and naming. *Journal of Memory and Language, 29*, 228–244.

Graves, R. E. (1997). The legacy of the Wernicke–Lichtheim model. *Journal of the History of the Neurosciences, 6*(1), 3–20.

Green, D. W., & Abutalebi, J. (2008). Understanding the link between bilingual aphasia and language control. *Journal of Neurolinguistics, 21*(6), 558–576.

Guion, S. G. (2003). The vowel systems of Quichua–Spanish bilinguals: An investigation into age of acquisition effects on the mutual influence of the first and second languages. *Phonetica, 60*, 98–128.

Harwerth, R., Smith, E., Duncan, G., Crawford, M., & von Noorden, G. (1986). Multiple sensitive periods in the development of the primate visual system. *Science, 232*, 235–238.

Heim, S., Opitz, B., & Friederici, A. D. (2002). Broca's area in the human brain is involved in the selection of grammatical gender for language production: Evidence from event-related functional magnetic resonance imaging. *Neuroscience Letters, 328*(2), 101–104.

Hernandez, A. E. (2013). *The bilingual brain.* New York, NY: Oxford University Press.

Hernandez, A. E., Bates, E., & Avila, L. X. (1996). Processing across the language boundary: A cross modal priming study of Spanish–English bilinguals. *Journal of Experimental Psychology: Learning, Memory, and Cognition, 22*, 846–864.

Hernandez, A. E., & Kohnert, K. (1999). Aging and language switching in bilinguals. *Aging, Neuropsychology and Cognition, 6*, 69–83.

Hernandez, A. E., Kotz, S. A., Hofmann, J., Valentin, V. V., Dapretto, M., & Bookheimer, S. Y. (2004). The neural correlates of grammatical gender decisions in Spanish. *Neuroreport, 15*(5), 863–866.

Hernandez, A. E., & Li, P. (2007). Age of acquisition: Its neural and computational mechanisms. *Psychological Bulletin, 133*(4), 638–650.

Hernandez, A. E., & Reyes, I. (2002). Within- and between-language priming differ: Evidence from repetition of pictures in Spanish-English bilinguals. *Journal of Experimental Psychology: Learning, Memory, and Cognition, 28*(4), 726–734. Available from: http://dx.doi.org/10.1037/0278-7393.28.4.726.

Huang, Z., Kirkwood, A., Pizzorusso, T., Porciatti, V., Morales, B., Bear, M., et al. (1999). BDNF regulates the maturation of inhibition and the critical period of plasticity in mouse visual cortex. *Cell, 98*, 739–755.

Hubel, D. H., & Wiesel, T. N. (1963). Receptive fields of cells in striate cortex of very young, visually inexperienced kittens. *Journal of Neurophysiology, 26*, 994–1002.

Huttenlocher, P. R. (1990). Morphometric study of human cerebral cortex development. *Neuropsychologia, 28*(6), 517–527.

Huttenlocher, P. R. (1994). Synaptogenesis in the human cerebral cortex. In G. Dawson, & K. W. Fischer (Eds.), *Human behavior and the developing brain* (pp. 137–152). New York, NY: Guilford Press.

Huttenlocher, P. R., & Dabholkar, A. S. (1997). Regional differences in synaptogenesis in human cerebral cortex. *The Journal of Comparative Neurology, 387*, 167–178.

Huttenlocher, P. R., & de Courten, C. (1987). The development of synapses in striate cortex of man. *Human Neurobiology, 6*(1), 1–9.

Huttenlocher, P. R., de Courten, C., Garey, L. J., & Van der Loos, H. (1982a). Synaptic development in human cerebral cortex. *International Journal of Neurology, 16–17*, 144–154.

Huttenlocher, P. R., de Courten, C., Garey, L. J., & Van der Loos, H. (1982b). Synaptogenesis in human visual cortex—evidence for synapse elimination during normal development. *Neuroscience Letters, 33*(3), 247–252.

Issa, N., Trachtenberg, J., Chapman, B., Zahs, K., & Stryker, M. (1999). The critical period for ocular dominance plasticity in the ferret's visual cortex. *Journal of Neuroscience, 19*, 6965–6978.

Johnson, J. S., & Newport, E. L. (1989). Critical period effects in second language learning: The influence of maturational state on the acquisition of English as a second language. *Cognitive Psychology, 21*(1), 60–99.

Kim, K. H. S., Relkin, N. R., Lee, K.-M., & Hirsch, J. (1997). Distinct cortical areas associated with native and second languages. *Nature, 388*(6638), 171–174.

Konishi, T. (1993). The semantics of grammatical gender: A cross-cultural study. *Journal of Psycholinguistic Research, 22*(5), 519–534.

Kuhl, P. K., Conboy, B. T., Coffey-Corina, S., Padden, D., Rivera-Gaxiola, M., & Nelson, T. (2008). Phonetic learning as a pathway to language: New data and native language magnet theory expanded (NLM-e). *Philosophical Transactions of the Royal Society of London Series B, Biological Sciences, 363*(1493), 979–1000.

Kuhl, P. K., Tsao, F. M., & Liu, H. M. (2003). Foreign-language experience in infancy: Effects of short-term exposure and social interaction on phonetic learning. *Proceedings of the National Academy of Sciences of the United States of America, 100*(15), 9096–9101.

Lenneberg, A. (1967). *Biological foundations of language.* New York, NY: John Wiley.

MacWhinney, B. (Ed.), (1999). *Emergence of language.* Hillsdale, NJ: Lawrence Erlbaum Associates.

MacWhinney, B. (2002a). Language emergence. In P. Burmeister, T. Piske, & A. Rohde (Eds.), *An integrated view of language development—Papers in honor of Henning Wode.* Trier: Wissenshaftliche Verlag.

MacWhinney, B. (2002b). The gradual emergence of language. In T. Givón, & B. Malle (Eds.), *The evolution of language from prelanguage.* Amsterdam: John Benjamins.

MacWhinney, B. (2004). A unified model of language acquisition. In J. K. A. D. Groot (Ed.), *Handbook of bilingualism: Psycholinguistic approaches.* New York, NY: Oxford University Press.

Meschyan, G., & Hernandez, A. E. (2002). Age of acquisition and word frequency: Determinants of object-naming speed and accuracy. *Memory and Cognition, 30*(2), 262–269.

Morgan, J. L., & Demuth, K. (Eds.), (1996). *Signal to syntax: Bootstrapping from speech to grammar in early acquisition.* Mahwah, NJ: Lawrence Erlbaum Associates Inc.

Morgan-Short, K., Sanz, C., Steinhauer, K., & Ullman, M. T. (2010). Second language acquisition of gender agreement in explicit and implicit training conditions: An event-related potential study. *Language Learning, 60*(1), 154–193. Available from: http://dx.doi.org/10.1111/j.1467-9922.2009.00554.x.

Morgan-Short, K., Steinhauer, K., Sanz, C., & Ullman, M. T. (2012). Explicit and implicit second language training differentially affect the achievement of native-like brain activation patterns. *Journal of Cognitive Neuroscience, 24*(4), 933–947. Available from: http://dx.doi.org/10.1162/jocn_a_00119.

Morrison, C. M., & Ellis, A. W. (1995). Roles of word frequency and age of acquisition in word naming and lexical decision. *Journal of Experimental Psychology: Learning, Memory, and Cognition, 21*(1), 116–133.

Norcia, A. M. (1996). Abnormal motion processing and binocularity: Infantile esotropia as a model system for effects of early interruptions of binocularity. *Eye, 10*, 259–265.

Oh, J. S., Au, T. K.-F., & Jun, S.-A. (2010). Early childhood language memory in the speech perception of international adoptees. *Journal of Child Language, 37*(5), 1123–1132.

Oh, J. S., Jun, S.-A., Knightly, L. M., & Au, T. K.-F. (2003). Holding on to childhood language memory. *Cognition, 86*(3), B53–b64.

Olson, C., & Freeman, R. (1980). Profile of the sensitive period for monocular deprivation in kittens. *Experimental Brain Research, 39*, 17–21.

Ostergaard, A. L. (1998). The effects on priming of word frequency, number of repetitions, and delay depend on the magnitude of priming. *Memory and Cognition, 26*(1), 40–60.

Osterhout, L., & Holcomb, P. J. (1992). Event-related brain potentials elicited by syntactic anomaly. *Journal of Memory and Language, 31* (6), 785–806.

Osterhout, L., Holcomb, P. J., & Swinney, D. A. (1994). Brain potentials elicited by garden-path sentences: Evidence of the application of verb information during parsing. *Journal of Experimental Psychology Learning, Memory, and Cognition, 20*(4), 786–803.

Pallier, C., Dehaene, S., Poline, J. B., LeBihan, D., Argenti, A. M., Dupoux, E., et al. (2003). Brain imaging of language plasticity in adopted adults: Can a second language replace the first? *Cerebral Cortex, 13*(2), 155–161.

Perani, D., Dehaene, S., Grassi, F., Cohen, L., Cappa, S., Dupoux, E., et al. (1996). Brain processing of native and foreign languages. *Neuroreport, 7*, 2439–2444.

Perani, D., Paulesu, E., Galles, N. S., Dupoux, E., Dehaene, S., Bettinardi, V., et al. (1998). The bilingual brain: Proficiency and age of acquisition of the second language. *Brain, 121*(10), 1841–1852.

Pinker, S. (1994). *The language instinct*. New York, NY: William Morrow.

Piske, T., MacKay, I. R. A., & Flege, J. E. (2001). Factors affecting degree of foreign accent in an L2: A review. *Journal of Phonetics, 29*(2), 191–215.

Poldrack, R. A., Wagner, A. D., Prull, M. W., Desmond, J. E., Glover, G. H., & Gabrieli, J. D. (1999). Functional specialization for semantic and phonological processing in the left inferior prefrontal cortex. *NeuroImage, 10*(1), 15–35.

Rossi, S., Gugler, M. F., Friederici, A. D., & Hahne, A. (2006). The impact of proficiency on syntactic second-language processing of German and Italian: Evidence from event-related potentials. *Journal of Cognitive Neuroscience, 18*(12), 2030–2048.

Schlaggar, B. L., Brown, T. T., Lugar, H. M., Visscher, K. M., Miezin, F. M., & Petersen, S. E. (2002). Functional neuroanatomical differences between adults and school-age children in the processing of single words. *Science, 296*(5572), 1476–1479.

Schlaug, G., Jancke, L., Huang, Y., Staiger, J. F., & Steinmetz, H. (1995). Increased corpus callosum size in musicians. *Neuropsychologia, 33*, 1047–1055.

Sera, M. D., Berge, C. H., & del Castillo Pintado, J. (1994). Grammatical and conceptual forces in the attribution of gender by English and Spanish speakers. *Cognitive Development, 9*(3), 261–292. Available from: http://dx.doi.org/10.1016/0885-2014(94)90007-8.

Sowell, E. R., Peterson, B. S., Thompson, P. M., Welcome, S. E., Henkenius, A. L., & Toga, A. W. (2003). Mapping cortical change across the human life span. *Nature Neuroscience, 6*, 309–315.

Trainor, L. J. (2005). Are there critical periods for musical development? *Developmental Psychobiology, 46*(3), 262–278.

Ullman, M. T. (2001). The neural basis of lexicon and grammar in first and second language: The declarative/procedural model. *Bilingualism: Language and Cognition, 4*, 105–122.

Ullman, M. T. (2005). A cognitive neuroscience perspective on second language acquisition: The declarative/procedural model. In C. Sanz (Ed.), *Mind and context in adult second language acquisition: Methods, theory, and practice* (pp. 141–178). Washington, DC: Georgetown University Press.

Ventureyra, V. A. G., Pallier, C., & Yoo, H.-Y. (2004). The loss of first language phonetic perception in adopted Koreans. *Journal of Neurolinguistics, 17*, 79–91.

Wagner, A. D., Maril, A., Bjork, R. A., & Schacter, D. L. (2001). Prefrontal contributions to executive control: fMRI evidence for functional distinctions within lateral prefrontal cortex. *NeuroImage, 14*(6), 1337–1347.

Wagner, A. D., Pare-Blagoev, E. J., Clark, J., & Poldrack, R. A. (2001). Recovering meaning: Left prefrontal cortex guides controlled semantic retrieval. *Neuron, 31*(2), 329–338.

Wartenburger, I., Heekeren, H. R., Abutalebi, J., Cappa, S. F., Villringer, A., & Perani, D. (2003). Early setting of grammatical processing in the bilingual brain. *Neuron, 37*(1), 159–170.

Watanabe, D., Savion-Lemieux, T., & Penhune, V. B. (2007). The effect of early musical training on adult motor performance: Evidence for a sensitive period in motor learning. *Experimental Brain Research, 176*, 332–340.

Weber-Fox, C., & Neville, H. J. (1996). Maturational constraints on functional specializations for language processing: ERP and behavioral evidence in bilingual speakers. *Journal of Cognitive Neuroscience, 8*, 231–256.

Werker, J. F., & Curtin, S. (2005). PRIMIR: A developmental framework of infant speech processing. *Language Learning and Development, 1*(2), 197–234.

Wiesel, T. N., & Hubel, D. H. (1963a). Effects of visual deprivation on morphology and physiology of cells in the cats lateral geniculate body. *Journal of Neurophysiology, 26*, 978–993.

Wiesel, T. N., & Hubel, D. H. (1963b). Single-cell responses in striate cortex of kittens deprived of vision in one eye. *Journal of Neurophysiology, 26*, 1003–1017.

CHAPTER

35

Bilingualism: Switching

Albert Costa[1,2], *Francesca M. Branzi*[1] *and Cesar Ávila*[3]

[1]Center for Brain and Cognition (CBC), Universitat Pompeu Fabra, Barcelona, Spain;
[2]Institució Catalana de Recerca i Estudis Avançats (ICREA), Barcelona, Spain;
[3]Dpt. Psicologia Bàsica, Clínica i Psicobiologia, Universitat Jaume I, Castelló de la Plana, Spain

35.1 INTRODUCTION

The ability to use two languages at will often marvels monolingual speakers and, at the same time, comes naturally to bilingual speakers who appear to do so without difficulty. The critical issue in this context refers to the cognitive processes and brain basis involved in controlling the two languages. In other words, what are the brain mechanisms that allow bilingual speakers to focus on one language while preventing interference from the non-response language?

This issue has been extensively explored by studying the ability of bilingual speakers to switch between languages. The ability to switch back and forth between languages according to the needs of the interlocutor is perhaps the activity that most often strikes monolingual speakers. In this chapter, we review some of the most relevant available evidence regarding the time course and the brain basis of language switching and how this information can inform models of language control. In doing so, we particularly focus on the studies exploring language switching in speech production and on how certain variables, such as language proficiency level, affect the brain network sustaining language switching.

35.2 LANGUAGE SWITCHING: INSTANTIATING THE PARADIGM

Language switching studies in language production usually involve participants naming aloud a series of stimuli (pictures, digits, etc.). There are two main instantiations of this paradigm. In the first one (trial-by-trial switching task), the language in which a given stimulus has to be named is signaled by a cue (e.g., the picture's color) and the response language is randomly assigned to the stimuli. Crucially, there are two main types of trials: trials in which a picture is named in the same language as the language used in the immediately preceding trial (nonswitch trials) and trials in which the target language is different from the one used in the immediately preceding trial (switch trials). The so-called "language switch cost" is computed by comparing the performance in switch versus non-switch trials. In the second instantiation (blocked switching task), the language in which a given picture has to be named is "blocked," that is, participants name a series of pictures in only one language, and then a new block starts in which the other language is used. In the second instantiation, the "language switch cost" is calculated by comparing naming latencies between the different blocks. The argument is that if speech production in a given language (e.g., language A) affects the successive naming in the other language (e.g., language B), then comparing the same language (language B) after and before the use of the other language (language A) should incur cognitive and neural costs. To investigate this question, some authors compared the behavioral performance and the neural effects elicited by naming in language B in a first block of tasks versus naming in language B in a second block of tasks (i.e., after naming in language A in the first block) (Branzi, Martin, Abutalebi, & Costa, 2014).

Both the first and the second instantiations of the language switching task allow computing the "after-effects" of naming in one language on the performance of the other language (Branzi et al., 2014; Guo, Liu, Misra, & Kroll, 2011). However, the majority of the studies on bilingual language control used the trial-by-trial

Neurobiology of Language. DOI: http://dx.doi.org/10.1016/B978-0-12-407794-2.00035-3

instantiation of the language switching[1] task, sometimes accompanied by entire blocks of naming in dominant (L1) and non-dominant (L2) language (blocked conditions) (Christoffels, Firk, & Schiller, 2007).

It has been argued that the use of these two different naming contexts (mixed and blocked) allowed for testing different types of language control ("local" versus "global" control) and various timing aspects of the control processes ("transient" versus "sustained" control). In fact, it has been proposed that language control might be implemented in different ways (Green, 1998). The first one would act through a control process that globally adapts the activation levels of all lemmas in both languages (increasing activation for lexical representations of the target language and decreasing those of the non-target language). The second mechanism would act by locally suppressing the activation of any (specific) non-target language lemmas that escape the global inhibition process (see De Groot & Christoffels, 2006). Thus, bilinguals may use a whole-language control process to suppress a complete language subsystem affecting all lexical representations in that language ("global control") and an additional control process that affects a restricted set of items ("local control").

Beyond the representational scope of bilingual language control, there are also different processes related to the timing of such control (Christoffels et al., 2007). "Transient control" refers to reactive trial-by-trial control applied during the continuous switching between languages/tasks, and it is generally measured through "switch costs." Conversely, "sustained control" refers to an proactive control influenced by the context in which a given language has to be produced. This type of control is applied for maintaining language/task sets throughout the whole task, and it is measured through the so-called "mixing cost", that is, by comparing non-switch trials[2] in mixed versus blocked naming contexts (Braver, Reynolds, & Donaldson, 2003; Christoffels et al., 2007). It has been suggested that both "sustained" and "transient" components of control might be crucial for bilingual language production. In fact, many studies reported these two components as being differently involved in the control of L1 and L2 production.

Before explaining these issues in detail, we first describe the two different instantiations of language switching paradigm and the main relevant findings.

At the behavioral level, trial-by-trial switching paradigm reveals slower response times (RTs) for switch trials than for non-switch trials, indicating that switching languages incurs a behavioral cost (Christoffels et al., 2007; Jackson, Swainson, Cunnington, & Jackson, 2001). Interestingly, however, several studies have revealed that the magnitude of the language switch cost depends, to some extent, on the relative proficiency of the two languages involved in a switching task. When there is a difference between the proficiency in the two languages, switch costs tend to be larger for the stronger language (i.e., L1) than for the weaker language (i.e., L2) (Meuter & Allport, 1999). That is, switching into the easy task (stronger language, L1) is more costly than switching into the difficult task (weaker language, L2). This pattern resembles that observed in domain-general switching tasks in which little language is involved (Martin, Barceló, Hernández, & Costa, 2011; Nagahama et al., 2001; Rubinstein, Meyer, & Evans, 2001) and has been taken to reveal the participation of inhibitory processes in bilingual language control. The Inhibitory Control model (see IC model by Green, 1998) explicitly predicts such inhibitory effects: because both languages are active even when naming in only one language (Colomé, 2001; Costa, Miozzo, & Caramazza, 1999; Kroll, Bobb, & Wodniecka, 2006), a control mechanism suppresses the activation of non-target lexical items, and this suppression is proportional to the amount of activation and potential interference of a given language. To be more specific, speaking in the weaker language (i.e., L2) requires engaging inhibitory control mechanisms that reduce or suppress the activation of the more ready representations of the stronger language (i.e., L1). Consequently, if in the next trial a response in the stronger language (i.e., L1) is required, then the speaker needs to overcome the lingering effects of the inhibition applied in the preceding trial, therefore delaying production (see *Task-set inertia* hypothesis in Allport, Styles, & Hsieh, 1994). Given that inhibitory processes are supposed to be proportional to strength of the language (Green, 1998), one would expect the stronger language (i.e., L1) to be more inhibited than the weaker language (i.e., L2), therefore leading to the asymmetrical switch cost (Meuter & Allport, 1999). Importantly, however, when participants are asked to switch between languages in which they have similar proficiency, then the asymmetrical switch cost goes away and the switch costs for both languages are of similar magnitude when switching not only between L1 and L2 but also between L1 and a much weaker L3 (e.g., Calabria, Hernández, Branzi, & Costa, 2012; Costa, Santesteban, & Ivanova, 2006).

[1]The second instantiation (i.e., the "blocked switching task") is very recent (Branzi et al., 2014; Guo et al., 2011; Misra et al., 2012; Strijkers et al., 2013).

[2]In the literature, "mixing costs" have been assessed by computing either the difference between mixed (switch and non-switch trials) and blocked conditions. (Koch, Prinz, & Allport, 2005; Los, 1996) or the difference between non-switch trials in mixed versus blocked conditions. (Rubin & Meiran, 2005).

These sorts of studies have been rather influential when trying to understand the cognitive processes behind bilingual language control and have highlighted how proficiency may affect such processes. The interpretation of the asymmetrical switch costs in terms of inhibition inflicted on the L1 has been one of the fundamental results used to support the IC model proposed by Green (1998). However, somewhat problematic for the IC account is that symmetrical switch costs have been reported not only for highly proficient bilinguals but also for unbalanced bilinguals during voluntary switching and with long preparation times (cue-stimulus intervals) (Christoffels et al., 2007; Gollan, & Ferreria, 2009; Verhoef, Roelofs, & Chwilla, 2009). Moreover, some other results (Runnqvist, Strijkers, Alario, & Costa, 2012) showed that semantic effects appear to survive language switching, suggesting that words from the non-used language are not inhibited during language production in the other language. Henceforth, alternative explanations that do not rely (or do not only rely) on inhibitory processes have been put forward (e.g., L2 overactivation account,[3] for a review see Koch, Gade, Schuch, & Philipp, 2010, but also "L1 repetition benefit hypothesis" in Verhoef et al., 2009). At any rate, this sort of language switching instantiation is the kind that has been used most often in neuroscientific studies on bilingual language control.

As previously hinted, the use of the other instantiation of the language switch paradigm is very recent and has given special attention to the "after-effects" of naming in one language on the successive use of the other language without mixing the two languages in the same block. In this paradigm, naming latencies are usually compared across groups of participants. For example, while group 1 starts naming a block of pictures in L1 and then names them in L2, another group starts in L2 and subsequently does so in L1. The critical issue here is whether naming performance in the second block departs from naming performance in the first block while keeping the languages constant. Responses in an L2 do not seem very affected by whether they are given in the first block or rather after having named the pictures in an L1 (hence, the second block). However, responses in L1 are actually affected by the order of presentation. L1 naming is hindered after having named the pictures in an L2 (Branzi et al., 2014; Misra, Guo, Bobb, & Kroll, 2012). This effect appears to find a ready explanation in terms of inhibitory processes (see IC model, Green, 1998), according to which naming the pictures in the weaker language entails a strong inhibition of the corresponding representations in the L1, inhibition that then affects negatively subsequent naming in the L1 (see also Meuter & Allport, 1999 for similar explanation in the trial-by-trial language switching task).

Regardless of whether the inhibitory account of these effects is the correct one, these results highlight two important factors to consider when exploring the brain basis of language switching and the corresponding links to bilingual language control. First, the neural correlates of language switching need to be considered in the context of the language proficiency of the bilingual speaker. That is, it is likely that the neural substrates of language switching and, consequently, those of bilingual language control depend on the proficiency level attained by the bilingual speaker in the two languages (Costa & Santesteban, 2004). Second, for cases in which language proficiency is very different in the two languages, special attention needs to be given to the direction of the language switch. That is, it could be that the brain networks involved in switching into the weaker language are different from those involved in switching into the stronger language. It is worth mentioning that besides language proficiency, other sociolinguistic variables such as language exposure and/or frequency of language switching have been shown to affect somehow bilingual language control (Christoffles et al., 2007; Perani et al., 2003; Prior & Gollan, 2011). Therefore, such factors also need to be taken into consideration when assessing the behavioral and neural effects related to bilingual language control in switching tasks.

In the next sections, we review some relevant studies that describe the time course and the neural correlates of bilingual language control.

35.3 EVIDENCE FROM ELECTROPHYSIOLOGY

During the past decade, the event-related potentials (ERPs) technique has been largely used to study the time course of bilingual language control. Behavioral differences between L1 and L2 processing can be explained by calling into account different control mechanisms (e.g., L1 inhibition, L2 overactivation). Hence, ERP studies were designed to reveal whether such behavioral differences between L1 and L2 were due to the

[3]According to "proactive interference" accounts, switch costs would primarily reflect the passive *after-effects* of previous active control processes (i.e., *task-set inertia*), which result in both positive and negative priming of task sets (Allport et al., 1994; Allport & Wylie, 1999; Wylie & Allport, 2000). Hence, regarding the origin of switch costs and related asymmetries, there are at least two equally good explanations within the "proactive interference" accounts (Allport et al., 1994; Allport & Wylie, 1999; Wylie & Allport, 2000). One is that switch costs origin because of previous task inhibition (Green, 1998; Meuter & Allport, 1999). The other possibility is that at the origin of switch costs and related effects, there is a carryover effect of the activation of the previous task on the successive one.

implementation of qualitatively different control mechanisms (as reflected by different effects at different ERP components) or rather to the recruitment of the same control mechanisms but to a different extent (as reflected by smaller or larger effects at the same ERP component).

The most common ERP component associated with language control mechanisms is the so-called N200 component (Christoffels et al., 2007; Jackson et al., 2001; Misra et al., 2012; Verhoef et al., 2009). This negative deflection (peaking at approximately 250–350 ms after stimulus onset) has been generally described as an index of general control processes (Nieuwenhuis, Yeung, & Cohen, 2004). Even though it is still unclear whether the N200 component reflects inhibitory control processes (Falkenstein, Hoormann, & Hohnsbein, 1999) or response conflict monitoring (Bruin, Wijers, & Van Staveren, 2001; Donkers & van Boxtel, 2004; Nieuwenhuis, Yeung, van den Wildenberg, & Ridderinkhof, 2003), in language switching studies it has been often attributed to inhibitory control applied during bilingual speech production (Christoffels et al., 2007; Jackson et al., 2001; Misra et al., 2012; Verhoef et al., 2009).

As previously described, at the behavioral level trial-by-trial language switching paradigm reveals slower RTs for switch trials than for non-switch trials, indicating that switching languages incurs a behavioral cost. An additional observation is that this switch cost becomes asymmetrical when the L2 proficiency is quite different from that of the L1 (Meuter & Allport, 1999). Hence, if the linguistic switch cost originates from the suppression of the non-target language (see IC model, Green, 1998), a greater ERP N200 negativity for switch than for non-switch trials would be expected. Regarding the asymmetries of switch costs, if the switch-related modulation of the N200 component is associated with response suppression during language switching, then there should be an increased modulation (switch–non-switch difference) of this negativity on L2 trials (when the L1 response needs to be inhibited) compared with L1 trials. Interestingly, this result was observed by Jackson et al. (2001), along with longer RTs to switch to L1 than to switch to L2.

In detail, Jackson et al. (2001) studied a group of unbalanced bilinguals who were presented with digits to be named in either their L1 or their L2 in a predictable fashion. The language to be uttered was signaled by a visual cue (the color of the digit). Behavioral results revealed an asymmetrical pattern of switch costs, with larger costs for switching into the L1 than into the L2.[4] ERP results showed an increased N200 negativity for switch versus non-switch trials only when responses were given in the L2.

These findings strongly suggest that inhibitory processes are involved in bilingual language control. However, this conclusion is undermined by the fact that subsequent ERP studies were not able to replicate these observations. In fact, these studies reported somewhat inconsistent results related to the N200 component. Sometimes the N200 effects have been found in the opposite direction, with more negative deflections for L1 non-switch versus switch conditions (Christoffels et al., 2007), whereas other studies failed to observe any modulation of the N200 component regarding switch cost effects (Martin et al., 2013; Verhoef et al., 2009).

In Christoffels et al. (2007), a group of moderately proficient German–Dutch bilinguals named pictures either in their L1 or in their L2 (blocked language conditions), or they switched between languages in an unpredictable manner (mixed language conditions). This experimental design allowed the dissociation of the effects of the "transient" and "sustained" components of language control. Behaviorally, "transient" processes for L1 and L2 were of the same magnitude, as revealed by symmetrical switch costs. ERP results relative to switch costs showed an increased N200 negativity for the L1 only (this effect was absent for the L2) in non-switch trials as compared with switch trials in mixed language conditions. Behavioral "mixing costs" reflecting "sustained" processes revealed an asymmetric pattern: naming latencies in L2 were not affected by naming contexts, whereas naming latencies in L1 were slower in the mixed condition. Furthermore, authors observed an enhanced N200 negativity in the ERPs for non-switch trials compared with blocked conditions, irrespective of the language in the first time-window (275–375 ms). Interestingly, in a second time-window (375–475 ms), authors found enlarged negative ERP amplitudes for blocked compared with non-switch trials (mixed language conditions) for L1, but not for L2. This "frontal negativity effect" found for L1, but not for L2 was taken to reflect that bilinguals control their two languages by adjusting selectively the availability of the L1 only. This result supported by behavioral mixing costs for the L1 only was taken as evidence of sustained inhibitory control (Green, 1998) applied to the L1 to favor L2 production. Conversely, the observation of symmetrical switch costs in RTs and relatively small effects of switching in the ERP data led authors to conclude that their results did not support the presence of reactive inhibition in bilingual language control (Green, 1998).

[4]Note that the asymmetrical pattern of switch costs was due to a difference between the L1 and the L2 for non-switch trials, with faster RTs for L1 than for L2. Instead, RTs for switch trials were identical between the L1 and the L2. Note that these data do not replicate exactly those reported previously by Meuter and Allport (1999). Specifically, Jackson et al. (2001) did not obtain the crossover interaction pattern (i.e., L1 RTs longer than L2 RTs for switch trials) obtained by Meuter and Allport (1999).

Verhoef et al. (2009) obtained a quite different pattern of results regarding the aforementioned studies. In this study, unbalanced Dutch–English bilinguals performed a language switching task where preparation time (cue-stimulus interval) was manipulated to test whether the asymmetries in switch costs were due to L1 slower RTs on switch trials (IC hypothesis, see Green, 1998) or rather to an L1 repeat benefit on nonswitch trials (L1 repeat benefit hypothesis). The results revealed that the occurrence of asymmetrical or symmetrical switch costs did not depend on language proficiency, but rather on preparation time (short preparation times elicited asymmetrical switch costs whether long preparation times elicited symmetrical switch costs), and that the modulation of the N200 ERP component, previously taken as the expression of language inhibition (Jackson et al., 2001), was only sensitive to preparation times. Based on these results, the authors concluded that bilinguals could use inhibition as a strategy during language switching but do not need inhibition to control their language use. In accord with this conclusion, Martin et al. (2013) tested two groups of bilinguals (early and late Catalan–Spanish bilinguals) and failed to reveal any modulation of the N200 component in relation to linguistic switch costs.

The evidence reviewed indicates a complex scenario. The behavioral patterns of switch costs and the ERP components related to inhibition seem to be modulated by other up-to-date unknown factors besides language proficiency. This observation suggests that some caution should be exercised when drawing strong conclusions from these nonsystematic observations.

At present, just a few ERP studies revealed the time course of the effects related to language control by using the other instantiation of language switching. Misra et al. (2012) used a "blocked switching task" to investigate how performance in one language is affected by the previous use of a different language. Hence, two groups of unbalanced bilinguals took part in the study. The first one named a set of pictures in the L1 (two blocks of naming) and, subsequently, the same set of pictures in the L2 (two blocks of naming). The second group instead named a set of pictures in the L2 (two blocks of naming), and then the same pictures in the L1 (two blocks of naming). Therefore, to assess whether naming in one language was affected by previous use of the other language, authors compared across the two groups the third versus the first block of naming in the same language. Results indicated that naming in L1 was slowed by previous naming in the L2 as compared with naming in L1 without previous use of any language (i.e., first block). Conversely, the L2 was not negatively affected by the previous use of the L1. This asymmetrical pattern of RTs, which resembles that of trial-by-trial language switching tasks, was accompanied by ERP effects in the N200 time-window (Figure 35.1). Specifically, it was found that naming in L1 after naming in L2 elicited an enhancement of the N200 component, as compared with naming in L1 in the first block. In contrast, naming in L2 after L1 did not modulate the N200 component. Overall, these results have been taken as evidence of inhibition of the L1 during naming in the L2, with this inhibition having a negative after-effect when naming the same pictures later in L1.

A recent study with highly proficient bilinguals (Branzi et al., 2014) revealed a pattern of RTs similar to that found by Misra et al. (2012), that is, naming in L1 was hindered by previous use of the L2. Interestingly, this was found for both pictures that had to be named in the two languages or only in one (i.e., repeated and unrepeated pictures), suggesting that language control is applied globally on the whole non-target language set. However, differently from what was reported by Misra et al. (2012), the ERP effects of language control occurred in an earlier time-window than the N200, that is, in the P200 time-window. The ERP P200 component has been associated with the ease of lexical access (Costa, Strijkers, Martin, & Thierry, 2009; Strijkers, Baus, Runnqvist, Fitzpatrick, & Costa, 2013; Strijkers, Costa, & Thierry, 2010; Strijkers, Holcomb, & Costa, 2011) rather than with inhibitory control processes. In accord with other recent findings (Strijkers et al., 2013), these results suggest that the way language control applies in bilingual speakers might be different for highly and low-proficients. Specifically, language control might not rely on inhibition in the first case.

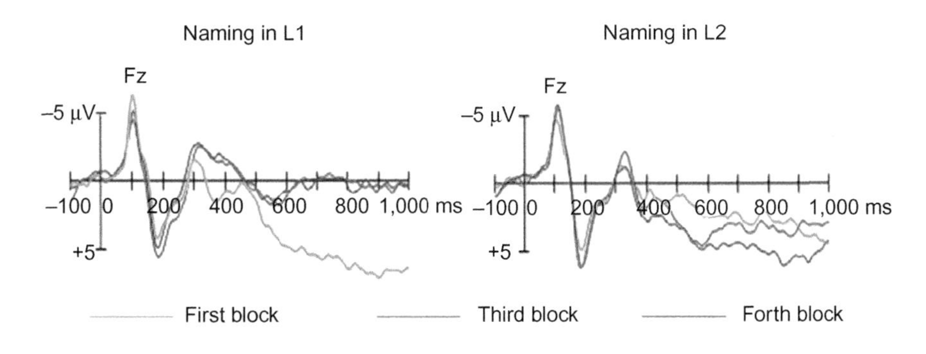

FIGURE 35.1 Grand average ERPs for naming pictures in L1 (left panel) and in L2 (right panel). First block (in green) represents the first time that pictures were named in L1 or in L2. Third and fourth blocks (blue and red) represent the third and the fourth times that pictures were seen. In these blocks pictures were named in L1 or in L2, after naming in L2 and in L1 (in the first two blocks), respectively. Note that negative is plotted up. *Adapted from Misra et al. (2012) with permission from the publisher.*

An interesting aspect of the studies presented refers to the fact that similar patterns of behavioral results may be related to qualitatively different control processes driven by L2 proficiency.

It is evident that when taking into account more variables, the image drawn from these reviewed studies becomes quite complex. As already mentioned, one such variable is language proficiency. Different patterns of behavioral and ERP effects between groups of bilinguals with different levels of L2 proficiency suggest that this factor may influence the way in which bilinguals control their languages. However, given that many of these studies were conducted with bilinguals with low or medium proficiency in the second language, much research is needed to address whether this is the case.

Besides proficiency, other variables may influence the way in which bilingual language control is applied. One of them is the "frequency of language switching" in daily life (Prior & Gollan, 2011). As previously hinted, Christoffels et al. (2007) revealed that bilinguals who were moderately proficient in their second language showed symmetrical switch costs between their L1 and their L2. This pattern, typically observable in highly proficient bilinguals, has been the argument for language-specific control mechanisms driven by L2 proficiency (Costa et al., 2004, 2006). Because bilingual participants in Christoffels et al. (2007) were only moderately proficient in L2, the authors hypothesized that the symmetrical pattern of switch costs was due to the fact that these participants were accustomed to switch frequently between languages in daily life. Hence, in addition to L2 proficiency and, daily experience with language switching may also boost the language control system and influence how language control mechanisms are applied.

It has also been proposed that some differences in the experimental designs could explain the different outcomes in these studies. For example, Christoffels et al. (2007) suggested that the different ERP results obtained in comparison with those of Jackson et al. (2001) could be explained by differences in the experimental design. That is, in Christoffels et al. (2007), participants had to switch between the L1 and the L2 in an unpredictable manner, whereas in Jackson et al. (2001) the occurrence of switch trials was totally predictable (e.g., every second trial). Even though the effect of predictability of language switches on the ERP components has yet to be established, the N200 component is particularly sensitive to context (Nieuwenhuis et al., 2003). Hence, this aspect could have influenced the way in which the N200 ERP component related to language control was modulated in these two studies (see also Martin et al., 2013).

Verhoef et al. (2009) provided an example of how experimental design manipulations could affect the pattern of behavioral and ERP results. In fact, authors revealed that preparation times influenced not only the pattern of switch costs but also the effects at the N200 time-window. These results suggested that the behavioral pattern of switch costs was not proficiency-dependent, unlike what Costa and Santesteban (2004) proposed, given that in a group of unbalanced bilinguals short–long preparation times elicited different patterns of switch costs. Even if this result suggests that high proficiency in an L2 is not a necessary prerequisite for symmetrical switch costs, the fact that the N200 component was sensitive to preparation times but not to language switching effects makes it difficult to establish whether this component has a key role in bilingual language control.

Overall, the current set of ERP findings does not provide consistent evidence and leaves open the question of the mechanisms involved in bilingual language control. Importantly, we believe there is still a need to address the weights of different factors (i.e., language proficiency, daily frequency of language switching) in modulating the ERP components and behavioral effects related to bilingual language control in language switching tasks.

The neuroimaging findings represent an important source of evidence to understand the functional significance of the ERP components related to language control. For example, the N200 component has been related to a specific neural generator: the anterior cingulate cortex (ACC) (Ladouceur, Dahl, & Carter, 2007; Van Veen & Carter, 2002). This brain structure has also been found to be involved in many language switching tasks and has been referred to as specific language control mechanisms (Wang, Xue, Chen, Xue, & Dong, 2007). In the next section, we review the most important functional magnetic resonance imaging (fMRI) studies in the language switching literature, which describe how proficiency level can affect the brain network sustaining language switching.

35.4 THE NEURAL CORRELATES OF LANGUAGE CONTROL: A FRONTAL, PARIETAL, AND SUBCORTICAL NETWORK

A number of neuropsychological studies have presented patients showing selective linguistic alterations and/or pathological language switching (Abutalebi, Miozzo, & Cappa, 2000; Fabbro, Skrap, & Aglioti, 2000; Paradis, 2001). These cases suggest the existence of specific brain areas involved in language control. With the complementary aid of neuroimaging techniques, brain areas such as the left caudate, the ACC, the lateral prefrontal cortex, and the left inferior parietal cortex have been proposed to conform the language control network (Abutalebi & Green, 2007). These brain areas were not specifically dedicated to this aim but acquired a specific role in language control as a part of its general function.

The brain lesion most frequently related to pathological language switching in bilinguals is that affecting the left basal ganglia (Abutalebi et al., 2000; Adrover-Roig et al., 2011; Aglioti & Fabbro, 1993). The general role of this area is to integrate information from multiple brain regions to shape motor learning. In the case of bilinguals, this area is required to establish the adequate motor language program including language planning, selection, and switching. A second area involved in language control is the ACC, which participates in the monitoring of different response alternatives during conflict processing (e.g., Botvinick, Braver, Barch, Carter, & Cohen, 2001; Botvinick, Cohen, & Carter, 2004; Braver, Barch, Gray, Molfese, & Snyder, 2001). In the case of bilinguals, the ACC may participate in the control process of L1 or L2, and in error detection and selective attention during language monitoring. Consistent with this role, Fabbro et al. (2000) reported a case of pathological language switching after a lesion in the left ACC. A third area involved in language control is the left lateral prefrontal cortex including the dorsolateral and ventrolateral parts. These parts exert general executive control functions over behavior in response to stimuli. For instance, when bilinguals named pictures or read words aloud in their L2, this area was more activated in highly proficient bilinguals than in monolinguals (Jones et al., 2012). Thus, the left lateral prefrontal cortex participates actively in the language control network by exerting a role in response selection and inhibition and in working memory. The last part of the brain proposed to participate in the network is the left inferior parietal lobe that participates in the maintaining of language representations in working memory.

The language control network described seems to be somehow affected by the language proficiency level in the second language. For example, bilinguals with high proficiency in the two languages (Garbin et al., 2011) engage different brain areas when switching between L1 and L2 as compared with low-proficient bilinguals (Wang et al., 2007). Note also that the language control network seems to be functionally influenced by sociolinguistic aspects other than language proficiency. For example, Perani et al. (2003) examined the role of age acquisition and language exposure/usage in early and highly proficient Catalan-Spanish bilinguals by means of a fluency task performed in the two languages. Authors revealed the lexical retrieval in the language acquired earlier in life was associated with less extensive activation in the brain. Language exposure/usage modulated the brain areas involved in lexical retrieval. In fact, in one of two groups of participants, those who were less exposed to the L2 showed larger activations during L2 lexical retrieval. These results indicated that in addition to language proficiency, also age of language acquisition and language exposure/usage are crucial factors in determining the neural pattern during lexical processing in bilinguals. This conclusion is in accord with recent ERP evidence suggesting that different linguistic experiences (early versus late acquisition of the L2) in bilinguals may affect the way in which the brain recruits language control (Martin et al., 2013).

Taken together, the network described, which consists of the prefrontal cortex, the ACC, the posterior parietal cortex, and the basal ganglia, constitutes an efficient brain network for language selection and control. In the next section, we describe how this network is specifically engaged when bilinguals have to switch between languages, because language switching relies heavily on cognitive control (Monsell, 2003) and how L2 proficiency can influence this activity.

35.4.1 The Neural Correlates of Language Switching

Neuroimaging techniques like positron emission tomography (PET) and fMRI have been used during the past 20 years to measure brain activity. They provide accurate information about time course of brain activity, and especially about brain localization of this activity. Initial studies on bilingualism were more focused on the representation problem, that is, in knowing whether both languages were represented in the same or different parts of the brain (Kim, Relkin, Lee, & Hirsch, 1997; Perani et al., 2003). Once the great overlapping in the brain activity for L1 and L2 representations was determined, the remaining question was to investigate how language control is exerted in the brain (i.e., how these overlapping brain areas are recruited for language selection, language inhibition, and language switching).

As in the ERP studies, the different fMRI experiments conducted to investigate language switching have been designed to investigate "sustained" and/or "transient" control mechanisms in language control. As previously introduced, both "sustained" and "transient" control processes are important for language control and both processes may be best characterized in a qualitatively different way and subserve different aspects of language control (Christoffels et al., 2007).

"Sustained" activity was studied by Hernandez, Martinez, and Kohnert (2000) and Hernandez, Dapretto, Mazziotta, and Bookheimer (2001) by comparing brain activation in early and highly proficient Spanish–English bilinguals during naming blocks in one language with naming in mixing blocks in either L1 or L2 (fixed order). Mixed compared with blocked naming conditions increased the activation of the left inferior frontal cortex and the bilateral dorsolateral prefrontal cortex for the two languages. These results suggest that switching between languages requires the extra participation of executive control areas. Similar results (bilateral activation of dorsolateral prefrontal cortex) were obtained by Wang, Kuhl, Chen, and Dong (2009) in late, low-proficient, Chinese–English bilinguals using a single digit naming task in mixed and blocked conditions. Also, they found an additional activation in the supplementary motor area (SMA) (see Guo et al., 2011 for similar results in low-proficient bilinguals). Interestingly, Wang et al. (2009) revealed some dissociation between the two languages for sustained control. The mixed condition elicited the activation of the left middle frontal gyrus and right precuneus relative to blocked naming in L1 (Chinese). The mixed condition as compared with the blocked naming in L2 (English) instead revealed the activation of a large network of brain areas: the bilateral middle frontal gyri, the cerebellum, the left inferior frontal gyrus, and the SMA. This recruitment of frontal areas for both languages may reflect proactive processes necessary to regulate the activation level of the two languages when the interference is high (i.e., mixed condition) (Braver et al., 2003). In accord with this observation, Ma et al. (2014) have explored sustained control in a group of Chinese–English unbalanced bilinguals, relatively highly proficient in their L2. In detail, authors revealed that switching between languages (mixed condition) as compared with naming in L1 (blocked L1 condition) elicited the activation of two large clusters. The first one included the left inferior frontal gyrus, the SMA bilaterally, the left insula, and the basal ganglia (including caudate and putamen portions). The second big cluster peaked into the left inferior parietal gyrus and extended to the left supramarginal gyrus and the angular gyrus and precuneus. Switching between languages (mixed condition) as compared with naming in L2 (blocked L2 condition) elicited activations mainly in the left inferior frontal gyrus, the bilateral precentral gyrus and SMA, the bilateral inferior parietal gyrus, the bilateral fusiform, the left lingual gyrus, the left inferior temporal gyrus, as well as the hippocampus bilaterally. These results revealed a neural dissociation that suggests that the sustained mechanisms for L1 and L2 involve different levels of control demands. This may be taken to reflect that when competition increases because of the need to alternate the two languages, the level of activation of the L1 needs to be reduced to favor L2 production. According to Ma et al. (2014), this might be achieved through the coupling between the frontal and basal ganglia brain regions.

These evidences suggest that language proficiency may affect the way in which sustained control processes are applied to regulate the availability of words of the two languages during speech production.

In a different study, Abutalebi et al. (2008) studied highly proficient bilinguals (university students of the Translation Department) using a similar procedure with a random presentation of cues to name in the L1 or in the L2. Besides the activation of left Broca's area, it was reported that naming in L1 in the bilingual context (where subjects had to select L1 or L2 nouns following a cue) compared with monolingual contexts (where subjects had to select L1 nouns or L1 verbs following a cue) induced an increased activation in the left prefrontal cortex and specifically engaged the left caudate and the ACC. Strikingly, this pattern of activity was absent for the same L1 nouns when the same subjects were placed in a monolingual context, therefore highlighting the crucial role of these neural structures in language control and particularly in language switching.

As in ERPs, different studies have used the first instantiation of the language switching paradigm designed to study "transient" effects. For example, Wang et al. (2007) applied this paradigm to a group of late Chinese–English bilinguals. The overall switching condition when compared with non-switching condition activated the language control network, including the left and right dorsolateral prefrontal cortex, the right ACC, and right caudate. When considering directional changes, the authors observed an asymmetry in behavioral switch costs, such as a larger cost present when switching to the L1 than when switching to the L2. The notion that there is competition between languages predicts that increased executive processes are recruited to allow L2 production compared with L1, especially in the case of low-proficient bilinguals. In line with this prediction, Wang et al. (2007) observed that switches from L1 to L2 ("forward switching") activated the left ACC/SMA, whereas the switches from L2 to L1 ("backward switching") did not activate any brain area within the language control network (Figure 35.2).

In a subsequent study, Wang et al. (2009) studied "transient" changes of language in low-proficient bilinguals using a single digit naming task. They replicated behavioral asymmetries of switch costs (larger switch cost to L1 than to L2) and that switching from L1 to L2 ("forward switching") activated the language control network (left SMA, left dorsolateral prefrontal cortex, and left inferior parietal lobe), but the results for overall switches and switches from L2 to L1 ("backward switching") did not show activations in areas of the language control network. In a further study using the first instantiation of language switching paradigm, Garbin et al. (2011) studied a group of early and highly proficient bilinguals and found a different pattern of results. Switches from L1 to L2 ("forward switching") activated the left caudate, whereas switches from L2 to L1 ("backward switching") activated the SMA/ACC (Figure 35.3).

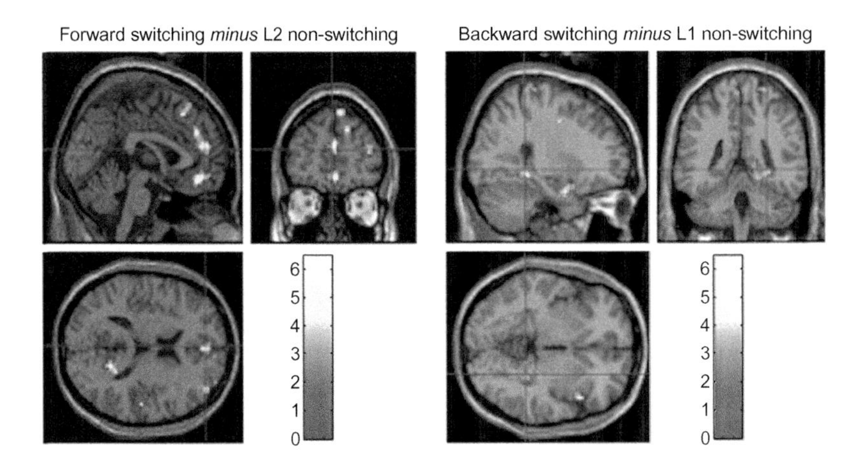

FIGURE 35.2 Brain regions involved in "forward switching" (from L1 to L2) relative to L2 non-switching (left panel) and in "backward switching" (from L2 to L1) relative to L1 non-switching (right panel).

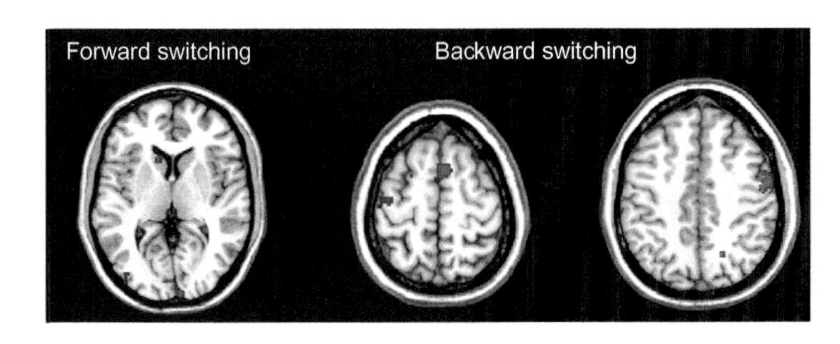

FIGURE 35.3 Brain regions involved in "forward switching" and "backward switching." *From Garbin et al. (2011) with permission from the publisher.*

At the neural level, different brain areas appear to be engaged during language switching and, interestingly, functional data indicate that the nature of this may alter with L2 proficiency, consistent with a change from controlled to more automatic L2 processing (Abutalebi & Green, 2007).

Abutalebi et al. (2012, 2013) used a different strategy based on the comparison of the process of language switching in bilinguals to a within-language switching task in monolinguals. The first study focused on the ACC and described this brain area as tuned by bilingualism to resolve cognitive and language conflicts (Abutalebi et al., 2012). The second study compared multilinguals with high proficiency in the L2 and poor proficiency in the L3 with monolinguals[5] while performing the first instantiation of the language switching paradigm twice (once with L1 and L2, and the other with L1 and L3). In this study, authors were particularly interested in evaluating how language proficiency modulates the brain network of bilingual language control during a language switching task. The crucial finding of this study is that the pre-SMA/ACC participated in language switching regardless of individual proficiency, whereas the left caudate only was active when switching from L1 to the less proficient language (L3). Contrasting with these results, no neural significant differences were found in this study when comparing direction of switches within the multilingual group.

These studies suggest that language proficiency has an important role in determining how brain areas are recruited for language control, both for the "sustained" and "transient" components.

Besides these adaptations of the switching behavioral paradigms, other studies have used perceptive tasks. Crinion et al. (2006) used a classical semantic priming procedure to compare neural activation when the prime and the target belong to the same language or to different languages. Results showed an activation of the left caudate during language switching compared with nonlanguage switching trials across three different groups of highly proficient bilinguals. A different perceptive task was used by Abutalebi et al. (2007) in a sample of early, highly proficient bilinguals with more experience in L2. They used narratives that included unpredictable changes of languages that might be regular or irregular depending on the respect or violation of the constituents of sentence structure. The comparison of switch and nonswitch trials activated more language areas, including the left inferior frontal gyrus and

[5]The switching task for monolinguals required to switch between two different categories of naming in the same language: nouns and verb forms.

middle temporal gyrus. Switches into L1 (the less exposed language) activated the left caudate and the ACC, whereas switches into L2 did not activate the language control network.

In general terms, the reviewed studies provide an acceptable uniformity of the results highly consistent with brain areas proposed to form the language control network. This uniformity is greater if we attend to the special role played by variables, such as the type of task and proficiency of bilinguals. The studies designed to investigate "sustained" processes in language control showed consistent activations in the dorsolateral prefrontal cortex and the SMA/ACC, whereas those more related to the "transient" process circumscribed the neural basis of language switching to the SMA/ACC. As in other conflict processes, the language conflict seems to require the participation of ACC that may exert the functions of monitoring and resolution. The dorsolateral prefrontal cortex would remain a global cognitive control mechanism of language conflicts. The role of the left caudate seems to be restricted to language in early and highly proficient bilinguals in tasks that involve language production and comprehension (see Ma et al., 2014 for involvement of caudate in late bilinguals more proficient in L2). Even though its specific role is not clear because in some studies it participated in "backward switching" and in others it appeared to be more related to "forward switching," the diverse studies were consistent in showing a role in detection of language switching when both languages were acquired early. Finally, the left inferior parietal lobe is the brain area of the language network less often detected, probably due to the low working memory demands of the used paradigms. The activation of this area has only been found in tasks using the same stimuli (i.e., numbers) and responses. Maybe its role would be more prominent in translation tasks (Price, Green, & Von Studnitz, 1999).

35.5 CONCLUSION

In this chapter we reviewed the main literature on bilingual language control, giving special attention to the studies that have made use of language switching paradigms. When combining behavioral, electrophysiological, and neuroimaging evidence, it is evident that some sociolinguistic variables, such as language proficiency, appear to influence the way in which language control is applied. This aspect is well-supported by both behavioral and neuroimaging evidence. However, more research is needed to determine how other sociolinguistic variables, besides language proficiency, can shape language control ability in bilinguals. Also, a distinction between different components of attentional control appears to be crucial within the linguistic domain. "Sustained" and "transient" components of attention show dissociable effects in the control of the two languages of a bilingual, as reflected by the recruitment of different brain areas.

References

Abutalebi, J., Annoni, J. M., Seghier, M., Zimine, I., Lee-Jahnke, H., Lazeyras, F., et al. (2008). Language control and lexical competition in bilinguals: An event-related fMRI study. *Cerebral Cortex, 18*, 1496–1505.

Abutalebi, J., Brambati, S. M., Annoni, J. M., Moro, A., Cappa, S. F., & Perani, D. (2007). The neural cost of the auditory perception of language switches: An event-related functional magnetic resonance imaging study in bilinguals. *The Journal of Neuroscience, 27*, 13762–13769.

Abutalebi, J., Della Rosa, P. A., Ding, G., Weekes, B. S., Costa, A., & Green, D. W. (2013). Language proficiency modulates the engagement of cognitive control areas in multilinguals. *Cortex, 49*, 905–911.

Abutalebi, J., Della Rosa, P. A., Green, D., Hernández, M., Scifo, P., Keim, R., et al. (2012). Bilingualism tunes the anterior cingulate cortex for conflict monitoring. *Cerebral Cortex, 22*, 2076–2086.

Abutalebi, J., & Green, D. (2007). Bilingual language production: The neurocognition of language representation and control. *Journal of Neurolinguistics, 20*, 242–275.

Abutalebi, J., Miozzo, A., & Cappa, S. F. (2000). Do subcortical structures control "language selection" in polyglots? Evidence from pathological language mixing. *Neurocase, 6*, 51–56.

Adrover-Roig, D., Galparsoro-Izagirre, N., Marcotte, K., Ferré, P., Wilson, M. A., & Inés Ansaldo, A. (2011). Impaired L1 and executive control after left basal ganglia damage in a bilingual Basque-Spanish person with aphasia. *Clinical Linguistics and Phonetics, 25*, 480–498.

Aglioti, S., & Fabbro, F. (1993). Paradoxical selective recovery in a bilingual aphasic following subcortical lesions. *Neuroreport, 4*, 1359–1362.

Allport, A., & Wylie, G. (1999). Task switching: Positive and negative priming of task-set. In G. W. Humphreys, J. Duncan, & A. M. Treisman (Eds.), *Attention, space and action: Studies in cognitive neuroscience* (pp. 273–296). Oxford, England: Oxford University Press.

Allport, D. A., Styles, E. A., & Hseih, S. (1994). Shifting intentional set: Exploring the dynamic control of tasks. In C. Umiltà, & M. Moscovitch (Eds.), *Attention and performance XV: Conscious and non conscious information processing* (pp. 421–452). Cambridge, MA: MIT Press.

Botvinick, M. M., Braver, T. S., Barch, D. M., Carter, C. S., & Cohen, J. D. (2001). Conflict monitoring and cognitive control. *Psychological Review, 108*, 624.

Botvinick, M. M., Cohen, J. D., & Carter, C. S. (2004). Conflict monitoring and anterior cingulate cortex: An update. *Trends in Cognitive Sciences, 8*, 539–546.

Branzi, F. M., Martin, C. D., Abutalebi, J., & Costa, A. (2014). The after-effects of bilingual language production. *Neuropsychologia, 52*, 102–116.

Braver, T. S., Barch, D. M., Gray, J. R., Molfese, D. L., & Snyder, A. (2001). Anterior cingulate cortex and response conflict: Effects of frequency, inhibition and errors. *Cerebral Cortex, 11*, 825–836.

Braver, T. S., Reynolds, J. R., & Donaldson, D. I. (2003). Neural mechanisms of transient and sustained cognitive control during task switching. *Neuron, 39*, 713–726.

Bruin, K. J., Wijers, A. A., & Van Staveren, A. S. J. (2001). Response priming in a go/nogo task: Do we have to explain the go/nogo N200 effect in terms of response activation instead of inhibition? *Clinical Neurophysiology, 112*, 1660–1671.

Calabria, M., Hernández, M., Branzi, F. M., & Costa, A. (2012). Qualitative differences between bilingual language control and executive control: Evidence from task-switching. *Frontiers in Psychology, 2*, 1–10.

Christoffels, I. K., Firk, C., & Schiller, N. O. (2007). Bilingual language control: An event-related potential study. *Brain Research, 1147*, 192–208.

Colomé, A. (2001). Lexical activation in bilinguals' speech production: Language-specific or language-independent?. *Journal of Memory and Language, 45*, 721–736.

Costa, A., Miozzo, M., & Caramazza, A. (1999). Lexical selection in bilinguals: Do words in the bilingual's two lexicons compete for selection? *Journal of Memory and Language, 41*, 365–397.

Costa, A., & Santesteban, M. (2004). Lexical access in bilingual speech production: Evidence from language switching in highly proficient bilinguals and L2 learners. *Journal of Memory and Language, 50*, 491–511.

Costa, A., Santesteban, M., & Ivanova, I. (2006). How do highly proficient bilinguals control their lexicalization process? Inhibitory and language-specific selection mechanisms are both functional. *Journal of Experimental Psychology: Learning, Memory, and Cognition, 32*, 1057–1074.

Costa, A., Strijkers, K., Martin, C. D., & Thierry, G. (2009). The time course of word retrieval revealed by event-related brain potentials during overt speech. *Proceedings of the National Academy of Sciences of the United States of America, 106*, 21442–21446.

Crinion, J., Turner, R., Grogan, A., Hanakawa, T., Noppeney, U., Devlin, J. T., et al. (2006). Language control in the bilingual brain. *Science, 312*, 1537–1540.

De Groot, A. M. B., & Christoffels, I. K. (2006). Language control in bilinguals: Monolingual tasks and simultaneous interpreting. *Bilingualism: Language and Cognition, 9*, 189–201.

Donkers, F. C. L., & van Boxtel, G. J. M. (2004). The N2 in go/no-go tasks reflects conflict monitoring not response inhibition. *Brain and Cognition, 56*, 165–176.

Fabbro, F., Skrap, M., & Aglioti, S. (2000). Pathological switching between languages after frontal lesions in a bilingual patient. *Journal of Neurology, Neurosurgery and Psychiatry, 68*, 650–652.

Falkenstein, M., Hoormann, J., & Hohnsbein, J. (1999). ERP components in Go/Nogo tasks and their relation to inhibition. *Acta Psychologica, 101*, 267–291.

Garbin, G., Costa, A., Sanjuan, A., Forn, C., Rodriguez-Pujadas, A., Ventura, N., et al. (2011). Neural bases of language switching in high and early proficient bilinguals. *Brain and Language, 119*, 129–135.

Gollan, T. H., & Ferreria, V. S. (2009). Should I stay or should I switch? A cost-benefit analysis of voluntary language switching in young and aging bilinguals. *Journal of Experimental Psychology: Learning, Memory, and Cognition, 35*, 640–665.

Green, D. W. (1998). Mental control of the bilingual lexico-semantic system. *Bilingualism: Language and Cognition, 1*, 67–81.

Guo, T., Liu, H., Misra, M., & Kroll, J. F. (2011). Local and global inhibition in bilingual word production: fMRI evidence from Chinese–English bilinguals. *NeuroImage, 56*, 2300–2309.

Hernandez, A. E., Dapretto, M., Mazziotta, J., & Bookheimer, S. (2001). Language switching and language representation in Spanish–English bilinguals: An fMRI study. *NeuroImage, 14*, 510–520.

Hernandez, A. E., Martinez, A., & Kohnert, K. (2000). In search of the language switch: An fMRI study of picture naming in Spanish–English bilinguals. *Brain and Language, 73*, 421–431.

Jackson, G. M., Swainson, R., Cunnington, R., & Jackson, S. R. (2001). ERP correlates of executive control during repeated language switching. *Bilingualism: Language and Cognition, 4*, 169–178.

Jones, Ō. P., Green, D. W., Grogan, A., Pliatsikas, C., Filippopolitis, K., Ali, N., et al. (2012). Where, when and why brain activation differs for bilinguals and monolinguals during picture naming and reading aloud. *Cerebral Cortex, 22*, 892–902.

Kim, K. H., Relkin, N. R., Lee, K. M., & Hirsch, J. (1997). Distinct cortical areas associated with native and second languages. *Nature, 388*, 171–174.

Koch, I., Gade, M., Schuch, S., & Philipp, A. M. (2010). The role of inhibition in task switching: A review. *Psychonomic Bulletin and Review, 17*, 1–14.

Koch, I., Prinz, W., & Allport, A. (2005). Involuntary retrieval in alphabet-arithmetic tasks: Task-mixing and task-switching costs. *Psychological Research, 69*(4), 252–261.

Kroll, J. F., Bobb, S., & Wodniecka, Z. (2006). Language selectivity is the exception, not the rule: Arguments against a fixed locus of language selection in bilingual speech. *Bilingualism: Language and Cognition, 9*, 119–135.

Ladouceur, C. D., Dahl, R. E., & Carter, C. S. (2007). Development of action monitoring through adolescence into adulthood: ERP and source localization. *Developmental Science, 10*, 874–891.

Los, S. A. (1996). On the origin of mixing costs: Exploring information processing in pure and mixed blocks of trials. *Acta Psychologica, 94*(2), 145–188.

Ma, H., Hu, J., Xi, J., Shen, W., Ge, J., Geng, F., et al. (2014). Bilingual cognitive control in language switching: An fMRI study of English–Chinese late bilinguals. *PLoS One, 9*(9), e106468.

Martin, C. D., Barceló, F., Hernández, M., & Costa, A. (2011). The time course of the asymmetrical "local" switch cost: Evidence from event-related potentials. *Biological Psychology, 86*, 210–218.

Martin, C. D., Strijkers, K., Santesteban, M., Escera, C., Hartsuiker, R. J., & Costa, A. (2013). The impact of early bilingualism on controlling a language learned late: An ERP study. *Frontiers in Psychology, 4*, 1–15.

Meuter, R. F. I., & Allport, A. (1999). Bilingual language switching in naming: Asymmetrical costs of language selection. *Journal of Memory and Language, 40*, 25–40.

Misra, M., Guo, T., Bobb, S., & Kroll, J. F. (2012). When bilinguals choose a single word to speak: Electrophysiological evidence for inhibition of the native language. *Journal of Memory and Language, 67*, 224–237.

Monsell, S. (2003). Task switching. *Trends in Cognitive Sciences, 7*, 134–140.

Nagahama, Y., Okada, T., Katsumi, Y., Hayashi, T., Yamauchi, H., Oyanagi, C., et al. (2001). Dissociable mechanisms of attentional control within the human prefrontal cortex. *Cerebral Cortex, 11*, 85–92.

Nieuwenhuis, S., Yeung, N., & Cohen, J. D. (2004). Stimulus modality, perceptual overlap, and the go/nogo N2. *Psychophysiology, 41*, 157–160.

Nieuwenhuis, S., Yeung, N., van den Wildenberg, W., & Ridderinkhof, K. R. (2003). Electrophysiological correlates of anterior cingulate function in a go/nogo task: Effects of response conflict and trial type frequency. *Cognitive, Affective, and Behavioral Neuroscience, 3*, 17–26.

Paradis, M. (2001). Bilingual and polyglot aphasia. In R. S. Berndt (Ed.), *Handbook of neuropsychology* (pp. 69–91). Amsterdam: Elsevier Science.

Perani, D., Abutalebi, J., Paulesu, E., Brambati, S., Scifo, P., Cappa, S. F., et al. (2003). The role of age of acquisition and language usage in early, high-proficient bilinguals: An fMRI study during verbal fluency. *Human Brain Mapping, 19*, 170–182.

Price, C. J., Green, D. W., & Von Studnitz, R. (1999). A functional imaging study of translation and language switching. *Brain, 122*, 2221–2235.

Prior, A., & Gollan, T. H. (2011). Good language-switchers are good task-switchers: Evidence from Spanish–English and Mandarin–

English bilinguals. *Journal of the International Neuropsychological Society, 17*, 682–691.
Rubin, O., & Meiran, N. (2005). On the origins of the task mixing cost in the cuing task-switching paradigm. *Journal of Experimental Psychology: Learning, Memory, and Cognition, 31*(6), 1477.
Rubinstein, J. S., Meyer, D. E., & Evans, J. E. (2001). Executive control of cognitive processes in task switching. *Journal of Experimental Psychology: Human Perception and Performance, 27*, 763–797.
Runnqvist, E., Strijkers, K., Alario, F., & Costa, A. (2012). Cumulative semantic interference is blind to language: Implications for models of bilingual speech production. *Journal of Memory and Language, 66*, 850–869.
Strijkers, K., Baus, C., Runnqvist, E., Fitzpatrick, I., & Costa, A. (2013). The temporal dynamics of first versus second language speech production. *Brain and Language, 127*, 6–11.
Strijkers, K., Costa, A., & Thierry, G. (2010). Tracking lexical access in speech production: Electrophysiological correlates of word frequency and cognate effects. *Cerebral Cortex, 20*, 912–928.
Strijkers, K., Holcomb, P., & Costa, A. (2011). Conscious intention to speak facilitates lexical access during overt object naming. *Journal of Memory and Language, 65*, 345–362.
Van Veen, V., & Carter, C. S. (2002). The timing of action-monitoring processes in the anterior cingulate cortex. *Journal of Cognitive Neuroscience, 14*, 593–602.
Verhoef, K. M. W., Roelofs, A., & Chwilla, D. J. (2009). Role of inhibition in language switching: Evidence from event-related brain potentials in overt picture naming. *Cognition, 110*, 84–99.
Wang, Y., Kuhl, P. K., Chen, C., & Dong, Q. (2009). Sustained and transient language control in the bilingual brain. *NeuroImage, 47*, 414–422.
Wang, Y., Xue, G., Chen, C., Xue, F., & Dong, Q. (2007). Neural bases of asymmetric language switching in second-language learners: An ER-fMRI study. *NeuroImage, 35*, 862–870.
Wylie, G., & Allport, A. (2000). Task switching and the measurement of "switch costs.". *Psychological Research, 63*, 212–233.

CHAPTER

36

Neurobiology of Sign Languages

David P. Corina[1,2] *and Shane Blau*[1]

[1]Department of Linguistics, Center for Mind and Brain, University of California, Davis, CA, USA; [2]Department of Psychology, Center for Mind and Brain, University of California, Davis, CA, USA

36.1 INTRODUCTION

Signed languages used in deaf communities are naturally occurring human languages that exhibit the full range of linguistic complexity found in spoken languages. Just as there are a multitude of spoken language communities around the world (e.g., speakers of Quechua, Farsi, Portuguese, and English), there are many different sign language communities (e.g., signers of Langue des Signes du Québec [LSQ], Deutsche Gebärdensprache [DGS], Taiwan Ziran Shouyu [TZS], and British Sign Language [BSL], to name but a few). Although the histories and geographies of signed languages are less well-documented, it is known that signed languages arise spontaneously, over several generations, from isolated communities that have a preponderance of deaf individuals. Such situations are not as rare as one might think because genetic influences on the transmission of deafness are well-attested (Groce, 1985; Sandler, Meir, Padden, & Aronoff, 2005).

The comparison of signed and spoken human languages provides a unique opportunity for investigating the biological substrates of human language. The existence of two fundamentally different forms of human language, one expressed by manual and body articulations and perceived by the visual system, and one expressed by the vocal system and perceived by the auditory system, allows researchers to identify neural systems that are common to all forms of human linguistic communication from those that reflect a languages modality of expression.

Neurolinguistic studies of life-long signers have documented patterns of sign language disturbances after acute brain damage and have helped to identify brain regions that are critical for sign language processing. These studies indicate remarkable parallels between left hemisphere regions that support signed language processing and left hemisphere regions that support spoken language processing. Collectively, the pattern of hemispheric asymmetries found in language holds independently of whether the language involved is signed or spoken. These studies point toward a common network of brain regions that support the human capacity for linguistic communication. Moreover, these studies have helped advance our understanding of the homologies between the linguistic properties of spoken and signed languages. Research using neuroimaging and electrophysiology has confirmed and extended our understanding of the neuroanatomical and functional relationships between signed and spoken languages and has raised new questions about the similarities and the modality-specific differences observed. More recently, neurolinguistic studies of signed language have helped us to evaluate the role of *mirror neuron*–inspired theories of language understanding, a central theme in embodied accounts of human language and cognition.

36.2 SIGN LANGUAGE APHASIA

Case studies of deaf signing individuals with acquired brain damage provide evidence for the contribution of left hemisphere perisylvian regions in the mediation of signed language. Deaf signers, like hearing speakers, exhibit distinct language disturbances when left hemisphere cortical regions are damaged, and these disturbances vary systematically depending on lesion locus (Hickok, Love-Geffen, & Klima, 2002; Marshall, Atkinson, Smulovitch, Thacker, & Woll, 2004; Poizner, Bellugi, & Klima, 1987; for a review see Corina, 1998a, 1998b). One of the most interesting facts about the realization of aphasic deficits in right-handed deaf signers is that language errors are often realized on the ipsilesional limb, (i.e., the left hand in

Neurobiology of Language. DOI: http://dx.doi.org/10.1016/B978-0-12-407794-2.00036-5

cases of left hemisphere damage). This indicates that the observed language errors are not an artifact of motoric weakness due to contralesional hemiplegia. Rather, these deficits reflect a central disorder of linguistic control mediated by the left hemisphere.

36.2.1 Broca-Like Signing

In spoken languages, language production impairments are typically associated with left hemisphere frontal anterior lesions that involve the lower posterior portion of the left frontal lobe (e.g., Broca's area, Brodmann area 44/45). These lesions often extend in depth to the periventricular white matter (Goodglass, 1993; Mohr et al., 1978). The anterior insula has also been implicated in chronic speech production problems (Dronkers, Redfern, & Knight, 2000). Two well-documented cases of sign language aphasia after left hemisphere frontal damage demonstrate this familiar structure–function relationship for signed languages. The case study of patient G.D., reported in Poizner et al. (1987), documents a profound expressive nonfluent sign language aphasia after large left frontal lobe lesion that encompassed Broca's territory including BA 44/45. The subject presented with halting dysfluent sign articulation and an agrammatic language profile—American Sign Language (ASL) sentence structure was greatly simplified (i.e., telegraphic) and signs did not show required movement modulations that signal morpho-syntactic inflection. In contrast to her expressive deficits, G.D.'s comprehension was well within normal limits.

Case study R.S. had a left inferior frontal lesion that was more restricted to classically defined Broca's territory (Hickok, Kritchevsky, Bellugi, & Klima, 1996). This lesion involved inferior motor cortex and extended anteriorly to involve most of the par opercularis and inferiorly involved anterior and superior insula. Subcortically, this lesion undercut most of the par triangularis and involved subcortical white matter deep to the lower motor cortex. During the acute phase of this stroke, the patient showed a marked expressive aphasia with nonfluent effortful production. Sign repetition was also affected. Sign language comprehension, however, was intact. Over time, the expressive aphasia largely resolved with lingering problems of word finding. As discussed, an unusual feature of R.S.'s signing was formational paraphasia. These impairments were not attributable to motor weakness, nor did this patient show signs of limb apraxia.

36.2.2 Wernicke-Like Signing

Spoken language comprehension deficits are well-attested after left hemisphere temporal lobe damage (Damasio, 1992; Dronkers & Baldo, 2009; Naeser, Helm-Estabrooks, Haas, Auerbach, & Srinivasan, 1987). For example, Wernicke's aphasia, which is expressed as impaired language comprehension accompanied with fluent but often paraphasic (semantic and phonemic) output is often associated with damage to the posterior regions of the left superior temporal gyrus. More recent work has suggested the contribution of posterior middle temporal gyrus in cases of chronic Wernicke's aphasia (Dronkers, Redfern, & Knight, 2000; Dronkers, Redfern, & Ludy, 1995). Signers with left hemisphere posterior lesions also evidence fluent sign aphasia with associated comprehension deficits. However, as discussed, there is some controversy regarding whether anatomical locations of lesions that result in sign language comprehension deficits are fully comparable with those seen in users of spoken languages.

Subject L.K. described in Chiarello, Knight, and Mandel (1982) and in Poizner et al. (1987) had a left anterior parietal lesion in the region of the supramarginal and angular gyri. The inferior frontal operculum (areas 44 and 45) and the posterior superior temporal plane (areas 41, 42, and 22) were spared by the lesion. She presented with motorically facile signing, but with phonemic paraphasia and severe anomia. She had difficulty in grasping test instructions and was unable to perform two- or three-part commands. She experienced profound and lasting sign comprehension impairment. Subject P.D., reported in Poizner et al. (1987), had a left subcortical lesion deep to Broca's area, extending posteriorly beneath the parietal lobe. His signing, while fluent, exhibited numerous grammatically inappropriate signs (paragrammatism). His expression of sentence-level grammatical roles was disturbed. ASL syntax relies heavily on the use of an imaginary horizontal plane in front of the signer, which the signer may use to locate nominal referents. The movements of verb and predicate sign forms between these locations can signal subject and object relations; pronominal reference also makes use of this signing space convention. P.D. showed a lack of consistency in spatial agreement required of ASL syntax (Bellugi, Poizner, & Klima, 1989). That is, he would establish nominal referents in space, but he was inconsistent in maintaining co-reference to these previously established locations.

36.2.2.1 Comprehension Deficits

Group-level analyses have helped assess the structure–function relationships implicated in sign language comprehension deficits. In two reports, one on ASL and one on BSL (a completely distinct language from ASL), damage to the left hemisphere was observed to factor in sign comprehension deficits

(Atkinson, Marshall, Woll, & Thacker, 2005; Hickok et al., 2002). These studies report an increasing gradation of impairments for single signs, verb meaning, and simple sentence understanding (e.g., sentence verification and token test commands) in patients with left hemisphere damage but not right hemisphere damage.

There is, however, controversy with respect to degree of anatomical overlap observed in comprehension problems in spoken and signed languages. Whereas the study by Hickok et al. (2002) highlights the role of the left posterior temporal lobe in sign comprehension deficits, others have remarked on the potentially unique role of the left inferior parietal region in sign processing (Chiarello et al., 1982; Corina, 1998a, 1998b; Corina, Lawyer, & Cates, 2013; Leischner, 1943; Poizner et al., 1987). For example, in the group study data presented by Hickok et al. (2002) comparing the sign language comprehension abilities of signers with damaged left and right hemispheres, signers with left hemisphere posterior temporal lobe damage were found to perform worse than any other group. Although the authors emphasize the involvement of the damaged temporal lobe in these comprehension deficits, in all cases lesions of these subjects additionally extend into the parietal lobe. It is noteworthy that in the cases of L.K. and P.D. described, neither appears have to temporal lobe lesions. Case study W.L. (Corina et al., 1992) exhibited fluent aphasia with severe comprehension deficits. Lesions once again did not occur in cortical Wernicke's area; rather, this lesion undercuts frontal and inferior parietal areas, including the supramarginal gyrus. These observations have led some to suggest that sign language comprehension may be more dependent than speech on left hemisphere inferior parietal areas traditionally associated with somatosensory and visual motor integration, whereas spoken language comprehension might weigh more heavily on posterior temporal lobe association regions whose input includes networks intimately involved with auditory speech processing.

In summary, although it is clear that sign language comprehension suffers after left hemisphere damage, the contributions of the temporal and parietal lobes warrant further investigation. This controversy illustrates a theoretical difference between a modality-influenced view of language comprehension and an a-modal account. Proponents of a modality-influenced model would claim that cortical regions such as the IPL that support associative visual and motoric properties of sign language processing are intimately connected with language comprehension, whereas the a-modal perspective views the posterior temporal lobe as a unifying computational stage in language understanding, independent of language modality (see Hickok et al., 2002 for some discussion). Additional studies are needed to fully assess these competing hypotheses.

36.2.2.2 Sign Language Paraphasia

Language breakdown after left hemisphere damage is not haphazard, but it affects independently motivated linguistic categories. An example of the systematicity in sign and spoken language breakdown is illustrated through consideration of parapahsic errors (Corina, 2000). Paraphasia refers to the substitution of an unexpected word for an intended target. Semantic paraphasias often have a clear semantic relationship to the desired word and represent the same part of speech (Goodglass, 1993). In contrast, phonemic or "literal" paraphasia refers to the production of unintended sounds or syllables in the utterance of a partially recognizable word (Blumstein, 1973; Goodglass, 1993). Sound substitutions in phonemic paraphasia may result in the production of a real word related in sound, but not in meaning (e.g., telephone becomes television). Also attested are cases in which the erroneous word shares both sound characteristics and meaning with the target (broom becomes brush; Goodglass, 1993).

Sign language paraphasias are produced by signers with aphasia and have clear parallel forms found in spoken languages. For example, subject P.D. (Poizner et al., 1987) produced semantic paraphasias, signing BED for CHAIR, DAUGHTER for SON, and QUIT for DEPART. ASL semantic paraphasias tend to overlap in meaning and lexical class with the intended target. Semantic paraphasias, whether in sign or speech, suggest that the mental lexicon is structured according to semantic principles, whereby similar semantic items share a representational proximity. In this view, coactivation of closely related representations and/or an absence of appropriate inhibition from competing lexical entries may lead to substitutions and blends.

Descriptions of formational or literal paraphasias in aphasic signing provide insight into the structural properties of signed language. ASL formational errors may encompass both *phonological* and *phonetic* levels of impairment (see Corina, 2000 for some discussion).[1] *A priori*, we may expect to find phonological errors affecting the four major formational parameters of ASL phonology: handshape, movement, location, and orientation. However, the distribution of these paraphasic errors appears to be unequal; handshape configuration errors are the most widely reported, whereas sign

[1]The use of the terms phonology and phonetics in the present context refer to the abstract structural and articulatory properties of a sign, respectively.

errors affecting movement and especially location, although attested, are far less frequent. The aphasic signer WL described produced numerous *phonemic* errors, nearly all of which were errors involving substitutions of handshape. For example, he produced the sign TOOTHBRUSH with a Y handshape rather than the required G handshape. He produced the sign SISTER with an F handshape rather than the required L handshape (Figure 36.1).

Data from rare clinical cases of cortical stimulation mapping (CSM) performed in deaf and signing neurosurgical patients provide important clues to neural regions involved in semantic and phonemic paraphasia. Corina et al. (1999) reports the effects of left hemisphere cortical stimulation on sign language production in a deaf individual, patient S.T., undergoing an awake CSM procedure. Stimulation to two anatomical sites, one in the frontal operculum and one in the parietal operculum, led to consistent naming disruption. The errors, however, were qualitatively different.

Stimulation to the frontal opercular region of Broca's area (BA 44) resulted in errors involving the motor execution of signs. These errors were characterized by a laxed articulation of the intended sign, with nonspecific movements (repeated tapping or rubbing) and a reduction in handshape configurations to a lax closed fist handshape. Interestingly, there was no effort on the part of S.T. to self-correct these imperfect forms. In addition, such errors were observed during trials of sign and nonsign repetition. These results suggest that the posterior portion of Broca's area is involved in the motoric execution of complex articulatory forms, especially those underlying the phonetic level of language structure.

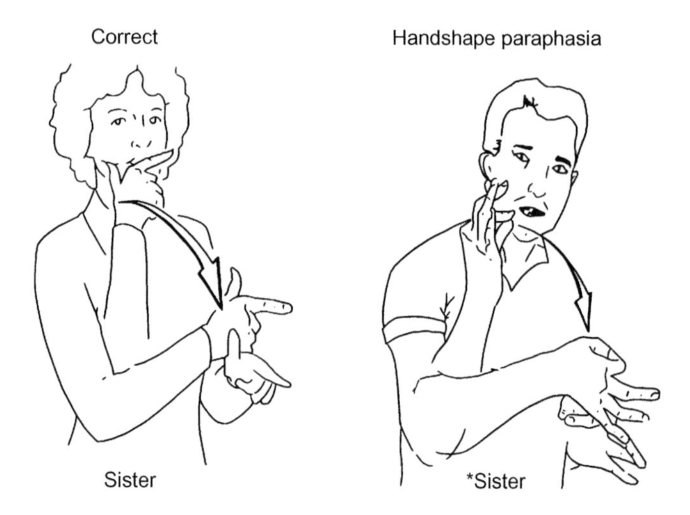

FIGURE 36.1 In ASL paraphasia, the correct sign Sister is made with an "L" handshape; in the error, the aphasic subject produced this form with an "F" handshape. *From Corina et al. (1992) with permission from the publisher.*

The sign errors observed with stimulation to the supramarginal gyrus (BA 40) resulted in both formational and semantic errors. Formational errors were characterized by repeated attempts to articulate the intended targets, and successive approximations of the target sign were common. For example, the sign PEANUT is normally signed with a closed fist with an outstretched thumb. During the movement of the sign, the thumb is flicked off the front of teeth. Under stimulation, this sign began as an incorrect, but clearly articulated, X handshape (closed fist with a protruding bent index finger) articulated at the correct location (in front of the mouth), but with an incorrect inward rotation movement. In two successive attempts to correct this error, S.T. first corrected the handshape and then went on to correct the movement as well. Notably, we do not find the laxed and reduced articulations characteristic of the signs seen with stimulation to Broca's area. Instead, these examples suggest problems involving the selection of the individual components of sign forms (i.e., handshape, movement, and, to a lesser extent, location). Moreover, although stimulation to the frontal operculum triggered errors in both sign and pseudosign production, pseudosign repetition was unaffected with stimulation to the supramarginal gyrus (SMG).

Semantic errors were also observed with stimulation to the SMG. Interestingly, many of these errors involve semantic substitutions that were formationally quite similar to the intended targets. For example, in one instance the stimulus picture "pig" elicited the sign FARM. The signs PIG and FARM differ in movement but share an identical articulatory location (the chin), and each is made with similar handshapes. This formational-semantic similarity is similar to the blended spoken paraphasia (e.g., *broom* → *brush*). Thus, in patient S.T., stimulation to the ventral portion of Broca's area had a global effect on the motor output of signing, whereas stimulation to the supramarginal gyrus affected the correct selection of the linguistic elements of a sign (including both phonological and semantic components).

The higher incidence of handshape errors is interesting, and linguistic analyses of ASL have argued that the parameter of handshape has autosegmental properties. Autosegmental representations are used to capture the independence of phonological features and posit representations in which features appear on separate tiers. This approach has been useful in capturing phonological phenomena in cases where phonological features span domains greater than the traditional segment, for example, in the cases of tone in tonal languages (Goldsmith, 1990). In many linguistic accounts of signed languages, handshape specifications are represented on an independent tier in a phonological representation (Sandler & Lillo-Martin, 2006). The independent vulnerability of handshape specifications in ASL formational paraphasia may be evidence for the psychological reality of this abstract linguistic representation. In spoken language phonemic paraphasias, a curious asymmetry exists; the majority of phonemic paraphasias involve consonant rather than vowel distortions. In sign language, handshape errors are far more common than path movement errors. What is striking is that in some analyses, handshape specifications are suggested to be more consonantal in nature, whereas movement components of ASL signs may be more analogous to vowels (see Corina & Sandler, 1993, for some discussion). Collectively, these errors demonstrate how functionally similar language categories (i.e., *consonants, vowels*) may be selectively vulnerable to impairment across signed and spoken languages. Although the surface manifestations differ, the underlying disruption may be related to a common abstract modality-independent level of representation.

Although compelling commonalities exist in the expression of paraphasic errors in spoken and signed language, one might also expect errors that reflect a language's modality of expression. For example, errors affecting the feature of voicing may not have any direct parallel in a signed language, so this type of error is likely modality-specific. The control of the articulators for speech and sign has different processing requirements. For signed languages, low-level motor control of the limbs is contralateral, whereas low-level motor control of the vocal tract is largely bilaterally. The processing requirements for two-handed articulations may place qualitatively different demands on the linguistic system and result in the expression of modality-conditioned errors.

Patient R.S., described previously, exhibited paraphasia restricted to two-handed signs. In signs that require two hands to assume different handshapes and/or move independently, R.S. would incorrectly move one of her hands. In other cases, R.S. would fail to preserve the required spatial relationships between the two hands. Moreover, during one-handed signing, R.S. mirrored the movements and handshapes of the dominant hand on the nondominant hand, but somewhat reduced in degree of movement. This mirroring was not seen during nonlinguistic movements and appears different from mirror movements seen in some cases of hemiparesis. This case is important for our understanding of neurobiology of language because it indicates that the modality of the linguistic system can place unique demands on the neural mediation and implementation of language. These errors may be taken as evidence of modality-dependent linguistic impairment.

36.3 RIGHT HEMISPHERE DAMAGE

Studies of deaf signers with right hemisphere damage often exhibit visual-spatial deficits, but evidence only subtle language problems, similar to those observed in hearing nonsigners. The case study of J.H., a deaf signer with a large right hemisphere lesion involving frontal, temporal, and parietal cortex, shows a remarkable sparing of sign language comprehension with severely compromised visual-spatial abilities (Corina, Kritchevsky, & Bellugi, 1996). J.H. exhibited persistent left-neglect, an attention disorder that causes patients to ignore contralesional spatial locations.

Formal testing showed that even in cases of face-to-face signing, where the perception of a two-handed sign such as DEER (the right and left hands, finger spread, are placed on each side of the forehead like antlers) could easily be perceived as the well-formed one-handed sign FATHER (only the dominant hand, fingers spread, is placed in the forehead). J.H. showed little to no evidence that this visual-spatial neglect affected his perception of ASL. In contrast, if two identical objects were presented for identification, one in J.H.'s left visual field and the other in the right visual field, J.H. would show consistent *extinction* for the physical object in his left visual field (see Corina et al., 1996 for a discussion).

36.3.1 Discourse Abilities

The disruption of discourse abilities is well-attested in hearing individuals with right hemisphere damage (Brownell, Simpson, Bihrle, Potter, & Gardner, 1990; Kaplan, Brownell, Jacobs, & Gardner, 1990; Rehak et al., 1992). Similar patterns of discourse impairment are seen in right hemisphere–damaged (RHD) users of signed language. Analysis of language use in subjects with right hemisphere lesions, J.H. and D.N. (Emmorey, 1993; Emmorey, Corina, & Bellugi, 1995; Poizner & Kegl, 1992), reveal contrasting disruptions. Lesion sites differed in these two patients; J.H.'s stroke involved cortical regions in the distribution of the right middle cerebral artery, whereas D.N.'s lesion was predominantly medial and involved the upper part of the occipital lobe and superior parietal lobule. In everyday signing and in picture description tasks, J.H. showed occasional nonsequiturs and abnormal attention to details, behaviors that are typically found in the discourse of hearing patients with right hemisphere lesions. Subject D.N. showed a different pattern of discourse disruption: one that affected the discourse cohesion. Spatial indexing is a commonly used device for pronominal reference in signed languages. Signers will point to the spatial location of a previously signed nominal, in essence re-referencing this previously introduced concept. Although her within-sentence spatial indexing was unimpaired, her use of space across sentences was inconsistent. That is, she did not consistently use the same index points to refer to individuals throughout a discourse.[2] To salvage intelligibility, D.N. used a compensatory strategy in which she restated the noun phrase in each sentence, resulting in an overly repetitive discourse style. Taken together, the cases of J.H. and D.N. suggest that right hemisphere lesions in signers can differentially disrupt discourse content (as in the case of JH) and discourse cohesion (as in the case of DN) (see also Hickok et al., 1999).

Signed languages are unique in their ability to spatially represent prepositional relationships between objects such as *on, above, under*, and others. These concepts are often conveyed via the depiction of the physical relation itself rather than encoded by a discrete lexical item. For example, an ASL translation of the English sentence "The pen rests on a book" may, in part, involve the use of the two hands, whereby one hand configuration with an outstretched finger (representing the pen) is placed on the back of a flat open hand (representing the book). This configuration encodes the spatial meaning "on" but without the need for an explicit lexical preposition.

Many signed languages express spatial relationships and events in this manner and have discrete inventories of highly productive grammatical forms, often referred to as "classifiers" or "classifier predicates," which participate in a wide range of depictive constructions. These highly productive forms play an important linguistic function in signed languages (Supalla, 1982; see also Emmorey, 2003 and papers therein) and their theoretical status remains a point of vibrant discussion and debate (for contrastive views see Dudis, 2007; Liddell, 2003; Sandler & Lillo-Martin, 2006).

The linguistic status of these classifier forms and their conventions of use have important implications for our understanding of neurolinguistics of signed languages. Several studies have found differential disruptions in the use and comprehension of sentences that involve these spatial-depictive forms. For example, Atkinson and colleagues conducted a group study of left-hemisphere-damaged and right-hemisphere-damaged signers of BSL (Atkinson et al., 2005). They devised comprehension tests that examined a wide range of sentence types including simple and complex argument structures and semantically reversible sentences. They also included a test of classifier placement, orientation, and rotation. Their findings indicated that left hemisphere damaged BSL signers relative to age-matched control subjects exhibited deficits on all comprehension tests. Right hemisphere damaged signers did not differ from controls on single sign and single predicate-verb constructions, nor on sentences that ranged in argument structure and semantic reversibility. However, RHD signers (like left hemisphere–damaged [LHD] signers) were impaired on tests of locative relationships expressed via classifier constructions, and on the test of classifier placement, orientation, and rotation (see also Hickok, Pickell, Klima, & Bellugi, 2009).

One interpretation offered for this pattern of responses is that the comprehension of classifier constructions requires not only intact left hemisphere resources but also intact right hemisphere visual-spatial processing mechanisms. Atkinson et al. states "the deficits stem from the disruption of processes which map nonarbitrary sign locations on the real-world's spatial position." That is, whereas both LHD and RHD signers show comprehension deficits, the RHD signer's difficulties stem from a more general *extralinguistic* visual-spatial deficit rather than linguistic malfunction *per se*. However, depictive forms are used to refer to not only real-world events but also imaginary and nonpresent events as well. Thus, a theoretical question arises as to how specific these visual-spatial deficits must be to be deemed *extralinguistic*. The broader point is whether aphasic deficits should solely be defined as those that have clear homologies to the left hemisphere maladies that are evidenced in spoken languages, or whether the existence of sign languages will force us to reconsider the conception of linguistics deficits. As discussed in the case of R.S., who showed evidence of modality-specific linguistic paraphasias in language production and in the realm of language comprehension, we may need to acknowledge modality-specific linguistic deficits. These sign language impairments may implicate neural regions such as left parietal areas and right hemisphere regions that lie well beyond traditionally defined perisylvian language areas that support properties of spoken languages.

In summary, aphasia studies to date provide ample evidence for the importance of the left hemisphere in mediation of sign language. After left hemisphere damage, sign language performance breaks down in a linguistically significant manner. In addition, there is growing evidence for the role of the right hemisphere in aspects of ASL discourse, classifier use, and syntactic comprehension. Case studies of deaf signers with aphasia have been useful in illuminating the nature of aphasic breakdown in signed languages as well as raising new questions concerning the neurobiology of linguistic processing.

[2]Contrast this impairment with left hemisphere subject P.D. who showed a lack of spatial agreement within single sentences.

36.4 NEUROIMAGING

Functional imaging studies of deaf signers provide additional evidence for left hemisphere specialization of signed language and provide an opportunity to explore more targeted questions regarding neurobiology of signed languages. Early studies often emphasized the overlap of activation between regions subserving sign processing and those regions traditionally thought to mediate speech processing (Bavelier et al., 1998; McGuire et al., 1997; Neville et al., 1998; Nishimura et al., 1999; Petitto et al., 2000; Söderfeldt et al., 1997). For example, Petitto et al. (2000) and McGuire et al. (1997) reported significant activation in left inferior frontal regions, specifically Broca's area (BA 44 and 45) and anterior insula during overt and covert tasks of sign generation. Subsequent work using more refined cytoarchitectonic references have indicated that BA 45 is activated during both sign and speech, and it can be differentiated from complex oral/laryngeal and limb movements, resulting in more ventral activation of BA 44. This interpretation accords well with the cortical stimulation finding reported in Corina et al. (1997). Overall, this research speaks to the modality independence of the pars triangularis (BA 45) (Horwitz et al., 2003), a region that has been traditionally considered a "speech-motor" region.

Studies have also reported that processing of signs in deaf signers can lead to activation in the temporal lobe. Petitto et al. (2000) and Nishimura et al. (1999) reported significant activation in left superior temporal cortex, a region often associated with auditory processing, in response to the perception of single signs. MacSweeney, Woll, Campbell, McGuire, et al. (2002) also observed the activation of auditory association areas in deaf and hearing native signers of BSL. Their findings revealed relatively greater activation for deaf native signers than hearing native signers in the left temporal auditory association cortex during the perception of signed sentences (see also Kassubek, Hickok, & Erhard, 2004). These findings are taken as evidence of cross-modal plasticity whereby auditory cortex and auditory association cortex may be modified in the absence of auditory input to become specialized for visual language input. Although this left hemisphere auditory activation is presumed to reflect linguistic processing, the functional significance of these activations is not well-understood. For example, several studies have reported activation of primary auditory cortex in deaf individual in response to low-level (nonlinguistic) visual stimulation (Fine, Finney, Boynton, & Dobkins, 2005; Finney, Clementz, Hickok, & Dobkins, 2003; Scott et al., 2014). The findings that auditory cortex is responsive to visual information is perhaps not that surprising given recent evidence that auditory cortex supports silent speech reading and audiovisual integration (Calvert et al., 1997; Calvert & Campbell, 2003; Kauramäki, 2010; Okada et al., 2013).

In studies of ASL verb generation in which subjects are required to generate a verb in response to a given noun, San José-Robertson, Corina, Ackerman, Guillemin, and Braun (2004) reported left-lateralized activation within perisylvian frontal and subcortical regions commonly observed in spoken language generation tasks. In an extension of this work, Corina, San Jose-Robertson, Guillemin, High, and Braun (2003) reported the left-lateralized pattern was not significantly different when the production of a repeat-generate task was conducted with a signer's right dominant or left nondominant hand. This finding is consistent with the fact that left-hemisphere-damaged aphasic signers will evidence linguistic errors while using their nondominant hand.

36.4.1 Sign Language Production

Neuroimaging studies of sign language production reveal further commonalities in the neural systems underlying core properties of language function in sign and speech. In a large-scale analysis of deaf signers and hearing nonsigners engaged in object naming tasks, Emmorey, Mehta, and Grabowski (2007) reported areas of overlap in neural activation for sign production in the deaf subjects and word production in the hearing subjects. This analysis identified regions supporting modality-independent lexical access. Common regions included the left mesial temporal cortex and the left inferior frontal gyrus. Emmorey et al. suggest that left temporal activation reflects conceptually driven lexical access (Indefrey & Levelt, 2004). In this view, for both speakers and signers, activation within the left inferior temporal gyrus may reflect prelexical conceptual processing of the pictures to be named, whereas activation within the more mesial temporal regions may reflect lemma selection, prior to phonological code retrieval. These results argue for a modality-independent frontotemporal network that subserves both sign and word production.

Differences in activated regions in speakers and signers were also observed. Within the left parietal lobe, two regions were more active for sign than for speech: the supramarginal gyrus and the superior parietal lobule. As discussed, parietal regions may be linked to modality-specific output parameters of sign language. Specifically, activation within left SMG may reflect aspects of phonological processing in ASL (e.g., selection of hand configuration and place of articulation features), whereas activation within the superior parietal lobule (SPL) may reflect proprioceptive monitoring of motoric output.

A study of discourse production in ASL-English native bilinguals further underscores the similarities between speech and sign. In this study, spontaneous generation of autobiographical narratives in ASL and English revealed complementary progression from early stages of concept formation and lexical access to later stages of phonological encoding and articulation. This progression proceeds from bilateral to left-lateralized representations, with posterior regions—especially posterior cingulate, precuneous, and basal-ventral temporal regions—activated during encoding of semantic information (Braun, Guillemin, Hosey, & Varga, 2001).

36.4.2 Sentence Comprehension

Studies of sentence processing in signed languages have repeatedly reported left-hemisphere activations that parallel those found for spoken languages (Figure 36.2). These activation patterns include inferior frontal gyrus (including Broca's area and insula), precentral sulcus, superior and middle temporal cortical regions, posterior superior temporal sulcus (STS), angular gyrus (AG), and SMG (Lambertz, Gizewski, de Greiff, & Forsting, 2005; MacSweeney, Woll, Campbell, Gemma, et al., 2002; MacSweeney et al., 2006; Newman, Bavelier, Corina, Jezzard, & Neville, 2002; Neville et al., 1998; Petitto et al., 2000; Sakai, Tatsuno, Suzuki, Kimura, & Ichida, 2005). The majority of functional imaging studies of sign language confirm the importance of the left hemisphere in sign processing and emphasize the similarity of patterns of activation for signed and spoken languages.

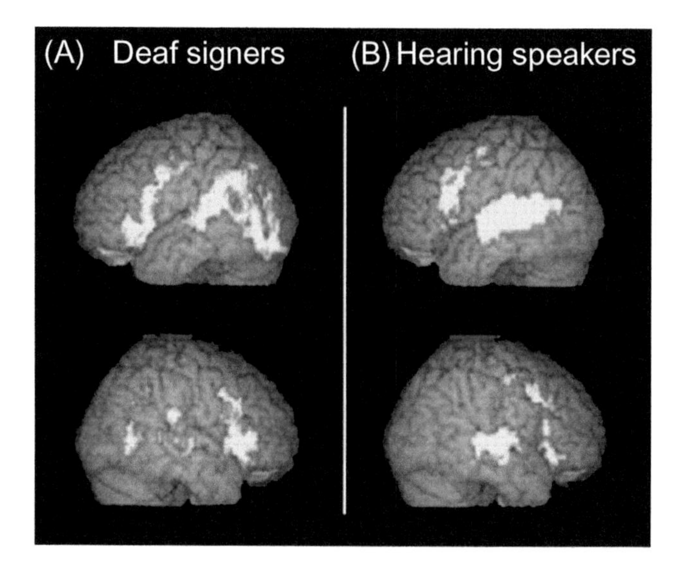

FIGURE 36.2 (A) Regions activated by BSL sentence comprehension in deaf native signers. (B) Regions activated by audiovisual English sentence comprehension in hearing nonsigners (voxelwise $p < 0.00005$). Both language inputs were contrasted with a low-level baseline: perception of the still model and a low-level target detection task (visual for deaf; auditory for hearing). Activation up to 5 mm beneath the cortical surface is displayed. *From MacSweeney, Capek, Campbell, and Woll (2008) with permission from the publisher.*

In addition to the more familiar left hemisphere activations, imaging studies of sentence processing in sign language have also noted significant right hemisphere activation. For example, activations in right hemisphere superior temporal, inferior frontal, and posterior parietal regions have been reported (MacSweeney, Woll, Campbell, McGuire, et al., 2002; MacSweeney et al., 2006; Neville et al., 1998; Newman et al., 2002). The question of whether these patterns of activation are unique to sign has been the topic of debate (Corina, Neville, & Bavelier, 1998; Hickok, Bellugi, & Klima, 1998). Mounting evidence suggests that some aspects of this right hemisphere activation may be attributable to the processing of facial information, which factors significantly in the sign signal. Studies of auditory and audiovisual speech have observed right hemisphere activations that appear similar to those reported in signing (Capek et al., 2004; Davis & Johnsrude, 2003; Schlosser, Aoyagi, Fulbright, Gore, & McCarthy, 1998).

However, as hinted, right hemisphere posterior parietal and temporal lobe regions may play a unique role in the mediation of signed languages. In a study by Newman et al. (2002), deaf and hearing native signers, hearing nonsigners, and hearing late learners of ASL viewed sign language sentences contrasted with sign gibberish. Deaf and hearing native signers showed significant activation in right hemisphere posterior parietal and posterior temporal regions. These activation patterns were not seen in nonsigners, nor were they observed in hearing late learners of sign language. A group analysis of hearing participants confirmed that only hearing native users of ASL, but not late learners who had learned ASL after puberty, recruited the right angular gyrus during this task. Newman and colleagues suggested that the activation of this neural region may be a neural signature of sign language being acquired during the critical period for language.

MacSweeney, Woll, Campbell, Gemma, et al. (2002) compared the role of left and right parietal cortices in an anomaly detection task in BSL. They tested deaf and hearing native signers in a paradigm that utilized BSL sentence contexts that either made use of spatial-topographic signing space or did not require spatial mapping. Similar to the findings reported by Newman et al. (2002), native deaf signers in the MacSweeney et al. study also showed activation in the right angular gyrus (BA 39). Parietal activation in the hearing native signers, however, was modulated by accuracy on the task, with more accurate subjects showing greater activation. This finding suggests that proficiency, rather than age of acquisition, may be a critical determinant of right hemisphere engagement. Importantly, activation in right hemisphere temporal-parietal regions was specific to BSL and was not observed in hearing nonsigners watching audiovisual English translations of the same sentences.

Neuroimaging studies of working memory processes in users of sign language have also reported modality-specific activations in response to the perception and maintenance of sign-based information. These activations typically include bilateral temporal-occipital activations and posterior parietal cortex (Buchsbaum et al., 2005; Pa, Wilson, Pickell, Bellugi, & Hickok, 2008; Ronnberg, Rudner, & Ingvar, 2004).

In an electrophysiological study of ASL sentence processing, Capek et al. (2009) reported left anterior negativity (i.e., LAN) syntactic verb-agreement violations in ASL. However, in a condition where the ungrammaticality was signaled by a spatially inflecting verb that lacked an overtly specified object, anterior negativity (200–360 ms) that was largest over the right lateral frontal sites was observed.

Many researchers have speculated that right hemisphere parietal activation in signers is associated with the linguistic use of space in sign language, and recent studies have sought to clarify the contributions of spatial processing in ASL signing to observed right hemisphere activations. A positron emission tomography (PET) study by Emmorey et al. (2002) is noteworthy in this regard. Emmorey et al. (2002) required deaf subjects to examine line drawings of two spatially arrayed objects and produce either a classifier description or a description of the spatial relationship using ASL lexical prepositions. This study found evidence for right hemisphere SMG activation for both prepositional forms and classifiers compared with object naming; however, the direct comparison between classifier constructions and lexical prepositions in sign revealed only left hemisphere inferior parietal lobule activation. This counterintuitive finding suggests that right hemisphere activation must be related to some common process, perhaps the spatial analysis of stimulus to be described, rather than a special spatial-linguistic property of ASL classifiers *per se*. Note this finding would support the contention of Atkinson et al. that restricted right hemisphere comprehension impairments seen in some deaf aphasics may reflect more generalized disruptions of core mechanisms needed for spatial analysis.

36.5 SIGN LANGUAGE AND THE MIRROR NEURON SYSTEM

Studies of the neurobiology sign language have been called on to inform motor simulation theories of language, particularly mirror neuron–based accounts of language evolution and language processing. Arbib (2005, 2008) and Rizzolatti and Arbib (1998), Rizzolatti and Craighero (2004) propose that a human analog of the bilateral frontoparietal macaque mirror neuron system that subserves action execution and action perception supports a variety of complex sociocognitive phenomena, most notably human action understanding and human language. Signed languages provide an important opportunity to examine hypotheses such as that sign language production requires the orchestration of complex manual gestural forms and its comprehension depends on the direct perception of these same gestures.

Data from lesion and neuroimaging studies place some limits on the feasibility of the mirror neuron account of sign language. For example, researchers have suggested that Broca's region is a likely homologue of Macaque F5, a frontal ventral region where neurons with mirroring properties have been observed (Gallese, Fadiga, Fogassi, & Rizzolatti, 1996). Yet, as discussed, studies of aphasic patients show that damage to left frontal ventral and opercular regions do not result in comprehension deficits in signers, as would be predicted from a mirror neuron account of human language. Recent group-level analyses of signing aphasics reinforce this point (Rogalsky et al., 2013). In addition, case studies of deaf signers have shown compelling dissociations between nonlinguistic pantomime and sign language abilities. Patient W.L., for example, produced and comprehended pantomime normally but demonstrated marked deficits in sign language production and comprehension (Corina et al., 1992). His gestures were clearly intended to convey symbolic information that he ordinarily would have imparted with sign language. Marshall et al. (2004) reported on the dissociation of sign and gesture in Charles, an aphasic user of BSL. His ability to use nonlinguistic gesture was intact while sign language expression showed significant impairment. These accounts are difficult to reconcile within a human mirror neuron system view of action execution and perception that fail to distinguish between different classes of human actions. Data from neuroimaging studies that have directly compared the perception of different classes of human actions and signs in deaf signers also show little evidence of overlap in the neural system mediating the perception of transitive and intransitive human actions, pantomimes, and linguistic signs (Corina et al., 2007; Emmorey, Xu, Gannon, Goldin-Meadow, & Braun, 2010 but see also MacSweeney et al. 2004). In these studies, activation patterns in response to sign language forms activate the now-familiar perisylvian language areas while nonlinguistic human actions and pantomimes show prominent activation in the inferior occipital-temporal regions in deaf signers. One explanation for these differences is that deaf signers may show a greater reliance on top-down processing in the recognition of signs, leading to a more automatic and efficient early visual processing of highly familiar linguistic features. In contrast, nonlinguistic gesture detection may be driven by bottom-up processing, in which preliminary visual analysis is crucial to interpretation of these forms.

Some support for the involvement of a unified action perception/action execution system for sign language comes from consideration of the functional role of the parietal lobes in sign language behaviors. Mirror neuron system accounts of language processing hold that one and the same region should participate both in the production and comprehension of language forms. In a review, Corina and Knapp (2006) compiled neuroimaging data from studies of sign language production and sign language comprehension and examined regions that showed significant overlap. One region that showed compelling overlap was in the left SMG of the inferior parietal lobe. Recall that lesion studies have shown evidence for sign comprehension deficits with damage to left inferior parietal regions, and both lesion and stimulation mapping studies have shown impairment in sign production implicating the SMG. This is a region that is often observed in studies sign language working memory (Buchsbaum et al., 2005; Pa et al., 2008; Ronnberg et al., 2004). In hearing individuals, this region is considered a polymodal association region with involvement in tasks that range from phonological short-term memory (Jacquemot & Scott, 2006), to judgments of hand gestures (Hermsdorfer et al., 2001), to motor attention to limb movements (Rushworth, Krams, & Passingham, 2001), and to phonological agraphia (Alexander, Friedman, Loverso, & Fischer, 1992). Given this multiplicity of functions of the SMG, it is perhaps not surprising that aspects of sign comprehension and sign production are subserved by this region. Whether individual neurons within this region exhibit mirror properties remains an important question (see Chong, Cunnington, Williams, Kanwisher, & Mattingley, 2008, for example).

In summary, functional imaging studies have largely replicated and helped extend the findings of lesion studies in deaf signers. These studies point to the importance of the left hemisphere in sign language production and comprehension, and have helped illuminate the modality-independent regions of the left hemisphere perisylvian language network. Functional imagining studies typically permit more directed inquiry than lesion studies. As such, these investigations have begun to address more subtle issues related to the modality-dependent contributions of the left and right parietal lobes in sign language processing and have proved to be an important source of data for constraining contemporary theories of biological basis of language.

36.5.1 Morphometric Studies

A small number of morphometric studies have been conducted to examine the presence of anatomical differences in the brains of deaf and hearing subjects. Two studies have reported reduced white matter in the left posterior superior temporal gyrus adjacent to language cortex in deaf subjects, but no difference in gray matter volume of temporal auditory and speech areas (Emmorey, Allen, Bruss, Schenker, & Damasio, 2003; Shibata, 2007). It is speculated that the reduced white matter volume may indicate a hypoplasia in the development of specific tracts related to speech. The finding that auditory cortices show no differences in gray matter volume has been taken as evidence for preserved functionality of these regions, perhaps in the form of cross-modal plasticity. As noted previously, several studies have shown activation of auditory cortex in response to the perception of visual stimuli and visual language. It is also interesting to note that in the Shibata study, deaf subjects showed trends for larger gray matter differences in superior frontal gyrus (BA 6) thought to reflect differences related to the use of manual language in this right-handed cohort of signers. In a study examining morphology of the insula in deaf and signing populations, Allen et al. reported volumetric differences attributed to both auditory deprivation and sign experience. Deaf subjects exhibited a significant increase in the amount of gray matter in the left posterior insular lobule, which may be related to dependence on lip-reading and articulatory (rather than auditory-based) representation. In contrast to nonsigners, both deaf and hearing signers exhibited increased volume in white matter in the right insula. This latter difference may be attributed to increased reliance on cross-modal sensory integration in sign compared with spoken languages (Allen, Emmorey, Bruss, & Damasio, 2008).

36.6 CONCLUSION

Patterns for deficits seen in deaf signers following left and right hemisphere lesions show great commonalities to the deficits seen in users of spoken languages. Functional imaging studies of sign language comprehension and production in deaf and hearing signers provide further evidence for the uniformity in the neural systems underlying spoken and signed languages. Taken together, these studies provide evidence for a core neurobiological system that underlies human language, regardless of the modality of expression (Fedorenko, Behr, & Kanwisher, 2011). At the same time, and as might be expected, studies of sign language have also revealed instances in which the properties of sign languages invoke neural processing resources that differ from those observed for spoken languages. These instances force us to acknowledge the possibility of language specificity of linguistics impairment (and whether these fall under the purview of aphasic deficits) and illuminate the intimate connections between language modality and brain structures for language. Neurolinguistics studies of signed languages have been used to motivate and constrain models of language evolution and language biology offered by mirror neuron system accounts of language understanding. Structural imaging studies of deaf signers have begun to observe and differentiate the roles of sensory and language experience in sculpting cortical connections. Studies of signed languages and the experiences of deaf individuals provide an important means to advance our understanding of the neurobiology of human language.

Acknowledgments

This work was support by grants NIH NIDCD R01 DC011538 and NSF SBE-1041725.

References

Alexander, M. P, Friedman, R. B., Loverso, F., & Fischer, R. S. (1992). Lesion localization of phonological agraphia. *Brain and Language*, *43*, 83–95.

Allen, J. S., Emmorey, K., Bruss, J., & Damasio, H. (2008). Morphology of the insula in relation to hearing status and sign language experience. *The Journal of Neuroscience*, *28*(46), 11900–11905.

Arbib, M. A. (2005). From monkey-like action recognition to human language: An evolutionary framework for neurolinguistics (with commentaries and author's response). *Behavioral and Brain Sciences*, *28*, 105–167.

Arbib, M. A. (2008). From grasp to language: Embodied concepts and the challenge of abstraction. *Journal de Physiologie Paris*, *102*, 4–20.

Atkinson, J, Marshall, J., Woll, B., & Thacker, A. (2005). Testing comprehension abilities in users of British Sign Language following CVA. *Brain and Language*, *94*, 233–248.

Bavelier, D., Corina, D. P., Jezzard, P., Clark, V., Karni, A., Lalwani, A., et al. (1998). Hemispheric specialization for English and ASL: Left invariance-right variability. *Neuroreport*, *9*(7), 1537–1542.

Bellugi, U., Poizner, H., & Klima, E. S. (1989). Language, modality and the brain. *Trends in Neuroscience*, *12*(10), 380–388.

Blumstein, S. E. (1973). *A phonological investigation of aphasic speech*. The Hague: Mouton.

Braun, A. R., Guillemin, A., Hosey, L., & Varga, M. (2001). The neural organization of discourse: An H2 15O-PET study of narrative production in English and American Sign Language. *Brain*, *124* (10), 2028–2044.

Brownell, H. H., Simpson, T. L., Bihrle, A. M., Potter, H. H., & Gardner, H. (1990). Appreciation of metaphoric alternative word meanings by left and right brain-damaged patients. *Neuropsychologia*, *28*(4), 375–383.

Buchsbaum, B., Pickell, B., Love, T., Hatrak, M., Bellugi, U., & Hickok, G. (2005). Neural substrates for verbal working memory in deaf signers: fMRI study and lesion case report. *Brain and Language*, *95*(2), 265–272.

Calvert, G. A., Bullmore, E. T., Brammer, M. J., Campbell, R., Williams, S. C. R., McGuire, P. K., et al. (1997). Activation of auditory cortex during silent lipreading. *Science*, *276*(5312), 593–596.

Calvert, G. A, & Campbell, R. (2003). Reading speech from still and moving faces: The neural substrates of visible speech. *Journal Cognitive Neuroscience*, *15*, 57–70.

Capek, C., Bavelier, D., Corina, D. P., Newman, A. J., Jezzard, P., & Neville, H. J. (2004). The cortical organization of audio-visual sentence comprehension: an fMRI study at 4Tesla. *Cognitive Brain Research*, *2*, 111–119.

Capek, C. M., Grossi, G., Newman, A. L., McBurney, S. L., Corina, D., Roeder, B., et al. (2009). Brain systems mediating semantic and syntactic processing in deaf native signers: Biological invariance and modality specificity. *Proceedings of the National Academy of Sciences of the United States of America*, *106*, 8784–8798.

Chiarello, C., Knight, R., & Mandel, M. (1982). Aphasia in a prelingually deaf woman. *Brain*, *105*(1), 29–51.

Chong, T. T, Cunnington, R., Williams, M. A, Kanwisher, N., & Mattingley, J. B. (2008). fMRI adaptation reveals mirror neurons in human inferior parietal cortex. *Current Biology: CB*, *18*, 1576–1580.

Corina, D., Neville, H. J., & Bavelier, D. (1998). Response from Corina, Neville and Bavelier. *Trends in Cognitive Sciences*, *2*(12).

Corina, D. P. (1998a). Aphasia in users of signed language. In P. Coppens, Y. Lebrun, & A. Basso (Eds.), *Aphasia in atypical populations* (pp. 261–309). Mahwah, NJ: Lawrence Erlbaum Associates.

Corina, D. P. (1998b). The processing of sign language: Evidence from aphasia. In H. Whitaker, & B. Stemmer (Eds.), *Handbook of neurology* (pp. 313–329). San Diego, CA: Academic Press.

Corina, D. P. (2000). Some observations regarding paraphasia in American Sign Language. In K Emmorey, & H. Lane (Eds.), *The signs of language revisited: An anthology to honor Ursula Bellugi and Edward Klima* (pp. 493–507). Mahwah, NJ: Lawrence Erlbaum Associates.

Corina, D. P., & Knapp, H. (2006). Sign language processing and the mirror neuron system. *Cortex*, *42*(4), 529–539. Available from: http://dx.doi.org/10.1016/S0010-9452(08)70393-70399.

Corina, D. P., Chiu, Y., Knapp, H., Greenwald, R., San Jose-Robertson, L., & Allen Braun, A. R. (2007). Neural correlates of human action observation in hearing and deaf subjects. *Brain Research*, *1152*, 111–129.

Corina, D. P., Kritchevsky, M., & Bellugi, U. (1996). Visual language processing and unilateral neglect: Evidence from American Sign Language. *Cognitive Neuropsychology*, *3*(13), 321–356.

Corina, D. P., Lawyer, L., & Cates, D. (2013). Cross-linguistic differences in the neural representation of human language: Evidence from users of signed languages. *Frontiers in Psychology*, *3*, 1–8. Available from: http://dx.doi.org/10.3389/fpsyg.2012.00587.

Corina, D. P., McBurney, S. L., Dodrill, C., Hinshaw, K., Brinkley, J., & Ojemann, G. (1999). Functional roles of Broca's area and SMG: Evidence from cortical stimulation mapping in a deaf signer. *NeuroImage*, *10*(5), 570–581.

Corina, D. P., Poizner, H., Bellugi, U., Feinberg, T., Dowd, D., & O'Grady-Batch, L. (1992). Dissociation between linguistic and nonlinguistic gestural systems: A case for compositionality. *Brain and Language*, *43*(3), 414–447.

Corina, D. P., & Sandler, W. (1993). On the nature of phonological structure in sign language. *Phonology*, *10*(2), 165–207.

Corina, D. P., San Jose-Robertson, L., Guillemin, A., High, J., & Braun, A. R. (2003). Language lateralization in a bimanual language. *Journal of Cognitive Neuroscience*, *15*(5), 718–730.

Damasio, A. R. (1992). Aphasia. *New England Journal of Medicine.*, *326* (8), 351–539.

Davis, M. H., & Johnsrude, I. S. (2003). Hierarchical processing in spoken language comprehension. *The Journal of Neuroscience, 23* (8), 3423–3431.

Dronkers, N., & Baldo, J. (2009). Language: Aphasia. In L. R. Squire (Ed.), *Encyclopedia of neuroscience 5* (pp. 343–348). Oxford: Academic Press.

Dronkers, N. F., Redfern, B. B., & Knight, R. T. (2000). The neural architecture of language disorders. In M. S. Gazzaniga (Ed.), *The new cognitive neurosciences* (pp. 949–958). Cambridge, MA: MIT Press.

Dronkers, N. F., Redfern, B. B., & Ludy, C. A. (1995). Lesion localization in chronic Wernicke's aphasia. *Brain and Language, 51*(1), 62–65.

Dudis, P. (2007). *Types of depiction in ASL.* Washington, DC: Manuscript Gallaudet University.

Emmorey, K. (1993). Processing a dynamic visual-spatial language: Psycholinguistic studies of American Sign Language. *Journal of Psycholinguistic Research, 22*(2), 153–188.

Emmorey, K. (2003). *Perspectives on classifier constructions in sign languages.* Mahwah, NJ: Lawrence Erlbaum Associates.

Emmorey, K., Allen, J. S., Bruss, J., Schenker, N., & Damasio, H. (2003). A morphometric analysis of auditory brain regions in congenitally deaf adults. *Proceedings of the National Academy of Sciences of the United States of America, 100*(17), 10049–10054.

Emmorey, K., Corina, D. P., & Bellugi, U. (1995). Differential processing of topographic and referential functions of space. In K. Emmorey, & J. Snitzer Reilly (Eds.), *Language, gesture, and space* (pp. 43–62). Hillsdale, NJ: Lawrence Erlbaum Associates.

Emmorey, K., Damasio, H., McCullough, S., Grabowski, T., Ponto, L. L. B., Hichwa, R. D., et al. (2002). Neural systems underlying spatial language in American Sign Language. *NeuroImage, 17*(2), 812–824.

Emmorey, K., Mehta, S., & Grabowski, T. J. (2007). The neural correlates of sign versus word production. *NeuroImage, 36*(1), 202–208.

Emmorey, K., Xu, J., Gannon, P., Goldin-Meadow, S., & Braun, A. (2010). CNS activation and regional connectivity during pantomime observation: No engagement of the mirror neuron system for deaf signers. *NeuroImage, 49*, 994–1005.

Fedorenko, E., Behr, M. K., & Kanwisher, N. (2011). Functional specificity for high-level linguistic processing in the human brain. *Proceedings for the National Academy of Sciences, 108*(39), 16428–16433. Available from: http://dx.doi.org/10.1073/pnas.1112937108.

Fine, I., Finney, E. M., Boynton, G. M., & Dobkins, K. R. (2005). Comparing the effects of auditory deprivation and sign language within the auditory and visual cortex. *Journal of Cognitive Neuroscience, 17*(10), 1621–1637.

Finney, E. M., Clementz, B. A., Hickok, G., & Dobkins, K. R. (2003). Visual stimuli activate auditory cortex in deaf subjects: Evidence from MEG. *Neuroreport, 14*(11), 1425–1427.

Gallese, V., Fadiga, L., Fogassi, L., & Rizzolatti, G. (1996). Action recognition in the premotor cortex. *Brain, 119*(Pt 2), 593–609.

Goodglass, H. (1993). *Understanding aphasia.* San Diego, CA: Academic Press.

Goldsmith, J. (1990). *Autosegmental and metrical phonology.* Oxford: Basil Blackwell.

Groce, N. E. (1985). *Everyone here spoke sign language: Hereditary deafness on Martha's Vineyard.* Cambridge, MA: Harvard University Press.

Hermsdorfer, J., Goldenberg, G., Wachsmuth, C., Conrad, B., Ceballos-Baumann, A., & Bartenstein, P. (2001). Cortical correlates of gesture processing: Clues to the cerebral mechanisms underlying apraxia during the imitation of meaningless gestures. *NeuroImage, 14*, 149–161.

Hickok, G., Bellugi, U., & Klima, E. S. (1998). What's right about the neural organization of sign language? A perspective on recent neuroimaging results. *Trends in Cognitive Science, 2*, 465–468.

Hickok, G., Kritchevsky, M., Bellugi, U., & Klima, E. (1996). The role of the left frontal operculum in sign language. *Neurocase, 2*, 373–380.

Hickok, G., Love-Geffen, T., & Klima, E. S. (2002). Role of the left hemisphere in sign language comprehension. *Brain and Language, 82*(2), 167–178.

Hickok, G., Pickell, H., Klima, E., & Bellugi, U. (2009). Neural dissociation in the production of lexical versus classifier signs in ASL: Distinct patterns of hemispheric asymmetry. *Neuropsychologia, 47* (2), 382–387 (for classifier production data)

Hickok, G., Wilson, M., Clark, K., Klima, E. S., Kritchevsky, & Bellug, U. (1999). Discourse deficits following right hemisphere damage in deaf signers. *Brain and Language, 66*(2), 233–248.

Horwitz, B., Amunts, K., Bhattacharyya, R., Patkin, D., Jeffries, K., Zilles, K., et al. (2003). Activation of Broca's area during the production of spoken and signed language: A combined cytoarchitectonic mapping and PET analysis. *Neuropsychologia, 41*(14), 1868–1876.

Indefrey, P., & Levelt, W. J. M. (2004). The spatial and temporal signatures of word production components. *Cognition, 92*, 101–144.

Jacquemot, C., & Scott, S. K. (2006). What is the relationship between phonological short-term memory and speech processing? *Trends in Cognitive Science, 10*(11), 480–486.

Kaplan, J. A., Brownell, H. H., Jacobs, J. R., & Gardner, H. (1990). The effects of right hemisphere damage on the pragmatic interpretation of conversational remarks. *Brain and Language, 38*(2), 315–333.

Kassubek, J., Hickok, G., & Erhard, P. (2004). Involvement of classical anterior and posterior language areas in sign language production, as investigated by 4T functional magnetic resonance imaging. *Neuroscience Letters, 364*(3), 168–172.

Kauramäki, J., Jääskeläinen, I. P., Hari, R, Möttönen, R., Rauschecker, J. P., & Sams, M. (2010). Lipreading and covert speech production similarly modulate human auditory-cortex responses to pure tones. *The Journal of Neuroscience, 30*, 1314–1321.

Lambertz, N., Gizewski, E. R., de Greiff, A., & Forsting, M. (2005). Cross-modal plasticity in deaf subjects dependent on the extent of hearing loss. *Cognitive Brain Research, 25*(3), 884–890.

Leischner, A. (1943). Die "apasie" der taubstummen. *Archiv fur Psychiatrie und Nervenkrankheiten, 115*, 469–548.

Liddell, S. K. (2003). *Grammar, gesture, and meaning in American Sign Language.* New York, NY: Cambridge University Press.

MacSweeney, M., Campbell, R., Woll, B., Brammer, M. J., Giampietro, V., David, A. S., et al. (2006). Lexical and sentential processing in British Sign Language. *Human Brain Mapping, 27*(1), 63–76.

MacSweeney, M., Campbell, R., Woll, B., Giampietro, V., David, A. S., McGuire, P. K., et al. (2004). Dissociating linguistic and nonlinguistic gestural communication in the brain. *NeuroImage, 22*(4), 1605–1618.

MacSweeney, M., Capek, C. M., Campbell, R., & Woll, R. (2008). The signing brain: The neurobiology of sign language. *Trends in Cognitive Sciences, 12*(11), 432–440.

MacSweeney, M., Woll, B., Campbell, R., Gemma, A., Calvert, G. A., Philip, K., et al. (2002). Neural correlates of British sign language comprehension: Spatial processing demands of topographic language. *Journal of Cognitive Neuroscience, 14*(7), 1064–1075.

MacSweeney, M., Woll, B., Campbell, R., McGuire, P. K., David, A. S., Williams, S. C., et al. (2002). Neural systems underlying British Sign Language and audio-visual English processing in native users. *Brain, 125*(7), 1583–1593.

Marshall, J., Atkinson, J., Smulovitch, E., Thacker, A., & Woll, B. (2004). Aphasia in a user of British Sign Language: Dissociation between sign and gesture. *Cognitive Neuropsychology, 21*(5), 537–554.

McGuire, P. K., Robertson, D., Thacker, A., David, A. S., Kitson, N., Frackowiak, R. S., et al. (1997). Neural correlates of thinking in sign language. *Neuroreport*, *8*(3), 695–698.

Mohr, J. P., Pessin, M. S., Finkelstein, S., Funkenstein, H. H., Duncan, G. W., & Davis, K. R. (1978). Broca aphasia: Pathologic and clinical. *Neurology*, *28*(4), 311–324.

Naeser, M. A., Helm-Estabrooks, N., Haas, G., Auerbach, S., & Srinivasan, M. (1987). Relationship between lesion extent in Wernicke's area on computed tomographic scan and predicting recovery of comprehension in Wernicke's aphasia. *Archives of Neurology*, *44*(1), 73–82.

Neville, H. J., Bavelier, D., Corina, D., Rauschecker, J. P., Karni, A., Lalwani, A., et al. (1998). Cerebral organization for language in deaf and hearing subjects: Biological constraints and effects of experience. *Proceedings of the National Academy of Sciences of the United States of America*, *95*(3), 922–929.

Newman, A. J., Bavelier, D., Corina, D., Jezzard, P., & Neville, H. J. (2002). A critical period for right hemisphere recruitment in American Sign Language processing. *Nature Neuroscience*, *5*(1), 76–80.

Nishimura, H., Hashikawa, K., Doi, K., Iwaki, T., Watanabe, Y., Kusuoka, H., et al. (1999). Sign language "heard" in the auditory cortex. *Nature*, *397*(6715), 116.

Pa, J., Wilson, S. M., Pickell, B., Bellugi, U., & Hickok, G. (2008). Neural organization of linguistic short-term memory is sensory modality-dependent: Evidence from signed and spoken language. *Journal of Cognitive Neuroscience*, *20*, 2198–2210.

Petitto, L. A., Zatorre, R. J., Gauna, K., Nikelski, E. J., Dostie, D., & Evans, A. C. (2000). Speech-like cerebral activity in profoundly deaf people processing signed languages: Implications for the neural basis of human language. *Proceedings of the National Academy of Sciences of the United States of America*, *97*(25), 13961–13966.

Poizner, H., Bellugi, U., & Klima, E. S. (1987). *What the hands reveal about the brain*. Cambridge, MA: MIT Press.

Poizner, H., & Kegl, J. (1992). Neural basis of language and motor behavior: Perspectives from American Sign Language. *Aphasiology*, *6*(3), 219–256.

Rehak, A., Kaplan, J. A., Weylman, S. T., Kelly, B., Brownell, H. H., & Gardner, H. (1992). Story processing in right-hemisphere brain-damaged patients. *Brain and Language*, *42*(3), 320–336.

Rizzolatti, G., & Arbib, M. A. (1998). Language within our grasp. *Trends in Neuroscience*, *21*, 188–194.

Rizzolatti, G., & Craighero, L. (2004). The mirror-neuron system. *Annual Review of Neuroscience*, *27*, 169–192.

Rogalsky, C, Kristin Raphel, K., Tomkovicz, V., O'Grady, L., Damasio, H., Bellugi, U., et al. (2013). Neural basis of action understanding: Evidence from sign language aphasia. *Aphasiology*, *27*(9), 1147–1158.

Ronnberg, J., Rudner, M., & Ingvar, M. (2004). Neural correlates of working memory for sign language. *Brain Research, Cognitive Brain Research*, *20*, 165–182.

Rushworth, M. F., Krams, M., & Passingham, R. E. (2001). The attentional role of the left parietal cortex: The distinct lateralization and localization of motor attention in the human brain. *Journal of Cognitive Neuroscience*, *13*, 698–710.

Sakai, K. L., Tatsuno, Y., Suzuki, K., Kimura, H., & Ichida, Y. (2005). Sign and speech: Amodal commonality in left hemisphere dominance for comprehension of sentences. *Brain*, *128*(6), 1407–1417.

Sandler, W., & Lillo-Martin, D. C. (2006). *Sign language and linguistic universals*. Cambridge, UK; New York, NY: Cambridge University Press.

Sandler, W., Meir, I., Padden, C., & Aronoff, A. (2005). The emergence of grammar: Systematic structure in a new language. *Proceedings of the National Academy of Sciences of the United States of America*, *7*(102), 2661–2665. Available from: http://dx.doi.org/10.1073/pnas.0405448102.

San José-Robertson, L., Corina, D. P., Ackerman, D., Guillemin, A., & Braun, A. R. (2004). Neural systems for sign language production: Mechanisms supporting lexical selection, phonological encoding, and articulation. *Human Brain Mapping*, *23*(3), 156–167.

Schlosser, M. J., Aoyagi, N., Fulbright, R. K., Gore, J. C., & McCarthy, G. (1998). Functional MRI studies of auditory comprehension. *Human Brain Mapping*, *6*(1), 1–13.

Scott, G. D., Karns, C. M., Dow, M. W., Stevens, C., & Neville, H. J. (2014). Enhanced peripheral visual processing in congenitally deaf humans is supported by multiple brain regions, including primary auditory cortex. *Frontiers in Human Neuroscience*, *8*, 177. Available from: http://dx.doi.org/10.3389/fnhum.2014.00177.

Shibata, D. K. (2007). Differences in brain structure in deaf persons on MR imaging studied with voxel-based morphometry. *American Journal of Neuroradiology*, *28*(2), 243–249.

Söderfeldt, B., Ingvar, M., Rönnberg, J., Eriksson, L., Serrander, M., & Stone-Elander, S. (1997). Signed and spoken language perception studied by positron emission tomography. *Neurology*, *49*(1), 82–87.

Supalla, T. (1982). *Structure and acquisition of verbs of motion and location in American Sign Language*. PhD thesis, Department of Psychology, University of California, San Diego, CA.

SECTION F

PERCEPTUAL ANALYSIS OF THE SPEECH SIGNAL

CHAPTER

37

Phoneme Perception

Jeffrey R. Binder

Department of Neurology, Medical College of Wisconsin, Milwaukee, WI, USA

Human vocalizations include the set of sounds called phonemes, which are the basic linguistic units of speech. Phonemes are approximately equivalent to the set of vowel and consonant sounds of a language. For example, the word *cat* comprises the three phonemes (here transcribed in International Phonetic Alphabet notation) /k/, /æ/, and /t/. Changing any one of these phonemes, such as substituting /p/ for /k/, results in a different word. Thus, a phoneme is often defined as the smallest contrastive speech unit that brings about a change of meaning. Phonemes are also the product of complex motor acts involving the lungs, diaphragm, larynx, tongue, palate, lips, and so on. The actual sounds produced by this complex apparatus vary from one instance to the next, even for a single speaker repeating the same phoneme. They are also altered by the context in which the act occurs, resulting in allophonic variations such as the difference in the middle "t" sound of *night rate* and *nitrate*. Phoneme realizations also show large variations across geographical and social groups, resulting in regional and class accents. Therefore, a phoneme is best viewed as an abstract representation of a set of related speech sounds that are perceived as equivalent within a particular language. The main task of phoneme perception systems in the brain is to enable this perceptual equivalence.

The acoustic analysis of phonemes is a relatively recent science, becoming systematic only with the development of acoustic spectral analysis in the 1940s using analog filter banks (Koenig, Dunn, & Lacy, 1946). Accordingly, the standard classification of phonemes refers mainly to the motor gestures that produce them. "Manner" of articulation refers to the general type of gesture performed. Examples of distinct manners of articulation include: transient complete closure of the vocal tract, called a stop or plosive (example: p, d, k); sustained constriction of the vocal tract resulting in turbulent air flow, called a fricative (example: f, th, s); lowering of the soft palate and relaxation of the upper pharynx during occlusion of the oral tract to allow air flow and resonance in the nasal cavity, called a nasal (example: m, n, ng); and relatively unobstructed phonation using the tongue and lips to shape the resonant properties of the oral tract, resulting in a vowel. "Place" of articulation refers to the location in the vocal tract at which maximal restriction occurs during a gesture. Examples include: restriction at the lips, referred to as bilabial (example: b, p, m); restriction with the tip of the tongue against the teeth or alveolar palate, referred to as coronal (example: th, d, s); and restriction with the body of the tongue against the soft palate, called velar (example: g, k). Finally, the consonants are divided into two "voicing" categories on the basis of whether the vocal folds in the larynx are made to vibrate during or very close to the time of maximal articulation. When phonatory muscles in the larynx contract, the vocal folds vibrate due to periodic build-up of air pressure from the lungs, resulting in an audible pitch or voice. When this vibration is present during maximal articulation, the consonant is "voiced" (example: b, z, g). When no vocal fold vibration is present, the consonant is "unvoiced" (example: p, s, k). For stop consonants followed by a vowel, voicing is largely determined by the time that elapses between release of the stop and the onset of vocal fold vibration, referred to as voice onset time (VOT).

Although derived originally from motor descriptions, each of these distinctions is associated with corresponding acoustic phenomena (Figure 37.1). Vocal cord vibrations produce a periodic sound wave at a particular rate, called the fundamental frequency of the voice (F0), as well as harmonics at multiples of the fundamental. The relative intensity of these harmonics is determined by the shape of the vocal tract, which changes dynamically due to movement of the tongue and other articulators, altering the resonant (and antiresonant) properties of the tract. Resonant harmonics form distinct bands of increased power in the spectrum, called formants (numbered F1, F2, etc.), and the center frequency and

Neurobiology of Language. DOI: http://dx.doi.org/10.1016/B978-0-12-407794-2.00037-7

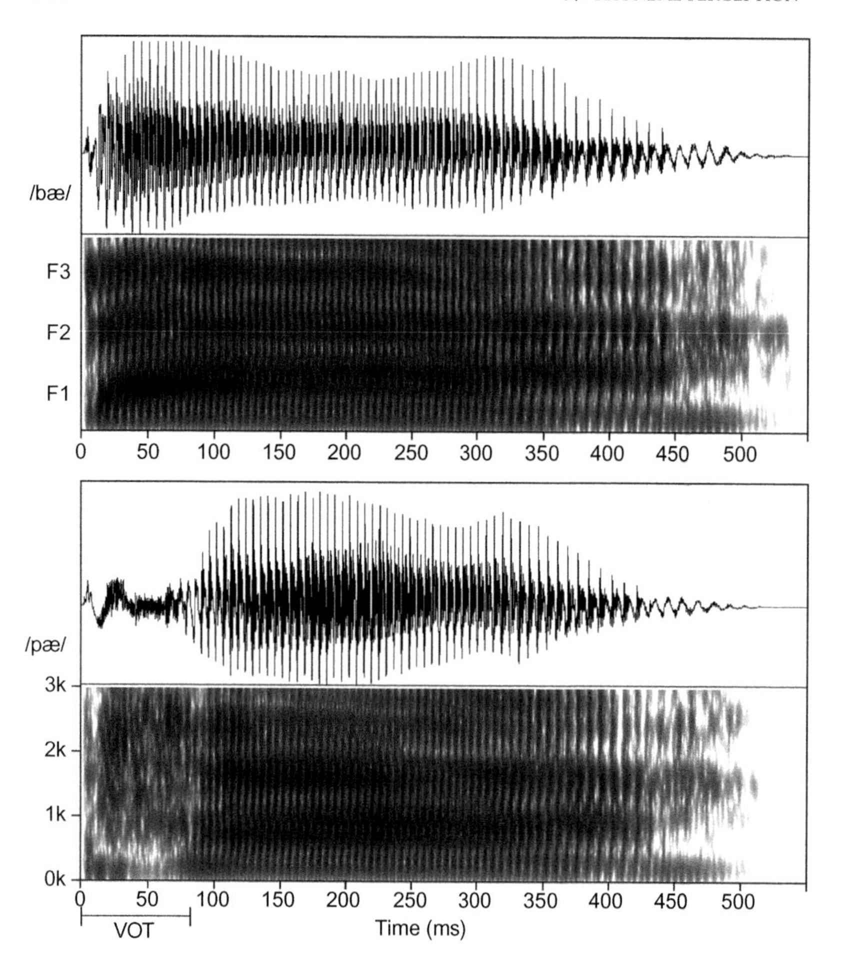

FIGURE 37.1 Acoustic waveforms (top row in each pair) and time-frequency displays ("spectrograms") showing spectral power at each point in time (bottom row in each pair) for the syllables /bæ/ and /pæ/ (as in "bat" and "pat"). Vertical striations seen in both displays are due to vocal cord vibrations at the fundamental frequency of the speaker's voice. Dark horizontal bands in the time-frequency displays are formants (labeled F1, F2, and F3 on the /bæ/ display) caused by resonances related to the shape of the oral cavity. The first 50–60 ms of /bæ/ is the period of formant transition, during which the jaw opens and the spectral position of the formants changes rapidly. Unvoiced /pæ/ differs from voiced /bæ/ in that the period of formant transition is replaced in /pæ/ by a period of aspiration noise (indicated by "VOT") during which there are no vocal cord vibrations. The absence of voicing also greatly shortens the formant transition duration and raises the starting frequency of F1 for /pæ/.

bandwidth of these formants change as the vocal tract changes shape. Stop consonants are characterized by abrupt changes in tract shape as the jaw, tongue, and lips open or close, resulting in rapid changes (typically occurring over 20–50 ms) in the center frequencies of the formants (called formant transitions), as well as rapid changes in overall intensity as air flow is released or stopped. Fricatives are characterized by sustained, high-frequency broadband noise produced by turbulent air flow at the point of vocal tract constriction. Nasal consonants are characterized by a prominent nasal formant at approximately 250–300 Hz (the "nasal murmur") and damping of higher-frequency oral formants.

The acoustic correlates of consonant place of articulation were a central focus of much early research in acoustic phonetics, in large part due to the difficulty in identifying invariant characteristics that distinguished stops in different place categories. Place distinctions among stops are associated with differences in the direction and slope of formant transitions, but it was quickly noticed that these parameters are context-dependent. For example, the second formant rises during articulation of syllable-initial /d/ when the following vowel is "ee," but it falls during articulation of the same segment when the following vowel is "ah" or "oo" (Delattre, Liberman, & Cooper, 1955). Such context dependency, which is ubiquitous in speech, proved challenging to explain and even led to a prominent theory proposing that phoneme identification depends on activation of an invariant motor representation (Liberman, Cooper, Shankweiler, & Studdert-Kennedy, 1967), although exactly how this could occur if the acoustic signal is uninformative was never fully specified. Later studies showed that overall spectral shape at the time of stop release and change in overall shape just after release provide a context-invariant acoustic cue for place of articulation (Kewley-Port, 1983; Stevens & Blumstein, 1981). Similarly, place of articulation for fricatives and nasal consonants is cued primarily by gross spectral shape (Hughes & Halle, 1956; Kurowski & Blumstein, 1987).

Phoneme perception is often based on multiple interacting acoustic cues. The cues signaling stop consonant voicing are a clear case in point. Although VOT is the most prominent cue, with VOT values longer than 25–30 ms generally creating a strong unvoiced percept, unvoiced stops (i.e., p, t, and k) are also characterized by aspiration (i.e., noise from the burst of air that accompanies release of the stop), a higher center frequency of the first formant at the onset of voicing, and a shorter formant transition duration, among other cues (Lisker, 1986) (Figure 37.1). Studies using synthesized speech show that these cues jointly contribute to voicing discrimination in that reduction of one cue can be compensated for by an increase in others (e.g., reduction of VOT can be "traded" for an increase in aspiration amplitude or first formant onset frequency to maintain the likelihood of an unvoiced percept) (Summerfield & Haggard, 1977).

As mentioned, a major task of the phoneme perception system is to detect these acoustic cues despite extensive variation in their realization. In addition to variation due to phonetic context (i.e., preceding or following phonemes), speaker accent, and a host of purely random factors, speakers vary in overall speed of production and in the length and size of their vocal tracts, producing variation in the absolute value of fundamental and formant center frequencies. Two types of mechanisms are thought to mitigate this problem. The first are normalization processes that adjust for variation in vocal tract characteristics and articulation speed. Evidence suggests, for example, that although absolute formant frequencies for any particular phoneme vary between speakers, the ratios of formants are more constant and therefore provide more invariant cues (Miller, 1989). Similarly, there is evidence that listeners automatically apply a temporal normalization process that adjusts for variation in speed of production within and across speakers (Sawusch & Newman, 2000).

A second mechanism for dealing with acoustic variation is known as categorical perception. This term refers to a relative inability to perceive variation within a set of items belonging to the same perceptual category. The phenomenon is most strikingly demonstrated using synthesized speech continua that vary in acoustically equidistant steps from one phoneme category to another, such as from /b/ to /p/ (Liberman, Harris, Hoffman, & Griffith, 1957). Listeners fail to perceive changes in VOT from 0 to 20 ms, for example, but can easily perceive a change from 20 to 40 ms. The 0-ms and 20-ms VOT sounds are both perceived as the same voiced stop consonant, whereas the 40-ms VOT sound is perceived as unvoiced and, therefore, a different phoneme. Importantly, the phenomenon is not simply attributable to response bias or limited availability of response labels, because it is observed using alternative forced-matching (e.g., ABX) tasks as well as labeling tasks. Categorical perception indicates that the phoneme perception system has an automatic (i.e., subconscious) mechanism for "suppressing" acoustic detail of a speech sound, leaving behind, in consciousness, mainly the abstract identity of the phoneme. Which specific details are suppressed, however, depends on the phoneme categories to which one is exposed during early linguistic experience. Speakers of English, for example, cannot appreciate distinctions between the dentoalveolar and (postalveolar) retroflex stops of Hindi and other Indo-Aryan languages, whereas these are distinct phonemes with categorical perceptual boundaries for Hindi listeners (Pruitt, Jenkins, & Strange, 2006). It thus appears that the phoneme perceptual system becomes "tuned" during language development through bottom-up statistical learning processes that reflect overrepresentation of category prototypes in the language environment (Kuhl, 2000; Werker & Tees, 2005). The phoneme categories that result from this tuning process can be thought of as "attractor states" (Damper & Harnad, 2000) or "perceptual magnets" (Kuhl, 1991) that "pull" the neural activity pattern toward a target position in phoneme perceptual space regardless of small variations in the acoustic input.

Before moving on to a consideration of relevant neurobiological data, two additional core aspects of phoneme perception should be mentioned. The first involves the integration of visual and auditory input. Many of the articulatory movements used in speaking are visible, and because this visual information is strongly correlated with the resulting acoustic phenomena, it is useful for phoneme perception. For people with normal hearing, visual information is particularly helpful in noisy environments (Sumby & Pollack, 1954). For people with little or no hearing, visual information alone can often provide enough information for phoneme recognition (Bernstein, Demorest, & Tucker, 2000). Auditory and visual information are "integrated" at a relatively early, preconscious level during phoneme perception, as shown by the McGurk illusion, in which the perceived identity of a heard phoneme is altered by changing the content of a simultaneous video display (McGurk & MacDonald, 1976).

A final point is that phoneme perception should not be thought of as an entirely bottom-up sensory process. As with all perceptual recognition tasks, identification of the sensory input is influenced by knowledge about what it is most *likely* to be. One illustration of this phenomenon is the influence of lexical status on phoneme category boundaries (Ganong, 1980; Pitt & Samuel, 1993). A sound that is acoustically near the VOT boundary between /b/ and /p/, for example, is likely to be heard as /b/ after the stem /dræ/, thereby forming a word (drab) rather than a nonword (drap), whereas the same sound is likely to be heard as /p/ after the stem /træ/ (again, favoring a real word over a nonword). The connectionist model of phoneme perception called TRACE instantiates this top-down influence using feedback connections from lexical to phoneme representations (McClelland & Elman, 1986). Although feedback from higher levels provides one biologically plausible account of lexical effects on phoneme perception, it may not be strictly necessary (Norris, McQueen, & Cutler, 2000). Another example of "top-down" effects in phoneme perception is the phenomenon of "perceptual restoration," in which a phoneme is perceived within a word despite the fact that the actual phoneme has been replaced by a segment of noise or other nonspeech sound (Samuel, 1997; Warren & Obusek, 1971). An extreme example of such "top-down restoration" is the successful perception of phonemes in sounds constructed of three or four pure tones that represent only the center frequency and intensity of the formants in the original speech. Naïve listeners initially perceive such "sine wave speech" as nonverbal "chirps" or "electronic sounds," but they are then immediately able to hear phonemes on being told that the sounds are speech (Remez, Rubin, Pisoni, & Carrell, 1981). Such phenomena vividly illustrate the general principle that what we perceive is determined as much by what we expect as by what is available from sensory input.

37.1 NEUROPSYCHOLOGICAL STUDIES

As described in the preceding section, phoneme perception in the healthy brain is influenced by stored lexical and phonetic knowledge. However, neuropsychological studies amply demonstrate that phoneme perception and higher lexical processes are often differentially affected by focal lesions. For example, patients with transcortical sensory aphasia have severe impairments of word comprehension but are perfectly able to repeat words they do not understand (Berthier, 1999). Intact repetition indicates preserved perception of phonemes, and thus the word comprehension deficit must be due to either an impairment in mapping the phonemes to a lexical representation or an impairment at the semantic level. Conversely, patients with "pure word deafness" (PWD) are unable to recognize or repeat spoken words but have normal understanding of written language and normal language production, suggesting a relatively specific impairment at the level of phoneme perception (Buchman, Garron, Trost-Cardamone, Wichter, & Schwartz, 1986). This well-documented double dissociation should be kept in mind when encountering ambiguous terms such as "speech comprehension" and "intelligibility," which unnecessarily conflate different levels of processing involving distinct kinds of information.

Early studies of PWD provided the first hard evidence that speech sounds are processed in the temporal lobe. The work of Salomon Henschen, in particular, distinguished lesions causing specific phoneme perception impairment from lesions causing other forms of "word deafness" (at the time, this term was used to designate any speech comprehension impairment), linking the former to damage in the superior temporal gyrus (STG) (Henschen, 1918–1919). Lesion data from patients with PWD strongly suggest some degree of bilateral phoneme processing, because the vast majority of cases have bilateral STG lesions (Buchman et al., 1986; Poeppel, 2001). Particularly compelling in this regard are those who have a unilateral STG lesion with no symptomatic speech perception impairment and then develop PWD only after the contralateral STG is affected (Buchman et al., 1986; Ulrich, 1978). Also relevant to the issue of bilateral representation is evidence from intracarotid amobarbital (Wada) testing that shows that patients maintain much of their ability to discriminate phonemes even after anesthetization of the language-dominant hemisphere (Boatman et al., 1998; Hickok et al., 2008). Taken as a whole, these data suggest that phoneme perceptual processes are much more bilaterally represented than are later stages of speech comprehension. Nevertheless, PWD after unilateral damage is occasionally reported, and the damage in these instances is nearly always on the left side, suggesting a degree of asymmetry in the bilateral representation (Buchman et al., 1986; Poeppel, 2001).

PWD is a misnomer for several reasons. The deficit involves perception of phonemes, not words; therefore, sublexical phoneme combinations and other nonword stimuli are affected at least as much as real words. More appropriate designations include "phoneme deafness" and "auditory verbal agnosia." The "purity" of the deficit is also highly variable. Although patients with PWD have normal pure tone hearing thresholds, most cases show impairments on more complex auditory perceptual tasks that do not involve speech sounds, such as recognition of nonspeech environmental sounds, pitch discrimination and other musical tasks, and various measures of temporal sequence perception (Buchman et al., 1986; Goldstein, 1974). The variety and severity of such nonverbal perceptual impairments are generally greater in patients with bilateral lesions than in those with unilateral left lesions. Two patients with unilateral left temporal lesions, for example, showed relatively circumscribed deficits involving perception of rapid spectral changes, including impaired consonant but spared vowel discrimination (Stefanatos, Gershkoff, & Madigan, 2005; Wang, Peach, Xu, Schneck, & Manry, 2000).

37.2 FUNCTIONAL IMAGING STUDIES

Phoneme perception has been the focus of many functional imaging studies. The loud noises produced by rapid gradient switching present a challenge for fMRI studies of auditory processing, because these noises not only mask experimental stimuli but also activate many of the brain regions of interest, limiting the dynamic range available for hemodynamic responses. Most auditory fMRI studies avoid these problems by separating successive image volume acquisitions with a silent period during which the experimental stimuli are presented (Hall et al., 1999). If the acquisitions are separated by a sufficient amount of time (i.e., >7 seconds), then the measured BOLD responses primarily reflect activation by the experimental stimuli rather than the noise produced by the previous image acquisition.

Compared with a silent baseline, speech sounds (whether syllables, words, pseudowords, or reversed speech) activate most of the STG bilaterally (Binder et al., 2000; Wise et al., 1991). Although this speech–silence contrast is sometimes advocated as a method of "mapping Wernicke's area" for clinical purposes (Hirsch et al., 2000), the resulting activation is typically symmetric and is unrelated to language lateralization as determined by Wada testing (Lehéricy et al., 2000). The activated regions include Heschl's gyrus and surrounding dorsal STG areas that contain primary and belt auditory cortex (Rauschecker & Scott, 2009), suggesting that much of the activation is due to low-level, general auditory processes. A variety of nonspeech control stimuli (frequency-modulated tones, amplitude-modulated noise) were introduced to identify areas that might be more speech-specific (Binder et al., 2000; Jäncke, Wüstenberg, Scheich, & Heinze, 2002; Mummery, Ashburner, Scott, & Wise, 1999; Zatorre, Evans, Meyer, & Gjedde, 1992). The main finding in these studies was that primary and belt regions on the dorsal STG respond similarly to speech and simpler nonspeech sounds, whereas areas in the ventral STG and superior temporal sulcus (STS) respond more to speech than to simpler sounds (see DeWitt & Rauschecker, 2012 for an activation likelihood estimation [ALE] meta-analysis). The pattern suggests a hierarchically organized processing stream running from the primary auditory cortex on the dorsal surface to higher-level areas in the lateral and ventral STG, a pattern very reminiscent of the core-belt-parabelt organization in monkey auditory cortex (Kaas & Hackett, 2000).

The tone and noise controls used in the studies just mentioned are far less acoustically complex than speech sounds, leaving open the possibility that the observed ventral STG/STS activation reflects general acoustic complexity rather than phoneme-specific processing. The degree to which speech perception requires "special" mechanisms has been a contentious topic for many decades (Diehl & Kluender, 1989; Liberman et al., 1967; Liberman & Mattingly, 1985; Lotto, Hickok, & Holt, 2009). This unresolved issue motivated functional imaging researchers to begin the search for auditory cortical areas that respond selectively to speech phonemes compared with nonphonemic sounds of equivalent spectrotemporal complexity.

Scott, Blank, Rosen, and Wise (2000) addressed this problem using spectrally "rotated" speech. In this manipulation, the instantaneous auditory spectrum at each point in time is rotated about a central frequency axis, such that high frequencies become low frequencies and vice versa (Blesser, 1972). The rotation preserves overall spectrotemporal complexity (though the long-term spectrum and spectral distribution of dynamic features are both greatly altered) while reducing phoneme intelligibility. The contrast between speech and rotated speech produced activation of the left middle and anterior STS. A number of subsequent studies using similar methods have replicated this initial finding (Davis & Johnsrude, 2003; Friederici, Kotz, Scott, & Obleser, 2010; Narain et al., 2003; Okada et al., 2010). One difficulty with interpreting these results arises from the fact that the stimuli were sentences and therefore contained considerable semantic and syntactic information. Phoneme perception would have been perfectly correlated with engagement of semantic and syntactic processes. Thus, what is sometimes referred to as an "acoustically invariant response to speech" may simply reflect semantic or syntactic processing independent of phoneme perception per se. This would explain the somewhat anterior location of the activation compared with prior speech studies, because semantic and syntactic integration at the sentence level is thought to particularly engage the anterior temporal lobe (DeWitt & Rauschecker, 2012; Humphries, Binder, Medler, & Liebenthal, 2006; Humphries, Swinney, Love, & Hickok, 2005; Pallier, Devauchelle, & Dehaene, 2011; Spitsyna, Warren, Scott, Turkheimer, & Wise, 2006; Vandenberghe, Nobre, & Price, 2002; Visser, Jefferies, & Lambon Ralph, 2010).

Subsequent work focused more specifically on phoneme perception using a variety of controls for spectrotemporal complexity. Liebenthal, Binder, Spitzer, Possing, and Medler (2005) created synthetic nonphonemic syllables by inverting the direction of the first formant (F1) transition in a series of stop consonants. Stops are characterized acoustically by a rising F1 trajectory that results from opening the closed vocal tract and sudden enlargement of the oral cavity (Fant, 1973). Inverting the F1 trajectory results in speech-like sounds that have no phoneme identity. Brain activity was measured during ABX discrimination along a synthetic continuum ranging in equal steps from /b/ to /d/ (accomplished by gradual shift of the F2 transition from rising to falling) and along an otherwise similar nonphonemic continuum with inverted F1 transitions. Participants showed typical categorical perception behavior (superior discrimination across the category boundary compared with within category) for the phoneme continuum, but a flat discrimination function for the nonphonemic items, verifying that the latter stimuli failed to evoke a phoneme percept. The conditions were matched on overall accuracy (i.e.,

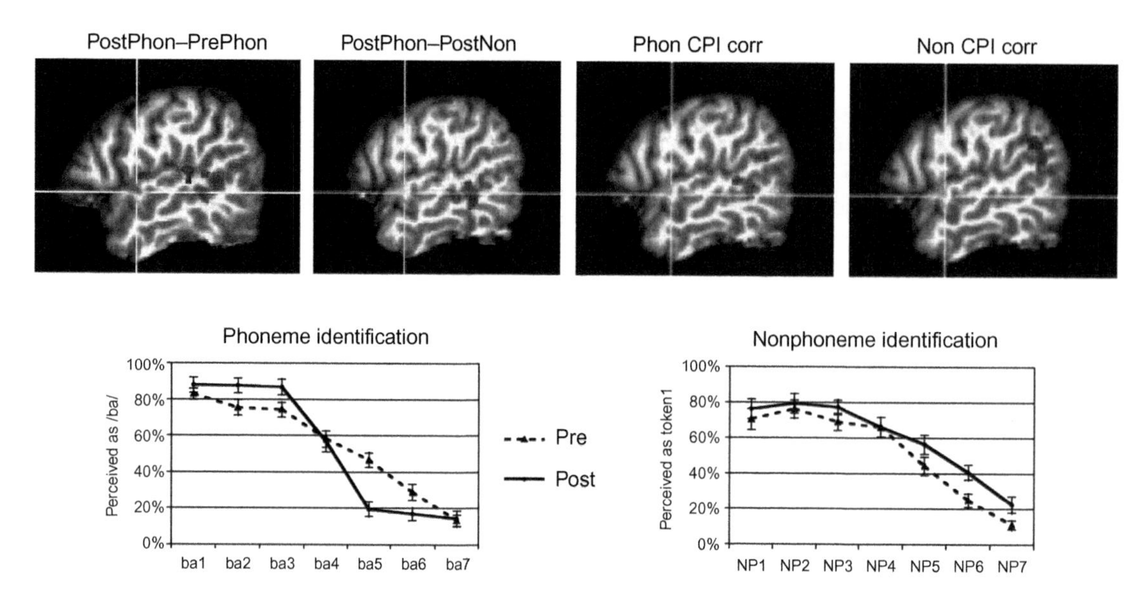

FIGURE 37.2 Left posterior temporal activation correlated with perception of phonemes in sine wave speech. The first upper panel (PostPhon–PrePhon) shows left posterior STS areas engaged only after the phonemic stimuli were perceived as /ba/ and /da/ as a result of training. The second upper panel (PostPhon–PostNon) shows selective activation of this region by phonemic relative to nonphonemic stimuli after training. Graphs in the lower panels show group average identification functions for the phonemic (left) and nonphonemic (right) continua, and the selective induction of categorical labeling for the phonemic stimuli after training. The third upper panel shows areas where increased activation for the phonemic stimuli from before (Pre) to after (Post) was correlated with individual change in slope of the phoneme identification function from Pre to Post (phonemic categorical perception index). The final upper panel shows areas where increased activation for the nonphonemic stimuli from Pre to Post was correlated with individual change in slope of the nonphonemic identification function from Pre to Post (nonphonemic CPI). *Adapted from Desai et al. (2008) with permission from the publisher.*

discriminability) and reaction time. A contrast between the phonemic and nonphonemic conditions showed activation localized in the mid-to-posterior left STS (peaks at Talairach coordinates −60, −8, −3 and −56, −31, 3). The spectrotemporally matched control condition effectively rules out auditory complexity as an explanation for this effect. This portion of the left STS thus appears to be a location where complex auditory information activates phoneme category representations.

Liebenthal et al. (2010) replicated this result using the same stimuli together with a labeling task. Event-related potentials data were obtained simultaneously with fMRI, showing that the difference between phonemic and nonphonemic processing begins at approximately 180 ms after stimulus onset and peaks at approximately 200 ms, manifesting as a larger P2 deflection in the phonemic condition.

Dehaene-Lambertz et al. (2005) and Möttönen et al. (2006) addressed the problem of spectrotemporal complexity confounds by comparing activation to sine wave speech presented before and after training subjects to perceive the sounds as phonemes. In both studies, participants performed a discrimination task in the naïve state during which they reported hearing the sounds as nonspeech. After explicit instruction that the sounds contained particular phonemes and after training in the scanner with explicit phoneme identification tasks, the participants were scanned again. Dehaene-Lambertz et al. included a manipulation of across-category versus within-category discrimination and demonstrated a specific improvement in across-category discrimination after training. In both studies, a left posterior STS region showed stronger activation after training compared with before training. The clusters were somewhat more posterior along the STS than in the Liebenthal studies (Dehaene-Lambertz et al. peaks: −60, −24, 4 and −56, −40, 0; Möttönen et al. peak: −61, −39, 2), although both overlap with the posterior aspect of the Liebenthal et al. clusters.

Desai, Liebenthal, Waldron, and Binder (2008) combined these approaches by comparing sine wave syllables and nonphonemic sine wave stimuli before and after training. The nonphonemic items were created by inverting the initial trajectory of the sine tone representing "F1," analogous to the approach followed by Liebenthal et al. with their speech sounds. The question the authors addressed was whether the expectation of hearing phonemes after training is sufficient in itself to activate the left STS, or if it is necessary for the stimulus to contain phonemic information. The main results (Figure 37.2) suggest that both top-down expectation and bottom-up phoneme input are necessary to activate the left STS region observed in the prior studies, and that the same region is activated whether comparing phoneme sounds after versus before training (replicating the Dehaene-Lambertz et al. and Möttönen et al. studies) or phonemic and nonphonemic sounds after training. Desai et al. also obtained labeling data for the phonemic and nonphonemic continua before and after training and used a logistic function to model the slope of the boundary region as an index of categorical perception. This index increased significantly for the phoneme continuum after training but not for the nonphonemic continuum (Figure 37.2, bottom), and activation in the left STS was correlated with individual variation in the size of this increase in categorical perception.

It is noteworthy that in all of these studies in which phonemic stimuli were contrasted with nonphonemic stimuli of equal spectrotemporal complexity, the phoneme-specific responses were strongly left-lateralized. In contrast to the data from Wada testing and patients with PWD, discussed previously, these fMRI studies suggest that, at least in the intact brain, the left auditory system performs most of the computation related specifically to phoneme perception. The issue of lateralization of phoneme perception is discussed further in the final section of this chapter.

A major unanswered question from these fMRI studies is the extent to which the left STS response is necessarily exclusive to phoneme sounds. Phonemes are, by definition, highly familiar sounds for which behaviorally relevant perceptual categories have developed as a result of extensive experience. Would the left STS respond similarly during categorical perception of nonphoneme sounds? Does the response depend on the particular acoustic properties of phonemes, or does it reflect a more general process of auditory category perception? Desai et al. addressed this issue using individual variation in the categorical perception index they derived from their nonphonemic labeling functions (Figure 37.2). Although the participants, on average, showed no increase in categorical perception of these sounds after training, the small variation in the categorical perception index across participants actually predicted the level of activation in a small region in the left supramarginal gyrus and posterior STS. This cluster was somewhat posterior and dorsal to the one activated by phoneme perception (Figure 37.2). The authors proposed that the left posterior STS serves a general function of auditory category perception, and that small differences in the location of activation for phonemic and nonphonemic sounds reflect differences in familiarity between these types of categories.

Desai et al. (2008) provided only minimal perceptual training with their nonphonemic continuum. In a follow-up study, Liebenthal et al. (2010) asked whether more extensive training that induced stronger categorical perception of nonphonemic sounds would result in a similar recruitment of left posterior STS regions by phonemic and nonphonemic sounds after training. Participants participated in four sessions in which they were taught to label items in the nonphonemic speech continuum developed by Liebenthal et al. (2005) using two (arbitrary) category names. Identification functions became more categorical after training (Figure 37.3, top). Activation in the left mid-STS, which had been stronger for phonemic sounds prior to training, showed no difference between phonemic and nonphonemic stimuli after training, and a large region in left posterior middle temporal gyrus (MTG) and STS showed stronger responses to the nonphonemic sounds after training compared with before training (Figure 37.3, bottom). An interaction analysis showed that the left posterior STS was recruited by training specifically for the nonphonemic sounds (Figure 37.3, bottom right).

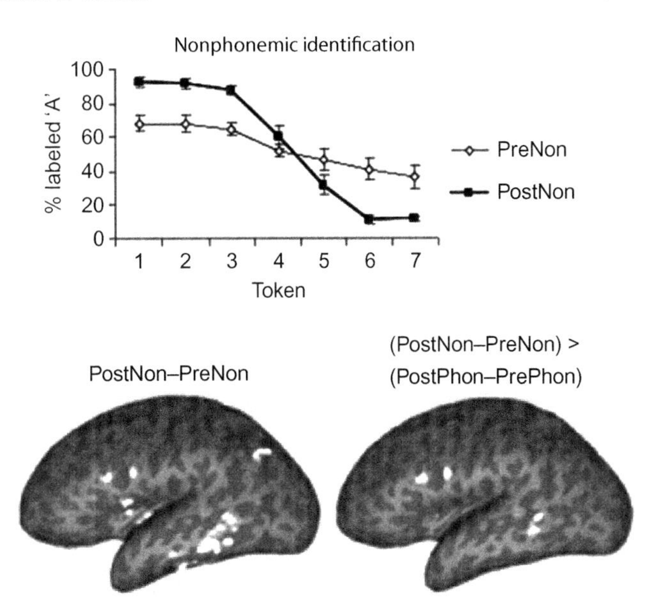

FIGURE 37.3 Coding of novel speech categories in the posterior left STS. The top graph shows induction of categorical labeling of nonphonemic consonant-like speech sounds after four training sessions. Training resulted in recruitment of a large left middle temporal region (lower left) and a more focal region of posterior left STS that was recruited more by learning novel nonphonemic categories than by similar training with familiar phonemic categories (/ba/ and /da/). *Adapted from Liebenthal et al. (2010) with permission from the publisher.*

These five studies paint a somewhat complex picture of phoneme processing in the left STS, summarized in Figure 37.4. Highly familiar phonemes, for which strong and overlearned perceptual categories exist, selectively engage the mid-portion of the left STS compared with closely matched acoustic control stimuli. Sine wave speech phonemes engage a slightly more posterior region of left STS compared with the same sounds heard as nonspeech. Finally, perception of newly learned nonphonemic speech categories engages still more posterior regions in the left STS, MTG, and supramarginal gyrus.

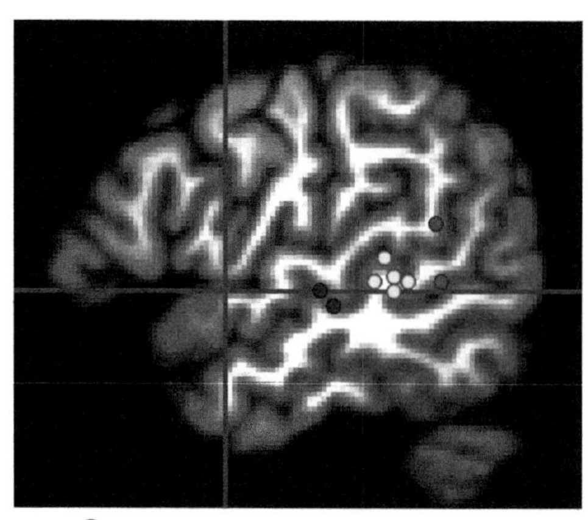

FIGURE 37.4 Summary of five controlled studies of phoneme and matched nonphoneme perception. Red markers indicate peaks in two studies comparing phonemic (/ba/-/da/) with spectrotemporally matched nonphonemic speech sounds (Liebenthal et al., 2005, 2010). Yellow markers indicate peaks from three studies examining induction of phoneme perception with sine wave speech (Dehaene-Lambertz et al., 2005; Desai et al., 2008; Möttönen et al., 2006). Blue markers indicate peaks in two studies examining acquisition of categorical perception with nonphonemic sine wave stimuli (Desai et al., 2008) and nonphonemic synthetic speech (Liebenthal et al., 2010).

One interpretation of this "familiarity gradient" is that it reflects a greater dependence on short-term auditory memory mechanisms in the case of less familiar sounds. As the mapping from spectrotemporal forms to categories becomes less automatic, the spectrotemporal information needs to be maintained in a short-term memory store. For speech sounds, this memory store likely constitutes the first stage in a supra-sylvian speech production pathway (Hickok & Poeppel, 2007; Wise et al., 2001), although the evidence obtained from nonphonemic categorization tasks (Desai et al., 2008; Liebenthal et al., 2010) indicates that these short-term auditory memory representations are not strictly speech-specific (Figure 37.4).

The perceptual integration of auditory and visual phoneme information has been examined in a number of studies, most of which manipulated the congruency or temporal synchrony of auditory and visual speech inputs (see Calvert, 2001 for an early review). These studies have generally implicated the left posterior STS and superior parietal regions such as the intraparietal sulcus in audiovisual integration of speech. Miller and D'Esposito (2005) used synchrony manipulation together with an event-related analysis of perceptual fusion judgments to separate brain responses due to successful perceptual fusion from responses reflecting the degree of synchrony. The left mid-STS (MNI coordinates −46, −28, 0) and left Heschl's gyrus responded more strongly when the audiovisual stop consonants were successfully fused than when they were perceived as separate events. A number of other regions, mainly nodes in the attention and cognitive control networks, showed the reverse pattern. The left STS response depended on successful perceptual fusion but showed no sensitivity to the degree of actual audiovisual synchrony.

Recent studies also implicate the left mid-STS in audiovisual integration processes underlying the McGurk illusion. Susceptibility to the McGurk effect is known to vary across individuals. Nath and Beauchamp (2012) divided their participant sample into low- and high-susceptibility McGurk perceivers, defined by their probability of perceiving "da" when presented with simultaneous auditory "ba" and visual "ga." High-susceptibility perceivers showed stronger responses to these stimuli in a functionally defined audiovisual left mid-STS region of interest (ROI) (mean center-of-mass Talairach coordinates −58, −28, 4) compared with low-susceptibility perceivers. Across all participants, activation in this ROI was significantly correlated with individual susceptibility scores. In a complementary study, Beauchamp, Nath, and Pasalar (2010) showed that single-pulse transcranial magnetic stimulation (TMS) applied to a functionally defined audiovisual left mid-STS region disrupted the McGurk illusion in high-susceptibility perceivers, causing participants to report only the auditory stimulus. Taken together, these results suggest that a focal region in the left mid-STS integrates auditory and visual phoneme information, but that the visual input to this region is more variable across individuals and more easily disrupted.

37.3 DIRECT ELECTROPHYSIOLOGICAL RECORDINGS

Intracranial cortical electrophysiological recording (electrocorticography, or ECog) offers a means of recording neural activity directly rather than indirectly via a hemodynamic response (as in positron emission tomography [PET] or fMRI). The technique provides much higher temporal resolution than fMRI and can also provide more precise spatial localization. Recent work has focused on induced changes in high gamma (>60 Hz) power, which is tightly correlated with local neural activity and the BOLD signal (Crone, Miglioretti, Gordon, & Lesser, 1998; Goense & Logothetis, 2008). The method usually involves a regularly spaced grid of electrodes draped on the surface convexity of the brain. One limitation of this approach is that the electrodes make contact with gyral crests and do not record from sulcal depths. Thus, most ECog studies of phoneme perception provide data regarding the lateral STG but little or none regarding the STS.

ECog studies suggest that overall spectrotemporal form is relatively well-preserved in the pattern of neural activity on the left lateral mid-STG. The recorded spectrotemporal neural activity is sufficient to reconstruct intelligible speech waveforms (Pasley et al., 2012), indicating that the represented information is primarily, if not exclusively, acoustic rather than linguistic or symbolic in nature. Mesgarani, Cheung, Johnson, and Chang (2014) demonstrated that this region contains patches selective for different kinds of phonemes and phoneme features, with selectivity perhaps most closely reflecting manner cues such as overall spectral shape and dynamic changes in spectral shape. Thus, distinct groups of electrodes respond selectively to the following: plosive consonants, which are characterized by rapid changes over a broad frequency spectrum; sibilant fricatives, which have less dynamic and much more focused high-frequency energy; vowels, which have less dynamic and grossly bimodal frequency spectra; and nasals, which are characterized by prominent low-frequency energy. The fricative manner category includes both sibilant and nonsibilant (anterior) phonemes, and it is notable that the anterior voiced fricatives /v/ and /ð/ (as in "vee" and "thee") did not pattern with the sibilant phonemes. This is further evidence that the patterns reflect spectrotemporal information, which is quite different for sibilant unvoiced fricatives and anterior voiced fricatives, rather than more abstract phonetic features like "fricative."

Another notable finding from these studies is the combinatorial nature of responses in the lateral STG. For example, Mesgarani et al. report that areas sensitive to vowels show negatively correlated sensitivity to first and second formant values, indicating that they integrate information about F1 and F2 and respond optimally to particular F1/F2 combinations. Such combinatorial encoding not only is expected given the essential acoustic properties that distinguish vowels (Peterson & Barney, 1952) but also provides a probable mechanism for perceptual normalization across speakers with different vocal tract properties (Sussman, 1986).

ECog data also provide novel information about the timing of activation across the STG and STS during phoneme perception. Canolty et al. (2007) compared spoken word stimuli with nonphonemic sounds created by removing formant detail from the auditory waveform. High gamma responses were larger for words than for the nonphonemic sounds, and these differences emerged at 120 ms after stimulus onset in the posterior STG, at 193 ms in the mid-STG, and at 268 ms in the mid-STS. This timing is in good agreement with the findings of a larger scalp P2 response for phonemic compared with nonphonemic sounds peaking at approximately 200 ms (Liebenthal et al., 2010). The ECog data furthermore show the spread of activation over time from earlier posterior auditory areas to higher-level anterior and lateral STG cortex and, finally, to mid-STS, consistent with the proposed hierarchical organization of this network (Binder et al., 2000; Kaas & Hackett, 2000).

These data add enormously to our understanding of spectrotemporal encoding in higher-level auditory cortex. They indicate a process of multidimensional acoustic feature representation from which subphonemic features arise through combinatorial processing. The extent to which such representations are specific to phonemes would require further testing with appropriate nonphonemic controls, but presumably identical subphonemic acoustic features would be represented by the same neurons in lateral STG regardless of whether they occurred in phonemes or nonphonemic sounds. As described in the preceding section, it appears likely that encoding of more abstract phonetic category percepts occurs at still higher levels of processing performed in the left STS.

37.4 THE ROLE OF ARTICULATORY REPRESENTATIONS IN PHONEME PERCEPTION

The role of motor speech codes in phoneme perception has been a long and heatedly debated topic. There are theories emphasizing that speech sounds arise from articulatory gestures, and that these gestures, or at the least the motor programs that underlie them, are less variable than the sounds they produce (Galantucci, Fowler, & Turvey, 2006; Liberman et al., 1967). According to such theories, motor programs provide an internal template (or "efference copy") of the phoneme against which the auditory input can be matched. On the other side of the debate are purely auditory theories of phoneme perception in which motor templates have no necessary role (Diehl & Kluender, 1989; Lotto et al., 2009). An intermediate view is that motor templates are increasingly engaged and become more necessary as the auditory signal is more unfamiliar or degraded (Callan, Jones, Callan, & Akahane-Yamada, 2004). There seems little doubt that strong connections exist between speech perception and articulation systems, enabling auditory feedback to guide articulation during speech production (Guenther, Hampson, & Johnson, 1998; Tourville, Reilly, & Guenther, 2008). It therefore appears reasonable that these connections might operate in reverse during phoneme perception tasks.

Numerous fMRI and PET studies show activation in the left precentral gyrus (ventral premotor cortex) and inferior frontal gyrus (IFG) pars opercularis during speech perception tasks (Binder et al., 2000; Burton, Small, & Blumstein, 2000; Démonet et al., 1992), and show common activation in these areas during speech perception and production tasks (Callan, Callan, Gamez,

Sato, & Kawato, 2010; Callan et al., 2006; Tremblay & Small, 2011; Wilson & Iacoboni, 2006; Wilson, Saygin, Sereno, & Iacoboni, 2004). A persistent difficulty with interpreting these posterior frontal activations is the possibility that they may in some cases reflect decision, working memory, or other cognitive control processes rather than activation of motor codes (Binder, Liebenthal, Possing, Medler, & Ward, 2004). A more specific activation of motor representations is suggested by TMS studies showing enhancement of motor-evoked potentials in articulatory muscles during passive listening to speech (Fadiga, Craighero, Buccino, & Rizzolatti, 2002; Watkins, Strafella, & Paus, 2003) (see Yuen, Davis, Brysbaert, & Rastle, 2010 for related evidence). These findings suggest that merely listening to speech sounds activates motor speech areas sufficiently to lower the threshold for eliciting an evoked potential. Pulvermuller et al. (2006) presented fMRI evidence that this activation is articulator-specific. Premotor areas controlling tongue and lip movements were determined using motor localizer tasks. The tongue area was differentially activated by hearing the coronal stop /t/ relative to hearing the bilabial stop /p/, and conversely the lip area was more activated by /p/ compared to /t/.

As pointed out by many authors, the common activation in premotor areas by production and auditory perception of speech gestures is analogous to the cross-modal responsiveness of mirror neurons to execution and visual observation of other actions (Rizzolatti & Arbib, 1998; Rizzolatti & Craighero, 2004). In both cases the performer is simultaneously a perceiver and uses perceptual information to guide performance, presumably creating strong functional connections between perceptual and motor networks.

The possibility that motor cortex activation might be merely an epiphenomenon unrelated to the perceptual process itself was countered by several studies showing that repetitive TMS delivered to the articulatory motor cortex to produce a functional "lesion" reduces phoneme discrimination performance (D'Ausilio et al., 2009; Meister, Wilson, Deblieck, Wu, & Iacoboni, 2007; Möttönen & Watkins, 2009). In two of these studies, the effects on speech perception were specific to the place of articulation affected by TMS: identification of bilabial stops ("ba" and "pa") was selectively affected by TMS to the lip representation, whereas identification of lingual stops ("da" and "ta") was selectively affected by TMS to the tongue representation (D'Ausilio et al., 2009; Möttönen & Watkins, 2009). Such results provide rather compelling evidence for at least some contribution from motor codes to phoneme perception.

At issue is the extent to which this contribution is critical under typical listening circumstances. The aforementioned TMS lesion studies all used relatively difficult phoneme discrimination tasks with intelligibility compromised by adding noise or by manipulating the distinctiveness of synthetic formant transitions. Two other motor cortex TMS lesion studies showed no effect of TMS on perception of highly intelligible phonemes (D'Ausilio, Bufalari, Salmas, & Fadiga, 2012; Sato, Tremblay, & Gracco, 2009), suggesting that motor codes may only contribute to phoneme perception under adverse listening conditions. A more recent TMS study by Möttönen, Dutton, and Watkins (2013), however, provides evidence against this interpretation. They examined effects of motor cortex TMS on auditory mismatch negativity (MMN) responses to ignored speech. The MMN is a relatively preattentive auditory cortex response to infrequent deviant stimuli appearing within a sequence of repeating stimuli. Repetitive TMS to the motor lip representation reduced the magnitude of the MMN to deviant "ba" or "ga" syllables in a train of "da" syllables. This effect was specific to lip cortex stimulation and did not appear when the hand motor area was stimulated, when the lip cortex was stimulated and tone trains with intensity or duration deviants were used, or when the lip cortex was stimulated and syllable trains with duration deviants were used. One wrinkle in the results was that stimulation of the lip cortex also reduced the MMN to syllables with intensity deviations, which would be unexpected if the input from motor cortex reflects phoneme identity. Although these results do not imply that activation of motor cortex is necessary for successful conscious phoneme perception, they do demonstrate that input from motor areas involved in speech articulation reduces the ability of auditory cortex to process changes in phoneme input under noise-free listening conditions and in the absence of any attentional task.

Another notable aspect of the Möttönen and Watkins MMN study concerns the locus of motor activation effects. Activation of articulatory representations could affect perception in at least two distinct ways. According to the original Motor Theory of speech perception, articulatory representations are the actual targets of perception (Liberman et al., 1967). That is, perceptual decisions are based on activation of these codes rather than on auditory phoneme representations, and there is no role for feedback from motor to auditory codes. In contrast, most recent work assumes that motor codes provide feedback that assists phoneme recognition by the auditory system through a mechanism variously described as priming, prediction, or constraint. Suppression of the preattentive auditory MMN by motor cortex TMS suggests that a feedback mechanism from motor to auditory cortex plays at least some role.

37.5 HEMISPHERIC SPECIALIZATION IN PHONEME PERCEPTION

Speech contains much more than just phoneme information. Speech provides rich information about the age, gender, and specific identity of the speaker (based on fundamental frequency, voice quality and timbre, speed of articulation, accent, and other cues); emotional state of the speaker and affective content of the message (derived from emphasis cues and prosodic variation in fundamental frequency); and a variety of nonphonemic linguistic content (syllabic emphasis, semantic emphasis, sentential prosody). To refer to phoneme perception as "speech perception" is therefore both illogical and confusing, yet this usage is rampant and has contributed to some of the confusion regarding hemispheric lateralization of phoneme perception. As noted, speech sounds induce relatively symmetric bilateral activation of the STG compared with nonspeech sounds (tones and noise), which has contributed to a general sense that phoneme perception is symmetrically represented. Much of this right STG activation from speech, however, likely reflects processing of indexical and prosodic cues (Baum & Pell, 1999; Belin, Zatorre, Lafaille, Ahad, & Pike, 2000; Bonte, Hausfeld, Scharke, Valente, & Formisano, 2014; Latinus, McAleer, Bestelmeyer, & Belin, 2013; Lattner, Meyer, & Friederici, 2005; Ross, Thompson, & Yenkosky, 1997; Van Lancker, Kreiman, & Cummings, 1989; von Kriegstein, Eger, Kleinschmidt, & Giraud, 2003; Zatorre, Belin, & Penhune, 2002; Zatorre et al., 1992). With adequate controls for these cues, activation associated specifically with phoneme perception is much more strongly lateralized to the left STS (Dehaene-Lambertz et al., 2005; Desai et al., 2008; Liebenthal et al., 2005, 2010; Meyer et al., 2005; Möttönen et al., 2006).

Phoneme perception, however, does not depend solely on the left hemisphere. As noted, data from patients with focal brain lesions and patients undergoing Wada testing indicate that the right hemisphere has the capacity to perceive phonemes when the left hemisphere is compromised. Again, however, it is useful to note that unilateral left STG lesions occasionally produce phoneme perception deficits (this could depend on the extent of left STG and STS damage), and phoneme perception by the right hemisphere during Wada testing is not perfect (Hickok et al., 2008). Moreover, Boatman and colleagues demonstrated unequivocal phoneme perception deficits during unilateral electrical stimulation of the left mid-posterior STG (Boatman, Lesser, & Gordon, 1995). Perception was more impaired for consonant than for vowel discriminations (Boatman, Hall, Goldstein, Lesser, & Gordon, 1997). These mixed findings are difficult to explain with models that propose either bilateral symmetric processing of phonemes or unilateral left hemisphere processing, despite the attractiveness of such models. As with many theories of brain function, the weight of the evidence suggests a less dichotomous state of affairs in which phoneme information is processed mainly but not exclusively by the left STG/STS, and lesions to this region cause varying degrees of phoneme perception deficit, depending on the type and extent of the lesion and the premorbid functional capacity of the individual's right hemisphere.

Zatorre et al. proposed a theory of auditory hemispheric specialization in which the left auditory system excels at temporal resolution and the right auditory system excels at spectral resolution (Zatorre & Belin, 2001). Analogous to the trade-off in temporal versus spectral resolution that characterizes any time-frequency analysis, the left auditory system is proposed to integrate information across wide spectral bandwidths but over short time windows, whereas the right auditory system integrates across longer time windows in narrower spectral bands. As a general concept, the theory provides considerable explanatory power, accounting for right hemisphere specializations for prosody, melody, and suprasegmental linguistic perception, and left hemisphere specialization for perception of short-segment (i.e., <50 ms) phoneme cues. Specialization of the right auditory system for frequency discrimination and melody processing is now well-documented in functional imaging studies (Belin, Zilbovicius, Crozier, Thivard, & Fontaine, 1998; Brechmann & Scheich, 2005; Jamison, Watkins, Bishop, & Matthews, 2006; Schonwiesner, Rubsamen, & von Cramon, 2005; Zatorre & Belin, 2001; Zatorre et al., 2002), although the evidence supporting left hemisphere specialization for high temporal resolution is more mixed (Belin et al., 1998; Boemio, Fromm, Braun, & Poeppel, 2005; Jamison et al., 2006; Overath, Kumar, von Kriegstein, & Griffiths, 2008; Schonwiesner et al., 2005; Zaehle, Wustenberg, Meyer, & Jancke, 2004; Zatorre & Belin, 2001). McGettigan and Scott (2012) criticized the proposal on the basis that phoneme perception requires extensive spectral analysis; however, this criticism misses the central fact that the spectral resolution needed for phoneme perception is still very coarse. Discriminating place of articulation for stop consonants, for example, requires only an overall representation of spectral shape across a range of >1,000 Hz. Similarly, discriminating fricative place of articulation and voicing, voiced from unvoiced stops, and nasal from oral stops require only fairly coarse information about spectral shape over bandwidths of several hundred Hz (see Shannon, Zeng, Kamath, Wygonski, & Ekelid, 1995 for related evidence). In contrast, listeners can discriminate changes in F0 of 5 Hz or less (Klatt, 1973), which is required for prosodic perception (and for singing in tune!) yet constitutes spectral resolution several orders of magnitude finer than is needed for phoneme discrimination. Changes of this magnitude in formant center frequencies have no effect on phoneme intelligibility (Ghitza & Goldstein, 1983). Vowels are a somewhat special case in which phoneme discrimination depends on a relatively fine-grained analysis of spectral shape (although still coarse by musical standards) and involves longer segments. This difference is consistent with broad evidence for relatively right-lateralized processing of vowel categories (Boatman et al., 1997; Bonte et al., 2014; Shankweiler & Studdert-Kennedy, 1970; Stefanatos et al., 2005; Wang et al., 2000).

Working from the principles articulated by Zatorre, Poeppel proposed a relationship between asymmetric time scales and corresponding brain oscillations (Poeppel, 2003). Gamma frequency oscillations (40 Hz and higher) were proposed as the physiological correlate of integration over short time windows (20–50 ms) and linked with processing of phonemes, whereas alpha–theta oscillations (4–10 Hz) were proposed as the physiological correlate of integration over longer time windows (150–250 ms) and linked with processing of syllables and suprasegmental frequency modulations. The theory is otherwise (to this reader, at least) identical to Zatorre's. More recent versions of the theory emphasize the role of endogenous theta oscillations, which are proposed to be entrained by amplitude modulations at the syllable level in ongoing speech (Giraud & Poeppel, 2012). Entrainment of an endogenous theta rhythm enhances processing of nested gamma oscillations that sample the auditory input on a finer time scale to support processing of phoneme cues. Asymmetry arises because entrainment of theta oscillations dominates in the right auditory system, whereas gamma oscillations are more prominent in the left auditory cortex. Some support for this linkage comes from studies showing relative asymmetries of theta and gamma power at rest (Giraud et al., 2007; Morillon et al., 2010). The emphasis on syllabic rhythms reflects a current renewed interest in the role of syllabic organization in speech perception (Ghitza, 2012; Peelle & Davis, 2012; Stevens, 2002), which in turn reflects the dominant rate of articulator movements during production. However, entrainment by a syllabic rhythm cannot be critical for phoneme perception because phonemes can be perceived in isolated monosyllabic stimuli, which do not allow enough time for such entrainment (Ghitza, 2013). At present, the links between acoustic phonetic perception and endogenous neural oscillations is a topic of ongoing investigation.

References

Baum, S., & Pell, M. (1999). The neural bases of prosody: Insights from lesion studies and neuroimaging. *Aphasiology, 13*, 581–608.

Beauchamp, M. S., Nath, A. R., & Pasalar, S. (2010). fMRI-guided transcranial magnetic stimulation reveals that the superior temporal sulcus is a cortical locus of the McGurk effect. *Journal of Neuroscience, 30*, 2414–2417.

Belin, P., Zatorre, R. J., Lafaille, P., Ahad, P., & Pike, B. (2000). Voice-selective areas in human auditory cortex. *Nature, 403*, 309–312.

Belin, P., Zilbovicius, M., Crozier, S., Thivard, L., & Fontaine, A. (1998). Lateralization of speech and auditory temporal processing. *Journal of Cognitive Neuroscience, 10*(4), 536–540.

Bernstein, L. E., Demorest, M. E., & Tucker, P. E. (2000). Speech perception without hearing. *Perception and Psychophysics, 62*(2), 233–252.

Berthier, M. L. (1999). *Transcortical aphasias*. Hove: Psychology Press.

Binder, J. R., Frost, J. A., Hammeke, T. A., Bellgowan, P. S. F., Springer, J. A., Kaufman, J. N., et al. (2000). Human temporal lobe activation by speech and nonspeech sounds. *Cerebral Cortex, 10*, 512–528.

Binder, J. R., Liebenthal, E., Possing, E. T., Medler, D. A., & Ward, B. D. (2004). Neural correlates of sensory and decision processes in auditory object identification. *Nature Neuroscience, 7*(3), 295–301.

Blesser, B. (1972). Speech perception under conditions of spectral transformation: I. Phonetic characteristics. *Journal of Speech and Hearing Research, 15*, 5–41.

Boatman, D., Hall, C., Goldstein, M. H., Lesser, R., & Gordon, B. (1997). Neuroperceptual differences in consonant and vowel discrimination as revealed by direct cortical electrical interference. *Cortex, 33*(1), 83–98.

Boatman, D., Hart, J., Lesser, R. P., Honeycutt, N., Anderson, N. B., Miglioretti, D., et al. (1998). Right hemisphere speech perception revealed by amobarbital injection and electrical interference. *Neurology, 51*(2), 458–464.

Boatman, D., Lesser, R., & Gordon, B. (1995). Auditory speech processing in the left temporal lobe: An electrical interference study. *Brain and Language, 51*, 269–290.

Boemio, A., Fromm, S., Braun, A., & Poeppel, D. (2005). Hierarchical and asymmetric temporal sensitivity in human auditory cortices. *Nature Neuroscience, 8*(3), 389–395.

Bonte, M., Hausfeld, L., Scharke, W., Valente, G., & Formisano, E. (2014). Task-dependent decoding of speaker and vowel identity from auditory cortical response patterns. *Journal of Neuroscience, 34*(13), 4548–4557.

Brechmann, A., & Scheich, H. (2005). Hemispheric shifts of sound representation in auditory cortex with conceptual listening. *Cerebral Cortex, 15*, 578–587.

Buchman, A. S., Garron, D. C., Trost-Cardamone, J. E., Wichter, M. D., & Schwartz, D. (1986). Word deafness: One hundred years later. *Journal of Neurology, Neurosurgery, and Psychiatry, 49*, 489–499.

Burton, M. W., Small, S., & Blumstein, S. E. (2000). The role of segmentation in phonological processing: An fMRI investigation. *Journal of Cognitive Neuroscience, 12*, 679–690.

Callan, D., Callan, A., Gamez, M., Sato, M. A., & Kawato, M. (2010). Premotor cortex mediates perceptual performance. *NeuroImage, 51*, 844–858.

Callan, D., Tsytsarev, V., Hanakawa, T., Callan, A., Katsuhara, M., Fukuyama, H., et al. (2006). Song and speech: Brain regions involved with perception and covert production. *NeuroImage, 31*, 1327–1342.

Callan, D. E., Jones, J. A., Callan, A. M., & Akahane-Yamada, R. (2004). Phonetic perceptual identification by native- and second-language speakers differentially activates brain regions involved with acoustic phonetic processing and those involved with articulatory-auditory/orosensory internal models. *NeuroImage, 22*(3), 1182–1194.

Calvert, G. A. (2001). Crossmodal processing in the human brain: Insights from functional neuroimaging studies. *Cerebral Cortex, 11*, 1110–1123.

Canolty, R. T., Soltani, M., Dalal, S. S., Edwards, E., Dronkers, N. F., Nagarajan, S. S., et al. (2007). Spatiotemporal dynamics of word processing in the human brain. *Frontiers in Neuroscience, 1*(1), 185–196.

Crone, N. E., Miglioretti, D. L., Gordon, B., & Lesser, R. P. (1998). Functional mapping of human sensorimotor cortex with electrocorticographic spectral analysis: II. Event-related synchronization in the gamma band. *Brain, 121*, 2301–2315.

Damper, R. I., & Harnad, S. R. (2000). Neural network models of categorical perception. *Perception and Psychophysics, 62*(4), 843–867.

D'Ausilio, A., Bufalari, I., Salmas, P., & Fadiga, L. (2012). The role of the motor system in discriminating normal and degraded speech sounds. *Cortex, 48*(7), 882–887.

D'Ausilio, A., Pulvermüller, F., Salmas, P., Bufalari, I., Begliomini, C., & Fadiga, L. (2009). The motor somatotopy of speech perception. *Current Biology, 19*, 381–385.

Davis, M. H., & Johnsrude, I. S. (2003). Hierarchical processing in spoken language comprehension. *Journal of Neuroscience, 23*(8), 3423–3431.

Dehaene-Lambertz, G., Pallier, C., Serniclaes, W., Sprenger-Charolles, L., Jobert, A., & Dehaene, S. (2005). Neural correlates of switching from auditory to speech perception. *NeuroImage, 24*, 21–33.

Delattre, P., Liberman, A. M., & Cooper, F. S. (1955). Acoustic loci and transitional cues for consonants. *Journal of the Acoustical Society of America, 27*, 769–773.

Démonet, J.-F., Chollet, F., Ramsay, S., Cardebat, D., Nespoulous, J.-L., Wise, R., et al. (1992). The anatomy of phonological and semantic processing in normal subjects. *Brain, 115*, 1753–1768.

Desai, R., Liebenthal, E., Waldron, E., & Binder, J. R. (2008). Left posterior temporal regions are sensitive to auditory categorization. *Journal of Cognitive Neuroscience, 20*, 1174–1188.

DeWitt, I., & Rauschecker, J. P. (2012). Phoneme and word recognition in the auditory ventral stream. *Proceedings of the National Academy of Sciences of the United States of America, 109*, E505–514.

Diehl, R., & Kluender, K. (1989). On the objects of speech perception. *Ecological Psychology, 1*, 121–144.

Fadiga, L., Craighero, L., Buccino, G., & Rizzolatti, G. (2002). Speech listening specifically modulates the excitability of tongue muscles: A TMS study. *European Journal of Neuroscience, 15*(2), 399–402.

Fant, G. (1973). *Speech sounds and features.* Cambridge, MA: MIT Press.

Friederici, A. D., Kotz, S. A., Scott, S. K., & Obleser, J. (2010). Disentangling syntax and intelligibility in auditory language comprehension. *Human Brain Mapping, 31*, 448–457.

Galantucci, B., Fowler, C. A., & Turvey, M. T. (2006). The motor theory of speech perception reviewed. *Psychonomic Bulletin Review, 13*(3), 361–377.

Ganong, W. F. (1980). Phonetic categorization in auditory word perception. *Journal of Experimental Psychology: Human Perception and Performance, 6*, 110–115.

Ghitza, O. (2012). On the role of theta-driven syllabic parsing in decoding speech: Intelligibility of speech with a manipulated modulation spectrum. *Frontiers in Psychology, 3*, 238.

Ghitza, O. (2013). The theta-syllable: A unit of speech information defined by cortical function. *Frontiers in Psychology, 4*, 138.

Ghitza, O., & Goldstein, J. L. (1983). JNDs for the spectral envelope parameters in natural speech. In R. Klinke, & R. Hartmann (Eds.), *Hearing—Physiological bases and psychophysics* (pp. 352–358). Berlin: Springer-Verlag.

Giraud, A.-L., Kleinschmidt, A., Poeppel, D., Lund, T. E., Frackowiak, R. S. J., & Laufs, H. (2007). Endogenous cortical rhythms determine cerebral specialization for speech perception and production. *Neuron, 56*(6), 1127–1134.

Giraud, A.-L., & Poeppel, D. (2012). Cortical oscillations and speech processing: Emerging computational principles and operations. *Nature Neuroscience, 15*(4), 511–517.

Goense, J. M., & Logothetis, N. K. (2008). Neurophysiology of the BOLD fMRI signal in awake monkeys. *Current Biology, 18*(9), 631–640.

Goldstein, M. (1974). Auditory agnosia for speech ("pure word deafness"): A historical review with current implications. *Brain and Language, 1*, 195–204.

Guenther, F. H., Hampson, M., & Johnson, D. (1998). A theoretical investigation of reference frames for the planning of speech movements. *Psychological Review, 105*(4), 611–633.

Hall, D. A., Haggard, M. P., Akeroyd, M. A., Palmer, A. R., Summerfield, A. Q., Elliott, M. R., et al. (1999). Sparse temporal sampling in auditory fMRI. *Human Brain Mapping, 7* 213–223.

Henschen, S. E. (1918). On the hearing sphere. *Acta Oto-laryngologica, 1*, 423–486.

Hickok, G., Okada, K., Barr, W., Pa, J., Rogalsky, C., Donnelly, K., et al. (2008). Bilateral capacity for speech sound processing in auditory comprehension: Evidence from Wada procedures. *Brain and Language, 107*, 179–184.

Hickok, G., & Poeppel, D. (2007). The cortical organization of speech processing. *Nature Reviews Neuroscience, 8*(5), 393–402.

Hirsch, J., Ruge, M. I., Kim, K. H. S., Correa, D. D., Victor, J. D., Relkin, N. R., et al. (2000). An integrated functional magnetic resonance imaging procedure for preoperative mapping of cortical areas associated with tactile, motor, language, and visual functions. *Neurosurgery, 47*(3), 711–722.

Hughes, G. W., & Halle, M. (1956). Spectral properties of fricative consonants. *Journal of the Acoustical Society of America, 28*, 303–310.

Humphries, C., Binder, J. R., Medler, D. A., & Liebenthal, E. (2006). Syntactic and semantic modulation of neural activity during auditory sentence comprehension. *Journal of Cognitive Neuroscience, 18*, 665–679.

Humphries, C., Swinney, D., Love, T., & Hickok, G. (2005). Response of anterior temporal cortex to syntactic and prosodic manipulations during sentence processing. *Human Brain Mapping, 26*, 128–138.

Jamison, H. L., Watkins, K. E., Bishop, D. V. M., & Matthews, P. M. (2006). Hemispheric specialization for processing auditory nonspeech stimuli. *Cerebral Cortex, 16*(9), 1266–1275.

Jäncke, L., Wüstenberg, T., Scheich, H., & Heinze, H. J. (2002). Phonetic perception and the temporal cortex. *NeuroImage, 15*, 733–746.

Kaas, J. H., & Hackett, T. A. (2000). Subdivisions of auditory cortex and processing streams in primates. *Proceedings of the National Academy of Sciences of the United States of America, 97*, 11793–11799.

Kewley-Port, D. (1983). Time-varying features as correlates of place of articulation in stop consonants. *Journal of the Acoustical Society of America, 73*, 322–335.

Klatt, D. H. (1973). Discrimination of fundamental frequency contours in synthetic speech: Implications for models of speech perception. *Journal of the Acoustical Society of America, 53*, 8–16.

Koenig, W., Dunn, H. K., & Lacy, L. Y. (1946). The sound spectrograph. *Journal of the Acoustical Society of America, 18*, 21–32.

Kuhl, P. K. (1991). Human adults and human infants show a "perceptual magnet effect" for the prototypes of speech categories, monkeys do not. *Perception and Psychophysics, 50*, 93–107.
Kuhl, P. K. (2000). A new view of language acquisition. *Proceedings of the National Academy of Sciences of the United States of America, 97* (22), 11850–11857.
Kurowski, K., & Blumstein, S. E. (1987). Acoustic properties for place of articulation in nasal consonants. *Journal of the Acoustical Society of America, 81*, 1917–1927.
Latinus, M., McAleer, P., Bestelmeyer, P. E., & Belin, P. (2013). Norm-based coding of voice identity in human auditory cortex. *Current Biology, 23*, 1075–1080.
Lattner, S., Meyer, M. E., & Friederici, A. D. (2005). Voice perception: Sex, pitch, and the right hemisphere. *Human Brain Mapping, 24*, 11–20.
Lehéricy, S., Cohen, L., Bazin, B., Samson, S., Giacomini, E., Rougetet, R., et al. (2000). Functional MR evaluation of temporal and frontal language dominance compared with the Wada test. *Neurology, 54*, 1625–1633.
Liberman, A. M., Cooper, F. S., Shankweiler, D. P., & Studdert-Kennedy, M. (1967). Perception of the speech code. *Psychological Review, 74*, 431–461.
Liberman, A. M., Harris, K. S., Hoffman, H. S., & Griffith, B. C. (1957). The discrimination of speech sounds within and across phoneme boundaries. *Journal of Experimental Psychology, 54*, 358–368.
Liberman, A. M., & Mattingly, I. G. (1985). The motor theory of speech perception revised. *Cognition, 21*, 1–36.
Liebenthal, E., Binder, J. R., Spitzer, S. M., Possing, E. T., & Medler, D. A. (2005). Neural substrates of phonemic perception. *Cerebral Cortex, 15*, 1621–1631.
Liebenthal, E., Desai, R., Ellingson, M. M., Ramachandran, B., Desai, A., & Binder, J. R. (2010). Specialization along the left superior temporal sulcus for phonemic and non-phonemic categorization. *Cerebral Cortex, 20*, 2958–2970.
Lisker, L. (1986). "Voicing" in English: A catalogue of acoustic features signaling /b/ versus /p/ in trochees. *Language and Speech, 29*, 3–11.
Lotto, A. J., Hickok, G. S., & Holt, L. L. (2009). Reflections on mirror neurons and speech perception. *Trends in Cognitive Sciences, 13*, 110–114.
McClelland, J. L., & Elman, J. L. (1986). The TRACE model of speech perception. *Cognitive Psychology, 18*, 1–86.
McGettigan, C., & Scott, S. K. (2012). Cortical asymmetries in speech perception: What's wrong, what's right and what's left? *Trends in Cognitive Sciences, 16*(5), 269–276.
McGurk, H., & MacDonald, J. (1976). Hearing lips and seeing voices. *Nature, 264*(5588), 746–748.
Meister, I. G., Wilson, S. M., Deblieck, C., Wu, A. D., & Iacoboni, M. (2007). The essential role of premotor cortex in speech perception. *Current Biology, 17*, 1692–1696.
Mesgarani, N., Cheung, C., Johnson, K., & Chang, E. F. (2014). Phonetic feature encoding in human superior temporal gyrus. *Science, 343*(6174), 1006–1010.
Meyer, M., Zaehle, T., Gountouna, V.-E., Barron, A., Jancke, L., & Turk, A. (2005). Spectro-temporal processing during speech perception involves left posterior temporal auditory cortex. *Neuroreport, 16*(18), 1985–1989.
Miller, J. D. (1989). Auditory-perceptual interpretation of the vowel. *Journal of the Acoustical Society of America, 85*, 2114–2134.
Miller, L. M., & D'Esposito, M. (2005). Perceptual fusion and stimulus coincidence in the cross-modal integration of speech. *Journal of Neuroscience, 25*(25), 5884–5893.
Morillon, B., Lehongre, K., Frackowiak, R. S., Ducorps, A., Kleinschmidt, A., Poeppel, D., et al. (2010). Neurophysiological origin of human brain asymmetry for speech and language. *Proceedings of the National Academy of Sciences of the United States of America, 107*, 18688–18693.
Möttönen, R., Calvert, G. A., Jaaskelainen, I. P., Matthews, P. M., Thesen, T., Tuomainen, J., et al. (2006). Perceiving identical sounds as speech or non-speech modulates activity in the left posterior superior temporal sulcus. *NeuroImage, 30*, 563–569.
Möttönen, R., Dutton, R., & Watkins, K. E. (2013). Auditory-motor processing of speech sounds. *Cerebral Cortex, 23*(5), 1190–1197.
Möttönen, R., & Watkins, K. E. (2009). Motor representations of articulators contribute to categorical perception of speech sounds. *Journal of Neuroscience, 29*, 9819–9825.
Mummery, C. J., Ashburner, J., Scott, S. K., & Wise, R. J. S. (1999). Functional neuroimaging of speech perception in six normal and two aphasic subjects. *Journal of the Acoustical Society of America, 106*, 449–457.
Narain, C., Scott, S. K., Wise, R. J. S., Rosen, S., Leff, A., Iversen, S. D., et al. (2003). Defining a left-lateralized response specific to intelligible speech using fMRI. *Cerebral Cortex, 13*, 1362–1368.
Nath, A. R., & Beauchamp, M. S. (2012). A neural basis for interindividual differences in the McGurk effect, a multisensory speech illusion. *NeuroImage, 59*, 781–787.
Norris, D., McQueen, J. M., & Cutler, A. (2000). Merging information in speech recognition: Feedback is never necessary. *Behavioral and Brain Sciences, 23*, 299–325.
Okada, K., Rong, F., Venezia, J., Matchin, W., Hsieh, I.-H., Saberi, K., et al. (2010). Hierarchical organization of human auditory cortex: Evidence from acoustic invariance in the response to intelligible speech. *Cerebral Cortex, 20*, 2486–2495.
Overath, T., Kumar, S., von Kriegstein, K., & Griffiths, T. D. (2008). Encoding of spectral correlation over time in auditory cortex. *Journal of Neuroscience, 28*(49), 13268–13273.
Pallier, C., Devauchelle, A.-D., & Dehaene, S. (2011). Cortical representation of the constituent structure of sentences. *Proceedings of the National Academy of Sciences of the United States of America, 108* (6), 2522–2527.
Pasley, B. N., David, S. V., Mesgarani, N., Flinker, A., Shamma, S. A., Crone, N. E., et al. (2012). Reconstructing speech from human auditory cortex. *Plos Biology, 10*(1), e1001251.
Peelle, J. E., & Davis, M. H. (2012). Neural oscillations carry speech rhythm through to comprehension. *Frontiers in Psychology, 3*, 320.
Peterson, G. E., & Barney, H. L. (1952). Control methods used in a study of vowels. *Journal of the Acoustical Society of America, 24*, 175–184.
Pitt, M. A., & Samuel, A. G. (1993). An empirical and meta-analytic evaluation of the phoneme identification task. *Journal of Experimental Psychology: Human Perception and Performance, 19*, 699–725.
Poeppel, D. (2001). Pure word deafness and the bilateral processing of the speech code. *Cognitive Science, 25*, 679–693.
Poeppel, D. (2003). The analysis of speech in different temporal integration windows: Cerebral lateralization as "asymmetric sampling in time". *Speech Communication, 41*, 245–255.
Pruitt, J. S., Jenkins, J. J., & Strange, W. (2006). Training the perception of Hindi dental and retroflex stops by native speakers of American English and Japanese. *Journal of the Acoustical Society of America, 119*(3), 1684–1696.
Pulvermuller, F., Huss, M., Kheri, F., Moscoso Del Prado Martin, F., Hauk, O., & Shtrov, Y. (2006). Motor cortex maps articulatory features of speech sounds. *Proceedings of the National Academy of Sciences of the United States of America, 103*, 7865–7870.
Rauschecker, J. P., & Scott, S. K. (2009). Maps and streams in the auditory cortex: Nonhuman primates illuminate human speech processing. *Nature Neuroscience, 12*(6), 718–724.

Remez, R. E., Rubin, P. E., Pisoni, D. B., & Carrell, T. D. (1981). Speech perception without traditional speech cues. *Science*, *212*, 947–950.

Rizzolatti, G., & Arbib, M. A. (1998). Language within our grasp. *Trends in Neurosciences*, *21*, 188–194.

Rizzolatti, G., & Craighero, L. (2004). The mirror-neuron system. *Annual Review of Neuroscience*, *27*, 169–192.

Ross, E. D., Thompson, R. D., & Yenkosky, J. (1997). Lateralization of affective prosody in brain and the callosal integration of hemispheric language functions. *Brain and Language*, *56*, 27–54.

Samuel, A. G. (1997). Lexical activation produces potent phonemic percepts. *Cognitive Psychology*, *32*, 97–127.

Sato, M., Tremblay, P., & Gracco, V. L. (2009). A mediating role of the premotor cortex in phoneme segmentation. *Brain and Language*, *111*(1), 1–7.

Sawusch, J. R., & Newman, R. S. (2000). Perceptual normalization for speaking rate II: Effects of signal discontinuities. *Perception and Psychophysics*, *62*(2), 285–300.

Schonwiesner, M., Rubsamen, R., & von Cramon, D. Y. (2005). Hemispheric asymmetry for spectral and temporal processing in the human antero-lateral auditory belt cortex. *European Journal of Neuroscience*, *22*(6), 1521–1528.

Scott, S. K., Blank, C., Rosen, S., & Wise, R. J. S. (2000). Identification of a pathway for intelligible speech in the left temporal lobe. *Brain*, *123*, 2400–2406.

Shankweiler, D. P., & Studdert-Kennedy, M. (1970). Hemispheric specialization for speech perception. *Journal of the Acoustical Society of America*, *48*(2), 579–594.

Shannon, R. V., Zeng, F.-G., Kamath, V., Wygonski, J., & Ekelid, M. (1995). Speech recognition with primarily temporal cues. *Science*, *270*, 303–304.

Spitsyna, G., Warren, J. E., Scott, S. K., Turkheimer, F. E., & Wise, R. J. S. (2006). Converging language streams in the human temporal lobe. *Journal of Neuroscience*, *26*(28), 7328–7336.

Stefanatos, G. A., Gershkoff, A., & Madigan, S. (2005). On pure word deafness, temporal processing and the left hemisphere. *Journal of the International Neuropsychological Society*, *11*(4), 456–470.

Stevens, K. (2002). Toward a model for lexical access based on acoustic landmarks and distinctive features. *Journal of the Acoustical Society of America*, *111*(4), 1872–1891.

Stevens, K. N., & Blumstein, S. E. (1981). The search for invariant acoustic correlates of phonetic features. In P. D. Eimas, & J. L. Miller (Eds.), *Perspectives on the study of speech* (pp. 1–38). Hillsdale, NJ: Erlbaum.

Sumby, W. H., & Pollack, I. (1954). Visual contribution to speech intelligibility in noise. *Journal of the Acoustical Society of America*, *26*, 212–215.

Summerfield, Q., & Haggard, M. (1977). On the dissociation of spatial and temporal cues to the voicing distinction in initial stop consonants. *Journal of the Acoustical Society of America*, *62*, 435–448.

Sussman, H. M. (1986). A neuronal model of vowel normalization and representation. *Brain and Language*, *28*, 12–23.

Tourville, J. A., Reilly, K. J., & Guenther, F. H. (2008). Neural mechanisms underlying auditory feedback control of speech. *NeuroImage*, *39*(3), 1429–1443.

Tremblay, P., & Small, S. L. (2011). On the context-dependent nature of the contribution of the ventral premotor cortex to speech perception. *NeuroImage*, *57*, 1561–1571.

Ulrich, G. (1978). Interhemispheric relationships in auditory agnosia: An analysis of the preconditions and a conceptual model. *Brain and Language*, *5*, 286–300.

Van Lancker, D. R., Kreiman, J., & Cummings, J. (1989). Voice perception deficits: Neuroanatomical correlates of phonagnosia. *Journal of Clinical and Experimental Neuropsychology*, *11*(5), 665–674.

Vandenberghe, R., Nobre, A. C., & Price, C. J. (2002). The response of left temporal cortex to sentences. *Journal of Cognitive Neuroscience*, *14*(4), 550–560.

Visser, M., Jefferies, E., & Lambon Ralph, M. A. (2010). Semantic processing in the anterior temporal lobes: A meta-analysis of the functional neuroimaging literature. *Journal of Cognitive Neuroscience*, *22* (6), 1083–1094.

von Kriegstein, K., Eger, E., Kleinschmidt, A., & Giraud, A. L. (2003). Modulation of neural responses to speech by directing attention to voices or verbal content. *Cognitive Brain Research*, *17*, 48–55.

Wang, E., Peach, R. K., Xu, Y., Schneck, M., & Manry, C. (2000). Perception of dynamic acoustic patterns by an individual with unilateral verbal auditory agnosia. *Brain and Language*, *73*(3), 442–455.

Warren, R. M., & Obusek, C. J. (1971). Speech perception and phonemic restorations. *Perception and Psychophysics*, *9*, 358–362.

Watkins, K., Strafella, A., & Paus, T. (2003). Seeing and hearing speech excites the motor system involved in speech production. *Neuropsychologia*, *41*(8), 989–994.

Werker, J. F., & Tees, R. C. (2005). Speech perception as a window for understanding plasticity and commitment in language systems of the brain. *Developmental Psychobiology*, *46*(3), 233–251.

Wilson, S., Saygin, A., Sereno, M., & Iacoboni, M. (2004). Listening to speech activates motor areas involved in speech production. *Nature Neuroscience*, *7*, 701–702.

Wilson, S. M., & Iacoboni, M. (2006). Neural responses to non-native phonemes varying in producibility: Evidence for the sensorimotor nature of speech perception. *NeuroImage*, *33*, 316–325.

Wise, R., Chollet, F., Hadar, U., Friston, K., Hoffner, E., & Frackowiak, R. (1991). Distribution of cortical neural networks involved in word comprehension and word retrieval. *Brain*, *114*, 1803–1817.

Wise, R. S. J., Scott, S. K., Blank, S. C., Mummery, C. J., Murphy, K., & Warburton, E. A. (2001). Separate neural subsystems within "Wernicke's area". *Brain*, *124*, 83–95.

Yuen, I., Davis, M. H., Brysbaert, M., & Rastle, K. (2010). Activation of articulatory information in speech perception. *Proceedings of the National Academy of Sciences of the United States of America*, *107*, 592–597.

Zaehle, T., Wustenberg, T., Meyer, M., & Jancke, L. (2004). Evidence for rapid auditory perception as the foundation of speech processing: A sparse temporal sampling fMRI study. *European Journal of Neuroscience*, *20*(9), 2447–2456.

Zatorre, R. J., & Belin, P. (2001). Spectral and temporal processing in human auditory cortex. *Cerebral Cortex*, *11*(10), 946–953.

Zatorre, R. J., Belin, P., & Penhune, V. B. (2002). Structure and function of auditory cortex: Music and speech. *Trends in Cognitive Sciences*, *6*(1), 37–46.

Zatorre, R. J., Evans, A. C., Meyer, E., & Gjedde, A. (1992). Lateralization of phonetic and pitch discrimination in speech processing. *Science*, *256*, 846–849.

CHAPTER

38

A Neurophysiological Perspective on Speech Processing in "The Neurobiology of Language"

Luc H. Arnal[1,2], David Poeppel[2,3] and Anne-Lise Giraud[1]

[1]Department of Neurosciences, Biotech Campus, University of Geneva, Geneva, Switzerland; [2]Department of Psychology, New York University, New York, NY, USA; [3]Max-Planck-Institute for Empirical Aesthetics, Frankfurt, Germany

38.1 OVERVIEW

Of all the signals the human auditory system has to process, the most relevant to the listener is arguably speech. Speech perception is learned and executed with automaticity and great ease even by very young children, but it is handled surprisingly poorly by even sophisticated automatic devices. Therefore, parsing and decoding speech can be considered one of the main challenges of the auditory system. This chapter focuses on how the human auditory cortex uses the temporal structure of the acoustic signal to extract phonemes and syllables, two types of events that need to be identified in connected speech.

Speech is a complex acoustic signal exhibiting quasiperiodic behavior at several timescales. The neural signals recorded from the auditory cortex using EEG or MEG also exhibit a quasiperiodic structure, whether in response to speech or not. In this chapter we present different models grounded on the assumption that the quasiperiodic structure of collective neural activity in auditory cortex represents the ideal mechanical infrastructure to solve the speech demultiplexing problem (i.e., the fractioning of speech into linguistic constituents of variable size). These models remain largely hypothetical. However, they constitute exciting theories that will hopefully lead to new research approaches and incremental progress on this foundational question about speech perception.

The chapter proceeds as follows. First, some of the essential features of natural and speech auditory stimuli are outlined. Next, the properties of auditory cortex that reflect its sensitivity to these features are reviewed, and current ideas about the neurophysiological mechanisms underpinning the processing of connected speech are discussed. The chapter closes with a summary of speech processing models on a larger scale, attempting to capture most of these phenomena in an integrated vision.

38.1.1 Timescales in Auditory Perception

Sounds are audible over a broad frequency range between 20 and 20,000 Hz. They enter the outer ear and travel through the middle ear to the inner ear, where they provoke the basilar membrane to vibrate at a specific location, depending on the sound frequency. Low and high frequencies induce vibrations of the apex and base of the basilar membrane, respectively. The deformation of the membrane on acoustic stimulation provokes the deflection of inner hair cells, *ciliae*, and the emission of a neural signal to cochlear neurons, subsequently transmitted to neurons of the cochlear nucleus in the brainstem. Each cochlear neuron is sensitive to a specific range of acoustic frequencies between 20 Hz and 20 kHz. Because of their regular position along the basilar membrane, the cochlear neurons ensure the place-coding of acoustic frequencies, also called "tonotopy," which is preserved up to the cortex (Moerel et al., 2013; Saenz & Langers, 2014).

Acoustic fluctuations <20 Hz are not audible. They do not elicit place-specific responses in the cochlea.

Neurobiology of Language. DOI: http://dx.doi.org/10.1016/B978-0-12-407794-2.00038-9

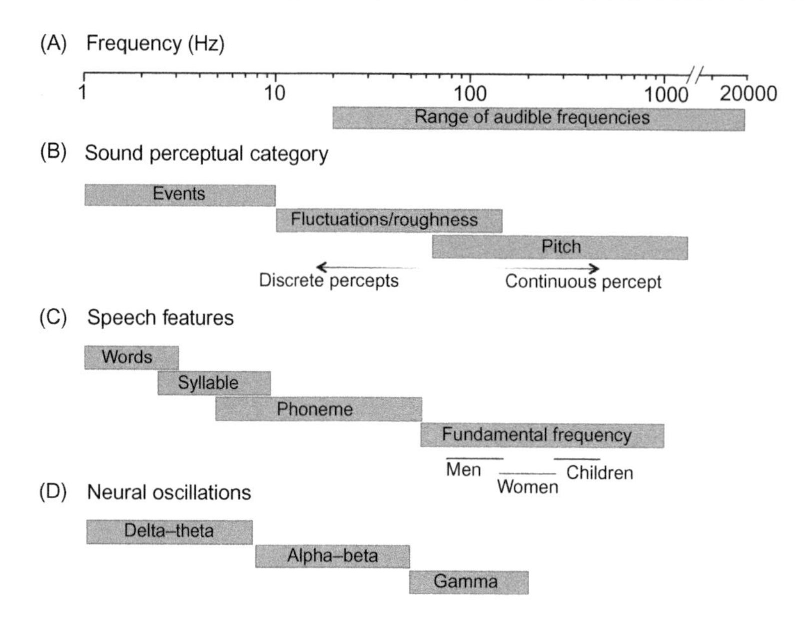

FIGURE 38.1 (A) Scale of perceived temporal modulation (modified from Joris, Schreiner, & Rees, 2004 and Nourski & Brugge, 2011). (B) Relevant psychophysical parameters (perceptual changes) of the spectrogram reflect the temporal constraints that superimpose on the structure of linguistic signals. (C) Temporal structure of linguistic features. (D) The length of linguistic features remarkably matches the frequency of oscillations that are observed at rest in the brain. Note that the frequency ranges at which auditory percepts switch from discrete (flutter) to continuous (pitch) roughly match the upper limit at which gamma rhythms can be entrained by the stimulus (~200 Hz).

Low frequencies <300 Hz are present in complex sounds, such as temporal fluctuations of audible frequencies, and are encoded through the discharge rate of cochlear neurons (Zeng, 2002). A range of frequencies from 20 to 300 Hz is both place-coded and rate-coded at the auditory periphery. Temporal modulation of sounds in these frequencies typically elicits a sensation of pitch. Figure 38.1A and 38.1B summarizes the correspondence between categories of perceptual attributes and the sound modulation frequency (see also Nourski & Brugge, 2011 for a review). When sounds are modulated at very slow rates <10 Hz, a sequence of distinct events is perceived. When modulations accelerate from 10 to approximately 100 Hz, distinct events merge into a single auditory stream, and the sensation evolves from fluctuating magnitude to a sensation of acoustic roughness (Figure 38.1B).

Speech sounds are complex acoustic signals that involve only the lower part of audible frequencies (20–8,000 Hz). They are called "complex" because both their frequency distribution and their magnitude vary strongly and quickly over time. In natural speech, amplitude modulations at slow (<20 Hz) and fast (>100 Hz) timescales are coupled (Figure 38.2), and slower temporal fluctuations modulate the amplitude of spectral fluctuations. Slow modulations (<5 Hz) signal word and syllable boundaries (Hyafil, Fontolan, Gutkin, & Giraud, 2012; Rosen, 1992), which are perceived as a sequence of distinct events. Phonemes (speech sounds) are signaled by fast spectrotemporal modulations (<30 Hz). They can be perceived as distinct events when being discriminated from each other. Faster modulations, such as those imposed by the glottal pulse (100–300 Hz), indicate the voice pitch (Figure 38.1C). Figure 38.1D shows how these perceptual events relate to the different frequency ranges of the EEG.

38.1.2 The Temporal Structure of Speech Sounds

Figure 38.2 illustrates two useful ways to visualize the signal: as a waveform (38.2A) and as a spectrogram (38.2B). The waveform represents energy variation over time—the input that the ear actually receives. The outlined "envelope" (thick line) reflects that there is a temporal regularity in the signal at relatively low modulation frequencies. These modulations of signal energy (in reality, spread out across the "cochlear" filterbank) are <20 Hz and peak at a rate of approximately 4–6 Hz (Elliott & Theunissen, 2009; Steeneken & Houtgast, 1980). From the perspective of what auditory cortex receives as input, namely the modulations at the output of each frequency channel of the filterbank that constitutes the auditory periphery, these energy fluctuations can be characterized by the modulation spectrum (Chi, Ru, & Shamma, 2005; Dau, Kollmeier, & Kohlrausch, 1997; Kanedera, Arai, Hermansky, & Pavel, 1999; Kingsbury, Morgan, &

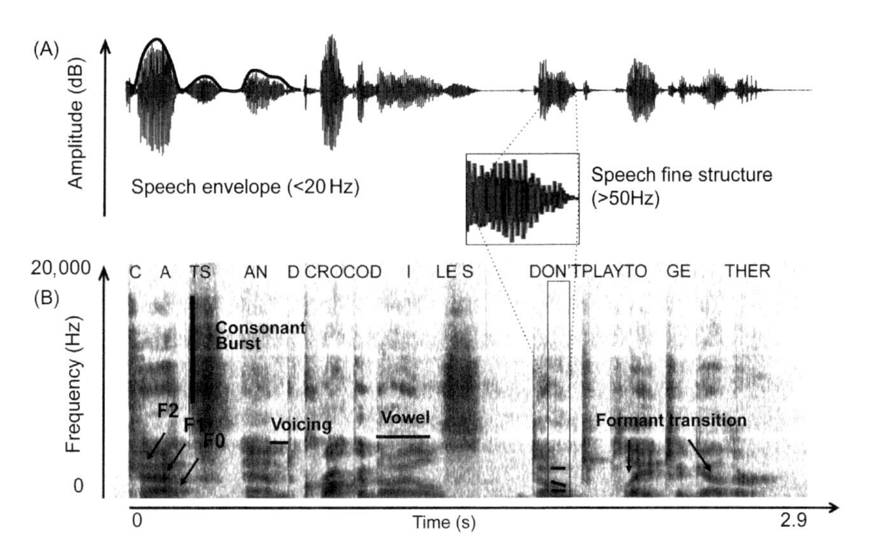

FIGURE 38.2 (A) Waveform and (B) spectrogram of the same sentence uttered by a male speaker. Some of the key acoustic cues in speech comprehension are highlighted in black.

Greenberg, 1998; McDermott & Simoncelli, 2011). At the shortest timescale (below 1 ms or equivalently above 1 kHz), the very fast temporal fluctuations are transformed into a spectral representation at the cochlea and the neural processing of these features is generally known as spectral processing. At an intermediate timescale (~70 Hz–1 kHz), the temporal fluctuations are usually referred to as the temporal fine structure. The temporal fine structure is critical to the perception of pitch and inter-aural time differences that are important cues for sound source localization (Grothe, Pecka, & McAlpine, 2010; Plack, Oxenham, Fay, & Popper, 2005). Temporal fluctuations on an even longer timescale (~1–10 Hz) are heard as a sequence of discrete events. Acoustic events occurring on this timescale include syllables and words in speech and notes and beats in music. Of course, there are no clear boundaries between these timescales; the timescales are divided here based on human auditory perception.

The second analytic representation, the spectrogram, decomposes the acoustic signal in the frequency, time, and amplitude domains (Figure 38.2B). Although the human auditory system captures frequency information between 20 Hz and 20 kHz (and such a spectrogram is plotted here), most of the information that is extracted for effective recognition is below 8 kHz. It is worth remembering that speech transmitted over telephone landlines contains a much narrower bandwidth (200–3600 Hz) and is comfortably understood by normal listeners. A number of critical acoustic features can be identified in the spectrogram. The faintly visible vertical stripes represent the glottal pulse, which reflects the speaker's fundamental frequency, F0. This can range from approximately 100 Hz (male adult) to 300 Hz (child; see Figure 38.1D). The horizontal bands of energy show where in frequency space a particular speech sound is carried. The spectral structure thus reflects the articulator configuration. These bands of energy include the formants (F1, F2, etc.) definitional of vowel identity; high-frequency bursts associated, for example, with frication in certain consonants (e.g., /s/, /f/); and formant transitions that signal the change from a consonant to a vowel or vice versa.

Notwithstanding the importance of the spectral fine structure, there is a big caveat: speech can be understood, in the sense of being intelligible in psychophysical experiments, when the spectral content is replaced by noise and only the envelope is preserved. Importantly, this manipulation is done in separate bands across the spectrum, for example, as few as four separate bands (Shannon, Zeng, Kamath, Wygonski, & Ekelid, 1995). Speech that contains only envelope but no fine structure information is called vocoded speech (Faulkner, Rosen, & Smith, 2000). Compelling demonstrations that exemplify this type of signal decomposition illustrate that the speech signal can undergo radical alterations and distortions and yet remain intelligible (Shannon et al., 1995; Smith, Delgutte, & Oxenham, 2002). Such findings have led to the idea that the temporal envelope, that is, temporal modulations of speech at relatively slow rates, is sufficient to yield speech comprehension (Drullman, Festen, & Plomp, 1994a, 1994b; Giraud et al., 2004; Loebach & Wickesberg, 2008; Rosen, 1992; Scott, Rosen, Lang, & Wise, 2006; Shannon et al., 1995; Souza & Rosen, 2009). When using stimuli in which the fine structure is compromised or not available at all, envelope modulations below 16 Hz appear to suffice for adequate intelligibility. The remarkable comprehension level reached by most patients with cochlear implants, in whom approximately 15–20 electrodes replace 3,000 hair cells, remains the best empirical demonstration that the spectral content of speech can be degraded with tolerable alteration of speech perception (Roberts, Summers, & Bailey, 2011). A related demonstration showing the resilience of speech comprehension in the face of radical signal impoverishment is provided by sine-wave speech (Remez, Rubin, Pisoni, & Carrell, 1981). In these stimuli both envelope and spectral content are degraded, but enough information is preserved to permit intelligibility. Typically, sine-wave speech preserves the modulations of the three first formants, which are replaced by sine-waves centered on F0, F1, and F2. In summary, dramatically impoverished stimuli remain intelligible insofar as enough information in the spectrum is available to convey temporal modulations at appropriate rates.

Based on this brief and selective summary, two concepts merit emphasis. First, the extended speech signal contains critical information that is modulated at rates below 20 Hz, with the modulation peaking at approximately 5 Hz (Edwards & Chang, 2013). This low-frequency information correlates closely with the syllabic structure of connected speech (Giraud & Poeppel, 2012). Second, the speech signal contains critical information at modulation rates higher than, for example, 50 Hz. This rapidly changing information is associated with fine spectral changes that carry information about the speaker's gender or identity and other relevant speech attributes (Elliott & Theunissen, 2009). Thus, two surprisingly different timescales are concurrently at play in the speech signal. This important issue is described in the text that follows. In this chapter, we discuss the timescales longer than 5 ms (<200 Hz) with a focus on the timescale between 100 ms and 1 s (1–10 Hz). Temporal features that contribute to the spatial localization of sounds are not discussed (see Grothe et al., 2010 for a review).

38.2 CORTICAL PROCESSING OF CONTINUOUS SOUNDS STREAMS

38.2.1 The Discretization Problem

In natural connected speech, speech information is embedded in a continuous acoustic flow, and sentences are not "presegmented" in perceptual units of analysis. Recent work on sentence-level stimuli (i.e., materials with a duration exceeding 1–2 s) using experimental tasks such as intelligibility, suggests that long-term temporal parameters of the acoustic signal are of major importance (Ghitza & Greenberg, 2009; Luo, Liu, & Poeppel, 2010; Luo & Poeppel, 2007; Peelle, Gross, & Davis, 2013). Online segmentation remains a major challenge to contemporary models of speech perception as well as automatic speech recognition.

Interestingly, a large body of psychophysical work studied speech perception and intelligibility using phrasal or sentential stimuli (see Miller, 1951 for a summary of many experiments and Allen, 2005 for a review of the influential work of Fletcher and others). Fascinating findings emerged from that work, emphasizing the role of signal-to-noise ratio in speech comprehension, but perhaps the most interesting feature is that connected speech has principled and useful temporal properties that may play a key role in the problem of speech parsing and decoding. Natural speech usually comes to the listener as a continuous stream and needs to be analyzed online and decoded by mechanisms that are unlikely to be continuous (Giraud & Poeppel, 2012). The parsing mechanism corresponds to the discretization of the continuous input signal into subsegments of speech information that are read out, to a certain extent, independently from each other. The notion that perception is discrete has been extensively discussed and generalized in numerous sensory modalities and contexts (Pöppel & Artin, 1988; VanRullen & Koch, 2003). Here, we discuss the hypothesis that neural oscillations constitute a possible mechanism for discretizing temporally complex sounds such as speech (Giraud & Poeppel, 2012).

38.2.2 Analysis at Mutiple Timescales

Speech is a multiplexed signal, that is, it interlinks several levels of complexity, and organizational principles and perceptual units of analysis exist at distinct timescales. Using data from linguistics, psychophysics, and physiology, Poeppel and colleagues proposed that speech is analyzed in parallel at multiple timescales (Boemio, Fromm, Braun, & Poeppel, 2005; Poeppel, 2001, 2003; Poeppel, Idsardi, & van Wassenhove, 2008). The central idea is that both local-to-global and global-to-local types of analyses are carried out concurrently (multitime-resolution processing). This assumption adds to the notion of reverse hierarchy (Hochstein & Ahissar, 2002; Nahum, Nelken, & Ahissar, 2008) and other hierarchical models in perception, which propose that the hierarchical complexification of sensory information (e.g., the temporal hierarchy) maps onto the anatomo-functional hierarchy of the brain (Giraud et al., 2000; Kiebel, Daunizeau, & Friston, 2008). The principal motivations for extending such a hypothesis are two-fold. First, a single, short temporal window that forms the basis for hierarchical processing, that is, increasingly larger temporal analysis units as one ascends the processing system, fails to account for the spectral and temporal sensitivity of the speech processing system and is difficult to reconcile with behavioral performance. Second, the computational strategy of analyzing information on multiple scales is widely used in engineering and biological systems, and the neuronal infrastructure exists to support multiscale computation (Canolty & Knight, 2010). According to the view summarized here, speech is chunked into segments of roughly featural or phonemic length, and then integrated into larger units, such as segments, diphones, syllables, and words. In parallel, there is a fast global analysis that yields coarse inferences about speech (akin to Stevens, 2002 "landmarks" hypothesis) and that subsequently refines segmental analysis. Here, we propose that segmental and suprasegmental analyses could be carried out concurrently and "packaged" for parsing and decoding by neuronal oscillations at different rates.

The notion that speech analysis occurs in parallel at multiple timescales justifies moving away from strictly hierarchical models of speech perception (Giraud & Price, 2001). Accordingly, the simultaneous extraction of different acoustic cues permits simultaneous high-order processing of different information from a unique input signal. That speech should be analyzed in parallel at different timescales derives, among other reasons, from the observation that articulatory–phonetic phenomena occur at different timescales. It was noted previously (Figure 38.1C and Figure 38.2) that the speech signal contains events of different durations: short energy bursts and formant transitions occur within a 20- to 80-ms timescale, whereas syllabic information occurs over 150–300 ms. The processing of both types of events could be accounted for either by a hierarchical model in which smaller acoustic units (segments) are concatenated into larger units (syllables) or by a parallel model in which both temporal units are extracted independently, and then combined. A degree of independence in the processing of long (slow modulation) and short (fast modulation) units is observed at the behavioral level. For instance, speech can be understood well when it is first segmented into units up to 60 ms and when these local units are temporally reversed (Greenberg & Arai, 2001; Saberi & Perrott, 1999). Because the correct extraction of short units is not a prerequisite for comprehension, this rules out the notion that speech processing relies solely on hierarchical processing of short and then larger units. Overall, there appears to be a grouping of psychophysical phenomena such that some cluster at thresholds of approximately 50 ms and below and others cluster at approximately 200 ms and above (a similar clustering is observed for temporal properties in vision; Holcombe, 2009). Importantly, nonspeech signals are subject to similar thresholds. For example, 15–20 ms is the minimal stimulus duration required for correctly identifying upward versus downward FM sweeps (Luo, Boemio, Gordon, & Poeppel, 2007). By comparison, 200-ms stimulus duration underlies loudness judgments. In summary, physiological events at related scales form the basis for processing at that level. Therefore, the neuronal oscillatory machinery (together with motor constraints related to speech production; Morillon et al., 2010) presumably imposed strong temporal constrains that might have shaped the size of acoustic features selected to carry speech information. This is consistent with the notion that perception is discrete and that the exogenous recruitment of neuronal populations is followed by refractory periods that temporarily reduce the ability to optimally extract sensory information (Ghitza, 2011; Ghitza & Greenberg, 2009). According to this hypothesis, the temporally limited capacity of gamma oscillations to integrate information over time possibly imposes a lower limit to the phoneme length. This also suggests that oscillatory constrains in the delta–theta range possibly constrained the size of syllables to be approximately the size of a delta–theta cycle. Considering that the average lengths of phoneme and syllable are approximately 25–80 and 150–300 ms, respectively (Figure 38.1 and Figure 38.2), the dual timescale segmentation requires two parallel sampling mechanisms, one at approximately 40 Hz (or, more broadly, in the low gamma range) and one at approximately 4 Hz (or in the theta range).

38.2.3 Neural Oscillations as Endogenous Temporal Constraints

Neural oscillations correspond to synchronous activity of neuronal assemblies that are both intrinsically coupled and coupled by a common input. It was proposed that these oscillations reflect modulations of neuronal excitability that temporally constrain the sampling of sensory information (Schroeder & Lakatos, 2009a). The intriguing correspondence between the size of certain speech temporal units and the frequency of oscillations in certain frequency bands (see Figure 38.1) has elicited the intuition that they might play a functional role in sensory sampling. Oscillations are evidenced by means of a spectrotemporal analysis of electrophysiological recordings (see Wang, 2010 for a review). The requirements for measuring oscillations and spiking activity are different. The presentation of an exogenous stimulus typically results in an increase of spiking activity (i.e., an increase of synaptic output relative to baseline spiking activity) in those brain areas that are functionally sensitive to such sensory inputs. Neural oscillations, however, can be observed in local field potential recordings (LFPs), which reflect synchronized dentritic inputs into the observed area, even in the absence of any external stimulation. Exogenous stimulation, however, typically modulates oscillatory activity, resulting either in a reset of their phase and/or in a change (increase or decrease) in the magnitude of these oscillations (Howard & Poeppel, 2012).

Cortical oscillations are proposed to shape spike-timing dynamics and to impose phases of high and low neuronal excitability (Britvina & Eggermont, 2007; Panzeri, Brunel, Logothetis, & Kayser, 2010; Schroeder & Lakatos, 2009a, 2009b). The assumption that it is oscillations that cause spiking to be temporally clustered is derived from the observation that spiking tends to occur in specific phases (i.e., the trough) of oscillatory activity (Womelsdorf et al., 2007). It is also assumed that spiking and oscillations do not reflect the same aspect of information processing. Whereas spiking reflects axonal activity, oscillations are said to reflect mostly dendritic synaptic activity (Wang, 2010). Although both measures are relevant to address how sensory information is encoded in the brain, we believe that the ability of neural oscillations to temporally organize spiking activity support the functional relevance of neural oscillations to solve the discretization problem and to permit the integration of complex sensory signals across time.

Neuronal oscillations are ubiquitous in the brain, but they vary in strength and frequency depending on their location and the exact nature of their neuronal generators (Mantini, Perrucci, Del Gratta, Romani, & Corbetta, 2007). The notion that neural oscillations shape the way the brain processes sensory information is supported by a wealth of electrophysiological findings in humans and animals. Stimuli that occur in the ideal excitability phase of slow oscillations (<12 Hz) are processed faster and with a higher accuracy (Busch, Dubois, & VanRullen, 2009; Henry & Obleser, 2012; Lakatos, Karmos, Mehta, Ulbert, & Schroeder, 2008; Ng, Schroeder, & Kayser, 2012; Wyart, de Gardelle, Scholl, & Summerfield, 2012). However, gamma-band 40 Hz activity (low gamma band) can be observed at rest in both monkey (Fukushima, Saunders, Leopold, Mishkin, & Averbeck, 2012) and human auditory cortex. In humans, it can be measured using EEG, MEG, concurrent EEG and fMRI (with more precise localization) (Morillon et al., 2010), and intracranial electroencephalographic recordings (sEEG, EcoG) in patients. Neural oscillations in this range are endogenous in the sense that one can observe a spontaneous spike clustering at approximately 40 Hz even in the absence of external stimulation. This gamma activity is thought to be generated by a ping-pong interaction between pyramidal cells and inhibitory interneurons (Borgers, Epstein, & Kopell, 2005, Borgers, Epstein, & Kopell, 2008), or even just among interneurons that are located in superficial cortical layers (Tiesinga & Sejnowski, 2009). Exogenous inputs usually increase gamma-band activity in sensory areas, presumably clustering spiking activity that is propagated to higher hierarchical processing stages (Arnal & Giraud, 2012; Arnal, Wyart, & Giraud, 2011; Bastos et al., 2012; Fontolan, Morillon, Liegeois-Chauvel, & Giraud, 2014). By analogy with the proposal of Elhilali, Fritz, Klein, Simon, and Shamma (2004) that slow responses gate faster ones, it is interesting to envisage this periodic modulation of spiking by oscillatory activity as an endogenous mechanism to optimize the extraction of relevant sensory input in time. Such integration could occur under the patterning of slower oscillations in the delta–theta range.

38.2.4 Alignment of Neuronal Excitability with Speech Timescales

Experimental exploration of how speech parsing and encoding is performed by the brain is nontrivial. One approach has been to explore how neural responses can discriminate different sentences, assuming that the features of neural signals that are sensitive to such differences (e.g., frequency band, amplitude, and phase) should reveal the features that are vital to sentence decoding. Using this approach, it was shown that the phase of theta-band neural activity reliably discriminates different sentences (Luo & Poeppel, 2007). Specifically, when one sentence is repeatedly presented to listeners, the phase of ongoing theta-band activity follows a consistent phase sequence. When different sentences are played, however, different phase sequences are observed. Because the theta-band (4–8 Hz) falls around the mean syllabic rate of speech (~5 Hz), the phase of theta-band activity likely tracks syllabic-level features of speech (Doelling, Arnal, Ghitza, & Poeppel, 2014; Edwards & Chang, 2013; Giraud & Poeppel, 2012; Hyafil et al., 2012; Luo & Poeppel, 2007; Peelle et al., 2013). These findings support the notion that the syllabic timescale has adapted to a preexisting cortical preference for temporal information in this frequency range. At this point, however, it is not clear whether the phase-locking between speech inputs and neural oscillations is necessary for speech intelligibility. Sentences played backward (and therefore unintelligible) can similarly be discriminated on the basis of their phase course, which tempers the interpretation that these oscillations play a causal role in speech perception (Howard & Poeppel, 2011). However, two recent studies using distinct ways of acoustically degrading speech intelligibility demonstrate that the temporal alignment between the stimulus and delta–theta band responses is higher when the stimulus is intelligible (Doelling et al., 2014; Peelle et al., 2013). This, again, supports the notion that those neural oscillations that match the slow (syllabic) speech timescales are useful (if not necessary) for the extraction of relevant speech information.

Neural oscillatory responses can also be entrained at much higher rates in the middle to high (40–200 Hz) gamma band (Brugge et al., 2009; Fishman, Reser, Arezzo, & Steinschneider, 2000). This could suggest that faster speech segments such as phonemic transitions could be extracted using the same encoding principle. High gamma responses in early auditory regions (Ahissar et al., 2001; Mesgarani & Chang, 2012; Morillon, Liegeois-Chauvel, Arnal, Benar, & Giraud, 2012; Nourski et al., 2009) reflect the fast temporal fluctuations in the speech envelope. A recent EcoG study succeeded at reconstructing the original speech input by using a combination of linear and nonlinear methods to decode neural responses from high gamma (70–150 Hz) activity recorded in auditory cortical regions (Pasley et al., 2012). Therefore, the decoding of auditory activity on a large spatial scale (at the population level) demonstrates that the auditory cortex maintains a high-fidelity representation of temporal modulations up to 200 Hz. However, according to psychophysiological findings described previously, speech intelligibility mostly relies on the preservation of the low-frequency (<50 Hz) temporal fluctuations rather than on higher-frequency information. Therefore, whether it is necessary to maintain a representation of such acoustic features to correctly perceive speech remains unclear. The following section aims at clarifying the putative neural mechanisms underpinning the segmentation and the integration of auditory speech signals into an intelligible percept.

38.2.5 Parallel Processing at Multiple Timescales

Schroeder & Lakatos (2009a, 2009b) have argued that oscillations correspond to the alternation of phases of high and low neuronal excitability, which temporally constrain sensory processing. This means that low gamma oscillations at 40 Hz, which have a period of approximately 25 ms, provide a 10- to 15-ms window for integrating spectrotemporal information (low spiking rate), followed by a 10- to 15-ms window for propagating the output (high spiking rate; see Figure 38.3A. for illustration). However, because the average length of a phoneme is approximately 50 ms on average, a 10- to 15-ms window might be too short for integrating

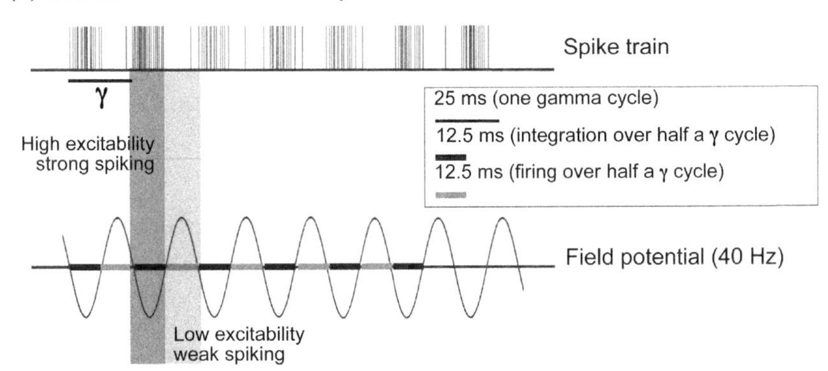

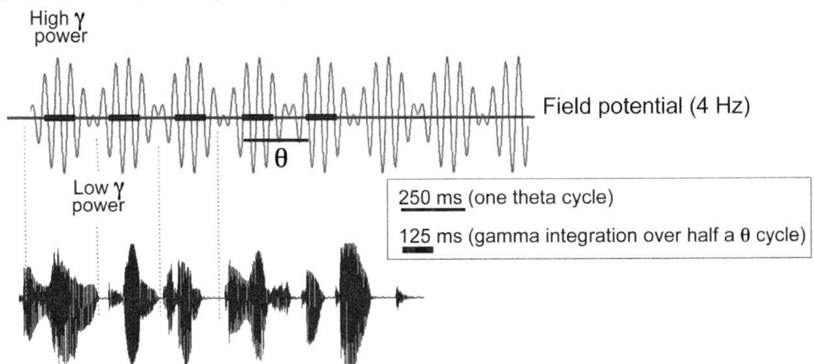

FIGURE 38.3 The temporal relationship between speech and brain oscillations. (A) Gamma oscillations periodically modulate neuronal excitability and spiking. The hypothesized mechanism is that neurons fire for approximately 12.5 ms and integrate for the rest of the 25-ms time window. Note that these values are approximate; we consider the relevant gamma range for speech to lie between 28 and 40 Hz. (B) Gamma power is modulated by the phase of the theta rhythm (approximately 4 Hz). Theta rhythm is reset by speech, resulting in maintaining the alignment between brain rhythms and speech bursts.

this information. Using a computational model of gamma oscillations generated by a pyramidal interneuron network (PING model; Borgers et al., 2005), Shamir, Ghitza, Epstein, and Kopell (2009) showed that the shape of a sawtooth input signal designed to have the typical duration and amplitude modulation of a diphone (~50 ms; typically a consonant–vowel or vowel–consonant transition) can correctly be represented by three gamma cycles, which act as a three-bit code. Such a code has the capacity required to distinguish different shapes of the stimulus and is therefore a plausible means to distinguish between phonemes. That 50-ms diphones could be correctly discriminated with three gamma cycles suggests that phonemes could be sampled with one or two gamma cycles. This issue is critical because the frequency of neural oscillations in the auditory cortex might constitute a strong biophysical determinant with respect to the size of the minimal acoustic unit that can be manipulated for linguistic purposes. In a recent extension of this model, the parsing and encoding capacity of coupled theta and gamma oscillating networks was studied (Hyafil et al., 2012). In combination, these modules succeed in signaling syllables boundaries and to orchestrate spiking within syllabic windows, so that online speech decoding becomes possible with a similar accuracy as experimental findings using intracortical recordings (Kayser, Ince, & Panzeri, 2012).

An important requirement of the computational model mentioned previously (Shamir et al., 2009) is that ongoing gamma oscillations are phase-reset, for example, by a population of onset excitatory neurons. In the absence of this onset signal, the performance of the model declines. Ongoing intrinsic oscillations appear to be effective as a segmenting tool only if they align with the stimulus. Several studies in humans and nonhuman primates have suggested that gamma and theta rhythms work together, and that the phase of theta oscillations determines the power and possibly also the phase of gamma oscillations (see Figure 38.3B; Canolty et al., 2006; Csicsvari, Jamieson, Wise, & Buzsaki, 2003; Lakatos et al., 2008, 2005; Schroeder, Lakatos, Kajikawa, Partan, & Puce, 2008). This cross-frequency relationship is referred to as "nesting." Electrophysiological recordings suggest that theta oscillations can be phase-reset by several means, such as through multimodal corticocortical pathways (Arnal, Morillon, Kell, & Giraud, 2009; Lakatos, Chen, O'Connell, Mills, & Schroeder, 2007; Thorne, De Vos, Viola, & Debener, 2011) or through predictive top-down modulations, but most probably by the stimulus onset itself. The largest cortical auditory-evoked response measured with EEG and MEG, approximately 100 ms after stimulus onset, corresponds to a phase-reset and magnitude increase of theta activity (Arnal et al., 2011; Howard & Poeppel, 2012; Sauseng et al., 2007; Sayers & Beagley, 1974). This phase-reset would align the speech signal and the cortical theta rhythm, the proposed instrument of speech segmentation into syllable/word units. Because speech is strongly amplitude-modulated at the theta rate, this would result in aligning neuronal excitability with those parts of the speech signals that are most informative in terms of energy and spectrotemporal content (Figure 38.3B). There remain critical computational issues, such as the means to get strong gamma activity at the moment of theta reset. Recent psychophysical research emphasizes the importance of aligning the acoustic speech signal with the brain's oscillatory/quasi-rhythmic activity. Ghitza and Greenberg (2009) demonstrated that while comprehension was drastically reduced by time-compressing speech signals by a factor of 3, comprehension was restored by artificially inserting periods of silence. The mere fact of adding silent periods to speech to restore an optimal temporal rate—which is equivalent to restoring "syllabicity"—improves performance even though the speech segments that remained available are not more intelligible. Optimal performance is obtained when 80-ms silent periods alternate with 40-ms time-compressed speech signals. These time constants allowed the authors to propose a phenomenological model involving three nested rhythms in the theta (5 Hz), beta, or low gamma (20–40 Hz) and gamma (80 Hz) domains (for an extended discussion, see Ghitza, 2011).

38.2.6 Parallel Processing in Bilateral Auditory Cortices

There is emerging consensus, based on neuropsychological and imaging data, that speech perception is mediated bilaterally. Poeppel (2003) attempted to integrate and reconcile several of the strands of evidence. First, speech signals contain information on at least two critical timescales, correlating with segmental and syllabic information. Second, many nonspeech auditory psychophysical phenomena fall in two groups, with integration constants of approximately 25–50 and 200–300 ms. Third, both patient and imaging data reveal cortical asymmetries such that both sides participate in auditory analysis but are optimized for different types of processing in left versus right. Fourth, crucial for the present chapter, neuronal oscillations might relate in a principled way to temporal integration constants of different sizes. Poeppel (2003) proposed that there are asymmetric distributions of neuronal ensembles between hemispheres with preferred shorter versus longer

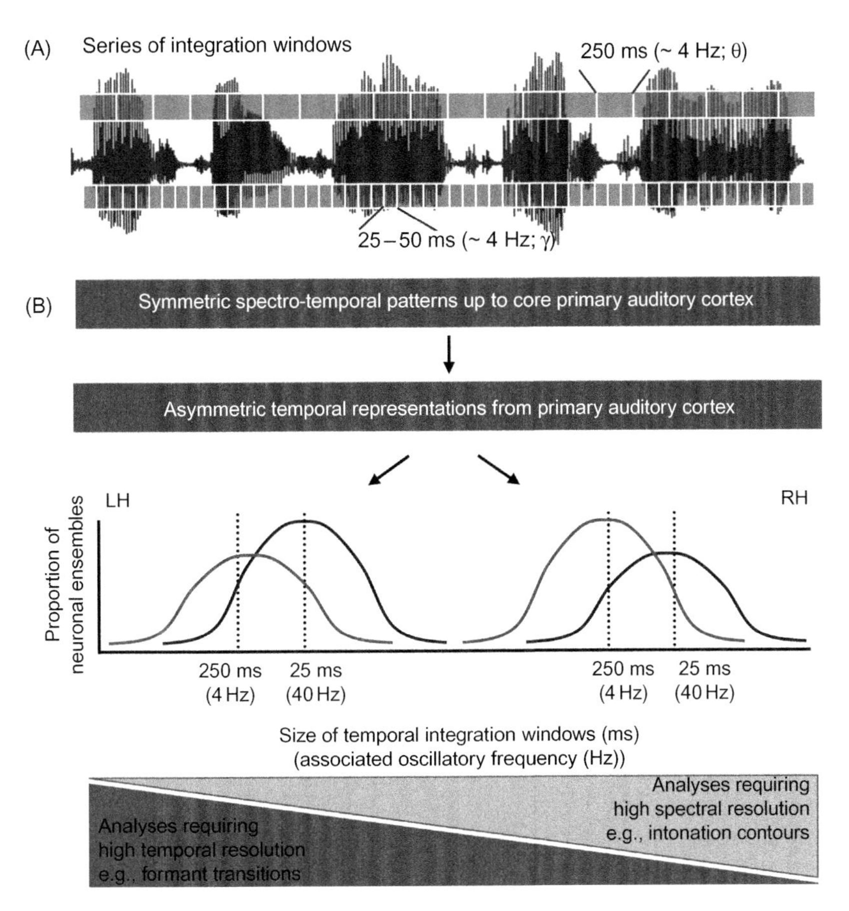

FIGURE 38.4 The AST hypothesis. (A) Temporal relationship between the speech waveform and the two proposed integration timescales (in milliseconds) and associated brain rhythms (in hertz). (B) Proposed mechanisms for asymmetric speech parsing: left auditory cortex (LH) contains a larger proportion of neurons able to oscillate at gamma frequency than the right one (RH).

integration constants; these cell groups "sample" the input with different sampling integration constants (Figure 38.4A). Specifically, left auditory cortex has a relatively higher proportion of short-term (gamma) integrating cell groups, whereas right auditory cortex has a larger long-term (theta) integrating proportion (Figure 38.4B). As a consequence, left hemisphere auditory cortex would be better equipped for parsing speech at the segmental timescale, and right auditory cortex would be better equipped for parsing speech at the syllabic timescale. This hypothesis, referred to as the asymmetric sampling in time (AST) theory, is summarized in Figure 38.4. It accounts for a variety of psychophysical and functional neuroimaging results that show that left temporal cortex responds better to many aspects of rapidly modulated speech content, whereas right temporal cortex responds better to slowly modulated signals, including music, voices, and other sounds (Warrier et al., 2009; Zatorre, Belin, & Penhune, 2002). A difference in the size of the basic integration window between left and right auditory cortices would explain speech functional asymmetry by better sensitivity of left auditory cortex to information carried in fast temporal modulations that convey, for example, phonetic cues. A specialization of right auditory cortex to slower modulations would grant it better sensitivity to slower and stationary cues, such as harmonicity and periodicity (Rosen, 1992), which are important to identify vowels, syllables, and thereby speaker identity. The AST theory is very close, in kind, to the spectrotemporal asymmetry hypothesis promoted by Zatorre (Zatorre et al., 2002; Zatorre & Gandour, 2008). Although many psychophysics and neurophysiological experiments seem to support this idea (see Giraud & Poeppel, 2012; Poeppel, 2003; Poeppel et al., 2008 for reviews on the topic), there is a lot of work in progress regarding this unresolved question.

38.2.7 Dysfunctional Oscillatory Sampling

Additional evidence to support the notion that neural oscillations play an instrumental role in speech processing would be to show that dysfunctional oscillatory mechanisms result in speech processing impairments.

Dyslexia, which is a phonological deficit (i.e., a deficit in processing speech sounds), presumably is a good candidate to test this hypothesis. Temporal sampling mediated by cortical oscillations has recently been proposed to be a central mechanism in several aspects of dyslexia (Goswami, 2011). This proposal suggests that a deficit involving theta oscillations might impair the tracking of low temporal modulations in the syllabic range. In a complementary way, it was proposed recently that gamma oscillations might play a role in yielding an auditory phonemic deficit.

Interestingly, at approximately 30 Hz, the left-dominant phase-locking profile of auditory responses in MEG (auditory steady-state responses) was only present in subjects with normal reading ability (Lehongre, Ramus, Villiermet, Schwartz, & Giraud, 2011). Because this response is absent in dyslexic participants, the authors suggested that the ability of their left auditory cortex to parse speech at the appropriate phonemic rate was altered. Those with dyslexia had a strong response at this frequency in right auditory cortex and therefore presented an abnormal asymmetry between left and right auditory cortices. Importantly, the magnitude of the anomalous asymmetry correlated with behavioral measures in phonology (such as non-word repetition and rapid automatic naming). Finally, it was also shown that dyslexic readers had stronger resonance than controls in both left and right auditory cortices at frequencies between 50 and 80 Hz. This supports the notion that these participants had a tendency to oversample information in the phonemic range, with this latter effect being positively correlated with a phonological memory deficit. As a consequence, if dyslexia induces speech parsing at a wrong frequency, then phonemic units would be sampled erratically, without necessarily inducing major perceptual deficits (Ramus & Szenkovits, 2008; Ziegler, Pech-Georgel, George, & Lorenzi, 2009). As a consequence, the phonological impairment could take different forms, with a stronger impact on the acoustic side for undersampling (insufficient acoustic detail per time unit) and on the memory side for oversampling (too many frames to be integrated per time unit).

Although important, the observation that oscillatory anomalies co-occur with atypical phonological representations remains insufficient to establish a causal role of dysfunctional oscillatory sampling. Causal evidence that auditory sampling depends on intrinsic oscillatory properties (and is, as a consequence, determined by cortical columnar organization) could be obtained from knockout animal models comparing neuronal activity with continuous auditory stimuli in sites with various degrees of columnar disorganization. However, such animal work can only indirectly address a specific relation to speech processing.

38.3 BROADENING THE SCOPE: FUNCTIONAL MODELS

Although the perceptual analysis of speech is rooted in the different anatomic subdivisions of auditory cortex in the temporal lobe, speech processing involves a large network that includes areas in parietal and frontal cortices, the relative activations of which strongly depend on the task performed. Several reviews have synthesized the state of the art of functional neuroanatomy of speech perception (Hickok & Poeppel, 2000, 2004, 2007; Rauschecker & Scott, 2009; Scott & Johnsrude, 2003). Figure 38.5A summarizes the main consensus based on functional neuroimaging (fMRI; positron emission tomography [PET], MEG/EEG) and lesion data.

Departing from the classical model in which both a posterior (Wernicke's) and an anterior (Broca's) area form the anatomic network, it is now argued that speech is processed in parallel in at least two streams, a ventral stream for speech-to-meaning mapping (a "what" stream) and a dorsal stream for speech-to-articulation mapping (a "how" stream). Both streams converge on prefrontal cortex, with a tendency for the ventral pathway to contact ventral prefrontal cortex (BA 44/45, also referred to as Broca's area) and for the dorsal pathway to contact dorsal premotor regions (Hickok & Poeppel, 2007; Rauschecker & Scott, 2009). The dual path network operates both in a feedforward (bottom-up) and feedback (top-down) manner, highlighting the need for neurophysiologically grounded theories that have appropriate primitives to permit such bidirectional processing in real time.

38.3.1 An Oscillation-Based Model of Speech Processing

This chapter emphasizes a neurophysiological perspective and especially the potential role of neuronal oscillations as "administrative mechanisms" to parse and decode speech signals. Does such a focus converge with the current functional anatomical models? Recent experimental research has begun developing a functional anatomic model solely derived from recordings of neuronal oscillations. Based on analyses of the sources of oscillatory activity, that is, brain regions showing asymmetric theta/gamma activity at rest and under linguistic stimulation, Morillon et al. (2010) proposed a new functional model of speech and language processing (Figure 38.5B) that links to the textbook anatomy (illustrated in Figure 38.5A). This model is grounded in a "core network" showing left oscillatory dominance at rest (no linguistic stimulation, no task), encompassing auditory, somatosensory, and motor

(A) Dual stream functional neuroanatomical model

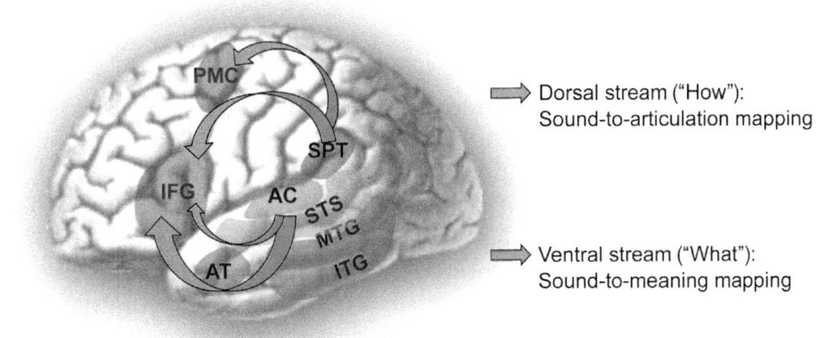

(B) Oscillation-based functional model

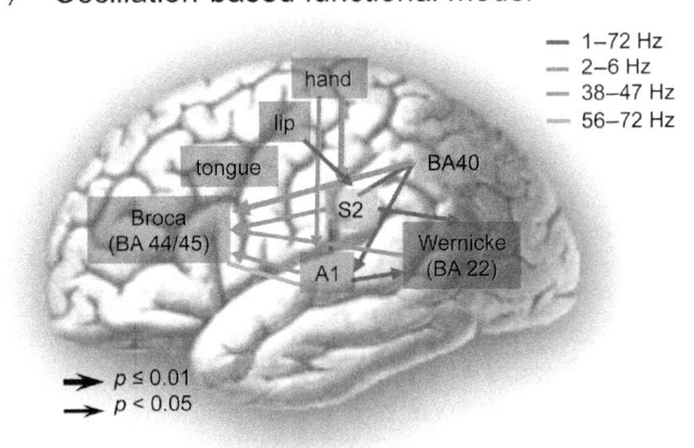

(C) Top-down predictions in the AC

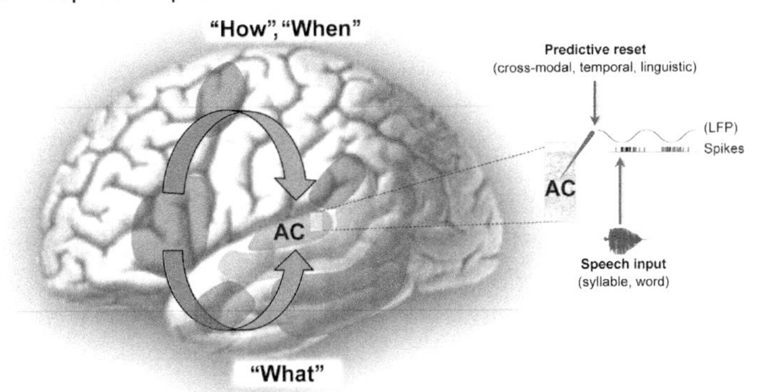

FIGURE 38.5 Three functional neuroanatomical models of speech perception. (A) Model based on neuropsychology and functional neuroimaging data (PET and fMRI; after Hickok & Poeppel, 2007). (B) Model based on the propagation of resting oscillatory asymmetry during an audiovisual linguistic stimulation (a spoken movie). (C) The Predictive (Bayesian) View on Speech Processing relies on the inversion of the dual-stream model to support the propagation of "what" and "when" predictions toward sensory regions.

cortices, and BA40 in inferior parietal cortex. The strongest asymmetries are observed in motor cortex and in BA40, which presumably play an important causal role in left hemispheric dominance during language processing. Critically, the proposed core network does not include Wernicke's (BA22) and Broca's (BA 44/45) areas, despite the fact that both are classically related to speech and language processing. Interestingly, whereas these areas show no sign of asymmetry at rest, they possibly "inherit" left-dominant oscillatory activity during linguistic processing from the putative core regions. The model argues that posterior superior temporal cortex (Wernicke's area) inherits its profile from auditory and somatosensory cortices, whereas Broca's area inherits its profile from all posterior regions including auditory, somatosensory, Wernicke, and BA40. This model specifies that posterior regions share their oscillatory activity over the whole range of frequencies examined (1–72 Hz), whereas Broca's area inherits only the gamma range of the posterior oscillatory activity. This might reflect that oscillatory activity in Broca's area does not exclusively pertain to language. Finally, an important feature of the model is the influence of the motor lip and hand areas on auditory cortex oscillatory activity on the delta/theta scale, which underlines the importance of syllable and co-speech gesture production rates, on the receptive auditory sampling, and its asymmetric implementation. This model is compatible with a hardwired alignment of speech perception and production capacities at a syllable but not at a phonemic timescale, suggesting that sensory/motor alignment at the phonemic timescale is presumably acquired. Using an approach entirely driven by oscillations, this model is largely consistent with the traditional one (Figure 38.5A), but it places a new emphasis on hardwired auditory–motor interactions and on a determinant role of BA40 in language lateralization, which remains to be clarified.

38.3.2 Predictive Models of Speech Processing

When processing continuous speech, the brain needs to simultaneously carry out acoustic and linguistic operations; at every instant there is both acoustic input to be processed and meaning to be calculated from the preceding input. Discretization using phases during which cortical neurons are either highly or weakly receptive to input is one computational principle that could ensure constant alternation between sampling the input and matching this input onto higher-level, more abstract representations. A Bayesian perspective on this issue would indicate that the brain decodes sounds by constantly generating descending (top-down) inferences about what is and will be said on the basis of the quickest and crudest neural representation it can make with an acoustic input (Poeppel et al., 2008). Consistent with this view, recent models of perception suggest that the brain hosts an internal model of the world that is used to generate, test, and update predictions about future sensory events. A proposal in very similar spirit is the "reverse hierarchy theory," a conceptualization developed to meet certain challenges in visual object recognition (Hochstein & Ahissar, 2002), and more recently extended to speech processing (Nahum et al., 2008).

Adding to the dual-stream model of perception that splits the processing of "what" and "how" information (Figure 38.5A), the predictive coding framework (Friston, 2005, 2010) posits that each stream also generates distinct types of predictions with regard to expected sensory inputs (Figure 38.5C). Top-down predictions that propagate throughout the ventral stream relate to the content of expected stimuli (the "what" aspect), whereas the dorsal stream generates predictions that pertain to the timing of events (the "when" aspect). Here we propose that in both cases, top-down expectations predictively modulate neural oscillations to facilitate the sampling of predicted sensory inputs.

Audiovisual speech perception provides an ideal paradigmatic situation to test how the brain can use information from one sensory (visual) stream to derive predictions about upcoming events in another sensory (auditory) stream. During face-to-face conversation, because the onset of visual cues (oro-facial movements) often precedes the auditory onset (syllable) by approximately 150 ms, visual information can be used to predict upcoming auditory syllables (van Wassenhove, Grant, & Poeppel, 2005), speeding-up speech perception. A series of recent works have proposed that this predictive cross-modal mechanism relies on a corticocortical phase-reset from visual regions to auditory ones (Arnal et al., 2009, 2011; Luo et al., 2010). According to this hypothesis, a cross-modal visual-to-auditory phase-reset aligns ongoing low-frequency delta–theta oscillations in the auditory cortex in an ideal phase allowing for optimal sampling of expected auditory inputs (Schroeder et al., 2008). Therefore, during the perception of a continuous audiovisual speech stream, visual inputs would predictively entrain ongoing oscillations in the auditory cortex, which in turn facilitates the encoding of syllabic information (Zion Golumbic, Cogan, Schroeder, & Poeppel, 2013).

When presented to a rhythmic stream of events, the brain can also predict the temporal occurrence of future sensory inputs (Arnal & Giraud, 2012). Because the speech signal exhibits quasiperiodic modulations at the syllabic rate, the syllable timing is relatively predictable in time and could be predictively encoded by the phase of ongoing delta–theta (2–8 Hz) (Lakatos et al., 2008). The alignment of incoming speech signals with slow endogenous cortical activity and the resonance with neocortical delta–theta oscillations represents a plausible way to automatize predictive timing at an early, prelexical, processing stage (Ding & Simon, 2012; Henry & Obleser, 2012). However, the observation that the dorsal pathway (Figure 38.5A and 38.5C) is recruited during the perception of temporally regular event streams (e.g., during beat perception) suggests that the motor system also plays a role in the predictive alignment of ongoing oscillation in the auditory cortex (Arnal, 2012; Arnal & Giraud, 2012; Fujioka, Trainor, Large, & Ross, 2012). Consistent with the concepts of active sensing or active inference, motor efferent signals that are generated in synchrony with the beat may predictively modulate the activity in the auditory cortex and facilitate the processing of incoming events (Arnal & Giraud, 2012). In other words, the active entrainment of slow endogenous cortical activity via a periodic resetting from the motor system represents a plausible way to facilitate the processing of expected auditory events at a low processing level. Such a mechanism is possibly at the origin of the attentional selection of the relevant streams in the cocktail party effect (Zion Golumbic, Ding, et al., 2013).

In summary, predictions play a major role in the encoding of temporal information in the auditory cortex. The ability to extract regularities and generate inferences about upcoming events primarily allows the periodic cortical activity to facilitate sensory processing, regardless of the informational content of forthcoming information. Additionally, when targeting specific neuronal populations, top-down signals could provide content-related priors. These two mechanisms are complementary at the computational level. Whereas temporal predictions ("when") align neuronal excitability by controlling the momentary phase of low-frequency oscillations relative to incoming stimuli, content predictions target neuronal populations specific to the representational content ("what") of forthcoming stimuli. The combination of these two types of mechanisms is again ideally illustrated by speech processing. Speech comprehension has long been argued to rely on cohort models in which each heard word preactivates a pool of other words with the same onset, until it reaches a point at which the word is uniquely identified. This model assumes that cognitive resources are used at the lexical level, where predictions are formed. Gagnepain and collaborators (Gagnepain, Henson, & Davis, 2012) recently demonstrated, however, that the predictive mechanisms in word comprehension involve segmental rather than lexical predictions, meaning that each segment is likely used to predict the next. Computationally, this observation supports the view that auditory cortex samples speech into segments using mechanisms that make them predictable in time, and that a representation of these segments is used to test specific predictions in a recurrent, predictable fashion.

38.3.3 Conclusion

Time is an essential feature of speech perception. No speech sound can be identified without integrating the acoustic input over time, and the temporal scale at which such integration operates determines whether we are hearing phonemes, syllables, or words. The central idea of this chapter is that, unlike subcortical processing that faithfully encodes speech sounds in their precise spectrotemporal structure, processing in primary and association auditory cortices results in the discretization of spectrotemporal patterns, using variable temporal integration scales. It is unlikely that speech representations are precise in both time and space. The limited phase-locking capacity of the auditory cortex thus appears a likely counterpart to its spatial integration properties (across cortical layers and functional regions). Speech processing through and across cortical columns containing complex recurrent circuits bears a cost on the temporal precision of speech representations, and integration at the gamma scale could be a direct consequence of processing at the cortical column scale. In this chapter we argue that the auditory cortex uses gamma oscillations to integrate the speech auditory stream at the phonemic timescale, and uses theta oscillations to signal syllable boundaries and orchestrate gamma activity. Although the generation mechanisms are less well-known for theta than for gamma oscillations, at present we see no alternative computational solution to the online speech segmentation and integration problem than invoking coupled theta and gamma activity. More research is needed to evaluate the detailed neural operations that are necessary to transform the acoustic input into linguistic representations, and it remains possible that other nonoscillatory mechanisms also contribute to these transformations.

References

Ahissar, E., Nagarajan, S., Ahissar, M., Protopapas, A., Mahncke, H., & Merzenich, M. M. (2001). Speech comprehension is correlated with temporal response patterns recorded from auditory cortex. *Proceedings of the National Academy of Sciences of the United States of America, 98*, 13367–13372.

Allen, J. B. (2005). Articulation and intelligibility. *Synthesis Lectures on Speech and Audio Processing, 1*(1), 1–124.

Arnal, L. H. (2012). Predicting "when" using the motor system's beta-band oscillations. *Frontiers in Human Neuroscience, 6*, 225.

Arnal, L. H., & Giraud, A. L. (2012). Cortical oscillations and sensory predictions. *Trends in Cognitive Sciences, 16*, 390–398.

Arnal, L. H., Morillon, B., Kell, C. A., & Giraud, A.-L. (2009). Dual neural routing of visual facilitation in speech processing. *The Journal of Neuroscience, 29*, 13445–13453.

Arnal, L. H., Wyart, V., & Giraud, A. L. (2011). Transitions in neural oscillations reflect prediction errors generated in audiovisual speech. *Nature Neuroscience, 14*, 797–801.

Bastos, A. M., Usrey, W. M., Adams, R. A., Mangun, G. R., Fries, P., & Friston, K. J. (2012). Canonical microcircuits for predictive coding. *Neuron, 76*, 695–711.

Boemio, A., Fromm, S., Braun, A., & Poeppel, D. (2005). Hierarchical and asymmetric temporal sensitivity in human auditory cortices. *Nature Neuroscience, 8*, 389–395.

Borgers, C., Epstein, S., & Kopell, N. J. (2005). Background gamma rhythmicity and attention in cortical local circuits: A computational study. *Proceedings of the National Academy of Sciences of the United States of America, 102*, 7002–7007.

Borgers, C., Epstein, S., & Kopell, N. J. (2008). Gamma oscillations mediate stimulus competition and attentional selection in a cortical network model. *Proceedings of the National Academy of Sciences of the United States of America, 105*, 18023–18028.

Britvina, T., & Eggermont, J. J. (2007). A markov model for interspike interval distributions of auditory cortical neurons that do not show periodic firings. *Biological Cybernetics, 96*, 245–264.

Brugge, J. F., Nourski, K. V., Oya, H., Reale, R. A., Kawasaki, H., Steinschneider, M., et al. (2009). Coding of repetitive transients by auditory cortex on Heschl's gyrus. *Journal of Neurophysiology, 102*, 2358–2374.

Busch, N. A., Dubois, J., & VanRullen, R. (2009). The phase of ongoing EEG oscillations predicts visual perception. *The Journal of Neuroscience, 29*, 7869–7876.

Canolty, R. T., Edwards, E., Dalal, S. S., Soltani, M., Nagarajan, S. S., Kirsch, H. E., et al. (2006). High gamma power is phase-locked to theta oscillations in human neocortex. *Science, 313*, 1626–1628.

Canolty, R. T., & Knight, R. T. (2010). The functional role of cross-frequency coupling. *Trends in Cognitive Sciences, 14*, 506–515.

Chi, T., Ru, P., & Shamma, S. A. (2005). Multiresolution spectrotemporal analysis of complex sounds. *The Journal of the Acoustical Society of America, 118*, 887–906.

Csicsvari, J., Jamieson, B., Wise, K. D., & Buzsaki, G. (2003). Mechanisms of gamma oscillations in the hippocampus of the behaving rat. *Neuron, 37*, 311–322.

Dau, T., Kollmeier, B., & Kohlrausch, A. (1997). Modeling auditory processing of amplitude modulation. II. Spectral and temporal integration. *The Journal of the Acoustical Society of America, 102*, 2906–2919.

Ding, N., & Simon, J. Z. (2012). Neural coding of continuous speech in auditory cortex during monaural and dichotic listening. *Journal of Neurophysiology, 107*, 78–89.

Doelling, K. B., Arnal, L. H., Ghitza, O., & Poeppel, D. (2014). Acoustic landmarks drive delta-theta oscillations to enable speech comprehension by facilitating perceptual parsing. *Neuroimage, 85*, 761–768.

Drullman, R., Festen, J. M., & Plomp, R. (1994a). Effect of reducing slow temporal modulations on speech reception. *The Journal of the Acoustical Society of America, 95*, 2670–2680.

Drullman, R., Festen, J. M., & Plomp, R. (1994b). Effect of temporal envelope smearing on speech reception. *The Journal of the Acoustical Society of America, 95*, 1053–1064.

Edwards, E., & Chang, E. F. (2013). Syllabic (~2–5 Hz) and fluctuation (~1–10 Hz) ranges in speech and auditory processing. *Hearing research, 305*, 113–134.

Elhilali, M., Fritz, J. B., Klein, D. J., Simon, J. Z., & Shamma, S. A. (2004). Dynamics of precise spike timing in primary auditory cortex. *The Journal of Neuroscience, 24*, 1159–1172.

Elliott, T. M., & Theunissen, F. E. (2009). The modulation transfer function for speech intelligibility. *PLoS Computational Biology, 5*, e1000302.

Faulkner, A., Rosen, S., & Smith, C. (2000). Effects of the salience of pitch and periodicity information on the intelligibility of four-channel vocoded speech: Implications for cochlear implants. *The Journal of the Acoustical Society of America, 108*, 1877–1887.

Fishman, Y. I., Reser, D. H., Arezzo, J. C., & Steinschneider, M. (2000). Complex tone processing in primary auditory cortex of the awake monkey. II. Pitch versus critical band representation. *The Journal of the Acoustical Society of America, 108*, 247–262.

Fontolan, L., Morillon, B., Liegeois-Chauvel, C., & Giraud, A.-L. (2014). The contribution of frequency-specific activity to hierarchical information processing in the human auditory cortex. *Nature Communications, 5*, 4694.

Friston, K. (2005). A theory of cortical responses. *Philosophical transactions of the Royal Society of London Series B, Biological sciences, 360*, 815–836.

Friston, K (2010). The free-energy principle: A unified brain theory? *Nature Reviews Neuroscience, 11*, 127–138.

Fujioka, T., Trainor, L. J., Large, E. W., & Ross, B. (2012). Internalized timing of isochronous sounds is represented in neuromagnetic beta oscillations. *The Journal of Neuroscience, 32*, 1791–1802.

Fukushima, M., Saunders, R. C., Leopold, D. A., Mishkin, M., & Averbeck, B. B. (2012). Spontaneous high-gamma band activity reflects functional organization of auditory cortex in the awake macaque. *Neuron, 74*, 899–910.

Gagnepain, P., Henson, R. N., & Davis, M. H. (2012). Temporal predictive codes for spoken words in auditory cortex. *Current Biology, 22*, 615–621.

Ghitza, O. (2011). Linking speech perception and neurophysiology: Speech decoding guided by cascaded oscillators locked to the input rhythm. *Front Psychology, 2*, 130.

Ghitza, O., & Greenberg, S. (2009). On the possible role of brain rhythms in speech perception: Intelligibility of time-compressed speech with periodic and aperiodic insertions of silence. *Phonetica, 66*, 113–126.

Giraud, A. L., Kell, C., Thierfelder, C., Sterzer, P., Russ, M. O., Preibisch, C., et al. (2004). Contributions of sensory input, auditory search and verbal comprehension to cortical activity during speech processing. *Cerebral Cortex, 14*, 247–255.

Giraud, A. L., Lorenzi, C., Ashburner, J., Wable, J., Johnsrude, I., Frackowiak, R., et al. (2000). Representation of the temporal envelope of sounds in the human brain. *Journal of Neurophysiology, 84*, 1588–1598.

Giraud, A. L., & Poeppel, D. (2012). Cortical oscillations and speech processing: Emerging computational principles. *Nature Neuroscience, 15*, 511–517.

Giraud, A. L., & Price, C. J. (2001). The constraints functional neuroimaging places on classical models of auditory word processing. *Journal of Cognitive Neuroscience, 13*, 754–765.

Goswami, U. (2011). A temporal sampling framework for developmental dyslexia. *Trends in Cognitive Sciences, 15*, 3–10.

Greenberg, S., & Arai, T. (2001). The relation between speech intelligibility and the complex modulation spectrum. *Proceedings of the 7th Eurospeech Conference on Speech Communication and Technology (Eurospeech-2001)* (pp. 473–476). Aalborg, Denmark.

Grothe, B., Pecka, M., & McAlpine, D. (2010). Mechanisms of sound localization in mammals. *Physiological Reviews, 90*, 983–1012.

Henry, M. J., & Obleser, J. (2012). Frequency modulation entrains slow neural oscillations and optimizes human listening behavior. *Proceedings of the National Academy of Sciences of the United States of America, 109*, 20095–20100.

Hickok, G., & Poeppel, D. (2000). Towards a functional neuroanatomy of speech perception. *Trends in Cognitive Sciences, 4*, 131–138.

Hickok, G., & Poeppel, D. (2004). Dorsal and ventral streams: A framework for understanding aspects of the functional anatomy of language. *Cognition, 92*, 67–99.

Hickok, G., & Poeppel, D. (2007). The cortical organization of speech processing. *Nature Reviews Neuroscience, 8*, 393–402.

Hochstein, S., & Ahissar, M. (2002). View from the top: Hierarchies and reverse hierarchies in the visual system. *Neuron, 36*, 791–804.

Holcombe, A. O. (2009). Seeing slow and seeing fast: Two limits on perception. *Trends in Cognitive Sciences, 13*, 216–221.

Howard, M. F., & Poeppel, D. (2011). Discrimination of speech stimuli based on neuronal response phase patterns depends on acoustics but not comprehension. *Journal of Neurophysiology, 104*, 2500–2511.

Howard, M. F., & Poeppel, D. (2012). The neuromagnetic response to spoken sentences: Co-modulation of theta band amplitude and phase. *NeuroImage, 60*, 2118–2127.

Hyafil, A., Fontolan, L., Gutkin, B., & Giraud, A.-L. (2012). A theoretical exploration of speech/neural oscillation alignment for speech parsing. *FENS Abstract, 6*, S4704.

Joris, P. X., Schreiner, C. E., & Rees, A. (2004). Neural processing of amplitude-modulated sounds. *Physiological Reviews, 84*, 541–577.

Kanedera, N., Arai, T., Hermansky, H., & Pavel, M. (1999). On the relative importance of various components of the modulation spectrum for automatic speech recognition. *Speech Communication, 28*, 43–55.

Kayser, C., Ince, R. A., & Panzeri, S. (2012). Analysis of slow (theta) oscillations as a potential temporal reference frame for information coding in sensory cortices. *PLoS Computational Biology, 8*, e1002717.

Kiebel, S. J., Daunizeau, J., & Friston, K. J. (2008). A hierarchy of time-scales and the brain. *PLoS Computational Biology, 4*, e1000209.

Kingsbury, B. E. D., Morgan, N., & Greenberg, S. (1998). Robust speech recognition using the modulation spectrogram. *Speech Communication, 25*, 117–132.

Lakatos, P., Chen, C. M., O'Connell, M. N., Mills, A., & Schroeder, C. E. (2007). Neuronal oscillations and multisensory interaction in primary auditory cortex. *Neuron, 53*, 279–292.

Lakatos, P., Karmos, G., Mehta, A. D., Ulbert, I., & Schroeder, C. E. (2008). Entrainment of neuronal oscillations as a mechanism of attentional selection. *Science, 320*, 110–113.

Lakatos, P., Shah, A. S., Knuth, K. H., Ulbert, I., Karmos, G., & Schroeder, C. E. (2005). An oscillatory hierarchy controlling neuronal excitability and stimulus processing in the auditory cortex. *Journal of Neurophysiology, 94*, 1904–1911.

Lehongre, K., Ramus, F., Villiermet, N., Schwartz, D., & Giraud, A. L. (2011). Altered low-gamma sampling in auditory cortex accounts for the three main facets of dyslexia. *Neuron, 72*, 1080–1090.

Loebach, J. L., & Wickesberg, R. E. (2008). The psychoacoustics of noise vocoded speech: A physiological means to a perceptual end. *Hearing Research, 241*, 87–96.

Luo, H., Boemio, A., Gordon, M., & Poeppel, D. (2007). The perception of FM sweeps by Chinese and English listeners. *Hearing Research, 224*, 75–83.

Luo, H., Liu, Z. X., & Poeppel, D. (2010). Auditory cortex tracks both auditory and visual stimulus dynamics using low-frequency neuronal phase modulation. *PLoS Biology, 8*, 13.

Luo, H., & Poeppel, D. (2007). Phase patterns of neuronal responses reliably discriminate speech in human auditory cortex. *Neuron, 54*, 1001–1010.

Mantini, D., Perrucci, M. G., Del Gratta, C., Romani, G. L., & Corbetta, M. (2007). Electrophysiological signatures of resting state networks in the human brain. *Proceedings of the National Academy of Sciences of the United States of America, 104*, 13170–13175.

McDermott, J. H., & Simoncelli, E. P. (2011). Sound texture perception via statistics of the auditory periphery: Evidence from sound synthesis. *Neuron, 71*, 926–940.

Mesgarani, N., & Chang, E. F. (2012). Selective cortical representation of attended speaker in multi-talker speech perception. *Nature, 485*, 233–236.

Miller, G. A. (1951). *Language and communication*. New York, NY: McGraw-Hill.

Moerel, M., De Martino, F., Santoro, R., Ugurbil, K., Goebel, R., Yacoub, E., et al. (2013). Processing of natural sounds: Characterization of multipeak spectral tuning in human auditory cortex. *The Journal of Neuroscience, 33*, 11888–11898.

Morillon, B., Lehongre, K., Frackowiak, R. S., Ducorps, A., Kleinschmidt, A., Poeppel, D., et al. (2010). Neurophysiological origin of human brain asymmetry for speech and language. *Proceedings of the National Academy of Sciences of the United States of America, 107*, 18688–18693.

Morillon, B., Liegeois-Chauvel, C., Arnal, L. H., Benar, C. G., & Giraud, A. L. (2012). Asymmetric function of theta and gamma activity in syllable processing: An intra-cortical study. *Front Psychology, 3*, 248.

Nahum, M., Nelken, I., & Ahissar, M. (2008). Low-level information and high-level perception: The case of speech in noise. *PLoS Biology, 6*, e126.

Ng, B. S., Schroeder, T., & Kayser, C. (2012). A precluding but not ensuring role of entrained low-frequency oscillations for auditory perception. *The Journal of Neuroscience, 32*, 12268–12276.

Nourski, K. V., & Brugge, J. F. (2011). Representation of temporal sound features in the human auditory cortex. *Reviews in the Neurosciences, 22*, 187–203.

Nourski, K. V., Reale, R. A., Oya, H., Kawasaki, H., Kovach, C. K., Chen, H., et al. (2009). Temporal envelope of time-compressed speech represented in the human auditory cortex. *The Journal of Neuroscience, 29*, 15564–15574.

Panzeri, S., Brunel, N., Logothetis, N. K., & Kayser, C. (2010). Sensory neural codes using multiplexed temporal scales. *Trends in Neurosciences, 33*, 111–120.

Pasley, B. N., David, S. V., Mesgarani, N., Flinker, A., Shamma, S. A., Crone, N. E., et al. (2012). Reconstructing speech from human auditory cortex. *PLoS Biology, 10*, e1001251.

Peelle, J. E., Gross, J., & Davis, M. H. (2013). Phase-locked responses to speech in human auditory cortex are enhanced during comprehension. *Cerebral Cortex, 23*, 1378–1387.

Plack, C. J., Oxenham, A. J., Fay, R. R., & Popper, A. N. (2005). *Pitch: Neural coding and perception*. New York, NY: Springer.

Poeppel, D. (2001). New approaches to the neural basis of speech sound processing: Introduction to special section on brain and speech. *Cognitive Science, 25*, 659–661.

Poeppel, D. (2003). The analysis of speech in different temporal integration windows: Cerebral lateralization as "asymmetric sampling in time". *Speech Communication, 41*, 245–255.

Poeppel, D., Idsardi, W. J., & van Wassenhove, V. (2008). Speech perception at the interface of neurobiology and linguistics. *Philosophical Transactions of the Royal Society of London Series B, Biological Sciences, 363*, 1071–1086.

Pöppel, E., & Artin, T. T. (1988). *Mindworks: Time and conscious experience*. Harcourt Brace Jovanovich.

Ramus, F., & Szenkovits, G. (2008). What phonological deficit? *Quarterly Journal of Experimental Psychology, 61*, 129–141.

Rauschecker, J. P., & Scott, S. K. (2009). Maps and streams in the auditory cortex: Nonhuman primates illuminate human speech processing. *Nature Neuroscience, 12*, 718–724.

Remez, R. E., Rubin, P. E., Pisoni, D. B., & Carrell, T. D. (1981). Speech perception without traditional speech cues. *Science, 212*, 947–949.

Roberts, B., Summers, R. J., & Bailey, P. J. (2011). The intelligibility of noise-vocoded speech: Spectral information available from across-channel comparison of amplitude envelopes. *Proceedings Biological Sciences/The Royal Society, 278*, 1595–1600.

Rosen, S. (1992). Temporal information in speech: Acoustic, auditory and linguistic aspects. *Philosophical Transactions of the Royal Society of London Series B, Biological Sciences, 336*, 367–373.

Saberi, K., & Perrott, D. R. (1999). Cognitive restoration of reversed speech. *Nature, 398*, 760.

Saenz, M., & Langers, D. R. (2014). Tonotopic mapping of human auditory cortex. *Hearing Research, 307*, 42–52.

Sauseng, P., Klimesch, W., Gruber, W. R., Hanslmayr, S., Freunberger, R., & Doppelmayr, M. (2007). Are event-related potential components generated by phase resetting of brain oscillations? A critical discussion. *Neuroscience, 146*, 1435–1444.

Sayers, B. M., & Beagley, H. A. (1974). Objective evaluation of auditory evoked EEG responses. *Nature, 251*, 608–609.

Schroeder, C. E., & Lakatos, P. (2009a). Low-frequency neuronal oscillations as instruments of sensory selection. *Trends in Neurosciences, 32*, 9–18.

Schroeder, C. E., & Lakatos, P. (2009b). The gamma oscillation: Master or slave? *Brain Topography, 22*, 24–26.

Schroeder, C. E., Lakatos, P., Kajikawa, Y., Partan, S., & Puce, A. (2008). Neuronal oscillations and visual amplification of speech. *Trends in Cognitive Sciences, 12*, 106–113.

Scott, S. K., & Johnsrude, I. S. (2003). The neuroanatomical and functional organization of speech perception. *Trends in Neurosciences, 26*, 100–107.

Scott, S. K., Rosen, S., Lang, H., & Wise, R. J. (2006). Neural correlates of intelligibility in speech investigated with noise vocoded speech—A positron emission tomography study. *The Journal of the Acoustical Society of America*, *120*, 1075–1083.

Shamir, M., Ghitza, O., Epstein, S., & Kopell, N. (2009). Representation of time-varying stimuli by a network exhibiting oscillations on a faster time scale. *PLoS Computational Biology*, *5*, e1000370.

Shannon, R. V., Zeng, F. G., Kamath, V., Wygonski, J., & Ekelid, M. (1995). Speech recognition with primarily temporal cues. *Science*, *270*, 303–304.

Smith, Z. M., Delgutte, B., & Oxenham, A. J. (2002). Chimaeric sounds reveal dichotomies in auditory perception. *Nature*, *416*, 87–90.

Souza, P., & Rosen, S. (2009). Effects of envelope bandwidth on the intelligibility of sine- and noise-vocoded speech. *The Journal of the Acoustical Society of America*, *126*, 792–805.

Steeneken, H. J., & Houtgast, T. (1980). A physical method for measuring speech-transmission quality. *The Journal of the Acoustical Society of America*, *67*, 318–326.

Stevens, K. N. (2002). Toward a model for lexical access based on acoustic landmarks and distinctive features. *The Journal of the Acoustical Society of America*, *111*, 1872–1891.

Thorne, J. D., De Vos, M., Viola, F. C., & Debener, S. (2011). Cross-modal phase reset predicts auditory task performance in humans. *The Journal of Neuroscience*, *31*, 3853–3861.

Tiesinga, P., & Sejnowski, T. J. (2009). Cortical enlightenment: Are attentional gamma oscillations driven by ING or PING? *Neuron*, *63*, 727–732.

VanRullen, R., & Koch, C. (2003). Is perception discrete or continuous? *Trends in Cognitive Sciences*, *7*, 207–213.

van Wassenhove, V., Grant, K. W., & Poeppel, D. (2005). Visual speech speeds up the neural processing of auditory speech. *Proceedings of the National Academy of Sciences of the United States of America*, *102*, 1181–1186.

Wang, X. J. (2010). Neurophysiological and computational principles of cortical rhythms in cognition. *Physiological Reviews*, *90*, 1195–1268.

Warrier, C., Wong, P., Penhune, V., Zatorre, R., Parrish, T., Abrams, D., et al. (2009). Relating structure to function: Heschl's gyrus and acoustic processing. *The Journal of Neuroscience*, *29*, 61–69.

Womelsdorf, T., Schoffelen, J. M., Oostenveld, R., Singer, W., Desimone, R., Engel, A. K., et al. (2007). Modulation of neuronal interactions through neuronal synchronization. *Science*, *316*, 1609–1612.

Wyart, V., de Gardelle, V., Scholl, J., & Summerfield, C. (2012). Rhythmic fluctuations in evidence accumulation during decision making in the human brain. *Neuron*, *76*, 847–858.

Zatorre, R. J., Belin, P., & Penhune, V. B. (2002). Structure and function of auditory cortex: Music and speech. *Trends in Cognitive Sciences*, *6*, 37–46.

Zatorre, R. J., & Gandour, J. T. (2008). Neural specializations for speech and pitch: Moving beyond the dichotomies. *Philosophical Transactions of the Royal Society of London Series B, Biological Sciences*, *363*, 1087–1104.

Zeng, F. G. (2002). Temporal pitch in electric hearing. *Hearing Research*, *174*, 101–106.

Ziegler, J. C., Pech-Georgel, C., George, F., & Lorenzi, C. (2009). Speech-perception-in-noise deficits in dyslexia. *Developmental Science*, *12*, 732–745.

Zion Golumbic, E., Cogan, G. B., Schroeder, C. E., & Poeppel, D. (2013a). Visual input enhances selective speech envelope tracking in auditory cortex at a "cocktail party". *The Journal of Neuroscience*, *33*, 1417–1426.

Zion Golumbic, E. M., Ding, N., Bickel, S., Lakatos, P., Schevon, C. A., McKhann, G. M., et al. (2013b). Mechanisms underlying selective neuronal tracking of attended speech at a "cocktail party". *Neuron*, *77*, 980–991.

CHAPTER

39

Direct Cortical Neurophysiology of Speech Perception

Matthew K. Leonard and Edward F. Chang

Department of Neurological Surgery, University of California, San Francisco, CA, USA

39.1 INTRODUCTION

Language is among the most unique and complex human behaviors. Understanding how the brain coordinates multiple representations (e.g., for speech: acoustic–phonetic, phonemic, lexical, semantic, and syntactic) to achieve a seemingly unified linguistic experience is among the most important questions in cognitive neuroscience. This is not a new question (Geschwind, 1974; Wernicke, 1874); however, methodological developments in the past decade have provided a new perspective on its biological underpinnings. Combined with theoretical advances in linguistics (Poeppel, Idsardi, & van Wassenhove, 2008; also refer Chapters 12–14) and important discoveries in the fundamental mechanisms of neural communication and representation (Friston, 2010; Simoncelli & Olshausen, 2001), we are at a critical moment in the study of the neurobiology of language.

One challenging consequence of language being unique to humans is that we are more limited in our ability to observe the brain during linguistic tasks. Whereas sensory, motor, and even some cognitive abilities like decision-making can be studied in detail in animal models using methods that afford high spatial and temporal resolution, there are only limited opportunities for investigations at the same level of detail in humans. In this chapter, we consider how data acquisition methods that provide unparalleled spatiotemporal resolution, combined with new applications of multivariate statistical and machine learning analysis tools, have significantly advanced our understanding of the neural basis of language. In particular, we focus on a relatively rare and invasive recording method known as electrocorticography (ECoG) and its use in the direct study of the human auditory, motor, and speech systems. ECoG has been in use for much longer than many noninvasive methods, including functional magnetic resonance imaging (fMRI), but recent advances in the design and manufacturing of electrode arrays, in addition to important discoveries on the frequency dynamics of neural signals recorded directly from the cortical surface, have enabled critical new insights into neural processing.

We argue that the ability to study language at the level of neural population dynamics (and in some cases, single neurons) is fundamentally necessary if we wish to understand how sensory input is transformed into the phenomenologically rich linguistic representations that are at the center of human communication and thought. Invasive electrophysiological methods are uniquely situated to examine these questions and, when interpreted in the context of the vast (and growing) knowledge gained from noninvasive approaches, allow us to move beyond the localization questions that have dominated the field for decades and toward an understanding of neural representations.

39.2 INVASIVE NEURAL RECORDING METHODS

39.2.1 Event-Related Neural Responses

Since soon after the discovery of the scalp electroencephalogram (EEG) 85 years ago (Berger, 1929), several aspects of neural population activity have been clear. The coordinated, stimulus-evoked responses of large groups of similarly oriented pyramidal neurons can be closely related to behavior. One of the most important examples in language research is the discovery of the N400 event-related potential (ERP) (Kutas & Federmeier, 2011). When

Neurobiology of Language. DOI: http://dx.doi.org/10.1016/B978-0-12-407794-2.00039-0

EEG subjects read or hear sentences that end with either contextually congruent ("I like my coffee with milk and *sugar*") or contextually incongruent ("I like my coffee with milk and *bolts*") endings, neural activity between approximately 200 and 600 ms after critical word onset is stronger for the incongruent condition. This finding, replicated hundreds of times in a variety of subject populations (Federmeier & Kutas, 2005) and stimulus modalities (Leonard et al., 2012; Marinkovic et al., 2003), suggests that activity in large groups of cells is sensitive to the linguistic features of the stimulus and to the context in which individual words occur.

Understanding how these effects arise from local neuronal firing requires more detailed information than is typically available from scalp recordings. Electrical currents from the brain are typically measured in microvolts, approximately 1.5 million-times weaker than the power of a standard AA battery. Furthermore, the electromagnetic inverse problem, where signals recorded at the scalp cannot unambiguously be localized in the brain, means that noninvasive recordings of electrical activity are unable to resolve separate neural populations (Pascual-Marqui, 1999). Compounding the lack of an analytic solution to the inverse problem, the biophysical properties of the head, including the various layers of parenchyma, corticospinal fluid, meninges, skull, and scalp, mean that there is up to approximately 15 mm of tissue between the signal source and the recording electrode (Beauchamp et al., 2011). This tissue further smears the signal spatially and also acts as a low-pass filter, preventing the observation of oscillatory activity at higher frequencies.

ECoG does not suffer from these problems to the same extent because the recording electrodes are placed directly on the surface of the cortex during a surgical procedure that involves exposing the brain through either a Burr hole or a craniotomy (Figure 39.1A) (Penfield & Baldwin, 1952). Obviously, the methods for implanting electrodes limit the populations that can be studied to patients who require such invasive surgery for clinical reasons. In a small portion of the epilepsy patient population, anti-seizure medications are ineffective, and the clinicians determine that the severity of the seizures is sufficiently disabling that the best course of treatment is to identify and resect the epileptogenic tissue.

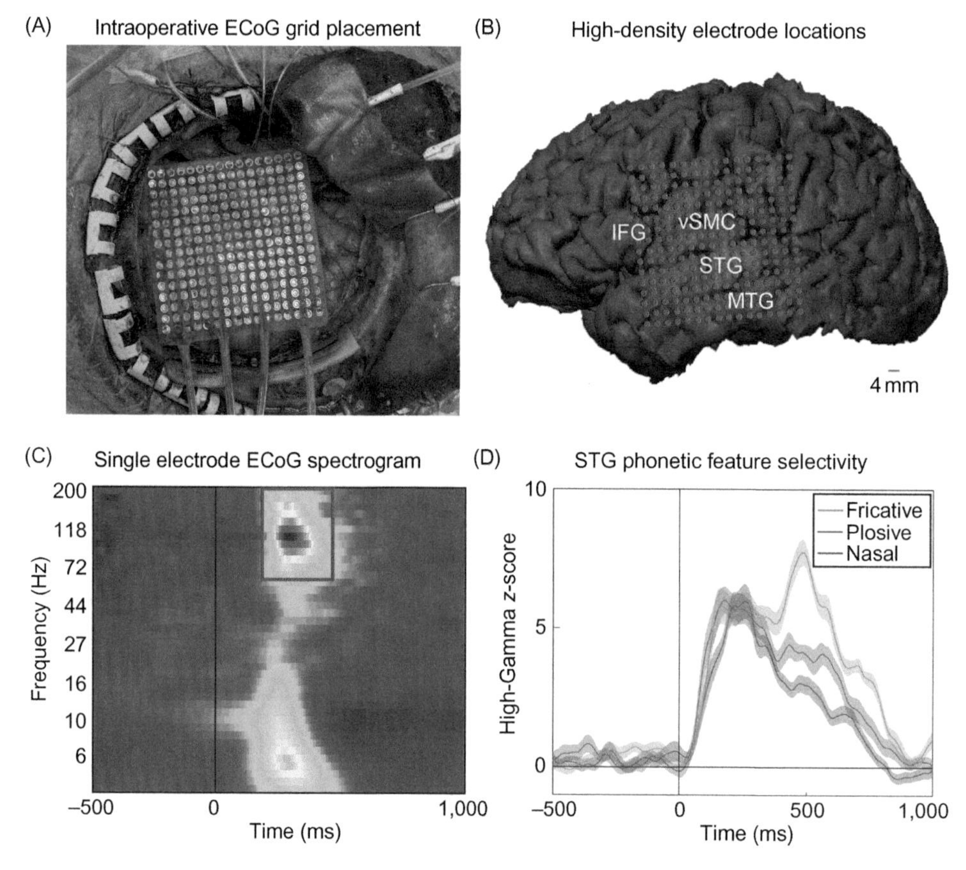

FIGURE 39.1 Invasive recording methods. (A) Intraoperative photo showing placement of high-density (4 mm pitch) ECoG grid on cortical surface. (B) Brain reconstruction showing coverage of ECoG grid. (C) Spectrogram of average speech-evoked neural activity from STG, with prominent high-gamma band responses (red box). (D) STG electrode showing high-gamma sensitivity to phonetic features (fricatives, plosives, nasals).

In many cases, localizing these foci can be done through a combination of behavioral and scalp EEG measures. However, in some cases, either because of the depth of the seizure focus or because of the inverse problem, it is only possible to identify the abnormal tissue through electrodes implanted directly in the brain (Crone, Sinai, & Korzeniewska, 2006). In these instances, the electrodes are placed according to the clinicians' best estimates of probable seizure foci, and the patient is kept in the epilepsy monitoring unit for a few days to several weeks, or however long it takes to record enough seizure activity to localize the source. If the patient has consented prior to surgery, then this hospitalization period provides a unique opportunity for researchers to collect neural data directly from the brain during a variety of experimental tasks.

The placement of ECoG electrodes often overlaps with peri-Sylvian areas of the brain that are critical for linguistic functions, particularly for speech. Electrode arrays can routinely obtain high-density coverage of the superior temporal gyrus (STG), middle temporal gyrus (MTG), ventral sensory-motor cortex (vSMC), and the inferior frontal gyrus (IFG), regions that are known to be involved in speech perception and production (Figure 39.1B).

Additionally, given that medial temporal lobe epilepsy is common in this patient population, activity is often recorded from the hippocampal and peri-hippocampal structures, which are also thought to be involved in the encoding of linguistic information (Halgren, Baudena, Heit, Clarke, & Marinkovic, 1994). These recordings are obtained by implanting penetrating depth electrodes, which reach these deep structures through basal or lateral cortical areas that are likely to be resected (Jerbi et al., 2009). Depth electrode recordings are difficult, both surgically and electrophysiologically; however, several research groups have pioneered these methods and obtained incredibly fruitful results.

It is also possible to record the activity of single neurons in the human brain. We present several examples of work that have examined the fundamental currency of neural computation, the spike. Although relatively less common, these studies provide a critical link between human cognitive neuroscience and animal models, where much more is known about the cellular and neurophysiological properties of neuronal function.

39.2.2 High-Frequency Oscillations

In addition to spatial (electrodes) and temporal (milliseconds) dimensions, neural population activity can be decomposed into the frequency dimension. One of Berger's initial observations in the scalp EEG was a dominant oscillation in the 8- to 12-Hz range, which he referred to as the "alpha" band (Berger, 1929). Over the next several decades, more frequency bands were discovered and analyzed using a variety of techniques including Fourier decomposition, wavelet analysis, and the Hilbert transform (Bruns, 2004). Different frequency ranges have been associated with a variety of behavioral states and stimulus-related responses. For example, in speech, the theta band ($\sim$4–7 Hz) is proposed to be a critical range for encoding the temporal structure of spoken input (Edwards & Chang, 2013; Poeppel, 2003).

However, the biophysics of the various media in the head prevents some of the higher frequencies from being observed at the scalp. Thus, one major advantage of ECoG is the ability to detect oscillations above approximately 70 Hz. Approximately 15 years ago, these frequencies were discovered in humans to contain highly coherent and reliable information about the stimulus (Crone, Boatman, Gordon, & Hao, 2001; Crone, Miglioretti, Gordon, & Lesser, 1998). Since then, the majority of studies that have examined speech processing using ECoG have focused on the high-gamma range, which is typically between 60 and 200 Hz (Figure 39.1C). Whereas the broadband local field potential ERP is dominated by low-frequency oscillations between approximately 2 and 20 Hz (in part due to the $1/f$ power law), the temporal resolution of high-gamma appears to carry critical information about speech, which is equally dynamic on a millisecond time scale. As we describe, it has been shown that high-gamma evoked activity correlates with acoustic–phonetic variability in primary and secondary auditory areas (Figure 39.1D) and with the coordinated movement of the articulators in motor regions during speech production. Perhaps of great interest to readers of this book is that high-gamma activity also carries information about higher level and abstract aspects of the speech signal, which suggests that it may be an important tool for understanding how both local and long distance neural networks give rise to rich linguistic representations and concepts from incoming sensory signals. Finally, high-gamma activity is well-suited to link animal models with human cognitive studies because it is strongly correlated with multiunit spiking activity (Ray & Maunsell, 2011; Steinschneider, Fishman, & Arezzo, 2008) and the blood oxygen level–dependent (BOLD) response in fMRI (Mukamel et al., 2005; Ojemann, Ojemann, & Ramsey, 2013).

39.2.3 Limitations of Invasive Methods

Before describing the scientific advancements that have been made using this set of tools, it is important to consider the limitations of invasive recording methods. First, the data are obtained from patient volunteers who are undergoing surgical procedures for serious medical conditions. Therefore, all recordings are dictated by the willingness of the patient to participate in the hospital setting and by the areas of the brain that the clinicians deem relevant to their treatment. A related limitation is that although electrode arrays are becoming more advanced in their density and coverage (Insel, Landis, & Collins, 2013; Viventi et al., 2011), invasive methods necessarily under-sample the brain. Except in cases when it is clinically necessary, recordings are typically obtained only from the exposed surface of the cortex, which limits access to the sulci and deep structures. Additionally, because most epilepsy cases have unilateral seizure foci, simultaneous recordings from both cerebral hemispheres are rare, although there have been some bilateral cases (Cogan et al., 2014).

Finally, it is important to consider that recordings are obtained from patients who have abnormal brains. Many of these surgeries involve removing much of the anterior temporal lobe and hippocampal/amygdala complex, sometimes with only limited behavioral effects on the patients (Tanriverdi, Ajlan, Poulin, & Olivier, 2009). This suggests that these areas had only limited functionality, perhaps due to plastic reorganization around epileptogenic tissue. In practice, electrodes that show any sort of abnormal neurophysiological activity are excluded from analyses; however, this concern warrants caution in interpreting results.

Despite all of these caveats, invasive recording methods offer numerous advantages for advancing our knowledge of the neurobiology of language. Limited sampling of select brain regions precludes an understanding of the dynamics of large-scale inter-regional neural networks, but the frequency and spatiotemporal resolutions of the data obtained from these areas are unparalleled compared with noninvasive tools. Furthermore, results from invasive methods consistently agree with and extend findings from other techniques, despite the nature of the patient populations. As we argue in the rest of this chapter, ECoG, depth electrode, and single unit recordings must be interpreted in the context of the current understanding from other methodologies. Ultimately, it is a rare opportunity and a privilege to work with these patients, and their contributions to neuroscience and the neurobiology of language have been critical over the past several years.

39.3 INTRACRANIAL CONTRIBUTIONS TO THE NEUROBIOLOGY OF LANGUAGE

In this section, we present examples of how invasive recording techniques have provided new insights into the neural basis of language. Although this is not an exhaustive review of the science, we hope to convey the contributions of these studies to the broader view of neurolinguistics. Thus, we consider these studies in the context of fMRI, EEG, and MEG studies, particularly those that are using analogous multivariate statistical and machine learning analysis methods. We also focus primarily on the sensory, acoustic–phonetic, and phonemic aspects of speech perception because implanted electrodes typically cover areas that are most closely associated with these functions. We also briefly argue that invasive methods are already proving useful for examining other aspects of language and linguistic encoding, particularly higher level abstract representations, such as those for lexical and semantic information (Leonard & Chang, 2014).

39.3.1 Sensory Encoding in Primary Auditory Cortex

We begin this section with a brief overview of sensory processing in the primary auditory cortex (A1). As shown in subsequent sections, downstream regions that encode acoustic–phonetic and phonemic information are highly sensitive to the spectrotemporal characteristics of the stimulus. Therefore, it is crucial to understand the inputs to these regions and to recognize that even the earliest cortical auditory areas show selective tuning that facilitates speech perception.

The ascending auditory system sends signals from the tonotopically organized cochlea through the lateral lemniscal pathway to an area on the posteromedial portion of Heschl's gyrus, which is the primary auditory cortex in humans (Hackett, 2011). Like the areas that precede it in the auditory hierarchy, A1 is tonotopically organized along at least one major axis, meaning that different neural populations are sensitive to different acoustic frequencies (Barton, Venezia, Saberi, Hickok, & Brewer, 2012; Baumann, Petkov, & Griffiths, 2013; also refer to Chapter 5). Following initial work by Celesia (Celesia, 1976) and Liégeois-Chauvel (Liegeois-Chauvel, Musolino, & Chauvel, 1991), researchers at the University of Iowa have provided a detailed description of human A1 using multicontact depth electrodes and have found that there are frequency-specific responses that are indicative of both rate and temporal coding of auditory stimuli both within and across neural populations (Brugge et al., 2009). These different coding mechanisms occur at different stimulus frequencies and may also contribute to the perception of pitch (Griffiths et al., 2010). Furthermore, narrow spectral tuning in human A1 is not just a feature of the population neural activity; it is also reflected in the selective responses of single neurons (Bitterman, Mukamel, Malach, Fried, & Nelken, 2008).

These depth electrode and single unit studies in Heschl's gyrus extend important work from noninvasive methods. Using machine learning analysis methods to relate the acoustic properties of natural sounds to changes in the BOLD response in fMRI, Moerel and colleagues have shown that A1 exhibits narrow spectral tuning in addition to tonotopic organization (Moerel, De Martino, & Formisano, 2012). The same researchers have also used high-field fMRI to show that individual neural populations have multipeak tuning at octave intervals, possibly facilitating the binding of complex sound features (Moerel et al., 2013). In agreement with temporal coding mechanisms giving rise to the perceptual phenomenon of pitch, it has also been shown with fMRI that A1 has spectro-temporal modulation transfer functions that track temporal modulations in the stimulus (Schönwiesner & Zatorre, 2009). The combined spatial organization of spectral and temporal modulation selectivity reveals multiple subfields of both primary and secondary auditory regions, suggesting a functional microstructure of the earliest cortical sensory areas that may contribute to the complex nonlinear representations of input like speech (Barton et al., 2012).

Sensitivity to spectral and temporal stimulus features in A1 most likely facilitates our ability to understand dynamic acoustic input like speech. However, it is not necessarily the case that this type of tuning reflects speech-specific or even speech-oriented specialization. In a recent study, Steinschneider and colleagues examined single-unit and multi-unit spiking activity in the primary auditory cortex of human epilepsy patients and macaques. For broadband auditory evoked potentials and high-gamma activity, both species showed similar spectrotemporal selectivity (Steinschneider, Nourski, & Fishman, 2013). Remarkably, this selectivity included acoustic aspects of speech, such as voice-onset-time and place of articulation (POA), which are critical for making phonetic distinctions. This suggests that speech-specific representations do not reside at the level of A1, although the tuning characteristics of this region allow us to parse spoken input efficiently.

39.3.2 Acoustic–Phonetic Representations in Lateral Superior Temporal Cortex

In humans, Heschl's gyrus is in close proximity and is densely connected to the lateral superior temporal cortex, in particular the STG (Hackett, 2011). In part because of the most common placement of ECoG electrode arrays, and also because of the functions of this area, STG is among the best characterized regions in the speech system. It is well known that STG is involved in phonological representation, because neural activation is observed for most speech tasks (DeWitt & Rauschecker, 2012; Hickok & Poeppel, 2007; Rauschecker & Scott, 2009). In fMRI, a very productive paradigm has contrasted BOLD responses to speech and nonspeech sounds that preserve important spectral or temporal aspects of the signal (Davis & Johnsrude, 2003; Scott, Blank, Rosen, & Wise, 2000). Numerous studies have demonstrated stronger STG activation for speech, suggesting that this region encodes aspects of the stimuli that are intelligible and behaviorally relevant to the listener (Obleser, Eisner, & Kotz, 2008). Recently, multivariate pattern analysis methods have been applied to fMRI data and have shown that STG (among other temporal lobe regions including superior temporal sulcus) are involved in representing meaningful aspects of the speech signal (Evans et al., 2013).

These results have been advanced using both invasive and noninvasive methods that afford high spatial and temporal resolution. Using MEG in healthy volunteers and ECoG in epilepsy patients, Travis and colleagues showed that STG responds preferentially to speech relative to a noise-vocoded control, which smoothes the speech spectrogram in the spectral axis, preserving the temporal and most of the frequency content but destroying intelligibility (Travis et al., 2012). This study also demonstrates that STG speech-selective responses observed in fMRI most likely reflect stimulus-evoked neural activity between approximately 50 and 150 ms, significantly earlier than activity in the same region that is associated with higher level linguistic encoding of lexical and semantic information. This demonstrates that at least at the level of nearby cortical columns, STG encodes multiple levels of acoustic–phonetic and abstract linguistic information.

What gives rise to this selectivity for meaningful aspects of the acoustic speech signal? Like A1, distinct STG neuronal populations encode the temporal structure of nonspeech acoustic input differently depending on the frequency content of the input (Nourski et al., 2013). Also like A1, populations are spectrally selective to ranges of frequencies, although they are more broadly tuned than in primary auditory regions (Nourski et al., 2012). These two ECoG studies with nonspeech tones and clicks are complemented by a set of speech ECoG studies using high-density electrode grids over STG. Spectrotemporal selectivity at single electrodes has important consequences for the representation of spoken words because it suggests that those acoustic inputs are represented as a complex pattern of activity across electrodes over time as the input unfolds. Using linear stimulus reconstruction methods, it is possible to generate a spectrogram from the population STG activity that corresponds closely to the original stimulus spectrogram (Pasley et al., 2012). This correspondence is particularly strong when the reconstruction model applies stronger weights to the spectrotemporal aspects of the acoustic input that are relevant for speech intelligibility, including temporal modulation rates that correspond to syllable onsets and offsets.

This spectrotemporal representation across electrodes indicates that separate neural populations may be contributing sensitivity to particular features of the speech stimulus. Chang and colleagues presented ECoG participants with hundreds of naturally spoken sentences that contained a large number of examples of all English phonemes (Mesgarani, Cheung, Johnson, & Chang, 2014). This allowed the researchers to examine activity at the closely spaced ECoG electrodes in response to each phoneme. They found that individual electrodes responded selectively to linguistically meaningful phonetic features, such as fricatives, plosives, and nasals. For vowels, there was a similar feature-based representation for low-back, low-front, and high-front vowels, which arose from the encoding of formant frequencies, particularly the ratio of F1 and F2. These results provide crucial information about the role of STG in speech perception. It has been suggested that STG is involved in spectrotemporal (Pasley et al., 2012), acoustic–phonetic (Boatman, 2004), and phonemic (Chang et al., 2010; DeWitt & Rauschecker, 2012; Molholm et al., 2013) processing; however, the functional and spatial specificity of these features have not been described previously. Chang and colleagues showed that STG does represent acoustic–phonetic features (rather than individual phonemes, as single electrodes responded to multiple phonemes sharing a particular feature), but that this selectivity arises from tuning to specific spectrotemporal cues.

Similar results have been obtained at the single neuron level. Chan and colleagues collected neural responses from an epilepsy patient implanted with a Utah array, which consists of 100 penetrating electrodes with 20 micron tips, and recorded spiking activity from approximately 150 single and multiunits in anterior STG (Chan et al., 2013). They found neurons that responded selectively to specific phonemes that shared particular phonetic features (e.g., high-front vowels), but not to individual phonemes. Remarkably, they also demonstrated that responses are similarly selective to phonemes when they are heard and when they are read. This suggests that written words activate phonological representations and also that single neurons in the human STG are tuned to phonetic features, independent of the nature of the sensory input.

39.3.3 Population Encoding of Phonemic Information in STG

Spectrotemporally based phonetic tuning at individual neural populations may also give rise to more complex representations of speech. Although single electrodes do not show sensitivity to specific phonemes (Mesgarani et al., 2014), the combined activity across electrodes generates a population code for complex sets of phonetic features. Two important higher level features that help distinguish phonemes are POA and voice onset time (VOT). POA refers to the configuration of the articulators and distinguishes different phonemes that share a particular feature, such as plosives (e.g., /ba/ vs. /ga/). VOT refers to the temporal delay between plosive closure and the onset of vibrations in the larynx, and distinguishes plosives that have the same POA (e.g., /ba/ vs. /pa/).

In an ECoG study, Steinschneider et al. (2011) examined how event-related high-gamma activity differed for sounds that differed parametrically in POA or VOT. They found that posterior STG electrodes correlated most strongly with changes in POA. VOT was also represented in both medial and posterior STG, and it tracked the timing of voice onset through latency shifts in the peaks of the average high-gamma waveforms. This demonstrates that understanding the encoding of complex stimulus features requires the ability to discriminate activity at local neural populations (spatial selectivity, achieved through high-density electrode arrays and high-gamma band filtering) and also to track neural activity on a millisecond level.

Another common method for examining the encoding of individual phonemes is through categorical phoneme perception. In this paradigm, a linear continuum of speech sounds varies in a particular acoustic parameter from one unambiguous extreme to another (e.g., F2 frequency, distinguishing POA for the plosives /b/, /d/, and /g/). The sounds are perceived nonlinearly, such that listeners hear one phoneme up to a particular point on the continuum, and then abruptly switch to perceiving a different phoneme; in other words, listeners fit acoustically ambiguous examples into established phoneme categories. This phenomenon has been localized using fMRI primarily to the lateral superior temporal cortex (Joanisse, Zevin, & McCandliss, 2007); however, these studies generally do not reveal how population neural activity enables this important perceptual effect.

ECoG has sufficient spatial and temporal resolution to address how nonlinear perceptual phenomena such as categorical perception are encoded in the brain. Using a continuum of synthesized speech sounds that varied linearly in F2 onset from /ba/ to /da/ to /ga/ (Figure 39.2A), it has been shown that high-gamma activity across electrodes can distinguish these sounds categorically (Chang et al., 2010). Distinct neural populations within the posterior STG generate neural activity patterns that correspond to each phoneme along

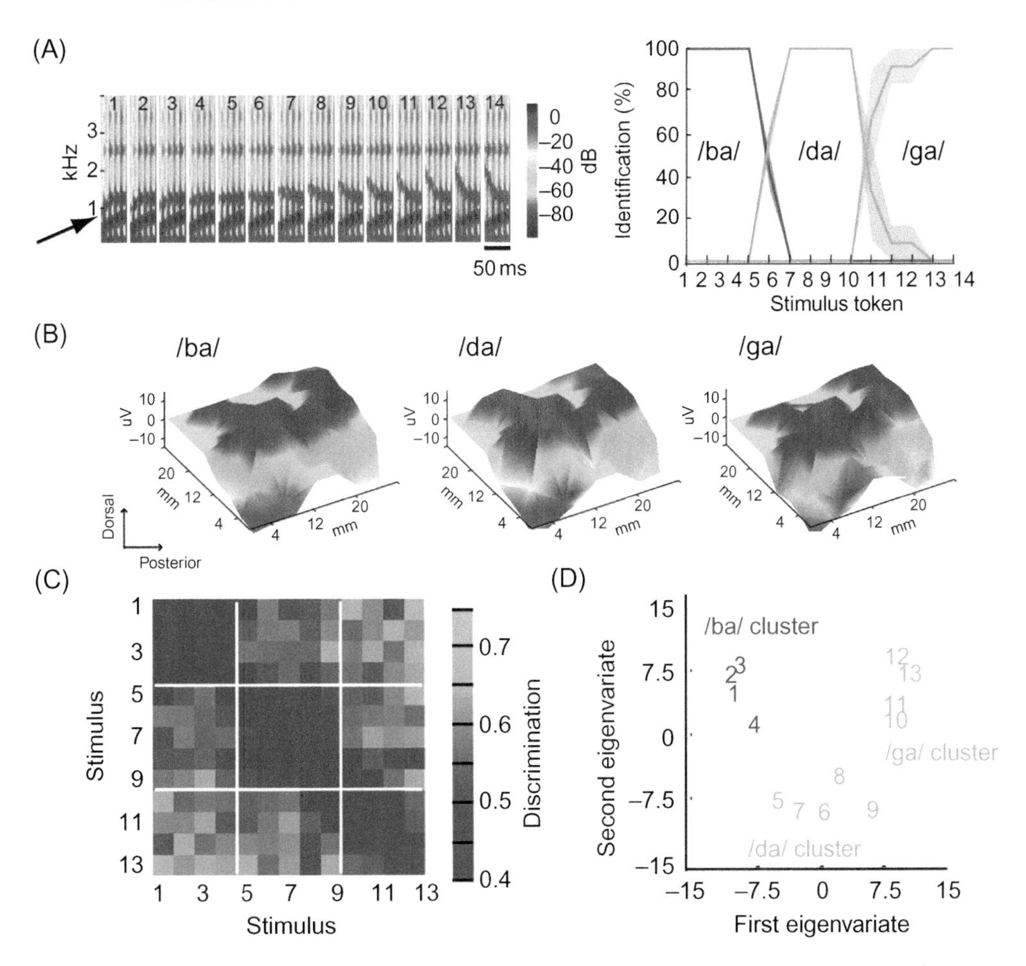

FIGURE 39.2 Speech representation in human STG. (A) Speech sounds synthesized along an acoustic continuum (stimuli 1–14, small increases in the second formant [arrow]) are perceived categorically rather than linearly in the behavior. (B) Spatial topography of evoked potentials recorded directly from the cortical surface of STG using ECoG is highly distributed and complex for each sound class. (C) Neural confusion matrix plots the pair-wise dissimilarity of neural patterns using a multivariate classifier. (D) Multidimensional scaling shows that response patterns are organized in discrete clustered categories along both acoustic and perceptual sensitivities. *Adapted from Chang et al. (2010) with permission from the publisher.*

the POA continuum, even within the space of only a couple of centimeters (Figure 39.2B). Using multivariate classification methods, stimulus-specific discriminability was observed in this activity (Figure 39.2C), which suggests that at the peak of neural pattern dissimilarity across categories (occurring at ~110 ms after stimulus onset), the perceptual effect arises from the activity at specific groups of electrodes. The timing of this discriminability is sufficiently early to suggest that categorical phoneme perception occurs *in situ* within the STG, rather than through top-down influences of other brain regions, although the limited sampling of brain areas with ECoG does not rule out the latter possibility. It is also important to note that, consistent with the evidence described for phonetic feature representations at individual STG electrodes (Mesgarani et al., 2014), categorical perception of the linear continuum in this study was organized along acoustic sensitivities. Specifically, the representations of speech tokens were ordered according to F2 in one dimension (see ordering along the *x*-axis in Figure 39.2D), but the overall pattern was categorical in two dimensions (Figure 39.2D). This is a clear demonstration that nonlinear perceptual phenomena are encoded nonlinearly in the brain.

39.3.4 Cognitive Influences on Speech in STG

Intracranial recording methods have also recently proven useful for understanding how spectrotemporal, phonetic, and phonemic representations in STG are modulated by cognitive processes. For example, these methods have renewed an interest in the "cocktail party" problem, where listeners can selectively filter out irrelevant auditory streams in a noisy environment (Cherry, 1953). Several recent studies have examined neural responses to overlapping, multispeaker acoustic stimuli, where listeners were directed to focus their attention on only one speaker (Ding & Simon, 2012; Kerlin, Shahin, & Miller, 2010; Sabri et al., 2008; Zion Golumbic et al., 2013). Using ECoG, one study used stimulus reconstruction methods (Mesgarani, David, Fritz, & Shamma, 2009) to investigate how attention modulates the spectrotemporal representation of the acoustic input in STG (Mesgarani & Chang, 2012). Participants listened to two speakers simultaneously (Figure 39.3A and B) and were asked to report the content of just one of the speech streams, thus attending to only part of the acoustic input. Remarkably, the spectrotemporal representation of the stimulus in STG only reflected the attended speech stream, as if the ignored speaker had not been heard at all (Figure 39.3C).

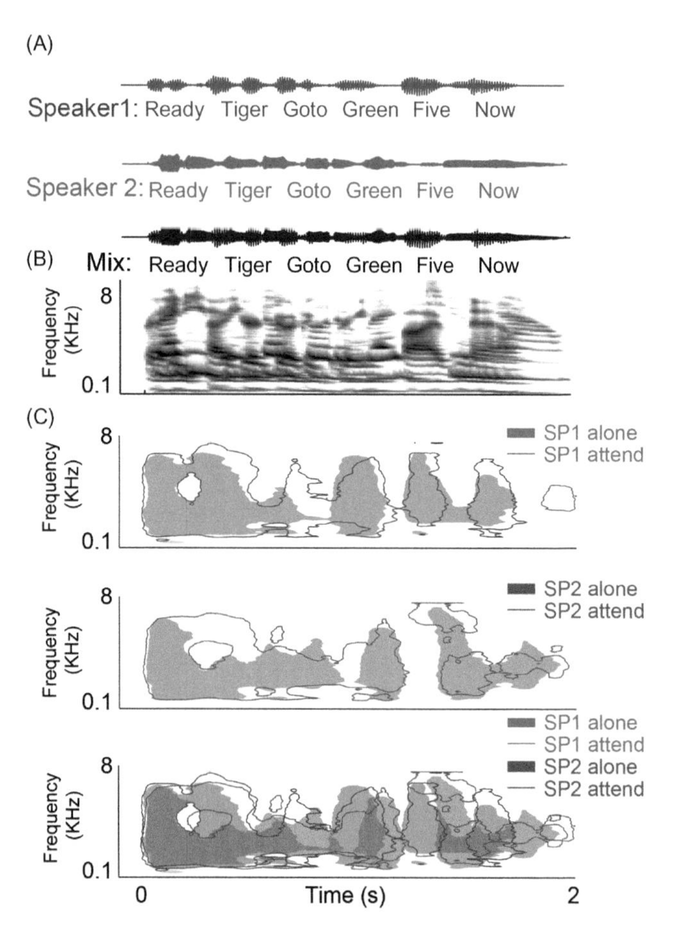

FIGURE 39.3 Attention modulates STG representations of spectrotemporal speech content. (A) ECoG participants listened to two speech streams either alone or simultaneously and were cued to focus on a particular call sign ("tiger" or "ringo") and to report the color/number combination (e.g., "green five") associated with that speaker. (B) The acoustic spectrogram of the mixed speech streams shows highly overlapping energy distributions across time. (C) Neural population-based reconstruction of the spectrograms for speaker 1 (blue) and speaker 2 (red), when participants heard each speaker alone (shaded area) or in the mixed condition (outline). Results demonstrate that in the mixed condition, attention to a particular speaker results in a spectrotemporal representation in STG as if that speaker were heard alone. *Adapted from Mesgarani and Chang (2012) with permission from the publisher.*

Using similar methods, Zion-Golumbic and colleagues extended these findings to show that this attentional modulation is reflected not only in high-gamma power changes but also in low-frequency phase over the course of several hundred milliseconds (Zion Golumbic et al., 2013). They also demonstrated that the extent to which the ignored speech stream is represented changes as a function of brain region, specifically that higher level frontal and pre-frontal areas show less robust representation. There was also a stronger correlation between the attended stimulus and the neural response in STG for high-gamma power, whereas frontal regions showed stronger correlations with the low-frequency phase, implicating different frequency-dependent encoding mechanisms in different brain regions.

As described, there have also been attempts to characterize higher order representations of linguistic input using intracranial recording methods (Sahin, Pinker, Cash, Schomer, & Halgren, 2009). The N400 response, which reflects the degree of semantic contextual mismatch, has been localized to left anterior, ventral, and superior temporal cortex (Halgren et al., 1994). Recent work has further shown that responses within STG are sensitive to semantic context, independent of lower level acoustic features (Travis et al., 2012). ECoG and depth electrode recordings have also revealed that ventral temporal regions (including perirhinal cortex) show semantic category selectivity for words, regardless of whether they are written or spoken, suggesting that abstract semantic and conceptual representations are resolvable using these methods (Chan et al., 2011). Halgren, Chan, and colleagues have also used penetrating laminar depth electrodes to record multiunit activity in individual cortical layers during these tasks and have shown that the timing of responses in input and output layers suggests an early ($\sim$130 ms) first-pass selective response, followed by a later ($\sim$400 ms) selective response that is consistent with the N400 recorded at the scalp (Chan et al., 2011; Halgren et al., 2006). This suggests that higher order linguistic representations not only are spatially distributed but also arise from a complex temporal pattern of neural activity.

Together, these studies demonstrate the context-sensitive nature of evoked neural responses. Crucially, they demonstrate that both fine spatial resolution and temporal resolution are necessary to understand how cognitive representations influence and modulate lower level sensory and perceptual neural responses. Differences across frequencies in the neural signal also encode important aspects of this modulation, which nicely complements work using noninvasive methods.

39.4 THE FUTURE OF INVASIVE METHODS IN LANGUAGE RESEARCH

We have argued that methodological advances in recording and analysis tools have allowed invasive neurophysiology to provide a window into the neural basis of speech and language that cannot be seen by other methods. The past several years have been particularly exciting because there has been a fusion of novel engineering approaches (machine learning and multivariate statistics) with linguistic and neuroscientific questions. In general, there have been two ways in which these fields have met. In the first, traditional experimental paradigms have been used with increasingly higher density neural recordings, which have in some cases provided new spatial information and often an entirely novel temporal dimension (Edwards et al., 2009, 2010; Molholm et al., 2013; Steinschneider et al., 2011; Travis et al., 2012). In the second approach, researchers have used data mining and machine learning techniques to understand how the activity observed in the paradigms from the first approach reflect the encoding and representation of stimulus and abstract information (Chan et al., 2013; Chang et al., 2010; Cogan et al., 2014; Mesgarani et al., 2014; Pasley et al., 2012; Zion Golumbic et al., 2013). Together, and in the context of noninvasive studies, these investigations are converging on the ultimate goal of neurolinguistics: to understand how neurons and neural networks generate the linguistic representations that are at the core of our daily experience.

Although the machine learning and data mining approaches have provided an unprecedented understanding of the underlying information representations in areas such as STG, there are many equally fundamental questions that remain that invasive recording methods may be well-suited to answer. A basic question for speech perception is how tuning and representations in A1 and STG are integrated over time to form more abstract word-level representations. Surprisingly, many of the studies cited neglect the temporal dimension of the data; it is common for input to linear classifiers to be an average over a large time window or even a single time point (Mesgarani et al., 2014). Although examining high-gamma band power inherently focuses on an aspect of the data involving fast time scales (e.g., acoustic onsets), there is a major question regarding how information is encoded over time in different brain regions.

One method for examining neural activity over time, which was originally developed for understanding the dynamics of populations of single units (Churchland et al., 2012), involves characterizing the shared variability across space and plotting neural activity as trajectories through a "state-space" over time. In a recent study on the organization of articulatory representations in vSMC, principal components analysis was used to reduce the dimensionality across ECoG electrodes and showed that activity closely tracks the articulatory features of consonant-vowel syllables on a millisecond time scale (Bouchard, Mesgarani, Johnson, & Chang, 2013). Given that abstract representations such as lexical and semantic features are known to be more spatially distributed (Huth, Nishimoto, Vu, & Gallant, 2012) than lower level sensory features, it may be possible to use high-density neurophysiological recordings to understand how speech is transformed from primary sensory areas to distributed cortical networks. Overall, the field of neurolinguistics faces an exciting period with a multitude of experimental approaches that contribute unique and complementary information to our understanding of language.

References

Barton, B., Venezia, J. H., Saberi, K., Hickok, G., & Brewer, A. A. (2012). Orthogonal acoustic dimensions define auditory field maps in human cortex. *Proceedings of the National Academy of Sciences, 109*(50), 20738–20743.

Baumann, S., Petkov, C. I., & Griffiths, T. D. (2013). A unified framework for the organization of the primate auditory cortex. *Frontiers in Systems Neuroscience, 7*.

Beauchamp, M. S., Beurlot, M. R., Fava, E., Nath, A. R., Parikh, N. A., Saad, Z. S., et al. (2011). The developmental trajectory of brain-scalp distance from birth through childhood: Implications for functional neuroimaging. *PLoS One, 6*(9), e24981.

Berger, H. (1929). Über das elektrenkephalogramm des menschen. *European Archives of Psychiatry and Clinical Neuroscience, 87*(1), 527–570.

Bitterman, Y., Mukamel, R., Malach, R., Fried, I., & Nelken, I. (2008). Ultra-fine frequency tuning revealed in single neurons of human auditory cortex. *Nature, 451*, 197–201.

Boatman, D. (2004). Cortical bases of speech perception: Evidence from functional lesion studies. *Cognition, 92*(1), 47–65.

Bouchard, K. E., Mesgarani, N., Johnson, K., & Chang, E. F. (2013). Functional organization of human sensorimotor cortex for speech articulation. *Nature, 495*(7441), 327–332.

Brugge, J. F., Nourski, K. V., Oya, H., Reale, R. A., Kawasaki, H., Steinschneider, M., et al. (2009). Coding of repetitive transients by auditory cortex on Heschl's gyrus. *Journal of Neurophysiology, 102*, 2358–2374.

Bruns, A. (2004). Fourier-, Hilbert-and wavelet-based signal analysis: Are they really different approaches? *Journal of Neuroscience Methods, 137*(2), 321–332.

Celesia, G. G. (1976). Organization of auditory cortical areas in man. *Brain: A Journal of Neurology, 99*(3), 403.

Chan, A. M., Baker, J. M., Eskandar, E., Schomer, D., Ulbert, I., Marinkovic, K., et al. (2011). First-pass selectivity for semantic categories in human anteroventral temporal lobe. *The Journal of Neuroscience, 31*(49), 18119–18129.

Chan, A. M., Dykstra, A. R., Jayaram, V., Leonard, M. K., Travis, K. E., Gygi, B., et al. (2013). Speech-specific tuning of neurons in human superior temporal gyrus. *Cerebral Cortex, 24*(10), 2679–2693.

Chang, E. F., Rieger, J. W., Johnson, K., Berger, M. S., Barbaro, N. M., & Knight, R. T. (2010). Categorical speech representation in human superior temporal gyrus. *Nature Neuroscience, 13*, 1428–1432.

Cherry, E. C. (1953). Some experiments on the recognition of speech, with one and with two ears. *Journal of the Acoustical Society of America, 25*(5), 975–979.

Churchland, M. M., Cunningham, J. P., Kaufman, M. T., Foster, J. D., Nuyujukian, P., Ryu, S. I., et al. (2012). Neural population dynamics during reaching. *Nature, 487*, 51–56.

Cogan, G. B., Thesen, T., Carlson, C., Doyle, W., Devinsky, O., & Pesaran, B. (2014). Sensory-motor transformations for speech occur bilaterally. *Nature, 507*, 94–98.

Crone, N. E., Boatman, D., Gordon, B., & Hao, L. (2001). Induced electrocorticographic gamma activity during auditory perception. *Clinical Neurophysiology, 112*, 565–582.

Crone, N. E., Miglioretti, D. L., Gordon, B., & Lesser, R. P. (1998). Functional mapping of human sensorimotor cortex with electrocorticographic spectral analysis. II. Event-related synchronization in the gamma band. *Brain, 121*, 2301–2315.

Crone, N. E., Sinai, A., & Korzeniewska, A. (2006). High-frequency gamma oscillations and human brain mapping with electrocorticography. *Progress in Brain Research, 159*, 275–295.

Davis, M. H., & Johnsrude, I. S. (2003). Hierarchical processing in spoken language comprehension. *The Journal of Neuroscience, 23*, 3423–3431.

DeWitt, I., & Rauschecker, J. P. (2012). Phoneme and word recognition in the auditory ventral stream. *Proceedings of the National Academy of Sciences, 109*, E505–E514.

Ding, N., & Simon, J. Z. (2012). Emergence of neural encoding of auditory objects while listening to competing speakers. *Proceedings of the National Academy of Sciences, 109*, 11854–11859.

Edwards, E., & Chang, E. F. (2013). Syllabic (~2–5 Hz) and fluctuation (~1–10 Hz) ranges in speech and auditory processing. *Hearing Research, 305*, 113–134.

Edwards, E., Nagarajan, S. S., Dalal, S. S., Canolty, R. T., Kirsch, H. E., Barbaro, N. M., et al. (2010). Spatiotemporal imaging of cortical activation during verb generation and picture naming. *Neuroimage, 50*(1), 291–301.

Edwards, E., Soltani, M., Kim, W., Dalal, S. S., Nagarajan, S. S., Berger, M. S., et al. (2009). Comparison of time–frequency responses and the event-related potential to auditory speech stimuli in human cortex. *Journal of Neurophysiology, 102*(1), 377.

Evans, S., Kyong, J., Rosen, S., Golestani, N., Warren, J., McGettigan, C., et al. (2013). The pathways for intelligible speech: Multivariate and univariate perspectives. *Cerebral Cortex, 24*(9), 2350–2361.

Federmeier, K. D., & Kutas, M. (2005). Aging in context: Age-related changes in context use during language comprehension. *Psychophysiology, 42*(2), 133–141.

Friston, K. (2010). The free-energy principle: A unified brain theory? *Nature Reviews Neuroscience, 11*, 127–138.

Geschwind, N. (1974). *Disconnexion syndromes in animals and man*. The Netherlands: Springer.

Griffiths, T. D., Kumar, S., Sedley, W., Nourski, K. V., Kawasaki, H., Oya, H., et al. (2010). Direct recordings of pitch responses from human auditory cortex. *Current Biology, 20*, 1128–1132.

Hackett, T. A. (2011). Information flow in the auditory cortical network. *Hearing Research, 271*, 133–146.

Halgren, E., Baudena, P., Heit, G., Clarke, M., & Marinkovic, K. (1994). Spatio-temporal stages in face and word processing. 1. Depth recorded potentials in the human occipital and parietal lobes. *Journal of Physiology—Paris, 88*(1), 1–50.

Halgren, E., Wang, C., Schomer, D. L., Knake, S., Marinkovic, K., Wu, J., et al. (2006). Processing stages underlying word recognition in the anteroventral temporal lobe. *Neuroimage, 30*(4), 1401–1413.

Hickok, G., & Poeppel, D. (2007). The cortical organization of speech processing. *Nature Reviews Neuroscience, 8*, 393–402.

Huth, A. G., Nishimoto, S., Vu, A. T., & Gallant, J. L. (2012). A continuous semantic space describes the representation of thousands of object and action categories across the human brain. *Neuron, 76* (6), 1210–1224.

Insel, T. R., Landis, S. C., & Collins, F. S. (2013). The NIH brain initiative. *Science, 340*(6133), 687–688.

Jerbi, K., Ossandón, T., Hamame, C. M., Senova, S., Dalal, S. S., Jung, J., et al. (2009). Task-related gamma-band dynamics from an intracerebral perspective: Review and implications for surface EEG and MEG. *Human Brain Mapping, 30*(6), 1758–1771.

Joanisse, M. F., Zevin, J. D., & McCandliss, B. D. (2007). Brain mechanisms implicated in the preattentive categorization of speech sounds revealed using fMRI and a short-interval habituation trial paradigm. *Cerebral Cortex, 17*, 2084–2093.

Kerlin, J. R., Shahin, A. J., & Miller, L. M. (2010). Attentional gain control of ongoing cortical speech representations in a "cocktail party". *The Journal of Neuroscience, 30*, 620–628.

Kutas, M., & Federmeier, K. D. (2011). Thirty years and counting: Finding meaning in the N400 component of the event-related brain potential (ERP). *Annual Review of Psychology, 62*, 621–647.

Leonard, M. K., & Chang, E. F. (2014). Dynamic speech representations in the human temporal lobe. *Trends in Cognitive Sciences, 18* (9), 472–479.

Leonard, M. K., Ramirez, N. F., Torres, C., Travis, K. E., Hatrak, M., Mayberry, R. I., et al. (2012). Signed words in the congenitally deaf evoke typical late lexicosemantic responses with no early visual responses in left superior temporal cortex. *The Journal of Neuroscience, 32*(28), 9700–9705.

Liegeois-Chauvel, C., Musolino, A., & Chauvel, P. (1991). Localization of the primary auditory area in man. *Brain, 114*(1), 139–153.

Marinkovic, K., Dhond, R. P., Dale, A. M., Glessner, M., Carr, V., & Halgren, E. (2003). Spatiotemporal dynamics of modality-specific and supramodal word processing. *Neuron, 38*(3), 487–497.

Mesgarani, N., & Chang, E. F. (2012). Selective cortical representation of attended speaker in multi-talker speech perception. *Nature, 485*, 233–236.

Mesgarani, N., Cheung, C., Johnson, K., & Chang, E. F. (2014). Phonetic feature encoding in human superior temporal gyrus. *Science, 343*(6174), 1006–1010.

Mesgarani, N., David, S. V., Fritz, J. B., & Shamma, S. A. (2009). Influence of context and behavior on stimulus reconstruction from neural activity in primary auditory cortex. *Journal of Neurophysiology, 102*(6), 3329–3339.

Moerel, M., De Martino, F., & Formisano, E. (2012). Processing of natural sounds in human auditory cortex: Tonotopy, spectral tuning, and relation to voice sensitivity. *The Journal of Neuroscience, 32*, 14205–14216.

Moerel, M., De Martino, F., Santoro, R., Ugurbil, K., Goebel, R., Yacoub, E., et al. (2013). Processing of natural sounds: Characterization of multipeak spectral tuning in human auditory cortex. *The Journal of Neuroscience, 33*, 11888–11898. Available from: http://dx.doi.org/doi:10.1523/jneurosci.5306-12.2013.

Molholm, S., Mercier, M. R., Liebenthal, E., Schwartz, T. H., Ritter, W., Foxe, J. J., et al. (2013). Mapping phonemic processing zones along human perisylvian cortex: An electro-corticographic investigation. *Brain Structure and Function*1–15.

Mukamel, R., Gelbard, H., Arieli, A., Hasson, U., Fried, I., & Malach, R. (2005). Coupling between neuronal firing, field potentials, and FMRI in human auditory cortex. *Science, 309*, 951–954.

Nourski, K. V., Brugge, J. F., Reale, R. A., Kovach, C. K., Oya, H., Kawasaki, H., et al. (2013). Coding of repetitive transients by auditory cortex on posterolateral superior temporal gyrus in humans: An intracranial electrophysiology study. *Journal of Neurophysiology, 109*, 1283–1295.

Nourski, K. V., Steinschneider, M., Oya, H., Kawasaki, H., Jones, R. D., & Howard, M. A. (2012). Spectral organization of the human lateral superior temporal gyrus revealed by intracranial recordings. *Cerebral Cortex, 24*(2), 340–352.

Obleser, J., Eisner, F., & Kotz, S. A. (2008). Bilateral speech comprehension reflects differential sensitivity to spectral and temporal features. *The Journal of Neuroscience, 28*, 8116–8123.

Ojemann, G. A., Ojemann, J., & Ramsey, N. F. (2013). Relation between functional magnetic resonance imaging (fMRI) and single neuron, local field potential (LFP) and electrocorticography (ECoG) activity in human cortex. *Frontiers in Human Neuroscience, 7*. Available from: http://dx.doi.org/10.3389/fnhum.2013.00034.

Pascual-Marqui, R. D. (1999). Review of methods for solving the EEG inverse problem. *International Journal of Bioelectromagnetism, 1*(1), 75–86.

Pasley, B. N., David, S. V., Mesgarani, N., Flinker, A., Shamma, S. A., Crone, N. E., et al. (2012). Reconstructing speech from human auditory cortex. *PLoS Biology, 10*, e1001251.

Penfield, W., & Baldwin, M. (1952). Temporal lobe seizures and the technic of subtotal temporal lobectomy. *Annals of Surgery, 136*(4), 625.

Poeppel, D. (2003). The analysis of speech in different temporal integration windows: Cerebral lateralization as "asymmetric sampling in time". *Speech Communication, 41*, 245–255.

Poeppel, D., Idsardi, W. J., & van Wassenhove, V. (2008). Speech perception at the interface of neurobiology and linguistics. *Philosophical Transactions of the Royal Society B: Biological Sciences, 363*, 1071–1086.

Rauschecker, J. P., & Scott, S. K. (2009). Maps and streams in the auditory cortex: Nonhuman primates illuminate human speech processing. *Nature Neuroscience, 12*, 718–724.

Ray, S., & Maunsell, J. H. (2011). Different origins of gamma rhythm and high-gamma activity in macaque visual cortex. *PLoS Biology, 9*, e1000610.

Sabri, M., Binder, J. R., Desai, R., Medler, D. A., Leitl, M. D., & Liebenthal, E. (2008). Attentional and linguistic interactions in speech perception. *Neuroimage, 39*, 1444–1456.

Sahin, N. T., Pinker, S., Cash, S. S., Schomer, D., & Halgren, E. (2009). Sequential processing of lexical, grammatical, and phonological information within Broca's area. *Science, 326*(5951), 445–449.

Schönwiesner, M., & Zatorre, R. J. (2009). Spectro-temporal modulation transfer function of single voxels in the human auditory cortex measured with high-resolution fMRI. *Proceedings of the National Academy of Sciences, 106*, 14611–14616.

Scott, S. K., Blank, C. C., Rosen, S., & Wise, R. J. (2000). Identification of a pathway for intelligible speech in the left temporal lobe. *Brain, 123*(12), 2400–2406.

Simoncelli, E. P., & Olshausen, B. A. (2001). Natural image statistics and neural representation. *Annual Review of Neuroscience, 24*, 1193–1216.

Steinschneider, M., Fishman, Y. I., & Arezzo, J. C. (2008). Spectrotemporal analysis of evoked and induced electroencephalographic responses in primary auditory cortex (A1) of the awake monkey. *Cerebral Cortex, 18*, 610–625.

Steinschneider, M., Nourski, K. V., & Fishman, Y. I. (2013). Representation of speech in human auditory cortex: Is it special? *Hearing Research, 305*, 57–73.

Steinschneider, M., Nourski, K. V., Kawasaki, H., Oya, H., Brugge, J. F., & Howard, M. A. (2011). Intracranial study of speech-elicited activity on the human posterolateral superior temporal gyrus. *Cerebral Cortex, 21*, 2332–2347.

Tanriverdi, T., Ajlan, A., Poulin, N., & Olivier, A. (2009). Morbidity in epilepsy surgery: An experience based on 2449 epilepsy surgery procedures from a single institution: Clinical article. *Journal of Neurosurgery, 110*(6), 1111–1123.

Travis, K. E., Leonard, M. K., Chan, A. M., Torres, C., Sizemore, M. L., Qu, Z., et al. (2012). Independence of early speech processing from word meaning. *Cerebral Cortex, 23*(10), 2370–2379.

Viventi, J., Kim, D.-H., Vigeland, L., Frechette, E. S., Blanco, J. A., Kim, Y.-S., et al. (2011). Flexible, foldable, actively multiplexed, high-density electrode array for mapping brain activity in vivo. *Nature Neuroscience, 14*(12), 1599–1605.

Wernicke, C. (1874). *Der aphasische Symptomencomplex: eine psychologische Studie auf anatomischer Basis*. Breslau: Cohn.

Zion Golumbic, E. M., Ding, N., Bickel, S., Lakatos, P., Schevon, C. A., McKhann, G. M., et al. (2013). Mechanisms underlying selective neuronal tracking of attended speech at a "cocktail party". *Neuron, 77*, 980–991.

C H A P T E R

40

Factors That Increase Processing Demands When Listening to Speech

Ingrid S. Johnsrude[1] *and Jennifer M. Rodd*[2]

[1]Department of Psychology and Centre for Neuroscience Studies, Queen's University, Kingston, ON, Canada;
[2]Department of Experimental Psychology, University College London, London UK

For many individuals, speech listening usually feels effortless. However, whenever background noise is present, or when the person is hearing-impaired, perception of speech can be difficult and tiring (Ivarsson & Arlinger, 1994). It has become increasingly evident to applied hearing researchers that "listening effort" is an essential concept (McGarrigle et al., 2014) because it can explain behavior. For example, two hearing-impaired individuals may experience similar improvements in speech perception with an auditory prosthesis, but one may also experience increased listening effort and choose not to wear a listening device, whereas the other one who does not feel like he/she must "listen harder" wears it habitually. Although listening effort is an appealing and intuitive concept, it has not been well-elaborated in the cognitive literature. In particular, it has not been differentiated from nonspecific factors such as arousal, vigilance, motivation, or selective attention. We suggest that listening effort can be productively considered as the interaction of two factors—the processing demands, or challenges, imposed by the utterance and the listening situation and cognitive resources that an individual brings to the situation, that they can deploy to compensate for the demand. Different types of processing load draw on different resources in different cognitive domains. Listening effort is only manifest when the resources available to an individual are only barely adequate, or are inadequate, given the demand. For example, when utterance and listening conditions are straightforward and demands are low, effort is necessarily low for everyone. As processing load increases, effort will begin to increase sooner for people with more limited resources in the cognitive domains relevant for that type of load compared with people with greater resources. Eventually, effort asymptotes, at which point the cognitive resources are no longer adequate for the conditions, and perception and comprehension suffer (Figure 40.1).

We focus on the different kinds of processing demands that arise for all listeners in different listening situations and discuss the kinds of perceptual and cognitive processes that may be required to meet these challenges. Individual differences in cognitive resources (such as memory, perceptual learning, processing speed, fluid intelligence, and control processes) that permit one person to cope more efficiently or more successfully than another with the challenges imposed by a listening situation (i.e., that allow them to cope with the processing demands) are outside the remit of this chapter.

The listening conditions of everyday life are highly variable. Sometimes speech is heard in quiet. More often, however, it is degraded or masked by other sounds. Such challenging situations increase processing demand (which we also call processing load) in three different ways. First, an increase in processing load can be driven by perceptual challenges. Processing load can increase, for example, when the stimulus is masked by interfering background noise or by speech from other talkers, or as a result of degradation of the stimulus caused by hearing loss.

Second, processing load depends strongly on the linguistic properties of the stimulus. For example, the processing load required to understand a sentence increases with the syntactic (i.e., grammatical) complexity (Gibson & Pearlmutter, 1998; Stewart & Wingfield, 2009), or when its meaning is difficult to compute due to the presence of homophones (Rodd, Davis, & Johnsrude, 2005; Rodd, Johnsrude, & Davis, 2012), or

Neurobiology of Language. DOI: http://dx.doi.org/10.1016/B978-0-12-407794-2.00040-7

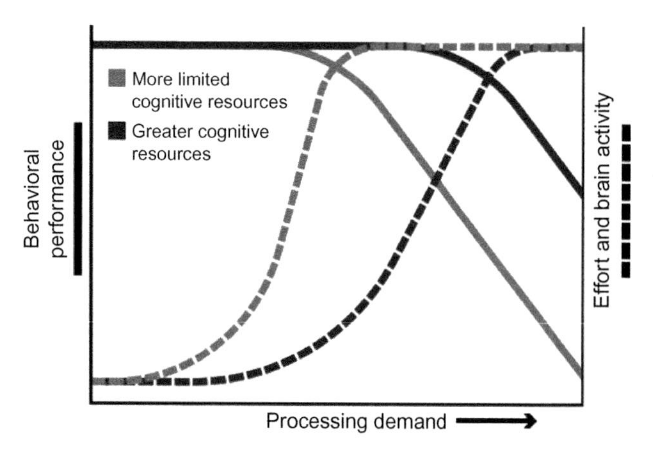

FIGURE 40.1 Schematic depiction of the interaction between processing demand and cognitive resources and how they manifest in behavioral performance, effort, and brain activity (as measured by, e.g., fMRI). The three different kinds of processing demand reviewed in this chapter will recruit somewhat different cognitive resources, but the general principle is that, for those with more limited cognitive resources, effort begins to be required at a lower level of processing demand and may grow more quickly as demand grows compared with those with greater cognitive resources. Performance is high when processing demand is low and declines as cognitive resources become insufficient to cope with the demand (i.e., as effort reaches an asymptotic maximum). Brain activity in areas sensitive to a particular type of processing demand (e.g., perceptual demand) will be low when that demand is low (e.g., speech is produced clearly, against a quiet background) because cognitive resources are not recruited to cope with the demand. As demand increases (as stimulus quality degrades), effort and brain activity increase. As demand exceeds available cognitive resources, effort and brain activity reach asymptotic levels.

when the structure of the utterance places demands on memory (Lewis, Vasishth, & Van Dyke, 2006).

Finally, the overall processing load placed on the listener can increase because of the concurrent demands of another extrinsic task performed at the same time as speech is being processed. These dual-task situations are relatively common in everyday life: we often listen to speech while doing something else (e.g., driving; Wild, Yusuf, et al., 2012). We first review the ways in which processing load is measured, and then we consider each of these different types of processing loads in turn.

The processing load associated with any given listening situation is difficult to measure directly: what can be measured is behavior or performance. Speech perception (and, by inference, comprehension) is often measured using word-report tasks, in which listeners report all the words they can understand from an utterance. Performance on such tasks not only depends on signal quality but also depends on the cognitive resources the individual can use to compensate for declines in signal quality. In other words, performance on a word report task is a function both of the signal quality and of the ability of the individual listener to overcome degradations in signal quality.

A particular problem in using behavioral measures of speech perception (like word-report scores) to measure processing load is that these often suffer from ceiling effects. Perfect or near-perfect performance (i.e., near 100% intelligibility) may result either from a listening situation that involves a low processing load (comprehension of clearly audible speech in quiet), with which cognitive resources can cope easily, or from a higher-load situation with greater recruitment of cognitive resources (see Figure 40.1). Methods used to measure the processing load associated with different listening situations therefore not only must be sensitive to listeners' overall performance on some measure of speech perception/comprehension, but also must be sensitive to the degree to which cognitive resources were taxed to achieve that level of performance. The degree to which cognitive resources are taxed has been, in the literature, operationally defined as listening effort. Effort is high if processing load is nonzero, and if the available cognitive resources are not easily or amply sufficient to cope with the processing load (Figure 40.1). Performance can be measured behaviorally; effort can be measured using either behavioral dual-task methods or physiological (including neuroimaging) methods. The constructs that interact to give rise to such measures (i.e., processing load, which is the topic of this chapter, and cognitive resources, which are not) cannot be measured directly.

Broadly speaking, there have been two approaches to the measurement of listening effort: (i) dual-task methods and (ii) methods that measure changes in physiological responses. Dual-task methods typically measure performance on a second (usually unrelated) task that is being performed at the same time as the speech is being presented. Reduced performance on this second task is assumed to reflect increased processing load of the first task (one involving speech perception/comprehension), resulting in a commensurate increase in listening effort. For example, researchers have shown that performance on a secondary task can change as a function either of the perceptual properties of the speech (e.g., the sensory quality; Gosselin & Gagné, 2011; Pals, Sarampalis, & Baskent, 2012) or of the linguistic properties of the speech (e.g., the presence of a semantic ambiguity; Rodd, Johnsrude, & Davis, 2010). Although dual-task approaches have been extensively used, these behavioral measures are, at best, an indirect test of effort required to process a speech stimulus. Furthermore, claims that dual-task effects are only observed when the two concurrent tasks overlap in the set of cognitive processes they recruit raise the possibility that such methods may be differentially sensitive to different forms of listening effort (Rodd, Johnsrude, et al., 2010).

Physiological responses have also been used to provide a more direct index of effort. For example, some researchers have suggested that pupillometry (i.e., changes in pupil diameter) can provide an index of cognitive and listening effort (Kahneman & Beatty, 1966; Piquado, Isaacowitz, & Wingfield, 2010; Sevilla, Maldonado, & Shalóm, 2014; Zekveld & Kramer, 2014; Zekveld, Kramer, & Festen, 2011). This approach is limited by difficulties in understanding exactly *how* and *why* pupil diameter changes in response to changes in the task being performed. It is still not clear whether this measure indexes effort per se or instead reflects correlated factors such as task engagement (Franklin, Broadway, Mrazek, Smallwood, & Schooler, 2013), aspects of perception/comprehension (Naber & Nakayama, 2013), or arousal (Bradshaw, 1967; Brown, van Steenbergen, Kedar, & Nieuwenhuis, 2014).

Another physiological measure of effort is brain activity. Researchers typically assume that an increase in brain activity in a region (measured, e.g., using electroencephalography [EEG], magnetoencephalography [MEG], or functional magnetic resonance imaging [fMRI]) indexes the extent to which a particular stimulus has recruited that region (i.e., the "effort" being expended within that region) (Figure 40.1). As with all measures of effort, processing load and cognitive resources interact to produce brain activity. Imaging measures have two key advantages over other approaches to measuring effort. First, they can be used for a wide range of listening situations and are potentially sensitive to changes in listening effort across all stages of speech processing, from early perceptual processes (i.e., analysis of sound in primary auditory cortex [PAC]) to higher-level aspects of semantic and syntactic processing involving anterior and posterior temporal, frontal, and parietal regions. Second, imaging measures can be used in conditions in which a participant is not exclusively attending to the critical speech stimulus and therefore can be used to measure changes in processing that arise as a function of attentional modulation. Behavioral reports for unattended speech may be unreliable, and being asked to report something out of the focus of attention may inadvertently reorient attention back to that stimulus. Imaging provides an excellent way to measure processing and how it varies with attention because imaging data can be acquired both when a participant is attending to a stimulus and when he or she is distracted from it (Heinrich, Carlyon, Davis, & Johnsrude, 2011; Wild, Yusuf, et al., 2012). If one assumes that effort requires attention, then imaging can be used to compare relatively load-free stimulus-driven processing (Wild, Yusuf, et al., 2012), with the more effortful processing observed under full attention, when a processing load is present.

40.1 TYPES OF PROCESSING DEMAND

40.1.1 Perceptual Demands

Perceptual demands are imposed by a target signal that is not perfectly audible, such as hearing an announcement over loudspeakers in a busy train station. The audibility of speech can be compromised in many different ways, through degradation or distortion of the signal, and/or through the presence of concurrent interfering sounds. These different demands are probably met through different cognitive mechanisms. For example, a degraded signal requires that missing bits be "filled in," whereas following a target in the presence of multiple competing sounds requires sound segregation processes and selective attention. In this section, we review the different types of perceptual demand and the cognitive mechanisms that may be required to compensate, as summarized in Table 40.1.

Taking the public announcement example, the speech signal itself is likely to be both degraded and distorted because of the limitations of most public-transport loudspeaker systems. (An audio signal is *degraded* if frequency components have been reduced or removed; it is *distorted* if components, not originally present, have been somehow introduced). Degradation places demands on processes that help to recover obliterated signal, whereas distortion introduces segregation demands—the listener must be able to perceptually distinguish the components of the original signal from the artefactual components of the distortion to recognize the signal. A common form of distortion is reverberation: the surfaces of indoor spaces, in which so much communication happens, reflect sound and the reflected sounds must be distinguished from the direct sound source for the sound to

TABLE 40.1 Different Types of Perceptual Demand and the Nature of the Cognitive Processes Required to Cope with Such Demands

Perceptual demand	Imposes a processing load on cognitive mechanisms required for
Target signal degradation, energetic masking of target	Perceptual closure, including linguistically guided perceptual closure Linguistically guided word-segmentation processes
Target signal distortion; and informational masking of target due to reverberation or competing, acoustically similar, sounds	Sound source segregation Selective attention Voice identity processing
Hearing impairment	All of the above

be understood. The presence of concurrent sounds (also introduced by the competing sounds, including concurrent speech, in that busy train station) requires listeners to recruit processes that serve to segregate the target signal from the background, to selectively attend to the target, and to recover obliterated signal (perceptual closure). Masking that physically interrupts or occludes a target signal so that it is effectively not transduced at the periphery is called *energetic masking*. When both target and masker are audible but are difficult to separate perceptually, *informational masking* occurs (Kidd, Mason, Richards, Gallun, & Durlach, 2007; Leek, Brown, & Dorman, 1991; Pollack, 1975).

Compensation for energetic masking (which is typically transient in the dynamic conditions of real-world communicative environments) requires that the occluded signal be "filled in" in some way. Accumulating evidence suggests that such perceptual closure comprises at least two, if not more, physiologically separable phenomena. First, if discrete frequencies are occluded for a brief period, then the brain can complete the partially masked sound so that it is heard as a complete whole (Carlyon, Deeks, Norris, & Butterfield, 2002; Ciocca & Bregman, 1987; Heinrich, Carlyon, Davis, & Johnsrude, 2008). For example, if a steady-state tone is interrupted by a gap filled with noise, the tone is "heard" to continue through the gap. Electrophysiological and MRI studies suggest that, in such cases, the illusion is complete at a relatively early stage of processing. Petkov and colleagues (Petkov, O'Connor, & Sutter, 2007) measured the responses of macaque A1 neurons to steady tones, to noise bursts, and to interrupted tones where the silent gap could optionally be filled by a noise. In this latter case, they found that the response of approximately half the neurons to the noise was more similar to that elicited by an uninterrupted tone than it was to that produced by an isolated noise burst. This suggests that the continuity illusion, at least for tones, is complete by the level of auditory cortex. Physiological investigations involving humans also suggest that continuity for tones is complete at relatively early stages of auditory processing and happens obligatorily in the absence of full attention (Heinrich et al., 2011; Micheyl et al., 2003).

The second type of perceptual closure also compensates for transient energetic masking but works to fill in missing speech sounds, or even missing words, on the basis of linguistic knowledge (Bashford, Riener, & Warren, 1992; Saija, Akyürek, Andringa, & Başkent, 2014; Shahin, Bishop, & Miller, 2009; Warren, 1970). For example, sounds that are distinguished by one phonetic feature such as /k/ and /g/ are acoustically similar and, if degraded, perhaps acoustically identical. Uncertainty about the identity of a phoneme (/k/ versus /g/) can be resolved through knowledge of words: it will be heard as /k/ in the context *iss, and /g/ in the context *ift, because "kiss" and "gift" are words and "giss" and "kift" are not (Ganong, 1980). Such perceptual ambiguity can also increase processing load on higher-level linguistic processes because there are more candidate linguistic representations ("bark" versus "park" for example). This load can be met through effective use of syntactic and semantic context; many studies demonstrate that speech materials embedded in a meaningful context are easier to report than those without a constraining context (Davis, Ford, Kherif, & Johnsrude, 2011a; Dubno, Ahlstrom, & Horwitz, 2000; Kalikow, Stevens, & Elliott, 1977; Miller, Heise, & Lichten, 1951; Nittrouer & Boothroyd, 1990). Degradation would also stress processes involved in segmenting adjacent words because degradation means that word boundaries are less clearly demarcated. Thus, linguistically guided perceptual closure phenomena draw on prior knowledge at multiple levels: such linguistic priors would include knowledge of the words and morphology of a language and how words can be combined. These phenomena would also draw on word frequency effects, contextually contingent prior probabilities, and knowledge of the world.

Evidence from studies in which speech has been perceptually degraded (e.g., using noise vocoding; Shannon, Zeng, Kamath, Wygonski, & Ekelid, 1995) suggests that processes involving the use of prior knowledge to enhance speech comprehension (i.e., linguistically guided perceptual closure phenomena) rely on regions of the left inferior frontal gyrus (LIFG) and/or regions along the superior and middle temporal gyri (Davis, Ford, Kherif, & Johnsrude, 2011b; Davis & Johnsrude, 2003; Hervais-Adelman, Carlyon, Johnsrude, & Davis, 2012; Obleser & Kotz, 2010; Scott, Blank, Rosen, & Wise, 2000; Wild, Yusuf, et al., 2012), and possibly also on left angular gyrus (Golestani, Hervais-Adelman, Obleser, & Scott, 2013; Wild, Davis, & Johnsrude, 2012). Even though stimuli in these studies are carefully controlled and energetic masking introduces uncertainty at the acoustic level, the widespread patterns of activity in response are probably due to the fact that uncertainty propagates across linguistic levels—lexical, morphological, syntactic and semantic.

Informational masking is defined as any decline in perception or comprehension of a target stimulus (therefore "masking") that cannot be attributed to energetic masking (Cooke, Garcia Lecumberri, & Barker, 2008; Kidd et al., 2007; Leek et al., 1991; Pollack, 1975). This includes everything from difficulty with segregation (i.e., due to perceptual similarity between target and masker) to attentional bias effects (due to the salience or "ignorability" of a target, or of a masker due to familiarity or meaningfulness). If knowledge-based factors can produce a "release from masking" (i.e., better performance on a listening task), then the masking was, by definition, informational. However, informational masking should be distinguished from general inattentiveness (Durlach et al., 2003). There may be good reason to consider processing demands resulting from at least some types of informational masking as very different from those resulting from the perceptual demands of energetic masking (Cooke et al., 2008; Mattys, Barden, & Samuel, 2014; Mattys, Brooks, & Cooke, 2009; Mattys & Wiget, 2011). We discuss this idea in the section on *Concurrent task demands*.

Importantly, familiarity and expertise can reduce informational masking. In a recent study, Johnsrude et al. (2013) observed that listeners obtained a substantial release from (informational) masking when they heard their spouse's voice in a two-voice mixture either when it was the target or when it was the masker, suggesting that familiar-voice information can facilitate segregation. Furthermore, although the ability to report novel-voice targets declined with age, there was no such decline for spouse-voice targets even though performance on spouse-voice targets was not at ceiling. The decline with age for novel targets could not be due to energetic masking (because this would apply equally to spouse-voice targets, which were acoustically perfectly matched, across participants, to the novel-voice targets) and thus must result from increased informational masking with age, which is reduced (or possibly eliminated) by familiar-voice information.

Perceptual demands imposed by degradation, distortion, and concurrent sounds are exacerbated by hearing impairment, including age-related hearing loss (presbycusis). Hearing impairment commonly involves both diminished sensitivity to sound (degrading the input signal) and diminished frequency acuity. Poorer frequency acuity means that acoustically similar sounds are less discriminable: the signal is degraded and harder to segregate from other sounds. Thus, hearing loss increases perceptual load and, like the other forms of perceptual load, hearing loss places greater demands on processes involved in perceptual closure, in selective attention, and in segregation (Gatehouse & Noble, 2004).

40.1.2 Linguistic Demands

Even in the case where all words are fully intelligible, the effort required to understand language can vary greatly. For example, cognitive models of sentence comprehension describe in detail the additional processes that are required to comprehend a sentence that has a complex grammatical structure or that contains words with more than one possible meaning (Duffy, Kambe, & Rayner, 2001; Duffy, Morris, & Rayner, 1988; Frazier, 1987; MacDonald, Pearlmutter, & Seidenberg, 1994). These cognitive models have been developed largely on the basis of reading experiments, in which a relatively direct measure of the amount of processing that is required by different types of sentences can be obtained by measuring the time taken to read them. In contrast, it is more difficult to measure the processing demand imposed by particular types of spoken sentences, because the rate at which words arrive is typically controlled by the speaker (and not the listener).

One form of linguistic challenge that has been relatively well-studied using spoken as well as written language is lexical ambiguity, also known as semantic ambiguity, which arises when words with multiple meanings are included within a sentence. For example, to understand the phrase "the bark of the dog," a listener must use the meaning of the word "dog" to work out that "bark" is probably referring to the noise made by that animal and not the outer covering of a tree. These forms of ambiguity are ubiquitous in language; at least 80% of the common words in a typical English dictionary have more than one definition (Rodd, Gaskell, & Marslen-Wilson, 2002), and most dictionaries list more than 40 different definitions for the word "run" (e.g., "an athlete runs a race," "a river runs to the sea," "a politician runs for office"). Each time one of these ambiguous words is encountered, the listener must select the appropriate meaning on the basis of its sentential context.

A dual-task experiment (Figure 40.2A) has revealed longer reaction times on an unrelated visual case-judgement task when participants are listening to high-ambiguity, compared with low-ambiguity, sentences (Rodd, Johnsrude, et al., 2010), indicating that high-ambiguity sentences are more effortful to understand. High-ambiguity sentences contained at least two ambiguous words (e.g., "There were *dates* and *pears* [homophonous with pairs] in the fruit bowl") and were compared with well-matched low-ambiguity control sentences (e.g., "There was beer and cider on the kitchen shelf"; Rodd, Johnsrude, et al., 2010). A concurrent case-judgement task required participants to indicate, as quickly as possible with a keypress, whether a cue letter on the screen was presented in UPPER case or lower case. Interestingly, this ambiguity effect varies as a function of the sentence structure. In cases when the ambiguous word is preceded by disambiguating sentence context (e.g., "The hunter thought that the HARE/HAIR ..."), there is no observed cost of the case-judgement task, suggesting

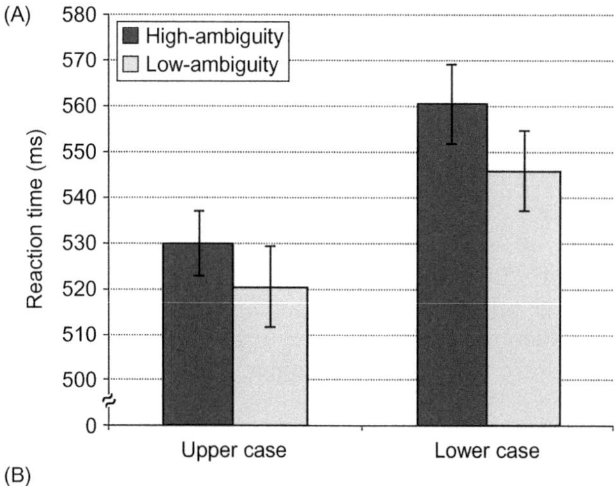

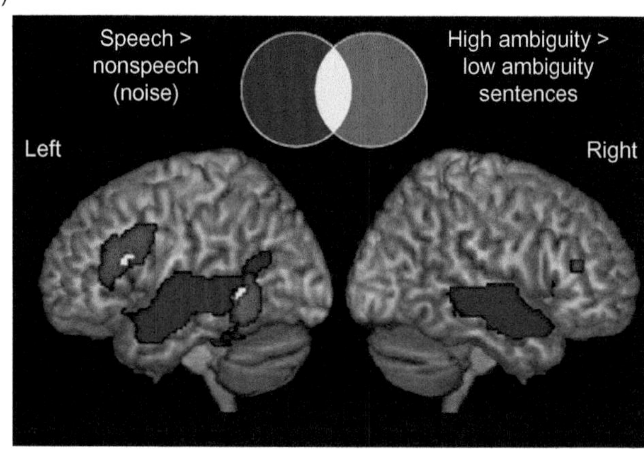

FIGURE 40.2 (A) Reaction times (in ms) for a case-judgment task presented while participants listened to high- and low-ambiguity sentences as a function of the case of the presented letter (UPPER case or lower case). Participants were asked to respond as quickly as possible when the letter cue was presented on the screen at the offset of the second homophone (or homologous word in the matched low-ambiguity sentence). At the end of each sentence, participants were also required to make a semantic relatedness judgment to a word presented on the screen. Case-judgment reaction times were significantly longer for high- compared with low-ambiguity sentences, suggesting that some cognitive process involved in resolving the meaning of the homophones in the high-ambiguity sentences is also involved in case-judgment. (B) Brain regions that are significantly more active for high-ambiguity sentences compared with low-ambiguity sentences in a group of normal young-adult right-handed participants are shown in red/yellow. These regions include the inferior frontal gyri bilaterally and the left inferior temporal gyrus. Brain regions that are significantly more active for all speech compared with unintelligible signal-correlated noise are shown in blue/yellow. These regions include the superior and middle temporal gyri bilaterally. *Data from Rodd et al. (2005) with permission from the publisher.*

that this process of selecting a contextually relevant word meaning on the basis of the information that is currently available is relatively undemanding. In contrast, when the disambiguating context is positioned *after* the ambiguous word (e.g., "The ecologist thought that the PLANT by the river should be closed down"), there is a considerable increase in the time taken to make the case judgements (Rodd, Johnsrude, et al., 2010). The most likely explanation is that, for these sentences, listeners are frequently misinterpreting the ambiguous word by selecting the more frequent, but contextually inappropriate, meaning, and that the dual-task effect arises because of the increased processing load associated with the reinterpretation process that is triggered by the final word in the sentence.

Several neuroimaging studies using a range of different spoken-sentence materials confirm that comprehension of sentences that contain ambiguous words is more demanding than comprehension of sentences that do not. Rodd et al. (2005) compared high-ambiguity sentences with well-matched low-ambiguity sentences and reported a large cluster of LIFG activation with its peak within the pars opercularis, as well as a cluster in the posterior portion of the left inferior temporal gyrus (LITG) (Figure 40.2B). This is broadly consistent with the results of subsequent studies, which consistently find ambiguity-related activation in both posterior and anterior subdivisions of the LIFG (Bekinschtein, Davis, Rodd, & Owen, 2011; Davis et al., 2007; Rodd et al., 2012; Rodd, Longe, Randall, & Tyler, 2010). The precise location of the LITG activation has been more variable, with some studies finding no significant effect (Rodd, Longe, et al., 2010). Again, as with the dual-task studies, the ambiguity-related increases in BOLD signal have been shown to be largest when listeners must reinterpret a sentence that was initially misunderstood (Rodd et al., 2012), suggesting that this aspect of semantic processing is particularly demanding.

These two key brain regions (LIFG and posterior temporal lobe) have also been activated for a number of other linguistic manipulations. Rodd, Longe, et al. (2010) have shown that *syntactic* ambiguities (e.g., "VISITING RELATIVES is/are boring") produce LIFG activation (compared with unambiguous sentences) in the same region as *semantic* ambiguities (within the same participants), as well as in posterior left middle temporal gyrus (MTG). Tyler et al. (2011) found similar effects of syntactic ambiguity that extended across both anterior and posterior LIFG, as well as smaller activations in homologous right frontal regions, posterior MTG, and parietal regions. Finally, similar frontal and posterior temporal regions have been shown to produce elevated activation for sentences that have a grammatical structure that is more difficult to process (Mack, Meltzer-Asscher, Barbieri, & Thompson, 2013; Meltzer, McArdle, Schafer, & Braun, 2010; Peelle, McMillan, Moore, Grossman, & Wingfield, 2004; Shetreet & Friedmann, 2014), although there is variability in the precise location of the posterior temporal lobe activations. In addition, studies have also found that increased demands on syntactic processes are associated with activation in the *anterior* temporal lobe (Brennan et al., 2012; Obleser et al., 2011) and temporo-parietal regions (Meyer et al., 2012).

Another linguistic challenge to comprehension is a relative lack of contextual constraint. For sentences that lack a strong determinative meaningful context (such as "Her good slope was done in carrot"), a region of LIFG appears more active when compared with sentences that have context (such as "Her new skirt was made of denim") (Davis et al., 2011b). This increased activity is consistent with the idea that a lack of supportive context increases processing demands, possibly because weaker semantic constraint renders speech less predictable.

Although the precise cognitive roles of these different brain regions associated with higher-level aspects of speech comprehension remains unclear (Rodd et al., 2012), results are consistent with the view that the LIFG contributes to the integrative processes that are required to combine the meanings and syntactic properties of individual words into higher-level linguistic structures (Hagoort, 2005; Rodd et al., 2005; Willems & Hagoort, 2009). The structures of the LIFG may be required for cognitive control processes required to select and maintain task-relevant information while inhibiting information that is not currently required (Novick, Kan, Trueswell, & Thompson-Schill, 2009; Thompson-Schill, D'Esposito, Aguirre, & Farah, 1997).

Activation of posterior (and not anterior) aspects of the temporal lobe provide clear support for the view that these regions form part of a processing stream that is critical for accessing the meaning of spoken words (Hickok & Poeppel, 2007), although other authors have suggested that this region, like the LIFG, is associated with control processes that are necessary to select and maintain aspects of meaning that are currently relevant, rather than the stored representation of semantic/syntactic information (Noonan, Jefferies, Visser, & Lambon Ralph, 2013).

In summary, higher-level aspects of speech comprehension, such as selecting contextually appropriate word meanings and assigning appropriate syntactic roles, are largely associated with frontal and posterior temporal brain regions. Interestingly, these overlap with areas that are recruited to cope with the "lower level" perceptual demands reviewed earlier, suggesting that similar cognitive mechanisms are required to cope with different kinds of processing demand.

Overlapping domain-general processing resources are recruited in the face of both perceptual and cognitive demands, raising the possibility that competition for these resources may result in a "double whammy" effect: when both forms of processing resources are required, the listener is disproportionately disadvantaged. For example, it is possible that when a listener needs to deploy additional resources to identify the sounds that are present in the speech stream, the resources that they have available to them for dealing with any semantic/syntactic demands may be reduced (and vice versa).

Because high-level cognitive processes operate on the output from lower-level perceptual processes, conditions in which the perceptual input is degraded may produce perceptual uncertainty that cascades to higher-level cognitive processes. For example, an "easy" sentence that, when presented with high perceptual clarity, would produce relatively low demand on semantic/syntactic processes might become more demanding at these higher cognitive levels when the input is degraded such that the listener is unsure (based on perceptual input alone) of the words that were actually present. As reviewed previously, demands on semantic/syntactic processes almost certainly increase as a function of the perceptual clarity of the input, even when semantic/syntactic properties are held constant.

In the face of perceptual degradation, higher-level cognitive resources make a key contribution to resolving perceptual ambiguity, such that information about the meaning, syntactic structure, or identity of a sentence (or a word) plays a key role in providing contextual cues that can be used to resolve perceptual ambiguities (Davis et al., 2011b; Sohoglu, Peelle, Carlyon, & Davis, 2012; Wild, Davis, et al., 2012). For example, Wild, Davis, et al. (2012) presented acoustically degraded speech that was preceded (200 ms) earlier by text primes that either matched each upcoming word or was incongruent. Preceding text primes enhanced the rated subjective clarity of heard speech, and activity within PAC was sensitive to this manipulation. A region-of-interest defined using cytoarchitectonic maps of PAC (Tahmasebi et al., 2009) revealed that activity was significantly higher for degraded speech preceded by matching compared with mismatching primes, and that this difference was significantly greater than the difference observed for matching and mismatching primes preceding either clear speech or unintelligible control sounds. Connectivity analyses suggested that sources of input to PAC are higher-order temporal, frontal, and motor regions, consistent with the extensive network of feedback connections in the auditory system (De la Mothe, Blumell, Kajikawa, & Hackett, 2006; Winer, 2006) that, in humans, may permit higher-level interpretations of speech to be tested against incoming auditory representations.

40.1.3 Concurrent Task Demands

A final type of processing load is that imposed by extrinsic tasks that are being performed while speech is heard. Although much research on speech perception and comprehension is conducted under ideal conditions (undivided attention, in quiet environments, and with carefully recorded material), real-world speech perception often involves hearing speech while in the middle of other tasks that divert attention or impose perceptual or cognitive loads (Mattys et al., 2009). Such concurrent processing loads, if they render target speech less intelligible, qualify as "informational masking" according to the definition offered in the previous section on *Perceptual demands*. To what extent is speech that is heard under such conditions processed for information, and how is processing affected by the nature of the secondary task and by the acoustic quality of the speech signal?

The degree to which speech is processed when attention is elsewhere is a difficult question to address behaviorally because it is hard to measure the perception of a stimulus to which a participant is not attending. Wild, Yusuf, et al. (2012) used fMRI to compare processing of degraded (noise-vocoded) and clear speech under full attention and when distracted by other tasks. In every trial, young listeners with normal hearing were directed to attend to one of three simultaneously presented stimuli: a sentence (at one of four acoustic clarity levels), an auditory distracter, or a visual distracter. When attending to the distracters, participants performed a target-monitoring task. A postscan recognition test showed that clear speech was processed even when not attended, but that attention greatly enhanced the processing of degraded speech. Speech-sensitive cortex could be fractionated according to how speech-evoked responses were modulated by attention, and these divisions appeared to map onto the hierarchical organization of the auditory system (Davis & Johnsrude, 2003; Hickok & Poeppel, 2007; Okada et al., 2010; Peelle, Johnsrude, & Davis, 2010). Critically, only in the region of the superior temporal sulcus (STS) and in frontal regions—regions corresponding to the highest stages of auditory processing—was the effect of speech clarity on BOLD activity influenced by attentional state (Figure 40.3). Frontal and temporal regions manifested this interaction

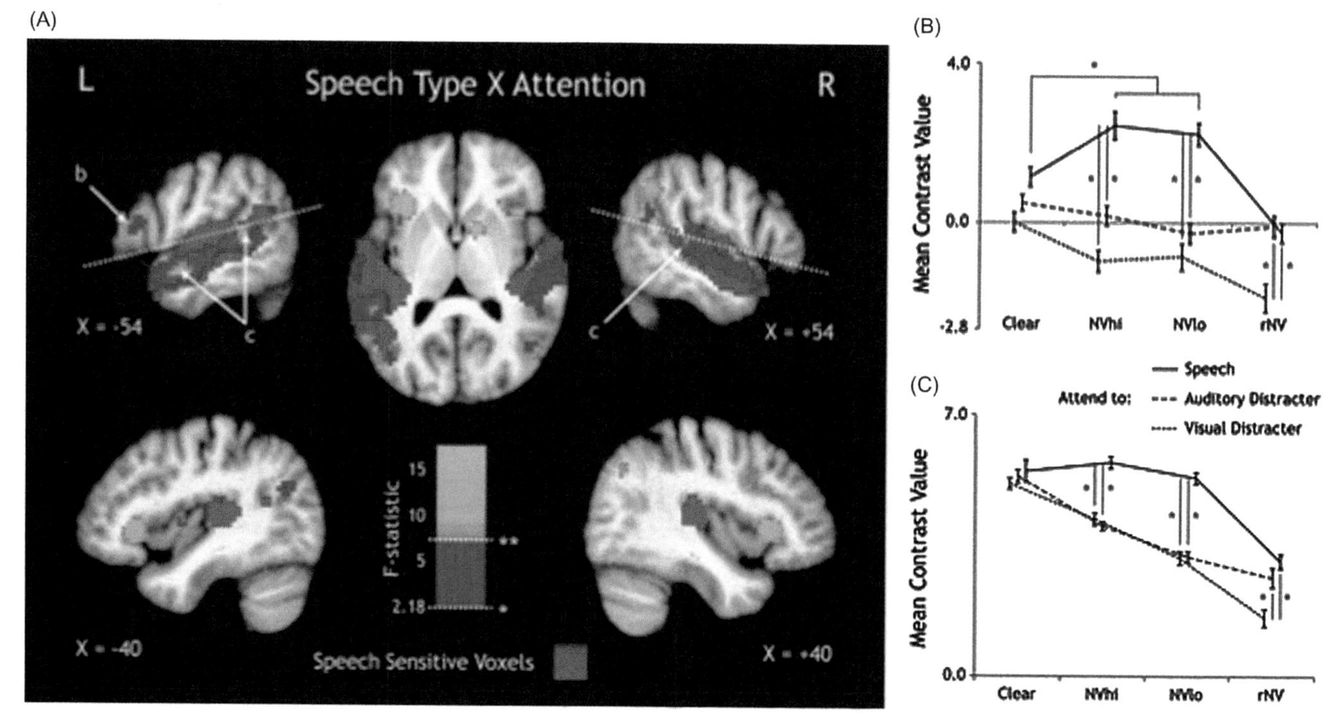

FIGURE 40.3 The interaction between Attention (three levels: attention to speech; attention to an auditory distracter, or attention to a visual distracter) and Speech Clarity (four levels: Clear Speech; 6-band noise vocoded speech [NVhi]; compressed 6-band noise vocoded speech [NVlo]; and spectrally rotated noise vocoded speech [rNV]). Materials were meaningful English sentences such as "His handwriting was very difficult to read." Whereas both NVhi and NVlow are potentially intelligible (word-report scores for the materials in a pilot group of participants were 88% and 68% respectively), rNV is completely unintelligible. (A) The Speech Clarity by Attentional State interaction F-contrast is shown in blue. *Indicates the threshold at $p < 0.05$ uncorrected for multiple comparisons; **indicates family-wise correction for multiple comparisons. Voxels in red are sensitive to the different levels of Speech Clarity (i.e., demonstrate a significant main effect of speech type) but show no evidence for an interaction with Attention ($p < 0.05$, uncorrected). (B) Contrast values (i.e., estimated signal relative to baseline; arbitrary units) are plotted for LIFG and (C) an average of the STS peaks. Error bars represent the SEM suitable for repeated-measures data (Loftus & Masson, 1994). Vertical lines with asterisks indicate significant pairwise comparisons ($p < 0.05$, Bonferroni corrected for 12 comparisons). *Figure taken from Wild, Yusuf, et al. (2012) with permission from the publisher.*

in different ways. In the LIFG, when speech was attended, activity was greater for degraded than for clear speech. However, when the listeners attended elsewhere, LIFG activity did not depend on speech type (Figure 40.3B). Increased activity for attended degraded speech may reflect effortful processing that is required to enhance intelligibility or additional cognitive processes (such as those required for perceptual learning) that are engaged uniquely when a listener is attending to speech (Davis, Johnsrude, Hervais-Adelman, Taylor, & McGettigan, 2005; Eisner, McGettigan, Faulkner, Rosen, & Scott, 2010; Huyck & Johnsrude, 2012).

The pattern of activity around the STS was somewhat different; when speech was attended, degraded and clear speech produced similar amounts of activation, consistent with attention enhancing the intelligibility of degraded speech. When distracted, activity in the STS appeared to be driven by the speech quality of the stimulus (Figure 40.3C). Together, these results suggest that the LIFG only responds to degraded speech when listeners are attending to it, whereas the STS responds to speech intelligibility, regardless of attention or how that intelligibility is achieved.

A series of studies by Mattys and colleagues (Mattys, Barden, & Samuel, 2014; Mattys et al., 2009; Mattys & Wiget, 2011) explored how additional extrinsic cognitive load alters the processing of simultaneously presented spoken words. They observed that the mechanisms involved in processing speech under conditions of divided attention are not the same as those involved in processing speech in conditions of undivided attention. When listeners were required to perform a concurrent task that involved divided attention (Mattys et al., 2009), a memory load (Mattys et al., 2009), or a visual search task (Mattys & Wiget, 2011), they relied more on lexical semantic structure to aid segmentation of words and on lexical knowledge to identify phonemes, rather than on acoustic cues conveyed in fine phonetic detail. In fact, the "lexical drift"—this increased reliance on lexical knowledge when distracted—appears to be driven by the reduced acuity for acoustic cues to phoneme identity and word boundaries (Mattys et al., 2014). This may at first seem surprising. As load on central cognitive resources increases, listeners rely more, not less, on knowledge-guided factors (that presumably rely on the same central cognitive resources) for speech perception. Mattys and colleagues (2014) attribute this to a re-weighting of perceptual cues, such that acoustic cues to the identity of words and phonemes are weighted less than linguistic cues. Alternatively, a concurrent task may alter the rate at which acoustic information is sampled, causing some samples to be missed while attention is elsewhere. The results of these experiments, indicating that concurrent tasks do not simply impair perception but rather alter it qualitatively by affecting perceptual decision criteria, highlight the importance of considering how different real-world situations might affect the effort required to understand speech.

40.2 SUMMARY

"Listening effort" is a unitary term that probably reflects several different challenges to speech comprehension met through a variety of cognitive mechanisms. Our position is that effort is due to an interaction between the perceptual, linguistic, or task challenges imposed by the situation and stimulus and the cognitive resources that the listener brings to bear. This framework allows us to separate different cognitive mechanisms that contribute to listening effort. Behavioral measures such as effort ratings, dual-task reaction time, and pupil dilation may not be able to differentiate among different kinds of processing demand because the measure is one-dimensional. In contrast, neuroimaging methods such as fMRI permit exploration of different kinds of processing demand because they may recruit different regions/networks.

Although this chapter has considered these demands separately, they are not independent. An increase in processing demand at a lower level of representation (e.g., perceptual) can affect processing at a higher level of representation (e.g., semantic/syntactic) in at least three ways: (i) by reducing the domain-general processing resources that are available; (ii) by increasing the ambiguity at later levels of representation; and (iii) by increasing the listeners' reliance on these representations to resolve lower-level ambiguities. Increases in processing demands at a *higher* level of representation may also have consequences for *lower*-level processing through the extensive network of feedback connections documented in the auditory system. These complex interactions mean that processing demands cannot be fully understood by studying them in isolation.

References

Bashford, J. A., Riener, K. R., & Warren, R. M. (1992). Increasing the intelligibility of speech through multiple phonemic restorations. *Perception and Psychophysics*, *51*(3), 211–217.

Bekinschtein, T. A., Davis, M. H., Rodd, J. M., & Owen, A. M. (2011). Why clowns taste funny: The relationship between humor and semantic ambiguity. *The Journal of Neuroscience: The Official Journal of the Society for Neuroscience*, *31*(26), 9665–9671.

Bradshaw, J. (1967). Pupil size as a measure of arousal during information processing. *Nature*, *216*(5114), 515–516.

Brennan, J., Nir, Y., Hasson, U., Malach, R., Heeger, D. J., & Pylkkänen, L. (2012). Syntactic structure building in the anterior temporal lobe during natural story listening. *Brain and Language, 120*(2), 163–173.

Brown, S. B., van Steenbergen, H., Kedar, T., & Nieuwenhuis, S. (2014). Effects of arousal on cognitive control: Empirical tests of the conflict-modulated Hebbian-learning hypothesis. *Frontiers in Human Neuroscience, 8*, 23.

Carlyon, R. P., Deeks, J., Norris, D., & Butterfield, S. (2002). The continuity illusion and vowel identification. *Acta Acustica United with Acustica, 88*, 408–415.

Ciocca, V., & Bregman, A. S. (1987). Perceived continuity of gliding and steady-state tones through interrupting noise. *Perception and Psychophysics, 42*(5), 476–484.

Cooke, M., Garcia Lecumberri, M. L., & Barker, J. (2008). The foreign language cocktail party problem: Energetic and informational masking effects in non-native speech perception. *The Journal of the Acoustical Society of America, 123*(1), 414–427.

Davis, M. H., Coleman, M. R., Absalom, A. R., Rodd, J. M., Johnsrude, I. S., Matta, B. F., et al. (2007). Dissociating speech perception and comprehension at reduced levels of awareness. *Proceedings of the National Academy of Sciences of the United States of America, 104*(41), 16032–16037.

Davis, M. H., Ford, M. A., Kherif, F., & Johnsrude, I. S. (2011a). Does semantic context benefit speech understanding through "top-down" processes? Evidence from time-resolved sparse fMRI. *Journal of Cognitive Neuroscience, 23*(12), 3914–3932.

Davis, M. H., Ford, M. A., Kherif, F., & Johnsrude, I. S. (2011b). Does semantic context benefit speech understanding through "top-down" processes? Evidence from time-resolved sparse fMRI. *Journal of Cognitive Neuroscience, 23*(12), 3914–3932.

Davis, M. H., & Johnsrude, I. S. (2003). Hierarchical processing in spoken language comprehension. *Journal of Neuroscience, 23*(8), 3423–3431.

Davis, M. H., Johnsrude, I. S., Hervais-Adelman, A., Taylor, K., & McGettigan, C. (2005). Lexical information drives perceptual learning of distorted speech: Evidence from the comprehension of noise-vocoded sentences. *Journal of Experimental Psychology: General, 134*(2), 222–241.

De la Mothe, L. A., Blumell, S., Kajikawa, Y., & Hackett, T. A. (2006). Cortical connections of the auditory cortex in marmoset monkeys: Core and medial belt regions. *The Journal of Comparative Neurology, 496*(1), 27–71.

Dubno, J. R., Ahlstrom, J. B., & Horwitz, A. R. (2000). Use of context by young and aged adults with normal hearing. *The Journal of the Acoustical Society of America, 107*(1), 538–546.

Duffy, S. A., Kambe, G., & Rayner, K. (2001). The effect of prior disambiguating context on the comprehension of ambiguous words: Evidence from eye movements. In D. S. Gorfein (Ed.), *On the consequences of meaning selection: Perspectives on resolving lexical ambiguity*. Washington DC: American Psychological Association.

Duffy, S. A., Morris, R. K., & Rayner, K. (1988). Lexical ambiguity and fixation times in reading. *Journal of Memory and Language, 27* (4), 429–446.

Durlach, N. I., Mason, C. R., Kidd, G., Arbogast, T. L., Colburn, H. S., & Shinn-Cunningham, B. G. (2003). Note on informational masking. *The Journal of the Acoustical Society of America, 113*(6), 2984–2987.

Eisner, F., McGettigan, C., Faulkner, A., Rosen, S., & Scott, S. K. (2010). Inferior frontal gyrus activation predicts individual differences in perceptual learning of cochlear-implant simulations. *Journal of Neuroscience, 30*(21), 7179–7186.

Franklin, M. S., Broadway, J. M., Mrazek, M. D., Smallwood, J., & Schooler, J. W. (2013). Window to the wandering mind: Pupillometry of spontaneous thought while reading. *Quarterly Journal of Experimental Psychology (2006), 66*(12), 2289–2294.

Frazier, L. (1987). Sentence processing: A tutorial review. In M. Coltheart (Ed.), *Attention and performance XII: The psychology of reading*. Lawrence Erlbaum Associates.

Ganong, W. F. (1980). Phonetic categorization in auditory word perception. *Journal of Experimental Psychology. Human Perception and Performance, 6*(1), 110–125.

Gatehouse, S., & Noble, W. (2004). The Speech, Spatial and Qualities of hearing scale (SSQ). *International Journal of Audiology, 43*(2), 85–99.

Gibson, E., & Pearlmutter, N. J. (1998). Constraints on sentence comprehension. *Trends in Cognitive Sciences, 2*(7), 262–268.

Golestani, N., Hervais-Adelman, A., Obleser, J., & Scott, S. K. (2013). Semantic versus perceptual interactions in neural processing of speech-in-noise. *NeuroImage, 79*, 52–61.

Gosselin, P. A., & Gagné, J.-P. (2011). Older adults expend more listening effort than young adults recognizing audiovisual speech in noise. *International Journal of Audiology, 50*(11), 786–792.

Hagoort, P. (2005). On Broca, brain, and binding: A new framework. *Trends in Cognitive Sciences, 9*(9), 416–423.

Heinrich, A., Carlyon, R. P., Davis, M. H., & Johnsrude, I. S. (2008). Illusory vowels resulting from perceptual continuity: A functional magnetic resonance imaging study. *Journal of Cognitive Neuroscience, 20*(10), 1737–1752.

Heinrich, A., Carlyon, R. P., Davis, M. H., & Johnsrude, I. S. (2011). The continuity illusion does not depend on attentional state: FMRI evidence from illusory vowels. *Journal of Cognitive Neuroscience, 23*(10), 2675–2689.

Hervais-Adelman, A., Carlyon, R., Johnsrude, I., & Davis, M. (2012). Brain regions recruited for the effortful comprehension of noise-vocoded words. *Language and Cognitive Processes, 28*, 1145–1166.

Hickok, G., & Poeppel, D. (2007). The cortical organization of speech processing. *Nature Reviews Neuroscience, 8*(5), 393–402.

Huyck, J. J., & Johnsrude, I. S. (2012). Rapid perceptual learning of noise-vocoded speech requires attention. *Journal of the Acoustical Society of America, 131*(3), EL236–42.

Ivarsson, U. S., & Arlinger, S. D. (1994). Speech recognition in noise before and after a work-day's noise exposure. *Scandinavian Audiology, 23*(3), 159–163.

Johnsrude, I. S., Mackey, A., Hakyemez, H., Alexander, E., Trang, H. P., & Carlyon, R. P. (2013). Swinging at a cocktail party: Voice familiarity aids speech perception in the presence of a competing voice. *Psychological Science, 24*(10), 1995–2004.

Kahneman, D., & Beatty, J. (1966). Pupil diameter and load on memory. *Science, 154*(3756), 1583–1585.

Kalikow, D. N., Stevens, K. N., & Elliott, L. L. (1977). Development of a test of speech intelligibility in noise using sentence materials with controlled word predictability. *Journal of the Acoustical Society of America, 61*(5), 1337–1351.

Kidd, G. J., Mason, C. R., Richards, V. M., Gallun, F. J., & Durlach, N. I. (2007). Informational masking. In W. A. Yost, & R. R. Fay (Eds.), *Auditory perception of sound sources* (pp. 143–190). Springer.

Leek, M. R., Brown, M. E., & Dorman, M. F. (1991). Informational masking and auditory attention. *Perception and Psychophysics, 50* (3), 205–214.

Lewis, R. L., Vasishth, S., & Van Dyke, J. A. (2006). Computational principles of working memory in sentence comprehension. *Trends in Cognitive Sciences, 10*(10), 447–454.

Loftus, G. R., & Masson, M. E. (1994). Using confidence intervals in within-subject designs. *Psychonomic Bulletin and Review, 1*(4), 476–490.

MacDonald, M. C., Pearlmutter, N. J., & Seidenberg, M. S. (1994). The lexical nature of syntactic ambiguity resolution [corrected]. *Psychological Review, 101*(4), 676–703.

Mack, J., Meltzer-Asscher, A., Barbieri, E., & Thompson, C. (2013). Neural correlates of processing passive sentences. *Brain Sciences, 3* (3), 1198–1214.

Mattys, S. L., Barden, K., & Samuel, A. G. (2014). Extrinsic cognitive load impairs low-level speech perception. *Psychonomic Bulletin and Review, 21*(3), 748–754.

Mattys, S. L., Brooks, J., & Cooke, M. (2009). Recognizing speech under a processing load: Dissociating energetic from informational factors. *Cognitive Psychology, 59*(3), 203–243.

Mattys, S. L., & Wiget, L. (2011). Effects of cognitive load on speech recognition. *Journal of Memory and Language, 65*(2), 145–160.

McGarrigle, R., Munro, K. J., Dawes, P., Stewart, A. J., Moore, D. R., Barry, J. G., et al. (2014). Listening effort and fatigue: What exactly are we measuring? A British society of audiology cognition in hearing special interest group "white paper.". *International Journal of Audiology*.

Meltzer, J. A., McArdle, J. J., Schafer, R. J., & Braun, A. R. (2010). Neural aspects of sentence comprehension: Syntactic complexity, reversibility, and reanalysis. *Cerebral Cortex, 20*(8), 1853–1864.

Meyer, L., Obleser, J., Kiebel, S. J., & Friederici, A. D. (2012). Spatiotemporal dynamics of argument retrieval and reordering: An FMRI and EEG study on sentence processing. *Frontiers in Psychology, 3*, 523.

Micheyl, C., Carlyon, R. P., Shtyrov, Y., Hauk, O., Dodson, T., & Pullvermüller, F. (2003). The neurophysiological basis of the auditory continuity illusion: A mismatch negativity study. *Journal of Cognitive Neuroscience, 15*(5), 747–758.

Miller, G. A., Heise, G. A., & Lichten, W. (1951). The intelligibility of speech as a function of the context of the test materials. *Journal of Experimental Psychology, 41*(5), 329–335.

Naber, M., & Nakayama, K. (2013). Pupil responses to high-level image content. *Journal of Vision, 13*(6).

Nittrouer, S., & Boothroyd, A. (1990). Context effects in phoneme and word recognition by young children and older adults. *The Journal of the Acoustical Society of America, 87*(6), 2705–2715.

Noonan, K. A., Jefferies, E., Visser, M., & Lambon Ralph, M. A. (2013). Going beyond inferior prefrontal involvement in semantic control: Evidence for the additional contribution of dorsal angular gyrus and posterior middle temporal cortex. *Journal of Cognitive Neuroscience, 25*(11), 1824–1850.

Novick, J. M., Kan, I. P., Trueswell, J. C., & Thompson-Schill, S. L. (2009). A case for conflict across multiple domains: Memory and language impairments following damage to ventrolateral prefrontal cortex. *Cognitive Neuropsychology, 26*(6), 527–567.

Obleser, J., & Kotz, S. A. (2010). Expectancy constraints in degraded speech modulate the language comprehension network. *Cerebral Cortex, 20*(3), 633–640.

Obleser, J., Meyer, L., & Friederici, A. D. (2011). Dynamic assignment of neural resources in auditory comprehension of complex sentences. *Neuroimage, 56*(4), 2310–2320.

Okada, K., Rong, F., Venezia, J., Matchin, W., Hsieh, I.-H., Saberi, K., et al. (2010). Hierarchical organization of human auditory cortex: Evidence from acoustic invariance in the response to intelligible speech. *Cerebral Cortex, 20*(10), 2486–2495.

Pals, C., Sarampalis, A., & Baskent, D. (2012). Listening effort with cochlear implant simulations. *Journal of Speech, Language, and Hearing Research*.

Peelle, J. E., Johnsrude, I. S., & Davis, M. H. (2010). Hierarchical processing for speech in human auditory cortex and beyond. *Frontiers in Human Neuroscience, 4*, 51.

Peelle, J. E., McMillan, C., Moore, P., Grossman, M., & Wingfield, A. (2004). Dissociable patterns of brain activity during comprehension of rapid and syntactically complex speech: Evidence from fMRI. *Brain and Language, 91*(3), 315–325.

Petkov, C. I., O'Connor, K. N., & Sutter, M. L. (2007). Encoding of illusory continuity in primary auditory cortex. *Neuron, 54*(1), 153–165.

Piquado, T., Isaacowitz, D., & Wingfield, A. (2010). Pupillometry as a measure of cognitive effort in younger and older adults. *Psychophysiology, 47*(3), 560–569.

Pollack, I. (1975). Auditory informational masking. *Journal of the Acoustical Society of America, 57*, S5.

Rodd, J., Gaskell, G., & Marslen-Wilson, W. (2002). Making sense of semantic ambiguity: Semantic competition in lexical access. *Journal of Memory and Language, 46*(2), 245–266.

Rodd, J. M., Davis, M. H., & Johnsrude, I. S. (2005). The neural mechanisms of speech comprehension: fMRI studies of semantic ambiguity. *Cerebral Cortex, 15*(8), 1261–1269.

Rodd, J. M., Johnsrude, I. S., & Davis, M. H. (2010). The role of domain-general frontal systems in language comprehension: Evidence from dual-task interference and semantic ambiguity. *Brain and Language, 115*(3), 182–188.

Rodd, J. M., Johnsrude, I. S., & Davis, M. H. (2012). Dissociating frontotemporal contributions to semantic ambiguity resolution in spoken sentences. *Cerebral Cortex, 22*(8), 1761–1773.

Rodd, J. M., Longe, O. A., Randall, B., & Tyler, L. K. (2010). The functional organisation of the fronto-temporal language system: Evidence from syntactic and semantic ambiguity. *Neuropsychologia, 48*(5), 1324–1335.

Saija, J. D., Akyürek, E. G., Andringa, T. C., & Başkent, D. (2014). Perceptual restoration of degraded speech is preserved with advancing age. *Journal of the Association for Research in Otolaryngology, 15*(1), 139–148.

Scott, S. K., Blank, C. C., Rosen, S., & Wise, R. J. (2000). Identification of a pathway for intelligible speech in the left temporal lobe. *Brain, 123*(Pt 12), 2400–2406.

Sevilla, Y., Maldonado, M., & Shalóm, D. E. (2014). Pupillary dynamics reveal computational cost in sentence planning. *Quarterly Journal of Experimental Psychology (2006), 67*(6), 1041–1452.

Shahin, A. J., Bishop, C. W., & Miller, L. M. (2009). Neural mechanisms for illusory filling-in of degraded speech. *NeuroImage, 44*(3), 1133–1143.

Shannon, R. V., Zeng, F. G., Kamath, V., Wygonski, J., & Ekelid, M. (1995). Speech recognition with primarily temporal cues. *Science, 270*(5234), 303–304.

Shetreet, E., & Friedmann, N. (2014). The processing of different syntactic structures: fMRI investigation of the linguistic distinction between wh-movement and verb movement. *Journal of Neurolinguistics, 27*(1), 1–17.

Sohoglu, E., Peelle, J. E., Carlyon, R. P., & Davis, M. H. (2012). Predictive top-down integration of prior knowledge during speech perception. *The Journal of Neuroscience, 32*(25), 8443–8453.

Stewart, R., & Wingfield, A. (2009). Hearing loss and cognitive effort in older adults' report accuracy for verbal materials. *Journal of the American Academy of Audiology, 20*(2), 147–154.

Tahmasebi, A. M., Abolmaesumi, P., Geng, X., Morosan, P., Amunts, K., Christensen, G. E., et al. (2009). A new approach for creating customizable cytoarchitectonic probabilistic maps without a template. *Medical Image Computing and Computer-Assisted Intervention, 12*(Pt 2), 795–802.

Thompson-Schill, S. L., D'Esposito, M., Aguirre, G. K., & Farah, M. J. (1997). Role of left inferior prefrontal cortex in retrieval of semantic knowledge: A reevaluation. *Proceedings of the National Academy of Sciences of the United States of America, 94*(26), 14792–14797.

Tyler, L. K., Marslen-Wilson, W. D., Randall, B., Wright, P., Devereux, B. J., Zhuang, J., et al. (2011). Left inferior frontal cortex and syntax: Function, structure and behaviour in patients with left hemisphere damage. *Brain, 134*(Pt 2), 415–431.

Warren, R. M. (1970). Perceptual restoration of missing speech sounds. *Science, 167*(3917), 392–393.

Wild, C. J., Davis, M. H., & Johnsrude, I. S. (2012). Human auditory cortex is sensitive to the perceived clarity of speech. *NeuroImage*, *60*(2), 1–13.

Wild, C. J., Yusuf, A., Wilson, D. E., Peelle, J. E., Davis, M. H., & Johnsrude, I. S. (2012). Effortful listening: The processing of degraded speech depends critically on attention. *Journal of Neuroscience*, *32*(40), 14010–14021.

Willems, R. M., & Hagoort, P. (2009). Broca's region: Battles are not won by ignoring half of the facts. *Trends in Cognitive Sciences*, *13* (3), 101 (author reply 102)

Winer, J. A. (2006). Decoding the auditory corticofugal systems. *Hearing Research*, *212*(1–2), 1–8.

Zekveld, A. A., & Kramer, S. E. (2014). Cognitive processing load across a wide range of listening conditions: Insights from pupillometry. *Psychophysiology*, *51*(3), 277–284.

Zekveld, A. A., Kramer, S. E., & Festen, J. M. (2011). Cognitive load during speech perception in noise: The influence of age, hearing loss, and cognition on the pupil response. *Ear and Hearing*, *32*(4), 498–510.

CHAPTER

41

Neural Mechanisms of Attention to Speech

Lee M. Miller

Center for Mind & Brain, and Department of Neurobiology, Physiology, & Behavior, University of California, Davis, CA, USA

41.1 OVERVIEW AND HISTORY

Speech perception rarely occurs under pristine acoustic conditions. Background noises, competing talkers, and reverberations often corrupt speech signals and confuse the basic organization of our auditory world. In modern life this perceptual challenge is inescapable, pervading the home, workplace, social situations (e.g., restaurants), and modes of transit (cars, subways). Fortunately, our brains have a remarkable ability to filter out undesired sounds and attend to a single talker, dramatically improving our comprehension (Kidd, Arbogast, Mason, & Gallun, 2005).

In this chapter we address *selective attention* to speech, as distinguished from general arousal or vigilance that is not directed at specific locations, objects, or features. William James described the essential phenomenology of selective attention as "the taking possession by the mind, in clear and vivid form, of one out of what seem several simultaneously possible objects or trains of thought... It implies withdrawal from some things in order to deal effectively with others" (James, 1890). Mechanistically, this comprises two separable components: *control* and *selection*. Attentional control refers to the cognitive processes that coordinate and direct our attention to specific aspects of our world, for instance, to the high-pitch female talker on the left. In contrast, attentional selection describes the biasing or filtering operation by which one object is highlighted or processed further and the rest is ignored. Selection is therefore the effect that attentional control exerts on the representations of speech content (e.g., acoustic, phonetic, semantic, affective). In the metaphor of the spotlight (Posner, Snyder, & Davidson, 1980), attentional control points the spotlight and attentional selection is evident by its illumination.

The modern study of attention to speech began in 1953 with Colin Cherry's behavioral work on speech comprehension in the presence of other distracting talkers, a situation he termed the "cocktail party problem" (Cherry, 1953). His observations and the experimental paradigms he pioneered, such as dichotic listening (different messages to left and right ears) and shadowing (repeating aloud), attracted many investigators to the topic over the ensuing decades. Attention-to-speech therefore became a driving force behind general cognitive theories of attention through the 1950s and 1960s, when the "early versus late selection" debate arose (Broadbent, 1958; Deutsch & Deutsch, 1963; Treisman, 1960). In contrast to the cognitive theory, the neuroscience of attention to speech began much later. In part, this delay was methodological, awaiting the widespread use of noninvasive human electrophysiology in the 1970s (electroencephalography [EEG]) and the advent of neuroimaging in the 1980s and 1990s (positron emission tomography [PET] and functional magnetic resonance imaging [fMRI]). Additionally, much early neuroscientific work on audition and attention eschewed speech for simpler stimuli such as isolated pure tones or noise bursts (Miller et al., 1972). Simple, isolated stimuli offer substantial experimental benefit because they are easily parameterized, have consistent neural responses that are straightforward to analyze, and provide a ready comparison between humans and nonhuman neurophysiology where simple stimuli prevail. However, especially when presented in isolation, they cannot approximate a "cocktail party" environment where selective attention is so crucial. Moreover, simple stimuli lack speech's rich spectro-temporal structure that may be indispensable to understand the mechanisms of attentional selection.

Consequently, the neuroscience of selective attention to speech—in the tradition of Cherry—began only in 1976, using traditional low-density (here, three

Neurobiology of Language. DOI: http://dx.doi.org/10.1016/B978-0-12-407794-2.00041-9

electrodes) EEG in humans (Hink & Hillyard, 1976), which showed increased neural response to speech in the attended ear with a latency of approximately 100 ms. In other words, attention can "select" a target speech stream by enhancing its neural processing at an early stage, a result that has been replicated many times since. We build on this basic idea, reviewing some of the evidence and conceptual advances that have brought us to our current understanding. Notably, recent years have witnessed a surge of interest and increased appreciation for realistic conditions with high perceptual load (Lavie, 2005), especially using continuous competing speech stimuli, degraded acoustics, and realistic spatial scenes. In the following sections we review the neural networks for attentional control, the levels of processing or representations that attention selects, and current neural theories about how mechanistically selective attention brings about such profound improvement in speech perception.

41.2 NEURAL NETWORKS FOR ATTENTIONAL CONTROL

Communicating in daily life requires that we direct the focus of our attention: maintain it on one talker in a noisy background, scan for a friend's voice in a crowd, listen for a certain word, or switch between different talkers in a conversation. All these are aspects of attentional control, whose functional neuroanatomy provides a framework for understanding attention-to-speech generally. Although attentional control to speech has been studied less than attentional selection, many aspects of control are cross-modal or supramodal (Krumbholz, Nobis, Weatheritt, & Fink, 2009; Macaluso, 2010; Salmi, Rinne, Degerman, Salonen, & Alho, 2007; Shomstein & Yantis, 2004), so several key principles observed in other sensory systems apply to speech as well. One influential model holds that two independent neural networks exert attentional control (Corbetta & Shulman, 2002). An "endogenous" dorsal fronto-parietal network guides top-down or volitional attention. This network consistently includes intraparietal sulcus (IPS) or superior parietal lobule and the frontal eye fields (FEF). A second, "exogenous" ventral fronto-temporo-parietal control network is proposed to handle reflexive or bottom-up reorienting of attention, as when triggered by stimulus saliency and target detection. This ventral control network consistently includes lateral cortical areas such as the temporoparietal junction (TPJ).

Adverse speech environments virtually always require top-down, volitional attention to particular talker locations, voice qualities, and speech content. Consistent with this requirement, "cocktail party" paradigms often recruit the dorsal fronto-parietal "endogenous" attention network: particularly IPS and superior parietal lobule, and the superior precentral sulcus in the characteristic location of FEF. Clear evidence for this dorsal network is observed for cued attention (Krumbholz et al., 2009), including when listeners deploy the attention to a desired pitch or location of a talker before the speech occurs (Hill & Miller, 2009). Dorsal network involvement is also evident through suppression of alpha ($\sim$10 Hz) EEG power over the parietal lobe contralateral to the attended hemifield (Kerlin, Shahin, & Miller, 2010) and later in the fronto-parietal network after perception of degraded speech (Obleser & Weisz, 2012). Notice that in speech studies, such superior fronto-parietal regions including likely FEF appear to be involved when attending not only to space but also to speech features such as pitch, and they are even recruited when attending to speech with no spatial acoustic cues whatsoever (as well as visual distractor stimuli; Wild et al., 2012). This supports the model that a supramodal dorsal fronto-parietal network handles volitional attention to speech (Corbetta & Shulman, 2002), in contrast to alternate models that postulate, for instance, a dorsal-spatial versus ventral nonspatial specialization (Hill & Miller, 2009; Lee et al., 2012). That said, some studies report little if any fronto-parietal activation associated with attention to speech (Alho et al., 2006). Though the reasons are unclear, in some cases this may be due to low perceptual interference (high speech clarity) or ease of the behavioral task, both of which would reduce the need for attentional control.

Another area robustly involved in attention to speech is the inferior frontal gyrus (IFG), especially the dorsal part and in some cases specifically the inferior frontal junction (IFJ; posterior part of the inferior frontal sulcus where it meets the precentral sulcus). Early suggestions for an attentional role of IFG came not from neuroimaging studies on attention per se, but from those that happened to require attention to phonetic attributes of clean speech. For instance, Zatorre reviewed a number of reports that used phonetic monitoring and discrimination, all of which show activation in dorsal posterior Broca's area (Zatorre, Meyer, Gjedde, & Evans, 1996). Although Broca's area is a large region heterogeneous in its cytoarchitectonics, anatomical connectivity, and function (Amunts & Zilles, 2012; Rogalsky & Hickok, 2011), evidence suggests that in addition to phonetic and other linguistic processing, this consistency in *dorsal* IFG is due to sustained active monitoring (i.e., attentional control). Subsequent studies using selective attention to one of multiple concurrent speech sounds also found left lateralized dorsal-posterior IFG activity when selective attention was required (Hashimoto, Homae, Nakajima, Miyashita, & Sakai, 2000; Lipschutz, Kolinsky, Damhaut,

Wikler, & Goldman, 2002), or bilateral inferior frontal involvement only when acoustic-phonetic cues could not easily distinguish the two competing speech streams (Nakai, Kato, & Matsuo, 2005). Another study approximated the challenge of a cocktail party using degraded speech and clever manipulation to distinguish controlled attention: subjects were first presented with noises, unaware that they could be heard as speech, and then the subjects learned to listen to the noises as degraded words (Giraud et al., 2004). The results showed that effortful "auditory search" recruits left dorsal posterior IFG without reflecting comprehension directly. Wild et al. (2012) also reported effects in IFG with similarly degraded sentences, in this case effects of auditory attention broadly but with an interaction such that a portion of dorsal IFG was only active when degraded but *potentially comprehensible* speech was attended. This is similar to the aforementioned "auditory search" in that listeners use internal expectations of speech to parse it, effortfully, from noise. In a rather different paradigm, left dorsal IFG/IFJ was specifically related to attentional control as opposed to selection of a competing speech stream, because it was shown to be active while directing attention before the speech occurred (Hill & Miller, 2009). Finally, although it is not universally the case (Alho et al., 2003), a majority of speech paradigms show this activation to have a left-hemisphere bias, likely reflecting linguistic hemispheric dominance (Giraud et al., 2004; Hill & Miller, 2009).

Even when attending to nonspeech noise bursts, IFG is active when listening in silence before the stimulus occurs (but here leading to a right hemisphere activation; Voisin, Bidet-Caulet, Bertrand, & Fonlupt, 2006). Moreover, specifically the dorsal posterior part of IFG is involved in a wide array of challenging tasks—both auditory, such as sequencing speech or nonspeech sounds (Gelfand & Bookheimer, 2003), and nonauditory, from math to cognitive control to working memory (Fedorenko, Duncan, & Kanwisher, 2012). All of these are consistent with a domain-general function of (usually left) dorsal IFG in attentional control or perhaps more broadly cognitive control (Novick, Trueswell, & Thompson-Schill, 2005). Interestingly, rather than being squarely part of a ventral reflexive attention network, IFJ may play an integrative role in both goal-directed and stimulus-driven attention (Asplund, Todd, Snyder, & Marois, 2010). This would clearly be important in a "cocktail party" environment, where momentary attentional goals would be continually updated by the varying salience of corrupted speech attributes.

In addition to IFG, attention to speech may recruit the hypothesized posterior member of the ventral "exogenous" control system—inferior parietal lobule and especially the TPJ—although here the evidence is not quite as consistent. Nevertheless, in the first neuroimaging study of concurrent *continuous* speech (as opposed to brief tokens), Alho et al. (2003) observed right lateralized TPJ activity regardless of whether subjects attended to the left or right ear. Similar activation is seen when attending to one of two concurrent streams of dichotic syllables (Lipschutz et al., 2002) and when switching and reorienting attention to one of two streams of spoken digits, whether simultaneously presented (Larson & Lee, 2012) or not (Krumbholz et al., 2009). Although bilateral TPJ involvement has been shown in a number of auditory attention studies (Huang, Belliveau, Tengshe, & Ahveninen, 2012), a right-hemisphere bias of a ventral attentional network is nevertheless consistent with visual attention and the phenomenon of hemispatial neglect (Corbetta & Shulman, 2011).

Thus, attentional control to speech relies on a supramodal, largely bilateral, dorsal fronto-parietal "volitional" network that directs attention to numerous speech attributes including but not limited to spatial location. Additionally, ventral prefrontal cortex (IFG) appears to be crucial for deploying attention flexibly to comprehend degraded or competing speech, whether it be attending to pitch or location, monitoring phonemes, or analyzing sound sequences. Like the dorsal network, left IFG appears to be supramodal in the sense that parts of it handle a wide array of attention-demanding perceptual and cognitive functions, both verbal and nonverbal. This agrees with the suggestion that the ventral frontal attention network, IFJ in particular, may be a kind of attentional nexus playing an integrative role in both goal-directed and stimulus-driven attention (Asplund et al., 2010; Huang et al., 2012). Finally, ventral posterior cortices consistent with a supramodal "orienting network," usually TPJ and particularly in the right hemisphere, appear to be important for attentional control and switching to speech in certain paradigms. Overall, the evidence shows that attentional control to speech is domain-general, sharing principles and functional neuroanatomy with other modalities.

41.3 LEVELS OF ATTENTIONAL SELECTION

Once attentional control "points the spotlight," its "illumination" then selects or modulates internal speech representations. Historically, since the time of Cherry, perception of competing speech has been central to the debate about attentional selection. Cherry and others observed that when attending to and shadowing one of two competing speech streams, listeners may not notice even dramatic changes in the unattended stream, such as a switch from English to German or backward speech (Cherry, 1953; Wood & Cowan, 1995). Such evidence inspired the "early selection" theory, championed by Broadbent (1958), which held that processing of unattended speech is thoroughly blocked at an early sensory stage. Other investigators advocated "late selection": that some speech information was still maintained through the entire sensory processing hierarchy, with attentional selection occurring just before entering working memory (Deutsch & Deutsch, 1963). Still others developed variants of these, with different stages of attenuation or filtering by attention (Treisman, 1960).

Today the debate is less heated, with most investigators recognizing that attention may act at different levels, and unattended speech may be processed to different depths, depending on the nature of the stimuli and task goals. The neural evidence shows that in some cases, attention can modulate the ascending pathway at the thalamus (Frith & Friston, 1996), brainstem including inferior colliculus (Galbraith & Arroyo, 1993; Rinne et al., 2008), or even the cochlea (Delano, Elgueda, Hamame, & Robles, 2007; Giraud et al., 1997). However, most evidence for attentional modulation of speech, based on latency and anatomical location, points to auditory cortex and higher-level speech-related cortex. The emerging pattern is that selective attention can modulate acoustic speech representations in primary or early nonprimary auditory cortical areas and at numerous higher levels, evidently depending on the task demands and on the degree of perceptual load or interference.

Our understanding of the neural selection of speech builds on and parallels a large body of research on auditory attention to nonspeech, especially simpler stimuli. Although this chapter does not aspire to thoroughly review auditory attention, a few notable studies provide some context. The earliest demonstration of auditory attentional selection occurred in cat auditory cortex, where Hubel observed cells that only responded when the animal paid attention to the stimulus (Hubel, Henson, Rupert, & Galambos, 1959). This appeared to be an early and strong form of gating or gain modulation due to auditory attentional selection, but little scientific work pursued the idea for many years. In humans, the "early versus late selection" debate was finally addressed with neural data using EEG; in a dichotic listening task, attention modulated the neural response with an enhanced negativity (at the top of the scalp) to tones in the attended ear at latencies 80–110 ms, corresponding to the auditory N1 wave (Hillyard, Hink, Schwent, & Picton, 1973). This observation has been replicated many times, and the modulation has been shown in some circumstances to occur earlier, from 20 to 50 ms and from 80 to 130 ms (Woldorff et al., 1993). Thus, early studies demonstrated a gain modulation of neural responses to attended sounds at latencies consistent with low-level auditory cortex.

These principles appeared to hold for speech sounds as well, with the aforementioned Hink and Hillyard dichotic speech study (Hink & Hillyard, 1976) that showed a larger N1 to stimuli in the attended ear. Subsequent EEG work approximated Cherry's shadowing paradigm, again reporting enhanced negativity to speech probe sounds and again co-temporal with the N1 wave (Woods, Hillyard, & Hansen, 1984). In a significant step toward ecologically valid "cocktail party" perception, Teder et al. first used continuous speech itself to elicit ERPs rather than independent probe sounds in the attended "channel," finding increased negativity beginning at approximately 40 ms and peaking at 80–90 ms latency (Teder, Kujala, & Naatanen, 1993). This study also used free-field speakers rather than dichotic presentation and highlighted the potential importance of realistic spatial cues, which may influence comprehension more as the scene becomes complex (Yost, Dye, & Sheft, 1996).

Neuroanatomically, the timing of effects in the electromagnetic studies suggest attentional modulation of primary and early nonprimary auditory cortices (Godey, Schwartz, de Graaf, Chauvel, & Liegeois-Chauvel, 2001). This conjecture was supported by a number of neuroimaging studies that used attention to speech in the absence of competing sounds or degradation (Grady et al., 1997; Jancke, Mirzazade, & Shah, 1999), or with brief dichotic speech stimuli spaced out in time (Jancke, Buchanan, Lutz, & Shah, 2001). Many early imaging studies using concurrent speech reported attentional modulation of large areas of the superior temporal lobe and give mixed results or lacked the spatial specificity to determine whether the modulation extended to presumed primary or early secondary auditory cortex (Alho et al., 2003; Hashimoto et al., 2000; Hugdahl et al., 2000; Lipschutz et al., 2002; Nakai et al., 2005; O'Leary et al., 1996). As a group, these leave unanswered the question of "early selection" but provide powerful evidence for attentional modulation at multiple higher stages of processing.

Meanwhile, studies using nonspeech sounds with more precise mapping techniques demonstrated that even for isolated (nonconcurrent) sounds, attentional modulation tended to be weak or absent in primary auditory cortex and stronger in secondary and tertiary fields (Petkov et al., 2004; Woods et al., 2009). Recent reports using more ecological "cocktail party" speech tasks also tend to report superior temporal attentional modulations that selectively spare primary auditory cortex but affect neighboring areas (Hill & Miller, 2009). On the basis of timing and source location, MEG demonstrated such modulations in planum temporale at approximately 100 ms latency but not earlier (Ding & Simon, 2012). This contrasts with attentional enhancements of response phase and coherence to nonspeech tonal stimuli, interpreted to occur in core or primary auditory cortex (Elhilali, Xiang, Shamma, & Simon, 2009). Another recent imaging study addressed selective attention to distorted speech in the presence of additional auditory and visual distractors, in other words, a very demanding and realistic task (Wild et al., 2012). Attentional modulation occurred throughout auditory cortex when subjects attended to audition as opposed to vision. However, attentional enhancement *specifically to distorted speech* occurred only in speech-sensitive anterior and posterior STS—high-level areas hierarchically distant from primary auditory cortex. In other words, much of auditory cortex showed some nonspecific modulation but only high-level speech processing benefitted from targeted, selective attentional enhancement.

Attentional selection of speech is thus remarkably flexible and can modulate processing at multiple levels of the cortical auditory and speech hierarchy, even during the same task. In all likelihood, this variable selection is as rapidly deployed and adaptive as observed in vision (Hopf et al., 2006). However, the specific modulation appears to be constrained by perceptual load or separability at the acoustic-phonetic level (Shinn-Cunningham, 2008). For sounds that differ on basic acoustic dimensions (or for sounds versus stimuli in another sensory modality), attention may be able to act earlier in the hierarchy. For competing speech that shares many physical characteristics or in cases of high perceptual load, attention may only be able to modulate higher-level, more abstract speech processing. This suggests that the levels of attentional selection are determined by both the processing level at which sensory representations can be readily distinguished and the amount of competition among those representations.

41.4 SPEECH REPRESENTATIONS THAT ATTENTION SELECTS

To fully understand the levels of attentional selection, we must also characterize the nature or content of speech representations that are actually selected. Here, manipulating task demands has been particularly informative, showing that selection works on a vast array of perceptually salient speech attributes depending on their momentary relevance. For this reason, attention to speech has been used most often *not to study attention* but as a tool to identify brain areas devoted to components of speech-perceptual and linguistic processing. The advantage of using attention in such paradigms is that one can equate low-level stimulus attributes among conditions, changing only the task goal, and with it the attentional focus. These studies range from the well-established "what versus where" streams distinguishing speech identity from location processing (Ahveninen et al., 2006; Alain, Arnott, Hevenor, Graham, & Grady, 2001; Maeder et al., 2001), to recognizing voices versus comprehending them (Alho et al., 2006; von Kriegstein, Eger, Kleinschmidt, & Giraud, 2003), to performing syntactic and semantic operations (Rogalsky & Hickok, 2009). However, although they use attentional selection as a tool, most of these approaches are not intended to reproduce the ecological conditions (e.g., high perceptual load) that may be necessary to strongly engage and reveal the attentional system (Lavie, 2005).

When listeners are challenged to understand degraded or competing speech, one of the key neural representations that correlates with comprehension is a low-frequency response (approximately 2–16 Hz, and especially 4–8 Hz) likely evoked or entrained by the speech amplitude envelope (Ahissar et al., 2001; Ding & Simon, 2013; Giraud & Poeppel, 2012; Luo & Poeppel, 2007; Peelle & Davis, 2012) or other salient acoustic features time-varying at a syllabic rate (Ghitza, Giraud, & Poeppel, 2012; Obleser, Herrmann, & Henry, 2012). That is, the neural response has substantial power at the same frequencies that speech acoustic power fluctuates across syllables. And it is phase-locked in the sense that neural fluctuations in that frequency range follow the acoustic speech events with a consistent latency. This particular response has a latency of approximately 100 ms, appears to be strongly represented in early auditory cortices, and is modulated by attention (Ding & Simon, 2012; Kerlin et al., 2010; Power, Lalor, & Reilly, 2011; Zion Golumbic et al., 2013). Recordings from the cortical surface in humans furthermore show that not only temporal envelope but also *spectro*-temporal envelope of the attended talker are preferentially modulated by attention, largely along the lateral superior temporal gyrus (STG) (Mesgarani & Chang, 2012). These observations are consistent with reports of nonhuman animals showing rapid, attention-dependent changes in spectro-temporal tuning of single cells and populations in auditory cortex (Fritz, Elhilali, & Shamma, 2007). As noted, other perceptually salient speech features such as pitch and spatial location are also clearly modulated by attention, but compared with spectro-temporal envelope the specific nature of these selected representations is relatively less apparent or less understood beyond a well-documented contralateral bias for spatial hemifield (Alho et al., 2003).

The prominent role of dynamic speech attributes for attention and comprehension highlights the unique importance of time. Unlike in vision, *all* auditory stimuli—and particularly speech—are distinguished by their temporal evolution. Thus, speech perception may rely on attention to specific moments even more than in other perceptual domains (Coull & Nobre, 1998; Lange & Roder, 2006). Behaviorally, temporal expectancy or attention in time has a powerful influence on speech comprehension. For instance, knowing both an expected word and the moment of a critical phoneme, but not one or the other, allows focused attention to alter the perceptual continuity of speech (Samuel & Ressler, 1986), even when the attentional state generally has no effect on continuity (Heinrich, Carlyon, Davis, & Johnsrude, 2011). This suggests that attention-in-time to speech may require an internal representation or prediction (e.g., lexical or syntactic) on which it operates (Shahin, Bishop, & Miller, 2009). In adverse environments such as context-guided, temporal expectations would be the rule rather than the exception. EEG shows that specific attention in time results in an enhanced N1 response (~100 ms) around the time of word onsets (Astheimer & Sanders, 2008), similar to the enhanced N1 often observed for nontemporal attention (Hillyard et al., 1973). These results complement mounting evidence that context and predictions, whether strictly attentional in nature or not, improve the neural processing of speech (Obleser, Wise, Alex, Dresner, & Scott, 2007; Sohoglu, Peelle, Carlyon, & Davis, 2012).

Notice that most of our understanding about the neural basis of attentional selection for speech regards specific speech features or attributes (e.g., pitch, temporal envelope, spectral attributes; Tuomainen, Savela, Obleser, & Aaltonen, 2013). Perceptually, rather than experiencing a collection of individual features, we perceive talkers as coherent, unified *objects* or *streams*. An important theoretical question, then, has been whether attention acts on neural speech representations as objects or as individual features. And if it acts on individual features, does attention then automatically "spread" to other features shared by the object? Although the question is not yet settled, most evidence is consistent with an object-based mechanism of attention to speech (Alain & Arnott, 2000; Ihlefeld & Shinn-Cunningham, 2008; Shinn-Cunningham, 2008). More specifically it seems that, similar to vision (Kravitz & Behrmann, 2011), auditory-speech object formation causally precedes but potentially still interacts with attentional selection. A recent MEG study supports this idea, showing that when the intensity of competing talkers is separately varied, the neural representation of attended speech adapts only to the intensity of that speaker and not to the background object even though they overlap spectro-temporally (Ding & Simon, 2012). These object representations, related to speech envelope and evident in the M100 response (~100 ms latency), were localized to planum temporale (i.e., in nonprimary auditory cortex). So even though we may attend to individual components of speech, naturalistic speech perception likely relies on bound object representations beginning early in the processing hierarchy.

41.5 NEURAL MECHANISMS AND TOP-DOWN/BOTTOM-UP INTERACTIONS

In the tradition of the "early versus late selection" debate, most inquiry into the neural bases of attention-to-speech has focused on neural timing and functional anatomy to determine the hierarchical level of speech processing where attention acts. Recent years, however, have brought an increased interest in the specific mechanisms by which attention modulates the speech representations. The historical, often implicit, concept of attentional selection has been that of a gain model: selection increases neural firing for the attended speech feature or object, and perhaps suppresses it for unattended ones. Prima facie, most empirical evidence is consistent with some type of gain modulation, including single-cell neurophysiology (beginning with Hubel's "attention units"; Hubel et al., 1959) as well as the majority of data from neuroimaging and EEG/MEG. One should of course keep in mind that different approaches may be more sensitive to different mechanisms (e.g., noninvasive neuroimaging techniques such as PET or fMRI might be blind to many non-gain temporal mechanisms), whereas in EEG and MEG increased neural synchrony (without more activity overall) may be interpreted as a "gain" effect.

Nevertheless, recent studies using a variety of methods continue to support selectively greater activity for attended speech representations. In several EEG or MEG reports, gain modulation has been observed for the phase-locked, syllable-rate responses noted, reflecting increased neural following of the attended speech envelope (approximately 4–8 Hz; Ding & Simon, 2012; Kerlin et al., 2010). The envelope response to unattended speech may also be somewhat suppressed (Kerlin et al., 2010) as it is with music (Choi, Rajaram, Varghese, & Shinn-Cunningham, 2013). Moreover, the selection of spectro-temporal speech modulations observed in cortical surface recordings (Mesgarani & Chang, 2012) appears to reflect a gain effect. Interestingly, the amount of gain modulation in the time-varying speech response, effectively the strength of attentional selection, is predicted by the strength of attentional control exerted. Specifically, gain in the speech representation correlates with the strength of *anticipatory* alpha power (~8–12 Hz) suppression over the parietal lobe contralateral to attended speaker hundreds of milliseconds earlier (Kerlin et al., 2010). Greater alpha suppression throughout the superior fronto-parietal control network furthermore predicts intelligibility of degraded speech (Obleser & Weisz, 2012). Thus, consistent with the visual system, the engagement of the attentional control network and consequent suppression of alpha oscillations may reflect the lifting of inhibition, followed by increased gain in attended speech representations (Jensen & Mazaheri, 2010). However, the details of such a "gating by inhibition" for speech, including any potential relationships between speech envelope and alpha oscillation phase, need further study.

In addition to modulating the intensity of activity in certain neural populations, attentional selection may also rapidly alter the spectrotemporal tuning of auditory cortical cells and populations to match the target sound features. Using nonspeech stimuli, Fritz and colleagues have shown such effects in nonhuman animals over multiple time scales (Elhilali, Fritz, Chi, & Shamma, 2007; Fritz, Shamma, Elhilali, & Klein, 2003). Evidence that attention can narrow spectral tuning in auditory cortex has also been observed in humans with MEG/EEG and fMRI (Ahveninen et al., 2011). Yet another proposed mechanism broadly compatible with increased gain and tuning is that attention aligns intrinsic, ongoing neural oscillations with the temporal evolution of attended speech (Besle et al., 2011; Lakatos, Karmos, Mehta, Ulbert, & Schroeder, 2008; Lakatos et al., 2009; Schroeder & Lakatos, 2009; Zion Golumbic et al., 2013). This "selective entrainment hypothesis" holds that spontaneous low-frequency neural oscillations (e.g., 1–8 Hz) reflect fluctuations in neural excitability (Bastos et al., 2012) and, therefore, preferred moments for sensory processing. In other words, the brain has natural rhythms and, during different moments in the rhythmic cycles, neurons are more or less excitable. Attention, using temporal predictions about precisely when speech information should occur, would align these endogenous oscillations to be in phase with the evoked sensory response, thereby selecting and boosting the attended speech representation. The idea that attention alters the temporal alignment of neural responses has further support from a study using nonspeech sounds showing greater temporal coherence for an attended rhythm across a bilateral auditory network (Elhilali et al., 2009). Therefore, attention may act through several mechanisms that complement simple gain modulation or may lead to it via different means. Much work remains to clarify both these additional possibilities and the behavioral circumstances in which they play a role.

41.6 INTERACTIONS BETWEEN ATTENTION, PERCEPTION, AND PREDICTION

As our mechanistic understanding of attention to speech grows, so does the recognition that attention may not simply modulate early sensory activity—boosting or shifting it but leaving the nature of the representations essentially intact. Rather, attention and similar mechanisms may interact with how our perceptual systems fundamentally parse and organize the world, a process known as auditory scene analysis (Bregman, 1990; McDermott, 2009; Shinn-Cunningham, 2008; Snyder, Gregg, Weintraub, & Alain, 2012). Most research on this topic has used nonspeech stimuli, but the principles may readily apply. For instance, it can take time, approximately 1 second, for our auditory systems to segregate or stream multiple objects in a complex scene. This "building up" of streaming may begin without attentional influence (Sussman, Horvath, Winkler, & Orr, 2007) but might require attention in some circumstances (Carlyon, Cusack, Foxton, & Robertson, 2001; Snyder, Alain, & Picton, 2006). These ideas are incorporated in a theory of auditory scene analysis that posits that the formation of auditory streams depends primarily on temporal coherence of neural responses to different object features (Shamma, Elhilali, & Micheyl, 2011). The theory furthermore holds that attention to one object feature is then *necessary* to bind the other temporally coherent features into a stream. All the other incoherent, unattended features are left unstreamed, as if you "wrap up all your garbage in the same bundle" (Bregman, 1990, p. 193). This is related to the idea, also shown in vision and cross-modally, that attention "spreads" to encompass the bound features of an object (Cusack, Deeks, Aikman, & Carlyon, 2004). However, this "temporal coherence" theory contradicts some evidence mentioned that attention acts on objects (Alain & Arnott, 2000; Shinn-Cunningham, 2008) and that object formation is largely pre-attentive or automatic (Bregman, 1990). Perhaps as with the levels of attentional selection, the degree to which attention interacts with basic object formation depends on the intrinsic competition or ambiguity among the sensory representations.

The notion that attention interacts fundamentally with speech processing is also consistent with a growing body of research on the perceptual role of context. We mentioned several instances where attention may guide sensory predictions or expectations (Astheimer & Sanders, 2008; Lakatos et al., 2008). Many more studies beyond the scope of this chapter show that the auditory system maintains a continuous, adaptive representation of context at multiple levels. Some expectations are largely automatic and depend on stimulus statistics, as indexed by the well-known MMN EEG response when patterns are violated (Naatanen, Gaillard, & Mantysalo, 1978; Sussman, Ritter, & Vaughan, 1998). However, even this so-called automatic response can be modulated by attention (Nyman et al., 1990; Sussman, Winkler, Huotilainen, Ritter, & Naatanen, 2002; Woldorff, Hillyard, Gallen, Hampson, & Bloom, 1998). Others, still largely unconscious, will be more linguistic in nature (Clos et al., 2014) and rely more on long-term memory. These linguistic contextual cues profoundly influence how well speech is processed, particularly when it is degraded (Baumgaertner, Weiller, & Buchel, 2002; Obleser & Kotz, 2009; Pichora-Fuller, Schneider, & Daneman, 1995) and when accompanied by age-related hearing loss, where peripheral impairment has been shown to interact with numerous cognitive and attentional functions (Humes et al., 2012). Still other expectations will be conscious, as when anticipating a specific word. Admittedly, not all of these phenomena will use attention-like mechanisms to select the predicted speech content, but some evidently do (e.g., with attention-in-time to emphasize representations of word onsets) (Astheimer & Sanders, 2008).

One increasingly prominent perceptual theory known as "predictive coding" relies inherently on the interaction between sensory input and expectation or context. Bayesian in nature, it holds that predictions formulated at higher levels are fed back to early sensory cortices, which act as comparators between prediction and sensory input, with the ultimate goal of minimizing prediction error. Thus, early sensory activity feeding forward reflects an error signal or mismatch between expectation and reality. Because the acoustic and linguistic statistics of speech support such rich predictions, this theory is a promising model to characterize attention to speech as well. Several recent results lend some empirical support to predictive coding. For instance, in a concurrent EEG + MEG study, increased prior knowledge about speech increased activity in IFG but *decreased* activity in STG, in line with predictive coding (Sohoglu et al., 2012). The timing of these effects suggested that an initial feed-forward draft of the speech allows IFG to activate likely predictions that are then fed back to sensory cortex. In another study, MEG responses localized to STG were consistent with predictive coding when perceiving newly learned words that were similar to known (predictable) words (Gagnepain, Henson, & Davis, 2012). Further clues about predictive coding used audiovisual speech, suggesting that "top-down" predictions based on vision may be fed back via lower-frequency oscillations (14–15 Hz) and, coupled with them, error signals propagated forward via high-frequency (60–80 Hz) activity (Arnal, Wyart, & Giraud, 2011). Typically "top-down" implies more abstract representations or rules originating at higher levels and flowing to lower levels of an anatomical or functional hierarchy, whereas bottom-up implies information, in this model an error signal, flowing up the hierarchy in the direction of an initial sensory volley. Such mechanistically different roles for different frequency bands agree with recent models of cortical information processing (Bastos et al., 2012; Wang, 2010). It also recalls a seminal nonhuman primate study of volitional versus stimulus-driven attentional control in vision, where volitional attention began in prefrontal cortex (FEF), exogenous attention initiated in the parietal lobe, and the coherence between them was greater in lower frequencies for volitional control and in higher frequencies for stimulus-driven attention (Buschman & Miller, 2007). Although results from different systems and paradigms should be taken with care, they are at least consistent with the notion that attention to speech could act within a predictive coding framework.

41.7 FUTURE DIRECTIONS

Our understanding of attention to speech has advanced considerably in the past half century, but many questions remain. Regarding basic functional neuroanatomy, for instance, how should attention to speech be reconciled with the influential dual-stream theory of speech perception (Hickok & Poeppel, 2007)? The dual stream model broadly conforms to a dorsal "where/how" pathway for sensorimotor transformation and perception-for-action and a ventral "what" pathway for identifying objects (Goodale & Milner, 1992; Hickok & Poeppel, 2007; O'Reilly, 2010; Scott, 2005). This is consistent with evidence from attentional selection, with multiple, flexible levels of modulation along the speech hierarchy. However, correspondence with the dual-stream model is less clear for the fronto-temporo-parietal attentional control network. For example, ventral areas for orienting attention overlap those thought to perform sensorimotor (acoustic-articulatory) integration and, particularly in Broca's area, other linguistic perceptual functions. Perhaps the supramodal dorsal fronto-parietal attentional network controls volitional attention to speech as an acoustic *object in the world* (e.g., to its spatial location) or when using other attributes to segregate the object from others. In contrast, perhaps the inferior frontal attentional system controls volitional attention as well as stimulus-driven orienting to identifying attributes of speech (i.e., those that contribute directly to its intelligibility). Perhaps function is lateralized in the ventral attention network such that right

TPJ is devoted to exogenous attention while left TPJ handles acoustic-motor mapping. At present, we lack a framework that integrates these necessary but somewhat incompatible roles.

In terms of mechanisms, most investigators would agree that gain modulation plays a central role in attentional selection of speech. But other plausible mechanisms have not yet been thoroughly vetted. Are some effects that we attribute to gain instead due to increased synchrony or coherence in neural activity, or are they due to altered receptive field tuning in individual auditory cortical cells? Does attentional control selectively entrain endogenous fluctuations to be in phase with incoming speech, thereby boosting it at the expense of distractors? Does alpha activity act as "gating inhibition" and, when suppressed, dis-inhibit the attended speech representations? If so, then how would alpha fluctuations relate to the well-known syllabic rate sensory activity or to behavioral performance (Zoefel & Heil, 2013)? Such interactions between top-down control and sensory selection also lead naturally to the relation between attention and perceptual segregation of speech streams. Are speech objects formed independently of attention (and then selected by it), or is attention required—and, if so, under which circumstances and at what levels of processing? Is attention organized hierarchically for features and objects so that attention to any feature always spreads to other features of that speech stream? Does attention to speech operate as "predictive coding," seeking to minimize the error between context-based expectations and reality? Far more empirical evidence is needed, but the proposal fits well with a growing appreciation that the brain often adheres to Bayesian principles (Bastos et al., 2012).

Still other possible attentional mechanisms have hardly been explored with speech, including several already demonstrated in the visual system. For instance, should we conceive of selective attention to speech as a form of biased competition (Desimone & Duncan, 1995; Shinn-Cunningham, 2008) and, if so, how should we think of the high-order "receptive field" within which different stimuli compete? As noted, at the single neuron and population level, we expect that attention might induce rapid adaptive receptive field changes in how auditory neurons respond to speech. But does it also decorrelate neural activity, which in visual cortex can explain behavioral performance more than gain modulation (Cohen & Maunsell, 2009)? Or does attention perform a kind of "normalizing" computation to yield a winner-take-all selected speech object in certain circumstances (Reynolds & Heeger, 2009)? Already we know that attentional control networks are neuroanatomically largely supramodal and so, too, could the underlying mechanisms of attention be universal.

As the abundance of open questions makes clear, great effort will be required to characterize the neural bases of attention to speech. But it also demonstrates how our understanding has become more informed, nuanced, and mechanistic over time. We are now at the threshold of characterizing how this profoundly important everyday ability works. After 60 years, we have returned to Cherry's cocktail party.

Acknowledgments

This work was supported by the National Institutes of Health/National Institute on Deafness and other Communication Disorders (NIH/NIDCD), grant R01-DC008171 (LMM).

References

Ahissar, E., Nagarajan, S., Ahissar, M., Protopapas, A., Mahncke, H., & Merzenich, M. M. (2001). Speech comprehension is correlated with temporal response patterns recorded from auditory cortex. *Proceedings of the National Academy of Sciences of the United States of America*, *98*(23), 13367–13372.

Ahveninen, J., Hamalainen, M., Jaaskelainen, I. P., Ahlfors, S. P., Huang, S., Lin, F. H., et al. (2011). Attention-driven auditory cortex short-term plasticity helps segregate relevant sounds from noise. *Proceedings of the National Academy of Sciences of the United States of America*, *108*(10), 4182–4187.

Ahveninen, J., Jaaskelainen, I. P., Raij, T., Bonmassar, G., Devore, S., Hamalainen, M., et al. (2006). Task-modulated "what" and "where" pathways in human auditory cortex. *Proceedings of the National Academy of Sciences of the United States of America*, *103*(39), 14608–14613.

Alain, C., & Arnott, S. R. (2000). Selectively attending to auditory objects. *Frontiers in Bioscience*, *5*, D202–212.

Alain, C., Arnott, S. R., Hevenor, S., Graham, S., & Grady, C. L. (2001). "What" and "where" in the human auditory system. *Proceedings of the National Academy of Sciences of the United States of America*, *98*(21), 12301–12306.

Alho, K., Vorobyev, V. A., Medvedev, S. V., Pakhomov, S. V., Roudas, M. S., Tervaniemi, M., et al. (2003). Hemispheric lateralization of cerebral blood-flow changes during selective listening to dichotically presented continuous speech. *Brain Research Cognitive Brain Research*, *17*(2), 201–211.

Alho, K., Vorobyev, V. A., Medvedev, S. V., Pakhomov, S. V., Starchenko, M. G., Tervaniemi, M., et al. (2006). Selective attention to human voice enhances brain activity bilaterally in the superior temporal sulcus. *Brain Research*, *1075*(1), 142–150.

Amunts, K., & Zilles, K. (2012). Architecture and organizational principles of Broca's region. *Trends in Cognitive Sciences*, *16*(8), 418–426.

Arnal, L. H., Wyart, V., & Giraud, A. L. (2011). Transitions in neural oscillations reflect prediction errors generated in audiovisual speech. *Nature Neuroscience*, *17*, 797–801.

Asplund, C. L., Todd, J. J., Snyder, A. P., & Marois, R. (2010). A central role for the lateral prefrontal cortex in goal-directed and stimulus-driven attention. *Nature Neuroscience*, *13*(4), 507–512.

Astheimer, L. B., & Sanders, L. D. (2008). Listeners modulate temporally selective attention during natural speech processing. *Biological Psychology*, *80*(1), 23–34.

Bastos, A. M., Usrey, W. M., Adams, R. A., Mangun, G. R., Fries, P., & Friston, K. J. (2012). Canonical microcircuits for predictive coding. *Neuron*, *76*(4), 695–711.

Baumgaertner, A., Weiller, C., & Buchel, C. (2002). Event-related fMRI reveals cortical sites involved in contextual sentence integration. *Neuroimage, 16*(3 Pt 1), 736–745.

Besle, J., Schevon, C. A., Mehta, A. D., Lakatos, P., Goodman, R. R., McKhann, G. M., et al. (2011). Tuning of the human neocortex to the temporal dynamics of attended events. *The Journal of Neuroscience, 31*(9), 3176–3185.

Bregman, A. S. (1990). *Auditory scene analysis*. Cambridge, MA: Bradford.

Broadbent, D. (1958). *Perception and communication*. London: Pergamon Press.

Buschman, T. J., & Miller, E. K. (2007). Top-down versus bottom-up control of attention in the prefrontal and posterior parietal cortices. *Science, 315*(5820), 1860–1862.

Carlyon, R. P., Cusack, R., Foxton, J. M., & Robertson, I. H. (2001). Effects of attention and unilateral neglect on auditory stream segregation. *Journal of Experimental Psychology Human Perception and Performance, 27*(1), 115–127.

Cherry, E. C. (1953). Some experiments on the recognition of speech with one and with two ears. *The Journal of the Acoustical Society of America, 25*, 975–979.

Choi, I., Rajaram, S., Varghese, L. A., & Shinn-Cunningham, B. G. (2013). Quantifying attentional modulation of auditory-evoked cortical responses from single-trial electroencephalography. *Frontiers in Human Neuroscience, 7*, 115.

Clos, M., Langner, R., Meyer, M., Oechslin, M. S., Zilles, K., & Eickhoff, S. B. (2014). Effects of prior information on decoding degraded speech: An fMRI study. *Human Brain Mapping, 35*(1), 61–74.

Cohen, M. R., & Maunsell, J. H. (2009). Attention improves performance primarily by reducing interneuronal correlations. *Nature Neuroscience, 12*(12), 1594–1600.

Corbetta, M., & Shulman, G. L. (2002). Control of goal-directed and stimulus-driven attention in the brain. *Nature Reviews Neuroscience, 3*(3), 201–215.

Corbetta, M., & Shulman, G. L. (2011). Spatial neglect and attention networks. *Annual Review of Neuroscience, 34*, 569–599.

Coull, J. T., & Nobre, A. C. (1998). Where and when to pay attention: The neural systems for directing attention to spatial locations and to time intervals as revealed by both PET and fMRI. *The Journal of Neuroscience, 18*(18), 7426–7435.

Cusack, R., Deeks, J., Aikman, G., & Carlyon, R. P. (2004). Effects of location, frequency region, and time course of selective attention on auditory scene analysis. *Journal of Experimental Psychology Human Perception and Performance, 30*(4), 643–656.

Delano, P. H., Elgueda, D., Hamame, C. M., & Robles, L. (2007). Selective attention to visual stimuli reduces cochlear sensitivity in chinchillas. *The Journal of Neuroscience, 27*(15), 4146–4153.

Desimone, R., & Duncan, J. (1995). Neural mechanisms of selective visual attention. *Annual Review of Neuroscience, 18*, 193–222.

Deutsch, J. A., & Deutsch, D. (1963). Attention: Some theoretical considerations. *Psychological Review, 70*, 80–90.

Ding, N., & Simon, J. Z. (2012). Emergence of neural encoding of auditory objects while listening to competing speakers. *Proceedings of the National Academy of Sciences of the United States of America, 109*(29), 11854–11859.

Ding, N., & Simon, J. Z. (2013). Adaptive temporal encoding leads to a background-insensitive cortical representation of speech. *The Journal of Neuroscience, 33*(13), 5728–5735.

Elhilali, M., Fritz, J. B., Chi, T. S., & Shamma, S. A. (2007). Auditory cortical receptive fields: Stable entities with plastic abilities. *The Journal of Neuroscience, 27*(39), 10372–10382.

Elhilali, M., Xiang, J., Shamma, S. A., & Simon, J. Z. (2009). Interaction between attention and bottom-up saliency mediates the representation of foreground and background in an auditory scene. *PLoS Biology, 7*(6), e1000129.

Fedorenko, E., Duncan, J., & Kanwisher, N. (2012). Language-selective and domain-general regions lie side by side within Broca's area. *Current Biology, 22*(21), 2059–2062.

Frith, C. D., & Friston, K. J. (1996). The role of the thalamus in "top down" modulation of attention to sound. *Neuroimage, 4*(3 Pt 1), 210–215.

Fritz, J., Shamma, S., Elhilali, M., & Klein, D. (2003). Rapid task-related plasticity of spectrotemporal receptive fields in primary auditory cortex. *Nature Neuroscience, 6*(11), 1216–1223.

Fritz, J. B., Elhilali, M., & Shamma, S. A. (2007). Adaptive changes in cortical receptive fields induced by attention to complex sounds. *Journal of Neurophysiology, 98*(4), 2337–2346.

Gagnepain, P., Henson, R. N., & Davis, M. H. (2012). Temporal predictive codes for spoken words in auditory cortex. *Current Biology, 22*(7), 615–621.

Galbraith, G. C., & Arroyo, C. (1993). Selective attention and brainstem frequency-following responses. *Biological Psychology, 37*(1), 3–22.

Gelfand, J. R., & Bookheimer, S. Y. (2003). Dissociating neural mechanisms of temporal sequencing and processing phonemes. *Neuron, 38*(5), 831–842.

Ghitza, O., Giraud, A. L., & Poeppel, D. (2012). Neuronal oscillations and speech perception: Critical-band temporal envelopes are the essence. *Frontiers in Human Neuroscience, 6*, 340.

Giraud, A. L., Garnier, S., Micheyl, C., Lina, G., Chays, A., & Chery-Croze, S. (1997). Auditory efferents involved in speech-in-noise intelligibility. *Neuroreport, 8*(7), 1779–1783.

Giraud, A. L., Kell, C., Thierfelder, C., Sterzer, P., Russ, M. O., Preibisch, C., et al. (2004). Contributions of sensory input, auditory search and verbal comprehension to cortical activity during speech processing. *Cerebral Cortex, 14*(3), 247–255.

Giraud, A. L., & Poeppel, D. (2012). Cortical oscillations and speech processing: Emerging computational principles and operations. *Nature Neuroscience, 15*(4), 511–517.

Godey, B., Schwartz, D., de Graaf, J. B., Chauvel, P., & Liegeois-Chauvel, C. (2001). Neuromagnetic source localization of auditory evoked fields and intracerebral evoked potentials: A comparison of data in the same patients. *Clinical Neurophysiology, 112*(10), 1850–1859.

Goodale, M. A., & Milner, A. D. (1992). Separate visual pathways for perception and action. *Trends in Neurosciences, 15*(1), 20–25.

Grady, C. L., Van Meter, J. W., Maisog, J. M., Pietrini, P., Krasuski, J., & Rauschecker, J. P. (1997). Attention-related modulation of activity in primary and secondary auditory cortex. *Neuroreport, 8*(11), 2511–2516.

Hashimoto, R., Homae, F., Nakajima, K., Miyashita, Y., & Sakai, K. L. (2000). Functional differentiation in the human auditory and language areas revealed by a dichotic listening task. *Neuroimage, 12*(2), 147–158.

Heinrich, A., Carlyon, R. P., Davis, M. H., & Johnsrude, I. S. (2011). The continuity illusion does not depend on attentional state: FMRI evidence from illusory vowels. *Journal of Cognitive Neuroscience, 23*(10), 2675–2689.

Hickok, G., & Poeppel, D. (2007). The cortical organization of speech processing. *Nature Reviews Neuroscience, 8*(5), 393–402.

Hill, K. T., & Miller, L. M. (2009). Auditory attentional control and selection during cocktail party listening. *Cerebral Cortex, 20*(3), 583–590.

Hillyard, S. A., Hink, R. F., Schwent, V. L., & Picton, T. W. (1973). Electrical signs of selective attention in the human brain. *Science, 182*(4108), 177–180.

Hink, R. F., & Hillyard, S. A. (1976). Auditory evoked potentials during selective listening to dichotic speech messages. *Perception & Psychophysics, 20*(4), 236–242.

Hopf, J. M., Luck, S. J., Boelmans, K., Schoenfeld, M. A., Boehler, C. N., Rieger, J., et al. (2006). The neural site of attention matches the spatial scale of perception. *The Journal of Neuroscience*, *26*(13), 3532–3540.

Huang, S., Belliveau, J. W., Tengshe, C., & Ahveninen, J. (2012). Brain networks of novelty-driven involuntary and cued voluntary auditory attention shifting. *PLoS One*, *7*(8), e44062.

Hubel, D. H., Henson, C. O., Rupert, A., & Galambos, R. (1959). Attention units in the auditory cortex. *Science*, *129*(3358), 1279–1280.

Hugdahl, K., Law, I., Kyllingsbaek, S., Bronnick, K., Gade, A., & Paulson, O. B. (2000). Effects of attention on dichotic listening: An 15O-PET study. *Human Brain Mapping*, *10*(2), 87–97.

Humes, L. E., Dubno, J. R., Gordon-Salant, S., Lister, J. J., Cacace, A. T., Cruickshanks, K. J., et al. (2012). Central presbycusis: A review and evaluation of the evidence. *Journal of the American Academy of Audiology*, *23*(8), 635–666.

Ihlefeld, A., & Shinn-Cunningham, B. (2008). Disentangling the effects of spatial cues on selection and formation of auditory objects. *The Journal of the Acoustical Society of America*, *124*(4), 2224–2235.

James, W. (1890). *The principles of psychology*. New York, NY: Dover Publications.

Jancke, L., Buchanan, T. W., Lutz, K., & Shah, N. J. (2001). Focused and nonfocused attention in verbal and emotional dichotic listening: An FMRI study. *Brain and Language*, *78*(3), 349–363.

Jancke, L., Mirzazade, S., & Shah, N. J. (1999). Attention modulates activity in the primary and the secondary auditory cortex: A functional magnetic resonance imaging study in human subjects. *Neuroscience Letters*, *266*(2), 125–128.

Jensen, O., & Mazaheri, A. (2010). Shaping functional architecture by oscillatory alpha activity: Gating by inhibition. *Frontiers in Human Neuroscience*, *4*, 186.

Kerlin, J. R., Shahin, A. J., & Miller, L. M. (2010). Attentional gain control of ongoing cortical speech representations in a "cocktail party". *The Journal of Neuroscience*, *30*(2), 620–628.

Kidd, G., Jr., Arbogast, T. L., Mason, C. R., & Gallun, F. J. (2005). The advantage of knowing where to listen. *The Journal of the Acoustical Society of America*, *118*(6), 3804–3815.

Kravitz, D. J., & Behrmann, M. (2011). Space-, object-, and feature-based attention interact to organize visual scenes. *Attention, Perception, & Psychophysics*, *73*(8), 2434–2447.

Krumbholz, K., Nobis, E. A., Weatheritt, R. J., & Fink, G. R. (2009). Executive control of spatial attention shifts in the auditory compared to the visual modality. *Human Brain Mapping*, *30*(5), 1457–1469.

Lakatos, P., Karmos, G., Mehta, A. D., Ulbert, I., & Schroeder, C. E. (2008). Entrainment of neuronal oscillations as a mechanism of attentional selection. *Science*, *320*(5872), 110–113.

Lakatos, P., O'Connell, M. N., Barczak, A., Mills, A., Javitt, D. C., & Schroeder, C. E. (2009). The leading sense: Supramodal control of neurophysiological context by attention. *Neuron*, *64*(3), 419–430.

Lange, K., & Roder, B. (2006). Orienting attention to points in time improves stimulus processing both within and across modalities. *Journal of Cognitive Neuroscience*, *18*(5), 715–729.

Larson, E., & Lee, A. K. (2012). The cortical dynamics underlying effective switching of auditory spatial attention. *Neuroimage*, *64*, 365–370.

Lavie, N. (2005). Distracted and confused?: Selective attention under load. *Trends in Cognitive Sciences*, *9*(2), 75–82.

Lee, A. K., Rajaram, S., Xia, J., Bharadwaj, H., Larson, E., Hamalainen, M. S., et al. (2012). Auditory selective attention reveals preparatory activity in different cortical regions for selection based on source location and source pitch. *Frontiers in Neuroscience*, *6*, 190.

Lipschutz, B., Kolinsky, R., Damhaut, P., Wikler, D., & Goldman, S. (2002). Attention-dependent changes of activation and connectivity in dichotic listening. *Neuroimage*, *17*(2), 643–656.

Luo, H., & Poeppel, D. (2007). Phase patterns of neuronal responses reliably discriminate speech in human auditory cortex. *Neuron*, *54* (6), 1001–1010.

Macaluso, E. (2010). Orienting of spatial attention and the interplay between the senses. *Cortex*, *46*(3), 282–297.

Maeder, P. P., Meuli, R. A., Adriani, M., Bellmann, A., Fornari, E., Thiran, J. P., et al. (2001). Distinct pathways involved in sound recognition and localization: A human fMRI study. *Neuroimage*, *14*(4), 802–816.

McDermott, J. H. (2009). The cocktail party problem. *Current Biology*, *19*(22), R1024–1027.

Mesgarani, N., & Chang, E. F. (2012). Selective cortical representation of attended speaker in multi-talker speech perception. *Nature*, *485* (7397), 233–236.

Miller, J. M., Sutton, D., Pfingst, B., Ryan, A., Beaton, R., & Gourevitch, G. (1972). Single cell activity in the auditory cortex of Rhesus monkeys: Behavioral dependency. *Science*, *177*(4047), 449–451.

Naatanen, R., Gaillard, A. W., & Mantysalo, S. (1978). Early selective-attention effect on evoked potential reinterpreted. *Acta Psychologica (Amsterdam)*, *42*(4), 313–329.

Nakai, T., Kato, C., & Matsuo, K. (2005). An FMRI study to investigate auditory attention: A model of the cocktail party phenomenon. *Magnetic Resonance in Medical Sciences*, *4*(2), 75–82.

Novick, J. M., Trueswell, J. C., & Thompson-Schill, S. L. (2005). Cognitive control and parsing: Reexamining the role of Broca's area in sentence comprehension. *Cognitive, Affective & Behavioral Neuroscience*, *5*(3), 263–281.

Nyman, G., Alho, K., Laurinen, P., Paavilainen, P., Radil, T., Reinikainen, K., et al. (1990). Mismatch Negativity (MMN) for sequences of auditory and visual stimuli: Evidence for a mechanism specific to the auditory modality. *Electroencephalography and Clinical Neurophysiology*, *77*(6), 436–444.

Obleser, J., Herrmann, B., & Henry, M. J. (2012). Neural oscillations in speech: Don't be enslaved by the envelope. *Frontiers in Human Neuroscience*, *6*, 250.

Obleser, J., & Kotz, S. A. (2009). Expectancy constraints in degraded speech modulate the language comprehension network. *Cerebral Cortex*, *20*, 633–640.

Obleser, J., & Weisz, N. (2012). Suppressed alpha oscillations predict intelligibility of speech and its acoustic details. *Cerebral Cortex*, *22*(11), 2466–2477.

Obleser, J., Wise, R. J., Alex Dresner, M., & Scott, S. K. (2007). Functional integration across brain regions improves speech perception under adverse listening conditions. *The Journal of Neuroscience*, *27*(9), 2283–2289.

O'Leary, D. S., Andreason, N. C., Hurtig, R. R., Hichwa, R. D., Watkins, G. L., Ponto, L. L., et al. (1996). A positron emission tomography study of binaurally and dichotically presented stimuli: Effects of level of language and directed attention. *Brain and Language*, *53*(1), 20–39.

O'Reilly, R. C. (2010). The what and how of prefrontal cortical organization. *Trends in Neurosciences*, *33*(8), 355–361.

Peelle, J. E., & Davis, M. H. (2012). Neural oscillations carry speech rhythm through to comprehension. *Front Psychology*, *3*, 320.

Petkov, C. I., Kang, X., Alho, K., Bertrand, O., Yund, E. W., & Woods, D. L. (2004). Attentional modulation of human auditory cortex. *Nature Neuroscience*, *7*(6), 658–663.

Pichora-Fuller, M. K., Schneider, B. A., & Daneman, M. (1995). How young and old adults listen to and remember speech in noise. *The Journal of the Acoustical Society of America*, *97*(1), 593–608.

Posner, M. I., Snyder, C. R., & Davidson, B. J. (1980). Attention and the detection of signals. *Journal of Experimental Psychology*, *109*(2), 160–174.

Power, A. J., Lalor, E. C., & Reilly, R. B. (2011). Endogenous auditory spatial attention modulates obligatory sensory activity in auditory cortex. *Cerebral Cortex*, *21*(6), 1223–1230.

Reynolds, J. H., & Heeger, D. J. (2009). The normalization model of attention. *Neuron*, *61*(2), 168–185.

Rinne, T., Balk, M. H., Koistinen, S., Autti, T., Alho, K., & Sams, M. (2008). Auditory selective attention modulates activation of human inferior colliculus. *Journal of Neurophysiology*, *100*(6), 3323–3327.

Rogalsky, C., & Hickok, G. (2009). Selective attention to semantic and syntactic features modulates sentence processing networks in anterior temporal cortex. *Cerebral Cortex*, *19*(4), 786–796.

Rogalsky, C., & Hickok, G. (2011). The role of Broca's area in sentence comprehension. *Journal of Cognitive Neuroscience*, *23*(7), 1664–1680.

Salmi, J., Rinne, T., Degerman, A., Salonen, O., & Alho, K. (2007). Orienting and maintenance of spatial attention in audition and vision: Multimodal and modality-specific brain activations. *Brain Structure & Function*, *212*(2), 181–194.

Samuel, A. G., & Ressler, W. H. (1986). Attention within auditory word perception: Insights from the phonemic restoration illusion. *Journal of Experimental Psychology Human Perception and Performance*, *12*(1), 70–79.

Schroeder, C. E., & Lakatos, P. (2009). Low-frequency neuronal oscillations as instruments of sensory selection. *Trends in Neurosciences*, *32*(1), 9–18.

Scott, S. K. (2005). Auditory processing—speech, space and auditory objects. *Current Opinion in Neurobiology*, *15*(2), 197–201.

Shahin, A. J., Bishop, C. W., & Miller, L. M. (2009). Neural mechanisms for illusory filling-in of degraded speech. *Neuroimage*, *44*(3), 1133–1143.

Shamma, S. A., Elhilali, M., & Micheyl, C. (2011). Temporal coherence and attention in auditory scene analysis. *Trends in Neurosciences*, *34*(3), 114–123.

Shinn-Cunningham, B. G. (2008). Object-based auditory and visual attention. *Trends in Cognitive Sciences*, *12*(5), 182–186.

Shomstein, S., & Yantis, S. (2004). Control of attention shifts between vision and audition in human cortex. *The Journal of Neuroscience*, *24*(47), 10702–10706.

Snyder, J. S., Alain, C., & Picton, T. W. (2006). Effects of attention on neuroelectric correlates of auditory stream segregation. *Journal of Cognitive Neuroscience*, *18*(1), 1–13.

Snyder, J. S., Gregg, M. K., Weintraub, D. M., & Alain, C. (2012). Attention, awareness, and the perception of auditory scenes. *Front Psychology*, *3*, 15.

Sohoglu, E., Peelle, J. E., Carlyon, R. P., & Davis, M. H. (2012). Predictive top-down integration of prior knowledge during speech perception. *The Journal of Neuroscience*, *32*(25), 8443–8453.

Sussman, E., Ritter, W., & Vaughan, H. G., Jr. (1998). Attention affects the organization of auditory input associated with the mismatch negativity system. *Brain Research*, *789*(1), 130–138.

Sussman, E., Winkler, I., Huotilainen, M., Ritter, W., & Naatanen, R. (2002). Top-down effects can modify the initially stimulus-driven auditory organization. *Brain Research Cognitive Brain Research*, *13* (3), 393–405.

Sussman, E. S., Horvath, J., Winkler, I., & Orr, M. (2007). The role of attention in the formation of auditory streams. *Perception & Psychophysics*, *69*(1), 136–152.

Teder, W., Kujala, T., & Naatanen, R. (1993). Selection of speech messages in free-field listening. *Neuroreport*, *5*(3), 307–309.

Treisman, A. (1960). Contextual cues in selective listening. *The Quarterly Journal of Experimental Psychology*, *12*(4), 242–248.

Tuomainen, J., Savela, J., Obleser, J., & Aaltonen, O. (2013). Attention modulates the use of spectral attributes in vowel discrimination: Behavioral and event-related potential evidence. *Brain Research*, *1490*, 170–183.

Voisin, J., Bidet-Caulet, A., Bertrand, O., & Fonlupt, P. (2006). Listening in silence activates auditory areas: A functional magnetic resonance imaging study. *The Journal of Neuroscience*, *26*(1), 273–278.

von Kriegstein, K., Eger, E., Kleinschmidt, A., & Giraud, A. L. (2003). Modulation of neural responses to speech by directing attention to voices or verbal content. *Brain Research Cognitive Brain Research*, *17*(1), 48–55.

Wang, X. J. (2010). Neurophysiological and computational principles of cortical rhythms in cognition. *Physiological Reviews*, *90*(3), 1195–1268.

Wild, C. J., Yusuf, A., Wilson, D. E., Peelle, J. E., Davis, M. H., & Johnsrude, I. S. (2012). Effortful listening: The processing of degraded speech depends critically on attention. *The Journal of Neuroscience*, *32*(40), 14010–14021.

Woldorff, M. G., Gallen, C. C., Hampson, S. A., Hillyard, S. A., Pantev, C., Sobel, D., et al. (1993). Modulation of early sensory processing in human auditory cortex during auditory selective attention. *Proceedings of the National Academy of Sciences of the United States of America*, *90*(18), 8722–8726.

Woldorff, M. G., Hillyard, S. A., Gallen, C. C., Hampson, S. R., & Bloom, F. E. (1998). Magnetoencephalographic recordings demonstrate attentional modulation of mismatch-related neural activity in human auditory cortex. *Psychophysiology*, *35*(3), 283–292.

Wood, N. L., & Cowan, N. (1995). The cocktail party phenomenon revisited: Attention and memory in the classic selective listening procedure of Cherry (1953). *Journal of Experimental Psychology General*, *124*(3), 243–262.

Woods, D. L., Hillyard, S. A., & Hansen, J. C. (1984). Event-related brain potentials reveal similar attentional mechanisms during selective listening and shadowing. *Journal of Experimental Psychology Human Perception and Performance*, *10*(6), 761–777.

Woods, D. L., Stecker, G. C., Rinne, T., Herron, T. J., Cate, A. D., Yund, E. W., et al. (2009). Functional maps of human auditory cortex: Effects of acoustic features and attention. *PLoS One*, *4*(4), e5183.

Yost, W. A., Dye, R. H., Jr., & Sheft, S. (1996). A simulated "cocktail party" with up to three sound sources. *Perception & Psychophysics*, *58*(7), 1026–1036.

Zatorre, R. J., Meyer, E., Gjedde, A., & Evans, A. C. (1996). PET studies of phonetic processing of speech: Review, replication, and reanalysis. *Cerebral Cortex*, *6*(1), 21–30.

Zion Golumbic, E. M., Ding, N., Bickel, S., Lakatos, P., Schevon, C. A., McKhann, G. M., et al. (2013). Mechanisms underlying selective neuronal tracking of attended speech at a "cocktail party". *Neuron*, *77*(5), 980–991.

Zoefel, B., & Heil, P. (2013). Detection of near-threshold sounds is independent of EEG phase in common frequency bands. *Front Psychology*, *4*, 262.

CHAPTER

42

Audiovisual Speech Integration: Neural Substrates and Behavior

Michael S. Beauchamp

Department of Neurosurgery and Core for Advanced MRI, Baylor College of Medicine, Houston, TX, USA

Speech is the most important form of human communication. Speech perception is multisensory, with both an auditory component (the talker's voice) and a visual component (the talker's face). When the auditory speech signal is compromised because of auditory noise or hearing loss, the visual speech signal gains in importance (Bernstein, Auer, & Takayanagi, 2004; Ross et al., 2007; Sumby & Pollack, 1954). In the case of profound deafness, lip-reading by itself can be sufficient for speech perception (Suh, Lee, Kim, Chung, & Oh, 2009). Visual speech information is also beneficial in cases of partial hearing loss; older adults with hearing impairments identify visual-only speech *better* than older adults with normal hearing (Tye-Murray, Sommers, & Spehar, 2007).

42.1 NEUROARCHITECTURE OF AUDIOVISUAL SPEECH INTEGRATION

The most popular technique for examining human brain function is blood-oxygen level-dependent functional magnetic resonance imaging (BOLD fMRI) (Friston, 2009), and neurophysiological studies of multisensory speech are no exception (Beauchamp, Lee, Argall, & Martin, 2004; Blank & von Kriegstein, 2013; Lee & Noppeney, 2011; Miller & D'Esposito, 2005; Nath & Beauchamp, 2011; Noppeney, Josephs, Hocking, Price, & Friston, 2008; Okada, Venezia, Matchin, Saberi, & Hickok, 2013; Wilson, Molnar-Szakacs, & Iacoboni, 2008). The multisensory speech network in a typical subject is easily identifiable with fMRI and includes regions of occipital, temporal, parietal, and frontal cortex (Figure 42.1A: left and right hemisphere in one subject). The most consistent activations (found in every hemisphere in every subject, example subject in Figure 42.1B) are in three regions: (i) visual cortex, especially lateral extrastriate motion-sensitive areas, including areas MT and MST; (ii) auditory cortex and auditory association areas on the superior temporal gyrus; and (iii) multisensory areas in the posterior superior temporal sulcus (STS) and adjacent superior temporal gyrus and middle temporal gyrus (pSTS/G).

A working hypothesis for the function of these areas is that extrastriate visual areas process the complex biological motion signals carried by the talker's facial motion and form a representation of visual speech. Auditory association areas process the complex auditory information necessary to form a representation of auditory speech sounds. These representations must abstract away from the physical stimulus properties to represent the key features of speech to be behaviorally useful; for instance, it is important to integrate auditory and visual speech from a talker even if we have never seen or heard them before. Multisensory areas in the STS then integrate the visual and auditory speech representations to decide the most likely speech percept produced by the talker. A critical feature of speech is that the amount of information carried by the auditory and visual modalities varies tremendously from moment to moment. According to Bayesian models of perception, ideal observers should weight each modality by its reliability (the inverse of its variance). For instance, in a conversation in a very loud room, the visual information is more reliable than the auditory information and should be given more weight; in a dark room, the auditory information is more reliable and should receive more weight. In a BOLD fMRI experiment, Nath and Beauchamp (2011) examined the neural mechanisms for this dynamic weighting of auditory and visual information. Auditory speech

Neurobiology of Language. DOI: http://dx.doi.org/10.1016/B978-0-12-407794-2.00042-0

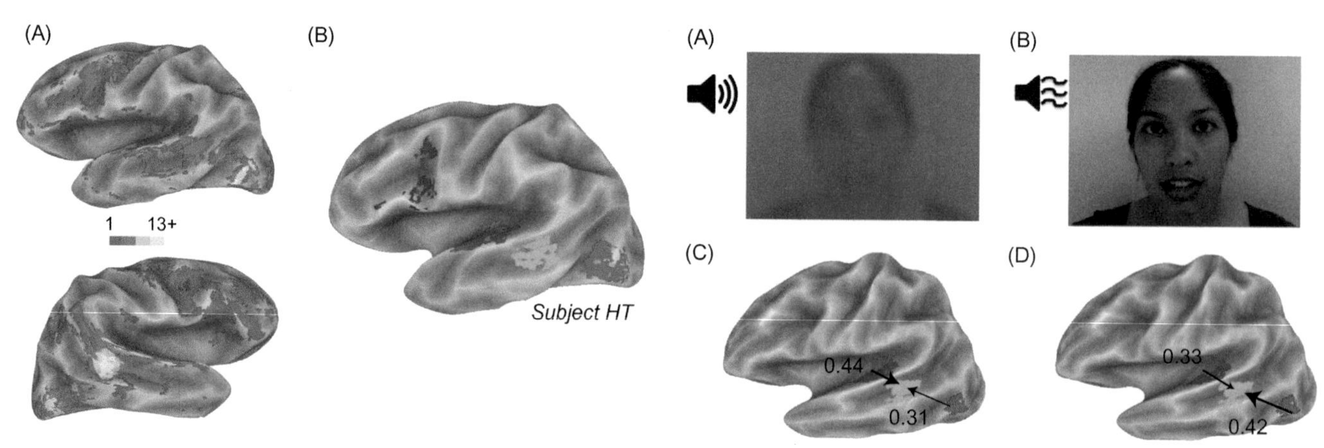

FIGURE 42.1 The neural network for processing audiovisual speech measured with BOLD fMRI. (A) Lateral views of the partially inflated left hemisphere (top) and right hemisphere (bottom). Dark gray shows sulcal depths and light gray shows gyral crowns. Colored brain regions showed a significant increase in BOLD fMRI signal to audiovisual speech (recordings of meaningless syllables) compared with fixation baseline. All activations thresholded at $F > 5.0$, $P < 0.05$, corrected for false discovery rate. Color scale shows t-values for contrast of all syllables versus fixation. (B) Anatomical-functional regions of interest (ROIs) created from activation map in (A). The STS ROI (green) contains voxels responsive to both auditory and visual speech greater than baseline in the posterior STS. The auditory cortex ROI (blue) contains voxels responsive to auditory speech greater than baseline within Heschl's gyrus. The visual cortex ROI (red) contains voxels responsive to visual speech greater than baseline within extrastriate lateral occipitotemporal cortex. The IFG ROI (purple) contains voxels responsive to both auditory and visual speech greater than baseline within the opercular region of the inferior frontal gyrus and the inferior portion of the precentral sulcus.

FIGURE 42.2 fMRI functional connectivity within the audiovisual speech network. (A) Auditory-reliable speech: undegraded auditory speech (loudspeaker icon) with degraded visual speech (single video frame shown). (B) Visual-reliable speech: undegraded visual speech with degraded auditory speech. (C) Functional connectivity during auditory-reliable speech in a representative subject. The green region on the cortical surface model shows the location of the pSTS/G ROI; the blue region shows the location of the auditory cortex ROI; and the red region shows the location of the lateral extrastriate visual cortex ROI. The top number is the path coefficient between pSTS/G and auditory cortex. The bottom number is the path coefficient between the pSTS/G and visual cortex. (D) Functional connectivity during visual-reliable speech (same subject and color scheme as in **C**).

information was degraded by adding masking noise, and visual speech information was degraded by blurring the stimulus (across experiments, different levels of degradation were used with similar results). To assess functional connectivity, the psychophysiological interaction (PPI) method was used. Functional connections between STS and auditory association areas and lateral extrastriate visual cortical areas were observed, but not between STS and primary auditory or primary visual cortex.

As shown in Figure 42.2, increased functional connectivity between the STS and auditory association cortex was observed when the auditory modality was more reliable, and increased functional connectivity between the STS and extrastriate visual cortex was observed when the visual modality was more reliable, even when the reliability changed rapidly during rapid event-related presentation of successive words. This finding suggests that changes in STS functional connectivity may be an important neural mechanism underlying the perception of noisy speech. Changes in functional connectivity may also be important for audiovisual learning (Driver & Noesselt, 2008; Powers, Hevey, & Wallace, 2012).

fMRI studies have provided a wealth of information about audiovisual speech perception, but they are constrained by the limitations of the BOLD signal. Most critically, BOLD is an indirect measure of neural function that relies on slow changes in the cerebral vasculature. The BOLD response to a stimulus consisting of a single syllable (duration ~0.5 s) extends for 15 s; most fMRI studies have a temporal resolution of only 2 s. This temporal resolution is thousands of times slower than the underlying neural information processing (duration of a sodium-based action potential <1 ms). Because speech is itself a rapidly changing signal, and because the speech information must be processed rapidly by the brain to be behaviorally useful, fMRI is far too slow for direct observation of the underlying neural events.

In contrast, studies that use electrical and magnetic measurements can directly measure neural activity with high temporal precision. In the first approach, event-related potentials are recorded. Transient responses to audiovisual syllables display peaks between 100 and 200 ms after stimulus onset (Bernstein, Auer, Wagner, & Ponton, 2008). In a second approach, induced oscillations are recorded. These have the advantage of allowing for the examination of sustained neuronal activity that is not phase-locked to the stimulus and can extend for several hundred milliseconds after stimulus onset (Schepers, Schneider, Hipp, Engel, & Senkowski, 2013).

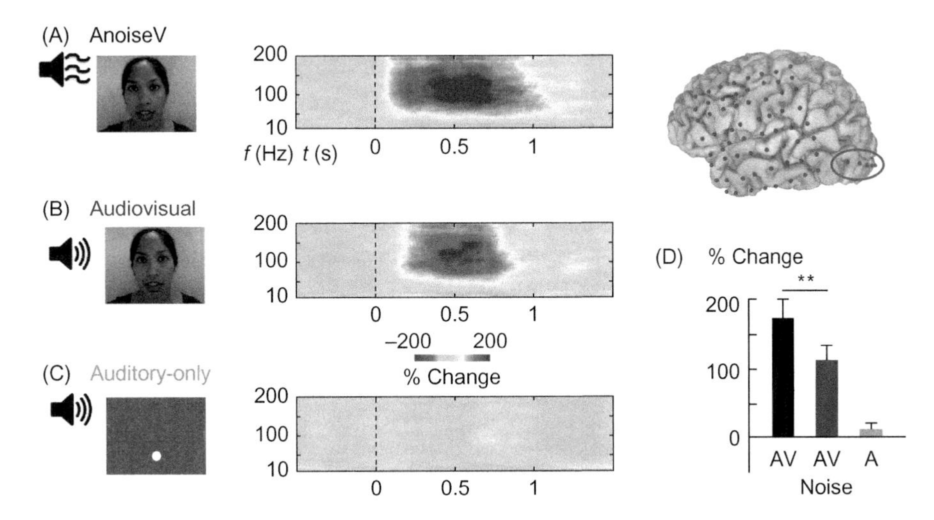

FIGURE 42.3 Electrocorticography of visual cortex responses to audiovisual speech. (A) A patient being treated for intractable epilepsy was implanted with subdural electrodes (red dots on cortical surface model, right). Brain activity was recorded from electrodes over visual cortex (red dots inside ellipse) as the subject was presented with noisy audiovisual speech (visual speech with auditory white noise, AnoiseV). A time-frequency analysis was used to determine the neural response at different frequencies (y-axis) at each time (x-axis) after stimulus onset (dashed vertical line). Color scale indicates amplitude of change from pre-stimulus baseline. (B) Time-frequency response of visual cortex to audiovisual speech (undegraded visual and auditory). (C). Response to auditory-only speech. (D) Amplitude of high-frequency responses to each stimulus condition averaged across the entire course of the response. **Significant difference between AnoiseV and AV speech, $P = 0.0016$.

EEG and MEG have provided important insights into the neural basis of multisensory speech perception. Another technique known as electrocorticography has two important advantages (Allison, Puce, Spencer, & McCarthy, 1999). First, electrocorticography offers superior spatial resolution: activity recorded from each approximately 1-mm electrode on the cortical surface is generated by neurons in close proximity to the electrode. In contrast, EEG and MEG suffer from the inverse problem: for any distribution of electrical or magnetic fields measured outside the head, there are an infinite number of patterns of brain electrical activity that could have resulted in the observed fields. Practically speaking, this limits the spatial resolution, making it difficult to untangle the location of sources that are in close proximity, as are the areas in the audiovisual speech network. Second, electrocorticography signals have much greater signal-to-noise than EEG/MEG, especially for high-frequency responses, because the electrode is much closer to the source, eliminating filtering by the cerebrospinal fluid and skull that lies between the cortex and the EEG/MEG sensors. High-frequency activity is ubiquitous in the human brain (Canolty et al., 2006; Crone, Miglioretti, Gordon, & Lesser, 1998; Schepers, Hipp, Schneider, Roder, & Engel, 2012) and is thought to be an excellent population-level measure of neuronal activity correlated with single-neuron spiking (for a recent review see Lachaux, Axmacher, Mormann, Halgren, & Crone, 2012; Logothetis, Kayser, & Oeltermann, 2007; Nir et al., 2007; Ray & Maunsell, 2011).

Most electrocorticography studies of speech perception have investigated auditory-only speech, either in a single brain area, such as auditory cortex (Chang et al., 2010), or in a more distributed network (Canolty et al., 2007; Pei et al., 2011; Towle et al., 2008). We collected electrocorticography data while subjects were presented with clear and noisy auditory and visual speech. Our hypothesis was that multisensory interactions between auditory and visual speech processing should be most pronounced when the visual speech information is most important, such as when noisy auditory speech information is present. We tested this hypothesis by examining the responses to visual speech in visual cortex.

In the noisy audiovisual and audiovisual speech conditions (Figure 42.3A and B), the subject viewed a video of a talker speaking a word. This is a powerful visual stimulus that evoked a strong response in visual electrodes. As is commonly reported for responses to visual stimuli, the response differed by frequency. For low frequencies (<20 Hz) there was a decrease in power relative to pre-stimulus baseline (blue colors at bottom of plot). For high frequencies (>40 Hz) there was a large increase in power relative to pre-stimulus baseline (yellow and red colors at middle of each plot). These high-frequency responses are thought to reflect spiking activity in neurons underlying the electrode and therefore served as our primary measure of interest. As is expected for visual cortex, there was a negligible response to stimuli without a visual component (auditory-only speech, Figure 42.3C). Responses to noisy audiovisual speech (video of a speaker accompanied by auditory white noise) were stronger than the responses to AV speech (Figure 42.3D).

This supports a simple model in which the goal of the cortical speech network is to extract meaning from AV speech as quickly and efficiently as possible. When the auditory component of the speech input is sufficient to extract meaning, the visual response is not needed and therefore not enhanced. When the auditory stimulus does not contain sufficient information to extract meaning (as is the case for visual-only speech or visual speech with auditory white noise), the visual response is needed to extract meaning and activity in visual cortex is upregulated.

42.2 BEHAVIORAL APPROACHES FOR STUDYING AUDIOVISUAL SPEECH INTEGRATION

There are a number of behavioral approaches to studying the contributions of the auditory and visual modalities to audiovisual speech perception. In the most straightforward method, noisy auditory speech is accompanied by clear visual speech, such as a video of a talker saying the words that match the auditory signal (Figure 42.4A). The information available from viewing the face of the talker improves the ability to understand speech. This improvement is equivalent to an approximately 10-dB increase in signal-to-noise ratio; at increased levels of auditory noise, this can correspond to a dramatic improvement in the number of words understood (Ross, Saint-Amour, Leavitt, Javitt, & Foxe, 2007b).

At a neural level, we might imagine that there are different pools of neurons that represent representations of different auditory and visual speech representations. When presented with a noisy auditory phoneme, many pools representing many different auditory phonological representations are weakly active, with no single pool reaching an activity level sufficient to form a perceptual decision, assuming a Bayesian evidence accumulation model (Beck et al., 2008). When visual information is added, pools of neurons representing the

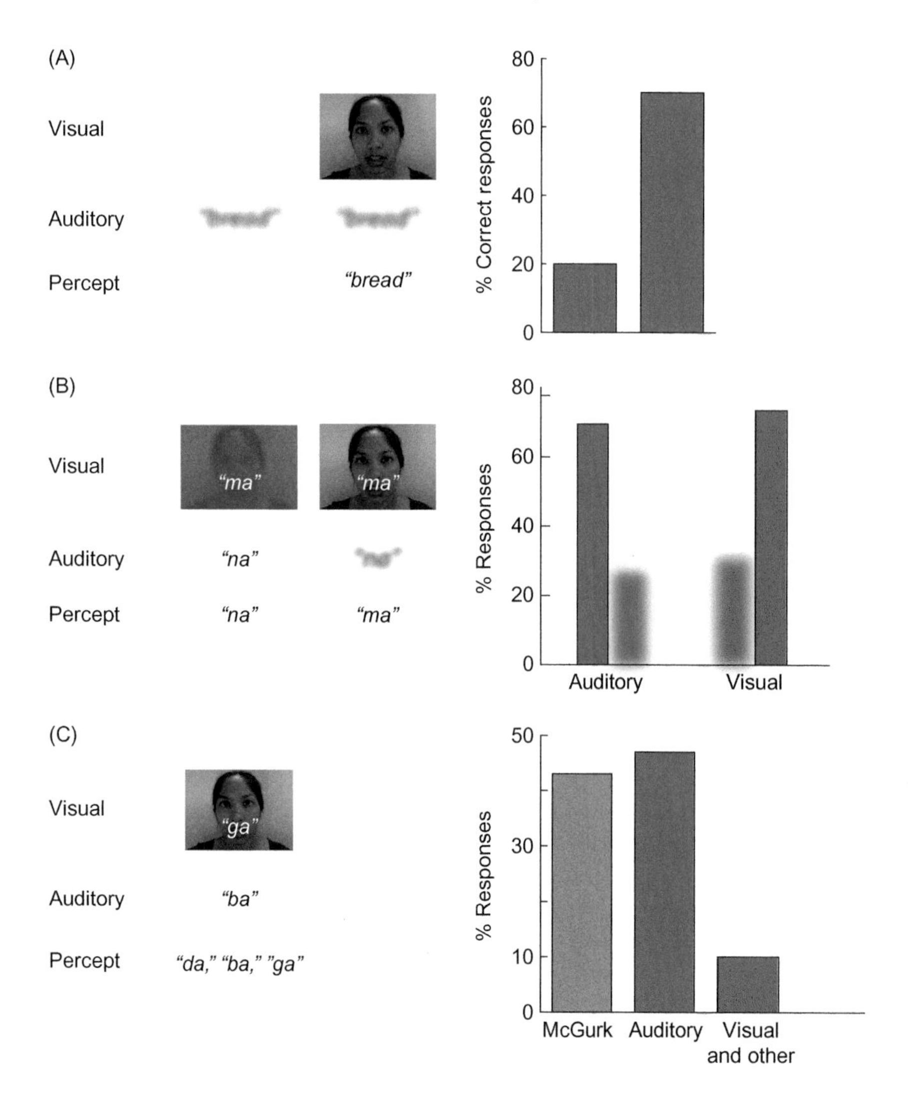

FIGURE 42.4 Behavioral responses to incongruent audiovisual speech. (A) Noisy auditory speech with and without congruent visual speech. If a degraded auditory word (e.g., bread; auditory filtering indicated by blurriness of printed word) is presented, then the subject only rarely is able to correctly identify it (left bar in plot). If the same degraded auditory word is accompanied by a video of a talker speaking the word, then the subject is usually able to identify it (right bar in plot). Data replotted from Ross, Saint-Amour, Leavitt, Javitt, and Foxe (2007a) with permission from the publisher. (B) Incongruent audiovisual speech consisting of a visual /ma/ and an auditory /na/ (or vice versa). If the visual component of speech is degraded by blurring (left column), then subjects report a percept corresponding to the auditory component of the stimulus (indicated in plot by solid blue bar and complementary blurry red bar). If the auditory component of speech is degraded by filtering (right column), then subjects report a percept corresponding to the visual component of the stimulus (indicated in plot by blurry blue bar and complementary solid red bar). (C) Incongruent audiovisual speech consisting of a visual /ga/ and an auditory /ba/ (or a visual /pa/ and an auditory /ka/). In 43% of trials (green bar in plot) subjects reported a McGurk fusion percept of /da/ (or /ta/), corresponding to neither the auditory nor the visual components of the stimulus.

visual speech information respond strongly. Combining the information available in the visual and auditory-responsive pools of neurons allows a perceptual decision to be reached.

In a more complex approach, auditory speech and visual speech are presented together, but the modalities are incongruent; the auditory modality signals one syllable at the same time as the visual modality signals a different one (Figure 42.4B). The advantage of this approach is that it pits the auditory and visual modalities against each other, allowing an evaluation of their relative contributions to perception. If the reported percept matches the auditory modality, then we infer that the auditory modality was most important; if the reported percept matched the visual modality, then we infer that the visual modality was most important. For instance, if a visual /ma/ is paired with an auditory /na/ that is heavily degraded, subjects will often report perceiving /ma/, implying that they used the visual information more than the auditory information in forming their decision (Nath & Beauchamp, 2011). However, it is not possible to directly quantify the contribution of each modality to each response as it is with other perceptual decisions. For continuous variables, such as location, the percept of an incongruent audiovisual stimulus (e.g., beep in one location and flash in a different location) can vary anywhere between the location of the two unisensory stimuli. A report of a percept at a location exactly midway between the locations of the two stimuli indicates that both modalities contributed equally to the percept, with locations closer to one or the other of the two stimuli indicating more influence of that modality. However, speech perception is categorical—any given speech sound is perceived as a particular phoneme, and subjects in general do not make intermediate responses. Therefore, it is only possible to state that one modality or the other was dominant. Subjects are often aware that the auditory and visual modalities are incongruent, but this does not change their conscious percept of a single syllable.

At a neural level, we could imagine that adding auditory noise weakens the response of the pool of neurons representing the auditory speech representation. At high levels of auditory noise, the response of the neurons representing the incongruent visual speech representation is stronger than the response of the neurons representing the auditory speech, and thus the reported percept corresponds to the visual speech representation.

A third approach to studying multisensory integration in speech also involves the presentation of incongruent audiovisual syllables (Figure 42.4C). For most incongruent audiovisual syllables, subjects perceive either the auditory component of the stimulus or the visual component. However, for a few incongruent audiovisual syllables, subjects perceive a third syllable that corresponds to neither the auditory nor the visual component of the stimulus. Two of these unusual incongruent syllables, first discovered serendipitously by McGurk and MacDonald (1976), consist of an auditory /ba/ paired with a visual /ga/ (perceived as /da/) and an auditory /pa/ paired with a visual /ka/ (perceived as /ta/). These illusory percepts have come to be known as the McGurk effect.

The McGurk effect is widely used in research for at least two reasons. First, because the percept corresponds to neither the auditory nor the visual component of the stimulus, the McGurk effect provides strong evidence that speech perception is multisensory. Second, the McGurk effect offers a quick and simple way to measure multisensory integration. A single trial in which the subject perceives the illusion demonstrates that the subject is integrating auditory and visual information. In contrast, measuring audiovisual integration with other approaches requires many trials at different levels of auditory noise, which is time-consuming for normal volunteers and may be impossible for clinical groups such as elderly subjects or children.

42.3 INTERSUBJECT VARIABILITY

The McGurk effect has also been used as a sensitive assay of individual differences in audiovisual integration. Although it is generally appreciated that the ability to see and hear varies across the population (hence the need for eye glasses and hearing aids), difference in central perceptual abilities are less well-appreciated. Healthy young subjects were tested with auditory-only stimuli and congruent audiovisual stimuli and identified them without errors. However, when presented with McGurk stimuli, there were dramatic differences between individuals. Across 14 different McGurk stimuli (recorded by many different speakers), some subjects never reported the illusion, whereas some subjects reported the illusory percept on every trial of every stimulus (Figure 42.5A). However, for most subjects, the illusion was not "all or nothing." Most subjects reported the illusion for some stimuli but not others, and, even for multiple presentations of the same stimulus, a subject might report the illusion on some trials but not others.

To better understand this variability, we created a simple model with three sets of parameters. One parameter described the strength of each stimulus (across subjects, how often this stimulus elicited an illusory percept). Another parameter described each subject's individual threshold for reporting the illusion (any stimulus perceived to lie above this threshold would evoke the illusion). A third parameter described the sensory noise in each subject's measurement of a

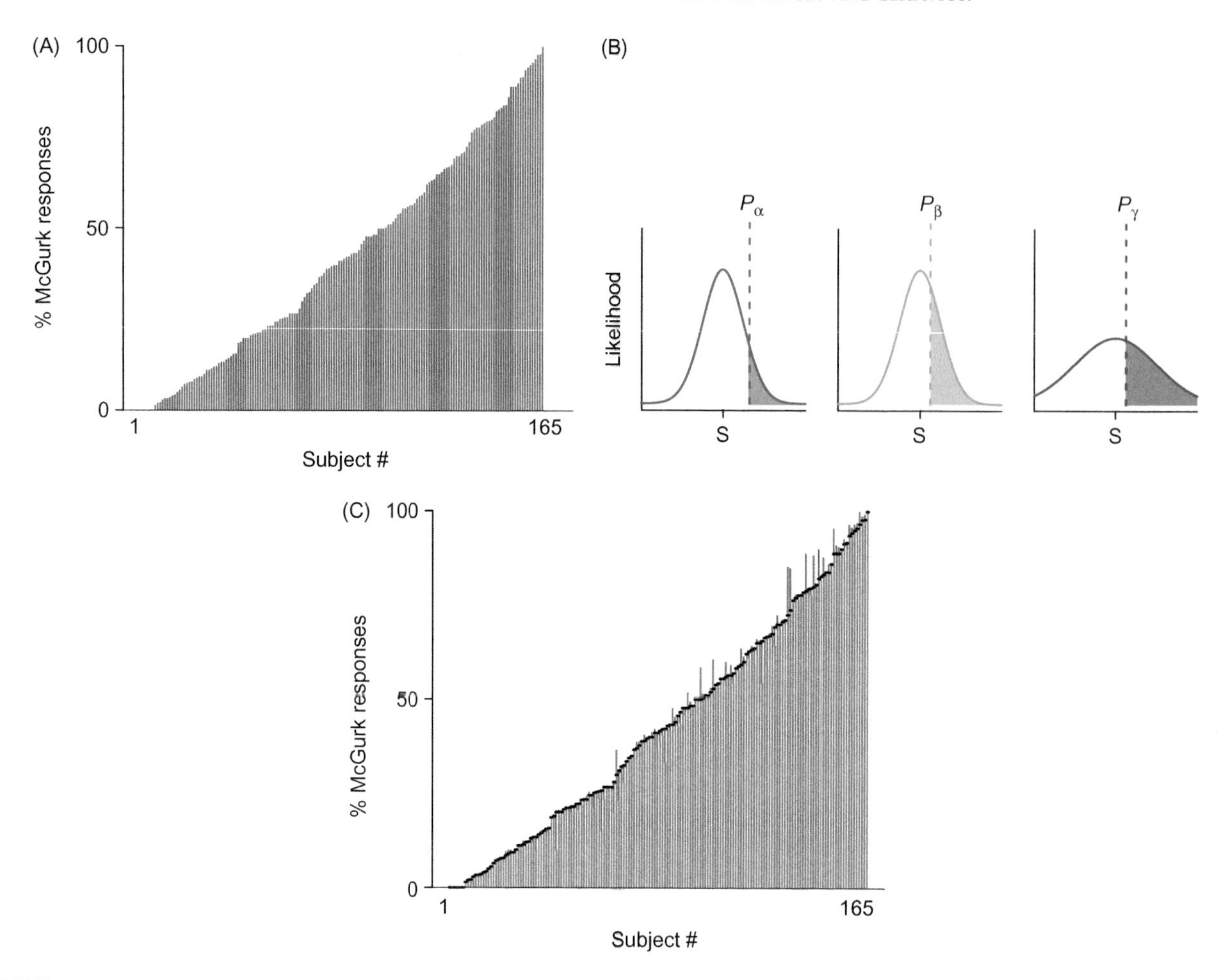

FIGURE 42.5 Intersubject variability in the McGurk effect. (A) Each bar represents a single subject ($n = 165$). The height of each bar corresponds to the proportion of McGurk fusion responses (e.g., percept of /da/ when presented with visual /ga/ and auditory /ba/) out of all responses when the subject was presented with 10 repetitions of up to 14 different McGurk stimuli. Subjects are ordered by rate of McGurk fusion responses. (B) A model of McGurk perception in which each stimulus is fit with a given strength (tick mark labeled S on x-axis) and each subject is fit with a given threshold (vertical dashed line). In each trial, the strength of the stimulus is measured by the subject with a process that includes sensory noise (width of Gaussian curve corresponds to amount of sensory noise). This leads to a distribution of measured strengths across trials. In those trials in which the measured strength exceeds threshold, the McGurk effect is perceived by the subject (shaded regions). The left plot shows a subject with low susceptibility (P_α): the subject threshold is greater than the stimulus strength, leading to few trials on which the illusion is perceived (small red shaded area). A second subject (P_β, middle plot) has the same sensory noise as P_α (width of yellow curve same as width of red curve) but a lower threshold (dashed yellow line to the left of dashed red line along x-axis), leading to increased perception of the McGurk effect (yellow shaded region) for the identical stimulus. A third subject (P_γ, right plot) has greater sensory noise (width of green curve greater) but the same threshold as P_β, leading to increased perception of the McGurk effect (green shaded region). (C) The model is fit to the behavioral data from each subject shown in (A). Each black point shows the behavioral data (mean proportion of fusion responses across all stimuli). Each gray bar shows the model prediction for that subject. Each red bar shows the prediction error for that subject.

stimulus's strength. In any given trial, the subject does not perceive the actual strength of the stimulus, only the subject's noisy internal representation of it. The sensory noise term describes how much variability is added to the (true, external) stimulus strength when creating this internal representation. The sensory noise term is critical to account for the observation that the exact same stimulus can evoke different reports on different trials. If subjects perceived the true stimulus strength, then for any given stimulus they should either always report the illusion (if the stimulus strength falls above their threshold) or never report it (if the stimulus strength falls below their threshold). Because subjects instead perceive a corrupted version of stimulus strength, in some trials, the perceived strength lies above the subject's fusion threshold, and the illusion is reported. In other trials, the perceived strength lies below the subject's fusion threshold and the illusion is not reported. Using these simple sets of parameters (two for each subject and one for each stimulus), we can accurately reproduce the pattern of observed reports of the illusion. Even more importantly, the model can predict responses to stimuli that subjects have never before seen.

42.4 NEURAL SUBSTRATES OF THE McGURK EFFECT

Different individuals with normal hearing and vision report very different percepts when presented with the same McGurk stimulus. The source of this variability must lie within the central nervous system. To examine the neural underpinnings of this behavioral variability, we conducted three parallel BOLD fMRI studies in young adults (Nath & Beauchamp, 2012), children (Nath, Fava, & Beauchamp, 2011), and older adults. Subjects were divided into those who only weakly perceived the McGurk effect and subjects who more strongly perceived it in a behavioral pre-test performed outside the scanner (Figure 42.6A). Then, the STS was identified using an independent set of fMRI data that used different stimuli (audiovisual words) to eliminate bias in voxel selection (Simmons, Bellgowan, & Martin, 2007).

After the STS was identified, responses in the STS to individual McGurk syllables were measured. Only passive viewing of the syllables was used; subjects did not make a behavioral response during MR scanning. Even though the physical stimulus presented to the two groups was identical, subjects who strongly

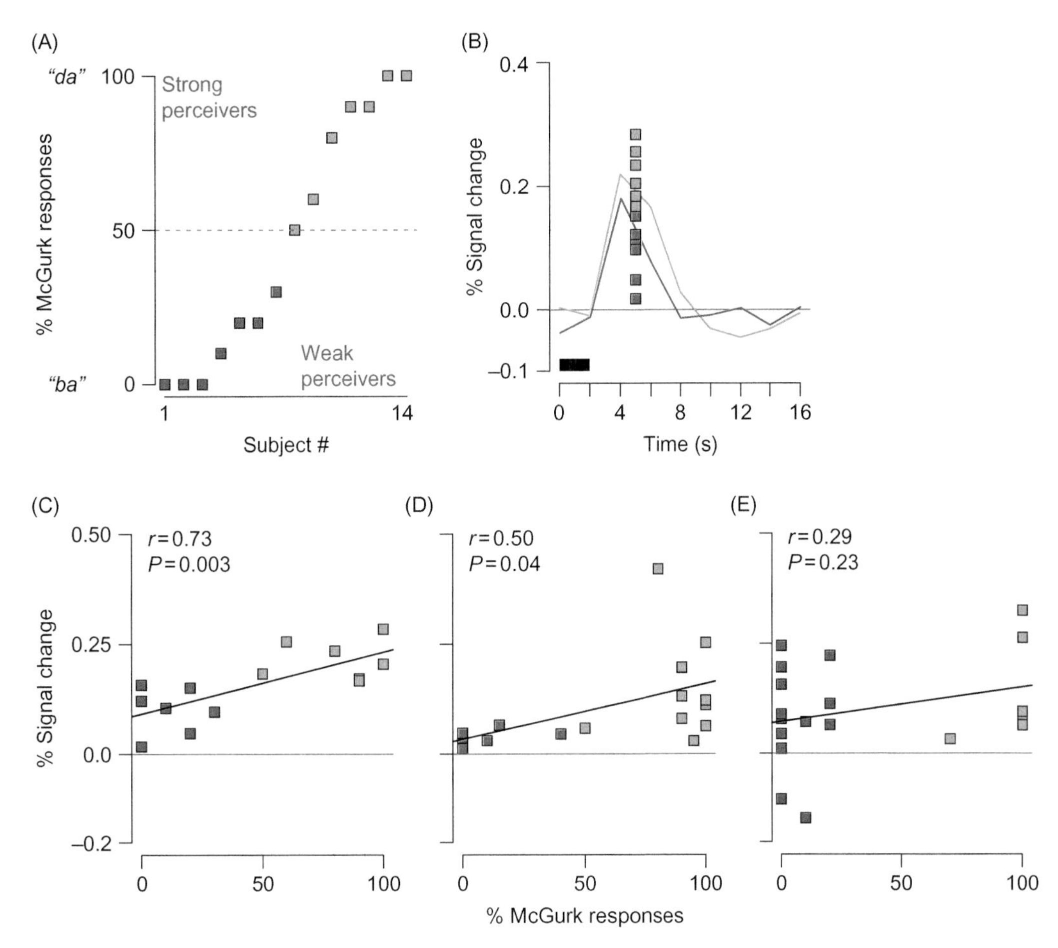

FIGURE 42.6 Intersubject variability in neural responses to the McGurk effect. (A) Young adult subjects ($n = 14$) were ordered by proportion of McGurk fusion responses and divided into strong perceivers of the McGurk effect (green squares, proportion of McGurk fusion responses ≥ 0.50) and weak perceivers of the McGurk effect (red squares, proportion of McGurk fusion responses <0.50). (B) The STS was localized in each young adult subject using independent stimuli (auditory and visual words). Then, the BOLD fMRI response to single McGurk syllables (black bar) was measured in the STS ROI. The average BOLD response across all strong perceivers (green curve) and across all weak perceivers (red curve) is shown. Each square shows the peak response of a single subject who strongly perceived the effect (green square) or weakly perceived it (red square). (C) Each young adult subject's McGurk fusion rate (x-axis) was correlated with the BOLD fMRI signal change in the STS ROI (y-axis). Error bars show the intersubject variability in the fMRI response across trials for each subject (green squares are strong perceivers, red squares are weak perceivers). (D) Children aged 5–12 years were tested using a similar protocol (the STS was localized with independent localizer; subjects were divided into strong and weak perceivers based on their behavioral responses to the McGurk effect; the fMRI signal in the STS ROI was plotted against the proportion of fusion responses; error bars show within-subject variability). (E) Older adult subjects aged 5–12 years were tested using a similar protocol (the STS was localized with independent localizer; subjects were divided into strong and weak perceivers based on their behavioral responses to the McGurk effect; the fMRI signal in the STS ROI was plotted against the proportion of fusion responses).

perceived the McGurk effect had greater BOLD responses to McGurk stimuli. There was a positive correlation between how often they perceived the illusion and the amplitude of the BOLD fMRI response of the STS. No other brain regions showed such a correlation.

A possible explanation for this finding is that neurons in the STS perform the computations necessary to integrate the auditory and visual components of the stimulus and create the McGurk illusion. If these neurons are not active, then the illusory percept is not created. It is important to emphasize that this does not indicate anything defective about audiovisual integration in subjects who do or do not perceive the McGurk illusion. In fact, the STS in both groups responded identically to congruent audiovisual syllables. Rather, subjects may be differentially sensitive to the incongruent nature of the McGurk stimuli. If subjects infer that the auditory and visual components of the stimulus are likely to have originated from different speakers, then the optimal solution is to respond with only one component of the stimulus (typically the auditory component, because the auditory modality is usually more reliable for speech perception). If subjects infer that the auditory and visual compoments of the stimulus are more likely to have originated from the same speaker, then the optimal solution is to fuse the components and respond with the illusory percept, because this percept is most consistent with both the auditory and visual components. This process of Bayesian causal inference may account for audiovisual speech perception under conditions in which the synchrony between the auditory and visual modalities varies (Magnotti, Ma, & Beauchamp, 2013).

In children aged 5 to 12 years, the correlation between STS responses and McGurk susceptibility was weaker than in adults but still significant (Figure 42.6B). In addition to the STS, the fusiform face area (FFA) on the ventral surface of the temporal lobe was also correlated with McGurk susceptibility.

In older adults aged 53 to 75 years, the correlation between STS responses and McGurk susceptibility was not significant. One possible explanation for the difference is that older adults had significantly greater intrasubject variability than younger adults. Each subject viewed the same McGurk stimuli multiple times, allowing for an estimate both of the mean BOLD fMRI response (used to correlate with the McGurk susceptibility for that individual) and the variability of the BOLD fMRI response. This variability was much larger in older adults, possibly obscuring a true correlation between the mean response and susceptibility. A second possible explanation is that most of the older subjects did not perceive the illusion. Because the number of weak and strong perceivers was unbalanced, this reduced the power available to detect differences between them. In the younger adult study, this imbalance was avoided by pre-screening the subjects so that equal numbers of strong and weak perceivers were included.

The observed positive correlation between left STS activity and perception of the McGurk effect (and no significant correlation in any other brain region) is provocative, but it does not demonstrate that activity in the STS is necessary for the illusion. To demonstrate this, we interfered with activity in the STS using transcranial magnetic stimulation (TMS).

The STS is found in every human brain, but there is substantial variability in its anatomical configuration, as well as in the anatomical location of the functionally defined subregion within the STS that responds to auditory and visual speech (although it is often found at the bend where the STS turns upward into posterior lobe). Most TMS studies stimulate the brain based on skull landmarks (e.g., 2 cm lateral to the occipital pole) or on functional landmarks (e.g., 2 cm anterior to motor cortex). However, the STS is not easily localized with either of these techniques. Therefore, we used fMRI to identify the multisensory region of the STS in each subject (Beauchamp, Nath, & Pasalar, 2010). Then, frameless stereotaxic registration was used to position the fMRI coil over the STS. Single pulses of TMS (<1 ms duration) were delivered as subjects viewed McGurk and control stimuli (Figure 42.7). Subjects were pre-screened so that only those who strongly perceived the illusion were tested. Without TMS, the McGurk percept was reported on nearly every trial. However, when TMS was delivered to the STS, the illusion was reported in only approximately half of trials. Perception of other auditory or visual stimuli was not affected. Stimulating cortex outside the STS did not produce any change in the illusion, demonstrating that it was a specific result of STS stimulation and not related to the auditory click produced by the TMS coil or the scalp tactile stimulation induced by the pulse. The "virtual lesions" introduced with TMS interfered with McGurk only in a temporal window just before and after onset of the auditory stimulus (Figure 42.7C), similar to the window in which electrocorticographic activity is observed. The impairment in perception of the McGurk effect caused by TMS stimulation of the STS lasted for <1 s. Across multiple trials of TMS, there were no appreciable changes: TMS interfered with McGurk perception as effectively at the end of the experiment as it did at the beginning. However, permanent damage to the STS might engage cortical mechanisms for plasticity.

To examine this possibility, we tested a patient, SJ, who experienced a stroke that ablated the entirety of her left posterior STS (Figure 42.8); her right hemisphere was unaffected (Baum, Martin, Hamilton, & Beauchamp, 2012). She was severely impaired immediately after the brain injury but underwent extensive

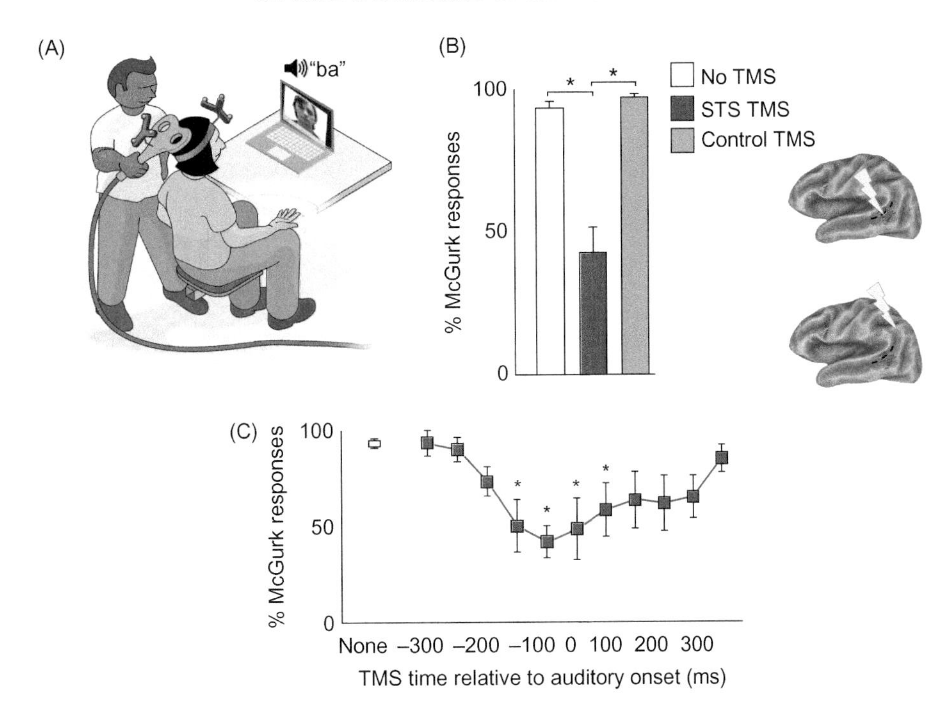

FIGURE 42.7 Disruption of the McGurk Effect with TMS. (A) A TMS coil was used to stimulate the STS or a control location as subjects viewed a McGurk stimulus and reported their percept. fMRI was used to identify target regions in each subject and frameless stereotaxy was used to position the TMS coil over the identified target. A single pulse of TMS was applied in each trial. (B) Behavioral responses to McGurk stimuli with no TMS, with STS TMS, and with control TMS. The lightning bold icon positioned over the inflated cortical surface model and fMRI activation map of a sample subject shows the location of the applied TMS. Figure adapted from Beauchamp et al. (2010) with permission from the publisher. (C) The time at which the single pulse of TMS was delivered was varied. With no TMS, the McGurk effect was always perceived. Maximal interference with the McGurk effect was observed when the TMS was delivered at times near the onset of the auditory stimulus. *Significant reduction in McGurk perception, $P < 0.05$.

daily rehabilitation in the 5 years after her stroke. After recovery, she reported good ability to understand speech but reported that it was very effortful. At testing, her ability to understand auditory syllables was similar to that of age-matched controls. Surprisingly, she also reported experiencing the McGurk illusion. Because her left STS was ablated, we performed a BOLD fMRI study to try and understand the brain circuits active when she was presented with audiovisual speech. No activity was observed in here left STS, but her right STS was highly active when she was presented with speech, including words, congruent syllables, and McGurk syllables. The volume of cortex active in her right STS exceeded that of any age-matched control. This suggests a scenario in which the right STS reorganized to perform the multisensory integration performed in the left STS by normal individuals. This, in turn, suggests substantial plasticity in the circuit for perceiving audiovisual speech. A promising direction for future research is exploring the training parameters that are most effective at modifying the multisensory speech perception network.

Another important direction for future research is a better understanding of how the auditory cortex, visual cortex, and STS interact with other brain areas important for speech perception. A subcortical structure, the superior and inferior colliculus (Wallace, Wilkinson, & Stein, 1996), is a brainstem hub for integrating auditory and visual information. Although the role of the colliculus in audiovisual speech perception is unknown, a lesion study suggested that damage to subcortical structures can impair audiovisual speech perception (Chamoun, Takashima, & Yoshor, 2014). Within the cortex, Broca's area is critical for linguistic processing (Nagy, Greenlee, & Kovacs, 2012; Sahin, Pinker, Cash, Schomer, & Halgren, 2009) and may play a role in audiovisual integration and speech comprehension, especially for noisy speech (Abrams et al., 2012; Noppeney, Ostwald, & Werner, 2010; but see Hickok, Costanzo, Capasso, & Miceli, 2011). A common finding is activity in Broca's area for producing (or repeating), but not perceiving, words (Towle et al., 2008). Another important brain area to consider is the FFA, which plays an important role in face processing (Engell & McCarthy, 2010; Grill-Spector & Malach, 2004) and has also been implicated in the processing of voices (von Kriegstein, Kleinschmidt, & Giraud, 2006). Recent evidence from probabilistic tractography suggests that there exist anatomical connections between FFA and STS (Blank, Anwander, & von Kriegstein, 2011). High-resolution fMRI and

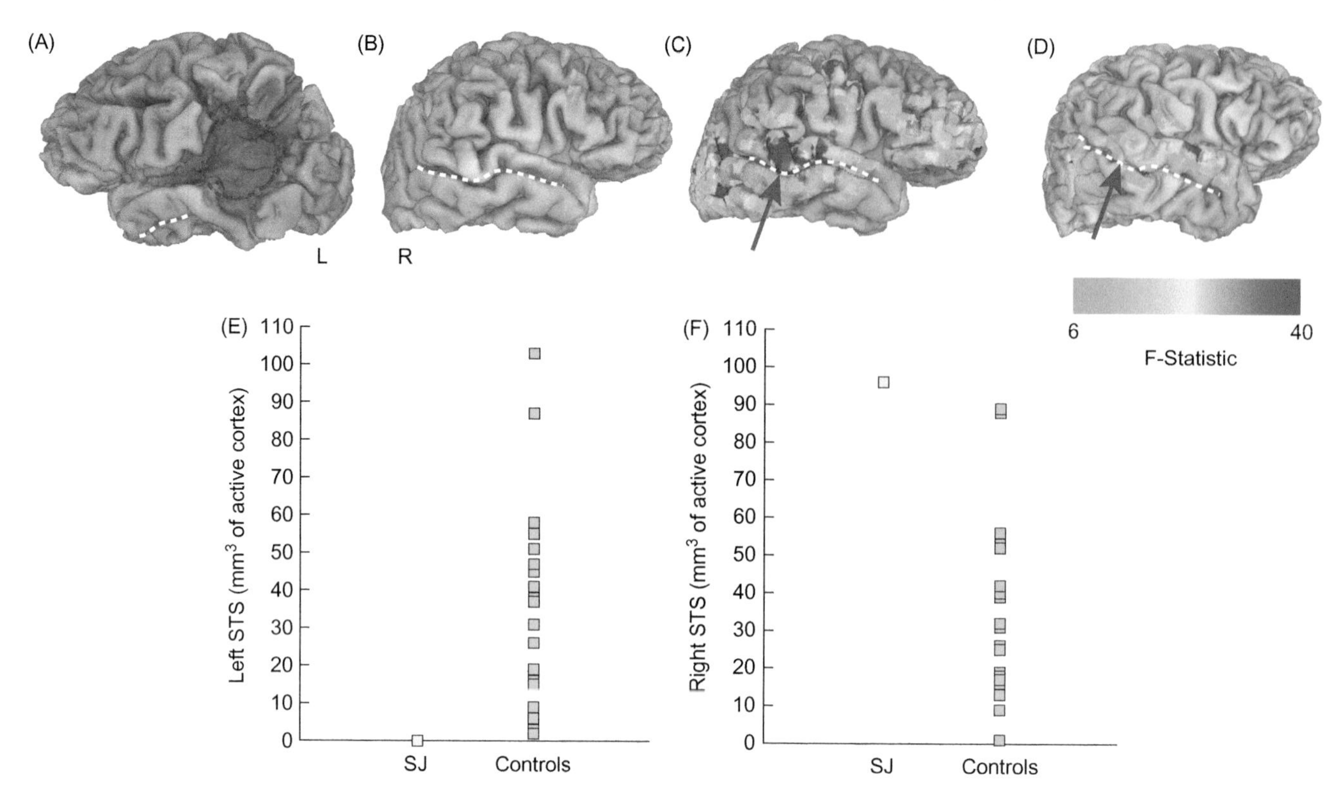

FIGURE 42.8 Plasticity of the neural substrates of audiovisual speech perception after brain injury. (A) Cortical surface model created from a T1-weighted MRI of patient SJ 5 years after stroke showing extensive lesion to left posterior STS (red circle). Dashed white line shows the location of STS; only the most anterior portion is spared. Adapted from Baum et al. (2012) with permission from the publisher. (B) Cortical surface model of right hemisphere of patient SJ showing undamaged STS (dashed white line). (C) Cortical surface model of right hemisphere of SJ showing BOLD fMRI activity in response to audiovisual speech. Color scale shows significance of activation. Red arrow highlights extensive activity in posterior STS. (D) BOLD fMRI responses to audiovisual speech in the right hemisphere of a healthy age-matched control. Red arrow highlight paucity of activity in posterior STS (same color scale for (C) and (D)). (E) Volume of cortex in left posterior STS responsive to audiovisual speech in patient SJ (yellow square) and 23 healthy age-matched controls (blue squares). (F) Volume of cortex in right posterior STS responsive to audiovisual speech in patient SJ (yellow square) and 23 healthy age-matched controls (blue squares).

electrocorticography may also make it possible to record responses from small pools of neurons that respond to auditory or visual speech (Chang et al., 2010), allowing a detailed understanding of the neural circuits underlying audiovisual speech perception.

Acknowledgments

This research was supported by NIH 5R01NS065395 and would not have been possible without the research subjects who provided the data and the scientists who made sense of it (in alphabetical order): Sarah Baum, Eswen Fava, Cris Hamilton, Haley Lindsay, Weiji Ma, John Magnotti, Debshila Basu Mallick, Randi Martin, Cara Miekka, Audrey Nath, Xiaomei Pei, Inga Schepers, Muge Ozker, Ping Sun, Siavash Pasalar, Nafi Yasar, and Daniel Yoshor.

References

Abrams, D. A., Ryali, S., Chen, T., Balaban, E., Levitin, D. J., & Menon, V. (2012). Multivariate activation and connectivity patterns discriminate speech intelligibility in Wernicke's, Broca's, and Geschwind's areas. *Cerebral Cortex*. Available from: http://dx.doi.org/doi:10.1093/cercor/bhs165.

Allison, T., Puce, A., Spencer, D. D., & McCarthy, G. (1999). Electrophysiological studies of human face perception. I: Potentials generated in occipitotemporal cortex by face and non-face stimuli. *Cerebral Cortex*, *9*(5), 415–430.

Baum, S. H., Martin, R. C., Hamilton, A. C., & Beauchamp, M. S. (2012). Multisensory speech perception without the left superior temporal sulcus. [Research Support, N.I.H., Extramural Research Support, U.S. Gov't, Non-P.H.S.]. *NeuroImage*, *62*(3), 1825–1832. Available from: http://dx.doi.org/doi:10.1016/j.neuroimage.2012.05.034.

Beauchamp, M. S., Lee, K. E., Argall, B. D., & Martin, A. (2004). Integration of auditory and visual information about objects in superior temporal sulcus. *Neuron*, *41*(5), 809–823.

Beauchamp, M. S., Nath, A. R., & Pasalar, S. (2010). fMRI-Guided transcranial magnetic stimulation reveals that the superior temporal sulcus is a cortical locus of the McGurk effect. *Journal of Neuroscience*, *30*(7), 2414–2417. Available from: http://dx.doi.org/10.1523/JNEUROSCI.4865-09.2010.

Beck, J. M., Ma, W. J., Kiani, R., Hanks, T., Churchland, A. K., Roitman, J., et al. (2008). Probabilistic population codes for Bayesian decision making. *Neuron*, *60*(6), 1142–1152.

Bernstein, L. E., Auer, E. T., Jr., Wagner, M., & Ponton, C. W. (2008). Spatiotemporal dynamics of audiovisual speech processing. [Research Support, N.I.H., Extramural Research Support, U.S. Gov't, Non-P.H.S.]. *NeuroImage*, *39*(1), 423–435. Available from: http://dx.doi.org/doi:10.1016/j.neuroimage.2007.08.035.

Bernstein, L. E., Auer, J. E. T., & Takayanagi, S. (2004). Auditory speech detection in noise enhanced by lipreading. *Speech Communication*, *44*(1–4), 5–18.

Blank, H., Anwander, A., & von Kriegstein, K. (2011). Direct structural connections between voice- and face-recognition areas. *Journal of Neuroscience*, *31*(36), 12906–12915. Available from: http://dx.doi.org/10.1523/JNEUROSCI.2091-11.2011.

Blank, H., & von Kriegstein, K. (2013). Mechanisms of enhancing visual-speech recognition by prior auditory information. *Neuroimage*, *65*, 109–118. Available from: http://dx.doi.org/doi:10.1016/j.neuroimage.2012.1009.1047. Epub 2012 Sep 1027.

Canolty, R. T., Edwards, E., Dalal, S. S., Soltani, M., Nagarajan, S. S., Kirsch, H. E., et al. (2006). High gamma power is phase-locked to theta oscillations in human neocortex. *Science*, *313*(5793), 1626–1628.

Canolty, R. T., Soltani, M., Dalal, S. S., Edwards, E., Dronkers, N. F., Nagarajan, S. S., et al. (2007). Spatiotemporal dynamics of word processing in the human brain. *Front Neuroscience*, *1*(1), 185–196.

Chamoun, R., Takashima, M., & Yoshor, D. (2014). Endoscopic extracapsular dissection for resection of pituitary macroadenomas: Technical note. *Journal of Neurological Surgery. Part A: Central European Neurosurgery*, *75*(1), 48–52. Available from: http://dx.doi.org/doi:10.1055/s-0032-1326940.

Chang, E. F., Rieger, J. W., Johnson, K., Berger, M. S., Barbaro, N. M., & Knight, R. T. (2010). Categorical speech representation in human superior temporal gyrus. [Research Support, N.I.H., Extramural Research Support, Non-U.S. Gov't]. *Nature Neuroscience*, *13*(11), 1428–1432. Available from: http://dx.doi.org/doi:10.1038/nn.2641.

Crone, N. E., Miglioretti, D. L., Gordon, B., & Lesser, R. P. (1998). Functional mapping of human sensorimotor cortex with electrocorticographic spectral analysis. II. Event-related synchronization in the gamma band. *Brain*, *121*(Pt 12), 2301–2315.

Driver, J., & Noesselt, T. (2008). Multisensory interplay reveals crossmodal influences on 'sensory-specific' brain regions, neural responses, and judgments. *Neuron*, *57*(1), 11–23.

Engell, A. D., & McCarthy, G. (2010). Selective attention modulates face-specific induced gamma oscillations recorded from ventral occipitotemporal cortex. *Journal of Neuroscience*, *30*(26), 8780–8786.

Friston, K. J. (2009). Modalities, modes, and models in functional neuroimaging. *Science*, *326*(5951), 399–403.

Grill-Spector, K., & Malach, R. (2004). The human visual cortex. *Annual Review of Neuroscience*, *27*, 649–677.

Hickok, G., Costanzo, M., Capasso, R., & Miceli, G. (2011). The role of Broca's area in speech perception: Evidence from aphasia revisited. [Research Support, N.I.H., Extramural Research Support, Non-U.S. Gov't]. *Brain and Language*, *119*(3), 214–220. Available from: http://dx.doi.org/doi:10.1016/j.bandl.2011.08.001.

Lachaux, J. P., Axmacher, N., Mormann, F., Halgren, E., & Crone, N. E. (2012). High-frequency neural activity and human cognition: Past, present and possible future of intracranial EEG research. *Progress in Neurobiology*, *98*(3), 279–301. Available from: http://dx.doi.org/10.1016/j.pneurobio.2012.06.008.

Lee, H., & Noppeney, U. (2011). Physical and perceptual factors shape the neural mechanisms that integrate audiovisual signals in speech comprehension. *Journal of Neuroscience*, *31*(31), 11338–11350.

Logothetis, N. K., Kayser, C., & Oeltermann, A. (2007). *In vivo* measurement of cortical impedance spectrum in monkeys: Implications for signal propagation. *Neuron*, *55*(5), 809–823.

Magnotti, J. F., Ma, W. J., & Beauchamp, M. S. (2013). Causal inference of asynchronous audiovisual speech. *Frontiers in Psychology*, *4*, 798. Available from: http://dx.doi.org/doi:10.3389/fpsyg.2013.00798.

McGurk, H., & MacDonald, J. (1976). Hearing lips and seeing voices. *Nature*, *264*(5588), 746–748.

Miller, L. M., & D'Esposito, M. (2005). Perceptual fusion and stimulus coincidence in the cross-modal integration of speech. *The Journal of Neuroscience*, *25*(25), 5884–5893.

Nagy, K., Greenlee, M. W., & Kovacs, G. (2012). The lateral occipital cortex in the face perception network: an effective connectivity study. *Frontiers in Psychology*, *3*, 141. Available from: http://dx.doi.org/doi:10.3389/fpsyg.2012.00141.

Nath, A. R., & Beauchamp, M. S. (2011). Dynamic changes in superior temporal sulcus connectivity during perception of noisy audiovisual speech. *The Journal of Neuroscience*, *31*(5), 1704–1714.

Nath, A. R., & Beauchamp, M. S. (2012). A neural basis for interindividual differences in the McGurk effect, a multisensory speech illusion. [Research Support, N.I.H., Extramural Research Support, U.S. Gov't, Non-P.H.S.]. *NeuroImage*, *59*(1), 781–787. Available from: http://dx.doi.org/doi:10.1016/j.neuroimage.2011.07.024.

Nath, A. R., Fava, E. E., & Beauchamp, M. S. (2011). Neural correlates of interindividual differences in children's audiovisual speech perception. [Research Support, N.I.H., Extramural Research Support, U.S. Gov't, Non-P.H.S.]. *Journal of Neuroscience*, *31*(39), 13963–13971. Available from: http://dx.doi.org/doi:10.1523/JNEUROSCI.2605-11.2011.

Nir, Y., Fisch, L., Mukamel, R., Gelbard-Sagiv, H., Arieli, A., Fried, I., et al. (2007). Coupling between neuronal firing rate, gamma LFP, and BOLD fMRI is related to interneuronal correlations. *Current Biology*, *17*(15), 1275–1285.

Noppeney, U., Josephs, O., Hocking, J., Price, C. J., & Friston, K. J. (2008). The effect of prior visual information on recognition of speech and sounds. *Cerebral Cortex*, *18*(3), 598–609.

Noppeney, U., Ostwald, D., & Werner, S. (2010). Perceptual decisions formed by accumulation of audiovisual evidence in prefrontal cortex. [Research Support, Non-U.S. Gov't]. *The Journal of Neuroscience: The Official Journal of the Society for Neuroscience*, *30*(21), 7434–7446. Available from: http://dx.doi.org/doi:10.1523/JNEUROSCI.0455-10.2010.

Okada, K., Venezia, J. H., Matchin, W., Saberi, K., & Hickok, G. (2013). An fMRI Study of audiovisual speech perception reveals multisensory interactions in auditory cortex. *PLoS*, *8*(6), e68959 Print 62013.

Pei, X., Leuthardt, E. C., Gaona, C. M., Brunner, P., Wolpaw, J. R., & Schalk, G. (2011). Spatiotemporal dynamics of electrocorticographic high gamma activity during overt and covert word repetition. *Neuroimage*, *54*(4), 2960–2972.

Powers, A. R., III, Hevey, M. A., & Wallace, M. T. (2012). Neural correlates of multisensory perceptual learning. *Journal of Neuroscience*, *32*(18), 6263–6274. Available from: http://dx.doi.org/10.1523/JNEUROSCI.6138-11.2012.

Ray, S., & Maunsell, J. H. (2011). Different origins of gamma rhythm and high-gamma activity in macaque visual cortex. *PLoS Biology*, *9*(4), e1000610.

Ross, L. A., Saint-Amour, D., Leavitt, V. M., Javitt, D. C., & Foxe, J. J. (2007a). Do you see what I am saying? Exploring visual enhancement of speech comprehension in noisy environments. *Cerebral Cortex*, *17*(5), 1147–1153. Available from: http://dx.doi.org/doi:10.1093/cercor/bhl024.

Ross, L. A., Saint-Amour, D., Leavitt, V. M., Javitt, D. C., & Foxe, J. J. (2007b). Do you see what I am saying? Exploring visual enhancement of speech comprehension in noisy environments. [Research Support, N.I.H., Extramural]. *Cerebral Cortex*, *17*(5), 1147–1153. Available from: http://dx.doi.org/doi:10.1093/cercor/bhl024.

Ross, L. A., Saint-Amour, D., Leavitt, V. M., Molholm, S., Javitt, D. C., & Foxe, J. J. (2007). Impaired multisensory processing in schizophrenia: Deficits in the visual enhancement of speech comprehension under noisy environmental conditions. [Research Support, N.I.H., Extramural]. *Schizophrenia Research*, *97*(1–3), 173–183. Available from: http://dx.doi.org/doi:10.1016/j.schres.2007.08.008.

Sahin, N. T., Pinker, S., Cash, S. S., Schomer, D., & Halgren, E. (2009). Sequential processing of lexical, grammatical, and phonological information within Broca's area. *Science*, *326*(5951), 445–449.

Schepers, I. M., Hipp, J. F., Schneider, T. R., Roder, B., & Engel, A. K. (2012). Functionally specific oscillatory activity correlates between visual and auditory cortex in the blind. *Brain*, *135*(Pt 3), 922–934.

Schepers, I. M., Schneider, T. R., Hipp, J. F., Engel, A. K., & Senkowski, D. (2013). Noise alters beta-band activity in superior temporal cortex during audiovisual speech processing. *Neuroimage*, *70*, 101–112. Available from: http://dx.doi.org/10.1016/j.neuroimage.2012.11.066.

Simmons, W. K., Bellgowan, P. S., & Martin, A. (2007). Measuring selectivity in fMRI data. *Nature Neuroscience*, *10*(1), 4–5.

Suh, M. W., Lee, H. J., Kim, J. S., Chung, C. K., & Oh, S. H. (2009). Speech experience shapes the speechreading network and subsequent deafness facilitates it. *Brain*, *132*(Pt 10), 2761–2771. Available from: http://dx.doi.org/10.1093/brain/awp159.

Sumby, W. H., & Pollack, I. (1954). Visual contribution to speech intelligibility in noise. *Journal of the Acoustical Society of America*, *26*(2), 212–215.

Towle, V. L., Yoon, H. A., Castelle, M., Edgar, J. C., Biassou, N. M., Frim, D. M., et al. (2008). ECoG gamma activity during a language task: Differentiating expressive and receptive speech areas. *Brain*, *131*(Pt 8), 2013–2027.

Tye-Murray, N., Sommers, M. S., & Spehar, B. (2007). Audiovisual integration and lipreading abilities of older adults with normal and impaired hearing. *Ear and Hearing*, *28*(5), 656–668. Available from: http://dx.doi.org/10.1097/AUD.0b013e31812f7185.

von Kriegstein, K., Kleinschmidt, A., & Giraud, A. L. (2006). Voice recognition and cross-modal responses to familiar speakers' voices in prosopagnosia. *Cerebral Cortex*, *16*(9), 1314–1322. Available from: http://dx.doi.org/10.1093/cercor/bhj073.

Wallace, M. T., Wilkinson, L. K., & Stein, B. E. (1996). Representation and integration of multiple sensory inputs in primate superior colliculus. *Journal of Neurophysiology*, *76*(2), 1246–1266.

Wilson, S. M., Molnar-Szakacs, I., & Iacoboni, M. (2008). Beyond superior temporal cortex: Intersubject correlations in narrative speech comprehension. *Cerebral Cortex*, *18*(1), 230–242. Available from: http://dx.doi.org/10.1093/cercor/bhm049.

CHAPTER

43

Neurobiology of Statistical Information Processing in the Auditory Domain

Uri Hasson[1] *and Pascale Tremblay*[2,3]

[1]Center for Mind and Brain Sciences (CIMeC), University of Trento, Mattarello (TN), Italy; [2]Centre de Recherche de l'Institut Universitaire en Santé Mentale de Québec, Québec City, QC, Canada; [3]Département de Réadaptation, Faculté de Médecine, Université Laval, Québec City, QC, Canada

43.1 INTRODUCTION

In this chapter, we provide an overview of the literature on the neurobiology of statistical information processing that is relevant for studies of language processing. We intend this chapter for those interested in understanding how the coding of statistics between sublexical (nonsemantic) speech elements may impact language acquisition and online speech processing. For this reason, we focus particularly on studies speaking to the sublexical level (syllables), but we also addresses the processing of statistics within auditory nonspeech stimuli.

As detailed in this chapter, there is increasing evidence that the human brain implements computations that allow it to understand the environment by extracting and analyzing its regularities to derive meaning, simplify the representation of inputs via categorization or compression, and predict upcoming events. The auditory environment is particularly complex, presenting humans with a seemingly infinite combination of sounds with an immensely rich spectral diversity, originating from a wide variety of sources. Among these sounds are speech sounds, that is, syllables and phonemes. Speech sounds occur within rapid and continuous streams that must be parsed into their constituent units for meaning to be extracted. Speech streams are sequentially organized based on complex language-specific (i.e., phonotactic) constraints. These constraints, whose delineation is the focus of linguistic approaches (e.g., Optimality theory; see McCarthy, 2001), are manifested, at the surface level, in language-specific *statistics*. One such statistic is "transition probability" (TP), which refers to the conditional probability of transitioning from one syllable to the immediately successive one, P(syllable2 | syllable1). Importantly, syllables with high TPs tend to form words, whereas those with lower TPs mark word boundaries. Because, unlike written language, the speech stream does not contain invariant cues marking word boundaries (Cole & Jakimik, 1980; Lehiste & Shockey, 1972), access to statistical information, in particular TP, is considered critical for learning how to segment continuous speech into syllables and words, especially during early childhood when word boundaries are still unknown. There is a large behavioral literature detailing the statistical abilities of infants and children, but also adults, all of whom are sensitive to TP in speech (Newport & Aslin, 2004; Pelucchi, Hay, & Saffran, 2009a, 2009b; Pena, Bonatti, Nespor, & Mehler, 2002; Saffran, Aslin, & Newport, 1996; Saffran, Johnson, Aslin, & Newport, 1999) as well as in nonspeech auditory and visual stimuli (Fiser & Aslin, 2001, 2002; Kirkham, Slemmer, & Johnson, 2002; Saffran et al., 1999; Saffran, Pollak, Seibel, & Shkolnik, 2007).

Importantly, statistics within the auditory domain support more than just language acquisition. In early and late adulthood, statistical information can be used for predicting upcoming syllables and words or for disambiguating partly heard words, especially in contexts in which the speech signal is degraded, such as in noisy environments and for people with hearing loss. It is well-known that during speech comprehension the speech system seeks to identify the target lexical item as quickly as possible (Gaskell & Marslen-Wilson, 1997). While processing the auditory input, the

Neurobiology of Language. DOI: http://dx.doi.org/10.1016/B978-0-12-407794-2.00043-2

cohort of lexical items that form possible lexical continuations becomes accessible, and the lexical item is identified as soon as the "uniqueness point" is reached, that is, the point at which only one continuation is possible and at which it can be discriminated from similar words in the "mental lexicon."[1] This process is further sensitive to the statistical distribution of potential complements: the uncertainty of possible word continuations (the potential cohort) near the recognition point, as quantified by the cohort's entropy, is negatively correlated with word identification, suggesting that the set of potential complements is made accessible during the word comprehension itself (Wurm, Ernestus, Schreude, & Baayen, 2006). Given that interpretations are sought prior to word completion, a probabilistic model of potential completions can optimize recognition by differentially activating more probable completions. Recent work supports the premise that such predictions are made prior to reaching the uniqueness point (Gagnepain, Henson, & Davis, 2012). Beyond accounting for online processing, it has been shown that adults with better capacities for representing input statistics have more efficient language processing capacities, suggesting a tight link between the two mechanisms (Conway, Bauernschmidt, Huang, & Pisoni, 2010; Misyak & Christiansen, 2012).

The studies we review here focus on the neurobiology of statistical information processing. These studies draw on three research traditions: information theory, artificial grammar learning (AGL), and perceptual learning. Therefore, they use different terminologies. Studies grounded in information theory differentiate "regular" from "irregular" or "random" inputs. Here, regular inputs are generated by a process with some level of constraints (e.g., a first- or second-order Markov process[2]) in which the transitions between tokens is not uniform in the ordered condition but is uniform in the random condition. For instance, given a grammar of N tokens, in the random condition transition probabilities between tokens will be $1/N$ uniformly for all token-pair transitions, whereas the regular conditions will increase the probability of some transitions and decrease the probability of others. Studies using AGL paradigms, which are grounded in formal learning theories, often refer to the generating process as a transition network and examine responses to sequences that could or could not be generated by the learned grammar. Many of the grammars used in the AGL literature are of the simplest type (regular grammars) that can be represented via a Markov process. The third body of literature, grounded in perceptual learning, uses manipulations in which random token-sequences are contrasted with sequences in which some tokens are always concatenated in a fixed order to form "words." It is important to note that both the random and word-forming conditions can be described by a first-order Markov process: the random condition has a uniform distribution of transitions whereas the word-forming condition contains several cells with a 100% TP. We further note that these latter studies are different from those grounded in AGL or information theory in that once the "words" are decoded, learners benefit from numerous points in the sequence that completely lack uncertainty (particularly when each word begins with a unique syllable). To illustrate, if words are four syllables long, then after the words have been learned, identifying a word-initial syllable will uniquely determine the next three syllables for the listener. That is, including "words" introduces a measure of determinism that is not present in other studies of statistical processing, and there is evidence that this results in chunking of the words into single units, with less attention given to those syllables in the sequence that are predictable with absolute certainty (Astheimer & Sanders, 2011; Buiatti, Pena, & Dehaene-Lambertz, 2009).

We also emphasize that the findings we review in this chapter derive from studies examining the acquisition or representation of statistical constraints, a field that only partly overlaps with studies of grammar acquisition and representation or rule learning. The study of transition-probability statistics takes as its model inputs that are generated by regular grammars, that is, grammars that can be described via finite state automata (also referred to as finite state systems or Markov processes), that are defined solely by an alphabet, states, and transitions. Finite state systems can model simple grammars, but not grammars that require a memory stack (e.g., $A^n B^n$ grammars). Studies that examine the representation of more complex grammars such as these, which necessitate memory stores, hierarchical representations, recursion, or that use of symbolisms such as phrase structures or unification-based constraints, are outside the scope of this review and may utilize different neural systems (see Fitch & Friederici, 2012, for recent neurobiological studies

[1]The mental lexicon is a concept used in linguistics and in psycholinguistics to refer to a speaker's lexical representations.

[2]Markov chains are stochastic processes that can be parameterized by estimating TP between discrete states in a system. In first-order Markov chains, each subsequent state depends only on the immediately preceding one. In second-order Markov chains, the next state depends on two or more preceding states.

comparing regular versus context free grammars and for representative experimental work; Bahlmann, Schubotz, & Friederici, 2008; Folia, Forkstam, Ingvar, Hagoort, & Petersson, 2011). Work on formal rule learning and evaluation of stimuli against rules is also outside the scope of the current chapter.

Importantly, whereas much of the data in the studies we discuss are interpreted in terms of the learning of statistical relations (transition probabilities) or use of systems that represent statistics, these interpretations often ignore the relation between processes associated with the coding of statistics and those related to chunking of tokens. There is a strong relation between sequence statistics, the process of chunking sequences, and the resulting segmentation of the input into a "mental alphabet" consisting of distinct units (Perruchet & Pacton, 2006). Computationally, if two or more tokens always appear jointly, it would be most optimal to chunk (represent) them as a single unit for the purpose of any statistical computation. Depending on interpretive traditions, some of the studies reviewed in this chapter address differences in brain activity for regular and random sequences, not in terms of statistical learning of relations between items, but in terms of chunking. It has been shown that chunking-based accounts can explain many of the findings in the statistical learning literature and whether the same systems mediate chunking and statistical learning is an ongoing issue (see Perruchet & Pacton, 2006 for an important discussion). Thus, from a neurobiological perspective, it is still difficult to determine whether sensitivity to statistical regularities is dependent on chunking-related processes or on processes more specifically related to statistical learning, such as the updating of associative information and the construction of predictions.

The structure of the rest of the chapter is as follows. Section 2 identifies and discusses the brain regions most commonly involved in processing statistical information. Section 2.1 focuses on the cortical regions, and section 2.2 focuses on the subcortical ones. Finally, section 2.3 discusses the connectivity between the cortical and subcortical structures involved in statistical information processing. We conclude by discussing the potential usefulness of statistical information processing in the context of auditory, language, and system neurosciences.

43.2 BRAIN SYSTEMS INVOLVED IN STATISTICAL INFORMATION PROCESSING

Not much is known about the anatomy and functional organization of the brain systems that process and use statistical information in the speech signal. From a neurobiological perspective, different systems might be involved in the representation and use of such statistical information. Some might represent associative links between speech elements (e.g., the distribution of all elements that follow what is currently perceived), whereas some may mediate more general auditory-attention processes that enable the use of statistical information. For instance, hearing the syllable "que" [/kwɛ/] can invoke a representation of a range of possible complements such as "est" or "estion." Processing of acoustic signals uses active "look-ahead" mechanisms that aim to identify target words given minimally sufficient acoustic information prior to the moment when a word has been presented in its entirety (Gaskell & Marslen-Wilson, 1997). This representation, which can be formally represented via a cumulative distribution of possible following syllables could be instantiated on the fly. The deployment of neural systems to construct and utilize these distributions may further depend on the "diagnosticity" of the auditory cue (i.e., the degree to which it constrains future inputs). Whereas a syllable such as /kwɛ/ is highly diagnostic because it can be followed by relatively few continuations, a syllable such as "bi" [/bɑɪ/] allows numerous ones (bite, bilateral, bicycle) in which case making predictions may be counterproductive and, instead, paying greater attention to bottom-up input (i.e., to the acoustic details themselves) may be more optimal. When the number of potential complements is large, systems associated with modulation of auditory attention likely play a role to favor bottom-up processing.

As discussed in sections 2.1 and 2.2, neuroimaging research has begun to uncover a complex and distributed system, including both cortical and subcortical components, which is important for processing auditory statistical information either by directly supporting statistical mechanisms, or by controlling more general mechanisms such as auditory segmentation, auditory attention, and working memory. The functional organization and specific contributions of the different components of this distributed system, however, remain largely unknown. A range of experimental paradigms has been used to study this question, which included either speech or nonspeech stimuli (such as pure tones), or sometimes both, presented in either auditory or visual modalities. Passive listening tasks have been used to focus on automatic and perhaps more naturally occurring processes, whereas other tasks have targeted deliberate/executive processes, including various forms of learning and pattern recognition paradigms. This range of experimental paradigms may explain the diverse network of regions that have been associated with statistical information processing, which includes part of temporal and frontal parietal lobes and subcortical structures. That said, in the present chapter we do not review research focusing on decision-making about future outcomes.

43.2.1 Cortical Systems Underlying Statistical Processing of Linguistic Inputs

At the cortical level, perhaps not surprisingly, sensitivity to statistical information in auditory input streams has been found mainly in the supratemporal plane (STP) of healthy adults, including the auditory cortices. This is generally consistent with the idea that representation of statistics in different modalities relies on separate circuits rather than on a general system for coding statistics (Bastos et al., 2012). However, several studies have also implicated the left inferior frontal gyrus (IFG), a region that may play a more domain-general role in this process.

43.2.1.1 Involvement of Temporal Regions

Several studies examining statistical information processing in auditory nonspeech streams, usually using pure-tone streams, have identified temporal and temporal-parietal regions whose activation is modulated by the statistical properties of the input signal. In a magnetoencephalographic (MEG) study (Furl et al., 2011), sensitivity to statistical information, quantified as transition probabilities (second-order Markov chains of tones comprising five possible pitches), was found in the right temporal-parietal junction, including the most posterior part of the STP. Using functional magnetic resonance imaging (fMRI), we found sensitivity to perceived changes in statistical regularity in auditory sequences of pure tones in a region of the left superior temporal gyrus (STG) as well as in several other regions, including the bilateral medial frontal gyrus (Tobia, Iacovella, Davis, & Hasson, 2012). Using near-infrared spectroscopy (NIRS), Abla and Okanoya (2008) examined the processing of tone sequences containing "tone-words" and random sequences. Prior to the experiment, participants were presented with the tone-words during a 10-minute training phase. During the experiment, participants were asked to try to recognize the word tones they had learned during the training phase, a task that requires continuous segmentation of the input and evaluation against the learned words. The results showed increased activation in both left STG and left IFG. Although the authors argued *against* the STG as being involved in statistical information processing, suggesting instead that the observed task-related modulation reflects auditory selective attention, the finding of greater activation in this area adds to a literature suggesting that STG is indeed involved in statistical information processing. Comparing auditory melodies to unstructured sequences of tones comparable in terms of pitch and rhythm revealed stronger fMRI signal in bilateral IFG and anterior and posterior STG, including the planum temporale (PT) (Minati et al., 2008). Interestingly, the PT has been shown to respond to the presentation of unexpected silences in auditory sequences of complex sounds (modulated scanner noise), further suggesting that it is sensitive to the structure of sound sequences (Mustovic et al., 2003). In sum, neuroimaging studies, including MEG, fMRI, and NIRS studies, focusing on the processing of auditory sequences robustly find activation along the supratemporal cortex, particularly at the level of the posterior STG and PT, that is modulated by the internal statistical structure of nonspeech auditory sequences.

Neuroimaging studies that have used speech instead of nonspeech stimuli lead to conclusions that are consistent with those found in nonspeech studies. Using fMRI, McNealy, Mazziotta, and Dapretto (2006) examined brain activity of healthy adults during passive listening to: (i) syllable streams from an artificial language containing statistical regularities; (ii) streams containing statistical regularities and speech cues (stressed syllables); and (iii) streams containing no statistical regularities and no speech cues. The comparison of streams containing regularities with those that did not revealed activity in the left posterior STG. Importantly, activity in this region was found to increase over time for the streams containing statistical regularities but not for the random streams. This finding is particularly important for understanding the potential role of the posterior STG in statistical processing. If the posterior STG were involved in the updating of statistical representations (i.e., statistical updating *per se*), then this process of updating probabilities should not differ in its dynamics for regular versus random series. In contrast, if it were implicated in segmentation of words from the stream, then it would show more activation for regular series, as was found by the authors. A subset of the participants in McNealy et al. study also completed a word discrimination task in which words, part-words, and nonwords taken from the speech streams were presented to the participants who were given no explicit task. Results show activity in the left posterior IFG and left middle frontal gyrus (MFG) for words compared with nonwords and part-words, altogether suggesting that posterior frontal regions are involved in word recognition once these words are learned, whereas statistical information and segmentation processes may be accomplished within the STP. This experimental paradigm was reapplied with a group of 10-year-old children; results show essentially the same BOLD activation pattern (McNealy, Mazziotta, & Dapretto, 2010).

In a related fMRI study (Karuza et al., 2013), participants were also exposed to an artificial language and tested on their ability to recognize words (three-syllable combinations with high transitional probabilities) and part-words that occurred during the exposure streams (learning). Exposure streams included forward and backward speech streams. Greater activation was found in the left STG, right posterior middle temporal gyrus (MTG), and right STG when comparing the forward speech stream with the backward speech control condition. Relating each participant's change in learning across postexposure test phases to changes in neural activity during the forward as compared with the backward exposure phase revealed activation in the left IFG pars triangularis and a small portion of the left IFG pars opercularis. Together with the findings of McNealy et al. (2006), this suggests that when statistical dependencies are found in speech stream, "chunks" of colinear elements are coded as single items ("words") in a representation accessible to left inferior frontal regions. That is, these regions may be specifically involved in word learning or word recognition, whereas more general auditory statistical information processing mechanisms involve the STP.

In our own fMRI work (Tremblay, Baroni, & Hasson, 2012), we used a slightly different approach to study statistical information processing in speech and nonspeech auditory sequences. We exposed participants to meaningless sequences of native syllables and spectrally complex natural nonspeech sounds (bird chirps) that varied in terms of their statistical structure, which was determined by their transition-probability matrices (participants monitored a visual stimulus in an incidental task). The auditory sequences ranged from random to highly structured in three steps: predictable sequences were associated with strong transition constraints between elements in a way that allowed making a strong prediction about the next element, whereas mid-level or random sequences were associated with weaker or no transition constraints. In contrast to prior work, we used a detailed anatomical ROI approach and focused on the STP, which we partitioned into 13 subregions in each hemisphere. Our results emphasize the importance of the posterior STP in the automatic processing of statistical information for both speech and nonspeech sequences. Several areas in the left STP were sensitive to statistical regularities in both speech and nonspeech inputs, including the PT, posterior STG, and primary auditory cortex in the medial transverse temporal gyrus (TTG). The bilateral lateral TTG and transverse temporal sulcus (TTS) were sensitive to statistical regularities in both speech and nonspeech, but patterns of activation were distinct with earlier activation for speech than nonspeech, suggesting easier segmentation for familiar than unfamiliar sounds. These findings for TTG are also in agreement with work showing sensitivity to regularities in that region using direct recordings in rats (Yaron, Hershenhoren, & Nelken, 2012). Recent data from our group (Deschamps, Hasson, & Tremblay, submitted) shows that the cortical thickness of left anterior STG correlates with the ability to perceive statistical structure in speech sequences. An additional exploratory whole-brain analysis revealed significant correlations in the superior parietal lobule (SPL) and intraparietal sulcus (IPS), regions more typically associated with attentional processes. These data suggest that the ability to process statistical information depends on auditory-specialized cortical networks in STP, as well as a network of cortical regions supporting executive functions.

43.2.1.2 Involvement of Left IFG

As indicated, the left IFG appears to play a role in the segmentation of word or word-like structures in an auditory stream. It is an outstanding question whether the left IFG mediates a modality-general function. In a nonspeech study, Petersson, Folia, and Hagoort (2012) introduced participants to written (visual) sequences generated by a seven-item artificial grammar. In a subsequent fMRI session, these participants were asked to determine whether sequences presented to them were grammatical. The authors found greater activation for accurate nongrammatical versus grammatical judgments in left IFG, suggesting that these regions generally evaluate the well-formedness of inputs by examining whether they are licensed given the grammar learned, even outside language contexts. However, other fMRI work (Nastase, Iacovella, & Hasson, 2014) argues that the left IFG may mediate low-level auditory statistical processing rather than a modality-independent process. The Nastase et al. study showed that activity in left IFG was sensitive to regular versus random series of pure-tone series, but it was not sensitive to regularity when the tokens were substituted with simple visual objects. One possibility is that participants in the Petersson et al. study represented the visual stimuli using a verbal code in preparation for the task required of them. In contrast, the short sequences, passive perception context, and rapid presentation rate used by Nastase et al. do not favor that strategy, which may explain the discrepant findings.

In sum, a review of the literature indicates that the posterior STP, particularly at the level of the PT and posterior STG, as well as the left IFG, are the cortical structures most frequently implicated in the processing of statistical information in the auditory domain. Further studies are needed, however, to delineate the specific role(s) of each region as well as the manner in which information flows between these different regions.

43.2.2 Subcortical Systems Underlying Statistical Information Processing

43.2.2.1 Involvement of Basal Ganglia in Statistical Learning

Relying mainly on neuroimaging and neurophysiology studies, we now consider the potential role of the basal ganglia (BG) in mediating statistically informed aspects of phonemic and syllable-level processing. The BG is involved in several cortico-striatal loops, including limbic, associative, and motor loops, each originating and terminating in distinct cortical areas, and connecting via different parts of the BG (Alexander, DeLong, & Strick, 1986). The basal motor loop connects primary and nonprimary motor areas including the supplementary motor area (SMA) to the posterior striatum, particularly the putamen, the external and internal pallidum, the subthalamic nucleus, and ventrolateral thalamic nuclei. This loop has been associated with the acquisition, storage, and execution of motor patterns and sequence learning (Graybiel, 1998). Functionally, it has been suggested that the BG is involved in the coding of context (Saint-Cyr, 2003) and in the generation of sequenced outputs via rule application (Lieberman, 2001). Neuropsychological, anatomical, and behavioral evidence for these notions are not discussed here; they have been discussed extensively elsewhere, particularly in relation to populations with BG dysfunction (see Alm, 2004 for dysfunctions related to stuttering; see Saint-Cyr, 2003 for a review of BG dysfunctions such as Parkinson's and Hungtington's diseases).

The potential role of BG in representing statistical constraints among short nonsemantic units has been examined in several studies. An early neuroimaging study (Lieberman, Chang, Chiao, Bookheimer, & Knowlton, 2004) provides a convincing demonstration of this point. In this study, participants were presented with letter strings constructed from an artificial grammar as part of a training stage; then, in a test stage, they were tested on novel grammatical and nongrammatical letter strings. Orthogonally to the grammaticality manipulation, the authors constructed test items that either strongly or weakly overlapped with the training items in "chunk strength," which is a measure not relating to the grammaticality of the string, but rather to the frequency in which the "alphabet" of the grammar tended to occur conjointly (co-occurrence frequency). In this way, the authors could separate regions implicated in retention of grammar rules from those involved in retention of chunks of letters. When chunk strength was low, offering little superficial overlap with the training items, the caudate nucleus was sensitive to grammaticality, showing more activity for grammatical items. Examination of the correlates of chunk strength revealed hippocampal and putamen activation, with hippocampal activation interpreted as being related to retrieval of training items.

The putamen has been linked to the coding of regularity in the study of McNealy et al. (2006) discussed in the previous section. That study reported stronger activity in the putamen during presentation of random syllable strings than during presentation of artificial languages. Furthermore, activation in the region was stronger for unstressed language-like streams than for stressed language-like streams (in the latter, word-strings were easier to disambiguate). Thus, the putamen showed a gradient of activation with the least activity during the condition in which words were easiest to segment. Interpreting such relatively focal regional effects in the context of more general approach to understanding BG-thalamic loops is an important goal for future work. The BG is internally connected via inhibitory connections so that increased activity in the striatum can inhibit or excite the pallidum and, via different pathways, result in either increased or decreased activity in thalamus. Future work with high temporal and spatial resolution (e.g., time-resolved fMRI or MEG) may identify such interactions. In any case, the involvement of BG in grammar learning is supported by work (De Diego-Balaguer et al., 2008) showing that individuals diagnosed with Huntington's disease, which affects the striatum (i.e., the caudate and the putamen), show deficits in word and morph-syntactic learning, with the former necessitating veridical memory for three-syllable combinations and the latter necessitating the extraction of morphological regularities (e.g., artificial words ending in "mi" always begin with "pa").

A role for the BG in representing statistical relationships between input elements, or their prediction, is also supported by research on auditory processing. For example, Leaver, Van Lare, Zielinski, Halpern, and Rauschecker (2009) linked the BG to auditory sequencing, a process that is closely related to the processing of statistical information, given that statistics information can only be computed and processed if the independent units in a signal have been properly segmented. In that study, participants were scanned while anticipating a familiar or unfamiliar melody. While silently anticipating a familiar melody, SMA, pre-SMA, left BG (globus pallidus/putamen), and several other regions showed more activity than when anticipating a recently learned melody, suggesting a role in auditory prediction. However, for the BG, this pattern was stronger during the early stages of familiarization with the "familiar" melody than in later stages, suggesting that the involvement of BG is related to the learning process itself ("learning stimulus associations," *ibid.*, p. 2482) rather than prediction.

Whether the BG mediates pattern learning or use of statistical information for prediction is more directly addressed by recent work on beat perception, stimulus anticipation, and music perception (Grahn & Rowe, 2013). In that study, the authors presented participants with auditory streams consisting of regular beats that either changed or were the same as the beat patterns in streams presented immediately before. Results showed that the putamen was most strongly active in the condition where the current pattern was the same as the previous one, and showed the least activity when the current pattern was new and did not bear a relation to the prior one. Mid-levels of activity were found for changes in speed. On the basis of these findings, the authors proposed that "the basal ganglia's response profile suggests a role in beat prediction not beat finding" (Grahn & Rowe, 2013, p. 913). Geiser, Notter, and Gabrieli (2012) also linked the putamen to processing of auditory regularities by demonstrating that it shows more activity in the context of regular (versus irregular) beat patterns (the opposite pattern found by McNealy et al., 2006) and, furthermore, across individuals, greater activity in the putamen was negatively correlated with activity in auditory cortex. We note that the studies of Grahn and Rowe and that of Geiser et al. address predictions given temporal regularities rather than regularities of content; establishing the relation between these two kinds of phenomena, in the domain of language, is a topic for future work.

It is still unclear whether activity in BG reflects the coding of regularity, predictive processes, or evaluation of input against predictions. Seger et al. (2013) found that the caudate head and body (though not the putamen) were sensitive to the degree of mismatch in musical progression of a chord series, suggesting that the caudate is involved in evaluation of predictions, whereas the putamen is involved in constructing predictions. However, even this interpretation of putamen activity necessitates further work, as Langners et al. (2011) have shown that the putamen shows higher activity when more specific versus less specific predictions are violated (see also Haruno & Kawato, 2006).

Furthermore, it is also unclear whether temporal or content-based predictions mediated by BG, such as those reviewed in the aforementioned studies, extend to the lexical level. Interestingly, some work suggests that the role of BG in the construction or evaluation of predictions is limited to the sublexical level. Wahl et al. (2008) examined responses to semantic or syntactic violations in spoken sentences. Deep brain recordings monitored activity in the thalamus and BG. Results showed that BG were not sensitive to either type of violation, leading the authors to propose that "syntactic and semantic language analysis is primarily realized within cortico-thalamic networks, whereas a cohesive BG network is not involved in these essential operations of language analysis" (p. 697). It is therefore possible that the BG mediates statistical learning, but only when learning applies to sublexical units (phonemic, syllabic, morphemic) rather than to word or clause-level phenomenon.

43.2.2.2 Debates on the Functional Role of Basal Ganglia

It is important to consider the empirical findings reviewed here in the context of the continuous discussion on the role of BG in language comprehension. Whether the BG's computations during language comprehension are intrinsically linguistic and related to grammatical computations or, alternatively, associated with more general functions is currently being debated. Arguing for the linguistic-computation view, Ullman (2001) has suggested that the BG subserves aspects of mental grammar and that it allows learning of grammars in various domains, including syntax, morphology, and even phonology. Given that grammar learning reflects, at minimum, extraction of regularities or invariants, this would entail sophisticated mechanisms for extracting regularities that can include phrase-structure rules, lexical-level constraints, or rules governing long-distance movements between constituents. This viewpoint has received empirical support from work (Nemeth et al., 2011) showing that in dual-task conditions where one task involves statistically driven motor sequence learning, a parallel sentence comprehension task diminishes learning, whereas mathematical or word-monitoring parallel tasks do not. In contrast, Crosson and colleagues have argued that the BG is not involved in primary language functions, but rather increases signal to noise during action selection. Thus, during speech comprehension, the region may be implicated in increasing the coding precision (activation sharpening) of those phonemes most appropriate in a given context (Ford et al., 2013, p. 6). Understanding the role of BG in statistical contexts containing both auditory and nonauditory language content is important for differentiating these different theoretical frameworks.

43.3 CONNECTIONAL ANATOMY OF THE STATISTICAL NETWORK

Our review suggests that the caudal part of STP, including the posterior STG and PT extending to TTG, as well as the striatum, and possibly the IFG (pars opercularis, triangularis), form the core regions involved in processing statistical information. However, whether these regions form a functional network for segmenting auditory inputs or processing embedded statistical information remains unclear, because no studies thus far have attempted to examine the connectivity patterns associated with processing auditory statistics. We briefly outline reasons for supposing these three areas (IFG, posterior-middle temporal cortex, BG) have the potential for integrative processing of statistics in the auditory stream.

The idea of some degree of anatomical connectivity between the human IFG and posterior temporal areas has a long history, but it is not without controversy (see Dick, Bernal, & Tremblay, 2013; Dick & Tremblay, 2012 for reviews on the subject and references therein for data in support of and against the presence of such connectivity). There is, however, evidence for functional connectivity between these regions (Hampson et al., 2006).

The other question of interest concerns the connectivity of the striatum with posterior STP and/or IFG. Although there is extended structural connectivity (monosynaptic) between the cortex and the striatum, evidence of anatomical connectivity between the IFG and the striatum, as well as between STP and the striatum, is scarce. Schmahmann and Pandya (2006), using an anterograde tract-tracer technique in the macaque monkey, reported some projections from the posterior STG to the caudate nucleus. In humans, a recent diffusion MRI study conducted on 13 healthy adults suggest that the temporal cortex (including posterior segment) projects to the posterior part of the caudate nucleus (Kotz, Anwander, Axer, & Knosche, 2013), consistent with the macaque evidence. Documentation of structural connectivity between the IFG and the striatum is scarce. A recent DTI study conducted on a relatively small sample ($N = 10$) and focusing on the IFG reported that both the opercular and triangular convolutions of the IFG connect with the rostral putamen (Ford et al., 2013).

Another way of assessing connectivity between regions is to examine their functional connectivity, that is, the tendency (in the statistical sense) of different regions to be active synchronously, which does not necessarily imply direct (monosynaptic) connectivity. The functional connectivity literature, however, provides a slightly less consistent account. A detailed study of the functional connectivity of the human striatum conducted on a large sample ($N = 1000$; Choi, Yeo, & Buckner, 2012) reported only weak connectivity between the posterior STP and the striatum and between the IFG and the striatum. Using dynamic causal modeling (DCM) during a phonological (rhyming judgment) task, Booth et al. (Booth, Wood, Lu, Houk, & Bitan, 2007) reported functional connectivity between the putamen and left IFG as well as between the putamen and the lateral temporal cortex (including both superior and middle temporal areas). Another study, using resting-state connectivity, documented functional connectivity between the putamen and posterior STP (Di Martino et al., 2008).

In sum, there is some evidence of functional connectivity but very limited evidence of anatomical connectivity between the IFG and posterior STP and the striatum. Additional research is needed to further current understanding of how these areas, which appear to form the core of a statistical information processing network, interact either directly or indirectly to extract and process that information.

43.4 RELATED WORK AND FURTHER AFIELD

There are a number of research lines that appear related to the processing of statistical information but address substantially different issues. The first defines statistical relations between inputs not in terms of formal properties such as transition probabilities, but in terms of the degree of physical (metric) changes in (sound) frequency space that occur between tones in more and less ordered sequences. In that literature, regularity is quantified via the serial autocorrelation of tones in the pitch sequence (Overath et al., 2007; Patel & Balaban, 2000) so that regularity is higher when low frequencies dominate or sample entropy is lower. These studies have implicated auditory cortical regions in regularities, but it is important to note that the term "regularity" is used differently and that the explanations for such effects could have a purely physiological basis; when "statistical regularity" is related to the magnitude of transitions in frequency space, the explanation for these effects could be due to low-level repetition suppression effects *per se* driven by the tuning curves of auditory neurons.

There is also a body of work examining neural responses to varying levels of uncertainty, but in which uncertainty originates from the degree of uniformity of the distribution from which the tokens are sampled (captured by Shannon's entropy of the variable generating the input), rather than to the strength of transition probabilities. Particularly relevant are studies that did not require directing attention to stimulus statistics (Harrison, Duggins, & Friston, 2006; Strange, Duggins, Penny, Dolan, & Friston, 2005; Tobia, Iacovella, & Hasson, 2012). The studies of Strange et al. (2005) and Harrison et al. (2006) used visual stimuli and linked the hippocampus to coding series uncertainty (quantified via Shannon's entropy). However, one study using passive listening to auditory stimuli (Tobia, Iacovella, & Hasson, 2012) has identified perisylvian regions including both inferior parietal and lateral temporal cortex as sensitive to the relative frequency distribution of auditory tokens without hippocampal involvement (see Tobia, Iacovella, & Hasson, 2012 for a discussion of the role of the hippocampus in statistical learning).

Another important and unresolved issue concerns the domain specificity of the neural processes involved in learning and representation of statistical information. Our own work (Nastase et al., 2014) suggests that rapid learning of sequence statistics within auditory or visual steams is largely mediated by separate neural systems, with little overlap and without hippocampal involvement. Others (Bubic, von Cramon, & Schubotz, 2011) have shown that violations of regularity-induced predictions evoke activities in different systems depending on whether the regularity is temporal-, location-, or feature-based, suggesting a domain-specific basis for the construction or evaluation of predictions. However, several studies (Cristescu, Devlin, & Nobre, 2006; Egner et al., 2008) suggest that there are domain-general systems that mediate predictions of items' perceptual features, location, and semantic features.

43.5 CONCLUSION AND FUTURE WORK

The study of the neurobiological substrates underlying the representation and use of statistical information is in its initial stages. Although statistical information may be crucial to language acquisition and online speech processing, experimental work has just begun identifying the neural mechanisms mediating these processes as well as their functional importance. Several core areas identified in this experimental work are regions of the STP and lateral temporal cortex that appear to be sensitive to the degree of statistical regularity in an input and potentially initial word segmentation and left IFG regions that play a role in word recognition but that may not be involved in the coding of statistics *per se*. The BG appears to have a role in online temporal predictions and anticipation, with a particularly important role for the putamen, a core part of the cortico-basal motor loop. Finally, although there is some evidence for functional and structural connectivity between IFG, STP, and BG, future work is needed to examine the relation between statistical processing mechanisms and the strength of functional and structural connectivity within this network.

References

Abla, D., & Okanoya, K. (2008). Statistical segmentation of tone sequences activates the left inferior frontal cortex: A near-infrared spectroscopy study. *Neuropsychologia*, *46*(11), 2787–2795. Available from: http://dx.doi.org/10.1016/j.neuropsychologia.2008.05.012.

Alexander, G. E., DeLong, M. R., & Strick, P. L. (1986). Parallel organization of functionally segregated circuits linking basal ganglia and cortex. *Annual Review of Neuroscience*, *9*, 357–381.

Alm, P. A. (2004). Stuttering and the basal ganglia circuits: A critical review of possible relations. *Journal of Communication Disorders*, *37* (4), 325–369. Available from: http://dx.doi.org/10.1016/j.jcomdis.2004.03.001.

Astheimer, L. B., & Sanders, L. D. (2011). Predictability affects early perceptual processing of word onsets in continuous speech. *Neuropsychologia*, *49*(12), 3512–3516. Available from: http://dx.doi.org/10.1016/j.neuropsychologia.2011.08.014.

Bahlmann, J., Schubotz, R. I., & Friederici, A. D. (2008). Hierarchical artificial grammar processing engages Broca's area. *Neuroimage*, *42*(2), 525–534. Available from: http://dx.doi.org/10.1016/j.neuroimage.2008.04.249.

Bastos, A. M., Usrey, W. M., Adams, R. A., Mangun, G. R., Fries, P., & Friston, K. J. (2012). Canonical microcircuits for predictive coding. *Neuron*, *76*(4), 695–711. Available from: http://dx.doi.org/10.1016/j.neuron.2012.10.038.

Booth, J. R., Wood, L., Lu, D., Houk, J. C., & Bitan, T. (2007). The role of the basal ganglia and cerebellum in language processing. *Brain Research*, *1133*(1), 136–144. Available from: http://dx.doi.org/10.1016/j.brainres.2006.11.074.

Bubic, A., von Cramon, D. Y., & Schubotz, R. (2011). Exploring the detection of associatively novel events using fMRI. *Human Brain Mapping*, *32*(3), 370–381. Available from: http://dx.doi.org/10.1002/hbm.21027.

Buiatti, M., Pena, M., & Dehaene-Lambertz, G. (2009). Investigating the neural correlates of continuous speech computation with frequency-tagged neuroelectric responses. *Neuroimage*, *44*(2), 509–519. Available from: http://dx.doi.org/10.1016/j.neuroimage.2008.09.015.

Choi, E. Y., Yeo, B. T., & Buckner, R. L. (2012). The organization of the human striatum estimated by intrinsic functional connectivity. *Journal of Neurophysiology*, *108*(8), 2242–2263. Available from: http://dx.doi.org/10.1152/jn.00270.2012.

Cole, R. A., & Jakimik, J. (1980). How are syllables used to recognize words? *The Journal of the Acoustical Society of America*, *67*(3), 965–970.

Conway, C. M., Bauernschmidt, A., Huang, S. S., & Pisoni, D. B. (2010). Implicit statistical learning in language processing: Word predictability is the key. *Cognition*, *114*(3), 356–371. Available from: http://dx.doi.org/10.1016/j.cognition.2009.10.009.

Cristescu, T. C., Devlin, J. T., & Nobre, A. C. (2006). Orienting attention to semantic categories. *NeuroImage*, *33*(4), 1178–1187. Available from: http://dx.doi.org/10.1016/j.neuroimage.2006.08.017.

De Diego-Balaguer, R., Couette, M., Dolbeau, G., Durr, A., Youssov, K., & Bachoud-Levi, A. C. (2008). Striatal degeneration impairs language learning: evidence from Huntington's disease. *Brain*, *131* (Pt 11), 2870–2881. Available from: http://dx.doi.org/10.1093/brain/awn242.

Deschamps, I., Hasson, U., & Tremblay, P. (submitted). Cortical thickness in perisylvian regions correlates with individuals' sensitivity to statistical regularities in auditory sequences.

Dick, A. S., Bernal, B., & Tremblay, P. (2013). The language connectome: New pathways, new concepts. *Neuroscientist*. Available from: http://dx.doi.org/10.1177/1073858413513502.

Dick, A. S., & Tremblay, P. (2012). Beyond the arcuate fasciculus: Consensus and controversy in the connectional anatomy of language. *Brain*, *135*(Pt 12), 3529–3550. Available from: http://dx.doi.org/10.1093/brain/aws222.

Di Martino, A., Scheres, A., Margulies, D. S., Kelly, A. M., Uddin, L. Q., Shehzad, Z., et al. (2008). Functional connectivity of human striatum: A resting state FMRI study. *Cerebral Cortex*, *18*(12), 2735–2747. Available from: http://dx.doi.org/10.1093/cercor/bhn041.

Egner, T., Monti, J. M. P., Trittschuh, E. H., Wieneke, C. A., Hirsch, J., & Mesulam, M. M. (2008). Neural integration of top-down spatial and feature-based information in visual search. *The Journal of Neuroscience, 28*(24), 6141–6151. Available from: http://dx.doi.org/10.1523/JNEUROSCI.3562-08.2008.

Fiser, J., & Aslin, R. N. (2001). Unsupervised statistical learning of higher-order spatial structures from visual scenes. *Psychological Science, 12*(6), 499–504.

Fiser, J., & Aslin, R. N. (2002). Statistical learning of new visual feature combinations by infants. *Proceedings of the National Academy of Sciences of the United States of America, 99*(24), 15822–15826. Available from: http://dx.doi.org/10.1073/pnas.232472899.

Fitch, W. T., & Friederici, A. D. (2012). Artificial grammar learning meets formal language theory: An overview. *Philosophical Transactions of the Royal Society of London Series B, Biological Sciences, 367*(1598), 1933–1955. Available from: http://dx.doi.org/10.1098/rstb.2012.0103.

Folia, V., Forkstam, C., Ingvar, M., Hagoort, P., & Petersson, K. M. (2011). Implicit artificial syntax processing: Genes, preference, and bounded recursion. *Biolinguistics, 5*, 105–132.

Ford, A. A., Triplett, W., Sudhyadhom, A., Gullett, J., McGregor, K., Fitzgerald, D. B., et al. (2013). Broca's area and its striatal and thalamic connections: A diffusion-MRI tractography study. *Frontiers in Neuroanatomy, 7*, 8. Available from: http://dx.doi.org/10.3389/fnana.2013.00008.

Furl, N., Kumar, S., Alter, K., Durrant, S., Shawe-Taylor, J., & Griffiths, T. D. (2011). Neural prediction of higher-order auditory sequence statistics. *Neuroimage, 54*(3), 2267–2277. Available from: http://dx.doi.org/10.1016/j.neuroimage.2010.10.038.

Gagnepain, P., Henson, R. N., & Davis, M. H. (2012). Temporal predictive codes for spoken words in auditory cortex. *Current Biology, 22*(7), 615–621. Available from: http://dx.doi.org/10.1016/j.cub.2012.02.015.

Gaskell, M. G., & Marslen-Wilson, W. D. (1997). Integrating form and meaning: A Distributed model of speech perception. *Language and Cognitive Processes, 12*(5–6), 613–656. Available from: http://dx.doi.org/10.1080/016909697386646.

Geiser, E., Notter, M., & Gabrieli, J. D. (2012). A corticostriatal neural system enhances auditory perception through temporal context processing. *The Journal of Neuroscience: The Official Journal of the Society for Neuroscience, 32*(18), 6177–6182. Available from: http://dx.doi.org/10.1523/JNEUROSCI.5153-11.2012.

Grahn, J. A., & Rowe, J. B. (2013). Finding and feeling the musical beat: Striatal dissociations between detection and prediction of regularity. *Cerebral Cortex, 23*(4), 913–921. Available from: http://dx.doi.org/10.1093/cercor/bhs083.

Graybiel, A. M. (1998). The basal ganglia and chunking of action repertoires. *Neurobiology of Learning and Memory, 70*(1–2), 119–136. Available from: http://dx.doi.org/10.1006/nlme.1998.3843.

Hampson, M., Tokoglu, F., Sun, Z., Schafer, R. J., Skudlarski, P., Gore, J. C., et al. (2006). Connectivity-behavior analysis reveals that functional connectivity between left BA39 and Broca's area varies with reading ability. *Neuroimage, 31*(2), 513–519. Available from: http://dx.doi.org/10.1016/j.neuroimage.2005.12.040.

Harrison, L. M., Duggins, A., & Friston, K. J. (2006). Encoding uncertainty in the hippocampus. *Neural Networks, 19*(5), 535–546.

Haruno, M., & Kawato, M. (2006). Different neural correlates of reward expectation and reward expectation error in the putamen and caudate nucleus during stimulus-action-reward association learning. *Journal of Neurophysiology, 95*(2), 948–959. Available from: http://dx.doi.org/10.1152/jn.00382.2005.

Karuza, E. A., Newport, E. L., Aslin, R. N., Starling, S. J., Tivarus, M. E., & Bavelier, D. (2013). The neural correlates of statistical learning in a word segmentation task: An fMRI study. *Brain and Language, 127*(1), 46–54. Available from: http://dx.doi.org/10.1016/j.bandl.2012.11.007.

Kirkham, N. Z., Slemmer, J. A., & Johnson, S. P. (2002). Visual statistical learning in infancy: Evidence for a domain general learning mechanism. *Cognition, 83*(2), B35–42. Available from: http://dx.doi.org/S0010027702000045 [pii].

Kotz, S. A., Anwander, A., Axer, H., & Knosche, T. R. (2013). Beyond cytoarchitectonics: The internal and external connectivity structure of the caudate nucleus. *PLoS One, 8*(7), e70141. Available from: http://dx.doi.org/10.1371/journal.pone.0070141.

Langner, R., Kellermann, T., Boers, F., Sturm, W., Willmes, K., & Eickhoff, S. B. (2011). Modality-specific perceptual expectations selectively modulate baseline activity in auditory, somatosensory, and visual cortices. *Cerebral Cortex, 21*(12), 2850–2862. Available from: http://dx.doi.org/10.1093/cercor/bhr083.

Leaver, A. M., Van Lare, J., Zielinski, B., Halpern, A. R., & Rauschecker, J. P. (2009). Brain activation during anticipation of sound sequences. *The Journal of Neuroscience: The Official Journal of the Society for Neuroscience, 29*(8), 2477–2485. Available from: http://dx.doi.org/10.1523/JNEUROSCI.4921-08.2009.

Lehiste, I., & Shockey, L. (1972). On the perception of coarticulation effects in English VCV syllables. *Journal of Speech and Hearing Research, 15*(3), 500–506.

Lieberman, M. D., Chang, G. Y., Chiao, J., Bookheimer, S. Y., & Knowlton, B. J. (2004). An event-related fMRI study of artificial grammar learning in a balanced chunk strength design. *Journal of Cognitive Neuroscience, 16*(3), 427–438. Available from: http://dx.doi.org/10.1162/089892904322926764.

Lieberman, P. (2001). On the subcortical bases of the evolution of language. In J. r. Trabant, & S. Ward (Eds.), *New essays on the origin of language* (pp. 1–20). Berlin/Hawthorne, NY: Mouton de Gruyter.

McCarthy, J. (2001). *A thematic guide to optimality theory*. Cambridge: Cambridge University Press.

McNealy, K., Mazziotta, J. C., & Dapretto, M. (2006). Cracking the language code: Neural mechanisms underlying speech parsing. *The Journal of Neuroscience: The Official Journal of the Society for Neuroscience, 26*(29), 7629–7639. Available from: http://dx.doi.org/10.1523/JNEUROSCI.5501-05.2006.

McNealy, K., Mazziotta, J. C., & Dapretto, M. (2010). The neural basis of speech parsing in children and adults. *Developmental Science, 13*(2), 385–406. Available from: http://dx.doi.org/10.1111/j.1467-7687.2009.00895.x.

Minati, L., Rosazza, C., D'Incerti, L., Pietrocini, E., Valentini, L., Scaioli, V., et al. (2008). FMRI/ERP of musical syntax: Comparison of melodies and unstructured note sequences. *Neuroreport, 19*(14), 1381–1385. Available from: http://dx.doi.org/10.1097/WNR.0b013e32830c694b.

Misyak, Jennifer B., & Christiansen, Morten H. (2012). Statistical learning and language: An individual differences study. *Language Learning, 62*(1), 302–331. Available from: http://dx.doi.org/10.1111/j.1467-9922.2010.00626.x.

Mustovic, H., Scheffler, K., Di Salle, F., Esposito, F., Neuhoff, J. G., Hennig, J., et al. (2003). Temporal integration of sequential auditory events: Silent period in sound pattern activates human planum temporale. *Neuroimage, 20*(1), 429–434.

Nastase, S., Iacovella, V., & Hasson, U. (2014). Uncertainty in visual and auditory series is coded by modality-general and modality-specific neural systems. *Human Brain Mapping*. Available from: http://dx.doi.org/10.1002/hbm.22238.

Nemeth, D., Janacsek, K., Csifcsak, G., Szvoboda, G., Howard, J. H., Jr., & Howard, D. V. (2011). Interference between sentence processing and probabilistic implicit sequence learning. *PLoS One, 6*(3), e17577. Available from: http://dx.doi.org/10.1371/journal.pone.0017577.

Newport, E. L., & Aslin, R. N. (2004). Learning at a distance I. Statistical learning of non-adjacent dependencies. *Cognitive Psychology*, *48*(2), 127–162. Available from: http://dx.doi.org/10.1016/S0010-0285(03)00128-2 [pii].

Overath, T., Cusack, R., Kumar, S., von Kriegstein, K., Warren, J. D., Grube, M., et al. (2007). An information theoretic characterisation of auditory encoding. *PLoS Biology*, *5*(11), e288. Available from: http://dx.doi.org/10.1371/journal.pbio.0050288.

Patel, A. D., & Balaban, E. (2000). Temporal patterns of human cortical activity reflect tone sequence structure. *Nature*, *404*(6773), 80–84. Available from: http://dx.doi.org/10.1038/35003577.

Pelucchi, B., Hay, J. F., & Saffran, J. R. (2009a). Learning in reverse: Eight-month-old infants track backward transitional probabilities. *Cognition*, *113*(2), 244–247. Available from: http://dx.doi.org/10.1016/j.cognition.2009.07.011.

Pelucchi, B., Hay, J. F., & Saffran, J. R. (2009b). Statistical learning in a natural language by 8-month-old infants. *Child development*, *80*(3), 674–685. Available from: http://dx.doi.org/10.1111/j.1467-8624.2009.01290.x.

Pena, M., Bonatti, L. L., Nespor, M., & Mehler, J. (2002). Signal-driven computations in speech processing. *Science*, *298*(5593), 604–607. Available from: http://dx.doi.org/10.1126/science.1072901.

Perruchet, P., & Pacton, S. (2006). Implicit learning and statistical learning: One phenomenon, two approaches. *Trends in Cognitive Sciences*, *10*(5), 233–238. Available from: http://dx.doi.org/10.1016/j.tics.2006.03.006.

Petersson, K. M., Folia, V., & Hagoort, P. (2012). What artificial grammar learning reveals about the neurobiology of syntax. *Brain and Language*, *120*(2), 83–95. Available from: http://dx.doi.org/10.1016/j.bandl.2010.08.003.

Saffran, J. R., Aslin, R. N., & Newport, E. L. (1996). Statistical learning by 8-month-old infants. *Science*, *274*(5294), 1926–1928.

Saffran, J. R., Johnson, E. K., Aslin, R. N., & Newport, E. L. (1999). Statistical learning of tone sequences by human infants and adults. *Cognition*, *70*(1), 27–52. Available from: http://dx.doi.org/S0010-0277(98)00075-4.

Saffran, J. R., Pollak, S. D., Seibel, R. L., & Shkolnik, A. (2007). Dog is a dog is a dog: Infant rule learning is not specific to language. *Cognition*, *105*(3), 669–680. Available from: http://dx.doi.org/10.1016/j.cognition.2006.11.004.

Saint-Cyr, J. A. (2003). Frontal-striatal circuit functions: Context, sequence, and consequence. *Journal of the International Neuropsychological Society*, *9*(1), 103–127.

Schmahmann, J. D., & Pandya, D. N. (2006). *Fiber pathways of the brain*. Oxford/New York, NY: Oxford University Press.

Seger, C. A., Spiering, B. J., Sares, A. G., Quraini, S. I., Alpeter, C., David, J., et al. (2013). Corticostriatal contributions to musical expectancy perception. *Journal of Cognitive Neuroscience*, *25*(7), 1062–1077. Available from: http://dx.doi.org/10.1162/jocn_a_00371.

Strange, B. A., Duggins, A., Penny, W., Dolan, R. J., & Friston, K. J. (2005). Information theory, novelty and hippocampal responses: Unpredicted or unpredictable? *Neural Networks*, *18*(3), 225–230.

Tobia, M. J., Iacovella, V., Davis, B., & Hasson, U. (2012). Neural systems mediating recognition of changes in statistical regularities. *Neuroimage*, *63*(3), 1730–1742. Available from: http://dx.doi.org/10.1016/j.neuroimage.2012.08.017.

Tobia, M. J., Iacovella, V., & Hasson, U. (2012). Multiple sensitivity profiles to diversity and transition structure in non-stationary input. *Neuroimage*, *60*(2), 991–1005. Available from: http://dx.doi.org/10.1016/j.neuroimage.2012.01.041.

Tremblay, P., Baroni, M., & Hasson, U. (2012). Processing of speech and non-speech sounds in the supratemporal plane: Auditory input preference does not predict sensitivity to statistical structure. *Neuroimage*, *66C*, 318–332. Available from: http://dx.doi.org/10.1016/j.neuroimage.2012.10.055.

Ullman, M. T. (2001). A neurocognitive perspective on language: The declarative/procedural model. *Nature Reviews Neuroscience*, *2*(10), 717–726. Available from: http://dx.doi.org/10.1038/35094573.

Wahl, M., Marzinzik, F., Friederici, A. D., Hahne, A., Kupsch, A., Schneider, G. H., et al. (2008). The human thalamus processes syntactic and semantic language violations. *Neuron*, *59*(5), 695–707. Available from: http://dx.doi.org/10.1016/j.neuron.2008.07.011.

Wurm, L. H., Ernestus, M., Schreude, R., & Baayen, R. H. (2006). Dynamics of the auditory comprehension of prefixed words: Cohort entropies and conditional root uniqueness points. *Mental Lexicon*, *1*(1), 125–146.

Yaron, A., Hershenhoren, I., & Nelken, I. (2012). Sensitivity to complex statistical regularities in rat auditory cortex. *Neuron*, *76*(3), 603–615. Available from: http://dx.doi.org/10.1016/j.neuron.2012.08.025.

SECTION G

WORD PROCESSING

CHAPTER

44

The Neurobiology of Lexical Access

Matthew H. Davis

Medical Research Council, Cognition and Brain Sciences Unit, Cambridge, UK

44.1 INTRODUCTION

By comparison with other animals, humans use an exceptionally large inventory of communicative sounds (i.e., words). Infants learn to recognize their first words in the first year of life (Kuhl, 2004; Werker & Tees, 1999), and from approximately 18 months of age they add words to their vocabulary at an astonishing rate, learning perhaps as many as 10 new words per day (McMurray, 2007). Estimates of the number of words known by a typical adult who is a native English speaker vary from 25,000 to 75,000, but they clearly depend on the degree of lexical redundancy assumed (Altmann, 1997).[1] The resulting lexicon of familiar words is key to effective communication. The neurobiological foundations of lexical knowledge are correspondingly central to understanding the human faculty of language.

Knowing a word involves several distinct forms of knowledge. Critically, a listener must know the specific sequence of speech sounds that comprise a word, and these word forms must be linked to representations of word meaning, syntactic properties, and orthographic form. The focus of the present chapter is to provide the neural basis of lexical access for spoken words. We consider three functional challenges in guiding our specification of the neurobiological foundations of lexical access. Together, these help us to account for the remarkable speed, robustness, and flexibility of human spoken word recognition.

44.2 THREE CHALLENGES FOR LEXICAL ACCESS IN SPEECH

A typical spoken word in connected speech lasts less than half a second, during which time we must recognize one specific word from the thousands that we know. Words are fleeting and transient, speech sounds unfold rapidly in time, and we cannot listen again to earlier speech sounds (unlike regressive eye movements in reading) to compensate for uncertain information heard at the start of the current spoken word (e.g., in distinguishing between *speaker* and *beaker*; Allopenna, Magnuson, & Tanenhaus, 1998). In connected speech there may be even longer delays before the arrival of critical information. For example, listeners may use subsequent words to resolve perceptual ambiguity when short words also match the start of longer words (Davis, Marslen-Wilson, & Gaskell, 2002), or to correctly interpret word-final coarticulation (Gaskell & Marslen-Wilson, 2001).

The first of our three challenges arises because spoken words unfold so rapidly, and yet the neurobiological system that supports spoken word recognition must use information that is spread out in time. Lexical access requires an internal memory that retains a record of prior speech to guide perception of current speech sounds or words (the TRACE in McClelland & Elman, 1986, or recurrent context units in Gaskell & Marslen-Wilson, 1997). In Section 44.3.2 of this chapter, we link this internal memory to a processing hierarchy in anterior regions of the superior

[1]It can be unclear, for example, whether inflections ("drives," "driving") and derivations ("driver") are counted as additional known words despite being predictably given knowledge of the stem ("drive"). Consideration of morphological relationships in the lexicon is of great importance but is beyond the scope of the present chapter, which concerns the recognition of monomorphemic words (see Chapter 13; Marslen-Wilson & Tyler, 2007).

Neurobiology of Language. DOI: http://dx.doi.org/10.1016/B978-0-12-407794-2.00044-4

and middle temporal gyrus (STG/MTG). Neurons in more anterior regions integrate information over progressively longer time periods (i.e., have longer temporal receptive fields) to support spoken word identification and recognition of longer prosodic elements.

A second challenge in recognizing spoken words is to identify a single spoken word despite noise and variability in the form of speech sounds. Human spoken word recognition is robust to the range of speakers that we converse with, each with characteristic fundamental and formant frequencies (Smith, Patterson, Turner, Kawahara, & Irino, 2005), speech rates (Summerfield, 1981), or accents (Clarke & Garrett, 2004). This chapter proposes that a key process during identification (particularly for degraded spoken words) is that listeners retrieve the speaker's intended articulatory gestures (or equivalently) and determine how they would say the words that they hear. This computational goal is supported by the sensorimotor or dorsal processing pathway in the temporoparietal junction (TPJ). This dorsal pathway is known to link the words that we hear to motor representations used in speaking (Hickok & Poeppel, 2000; Scott & Johnsrude, 2003). However, the claim that this pathway contributes to speech perception remains controversial.

In accessing the meaning of spoken words, a third computational requirement is that we deal flexibly with the lexical and semantic ambiguity that is ubiquitous in spoken language (Piantadosi, Tily, & Gibson, 2012; Rodd, Gaskell, & Marslen-Wilson, 2002). Ambiguity occurs because spoken language fails to distinguish words that differ orthographically (e.g., *knight/night*) and/or semantically (e.g., *bark*) or due to the flexible meanings of polysemous words (e.g., *run*; Rodd et al., 2002). In such cases, additional neural processes are recruited to support retrieval of multiple meanings and for the selection of the contextually appropriate meaning. Functional imaging links these processes to posterior inferior temporal regions (Rodd, Davis, & Johnsrude, 2005; Rodd, Johnsrude, & Davis, 2012) that contribute to accessing the meaning of spoken words.

44.3 MAPPING LEXICAL COMPUTATIONS ONTO NEUROBIOLOGY

A key method for isolating the brain regions that make distinct contributions to lexical access for spoken words comes from studies contrasting hemodynamic responses to familiar words to unfamiliar word-like pseudowords. This contrast has been assessed in a number of published functional imaging studies. We review these studies by summarizing a meta-analysis of 11 published PET and fMRI studies including data from 184 healthy adult participants (Davis & Gaskell, 2009). These studies report significant differential responses to spoken words and pseudowords with substantial inter-study agreement in the location and direction of neural response differences between words and pseudowords, as shown in Figure 44.1.

However, a significant challenge in interpreting the results of the Davis and Gaskell (2009) meta-analysis is that differential responses are observed in both directions: additional responses for words compared with pseudowords are observed in anterior, posterior, and inferior regions of the lateral temporal lobe and for pseudowords greater than words in peri-auditory regions of the STG. These bidirectional findings are not indicative of unreliable findings in functional imaging—the activation likelihood estimation method confirms that both response differences are statistically significant when corrected for multiple comparisons. Furthermore, bidirectional response differences were

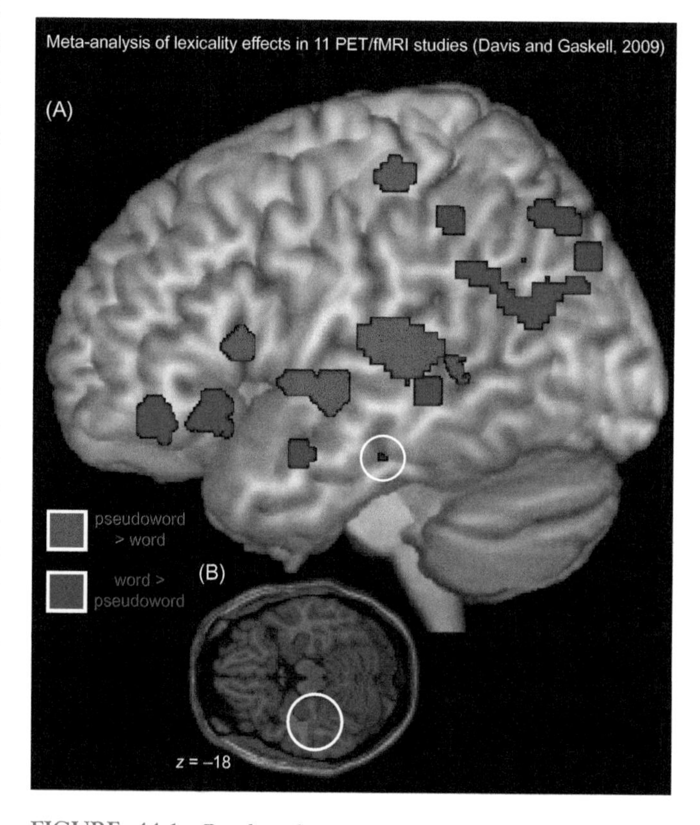

FIGURE 44.1 Results of an Activation Likelihood Estimation (ALE) meta-analysis of 11 functional imaging studies comparing neural responses to spoken words and pseudowords (from Davis & Gaskell, 2009). (A) ALE maps are thresholded at $P < 0.05$ FDR corrected with only clusters larger than 100 mm^3 displayed rendered onto a canonical brain. Greater activation for pseudowords compared with words (red) is seen primarily in peri-auditory regions of the STG. Additional activation for words compared with pseudowords (blue) is seen in anterior, posterior, and inferior regions of the lateral temporal lobe and adjacent parietal regions. Inset (B) shows a single axial slice at $z = -18$ to more clearly illustrate word activation in inferior temporal regions (circled).

observed in 4 of the 11 single studies when considered individually. Hence, before we can draw conclusions concerning the neurobiological systems supporting lexical access, we must first consider how these two directions of neural difference arise from brain regions supporting the recognition of spoken words.

The standard, subtractive assumption for functional imaging studies suggests that lexical processing should be localized to brain regions that respond to familiar words more than to pseudowords. However, when a similar subtraction is used to isolate stages of language processing (e.g., syntactic or semantic computations), potentially contradictory findings are observed. Many studies report additional activation for syntactically or semantically anomalous compared with well-formed sentences in regions that contribute to higher-level sentence comprehension (see Davis & Rodd, 2011, for discussion). These findings suggest that unsuccessful lexical processing could provide additional activation compared with successful recognition, and that the reverse contrast of pseudowords versus words would also highlight lexical processes. These conflicting assumptions suggest that explicit linking assumptions are required to mediate between computational and neural explanations when interpreting word versus pseudoword differences.

A recent review article on visual word recognition (Taylor, Rastle, & Davis, 2013) proposed two linking principles by which cognitive models can inform interpretation of differential activity in functional imaging studies. First, neural engagement can lead to additional responses for stimuli that are represented in a specific brain region compared with stimuli that are not (explaining standard cognitive subtractions). However, if both classes of stimuli are represented in a specific brain region, then differential processing effort leads to an increased neural response for items with less familiar representations because they are more difficult to process (explaining increased responses to anomalous sentences). In combination, these two principles combine to produce an inverted U-shape linking function depicted in Figure 44.2. The Taylor et al. (2013) article offers further background and simulations to justify these linking principles.

Predictions due to differential neural engagement are easily applied to localist models of spoken word recognition such as TRACE (McClelland & Elman, 1986) or NAM (Luce & Pisoni, 1998). These models propose a lexicon of localist nodes that represent words but not pseudowords. Thus, engagement of lexical representations would predict additional activation for words versus pseudowords. Based on these models, then, we must explain why multiple brain regions (at least three in the left temporal lobe alone) show this profile in the Davis and Gaskell (2009)

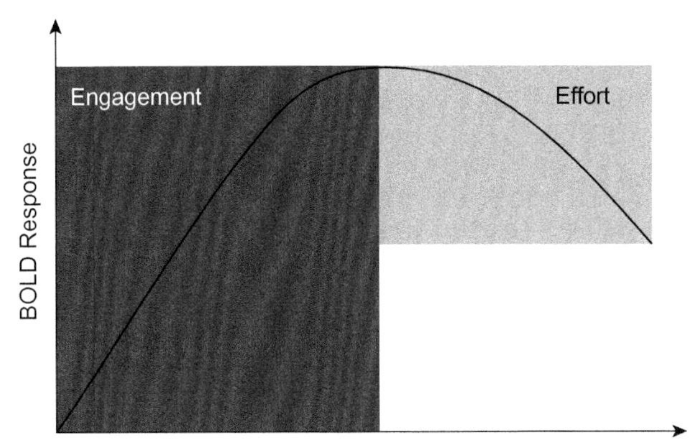

FIGURE 44.2 Illustration of an inverted U-shape function proposed to relate the goodness of fit between the current speech input and neural representations and the observed BOLD response measured with functional MRI (following Taylor et al., 2013). This inverted U function arises due to a large increase in BOLD response for stimuli that engage a specific neural system (engagement) and a reduced neural response for stimuli that are processed more easily (effort). For example, a system representing phonetic form (such as the STG) will produce a minimal response to nonspeech stimuli (no engagement), a large response to pseudowords (engaged with maximal processing effort), and an intermediate response for highly familiar words (engaged with minimal effort). Conversely, a brain region representing the meaning of familiar words (such as the ITG) will respond minimally to pseudowords and maximally to low-familiarity or ambiguous words, and will produce a reduced response to words that are high-familiarity, primed, or that can be processed with reduced effort. A similar inverted U-shape profile is described by Price and Devlin (2011) motivated by predictive coding principles.

meta-analysis. Furthermore, although additional activation for words is straightforward, localist models struggle to explain the reverse observation in the STG. Feedforward accounts of spoken word recognition (such as Shortlist; Norris, 1994) provide no mechanism by which lexical knowledge can modulate activity in prelexical processes. Even interactive models such as TRACE propose top-down excitatory connections, which seem likely to increase rather than decrease responses to familiar words compared with pseudowords.

Conversely, models that include distributed representations of the lexical form (such as the Distributed Cohort Model of Gaskell & Marslen-Wilson, 1997, hereafter DCM) assume that representations of the phonetic form are common to words and pseudowords. In these distributed models, neural responses associated with sublexical processing would be increased for pseudowords relative to words because increased familiarity for words leads to reduced processing effort. Therefore, these models have a ready explanation for increased activation for pseudowords in the STG. The following section assesses a further prediction made by the DCM that additional responses for pseudowords should overlap with brain regions in which neural activity differs for classes of words (e.g., high versus low neighborhood items) that require more or less processing effort for recognition.

44.3.1 Processing Effort and STG Responses to Familiar and Unfamiliar Words

A range of functional imaging studies has localized processing effort during the recognition of spoken words to regions of the STG that respond more strongly to pseudowords than to words. For example, Gagnepain et al. (2008) contrasted neural responses to degraded spoken words and pseudowords, half of which were primed in a study phase prior to participants entering the scanner. They report neural interactions between lexicality and priming in bilateral regions of the STG; additional responses for pseudowords clearly overlap with regions showing repetition suppression for spoken words. Other studies have similarly localized immediate repetition suppression for spoken words to peri-auditory regions of the STG (Cohen, Jobert, LeBihan, & Dehaene, 2004; Kouider, de Gardelle, Dehaene, Dupoux, & Pallier, 2010). Studies using semantic priming with spoken words also localize reduced neural responses for primed words to regions of the STG that respond more to pseudowords (Kotz, Cappa, von Cramon, & Friederici, 2002; Rissman, Eliassen, & Blumstein, 2003; Wible et al., 2006). Finally, reduced activity for spoken words with phonological forms that are more easily recognized (due to having fewer lexical competitors) localizes to the same regions of the STG. This finding has been reported for comparisons of words with smaller and larger phonological neighborhoods (Okada & Hickok, 2006), for words without versus with onset-embedded competitors (such as *claim*, which contains the word *clay*; Bozic, Tyler, Ives, Randall, & Marslen-Wilson, 2010), and for pseudowords that diverge earlier from existing words and are thus more phonologically distinct from real words (Zhuang, Tyler, Randall, Stamatakis, & Marslen-Wilson, 2014). Additional processing effort in recognition of spoken words explain additional responses to pseudowords in the STG and all these observations of differential responses during recognition of familiar words.

Despite this compelling correspondence between neural data and computational predictions, we must still explain additional responses to real words compared with pseudowords in several left[2] temporal regions: anterior STG/MTG, anterior fusiform, and posterior MTG, and inferior temporal gyrus (ITG). Outside the temporal lobe, additional responses to real words are also observed in the TPJ (including supramarginal gyrus [SMG] and adjacent regions of posterior STG) and the ventral and anterior portions of the inferior frontal gyrus (IFG orbitalis). For these temporal lobe regions, we invoke the first linking principle described by Taylor et al. (2013) and propose that these activations reflect neural representations that are uniquely engaged for familiar words. However, although it might appear self-evident that these responses delineate the neural basis for lexical processing, we must still explain why several different neural systems show an elevated response to familiar words. This redundancy is not anticipated by typical cognitive models that propose a unitary lexical system or domain-specific lexicons for reading, listening, and speaking. However, these three systems might serve to address the three computational challenges involved in spoken word recognition. We assess this proposal by considering other imaging contrasts that also activate each of these temporal lobe regions.

44.3.2 Anterior Temporal Lobe Contributions to Integrating Speech Over Time

The first system to consider is the anterior temporal region that responds more strongly to spoken words than to pseudowords (Davis & Gaskell, 2009). This ventral processing pathway for spoken language comprehension has been described in detail by Scott and colleagues (Rauschecker & Scott, 2009; Scott & Johnsrude, 2003). They designate the anterior STG/MTG as an auditory association area by analogy with homologous regions in the auditory system of macaque monkeys (Scott & Johnsrude, 2003). It seems reasonable to assume that associative knowledge is similarly critical for the recognition of familiar spoken words. However, other authors (notably Hickok & Poeppel, 2007) have cast doubt on these claims, instead proposing that anterior regions of the temporal lobe support combinatorial processing of sentences.

Consistent with the claim that anterior temporal regions contribute to combinatorial processing, functional imaging studies using spoken sentences are more likely to report activation of anterior temporal regions than studies using single words or syllables (see a meta-analysis by DeWitt & Rauschecker, 2012). Among the contrasts that show reliable anterior temporal activity are comparisons of more and less

[2]In summarizing this meta-analysis, we focus on responses in the left hemisphere; however, many of the differential responses reported are similarly observed in homologous regions of the left hemisphere. Some authors—most notably Bozic et al. (2010)—have argued that lexical access processes are supported by the temporal lobe bilaterally. However, the extended lexical network (including associated frontal regions) appears to be largely left-lateralized.

intelligible sentences (Scott, Blank, Rosen, & Wise, 2000). Those studies that compare responses to different acoustic forms of intelligible speech confirm that more anterior regions of the STG and MTG respond to intelligible speech in an acoustically invariant manner, consistent with higher-level linguistic processes (Davis & Johnsrude, 2003; Evans et al., 2014; but see Okada et al., 2010). Studies that manipulate sentence coherence and speech intelligibility report an interaction between these factors in anterior regions of the MTG (Davis, Ford, Kherif, & Johnsrude, 2011) with delayed responses in more anterior MTG regions, consistent with temporal integration. Similar findings are apparent in analysis of the phase of the BOLD responses to sentences in anterior STG regions (Dehaene-Lambertz et al., 2006), or in studies using dynamic causal modeling (Leff et al., 2008). For clearly spoken materials, studies contrasting sentences versus word lists consistently show anterior STG activation (Mazoyer et al., 1993) with additional activation in overlapping regions due to the presence of either sentential prosody or syntactic structure (Humphries, Binder, Medler, & Liebenthal, 2006). Another fMRI study shows activity in anterior temporal regions during narrative comprehension is correlated over subjects and perturbed by reordering larger units in progressively more anterior regions (Lerner, Honey, Silbert, & Hasson, 2011).

These functional imaging findings point to a hierarchy of temporal receptive fields in more anterior regions of the superior and middle temporal gyri. More anterior regions respond preferentially during sentence comprehension, particularly for connected speech containing familiar lexical items and supra-lexical elements (e.g., prosodically well-formed phrases). Within this temporal hierarchy, then, additional responses to familiar words would be expected at the relevant level of temporal structure (the 300- to 500-ms duration of typical spoken words in connected speech). Such regions serve as an internal memory for prior speech to guide perception of current speech sounds and words (TRACE in McClelland & Elman, 1986, or recurrent context units in Gaskell & Marslen-Wilson, 1997). Interestingly, a similarly anterior temporal system has been proposed as the neural basis of auditory echoic memory for speech in short-term memory experiments (Buchsbaum, Olsen, Koch, & Berman, 2005). We argue that this is a critical part of the lexical system and retains a representation of recently heard spoken words.

44.3.3 Temporoparietal Regions Link Auditory and Motor Representations of Spoken Words

The second neural system that responds more to familiar words than to pseudowords is the TPJ. This is the largest cluster highlighted in the Davis and Gaskell (2009) meta-analysis and extends over a number of cortical areas, including posterior portions of the STG (classically, Wernicke's area), and adjacent parietal regions in the supramarginal and angular gyri. This anatomy broadly corresponds to the temporoparietal portion of the dorsal processing pathway, which is the main point of agreement between dual-pathway accounts by Scott (Rauschecker & Scott, 2009; Scott & Johnsrude, 2003), and Hickok and Poeppel (2007). Both have proposed that this temporoparietal pathway forms an auditory-motor interface, mapping heard speech onto the premotor and motor representations used to control speech production. However, whereas others have emphasized nonlexical contributions of the dorsal stream (e.g., in supporting nonword repetition; Hickok, Buchsbaum, Humphries, & Muftuler, 2003), this chapter follows Gow (2012) in proposing this as a "dorsal lexicon" critical for mapping heard words onto motor representations involved in producing these words aloud. We further propose that this system supports processes that contribute to the recognition of degraded or acoustically ambiguous spoken words and that guide perceptual learning.

The clearest evidence that TPJ regions contribute to lexical processing is the observation of increased responses to spoken words versus pseudowords (Davis & Gaskell, 2009). This cannot be explained by additional processing effort or task difficulty (which would typically be greater for pseudowords), but instead it suggests neural representations that are specifically engaged by spoken words. Further, evidence for lexical representations in the TPJ is that exactly the same region is activated for words versus pseudowords in visual word recognition (Taylor et al., 2013) and in short-term memory tasks with written and spoken words (Buchsbaum et al., 2005). This is in contrast to the anterior temporal system, which is seldom activated in studies using single written words (although activation is observed for written sentences and narratives; Spitsyna, Warren, Scott, Turkheimer, & Wise, 2006).

The cross-modal convergence of lexicality effects for spoken and written words in TPJ is compatible with two distinct functional proposals discussed by Taylor et al. (2013): (i) TPJ contributes to semantic processing (Binder, Desai, Graves, & Conant, 2009) or (ii) TPJ supports a multimodal phonological lexicon shared between speech and print (the sensorimotor account proposed by Hickok & Poeppel, 2007). Functional imaging evidence is more consistent with the second of these proposals because purely phonological manipulations lead to differential TPJ activation. For example, repetition suppression effects for pseudowords and isolated syllables localize to posterior STG and adjacent parietal regions such as the SMG

(Jacquemot, Pallier, LeBihan, Dehaene, & Dupoux, 2003). The SMG has been associated with phonological change responses for sine-wave stimuli perceived as speech, but not equivalent nonspeech (Dehaene-Lambertz et al., 2005). Furthermore, changes before and after training with sine-wave speech are correlated with individual differences in the strength of categorical responses in the SMG (Desai, Liebenthal, Waldron, & Binder, 2008).

Studies that manipulate the difficulty of lexical processing provide further evidence that phonological processes in the TPJ contribute to lexical access in speech perception and speech production. For instance, additional neural responses in posterior superior temporal regions are seen during perception and production of words with greater phonological complexity (due to more consonant clusters; Tremblay & Small, 2011). Similarly, words with higher neighborhood density (i.e., that have more similar sounding competitors) require greater phonological specificity for identification and are associated with increased activation of the SMG (Prabhakaran, Blumstein, Myers, Hutchison, & Britton, 2006) during speech perception and similarly in the SMG during speech production (Peramunage, Blumstein, Myers, Goldrick, & Baese-Berk, 2011). These results are distinct from the findings reviewed in Section 44.3.2 because mid-STG activity is associated with the difficulty of lexical access during spoken word recognition (Bozic et al., 2010; Okada & Hickok, 2006), but not during speech production. However, there are inconsistencies in lexical difficulty effects observed in spoken word recognition (see Vaden, Piquado, & Hickok, 2011). These may be resolved by using computationally informed measures of processing difficult (such as prediction error) combined with time-locked measures of neural activity (MEG or EEG; see Gagnepain, Henson, & Davis, 2012).

How, then, do these TPJ regions—and the dorsal auditory-motor pathway more broadly—contribute to the comprehension of spoken words? The introduction to this chapter highlighted the problems faced by listeners who must understand words despite variation in their acoustic form (due to different speakers, accents, acoustic degradation, etc.). For a listener attempting to repeat back single words, many different acoustic forms should lead to the same motor response (assuming that words are correctly perceived). In the context of a repetition task, then, the computational goal of the auditory-motor pathway should be to take variable or degraded speech as an input, to abstract from this acoustic or phonetic variability, and to extract the correct motor commands or articulatory gestures that the listener should use to repeat the speech that was heard. Although normalization for variability in visual processing is traditionally assumed to be part of the ventral visual processing stream (Ungerleider & Mishkin, 1982), speech is different from vision because we not only hear speech but also produce it.[3] For verbal communication to be successful, listeners must be able to produce sounds that lead to essentially the same phonological percept as the speech that they hear (Liberman & Whalen, 2000). That is, listeners in a repetition task should be able to retrieve the same articulatory gestures as those produced by the person speaking. In this way, the goal of the auditory-motor pathway during repetition (though perhaps not the goal of the perceptual system as a whole or during all tasks) embodies a key proposal of the motor theory of speech perception that the similarity of heard and spoken necessitates a direct mapping between sensory and motor representations of speech (Liberman & Whalen, 2000; Chapter 15, but see Chapter 16).

Accessing the intended articulatory gesture for a heard spoken word is thus one key outcome in the perceptual identification of degraded speech (Davis & Johnsrude, 2007). Evidence for activation of articulatory representations comes from imaging studies showing increased inferior frontal and motor recruitment during perception of degraded speech compared with clear speech (Adank & Devlin, 2010; Davis & Johnsrude, 2003), even if the current task does not require access to articulatory representations (Hervais-Adelman, Carlyon, Johnsrude, & Davis, 2012). Recent fMRI studies using multivoxel pattern analysis methods show that (particularly for degraded or ambiguous syllables) frontal and motor regions represent the phonological content of heard speech (Du, Buchsbaum, Grady, & Alain, 2014), although this study included overt phoneme categorization tasks. Venezia, Saberi, Chubb, and Hickok (2012) showed that activity in these frontal and motor regions is correlated with response bias in phoneme decision tasks and, hence, argue that frontal regions are involved in response selection but do not contribute to perception *per se* (Binder, Liebenthal, Possing, Medler, & Ward, 2004). However, other multivoxel pattern analysis studies have shown that inferior frontal (Lee, Turkeltaub, Granger, & Raizada, 2012) and motor areas (Evans & Davis, submitted) encode the perceptual content of speech even when no response is

[3]One exception is that skilled artists can generate images (e.g., using a pencil on paper) that reproduce essential aspects of their subjective visual experience. However, this ordinarily requires substantial training and many (including this author) never achieve even limited proficiency at this. Note that the same is not true for speech. One of the earliest foundations for spoken language is the ability to reproduce (with the voice) the auditory patterns that one hears (i.e., verbal repetition).

required to indicate this perceptual content. We take these findings as evidence against the view that frontal and motor responses are only involved in response selection. At the same time, we acknowledge that there is less functional imaging evidence for motor responses to clear speech or in tasks focused on natural comprehension (see Scott, Mcgettigan, & Eisner, 2009 for a review). We have previously shown that frontal activation is both specific to degraded speech and linked to attentive perception because it is absent during distraction (Wild et al., 2012). Similarly, it is only in challenging listening situations—such as when the speech signal is perturbed by noise (Meister, Wilson, Deblieck, Wu, & Iacoboni, 2007) or made perceptually ambiguous (Möttönen & Watkins, 2009)—that TMS disruption of motor regions leads to perceptual impairments (though, again, Venezia et al., 2012 have argued against this).

One possible contribution of the dorsal auditory-motor pathway to speech perception that is consistent with the evidence reviewed is to support perceptual learning or adaptation. Perceptual learning is apparent in a range of circumstances in which listeners hear spoken words that are artificially degraded (by noise-vocoding, Hervais-Adelman, Davis, Johnsrude, & Carlyon, 2008; or by sine-wave synthesis, Remez et al., 2011). For both these forms of artificial distortion, prior exposure enhances perception of trained and untrained words (Dahan & Mead, 2010; Hervais-Adelman, Davis, Johnsrude, Taylor, & Carlyon, 2011). For vocoded speech, perceptual learning is enhanced by prior knowledge of speech content (Hervais-Adelman et al., 2008) and feedback from lexical representations (Davis, Johnsrude, Hervais-Adelman, Taylor, & McGettigan, 2005) or visual speech (Wayne & Johnsrude, 2012). Evidence to tie perceptual learning to the dorsal auditory-motor pathway is that lexical feedback is equally successful in enhancing perceptual learning when it is supplied in written or spoken form (Davis et al., 2005). Thus, perceptual learning is plausibly linked to convergence of written and spoken lexical representations in TPJ regions. However, existing functional imaging evidence has inconsistently highlighted frontal (Eisner, McGettigan, Faulkner, Rosen, & Scott, 2010), motor (Hervais-Adelman et al., 2012), and thalamus (Erb, Henry, Eisner, & Obleser, 2013) as supporting perceptual learning. A recent tDCS study provides further evidence for inferior frontal contributions to perceptual learning (Sehm et al., 2013).

A link between lexical knowledge and frontal/TPJ activity is also apparent for a second perceptual learning paradigm first reported by Norris, McQueen, and Cutler (2003). In this paradigm, phonetic category boundaries are systematically altered after exposure to ambiguous segments (e.g., a segment intermediate between /f/ and /s/) in the context of disambiguating familiar words ("peace" or "beef"), but not in pseudowords ("dreace" or "dreef"; Norris et al., 2003). Regarding degraded speech, perceptual learning generalizes to untrained words consistent with a prelexical locus for learning (McQueen, Cutler, & Norris, 2006). Functional imaging evidence links perceptual learning to auditory-motor pathways because ambiguous segments in contexts that support learning activate posterior temporal and adjacent parietal regions (Myers & Blumstein, 2008) in addition to the inferior frontal regions activated for nonlexical segment disambiguation (Blumstein, Myers, & Rissman, 2005). MEG data suggests top-down causal influences from SMG to posterior STG regions during perception of Ganong stimuli (Gow, Segawa, Ahlfors, & Lin, 2008) similar to those that induce perceptual learning. Functional imaging during periods of audio-visually induced phoneme category boundary changes has also been shown to be associated with activation of inferior frontal and parietal regions (angular gyrus; Kilian-Hütten, Vroomen, & Formisano, 2011).

Another situation in which perceptual learning enhances comprehension is when listeners hear an unfamiliar or artificial accent (Clarke & Garrett, 2004). As previously described for vocoded speech, parallel presentation of a written transcription supports perceptual learning of accented speech (e.g., movie subtitles; Mitterer & McQueen, 2009). Furthermore, requiring listeners to imitate the novel accent (i.e., engaging audio-motor pathways) also enhances learning (Adank, Hagoort, & Bekkering, 2010). Functional imaging evidence to link perceptual learning of accented speech to TPJ regions comes from the observation that posterior STG is activated during perception of accented speech relative to speech in noise of equivalent intelligibility (Adank, Davis, & Hagoort, 2012) and that accent repetition leads to suppression of activity in posterior STG and SMG (Adank, Noordzij, & Hagoort, 2012). These results suggest a role for auditory-motor representations during accent perception that parallels the involvement of this same system in other forms of lexically driven perceptual learning. Thus, evidence from three different forms of perceptual adaptation is consistent with auditory-motor mappings for familiar words in the dorsal auditory pathway being key to the accurate perceptions of the degraded or variable speech that are part of the everyday experience of spoken words for listeners.

44.3.4 Posterior Middle and Inferior Temporal Regions Map Spoken Words onto Meaning

The ultimate goal of listeners hearing spoken words is surely to access the meaning that the speaker intended. Determining the anatomical and functional organizations of neural systems that support access to meaning from spoken words and distinguishing this process from semantic representation *per se* has proved challenging. For a time it was unclear whether any single neural system supports semantic memory: recent reviews on semantic processing of language stimuli argued for the involvement of multiple cortical areas. For example, a functional imaging meta-analysis of semantic processing of spoken and written words (Binder et al., 2009) highlighted a semantic system that encompasses most of the lateral temporal lobe and inferior parietal lobe, including both the anterior and posterior pathways reviewed previously. This is perhaps because Binder included the contrast of spoken and written words versus pseudowords as a means of identifying semantic regions. This may highlight neural systems involved in processing the form of familiar words as well as their meaning. Other semantic contrasts assessed by Binder et al. (2009) involve comparing semantic and phonological tasks or more/less meaningful utterances. These comparisons may be less sensitive to form-based processes, although they may instead highlight brain regions that are sensitive to effects of task or comprehension difficulty (see Davis & Rodd, 2011 for discussion).

A more focal localization of semantic processing is suggested by meta-analyses that contrast the neural representation of different types of semantic information also reported by Binder et al. (2009). Binder and colleagues showed activation during semantic processing of words referring to actions in posterior portions of the MTG. However, inferior temporal and fusiform regions are activated in tasks that require access to perceptual attributes of the referents of spoken or written words (e.g., in responding to words with concrete meanings). These findings may explain some of the other regions activated for word versus pseudoword contrasts in Davis and Gaskell (2009) and Taylor et al. (2013): many of the studies reported use of tasks that involve access to word meaning (e.g., lexical decision; Plaut, 1997) and use of words with object and action associations. Thus, activation of posterior middle and inferior temporal regions for familiar words compared with pseudowords might reflect responses to perceptual or action semantic attributes.

The semantic feature specificity reported by Binder et al. (2009) is consistent with a distributed, embodied view of semantic processing in which some forms of conceptual information (visual appearance, associated actions, etc.) are grounded in sensory and motor cortices (see Martin, 2007; Chapter 62 for reviews). Although these data rule out purely symbolic accounts of semantic cognition, strong and weak embodied theories remain to be distinguished (Meteyard, Cuadrado, Bahrami, & Vigliocco, 2012). One key question concerns whether embodied semantic representations are accessed directly from relevant perceptual representations (e.g., from superior temporal regions, in the case of spoken words; Martin, 2007) or whether semantic representations are accessed via an amodal semantic hub localized in the ventral anterior temporal lobe (Chapter 61; Patterson, Nestor, & Rogers, 2007).

Considerable evidence from neuroimaging suggests that these anterior temporal regions are activated during semantic processing (Visser, Jefferies, & Lambon Ralph, 2009), particularly in studies that overcome the susceptibility artifacts that lead to loss of signal in conventional fMRI acquisitions (Devlin et al., 2000; Visser & Lambon Ralph, 2011). Evidence for anterior temporal lobe contributions to semantic processing also comes from neuropsychological deficits in individuals with focal neurodegeneration of ventral temporal regions (specifically, semantic dementia; see Patterson et al., 2007) and from semantic impairments observed after rTMS to similar regions (Pobric, Jefferies, & Ralph, 2007). However, one striking aspect of all these observations is that the semantic contribution observed is domain general—impacting all modalities of input (written and spoken language, pictures, etc.) and multiple output modalities (naming, drawing, familiarity judgment—Patterson et al., 2007). Thus, this semantic hub may be responsible for binding together (and maintaining the links between) distributed representations that themselves code for the form and meaning of spoken words. In other work, we have proposed that the semantic hub may play a mnemonic function in acquiring and maintaining lexical representations (perhaps due to projections to medial temporal regions) rather than a purely linguistic function (Davis & Gaskell, 2009).

In contrast, more posterior portions of the MTG and ITG are more specifically associated with semantic processing of spoken words (Gow, 2012; Lau, Phillips, & Poeppel, 2008) and may explain the observation of increased activation for spoken words compared with pseudowords in the Davis and Gaskell (2009) meta-analysis. Evidence from imaging studies using semantic priming manipulations also show reductions in neural activity in the same posterior temporal regions for spoken word pairs that are semantically related (Rissman et al., 2003; Wible et al., 2006). Other manipulations of the ease of meaning access—such as variations in word frequency—also modulate activity in these regions (e.g., reduced activity for higher-frequency words in Hauk, Davis, Kherif, & Pulvermüller, 2008). This inverted U-shape response—neural engagement for words compared with pseudowords and a decreased response to words that are less-effortfully processed—corresponds closely with expectations for a region involved in accessing semantic representations given the inverted U-shape in Figure 44.2. These findings provide further evidence that neural processes involved in accessing word meaning localize to posterior inferior temporal regions.

Another semantic challenge that activates posterior middle and inferior temporal regions during comprehension of spoken language is seen in studies exploring neural responses to lexical-semantic ambiguities in connected speech. High-ambiguity sentences (such as "there were *dates* and *pears* in the fruit bowl") require additional semantic processes to activate multiple possible meanings of ambiguous words (e.g., dates/pears) and select appropriate meanings in context. By comparison with matched, low-ambiguity sentences ("there was beer and cider on the kitchen shelf"), comprehension of these sentences produces additional activation in posterior middle and inferior temporal gyri and adjacent fusiform regions (Rodd et al., 2005) as well as in inferior frontal regions. Similar combinations of inferior frontal and posterior temporal responses have been obtained for sentences in which lexical ambiguities lead to syntactic as well as semantic selection (Rodd, Longe, Randall, & Tyler, 2010) and for sentences in which the timing of the initial ambiguity and subsequent disambiguation is varied (Rodd, Johnsrude, & Davis, 2012).

Neuropsychological data from brain-injured patients also suggest impaired spoken word comprehension following lesions to posterior portions of the MTG and ITG following stroke (Bates et al., 2003; Dronkers, Wilkins, Van Valin, Redfern, & Jaeger, 2004) and in some, but not all, studies of neurodegenerative conditions such as semantic dementia (Peelle et al., 2008). Further evidence to link posterior activation to successful semantic processing comes from activation during graded anesthetic sedation that disrupts comprehension. Propofol sedation attenuates activation in posterior temporal regions (Davis et al., 2007) in a semantic ambiguity paradigm. Yet, posterior ITG activation remains reliable for partially sedated participants who are able to make accurate semantic judgments on spoken words (Adapa, Davis, Stamatakis, Absalom, & Menon, 2014). Such findings further implicate posterior inferior regions of the temporal lobe in accessing semantic representations of heard words (consistent with the proposal made by Hickok & Poeppel, 2007).

44.4 FUNCTIONAL SEGREGATION AND CONVERGENCE IN LEXICAL PROCESSING

This chapter has presented the functional organization of three temporal lobe pathways that contribute to lexical access for spoken words. A sketch of the anatomical and functional organization of this tripartite model is shown in Figure 44.3. This account suggests considerable functional segregation, with three distinct pathways contributing to recognition and comprehension of spoken words.

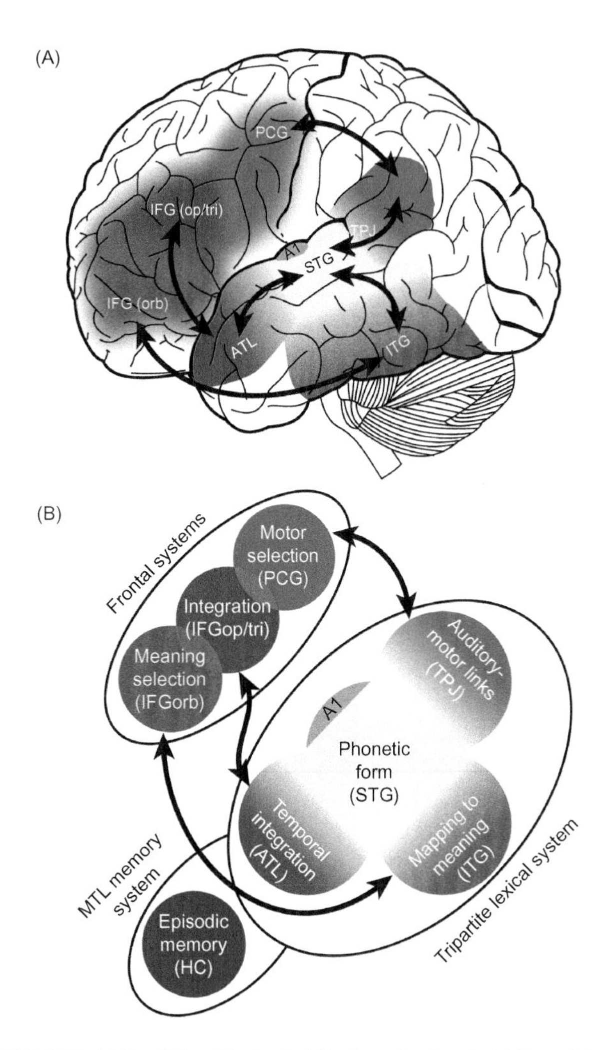

FIGURE 44.3 Tripartite Lexical System. Anatomical (A) and functional (B) organizations of the Tripartite Lexical System and associated frontal and medial temporal systems. The sounds of spoken words are coded in primary regions (A1, gray) before processing of phonetic forms in the STG (yellow). Representations of speech sounds in the STG are combined with three other systems during lexical processing: (i) they are temporally integrated with hierarchical representations of prior segmental and lexical context in anterior regions of the STG/MTG (yellow–red, ATL); (ii) they are mapped onto articulatory representations via auditory-motor links in the TPJ (yellow–blue); and (iii) they are mapped onto meaning representations in posterior and inferior portions of the temporal lobe (yellow–green, ITG). Associated frontal systems show a graded form of the same specialization based on convergent connectivity: precentral gyrus (PCG; blue), inferior frontal gyrus opercularis/triangularis (IFGop/tri; red), and inferior frontal gyrus orbitalis (IFGorb; green). Convergent connectivity among the three parts of the lexical system is apparent in two other systems: re-entrant connections to the STG (in A) and in MTL/HC regions (purple in B, not shown in A) that code episodic representations of words and associated contextual information during lexical learning.

One challenge for this tripartite account is to specify how particular tasks or behavioral goals are achieved. This is particularly true for listening situations that challenge one or more processing streams (e.g., degraded speech sounds, ambiguous word). Listeners hearing an unexpected word in a sentence must determine whether this is the result of mishearing the current word, erroneous segmental or lexical predictions, an unfamiliar accent, ambiguity in word meaning, or some other error. Each of these causes of misidentification brings specific computational challenges or repair operations associated with one or more processing pathway.

One common observation across a range of different experimental situations that involve effortful listening to speech is that inferior frontal, insula, and motor regions show additional activation compared with easy or effortless comprehension conditions (see Chapter 40). Anatomical evidence would suggest that these prefrontal regions receive convergent projections from the three posterior pathways (Scott & Johnsrude, 2003) and, in turn, provide re-entrant connections back to posterior regions. A key function of frontal connections may be to support learning processes that guide interpretation of noisy or ambiguous speech input (Davis & Johnsrude, 2007). However, frontal mechanisms can also support other processes that contribute to lexical access—such as competitive selection of single lexical items (Bozic et al., 2010; Prabhakaran et al., 2006), maintenance of recognized words in working memory (Rodd et al., 2012), combining multiple words to infer sentence level meaning (Hagoort, 2005; Chapter 28), and others. These and other higher-level processes will be actively engaged during effortful comprehension of speech (see Chapter 40; Guediche, Blumstein, Fiez, Holt, & Peelle, 2014).

Another form of functional convergence that may be critical for successful recognition of spoken words are top-down projections from frontal and lateral temporal regions back to earlier auditory and peri-auditory regions. Direct evidence for top-down connections has been difficult to obtain from fMRI data (Davis et al., 2011; Guediche, Salvata, & Blumstein, 2013). However, evidence from MEG/EEG (Arnal, Wyart, & Giraud, 2011; Sohoglu, Peelle, Carlyon, & Davis, 2012) is more clearly consistent with the top-down mechanism as proposed in predictive coding theories (Rao & Ballard, 1999). One such proposal is that spoken word recognition is achieved by comparing lexically informed predictions for upcoming segments with incoming speech sounds (Gagnepain et al., 2012). Prediction errors are propagated forward from early auditory regions and provide a bottom-up signal to support or reject previously activated lexical candidates. This predictive coding view is supported by evidence that phonological mismatch responses in EEG studies are time-locked to deviating or absent segments in expected spoken words (Bendixen, Scharinger, Strauß, & Obleser, 2014; Desroches, Newman, & Joanisse, 2009). Furthermore, MEG responses to spoken words and pseudowords in the STG and changes due to learning and consolidation study are uniquely explained by computational simulations in which word recognition is achieved using prediction error signals (Gagnepain et al., 2012).

A final form of convergence involves projections from lateral to medial temporal lobe (MTL) regions including the hippocampus (HC) (Davis & Gaskell, 2009). Connections to the MTL are proposed to play a key role in initial learning of phonological forms (Davis, Di Betta, Macdonald, & Gaskell, 2009) or form-meaning links (Breitenstein et al., 2005). MTL regions are commonly co-activated with lateral temporal regions (such as the STG) when recognizing recently learned spoken words (Takashima, Bakker, van Hell, Janzen, & McQueen, 2014), and the STG shows enhanced coupling with the hippocampus during memory encoding for spoken words (Gagnepain et al., 2011). Involvement of the MTL in initial acquisition of novel spoken words is consistent with complementary learning system theories of lexical learning (Davis & Gaskell, 2009). One proposal of the CLS theory is that overnight sleep after learning plays a key role in consolidating neocortical representations of newly acquired words. This has support from overnight changes in cortical responses to novel words (Davis et al., 2009; Gagnepain et al., 2012; Takashima et al., 2014) and correlations between sleep-spindles measured during postlearning slow-wave sleep and individual differences in lexical integration of novel words (Tamminen, Payne, Stickgold, Wamsley, & Gaskell, 2010). One question that remains, however, is whether these MTL regions have functional overlap with, or are distinct from, the anterior temporal semantic hub proposed to bind the form and meaning of familiar spoken words (Patterson et al., 2007).

44.5 CONCLUSION

This chapter has presented a tripartite account of the brain regions that support lexical processing of spoken words. The functional goals of three temporal lobe systems have been introduced in the context of key computational challenges associated with spoken word recognition. First, listeners must integrate current speech sounds with previously heard speech in recognizing words. This motivates a hierarchy of representations that temporally integrate speech signals over time localized to anterior regions of the STG. A second challenge is that for listeners to repeat degraded words correctly requires that they recover the speakers' intended articulatory gestures. The tripartite account localizes this process to auditory-motor links mediated by TPJ regions and proposes that these links play a key role in supporting robust identification and perceptual learning of degraded spoken words. The third challenge relates to extracting meaning from spoken words, which is proposed to be supported by cortical areas in posterior ITG and surrounding regions (MTG and fusiform). Despite the three-way functional segregation that is at the heart of this triparate account, this chapter also acknowledges that reliable recognition of familiar words, optimally efficient processing of incoming speech, and learning of novel spoken words all depend on combining information between these processing pathways. This is achieved through convergent connectivity within the lateral temporal lobe, in frontal or medial temporal regions, and through top-down predictions mapped back to peri-auditory regions of the STG. A key goal for future research must be to specify these convergent mechanisms in more detail and derive precise computational proposals for how the tripartite lexical system supports the everyday demands of human speech comprehension.

Acknowledgment

This research was funded by the Medical Research Council MC-A060-5PQ80.

References

Adank, P., Davis, M. H., & Hagoort, P. (2012). Neural dissociation in processing noise and accent in spoken language comprehension. *Neuropsychologia*, *50*(1), 77–84.

Adank, P., & Devlin, J. T. (2010). On-line plasticity in spoken sentence comprehension: Adapting to time-compressed speech. *NeuroImage*, *49*(1), 1124–1132. Available from: http://dx.doi.org/10.1016/j.neuroimage.2009.07.032.

Adank, P., Hagoort, P., & Bekkering, H. (2010). Imitation improves language comprehension. *Psychological Science*, *21*(12), 1903–1909. Available from: http://dx.doi.org/10.1177/0956797610389192.

Adank, P., Noordzij, M. L., & Hagoort, P. (2012). The role of planum temporale in processing accent variation in spoken language comprehension. *Human Brain Mapping*, *33*(2), 360–372. Available from: http://dx.doi.org/10.1002/hbm.21218.

Adapa, R. M., Davis, M. H., Stamatakis, E. A, Absalom, A. R., & Menon, D. K. (2014). Neural correlates of successful semantic processing during propofol sedation. *Human Brain Mapping*, *35*(7), 2935–2942. Available from: http://dx.doi.org/10.1002/hbm.22375.

Allopenna, P. D., Magnuson, J. S., & Tanenhaus, M. K. (1998). Tracking the time course of spoken word recognition using eye movements: Evidence for continuous mapping models. *Journal of Memory and Language*, *38*(4), 419–439. Available from: http://dx.doi.org/10.1006/jmla.1997.2558.

Altmann, G. T. M. (1997). *The Ascent of Babel: An exploration of language, mind, and understanding*. Oxford University Press.

Arnal, L. H., Wyart, V., & Giraud, A.-L. (2011). Transitions in neural oscillations reflect prediction errors generated in audiovisual speech. *Nature Neuroscience*, *14*(6), 797–801. Available from: http://dx.doi.org/10.1038/nn.2810.

Bates, E., Wilson, S. M., Saygin, A. P., Dick, F., Sereno, M. I., Knight, R. T., et al. (2003). Voxel-based lesion-symptom mapping. *Nature Neuroscience*, *6*(5), 448–450. Available from: http://dx.doi.org/10.1038/nn1050.

Bendixen, A., Scharinger, M., Strauß, A., & Obleser, J. (2014). Prediction in the service of comprehension: Modulated early brain responses to omitted speech segments. *Cortex*, *53*, 9–26. Available from: http://dx.doi.org/10.1016/j.cortex.2014.01.001.

Binder, J. R., Desai, R. H., Graves, W. W., & Conant, L. L. (2009). Where is the semantic system? A critical review and meta-analysis of 120 functional neuroimaging studies. *Cerebral Cortex*, *19*(12), 2767–2796. Available from: http://dx.doi.org/10.1093/cercor/bhp055.

Binder, J. R., Liebenthal, E., Possing, E. T., Medler, D. A., & Ward, B. D. (2004). Neural correlates of sensory and decision processes in auditory object identification. *Nature Neuroscience*, *7*(3), 295–301. Available from: http://dx.doi.org/10.1038/nn1198.

Blumstein, S. E., Myers, E. B., & Rissman, J. (2005). The perception of voice onset time: An fMRI investigation of phonetic category structure. *Journal of Cognitive Neuroscience*, *17*(9), 1353–1366. Available from: http://dx.doi.org/10.1162/0898929054985473.

Bozic, M., Tyler, L. K., Ives, D. T., Randall, B., & Marslen-Wilson, W. D. (2010). Bihemispheric foundations for human speech comprehension. *Proceedings of the National Academy of Sciences of the United States of America*, *107*(40), 17439–17444. Available from: http://dx.doi.org/10.1073/pnas.1000531107.

Breitenstein, C., Jansen, A., Deppe, M., Foerster, A.-F., Sommer, J., Wolbers, T., et al. (2005). Hippocampus activity differentiates good from poor learners of a novel lexicon. *NeuroImage*, *25*(3), 958–968. Available from: http://dx.doi.org/10.1016/j.neuroimage.2004.12.019.

Buchsbaum, B. R., Olsen, R. K., Koch, P., & Berman, K. F. (2005). Human dorsal and ventral auditory streams subserve rehearsal-based and echoic processes during verbal working memory. *Neuron*, *48*(4), 687–697. Available from: http://dx.doi.org/10.1016/j.neuron.2005.09.029.

Clarke, C. M., & Garrett, M. F. (2004). Rapid adaptation to foreign-accented English. *The Journal of the Acoustical Society of America*, *116*(6), 3647–3658. Available from: http://dx.doi.org/10.1121/1.1815131.

Cohen, L., Jobert, A., LeBihan, D., & Dehaene, S. (2004). Distinct unimodal and multimodal regions for word processing in the left temporal cortex. *NeuroImage*, *23*(4), 1256–1270. Available from: http://dx.doi.org/10.1016/j.neuroimage.2004.07.052.

Dahan, D., & Mead, R. L. (2010). Context-conditioned generalization in adaptation to distorted speech. *Journal of Experimental Psychology. Human Perception and Performance*, *36*(3), 704–728. Available from: http://dx.doi.org/10.1037/a0017449.

Davis, M. H., Coleman, M. R., Absalom, A. R., Rodd, J. M., Johnsrude, I. S., Matta, B. F., et al. (2007). Dissociating speech perception and comprehension at reduced levels of awareness. *Proceedings of the National Academy of Sciences of the United States of America*, *104*(41), 16032–16037. Available from: http://dx.doi.org/10.1073/pnas.0701309104.

Davis, M. H., Di Betta, A. M., Macdonald, M. J. E., & Gaskell, M. G. (2009). Learning and consolidation of novel spoken words. *Journal of Cognitive Neuroscience*, *21*(4), 803–820. Available from: http://dx.doi.org/10.1162/jocn.2009.21059.

Davis, M. H., Ford, M. A., Kherif, F., & Johnsrude, I. S. (2011). Does semantic context benefit speech understanding through "top-down" processes? Evidence from time-resolved sparse fMRI. *Journal of Cognitive Neuroscience*, *23*(12), 3914–3932. Available from: http://dx.doi.org/10.1162/jocn_a_00084.

Davis, M. H., & Gaskell, M. G. (2009). A complementary systems account of word learning: Neural and behavioural evidence. *Philosophical Transactions of the Royal Society B, Biological Sciences*, *364*(1536), 3773–3800. Available from: http://dx.doi.org/10.1098/rstb.2009.0111.

Davis, M. H., & Johnsrude, I. S. (2003). Hierarchical processing in spoken language comprehension. *The Journal of Neuroscience, 23*, 3423–3431. Available from: http://dx.doi.org/23/8/3423 [pii].

Davis, M. H., & Johnsrude, I. S. (2007). Hearing speech sounds: Top-down influences on the interface between audition and speech perception. *Hearing Research, 229*(1–2), 132–147. Available from: http://dx.doi.org/10.1016/j.heares.2007.01.014.

Davis, M. H., Johnsrude, I. S., Hervais-Adelman, A., Taylor, K., & McGettigan, C. (2005). Lexical information drives perceptual learning of distorted speech: Evidence from the comprehension of noise-vocoded sentences. *Journal of Experimental Psychology: General, 134*(2), 222–241. Available from: http://dx.doi.org/10.1037/0096-3445.134.2.222.

Davis, M. H., Marslen-Wilson, W. D., & Gaskell, M. G. (2002). Leading up the lexical garden path: Segmentation and ambiguity in spoken word recognition. *Journal of Experimental Psychology: Human Perception and Performance, 28*(1), 218–244. Available from: http://dx.doi.org/10.1037//0096-1523.28.1.218.

Davis, M. H., & Rodd, J. M. (2011). Brain structures underlying lexical processing of speech: Evidence from brain imaging. In P. Zwitserlood & M. G. Gaskell (Eds.), *Lexical representation: A multidisciplinary approach*. Berlin, Germany: De Gruyter Mouton.

Dehaene-Lambertz, G., Dehaene, S., Anton, J.-L., Campagne, A., Ciuciu, P., Dehaene, G. P., et al. (2006). Functional segregation of cortical language areas by sentence repetition. *Human Brain Mapping, 27*(5), 360–371. Available from: http://dx.doi.org/10.1002/hbm.20250.

Dehaene-Lambertz, G., Pallier, C., Serniclaes, W., Sprenger-Charolles, L., Jobert, A., & Dehaene, S. (2005). Neural correlates of switching from auditory to speech perception. *NeuroImage, 24*(1), 21–33. Available from: http://dx.doi.org/10.1016/j.neuroimage.2004.09.039.

Desai, R., Liebenthal, E., Waldron, E., & Binder, J. R. (2008). Left posterior temporal regions are sensitive to auditory categorization. *Journal of Cognitive Neuroscience, 20*(7), 1174–1188. Available from: http://dx.doi.org/10.1162/jocn.2008.20081.

Desroches, A. S., Newman, R. L., & Joanisse, M. F. (2009). Investigating the time course of spoken word recognition: Electrophysiological evidence for the influences of phonological similarity. *Journal of Cognitive Neuroscience, 21*(10), 1893–1906. Available from: http://dx.doi.org/10.1162/jocn.2008.21142.

Devlin, J. T., Russell, R. P., Davis, M. H., Price, C. J., Wilson, J., Moss, H. E., et al. (2000). Susceptibility-induced loss of signal: Comparing PET and fMRI on a semantic task. *NeuroImage, 11*(6 Pt 1), 589–600. Available from: http://dx.doi.org/10.1006/nimg.2000.0595.

DeWitt, I., & Rauschecker, J. P. (2012). Phoneme and word recognition in the auditory ventral stream. *Proceedings of the National Academy of Sciences of the United States of America, 109*(8), E505–E514. Available from: http://dx.doi.org/10.1073/pnas.1113427109.

Dronkers, N. F., Wilkins, D. P., Van Valin, R. D., Redfern, B. B., & Jaeger, J. J. (2004). Lesion analysis of the brain areas involved in language comprehension. *Cognition, 92*(1–2), 145–177. Available from: http://dx.doi.org/10.1016/j.cognition.2003.11.002.

Du, Y., Buchsbaum, B. R., Grady, C. L., & Alain, C. (2014). Noise differentially impacts phoneme representations in the auditory and speech motor systems. *Proceedings of the National Academy of Sciences of the United States of America, 111*(19), 7126–7131. Available from: http://dx.doi.org/10.1073/pnas.1318738111.

Eisner, F., McGettigan, C., Faulkner, A., Rosen, S., & Scott, S. K. (2010). Inferior frontal gyrus activation predicts individual differences in perceptual learning of cochlear-implant simulations. *The Journal of Neuroscience, 30*(21), 7179–7186. Available from: http://dx.doi.org/10.1523/JNEUROSCI.4040-09.2010.

Erb, J., Henry, M. J., Eisner, F., & Obleser, J. (2013). The brain dynamics of rapid perceptual adaptation to adverse listening conditions. *The Journal of Neuroscience, 33*(26), 10688–10697. Available from: http://dx.doi.org/10.1523/JNEUROSCI.4596-12.2013.

Evans, S., & Davis, M. H. (submitted). Hierarchical organisation of auditory and motor representations in speech perception: Evidence from searchlight similarity analysis.

Evans, S., Kyong, J. S., Rosen, S., Golestani, N., Warren, J. E., McGettigan, C., et al. (2014). The pathways for intelligible speech: Multivariate and univariate perspectives. *Cerebral Cortex, 24*(9), 2350–2361. Available from: http://dx.doi.org/10.1093/cercor/bht083.

Gagnepain, P., Chételat, G., Landeau, B., Dayan, J., Eustache, F., & Lebreton, K. (2008). Spoken word memory traces within the human auditory cortex revealed by repetition priming and functional magnetic resonance imaging. *The Journal of Neuroscience, 28*(20), 5281–5289. Available from: http://dx.doi.org/10.1523/JNEUROSCI.0565-08.2008.

Gagnepain, P., Henson, R., Chételat, G., Desgranges, B., Lebreton, K., & Eustache, F. (2011). Is neocortical-hippocampal connectivity a better predictor of subsequent recollection than local increases in hippocampal activity? New insights on the role of priming. *Journal of Cognitive Neuroscience, 23*(2), 391–403. Available from: http://dx.doi.org/10.1162/jocn.2010.21454.

Gagnepain, P., Henson, R. N., & Davis, M. H. (2012). Temporal predictive codes for spoken words in auditory cortex. *Current Biology, 22*(7), 615–621.

Gaskell, M. G., & Marslen-Wilson, W. D. (1997). Integrating form and meaning: A distributed model of speech perception. *Language and Cognitive Processes, 12*, 613–656.

Gaskell, M. G., & Marslen-wilson, W. D. (2001). Lexical ambiguity resolution and spoken word recognition: Bridging the gap. *Journal of Memory and Language, 44*, 325–349. Available from: http://dx.doi.org/10.1006/jmla.2000.2741.

Gow, D. W. (2012). The cortical organization of lexical knowledge: A dual lexicon model of spoken language processing. *Brain and Language, 121*(3), 273–288. Available from: http://dx.doi.org/10.1016/j.bandl.2012.03.005.

Gow, D. W., Segawa, J. a, Ahlfors, S. P., & Lin, F.-H. (2008). Lexical influences on speech perception: A granger causality analysis of MEG and EEG source estimates. *NeuroImage, 43*(3), 614–623. Available from: http://dx.doi.org/10.1016/j.neuroimage.2008.07.027.

Guediche, S., Blumstein, S. E., Fiez, J. A., Holt, L. L., & Peelle, J. E. (2014). Speech perception under adverse conditions: Insights from behavioral, computational, and neuroscience research. *Frontiers in Systems Neuroscience, 7*, 1–16. Available from: http://dx.doi.org/10.3389/fnsys.2013.00126.

Guediche, S., Salvata, C., & Blumstein, S. E. (2013). Temporal cortex reflects effects of sentence context on phonetic processing. *Journal of Cognitive Neuroscience, 25*(5), 706–718. Available from: http://dx.doi.org/10.1162/jocn.

Hagoort, P. (2005). On Broca, brain, and binding : A new framework. *Trends in Cognitive Sciences, 9*(9). Available from: http://dx.doi.org/10.1016/j.tics.2005.07.004.

Hauk, O., Davis, M. H., Kherif, F., & Pulvermüller, F. (2008). Imagery or meaning? Evidence for a semantic origin of category-specific brain activity in metabolic imaging. *The European Journal of Neuroscience, 27*(7), 1856–1866. Available from: http://dx.doi.org/10.1111/j.1460-9568.2008.06143.x.

Hervais-Adelman, A., Davis, M. H., Johnsrude, I. S., & Carlyon, R. P. (2008). Perceptual learning of noise vocoded words: Effects of feedback and lexicality. *Journal of Experimental Psychology. Human Perception and Performance, 34*(2), 460–474. Available from: http://dx.doi.org/10.1037/0096-1523.34.2.460.

Hervais-Adelman, A. G., Carlyon, R. P., Johnsrude, I. S., & Davis, M. H. (2012). Brain regions recruited for the effortful comprehension of noise-vocoded words. *Language and Cognitive Processes, 27*(7/8), 1145–1166.

Hervais-Adelman, A. G., Davis, M. H., Johnsrude, I. S., Taylor, K. J., & Carlyon, R. P. (2011). Generalization of perceptual learning of vocoded speech. *Journal of Experimental Psychology. Human Perception and Performance*, *37*(1), 283–295. Available from: http://dx.doi.org/10.1037/a0020772.

Hickok, G., Buchsbaum, B., Humphries, C., & Muftuler, T. (2003). Auditory-motor interaction revealed by fMRI: Speech, music, and working memory in area Spt. *Journal of Cognitive Neuroscience*, *15*(5), 673–682. Available from: http://dx.doi.org/10.1162/089892903322307393.

Hickok, G., & Poeppel, D. (2000). Towards a functional neuroanatomy of speech perception. *Trends in Cognitive Sciences*, *4*(4), 131–138.

Hickok, G., & Poeppel, D. (2007). The cortical organization of speech processing. *Nature Reviews Neuroscience*, *8*, 393–402.

Humphries, C., Binder, J. R., Medler, D. A., & Liebenthal, E. (2006). Syntactic and semantic modulation of neural activity during auditory sentence comprehension. *Journal of Cognitive Neuroscience*, *18*(4), 665–679. Available from: http://dx.doi.org/10.1162/jocn.2006.18.4.665.

Jacquemot, C., Pallier, C., LeBihan, D., Dehaene, S., & Dupoux, E. (2003). Phonological grammar shapes the auditory cortex: A functional magnetic resonance imaging study. *The Journal of Neuroscience*, *23*(29), 9541–9546.

Kilian-Hütten, N., Vroomen, J., & Formisano, E. (2011). Brain activation during audiovisual exposure anticipates future perception of ambiguous speech. *NeuroImage*, *57*(4), 1601–1607. Available from: http://dx.doi.org/10.1016/j.neuroimage.2011.05.043.

Kotz, S., Cappa, S. F., von Cramon, D. Y., & Friederici, A. D. (2002). Modulation of the lexical–semantic network by auditory semantic priming: An event-related functional MRI study. *NeuroImage*, *17*(4), 1761–1772. Available from: http://dx.doi.org/10.1006/nimg.2002.1316.

Kouider, S., de Gardelle, V., Dehaene, S., Dupoux, E., & Pallier, C. (2010). Cerebral bases of subliminal speech priming. *NeuroImage*, *49* (1), 922–929. Available from: http://dx.doi.org/10.1016/j.neuroimage.2009.08.043.

Kuhl, P. K. (2004). Early language acquisition: Cracking the speech code. *Nature Reviews. Neuroscience*, *5*(11), 831–843. Available from: http://dx.doi.org/10.1038/nrn1533.

Lau, E. F., Phillips, C., & Poeppel, D. (2008). A cortical network for semantics: (De)constructing the N400. *Nature Reviews Neuroscience*, *9*(12), 920–933. Available from: http://dx.doi.org/10.1038/nrn2532.

Lee, Y.-S., Turkeltaub, P., Granger, R., & Raizada, R. D. S. (2012). Categorical speech processing in Broca's area: An fMRI study using multivariate pattern-based analysis. *Journal of Neuroscience*, *32*(11), 3942–3948. Available from: http://dx.doi.org/10.1523/JNEUROSCI.3814-11.2012.

Leff, A. P., Schofield, T. M., Stephan, K. E., Crinion, J. T., Friston, K. J., & Price, C. J. (2008). The cortical dynamics of intelligible speech. *The Journal of Neuroscience: The Official Journal of the Society for Neuroscience*, *28*(49), 13209–13215. Available from: http://dx.doi.org/10.1523/JNEUROSCI.2903-08.2008.

Lerner, Y., Honey, C. J., Silbert, L. J., & Hasson, U. (2011). Topographic mapping of a hierarchy of temporal receptive windows using a narrated story. *The Journal of Neuroscience*, *31*(8), 2906–2915. Available from: http://dx.doi.org/10.1523/JNEUROSCI.3684-10.2011.

Liberman, A., & Whalen, D. (2000). On the relation of speech to language. *Trends in Cognitive Sciences*, *4*(5), 187–196.

Luce, P. A., & Pisoni, D. (1998). Recognizing spoken words: The neighborhood activation model. *Ear and Hearing*, *19*, 1–36.

Marslen-Wilson, W. D., & Tyler, L. K. (2007). Morphology, language and the brain: The decompositional substrate for language comprehension. *Philosophical Transactions of the Royal Society of London. Series B, Biological Sciences*, *362*(1481), 823–836. Available from: http://dx.doi.org/10.1098/rstb.2007.2091.

Martin, A. (2007). The representation of object concepts in the brain. *Annual Review of Psychology*, *58*, 25–45. Available from: http://dx.doi.org/10.1146/annurev.psych.57.102904.190143.

Mazoyer, B. M., Tzourio, N., Frak, V., Syrota, A., Murayama, N., Levrier, O., et al. (1993). The cortical representation of speech. *Journal of Cognitive Neuroscience*, *5*(4), 467–479. Available from: http://dx.doi.org/10.1162/jocn.1993.5.4.467.

McClelland, J. L., & Elman, J. L. (1986). The TRACE model of speech perception. *Cognitive Psychology*, *18*, 1–86.

McMurray, B. (2007). Defusing the childhood vocabulary explosion. *Science*, *317*(3 August), 631.

Okada, K., & Hickok, G. S. (2006). Identification of lexical-phonological networks in the superior temporal sulcus using functional magnetic resonance imaging. *Neuroreport*, *17*(12), 1293–1296.

McQueen, J. M., Cutler, A., & Norris, D. (2006). Phonological abstraction in the mental lexicon. *Cognitive Science*, *30*, 1113–1126.

Meister, I. G., Wilson, S. M., Deblieck, C., Wu, A. D., & Iacoboni, M. (2007). The essential role of premotor cortex in speech perception. *Current Biology*, *17*(19), 1692–1696. Available from: http://dx.doi.org/10.1016/j.cub.2007.08.064.

Meteyard, L., Cuadrado, S. R., Bahrami, B., & Vigliocco, G. (2012). Coming of age: A review of embodiment and the neuroscience of semantics. *Cortex*, *48*(7), 788–804. Available from: http://dx.doi.org/10.1016/j.cortex.2010.11.002.

Mitterer, H., & McQueen, J. M. (2009). Foreign subtitles help but native-language subtitles harm foreign speech perception. *PLoS One*, *4*(11), e7785. Available from: http://dx.doi.org/10.1371/journal.pone.0007785.

Möttönen, R., & Watkins, K. E. (2009). Motor representations of articulators contribute to categorical perception of speech sounds. *The Journal of Neuroscience*, *29*(31), 9819–9825. Available from: http://dx.doi.org/10.1523/JNEUROSCI.6018-08.2009.

Myers, E. B., & Blumstein, S. E. (2008). The neural bases of the lexical effect: An fMRI investigation. *Cerebral Cortex.*, *18*(2), 278–288. Available from: http://dx.doi.org/10.1093/cercor/bhm053.

Norris, D. (1994). Shortlist: A connectionist model of continuous speech recognition. *Cognition*, *52*, 189–234.

Norris, D., McQueen, J. M., & Cutler, A. (2003). Perceptual learning in speech. *Cognitive Psychology*, *47*(2), 204–238. Available from: http://dx.doi.org/10.1016/S0010-0285(03)00006-9.

Okada, K., & Hickok, G. S. (2006). Identification of lexical-phonological networks in the superior temporal sulcus using functional magnetic resonance imaging. *Neuroreport*, *17*(12), 1293–1296.

Okada, K., Rong, F., Venezia, J., Matchin, W., Hsieh, I.-H., Saberi, K., et al. (2010). Hierarchical organization of human auditory cortex: Evidence from acoustic invariance in the response to intelligible speech. *Cerebral Cortex*, *20*(10), 2486–2495. Available from: http://dx.doi.org/10.1093/cercor/bhp318.

Patterson, K., Nestor, P. J., & Rogers, T. T. (2007). Where do you know what you know? The representation of semantic knowledge in the human brain. *Nature Reviews. Neuroscience*, *8*(12), 976–987. Available from: http://dx.doi.org/10.1038/nrn2277.

Peelle, J. E., Troiani, V., Gee, J., Moore, P., McMillan, C., Vesely, L., et al. (2008). Sentence comprehension and voxel-based morphometry in progressive nonfluent aphasia, semantic dementia, and nonaphasic frontotemporal dementia. *Journal of Neurolinguistics*, *21*(5), 418–432. Available from: http://dx.doi.org/10.1016/j.jneuroling.2008.01.004.

Peramunage, D., Blumstein, S. E., Myers, E. B., Goldrick, M., & Baese-Berk, M. (2011). Phonological neighborhood effects in spoken word production: An fMRI study. *Journal of Cognitive Neuroscience*, *23*(3), 593–603. Available from: http://dx.doi.org/10.1162/jocn.2010.21489.

Piantadosi, S. T., Tily, H., & Gibson, E. (2012). The communicative function of ambiguity in language. *Cognition*, *122*(3), 280–291. Available from: http://dx.doi.org/10.1016/j.cognition.2011.10.004.

Plaut, D. C. (1997). Structure and function in the lexical system: Insights from distributed models of word reading and lexical decision. *Language and Cognitive Processes*, *12*(5–6), 765–806. Available from: http://dx.doi.org/10.1080/016909697386682.

Pobric, G., Jefferies, E., & Ralph, M. A. L. (2007). Anterior temporal lobes mediate semantic representation: Mimicking semantic dementia by using rTMS in normal participants. *Proceedings of the National Academy of Sciences of the United States of America*, *104*(50), 20137–20141. Available from: http://dx.doi.org/10.1073/pnas.0707383104.

Prabhakaran, R., Blumstein, S. E., Myers, E. B., Hutchison, E., & Britton, B. (2006). An event-related fMRI investigation of phonological-lexical competition. *Neuropsychologia*, *44*(12), 2209–2221. Available from: http://dx.doi.org/10.1016/j.neuropsychologia.2006.05.025.

Price, C. J., & Devlin, J. T. (2011). The interactive account of ventral occipitotemporal contributions to reading. *Trends in Cognitive Sciences*, *15*(6), 246–253. Available from: http://dx.doi.org/10.1016/j.tics.2011.04.001.

Rao, R. P., & Ballard, D. H. (1999). Predictive coding in the visual cortex: A functional interpretation of some extra-classical receptive-field effects. *Nature Neuroscience*, *2*(1), 79–87. Available from: http://dx.doi.org/10.1038/4580.

Rauschecker, J. P., & Scott, S. K. (2009). Maps and streams in the auditory cortex: Nonhuman primates illuminate human speech processing. *Nature Neuroscience*, *12*(6), 718–724. Available from: http://dx.doi.org/10.1038/nn.2331.

Remez, R. E., Dubowski, K. R., Broder, R. S., Davids, M. L., Grossman, Y. S., Moskalenko, M., et al. (2011). Auditory-phonetic projection and lexical structure in the recognition of sine-wave words. *Journal of Experimental Psychology. Human Perception and Performance*, *37*(3), 968–977. Available from: http://dx.doi.org/10.1037/a0020734.

Rissman, J., Eliassen, J. C., & Blumstein, S. E. (2003). An event-related FMRI investigation of implicit semantic priming. *Journal of Cognitive Neuroscience*, *15*(8), 1160–1175. Available from: http://dx.doi.org/10.1162/089892903322598120.

Rodd, J., Gaskell, G., & Marslen-Wilson, W. (2002). Making sense of semantic ambiguity: Semantic competition in lexical access. *Journal of Memory and Language*, *46*(2), 245–266. Available from: http://dx.doi.org/10.1006/jmla.2001.2810.

Rodd, J. M., Davis, M. H., & Johnsrude, I. S. (2005). The neural mechanisms of speech comprehension: fMRI studies of semantic ambiguity. *Cerebral Cortex*, *15*(8), 1261–1269.

Rodd, J. M., Johnsrude, I. S., & Davis, M. H. (2012). Dissociating frontotemporal contributions to semantic ambiguity resolution in spoken sentences. *Cerebral Cortex*, *22*(8), 1761–1773. Available from: http://dx.doi.org/10.1093/cercor/bhr252.

Rodd, J. M., Longe, O., Randall, B., & Tyler, L. K. (2010). The functional organisation of the fronto-temporal language system: Evidence from syntactic and semantic ambiguity. *Neuropsychologia*, *48*(5), 1324–1335. Available from: http://dx.doi.org/10.1016/j.neuropsychologia.2009.12.035.

Scott, S. K., Blank, C. C., Rosen, S., & Wise, R. J. (2000). Identification of a pathway for intelligible speech in the left temporal lobe. *Brain*, *123*(Pt 12), 2400–2406. Retrieved from: http://www.ncbi.nlm.nih.gov/pubmed/11099443.

Scott, S. K., & Johnsrude, I. S. (2003). The neuroanatomical and functional organization of speech perception. *Trends in Neurosciences*, *26*(2), 100–107. Available from: http://dx.doi.org/10.1016/S0166-2236(02)00037-1.

Scott, S. K., Mcgettigan, C., & Eisner, F. (2009). A little more conversation, a little less action – candidate roles for the motor cortex in speech perception. *Nature Reviews Neuroscience*, *10*(April), 295–302.

Sehm, B., Schnitzler, T., Obleser, J., Groba, A., Ragert, P., Villringer, A., et al. (2013). Facilitation of inferior frontal cortex by transcranial direct current stimulation induces perceptual learning of severely degraded speech. *The Journal of Neuroscience*, *33*(40), 15868–15878. Available from: http://dx.doi.org/10.1523/JNEUROSCI.5466-12.2013.

Smith, D. R. R., Patterson, R. D., Turner, R., Kawahara, H., & Irino, T. (2005). The processing and perception of size information in speech sounds. *The Journal of the Acoustical Society of America*, *117*(1), 305. Available from: http://dx.doi.org/10.1121/1.1828637.

Sohoglu, E., Peelle, J. E., Carlyon, R. P., & Davis, M. H. (2012). Predictive top-down integration of prior knowledge during speech perception. *The Journal of Neuroscience*, *32*(25), 8443–8453. Available from: http://dx.doi.org/10.1523/JNEUROSCI.5069-11.2012.

Spitsyna, G., Warren, J. E., Scott, S. K., Turkheimer, F. E., & Wise, R. J. S. (2006). Converging language streams in the human temporal lobe. *The Journal of Neuroscience*, *26*(28), 7328–7336. Available from: http://dx.doi.org/10.1523/JNEUROSCI.0559-06.2006.

Summerfield, A. Q. (1981). Articulatory rate and perceptual constancy in phonetic perception. *Journal of Experimental Psychology: Human Perception and Performance*, *7*, 1074–1095.

Takashima, A., Bakker, I., van Hell, J. G., Janzen, G., & McQueen, J. M. (2014). Richness of information about novel words influences how episodic and semantic memory networks interact during lexicalization. *NeuroImage*, *84*, 265–278. Available from: http://dx.doi.org/10.1016/j.neuroimage.2013.08.023.

Tamminen, J., Payne, J. D., Stickgold, R., Wamsley, E. J., & Gaskell, M. G. (2010). Sleep spindle activity is associated with the integration of new memories and existing knowledge. *The Journal of Neuroscience: The Official Journal of the Society for Neuroscience*, *30*(43), 14356–14360. Available from: http://dx.doi.org/10.1523/JNEUROSCI.3028-10.2010.

Taylor, J. S. H., Rastle, K., & Davis, M. H. (2013). Can cognitive models explain brain activation during word and pseudoword reading? A meta-analysis of 36 neuroimaging studies. *Psychological Bulletin*, *139*(4), 766–791. Available from: http://dx.doi.org/10.1037/a0030266.

Tremblay, P., & Small, S. L. (2011). On the context-dependent nature of the contribution of the ventral premotor cortex to speech perception. *NeuroImage*, *57*(4), 1561–1571. Available from: http://dx.doi.org/10.1016/j.neuroimage.2011.05.067.

Ungerleider, L. G., & Mishkin, M. (1982). Two cortical visual systems. In D. J. Ingle, M. A. Goodale, & R. J. W. Mansfield (Eds.), *Analysis of visual behavior* (pp. 549–586). Cambridge, MA: MIT Press.

Vaden, K. I., Piquado, T., & Hickok, G. (2011). Sublexical properties of spoken words modulate activity in Broca's area but not superior temporal cortex: Implications for models of speech recognition. *Journal of Cognitive Neuroscience*, *23*(10), 2665–2674.

Venezia, J. H., Saberi, K., Chubb, C., & Hickok, G. (2012). Response bias modulates the speech motor system during syllable discrimination. *Frontiers in Psychology*, *3*(May), 157. Available from: http://dx.doi.org/10.3389/fpsyg.2012.00157.

Visser, M., Jefferies, E., & Lambon Ralph, M. A. (2009). Semantic processing in the anterior temporal lobes: A meta-analysis of the functional neuroimaging Literature. *Journal of Cognitive Neuroscience*, *22*, 1083–1094.

Visser, M., & Lambon Ralph, M. A. (2011). Differential contributions of bilateral ventral anterior temporal lobe and left anterior superior temporal gyrus to semantic processes. *Journal of Cognitive Neuroscience*, *23*, 3121–3131.

Wayne, R. V., & Johnsrude, I. S. (2012). The role of visual speech information in supporting perceptual learning of degraded speech. *Journal of Experimental Psychology: Applied, 18*, 419–435.

Werker, J. F., & Tees, R. C. (1999). Influences on infant speech processing: Toward a new synthesis. *Annual Review of Psychology, 50*, 509–535. Available from: http://dx.doi.org/10.1146/annurev.psych.50.1.509.

Wible, C. G., Han, S. D., Spencer, M. H., Kubicki, M., Niznikiewicz, M. H., Jolesz, F. A., et al. (2006). Connectivity among semantic associates: An fMRI study of semantic priming. *Brain and Language, 97*(3), 294–305. Available from: http://dx.doi.org/10.1016/j.bandl.2005.11.006.

Wild, C. J., Yusuf, A., Wilson, D. E., Peelle, J. E., Davis, M. H., & Johnsrude, I. S. (2012). Effortful listening: The processing of degraded speech depends critically on attention. *The Journal of Neuroscience, 32*(40), 14010–14021. Available from: http://dx.doi.org/10.1523/JNEUROSCI.1528-12.2012.

Zhuang, J., Tyler, L. K., Randall, B., Stamatakis, E. A., & Marslen-Wilson, W. D. (2014). Optimally efficient neural systems for processing spoken language. *Cerebral Cortex, 24*(4), 908–918. Available from: http://dx.doi.org/10.1093/cercor/bhs366.

CHAPTER

45

A Common Neural Progression to Meaning in About a Third of a Second

Kara D. Federmeier[1,2,3], *Marta Kutas*[4,5,6] *and Danielle S. Dickson*[1]

[1]Department of Psychology, University of Illinois, Urbana, IL, USA; [2]Program in Neuroscience, University of Illinois, Urbana, IL, USA; [3]Beckman Institute for Advanced Science and Technology, University of Illinois, Urbana, IL, USA; [4]Department of Cognitive Science, University of California, San Diego, CA, USA; [5]Department of Neurosciences, University of California, San Diego, CA, USA; [6]Center for Research in Language, University of California, San Diego, CA, USA

Language is a system that allows an interlocutor to effectively and efficiently create mental states in a communication partner. Words and phrases are memory cues that are often embedded in sentences and larger language units that provide directions for combining those cues into complex, structured mental representations. At its heart, therefore, language comprehension involves a temporally extended process of accessing and updating semantic memory, the brain's repository of general experience and implicit and explicit knowledge. Semantic memory consists of stored experiences that can be accessed without making reference to specific details of the episodic context from which those experiences were derived. It includes information about people (and their faces, voices, and biographies), objects (attributes, functions), places, words (features, meanings, and combinatorial statistics), and general world knowledge of all types (facts, events, social norms, beliefs, etc.). Semantic access, then, is a protracted process whereby information in semantic memory comes to be activated via links to an internal or external stimulus. The specific nature and structure of semantic memory remains a topic of debate. In particular, there is disagreement about the extent to which semantics is abstract/symbolic in nature versus directly built from perceptual stimulations, amodal versus modality-specific, spatially organized according to features versus discrete categories, and so forth (for a variety of views see, e.g., Barsalou, 2008; Binder & Desai, 2011; Mahon & Caramazza, 2008; Martin, 2007; Patterson, Nestor, & Rogers, 2007). However, there is a general consensus on some aspects of semantic memory, and these pose important challenges for understanding the neural and cognitive mechanisms involved in semantic access.

First, a highly distributed (albeit not exactly the same) brain network is engaged in association with semantic memory access. This network comprises a variety of heteromodal areas, including much of the temporal lobe and inferior parietal cortex, coupled with higher-order sensory and motor processing areas and control structures in the left prefrontal lobe (e.g., review in Binder & Desai, 2011). Understanding how representations and processes in such an extensive set of brain areas are coordinated to mediate semantic access would seem to call for a fundamentally different framework for thinking about the relationship between cognitive functions and the brain than has been typical in much of the cognitive neuroscience literature to date, wherein a common approach to pinpointing the neurobiological roots of a particular cognitive process is to locate that process in a particular brain region or small network of areas. For semantic memory, however, spatial localization in isolation is unarguably an inadequate anchor for function, if not also representation. Instead, the process (and not moment) of semantic access requires the dynamic creation of a functional circuit, with *time* as a critical component (see, e.g., Federmeier & Laszlo, 2009).

Second, access to semantic memory manifests a jarring juxtaposition of consistency and flexibility. Semantic access must be sufficiently stable that the mental states elicited by physically different stimuli

Neurobiology of Language. DOI: http://dx.doi.org/10.1016/B978-0-12-407794-2.00045-6

(e.g., hearing the word "brain" and seeing a picture of a brain) as well as by the same type of stimulus encountered at different times and in different contexts are all recognizable as (linking to) the "same" concept. Such stability is necessary for learning and generalization and also critically enables human communication. Moreover, for any given instance of semantic access, only a subset of all the information linked to a particular concept is likely to be relevant, and, correspondingly, the exact configuration of conceptual features accessed for a given input (word, picture, etc.) is highly context-sensitive. In fact, although in the laboratory semantic access has often been examined for isolated stimuli (a single face, object, or word), in everyday life, and especially in the context of language comprehension, semantic access for any given stimulus occurs in a rich context, including information that has come in (and will continue to come in) over time.

Time and timing thus play a critical role in semantic access (indeed, in most cognitive processes and their neural mechanisms!). In this chapter, we therefore examine the time course of semantic access across two temporal scales: first, that associated with accessing long-term memory from an individual stimulus and, second, the scale associated with building meaning across multiple inputs in a stimulus stream (e.g., an unfolding sentence or discourse).

45.1 PART 1: THE TIMECOURSE OF SEMANTIC ACCESS OUT OF CONTEXT

Access to semantic memory is not exclusive to language stimuli. Linking perceptual inputs to meaning is arguably one of the brain's most fundamental tasks, because this allows more flexible, well-tuned, and complex responses to stimuli. Meaning is accessed from all modalities and from many sundry representation types: environmental sounds; somatosensory stimuli; visual objects, faces, and scenes; tastes and smells; and nonlinguistic symbolic systems (such as numbers); as well as spoken, written, and signed language stimuli. However, each of these stimulus types present different challenges for semantic access. Comparisons across modality/representation type have played a critical role in investigations of the *structure* of semantic memory (see, e.g., discussion in Federmeier & Kutas, 2001), offering important insights into the factors that govern how semantic information is stored and organized. Surprisingly, however, the literature looking at the *process* of semantic access from different types of inputs (faces, objects, printed words, spoken words, etc.) have not been similarity integrated. Such comparisons afford an understanding of the extent to which the timing—and, by inference, the mechanism—of semantic access does or does not differ as a function of factors such as input modality, perceptual expertise (greater for faces and written or spoken words than objects and environmental sounds), developmental trajectory (earlier for faces and objects than written words), evolutionary precedence (earlier for nonlinguistic than linguistic inputs overall; earlier for auditory than for visual words), and iconicity (more for faces and objects than words), among others.

Moreover, debates about the timecourse of semantic access from linguistic stimuli (e.g., Laszlo & Federmeier, 2014; Rayner & Clifton, 2009) can be importantly informed by knowledge about the timecourse of access from nonlinguistic ones. For example, the ability to rapidly analyze a face stimulus and link it with biographical and emotion-related information is clearly a critical social skill with implications for an organism's health and survival. Accessing memory from faces is also a phylogenetically and ontologically earlier ability than is accessing memory from words (especially printed words). The timecourse of semantic access from faces can thus provide an important point of reference for understanding semantic access from words, serving as a possible upper limit on the speed with which perceptual stimuli can be linked with information in long-term memory. Therefore, we review the timecourse of semantic processing, as revealed by event-related brain potentials (ERPs), for faces and visual objects, visual and auditory words, and numbers.

45.1.1 Face and Object Processing

A large body of literature has used techniques with high temporal resolution, especially ERPs, to track the timecourse of face and object processing, from extracting basic features, isolating the figure from the ground, identifying whole complex structures, to ultimately accessing meaning. The timecourse thus obtained is particularly well-attested because, unlike for language stimuli, it can be validated through work in nonhuman animals using more invasive techniques. Recordings in monkeys have revealed onset latencies of about 100 ms for face-selective neurons in the inferior temporal cortex (Kiani, Esteky, & Tanaka, 2005), and intracranial recordings in humans (who have bigger brains, meaning generally longer neural response latencies) first find face-specific responses beginning at approximately 160 ms in posterior fusiform gyrus (Halgren, Baudena, Heit, Clarke, & Marinkovic, 1994). This accords well with the timecourse inferred from scalp-recorded ERPs, for which the first reliable categorization of visual stimuli is seen in amplitude modulations on the N1 component, part of the normal evoked response to all visual onsets, offsets, or changes (see Di Russo, Martínez, Sereno, Pitzalis, & Hillyard, 2002 for a description of onset latencies and hypothesized localizations of early visual components). This neural response begins at approximately 130 ms after stimulus onset and peaks between 150 and 170 ms. In the object and face recognition literatures, the N1 is often referred to as the "N170," with (what is likely) the positive end of this dipole seen over central electrode sites and referred to as the Vertex Positive Potential or the P150 (see Jeffreys, 1989).

The N170/P150 to images containing objects reliably differs from those to nonsense images generated by phase-scrambling their intact counterparts (while maintaining low-level cues including spatial frequency; Schendan & Lucia, 2010). In this same time window, visual stimuli with which people have special expertise, such as faces (Bentin, Allison, Puce, Perez, & McCarthy, 1996) and word-like strings (Schendan, Ganis, & Kutas, 1998), elicit enhanced responses compared with responses to objects. Critically, N170/P150 amplitude modulations do not simply track the presence of particular visual features, but seem to reflect the *detection* of a particular type of stimulus. For example, studies using Mooney images—which, when presented in one orientation (rightside up) are generally perceived as faces but, presented upside-down, are not—have shown that N170 enhancements occur only when a visual stimulus is perceived as a face (George, Jemel, Fiori, Chaby, & Renault, 2005). However, faces need not have typical or even coherent structure to elicit enhanced N170s, which have been seen even when faces are upside-down or cut in half, or when facial features are scrambled or even missing (e.g., Bentin et al., 1996). Similarly, P150 enhancements to word-like strings are obtained even when the elements comprising the string are unfamiliar ("pseudofonts"; Schendan et al., 1998), and P150 sensitivity to repetition is independent of the familiarity or meaningfulness of the eliciting stimuli (Voss, Schendan, & Paller, 2010). The initial categorization of a stimulus as "object-like," "face-like," or "word-like" at a basic level thus seems to begin in the time window of the N170/P150 potential.

Responses in the late part of the N170 time window (160–190 ms) and on the subsequent P2 component have been tied to perceptual grouping and structural encoding processes, which create a structured representation of an object or face and enable, for example, the matching of different instantiations and viewpoints of the same form. Studies using fragmented line drawings have shown repetition effects on the posterior P2 for globally similar forms (e.g., drawings of the same object) despite differences in lower-level features (e.g., local contours) (Schendan & Kutas, 2007a, 2007b). Critically, P2 reductions in this design only occur when images on both first and second presentations are fragmented or incomplete (not when images are intact)—in other words, presumably only when both presentations pose a challenge to perceptual grouping processes. P2 amplitudes are also larger for phase-scrambled than for intact images, likely reflecting grouping difficulties for images without perceivable objects (Schendan & Lucia, 2010).

Structural encoding of faces has further been associated with an ERP response known as the N250r. The N250r is a negative-going potential seen between about 230 and 300 ms over (right) inferior temporal electrode sites (thought to have a neural generator in the fusiform gyrus; Schweinberger, Pickering, Jentzsch, Burton, & Kaufmann, 2002) and is selectively enhanced to immediate repetitions of faces (compared with nonface objects). These repetition effects are transient and do not persist beyond more than a few intervening stimuli (Schweinberger, Pickering, Burton, & Kaufmann, 2002). Repetition effects for intact objects appear a little later on a component known as the frontal N300 (sometimes N350). This potential is thought to reflect global object processing after perceptual grouping has occurred and independent of local features (Schendan & Kutas, 2007a); it has been linked to occipitotemporal generators, including the object-selective lateral occipital cortex (Schendan & Kutas, 2002, 2003).

Critically, these components observed in the first 300 ms of processing have been argued to precede access to long-term semantic memory. The N170/P150 does not differentiate between unfamiliar and familiar or famous faces (Bentin & Deouell, 2000; Eimer, 2000) or between familiar objects and unfamiliar pseudo-objects (Schendan et al., 1998). The N250r is reduced for novel faces compared with those that are perceptually familiar (Begleiter, Porjesz, & Wang, 1995; Schweinberger, Pfütze, & Sommer, 1995). It has been linked to the creation of a transient, individuated facial representation as a necessary precursor to recognition and semantic access, and prior experience creating that structural representation seems to facilitate the creation process. The N250r, however, is unaffected by the degree to which a face has any learned associations (Paller, Gonsalves, Grabowecky, Bozic, & Yamada, 2000). The N300 has similarly been linked to the creation of higher-level object representations. Repetition effects on the N300 are reduced if the object in the two presentations differs in viewpoint (Schendan & Kutas, 2003) and the component is larger for unusual views of objects than for canonical ones and for unidentified as opposed to identified objects (Schendan & Kutas, 2002, 2003; Schendan & Lucia, 2010).

Effects of semantic variables are first consistently reported on an ERP component known as the N400. The N400 is a negative-going response seen to all potentially meaningful stimuli, including faces, objects, scenes, environmental sounds, and auditory and visual words, among others. It begins at approximately 250 to 300 ms and peaks just before 400 ms in healthy young adults. The scalp-recorded N400 seems to arise from a highly distributed network of brain areas, with important sources in the medial and anterior temporal lobes, but also reflects activity in higher-level perceptual areas, the superior temporal lobe, and dorsolateral frontal regions (Halgren et al., 1994, 2002; Lau, Phillips, & Poeppel, 2008; Tse et al., 2007). The widespread nature of these neural sources is consistent with the well-attested functional link between the N400 and semantic memory: the amplitude (but, strikingly, not the timing) of this potential is highly sensitive to a wide range of factors related to the ease of semantic access (for review, see Kutas & Federmeier, 2011).

In the face processing literature, the first effects of delayed repetition—that is, repetition effects that clearly involve long-term memory—present as N400 amplitude reductions (e.g., Joyce & Kutas, 2005). The N400 window also encompasses the first reliable differentiation of unfamiliar faces from those that are familiar and associated with biographical information, such as faces that have been learned along with names (Paller et al., 2000) or famous faces (Bentin & Deouell, 2000), and first reliable effects of semantic association, such as between a face and a name (Huddy, Schweinberger, Jentzsch, & Burton, 2003). Similarly, in the literature on object/picture processing, semantic effects of a variety of kinds all consistently manifest within the N400 time window. This includes, but is not limited to, semantic priming effects between pictures and effects from words onto pictures, in which semantically primed pictures elicit smaller N400s, as well as congruity effects for a pictured object in a sentence, in a visual scene, in a picture story, or in a video, wherein congruous objects elicit reduced N400s compared to incongruous or unexpected ones (e.g., Federmeier & Kutas, 2001; Ganis & Kutas, 2003; Ganis, Kutas, & Sereno, 1996; Holcomb & McPherson, 1994; Sitnikova, Kuperberg, & Holcomb, 2003; West & Holcomb, 2002). Additional evidence that conceptual processing of objects occurs during the N400 time window comes from a study using novel, meaningless "squiggles," which each participant rated for meaningfulness. Only squiggles rated high in meaningfulness showed a repetition effect between 300 and 500 ms after stimulus onset (Voss et al., 2010). Because ratings were idiosyncratic, the same visual forms were assigned to the low and high meaningful categories across participants; thus, the repetition effect could not have been driven by perceptual features, but rather reflects the conceptual representations and associated semantic information evinced by the stimuli for each participant.

In summary, the majority of studies examining the timecourse of processing for nonlinguistic visual stimuli suggest that it takes about 150 ms for the brain to be able to extract sufficient information to reliably categorize a stimulus at a basic level (as an object, string, or face) and an additional 150 ms to exercise perceptual grouping processes and to create a (transient) structured representation of an individual object or face. Beginning at approximately 300 ms, this representation makes links to associated information in semantic memory.

45.1.2 Visual Word and Number Processing

With the timecourse of semantic access from nonlinguistic stimuli as a basis for comparison, we next examine ERP data regarding the processing timecourse for written words and numeric symbols. As already discussed, early differentiation of word-like stimuli (strings) from nonword-like stimuli appears in the range of 150 ms. Potentials in this timeframe (the N/P150 component; see review by Grainger & Holcomb, 2009) are sensitive to the repetition of single letters and words in a manner that is independent of size but sensitive to both font (Chauncey, Holcomb, & Grainger, 2008) and letter case (Petit, Midgley, Holcomb, & Grainger, 2006). This repetition effect is eliminated by changes in the position of the words by a shift of even a single letter, suggesting that at this point in processing representations are still retinotopic (Dufau, Grainger, & Holcomb, 2008). This aligns well with what is known about the rate of information flow through visual cortex, where single unit recordings in higher visual areas with complex receptive fields (e.g., V4) tend to show activation peaks between 100 and 200 ms after stimulus onset (e.g., Luck, Chelazzi, Hillyard, & Desimone, 1997).

The N/P150 is followed by a negativity peaking at approximately 250 ms, known as the N250 (Holcomb & Grainger, 2006). Like the similarly named potential to faces, the N250 is most robustly characterized in the context of repetition (in this case, primarily masked repetition). The N250 to letter strings, whether familiar (words) or unfamiliar (pseudowords), is larger (more negative) for unrepeated than for repeated items and has a widespread distribution maximal over (left) midline anterior sites. (*Note*: face processing studies often use an average reference, whereas word processing studies generally use an average mastoid reference, making comparisons of waveform morphology and scalp topography difficult; using an average reference, Pickering and Schweinberger reported an N250r to names, which is similar in form to the face N250r but is more prominent over left rather than right posterior electrode sites.) The N250 is sensitive to the degree of orthographic overlap between visually presented primes and targets but is not sensitive to font changes or positional shifts (Chauncey et al., 2008; Dufau et al., 2008; Grainger, Kiyonaga, & Holcomb, 2006; see also Pickering & Schweinberger, 2003). The N250 response is not domain general, because it is not seen for spoken words or pictures or in cross-modal priming onto visual words (Kiyonaga, Grainger, Midgley, & Holcomb, 2007; Pickering & Schweinberger, 2003). Like the face N250r, the N250 to words is transient. Studies varying the time interval between prime and target have observed the N250 with stimulus-onset asynchronies of 60 and 180 ms, but not 300 ms or longer (Holcomb & Grainger, 2007). Grainger and Holcomb (2009) suggest that the neural activity indexed by the N250 arises from the processing of sublexical orthographic representations important for building whole word representations from letters.

Following the N250 is a positivity, known as the P325, that manifests larger repetition effects for words than for pseudowords. These effects are obtained only when prime and target are identical, whereas strings with partial overlap produce graded responses on the N250. Thus, the P325, unlike components before it, seems to be sensitive to whole item identity and familiarity (see Friedrich, Kotz, Friederici, & Gunter, 2004; King & Kutas, 1998; Pylkkänen, Stringfellow, & Marantz, 2002 for other reports of word-level effects in this time window). This also seems to be a time window during which phonological information is being associated with written words, as Grainger et al. (2006) report pseudohomophone priming effects (e.g., "bacon" primed by "bakon" more than "bafon") on the P325. Holcomb and Grainger (2006) thus link the P325 to processes at the whole word—but not yet semantic—level. Semantic processing effects are then seen in the form of amplitude changes on the N400, which, as for faces and objects, peaks just before 400 ms.

This timecourse, established primarily using component-based analyses on average data and captured in a language processing architecture known as the Bimodal Interactive Activation Model (see review in Grainger & Holcomb, 2009) has also been validated using a multiple regression approach on single-item ERP data, including responses to words, pseudowords, familiar acronyms, and illegal letter strings collected from a large set of participants (Laszlo & Federmeier, 2014). This regression approach revealed effects of bigram frequency and other orthographic variables (orthographic neighborhood size and frequency) beginning at approximately 130 ms (in the range of the N/P150) and continuing throughout most of the first half-second of processing. Whole word frequency began to have an independent effect on the ERP waveform beginning at approximately 270 ms (i.e., in the late N250/early P325 timewindow). In contrast, semantic variables (number of lexical associates, concreteness, imageability, number of senses, and noun–verb ambiguity) did not begin to influence the waveform until after 300 ms in the N400 time window.

The impact of a wide range of semantic variables on N400 amplitudes to visual words is well-attested (see Kutas & Federmeier, 2011, for a review). In particular, the N400 is sensitive to immediate and long-term repetitions (e.g., Laszlo & Federmeier, 2007) and to semantic priming of all types: association-based, graded by association strength (e.g., Kandhadai & Federmeier, 2010), and category-based, graded by typicality (e.g., Federmeier, Kutas, & Schul, 2010). The N400 is also sensitive to the semantic fit between a word and its phrasal, sentential, or discourse context, based on the structure of semantic memory (e.g., Federmeier & Kutas, 1999b), event knowledge (Metusalem et al., 2012), and world knowledge (Hagoort, Hald, Bastiaansen, & Petersson, 2004). Repetition, priming, and contextual fit all decrease N400 amplitude while leaving latency relatively unchanged. As Federmeier and Laszlo (2009) reviewed, the N400 has all of the properties one would expect of a neural marker of access to semantic memory (see also Lau et al., 2008).

Despite readers having relatively less experience with numerals than with other visual forms, the extraction of meaning from numbers appears to proceed with little to no differences compared with that from faces, objects, and visual words. For example, when alphanumeric symbols are rendered predictable by contextual cues (e.g., "20" when preceded by the simple arithmetical formula of "5 × 4"), correct answers elicit smaller (facilitated) N400s relative to incorrect answers (Niedeggen & Rösler, 1999; Niedeggen, Rösler, & Jost, 1999).

Analogous findings across the face, object, visual word, and number processing literatures thus reveal a reliable timecourse of processes leading to the access of semantic information, despite the fact that these stimulus types are associated with different degrees of expertise, engender different processing challenges, and have been linked to partially distinct neural pathways. Processing of these stimuli all culminate in a similar response during the same time window: the N400. The consistency in when and how semantic information comes to be linked with a perceptual stimulus becomes even more striking when the processing timecourse for these meaningful visual stimuli is compared with that for auditory words and other meaningful sounds.

45.1.3 Auditory Word Processing

Whereas "the time locking of neural activity in the visual pathway is poor, its timing is sluggish, and its ability to follow fast transitions is limited"

(Pratt, 2012), the auditory system is fast and exquisitely sensitive to timing. Nevertheless, semantic access seems to proceed with a remarkably similar timecourse in the two modalities.

One of the earliest language-related, albeit not language-specific, effects is an enhancement of the amplitude of the auditory N100 component, which peaks at approximately 100 ms and is part of the normal evoked response to acoustic onsets, offsets, and deviations. Although not acoustic onsets, word onsets in continuous speech elicit larger N100s than matched, non-onset sounds (Sanders & Neville, 2003). This effect seems to reflect increased temporally directed selective attention to less predictable information in an auditory stream (Astheimer & Sanders, 2011).

Paralleling the timecourse of orthographic processing for written words, phonological—but not yet semantic—processing of spoken words has been associated with ERP amplitude modulations between 250 and 300 ms. The Phonological Mapping Negativity (PMN) has been attributed to the detection of a mismatch between expected and realized phonological information, such as when an incoming phoneme violates a context-based or task-induced auditory/phonological expectation (e.g., Connolly & Phillips, 1994; Desroches, Newman, & Joanisse, 2009; Newman & Connolly, 2009). For example, in a task wherein participants were given an input to transform (e.g., "telk" or "hat") and a target transformation of the initial phoneme (e.g., "m"), PMNs were larger when the onset phoneme of the probe stimulus mismatched the resulting expectation (e.g., for "melk" or "mat"). Importantly, PMN effects are obtained for both words and pseudowords (e.g., Newman & Connolly, 2009). The PMN has therefore been proposed to reflect phonological mapping processes that precede semantic access.

Effects associated with word level processing have been reported beginning at approximately 300 ms. Using a cross-modal word fragment priming task, Friedrich and colleagues (Friedrich, Kotz, Friederici, & Alter, 2004; Friedrich, Kotz, Friederici, & Gunter, 2004; Friedrich, Schild, & Röder, 2009) found larger positivities peaking at approximately 350 ms (P350) over left frontal electrode sites for words that mismatched (versus matched) the visual fragment prime. Partially mismatching targets elicited an intermediate response. The authors interpret their results as reflecting activation of abstract word forms. Effects of word repetition and word frequency have also been reported in this time window in studies using MEG (the M350; Pylkkänen & Marantz, 2003) and have been linked to sources in the left superior temporal lobe (Pylkkänen et al., 2002).

P/M350 modulations are coincident in time with (but are of opposite polarity from) effects of repetition, frequency, and phonological priming on the N400. The N400 time window is also when effects of semantic manipulations of many kinds (semantic priming, context effects, etc.) are first observed for auditory words (see review in Kutas & Federmeier, 2011). As with the visual modality, the N400 elicited by auditory stimuli is not specific to linguistically relevant inputs (syllables or words), because meaningful nonlinguistic environmental sounds also elicit N400s and N400 effects (Van Petten & Rheinfelder, 1995).

45.1.4 Summary

Comparisons of N400 responses to spoken and written words or for linguistic and nonlinguistic stimuli (e.g., pictures and visual words; auditory words and environmental sounds) assessed for the same stimuli and/or under the same task conditions (e.g., Federmeier & Kutas, 1999b, 2001; Federmeier, McLennan, De Ochoa, & Kutas, 2002; Ganis et al., 1996; Holcomb & Neville, 1990; Van Petten & Rheinfelder, 1995) reveal striking similarities: similar peak latencies, similar response characteristics (amplitude reductions—but not latency shifts—to repeated, primed, or contextually more predictable items), and similar patterns of sensitivity to many factors that are taken to affect semantic access. Remarkably, then, different neural pathways with different functional and anatomical constraints ultimately culminate in a similar brain response, with consistent and stable timing. Federmeier and Laszlo (2009) argued that the collective N400 literature highlights the importance of time as a mechanism for linking incoming stimuli (of all types and in all modalities) to distributed information in long-term semantic memory. Although there are brain areas that are likely to be common across specific instances of the N400, neither the neural process nor the cognitive function the N400 indexes is localized. Instead, semantic access arises from a dynamically configured neural circuit, for which time and timing are critical and which implements a shared function, even in the absence of a spatial anchor. The dynamic nature of the mechanism affords it the necessary speed and flexibility to accrue information from stimuli over time and thereby construct complex messages, which, in turn, shape the processing of subsequent meaningful stimuli.

45.2 PART 2: CONTEXT AND THE TIMECOURSE OF SEMANTIC ACCESS

Making sense clearly requires a stable (if transient) linking between perceived sensory input and information in long-term memory (i.e., semantic access), and, as we have described, the timing of semantic access so defined is relatively reliable, occurring between approximately 300 and 500 ms, whether words are encountered in isolation or embedded in a sentence and/or a larger discourse. However, this reliability belies the fact that the availability of context information allows word-related information to be activated—via prediction—before the word is actually encountered, and, to the extent that it is, processing (identification, access, integration) of that word's "word form" and the perceptual and semantic information linked to it are eased. The past decade of ERP research attests to the pervasive presence of predictive processing with graded facilitation at multiple levels (semantic, morphosyntactic, phonological, and orthographic) during natural language comprehension, at least in young adult readers and listeners (DeLong, Urbach, & Kutas, 2005; Federmeier & Kutas, 1999b; Van Berkum, Brown, Zwitserlood, Kooijman, & Hagoort, 2005; Wicha, Moreno, & Kutas, 2004).

As previously noted, N400 amplitude modulations provide a sensitive measure of the state of activation of the probing item in semantic memory and of the various factors that influence that state. As such, N400 data have proven especially useful in determining the extent to which there are predictive processes at play during routine language comprehension. Through clever designs, several researchers have found evidence of graded contextually driven facilitation of semantic, morphosyntactic, phonological, and orthographic features of likely upcoming words or pictures (DeLong et al., 2005; Federmeier & Kutas, 1999b; Laszlo & Federmeier, 2009; Van Berkum et al., 2005; Wicha, Bates, Moreno, & Kutas, 2003).

Federmeier and Kutas (1999b), for example, examined the ERPs to sentence final words as young adults read two sentences for comprehension; final words were moderately to highly expected members of a particular category, unexpected members of the same category, or unexpected members of a different category. N400 amplitudes were reduced for the unexpected endings that shared features with the expected endings relative to those that did not. The fact that this pattern of N400 effects came primarily from the strongly (versus weakly) constraining sentence frames, wherein the unexpected categorically related items were highly implausible if not downright anomalous, evidenced that processing of the accruing context preactivated semantic features of likely upcoming words and related ones. The interpretation of this effect as demonstrating that prediction plays an important role in normal language comprehension was bolstered by the fact that the pattern replicated for natural speech for young adult comprehenders (Federmeier et al., 2002). However, this study also showed that prediction is not the only mechanism for context-based facilitation. Although healthy older adults showed clear effects of contextual congruency on the N400, they did not show the pattern indicative of predictive processing. The tendency for (most) older adults to rely more heavily on stimulus-driven rather than predictive processing comprehension mechanisms has been seen in a number of studies (see review by Wlotko, Lee, & Federmeier, 2010).

The availability of different mechanisms by which context can shape language comprehension is also attested by studies of hemispheric differences. For example, a (young adult) visual half-field version of the Federmeier and Kutas study, with the sentence final words randomly lateralized to the left or right visual field (LVF/RVF), revealed asymmetric patterns of context use during online sentence processing (Federmeier & Kutas, 1999a). The left hemisphere (based on RVF initiated processing) seems to use context to preactivate information and compare the actual input with the prediction, whereas the right hemisphere (based on LVF initiated processing) seems to adopt a more passive (what we have called "integrative") strategy that does not involve prediction (for a review, see Federmeier, 2007). Notably, the local scale timing of semantic effects retains its typical stable form in the two hemispheres, with context effects manifesting between about 300 and 500 ms, in the form of N400 amplitude reductions. This further emphasizes our point that semantic access is not a unitary or localized process, but rather is a temporally constrained one. However, at the larger scale, the effective timecourse of availability of information is different in the two hemispheres, creating an emergent mechanism by which the brain can both benefit from the use of context information to constrain processing via predictions while still maintaining a more veridical representation of the actual bottom-up stimulus stream to allow for recovery when predictions are incomplete or misleading.

The Federmeier and Kutas studies provide evidence that semantic feature information, essential for semantic categorization, can be predicted (i.e., activated in advance) during sentence comprehension (but see Van Petten & Luka, 2012, for a proposed distinction between expectation of semantic features and prediction of word forms). A growing body of evidence further attests to the ability of the language comprehension system to predict a wide range of information, including morpho-syntactic, phonological, and orthographic features of upcoming words. Investigations of gender-marked languages, such as Spanish (Wicha, Bates, et al., 2003; Wicha, Moreno, & Kutas, 2003; Wicha et al., 2004) and Dutch (Otten, Nieuwland, & Van Berkum, 2007; Van Berkum et al., 2005), have reported ERP differences to articles (Wicha, Moreno, et al., 2003; Wicha et al., 2004) or adjectives (Otten et al., 2007) depending on whether they agree in gender with the predicted (but not yet presented) noun that they modify. DeLong et al. (2005) found similar results by exploiting the A/AN alternation in English, which varies with the onset phonology of the upcoming word (beginning with a consonant versus vowel sound, respectively); N400 amplitudes to the article closely tracked the predictability of the not yet presented noun.

Clearly, then, sentence contextual information can lead to expectancies for specific words. Moreover, Laszlo and Federmeier (2009) demonstrated that information brought online by prediction is specific enough to affect processing based on orthography: nonsensical orthographic neighbors of expected sentence endings had smaller N400s than nonsensical non-neighbors, regardless of whether the item was a word, a meaningless pseudoword, or an orthographically illegal string of letters. Critically, this means that prediction affects the availability of specific orthographic features of likely upcoming items and begins to influence the bottom-up processing of inputs before what might be thought of as "word recognition," manifesting through clear downstream consequences on N400 amplitude and reflecting the mapping of the stimulus form to its meaning.

As these N400 findings demonstrate, processing of contextual information (or experience or task goals) can allow predictions to be made about specific stimulus features; consequently, any perceptual information accrued about incoming stimuli—available before the stimulus can be linked with semantic memory—should provide early indicators about what the stimulus is likely to be. In these cases, then, relatively early effects of predictive processes are evident in modulations of sensory ERP components that precede the N400. For example, a number of ERP investigations have also observed semantic context-driven modulation of the amplitude of a frontally maximal sensory-evoked P2 potential (180–250 ms) that has been linked to higher-order visual processing, including perceptual analysis and the allocation of visuo-spatial attention (e.g., Luck & Hillyard, 1994). Words in highly constraining sentence contexts have been found to elicit enhanced P2s, presumably because contextually driven expectations for a particular form affords the focusing of attentional resources. Federmeier and colleagues have attributed this P2 effect to predictive mechanisms in the left hemisphere, based on findings of selective enhancement of P2 amplitudes to words in highly (as opposed to weakly) constraining contexts when those words were presented to the RVF/LH but not when presented to the LVF/RH (Federmeier & Kutas, 2002; Federmeier, Mai, & Kutas, 2005; Huang, Lee, & Federmeier, 2010; Kandhadai & Federmeier, 2010; Wlotko & Federmeier, 2007).

Predictive processing mechanisms are also reflected on even earlier components under the right conditions. For example, when the sentence final words in the Federmeier and Kutas experiments were replaced by line drawings, either unfamiliar or rendered familiar via pre-exposure, the N400 was qualitatively similar in its sensitivity to both contextual fit and category-based semantic feature overlap with the predicted item (Federmeier & Kutas, 2001). Critically, the results also revealed that context also interacted with perceptual predictability, affecting both perceptual and semantic processing. When perceptual features of expected pictures were predictable based on pre-exposure, as manifest in smaller early sensory components (anterior N1, posterior P1, and P2) linked to visual processing and the allocation of visuospatial attention, contextual constraint did not provide any additional facilitation (i.e., no effect on visual components or N400 amplitude). By contrast, when pictures were unfamiliar and thus less perceptually predictable, more constraining contexts allowed better predictions for perceptual features of expected pictures, easing perceptual processing (as seen on the anterior N1 and P2) and creating downstream benefits for semantic access (N400).

Evidence for the prediction of visual word form properties has come from studies of the so-called early Left Anterior Negativity (eLAN), originally characterized as a response to word category violations (Friederici & Kotz, 2003). Having noticed that eLANs were observed almost exclusively in sentences wherein a specific word category was highly predictable, Lau, Stroud, Plesch, and Phillips (2006) proposed an alternative account on which it is the violation of expectations for word forms derived from syntactic regularities, rather than the word category violation as such, that is critical for eLAN elicitation. Dikker and colleagues likewise argued that the eLAN or the early negativity (EN, renamed to acknowledge the more broad bilateral distribution of this class of negativities across the scalp) is a response to the presence of unexpected sensory information in the context of a predicted word and/or word category. Dikker, Rabagliati, and Pylkkänen (2009) tested their hypothesis by comparing word category violations that contained an unexpected morpheme or were missing an expected morpheme. They observed an M100 (the presumed magnetic counterpart of the eLAN/EN) to the unexpected morpheme but not to the absence of an expected one (e.g., *The discovery was REPORT). Definitive interpretation of the latter finding, however, is confounded by the word category ambiguity of the critical word—it could be experienced as a noun (word category violation) or a verb (agreement violation). In a follow-up study, Dikker, Rabagliati, Farmer, and Pylkkänen (2010) further tested their hypothesis by comparing responses to two types of nouns following an adverb that led to an expectation for an upcoming verb instead. Members of one class of nouns had the phonotactics of typical nouns (*The beautifully SOFA), whereas those of the other had phonotactics equally consistent with nouns or verbs (*The beautifully WINDOW). An EN, with a left occipital source, was elicited only by the unexpected noun that clearly violated the phonotactics of the expected category (verb). Whether or not these negativities are all the same, it does appear that the language system is sensitive to expectations based on syntactic constraints and violations thereof, which are seen in ERP modulations within the first 200 ms after word onset and clearly can have downstream repercussions for semantic processing.

The language comprehension system thus takes advantage of context information to make predictions about features of likely upcoming stimuli, and, in turn, those predictions shape attention and affect how and how easily a stimulus is perceived and then linked to aspects of meaning in the course of interpretation. Perhaps surprisingly, context-based facilitation does not seem to speed the timecourse with which contact is established between perception and long-term semantic memory. However, at least in some cases, indications that an incoming stimulus matches or mismatches predictions at a perceptual level can provide early cues about whether that stimulus is likely to have particular semantic features as well—and the brain does a remarkable job of making use of this information in the service of preparing rapid responses (e.g., to initiate/inhibit a saccade or other motor response). Thus, although context may not actually speed semantic access as such, it can shift the apparent timecourse of semantic processing, as is revealed in brain responses and unfolds in behavior.

45.3 CONCLUSIONS

Information of all types dynamically accrues via context and experience, allowing comprehenders to construct meaning from past and present inputs and to anticipate future ones as a function of their likelihood in the ongoing context. The process of semantic access—and an extended process it is—takes time because it requires the brain to establish a stable and reliable, yet ultimately idiosyncratic, neural linking between the perceptual form constructed from an incoming stimulus and distributed, stored representations of relevant prior experience. Moreover, meaningful stimuli typically accrue rapidly, and the brain must then collate and integrate the information it collects from, for example, a continuous auditory signal or a sequence of saccades over text or a scene. Constructing meaning is thus arguably one of the most challenging tasks the brain must perform. However, it is also what the brain evolved to do, and to do quickly and well enough, even under less than ideal circumstances. Semantic constraints may not speed aspects of brain processing (e.g., N400 latency) as they do offline reaction times, but they can (re)direct attentional focus and shape current activation states to facilitate sensory and mapping processes for the future—allowing the brain to effectively bypass its own temporal constraints.

Acknowledgments

We gratefully acknowledge funding from the McDonnell Foundation and NIA grant 26308 (to K.D.F.) and NICHD Grant 22614 (to M.K.). We also thank Katherine Mimnaugh for help with document preparation.

References

Astheimer, L. B., & Sanders, L. D. (2011). Predictability affects early perceptual processing of word onsets in continuous speech. *Neuropsychologia*, *49*(12), 3512–3516.

Barsalou, L. W. (2008). Cognitive and neural contributions to understanding the conceptual system. *Current Directions in Psychological Science*, *17*(2), 91–95.

Begleiter, H., Porjesz, B., & Wang, W. (1995). Event-related brain potentials differentiate priming and recognition to familiar and unfamiliar faces. *Electroencephalography and Clinical Neurophysiology*, *94*(1), 41–49.

Bentin, S., Allison, T., Puce, A., Perez, E., & McCarthy, G. (1996). Electrophysiological studies of face perception in humans. *Journal of Cognitive Neuroscience*, *8*(6), 551–565.

Bentin, S., & Deouell, L. Y. (2000). Structural encoding and identification in face processing: ERP evidence for separate mechanisms. *Cognitive Neuropsychology*, *17*(1–3), 35–55.

Binder, J. R., & Desai, R. H. (2011). The neurobiology of semantic memory. *Trends in Cognitive Sciences*, *15*(11), 527–536.

Chauncey, K., Holcomb, P. J., & Grainger, J. (2008). Effects of stimulus font and size on masked repetition priming: An event-related potentials (ERP) investigation. *Language and Cognitive Processes*, *23* (1), 183–200.

Connolly, J. F., & Phillips, N. A. (1994). Event-related potential components reflect phonological and semantic processing of the terminal word of spoken sentences. *Journal of Cognitive Neuroscience*, *6*(3), 256–266.

DeLong, K. A., Urbach, T. P., & Kutas, M. (2005). Probabilistic word pre-activation during language comprehension inferred from electrical brain activity. *Nature Neuroscience*, *8*(8), 1117–1121.

Desroches, A. S., Newman, R. L., & Joanisse, M. F. (2009). Investigating the time course of spoken word recognition: Electrophysiological evidence for the influences of phonological similarity. *Journal of Cognitive Neuroscience*, *21*(10), 1893–1906.

Dikker, S., Rabagliati, H., Farmer, T. A., & Pylkkänen, L. (2010). Early occipital sensitivity to syntactic category is based on form typicality. *Psychological Science*, *21*(5), 629–634.

Dikker, S., Rabagliati, H., & Pylkkänen, L. (2009). Sensitivity to syntax in visual cortex. *Cognition*, *110*(3), 293–321.

Di Russo, F., Martínez, A., Sereno, M. I., Pitzalis, S., & Hillyard, S. A. (2002). Cortical sources of the early components of the visual evoked potential. *Human Brain Mapping*, *15*(2), 95–111.

Dufau, S., Grainger, J., & Holcomb, P. J. (2008). An ERP investigation of location invariance in masked repetition priming. *Cognitive, Affective, & Behavioral Neuroscience*, *8*(2), 222–228.

Eimer, M. (2000). Event-related brain potentials distinguish processing stages involved in face perception and recognition. *Clinical Neurophysiology*, *111*(4), 694–705.

Federmeier, K. D. (2007). Thinking ahead: The role and roots of prediction in language comprehension. *Psychophysiology*, *44*(4), 491–505.

Federmeier, K. D., & Kutas, M. (1999a). Right words and left words: Electrophysiological evidence for hemispheric differences in meaning processing. *Cognitive Brain Research*, *8*(3), 373–392.

Federmeier, K. D., & Kutas, M. (1999b). A rose by any other name: Long-term memory structure and sentence processing. *Journal of Memory and Language*, *41*(4), 469–495. Available from: http://dx.doi.org/10.1006/Jmla.1999.2660.

Federmeier, K. D., & Kutas, M. (2001). Meaning and modality: Influences of context, semantic memory organization, and perceptual predictability on picture processing. *Journal of Experimental Psychology: Learning, Memory, and Cognition*, *27*(1), 202.

Federmeier, K. D., & Kutas, M. (2002). Picture the difference: Electrophysiological investigations of picture processing in the two cerebral hemispheres. *Neuropsychologia*, *40*(7), 730–747.

Federmeier, K. D., Kutas, M., & Schul, R. (2010). Age-related and individual differences in the use of prediction during language comprehension. *Brain and Language*, *115*(3), 149–161. Available from: http://dx.doi.org/10.1016/j.bandl.2010.07.006.

Federmeier, K. D., & Laszlo, S. (2009). Chapter 1 Time for meaning: Electrophysiology provides insights into the dynamics of representation and processing in semantic memory. *Psychology of Learning and Motivation*, *51*, 1–44. Available from: http://dx.doi.org/10.1016/S0079-7421(09)51001-8.

Federmeier, K. D., Mai, H., & Kutas, M. (2005). Both sides get the point: Hemispheric sensitivities to sentential constraint. *Memory & Cognition*, *33*(5), 871–886.

Federmeier, K. D., McLennan, D. B., De Ochoa, E., & Kutas, M. (2002). The impact of semantic memory organization and sentence context information on spoken language processing by younger and older adults: An ERP study. *Psychophysiology*, *39*(2), 133–146. Available from: http://dx.doi.org/10.1017/s0048577202001373.

Friederici, A. D., & Kotz, S. A. (2003). The brain basis of syntactic processes: Functional imaging and lesion studies. *Neuroimage*, *20*, S8–S17.

Friedrich, C. K., Kotz, S. A., Friederici, A. D., & Alter, K. (2004). Pitch modulates lexical identification in spoken word recognition: ERP and behavioral evidence. *Cognitive Brain Research*, *20*(2), 300–308.

Friedrich, C. K., Kotz, S. A., Friederici, A. D., & Gunter, T. C. (2004). ERPs reflect lexical identification in word fragment priming. *Journal of Cognitive Neuroscience*, *16*(4), 541–552.

Friedrich, C. K., Schild, U., & Röder, B. (2009). Electrophysiological indices of word fragment priming allow characterizing neural stages of speech recognition. *Biological Psychology*, *80*(1), 105–113.

Ganis, G., & Kutas, M. (2003). An electrophysiological study of scene effects on object identification. *Cognitive Brain Research*, *16*(2), 123–144.

Ganis, G., Kutas, M., & Sereno, M. I. (1996). The search for "common sense": An electrophysiological study of the comprehension of words and pictures in reading. *Journal of Cognitive Neuroscience*, *8*(2), 89–106.

George, N., Jemel, B., Fiori, N., Chaby, L., & Renault, B. (2005). Electrophysiological correlates of facial decision: Insights from upright and upside-down Mooney-face perception. *Cognitive Brain Research*, *24*(3), 663–673.

Grainger, J., & Holcomb, P. J. (2009). Watching the word go by: On the time-course of component processes in visual word recognition. *Language and Linguistics Compass*, *3*(1), 128–156.

Grainger, J., Kiyonaga, K., & Holcomb, P. J. (2006). The time course of orthographic and phonological code activation. *Psychological Science*, *17*(12), 1021–1026.

Hagoort, P., Hald, L., Bastiaansen, M., & Petersson, K. M. (2004). Integration of word meaning and world knowledge in language comprehension. *Science*, *304*(5669), 438–441.

Halgren, E., Baudena, P., Heit, G., Clarke, M., & Marinkovic, K. (1994). Spatio-temporal stages in face and word processing. 1. Depth recorded potentials in the human occipital and parietal lobes. *Journal of Physiology-Paris*, *88*(1), 1–50.

Halgren, E., Dhond, R. P., Christensen, N., Van Petten, C., Marinkovic, K., Lewine, J. D., et al. (2002). N400-like magnetoencephalography responses modulated by semantic context, word frequency, and lexical class in sentences. *Neuroimage*, *17*(3), 1101–1116.

Holcomb, P. J., & Grainger, J. (2006). On the time course of visual word recognition: An event-related potential investigation using masked repetition priming. *Journal of Cognitive Neuroscience*, *18*(10), 1631–1643.

Holcomb, P. J., & Grainger, J. (2007). Exploring the temporal dynamics of visual word recognition in the masked repetition priming paradigm using event-related potentials. *Brain Research*, *1180*, 39–58.

Holcomb, P. J., & McPherson, W. B. (1994). Event-related brain potentials reflect semantic priming in an object decision task. *Brain and Cognition*, *24*(2), 259–276.

Holcomb, P. J., & Neville, H. J. (1990). Auditory and visual semantic priming in lexical decision: A comparison using event-related brain potentials. *Language and Cognitive Processes*, *5*(4), 281–312.

Huang, H.-W., Lee, C.-L., & Federmeier, K. D. (2010). Imagine that! ERPs provide evidence for distinct hemispheric contributions to the processing of concrete and abstract concepts. *Neuroimage*, *49*(1), 1116–1123.

Huddy, V., Schweinberger, S. R., Jentzsch, I., & Burton, A. M. (2003). Matching faces for semantic information and names: An event-related brain potentials study. *Cognitive Brain Research*, *17*(2), 314–326.

Jeffreys, D. A. (1989). A face-responsive potential recorded from the human scalp. *Experimental Brain Research*, *78*(1), 193–202.

Joyce, C. A., & Kutas, M. (2005). Event-related potential correlates of long-term memory for briefly presented faces. *Journal of Cognitive Neuroscience*, *17*(5), 757–767.

Kandhadai, P., & Federmeier, K. D. (2010). Automatic and controlled aspects of lexical associative processing in the two cerebral hemispheres. *Psychophysiology*, *47*, 774–785 (United States).

Kiani, R., Esteky, H., & Tanaka, K. (2005). Differences in onset latency of macaque inferotemporal neural responses to primate and non-primate faces. *Journal of Neurophysiology*, *94*(2), 1587–1596.

King, J. W., & Kutas, M. (1998). Neural plasticity in the dynamics of human visual word recognition. *Neuroscience Letters*, *244*(2), 61–64.

Kiyonaga, K., Grainger, J., Midgley, K., & Holcomb, P. J. (2007). Masked cross-modal repetition priming: An event-related potential investigation. *Language and Cognitive Processes*, *22*(3), 337–376.

Kutas, M., & Federmeier, K. D. (2011). Thirty years and counting: Finding meaning in the N400 component of the event-related brain potential (ERP). *Annual Review of Psychology*, *62*, 621–647. Available from: http://dx.doi.org/10.1146/annurev.psych.093008.131123.

Laszlo, S., & Federmeier, K. D. (2007). Better the DVL you know acronyms reveal the contribution of familiarity to single-word reading. *Psychological Science*, *18*(2), 122–126.

Laszlo, S., & Federmeier, K. D. (2009). A beautiful day in the neighborhood: An event-related potential study of lexical relationships and prediction in context. *Journal of Memory and Language*, *61*(3), 326–338.

Laszlo, S., & Federmeier, K. D. (2014). Never seem to find the time: Evaluating the physiological time course of visual word recognition with regression analysis of single item ERPs. *Language, Cognition, and Neuroscience*, *29*, 642–661.

Lau, E. F., Phillips, C., & Poeppel, D. (2008). A cortical network for semantics: (de) constructing the N400. *Nature Reviews Neuroscience*, *9*(12), 920–933.

Lau, E., Stroud, C., Plesch, S., & Phillips, C. (2006). The role of structural prediction in rapid syntactic analysis. *Brain and Language*, *98*(1), 74–88.

Luck, S. J., Chelazzi, L., Hillyard, S. A., & Desimone, R. (1997). Neural mechanisms of spatial selective attention in areas V1, V2, and V4 of macaque visual cortex. *Journal of Neurophysiology, 77*(1), 24–42.

Luck, S. J., & Hillyard, S. A. (1994). Electrophysiological correlates of feature analysis during visual search. *Psychophysiology, 31*(3), 291–308.

Mahon, B. Z., & Caramazza, A. (2008). A critical look at the embodied cognition hypothesis and a new proposal for grounding conceptual content. *Journal of Physiology-Paris, 102*(1), 59–70.

Martin, A. (2007). The representation of object concepts in the brain. *Annual Review of Psychology, 58*, 25–45.

Metusalem, R., Kutas, M., Urbach, T. P., Hare, M., McRae, K., & Elman, J. L. (2012). Generalized event knowledge activation during online sentence comprehension. *Journal of Memory and Language, 66*(4), 545–567.

Newman, R. L., & Connolly, J. F. (2009). Electrophysiological markers of pre-lexical speech processing: Evidence for bottom–up and top–down effects on spoken word processing. *Biological Psychology, 80*(1), 114–121.

Niedeggen, M., & Rösler, F. (1999). N400 effects reflect activation spread during retrieval of arithmetic facts. *Psychological Science, 10*(3), 271–276.

Niedeggen, M., Rösler, F., & Jost, K. (1999). Processing of incongruous mental calculation problems: Evidence for an arithmetic N400 effect. *Psychophysiology, 36*(3), 307–324.

Otten, M., Nieuwland, M. S., & Van Berkum, J. J. A. (2007). Great expectations: Specific lexical anticipation influences the processing of spoken language. *BMC Neuroscience, 8*(1), 89.

Paller, K. A., Gonsalves, B., Grabowecky, M., Bozic, V. S., & Yamada, S. (2000). Electrophysiological correlates of recollecting faces of known and unknown individuals. *Neuroimage, 11*(2), 98–110.

Patterson, K., Nestor, P. J., & Rogers, T. T. (2007). Where do you know what you know? The representation of semantic knowledge in the human brain. *Nature Reviews Neuroscience, 8*(12), 976–987.

Petit, J.-P., Midgley, K. J., Holcomb, P. J., & Grainger, J. (2006). On the time course of letter perception: A masked priming ERP investigation. *Psychonomic Bulletin & Review, 13*(4), 674–681.

Pickering, E. C., & Schweinberger, S. R. (2003). N200, N250r, and N400 event-related brain potentials reveal three loci of repetition priming for familiar names. *Journal of Experimental Psychology: Learning, Memory, and Cognition, 29*(6), 1298.

Pratt, H. (2012). Sensory ERP components. In S. Luck, & E. S. Kappenman (Eds.), *The Oxford handbook of event-related potential components* (pp. 89–114). New York, NY: Oxford University Press.

Pylkkänen, L., & Marantz, A. (2003). Tracking the time course of word recognition with MEG. *Trends in Cognitive Sciences, 7*(5), 187–189.

Pylkkänen, L., Stringfellow, A., & Marantz, A. (2002). Neuromagnetic evidence for the timing of lexical activation: An MEG component sensitive to phonotactic probability but not to neighborhood density. *Brain and Language, 81*(1), 666–678.

Rayner, K., & Clifton, C., Jr. (2009). Language processing in reading and speech perception is fast and incremental: Implications for event-related potential research. *Biological Psychology, 80*(1), 4–9.

Sanders, L. D., & Neville, H. J. (2003). An ERP study of continuous speech processing: I. Segmentation, semantics, and syntax in native speakers. *Cognitive Brain Research, 15*(3), 228–240.

Schendan, H. E., Ganis, G., & Kutas, M. (1998). Neurophysiological evidence for visual perceptual categorization of words and faces within 150 ms. *Psychophysiology, 35*(3), 240–251.

Schendan, H. E., & Kutas, M. (2002). Neurophysiological evidence for two processing times for visual object identification. *Neuropsychologia, 40*(7), 931–945.

Schendan, H. E., & Kutas, M. (2003). Time course of processes and representations supporting visual object identification and memory. *Journal of Cognitive Neuroscience, 15*(1), 111–135.

Schendan, H. E., & Kutas, M. (2007a). Neurophysiological evidence for the time course of activation of global shape, part, and local contour representations during visual object categorization and memory. *Journal of Cognitive Neuroscience, 19*(5), 734–749.

Schendan, H. E., & Kutas, M. (2007b). Neurophysiological evidence for transfer appropriate processing of memory: Processing versus feature similarity. *Psychonomic Bulletin & Review, 14*(4), 612–619.

Schendan, H. E., & Lucia, L. C. (2010). Object-sensitive activity reflects earlier perceptual and later cognitive processing of visual objects between 95 and 500 ms. *Brain Research, 1329*, 124–141.

Schweinberger, S. R., Pfütze, E.-M., & Sommer, W. (1995). Repetition priming and associative priming of face recognition: Evidence from event-related potentials. *Journal of Experimental Psychology: Learning, Memory, and Cognition, 21*(3), 722.

Schweinberger, S. R., Pickering, E. C., Burton, A. M., & Kaufmann, J. M. (2002). Human brain potential correlates of repetition priming in face and name recognition. *Neuropsychologia, 40*(12), 2057–2073.

Schweinberger, S. R., Pickering, E. C., Jentzsch, I., Burton, A. M., & Kaufmann, J. M. (2002). Event-related brain potential evidence for a response of inferior temporal cortex to familiar face repetitions. *Cognitive Brain Research, 14*(3), 398–409.

Sitnikova, T., Kuperberg, G., & Holcomb, P. J. (2003). Semantic integration in videos of real-world events: An electrophysiological investigation. *Psychophysiology, 40*(1), 160–164. Available from: http://dx.doi.org/10.1111/1469-8986.00016.

Tse, C.-Y., Lee, C.-L., Sullivan, J., Garnsey, S. M., Dell, G. S., Fabiani, M., et al. (2007). Imaging cortical dynamics of language processing with the event-related optical signal. *Proceedings of the National Academy of Sciences, 104*(43), 17157–17162.

Van Berkum, J. J. A., Brown, C. M., Zwitserlood, P., Kooijman, V., & Hagoort, P. (2005). Anticipating upcoming words in discourse: Evidence from ERPs and reading times. *Journal of Experimental Psychology: Learning, Memory, and Cognition, 31*(3), 443.

Van Petten, C., & Luka, B. J. (2012). Prediction during language comprehension: Benefits, costs, and ERP components. *International Journal of Psychophysiology, 83*(2), 176–190. Available from: http://dx.doi.org/10.1016/j.ijpsycho.2011.09.015.

Van Petten, C., & Rheinfelder, H. (1995). Conceptual relationships between spoken words and environmental sounds: Event-related brain potential measures. *Neuropsychologia, 33*(4), 485–508.

Voss, J. L., Schendan, H. E., & Paller, K. A. (2010). Finding meaning in novel geometric shapes influences electrophysiological correlates of repetition and dissociates perceptual and conceptual priming. *Neuroimage, 49*(3), 2879–2889.

West, W. C., & Holcomb, P. J. (2002). Event-related potentials during discourse-level semantic integration of complex pictures. *Cognitive Brain Research, 13*(3), 363–375.

Wicha, N. Y. Y., Bates, E. A., Moreno, E. M., & Kutas, M. (2003). Potato not pope: Human brain potentials to gender expectation and agreement in Spanish spoken sentences. *Neuroscience Letters, 346*(3), 165–168.

Wicha, N. Y. Y., Moreno, E. M., & Kutas, M. (2003). Expecting gender: An event related brain potential study on the role of grammatical gender in comprehending a line drawing within a written sentence in Spanish. *Cortex, 39*(3), 483–508.

Wicha, N. Y. Y., Moreno, E. M., & Kutas, M. (2004). Anticipating words and their gender: An event-related brain potential study of semantic integration, gender expectancy, and gender agreement in Spanish sentence reading. *Journal of Cognitive Neuroscience, 16*(7), 1272–1288.

Wlotko, E. W., & Federmeier, K. D. (2007). Finding the right word: Hemispheric asymmetries in the use of sentence context information. *Neuropsychologia, 45*(13), 3001–3014.

Wlotko, E. W., Lee, C. L., & Federmeier, K. D. (2010). Language of the aging brain: Event-related potential studies of comprehension in older adults. *Language and Linguistics Compass, 4*(8), 623–638.

CHAPTER

46

Left Ventrolateral Prefrontal Cortex in Processing of Words and Sentences

Nazbanou Nozari[1,2] *and Sharon L. Thompson-Schill*[3]

[1]Department of Neurology, Johns Hopkins University, Baltimore, MD, USA;
[2]Department of Cognitive Science, Johns Hopkins University, Baltimore, MD, USA;
[3]Department of Psychology, University of Pennsylvania, Philadelphia, PA, USA

46.1 INTRODUCTION

Left ventrolateral prefrontal cortex (VLPFC) has attracted much attention in cognitive neuroscience, not only for its involvement in numerous cognitive operations but also for its historical significance as one of the first brain regions to be formally linked to a specific function. In the mid 19th century, Pierre Paul Broca called this area (later to be called "Broca's area") the locus for production of articulate language and the seat of motor speech (Broca, 1861, 1865).

Although modern theories of language production (i.e., the dual stream model) have revived the notion that this area might be involved in motor production by proposing its involvement in storing articulatory representations, phonetic encoding, and retrieving or generating the articulatory codes (Hickok & Poeppel, 2004, 2007), much more has come to light about VLPFC in the past few decades. These new findings do not negate the possible role of this region in motor production, but rather call for an expansion of its role in the processing of language and perhaps even in nonlinguistic tasks. For instance, VLPFC has been implicated in semantic processing (Buckner et al., 1995; Demb et al., 1995; Démonet et al., 1992; Fiez, 1997; Kapur et al., 1994; Martin, Haxby, Lalonde, Wiggs, & Ungerleider, 1995; McCarthy, Blamire, Rothman, Gruetter, & Shulman, 1993; Petersen, Fox, Snyder, & Raichle, 1990; Raichle et al., 1994), phonological/phonetic processing, especially when phonological segmentation and sequencing is required (Démonet et al., 1992; Démonet, Price, Wise, & Frackowiak, 1994; Fiez et al., 1995; Newman, Twieg, & Carpenter, 2001; Paulesu, Frith, & Frackowiak, 1993; Price et al., 1994; Shaywitz et al., 1995; Zatorre, Evans, Meyer, & Gjedde, 1992), phoneme-to-grapheme conversional processes (Fiebach, Friederici, Müller, & Von Cramon, 2002), syntactic processing (Ben-Shachar, Hendler, Kahn, Ben-Bashat, & Grodzinsky, 2003; Embick, Marantz, Miyashita, O'Neil, & Sakai, 2000; Grodzinsky, 2000), and domain-general processes such as temporal sequencing regardless of the specific stimulus type (Gelfand & Bookheimer, 2003). Also, the term "working memory" appears frequently in the VLPFC literature (Awh et al., 1996; Paulesu et al., 1993).

Moreover, within each domain there is more than one view. For example, even among researchers who agree on VLPFC's involvement in semantic processing, there has been disagreement about whether it is involved in semantic retrieval (Demb et al., 1995; Démonet et al., 1992; Martin et al., 1995), conflict resolution (Thompson-Schill, D'Esposito, Aguirre, & Farah, 1997), or controlled semantic processing (Wagner, Paré-Blagoev, Clark, & Poldrack, 2001). Similarly, there have been debates among proponents of VLPFC's role in syntactic processing, some of whom had a strong hypothesis for the region's involvement in a specific syntactic operation (Grodzinsky, 2000; Musso et al., 2003) whereas others argued for multiple syntactic functions being mediated by the region (Friederici, Meyer, & von Cramon, 2000). The discussion of all these functions is obviously beyond the scope of this chapter. Therefore, we focus on the role of VLPFC in semantic processing, especially in the context of language. In most of what we discuss, the nature of the tasks addresses semantic-lexical mapping; however,

Neurobiology of Language. DOI: http://dx.doi.org/10.1016/B978-0-12-407794-2.00046-8

some of the concepts could apply to semantic processing without lexical retrieval.

Because it is our goal to discuss not only single-word but also sentence-level processing, it is inevitable to discuss syntactic theories of VLPFC as well. We review evidence from language comprehension and production studies and, whenever possible, present converging evidence from multiple sources (neuroimaging studies, patient studies, and transcranial magnetic stimulation (TMS) studies) to build a complete picture of the circumstances that lead to VLPFC recruitment and operations that are crucially dependent on this region. The chapter begins with an overview of the anatomy of VLPFC, followed by two main sections, discussing the debates on the region's role in (semantic) the processing of single words and (semantic–syntactic) the processing of sentences. We close by offering a unifying account that best summarizes the body of evidence in the earlier sections, with some future questions to consider.

The terms Broca's area, left inferior frontal gyrus (LIFG), left inferior prefrontal cortex (LIPC), and VLPFC have sometimes been used interchangeably. It is generally accepted that Brodmann area (BA) 44 and BA 45 (corresponding roughly to pars opercularis and pars triangularis) are the cytoarchitectonic correlates of Broca's region (Aboitiz & García, 1997; Uylings, Malofeeva, Bogolepova, Amunts, & Zilles, 1999; Figure 46.1), although there are finer subdivisions in these areas as well, with the dorsal part of BA45 (BA45B in Figure 46.1) resembling BA44 more closely than its ventral part (BA45A in Figure 46.1; Amunts et al., 2004). VLPFC also includes BA47 (pars orbitalis), the ventral cortical area inferior and anterior to the horizontal ramus of the lateral fissure. Most of the findings in this chapter concern BA44 and/or BA45.

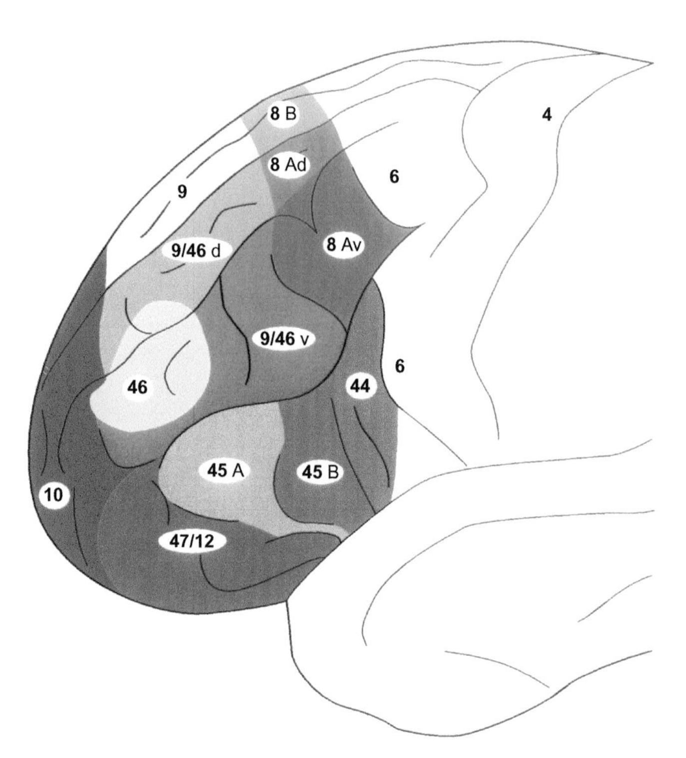

FIGURE 46.1 Anatomy of the VLPFC. *Figure adapted from Petrides and Pandya (2002). Reproduced with permission from Michael Petrides and Wiley-Blackwell publishing.*

46.2 VLPFC IN SINGLE-WORD PROCESSING

Although classic psycholinguistic studies primarily conceived of VLPFC as involved in either motor processing (following Broca's suggestion) or syntactic processing, memory researchers had an altogether different idea. In PET and fMRI studies, VLPFC activation has been consistently found in a variety of tasks requiring semantic processing. Among these are living/nonliving classification (Demb et al., 1995; Gabrieli et al., 1996; Kapur et al., 1994), feature-based similarity judgment (Thompson-Schill et al., 1997; Whitney, Kirk, O'Sullivan, Ralph, & Jefferies, 2011), global similarity judgment (Thompson-Schill et al., 1997; Wagner et al., 2001), and category-based verbal fluency (Basho, Palmer, Rubio, Wulfeck, & Müller, 2007; Birn et al., 2010; Gurd et al., 2002; Paulesu et al., 1997). VLPFC activation in a wide variety of tasks requiring semantic access points strongly to its role in some aspect of semantic processing. Initially, this was hypothesized to be semantic retrieval (Demb et al., 1995; Martin et al., 1995). Starting in the late 1990s, a series of experiments pointed out the region's particular sensitivity to the *control demands* of semantic retrieval as opposed to retrieval *per se*.

The first in this series was the study by Thompson-Schill et al. (1997) in which, through three experiments, the authors showed that activation of VLPFC was much more prominent in conditions with high-selection demands. Two of these experiments concerned single-word comprehension. In the first task, subjects had to judge whether a word matched a picture either in its identity (e.g., the word CAR matching the picture of a car; low selection) or in a feature belonging to it (e.g., the word EXPENSIVE matching the picture of a car; high selection). In the second task, subjects had to judge the similarity of a probe word to a number of alternatives. In the low-selection condition, judgment was to be based on global similarity (e.g., TICK–FLEA/WELL), whereas in the high-selection condition, selection based on a specific feature—and ignoring other attributes—was necessary for making the correct choice (e.g., TOOTH–BONE/TONGUE). The third experiment was a production task. Subjects had to generate verbs in response to nouns that induced either low competition (e.g., SCISSORS, which strongly evokes the verb CUT) or high competition (e.g., CAT, which could be associated with a number of verbs, like MEOW, PLAY, EAT, etc.) between response alternatives. In the same group of subjects, in all three tasks an overlapping area of VLPFC was found to respond more to conditions that placed higher demands on selection. Wagner et al. (2001) extended these findings by showing that it was possible to manipulate control demands even within the global similarity judgment task. Increased VLPFC activation was found when global similarity was probed for low-association items (e.g., CANDLE-HALO) compared with high-association items (e.g., CANDLE-FLAME).

Although the pattern of activation in these neuroimaging studies was suggestive of a role for VLPFC in semantic control, it remained to be seen if such control depended crucially on this prefrontal region. It was Thompson-Schill et al. (1998) who showed that patients with lesions encompassing VLPFC had trouble with the verb generation task described, but only when the noun was not strongly associated with a unique verb. They further showed that the percentage of damage to BA 44 (but not overall lesion volume or damage to dorsolateral prefrontal regions) was a significant predictor of errors in the trials with high-selection demand.

These early studies convincingly demonstrated that the picture was incompatible with VLPFC's role in pure semantic retrieval but pointed to a role in enforcing top-down control when stimulus-response associations were weak (Miller & Cohen, 2001; Norman & Shallice, 1986). Following Desimone and Duncan (1995), it was proposed that when bottom-up association is not strong enough to pick a clear "winner" among the alternatives, competition must be resolved by top-down control to bias selection toward a single representation, and VLPFC was a likely candidate for implementing this bias (Kan & Thompson-Schill, 2004; Miller & Cohen, 2001; Wagner et al., 2001). Since that time, this idea has been tested using various tasks and paradigms; some of which that are related to word processing are reviewed later.

46.2.1 Deciphering Words with Multiple Meanings

Every tongue contains words that carry more than one meaning, and efficient processing of language requires that speakers and listeners would be able to handle this ambiguity by flexibly retrieving the relevant meaning and suppressing the irrelevant meaning in different contexts. As such, these types of words provide a good opportunity for investigating which brain region is involved in biasing selection. Bedny, McGill, and Thompson-Schill (2008) had participants judge the relatedness of word pairs, with some subsequent pairs containing ambiguous words (e.g., SUMMER-FAN → CEILING-FAN [same meaning]; ADMIRER-FAN → CEILING-FAN [different meaning]). Whereas posterior temporal cortex proved to be sensitive to semantic overlap regardless of ambiguity, VLPFC activity depended on the amount of semantic ambiguity.

This finding is mirrored by VLPFC patients' difficulty in efficiently selecting the appropriate meaning of ambiguous words. Bedny, Hulbert, and Thompson-Schill (2007) chose three groups of participants, patients with VLPFC damage, patients with frontal damage sparing VLPFC, and matched controls, and asked them to determine the lexicality of the third word in a triplet of words, with the second word being either a homonym or a polysemous word. Both homonyms and polysemous words have more than one meaning, although these meanings are unrelated in the case of homonyms (e.g., ceiling FAN, vs. football FAN) and related in the case of polysemous words (e.g., live CHICKEN, vs. food CHICKEN). In the triplet, the first and the third words were related to either the same meaning of the second word (e.g., BACK, PACK, BAG; consistent condition) or a different meaning (WOLF, PACK, BAG; inconsistent condition). Unlike controls and patients with non-VLPFC frontal lobe lesions, VLPFC patients' performance did not show a reliable difference between the consistent and inconsistent conditions. These patients did, however, show priming in the consistent condition compared with an unrelated baseline. This finding suggests that, in all likelihood, both meanings of the words were activated, and the activation of the context-relevant meaning benefited performance. However, the context-irrelevant meaning was not properly suppressed. This is consistent with the distinction we made earlier between semantic retrieval versus biasing competition; only the latter of which seems to depend critically on VLPFC.

Event-related potentials (ERP) findings are consistent with the results of Bedny et al. (2007). Swaab, Brown, and Hagoort (1998) presented participants with sentences that primed either the dominant or the subordinate meaning of a homonym, or was unrelated to the target (e.g., target word: RIVER; consistent prime: "The man planted a tree on the bank"; inconsistent prime: "The man made a phone call to the bank," and unrelated prime: "The boy petted the dog on the head"). The target word (RIVER) was always related to the subordinate meaning of the homonym (BANK) and followed the sentence with either a short (100 ms) or a long (1,250 ms) ISI. N400 (as a measure of violation of semantic expectancy) was measured in response to the target word. The logic was that proper priming of the meaning should decrease the amplitude of N400 in the consistent, but not in the inconsistent, condition. Healthy controls showed this pattern. Broca's patients, the majority of whom had prefrontal lesions, however, showed evidence of reduced N400 under *both* conditions for the short ISI, meaning that context had not successfully abolished the irrelevant meaning. In the longer ISI, some of the patients no longer exhibited this abnormal pattern, implying that given enough processing time, suppression of the irrelevant meaning was slowly achieved, perhaps through a complementary/compensatory network. More recently, Vuong and Martin (2011) supported this position by showing that a patient with damage that included VLPFC damage was considerably slower than two patients with non-VLPFC lesions in using context to bias toward the subordinate meaning of an ambiguous word. However, when tested using balanced ambiguous words, this patient, similar to the other two patients and healthy controls, was unimpaired in using context to resolve ambiguity.

46.2.2 Verbal Fluency

Verbal fluency tasks are among the oldest neuropsychological tasks for assessing the integrity of memory and language. There are two categories of verbal fluency tasks. In the *semantic* verbal fluency task a semantic cue is provided, based on which the individual must search for as many words as possible in a short duration of time (e.g., "Name all the animals you can."). In the *phonological* variant, the cue is phonological (e.g., "Name all the words you can that start with B."). Given the very different nature of the search in these two tasks, it is not surprising that semantic and phonological verbal fluency tasks rely, at least in part, on different cognitive architectures. For example, Martin, Wiggs, Lalonde, and Mack (1994) found that the semantic verbal fluency task is subject to interference from object recognition, whereas phonological verbal fluency task performance is sensitive to motor sequence tasks. Likewise, each variant of the task induces preferential activation in a certain frontal region (Birn et al., 2010). However, despite their unique elements, both tasks have been shown to activate VLPFC, although there is disagreement on whether they activate the same or different subpopulation of neurons in this region (Frith, Friston, Liddle, & Frackowiak, 1991; Mummery, Patterson, Hodges, & Wise, 1996; Paulesu et al., 1997).

The verbal fluency task provides a unique opportunity to study two cognitive operations within the verbal fluency task. At any given point during this task it is possible to continue retrieving names from the same subcategory as the previous item or to switch to a new subcategory. For example, when prompted to name as many animals as you can, you may cue yourself by starting with the subcategory of farm animals. The ability to retrieve many names in one subcategory is called *Clustering*. Even when you have good clustering abilities, at some point you will run out of names of farm animals, and to maximize your output you must switch to a new subcategory, for example, wild animal. The ability to switch from one subcategory to another is called *Switching*. It has been proposed that clustering reflects the intactness of semantic knowledge, whereas switching reflects the biasing ability. In tandem with this proposition, damage to the temporal cortex typically causes clustering deficits and damage to the left prefrontal cortex is associated with switching deficits (Chertkow & Bub, 1990; Troyer, Moscovitch, & Winocur, 1997). In two experiments, Hirshorn and Thompson-Schill (2006) showed that it was the switching aspects of the task that elicited VLPFC activation. In Experiment 1, subjects were told either to switch on every trial or to freely produce words given a category cue. In Experiment 2, they were given instructions for free category-based name generation while pushing a button but were asked to push a different button whenever they switched to a new category. In both cases, VLPFC activation was linked directly to switching.

On the lesion side, lesions in the frontal cortex have long been known to cause impairments in performing verbal fluency tasks (Baldo & Shimamura, 1998; Janowsky, Shimamura, Kritchevsky, & Squire, 1989; Owen, Downes, Sahakian, Polkey, & Robbins, 1990; Perret, 1974; Robinson, Shallice, Bozzali, & Cipolotti, 2012; Stuss et al., 1998; Troyer, Moscovitch, Winocur, Alexander, & Stuss, 1998), and some have specifically pinned the effect down to the VLPFC (Novick, Kan, Trueswell, & Thompson-Schill, 2009).

46.2.3 Picture Naming and the Need for Control

This discussion raises the question of where the boundary is for the need for top-down control. Does naming pictures, for example, require top-down control? This question is particularly important because picture naming is the single most important neuropsychological test in localizing deficits in lexical retrieval. Furthermore, in many cases a patient's response pattern in picture naming allows for predictions to be made about their other production abilities, such as word repetition (Nozari & Dell, 2013; Nozari, Kittredge, Dell, & Schwartz, 2010). But is picture naming measuring lexical-semantic retrieval or controlled selection? The reports are mixed. Although some have found activation of VLPFC in picture naming (Murtha, Chertkow, Beauregard, Dixon, & Evans, 1996; Murtha, Chertkow, Beauregard, & Evans, 1999), some have not (Etard et al., 2000). We briefly discuss two factors, name agreement and context, that speak directly to the control aspect of picture naming, and help reconcile these seemingly disparate findings. Novick et al. (2009; Experiment 2) showed that a patient with VLPFC lesion was significantly impaired in naming pictures with low name agreement (e.g., COUCH/SOFA) compared with controls and patients with frontal lesions sparing VLPFC. However, the same patient showed no marked deficits in naming pictures with high name agreement. Given what has been discussed earlier in this chapter, the explanation for this finding should be clear. When bottom-up cues are not strong enough to unequivocally select a unique representation, VLPFC is needed to implement top-down control and help with selection.

But even when the picture itself is associated with a unique label, the context in which the name is to be retrieved can modulate retrieval demands. When pictures are named in the context of same-category items (e.g., DOG, HORSE, LION), they are named more slowly in the neurologically intact adults (Belke, Meyer, & Damian, 2005; Damian, Vigliocco, & Levelt, 2001; Kroll & Stewart, 1994) and with more errors in aphasic patients with left frontal lesions (McCarthy & Kartsounis, 2000; Schnur, Schwartz, Brecher, & Hodgson, 2006). These tasks control for pure lexical retrieval because picture naming in a semantically heterogeneous context (e.g., DOG, TRUCK, APPLE) is also assessed and acts as a baseline for comparison. While the exact mechanisms of this *semantic blocking effect* are disputed (Howard, Nickels, Coltheart, & Cole-Virtue, 2006; Oppenheim, Dell, & Schwartz, 2010), there is consensus that competition is higher in the semantically homogenous context, and both neuroimaging and lesion studies link the effect to VLPFC (Schnur et al., 2009). Incidentally, the Murtha et al. (1996, 1999) picture naming study administered only a homogenous block of animals to subjects to name, whereas Etard et al. (2000) used a heterogenous block. Given the well-established semantic blocking effect, it is not surprising that picture naming designs that strongly tap into the effect activate VLPFC. In summary, picture naming seems to require top-down support from VLPFC primarily under conditions of high competition.

46.2.4 TMS Studies

So far, we have shown that neuroimaging and patient studies converge on the biasing role of VLPFC in semantic/lexical retrieval. We end this section by providing a brief review of the TMS studies related to this issue. Devlin, Matthews, and Rushworth (2003; Experiment 2) presented subjects with single words and asked them to make a natural/man-made judgment for each word. In a control perceptual task, subjects judged whether the horizontal line above the word was shorter than the word. They showed that TMS to the anterior portion of VLPFC interfered with the semantic, but not the perceptual, decision. This finding, along with higher activation of anterior VLPFC in making semantic versus phonological judgments (Devlin et al., 2003; Experiment 1), was taken as evidence for this region's contribution to making semantic judgments. Gough, Nobre, and Devlin (2005) provided additional support for this finding by showing that TMS over the anterior VLPFC caused a selective impairment in a semantic (synonym judgment) as opposed to a phonological (homophone judgment) task, whereas the opposite pattern was found when posterior VLPFC was stimulated (see also Wig, Grafton, Demos, & Kelley, 2005) for a demonstration of eliminated priming as a function of encoding under TMS).

The direct involvement of VLPFC in biasing semantic selection was also demonstrated in a number of recent TMS studies. Hindy, Hamilton, Houghtling, Coslett, and Thompson-Schill (2009) used TMS and computer-mouse tracking to examine the role of VLPFC in semantic processing. In each trial, two words appeared on the screen, one of which was better matched to a probe word that appeared with some delay, and participants had to move the mouse toward the correct response. Either the probe was strongly associated with the response (e.g., KING-HOOK; probe = QUEEN) or this association was weak (e.g., CARDS-HOOK; probe = QUEEN). Repetitive TMS was delivered soon after the onset of the first two words and before the appearance of the probe item. The results showed greater deviation of mouse movement trajectories toward the incorrect response when the association was weak, compatible with the hypothesized role of VLPFC in biasing competition toward the correct meaning (however, the effect was not found when the congruency between the stimuli and the probe item was manipulated). Importantly, when the delay between the response and target words was removed (and TMS was delivered afterwards), the effect disappeared. Hindy et al. (2009) suggested that the disappearance of the effect with this timing manipulation implies that the contextually appropriate association is formed, and conflict is already resolved before receiving TMS. In a conceptually similar study, Whitney et al. (2011) used a design similar to that of Wagner et al. (2001), in which subjects had to select a word that had either a strong (e.g., PEPPER) or a weak (e.g., GRAIN) association with the probe word (SALT). Although performance in the strong-association condition was unaffected by TMS over VLPFC, stimulation disrupted performance in the low-association condition. Similar to the results of Devlin et al. (2003), a control (nonsemantic) judgment task was insensitive to TMS effects in this area (see also Whitney, Kirk, O'Sullivan, Ralph, & Jefferies, 2012).

To summarize, we reviewed evidence for the possible role of VLPFC in comprehension and production of single words and showed that while this region does have an involvement in semantic/lexical retrieval, this involvement appears to be specific to situations in which there is a need for biasing competition through top-down control.

46.3 VLPFC IN SENTENCE PROCESSING

Although the role of VLPFC was being investigated in semantic processing mostly by memory researchers, many linguists and psycholinguists were attempting to understand the role that VLPFC played in processing syntax. The initial motivation for the various syntactic hypotheses might have stemmed from the clinical profile of Broca's aphasia (Grodzinsky, 2000), an impairment that is characterized by agrammatic speech without marked semantic difficulties. There was a major problem with this inference. Damage to Broca's area is neither sufficient nor necessary for generating symptoms constituting Broca's aphasia. Damage restricted to Broca's area leads only to a transient mutism with spontaneous recovery (Levine & Mohr, 1979; Mohr et al., 1978). However, Broca's aphasics' lesions often extend beyond BA 44 and BA 45 to involve some parts of BA 6, BA 8, BA 9, BA 10, and BA 46, as well as the underlying white matter and basal ganglia (Damasio, 1992; Dronkers, Plaisant, Iba-Zizen, & Cabanis, 2007).

However, regardless of whether the profile of Broca's aphasia is informative about the role of Broca's area, a large number of neuroimaging studies have also implicated VLPFC in "syntactic processing" (see Kaan & Swaab, 2002, for an excellent review). Here, we use the general framework used by Kaan and Swaab (2002) to review the evidence for a syntactic role of VLPFC.

46.3.1 Syntactic Complexity

The first group of studies constitutes experiments comparing syntactically simple versus syntactically complex sentences using a variety of manipulations (Caplan, Alpert, & Waters, 1998, 1999; Cooke et al., 2002; Dapretto & Bookheimer, 1999; Fiebach, Schlesewsky, & Friederici, 2001; Inui et al., 1998; Keller, Carpenter, & Just, 2001; Michael, Keller, Carpenter, & Just, 2001; Stowe et al., 1998; Stromswold, Caplan, Alpert, & Rauch, 1996). For example, Caplan et al. (1999) had subjects judge the plausibility of cleft object and cleft subject sentences (e.g., "It was the actress that the award thrilled" vs. "It was the award that thrilled the actress") and found increased VLPFC activation when subjects processed the cleft object sentences. Although these two sentences are close in meaning in the sense that they both convey that the actress was thrilled by the award, they are different in their syntactic forms, making VLPFC a suspect in processing syntactic complexity. However, Cooke et al. (2002) failed to find increased VLPFC activation in object-relative clauses with short antecedent gaps (e.g., "The flower girl who Andy punched in the arm was 5 years old"), but they did find it when the gap was long (e.g., "The messy boy who Janet, the very popular hairdresser, grabbed was extremely hairy."). In line with this, Fiebach et al. (2001) showed that when the case marker disambiguated the thematic roles in sentences containing relative clauses (as is the case in German), it was only the long-distance dependencies that induced VLPFC activation. This led to the proposal that VLPFC mediates *syntactic working memory*. However, it is unclear how the finding of Caplan et al. (1999) would fit into this account.

Moreover, Chen, West, Waters, and Caplan (2006) showed that when length and syntactic structure were kept constant, higher activation of VLPFC was found when the subject of the relative clause was inanimate (e.g., "The golfer that the lightning struck survived the incident") compared with when it was animate (e.g., "The wood that the man chopped heated the cabin."). Neither syntactic ambiguity nor syntactic working memory can be blamed for this difference. Corroborating the antisyntactic evidence were findings of Keller et al. (2001), who showed that syntactically complex sentence with object-relative clauses (e.g., "The boy who the doctor visited had contracted pneumonia") actually elicited *less* activity in VLPFC compared with syntactically simple sentences that contained a temporarily ambiguous word (e.g., "The desert trains [= noun] usually are late"; "The desert trains [= verb] its inhabitants to conserve their resources"). An innovative study in this genre was conducted by Dapretto and Bookheimer (1999), who claimed that their design allowed for teasing apart semantic from syntactic processing. They asked subjects to make same/different judgments for sentences such as "The bridge is west of the airport" paired with what the authors considered either a semantically different sentence (e.g., "The bridge is west of the river") or a syntactically different sentence ("West of the bridge is the airport"). VLPFC was selectively implicated for the syntactically different sentences. However, the syntactically different sentences also perform different meanings ("The bridge is west of the airport" does not mean that "West of the bridge is the airport"). It is therefore unlikely that this design can single-out syntactic processing.

One of the most famous syntactic accounts of VLPFC, based on syntactic complexity, is its role in syntactic (or *transformational*) movements (Grodzinsky, 1995, 2000). This account proposes that when faced with a sentence like "Which man did the woman like?", the parser would create a placeholder (trace) to the right-hand side of the verb to simulate an active sentence structure: "[Which man] did the woman like {trace}?". By establishing a common index between the trace and its antecedent (which man), the parser recognizes the antecedent as the theme of the verb "like." Following this proposal, Grodzinsky (2000) claimed that in Broca's aphasia, all traces of movements are deleted and, primarily based on this assumption, concluded that Broca's area handles exclusively intrasentential dependency relations. Grodzinsky's theory has received criticism not only for equating Broca's aphasia with Broca's area, an issue that we touched on earlier in this chapter but also because of its oversight in explaining the full picture of Broca's aphasia that it targets (see Grodzinsky, 2000 for full commentaries). For one thing, there are Broca's patients who perform better than chance in comprehension of certain transformationally derived passives (Balogh & Grodzinsky, 1996; Druks & Marshall, 1995; Hickok & Avrutin, 1995; Saddy, 1995). The most serious criticism for the theory, though, is the finding that the deficits in such patients is not limited to processing sentences with syntactic structures requiring transitional movements. Agrammatic patients can have problems with active sentences, too, as long as the semantic roles are reversible. Schwartz, Saffran, and Marin (1980), among others, have reported such deficits when semantic symmetry is possible around a verb (e.g., "The dancer applauds the clown") or around a spatial preposition (e.g., "The square is above the circle").

To summarize, VLPFC is recruited for processing of certain syntactic complexities, but its activation does not seem to be either necessary for or limited to all syntactically complex sentences. It is worth mentioning that the activation of VLPFC in difficult sentences has been attributed to subvocal rehearsal, which presumably helps with parsing (Paulesu et al., 1993; Smith & Jonides, 1999). To test this hypothesis specifically, Caplan, Alpert, Waters, and Olivieri (2000) investigated VLPFC activation in subject-relative and object-relative sentences under concurrent articulation conditions (which seriously reduce the chance of subvocal rehearsal) and showed that the differential activation of VLPFC for object-relative sentences survived this manipulation. It is therefore unlikely that this explanation sufficiently justifies the nature of VLPFC's involvement in these cases.

46.3.2 Anomalous Sentences

Another category of studies aiming at semantic–syntactic comparison of VLPFC involves presenting anomalous sentences. Many such studies have failed to show a syntax-specific role for this cortical region (Indefrey, Hagoort, Herzog, Seitz, & Brown, 2001; Kuperberg et al., 2000; Newman, Pancheva, Ozawa, Neville, & Ullman, 2001; Ni et al., 2000; Nichelli et al., 1995; cf., Embick et al., 2000). In some cases VLPFC was not recruited when detecting syntactic anomalies such as "Trees can grew," and in some other cases detection of syntactic and semantic anomalies was not dissociable. Also, the choice of baseline should be taken into account when interpreting claims about the exclusivity of VLPFC's processing to syntax. For example, Moro et al. (2001) showed VLPFC's selective activation during detection of syntactic anomalies, but the baseline for comparison was detection of phonotactic and orthographic anomalies. To attribute greater VLPFC activation in detecting syntactic over, for example, phonotactic anomalies, one must assume that the cognitive processes required for detecting these two types of violation are the same, except for the materials. This is very unlikely to be the case because spotting syntactic errors, for example, noun–verb agreement errors, requires keeping track of earlier parts of the utterance, whereas phonotactic violations can be detected without any memory of what has been stated earlier. In fact, there is now ample evidence that semantic and pragmatic violations do recruit VLPFC (Hagoort, Hald, Bastiaansen, & Petersson, 2004; Kiehl, Laurens, & Liddle, 2002; Kuperberg, Holcomb, et al., 2003; Kuperberg, Sitnikova, Caplan, & Holcomb, 2003; Newman et al., 2001; Ni et al., 2000; see also Marini & Urgesi, 2012 for a TMS study). Baumgaertner, Weiller, and Büchel (2002) took this one step further by showing that unexpected sentence endings, even if they do not qualify as violations, elicit activation in VLPFC.

46.3.3 Other Semantic–Syntactic Comparisons

Another category of studies speaking to a semantic–syntactic differentiation in the role of VLPFC compared either word lists to sentences or meaningful sentences to jabberwocky and syntactic prose. Kaan and Swaab (2002) point out that the studies in this category that have failed to identify VLPFC (Kuperberg et al., 2000; Mazoyer et al., 1993; Stowe et al., 1998; 1999) outnumber the studies that have reported a positive effect (Bottini et al., 1994) and point to task-specific demands for recruiting VLPFC in the latter cluster.

46.3.4 Ambiguity

Although there is simply too much evidence to deny the involvement of VLPFC in some aspect of semantic processing, it is also clear that there are some syntactic structures that elicit VLPFC activation more than others, as pointed out in the *syntactic complexity* section. Here, we separately review a subset of these studies, focusing on garden path sentences (temporarily ambiguous sentences that initially lead to an incorrect interpretation), because they have been particularly useful in reconciling the semantic–syntactic debate over the role of VLPFC. We mentioned the Keller et al. (2001) study showing VLPFC activation when subjects encountered sentences containing ambiguous words like TRAIN. Minimizing the differences between surface forms of sentences, Mason, Just, Keller, and Carpenter (2003) used a similar logic by using ambiguous words like WARNED in the following sentences and observed the following activation pattern in VLPFC: a > b > c.

a. *The experienced soldiers warned about the dangers conducted the midnight raid. (ambiguous verb, subordinate form)*
b. *The experienced soldiers warned about the dangers before the midnight raid. (ambiguous verb, dominant form)*
c. *The experienced soldiers spoke about the dangers before the midnight raid. (unambiguous verb)*

When there was a single possible interpretation of the verb (i.e., spoke = past tense, active), VLPFC's role in parsing was minimal. When there were alternative meanings to be considered, VLPFC activation increased to bias toward one of the two interpretations (i.e., warned = past tense, active or passive). The maximum involvement of VLPFC was observed when the correct interpretation required biasing toward the subordinate meaning of an ambiguous verb (i.e., warned = passive; see Garnsey, Pearlmutter, & Myers, 1997, for a discussion of verb bias).

A few years after the Mason et al. study, January, Trueswell, and Thompson-Schill (2009) presented the first clear demonstration for the involvement of the same cortical region in processing both syntactic and nonsyntactic high-conflict tasks. These authors demonstrated within-subject overlap in neural substrates of processing garden path sentences and the word Stroop task, localizing both effects to VLPFC. Studies of patients with VLPFC lesions confirm that they have problems recovering from the wrong interpretation of sentences that initially induce a bias toward the incorrect meaning (Novick et al., 2009; Novick, Trueswell, & Thompson-Schill, 2005). The same patients show selective impairment in suppressing interference in a memory task (Novick et al., 2009; see also Hamilton & Martin, 2005).

But does VLPFC respond to the ambiguous word itself, or to the need for revision (i.e., the disambiguating information)? Fiebach, Schlesewsky, Lohmann, Von Cramon, and Friederici (2005) found greater VLPFC activation when disambiguating information came later, as opposed to earlier, in the sentence. This is expected if VLPFC starts the biasing process at the moment the parser encounters an ambiguity and continues to update the bias as more cues accumulate. If the initial commitment to a meaning is incorrect, then the later the disambiguating information comes in, the farther the bias in the incorrect direction, the more work needed to shift the competition in favor of the alternative meaning, and, hence, the greater activation in the delayed disambiguation condition. Compatible with this interpretation, Rodd, Johnsrude, and Davis (2012) manipulated the relative timing of the ambiguous word and the disambiguating information and showed VLPFC activation by the ambiguous word and by the subsequent disambiguating information. In contrast, left inferior temporal gyrus responded only to the disambiguating information.

The evidence strongly suggests that when there is need for biasing interpretation toward one of the two meanings, VLPFC is activated, especially if the favored meaning turns out to be incorrect. If this is really the role of this region, then the bias need not be induced by syntax; semantics also should be able to create the incorrect bias. Recently, Thothathiri, Kim, Trueswell, and Thompson-Schill (2012) demonstrated that this is true. By keeping the verb constant in an unambiguous syntactic structure and changing the content nouns around the verb, they evaluated parsing of three types of sentences (a–c in the following sentences). The pattern of VLPFC activation was as follows: a > b > c.

a. *The journalist was interviewed by the undergraduate. (violation of the usual role of journalist)*
b. *The patient was interviewed by the attractive man. (neutral roles)*
c. *The celebrity was interviewed by a reporter. (congruent with the usual role of reporter)*

Although there is no ambiguity in these sentences, violation of the usual semantic roles recruits VLPFC proportionally to the degree of violation. Similarly, Saffran, Schwartz, and Linebarger (1998) reported that patients with agrammatic aphasia exhibited high error rates even with syntactically simple sentences when semantic information contradicted the correct thematic role assignment (e.g., "The deer shot the hunter"), but the exact site of lesion was not specified for the patients in this study.

46.4 SUMMARY

46.4.1 Against a Rigid Semantic–Syntactic Distinction in VLPFC

Our goal for organizing this chapter into two sections was to better classify studies pertaining to the role of VLPFC in single-word versus sentence proessing. In so doing, we also captured the spirit of a fundamental debate over the role of VLPFC, namely the semantic–syntactic debate. In the first section, we discussed an abundance of evidence from neuroimaging, TMS, and lesion studies converging on the role of VLPFC in biasing competition during semantic/lexical selection in the absence of any syntax-like structure. In the second section, we discussed studies that directly pitted semantic processing against syntactic processing in VLPFC and showed that the evidence for a pure syntactic account is sparse.

Our contention is that drawing a hard line between semantic and syntactic processing in understanding the role of VLPFC in language processing is not very useful. By this assertion, we do not mean to deny that semantic and syntactic aspects of language processing are distinct and can be teased apart, but that given the empirical evidence, this distinction does not seem critical to VLPFC operations. ERP studies have shown that both syntactic violations (e.g., "at breakfast the boys would eats...") and certain semantic violations in the absence of syntactic violations ("at breakfast the eggs would eat...") evoke a positivity called P600 (Kuperberg, 2007; Kuperberg, Caplan, Sitnikova, Eddy, & Holcomb, 2006; Kuperberg, Holcomb, et al., 2003; Kuperberg, Sitnikova, et al., 2003). Interestingly, even though these two sentence types are different from violations of world knowledge, such as "The Dutch trains are white..." (Hagoort et al., 2004), or expected events "...at breakfast the boys would plant...," which elicit N400 instead, all of these violations activate VLPFC (Kuperberg, Sitnikova, & Lakshmanan, 2008).

What is the cognitvie explanation for this? Modern theories of language comprehension (MacDonald, Pearlmutter, & Seidenberg, 1994) propose that on encountering a sentence, multiple sources of information (syntactic, semantic, pragmatic, etc.) are triggered and collaborate to derive the meaning. Activation of information in each domain is probabilistic and frequency-dependent, and convergence of information from these multiple domains leads to proper comprehension of a sentence. Because the ultimate goal is to arrive at a coherent meaning supported by all cues, any type of information, if it creates a bias toward an interpretation that clashes with other types of information, leads to conflict, and this conflict requires top-down control to be resolved. The source of initial incorrect bias could be syntactic (e.g., in the case of less frequent object-relative clauses without strong semantic cues), semantic (e.g., when the world knowledge is incongruent with the thematic roles, like celebraties photographing paparazzi), or perhaps other (e.g., pragmatic and prosodic). Regardless, on encountering potential for multiple interpretations, VLPFC is activated to use the existing information to create the bias and continues to do so until conflict is minimized. If VLPFC fails, due to lesion or temporary deactivation via TMS, then top-down control is significantly reduced and processing would follow bottom-up cues, giving rise to the difficulty that VLPFC patients have with sentences, the correct interpretation of which requires overriding a strong semantic–syntactic cue. Complementary to this picture are cases where bottom-up cues are limited. Although a sentence bombards the comprehension system with multiple external cues, production is much more internally driven. Thus, the speaker must use top-down control to successfully initiate (and to fluently continue) the generation of concept/linguistic materials, except for cases when a strong bottom-up cue is presented (e.g., a picture of an object with a unique label), eliminating the need for top-down biasing.

46.4.2 A New "Broca's" Aphasia

Earlier in this chapter we alluded to the fact that Broca's aphasia does not necessarily correspond to pathology in Broca's area. In this section, we present other clinical profiles of aphasia that have stronger ties to VLPFC lesions. First, we discuss the revival of an old profile described by Alexander Luria, called "dynamic aphasia" (Luria, 1970, 1973; Luria & Tsvetkova, 1968), which has been directly linked to VLPFC damage. Next, we discuss a syndrome called Semantic Aphasia (SA), also seen in VLPFC patients, although certain temporoparietal regions can induce similar symptoms. Our goal is not to argue that these two are distinct syndromes. In fact, it is quite likely that the patients reported under these two labels have similar clinical deficits; however, to date they have been studied and discussed under different literature.

A typical dynamic aphasic profile is portrayed by a patient described by Robinson, Blair, and Cipolotti (1998). This patient had a frontal meningioma that impinged on BA 45 and presented with no impairment in simple picture naming, word repetition, comprehension, or reading, but who did have markedly decreased spontaneous or propositional speech. In multiple experiments, the authors showed that the patient was unimpaired in generating words or phrases given a strong cue that limited the possible responses, and also was severely impaired under conditions where the cue was not strongly associated with a unique response (see Robinson, Shallice, & Cipolotti, 2005, for a similar report). It is worth mentioning that the patient reported in Robinson et al. (1998) had a lesion that also impacted the dorsolateral PFC (DLPFC). Alexander (2006) has also considered lesions of DLPFC to be relevant to symptoms of dynamic aphasia. To investigate exactly which aspect of the impairment in dynamic aphasia was linked to the VLPFC, Robinson,

Shallice, Bozzali, and Cipolotti (2010) extended their case study to a group study. In this new study, patients with VLPFC damage were compared with patients with frontal lesions sparing VLPFC and patients with posterior lesions. Compared with the other two groups, VLPFC patients showed selective impairment in sentence generation tasks only when stimuli activated multiple conceptual propositions that competed with each other for selection. For example, VLPFC patients were impaired in generating sentences from high-frequency words, but not from low-frequency words and proper nouns, because the latter two are more constraining in their associations. In contrast, when the lesion spared VLPFC (i.e., non-VLPFC and the left temporal groups), the patients did not show sensitivity to the number of possible propositions.

Impairment in this experimental task was shown to be predictive of the clinical problem. VLPFC patients' scores in the high-frequency minus low-frequency and in the high-frequency minus proper noun conditions showed a reliable correlation with their spontaneous speech rate. In keeping with this, Blank, Scott, Murphy, Warburton, and Wise (2002) found that BA 44 showed greater activation under conditions of propositional compared with nonpropositional speech (i.e., counting and overlearned nursery rhymes). However, it must be pointed out that based on these findings, Robinson et al. (2010) concluded that VLPFC has a fundamental role in selection for conceptual propositions, as opposed to selection during lexical retrieval. Given the evidence reviewed in earlier sections, we are skeptical about the exclusivity of the role of VLPFC to conceptual biasing. For instance, under this account, it is unclear why VLPFC lesions would cause impairment in naming a picture COUCH/SOFA when the "concept" is right in front of the patient. However, we concur that profile of dynamic aphasia is compatible with the perspective taken in this chapter on the role of VLPFC.

Another clinical syndrome that has been recently linked to VLPFC is SA (Jefferies & Ralph, 2006), a deficit that is meant to be contrasted with Semantic Dementia (SD). SD, which generally results from bilateral damage (often of the atrophic type) to anterior temporal lobes, affects conceptual knowledge in verbal and nonverbal domains across different sensory modalities (Bozeat, Lambon Ralph, Patterson, Garrard, & Hodges, 2000; Coccia, Bartolini, Luzzi, Provinciali, & Lambon Ralph, 2004; Luzzi et al., 2007). Thus, the hallmark of SD is that failure to activate a concept is stable across tasks and processing modalities (Bozeat et al., 2000; Jefferies & Ralph, 2006). TMS studies on healthy controls corroborate this assertion (Pobric, Jefferies, & Ralph, 2007; Ralph, Pobric, & Jefferies, 2009). In contrast to SD, SA patients have intact conceptual knowledge, but controlling that knowledge has become dysregulated. These patients usually have poststroke lesions in the VLPFC, although it has been reported that lesions in inferior parietal cortex (i.e., BA 39/40) and posterior temporal (especially temporoparietal junction) can create similar symptoms, suggesting that semantic control is achieved through a distributed network (Whitney et al., 2012). Importantly, SA patients differ from SD patients in their inconsistent performance on the same items when the control demands of the task change, and their picture naming performance shows great sensitivity to constraining or distracting cues.

Deficits of SA patients closely mirror what we discussed as VLPFC's function. For example, Noonan, Jefferies, Corbett, and Ralph (2010) performed four experiments in a group of SA patients, three of which are in line with the studies previously discussed in this chapter. In Experiment 1, participants had to judge which of three alternatives was closest in meaning to a probe item. The semantic distance between the probe item (CHIPMUNK) and the target was manipulated to be low (e.g., SQUIRREL) in half of the trials and high (e.g., BEE) in the other half, whereas the two distractors were kept the same (and unrelated to both the probe and the target). In Experiment 2, they examined synonym and antonym judgments when the associative strength between the probe (e.g., HAPPY/NEAT) and one of the distractors was either stronger or weaker than the association between the probe and the target (e.g., HAPPY–SAD [distractor] > HAPPY–CHEERFUL [target]; NEAT–MESSY [distractor] < NEAT–TIDY [target]). Consistent with the past findings, SA patients showed selective impairment when the correct response required biasing towards the weak association. In Experiment 4, participants were asked to select which of four words was related in meaning to a probe word presented at the top of the page. In half of the trials, the target referred to the dominant meaning of the probe word (FIRE → HOT). In the remaining trials, the target word was related to the subordinate meaning of the probe word (FIRE → RIFLE) while distractors remained the same. Three cue conditions preceded these trials: no cues, correct cues (e.g., "I lit a fire" → HOT), or miscues (e.g., "Fire at will" → HOT). SA patients showed great difficulty activating the less frequent meaning, and this was the condition that specifically benefitted from cueing.

On the production side, the cue/miscue paradigm has been used to probe the sensitivity of SA patients to constraining and distracting information. Jefferies, Patterson, and Ralph (2008) showed that phonological onset cues (e.g., /k/) when naming pictures (e.g., a cup) were very useful in helping SA patients overcome their difficulty in suppressing competing names (e.g., tea). Noonan et al. (2010; Experiment 3) showed that the opposite manipulation in the cuing paradigm of Jeffries et al. (2008) does yield the opposite effect. Miscuing the picture name (e.g., /t/ for the picture of a cup) impaired the SA patients' ability to name the pictures correctly and elicited additional semantic errors.

It is noteworthy that semantic control deficits in SA patients are not restricted to the verbal domain. Corbett, Jefferies, Ehsan, and Ralph (2009) compared object use in SD and SA patients and found that SD, but not SA, patients' performance was sensitive to item frequency and was consistent irrespective of task demands. Straightforward object use demonstration, for example, was relatively intact in SA patients in comparison with much poorer performance on an executively demanding, mechanical-puzzles task (see also Corbett, Jefferies, & Ralph, 2009).

46.5 CONCLUDING REMARKS AND FUTURE AVENUES

Although left temporal cortex is consistently implicated in storing long-term representations of knowledge (Binder, Desai, Graves, & Conant, 2009; Indefrey & Levelt, 2004; Vigneau et al., 2006), VLPFC seems to mediate processes necessary for controlling the use of this knowledge. We reviewed evidence from multiple sources in single-word and sentence production and comprehension consistent with a crucial role of this region in implementing top-down biasing at the semantic/lexical level. It is important to point out that this is not restricted to the language domain. VLPFC's involvement has been shown in selecting the target among nontarget items in target detection tasks, even when items are complex symbols without known lexical labels (Hampshire, Duncan, & Owen, 2007). More generally, VLPFC has been proposed as a critical part of a system involved in processing hierarchical structure of goal-directed behavior (Koechlin & Jubault, 2006).

Although we have made considerable progress in understanding the role of VLPFC in processing language, much remains to be explored. We close by posing four main questions that we consider to be excellent avenues for promoting our understanding of the role of VLPFC in language:

1. How many stages are there in (controlled) lexicosemantic retrieval, and which stage/stages requires VLPFC? Badre and Wagner (2007) proposed that VLPFC implements semantic control in two steps. Step 1 constitutes controlled access to stored representations when bottom-up input is not enough. Step 2 operates postretrieval and is thought to bias competition among representations that have been activated during Step 1. A similar idea has been expressed by Thompson-Schill and Botvinick (2006) using a Bayesian framework. According to Badre and Wagner (2007), both steps recruit VLPFC, albeit different parts of it (BA 47 and BA 45, respectively; Whitney et al., 2012). The exact computations by which this is achieved are not clear.
2. Are subdivisions of VLPFC specialized based on different materials or different processes? An example of a material-based parcellation in the context of language is the semantic-phonological distinction, born out of the differential sensitivity of the posterior parts of VLPFC to phonological, and the anterior parts to semantic processing (Devlin et al., 2003; Gold, Balota, Kirchhoff, & Buckner, 2005; Gough et al., 2005; Poldrack & Wagner, 2004). Studies of statistical mapping between speech errors and lesion sites using Voxel-based Lesion-Symptom Mapping also link damage to the posterior aspects of VLPFC to phonological errors (Schwartz, Faseyitan, Kim, & Coslett, 2012) and its anterior aspects to semantic errors (Schwartz, Kimberg, Walker, Faseyitan, & Brecher, 2009). Similarly, using a version of the technique that maps lesions to parameters in a computational model of language production, Dell, Schwartz, Nozari, Faseyitan, and Branch Coslett (2013) showed that lesions in the more posterior parts of VLPFC correspond to the phonological parameter of the model, whereas lesions in the more anterior parts of VLPFC are associated with semantic errors.

   The semantic-phonological distinction, however, does not negate a process-based organization. Semantic and phonological processing may differ not only in the materials they use but also in the operations performed on those materials. One dominant process-based view is that of a rostrocaudal functional gradient of abstraction, with more anterior regions processing more abstract information (Badre, 2008; Badre & D'Esposito, 2009). This view has found support in the dorsolateral prefrontal cortex (Badre & D'Esposito, 2007; Koechlin & Jubault, 2006; Koechlin, Ody, & Kouneiher, 2003) and, more recently, also in the VLPFC (Race, Shanker, & Wagner, 2009). It remains to be seen if such a view is sufficient for explaining the separation of semantic and phonological processing in VLPFC.
3. Do the same regions in the prefrontal cortex that support domain-general mechanisms also support language processing? Although some have argued for domain-generality (January et al., 2009), some have contested this view (Fedorenko, Behr, &

Kanwisher, 2011) and some have proposed a middle ground, identifying both language-specific and domain-general regions in the PFC (Sakai & Passingham, 2002). A recent study tested whether the rostrocaudal gradient of abstraction is sensitive to the nature of the representations (spatial versus verbal). Bahlmann, Blumenfeld, and D'Esposito (2014) found an indistinguishable pattern of activity for the two stimulus types along this rostrocaudal axis. However, a task-sensitive topographic segregation was also found in the dorsolateral axis, such that processing of spatial information was localized to more dorsal areas, whereas processing of verbal information activated more ventral regions.

4. What other regions are in the control network that VLPFC is part of, and what role do they play? As pointed out, semantic control deficits of the SA type can also arise from lesions to posterior temporal or inferior parietal lobes. These two areas are heavily connected to the prefrontal cortex via arcuate and longitudinal fasciculi (Parker et al., 2005) and have both been previously implicated in tasks requiring controlled semantic access (Badre, Poldrack, Paré-Blagoev, Insler, & Wagner, 2005; Gennari, MacDonald, Postle, & Seidenberg, 2007; Lee & Dapretto, 2006; Rodd, Davis, & Johnsrude, 2005; Thompson-Schill et al., 1997; Wagner et al., 2001; Whitney et al., 2012; Zempleni, Renken, Hoeks, Hoogduin, & Stowe, 2007). It remains to be demonstrated how labor is divided in this frontotemporoparietal network that supports semantic and lexical retrieval.

Acknowledgments

We thank Gary Dell and Tilbe Göksun for their valuable comments.

References

Aboitiz, F., & García, V. R. (0030). The evolutionary origin of the language areas in the human brain. A neuroanatomical perspective. *Brain Research Reviews*, *25*(3), 381–396.

Alexander, M. P. (2006). Impairments of procedures for implementing complex language are due to disruption of frontal attention processes. *Journal of the International Neuropsychological Society*, *12* (02), 236–247.

Amunts, K., Weiss, P. H., Mohlberg, H., Pieperhoff, P., Eickhoff, S., Gurd, J. M., et al. (2004). Analysis of neural mechanisms underlying verbal fluency in cytoarchitectonically defined stereotaxic space—The roles of Brodmann areas 44 and 45. *NeuroImage*, *22*(1), 42–56.

Awh, E., Jonides, J., Smith, E. E., Schumacher, E. H., Koeppe, R. A., & Katz, S. (1996). Dissociation of storage and rehearsal in verbal working memory: Evidence from positron emission tomography. *Psychological Science*, 25–31.

Badre, D. (2008). Cognitive control, hierarchy, and the rostro–caudal organization of the frontal lobes. *Trends in Cognitive Sciences*, *12* (5), 193–200.

Badre, D., & D'Esposito, M. (2007). Functional magnetic resonance imaging evidence for a hierarchical organization of the prefrontal cortex. *Journal of Cognitive Neuroscience*, *19*(12), 2082–2099.

Badre, D., & D'Esposito, M. (2009). Is the rostro-caudal axis of the frontal lobe hierarchical? *Nature Reviews Neuroscience*, *10*(9), 659–669.

Badre, D., Poldrack, R. A., Paré-Blagoev, E. J., Insler, R. Z., & Wagner, A. D. (2005). Dissociable controlled retrieval and generalized selection mechanisms in ventrolateral prefrontal cortex. *Neuron*, *47*(6), 907–918.

Badre, D., & Wagner, A. D. (2007). Left ventrolateral prefrontal cortex and the cognitive control of memory. *Neuropsychologia*, *45*(13), 2883–2901.

Bahlmann, J., Blumenfeld, R. S., & D'Esposito, M. (2014). The rostro-caudal axis of frontal cortex is sensitive to the domain of stimulus information. *Cerebral Cortex*. Available from http://dx.doi.org/10.1093/cercor/bht419.

Baldo, J. V., & Shimamura, A. P. (1998). Letter and category fluency in patients with frontal lobe lesions. *Neuropsychology*, *12* (2), 259.

Balogh, J., & Grodzinsky, Y. (1996). Varieties of passives in agrammatic Broca's aphasia: Upsilon-grids, arguments, and referentiality. *Brain and Language*, *55*, 54–56.

Basho, S., Palmer, E. D., Rubio, M. A., Wulfeck, B., & Müller, R.-A. (2007). Effects of generation mode in fMRI adaptations of semantic fluency: Paced production and overt speech. *Neuropsychologia*, *45*(8), 1697–1706.

Baumgaertner, A., Weiller, C., & Büchel, C. (2002). Event-related fMRI reveals cortical sites involved in contextual sentence integration. *NeuroImage*, *16*(3), 736–745.

Bedny, M., Hulbert, J. C., & Thompson-Schill, S. L. (2007). Understanding words in context: The role of Broca's area in word comprehension. *Brain Research*, *1146*, 101–114.

Bedny, M., McGill, M., & Thompson-Schill, S. L. (2008). Semantic adaptation and competition during word comprehension. *Cerebral Cortex*, *18*(11), 2574–2585.

Belke, E., Meyer, A. S., & Damian, M. F. (2005). Refractory effects in picture naming as assessed in a semantic blocking paradigm. *The Quarterly Journal of Experimental Psychology*, *58*(4), 667–692.

Ben-Shachar, M., Hendler, T., Kahn, I., Ben-Bashat, D., & Grodzinsky, Y. (2003). The neural reality of syntactic transformations evidence from functional magnetic resonance imaging. *Psychological Science*, *14*(5), 433–440.

Binder, J. R., Desai, R. H., Graves, W. W., & Conant, L. L. (2009). Where is the semantic system? A critical review and meta-analysis of 120 functional neuroimaging studies. *Cerebral Cortex*, *19*(12), 2767–2796.

Birn, R. M., Kenworthy, L., Case, L., Caravella, R., Jones, T. B., Bandettini, P. A., et al. (2010). Neural systems supporting lexical search guided by letter and semantic category cues: A self-paced overt response fMRI study of verbal fluency. *NeuroImage*, *49*(1), 1099–1107.

Blank, S. C., Scott, S. K., Murphy, K., Warburton, E., & Wise, R. J. (2002). Speech production: Wernicke, Broca and beyond. *Brain*, *125*(8), 1829–1838.

Bottini, G., Corcoran, R., Sterzi, R., Paulesu, E., Schenone, P., Scarpa, P., et al. (1994). The role of the right hemisphere in the interpretation of figurative aspects of language A positron emission tomography activation study. *Brain*, *117*(6), 1241–1253.

Bozeat, S., Lambon Ralph, M. A., Patterson, K., Garrard, P., & Hodges, J. R. (2000). Non-verbal semantic impairment in semantic dementia. *Neuropsychologia*, *38*(9), 1207–1215.

Broca, P. (1861). Remarques sur le siège de la faculté du langage articulé; suivies d'une observation d'aphémie(perte de la parole). *Bulletins de la Société Anatomique (Paris)*, *6*, 330–357, 398–407.

Broca, P. (1865). Sur le siège de la faculté du langage articulé. *Bulletins de la Société d'Anthropologie*, *6*, 337–393.

Buckner, R. L., Petersen, S. E., Ojemann, J. G., Miezin, F. M., Squire, L. R., & Raichle, M. E. (1995). Functional anatomical studies of explicit and implicit memory retrieval tasks. *The Journal of Neuroscience*, *15*(1), 12–29.

Caplan, D., Alpert, N., & Waters, G. (1998). Effects of syntactic structure and propositional number on patterns of regional cerebral blood flow. *Journal of Cognitive Neuroscience*, *10*(4), 541–552.

Caplan, D., Alpert, N., & Waters, G. (1999). PET studies of syntactic processing with auditory sentence presentation. *NeuroImage*, *9*(3), 343–351.

Caplan, D., Alpert, N., Waters, G., & Olivieri, A. (2000). Activation of Broca's area by syntactic processing under conditions of concurrent articulation. *Human Brain Mapping*, *9*(2), 65–71.

Chen, E., West, W. C., Waters, G., & Caplan, D. (2006). Determinants of BOLD signal correlates of processing object-extracted relative clauses. *Cortex*, *42*(4), 591–604.

Chertkow, H., & Bub, D. (1990). Semantic memory loss in dementia of Alzheimer's type: What do various measures measure? *Brain*, *113*(2), 397–417.

Coccia, M., Bartolini, M., Luzzi, S., Provinciali, L., & Lambon Ralph, M. A. (2004). Semantic memory is an amodal, dynamic system: Evidence from the interaction of naming and object use in semantic dementia. *Cognitive Neuropsychology*, *21*(5), 513–527.

Cooke, A., Zurif, E. B., DeVita, C., Alsop, D., Koenig, P., Detre, J., et al. (2002). Neural basis for sentence comprehension: Grammatical and short-term memory components. *Human Brain Mapping*, *15*(2), 80–94.

Corbett, F., Jefferies, E., Ehsan, S., & Ralph, M. A. L. (2009). Different impairments of semantic cognition in semantic dementia and semantic aphasia: Evidence from the non-verbal domain. *Brain*, *132*(9), 2593–2608.

Corbett, F., Jefferies, E., & Ralph, M. A. L. (2009). Exploring multimodal semantic control impairments in semantic aphasia: Evidence from naturalistic object use. *Neuropsychologia*, *47*(13), 2721–2731.

Damasio, A. R. (1992). Aphasia. *New England Journal of Medicine*, *326* (8), 531–539.

Damian, M. F., Vigliocco, G., & Levelt, W. J. (2001). Effects of semantic context in the naming of pictures and words. *Cognition*, *81*(3), B77–B86.

Dapretto, M., & Bookheimer, S. Y. (1999). Form and content: Dissociating syntax and semantics in sentence comprehension. *Neuron*, *24*(2), 427–432.

Dell, G. S., Schwartz, M. F., Nozari, N., Faseyitan, O., & Branch Coslett, H. (2013). Voxel-based lesion-parameter mapping: Identifying the neural correlates of a computational model of word production. *Cognition*, *128*(3), 380–396.

Demb, J. B., Desmond, J. E., Wagner, A. D., Vaidya, C. J., Glover, G. H., & Gabrieli, J. D. (1995). Semantic encoding and retrieval in the left inferior prefrontal cortex: A functional MRI study of task difficulty and process specificity. *Journal of Neuroscience*, *15*(9), 5870–5878.

Démonet, J.-F., Chollet, F., Ramsay, S., Cardebat, D., Nespoulous, J.-L., Wise, R., et al. (1992). The anatomy of phonological and semantic processing in normal subjects. *Brain*, *115*(6), 1753–1768.

Démonet, J.-F., Price, C., Wise, R., & Frackowiak, R. S. J. (1994). Differential activation of right and left posterior sylvian regions by semantic and phonological tasks: A positron-emission tomography study in normal human subjects. *Neuroscience Letters*, *182* (1), 25–28.

Desimone, R., & Duncan, J. (1995). Neural mechanisms of selective visual attention. *Annual Review of Neuroscience*, *18*(1), 193–222.

Devlin, J. T., Matthews, P. M., & Rushworth, M. F. (2003). Semantic processing in the left inferior prefrontal cortex: A combined functional magnetic resonance imaging and transcranial magnetic stimulation study. *Journal of Cognitive Neuroscience*, *15*(1), 71–84.

Dronkers, N. F., Plaisant, O., Iba-Zizen, M. T., & Cabanis, E. A. (2007). Paul Broca's historic cases: High resolution MR imaging of the brains of Leborgne and Lelong. *Brain*, *130*(5), 1432–1441.

Druks, J., & Marshall, J. C. (1995). When passives are easier than actives: Two case studies of aphasic comprehension. *Cognition*, *55* (3), 311–331.

Embick, D., Marantz, A., Miyashita, Y., O'Neil, W., & Sakai, K. L. (2000). A syntactic specialization for Broca's area. *Proceedings of the National Academy of Sciences*, *97*(11), 6150–6154.

Etard, O., Mellet, E., Papathanassiou, D., Benali, K., Houdé, O., Mazoyer, B., et al. (2000). Picture naming without Broca's and Wernicke's area. *Neuroreport*, *11*(3), 617–622.

Fedorenko, E., Behr, M. K., & Kanwisher, N. (2011). Functional specificity for high-level linguistic processing in the human brain. *Proceedings of the National Academy of Sciences*, *108*(39), 16428–16433.

Fiebach, C. J., Friederici, A. D., Müller, K., & Von Cramon, D. Y. (2002). fMRI evidence for dual routes to the mental lexicon in visual word recognition. *Journal of Cognitive Neuroscience*, *14*(1), 11–23.

Fiebach, C. J., Schlesewsky, M., & Friederici, A. D. (2001). Syntactic working memory and the establishment of filler-gap dependencies: Insights from ERPs and fMRI. *Journal of Psycholinguistic Research*, *30*(3), 321–338.

Fiebach, C. J., Schlesewsky, M., Lohmann, G., Von Cramon, D. Y., & Friederici, A. D. (2005). Revisiting the role of Broca's area in sentence processing: Syntactic integration versus syntactic working memory. *Human Brain Mapping*, *24*(2), 79–91.

Fiez, J. A. (1997). Phonology, semantics, and the role of the left inferior prefrontal cortex. *Human Brain Mapping*, *5*(2), 79–83.

Fiez, J. A., Raichle, M. E., Miezin, F. M., Petersen, S. E., Tallal, P., & Katz, W. F. (1995). PET studies of auditory and phonological processing: Effects of stimulus characteristics and task demands. *Journal of Cognitive Neuroscience*, *7*(3), 357–375.

Friederici, A. D., Meyer, M., & von Cramon, D. Y. (2000). Auditory language comprehension: An event-related fMRI study on the processing of syntactic and lexical information. *Brain and Language*, *74*(2), 289–300.

Frith, C. D., Friston, K. J., Liddle, P. F., & Frackowiak, R. S. J. (1991). A PET study of word finding. *Neuropsychologia*, *29*(12), 1137–1148.

Gabrieli, J. D., Desmond, J. E., Demb, J. B., Wagner, A. D., Stone, M. V., Vaidya, C. J., et al. (1996). Functional magnetic resonance imaging of semantic memory processes in the frontal lobes. *Psychological Science*, *7*(5), 278–283.

Garnsey, S. M., Pearlmutter, N. J., Myers, E., & Lotocky, M. A. (1997). The contributions of verb bias and plausibility to the comprehension of temporarily ambiguous sentences. *Journal of Memory and Language*, *37*(1), 58–93.

Gelfand, J. R., & Bookheimer, S. Y. (2003). Dissociating neural mechanisms of temporal sequencing and processing phonemes. *Neuron*, *38*(5), 831–842.

Gennari, S. P., MacDonald, M. C., Postle, B. R., & Seidenberg, M. S. (2007). Context-dependent interpretation of words: Evidence for interactive neural processes. *NeuroImage*, *35*(3), 1278–1286.

Gold, B. T., Balota, D. A., Kirchhoff, B. A., & Buckner, R. L. (2005). Common and dissociable activation patterns associated with controlled semantic and phonological processing: Evidence from fMRI adaptation. *Cerebral Cortex*, *15*(9), 1438–1450.

Gough, P. M., Nobre, A. C., & Devlin, J. T. (2005). Dissociating linguistic processes in the left inferior frontal cortex with transcranial magnetic stimulation. *The Journal of Neuroscience, 25*(35), 8010–8016.

Grodzinsky, Y. (1995). Trace deletion, [Theta]-roles, and cognitive strategies. *Brain and Language, 51*(3), 469–497.

Grodzinsky, Y. (2000). The neurology of syntax: Language use without Broca's area. *Behavioral and Brain Sciences, 23*(01), 1–21.

Gurd, J. M., Amunts, K., Weiss, P. H., Zafiris, O., Zilles, K., Marshall, J. C., et al. (2002). Posterior parietal cortex is implicated in continuous switching between verbal fluency tasks: An fMRI study with clinical implications. *Brain, 125*(5), 1024–1038.

Hagoort, P., Hald, L., Bastiaansen, M., & Petersson, K. M. (2004). Integration of word meaning and world knowledge in language comprehension. *Science, 304*(5669), 438–441.

Hamilton, A. C., & Martin, R. C. (2005). Dissociations among tasks involving inhibition: A single-case study. *Cognitive, Affective, and Behavioral Neuroscience, 5*(1), 1–13.

Hampshire, A., Duncan, J., & Owen, A. M. (2007). Selective tuning of the blood oxygenation level-dependent response during simple target detection dissociates human frontoparietal subregions. *The Journal of Neuroscience, 27*(23), 6219–6223.

Hickok, G., & Avrutin, S. (1995). Representation, referentiality, and processing in agrammatic comprehension: Two case studies. *Brain and Language, 50*(1), 10–26.

Hickok, G., & Poeppel, D. (2004). Dorsal and ventral streams: A framework for understanding aspects of the functional anatomy of language. *Cognition, 92*(1–2), 67–99. Available from http://dx.doi.org/doi:10.1016/j.cognition.2003.10.011.

Hickok, G., & Poeppel, D. (2007). The cortical organization of speech processing. *Nature Reviews Neuroscience, 8*(5), 393–402. Available from http://dx.doi.org/doi:10.1038/nrn2113.

Hindy, N. C., Hamilton, R., Houghtling, A. S., Coslett, H. B., & Thompson-Schill, S. L. (2009). Computer-mouse tracking reveals TMS disruptions of prefrontal function during semantic retrieval. *Journal of Neurophysiology, 102*(6), 3405–3413.

Hirshorn, E. A., & Thompson-Schill, S. L. (2006). Role of the left inferior frontal gyrus in covert word retrieval: Neural correlates of switching during verbal fluency. *Neuropsychologia, 44*(12), 2547–2557.

Howard, D., Nickels, L., Coltheart, M., & Cole-Virtue, J. (2006). Cumulative semantic inhibition in picture naming: Experimental and computational studies. *Cognition, 100*(3), 464–482.

Indefrey, P., Hagoort, P., Herzog, H., Seitz, R. J., & Brown, C. M. (2001). Syntactic processing in left prefrontal cortex is independent of lexical meaning. *NeuroImage, 14*(3), 546–555.

Indefrey, P., & Levelt, W. J. (2004). The spatial and temporal signatures of word production components. *Cognition, 92*(1), 101–144.

Inui, T., Otsu, Y., Tanaka, S., Okada, T., Nishizawa, S., & Konishi, J. (1998). A functional MRI analysis of comprehension processes of Japanese sentences. *NeuroReport, 9*(14), 3325–3328.

Janowsky, J. S., Shimamura, A. P., Kritchevsky, M., & Squire, L. R. (1989). Cognitive impairment following frontal lobe damage and its relevance to human amnesia. *Behavioral Neuroscience, 103*(3), 548.

January, D., Trueswell, J. C., & Thompson-Schill, S. L. (2009). Co-localization of Stroop and syntactic ambiguity resolution in Broca's area: Implications for the neural basis of sentence processing. *Journal of Cognitive Neuroscience, 21*(12), 2434–2444.

Jefferies, E., Patterson, K., & Ralph, M. A. L. (2008). Deficits of knowledge versus executive control in semantic cognition: Insights from cued naming. *Neuropsychologia, 46*(2), 649–658.

Jefferies, E., & Ralph, M. A. L. (2006). Semantic impairment in stroke aphasia versus semantic dementia: A case-series comparison. *Brain, 129*(8), 2132–2147.

Kaan, E., & Swaab, T. Y. (2002). The brain circuitry of syntactic comprehension. *Trends in Cognitive Sciences, 6*(8), 350–356.

Kan, I. P., & Thompson-Schill, S. L. (2004). Selection from perceptual and conceptual representations. *Cognitive, Affective, and Behavioral Neuroscience, 4*(4), 466–482.

Kapur, S., Craik, F. I., Tulving, E., Wilson, A. A., Houle, S., & Brown, G. M. (1994). Neuroanatomical correlates of encoding in episodic memory: Levels of processing effect. *Proceedings of the National Academy of Sciences, 91*(6), 2008–2011.

Keller, T. A., Carpenter, P. A., & Just, M. A. (2001). The neural bases of sentence comprehension: A fMRI examination of syntactic and lexical processing. *Cerebral Cortex, 11*(3), 223–237.

Kiehl, K. A., Laurens, K. R., & Liddle, P. F. (2002). Reading anomalous sentences: an event-related fMRI study of semantic processing. *NeuroImage, 17*(2), 842–850.

Koechlin, E., & Jubault, T. (2006). Broca's area and the hierarchical organization of human behavior. *Neuron, 50*(6), 963–974.

Koechlin, E., Ody, C., & Kouneiher, F. (2003). The architecture of cognitive control in the human prefrontal cortex. *Science, 302*(5648), 1181–1185.

Kroll, J. F., & Stewart, E. (1994). Category interference in translation and picture naming: Evidence for asymmetric connections between bilingual memory representations. *Journal of Memory and Language, 33*(2), 149–174.

Kuperberg, G. R. (2007). Neural mechanisms of language comprehension: Challenges to syntax. *Brain Research, 1146*, 23–49.

Kuperberg, G. R., Caplan, D., Sitnikova, T., Eddy, M., & Holcomb, P. J. (2006). Neural correlates of processing syntactic, semantic, and thematic relationships in sentences. *Language and Cognitive Processes, 21*(5), 489–530.

Kuperberg, G. R., Holcomb, P. J., Sitnikova, T., Greve, D., Dale, A. M., & Caplan, D. (2003). Distinct patterns of neural modulation during the processing of conceptual and syntactic anomalies. *Journal of Cognitive Neuroscience, 15*(2), 272–293.

Kuperberg, G. R., McGuire, P. K., Bullmore, E. T., Brammer, M. J., Rabe-Hesketh, S., Wright, I. C., et al. (2000). Common and distinct neural substrates for pragmatic, semantic, and syntactic processing of spoken sentences: An fMRI study. *Journal of Cognitive Neuroscience, 12*(2), 321–341.

Kuperberg, G. R., Sitnikova, T., Caplan, D., & Holcomb, P. J. (2003). Electrophysiological distinctions in processing conceptual relationships within simple sentences. *Cognitive Brain Research, 17*(1), 117–129.

Kuperberg, G. R., Sitnikova, T., & Lakshmanan, B. M. (2008). Neuroanatomical distinctions within the semantic system during sentence comprehension: Evidence from functional magnetic resonance imaging. *NeuroImage, 40*(1), 367–388.

Lee, S. S., & Dapretto, M. (2006). Metaphorical vs. Literal word meanings: fMRI evidence against a selective role of the right hemisphere. *NeuroImage, 29*(2), 536–544.

Levine, D. N., & Mohr, J. P. (1979). Language after bilateral cerebral infarctions role of the minor hemisphere in speech. *Neurology, 29*(7), 927–938.

Luria, A. R. (1970). *Traumatic aphasia: Its syndromes, psychology and treatment* (D. Bowden, Trans.). The Hague: Mouton.

Luria, A. R. (1973). *The working brain: An introduction to neuropsychology.* New York, NY: Basic Books.

Luria, A. R., & Tsvetkova, L. S. (1968). The mechanism of 'dynamic aphasia'. *Foundations of Language, 4*(3), 296–307.

Luzzi, S., Snowden, J. S., Neary, D., Coccia, M., Provinciali, L., & Lambon Ralph, M. A. (2007). Distinct patterns of olfactory impairment in Alzheimer's disease, semantic dementia, frontotemporal dementia, and corticobasal degeneration. *Neuropsychologia, 45*(8), 1823–1831.

MacDonald, M. C., Pearlmutter, N. J., & Seidenberg, M. S. (1994). The lexical nature of syntactic ambiguity resolution. *Psychological Review, 101*(4), 676.

Marini, A., & Urgesi, C. (2012). Please get to the point! A cortical correlate of linguistic informativeness. *Journal of Cognitive Neuroscience*, *24*(11), 2211–2222.

Martin, A., Haxby, J. V., Lalonde, F. M., Wiggs, C. L., & Ungerleider, L. G. (1995). Discrete cortical regions associated with knowledge of color and knowledge of action. *Science*, *270*(5233), 102–105.

Martin, A., Wiggs, C. L., Lalonde, F., & Mack, C. (1994). Word retrieval to letter and semantic cues: A double dissociation in normal subjects using interference tasks. *Neuropsychologia*, *32*(12), 1487–1494.

Mason, R. A., Just, M. A., Keller, T. A., & Carpenter, P. A. (2003). Ambiguity in the brain: What brain imaging reveals about the processing of syntactically ambiguous sentences. *Journal of Experimental Psychology: Learning, Memory, and Cognition*, *29*(6), 1319.

Mazoyer, B. M., Tzourio, N., Frak, V., Syrota, A., Murayama, N., Levrier, O., et al. (1993). The cortical representation of speech. *Journal of Cognitive Neuroscience*, *5*(4), 467–479.

McCarthy, G., Blamire, A. M., Rothman, D. L., Gruetter, R., & Shulman, R. G. (1993). Echo-planar magnetic resonance imaging studies of frontal cortex activation during word generation in humans. *Proceedings of the National Academy of Sciences*, *90*(11), 4952–4956.

McCarthy, R. A., & Kartsounis, L. D. (2000). Wobbly words: Refractory anomia with preserved semantics. *Neurocase*, *6*(6), 487–497.

Michael, E. B., Keller, T. A., Carpenter, P. A., & Just, M. A. (2001). fMRI investigation of sentence comprehension by eye and by ear: Modality fingerprints on cognitive processes. *Human Brain Mapping*, *13*(4), 239–252.

Miller, E. K., & Cohen, J. D. (2001). An integrative theory of prefrontal cortex function. *Annual Review of Neuroscience*, *24*(1), 167–202.

Mohr, J. P., Pessin, M. S., Finkelstein, S., Funkenstein, H. H., Duncan, G. W., & Davis, K. R. (1978). Broca aphasia pathologic and clinical. *Neurology*, *28*(4), 311–324.

Moro, A., Tettamanti, M., Perani, D., Donati, C., Cappa, S. F., & Fazio, F. (2001). Syntax and the brain: Disentangling grammar by selective anomalies. *NeuroImage*, *13*(1), 110–118.

Mummery, C. J., Patterson, K., Hodges, J. R., & Wise, R. J. (1996). Generating "tiger" as an animal name or a word beginning with T: Differences in brain activation. *Proceedings of the Royal Society of London. Series B: Biological Sciences*, *263*(1373), 989–995.

Murtha, S., Chertkow, H., Beauregard, M., Dixon, R., & Evans, A. (1996). Anticipation causes increased blood flow to the anterior cingulate cortex. *Human Brain Mapping*, *4*(2), 103–112.

Murtha, S., Chertkow, H., Beauregard, M., & Evans, A. (1999). The neural substrate of picture naming. *Journal of Cognitive Neuroscience*, *11*(4), 399–423.

Musso, M., Moro, A., Glauche, V., Rijntjes, M., Reichenbach, J., Büchel, C., et al. (2003). Broca's area and the language instinct. *Nature Neuroscience*, *6*(7), 774–781. Available from http://dx.doi.org/doi:10.1038/nn1077.

Newman, A. J., Pancheva, R., Ozawa, K., Neville, H. J., & Ullman, M. T. (2001). An event-related fMRI study of syntactic and semantic violations. *Journal of Psycholinguistic Research*, *30*(3), 339–364.

Newman, S. D., Twieg, D. B., & Carpenter, P. A. (2001). Baseline conditions and subtractive logic in neuroimaging. *Human Brain Mapping*, *14*(4), 228–235.

Ni, W., Constable, R. T., Mencl, W. E., Pugh, K. R., Fulbright, R. K., Shaywitz, S. E., et al. (2000). An event-related neuroimaging study distinguishing form and content in sentence processing. *Journal of Cognitive Neuroscience*, *12*(1), 120–133.

Nichelli, P., Grafman, J., Pietrini, P., Clark, K., Lee, K. Y., & Miletich, R. (1995). Where the brain appreciates the moral of a story. *Neuroreport*, *6*(17), 2309–2313.

Noonan, K. A., Jefferies, E., Corbett, F., & Ralph, M. A. L. (2010). Elucidating the nature of deregulated semantic cognition in semantic aphasia: Evidence for the roles of prefrontal and temporo-parietal cortices. *Journal of Cognitive Neuroscience*, *22*(7), 1597–1613.

Norman, D. A., & Shallice, T. (1986). Attention to action. In R. Davidson, R. Schwartz, & D. Shapiro (Eds.), *Consciousness and self-regulation: Advances in research and theory IV* (pp. 1–18). New York, NY: Plenum Press.

Novick, J. M., Kan, I. P., Trueswell, J. C., & Thompson-Schill, S. L. (2009). A case for conflict across multiple domains: Memory and language impairments following damage to ventrolateral prefrontal cortex. *Cognitive Neuropsychology*, *26*(6), 527–567.

Novick, J. M., Trueswell, J. C., & Thompson-Schill, S. L. (2005). Cognitive control and parsing: Reexamining the role of Broca's area in sentence comprehension. *Cognitive, Affective, and Behavioral Neuroscience*, *5*(3), 263–281.

Nozari, N., & Dell, G. S. (2013). How damaged brains repeat words: A computational approach. *Brain and Language*, *126*(3), 327–337.

Nozari, N., Kittredge, A. K., Dell, G. S., & Schwartz, M. F. (2010). Naming and repetition in aphasia: Steps, routes, and frequency effects. *Journal of Memory and Language*, *63*(4), 541–559.

Oppenheim, G. M., Dell, G. S., & Schwartz, M. F. (2010). The dark side of incremental learning: A model of cumulative semantic interference during lexical access in speech production. *Cognition*, *114*(2), 227–252.

Owen, A. M., Downes, J. J., Sahakian, B. J., Polkey, C. E., & Robbins, T. W. (1990). Planning and spatial working memory following frontal lobe lesions in man. *Neuropsychologia*, *28*(10), 1021–1034.

Parker, G. J., Luzzi, S., Alexander, D. C., Wheeler-Kingshott, C. A., Ciccarelli, O., & Lambon Ralph, M. A. (2005). Lateralization of ventral and dorsal auditory-language pathways in the human brain. *NeuroImage*, *24*(3), 656–666.

Paulesu, E., Frith, C. D., & Frackowiak, R. S. (1993). The neural correlates of the verbal component of working memory. *Nature*342–345.

Paulesu, E., Goldacre, B., Scifo, P., Cappa, S. F., Gilardi, M. C., Castiglioni, I., et al. (1997). Functional heterogeneity of left inferior frontal cortex as revealed by fMRI. *Neuroreport*, *8*(8), 2011–2016.

Perret, E. (1974). The left frontal lobe of man and the suppression of habitual responses in verbal categorical behaviour. *Neuropsychologia*, *12*(3), 323–330.

Petersen, S. E., Fox, P. T., Snyder, A. Z., & Raichle, M. E. (1990). Activation of extrastriate and frontal cortical areas by visual words and word-like stimuli. *Science*, *249*(4972), 1041–1044.

Petrides, M., & Pandya, D. N. (2002). Comparative cytoarchitectonic analysis of the human and the macaque ventrolateral prefrontal cortex and corticocortical connection patterns in the monkey. *European Journal of Neuroscience*, *16*(2), 291–310.

Pobric, G., Jefferies, E., & Ralph, M. A. L. (2007). Anterior temporal lobes mediate semantic representation: Mimicking semantic dementia by using rTMS in normal participants. *Proceedings of the National Academy of Sciences*, *104*(50), 20137–20141.

Poldrack, R. A., & Wagner, A. D. (2004). What can neuroimaging tell us about the mind? Insights from prefrontal cortex. *Current Directions in Psychological Science*, *13*(5), 177–181.

Price, C. J., Wise, R. J. S., Watson, J. D., Patterson, K., Howard, D., & Frackowiak, R. S. (1994). Brain activity during reading. The effects of exposure duration and task. *Brain*, *117*(6), 1255–1269.

Race, E. A., Shanker, S., & Wagner, A. D. (2009). Neural priming in human frontal cortex: Multiple forms of learning reduce demands on the prefrontal executive system. *Journal of Cognitive Neuroscience*, *21*(9), 1766–1781.

Raichle, M. E., Fiez, J. A., Videen, T. O., MacLeod, A.-M. K., Pardo, J. V., Fox, P. T., et al. (1994). Practice-related changes in human brain functional anatomy during nonmotor learning. *Cerebral Cortex*, *4*(1), 8–26. Available from http://dx.doi.org/doi:10.1093/cercor/4.1.8.

Ralph, M. A. L., Pobric, G., & Jefferies, E. (2009). Conceptual knowledge is underpinned by the temporal pole bilaterally: Convergent evidence from rTMS. *Cerebral Cortex*, *19*(4), 832–838.

Robinson, G., Blair, J., & Cipolotti, L. (1998). Dynamic aphasia: An inability to select between competing verbal responses? *Brain*, *121* (1), 77–89.

Robinson, G., Shallice, T., Bozzali, M., & Cipolotti, L. (2010). Conceptual proposition selection and the LIFG: Neuropsychological evidence from a focal frontal group. *Neuropsychologia*, *48*(6), 1652–1663.

Robinson, G., Shallice, T., Bozzali, M., & Cipolotti, L. (2012). The differing roles of the frontal cortex in fluency tests. *Brain*, *135*(7), 2202–2214.

Robinson, G., Shallice, T., & Cipolotti, L. (2005). A failure of high level verbal response selection in progressive dynamic aphasia. *Cognitive Neuropsychology*, *22*(6), 661–694.

Rodd, J. M., Davis, M. H., & Johnsrude, I. S. (2005). The neural mechanisms of speech comprehension: fMRI studies of semantic ambiguity. *Cerebral Cortex*, *15*(8), 1261–1269.

Rodd, J. M., Johnsrude, I. S., & Davis, M. H. (2012). Dissociating frontotemporal contributions to semantic ambiguity resolution in spoken sentences. *Cerebral Cortex*, *22*(8), 1761–1773.

Saddy, J. D. (1995). Variables and events in the syntax of agrammatic speech. *Brain and Language*, *50*(2), 135–150.

Saffran, E. M., Schwartz, M. F., & Linebarger, M. C. (1998). Semantic influences on thematic role assignment: Evidence from normals and aphasics. *Brain and Language*, *62*(2), 255–297.

Sakai, K., & Passingham, R. E. (2002). Prefrontal interactions reflect future task operations. *Nature Neuroscience*, *6*(1), 75–81.

Schnur, T. T., Schwartz, M. F., Brecher, A., & Hodgson, C. (2006). Semantic interference during blocked-cyclic naming: Evidence from aphasia. *Journal of Memory and Language*, *54*(2), 199–227.

Schnur, T. T., Schwartz, M. F., Kimberg, D. Y., Hirshorn, E., Coslett, H. B., & Thompson-Schill, S. L. (2009). Localizing interference during naming: Convergent neuroimaging and neuropsychological evidence for the function of Broca's area. *Proceedings of the National Academy of Sciences*, *106*(1), 322–327.

Schwartz, M. F., Faseyitan, O., Kim, J., & Coslett, H. B. (2012). The dorsal stream contribution to phonological retrieval in object naming. *Brain*, *135*(12), 3799–3814.

Schwartz, M. F., Kimberg, D. Y., Walker, G. M., Faseyitan, O., & Brecher, A. D. (2009). Anterior temporal involvement in semantic word retrieval: VLSM evidence from aphasia. *Brain*, *132*(12), 3411–3427.

Schwartz, M. F., Saffran, E. M., & Marin, O. S. (1980). The word order problem in agrammatism: I. Comprehension. *Brain and Language*, *10*(2), 249–262.

Shaywitz, B. A., Shaywltz, S. E., Pugh, K. R., Constable, R. T., Skudlarski, P., Fulbright, R. K., et al. (1995). Sex differences in the functional organization of the brain for language. *Nature*, *373*, 607–609.

Smith, E. E., & Jonides, J. (1999). Storage and executive processes in the frontal lobes. *Science*, *283*(5408), 1657–1661.

Stowe, L. A., Broere, C. A., Paans, A. M., Wijers, A. A., Mulder, G., Vaalburg, W., et al. (1998). Localizing components of a complex task: Sentence processing and working memory. *Neuroreport*, *9* (13), 2995–2999.

Stowe, L. A., Paans, A. M., Wijers, A. A., Zwarts, F., Mulder, G., & Vaalburg, W. (1999). Sentence comprehension and word repetition: A positron emission tomography investigation. *Psychophysiology*, *36*(06), 786–801.

Stromswold, K., Caplan, D., Alpert, N., & Rauch, S. (1996). Localization of syntactic comprehension by positron emission tomography. *Brain and Language*, *52*(3), 452–473.

Stuss, D. T., Alexander, M. P., Hamer, L., Palumbo, C., Dempster, R., Binns, M., et al. (1998). The effects of focal anterior and posterior brain lesions on verbal fluency. *Journal of International Neuropsychological Society*, *4*, 265–278.

Swaab, T. Y., Brown, C., & Hagoort, P. (1998). Understanding ambiguous words in sentence contexts: Electrophysiological evidence for delayed contextual selection in Broca's aphasia. *Neuropsychologia*, *36*(8), 737–761.

Thompson-Schill, S. L., & Botvinick, M. M. (2006). Resolving conflict: A response to Martin and Cheng (2006). *Psychonomic Bulletin and Review*, *13*(3), 402–408.

Thompson-Schill, S. L., D'Esposito, M., Aguirre, G. K., & Farah, M. J. (1997). Role of left inferior prefrontal cortex in retrieval of semantic knowledge: A reevaluation. *Proceedings of the National Academy of Sciences*, *94*(26), 14792–14797.

Thompson-Schill, S. L., Swick, D., Farah, M. J., D'Esposito, M., Kan, I. P., & Knight, R. T. (1998). Verb generation in patients with focal frontal lesions: A neuropsychological test of neuroimaging findings. *Proceedings of the National Academy of Sciences*, *95* (26), 15855–15860.

Thothathiri, M., Kim, A., Trueswell, J. C., & Thompson-Schill, S. L. (2012). Parametric effects of syntactic–semantic conflict in Broca's area during sentence processing. *Brain and Language*, *120*(3), 259–264.

Troyer, A. K., Moscovitch, M., & Winocur, G. (1997). Clustering and switching as two components of verbal fluency: Evidence from younger and older healthy adults. *Neuropsychology*, *11*(1), 138.

Troyer, A. K., Moscovitch, M., Winocur, G., Alexander, M. P., & Stuss, D. (1998). Clustering and switching on verbal fluency: The effects of focal frontal- and temporal-lobe lesions. *Neuropsychologia*, *36*(6), 499–504.

Uylings, H. B. M., Malofeeva, L. I., Bogolepova, I. N., Amunts, K., & Zilles, K. (1999). Broca's language area from a neuroanatomical and developmental perspective. In C. M. Brown & P. Hagoort (Eds.), *The neurocognition of language* (pp. 319–336). Oxford: Oxford University Press.

Vigneau, M., Beaucousin, V., Herve, P.-Y., Duffau, H., Crivello, F., Houde, O., et al. (2006). Meta-analyzing left hemisphere language areas: Phonology, semantics, and sentence processing. *NeuroImage*, *30*(4), 1414–1432.

Vuong, L. C., & Martin, R. C. (2011). LIFG-based attentional control and the resolution of lexical ambiguities in sentence context. *Brain and Language*, *116*(1), 22–32.

Wagner, A. D., Paré-Blagoev, E. J., Clark, J., & Poldrack, R. A. (2001). Recovering meaning: Left prefrontal cortex guides controlled semantic retrieval. *Neuron*, *31*(2), 329–338.

Whitney, C., Kirk, M., O'Sullivan, J., Ralph, M. A. L., & Jefferies, E. (2011). The neural organization of semantic control: TMS evidence for a distributed network in left inferior frontal and posterior middle temporal gyrus. *Cerebral Cortex*, *21*(5), 1066–1075.

Whitney, C., Kirk, M., O'Sullivan, J., Ralph, M. A. L., & Jefferies, E. (2012). Executive semantic processing is underpinned by a large-scale neural network: Revealing the contribution of left prefrontal, posterior temporal, and parietal cortex to controlled retrieval and selection using TMS. *Journal of Cognitive Neuroscience*, *24*(1), 133–147.

Wig, G. S., Grafton, S. T., Demos, K. E., & Kelley, W. M. (2005). Reductions in neural activity underlie behavioral components of repetition priming. *Nature Neuroscience*, *8*(9), 1228–1233.

Zatorre, R. J., Evans, A. C., Meyer, E., & Gjedde, A. (1992). Lateralization of phonetic and pitch discrimination in speech processing. *Science*, *256*(5058), 846–849.

Zempleni, M.-Z., Renken, R., Hoeks, J. C., Hoogduin, J. M., & Stowe, L. A. (2007). Semantic ambiguity processing in sentence context: Evidence from event-related fMRI. *NeuroImage*, *34*(3), 1270–1279.

SECTION H

SENTENCE PROCESSING

CHAPTER

47

The Role of the Anterior Temporal Lobe in Sentence Processing

Corianne Rogalsky

Department of Speech and Hearing Science, Arizona State University, Tempe, AZ, USA

There are numerous computations necessary for successful sentence comprehension; phonological, semantic, syntactic, and combinatorial processes are all critical. This chapter focuses on the syntactic and combinatorial operations involved in sentence comprehension. The neuroanatomy supporting these processes, and perhaps even the definitions of these terms, are hotly debated (Grodzinsky & Santi, 2008; Hickok & Poeppel, 2004, 2007; Rogalsky & Hickok, 2011; Scott, Blank, Rosen, & Wise, 2000). Although Broca's area has long been the focus of these debates, the anterior temporal lobe (ATL) has emerged as a strong candidate for supporting sentence-level computations because there is strong evidence that the ATL is sensitive to the presence of sentence structure (Brennan et al., 2012; Humphries, Binder, Medler, & Liebenthal, 2006; Humphries, Love, Swinney, & Hickok, 2005; Humphries, Willard, Buchsbaum, & Hickok, 2001; Mazoyer et al., 1993; Rogalsky & Hickok, 2009; Rogalsky, Rong, Saberi, & Hickok, 2011). Two hypotheses regarding the ATL's responsiveness to sentence structure are: (i) the ATL supports basic syntactic operations or (ii) the ATL supports combinatorial semantic operations. In this chapter, we examine the evidence in support of both of these hypotheses. First, we discuss the historical framework and evidence for the ATL being sensitive to sentence structure.

47.1 WHAT ABOUT BROCA'S AREA?

Broca's area has traditionally been the focus of sentence-level investigations, particularly regarding syntactic processes (Bradley, Garrett, & Zurif, 1980; Caramazza & Zurif, 1976; Grodzinsky, 2000). This view originates from neuropsychological findings involving individuals with Broca's aphasia who often produce syntactically simple constructions and exhibit agrammatic production (i.e., speech that lacks function words and inflections) (Gleason, Goodglass, Green, Ackerman, & Hyde, 1975; Goodglass, 1968, 1976; Goodglass & Berko, 1960). Those with Broca's aphasia also demonstrate sentence comprehension deficits that similarly appear to implicate syntactic processes because comprehension failures are common for semantically reversible sentences that do not follow canonical word order (which in English is subject–verb–object; e.g., "It was the boy that the girl kissed") (Bradley et al., 1980; Caramazza & Zurif, 1976). However, there is also evidence that those with Broca's aphasia do have access to syntactic knowledge to some degree. For example, those with Broca's aphasia are able to make grammaticality judgments of sentences that they cannot completely comprehend (Linebarger, Schwartz, & Saffran, 1983; Wulfeck, 1988). It is also important to note that those with Broca's aphasia do not always have damage to Broca's area (Dronkers, Shapiro, Redfern, & Knight, 1992), and damage to Broca's area is not sufficient for Broca's aphasia (Mohr, 1976; Mohr et al., 1978). This lack of correspondence further complicates the claims that Broca's area plays a fundamental role in syntactic processing.

Functional imaging studies implicate Broca's area in aspects of syntactic processing, particularly in the comprehension of complex syntactic structures, such as noncanonical compared with canonical structures (Caplan, Alpert, & Waters, 1998, 1999; Dapretto & Bookheimer, 1999; Just, Carpenter, Keller, Eddy, & Thulborn, 1996; Stromswold, Caplan, Alpert, & Rauch, 1996). However, other functional imaging studies have found that activation in Broca's area does not track

Neurobiology of Language. DOI: http://dx.doi.org/10.1016/B978-0-12-407794-2.00047-X

with the presence or absence of syntactic information (Humphries et al., 2001, 2005, 2006; Mazoyer et al., 1993; Rogalsky et al., 2011; Stowe et al., 1998), suggesting that Broca's area plays a restricted role in sentence processing rather than a fundamental role in structure building or combinatorial processes. This has led researchers interested in syntactic processing to explore the response properties of other brain areas, including the ATL.

47.2 WHERE IS THE ATL?

The term "anterior temporal lobe" in this chapter is used to refer to cortex anterior to Heschl's gyrus in the lateral temporal lobe. The ATL regions implicated in sentence comprehension are in both the superior temporal gyrus and the middle temporal gyrus primarily, including in the superior temporal sulcus (e.g., the green areas in Figure 47.1). The approximate Brodmann areas corresponding to these functionally defined ATL regions include anterior BA 21 and BA 22 and posterior BA 38 (Brodmann, 1909; Wong & Gallate, 2012). It is worth noting that these ATL regions extend more posteriorly than the temporal pole (approximately BA 38), which has been proposed to be a multimodal hub supporting processes such as semantic memory, abstract conceptualizations, object concepts, personal narratives, and social concepts (Bonner & Price, 2013; Olson, McCoy, Klobusicky, & Ross, 2013; Simmons & Martin, 2009; Wong & Gallate, 2012).

47.3 DOMAIN-GENERAL SEMANTICS

Across the field of cognitive neuroscience, the term "anterior temporal lobe" is perhaps most associated with semantic memory (Simmons & Martin, 2009). Semantic dementia (SD), related to atrophy originating in the temporal pole(s), is associated with domain-general semantic deficits seen in naming, categorizing, and discriminating objects (Hodges, Patterson, Oxbury, & Funnell, 1992; Hodges et al., 1999; Mummery et al., 2000). Semantic deficits are also

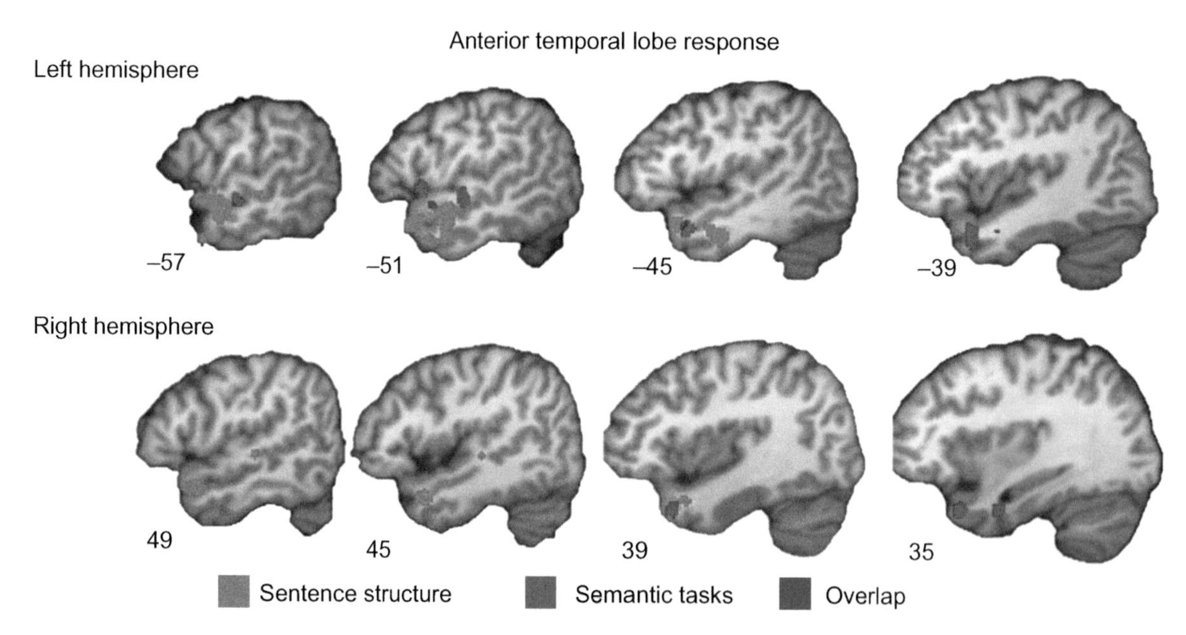

FIGURE 47.1 Meta-analysis of neuroimaging studies implicating the ATL in sentence structure and semantic processing. Activation likelihood estimations (ALEs; Eickhoff et al., 2009; Turkeltaub et al., 2012) derived from 17 studies reporting ATL activation to sentence structure (green), and 18 studies reporting ATL activation during single-word or object semantic tasks (red) (totaling 569 subjects and 157 peak coordinates); $p < 0.05$, false discovery rate (FDR)-corrected. The overlap between the sentence and semantic ALEs is shown in blue. The semantic task studies are those described in the review by Wong and Gallate (2012) of ATL function that reported peak activation coordinates (Wong & Gallate's Table 1, labeled as domain-general and nondomain general semantic functions) and in articles referencing or referenced by these studies. The sentence structure studies were identified by a PubMed.gov search using the terms "fMRI" or positron emission tomography ("PET") or "neuroimaging" and "sentence," and then by including all studies reporting ATL coordinates in response to sentences compared to stimuli without sentence structure (e.g., word lists). In addition, a dataset of Rogalsky et al. (submitted) was included. Four representative slices for each hemisphere are shown (left to right = lateral to medial). Only coordinates in the temporal lobes anterior to the posterior boundary of Heschl's gyrus were included. Included studies are denoted in the references by "*" for sentence structure and "**" for semantic tasks. This figure is not intended to portray an exhaustive meta-analysis of ATL function, but rather the paucity of overlap between the significant sentence structure and semantic activation estimates suggests that the ATL's response to sentence structure cannot be accounted for by semantic processing alone.

reported in temporal lobe epilepsy patients who have undergone anterior temporal lobectomies (Drane et al., 2009; Ellis, Young, & Critchley, 1989). Functional imaging data suggest that the ATL is engaged by a variety of lexical and semantic tasks, including categorization, naming, lexical decisions, and semantic knowledge decisions (Bright, Moss, & Tyler, 2004; Devlin et al., 2002; Ellis, Burani, Izura, Bromiley, & Venneri, 2006; Kellenbach, Hovius, & Patterson, 2005; Noppeney & Price, 2002; Peelen & Caramazza, 2012). The ATL is also more activated by content versus function words, and more by concrete versus abstract words (Diaz & McCarthy, 2009; Kiehl et al., 1999). The "semantic hub" hypothesis has recently been modified to include the idea that the ATL is not amodal, but rather there are modality-specific subregions (Olson et al., 2013; Skipper, Ross, & Olson, 2011). Nevertheless, as is discussed and as can be seen in Figure 47.1, these "semantic" ATL regions are physically dissociated from the ATL's activation during sentence comprehension. A meta-analysis of neuroimaging data by Visser, Jeffries, and Lambon Ralph (2010) also indicates that the activations in the ATL, across a variety of tasks, for auditory sentences are more dorsal and lateral than the ATL responses to pictures.

47.4 THE ATL RESPONDS TO SENTENCE STRUCTURE

Functional imaging first implicated the ATL in sentence processing. The Mazoyer et al. (1993) positron emission tomography (PET) study found that the ATL, and not Broca's area, reliably activated to sentence structure. Mazoyer et al. presented several types of speech stimuli to participants: sentences in the participant's native language; sentences in an unfamiliar language; pseudoword sentences (sentences in which the content words were replaced with nonwords); semantically anomalous sentences; and word lists. The three stimuli types containing syntactic information (native language sentences, pseudoword sentences, and semantically anomalous sentences) reliably activated the ATL, whereas the remaining stimuli did not. Several subsequent neuroimaging studies also found the ATL to track sentence structure; sentences and pseudoword sentences reliably activate the bilateral ATL significantly more than scrambled sentences, semantically related and unrelated word lists, spectrally rotated speech, and environmental sounds sequences (Friederici, Kotz, Scott, & Obleser, 2010; Humphries et al., 2001, 2005, 2006; Rogalsky & Hickok, 2009; Rogalsky et al., 2011; Spitsyna, Warren, Scott, Turkheimer, & Wise, 2006; Xu, Kemeny, Park, Frattali, & Braun, 2005). The ATL's response to sentence structure is independent of input modality; both listening to and reading sentences elicit greater responses in the ATL than word lists in each modality, respectively (Bemis & Pylkkanen, 2013a; Jobard, Vigneau, Mazoyer, & Tzourio-Mazoyer, 2007; Stowe et al., 1998; Vandenberghe, Nobre, & Price, 2002).

Lesion evidence also indicates that the ATL is engaged in sentence-level processing. ATL lesions are associated with sentence comprehension deficits for all but the simplest of sentence structures (Dronkers, Wilkins, Van Valin, Redfern, & Jaeger, 2004). Damage to the left ATL has also been implicated in complex syntactic processing deficits. Specific impairment of noncanonical sentence comprehension was associated with left ATL damage, whereas the comprehension of canonically structured sentences implicated a large temporoparietaloccipital network (Magnusdottir et al., 2013). It is noteworthy that there are very few lesion studies specifically addressing ATL function. This may, in part, be due to the rarity of lesions from stroke in the ATL. Typically, ATL damage from stroke coincides with overall large left hemisphere damage and strokes restricted to the ATL are rare (Holland & Lambon Ralph, 2010).

In summary, the ATL responds to sentence structure. But is the ATL responding to sentence structure *per se* or hierarchical structure more generally? The ATL does respond significantly more to sentences than environmental sound sequences portraying the same event (Humphries et al., 2001), suggesting that the ATL is not responding to auditory sound sequences *per se*. However, a question still remains regarding whether the ATL is responding to speech-specific hierarchical structures or if it is involved in hierarchical processing more generally. One way to test these possibilities is with music. Melodies, like sentences, can be complex acoustic stimuli, with hierarchical structure, tone, and rhythm, requiring combinatorial representations (Lerdahl & Jackendoff, 1983; Patel, 2007). There are also behavioral findings of interactions between structured speech and music processes, suggesting that they may share neural resources (Fedorenko, Patel, Casanato, Winawer, & Gibson, 2009; Patel, 2003). Regarding the specificity of the ATL's response to sentence structure, one PET study does report that ATL activity is modulated by structural complexity of the melodies presented (Griffiths, Buchel, Frackowiak, & Patterson, 1998). However, the only work, to our knowledge, to directly compare sentences and melodies has found that the ATL responds significantly more to the sentences than melodies, and that the ATL does not significantly activate for melodies compared with rest (Rogalsky et al., 2011). This suggests that it is not structural processing *per se* that is driving the ATL response to sentences.

47.5 SYNTAX

Now, we turn to the question of *why* the ATL is responsive to sentence structure: is the ATL's response to sentences driven by compositional semantic properties (i.e., integrating lexical semantic information to extract sentence meaning) or more basic syntactic (i.e., parsing) processes? There is evidence for both hypotheses and, in fact, there is evidence to suggest that both processes may be preferentially engaging subregions of the ATL (Rogalsky & Hickok, 2009; Vandenberghe et al., 2002). In this section, we discuss the evidence for a syntactic account of the ATL's role in sentence processing. The three main lines of evidence that support the syntactic account are: (i) the ATL is activated by sentences lacking lexical semantic content; (ii) the ATL is more responsive to syntactic errors than semantic errors; and (iii) different grammatical constructions elicit different activation patterns in the ATL. These three points are described in more detail here.

As described in the previous section, the ATL shows a greater response to sentences compared with scrambled sentences and word lists, suggesting that the ATL is involved in sentence processing beyond lexical-semantics. However, this finding does not rule out the possibility that the ATL is involved in sentence-level semantic processes, commonly referred to as combinatorial semantics (Fodor, 1995; Partee, 1995; Vandenberghe et al., 2002). Perhaps the strongest argument against this possibility, and for the hypothesis that syntactic information drives ATL activation, is how the ATL responds to pseudoword sentences (sentences in which the content words have been replaced by phonologically plausible nonwords). For example, "The klib were frimming in swak," or "A klinder ghasted the nederopit." The ATL consistently and robustly activates to pseudoword sentences compared with scrambled pseudowords and pseudoword lists, suggesting that the ATL's response to sentence structure is not necessarily driven by semantic information, but rather it is in response to the syntactic structure, which is preserved in pseudoword sentences (Humphries et al., 2006; Rogalsky et al., 2011). However, it could be argued that semantic combinatorial processes are driving this effect. Perhaps combinatorial analyses are still being attempted for these pseudoword sentences despite the lack of semantic content because typically the presence of syntactic structure correctly predicts the need to build a sentence-level semantic representation, thereby engaging the ATL.

Evidence that the ATL is responding to syntactic information also comes from error detection paradigms; sentences containing syntactic errors elicit more of a response in the ATL than correct sentences or those containing semantic errors (Friederici, Ruschemeyer, Hahne, & Fiebach, 2003; Herrmann, Maess, Hahne, Schroger, & Friederici, 2011; Herrmann, Obleser, Kalberlah, Haynes, & Friederici, 2012). Syntactic errors include case, number, or gender disagreement, whereas semantic errors typically involve thematic role incompatibility or semantic anomalies. For example, an event-related potential study found that syntactic (but not semantic) errors elicit an early left anterior negativity, which is proposed to be associated with anterior superior temporal and inferior frontal regions (Friederici & Kotz, 2003). fMRI data also implicate the ATL in syntactic error detection; compared with correct sentences, syntactic errors activated a network of regions including the ATL, whereas semantic errors did not activate the ATL (Friederici et al., 2003; Herrmann et al., 2012).

Further evidence for the ATL supporting syntactic processing is that under some conditions, ATL regions respond differently to different grammatical constructions, even when semantic content and complexity are controlled (Allen, Pereira, Botvinick, & Goldberg, 2012; Hammer, Jansma, Tempelmann, & Munte, 2011; Newman, Supalla, Hauser, Newport, & Bavelier, 2010). For example, the Allen et al. fMRI study found differing patterns of activation in the ATL for the English dative construction (e.g., Mark gave the gift to Felix) versus the English ditransitive (e.g., Mark gave Felix the gift). The pattern of activation in the left ATL (in conjunction with inferior frontal areas) was different for these two grammatical constructions despite containing essentially the same semantic content, thereby suggesting that the syntactic structure is driving the patterns of activation.

One piece of evidence complicating the syntactic account of the ATL's role in sentence comprehension comes from work with SD patients. SD patients have progressive atrophy originating in the ATLs and present with progressive, selective deficits for lexical semantic knowledge (Wilson, Galantucci, Tartaglia, & Gorno-Tempini, 2012). Several studies report that SD patients do not exhibit syntactic deficits once lexical semantic deficits are accounted for (Gorno-Tempini et al., 2004; Hodges et al., 1992; Warrington, 1975; Wilson et al., 2011). However, Wilson et al. (2014) found SD patients to have impaired performance on a sentence comprehension task with minimal lexical semantic demands, and to have different levels of activation than controls in their atrophied ATLs for the sentence task compared with rest. This result is complicated by the SD patients also having posterior temporal atrophy in regions activated in controls during the sentence task, but this study suggests that the link between SD and lack of syntactic deficits is not as straightforward as previous behavioral studies would suggest.

As discussed previously, the ATL is activated more for sentences than for a variety of control conditions (e.g., word lists, scrambled sentences, and acoustically matched auditory stimuli). One may then wonder whether the kinds of sentence comprehension manipulations that implicate Broca's area also implicate the ATL (e.g., noncanonical versus canonical syntactic constructions). In fact, these types of "syntactic complexity" comparisons that consistently implicate inferior frontal regions do not activate ATL regions (Ben-Shachar, Hendler, Kahn, Ben-Bashat, & Grodzinsky, 2003; Ben-Shachar, Palti, & Grodzinsky, 2004; Caplan et al., 1998; Rogalsky & Hickok, 2011; Rogalsky, Matchin, & Hickok, 2008; Stromswold et al., 1996; see Santi & Grodzinsky, 2012 for one notable exception). The ATL's lack of sensitivity to different syntactic constructions could be interpreted as evidence against the ATL's involvement in syntactic processing. However, given the ATL's robust response to sentence structure in general, it is more likely that the ATL is supporting syntactic processes in both the canonical and noncanonical sentences, and that additional resources (such as in the inferior frontal lobe) are being recruited for the more difficult constructions (Brennan et al., 2012; Fiebach, Schlesewsky, Lohmann, von Cramon, & Friederici, 2005; Rogalsky & Hickok, 2011).

47.6 COMBINATORIAL SEMANTICS

The ATL and temporal pole broadly speaking are clearly involved in building semantic representations across modalities and for a variety of stimuli, as findings from SD and neuroimaging studies (Figure 47.1) attest. Accordingly, semantic processing at the sentence level has been suggested as a possible role of the ATL in sentence comprehension (Wilson et al., 2014; Wong & Gallate, 2012). Indeed, one could argue that many of the observations that have linked ATL activity to syntactic structure are attributable to combinatorial processes, for example, the difference in activation between sentences and word lists.

There are several pieces of evidence to suggest that the ATL's response to sentence structure is due to its involvement in combinatorial semantics (i.e., combining the meanings of the words in a sentence into a cohesive, meaningful unit) (Pallier, Devauchelle, & Dehaene, 2011; Vandenberghe et al., 2002; but for a notable exception, see Humphries et al., 2006). The evidence supporting this claim mostly comes from the fact that manipulating various semantic features of sentences modulates ATL activity. Semantic anomalies (e.g., "the excited students rode the *trampoline* to school") consistently modulate ATL activity, suggesting that the unexpected word is making the formation of a global meaning more difficult to construct and thereby increasing activity in the ATL (Dapretto & Bookheimer, 1999; Ferstl, Rinck, & von Cramon, 2005; Vandenberghe et al., 2002). Even correct sentences in a semantic anomaly detection task drive ATL activation more than correct sentences in a syntactic error detection task, indicating that ATL regions are involved in building the sentence-level semantic representations necessary to complete the semantic anomaly detection task (Rogalsky & Hickok, 2009).

A variety of other paradigms besides semantic anomalies also implicate the ATL in combinatorial semantic processes in sentence comprehension. Idioms activate the ATL significantly more than literal sentences (Lauro, Tettamanti, Cappa, & Papagno, 2008), and narratives elicit greater activation in the ATL than sentences (Xu et al., 2005). ATL activity has also been found to increase as a function of combinatorial semantic load; ATL activity increases as the constituent size within word streams increases (Pallier et al., 2011). This effect was not found for pseudoword streams, suggesting that the ATL is responding to the presence of lexical semantic information but not the syntactic structure *per se*. Similarly, adaptation paradigms produce effects in the ATL for sentences with different syntactic structures but the same semantic content (Devauchelle, Oppenheim, Rizzi, Dehaene, & Pallier, 2009), and ATL activity is evoked more by minimal phrases (e.g., "blue car") than comparable word lists (e.g., "car, blue") (Bemis & Pylkkanen, 2011, 2013a, 2013b).

47.7 PROSODY

Another component of the ATL's contribution to sentence comprehension may be prosody. Variations in prosody and sentence structure are highly correlated in natural environments (Herrmann, Maess, & Friederici, 2011); the human brain has an expectation that speech will contain both syntactic and acoustic/prosodic information that will be informative regarding the overall meaning and tone of the speech stimulus (Frazier, Carlson, & Clifton, 2006; Herrmann, Maess, & Friederici, 2011; Shattuck-Hufnagel & Turk, 1996; Weintraub, Mesulam, & Kramer, 1981). One complication of the studies demonstrating that the ATL is sensitive to sentence structure and/or combinatorial semantics is that many of the auditory stimuli used were spoken by talkers using sentence-like prosody and compared with word lists or other baseline conditions lacking sentence-like prosody (e.g., word lists with list-like prosody; Humphries et al., 2001; Meyer, Steinhauer, Alter, Friederici, & von Cramon, 2004; Rogalsky & Hickok, 2009; Spitsyna et al., 2006). One attempt to control for prosody was a study that applied "list prosody" to sentences and compared activation to lists of words with list prosody (Humphries et al., 2005). Sentences still activated portions of the ATL more robustly, but it is possible that the prosodic chimera sentences created a prosodic anomaly response that drove the activation. So, is prosody, not structure, driving ATL involvement?

Both functional imaging and lesion studies of prosody clearly implicate the ATLs (Adolphs, Damasio, & Tranel, 2002; Johnstone, van Reekum, Oakes, & Davidson, 2006; Phillips et al., 1998). The right ATL is implicated in perceiving a variety of emotional prosodic manipulations (Gorelick & Ross, 1987; Heilman, Scholes, & Watson, 1975; Hoekert, Bais, Kahn, & Aleman, 2008; Phillips et al., 1998), for example, angry, sad, or anxious tones of voice versus neutral tone (Hoekert et al., 2008). The left ATL has not been implicated in emotional prosody, but it does have subregions that are responsive to normal sentence prosody. The Humphries et al. (2005) fMRI study presented subjects with sentences, scrambled sentences, and lists of content words. All three stimulus types were presented in two ways, with sentence prosody and then also with flat list prosody. A left ATL subregion showed a main effect for syntactic structure, but both left and right ATL subregions exhibited main effects for sentence prosody (i.e., was activated by sentence prosody even for the scrambled and list stimuli). These findings may reflect the ATL's involvement in prosody, but it is also possible that the ATL is responding because of its role in processing socio-emotional content in general (Olson et al., 2013; Wong & Gallate, 2012). The ATL and temporal pole are responsive to a variety of social and emotional parameters (Burnett & Blakemore, 2009; Reiman et al., 1997; Shin et al., 2000; Zahn et al., 2007), so it may be that the ATL is engaged by sentences, at least in part, to process the socio-emotional information that prosody provides as one component of a sentence-level representation.

47.8 THE ATL IS PART OF A LARGE LANGUAGE NETWORK

This chapter focused on how the ATL is sensitive to the various elements of sentence comprehension. However, it is critical to understand that these elements (syntactic structure, prosody, combinatorial semantics, lexical-semantics) engage large frontal–temporal–parietal networks, of which the ATL is one component (Hickok & Poeppel, 2000, 2004, 2007). Notably, posterior temporal regions are also sensitive to syntactic structure and sentence-level semantic information (Griffiths, Marslen-Wilson, Stamatakis, & Tyler, 2013; Humphries et al., 2006; Wilson et al., 2014). Broca's area, as discussed, is recruited during sentence comprehension in high-load, difficult situations. The point of this chapter is not to identify the ATL as the "site" of sentence processing, but rather the point is to discuss how the ATL is sensitive to sentence-level structure and semantics, and how the ATL may be contributing to sentence comprehension as part of a large network.

47.9 SUMMARY

The ATL is not a homogenous region with only one specific function. Overall, the functionally diverse ATL is involved in integrating information to form meaningful representations from all modalities (Visser et al., 2010; Wong & Gallate, 2012). The reliable activation of ATL subregions to syntactic structure suggests that portions of the ATL are particularly tuned to form representations from a speech signal. Although it is difficult to pinpoint the exact computations that the ATL is providing for sentence comprehension, it is clear that the ATL is engaged during the comprehension of sentences and is sensitive to syntactic, combinatorial semantic, and prosodic information. The ATL is one part of a large frontotemporoparietal network engaged in sentence comprehension.

Acknowledgments

The author thanks Colin Humphries and Greg Hickok for their help in developing this chapter, and Alvaro Diaz for his assistance in generating the figure.

References

Adolphs, R., Damasio, H., & Tranel, D. (2002). Neural systems for recognition of emotional prosody: A 3-D lesion study. *Emotion, 2*(1), 23–51.

Allen, K., Pereira, F., Botvinick, M., & Goldberg, A. E. (2012). Distinguishing grammatical constructions with fMRI pattern analysis. *Brain and Language, 123*(3), 174–182.

Baron, S. G., & Osherson, D. (2011). Evidence for conceptual combination in the left anterior temporal lobe. *NeuroImage, 55*(4), 1847–1852.

Bemis, & Pylkkanen (2011). Simple composition: A magnetoencephalography investigation into the comprehension of minimal linguistic phrases. *Journal of Neuroscience, 31*(8), 2801–2814.

Bemis, & Pylkkanen (2013a). Basic linguistic composition recruits the left anterior temporal lobe and left angular gyrus during both listening and reading. *Cerebral Cortex, 23*(8), 1859–1873.

Bemis, & Pylkkanen (2013b). Flexible composition: MEG evidence for the deployment of basic combinatorial mechanisms in response to task demands. *PLoS One, 8*(9), e73949.

Ben-Shachar, M., Hendler, T., Kahn, I., Ben-Bashat, D., & Grodzinsky, Y. (2003). The neural reality of syntactic transformations: Evidence from functional magnetic resonance imaging. *Psychological Science, 14*(5), 433–440.

Ben-Shachar, M., Palti, D., & Grodzinsky, Y. (2004). Neural correlates of syntactic movement: Converging evidence from two fMRI experiments. *NeuroImage*, *21*(4), 1320–1336.

Bonner, M. F., & Price, A. R. (2013). Where is the anterior temporal lobe and what does it do? *The Journal of Neuroscience*, *33*(10), 4213–4215.

Bradley, D. C., Garrett, M. E., & Zurif, E. B. (1980). Syntactic deficits in Broca's aphasia. In D. Caplan (Ed.), *Biological studies of mental processes*. Cambridge, MA: MIT Press.

*Brennan, J., Nir, Y., Hasson, U., Malach, R., Heeger, D. J., & Pylkkanen, L. (2012). Syntactic structure building in the anterior temporal lobe during natural story listening. *Brain and Language*, *120*(2), 163–173.

*Bright, P., Moss, H., & Tyler, L. K. (2004). Unitary vs multiple semantics: PET studies of word and picture processing. *Brain and Language*, *89*(3), 417–432.

Brodmann, K. (1909). *Vergleichende Lokalisationslehre der Grobhirnrinde in ihren Prinzipien dargestellt aufgrund des Zellenbaues*. Leipzig: Johann Ambrosius Barth Verlag.

Burnett, S., & Blakemore, S. J. (2009). Functional connectivity during a social emotion task in adolescents and in adults. *The European Journal of Neuroscience*, *29*(6), 1294–1301.

Caplan, D., Alpert, N., & Waters, G. (1998). Effects of syntactic structure and propositional number on patterns of regional cerebral blood flow. *Journal of Cognitive Neuroscience*, *10*, 541–552.

Caplan, D., Alpert, N., & Waters, G. (1999). PET studies of syntactic processing with auditory sentence presentation. *NeuroImage*, *9*, 343–351.

Caramazza, A., & Zurif, E. B. (1976). Dissociation of algorithmic and heuristic processes in sentence comprehension: Evidence from aphasia. *Brain and Language*, *3*, 572–582.

*Cutting, L. E., Clements, A. M., Courtney, S., Rimrodt, S. L., Schafer, J. G., Bisesi, J., et al. (2006). Differential components of sentence comprehension: Beyond single word reading and memory. *NeuroImage*, *29*, 429–438.

Dapretto, M., & Bookheimer, S. Y. (1999). Form and content: Dissociating syntax and semantics in sentence comprehension. *Neuron*, *24*, 292–293.

Devauchelle, A., Oppenheim, C., Rizzi, L., Dehaene, S., & Pallier, C. (2009). Sentence syntax and content in the human temporal lobe: An fMRI adaptation study in auditory and visual modalities. *Journal of Cognitive Neuroscience*, *21*(5), 1000–1012.

**Devlin, J. T., Russell, R. P., Davis, M. H., Price, C. J., Moss, H. E., Fadili, M. J., et al. (2002). Is there an anatomical basis for category-specificity? Semantic memory studies in PET and fMRI. *Neuropsychologia*, *40*(1), 54–75.

**Diaz, M. T., & McCarthy, G. (2009). A comparison of brain activity evoked by single content and function words: An fMRI investigation of implicit word processing. *Brain Research*, *1282*, 38–49.

Drane, D. L., Ojemann, G. A., Ojemann, J. G., Aylward, E., Silbergeld, D. L., Miller, J. W., et al. (2009). Category-specific recognition and naming deficits following resection of a right anterior temporal lobe tumor in a patient with atypical language lateralization. *Cortex*, *45*(5), 630–640.

Dronkers, N. F., Shapiro, J. K., Redfern, B., & Knight, R. T. (1992). The role of Broca's area in Broca's aphasia. *Journal of Clinical and Experimental Neuropsychology*, *14*, 52–53.

Dronkers, N. F., Wilkins, D. P., Van Valin, R. D., Jr., Redfern, B. B., & Jaeger, J. J. (2004). Lesion analysis of the brain areas involved in language composition. *Cognition*, *92*(1–2), 145–177.

Eickhoff, S. B., Laird, A. R., Grefkes, C., Wang, L. E., Zilles, K., & Fox, P. T. (2009). Coordinate-based activation likelihood estimation meta-analysis of neuroimaging data: A random-effects approach based on empirical estimates of spatial uncertainty. *Human Brain Mapping*, *30*, 2907–2926.

**Ellis, A. W., Burani, C., Izura, C., Bromiley, A., & Venneri, A. (2006). Traces of vocabulary acquisition in the brain: Evidence from covert object naming. *NeuroImage*, *33*(3), 958–968.

Ellis, A. W., Young, A. W., & Critchley, E. M. (1989). Loss of memory for people following temporal lobe damage. *Brain*, *112*(Pt 6), 1469–1483.

**Emerton, B. C., Gansler, D. A., Sandberg, E. H., & Jerram, M. (2013). Functional anatomical dissociation of description and picture naming in the left temporal lobe. *Brain Imaging and Behavior*, *8*, 570–578.

Fedorenko, E., Patel, A., Casanato, D., Winawer, J., & Gibson, E. (2009). Structural integration in language and music: Evidence for a shared system. *Memory and Cognition*, *37*(1), 1–9.

Ferstl, E. C., Rinck, M., & von Cramon, D. Y. (2005). Emotional and temporal aspects of situation model processing during text comprehension: An event-related fMRI study. *Journal of Cognitive Neuroscience*, *17*(5), 724–739.

Fiebach, C. J., Schlesewsky, M., Lohmann, G., von Cramon, D. Y., & Friederici, A. D. (2005). Revisiting the role of Broca's area in sentence processing: Syntactic integration versus syntactic working memory. *Human Brain Mapping*, *24*, 79–91.

Fodor, J. (1995). Comprehending sentence structure. In D. Osherson, & M. Liberman (Eds.), *Language: An invitation to cognitive science*. Cambridge, MA: MIT Press.

Frazier, L., Carlson, K., & Clifton, C., Jr. (2006). Prosodic phrasing is central to language comprehension. *Trends in Cognitive Sciences*, *10*(6), 244–249.

Friederici, A. D., & Kotz, S. A. (2003). The brain basis of syntactic processes: Functional imaging and lesion studies. *NeuroImage*, *20* (Suppl. 1), S8–S17.

*Friederici, A. D., Kotz, S. A., Scott, S. K., & Obleser, J. (2010). Disentangling syntax and intelligibility in auditory language comprehension. *Human Brain Mapping*, *31*(3), 448–457.

**Friederici, A. D., Opitz, B., & von Cramon, D. Y. (2000). Segregating semantic and syntactic aspects of processing in the human brain: An fMRI investigation of different word types. *Cerebral Cortex*, *10*(7), 698–705.

*Friederici, A. D., Ruschemeyer, S. A., Hahne, A., & Fiebach, C. J. (2003). The role of left inferior frontal and superior temporal cortex in sentence comprehension: Localizing syntactic and semantic processes. *Cerebral Cortex*, *13*, 170–177.

Gleason, J. B., Goodglass, H., Green, E., Ackerman, N., & Hyde, M. (1975). The retrieval of syntax in Broca's aphasia. *Brain and Language*, *2*, 451–471.

Goodglass, H. (1968). Studies in the grammar of aphasics. In S. Rosenberg, & J. Koplin (Eds.), *Developments in applied psycholinguistic research*. New York, NY: MacMillan.

Goodglass, H. (1976). Agrammatism. In H. Whitaker, & H. A. Whitaker (Eds.), *Studies in neurolinguistics* (Vol. 1). New York, NY: Academic Press.

Goodglass, H., & Berko, J. (1960). Agrammatism and inflectional morphology in English. *Journal of Speech and Hearing Research*, *3*, 257–267.

Gorelick, R. B., & Ross, E. D. (1987). The aprosodias: Further functional-anatomical evidence for the organisation of affective language in the right hemisphere. *Journal of Neurology, Neurosurgery, and Psychiatry*, *50*(5), 553–560.

Gorno-Tempini, M. L., Dronkers, N. F., Rankin, K. P., Ogar, J. M., Phengrasamy, L., Rosen, H. J., et al. (2004). Cognition and anatomy in the three variants of primary progressive aphasia. *Annals of Neurology*, *55*(3), 335–346.

**Grabowski, T. J., Damasio, H., Tranel, D., Ponto, L. L., Hichwa, R. D., & Damasio, A. R. (2001). A role for left temporal pole in the retrieval of words for unique entities. *Human Brain Mapping*, *13*(4), 199–212.

Griffiths, J. D., Marslen-Wilson, W. D., Stamatakis, E. A., & Tyler, L. K. (2013). Functional organization of the neural language system: Dorsal and ventral pathways are critical for syntax. *Cerebral Cortex, 23*(1), 139–147.

Griffiths, T. D., Buchel, C., Frackowiak, R. S. J., & Patterson, R. D. (1998). Analysis of temporal structure in sound by the human brain. *Nature Neuroscience, 1*(5), 422–427.

Grodzinsky, Y. (2000). The neurology of syntax: Language use without Broca's area. *Behavioral and Brain Sciences, 23*, 1–21.

Grodzinsky, Y., & Santi, A. (2008). The battle for Broca's region. *Trends in Cognitive Sciences, 12*(12), 474–480.

**Hamberger, M. J., Habeck, C. G., Pantazatos, S. P., Williams, A. C., & Hirsch, J. (2014). Shared space, separate processes: Neural activation patterns for auditory description and visual object naming in healthy adults. *Human Brain Mapping, 35*(6), 2507–2520.

Hammer, A., Jansma, B. M., Tempelmann, C., & Munte, T. F. (2011). Neural mechanisms of anaphoric reference revealed by fMRI. *Frontiers in Psychology, 2*, 32.

Heilman, K. M., Scholes, R., & Watson, R. T. (1975). Auditory affective agnosia: Disturbed comprehension of affective speech. *Journal of Neurology, Neurosurgery and Psychiatry, 38*(1), 69–72.

Herrmann, B., Maess, B., & Friederici, A. D. (2011). Violation of syntax and prosody-disentangling their contributions to the early left anterior negativity (ELAN). *Neuroscience Letters, 490*(2), 116–120.

Herrmann, B., Maess, B., Hahne, A., Schroger, E., & Friederici, A. D. (2011). Syntactic and auditory spatial processing in the human temporal cortex: An MEG study. *NeuroImage, 57*(2), 624–633.

*Herrmann, B., Maess, B., Hasting, A. S., & Friederici, A. D. (2009). Localization of the syntactic mismatch negativity in the temporal cortex: An MEG study. *NeuroImage, 48*, 590–600.

*Herrmann, B., Obleser, J., Kalberlah, C., Haynes, J., & Friederici, A. D. (2012). Dissociable neural imprints of perception and grammar in auditory functional imaging. *Human Brain Mapping, 33*(3), 584–595.

Hickok, G., & Poeppel, D. (2000). Towards a functional neuroanatomy of speech perception. *Trends in Cognitive Sciences, 4*, 131–138.

Hickok, G., & Poeppel, D. (2004). Dorsal and ventral streams: A framework for understanding aspects of the functional anatomy of language. *Cognition, 92*, 67–99.

Hickok, G., & Poeppel, D. (2007). The cortical organization of speech processing. *Nature Reviews Neuroscience, 8*, 393–402.

Hodges, J. R., Patterson, K., Oxbury, S., & Funnell, E. (1992). Semantic dementia. Progressive fluent aphasia with temporal lobe atrophy. *Brain: A Journal of Neurology, 115*(Pt 6), 1783–1806.

Hodges, J. R., Patterson, K., Ward, R., Garrard, P., Bak, T., Perry, R., et al. (1999). The differentiation of semantic dementia and frontal lobe dementia (temporal and frontal variants of frontotemporal dementia) from early Alzheimer's disease: A comparative neuropsychological study. *Neuropsychology, 13*(1), 31–40.

Hoekert, M., Bais, L., Kahn, R. S., & Aleman, A. (2008). Time course of the involvement of the right anterior superior temporal gyrus and the right fronto-parietal operculum in emotional prosody perception. *PLoS One*, 3(5), e2244.

Holland, R., & Lambon Ralph, M. A. (2010). The anterior temporal lobe semantic hub is a part of the language neural network: Selective disruption of irregular past tense verbs by rTMS. *Cerebral Cortex, 20*(12), 2771–2775.

Humphries, C., Binder, J. R., Medler, D. A., & Liebenthal, E. (2006). Syntactic and semantic modulation of neural activity during auditory sentence comprehension. *Journal of Cognitive Neuroscience, 18*, 665–679.

Humphries, C., Love, T., Swinney, D., & Hickok, G. (2005). Response of anterior temporal cortex to syntactic and prosodic manipulations during sentence processing. *Human Brain Mapping, 26*, 128–138.

Humphries, C., Willard, K., Buchsbaum, B., & Hickok, G. (2001). Role of anterior temporal cortex in auditory sentence comprehension: An fMRI study. *NeuroReport, 12*, 1749–1752.

*Jobard, G., Vigneau, M., Mazoyer, B., & Tzourio-Mazoyer, N. (2007). Impact of modality and linguistic complexity during reading and listening tasks. *NeuroImage, 34*(2), 784–800.

Johnstone, T., van Reekum, C. M., Oakes, T. R., & Davidson, R. J. (2006). The voice of emotion: An fMRI study neural responses to angry and happy vocal expressions. *Social Cognitive and Affective Neuroscience, 1*(3), 242–249.

Just, M. A., Carpenter, P. A., Keller, T. A., Eddy, W. F., & Thulborn, K. R. (1996). Brain activation modulated by sentence comprehension. *Science, 274*, 114–116.

**Kellenbach, M. L., Hovius, M., & Patterson, K. (2005). A pet study of visual and semantic knowledge about objects. *Cortex, 41*(2), 121–132.

**Kiehl, K. A., Liddle, P. F., Smith, A. M., Mendrek, A., Forster, B. B., & Hare, R. D. (1999). Neural pathways involved in the processing of concrete and abstract words. *Human Brain Mapping, 7*(4), 225–233.

Lauro, L. J. R., Tettamanti, M., Cappa, S. F., & Papagno, C. (2008). Idiom comprehension: A prefrontal task? *Cerebral Cortex, 18*(1), 162–170.

Lerdahl, F., & Jackendoff, R. (1983). An overview of hierarchical structure in music. *Music Percept, 1*, 229–252.

Linebarger, M. C., Schwartz, M., & Saffran, E. (1983). Sensitivity to grammatical structure in so-called agrammatic aphasics. *Cognition, 13*, 361–393.

Magnusdottir, S., Fillmore, P., den Ouden, D. B., Hjaltson, H., Rorden, C., Kjartansson, O., et al. (2013). Damage to left anterior temporal cortex predicts impairment of complex syntactic processing: A lesion-symptom mapping study. *Human Brain Mapping, 34*(10), 2715–2723.

Mazoyer, B. M., Tzourio, N., Frak, V., Syrota, A., Murayama, N., Levrier, O., et al. (1993). The cortical representation of speech. *Journal of Cognitive Neuroscience, 5*, 467–479.

Meyer, M., Steinhauer, K., Alter, K., Friederici, A. D., & von Cramon, D. Y. (2004). Brain activity varies with modulation of dynamic pitch variance in sentence melody. *Brain and Language, 89*(2), 277–289.

Mohr, J. P. (1976). Broca's area and Broca's aphasia. In H. Whitaker, & H. A. Whitaker (Eds.), *Studies in neurolinguistics* (Vol. 1, pp. 201–235). New York, NY: Academic Press.

Mohr, J. P., Pessin, M. S., Finkelstein, S., Funkenstein, H. H., Duncan, G. W., & Davis, K. R. (1978). Broca's aphasia: Pathologic and clinical. *Neurology, 28*, 311–324.

Mummery, C. J., Patterson, K., Price, C. J., Ashburner, J., Frackowiak, R. S., & Hodges, J. R. (2000). A voxel-based morphometry study of semantic dementia: Relationship between temporal lobe atrophy and semantic memory. *Annals of Neurology, 47*(1), 36–45.

**Mummery, C. J., Shallice, T., & Price, C. J. (1999). Dual-process model in semantic priming: A functional imaging perspective. *NeuroImage, 9*(5), 516–525.

**Murtha, S., Chertkow, H., Beauregard, M., & Evans, A. (1999). The neural substrate of picture naming. *Journal of Cognitive Neuroscience, 11*(4), 399–423.

Newman, A. J., Supalla, T., Hauser, P., Newport, E. L., & Bavelier, D. (2010). Dissociating neural subsystems for grammar by contrasting word order and inflection. *Proceedings of the National Academy of Sciences of the United States of America, 107*(6), 7539–7544.

**Noppeney, U., & Price, C. J. (2002). Retrieval of visual, auditory, and abstract semantics. *NeuroImage, 15*(4), 917–926.

Olson, I. R., McCoy, D., Klobusicky, E., & Ross, L. A. (2013). Social cognition and the anterior temporal lobes: A review and theoretical framework. *Social Cognitive and Affective Neuroscience, 8*(2), 123–133.

Pallier, C., Devauchelle, A., & Dehaene, S. (2011). Cortical representations of the constituent structure of sentences. *Proceedings of the National Academy of Sciences of the United States of America*, *108*(6), 2522–2527.

Partee, B. (1995). Lexical semantics and compositionality. In D. Osherson, & M. Liberman (Eds.), *Language: An invitation to cognitive science*. Cambridge, MA: MIT Press.

Patel, A. (2003). Language, music, syntax and the brain. *Nature Neuroscience*, *6*(7), 674–681.

Patel, A. D. (2007). *Music, language, and the brain*. Oxford: Oxford UP.

Peelen, M. V., & Caramazza, A. (2012). Conceptual object representations in human anterior temporal cortex. *Journal of Neuroscience*, *32*(45), 15728–15736.

Phillips, M. L., Young, A. W., Scott, S. K., Calder, A. J., Andrew, C., Giampietro, V., et al. (1998). Neural responses to facial and vocal expressions of fear and disgust. *Proceedings of the Royal Society B: Biological Sciences*, *265*(1408), 1809–1817.

**Price, C. J., Moore, C. J., Humphreys, G. W., & Wise, R. J. S. (1997). Segregating semantic from phonological processes during reading. *Journal of Cognitive Neuroscience*, *9*(6), 727–733.

Reiman, E. M., Lane, R. D., Ahern, G. L., Schwartz, G. E., Davidson, R. J., Friston, K. J., et al. (1997). Neuroanatomical correlates of externally and internally generated human emotion. *The American Journal of Psychiatry*, *154*(7), 918–925.

*Rogalsky, C., & Hickok, G. (2009). Selective attention to semantic and syntactic features modulates sentence processing networks in anterior temporal cortex. *Cerebral Cortex*, *19*, 786–796.

Rogalsky, C., & Hickok, G. (2011). The role of Broca's area in sentence comprehension. *Journal of Cognitive Neuroscience*, *23*(7) 1664–1680.

Rogalsky, C., Matchin, W., & Hickok, G. (2008). Broca's area, sentence comprehension, and working memory: An fMRI study. *Frontiers in Human Neuroscience*, *2*, 14.

*Rogalsky, C., Rong, F., Saberi, K., & Hickok, G. (2011). Functional anatomy of language and music perception: Temporal and structural factors investigated using functional magnetic resonance imaging. *Journal of Neuroscience*, *31*(10), 3843–3852.

Santi, A., & Grodzinsky, Y. (2012). Broca's area and sentence comprehension: A relationship parasitic on dependency, displacement or predictability. *Neuropsychologia*, *50*(5), 821–832.

Scott, S. K., Blank, C. C., Rosen, S., & Wise, R. J. S. (2000). Identification of a pathway for intelligible speech in the left temporal lobe. *Brain*, *123*, 2400–2406.

Shattuck-Hufnagel, S., & Turk, A. E. (1996). A prosody tutorial for investigators of auditory sentence processing. *The Journal of Psycholinguistic Research*, *25*(2), 193–247.

Shin, L. M., Dougherty, D. D., Orr, S. P., Pitman, R. K., Lasko, M., Macklin, M. L., et al. (2000). Activation of anterior paralimbic structures during guilt-related script-driven imagery. *Biological Psychiatry*, *48*(1), 43–50.

Simmons, W. K., & Martin, A. (2009). The anterior temporal lobes and the functional architecture of semantic memory. *Journal of the International Neuropsychological Society*, *15*(5), 645–649.

Skipper, L. M., Ross, L. A., & Olson, I. R. (2011). Sensory and semantic category subdivisions within the anterior temporal lobes. *Neuropsychologia*, *49*(12), 3419–3429.

*Snijders, T. M., Vosse, T., Kempen, G., Van Berkum, J. J., Petersson, K. M., & Hagoort, P. (2009). Retrieval and unification of syntactic structure in sentence comprehension: An fMRI study using word-category ambiguity. *Cerebral Cortex*, *19*, 1493–1503.

Spitsyna, G., Warren, J. E., Scott, S. K., Turkheimer, F. E., & Wise, R. J. (2006). Converging language streams in the human temporal lobe. *Journal of Neuroscience*, *26*(28), 7328–7336.

*Stowe, L. A., Broere, C. A., Paans, A. M., Wijers, A. A., Mulder, G., Vaalburg, W., et al. (1998). Localizing components of a complex task: Sentence processing and working memory. *NeuroReport*, *9*, 2995–2999.

*Stowe, L. A., Paans, A. M. J., Wijers, A. A., Zwarts, F., Mulder, G., & Vaalburg, W. (1999). Sentence comprehension and word repetition: A positron emission tomography investigation. *Psychophysiology*, *36* (6), 786–801.

Stromswold, K., Caplan, D., Alpert, N., & Rauch, S. (1996). Localization of syntactic comprehension by positron emission tomography. *Brain and Language*, *52*, 452–473.

**Thompson-Schill, S. L., Aguirre, G. K., D'Esposito, M., & Farah, M. J. (1999). A neural basis for category and modality specificity of semantic knowledge. *Neuropsychologia*, *37*(6), 671–676.

**Tomaszewki Farias, S., Harrington, G., Broomand, C., & Seyal, M. (2005). Differences in functional MR imaging activation patterns associated with confrontation naming and responsive naming. *American Journal of Neuroradiology*, *26*(10), 2492–2499.

Turkeltaub, P. E., Eickhoff, S. B., Laird, A. R., Fox, M., Wiener, M., & Fox, P. (2012). Minimizing within-experiment and within-group effects in activation likelihood estimation meta-analyses. *Human Brain Mapping*, *33*, 1–13.

*Vandenberghe, R., Nobre, A. C., & Price, C. J. (2002). The response of left temporal cortex to sentences. *Journal of Cognitive Neuroscience*, *14*, 550–560.

**Visser, M., Jeffries, E., & Lambon Ralph, M. A. (2010). Semantic processing in the anterior temporal lobes: A meta-analysis of the functional neuroimaging literature. *Journal of Cognitive Neuroscience*, *22*(6), 1083–1094.

Warrington, E. K. (1975). The selective impairment of semantic memory. *Quarterly Journal of Experimental Psychology*, *27*(4), 635–657.

Weintraub, S., Mesulam, M. M., & Kramer, L. (1981). Disturbances in prosody. A right-hemisphere contribution to language. *Archives of Neurology*, *38*(12), 742–744.

Wilson, S. M., DeMarco, A. T., Henry, M. L., Gesierich, B., Babiak, M., Mandelli, M. L., et al. (2014). What role does the anterior temporal lobe play in sentence-level processing? Neural correlates of syntactic processing in semantic variant primary progressive aphasia. *Journal of Cognitive Neuroscience*, *26*(5), 970–985.

Wilson, S. M., Galantucci, S., Tartaglia, M. C., & Gorno-Tempini, M. L. (2012). The neural basis of syntactic deficits in primary progressive aphasia. *Brain and Language*, *122*(3), 190–198.

Wilson, S. M., Galantucci, S., Tartaglia, M. C., Rising, K., Patterson, D. K., Henry, M. L., et al. (2011). Syntactic processing depends on dorsal language tracts. *Neuron*, *72*(2), 397–403.

Wong, C., & Gallate, J. (2012). The function of the anterior temporal lobe: A review of the empirical evidence. *Brain Research*, *1449*, 94–116.

Wulfeck, B. (1988). Grammaticality judgments and sentence comprehension in agrammatic aphasia. *Journal of Speech and Hearing Research*, *31*, 72–81.

*Xu, J., Kemeny, S., Park, G., Frattali, C., & Braun, A. (2005). Language in context: Emergent features of word, sentence, and narrative comprehension. *NeuroImage*, *25*(3), 1002–1015.

Zahn, R., Moll, J., Krueger, F., Huey, E. D., Garrido, G., & Grafman, J. (2007). Social concepts are represented in the superior anterior temporal cortex. *Proceedings of the National Academy of Sciences of the United States of America*, *104*(15), 6430–6435.

CHAPTER

48

Neural Systems Underlying the Processing of Complex Sentences

Lars Meyer and Angela D. Friederici

Deparment of Neuropsychology, Max Planck Institute for Human Cognitive and Brain Sciences, Leipzig, Germany

48.1 INTRODUCTION

The functional neuroanatomy of sentence processing is a major and, at the same time, one of the most enduring topics of the cognitive neuropsychology of speech and language processing. The present chapter serves three purposes. First, we define the cognitive processes necessitated by the processing of the two sentence types most extensively studied by neuroimaging research: noncanonical and embedded sentences. To this end, we review extensive behavioral evidence for the involved cognitive processes. In the second step, we perform a meta-analysis on all neuroimaging studies that directly contrasted either canonical with noncanonical or nonembedded with embedded sentences, identifying the brain regions that most likely activate during the processing of noncanonical and embedded sentences. Based on our meta-analysis, we explain why the functional neuroanatomy of complex sentence processing involves two core areas: the left inferior frontal cortex and the left middle and superior posterior temporal gyri. Finally, based on the resemblance in brain activations for noncanonical and embedded sentences in our meta-analysis, we discuss a possible common definition of the processing difficulty associated with complex sentences.

48.2 WHY ARE WORD-ORDER DEVIATIONS DIFFICULT TO PROCESS?

The increased processing difficulty of sentences with word orders deviating from the canonical order is a classical finding in psycholinguistics and neurolinguistics across languages—such as Chinese (Hsiao & Gibson, 2003), Dutch (Frazier, 1987), English (King & Just, 1991), Finnish (Hyönä & Hujanen, 1997), French (Holmes & O'Regan, 1981), German (Friederici, Fiebach, Schlesewsky, Bornkessel, & von Cramon, 2006), Hindi (Vasishth & Lewis, 2006), Hungarian (MacWhinney & Pléh, 1988), Japanese (Mazuka, Itoh, & Kondo, 2002), and Kaqchikel (Ohta, Koizumi, & Sakai, 2014). Theoretical linguistics has long derived the increased difficulty of word-order deviations from their increased level of hierarchy and the according necessity to reorder a noncanonical order of object and subject into a canonical order of subject and object for the integration with their verb (Bever, 1970; Chomsky, 1955; Fodor, 1978; see Figure 48.1).

Psycholinguistic research inspired by this theoretical account has accumulated strong evidence that reordering can be translated into real-time processing reflections. Seminal work found that grammaticality judgment slows not only for object-initial sentences but also when transitive verbs are not followed by a prepositional phrase instead of their object, suggesting that objects are expected at a fixed position (Clifton, Frazier, & Connine, 1984; Holmes, 1987). In line with this interpretation, Trueswell, Tanenhaus, and Kello (1993) observed that in English, postverbal nouns are preferentially interpreted as objects—even if they are the subjects of a subsequent sentence. Complementary evidence is provided by cross-modal priming work. Tanenhaus, Carlson, and Seidenberg (1985) presented words that rhymed with an object that occurred early in the sentence at the canonical object position of an English sentence. Processing of the rhyming words was facilitated compared with control words, suggesting that the remote object had been retrieved at the canonical object position—a finding that was replicated numerous times (for review, see Nicol, Fodor, & Swinney, 1994). Evidence that such priming effects are

Neurobiology of Language. DOI: http://dx.doi.org/10.1016/B978-0-12-407794-2.00048-1

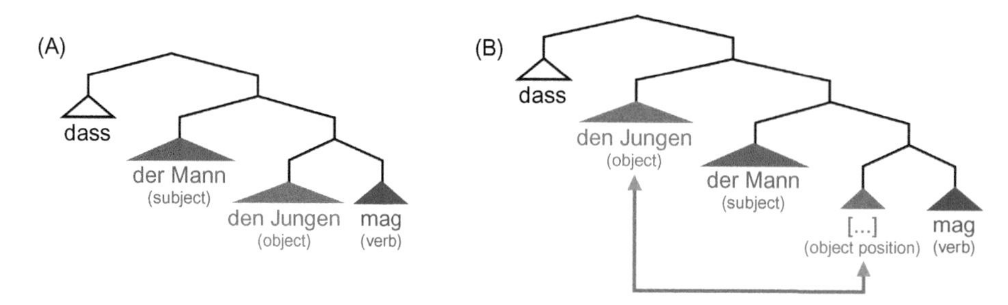

FIGURE 48.1 Simplified examples of (A) a canonical and (B) a noncanonical German subordinate clause. While the order of the subject (red) preceding the object (green) is the canonical order in German sentences, the order of the object (green) preceding the subject (red) is the noncanonical order in German sentences. Note also that the noncanonical sentence involves an additional level of hierarchy relative to the canonical sentence, and that the dependency between the object and its verb in the noncanonical structure crosses the subject (indicated by the arrow). Both sentences have the same meaning, translating to *that the man likes the boy*.

restricted to subject positions for subjects and to object positions for objects was presented by Osterhout and Swinney (1993). Besides the cross-modal priming studies, the evidence for fixed positions of subject and object is extended by syntactic interference studies. The work by Van Dyke (2007) used sentences with additional nouns whose syntactic features overlapped with those of a subject that occurred early in the sentence and needed integration with its verb late in the sentence. The results showed that processing at the verb's canonical subject position was exacerbated by feature overlap.

As a recurrent motif, an alternative explanation to the reordering mechanism is put forward time and again (for a recent instance, see Rogalsky & Hickok, 2010). This explanation is based on the fact that English noncanonical sentences are not only noncanonical but also place the object further away from its object position than the subject from its subject position. The time interval between subject or object and verb impedes processing (Behaghel, 1932; Frazier, Clifton, & Randall, 1983; Yngwe, 1960). This is most likely because of the decay of the subject or object in working memory prior to verb encounter (Cowper, 1976; Lewis, Vasishth, & Van Dyke, 2006) and collaterally induced working memory storage and rehearsal demands. Still, while some of the English results may thus be ambiguous between a reordering and a working memory explanation, this is not the case for all languages that show noncanonicity effects. In German, recent work has compared canonical and noncanonical orders, keeping the time interval between either subject or object and their verb constant, finding that noncanonicity still impedes processing over canonicity (Friederici et al., 2006; Meyer, Obleser, Anwander, & Friederici, 2012). In summary, the available evidence supports the notion that sentences with noncanonical orders are difficult to process because they necessitate reordering into the canonical order.

48.3 WHY ARE EMBEDDED SENTENCES DIFFICULT TO PROCESS?

When a subordinate clause is embedded into a superordinate clause, the resulting sentence has a hierarchical structure, and its processing becomes difficult (Figure 48.2). Such effects have been reported for Dutch (Bach, Brown, & Marslen-Wilson, 1986; Kaan & Vasić, 2004), English (Miller & Isard, 1964), German (Bach et al., 1986), Hindi (Vasishth, 2003), Japanese (Babyonyshev & Gibson, 1999), and Russian (Fedorenko, Babyonyshev, & Gibson, 2004). The associated difficulty can have drastic consequences. When one of the three required verbs of a double-embedded sentence is experimentally removed, participants do not even realize the absence (Gibson & Thomas, 1999; Gimenes, Rigalleau, & Gaonach, 2009).

One of the main cognitive challenges associated with the processing of embedded sentences lies in their hierarchy of multiple parallel syntactic dependencies between multiple subjects and objects and their respective verbs (Bever, 1974; Gibson, 1998; Makuuchi, Bahlmann, Anwander, & Friederici, 2009). Processing of an embedded sentence temporarily requires the parallel storage of two or more subjects or objects until their verbs occur. Although embedded sentences whose multiple subjects or objects are highly distinct along syntactic lines are processed relatively easily, recent work has reported that an overlap of syntactic features between the multiple subjects or objects increases processing difficulty, likely as a result of memory interference (Badecker & Kuminiak, 2007; Gibson, 1998; Glaser, Martin, Van Dyke, Hamilton, & Tan, 2013; Gordon, Hendrick, & Johnson, 2001; Lewis, 1996; Van Dyke & McElree, 2006). Initial behavioral evidence for this explanation has been found in Japanese. Babyonyshev and Gibson (1999) presented embedded sentences in which the form of the case markings of three subjects matched and contrasted

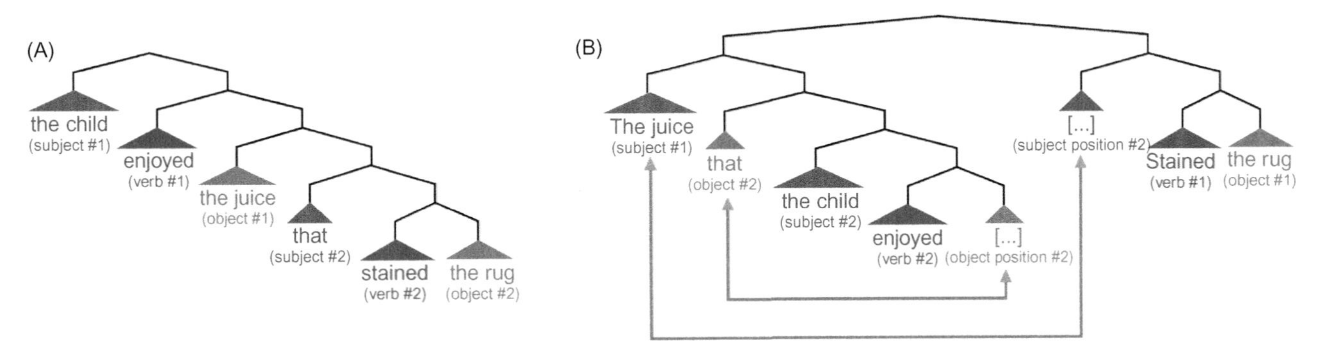

FIGURE 48.2 Simplified examples of (A) a nonembedded and (B) an embedded English sentence. In the nonembedded sentence, all subjects (red) and objects (green) are in their canonical positions and can be linked to their verbs (blue) directly, whereas in the embedded sentence, both the subject (red) of the superordinate clause and the object (green) of the subordinate clause form a dependency with their respective canonical positions (indicated by the arrows). Note that the embedded sentence is hierarchically more complex, and that the dependency between the superordinate-clause subject and its canonical position spans the subject and object of the subordinate clause. Both sentences have the same meaning. *Adapted from Caplan et al., 2000, with permission from the publisher.*

these with embedded sentences in which only two subjects' case markings matched. The result was that the sentences with three matching subjects were more difficult to process, pointing to interference-induced processing difficulty. Similar evidence has been found in Russian. Fedorenko et al. (2004) used embedded sentences in which the case marking of two objects either matched or did not match, finding that sentences with matching objects increased processing demands. In summary, there is good behavioral evidence that the difficulty of embeddings results from their hierarchical nature and the involvement of multiple dependencies between multiple subjects or objects and their respective multiple verbs.

48.4 WHICH BRAIN REGIONS ARE INVOLVED IN PROCESSING COMPLEX SENTENCES?

The functional neuroanatomy of noncanonicity and embedding has been extensively investigated by neuroimaging research. For this review, we conducted an activation likelihood estimation (ALE) meta-analysis on the left-hemispheric peak activation coordinates from all published studies that tested for a main effect of either noncanonicity or embedding, did not report an interaction with other experimental factors, and did report Talairach (Talairach & Tournoux, 1988) or Montreal Neurological Institute (MNI) coordinates (Table 48.1 and Figure 48.3A). We first converted all non-MNI coordinates into MNI space (Brett, Johnsrude, & Owen, 2002). The ALE analysis then generated a smoothed brain map from each study's peak coordinate map, applying an individual smoothing kernel whose full width at half maximum was calculated from each study's sample size (Eickhoff et al., 2009). The overlap of the individual study maps results in a voxel-wise ALE value (Turkeltaub, Eden, Jones, & Zeffiro, 2002; Turkeltaub et al., 2012), which is then tested against the null distribution ($P < 0.001$). The resulting whole-brain significance map is corrected for multiple comparisons using a cluster statistic (Eickhoff, Bzdok, Laird, Kurth, & Fox, 2012), which we ran at 1,000 permutations and a conservative cluster-level threshold of $P < 0.001$. The result of the ALE procedure (Table 48.2 and Figure 48.3B) showed a striking likelihood for noncanonicity and embedding to activate two core regions: Brodmann area (BA) 44 of the left inferior frontal gyrus (IFG), MNI peak coordinates at maximum $X = -52$, $Y = 10$, $Z = 16$; and, less consistently reported across studies, BA 22 of the middle and superior temporal gyri (MTG/STG), MNI peak coordinates at maximum $X = -54$, $Y = -52$, $Z = -4$ (Figure 48.3 and Table 48.2).

Representative of the coordinate overlap in BA 44 among studies are seven functional magnetic resonance imaging (fMRI) studies from languages that allow for testing the factor syntactic complexity while maximally controlling for confounding working memory storage and rehearsal demands—that is, keeping the temporal interval between either the subject or the object and their verb constant across experimental conditions. These are languages as diverse as German (Friederici et al., 2006; Meyer et al., 2012; Obleser, Meyer, & Friederici, 2011; Röder, Stock, Neville, Bien, & Rösler, 2002), Hebrew (Ben-Shachar, Hendler, Kahn, Ben-Bashat, & Grodzinsky, 2003), and Japanese (Kim et al., 2009; Kinno, Kawamura, Shioda, & Sakai, 2008). First, despite the linguistic heterogeneity of these languages, all these studies used either a factorial or a parametric design comparing sentences involving

TABLE 48.1 Overview of all Left-Hemispheric Peak Activation Coordinates from Main Effects of Noncanonicity or Embedding in Published Neuroimaging Studies on Reordering and Embedding

Study	Contrast[a]	Coordinate (MNI)			Coordinate (Talairach)			Region[b]
		X	Y	Z	X	Y	Z	
Amici et al. (2007)		−34	39	6	−34	38	4	IFG/IFS
Bahlmann et al. (2007)	OF > SF	−40	−52	28	−40	−49	28	SMG
		−44	24	28	−44	25	25	IFG
Ben-Shachar et al. (2003)	OR > SC	−45	23	9	−45	23	7	IFG
		−54	−44	5	−53	−42	7	pSTS
Ben-Shachar et al. (2004) (experiment 1)	OF > SF	−43	21	9	−43	21	7	IFG
		−41	10	30	−41	11	27	vPCS
		−57	−44	5	−56	−42	7	pSTS
		−55	−19	10	−54	−18	10	HC
Ben-Shachar et al. (2004) (experiment 2)	OWh > SWh	−44	21	10	−44	21	8	IFG
		−45	7	28	−45	8	25	vPCS
		−56	−43	4	−55	−41	6	pSTS
Bornkessel et al. (2005)	OF > SF	−43	13	20	−43	14	18	IFG
		−53	−45	17	−52	−43	18	pSTG
		−47	−61	23	−47	−58	24	pSTS
		−35	3	38	−35	5	35	IFJ
		−35	−4	54	−35	−1	50	vPMC
		−28	−62	38	−28	−58	38	IPS
		−40	−60	−3	−40	−58	0	OTS
Bornkessel-Schlesewsky, Schlesewsky, and von Cramon (2009)	OF > SF	−54	11	6	−53	11	5	IFG
		−34	23	7	−34	23	5	AI
		−40	3	35	−40	5	32	IFJ
		−7	22	46	−7	23	41	pre-SMA
		−32	−62	41	−32	−58	41	IPS
		−14	0	20	−14	1	18	CN
Caplan et al. (2000)	CE > RB	−46	37	6	−46	36	4	BA
		−14	−21	3	−14	−20	4	Th
		−10	−39	41	−10	−36	40	CG
		−44	5	17	−44	6	15	IFG
Constable et al. (2004)	OR > SR	−52	−60	0	−51	−58	3	PP
		−49	11	15	−49	11	13	IF
		−36	−67	30	−36	−64	31	ST/P
		−36	2	50	−36	4	46	pre-M
		−2	4	36	−2	6	33	AC
		−3	−25	15	−3	−24	15	Th
Dapretto and Bookheimer (1999)	P/pre-PP > A/post-PP	−53	9	31	−52	10	28	IFG
		−40	30	17	−40	30	14	IFG
		−59	−60	12	−58	−58	14	STG
		−61	−38	15	−60	−36	16	TTG

(Continued)

TABLE 48.1 (Continued)

Study	Contrast[a]	Coordinate (MNI)			Coordinate (Talairach)			Region[b]
		X	Y	Z	X	Y	Z	
		−48	21	−18	−48	20	−16	TP
		−42	−60	38	−42	−56	38	SMG
Friederici et al. (2006)	OF > SF	−50	10	5	−49	10	4	PO/IFG
Grewe et al. (2005)	OF > SF	−32	20	4	−32	20	3	FO/AI
		−53	14	17	−52	14	15	IFG
		−2	31	34	−2	32	30	pre-SMA
		−38	6	42	−38	8	38	IFJ
Kim et al. (2009)	OF > SF	−42	−3	54	−42	0	50	dPFC
		−55	26	23	−54	26	20	IFG
Kinno et al. (2008) (comparison 1)	OF > SF	−39	0	45	−39	2	41	lPMC
		−52	21	21	−51	21	18	IFG
		−54	−54	3	−53	−52	5	pSTG/MTG
Kinno et al. (2008) (comparison 2)	(OF > SF) > (SF > CS)	−51	21	18	−50	21	16	IFG
		−51	−51	3	−50	−49	5	pSTG/MTG
Kinno et al. (2008) (comparison 3)	P > A	−48	24	21	−48	24	18	IFG
Makuuchi et al. (2009)	CE > SF	−45	6	24	−45	7	22	PO
Meyer et al. (2012)	OF > SF	−54	10	18	−53	11	16	IFG
Michael et al. (2001) (visual)	OR > CA	−53	−37	3	−52	−36	5	T
		−42	10	29	−42	11	26	IFG
Michael et al. (2001) (auditory)	OR > CA	−53	−30	4	−52	−29	5	T
		−43	14	27	−43	15	24	IFG
Newman et al. (2010) (phase 1)	OR > CA	−40	14	24	−40	15	21	IFG
		−58	−36	2	−57	−35	4	T
Newman et al. (2010) (phase 2)	OR > CA	−54	8	16	−53	8	14	IFG
Obleser et al. (2011) (experiment 1)	OF > SF	−50	16	−20	−50	15	−18	STG
		−64	−54	10	−63	−52	12	IFG
		−52	12	14	−51	12	12	STS
Obleser et al. (2011) (experiment 2)	OF > SF	−48	10	18	−48	11	16	IFG
Santi and Grodzinsky (2010) (comparison 1)	OF > SF	−48	19	17	−48	19	15	IFG
		−41	9	34	−41	10	31	IFG/IPCG
		−53	−35	0	−52	−34	2	STG
Santi and Grodzinsky (2010) (comparison 2)	CE > RB	−41	9	34	−41	10	31	IFG/IPCG
		−53	−35	0	−52	−34	2	STG
Stromswold, Caplan, Alpert, and Rauch (1996)	CE > RB	−47	10	5	−47	10	4	PO

[a] *A, active; CA, conjoined active; CE, center-embedded; CS, conjoined-subject; OF, object-first; OR, object-relative; OWh, object-wh; P, passive; post-PP, postposed-prepositional-phrase; pre-PP, preposed-prepositional-phrase; RB, right-branching; SC, sentential-complement; SF, subject-first; SR, subject-relative; SWh, subject-wh.*

[b] *AC, anterior cingulate; AI, anterior insula; BA, Broca's area; CG, cingulate gyrus; CN, caudate nucleus; dPFC, dorsal prefrontal cortex; FO, frontal operculum; HC, Heschl's cortex; IF, inferior frontal; IFG, inferior frontal gyrus; IPCG, inferior precentral gyrus; IFJ, inferior frontal junction; IPS, intraparietal sulcus; lPMC, lateral premotor cortex; OTS, occipito-temporo sulcus; PO, pars opercularis; PP, posterior parietal; pre-M, premotor; pre-SMA, pre-supplementary-motor area; pSTG, posterior superior temporal gyrus; MTG, middle temporal gyrus; pSTS, posterior superior temporal sulcus; SMG, supramarginal gyrus; ST/P, superior temporal/parietal; STS, superior temporal sulcus; T, temporal; Th, Thalamus; TP, temporal pole; TTG, transverse temporal gyrus; vPCS, ventral precentral gyrus; vPMC, ventral premotor cortex.*

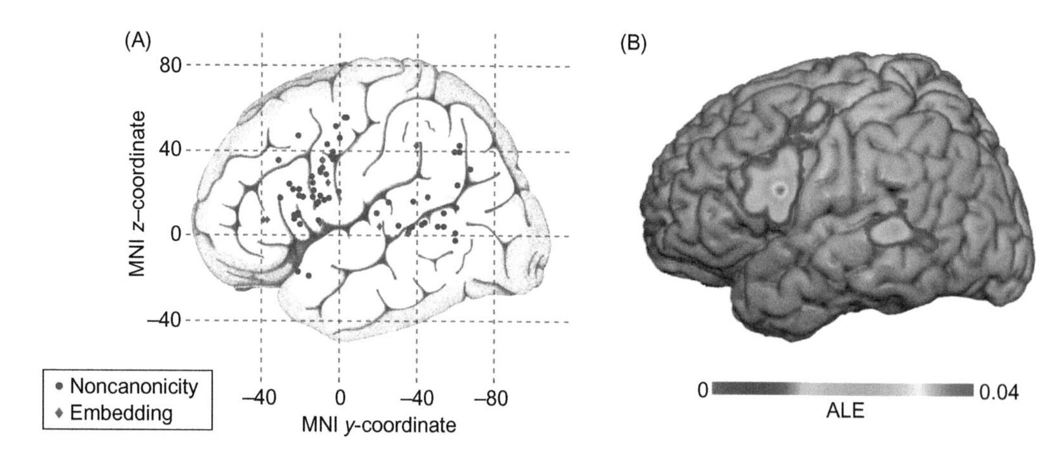

FIGURE 48.3 (A) Overview of all left-hemispheric peak activation coordinates from main effects of noncanonicity or embedding in published neuroimaging studies on noncanonicity (blue) and embedding (red). (B) Results of the ALE analysis (cluster-level $P < 0.001$; voxel-level $P < 0.001$; 1,000 permutations) across the plotted coordinates, rendered onto the Colin 27 template brain (Holmes et al., 1998). It is visible that two major regions are most likely activated during complex sentence processing, one in BA 44 of the IFG and one in the SMG/STG.

TABLE 48.2 Overview of Significant ALE Clusters Across Peak Coordinates from Neuroimaging Studies on Reordering and Embedding (Cluster-Level $p < 0.001$; Voxel-Level $p < 0.001$; 1,000 Permutations); Coordinates Are in MNI Space

	Coordinate (center)			Coordinate (peak)				
Anatomical label	*X*	*Y*	*Z*	*X*	*Y*	*Z*	**Volume (mm³)**	**ALE (extremum)**
Inferior frontal gyrus (BA 44)	− 45	12	23	−52	10	16	13,128	0.037
Inferior frontal gyrus (BA 9)				−44	8	28		0.025
Inferior frontal gyrus (BA 45)				−44	22	10		0.021
Insula (BA 13)				−34	22	6		0.015
Middle frontal gyrus (BA 6)				−38	0	50		0.015
Middle temporal gyrus (BA 22)	− 54	− 42	4	−54	−42	4	3,184	0.024
Middle temporal gyrus (BA 37)				−52	−58	2		0.011
Superior temporal gyrus (BA 22)				−54	−44	16		0.008

canonical and noncanonical orders of subject and objects. Note also that all of these studies rigorously controlled their experimental paradigms for the confounding influence of working memory storage and rehearsal demands, which is crucial because BA 44 activity has been observed for working memory tasks outside of the sentence processing domain as well (Gerton et al., 2004; Paulesu, Frith, & Frackowiak, 1993). Conflicting working memory demands may render ambiguous some of the results of prior noncanonicity studies conducted in English (Ben-Shachar, Palti, & Grodzinsky, 2004; Just, Carpenter, Keller, Eddy, & Thulborn, 1996; Rogalsky, Matchin, & Hickok, 2008), where noncanonicity induced a collateral increase in working memory storage and rehearsal demands. However, the credibility of the results from the nonconfounded paradigms is increased by a line of clinical research that has demonstrated that reordering abilities can be impaired independently of verbal working memory as tested by standard working memory tests (Caplan & Waters, 1999; Martin & Romani, 1994). Although collateral working memory demands cannot be fully ruled out as a confound in the three imaging studies that compared embedded and nonembedded sentences (Caplan, Alpert, Waters, & Olivieri, 2000; Makuuchi et al., 2009; Santi & Grodzinsky, 2010), an explanation of the observed IFG activations by working memory demands is unlikely. First, the study by Rogalsky et al. (2008) on English crossed a canonicity manipulation with an overt rehearsal task, but it did not find activation in BA 44 under concurrent articulation to selectively increase during the processing of noncanonical sentences. Rather, BA 44 activation for both canonical and noncanonical sentences increased under concurrent articulation, suggesting that sentence comprehension may interfere with a concurrent articulation task in a canonicity-insensitive manner. Second, the recent study by Ohta et al. (2014) on Kaqchikel reports BA 44 activity to increase during the processing of noncanonical sentences even when these exhibit a lower working memory storage and rehearsal demand than their canonical counterparts. We could not include the study by Ohta et al. (2014) in our ALE analysis because no activation coordinates were reported; only anatomical labels were reported.

Is this strong position on the involvement of BA 44 in the processing of noncanonicity and embedding warranted by causal, that is, clinical, evidence? A strong relationship between stroke lesions involving BA 44 and processing deficits on noncanonical sentences has been proposed (Grodzinsky, 2001), but stroke lesions are relatively large, associated comprehension patterns are diverse, and a number of methodological concerns about the studies have been raised (Bastiaanse & van Zonneveld, 2006). However, voxel-based lesion-symptom mapping work with glioma patients, lesion patients, and nonfluent primary progressive aphasics reported a processing decline on noncanonical sentences after damage to the left IFG. Two studies involving Japanese glioma patients by Kinno et al. (2009), Kinno, Ohta, Muragaki, Maruyama, and Sakai (2014) indicated that glioma in BA 44, but also in superior regions, predicts decreased performance on noncanonical sentences. In line with this, a study involving Icelandic stroke patients by Magnusdottir et al. (2013) indicates that damage to the IFG, among other left-hemispheric regions, is significantly associated with decreased performance on noncanonical sentences. Additionally, in recent work involving nonfluent primary progressive aphasics, Wilson et al. (2010) further reduced the localization variance, reporting both tissue degeneration in the left posterior IFG and a focal lack of activation increase in this region for noncanonical sentences (for review, see Leyton et al., 2011). Thus, the patient data do support the primary role of BA 44 in the processing of noncanonicity and embedding.

In addition to BA 44, the present ALE analysis isolated the MTG/STG as a second region to most likely activate during the processing of noncanonicity and embedding (Bahlmann, Rodriguez-Fornells, Rotte, & Munte, 2007; Ben-Shachar et al., 2003, 2004; Bornkessel, Zysset, Friederici, von Cramon, & Schlesewsky, 2005; Constable et al., 2004; Dapretto & Bookheimer, 1999; Friederici et al., 2006; Kinno et al., 2008; Michael, Keller, Carpenter, & Just, 2001; Newman, Ikuta, & Burns, 2010; Obleser et al., 2011; Santi & Grodzinsky, 2010). According to a meta-study by Vigneau et al. (2006), the MTG/STG is involved in semantic integration across linguistic tasks. In the case of noncanonicity and embedding, increasing brain activity in a semantic integration region may result from increasing demands for sentence-level semantic information, most likely in accessing verb–argument structure representations (Friederici, Rueschemeyer, Hahne, & Fiebach, 2003). This interpretation is plausible considering a study by Shetreet, Palti, Friedmann, and Hadar (2007), who assessed the brain activation associated with increasing numbers of possible verb–argument structures in which a verb could be involved. The results show that the number of possible verb–argument structures associated with a verb parametrically increases activity in the MTG/STG region, in close proximity to the MTG/STG ALE maximum in the current meta-analysis. Our comparably agnostic position regarding the involvement of the MTG/STG with respect to the primary mechanisms involved in the processing of noncanonicity and embedding is in line with the studies enrolled in our meta-analysis. Although all studies report at least one inferior frontal activation, not all studies report posterior temporal activations.

48.5 WHAT DO WORD-ORDER DEVIATIONS AND EMBEDDING HAVE IN COMMON?

Our meta-analysis suggests that the processing difficulty associated with the processing of noncanonicity and embedding rely on a similar brain substrate. But what is the similarity between the involved cognitive processes? First, the difficulty associated with noncanonicity lies in a mismatch between the incoming order of object and the underlying canonical order of the subject and object positions at their verb. The resolution of this mismatch requires an assumed process that we defined as reordering. This process, independent of working memory storage and rehearsal, needs to ensure that subject and object of a verb are retrieved in the canonical order once the verb is reached. Second, the difficulty associated with embedded sentences lies in the multiple dependencies between multiple subjects and objects and their multiple verbs. This requires a process independent of working memory storage and rehearsal to keep items distinct, enabling the retrieval of each item only at their appropriate canonical position in relation to their appropriate verb.

Word-order deviations and embeddings are similar along two isomorphic lines. Both sentence types involve a syntactic structure that is more complex than the structure of their canonical or nonembedded counterparts—in formal terms—and both noncanonical and embedded sentences involve an additional level of hierarchy in the syntactic tree. However, both sentence types involve the temporary storage of at least a first

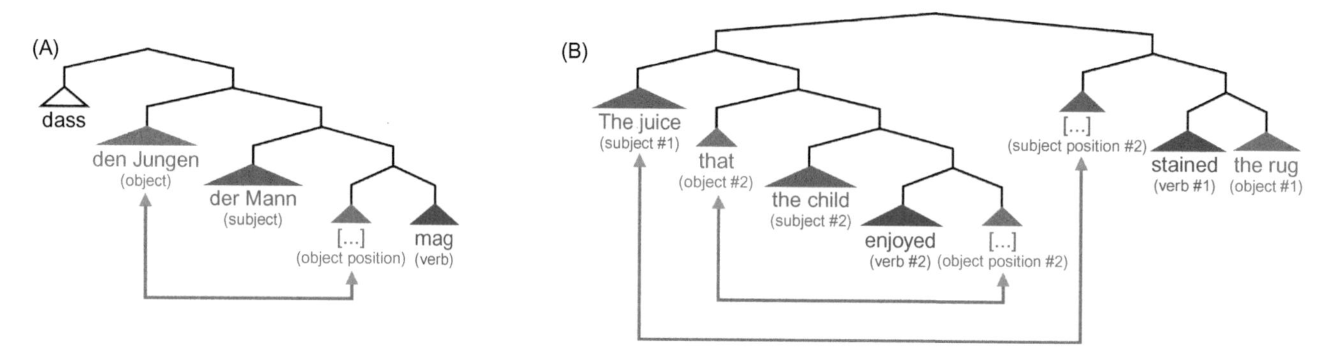

FIGURE 48.4 Comparison of the sentence structures associated with (A) a German noncanonical subordinate clause and (B) an English embedded sentence. The remote positioning of the object in the noncanonical sentence and the remote positioning of the superordinate subject and subordinate object in the embedded sentence result in an additional level of hierarchy (see Figure 48.1 and 48.2 and compare the number of nodes/triangles for sentence types (A) and (B), respectively) and a dependency that crosses at least either a second subject or object (consult Figure 48.1 and 48.2 and compare the number of arrows for sentence types (A) and (B), respectively).

noun phrase (subject or object) across at least a second noun phrase (subject or object) that may interfere—in functional terms—and noncanonical and embedded sentences necessitate the avoidance of an erroneous premature subject or object retrieval (Figure 48.4).

We consider that the establishment of syntactic hierarchy and the possible interference of elements in working memory may contribute to the processing difficulty of noncanonical and embedded sentences. This interpretation is in line with both classical assumptions of syntactic theory (Chomsky, 1955), more recent proposals of psycholinguistic sentence-processing research (Gibson et al., 2013; Lewis, 1996), and a recent fMRI study by Glaser et al. (2013) that reported BA 44 in the IFG activates more strongly for the processing of embeddings whose relative-clause subject interfered with the matrix-clause subject as compared with noninterfering embeddings. Thus, the available neurobiological literature suggests that the main function of BA 44 during sentence processing is to establish the hierarchical dependencies between subjects and objects and their verbs. There is now initial evidence for the neural basis of the second processing aspect, that is, the distinctive temporary storage and retrieval of possibly interfering subjects and objects from working memory; the interference account needs further neurolinguistic substantiation.

48.6 SUMMARY

The current chapter has outlined the cognitive processes involved in the processing of noncanonical and embedded sentences, for which cross-linguistic psycholinguistic research has reported increased processing difficulty relative to their canonical or nonembedded counterparts, respectively. Likely, the increased processing difficulty of noncanonicity results from the need to reorder the noncanonical order of object and subject into the canonical order of subject and object at the verb; the increased processing difficulty of embedding results from its hierarchical nature and the need to distinctly track multiple dependencies between multiple subjects or objects and their respective multiple verbs. Our ALE meta-analysis provides strong evidence that both processes rely on a left-hemispheric brain substrate consisting of BA 44 in the IFG and the MTG/STG. Because only inferior frontal regions are found to be active in all studies, including in our meta-analysis, we suggest that BA 44 is primarily involved in the processing of noncanonicity and embedding. From our literature review and our meta-analysis, we synthesized the common underlying mechanism of BA 44 in complex sentence processing to lie in the tracking of the multiple hierarchical dependencies between subjects and objects and their verbs, avoiding their interference to ensure that subjects and objects are only retrieved from working memory at their appropriate positions.

References

Amici, S., Brambati, S. M., Wilkins, D. P., Ogar, J., Dronkers, N. L., Miller, B. L., et al. (2007). Anatomical correlates of sentence comprehension and verbal working memory in neurodegenerative disease. *Journal of Neuroscience*, *27*, 6282–6290.

Babyonyshev, M., & Gibson, E. (1999). The complexity of nested structures in Japanese. *Language*, *75*, 423–450.

Bach, E., Brown, C., & Marslen-Wilson, W. (1986). Crossed and nested dependencies in German and Dutch: A psycholinguistic study. *Language and Cognitive Processes*, *1*, 249–262.

Badecker, W., & Kuminiak, F. (2007). Morphology, agreement and working memory retrieval in sentence production: Evidence from gender and case in Slovak. *Journal of Memory and Language*, *56*, 65–85.

Bahlmann, J., Rodriguez-Fornells, A., Rotte, M., & Munte, T. F. (2007). An fMRI study of canonical and noncanonical word order in German. *Human Brain Mapping*, *28*, 940–949.

Bastiaanse, R., & van Zonneveld, R. (2006). Comprehension of passives in Broca's aphasia. *Brain and Language, 96*, 135–142.

Behaghel, O. (1932). *Deutsche Syntax: Eine geschichtliche Darstellung. Band IV. Wortstellung, Periodenbau.* Heidelberg: Winter.

Ben-Shachar, M., Hendler, T., Kahn, I., Ben-Bashat, D., & Grodzinsky, Y. (2003). The neural reality of syntactic transformations. *Psychological Science, 14*, 433–440.

Ben-Shachar, M., Palti, D., & Grodzinsky, Y. (2004). Neural correlates of syntactic movement: Converging evidence from two fMRI experiments. *NeuroImage, 21*, 1320–1336.

Bever, T. G. (1970). The cognitive basis for linguistic structures. In R. Hayes (Ed.), *Cognition and language development* (pp. 279–362). New York, NY: Wiley & Sons, Inc.

Bever, T. G. (1974). The ascent of the specious, or there's a lot we don't know about mirrors. In D. Cohen (Ed.), *Explaining linguistic phenomena* (pp. 173–200). Washington, DC: Hemisphere.

Bornkessel, I., Zysset, S., Friederici, A. D., von Cramon, D., & Schlesewsky, M. (2005). Who did what to whom? The neural basis of argument hierarchies during language comprehension. *NeuroImage, 26*, 221–233.

Bornkessel-Schlesewsky, I., Schlesewsky, M., & von Cramon, D. Y. (2009). Word order and Broca's region: Evidence for a supra-syntactic perspective. *Brain and Language, 111*, 125–139.

Brett, M., Johnsrude, I. S., & Owen, A. M. (2002). The problem of functional localization in the human brain. *Nature Reviews Neuroscience, 3*, 243–249.

Caplan, D., Alpert, N., Waters, G., & Olivieri, A. (2000). Activation of Broca's area by syntactic processing under conditions of concurrent articulation. *Human Brain Mapping, 9*, 65–71.

Caplan, D., & Waters, G. (1999). Verbal working memory and sentence comprehension. *Behavioral and Brain Sciences, 22*, 77–94.

Chomsky, N. (1955). *The logical structure of linguistic theory*. New York, NY: Plenum Press.

Clifton, C., Frazier, L., & Connine, C. (1984). Lexical expectations in sentence comprehension. *Journal of Verbal Learning and Verbal Behavior, 23*, 696–708.

Constable, R., Pugh, K., Berroya, E., Mencl, W., Westerveld, M., Ni, W., et al. (2004). Sentence complexity and input modality effects in sentence comprehension: An fMRI study. *NeuroImage, 22*, 11–21.

Cowper, E. A. (1976). *Constraints on sentence complexity: A model for syntactic processing*. PhD Thesis, Brown University, Providence, RI.

Dapretto, M., & Bookheimer, S. Y. (1999). Form and content: Dissociating syntax and semantics in sentence comprehension. *Neuron, 24*, 427–432.

Eickhoff, S. B., Bzdok, D., Laird, A. R., Kurth, F., & Fox, P. T. (2012). Activation likelihood estimation meta-analysis revisited. *NeuroImage, 59*, 2349–2361.

Eickhoff, S. B., Laird, A. R., Grefkes, C., Wang, L. E., Zilles, K., & Fox, P. T. (2009). Coordinate-based activation likelihood estimation meta-analysis of neuroimaging data: A random-effects approach based on empirical estimates of spatial uncertainty. *Human Brain Mapping, 30*, 2907–2926.

Fedorenko, E., Babyonyshev, M., & Gibson, E. (2004). The nature of case interference in on-line sentence processing in Russian. *NELS 34 conference proceedings*, Stony Brook, New York.

Fodor, J. (1978). Parsing strategies and constraints on transformations. *Linguistic Inquiry, 9*, 427–473.

Frazier, L. (1987). Syntactic processing: Evidence from Dutch. *Natural Language and Linguistic Theory, 5*, 519–559.

Frazier, L., Clifton, C., & Randall, J. (1983). Filling gaps: Decision principles and structure in sentence comprehension. *Cognition, 13*, 187–222.

Friederici, A. D., Fiebach, C., Schlesewsky, M., Bornkessel, I., & von Cramon, D. Y. (2006). Processing linguistic complexity and grammaticality in the left frontal cortex. *Cerebral Cortex, 16*, 1709–1717.

Friederici, A. D., Rueschemeyer, S.-A., Hahne, A., & Fiebach, C. (2003). The role of left inferior frontal and superior temporal cortex in sentence comprehension: Localizing syntactic and semantic processes. *Cerebral Cortex, 13*, 170–177.

Gerton, B. K., Brown, T. T., Meyer-Lindenberg, A., Kohn, P., Holt, J. L., Olsen, R. K., et al. (2004). Shared and distinct neurophysiological components of the digits forward and backward tasks as revealed by functional neuroimaging. *Neuropsychologia, 42*, 1781–1787.

Gibson, E. (1998). Linguistic complexity: Locality of syntactic dependencies. *Cognition, 68*, 1–76.

Gibson, E., Piantadosi, S. T., Brink, K., Bergen, L., Lim, E., & Saxe, R. (2013). A noisy-channel account of crosslinguistic word-order variation. *Psychological Science, 24*, 1079–1088.

Gibson, E., & Thomas, J. (1999). Memory limitations and structural forgetting: The perception of complex ungrammatical sentences as grammatical. *Language and Cognitive Processes, 14*, 225–248.

Gimenes, M., Rigalleau, F., & Gaonach, D. (2009). When a missing verb makes a French sentence more acceptable. *Language and Cognitive Processes, 24*, 440–449.

Glaser, Y. G., Martin, R. C., Van Dyke, J. A., Hamilton, A. C., & Tan, Y. (2013). Neural basis of semantic and syntactic interference in sentence comprehension. *Brain and Language, 126*, 314–326.

Gordon, P., Hendrick, R., & Johnson, M. (2001). Memory interference during language processing. *Journal of Experimental Psychology: Learning, Memory and Cognition, 27*, 1411–1423.

Grewe, T., Bornkessel, I., Zysset, S., Wiese, R., von Cramon, D. Y., & Schlesewsky, M. (2005). The emergence of the unmarked: A new perspective on the language-specific function of Brocaís area. *Human Brain Mapping, 26*, 178–190.

Grodzinsky, Y. (2001). The neurology of syntax: Language use without Broca's area. *Behavioral and Brain Sciences, 23*, 1–71.

Holmes, C. J., Hoge, R., Collins, L., Woods, R., Toga, A. W., & Evans, A. C. (1998). Enhancement of MR images using registration for signal averaging. *Journal of Computer Assisted Tomography, 22*, 324–333.

Holmes, V. (1987). Syntactic parsing: In search of the garden path. In M. Coltheart (Ed.), *Attention and performance XII: The psychology of reading* (pp. 587–599). London: Lawrence Erlbaum Associates.

Holmes, V. M., & O'Regan, J. K. (1981). Eye fixation patterns during the reading of relative-clause sentences. *Journal of Verbal Learning and Verbal Behavior, 20*, 417–430.

Hsiao, F., & Gibson, E. (2003). Processing relative clauses in Chinese. *Cognition, 90*, 3–27.

Hyönä, J., & Hujanen, H. (1997). Effects of case marking and word order on sentence parsing in Finnish: An eye fixation analysis. *Quarterly Journal of Experimental Psychology, 50*, 841–858.

Just, M., Carpenter, P., Keller, T., Eddy, W., & Thulborn, K. (1996). Brain activation modulated by sentence comprehension. *Science, 274*, 114–116.

Kaan, E., & Vasić, N. (2004). Cross-serial dependencies in Dutch: Testing the influence of NP type on processing load. *Memory and Cognition, 32*, 175.

Kim, J., Koizumi, M., Ikuta, N., Fukumitsu, Y., Kimura, N., Iwata, K., et al. (2009). Scrambling effects on the processing of Japanese sentences: An fMRI study. *Journal of Neurolinguistics, 22*, 151–166.

King, J., & Just, M. (1991). Individual differences in syntactic processing: The role of working memory. *Journal of Memory and Language, 30*, 580–602.

Kinno, R., Kawamura, M., Shioda, S., & Sakai, K. (2008). Neural correlates of noncanonical syntactic processing revealed by a picture sentence matching task. *Human Brain Mapping, 29*, 1015–1027.

Kinno, R., Muragaki, Y., Hori, T., Maruyama, T., Kawamura, M., & Sakai, K. (2009). Agrammatic comprehension caused by a glioma in the left frontal cortex. *Brain and Language, 110*, 71–80.

Kinno, R., Ohta, S., Muragaki, Y., Maruyama, T., & Sakai, K. L. (2014). Differential reorganization of three syntax-related networks induced by a left frontal glioma. *Brain, 137*, 1193–1212.

Lewis, R. (1996). Interference in short-term memory: The magical number two (or three) in sentence processing. *Journal of Psycholinguistic Research, 25*, 93–115.

Lewis, R., Vasishth, S., & Van Dyke, J. (2006). Computational principles of working memory in sentence comprehension. *Trends in Cognitive Sciences, 10*, 447–454.

Leyton, C. E., Villemagne, V. L., Savage, S., Pike, K. E., Ballard, K. J., Piguet, O., et al. (2011). Subtypes of progressive aphasia: Application of the International Consensus Criteria and validation using beta-amyloid imaging. *Brain: A Journal of Neurology, 134*, 3030–3043.

MacWhinney, B., & Pléh, C. (1988). The processing of restrictive relative clauses in Hungarian. *Cognition, 29*, 95–141.

Magnusdottir, S., Fillmore, P., den Ouden, D., Hjaltason, H., Rorden, C., Kjartansson, O., et al. (2013). Damage to left anterior temporal cortex predicts impairment of complex syntactic processing: A lesion-symptom mapping study. *Human Brain Mapping, 34*, 2715–2723.

Makuuchi, M., Bahlmann, J., Anwander, A., & Friederici, A. D. (2009). Segregating the core computational faculty of human language from working memory. *Proceedings of the National Academy of Sciences of the United States of America, 106*, 8362–8367.

Martin, R., & Romani, C. (1994). Verbal working memory and sentence comprehension: A multiple-components view. *Neuropsychology, 8*, 506–523.

Mazuka, R., Itoh, K., & Kondo, T. (2002). Costs of scrambling in Japanese sentence processing. In M. Nakayama (Ed.), *Sentence processing in East-Asian languages* (pp. 131–166). Stanford, CA: CSLI Publications.

Meyer, L., Obleser, J., Anwander, A., & Friederici, A. D. (2012). Linking ordering in Broca's area to storage in left temporo-parietal regions: The case of sentence processing. *NeuroImage, 62*, 1987–1998.

Michael, E. B., Keller, T. A., Carpenter, P. A., & Just, M. A. (2001). fMRI investigation of sentence comprehension by eye and by ear: Modality fingerprints on cognitive processes. *Human Brain Mapping, 13*, 239–252.

Miller, G. A., & Isard, S. (1964). Free recall of self-embedded English sentences. *Information and Control, 7*, 292–303.

Newman, S., Ikuta, T., & Burns, T. (2010). The effect of semantic relatedness on syntactic analysis: An fMRI study. *Brain and Language, 113*, 51–58.

Nicol, J., Fodor, J., & Swinney, D. (1994). Using cross-modal lexical decision tasks to investigate sentence processing. *Journal of Experimental Psychology: Learning, Memory, and Cognition, 20*, 1229–1238.

Obleser, J., Meyer, L., & Friederici, A. D. (2011). Dynamic assignment of neural resources in auditory comprehension of complex sentences. *NeuroImage, 4*, 2310–2320.

Ohta, S., Koizumi, M., & Sakai, K. (2014). Activation modulation in the left inferior frontal gyrus caused by scrambled word orders: An fMRI study in Kaqchikel Maya. Poster presented at *annual meeting of the society for the neurobiology of language*. Amsterdam, The Netherlands.

Osterhout, L., & Swinney, D. (1993). On the temporal course of gap-filling during comprehension of verbal passives. *Journal of Psycholinguistic Research, 22*, 273–286.

Paulesu, E., Frith, C., & Frackowiak, R. (1993). The neural correlates of the verbal component of working memory. *Nature, 362*, 342–345.

Röder, B., Stock, O., Neville, H., Bien, S., & Rösler, F. (2002). Brain activation modulated by the comprehension of normal and pseudo-word sentences of different processing demands: A functional magnetic resonance imaging study. *NeuroImage, 15*, 1003–1014.

Rogalsky, C., & Hickok, G. (2010). The role of Broca's area in sentence comprehension. *Journal of Cognitive Neuroscience, 23*, 1664–1680.

Rogalsky, C., Matchin, W., & Hickok, G. (2008). Broca's area, sentence comprehension, and working memory: An fMRI study. *Frontiers in Human Neuroscience, 2*, 1–13.

Santi, A., & Grodzinsky, Y. (2010). fMRI adaptation dissociates syntactic complexity dimensions. *NeuroImage, 51*, 1285–1293.

Shetreet, E., Palti, D., Friedmann, N., & Hadar, U. (2007). Cortical representation of verb processing in sentence comprehension: Number of complements, subcategorization, and thematic frames. *Cerebral Cortex, 17*, 1958–1969.

Stromswold, K., Caplan, D., Alpert, N., & Rauch, S. (1996). Localization of syntactic comprehension by positron emission tomography. *Brain and Language, 52*, 452–473.

Talairach, J., & Tournoux, P. (1988). *Co-planar stereotaxic atlas of the human brain (3-dimensional proportional system: an approach to cerebral imaging)*. Stuttgart: Thieme.

Tanenhaus, M., Carlson, G., & Seidenberg, M. (1985). Do listeners compute linguistic representations. In D. R. Dowty, L. Karttunen, & A. M. Zwicky (Eds.), *Natural language parsing: Psychological, computational, and theoretical perspectives* (pp. 359–408). New York, NY: Cambridge University Press.

Trueswell, J., Tanenhaus, M., & Kello, C. (1993). Verb-specific constraints in sentence processing: Separating effects of lexical preference from garden-paths. *Journal of Experimental Psychology: Learning, Memory, and Cognition, 19*, 528–533.

Turkeltaub, P., Eden, G. F., Jones, K. M., & Zeffiro, T. A. (2002). Meta-analysis of the functional neuroanatomy of single-word reading: Method and validation. *NeuroImage, 16*, 765–780.

Turkeltaub, P., Eickhoff, S., Laird, A., Fox, M., Wiener, M., & Fox, P. (2012). Minimizing within-experiment and within-group effects in activation likelihood estimation meta-analyses. *Human Brain Mapping, 33*, 1–13.

Van Dyke, J. (2007). Interference effects from grammatically unavailable constituents during sentence processing. *Journal of Experimental Psychology: Learning, Memory and Cognition, 33*, 407–430.

Van Dyke, J., & McElree, B. (2006). Retrieval interference in sentence comprehension. *Journal of Memory and Language, 55*, 157–166.

Vasishth, S. (2003). *Working memory in sentence comprehension: Processing Hindi center embeddings*. New York, NY: Routledge.

Vasishth, S., & Lewis, R. (2006). Argument-head distance and processing complexity: Explaining both locality and antilocality effects. *Language, 82*, 767–794.

Vigneau, M., Beaucousin, V., Herve, P. Y., Duffau, H., Crivello, F., Houde, O., et al. (2006). Meta-analyzing left hemisphere language areas: Phonology, semantics, and sentence processing. *NeuroImage, 30*, 1414–1432.

Wilson, S. M., Dronkers, N. F., Ogar, J. M., Jang, J., Growdon, M. E., Agosta, F., et al. (2010). Neural correlates of syntactic processing in the nonfluent variant of primary progressive aphasia. *Journal of Neuroscience, 30*, 16845–16854.

Yngwe, V. (1960). A model and a hypothesis for language structure. *Proceedings of the American Philosophical Society, 104*, 444–466.

CHAPTER

49

The Timecourse of Sentence Processing in the Brain

Ina Bornkessel-Schlesewsky[1,2], *Adrian Staub*[3] *and Matthias Schlesewsky*[4]

[1]Cognitive Neuroscience Laboratory, School of Psychology, Social Work and Social Policy, University of South Australia, Adelaide, SA, Australia; [2]Department of Germanic Linguistics, University of Marburg, Marburg, Germany; [3]Department of Psychological and Brain Sciences, University of Massachusetts, Amherst, MA, USA; [4]Department of English and Linguistics, Johannes Gutenberg-University Mainz, Mainz, Germany

49.1 PRELIMINARIES: CHALLENGES TO A NEUROBIOLOGICAL PERSPECTIVE ON THE TIMECOURSE OF SENTENCE PROCESSING

Timecourse information has long held a special significance for researchers interested in the cognitive architecture of language processing. At the level of sentence processing in particular, insights about timing have played a central role in shaping the defining debates of the field, such as the question of how different information sources are utilized during the comprehension of a sentence. In this respect, modular models maintain that distinct linguistic information sources are processed in succession, typically with at least some aspects of syntactic processing thought to take place first and in an informationally encapsulated manner (Frazier, 1987; Frazier & Clifton, 1996; Rayner, Carlson, & Frazier, 1983). Other information sources (semantics, plausibility, context, etc.) are not utilized until a later point during processing. In contrast, (strongly) interactive models posit that there is no priority for one particular information source over others and, thereby, that all available sources of information interact from the earliest stages of processing (MacDonald, Pearlmutter, & Seidenberg, 1994; Trueswell & Tanenhaus, 1994). Clearly, an evaluation of these competing architectural concepts requires detailed timecourse measures.

A problem arises, however, when we attempt to translate this classic cognitive view into a neurobiologically plausible perspective. In particular, what are the relevant units and mechanisms to consider when describing the timecourse of the neurobiology of sentence processing? The heritage of cognitive models has typically led us to pose this question in a particular way: we tend to think of the relevant units as being part of linguistic subdomains such as phonology, syntax, and semantics, and of the relevant mechanisms as being tied to cognitive metaphors such as the notion of a mental lexicon (e.g., "lexical access," "prelexical," or "postlexical" processing). However, the utility of such concepts for the neurobiology of language is considerably less clear. Consider the intuitively appealing and cognitively plausible mental lexicon metaphor as an example. It is generally accepted that conceptual information is represented in the brain in a distributed fashion (e.g., Patterson, Nestor, & Rogers, 2007, for a review). Thus, the notion of a lexical "entry" that can be activated and accessed cannot be translated straightforwardly into neurobiological terms. Moreover, the mechanisms suggested by terms such as "lexical access" are unrevealing with respect to the question of how the human brain actually *implements* the processing of information thought to be encoded in a lexical entry. In their recent review of the N400 event-related brain potential (ERP) component in language processing, Kutas and Federmeier (2011) also conclude that the notion of a mental lexicon and the mechanisms suggested by it (such as prelexical or postlexical processing) are empirically inadequate for explaining the existing range of N400 results. In addition, the dissociation between the cognitive view and a neurobiological perspective is underscored by the discrepancy between timecourse estimates obtained using

Neurobiology of Language. DOI: http://dx.doi.org/10.1016/B978-0-12-407794-2.00049-3

ERPs as opposed to eye movements during reading (see Bornkessel & Schlesewsky, 2006; Rayner & Clifton, 2009; Sereno & Rayner, 2003; and Sections 49.3.1 and 49.4 in this chapter). In summary, advances in understanding the neurobiology of language will likely require a new, neurobiologically plausible conception of the timecourse of sentence processing.

The remainder of this chapter is organized as follows. In Section 49.2, we discuss some neurobiological considerations with respect to timecourse before turning to the major current perspectives on the timecourse of sentence processing in the brain in Section 49.3. Section 49.4 then attempts to relate these approaches to behavioral timecourse measures obtained using eye-tracking during natural reading. Finally, Section 49.5 offers some conclusions and also discusses open questions for future research.

49.2 NEUROBIOLOGICAL CONSIDERATIONS

So far, we have suggested that it is not trivial to map traditional cognitive perspectives on the timecourse of sentence processing directly onto the neurobiology of language. In this section, we briefly discuss some neurobiological considerations related to timecourse information.

With respect to the nature of the "units" or "domains" of processing, it is important to keep in mind that most neuroscientists do not consider the purpose of the human brain to lie in the construction of elaborate cognitive representations. Rather, there is a relatively broad consensus that the brain implements perception–action cycles (Fuster, 1997) and, hence, that "the purpose of the human brain is to use sensory representations to determine future actions" (Wolpert, Doya, & Kawato, 2003, p. 593). Ultimately, a neurobiologically plausible perspective on language (including sentence) processing should be compatible with these insights. Perhaps somewhat reassuringly, however, a commonality between these neurobiological considerations and the classic cognitive perspective lies in the notion of hierarchical organization as the basis for stable complex systems (Simon, 1962).

In terms of neurobiological organization, neither the strictly serial and encapsulated perspective of modular models nor the "everything is connected to everything" perspective of strongly interactive models (often with connectionist implementations) is neurobiologically plausible. Rather, perception–action cycles in vision and audition (including the processing of speech and language) involve information transfer along multiple streams, with a hierarchical organization within streams (Rauschecker & Scott, 2009; see also Bornkessel-Schlesewsky & Schlesewsky, 2013 for discussion). Hierarchical processing implies properties of both serial and parallel organization. Ultimately, however, connectivity is considerably more complicated. First, because streams of information processing are inherently bidirectional, anatomical feedback connections must be considered in addition to feedforward connections.[1] Second, hierarchical organization does not imply "serial" connectivity in the sense that a level n of the hierarchy is only connected to levels $n+1$ and $n-1$. Rather, long-range (feedforward and feedback) connections can "skip" individual levels of the hierarchy (for possible implications, see Bar, 2003). Implications of these considerations for associated timecourse scenarios include the following: (i) information processing proceeds in a cascaded manner incorporating both serial and parallel aspects (consequence of hierarchical organization); and (ii) the interpretation of observable effects must take into account that processing comprises more than a feedforward sweep, with feedback connections and top-down modulations also playing an essential role (consequence of bidirectional and long-range connectivity).

A further complication arises from the methods currently available to study the timecourse of language processing in the brain. Electroencephalography (EEG) and magnetoencephalography (MEG) have long been the methods of choice for this purpose but, despite their exquisite temporal resolution, provide rather macroscopic measures of brain activity. Functional magnetic resonance imaging (fMRI), by contrast, does not offer the temporal resolution required for the question under consideration. Electrocorticography (ECoG) (i.e., the recording of electrophysiological activity directly from the cortex using subdural electrodes) (see Chang et al., 2010, for a recent application to language) is a highly promising approach because it offers both high temporal and high spatial resolution. However, ECoG studies can only be performed with patients and the electrode grids used typically only offer limited coverage because electrode placement is determined by clinical motivations.

[1]Feedback and feedforward connections may further be asymmetrical (i.e., connectivity between regions A and B may be bidirectional but connectivity between regions A and C may not). For a detailed demonstration of this property for visual processing streams, see recent work by Kravitz and colleagues (Kravitz, Saleem, Baker, & Mishkin, 2011; Kravitz, Saleem, Baker, Ungerleider, & Mishkin, 2013).

These difficulties, which currently provide significant challenges to a neurobiological perspective on the timecourse of language processing, apply to all aspects of language. They are, however, aggravated further by the particular demands of sentence processing, which necessarily involves more complex units[2] than, for example, sound or word processing and a more complex interplay of different information sources.

In view of all of these considerations, it is virtually impossible to provide a comprehensive picture of the timecourse of sentence processing from a truly neurobiological perspective at this point. However, there exist a number of approaches that have set out to tackle these challenges from a variety of perspectives. In what follows, we aim to provide an overview of the different (types of) approaches that have been put forward and their respective claims about the timecourse of sentence processing in the brain.

49.3 DIFFERING PERSPECTIVES ON THE TIMECOURSE OF SENTENCE PROCESSING IN THE BRAIN

As is already the case in the cognitive domain, the timecourse of sentence processing in the brain has been debated intensely and it is not yet possible to outline a consensus view. Accordingly, we discuss a variety of different perspectives primarily based on ERP data because this remains the most widely used method by far for the examination of timecourse questions from a neurobiological perspective.

49.3.1 Component Mapping/Absolute Timing

Originally, ERP-based research into the timecourse of sentence processing closely mirrored the cognitive perspective that had been developed on the basis of behavioral results. Thus, ERP components such as the N400 and P600 were viewed as correlates of linguistic domains (semantic and syntactic processing, respectively), and the relative timing of these components as well as their interaction or mutual independence were thus thought to shed light on the organization of the language architecture (e.g., Friederici, Pfeifer, & Hahne, 1993; Hagoort, Brown, & Osterhout, 1999; Osterhout & Nicol, 1999, among many others).

This perspective forms part of two highly influential models of sentence processing in the brain, Friederici's *Neurocognitive Model of Auditory Sentence Processing* (Friederici, 1995, 1999, 2002, 2009, 2011) and Hagoort's *Memory, Unification, and Control (MUC)* Model (Hagoort, 2003, 2005, 2013). In the spirit of the classic modularity versus interactivity debate, the Neurocognitive Model posits that different information sources are processed in several successive processing stages that are organized serially, whereas the MUC model advocates an interactive processing architecture. In each case, evidence for the respective view was provided by an assessment of ERP component timing. According to Friederici (see Friederici, 2011, for a recent review), acoustic–phonetic processing is followed by an initial stage of local syntactic structure building (150–250 ms, indexed by the early left anterior negativity [ELAN]). In a second processing stage, semantic and morphosyntactic relations are processed independently but in parallel, as evidenced by the N400 and the left anterior negativity (LAN), respectively. Finally, these separate information processing streams are integrated with one another in stage 3 of processing (correlating with the P600), which also comprises processes of reanalysis and/or repair if necessary. In this view, crucial evidence for the seriality of processing stems from the observation that a disruption of early syntactic processing (i.e., a word category violation that engenders an ELAN, example 1a) can "block" semantic processing (example 1b) because a double (both word category and semantic) violation (example 1c) engenders an ELAN-P600 pattern with no additional N400 (Friederici, 2002; Hahne & Friederici, 2002).

(1) Examples of the syntactic, semantic, and double violations used by Friederici and colleagues to infer a staged processing timecourse (examples and ERP responses from Hahne & Friederici, 2002)

a. Syntactic (word category) violation (ELAN-P600 response in comparison with 1d)
*Das Eis wurde im gegessen
the ice cream was in-the eaten

b. Semantic violation (N400 response in comparison with 1d)
Der Vulkan wurde gegessen
the volcano was eaten

c. Double violation (ELAN-P600 response in comparison with 1d)
*Das Türschloss wurde im gegessen
the door lock was in-the eaten

d. Control condition
Das Brot wurde gegessen
the bread was eaten

[2] In view of the problem of distributed representations (see the discussion of the mental lexicon in Section 49.1), it is currently not at all clear how anatomical considerations such as hierarchical organization actually map onto the kinds of cognitively inspired units that (at least currently) appear to be unavoidable in the characterization of sentence processing (e.g., words, morphemes).

In contrast, while also subscribing to the view that the N400 and LAN/P600 are related to semantic and syntactic aspects of processing, respectively, the MUC model assumes that there is no priority for any one information source over another during sentence interpretation. Rather, the relative timing of when a particular information source is used depends on its availability in the input signal (Hagoort, 2008). In addition, the model posits a "one-step" interpretation process in which all available information sources interact directly irrespective of whether they are sentence-internal, sentence-external (e.g., discourse context), or even external to language (e.g., co-speech gestures, identity of the speaker) (Hagoort & van Berkum, 2007). It thus rejects any notion of stages during the comprehension process, including word recognition; there is no separate stage of "lexical access" that needs to precede integration into the current sentence (Hagoort, 2008). Evidence for timing based on availability stems, for example, from a study by Van Den Brink and Hagoort (2004) that used logic similar to that employed by Hahne and Friederici (2002) but changed the temporal availability of the syntactic and semantic information sources by incorporating the semantic violation into the word stem and the word category violation into the suffix. In contrast to Friederici and colleagues, they observed an N400 relative to word onset and an early anterior negativity relative to the category violation point (CVP), thus leading them to reject the stages posited in Friederici's model.

(2) Examples of the stimuli used by Van Den Brink and Hagoort (2004) to infer that the relative timing of syntactic and semantic processing depends on temporal availability within the acoustic signal

a. Double violation (N400 followed by an ELAN relative to the CVP in comparison with 2b)

*Het vrouwtje veegde de vloer met een oude kliederde gemaakt van twijgen
The woman swept the floor with an old messed made of twigs

b. Control condition

Het vrouwtje veegde de vloer met een oude bezem gemaakt van twijgen
The woman swept the floor with an old broom made of twigs

The second crucial assumption of the MUC model (one-step interpretation) is supported by results showing that sentence-internal and sentence-external information all modulate the N400 component, with no evidence for a different timecourse depending on the source of the information (see Hagoort & van Berkum, 2007 for a detailed review of the relevant studies).

As in the cognitive domain, the debate between modular and interactive neurocognitive models of sentence processing remains unresolved even after several decades of work on this subject. Furthermore, continuing progress in the field has revealed challenges to some of the basic assumptions underlying the component mapping approach:

(a) *Absolute timing*: Several observations have called into question the assumption that ERP component latencies provide absolute estimates of timing information. The first stems from measures of eye movements during reading, which, as briefly noted, sometimes provide different timing estimates to ERPs. This is perhaps most readily apparent with regard to the N400 component, which typically shows an onset latency of a little more than 200 ms in young adults (Kutas & Iragui, 1998 report an average onset latency of 236 ms for adults in their 20s) and is highly stable in terms of its timing (Federmeier & Laszlo, 2009). The N400 is reliably modulated by word frequency (at least for words occurring early on in a sentence; Van Petten & Kutas, 1990), a factor which is also known to affect the duration of the first fixation on a word during reading (Inhoff & Rayner, 1986). Because fixations last approximately 200–250 ms on average for English (Rayner, 1998), and because saccade programming is thought to require approximately 150–175 ms (Rayner, Slowiaczek, Clifton, & Bertera, 1983), word frequency appears to influence eye movements at a point in time that *precedes* the onset of the N400 (Rayner & Clifton, 2009; Sereno & Rayner, 2003; for discussion, see Bornkessel & Schlesewsky, 2006). This suggests that latencies of components such as the N400 and P600 may not be reliable indicators of when a particular information source first becomes available or when it is first utilized during sentence processing.

A second and related observation stems from the ERP domain itself. If components such as the N400 do not provide absolute timing information, perhaps markers of this information should be sought earlier in the ERP record. There have been a number of reports of language-related ERP effects "before the N400" (Dien, 2009). These include word frequency-based modulations of the N100 (Sereno, Rayner, & Posner, 1998) and modulations of the mismatch negativity (MMN), which peaks between approximately

100 and 250 ms poststimulus onset, by morphosyntactic violations presented as part of an oddball design (Hasting, Kotz, & Friederici, 2007). Recent MEG studies have also provided converging evidence for early effects by demonstrating a dissociation of brain responses to words and pseudowords from as early as 50 ms (MacGregor, Pulvermuller, van Casteren, & Shtyrov, 2012) or a modulation of the visual M100 by syntactic (Dikker, Rabagliati, & Pylkkänen, 2009) and lexical-semantic violations (Dikker & Pylkkänen, 2011) in highly predictive contexts. It is possible that findings such as these should be taken to indicate that ERPs and event-related magnetic fields (ERFs) *do* provide absolute timing information once early components—rather than "late" components such as the N400—are taken into account. We return to this perspective in Section 3.2.

(b) *1:1 component mapping*: Although highly appealing, the notion that language-related ERP components can be used as diagnostic indicators of particular linguistic or cognitive domains, such as the N400 as a correlate of lexical-semantic processing or declarative memory contributions to language (Ullman, 2001) versus the LAN/P600 as correlates of morphosyntactic processing or procedural memory, is now challenged by a continually increasing number of findings. The occurrence of monophasic P600 effects for certain types of semantic violations—particularly violations that a plausible sentence would obtain if thematic role assignments to arguments were reversed (example 3)—has been discussed intensely in the literature on the electrophsyiology of language over the past decade (e.g., Hoeks, Stowe, & Doedens, 2004; Kim & Osterhout, 2005; Kolk, Chwilla, van Herten, & Oor, 2003; Kuperberg, Sitnikova, Caplan, & Holcomb, 2003, among many others; see Bornkessel-Schlesewsky & Schlesewsky, 2008; Brouwer, Fitz, & Hoeks, 2012; van de Meerendonk, Kolk, Chwilla, & Vissers, 2009, for reviews; for an account of "semantic P600" effects that subscribes to a traditional functional interpretation of language-related ERP components, see Kim & Osterhout, 2005). In addition, many studies using "traditional" semantic violations in the style of Kutas and Hillyard's famous "He spread the warm bread with *socks*" (Kutas & Hillyard, 1980) have observed N400s followed by P600 effects, and P600 effects have also been observed in response to other types of incongruencies such as orthographic violations (Münte, Heinze, Matzke, Wieringa, & Johannes, 1998; Vissers, Chwilla, & Kolk, 2006), thus leading several groups to suggest more general interpretations of the P600 (e.g., as a correlate of "conflict monitoring"; Kolk et al., 2003; van de Meerendonk et al., 2009) or an instance of the domain-general P300 (Bornkessel-Schlesewsky et al., 2011; Coulson, King, & Kutas, 1998; Gunter, Stowe, & Mulder, 1997; Sassenhagen, Schlesewsky, & Bornkessel-Schlesewsky, 2014).

(3) Semantic reversal anomaly (from Kim & Osterhout, 2005)

The hearty meals were *devouring*...

Although less well-known than semantic P600s, syntax-related or morphosyntax-related N400 effects have also been reported in a range of studies, such as for case violations (Choudhary, Schlesewsky, Roehm, & Bornkessel-Schlesewsky, 2009; Frisch & Schlesewsky, 2001; Mueller, Hahne, Fujii, & Friederici, 2005), gender incongruencies (Wicha, Moreno, & Kutas, 2004), object case ambiguities (Hopf, Bayer, Bader, & Meng, 1998), and word order variations (Bornkessel, McElree, Schlesewsky, & Friederici, 2004; Haupt, Schlesewsky, Roehm, Friederici, & Bornkessel-Schlesewsky, 2008). In summary, all of these results suggest that ERP components do not reflect a clear-cut functional divide between linguistic subdomains and that somewhat more general explanations may be required.

(c) *Data interpretability*: A final problem concerns the ability of ERP or ERF data to conclusively arbitrate between competing models. Consider the ELAN/N400 debate, which we discussed in relation to examples (1) and (2). As is apparent from the discussion, these results have been interpreted differently by proponents of alternative models rather than helping to conclusively settle the modularity versus interactivity debate with respect to the neurocognition of language.

In view of these conflicting interpretations, some of us have previously proposed that the data may require a third type of explanation that shares characteristics with both a modular and an interactive view (Bornkessel-Schlesewsky & Schlesewsky, 2008, 2009). From this perspective, the processing architecture is organized in a hierarchical and cascaded manner (McClelland, 1979): whereas there are hierarchically organized processing "stages", processing stage n need not be completed when processing stage $n+1$ begins, thus allowing for a certain degree of parallelism.

Applied to the ELAN/N400 discussion, this proposal assumes that although word category processing does take hierarchical precedence vis à vis semantic processing, this does not mean that semantic processing cannot be initiated until unambiguous word category information is encountered. Rather, because the critical word in sentence example (2a) occurs in a sentence context that is highly predictive of a noun (a determiner + adjective sequence) and the initial several hundred milliseconds of the acoustic signal from word onset are compatible with this prediction, information for a "noun" analysis quickly accrues and processing can proceed. Accordingly, the semantic incongruity leads to an N400 just as would be expected if there were no additional word category violation. In contrast, when the word category violation is encountered several hundred milliseconds later, it nevertheless elicits an anterior negativity. An account along these lines thus explains the asymmetry between the ERP patterns for examples (1) and (2): when the word category violation precedes a semantic violation in the input signal, the ELAN "blocks" the N400, but not vice versa (i.e., an N400 for a semantic incongruity that temporally precedes a word category violation does not block an ELAN). Note also in this context, however, that the experimental paradigms that have traditionally been used to elicit ELAN effects have recently been criticized on a number of grounds (Steinhauer & Drury, 2012). Thus, some of the claims based on the ELAN literature may eventually require reevaluation.

Perhaps most importantly, irrespective of whether one is willing to accept the cascaded view as a viable alternative, these considerations at least highlight the need to take into account issues such as prediction and the interaction between top-down and bottom-up information sources when attempting to account for the timecourse of sentence processing in the brain. These concepts play an important role in the accounts discussed in the following sections.

49.3.2 Absolute Timing Revisited: Virtual Parallelism and an Early Cascade

As already discussed in some detail in the preceding section, there is considerable evidence to suggest that the latency of classic language-related ERP components such as the N400, LAN, and P600 does not provide an adequate estimate of absolute timecourse information. Within the neurocognitive domain, this perspective has been advocated for some time by Pulvermüller and colleagues, drawing primarily on experimental paradigms using the MMN. The MMN typically occurs in response to deviant stimuli in oddball paradigms, such as stimuli that occur infrequently within sequences of high probability "standards." The processes underlying the MMN are thought to be preattentive because it can be elicited even when participants are attending to something other than the critical stimuli (see Kujala, Tervaniemi, & Schröger, 2007 for a review of the MMN). Traditionally, MMN effects are elicited using relatively simple stimuli (e.g., sequences of tones) and their occurrence is explained in terms of sensory memory: the sequence of standard stimuli is assumed to comprise a sensory memory trace and MMN results when a (deviant) stimulus produces a mismatch with this memory representation (Näätänen, Paavilainen, Alho, Reinikainen, & Sams, 1989).

Pulvermüller and colleagues have extended this line of research to considerably more complex stimuli, including minimal phrases such as "we come" or, as a violation condition, "we comes" (Pulvermüller & Shtyrov, 2003; Shtyrov & Pulvermüller, 2002). In comparison with the same word presented out of context (i.e., "comes" preceded by a nonlinguistic noise mimicking the spectral envelope of the word "we"), the ungrammatical string elicited a larger MMN than the grammatical string between 100 and 150 ms after the acoustic divergence point between the two strings. MMN effects related to morphosyntactic information have also been reported for German using both subject–verb agreement violations and word category violations (Hasting et al., 2007). Semantic effects in MMN paradigms have also been observed (Menning et al., 2005; Shtyrov & Pulvermüller, 2007).

On the basis of these results and other findings on the processing of single written words, Pulvermüller and colleagues argue for early "near simultaneity," with initial processing of different linguistic information sources (phonological, lexical, syntactic, and semantic information) taking place before 200 ms after stimulus onset and in an almost simultaneous but nevertheless cascaded fashion. The single word processing studies contributing to this argument contrasted various parameters (e.g., word frequency, word length, phonological/orthographic typicality, word category [considered a syntactic factor], semantic categories, and affective parameters) and all have in common that they observed effects that occurred considerably earlier than the classic language-related ERP components (see Pulvermüller, Shtyrov, & Hauk, 2009 for review and references to the individual studies). Pulvermüller and colleagues suggest that these very early effects may have been missed in many studies because they are focal and transient, thus requiring very tightly controlled stimuli and appropriately narrow time windows. In their account, the classic language-related ERP effects such as N400, LAN, and P600 reflect either a second stage of information processing or some type of "postprocessing" that is not specified further. The early effects, in contrast, that are thought to reflect the initial processing of the relevant information sources are viewed as resulting from the activation of different cortical circuits depending on the exact type of information being processed. Although these circuits are activated with similar latencies (i.e., nearly instantaneously), thus accounting for the near simultaneity of the differing information sources, their internal conduction delays are assumed to differ, thus accounting for the "fine-grained activation delays in the millisecond range" (Pulvermüller et al., 2009, p. 89).

One problem that arises in this respect is that it is not clear whether the effects observed under these circumstances (i.e., single word presentation, MMN paradigms) generalize to more typical sentence processing settings. With respect to the single word paradigms, it is well-known, for example, that word frequency effects interact with sentence context in EEG studies (Van Petten & Kutas, 1990) and that word category effects vary depending on whether words are presented in isolation or in sentence context (Vigliocco, Vinson, Druks, Barber, & Cappa, 2011). Thus, although the observation of ERP modulations "before the N400" in response to various information sources is clearly an important finding that needs to be pursued further, it is not yet clear what conclusions should be drawn from such findings for the timecourse of sentence processing in the brain (for some initial reports of early semantic context effects, see Penolazzi, Hauk, & Pulvermüller, 2007; Sereno, Brewer, & O'Donnell, 2003; Tse et al., 2007).

Evidence from MMN paradigms is also sometimes viewed as problematic because of the highly repetitive nature of the stimuli and the lack of lexical variation (i.e., an experiment may consist simply of a large number of repetitions of the two stimuli "he walks" and "he walk," with one the standard and one the deviant). However, because syntactic or semantic violation deviants are always contrasted against grammatical deviants in these paradigms, the linguistic properties of the stimuli obviously play at least some role in engendering the larger MMN response to incongruous stimuli. Perhaps the first word in a minimal phrase such as "he walks" creates certain expectations with regard to possible continuations that are commensurate with the prior experience of the language processing system.

The involvement of predictive processes in MMN paradigms in language processing is supported by at least two observations. First, context influences MMN magnitude, as shown by larger MMN responses when a deviant syllable or phoneme occurs in the context of a real word as opposed to a pseudoword (see Pulvermüller et al., 2009). Second, as argued by Friston and colleagues (Friston, 2005; Garrido, Kilner, Stephan, & Friston, 2009), the MMN response can be derived very plausibly within a predictive coding account of brain architecture. This view assumes a hierarchically organized cortical architecture that attempts to minimize prediction error for sensory stimuli by using internal models of the world to infer the causes of sensory events and, thereby, to formulate appropriate top-down predictions. The MMN is thought to arise from the prediction error that arises after sensory learning has taken place within the context of an oddball paradigm and the deviant stimulus does not match the prediction for the upcoming stimulus. The prediction error is subsequently propagated up the cortical hierarchy and thereby leads to model adjustment.

In summary, a perspective advanced most prominently by Pulvermüller and colleagues (among other researchers) assumes that ERP/ERF components are informative with regard to absolute timecourse information but proposes that the relevant effects occur in a time range prior to 200 ms and, therefore, precede typical language-related components such as the N400. Because these early effects are thought to arise from the near-simultaneous activation of word-specific cortical circuits, with differences in conduction delays (i.e., speeds of information transfer) giving rise to the apparent cascade of effects, this account essentially abolishes the idea of component mapping. As we suggested here, however, predictive processing may also play an important role in engendering—or at least modulating—the effects under discussion, and this aspect may help to link the relevant explanations to sentence-level processing.

49.3.3 The Crucial Role of Top-Down Predictions

The importance of top-down predictions is emphasized more strongly in an approach that has been championed by Dikker, Pylkkänen, and colleagues. Using MEG, these researchers have repeatedly shown that sentence-level (syntactic and semantic) incongruities engender modulations of the visual M100, a component that is typically localized to visual cortex (Dikker & Pylkkänen, 2011; Dikker et al., 2009; Dikker, Rabagliati, Farmer, & Pylkkänen, 2010; for an early report of auditory N100 differences in response to semantic incongruity in sentence processing, see Besson & Macar, 1987). In the syntactic domain, these early effects have been observed for word category violations such as those known from the ELAN literature (see Section 3.1), such as "The boys heard Joe's *about...*" or "The tastelessly *soda...*". In the semantic domain, they have been demonstrated using a picture–word matching paradigm (e.g., the noun phrase "the apple" following an image of an apple (+predictive, +match), a picture of a banana

(+ predictive, − match), a picture of a shopping bag with many items (− predictive, + match), or a picture of Noah's ark with many animals (− predictive, − match)).

The early latency of the M100 and its likely source localization suggest that the effects in question may be indexing prediction mismatches at a rather early sensory level. In other words, if the predictive context is strong enough, predictions extend not only to abstract features of an upcoming linguistic item (e.g., word class, semantic category) but also to its sensory properties (i.e., a concrete word form). Similar effects have been observed in the auditory domain, as demonstrated by the so-called phonological mapping negativity (PMN). The PMN, previously labeled phonological mismatch negativity and analyzed by some as an instance of the N200 component (van den Brink, Brown, & Hagoort, 2001, 2006), is a negativity with a peak latency of approximately 250–300 ms that occurs in response to phonologically unexpected input (Connolly & Phillips, 1994). As recently demonstrated by Newman and Connolly (2009), the PMN is independent of whether the current input item is a word or a nonword, thus suggesting that it reflects a purely form-based expectation mismatch. At least to some degree, an account of language-related ERP components that focuses on sensory predictability shifts the burden of explanation for observed ERP or ERF effects from the current input item to the preceding linguistic context. In other words, when evaluating timecourse information, the properties of the current stimulus item may be less important than the predictive capacity of the context in which it is encountered.

This perspective is highly compatible with a hierarchically organized predictive coding architecture in which predictions are passed from higher to lower areas within the cortical hierarchy and, depending on their specificity, can lead to prediction errors at any hierarchical level. Thus, the prediction error signal itself is not necessarily diagnostic of any particular functional interpretation, but rather varies as a function of where within the overall architecture the prediction error is registered.

49.3.4 Integrating Multiple Information Sources During Multimodal Language Processing

The notion of prediction also plays a crucial role in an account of language-related ERP components—and specifically the N400—that abstracts away from timecourse information to a relatively large degree (Kutas & Federmeier, 2011). This perspective builds on the observation that the N400 is not parsimoniously described in relation to lexical (and prelexical, postlexical, etc.) processing because, for example, it also responds to nonword stimuli as long as they are meaningful in a given context (Federmeier & Laszlo, 2009, for an overview), and that the N400 is remarkably stable with respect to its latency. Accordingly, Federmeier, Kutas and colleagues argue that the N400 should be viewed as indexing a period of convergence between multiple, multimodal information sources, during which meaning is constructed in a dynamic and context-dependent manner. In other words, the N400 indexes the "transition from unimodal to multimodal processing (e.g., from word form to the concept that word brings to mind)" and "reflect[s] the activity in a multimodal long-term memory system that is induced by a given input stimulus during a delimited time window as meaning is dynamically constructed" (Kutas & Federmeier, 2011, p. 640).

This perspective, too, emphasizes the crucial role of prediction during language processing, because linguistic context is assumed to preactivate particular linguistic features as well as their combination. This preactivation process modulates the dynamic construction of meaning as described and hence impacts N400 amplitude. Like the perspectives discussed in the preceding sections, it thus de-emphasizes the notion of component mapping and, although it does not focus on early language-related ERP components, it shares with Pylkkänen and colleagues the assumption that predictive processing plays a crucial role in language comprehension. In view of the focus on the N400 and the fixed latency of this component, however, detailed timecourse considerations do not feature as part of this account.

49.3.5 The Interplay Between Top-Down and Bottom-Up Information Sources

In addition to the focus on prediction that has been central to the discussion in the preceding subsections, a range of findings now demonstrate that top-down predictability or "preactivation" is not the sole determinant of neurophysiological effects during sentence comprehension. Rather, the "balance" between top-down and bottom-up influences needs to be taken into account. Once again, a great deal of the relevant evidence stems from the N400 ERP component. A recent finding that shows that N400 amplitude is modulated by properties of the current stimulus rather than the preceding context was reported by Laszlo and Federmeier (2011). They showed that N400 amplitudes are higher for input items—both words and letter strings—with more or higher frequency orthographic neighbors and lexical associates. This result also provides compelling evidence against a view of the N400 as a correlate of lexical access or postlexical processing because accounts along these lines should predict differences between word and nonword stimuli.

The need to consider the integration of top-down and bottom-up information was further emphasized by Lotze, Tune, Schlesewsky, and Bornkessel-Schlesewsky (2011), who found that an unpredictable physical change from normal orthography to a capitalized sentence-final word almost completely neutralized an implausibility-based N400 effect on that word, whereas a change from all capitalized input to a lower case sentence-final word had no effect on the N400. Clearly, the attenuation of the N400 must be attributable to bottom-up stimulus features in this study, because top-down factors did not change between conditions. A somewhat similar result was reported by Delaney-Busch and Kuperberg (2013), who found that N400 effects for incongruous scenarios were neutralized when the critical word was an emotion word (of either positive or negative valence). A possible explanation for these results is that, in line with Bar's assumptions concerning visual object recognition (Bar, 2003 and see Section 2), very basic and salient stimulus features (e.g., sudden capitalization as a visual "pop out" or high emotional valence) are rapidly projected to prefrontal cortex (or other higher-level regions within the cortical hierarchy) via long-range feedforward connections, thus enabling these features to lead to a top-down modulation of plausibility or congruity effects.

If this proposal turns out to be viable, it could provide the basis for a new approach for inferring the organization of the neural language architecture. Rather than basing arguments on temporal precedence, investigations should focus more closely on *functional* precedence (see also our interpretation of the ELAN/N400 debate in Section 3.1): if an information source A can "block" the effects typically associated with a second information source B, this suggests that A takes functional (hierarchical) precedence over B. Depending on the particular type of manipulation examined, further inferences can be drawn regarding the likely nature of the blocking effect, that is, whether it results from a feedforward influence (as is likely the case for concurrent word category and semantic violations) or from a combination of long-range feedforward and feedback connections (as may be the case for the orthographic and emotional influences on semantic congruity effects).

49.3.6 The Emerging Picture

Although, as already noted at the outset of this chapter, our understanding of the timecourse of sentence processing in the brain is still woefully incomplete, and although we have a long way to go in advancing toward a truly neurobiologically grounded perspective, the preceding discussion shows that substantial progress has been made. Here, we attempt to summarize some of the insights emerging from this overall picture:

(a) ERP components appear to be less specific for particular cognitive—and, specifically, particular linguistic—representations and/or operations than was once thought. Accordingly, more and more accounts are adopting more general mechanistic interpretations of the components and are no longer emphasizing the distinction between them as much as used to be common.

(b) The importance of top-down versus bottom-up information has emerged as a major topic in the literature on sentence processing (Federmeier, 2007; Federmeier & Laszlo, 2009; Kutas & Federmeier, 2011; Lotze et al., 2011; Newman & Connolly, 2009; Tune et al., 2014). This view includes a critical role for prediction (DeLong, Urbach, & Kutas, 2005; Dikker & Pylkkänen, 2011; Dikker et al., 2010; Federmeier, Kutas, & Schul, 2010) and is also highly compatible with the assumption of a hierarchically organized neural (cortical) architecture with predictive coding and asymmetric forward and backward connections (Friston, 2005, 2010).

What are the consequences of these insights for our understanding of the timecourse of sentence processing in the brain? While the question remains difficult, we can draw at least the following minimal conclusion: if we want to think in cognitive terms, there is a high likelihood of a cascaded architecture (although an "imperfect" one due to the feedback connections and long-range feedforward connections described throughout this chapter). This also means that specific information sources are less important than was long thought for the processing of the current input item in comparison with their predictive capacity for upcoming input. This view essentially allows for a new synthesis of the classic modular and interactive views: it suggests that modular models placed too much emphasis on bottom-up information and a feedforward information flow, whereas interactive models placed too much emphasis on top-down information and a parallel information flow between all available information sources or representations.

49.4 BEHAVIORAL INSIGHTS

Before turning to some open questions arising from synthesis, we first briefly discuss how the neurophysiologically determined picture outlined relates to behavioral insights. To this end, we focus particularly on measures of eye movements during natural reading for several reasons: eye-tracking has long been the most influential behavioral method for inferring timecourse information during sentence processing; like neurophysiological measures, it does not require an ancillary task; and eye movement measures have challenged timecourse estimates provided by ERP components such as the N400 by demonstrating earlier influences of parameters such as word frequency. Furthermore, from a neurobiological perspective, it has been suggested that eye movements play a crucial role in probing our environment to test hypotheses arising from the current internal model about that environment (Friston, Adams, Perrinet, & Breakspear, 2012). It appears plausible that eye movements during reading may play a similar role. Accordingly, eye movement data appear highly relevant for informing our understanding of the timecourse of sentence processing, and for allowing us to gain a better understanding of the link between brain and behavior during language processing.

Recent results appear to emphasize the problems regarding absolute timing and ERP/ERF effects briefly discussed in Section 49.3.1. A range of recent studies has investigated in fine detail the question of when exactly factors such as frequency, predictability, and stimulus quality influence the eye movement record (Reingold, Reichle, Glaholt, & Sheridan, 2012; Sheridan & Reingold, 2012a, 2012b; Staub, 2011; Staub, White, Drieghe, Hollway, & Rayner, 2010; White & Staub, 2012) and has led to the conclusion that factors such as frequency and predictability manifest themselves even earlier than previously thought. For example, effects of word frequency were found in fixations as short as 140 ms (Reingold et al., 2012), as were effects of predictability (Sheridan & Reingold, 2012a, 2012b) and of sentence context on the processing of homographs such as *bank* (Sheridan & Reingold, 2012a, 2012b). Recall from Section 49.3.1 that, for fixation effects, the time needed for saccade planning must also be taken into account. Thus, these effects are exceedingly early, even from the perspective of possible neurophysiological effects "before the N400" (see Section 49.3.2).

However, the fact that these different information sources appear to have an equally early influence on the eye movement record during natural sentence reading does not necessarily mean that they should be viewed as playing functionally identical roles in the language processing architecture. As demonstrated by Staub et al. (2010) using distributional analyses of fixation durations, word frequency affects both the mean of the distribution and its skew, with low frequency words shifting the distribution to the right but also leading to a more pronounced right tail (Staub et al., 2010). However, Staub (2011) found that low predictability shifts the distribution of first-fixation and single-fixation durations to the right but has little effect on the shape of the distribution. This may amount to a "head start" in the word recognition process for predictability (Balota, Yap, Cortese, & Watson, 2008), which also affects skipping rates for a critical word and thus has an impact even before that word is fixated. In this regard, recent eye movement results are consistent with the conclusions drawn regarding the crucial role of prediction (or, more broadly speaking, top-down expectation) for our understanding of the timecourse of sentence processing. Intriguingly, however, the top-down/bottom-up balance manifests differently in the eye movement record than in ERP measures: although the effect of word frequency on the N400 is correlated negatively with the position of the critical word in the sentence (Van Petten & Kutas, 1990), frequency effects on fixation durations are reliably observed even in late positions within a sentence and many eye-tracking studies have failed to find a frequency by predictability interaction (Kennedy, Pynte, Murray, & Paul, 2013; Rayner, Ashby, Pollatsek, & Reichle, 2004). This suggests that eye movements reflect different aspects of the timecourse of sentence processing than neurophysiological measures, placing a stronger emphasis on bottom-up factors such as word frequency.

Regarding this possible functional discrepancy, we advance the following speculation. If Friston and colleagues are correct in their assumption that eye movements reflect experiments designed to test hypotheses arising from the current internal model about the external world (Friston et al., 2012), and if this perspective is also applicable to reading, then it appears entirely plausible that fixations during reading should be more sensitive to bottom-up characteristics of the input. If this were not the case, eye movements would be poorly suited to hypothesis testing. This assumption appears compatible with recent claims that eye movements during reading should be considered an "input system" rather than directly reflecting the cognitive processes associated with language comprehension (Kretzschmar, 2010; Kretzschmar, Bornkessel-Schlesewsky, & Schlesewsky, 2009). Clearly, however, this currently speculative hypothesis will need to be tested explicitly in future research.

Regarding absolute timecourse measures and the dissociation between fixation durations and ERP latencies, a few considerations are also in order. First, the very early effects for parameters such as frequency are only applicable with a valid parafoveal preview of (the) upcoming word(s). As shown by Reingold and colleagues (2012), without a valid preview, the earliest effects of frequency on fixation times were observed in fixations of 250 ms in length. Because parafoveal preview is not available in most visual EEG or MEG studies, which typically use rapid serial visual presentation (RSVP), this reduces the discrepancy between the methods somewhat. As demonstrated (Kretzschmar, 2010; Kretzschmar et al., 2009) using concurrent EEG and eye-tracking measures during natural reading, ERP patterns appear, under certain circumstances, to be qualitatively different during natural reading as opposed to RSVP, and this difference appears directly attributable to the presence versus absence of parafoveal preview. Second, as first proposed in Bornkessel and Schlesewsky (2006), if ERPs are viewed as resulting from phase resets within ongoing oscillatory activity in different frequency bands (Dogil, Frese, Haider, Roehm, & Wokurek, 2004; Makeig et al., 2002; Roehm, Bornkessel-Schlesewsky, & Schlesewsky, 2007), then it is possible that eye movements reflect (or are in some sense "triggered by") the initial point of reset, whereas the latency of the corresponding ERP component will depend on the frequency band involved. This proposal, while potentially promising, requires integration within a more comprehensive neurobiologically grounded model of sentence reading for it to generate testable predictions. Such a model is currently lacking, with all existing neurobiological models of reading focusing on the single word level (Dehaene & Cohen, 2011; Price & Devlin, 2011; see also Carreiras, Armstrong, Perea, & Frost, 2014).

49.5 OPEN QUESTIONS/PERSPECTIVES FOR FUTURE RESEARCH

This chapter has perhaps raised more questions than it has answered. Nevertheless, we hope to have provided some insights into what we view as the current state of the art with regard to the timecourse of sentence processing in the brain.

One of the central points discussed in this regard is that we are currently unable to make absolute claims about the timecourse of sentence processing from a neurobiological perspective; rather than being able to label particular information sources or mechanisms as becoming accessible or applying at *x* ms, we have argued that we can currently only make assumptions about the overall structure of the processing architecture and the relative functional role of different information types or mechanisms within that architecture. A central goal for future research thus appears to lie in the development of appropriate experimental approaches to map out absolute timecourse information. However, in view of the fact that this may not be possible with noninvasive methods, it is not clear at present whether this is an entirely realistic goal.

A second crucial point within this chapter concerned the role of predictions and the interaction of top-down and bottom-up information sources. Again, some of the crucial pieces to the puzzle are still missing in our understanding of these influences. First, future research must determine more definitively why some cues appear to be more highly predictive (top-down) or salient (bottom-up) than others, and under which circumstances bottom-up cues can "override" top-down cues. We have suggested here that the key to explaining at least the latter phenomenon may lie in the hierarchical organization of the neural language processing architecture, but this remains to be tested in detail. Second, with regard to the top-down/bottom-up balance, a number of influencing factors appear to contribute, including the properties of the specific language under consideration (Bornkessel-Schlesewsky et al., 2011; Tune et al., 2014). How these system-wide properties interact with more phasic influences—and how they are learned during development—is currently unknown.

Finally, given that top-down influences from previous input appear to play a particularly important role in influencing the timecourse of sentence processing in the brain, we clearly need to develop an understanding of the appropriate "cycles" of processing, which determine the influence of top-down processing from one input item to the next. Although words appear to be a natural and intuitive choice, they may not necessarily be the appropriate one. For example, approaches to sentence parsing in computational linguistics suggest that, in morphologically rich languages such as Turkish or Hindi, the use of morphologically based units smaller than the word improves parsing performance (Eryiğit, Nivre, & Oflazer, 2008). This observation raises the intriguing complication that the relevant units may prove to be language-specific. From a neurobiological perspective, the notion of integration windows has also played an important role in several approaches. For example, Pulvermüller and colleagues argue for a window of approximately 200 ms as a "discrete processing step [...] in perceptual, motor, and cognitive processes" (Pulvermüller et al., 2009, p. 87). This claim is partially compatible with the long temporal integration window (approximately 150–250 ms) proposed by Poeppel (2003) as the basis for right hemisphere auditory processing (in contrast to the short temporal integration window of approximately 20–40 ms, which, according to this account, better describes auditory processing in the left hemisphere). The longer window, which aligns with theta oscillations, represents a timescale at approximately the syllable level (Giraud & Poeppel, 2012), an idea that appears potentially compatible with the assumption that units at a level below the word may ultimately prove appropriate for a more comprehensive understanding of the timecourse of sentence processing from a truly neurobiological perspective.

References

Balota, D. A., Yap, M. J., Cortese, M. J., & Watson, J. M. (2008). Beyond mean response latency: Response time distributional analyses of semantic priming. *Journal of Memory and Language*, *59*(4), 495–523.

Bar, M. (2003). A cortical mechanism for triggering top-down facilitation in visual object recognition. *Journal of Cognitive Neuroscience*, *15*(4), 600–609.

Besson, M., & Macar, F. (1987). An event-related potential analysis of incongruity in music and other non-linguistic contexts. *Psychophysiology*, *24*, 14–25.

Bornkessel, I., McElree, B., Schlesewsky, M., & Friederici, A. D. (2004). Multi-dimensional contributions to garden path strength: Dissociating phrase structure from case marking. *Journal of Memory and Language*, *51*, 495–522.

Bornkessel, I., & Schlesewsky, M. (2006). The extended argument dependency model: A neurocognitive approach to sentence comprehension across languages. *Psychological Review*, *113*, 787–821.

Bornkessel-Schlesewsky, I., Kretzschmar, F., Tune, S., Wang, L., Genç, S., Philipp, M., et al. (2011). Think globally: Cross-linguistic variation in electrophysiological activity during sentence comprehension. *Brain and Language*, *117*(3), 133–152.

Bornkessel-Schlesewsky, I., & Schlesewsky, M. (2008). An alternative perspective on "semantic P600" effects in language comprehension. *Brain Research Reviews*, *59*, 55–73.

Bornkessel-Schlesewsky, I., & Schlesewsky, M. (2009). *Processing syntax and morphology: A neurocognitive perspective*. Oxford: Oxford University Press.

Bornkessel-Schlesewsky, I., & Schlesewsky, M. (2013). Reconciling time, space and function: A new dorsal-ventral stream model of sentence comprehension. *Brain and Language*, *125*, 60–76.

Brouwer, H., Fitz, H., & Hoeks, J. C. (2012). Getting real about semantic illusions: Rethinking the functional role of the P600 in language comprehension. *Brain Research*, *1446*, 127–143.

Carreiras, M., Armstrong, B. C., Perea, M., & Frost, R. (2014). The what, when, where, and how of visual word recognition. *Trends in Cognitive Sciencesognitive Sciences*, *18*(2), 90–98.

Chang, E. F., Rieger, J. W., Johnson, K., Berger, M. S., Barbaro, N. M., & Knight, R. T. (2010). Categorical speech representation in human superior temporal gyrus. *Nature Neuroscience*, *13*(11), 1428–1432.

Choudhary, K. K., Schlesewsky, M., Roehm, D., & Bornkessel-Schlesewsky, I. (2009). The N400 as a correlate of interpretively-relevant linguistic rules: Evidence from Hindi. *Neuropsychologia*, *47*, 3012–3022.

Connolly, J. F., & Phillips, N. A. (1994). Event-related potential components reflect phonological and semantic processing of the terminal word of spoken sentences. *Journal of Cognitive Neuroscience*, *6*, 256–266.

Coulson, S., King, J. W., & Kutas, M. (1998). Expect the unexpected: Event-related brain response to morphosyntactic violations. *Language and Cognitive Processes*, *13*, 21–58.

Dehaene, S., & Cohen, L. (2011). The unique role of the visual word form area in reading. *Trends in Cognitive Sciences*, *15*(6), 254–262.

Delaney-Busch, N., & Kuperberg, G. R. (2013). Friendly drug-dealers and terrifying puppies: Affective primacy can attenuate the N400 effect in emotional discourse contexts. *Cognitive, Affective and Behavioral Neuroscience*, *13*, 473–490.

DeLong, K. A., Urbach, T. P., & Kutas, M. (2005). Probabilistic word pre-activation during language comprehension inferred from electrical brain activity. *Nature Neuroscience*, *8*, 1117–1121.

Dien, J. (2009). Foreword to the special issue "Before the N400: Early Latency Language ERPs". *Biological Psychology*, *80*, 1–3.

Dikker, S., & Pylkkänen, L. (2011). Before the N400: Effects of lexical–semantic violations in visual cortex. *Brain and Language*, *118*, 23–28.

Dikker, S., Rabagliati, H., Farmer, T. A., & Pylkkänen, L. (2010). Early occipital sensitivity to syntactic category is based on form typicality. *Psychological Science*, *21*(5), 629–634. Available from: http://dx.doi.org/10.1177/0956797610367751.

Dikker, S., Rabagliati, H., & Pylkkänen, L. (2009). Sensitivity to syntax in visual cortex. *Cognition*, *110*, 293–321.

Dogil, G., Frese, I., Haider, H., Roehm, D., & Wokurek, W. (2004). Where and how does grammatically geared processing take place—And why is Broca's area often involved. A coordinated fMRI/ERBP study of language processing. *Brain and Language*, *89*, 337–345.

Eryiğit, G., Nivre, J., & Oflazer, K. (2008). Dependency parsing of Turkish. *Computational Linguistics*, *34*, 357–389.

Federmeier, K. D. (2007). Thinking ahead: The role and roots of prediction in language comprehension. *Psychophysiology*, *44*, 491–505.

Federmeier, K. D., Kutas, M., & Schul, R. (2010). Age-related and individual differences in the use of prediction during language comprehension. *Brain and Language*, *115*, 149–161.

Federmeier, K. D., & Laszlo, S. (2009). Time for meaning: Electrophysiology provides insights into the dynamics of representation and processing in semantic memory. In B. H. Ross (Ed.), *Psychology of learning and motivation* (Vol. 51, pp. 1–44). Burlington: Academic Press.

Frazier, L. (1987). Sentence processing: A tutorial review. In M. Coltheart (Ed.), *Attention and performance, vol. 12: The psychology of reading* (pp. 559–586). Hove: Erlbaum.

Frazier, L., & Clifton, C. (1996). *Construal*. Cambridge, MA: MIT Press.

Friederici, A. D. (1995). The time course of syntactic activation during language processing: A model based on neuropsychological and neurophysiological data. *Brain and Language*, *50*, 259–281.

Friederici, A. D. (1999). The neurobiology of language comprehension. In A. D. Friederici (Ed.), *Language comprehension: A biological perspective* (pp. 263–301). Berlin/Heidelberg/New York: Springer.

Friederici, A. D. (2002). Towards a neural basis of auditory sentence processing. *Trends in Cognitive Sciences*, *6*(2), 78–84.

Friederici, A. D. (2009). Pathways to language: Fiber tracts in the human brain. *Trends in Cognitive Sciences*, *13*, 175–181.

Friederici, A. D. (2011). The brain basis of language processing: From structure to function. *Physiological Reviews*, *91*(4), 1357–1392.

Friederici, A. D., Pfeifer, E., & Hahne, A. (1993). Event-related brain potentials during natural speech processing: Effects of semantic, morphological, and syntactic violations. *Cognitive Brain Research*, *1*, 183–192.

Frisch, S., & Schlesewsky, M. (2001). The N400 indicates problems of thematic hierarchizing. *Neuroreport*, *12*, 3391–3394.

Friston, K. (2005). A theory of cortical responses. *Philosophical Transactions of the Royal Society B: Biological Sciences*, *360*(1456), 815–836.

Friston, K., Adams, R. A., Perrinet, L., & Breakspear, M. (2012). Perceptions as hypotheses: Saccades as experiments. *Frontiers in Psychology*, *3*, 151. Available from: http://dx.doi.org/10.3389/fpsyg.2012.00151.

Friston, K. J. (2010). The free-energy principle: A unified brain theory? *Nature Reviews Neuroscience*, *11*, 127–138.

Fuster, J. M. (1997). Network memory. *Trends in Neurosciences*, *20*(10), 451–459.

Garrido, M. I., Kilner, J. M., Stephan, K. E., & Friston, K. J. (2009). The mismatch negativity: A review of underlying mechanisms. *Clinical Neurophysiology*, *120*, 453–463.

Giraud, A.-L., & Poeppel, D. (2012). Cortical oscillations and speech processing: Emerging computational principles and operations. *Nature Neuroscience*, *15*, 511–517.

Gunter, T. C., Stowe, L. A., & Mulder, G. (1997). When syntax meets semantics. *Psychophysiology*, *34*, 660–676.

Hagoort, P. (2003). How the brain solves the binding problem for language: A neurocomputational model of syntactic processing. *NeuroImage, 20*, S18–S29.
Hagoort, P. (2005). On Broca, brain, and binding: A new framework. *Trends in Cognitive Sciences, 9*, 416–423.
Hagoort, P. (2008). The fractionation of spoken language understanding by measuring electrical and magnetic brain signals. *Philosophical Transactions of the Royal Society B, 363*, 1055–1069.
Hagoort, P. (2013). MUC (Memory, Unification, Control) and beyond. *Frontiers in Psychology, 4*, 416.
Hagoort, P., Brown, C., & Osterhout, L. (1999). The neurocognition of syntactic processing. In C. Brown, & P. Hagoort (Eds.), *The neurocognition of language* (pp. 273–316). Oxford: Oxford University Press.
Hagoort, P., & van Berkum, J. J. (2007). Beyond the sentence given. *Philosophical Transactions of the Royal Society B, 362*, 801–811.
Hahne, A., & Friederici, A. D. (2002). Differential task effects on semantic and syntactic processes as revealed by ERPs. *Cognitive Brain Research, 13*, 339–356.
Hasting, A. S., Kotz, S. A., & Friederici, A. D. (2007). Setting the stage for automatic syntax processing: The mismatch negativity as an indictor of syntactic priming. *Journal of Cognitive Neuroscience, 19*, 386–400.
Haupt, F. S., Schlesewsky, M., Roehm, D., Friederici, A. D., & Bornkessel-Schlesewsky, I. (2008). The status of subject-object reanalyses in the language comprehension architecture. *Journal of Memory and Language, 59*, 54–96.
Hoeks, J. C., Stowe, L. A., & Doedens, G. (2004). Seeing words in context: The interaction of lexical and sentence level information during reading. *Cognitive Brain Research, 19*, 59–73.
Hopf, J.-M., Bayer, J., Bader, M., & Meng, M. (1998). Event-related brain potentials and case information in syntactic ambiguities. *Journal of Cognitive Neuroscience, 10*, 264–280.
Inhoff, W. A., & Rayner, K. (1986). Parafoveal word processing during eye fixations in reading: Effects of word frequency. *Perception and Psychophysics, 40*, 431–439.
Kennedy, A., Pynte, J., Murray, W. S., & Paul, S.-A. (2013). Frequency and predictability effects in the Dundee Corpus: An eye movement analysis. *The Quarterly Journal of Experimental Psychology, 66*(3), 601–618.
Kim, A., & Osterhout, L. (2005). The independence of combinatory semantic processing: Evidence from event-related potentials. *Journal of Memory and Language, 52*, 205–225.
Kolk, H. H., Chwilla, D. J., van Herten, M., & Oor, P. J. (2003). Structure and limited capacity in verbal working memory: A study with event-related potentials. *Brain and Language, 85*, 1–36.
Kravitz, D. J., Saleem, K. S., Baker, C. I., & Mishkin, M. (2011). A new neural framework for visuospatial processing. *Nature Reviews Neuroscience, 12*(4), 217–230. Available from: http://dx.doi.org/10.1038/nrn3008.
Kravitz, D. J., Saleem, K. S., Baker, C. I., Ungerleider, L. G., & Mishkin, M. (2013). The ventral visual pathway: An expanded neural framework for the processing of object quality. *Trends in Cognitive Sciences, 17*(1), 26–49.
Kretzschmar, F. (2010). *The electrophysiological reality of parafoveal processing: On the validity of language-related ERPs in natural reading*. Unpublished Doctoral dissertation, University of Marburg.
Kretzschmar, F., Bornkessel-Schlesewsky, I., & Schlesewsky, M. (2009). Parafoveal vs. foveal N400s dissociate spreading activation from contextual fit. *NeuroReport, 20*, 1613–1618.
Kujala, T., Tervaniemi, M., & Schröger, E. (2007). The mismatch negativity in cognitive and clinical neuroscience: Theoretical and methodological considerations. *Biological Psychology, 74*(1), 1–19.
Kuperberg, G. R., Sitnikova, T., Caplan, D., & Holcomb, P. (2003). Electrophysiological distinctions in processing conceptual relationships within simple sentences. *Cognitive Brain Research, 17*, 117–129.
Kutas, M., & Federmeier, K. D. (2011). Thirty years and counting: Finding meaning in the N400 component of the event-related brain potential (ERP). *Annual Review of Psychology, 62*, 621–647.
Kutas, M., & Hillyard, S. A. (1980). Reading senseless sentences: Brain potentials reflect semantic incongruity. *Science, 207*, 203–205.
Kutas, M., & Iragui, V. (1998). The N400 in a semantic categorization task across 6 decades. *Electroencephalography and Clinical Neurophysiology, 108*, 456–471.
Laszlo, S., & Federmeier, K. D. (2011). The N400 as a snapshot of interactive processing: Evidence from regression analyses of orthographic neighbor and lexical associate effects. *Psychophysiology, 48*, 176–186.
Lotze, N., Tune, S., Schlesewsky, M., & Bornkessel-Schlesewsky, I. (2011). Meaningful physical changes mediate lexical-semantic integration: Top-down and form-based bottom-up information sources interact in the N400. *Neuropsychologia, 49*, 3573–3582.
MacDonald, M. C., Pearlmutter, N. J., & Seidenberg, M. S. (1994). The lexical nature of syntactic ambiguity resolution. *Psychological Review, 101*, 676–703.
MacGregor, L. J., Pulvermuller, F., van Casteren, M., & Shtyrov, Y. (2012). Ultra-rapid access to words in the brain. *Nature Communications, 3*, 711.
Makeig, S., Westerfield, M., Jung, T. P., Enghoff, S., Townsend, J., Courchesne, E., et al. (2002). Dynamic brain sources of visual evoked responses. *Science, 295*, 690–694.
McClelland, J. L. (1979). On the time relations of mental processes: An examination of systems of processes is cascade. *Psychological Review, 86*, 287–330.
Menning, H., Zwitserlood, P., Schöning, S., Hihn, H., Bölte, J., Dobel, C., et al. (2005). Pre-attentive detection of syntactic and semantic errors. *Neuroreport, 16*(1), 77–80.
Mueller, J. L., Hahne, A., Fujii, Y., & Friederici, A. D. (2005). Native and nonnative speakers' processing of a miniature version of Japanese as revealed by ERPs. *Journal of Cognitive Neuroscience, 17*, 1229–1244.
Münte, T. F., Heinze, H. J., Matzke, M., Wieringa, B. M., & Johannes, S. (1998). Brain potentials and syntactic violations revisited: No evidence for specificity of the syntactic positive shift. *Neuropsychologia, 36*, 217–226.
Näätänen, R., Paavilainen, P., Alho, K., Reinikainen, K., & Sams, M. (1989). Do event-related potentials reveal the mechanism of the auditory sensory memory in the human brain? *Neuroscience Letters, 98*(2), 217–221.
Newman, R. L., & Connolly, J. F. (2009). Electrophysiological markers of pre-lexical speech processing: Evidence for bottom-up and top-down effects on spoken word processing. *Biological Psychology, 80*, 114–121.
Osterhout, L., & Nicol, J. (1999). On the distinctiveness, independence, and time course of the brain response to syntactic and semantic anomalies. *Language and Cognitive Processes, 14*, 283–317.
Patterson, K., Nestor, P. J., & Rogers, T. T. (2007). Where do you know what you know? The representation of semantic knowledge in the human brain. *Nature Reviews Neuroscience, 8*, 976–988.
Penolazzi, B., Hauk, O., & Pulvermüller, F. (2007). Early semantic context integration and lexical access as revealed by event-related brain potentials. *Biological Psychology, 74*(3), 374–388.
Poeppel, D. (2003). The analysis of speech in different temporal integration windows: Cerebral lateralizations as "asymmetric sampling in time". *Speech Communication, 41*, 245–255.
Price, C. J., & Devlin, J. T. (2011). The interactive account of ventral occipitotemporal contributions to reading. *Trends in Cognitive sciences, 15* (6), 246–253. Available from: http://dx.doi.org/10.1016/j.tics.2011.04.001.

Pulvermüller, F., & Shtyrov, Y. (2003). Automatic processing of grammar in the human brain as revealed by the mismatch negativity. *NeuroImage*, *20*(1), 159–172.

Pulvermüller, F., Shtyrov, Y., & Hauk, O. (2009). Understanding in an instant: Neurophysiological evidence for mechanistic language circuits in the brain. *Brain and Language*, *110*, 81–94.

Rauschecker, J. P., & Scott, S. K. (2009). Maps and streams in the auditory cortex: Nonhuman primates illuminate human speech processing. *Nature Neuroscience*, *12*(6), 718–724.

Rayner, K. (1998). Eye movements in reading and information processing: 20 years of research. *Psychological Bulletin*, *124*, 372–422.

Rayner, K., Ashby, J., Pollatsek, A., & Reichle, E. D. (2004). The effects of frequency and predictability on eye fixations in reading: Implications for the EZ Reader model. *Journal of Experimental Psychology: Human Perception and Performance*, *30*(4), 720.

Rayner, K., Carlson, G., & Frazier, L. (1983). The interaction of syntax and semantics during sentence processing: Eye movements in the analysis of semantically biased sentences. *Journal of Verbal Learning and Verbal Behavior*, *22*, 657–673.

Rayner, K., & Clifton, C. (2009). Language processing in reading and speech perception is fast and incremental: Implications for event-related potential research. *Biological Psychology*, *80*(1), 4–9.

Rayner, K., Slowiaczek, M. L., Clifton, C., & Bertera, J. H. (1983). Latency of sequential eye movements: Implications for reading. *Journal of Experimental Psychology: Human Perception and Performance*, *9*(6), 912.

Reingold, E. M., Reichle, E. D., Glaholt, M. G., & Sheridan, H. (2012). Direct lexical control of eye movements in reading: Evidence from a survival analysis of fixation durations. *Cognitive Psychology*, *65*(2), 177–206.

Roehm, D., Bornkessel-Schlesewsky, I., & Schlesewsky, M. (2007). The internal structure of the N400: Frequency characteristics of a language-related ERP component. *Chaos and Complexity Letters*, *2*, 365–395.

Sassenhagen, J., Schlesewsky, M., & Bornkessel-Schlesewsky, I. (2014). The P600-as-P3 hypothesis revisited: Single-trial analyses reveal that the late EEG positivity following linguistically deviant material is reaction time aligned. *Brain and Language*, *137*, 29–39.

Sereno, S. C., Brewer, C. C., & O'Donnell, P. J. (2003). Context effects in word recognition: Evidence for early interactive processing. *Psychological Science*, *14*, 328–333.

Sereno, S. C., & Rayner, K. (2003). Measuring word recognition in reading: Eye movements and event-related potentials. *Trends in Cognitive Sciences*, *7*, 489–493.

Sereno, S. C., Rayner, K., & Posner, M. I. (1998). Establishing a time-line of word recognition: Evidence from eye movements and event-related potentials. *Neuroreport*, *9*, 2195–2200.

Sheridan, H., & Reingold, E. M. (2012a). The time course of contextual influences during lexical ambiguity resolution: Evidence from distributional analyses of fixation durations. *Memory and Cognition*, *40*(7), 1122–1131.

Sheridan, H., & Reingold, E. M. (2012b). The time course of predictability effects in reading: Evidence from a survival analysis of fixation durations. *Visual Cognition*, *20*(7), 733–745.

Shtyrov, Y., & Pulvermüller, F. (2002). Memory traces for inflectional affixes as shown by mismatch negativity. *European Journal of Neuroscience*, *15*, 1085–1091.

Shtyrov, Y., & Pulvermüller, F. (2007). Early MEG activation dynamics in the left temporal and inferior frontal cortex reflect semantic context integration. *Journal of Cognitive Neuroscience*, *19*(10), 1633–1642.

Simon, H. A. (1962). The architecture of complexity. *Proceedings of the American Philosophical Society*, *106*, 467–482.

Staub, A. (2011). The effect of lexical predictability on distributions of eye fixation durations. *Psychonomic Bulletin and Review*, *18*(2), 371–376.

Staub, A., White, S. J., Drieghe, D., Hollway, E. C., & Rayner, K. (2010). Distributional effects of word frequency on eye fixation durations. *Journal of Experimental Psychology: Human Perception and Performance*, *36*(5), 1280.

Steinhauer, K., & Drury, J. E. (2012). On the early left-anterior negativity (ELAN) in syntax studies. *Brain and Language*, *120*(2), 135–162.

Trueswell, J. C., & Tanenhaus, M. K. (1994). Toward a lexicalist framework for constraint-based syntactic ambiguity resolution. In C. Clifton, L. Frazier, & K. Rayner (Eds.), *Perspectives in sentence processing*. Hillsdale, NJ: Erlbaum.

Tse, C.-Y., Lee, C.-L., Sullivan, J., Garnsey, S. M., Dell, G. S., Fabiani, M., et al. (2007). Imaging cortical dynamics of language processing with the event-related optical signal. *Proceedings of the National Academy of Sciences of the United States of America*, *104*(43), 17157–17162.

Tune, S., Schlesewsky, M., Small, S. L., Sanford, A. J., Bohan, J., Sassenhagen, J., et al. (2014). Cross-linguistic variation in the neurophysiological response to semantic processing: Evidence from anomalies at the borderline of awareness. *Neuropsychologia*, *56*, 147–166.

Ullman, M. T. (2001). A neurocognitive perspective on language: The declarative/procedural model. *Nature Reviews Neuroscience*, *2*, 717–726.

van de Meerendonk, N., Kolk, H. H., Chwilla, D. J., & Vissers, C. T. W. M. (2009). Monitoring in language perception. *Language and Linguistics Compass*, *3*, 1211–1224.

van den Brink, D., Brown, C. M., & Hagoort, P. (2001). Electrophysiological evidence for early contextual influences during spoken-word recognition: N200 versus N400 effects. *Journal of Cognitive Neuroscience*, *13*(7), 967–985.

van den Brink, D., Brown, C. M., & Hagoort, P. (2006). The cascaded nature of lexical selection and integration in auditory sentence processing. *Journal of Experimental Psychology Learning, Memory, and Cognition*, *32*(2), 364–372.

Van Den Brink, D., & Hagoort, P. (2004). The influence of semantic and syntactic context constraints on lexical selection and integration in spoken-word comprehension as revealed by ERPs. *Journal of Cognitive Neuroscience*, *16*(6), 1068–1084.

Van Petten, C., & Kutas, M. (1990). Interactions between sentence context and word frequency in event-related brain potentials. *Memory and Cognition*, *18*(4), 380–393.

Vigliocco, G., Vinson, D. P., Druks, J., Barber, H., & Cappa, S. F. (2011). Nouns and verbs in the brain: A review of behavioural, electrophysiological, neuropsychological and imaging studies. *Neuroscience and Biobehavioral Reviews*, *35*, 407–426.

Vissers, C. T. W. M., Chwilla, D. J., & Kolk, H. H. J. (2006). Monitoring in language perception: The effect of misspellings of words in highly constrained sentences. *Brain Research*, *1106*, 150–163.

White, S. J., & Staub, A. (2012). The distribution of fixation durations during reading: Effects of stimulus quality. *Journal of Experimental Psychology: Human Perception and Performance*, *38*(3), 603.

Wicha, N. Y. Y., Moreno, E. M., & Kutas, M. (2004). Anticipating words and their gender: An event-related brain potential study of semantc integration, gender expectancy, and gender agreement in Spanish sentence reading. *Journal of Cognitive Neuroscience*, *16*, 1271–1288.

Wolpert, D. M., Doya, K., & Kawato, M. (2003). A unifying computational framework for motor control and social interaction. *Philosophical Transactions of the Royal Society B*, *358*(1431), 593–602.

CHAPTER

50

Composition of Complex Meaning: Interdisciplinary Perspectives on the Left Anterior Temporal Lobe

Liina Pylkkänen

Department of Linguistics, Department of Psychology, New York University, New York, NY, USA; NYUAD Institute, New York University Abu Dhabi, Abu Dhabi, United Arab Emirates

The ultimate goal of the cognitive neuroscience of language is to fully characterize the computations by which complex meanings are constructed from sensory input in comprehension and the way such meanings are externalized as motor commands in language production. Thus, a complex semantic representation is the end product of comprehension and the starting point of production. What are the neural computations that construct these complex semantic representations from smaller, stored units of meaning? Addressing this question must begin with a definition of the relevant computations.

The simplest way to define semantic composition is as the process by which the meaning of a lexical item combines with its context. In a simple case, the comprehension of the noun phrase *black cat* requires us to create a mental representation of an entity that is both black and a cat. However, at least descriptively, composition clearly has a variety of forms, for example, to interpret the verb phrase *banks money*, we do not construct a mental representation of an entity that is both a bank and money, but rather take money to be an object affected by banking. Even if we keep the syntactic categories of the composing items constant, composition plays out in many different ways. For example, although a black cat is an individual that is both black and a cat, a possible murderer is not an individual who is both "possible" and a murderer—in fact, that person may not be a murderer at all. Or, if your colleague reports that an occasional student walked into their office, you do not take this to mean that a person who is occasionally a student visited the office, but rather that the office was occasionally visited by students. In sum, the composition of natural language meanings is complicated. Nevertheless, our formal understanding of the nature of the relevant representations is highly detailed, thanks to decades of research within formal semantics, a sub-branch of linguistics with origins in philosophical logic and generative grammar (Portner & Partee, 2008).

If the empirical phenomena within natural language semantics are this complex and multifaceted, then we should consequently expect the brain profile of semantic processing to exhibit similar intricacies. However, the state of the art within the cognitive neuroscience of sentence level semantics is that we barely understand the composition of even the simplest of cases, such as *black cat*. How can this be at a time when a search for "brain" and "semantic integration" yields thousands of Google Scholar hits?

At least partly, the answer lies in the fact that instead of starting with the simplest cases, such as the composition of two simple elements, a major focus in the field has been to investigate rather complex phenomena, such as the processing of semantically anomalous expressions or ones containing complex embeddings. These large bodies of research have been inspired by the seminal and highly replicable findings of Kutas and Hillyard (1980) and Stromswold and colleagues (1996), who reported the electrophysiological N400 as a response sensitive to semantic incongruity and Broca's area as a region supporting the comprehension of center-embedded structures, respectively. Although this work has provided invaluable insight to

Neurobiology of Language. DOI: http://dx.doi.org/10.1016/B978-0-12-407794-2.00050-X

our understanding of the interface between linguistic meaning and neighboring components of the mind, such as world knowledge and working memory, these rather complex phenomena have been addressed before developing a basic understanding of how words compose into larger meanings within simple well-formed expressions.

A second limiting factor for our understanding of the brain basis of meaning representation has been that the cognitive neuroscience of semantics, especially at the sentence level, has not been tightly connected with linguistic theory or psycholinguistics; thus, the brain science of meaning has not substantially benefitted from the large bodies of results within these older disciplines. In fact, among the various representational and processing levels within language, semantics arguably exemplifies the starkest disconnect between brain research and corresponding theoretical work. Although sizeable bodies of cognitive neuroscience research have investigated the brain basis of theoretically motivated constructs such as phonological categories (Dehaene-Lambertz, 1997; Näätänen et al., 1997) or syntactic displacement (Grodzinsky & Friederici, 2006), terms from the semanticist's basic tool kit such as "function application" or "variable binding" (Heim & Kratzer, 1998) do not figure within the cognitive neuroscience of semantics.

In this chapter I first describe in more detail what "semantics" is typically taken to mean in cognitive neuroscience on the one hand, and in linguistics on the other. Within mainstream hemodynamic research on sentence processing, one popular paradigm, the so-called sentence versus list paradigm, arises as the most relevant for identifying brain activity related to basic composition; I discuss this experimental design from the point of view of formal semantics. Although it is unlikely that any type of composition would be accomplished by a single brain region—even experiments examining the composition of just two words have implicated at least three regions as sensitive to such composition, namely the left anterior temporal lobe (LATL), the ventromedial prefrontal cortex (vmPFC), and the angular gyrus (Bemis & Pylkkänen, 2011, 2012, 2013; Pylkkänen, Bemis, & Blanco Elorrieta, 2014; for relevant reviews, see also Binder & Desai, 2011; Friederici, 2012; Hagoort & Indefrey, 2014; Lau, Phillips, & Poeppel, 2008)—there is a single region, namely the LATL, within which combinatory effects have had a very high rate of replicability across studies, labs, and techniques. This has enabled a productive incremental research program aimed at characterizing the computational contribution of LATL activity in a way that has been harder for other regions whose contribution has been less systematic (Bemis & Pylkkänen, 2011 vs. 2012)—clearly, the absence of an effect even in an interaction setting can only be meaningful if the activity in question is known to generate few false negatives. At large, a combined body of hemodynamic and magnetoencephalography (MEG) results strongly implicate the LATL as a basic combinatory region supporting incremental sentence interpretation. In particular, MEG studies have enabled the characterization of the temporal profile of this activity, demonstrating that combinatory effects in the LATL peak relatively early, at 200–250 ms. However, in the latter half of the chapter, I review a body of recent results suggesting that this activity, despite its timing, is fundamentally semantic as opposed to syntactic in nature.

Given that this chapter mainly discusses MEG research, in which spatial resolution is fuzzier than in PET or fMRI, my primary focus is on time rather than precise spatial details. The MEG localization of LATL combinatory effects has varied somewhat, most consistently implicating the temporal pole and immediately adjacent tissue (Bemis & Pylkkänen, 2011, 2012, 2013; Del Prato & Pylkkänen, 2014; Leffel, Lauter, Westerlund, & Pylkkänen, 2014; Pylkkänen et al., 2014), conforming to relevant hemodynamic data (Baron & Osherson, 2011), but sometimes also localizing on the more lateral surface of left anterior temporal cortex (Westerlund & Pylkkänen, 2014). A similar and somewhat variable picture arises from hemodynamic LATL research on single concepts (see Section 50.6.1), where semantic effects have occurred at least within the temporal pole (Gauthier, Anderson, Tarr, Skudlarski, & Gore, 1997), on the ventral surface of the anterior temporal lobe (Binney, Embleton, Jefferies, Parker, & Ralph, 2010; Visser et al., 2012), and more medially (Clarke, Taylor, & Tyler, 2011; Tyler et al., 2013). For the purpose of this chapter, the reader can assume that LATL refers to a somewhat large region covering the temporal pole (BA 38) and anterior portions of middle and inferior temporal gyri (BAs 20 and 21), because this is the area within which the LATL combinatory response systematically localizes in MEG. Thus, there is obviously more to be said about the internal organization of this large area, but such an understanding will likely come from techniques other than MEG.

50.1 "SEMANTICS" IN THE BRAIN SCIENCES VERSUS LINGUISTICS

The word "semantics" can be used in many different ways and, in particular, this term means something different for most cognitive neuroscientists than what it does for linguists. Traditionally, within the brain sciences, "semantics" has meant more or less what the "man-on-the-street" might mean by it: the conceptual content of words and their combinations. Although one would not usually find this definition in publications, it is deducible from the nature of the stimulus manipulations that have been considered "semantic." At the word level, relevant distinctions have included contrasts such as tools versus animals (Devlin et al., 2002; Ilmberger, Rau, Noachtar, Arnold, & Winkler, 2002; Vitali et al., 2005), living versus nonliving (Tyler et al., 2003), and abstract versus concrete (Noppeney & Price, 2004). Commonly, this literature has also aimed to modulate "semantics" by keeping the stimuli constant and varying the task; for example, an abstract versus concrete judgment might count as a semantic task and syllable counting might be considered as a phonological task (Poldrack et al., 1999). At the sentence level, semantic manipulations have mostly varied factors such as coherence—word salad sentences such as *the freeway on a pie watched the house and a window* counting as incoherent expressions (Humphries, Binder, Medler, & Liebenthal, 2006)—or plausibility, a similar variable that typically refers to the plausibility of a single lexical item in its context (Kutas & Federmeier, 2000). The origin of this approach to sentence semantics straightforwardly dates to Kutas and Hillyard's (1980) pioneering discovery of the electrophysiological N400, a brain response to anomalous sentence endings such as *socks* as in their much-cited example *he spread the warm bread with socks*.

While this lexical-level literature has significantly advanced our understanding of the neural underpinnings of long-term semantic memory (in the sense of Tulving, 1972; for reviews see Binder & Desai, 2011; Kutas & Federmeier, 2000) and the sentence-level research has significantly advanced our understanding of the interface between sentence meaning and world knowledge (Hagoort, Hald, Bastiaansen, & Petersson, 2004; Pylkkänen, Martin, McElree, & Smart, 2009; Pylkkänen, Brennan, & Bemis, 2010), there is a crucial level of processing that has remained largely unaddressed by the majority of brain research on semantics: the computations that create the complex meanings of well-formed sentences. Characterizing the rule system that enables such computations is the primary subject matter of formal semantics in linguistics; thus, linguists typically find little of relevance to their research in the vast cognitive neuroscience literature on semantics. Specifically, the distinctions that dominate the lexical level literature in the brain sciences (i.e., differences such as the contrast between fruits and vegetables, or tools and animals) are precisely the types of factors that are thought *not* to matter for the generative procedures that compose meanings together. Rather, the variables that do matter for these routines include contrasts such as the mass/count distinction (Chierchia, 1998; Link, 1983), the relationality of nouns (Barker, 1995), the internal temporal structure of the events described by verbs (Dowty, 1979), and the gradability of adjectives (Kennedy, 2007) (i.e., factors that matter for the possible distributions of lexical items within well-formed expressions). The majority of such variables have no corresponding cognitive neuroscience associated with them, with the mass/count distinction perhaps standing out as somewhat of an exception, having inspired several brain investigations (Chiarelli et al., 2011; Fieder, Nickels, Biedermann, & Best, 2014; Steinhauer, Pancheva, Newman, Gennari, & Ullman, 2001; Taler, Jarema, & Saumier, 2005).

In sum, the brain sciences have focused on the more conceptual aspects of meaning and formal semantics on the more grammatical ones. Consequently, the more grammatical aspects of meaning constitute a large and mostly uninvestigated terrain for the neurobiology of language.

50.2 THE SENTENCE VERSUS LIST PARADIGM

To characterize the brain basis of the computations that create complex meanings, one should vary the presence, number, or difficulty of those computations. One line of hemodynamic literature on sentence processing, situated outside the majority paradigms focused on anomaly, has in fact directly aimed to do this, although from the perspective of formal semantics; it is not entirely obvious whether and how the manipulation in question achieves this. In these studies, the critical contrast is always between a fully structured sentence and an unstructured list of words (Friederici, Meyer, & von Cramon, 2000; Humphries, Love, Swinney, & Hickok, 2005; Humphries et al., 2006; Jobard, Vigneau, Mazoyer, & Tzourio-Mazoyer, 2007; Mazoyer et al., 1993; Rogalsky & Hickok, 2009; Snijders et al., 2009; Stowe et al., 1998; Vandenberghe et al., 2002; Xu, Kemeny, Park, Frattali, & Braun, 2005), with the idea being that the sentence engages all the computations involved in sentence processing whereas the list does not, which of course should, at least in most cases, be true[1] (for discussion, see Brennan & Pylkkänen, 2012).

The version of this paradigm that is most relevant for *semantic* composition is one that also includes

[1]The only hesitation arising from possibly divergent processing strategies when encountering a stimulus such as *the do the accept shopkeeper napkin* (from Vandenberghe, Nobre, & Price, 2002); it seems at least possible that the participant might still attempt to construct some kind of meaning from this.

conditions in which all the open class words within the stimuli are replaced by pseudowords, intended as a way to remove semantics from the experimental materials (Friederici et al., 2000; Humphries et al., 2006; Mazoyer et al., 1993; Pallier, Devauchelle, & Dehaene, 2011). This results in a contrast between pseudoword sentences and pseudoword lists, such as *the solims on a sonting grilloted a yome and a sovir* (pseudoword sentence) versus *rooned the sif into lilf the and the foig aurene to* (pseudoword list) from Humphries et al. (2006). On a descriptive level, this particular pseudoword sentence of course includes many constituents, and the first five words of the stimulus have the structure of a well-formed sentence in an informal register of English with subject dropping (e.g., *turned the water into wine*). This is common in the sentence versus word list literature; the list stimuli are usually not completely void of structure (though see Brennan & Pylkkänen, 2012) and therefore should be thought of as the stimulus that involves *less* structure (as opposed to no structure). More relevant for the current discussion is the extent to which the pseudoword sentence is free of meaning. If it does not engage the semantic combinatory machinery, which could be true if that machinery only took as its input existing words, then any effects elicited for the real-word sentence versus list contrast that were absent for pseudoword sentences would be good candidates for neural correlates of semantic composition.

However, formal semantic theories of the generative tradition make almost the opposite prediction: the semantic combinatory system operates to a large extent without caring much about the conceptual details of its input items. In other words, the meanings of the sentences *the cat chased the mouse* and *the mouse ate the cheese* compose exactly in the same way even though the direct objects in the two sentences differ in animacy. Further, the same routines could just as easily be applied to *the solim grilloted the yome*, yielding an assertion of the existence of a grilloting event with the solim as a likely agent (and less likely experiencer) and the yome as the affected party. This pseudoword sentence is void of actual conceptual content, but it is not void of meaning. It asserts the existence of an eventuality in the past and relates two event participants to that eventuality (Parsons, 1990).

50.3 AN EMPIRICAL QUESTION: DO CONCEPTS MATTER FOR COMPOSITION?

As the previous discussion illustrates, semantic composition within the formal semantics tradition refers to the logical aspects of meaning construction, accomplishing computations such as the saturation of arguments (as in when a verb takes its direct object as an argument), the intersection of properties (the *black cat* type examples), or the binding of variables (assuring that in expressions such as *every diver defended himself*, "himself" refers to the divers) (Heim & Kratzer, 1998). The empirical question then is: is the brain as a composer of meaning like the linguist's combinatory system, abstract and indifferent to the conceptual content of its input items, or do the types of contrasts traditionally used in the brain imaging literature in fact tap into the construction of complex meaning, contrary to what most formal semantic theories would necessarily predict? Although we do not have a definitive answer to this, a recent line of research approaching the brain basis of semantic composition from the linguist's perspective has uncovered that the brain activity that is affected by composition, as linguists would define this notion, *is* in fact remarkably sensitive to the conceptual content of the input words. In what follows, I first define some prerequisites for a research program addressing composition and then summarize the main results of this body of work, relating it to the prior hemodynamic literature as well as neuropsychological research on semantic dementia (SD).

50.4 METHODOLOGICAL STARTING POINTS FOR THE COGNITIVE NEUROSCIENCE OF SEMANTIC COMPOSITION

Isolating brain activity that specifically reflects semantic composition is highly challenging given that semantic composition correlates with many other computations, all of which occur within a few hundred milliseconds on the introduction of a new word into an expression. Lexical access, syntactic composition, and pragmatic inferencing is a short list of such computations, and all of these are themselves complex processes.

Given this challenge, one can imagine two rather different starting points for an investigation aimed at revealing the brain basis of semantic composition: either one could design clever experiments that somehow only vary semantic composition or one could initially conduct simpler, more confounded experiments to extract a broader network of regions and then proceed to try to unpack that network. The sentence versus word list literature discussed could exemplify the latter approach, although few studies have taken steps toward the "unpacking" of the implicated network. One line of research pursuing this goal has focused on delineating the network of "sentence processing regions" into domain general versus more language-specific regions (Fedorenko, Duncan, & Kanwisher,

2012, 2013). More relevantly for semantic composition, Rogalsky and Hickok (2009) used the sentence versus list manipulation as an initial localizer and then examined what activity within the regions sensitive to this contrast showed an enhanced signal when subjects attended to either syntactic or semantic aspects of well-formed sentences. Their focus was on the LATL, where sentence over word list effects have localized the most consistently, but the results indicated only a subtle task effect. Although most of the left anterior temporal cortex responded similarly during the syntactic and semantic attentional demands, an anterior subregion within the general region of interest responded more when semantics was attended to. As acknowledged by these authors, this experimental design clearly did not vary the presence of syntactic or semantic computations; rather, both types of composition played out fully for all stimuli. Thus, the rather subtle effects are perhaps not surprising.

Importantly, all the literature cited so far has used hemodynamic brain imaging techniques such as PET or fMRI, which only have temporal resolution in the order of seconds. This is clearly too slow for characterizing the rapid dynamics of linguistic computations, which happen at the millisecond scale. For example, it is not inconceivable that a single combinatory region could combine both syntactic and semantic inputs but at slightly different times. This type of result would not be obtainable without both detailed timing and spatial information. Of currently available noninvasive cognitive neuroscience techniques, the best combination of both is offered by MEG, which measures the magnetic fields associated with neural activity. Although its spatial resolution is worse than that of PET or fMRI, it is at the centimeter scale and, crucially, MEG offers millisecond temporal resolution. This makes MEG ideal for research aimed at characterizing language processing not simply as a static brain image but also as a dynamic process unfolding over time.

50.5 THE LATL AS A COMBINATORY REGION: EVIDENCE FROM MEG

Decades of behavioral research have shown that sentence interpretation proceeds incrementally (Altmann & Kamide, 1999; Altmann & Steedman, 1988; Chambers, Tanenhaus, Eberhard, Filip, & Carlson, 2002; Kamide, Altmann, & Haywood, 2003; Tanenhaus, Spivey-Knowlton, Eberhard, & Sedivy, 1995). Thus, we should expect syntactic and semantic combinatory operations to occur word-by-word as a sentence unfolds. Electrophysiological research using violations at various representational levels has conformed to this prediction (Friederici, 2002; Kutas & Hillyard, 1980, 1984; Neville, Nicol, Barss, Forster, & Garrett, 1991), but measuring responses to violations does not necessarily address the computations that serve to build well-formed structure. As already discussed, from hemodynamic research we have learned that when the brain activity elicited by sentences and unstructured lists is compared, the most consistent activity increase for sentences systematically localizes to the LATL. To address whether this activity reflects some aspect of incremental combinatory operations, as opposed to some computation occurring at a different time scale (e.g., clause-final wrap-up effects; Aaronson & Scarborough, 1976; Just & Carpenter, 1980; Rayner, Sereno, Morris, Schmauder, & Clifton, 1989), it is crucial to address whether LATL effects of sentential structure are a component of the usual brain response to words within a structured context. To investigate this, Brennan and Pylkkänen (2012) imported the sentence versus word list manipulation into a MEG setting and found that the LATL effect was reliable at the word level, alongside increases in a broad distributed network covering much of left temporal and frontal cortex. Thus, the LATL does in some way appear to participate in incremental sentence comprehension.

The distributed nature of the sentence versus list effect is not surprising given the coarseness of this contrast, which manipulates more or less every process-related sentence interpretation. As a more focused approach, a series of recent MEG studies has adopted a more controlled experimental paradigm aimed at contrasting conditions only in the presence of a single phrasal combinatory step (Bemis & Pylkkänen, 2011, 2012, 2013; Del Prato & Pylkkänen, 2014; Pylkkänen et al., 2014, Westerlund & Pylkkänen, 2014). As an initial case study, a simple manipulation of adjectival modification was chosen, with color adjectives modifying nouns denoting concrete objects, as in *red boat* (Bemis & Pylkkänen, 2011). MEG activity was always measured at the noun, which in the control conditions was either presented without a linguistic context (the color word was replaced by an unpronounceable consonant string) or preceded by a noun that could not combine with it semantically, given the nature of the experimental task. Specifically, subjects engaged in both a combinatory task, where they matched the just presented verbal stimulus (*red boat* or *xtp boat*) to a picture of a colored object, eliciting natural combinatory processing, and a noncombinatory list task, where the stimulus was either a pair of nouns (*cup, boat*) or a single noun (*xtp boat*), and the participants indicated whether a subsequent picture of a colored object was a match to any noun within the verbal stimulus. The intention in this latter task was to elicit a maximally noncombinatory list-like interpretation (despite the logical possibility for interpreting any noun–noun sequence as a noun–noun compound), and thus the design was in a sense a mini-version of the traditional sentence versus list contrast. Given this design, any neural effect that was unique to the adjective–noun phrases was considered a potential correlate of basic combinatory processes. The LATL showed such an effect, with responses showing a reliable increase only for the phrasal condition. The effect occurred relatively early, at approximately 200–250 ms after noun onset; according to a follow-up study, it was automatic in the sense that it was elicited even when adjective–noun combinations were presented within a task that discouraged composition (Bemis & Pylkkänen, 2013). The same effect in a somewhat later time window was elicited auditorily (Bemis & Pylkkänen, 2012) as well as during language production (Pylkkänen et al., 2014), conforming to the hypothesis that this activity reflects a modality general stage of processing where the basic building blocks of language are combined into a complex, structured representation.

50.6 DELVING DEEPER: WHAT TYPES OF REPRESENTATIONS DOES THE LATL COMBINE?

50.6.1 Composition Versus Conceptual Specificity

None of the MEG results discussed yet speak to the computational details of LATL activity, such as whether it reflects syntactic or semantic composition, given that the processing of adjective–noun combinations obviously involves both. At the heart of a syntactic hypothesis is the prediction that for syntactically uniform structures, LATL effects should pattern similarly. However, this prediction has been disconfirmed by a study aimed at relating the LATL literature on composition to a thus far disconnected literature on the LATL and SD (Westerlund & Pylkkänen, 2014). SD is a progressive neurodegenerative disorder characterized by loss of semantic memory and, crucially, the most common locus of cerebral atrophy in this population is in the LATL (Chan et al., 2001). SD patients have trouble recognizing and naming concepts (Hodges, Graham, & Patterson, 1995; Hodges, Patterson, Oxbury, & Funnell, 1992; Mummery et al., 1999, 2000; Patterson et al., 2006; Patterson, Nestor, & Rogers, 2007; Rogers et al., 2004) and, in particular, more "specific" concepts (e.g., swallow as compared to bird) are disproportionately affected (Done & Gale, 1997; Hodges et al., 1995; Rogers et al., 2004, 2006; Warrington, 1975). Additionally, when the types of tasks that SD patients struggle with have been conducted on healthy participants using fMRI or PET, the LATL has shown increased activation for more specific category labels (Bright, Moss, Stamatakis, & Tyler, 2005; Moss, Rodd, Stamatakis, Bright, & Tyler, 2005; Rogers et al., 2006; Tyler et al., 2004). Because these results implicate the LATL for the processing of individual concepts, with no obvious combinatory components, the question arises whether the phrasal LATL effects could in fact be explained as conceptual specificity effects, given that adjectival modification typically increases the conceptual specificity of the noun. If this was the case, then the role of the LATL would clearly not be syntactic.

To test this, Westerlund and Pylkkänen (2014) investigated the relationship between the composition and single word specificity effects in MEG by combining the basic composition paradigm of Bemis and Pylkkänen (2011) with a manipulation of the conceptual specificity of the nouns. As before, participants read nouns with and without adjectival modifiers, but now the nouns were further divided into more and less conceptually specific classes (e.g., *blue canoe/boat*). If the previously observed composition effects in the LATL reflected the increased conceptual specificity associated with adjectival modification, then a similar conceptual specificity effect should have been elicited for the single nouns and, potentially, the combinatory and single word specificity effects could have been expected to add up in a somewhat linear fashion (causing the highest amplitudes for the adjectivally modified specific nouns). Alternatively, the conceptual specificity effect of the hemodynamic literature could be temporally and/or spatially distinct from the LATL composition effect; given that neither PET nor fMRI have sufficient time resolution to attribute an effect to a particular subpart of relatively quick trials, the prior literature does not yet tell us whether the specificity effects are associated with the processing of the noun itself or some other part of the trial, such as the match/mismatch judgment on the subsequently presented pictures. Although the spatial resolution of MEG is not sufficient to detect subtle differences in localizations, it is an ideal technique for assessing whether the two effects are timed similarly.

Single word specificity did affect LATL amplitudes at the nonmodified nouns in a way consistent with the hemodynamic literature, but statistically this was not a very robust finding. Instead, the strongest result was an interaction between the two variables: adjectival modification reliably increased LATL amplitudes only when the modified noun was less specific. In other words, adding a feature to a noun that was already conceptually quite specific, such a *canoe*, did not result in a boost in LATL activity; rather, an amplitude increase was only elicited when the noun itself did not convey a very specific meaning.

Although this result does not yet definitively settle the question of whether the composition effect might "reduce" to a specificity effect—the single word specificity effect was too weak for this to follow straightforwardly—it does demonstrate that the two variables are fundamentally connected: as a composer of new concepts, the LATL is not insensitive to the conceptual make-up of its input items. This simple finding strongly rules out a syntactic account of the LATL composition effect: no theory of phrase structure building would draw a difference between *blue boat* and *blue canoe*. More interestingly, no theory of semantic composition within formal semantics would either. In these systems, the composition of *blue* with either *boat* or *canoe* would yield an intersection of the two properties, picking out a set of entities in one's mental model that have both the properties of being blue and being a boat/canoe (Heim & Kratzer, 1998). Thus, there is no combinatory routine within contemporary linguistic theories that would adequately model the function of the LATL, at least as it is implicated by the currently available MEG data.

50.6.2 Composition Versus Conceptual Combination

Although the sensitivity of the LATL's combinatory response to conceptual specificity does not straightforwardly fit into theories of formal semantics, this functional profile of the LATL is potentially less surprising if approached from the perspective of classic psychological research on so-called "conceptual combination" (Hampton, 1997; Murphy, 1990; Smith & Osherson, 1984). Like work in formal semantics, research on conceptual combination has aimed to explain the meanings of linguistically complex expressions; unfortunately, these two approaches to meaning composition have had little contact [with the notable exception of Kamp and Partee's (1995) extensive response to Osherson and Smith (1981)]. Compared with research in formal semantics, which explicitly aims to address the entire grammar (Bach, 1989; Chierchia & McConnell-Ginet, 1990; Gamut, 1991; Heim & Kratzer, 1998; Montague, 1970a, 1970b, 1973) with hopefully substantial crosslinguistic coverage (Bittner, 1994; Matthewson, 2001), work on conceptual combination has been quite focused on one particular domain, the modification of nouns (Costello & Keane, 2000; Hampton, 1997; Medin & Shoben, 1988; Murphy, 1990; Wisniewski, 1996), with little discussion on exactly what phenomena should more generally fall under the notion of "conceptual combination" (but see Hampton, 2011). Nonetheless, it is possible that the psychologist's notion of conceptual combination may model the function of the LATL better than the linguist's notion of semantic composition, although this is still nothing but the crudest starting point. Importantly, the term conceptual combination has already been linked to the LATL in an fMRI study showing that the LATL activation elicited by a concept such as "boy" is correlated with the product of activations for concepts representing features that contribute to the meaning of "boy," such as "male" and "child" (Baron & Osherson, 2011). Whereas the methodology of this work was very different from the MEG studies discussed here, it is clear that, like Westerlund and Pylkkänen's (2014) findings, Baron & Osherson's results are also neither syntactic in nature nor accountable in terms of semantic composition of the formal, conceptually blind kind.

If the input to the LATL function is something like a "concept" as opposed to a syntactic category or a semantic type, modulations of its activity by variables such as typicality or feature diagnosticity obviously become more expected (Costello & Keane, 2000; Medin & Shoben, 1988). For example, the lesser LATL effect of adjectival modification when combined with more specific nouns may relate to a lesser diagnostic value of adjectival modifiers in this context (Westerlund & Pylkkänen, 2014). However, for this functional hypothesis of the LATL to actually gain some predictive power, we must develop some understanding of the bounds and generality of the relevant LATL-housed computation: exactly what types of representations does it combine and how?

One task is to significantly expand the range of constructions under investigation. Initial steps toward this were taken by Westerlund, Kastner, Al Kaabi, and Pylkkänen (under revision), who tested for LATL composition effects in multiple different types of two-word phrases covering both cases of modification and argument saturation, results suggesting robust generality not only across constructions but also crosslinguistically. In contrast, computational limits on the function of the LATL have been established by a simple extension of the basic composition paradigm contrasting color modifications with numeral quantifications as in *red cups* versus *two cups* (Del Prato & Pylkkänen, 2014). Given that combinatory effects in the LATL had already been shown to extend to language production (Pylkkänen et al., 2014), the study was conducted as a production task because this allowed a manipulation of composition type while keeping the physical stimulus constant: subjects named pluralities of colored objects (e.g., a picture of two red cups) using either color modifiers (e.g., *red cups*) or numeral quantifiers (*two cups*), depending on task instruction. Noncombinatory list productions served as control conditions. It is clear that any version of semantic composition as construed in formal semantics would predict both the color and number phrases to elicit a composition effect, because clearly both types of phrases involve semantic composition. However, whether *two cups* is an instance of conceptual combination because this notion is construed within the psychology literature is much less clear. The MEG results indicated that the LATL discriminates between the two cases: only adjectival modification, and not numeral quantification, increased LATL amplitudes. Thus, whereas the addition of a color feature to an object concept increased LATL amplitude, the enumeration of tokens belonging to a set did not. This provides further evidence that the LATL's combinatory role may be limited to some version of "conceptual combination," the details of which are yet to be defined.

50.6.3 Absence of LATL Effects in Semantic Mismatch Configurations: Studies on Coercion

One notable manipulation that is semantic in nature but has systematically not elicited LATL effects is studies on coercion (i.e., configurations where a type mismatch triggers an inference for an implicit meaning). For example, although verbs such as *begin* describe the beginnings of eventualities, they do not necessarily require a direct object describing an event, as in *begin to write an article*, but rather they are perfectly well-formed when combined directly with an entity-describing noun phrase: *the author began an article*. When the object describes an entity, there is a strong intuition, supported by a sizeable psycholinguistic literature (Frisson & McElree, 2008; McElree, Pylkkänen, Pickering, & Traxler, 2006; Traxler, Pickering, & McElree, 2002), that the interpretation of the sentence nevertheless involves a covert activity (i.e., "the author began some activity involving a book"). Because the computation of the covert meaning is a combinatory semantic process not corresponding to steps in syntactic processing, a series of MEG studies used constructions such as this one as a starting point for identifying activity related to semantic composition (Brennan & Pylkkänen, 2008, 2010; Pylkkänen, 2008; Pylkkänen & McElree, 2007; Pylkkänen Oliveri, & Smart, 2009). A systematic finding arose from these studies implicating MEG activity at approximately 400 ms localizing in vmPFC as sensitive to the construction of covert meaning. The LATL, however, showed no such effects, a finding that is somewhat at odds with the hypothesis that the LATL is a semantic combinatory site. As for frameworks that would account for this null finding, one possibility is that semantic composition proceeds in several stages, with the LATL at approximately 200 ms representing a very early stage during which only the most readily available features combine, whereas the computations at the vmPFC stage are more clearly post-lexical. Given that coercion configurations always involve an initial mismatch between the semantic requirements of two composing constituents, it is then conceivable that an early stage of composition would not be able to operate on such input; rather, composition would only succeed at a later stage, after the semantic mismatch has been resolved. Crucially, even when LATL effects are observed for seemingly simple expressions such as *red boat*, they are typically accompanied by later vmPFC effects, suggesting that, in the usual case, shallow early composition is followed by a later stage potentially incorporating more semantic detail. In production, we would expect the reverse, because the detailed message to be conveyed at least logically precedes more abstract levels of linguistic computation, a prediction that was born out in the production results of Pylkkänen et al. (2014). This type of proposal conforms to many extant theories, including old schema models of conceptual combination where an initial filling of feature slots is followed by an elaboration of the composed representation on the basis of world knowledge[2] (Cohen & Murphy, 1984; Murphy, 1988), good-enough parsing theories in psycholinguistics (Ferreira et al., 2002; Ferreira & Patson, 2007), as well as recent models within cognitive neuroscience proposing semantic activation to be gradient (Binder & Desai, 2011).

50.7 CLOSING REMARKS

In this brief chapter, I have aimed to illustrate that identifying a brain region as sensitive to composition is only a starting point. The real work lies in trying to characterize the details of the computation performed by the region. The rather vast hypothesis space capable of explaining any initial "combinatory effect" can only be narrowed by systematic, incremental investigation replicating the initial, basic finding over and over again. In this endeavor, temporal information is crucial because, in the absence of a motivating hypothesis, one would not want to call, say, a LATL effect occurring at 200 ms the "same" as one occurring at 600 ms. Further, in the spirit of the initial mission statement of cognitive neuroscience, our hypothesis generation should take advantage of the vast bodies of results developed in older disciplines addressing the processing and representation of language. Without this

[2]Thanks to Masha Westerlund for pointing out this connection.

knowledge base to help us carve out the space of possibilities, characterizing the neural underpinnings of something as complex as semantics would simply be too hard.

Acknowledgments

The writing of this chapter and much of the research discussed in it were supported by National Science Foundation Grant BCS-1221723 and grant G1001 from the NYUAD Institute, New York University Abu Dhabi.

References

Aaronson, D., & Scarborough, H. S. (1976). Performance theories for sentence coding: Some quantitative evidence. *Journal of Experimental Psychology: Human Perception and Performance*, 2(1), 56.

Altmann, G., & Kamide, Y. (1999). Incremental interpretation at verbs: Restricting the domain of subsequent reference. *Cognition, 73*, 247–264.

Altmann, G., & Steedman, M. (1988). Interaction with context during human sentence processing. *Cognition, 30*, 191.

Bach, E. (1989). *Informal lectures on formal semantics*. New York, NY: State University of New York Press.

Barker, C. (1995). *Possessive descriptions*. Stanford: CSLI publications.

Baron, S. G., & Osherson, D. (2011). Evidence for conceptual combination in the left anterior temporal lobe. *Neuroimage, 55*(4), 1847–1852.

Bemis, D. K., & Pylkkänen, L. (2011). Simple composition: An MEG investigation into the comprehension of minimal linguistic phrases. *Journal of Neuroscience, 31*(8), 2801–2814.

Bemis, D. K., & Pylkkänen, L. (2012). Basic linguistic composition recruits the left anterior temporal lobe and left angular gyrus during both listening and reading. *Cerebral Cortex, 23*(8), 1859–1873.

Bemis, D. K., & Pylkkänen, L. (2013). Flexible composition: MEG evidence for the deployment of basic combinatorial linguistic mechanisms in response to task demands. *PLoS One, 8*(9), e73949.

Binder, J. R., & Desai, R. H. (2011). The neurobiology of semantic memory. *Trends in Cognitive Sciences, 15*(11), 527–536.

Binney, R. J., Embleton, K. V., Jefferies, E., Parker, G. J., & Ralph, M. A. L. (2010). The ventral and inferolateral aspects of the anterior temporal lobe are crucial in semantic memory: Evidence from a novel direct comparison of distortion-corrected fMRI, rTMS, and semantic dementia. *Cerebral Cortex, 20*(11), 2728–2738.

Bittner, M. (1994). Cross-linguistic semantics. *Linguistics and Philosophy, 17*(1), 53–108.

Brennan, J., & Pylkkänen, L. (2008). Processing events: Behavioral and neuromagnetic correlates of aspectual coercion. *Brain and Language, 106*(2), 132–143.

Brennan, J., & Pylkkänen, L. (2010). Processing psych verbs: Behavioural and MEG measures of two different types of semantic complexity. *Language and Cognitive Processes, 25*(6), 777–807.

Brennan, J., & Pylkkänen, L. (2012). The time-course and spatial distribution of brain activity associated with sentence processing. *Neuroimage, 60*(2), 1139–1148.

Bright, P., Moss, H. E., Stamatakis, E. A., & Tyler, L. K. (2005). The anatomy of object processing: The role of anteromedial temporal cortex. *The Quarterly Journal of Experimental Psychology Section B, 58*(3–4), 361–377.

Chambers, C. G., Tanenhaus, M. K., Eberhard, K. M., Filip, H., & Carlson, G. N. (2002). Circumscribing referential domains during real-time language comprehension. *Journal of Memory and Language, 47*, 30–49.

Chan, D., Fox, N. C., Scahill, R. I., Crum, W. R., Whitwell, J. L., Leschziner, G., et al. (2001). Patterns of temporal lobe atrophy in semantic dementia and Alzheimer's disease. *Annals of Neurology, 49*(4), 433–442.

Chiarelli, V., El Yagoubi, R., Mondini, S., Bisiacchi, P., & Semenza, C. (2011). The syntactic and semantic processing of mass and count nouns: An ERP study. *PLoS One, 6*(10), e25885.

Chierchia, G. (1998). Reference to kinds across language. *Natural Language Semantics, 6*(4), 339–405.

Chierchia, G., & McConnell-Ginet. (1990). Meaning and grammar: An introduction to semantics. Cambridge, MA.

Clarke, A., Taylor, K. I., & Tyler, L. K. (2011). The evolution of meaning: Spatio-temporal dynamics of visual object recognition. *Journal of Cognitive Neuroscience, 23*(8), 1887–1899.

Cohen, B., & Murphy, G. L. (1984). Models of concepts. *Cognitive Science, 8*(1), 27–58.

Costello, F. J., & Keane, M. T. (2000). Efficient creativity: Constraint-guided conceptual combination. *Cognitive Science, 24*(2), 299–349.

Dehaene-Lambertz, G. (1997). Electrophysiological correlates of categorical phoneme perception in adults. *Neuroreport, 8*(4), 919–924.

Del Prato, P., & Pylkkänen, L. (2014). MEG evidence for conceptual combination but not numeral quantification in the left anterior temporal lobe during language production. *Frontiers in Psychology, 5*(524).

Devlin, J. T., Russell, R. P., Davis, M. H., Price, C. J., Moss, H. E., Fadili, M. J., et al. (2002). Is there an anatomical basis for category-specificity? Semantic memory studies in PET and fMRI. *Neuropsychologia, 40*(1), 54–75.

Done, D. J., & Gale, T. M. (1997). Attribute verification in dementia of Alzheimer type: Evidence for the preservation of distributed concept knowledge. *Cognitive Neuropsychology, 14*(4), 547–571.

Dowty, D. R. (1979). *Word meaning and Montague grammar: The semantics of verbs and times in generative semantics and in Montague's PTQ* (Vol. 7). Reidel: Springer.

Fedorenko, E., Duncan, J., & Kanwisher, N. (2012). Language-selective and domain-general regions lie side by side within Broca's area. *Current Biology, 22*(21), 2059–2062.

Fedorenko, E., Duncan, J., & Kanwisher, N. (2013). Broad domain generality in focal regions of frontal and parietal cortex. *Proceedings of the National Academy of Sciences, 110*(41), 16616–16621.

Ferreira, F., Ferraro, V., & Bailey, K. G. D. (2002). Good-enough representations in language comprehension. *Current Directions in Psychological Science, 11*, 11–15.

Ferreira, F., & Patson, N. D. (2007). The 'good enough' approach to language comprehension. *Language and Linguistics Compass, 1* (1–2), 71–83.

Fieder, N., Nickels, L., Biedermann, B., & Best, W. (2014). From "some butter" to "a butter": An investigation of mass and count representation and processing. *Cognitive Neuropsychology*, 1–37.

Friederici, A. D. (2002). Towards a neural basis of auditory sentence processing. *Trends in Cognitive Sciences, 6*(2), 78–84.

Friederici, A. D. (2012). The cortical language circuit: From auditory perception to sentence comprehension. *Trends in Cognitive Sciences, 16*(5), 262–268.

Friederici, A. D., Meyer, M., & von Cramon, D. Y. (2000). Auditory language comprehension: An event-related fMRI study on the processing of syntactic and lexical information. *Brain and Language, 74*(2), 289–300.

Frisson, S., & McElree, B. (2008). Complement coercion is not modulated by competition: Evidence from eye movements. *Journal of Experimental Psychology: Learning, Memory, and Cognition, 34*(1), 1.

Gamut, L. T. F. (1991). *Logic, language, and meaning*. Chicago, IL: University of Chicago Press.

Gauthier, I., Anderson, A. W., Tarr, M. J., Skudlarski, P., & Gore, J. C. (1997). Levels of categorization in visual recognition studied using functional magnetic resonance imaging. *Current Biology, 7* (9), 645–651.

Grodzinsky, Y., & Friederici, A. D. (2006). Neuroimaging of syntax and syntactic processing. *Current Opinion in Neurobiology, 16*(2), 240–246.

Hagoort, P., Hald, L., Bastiaansen, M., & Petersson, K. M. (2004). Integration of word meaning and world knowledge in language comprehension. *Science, 304*(5669), 438–441.

Hagoort, P., & Indefrey, P. (2014). The neurobiology of language beyond single words. *Annual Review of Neuroscience, 37*, 1.

Hampton, J. A. (1997). Conceptual combination. In K. Lamberts, & D. Shanks (Eds.), *Knowledge, concepts, and categories* (pp. 133–159). London: UCL Press.

Hampton, J. A. (2011). Conceptual combinations and fuzzy logic. *Concepts and Fuzzy Logic, 209*.

Heim, I., & Kratzer, A. (1998). *Semantics in generative grammar*. London: Blackwell.

Hodges, J. R., Graham, N., & Patterson, K. (1995). Charting the progression in semantic dementia: Implications for the organization of semantic memory. *Semantic Knowledge and Semantic Representations, 3*(3–4), 463–495.

Hodges, J. R., Patterson, K., Oxbury, S., & Funnell, E. (1992). Progressive fluent aphasia with temporal lobe atrophy. *Brain, 115* (6), 1783–1806.

Humphries, C., Love, T., Swinney, D., & Hickok, G. (2005). Response of anterior temporal cortex to syntactic and prosodic manipulations during sentence processing. *Human brain mapping, 26*(2), 128–138.

Humphries, C., Binder, J. R., Medler, D. A., & Liebenthal, E. (2006). Syntactic and semantic modulation of neural activity during auditory sentence comprehension. *Journal of Cognitive Neuroscience, 18* (4), 665–679.

Ilmberger, J., Rau, S., Noachtar, S., Arnold, S., & Winkler, P. (2002). Naming tools and animals: Asymmetries observed during direct electrical cortical stimulation. *Neuropsychologia, 40* (7), 695–700.

Jobard, G., Vigneau, M., Mazoyer, B., & Tzourio-Mazoyer, N. (2007). Impact of modality and linguistic complexity during reading and listening tasks. *Neuroimage, 34*, 784–800.

Just, M. A., & Carpenter, P. A. (1980). A theory of reading: From eye fixations to comprehension. *Psychological Review, 87*(4), 329.

Kamide, Y., Altmann, G. T. M., & Haywood, S. L. (2003). The time-course of prediction in in-cremental sentence processing: Evidence from anticipatory eye movements. *Journal of Memory and Language, 49*, 133–156.

Kamp, H., & Partee, B. (1995). Prototype theory and compositionality. *Cognition, 57*(2), 129–191.

Kennedy, C. (2007). Vagueness and grammar: The semantics of relative and absolute gradable adjectives. *Linguistics and Philosophy, 30*(1), 1–45.

Kutas, M., & Federmeier, K. D. (2000). Electrophysiology reveals semantic memory use in language comprehension. *Trends in Cognitive Sciences, 4*(12), 463–470.

Kutas, M., & Hillyard, S. A. (1980). Reading senseless sentences: Brain potentials reflect semantic incongruity. *Science, 207*(4427), 203–205.

Kutas, M., & Hillyard, S. A. (1984). Brain potentials during reading reflect word expectancy and semantic association. *Nature, 307*, 161–162.

Lau, E. F., Phillips, C., & Poeppel, D. (2008). A cortical network for semantics:(de) constructing the N400. *Nature Reviews Neuroscience, 9*(12), 920–933.

Leffel, T., Lauter, M., Westerlund, M., & Pylkkänen, L. (2014). Restrictive vs. non-restrictive composition: An MEG study. *Language, Cognition and Neuroscience, 29*(10), 1191–1204.

Link, G. (1983). The logic analysis of plural and mass terms: A lattice theoretical approach. In R. Bäuerle, C. Schwarze, & A. v. Stechow (Eds.), *Meaning, use and interpretation of language* (pp. 302–323). Berlin: Walter de Gruiter.

Matthewson, L. (2001). Quantification and the nature of crosslinguistic variation. *Natural Language Semantics, 9*(2), 145–189.

Mazoyer, B. M., Tzourio, N., Frak, V., Syrota, A., Murayama, N., Levrier, O., et al. (1993). The cortical representation of speech. *Journal of Cognitive Neuroscience, 5*(4), 467–479.

McElree, B., Pylkkänen, L., Pickering, M. J., & Traxler, M. J. (2006). A time course analysis of enriched composition. *Psychonomic Bulletin & Review, 13*(1), 53–59.

Medin, D. L., & Shoben, E. J. (1988). Context and structure in conceptual combination. *Cognitive Psychology, 20*(2), 158–190.

Montague, R. (1970a). English as a Formal Language. In Linguaggi nella Società e nella Tecnica, ed. Bruno Visentini et al., 189–224. Milan: Edizioni di Comunità. Reprinted in Montague 1974, 188–221.

Montague, R. (1970b). Universal grammar. *Theoria, 36*, 373–398.

Montague, R. (1973). The proper treatment of quantification in ordinary English. In K. J. J. Hintikka, J. M. E. Moravcsik, & P. Suppes (Eds.), *Approaches to natural language* (pp. 221–242). Dordrecht: Reidel.

Moss, H. E., Rodd, J., Stamatakis, E., Bright, P., & Tyler, L. K. (2005). Anteromedial temporal cortex supports fine-grained differentiation among objects. *Cerebral Cortex, 15*(5), 616–627.

Mummery, C., Patterson, K., Price, C., Ashburner, J., Frackowiak, R., & Hodges, J. R. (2000). A voxel-based morphometry study of semantic dementia: Relationship between temporal lobe atrophy and semantic memory. *Annals of Neurology, 47*(1), 36–45.

Mummery, C., Patterson, K., Wise, R., Vandenbergh, R., Price, C., & Hodges, J. (1999). Disrupted temporal lobe connections in semantic dementia. *Brain, 122*(1), 61.

Murphy, G. L. (1988). Comprehending complex concepts. *Cognitive Science, 12*(4), 529–562.

Murphy, G. L. (1990). Noun phrase interpretation and conceptual combination. *Journal of Memory and Language, 29*(3), 259–288.

Näätänen, R., Lehtokoski, A., Lennest, M., Luuki, A., Alliki, J., Sinkkonen, J., et al. (1997). Language-specific phoneme representations revealed by electric and magnetic brain responses. *Nature, 385*, 432–4.

Neville, H., Nicol, J., Barss, A., Forster, K., & Garrett, M. (1991). Syntactically based sentence processing classes: Evidence from event-related brain potentials. *Journal of Cognitive Neuroscience, 3* (2), 151–165.

Noppeney, U., & Price, C. J. (2004). Retrieval of abstract semantics. *Neuroimage, 22*(1), 164–170.

Osherson, D. N., & Smith, E. E. (1981). On the adequacy of prototype theory as a theory of concepts. *Cognition, 9*(1), 35–58.

Pallier, C., Devauchelle, A. D., & Dehaene, S. (2011). Cortical representation of the constituent structure of sentences. *Proceedings of the National Academy of Sciences, 108*(6), 2522–2527.

Parsons, T. (1990). *Events in the semantics of English* (Vol. 5). Cambridge, MA: MIT Press.

Patterson, K., Ralph, M. A. L., Jefferies, E., Woollams, A., Jones, R., Hodges, J. R., et al. (2006). "Presemantic" cognition in semantic dementia: Six deficits in search of an explanation. *Journal of Cognitive Neuroscience, 18*(2), 169–183.

Patterson, K., Nestor, P. J., & Rogers, T. T. (2007). Where do you know what you know? The representation of semantic knowledge in the human brain. *Nature Reviews Neuroscience, 8*(12), 976–987.

Poldrack, R. A., Wagner, A. D., Prull, M. W., Desmond, J. E., Glover, G. H., & Gabrieli, J. D. (1999). Functional specialization for semantic and phonological processing in the left inferior prefrontal cortex. *Neuroimage, 10*(1), 15–35.

Portner, P. H., & Partee, B. H. (Eds.), (2008). *Formal semantics: The essential readings* (Vol. 7). Oxford: John Wiley & Sons.

Pylkkänen, L. (2008). Mismatching meanings in brain and behavior. *Language and Linguistics Compass, 2*(4), 712–738.

Pylkkänen, L., Bemis, D. K., & Blanco Elorrieta, E. (2014). Building phrases in language production: An MEG study of simple composition. *Cognition, 133*(2), 371–384.

Pylkkänen, L., Brennan, J., & Bemis, D. K. (2010). Grounding the cognitive neuroscience of semantics in linguistic theory. *Language and Cognitive Processes, 26*(9), 1317–1337.

Pylkkänen, L., Martin, A. E., McElree, B., & Smart, A. (2009). The anterior midline field: Coercion or decision making? *Brain and Language, 108*(3), 184–190.

Pylkkänen, L., & McElree, B. (2007). An MEG study of silent meaning. *Journal of Cognitive Neuroscience, 19*, 1905–1921.

Pylkkänen, L., Oliveri, B., & Smart, A. (2009). Semantics vs. World knowledge in prefrontal cortex. *Language and Cognitive Processes, 24*, 1313–1334.

Rayner, K., Sereno, S. C., Morris, R. K., Schmauder, A. R., & Clifton, C., Jr. (1989). Eye movements and on-line language comprehension processes. *Language and Cognitive Processes, 4*(3–4), SI21–SI49.

Rogalsky, C., & Hickok, G. (2009). Selective attention to semantic and syntactic features modulates sentence processing networks in anterior temporal cortex. *Cerebral Cortex, 19*, 786–796.

Rogers, T. T., Hocking, J., Noppeney, U., Mechelli, A., Gorno-Tempini, M. L., Patterson, K., et al. (2006). Anterior temporal cortex and semantic memory: Reconciling findings from neuropsychology and functional imaging. *Cognitive, Affective, & Behavioral Neuroscience, 6*(3), 201–213.

Rogers, T. T., Lambon Ralph, M. A., Garrard, P., Bozeat, S., McClelland, J. L., Hodges, J. R., et al. (2004). Structure and deterioration of semantic memory: A neuropsychological and computational investigation. *Psychological Review, 111*(1), 205.

Smith, E. E., & Osherson, D. N. (1984). Conceptual combination with prototype concepts. *Cognitive Science, 8*(4), 337–361.

Snijders, T. M., Vosse, T., Kempen, G., Van Berkum, J. J. A., Petersson, K. M., & Hagoort, P. (2009). Retrieval and unification of syntactic structure in sentence comprehension: An FMRI study using word-category ambiguity. *Cerebral Cortex, 19*, 1493–1503.

Steinhauer, K., Pancheva, R., Newman, A. J., Gennari, S., & Ullman, M. T. (2001). How the mass counts: An electrophysiological approach to the processing of lexical features. *Neuroreport, 12*(5), 999–1005.

Stowe, L. A., Broere, C. A., Paans, A. M., Wijers, A. A., Mulder, G., Vaalburg, W., et al. (1998). Localizing components of a complex task: Sentence processing and working memory. *Neuroreport, 9*, 2995–2999.

Stromswold, K., Caplan, D., Alpert, N., & Rauch, S. (1996). Localization of syntactic comprehension by positron emission tomography. *Brain and language, 52*(3), 452–473.

Taler, V., Jarema, G., & Saumier, D. (2005). Semantic and syntactic aspects of the mass/count distinction: A case study of semantic dementia. *Brain and Cognition, 57*(3), 222–225.

Tanenhaus, M., Spivey-Knowlton, M., Eberhard, K., & Sedivy, J. (1995). Integration of visual and linguistic information in spoken language comprehension. *Science, 268*, 1632–1634.

Traxler, M. J., Pickering, M. J., & McElree, B. (2002). Coercion in sentence processing: Evidence from eye-movements and self-paced reading. *Journal of Memory and Language, 47*(4), 530–547.

Tulving, E. (1972). *Episodic and semantic memory 1, Organization of memory* (Vol. 381, p. e402). London: Academic.

Tyler, L. K., Bright, P., Dick, E., Tavares, P., Pilgrim, L., Fletcher, P., et al. (2003). Do semantic categories activate distinct cortical regions? Evidence for a distributed neural semantic system. *Cognitive Neuropsychology, 20*(3–6), 541–559.

Tyler, L. K., Chiu, S., Zhuang, J., Randall, B., Devereux, B. J., Wright, P., et al. (2013). Objects and categories: Feature statistics and object processing in the ventral stream. *Journal of Cognitive Neuroscience, 25*(10), 1723–1735.

Tyler, L. K., Stamatakis, E. A., Bright, P., Acres, K., Abdallah, S., Rodd, J., et al. (2004). Processing objects at different levels of specificity. *Journal of Cognitive Neuroscience, 16*(3), 351–362.

Vandenberghe, R., Nobre, A. C., & Price, C. J. (2002). The response of left temporal cortex to sentences. *Journal of Cognitive Neuroscience, 14*(4), 550–560.

Visser, M., Jefferies, E., Embleton, K. V., & Ralph, M. A. L. (2012). Both the middle temporal gyrus and the ventral anterior temporal area are crucial for multimodal semantic processing: Distortion-corrected fMRI evidence for a double gradient of information convergence in the temporal lobes. *Journal of Cognitive Neuroscience, 24*(8), 1766–1778.

Vitali, P., Abutalebi, J., Tettamanti, M., Rowe, J., Scifo, P., Fazio, F., et al. (2005). Generating animal and tool names: An fMRI study of effective connectivity. *Brain and Language, 93*(1), 32–45.

Warrington, E. K. (1975). The selective impairment of semantic memory. *The Quarterly Journal of Experimental Psychology, 27*(4), 635–657.

Westerlund, M., Kastner, I., Al Kaabi, M., & Pylkkänen, L. (under revision). The LATL as locus of composition: MEG evidence from English and Arabic. *Brain and Language*

Westerlund, M., & Pylkkänen, L. (2014). The role of the left anterior temporal lobe in semantic composition vs. semantic memory. *Neuropsychologia, 57*, 59–70.

Wisniewski, E. J. (1996). Construal and similarity in conceptual combination. *Journal of Memory and Language, 35*(3), 434–453.

Xu, J., Kemeny, S., Park, G., Frattali, C., & Braun, A. (2005). Language in context: Emergent features of word, sentence, and narrative comprehension. *Neuroimage, 25*, 1002–1015.

CHAPTER

51

Working Memory and Sentence Comprehension

David Caplan

Neuropsychology Laboratory, Department of Neurology, Massachusetts General Hospital, Boston, MA, USA

Sentence comprehension requires memory over time scales ranging from deciseconds through several seconds and more. Even adjacent words must be assigned a syntactic relationship, and syntactic relations often span many words. For instance, in (1), the subject and object of *grabbed*, the subject of *lost*, and the antecedent of *his* must be retrieved at the points at which *grabbed*, *lost*, and *his* are encountered (or later, if parsing and interpretation is deferred):

1. The boy who the girl who fell down the stairs grabbed lost his balance.

Historically, the memory system that has been most often connected to parsing and interpretation is "short-term memory (STM)." The hypothesis that the memory system that supports parsing and interpretation utilizes STM is intuitively appealing because the temporal intervals over which parsing and interpretation usually apply are approximately the same as those over which STM operates. In addition, STM is an appealing construct to apply to sentence memory because it is thought to have capacity and temporal limitations that might account for the difficulty of comprehending certain sentences. This chapter reviews research on the relation of STM and comprehension, focusing on syntactic processing, beginning with theories that relate components of models of STM to this process, then turning to studies of the retrieval mechanisms in STM tasks and STM capacity limits and their relation to parsing and interpretation. Throughout the presentation, I critique the evidence supporting the view that STM is the memory system that supports parsing and interpretation. I end this chapter by presenting an outline of an alternative model.

51.1 EARLY STUDIES OF STM/WM AND ITS RELATION TO COMPREHENSION

Baddeley's model of ST-WM, an influential model of short-term and working memory, was one of the first that was related to comprehension. Baddeley's initial model of ST-WM (Baddeley, 1986) contained two major components. A "Central Executive (CE)" maintained multidimensional representations and was also considered to have some computational functions. Some authors suggested these computational functions consisted, in part, of operations needed to activate and transform representations, particularly functional domains (such as syntactic proceedures; Just & Carpenter, 1992); others suggested that these functions schedule the entry and removal of information in the limited-capacity central store (Engle & Kane, 2004). Visuo-spatial and verbal "slave systems" maintained domain-specific representations. The verbal slave system, the "Phonological Loop (PL)," consists of two components—a "Phonological Store (PS)" that maintains information in phonological form subject to rapid decay and an articulatory mechanism that rehearses items in the PS and transcodes written verbal stimuli into phonological form. Baddeley (2000) introduced a third type of store—the episodic buffer (EB)—that retained integrated units of visual, spatial, and verbal information marked for temporal occurrence. The slave systems and the EB had no computational functions themselves.

From approximately 1980 to 2000, a number of researchers sought to relate the memory system that supports aspects of sentence comprehension to components of the model of ST-WM developed by Baddeley and his colleagues. The main question that was considered was what component(s) of the ST-WM system,

Neurobiology of Language. DOI: http://dx.doi.org/10.1016/B978-0-12-407794-2.00051-1

if any, supported aspects of sentence memory; as noted, our focus is on the sentence memory that supports syntactic comprehension. Either the CE or the PL could play this role (the role of the EB, which had not been introduced at the time when much of this work was done, has not been investigated). The CE maintains abstract representations that could include syntactic and semantic information, making it suitable for this purpose. The PL maintains phonological representations that are linked to lexical items that contain the needed semantic and syntactic information, and it might be easier for the memory system to maintain such representations in the PL as pointers to the needed information than to maintain what are arguably more complex representations in the CE.[1]

Data regarding these possibilities come from several sources: interference effects of concurrent tasks that required the CE or the PL on parsing and interpretation (King & Just, 1991; Waters, Caplan, & Hildebrandt, 1987); tongue twister effects (Acheson & MacDonald, 2011; Ayres, 1984; Keller, Carpenter, & Just, 2003; Kennison, 2004; McCutchen, Bell, France, & Perfetti, 1991; McCutchen & Perfetti, 1982; Zhang & Perfetti, 1993); effects of homophones on comprehension (Coltheart, Patterson, & Leahy, 1994; Waters, Caplan, & Leonard, 1992); correlations between measures of the capacity of the CE and performance in parsing and interpretation (Just & Carpenter, 1992; King & Just, 1991; MacDonald, Just, & Carpenter, 1992; Miyake, Carpenter, & Just, 1994; Waters & Caplan, 1996); associations and dissociations of effects of brain damage on the CE or the PL and parsing and interpretation (Caplan & Waters, 1996; Emery, 1985; Grossman et al., 1991; Grossman, Carvell, Stern, Gollomp, & Hurtig, 1992; Kontiola, Laaksonen, Sulkava, & Erkinjuntti, 1990; Lalami et al., 1996; Lieberman, Friedman, & Feldman, 1990; Martin, 1990; Natsopoulos et al., 1991; Rochon & Saffran, 1995; Rochon, Waters, & Caplan, 1994; Stevens, Kempler, Andersen, & MacDonald, 1996; Tomoeda, Bayles, Boone, Kaszniak, & Slauson, 1990; Waters, Caplan, & Hildebrandt, 1991; Waters, Caplan, & Rochon, 1995); and neural evidence in the form of differences in neural activation in high-span and low-span participants during parsing and interpretation and overlap or nonoverlap of brain areas activated in tasks that involve the CE and ones that involve parsing and interpretation (Fiebach, Schlesewsky, & Friederici, 2001; Fiebach, Vos, & Friederici, 2004).

Caplan and Waters (1990) argued that these sources of data suggested that the PL did not support the memory requirements of the initial assignment of the preferred structure and interpretation of a sentence (what we called "first pass" parsing and interpretation), but they may play a role in supporting review and reanalysis of previously encountered information and in using the products of the comprehension process to accomplish a task (e.g., in maintaining a sentence in memory to later answer a question about its meaning). There is a general consensus that this view is correct; in an article relating the memory system that supports syntactic comprehension to ST-WM, Just and Carpenter (1992) explicitly say they are not advocating a role for the PL, only the CE, in this process.

The role of the CE in parsing and interpretation has been more controversial. Caplan and Waters (1999) argued that evidence of the sort described indicated that, like the PL, the CE does not support first-pass parsing and interpretation, but other researchers disagreed. Just and Carpenter (1992), for instance, developed a parsing model that included a limit on combined storage and processing capacity that they argued was derived from the CE.

51.2 CHANGES IN MODELS OF STM/WM

As noted, Baddeley's model of the CE includes both a limited-capacity memory store of multidimensional representations and a computational capacity. However, aside from work by Engle and his colleagues on how representations are entered and removed from ST-WM, recent work in ST-WM has largely ignored the computational functions of the CE and has focused on its storage capacity. A second line of research has studied the nature of retrieval of items in STM tasks, a topic that had been studied prior to the Baddeley model but that had become less the focus of work since the advent of that model.

Seen as a storage system devoid of a computational function, the CE generally goes by the name "Central Store (CS)" (Cowan, 2000; Ricker, AuBuchon, & Cowan, 2010). The capacity of the CS has been estimated based on accuracy and speed of recall data. Using tasks that prevent the persistence of sensory stores and the use of rehearsal, the capacity of the CS based on accuracy has been estimated at between three

[1]A store that is not part of Baddeley's formulation of ST-WM but that has been argued to play a role in sentence comprehension is a "short-term semantic store" postulated by Martin, Shelton, and Yaffee (1994). Caplan, DeDe, Waters, Michaud, and Tripodis (2011) raise questions about the existence of this store. If it exists, then its role in generating STM phenomena has not been examined in detail. Martin and He (2004) have argued that the "short-term semantic store" is used when items in a sentence are not assigned propositional meaning, as in prenominal sequences of adjectives (*the old, rusty, twisted...*N). If this store exists, then this would be, at most, a minor role in sentence comprehension, given the incremental nature of comprehension.

and five items (see Cowan, 2000 for review). For instance, Saults and Cowan (2007) exposed participants to four spoken digits presented in four different voices by loudspeakers in four spatial positions, and four colored spots presented at four different locations in a recognition task. Participants recognized three to four items regardless of whether only the visually presented colors or both the visually presented colors and the auditorily presented digits had to be remembered. Temporal data result in a similar, but more subtle, picture. Based on data from Verhaeghen and Basak (2005) regarding reaction times (RTs) in the *n*-back task, Verhaeghen et al. (2007) argued one item was maintained in the focus of attention but that this could be expanded to four items in highly practiced tasks.

The capacity of ST-WM has also been studied through the analysis of speed–accuracy trade-offs (SAT), which also provide information about the nature of retrieval. In the SAT approach, participants are presented a stimulus followed by a probe at a variable lag and are required to make a two-alternative forced new–old choice within a very short time (usually less than 200 ms; responses less than 100 ms are discarded as anticipations). Asymptotic d′ is considered to be a measure of the availability of an item, the temporal point at which d′ rises above 0 (the intercept), and the slope of the rise of d′ from the point at which it rises above 0 to its asymptotic level to reflect how an item is accessed. McElree and Dosher (1989) found that set size (3 and 5) affected asymptotic accuracy but not temporal dynamics, except for the set-final item, even when the probe was presented in a different font. McElree (1998) found that there were different retrieval speeds for probes drawn from the last and all other categories when triads of words in different semantic categories were presented. McElree (1996) found that recognizing rhyming words and semantically related words resulted in uniformly slower dynamics than recognizing previously presented words as wholes. McElree (2006) argued that ST-WM consisted of a focus of attention with one item, which was accessed by a "matching" procedure based on abstract (nonsensory) features of the retrieval cue, and items outside the focus of attention, in long term memory (LTM), that were accessed by a content-addressable retrieval mechanism, with "chunking" of the memory set in accordance with semantic features. Based on an analysis of the distribution of RTs in the *n*-back task, Verhaeghen et al. (2007) also argued that items outside the focus of attention are content-addressable, although only under what they called "ideal" circumstances—precisely predictable switching out of the focus of attention on the part of high-functioning individuals.

These two features of memory mechanisms in modern models of STM/WM have been considered for their possible relation to sentence comprehension: (i) the possibility that retrieval of information in parsing and interpretation is content-addressable and (ii) the possibility that a small number of items—at most four or five—are maintained in a highly accessible form and that this number of available items sets a capacity limit in parsing and interpretation. I review the work on these topics in turn.

51.3 RETRIEVAL MECHANISMS IN PARSING

A productive approach to modeling parsing is "retrieval-based parsing." In retrieval-based parsing, words serve as retrieval cues for previous portions of a sentence. In one well-known model (Lewis & Vasishth, 2005), parsing begins with lexical access, which places a new word and its features in a lexical buffer. The syntactic features of the current lexical entry combine with the structure in a control buffer that contains the categories that are needed to complete phrase markers already constructed to form retrieval cues, which may retrieve previously constructed constituents or symbols in long-term memory. Based on the retrieved item and the current lexical content, a production creates a new syntactic structure and attaches it to the retrieved constituent. This structure replaces the one in a problem state buffer that contains the existing constructed syntactic representation, the categories that are needed to complete phrase markers in this buffer are updated in a control buffer, and a production rule guides attention to the next word.

In this model, the retrieval process for lexical items and syntactic symbols is content-addressable. It consists of matching features of a retrieval cue to features of items in the memory set. Retrieval success is determined by the ratio of the match of the retrieval cue with the retrieved item divided by its match with other items in the memory set (Nairne, 1990a; Ratcliff, 1978), subject to noise. This model has two features: (i) items in the memory set that share features with a retrieval cue interfere with retrieval of a target and (ii) shared features of items in the memory set do not affect retrieval unless these features are part of the retrieval cue.

Evidence for this model comes from interference effects in online measures of parsing and interpretation, and from the SAT paradigm applied to sentence comprehension.

If retrieval success is determined by the ratio of the match of the retrieval cue with the retrieved item divided by its match with other items in the memory set, then retrieval difficulty will increase as a function of the extent to which nontarget and target items share features of the retrieval cue. This has been found for semantic and syntactic features of items.

Effects of the semantic similarity of noun phrases on online sentence processing measures have been described for both sentence-internal and sentence-external noun phrases. An example of sentence-internal interference is the report of Gordon, Hendrick, and Johnson (2001) regarding self-paced reading times in subject-extracted and object-extracted relative clauses (2a, b) and clefts (3a, b) with common nouns (definite descriptions), proper nouns, or pronouns:

2a. The banker that praised the barber/Sue/you climbed the mountain...
b. The banker that the barber/Sue/you praised climbed the mountain...
3a. It was the banker/Sue that praised the barber/Dee...
b. It was the banker/Sue that the barber/Dee praised...

The verb of an object relative creates a retrieval cue for its subject and object, and the verb of a subject relative creates a retrieval cue for its subject only, leading to more interference at the verb of the object than the subject relative clause. This was confirmed. In the relative clauses (2a, b), the sentence type effect was reported as being greater in the definite description (*the barber*) than either the pronoun (*you*) or the proper name (*Sue*) conditions. In clefts (3a, b), there was a greater sentence type effect in sentences in which the noun phrases (NPs) were matched for noun type (both definite descriptions or both proper nouns) than in sentences in which they differed.

An example of sentence-external interference is shown in the Van Dyke and McElree (2006) report of self-paced reading times for object clefts with or without concurrent recall of sets of three words that could or could not be integrated into the sentence, as in (4):

4: Word list: table-sink-truck
Sentence: It was the boat that the guy who lived by the sea sailed/fixed in two sunny days.

If verbs create retrieval cues that include information about semantic features of their arguments, there should be more interference from the integrated than the nonintegrated sets of words. The authors found that reading times were longer in the integrated than the unintegrated conditions.

Syntactic interference effects would be expected to arise when intervening items have syntactic properties shared by the retrieval cue and the to-be-retrieved item. Lewis and his colleagues (Lewis, 1996, 2000; Lewis & Vasishth, 2005; Lewis, Vasishth, & Van Dyke, 2006; Van Dyke & Lewis, 2003; Vasishth & Lewis, 2006) have shown such effects in online processing. For instance, Van Dyke and Lewis (2003) presented sentences such as (5) and (6). Words in parentheses were omitted in half the presentations of the sentences to produce ambiguities; we have annotated some words with subscripts for ease of reference.

5. Low interference
The secretary forgot (that) the student who was waiting for the exam was standing in the hallway.
6. High interference
The secretary forgot (that) the student who knew (that) the exam was important was standing in the hallway.

If *was standing* is a retrieval cue for a subject, then there should be more syntactic interference with retrieval of *the student* in (6) than in (5) because *the exam* is a subject in (6) and an object in (5). Van Dyke and Lewis presented four studies of accuracy and self-paced reading times in an acceptability judgment task with these sentences. In three of the four studies in which unambiguous versions of these sentences were presented, there was greater acceptance of well-formedness of (5) than (6) and longer self-paced reading times for the critical segment *was standing* in (6) than in (5).

Retrieval of information in sentences has also been studied using the speed-accuracy trade-off (SAT) technique. In applying paradigm to sentences, participants make judgments about the acceptability of sentences. The last word in the experimental sentence serves as a retrieval cue for an earlier word, and the position of the retrieved word and the number and type of potentially interfering items are varied across sentence types. Asymptotic d′ is taken as a measure of the availability of the earlier word, and d′ dynamics are taken as the measure of its accessability.

McElree and his colleagues have used the SAT technique to study many types of sentences and found that d′ measures are not affected by the number of potentially distracting items. For instance, Martin and McElree (2008, 2009) varied the distance between a verb phrase ellipsis and its antecedent (6a/b), the length and complexity of the antecedent (7c/d), and the amount of proactive or retroactive interference (7e/f).

7a. The editor admired the author's writing but the critics did not.
b. The editor admired the author's writing but everyone at the publishing house was shocked to hear that the critics did not.
c. The history professor understood Roman mythology but the principal was displeased to learn that the over-worked students attending the summer school did not.
d. The history professor understood Rome's swift and brutal destruction of Carthage but the principal knew that the over-worked students attending the summer school did not.
e. Even though Claudia was not particularly angry, she filed a complaint. Ron did too.
f. Claudia filed a complaint and she also wrote an angry letter. Ron did too.

None of these variables affected SAT dynamics, indicating that retrieval is unaffected by the amount of proactive or retroactive interference or the complexity of the retrieved item. This implies that retrieval uses a content-addressable mechanism, similar to findings in ST-WM reviewed.

Despite these results, the effort to relate retrieval-based parsing models of STM/WM faces a number of challenges. One issue is how certain results are to be interpreted. For instance, the semantic similarity effects in studies such as those of Gordon et al. (2001) cannot be due to the interference-generating mechanism outlined, because the verb of an object relative clause cannot generate a cue to retrieve two noun phrases that share a common semantic property or an aspect of form such as being a common or proper noun. An alternative, suggested by Lewis et al. (2006), is that the semantic similarity effects in the studies by Gordon and his colleagues arose during encoding. If so, then encoding of words into memory during sentence processing appears to differ from encoding of words in ST-WM, where similarity effects appear to arise during retrieval (Baddeley, 1968; Neath, 2000).

The comparison of the effects of semantic similarity in parsing/interpretation and in SAT studies also leads to a problem. Similarity of list items does not affect, and may slightly improve, retrieval of item information (Baddeley & Hitch, 1974; Nairne, 1990b; Nimmo & Roodenrys, 2004; Poirier & Saint-Aubin, 1996; Postman & Keppel, 1977, exp 2; Underwood & Ekstrand, 1967; Watkins, Watkins, & Crowder, 1974; Wickelgren, 1965, 1966). Therefore, if semantic similarity effects arise at retrieval, then what must be retrieved is item and order information, as Gordon et al. (2001) suggest. However, the content-addressable retrieval suggested by the SAT dynamics results is consistent only with retrieval of item information in ST-WM, not order information (Hockley, 1984, exp 3; McElree & Dosher, 1993). Because both the interference and SAT studies involve retrieval of NPs in the same structures (relative clauses), they must be characterizing the same retrieval process. If both sets of results arise during retrieval, then the difference between the two studies reflects strategic effects of task or other factors on retrieval mechanisms.

An issue that has been emphasized in recent literature is the absence of expected effects of potentially interfering items in situations where such effects would be expected if content-addressable retrieval occurs. One phenomenon that has been studied is the selective absence of illusions of grammaticality. For instance, many readers initially find sentence (8) grammatical (Eberhard, Cutting, & Bock, 2005), but readily report the ungrammaticality of (9) (Dillon, Mishler, Sloggett, & Phillips, under review, cited in Phillips, Wagers, & Lau, 2011).

*8. The key to the cabinets are on the table

*9. The diva that accompanied the harpists on stage presented themselves with lots of fanfare.

Phillips et al. (2011) suggested that the difference between (8) and (9) is due to the nature of the retrieval cues established by the agreement markers on verbs and reflexives. They proposed that reflexives retrieve their antecedents using only structural cues, whereas verbs create retrieval cues that specify the features of the subject. In their view, this difference is due to the fact that a sentence subject reliably predicts a verb and its agreement features, whereas a reflexive cannot be reliably predicted.

Items in certain syntactic positions appear not to be contacted during the retrieval process even when the retrieval cue specifies their features. For instance, NPs in certain positions do not appear to be contacted when a verb of an object relative clause retrieves its object. Traxler and Pickering (1996) found evidence for a retrieval cue that specified semantic features of the object of the verb of a relative clause (similar to Van Dyke & McElree, 2006) in (10) in the form of longer reading times for *shot* in the anomalous version of that sentence (with the NP *garage*), but no prolongation of reading times for *wrote* in the version of (11) with the NP *city*:

10. That's the pistol/garage with which the heartless killer shot the hapless man.
11. We liked the book/city that the author who wrote unceasingly saw while waiting for a contract.

This suggests that *the city/book* are not in the list of items contacted by the retrieval cue established by *wrote* in (11). It has been suggested that this is because of the syntactic environment in which *the city/book* and *wrote* occur; *the book/city* is outside a syntactic "island" that contains *wrote*.

The implication of the last two sets of results is that there are several retrieval mechanisms in sentence comprehension. At times, items are retrieved on a content-addressable basis. At other times, a search process applies. In addition, there are mechanisms that are reasonably well-established in the ST-WM literature, such as Weber-compressed temporal intervals from retrieved item to retrieval cue or response (Brown, Neath, & Chater, 2007), positional activation gradients measured from both list-initial and list-terminal items (Henson, 1998), temporal oscillators that estimate list length (Brown, Preece, & Hulme, 2000; Henson & Burgess, 1997), and others (for review, see Henson, 1998) that are not plausible candidates for mechanisms that support the memory requirements of parsing and interpretation. If there are mechanisms found in ST-WM and not in parsing and interpretation, then parsing and interpretation at most select from memory mechanisms found in ST-WM.

Altogether, the relation between the memory mechanisms that have been documented in sentence comprehension and those that apply in STM/WM tasks is not clear.

51.4 CAPACITY LIMITS IN STM/WM AND SENTENCE COMPREHENSION

The idea that parsing is capacity-limited has been part of thinking about sentence comprehension for decades (Miller & Chomsky, 1963). If it is correct that capacity limitations of the parser/interpreter are those found in the capacity-limited portion of ST-WM, then this would tie a major feature of parsing and interpretation to a property of ST-WM. Lewis and his colleagues have proposed several such relations.

Lewis (1996) developed a model in which lexical items can be either heads or dependents (or both) in structural relations. Parsing consists of connecting heads and dependents under higher nodes. Lewis (1996) argued that capacity limits on parsing arise because the buffer can contain only two constituents of the same type. For instance, in an object relative clause such as (1b), when *the clerk* is encountered, the buffer contains the structure [DEPENDENT—SPEC-IP—*the manager*; *the clerk*], which is within the capacity of the buffer. However, in the double-center embedded structure (12):

12. It was the book that the editor who the secretary who quit married enjoyed at *the secretary*, the buffer contains the structure [DEPENDENT—SPEC-IP—*the book*; *the editor*; *the secretary*] and the sentence becomes difficult because three NPs in the buffer are assigned the value [DEPENDENT—SPEC-IP].

In a related vein, Lewis and Vasishth (2005) suggested that the structure of their retrieval-based parsing model with three buffers (a control buffer, a problem state buffer, and a retrieval buffer) "has much in common with conceptions of working memory and STM that posit an extremely limited focus of attention of one to three items, with retrieval processes required to bring items into focus for processing. (Lewis & Vasishth, 2005, p. 380)."

The relation of capacity and temporal limits in parsing and interpretation to ST-WM is also subject to question. Lewis's (1996) limit of two on the number of items with the same relation that can be maintained in a buffer does not correspond to any suggestion about the size of a fixed-capacity, capacity-limited portion of ST-WM, which has been estimated as one by McElree (2006) and as four or five by Cowan (2000). Lewis and Vasishth's three buffers differ from the ST-WM concept of a single CS, and triple the capacity limit of the CS.

Another consideration is that Lewis' (1996) model requires that the contents of buffers remain accessible, because new items must be able to attach within them. The transparency of chunks in Lewis' model differs from applications of models of capacity limits of the CS to other cognitive phenomena. For instance, Halford, Cowan, and Andrews (2007) argued that the limits of the CS restrict human reasoning to problems with relational complexity ("arity") of no more than four, but their model requires that the reasoning process cannot access items within a chunk. If the capacity limits in processing in various domains are to be explained because of capacity limits in the CS, then the characterization of how chunks are processed must apply universally.

A third issue is that capacity limits and content-addressable retrieval apply to different stores in models of ST-WM; capacity limits are features of the CS or an expanded focus of attention, and content-addressable retrieval applies to items outside the capacity-limited portion of ST-WM (i.e., to items in LTM). If capacity limits exist in parsing and interpretation, then this would require that they are due to items being stored in a store from which they are not retrieved by a content-addressable mechanism, a noticeable discrepancy across the models we have been discussing. A model of STM/WM that recognizes only a focus of attention, such as McElree's, cannot maintain that parsing and interpretation retrieves items in STM/WM (the focus of attention), but requires that memory limits in parsing and interpretation arise because of interference effects during retrieval of information from LTM (see Van Dyke & Johns, 2012, discussion).

51.5 AN ALTERNATIVE FRAMEWORK FOR VIEWING THE MEMORY SYSTEM FOR PARSING AND INTERPRETATION

A central feature of the memory system that supports parsing and interpretation that emerges from this discussion of the recent literature is the domain-specificity of many aspects of memory in parsing and interpretation. Retrieval cues, aspects of retrieval processes, and any capacity limits that may apply in parsing and interpretation all have important domain-specific features. Domain-specific aspects of memory are characteristic features of cognitive skills that have been related to a different postulated memory system—"long-term working memory (LT-WM)."

The concept of LT-WM was proposed and developed by Ericsson and Kintsch (1995) as a response to two phenomena: (i) the greatly expanded working memory capacity of experts and skilled performers and (ii) the fact that skilled activities can be interrupted and later resumed without major effects on performance. Examples of activities supported by LT-WM in individuals with established expertise include using an abacus, mental calculation, memory of servers for orders, and memory of chess players for chessboard arrays, and examples of activities supported by LT-WM in individuals after training include expansion of span (to as much as 80 items). Considerable resistance of these memory performances to interruption has been documented in studies such as Chase and Ericsson (1982), who found only a small decrement in recall of 30 item lists due to proactive interference from other lists in a trained subject. Ericsson and Kintsch argue that several capacities are required for these memory performances. The individuals who show these memory skills must have a large body of relevant knowledge that allows them to store the incoming information in LTM, the activity must be familiar, and the encoded information must be associated with a "retrieval structure"—a domain-specific complex set of retrieval cues.

The phenomena that led to the concept of LT-WM are seen in parsing and interpretation: a large capacity for memory of items; domain-specific knowledge; and a high degree of familiarity on the part of performers. The imperviousness of parsing and interpretation to temporal interruption is unknown; the few available results (Wanner & Maratsos, 1978) suggest it is unexpectedly robust. Applying Ericsson and Kintsch's model to parsing and interpretation would lead to a model in which input activates items in long-term memory; these items are related by knowledge-based associations, patterns, and schemas to establish an integrated memory representation, and what is maintained in a STM system are retrieval cues for these items. It is of interest to reexamine the model of Lewis and colleagues with this perspective in mind. The representation of syntactic knowledge in declarative memory in the model of Lewis and colleagues corresponds to the "knowledge-based associations, patterns, and schemas" that are the basis for establishing an "integrated memory representation" in Ericsson and Kintsch's model. Both Lewis and Ericsson and Kintsch maintain that the application of "knowledge-based associations, patterns, and schemas" to the creation of an "integrated memory representation" is a domain-specific process. As noted, the operations of Lewis' model involve procedural memory and declarative long-term memory; ST-WM is only invoked, questionably (I have argued), to account for capacity limits on parsing. Van Dyke and Lewis (2003) say that "cues in the retrieval context are combined multiplicatively to produce a single retrieval probe...all cues are combined into a retrieval probe, which gives the strength of the relationship between each possible probe cue ($Q_1, \ldots, Q_m$) and the memory trace." This process has many similarities to the formation of cues in LT-WM, as described by Ericsson and Kintsch.

Placing the memory system that supports parsing and interpretation within the LT-WM/skilled performance framework captures the facts that much of encoding, storage, and retrieval of information in parsing and interpretation is unconscious and subjectively undemanding, and that the retrieval cues used in parsing and interpretation are domain-specific. The domain specificity of retrieval cues and aspects of the retrieval process contributes toward the degree of independence of skills in individual domains from skills in others (great servers are not particularly likely to be great chess players). This predicts that the skill of parsing and interpretation will not necessarily covary with other skills, and it accounts for the findings that measures of memory use in parsing and interpretation—such as online processing times at points at which parsing and interpretation require retrieval from memory—correlate poorly or not at all with memory performance on ST-WM tasks (Caplan & Waters, 1999). However, overlap of components of skills or their utilization of a common functional architecture for memory might lead to correlations between measures of online effects in parsing and interpretation and online implicit memory effects in selected implicit memory tasks, as has been shown for statistical learning tasks (Conway, Karpicke, & Pisoni, 2007; Conway & Pisoni, 2008; Misyak, Christiansen, & Tomblin, 2010) and to correlation of measures of online effects in parsing and interpretation with elementary speed of processing (Caplan et al., 2011), which is common to many skills.

I have suggested that LT-WM supports skilled parsing and interpretation. However, skilled parsing and interpretation occasionally breaks down. Caplan and Waters (2013) suggested that ST-WM plays a role in supporting memory demands of processes that occur at points of incremental comprehension failure. At these points, comprehenders at times review previously presented material that is held in memory. In spoken language comprehension and noncumulative self-paced or externally paced reading, the verbatim representation of the input must be retrieved from memory; no matter what the presentation method, the constructed representations exist only in memory (they are not part of the literal input), where they must be accessed and manipulated. Even when the input is available in written form, items appear to be accessed in memory at times, rather than by visual reinspection of the text. Regressive eye movements that are targeted to within one word of an item that is relevant to the clarification of the structure and meaning of the sentence (Frazier & Rayner, 1982; Inhoff & Weger, 2005; Meseguer, Carreiras, & Clifton, 2002; Mitchell, Shen, Green, & Hodgson, 2008; von der Malsburg & Vasishth, 2011; Weger & Inhoff, 2007) have been taken as being controlled by a representation in memory of the words in a sentence that is linked to memory of their spatial coordinates (Inhoff & Weger, 2005; Weger & Inhoff, 2007).

Caplan and Waters (1999, 2013) argued that retrieval of items under these circumstances directly utilizes mechanisms that apply to STM/WM tasks involving retrieval of items in lists. One argument that this is the case is that immediate serial recall of words in sentences with two clauses shows serial position effects on word recall in each clause (Marslen Wilson & Tyler, 1976), pointing to an interaction of recall based on linguistic structures (reconstruction from a conceptual representation; Lombardi & Potter, 1992; Potter & Lombardi, 1990) and mechanisms that support recall of words in lists. Theoretically, retrieval in the service of reanalysis occurs when the input is incompletely structured syntactically and/or semantically, and therefore has structural properties more akin to those of words in lists, which is the case during successful comprehension.

Assuming this framework is basically correct, an important question is, "when does the specialized LT-WM memory support comprehension and when do mechanisms directly tied to retrieval of information in lists apply?" In Caplan and Waters (1999), we suggested that ST-WM supported retrieval of information when syntactic structures had to be revised. This suggests that one could identify points in sentences at which ST-WM supports retrieval on the basis of syntactic features and parsing considerations. The literature provides many hypotheses about structural and/or processing determinants of points of possible incremental parsing failure, such as very high surprisal values (Hale, 2001; Levy, 2008; Staub 2010), high integration costs (Gibson, 1998), the need for nonmonotonic parsing during revisions (Sturt & Crocker, 1996), memory capacity limits (as discussed), and others. However, Caplan and Waters (2013) argued that no purely structural or processing theory can determine points at which ST-WM may support retrieval because these points are determined by incremental *comprehension* failure, not incremental *parsing and interpretation* failure. An important determinant of the success of sentence comprehension is intentional context—the comprehender's task and the criterion she/he sets for accomplishing the task. Tasks that require superficial comprehension or low criteria for task performance can allow comprehension to succeed even if the parser/interpreter fails to generate a grammatically licensed meaning (e.g., if all that the comprehender needs to know is who participated in an action and not what thematic roles different participants played), and, conversely, comprehenders may not have achieved adequate comprehension when the parser–interpreter generates a well-formed, semantically coherent representation if criteria for successful comprehension are set high (as seen in "wrap-up" effects at clause and sentence boundaries). This view is closely related to Lewis's idea of bounded optimality (Howes, Lewis, & Vera, 2009) and Ferreira's views on "good enough parsing" (Ferreira, Engelhardt, & Jones, 2009; Ferreira & Patson, 2007). Models of structural and processing load can provide an upper bound to what can be comprehended by most individuals, but this only partially determines incremental comprehension failure.

To investigate the possible role of ST-WM in supporting processing that arises after incremental comprehension failure, it is therefore necessary to identify empirical markers of incremental parsing success and failure. Because comprehenders have great freedom with respect to their behaviors, it is not clear that there are any behaviors that are inevitably triggered by either of these two states. However, the converse may be the case: certain behaviors, if present, may reliably indicate incremental comprehension failure. Caplan and Waters (2013) suggested two such behaviors: (i) in whole sentence reading, regressive eye movements "targeted" to within one word of an item that is relevant to the clarification of the structure and meaning of the sentence (discussed above) and (ii) in noncumulative self-paced reading and listening, positive correlations between reading and listening times for segments and task performance mark points at which comprehension is incrementally unsuccessful and controlled processing occurs. The basis for this latter suggestion is that longer self-paced reading and listening times result from the reader doing more "work" at a point in a sentence and therefore will occur in less efficient parsers, leading to a negative correlation of self-paced reading and listening times and comprehension accuracy across individuals if the extra work done by less efficient processors does not fully compensate for their inefficiency, or no relation between self-paced reading and listening processing times and task performance if the extra work done by less efficient processors does fully compensate for their inefficiency. If longer self-paced reading and listening times are positively correlated with task performance across individuals, then some process other than less efficient processors working harder to accomplish the comprehension task to the average level must have applied. These processes include controlled problem-solving operations such as deliberate application of "rules of grammar" that a comprehender has learned, readers' revisions of mental intonation contours to help determine constituent boundaries, and others that require accessing representations of a presented sentence in memory (for discussion and relevant results, see Caplan et al., 2011).

51.6 A COMMENT ON THE NEURAL BASIS OF PROCEDURAL (LT-WM) MEMORY MECHANISMS UNDERLYING SENTENCE COMPREHENSION

The suggestion that parsing and interpretation rely on LT-WM has much in common with the suggestion that parsing utilizes procedural memory (Ullman, 2004). There is evidence that neural structures that support procedural memory include a network that includes the basal ganglia and cerebellum and likely frontal lobe structures (Mizumori, Puryear, & Martig, 2009; Nagao & Kitazawa, 2008), and Ullman and others have suggested that these structures—especially the frontal-basal ganglionic portion of the network—support retrieval of information from memory in skilled parsing and interpretation.

The appeal of this hypothesis on theoretical grounds is tempered by the fact that, as reviewed, many features of memory in skilled parsing and interpretation are "domain-specific." Rick Lewis (personal communication) has suggested that it might be most profitable to view the memory mechanisms that support parsing and interpretation as being tokens of types of mechanisms used more generally, pointing to the many memory phenomena to which the ACT-R architecture has been applied (list memory: Anderson & Matessa, 1997; sentence memory: Budiu & Anderson, 2004; expertise: Anderson & Fincham, 1994; Anderson, Fincham, & Douglass, 1999). However, even if memory mechanisms that support different domain-specific skills are variants and specializations of a common functional architecture, there are substantial differences in the memory systems that support different tasks and skills. In Lewis' terms, the "tokens" of a common memory system "type" differ significantly because of domain-specific representations in the set of to-be-recalled items, the nature of the task that creates the need for retrieval, the extent to which encoding, storage, and rehearsal are performed in isolation or while other operations (such as comprehension or applying the products of comprehension to a task) are performed (Humphreys, Bain, & Pike, 1989). Domain-specific procedural memory systems arise because different sets of mechanisms are used to satisfy the memory needs of performance under the conditions found in different domains. Skilled, incrementally successful parsing and interpretation have a unique set of properties that are supported by one such domain-specific adaptation of memory mechanisms. It is therefore theoretically conceivable that aspects of the memory processes that are deployed during comprehension involve specialized areas of the brain, such as those that store lexical syntactic information in LTM or maintain transiently activated phrase markers during online comprehension.

The neural structures that support the memory processes that apply during skilled sentence comprehension need to be investigated empirically. To date, studies of the localization of comprehension of sentences that require parsing and interpretation have yielded contradictory results. For one structure—the contrast of object and subject relative clauses—a wide variety of patterns of activation have been reported: L inferior frontal gyrus (IFG) (Stromswold, Caplan, Alpert, & Rosch, 1996), Broca's area and Wernicke's area, and, to a lesser degree, in their right hemisphere homologues (Just, Carpenter, Keller, Eddy, & Thulborn, 1996), bilateral inferior temporal lobe (Cooke et al., 2002), medial anterior structures (cingulate, middle frontal and superior frontal gyri) and left thalamus, left superior parietal lobe, and right inferior frontal lobe (Caplan, Alpert, & Waters 1998, 1999; Caplan, Alpert, Waters, Olivieri, 2000; Chen, West, Waters, & Caplan, 2006; Waters, Caplan, & Yampolsky, 2003), and superior temporal gyrus (STG) (Ben-Shachar, Hendler, Kahn, Ben-Bashat, & Grodzinsky, 2003; Just et al., 1996; see Grodzinsky and Freiderici, 2006; and Caplan 2006, 2007 for review). Chen et al. (2006) found that some effects attributed to processing syntactic structure were, in reality, due to semantic factors, and Caplan, Chen, and Waters (2008) and Caplan (2010) have provided evidence showing that how syntactic structure is processed depends on the task, which influences neural responses. The areas that are activated by encoding, storing, and retrieving information in memory during the process of skilled sentence comprehension, as well as controlled processing that applies when skilled comprehension fails, require more study.

Acknowledgments

This work was supported by grants from NIH to David Caplan (DC00942, DC02146, DC003108, DC011032).

References

Acheson, D. J., & MacDonald, M. C. (2011). The rhymes that the reader perused confused the meaning: Phonological effects during on-line sentence comprehension. *Journal of Memory and Language*, *65*, 193–207.

Anderson, J. R., & Fincham, J. M. (1994). Acquisition of procedural skills from examples. *Journal of Experimental Psychology: Learning, Memory, and Cognition*, *20*, 1322–1340.

Anderson, J. R., Fincham, J. M., & Douglass, S. (1999). Practice and retention: A unifying analysis. *Journal of Experimental Psychology: Learning, Memory, and Cognition, 25*, 1120–1136.

Anderson, J. R., & Matessa, M. P. (1997). A production system theory of serial memory. *Psychological Review, 104*, 728–748.

Ayres, T. J. (1984). Silent reading time for tongue-twister paragraphs. *American Journal of Psychology, 97*, 605–609.

Baddeley, A. D. (1968). How does acoustic similarity influence short-term memory? *The Quarterly Journal of Experimental Psychology, 20*, 249–263.

Baddeley, A. D. (1986). *Working memory*. New York, NY: Oxford University Press.

Baddeley, A. D. (2000). The episodic buffer: A new component of working memory? *Trends in Cognitive Sciences, 4*, 417–423.

Baddeley, A. D., & Hitch, G. (1974). Working memory. In G. H. Bower (Ed.), *The psychology of learning and motivation: Advances in research and theory* (Vol. 8, pp. 47–89). New York, NY: Academic Press.

Ben-Shachar, M., Hendler, T., Kahn, I., Ben-Bashat, D., & Grodzinsky, Y. (2003). The neural reality of syntactic transformations: Evidence from functional magnetic resonance imaging. *Psychological Science, 14*, 433–440.

Brown, G. D. A., Neath, I., & Chater, N. (2007). A temporal ratio model of memory. *Psychological Review, 114*, 539–576.

Brown, G. D. A., Preece, T., & Hulme, C. (2000). Oscillator-based memory for serial order. *Psychological Review, 107*, 127–181.

Budiu, R., & Anderson, J. R. (2004). Interpretation-based processing: A unified theory of semantic sentence processing. *Cognitive Science, 28*, 1–44.

Caplan, D. (2006). Functional Neuroimaging of Syntactic Processing: New Claims and Methodological Issues. *Current Reviews in Medical Imaging, 2*, 443–451.

Caplan, D. (2007). Functional Neuroimaging Studies of Syntactic Processing in Sentence Comprehension: A Critical Selective Review. *Language and Linguistic Compass, 1*, 32–47.

Caplan, D. (2010). Task effects on BOLD signal correlates of implicit syntactic processing. *Language and Cognitive Processes, 25*, 866–901.

Caplan, D., Alpert, N., & Waters, G. S. (1998). Effects of syntactic structure and propositional number on patterns of regional cerebral blood flow. *Journal of Cognitive Neuroscience, 10*(4), 80–94.

Caplan, D., Alpert, N., & Waters, G. S. (1999). PET Studies of Sentence Processing with Auditory Sentence Presentation. *Neuroimage, 9*, 343–351.

Caplan, D., Alpert, N., Waters, G., & Olivieri, A. (2000). Activation of Broca's Area by Syntactic Processing under Conditions of Concurrent Articulation. *Human Brain Mapping, 9*, 65–71.

Caplan, D., Chen, E., & Waters, G. (2008). Task-dependent and task-independent neurovascular responses to syntactic processing. *Cortex, 44*, 257–275.

Caplan, D., DeDe, G., Waters, G., Michaud, J., & Tripodis, Y. (2011). Effects of age, speed of processing and working memory on comprehension of sentences with relative clauses. *Psychology and Aging, 26*, 439–450.

Caplan, D., & Waters, G. (1996). Syntactic processing in sentence comprehension under dual-task conditions in aphasic patients. *Language and Cognitive Processes, 11*, 525–551.

Caplan, D., & Waters, G. S. (1990). Short-term memory and language comprehension: A critical review of the neurophyschological literature. In G. Vallar, & T. Shallice (Eds.), *Neuropsychological impairments of short-term memory* (pp. 337–389). Cambridge: Cambridge University Press.

Caplan, D., & Waters, G. S. (1999). Verbal working memory and sentence comprehension. *Behavioral and Brain Sciences, 22*(1), 77–94. Available from: http://dx.doi.org/10.3758/s13423-012-0369-9.

Caplan, D., & Waters, G. S. (2013). Memory systems underlying syntactic comprehension. *Psychonomic Bulletin and Review, 20*(2), 243–268.

Chase, W. G., & Ericsson, K. A. (1982). Skill and working memory. In G. H. Bower (Ed.), *The psychology of learning and motivation* (Vol. 16, pp. 1–58). New York, NY: Academic Press.

Chen, E., West, C., Waters, G., & Caplan, D. (2006). Determinants of BOLD signal correlates of processing object-extracted relative clauses. *Cortex, 42*, 591–604.

Coltheart, V., Patterson, K., & Leahy, J. (1994). When a ROWS is a ROSE: Phonological effects in written word comprehension. *Quarterly Journal of Experimental Psychology: Learning, Memory, and Cognition, 15*, 824–845.

Conway, C. M., Karpicke, J., & Pisoni, D. B. (2007). Contribution of implicit sequence learning to spoken language processing: Some preliminary findings from normal-hearing adults. *Journal of Deaf Studies and Deaf Education, 12*, 317–334.

Conway, C. M., & Pisoni, D. B. (2008). Neurocognitive basis of implicit learning of sequential structure and its relation to language processing. *Annals of the New York Academy of Sciences, 1145*, 113–131.

Cooke, A., Zurif, E. B., DeVita, C., Alsop, D., Koenig, P., Detre, J., et al. (2002). Neural basis for sentence comprehension: grammatical and short-term memory components. *Hum. Brain Mapp, 15*, 80–94.

Cowan, N. (2000). The magical number 4 in short-term memory: A reconsideration of mental storage capacity. *Behavioral and Brain Sciences, 24*, 87–185.

Dillon, B., Mishler, A., Sloggett, S., & Phillips, C. (under review). Contrasting interference profiles for agreement and anaphora: Experimental and modeling evidence.

Eberhard, K. M., Cutting, J. C., & Bock, J. K. (2005). Making syntax of sense: Number agreement in sentence production. *Psychological Review, 112*, 531–559.

Emery, O. B. (1985). Language and aging. *Experimental Aging Research (Monograph), 11*, 3–60.

Engle, R. W., & Kane, M. J. (2004). Exectutive attention, working memory capactiy, and a two-factor theory of cognitive control. In B. Ross (Ed.), *The psychology of learning and motivation* (Vol. 44, pp. 145–199). New York, NY: Elsevier.

Ericsson, K. A., & Kintsch, W. (1995). Long-term working memory. *Psychological Review, 102*(2), 211–245.

Ferreira, F., Engelhardt, P. E., & Jones, M. W. (2009). Good enough language processing: A satisficing approach. In N. Taatgen, H. Rijn, J. Nerbonne, & L. Schomaker (Eds.), *Proceedings of the 31st annual conference of the cognitive science society*. Austin, TX: Cognitive Science Society.

Ferreira, F., & Patson, N. (2007). The good enough approach to language comprehension. *Language and Linguistics Compass, 1*, 71–83.

Fiebach, C. J., Schlesewsky, M., & Friederici, A. D. (2001). Syntactic working memory and the establishment of filler-gap dependencies: Insights from ERPs and fMRI. *Journal Psycholing Research, 30*, 321–338.

Fiebach, J., Vos, S. H., & Friederici, A. D. (2004). Neural correlates of syntactic ambiguity in sentence comprehension for low and high span readers. *Journal of Cognitive Neuroscience, 16*, 1562–1575.

Frazier, L., & Rayner, K. (1982). Making and correcting errors during sentence comprehension: Eye movements in the analysis of structurally ambiguous sentences. *Cognitive Psychology, 14*, 178–210.

Gibson, E. (1998). Linguistic complexity: Locality of syntactic dependencies. *Cognition, 68*, 1–76.

Gordon, P. C., Hendrick, R., & Johnson, M. (2001). Memory interference during language processing. *Journal of Experimental Psychology: Learning, Memory and Cognition, 27*(6), 1411–1423.

Grodzinsky, Y., & Friederici, A. (2006). Neuroimaging of syntax and syntactic processing. *Current Opinion in Neurobiology, 16*(2), 240–246.

Grossman, M., Carvell, S., Gollomp, S., Stern, M. B., Vernon, G., & Hurtig, H. I. (1991). Sentence comprehension and praxis deficits in Parkinson's disease. *Neurology, 41*, 1620–1628.

Grossman, M., Carvell, S., Stern, M. B., Gollomp, S., & Hurtig, H. (1992). Sentence comprehension in Parkinson's disease: The role of attention and memory. *Brain and Language, 42*, 347–384.

Hale, J. (2001). A probabilistic Earley parser as a psycholinguistic model. In *Proceedings of NAACL* (Vol. 2, pp. 159–166).

Halford, G. S., Cowan, N., & Andrews, G. (2007). Separating cognitive capacity from knowledge: A new hypothesis. *Trends in Cognitive Sciences, 11*(6), 236–242.

Henson, R. N., & Burgess, N. (1997). Representations of serial order. In J. A. Bullinaria, D. W. Glasspool, & G. Houghton (Eds.), *4th neural computation and psychology workshop* (pp. 283–300). London: Springer.

Henson, R. N. A. (1998). Short-term memory for serial order. The start-end model. *Cognitive Psychology, 36*, 73–137.

Hockley, W. E. (1984). Analysis of response time distributions in the study of cognitive processes. *Journal of Experimental Psychology: Learning, Memory and Cognition, 10*, 598–615.

Howes, A., Lewis, R. L., & Vera, A. H. (2009). Rational adaptation under task and processing constraints: Implications for testing theories of cognition and action. *Psychological Review, 116*(4), 717–751.

Humphreys, M. S., Bain, J. D., & Pike, R. (1989). Different ways to cue a coherent memory system: A theory for episodic, semantic and procedural tasks. *Psychological Review, 96*, 208–233.

Inhoff, A. W., & Weger, U. W. (2005). Memory for word location during reading: Eye movements to previously read words are spatially selective but not precise. *Memory & Cognition, 33*, 447–461.

Just, M. A., & Carpenter, P. A. (1992). A capacity theory of comprehension: Individual differences in working memory. *Psychological Review, 99*(1), 122–149.

Just, M. A., Carpenter, P. A., Keller, T. A., Eddy, W. F., & Thulborn, K. R. (1996). Brain activation modulated by sentence comprehension. *Science, 274*, 114–116.

Keller, T. A., Carpenter, P. A., & Just, M. A. (2003). Brain imaging of tongue twister sentence comprehension: Twisting the tongue and the brain. *Brain and Language, 84*(2), 189–203.

Kennison, S. M. (2004). The effect of phonemic repetition on syntactic ambiguity resolution: Implications for models of working memory. *Journal of Psycholinguistic Research, 33*, 493–516.

King, J., & Just, M. A. (1991). Individual differences in syntactic processing: The role of working memory. *Journal of Memory and Language, 30*, 580–602.

Kontiola, P., Laaksonen, R., Sulkava, R., & Erkinjuntti, T. (1990). Pattern of language impairment is different in Alzheimer's disease and multi-infarct dementia. *Brain and Language, 38*, 364–383.

Lalami, L., Marblestone, K., Schuster, S., Andersen, E., Kempler, D., Tyler, L., et al. (1996). On-line vs. off-line sentence processing in Alzheimer's disease. Poster presented at the meeting of cognitive aging, Atlanta, GA.

Levy, R. (2008). Expectation-based syntactic comprehension. *Cognition, 106*(3), 1126–1127.

Lewis, R. L. (1996). Interference in short-term memory: The magical number two (or three) in sentence processing. *Journal of Psycholinguistic Research, 25*(1), 93–115.

Lewis, R. L. (2000). Specifying architectures for language processing: Process, control, and memory in parsing and interpretation. In M. W. Crocker, M. Pickering, & C. Clifton, Jr. (Eds.), *Architectures and mechanisms for language processing*. Cambridge, UK: Cambridge University Press.

Lewis, R. L., & Vasishth, S. (2005). An activation-based model of sentence processing as skilled memory retrieval. *Cognitive Science, 29*, 375–419.

Lewis, R. L., Vasishth, S., & Van Dyke, J. A. (2006). Computational principles of working memory in sentence comprehension. *Trends in Cognitive Sciences, 10*, 44–54.

Lieberman, P., Friedman, J., & Feldman, L. S. (1990). Syntax comprehension in Parkinson's disease. *Journal of Nervous and Mental Diseases, 178*, 360–366.

Lombardi, L., & Potter, M. C. (1992). The regeneration of syntax in short term memory. *Journal of Memory and Language, 31*, 713–733.

MacDonald, M. C., Just, M. A., & Carpenter, P. A. (1992). Working memory constraints on the processing of syntactic ambiguity. *Cognitive Psychology, 24*, 56–98.

Marslen Wilson, W., & Tyler, L. (1976). Memory and levels of processing in a psycholinguistic context. *Journal of Experimental Psychology: Human Learning and Memory, 2*, 112–119.

Martin, A. E., & McElree, B. (2008). A content-addressable pointer mechanism underlies comprehension of verb-phrase ellipsis. *Journal of Memory and Language, 58*, 879–906.

Martin, A. E., & McElree, B. (2009). Memory operations that support language comprehension: Evidence from verb-phrase ellipsis. *Journal of Experimental Psychology: Learning Memory and Cognition, 35*, 1231–1239.

Martin, R. (1990). The consequences of reduced memory span for the comprehension of semantic versus syntactic information. *Brain and Language, 38*, 1–20.

Martin, R. C., & He, T. (2004). Semantic short-term memory and its role in sentence processing: A replication. *Brain and Language, 89*, 76–82.

Martin, R. C., Shelton, J., & Yaffee, L. (1994). Language processing and working memory: Neuropsychological evidence for separate phonological and semantic capacities. *Journal of Memory and Language, 33*, 83–111.

McCutchen, D., Bell, L. C., France, I. M., & Perfetti, C. A. (1991). Phoneme-specific interference in reading: The visual tongue-twister effect revisited. *Reading Research Quarterly, 26*, 87–103.

McCutchen, D., & Perfetti, C. A. (1982). The visual tongue-twister effect: Phonological activation in silent reading. *Journal of Verbal Learning and Verbal Behavior, 21*, 672–687.

McElree, B. (1996). Accessing short-term memory with semantic and phonological information: A time-course analysis. *Memory and Cognition, 24*, 173–187.

McElree, B. (1998). Attended and non-attended states in working memory: Accessing categorized structures. *Journal of Memory and Language, 38*, 225–252.

McElree, B. (2006). Accessing recent events. In B. H. Ross (Ed.), *The psychology of learning and motivation* (Vol. 46). San Diego, CA: Academic Press.

McElree, B., & Dosher, B. A. (1989). Serial position and set size in short-term memory: Time course of recognition. *Journal of Experimental Psychology: General, 18*, 346–373.

McElree, B., & Dosher, B. A. (1993). Serial retrieval processes in the recovery of order information. *Journal of Experimental Psychology: General, 122*, 291–315.

Meseguer, E., Carreiras, M., & Clifton, C. (2002). Overt reanalysis strategies during the reading of mild garden path sentences. *Memory and Cognition, 30*, 551–561.

Miller, G. A., & Chomsky, N. (1963). Finitary models of language users. In R. D. Luce, R. R. Bush, & E. Galanter (Eds.), *Handbook of mathematical psychology* (Vol. 2, pp. 419–491). New York, NY: Wiley.

Misyak, J. B., Christiansen, M. H., & Tomblin, J. B. (2010). On-line individual differences in statistical learning predict language processing. *Frontiers in Psychology, 1*, 31.

Mitchell, D. C., Shen, X., Green, M. J., & Hodgson, T. (2008). Accounting for regressive eye-movements in models of sentence processing: A reappraisal of the Selective Reanalysis hypothesis. *Journal of Memory and Language, 59*, 266–293.

Miyake, A., Carpenter, P., & Just, M. (1994). A capacity approach to syntactic comprehension disorders: Making normal adults perform like aphasic patients. *Cognitive Neuropsychology, 11*, 671–717.

Mizumori, S. J., Puryear, C. B., & Martig, A. K. (2009). Basal ganglia contributions to adaptive navigation. *Behavioural Brain Research, 199*, 32–42.

Nagao, S., & Kitazawa, H. (2008). Role of the cerebellum in the acquisition and consolidation of motor memory. *Brain and Nerve, 60*(7), 783–790.

Nairne, J. S. (1990a). A feature model of immediate memory. *Memory and Cognition, 18*, 251–269.

Nairne, J. S. (1990b). Similarity and long-term memory for order. *Journal of Memory and Language, 29*, 733–746.

Natsopoulos, D., Katsarou, Z., Bostantzopoulo, S., Grouios, G., Mentanopoulos, G., & Logothetis, J. (1991). Strategies in comprehension of relative clauses in Parkinsonian patients. *Cortex, 27*, 255–268.

Neath, I. (2000). Modeling the effects of irrelevant speech on memory. *Psychonomic Bulletin and Review, 7*, 403–423.

Nimmo, L., & Roodenrys, S. (2004). Investigating the phonological similarity effect: Syllable structure and the position of common phonemes. *Journal of Memory and Language, 50*, 245–258.

Phillips, C., Wagers, M. W., & Lau, E. F. (2011). Grammatical illusions and selective fallibility in real-time language comprehension. In J. Runner (Ed.), *Experiments at the Interfaces, Syntax and Semantics* (Vol. 37, pp. 153–186). Bingley, UK: Emerald Publications.

Poirier, M., & Saint-Aubin, J. (1996). Immediate serial recall, word frequency, item identity and item position. *Canadian Journal of Experimental Psychology, 50*(4), 408–412.

Postman, L., & Keppel, G. (1977). Conditions of cumulative proactive inhibition. *Journal of Experimental Psychology: General, 106*, 376–403.

Potter, M. C., & Lombardi, L. (1990). Regeneration in the short-term recall of sentences. *Journal of Memory and Learning, 29*, 633–654.

Ratcliff, R. (1978). A theory of memory retrieval. *Psychological Review, 85*, 59–108.

Ricker, T., AuBuchon,, A. M., & Cowan, N. (2010). Working memory. In L. Nadel (Ed.), *Wiley interdisciplinary reviews: Cognitive science* (Vol. 1, pp. 573–585).

Rochon, E., & Saffran, E. M. (1995). A semantic contribution to sentence comprehension impairments in Alzheimer's disease. Poster presented at the academy of aphasia, San Diego, CA.

Rochon, E., Waters, G. S., & Caplan, D. (1994). Sentence comprehension in patients with Alzheimer's Disease. *Brain and Language, 46*, 329–349.

Saults, J. S., & Cowan, N. (2007). A central capacity limit to the simultaneous storage of visual and auditory arrays in working memory. *Journal of Experimental Psychology: General, 136*(4), 663–684.

Staub, A. (2010). Eye movements and processing difficulty in object relative clauses. *Cognition, 116*, 71–86.

Stevens, K., Kempler, D., Andersen, E., & MacDonald, M. C. (1996). Preserved use of semantic and syntactic context in Alzheimer's disease. Poster presented at the meeting of cognitive aging, Atlanta, GA.

Stromswold, K., Caplan, D., Alpert, N., & Rosch, S. (1996). Localization of syntactic comprehension by pos- itron emission tomography. *Brain and Language, 52*, 452–473.

Sturt, P., & Crocker, M. (1996). Monotonic syntactic processing: A cross-linguistic study of attachment. *Language and Cognitive Processes, 11*(5), 449–494.

Tomoeda, C. K., Bayles, K. A., Boone, D. R., Kaszniak, A. W., & Slauson, T. J. (1990). Speech rate and syntactic complexity effects on the auditory comprehension of Alzheimer patients. *Journal of Communication Disorders, 23*, 151–161.

Traxler, M. J., & Pickering, M. J. (1996). Plausibility and the processing of unbounded dependencies: An eye-tracking study. *Journal of Memory and Language, 35*, 454–475.

Ullman, M. T. (2004). Contributions of memory circuits to language: The declarative/procedural model. *Cognition, 92*, 231–270.

Underwood, B. J., & Ekstrand, B. R. (1967). Effect of distributed practice on paired-associate learning. *Journal of Experimental Psychology, 73*, 1–21.

Van Dyke, J., & Lewis, R. L. (2003). Distinguishing effects of structure and decay on attachment and repair: A cue-based parsing account of recovery from misanalysed ambiguities. *Journal of Memory and Language, 49*, 285–316.

Van Dyke, J. A., & Johns, C. L. (2012). Memory interference as a determinant of language comprehension. *Language and Linguistics Compass, 6*(4), 193–211.

Van Dyke, J. A., & McElree, B. (2006). Retrieval interference in sentence comprehension. *Journal of Memory and Language, 55*, 157–166.

Vasishth, S., & Lewis, R. L. (2006). Argument-head distance and processing complexity: Explaining both locality and anti-locality effects. *Language, 82*, 767–794.

Verhaeghen, P., & Basak, C. (2005). Aging and switching of the focus of attention in working memory: Results from a modified n-back task. *Quarterly Journal of Experimental Psychology (A), 58*, 134–154.

Verhaeghen, P., Cerella, J., Basak, C., Bopp, K. L., Zhang, Y., & Hoyer, W. J. (2007). The ins and outs of working memory: Dynamic processes associated with focus switching and search. In N. Osaka, R. Logie, & M. D'Esposito (Eds.), *Working memory: Behavioural and neural correlates* (pp. 81–98). Oxford: Oxford University Press.

von der Malsburg, T., & Vasishth, S. (2011). What is the scanpath signature of syntactic reanalysis? *Journal of Memory and Language, 65*(2), 109–127.

Wanner, E., & Maratsos, M. P. (1978). An ATN approach to comprehension. In M. Halle, J. Bresnan, & G. A. Miller (Eds.), *Linguistic theory and psychological reality*. Cambridge, MA: MIT Press.

Waters, G., & Caplan, D. (1996). Processing resource capacity and the comprehension of garden path sentences. *Memory and Cognition, 24*, 342–355.

Waters, G., Caplan, D., & Hildebrandt, N. (1987). Working memory and written sentence comprehension. In M. Coltheart (Ed.), *Attention and performance XII: The psychology of reading* (pp. 531–555). London: Lawrence Erlbaum.

Waters, G. S., Caplan, D., & Hildebrandt, N. (1991). On the structure of verbal short-term memory and its functional role in sentence comprehension: Evidence from neuropsychology. *Cognitive Neuropsychology, 8*(2), 81–126.

Waters, G. S., Caplan, D., & Leonard, C. (1992). The role of phonology in reading comprehension: Implications of the effects of homophones on processing sentences with referentially dependent categories. *Quarterly Journal of Experimental Psychology A, 44*, 343–372.

Waters, G. S., Caplan, D., & Rochon, E. (1995). Processing capacity and sentence comprehension in patients with Alzheimer's Disease. *Cognitive Neuropsychology, 12*, 1–30.

Waters, G. S., Caplan, D., & Yampolsky, S. (2003). On-line syntactic processing under concurrent memory load. *Psychonomic Bulletin & Review, 10*(1), 88–95.

Watkins, M. J., Watkins, O. C., & Crowder, R. G. (1974). The modality effect in free and serial recall as a function of phonological similarity. *Journal of Verbal Learning and Verbal Behavior, 13*, 430–447.

Weger, U. W., & Inhoff, A. W. (2007). Long-range regressions to previously read words are guided by spatial and verbal memory. *Memory and Cognition, 35*, 1293–1306.

Wickelgren, W. A. (1965). Short-term memory for phonemically similar lists. *American Journal of Psychology, 78*, 567–574.

Wickelgren, W. A. (1966). Associative intrusions in short-term recall. *Journal of Experimental Psychology, 72*, 853–858.

Zhang, S., & Perfetti, C. A. (1993). The tongue twister effect in reading Chinese. *Journal of Experimental Psychology: Learning, Memory, and Cognition, 19*, 1082–1093.

Further Reading

Basak, C. (2005). *Capacity limits of the focus of attention and dynamics of the focus switch cost in working memory* (unpublished doctoral dissertation). Syracuse, NY: Syracuse University.

Caplan, D., Waters, G. S., & Howard, D. (2012). Slave systems in verbal working memory. *Aphasiology: Special Issue on Neuropsychology of Short Term Memory, 26*, 279–316.

Colle, H. A. (1980). Auditory encoding in visual short-term recall: Effects of noise intensity and spatial location. *Journal of Verbal Learning and Verbal Behavior, 19*, 722–735.

Colle, H. A., & Welsh, A. (1976). Acoustic masking in primary memory. *Journal of Verbal Learning and Verbal Behavior, 15*, 17–32.

Conrad, R. (1963). Acoustic confusions and memory span for words. *Nature, 197*, 1029–1030.

Conrad, R. (1964). Acoustic confusions in immediate memory. *British Journal of Psychology, 55*, 75–84.

Craik, F. I. M. (1970). The fate of primary memory items in free recall. *Journal of Verbal Learning and Verbal Behavior, 9*, 143–148.

Craik, F. I. M., & Levy, B. A. (1970). Semantic and acoustic information in primary memory. *Journal of Experimental Psychology, 86*, 77–82.

Henson, R. N. A., Page, M. P., Norris, D., & Baddeley, A. D. (1996). Unchained memory: Error patterns rule out chaining models of immediate serial recall. *The Quarterly Journal of Experimental Psychology, 49A*(8), 0–115.

James, W. (1890). *The principles of psychology*. New York, NY: Holt.

Johnson, M. L., Lowder, M. W., & Gordon, P. C. (2011). The sentence composition effect: Processing of complex sentences depends on the configuration of common versus unusual noun phrases. *Journal of Experimental Psychology: General, 140*, 707–724.

McElree, B. (2001). Working memory and focal attention. *Journal of Experimental Psychology: Learning, Memory and Cognition, 27*, 817–835.

Shallice, T. (1975). On the contents of primary memory. In P. M. A. Rabbit, & S. Domic (Eds.), *Attention and performance* (Vol. 5, pp. 269–280). London: Academic Press.

Shallice, T., & Vallar, G. (1990). The impairment of auditory-verbal short-term storage. In G. Vallar, & T. Shallice (Eds.), *Neuropsychological impairments of short-term memory* (pp. 11–53). Cambridge: Cambridge University Press.

Verhaeghen, P., Cerella, J., & Basak, C. (2004). A working memory workout: How to change to size of the focus of attention from one to four in ten hours or less. *Journal of Experimental Psychology: Learning, Memory, and Cognition, 30*, 1322–1337.

CHAPTER

52

Grounding Sentence Processing in the Sensory-Motor System

Marta Ghio[1] *and Marco Tettamanti*[2]

[1]Institute for Experimental Psychology, Heinrich-Heine-University, Dusseldorf, Germany; [2]Department of Nuclear Medicine and Division of Neuroscience, San Raffaele Scientific Institute, Milano, Italy

52.1 INTRODUCTION

Current neuroscientific models of sentence processing recognize the coordinate labor between perisylvian language areas in the frontal and temporo-parietal cortices and supportive brain networks supplying short-term memory and other executive functions. Together, they subserve the reversible mapping of phonological lexical forms onto semantic and syntactic information in language comprehension and production, respectively (Friederici, 2012; Kemmerer, 2015). The involvement of left-hemispheric middle and superior temporal, inferior-posterior parietal, as well as inferior frontal brain regions in various sentence comprehension tasks has been revealed by a host of neuropsychological and neuroimaging studies (Dronkers, Wilkins, Van Valin, Redfern, & Jaeger, 2004; Pallier, Devauchelle, & Dehaene 2011; Vigneau et al., 2006). Part of this supportive brain network is also the left-hemispheric lateral premotor cortex, sometimes extending more posteriorly into the primary motor area and more anteriorly into the middle frontal gyrus (Kemmerer, 2015; Vigneau et al., 2006). The involvement of the motor system in sentence processing is not only due to phonological and articulatory mapping (Pulvermueller & Fadiga, 2010)—aspects that are not dealt with in this chapter—because it also provides a grounding node for certain kinds of conceptual-semantic information. The motor system involvement in conceptual-semantic language processing has particularly intrigued researchers over the past decade and led to hotly debated divergent positions in the field. Activations in the sensory-motor system during processing of action-related conceptual knowledge have been interpreted in epiphenomenal terms or in "embodied" terms. Although several variants of embodiment have been formulated (Meteyard, Cuadrado, Bahrami, & Vigliocco, 2012), here we refer to a more general assumption formalized within the theoretical framework of grounded cognition, namely that concepts are represented in multiple, distributed brain networks reflecting the quality of experience that is characteristic for the concepts' referents. In contrast to other accounts, the grounded cognition framework makes explicit that evidence about action-related conceptual representations is just one instance regarding the much broader domain of conceptual-semantic representations, including object-related, emotion-related, and other types of even more abstract concepts. Following this more general perspective, the theoretical construct of "grounded cognition" incorporates not only bodily (embodied) states but also the physical, social, and linguistic environment, affective, and internal states (Barsalou, 2010; Pezzulo et al., 2013).

Grounded cognition provides a coherent and scientifically testable theoretical framework to account for the involvement of the sensory-motor and other experiential brain systems in conceptual-semantic processing. Research on grounded cognition has provided an increasing amount of empirical data in favor of the involvement of these brain systems in conceptual-semantic processing (Kiefer & Pulvermueller, 2012; Meteyard et al., 2012). Most studies focused on processing words in isolation, either nouns or action verbs (Hauk & Tschentscher, 2013), whereas much less attention has been given to multi-word utterances such as phrases and sentences. In this respect, it has been observed that the meaning of a word presented in isolation is often underspecified and is, in many cases, prone to different interpretations. Desai, Binder,

Neurobiology of Language. DOI: http://dx.doi.org/10.1016/B978-0-12-407794-2.00052-3

Conant, and Seidenberg (2010) claimed that the meaning of a verb like "to use" is closely related to its arguments, with "use the hammer" and "use the opportunity" conveying an action-related and an abstract meaning, respectively. Moreover, processing a verb in isolation involves not only its meaning but also the information conveyed by its syntactic structure, such as the arguments it takes and their typical meanings. Ferretti, McRae, and Hatherell (2001) found that verbs prime typical agents and patients (to arrest—cop, criminal), or typical instruments (to stir—spoon), but not location (to swim—ocean). Similar results have been found for nouns (McRae & Jones, 2013). Providing verbs and nouns within a sentence structure can thus be considered as a linguistically more constrained access to conceptual knowledge that disambiguates the meaning of a word by the sentential context and that, by keeping the syntactic form constant, allows for controlling the information conveyed by the syntactic structure of a word.

A further important aspect to be considered is that only at the multi-word level do certain aspects emerge, such as how the meanings of individual words embedded within a phrase or a sentence are compositionally combined into complex meanings. Only recently has the question of the compositional mechanisms of language understanding been addressed from a grounded perspective. A first, still poorly investigated issue regards the role of grammatical/morpho-syntactic information, not only in assembling the meanings of lexical-semantic units but also in modulating grounded representations of action meanings by giving focus to different aspects of the linguistically described action (e.g., verb tense and aspect, or affirmative versus negative sentences). A second issue regards the role of the sentential semantic context in which action-related words are embedded in modulating motor resonance, such as literal (e.g., "Grasp the hammer") versus figurative metaphorical (e.g., "Grasp the idea") or idiomatic (e.g., "Grasp the nettle") language.

Thus, the experimental manipulation of multi-word compositional parameters cannot but extend the discussion on the flexible nature of conceptual-semantic grounded representations (Kiefer & Pulvermueller, 2012) by emphasizing the modulatory role of the sentential linguistic context.

In this chapter, we cover and review the fundamental observations that have demonstrated the grounding of action-related sentence processing in the sensory-motor system and the debate on its causal, necessary role (see Section 52.2). Furthermore, we address increasing evidence about context-dependent flexible modulations with respect to the modulatory role of grammar/syntax in grounded conceptual-semantic representations at the multi-word level (see Section 52.3) and about action-related figurative language (see Section 52.4). Finally, we deal with attempts to provide a more overarching framework for the neural mechanisms underlying the conceptual-semantic processing of emotion-related (see Section 52.5) and other abstract concepts (see Section 52.6) that may be grounded not only in action and perception but also in other experiential brain systems.

52.2 GROUNDING OF ACTION-RELATED SENTENCE PROCESSING IN THE SENSORY-MOTOR SYSTEM

The idea rooted in philosophy and psychology that the coding of conceptual information in semantic memory is tied to inherent sensory-motor properties of real-world objects and to the kind of sensory-motor experience we maintain with those objects eventually begun being supported by neuroscientific observations, particularly with seminal studies using neuropsychological patients (Patterson, Nestor, & Rogers 2007; Warrington & McCarthy, 1987). Neuroimaging studies using healthy normal subjects further demonstrated that passive viewing or naming of manipulable tools activates the lateral frontal cortex in or in proximity of the premotor cortex (Martin, 2007). From the point of view of language processing, one of the first ideas to be tested with functional magnetic resonance imaging (fMRI) was that linguistic descriptions of motor actions evoke activations of the neural motor system (Hauk, Johnsrude, & Pulvermueller, 2004): somatotopically arranged activations in the primary motor and premotor cortex were found for reading mouth-, hand-, and leg-related motor verbs presented in isolated, infinitive forms. Other fMRI studies closely replicated these findings at the multi-word level, using either Italian subject-verb-object sentences (Tettamanti et al., 2005) or English verb-object phrases (Aziz-Zadeh, Wilson, Rizzolatti, & Iacoboni, 2006). Several methodological differences confound the comparison between the results of these three studies, such as reading (Aziz-Zadeh et al., 2006; Hauk et al., 2004) versus passive listening tasks (Tettamanti et al., 2005), and the syntactic structure of action-related linguistic stimuli. These ranged from having both the predicate's subject and object unspecified (Hauk et al., 2004), to only the object specified (Aziz-Zadeh et al., 2006), to both subject and object specified (Tettamanti et al., 2005). This may, at least in part, explain why only in the latter study the activations in motor areas were clearly accompanied by the activation of a broader left-hemispheric network, including the supramarginal and the posterior middle temporal gyri. The involvement of a left-hemispheric premotor-parieto-temporal network for listening to action-related, subject-verb-object sentences was confirmed by other studies (Boulenger, Hauk, & Pulvermueller, 2009; Desai et al., 2010), with some also showing effective connectivity interactions between these brain regions (Ghio & Tettamanti, 2010; Tettamanti et al., 2008). One likely advanced possibility is that this premotor-parieto-temporal network reflects the involvement of action representation processes known to involve the same brain regions. In this view, the linguistic specification of both the subject and the object of a goal-directed, action-related verb predicate appears to lead to functional interactions between premotor areas, coding for the action motor program, anterior parietal regions, coding for the object's affordance features, and posterior temporal regions close to visual areas coding for biological motion (Arbib, 2012). It is possible that using isolated words or simple phrases, leaving one or more thematic roles unspecified, leads to a more restricted activation spread within the action representation network. This may be either because some properties of the action's agents and recipients are not being represented or because activity reverberation within the broader network is hindered by compensatory activation in some network nodes for reconstructing the unspecified thematic role information. Compatibly with these lines of reasoning, an earlier fMRI study on action observation (Buccino et al., 2001) reported premotor-parietal activations for viewing transitive actions (i.e., directed toward a visible object goal), but only premotor activations and no parietal activations for viewing intransitive actions (i.e., pantomimes, lacking an object goal).

Whereas all these studies have regarded sentence processing as an integrated unitary process over time, mainly due to the coarse temporal resolution of fMRI, other techniques with higher temporal resolution may allow the breakdown of the specific neural dynamics that are instantiated as the processing of the sentence grammatical and thematic roles unfolds. Along these lines, Scorolli et al. (2012) delivered transcranial magnetic stimulation (TMS) to the left primary motor cortex while participants read action-related verbs followed by concrete nouns, together forming verb-object phrases. Semantic acceptability response times for experimental phrases, compared with nonsense control phrases, were faster when the TMS pulse was delivered during verb than during noun presentation. This facilitation effect suggests that the semantic processing of action-related verbs, but not that of the associated concrete object nouns, rapidly activates the left-hemispheric motor system. A limitation of this study is the lack of additional TMS target anatomical loci, for instance, to test the hypothesis of parietal involvement in the processing of manipulable object nouns in verb predicates.

More indirect (lacking any measurements or modulations of neural activity targeted anatomically to the motor cortex) evidence of the involvement of the motor system in sentence processing at the conceptual-semantic level came from behavioral studies testing the so-called action-sentence compatibility effect (ACE). Because the processing of action-related word meanings and congruent motor actions are thought to be subserved by partially overlapping neural networks, several experimental studies have tested whether the temporal proximity/overlap between language processing and action execution tasks can lead to facilitatory/interference effects. For example, Glenberg and Kaschak (2002) showed that hand movements toward or away from the body were facilitated by sentences describing a congruent (e.g., "He opened/closed the drawer," respectively) compared with an incongruent action. Boulenger et al. (2006) found that the processing of action-related verbs presented before a signal prompting for an upper-limb grasping movement facilitated movement kinematics, an effect that was ascribed to residual activation of motor areas by verb processing that lowered the amount of activation required by the subsequent grasping movement to reach threshold. In turn, when the action-related verbs were presented simultaneously to the start of the grasping movement, interference in kinematic parameters was observed; this interference effect was ascribed to language and action processing simultaneously competing for the same neural resources.

However, whether sensory and motor representations play a causal (i.e., necessary) role in conceptual processing is still a matter of debate between defenders of the epiphenomenal (Mahon & Caramazza, 2008) and strong embodiment views (Gallese & Lakoff, 2005). Somewhat in between, moderate and weak versions of embodiment proposed that semantic representations are at least partly constituted by sensory-motor information, although with a certain degree of abstraction (Binder & Desai, 2011; Meteyard et al., 2012) and with flexible dynamic modulations according to tasks, context, and situations (Kiefer & Pulvermueller, 2012).

The choice of experimental approaches to test the necessary role of sensory-motor representation is controversial. fMRI and electroencephalography/magnetoencephalography evidence of fast and automatic activations of grounded systems during semantic processing (Kiefer & Pulvermueller, 2012; Pulvermueller & Fadiga, 2010) is not considered to be conclusive, because these methodologies do not allow establishing whether the observed correlations between brain activities and cognitive functions are causal. More decisive evidence is generally thought to be provided from neuropsychological studies involving patients with various disorders involving the motor system.

For instance, a sentence processing study by Ibáñez et al. (2013) showed that, relative to healthy control subjects, the ACE effect is reduced in Parkinson's disease patients. Similarly, Fernandino et al. (2013) showed that Parkinson's disease patients were slower than healthy subjects in a comprehension task with hand action–related sentences. However, motor and premotor brain regions do not appear to play a necessary role in sentence comprehension because their damage does not incontrovertibly lead to massive comprehension deficits in large patient cohorts (Kemmerer, 2015). Altogether, neuropsychological studies have provided conflicting results, leaving the causality debate from this perspective still open.

An alternative approach to argue against a merely epiphenomenal influence of motor system activity on language processing, but one that does not deny its flexible modulatory nature, is to demonstrate that the processing of word meaning involves the activation of sensory-motor brain areas at a varying degree that determines the efficiency of conceptual-semantic language understanding. Two studies on sentence processing have substantially contributed advances in this direction. Beilock, Lyons, Mattarella-Micke, Nusbaum, and Small (2008), in a combined fMRI and behavioral study, showed that individual sensory-motor expertise can improve the comprehension of related concepts in a semantic language task. Expertise was positively correlated with higher activation of the left dorsal premotor cortex, a brain region supporting the selection of well-learned action plans. Locatelli, Gatti, and Tettamanti (2012) showed that sensory-motor expertise gained by training naive subjects to perform specific manual dexterity actions (e.g., origami, knot-tying) can lead to an improvement of cognitive-linguistic skills related to the specific conceptual-semantic domain associated with the trained actions. More specifically, the authors observed, after manual dexterity training, a speeding of reaction times in a sentence-picture semantic congruency judgment task that was selective for sentence-picture pairs semantically related to the trained manual actions versus unrelated sentence-picture pairs.

In sum, the investigation of sentence processing can substantially contribute to the debate on the grounding of action-related semantic processing in the sensory-motor system. Specifically, action-related sentences may help clarify the specific neural mechanisms for representing the subject and object thematic roles of an action's verb predicate. The debate regarding whether the involvement of the motor system in action-related sentence processing is obligate or optional continues, but researchers increasingly recognize—against a strictly epiphenomenal view—that the most relevant question is to clarify when and how this flexible involvement occurs (Willems & Francken, 2012) and whether this involvement can be beneficial to language processing.

52.3 FLEXIBLE MODULATIONS OF SENSORY-MOTOR GROUNDING BY GRAMMATICAL AND SYNTACTIC ASPECTS

As discussed, the investigation of conceptual-semantic processing at the sentence level can help gain a better control over the polysemous and thematic ambiguities that individual words in isolation leave unspecified. However, when it comes to sentence-level language processing, as opposed to single words, a set of nonlexical variables markedly comes into play, including combinatorial and compositional semantics, verb tense and aspect, and syntactic structure. These variables could, in principle, influence the kind of grounded conceptual-semantic representations reviewed (Bergen & Wheeler, 2010). In this respect, some behavioral evidence has been provided showing that grammatical parameters, such as verb aspect or temporal markers, affect the ACE. Zwaan, Taylor, and de Boer (2010) asked subjects to read action sentences embedded in a text describing an event as ongoing (e.g., "He opened the bottle"), about to happen (e.g., "He was about to open the bottle"), or completed (e.g., "He had opened the bottle") while performing a knob rotation task. Results revealed ACEs (longer reading and longer knob rotating times) for both ongoing and completed action sentences, but not for about to happen sentences. Bergen and Wheeler (2010) investigated the role of verb aspect in modulating the motor representations in language comprehension. Based on simulations, they argued that progressive sentences (e.g., "Ashley is stretching her arms") result in the detailed simulation of the described action (and should yield ACEs), whereas perfect sentences (e.g., "Ashley stretched her arms") result in the simulation of the action end-state (and should not yield ACEs). Accordingly, they found evidence for ACEs in progressive sentences, but not in perfect sentences. Thus, it seems that the activation of the motor system can be modulated by grammatical/syntactic information related to verb tense and aspect.

Among the more straightforward instances of compositional semantic integration at the sentence level that have contributed to the view of the flexible nature of conceptual-semantic sensory-motor representations is the case of sentential negation. In sentential negation, an interaction occurs between a syntactic element (i.e., the negation operator *not*) and conceptual-semantic information (i.e., the scope of sentential negation containing the negated information) (Tettamanti & Moro, 2012). Building on psycholinguistic evidence (Kaup, Zwaan, & Lüdtke, 2007) indicating reduced speed in the access to lexical-semantic information under the scope of sentential negation, the initial experimental question addressed by Tettamanti et al. (2008) in an fMRI study was the following: given that the semantic processing of action-related sentences like "I grasp an apple" activates the left-hemispheric action representation system, can we measure any kind of modulations of such an activation when processing sentences with reversed negation polarity like "I do not grasp an apple"? They observed reduced activation and reduced effective connectivity within the left-hemispheric premotor-parieto-temporal action representation system for negative versus affirmative action-related sentences. This effect was interpreted as a reduced grounding in the sensory-motor system for the conceptual-semantic processing of negative action-related sentences, possibly driven by a drain of neuro-computational resources in favor of syntactic processing, given the greater syntactic and computational complexity of negative versus affirmative sentences (Christensen, 2009; Kaup et al., 2007).

Several other studies using different experimental techniques have yielded compatible observations. Tomasino, Weiss, and Fink (2010) showed that fMRI activations in the hand region of the primary motor and premotor cortices were reduced for negative hand action–related imperatives (e.g., "Don't grasp!") compared with affirmative counterparts (e.g., "Grasp!"). By means of TMS of the hand motor cortex and a concurrent reading task, Liuzza, Candidi, and Aglioti (2011) showed that the suppression of motor-evoked potentials (MEPs) from hand muscles observed for affirmative hand action–related sentences was reduced for negative sentences. A similar reduction of motor cortex activity for negative sentences was observed by Aravena et al. (2012) in a study in which they measured hand grip force continuously after target verb presentation; they found an enhancement of grip force for affirmative but not for negative hand action–related sentences.

On the basis of this motor cortex activity reduction due to sentential negation, Bartoli et al. (2013) postulated a "disembodiment" mechanism of sentential negation, by which the reduced grounding of conceptual-semantic processing in the sensory-motor system for negative action-related sentences leads to a computational load reduction in these brain regions, yielding a reduced interference in concurrent motor tasks. This mechanism was verified in a kinematic study in which participants listened to affirmative or negative action-related subject-verb phrases and immediately afterward performed a congruent upper-limb movement (Bartoli et al., 2013). Negative sentences were found to interfere less with the movement kinematic parameters than matching affirmative sentences.

Foroni and Semin (2013) measured electromyographic recordings from zygomatic muscles continuously after the presentation of face action-related subject-verb phrases (e.g., "I am smiling"/"I am not smiling") and found muscle activation for affirmative sentences and muscle inhibition for negative sentences. These results suggest a possible neurophysiological mechanism by which the semantic comprehension of negative action-related sentences is encoded in terms of an inhibition of the muscles whose activation is negated.

Another sentence-level linguistic construction that is related to but distinct from sentential negation is that of counterfactuals—a linguistic combination of semantic and grammatical elements to express a given modality of the declarative state of affairs, namely that something does not occur but could have occurred under some hypothetical plausible circumstances. Evidence limited to one fMRI study (Urrutia, Gennari, & de Vega, 2012) using complex action-related sentences such as "If Pedro had decided to paint the room, he would have moved the photograph" showed higher activation of the fronto-medial cortex extending into the supplementary motor area for counterfactuals versus factuals (e.g., "Since Pedro decided to paint the room, he is moving the photograph"). Thus, the conceptual-semantic comprehension of counterfactuals appears to be grounded in the motor system in a manner that differs from that of sentential negation with respect to direction (higher versus lower activation) and anatomical location (fronto-medial versus premotor and primary motor cortex). The fronto-medial cortex has been implicated in the programming of actions that are withheld or not executed (Brass & Haggard, 2007).

All these instances of multi-word grammaticality underline the fact that future studies on sentence processing shall more and more disentangle relevant dimensions of sentence structure with reference to sentence processing anatomo-functional models and not treat anymore sentences as unitary blocks. As already discussed in the previous section, time-resolved neuroimaging techniques can provide a crucial complementary contribution in this respect.

52.4 FIGURATIVE LANGUAGE AS AN ABSTRACT SENTENTIAL-SEMANTIC CONTEXT FOR ACTION-RELATED VERBS

Figurative language forms still another traditional research topic within the domain of grounded cognition having a focus on sentence processing, because figurative language is typically instantiated at the sentence level. Metaphors are linguistic utterances in which the intended conceptual meaning differs from the meaning conveyed by the literal combination of the individual lexical-semantic constituents, thus forming a conceptual-semantic representation at a somewhat more "abstract" level. Some proponents of embodied cognition theories (Gallese & Lakoff, 2005; Gibbs, 2006) have proposed that these more abstract concepts, particularly when formed by lexical-semantic constituents with a physical, sensory-motor meaning (e.g., "Grasp the idea"), are grounded in the sensory-motor, premotor-parietal system. In metaphoric expressions, the motivations for using a certain literal image or meaning are generally manifest to the communication recipients. On the contrary, more conventionalized figurative expressions, such as idioms, can present varying degrees of semantic transparency, ranging from transparent ones (e.g., "Smell a rat") to more opaque ones (e.g., "Kick the bucket"). It is debated whether the comprehension of idioms compositionally entails the literal comprehension of the individual lexical-semantic constituents until some (e.g., contextual) cues prompt the idiomatic interpretation, or whether it is directly gained by retrieving a morphologically complex entry in the mental lexicon (Papagno & Caporali, 2007). Most empirical evidence suggests that idiom processing involves, to some extent, a decomposition into individual lexical units, even in the case of more opaque idioms (Papagno & Caporali, 2007; Romero Lauro, Tettamanti, Cappa, & Papagno, 2008). If so, then we should expect that figurative expressions of different sorts containing action-related verbs activate motor areas, and this should reveal the extent to which the motor system involvement in the processing of action-related words is lexically bound, independently of the global conceptual-semantic representation formed, or flexibly modulated depending on the type of literal versus figurative sentential context.

Evidence in support of either hypothesis until now has been somewhat inconsistent, although overall the number of studies that found an association between the motor system and specific types of figurative action-related language probably exceeds the number of those that did not (for an earlier review, including a coordinate meta-analysis, see Rapp, Mutschler, & Erb, 2012). An early fMRI study (Aziz-Zadeh et al., 2006) found no significant motor system activations for idiomatic phrases related to mouth, hand, or foot actions, despite a somatotopic effect for literal sentences. Negative fMRI evidence for idiomatic, but not for literal, sentences were also supplied by Raposo, Moss, Stamatakis, and Tyler (2009). Other fMRI studies, however, found a significant involvement of the motor and/or premotor cortex in the processing of figurative action-related sentences. Boulenger et al. (2009) found arm/leg somatotopic effects for both idiomatic and literal sentences, effects that were confirmed in a magnetoencephalography study (Boulenger, Shtyrov, & Pulvermueller, 2012), showing a fast spread of activity in the motor cortex within 250 ms of latency. Desai, Binder, Conant, Mano, and Seidenberg (2011) found larger activations for idiomatic versus literal sentences in the motor cortex, as well as in the posterior superior temporal sulcus coding for biological motion; these effects displayed an inverse correlation with familiarity, such that unfamiliar sentences showed the strongest activation in these brain regions. An fMRI study by Romero Lauro, Mattavelli, Papagno, and Tettamanti (2013) included metaphoric and fictive motion (e.g., "The road runs through the valley") sentences in addition to idiomatic and literal ones. Left premotor cortex activations were found for processing both literal (arm and leg somatotopy) and fictive motion sentences (only leg, because fictive motion cannot involve upper-limb verbs by definition); arm-related but not leg-related metaphoric sentences induced significant activations in the left premotor cortex, whereas idiomatic sentences only presented a trend toward significant left premotor activations when an upper-limb verb was involved. This gradient of the motor effects from literal to figurative conditions was also confirmed by another fMRI study (Desai, Conant, Binder, Park, & Seidenberg, 2013) and by an earlier TMS study (Cacciari et al., 2011) that found significant effects for literal, fictive motion, and metaphoric sentences, but not for idiomatic sentences. An ACE study by Santana and de Vega (2011) showed faster reaction times in the action-sentence matching condition for both literal and metaphoric sentences. Finally, a study on action-related figurative language in Parkinson's disease patients (Fernandino et al., 2013, already discussed in Section 52.2) reported slower reaction times compared with healthy control subjects for literal and idiomatic sentences, as well as worse accuracy for metaphoric sentences.

Overall, multiple complementary techniques provide some convergent positive evidence for an involvement of the motor system in figurative action-related language processing. This conclusion can be drawn with somewhat greater confidence for figurative expressions that, as we have seen, undergo decomposition into individual lexical-semantic components, such as metaphors and fictive motion expressions. Idioms, although at least partially undergoing lexical decomposition, appear to activate motor areas in a less consistent manner, possibly inversely correlated with idiom familiarity. Therefore, at least to some extent, the literal versus figurative contexts in which action-related verbs appear flexibly modulates the degree of grounding in the sensory-motor system.

As already discussed in relation to Desai et al. (2011) study, a quite consistent finding of action-related language fMRI studies, distinct from the activation of the motor cortex, is the involvement of the posterior temporal cortex coding for biological motion. Accordingly, fMRI studies on action-related figurative expressions, whether idiomatic, metaphoric, or more frequently involving fictive motion, have found consistent activation in posterior temporal brain regions (Romero Lauro et al., 2013; Saygin, McCullough, Alac, & Emmorey, 2010; Wallentin, Lund, Ostergaard, Ostergaard, & Roepstorff, 2005; Wallentin et al., 2011). Chen, Widick, and Chatterjee (2008) reported a similar result for predicate metaphors (e.g., "The man fell under her spell") but importantly distinguished this activation effect as being located more anteriorly along the posterior middle temporal cortex compared with the activation elicited by literal motion sentences. This gradient was interpreted as evidence that increasingly abstract motion knowledge progressively involves more anterior portions of the lateral temporal cortex.

The recent neuroscientific literature has shown that the tight functional link between sensory-motor conceptual knowledge conveyed by language and neural circuits mediating sensory-motor experience related to the concepts' referents extends more in general to virtually every aspect of the perceptual domain (Kiefer & Pulvermueller, 2012; Patterson et al., 2007). Within the realm of figurative language studies, a notable example of these more general grounded associations is an fMRI study on metaphors based on concepts related to object texture (e.g., "She had a rough day"), which showed a specific activation in the texture-selective somatosensory cortex in the parietal operculum (Lacey, Stilla, & Sathian, 2012).

52.5 EMOTION-RELATED LANGUAGE: ABSTRACT BUT PARTIALLY GROUNDED IN THE SENSORY-MOTOR SYSTEM

The use of multi-word expressions opens the possibility of combining concrete words to convey more abstract meanings, as we have illustrated. Abstract meanings, however, can also be conveyed by words or sentences referring to entities, events, or states that cannot be experienced through senses in the external world. The controversial possibility of grounding the processing of abstract knowledge in the sensory-motor system is best illustrated by emotion-related knowledge.

Emotion meanings (e.g., happiness, sadness) are generally considered abstract concepts because they mainly refer to internal states and have no obvious referents in sensory-motor experience. From a grounded perspective, internal states can be as much the object of experience as external states and actions (Kousta, Vigliocco, Vinson, Andrews, & Del Campo, 2011). In addition to internal experiences—including affective states, introspection, interoceptions, and mentalizing—emotion concepts are also believed to be grounded in situations in which people experienced them—including settings, agents, objects, events, and even motor actions (Barsalou, 2010; Wilson-Mendenhall, Barrett, Simmons, & Barsalou, 2011). The relative contribution of the internal and external sensory-motor experiences associated with a given emotion concept may vary (Wilson-Mendenhall et al., 2011). Accordingly, retrieving and processing abstract emotion knowledge conveyed by language should activate a distributed brain circuitry including (in addition to language areas) neural networks processing emotion and introspection and the sensory-motor system in a context-dependent manner (Binder & Desai, 2011; Kiefer & Pulvermueller, 2012). As far as motor representations are concerned, it has been observed (Winkielman, Niedenthal, & Oberman, 2008), for example, that covert emotional states (e.g., anger) are often associated with overt motor behaviors, such as facial expressions (e.g., frown) and emotional hand/arm gestures (e.g., clenched fists).

Neuroimaging studies (for a review, see Citron, 2012) have provided evidence of activations in emotion brain networks during the processing of linguistic utterances conveying an emotional content, be it words denoting a specific emotion (e.g., sadness) or characterized by some emotional connotations (e.g., flower, war). On the contrary, data about the involvement of the sensory-motor system in emotion-related language are scant. Initial evidence came from an fMRI study (Moseley, Carota, Hauk, Mohr, & Pulvermueller, 2012) in which participants silently read single words denoting either emotions (e.g., "mock") or sensory-motor emotional actions (e.g., "frown"), plus mouth- and hand-related words (e.g., "bite," "pinch") as a control. Both types of emotional words activated (in addition to emotion circuitries) the same motor areas activated by hand- and mouth-related verbs. These findings are compatible with the idea that processing emotion-related meanings is grounded in motor areas, even if words denote emotional states ("mock"), without explicitly referring to emotionally charged objects and actions (as in "frown"). At the sentence level, neuroimaging data are still lacking. In a rating study, we provided initial evidence that emotion-related sentences (e.g., "She mocks the disappointment"), although judged as abstract on a concreteness scale, exhibited a specific involvement of body parts and were more associated with mouth, hand, and leg movements relative to other types of abstract-related meanings (Ghio, Vaghi, & Tettamanti, 2013).

One alternative (but partially compatible) view stresses the grounding of emotion meanings in sensory-motor systems, arguing that language and emotion are mutually related through action (Glenberg, Webster, Mouilso, Havas, & Lindeman, 2009; Niedenthal, 2007). Glenberg et al. (2009) showed that language understanding induces emotion-specific changes in the facial muscles. Participants read sentences describing angry, sad, and happy situations while recording electromyographic activity from the facial muscles involved in producing expressions of happiness (*zygomatic major* and *orbicularis oculi*) versus anger and sadness (*corrugator supercilii*). The muscles used in smiling were more active during reading of happy sentences, whereas the frowning muscle was more active during reading of sad and angry sentences. Havas, Glenberg, Gutowski, Lucarelli, and Davidson (2010) tested whether action simulations play a causal role in emotional sentence processing by evoking a reversible paralysis of the *corrugator supercilii* muscle used in expressing negative emotions through the injection of the botulinum toxin. During sessions before and after injections, participants read sentences describing angry, happy, and sad situations and then pressed a button. A sentence-by-session interaction was found, with reading times for sad and angry sentences being longer during sessions before injections than after injection, and there were no differences for happy sentences.

In sum, limited evidence supports the view of a grounding of emotion language processing in the sensory-motor system. More clearly, the broadening of the grounded cognition perspective to account for the linguistic processing of "abstract" emotion-related conceptual knowledge seems to require the involvement of distributed brain networks beyond the sensory-motor system.

52.6 ABSTRACT SENTENCE PROCESSING IS GROUNDED IN EXPERIENTIAL NEUROCOGNITIVE SYSTEMS

Contrary to traditional accounts positing that abstract knowledge relies on verbal representations only (Paivio, 1971; Schwanenflugel, Harnishfeger, & Stowe, 1988), grounded accounts postulate that, also in the abstract domain, the storage of conceptual knowledge may reflect the type of experience maintained with the concepts' referents (Barsalou, 2010; Kiefer & Pulvermueller, 2012), as exemplified by emotion concepts. Among grounded accounts, different nuances exist with respect to the type of experience considered characteristic for abstract concepts (Meteyard et al., 2012). According to a strong embodied view, abstract concepts are grounded in sensory-motor experience via conceptual metaphors (Gallese & Lakoff, 2005; Gibbs, 2006). However, criticisms have been raised about the generalizability of conceptual metaphors as foundational mechanisms to all abstract concepts (Barsalou, 1999; Vigliocco, Meteyard, Andrews, & Kousta, 2009; Tettamanti & Moro, 2012), and empirical evidence is limited to a set of specific conceptual domains (Boroditsky & Ramscar, 2002; Casasanto & Boroditsky, 2008). Another influential embodied view proposed that, in addition to sensory-motor information, abstract meanings rely on emotional and introspective information about internal states, such as interoception, mentalizing, beliefs, affects, self-thoughts, and intention recognition (Barsalou, 2008; Ghio & Tettamanti, 2010; Kiefer & Pulvermueller, 2012; Wilson-Mendenhall et al., 2011; Wilson-Mendenhall, Simmons, Martin & Barsalou, 2013). Also, linguistic information has been considered relevant for abstract meanings, either in the form of word associations resulting from co-occurrence patterns and syntactic information (Simmons, Hamann, Harenski, Hu, & Barsalou, 2008; Vigliocco et al., 2009) or in the form of social and normative linguistic boundaries in which words may express previous experiences (Prinz, 2002) but may also constitute actions/experiences in their own right (Borghi & Cimatti, 2012; Sakreida et al., 2013).

In general, a shared assumption of grounded accounts is that multiple experiential representations contribute to the processing of both concrete and abstract meanings, although in different proportions. Concrete meanings may more heavily rely on sensory-motor information, whereas abstract meanings may rely on affective and linguistic information, depending on context and tasks (Barsalou, 2008; Kiefer & Pulvermueller, 2012; Vigliocco et al., 2009).

Despite the existing variety of abstract meanings, the abstract domain has generally been regarded as an undifferentiated whole (Cappa, 2008; Wang, Conder, Blitzer, & Shinkareva, 2010). In this respect, grounded accounts seem to have a great potential in explaining such variety of abstract meanings, encompassing various entities and processes such as social relationships or facts, events, and introspective states (Cappa, 2008). Depending on the particular abstract concept, a specific combination of modalities and systems that process perception, action, language, emotions, and internal states may be more or less relevant. By extending this line of reasoning, recent theoretical advancements proposed that distributed neural representations of experiential information related to the concepts' referents might distinguish concepts with fine-grained specificity in the abstract domain by analogy to what has been demonstrated for action-related and object knowledge (Ghio et al., 2013; Wilson-Mendenhall et al., 2013). For example, as the processing of affective concepts conveyed by language appears to involve the emotion processing network—as discussed in the previous section—similarly, the processing of introspective concepts referring to mental states might activate the mentalizing neural network, and the processing of abstract social meanings (i.e., meanings referring to social behaviors like "convince" or psychological traits like "ambitious") might activate brain regions underlying mentalizing and social cognition. Consistently, fMRI studies using either single words (e.g., Zahn et al., 2007) or sentences (e.g., Simmons, Reddish, Bellgowan, & Martin, 2010) referring to abstract social meanings have found comparable activations in brain regions typically implicated in social cognition (particularly the anterior temporal lobe, but also the temporal poles, the medial prefrontal cortex, the posterior superior temporal sulcus, and the precuneus/posterior cingulate).

Similarly, for the semantic processing of mathematical-related concepts, the grounded framework posited an involvement of the same brain areas involved in the actual processing of numbers and quantities, such as the horizontal intraparietal sulci (Piazza, Pinel, Le Bihan, & Dehaene, 2007). In an fMRI study, Wilson-Mendenhall et al. (2013) showed that the processing of the concept "arithmetic" compared with "convince" relies on the same brain regions activated by a numerical localizer task. Another line of research suggests that number cognition is grounded in the sensory-motor system, with number processing involving hand representations related to finger-counting, for example. Developmental, behavioral, and neuropsychological studies seem to indicate that finger-counting habits have an effect on numerical and arithmetic processing (Hauk & Tschentscher, 2013). At the neural level, several neuroimaging studies provided evidence of an anatomical overlap between activations in parietal and precentral areas for numerical and arithmetic processing and activations for grasping and pointing movements (Tschentscher, Hauk, Fischer, & Pulvermueller, 2012). Such overlap, however, has been differently interpreted as resulting either from the invasion of evolutionary older brain circuits by the more recent cultural invention of numeracy (Dehaene & Cohen, 2007) or from hebbian learning mechanisms (Tschentscher et al., 2012).

Altogether, the empirical evidence of grounding abstract knowledge expressed linguistically is still quite limited. What seems to emerge, however, is a more general grounded framework in which (in addition to sensory-motor representations) different kinds of experience-based representations tied in distributed neural networks may contribute to abstract semantic processing to different degrees.

52.7 CONCLUDING REMARKS

Convergent experimental evidence has demonstrated that conceptual-semantic sentence processing is flexibly grounded in sensory-motor and experience-based brain networks. Further research should better clarify how the neural activity in experience-related brain systems is functionally coordinated with that of language perisylvian and other supportive brain regions and the impact of inter-individual differences in the quality and quantity of experience in modulating experience-based neural representations.

References

Aravena, P., Delevoye-Turrell, Y., Deprez, V., Cheylus, A., Paulignan, Y., Frak, V., et al. (2012). Grip force reveals the context sensitivity of language-induced motor activity during "Action Words" processing: Evidence from sentential negation. *PLoS One*, *7*(12), e50287.

Arbib, M. A. (2012). *How the brain got language: The mirror system hypothesis*. New York, NY: Oxford University Press.

Aziz-Zadeh, L., Wilson, S. M., Rizzolatti, G., & Iacoboni, M. (2006). Congruent embodied representations for visually presented actions and linguistic phrases describing actions. *Current Biology*, *16*(18), 1818–1823.

Barsalou, L. W. (1999). Perceptual symbol systems. *Behavioral and Brain Sciences*, *22*(4), 577–609.

Barsalou, L. W. (2008). Grounded cognition. *Annual Review of Psychology*, *59*, 617–645.

Barsalou, L. W. (2010). Grounded cognition: Past, present, and future. *Topics in Cognitive Science*, *2*(4), 716–724.

Bartoli, E., Tettamanti, A., Farronato, P., Caporizzo, A., Moro, A., Gatti, R., et al. (2013). The disembodiment effect of negation: Negating action-related sentences attenuates their interference on congruent upper limb movements. *Journal of Neurophysiology*, *109* (7), 1782–1792.

Beilock, S. L., Lyons, I. M., Mattarella-Micke, A., Nusbaum, H. C., & Small, S. L. (2008). Sports experience changes the neural processing of action language. *Proceedings of the National Academy of Sciences of the United States of America*, *105*(36), 13269–13273.

Bergen, B., & Wheeler, K. (2010). Grammatical aspect and mental simulation. *Brain and Language*, *112*(3), 150–158.

Binder, J. R., & Desai, R. H. (2011). The neurobiology of semantic memory. *Trends in Cognitive Science*, *15*(11), 527–536.

Borghi, A. M., & Cimatti, F. (2012). Words are not just words: The social acquisition of abstract words. *Rivista Italiana di Filosofia del Linguaggio*, *5*, 22–37.

Boroditsky, L., & Ramscar, M. (2002). The roles of body and mind in abstract thought. *Psychological Science*, *13*(2), 185–189.

Boulenger, V., Hauk, O., & Pulvermueller, F. (2009). Grasping ideas with the motor system: Semantic somatotopy in idiom comprehension. *Cerebral Cortex*, *19*(8), 1905–1914.

Boulenger, V., Roy, A. C., Paulignan, Y., Deprez, V., Jeannerod, M., & Nazir, T. A. (2006). Cross-talk between language processes and overt motor behavior in the first 200 msec of processing. *Journal of Cognitive Neuroscience*, *18*(10), 1607–1615.

Boulenger, V., Shtyrov, Y., & Pulvermueller, F. (2012). When do you grasp the idea? MEG evidence for instantaneous idiom understanding. *Neuroimage*, *59*(4), 3502–3513.

Brass, M., & Haggard, P. (2007). To do or not to do: The neural signature of self-control. *Journal of Neuroscience*, *27*(34), 9141–9145.

Buccino, G., Binkofski, F., Fink, G. R., Fadiga, L., Fogassi, L., Gallese, V., et al. (2001). Action observation activates premotor and parietal areas in a somatotopic manner: An fMRI study. *European Journal of Neuroscience*, *13*, 400–404.

Cacciari, C., Bolognini, N., Senna, I., Pellicciari, M. C., Miniussi, C., & Papagno, C. (2011). Literal, fictive and metaphorical motion sentences preserve the motion component of the verb: A TMS study. *Brain and Language*, *119*(3), 149–157.

Cappa, S. F. (2008). Imaging studies of semantic memory. *Current Opinion in Neurology*, *21*(6), 669–675.

Casasanto, D., & Boroditsky, L. (2008). Time in the mind: Using space to think about time. *Cognition*, *106*(2), 579–593.

Chen, E., Widick, P., & Chatterjee, A. (2008). Functional–anatomical organization of predicate metaphor processing. *Brain and Language*, *107*(3), 194–202.

Christensen, K. R. (2009). Negative and affirmative sentences increase activation in different areas in the brain. *Journal of Neurolinguistics*, *22*(1), 1–17.

Citron, F. M. M. (2012). Neural correlates of written emotion word processing: A review of recent electrophysiological and hemodynamic neuroimaging studies. *Brain and Language*, *122*(3), 211–226.

Dehaene, S., & Cohen, L. (2007). Cultural recycling of cortical maps. *Neuron*, *56*(2), 384–398.

Desai, R. H., Binder, J. R., Conant, L. L., Mano, Q. R., & Seidenberg, M. S. (2011). The neural career of sensory-motor metaphors. *Journal of Cognitive Neuroscience*, *23*(9), 2376–2386.

Desai, R. H., Binder, J. R., Conant, L. L., & Seidenberg, M. S. (2010). Activation of sensory-motor areas in sentence comprehension. *Cerebral Cortex*, *20*(2), 468–478.

Desai, R. H., Conant, L. L., Binder, J. R., Park, H., & Seidenberg, M. S. (2013). A piece of the action: Modulation of sensory-motor regions by action idioms and metaphors. *NeuroImage*, *83*, 862–869.

Dronkers, N. F., Wilkins, D. P., Van Valin, R. D., Redfern, B. B., & Jaeger, J. J. (2004). Lesion analysis of the brain areas involved in language comprehension. *Cognition*, *92*(1–2), 145–177.

Fernandino, L., Conant, L. L., Binder, J. R., Blindauer, K., Hiner, B., Spangler, K., et al. (2013). Where is the action? Action sentence processing in Parkinson's disease. *Neuropsychologia*, *51*(8), 1510–1517.

Ferretti, T. R., McRae, K., & Hatherell, A. (2001). Integrating verbs, situation schemas, and thematic role concepts. *Journal of Memory and Language*, *44*(4), 516–547.

Foroni, F., & Semin, G. R. (2013). Comprehension of action negation involves inhibitory simulation. *Frontiers in Human Neuroscience*, *7*, 209.

Friederici, A. D. (2012). The cortical language circuit: From auditory perception to sentence comprehension. *Trends in Cognitive Science*, *16*(5), 262–268.

Gallese, V., & Lakoff, G. (2005). The brain's concepts: The role of the sensory-motor system in conceptual knowledge. *Cognitive Neuropsychology*, *22*, 455–479.

Ghio, M., & Tettamanti, M. (2010). Semantic domain-specific functional integration for action-related vs. abstract concepts. *Brain and Language*, *112*(3), 223–232.

Ghio, M., Vaghi, M. M. S., & Tettamanti, M. (2013). Fine-grained semantic categorization across the abstract and concrete domains. *PLoS One*, *8*(6), e67090.

Gibbs, R. W. (2006). Metaphor interpretation as embodied simulation. *Mind & Language*, *21*(3), 434–458.

Glenberg, A. M., & Kaschak, M. P. (2002). Grounding language in action. *Psychonomic Bulletin & Review*, *9*(3), 558–565.

Glenberg, A. M., Webster, B. J., Mouilso, E., Havas, D., & Lindeman, L. M. (2009). Gender, emotion, and the embodiment of language comprehension. *Emotion Review*, *1*(2), 151–161.

Hauk, O., Johnsrude, I., & Pulvermueller, F. (2004). Somatotopic representation of action words in human motor and premotor cortex. *Neuron*, *41*(2), 301–307.

Hauk, O., & Tschentscher, N. (2013). The body of evidence: What can neuroscience tell us about embodied semantics? *Frontiers in Psychology*, *4*, 50.

Havas, D. A., Glenberg, A. M., Gutowski, K. A., Lucarelli, M. J., & Davidson, R. J. (2010). Cosmetic use of botulinum toxin-a affects processing of emotional language. *Psychological Science*, *21*(7), 895–900.

Ibáñez, A., Cardona, J. F., Dos Santos, Y. V., Blenkmann, A., Aravena, P., Roca, M., et al. (2013). Motor-language coupling: Direct evidence from early Parkinson's disease and intracranial cortical recordings. *Cortex*, *49*(4), 968–984.

Kaup, B., Zwaan, R. A., & Lüdtke, J. (2007). The experiential view of language comprehension: How is negation represented? In F. A. Schmalhofer, & C. A. Perfetti (Eds.), *Higher level language processes in the brain: Inference and comprehension processes* (pp. 255–288). Mahwah, NJ: Erlbaum.

Kemmerer, D. (2015). *Sentence comprehension. Cognitive neuroscience of language*. New York, NY: Psychology Press.

Kiefer, M., & Pulvermueller, F. (2012). Conceptual representations in mind and brain: Theoretical developments, current evidence and future directions. *Cortex*, *48*(7), 805–825.

Kousta, S. T., Vigliocco, G., Vinson, D. P., Andrews, M., & Del Campo, E. (2011). The representation of abstract words: Why emotion matters. *Journal of Experimental Psychology: General*, *140*(1), 14–34.

Lacey, S., Stilla, R., & Sathian, K. (2012). Metaphorically feeling: Comprehending textural metaphors activates somatosensory cortex. *Brain and Language*, *120*(3), 416–421.

Liuzza, M. T., Candidi, M., & Aglioti, S. M. (2011). Do not resonate with actions: Sentence polarity modulates cortico-spinal excitability during action-related sentence reading. *PLoS One*, *6*(2), e16855.

Locatelli, M., Gatti, R., & Tettamanti, M. (2012). Training of manual actions improves language understanding of semantically related action sentences. *Frontiers in Cognitive Science*, *3*, 547.

Mahon, B. Z., & Caramazza, A. (2008). A critical look at the embodied cognition hypothesis and a new proposal for grounding conceptual content. *Journal of Physiology-Paris*, *102*(1–3), 59–70.

Martin, A. (2007). The representation of object concepts in the brain. *Annual Review of Psychology*, *58*(1), 25–45.

McRae, K., & Jones, M. N. (2013). Semantic memory. In D. Reisberg (Ed.), *The Oxford handbook of cognitive psychology*. New York, NY: Oxford University Press.

Meteyard, L., Cuadrado, S. R., Bahrami, B., & Vigliocco, G. (2012). Coming of age: A review of embodiment and the neuroscience of semantics. *Cortex*, *48*(7), 788–804.

Moseley, R., Carota, F., Hauk, O., Mohr, B., & Pulvermueller, F. (2012). A role for the motor system in binding abstract emotional meaning. *Cerebral Cortex*, *22*(7), 1634–1647.

Niedenthal, P. M. (2007). Embodying emotion. *Science*, *316*(5827), 1002–1005.

Paivio, A. (1971). *Imagery and verbal processes*. New York, NY: Holt, Rinehart & Winston.

Pallier, C., Devauchelle, A.-D., & Dehaene, S. (2011). Cortical representation of the constituent structure of sentences. *Proceedings of the National Academy of Sciences of the United States of America, 108*(6), 2522–2527.

Papagno, C., & Caporali, A. (2007). Testing idiom comprehension in aphasic patients: The effects of task and idiom type. *Brain and Language, 100*(2), 208–220.

Patterson, K., Nestor, P. J., & Rogers, T. T. (2007). Where do you know what you know? The representation of semantic knowledge in the human brain. *Nature Reviews Neuroscience, 8*(12), 976–987.

Pezzulo, G., Barsalou, L. W., Cangelosi, A., Fischer, M. H., McRae, K., & Spivey, M. J. (2013). Computational grounded cognition: A new alliance between grounded cognition and computational modeling. *Frontiers in Psychology, 3*, 612.

Piazza, M., Pinel, P., Le Bihan, D., & Dehaene, S. (2007). A magnitude code common to numerosities and number symbols in human intraparietal cortex. *Neuron, 53*(2), 293–305.

Prinz, J. J. (2002). *Furnishing the mind: Concepts and their perceptual basis*. Cambridge, MA: MIT Press.

Pulvermueller, F., & Fadiga, L. (2010). Active perception: Sensorimotor circuits as a cortical basis for language. *Nature Reviews Neuroscience, 11*(5), 351–360.

Raposo, A., Moss, H. E., Stamatakis, E. A., & Tyler, L. K. (2009). Modulation of motor and premotor cortices by actions, action words and action sentences. *Neuropsychologia, 47*(2), 388–396.

Rapp, A. M., Mutschler, D. E., & Erb, M. (2012). Where in the brain is nonliteral language? A coordinate-based meta-analysis of functional magnetic resonance imaging studies. *NeuroImage, 63*(1), 600–610.

Romero Lauro, L. J., Mattavelli, G., Papagno, C., & Tettamanti, M. (2013). She runs, the road runs, my mind runs, bad blood runs between us: Literal and figurative motion verbs: An fMRI study. *NeuroImage, 83*, 361–371.

Romero Lauro, L. J., Tettamanti, M., Cappa, S. F., & Papagno, C. (2008). Idiom comprehension: A prefrontal task? *Cerebral Cortex, 18*(1), 162–170.

Sakreida, K., Scorolli, C., Menz, M. M., Heim, S., Borghi, A. M., & Binkofski, F. (2013). Are abstract action words embodied? An fMRI investigation at the interface between language and motor cognition. *Frontiers in Human Neuroscience, 7*, 125.

Santana, E., & de Vega, M. (2011). Metaphors are embodied, and so are their literal counterparts. *Frontiers in Psychology, 2*, 90.

Saygin, A. P., McCullough, S., Alac, M., & Emmorey, K. (2010). Modulation of BOLD response in motion-sensitive lateral temporal cortex by real and fictive motion sentences. *Journal of Cognitive Neuroscience, 22*(11), 2480–2490.

Schwanenflugel, P. J., Harnishfeger, K. K., & Stowe, R. W. (1988). Context availability and lexical decisions for abstract and concrete words. *Journal of Memory and Language, 27*, 499–520.

Scorolli, C., Jacquet, P. O., Binkofski, F., Nicoletti, R., Tessari, A., & Borghi, A. M. (2012). Abstract and concrete phrases processing differentially modulates cortico-spinal excitability. *Brain Research, 1488*, 60–71.

Simmons, W. K., Hamann, S. B., Harenski, C. L., Hu, X. P., & Barsalou, L. W. (2008). fMRI evidence for word association and situated simulation in conceptual processing. *Journal of Physiology-Paris, 102*(1), 106–119.

Simmons, W. K., Reddish, M., Bellgowan, P. S. F., & Martin, A. (2010). The selectivity and functional connectivity of the anterior temporal lobes. *Cerebral Cortex, 20*(4), 813–825.

Tettamanti, M., Buccino, G., Saccuman, M. C., Gallese, V., Danna, M., Scifo, P., et al. (2005). Listening to action-related sentences activates fronto-parietal motor circuits. *Journal of Cognitive Neuroscience, 17*(2), 273–281.

Tettamanti, M., Manenti, R., Rosa, P. A. D., Falini, A., Perani, D., Cappa, S. F., et al. (2008). Negation in the brain: Modulating action representations. *NeuroImage, 43*(2), 358–367.

Tettamanti, M., & Moro, A. (2012). Can syntax appear in a mirror (system)? *Cortex, 48*(7), 923–935.

Tomasino, B., Weiss, P. H., & Fink, G. R. (2010). To move or not to move: Imperatives modulate action-related verb processing in the motor system. *Neuroscience, 169*(1), 246–258.

Tschentscher, N., Hauk, O., Fischer, M. H., & Pulvermueller, F. (2012). You can count on the motor cortex: Finger counting habits modulate motor cortex activation evoked by numbers. *NeuroImage, 59-318*(4–12), 3139–3148.

Urrutia, M., Gennari, S. P., & de Vega, M. (2012). Counterfactuals in action: An fMRI study of counterfactual sentences describing physical effort. *Neuropsychologia, 50*(14), 3663–3672.

Vigliocco, G., Meteyard, L., Andrews, M., & Kousta, S. (2009). Toward a theory of semantic representation. *Language and Cognition, 1*(2), 219–247.

Vigneau, M., Beaucousin, V., Hervé, P. Y., Duffau, H., Crivello, F., Houdé, O., et al. (2006). Meta-analyzing left hemisphere language areas: Phonology, semantics, and sentence processing. *NeuroImage, 30*(4), 1414–1432.

Wallentin, M., Lund, T. E., Ostergaard, S., Ostergaard, L., & Roepstorff, A. (2005). Motion verb sentences activate left posterior middle temporal cortex despite static context. *Neuroreport, 16*(6), 649–652.

Wallentin, M., Nielsen, A. H., Vuust, P., Dohn, A., Roepstorff, A., & Lund, T. E. (2011). BOLD response to motion verbs in left posterior middle temporal gyrus during story comprehension. *Brain and Language, 119*(3), 221–225.

Wang, J., Conder, J. A., Blitzer, D. N., & Shinkareva, S. V. (2010). Neural representation of abstract and concrete concepts: A meta-analysis of neuroimaging studies. *Human Brain Mapping, 31*(10), 1459–1468.

Warrington, E. K., & McCarthy, R. A. (1987). Categories of knowledge. Further fractionations and an attempted integration. *Brain, 110*(5), 1273–1296.

Willems, R. M., & Francken, J. C. (2012). Embodied cognition: Taking the next step. *Frontiers in Cognitive Science, 3*, 582.

Wilson-Mendenhall, C. D., Barrett, L. F., Simmons, W. K., & Barsalou, L. W. (2011). Grounding emotion in situated conceptualization. *Neuropsychologia, 49*(5), 1105–1127.

Wilson-Mendenhall, C. D., Simmons, W. K., Martin, A., & Barsalou, L. W. (2013). Contextual processing of abstract concepts reveals neural representations of nonlinguistic semantic content. *Journal of Cognitive Neuroscience, 25*(6), 920–935.

Winkielman, P., Niedenthal, P. M., & Oberman, L. (2008). The embodied emotional mind. In G. R. Semin, & E. R. Smith (Eds.), *Embodied grounding: Social, cognitive, affective, and neuroscientific approaches* (pp. 263–288). New York, NY: Cambridge University Press.

Zahn, R., Moll, J., Krueger, F., Huey, E. D., Garrido, G., & Grafman, J. (2007). Social concepts are represented in the superior anterior temporal cortex. *Proceedings of the National Academy of Sciences of the United States of America, 104*(15), 6430–6435.

Zwaan, R. A., Taylor, L. J., & de Boer, M. (2010). Motor resonance as a function of narrative time: Further tests of the linguistic focus hypothesis. *Brain and Language, 112*(3), 143–149.

SECTION I

DISCOURSE PROCESSING AND PRAGMATICS

CHAPTER

53

Discourse Comprehension

Jeffrey M. Zacks[1] *and Evelyn C. Ferstl*[2]

[1]Department of Psychology, Washington University, Saint Louis, MO, USA; [2]Institute for Informatics and Society, Centre of Cognitive Science, Albert-Ludwigs-University, Freiburg, Germany

Here is the opening to a story no one would ever write:

> "Over to the Indian camp." The Indian who was rowing them was working very hard, but the other boat moved further ahead in the mist all the time. "Oh," said Nick. Uncle George sat in the stern of the camp rowboat. It was cold on the water. "There is an Indian lady very sick." Nick lay back with his father's arm around him. The Indians rowed with quick choppy strokes. Nick and his father got in the stern of the boat and the Indians shoved it off and one of them got in to row. Nick heard the oarlocks of the other boat quite a way ahead of them in the mist. The young Indian shoved the camp boat off and got in to row Uncle George. The two boats started off in the dark. The two Indians stood waiting. "Where are we going, Dad?" Nick asked. At the lake shore there was another rowboat drawn up.

What is wrong with this passage? All the sentences were written by Ernest Hemingway in his iconic spare style. The vocabulary should be accessible to most readers. However, because we have rearranged the order of the sentences, the passage as a whole is probably quite difficult to comprehend.

Here is the original as Hemingway wrote it (Hemingway, 2007):

> At the lake shore there was another rowboat drawn up. The two Indians stood waiting. Nick and his father got in the stern of the boat and the Indians shoved it off and one of them got in to row. Uncle George sat in the stern of the camp rowboat. The young Indian shoved the camp boat off and got in to row Uncle George. The two boats started off in the dark. Nick heard the oarlocks of the other boat quite a way ahead of them in the mist. The Indians rowed with quick choppy strokes. Nick lay back with his father's arm around him. It was cold on the water. The Indian who was rowing them was working very hard, but the other boat moved further ahead in the mist all the time. "Where are we going, Dad?" Nick asked. "Over to the Indian camp. There is an Indian lady very sick." "Oh," said Nick.

In Hemingway's opening to "Indian Camp," a situation and a sequence of events are described. A skilled reader appreciates the setting and the events without apparent effort. The rearranged version preserves the words, grammatical structure, and sentence-level organization, but it is much less comprehensible. In this chapter we give an account of those processes that allow readers to make sense of discourse not just at the level of the sentence but also at the level of a connected sequence of sentences that describe settings and events.

When readers and listeners comprehend a narrative prose passage, van Dijk and Kintsch (1983) proposed that they construct three types of representation. The *surface* representation is the reader's mental representation of the specific wording and structure of the text. The *textbase* is the reader's representation of the items of information, called propositions, that the text asserts, independent of their form. The *situation model* is a structured mental representation of a situation and set of events. For example, consider the sentence "Nick heard the oarlocks of the other boat quite a way ahead of them in the mist." If the end was changed to "in the mist quite a way ahead of them" or "quite a way ahead of them in the fog," then this would probably change the surface representation but not affect the textbase or the situation model; neither of those representations depends substantively on exact word choices or word order. If "heard the oarlocks" were changed to "heard the oars," then this would likely change the surface representation and also the textbase, because the object of the proposition is "oarlocks" in the former case but "oars" in the second. However, this probably would not change the situation model because the sound in both cases comes from the oars working in the oarlocks, so the two propositions give alternate descriptions of the same

Neurobiology of Language. DOI: http://dx.doi.org/10.1016/B978-0-12-407794-2.00053-5

situation. Finally, changing "quite a way ahead of them" to "quite a way behind them" would change all three representations, because it changes the word itself, the information asserted, and the situation described by that assertion.

These three types of representation differ in the mechanisms that produce them, in their fidelity, and in the timecourse with which they are forgotten. Surface representations are notably incomplete and fleeting. After short delays, readers are quite poor at answering questions such as whether "in the mist" or "in the fog" was Hemingway's choice of words (for a review, see Fletcher, 1994). Surface representations of the sentences in Hemingway's opening should not be much affected by the scrambling we did in the example. However, scrambling sentence order should substantially disrupt the textbase and the situation model.

In this chapter, we consider mechanisms by which comprehenders construct representations of discourse, and, in particular, how comprehenders establish connections among components of a text at each of the three levels. We take a cognitive neuroscience perspective in which we view the construction of discourse representations as resulting from mechanisms that have both information-processing aspects and neurophysiological aspects, and we try to present both aspects in an integrated fashion. Neuroscientific methods lend themselves to the study of discourse processes because they enable the researcher to disentangle linguistic and cognitive components of comprehension. Furthermore, because they do not always require an explicit experimental task, they allow researchers to study language comprehension more naturalistically.

The most important methods are event-related potentials (ERPs) and functional magnetic resonance imaging (fMRI). ERPs are averaged electro-encephalograms, time-locked at the onset of stimuli of interest. They provide excellent temporal resolution and a dissociation of qualitatively distinct processes (see Kaan, 2007 for an introduction.). For example, it has been shown that the N400 component is sensitive to the semantic fit of a word in the context, whereas the P600, a later positivity, can be found, for example, when the integration or reanalysis of syntactically difficult materials is needed (Kutas & Federmeier, 2011). These components are an important tool in language processing research, which has been successfully applied to research on discourse comprehension and contextual integration (van Berkum, 2004).

Neuroimaging methods, in contrast, are useful for specifying the brain regions and networks involved during comprehension. Using quite extensive knowledge of the functional neuroanatomy accumulated to date, it is possible to associate activation patterns with specific cognitive or linguistic processes. Although neuroimaging of discourse comprehension is a young field, several reviews are available (Bornkessel-Schlesewsky & Friederici, 2007; Ferstl, 2007, 2010; Mar, 2004; Mason & Just, 2006). In a meta-analysis, Ferstl, Neumann, Bogler, and von Cramon (2008) identified an extended language network of candidate regions involved during discourse comprehension, which is discussed in depth later in the chapter; however, briefly, in addition to the perisylvian language regions (left inferior frontal lobe and posterior temporal/inferior parietal lobe), bilateral anterior temporal lobe, and dorso-lateral prefrontal cortex play a role. Most importantly, the dorso-medial prefrontal and parieto-medial cortex have been shown to be active during processing of coherent language.

Although not included in the meta-analysis at the time (due to a relatively small database of studies), the question of right-hemisphere contributions to language comprehension is also crucial. Based on observations from clinical linguistics, it has been proposed that discourse processing requires the language-dominant left hemisphere right hemisphere homologues (Brownell & Martino, 1998; Jung-Beeman, 2005).

Although there are still open questions regarding the processes involved (Goldman, Golden, & van den Broek, 2007), the key psycholinguistic concepts used to organize this chapter are useful for describing the processes involved in comprehending discourse beyond the sentence level. It should be noted, however, that the comprehension of nonliteral language (e.g., idioms, metaphors, irony or verbal humor) is related to discourse comprehension but falls outside the scope of what we can cover in this chapter (for further readings see Papagno & Lauro, 2010; Schmidt & Seger, 2009).

53.1 COHESION

Connections between text elements at the surface level produce *cohesion* (Halliday & Hasan, 1976). Cohesion can be distinguished *coherence*, which refers to connections at the level of the text base or situation model. Cohesive markers, or "cohesive ties," are words that directly signal the relationship between parts of sentences or between successive sentences. For example, in "The birthday girl unwrapped *her* presents *before* the guests ate the cake," both the possessive pronoun *her* as well as the conjunction *before* help the reader to understand the relationships between the two actions (eating and opening), as well as between the concepts *girl* and *presents*. In addition, even the definite article "the" in "the cake" provides cohesion, because it presupposes a given cake (the one that is usually part of a birthday celebration)—rather than introducing "a cake" as a novel concept in the discourse. Note that these relationships can easily be inferred in the absence of lexical markers just by using the order of the statements or by drawing on general world knowledge: presents, cakes, and guests are usually parts of birthday parties.

Psycholinguistic studies confirm that lexical cohesion facilitates comprehension when it is used appropriately. One of the first investigations of the comprehension of cohesion using neuroimaging was a study by Robertson et al. (2000). Short paragraphs with indefinite articles were compared with the same paragraphs with definite articles, which rendered the sentences more cohesive (i.e., more story-like). The indefinite article sentences included the following: "A grandmother sat at the table. A child played in the garden...A grandmother promised to bake cookies." The corresponding definite article sentences were included: "The grandmother sat at the table. The child played in the garden...The grandmother promised to bake cookies." Despite the subtlety of this manipulation, there were differential activations in the frontal lobes. The cohesive condition elicited more activation in right prefrontal regions. Ferstl and von Cramon (2001) used a 2 × 2-design, crossing cohesion (connections at the surface level) with coherence (connections at the level of the text base). When cohesive ties were falsely used in incoherent sentence pairs, thus rendering the sentences infelicitous, activation in bilateral inferior prefrontal regions was observed. This result was interpreted as reflecting task management processes, which are required for resolving the inconsistency between knowledge-based coherence and lexically signaled cohesion. In this study, both pronouns and conjunctions were used as cohesive ties.

Using electroencephalographic ERPs, Münte, Schiltz, and Kutas (1998) studied the use of conjunctions during sentence processing. Connecting two events with the conjunctions *before* and *after* renders the temporal order either consistent ("After x happens, y happened") or inconsistent ("Before y, x happened") with the order of mention. The latter sentence type elicited a slow negative shift, indicating higher cognitive demands. Converging evidence was provided by Ye, Habets, Jansma, and Münte (2011) in an fMRI study. Using similar sentences in a production task, they found extensive left middle temporal activations for *before* compared with *after* sentences. Prefrontal and hippocampal activations suggested that the cognitive demands are related to memory processes needed for rearranging the temporal sequence. These results show that cohesion plays a role both during comprehension and production.

The processing of pronouns has been investigated in a number of imaging studies. Very similar to the manipulation used by Robertson et al. (2000), Almor, Smith, Bonilha, Fridriksson, and Rorden (2007) compared short sentence sequences in which the discourse referent was repeated (e.g., "Anna went shopping. Anna wore a scarf. Anna liked ice cream") with the same sentences using pronouns after the first mention (e.g., "Anna went shopping. She wore..."). The pronouns clearly signal continuity of reference, whereas repetition is less felicitous in natural discourse. A rather extended temporo-parietal network showed higher levels of activation for the repetition condition compared with the pronoun condition. This result confirms the sensitivity of the language processing system to these subtle pragmatic cues.

Using a distinction between grammatical and natural gender, Hammer, Goebel, Schwarzbach, Münte, and Jansma (2007) found that a mismatch of grammatical gender induced activation increases in left perisylvian regions including the left inferior frontal gyrus (IFG) and the superior temporal sulcus. In contrast, when both natural and grammatical gender were violated (e.g., "The woman was popular because he was attractive"), extensive right hemisphere activation emerged, particularly in the IFG and the inferior parietal lobe. Interestingly, very similar results were obtained in a study using gender stereotype nouns in English (Ferstl, Manouilidou, & Garnham, 2010), strengthening the interpretation that these regions are important for the processing of natural gender information, rather than for the integration of explicit grammatical gender marking.

Similarly, Nieuwland, Petersson, and van Berkum (2007) used simple sentences of the form "X told Y that he/she...". Varying the gender of the two noun phrases, they created coherent (one unique referent), ambiguous (two possible referents—e.g., two women followed by "she"), and violation conditions (no possible referents—e.g., two women followed by "he"). Activations for the referential failure condition included bilateral inferior prefrontal regions, whereas ambiguity elicited activations in the posterior cingulate cortex (PCC) and the medial prefrontal cortex. This latter result suggests increased inference demands for the ambiguous sentences when pronoun reference is attempted in the absence of lexical cues (see next section).

Some evidence for a successful interaction of semantic and syntactic cues to reference comes from ERP studies on gender stereotype referents of reflexive pronouns (Osterhout, Bersick, & McLaughlin, 1997). Consider a sentence containing a cataphoric reference—a reference to a word that has not yet been mentioned—such as "After *she* thought about the letter, the *minister* left London." Readers encountering the word "she" in this sentence incur a processing cost when they are unable to find a referent yet mentioned for the pronoun. However, reflexive pronouns such as *himself* or *herself* are syntactically bound to the sentence subject, as in "After reminding *herself* about the letter, the *minister* left London." This syntactic binding eliminates the need to search based on semantic fit for a referent, and thus when readers encounter them they do not incur processing costs (Kreiner, Sturt, & Garrod, 2008).

Taken together, these and similar findings suggest that cohesive ties are an important cue to discourse coherence. Articles, pronouns, and conjunctions have been shown to influence processing immediately. Because these different ties have dissociable functions, there is no unique set of brain regions specialized for establishing cohesion. However, it has been shown that fronto-parietal regions are important for the processing of gender information, and the inferior frontal cortex is important for the sequencing of events based on conjunctions in sentences. Furthermore, the use of cohesive markers directly influences the inference demands in text comprehension, which are required to derive semantic coherence.

53.2 COHERENCE

Even in the absence of cohesive ties, readers and listeners attempt to establish *coherence.* Coherence is a content-based connection between successive phrases, sentences, or utterances, manifest at the level of the text base or situation model. For example, readers might assume co-occurrence in time or space, as in the sentences from Robertson and colleagues described previously: After reading "A grandmother sat at the table...A grandmother promised to bake cookies," readers might represent these as occurring continuously in time at the same location. Readers also might infer a causal relationship, as in the following example: "Dorothy poured the bucket of water on the fire. The fire went out" (Singer, Halldorson, Lear, & Andrusiak, 1992). These inference processes have been extensively studied in psycholinguistic research on text comprehension. Questions of interest are whether inferences are drawn automatically and online, if and how they depend on the comprehenders' goals, and whether inference types can be classified into qualitatively different categories (such as bridging, causal, elaborative, or predictive inferences).

Neuroscientific studies on inferencing are usually less concerned with the fine-grained temporal resolution of these processes. Here, the assumption is that, due to the human drive to make sense of the world, inferences are a core component of comprehension. Interestingly, specific deficits have been observed in patients with right hemisphere or frontal brain lesions, rather than in aphasic patients (e.g., Baumgärtner, Weiller, & Büchel, 2002; Beeman, 1993; Ferstl, Guthke, & von Cramon, 2002). Thus, one of the most important issues is the question of domain specificity. Do language-based inferences elicit activation in the perisylvian language cortex of the left hemisphere? Do they engage prefrontal regions related to problem-solving and executive functions? What is the role of the right hemisphere? Do the neuronal signatures of inferences during language comprehension resemble those of processes such as evaluation, reasoning, or social judgments?

To investigate these questions, the aforementioned study by Ferstl and von Cramon (2001) used sentence pairs a coherence judgment task. Participants evaluated whether the sentences were connected or unrelated. Coherent compared with incoherent sentence pairs elicited activations not in fronto-temporal language regions, but in left medial areas, particularly the dorso-medial prefrontal cortex (dmPFC) and the PCC, reaching into the precuneus (pCC/prec). This finding has since been replicated numerous times (e.g., Chow, Kaup, Raabe, & Greenlee, 2008; Ferstl, 2010; Ferstl et al., 2008; Friese, Rutschmann, Raabe, & Schmalhofer, 2008), but the dmPFC has also been implicated in a number of other cognitive functions. Most notably, it has been considered part of the "default" network that is active in the absence of an overt task, or an engaging activity (Raichle et al., 2001). Of course, this overlap might be due to self-guided thinking when task demands are low. Other functions include Theory-of-Mind, that is, the process of making sense of other people by inferring their intentions and emotions (cf. Ferstl & von Cramon, 2002), aesthetic judgments, or the processing of self-relevant stimuli (see Ferstl et al., 2008 for more detail). Although it is beyond the scope of this overview to attempt to dissociate these proposals, it is clear that the dmPFC cannot be specific for language processes, but that its function is rather general. The most parsimonious account is that it is involved whenever someone needs to make a judgment based on their own background knowledge and self-calibrated evaluation criteria.

Several studies manipulated the degree of relatedness to vary inference demands gradually, rather than in an all-or-none fashion. Adopting a classical paradigm, Mason and Just (2004) used three conditions in which causal relationships were varied. For example, a target sentence "The next day he was covered in bruises" was preceded by a direct cause "Jane punched Larry," by an intermediate condition ("Jane got angry at Larry"), or by a distantly related one

("Larry went to Jane's house"). Using region-of-interest analyses, they reported right-sided activation for the intermediate condition and bilateral frontal activation with decreasing relatedness of the sentences. Kuperberg, Lakshmanan, Caplan, and Holcomb (2006), again using an explicit judgment, showed an extended network of regions to vary with inference difficulty, including both medial and right hemisphere regions. With an individually defined empirical criterion (using participants' ratings of relatedness), Siebörger, Ferstl, and von Cramon (2007) also confirmed that dmPFC activation was related to the inference process, rather than to stimulus properties.

Finally, a number of studies used mid-length stories and tested activation patterns at certain inference points within the stories. For example, a story about someone getting ready for a wedding contained the sentences "The shirt was all wrinkled. He started work/ironing...The shirt was all smooth" (Virtue, Haberman, Clancy, Parrish, & Jung-Beeman, 2006). When the specific term "ironing" is used, inference demands are low. However, when the vague term "work" is used, the reader needs to infer that the person is probably ironing. If this inference is made, then the subsequent "smooth" is as easy as in the first condition, whereas it again elicits a backward inference when the earlier inference is missing from the discourse representation. Brain regions involved in these inferences included left-sided and right-sided temporal regions. With these longer stories, working memory demands increase. In a follow-up study (Virtue, Parrish, & Jung-Beeman, 2008), comprehenders with a high working memory capacity engaged the right-sided regions to a larger extent than readers with low working memory capacity.

Distinguishing the content of inferences (emotional, physical, intentional), Mason and Just (2009, see also 2006) argued that the dmPFC activation reflects a protagonist monitoring process during narrative comprehension. Inference processes are important for establishing local coherence. In the next section, we go beyond local coherence, that is, the connections between the current sentence or utterance and the immediately preceding context. Global coherence, that is, the overall structure and gist of connected text, requires setting up a situation model of the text.

53.3 SITUATION MODEL CONSTRUCTION

When one hears or reads a moderately coherent prose passage, one can construct a situation model representing the setting and events described by the passage. To the extent the text makes it easy to create a situation model reading is faster, comprehension is easier and memory is better. These effects were first established in a famous series of studies of long-term memory conducted by Bransford and his colleagues. In one set of experiments, the memory of readers for sentences was tested using a recognition memory test that included previously read sentences and sentences that were altered to affect only the surface structure or the situation model (and textbase) along with the surface structure (Bransford, Barclay, & Franks, 1972). For example, a reader might encounter the sentence, "Three turtles rested on a floating log, and a fish swam beneath them," and then be tested with "Three turtles rested on a floating log, and a fish swam beneath it." This lure is very difficult to reject, presumably because changing "them" to "it" does not change the situation described. However, if the encoded sentence began "Three turtles rested *beside* a floating long...", then changing "them" to "it" changes the situation—and such a change is more easily detected during the memory test. (Of course, none of the words were italicized in the experiments.)

Another set of experiments manipulated the ability of the readers to construct a situation model during reading (Bransford & Johnson, 1972; see also Dooling & Lachman, 1971). For these studies, the researchers created passages for which it was difficult to form a situation model without the addition of a key element of context. For example:

> If the balloons popped, the sound wouldn't be able to carry since everything would be too far away from the correct floor. A closed window would also prevent the sound from carrying, since most buildings tend to be well insulated. Since the whole operation depends on a steady flow of electricity, a break in the middle of the wire would also cause problems. Of course, the fellow could shout, but the human voice is not loud enough to carry that far. An additional problem is that a string could break on the instrument. Then there could be no accompaniment to the message. It is clear that the best situation would involve less distance. Then there would be fewer potential problems. With face to face contact, the least number of things could go wrong. (p. 719).

This passage makes little sense without the picture provided in Figure 53.1. As expected, readers found the passage much easier to read if it was preceded by the picture. Moreover, pre-exposure to the picture dramatically increased the amount that readers were able to recall from the passage. Interestingly, presenting the picture *after* reading had little effect.

How do situation models influence online reading? One thing they do is render relevant information more accessible for further processing. This was shown vividly in a study by Glenberg, Meyer, and Lindem (1987). Glenberg et al. asked participants to read

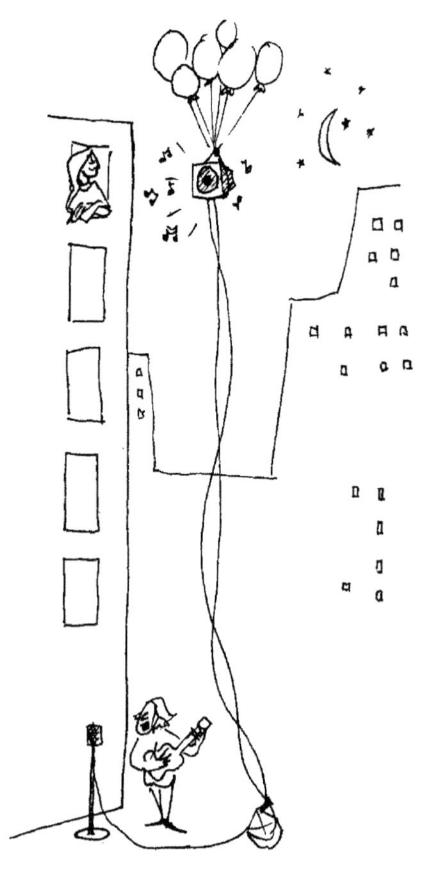

FIGURE 53.1 Picture specifying the appropriate context for the passage in Experiment 1 of Bransford and Johnson (1972).

passages that could be manipulated to change the situation such that a critical item was either associated with the main character or dissociated from her or him while controlling how recently the item was mentioned. An example passage follows: "John was preparing for a marathon in August. After doing a few warm-up exercises, he put on his sweatshirt and went jogging. He jogged halfway around the lake without difficulty. Further along his route, however, John's muscles began to ache." In this passage, the critical word is "sweatshirt," and it is *associated* with John. In the *dissociated* condition, "put on his sweatshirt" was replaced with "took off his sweatshirt." Memory for the critical object was tested using a yes–no recognition test, with probes placed immediately right after the sentence containing the critical word, or one or two sentences later. When tested immediately after presentation, recognition was fast independent of the associated/dissociated manipulation. After one or two sentences, however, responses were faster for words in the associated condition than in the dissociated condition.

As a narrative evolves, information about various dimensions of the situation changes. Comprehenders track these changes, incorporating the new information into their situation model. The event-indexing model (Zwaan, Langston, & Graesser, 1995) gives an account of how readers monitor dimensions of the situation described by the discourse and update their situation models when these dimensions change. According to the model, updating a model takes time and cognitive resources, and information that is no longer in a situation model as a result of updating is less accessible. As a result, signatures of updating can be seen in reading time and in memory accessibility. For example, Scott Rich and Taylor (2000) asked participants to read short narrative passages that contained changes in the main character, in the spatial location, or in temporal location. In the first experiment, readers were asked about the overall coherence of the narrative. In the second experiment, readers were asked about the local cohesion of the sentences ("how well the sentence fits with the previous sentence"). In the third experiment, participants' memory for recently presented actions was probed using a recognition test. In that experiment, it was also possible to measure reading time for the sentences that contained shifts and for control sentences. Changes in character, space, or time led readers to rate the text as less coherent and less cohesive. Actions from before a shift were responded to less quickly and less accurately than actions from since a shift. The three shifts did not function equivalently. Character changes produced the largest reductions in coherence and cohesion ratings, and led to memory responses that were more accurate but slower than the other changes. These results suggest that participants were putting more effort into tracking character information and incorporating it into long-term memory. This is surely not always the case, but tracking characters may be important under many reading conditions. Causality is another dimension that appears to be important for discourse comprehension under many conditions (van den Broek, Young, Tzeng, & Linderholm, 1999). For example, when the current sentence being read is more causally connected to the previous sentences, it is read more quickly (Myers, Shinjo, & Duffy, 1987). In contrast to characters and space, there is good evidence that comprehenders pay relatively less attention to the spatial dimension during normal reading for comprehension (see Radvansky & Zacks, 2014, Chapter 4 for a review).

Situation model updating can be conceptualized using a propositional—or language—like representational format, as originally proposed by Kintsch and van Dijk (1978). However, as van Dijk and Kintsch (1983) stressed, other domain-dependent representational formats (e.g., a map encoding spatial relationships) are perfectly plausible. Thus, approaches to situation model updating including the event-indexing model are commensurable with approaches to representation that emphasize the embodied nature of conceptual representations. Embodied cognition theories propose that comprehension depends on representations that use the same format and representational elements as those for basic perception and motor control (Barsalou, 2008; Glenberg, 2007; Zwaan, 2004).

What mechanisms does the brain bring to bear to construct a situation model? Neurophysiological data have helped to answer this question. First, when a comprehender can construct a situation model, mechanisms that build coherence may be more active or more effective. This is associated with activity in the medial prefrontal cortex, consistent with the previous section. Second, the data suggest that constructing a situation model brings online unique processing, which is associated with increased activity in posterior medial regions, the PCC and the precuneus. An early proposal stemming from neuropsychological case studies was that the right hemisphere is selectively involved in situation model construction. The strongest support for this proposal comes from the Robertson et al. (2000) study described previously (see *Cohesion* noted previously), which varied whether sentences began with the indefinite article (e.g., "A grandmother") or the definite article ("The grandmother"). Because the definite article indicates that the same grandmother is referred to in both sentences, this should encourage construction of an integrated situation model, whereas the indefinite article discourages this. Comparing these two conditions led to greater activation in the right IFG and inferior frontal sulcus. However, the right-hemisphere hypothesis has received mixed support in other studies (Ferstl, 2010).

One of the earliest neuroimaging studies to compare conditions that did or did not afford situation model construction was aimed at studying the comprehension of social information, or "theory of mind" (Fletcher et al., 1995). In this experiment, participants read paragraphs comprising unconnected sentences, narratives describing physical interactions among objects, or narratives describing social interactions. Comparing the social interaction paragraphs with the unconnected sentences revealed increased activity in the left dmPFC, the PCC, the left superior temporal gyrus, and the temporal poles bilaterally. A more recent study used a similar design to specifically investigate components of discourse comprehension. Yarkoni, Speer, and Zacks (2008) asked participants to read paragraphs interspersed with periods of rest. The paragraphs either were intact narratives or were composed by combining sentences from multiple narratives selected at random. Activity in the dmPFC bilaterally increased for the intact narratives but not at all for the scrambled narratives. This pattern was also observed in the posterior cerebellum. Compared to the resting baseline, the PCC and precuneus were observed to decrease for both scrambled sentences and for intact stories, and were observed to do so more for the stories (note that this effect in the opposite direction as that reported by Fletcher et al.).

Several studies have used the confusing passage developed by Bransford and Johnson (1972). The first used ERPs (St. George, Mannes, & Hoffman, 1994). In this study, participants read confusing paragraphs with or without titles while the ERP to each word was recorded. The authors found that the N400 associated with processing each content word was larger when reading the untitled paragraphs, and they concluded from this that the readers had more difficulty integrating the words into the ongoing discourse.

An fMRI study using the same design found greater activity in the right temporal pole when reading paragraphs with titles (St. George, Kutas, Martinez, & Sereno, 1999). However, this work reported analysis only of inferior parts of the brain, so neither the dmPFC nor the precuneus/PPC was examined. Another fMRI study used the picture version of the Bransford and Johnson paradigm (Maguire, Frith, & Morris, 1999). Comparing reading a story preceded by a relevant picture to reading the same story without the picture led to greater activity in the precuneus/PCC.

In summary, studies comparing intact stories with unconnected sentences and studies comparing confusing stories with disambiguated stories converge in implicating medial frontal and posterior regions as important for constructing a situation model. The ERP results flesh out this picture, suggesting that word-by-word integration is easier when one can construct a situation model.

We have seen that local inconsistencies in language processing produce time-locked electrophysiological responses—a larger N400—and corresponding responses in fMRI. What about language that is locally consistent but inconsistent at the level of the larger situation? Such inconsistencies also produce N400-like responses. van Berkum, Hagoort, and Brown (1999) had participants listen to the beginning of a short narrative and then read the last sentence word-by-word. The final sentences were set up to contain a word that could be either consistent or inconsistent with the preceding story context. For example, in one story the context established that a boy had gotten out of bed and dressed earlier than expected; the final sentence then had his sister telling him he was either *quick* (consistent) or *slow* (inconsistent). Processing the inconsistent words led to an N400-like ERP. Subsequent studies found similar results when the critical sentences were heard rather than read (van Berkum, Zwitserlood, Hagoort, & Brown,

2003), and suggested that this effect reflects a specific influence of the discourse-level constraints rather than just priming of words related to the critical word (Otten & van Berkum, 2007).

Bicknell, Elman, Hare, McRae, and Kutas (2010) developed a paradigm to examine situation-level violations of expectation within a single sentence. They used agent/verb/patient triples such that the verb was held constant and the patient was either expected or unexpected given the noun. For example, following "journalist checked," the noun "spelling" would be expected and "brakes" would be unexpected, but following "mechanic checked" the reverse would be true. They found that the unexpected nouns were read more slowly and produced a larger N400.

One important method to study situation model building is the inconsistency paradigm, in which the global fit of a locally acceptable word is varied. For example, Ferstl, Rinck, and von Cramon (2005) generated stories for an fMRI study that could be manipulated such that changing a word produced either a temporal inconsistency (e.g., "early" would be expected but "late" would be unexpected) or an emotional inconsistency (e.g., "happy" would be expected but "sad" would be unexpected). Activity time-locked to the inconsistent words was greater, compared with consistent words, in the right anterior temporal lobe. They also observed a sustained response in the dmPFC from the onset of the inconsistency until the end of the story as participants read the conclusions of the inconsistent emotional stories. This suggests that participants attempted to "fix" the inconsistency by an inference; a conclusion was strengthened by the fact that activity in the dmPFC was greater for those participants whose ratings agreed less well with the consensus of the group, suggesting they were working harder to integrate the inconsistent information. Using a similar paradigm, Hasson, Nusbaum, and Small (2007) showed that the extent of the medial activations was related to subsequent recognition memory for the stories.

More recent studies investigated the interactions between consistency on the local and global levels (Egidi & Caramazza, 2013). In an ERP study, Boudewyn, Long, and Swaab (2013) showed that the local effects of lexical association (a reduced N400) were delayed by an inconsistent global context.

Together, studies using the inconsistency paradigm suggest that how well a word fits into the larger discourse contributes uniquely to how much work it takes to process that word and to update the situation model accordingly. The fMRI results reinforce the importance of the dmPFC for this level of integration and also provide some support for a right-hemisphere role. Some studies also speak to how the process of assembling a situation model makes contact with neural systems that are specialized for representing different kinds of knowledge. In the study by Ferstl, Rinck, and von Cramon (2005), emotional stories elicited activations in limbic regions, including the amygdala, and temporal stories were processed in a bilateral fronto-parietal network. In another study, Ferstl and von Cramon (2007) once more found evidence for content-specific activations. With shorter stories consisting of two sentences only, they again found that temporal information selectively activated frontal and parietal areas, whereas emotional stories selectively activated the left anterior temporal lobes. In addition, selective activation for spatial information was found in the parahippocampal cortex. Such results support the view that situation model construction depends on semantic networks that localize different types of knowledge in different brain areas.

53.4 SHIFTING AND MAPPING

Discourses are dynamic structures, not static scenes, and the processes of discourse comprehension reflect this. As one proceeds through a narrative, one constructs a series of situation models corresponding to the events in the story (Zwaan & Radvansky, 1998). Gernsbacher (1990) distinguished between two qualitatively different phases in this dynamic process. During the *mapping* phase, the comprehender has an established situation model that provides a good fit to the incoming information. New information is incrementally incorporated into the situation model, resulting in a series of small changes. For example, suppose you were reading a biography of the pianist Glenn Gould and found yourself in the middle of a description of a recording session. As you incorporate information about Gould's repeated takes and conversations with the recording engineer, you would likely map the descriptive features and dialog into your current situation model. However, when the situation in the narrative changes substantially, it would be more adaptive to abandon one's current model and construct a new one; this is called *shifting*. For example, if you were to read that Gould left the studio, this would be likely to prompt a shift. The dynamics of comprehension, then, consist of relatively extended periods of mapping punctuated by briefer periods of shifting. If a narrative presents a sequence of event descriptions, then this is an efficient processing strategy. Zacks and colleagues have argued that this pattern of stable situation model maintenance punctuated by new model construction not only is a feature of discourse processing but also is carried over from real-life event comprehension

(Zacks, Speer, Swallow, Braver, & Reynolds, 2007; Zacks & Tversky, 2001).

Ditman, Holcomb, and Kuperberg (2008) studied shifting using a time-shift paradigm. Consider the following excerpt: "Kelly scolded the child. After one second, the child whimpered to his mother." The phrase "one second" denotes essentially continuous passage of time and is plausible in this context. In the sentence context, "one second" produced a larger N400 than "one day." Ditman and colleagues also were able to test for the immediate consequences of shifting. They predicted that if readers were establishing a new situation model at "one day later," the second instance of the direct noun anaphor "the child" would be pragmatically awkward, resulting in a larger N400 at the onset of "child." That is exactly what they found.

Two studies have studied shifting using event-related fMRI. Whitney et al. (2009) asked participants to listen to a short novella during MRI scanning. They coded the text for changes in characters, time, spatial location, and action. They hypothesized that at such changes, participants would be likely to update their situation models. Comparing sentences with any of these changes to sentences without changes revealed activity in midline posterior cortical regions—specifically, the PCC and precuneus.

Speer, Reynolds, and Zacks (2007) asked participants to read short narratives presented one word at a time during scanning. The narratives were coded for changes in spatial location, objects, characters, causes, and goals. (The narratives used did not have changes in time, but the mention of time also was coded.) Following the scanning session, each participant was shown the narratives again and asked to segment them into meaningful events. Each participant segmented the narratives twice, once to identify fine-grained event boundaries and once to identify coarse-grained boundaries. Speer and colleagues, like Whitney et al., hypothesized that readers would be likely to shift to building a new situation model when changes in space, objects, characters, causes, and goals occurred. Behaviorally, this was evident in readers' segmentation: event boundaries were more likely when more features changed (see also Zacks, Speer, & Reynolds, 2009). The fMRI results also provided evidence for shifting. Event boundaries were associated with increases in the precuneus and posterior cingulate, as reported by Whitney et al. (2009), and also in the right lateral temporal lobe (posterior superior temporal gyrus and anterior middle temporal gyrus) and in the right middle frontal gyrus. These regions also responded to changes on two or more of the situation dimensions, and the posterior cingulate/precuneus regions showed parametric increases such that they responded more strongly when more features changed (Speer, Reynolds, Swallow, & Zacks, 2009).

Like the data from the inconsistency paradigms, data on situation model shifting suggest that situation model construction depends on semantic representations that localize different sorts of information in different brain areas. For example, in the Speer et al. (2009) study, changes in characters selectively activated the left temporo-parietal junction, left anterior temporal lobe, and the dmPFC, changes in goals selectively activated the lateral middle frontal gyrus bilaterally, changes in objects selectively activated the left premotor cortex, and changes in time selectively activated the left insula.

The ERP and fMRI results both provide evidence for transient neural processing at those points during comprehension at which shifting would be predicted to occur. The fMRI data suggest that midline posterior cortical regions may be particularly important for this processing. However, they leave an important question unanswered: Do these transient effects reflect shifting or mapping? Recall that shifting is proposed to be a global updating mechanism that results in the creation of a new situation model, whereas mapping is proposed to reflect the incremental incorporation of new information as features in the narrative change. The effects reported in these studies could reflect global shifting, happening concurrent with some or all of the feature changes coded, incremental mapping, or a mixture of both. In one behavioral study, Kurby and Zacks (2012) tested for both shifting and mapping using a think-aloud task. Participants read short narratives and paused after each sentence to describe what they were thinking. Their responses indicated that when a particular feature changed (e.g., a change of spatial location), readers were more likely to mention that feature dimension. Above and beyond this, at event boundaries readers were more likely to mention features that changed and features that did not change. This result argues that both shifting and mapping can occur in response to feature changes. An important question for future research is whether the neurophysiological responses correspond to shifting, mapping, or both.

53.5 CONCLUSION

The short review shows promising first steps toward a neurobiological theory of language processing at the discourse level. Based on a well-founded tradition of psycholinguistic models (Kintsch, 1998) and the formulation of novel approaches (e.g., Zwaan, 2004), a creative set of experimental tools for constructing and manipulating narrative structure is available.

Supplementing the rich set of behavioral data, the beginnings of an empirical database of neurophysiological results are emerging. Let us summarize the major theoretical and empirical results.

Cohesion and local coherence are the most basic means to connect successive utterances. Both fMRI and ERP experiments have provided evidence for the use of even subtle cohesive cues during comprehension. The findings confirm the comprehenders' sensitivity to these cues. However, it is less clear whether the activation differences reflect the processing of the function words (e.g., in left frontal lobe) or the result of the cohesive markers, that is, the increase in coherence induced by them (e.g., right temporal activations).

Inference processes required for setting up a locally coherent text representation engage regions extending beyond the left perisylvian language cortex (see, e.g., Ferstl et al., 2008). In particular, right-hemisphere areas and medial regions—particularly the dmPFC—are important. In line with studies on situation model processing, inference studies suggest that the right hemisphere activations are observed whenever local, semantically based connections are set up, whereas medial regions become important whenever nonautomatic, evaluation-based inferences are drawn. Furthermore, there is evidence for individual differences with respect to right hemisphere involvement. Further research is needed to evaluate these hypotheses.

Both behavioral measures and neurophysiological measures support the idea that situation model construction is a crucial component during the comprehension of narrative language. Constructing a situation model appears to depend particularly on the medial prefrontal cortex and the medial PCC and precuneus. It may sometimes be associated with differential activity in the right hemisphere; the data on this point are mixed. When a text affords construction of a situation model, this facilitates sentence-by-sentence reading. This can be seen in reductions in the N400 ERP component. Situation model construction is valuable—it supports better subsequent memory for the contents of the text.

The distinction we have made between cohesion, coherence, and situation model construction is a substantive one, and it reflects how the field has parsed the computations the brain performs during discourse comprehension. However, it is important to note that cohesion, coherence, and situation model construction are closely interwoven: inferences are needed establishing local coherence, and also for setting up a situation model, and the prior discourse model facilitates later inference processes. Furthermore, not only long texts but also even short sentences can elicit a situation model representation.

When we read longer texts, a single situation model is not enough to represent the contents of the narrative. In this case, we need to construct a series of models, mapping information into the existing model when we can and shifting to a new model when we must. Shifting to a new model takes cognitive work, and we can see its signature in slowing in reading and in increased activity in multiple cortical areas. After we shift to a new model, some information that was maintained in the previous model becomes less accessible.

One important conclusion from the neuroscientific studies reported here is that theoretical approaches of text comprehension need to encompass cognitive mechanisms that are not specific to language. Whereas psycholinguistic models traditionally focus on the linguistic properties of text and the processes needed for interpreting them, the overlap of activation patterns with results from social cognitive neuroscience suggests that we need a more holistic approach to communication. Intentions, implications, Theory-of-Mind, and emotional connotations all need to be taken into account when studying language comprehension in context (see also Hasson and Tremblay, Chapter 43).

What will it take to move the field from this impressive but incomplete set of empirical regularities to a real theory? One useful approach is formal modeling that can make predictions about online reading processes and subsequent memory *at the discourse level*, sentence by sentence, clause by clause, and perhaps even word by word (Kintsch, 1988; van den Broek et al., 1999). Building such models is a great scientific challenge because, as the other sections of this volume illustrate, "lower-level" mechanisms of language comprehension are fantastically complex in their own right. However, a number of frameworks are available to support developing such models, including production systems (Anderson & Lebiere, 1998; Goldman, Varma, & Coté, 1996) and connectionist modeling systems (Christiansen & Chater, 1999; O'Reilly & Munakata, 2000). It will be a challenge to utilize such models for making specific predictions about the functional neuroanatomy of discourse comprehension (cf. Anderson et al., 2008).

A second means to make considerable progress is the development of experimental designs that allow us to study reading closer to how it occurs "in the wild" (Bailley & Zacks, 2011). Many of the studies described in this chapter utilized what Graesser and colleagues have dubbed "textoids"—brief, contrived texts that allow an experimenter to manipulate discourse structure at a fine grain (Graesser, Millis, & Zwaan, 1997).

Such studies are valuable and, indeed, necessary. We agree with Graesser et al. that this approach needs to be complemented by approaches that use longer, more natural texts. In short, what is needed is data on situations in which people read narratives that are like the ones they read in real life, in a manner similar to how they read in real life. There is a major downside to using naturalistic materials and tasks: variables of interest are confounded with each other and with other variables. For example, at the single word level the number of meanings associated with a word (e.g., "box" can mean a container or what one does with gloves in a ring) is confounded with its length and frequency (Yap & Balota, 2009). At the discourse level, changes in characters' goals are confounded with changes in spatial and temporal location (Zacks et al., 2009). The good news is that these confounds can be dealt with statistically—provided one has enough data. This suggests to us that a productive way forward will be to construct corpora of data from naturalistic comprehension tasks that are larger than the datasets typically collected to test individual hypotheses. Thankfully, naturalistic materials and tasks lend themselves to reuse. In our own research, we have been encouraged by the ability to mine naturalistic comprehension data to test novel hypotheses and language comprehension and perception (Magliano & Zacks, 2011; Speer et al., 2009; Yarkoni, Speer, Balota, McAvoy, & Zacks, 2008).

In short, processing discourse is a fascinating challenge. Our understanding of how the brain does so has benefited from a generation of behavioral research, which is now being complemented by neurophysiological measures. We look forward to much progress in the years to come.

References

Almor, A., Smith, D. V., Bonilha, L., Fridriksson, J., & Rorden, C. (2007). What is in a name? Spatial brain circuits are used to track discourse references. *Neuroreport*, *18*, 1215–1219.

Anderson, J. R., Carter, C. S., Fincham, J. M., Qin, Y., Ravizza, S. M., & Rosenberg-Lee, M. (2008). Using fMRI to test models of complex cognition. *Cognitive Science*, *32*, 1323–1348.

Anderson, J. R., & Lebiere, C. (1998). *The atomic components of thought.* Mahwah, NJ: Lawrence Erlbaum Associates.

Bailey, H. R., & Zacks, J. M. (2011). Literature and event understanding. *Scientific Study of Literature*, *1*(1), 72–78.

Barsalou, L. (2008). Grounded cognition. *Annual Review of Psychology*, *59*, 617–645.

Baumgärtner, A., Weiller, C., & Büchel, C. (2002). Event-related fMRI reveals cortical sites involved in contextual sentence integration. *Neuroimage*, *16*, 736–745.

Beeman, M. (1993). Semantic processing in the right hemisphere may contribute to drawing inferences from discourse. *Brain and Language*, *44*, 80–120.

Bicknell, K., Elman, J. L., Hare, M., McRae, K., & Kutas, M. (2010). Effects of event knowledge in processing verbal arguments. *Journal of Memory and Language*, *63*, 489–505.

Bornkessel-Schlesewsky, I. D., & Friederici, A. D. (2007). Neuroimaging studies of sentence and discourse comprehension. In M. G. Gaskell (Ed.), *Oxford handbook of psycholinguistics* (pp. 407–424). New York, NY: Oxford University Press.

Boudewyn, M. A., Long, D. L., & Swaab, T. Y. (2013). Effects of working memory span on processing of lexical associations and congruence in spoken discourse. *Frontiers in Psychology*, *4*.

Bransford, J. D., Barclay, J. R., & Franks, J. J. (1972). Sentence memory: a constructive versus interpretive approach. *Cognitive Psychology*, *3*, 193–209.

Bransford, J. D., & Johnson, M. K. (1972). Contextual prerequisites for understanding: some investigations of comprehension and recall. *Journal of Verbal Learning & Verbal Behavior*, *11*, 717–726.

Brownell, H. H., & Martino, G. (1998). Deficits in inference and social cognition: the effects of right hemisphere brain damage on discourse. In M. Beeman, & C. Chiarello (Eds.), *Right hemisphere language comprehension: Perspectives from cognitive neuroscience* (pp. 309–328). Mahwah, NJ: Lawrence Erlbaum Associates.

Chow, H. M., Kaup, B., Raabe, M., & Greenlee, M. W. (2008). Evidence of fronto-temporal interactions for strategic inference processes during language comprehension. *Neuroimage*, *40*, 940–954.

Christiansen, M. H., & Chater, N. (1999). Connectionist natural language processing: the state of the art. *Cognitive Science*, *23*, 417–437.

Ditman, T., Holcomb, P. I., & Kuperberg, G. F. (2008). Time travel through language: temporal shifts rapidly decrease information accessibility during reading. *Psychonomic Bulletin & Review*, *14*, 750–756.

Dooling, D. J., & Lachman, R. (1971). Effects of comprehension on retention of prose. *Journal of Experimental Psychology*, *88*, 216–222.

Egidi, G., & Caramazza, A. (2013). Cortical systems for local and global integration in discourse comprehension. *Neuroimage*, *71*, 59–74.

Ferstl, E. C. (2007). The functional neuroanatomy of text comprehension: what's the story so far? In F. Schmalhofer, & C. A. Perfetti (Eds.), *Higher level language processes in the brain: Inference and comprehension processes* (pp. 53–102). Mahwah, NJ: Lawrence Erlbaum Associates.

Ferstl, E. C. (2010). Neuroimaging of text comprehension: where are we now? *Italian Journal of Linguistics*, *22*, 61–88.

Ferstl, E. C., Guthke, T., & von Cramon, D. Y. (2002). Text comprehension after brain injury: left prefrontal lesions affect inference processes. *Neuropsychology*, *16*, 292–308.

Ferstl, E. C., Manouilidou, C., & Garnham, A. (2010). Gender stereotypes and implicit verb causality in language comprehension: an fMRI study. *Journal of Cognitive Neuroscience (Supplement)*, 127.

Ferstl, E. C., Neumann, J., Bogler, C., & von Cramon, D. Y. (2008). The extended language network: a meta-analysis of neuroimaging studies on text comprehension. *Human Brain Mapping*, *29*, 581–593.

Ferstl, E. C., Rinck, M., & von Cramon, D. Y. (2005). Emotional and temporal aspects of situation model processing during text comprehension: an event-related fMRI study. *Journal of Cognitive Neuroscience*, *17*, 724–739.

Ferstl, E. C., & von Cramon, D. Y. (2001). The role of coherence and cohesion in text comprehension: an event-related fMRI study. *Cognitive Brain Research*, *11*, 325–340.

Ferstl, E. C., & von Cramon, D. Y. (2002). What does the frontomedian cortex contribute to language processing: coherence or theory of mind? *Neuroimage, 17*, 1599–1612.

Ferstl, E. C., & von Cramon, D. Y. (2007). Time, space and emotion: fMRI reveals content-specific activation during text comprehension. *Neuroscience Letters, 427*, 159–164.

Fletcher, C. R. (1994). Levels of representation in memory for discourse. In M. A. Gernsbacher (Ed.), *Handbook of psycholinguistics* (pp. 589–607). New York, NY: Academic Press.

Fletcher, P. C., Happé, F., Frith, U., Baker, S. C., Dolan, R. J., Frackowiak, R. S. J., et al. (1995). Other minds in the brain: a functional imaging study of "theory of mind" in story comprehension. *Cognition, 57*, 109–128.

Friese, U., Rutschmann, R., Raabe, M., & Schmalhofer, F. (2008). Neural indicators of inference processes in text comprehension: an event-related functional magnetic resonance imaging study. *The Journal of Cognitive Neuroscience, 20*, 2110–2124.

Gernsbacher, M. A. (1990). *Language comprehension as structure building*. Hillsdale, NJ: Lawrence Erlbaum Associates.

Glenberg, A. M. (2007). Language and action: creating sensible combinations of ideas. In G. Gaskell (Ed.), *The Oxford handbook of psycholinguistics* (pp. 361–370).

Glenberg, A. M., Meyer, M., & Lindem, K. (1987). Mental models contribute to foregrounding during text comprehension. *Journal of Memory & Language, 26*, 69–83.

Goldman, S. R., Golden, R. M., & van den Broek, P. (2007). Why are computational models of text comprehension useful? In *Higher level language processes in the brain: Inference and comprehension processes* (pp. 9–26). Mahwah, NJ: Lawrence Erlbaum Associates.

Goldman, S. R., Varma, S., & Coté, N. (1996). Extending capacity-constrained construction integration: toward "smarter" and flexible models of text comprehension. In B. K. Britton, & A. C. Graesser (Eds.), *Models of understanding text* (pp. 73–113). Mahwah, NJ: Lawrence Erlbaum Associates.

Graesser, A. C., Millis, K. K., & Zwaan, R. A. (1997). Discourse comprehension. *Annual Review of Psychology, 48*, 163–189.

Halliday, M. A., & Hasan, R. (1976). *Cohesion in English*. London: Pearson Longman.

Hammer, A., Goebel, R., Schwarzbach, J., Münte, T. F., & Jansma, B. M. (2007). When sex meets syntactic gender on a neural basis during pronoun processing. *Brain Research, 1146*, 185–198.

Hasson, U., Nusbaum, H. C., & Small, S. L. (2007). Brain networks subserving the extraction of sentence information and its encoding to memory. *Cerebral Cortex, 17*, 2899–2913.

Hemingway, E. (2007). *The complete short stories of Ernest Hemingway: The Finca Vigia edition*. New York, NY: Simon and Schuster.

Jung-Beeman, M. (2005). Bilateral brain processes for comprehending natural language. *Trends in Cognitive Science, 9*, 512–518.

Kaan, E. (2007). Event-related potentials and language processing: a brief overview. *Language and Linguistics Compass, 1*, 571–591.

Kintsch, W. (1988). The role of knowledge in discourse comprehension: a construction-integration model. *Psychological Review, 95*, 163–182.

Kintsch, W. (1998). *Comprehension: a paradigm for cognition*. Cambridge: Cambridge University Press.

Kintsch, W., & van Dijk, T. A. (1978). Toward a model of text comprehension and production. *Psychological Review, 85*, 363–394.

Kreiner, H., Sturt, P., & Garrod, S. (2008). Processing definitional and stereotypical gender in reference resolution: evidence from eye-movements. *Journal of Memory and Language, 58*, 239–261.

Kuperberg, G. R., Lakshmanan, B. M., Caplan, D. N., & Holcomb, P. J. (2006). Making sense of discourse: an fMRI study of causal inferencing across sentences. *Neuroimage, 33*, 343–361.

Kurby, C. A., & Zacks, J. M. (2012). Starting from scratch and building brick by brick in comprehension. *Memory & Cognition*.

Kutas, M., & Federmeier, K. D. (2011). Thirty years and counting: finding meaning in the N400 component of the event-related brain potential (ERP). *Annual Review of Psychology, 62*, 621–647.

Magliano, J. P., & Zacks, J. M. (2011). The impact of continuity editing in narrative film on event segmentation. *Cognitive Science, 35*, 1489–1517.

Maguire, E. A., Frith, C. D., & Morris, R. G. (1999). The functional neuroanatomy of comprehension and memory: the importance of prior knowledge. *Brain, 122*, 1839–1850.

Mar, R. A. (2004). The neuropsychology of narrative: story comprehension, story production, and their interaction. *Neuropsychologia, 42*, 141–1434.

Mason, R. A., & Just, M. A. (2004). How the brain processes causal inferences in text: a theoretical account of generation and integration component processes utilizing both cerebral hemispheres. *Psychological Science, 15*, 1–7.

Mason, R. A., & Just, M. A. (2006). Neuroimaging contributions to the understanding of discourse processes. In M. J. Traxler, & M. A. Gernsbacher (Eds.), *Handbook of psycholinguistics* (pp. 765–799). Amsterdam: Elsevier.

Mason, R. A., & Just, M. A. (2009). The role of the theory-of-mind cortical network in the comprehension of narratives. *Linguistics and Language Compass, 3*, 157–174.

Münte, T. F., Schiltz, K., & Kutas, M. (1998). When temporal terms belie conceptual order. *Nature, 395*, 71–73.

Myers, J. L., Shinjo, M., & Duffy, S. A. (1987). Degree of causal relatedness and memory. *Journal of Memory and Language, 26*, 453–465.

Nieuwland, M. S., Petersson, K. M., & van Berkum, J. J. A. (2007). On sense and reference: examining the functional neuroanatomy of referential processing. *Neuroimage, 37*, 993–1004.

O'Reilly, R. C., & Munakata, Y. (2000). *Computational explorations in cognitive neuroscience: Understanding the mind by simulating the brain*. Cambridge, MA: MIT Press.

Osterhout, L., Bersick, M., & McLaughlin, J. (1997). Brain potentials reflect violations of gender stereotypes. *Memory & Cognition, 25*, 273–285.

Otten, M., & van Berkum, J. J. A. (2007). What makes a discourse constraining? Comparing the effects of discourse message and scenario fit on the discourse-dependent N400 effect. *Brain Research, 1153*, 166–177.

Papagno, C., & Lauro, L. J. R. (2010). The neural basis of idiom processing: neuropsychological, neurophysiological and neuroimaging evidence. *Italian Journal of Linguistics, 22*, 21–40.

Radvansky, G. A., & Zacks, J. M. (2014). *Event cognition*. New York, NY: Oxford University Press.

Raichle, M. E., MacLeod, A. M., Snyder, A. Z., Powers, W. J., Gusnard, D. A., & Shulman, G. L. (2001). A default mode of brain function. *Proceedings of the National Academy of Science of the United States, 98*, 676–692.

Robertson, D. A., Gernsbacher, M. A., Guidotti, S. J., Robertson, R. R., Irwin, W., Mock, B. J., et al. (2000). Functional neuroanatomy of the cognitive process of mapping during discourse comprehension. *Psychological Science, 11*, 255–260.

Schmidt, G. L., & Seger, C. A. (2009). Neural correlates of metaphor processing: the roles of figurativeness, familiarity and difficulty. *Brain and Cognition, 71*, 375–386.

Scott Rich, S., & Taylor, H. (2000). Not all narrative shifts function equally. *Memory & Cognition, 28*, 1257–1266.

Siebörger, F. T., Ferstl, E. C., & von Cramon, D. Y. (2007). Making sense of nonsense: an fMRI study of task induced inference processes during discourse comprehension. *Brain Research, 1166*, 77–91.

Singer, M., Halldorson, M., Lear, J. C., & Andrusiak, P. (1992). Validation of causal bridging inferences in discourse understanding. *Journal of Memory and Language, 31*, 507–524.

Speer, N. K., Reynolds, J. R., Swallow, K. M., & Zacks, J. M. (2009). Reading stories activates neural representations of perceptual and motor experiences. *Psychological Science, 20*, 989–999.

Speer, N. K., Reynolds, J. R., & Zacks, J. M. (2007). Human brain activity time-locked to narrative event boundaries. *Psychological Science, 18*, 449–455.

St. George, M., Kutas, M., Martinez, A., & Sereno, M. I. (1999). Semantic integration in reading: engagement of the right hemisphere during discourse processing. *Brain, 122*, 1317–1325.

St. George, M., Mannes, S., & Hoffman, J. E. (1994). Global semantic expectancy and language comprehension. *Journal of Cognitive Neuroscience, 6*, 70–83.

van Berkum, J. J. A. (2004). Sentence comprehension in a wider discourse: can we use ERPs to keep track of things. In M. Carreiras, & C. J. Clifton (Eds.), *The on-line study of sentence comprehension: Eyetracking, ERPs and beyond* (pp. 229–270). New York, NY: Psychology Press.

van Berkum, J. J. A., Hagoort, P., & Brown, C. M. (1999). Semantic integration in sentences and discourse: evidence from the N400. *Journal of Cognitive Neuroscience, 11*, 657–671.

van Berkum, J. J. A., Zwitserlood, P., Hagoort, P., & Brown, C. M. (2003). When and how do listeners relate a sentence to the wider discourse? Evidence from the N400 effect. *Cognitive Brain Research, 17*, 701–718.

van den Broek, P., Young, M., Tzeng, Y., & Linderholm, T. (1999). The landscape model of reading: inferences and the online construction of a memory representation. In H. van Oostendorp, & S. R. Goldman (Eds.), *The construction of mental representations during reading* (pp. 71–98). Mahwah, NJ: Lawrence Erlbaum Associates.

van Dijk, T. A., & Kintsch, W. (1983). *Strategies of discourse comprehension*. New York, NY: Academic Press.

Virtue, S., Haberman, J., Clancy, Z., Parrish, T., & Jung-Beeman, M. (2006). Neural activity of inferences during story comprehension. *Brain Research, 1084*, 104–114.

Virtue, S., Parrish, T., & Jung-Beeman, M. (2008). Inferences during story comprehension: cortical recruitment affected by predictability of events and working memory capacity. *Journal of Cognitive Neuroscience, 20*, 2274–2284.

Whitney, C., Huber, W., Klann, J., Weis, S., Krach, S., & Kircher, T. (2009). Neural correlates of narrative shifts during auditory story comprehension. *Neuroimage, 47*, 360–366.

Yap, M. J., & Balota, D. A. (2009). Visual word recognition of multisyllabic words. *Journal of Memory and Language, 60*, 502–529.

Yarkoni, T., Speer, N., Balota, D., McAvoy, M., & Zacks, J. (2008). Pictures of a thousand words: investigating the neural mechanisms of reading with extremely rapid event-related fMRI. *Neuroimage, 42*, 973–987.

Yarkoni, T., Speer, N., & Zacks, J. (2008). Neural substrates of narrative comprehension and memory. *Neuroimage, 41*, 1408–1425.

Ye, Z., Habets, B., Jansma, B. M., & Münte, T. F. (2011). Neural basis of linearization in speech production. *Journal of Cognitive Neuroscience, 23*, 3694–3702.

Zacks, J. M., Speer, N. K., & Reynolds, J. R. (2009). Segmentation in reading and film comprehension. *Journal of Experimental Psychology: General, 138*, 307–327.

Zacks, J. M., Speer, N. K., Swallow, K. M., Braver, T. S., & Reynolds, J. R. (2007). Event perception: a mind-brain perspective. *Psychological Bulletin, 133*, 273–293.

Zacks, J. M., & Tversky, B. (2001). Event structure in perception and conception. *Psychological Bulletin, 127*, 3–21.

Zwaan, R. A. (2004). The immersed experiencer: toward an embodied theory of language comprehension. In B. H. Ross (Ed.), *The psychology of learning and motivation* (Vol. 44, pp. 35–62). New York, NY: Academic Press.

Zwaan, R. A., Langston, M. C., & Graesser, A. C. (1995). The construction of situation models in narrative comprehension: an event-indexing model. *Psychological Science, 6*, 292–297.

Zwaan, R. A., & Radvansky, G. A. (1998). Situation models in language comprehension and memory. *Psychological Bulletin, 123*, 162–185.

CHAPTER

54

At the Core of Pragmatics

The Neural Substrates of Communicative Intentions

Bruno G. Bara[1,2], Ivan Enrici[1,2,3] and Mauro Adenzato[1,2]

[1]Center for Cognitive Science, Department of Psychology, University of Torino, Italy; [2]Neuroscience Institute of Turin, Italy; [3]Department of Philosophy and Educational Sciences, University of Torino, Italy

54.1 COMMUNICATIVE INTENTION: THE CORE FEATURE OF PRAGMATIC PHENOMENA

Cognitive theories of human communication are often theories of competence, that is, theories specified in terms of the mental states involved in communication independently of the means used to communicate. Examples of such theories were proposed by Strawson (1964), Schiffer (1972), Sperber and Wilson (1986), Airenti, Bara, and Colombetti (1993), Clark (1996), Tirassa (1999), and Bara (2010, 2011). Although different in many respects, these theories share the same logical assumptions as the pioneering work of Grice (1957) and Austin (1962), who proposed that human communication should be analyzed in terms of the role of mental states, such as intentions, beliefs, emotions, and desires, and the cognitive dynamics leading from one mental state to another. In particular, Grice (1957) described successful communication as the recognition of a specific set of mental states, including the intention to affect the communication partner and the higher-order intention that this intention is recognized. Although such conditions were subsequently strengthened by Strawson (1964) and Schiffer (1972), they remain at the core of the investigation of pragmatic phenomena (i.e., understanding the speaker's intention within a communicative exchange).

The key aspect of the approach proposed by Grice (1957, 1975) is that in communication, sentences are often underspecified with respect to what speakers intend to convey and, in a variety of pragmatic phenomena, there is a discrepancy between the literal and intended meanings, such as in indirect speech acts, conversational implicatures, sarcasm, metaphor, and irony. In other words, successful pragmatic understanding of an utterance depends on recognizing the intention of the speaker within a social context (Grice, 1975). Starting from this theoretical standpoint, Airenti et al. (1993) and, more recently, Bara (2010) proposed that the key concept of communicative intention is a *primitive* mental state (i.e., communicative intention that is not reducible to simple intentions).

The importance of communicative intention for the study of pragmatics lies in the fact that this concept permits the clear distinction between two separate phenomena: communication in and of itself, and information extraction (or information attribution). Communication is a social activity that requires the combined effort of at least two participants who consciously and intentionally cooperate to construct the meaning of their interaction. When people use language (spoken or written), they intrinsically convey communicative intentions simply by using this expressive mean. Likewise, a key concept in the process of information extraction is the *sign* (Hauser, 1996). A sign is a parameter that can take on different values. It is produced by the individual, sometimes with a precise aim, but with no communicative goal. The concept of signs is intrinsically ambiguous, given that any evidence of human activity may become communication. For example, an unmade bed may be a sign that a person has slept in it. However, in certain circumstances, this sign could be fully communicative, that is, a symbol that has been deliberately left to let the observer know that the bed has been slept in. Thus, signs may easily become signals, as long as they are left intentionally.

Accordingly, genuine communication and information extraction are two distinct but equally important phenomena. Confusing these phenomena and, hence,

Neurobiology of Language. DOI: http://dx.doi.org/10.1016/B978-0-12-407794-2.00054-7

being unable to distinguish between them has a long tradition in communication sciences. The well-known first axiom of communication proposed by Watzlawick, Bevin, and Jackson (1967)—"one cannot not communicate"—is the manifesto of this confusion, because it denies the possibility of information extraction in which an agent influences the mental states of another agent through his/her behavior or by manipulating the material surroundings *without any intention to do so*. In sum, we speak of fully communicative phenomena when reciprocal intentionality is involved and of extracting information when one of the actors does not possess the intention to communicate but is simply moving and acting in the world based on private intentions.

54.2 NEURAL SUBSTRATES OF COMMUNICATIVE INTENTION: THE INTENTION PROCESSING NETWORK

According to our perspective, human communication is a cooperative activity among social agents who jointly and intentionally construct the meaning of their interaction within a shared context. Communication is a social action used to affect and modify the mental states of others, and communicative intention is the primary mental state involved in explaining other people's communicative actions. For the study of the neurobiology of pragmatics, this means that full analysis of the processes involved in a communicative exchange cannot be limited to the study of the cognitive and neural mechanisms for coding and decoding words and phrases generated by actors, but should also include the context-based inferential processes involved in comprehending the agents' communicative intentions. Thus, the process involved in understanding this form of intentions connects human communication and a more general type of social competence, such as Theory of Mind (ToM) ability (Baron-Cohen, 1995; Premack & Woodruff, 1978). ToM is the ability to: (i) acknowledge that human beings possess mental states; (ii) attribute mental states to one's self and to others; (iii) recognize that the mental states of others do not necessarily correspond to one's own; (iv) understand that mental states can determine external behavior, such as decisions and actions; and (v) predict, describe, and explain behavior on the basis of these mental states.

Considering the role that the ToM neurocognitive system plays in understanding the communicative intentions underlying pragmatic phenomena, it is surprising that relatively few studies have investigated the neural substrates of the human ability to process this kind of social intention. In contrast, the study of the neural correlates of ToM for individual mental states, such as beliefs and desires, is a prosperous area of research. To bridge this gap, in the past decade we have performed a series of neuroimaging studies and proposed the Intention Processing Network (IPN) model, according to which a set of brain areas are differentially involved in comprehending different types of intention, including communicative intentions. The IPN model introduces a theoretical distinction that differentiates private versus social dimensions of intention and the temporal dimension (present or prospective) of the social interaction. *Private intentions* only involve the actor satisfying a particular goal (e.g., picking a bunch of grapes to eat them). Conversely, in *social intentions*, the goal of the actor is satisfied only if at least one other person is involved. For social intentions, we can distinguish between present and prospective (future) interactions. When two agents interact, the social intention is shared in the present (e.g., asking the waiter for a mojito). The prototypical example of a social intention shared in the present is a *communicative intention*. When a given social interaction is not present at the moment but the action of a single agent preludes to it, the social intention is potentially shared in the future (e.g., preparing a flower bouquet to give it to someone). We define this type of social intention as *prospective social intention*.

In two fMRI studies (Ciaramidaro et al., 2007; Walter et al., 2004), we used a story completion task presented in the form of a comic strip to show the progressive recruitment of the ToM network according to these theoretical dimensions. The brain areas associated with the IPN are shown in Figure 54.1 and include the complex formed by the medial prefrontal cortex (MPFC) and precuneus, and by the bilateral posterior superior temporal sulcus (pSTS) and adjacent temporoparietal junctions (brain areas referred to collectively as TPJ in this work, for simplicity, but see Gobbini, Koralek, Bryan, Montgomery, & Haxby, 2007; Saxe, 2006 for the dissociation between the pSTS as a part of the action understanding system and the TPJ as a part of the ToM system). In particular, we demonstrated that the whole IPN was only activated during communicative intention processing, when two characters were depicted in a communicative interaction within a pragmatic context. In contrast, the activated network was limited to the right TPJ and precuneus, when the depicted character was acting according to a private intention, or when two characters were acting independently according to their own private intentions without any shared context. For prospective social intentions, our data show that although both private and prospective social intentions involve single agents acting by themselves, the recruitment of the right TPJ and precuneus does not suffice when an agent is manifesting a social intention to be shared in the future. We explain this result by the widely recognized fact that MPFC activation is observed whenever a social

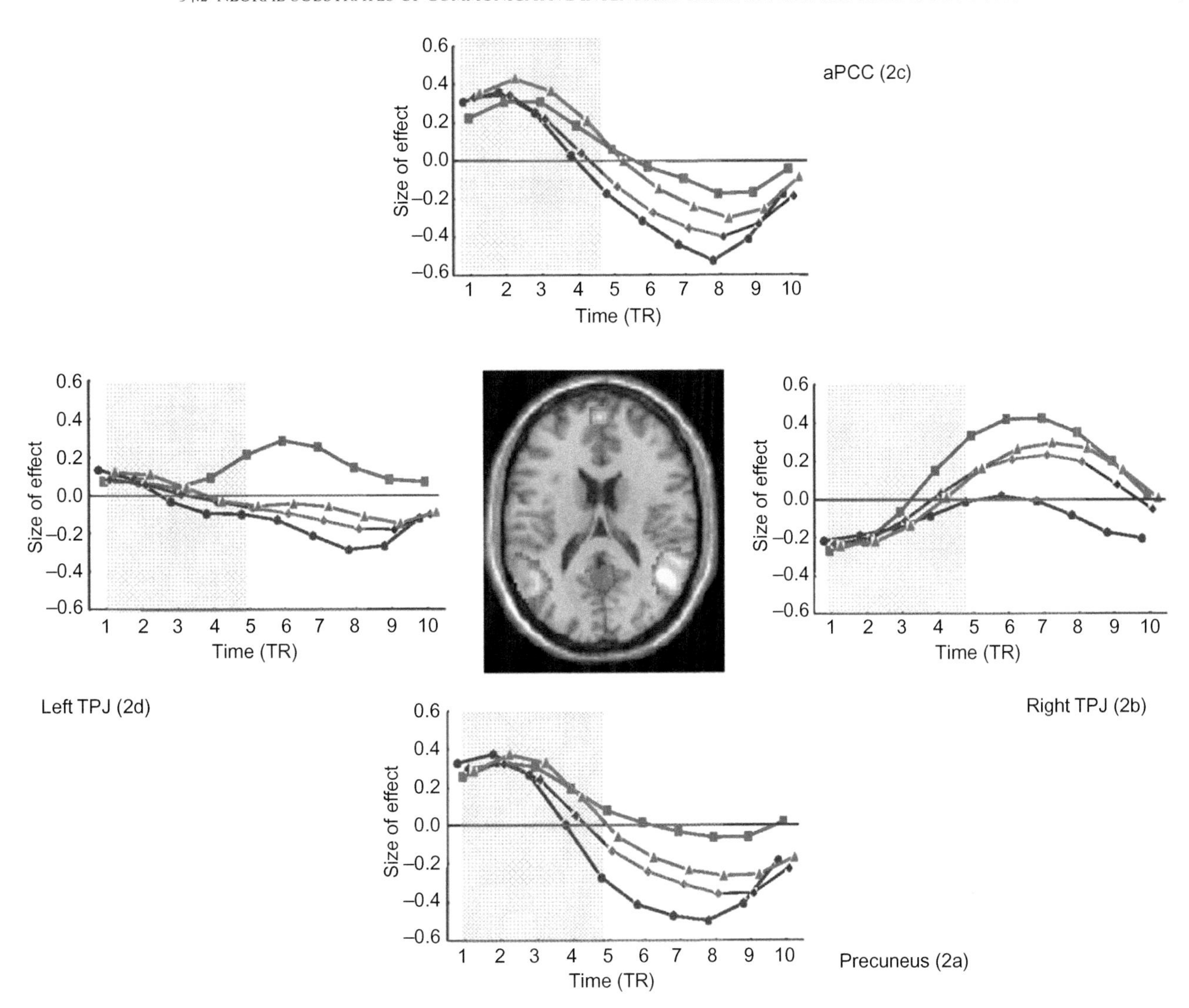

FIGURE 54.1 Signal time course for the three experimental conditions, private, prospective social, and communicative intention, and for control condition (physical causality) relative to rest for the IPN brain regions. The brain picture depicts the contrast communicative intention versus physical causality condition showing the four regions of interest, the aPCC (2c), precuneus (2a), and right and left TPJ (2b, 2d). Time courses were calculated by averaging across conditions within each participant (red curve for communicative intention, violet curve for prospective social intention, green curve for private intention, and blue curve for physical causality). The pink area represents the story phase and the grey area represents the choice phase of the comic strips task. aPCC, anterior paracingulate cortex (part of the MPFC); TPJ, temporoparietal junction. *From Ciaramidaro et al. (2007) with permission from the publisher.*

dimension is involved. For these reasons, we define the IPN model as the progressive recruitment of different brain areas (i.e., the precuneus and right TPJ, MPFC, left TPJ) according to, respectively, private, prospective social, and communicative intentions.

Different studies have provided independent evidence for the specific role of the IPN brain areas in social intention recognition and comprehension. For example, a meta-analysis (Van Overwalle & Baetens, 2009) suggested that the precuneus is crucial for elaboration of contextual information and identification of situational structure, and the TPJ is involved during the identification of end state behaviors. In particular, the TPJ along with the precuneus and MPFC take part in a larger process of goal identification in a social context. Van Overwalle (2009) showed impressive consistency in the empirical evidence demonstrating MPFC engagement in social inferences, specifically its crucial role in understanding social scripts that do not concern a single actor, but that describe adequate social actions for all the actors involved in a particular context.

More interesting for the purpose of the present discussion, the IPN model has received support from neuroimaging studies aimed at investigating the neural correlates of processing different pragmatic phenomena. Specifically, phenomena in which the literal

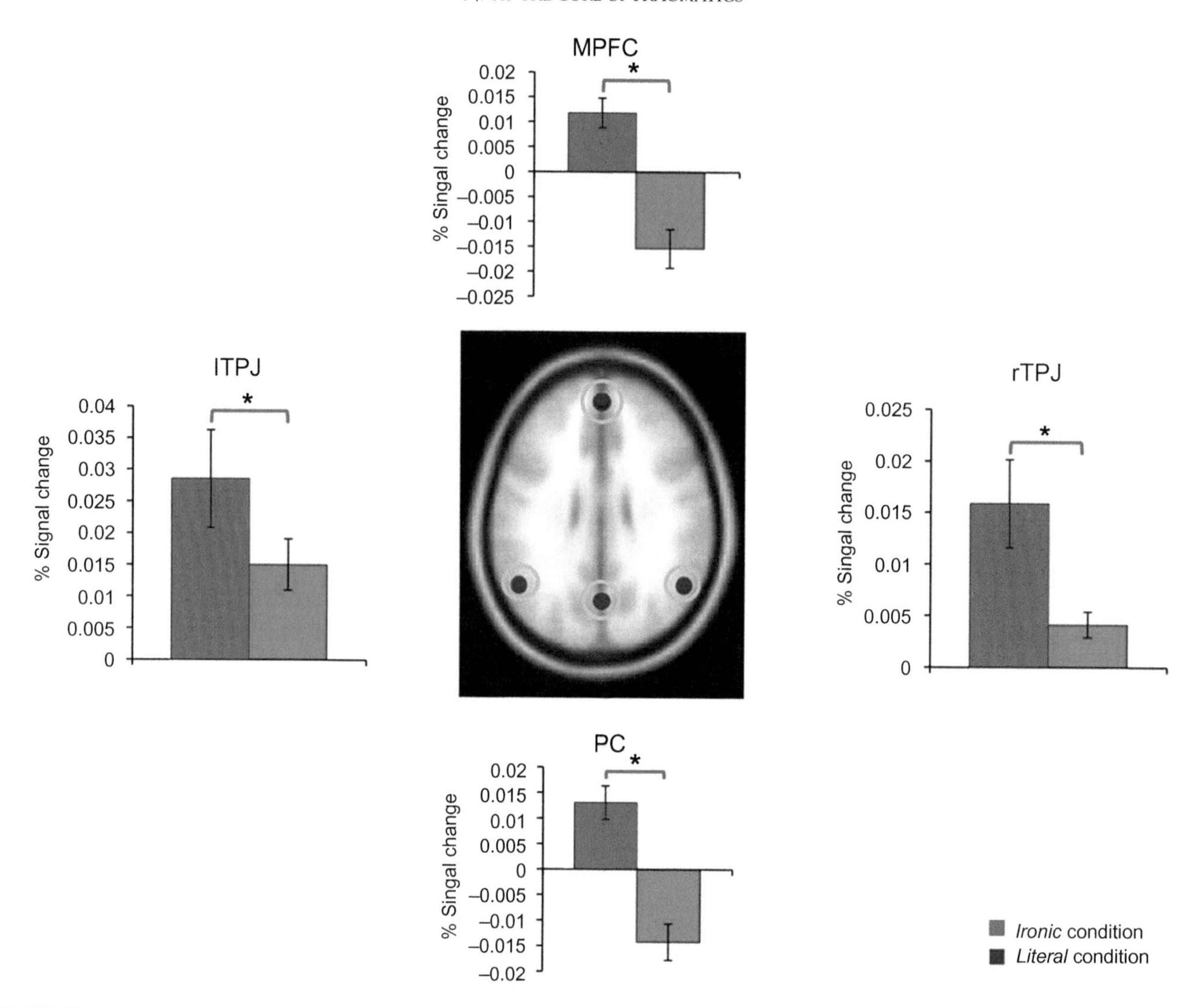

FIGURE 54.2 Regions of interest of the four main areas with significantly greater activity in the ironic than in the literal condition. MPFC, medial prefrontal cortex; lTPJ, left temporoparietal junction; rTPJ, right temporoparietal junction; PC, precuneus. *From Spotorno et al. (2012) with permission from the publisher.*

and intended meanings do not coincide, and inferences about the speaker's intended meaning are required.

It is unanimously agreed on that ironic remarks are typical examples of a high-level, pragmatic form of communication in which the literal and intended meaning do not coincide and, in fact, the communicative intention often corresponds to the opposite of literal meaning. Recently, Spotorno, Koun, Prado, van der Henst, and Noveck (2012) reviewed the neuroimaging work on irony and proposed a new methodological approach that overcomes the experimental problems in these studies, which show surprisingly little evidence for ToM network involvement in irony comprehension. In fact, although all of the studies reviewed found MPFC activation, none of them found TPJ activation, and only one reported precuneus activity. The main aim of Spotorno et al. (2012) was to uncover a direct, online link between language processing and ToM through neuroimaging, and they chose to test irony for the reasons stated previously. The authors compared participants' comprehension of brief stories in which a target sentence (e.g., "Tonight we gave a superb performance") was made either ironically or literally based on the context in which it was expressed (e.g., a terrible performance in the ironic condition and an impressive performance in the literal condition). They found that the ToM brain areas of the IPN model were active when participants understood verbal irony (Figure 54.2).

The key role that both the anterior (particularly the MPFC) and posterior (particularly the right TPJ) cortices play in understanding a speaker's intended meaning was recently confirmed in other pragmatic phenomena. For example, Prat, Mason, and Just (2012) demonstrated the involvement of these brain areas in understanding metaphors, and Bašnáková, Weber, Petersson, van Berkum, and Hagoort (2014) demonstrated their involvement in understanding indirect replies in spoken dialogue. In particular, Prat et al. (2012) manipulated figurativeness in metaphor comprehension (using three figurative conditions of increasing difficulty); they showed increased right TPJ and superior medial frontal activation for all figurative conditions compared with a literal condition. Bašnáková et al. (2014) demonstrated that compared with direct control utterances (in which the speaker's meaning was explicitly stated and corresponded to the literal meaning), indirect replies (in which the speaker's meaning was implicit and a pragmatic inference was necessary) engage a set of brain areas including mainly the MPFC and right TPJ.

Although neuroimaging evidence seems to support the IPN model as a neural substrate for the pragmatic comprehension of a speaker's intended meaning, it is important to note that neuroimaging findings are not a "smoking gun" when we want to ascribe a functional role to a specific brain area or brain network, such as the IPN. Neuroimaging techniques provide relevant information about the *involvement* of a brain area in a given task, but these methods are silent with respect to whether the brain structure is *necessary* for the task. Despite the inherent weaknesses of the neuropsychological approach, it continues to play an important role in complementing functional imaging techniques and in testing theoretical hypotheses about cognitive architecture (Rorden & Karnath, 2004).

A recent study by Cavallo et al. (2011) provided converging evidence for the role of the IPN in comprehending communicative intentions. The main aim of this study was to find neuropsychological evidence to complement neuroimaging findings that the MPFC, a key node of the IPN brain network, plays a crucial role in comprehending social situations in general and in comprehending communicative intentions in particular. To this end, they investigated how patients with amyotrophic lateral sclerosis (ALS) performed in an experimental protocol previously used in neuroimaging settings (Bara, Ciaramidaro, Walter, & Adenzato, 2011; Ciaramidaro et al., 2007; Walter et al., 2004, 2009). Although ALS was traditionally considered a neurodegenerative condition that exclusively affects the motor system with no cognitive repercussions, numerous studies have challenged this view demonstrating the presence of cognitive impairment in significant proportion of ALS patients (Abrahams, Leigh, & Goldstein, 2005). Accordingly, neuropathological investigations have shown the pathological involvement of prefrontal cortices (Maekawa et al., 2004). For this reason, the hypothesis driving the work by Cavallo et al. (2011) was that ALS patients should perform significantly worse than healthy controls on tasks requiring comprehension of social and communicative intentions, whereas performance should be comparable for patients and healthy controls on tasks that do not require comprehension of social intentions (i.e., private intentions). The results confirmed this hypothesis was in line with the following neuropsychological findings: (i) people with MPFC lesions show deficits in inferring speaker intentions (Lee et al., 2010) and (ii) people with neurodegenerative diseases that affect the functioning of the frontal cortex, such as the frontotemporal dementia, Tourette syndrome, and progressive supranuclear palsy (Eddy, Mitchell, Beck, Cavanna, & Rickards, 2010; Ghosh et al., 2012; Shany-Ur et al., 2012, for a review see Adenzato & Poletti, 2013; Poletti, Enrici, & Adenzato, 2012), show impaired comprehension of nonliteral language, such as sarcasm, metaphors, and indirect requests.

Other convergent evidence comes from patients with lexico-semantic impairments, such as in aphasia. These patients present profound language impairments with extensive damage to the traditional fronto-temporal language network, but without specific ToM deficits (see Willems & Varley, 2010, for a review). Accordingly, using alternative communicative resources, such as drawing, facial expression, and gesture, these patients are able to convey quite sophisticated messages (Siegal & Varley, 2006; Varley & Siegal, 2006). If language is crucially involved in communicative intention generation, then aphasic patients should not perform well on pragmatic tasks. However, patients with aphasia exhibited communication strategies that were comparable with those observed in the neurologically healthy population (Willems, Benn, Hagoort, Toni, & Varley, 2011).

54.3 COMMUNICATION IS MORE THAN LANGUAGE

To explain other people's communicative actions, communicative intention processing relies on different sources of information (i.e., linguistic or extralinguistic gestural means) to infer the underlying intended meaning. Viewed in this light, communication does not correspond with language competence, and lexico-semantic processing can be distinguished from communicative intention processing (Enrici, Adenzato, Cappa, Bara, & Tettamanti, 2011; Noordzij et al., 2009; Vicari & Adenzato, 2014; Willems et al., 2010; for a review see Willems & Varley, 2010). Thus, information acquired by different communicative modalities is equivalent from a mental processing standpoint, particularly when the actor's communicative intention has to be reconstructed.

Although different aspects of linguistic encoding (e.g., semantic and syntactic processes; see Bookheimer, 2002; Kaan & Swaab, 2002) and gestural encoding processes (Goldin-Meadow, 1999) have been widely investigated, few studies have focused on the level at which a communicative intention is reconstructed within a specific pragmatic context. Moreover, although studies have investigated the reciprocal influences and cross-modal interactions between speech and emblem gestures (where the prevailing view is that speech and gesture share a unified decoding process at the semantic level; see Willems & Hagoort, 2007 for a review), very few studies have addressed whether speech and gestures involve common processes at the pragmatic level, such as in communicative intention processing.

To answer this question, Enrici et al. (2011) performed an fMRI study that combined factorial and conjunction analyses to test two sets of predictions: (i) a common brain network is recruited for the comprehension of communicative intentions independently of the modality through which they are conveyed and (ii) additional brain areas are specifically recruited depending on the communicative modality, reflecting distinct sensorimotor gateways. Results clearly showed that the IPN is engaged in communicative intention processing, independently of the modality used, and that additional brain areas are specifically engaged by the particular communicative modality, for example, the peri-sylvian language network for the linguistic modality and the sensorimotor network for the extralinguistic gestural modality. Thus, communicative intention constitutes a shared pragmatic representation accessed by modality-specific gateways, which are distinct for linguistic versus extralinguistic expressive means.

Obviously, the IPN model does not exclude other levels of communicative processes (e.g., the lexico-semantic level) or other forms of shared representations (e.g., semantic representations). For example, Xu, Gannon, Emmorey, Smith, and Braun (2009) demonstrated that a shared semantic representation is accessed by distinct modality-specific sensory gateways by showing that communicative symbolic gestures and corresponding spoken glosses proposed by a single agent activated distinct modality-specific areas, but both engaged the left inferior frontal cortex and the posterior temporal cortex bilaterally to the same extent. The authors suggested that the peri-sylvian areas might function as a modality-independent semantic network linking *meaning* with *symbols* (both words and gestures). Similarly, we argue that the IPN might function as a modality-independent network, linking *shared meaning* with the actor's *intended meaning*.

Ferstl, Neumann, Bogler, and von Cramon (2008) have proposed an extended language network over and above the linguistic processes realized in left peri-sylvian areas. Their meta-analysis found that brain areas belonging to the IPN take part to an extend language network (i.e., a variety of additional cognitive processes recruited in language comprehension in written form). Even though we agree that language entails more than the traditional semantic, syntactic, and phonological processes, we are sympathetic with Willems and Varley (2010) who reported that incorporating ToM abilities into an extended language network is not a helpful conceptualization: "the separation of linguistic and communicative abilities seems a more fruitful characterization rather than calling both language, and allows for some forms of communication that are not linguistic" (Willems & Varley, 2010, p. 6).

Another important question concerns the roles of IPN brain areas and the mirror neuron system (MNS)—a system that allows action understanding without any explicit reflective mediation (Gallese, Keysers, & Rizzolatti, 2004)—in intention processing. Work to date suggests a dual-process model of the brain systems underlying action understanding and social cognition, where the MNS supports a bottom-up process for automatic behavior identification and the ToM system supports top-down processes for controlled social causal attribution (Keysers & Gazzola, 2007; Spunt & Lieberman, 2013). For example, in an interesting study, the kind of request the experimental subject had to respond to—namely *what*, *how*, or *why* an observed behavior was pursued (e.g., describing what an actor is doing, why he is doing it, or how he is doing it)—was manipulated. The results showed parametric variation in MNS and ToM network involvement in action identification (Spunt, Satpute, & Lieberman, 2011). These results suggest that the complex actions of others can be represented at multiple levels of an action hierarchy, and that ToM involvement varies parametrically: as the requested explanation for why an action is being performed becomes more plausible, the more ToM brain areas are recruited. Conversely, as the requested explanation for what or how an action is being performed becomes more plausible, the more MNS brain areas are recruited. Similarly, Van Overwalle (2009) and Van Overwalle and Baetens (2009) discussed which brain areas are responsible for understanding the actions of others and their underlying goals. These authors proposed a model in which actions and goals are organized hierarchically according to level of abstractness and that distinguishes between immediate goals that reflect the understanding of basic actions and long-term intentions that reflect the *why* of an action in a social context. The results of their meta-analysis provided additional evidence consistent with the role of the

IPN: although understanding basic actions requires the MNS, understanding social actions requires concurrent activation of the pSTS and TPJ areas as well as the precuneus and MPFC.

Different neuroimaging data support the assumption that both the MNS and IPN brain areas contribute to communicative comprehension particularly when extralinguistic communication ranging from imitation of social gesture (i.e., emblematic gesture; Mainieri, Heim, Straube, Binkofski, & Kircher, 2013) to hand sign recognition (Nakamura et al., 2004) is involved. Using magnetoencephalography, Nakamura et al. (2004) clearly showed that different systems are involved in symbolic (emblematic) hand sign recognition. These authors emphasized well-orchestrated regional electrical activity in multiple brain areas from the primary visual system to the object recognition system, as well as the MNS and IPN, indicating the involvement of a series of complex processes starting from visual gesture recognition and ending with inference of the underlying intended meaning. Different studies (Andric et al., 2013; Liew, Han, & Aziz-Zadeh, 2011; Noordzij et al., 2009, Schippers, Gazzola, Goebel, & Keysers, 2009; Spunt & Lieberman, 2013; Willems et al., 2010) found similar results suggesting that 1) the observation of communicative gestures relies on a combination of both MNS and IPN, 2) that there is a clear distinction between these systems on both neural and cognitive levels, and 3) that actual data do not support the view that the MNS alone provides the foundations for human communication.

Finally, Ciaramidaro, Becchio, Colle, Bara, and Walter (2014) investigated how mirror and ToM regions contribute to the implicit encoding of communicative intentions and whether activity in these regions is shaped and modulated by self-involvement. Their results revealed that the MPFC and bilateral premotor cortex are more active for second-person communicative intention (directed towards us) than for third-person communicative intention (directed toward others). Most importantly, they indicate that self-involvement may result in changes in functional connectivity between MNS and IPN regions.

In sum, data from recent neuroimaging research show that comprehending and generating communicative action is cognitively and neurally distinct from core linguistic processes (Willems & Varley, 2010). The information acquired by different communicative modalities is equivalent from a mental processing standpoint, particularly when the actor's communicative intention has to be reconstructed. Thus, communicative intention processing may build on different sources of information, such as language and gestures, to infer the underlying intended meaning.

54.4 COMMUNICATIVE EXCHANGE

Hamlet: Now, mother, what's the matter?
Queen Gertrude: Hamlet, thou hast thy father much offended.
Hamlet: Mother, you have my father much offended.

Hamlet, *W. Shakespeare, Act III, Scene IV.*

Communicative exchange involves constructing an acceptable interpretation of the reciprocal communication acts at all levels that participants consider significant. Although we may speak of a single agent when we refer to actions in general, we must always refer to at least one actor and a partner to whom the act is directed in the domain of communication. As well-known by Shakespeare, in this domain the meaning of a communicative exchange emerges from the mutual interplay of interactive agents embedded in their environment (Adenzato & Bucciarelli, 2008; Adenzato & Garbarini, 2006).

Very few neuroimaging studies have examined full communicative exchanges that include the action proposed by an actor and the concurrent reaction from a partner. Vice versa, most studies have focused on the comprehension of the early signals that convey an intention to communicate (Materna, Dicke, & Thier, 2008) (e.g., following someone's eye gaze or a pointing finger directed to the subject) or attempts to engage a communicative interaction (Kampe, Frith, & Frith, 2003) (e.g., looking directly at someone or calling their name). Moreover, other studies asked subjects to passively observe communicative actions (e.g., a gesture directed toward someone) performed by a single actor (Andric et al., 2013; Tylén, Allen, Hunter, & Roepstorff, 2012), such as placing a cup in front of them. Interestingly, in these studies, different areas of the IPN (such as MPFC or right TPJ)—but not necessarily the whole network—were recruited according to the kind of signal used to trigger a communicative interaction. It is important to underscore that within these experimental situations a communicative exchange between agents is merely put forward by the actor or character but does not actually take place.

Other studies focused on experimental tasks where the interaction between subjects consisted of different kinds of interaction by imitation, particularly gestural imitation (Nagy et al., 2010). For example, a recent study (Mainieri et al., 2013) required motor imitation of social (e.g., a "well-done" gesture with the thumbs up, or "wave a greeting"), nonsocial (e.g., gestures/mimes that depict common actions in the physical world, such as "take a note" or "cut something with scissors"), or meaningless gestures shown in short video clips with no speech. Neuroimaging results during both observation and execution (imitation) revealed activation in the core areas of the MNS, as well as in areas related to the IPN (e.g., middle frontal regions and TPJ), suggesting that imitating social gesture requires both MNS and IPN. According to Maineri and colleagues, the representation of communicative intentions might require a capacity mediated by the MPFC and related to ToM, although other networks, such as the MNS, might play an additional role in extracting motoric meaning.

Many studies have used interactive paradigms that represent a communicative interaction situation, for example, competitive games in which participants are directly involved in social interaction. McCabe, Houser, Ryan, Smith, and Trouard (2001) scanned participants while they played two-person decision-making games in which they could either cooperate or compete with human or computer opponents. Participants who cooperated with their opponent showed increased MPFC activity when playing against a human compared with the computer, a differentiation that was not seen when participants did not cooperate with their opponent. Gallagher, Jack, Roepstorff, and Frith (2002) used a version of the rock-paper-scissors game where participants played against three different opponents: a human competitor, a computer following a simple rule, and a computer making random choices. The human condition evoked activity in frontal regions different from both computer conditions, including the MPFC. Kircher et al. (2009) applied an adapted variation of the Prisoner's Dilemma Game, with subjects instructed to play either a putative human or computer partner (in actuality both were programmed to use a random sequence). Neuroimaging results showed MPFC and right TPJ activations in both conditions compared with a low-level baseline (relax condition), although activation was significantly stronger when participants thought they were playing with a human partner. Noordzij et al. (2009) required experimental subjects to engage in a communicative interaction using the tacit communication game, where two players, a sender and receiver, move a token on a game board displayed on a computer monitor. Participants communicated the position and orientation of a visual token to another individual using limited visuo-spatial means. Neuroimaging data were collected for both sender and receiver. Interestingly, results showed that activation in right TPJ increased when the sender designed a communicative act for another person, as well as when the receiver interpreted the communicative act; moreover, TPJ was more co-activated with MPFC compared with a control condition (asking the sender to completely ignore the receiver's token). Finally, Schippers et al. (2009) required experimental subjects to play Charades. In this game, participants were presented with a word on the screen and instructed to communicate this word to his or her partner using gestures. Activation in a combination of MNS and ToM brain areas was found.

Lastly very few studies have used real interactive communicative situations in which a full communicative exchange was presented. Calarge, Andreasen, and O'Leary (2003) proposed an imaginative communicative scenario and asked participants to invent and say aloud stories describing imaginary encounters with strangers. The results showed that compared with a control condition in which participants read stories requiring no mental state attribution aloud, the imaginative communicative scenario was associated with brain activations quite similar to those of the IPN. The fMRI work by Sassa et al. (2007) was the first to use a linguistic interaction paradigm, even though the interaction was one-sided. The authors used short video clips of daily actions in which an actor who was using a tool or handling an object (e.g., playing guitar) glanced at the camera. Participants responded to these movie clips in two conditions as follows: in the communicative condition, they talked to the person on the screen in a casual communicative manner; and in the descriptive condition, they verbally described the actor's situation. It is important to note that both conditions involved speech production, but there was only specific and explicit communicative intent directed to the actor in the communicative condition. Comparison of communicative and descriptive trials showed increased activation in MPFC, and bilateral TPJ and temporal poles in the communicative condition (Figure 54.3). Moreover, MPFC and precuneus activations were stronger for video clips in which the actor was a close friend of the subject compared with clips in which the actor was unfamiliar. Finally, compared with baseline (where a picture of a tool or object that appeared in the action video clips was presented on a moving mosaic), both experimental conditions led to activation in parts of the traditional language production network (i.e., the left inferior frontal gyrus) but, interestingly, these regions were not sensitive to the communicative/descriptive manipulation.

To the best of our knowledge, the study by Willems et al. (2010) is the only study that considered a full communicative exchange in which both the communicative action proposed by an actor and the concurrent reaction from a partner are considered. Participants were engaged in a communicative interaction through an interactive game (Taboo). The subject inside the MR scanner generated verbal descriptions of target words without using predetermined "taboo" words, whereas the other player outside the MR scanner listened to these descriptions and guessed the target word. Interestingly, manipulating communicative intent, that is, changing whether the listener already knew the target word or not, and linguistic difficulty, that is, modifying the semantic relationship between the Taboo words and the target word, influenced activation

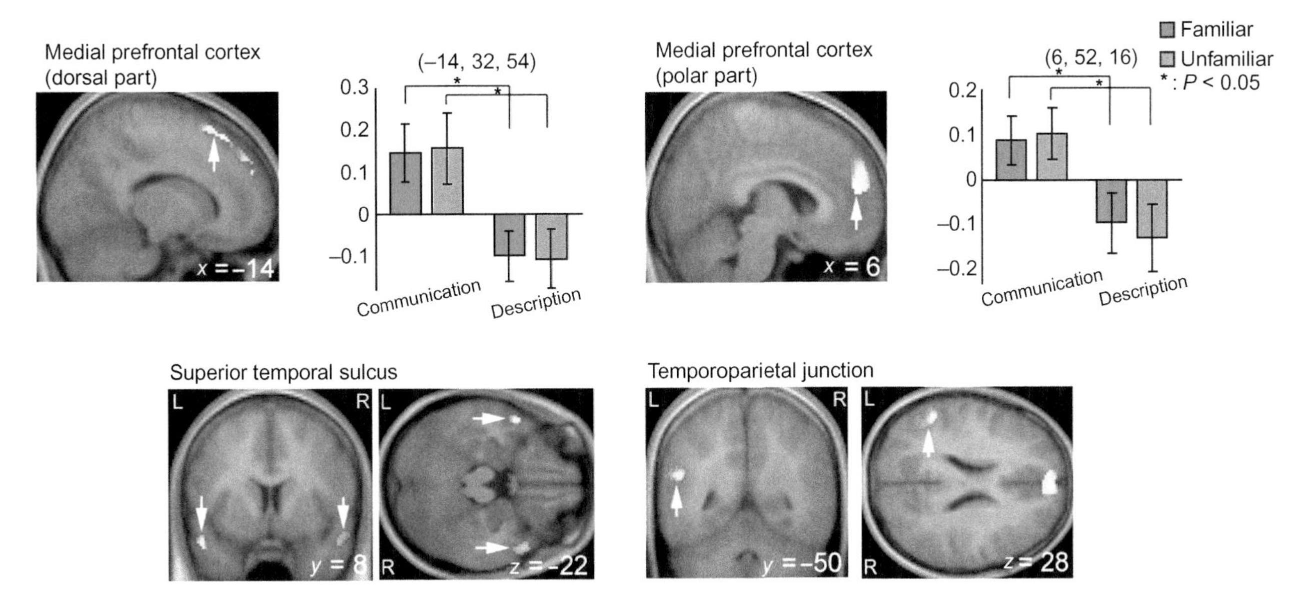

FIGURE 54.3 Regions of interest of the main areas with significantly greater activity during the communication relative to the description condition. *From Sassa et al. (2007) with permission from the publisher.*

patterns in different brain regions. Although MPFC activation was sensitive to the communicative intention manipulation, irrespective of linguistic difficulty, activation of the left inferior frontal cortex, a key region of language network, was sensitive to manipulations of linguistic demands but not communicative intention. In accordance with the aim of this work, this study provides an interesting neural evidence for a dissociation between communicative message generation and lexico-semantic language processes.

Taken together, we argue that these results reflect an experimental setting where a real interactive communicative situation occurred. Considering the complexity of the pragmatic dimension being investigated—communicative exchange between at least two people using different expressive modalities—we think that the more ecologically valid the study, the more pragmatic ability that is recruited. Hence, we argue that the most interesting findings in terms of the neurobiological bases of communicative comprehension competence will come from a more careful analysis of and ecological approach to the experimental setting.

54.5 STEPS TOWARD AN ECOLOGY OF COMMUNICATION

Communicative intentions are a core feature of any pragmatic phenomenon and, as mental states, they are irreducibly subjective entities (Searle, 1992). The intrinsic subjective nature of communicative intentions has relevant implications for the study of the neurobiology of pragmatics: the only judge of the intentionality of a behavioral act is the actor who performs that act; sometimes the observer has no way of determining whether an act was intentional and, thus, communicative or not. Searle (1983) distinguished between prior intention and intention-in-action: prior intention represents the aim of the action, whereas intention-in-action is the cause of bodily movement. Bearing this difference in mind, we note that in typical ToM experimental paradigms the causal order of prior and motor intention is reversed from a first-person to a third-person perspective. A prior intention logically and temporally precedes the motor intention in the first-person perspective (e.g., in Dante's Inferno, Minos intends to send the damned to their circle of Hell and, hence, girds himself with his tail), whereas prior intentions in the third-person perspective are inferred after an action has been observed (e.g., based on the fact that Minos girds himself with his tail, Dante infers that Minos intends to send the damned to their proper circle).

These considerations require overcoming the artificial separation between speaker/hearer or actor/partner to move toward a communal construction of meaning (Chatel-Goldman, Schwartz, Jutten, & Congedo, 2013; Konvalinka & Roepstorff, 2012; Pickering & Garrod, 2004). The first revolution in the study of the neurobiology of pragmatics should involve devising experimental paradigms in which the agent involved is constantly changing roles, and where the perspective continuously switches between first person (I am saying something to you) and second person (you are saying something to me). The second ecological revolution will be the study of normal conversation, where each participant voluntarily takes his/her turn and freely expresses goals through cooperative sharing of reciprocal mental states; only in this way will researchers gain access to the capacity for generating and comprehending effective communicative intentions. A technology that is able to monitor the brains of two agents online while they spontaneously interact is necessary for pragmatics concerned with ecological contexts.

Acknowledgments

This work was supported by a MIUR grant (FIRB 2012, protocol number: RBFR12F0BD) and by the University of Torino (Ricerca scientifica finanziata dall'Università "Cognizione sociale e attaccamento in popolazioni cliniche e non cliniche").

References

Abrahams, S., Leigh, P. N., & Goldstein, L. H. (2005). Cognitive change in ALS: A prospective study. *Neurology, 64*, 1222–1226.

Adenzato, M., & Bucciarelli, M. (2008). Recognition of mistakes and deceits in communicative interactions. *Journal of Pragmatics, 40*, 608–629.

Adenzato, M., & Garbarini, F. (2006). The As if in cognitive science, neuroscience and anthropology: A journey among robots, blacksmiths, and neurons. *Theory & Psychology, 16*, 747–759.

Adenzato, M., & Poletti, M. (2013). Theory of mind abilities in neurodegenerative diseases: An update and a call to introduce mentalizing tasks in standard neuropsychological assessments. *Clinical Neuropsychiatry, 10*, 226–234.

Airenti, G., Bara, B. G., & Colombetti, M. (1993). Conversation and behavior games in the pragmatics of dialogue. *Cognitive Science, 17*, 197–256.

Andric, M., Solodkin, A., Buccino, G., Goldin-Meadow, S., Rizzolatti, G., & Small, S. L. (2013). Brain function overlaps when people observe emblems, speech, and grasping. *Neuropsychologia, 51*, 1619–1629.

Austin, J. L. (1962). *How to do things with words*. Oxford: Oxford University Press.

Bara, B. G. (2010). *Cognitive pragmatics*. Cambridge, MA: MIT Press.

Bara, B. G. (2011). Cognitive pragmatics: The mental processes of communication. *Intercultural Pragmatics, 8*, 443–485.

Bara, B. G., Ciaramidaro, A., Walter, H., & Adenzato, M. (2011). Intentional minds: A philosophical analysis of intention tested through fMRI experiments involving people with schizophrenia, people with autism, and healthy individuals. *Frontiers in Human Neuroscience, 5*(7), 1–11. Available from: http://dx.doi.org/10.3389/fnhum.2011.00007.

Baron-Cohen, S. (1995). *Mindblindness: An essay on autism and Theory of Mind*. Cambridge, MA: MIT Press.

Bašnáková, J., Weber, K., Petersson, K. M., van Berkum, J., & Hagoort, P. (2014). Beyond the language given: The neural correlates of inferring speaker meaning. *Cerebral Cortex*. Available from: http://dx.doi.org/10.1093/cercor/bht112.

Bookheimer, S. (2002). Functional MRI of language: New approaches to understanding the cortical organization of semantic processing. *Annual Review of Neuroscience, 25*, 151–188.

Calarge, C., Andreasen, N. C., & O'Leary, D. S. (2003). Visualizing how one brain understands another: A PET study of Theory of Mind. *American Journal of Psychiatry, 160*, 1954–1964.

Cavallo, M., Adenzato, M., MacPherson, S. E., Karwig, G., Enrici, I., & Abrahams, S. (2011). Evidence of social understanding impairment in patients with Amyotrophic Lateral Sclerosis. *PLoS One, 6*(10), e25948.

Chatel-Goldman, J., Schwartz, J. L., Jutten, C., & Congedo, M. (2013). Non-local mind from the perspective of social cognition. *Frontiers in Human Neuroscience, 2*(7), 107. Available from: http://dx.doi.org/10.3389/fnhum.2013.00107.

Ciaramidaro, A., Adenzato, M., Enrici, I., Erk, S., Pia, L., Bara, B. G., et al. (2007). The intentional network: How the brain reads varieties of intentions. *Neuropsychologia, 45*, 3105–3113.

Ciaramidaro, A., Becchio, C., Colle, L., Bara, B. G., & Walter, H. (2014). Do you mean me? Communicative intentions recruit the mirror and the mentalizing system. *Social Cognitive and Affective Neuroscience, 9*, 909–916.

Clark, H. H. (1996). *Using language*. Cambridge, MA: Cambridge University Press.

Eddy, C. M., Mitchell, I. J., Beck, S. R., Cavanna, A. E., & Rickards, H. E. (2010). Impaired comprehension of nonliteral language in Tourette syndrome. *Cognitive and Behavioral Neurology, 23*, 178–184.

Enrici, I., Adenzato, M., Cappa, S., Bara, B. G., & Tettamanti, M. (2011). Intention processing in communication: A common brain network for language and gestures. *Journal of Cognitive Neuroscience, 23*, 2415–2431.

Ferstl, E. C., Neumann, J., Bogler, C., & von Cramon, D. Y. (2008). The extended language network: A meta-analysis of neuroimaging studies on text comprehension. *Human Brain Mapping, 29*, 581–593.

Gallagher, H. L., Jack, A. I., Roepstorff, A., & Frith, C. D. (2002). Imaging the intentional stance in a competitive game. *Neuroimage, 16*, 814–821.

Gallese, V., Keysers, C., & Rizzolatti, G. (2004). A unifying view of the basis of social cognition. *Trends in Cognitive Sciences, 8*, 396–403.

Ghosh, B. C. P., Calder, A. J., Peers, P. V., Lawrence, A. D., Acosta-Cabronero, J., Pereira, J. M., et al. (2012). Social cognitive deficits and their neural correlates in progressive supranuclear palsy. *Brain, 135*, 2089–2102.

Gobbini, M. I., Koralek, A. C., Bryan, R. E., Montgomery, K. J., & Haxby, J. V. (2007). Two takes on the social brain: A comparison of Theory of Mind tasks. *Journal of Cognitive Neuroscience, 19*, 1803–1814.

Goldin-Meadow, S. (1999). The role of gesture in communication and thinking. *Trends in Cognitive Sciences, 3*, 419–429.

Grice, H. P. (1957). Meaning. *Philosophical Review, 67*, 377–388.

Grice, H. P. (1975). Logic and conversation. In P. Cole, & J. L. Morgan (Eds.), *Syntax and semantics: Speech acts* (pp. 41–58). New York, NY: Academic Press.

Hauser, M. D. (1996). *The evolution of communication*. Cambridge, MA: MIT Press.

Kaan, E., & Swaab, T. Y. (2002). The brain circuitry of syntactic comprehension. *Trends in Cognitive Sciences, 6*, 350–356.

Kampe, K. K. W., Frith, C. D., & Frith, U. (2003). "Hey John": Signals conveying communicative intention toward the self activate brain regions associated with "mentalizing," regardless of modality. *Journal of Neuroscience, 23*, 5258–5263.

Keysers, C., & Gazzola, V. (2007). Integrating simulation and Theory of Mind: From self to social cognition. *Trends in Cognitive Sciences, 11*, 194–196.

Kircher, T., Straube, B., Leube, D., Weis, S., Sachs, O., Willmes, K., et al. (2009). Neural interaction of speech and gesture: Differential activations of metaphoric co-verbal gestures. *Neuropsychologia, 47*, 169–179.

Konvalinka, I., & Roepstorff, A. (2012). The two-brain approach: How can mutually interacting brains teach us something about social interaction? *Frontiers in Human Neuroscience, 6*, 215. Available from: http://dx.doi.org/10.3389/fnhum.2012.00215.

Lee, T. M. C., Ip, A. K. Y., Wang, K., Xi, C.-H., Hu, P.-P., Mak, H. K., et al. (2010). Faux pas deficits in people with medial frontal lesions as related to impaired understanding of a speaker's mental state. *Neuropsychologia, 48*, 1670–1676.

Liew, S. L., Han, S., & Aziz-Zadeh, L. (2011). Familiarity modulates mirror neuron and mentalizing regions during intention understanding. *Human Brain Mapping, 32*, 1986–1997.
Maekawa, S., Al-Sarraj, S., Kibble, M., Landau, S., Parnavelas, J., Cotter, D., et al. (2004). Cortical selective vulnerability in motor neuron disease: A morphometric study. *Brain, 127*, 1237–1251.
Mainieri, A. G., Heim, S., Straube, B., Binkofski, F., & Kircher, T. (2013). Differential role of the mentalizing and the mirror neuron system in the imitation of communicative gestures. *Neuroimage, 81*, 294–305.
Materna, S., Dicke, P. W., & Thier, P. (2008). The posterior superior temporal sulcus is involved in social communication not specific for the eyes. *Neuropsychologia, 46*, 2759–2765.
McCabe, K., Houser, D., Ryan, L., Smith, V., & Trouard, T. (2001). A functional imaging study of cooperation in two-person reciprocal exchange. *Proceedings of the National Academy of Sciences of the United States of America, 98*, 11832–11835.
Nagy, E., Liotti, M., Brown, S., Waiter, G., Bromiley, A., Trevarthen., et al. (2010). The neural mechanisms of reciprocal communication. *Brain Research, 1353*, 159–167.
Nakamura, A., Maess, B., Knösche, T. R., Gunter, T. C., Bach, P., & Friederici, A. D. (2004). Cooperation of different neuronal systems during hand sign recognition. *Neuroimage, 23*, 25–34.
Noordzij, M. L., Newman-Norlund, S. E., de Ruiter, J. P., Hagoort, P., Levinson, S. C., & Toni, I. (2009). Brain mechanisms underlying human communication. *Frontiers in Human Neuroscience, 3*, 14. Available from: http://dx.doi.org/doi:10.3389/neuro.09.014.2009.
Pickering, M. J., & Garrod, S. (2004). Toward a mechanistic psychology of dialogue. *Behavioral and Brain Sciences, 27*, 169–226.
Poletti, M., Enrici, I., & Adenzato, M. (2012). Cognitive and affective Theory of Mind in neurodegenerative diseases: Neuropsychological, neuroanatomical and neurochemical levels. *Neuroscience and Biobehavioral Reviews, 36*, 2147–2164.
Prat, C. S., Mason, R. A., & Just, M. A. (2012). An fMRI investigation of analogical mapping in metaphor comprehension: The influence of context and individual cognitive capacities on processing demands. *Journal of Experimental Psychology: Learning, Memory, and Cognition, 38*, 282–294.
Premack, D., & Woodruff, G. (1978). Does the chimpanzee have a Theory of Mind? *Behaviour & Brain Science, 1*, 515–526.
Rorden, C., & Karnath, H. O. (2004). Using human brain lesions to infer function: A relic from a past era in the fMRI age? *Nature Reviews Neuroscience, 5*, 813–819.
Sassa, Y., Sugiura, M., Jeong, H., Horie, K., Sato, S., & Kawashima, R. (2007). Cortical mechanism of communicative speech production. *Neuroimage, 37*, 985–992.
Saxe, R. (2006). Uniquely human social cognition. *Current Opinion in Neurobiology, 16*, 235–239.
Schiffer, S. R. (1972). *Meaning*. Oxford: Oxford University Press.
Schippers, M. B., Gazzola, V., Goebel, R., & Keysers, C. (2009). Playing charades in the fMRI: Are mirror and/or mentalizing areas involved in gestural communication? *PLoS One, 4*, e6801.
Searle, J. R. (1983). *Intentionality*. Cambridge, MA: Cambridge University Press.
Searle, J. R. (1992). *The rediscovery of mind*. Cambridge, MA: MIT Press.
Shany-Ur, T., Poorzand, P., Grossman, S. N., Growdon, M. E., Jang, J. Y., Ketelle, R. S., et al. (2012). Comprehension of insincere communications in neurodegenerative diseases: Lies, sarcasm and Theory of Mind. *Cortex, 48*, 1329–1341.
Siegal, M., & Varley, R. (2006). Aphasia, language, and Theory of Mind. *Social Neuroscience, 1*, 167–174.
Sperber, D., & Wilson, D. (1986). *Relevance. Communication and cognition* (2nd ed., 1995). Oxford: Blackwell.
Spotorno, N., Koun, E., Prado, J., van der Henst, J.-B., & Noveck, I. A. (2012). Neural evidence that utterance-processing entails mentalizing: The case of irony. *Neuroimage, 63*, 25–39.
Spunt, R. P., & Lieberman, M. D. (2013). The busy social brain: Evidence for automaticity and control in the neural systems supporting social cognition and action understanding. *Psychological Science, 24*, 80–86.
Spunt, R. P., Satpute, A. B, & Lieberman, M. D. (2011). Identifying the what, why, and how of an observed action: An fMRI study of mentalizing and mechanizing during action observation. *Journal of Cognitive Neuroscience, 23*, 63–74.
Strawson, P. (1964). Intention and convention in speech acts. *Philosophical Review, 73*, 439–469.
Tirassa, M. (1999). Communicative competence and the architecture of the mind/brain. *Brain and Language, 68*, 419–441.
Tylén, K., Allen, M., Hunter, B. K., & Roepstorff, A. (2012). Interaction vs. observation: Distinctive modes of social cognition in human brain and behavior? A combined fMRI and eye-tracking study. *Frontiers in Human Neuroscience, 6*, 331. Available from: http://dx.doi.org/10.3389/fnhum.2012.00331.
Van Overwalle, F. (2009). Social cognition and the brain: A meta-analysis. *Human Brain Mapping, 30*, 829–858.
Van Overwalle, F., & Baetens, K. (2009). Understanding others' actions and goals by mirror and mentalizing systems: A meta-analysis. *Neuroimage, 48*, 564–584.
Varley, R., & Siegal, M. (2006). Evidence for cognition without grammar from causal reasoning and 'Theory of Mind' in an agrammatic aphasic patient. *Current Biology, 10*, 723–726.
Vicari, G., & Adenzato, M. (2014). Is recursion language-specific? Evidence of recursive mechanisms in the structure of intentional action. *Consciousness and Cognition, 26*, 169–188.
Walter, H., Adenzato, M., Ciaramidaro, A., Enrici, I., Pia, L., & Bara, B. G. (2004). Understanding intentions in social interactions: The role of the anterior parancingulate cortex. *Journal of Cognitive Neuroscience, 16*, 1854–1863.
Walter, H., Ciaramidaro, A., Adenzato, M., Vasic, N., Ardito, R. B., Erk, S., et al. (2009). Dysfunction of the social brain in schizophrenia is modulated by intention type: An fMRI study. *Social Cognitive and Affective Neuroscience, 4*, 166–176.
Watzlawick, P., Bevin, J., & Jackson, D. (1967). *Pragmatics of human communication*. New York, NY: Norton.
Willems, R. M., Benn, Y., Hagoort, P., Toni, I., & Varley, R. (2011). Communicating without a functioning language system: Implications for the role of language in mentalizing. *Neuropsychologia, 49*, 3130–3135.
Willems, R. M., de Boer, M., de Ruiter, J. P., Noordzij, M. L., Hagoort, P., & Toni, I. (2010). A cerebral dissociation between linguistic and communicative abilities in humans. *Psychological Science, 21*, 8–14.
Willems, R. M., & Hagoort, P. (2007). Neural evidence for the interplay between language, gesture, and action: A review. *Brain and Language, 101*, 278–289.
Willems, R. M., & Varley, R. (2010). Neural insights into the relation between language and communication. *Frontiers in Human Neuroscience, 25*, 4. Available from: http://dx.doi.org/10.3389/fnhum.2010.00203.
Xu, J., Gannon, P. J., Emmorey, K., Smith, J. F., & Braun, A. R. (2009). Symbolic gestures and spoken language are processed by a common neural system. *Proceedings of the National Academy of Sciences of the United States of America, 106*, 20664–20669.

S E C T I O N J

SPEAKING

CHAPTER

55

Neurobiology of Speech Production: Perspective from Neuropsychology and Neurolinguistics

Sheila E. Blumstein[1] *and Shari R. Baum*[2]

[1]Department of Cognitive Linguistic and Psychological Sciences, Brown University and the Brown Institute for Brain Sciences, Providence, RI, USA; [2]School of Communication Sciences and Disorders and Centre for Research on Brain, Language & Music, McGill University, Montréal, QC, Canada

55.1 INTRODUCTION

For most of us, the production of speech appears to be effortless. And, yet, it is a multifaceted and complex process that requires access to different components of the language processing system. Although articulation of a word (i.e., patterns of articulatory movements of the vocal tract) may be the endpoint, the process begins with the conceptualization of a word, the selection of its lexical representation and corresponding sound properties from the mental lexicon, and the mapping of these phonological representations onto articulatory planning and implementation stages. Additionally, the speaker does not simply passively articulate speech, but rather he or she constantly monitors it on a moment-to-moment basis to adapt his/her output in either the short-term or the long-term based on both auditory and somatosensory feedback received from the production itself.

Although the details may vary, current models of the functional architecture of speech production typically identify these different aspects of speech production in terms of different stages of production. These stages are considered to be functionally distinct and hierarchically organized, with both feedforward and feedback mechanisms at each stage of processing (Figure 55.1). It is the goal of this chapter to examine current data and models of the speech production process, drawing from both neuropsychological and neurophysiological studies. To this end, we first provide a brief historical perspective drawn from the aphasia literature that has served as a framework for much of the neurophysiological research. We then consider the proposed stages of production outlined, first reviewing phonological stages of production and then phonetic/articulatory stages of production, integrating results based on lesion and neurophysiological studies. Together, they indicate that speech production recruits a neural network that encompasses the language network, including temporal, parietal, and frontal structures.

55.2 HISTORICAL PERSPECTIVE: SPEECH PRODUCTION DEFICITS IN APHASIA

Early research from the 1970s and 1980s suggested that the patterns of production in patients with aphasia reflected different stages of the speech production process. Patients with lesions to frontal structures displayed articulatory planning and implementation deficits, and patients with lesions to posterior structures displayed phonological deficits. One of the presenting clinical features long recognized in defining the aphasia syndromes is the nature of speech production output (Goodglass & Kaplan, 1972) (for discussion of clinical syndromes of aphasia, see Chapter 73). In particular, patients presenting with Broca's aphasia or apraxia of speech typically have nonfluent speech output characterized by distorted speech production, sound errors of various types, and a diminution of the prosodic range of speech. In contrast,

 DOI: http://dx.doi.org/10.1016/B978-0-12-407794-2.00055-9

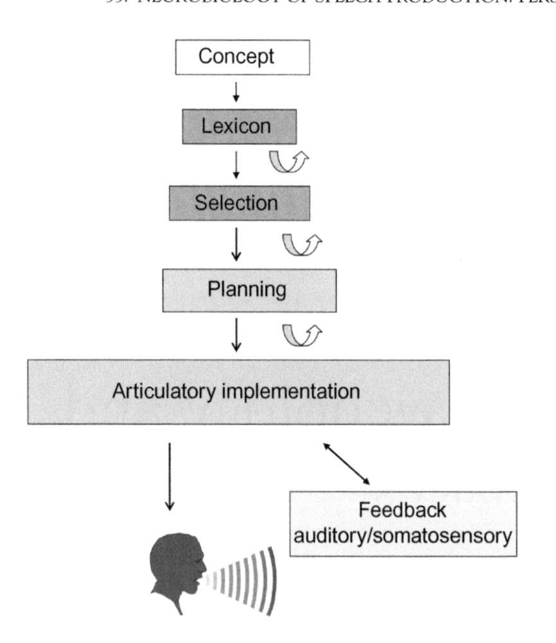

FIGURE 55.1 Functional architecture of stages of speech production processes. Colors in pink indicate phonological processes; those in blue indicate phonetic processes.

patients presenting with Wernicke's and conduction aphasia show fluent, well-articulated speech output. Nonetheless, these patients, and particularly those clinically diagnosed with conduction aphasia, make speech production errors characterized by sound substitutions and the misordering and transposition of sounds within and across words.

Studies examining the acoustic patterns of speech production of these patients largely supported this dichotomy. Acoustic measures of a number of parameters of speech, including voice-onset time (VOT) in stop consonants (Blumstein, Cooper, Goodglass, Statlender, & Gottlieb, 1980; Gandour & Dardarananda, 1984; Kent & Rosenbek, 1983), amplitude of voicing in fricative consonants (Kurowski, Hazen, & Blumstein, 2003), murmur duration, amplitude of the first harmonic at consonant release in nasal consonants (Kurowski, Blumstein, Palumbo, Waldstein, & Burton, 2007), and temporal parameters of consonant and vowel production (Baum, Blumstein, Naeser, & Palumbo, 1990) showed that patients with frontal lesions displayed deficits. They also displayed deficits in laryngeal control as shown by impairments in amplitude properties of nasals, fricatives, and prosody. In contrast, patients with posterior lesions showed normal patterns of production in the analysis of these parameters.

Although this dichotomy between phonological and phonetic stages of speech production is supported by more recent neuropsychological and neurophysiological studies, early on, studies of speech production in aphasia indicated that the picture was more complicated. These studies showed that in addition to phonetic (articulatory) deficits, those with Broca's aphasia and patients with apraxia of speech produced phonological errors similar to those made by those with fluent aphasia with posterior lesions (Blumstein, 1973; Canter, Trost, & Burns, 1985; Haley, Jacks, & Cunningham, 2013). For example, the patterns of phoneme substitution errors (e.g., pipe → bipe) made by those with Broca's, conduction, and Wernicke's aphasia were similar across groups, as were the distributions of different error types such as phoneme substitutions (e.g., teams → keams), deletions (e.g., brown → /bawn/), additions (e.g., papa → /paprə/), and transpositions (e.g., degree → /gədri/). Acoustic analysis revealed that both Wernicke's and conduction aphasics showed increased variability in the implementation of a number of acoustic parameters of speech, despite the fact that they showed normal patterns of articulatory implementation (Baum et al., 1990; Ryalls, 1986; Vijayan & Gandour, 1995). Taken together, these findings indicate that speech production recruits an integrated functional and neural system, one in which damage to one area has consequences throughout the system affecting stages of production both upstream and downstream from it.

55.3 PHONOLOGICAL PROCESSES IN SPEECH PRODUCTION

There is a rich literature that has examined the representations and processes involved in accessing words from the lexicon in spoken word production. Results from this research generally support the view that words are represented in terms of abstract, context-independent, phonological (phonemic and/or syllable-sized) units and that the mental lexicon and access to it is influenced by the phonological properties of individual words and the network of word representations that share structural properties with the word candidate (Dell, 1986; Gaskell & Marslen-Wilson, 1999). Information flow throughout the speech production system is considered to be interactive. Activation of a lexical candidate activates its associated phonological units, and these phonological units, in turn, boost the activation of the word candidate as well as those shared phonological units of other words in the lexicon. With this view, the selection of the phonological representation of a word is modulated by the number of phonological "neighbors" a word may have (Dell & Gordon, 2003), and this has a cascading effect on its articulatory implementation (Baese-Berk & Goldrick, 2009; Goldrick & Blumstein, 2006).

Evidence from aphasia has generally supported this view. Although these studies provide evidence in support of many of the theoretical claims about the functional architecture of spoken word production, most fail to provide information about or focus on the underlying neural substrates. Additionally, there have been surprisingly few functional neuroimaging studies examining these issues. With this limitation in mind, we now turn to a brief review of the extant neuropsychological and neurophysiological findings examining phonological processes in speech production.

55.3.1 Nature of Representations

Similar to individuals without brain damage, patients with aphasia show sensitivity to the sound structure properties of the lexicon in spoken word production. In a study of the production of speech errors in picture description and picture naming tasks in 43 aphasic patients (unselected for syndrome or lesion), Gordon (2002) showed that there were fewer errors produced for word targets that had many lexical neighbors than words that did not. These facilitative effects occur presumably because words from dense neighborhoods share phonological structure with the target, and the phonological similarities among the target word and its neighbors provide a further boost to the activation of the lexical target. That the effect is driven by lexical factors is further supported by the fact that errors were more likely to result in words (of similar sound shape) than in nonwords (see also Laganaro, Chetelat-Mabillard, & Frauenfelder, 2013).

As indicated, the general consensus in the spoken word production literature is that words are represented in terms of abstract phonological units. These units may include both sound segments and properties of sound segments (i.e., phonological/phonetic features). Evidence suggesting that these units are segmental comes from detailed analyses of the patterns of phonological errors produced by aphasic patients in spontaneous speech as well as in more controlled tasks such as repetition and naming (Blumstein, 1973; Lecours & Lhermitte, 1969). Results show that phoneme substitutions, the most common error produced by those with Broca's, conduction, and Wernicke's aphasia involve the replacement of a single phoneme by another; whether this happens within a word, such as teams → /kimz/, is the result of misordering of sounds within a word, such as "degrees" → /gədriz/, or is a sound substitution between words, such as roast beef → /rof bif/.

Additionally, the pattern of substitution errors suggests that phonological features of sounds are fundamental representational units. In particular, the large majority of substitution errors reflect the replacement of a single phonetic feature, with very few errors occurring between sounds distinguished by two or more features. For example, there are more voicing, place of articulation, or manner errors than combinations of these features. These findings are consistent with early work by Jakobson, Halle, and Fant (1967), who proposed that a finite set of distinctive features can be used to characterize the phonological systems of language and the acquisition or dissolution of language (Jakobson, 1968).

Although it may be the case that phonological units may be abstract and are represented independent of the phonetic context in which they occur, there must be a processing stage where the abstract sound shape is coded into fine "episodic" details that take into account the context in which the sounds appear. These details are ultimately used for motor planning and execution. Evidence from aphasia supports this two-stage process (Blumstein, 1973).

Consider the words "pill" and "spill" in English. The abstract representation of the /p/ in these two words is the same. However, their phonetic realization differs; /p/ in initial position in English is aspirated and has a long VOT (i.e., [p^h]), whereas /p/, following /s/ is unaspirated and has a short VOT (i.e., [p]). It is of interest to determine what happens when an aphasic patient simplifies a consonant cluster, particularly what happens when a patient deletes the /s/ in a word like "spill." Will the resultant production of the /p/ be produced with an aspirated and long VOT, consistent with the notion that the deletion of the /s/ occurred during phonological processing (i.e., before motor plans for a cluster are implemented), or will it be produced with an unaspirated and short VOT, consistent with the notion that the production reflected the contextually determined phonetic realization of the cluster /sp/ followed by the deletion of the initial fricative? A study by Buchwald and Miozzo (2011) measured the VOT productions of two aphasic patients who deleted /s/ in /sp/, /st/, and /sk/ clusters and compared these with the phonetic realization of correctly produced stop consonants in both singleton and cluster environments (i.e., "pill" and "spill"). Results showed two different patterns of production, with one patient producing the initial stop consonant of the word with a long VOT (similar to the realization of /p/ in "pill") and the other producing it with a short VOT (similar to the realization of /p/ in "spill"). These findings suggest that the errors of the former patient were phonologically based and the errors of the latter patient were phonetically based. Similar findings were shown examining duration properties of nasal consonants when deleted in /sn/ and /sm/ clusters (Buchwald & Miozzo, 2012).

55.3.2 Cascading Activation

During the stages of speech production, information flows from the more abstract properties of speech to more detailed phonetic representations required for articulatory planning and articulation. Recent research with normal individuals has suggested that there is a cascading flow of information with phonological properties of words influencing their fine-grained phonetic realization (Goldrick, 2006). For example, it has been shown that the implementation of voicing in stop consonants is influenced by whether the words have a voicing minimal pair (Baese-Berk & Goldrick, 2009; Goldrick & Blumstein, 2006). There is a longer VOT for initial voiceless stop consonants in words having a voiced minimal pair (e.g., "tart" versus "dart") compared with words that do not (e.g., tar versus *dar).

A recent neuroimaging study (Peramunage, Blumstein, Myers, Goldrick, & Baese-Berk, 2011) examined the neural substrates of this lexically conditioned phonetic effect. While in the scanner, participants read words with initial voiceless stop consonants that either had or did not have a voiced minimal pair. None of the voiced minimal pairs were included in the set of words produced by the participants. Filler words beginning with a variety of consonants were also included. Acoustic analysis of the productions replicated the behavioral effects showing longer VOTs for voiceless stop consonants in words that had a minimal pair compared with words that did not. fMRI results revealed reduced activation for minimal pair words compared with nonminimal words in a neural network, including the left posterior superior temporal gyrus (STGp), supramarginal gyrus (SMG), inferior frontal gyrus, and precentral gyrus.

These findings suggest that lexically driven differences in the activation of phonological representations in temporoparietal areas (posterior superior temporal and supramarginal gyri) modulate subsequent articulatory processes in frontal areas (the inferior frontal and precentral gyri). Moreover, they are consistent with the view that reduced activation for minimal pair words is a facilitatory process reflecting the overlap in shared phonological properties between the target word and its minimal pair (MP) neighbor. Of interest, a subsequent study (Bullock-Rest et al., 2013) examining the effects of lesions to the SMG extending into the STGp or to the inferior frontal gyrus on the production of minimal pair and nonminimal words using the Peramunage et al. (2011) stimuli showed that both groups of patients demonstrated the minimal pair effect. Of importance, this effect emerged for correct productions, suggesting that despite damage to the network, Broca's and conduction aphasic participants were still sensitive to the lexical properties of the target words and this, in turn, affected their phonetic output. Nonetheless, both groups showed a pattern of *errors* consistent with the view that they had a deficit in the spoken word production system. They made more errors on nonminimal pair words than minimal pair words and made phoneme substitution errors on the target words that were largely single-feature substitutions, and when such errors occurred, the production was more likely to be a word than a nonword.

Perhaps more compelling evidence comes from a recent study examining the origin of phonemic paraphasias (i.e., phoneme substitution errors) in aphasic patients (Kurowski & Blumstein, in preparation). The question asked here was whether phonemic substitution errors (e.g., [s] → [z]) reflect a planning error, where the wrong phoneme is selected and ultimately implemented correctly, or whether it reflects the simultaneous selection and activation of a mis-selected phoneme and the original target phoneme. In this latter case, the resulting production, although perceived by the listener as a phonemic paraphasia (i.e., a clear-cut substitution error), is a hybrid, realized acoustically as a phoneme substitution but with an acoustic trace of the original target phoneme. To examine this question, acoustic measures were performed of voicing in the fricative consonants [s z] produced in a consonant–vowel (CV) environment by seven aphasic participants, three clinically diagnosed with Broca's aphasia, three with conduction aphasia, and one with Wernicke's aphasia. Results showed that the phonemic paraphasias of the patients left traces of the original target phoneme. These findings suggest that the production of phonemic paraphasias reflect the simultaneous selection and activation of competing phonemes that subsequently influence articulatory planning and implementation stages downstream from it (see also Baum & Slatkovsky, 1993; Buckingham, 1992; Buckingham & Yule, 1987).

55.4 PHONETIC PROCESSES IN PRODUCTION

55.4.1 Articulation

Identifying the neurostructural and neurofunctional correlates of speech production at the articulatory phonetic level requires an understanding of both speech motor programming and sensorimotor integration. Until relatively recently, the majority of work on the neural networks underlying overt speech production relied primarily on evidence from studies of patient populations. In early years, symptomatology exhibited by stroke patients was coarsely correlated with lesion sites based on CT scans (Baum & Boyczuk, 1999; Blumstein et al., 1980; Blumstein, Cooper, Zurif, & Caramazza, 1977). As imaging technology improved, so did the accuracy with which symptoms were associated with lesion sites (Baum et al., 1990). In what has become a landmark study, Dronkers (1996) compared the lesions of 25 individuals diagnosed with apraxia of speech (in a chronic state) in an effort to identify a common area of infarct responsible for the phonetic implementation deficit. The only lesion site common to all patients was in the precentral gyrus of the insula, leading to the suggestion that this region was important in articulatory coordination and control (Dronkers, 1996; see also Dogil et al., 2002; Wise, Greene, Buchel, & Scott, 1999). However, more recent studies have implicated the pars opercularis, suggesting that the insula is part of a network involved in articulatory control (Hillis et al., 2004; Leonardo et al., 2006). Easier access to high-quality functional neuroimaging has permitted a far more detailed and broadly encompassing understanding of the neural substrates of speech production to emerge.

It is self-evident that the networks implicated in speech production include the cortical and subcortical regions involved in motor control more generally (i.e., primary motor cortex, supplementary motor area [SMA], regions within the basal ganglia, and the cerebellum). Numerous investigations over the past 15 years have attempted to identify the specific roles of each of these regions and others, including areas within auditory and somatosensory cortices, because each contributes to the neural network subserving speech production. We first review studies focused on the production of syllables and syllable sequences, and then we turn to sensorimotor integration in speech production.

In comparison with the vast literature on word production (e.g., as reviewed in Indefrey & Levelt, 2000) and a growing literature on covert speech production (e.g., Price, 2012, for review), far fewer neuroimaging studies have investigated overt articulation. Based on his well-developed DIVA model of speech production (Guenther, 1994, 1995), Guenther and colleagues have proposed and begun testing the neural bases of certain aspects of the model (e.g., Bohland & Guenther, 2006; Guenther & Ghosh, 2003; Guenther, Ghosh, & Tourville, 2006; and others). Broadly speaking, the DIVA model consists of two primary control systems—a feedforward system including areas involved in articulatory implementation (speech sound system and stored motor programs, hypothesized to be represented in left ventral premotor cortex [PMC]), articulatory positioning commands generated in motor cortex, and timing coordinated in the cerebellum, and a feedback system that allows for error detection and monitoring and sensorimotor integration based on both auditory (temporal cortex) and somatosensory (inferior parietal cortex) input, with bidirectional links between the two control systems (Guenther et al., 2006). In one of the first tests of the hypothesized neural correlates of the model's components, Guenther et al. (2006) conducted an event-related sparse sampling fMRI study in which 10 speakers produced monosyllabic or bisyllabic stimuli based on orthographic representations (i.e., reading aloud). Scans were acquired during silent intervals after syllable production, timed to co-occur with the idealized peak of the associated hemodynamic response function (HRF), and activation was compared with passive viewing of a string of Xs. Activation was found in a widespread bilateral network including primary motor cortex, primary auditory and sensory cortices, the cerebellum, left inferior prefrontal gyrus, SMA, SMG, and area Spt (Hickok & Poeppel, 2004)—all areas in keeping with the predictions of the model (see also Blank, Scott, Murphy, Warburton, & Wise, 2002; Price, Crinion, & MacSweeney, 2011; Price, Moore, & Frackowiak, 1996).

In a related study focused on the production of syllable sequences, Bohland and Guenther (2006) compared patterns of activation during production of syllables varying in complexity based on number of onset consonants (in this case, syllable length was also modified) and syllable sequences of varying complexity (i.e., the same CV repeated three times or three different CVs appended to one another). Subjects were required to overtly produce the stimuli for a subset of trials and to prepare to produce (but not actually do so) in others (through use of a go/no go task). A conjunction analysis of all speech conditions again yielded a broad bilateral network, including the bilateral central sulcus, precentral and postcentral gyri, anterior insula, superior temporal cortex, medial premotor areas (SMA and pre-SMA), the basal ganglia, thalamus, cerebellum, and left frontal operculum (Bohland & Guenther, 2006; see also Peeva et al., 2010). The increase in syllable complexity (which inherently entailed an increase in length) yielded increased activation in medial premotor regions, the frontal operculum bilaterally, and the left posterior parietal lobe. The increase in sequence complexity yielded increased activation in bilateral medial premotor regions, left frontal cortex, anterior insula bilaterally, left posterior parietal lobe, bilateral inferior posterior temporal lobe, bilateral cerebellum, anterior thalamus, and caudate nucleus (Bohland & Guenther, 2006; but see Dogil et al., 2002 for data suggesting that increased complexity yields reduced activation).

Based on these findings, the authors suggest that the inferior frontal sulcus may be the region where the motor programs required for the production of phonological units or syllables are represented. Given the increase in activation during complex sequences in the left IFGpo (see also Papoutsi et al., 2009; Park, Iverson, & Park, 2011; Soros et al., 2006), the authors suggest that this region houses the "speech sound map"—the phonetic codes for specific phonemes or syllables. The anterior insula—the region identified by Dronkers (1996) as common to all her patients who presented with apraxia of speech—was active in all overt speech tasks, but activation was not modulated by sequence or syllable complexity (Bohland & Guenther, 2006; see also Ackermann, Mathiak, & Ivry, 2004, but see Frenck-Mestre, Anton, Roth, Vaid, & Viallet, 2005; Riecker et al., 2000; Soros et al., 2006 for studies showing an absence of anterior insula activation during speech production). According to Bohland and Guenther (2006), the anterior insula may have a role in speech initiation (see also Shuren, 1993) but is not likely to be involved in the representation or implementation of syllable sequences (for yet a different perspective, see Moser et al. (2009) for data suggesting that the anterior insula has a role in speech implementation for *novel* motor programs). A nearby region that did show sensitivity to complexity was between the anterior insula and the frontal operculum; the authors conclude that this region is involved in syllable sequence representation, perhaps as a site of integration of lower and higher level representations of speech (Bohland & Guenther, 2006).

Other investigators have made similar but slightly different claims about the role of the IFGpo. For instance, based on a positron emission tomography (PET) study comparing silent speech to oral (nonspeech) movements, Price et al. (2011) suggest that the IFGpo (along with posterior superior temporal sulcus [STS]) is involved in predicting acoustic and motor consequences of speech articulation, that is, in the feedforward (internal model) system—not altogether different from Bohland and Guenther's (2006) speech sound map, but with a somewhat different precise localization. This predictive coding would then, hypothetically, be passed on to temporal regions (posterior STS and planum temporale) where it could be used in an optimization loop to limit both perceptual and production errors. The authors admit that functional connectivity analyses would be required to confirm this hypothesis. In this study, activation of the superior temporal plane region was not found to be specific to speech production (i.e., it emerged during oral, nonspeech movements as well), suggesting this region may play a more general role in sensorimotor integration (Price et al., 2011; see Hickok & Poeppel, 2004; Hickok, Okada, & Serences, 2009; Peschke, Ziegler, Eisenberger, & Baumgaertner, 2012).

In a meta-analysis of 19 fMRI and PET studies of overt speech production, Eickhoff, Heim, Zilles, and Amunts (2009) reported consistent activation in left inferior frontal gyrus (IFG), the face region of primary motor cortex, the anterior insula, lateral premotor cortex, the cerebellum, and the basal ganglia. They propose that the anterior insula is involved in preparation for speech, possibly in translating a phonetic "concept" obtained from left IFG into articulatory motor patterns, whereas the left IFG represents the site of the final stages of lexical retrieval that initiates the articulatory process. They further suggest that the basal ganglia and cerebellum are involved in the selection and implementation of motor programs, including timing and the incorporation of sensory feedback, whereas the premotor cortex combines the information from these two sources (basal ganglia and cerebellum) to convert into a specific movement (muscle contraction) plan. The involvement of primary motor cortex relates to actual execution via connections to lower motor neurons (Eickhoff et al., 2009).

As suggested in the description of the meta-analysis reported by Eickhoff and colleagues, additional regions that have been the focus of intense investigation regarding their role in speech production include several premotor regions, notably the pre-SMA, SMA proper, and dorsal and ventral premotor regions (Ackermann & Ziegler, 2010; Guenther et al., 2006; Hartwigsen et al., 2013; Peeva et al., 2010; Saur et al., 2008; Soros et al., 2006; Tremblay & Gracco, 2006). The pre-SMA has long been claimed to be important for speech initiation, based, in part, on evidence from lesion studies (Ackermann & Ziegler, 2010; Ziegler, Kilian, & Deger, 1997) and data from investigations of nonspeech motor control highlighting its role in movement initiation (Shima & Tanji, 2000). Dorsal and ventral premotor areas have frequently been found to be activated during speech production and are presumed to be implicated in the conversion of phonetic codes to articulatory implementation or execution (Guenther et al., 2006; Hickok, Houde, & Rong, 2011). Using dynamic causal modeling (DCM) to explore effective connectivity and diffusion tensor imaging (DTI) to examine anatomical connectivity, Hartwigsen et al. (2013) investigated the roles of these frontal premotor areas during real-word and pseudoword repetition. The investigators hypothesized that if pre-SMA is truly important in speech initiation, then the DCM model with the best fit to the data should include "driving input" into pre-SMA, with a strong influence of pre-SMA on dorsal and ventral premotor areas. The effective connectivity analyses did demonstrate such a pattern, with a stronger influence of pre-SMA on the dorsal premotor region for pseudowords relative to real words. DTI analyses revealed direct fiber connections between pre-SMA and both dorsal and ventral premotor cortex. Based on their findings, the authors propose that the pre-SMA is involved in the sequencing of motor programs for articulatory initiation, the dorsal premotor cortex is implicated in conversion of an abstract sequencing plan to an actual motor-articulatory plan (see also Alario, Chainay, Lehericy, & Cohen, 2006), and the ventral premotor cortex provides predictions of the auditory representations of common speech sounds and syllables (Hartwigsen et al., 2013; see also Guenther et al., 2006).

One final brain region that clearly plays an important role in speech production is the cerebellum, which has connections to both basal ganglia and sensorimotor cortex, as well as to premotor regions and the insula (Ackermann, 2008). Unfortunately, it has received substantially less attention in the literature. However, based on studies of patients with cerebellar lesions, some recent neuroimaging data, and studies of nonspeech motor control, the cerebellum has been argued to be involved in the temporal organization and sequencing of syllables, as well as the control of speech rate (beyond a certain speed) (e.g., Ackermann, 2008; Ackermann, Mathiak, & Riecker, 2007; see also Ivry & Keele, 1989; Ivry, Spencer, Zelznik, & Diedrichsen, 2002). Interestingly, parts of the cerebellum have also shown to be sensitive to speech perceptual demands, particularly in temporal discrimination (Ackermann, 2008; Mathiak, Hertrich, Grodd, & Ackermann, 2002, 2004; Petacchi, Laird, Fox, & Bower, 2005). The precise integration of cerebellar processing with cortical and subcortical processing for speech production remains to be determined.

55.4.2 Sensorimotor Integration

The integration of sensory and motor signals is a crucial component in sensorimotor learning and the establishment of perceptuo-motor representations. Speech production is, by nature, a sensorimotor behavior. Such behaviors require monitoring of input from one or more sensory systems to update or modify motor plans, either on a moment-by-moment compensatory basis (online sensorimotor control) or via longer-term adaptation (sensorimotor learning). In most models of speech production, sensorimotor feedback is compared with a predicted outcome to control ongoing movement and program future articulation; this process of internal feedforward models (efference copy) is proposed in one form or another in the DIVA model of Guenther (1995) and in the state feedback control model of Houde and Nagarajan (2011), among others (Price et al., 2011).

One means of addressing questions of sensorimotor integration is through empirical studies of imposed sensory-based manipulations. For example, it has frequently been demonstrated that the speech production system is able to compensate for many perturbations to the vocal tract—both predictable and unexpected—to achieve phonetic goals (e.g., Abbs, 1986; Lindblom, Lubker, & Gay, 1979, among many others); nonetheless, the extent of compensation varies depending on the nature of the perturbation and the specific speech sounds targeted (Aasland, Baum, & McFarland, 2006; Baum & McFarland, 1997, 2000). These findings suggest that speakers are generally able to adapt to modifications of somatosensory (kinesthetic or proprioceptive) feedback to produce perceptually adequate speech. Recently, a great deal of attention has been focused on speech adaptation to alterations of *auditory* feedback as well. As in the visual modality in studies of prism adaptation (Redding, Rossetti, & Wallace, 2005; Redding & Wallace, 1996; Rossetti, Koga, & Mano, 1993), investigations have demonstrated adaptive speech motor output in the face of auditory feedback that has been modified to yield what is perceived as a production error on the part of the speaker (Houde & Jordan, 1998, 2002; Jones & Munhall, 2005; Larson, Altman, Liu, & Hain, 2008; Larson, Burnett, Kiran, & Hain, 2000; Purcell & Munhall, 2006; Shiller, Sato, Gracco, & Baum, 2009). Adaptation in this context is not complete, in that the speech output does not fully compensate for the perceived target error (Jones & Munhall, 2005). The reasons for incomplete adaptation are complex (Shiller et al., 2009), but the limited adaptive response indicates online sensorimotor integration and suggests a recalibration of perceptual and motor responses (Baum & McFarland, 1997, 2000; Houde & Jordan, 1998, 2002; Jones & Munhall, 2005; Nasir & Ostry, 2009). Functional neuroimaging studies that investigate the neural mechanisms involved in such sensorimotor adaptation for speech are only beginning to emerge (Baciu, Abry, & Segebarth, 2000; Guenther et al., 2006; Ito, Kimura, & Gomi, 2005; Tourville, Reilly, & Guenther, 2008); thus far, they support a broad cortical and subcortical network, including both left and right hemisphere regions underlying the adaptation process, as detailed here. Other approaches to exploring the integration of perceptual and production processes have also been undertaken. For instance, Hickok and colleagues (Buchsbaum, Hickok, & Humphries, 2001; Hickok, Buchsbaum, Humphries, & Muftuler, 2003; Hickok et al., 2009, among others) have used nonword repetition tasks to tap into sensorimotor integration processes. These studies have identified the Spt, a region in the planum temporale, as a critical hub to this process (see also Hickok, 2014, for an overview).

In one of the earliest neuroimaging investigations to use feedback manipulations to examine sensorimotor integration for speech, Tourville et al. (2008) made use of an auditory feedback manipulation in the production of a series of consonant–vowel–consonant (CVC) stimuli, all of which included the vowel /ɛ/. Within a string of productions, an unexpected shift (up or down) of the F1 frequency in the vowel was induced (following Purcell & Munhall, 2006; Houde & Jordan, 1998). Speakers compensated for the unexpected perturbations, as expected. The neural regions that demonstrated increased activation under conditions of auditory perturbation and resulting compensation included bilateral STGp, planum temporale, left ventral motor and premotor cortex, and antero-medial cerebellum. Activation in these regions is consistent with a role for both secondary auditory and motor areas in the compensatory process. In another recent fMRI study investigating the neural substrates invoked during speech adaptation (i.e., speech motor learning) to consistent auditory feedback perturbation, Gracco, Sato, Shiller, and Baum (in preparation) found increased activation during the adaptation (hold) phase in bilateral IFG, premotor gyrus (PM), STS, posterior STG, posterior inferior temporal gyrus (pITG), right angular gyrus/posterior parietal gyrus (rAG/pPG), and left inferior parietal lobule/supramarginal gyrus (IPL/SMG).

Based on these findings, coupled with results in prism adaptation studies implicating a role for inferior parietal cortex in the control of adaptive movements (Redding et al., 2005), Gracco and colleagues (Dangler, Shiller, Baum, & Gracco, in prep; Shum, Shiller, Baum, & Gracco, 2011) conducted two rTMS studies stimulating this region prior to an auditory feedback manipulation/speech adaptation task. In a first study in which inhibitory stimulation was applied to left SMG (Shum et al., 2011), results revealed that those participants who had undergone transcranial magnetic stimulation (TMS) to this region produced a significantly reduced adaptive response as compared with individuals who had undergone sham stimulation. In contrast, Dangler et al. (in preparation) applied inhibitory TMS to the right hemisphere IPL and found a faster adaptive response subsequent to stimulation that was, however, of comparable magnitude in both experimental and control (sham) participant groups. The findings of both experiments were interpreted to suggest that the left inferior parietal region plays an important role in sensorimotor adaptation and integration, whereas the right hemisphere homologue seems to normally inhibit activity in the left IPL; in other words, when right IPL was inhibited, adaptation occurred more quickly (Dangler et al., in preparation; Gracco et al., 2012; Shum et al., 2011).

There have also been several recent investigations of the neural bases of sensorimotor integration with a focus on alteration of somatosensory feedback, rather than auditory feedback. For instance, Baciu et al. (2000), as reviewed in Golfinopoulos et al. (2011), made use of a lip tube to inhibit lip movement during rounded vowel production in an fMRI study. The static perturbation yielded increased activation in a bilateral distributed network that included frontal, parietal, and temporal areas, as well as the cerebellum; interestingly, there was an increase in the role of the right hemisphere during the perturbed trials. Using TMS to the left primary motor cortex, Ito et al. (2005) examined responses to unexpected jaw perturbation during speech and, not surprisingly, found an important role for that region in the compensatory response. Also, examining compensation to unexpected blocking of jaw movement via a pneumatic device, Golfinopoulos et al. (2011) reported increased activation in right anterior SMG, pre-SMA, SMA, IFG, precentral gyrus, thalamus, and cerebellum under perturbed speaking conditions. They also investigated effective connectivity and found increased connectivity between left and right anterior SMG, as well as between left anterior SMG and right ventral premotor cortex in the perturbed relative to the unperturbed conditions. Similarly, increased connectivity was found between right anterior SMG and right IFGpt, as well as bidirectionally between right ventral premotor cortex and right ventral primary motor cortex. The authors conclude that bilateral anterior SMG is involved in somatosensory error detection and correction (in keeping with the DIVA model of Guenther (1995); see also Golfinopoulos, Tourville, & Guenther, 2010, and a recent study by Schwartz, Faseyitan, Kim, and Coslett (2012), a voxel-based lesion symptom mapping analysis supporting the role of SMG in phonological error production in aphasia). Golfinopoulos et al. (2011) further suggest that whereas left IFG may be important for prelearned or automatized phonetic coding, right IFGpt may serve to maintain sensorimotor representations and assist in the adjustment of ongoing speech movements in response to sensory feedback (Golfinopoulos et al., 2011). This latter suggestion appears inconsistent with that proposed by Gracco and colleagues (2012); however, it must be borne in mind that the experiments were investigating different types of compensatory responses, in one case in response to unexpected somatosensory perturbation (Golfinopoulos et al., 2011) and in the other in response to static auditory feedback perturbation (Gracco et al., 2012; Shum et al., 2011).

55.5 SUMMARY

It has been the goal of this chapter to review neuropsychological and neurophysiological studies of speech production. Together, this research has shown a complex tapestry that includes different stages of production that ultimately map abstract phonological structures onto articulatory commands. This production system is integrated with both auditory and somatosensory systems, allowing it to maintain stability, and to adapt, as necessary, based on both immediate and longer-term perturbations. Speech production then requires both feedforward and feedback mechanisms where information flow is bidirectional. Not surprisingly, then, the speech production system recruits a large frontotemporoparietal cortical and subcortical network of left and right hemisphere regions, many of which have also been identified in language processing more generally. Future research holds much promise to provide further insights into the processes and mechanisms involved in speech production and their underlying neural mechanisms as ever-increasing sophisticated techniques become available for mapping lesions in neuropsychological studies and for examining the neural substrates of speech production in participants with and without lesions.

Acknowledgments

This research was supported in part by grants NIH R01 DC 006220 and R21DC013100, by the Dana Foundation to Brown University, and by the Natural Sciences and Engineering Council of Canada to McGill University. The content is solely the responsibility of the authors and does not necessarily represent the official views of the National Institute on Deafness and Other Communication Disorders or the National Institutes of Health.

References

Aasland, W., Baum, S., & McFarland, D. (2006). Electropalatographic, acoustic, and perceptual data on adaptation to a palatal perturbation. *Journal of the Acoustical Society of America*, *119*, 2372–2381.

Abbs, J. (1986). Invariance and variability in speech production: A distinction between linguistic intent and its neuromotor implementation. In J. Perkell, & D. Klatt (Eds.), *Invariance and variability in speech processes* (pp. 202–225). Hillsdale, NJ: Erlbaum.

Ackermann, H. (2008). Cerebellar contributions to speech production and speech perception: Psycholinguistic and neurobiological perspectives. *Trends in Neurosciences*, *31*, 265–272.

Ackermann, H., Mathiak, K., & Riecker, A. (2007). The contribution of the cerebellum to speech production and speech perception: Clinical and functional imaging data. *The Cerebellum*, *6*, 202–213.

Ackermann, H., & Ziegler, W. (2010). Brain mechanisms underlying speech. In W. J. Hardcastle, & J. Laver (Eds.), *The handbook of phonetic sciences* (2nd ed., pp. 202–250). Chichester: Wiley-Blackwell.

Ackermann, H., Mathiak, K., & Ivry, R. B. (2004). Temporal organization of "internal speech" as a basis for cerebellar modulation of cognitive functions. *Behavioral and Cognitive Neuroscience Reviews*, *3*, 14–22.

Alario, F., Chainay, H., Lehericy, S., & Cohen, L. (2006). The role of the supplementary motor area (SMA) in word production. *Brain Research*, *1076*, 129–143.

Baciu, M., Abry, C., & Segebarth, C. (2000). Équivalence motrice et dominance hémisphérique. Le cas de la voyelle (u). *Actes des journees d'etude sur la parole, Étude IRMf. Actes JEP*, *23*, 213–216.

Baese-Berk, M., & Goldrick, M. (2009). Mechanisms of interaction in speech production. *Language and Cognitive Processes*, *24*, 527–554.

Baum, S., Blumstein, S., Naeser, M., & Palumbo, C. (1990). Temporal dimensions of consonant and vowel production: An acoustic and CT scan analysis of aphasic speech. *Brain and Language*, *39*, 33–56.

Baum, S., & Boyczuk, J. (1999). Speech timing subsequent to brain damage: Effects of utterance length and complexity. *Brain and Language*, *67*, 30–45.

Baum, S., & McFarland, D. (1997). The development of speech adaptation to an artificial palate. *Journal of the Acoustical Society of America*, *102*, 2353–2359.

Baum, S., & McFarland, D. (2000). Individual differences in speech adaptation to an artificial palate. *Journal of the Acoustical Society of America*, *107*, 3572–3575.

Baum, S., & Slatkovsky, K. (1993). Phonemic false evaluation? Preliminary data from a conduction aphasia patient. *Clinical Linguistics and Phonetics*, *7*, 207–218.

Blank, S. C., Scott, S., Murphy, K., Warburton, E., & Wise, R. (2002). Speech production: Wernicke, Broca and beyond. *Brain*, *125*, 1829–1838.

Blumstein, S. (1973). *A phonological investigation of aphasic speech*. The Hague: Mouton.

Blumstein, S. E., Cooper, W., Goodglass, H., Statlender, S., & Gottlieb, J. (1980). Production deficits in aphasia: A voice onset time analysis. *Brain and Language*, *9*, 153–170.

Blumstein, S. E., Cooper, W., Zurif, E., & Caramazza, A. (1977). The perception and production of voice onset time in aphasia. *Neuropsychologia*, *15*, 371–383.

Bohland, J., & Guenther, F. (2006). An fMRI investigation of syllable sequence production. *NeuroImage*, *32*, 821–841.

Buchsbaum, B., Hickok, G., & Humphries, C. (2001). Role of left posterior superior temporal gyrus in phonological processing for speech perception and production. *Cognitive Science*, *25*, 663–678.

Buchwald, A., & Miozzo, M. (2011). Finding levels of abstraction in speech production: Evidence from sound production errors. *Psychological Science*, *22*, 1113–1119.

Buchwald, A., & Miozzo, M. (2012). Phonological and motor errors in individuals with acquired sound production impairment. *Journal of Speech, Language and Hearing Research*, *55*, 1573–1586.

Buckingham, H., & Yule, G. (1987). Phonemic false evaluation: Theoretical and clinical aspects. *Clinical Linguistics and Phonetics*, *1*, 113–125.

Buckingham, H. W. (1992). The mechanisms of phonemic paraphasia. *Clinical Linguistics and Phonetics*, *6*, 41–63.

Bullock-Rest, N., Cerny, A., Sweeney, C., Palumbo, C., Kurowski, K., & Blumstein, S. E. (2013). Neural systems underlying the influence of sound shape properties of the lexicon on spoken word production: Do fMRI findings predict effects of lesions in aphasia? *Brain and Language*, *126*, 159–168.

Canter, G. J., Trost, J. E., & Burns, M. S. (1985). Contrasting speech patterns in apraxia of speech and phonemic paraphasia. *Brain and Language*, *24*, 204–222.

Dangler, L., Shiller, D., Baum, S., & Gracco, V. (in preparation). Left but not right hemisphere inferior parietal lobe contributes to sensorimotor adaptation for speech.

Dell, G. S. (1986). A spreading activation theory of retrieval in sentence production. *Psychological Review*, *93*, 283–321.

Dell, G. S., & Gordon, J. K. (2003). Neighbors in the lexicon: Friends or foes? In N. O. Schiller, & A. S. Meyer (Eds.), *Phonetics and phonology in language comprehension and production: Differences and similarities* (pp. 9–37). New York, NY: Mouton de Gruyter.

Dogil, G., Ackermann, H., Grodd, W., Haider, H., Kamp, H., Mayer, J., et al. (2002). The speaking brain: A tutorial introduction to fMRI experiments in the production of speech, prosody and syntax. *Journal of Neurolinguistics*, *15*, 59–90.

Dronkers, N. (1996). A new brain region for coordinating speech articulation. *Nature*, *384*, 159–161.

Eickhoff, S., Heim, S., Zilles, K., & Amunts, K. (2009). A systems perspective on the effective connectivity of overt speech production. *Philosophical Transactions of the Royal Society A*, *367*, 2399–2421.

Frenck-Mestre, C., Anton, J. L., Roth, M., Vaid, J., & Viallet, F. (2005). Speech production in bilinguals. *NeuroReport*, *16*, 761–765.

Gandour, J., & Dardarananda, R. (1984). Voice-onset time in aphasia: Thai, II: Production. *Brain and Language*, *18*, 389–410.

Gaskell, M. G., & Marslen-Wilson, W. D. (1999). Ambiguity, competition, and blending in spoken word recognition. *Cognitive Science*, *23*, 439–462.

Goldrick, M. (2006). Limited interaction in speech production: Chronometric, speech error, and neuropsychological evidence. *Language and Cognitive Processes*, *21*, 817–855.

Goldrick, M., & Blumstein, S. E. (2006). Cascading activation from phonological planning to articulatory processes: Evidence from tongue twisters. *Language and Cognitive Processes*, *21*, 649–683.

Golfinopoulos, E., Tourville, J., Bohland, J., Ghosh, S., Nieto-Castanon, A., & Guenther, F. (2011). fMRI investigation of unexpected somatosensory feedback perturbation during speech. *NeuroImage*, *55*, 1324–1338.

Golfinopoulos, E., Tourville, J., & Guenther, F. (2010). The integration of large-scale neural network modeling and functional brain imaging in speech motor control. *NeuroImage*, *52*, 862–874.

Goodglass, H., & Kaplan, E. (1972). *The assessment of aphasia and related disorders*. Philadelphia, PA: Lea & Febiger.

Gordon, J. K. (2002). Phonological neighborhood effects in aphasic speech errors: Spontaneous and structured contexts. *Brain and Language*, *82*, 113–145.

Gracco, V., Deschamps, I., Shiller, D., Dangler, L., Elgie, B., & Baum, S. (2012). The role of the inferior parietal cortex in sensorimotor representations for speech, Paper presented at *Society for the Neurobiology of Language*, San Sebastian, Spain.

Gracco, V., Sato, M., Shiller, D., & Baum, S. (in prep). The neural networks underlying sensorimotor representations for speech.

Guenther, F. (1994). A neural network model of speech acquisition and motor equivalent speech production. *Biological Cybernetics*, *72*, 43–53.

Guenther, F. (1995). Speech sound acquisition, coarticulation, and rate effects in a neural network model of speech production. *Psychological Review*, *102*, 594–621.

Guenther, F., & Ghosh, S. (2003). A model of cortical and cerebellar function in speech. In: *Proceedings of the 15th international congress of phonetic sciences* (pp. 169–173) Barcelona, Spain.

Guenther, F., Ghosh, S., & Tourville, J. (2006). Neural modeling and imaging of the cortical interactions underlying syllable production. *Brain and Language*, *96*, 280–301.

Haley, K. L., Jacks, A., & Cunningham, K. T. (2013). Error variability and the differentiation between apraxia of speech and aphasia with phonemic paraphasia. *Journal of Speech Language and Hearing Research*, *56*, 891–905.

Hartwigsen, G., Saur, D., Price, C., Baumgaertner, A., Ulmer, S., & Siebner, H. (2013). Increased facilitatory connectivity from the pre-SMA to the left dorsal premotor cortex during pseudoword repetition. *Journal of Cognitive Neuroscience*, *25*, 580–594.

Hickok, G. (2014). The architecture of speech production and the role of the phoneme in speech processing. *Language, Cognition and Neuroscience*, *29*, 2–20.

Hickok, G., Buchsbaum, S., Humphries, C., & Muftuler, T. (2003). Auditory-motor interaction revealed by fMRI: Speech, music, and working memory in area Spt. *Journal of Cognitive Neuroscience*, *15*, 673–682.

Hickok, G., Houde, J., & Rong, F. (2011). Sensorimotor integration in speech processing: Computational basis and neural organization. *Neuron*, *69*, 407–422.

Hickok, G., Okada, K., & Serences, J. (2009). Area Spt in the human planum temporale supports sensory-motor integration for speech processing. *Journal of Neurophysiology*, *101*, 2725–2732.

Hickok, G., & Poeppel, D. (2004). Dorsal and ventral streams: A framework for understanding aspects of the functional anatomy of language. *Cognition*, *92*, 67–99.

Hillis, A., Work, M., Barker, P., Jacobs, M., Breese, E., & Maurer, K. (2004). Re-examining the brain regions crucial for orchestrating speech articulation. *Brain*, *127*, 1479–1487.

Houde, J., & Jordan, M. (1998). Sensorimotor adaptation in speech production. *Science*, *279*, 1213–1216.

Houde, J., & Jordan, M. (2002). Sensorimotor adaptation of speech I: Compensation and adaptation. *Journal of Speech, Language, and Hearing Research*, *45*, 295–310.

Houde, J., & Nagarajan, S. (2011). Speech production as state feedback control. *Frontiers of Human Neuroscience*, *5*, 82.

Indefrey, P., & Levelt, W. J. M. (2000). The neural correlates of language production. In M. Gazzaniga (Ed.), *The new cognitive neurosciences* (2nd ed., pp. 845–865). Cambridge, MA: MIT Press.

Ito, T., Kimura, T., & Gomi, H. (2005). The motor cortex is involved in reflexive compensatory adjustment of speech articulation. *NeuroReport*, *16*, 1791–1794.

Ivry, R., & Keele, W. (1989). Timing functions of the cerebellum. *Journal of Cognitive Neuroscience*, *1*, 136–152.

Ivry, R., Spencer, R., Zelznik, H., & Diedrichsen, J. (2002). The cerebellum and event timing. *Annals of the New York Academy of Sciences*, *978*, 302–317.

Jakobson, R. (1968). *Child language, aphasia, and phonological universals* (A.R. Keiler, Trans.). The Hague: Mouton.

Jakobson, R., Halle, M., & Fant, G. (1967). *Preliminaries to speech analysis: Distinctive features and their correlates*. Cambridge, MA: MIT Press.

Jones, J., & Munhall, K. (2005). Remapping auditory-motor representations in voice production. *Current Biology*, *15*, 1768–1772.

Kent, R., & Rosenbek, J. (1983). Acoustic patterns of apraxia of speech. *Journal of Speech and Hearing Research*, *26*, 231–248.

Kurowski, K., Blumstein, S. E., Palumbo, C. L., Waldstein, R., & Burton, M. W. (2007). Nasal production in anterior and posterior aphasics: Speech deficits and neuroanatomical correlates. *Brain and Language*, *100*, 262–275.

Kurowski, K., Hazen, E., & Blumstein, S. E. (2003). The nature of speech production impairments in anterior aphasics: An acoustic analysis of voicing in fricative consonants. *Brain and Language*, *84*, 353–371.

Laganaro, M., Chetelat-Mabillard, D., & Frauenfelder, U. H. (2013). Facilitatory and interfering effects of neighbourhood density on speech production: Evidence from aphasic errors. *Cognitive Neuropsychology*, *30*, 127–146.

Larson, C., Altman, K., Liu, H., & Hain, T. (2008). Interactions between auditory and somatosensory feedback for voice F0 control. *Experimental Brain Research*, *187*, 613–621.

Larson, C., Burnett, T., Kiran, S., & Hain, T. (2000). Effects of pitch-shift velocity on voice F0 responses. *Journal of the Acoustical Society of America*, *107*, 559–564.

Lecours, A. R., & Lhermitte, F. (1969). Phonemic paraphasias: Linguistic structures and tentative hypotheses. *Cortex*, *5*, 193–228.

Leonardo, B., Moser, D., Rorden, C., Baylis, C., Gordon, C., & Fridriksson, J. (2006). Speech apraxia without oral apraxia: Can normal brain function explain the physiopathology? *NeuroReport*, *17*, 1027–1031.

Lindblom, B., Lubker, J., & Gay, T. (1979). Formant frequencies of some fixed-mandible vowels and a model of speech motor

programming by predictive simulation. *Journal of Phonetics, 7*, 147–161.
Mathiak, K., Hertrich, I., Grodd, W., & Ackermann, H. (2002). Cerebellum and speech perception: A functional magnetic resonance imaging study. *Journal of Cognitive Neurosicnece, 14*, 902–912.
Mathiak, K., Hertrich, I., Grodd, W., & Ackermann, H. (2004). Discrimination of temporal information at the cerebellum: Functional magnetic resonance imaging of nonverbal auditory memory. *NeuroImage, 21*, 154–162.
Moser, D., Fridriksson, J., Bonilha, L., Healy, E., Baylis, G., Baker, J., et al. (2009). Neural recruitment for the production of native and novel speech sounds. *NeuroImage, 46*, 549–557.
Nasir, S., & Ostry, D. (2009). Auditory plasticity and speech motor learning. *Proceedings of the National Academy of Sciences of the United States of America, 106*, 20470–20475.
Papoutsi, M., de Zwart, J., Jansma, J., Pickering, M., Bednar, J., & Horwitz, B. (2009). From phonemes to articulatory codes: An fMRI study of the role of Broca's area in speech production. *Cerebral Cortex, 19*, 2156–2165.
Park, H., Iverson, G., & Park, H.-J. (2011). Neural correlates in the processing of phoneme-level complexity in vowel production. *Brain and Language, 119*, 158–166.
Peeva, M., Guenther, F., Tourville, J., Nieto-Castanon, A., Anton, J., Nazarian, B., et al. (2010). Distinct representations of phonemes, syllables, and supra-syllabic sequences in the speech production network. *NeuroImage, 50*, 626–638.
Peramunage, D., Blumstein, S. E., Myers, E. B., Goldrick, M., & Baese-Berk, M. (2011). Phonological neighborhood effects in spoken word production: An fMRI study. *Journal of Cognitive Neuroscience, 23*, 593–603.
Peschke, C., Ziegler, W., Eisenberger, J., & Baumgaertner, A. (2012). Phonological manipulation between speech perception and production activates a parieto-frontal circuit. *NeuroImage, 59*, 788–799.
Petacchi, A., Laird, A., Fox, P., & Bower, J. (2005). Cerebellum and auditory function: An ALE meta-analysis of functional neuroimaging studies. *Human Brain Mapping, 25*, 118–128.
Price, C. (2012). A review and synthesis of the first 20 years of PET and FMRI studies of heard speech, spoken language and reading. *Neuroimage, 62*, 816–847.
Price, C., Crinion, J., & MacSweeney, M. (2011). A generative model of speech production in Broca's and Wernicke's areas. *Frontiers in Psychology, 2*, 1–9.
Price, C., Moore, C., & Frackowiak, R. (1996). The effect of varying stimulus rate and duration on brain activity during reading. *NeuroImage, 3*, 40–52.
Purcell, D., & Munhall, K. (2006). Adaptive control of vowel formant frequency: Evidence from real-time formant manipulation. *Journal of the Acoustical Society of America, 120*, 966–977.
Redding, G., Rossetti, Y., & Wallace, B. (2005). Applications of prism adaptation: A tutorial in theory and method. *Neuroscience and Biobehavioral Reviews, 29*, 431–444.
Redding, G. M., & Wallace, B. (1996). Adaptive spatial alignment and strategic perceptualmotor control. *Journal of. Experimental Psychology: Human Perception and Performance, 22*, 379–394.
Riecker, A., Ackermann, H., Wildgruber, D., Mayer, J., Dogil, G., Haider, H., et al. (2000). Articulatory/phonetic sequencing at the level of the anterior perisylvian cortex: A functional magnetic resonance imaging (fMRI) study. *Brain and Language, 75*, 259–276.
Rossetti, Y., Koga, K., & Mano, T. (1993). Prismatic displacement of vision induces transient changes in the timing of eye-hand coordination. *Perception and Psychophysics, 54*, 355–364.
Ryalls, J. (1986). An acoustic study of vowel production in aphasia. *Brain and Language, 29*, 48–67.
Saur, D., Kreher, B., Schnell, S., Kummerer, D., Kellmeyer, P., Vry, M., et al. (2008). Ventral and dorsal pathways for language. *Proceedings of the National Academy of Sciences of the United States of America, 105*, 18035–18040.
Schwartz, M., Faseyitan, O., Kim, J., & Coslett, H. B. (2012). The dorsal stream contribution to phonological retrieval in object naming. *Brain, 135*, 3799–3814.
Shiller, D., Sato, M., Gracco, V., & Baum, S. (2009). Perceptual recalibration of speech sounds following speech motor learning. *Journal of the Acoustical Society of America, 125*, 1103–1113.
Shima, K., & Tanji, J. (2000). Neuronal activity in the supplementary and presupplementary motor areas of temporal organization of multiple movements. *Journal of Neurophysiology, 84*, 2148–2160.
Shum, M., Shiller, D., Baum, S., & Gracco, V. (2011). Sensorimotor integration for speech motor learning involves the inferior parietal cortex. *European Journal of Neuroscience, 34*, 1817–1822.
Shuren, J. (1993). Insula and aphasia. *Journal of Neurology, 240*, 216–218.
Soros, P., Sokologg, L., Bose, A., McIntosh, A., Graham, S., & Stuss, D. (2006). Clustered functional MRI of overt speech production. *NeuroImage, 32*, 376–387.
Tourville, J., Reilly, K., & Guenther, F. (2008). Neural mechanisms underlying auditory feedback control of speech. *NeuroImage, 39*, 1429–1443.
Tremblay, P., & Gracco, V. (2006). Contribution of the frontal lobe to externally and internally specified verbal responses: fMRI evidence. *NeuroImage, 33*, 947–957.
Vijayan, A., & Gandour, J. (1995). On the notion of a "subtle phonetic deficit" in fluent/posterior aphasia. *Brain and Language, 48*, 106–119.
Wise, R., Greene, J., Buchel, C., & Scott, S. (1999). Brain regions involved in articulation. *Lancet, 353*, 1057–1061.
Ziegler, W., Kilian, B., & Deger, K. (1997). The role of the left mesial frontal cortex in fluent speech: Evidence from a case of left supplementary motor area hemorrhage. *Neuropsychologia, 35*, 1197–1208.

CHAPTER

56

Word Production from the Perspective of Speech Errors in Aphasia

Myrna F. Schwartz[1] *and Gary S. Dell*[2]

[1]Moss Rehabilitation Research Institute, Elkins Park, PA, USA; [2]University of Illinois, Urbana-Champaign, Beckman Institute, Urbana, IL, USA

A common strategy for understanding the inner workings of a complex system is to study the ways in which it breaks down. Research in naturally occurring speech errors ("slips of the tongue") and research in aphasia both take advantage of this strategy. Language scientists study speech errors for insight into the workings of the cognitive system that enables the skilled production of words and sentences. Neuroscientists study language deficits in aphasia to learn how the mechanisms of language processing are represented in the brain. This chapter presents research that falls at the intersection of these two historical traditions. The studies we highlight use computational models to simulate the cognitive mechanisms responsible for speech errors in aphasia and large-scale lesion-symptom mapping to link those cognitive mechanisms with brain regions.

Naturally occurring speech errors are the product of momentary, functional disruption within a normal language system. Aphasia causes structural damage that has persistent functional consequences. The most important of these, for the present purposes, is exacerbation of the normal tendency to occasionally select the wrong word or the wrong phonological segments when speaking (Dell, Schwartz, Martin, Saffran, & Gagnon, 1997; Ellis, 1985; Freud, 1953). Aphasia creates vulnerability to speech errors in spontaneous speech and in single word production tasks, such as spoken object naming and auditory word repetition. The high degree of experimental control afforded by naming and word repetition tasks has made them the vehicle of choice for studying how patients' error types and error frequencies vary in relation to target properties (Cuetos, Aguado, Izura, & Ellis, 2002; Kittredge, Dell, Verkuilen, & Schwartz, 2008; Nickels & Howard, 1994, 1995) and clinical profiles (Blumstein, 1973; Jefferies & Lambon Ralph, 2006; Romani & Galluzzi, 2005). Furthermore, computational models of word production have been used to simulate lesions in aphasia and test competing theories of how errors arise. We exemplify this strategy with our own computational studies of naming and repetition.

56.1 SPEECH ERRORS IN APHASIA: THE NEUROLOGICAL TRADITION

The long history of neurological research in aphasia has been dominated by the classical syndromes or subtypes. These include, but are not limited to, Broca's aphasia (BA), Wernicke's aphasia (WA), Conduction aphasia (CA), and Anomic aphasia (AA). Patients are assigned to subtypes based primarily on how they perform on clinical examination of expressive language (fluency, grammar, naming), receptive language (comprehension of spoken words and sentences), and repetition (of words and sentence). Within this framework, BA is considered primarily a disorder of expressive speech, featuring dysfluency, altered and inconsistent articulation ("apraxia of speech"), and simplified grammar. WA is primarily a disorder of speech comprehension, CA is primarily a disorder of repetition, and AA is primarily a disorder of word retrieval.

Speech error tendencies also enter into the diagnosis of WA and CA. Patients with WA characteristically produce speech that is fluent but replete with errors (historically, "paraphasias") that particularly distort the semantic and also phonological content, sometimes to a degree that obscures the intended meaning. CA patients are less compromised in their functional

Neurobiology of Language. DOI: http://dx.doi.org/10.1016/B978-0-12-407794-2.00056-0

communication. Their comprehension is quite good and their speech is more semantically coherent. Their primary deficits are inability to repeat spoken speech, reduced auditory-verbal short-term memory span, and vulnerability to phonemic errors in all types of production tasks. Whereas WA patients are generally unaware of their errors, errorful productions in CA are often accompanied by successive repair attempts, referred to as "conduit d'approache."

The following speech samples exemplify paraphasic speech production in WA (Ex. 1) and CA (Ex. 2). The patients were asked to describe the *Cookie Theft* picture, which shows a woman washing dishes at an overflowing sink while a girl and boy, the latter perched on a tipping stool, grab cookies from a cabinet (Goodglass & Kaplan, 1983).

(1) *"So the two boys work together an one is sneakin' around here, making his. . .work an' his further funnas his time he had."*

(2) *"Well this um. . .somebody's. . .ah mather is takin the. . .washin' the dayshes* [material deleted] *and there's a. . .then the girl. . .not the girl. . .the boy who's getting the cooking is on this ah. . .strool and' startin' to fall off."*

The aphasia syndrome classification is rooted in the 19th century Broca–Wernicke–Lichtheim model of language and brain. The well-known claims of the model are that motor and auditory speech engrams localize to the left inferior frontal gyrus (IFG) and left posterior superior temporal gyrus (pSTG), respectively; these sensory and motor engrams are interconnected through the arcuate fasciculus. IFG lesions give rise to the expressive deficit in BA, pSTG lesions to the comprehension deficit in WA, and arcuate lesions to the repetition deficit in CA. The paraphasic speech patterns of WA and CA patients arise because the intact motor speech engrams are deprived of governing input from the speech comprehension center due to direct damage (WA) or disconnection (CA) (Compston, 2006; Geschwind, 1965).

The aphasia syndromes and classical model have an enduring legacy. The contemporary Dual Stream theory of auditory language processing (Hickok & Poeppel, 2004, 2007) bases its characterization of the dorsal stream on neuroimaging data and evidence concerning CA. The theory assigns to the dorsal stream the function of mapping between auditory and articulatory representations in speech. It identifies an area in the posterior Sylvian fissure at the temporoparietal boundary (area Spt) that plays a central role in this dorsal stream specialization and postulates that lesions here give rise to the CA symptom complex. Phonemic paraphasia in the speech of CA patients features importantly in the theory, providing evidence that normal speech depends on the integrity of the auditory-articulatory mapping served by the dorsal route (for discussion and evidence, see Buchsbaum et al., 2011).

Apart from the Dual Stream theory, the legacy of the classical aphasia model is seen in modern empirical studies of the relationship between aphasia syndromes and speech error patterns. This issue was featured in Blumstein's (1973) seminal monograph, "A phonological investigation of aphasic speech." The monograph's linguistic analysis of phonemic paraphasias in the speech of those with Broca's, Wernicke's and Conduction aphasia revealed surprising uniformity in the patterns of phonological breakdown. For example, in all three groups, phonemes were substituted more often than they were added or deleted, substitutions were strongly constrained by similarity in acoustic-articulatory features, and errors tended to replace marked structures (i.e., more complex, less frequent, and later acquired) with less marked ones.

Blumstein (1973) recognized that the mechanism for paraphasia might well be different in the three aphasic groups, notwithstanding their linguistic similarities. In line with this, researchers who study phonemic errors in WA often start from the assumption that these patients suffer from a central lexical deficit that impacts the retrieval of phonological and semantic information in language production, as it does in comprehension. In contrast, researchers who study phonemic errors in CA tend to emphasize that such errors occur even in tasks with an optional or obligatory "direct" route that bypasses the lexicon (e.g., repeating and reading words). On this basis (and others), it has been argued that phonemic paraphasia in at least some CA patients arises subsequent to lexical processing (i.e., at a postlexical phonological or phonetic stage of word production) (Caplan, Vanier, & Baker, 1986; Garrett, 1984; Goldrick & Rapp, 2007; Pate, Saffran, & Martin, 1987).

The notion that different mechanisms underlie the phonemic paraphasias of CA, WA, and BA patients is difficult to reconcile with the homogenizing impact of partial recovery. Experience suggests that some patients who are diagnosed in the acute phase with WA or BA in time evolve either to the CA profile, with its characteristic pattern of phonemic paraphasia and *conduit d'approache*, or to AA. This reminds us that what looks to be a difference in kind (e.g., phonemic errors, which resemble a plausible target, vs. neologisms, which do not) may actually be one of degree. Moreover, a particular diagnostic feature may become salient only on resolution of another symptom that initially masked or distorted its expression, as when the resolution of apraxia of speech reveals the presence of phonemic paraphasia in BA.

In our own cognitive studies of phonological and other errors in aphasia, we classify and interpret errors without regard to the speaker's aphasia subtype by applying a theoretical model that ascribes word-level substitutions to the lexical-semantic stage of lexical access and sublexical deviations to the lexical-phonological stage. The following section explains this further.

56.2 TWO STAGES OF LEXICAL ACCESS IN PRODUCTION

In the 1970s and 1980s, linguistic analysis of normal speech errors gave rise to a seminal psycholinguistic theory of sentence production (Dell, 1980; Fromkin, 1971; Garrett, 1975, 1980; MacKay, 1972; Shattuck-Hufnagel, 1979; Stemberger, 1985). The theory holds that planning a sentence involves the construction of multiple successive representations. A conceptual semantic representation is constructed first, followed by two linguistic levels that use syntactic and phonological information, respectively. Construction of the two linguistic levels involves building structure-representing frames and inserting content that is retrieved from the mental lexicon. The lexical units that fill slots in the syntactic frame are semantically and syntactically specified *words*. The units that fill slots in the phonological frame are phonological *segments*. It follows that entries in the mental lexical have multilevel representations and are retrieved in stages. Speech errors, according to the theory, arise when the wrong word/segment is retrieved from the lexicon, or the right word/segment is retrieved but inserted into the wrong slot.

The application of this theory to aphasia followed quickly. Researchers used the sentence planning framework to explain grammatical deficits (Saffran, 1982; Schwartz, 1987) and the multistage lexical retrieval model to explain aphasics' errors in word production (Buckingham, 1980, 1987; Butterworth, 1979; Ellis, Miller, & Sin, 1983; Garrett, 1984; Miller & Ellis, 1987; Pate et al., 1987; Schwartz, Saffran, Bloch, & Dell, 1994). The incorporation of interactive activation principles into multistage lexical models added to their range and power for explaining aphasic production errors. An early, excellent example is Ellis' (1985) proposal for how disruption of lexical dynamics could give rise to phonemic errors.

Ellis's informal (i.e., nonimplemented) model of word retrieval featured three cognitive components: a conceptual semantic system; a speech output lexicon

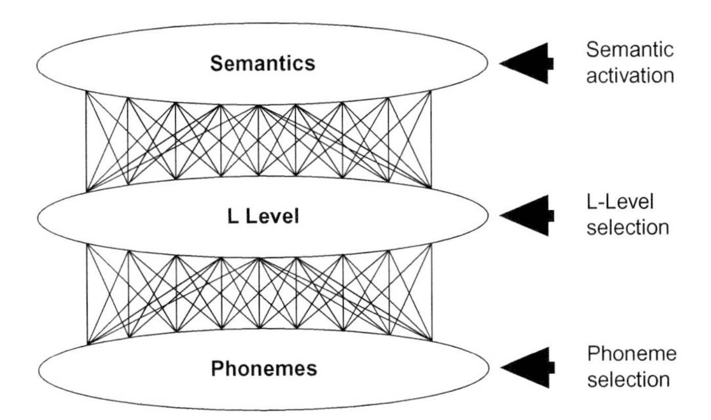

FIGURE 56.1 The generic two-stage account of spoken naming. Distributed semantic concepts map onto lexical nodes, which, in turn, map onto phonemes. *(From Rapp and Goldrick (2000). Reprinted with permission from the publisher (American Psychological Association)).*

with units or nodes representing known words[1]; and a component that represents articulable phonemes. He hypothesized that damage-induced reduction in the flow of activation between the lexical-semantic system and the speech output lexicon would compromise the competitive dynamics among nodes in the speech output lexicon. This would impact the spread of activation down to the phoneme level, resulting in utterances comprising "correct phonemes intermingled with inappropriate ones" (p. 132). Addressing the sources that contribute to the activation of inappropriate phonemes, Ellis discussed top-down, cascading activation from the lexical-semantic layer, feedback loops between phonemes and output lexicon, and persistence of activation from prior utterances. Addressing why some neologisms have strong overlap with the target while others are more remote, he speculated that the mix of inappropriate and appropriate phonemes might be related to patient factors (i.e., severity of the lexical access deficit) as well as target factors (long, infrequent targets more likely to yield remote neologisms).

Ellis' informal model laid the groundwork for later computational models that implemented many of its features. The basic architecture of several such models is shown in Figure 56.1 (from Rapp & Goldrick, 2000). This is a generic two-stage account of spoken naming, in which distributed semantic concepts map onto lexical nodes, and these, in turn, map onto phonemes. The first stage of naming is about selecting the lexical node that corresponds to the semantic concept; the second stage is about selecting the phonemes. The many-to-one mapping from distributed semantics to lexical units creates competition among semantic neighbors

[1]Ellis (1985) chose to remain agnostic on the question of whether the units in the "speech output" lexicon are the same lexical units that are accessed in comprehension. This was an important issue at the time (Allport, 1984) and it remains so today (Gow, 2012; Hickok, 2012).

and sets the stage for semantic and other lexical substitution errors to arise during step 1. Competition at the phoneme level, induced by processes described by Ellis (1985), invites phoneme substitution errors, most of which create nonwords. Models that use a noisy activation function can produce errors even in the unlesioned (default) state. Models that simulate aphasia do so by altering parameters of the model to reduce the efficiency of activation spreading and thus increase the impact of noise.

56.2.1 The Interactive Two-Step Model of Lexical Access in Naming

The generic model glosses over important differences in how alternative models formalize the semantic level and the lexical level, and how much they integrate or separate the processing that goes on at each level (for discussion, see Rapp & Goldrick, 2000). Dell's interactive two-step model of normal and aphasic naming (Dell & O'Seaghdha, 1991; Dell et al., 1997; Foygel & Dell, 2000) occupies a middle ground between discrete-stage theories, which do not allow any cross-stage influences (Levelt, Roelofs, & Meyer, 1999) and highly interactive models in which the mapping between meaning and the output of lexical forms is achieved in a single settling step (Plaut & Shallice, 1993).

TABLE 56.1 Taxonomy of Error Types[a,b]

Semantic: Real word response that is a synonym, category coordinate, superordinate, subordinate, or strong associate of the target (e.g., *bus* for van; *leash* for dog)
Mixed: Real word response that qualifies as a semantic error and that meets the criterion for phonological similarity (e.g., *snake* for snail)
Formal: Real word response that meets the criterion for phonological similarity (e.g., *shaft* for fish)
Unrelated: Real word response that is neither semantically nor phonologically similar to the target (e.g., *camp* for banana)
Nonword: String of phonemes that does not constitute a word in the language. Most such errors pass the phonological similarity criterion, and, depending on the study, this can be a requirement for inclusion (e.g., *goath* for goat (phonologically similar); *tuss* for cane (dissimilar)

[a]*Roach et al. (1996).*
[b]*The criterion for phonological similarity is that response and target share at least one phoneme in corresponding syllable or word position or two phonemes in any position, not counting unstressed vowels.*

In its current form, the interactive two-step model assumes that aphasia affects naming by weakening connections between semantics and L-level (s-weight lesion) and/or those between L-level and phonemes (*p*-weight lesion). S-weight lesions instantiate a lexical-semantic disorder; *p*-weight lesions instantiate a lexical-phonological disorder. Most patients have both, but the severity of each lesion can vary independently of the other, and the two together determine what types of errors are seen and in what proportions.

Our evaluations of the model have mostly used computational case series methods (Dell et al., 1997; Foygel & Dell, 2000; Schwartz, Dell, Martin, Gahl, & Sobel, 2006). This starts with systematic collection of behavioral measures from a large sample of chronic left hemisphere stroke survivors representing all the major subtypes. All have self-reported language deficits confirmed by the *Western Aphasia Battery* (WAB; Kertesz, 1982) or other language evaluation. Among the battery of tests that each participant performs is the 175-item test *Philadelphia Naming Test* (PNT) (Roach, Schwartz, Martin, Grewal, & Brecher, 1996), which assesses basic-level naming of line-drawn objects. PNT responses are categorized into six categories: correct responses and the five error types shown in Table 56.1, which are expressed as proportions (relative to all responses). For each patient in the series, the model is fit by finding the values of the s-weight and *p*-weight parameters that give the best match to the patient's actual response proportions (see Table 56.2 for example). The model is evaluated for the goodness of the quantitative fits and whether there are patients with particular response patterns that the model cannot fit well (called "deviating patterns").

A study performed in 2006 fits the interactive two-step model to the naming response proportions of 94 patients with diverse aphasia presentations

TABLE 56.2 Naming Response—Category Proportions from a Sample Patient from the Moss Database[a] and the Model's Simulated Proportions Generated When Best Fitting Parameters Are Chosen

Response categories	Correct	Semantic	Formal	Mixed	Unrelated	Nonwords
Examples	*"cat"*	*"dog"*	*"cabin," "mat"*	*"rat"*	*"log"*	*"cag," "gat"*
Patient naming	0.49	0.07	0.04	0.05	0.04	0.30
Model naming	0.52	0.07	0.08	0.02	0.04	0.26

[a]*Mirman et al. (2010).*
Best fitting parameter values: $s = 0.020$; $p = 0.016$ (root mean squared deviation = 0.03).
The example responses assume that the target is a picture of a cat.

(Schwartz et al., 2006). The model explained 94.4% of the total variance in naming proportions. There were two deviating patterns. One was attributable to the model's failure to account for perseverations. The other deviant pattern featured low rates of correctness with errors largely restricted to the categories of semantic errors and omissions. This "pure semantics" pattern of deviation highlights limitations in the model's treatment of semantic errors (Dell et al., 1997; Foygel & Dell, 2000; Rapp & Goldrick, 2000). We consider this further in the following section.

56.3 MODEL-INSPIRED LESION ANALYSIS OF SEMANTIC ERRORS

The case series methods we used in the model-fitting studies is well-suited to group-level voxel-based lesion-symptom mapping (VLSM: Bates et al., 2003; for discussion of case series methods, see Schwartz & Dell, 2010). VLSM aims to identify brain voxels anywhere in the brain that carry a correlation between lesion status and symptom severity of sufficient strength to pass a statistical threshold that corrects for the thousands of voxels tested. VLSM was inspired by functional neuroimaging, and the resulting brain maps resemble those produced in fMRI studies. Where fMRI maps tell us which areas activate during a particular cognitive task or process, VLSM maps tell us where lesions cause derailments in that task or function. We used this technique to address controversies surrounding the cognitive and neural basis of semantic error production.

The interactive two-step model associates all semantic errors with the transmission of activation from semantic to lexical representations (i.e., the error locus is *postsemantic* and *production-specific*). Many, ourselves included, have questioned the adequacy of this account. Studies have shown that some patients, including some with the "pure semantics" naming pattern, exhibit semantic difficulties in comprehension as well as production (Cloutman et al., 2009; Gainotti, Miceli, Caltagirone, Silveri, & Masullo, 1981; Hillis, Rapp, Romani, & Caramazza, 1990; Rapp & Goldrick, 2000; Schwartz et al., 2006). This raises the possibility that in stroke aphasia, as in some degenerative dementias, semantic naming errors might be due to compromised semantic representations. A second possibility is that aphasics' semantic errors result from an executive function disorder that affects the regulation of competition in the semantic system or more generally (Badre & Wagner, 2007; Corbett, Jefferies, & Lambon Ralph, 2011; Hamilton, Martin, & Burton, 2009; Jefferies & Lambon Ralph, 2006; Noonan, Jefferies, Corbett, & Lambon Ralph, 2010; Schnur, Schwartz, Brecher, & Hodgson, 2006; Schnur et al., 2009; Thompson-Schill, D'Esposito, Aguirre, & Farah, 1997; Whitney, Kirk, O'Sullivan, Lambon Ralph, & Jefferies, 2012).

Although most of the evidence for an executive-deficit account of semantic errors comes from experimental paradigms that exaggerate semantic competition and, hence, the need for executive control, Jefferies and Lambon Ralph (2006) argued that an executive deficit may be responsible for semantic errors in conventional naming as well. They supported this with evidence that the naming errors of stroke aphasics included semantic errors like apple → "worm," where target and error are related to one another thematically rather than taxonomically, as is typical of most semantic errors (e.g., apple → pear). Jeffries and Lambon Ralph hypothesized that thematic error production in conventional naming is symptomatic of failure to access semantic representations in accordance with task goals—an executive function deficit they attributed to damage in left frontal and/or temporoparietal regions (Corbett et al., 2011; Noonan et al., 2010).

We examined these alternative accounts in a study that used VLSM to map the lesions that correlated with rates of taxonomic and thematic semantic errors on the PNT (Schwartz et al., 2011). As noted, taxonomic errors are the most common type of semantic error in naming; such errors are categorically related to the target as coordinate, superordinate, or subordinate. Much rarer is the thematic semantic error in which the target and error are from different categories but often play complementary roles in events and sentences. Most error-coding schemes combine taxonomic and thematic errors into a general semantic error category; this is also true of the PNT coding scheme (Table 56.1). For the 2011 study, we subdivided the semantic error corpus by means of expert and normative judgments.

Separate counts of taxonomic and thematic errors were derived for the 86 aphasic individuals who participated, and their shared variance was regressed out of each measure. For both measures, we also regressed out the shared variance attributable to performance on a semantic comprehension test with a high requirement for semantic control (*Camel and Cactus Test* or CCT; Bozeat, Lambon Ralph, Patterson, Garrard, & Hodges, 2000; Jefferies & Lambon Ralph, 2006). We then conducted separate VLSM analyses on each of the dependent measures.

As expected, taxonomic errors ($N = 645$) predominated over thematic errors ($N = 134$). Many patients made all or mostly taxonomic errors, but none did the reverse. Anatomically, however, there was a clear taxonomic-thematic double dissociation (Figure 56.2). Taxonomic errors localized to the left anterior temporal lobe, including clusters of supra-threshold voxels in

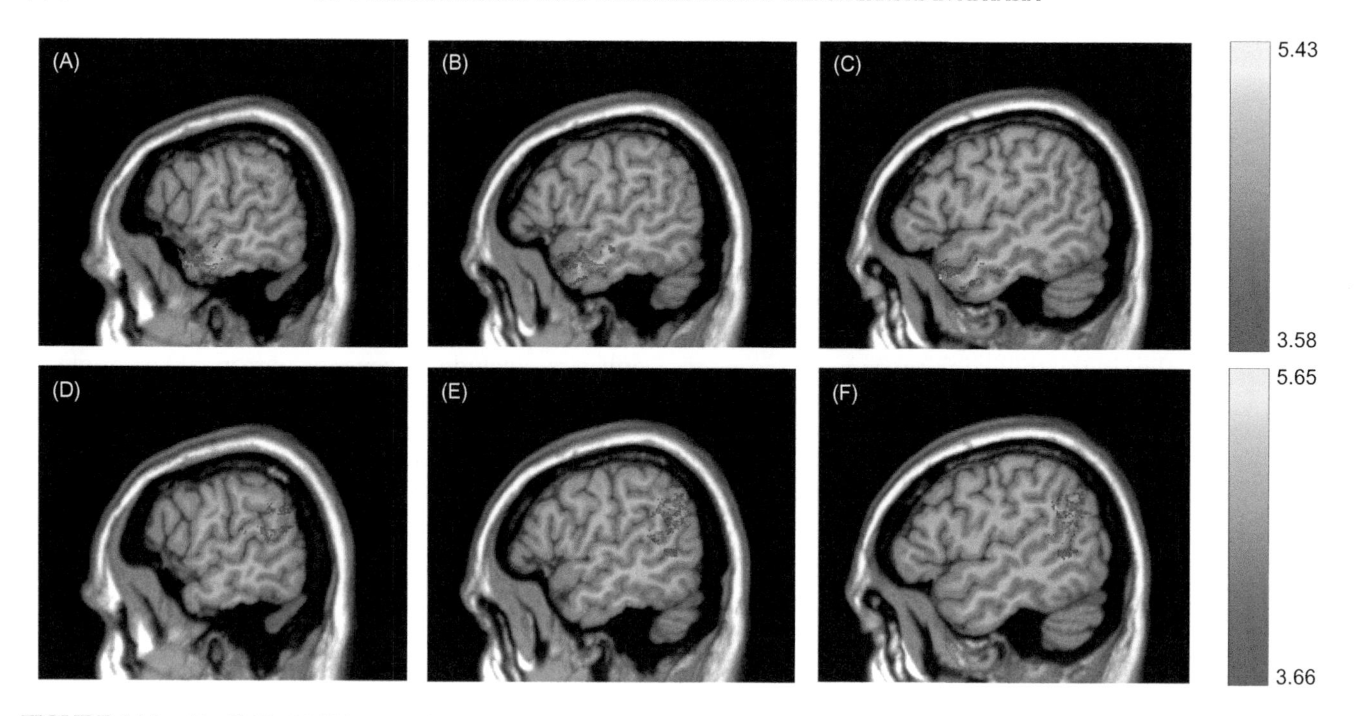

FIGURE 56.2 (A–C) The VLSM t-maps for *taxonomic semantic errors*, residualized for thematic semantic errors and for a measure of semantic comprehension. (D–F) The map for *thematic semantic errors*, residualized for taxonomic errors and semantic comprehension. Both maps were thresholded at a false discovery rate correction, $q = 0.02$; critical t-value for taxonomic errors was 3.58 and critical t-value for thematic errors was 3.66. Maps are rendered on the MNI-space Colin27 template, at x-coordinates $X = -60$ (A, D); $X = -56$ (B, E); $X = -52$ (C, F). *(Redrawn from Schwartz et al. (2011)).*

temporal pole (BA 38) and anterior portions of the middle and inferior temporal gyri (BA 20 and BA 21). Thematic errors, in contrast, localized to left temporoparietal cortices (TPCs), with the largest concentration of supra-threshold voxels in the angular gyrus (BA 39).

The ATL and TPC regions identified in these analyses are known from both neuropsychological and neuroimaging research to play important roles in multimodality semantic processing. Both have been cited as possible semantic "hubs" or high-level "convergence zones" (Binder & Desai, 2011; Binder, Desai, Graves, & Conant, 2009; Damasio, Tranel, Grabowski, Adolphs, & Damasio, 2004; Patterson, Nestor, & Rogers, 2007). However, because the analyses by Schwartz et al. (2011) statistically controlled for semantic comprehension, the implication for cognitive theory would seem to be that taxonomic and thematic semantic errors arise at a level of the production system that does not share resources with semantic comprehension. This goes along with the model's claim that semantic errors arise subsequent to semantic processing in the course of retrieving the correct word.

Schwartz et al. (2011) suggested that the anatomical double dissociation between taxonomic and thematic errors might be evidence of complementary semantic hubs in ATL and TPC, the former specialized for taxonomic (similarity) relations and the latter specialized for thematic (contiguity) relations. Although there is some supporting evidence for this position (Kalénine et al., 2009; Mirman & Graziano, 2012), much more research is required to establish whether the ATL and TPC process complementary semantic information, and whether these regions are best conceived as multimodality convergence zones or amodal hubs (Binder & Desai, 2011; Patterson et al., 2007). Schwartz et al.'s (2011) finding that regions within the left ATL and left TPC are uniquely or disproportionately concerned with semantically driven production would seem to argue for modality specificity, at least within left hemisphere sectors of these bilateral networks (for related evidence and discussion, Schwartz et al., 2009; Walker et al., 2011). However, Ueno, Saito, Rogers, and Lambon Ralph (2011) have proposed an alternative account of the ATL locus of semantic errors in naming that may be compatible with the amodal account of ATL semantics.

Finally, we return to the suggestion that the TPC is involved in executive control, and that the production of thematic errors in naming is symptomatic of an inability to access semantic processing in accordance with task demands. Two pieces of evidence in the Schwartz et al. (2011) study argue against this interpretation. First, regressing out the shared variance with CCT scores controlled for deficits in task-appropriate semantic selection affecting comprehension. Second, the study also included a VLSM of semantic circumlocution errors (e.g., apple → "they're crunchy to eat"). From a semantic control perspective, circumlocutions and thematic errors are both "off task" (circumlocutions describe rather than name the object; thematic errors express a frequent associate). Nevertheless, the VLSM findings indicate that circumlocutions are cognitively closer to taxonomic errors in that they strongly localize to the ATL. These data oppose the view that the left TPC subserves semantic control in a task general manner. Either it plays no role in semantic control or its role is modulated by task (production rather than comprehension) and the type of semantic information that is regulated.

As we noted, the study by Schwartz et al. (2011) of semantic errors confirmed the model's claim that these errors arise at a postsemantic, production-specific stage of lexical access. It reached this conclusion by controlling for alternative sources of semantic errors. However, insofar as all of these alternative sources contribute to observed semantic error counts, we would expect them to be absorbed into the fitted *s*-weight (hereafter, *s*) parameter and reflected in lesion sites that correlate with *s*. We examined this as part of a recent investigation of the lesions that correlate with the model parameters (Dell, Schwartz, Nozari, Faseyitan, & Coslett, 2013). We found that the voxels that correlated with *s* were widely distributed in the left hemisphere; in addition to ATL and TPC, there was a large concentration of such voxels in the prefrontal cortex (middle and inferior frontal gyri). Whether these prefrontal regions are part of a large distributed semantic-feature network or play a specific role in semantic control is a question for future research. What is clear, though, is that *s* itself is affected by processing at the semantic level in addition to the production-specific mapping from semantics to words.

56.4 SUMMATION DUAL-ROUTE MODEL OF REPETITION

This section takes a detailed look at the model's account of phonological errors in naming and its close relative, repetition. As in the discussion of semantic errors, we start by describing the relevant components of the model and how the implemented model was used to fit data from patients, followed by examination of the neural loci for the parameters fit by the model.

In the development of the interactive two-step model, naming and repetition have been linked through the assumption that the model's second step (lexical-phonological retrieval) is used in both; therefore, for a given patient, the status of lexical-phonological retrieval in naming (as indexed by the proportion of phonological errors, or by the fitted *p-weight*, hereafter, *p*) should be predictive of his or her accuracy in repeating words. This claim has been tested in multiple single case and case series model-fitting studies (Dell et al., 1997; Hanley, Dell, Kay, & Baron, 2004; Martin, Dell, Saffran, & Schwartz, 1994; Nozari, Kittredge, Dell, & Schwartz, 2010), resulting in successive refinements of the model and the theory of repetition that frames it.

The current model instantiates a *summation dual-route* theory of repetition. Words are repeated via a lexical route and a nonlexical route. The lexical route corresponds to step 2 in the naming model; the nonlexical route is contained in the connections from auditory input directly to the output phoneme units. Activation generated over both routes combines in the phoneme units to generate the final response.

An example will make the summation dual-route model's mechanisms and parameters more concrete. Let us start with the patient whose error pattern in naming is given in Table 56.2. This individual makes many nonword errors in naming (0.30). In the model, these errors can only arise during the retrieval of phonological units; therefore, a high rate of such errors suggests difficulty in lexical-phonological retrieval or in other processes involving sublexical units. The low value of the fitted *p* parameter (0.016, where more than 0.040 is normal) captures this difficulty in the model. If word repetition were performed using the lexical route alone, then we would expect that this individual would make a comparable number of nonword errors in that task as well, because lexical-route word repetition is, in the model, nothing more than the second step of naming. We can use the model to make this prediction precise by running just that step of the model using the *s* and *p* parameters determined from the person's naming performance:

(3) *Predicted word repetition from lexical route model for sample patient:*

Correct	*Semantic*	*Formal*	*Mixed*	*Unrelated*	*Nonwords*
0.65	0.00	0.07	0.01	0.00	0.27

Notice that the predicted proportion of nonwords for this patient from this model of repetition is similar to what it was for naming (Table 56.2). It turns out, though, that a pure lexical-route approach to repetition does not work in this case. The individual's actual word repetition is considerably better than predicted:

(4) *Obtained word repetition:*

Correct	*Semantic*	*Formal*	*Mixed*	*Unrelated*	*Nonwords*
0.95	0.00	0.01	0.00	0.00	0.04

Thus, it is possible that a nonlexical route to repetition is contributing to word repetition. We can show this by first measuring the effectiveness of this person's nonlexical route, and then by using the model to see if that route can explain the good repetition performance.

To measure the nonlexical route, we test the individual's ability to repeat nonword stimuli that are of similar phonological complexity as the naming targets. This yields:

(5) *Obtained nonword repetition:*

Correct	*Lexicalization Errors*	*Nonword Errors*
0.37	*0.33*	*0.30*

Notice that repetition of nonwords is considerably worse (0.37 correct) than words (0.95 correct), a consistent finding in aphasia that demonstrates at least some lexical contribution to word repetition. Because of this fact, we know that the nonlexical route cannot alone explain word repetition ability.

Next, we set up the model so that it has a nonlexical route to repetition, that is, connections between the auditory representation of the stimulus and phoneme nodes that mediate its production (e.g., as in Hanley et al., 2004). Specifically, we make the strength of these connections (the model's *nl* parameter) just as strong as necessary to simulate the patient's nonword repetition ability:

(6) *Predicted nonword repetition* ($nl = 0.026$; $s = 0.020$; $p = 0.016$):

Correct	*Lexicalization Errors*	*Nonword Errors*
0.36	*0.22*	*0.42*

Now, we can finally simulate dual-route word repetition by running the model so that it combines the activation generated by the lexical route and that coming from the nonlexical route:

(7) *Predicted word repetition from summation dual-route model* ($nl = 0.026$; $s = 0.020$; $p = 0.016$):

Correct	*Semantic*	*Formal*	*Mixed*	*Unrelated*	*Nonwords*
0.91	*0.00*	*0.02*	*0.00*	*0.00*	*0.07*

In this case, the predicted repetition (0.91 correct) is quite close to what was obtained (0.95), thus supporting the dual-route approach to repetition.

The dual-route model of word repetition has been tested in studies that have duplicated the steps that we just illustrated with actual patient samples (Abel, Huber, & Dell, 2009; Dell, Martin, & Schwartz, 2007). In these studies, word repetition performance is often better predicted by the dual-route than a pure lexical route, or pure nonlexical route model. Moreover, the assumption that the two routes sum to create the output (Hillis & Caramazza, 1991) appears necessary to explain the findings of Nozari et al. (2010). They found that in a large group of patients, the contribution of the nonlexical route was strong enough to reduce the rate of nonword errors in repetition (in comparison to naming), yet its contribution did not detract from the strength of the lexical frequency effect. By this we mean that in repetition, just as in naming, low-frequency targets elicited more nonword errors than high-frequency targets; the difference for high- versus low-frequency targets was as large for repetition as for naming.

56.4.1 Behavioral and Neural Predictors of Dual-Route Model Parameters

The essential claim of the dual-route model is that word repetition is achieved by the sum of activations over the lexical route (largely determined by parameter *p*) and the nonlexical route (parameter *nl*). Thus, aphasics' success in repeating words should be predicted primarily by *nl* and *p*. Dell et al. (2013) tested this prediction with data collected from 103 patients. They developed a regression model with word repetition accuracy as the dependent variable, and parameters, *p*, *nl*, and a variety of neuropsychological measures (nonverbal semantic comprehension, verbal semantic comprehension, short-term memory, apraxia of speech) as independent variables.[2] The resulting model had adjusted $R^2 = 0.61$, with strong and significant contributions from both model parameters, *nl* and *p*. *nl* independently contributed 14% of the variance explained; *p* independently contributed 5%, a smaller but significant value. A composite measure of verbal semantic comprehension also contributed a small amount to the model, demonstrating an indirect influence of semantics on repetition, a result expected from the interactive property of the model. None of the other variables was significant.

When Dell et al. (2013) used the same dataset to explore the relationship between *p* and *nl*, they found an association between the parameters ($r = 0.46$) that was not expected. The model assumes that each parameter indexes the strength of a different set of

[2]All of the behavioral and lesion-mapping analyses performed by Dell et al. (2013) involving fitted parameters used the square-root transform of the values, which made parameter variation more equal across the scale and also made the distributions more normal.

weights, and that each set of weights can be damaged independently of the others. This assumption holds for *s* and *p*: these parameters were indeed uncorrelated in this data set. However, the scatterplot relating *p* and *nl* revealed that patients with weak *p*-weights also tended to have weak *nl* weights, and there was a notable lack of patients with low values of *nl* (reflecting poor nonword repetition) who had high values of *p*.

To further explore this association between the parameters, as well as the contributions of other psycholinguistic and neuropsychological measures in our test battery, Dell et al. (2013) conducted regressions with each parameter as the dependent variable. They found that the sole (positive) predictors of *nl* were parameter *p* and auditory discrimination. *p* was predicted positively by *nl* and negatively by the presence of speech apraxia (i.e., presence of apraxia predicts low values of *p* [weak *p*-weights] reflecting many nonword errors). These results suggest that the nonlexical repetition route, whose strength is indexed by *nl*, reflects a production ability that is shared with parameter *p*, along with the ability to process auditory input. The association of low *p* with speech apraxia indicates the involvement of some sort of articulatory motor process in the mapping from words to phonemes.

These relationships have been further studied with VLSM (Schwartz, 2014; Schwartz, Faseyitan, Kim, & Coslett, 2012) and voxel-based lesion-*parameter* mapping (VLPM; Dell et al., 2013). "Symptom" in this context refers to one of the response proportions generated from naming or repetition. "Parameter" refers to the values of s, *p*, or *nl* that the model assigns to individuals based on their response proportions. Parameters reflect the overall response distribution in ways that can be quite complex, so it is of interest to know how the two sets of results compare. The VLSM and VPLM analyses of phonological processes tell a similar story, and so we restrict the present discussion to the VLPM results reported by Dell et al. (2013).

Figure 56.3 shows the voxels whose damaged status predicted the values of the *p* and *nl* parameters in the study by Dell et al. (2013). The two lesion maps overlap substantially, with most of the shared voxels occupying the anterior, inferior parietal lobe (supramarginal and postcentral gyri). Voxels unique to *nl* were found in superior temporal auditory regions, specifically STG, the posterior third of the planum temporale, and the cortex at the juncture of the parietal and temporal lobes, including area Spt. Voxels unique to *p* occupied more superior regions of the parietal lobe along with portions of the insula, which is considered important for speech articulation (Dronkers, 1996). Notice how these findings accord with the previously presented behavioral data in which the values of *nl* and *p* were positively correlated, but *nl* was uniquely predicted by auditory discrimination and *p* was uniquely predicted by the presence/absence of apraxia. Dell et al. (2013) also identified the voxels associated with word repetition, and these were found to be closely associated with the *nl* and *p* voxels, as expected from the dual-route approach to word repetition. To be specific, the dual-route model predicts that word repetition should be most closely related to the sum of the *nl* and *p* parameters, because the model's output sums nonlexical and lexical sources of activation. Figure 56.4 illustrates the close similarity between

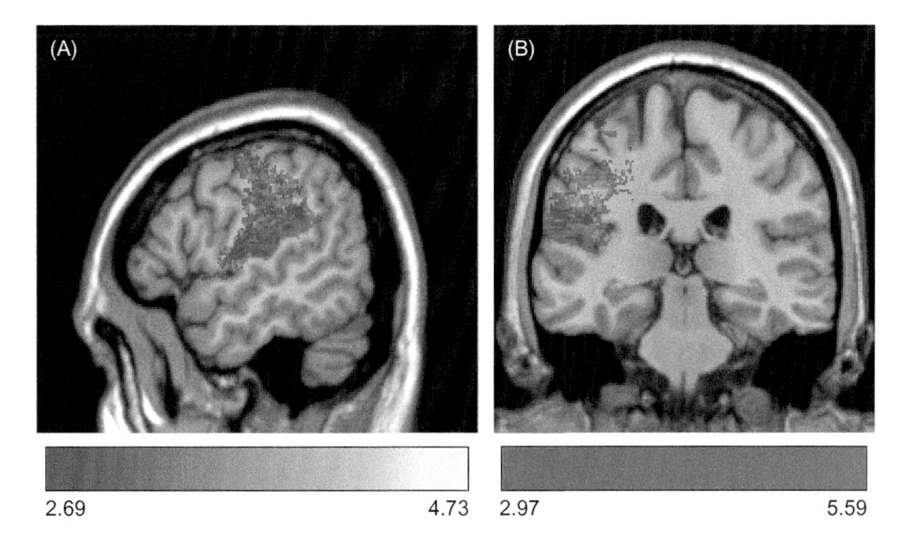

FIGURE 56.3 VLPM analyses of *p*-weight (red-yellow scale) and *nl*-weight (light blue to dark blue), both thresholded at a false discovery rate correction ($q = 0.05$), rendered on the MNI-space Colin27 template. The critical *t*-value for *p*-weight is 2.69, and for *nl*-weight it is 2.97. (A) A sagittal slice at MNI coordinate $x = -54$. (B) A coronal slice at MNI coordinate $y = -28$. *(Based on data reported in Dell et al. (2013) with permission from the publisher (Elsevier)).*

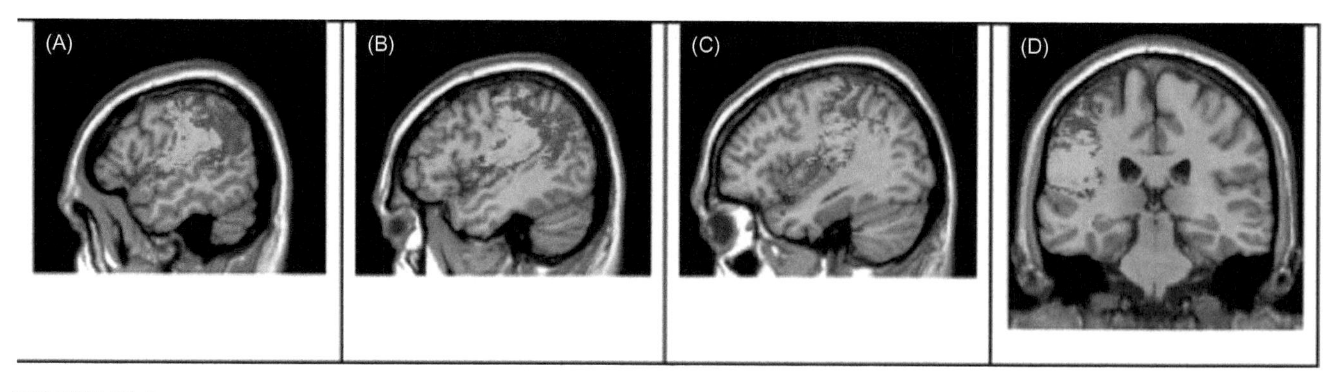

FIGURE 56.4 Lesion masks derived from the VLSM analyses of repetition accuracy (in red) and the VLSM analysis of *sum nl* + *p* (in blue; overlap shown in green). Statistical maps used to create masks were thresholded at a false discovery rate correction ($q = 0.05$) and rendered on the MNI-space Colin27 template. The critical *t*-value for repetition accuracy was 2.42 and for *sum nl* + *p* it was 2.69. Sagittal slices are at MNI coordinate $x = -54$, $x = -46$, $x = -38$; coronal slices are at MNI coordinate $y = -28$. *(From Dell et al. (2013) with permission from the publisher (Elsevier)).*

repetition voxels and those that associated with the *nl* + *p* sum.

Parameter *p* is not just associated with word repetition, though. Recall that it is derived from naming, not repetition. Specifically, in the model, *p* indexes step 2 of the lexical access process in naming, as determined largely on the basis of phonemic errors. It should come as no surprise, then, that a VLSM of phonemic errors in naming yielded a map very similar to the one shown here for *p* (i.e., centered on frontoparietal sensory-motor cortices) (Schwartz et al., 2012). What is puzzling, though, is that *p* was not found to be associated with Wernicke's area and surrounding posterior temporal and TPCs that are the long-hypothesized locus for lexical-phonological forms (Wernicke, 1874/1969; and more recently, Graves, Grabowski, Mehta, & Gupta, 2008; Indefrey & Levelt, 2004; Wilson, Isenberg, & Hickok, 2009). We explore this again in the following section of this chapter. First, though, we consider the implication of these findings for the model itself.

We learned from the regression and lesion mapping analyses that parameters *nl* and *p* are not as separate as the model originally claimed. On this basis, Dell et al. (2013) suggested a revision of the model that expands both *p*'s and *nl*'s functions such that they have a common function and distinct ones. The common function includes phonological representation. In the implemented model, the units of phonological representations were phonemes. However, finding that the *p*–*nl* overlap localizes to parietal cortices (supermarginal and postcentral) generally associated with sensory-motor processes invites speculation that phonological representations have a sensory motoric character (Gow, 2012). This is not inconsistent with linguistic theory, where the role of audition and articulation in shaping phonological generalizations is readily acknowledged (Cole & Hualde, 2003). Moreover, the theory of articulatory phonology (Browman & Goldstein, 1992) assumes that phonological forms consist of temporally coordinated gestures rather than abstract discrete segments. This approach to phonology is increasingly being used to interpret both linguistic and psycholinguistic data (Goldstein, Pouplier, Chen, Saltzman, & Byrd, 2007).

The distinct functions of *nl* and *p* also appear to be grounded in sensory-motor processes: *nl*'s distinct function includes the auditory processing of speech and its translation into phonological units, whereas *p*'s distinct functions include the mapping from lexical forms to phonemes (i.e., the original *p*-weights) and aspects of articulation. The partial sensory-motor characterization is supported by *nl*'s positive association with auditory discrimination ability and the association of apraxia of speech with low values of *p*.

Dell et al. (2013) also discussed the necessity of revising the model's treatment of the *s* parameter in light of the broadly dispersed temporal and frontal areas comprising the lesion map for *s* (see their Figure 7). They suggested that *s* not only is the lexical-semantic connections but also includes semantic representations and processes that control them. This treatment of *s* retains the original model's claim that *s* is a separate parameter from *p*, as supported by the finding of no correlation between the parameter values and little similarity in their brain maps.

56.5 IMPLICATIONS FOR NEUROCOGNITIVE MODELS OF LANGUAGE

The model's account of cognitive processes in naming and repetition and the lesions that compromise those processes have much in common with the Dual Stream theory (Hickok & Poeppel, 2004). In particular, the distribution of *nl* and *p* is consistent with the dorsal route's central role in the repetition of non-words (here, represented by *nl*) and words (both *nl* and *p*). More specifically, we propose that *p* and *nl* together represent the action of the dorsal stream and its role in the repetition of verbal stimuli (Baldo, Katseff, & Dronkers, 2012; Buchsbaum et al., 2011; Fridriksson et al., 2010; Hickok & Poeppel, 2004). The dorsal stream is largely distinct from the processing associated with parameter *s*, which indexes semantic processes (in the ventral stream) and their use during production. Given that parameter *p* is derived solely from performance in the naming task, this means that the part of the dorsal stream associated with *p* plays an important role during language production from meaning.

To flesh out this proposal, we now turn to two recent models that extend the Dual Stream framework in ways that make useful contact with the present approach. The *Lichtheim2 model* of Ueno et al. (2011) and Hickok's (2012, 2014) *hierarchical state feedback control* (HSFC) model are computational, neurally specified models that deal with both word retrieval from meaning and repetition of phonological forms. Each has points of convergence with the models and data presented in this chapter.

The *Lichtheim 2 model* of Ueno et al. (2011) is a multi-leveled parallel-distributed processing (PDP) model of language that links model processes with dorsal and ventral pathways in the brain. The name, Lichtheim 2, reflects the fact that this model, like its namesake, is concerned with the major aphasia syndromes. Ueno et al. trained the model to name, comprehend, and repeat a large set of words, after which they lesioned it at different points along its dorsal and ventral route to simulate the qualitative profiles of impaired and spared task performance typical of BA, WA, and CA.

To appreciate how Lichtheim 2 relates to our model, some further details are necessary. In Ueno et al.'s implementation, the dorsal path includes auditory cortex and surrounding areas, the inferior SMG, and insular-motor cortex. The ventral path also links auditory and insular-motor cortex, but through the temporal and frontal lobes. The model's connections link the various layers of these pathways, and their strengths are learned through training. To repeat words, the model learned to map from the auditory input to motor output layers; to produce them, it learned to map from the semantic layer (associated with the ventral anterior temporal lobe) to motor output. Because of the model's learning algorithm and its interactive architecture, which allows for activation to flow bidirectionally, both the dorsal and ventral paths contribute to both repetition and naming. Nonetheless, there is some specialization in the paths. The dorsal path is more important for the systematic mapping between sound and articulation, whereas the ventral path is specialized for the unsystematic mapping between word meaning and word form.

Lichteim 2's dorsal pathway specialization makes it well-suited for explaining our finding that *nl* and *p* mapped to dorsal stream areas, and particularly the SMG, which the Lichtheim 2 model associates with extracting and representing the statistical structure shared between speech sounds and phonotactics. Moreover, the role of *p* in naming is expected in that model from the fact that the dorsal path also makes a contribution to naming. The model also provides a good account of our finding that parameter *s* is strongly associated with temporal and frontal cortex. Finally, the fact that verbal semantic ability has a positive effect on word repetition is expected from the interactive property of the model.

Hickok's (2012, 2014) HSFC model links psycholinguistic approaches to production theories of motor control in speech. Word forms are retrieved and spoken through a control network involving phonological targets at both the syllable and phoneme level; the control network involves corresponding motor programs for these units and acoustic and somatosensory feedback to the target representations (Guenther, 2006). The crucial part of the HSFC model for our purposes is that it hypothesizes different brain circuits for programming syllable and phoneme-level units. The retrieval of whole syllable units involves a mapping between Wernicke's area (pSTG), which contains auditory syllable targets, and BA44, which is part of Broca's area containing syllable motor programs. Retrieval of phoneme units proceeds through the anterior supramarginal gyrus, which contains somatosensory phoneme targets, to vBA6-M1 for phoneme motor programs.

The HSFC model's phonemic control circuit corresponds well with our lesion map for *p* and with the overlap of *p* and *nl*. Parameters *p* and *nl* derive largely from phonemic errors, that is, responses such as "cap" or "cag" for CAT. So the fact that these parameters are more in line with the phonemic control circuits than with the syllabic control circuits supports this fundamental division in the HSFC model and associates its sensorimotor representations of phoneme targets with phonological error production. This allows us to explain why these parameters mapped to parietal regions rather than to Wernicke's area.

56.6 CONCLUSION

The central notion of this chapter is that data from aphasia provide a key link between behavioral studies of speech errors and neurocognitive models of language production. We reviewed how patterns of slips by unimpaired speakers led to models that identified representations and processing mechanisms in production. These models were then made concrete by developing computational implementations, such as our interactive two-step model. Once these computational models had been set up to explain speech error patterns, they could be altered (or "lesioned") to explain aphasic error patterns. Our model incorporates a two-step lexical access process and a nonlexical process for accessing phonology directly from auditory input. It attributes the variety of aphasic error patterns to variation in model parameters: the semantic (*s*), phonological (*p*), and nonlexical (*nl*) parameters. In a final step, we have mapped the lesions that correlate with error patterns and parameters in large numbers of individual patients to link model processes with brain areas.

Results of these studies have ramifications for contemporary neurocognitive accounts of language and brain. For example, we identified regions of the left anterior temporal and temporoparietal regions where lesions disrupt semantically driven word production (step 1 retrieval) over and above their impact on semantic comprehension. Lesions in these regions render the system vulnerable to mis-selection of words that share a categorical (anterior temporal region) or thematic (temporoparietal region) relation to the target. These regions may be part of complementary semantic hubs for language. They are also part of the ventral pathway for language, showing that, in addition to its well-established role in comprehension, the ventral route makes an essential contribution to the retrieval of words based on their meaning.

Lesions in the dorsal brain pathway also compromise the production of words from meaning (e.g., naming), but here the disruptive effect of lesions centers on step 2 processes. We found that parameter *p*, which indexes step 2 processes in naming and repetition, correlated with lesions in the left frontoparietal cortex and insula. The lesion map for *nl*, which indexes the nonlexical contribution to repetition, overlapped with *p* in postcentral and supramarginal gyri and extended into posterior dorsal route sectors in Spt and STG. We propose that the area of overlap identifies the brain's representation of phonological units accessed in production and that its anterior parietal distribution indicates that these representations might have a sensory-motor character.

Like all methods, ours have their limitations. From one perspective, the case series on which we based the analyses are too restrictive. Limiting the sample to patients with stroke aphasia guaranteed poor lesion coverage in areas such as the inferior temporal lobe and fusiform gyrus. These areas have been implicated in lexical-phonological retrieval (DeLeon et al., 2007; Lüders et al., 1991) but are generally spared in aphasia-producing strokes. Thus, our methods may underestimate the contribution to aphasic speech errors from these (and other) underpowered brain regions.

From another perspective, our case series samples may suffer from being overly inclusive. By grouping together evidence from patients of all different types, our analyses may have obscured differences in the mechanisms and corresponding lesion sites of phonological errors produced by patients with CA versus BA, for example (Romani & Galluzzi, 2005). This goes back to the long-standing debates about the relevance of the clinico-anatomical model, with which we began the chapter. However, there is good reason to be optimistic that the resolution to these debates is within reach. Case series methods can accommodate increases in *both* breadth and depth of analysis (i.e., broader lesion coverage and examination of effects in different subgroups; Schwartz & Dell, 2010). The limiting factor is sample size, but the expansion of institutional recruitment infrastructures and multisite collaborations makes this a surmountable problem. The most important reason for optimism is the range of sophisticated tools that are available to aphasia researchers today. Our group's work exemplifies the application of computational modeling and voxelwise mapping of model-defined symptoms and processing parameters. Others are correlating behavioral deficits with fMRI and with a multimodal characterization of tissue damage (e.g., structural damage, hypoperfusion, and/or white matter connectivity: Crinion, Warburton, Lambon Ralph, Howard, & Wise, 2006; Fridriksson et al., 2010; Han et al., 2013; Hillis et al., 2001; Turken & Dronkers, 2011). These advances in the neurocognitive characterization of impaired language complement neuroimaging studies of normal performance, making aphasia research as central to theory development today as it has been historically.

Acknowledgments

Preparation of this chapter was supported by grant RO1 DC000191-32 from the National Institute on Deafness and Other Communication Disorders (NIDCD), National Institutes of Health (NIH).

References

Abel, S., Huber, W., & Dell, G. S. (2009). Connectionist diagnosis of lexical disorders in aphasia. *Aphasiology*, *23*, 1353–1378.

Allport, D. A. (1984). Speech production and comprehension: One lexicon or two? In W. Prinz, & A. F. Sanders (Eds.), *Cognition and motor processes*. Berlin: Springer-Verlag.

Badre, D., & Wagner, A. D. (2007). Left ventrolateral prefrontal cortex and the cognitive control of memory. *Neuropsychologia, 45*, 2883–2901.

Baldo, J. V., Katseff, S., & Dronkers, N. F. (2012). Brain regions underlying repetition and auditory-verbal short-term memory deficits in aphasia: Evidence from voxel-based lesion symptom mapping. *Aphasiology, 26*, 338–354. Available from: http://dx.doi.org/doi:10.1080/02687038.2011.602391.

Bates, E., Wilson, S. M., Saygin, A. P., Dick, F., Sereno, M. I., Knight, R. T., et al. (2003). Voxel-based lesion-symptom mapping. *Nature Neuroscience, 6*, 448–450.

Binder, J. R., & Desai, R. H. (2011). The neurobiology of semantic memory. *Trends in Cognitive Sciences, 15*(11), 527–536. Available from: http://dx.doi.org/doi:10.1016/j.tics.2011.10.001.

Binder, J. R., Desai, R. H., Graves, W. W., & Conant, L. L. (2009). Where is the semantic system? A critical review and meta-analysis of 120 functional neuroimaging studies. *Cerebral Cortex, 19*(12), 2767–2796.

Blumstein, S. E. (1973). *A phonological investigation of aphasic speech*. The Hague: Mouton.

Bozeat, S., Lambon Ralph, M. A., Patterson, K., Garrard, P., & Hodges, J. R. (2000). Non-verbal semantic impairment in semantic dementia. *Neuropsychologia, 38*, 1207–1215.

Browman, C. P., & Goldstein, L. (1992). Articulatory phonology: An overview. Haskins Laboratories Status Report on Speech Research, Report no. 111/112, pp. 23–42.

Buchsbaum, B. R., Baldo, J. V., Okada, K., Berman, K. F., Dronkers, N., D'Esposito, M., et al. (2011). Conduction aphasia, sensory-motor integration, and phonological short-term memory—An aggregate analysis of lesion and fMRI data. *Brain and Language, 119*(3), 119–128.

Buckingham, H. D. (1980). On correlating aphasic errors with slips-of-the-tongue. *Applied Psycholinguistics, 1*, 199–220.

Buckingham, H. D. (1987). Phonemic paraphasias and psycholinguistic production models for neologistic jargon. *Aphasiology, 1*, 381–400.

Butterworth, B. (1979). Hesitation and the production of verbal paraphasias and neologisms in jargon aphasia. *Brain and Language, 8*, 133–161.

Caplan, D., Vanier, M., & Baker, E. (1986). A case study of reproduction conduction aphasia: 1. Word production. *Cognitive Neuropsychology, 3*, 99–128.

Cloutman, L., Gottesman, R., Chaudhry, P., Davis, C., Kleinman, J. T., Pawlak, M., et al. (2009). Where (in the brain) do semantic errors come from? *Cortex, 45*(5), 641–649.

Cole, J. S., & Hualde, J. I. (2003). *Papers in laboratory phonology 9*. New York, NY: Mouton de Greyter.

Compston, A. (2006). From the archives: On aphasia. By L. Lichtheim, *Brain, 129*(6), 1347–1350. Available from: http://dx.doi.org/10.1093/brain/awl134.

Corbett, F., Jefferies, E., & Lambon Ralph, M. A. (2011). Deregulated semantic cognition follows prefrontal and temporo-parietal damage: Evidence from the impact of task constraint on nonverbal object use. *Journal of Cognitive Neuroscience, 23*, 1125–1135.

Crinion, J. T., Warburton, E. A., Lambon Ralph, M. A., Howard, D., & Wise, R. J. S. (2006). Listening to narrative speech after aphasic stroke: The role of the left anterior temporal lobe. *Cerebral Cortex, 16*, 1116–1125.

Cuetos, F., Aguado, G., Izura, C., & Ellis, A. W. (2002). Aphasic naming in Spanish: Predictors and errors. *Brain and Language, 82*, 344–365.

Damasio, H., Tranel, D., Grabowski, T., Adolphs, R., & Damasio, A. (2004). Neural systems behind word and concept retrieval. *Cognition, 92*(1–2), 179–229.

DeLeon, J., Gottesman, R. F., Kleinman, J. T., Newhart, M., Davis, C., Heidler-Gary, J., et al. (2007). Neural regions essential for distinct cognitive processes underlying picture naming. *Brain, 130*, 1408–1422.

Dell, G.S. (1980). *Phonological and lexical encoding in speech production: An analysis of naturally occurring and experimentally elicited slips of the tongue*. Unpublished doctoral dissertation, University of Toronto.

Dell, G. S., Martin, N., & Schwartz, M. F. (2007). A case-series test of the interactive two-step model of lexical access: Predicting word repetition from picture naming. *Journal of Memory and Language, 56*, 490–520.

Dell, G. S., & O'Seaghdha, P. G. (1991). Mediated and convergent lexical priming in language production: A comment on Levelt et al (1991). *Psychological Review, 98*(4), 604–614.

Dell, G. S., Schwartz, M. F., Martin, N., Saffran, E. M., & Gagnon, D. A. (1997). Lexical access in aphasic and nonaphasic speakers. *Psychological Review, 104*, 801–838.

Dell, G. S., Schwartz, M. F., Nozari, N., Faseyitan, O., & Coslett, H. B. (2013). Voxel-based lesion-parameter mapping: Identifying the neural correlates of a computational model of word production in aphasia. *Cognition, 128*, 380–396.

Dronkers, N. F. (1996). A new brain region for coordinating speech articulation. *Nature, 384*, 159–161.

Ellis, A. W. (1985). The production of spoken words: A cognitive neuropsychological perspective. In A. W. Ellis (Ed.), *Progress in the psychology of language* (pp. 107–145). Hillsdale, NJ: Erlbaum.

Ellis, A. W., Miller, D., & Sin, G. (1983). Wernicke's aphasia and normal language processing: A case study in cognitive neuropsychology. *Cognition, 15*, 111–114.

Foygel, D., & Dell, G. S. (2000). Models of impaired lexical access in speech production. *Journal of Memory and Language, 43*, 182–216.

Freud, S. (1953). *On aphasia* (E. Stengel, Trans.). New York, NY: International University Press, Inc.

Fridriksson, J., Kiartansson, O., Morgan, P. S., Hialtason, H., Magnusdottir, S., Bonilha, L., et al. (2010). Imparied speech repetition and left parietal lobe damage. *Journal of Neuroscience, 30*, 11057–11061.

Fromkin, V. A. (1971). The non-anomalous nature of anomalous utterances. *Language, 47*(1), 26–52.

Gainotti, G., Miceli, G., Caltagirone, C., Silveri, M. C., & Masullo, C. (1981). The relationship between type of naming error and semantic-lexical discrimination in aphasic patients. *Cortex, 17*, 401–410.

Garrett, M. F. (1975). The analysis of sentence production. In G. H. Bower (Ed.), *The psychology of learning and motivation* (pp. 133–175). London: Academic Press.

Garrett, M. F. (1980). Levels of processing in sentence production. In B. Butterworth (Ed.), *Language production* (Vol. 1, pp. 177–220). London: Academic Press.

Garrett, M. F. (1984). *The organization of processing structure of language production: Application to aphasic speech*. Cambridge, MA: MIT Press.

Geschwind, N. (1965). Disconnection syndromes in animals and man. Part II. *Brain, 88*, 585–644.

Goldrick, M., & Rapp, B. (2007). Lexical and post-lexical phonological representations in spoken production. *Cognition, 102*(2), 219–260.

Goldstein, L., Pouplier, M., Chen, L., Saltzman, E., & Byrd, D. (2007). Dynamic action units slip in speech production errors. *Cognition, 103*, 386–412.

Goodglass, H., & Kaplan, E. (1983). *The assessment of aphasia and related disorders* (2nd ed.). Philadelphia, PA: Lea & Febiger.

Gow, D. W. (2012). The cortical organization of lexical knowledge: A dual lexicon model of spoken language processing. *Brain and Language, 121*, 273–288.

Graves, W. W., Grabowski, T. J., Mehta, S., & Gupta, P. (2008). The left posterior superior temporal gyrus participates specifically in accessing lexical phonology. *Journal of Cognitive Neuroscience, 20* (9), 1698–1710.

Guenther, F. H. (2006). Cortical interactions underlying the production of speech sounds. *Journal of Communication Disorders, 39*, 350–365.

Hamilton, A. C., Martin, R. C., & Burton, P. C. (2009). Converging fMRI evidence for a role of the left inferior frontal lobe in semantic retention during language comprehension. *Cognitive Neuropsychology, 26*, 685–704.

Han, Z., Ma, Y., Gong, G., He, Y., Caramazza, A., & Bi, Y. (2013). White matter structural connectivity underlying semantic processing: Evidence from brain damaged patients. *Brain, 136*(10), 2952–2965. Available from: http://dx.doi.org/doi:10.1093/brain/awt205.

Hanley, J. R., Dell, G. S., Kay, J., & Baron, R. (2004). Evidence for the involvement of a nonlexical route in the repetition of familiar words: A comparison of single and dual route models of auditory repetition. *Cognitive Neuropsychology, 21*(2–4), 147–158.

Hickok, G. (2012). Computational neuroanatomy of speech production. *Nature Reviews Neuroscience, 13*, 135–145.

Hickok, G. (2014). The architecture of speech production and the role of the phoneme in speech processing. *Language and Cognitive Processes, 29*, 2–20.

Hickok, G., & Poeppel, D. (2004). Dorsal and ventral streams: A framework for understanding aspects of the functional anatomy of language. *Cognition, 92*(1–2), 67–99.

Hickok, G., & Poeppel, D. (2007). The cortical organization of speech processing. *Nature Reviews Neuroscience, 8*(5), 393–402.

Hillis, A. E., & Caramazza, A. (1991). Mechanisms for accessing lexical representations for output: Evidence from a category specific semantic deficit. *Brain and Language, 40*, 106–144.

Hillis, A. E., Rapp, B. C., Romani, D., & Caramazza, A. (1990). Selective impairment of semantics in lexical processing. *Cognitive Neuropsychology, 7*(3), 191–243.

Hillis, A. E., Wityk, R. J., Tuffiash, E., Beauchamp, N. J., Jacobs, M. A., Barker, P. B., et al. (2001). Hypoperfusion of Wernicke's area predicts severity of semantic deficit in acute stroke. *Annals of Neurology, 50*(5), 561–566.

Indefrey, P., & Levelt, W. J. M. (2004). The spatial and temporal signatures of word production components. *Cognition, 92*(1–2), 101–144.

Jefferies, E., & Lambon Ralph, M. A. (2006). Semantic impairment in stroke aphasia versus semantic dementia: A case-series comparison. *Brain, 129*, 2132–2147.

Kalénine, S., Peyrin, C., Pichat, C., Segebarth, C., Bonthoux, F., & Baciu, M. (2009). The sensory-motor specificity of taxonomic and thematic conceptual relations: A behavioral and fMRI study. *NeuroImage, 44*, 1152–1162.

Kertesz, A. (1982). *Western aphasia battery test manual* (2nd ed.). New York, NY: Grune & Stratton.

Kittredge, A. K., Dell, G. S., Verkuilen, J., & Schwartz, M. F. (2008). Where is the effect of frequency in word production? Insights from aphasic picture naming errors. *Cognitive Neuropsychology, 25* (4), 463–492.

Levelt, W. J. M., Roelofs, A., & Meyer, A. S. (1999). A theory of lexical access in speech production. *Behavioral and Brain Sciences, 22*, 1–75.

Lüders, H., Lesser, R. P., Hahn, J., Dinner, D. S., Morris, H. H., Wyllie, E., et al. (1991). Basal temporal language area. *Brain, 114*, 743–754.

MacKay, D. G. (1972). The structure of words and syllables: Evidence from errors in speech. *Cognitive Psychology, 3*, 210–227.

Martin, N., Dell, G. S., Saffran, E. M., & Schwartz, M. F. (1994). Origins of paraphasias in deep dysphasia: Testing the consequences of a decay impairment to an interactive spreading activation model of lexical retrieval. *Brain and Language, 47*, 609–660.

Miller, D., & Ellis, A. W. (1987). Speech and writing errors in "neologistic jargonaphasia": A lexical activation hypothesis. In M. Coltheart, G. Santori, & R. Job (Eds.), *The cognitive neuropsychology of language* (pp. 253–271). London: Lawrence Erlbaum.

Mirman, D., & Graziano, K. M. (2012). Damage to temporo-parietal cortex decreases incidental activation of thematic relations during spoken word comprehension. *Neuropsychologia, 50*(8), 1990–1997. Available from: http://dx.doi.org/doi:10.1016/j.neuropsychologia.2012.04.024.

Mirman, D., Strauss, T. J., Brecher, A., Walker, G. M., Sobel, P., Dell, G. S., et al. (2010). A large, searchable, web-based database of aphasic performance on picture naming and other tests of cognitive function. *Cognitive Neuropsychology, 27*, 495–504.

Nickels, L., & Howard, D. (1994). A frequent occurrence? Factors accepting the production of semantic errors in aphasic naming. *Cognitive Neuropsychology, 11*, 289–320.

Nickels, L., & Howard, D. (1995). Aphasic naming: What matters? *Neuropsychologia, 33*, 1281–1303.

Noonan, K. A., Jefferies, E., Corbett, F., & Lambon Ralph, M. A. (2010). Elucidating the nature of deregulated semantic cognition in semantic aphasia: Evidence for the roles of prefrontal and temporo-parietal cortices. *Journal of Cognitive Neuroscience, 22*, 1597–1613.

Nozari, N., Kittredge, A. K., Dell, G. S., & Schwartz, M. F. (2010). Naming and repetition in aphasia: Steps, routes, and frequency effects. *Journal of Memory and Language, 63*, 541–559.

Pate, D. S., Saffran, E. M., & Martin, N. (1987). Specifying the nature of the production impairment in a conduction aphasic: A case study. *Language and Cognitive Processes, 2*, 43–84.

Patterson, K., Nestor, P. J., & Rogers, T. T. (2007). Where do you know what you know? The representation of semantic knowledge in the human brain. *Nature Reviews Neuroscience, 8*, 976–987.

Plaut, D. C., & Shallice, T. (1993). Deep dyslexia: A case study of connectionist neuropsychology. *Cognitive Neuropsychology, 10*, 377–500.

Rapp, B., & Goldrick, M. (2000). Discreteness and interactivity in spoken word production. *Psychological Review, 107*(3), 460–499.

Roach, A., Schwartz, M. F., Martin, N., Grewal, R. S., & Brecher, A. (1996). The Philadelphia naming test: Scoring and rationale. *Clinical Aphasiology, 24*, 121–133.

Romani, C., & Galluzzi, C. (2005). Effects of syllabic complexity in predicting accuracy of repetition and direction of errors in patients with articulatory and phonological difficulties. *Cognitive Neuropsychology, 22*, 817–850.

Saffran, E. M. (1982). Neuropsychological approaches to the study of language. *British Journal of Psychology, 73*, 317–337.

Schnur, T. T., Schwartz, M. F., Brecher, A., & Hodgson, C. (2006). Semantic interference during blocked-cyclic naming: Evidence from aphasia. *Journal of Memory and Language, 54*, 199–227.

Schnur, T. T., Schwartz, M. F., Kimberg, D. Y., Hirshorn, E., Coslett, H. B., & Thompson-Schill, S. L. (2009). Localizing interference during naming: Convergent neuroimaging and neuropsychological evidence for the function of Broca's area. *Proceedings of the National Academy of Sciences of the United States of America, 106*(1), 322–327.

Schwartz, M. F. (1987). Patterns of speech production deficit within and across aphasia syndromes: Application of a psycholinguistic model. In M Coltheart, G. Sartori, & R. Job (Eds.), *The cognitive neuropsychology of language* (pp. 163–199). London: Erlbaum.

Schwartz, M. F. (2014). Theoretical analysis of word production deficits in adult aphasia. *Philosophical Transactions of the Royal Society B: Biological Sciences, 369*, 20120390.

Schwartz, M. F., & Dell, G. S. (2010). Case series investigations in cognitive neuropsychology. *Cognitive Neuropsychology, 27*, 477–494.

Schwartz, M. F., Dell, G. S., Martin, N., Gahl, S., & Sobel, P. (2006). A case-series test of the interactive two-step model of lexical access: Evidence from picture naming. *Journal of Memory and Language, 54*, 228–264.

Schwartz, M. F., Faseyitan, O., Kim, J., & Coslett, H. B. (2012). The dorsal stream contribution to phonological retrieval in object naming. *Brain, 135*, 3799–3814.

Schwartz, M. F., Kimberg, D. Y., Walker, G. M., Brecher, A., Faseyitan, O., Dell, G. S., et al. (2011). Neuroanatomical dissociation for taxonomic and thematic knowledge in the human brain. *Proceedings of the National Academy of Sciences of the United States of America, 108*, 8520–8524.

Schwartz, M. F., Kimberg, D. Y., Walker, G. M., Faseyitan, O., Brecher, A., Dell, G. S., et al. (2009). Anterior temporal involvement in semantic word retrieval: VLSM evidence from aphasia. *Brain, 132*(12), 3411–3427.

Schwartz, M. F., Saffran, E. M., Bloch, D. E., & Dell, G. S. (1994). Disordered speech production in aphasic and normal speakers. *Brain and Language, 47*, 52–88.

Shattuck-Hufnagel, S. (1979). Speech errors as evidence for a serial ordering mechanism in speech production. In W. E. Cooper, & E. C. T. Walker (Eds.), *Sentence processing: Psycholinguistic studies presented to Merrill Garrett* (pp. 295–342). Hillsdale, NJ: Erlbaum.

Stemberger, J. P. (1985). An interactive activation model of language production. In A. W. Ellis (Ed.), *Progress in the psychology of language* (Vol. 1, pp. 143–186). Hillsdale, NJ: Erlbaum.

Thompson-Schill, S. L., D'Esposito, M., Aguirre, G. K., & Farah, M. J. (1997). Role of left inferior prefrontal cortex in retrieval of semantic knowledge: A reevaluation. *Proceedings of the National Academy of Sciences of the United States of America, 94*(26), 14792–14797.

Turken, A. U., & Dronkers, N. F. (2011). The neural architecture of the language comprehension network: Converging evidence from lesion and connectivity analysis. *Frontiers in Systems Neuroscience, 5*, 1–20.

Ueno, T., Saito, S., Rogers, T. T., & Lambon Ralph, M. A. (2011). Lichtheim 2: Synthesising aphasia and the neural basis of language in a neurocomputational model of the dual dorsal-ventral language pathways. *Neuron, 72*(2), 385–396.

Walker, G. M., Schwartz, M. F., Kimberg, D. Y., Faseyitan, O., Brecher, A., Dell, G. S., et al. (2011). Support for anterior temporal involvement in semantic error production in aphasia: New evidence from VLSM. *Brain and Language, 117*, 110–122.

Wernicke, C. (1874/1969). The aphasic symptom complex: A psychological study on a neurological basis. Breslau: Cohn and Weigert. Reprinted in R.S. Cohen & M.W. Wartofsky (Eds.), *Boston studies in the philosophy of science* (Vol. 4). Boston, MA: Reidel.

Whitney, C., Kirk, M., O'Sullivan, J., Lambon Ralph, M. A., & Jefferies, E. (2012). Executive semantic processing is underpinned by a large-scale neural network: Revealing the contribution of left prefrontal, posterior temporal, and parietal cortex to controlled retrieval and selection using TMS. *Journal of Cognitive Neuroscience, 24*, 133–147.

Wilson, S. M., Isenberg, A. L., & Hickok, G. (2009). Neural correlates of word production stages delineated by parametric modulation of psycholinguistic variables. *Human Brain Mapping, 30*(11), 3596–3608.

CHAPTER

57

Motor-Timing and Sequencing in Speech Production

A General-Purpose Framework

Sonja A. Kotz[1,2] *and Michael Schwartze*[1]

[1]School of Psychological Sciences, University of Manchester, Manchester, UK;
[2]Department of Neuropsychology, Max Planck Institute for Human Cognitive and Brain Sciences, Leipzig, Germany

57.1 FORMAL AND TEMPORAL PREDICTION: FUNDAMENTALS IN SPEECH PROCESSING

The primary purpose of language in general and speech in particular is to communicate meaning. Yet, whatever constitutes the meaning of an utterance, it is conveyed in the form of an acoustic signal that uses spectral and rhythmic variations of sound to transmit respective information. Speech rhythm is a trivial phenomenon in this regard. However, speech rhythm is a much more complex phenomenon in every other respect. Accordingly, just as there are different concepts of meaning in linguistics, there are different notions of rhythm, which already diverge in the way they reflect on a rhythm's inherent temporal structure (Martin, 1972). In the first instance, basic forms of rhythm in speech should be differentiated from meter (i.e., variations at the syllabic level such as the alternation of stressed and unstressed syllables). Further, rhythm is not synonymous with regularity because any temporal pattern—even a random one—may be considered rhythmic as long as it does not prevent perceptual grouping (Repp, 2000). Somewhat avoidant of the ubiquitous discussion of temporal regularity, Fraisse (1982) simply noted that there is rhythm if we can predict on the basis of what is perceived. In other words, predictions concerning future events, based on past events, are considered inherent to the concept of rhythm (Martin, 1972). This line of thought shifts the perspective of rhythm and timing from a descriptive level to psychological and functional interpretations of rhythmic variation, including a neurofunctional assessment of rhythm and prediction in speech processing.

Generally speaking, the ability to form predictions based on rhythmic features is equally important for both speech production and perception. It bears the potential for allowing an individual to anticipate what will happen and when it will happen, and to adapt its behavior not only appropriately but also in an anticipatory fashion. As such, prediction is more and more recognized as one of the primitives of brain function and in opposition to a reflexive operating mode (Arnal & Giraud, 2012; Friston, 2012; Friston, Kilner, & Harrison, 2006; Raichle, 2010).

Prediction is crucial for the adequate selection and identification of behavior ("what dimension"), as well as for the optimal timing of behavior ("when dimension"). However, although prediction plays a role in both dimensions, it is necessary to accentuate the fundamental difference between them, such as the fact that it is entirely possible to generate predictions that specify the type of future events independent of predictions that specify the timing of the very same events. For example, when an utterance such as "Good morning, ladies and..." is expressed, a listener can use past acquired knowledge to predict the completion of the utterance by "gentlemen." This specific type of prediction has become a standard in language research in the form of cloze probability testing (Taylor, 1953). Yet, at the same time, the listener can rely on past knowledge to predict when this particular speech event will occur. Dependent on intentional and/or contextual parameters that ultimately impact

Neurobiology of Language. DOI: http://dx.doi.org/10.1016/B978-0-12-407794-2.00057-2

on speech tempo, the respective time point when this speech event will occur may differ considerably. In other words, predictive mechanisms can build on prior knowledge and coherent information about the type or form of speech events (i.e., the formal structure of events) and/or prior knowledge and coherent information about their temporal arrangement (i.e., the temporal structure of events). Naturally, these processes should go hand-in-hand to achieve successful speech processing.

The example illustrates that both formal and temporal predictions are crucial in speech processing. Research along these lines led to detailed knowledge about the role of prediction, for example, in the identification of lexical speech events under normal or adverse listening conditions (Gagnepain, Henson, & Davis, 2012; Sohoglu, Pelle, Carlyon, & Davis, 2012). The underlying processes include a context-sensitive, predictive top-down component that guides lexical speech perception (Hannemann, Obleser, & Eulitz, 2007; Sohoglu et al., 2012; Strauß, Kotz, & Obleser, 2013). This predictive component implies prior knowledge that, to provide any substantial behavioral benefit, has to exploit some form of regularity extracted from the speech signal. However, the speech signal is flexible enough to convey various types of formal and temporal regularity at once. To infer meaning, speech processing may reveal itself equally flexible and capable of using different kinds of information as it becomes available.

57.2 A SYNCHRONIZED SPEECH PROCESSING MODE

Regularized formal structure may give rise to predictions at several levels of description, including segmental, semantic, or syntactic relations. In a similar vein, temporal structure may give rise to predictions at several timescales. In addition to the issue of hierarchical versus strictly serial order, this fact inevitably introduces the dimensions of weighting and granularity (i.e., leading to questions of whether predictions on one level or scale are more important than others and whether predictions are generated simultaneously at all of these levels). Moreover, as indicated, formal and temporal structure may interact across different levels to guarantee smooth processing and optimal behavior. This rather complex background rules out simplistic mappings between the two primary areas of interest, the speech signal and neural processes. Accordingly, recent years have seen a continuous increase in refined theoretical and empirical work that aims at linking these domains on the basis of oscillatory mechanisms (Ghitza & Greenberg, 2009; Giraud et al., 2007; Peelle & Davis, 2012). This work provides support for functional links between specific properties of the speech signal and neuronal oscillations across different frequency bands. For example, it is considered that oscillations in the beta and gamma bands play a role in the processing of phonetic features, whereas the processing of syllables and words has been linked to oscillations in the theta band (Ghitza, 2012).

One of the novel concepts that emerge from this work is the so-called "theta-syllable" (Ghitza, 2012, 2013). The theta-syllable is defined as "a theta-cycle long speech segment located between two successive nuclei" (Ghitza, 2013, p. 5). The theta-syllable has been suggested to assume a function in speech decoding as part of a hierarchical window structure, which uses cascaded oscillations to synchronize with the input signal (Ghitza, 2013). As such, the concept of the theta-syllable seems largely compatible with the notion of a constant attempt to synchronize internal and external oscillations and to form predictions on the basis of "speech events such as the rhythmic succession of vocalic nuclei" (Kotz & Schwartze, 2010). Thus, in addition to the dimensions of weighting and granularity, the question arises regarding whether one particular rhythmic level may guide different processes, such as memory access or word segmentation (Ghitza, 2013), and/or the allocation of limited resources, such as attention during the perceptual integration of information retrieved from memory (Kotz & Schwartze, 2010; Schwartze & Kotz, 2013). Furthermore, one may speculate that the dissociation of formal and temporal structure also holds true in this specific case. That is, the proposed function of the theta-syllable embedded in a cascade of oscillations in speech decoding may be related to the processing of formal structure and the identification of specific elements in memory, whereas the more isolated oscillatory function in the allocation of attention may be related to the processing of temporal structure and the temporally predictive adaptation of behavior (Schwartze, Tavano, Schröger, & Kotz, 2012). However, although these accounts assign different functions to the same phenomenon, they both ground their reasoning in the multitier, syllable-centric perspective of speech processing proposed by Greenberg, Carvey, Hitchcock, and Chang (2003; Greenberg, 2006). This theory developed around the "STeP" (spectrotemporal profile), which depicts the fluctuation of acoustic energy across time and frequency that is associated with the syllable and the articulatory circle (i.e., the opening and the closing movements of the mouth during speech production). The energy level typically increases to a peak close to the syllabic nucleus, reflecting maximal oral aperture. Thus, the contour of the energy fluctuation approximates an arc, which extends from the syllabic onset

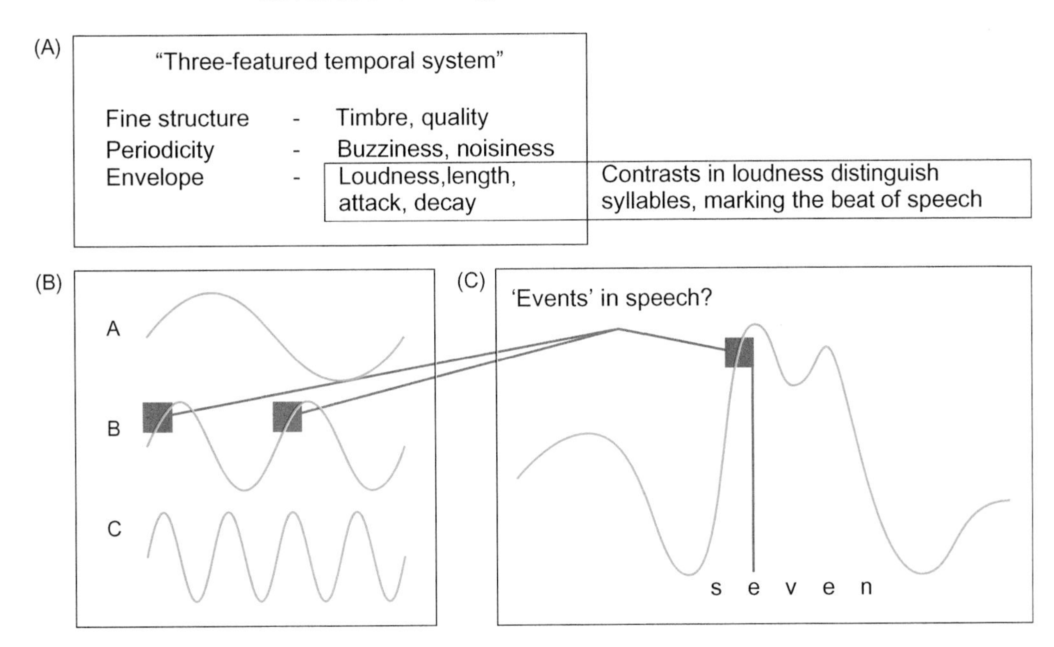

FIGURE 57.1 Different timescales in speech. Successive speech events such as phrases, words, syllables, or phonemes generate temporal structure at different timescales ranging from milliseconds to seconds. Physiological limits in temporal resolution set aside, the informational content of temporal structure may still be neglected at some scales, but it may contribute to successful speech processing on others. The value of temporal structure in this context probably varies with situational constraints. For example, the use of temporal structure may be more relevant under adverse conditions or, when perceived, regularity of temporal structure can be used to establish a predictive processing mode. Such a predictive mode may reflect a more general strive for optimal adaptation to a dynamic environment. It stands to reason whether there is a "default" scale that proves to be behaviorally most beneficial and, furthermore, whether auditory processing interacts with dedicated temporal processing at such a scale. To address these issues, the speech signal has to be decomposed on the basis of its temporal characteristics. Rosen (1992) suggests a "three-featured temporal system" (A) based on dominant fluctuation rates, which comprises fine structure, periodicity, and envelope levels. The envelope level (2–50 Hz) reflects acoustic features such as intensity, duration, rise and fall time, and their respective auditory correlates (loudness, length, attack, decay). A relative contrast in loudness distinguishes rhythmically prominent syllables, thus marking the "beat" of speech (Kochanski & Orphanidou, 2008). In a similar vein, Tillmann and Mansell (1980) propose a three-fold dissociation based on modulation rates into "ABC-prosody" (B). Slow modulations (A, melodic contour) and also fast modulations (C, intrinsic sound characteristics) convey a continuous impression, whereas intermediate rates are perceived as a series of pulses (B, rhythmic succession of syllables). This impression may be likened to a "perceptual beat" (Port, 2003) that occurs near the onset of vowels and that establishes a particularly salient event within a continuous speech stream. This notion is also central to the multitier framework introduced by Greenberg (2006), in which the nucleus of a syllable (especially stressed ones) sets the interpretational register for the whole syllable (Greenberg et al., 2003). Figure (C) is based on the time course of the so-called STeP (spectrotemporal profile; adapted from Greenberg et al., 2003), which has been derived from the energy contour of hundreds of individual realizations of the digit "seven." The relative increase in energy on or just prior to the peak in the vocalic nucleus may instantiate a distinct "event in speech" (red). Such events are the prime candidates for a nonlinear representation of the speech signal, and their scale and temporal structure may form the basis for an interaction between speech processing and dedicated temporal processing.

via the nucleus to the coda. The relative prominence of the nucleus, usually a vowel, is supposed to set a "register" for the decoding of other constituents of the speech signal. This functional interpretation of the STeP combines fundamental characteristics of speech production and speech perception in one cohesive conceptual framework. Accordingly, numerous links to other theories in both domains can be established.

Perhaps two of the more obvious points of contact exist in terms of the frame/content (f/c) theory of evolution of speech production (MacNeilage, 1998; MacNeilage & Davis, 2001) and in work related to the so-called perceptual-center phenomenon (p-center; Marcus, 1981; Morton, Marcus, & Frankish, 1976). Based on the dissociation of syllabic "frames" and segmental "content" elements, the former discusses the role of a simple biomechanical mechanism, that is, the mandibular cycle, in the acquisition and evolution of speech (MacNeilage & Davis, 2000). Essentially, this "register" versus "other constituents" dichotomy and the "frame" versus "content" dichotomy suggest hierarchies that may complement each other in speech processing, whereas they are anchored at different temporal levels or scales (Figure 57.1). In other words, whereas the notion of a register implies one event per syllable, a syllabic frame is defined by two events. However, the critical aspect in terms of the current discussion is the fact that temporal predictions based on the former scale would be predictive of intersyllabic temporal relations in the sense of the theta-syllable, whereas the latter would be predictive of intrasyllabic temporal relations (i.e., the duration of a syllable).

In principle, such a distinction may be relevant for different concrete realizations of speech, for example, the dissociation of stress-timed and syllable-timed languages. At a more abstract level, however, this difference raises the fundamental question regarding when a speech event such as the syllable becomes perceptually salient—is it at the point of the physical onset, corresponding to the starting point of an articulatory circle, its end, or at an intermediate point in time? At least for English, the aforementioned p-center phenomenon can be interpreted as pointing to the latter explanation.

The term p-center has been coined to describe "that what is regular in a perceptually regular sequence of speech sounds" (Marcus, 1981). The p-center marks the subjective moment, or point of occurrence, of an acoustic stimulus (Scott, 1998; Scott & McGettigan, 2012). Crucially, even for simple digit sequences, these points in time do not correspond to physical onset regularity (Morton et al., 1976). It is thus the perceived p-center regularity, rather than onset regularity, which may give rise to temporal predictions. In line with this proposal, Port (2003) discusses the p-center in the context of a neurocognitive system in which stimulus-driven internal oscillations create attractors in time for salient events. These, in turn, establish a framework for the processing of phonological units such as syllables and segments. One of the common denominators between such a system and the multitier framework based on the STeP is once more a hierarchical perspective of the speech signal and speech processing in which a particularly salient event sets the stage for the perceptual integration of other events. In terms of temporal structure, this perspective corresponds to a dissociation of a slower timescale or slower dynamic that guides processing on a faster dynamic timescale. A basic mechanism like this is probably not limited to speech perception. Likewise, it may be relevant in speech production in the sense that a speaker may involuntarily or voluntarily adjust slower dynamics to qualitatively adapt to the processing capacity of the listener. For example, regarding the p-center, speakers have been found to produce the asynchronies that are necessary to perceive a temporally regular sequence of speech events (Fowler, 1979). Moreover, on this basis, fundamental aspects of speech production and perception can be reflected on within a broader context (i.e., a communicative, general-purpose framework).

In previous work, we have discussed how the perceptual prominence of particular speech events, including the p-center, may serve to guide the predictive allocation of attention in time, which is expected to establish a "synchronized" and at the same time optimized speech processing mode (Kotz & Schwartze, 2010). From this integrative perspective, speech processing interacts with dedicated temporal processing mechanisms in an extended subcorticocortical network to efficiently exploit the temporal structure of speech (Kotz & Schwartze, 2010; Kotz, Schwartze, & Schmidt-Kassow, 2009; Schwartze, Tavano, et al., 2012). However, although the conceptual premise of this framework is equally applicable to perception and production, the primary focus, so far, has been on speech perception. In the following, we elaborate on its implications for speech production.

57.3 TIMING SPEECH: SUBCORTICO-CORTICAL INTERACTIONS

In this context, "temporal processing" refers broadly to the neural mechanisms that underlie the encoding, decoding, and evaluation of temporal structure. Neuroimaging research in this domain suggests several core anatomical substrates for temporal processing, including prefrontal cortical regions, the supplementary motor area (SMA), the basal ganglia (BG) and associated thalamocortical circuits, as well as the cerebellum (for recent reviews and a meta-analysis, see Allman, Teki, Griffiths, & Meck, 2014; Coull, Cheng, & Meck, 2010; Merchant, Harrington, & Meck, 2013; Spencer & Ivry, 2013; Wiener, Turkeltaub, & Coslett, 2010). It is important to note that these classical motor brain areas play a role in both the production and perception of temporal structure. A basic assumption in most of this research is that temporal processing engages some form of a specialized or dedicated system that is capable of representing the temporal relationship between events (Ivry & Schlerf, 2008). Another key aspect concerns the dissociation of event-based and interval-based mechanisms that have been proposed to rely primarily on the cerebellum and cortico-striato-thalamo-cortical loops, respectively (Buhusi & Meck, 2005; Spencer, Zelaznik, Diedrichsen, & Ivry, 2003). However, although temporal processing, articulatory processing, and auditory processing are naturally overlapping, the corresponding link is not particularly well-defined in terms of speech production.

As indicated, focusing on the temporal and essentially rhythmic nature of speech is more or less trivial if temporal structure is considered a mere by-product of "processing in time." Yet, the preceding sections have provided strong arguments in favor of a different viewpoint. Thus, temporal structure is considered to represent a valuable source of information in its own right. Moreover, the use of temporal structure serves the overarching goal to optimize predictive timing of behavior. The consequences arising from this perspective pose a serious challenge to neurofunctional models of speech processing. Any such model is incomplete unless it provides an adequate explanation with respect to how the temporal structure of the speech signal is generated, how it maps onto the temporal structure of neural processes, and how temporal information is used in predictive adaptation. Moreover, the immanent challenge of adding an explicit temporal dimension to the coordination between two dynamic systems (i.e., the acoustic signal and the brain) is further aggravated by the (at times) unspecific use of the term "temporal" in the context of speech processing, despite the fact that the time axis is the only common denominator across the various visual approximations that are typically used to capture the fleeting impression of the speech signal (McGettigan & Scott, 2012). However, there are some straightforward accounts (Schirmer, 2004) that discuss the putative connection between temporal processing and speech processing in a more detailed manner, for example, in terms of the cerebellar temporal processing system (Ackermann, Mathiak, & Riecker, 2007). The cerebellum and, more specifically, a right-cerebellar left-prefrontal loop have been proposed to play a specific role in the decoding of durational patterns in speech and nonspeech input in perception, and it appears to also engage in the temporal "shaping" of syllable templates in production (Ackermann et al., 2007). However, interactions between the cortico-striato-thalamo-cortical temporal processing system and speech processing are less specified, although it has been generally acknowledged as important to the ability to temporally structure the use of language and speech processing in particular (Allman et al., 2014; Buhusi & Meck, 2005).

The ability to temporally structure speech is, even though undoubtedly crucial, just a specific aspect of speech production. However, instead of summarizing in detail what is known about the neural correlates of speech production from neuroimaging research (for reviews and a meta-analysis, see Poeppel, Emmorey, Hickok, & Pylkkänen, 2012; Price, 2012; Scott, 2012; and Indefrey & Levelt, 2004, respectively), we focus on the proposed contribution of the dedicated temporal processing system to speech production. Likewise, we also refrain from an in-depth discussion of the well-documented role of the BG in motor control. Rather, on the basis of the preceding considerations, the goal here is to establish a functional link between theories in different domains that base their rationale on the same, or at least on overlapping, cortical and subcortical anatomical substrates (Figure 57.2).

Speech production is a complex multistage process linking intentions and conceptual ideas to articulation (Price, 2010). According to Levelt (2001), speech production involves processes of lemma selection, syllabification, and prosodification. These operate on the basis of a phonological code, which, in turn, comprises phonological segments that are incrementally inserted into a "morphological target frame." Concatenated frames are then expressed by means of appropriate articulatory gestures. However, to work efficiently, such processes require precise temporal coordination, because frames have to be "opened" and potentially "closed" or "filled" with segments differing in type as well as in number and "implemented." Moreover, speech production is essentially a sensorimotor process; speakers constantly operate a dual system (Indefrey & Levelt, 2004). This means that production and perception develop at the same time, thus adding another dimension of spatial, that is, (neuro)anatomical as well as temporal complexity. Hence, it is important to again emphasize why two neighboring domains such as speech processing and temporal processing need to be linked while acknowledging the fact that the picture is far from complete and that ultimately a much more fine-grained dissociation will be necessary. However, whereas the ability to generate predictions to optimize behavior already establishes a common ground in terms of conceptual development in both domains, potential cross-fertilization is also evident at the physiological level. Notwithstanding the vast literature on functional and anatomical differentiation in classic speech processing areas such as Broca's area or Wernicke's area, this aspect can be illustrated using the example of the SMA, where the relationship to dedicated temporal processing is perhaps most obvious.

The SMA comprises at least two subareas, the more rostral pre-SMA and the more caudal SMA proper (Matsuzaka, Aizawa, & Tanji, 1992; Picard & Strick, 2001). The SMA proper is connected to the primary motor cortex, whereas the pre-SMA is densely connected with prefrontal regions (Akkal, Dum, & Stick, 2007). Activation of the SMA in speech production can be distinguished along a rostrocaudal gradient associated with lexical selection, linear sequence encoding, and the control of motor output, respectively (Alario, Chainay, Lehericy, & Cohen, 2006). In the context of the f/c theory, the SMA engages in the primary control of the speech frame (MacNeilage & Davis, 2001). This basic function in linear sequencing can be further differentiated with respect to more explicit temporal characteristics as well as in relation to other structures, including the BG. For example, Riecker, Kassubek, Gröschel, Grodd, and Ackermann (2006) discussed the role of the SMA as the "starting mechanism" of speech production. In a similar vein, the Directions Into Velocities of Articulators (DIVA) model of speech production (Guenther, 2006; Guenther, Ghosh, & Tourville, 2006) suggests that the SMA and the BG provide a "GO signal" that controls the speaking rate (Guenther & Vladusich, 2012). More specifically, Bohland, Bullock, and Guenther (2009) assign the encoding of structural frames at an abstract level to the pre-SMA, whereas the SMA proper plays a role in the initiation and release of planned speech

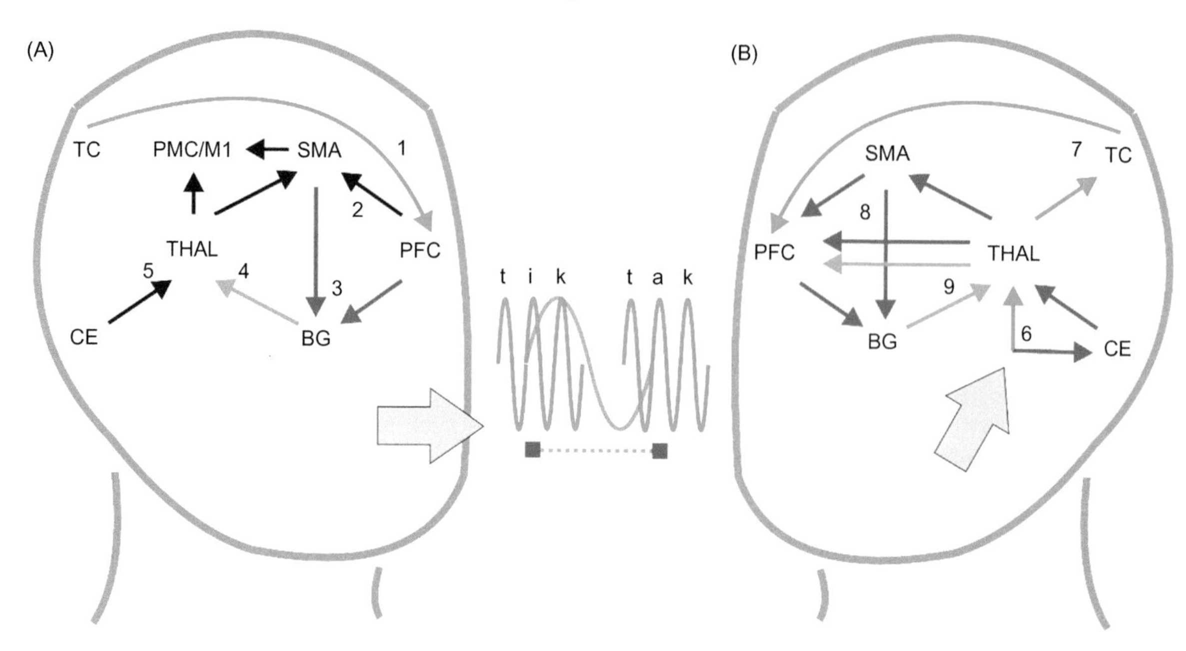

FIGURE 57.2 Schematic of the interaction of speech processing and dedicated temporal processing. (A) Once the decision to produce an utterance, that is, a time-varying sequence of speech events has been made (prefrontal cortex, PFC), speech production interacts with dedicated temporal processing in an extended subcorticocortical network (for further discussion, see Kotz and Schwartze, 2010). In accordance with the intentions of the speaker, elements from memory have to be retrieved, sequenced, and produced. The retrieval of memory elements (1) engages connections between PFC and temporal cortices (TC, blue). Hierarchical sequencing mechanisms in the sense of (morpho-)syntactic operations specify the combination of these elements, thus generating the formal structure of the utterance. Dedicated temporal processing is part of linear sequencing mechanisms that specify the temporal structure of the utterance. For this purpose, the PFC calls on the SMA (2) and, more specifically, its rostral portion, the pre-SMA, as well as the BG and associated thalamocortical circuits. In concert, these structures establish a temporal "grid" (3) that determines the global temporal structure of an utterance (green). In other words, the dedicated temporal processing system functions as a "pacemaker," that is, it provides a slower dynamic that serves as a reference for the subsequent, temporally precise shaping of faster dynamics and their actual implementation (4). The latter function engages the cerebellum and its connections to premotor (PMC) and primary motor (M1) cortices (5), and potentially also indirect connections to these areas via the caudal portion of the SMA (SMA proper). In line with the notion of a "spectrotemporal register" that holds the constituents of a syllable together (Greenberg et al., 2003), the pacemaker function maps onto the temporal structure expressed in the rhythmic succession of syllabic nuclei in the speech signal (i.e., the relation between "i" and "a" in the example). Similar mechanisms may be recruited during speech perception (B). Via its connections to the earliest stages of auditory processing, the cerebellar temporal processing system is expected to encode a temporally precise, event-based, nonlinear representation of temporal structure (6) that it rapidly transmits to frontal cortices, including the SMA, passing the thalamus (red). The ascending auditory pathway via the thalamus to the TC is expected to preserve a detailed representation of formal structure to guarantee stable memory access (blue). The nonlinear stimulus representation establishes a register for the perceptual integration (PFC) of elements retrieved from memory (7). However, it also serves as a trigger for oscillatory mechanisms (8) that explicitly encode the temporal relation between successive events, including the dedicated cortico-striato-thalamo-cortical temporal processing system (9, green). If the temporal relation between these events is perceived as regular, then the system may start to operate in a synchronized, predictive mode. On this basis (i.e., the separation of slower and faster dynamics), speech processing and dedicated temporal processing may interact not only to control the encoding of spatially distributed memory representations into a temporal sequence during speech production (Lashley, 1951) but also to optimize the processing of a sequential signal during speech perception.

acts. The pre-SMA then provides input to a planning loop involving the BG. Taken into a broader context, these functions may, to some extent, reflect the differential engagement of pre-SMA and SMA proper in temporal processing within, but also outside, the speech domain (Kotz & Schwartze, 2011; Schwartze, Rothermich, & Kotz, 2012).

57.4 CONCLUSION

In the most general sense, the proposed role of the pre-SMA as part of a corticostriatal "pacemaker" relates to establishing a slower temporal dynamic that guides the generation of faster temporal dynamics. While this function links speech processing and temporal processing, it also implies that speech processing builds on mechanisms that are essentially not domain-specific. Dedicated temporal processing is, per definition, a universal mechanism most likely serving the goal of optimizing predictive behavior. However, the integrative perspective discussed here offers a chance to ground speech processing in progressive general theories about brain function such as "predictive coding" (Friston, 2005, 2012) that build around a similar distinction and functional interaction of dynamics anchored at different timescales to recover information from time-varying signals. In this context, speech processing may reveal itself as the most sophisticated time-varying signal that recruits temporal processing mechanisms to optimize production and perception.

References

Ackermann, H., Mathiak, K., & Riecker, A. (2007). The contribution of the cerebellum to speech production and speech perception: Clinical and functional imaging data. *The Cerebellum, 6*, 202–213.

Akkal, D., Dum, R. P., & Stick, P. L. (2007). Supplementary motor area and presupplementary motor area: Targets of basal ganglia and cerebellar output. *The Journal of Neuroscience, 27*, 10659–10673.

Alario, F., Chainay, H., Lehericy, S., & Cohen, L. (2006). The role of the supplementary motor area (SMA) in word production. *Brain Research, 1076*, 129–143.

Allman, M. J., Teki, S., Griffiths, T. D., & Meck, W. H. (2014). Properties of the internal clock: First- and second-order principles of subjective time. *Annual Reviews of Psychology, 65*, 21.1–21.29.

Arnal, L. H., & Giraud, A. (2012). Cortical oscillations and sensory predictions. *Trends in Cognitive Sciences, 16*, 390–398.

Bohland, J. W., Bullock, D., & Guenther, F. H. (2009). Neural representations and mechanisms for the performance of simple speech sequences. *Journal of Cognitive Neuroscience, 22*, 1504–1529.

Buhusi, C. V., & Meck, W. H. (2005). What makes us tick? Functional and neural mechanisms of interval timing. *Nature Reviews Neuroscience, 6*, 755–765.

Coull, J. T., Cheng, R., & Meck, W. H. (2010). Neuroanatomical and neurochemical substrates of timing. *Neuropsychopharmacology, 36*, 1–25.

Fowler, C. A. (1979). "Perceptual centers" in speech production and perception. *Perception & Psychophysics, 25*, 375–388.

Fraisse, P. (1982). Rhythm and tempo. In D. Deutsch (Ed.), *The psychology of music* (pp. 149–180). New York, NY: Academic Press.

Friston, K. (2005). A theory of cortical responses. *Philosophical Transactions of the Royal Society B, 360*, 815–836.

Friston, K. (2012). Prediction, perception and agency. *International Journal of Psychophysiology, 83*, 248–252.

Friston, K., Kilner, J., & Harrison, L. (2006). A free energy principle for the brain. *Journal of Physiology, 100*, 70–87.

Gagnepain, P., Henson, R. N., & Davis, M. H. (2012). Temporal predictive codes for spoken words in auditory cortex. *Current Biology, 22*, 615–621.

Ghitza, O. (2012). On the role of theta-driven syllabic parsing in decoding speech: Intelligibility of speech with a manipulated modulation spectrum. *Frontiers in Psychology, 3*, 238.

Ghitza, O. (2013). The theta-syllable: A unit of speech information defined by cortical function. *Frontiers in Psychology, 4*, 138.

Ghitza, O., & Greenberg, S. (2009). On the possible role of brain rhythms in speech perception: Intelligibility of time-compressed speech with periodic and aperiodic insertions of silence. *Phonetica, 66*, 113–126.

Giraud, A., Kleinschmidt, A., Poeppel, D., Lund, E. L., Frackowiak, R. S. J., & Laufs, H. (2007). Endogenous cortical rhythms determine cerebral specialization for speech perception and production. *Neuron, 56*, 1127–1134.

Greenberg, S. (2006). A multi-tier framework for understanding spoken language. In S. Greenberg, & W. A. Ainsworth (Eds.), *Listening to speech: An auditory perspective* (pp. 411–434). Mahwah, NJ: Erlbaum.

Greenberg, S., Carvey, H., Hitchcock, L., & Chang, S. (2003). Temporal properties of spontaneous speech: A syllable-centric perspective. *Journal of Phonetics, 31*, 465–485.

Guenther, F. H. (2006). Cortical interactions underlying the production of speech sounds. *Journal of Communication Disorders, 39*, 350–365.

Guenther, F. H., Ghosh, S. S., & Tourville, J. A. (2006). Neural modeling and imaging of the cortical interactions underlying syllable production. *Brain and Language, 96*, 280–301.

Guenther, F. H., & Vladusich, T. (2012). A neural theory of speech acquisition and production. *Journal of Neurolinguistics, 25*, 408–422.

Hannemann, R., Obleser, J., & Eulitz, C. (2007). Top-down knowledge supports the retrieval of lexical information from degraded speech. *Brain Research, 1153*, 134–143.

Indefrey, P., & Levelt, W. J. M. (2004). The spatial and temporal signatures of word production components. *Cognition, 92*, 101–144.

Ivry, R. B., & Schlerf, J. E. (2008). Dedicated and intrinsic models of time perception. *Trends in Cognitive Sciences, 12*, 273–280.

Kochanski, G., & Orphanidou, C. (2008). What marks the beat of speech? *Journal of the Acoustical Society of America, 123*, 2781–2791.

Kotz, S. A., & Schwartze, M. (2010). Cortical speech processing unplugged: A timely subcortico-cortical framework. *Trends in Cognitive Sciences, 14*, 392–399.

Kotz, S. A., & Schwartze, M. (2011). Differential input of the supplementary motor area to a dedicated temporal processing network: Functional and clinical implications. *Frontiers in Integrative Neuroscience, 5*, 86.

Kotz, S. A., Schwartze, M., & Schmidt-Kassow, M. (2009). Non-motor basal ganglia functions: A review and proposal for a model of sensory predictability in auditory language perception. *Cortex, 45*, 982–990.

Lashley, K. S. (1951). The problem of serial order in behavior. In L. A. Jeffress (Ed.), *Cerebral mechanisms in behavior* (pp. 112–136). New York: Wiley.

Levelt, W. J. M. (2001). Spoken word production: A theory of lexical access. *Proceedings of the National Academy of Sciences of the United States of America, 98*, 13464–13471.

MacNeilage, P. F. (1998). The frame/content theory of evolution of speech production. *Behavioral and Brain Sciences, 21*, 499–546.

MacNeilage, P. F., & Davis, B. L. (2000). On the origin of internal structure of word forms. *Science, 288*, 527–531.

MacNeilage, P. F., & Davis, B. L. (2001). Motor mechanisms in speech ontogeny: Phylogenetic, neurobiological and linguistic implications. *Current Opinion in Neurobiology, 11*, 696–700.

Marcus, S. M. (1981). Acoustic determinants of perceptual center (P-center) location. *Perception & Psychophysics, 30*, 247–256.

Martin, J. G. (1972). Rhythmic (hierarchical) versus serial structure in speech and other behavior. *Psychological Review, 79*, 487–509.

Matsuzaka, Y., Aizawa, H., & Tanji, J. (1992). A motor area rostral to the supplementary motor area (presupplementary motor area) in the monkey: Neuronal activity during a learned motor task. *Journal of Neurophysiology, 68*, 653–662.

McGettigan, C., & Scott, S. K. (2012). Cortical asymmetries in speech perception: What's wrong, what's right and what's left? *Trends in Cognitive Sciences, 16*, 269–276.

Merchant, H., Harrington, D. L., & Meck, W. H. (2013). Neural basis of the perception and estimation of time. *Annual Reviews of Neuroscience, 36*, 313–336.

Morton, J., Marcus, S. M., & Frankish, C. R. (1976). Perceptual centers (P-centers). *Psychological Review, 83*, 405–408.

Peelle, J. E., & Davis, M. H. (2012). Neural oscillations carry speech rhythm through to comprehension. *Frontiers in Psychology, 3*, 320.

Picard, N., & Strick, P. L. (2001). Imaging the premotor areas. *Current Opinion in Neurobiology, 11*, 663–672.

Poeppel, D., Emmorey, K., Hickok, G., & Pylkkänen, L. (2012). Towards a new neurobiology of language. *The Journal of Neuroscience, 32*, 14125–14131.

Port, R. F. (2003). Meter and speech. *Journal of Phonetics, 31*, 599–611.

Price, C. J. (2010). The anatomy of language: A review of 100 fMRI studies published in 2009. *Annals of the New York Academy of Sciences, 1191*, 62–68.

Price, C. J. (2012). A review and synthesis of the first 20 years of PET and fMRI studies of heard speech, spoken language and reading. *NeuroImage*, 816–847.

Raichle, M. E. (2010). Two views of brain function. *Trends in Cognitive Sciences, 14*, 180–190.

Repp, B. H. (2000). Introduction. In P. Windsor, & P. Desain (Eds.), *Rhythm perception and production* (pp. 235–237). Lisse: Swets & Zeitlinger.

Riecker, A., Kassubek, J., Gröschel, K., Grodd, W., & Ackermann, H. (2006). The cerebral control of speech tempo: Opposite relationship between speaking rate and BOLD signal changes at striatal and cerebellar structures. *NeuroImage, 29*, 46–53.

Rosen, S. (1992). Temporal information in speech: Acoustic, auditory and linguistic aspects. *Philosophical Transactions of the Royal Society London B, 336*, 367–373.

Schirmer, A. (2004). Timing speech: A review of lesion and neuroimaging findings. *Cognitive Brain Research, 21*, 269–287.

Schwartze, M., & Kotz, S. A. (2013). A dual-pathway neural architecture for specific temporal prediction. *Neuroscience and Biobehavioral Reviews, 37*, 2587–2596.

Schwartze, M., Rothermich, K., & Kotz, S. A. (2012). Functional dissociation of pre-SMA and SMA-proper in temporal processing. *NeuroImage, 60*, 290–298.

Schwartze, M., Tavano, A., Schröger, E., & Kotz, S. A. (2012). Temporal aspects of prediction in audition: Cortical and subcortical neural mechanisms. *International Journal of Psychophysiology, 83*, 200–207.

Scott, S. K. (1998). The point of P-centres. *Psychological Research, 61*, 4–11.

Scott, S. K. (2012). The neurobiology of speech perception and production: Can functional imaging tell us anything we did not already know? *Journal of Communication Disorders, 45*, 419–425.

Scott, S. K., & McGettigan, C. (2012). Amplitude onsets and spectral energy in perceptual experience. *Frontiers in Psychology, 3*, 80.

Sohoglu, E., Pelle, J., Carlyon, R. P., & Davis, M. H. (2012). Predictive top-down integration of prior knowledge during speech perception. *The Journal of Neuroscience, 32*, 8443–8453.

Spencer, R. M. C., & Ivry, R. B. (2013). Cerebellum and timing. In M. Manto, D. L. Gruol, J. D. Schmahmann, N. Koibucchi, & F. Rossi (Eds.), *Handbook of the cerebellum and cerebellar disorders*. Dordrecht: Springer.

Spencer, R. M. C., Zelaznik, H. N., Diedrichsen, J., & Ivry, R. B. (2003). Disrupted timing of discontinuous but not continuous movements by cerebellar lesions. *Science, 300*, 1437–1439.

Strauß, A., Kotz, S. A., & Obleser, J. (2013). Narrowed expectancies under degraded speech: Revisiting the N400. *Journal of Cognitive Neuroscience, 15*, 1383–1395.

Taylor, W. L. (1953). "Cloze procedure": A new tool for measuring readability. *Journalism Quaterly, 30*, 415–433.

Tillmann, H. G., & Mansell, P. (1980). *Phonetik. Lautsprachliche Zeichen, Sprachsignale und lautsprachlicher Kommunikationsprozeß*. Stuttgart: Klett-Cotta.

Wiener, M., Turkeltaub, P., & Coslett, H. B. (2010). The image of time: A voxel-wise meta-analysis. *NeuroImage, 49*, 1728–1740.

CHAPTER

58

Neural Models of Motor Speech Control

Frank H. Guenther[1] *and Gregory Hickok*[2]

[1]Department of Speech, Language, and Hearing Sciences, Department of Biomedical Engineering, Boston University, Boston, MA, USA; [2]Department of Cognitive Sciences, Center for Language Science, Center for Cognitive Neuroscience, University of California Irvine, Irvine, CA, USA

58.1 INTRODUCTION

Speech production is a highly complex motor act involving the finely coordinated activation of approximately 100 muscles in the respiratory, laryngeal, and oral motor systems. To achieve this task, speakers utilize a large network of brain regions. This network includes regions involved in other motor tasks, such as the motor and somatosensory cortical areas, cerebellum, basal ganglia, and thalamus, as well as regions that are more specialized for speech and language, including inferior and middle prefrontal cortex and superior and middle temporal cortex. Our goal in this chapter is to describe the critical role of the auditory system in speech production. We first discuss the role of sensory systems in motor control broadly and summarize the long history of ideas and research on the interaction between auditory and motor systems for speech. We then describe current research on speech planning, which strongly implicates the auditory system in this process. Two large-scale neurocomputational models of speech production are then discussed. Finally, we highlight some future directions for research on speech production.

Movement is absolutely dependent on sensory information. We know where and how to reach for an object because we *see* its location and shape; we know how much force to exert while we are holding the object because we *feel* the pressure of the object on our hand and the weight on our limb; and we know how to initiate any of these movements because our sensory systems tell us where our limb is in relation to our body and the object. British neurologist and physiologist Henry Charlton Bastian (1837–1915) wrote on the topic of movement control in 1887, stating, "It may be regarded as a physiological axiom, that all purposive movements of animals are guided by sensations or by afferent impressions of some kind" (Bastian, 1887, p. 1). Experimental work over the decades backs these claims. This work has found, for example, that blocking somatosensory feedback from a monkey's limb (while leaving motor fibers intact) causes the limb to go dead. With training, the monkey can learn to reuse it clumsily, but only with visual feedback; blindfold the animal and motor control degrades dramatically (Sanes, Mauritz, Evarts, Dalakas, & Chu, 1984). Similar symptomology can be found in humans suffering from large-fiber sensory neuropathy, which deafferents the body sense while leaving motor fibers intact (Sanes et al., 1984).

Speech is no different. Without the auditory system, as in prelingual-onset peripheral deafness, normal speech development cannot occur. Importantly, it is not just during development that auditory information is critical. Experimental or naturally caused manipulations of acoustic input can have dramatic effects on speech production. For example, delayed auditory feedback induces nonfluency (Yates, 1963), altering feedback in the form of pitch or the formant frequency structure results in automatic and largely unconscious compensation in speech articulation (Burnett, Freedland, Larson, & Hain, 1998; Houde & Jordan, 1998; Larson, Burnett, Bauer, Kiran, & Hain, 2001), and exposure to a different linguistic environment can induce changes in the listener-speaker's articulation (picking up accents; Sancier & Fowler, 1997). Furthermore, although individuals who become deaf as adults can remain intelligible for years after they lose hearing, they show some speech output impairments immediately, including impaired ability to adjust pitch loudness in different listening conditions; over time, their phonetic contrasts become reduced (Perkell et al., 2000) and they exhibit articulatory decline (Waldstein, 1989).

The speech research literature contains numerous theoretical proposals that strongly link speech perception

Neurobiology of Language. DOI: http://dx.doi.org/10.1016/B978-0-12-407794-2.00058-4

and speech production. Notable examples include the motor theory of speech perception (Liberman, Cooper, Shankweiler, & Studdert-Kennedy, 1967; Liberman & Mattingly, 1985), which posits that speech perception involves translating acoustic signals into the motor gestures that produce them, as well as acoustic theories of speech production (Fant, 1960; Stevens, 1998), which highlight the importance of acoustic or auditory targets in the speech production process. In the following sections we elaborate on the roles of auditory information in recent neural models of speech planning and execution.

58.2 THE PLANNING OF SPEECH MOVEMENTS

Although it may seem obvious that auditory information is in some way involved in speech production—the lack of normal speech development in the absence of auditory feedback provides incontrovertible evidence of this—there has been much debate in the speech motor control literature over exactly what role(s) auditory feedback plays. Much of this debate revolves around the following central question: what exactly are the goals, or *targets*, of the speech production planning process?

To produce a goal-directed movement, the nervous system must generate a complex muscle activation pattern that satisfies the goals of the movement. For example, reaching to a target position in three-dimensional (3-D) space requires coordinated activation of muscles in the arm, wrist, and hand that move the hand in space to the target position. Various explanations have been put forth regarding how this complex task may be solved by the nervous system, ranging from a *motoric planning* extreme to a *task space planning* extreme. In a motoric planning solution, the nervous system translates the desired spatial position of the hand into a corresponding target in motoric terms, such as a set of joint angles or muscle lengths that place the hand at the appropriate position in 3-D space. Movement planning is then performed simply by interpolating between the current set of muscle lengths and the target set (Bullock & Grossberg, 1988; Rosenbaum, Engelbrecht, Bushe, & Loukopoulos, 1993; Rosenbaum, Loukopoulos, Meulenbroek, Vaughan, & Engelbrecht, 1995). In a task space planning scheme, the movement is planned in a reference frame that is much further removed from the musculature. In reaching, for example, a 3-D spatial reference frame may be used to plan a hand path to the target. Muscle activation patterns must still be generated to move the arm and hand, and different proposals for this transformation have been put forth. In one solution, the spatial movement *direction* needed to keep the hand moving along the planned spatial trajectory is transformed into appropriate muscle *velocities* during the reach (Bullock, Grossberg, & Guenther, 1993; Guenther, 1992). An important consequence of this form of planning is that it does not involve a specific target configuration of the arm for a given spatial position of the hand.

To understand the planning of speech movements, the relationship between articulator configuration and acoustic signal must be understood. At any point in time, the 3-D shape of the vocal tract (i.e., the shape of the "air tube" that exists from the vocal folds to the lips) is determined by the locations of speech articulators such as the tongue, lips, jaw, larynx, and the soft palate (velum). This shape manifests itself in the speech signal as a set of peaks in the envelope of the frequency spectrum of the acoustic signal; these peaks are referred to as *formants*. Roughly speaking, the ***first formant frequency, or F1***, corresponds to the degree of constriction formed by the tongue in the vocal tract, which in turn is closely related to tongue height. The ***second formant frequency, F2***, is related to the location of the tongue constriction along the length of the vocal tract, with higher F2 values corresponding to constrictions closer to the lips. These characterizations are rather crude "rules of thumb" that do not apply to all vocal tract configurations and leave out many important aspects of the relationship between speech articulation and acoustics. The relationships between the third and higher formant frequencies and vocal tract shape become increasingly complex and are not easily related to articulator positions.

Evidence from studies of speech movements indicate that speech sounds (e.g., phonemes or syllables) are not encoded as muscle length or articulator configuration targets (i.e., the speech motor system does not use a purely motoric planning strategy as defined). Instead, a large amount of variability is seen in the vocal tract configurations used to realize a particular phoneme. A stark example is the American English phoneme /r/. This phoneme is characterized acoustically by a deep dip in the third formant frequency; this formant frequency transition is highly consistent across phonetic contexts (Boyce & Espy-Wilson, 1997). Although the acoustic signal characterizing /r/ remains nearly constant, completely different articulatory gestures are used in the different contexts (Delattre, Freeman, 1968; Guenther et al., 1999; Nieto-Castanon, Guenther, Perkell, & Curtin, 2005). Figure 58.1 illustrates this articulatory variability. The top row shows tongue configurations at the start of /r/ production (dashed lines) and at the acoustic "center" of the /r/ (bold lines) when spoken by an American English participant in three different phonetic contexts (/ar/, /dr/, and /gr). In the /ar/ context, the participant raises the tongue tip to produce the /r/. In the /dr/ context, the speaker uses a backward movement of the tongue. In the /gr/ context, the speaker uses a downward movement of the tongue.

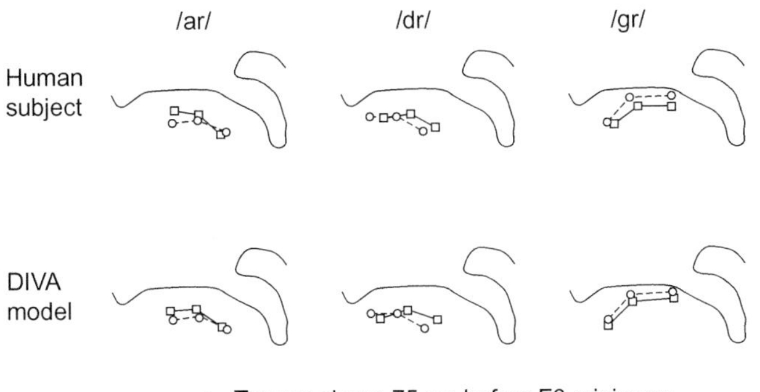

FIGURE 58.1. Articulatory variability during American English /r/ production. Measured (top) and simulated (bottom) tongue movements during production of the phoneme /r/ when preceded by the phoneme /a/ (left column), /d/ (center), and /g/ (right), shown in the midsagittal plane. Outline of the palate is shown for reference. Each connected line (bold or dashed) represents a tongue configuration; each circular or square data point corresponds to the position of a movement transducer on the tongue (tongue tip on the left; dorsum in the center; back on the right).The gesture used to produce /r/ is represented by the transition between the dashed and bold tongue configurations. Entirely different tongue gestures are used by both the human subject and the model in the three contexts. Simulations of the DIVA model utilizing a common auditory target for the three contexts capture this articulatory variability. *Adapted from Nieto-Castanon et al. (2005).*

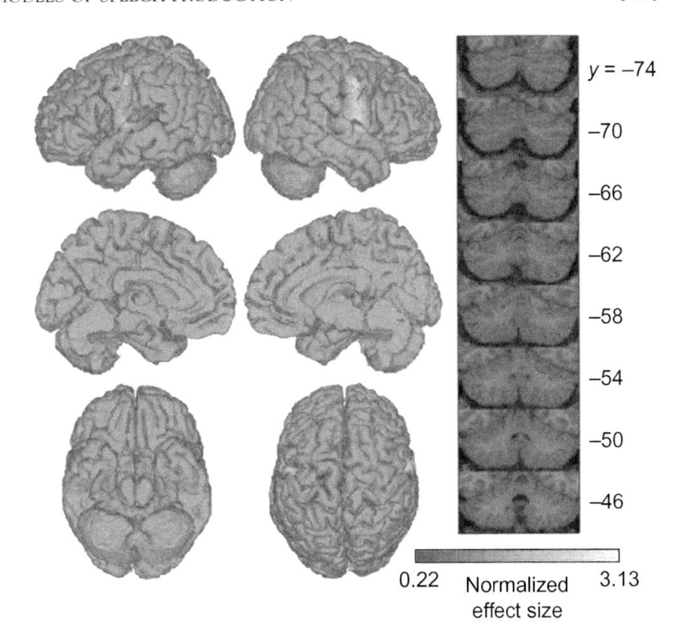

FIGURE 58.2. Neural activity in the cerebral cortex and cerebellum during production of monosyllabic words contrasted with a silent baseline. In the lateral views of the cerebral cortex (top panels), bilateral activity can be seen in the precentral gyrus of the frontal lobe, the postcentral gyrus of the parietal lobe, and the STG of the temporal lobe. Medial surfaces (middle panels) reveal activity in the SMA. Slices through the cerebellum (right column) indicate activity in the superior paravermal cerebellum, primarily in lobules V and VI. Additional activity in the basal ganglia and thalamus is not visible in this figure.

The articulatory variability seen for /r/ can be accounted for in a straightforward manner by a task space planning scheme in which the targets of speech are auditory rather than articulatory. The bottom row of Figure 58.1 shows the results of computer simulations of the Directions Into Velocities of Articulators (DIVA) model of speech motor control (Guenther, 1995a; Guenther, Ghosh, & Tourville, 2006; Guenther, Hampson, & Johnson, 1998; Guenther & Vladusich, 2012; Tourville & Guenther, 2011; described further) when producing /r/ in the three phonetic contexts. The same auditory target is used for /r/ in the three contexts, and articulator movements are planned in auditory space (see Nieto-Castanon et al., 2005 for details). This simple planning scheme accounts for the disparate tongue gestures seen in the three phonetic contexts, providing strong evidence for the use of an auditory planning space.

58.3 BRAIN REGIONS INVOLVED IN SPEECH ARTICULATION

Figure 58.2, adapted from Tourville, Reilly, and Guenther (2008), illustrates neural activity in the cerebral cortex and cerebellum during a simple speech task (monosyllabic word production) contrasted with a baseline task of silently viewing letters. As in most speech production neuroimaging studies, activity is seen in the ventral precentral gyrus (motor and premotor cortex), ventral postcentral gyrus (somatosensory cortex), superior temporal gyrus (STG) (auditory cortex), supplementary motor area (SMA), and superior paravermal cerebellum (primarily lobules V and VI). Although not visible in Figure 58.2, the basal ganglia and thalamus are also active during speech. In the following sections we discuss the roles of these regions within the context of neurocomputational models of speech production.

58.4 NEUROCOMPUTATIONAL MODELS OF SPEECH PRODUCTION

The extensive library of results from neuroimaging studies provides important insights into the roles of a large number of cortical and subcortical areas in speech production. In isolation, however, these results do not provide an integrated, mechanistic view of how the neural circuits engaged by speech tasks interact to produce fluent speech. To this end, computational models that both suggest the neural computations performed within specific modules and across pathways linking modules and propose specific neural correlates for these computational elements bridge this critical gap. Furthermore, these *neurocomputational* models lead to specific hypotheses that may be tested in both imaging and behavioral experiments, leading to a continuous cycle of model refinement that provides a unifying account of a wide range of experimental findings. We describe the DIVA model of speech motor control, which has been developed, tested, and refined by Guenther and colleagues over the past two decades, as well as an extension to that model, termed GODIVA, which begins to account for higher-level phonological processes involved in speech planning. Our discussion of neurocomputational models ends with a treatment of the Hierarchical State Feedback Control (HSFC) model of Hickok (2012), which addresses data from studies of conduction aphasia and internal correction of upcoming phonemic errors before they are articulated.

58.5 THE DIVA MODEL

The DIVA model, schematized in Figure 58.3, provides the most detailed and thoroughly tested account of the neural processes underlying speech motor control. Each box in the diagram corresponds to a set of neurons, or *map*, and arrows correspond to synaptic projections that transform one type of neural representation into another. The model learns to control movements of a simulated vocal tract that produces an acoustic signal (Maeda, 1990). The results of numerous computer simulations of the model have been shown to account for a wide range of findings regarding the brain activity, sound output, and movements produced by human speakers, as detailed elsewhere (Callan, Kent, Guenther, & Vorperian, 2000; Ghosh, Tourville, & Guenther, 2008; Guenther, 1995b; Guenther et al., 1999, 1998; Nieto-Castanon et al., 2005; Peeva et al., 2010; Perkell, Guenther, et al., 2004; Perkell, Matthies, et al., 2004; Tourville et al., 2008).

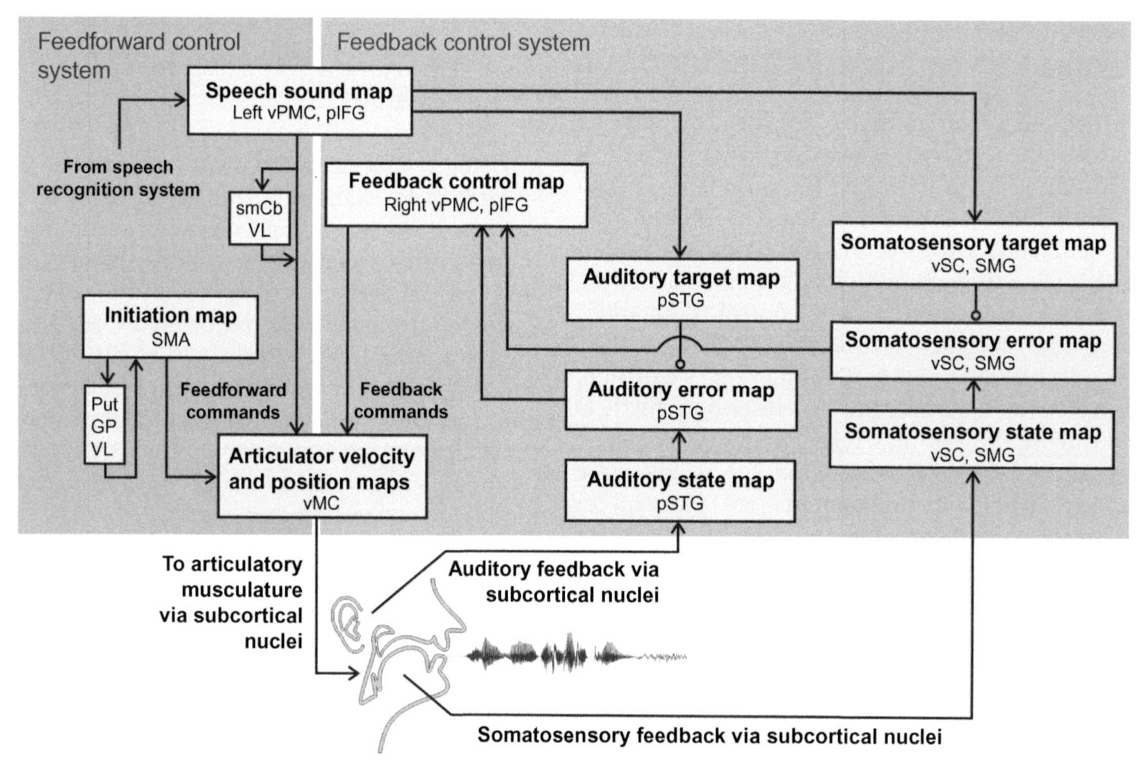

FIGURE 58.3. Schematic overview of the DIVA model of speech production. Each box in the diagram corresponds to a group of neurons; arrows represent excitatory axonal projections and empty circles represent inhibitory projections. The model breaks the control of speech production into a feedforward control system and a feedback control system, with the latter further broken into auditory and somatosensory feedback control subsystems. See text for details. GP, globus pallidus; Put, putamen; pIFG, posterior inferior frontal gyrus; pSTG, posterior superior temporal gyrus; smCb, superior medial cerebellum; SMA, supplementary motor area; SMG, supramarginal gyrus; VL, ventral lateral nucleus of the thalamus; vMC, ventral motor cortex; vPMC, ventral premotor cortex; vSC, ventral somatosensory cortex.

The production of a speech sound in the DIVA model starts with activation of neurons associated with that sound in the model's speech sound map. A "speech sound" in this map is defined as a segment of speech that has its own motor program, with the syllable being the most typical unit represented. The term "sound map" is used to highlight the fact that speech movement planning appears to aim at auditory targets rather than articulatory targets, as described in the preceding section. In the computer implementation of the model, each neuron in the map represents a different speech sound. When a new sound is encountered, a new neuron that codes for that sound is added to the speech sound map.

Activation of the speech sound map leads to articulator movement commands that arrive in primary motor cortex via two control subsystems. The feedforward control system projects directly, as well as indirectly, via a side loop through the cerebellum, from the speech sound map to primary motor cortex. The feedback control system, including both an auditory feedback control subsystem and a somatosensory feedback control subsystem, involves indirect projections passing through sensory brain areas.

The DIVA model provides a description of how an infant learns to produce speech sounds through babbling and imitation processes. According to the model, a combination of motor, auditory, and somatosensory information generated during early random and reduplicated babbling is used to tune the synaptic projections of the feedback control system. The sensory-motor transformations learned during babbling allow detected sensory errors to be mapped into corrective motor commands during the subsequent imitation stage, when feedforward commands are learned for the speech sounds of the infant's native language.

The imitation stage describes how syllable-specific learning occurs when an infant is presented with a new speech sound to learn, corresponding to an infant learning a new syllable from his/her parent, for example. Detection of a novel sound leads to activation of a new speech sound map neuron that will represent that sound. This is presumed to occur via projections from the speech recognition system in the auditory cortical areas to the left ventral premotor cortex (vPMC). Although not specified in the model, these projections may be part of the dorsal auditory stream (involving the superior longitudinal fasciculus or arcuate fasciculus) or the ventral stream (involving the uncinate fasciculus).

The model first learns an auditory target for the new sound from sound samples that it is presented. This auditory target is encoded in the synaptic projections from the speech sound map to the auditory target map in Figure 58.3. The model then repeatedly attempts to produce the sound, improving the feedforward command for the sound with each iteration. The functions of the feedforward and feedback control systems are further detailed in the following sections.

Since its inception (Guenther, 1992), the DIVA model has hypothesized the existence of a speech sound map in premotor cortex that is activated both during speech perception (when it is necessary for learning sensory targets for speech sounds) and speech production (when it is necessary for driving speech articulator movements via the feedforward and feedback control systems). Cells in premotor cortical areas that are active during both action perception and production of a motor act have since been identified in numerous electrophysiological studies in animals (see Rizzolatti & Craighero, 2004 for a review) and in fMRI studies showing activity while listening to speech in ventral premotor cortical areas involved in speech production (Buchsbaum, Hickok, & Humphries, 2001; Hickok, Buchsbaum, Humphries, & Muftuler, 2003; Wilson, Saygin, Sereno, & Iacoboni, 2004).

58.5.1 Auditory Feedback Control

As described, auditory feedback plays an important role in tuning the speech motor control system. According to the DIVA model, axonal projections from speech sound map cells in the left inferior frontal gyrus (IFG)/vPMC to the higher-order auditory cortical areas (via the dorsal auditory stream; Hickok & Poeppel, 2004) embody the auditory target region for the speech sound currently being produced. That is, they represent the auditory feedback that should arise when the speaker hears himself/herself producing the sound. This representation of expected auditory consequences is referred to as the *auditory target map*, and the model's auditory target consists of a temporally varying region of auditory space that encodes acceptable auditory variation for the current sound. The use of target regions, rather than points, is an important aspect of the DIVA model that provides a unified explanation for a wide range of speech production phenomena (see Guenther, 1995b for details). Projections such as those between the speech sound map and auditory state map in Figure 58.3, which predict the sensory state resulting from a movement, are often referred to as *forward models* (Davidson & Wolpert, 2005; Desmurget & Grafton, 2000; Guenther, 1995a; Guenther et al., 1998; Kawato, 1999; Miall & Wolpert, 1996), and they are closely related to the much older concepts of efference copy (originally proposed by von Helmholtz in the mid-19th century; see von Helmholtz, 1925) and corollary discharge (a term coined by Sperry, 1950).

The model's *auditory state map* represents the incoming auditory signal of one's own speech. This map is hypothesized to include a combination of relatively low-level auditory features such as pitch represented in the primary auditory cortex (Bendor & Wang, 2005; Kumar & Schönwiesner, 2012), located in Heschl's gyrus, as well as more complex auditory representations such as a speaker-normalized formant representation hypothesized to reside in higher-order auditory cortical areas in the planum temporale and posterior STG (including the superior bank of the superior temporal sulcus).

The auditory feedback control system detects speech errors, represented in the *auditory error map*, by comparing the auditory target map and auditory state map. Specifically, the auditory target map inhibits auditory error map cells, whereas the auditory state map excites auditory error map cells. If the incoming auditory signal is within the target region, then the inhibition from the auditory target cancels the excitatory effects of the auditory state map. If the incoming auditory signal is outside the target region, then the inhibitory target region will not completely cancel the excitatory input from the auditory periphery, resulting in activation of auditory error map cells. Evidence of inhibition in auditory cortical areas in the STG during one's own speech comes from several different sources, including recorded neural responses during open brain surgery (Creutzfeldt, Ojemann, & Lettich, 1989), MEG measurements (Houde, Nagarajan, Sekihara, & Merzenich, 2002; Numminen & Curio, 1999; Numminen, Salmelin, & Hari, 1999), and PET measurements (Wise, Greene, Buchel, & Scott, 1999). The auditory error map is hypothesized to reside in the same cortical areas as the auditory state and target maps; these regions include the Sylvian parietal temporal (Spt) area, which has been shown to become active during both speech perception and production (Buchsbaum, Hickok, & Humphries, 2001; Hickok & Poeppel, 2004).

Activation of the auditory error map during speech leads to corrective motor signals in the model via projections to the *feedback control map* in right vPMC. This map is responsible for transforming auditory errors into movements that correct these errors; such a transformation is often referred to as an *inverse model* (Kawato, 1999). The DIVA model derives its name from the particular form of inverse model used: a mapping from Directions (in sensory space) Into Velocities of Articulators. Such a mapping, which approximates the pseudoinverse of the Jacobian matrix relating changes in the auditory signal to the velocities of the speech articulators, results in many desirable properties, including motor equivalence (Guenther et al., 1998, 1999, 2006; see simulations of /r/ production in previous section) and efficient movement trajectories (Guenther et al., 1998).

Once the model has learned appropriate feedforward commands for a speech sound, it can successfully produce the sound using just those feedforward commands. That is, no auditory error will arise during production, and thus the auditory feedback control subsystem will not be activated. However, if an externally imposed perturbation occurs, such as a real-time "warping" of the subject's auditory feedback so that he hears himself producing the wrong sound (Houde & Jordan, 1998), then the auditory error cells will become active and attempt to correct for the perturbation through projections to the feedback control map in the right inferior frontal cortex. These predictions were verified by Tourville et al. (2008) in an fMRI study involving unexpected perturbations of auditory feedback during speech—specifically, upward or downward shifts of F1 by 30% on randomly dispersed trials. As illustrated in Figure 58.4, perturbed speech led to increased activity in the posterior STG bilaterally and in the right frontal cortex. These regions correspond to the auditory feedback control system in the schematic of the DIVA model (Figure 58.3).

58.5.2 Somatosensory Feedback Control

Like auditory information, somatosensory information has long been known to be important for speech production (Lindblom, Lubker, & Gay, 1979). The DIVA model posits a somatosensory feedback control subsystem operating in parallel with the auditory feedback control subsystem. The model's *somatosensory state map* corresponds to the representation of tactile and proprioceptive information from the speech articulators in primary and higher-order somatosensory cortical areas in the postcentral gyrus and supramarginal gyrus (SMG). The model's *somatosensory target map* and *somatosensory error map*, which play roles analogous to their auditory counterparts, are hypothesized to reside primarily in the SMG, a region that has been implicated in phonological processing for speech perception (Caplan, Gow, & Makris, 1995; Celsis et al., 1999), and speech production (Damasio & Damasio, 1980; Geschwind, 1965). This hypothesized role for inferior parietal cortex in the integration of motor commands and sensory feedback during speech production is analogous to the visual-motor integration role associated with more dorsal parietal regions during limb movements (Andersen, 1997; Rizzolatti, Fogassi, & Gallese, 1997).

According to the model, cells in the somatosensory error map become active during speech if the speaker's tactile and proprioceptive feedback from the vocal tract deviates from the somatosensory target region for the sound being produced; this prediction is supported

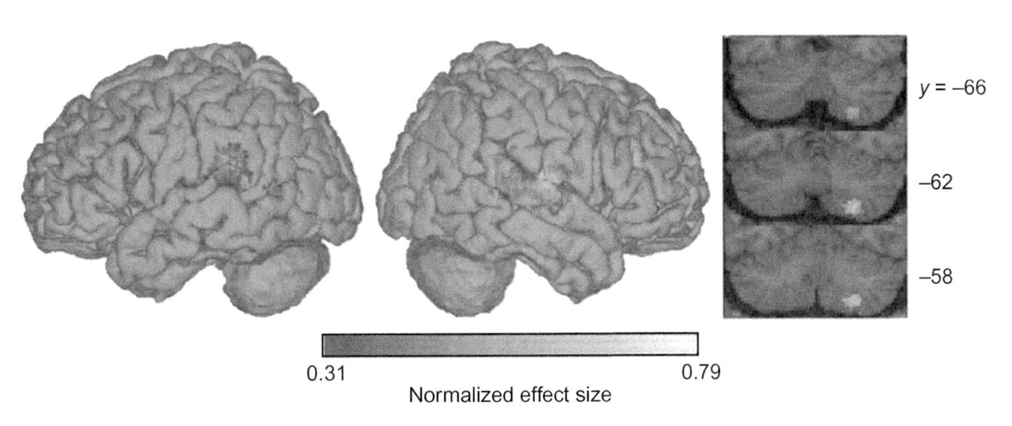

FIGURE 58.4. Auditory feedback control network for speech production. Areas of increased brain activity for auditorily perturbed speech compared with normal speech are indicated on the left and right lateral cortical hemispheres and coronal slices through the cerebellum (see Tourville et al., 2008 for details). The auditory perturbation consisted of a 30% increase or decrease of the first formant frequency (F1), which has the effect of making the vowel in a word like "bet" sound more like "bat" or "bit." This perturbation occurred in real time (i.e., the perturbed speech signal was delivered to the subject over headphons within 16 ms of the actual acoustic signal from his/her own voice), and it was applied in a random subset of trials. Subjects partially compensated for this perturbation by shifting their produced F1 in the direction opposite the perturbation. Increased activity was found bilaterally in the auditory cortical areas of the posterior STG during perturbed speech, supporting the DIVA model prediction of auditory error neurons in these areas. Increased activity was also seen in inferior frontal regions, with a right hemisphere bias; according to the DIVA model, these regions are involved in generating corrective motor commands.

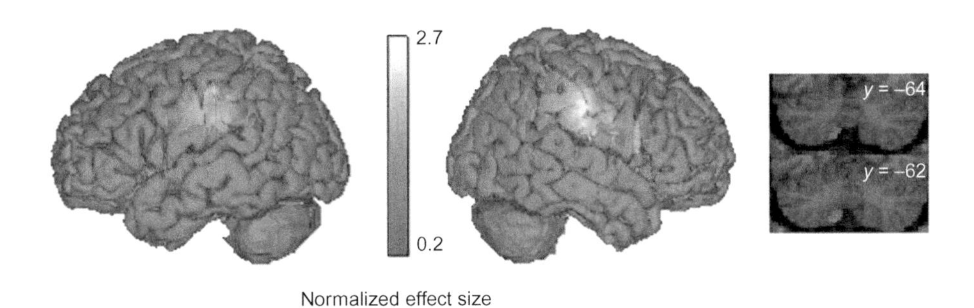

FIGURE 58.5. Somatosensory feedback control network for speech production. Areas of increased brain activity for speech produced with an unexpected blockage of the jaw contrasted with normal speech are indicated on the left and right lateral cortical hemispheres and coronal slices through the cerebellum. The unexpected jaw constraint caused bilateral activity in the postcentral gyrus and SMG, supporting the DIVA model prediction of a somatosensory error map in these regions. Increased activity was also seen in inferior frontal regions, with a right hemisphere bias; according to the DIVA model, these regions are involved in generating corrective motor commands.

by brain activity measured with fMRI during unexpected jaw perturbations as illustrated in Figure 58.5, which is adapted from Golfinopoulos et al. (2011). This study found increased activity bilaterally in the SMG and postcentral gyrus when contrasting the perturbed and unperturbed speaking conditions. The output of the somatosensory error map then propagates to the feedback control map in right vPMC, which transforms somatosensory errors into motor commands that correct those errors in a manner analogous to the auditory feedback control system (i.e., using an approximation to the pseudoinverse of the Jacobian matrix relating changes in the somatosensory state to velocities of the speech articulators).

Although the auditory and somatosensory feedback subsystems are implemented separately in computer simulations of the DIVA model, the existence of multisensory neurons that encode both auditory and somatosensory information during speech has been hypothesized (Guenther et al., 2006). Interactions between different sensory modalities involved in speech have been noted in a number of studies; for example, it has been shown that somatosensory stimulation can modulate auditory perceptual responses under some conditions (Ito, Tiede, & Ostry, 2009), auditory stimulation can alter skin sensation (Ito & Ostry, 2012), and viewing speech articulation can alter auditory perception, a phenomenon referred to as the McGurk Effect (McGurk & MacDonald, 1976).

58.5.3 Feedforward Control

According to the DIVA model, projections from left vPMC to primary motor cortex, supplemented by a cerebellar side loop between these two regions involving the superior paravermal region of the cerebellar cortex, constitute the feedforward motor commands for speech production. The primary motor and premotor cortices are well-known to be strongly interconnected (Krakauer & Ghez, 1999; Passingham, 1993). Furthermore, the cerebellum is known to receive input via the pontine nuclei from premotor cortical areas, as well as higher-order auditory and somatosensory areas that can provide state information important for choosing motor commands (Schmahmann & Pandya, 1997), and it projects heavily to the primary motor cortex via the thalamus (Middleton & Strick, 1997).

The DIVA model's feedforward commands constitute a form of motor program and, as such, they are closely related to the concepts of a *gestural score* (Browman & Goldstein, 1989) and *mental syllabary* (Levelt & Wheeldon, 1994). A gestural score is a stored representation of properly timed vocal tract gestures (defined primarily as constrictions in the vocal tract) needed to produce a phonological message. The mental syllabary is a repository of motor programs (or gestural scores) for the most frequently used syllables in a language. The DIVA model's conception of feedforward commands differs from these characterizations primarily in the choice of representation for the motor programs (in the DIVA model these programs command articulator movements to hit auditory targets rather than constriction targets) and size of the motor programs (the DIVA model allows for phonemic motor programs as well as syllabic and multisyllabic motor programs for frequently used phoneme sequences).

Prior to the development of speech, infants do not possess fine-tuned motor programs for producing the sounds of their language; in other words, their feedforward control system is not yet tuned. Only by attempting to produce the sounds of their language can accurate feedforward commands be learned. In the DIVA model, feedforward commands for a syllable are tuned on each production attempt. On the first attempt to produce a new sound, the model relies very heavily on auditory feedback control to produce the sound because its feedforward commands are inaccurate, thus resulting in auditory errors that are detected by comparing the stored auditory target for the sound with the auditory feedback generated by the production attempt. These auditory error signals activate the feedback control subsystem, which leads to corrective motor commands via the feedback control map. The corrective commands issued by the auditory feedback control subsystem during the current attempt to produce the sound become stored in the feedforward command for use on the next attempt. The superior paravermal region of the cerebellum is hypothesized to be involved in this process (see Guenther et al., 2006 for details). Each subsequent attempt to produce the sound results in a better feedforward command and less auditory error, until the feedforward command is capable of producing the sound fluently (i.e., without producing any auditory error), at which point the auditory feedback subsystem no longer contributes to production unless speech is perturbed in some way or the sizes or shapes of the articulators change. As the speech articulators get larger with growth, the auditory feedback control subsystem continues to provide corrective commands that are subsumed into the feedforward controller, thus allowing the feedforward controller to stay properly tuned despite dramatic changes in the sizes and shapes of the speech articulators over the course of a lifetime.

A second component of the feedforward control system schematized in Figure 58.3 is a basal ganglia-thalamo-cortical loop involving an Initiation Map in SMA. The SMA is strongly interconnected with lateral motor and premotor cortex and the basal ganglia (Jurgens, 1984; Lehericy et al., 2004; Luppino, Matelli, Camarda, & Rizzolatti, 1993; Matsumoto et al., 2004; Matsumoto et al., 2007), and a number of investigators have proposed that the SMA plays a critical role in controlling the initiation of speech motor commands (Alario, Chainay, Lehericy, & Cohen, 2006; Bohland & Guenther, 2006; Jonas, 1987; Ziegler, Kilian, & Deger, 1997). Each speech sound is represented by a different neuron in the model's *initiation map*, and activation of a sound's initiation map cell leads to readout of the feedforward commands for that sound via projections to ventral premotor and primary motor cortex (Tourville & Guenther, 2011). The basal ganglia, which receive input from a wide range of sensory, motor, and association areas of the cerebral cortex (not shown in Figure 58.3), are ideally suited for determining the proper sensorimotor and behavioral context for launching the feedforward commands for the next sound to be produced. This is hypothesized to occur via the basal ganglia-thalamo-cortical loop with SMA in Figure 58.3.

The feedforward control system of the DIVA model provides a mechanistic account for three disorders of speech motor control: apraxia of speech, ataxic dysarthria, and stuttering. Apraxia of speech is hypothesized to occur with damage to the speech sound map and/or projections from this map to primary motor cortex; such damage would impair the readout of speech motor programs, resulting in speech errors such as groping and syllable reduction that are associated with apraxia of speech (Ames, 2009). This account is consistent with the finding that damage to the brain regions thought to contain the speech sound map, left IFG and vPMC, is strongly associated with apraxia of speech (Hillis et al., 2004). Damage to the superior paravermal region of the cerebellar cortex results in ataxic dysarthria, a motor speech disorder characterized by slurred, poorly coordinated speech (Ackermann, Vogel, Petersen, & Poremba, 1992). This finding is in accord with the model's inclusion of this region for providing precisely timed feedforward commands necessary for fluent speech. Stuttering is hypothesized to arise from impairment of the basal ganglia-thalamo-cortical loop involving the initiation map in SMA (Civier, Bullock, Max, & Guenther, 2013). Alm (2004) provides a detailed treatment of the wide range of evidence implicating the basal ganglia in stuttering.

Interestingly, a number of auditory feedback manipulations have been shown to improve fluency, at least temporarily, in people who stutter. These include noise-masked auditory feedback (MacCulloch, Eaton, & Long, 1970), delayed auditory feedback (Van Riper, 1982), and frequency-altered feedback (Hargrave, Kalinowski, Stuart, Armson, & Jones, 1994). Alm (2004) proposes that such manipulations decrease the automaticity of speech and thereby reduce dependency on the impaired basal ganglia circuit; Civier, Tasko, and Guenther (2010) propose that such alterations prevent the speaker from hearing his/her own auditory errors, which when perceived may prohibit the basal ganglia from recognizing the proper sensory state for launching the next sound's feedforward commands.

58.6 THE GODIVA MODEL OF SPEECH SOUND SEQUENCING

The DIVA model accounts for production of individual speech motor programs, each corresponding to a different speech sound, with the typical unit of motor programming being the syllable. Additional brain areas, particularly in the left prefrontal cortex, become involved for longer speech utterances, which require the generation of an appropriately timed sequence of phonemes and syllables corresponding to a phrase or sentence. Here, we briefly treat some of the key areas involved in speech sound sequencing and introduce an extension to the DIVA model, called gradient order DIVA (GODIVA), that addresses the neural computations underlying speech sound sequencing.

The pre-SMA, like the SMA that lies immediately posterior to it in the medial frontal cortex, has been implicated in movement sequencing (Clower & Alexander, 1998; Shima & Tanji, 2000), including the sequencing of speech movements (Bohland & Guenther, 2006). Shima and Tanji (2000) found cells in pre-SMA that code for an entire sequence to be produced and other cells in pre-SMA that fired for the *n*th movement (e.g., the third movement) of a sequence regardless of the type of movement. Bohland and Guenther (2006) found increased activity in pre-SMA when the number of unique syllables in an utterance was increased, as well as when the complexity of the individual syllables was increased.

The lateral prefrontal cortex has been implicated in many studies of language and working memory (D'Esposito et al., 1998; Fiez et al., 1996; Gabrieli, Poldrack, & Desmond, 1998; Kerns, Cohen, Stenger, & Carter, 2004) and in serial order processing (Averbeck, Chafee, Crowe, & Georgopoulos, 2002, 2003; Petrides, 1991). Bohland and Guenther (2006) noted increased activity in lateral prefrontal cortex, specifically the region around the inferior frontal sulcus (IFS), with increased syllable sequence complexity of a spoken utterance.

The basal ganglia, particularly interactions between the cortex and the basal ganglia, are organized into multiple loop circuits (Alexander & Crutcher, 1990; Alexander, DeLong, & Strick, 1986; Middleton & Strick, 2000), including the *sensorimotor loop* described. The basal ganglia are known to be involved in sequencing motor acts (Harrington & Haaland, 1998), and abnormalities in the basal ganglia and/or associated circuitry can impact speech production (Kent, 2000; Murdoch, 2001), with some patients having particular difficulty fluently sequencing articulatory movements (Ho, Bradshaw, Cunnington, Phillips, & Iansek, 1998; Pickett, Kuniholm, Protopapas, Friedman, & Lieberman, 1998).

The GODIVA model (Bohland, Bullock, & Guenther, 2010) provides an account of how the pre-SMA, IFS, and basal ganglia interact with the speech motor control network described by the DIVA model to produce utterances containing multiple syllables. Briefly, left IFS acts as a phonological working memory that buffers the phonemes in an upcoming utterance, organized by their location within the syllable. Projections from left IFS to vPMC are responsible for choosing the speech sound map representation for the next syllable. Left pre-SMA is responsible for the sequencing of items within a syllable; specifically, pre-SMA represents the syllabic frame structure of the upcoming syllables (MacNeilage, 1998) without regard for the particular phonemes involved. Projections from pre-SMA to SMA activate, in the proper sequence, the appropriate neurons in the initiation map, thus leading to the readout of the feedforward commands for the next syllable via projections to the speech sound map. Loops through the basal ganglia are proposed to play central roles in the activation of pre-SMA, IFS, SMA, and vPMC neurons.

58.7 THE HSFC MODEL

DIVA is by far the most detailed and explicit model of speech motor control as well as the most influential

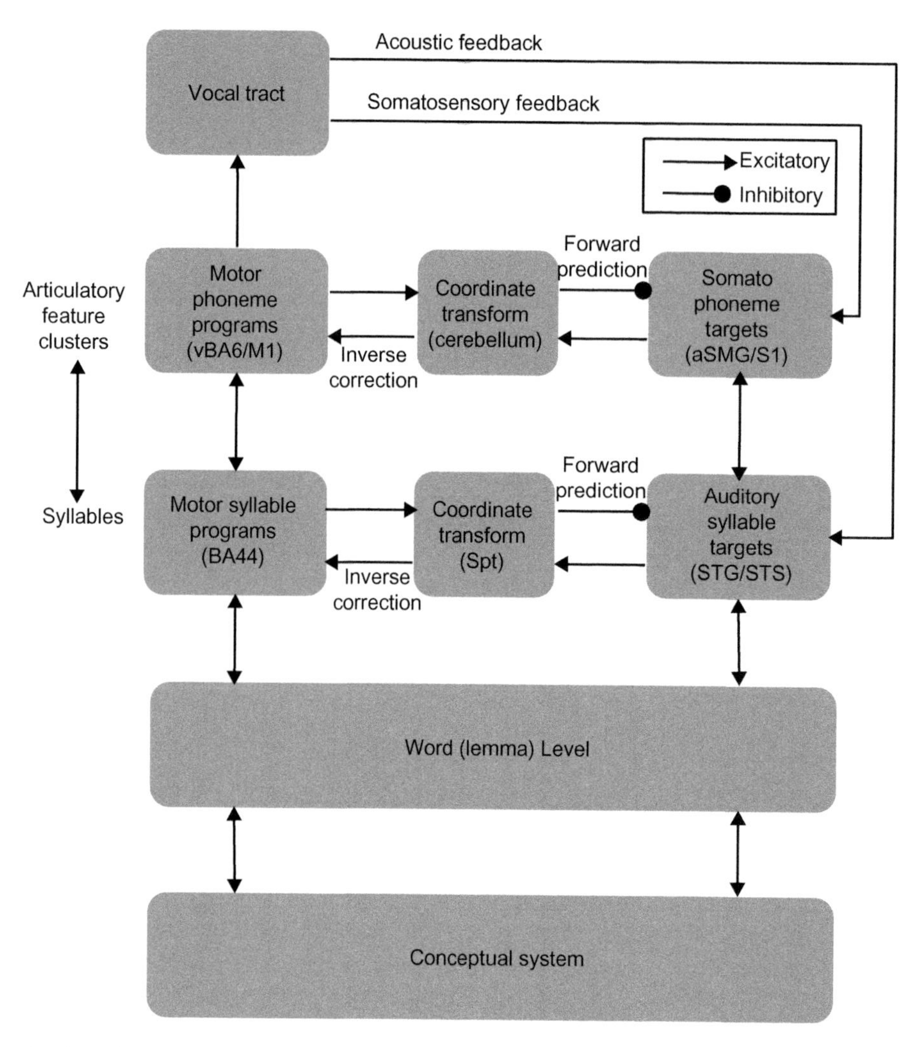

FIGURE 58.6. The HSFC model. The HSFC model includes two hierarchical levels of feedback control, each with its own internal and external sensory feedback loops. As in psycholinguistic models, the input to the HSFC model starts with the activation of a conceptual representation that, in turn, excites a corresponding word (lemma) representation. The word level projects in parallel to sensory and motor sides of the highest, fully cortical level of feedback control, the auditory–Spt–BA44 loop. This higher-level loop, in turn, projects, also in parallel, to the lower-level somatosensory–cerebellum–motor cortex loop. Direct connections between the word level and the lower-level circuit may also exist, although they are not depicted here. The HSFC model differs from the state feedback control (SFC) model in two main respects. First, "phonological" processing is distributed over two hierarchically organized levels, implicating a higher-level cortical auditory-motor circuit and a lower-level somatosensory-motor circuit, which approximately map onto syllabic and phonemic levels of analysis, respectively. Second, a true efference copy signal is not a component of the model. Instead, the function served by an efference copy is integrated into the motor planning process. BA, Brodmann area; M1, primary motor cortex; S1, primary somatosensory area; aSMG, anterior supramarginal gyrus; STG, superior temporal gyrus; STS, superior temporal sulcus; vBA6, ventral BA6. The HSFC model is squarely within the tradition of the DIVA model in that it assumes that the targets of speech gestures are coded in auditory space and that feedback control is a key computational operation of the network. HSFC differs from DIVA in three respects: (i) it assumes an internal as well as an external feedback detection/correction mechanism; (ii) it situates auditory and somatosensory feedback loops in a hierarchical arrangement (auditory loop being higher-level and somatosensory loop being lower-level); and (iii) it assumes a modified computational architecture for the feedback loops.

to date. As noted, DIVA uses feedback control architecture to detect and correct overtly produced errors. However, there is evidence in the motor control literature generally (Shadmehr & Mussa-Ivaldi, 1994; Tian & Poeppel, 2010; Wolpert, Ghahramani, & Jordan, 1995) and in the speech production literature more specifically for *internal* feedback control, that is, detecting and correcting internal coding errors prior to overt speech production. We describe one such model here, the HSFC model illustrated in Figure 58.6.

The empirical motivation for an internal feedback loop in the speech domain comes from three sources. One is simply that we can imagine speaking and hear ourselves in our "mind's ear." Experimental research on such inner speech has shown that imagined speech can contain inadvertent errors that are internally detected. Further, the types and distribution of such errors are similar to what is observed in overt speech (e.g., phonemic errors show a "lexical bias"0 (Nozari, Dell, & Schwartz, 2011; Oppenheim & Dell, 2008). A second source is that talkers correct partially articulated speech errors faster than they should be able to if they were relying on overt auditory feedback alone (Nozari et al., 2011); it is an open question whether somatosensory feedback may explain this phenomenon. The third source is conduction aphasia, a syndrome characterized by fluent speech output and intact speech perception abilities but with a higher than normal rate of phonemic errors in production. Crucially, affected individuals can readily detect their own errors once spoken but have trouble correcting them, even after they have been overtly detected (Goodglass, 1992). This pattern of speech behavior can be explained by a damaged internal error detection and correction loop (leading to the higher error rate) with an intact external feedback loop (allowing for detection of overtly produced errors).

The HSFC model, like DIVA, assumes that a basic planning unit in auditory space is approximately at the syllable level. The somatosensory circuit, however, is hypothesized to code sensory targets at a lower level. The basic idea is that speech production involves a cyclic opening and closing of the vocal tract (approximately corresponding to vowels and consonants respectively) and that the somatosensory system defines the targets of these opening or closing gestures (Gracco, 1994; Gracco & Lofqvist, 1994) similar to the somatosensory target map in the DIVA model but involving targets for individual phonemes rather than a single target for a whole syllable. The HSFC model holds that the internal auditory loop comprises a fully cortical network including auditory regions in the STG, motor regions in the IFG, and an auditory-motor interface network, Spt, in the posterior planum temporale region (of course, the external feedback loop involves noncortical structures). The somatosensory loop comprises somatosensory regions in the inferior parietal lobe, lower-level motor regions in primary motor cortex and/or Brodmann area 6, and a somatosensory-motor interface in the cerebellum. The hypothesis that the cerebellum is part of the lower-level sensory-motor circuit is motivated by the nature of speech deficits after cerebellar damage, which tend to be fairly low-level dysarthrias compared with the higher-level phonological deficits found in conduction aphasics with cortical damage (Ackermann, Mathiak, & Riecker, 2007; Baldo, Klostermann, & Dronkers, 2008; Buchsbaum et al., 2011; Kohn, 1984).

Architecturally and computationally, the HSFC (Figure 58.6) differs somewhat from DIVA (Figure 58.3). In the HSFC there are two sensory-motor feedback loops, both of which involve three components, sensory target representations, motor codes for "hitting" those targets (learned via external feedback), and a sensory-motor coordinate transform network. The latter is assumed to compute the relation between sensory and motor representations of speech units. In DIVA, speech production begins with the activation of a speech sound map in the frontal lobe; in HSFC, production begins with the activation of both auditory and motor units corresponding to the intended word. The auditory units comprise the target for motor unit selection in the same sense that a visually perceived object might comprise the target for activating motor units to execute a reaching action. The difference with speech is that the sensory (auditory) target is not physically present in the environment but is instead a (re-)activated mental representation of the sensory target (i.e., a sound pattern). During motor activation, the accuracy of motor unit selection is checked, internally, prior to speech articulation. If motor and sensory units match, then articulation proceeds. If there is a mismatch, then a correction signal can be generated prior to articulation. Computationally, this internal "checking" mechanism is instantiated via excitatory connections from auditory target units to their previously learned corresponding motor units (via the interface network) and via inhibitory feedback connections from motor units to their corresponding auditory units. When the motor and auditory units match, the motor units will inhibit the auditory target units and carry on with their activation pattern. When there is a mismatch, motor units will inhibit nontarget units in the auditory network, allowing the target auditory units to persist in exciting the correct motor units; this is the "error signal." Although a full-scale implementation of this architecture has not yet been demonstrated, a small-scale computational simulation has shown the feasibility of this architecture for internal error detection and correction (Hickok, 2012). A similar mechanism is assumed to hold at the various levels of the sensory-motor control hierarchy.

This architecture explains conduction aphasia as damage to the cortical interface network. Production is fluent because motor units are fully intact and can be activated directly from higher-level word representations. However, because of damage to the auditory-motor interface, motor unit activation cannot be checked against their auditory targets and an increase in error rates is observed. Once the conduction aphasic overtly produces an error, it is immediately detected via external feedback because the auditory target network is intact and the appropriate units are activated. However, subsequent attempts to correct such errors often fail, again because of the damage to the auditory-motor interface. Analysis of the relation between the lesions that typically cause conduction aphasia and the anatomical location of Spt, the auditory-motor interface, has shown good correspondence, lending further support for the proposed model (Buchsbaum et al., 2011).

The advantage of the HSFC model is that it incorporates an internal feedback loop that has some explanatory power regarding a speech disorder that has proven difficult to explain. It also includes a computational architecture that integrates auditory target activation, error detection, and error correction into a single process, which has some appeal from a parsimony standpoint. However, the model is far less developed than DIVA.

Although the models described herein can account for a wide range of experimental phenomena, it is important to note that these models are incomplete in their characterization of the vastly complex neural processes involved in speech production. An iterative process of generating testable predictions from a neurocomputational model, experimentally testing that prediction, and modifying the model as necessary to account for the experimental findings will lead to increasingly accurate accounts of the neural computations underlying speech.

58.8 FUTURE DIRECTIONS

The advent of noninvasive functional neuroimaging techniques such as PET and fMRI in the late 20th century has led to greatly accelerated progress toward understanding the neural mechanisms underlying speech production and perception. In more recent years, investigators have used multiple neuroimaging techniques in the same subject pool to overcome limitations in any given technique. For example, fMRI data, which have high spatial resolution but low temporal resolution, can be combined with either EEG or MEG, which have high temporal resolution but relatively low spatial resolution. New structural neuroimaging techniques such as diffusion-weighted imaging provide another dimension, namely the ability to correlate differences in brain function with differences in brain anatomy. We anticipate a continued increase in the use of multiple neuroimaging modalities as these neuroimaging technologies become more widely available and tools for analyzing the combined datasets improve.

The recent literature has seen a rapid increase in the number of publications that utilize electrocorticography (ECoG) to study the functioning human brain. ECoG involves electrodes placed on the surface of the cerebral cortex (either subdural or epidural); this technique is currently limited almost exclusively to epilepsy patients undergoing presurgical functional mapping to guide the surgical intervention. Despite the relatively limited opportunities for ECoG use, ECoG during speech production has already been used to investigate temporal lobe activity in response to one's own voice (Creutzfeldt et al., 1989), to compare spoken and sign language (Crone et al., 2001), to identify the time course of activity across the speech network during overt (Leuthardt et al., 2012; Pei et al., 2011) and covert (Pei et al., 2011) word repetition, and to identify the spatiotemporal pattern of activity in articulatory representations in sensorimotor cortex (Bouchard, Mesgarani, Johnson, & Chang, 2013). We expect this trend to continue due to ECoG's unique combination of high spatial and temporal resolution.

Early EEG studies identified correlations between mental state or behavior and brain rhythms (e.g., alpha rhythm, delta rhythm, etc.), which are quasi-oscillatory neural activity patterns that can be identified by peaks in the frequency spectrum of the recorded signals. The higher spatial resolution afforded by MEG and ECoG compared with EEG has led to the development of computational methods aimed at analyzing interactions between large-scale brain rhythms in different cortical areas, or in different frequency bands within the same cortical area. Among other things, these techniques can identify functional connectivity between different brain regions or functional coupling across frequency bands within a brain region. An increasing number of studies are now reporting relationships between brain rhythms and speech processing (Giraud et al., 2007; Leuthardt et al., 2011; Luo & Poeppel, 2007). We expect this trend to continue as MEG and ECoG data become more widely available.

The development of speech prostheses involving brain–computer interfaces is another promising area of future research in speech production. Guenther et al. (2009) utilized intracranial electrodes in the speech motor cortex to produce the first real-time speech prosthesis, a system that allowed an individual suffering from locked-in syndrome to produce vowels with a speech synthesizer using auditory feedback from the synthesizer to guide speech "movements." In this study, the DIVA model prediction of a spatiotemporal formant frequency trajectory in the speech premotor/motor cortex was first verified from electrophysiological recordings during attempted speech. A decoder was then built to translate brain activity into predicted formant frequencies, which were then synthesized into a speech signal using a formant synthesizer (Klatt, 1980). The delay between neural firing and corresponding sound output was 50 ms, which is approximately the delay from motor cortical activity to sound output in neurologically normal individuals. This real-time audio feedback allowed the users to improve their performance substantially with a small amount of practice. ECoG has also shown preliminary promise for brain–computer interfaces; for example, Leuthardt et al. (2011) developed an ECoG-based brain–computer interface that used activity in the speech network to control 1-D movement of a cursor.

Acknowledgments

Supported by the National Institute on Deafness and other Communication Disorders grants R01 DC007683 (FHG, PI), R01 DC002852 (FHG, PI), R01 DC03681 (GH, PI), and R01 DC009659 (GH, PI). We thank Barbara Holland and Jason Tourville for assistance with manuscript preparation.

References

Ackermann, H., Mathiak, K., & Riecker, A. (2007). The contribution of the cerebellum to speech production and speech perception: Clinical and functional imaging data. *Cerebellum, 6*, 202–213.

Ackermann, H., Vogel, M., Petersen, D., & Poremba, M. (1992). Speech deficits in ischaemic cerebellar lesions. *Journal of Neurology, 239*, 223–227.

Alario, F. X., Chainay, H., Lehericy, S., & Cohen, L. (2006). The role of the supplementary motor area (SMA) in word production. *Brain Research, 1076*(1), 129–143.

Alexander, G. E., & Crutcher, M. D. (1990). Functional architecture of basal ganglia circuits: Neural substrates of parallel processing. *Trends Neuroscience, 13*, 266–271.

Alexander, G. E., DeLong, M. R., & Strick, K. L. (1986). Parallel organization of functionally segregated circuits linking basal ganglia and cortex. *Annual Review of Neuroscience, 9*, 357–381.

Alm, P. (2004). Stuttering and the basal ganglia circuits: A critical review of possible relations. *Journal of Communication Disorders, 37*, 325–369.

Ames, H. (2009). *Neural dynamics of speech perception and production: From speaker normalization to apraxia of speech* (Ph.D. dissertation). Boston University.

Andersen, R. A. (1997). Multimodal integration for the representation of space in the posterior parietal cortex. *Philosophical Transactions of the Royal Society of London Series B: Biological Sciences, 352*(1360), 1421–1428.

Averbeck, B. B., Chafee, M. V., Crowe, D. A., & Georgopoulos, A. P. (2003). Neural activity in prefrontal cortex during copying geometrical shapes: I. single cells encode shape, sequence, and metric parameters. *Experimental Brain Research, 150*(2), 127–141.

Averbeck, B. E., Chafee, M. V., Crowe, D. A., & Georgopoulos, A. P. (2002). Parallel processing of serial movements in prefrontal cortex. *Proceedings of the National Academy of Sciences of the United States of America, 99*(20), 13172–13177.

Baldo, J. V., Klostermann, E. C., & Dronkers, N. F. (2008). It's either a cook or a baker: Patients with conduction aphasia get the gist but lose the trace. *Brain and Language, 105*, 134–140.

Bastian, H. C. (1887). The "muscular sense": Its nature and cortical localisation. *Brain, 10*, 1–89.

Bendor, D., & Wang, X. (2005). The neuronal representation of pitch in primate auditory cortex. *Nature, 436*(7054), 1161–1165.

Bohland, J. W., Bullock, D., & Guenther, F. H. (2010). Neural representations and mechanisms for the performance of simple speech sequences. *Journal of Cognitive Neuroscience, 22*, 1504–1529.

Bohland, J. W., & Guenther, F. H. (2006). An fMRI investigation of syllable sequence production. *NeuroImage, 32*, 821–841.

Bouchard, K. E., Mesgarani, N., Johnson, K., & Chang, E. F. (2013). Functional organization of human sensorimotor cortex for speech articulation. *Nature, 495*, 327–332.

Boyce, S., & Espy-Wilson, C. Y. (1997). Coarticulatory stability in American English /r/. *The Journal of the Acoustical Society of America, 101*, 3741–3753.

Browman, C. P., & Goldstein, L. (1989). Articulatory gestures as phonological units. *Phonology, 6*, 201–251.

Buchsbaum, B. R., Baldo, J., Okada, K., Berman, K. F., Dronkers, N., D'Esposito, M., et al. (2011). Conduction aphasia, sensory-motor integration, and phonological short-term memory—An aggregate analysis of lesion and fMRI data. *Brain and Language, 119*, 119–128.

Buchsbaum, B. R., Hickok, G., & Humphries, C. (2001). Role of left posterior superior temporal gyrus in phonological processing for speech perception and production. *Cognitive Science, 25*, 663–678.

Bullock, D., & Grossberg, S. (1988). Neural dynamics of planned arm movements: Emergent invariants and speed-accuracy properties during trajectory formation. *Psychological Review, 95*, 49–90.

Bullock, D., Grossberg, S., & Guenther, F. H. (1993). A self-organizing neural network model for redundant sensory-motor control, motor equivalence, and tool use. *Journal of Cognitive Neuroscience, 5*, 408–435.

Burnett, T. A., Freedland, M. B., Larson, C. R., & Hain, T. C. (1998). Voice F0 responses to manipulations in pitch feedback. *The Journal of the Acoustical Society of America, 103*, 3153–3161.

Callan, D. E., Kent, R. D., Guenther, F. H., & Vorperian, H. K. (2000). An auditory-feedback-based neural network model of speech production that is robust to developmental changes in the size and shape of the articulatory system. *Journal of Speech, Language, and Hearing Research, 43*, 721–736.

Caplan, D., Gow, D., & Makris, N. (1995). Analysis of lesions by MRI in stroke patients with acoustic-phonetic processing deficits. *Neurology, 45*, 293–298.

Celsis, P., Boulanouar, K., Doyon, B., Ranjeva, J. P., Berry, I., Nespoulous, J. L., et al. (1999). Differential fMRI responses in the left posterior superior temporal gyrus and left supramarginal gyrus to habituation and change detection in syllables and tones. *NeuroImage, 9*, 135–144.

Civier, O., Bullock, D., Max, L., & Guenther (2013). Computational modeling of stuttering caused by impairments in a basal ganglia thalamo-cortical circuit involved in syllable selection and initiation. *Brain and Language*, 126, 263–278

Civier, O., Tasko, S. M., & Guenther, F. H. (2010). Overreliance on auditory feedback may lead to sound/syllable repetitions: Simulations of stuttering and fluency-inducing conditions with a neural model of speech production. *Journal of Fluency Disorders, 35*, 246–279.

Clower, W. T., & Alexander, G. E. (1998). Movement sequence-related activity reflecting numerical order of components in supplementary and presupplementary motor areas. *Journal of Neurophysiology, 80*, 1562–1566.

Creutzfeldt, O., Ojemann, G., & Lettich, E. (1989). Neuronal-activity in the human lateral temporal-lobe.2. Responses to the subjects own voice. *Experimental Brain Research, 77*, 476–489.

Crone, N. E., Hao, L., Hart, J., Jr, Boatman, D., Lesser, R. P., Irizarry, R., et al. (2001). Electrocorticographic gamma activity during word production in spoken and sign language. *Neurology, 57*(11), 2045–2053.

Damasio, H., & Damasio, A. R. (1980). The anatomical basis of conduction aphasia. *Brain, 103*, 337–350.

Davidson, P. R., & Wolpert, D. M. (2005). Widespread access to predictive models in the motor system: A short review. *Journal of Neural Engineering*, 2(3), S313–S319.

Delattre, P., & Freeman, D. (1968). A dialect study of American R's by X-ray motion picture. *Linguistics, 44*, 29–68.
Desmurget, M., & Grafton, S. (2000). Forward modeling allows feedback control for fast reaching movements. *Trends in Cognitive Sciences, 4*(11), 423–431.
D'Esposito, M., Aguirre, G. K., Zarahn, E., Ballard, D., Shin, R. K., & Lease, J. (1998). Functional MRI studies of spatial and nonspatial working memory. *Cognitive Brain Research, 7*(1), 1–13.
Fant, G. (1960). *Acoustic theory of speech production*. The Hague: Mouton and Co.
Fiez, J. A., Raife, E. A., Balota, D. A., Schwarz, J. P., Raichle, M. E., & Petersen, S. E. (1996). A positron emission tomography study of the short-term maintenance of verbal information. *The Journal of Neuroscience, 16*(2), 808–822.
Gabrieli, J. D. E., Poldrack, R. A., & Desmond, J. E. (1998). The role of left prefrontal cortex in language and memory. *Proceedings of the National Academy of Sciences of the United States of America, 95*, 906–913.
Geschwind, N. (1965). Disconnexion syndromes in animals and man II. *Brain, 88*, 585–644.
Ghosh, S. S., Tourville, J. A., & Guenther, F. H. (2008). A neuroimaging study of premotor lateralization and cerebellar involvement in the production of phonemes and syllables. *Journal of Speech, Language, and Hearing Research, 51*, 1183–1202.
Giraud, A. L., Kleinschmidt, A., Poeppel, D., Lund, T. E., Frackowiak, R. S., & Laufs, H. (2007). Endogenous cortical rhythms determine cerebral specialization for speech perception and production. *Neuron, 56*, 1127–1134.
Golfinopoulos, E., Tourville, J. A., Bohland, Ghosh, S. S., Nieto-Castanon, A., & Guenther, F. H. (2011). fMRI investigation of unexpected somatosensory feedback perturbation during speech. *NeuroImage, 55*, 1324–1338.
Goodglass, H. (1992). Diagnosis of conduction aphasia. In S. E. Kohn (Ed.), *Conduction aphasia* (pp. 39–49). Hillsdale, NJ: Lawrence Erlbaum Associates.
Gracco, V. L. (1994). Some organizational characteristics of speech movement control. *Journal of Speech and Hearing Research, 37*, 4–27.
Gracco, V. L., & Lofqvist, A. (1994). Speech motor coordination and control: Evidence from lip, jaw, and laryngeal movements. *The Journal of Neuroscience, 14*, 6585–6597.
Guenther, F. H. (1992). *Neural models of adaptive sensory-motor control for flexible reaching and speaking* (Ph.D. dissertation). Boston University.
Guenther, F. H. (1995a). A modeling framework for speech motor development and kinematic articulator control. In K. Elenius, & P. Branderud (Eds.), *Proceedings of the XIIIth international congress of phonetic sciences, Stockholm, Sweden, 13–19 August, 1995* (Vol. 2, pp. 92–99). Stockholm, Sweden: KTH and Stockholm University.
Guenther, F. H. (1995b). Speech sound acquisition, coarticulation, and rate effects in a neural network model of speech production. *Psychological Review, 102*, 594–621.
Guenther, F. H., Brumberg, J. S., Wright, E. J., Nieto-Castanon, A., Tourville, J. A., Panko, M., et al. (2009). A wireless brain–machine interface for real-time speech synthesis. *PLoS ONE, 4*(12), e8218+.
Guenther, F. H., Espy-Wilson, C. Y., Boyce, S. E., Matthies, M. L., Zandipour, M., & Perkell, J. S. (1999). Articulatory tradeoffs reduce acoustic variability during American English /r/ production. *The Journal of the Acoustical Society of America, 105*, 2854–2865.
Guenther, F. H., Ghosh, S. S., & Tourville, J. A. (2006). Neural modeling and imaging of the cortical interactions underlying syllable production. *Brain and Language, 96*, 280–301.
Guenther, F. H., Hampson, M., & Johnson, D. (1998). A theoretical investigation of reference frames for the planning of speech movements. *Psychological Review, 105*, 611–633.
Guenther, F. H., & Vladusich, T. (2012). A neural theory of speech acquisition and production. *Journal of Neurolinguistics, 25*, 408–422.
Hargrave, S., Kalinowski, J. S., Stuart, A., Armson, J., & Jones, K. (1994). Effect of frequency-altered feedback on stuttering frequency at normal and fast speech rates. *Journal of Speech and Hearing Research, 37*, 1313–1319.
Harrington, D. L., & Haaland, K. Y. (1998). Sequencing and timing operations of the basal ganglia. In D. A. Rosenbaum, & C. E. Collyer (Eds.), *Timing of behavior: Neural, psychological, and computational perspectives* (pp. 35–61). Cambridge, MA: MIT Press.
Hickok, G. (2012). Computational neuroanatomy of speech production. *Nature Reviews Neuroscience, 13*, 135–145.
Hickok, G., Buchsbaum, B., Humphries, C., & Muftuler, T. (2003). Auditory-motor interaction revealed by fMRI: Speech, music, and working memory in area Spt. *Journal of Cognitive Neuroscience, 15*, 673–682.
Hickok, G., & Poeppel, D. (2004). Dorsal and ventral streams: A framework for understanding aspects of the functional anatomy of language. *Cognition, 92*, 67–99.
Hillis, A. E., Work, M., Barker, P. B., Jacobs, M. A., Breese, E. L., & Maurer, K. (2004). Reexamining the brain regions crucial for orchestrating speech articulation. *Brain, 127*(7), 1479–1487.
Ho, A. K., Bradshaw, J. L., Cunnington, R., Phillips, J. G., & Iansek, R. (1998). Sequence heterogeneity in Parkinsonian speech. *Brain and Language, 64*, 122–145.
Houde, J. F., & Jordan, M. I. (1998). Sensorimotor adaptation in speech production. *Science, 279*(5354), 1213–1216.
Houde, J. F., Nagarajan, S. S., Sekihara, K., & Merzenich, M. M. (2002). Modulation of the auditory cortex during speech: An MEG study. *Journal of Cognitive Neuroscience, 14*, 1125–1138.
Ito, T., & Ostry, D. J. (2012). Speech sounds alter facial skin sensation. *Journal of Neurophysiology, 107*(1), 442–447.
Ito, T., Tiede, M., & Ostry, D. J. (2009). Somatosensory function in speech perception. *Proceedings of the National Academy of Sciences of the United States of America, 106*(4), 1245–1248.
Jonas, S. (1987). The supplementary motor region and speech. In E. Perecman (Ed.), *The frontal lobes revisited* (pp. 241–250). New York, NY: IRBN Press.
Jurgens, U. (1984). The efferent and efferent connections of the supplementary motor area. *Brain Research, 300*, 63–81.
Kawato, M. (1999). Internal models for motor control and trajectory planning. *Current Opinion in Neurobiology, 9*(6), 718–727.
Kent, R. D. (2000). Research on speech motor control and its disorders: A review and prospective. *Journal of Communication Disorders, 33*, 391–427.
Kerns, J. G., Cohen, J. D., Stenger, V. A., & Carter, C. S. (2004). Prefrontal cortex guides context-appropriate responding during language production. *Neuron, 43*, 283–291.
Klatt, D. H. (1980). Software for a cascade/parallel formant synthesizer. *The Journal of the Acoustical Society of America, 67*, 971–995.
Kohn, S. E. (1984). The nature of the phonological disorder in conduction aphasia. *Brain and Language, 23*, 97–115.
Krakauer, J., & Ghez, C. (1999). Voluntary movement. In E. R. Kandel, J. H. Schwartz, & T. M. Jessell (Eds.), *Principles of neural science* (4th ed., pp. 756–781). New York, NY: McGraw Hill.
Kumar, S., & Schönwiesner, M. (2012). Mapping human pitch representation in a distributed system using depth-electrode recordings and modeling. *The Journal of Neuroscience, 32*, 13348–13351.
Larson, C. R., Burnett, T. A., Bauer, J. J., Kiran, S., Hain, T. C., & Bauer, J. J. (2001). Comparison of voice F0 responses to pitch-shift onset and offset conditions. *The Journal of the Acoustical Society of America, 110*, 2845–2848.
Lehericy, S., Ducros, M., Krainik, A., Francois, C., Van de Moortele, P. F., Ugurbil, K., et al. (2004). 3-D diffusion tensor axonal tracking

shows distinct SMA and pre-SMA projections to the human striatum. *Cerebral Cortex, 14*(12), 1302–1309.

Leuthardt, E. C., Gaona, C., Sharma, M., Szrama, N., Roland, J., Freudenberg, Z., et al. (2011). Using the electrocorticographic speech network to control a brain–computer interface in humans. *Journal of Neural Engineering, 8*, 036004.

Leuthardt, E. C., Pei, X. M., Breshears, J., Gaona, C., Sharma, M., Fredenberg, Z., et al. (2012). Temporal evolution of gamma activity in human cortex during an overt and covert word repetition task. *Frontiers in Human Neuroscience, 6*, 99.

Levelt, W. J., & Wheeldon, L. (1994). Do speakers have access to a mental syllabary? *Cognition, 50*, 239–269.

Liberman, A. M., Cooper, F. S., Shankweiler, D. P., & Studdert-Kennedy, M. (1967). Perception of the speech code. *Psychological Review, 74*(6), 431–461.

Liberman, A. M., & Mattingly, I. G. (1985). The motor theory of speech perception revised. *Cognition, 21*, 1–36.

Lindblom, B., Lubker, J., & Gay, T. (1979). Formant frequencies of some fixed-mandible vowels and a model of speech motor programming by predictive simulation. *Journal of Phonetics, 7*, 147–161.

Luo, H., & Poeppel, D. (2007). Phase patterns of neuronal responses reliably discriminate speech in human auditory cortex. *Neuron, 54*, 1001–1010.

Luppino, G., Matelli, M., Camarda, R., & Rizzolatti, G. (1993). Corticocortical connections of area F3 (SMA-proper) and area F6 (pre-SMA) in the macaque monkey. *The Journal of Comparative Neurology, 338*(1), 114–140.

MacCulloch, M. J., Eaton, R., & Long, E. (1970). The long term effect of auditory masking on young stutterers. *The British Journal of Disorders of Communication, 5*, 165–173.

MacNeilage, P. F. (1998). The frame/content theory of evolution of speech production. *The Behavioral and Brain Sciences, 21*, 499–511.

Maeda, S. (1990). Compensatory articulation during speech: Evidence from the analysis and synthesis of vocal tract shapes using an articulatory model. In W. J. Hardcastle, & A. Marchal (Eds.), *Speech production and speech modeling* (pp. 131–149). Boston, MA: Kluwer Academic.

Matsumoto, R., Nair, D. R., LaPresto, E., Bingaman, W., Shibasaki, H., & Luders, H. O. (2007). Functional connectivity in human cortical motor system: A cortico-cortical evoked potential study. *Brain, 130*(1), 181–197.

Matsumoto, R., Nair, D. R., LaPresto, E., Najm, I., Bingaman, W., Shibasaki, H., et al. (2004). Functional connectivity in the human language system: a cortico-cortical evoked potential study. *Brain, 127*, 2316–2330.

McGurk, H., & MacDonald, J. (1976). Hearing lips and seeing voices. *Nature, 264*(5588), 746–748.

Miall, R. C., & Wolpert, D. M. (1996). Forward models for physiological motor control. *Neural Networks, 9*(8), 1265–1279.

Middleton, F. A., & Strick, P. L. (1997). Cerebellar output channels. *International Review of Neurobiology, 41*, 61–82.

Middleton, F. A., & Strick, P. L. (2000). Basal ganglia and cerebellar loops: Motor and cognitive circuits. *Brain Research Reviews, 31*, 236–250.

Murdoch, B. E. (2001). Subcortical brain mechanisms in speech and language. *Folia Phoniatrica et Logopaedica, 53*, 233–251.

Nieto-Castanon, A., Guenther, F. H., Perkell, J. S., & Curtin, H. D. (2005). A modeling investigation of articulatory variability and acoustic stability during American English /r/ production. *The Journal of the Acoustical Society of America, 117*, 3196–3212.

Nozari, N., Dell, G. S., & Schwartz, M. F. (2011). Is comprehension necessary for error detection? A conflict-based account of monitoring in speech production. *Cognitive Psychology, 63*, 1–33.

Numminen, J., & Curio, G. (1999). Differential effects of overt, covert and replayed speech on vowel-evoked responses of the human auditory cortex. *Neuroscience Letters, 272*, 29–32.

Numminen, J., Salmelin, R., & Hari, R. (1999). Subject's own speech reduces reactivity of the human auditory cortex. *Neuroscience Letters, 265*, 119–122.

Oppenheim, G. M., & Dell, G. S. (2008). Inner speech slips exhibit lexical bias, but not the phonemic similarity effect. *Cognition, 106*, 528–537.

Passingham, R. E. (1993). *The frontal lobes and voluntary action*. Oxford: Oxford University Press.

Peeva, M. G., Guenther, F. H., Tourville, J. A., Nieto-Castanon, A., Anton, J. L., Nazarian, B., et al. (2010). Distinct representations of phonemes, syllables, and supra-syllabic sequences in the speech production network. *NeuroImage, 50*, 626–638.

Pei, X., Leuthardt, E. C., Gaona, C. M., Brunner, P., Wolpaw, J. R., & Schalk, G. (2011). Spatiotemporal dynamics of electrocorticographic high gamma activity during overt and covert word repetition. *NeuroImage, 54*, 2960–2972.

Perkell, J. S., Guenther, F. H., Lane, H., Matthies, M. L., Perrier, P., Vick, J., et al. (2000). A theory of speech motor control and supporting data from speakers with normal hearing and profound hearing loss. *Journal of Phonetics, 28*, 233–272.

Perkell, J. S., Guenther, F. H., Lane, H., Matthies, M. L., Stockmann, E., Tiede, M., et al. (2004). The distinctness of speakers' productions of vowel contrasts is related to their discrimination of the contrasts. *The Journal of the Acoustical Society of America, 116*(4), 2338–2344.

Perkell, J. S., Matthies, M. L., Tiede, M., Lane, H., Zandipour, M., Stockmann, E., et al. (2004). The distinctness of speakers' /s-sh/ contrast is related to their auditory discrimination and use of an articulatory saturation effect. *Journal of Speech Language and Hearing Research, 47*, 1259–1269.

Petrides, M. (1991). Functional specialization within the dorsolateral frontal cortex for serial order memory. *Proceedings of the Royal Society of London, B Biological Sciences, 246*(1317), 299–306.

Pickett, E. R., Kuniholm, E., Protopapas, A., Friedman, J., & Lieberman, P. (1998). Selective speech motor, syntax and cognitive deficits associated with bilateral damage to the putamen and the head of the caudate nucleus: A case study. *Neuropsychologia, 36*, 173–188.

Rizzolatti, G., & Craighero, L. (2004). The mirror-neuron system. *Annual Reviews in the Neurosciences, 27*, 169–192.

Rizzolatti, G., Fogassi, L., & Gallese, V. (1997). Parietal cortex: From sight to action. *Current Opinion in Neurobiology, 7*(4), 562–567.

Rosenbaum, D. A., Engelbrecht, S. E., Bushe, & Loukopoulos, L. D. (1993). Knowledge model for selecting and producing reaching movements. *Journal of Motor Behavior, 25*, 217–227.

Rosenbaum, D. A., Loukopoulos, L. D., Meulenbroek, R. G. J., Vaughan, J., & Engelbrecht, S. E. (1995). Planning reaches by evaluating stored postures. *Psychological Review, 192*, 28–67.

Sancier, M. L., & Fowler, C. A. (1997). Gestural drift in a bilingual speaker of Brazilian Portuguese and English. *Journal of phonetics, 25*, 421–436.

Sanes, J. N., Mauritz, K. H., Evarts, E. V., Dalakas, M. C., & Chu, A. (1984). Motor deficits in patients with large-fiber sensory neuropathy. *Proceedings of the National Academy of Sciences of the United States of America, 81*, 979–982.

Schmahmann, J. D., & Pandya, D. N. (1997). The cerebrocerebellar system. *International Review of Neurobiology, 41*, 31–60.

Shadmehr, R., & Mussa-Ivaldi, F. A. (1994). Adaptive representation of dynamics during learning of a motor task. *The Journal of Neuroscience, 14*, 3208–3224.

Shima, K., & Tanji, J. (2000). Neuronal activity in the supplementary and presupplementary motor areas for temporal organization of multiple movements. *Journal of Neurophysiology, 84*, 2148–2160.

Sperry, R. W. (1950). Neural basis of the spontaneous optokinetic response produced by visual inversion. *Journal of Comparative and Physiological Psychology*, *43*(6), 482–489.

Stevens, K. (1998). *Acoustic phonetics*. Cambridge, MA: MIT Press.

Tian, X., & Poeppel, D. (2010). Mental imagery of speech and movement implicates the dynamics of internal forward models. *Frontiers in Psychology*, *1*, 166.

Tourville, J. A., & Guenther, F. H. (2011). The DIVA model: A neural theory of speech acquisition and production. *Language and Cognitive Processes*, *26*, 952–981.

Tourville, J. A., Reilly, K. J., & Guenther, F. H. (2008). Neural mechanisms underlying auditory feedback control of speech. *NeuroImage*, *39*, 1429–1443.

Van Riper, C. (1982). *The nature of stuttering* (2nd ed.). Englewood Cliffs, NJ: Prentice-Hall.

von Helmholtz, H. (1925). *Helmholtz's treatise on physiological optics* (3rd ed., 1910) (J.P.C. Southall, Trans.). New York, NY: Optical Society of America.

Waldstein, R. S. (1989). Effects of postlingual deafness on speech production: Implications for the role of auditory feedback. *The Journal of the Acoustical Society of America*, *88*, 2099–2144.

Wilson, S. M., Saygin, A. E. P., Sereno, M. I., & Iacoboni, M. (2004). Listening to speech activates motor areas involved in speech production. *Nature Neuroscience*, *7*(7), 701–702.

Wise, R. J., Greene, J., Buchel, C., & Scott, S. K. (1999). Brain regions involved in articulation. *Lancet*, *353*, 1057–1061.

Wolpert, D. M., Ghahramani, Z., & Jordan, M. I. (1995). An internal model for sensorimotor integration. *Science*, *269*, 1880–1882.

Yates, A. J. (1963). Delayed auditory feedback. *Psychological Bulletin*, *60*, 213–251.

Ziegler, W., Kilian, B., & Deger, K. (1997). The role of the left mesial frontal cortex in fluent speech: Evidence from a case of left supplementary motor area hemorrhage. *Neuropsychologia*, *35*(9), 1197–1208.

CHAPTER

59

Neurobiology of Speech Production: A Motor Control Perspective

Pascale Tremblay[1,2], *Isabelle Deschamps*[1,2] and *Vincent L. Gracco*[3,4,5]

[1]Centre de Recherche de l'Institut Universitaire en Santé Mentale de Québec, Québec City, QC, Canada; [2]Département de Réadaptation, Faculté de Médecine, Université Laval, Québec City, QC, Canada; [3]Centre for Research on Brain, Language and Music, McGill University, Montreal, QC, Canada; [4]School of Communication Sciences and Disorders, McGill University, Montreal, QC, Canada; [5]Haskins Laboratories, New Haven, CT, USA

59.1 INTRODUCTION

Speech is one of the most distinguishing human traits. It represents a model neural system for studying a range of human characteristics from sensorimotor control to cognition. It uses a complex control system optimized for sequential output and is used for both self-expressive and interactive communication. The production of speech reflects a complex and dynamic process dependent on the interaction among multiple cortical and subcortical regions for the fine control of more than 100 muscles located in the oral cavity, neck, and abdomen (see Figure 59.1 for an overview). In the following, we identify the set of processes involved in speech production as well as their neural substrate.

59.2 NEUROBIOLOGY OF SPEECH MOTOR CONTROL

59.2.1 Speech Representations

From a linguistic perspective, a number of potential candidate constructs may be represented in the neural processes associated with the production of speech. One view in the psycholinguistic literature is that grammatical encoding, or the creation of lexical items within a syntactic frame, and phonological encoding, including the specification of prosodic structure, are the two fundamental processes that create the phonetic plan (Garrett, 1993; Levelt, 1992, 1993). The phonetic plan interfaces seamlessly with speech motor processes that generate the sequence of sounds specified in the plan.

One approach to associate these broad psycholinguistic processes with their neural substrates comes from studies using speech errors. Speech errors can provide valuable insights regarding the linguistic principles that are involved in the production of speech because these errors are generally consistent with language-specific phonological rules (Goldrick & Daland, 2009). Based on speech error analyses, some researchers have postulated that the units of speech planning are individual phonological features (Mowrey & MacKay, 1990), whereas others propose that these units are bigger [e.g., phonemes (Roelofs, 1997, 1999; Stemberger, 1982), syllables (Levelt, 1999), or words]. Recently, Peeva and colleagues (2010) used functional magnetic resonance imaging (fMRI) and a repetition–suppression (RS) paradigm (Grill-Spector, Henson, & Martin, 2006; Grill-Spector & Malach, 2001) to study speech representations. Capitalizing on the RS phenomenon in which the repetition of a stimulus leads to reduced neural activity (Henson & Rugg, 2003), the authors varied the repetition rate of phonemes, syllables, and pseudowords expecting that areas sensitive to the processing of a specific type of phonological unit would show specific RS effects. Sensitivity to phonemic information was found in the supplementary motor area (SMA), the left palladium (in the basal ganglia or BG), the left posterior superior temporal gyrus (STG), and the left superior posterior lateral cerebellum. Sensitivity to syllable

Neurobiology of Language. DOI: http://dx.doi.org/10.1016/B978-0-12-407794-2.00059-6

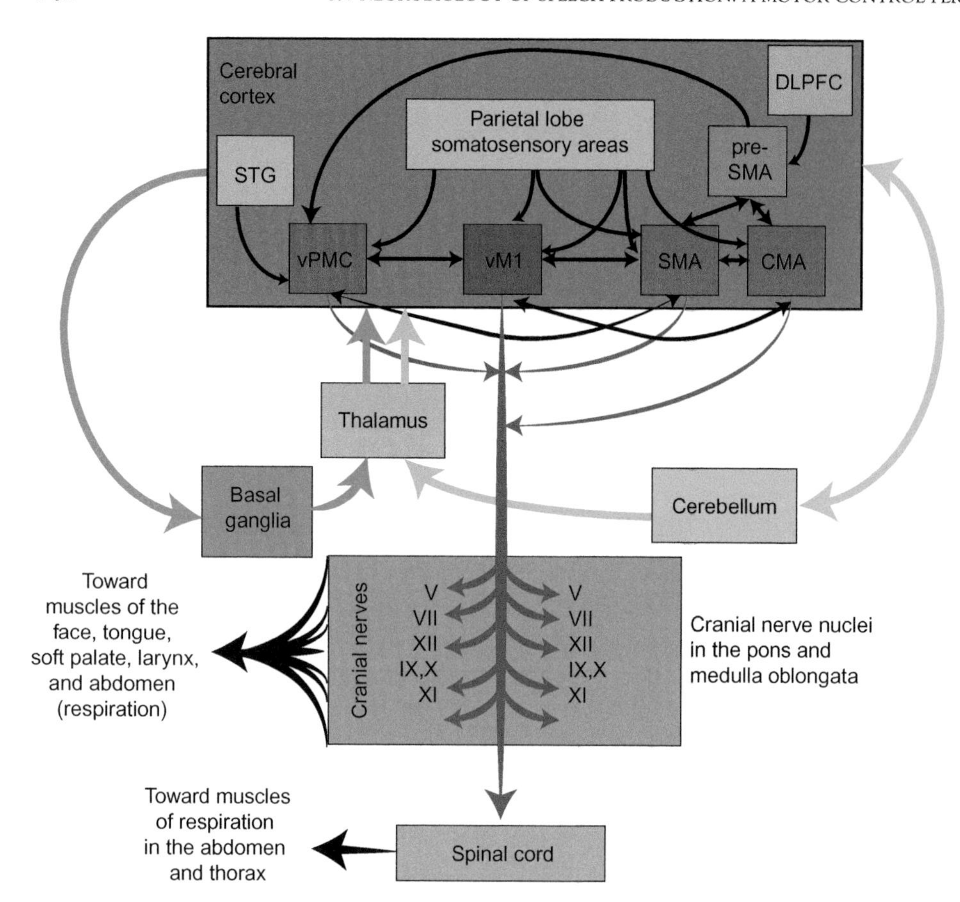

FIGURE 59.1 Simplified illustration of the motor speech system including main cortical components and their connectivity, subcortical loops (cortico-striatal-thalamic loops and cortico-cerebellar-thalamic loops), as well as peripheral components (cranial and spinal nerve innervations to speech-related structures in the face, oral cavity, neck, and abdomen).

level information was found in the ventral premotor cortex (vPMC), which was also sensitive to phonemic information. Finally, sensitivity to supra-syllabic information was found in the right superior posterior lateral cerebellum. These results suggest that multiple levels of representations, including phonemic and syllabic, are involved in the production of speech sounds.

59.2.2 Speech Motor Planning and Programming

During speech production, phonological encoding is the retrieval of the phonological code that consists of segmental (phonemes, syllables) and suprasegmental information (such as stress). This information is used to build a representation of the syllabified word form. The syllabified word form provides the framework for the planning of a motor act. Although models of speech production converge on the notion that the output of phonological encoding is a phonological word in which metrical, syllabic, and segmental properties are fully specified, models of speech production differ with regard to whether the retrieval of the phonological code is presyllabified. For some, the syllabification is computed online depending on the context (Levelt, Roelofs, & Meyer, 1999), whereas for others the phonological code is presyllabified (Dell, 1988). Regardless of the theoretical perspective, phonological encoding is associated with the process of speech motor preparation, which involves the activation and translation of phonological representations into multiple domain-general mechanisms, including response selection, response sequencing, and movement initiation. These important mechanisms, often referred to as "supra-motor functions" or "motor cognition" (Freund, Jeannerod, Hallett, & Leiguarda, 2005), are not specific to speech production but instead are part of the planning of all voluntary actions. Speech production builds on common action control mechanisms consistent with the notion that the speech system is an overlaid functional system that "*[...] gets what service it can out of organs and functions, nervous and muscular, that have come into being and are maintained for very different ends than its own*" (Sapir, 1921).

59.2.2.1 *Response Selection*

Response selection in spoken language production is the process by which a set of lexical units forming a message is transformed into motor programs, that is, stored motor routines. Several neuroimaging studies have examined the process of selecting nonspeech motor responses (such as finger and hand movements) and revealed activation in the presupplementary motor area (pre-SMA) (Brodmann's medial area 6) in which the increase in activation is commensurate with demands on response selection. For instance, activation in pre-SMA is enhanced when participants are free to choose a motor response from among several alternatives compared with when they are required to execute a specific, stimulus-driven, motor response (Deiber, Ibanez, Sadato, & Hallett, 1996; Lau, Rogers, & Passingham, 2006; Weeks, Honda, Catalan, & Hallett, 2001). Consistent with the nonspeech literature, several fMRI studies have shown that manipulating response selection during single word production modulates distributed brain networks including the pre-SMA, but also the adjacent cingulate motor area (CMA) and the vPMC (Crosson et al., 2001; Nagel, Schumacher, Goebel, & D'Esposito, 2008; Tremblay & Gracco, 2006; Tremblay & Small, 2011). Importantly, the pre-SMA is involved in selecting single words (Alario, Chainay, Lehericy, & Cohen, 2006; Tremblay & Gracco, 2006) but also noncommunicative oral motor gestures (Braun, Guillemin, Hosey, & Varga, 2001; Tremblay & Gracco, 2010), revealing a domain-general selection mechanism. Moreover, transcranial magnetic stimulation (TMS) to the pre-SMA leads to impaired voluntary selection of actions, including words and noncommunicative oral motor gestures (Tremblay & Gracco, 2009), supporting the notion of a domain-general selection process. Taken together, these results suggest that the pre-SMA plays a central role in selecting motor responses for speech production. The pre-SMA has a connectivity pattern that is ideal for linking higher-level cognitive (including linguistic) and motor processes, a *sine qua non* for the implementation of response selection, with important projections from the prefrontal cortex, particularly the dorsolateral prefrontal cortex or DLPFC (Lu, Preston, & Strick, 1994; Luppino, Matelli, Camarda, & Rizzolatti, 1993), and connections with several nonprimary motor areas, such as the SMA-proper and the PMC (Luppino & Rizzolatti, 2000), for controlling motor output. Recent fMRI evidence suggests a role for the caudate nucleus in response selection for speech production (Argyropoulos, Tremblay, & Small, 2013), consistent with evidence on the anatomical connectivity of the caudate, which connects with the prefrontal as well as the SMA/pre-SMA (Di Martino et al., 2008; Lehericy et al., 2004), suggesting that response selection is implemented through cortico-striatal connections between the pre-SMA and the caudate nucleus.

59.2.2.2 Response Sequencing

In his classic article on serial order, Lashley described the problem of organizing component parts of an action into movement sequences as the *action syntax problem* (Lashley, 1951). The manifestation of action syntax can be seen in a multitude of behaviors ranging from human thought (Marsden, 1984) to grooming behavior in rats (Aldridge, Berridge, & Rosen, 2004). For speech, action sequences can be organized around multiple components (phonemes, syllables, words, phrases, etc.); without appropriate timing, in terms of either initiating the action or sequencing the action units, communication would be difficult. fMRI studies have shown that motor sequencing is implemented in a network of regions organized around nonprimary motor areas (SMA-proper, PM), the cerebellum, and the BG (Bengtsson, Ehrsson, Forssberg, & Ullen, 2005; Gerloff, Corwell, Chen, Hallett, & Cohen, 1997; Macar et al., 2002). Repetitive TMS of the SMA-proper results in sequential timing disruptions in a complex finger movement task (Gerloff et al., 1997); SMA-proper activation accompanies tasks requiring the processing of temporal patterns (Macar et al., 2002). Using fMRI, Bohland and Guenther (2006) showed a bilateral network including the SMA, the anterior insula, and the superior cerebellum that was more strongly recruited for the production of complex sequences of syllables (ka-ru-ti) compared with the production of simpler sequences, which consisted of repeating the same sound three times (ta-ta-ta), consistent with the nonspeech literature. Although there remains a number of issues regarding the implementation of selection and sequencing mechanisms for speech, the available empirical evidence, though limited, suggests that speech production relies on common action control mechanisms centered on the pre-SMA, SMA, and vPMC.

59.2.2.3 Motor Programming

Preparing speech production also involves fine-tuning of the planned motor routines, including adjustments of velocity, muscle tone, movement range, and direction. Motor programming is necessary because even though a closed set of syllables is available in each language and probably stored as a set of motor routines, syllables and words are never produced identically, they are co-articulated and modulated as a function of the linguistic, environmental, emotional, and social contexts. Motor programming is usually believed to involve both online feedback-based and feedforward control systems. According to Van der Merwe (2009), regions involved in programming include the cerebellum, SMA-proper, M1, and the BG, but experimental evidence is lacking. The issue of feedback-based motor control is discussed in Section 59.3.

59.2.2.4 Movement Initiation

The initiation and termination of an action is fundamental to all voluntary motor behaviors. For speech production, starting and stopping speech movements is associated with a diverse range of communicative actions such as turn-taking, producing a list of words, and the insertion of pauses for emphasis. The SMA-proper has been previously identified as contributing to speech timing (Brendel et al., 2010; Gracco, 1997) and, more generally, has been associated with sequence timing as well as the perception of time. Recently, using functional connectivity analysis, and evaluating the temporal dynamics of the BOLD signal, two separately organized networks for speech production have been proposed (Brendel et al., 2010; Riecker et al., 2005) with the SMA and the insula identified as network components contributing to motor timing. For example, TMS to the SMA results in sequential timing disruptions (Gerloff et al., 1997), varying rate of stimulus presentation during reading results in modulation of SMA-proper activity (Price, Moore, Humphreys, Frackowiak, & Friston, 1996), and SMA activity accompanies tasks requiring the processing of temporal patterns (Bengtsson et al., 2005; Macar et al., 2002). Similarly, damage to the insula results in speech initiation impairments (Shuren, 1993) and apraxia of speech (Dronkers, 1996), a disorder of temporal sequencing, although there is controversy surrounding the role of insula in motor sequencing (Hillis et al., 2004). One possibility is that the SMA-proper and insula may be working to coordinate and time sequential actions, possibly through priming and then triggering motor cortex output. Speech and oral movements are localized around the central sulcus of the insula in an area that does not have direct projections onto lower motor neurons but does connect to frontal regions, including the DLPFC as well as the SMA and the sensorimotor portions of the striatum (Augustine, 1996). The DLPFC on the right hemisphere is known to modulate lower level systems (Shallice, 2004) and activity in the right DLPFC may be contributing to speech timing and/or temporal processing of action sequences (Coull, Frackowiak, & Frith, 1998; Vallesi, Shallice, & Walsh, 2007). It appears that the insula and SMA-proper form an integrated network component, operating in concert with peripheral feedback systems, to time sequential speech motor output. When there is a need for explicit timing control, prefrontal cortex participation is recruited.

Another way to study movement initiation is to compare the manner in which movements are triggered, whether externally by sensory events or at will. Movements initiated by external stimuli produce reliable activity in SMA-proper (Lee, Chang, & Roh, 1999; Thickbroom et al., 2000; Wiese et al., 2004) as well as in the left dorsal PMC (Krams, Rushworth, Deiber, Frackowiak, & Passingham, 1998; Lepage et al., 1999; Weeks et al., 2001), suggesting that these areas are involved in initiating actions based on external sensory triggers. In humans, TMS to the left PMC results in a response delay and a disruption in the early stage of reaching and grasping (before movement execution), suggesting a role in the onset of movement (Schluter, Rushworth, Passingham, & Mills, 1998). Importantly, the contrast of self-initiated and externally triggered movements reveals activation in the pre-SMA (Deiber et al., 1996; Jenkins, Jahanshahi, Jueptner, Passingham, & Brooks, 2000; Tsujimoto, Ogawa, Tsukada, Kakiuchi, & Sasaki, 1998), suggesting a role for this region in the generation of an internal trigger to move, which supports a role for the SMA/pre-SMA in the timing and initiation of actions. Another potentially important region for the timing of actions and the generation of a movement trigger is the BG. In patients with Parkinson's disease (PD), a disorder of BG, there is a clear decline in the ability to initiate movements at will without a concomitant reduction or slowing of externally triggered actions (Cunnington, Iansek, & Bradshaw, 1999; Freeman, Cody, & Schady, 1993; Praamstra, Stegeman, Cools, Meyer, & Horstink, 1998). Other BG dysfunctions lead to difficulty starting, stopping, or sustaining movements including speech (Speedie, Wertman, Ta'ir, & Heilman, 1993), as well as abnormal rate, regularity, and temporal ordering of speech movements (Ludlow, Connor, & Bassich, 1987; Skodda, 2011; Skodda & Schlegel, 2008; Volkmann, Hefter, Lange, & Freund, 1992), demonstrating the importance of BG for the timing of speech actions.

59.3 SPEECH MOVEMENT EXECUTION

The final output for speech comes mainly from the ventral part of the primary motor cortex (vM1), which contains the neurons controlling the vocal tract (Penfield & Boldrey, 1937). It has been estimated that approximately 100 striated and visceral muscles, distributed across the abdomen, neck, larynx, pharynx, and oral cavity, are involved in the production of speech, reflecting the immense complexity of this functional system, which, in mature speakers, may produce as many as 14 phonemes per second (i.e., between six and nine syllables per second) (Kent, 2000).

The pyramidal system, which includes the corticospinal and corticobulbar tracts, is one of the most important efferent pathways for the control of voluntary muscle contractions. It connects neurons in the cortex (upper motor neurons, UMN) to alpha (lower) motor neurons (LMN) located in the brainstem and spinal cord. LMN innervate the muscle fibers located in the face, neck, and abdomen. M1 is the cortical area that contains the largest number of pyramidal fibers (Kuypers, 1973; Murray & Coulter, 1981; Ralston & Ralston, 1985), particularly the giant Betz cells located in cortical layer V.

However, anatomical studies have shown that M1 is connected through long and short association fibers to multiple nonprimary motor areas, including the SMA (Dum & Strick, 1991, 1996; Muakkassa & Strick, 1979) the CMA located just beneath the SMA on the dorsal and ventral banks of the cingulate sulcus (Dum & Strick, 1991; Muakkassa & Strick, 1979) and the dorsal and ventral PMC (Barbas & Pandya, 1987; Dum & Strick, 1991; He, Dum, & Strick, 1993). Importantly, these nonprimary motor areas contain a high density of corticospinal and corticobulbar neurons, directly projecting to the spinal cord through the pyramidal tract (for a review of the connectivity of nonprimary motor areas, see Picard & Strick, 1996, 2001), and thus each has the potential to influence the generation and control of movement independently of M1 (Dum & Strick, 1991). Electrical stimulation of the SMA (Fried et al., 1991; Morris, Dinner, Luders, Wyllie, & Kramer, 1988; Penfield & Welch, 1951; Talairach & Bancaud, 1966) and CMA (von Cramon & Jurgens, 1983) induce vocalization and speech arrests in humans, suggesting a role in the control of phonation and articulation for these regions.

The corticospinal tracts innervate motor nuclei located in the spinal cord, whereas corticobulbar fibers innervate motor nuclei located in the brainstem. Because the motor nuclei involved in the control of respiration (mainly expiration), phonation, and articulation are located in the pons, down to the lumbar portion of the spinal cord, the production of speech depends on the integrity of both the corticospinal tract, for the innervations of the muscles of respiration in the abdomen, neck, and shoulder, and the corticobulbar tract, for the sensorimotor innervations of laryngeal and supralaryngeal muscles (for reviews, see Jurgens, 2002, 2009) through six pairs of cranial nerves (CN V: trigeminal; CN VII: facial; CN IX: glossopharyngeal; CN X: vagus; CN XI: accessory; CN XII: hypoglossal).

All cortical axons (originating from M1, SMA, CMA, and PMC) forming the corticobulbar and corticospinal tracts converge into the internal capsule, located between the thalamus and BG, with fibers originating from ventral areas located rostrally to those originating from more dorsal areas (Beevor & Horsley, 1890; Dejerine, 1901). Most pyramidal fibers cross from one side to the other before entering the spinal cord at the level of the medulla oblongata (i.e., the pyramidal decussation); corticobulbar fibers cross at the level of the brainstem, although there are substantial bilateral innervations of the CN motor nuclei. The exceptions include contralateral innervations of ventral cell groups of the motor nucleus of the facial nerve (CN VII), which supply muscles of the lower quadrants of the face (e.g., the orbicularis oris muscle), and the hypoglossal nucleus (CN XII), which supplies the intrinsic and extrinsic muscles of the tongue.

It has been suggested that vocalizations are controlled through two distinct cortical-subcortical pathways, one involving a circuit formed by the CMA, the periaqueductal gray matter (PAG), and the reticular formation for the control of innate vocal patterns (e.g., crying, laughing, and moaning), and another connecting M1 to the phonatory motoneurons through the reticular formation for the control of patterned speech and singing (Hsieh, Petrosyan, Goncalves, Hickok, & Saberi, 2011; Jurgens, 2002). This second circuit involves the cortico-striatal motor loop. Thus, M1 is connected not only to multiple nonprimary motor areas in the frontal lobe but also to the BG and cerebellum through the thalamus, and also to the reticular formation in the brainstem, controlling multiple aspects of speech production including respiration, vocalization, and articulation.

59.4 FEEDBACK PROCESSING AND SENSORY-MOTOR INTEGRATION

Early in the developmental process, the functional connection between speech perception and speech production is established and the ability to modify this coupling reflects the neural plasticity that continues throughout the life span. The resultant sensorimotor learning is the substrate on which developmental stages of speech and language develop, and one in which sensory feedback plays a crucial role (Mowrer, 1952). Somatosensory information from the lips and jaw have real-time access to modulate the spatial (Abbs & Gracco, 1983; Abbs, Gracco, & Cole, 1984; Gracco & Abbs, 1985; Shaiman & Gracco, 2002) and temporal aspects of speech sequences (Gracco & Abbs, 1989; Saltzman, Lofqvist, Kay, Kinsella-Shaw, & Rubin, 1998). Similar kinds of results are obtained from unanticipated alteration of auditory feedback for pitch (Burnett, Freedland, Larson, & Hain, 1998; Jones & Munhall, 2000) and formants (Purcell & Munhall, 2006a, 2006b). The overarching conclusion is that sensory and motor systems for speech are in a constant state of interaction and integration and, most importantly, the sensorimotor integration forms the basis for successful and efficient speech production (Gracco, 1991).

From a control perspective, speech production can be conceptualized as representing a hybrid control scheme consisting of feedforward and feedback-like neural processes (Abbs et al., 1984; Guenther, Ghosh, & Tourville, 2006; Hickok, 2012; Houde & Nagarajan, 2011; Tourville, Reilly, & Guenther, 2008). Feedforward control is used to compute, before movement onset, the necessary motor commands that will achieve generally a desired movement goal given the system's current state. That is, the feedforward controller assembles a basic motor plan prior to movement onset and sends the commands to the appropriate musculature for execution. In contrast, feedback processes are used to adjust and correct motor commands that are planned or executed by the feedforward controller. However, if such adjustments would depend solely on afferent input signals, there would be an unavoidable delay that may be too long for movements as fast as those involved in many skilled actions, including speech production. Feedback control processes can also be used to *predict* the sensory consequences of movements by making use of a copy of the prepared motor commands (efference copy or corollary discharge) (Sperry, 1950; Von Holst & Mittelstaedt, 1973). As such, information from somatosensory and auditory systems contributes in multiple ways. First, as part of the feedforward process, the sensory systems provide information about the initial conditions such that the motor commands for a desired outcome can be successfully achieved given the state of the vocal tract. Second, as part of the predictive process, they interact with the control signals to estimate the consequences of the planned action. Finally, as part of the feedback process, they modulate, in real time, adjustments to the motor commands based on re-afferent input during movement execution as well as signaling the achievement of the desired action.

The neural substrate associated with the sensorimotor aspects of speech production involves a mostly bilateral network of brain regions, including vM1 and sensory areas (somatosensory cortex, STG), nonprimary motor areas (vPMC, SMA-proper, CMA, and the insula), and subcortical regions associated with sensorimotor control (putamen, cerebellum, thalamus) (Ackermann & Riecker, 2004; Argyropoulos et al., 2013; Grabski, Tremblay, Gracco, Girin, & Sato, 2013; Riecker et al., 2004, 2005; Riecker, Wildgruber, Dogil, Grodd, & Ackermann, 2002; Tremblay, Deschamps, & Gracco, 2013; Tremblay & Gracco, 2006, 2009; Wise, Greene, Büchel, & Scott, 1999), areas that are known to receive afferent input from auditory and somatosensory areas. For example, in the macaque, the vPMC receives projection from sensory areas, including associative somatosensory area SII (Matelli, Camarda, Glickstein, & Rizzolatti, 1986) and the posterior STG (Chavis & Pandya, 1976; Schmahmann et al., 2007), whereas the SMA-proper receives important projection from the superior parietal lobule (area PEci, in the cingulate sulcus), which contains a complete somatosensory map of the body (Pandya & Seltzer, 1982) as well as from areas SII and SI (Luppino et al., 1993; McGuire, Bates, & Goldman-Rakic, 1991a, 1991b). Projections to the putamen have been reported from regions within the supratemporal plane and the STG (Yeterian & Pandya, 1998). The rostral and medial parts of STG project to rostroventral and caudoventral portions of the putamen, whereas the caudal portion of STG projects to caudal putamen. Recently, using resting state functional connectivity in humans, the dorsal portion of the putamen has been shown to connect with regions of the temporal cortex (Di Martino et al., 2008). As such, reafference may be an important source of information to assist in both the spatial and timing adjustments for the dynamic modulation of speech motor output as well as signaling successful achievement of speech motor goals (Gracco & Abbs, 1989).

For the cerebellum, bilateral posterior lobe activation in the vicinity of hemisphere lobule VI (Schmahmann et al., 1999) has been consistently reported during speech production (Ackermann, Mathiak, & Riecker, 2007; Riecker et al., 2002; Wise et al., 1999), most likely reflecting cortico-ponto-cerebellar projections from and to M1 (Kelly & Strick, 2003), possibly as part of an efference copy signal. A second area of activation on the inferior portion of the cerebellar hemisphere lobule VIIIA has been associated with auditory (Tourville et al., 2008) and somatosensory (Golfinopoulos et al., 2011) perturbations, as well as with rhythmic orofacial movements (Corfield et al., 1999) and sequencing nonmeaningful syllables (Bohland & Guenther, 2006; Riecker, Kassubek, Groschel, Grodd, & Ackermann, 2006). The posterior lobe of the cerebellum receives sensory input from the trigeminal nerve (which provides sensorimotor innervations of the muscles of mastication) as well as the auditory system (Huerta, Frankfurter, & Harting, 1983; Ikeda & Matsushita, 1992; Pastor et al., 2002), and this area of the cerebellum may be implicated in multisensory rather than motor processing (Stoodley & Schmahmann, 2009; Thickbroom, Byrnes, & Mastaglia, 2003). Hence, sensory information from the dynamics of speech articulation has access to multiple brain regions through cortico-cortico, cortico-striatal, and cortico-cerebellar control loops.

59.5 CONCLUSION

In this chapter, we have shown that the neural system that controls speech production is immensely complex at all levels of the nervous system, involving multiple sensorimotor regions for motor planning and execution including M1, PMC, SMA, pre-SMA, CMA, the insula, and the supratemporal and inferior parietal cortices. Loops of internal control involving the BG, thalamus, and cerebellum are also involved in several aspects of speech movement preparation, including sequencing and temporal ordering. All these regions work in concert to assemble complex, temporally ordered, and co-articulated sequences of speech movements; motor commands are sent through corticospinal and corticobulbar tracts involving seven cranial nerves, multiple spinal nerves, and more than 100 striatal and visceral nerves. Despite this remarkable complexity, the chain of events that leads to the production of speech occurs within several hundreds of milliseconds.

Acknowledgments

This project was funded by grants from the National Science and Engineering Research Council of Canada to PT and VLG, and from the National Institutes of Health Research (NIDCD—R01DC012502) to VLG.

References

Abbs, J. H., & Gracco, V. L. (1983). Sensorimotor actions in the control of multimovement speech gestures. *Trends in Neurosciences*, *6*(9), 391–395.

Abbs, J. H., Gracco, V. L., & Cole, K. J. (1984). Control of multimovement coordination: Sensorimotor mechanisms in speech motor programming. *Journal of Motor Behavior*, *16*(2), 195–231.

Ackermann, H., Mathiak, K., & Riecker, A. (2007). The contribution of the cerebellum to speech production and speech perception: Clinical and functional imaging data. *Cerebellum*, *6*(3), 202–213. Available from: http://dx.doi.org/10.1080/14734220701266742.

Ackermann, H., & Riecker, A. (2004). The contribution of the insula to motor aspects of speech production: A review and a hypothesis. *Brain and Language*, *89*(2), 320–328. Available from: http://dx.doi.org/10.1016/S0093-934X(03)00347-X.

Alario, F. X., Chainay, H., Lehericy, S., & Cohen, L. (2006). The role of the supplementary motor area (SMA) in word production. *Brain Research*, *1076*(1), 129–143.

Aldridge, J. W., Berridge, K. C., & Rosen, A. R. (2004). Basal ganglia neural mechanisms of natural movement sequences. *Canadian Journal of Physiology and Pharmacology*, *82*(8–9), 732–739.

Argyropoulos, G. P., Tremblay, P., & Small, S. L. (2013). The neostriatum and response selection in overt sentence production: An fMRI study. *NeuroImage*, *82C*, 53–60. Available from: http://dx.doi.org/10.1016/j.neuroimage.2013.05.064.

Augustine, J. R. (1996). Circuitry and functional aspects of the insular lobe in primates including humans. *Brain Research Brain Research Reviews*, *22*(3), 229–244.

Barbas, H., & Pandya, D. N. (1987). Architecture and frontal cortical connections of the premotor cortex (area 6) in the rhesus monkey. *Journal of Comparative Neurology*, *256*(2), 211–228.

Beevor, C. E., & Horsley, V. (1890). An experimental investigation into the arrangement of the excitable fibres of the internal capsule of the Bonnet Monkey (*Macacus sinicus*). *Philosophical Transactions of the Royal Society of London B.*, *181*, 49–88.

Bengtsson, S. L., Ehrsson, H. H., Forssberg, H., & Ullen, F. (2005). Effector-independent voluntary timing: Behavioural and neuroimaging evidence. *The European Journal of Neuroscience*, *22*(12), 3255–3265. Available from: http://dx.doi.org/10.1111/j.1460-9568.2005.04517.x.

Bohland, J. W., & Guenther, F. H. (2006). An fMRI investigation of syllable sequence production. *NeuroImage*, *32*(2), 821–841. Available from: http://dx.doi.org/10.1016/j.neuroimage.2006.04.173.

Braun, A. R., Guillemin, A., Hosey, L., & Varga, M. (2001). The neural organization of discourse: An H2 15O-PET study of narrative production in English and American sign language. *Brain*, *124*(10), 2028–2044.

Brendel, B., Hertrich, I., Erb, M., Lindner, A., Riecker, A., Grodd, W., et al. (2010). The contribution of mesiofrontal cortex to the preparation and execution of repetitive syllable productions: An fMRI study. *NeuroImage*, *50*(3), 1219–1230. Available from: http://dx.doi.org/10.1016/j.neuroimage.2010.01.039.

Burnett, T. A., Freedland, M. B., Larson, C. R., & Hain, T. C. (1998). Voice F0 responses to manipulations in pitch feedback. *The Journal of the Acoustical Society of America*, *103*(6), 3153–3161.

Chavis, D. A., & Pandya, D. N. (1976). Further observations on corticofrontal connections in the rhesus monkey. *Brain Research*, *117* (3), 369–386 . Available from: http://dx.doi.org/0006-8993(76)90089-5 [pii].

Corfield, D. R., Murphy, K., Josephs, O., Fink, G. R., Frackowiak, R. S., Guz, A., et al. (1999). Cortical and subcortical control of tongue movement in humans: A functional neuroimaging study using fMRI. *Journal of Applied Physiology*, *86*(5), 1468–1477.

Coull, J. T., Frackowiak, R. S., & Frith, C. D. (1998). Monitoring for target objects: Activation of right frontal and parietal cortices with increasing time on task. *Neuropsychologia*, *36*(12), 1325–1334.

Crosson, B., Sadek, J. R., Maron, L., Gökçay, D., Mohr, C., Auerbach, E. J., et al. (2001). Relative shift in activity from medial to lateral frontal cortex during internally versus externally guided word generation. *Journal of Cognitive Neuroscience*, *13*(2), 272–283.

Cunnington, R., Iansek, R., & Bradshaw, J. L. (1999). Movement-related potentials in Parkinson's disease: External cues and attentional strategies. *Movement Disorders*, *14*(1), 63–68.

Deiber, M. P., Ibanez, V., Sadato, N., & Hallett, M. (1996). Cerebral structures participating in motor preparation in humans: A positron emission tomography study. *Journal of Neurophysiology*, *75*(1), 233–247.

Dejerine, J. J. (1901). *Anatomie des centres nerveux* (Vol. 2). Paris: Rueff et Cie.

Dell, G. S. (1988). The retrieval of phonological forms in production: Tests of predictions from a connectionist model. *Journal of Memory and Language*, *27*(2), 124–142.

Di Martino, A., Scheres, A., Margulies, D. S., Kelly, A. M., Uddin, L. Q., Shehzad, Z., et al. (2008). Functional connectivity of human striatum: A resting state FMRI study. *Cerebral Cortex*, *18*(12), 2735–2747. Available from: http://dx.doi.org/10.1093/cercor/bhn041.

Dronkers, N. F. (1996). A new brain region for coordinating speech articulation. *Nature*, *384*(6605), 159–161. Available from: http://dx.doi.org/10.1038/384159a0.

Dum, R. P., & Strick, P. L. (1991). The origin of corticospinal projections from the premotor areas in the frontal lobe. *Journal of Neuroscience*, *11*(3), 667–689.

Dum, R. P., & Strick, P. L. (1996). Spinal cord terminations of the medial wall motor areas in macaque monkeys. *The Journal of Neuroscience*, *16*(20), 6513–6525.

Freeman, J. S., Cody, F. W., & Schady, W. (1993). The influence of external timing cues upon the rhythm of voluntary movements in Parkinson's disease. *Journal of Neurology, Neurosurgery and Psychiatry*, *56*(10), 1078–1084.

Freund, H. J., Jeannerod, M., Hallett, M., & Leiguarda, R. (2005). *Higher-order motor disorders: From neuroanatomy and neurobiology to clinical neurology*. Oxford: Oxford University Press.

Fried, I., Katz, A., McCarthy, G., Sass, K. J., Williamson, P., Spencer, S. S., et al. (1991). Functional organization of human supplementary motor cortex studied by electrical stimulation. *The Journal of Neuroscience*, *11*(11), 3656–3666.

Garrett, M. F. (1993). Errors and their relevance for models of language production. In G. Blanken, J. Dittman, H. Grim, J. Marshall, & C. Wallesch (Eds.), *Linguistic disorders and pathologies* (pp. 69–96). Berlin: de Gruyter.

Gerloff, C., Corwell, B., Chen, R., Hallett, M., & Cohen, L. G. (1997). Stimulation over the human supplementary motor area interferes with the organization of future elements in complex motor sequences. *Brain*, *120*(Pt 9), 1587–1602.

Goldrick, M., & Daland, R. (2009). Linking speech errors and phonological grammars: Insights from Harmonic Grammar networks. *Phonology, 26*(01), 147–185. Available from: http://dx.doi.org/10.1017/S0952675709001742.

Golfinopoulos, E., Tourville, J. A., Bohland, J. W., Ghosh, S. S., Nieto-Castanon, A., & Guenther, F. H. (2011). fMRI investigation of unexpected somatosensory feedback perturbation during speech. *NeuroImage, 55*(3), 1324–1338. Available from: http://dx.doi.org/10.1016/j.neuroimage.2010.12.065.

Grabski, K., Tremblay, P., Gracco, V. L., Girin, L., & Sato, M. (2013). A mediating role of the auditory dorsal pathway in selective adaptation to speech: A state-dependent transcranial magnetic stimulation study. *Brain Research, 1515*, 55–65. Available from: http://dx.doi.org/10.1016/j.brainres.2013.03.024.

Gracco, V. L. (1991). Sensorimotor mechanisms in speech motor control. In W. H. C. W. S. e. H. Peters (Ed.), *Speech motor control and stuttering* (pp. 53–78). Amsterdam: Elsevier.

Gracco, V. L. (1997). A neuromotor perspective on speech production. In W. Hulstijn, H. F. M. Peters, & P. H. H. M. Van Lieshout (Eds.), *Speech production: Motor control, brain research and fluency disorders* (pp. 37–56). Amsterdam: Elsevier.

Gracco, V. L., & Abbs, J. H. (1985). Dynamic control of the perioral system during speech: Kinematic analyses of autogenic and nonautogenic sensorimotor processes. *Journal of Neurophysiology, 54*(2), 418–432.

Gracco, V. L., & Abbs, J. H. (1989). Sensorimotor characteristics of speech motor sequences. *Experimental Brain Research, 75*(3), 586–598.

Grill-Spector, K., Henson, R., & Martin, A. (2006). Repetition and the brain: Neural models of stimulus-specific effects. *Trends in Cognitive Sciences, 10*(1), 14–23. Available from: http://dx.doi.org/10.1016/j.tics.2005.11.006.

Grill-Spector, K., & Malach, R. (2001). fMR-adaptation: A tool for studying the functional properties of human cortical neurons. *Acta Psychologica, 107*(1–3), 293–321.

Guenther, F. H., Ghosh, S. S., & Tourville, J. A. (2006). Neural modeling and imaging of the cortical interactions underlying syllable production. *Brain and Language, 96*(3), 280–301. Available from: http://dx.doi.org/10.1016/j.bandl.2005.06.001.

He, S. Q., Dum, R. P., & Strick, P. L. (1993). Topographic organization of corticospinal projections from the frontal lobe: Motor areas on the lateral surface of the hemisphere. *The Journal of Neuroscience, 13*(3), 952–980.

Henson, R. N., & Rugg, M. D. (2003). Neural response suppression, haemodynamic repetition effects, and behavioural priming. *Neuropsychologia, 41*(3), 263–270.

Hickok, G. (2012). The cortical organization of speech processing: Feedback control and predictive coding the context of a dual-stream model. *Journal of Communication Disorders*. Available from: http://dx.doi.org/10.1016/j.jcomdis.2012.06.004.

Hillis, A. E., Work, M., Barker, P. B., Jacobs, M. A., Breese, E. L., & Maurer, K. (2004). Re-examining the brain regions crucial for orchestrating speech articulation. *Brain, 127*(Pt 7), 1479–1487. Available from: http://dx.doi.org/10.1093/brain/awh172awh172[pii].

Houde, J. F., & Nagarajan, S. S. (2011). Speech production as state feedback control. *Frontiers in Human Neuroscience, 5*, 82. Available from: http://dx.doi.org/10.3389/fnhum.2011.00082.

Hsieh, I. H., Petrosyan, A., Goncalves, O. F., Hickok, G., & Saberi, K. (2011). Observer weighting of interaural cues in positive and negative envelope slopes of amplitude-modulated waveforms. *Hearing Research, 277*(1-2), 143–151. Available from: http://dx.doi.org/10.1016/j.heares.2011.01.008.

Huerta, M. F., Frankfurter, A., & Harting, J. K. (1983). Studies of the principal sensory and spinal trigeminal nuclei of the rat: Projections to the superior colliculus, inferior olive, and cerebellum. *The Journal of Comparative Neurology, 220*(2), 147–167. Available from: http://dx.doi.org/10.1002/cne.902200204.

Ikeda, M., & Matsushita, M. (1992). Trigeminocerebellar projections to the posterior lobe in the cat, as studied by anterograde transport of wheat germ agglutinin-horseradish peroxidase. *The Journal of Comparative Neurology, 316*, 221–237.

Jenkins, I. H., Jahanshahi, M., Jueptner, M., Passingham, R. E., & Brooks, D. J. (2000). Self-initiated versus externally triggered movements. II. The effect of movement predictability on regional cerebral blood flow. *Brain, 123*, 1216–1228.

Jones, J. A., & Munhall, K. G. (2000). Perceptual calibration of F0 production: evidence from feedback perturbation. *The Journal of the Acoustical Society of America, 108*(3 Pt 1), 1246–1251.

Jurgens, U. (2002). Neural pathways underlying vocal control. *Neuroscience and Biobehavioral Reviews, 26*(2), 235–258.

Jurgens, U. (2009). The neural control of vocalization in mammals: A review. *Journal of Voice: Official Journal of the Voice Foundation, 23* (1), 1–10. Available from: http://dx.doi.org/10.1016/j.jvoice.2007.07.005.

Kelly, R. M., & Strick, P. L. (2003). Cerebellar loops with motor cortex and prefrontal cortex of a nonhuman primate. *The Journal of Neuroscience, 23*(23), 8432–8444. Available from: http://dx.doi.org/23/23/8432 [pii].

Kent, R. D. (2000). Research on speech motor control and its disorders: A review and prospective. *Journal of Communication Disorders, 33*(5), 391–427. quiz 428. Available from: http://dx.doi.org/S0021-9924(00)00023-X[pii].

Krams, M., Rushworth, M. F., Deiber, M. P., Frackowiak, R. S., & Passingham, R. E. (1998). The preparation, execution and suppression of copied movements in the human brain. *Experimental Brain Research, 1203*, 386–398.

Kuypers, H. G. J. M. (1973). The anatomical organization of the descending pathways and their contribution to motor control especially in primates. In J. Desmedt (Ed.), *New developments in EMG and clinical neurophysiology* (pp. 38–68). Basel, Switzerland: Karger.

Lashley, K. S. (1951). The problem of serial order in behavior. In L. A. Jeffress (Ed.), *Cerebral mechanisms in behavior* (pp. 112–131). New York, NY: Wiley.

Lau, H., Rogers, R. D., & Passingham, R. E. (2006). Dissociating response selection and conflict in the medial frontal surface. *NeuroImage, 29*(2), 446–451.

Lee, K. M., Chang, K. H., & Roh, J. K. (1999). Subregions within the supplementary motor area activated at different stages of movement preparation and execution. *NeuroImage, 9*, 117–123.

Lehericy, S., Ducros, M., Krainik, A., Francois, C., Van de Moortele, P. F., Ugurbil, K., et al. (2004). 3-D diffusion tensor axonal tracking shows distinct SMA and pre-SMA projections to the human striatum. *Cerebral Cortex, 14*(12), 1302–1309. Available from: http://dx.doi.org/10.1093/cercor/bhh091.

Lepage, M., Beaudoin, G., Boulet, C., O'Brien, I., Marcantoni, W., Bourgouin, P., et al. (1999). Frontal cortex and the programming of repetitive tapping movements in man: Lesion effects and functional neuroimaging. *Brain Research. Cognitive Brain Research, 8*(1), 17–25.

Levelt, W. J. (1992). Accessing words in speech production: Stages, processes and representations. *Cognition, 42*(1–3), 1–22.

Levelt, W. J. (1993). Timing in speech production with special reference to word form encoding. *Annals of the New York Academy of Sciences, 682*, 283–295.

Levelt, W. J. (1999). Models of word production. *Trends in Cognitive Sciences, 3*(6), 223–232. Available from: http://dx.doi.org/S1364-6613(99)01319-4 [pii].

Levelt, W. J., Roelofs, A., & Meyer, A. S. (1999). A theory of lexical access in speech production. *The Behavioral and Brain Sciences, 22* (1), 1–38 (discussion 38–75).

Lu, M. T., Preston, J. B., & Strick, P. L. (1994). Interconnections between the prefrontal cortex and the premotor areas in the frontal lobe. *Journal of Comparative Neurology, 341*(3), 375–392.

Ludlow, C. L., Connor, N. P., & Bassich, C. J. (1987). Speech timing in Parkinson's and Huntington's disease. *Brain and Language, 32* (2), 195–214.

Luppino, G., Matelli, M., Camarda, R., & Rizzolatti, G. (1993). Corticocortical connections of area F3 (SMA-proper) and area F6 (pre-SMA) in the macaque monkey. *Journal of Comparative Neurology, 338*(1), 114–140.

Luppino, G., & Rizzolatti, G. (2000). The organization of the frontal motor cortex. *News and Views, 15*, 219–224.

Macar, F., Lejeune, H., Bonnet, M., Ferrara, A., Pouthas, V., Vidal, F., et al. (2002). Activation of the supplementary motor area and of attentional networks during temporal processing. *Experimental Brain Research, 142*(4), 475–485. Available from: http://dx.doi.org/10.1007/s00221-001-0953-0.

Marsden, C. D. (1984). Motor disorders in basal ganglia disease. *Human Neurobiology, 2*(4), 245–250.

Matelli, M., Camarda, R., Glickstein, M., & Rizzolatti, G. (1986). Afferent and efferent projections of the inferior area 6 in the macaque monkey. *Journal of Comparative Neurology, 251*(3), 281–298.

McGuire, P. K., Bates, J. F., & Goldman-Rakic, P. S. (1991a). Interhemispheric integration: I. Symmetry and convergence of the corticocortical connections of the left and the right principal sulcus (PS) and the left and the right supplementary motor area (SMA) in the rhesus monkey. *Cerebral Cortex, 1*(5), 390–407.

McGuire, P. K., Bates, J. F., & Goldman-Rakic, P. S. (1991b). Interhemispheric integration: II. Symmetry and convergence of the corticostriatal projections of the left and the right principal sulcus (PS) and the left and the right supplementary motor area (SMA) of the rhesus monkey. *Cerebral Cortex, 1*(5), 408–417.

Morris, H. H., III, Dinner, D. S., Luders, H., Wyllie, E., & Kramer, R. (1988). Supplementary motor seizures: Clinical and electroencephalographic findings. *Neurology, 38*(7), 1075–1082.

Mowrer, O. H. (1952). Speech development in the young child. I. The autism theory of speech development and some clinical applications. *The Journal of Speech and Hearing Disorders, 17*(3), 263–268.

Mowrey, R. A., & MacKay, I. R. (1990). Phonological primitives: Electromyographic speech error evidence. *The Journal of the Acoustical Society of America, 88*(3), 1299–1312.

Muakkassa, K. F., & Strick, P. L. (1979). Frontal lobe inputs to primate motor cortex: Evidence for four somatotopically organized "premotor" areas. *Brain Research, 177*(1), 176–182.

Murray, E. A., & Coulter, J. D. (1981). Organization of corticospinal neurons in the monkey. *The Journal of Comparative Neurology, 195* (2), 339–365. Available from: http://dx.doi.org/10.1002/cne.901950212.

Nagel, I. E., Schumacher, E. H., Goebel, R., & D'Esposito, M. (2008). Functional MRI investigation of verbal selection mechanisms in lateral prefrontal cortex. *NeuroImage, 43*(4), 801–807. Available from: http://dx.doi.org/10.1016/j.neuroimage.2008.07.017.

Pandya, D. N., & Seltzer, B. (1982). Intrinsic connections and architectonics of posterior parietal cortex in the rhesus monkey. *The Journal of Comparative Neurology, 204*, 196–210.

Pastor, M. A., Artieda, J., Arbizu, J., Marti-Climent, J. M., Penuelas, I., & Masdeu, J. C. (2002). Activation of human cerebral and cerebellar cortex by auditory stimulation at 40 Hz. *The Journal of Neuroscience, 22*(23), 10501–10506. Available from: http://dx.doi.org/22/23/10501[pii].

Peeva, M. G., Guenther, F. H., Tourville, J. A., Nieto-Castanon, A., Anton, J. L., Nazarian, B., et al. (2010). Distinct representations of phonemes, syllables, and supra-syllabic sequences in the speech production network. *NeuroImage, 50*(2), 626–638. Available from: http://dx.doi.org/10.1016/j.neuroimage.2009.12.065.

Penfield, W., & Boldrey, E. (1937). Somatic motor and sensory representation in the cerebral cortex of man as studied by electrical stimulation. *Brain: A Journal of Neurology, 60*, 389–443.

Penfield, W., & Welch, K. (1951). The supplementary motor area of the cerebral cortex; a clinical and experimental study. *A.M.A. Archives of Neurology and Psychiatry, 66*(3), 289–317.

Picard, N., & Strick, P. L. (1996). Motor areas of the medial wall: A review of their location and functional activation. *Cerebral Cortex, 6*, 342–353.

Picard, N., & Strick, P. L. (2001). Imaging the premotor areas. *Current Opinion in Neurobiology, 11*(6), 663–672.

Praamstra, P., Stegeman, D. F., Cools, A. R., Meyer, A. S., & Horstink, M. W. (1998). Evidence for lateral premotor and parietal overactivity in Parkinson's disease during sequential and bimanual movements. A PET study. *Brain, 121*, 769–772.

Price, C. J., Moore, C. J., Humphreys, G. W., Frackowiak, R. S., & Friston, K. J. (1996). The neural regions sustaining object recognition and naming. *Proceedings Biological Sciences/The Royal Society, 263*(1376), 1501–1507. Available from: http://dx.doi.org/10.1098/rspb.1996.0219.

Purcell, D. W., & Munhall, K. G. (2006a). Adaptive control of vowel formant frequency: Evidence from real-time formant manipulation. *The Journal of the Acoustical Society of America, 120*(2), 966–977.

Purcell, D. W., & Munhall, K. G. (2006b). Compensation following real-time manipulation of formants in isolated vowels. *The Journal of the Acoustical Society of America, 119*(4), 2288–2297.

Ralston, D. D., & Ralston, H. J., III (1985). The terminations of corticospinal tract axons in the macaque monkey. *The Journal of Comparative Neurology, 242*(3), 325–337. Available from: http://dx.doi.org/10.1002/cne.902420303.

Riecker, A., Gerloff, C., Wildgruber, D., Nagele, T., Grodd, W., Dichgans, J., et al. (2004). Transient crossed aphasia during focal right-hemisphere seizure. *Neurology, 63*(10), 1932. Available from: http://dx.doi.org/63/10/1932 [pii].

Riecker, A., Kassubek, J., Groschel, K., Grodd, W., & Ackermann, H. (2006). The cerebral control of speech tempo: Opposite relationship between speaking rate and BOLD signal changes at striatal and cerebellar structures. *NeuroImage, 29*(1), 46–53. Available from: http://dx.doi.org/10.1016/j.neuroimage.2005.03.046.

Riecker, A., Mathiak, K., Wildgruber, D., Erb, M., Hertrich, I., Grodd, W., et al. (2005). fMRI reveals two distinct cerebral networks subserving speech motor control. *Neurology, 64*, 700–706.

Riecker, A., Wildgruber, D., Dogil, G., Grodd, W., & Ackermann, H. (2002). Hemispheric lateralization effects of rhythm implementation during syllable repetitions: An fMRI study. *NeuroImage, 16*, 169–176.

Roelofs, A. (1997). The WEAVER model of word-form encoding in speech production. *Cognition, 64*(3), 249–284.

Roelofs, A. (1999). Phonological segments and features as planning units in speech production. *Language and Cognitive Processes, 14*(2), 173–200. Available from: http://dx.doi.org/10.1080/016909699386338.

Saltzman, E., Lofqvist, A., Kay, B., Kinsella-Shaw, J., & Rubin, P. (1998). Dynamics of intergestural timing: A perturbation study of lip-larynx coordination. *Experimental Brain Research, 123*(4), 412–424.

Sapir, E. (1921). *Language: An introduction to the study of speech*. New York, NY: Harcourt, Brace and Company.

Schluter, N. D., Rushworth, M. F., Passingham, R. E., & Mills, K. R. (1998). Temporary interference in human lateral premotor cortex suggests dominance for the selection of movements. A study using transcranial magnetic stimulation. *Brain, 121*(5), 785–799.

Schmahmann, J. D., Doyon, J., McDonald, D., Holmes, C., Lavoie, K., Hurwitz, A. S., et al. (1999). Three-dimensional MRI atlas of the human cerebellum in proportional stereotaxic space. *NeuroImage, 10*(3 Pt 1), 233–260. Available from: http://dx.doi.org/10.1006/nimg.1999.0459.

Schmahmann, J. D., Pandya, D. N., Wang, R., Dai, G., D'Arceuil, H. E., de Crespigny, A. J., et al. (2007). Association fibre pathways of the brain: Parallel observations from diffusion spectrum imaging and autoradiography. *Brain, 130*(Pt 3), 630–653. Available from: http://dx.doi.org/10.1093/brain/awl359.

Shaiman, S., & Gracco, V. L. (2002). Task-specific sensorimotor interactions in speech production. *Experimental Brain Research, 146*(4), 411–418. Available from: http://dx.doi.org/10.1007/s00221-002-1195-5.

Shallice, T. (2004). The fractionation of supervisory control. In M. S. Gazzaniga (Ed.), *The cognitive neurosciences* (pp. 943–956). Cambridge, MA: MIT Press.

Shuren, J. (1993). Insula and aphasia. *Journal of Neurology, 240*(4), 216–218.

Skodda, S. (2011). Aspects of speech rate and regularity in Parkinson's disease. *Journal of the Neurological Sciences, 310*(1-2), 231–236. Available from: http://dx.doi.org/10.1016/j.jns.2011.07.020.

Skodda, S., & Schlegel, U. (2008). Speech rate and rhythm in Parkinson's disease. *Movement Disorders: Official Journal of the Movement Disorder Society, 23*(7), 985–992. Available from: http://dx.doi.org/10.1002/mds.21996.

Speedie, L. J., Wertman, E., Ta'ir, J., & Heilman, K. M. (1993). Disruption of automatic speech following a right basal ganglia lesion. *Neurology, 43*(9), 1768–1774.

Sperry, R. W. (1950). Neural basis of the spontaneous optokinetic response produced by visual inversion. *Journal of Comparative Physiology and Psychology, 43*, 482–489.

Stemberger, J. P. (1982). The nature of segments in the lexicon: Evidence from speech errors. *Lingua, 56*(3–4), 235–259. Available from: http://dx.doi.org/10.1016/0024-3841(82)90012-2.

Stoodley, C. J., & Schmahmann, J. D. (2009). The cerebellum and language: Evidence from patients with cerebellar degeneration. *Brain and Language, 110*(3), 149–153. Available from: http://dx.doi.org/10.1016/j.bandl.2009.07.006.

Talairach, J., & Bancaud, J. (1966). The supplementary motor area in man. *International Journal of Neurology*, 5330–5347.

Thickbroom, G. W., Byrnes, M. L., & Mastaglia, F. L. (2003). Dual representation of the hand in the cerebellum: Activation with voluntary and passive finger movement. *NeuroImage, 18*(3), 670–674.

Thickbroom, G. W., Byrnes, M. L., Sacco, P., Ghosh, S., Morris, I. T., & Mastaglia, F. L. (2000). The role of the supplementary motor area in externally timed movement: The influence of predictability of movement timing. *Brain Research, 874*(2), 233–241.

Tourville, J. A., Reilly, K. J., & Guenther, F. H. (2008). Neural mechanisms underlying auditory feedback control of speech. *NeuroImage, 39*(3), 1429–1443. Available from: http://dx.doi.org/10.1016/j.neuroimage.2007.09.054.

Tremblay, P., Deschamps, I., & Gracco, V. L. (2013). Regional heterogeneity in the processing and the production of speech in the human planum temporale. *Cortex; A Journal Devoted to the Study of the Nervous System and Behavior, 49*(1), 143–157. Available from: http://dx.doi.org/10.1016/j.cortex.2011.09.004.

Tremblay, P., & Gracco, V. L. (2006). Contribution of the frontal lobe to externally and internally specified verbal responses: fMRI evidence. *NeuroImage, 33*(3), 947–957.

Tremblay, P., & Gracco, V. L. (2009). Contribution of the pre-SMA to the production of words and non-speech oral motor gestures, as revealed by repetitive transcranial magnetic stimulation (rTMS). *Brain Research, 1268*, 112–124. Available from: http://dx.doi.org/10.1016/j.brainres.2009.02.076.

Tremblay, P., & Gracco, V. L. (2010). On the selection of words and oral motor responses: Evidence of a response-independent fronto-parietal network. *Cortex; A Journal Devoted to the Study of the Nervous System and Behavior, 46*(1), 15–28. Available from: http://dx.doi.org/10.1016/j.cortex.2009.03.003.

Tremblay, P., & Small, S. L. (2011). Motor response selection in overt sentence production: A functional MRI study. *Frontiers in Psychology, 2*, 253. Available from: http://dx.doi.org/10.3389/fpsyg.2011.00253.

Tsujimoto, T., Ogawa, M., Tsukada, H., Kakiuchi, T., & Sasaki, K. (1998). Activation of the ventral and mesial frontal cortex of the monkey by self-initiated movement tasks as revealed by positron emission tomography. *Neuroscience Letters, 258*(2), 117–120.

Vallesi, A., Shallice, T., & Walsh, V. (2007). Role of the prefrontal cortex in the foreperiod effect: TMS evidence for dual mechanisms in temporal preparation. *Cerebral Cortex, 17*(2), 466–474. Available from: http://dx.doi.org/10.1093/cercor/bhj163.

Van der Merwe, A. (2009). A theoretical framework for the characterization of pathological speech sensorimotor control. In M. R. McNeil (Ed.), *Clinical management of sensorimotor speech disrders* (pp. 3–18). New York, NY: Thieme.

Volkmann, J., Hefter, H., Lange, H. W., & Freund, H. J. (1992). Impairment of temporal organization of speech in basal ganglia diseases. *Brain and Language, 43*(3), 386–399. Available from: http://dx.doi.org/0093-934X(92)90108-Q [pii].

von Cramon, D., & Jurgens, U. (1983). The anterior cingulate cortex and the phonatory control in monkey and man. *Neuroscience and Biobehavioral Reviews, 7*(3), 423–425.

Von Holst, E., & Mittelstaedt, H. (1973). The reafference principle. In R. Martin (Ed.), *The behavioral physiology of animals and man. The collected papers of Erich von Holst* (Vol. 1, pp. 139–173). Coral Gables, FL: University of Miami Press.

Weeks, R. A., Honda, M., Catalan, M. J., & Hallett, M. (2001). Comparison of auditory, somatosensory, and visually instructed and internally generated finger movements: A PET study. *NeuroImage, 14*(1), 219–230.

Wiese, H., Stude, P., Nebel, K., de Greiff, A., Forsting, M., Diener, H. C., et al. (2004). Movement preparation in self-initiated versus externally triggered movements: An event-related fMRI-study. *Neuroscience Letters, 371*(2–3), 220–225.

Wise, R. J., Greene, J., Büchel, C., & Scott, S. K. (1999). Brain regions involved in articulation. *Lancet, 353*, 1057–1061.

Yeterian, E. H., & Pandya, D. N. (1998). Corticostriatal connections of the superior temporal region in rhesus monkeys. *The Journal of Comparative Neurology, 399*(3), 384–402.

CHAPTER

60

Sentence and Narrative Speech Production: Investigations with PET and fMRI

Richard J.S. Wise and Fatemeh Geranmayeh

Computational, Cognitive and Clinical Neuroimaging Laboratory (C^3NL), Imperial College London, Hammersmith Hospital, London, UK

60.1 INTRODUCTION

An interview with Sir Christopher Woodhead, a former (and somewhat controversial) Chief Inspector of Schools for the English education system was recorded recently (http://www.bbc.co.uk/iplayer/episode/b040hx66/No_Triumph_No_Tragedy_Chris_Woodhead/). In 2000, he developed the first symptoms of motor neuron disease, a form that has resulted in slow but progressive weakness of the limbs; until now it has largely spared his axial musculature. Therefore, he is able to speak normally while confined to a wheelchair with tetraplegia. Although without the ability to walk and climb (he was formerly an enthusiastic rock climber), he remains able to control the many muscles that allow fluent speech production: he can control expiration, essential to the production of connected speech; he can control his larynx to produce sounds of appropriate pitch and loudness; and he can filter this sound by rapid, accurate sequential movements of his articulators remains unimpaired, with no hint of motor imprecision (i.e., dysarthria). Most importantly, his cognitive functions essential for producing an eloquent account of his plight, including quotations from literature and poetry, remain intact—there remains nothing wrong with his ability to retrieve words and construct sentences to express concepts about disability, the medical profession, and assisted dying. Then, a point was reached in the interview when Sir Christopher said that the time when he no longer wishes to go on living will come when he loses the power to speak.

Daniel Wolpert has famously said, in a recording for a TED broadcast, that the one and only reason we have a brain is to produce adaptable and complex movements (https://www.ted.com/talks/daniel_wolpert_the_real_reason_for_brains)—and the ability to move the muscles that result in fluent speech must rank as one of the most important functions of the human brain. That is clearly Sir Christopher's view, and it is seen in the frustration and despair expressed in angry gestures by patients with nonfluent progressive aphasia that has progressed to near mutism.

This preamble is to signal the conviction that one of the more important contributions functional neuroimaging can make to "systems neuroscience" is the study of speech production. This chapter addresses our views of the progress that has been made in the attempts to reveal the functional anatomy of "normal" propositional speech.

60.1.1 The Limitations

To date, progress has been slow. One reason is the unacceptable levels of noise introduced by articulation extended over time. It is evident that muscle artifact must intrude on electroencephalographic (EEG) and magnetoencephalographic (MEG) recordings, creating problems for analyzing recordings made during continuous speech. Therefore, MEG studies have largely been confined to exploring the responses to utterances of syllables or single words (Salmelin, 2007), one active area of research being the modulation of the response of auditory cortex to the sound of the participant's own utterances, heard unaltered or with distortion, such as changed pitch (Curio, Neuloh, Numminen, Jousmäki, & Hari, 2000; Kort, Nagarajan, & Houde, 2014; Niziolek, Nagarajan, & Houde, 2013). Artifacts generated by overt

Neurobiology of Language. DOI: http://dx.doi.org/10.1016/B978-0-12-407794-2.00060-2

speech also have a major impact on magnetic resonance images (MRI) using conventional blood-oxygen level-dependent contrast imaging (BOLD fMRI) (Gracco, Tremblay, & Pike, 2005; Mehta, Grabowski, Razavi, Eaton, & Bolinger, 2006), although the technique of fMRI, which uses arterial spin labeling (ASL), offers promise, albeit with less sensitivity (Kemeny, Ye, Birn, & Braun, 2005). Previous studies of overt single word production using BOLD fMRI demonstrated that whole-head movements contribute relatively little to the noise in the functional images (Huang, Carr, & Cao, 2002; Palmer et al., 2001). The study by Kemeny and colleagues (2005), which used overt sentence production, concluded that the major source of artifact using BOLD fMRI was the susceptibility effect generated by inhomogeneities in the magnetic field because of continuous changes in airflow and movements of the tongue and jaw. This speech-related artifact was greatest in anterior temporal regions bilaterally, impinging on ventral frontal and anterior insular cortices. The authors stressed the difference between these changes and the susceptibility artifacts related to air–bone interfaces above the nasal sinuses and the petrous temporal bone, which are always present and do not change across behavioral conditions. Speech-related signal obtained with ASL was not marred by anterior temporal artifact. On a personal note, when we attempted continuous data acquisition during narrative speech production using BOLD fMRI, conventional contrast between conditions—overt propositional speech production (we used picture description) and nonpropositional speech (counting)—was marred by unacceptable levels of noise (unpublished results), although this was rather more widespread than that observed by Kemeny and colleagues (2005). Rims of artifact were included around the edge of the brain and within the ventricular system. We attempted to clean the images using independent component analysis (ICA), which was successful in the sense that the multivariate analysis separated much of the noise from neural signal (Beckmann & Smith, 2004). However, those few components containing signal that was plausibly largely neural in origin were confined to lateral and midline premotor cortices, primary sensory-motor cortices, and paravermal cerebellum; in other words, signal from higher-order cortices involved in the selection and formation of a concept, which involves access to both episodic and semantic memory representations, and its transformation into a verbal message prior to actual articulation, were not visualized. However, improvements in denoising imaging datasets (Griffanti et al., 2014) may allow greater recovery of lost signal in future similar studies.

In contrast, artifact during single syllable or single word production, movements that only occur over a few hundreds of milliseconds, may be acceptably low; therefore, these tasks can be used with continuous data acquisition to visualize activity associated with motor-sensory and lexical retrieval processes, both cortical and subcortical, involved in speech production (Parker-Jones et al., 2014; Peeva et al., 2010). However, self-evidently these tasks bear a limited relationship to normal conversational speech production. An alternative for sentence-level speech production, and one that we and others have used, is "sparse" temporal sampling. Originally introduced so that participants could listen to auditory stimuli without interference from scanner noise (Hall et al., 1999), this technique relies on acquiring single volumes of brain images only at the end of stimulus delivery or speech response. Thus, in a typical study investigating speech production, a participant might be required to speak for ~7 s before being prompted to stop. Then, ~1 s later, a single T2*-weighted, gradient echo, echoplanar imaging (EPI) sequence with whole-brain coverage is performed over 2 s. The sequence can then be repeated as often as required. Although "sparse" temporal sampling can be performed at a faster rate, this is only appropriate for single-word rather than sentence-level speech production. This "slow" event-related design, which relies on the temporally extended hemodynamic response function (HRF) over ~15 s to obtain signal, is not as sensitive as continuous image acquisition, because it misses the peak of the HRF. It also requires a longer study to obtain sufficient functional imaging volumes and, therefore, more stoical participants. The gain is avoidance of some (but by no means all) of the artifacts. In fMRI studies using "sparse" temporal sampling that contrasted sentence production with various baseline conditions, regions active only during sentence production were revealed in midline and left frontotemporal cortical regions (Dhanjal, Handunnetthi, Patel, & Wise, 2008) and anterior striatum (Argyropoulos, Tremblay, & Small, 2013). Therefore, this data acquisition technique allows the visualization of signal associated with sentence-level speech production that is otherwise obscured by noise using BOLD fMRI and continuous data acquisition.

However, there is a downside when speech is elicited in short epochs when using "sparse" temporal sampling. Regular periodic speech controlled by external stimuli to start and stop the flow of speech is quite unlike normal speech production. For example, in the study of Dhanjal et al. (2008), it was deemed too "unnatural" to expect the participants to produce continuous narrative speech that was interrupted by these externally directed pauses during image acquisition. It was predicted that the results would have revealed activity dominated by "go"/"stop" cognitive control (it is easier to interrupt counting, the baseline task, periodically than the flow of a good narrative). Instead, names of objects were used as stimuli to prompt speech, and the participants were required to describe in one short sentence the attributes of each depicted object. Thus, the word "car" might elicit *"It's a way of getting around, it has four wheels, it runs on petrol/gas,"* a verbal-semantic task that requires speeded selection from among all the properties that a participant associates with a car. It does not feel like spontaneous speech, but more akin to a verbal fluency task, such as, "Think of action words (verbs) that you associate with the word "apple," which might elicit the response *"buy, eat, peel, cut, slice"*.

It might be considered that this may not matter much in terms of visualizing activity associated with sentence production. However, prior knowledge would predict a different distribution of activity associated with this task than may be present with normal narrative speech. This was evident from a study of continuous overt narrative speech production using positron emission tomography (PET) and radiolabeled water to assess regional cerebral blood flow as a marker of underlying net synaptic activity (Blank, Scott, Murphy, Warburton, & Wise, 2002). Artifact associated with continuous speech is less of a problem with PET; some artifacts are of course there, but they are disguised by the lower spatial and temporal resolution of this technique and the absence encountered with EPI fMRI of local magnetic field inhomogeneities. It has proven to be a technique that is capable of revealing relatively noise-free activity throughout widely distributed regions during the production of free narrative speech (Awad, Warren, Scott, Turkheimer, & Wise, 2007; Blank et al., 2002; Braun, Guillemin, Hosey, & Varga, 2001; Brownsett & Wise, 2010). In the study of Blank and colleagues (2002), the results acquired during narrative speech were contrasted with those from an earlier PET study in which participants had been required to generate verbs in response to noun prompts (Warburton et al., 1996). Activity in much of the left dorsolateral prefrontal cortex and in the dorsal anterior cingulate cortex (dACC) was present during verb generation but not during narrative speech production. This example demonstrates that word retrieval during a fluency task and during normal speech are not the "same thing" when it comes to frontal executive functions. Therefore, even when the data are reliable and (relatively) noise-free, there is always the problem of interpretation.

This is further exemplified in a study by Tremblay and Small (2011). Multiple conditions were included, but an important contrast was between sentence generation in response to object picture stimuli and repetition of heard sentences. Although increased activity in the midline dorsal prefrontal cortex, in the presupplementary area (pre-SMA) and adjacent cingulate sulcus, was interpreted as motor response selection during overt speech production, the authors discussed differences in the demands on lexical and semantic selection between the two tasks and how this may have influenced the results. Even then, the discussion remained within the limitations of speech-specific and language-specific processes. Those working on the systems neuroscience of domain-general attention and cognitive control could point out that this midline region, or at least a closely overlapping component of it, is involved during the performance of many cognitive tasks (Fedorenko, Duncan, & Kanwisher, 2013); task-dependent sentence generation is just one more "problem-solving" task that will engage domain-general as well as domain-specific processes.

Of course, the visualization of cognitive control during speech production may be a goal in itself. We have argued that at least some of the frontal activity elicited by language tasks observed in recovering aphasic stroke patients relative to normal control subjects is as likely to be the consequence of more domain-general cognitive control as domain-specific language processing (Geranmayeh, Brownsett, & Wise, 2014a), on the basis that the patients find the language activation tasks more difficult than the normal controls. In one study (Brownsett et al., 2014), we showed that reducing performance in the normal controls by degrading the language stimuli resulted in increased activity in midline frontal cortex, and to the same level as aphasic patients performing the task with normal stimuli. The proposal that functional imaging studies may reveal parallel activity in systems involved in more domain-general cognitive control during speech comprehension is also being considered (Fedorenko, 2014).

Therefore, the second difficulty with functional imaging studies of speech, and with language studies in general, is separating language from other processes, such as cognitive control in the examples cited. As importantly, communication in sentences and narratives is only a "normally" executed act if there is access to declarative memory systems, both episodic and semantic. The inevitable visualization of these systems in many studies will become evident in later examples. A univariate analysis comprising a contrast between conditions will reveal, to an extent determined by the nature of the activation and baseline conditions, both domain-specific and domain-general systems—language, memory, attention, and cognitive control—as an undifferentiated single "system," with interactions between them that may influence the effect size of signal observed (Friston et al., 1996). However, this is not all bad, because the composite reveals the many parallel processes, linguistic and nonlinguistic, that support meaningful human communication.

Perhaps it would be helpful at this stage to illustrate the previous arguments with one error of interpretation that our group made in the past; in hindsight, we might consider that the reinterpretation is potentially more interesting than the original interpretation. Although some have criticized the statistical power of functional neuroimaging studies in general (Button et al., 2013), we consider this is less of a problem than the manner of interpretation of what is, in fact, genuine signal. This means that past work can be read for the value of their results even if the original interpretation of the results might benefit from revision. Thus, Blank, Bird, Turkheimer, and Wise (2003) (with the first author of this chapter as senior author on the work) published a PET study that showed greatly increased activity in the right homologue of Broca's area during narrative speech production in patients who had partially recovered from a left posterior frontal stroke. Since the time of Paul Broca it has been dogma that recovery from a nonfluent aphasic stroke depends on a "laterality shift" of expressive language function from left to right frontal operculum (and currently the adjacent anterior insula would be included—FOp/aI). This was our interpretation (and confirming dogma gives one an easy time with reviewers); however, the one (major) inconsistency was that activity in the right FOp/aI did not correlate with the rate at which the patients were able to generate narrative speech, which remained more or less impaired in all subjects. Since then, there has been growing literature on the "cingulo-opercular" network, comprising both left and right FOp/aI and the dACC and the adjacent pre-SMA. Although there is debate about the precise function of this dorsal frontal network (Aron, Robbins, & Poldrack, 2014; Dosenbach et al., 2007; Hampshire, Highfield, Parkin, & Owen, 2012; Menon & Uddin, 2010), no one thinks that it is a domain-specific language system; rather, it is involved in aspects of domain-general cognitive control. This is not to deny that a component of the *left-lateralized* FOp/aI (Broca's area) is language-specific, although even in this "classic" language area it would appear that other anatomically overlapping components are involved in the more domain-general control of non-language tasks (and, quite possibly, the cognitive control of language in addition to language processing *per se*) (Fedorenko, Duncan, & Kanwisher, 2012). One might speculate on this evidence that evolution has resulted in part of a domain-general cognitive control system, namely Broca's area, developing a local language-specific component, but not in right Broca's area. In light of these insights from "non-language" systems neuroscience, the inference that the right FOp/aI signal in the patient population assembled by Blank and colleagues (2003) represented a "laterality shift" of language function looks much less secure; just as likely, when speech production becomes difficult as the result of a stroke, the observed signal may reflect compensatory upregulation of cognitive control networks rather than activity in "reorganized" language processors. Fortunately, this reinterpretation does not demean the original observation as a valueless epiphenomenon; as discussed, we now propose that the function of the cingulo-opercular system may contribute to recovery from aphasic stroke (Brownsett et al., 2014; Geranmayeh, Brownsett, & Wise, 2014).

The following sections attempt, as much as possible, to separate the different systems—linguistic, declarative memory retrieval (both semantic and episodic), and cognitive control (particularly the cingulo-opercular network)—involved during sentence and narrative production. As will become apparent, the context in which the participants are required to speak can have a major impact on the observed distribution of activity in univariate statistical analyses. When interpretation cannot be achieved through prior knowledge of which functions are located where, a more recent analytical method, which separates overlapping signal from multiple functional systems, is the introduction of multivariate analyses, such as ICA, to functional neuroimaging (Beckmann & Smith, 2004). A whole-brain ICA analysis can help distinguish between the different systems as they function together in a task-dependent manner (Geranmayeh, Wise, Mehta, & Leech, 2014b). In addition, ICA can reveal multiple separable spatiotemporal signals within any one brain region, each with different functional roles (for examples of this, see Leech, Braga, & Sharp, 2012; Simmonds et al., 2014). This overlap could be the result of the presence of spatially adjacent but functionally different neurons in that region, or of neurons that are flexibly involved in different functional networks. Lack of activity in a brain region on a univariate contrast does not necessarily mean that the region is not involved in the task.

60.2 WHAT HAVE WE LEARNED FROM META-ANALYSES OF LANGUAGE STUDIES

We should start with a review of some of the meta-analyses of functional neuroimaging studies of language, not least to determine whether we have anything further to add to their conclusions. Therefore, it is worth first considering the manner in which they presented their findings before discussing specific examples of speech production in later sections of this chapter. One such meta-analysis included 129 PET and fMRI studies and identified 730 activation peaks distributed over the neocortex of the left cerebral hemisphere based on univariate statistical analyses (Vigneau et al., 2006). These peaks were sorted into those that could broadly be interpreted as being associated with phonology, semantics, or sentence processing. For each of these three processes, the peaks were distributed across posterior frontal, lateral temporal (from posterior to anterior), and inferior parietal cortices, with considerable overlap. These peaks were then transformed into a smaller number of clusters that spatially separated one process from another. However, this did not alter the observation that all three processes appeared to be distributed across all three lobes of the left hemisphere. Most of the studies included were designed to investigate speech perception and/or comprehension, at the syllable, lexical, and sentential levels, with fewer studies available that had investigated speech production; however, what was apparent was that no one language function was restricted to one lobe of the left hemisphere. In the words of the authors of the meta-analysis, "these results argue for large-scale architecture networks rather than modular organization of language in the left hemisphere."

A more recent meta-analysis (Price, 2012) investigating both modalities, auditory and visual, of language perception and comprehension and speech production resulted in a figure that depicted the left cerebral hemisphere as a detailed patchwork of cortical regions, each associated with processing at different levels of language processing. This was somewhat reminiscent of a much earlier meta-analysis (Indefrey & Levelt, 2004), when PET was the dominant imaging methodology, that attempted to relate patches of left cerebral hemisphere cortex with processes that originated from an influential model of lexical access during speech production, largely based on evidence from chronometric experiments (Levelt, Roelofs, & Meyer, 1999). Regions associated with processes termed lemma selection and retrieval, syllabification, and others were also labeled with times at which these processes occurred in relation to one another (gleaned from chronometric data using other methodologies, because PET has a temporal resolution of many seconds). Price's map of language was much more detailed in its boundaries and distributions and used broader, less theoretically derived processing terms; but in many ways it looks superficially like a similar "neophrenology" of language processing. However, the author was careful to emphasize that "a distinction can be made between processes that are localized to specific structures (e.g., sensory and motor processing) and processes where specialization arises in the distributed pattern of activation over many different areas that each participate in multiple functions. Future studies will undoubtedly be able to improve the spatial precision with which functional regions can be dissociated, but the greatest challenge will be to understand how different brain regions interact with one another in their attempts to comprehend and produce language." We entirely agree and might argue that there is somewhat of a discord between these conclusions and detailed mapping of language functions as specific, tightly delineated patches of cortex. However, the author was only presenting the results of several decades of functional neuroimaging research on language as it had been presented in the original publications, as activated "blobs" with precise coordinates in anatomical stereotactic space.

One obvious problem with these meta-analyses is knowing whether the resulting composite maps have allowed activation sites to be included that do not represent a language-specific function. We have already discussed how this might occur in a patient population, but results from studies on normal participants included in meta-analyses are equally vulnerable to misinterpretation. Functional neuroimaging studies are performed almost exclusively on literate adults who, from the age of 2 years, have spent much of their daily lives comprehending and producing narrative language, both spoken and written. As a consequence, the distributed systems that support the many linguistic components of language, their dependent relationship with declarative memory systems (without which there is nothing to say), and the cognitive control systems that are variably engaged depending on the communicative context must be strongly "hard-wired" together. Under these constraints, even dissociating domain-specific language networks from these others may prove as uncertain as trying to identify a "module" for phonology or syntax.

And even after 150 years of localizing language, there remain disagreements on fundamentals. Years ago at an editorial board meeting of *Brain and Language*, one of the editors of this volume, Steven Small, made a comment about research on the functional anatomy of language—"we're still arguing about it 150 years after Broca, but at least we are certain about one thing: the left parietal lobe is involved." And yet a widely cited review of the functional anatomy of language, with the other editor of this volume, Greg Hickok, as first author, did not include any node in left inferior parietal cortex associated with any linguistic aspect of language. The authors of this chapter were initially in accord with the Hickok and Poeppel (2007) view on this issue, and we were inclined to relate parietal cortical functions (of which, of course, there are many) to processes associated with memory (working and declarative) rather than language *per se*. Indefrey and Levelt (2004) also made no mention of a function for left parietal cortex in word production. However, the meta-analyses of Vigneau et al. (2006) and Price (2012) have come to different conclusions. Unfortunately, lesion-deficit analyses are not a means to settle this dispute, because strokes restricted exclusively to the left supramarginal and/or angular gyrus are vanishingly rare, and almost invariably there will be associated infarction of the adjacent posterior temporal lobe or the frontal lobe (for example, although a study of language comprehension rather than production, see the patient series described in Dronkers, Wilkins, Van Valin, Redfern, & Jaeger, 2004). In our clinical practice it has been very rare to come across a patient with an isolated left parietal lesion and anything other than a transient aphasia. However, this is an anecdotal opinion. Large databases analyzed with appropriate statistical methods that relate lesion site and extent on anatomical images to behavioral measures of persistent aphasic deficits will have the power to confirm or refute this anecdotal impression (Mah et al., 2014; Price, Seghier, & Leff, 2010). Even then, a confound will be that strokes damage white matter tracts as well as gray matter, and the behavioral effects of anatomically disconnecting intact cortical and subcortical regions remote from the stroke may not be readily apparent. In view of these uncertainties, there was no good evidence, at least in our minds, about language processes that were dependent on left inferior parietal function—notwithstanding Geschwind's speculations that the angular gyrus is the interface between words (or, at least, object words, both heard and spoken) and the mental representations of their meaning, a hypothesis apparently supported by modern imaging of white matter tracts (Catani & ffytche, 2005). However, the advent of newer studies subjected to ICA (an example of which is shown later) has led us to be more inclined to believe Steven Small's assertion.

60.3 NARRATIVE SPEECH PRODUCTION

Braun et al. (2001) published a PET study that investigated narrative language production in speech and in sign language using the same group of participants for the two modalities. These were adults with normal hearing but who were fluent in sign language as the result of being the children of deaf parents. The baseline control conditions were non-communicative movements of the articulators and the limbs. The two communicative conditions activated a common system that consisted of "classic" Broca's and Wernicke's areas in the left cerebral hemisphere and, in addition, a small region of activity in the homologue of Wernicke's area in the right hemisphere and in the left superior frontal gyrus. Notably, there was activity in midline posterior cortex, centered on posterior cingulate cortex, but with adjacent retrosplenial cortex and anterior precuneus possibly included, and both left and right angular gyri.

A further PET study of narrative speech production was published the next year, but one that used the recital of overlearned nursery rhymes and counting as baseline tasks (Blank et al., 2002). In many respects this reproduced the results of Braun and colleagues (2001). The conjunction of activity for all three speaking conditions contrasted with a nonspeech condition revealed the expected activity in bilateral premotor, primary sensorimotor, and auditory cortices, with left-lateralized peaks in the pars opercularis, anterior insula, and medial planum temporale, and bilateral peaks in subcortical nuclei and in paravermal cerebellum. This distributed, largely motor-sensory, system controlling overt articulation is covered in detail in other chapters. The main interest in this study was the distributed activity associated with free narrative speech relative to both overlearned automatic speech (reciting nursery rhymes) and nonpropositional speech (counting). The distribution of this activity is shown in Figure 60.1. Left-lateralized activity was distributed along the superior frontal gyrus, from the presupplementary area forward, the posterior middle frontal gyrus, lateral anterior temporal cortex, and the mid-fusiform gyrus. There was bilateral activity in the angular gyri and posterior midline cortical activity, which again was described as being located in posterior cingulate cortex, but it may have included retrosplenial cortex. Additional activity was also observed in the ventral left temporal lobe and right lateral cerebellum, possibly regions that were outside the field of view in the study of Braun and colleagues (2001). The only subcortical region that survived the statistical threshold and spatial resolution of the technique was the right lateral cerebellum, so no activity was observed in the basal ganglia or thalami.

Uncontroversially, the results from these two studies could be interpreted as demonstrating that Broca's and Wernicke's areas are core linguistic nodes, because activity in these regions was evident during speech but was matched by the activity generated when narrative production was the result of reciting familiar nursery rhymes. So, what to make of all the extrasylvian activity depicted in Figure 60.1 and observed across both PET studies? A further PET study that compared narrative speech comprehension with narrative speech production demonstrated conjunctions of activity for both tasks in the same extrasylvian regions (Awad et al., 2007); clear evidence that the comprehension and production of narrative speech depend on many of these same, or closely overlapping, cortical regions.

In all three studies, the production of narratives was based around personal past experiences (e.g., "describe where you lived as a child"). The participants spoke for ~1 min, beginning a few seconds before the intravenously injected bolus of positron emitting water ($H_2{}^{15}O$) arrived at the brain and continuing as the head counts

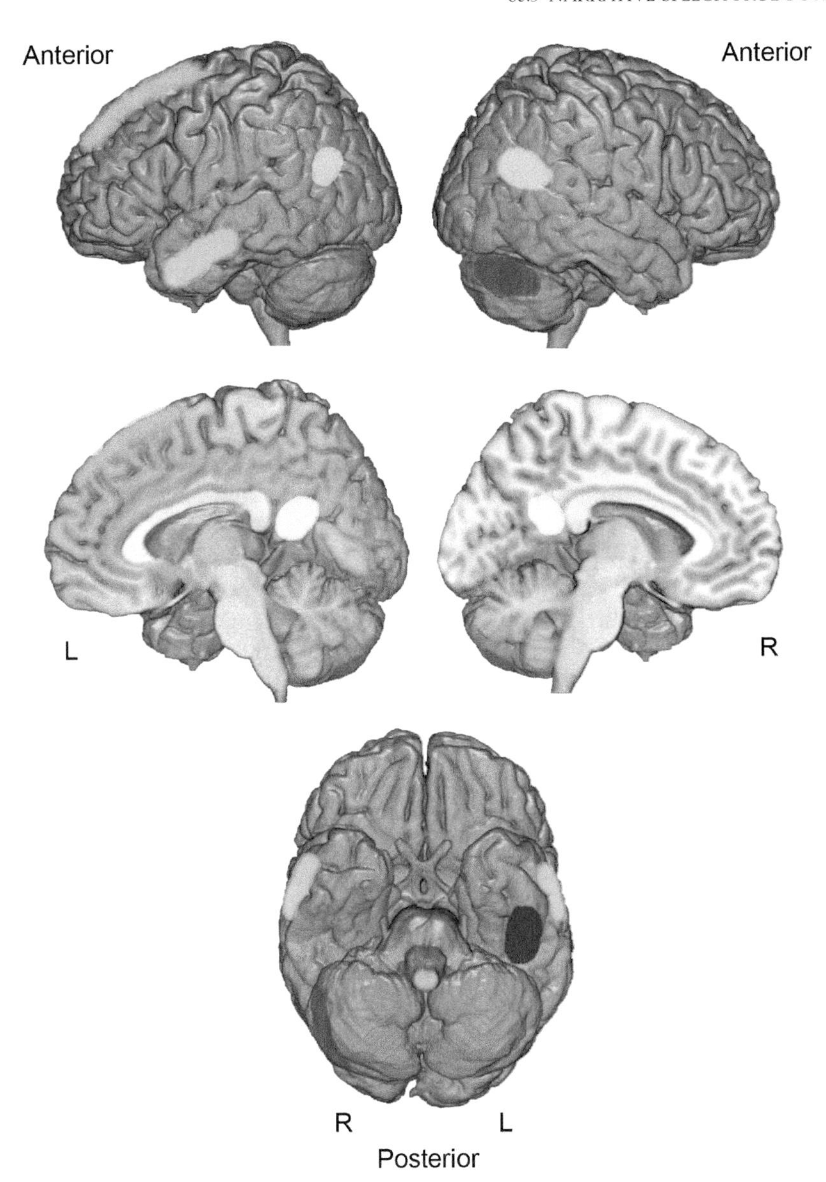

FIGURE 60.1 Schematic drawing of activity observed during a PET study of narrative speech production contrasted against shared activity during automatic and non-propositional speech baseline tasks (Blank et al., 2002). Top and middle panels show the lateral and medial views of both hemispheres, respectively. Bottom panel shows the ventral surface of the brain. Left-lateralized activity was seen in the superior frontal gyrus (orange), lateral anterior temporal cortex (light blue), and the mid-fusiform gyrus (dark blue). There was bilateral activity in the angular gyri (green) and posterior cingulate cortex/retrosplenial cortex (yellow). Additional activity was also observed in the right lateral cerebellum (red).

peaked and then began to decline due to washout of the radiolabeled water and its decay (half-life, 122.2 s). This relatively slow temporal resolution (although it is not as slow as it may seem, as the estimate of cerebral blood flow is weighted toward the period when the build-up of head counts is most rapid, over ~10 s) meant that the noise from physiological variables (such as small fluctuations of arterial carbon dioxide tension as the result of breathing-for-speech) could be ignored, as could (as mentioned previously) noise from small head movements "masked" by the technique's relatively low spatial resolution (~5–6 mm).

Although the studies were about speech production, important in terms of interpreting the results was not minimizing that narrative speech production depended on the continuing retrieval of both personal (autobiographical) and knowledge-based (semantic) meanings over many seconds. The posterior activity (the left and right angular gyri and the posterior midline cortex) forms the posterior components of the default mode network (DMN). The DMN is often considered as a single system, responsible, among other things, for internally generated thoughts and ruminations, which, of course, include thinking about past events, planning future events, and concerning ourselves about how we stand in relation to others in our personal and professional lives (for review, see Buckner, Andrews-Hanna, & Schacter, 2008). Reminiscing aloud is cognitively similar to reminiscing silently; therefore, it is no surprise that much of this network is active during narrating past events in one's life. By some accounts, it is also one component in the generation of creative narratives that are, of course, not created out of a void but depend on reformulations of one's own knowledge and past experiences (Shah et al., 2013). Of course, some of the accounts told by the participants in the PET studies may have been more fictional than factual.

Components of the DMN are the most richly connected "hubs" in the brain (Braga, Sharp, Leeson, Wise, & Leech, 2013), and the DMN probably has many diverse roles. Importantly, for the purposes of this discussion, much distributed activity that conforms to the known boundaries of the DMN appear in a meta-analysis performed on functional imaging studies that were considered to rely on semantic memory processes (Binder, Desai, Graves, & Conant, 2009). Other than inclusion of more ventral anterior midline cortex, the distribution of activity described in this meta-analysis was closely similar to that observed in the three PET studies under discussion. Components of the DMN are strongly associated with processes involving episodic memory, such as retrosplenial cortex (Aggleton, 2010), but the distinction between the classic *what?*, *when?*, and *where?* of episodic memory and fact-based general knowledge are only two extremes of a declarative memory continuum. It may be more accurate to refer to at least partially dissociable subcomponents of the angular gyri and posterior midline cortex when considering distinctions between autobiographical and semantic memory retrieval. Therefore, there is not necessarily a "disconnect" between those studies that relate these regions to episodic memory retrieval (Sestieri, Corbetta, Romani, & Shulman, 2011) and their inclusion as part of the semantic memory system. This issue is discussed in detail by Binder and his colleagues in their two publications (2009, 2011). To reemphasize, neophrenological interpretations should not enter into discussions of the functions of anatomically defined cortical regions such as "classic" language areas like Broca's area or the left angular gyrus, which may contain subcomponents that are both separately and conjointly involved in domain-specific processes such as language and memory and more domain-general attentional and control processes (Fedorenko et al., 2012; Seghier, Fagan, & Price, 2010).

60.4 FUNCTIONAL MRI STUDIES OF SENTENCE PRODUCTION

An example is illustrated in Figures 60.2–60.4, and these exemplify many of the points already made. They originate from an fMRI study previously published in *Brain and Language* (Geranmayeh et al., 2012). The univariate contrasts with sentence production (viewing the name of an object and describing it in ~7 s) contrasted with two different baseline conditions: noncommunicative repetitive movements of one of the articulators, the tongue, which is one of the control conditions used in the study by Braun et al. (2001), and a rest condition. In a sense, this is a rather "disappointing" result, showing much frontal activity, temporal lobe activity that can largely be attributed to auditory processing of own speech, and no activity in more posterior cortex. It looks most unlike what one would expect from the PET studies (Awad et al., 2007; Blank et al., 2002; Braun et al., 2001). Further, there is prominent activity in the cingulo-opercular network, which implies that the sentence production task depended on appreciable cognitive control. The sentence production task was similar to that previously mentioned in relation to the study by Dhanjal et al. (2008), namely describing attributes elicited by viewing a written text of an object, which is a task we consider more akin to verbal fluency than free narrative speech production, but a necessary constraint because of the requirement for "sparse" temporal sampling.

The ICA analysis illustrated in Figure 60.3 offers quite a different picture. One component includes bilateral auditory cortex along with bilateral lateral premotor cortex and the cingulo-opercular network. This can be reasonably interpreted as temporal coherence between a cognitive control network, higher-order premotor planning of the complex movements required for connected speech as opposed to repetitive tongue movements, and the auditory cortical response to own speech. Other processes, such as lexical selection and retrieval, may also be functions of subcomponents of this network, but this cannot be determined given the cognitive "differences" between the two tasks; attempts to further subdivide this network by increasing the number of components in the ICA were not successful. The other component in which sentence production was greater than tongue movements revealed an extensive left-lateralized dorsolateral prefrontal, posterior inferolateral temporal, and inferior parietal network, with small regions of signal in the right inferior parietal cortex and posterior midline cortex.

The left fronto-temporo-parietal network has considerable correspondence with the results of the meta-analysis by Vigneau et al. (2006) of many language studies. A further analysis, illustrated in Figure 60.4, is a schematic drawing of the univariate contrast of the participants making repetitive tongue movements—a monotonous task and one that can be expected to not deactivate the DMN (Buckner et al., 2008)—with the sentence generation task. In this contrast, there was greater activity in the posterior midline cortex and in the posterior lateral parietal lobe on the right. Of note, however, was the absence of a significant difference of activity in the left inferior parietal cortex.

Therefore, it was evident that the sentence production task was quite different from the narrative speech conditions used in the earlier PET studies. The reduced activity in posterior regions can be attributed to an absence of autobiographical memory recall, although

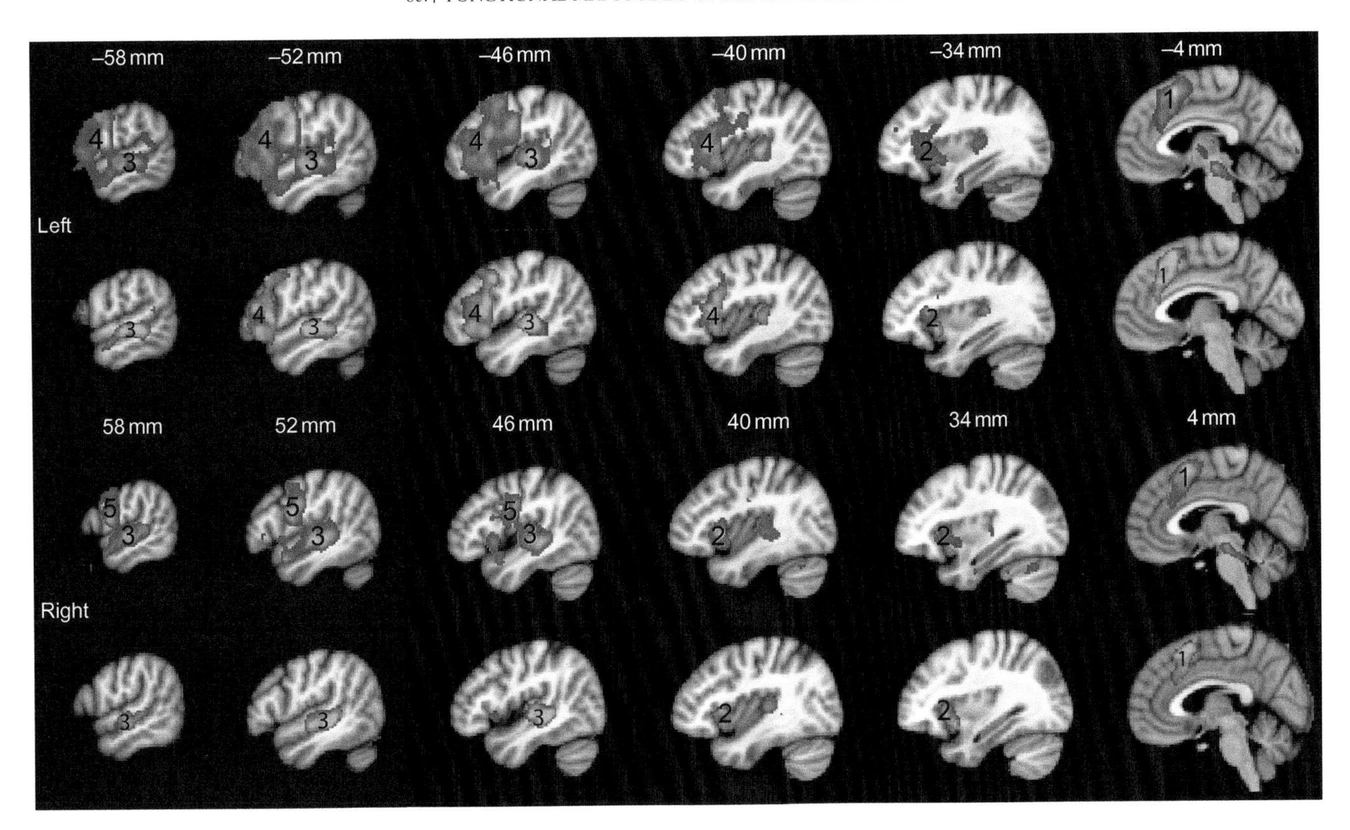

FIGURE 60.2 Standard sagittal T1-weighted anatomical slices through the left and right cerebral hemispheres overlaid with activity from the contrast of spoken language (sentence) production (a rest condition; red/yellow) and spoken language production (noncommunicative repetitive tongue movements; green). The statistical threshold was set at $Z > 2.3$, cluster-corrected. Anterior is to the left. The MNI coordinates are along the x-axis, with negative being to the left. The greater the number, the more lateral the sagittal slice. Regions of activity were located in: 1, pre-SMA and dACC; 2, bilateral anterior insula; 3, bilateral superior temporal cortex including left and right medial planum temporale; 4, left posterior inferior frontal gyrus (incorporating Broca's area) and extending dorsally into posterior middle frontal gyrus; and 5, right inferior frontal gyrus.

subcomponents of the posterior DMN, apparent from the ICA analysis shown in Figure 60.3, may have been responsible for the semantic demands of the task. The added advantage of the ICA analysis over and above the univariate analyses was demonstrating that subcomponents of left inferior parietal cortex and posterior temporal cortex were contributing to speech production, even though the overall effect size did not increase significantly.

Another study using very much the same experimental design and analyzed with 55-component ICA has demonstrated multiple overlapping networks, particularly in parietal cortex (Geranmayeh, Wise, et al., 2014). Although these many overlapping networks are a challenge to interpret, they confirm the left fronto-temporo-parietal network of the earlier study and the "deactivation" of much of the posterior DMN by the particular picture description task. This would seem to indicate that the activity in the angular gyri and the posterior midline cortex visualized in the earlier PET studies was predominantly related to the rich declarative memory recall associated with "storytelling" rather than a strict recall of factual knowledge about objects.

There are clearly other ways that can be used to elicit speech production from participants while they are encased in a scanner. An alternative method used picture description but put constraints on what words may be used to describe the scene (Grande et al., 2012). This inevitably led to periods when the participants formulated concepts to describe the scene, pauses as they searched for synonyms rather than uttering the "banned" words, and succeeded or failed at completing clause-like units. Analyzing the acquired data was performed in brief epochs so that signal could be related to particular episodes during speech production, such as successful lexical retrieval contrasted with a pause because of an initial failure of lexical retrieval. This required continuous data acquisition, which seems to have presented the authors with fewer problems with artifacts than has been our experience. Also, it has to be said that the task is quite unlike the seemingly effortless flow of normal conversational speech with no constraints (within the limits of polite discourse) on the vocabulary used.

Finally, it is apparent that we have made no mention of the functions of the left superior frontal gyrus,

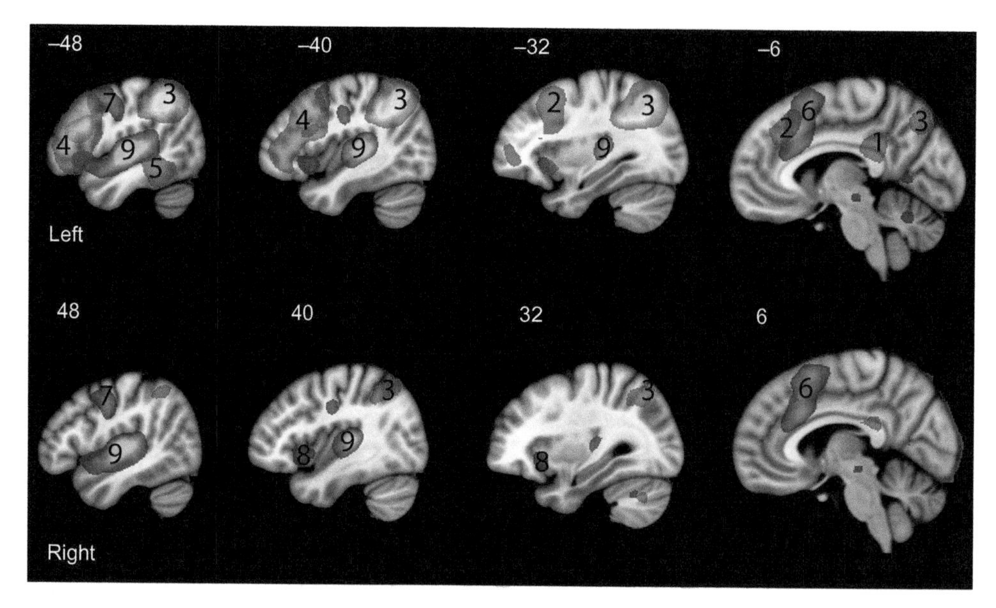

FIGURE 60.3 Standard T1-weighted anatomical slices overlaid with components from an ICA analysis of the study illustrated as univariate contrasts in Figure 60.2. Components 1 (shown in blue) and 24 (shown in red/yellow) demonstrated correlated activity for production of sentences that was significantly greater than for noncommunicative repetitive tongue movements. The statistical threshold was set at $Z > 4$. The images are sagittal views (MNI coordinates in the *x*-axis, with negative being to the left) with, in addition, one coronal slice (MNI coordinate in the *y*-axis) and one axial slice (MNI coordinate in the *z*-axis). The results from the ICA analysis demonstrated a widely distributed network for sentence production, including prominent left parietal activity not apparent in the univariate contrasts. The numbered regions show correlated activity for sentence production greater than for tongue moments in the following regions: 1, posterior cingulate cortex; 2, dACC; 3, left and right inferior parietal cortex; 4, left dorsolateral prefrontal cortex, including Broca's area; 5, left inferolateral temporal cortex; 6, presupplementary area and dACC; 7, lateral premotor cortex; 8, anterior insula; and 9, left and right superior temporal gyri.

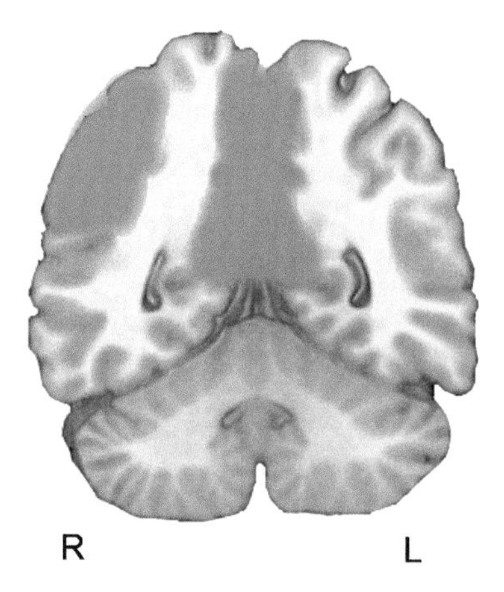

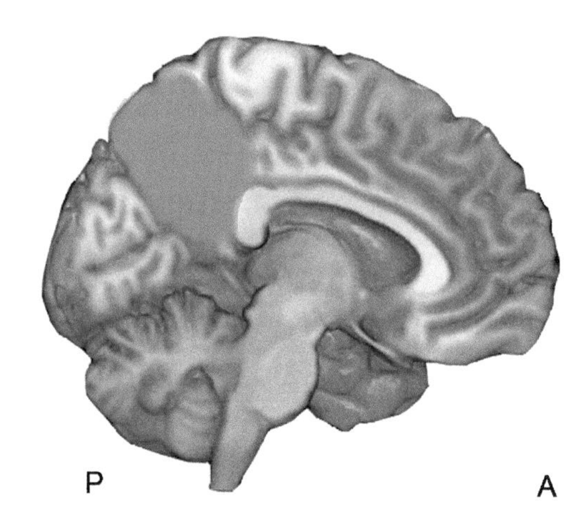

FIGURE 60.4 Schematic drawing of activity observed during an fMRI study contrasting non-communicative repetitive tongue movements with spoken language production (Geranmayeh et al., 2012). In addition to sensory cortices (not shown), activity was predominantly distributed in the posterior half of the DMN in the posterior midline cortex and in the posterior lateral parietal lobe on the right.

rostral to the pre-SMA, and the right lateral cerebellum, apparent in most of the studies on sentence-level and narrative-level speech production. These regions are clearly involved, but their role in the higher-order cognitive processes involved in speech generation remains speculative. Further, the basal ganglia must be involved at the cognitive as well as the motor-sensory level of speech production, but these are difficult regions to image during speech production even with the benefit of "sparse" temporal sampling. However, one study has demonstrated anterior striatal activity in relation to sentence production, with posterior striatal activity related to the articulatory processes (Argyropoulos et al., 2013). Future high-resolution scanning with meticulous attention to denoising the data is likely to be productive in revealing the roles of different corticostriatal loops during speech production.

60.5 CONCLUSION

This is an account of our views, limited though they are, on the functional anatomy and the neurobiology of sentence-level speech generation. The difficulties associated with data acquisition during sentence and narrative production have made it an under-researched area in functional neuroimaging. Further attempts to relate measures of fluctuations in local cerebral blood flow with poor temporal resolution to refined linguistic theory (and even more to the creative narratives of playwrights and novelists) make the whole enterprise seem distinctly inadequate. The generation of speech is strongly dependent on context, and the repetition of a sentence to the generation of a description of an object through memory-related reminiscing add increasing complexity in terms of both domain-specific and domain-general processing being embedded within functional neuroimaging data. Nevertheless, the ability to speak and to be able to express one's needs, thoughts, and desires is pre-eminent in humans, and its loss is so devastating to patients with brain disease that, in many ways, it can be considered the most important area of systems neuroscience research. If we could find a way reliably to improve speech after aphasic stroke over and above what natural recovery can offer, then that would be a very important achievement. Whether functional neuroimaging research in normal participants and in patients with brain injury or neurodegeneration can contribute to this goal remains to be seen. However, it is a fascinating endeavor. Although some may wish to try and dissect out the detail or to differentiate processes according to theoretical constructs of linguistic hierarchies, we ultimately have to accept that language is an anatomically and functionally interconnected system that lesion-deficit analyses have long demonstrated to be a largely left-lateralized network and one that is not well-represented by a "language neophrenology." Its function in normal communication is dependent on interactions with representations of memories, both autobiographical and semantic, and, depending on goals and contexts, is controlled by systems that also exert control over other cognitive functions. Above all, the language system interacts with the auditory system (during speech comprehension), the visual system (during reading), and the motor system (during speech production). As we image adults, we have to accept that we are making observations of brains that have "hard-wired" these disparate processes together over decades. Visualizing the "big picture," which in its entirety supports our ability to put what is in our head into the head of another through the medium of standing airwaves, is something on which we and others have made a modest start.

References

Aggleton, J. P. (2010). Understanding retrosplenial amnesia: Insights from animal studies. *Neuropsychologia*, *48*, 2328–2338.

Argyropoulos, G. P., Tremblay, P., & Small, S. L. (2013). The neostriatum and response selection in overt sentence production: An fMRI study. *NeuroImage*, *82*, 53–60.

Aron, A. R., Robbins, T. W., & Poldrack, R. A. (2014). Inhibition and the right inferior frontal cortex: One decade on. *Trends in Cognitive Sciences*, *18*, 177–185.

Awad, M., Warren, J. E., Scott, S. K., Turkheimer, F. E., & Wise, R. J. (2007). A common system for the comprehension and production of narrative speech. *Journal of Neuroscience*, *27*, 11455–11464.

Beckmann, C. F., & Smith, S. M. (2004). Probabilistic independent component analysis for functional magnetic resonance imaging. *IEEE Transactions in Medical Imaging*, *2*, 137–152.

Binder, J. R., & Desai, R. H. (2011). The neurobiology of semantic memory. *Trends in Cognitive Sciences*, *15*, 527–536.

Binder, J. R., Desai, R. H., Graves, W. W., & Conant, L. L. (2009). Where is the semantic system? A critical review and meta-analysis of 120 functional neuroimaging studies. *Cerebral Cortex*, *19*, 2767–2796.

Blank, S. C., Bird, H., Turkheimer, F., & Wise, R. J. (2003). Speech production after stroke: The role of the right pars opercularis. *Annals of Neurology*, *54*, 310–320.

Blank, S. C., Scott, S. K., Murphy, K., Warburton, E., & Wise, R. J. (2002). Speech production: Wernicke, Broca and beyond. *Brain*, *125*, 1829–1838.

Braga, R. M., Sharp, D. J., Leeson, C., Wise, R. J., & Leech, R. (2013). Echoes of the brain within default mode, association, and heteromodal cortices. *Journal of Neuroscience*, *33*, 14031–14039.

Braun, A. R., Guillemin, A., Hosey, L., & Varga, M. (2001). The neural organization of discourse: An $H_2{}^{15}O$ study of narrative production in English and American sign language. *Brain*, *124*, 2028–2044.

Brownsett, S. L., Warren, J. E., Geranmayeh, F., Woodhead, Z., Leech, R., & Wise, R. J. (2014). Cognitive control and its impact on recovery from aphasic stroke. *Brain*, *137*, 242–254.

Brownsett, S. L., & Wise, R. J. (2010). The contribution of the parietal lobes to speaking and writing. *Cerebral Cortex*, *20*, 517–523.

Buckner, R. L., Andrews-Hanna, J. R., & Schacter, D. L. (2008). The brain's default network: Anatomy, function, and relevance to disease. *Annals of the New York Academy of Sciences*, *1124*, 1–38.

Button, K. S., Ioannidis, J. P., Mokrysz, C., Nosek, B. A., Flint, J., Robinson, E. S., et al. (2013). Power failure: Why small sample size undermines the reliability of neuroscience. *Nature Reviews Neuroscience*, *14*, 365–376.

Catani, M., & ffytche, D. H. (2005). The rises and falls of disconnection syndromes. *Brain*, *128*, 2224–2239.

Curio, G., Neuloh, G., Numminen, J., Jousmäki, V., & Hari, R. (2000). Speaking modifies voice-evoked activity in the human auditory cortex. *Human Brain Mapping*, *9*, 183–191.

Dhanjal, N. S., Handunnetthi, L., Patel, M. C., & Wise, R. J. (2008). Perceptual systems controlling speech production. *Journal of Neuroscience*, *28*, 9969–9975.

Dosenbach, N. U., Fair, D. A., Miezin, F. M., Cohen, A. L., Wenger, K. K., Dosenbach, R. A., et al. (2007). Distinct brain networks for adaptive and stable task control in humans. *Proceedings of the National Academy of Sciences of the United States of America*, *104*, 11073–11078.

Dronkers, N. F., Wilkins, D. P., Van Valin, R. D., Jr., Redfern, B. B., & Jaeger, J. J. (2004). Lesion analysis of the brain areas involved in language comprehension. *Cognition*, *92*, 145–177.

Fedorenko, E. (2014). The role of domain-general cognitive control in language comprehension. *Frontiers in Psychology*, *5*, 335.

Fedorenko, E., Duncan, J., & Kanwisher, N. (2012). Language-selective and domain-general regions lie side by side within Broca's area. *Current Biology*, *22*, 2059–2062.

Fedorenko, E., Duncan, J., & Kanwisher, N. (2013). Broad domain generality in focal regions of frontal and parietal cortex. *Proceedings of the National Academy of Sciences of the United States of America*, *110*, 16616–16621.

Friston, K. J., Price, C. J., Fletcher, P., Moore, C., Frackowiak, R. S., & Dolan, R. J. (1996). The trouble with cognitive subtraction. *NeuroImage*, *4*, 97–104.

Geranmayeh, F., Brownsett, S. L., Leech, R., Beckmann, C. F., Woodhead, Z., & Wise, R. J. (2012). The contribution of the inferior parietal cortex to spoken language production. *Brain and Language, 121*, 47–57.

Geranmayeh, F., Brownsett, S. L., & Wise, R. J. (2014). Task-induced brain activity in aphasic stroke: What is driving recovery? *Brain, 137*, 2632–2648

Geranmayeh, F., Wise, R. J. S., Mehta, A., & Leech, R. (2014). Overlapping networks engaged during spoken language production, and its cognitive control. *Journal of Neuroscience, 34*, 8728–8740.

Gracco, V. L., Tremblay, P., & Pike, B. (2005). Imaging speech production using fMRI. *NeuroImage, 26*, 294–301.

Grande, M., Meffert, E., Schoenberger, E., Jung, S., Frauenrath, T., Huber, W., et al. (2012). From a concept to a word in a syntactically complete sentence: An fMRI study on spontaneous language production in an overt picture description task. *NeuroImage, 61*, 702–714.

Griffanti, L., Salimi-Khorshidi, G., Beckmann, C. F., Auerbach, E. J., Douaud, G., Sexton, C. E., et al. (2014). ICA-based artefact removal and accelerated fMRI acquisition for improved resting state network imaging. *NeuroImage, 95C*, 232–247.

Hall, D. A., Haggard, M. P., Akeroyd, M. A., Palmer, A. R., Summerfield, A. Q., Elliott, M. R., et al. (1999). "Sparse" temporal sampling in auditory fMRI. *Human Brain Mapping, 7*, 213–223.

Hampshire, A., Highfield, R. R., Parkin, B. L., & Owen, A. M. (2012). Fractionating human intelligence. *Neuron, 76*, 1225–1237.

Hickok, G., & Poeppel, D. (2007). The cortical organization of speech processing. *Nature Reviews Neuroscience, 8*, 393–402.

Huang, J., Carr, T. H., & Cao, Y. (2002). Comparing cortical activations for silent and overt speech using event-related fMRI. *Human Brain Mapping, 15*, 39–53.

Indefrey, P., & Levelt, W. J. (2004). The spatial and temporal signatures of word production components. *Cognition, 92*, 101–144.

Kemeny, S., Ye, F. Q., Birn, R., & Braun, A. R. (2005). Comparison of continuous overt speech fMRI using BOLD and arterial spin labeling. *Human Brain Mapping, 24*, 173–183.

Kort, N. S., Nagarajan, S. S., & Houde, J. F. (2014). A bilateral cortical network responds to pitch perturbations in speech feedback. *NeuroImage, 86*, 525–535.

Leech, R., Braga, R., & Sharp, D. J. (2012). Echoes of the brain within the posterior cingulate cortex. *Journal of Neuroscience, 32*, 215–222.

Levelt, W. J. M., Roelofs, A., & Meyer, A. S. (1999). A theory of lexical access in speech production. *Behavioral and Brain Sciences, 22*, 1–38.

Mah, Y. H., Husain, M., Rees, G., & Nachev, P. (2014). Human brain lesion-deficit inference remapped. *Brain, 137*, 2522–2531.

Mehta, S., Grabowski, T. J., Razavi, M., Eaton, B., & Bolinger, L. (2006). Analysis of speech-related variance in rapid event-related fMRI using a time-aware acquisition system. *NeuroImage, 29*, 1278–1293.

Menon, V., & Uddin, L. Q. (2010). Saliency, switching, attention and control: A network model of insula function. *Brain Structure and Function, 214*, 655–667.

Niziolek, C. A., Nagarajan, S. S., & Houde, J. F. (2013). What does motor efference copy represent? Evidence from speech production. *Journal of Neuroscience, 33*, 16110–16116.

Palmer, E. D., Rosen, H. J., Ojemann, J. G., Buckner, R. L., Kelley, W. M., & Petersen, S. E. (2001). An event-related fMRI study of overt and covert word stem completion. *NeuroImage, 14*, 182–193.

Parker-Jones, O., Prejawa, S., Hope, T. M., Oberhuber, M., Seghier, M. L., Leff, A. P., et al. (2014). Sensory-to-motor integration during auditory repetition: A combined fMRI and lesion study. *Frontiers in Human Neuroscience, 8*, 24.

Peeva, M. G., Guenther, F. H., Tourville, J. A., Nieto-Castanon, A., Anton, J. L., Nazarian, B., et al. (2010). Distinct representations of phonemes, syllables, and supra-syllabic sequences in the speech production network. *NeuroImage, 50*, 626–638.

Price, C. J. (2012). A review and synthesis of the first 20 years of PET and fMRI studies of heard speech, spoken language and reading. *NeuroImage, 62*, 816–847.

Price, C. J., Seghier, M. L., & Leff, A. P. (2010). Predicting language outcome and recovery after stroke: The PLORAS system. *Nature Reviews Neurology, 6*, 202–210.

Salmelin, R. (2007). Clinical neurophysiology of language: The MEG approach. *Clinical Neurophysiology, 118*, 237–254.

Seghier, M. L., Fagan, E., & Price, C. J. (2010). Functional subdivisions in the left angular gyrus where the semantic system meets and diverges from the default network. *Journal of Neuroscience, 30*, 16809–16817.

Sestieri, C., Corbetta, M., Romani, G. L., & Shulman, G. L. (2011). Episodic memory retrieval, parietal cortex, and the default mode network: Functional and topographic analyses. *Journal of Neuroscience, 31*, 4407–4420.

Shah, C., Erhard, K., Ortheil, H. J., Kaza, E., Kessler, C., & Lotze, M. (2013). Neural correlates of creative writing: An fMRI study. *Human Brain Mapping, 34*, 1088–1101.

Simmonds, A. J., Wise, R. J., Collins, C., Redjep, O., Sharp, D. J., Iverson, P., et al. (2014). Parallel systems in the control of speech. *Human Brain Mapping, 35*, 1930–1943.

Tremblay, P., & Small, S. L. (2011). Motor response selection in overt sentence production: A functional MRI study. *Frontiers in Psychology, 2*, 253.

Vigneau, M., Beaucousin, V., Hervé, P. Y., Duffau, H., Crivello, F., Houdé, O., et al. (2006). Meta-analyzing left hemisphere language areas: Phonology, semantics, and sentence processing. *NeuroImage, 30*, 1414–1432.

Warburton, E., Wise, R. J., Price, C. J., Weiller, C., Hadar, U., Ramsay, S., et al. (1996). Noun and verb retrieval by normal subjects. Studies with PET. *Brain, 119*, 159–179.

SECTION K

CONCEPTUAL SEMANTIC KNOWLEDGE

CHAPTER

61

The Hub-and-Spoke Hypothesis of Semantic Memory

Karalyn Patterson[1,2] *and Matthew A. Lambon Ralph*[3]

[1]Neurology Unit, Department of Clinical Neurosciences, University of Cambridge, Cambridge, UK; [2]MRC Cognition and Brain Sciences Unit, Cambridge, UK; [3]Neuroscience and Aphasia Research Unit, School of Psychological Sciences, University of Manchester, Manchester, UK

61.1 INTRODUCTION

The quantity of, quality of, and speed of access to semantic knowledge in the human brain is astonishing. Given any fragment of information, such as the word "camel," a healthy adult human can almost instantly generate a vast amount of other related information, such as the ways in which camels are and are not like other animals, how they look and move, where they live, how they interact with people, that it is the name of an American brand of cigarettes, that they are the subject of a famous story by Rudyard Kipling (*How the Camel Got His Hump*), and so forth. Curiously, this kind of conceptual knowledge—although often discussed by philosophers—did not become a major topic of research in cognitive science and neuroscience until relatively recently. Endel Tulving put it on the psychological map in 1972, when he proposed that semantic memory was a qualitatively different kind of knowledge from the episodic memory typically studied by experimental psychologists at that time. Elizabeth Warrington put it on the neuropsychological map in 1975, when she demonstrated that brain disease could selectively impair semantic memory. Now, approximately 40 years later, semantic memory is a commonplace object of behavioral study, brain research, and theorizing.

Over the past 15–20 years, we and our colleagues have been developing a story about the structure and neural basis of semantic memory that we call the hub-and-spoke hypothesis (Lambon Ralph, 2014; Lambon Ralph & Patterson, 2008; Lambon Ralph, Sage, Jones, & Mayberry, 2010; Patterson, Nestor, & Rogers, 2007; Rogers et al., 2004). The basic plot of this story comes in two parts (Figure 61.1). First, many different modality-specific cortical regions represent the aspects of conceptual knowledge that *are* modality-specific: the color of a camel in color regions, its shape in visual-form regions, the way it moves in regions that code visual movement, its name in language-specific cortex, and so on. Second, all of these regions send information to and receive information from a hub component of the semantic network, which codes semantic similarity structure and represents concepts in a manner that abstracts away from the specific features of how they look or sound or move or what they are called. The links between each modality-specific region and the hub are called spokes. Because it is the hub part of this story that is (somewhat) novel, other semantic-memory researchers—especially those critical of this hypothesis—occasionally describe our theory as one in which concepts have a completely abstract form of representation and reside in the hub. For example, Gainotti (2014) included our work in a short list of models "that assume that semantic representations... are stored in an abstract and propositional format" (Gainotti, 2014, p. 7). We acknowledge that some earlier expositions of our incompletely formed view may have sounded a bit like this (Patterson & Hodges, 2000), but it is certainly not an accurate description of our current hub-and-spoke hypothesis about semantic memory. We therefore begin the account of our position with an explanation of the crucial spoke components of the theory, called "the importance of the spokes and the regions from which they emanate" (Sections 61.2). Sections 61.3 argues that, although essential, the spokes and their modality-specific sources are insufficient and

Neurobiology of Language. DOI: http://dx.doi.org/10.1016/B978-0-12-407794-2.00061-4

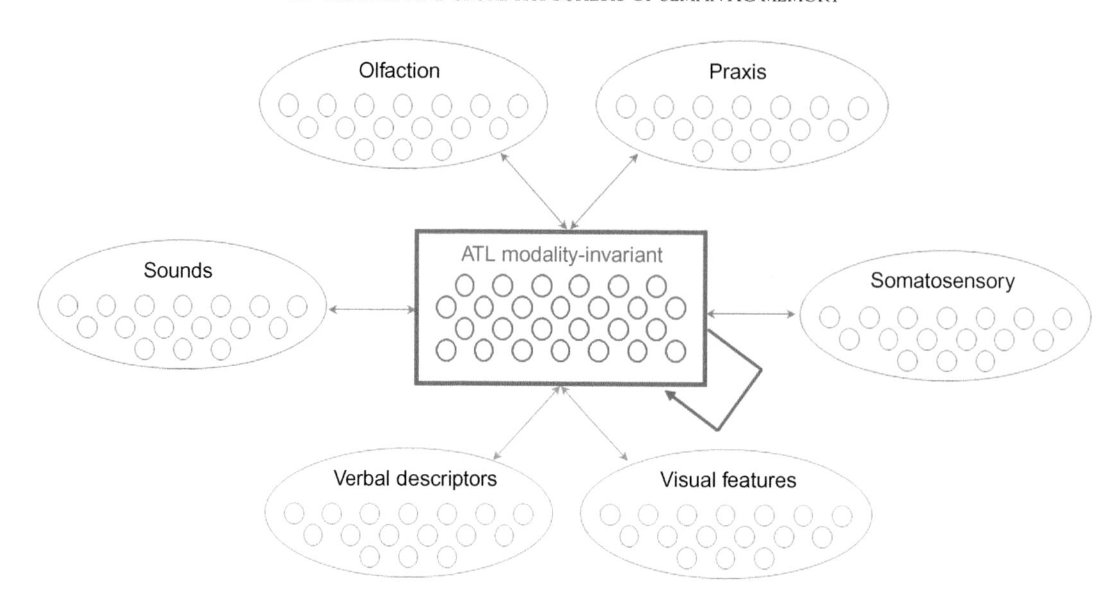

FIGURE 61.1 The "hub-and-spoke" model of conceptualization. Under this hypothesis, concepts are formed by the mutual interaction of modality-specific sources of information (the "spokes") with a central representational hub (red) that provides additional modality-invariant representational resource. The spokes are assumed to rely on modality-specific secondary association cortical regions while regions within the anterior temporal region (and, perhaps, other tertiary association areas) underpin the transmodal hub. Conceptualization follows from the joint action of the hub and spokes (see text for further details). Various computational implementations of this framework have been described in detail and demonstrate how concepts can be coded in this framework and how data from SD and patients with unilateral temporal lobe damage can be mimicked. *From Rogers et al., 2004 and Schapiro et al., 2013.*

that a transmodal hub is also necessary. Section 61.4 addresses reasons for locating the proposed hub in the anterior temporal lobe (ATL). Section 61.5 offers evidence for the bilateral nature of the ATL hub. Section 61.6, on the graded hub hypothesis, provides some more specific suggestions about different roles for various areas of the ATL.

61.2 THE IMPORTANCE OF THE SPOKES AND THE REGIONS FROM WHICH THEY EMANATE

It now seems almost incontrovertible that our knowledge about the modality-specific sensory and motor features of objects and other concepts is represented in or very near the same brain regions that process such information when it is encountered in the real world (Barsalou, 2008). This view, usually called either *grounded* or *embodied* cognition, is endorsed in one form or another by so many researchers that if we were to try to mention all of them, the rest of this section might consist of nothing but a list of references. Several different forms of evidence converge to support this idea, principally: (i) regionally specific patterns of brain activation, observed in healthy participants, that differ as a function of the modality or category of the stimulus and/or response; (ii) patterns of cognitive impairment, resulting from brain lesions in different locations, that also differ in principled ways depending on the sensory or motor features of the stimulus or response; and (iii) data from transient disruption of neural function in different regions via transcranial magnetic stimulation in neurologically intact participants (Pobric, Jefferies, & Lambon Ralph, 2010b). To be sure, some specific claims about modality or category-specific effects are open to debate (see Chen & Rogers, 2014, for a good discussion). To be even more sure, brain function is almost infinitely more subtle and complex than our limited ability to study it and think about it (Patterson & Plaut, 2009). This means that any claim for a simple relationship between a specific category and a specific brain region (e.g., "the fusiform face area [FFA] in the right posterior temporal lobe is completely and uniquely dedicated to processing faces") is unlikely to characterize the true state of things in the brain. We already know that areas other than the FFA respond to faces (Pitcher, Walsh, Yovel, & Duchaine, 2007) and that damage to the FFA resulting in prosopagnosia typically disrupts some abilities beyond face processing (Behrmann & Plaut, 2012). Nevertheless, there is a considerable degree of sensory-feature and motor-feature specialization in both processing and representation of different kinds of things that no theory of semantic representation can ignore. And it is crucial to emphasize that, according to embodied/grounded theories, this specialization does not just reflect surface features of stimuli: *words* referring to concepts of different types can also produce both brain activations in modality-relevant sensory or motor regions (Hauk, Shtyrov, & Pulvermüller, 2008; Pulvermuller, 2005; Simmons et al., 2007) (though see de Zubicaray, Arciuli, & McMahon, 2013 for an alternative interpretation of some results that might seem to support necessary activation in motor cortex when participants process action words) and distinctive patterns of impairment in patients with lesions in or near such sensory/motor areas (Boulenger et al., 2008; Pulvermüller et al., 2010). That is, a degree of specialization seems to apply to the concepts or mental representations of objects, not just objects in the real world.

In a useful review of "embodiment and the neuroscience of semantics," Meteyard, Cuadrado, Bahrami, and Vigliocco (2012) placed various theories about conceptual knowledge on an embodiment continuum, from "strong" to "weak" to "secondary" to "absent." The previous paragraph summarizes our reasons for rejecting theories that argue for purely abstract semantic representations with no element of embodiment. However, we also reject the most strongly embodied theories for reasons described in the next section. To summarize this first section, the hub-and-spoke hypothesis suggests that semantic knowledge is represented in a widely distributed brain network including modality-specific regions of cortex from which connections (spokes) emanate, sending activation to and receiving activation back from another essential component of the network: a transmodal semantic hub.

61.3 THE INSUFFICIENCY OF THE SPOKES AND THEIR SOURCES: WHY WE NEED A HUB

Strongly embodied theories of conceptual knowledge "effectively push semantics out into primary cortical areas and make it completely dependent on sensory and motor systems" (Meteyard et al., 2012, p. 793). In other words, according to such theories, the brain network for your knowledge of a concept such as *camel* consists only of the specific sensory/motor/linguistic features that you have experienced concerning camels. How do strongly embodied theories account for the fact that, on hearing the word "camel," you can almost instantly generate all that other information mentioned in the introduction? They assume that all of the feature representations in the modality-specific regions of cortex are connected to one another. Hearing the word "camel" initially produces selective activation of an auditory/linguistic representation; then, in a pattern-completion fashion, activation rapidly spreads via the inter-area connections to other modality-specific aspects of your camel knowledge. Straightforward enough, so why reject this hypothesis?

First, consider semantic dementia (SD), a neurodegenerative condition in the spectrum of frontotemporal dementia. The characteristics of SD especially germane to the current discussion—one neuroanatomical and four behavioral/cognitive—are as follows:

1. All clinical diagnoses of neurological origin are associated with a degree, larger or smaller, of variability across individual patients in "the three S's": site, size, and/or side of lesion. This fact, of course, makes it somewhat tricky to be confident about the precise neuroanatomical basis of the deficits. Such variability inevitably applies to SD, but it is widely agreed among researchers of neurodegenerative diseases that SD falls at the lower end of this spectrum of variability, especially in its early-middle phases (Acosta-Cabronero et al., 2011; Guo et al., 2013; Mion et al., 2010). To quote from Hodges and Patterson (2007), "Patients with even early-stage SD show bilateral, though typically asymmetrical, atrophy of the ATLs; as the disease progresses, the degeneration extends either caudally into the posterior temporal lobes or rostrally into the posterior, inferior frontal lobes, or both. Quantitative MRI studies. . .have refined these observations by showing consistent and extreme atrophy (commonly 50–80% gray matter loss) of the polar and perirhinal cortices and the anterior fusiform gyri (Davies, Graham, Xuereb, Williams, & Hodges, 2004; Du, Schuff, & Kramer, 2007; Gorno-Tempini, Dronkers, & Rankin, 2004)" (p. 1009). It should be noted that the hippocampus and entorhinal cortex are also substantially abnormal in SD (Chan, Fox, & Rossor, 2002; Williams, Nestor, & Hodges, 2005). Unlike the case of Alzheimer's disease (AD), however, the medial temporal atrophy in SD is much more severe in the rostral than in the caudal region. Why SD patients, despite their hippocampal abnormality, do not have the profoundly impaired anterograde amnesia characteristic of AD is a fascinating and evolving story (Nestor, Fryer, & Hodges, 2006; Nestor, Fryer, Ikeda, & Hodges, 2003), but this is not especially germane to this chapter on semantic memory.

   Results of the multimodal imaging study of SD patients and healthy controls by Guo et al. (2013), in agreement with many other investigations of SD, established that focal atrophy was confined to the ATL bilaterally. The critical finding of Guo et al., however, is that this focal structural damage was accompanied by reduced physiological integrity in many or all of the modality-specific regions demonstrated to be connected to the ATL in the control participants. Furthermore, the patients' semantic deficits correlated with both the ATL atrophy and the reduced connectivity to more specific regions. This one highly salient piece of research thus supports the existence and necessity of both the hub and the spokes. Guo et al. concluded that their evidence favors a model of the semantic network in which "a critical transmodal ATL semantic hub [is] positioned to integrate functionally relevant and topographically organized links to distributed modality-specific regions" (p. 2,988).

2. At least in the prefinal stages of the disease, the cognitive consequences of SD are largely confined to disruption of semantic knowledge per se (Guo et al., 2013; Hodges, Patterson, Oxbury, & Funnell, 1992; Snowden, Goulding, & Neary, 1989; Warrington, 1975) and of abilities in which semantic memory plays an important interactive role (Patterson et al., 2006).
3. The deterioration of semantic memory in SD is pan-category and transmodal: different kinds of concepts (e.g., concrete/abstract; living/human-made) are all impaired, and the semantic deficits are apparent no matter what format of stimulus (object, picture, spoken word, written word, environmental sound, taste, touch, etc.) is used to probe for conceptual knowledge and no matter what modality of response (pointing, speaking, writing, drawing, etc.) is used to measure performance (Bozeat, Lambon Ralph, Patterson, Garrard, & Hodges, 2000; Rogers et al., 2004).
4. The pattern of the cognitive deficits in SD invariably reflects an interactive impact of the familiarity or frequency of the stimulus concept (i.e., how often a concept is experienced) and the typicality of the stimulus in its domain (i.e., how similar each exemplar is to the category average or "prototype"). In many diverse tests and tasks that have manipulated the frequency and typicality of the stimuli, significantly impaired and poorest performance by SD patients is always observed for atypical and less familiar stimuli/concepts (Lambon Ralph et al., 2011; Patterson et al., 2006; Rogers et al., 2004; Woollams, Cooper-Pye, Hodges, & Patterson, 2008).
5. The nature of the errors made by SD patients is anything but random and, in virtually any task, their errors can be described as resulting from failures of appropriate generalization and discrimination (see Lambon Ralph & Patterson, 2008, for an extended discussion of such errors in both receptive and expressive experimental tasks, as well as in everyday life, and in both verbal and nonverbal contexts). In object naming, for example, instances of a category that are of moderate familiarity and typicality can no longer be discriminated correctly, resulting in errors such as naming a fox as "dog" or a zebra as "horse" or calling them both "animal." But less familiar and atypical animals such as a seahorse or a snail even fail to be generalized to the animal category: they usually result in content-less naming responses such as "a little thing" or "I don't know" or "what do you do with that?" (Patterson et al., 2007; Woollams et al., 2008). The same is true for object and word categorization: typical things are overgeneralized and atypical things are undergeneralized (Mayberry, Sage, & Lambon Ralph, 2011; Snowden, Griffiths, & Neary, 1996).

The first neuroanatomical characteristic indicates that the brain dysfunction in SD is typically centered on the ATL, with consequent disruption to the integrity of connections between the ATL and modality-specific primary and association cortices. The last four neuropsychological characteristics indicate that the cognitive abnormality is relatively restricted to semantic memory (and its impact on other tasks) and follows a highly consistent pattern. Given this set of well-established observations, it seems difficult to construct an account of SD that is compatible with a semantic network consisting only of modality-specific sensory, motor, and language regions of the brain. To our knowledge, there are no reports of significant structural degradation of modality-specific regions in SD. Furthermore, if SD did involve widespread degeneration of modality-specific regions, then it should yield cognitive deficits well beyond semantic memory, as seen in AD. For example, visuospatial processing depends substantially on parietal lobe regions; AD patients have parietal hypometabolism/atrophy with disrupted visuospatial ability; SD patients have neither (Pengas et al., 2010). Finally, if the transmodal semantic disorder in SD did arise from widespread abnormalities to modality-specific regions, then why would the pattern of disorder (the interaction between familiarity and typicality and the nature of the errors) be so consistent across all modalities and tasks? It seems much more plausible that the deficits in SD all emerge from disruption to a part of the semantic network that codes conceptual space independent of any modality-specific features of concepts in which the structure is defined by similarity of instances within domains and strongly influenced by familiar and typical instances (Rogers et al., 2004).

SD provides a clear, potentially definitive line of evidence against the strongest embodied theories of conceptualization, but it should also be noted that some researchers in related fields (such as philosophy, e.g., Wittgenstein, 2001, and cognitive science, e.g., Rosch, 1975; Smith & Medin, 1981) have also argued that models based on modality-specific features alone might be inadequate for the formation of coherent concepts and categories. A full exposition of the various arguments is beyond the scope of this chapter (these can be found in Lambon Ralph, 2014; Lambon Ralph, Sage, et al., 2010; Patterson et al., 2007; Rogers & McClelland, 2004), but the kernel of the issue is that the relationship between the abundance of modality-specific information and the plethora of concepts that emerge from them is highly complex and nonlinear. Our key working hypothesis, implemented computationally (Rogers et al., 2004; Schapiro, McClelland, Welbourne, Rogers, & Lambon Ralph, 2013) and supported neuroanatomically (Binney, Parker, & Lambon Ralph, 2012; Guo et al., 2013), is that concepts represent the *joint* action of distributed modality-specific regions (the "spokes") with an intermediate transmodal representational region (the "hub"). The hub provides the key additional computational capacity to draw the multiple and complex sources of information together, thereby deriving coherent and generalizable concepts.

61.4 WHY SHOULD THE HUB BE CENTERED ON THE ATL?

The principal argument here is, unsurprisingly, that the SD patients whose behavior strongly suggests the existence of a semantic hub have atrophy consistently centered on the ATL. It is worth noting that the best hypothesis about the location of a semantic hub in the ATL has changed somewhat since the earliest proposals. Because damage (e.g., as measured by VBM analyses on structural MR images in SD) appears to be maximal at the temporal pole (Brodmann area 38), that was the originally suggested hub region (Hodges et al., 1992; Patterson & Hodges, 2000). These earlier studies did not have sufficient power—because of relatively low patient numbers—to support correlations between lesion location and semantic performance. With improved diagnostic recognition of and further research on SD, however, that limitation has begun to recede, and the best evidence points to a region in the ATL posterior to the temporal pole, the anterior portion of the fusiform gyrus, which is subjacent to the head and body of the hippocampus (Brodmann areas 20 and 36). That is, atrophy and hypometabolism are severe across the entire anterior temporal region including both the temporal pole and this slightly more posterior and inferior region; however, the abnormality in the anterior fusiform correlates best with semantic scores (Libon et al., 2013; Mion et al., 2010).

Because no one is satisfied with scientific proposals/conclusions based on a single form of evidence, it is fortunate that there are populations other than SD patients and techniques other than the ones already mentioned that are compatible with a transmodal component of the semantic network located in the ATL. In this context, it is vital to note that standard fMRI—the source of most evidence concerning localization of function in the healthy brain—has been a rather inadequate source of evidence concerning the ATL because of various methodological limitations, including MRI field inhomogeneities, limited field-of-view, and the nature of the baseline task against which semantic task performance is compared (Devlin et al., 2000; Visser, Jefferies, & Lambon Ralph, 2010). Functional imaging studies using full field-of-view PET have often reported ATL activations when healthy participants perform semantic tasks (Sharp, Scott, & Wise, 2004; Vandenberghe, Price, Wise, Josephs, & Frackowiak, 1996). Furthermore, the relevance of the ATL to semantic knowledge is supported by recent results from: (i) distortion-corrected fMRI of this region (Visser, Jefferies, Embleton, & Lambon Ralph, 2012; Visser & Lambon Ralph, 2011); (ii) multivoxel pattern analyses of fMRI data (Peelen & Caramazza, 2012); (iii) the application of rTMS to the lateral part of this region (Lambon Ralph, Pobric, & Jefferies, 2009; Pobric, Jefferies, & Lambon Ralph, 2010a; Pobric et al., 2010b; Pobric, Jefferies, & Lambon Ralph, 2007); and (iv) source-based analyses of MEG signals (Marinkovic et al., 2003).

Three kinds of evidence combine to suggest that the ATL might constitute a reasonable location for a transmodal semantic hub. First, findings from nonhuman primate physiology (Gloor, 1997) and human functional and structural connectivity studies (Binney et al., 2012; Guo et al., 2013) indicate that a number of primary sensory and motor areas, along with their related association cortices, connect to and merge information in the ATL. Second, in addition to the rather factual nature of much conceptual information, semantic memory is often "flavored" by aspects of emotion and reward. Brain regions vital to these aspects of experience, including the amygdala and other limbic structures as well as the orbitofrontal cortex, are also close and well-connected to the ATL. Finally, the temporal lobe regions apparently important in semantic memory are immediately adjacent to the anterior parts of the medial temporal lobe memory system critical for rapid learning of new episodic information. Conceptual knowledge, although mostly, and for good reasons, acquired gradually (McClelland, McNaughton, & O'Reilly, 1995), is ultimately based on information experienced in episodes.

As a final note on this topic, it is potentially interesting and important that the center of the ATL transmodal hub (anterior fusiform/ITG) aligns with previously separate literature on neurophysiological evidence from patients who have depth or grid electrodes surgically implanted as part of investigation and treatment for long-term epilepsy. Ventral ATL regions are often considered to be visually specific, reflecting the apex of a visual "ventral" stream of processing (Albright, 2012); however, the same area also apparently plays a role in language processing. In their seminal studies, Lüders et al. (1986, 1991) demonstrated that stimulation of this region gave rise to transient language impairment (of reading and naming) and, consequently, they coined the term "basal temporal language area." Direct measurement of local field potentials confirmed its role in naming (Nobre, Allison, & McCarthy, 1994). More recent studies have demonstrated that these regions show sensitivity to semantic categories, but not low-level variations in visual characteristics (e.g., orientation and size). Strikingly, these semantic effects emerge very rapidly ($\sim$120 ms) after the onset of the stimulus (Chan et al., 2011; Liu, Agam, Madsen, & Kreiman, 2009). Although further direct comparisons with SD patients and functional neuroimaging are required, these data from the cognitive neurosurgery literature appear to support the hypothesis that this region is the centerpoint of a transmodal semantic hub.

61.5 EVIDENCE FOR AND POSSIBLE REASONS FOR A BILATERAL ATL HUB

So far, our attention has been on the ATL as a whole and in the singular. In the next two sections we consider emerging evidence for subdivisions both within the ATL in each hemisphere, from dorsal to ventral (see Section 61.6) and (in this section) across the two hemispheres. The ATL abnormality in SD patients is always bilateral (although often very asymmetrical, at least until late in the disease), suggesting that both left and right regions may contribute to conceptual knowledge. Other patient groups with clinically notable impairment of semantic memory typically have bilateral ATL damage [e.g., AD (Hodges & Patterson, 1995) and herpes simplex virus encephalitis (Lambon Ralph, Lowe, & Rogers, 2007; Noppeney et al., 2007)]. Likewise, formal meta-analyses of the functional neuroimaging literature indicate that both ATLs are implicated in semantic function in neurologically intact participants (Binder, Desai, Graves, & Conant, 2009; Visser et al., 2010). These neuropsychological and neuroimaging results prompt some obvious but important questions. First, does semantic impairment require bilateral ATL damage? And, if so, then what are the roles of the left versus right ATL?

Recent studies that have explored semantic function in patients with unilateral ATL damage indicate that semantic performance is generally much better than in patients with bilateral ATL diseases, especially SD. With more sensitive assessments, however, expressive and receptive semantic deficits can be observed after unilateral ATL lesions (Bi et al., 2011; Lambon Ralph, Cipolotti, Manes, & Patterson, 2010; Lambon Ralph, Ehsan, Baker, & Rogers, 2012). These contemporary studies fit with the older comparative neurology literature that consistently reported chronic multimodal (visual and auditory) "semantic" impairment in primates and monkeys after bilateral ATL resection, but only a milder and transient form of the same impairment after unilateral resection (Brown & Schafer, 1888; Klüver & Bucy, 1939). Intriguingly, and relevant to the cross-methodology data noted at the end of their seminal paper, Klüver and Bucy highlighted striking similarities between primates following bilateral ATL resection and the patients previously described by Pick with frontotemporal dementia. The same pattern as that observed in the primate studies was replicated in a rare human single-case neurosurgery study (Terzian & Dalle Ore, 1955). In an (ultimately unsuccessful) attempt to control the patient's epilepsy, these investigators performed sequential full-depth ATL resections. As in the previous primate studies, after initial unilateral removal, the patient exhibited transient deficits of language, recognition and memory. However, after bilateral resection, a chronic and severe comprehension deficit emerged, along with other features of what is now referred to as Klüver-Bucy syndrome. Taken together, the full (although admittedly still rather limited) set of neuropsychological and neurosurgery findings can be summarized as follows: unilateral ATL damage generates semantic impairment in the acute stage that often diminishes rapidly, leaving patients in the chronic phase with a mild comprehension deficit. This impairment can be detected with sensitive neuropsychological assessments but is nothing like the level of semantic impairment found in SD and other bilateral ATL diseases.

Before considering the explanation for these findings in more detail, a brief aside to avoid potential confusion is merited. By far the most famous example of bilateral temporal resection is that of patient HM. After surgery, he and the other patients reported in the seminal case-series study of Scoville and Milner (Scoville & Milner, 1957) presented with profound anterograde amnesia but not with other impairments of higher cognition, including language and semantic memory. At face value, this might appear to contradict the hypothesis that semantic representation is reliant on a bilateral ATL system; in fact, a closer consideration of the location of the resections adds further weight to our hypothesis that the ventrolateral aspects of the ATL are crucial to semantic memory. Because of the fact that Scoville used an orbital approach to the medial temporal regions when performing the bilateral selective hippocampectomy (rather than the lateral approach commonly used in most modern neurosurgical procedures), the ventral and lateral aspects of the ATL were preserved in patient HM and the other patients in the original case series.

The notion that both left and right ATL contribute to semantic representation is reinforced by recent studies applying rTMS to neurologically intact participants. The technique is complementary to neuropsychological studies in that one can assess performance before stimulation and then after it during the transient refractory period. Its advantage over lesion studies is that the location and timing of the stimulation are under experimental control; however, the behavioral effect of stimulation is (happily!) not only transient but also much smaller than that of neurological damage. Careful, sensitive assessment methods are thus required. As noted, SD patients always have bilateral temporal abnormalities, making it difficult in SD to separate and establish a selective role, if any, of left versus right ATL. It is possible, however, to contrast the effect of rTMS to left versus right ATL, and this has been done for verbal and nonverbal comprehension tasks in a series of studies (Lambon Ralph, Pobric, & Jefferies, 2009; Pobric et al., 2010a). In these studies, rTMS to either ATL generated an equivalent effect on both verbal and nonverbal semantic tasks, thus supporting our working hypothesis that both ATLs contribute to a transmodal semantic hub.

A potential explanatory framework for the patient (SD and unilateral resection) and rTMS results was explored in a recent computational model of bilateral ATL semantic representation (Schapiro et al., 2013). This was based on a direct extension of the original Rogers "hub-and-spoke" computational model (Rogers et al., 2004). Instead of a single transmodal representational hub, however, Schapiro and colleagues split it into two separate "demi-hubs" with only partial connectivity between them. The model, which was trained in exactly the same manner as the Rogers single-hub variant, was able to learn all of the training patterns despite having a lower number of connections overall. When considerable unilateral damage was applied to the simulation, a small semantic impairment resulted that could be reduced by allowing some postdamage weight adjustment (to mimic spontaneous recovery: Keidel, Welbourne, & Lambon Ralph, 2010; Welbourne & Lambon Ralph, 2007; Welbourne, Woollams, Crisp, & Lambon Ralph, 2011). In contrast, when the same total amount of damage was distributed across both demi-hubs, a substantial and considerably greater semantic deficit emerged—replicating the difference between patients with unilateral versus bilateral ATL lesions. Detailed analysis and mathematical exploration demonstrated that the difference between unilateral and bilateral damage in the model was related to the differential propagation of noisy activation. Damage in any computational model not only diminishes the level of activation but also adds noise to its processing. When damage is focussed within one (only partly interconnected) demi-hub, the unaffected system generates strong and relatively noise-free activations that are able to dominate the response from the model; thus, it performs generally accurately (albeit slightly less efficiently than before). In contrast, bilateral damage across both demi-hubs results in weakened and noisy representations, leading to inaccurate responses.

Although it seems clear that both left and right ATLs are important to conceptual knowledge, the current literature contains mixed results and conclusions regarding the nature of the contribution from each side. Based on results from a variety of techniques, some researchers argue for a left–right dissociation in which the left ATL is primarily responsible for verbal semantic processing, whereas the right is much more important in processing faces, objects, and nonverbal sounds (Gainotti, 2012, 2014; Mesulam et al., 2013; Snowden, Thompson, & Neary, 2004; Thierry & Price, 2006). Other research groups report results more compatible with either similar contributions from the two ATL sides or at least only graded dissociations. Furthermore, even results indicating left–right dissociation are open to a different and more subtle interpretation than a categorical "left = verbal, right = nonverbal." There is no doubt that cortical regions posterior to the ATL have significantly different involvement in the processing of verbal versus nonverbal information. Given that the great majority of corticocortical connections are very local and within rather than across hemisphere, this means that as information moves from caudal (sensory) regions to rostral (conceptual) regions, verbal input will activate the left ATL more than the right, and nonverbal input such as faces and objects will preferentially activate the right ATL. This account of left/right differences in semantic processing, including two implemented computational models, has been advanced by Ikeda, Patterson, Graham, Lambon Ralph, and Hodges (2006); Lambon Ralph, McClelland, Patterson, Galton, and Hodges (2001); Mion et al. (2010), and Schapiro et al. (2013).

In conclusion, the current literature provides only a partial answer to the nature of the contribution of left versus right ATL to semantic processing. There is broad agreement (across methods and patient groups) that both left and right ATL contribute in some fashion and that the left ATL is more involved than the right in generating names and speech from semantic memory. The issues that are subject to considerable current research activity center around two alternative hypotheses: (i) graded connectivity-related differences in the contribution of each ATL to a primarily bilateral neural system or (ii) a more modular distinction in which each ATL is relatively specialized for discrete subdivisions of semantic processing.

61.6 THE GRADED HUB HYPOTHESIS

In this final section, we consider potential functional variations within each ATL. As noted, although the robust and consistent data from SD patients have been crucial in linking bilateral ATLs to semantic representation, these findings are a less powerful source of information regarding the contributions of left versus right ATL. In a similar vein, although the atrophy in SD is generally agreed to be more profound in the ventral than the dorsal ATL, most of the ATL is abnormal by the time that SD patients come to clinical attention, with the result that the condition of SD is insufficient to inform us about potential variations in representation of conceptual knowledge from dorsal to ventral regions. There is also the question of lateral versus medial aspects of the ATL. All of these "unknowns" constitute yet another reason for adopting multiple methods, because limitations in one approach can often be compensated by advantages conferred by other techniques.

Ever since the seminal work of Brodmann (1909), it has been known that there are variations in cytoarchitecture that, in turn, imply different types of neurocomputation. Two major distinctions in the ATL region noted by Brodmann were a shift from granular through dysgranular to agranular cortex in moving from lateral to medial structures around the temporal pole and a significant change in cytoarchitecture from BA22 (approximately equivalent to the superior temporal gyrus) to the neighboring areas (pole, BA38; middle and inferior temporal gyri, BA21 and BA20, respectively). Although noting that there were also differences between the other ATL regions (sufficient to give them different labels), Brodmann commented that they were graded rather than absolute differences; in fact, it is possible to consider them to be one single larger cytoarchitectural region as some of his contemporaries did. Modern, sophisticated studies (Ding, Van Hoesen, Cassell, & Poremba, 2009) have increased the number of cytoarchitectural divisions proposed for the ATL but, echoing Brodmann's original observations, these divisions are thought to be graded rather than absolute in nature. Although local cytoarchitecture is only one factor in neurocomputation (others—like connectivity—are also crucial), these classic and contemporary findings might lead us to expect some functional variations across the ATL, albeit they are probably graded in nature.

Recent (distortion-corrected) neuroimaging studies have provided new insights about white-matter connectivity within the ATL and also the location of semantically related functional activations. With regard to white-matter connectivity, Binney et al. (2012) observed connections both along gyri and between neighboring gyri, providing a neural basis for blending of information in a rostral and lateral direction along the temporal cortex. In addition, there is a more regionally specific pattern of connectivity from the temporal lobe to other areas: the temporopolar cortex is primarily connected via the uncinate fasciculus to orbitofrontal and ventral prefrontal regions; posterior temporal areas to inferior parietal regions; and superior temporal areas to inferior prefrontal regions. Very similar patterns of connectivity have been revealed by a recent analysis of resting-state functional connectivity seeded from different parts of the human temporopolar cortex (Pascual et al., 2013). These long-range connections are likely to be germane not only for considering how various forms of modality-specific information interact with a transmodal ATL hub but also for considering how semantic representations interface with other cognitive mechanisms, such as executive functions and working memory (Binney et al., 2012; Jefferies & Lambon Ralph, 2006).

With the benefit of methods and techniques to allow effective probing of activations across the entire ATL (including ventral regions), recent fMRI studies have started to reveal various graded patterns of semantic activation, which seem to fit with the graded differences in connectivity and cytoarchitecture in this area (Binney et al., 2012; Ding et al., 2009). These neuroimaging studies point to the bilateral ventrolateral subregion as the centerpoint of a graded ATL semantic hub. Across studies, this region is consistently activated, irrespective of task, input modality, or category of stimuli (Sharp et al., 2004; Vandenberghe et al., 1996; Visser et al., 2012; Visser & Lambon Ralph, 2011). Moving away from this neural location, there seem to be gradual shifts in the semantic function dependent on the proximity/connectivity to different primary inputs (Visser et al., 2012; Visser & Lambon Ralph, 2011). Thus, activation is more sensitive to auditory-related stimuli (verbal and nonverbal) in lateral superior regions and to visual inputs in posterior ventral temporal areas. These early explorations of the ATL region are consistent with variants of the hub-and-spoke model that include graded variations of connectivity (Plaut, 2002).

In summary, recent evidence has made it clear that the ATL does not function as one undifferentiated mass but rather with important graded variations of function across the region. Given the pattern of its structural and functional connectivity, these graded changes in ATL function appear to reflect the varying influence of different inputs and outputs on the processing in local ATL subregions.

61.7 CONCLUDING COMMENT

Semantic memory/conceptual knowledge is central to much of human life. It may not be involved when you walk into a room to sit down, but recognizing that an unfamiliar chair is something on which to sit requires semantic memory. It may not be involved when you move your tongue and lips, but it is the basis of every novel utterance that you make. A huge amount has been learned about this central aspect of human function in the relatively short time in which cognitive neuroscientists have been addressing it in their experimental work and theorizing, but there is so far to go that future models of the organization and neural basis of semantic memory may look like "objects" that we have never encountered before. It would be a foolhardy researcher who tries to predict which of the field's current interpretations of data will still be alive one or several decades from now. Maybe the proposal of a semantic hub will be here today and gone tomorrow. All we can say is that today, we find it difficult to conceive of a hub-less framework that would account for existing data regarding semantic memory.

References

Acosta-Cabronero, J., Patterson, K., Fryer, T. D., Hodges, J. R., Pengas, G., Williams, G. B., et al. (2011). Atrophy, hypometabolism and white matter abnormalities in semantic dementia tell a coherent story. *Brain, 134*, 2025–2035.

Albright, T. D. (2012). On the perception of probable things: Neural substrates of associative memory, imagery, and perception. *Neuron, 74*(2), 227–245.

Barsalou, L. W. (2008). Grounded cognition. *Annual Review of Psychology, 59*(1), 617–645. Available from: http://dx.doi.org/10.1146/annurev.psych.59.103006.093639.

Behrmann, M., & Plaut, D. C. (2012). Bilateral hemispheric processing of words and faces: Evidence from word impairments in prosopagnosia and face impairments in pure alexia. *Cerebral Cortex, 24*, 1102–1118. Available from: http://dx.doi.org/10.1093/cercor/bhs390.

Bi, Y., Wei, T., Wu, C., Han, Z., Jiang, T., & Caramazza, A. (2011). The role of the left anterior temporal lobe in language processing revisited: Evidence from an individual with ATL resection. *Cortex, 47*(5), 575–587.

Binder, J. R., Desai, R. H., Graves, W. W., & Conant, L. L. (2009). Where is the semantic system? A critical review and meta-analysis of 120 functional neuroimaging studies. *Cerebral Cortex, 19*(12), 2767–2796. Available from: http://dx.doi.org/10.1093/cercor/bhp055.

Binney, R. J., Parker, G. J. M., & Lambon Ralph, M. A. (2012). Convergent connectivity and graded specialization in the rostral human temporal lobe as revealed by diffusion-weighted imaging probabilistic tractography. *Journal of Cognitive Neuroscience, 24*, 1998–2014.

Boulenger, V., Mechtouff, L., Thobois, S., Broussolle, E., Jeannerod, M., & Nazir, T. A. (2008). Word processing in Parkinson's disease is impaired for action verbs but not concrete nouns. *Neuropsychologia, 46*(2), 743–756.

Bozeat, S., Lambon Ralph, M. A., Patterson, K., Garrard, P., & Hodges, J. R. (2000). Non-verbal semantic impairment in semantic dementia. *Neuropsychologia, 38*, 1207–1215.

Brodmann, K. (1909). *Vergleichende lokalisationslehre der Grosshirnrinde*. Leipzig: Barth.

Brown, S., & Schafer, E. A. (1888). An investigation into the functions of the occipital and temporal lobes of the monkey's brain. *Philosophical Transactions of the Royal Society of London. B, 179*, 303–327.

Chan, A. M., Baker, J. M., Eskandar, E., Schomer, D., Ulbert, I., Marinkovic, K., et al. (2011). First-Pass selectivity for semantic categories in human anteroventral temporal lobe. *The Journal of Neuroscience, 31*(49), 18119–18129. Available from: http://dx.doi.org/10.1523/jneurosci.3122-11.2011.

Chan, D., Fox, N., & Rossor, M. (2002). Differing patterns of temporal atrophy in Alzheimer's disease and semantic dementia. *Neurology, 58*, 838.

Chen, L., & Rogers, T. T. (2014). Revisiting domain-general accounts of category specificity in mind and brain. *Wiley Interdisciplinary Reviews: Cognitive Science, 5*(3), 327–344. Available from: http://dx.doi.org/10.1002/wcs.1283.

Davies, R., Graham, K. S., Xuereb, J. H., Williams, G. B., & Hodges, J. R. (2004). The human perirhinal cortex and semantic memory. *European Journal of Neuroscience, 20*, 2441–2446.

Devlin, J. T., Russell, R. P., Davis, M. H., Price, C. J., Wilson, J., Moss, H. E., et al. (2000). Susceptibility-induced loss of signal: Comparing PET and fMRI on a semantic task. *NeuroImage, 11*(6; Pt 2), 589–600.

de Zubicaray, G., Arciuli, J., & McMahon, K. (2013). Putting an "end" to the motor cortex representations of action words. *Journal of Cognitive Neuroscience, 25*(11), 1957–1974. Available from: http://dx.doi.org/10.1162/jocn_a_00437.

Ding, S., Van Hoesen, G. W., Cassell, M. D., & Poremba, A. (2009). Parcellation of human temporal polar cortex: A combined analysis of multiple cytoarchitectonic, chemoarchitectonic, and pathological markers. *Journal of Comparative Neurology, 514*, 595–623.

Du, A. T., Schuff, N., Kramer, J. H., Rosen, H. J., Gorno-Tempini, M. L., Rankin, K., et al. (2007). Different regional patterns of cortical thinning in Alzheimer's disease and frontotemporal dementia. *Brain, 130*, 1159–1166.

Gainotti, G. (2012). The format of conceptual representations disrupted in semantic dementia: A position paper. *Cortex, 48*(5), 521–529.

Gainotti, G. (2014). Why are the right and left hemisphere conceptual representations different? *Behavioural Neurology, 2014*, 10. Available from: http://dx.doi.org/10.1155/2014/603134.

Gloor, P. (1997). *The temporal lobe and the limbic system*. Oxford: Oxford University Press.

Gorno-Tempini, M. L., Dronkers, N. F., Rankin, K. P., et al. (2004). Cognition and anatomy in three variants of primary progressive aphasia. *Annals of Neurology, 55*, 335–346.

Guo, C. C., Gorno-Tempini, M. L., Gesierich, B., Henry, M., Trujillo, A., Shany-Ur, T., et al. (2013). Anterior temporal lobe degeneration produces widespread network-driven dysfunction. *Brain, 136*(10), 2979–2991. Available from: http://dx.doi.org/10.1093/brain/awt222.

Hauk, O., Shtyrov, Y., & Pulvermüller, F. (2008). The time course of action and action-word comprehension in the human brain as revealed by neurophysiology. *Journal of Physiology Paris*, *102*(1–3), 50–58.

Hodges, J. R., & Patterson, K. (1995). Is semantic memory consistently impaired early in the course of Alzheimer's disease? Neuroanatomical and diagnostic implications. *Neuropsychologia*, *33*(4), 441–459.

Hodges, J. R., & Patterson, K. (2007). Semantic dementia: A unique clinico-pathological syndrome. *Lancet Neurology*, *6*, 1004–1014.

Hodges, J. R., Patterson, K., Oxbury, S., & Funnell, E. (1992). Semantic dementia: Progressive fluent aphasia with temporal lobe atrophy. *Brain*, *115*(Pt 6), 1783–1806.

Ikeda, M., Patterson, K., Graham, K. S., Lambon Ralph, M. A., & Hodges, J. R. (2006). A horse of a different colour: Do patients with semantic dementia recognise different versions of the same object as the same? *Neuropsychologia*, *44*(4), 566–575.

Jefferies, E., & Lambon Ralph, M. A. (2006). Semantic impairment in stroke aphasia vs. semantic dementia: A case-series comparison. *Brain*, *129*, 2132–2147.

Keidel, J. L., Welbourne, S. R., & Lambon Ralph, M. A. (2010). Solving the paradox of the equipotential and modular brain: A neurocomputational model of stroke vs. slow-growing glioma. *Neuropsychologia*, *48*(6), 1716–1724.

Klüver, H., & Bucy, P. (1939). Preliminary analysis of functions of the temporal lobes in monkeys. *Archives of Neurology and Psychiatry*, *42*(6), 979–1000.

Lambon Ralph, M. A. (2014). Neurocognitive insights on conceptual knowledge and its breakdown. *Philosophical Transactions of the Royal Society B: Biological Sciences*, *369*(1634), 20120392. Available from: http://dx.doi.org/10.1098/rstb.2012.0392.

Lambon Ralph, M. A., Cipolotti, L., Manes, F., & Patterson, K. (2010). Taking both sides: Do unilateral anterior temporal lobe lesions disrupt semantic memory? *Brain*, *133*(11), 3243–3255. Available from: http://dx.doi.org/10.1093/brain/awq264.

Lambon Ralph, M. A., Ehsan, S., Baker, G. A., & Rogers, T. T. (2012). Semantic memory is impaired in patients with unilateral anterior temporal lobe resection for temporal lobe epilepsy. *Brain*, *135*(1), 242–258. Available from: http://dx.doi.org/10.1093/brain/awr325.

Lambon Ralph, M. A., Lowe, C., & Rogers, T. T. (2007). Neural basis of category-specific semantic deficits for living things: Evidence from semantic dementia, HSVE and a neural network model. *Brain*, *130*, 1127–1137.

Lambon Ralph, M. A., McClelland, J. L., Patterson, K., Galton, C. J., & Hodges, J. R. (2001). No right to speak? The relationship between object naming and semantic impairment: Neuropsychological evidence and a computational model. *Journal of Cognitive Neuroscience*, *13*, 341–356.

Lambon Ralph, M. A., & Patterson, K. (2008). Generalisation and differentiation in semantic memory: Insights from semantic dementia. *Annals of the New York Academy of Science*, *1124*, 61–76.

Lambon Ralph, M. A., Pobric, G., & Jefferies, E. (2009). Conceptual knowledge is underpinned by the temporal pole bilaterally: Convergent evidence from rTMS. *Cerebral Cortex*, *19*(4), 832–838. Available from: http://dx.doi.org/10.1093/cercor/bhn131.

Lambon Ralph, M. A., Sage, K., Green Heredia, C., Berthier, M. L., Martínez-Cuitiño, M., Torralva, T., et al. (2011). El-La: The impact of degraded semantic representations on knowledge of grammatical gender in semantic dementia. *Acta Neuropsychologica*, *9*, 115–132.

Lambon Ralph, M. A., Sage, K., Jones, R. W., & Mayberry, E. J. (2010). Coherent concepts are computed in the anterior temporal lobes. *Proceedings of the National Academy of Sciences of the United States of America*, *107*(6), 2717–2722. Available from: http://dx.doi.org/10.1073/pnas.0907307107.

Libon, D. J., Rascovsky, K., Powers, J., Irwin, D. J., Boller, A., Weinberg, D., et al. (2013). Comparative semantic profiles in semantic dementia and Alzheimer's disease. *Brain*, *136*(8), 2497–2509. Available from: http://dx.doi.org/10.1093/brain/awt165.

Liu, H., Agam, Y., Madsen, J. R., & Kreiman, G. (2009). Timing, timing, timing: Fast decoding of object information from intracranial field potentials in human visual cortex. *Neuron*, *62*(2), 281–290.

Lüders, H., Lesser, R. P., Hahn, J., Dinner, D. S., Morris, H., Resor, S., et al. (1986). Basal temporal language area demonstrated by electrical stimulation. *Neurology*, *36*(4), 505–510.

Lüders, H., Lesser, R. P., Hahn, J., Dinner, D. S., Morris, H. H., Wyllie, E., et al. (1991). Basal temporal language area. *Brain*, *114* (Pt 2), 743–754.

Marinkovic, K., Dhond, R. P., Dale, A. M., Glessner, M., Carr, V., & Halgren, E. (2003). Spatiotemporal dynamics of modality-specific and supramodal word processing. *Neuron*, *38*, 487–497.

Mayberry, E. J., Sage, K., & Lambon Ralph, M. A. (2011). At the edge of semantic space: The breakdown of coherent concepts in semantic dementia is constrained by typicality and severity but not modality. *Journal of Cognitive Neuroscience*, *23*(9), 2240–2251. Available from: http://dx.doi.org/10.1162/jocn.2010.21582.

McClelland, J. L., McNaughton, B. L., & O'Reilly, R. C. (1995). Why there are complementary learning-systems in the hippocampus and neocortex: Insights from the successes and failures of connectionist models of learning and memory. *Psychological Review*, *102*(3), 419–457.

Mesulam, M.-M., Wieneke, C., Hurley, R., Rademaker, A., Thompson, C. K., Weintraub, S., et al. (2013). Words and objects at the tip of the left temporal lobe in primary progressive aphasia. *Brain*, *136*(2), 601–618. Available from: http://dx.doi.org/10.1093/brain/aws336.

Meteyard, L., Cuadrado, S. R., Bahrami, B., & Vigliocco, G. (2012). Coming of age: A review of embodiment and the neuroscience of semantics. *Cortex*, *48*(7), 788–804.

Mion, M., Patterson, K., Acosta-Cabronero, J., Pengas, G., Izquierdo-Garcia, D., Hong, Y. T., et al. (2010). What the left and right anterior fusiform gyri tell us about semantic memory. *Brain*, *133* (11), 3256–3268. Available from: http://dx.doi.org/10.1093/brain/awq272.

Nestor, P. J., Fryer, T. D., & Hodges, J. R. (2006). Declarative memory impairments in Alzheimer's disease and semantic dementia. *NeuroImage*, *30*, 1010–1020.

Nestor, P. J., Fryer, T. D., Ikeda, M., & Hodges, J. R. (2003). Retrosplenial cortex—B29/30—hypometabolism in mild cognitive impairment (prodromal Alzheimer's disease). *The European Journal of Neuroscience*, *18*, 2663–2667.

Nobre, A. C., Allison, T., & McCarthy, G. (1994). Word recognition in the human inferior temporal lobe. *Nature*, *372*(6503), 260–263. Available from: http://dx.doi.org/10.1038/372260a0.

Noppeney, U., Patterson, K., Tyler, L. K., Moss, H., Stamatakis, E. A., Bright, P., et al. (2007). Temporal lobe lesions and semantic impairment: A comparison of herpes simplex virus encephalitis and semantic dementia. *Brain*, *130*, 1138–1147.

Pascual, B., Masdeu, J. C., Hollenbeck, M., Makris, N., Insausti, R., Ding, S.-L., et al. (2013). Large-scale brain networks of the human left temporal pole: A functional connectivity MRI study. *Cerebral Cortex* [Epub ahead of print September 2013]. Available from: http://dx.doi.org/10.1093/cercor/bht260.

Patterson, K., & Hodges, J. R. (2000). Semantic dementia: One window on the structure and organization of semantic memory. In L. Cermak (Ed.), *Handbook of neuropsychology* (Vol. 2, pp. 313–333). Amsterdam: Elsevier.

Patterson, K., Lambon Ralph, M. A., Jefferies, E., Woollams, A., Jones, R. W., Hodges, J. R., et al. (2006). "Presemantic" cognition in semantic dementia: Six deficits in search of an explanation. *Journal of Cognitive Neuroscience, 18*(2), 169–183.

Patterson, K., Nestor, P. J., & Rogers, T. T. (2007). Where do you know what you know? The representation of semantic knowledge in the human brain. *Nature Reviews Neuroscience, 8*(12), 976–987. Available from: http://dx.doi.org/10.1038/nrn2277.

Patterson, K., & Plaut, D. (2009). "Shallow draughts intoxicate the brain": Lessons from cognitive science for cognitive neuropsychology. *Topics in Cognitive Science, 1*, 39–58.

Peelen, M. V., & Caramazza, A. (2012). Conceptual object representations in human anterior temporal cortex. *The Journal of Neuroscience, 32*(45), 15728–15736. Available from: http://dx.doi.org/10.1523/jneurosci.1953-12.2012.

Pengas, G., Patterson, K., Arnold, R. J., Bird, C. M., Burgess, N., & Nestor, P. J. (2010). Lost and found: Bespoke memory testing for Alzheimer's disease and semantic dementia. *Journal of Alzheimer's Disease, 21*(4), 1347–1365. Available from: http://dx.doi.org/10.3233/jad-2010-100654.

Pitcher, D., Walsh, V., Yovel, G., & Duchaine, B. (2007). TMS evidence for the involvement of the right occipital face area in early face processing. *Current Biology, 17*(18), 1568–1573. Available from: http://dx.doi.org/10.1016/j.cub.2007.07.063.

Plaut, D. C. (2002). Graded modality-specific specialization in semantics: A computational account of optic aphasia. *Cognitive Neuropsychology, 19*, 603–639.

Pobric, G., Jefferies, E., & Lambon Ralph, M. A. (2010a). Amodal semantic representations depend on both anterior temporal lobes: Evidence from repetitive transcranial magnetic stimulation. *Neuropsychologia, 48*(5), 1336–1342.

Pobric, G., Jefferies, E., & Lambon Ralph, M. A. (2010b). Category-specific versus category-general semantic impairment induced by transcranial magnetic stimulation. *Current Biology, 20*(10), 964–968.

Pobric, G. G., Jefferies, E., & Lambon Ralph, M. A. (2007). Anterior temporal lobes mediate semantic representation: Mimicking semantic dementia by using rTMS in normal participants. *Proceedings of the National Academy of Sciences of the United States of America, 104*, 20137–20141.

Pulvermuller, F. (2005). Brain mechanisms linking language and action. *Nature Reviews Neuroscience, 6*(7), 576–582.

Pulvermüller, F., Cooper-Pye, E., Dine, C., Hauk, O., Nestor, P. J., & Patterson, K. (2010). The word processing deficit in semantic dementia: All categories are equal, but some categories are more equal than others. *Journal of Cognitive Neuroscience, 22*(9), 2027–2041. Available from: http://dx.doi.org/10.1162/jocn.2009.21339.

Rogers, T. T., Lambon Ralph, M. A., Garrard, P., Bozeat, S., McClelland, J. L., Hodges, J. R., et al. (2004). The structure and deterioration of semantic memory: A neuropsychological and computational investigation. *Psychological Review, 111*, 205–235.

Rogers, T. T., & McClelland, J. L. (2004). *Semantic cognition: A parallel distributed processing approach*. Cambridge, MA: The MIT Press.

Rosch, E. (1975). Cognitive representations of semantic categories. *Journal of Experimental Psychology: General, 104*, 192–233.

Schapiro, A. C., McClelland, J. L., Welbourne, S. R., Rogers, T. T., & Lambon Ralph, M. A. (2013). Why bilateral damage is worse than unilateral damage to the brain. *Journal of Cognitive Neuroscience, 25*(12), 2107–2123.

Scoville, W. B., & Milner, B. (1957). Loss of recent memory after bilateral hippocampal lesions. *Journal of Neurology, Neurosurgery and Psychiatry, 20*, 11–21.

Sharp, D. J., Scott, S. K., & Wise, R. J. S. (2004). Retrieving meaning after temporal lobe infarction: The role of the basal language area. *Annals of Neurology, 56*(6), 836–846.

Simmons, W. K., Ramjee, V., Beauchamp, M. S., McRae, K., Martin, A., & Barsalou, L. W. (2007). A common neural substrate for perceiving and knowing about colour. *Neuropsychologia, 45*(12), 2802–2810.

Smith, E. E., & Medin, D. L. (1981). *Categories and concepts*. Cambridge, MA: Harvard University Press.

Snowden, J. S., Goulding, P. J., & Neary, D. (1989). Semantic dementia: A form of circumscribed cerebral atrophy. *Behavioural Neurology, 2*(3), 167–182.

Snowden, J. S., Griffiths, H. L., & Neary, D. (1996). Semantic-episodic memory interactions in semantic dementia: Implications for retrograde memory function. *Cognitive Neuropsychology, 13*(8), 1101–1137.

Snowden, J. S., Thompson, J. C., & Neary, D. (2004). Knowledge of famous faces and names in semantic dementia. *Brain, 127*, 860–872.

Terzian, H., & Dalle Ore, G. (1955). Syndrome of Kluver-Bucy reproduced in man by bilateral removal of the temporal lobes. *Neurology, 5*, 373–380.

Thierry, G., & Price, C. J. (2006). Dissociating verbal and nonverbal conceptual processing in the human brain. *Journal of Cognitive Neuroscience, 18*(6), 1018–1028. Available from: http://dx.doi.org/10.1162/jocn.2006.18.6.1018.

Tulving, E. (1972). Episodic and semantic memory. In E. Tulving, & W. Donaldson (Eds.), *Organisation of memory*. London: Academic Press.

Vandenberghe, R., Price, C., Wise, R., Josephs, O., & Frackowiak, R. S. J. (1996). Functional anatomy of a common semantic system for words and pictures. *Nature, 383*(6597), 254–256.

Visser, M., Jefferies, E., Embleton, K. V., & Lambon Ralph, M. A. (2012). Both the middle temporal gyrus and the ventral anterior temporal area are crucial for multimodal semantic processing: Distortion-corrected fMRI evidence for a double gradient of information convergence in the temporal lobes. *Journal of Cognitive Neuroscience, 24*(8), 1766–1778.

Visser, M., Jefferies, E., & Lambon Ralph, M. A. (2010). Semantic processing in the anterior temporal lobes: A meta-analysis of the functional neuroimaging literature. *Journal of Cognitive Neuroscience, 22*, 1083–1094.

Visser, M., & Lambon Ralph, M. A. (2011). Differential contributions of bilateral ventral anterior temporal lobe and left anterior superior temporal gyrus to semantic processes. *Journal of Cognitive Neuroscience, 23*(10), 3121–3131.

Warrington, E. K. (1975). The selective impairment of semantic memory. *Quarterly Journal of Experimental Psychology, 27*(4), 635–657.

Welbourne, S. R., & Lambon Ralph, M. A. (2007). Using PDP models to simulate phonological dyslexia: The key role of plasticity-related recovery. *Journal of Cognitive Neuroscience, 19*, 1125–1139.

Welbourne, S. R., Woollams, A. M., Crisp, J., & Lambon Ralph, M. A. (2011). The role of plasticity-related functional reorganization in the explanation of central dyslexias. *Cognitive Neuropsychology, 28*(2), 65–108.

Williams, G. B., Nestor, P. J., & Hodges, J. R. (2005). The neural correlates of semantic and behavioural deficits in frontotemporal dementia. *NeuroImage, 24*, 1042–1051.

Wittgenstein, L. (2001). *Philosophical investigations: The German text, with a revised English translation 50th anniversary commemorative edition*. Oxford: Wiley-Blackwell.

Woollams, A. M., Cooper-Pye, E., Hodges, J. R., & Patterson, K. (2008). Anomia: A doubly typical signature of semantic dementia. *Neuropsychologia, 46*(10), 2503–2514.

CHAPTER

62

What Does It Mean? A Review of the Neuroscientific Evidence for Embodied Lexical Semantics

Olaf Hauk

MRC Cognition and Brain Sciences Unit, Cambridge, UK

62.1 INTRODUCTION

We often have the impression that we create a "mental image" of concepts when we refer to them in discourse, remember them, or reason about them. This seems to suggest that concepts can evoke similar processes as those engaged during the acquisition of this concept. Philosophers have entertained the idea of mental images, in the sense of re-experienced perceptions or actions, for a long time. David Hume stated that the idea of the color red and the actual perception of red differ only in degree and not in nature (Hume, 1985). Immanuel Kant claimed that we cannot think about a line without drawing it in our mind (Kant, 1995). However, the delineation of unconscious and possibly automatic semantic processing from more deliberate and effortful conceptual processing and mental imagery is still a major issue in cognitive science. In the neuroscience of language, the degree to which representations and processing of semantics relies on brain systems for perception and action is still a matter of great controversy.

Most authors will agree that lexical semantics is about the relationship between symbols and what they represent, as well as their relationship to other symbols and their meanings. Unfortunately, relating specific experimental results to such a general definition is not straightforward. Any single study will need to make a choice of particular symbols and particular things they stand for, modes of presentation, contexts, task instructions, and measurements. This chapter introduces the major theoretical approaches in research on embodied semantics, describes the most popular neuroscientific methods used to test them, and provides a selective overview of the empirical literature. The main focus is on lexical semantics (i.e., the meaning of words and symbols), but references to other conceptual domains (e.g., sentence-level processing) will be made where appropriate.

62.2 MODELS OF EMBODIED SEMANTICS

Several authors have pointed out that a clear definition of "embodiment" or "grounding" is lacking (Chatterjee, 2010; Hauk & Tschentscher, 2013; Mahon & Caramazza, 2009; Meteyard, Cuadrado, Bahrami, & Vigliocco, 2012; Pulvermuller, 2013b; Wilson, 2002). Here, I review the literature in light of the empirically tractable question: which brain systems subserve specific perceptual and motor functions and contribute to semantic representations and processes? Answering this question requires researchers to define the brain systems that subserve specific perceptual and motor functions, as well as criteria for semantic representations and processes. Disentangling representations and processes may be impossible. Representations can be measured only if they affect the outcome of processes in both behavioral and neuroscientific studies (Anderson, 1978).

The "grounding problem," for example, was formulated by Harnad (1990). In a system that only consists of symbols and relationships among symbols, no symbol can have "meaning," because to get to the meaning of one symbol, one would have to follow a

Neurobiology of Language. DOI: http://dx.doi.org/10.1016/B978-0-12-407794-2.00062-6

path of relationships among symbols that do not have meaning either. Therefore, at least some concepts must be grounded in nonsymbolic representations. Harnad suggests a hybrid model of the conceptual system that contains symbolic and nonsymbolic representations. It is not obvious where these nonsymbolic representations should be located. The idea that semantics is widely distributed across the brain was articulated well before the neuroimaging days (Gage & Hickok, 2005). Does a concept have to reactivate the retina, LGN, V1, or higher visual association areas? Does it matter whether an action-word activates primary or premotor cortex or SMA? Unless we can specify a clear boundary of what is just periphery and preprocessing, and what is essential for the generation of meaning, we cannot come to a conclusion here—but this is exactly the question "what is meaning?"

Some of the major theoretical approaches dealing with the grounding problem are illustrated in Figure 62.1 and share some important features. They postulate that semantics emerges from an interplay of distributed sensory-motor areas that represent visual, auditory, and tactile information, and amodal or polymodal "core language areas," "hubs," or "higher-order association cortices" that bind these distributed areas together. These models have been developed for different purposes and on the basis of different types of empirical data, and it is unclear whether neuroscience can currently provide data that can clearly distinguish between these theories. In the following, the main features of these models are briefly highlighted.

Arguably the first empirical support for the view that semantic categories are differentially represented in the brain came from neuropsychological patients who were differentially impaired regarding concepts relating to living and nonliving things (Warrington & Shallice, 1984). This led to the "sensory-functional hypothesis" that semantic knowledge is structurally organized according to its reliance on perceptual (such as for animals) or functional (such as for tools) features (Martin, 2007).

The involvement of perceptual-motor systems is an essential part of Barsalou's theory of "Perceptual Symbol Systems" (PSS; Figure 62.1A) (Barsalou, 1999). Association areas of the brain are formed during perceptual experience in a bottom-up manner and are sensitive to patterns in sensory-motor systems. These association areas have the capacity to reactivate sensory-motor areas to form modal and analogical perceptual symbols. PSS theory later led to the "Language and Situated Simulation" theory (Barsalou, 2009; Simmons, Hamann, Harenski, Hu, & Barsalou, 2008), which assumes that conceptual processing can be broken down into two stages: an early "linguistic" stage, which comprises basic and possibly unconscious word recognition processes as well as retrieval of word associations, and a later "situated simulation" stage, which represents "deep conceptual information" that can be conscious and deliberate, which is the basis for "symbolic processes such as predication, conceptual combination, and recursion" (Barsalou, 2008).

The idea of hierarchically organized convergence zones is also a key part of Damasio's framework of "convergence and divergence zones" (CDZs; Figure 62.1B) (Damasio, 1989; Meyer & Damasio, 2009). Convergence zones develop on the basis that "temporally coincident activity at the separate sites modifies the connectivity patterns to, from, and within a shared CDZ downstream" and, therefore, "CDZs register linkages among knowledge fragments" (Meyer & Damasio, 2009). The authors point out that CDZs are not the representations themselves, "but, rather, establish meaning via time-locked multiregional retroactivation of early cortices."

The formation of connections between sensory-motor and perisylvian core language areas on the basis of associative learning is postulated in Pulvermüller's theory of semantics (Pulvermuller, 1999, 2013a; Figure 62.1C). During language acquisition, the co-occurrence of word forms such as "kick" or "grass" with the execution of kicking movements or the perception of grass, respectively, forms automatic associations between core language areas in perisylvian brain regions and the corresponding sensory-motor regions. This theory has led to the specific "somatotopy of action-words" (SAW) model, which makes the specific prediction that different types of action-words ("pick, kick, lick") should activate motor cortex in a somatotopic manner in accordance with the associated effectors (Pulvermüller, 2005).

Connectionist models of semantics attempt to model semantics as emerging from distributed neuronal networks, where semantic and word form features are represented as distributed activation patterns in different layers of a neuronal network. In a neuropsychologically inspired implementation, the connection between written and spoken word forms and distributed semantic features in sensory-motor areas are accomplished by a "semantic hub" in the anterior temporal lobes (Figure 62.1D; Patterson, Nestor, & Rogers, 2007; Rogers et al., 2004).

A number of authors have focused on amodal or polymodal brain areas involved in semantics but allow for the possibility that aspects of semantics may be distributed across larger brain systems (Binder & Desai, 2011; Poeppel & Hickok, 2004; Price, 2000). Some authors have concluded that

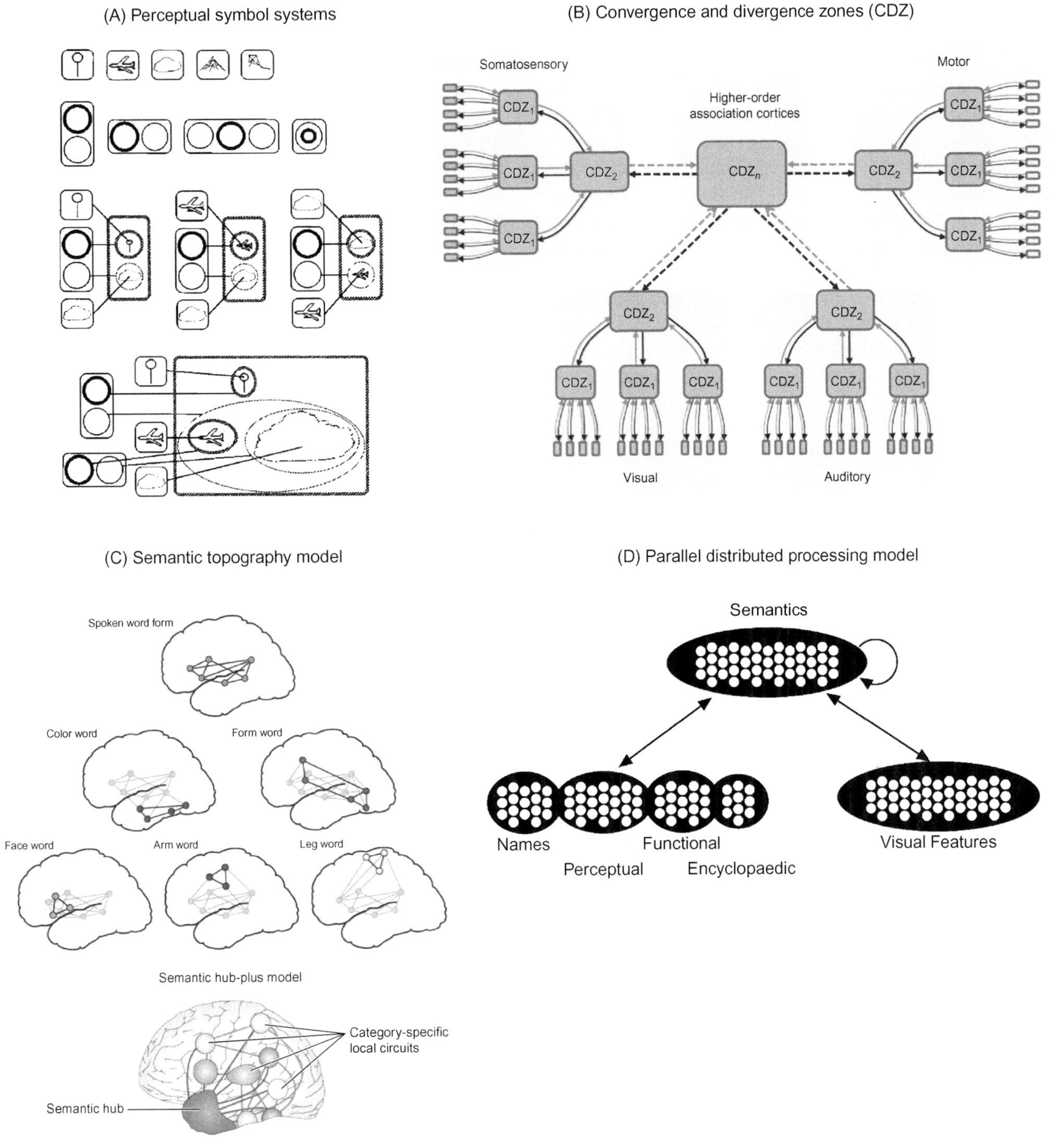

FIGURE 62.1 Illustration of models for conceptual-semantic representations that assume a link to sensory-motor systems. (A) Perceptual symbol systems and how they describe (i) object categorization, (ii) spatial relationships, (iii) combinatorial, and (iv) recursive processing. (B) Convergence and divergence zones. (C) Semantic topography model. (D) Parallel distributed processing model. *(A) From Meyer and Damasio (2009); (B) from Barsalou (1999); (C) from Pulvermuller et al. (2010); (D) modified from Rogers et al. (2004) with permission from the publisher.*

"conceptual knowledge is represented within a non-differentiated distributed system" (Tyler et al., 2003). Others have argued against the idea of strong embodied semantics and argue that "concepts are, at some level, 'abstract' and 'symbolic,' with the idea that sensory and motor information may 'instantiate' online conceptual processing" (Mahon & Caramazza, 2008).

62.3 METHODS FOR NEUROSCIENTIFIC RESEARCH ON EMBODIED SEMANTICS

A large number of behavioral studies have shown effects of congruency between semantic features of a stimulus (word, sentence, picture, etc.) and the type of response execution (finger response, foot response, pull versus push a lever, etc.). Results of this kind are interesting and useful because they demonstrate how task irrelevant semantic stimulus features can affect task performance. However, they are not direct evidence for the neuronal overlap of the representations and processes for action concepts and movements. Such interference could still occur at a higher level, where information from these two systems converges, or it could be caused by spreading activation (i.e., a passive "leakage" of activation from one brain system to another) without a functional contribution.

It is generally acknowledged that neuroimaging tools such as functional magnetic resonance imaging (fMRI), positron emission tomography (PET), electroencephalography (EEG), and magnetoencephalography (MEG) are correlational (Henson, 2005), that is, an effect obtained with either of these measures can be either cause or consequence of the perceptual or cognitive processes assumed to elicit it. Arguably the most extensively used method in the neuroscience of embodied semantics is fMRI, whose main appeal lies in its high anatomical precision in the millimeter range (Brett, Johnsrude, & Owen, 2002). Unfortunately, the BOLD response—which is the basis for most standard fMRI results—only provides snapshots of brain activity with an exposure duration of several seconds. Interpreting these data therefore requires a model for the relationship between stimulus and brain activity for at least 2 s or more.

One remedy for this problem is using the millisecond temporal resolution of EEG and MEG, which can provide upper limits for the earliest latency at which a stimulus feature affects brain responses. Psychophysiological measures such as EEG and MEG can be analyzed as event-related potentials or fields (ERPs and ERFs), often characterized in terms of "components" (P100, N400, etc.), or as time-frequency representations of the brain signal, which provide information about brain responses in specific frequency bands (such as alpha, beta, gamma, theta). However, a clear one-to-one relationship between components or frequency bands and brain processes cannot be assumed. The spatial resolution of EEG and MEG is limited (Hauk, Wakeman, & Henson, 2011), but it is reasonable to assume that high-density EEG and MEG recordings can distinguish different sensory-motor modalities and resolve the somatotopy of motor cortex. The most direct—although invasive—way to obtain signals from specific brain areas is to measure the electrocorticogram (ECoG) from the cortical surface (Edwards et al., 2010) or to record from intracranial electrodes.

A method that has the potential to reveal causal involvement of brain areas in cognitive processes is transcranial magnetic stimulation (Devlin & Watkins, 2006). Brief magnetic pulses, either applied during stimulus processing ("single-pulse") or as a sequence of pulses before testing for repetitive TMS ("rTMS"), are applied to specific brain sites. The effect of TMS is often described as "temporary lesions" but, depending on stimulus parameter and experimental paradigm, facilitatory effects can also be obtained. TMS can also be used to elicit motor-evoked potentials (MEPs), such as evoked electrical voltage deflections measured along a muscle (e.g., of the thumb). However, this method is also only correlational because it can detect an effect of stimulus or task features on motor cortex activity but not a causal role of motor cortex activity on behavioral performance.

Lesion studies can also provide evidence for causal relationships between brain areas and cognitive processes. It is not straightforward to determine how secondary effects of brain lesions, such as inability to move specific effectors and neuronal plasticity, may affect performance in experimental paradigms. Nevertheless, a clear double dissociation in patient groups, for example, demonstrating impairment of action-knowledge after impairment of motor cortex in one group and impaired visual knowledge in patients with impaired visual cortices, would be strong evidence for involvement of sensory-motor brain areas in semantics.

As with all empirical studies, there are general methodological issues with respect to replicability and publication bias (Ioannidis, 2005; Vul, Harris, Winkielman, & Pashler, 2009). For example, for a given result, it is worth asking how many unpublished null effects there may be or whether the results were obtained using a prespecified analysis pathway (Wagenmakers, Wetzels, Borsboom, van der Maas, & Kievit, 2012).

62.4 REVIEW OF THE EMPIRICAL LITERATURE

Some previous reviews of the empirical literature on embodied language have argued in favor of an embodied view on language and semantics (Fischer & Zwaan, 2008; Kiefer & Pulvermuller, 2011; Pulvermuller, 2013a). Others have focused on the difficulties in defining the concept of embodiment (Chatterjee, 2010; Hauk & Tschentscher, 2013;

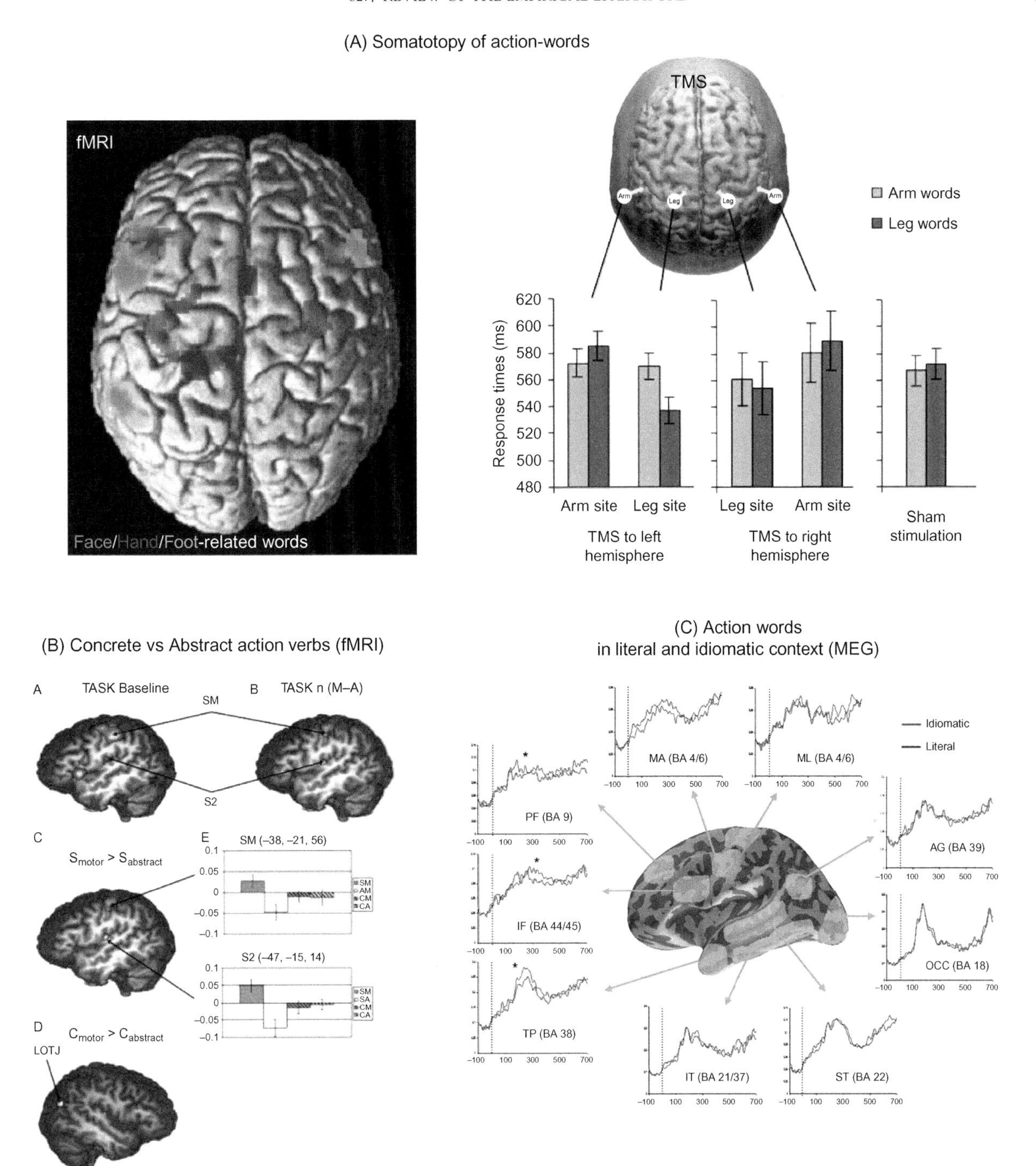

FIGURE 62.2 Examples of neuroscientific studies investigating the involvement of sensory-motor systems in action-word processing. (A) Left: fMRI results for somatotopic activation of cortical motor areas in response to face-, hand-, and foot-related words, respectively. Right: TMS evidence that stimulation of left-hemispheric hand-motor and foot-motor cortex facilitates hand-word and foot-word processing, respectively. (B) Results from an fMRI study that reported motor cortex activation to concrete action verbs, but not abstract verbs based on action concepts. SC/MA: simple/complex motor/abstract verbs. (C) Time course of brain activation for action-words in idiomatic and literal contexts as revealed by MEG source estimation. *(A) Left: modified from Hauk et al. (2004); right: from Pulvermuller, Hauk, Nikulin, and Ilmoniemi (2005); (B) from Ruschemeyer et al. (2007); (C) from Boulenger et al. (2012) with permission from the publisher.*

Mahon & Caramazza, 2009; Meteyard et al., 2012), and some have adopted a critical stance and argue that embodiment plays a minor role in language processing (Mahon & Caramazza, 2008). This chapter presents some typical studies that illustrate the experimental approaches (as shown in Figure 62.2), methodological challenges, and possible interpretations of the evidence as well as recent developments.

62.4.1 Concrete Lexical Semantics

A classic finding in support of embodied language processing is the Action-Sentence-Compatibility-Effect (ACE). Participants performed hand movements faster when the direction of movement was congruent with the direction of movement described in a preceding sentence compared with when it was incongruent (Glenberg & Kaschak, 2002; Zwaan, van der Stoep, Guadalupe, & Bouwmeester, 2012). Similar interference effects have been observed with visual and auditory motion paradigms, as well as in studies on movement kinematics (Boulenger et al., 2006; Kaschak, Zwaan, Aveyard, & Yaxley, 2006; Meteyard, Zokaei, Bahrami, & Vigliocco, 2008; Mirabella, Iaconelli, Spadacenta, Federico, & Gallese, 2012).

A number of fMRI and PET studies have shown that sensory-motor areas become active during language comprehension, mostly in the action domain (e.g., Hauk, Johnsrude, & Pulvermuller, 2004; for overviews see Kemmerer, Castillo, Talavage, Patterson, & Wiley, 2008; Postle, McMahon, Ashton, Meredith, & de Zubicaray, 2008; Pulvermüller, 2005), but also for auditory (Kiefer, Sim, Herrnberger, Grothe, & Hoenig, 2008) and visual concepts (Hauk, Davis, Kherif, & Pulvermuller, 2008; Pulvermuller & Hauk, 2006; Simmons et al., 2007), a mixture of those (Kiefer et al., 2012; Noppeney & Price, 2003), and other domains (Barros-Loscertales et al., 2012; van der Laan, de Ridder, Viergever, & Smeets, 2011). Although a number of studies have reported sensory-motor activation in language processing, null effects have been reported as well (Edwards et al., 2010; Postle et al., 2008), and some studies have illustrated significant variability of action-word activation across studies (Kemmerer et al., 2008) and across participants (Willems, Hagoort, & Casasanto, 2010).

Only a few fMRI studies have directly addressed the ambiguity of fMRI data with respect to semantic or postsemantic origins of the observed effects. Tomasino, Werner, Weiss, and Fink (2007) suggested that motor cortex activation for action-words is mainly caused by mental imagery processes. Willems, Hagoort, et al. (2010) reported that action-words in a lexical decision task produced activation patterns in motor areas that were nonoverlapping with activation patterns in a mental imagery task, suggesting that motor areas may play different roles in imagery and semantics. Wheatley, Weisberg, Beauchamp, and Martin (2005) concluded from priming effects in inferior temporal cortex that this area is involved in automatic object processing. Hauk et al. (2008) studied the effect of semantic category on the word frequency effect and provided evidence that differential effects for semantic word categories in posterior middle temporal cortex occur at a semantic rather than imagery stage.

The exact time course of semantic word processing is still a matter of debate, but it is plausible to assume that effects within the first 400 ms after stimulus presentation do not reflect deliberate mental imagery processes (Hauk et al., 2008). With respect to the embodiment of action-word semantics, Pulvermüller, Härle, and Hummel (2001) and Hauk and Pulvermuller (2004) reported differences between action-word types in the ERP at approximately 200 ms that were consistent with somatotopy. Similarly, differences between words with and without acoustic features occurred in ERPs at approximately 200 ms. In a parallel fMRI study, activation to words with acoustic features overlapped with areas activated during listening to sounds (Kiefer et al., 2008). Around the same latency, Moscoso del Prado Martin, Hauk, and Pulvermuller (2006) observed differences between color-related and form-related words. For auditory stimuli, the brain responses measured by MEG at approximately 200 ms differed depending on the action-word category, even when participants were distracted by watching a silent movie, suggesting early automatic activation of semantic cell assemblies (Pulvermüller, Shtyrov, & Ilmoniemi, 2005). With respect to brain oscillations, van Elk, van Schie, Zwaan, and Bekkering (2010) compared action verbs presented in sentence context related to human or animal actions. They observed stronger desynchronization in beta as well as mu frequency bands, with mu effects starting at approximately 160 ms, and starting in the beta band at approximately 500 ms. The estimated neuronal sources were consistent with motor cortex activation. One study investigated time-frequency ECoG responses to auditorily presented hand-verbs and mouth-verbs in four patients but did not find evidence for a somatotopic activation pattern (Canolty et al., 2007). TMS studies using MEPs have also provided evidence that motor areas are modulated by action-semantics within a few hundred milliseconds of word presentation (Buccino et al., 2005; Labruna, Fernandez-del-Olmo, Landau, Duque, & Ivry, 2011).

The main advantage of TMS, compared with correlational neuroimaging, is that it allows interfering with activation in some brain areas and measurement of the effect of this interference on behavioral performance. Pulvermüller et al. (2005) investigated the effects of TMS pulses delivered at 150 ms after hand-word and leg-word presentation to hand and leg motor cortex on performance in a lexical decision task, and they found an interaction of stimulation site and word type (i.e., responses to arm-related words were faster when hand motor was stimulated) and vice versa for leg-words after leg motor cortex stimulation. Similarly, Willems, Labruna, D'Esposito, Ivry, and Casasanto (2011) found facilitated action-word processing after premotor cortex rTMS. Tomasino, Fink, Sparing, Dafotakis, and Weiss (2008) studied effects of TMS on hand-action-verb processing at different stimulation latencies (150 and 750 ms after word presentation) and in different tasks. The main result was a facilitatory effect of hand motor cortex stimulation at all stimulation latencies, but only in the imagery task and not in lexical or silent reading tasks, suggesting a postsemantic origin of this effect. Interestingly, no disruption of action-word processing after (r)TMS to motor areas has been reported yet.

An interesting development in behavioral research is the use of motor or movement priming paradigms, where a particular movement type is used as a prime (presumably leading to preactivation of the corresponding brain system), and the effect on behavior is assessed. Glenberg, Sato, and Cattaneo (2008) asked their participants to move beans along a particular direction; they found an interference effect for this direction and the direction implied by concrete and abstract sentences. Hand and foot movements have also been reported to differentially affect working memory performance for action-word types (Shebani & Pulvermuller, 2013). Effects of motor priming on early brain responses have been reported in an EEG/MEG study (Mollo, Pulvermuller, & Hauk, 2011).

Evidence for impaired semantics after damage to sensory-motor brain systems from clinical populations has been provided from Parkinson's disease (Boulenger et al., 2008; Herrera, Rodriguez-Ferreiro, & Cuetos, 2012), stroke patients (Neininger & Pulvermüller, 2003; Trumpp, Kliese, Hoenig, Haarmeier, & Kiefer, 2012), and semantic dementia (Pulvermuller et al., 2010). Kemmerer, Rudrauf, Manzel, and Tranel (2010) reported that lesions in several brain areas including precentral gyrus, possibly extending into hand-related motor areas, as well as ventral postcentral gyrus, predicted performance in tasks such as word-picture matching or word comprehension for action verbs. However, using a similar approach but with smaller sample size, Arevalo, Baldo, and Dronkers (2012) did not find evidence for somatotopic effects of different action-word categories. Some authors have argued that neuropsychological studies so far have, at best, demonstrated subtle effects of deficits in patients with sensory-motor impairments on semantic processing (Binder & Desai, 2011; Mahon & Caramazza, 2009).

62.4.2 Abstract Lexical Semantics

A particularly challenging case for theories of embodiment is abstract concepts, because they have no obvious referents in sensory-motor experience. Possible approaches to the incorporation of abstract semantics in frameworks of embodiments are summarized by Glenberg et al. (2008) and Pecher, Boot, and Van Dantzig (2011). For example, it has been suggested that abstract semantics relies on concrete concepts by means of metaphor or image schemas (Gibbs & Steen, 1999; Lakoff, 1987), that some abstract concepts can be based on generalizations from situated simulations (Barsalou, 1999; Wilson-Mendenhall, Barrett, Simmons, & Barsalou, 2011), and that abstract concepts can be acquired as abstractions of concrete concepts (Glenberg & Robertson, 1999). It has also been pointed out that affective semantic features are important for abstract concepts (Kousta, Vigliocco, Vinson, Andrews, & Del Campo, 2011), and that abstract semantics cannot be explained within a single system (Shallice & Cooper, 2013).

Behavioral evidence for embodied semantics has mostly been provided by studies on sentence processing. For example, the action-sentence compatibility effect (ACE) has also been observed for abstract sentences, where direction of movement was implied by abstract sentence contents (e.g., "reading to" versus "being read to") (Glenberg et al., 2008). Similar effects have been observed with metaphors (Santana & de Vega, 2011).

Several fMRI studies have investigated the embodiment of abstract sentences. For example, Boulenger, Hauk, and Pulvermuller (2009) reported somatotopic activation for idiomatic sentences. Two studies failed to find such effects for abstract sentences (Aziz-Zadeh, Wilson, Rizzolatti, & Iacoboni, 2006; Raposo, Moss, Stamatakis, & Tyler, 2009), whereas others found effects of metaphor familiarity and sentence type in motor, motion, or somatosensory regions (Desai, Binder, Conant, Mano, & Seidenberg, 2011; Desai, Conant, Binder, Park, & Seidenberg, 2013; Lacey, Stilla, & Sathian, 2012). Citron and Goldberg (2014) found more activation for metaphorical sentences related to taste compared with their literal counterparts in emotion-related areas amygdala and anterior hippocampus, as well as in gustatory areas, suggesting a role of emotion in metaphor comprehension.

Only few fMRI studies have investigated the embodiment of abstract words in isolation. Ruschemeyer, Brass, and Friederici (2007) did not find motor cortex activation for abstract words based on action concepts. Moseley, Carota, Hauk, Mohr, and Pulvermuller (2012) found stronger motor cortex activation to emotion words compared with nonaction-related words, which was interpreted as embodied abstract semantics. Wilson-Mendenhall, Simmons, Martin, and Barsalou (2013) found more activation in frontal and temporal cortex for social concepts, and more activation in intraparietal sulcus for arithmetic concepts, indicating that even abstract concepts are represented in concept-specific distributed networks.

In a MEG version of their previous fMRI experiment, Boulenger, Shtyrov, and Pulvermuller (2012) presented literal and idiomatic sentences and analyzed the time course of brain responses after the critical word (e.g., "habit" in "she kicked the habit"). Literal and idiomatic sentences differed in their brain responses after approximately 200 ms, with some evidence for somatotopic activation for arm-related and leg-related idioms in this latency range.

Evidence on lexical embodied semantics from TMS studies is also scarce. Glenberg et al. (2008) measured TMS-induced MEPs in their study on the abstract ACE described and found that MEP amplitudes were greater in transfer sentences than in no-transfer sentences, and that there was little difference between concrete and abstract sentences. Scorolli et al. (2012) found MEP effects for both concrete and abstract words stimulating at 250 ms after word onset. Fernandino et al. (2013) reported differential processing of literal, idiomatic, metaphoric, and abstract action-related sentences in Parkinson's patients.

Considering the scarcity of evidence and the variability in stimuli, tasks, contexts, and measurement techniques, it is currently impossible to come to firm conclusions about abstract lexical semantics. Most studies reviewed used sentence stimuli, raising the question as to what degree does lexical semantics interact with context.

62.5 THE INFLUENCE OF TASK, CONTEXT, AND INDIVIDUAL EXPERIENCE

The proverb "If you have a hammer, everything looks like a nail" suggests that our environment and goals shape the way we assign meaning to objects around us. The way we use the meaning of words flexibly in different contexts and situations is a formidable challenge for theories of semantics. It is still unclear which levels of processing are affected by top-down processing, for example, depending on task and context. Some authors have pointed out that the field of word recognition suffers from the "curse of automaticity," and that lexical processing should be considered as flexible (Balota & Yap, 2006). Even masked semantic priming effects on N400 amplitudes have been shown to depend on task context (Kiefer & Martens, 2010). However, it is still unclear whether top-down control affects information retrieval during word recognition, or only later selection and decision processes (Chen, Davis, Pulvermuller, & Hauk, 2013).

Surprisingly few neuroscientific studies have systematically investigated the effects of task modulation on embodied semantic word processing (some examples are shown in Figure 62.3). A recent study looked at the specific effect of different semantic tasks. In an fMRI study, van Dam, van Dijk, Bekkering, and Rueschemeyer (2012) investigated brain activation to words that had to be judged either for color or for action attributes. Areas in the left parietal lobes activated more for action-words than for abstract words, but only during action-related judgments. This was interpreted as evidence for flexible and context-dependent semantic processing.

The degree to which semantic representations are embodied may depend on the experience of the individual (e.g., the context and frequency with which a concept is encountered and processed). An interesting test case involves left- and right-handed individuals: does the way we usually perform an action shape the way we represent it semantically? Two studies on this issue have led to different results. Willems, Toni, Hagoort, and Casasanto (2010) reported more activation in left motor cortex when right-handed individuals read unimanual action-related words (e.g., "throw"), whereas left-handed individuals showed more activation in right motor areas. Hauk and Pulvermuller (2011) found that unimanual words ("throw") activated left motor cortex in both left-handed and right-handed individuals, whereas bimanual words ("clap") activated motor cortex bilaterally in both groups. These seemingly contradictory findings may be explained by the use of different tasks. Willems, Toni, et al. (2010) used a lexical decision task that engages motor areas and may shift participants' attention to action-related aspects of the stimuli. Hauk and Pulvermuller (2011) used a silent reading task that did not require an explicit response. This highlights the importance of investigating effects of task demands on brain activation during semantic processing.

With respect to abstract concepts, it has been reported that brain responses to the same emotions differed depending on the situation in which they were situated (Wilson-Mendenhall et al., 2011). Recent studies have reported a "disembodiment effect of negation," that is, less motor cortex activation to action-words in negated ("He doesn't throw the ball") compared with affirmative context, for example, measuring grip force (Aravena et al., 2012), movement kinematics (Bartoli et al., 2013), and fMRI (Tettamanti et al., 2008). This suggests that the involvement of sensory-motor systems may not be necessary for semantic processing in general, but rather are dependent on context and task demands. Future research should address the question whether task effects occur at early and possibly unconscious stages of semantic processing, or at later stages that require combinatorial processes or deliberate mental imagery.

(A) Action-word processing in left- and right-handers

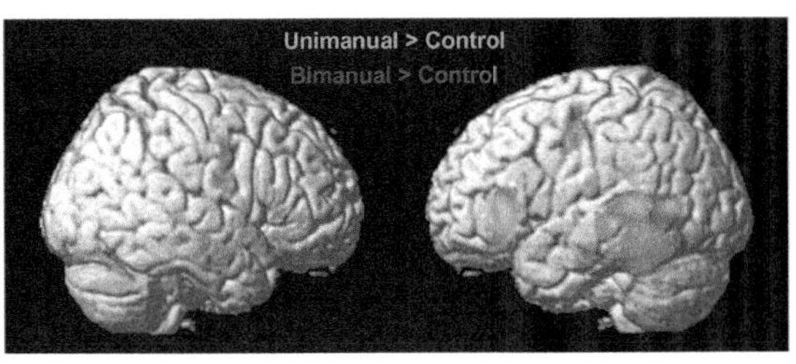

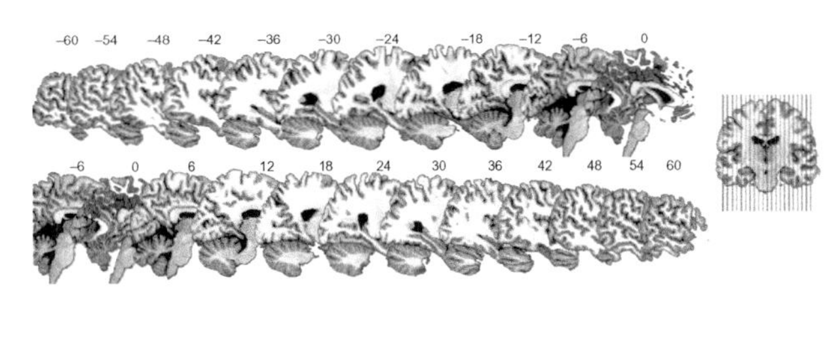

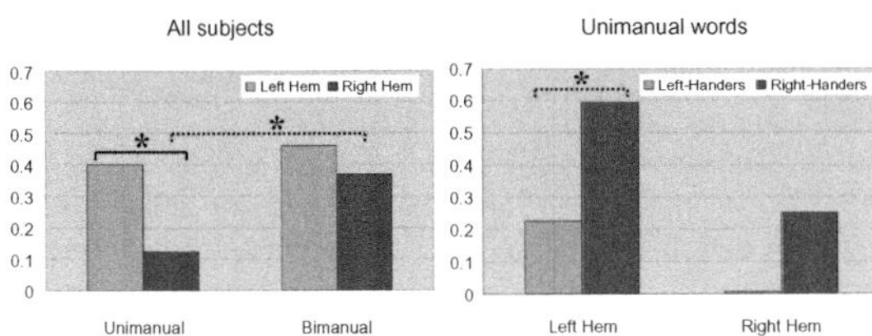

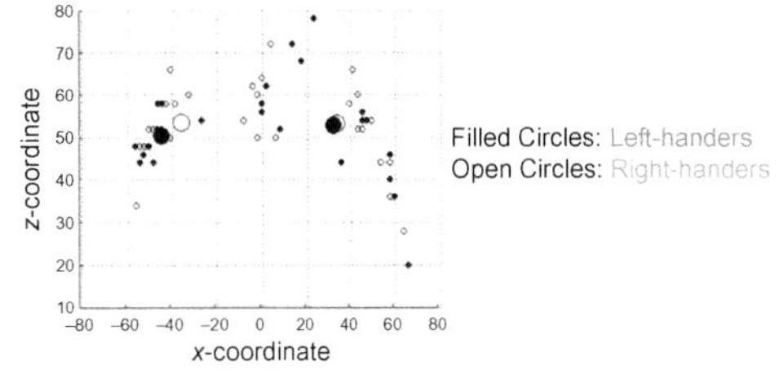

(B) Modulation of action-word processing by context

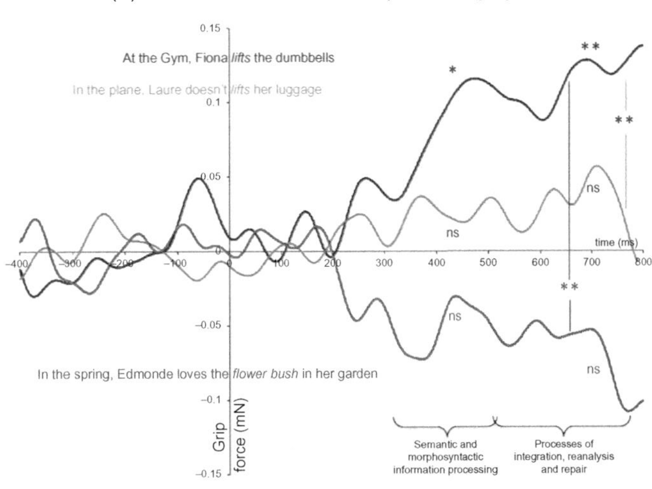

FIGURE 62.3 Examples of neuroscientific studies that investigated the influence of experience and task demands on action-word processing. (A) Left: fMRI results that indicate different lateralization of brain activation in motor areas for unimanual and bimanual action-words (e.g., "throw" and "clap"), but independently of the handedness of the participants. Right: fMRI results suggesting different lateralization of brain activation between left-handed and right-handed individuals in response to hand-related action-words. (B) Grip force amplitude is larger after action-words in affirmative (dark blue) compared with negative (light blue) context as well as with a nonaction control condition. *(A) Left: modified from Hauk and Pulvermuller (2011); right: modified from Willems, Hagoort, et al. (2010); (B) modified from Aravena et al. (2012) with permission from the publisher.*

62.6 CONCLUSION

Although most authors—especially those behind the models illustrated in Figure 62.1—acknowledge that sensory-motor systems may contribute to semantic processing at some stage, the main controversy surrounds the issue about how relevant or essential these contributions really are, and whether the existing evidence tells us anything interesting about how we represent and process meaning (Chatterjee, 2010; Hauk & Tschentscher, 2013; Mahon & Caramazza, 2008; Meteyard et al., 2012; Pulvermuller, 2013b).

The reasons for the grounding of concepts in nonsymbolic representations, such as those put forward by Harnad (1990), do not constitute a compelling theoretical argument for the embodiment of semantics in sensory-motor brain systems: As long as we do not know how the patterns of trees on our retinas evoke the concept of a "forest" in us, we also do not understand how reactivation of early visual brain areas can do this. Independently of the theoretical standpoint, one can address empirical questions regarding which brain areas contribute to the processing of semantic feature x in task y in participant group z. Even with respect to this dry formulation of the problem, the evidence is not yet conclusive. Most evidence stems from correlational studies predominantly reporting behavioral interference effects (which do not allow direct inferences about interaction at the neuronal level) and fMRI results (which are ambiguous with respect to semantic and postsemantic origins of effects). Time course information from EEG/MEG and TMS-MEP studies exists, but it is scarce and still susceptible to spreading-activation-criticism. The most direct evidence, that is, clearly replicable impaired performance in a semantic task after disruption of sensory-motor brain activity (using TMS or lesion data), has not yet been established. Studies investigating perceptual-motor priming on behavioral and brain responses in semantic tasks, possibly in combination with measures of brain connectivity, may be able to reveal links between specific perceptual-motor functions and brain systems in the future.

An inconclusive empirical basis is by no means specific to research on embodied semantics. Even the roles of the classical language areas, such as Broca's or Wernicke's, or the locations of semantic hubs, are still a matter of debate (Binder & Desai, 2011; Patterson et al., 2007; Pulvermuller, 2013a), and semantic effects in sensory-motor systems are unlikely to be more reliable. Clarifying this situation would be a prerequisite for the investigation of connectivity among the hubs and a distributed system, which is the logical next step in research of embodied semantics. A solid foundation in a "simple" field such as lexical semantics would also be beneficial for researchers attempting to address questions about polysemy or homonymy, combinatorial semantics, pragmatics, and others. In order to jump high, one should stand on firm ground. The empirical evidence base for embodied theories of semantics still needs solidification, and it is difficult to find neuroscientific benchmark data that all theorists agree have to be explained. Future research should attempt to make large-scale standardized data sets available and make use of the possibility to preregister acquisition and analysis parameters.

References

Anderson, J. R. (1978). Arguments concerning representations for mental imagery. *Psychological Review, 85*, 249–277.

Aravena, P., Delevoye-Turrell, Y., Deprez, V., Cheylus, A., Paulignan, Y., Frak, V., et al. (2012). Grip force reveals the context sensitivity of language-induced motor activity during "action words" processing: Evidence from sentential negation. *PLoS One, 7*, e50287.

Arevalo, A. L., Baldo, J. V., & Dronkers, N. F. (2012). What do brain lesions tell us about theories of embodied semantics and the human mirror neuron system? *Cortex, 48*, 242–254.

Aziz-Zadeh, L., Wilson, S. M., Rizzolatti, G., & Iacoboni, M. (2006). Congruent embodied representations for visually presented actions and linguistic phrases describing actions. *Current Biology, 16*, 1818–1823.

Balota, D. A., & Yap, M. J. (2006). Attentional control and flexible lexical processing: Explorations of the magic moment of word recognition. In S. Andrews (Ed.), *From inkmarks to ideas: Current issues in lexical processing* (pp. 229–258). New York, NY: Psychology Press.

Barros-Loscertales, A., Gonzalez, J., Pulvermuller, F., Ventura-Campos, N., Bustamante, J. C., Costumero, V., et al. (2012). Reading salt activates gustatory brain regions: fMRI evidence for semantic grounding in a novel sensory modality. *Cerebral Cortex, 22*, 2554–2563.

Barsalou, L. W. (1999). Perceptual symbol systems. *The Behavioral and Brain Sciences, 22*, 577–609 [discussion 610–560].

Barsalou, L. W. (2008). Grounded cognition. *Annual Review of Psychology, 59*, 617–645.

Barsalou, L. W. (2009). Simulation, situated conceptualization, and prediction. *Philosophical Transactions of the Royal Society of London Series B, Biological Sciences, 364*, 1281–1289.

Bartoli, E., Tettamanti, A., Farronato, P., Caporizzo, A., Moro, A., Gatti, R., et al. (2013). The disembodiment effect of negation: Negating action-related sentences attenuates their interference on congruent upper limb movements. *Journal of Neurophysiology, 109*, 1782–1792.

Binder, J. R., & Desai, R. H. (2011). The neurobiology of semantic memory. *Trends in Cognitive Sciences, 15*, 527–536.

Boulenger, V., Hauk, O., & Pulvermuller, F. (2009). Grasping ideas with the motor system: Semantic somatotopy in idiom comprehension. *Cerebral Cortex, 19*, 1905–1914.

Boulenger, V., Mechtouff, L., Thobois, S., Broussolle, E., Jeannerod, M., & Nazir, T. A. (2008). Word processing in Parkinson's disease is impaired for action verbs but not for concrete nouns. *Neuropsychologia, 46*, 743–756.

Boulenger, V., Roy, A. C., Paulignan, Y., Deprez, V., Jeannerod, M., & Nazir, T. A. (2006). Cross-talk between language processes and overt motor behavior in the first 200 msec of processing. *Journal of Cognitive Neuroscience, 18*, 1607–1615.

Boulenger, V., Shtyrov, Y., & Pulvermuller, F. (2012). When do you grasp the idea? MEG evidence for instantaneous idiom understanding. *NeuroImage, 59*, 3502–3513.

Brett, M., Johnsrude, I. S., & Owen, A. M. (2002). The problem of functional localization in the human brain. *Nature Reviews Neuroscience, 3*, 243–249.

Buccino, G., Riggio, L., Melli, G., Binkofski, F., Gallese, V., & Rizzolatti, G. (2005). Listening to action-related sentences modulates the activity of the motor system: A combined TMS and behavioral study. *Brain Research Cognitive Brain Research, 24*, 355–363.

Canolty, R. T., Soltani, M., Dalal, S. S., Edwards, E., Dronkers, N. F., Nagarajan, S. S., et al. (2007). Spatiotemporal dynamics of word processing in the human brain. *Frontiers in Neuroscience, 1*, 185–196.

Chatterjee, A. (2010). Disembodying cognition. *Language and Cognitive Processes, 2*, 79–116.

Chen, Y., Davis, M. H., Pulvermuller, F., & Hauk, O. (2013). Task modulation of brain responses in visual word recognition as studied using EEG/MEG and fMRI. *Frontiers in Human Neuroscience, 7*, 376.

Citron, F. M., & Goldberg, A. E. (2014). Metaphorical sentences are more emotionally engaging than their literal counterparts. *Journal of Cognitive Neuroscience, 26*, 2585–2595.

Damasio, A. R. (1989). Time-locked multiregional retroactivation: A systems-level proposal for the neural substrates of recall and recognition. *Cognition, 33*, 25–62.

Desai, R. H., Binder, J. R., Conant, L. L., Mano, Q. R., & Seidenberg, M. S. (2011). The neural career of sensory-motor metaphors. *Journal of Cognitive Neuroscience, 23*, 2376–2386.

Desai, R. H., Conant, L. L., Binder, J. R., Park, H., & Seidenberg, M. S. (2013). A piece of the action: Modulation of sensory-motor regions by action idioms and metaphors. *NeuroImage, 83C*, 862–869.

Devlin, J. T., & Watkins, K. E. (2006). Stimulating language: Insights from TMS. *Brain, 130*, 610–622.

Edwards, E., Nagarajan, S. S., Dalal, S. S., Canolty, R. T., Kirsch, H. E., Barbaro, N. M., et al. (2010). Spatiotemporal imaging of cortical activation during verb generation and picture naming. *NeuroImage, 50*, 291–301.

Fernandino, L., Conant, L. L., Binder, J. R., Blindauer, K., Hiner, B., Spangler, K., et al. (2013). Where is the action? Action sentence processing in Parkinson's disease. *Neuropsychologia, 51*, 1510–1517.

Fischer, M. H., & Zwaan, R. A. (2008). Embodied language: A review of the role of the motor system in language comprehension. *The Quarterly Journal of Experimental Psychology (Hove), 61*, 825–850.

Gage, N., & Hickok, G. (2005). Multiregional cell assemblies, temporal binding and the representation of conceptual knowledge in cortex: A modern theory by a "classical" neurologist, Carl Wernicke. *Cortex, 41*, 823–832.

Gibbs, R., & Steen, G. (1999). *Metaphor in cognitive linguistics*. Amsterdam: John Benjamins.

Glenberg, A. M., & Kaschak, M. P. (2002). Grounding language in action. *Psychonomic Bulletin and Review, 9*, 558–565.

Glenberg, A. M., & Robertson, D. A. (1999). Indexical understanding of instructions. *Discourse Processes, 28*, 1–26.

Glenberg, A. M., Sato, M., & Cattaneo, L. (2008). Use-induced motor plasticity affects the processing of abstract and concrete language. *Current Biology, 18*, R290–R291.

Glenberg, A. M., Sato, M., Cattaneo, L., Riggio, L., Palumbo, D., & Buccino, G. (2008). Processing abstract language modulates motor system activity. *The Quarterly Journal of Experimental Psychology (Hove), 61*, 905–919.

Harnad, S. (1990). The symbol grounding problem. *Physica D Nonlinear Phenomena, 42*, 335–346.

Hauk, O., Davis, M. H., Kherif, F., & Pulvermuller, F. (2008). Imagery or meaning? Evidence for a semantic origin of category-specific brain activity in metabolic imaging. *The European Journal of Neuroscience, 27*, 1856–1866.

Hauk, O., Johnsrude, I., & Pulvermuller, F. (2004). Somatotopic representation of action words in human motor and premotor cortex. *Neuron, 41*, 301–307.

Hauk, O., & Pulvermuller, F. (2004). Neurophysiological distinction of action words in the fronto-central cortex. *Human Brain Mapping, 21*, 191–201.

Hauk, O., & Pulvermuller, F. (2011). The lateralization of motor cortex activation to action-words. *Frontiers in Human Neuroscience, 5*, 149.

Hauk, O., & Tschentscher, N. (2013). The body of evidence: What can neuroscience tell us about embodied semantics? *Frontiers in Psychology, 4*, 50.

Hauk, O., Wakeman, D. G., & Henson, R. (2011). Comparison of noise-normalized minimum norm estimates for MEG analysis using multiple resolution metrics. *NeuroImage, 54*, 1966–1974.

Henson, R. N. (2005). What can functional neuroimaging tell the experimental psychologist? *Quarterly Journal of Experimental Psychology Section A—Human Experimental Psychology, 58*, 193–233.

Herrera, E., Rodriguez-Ferreiro, J., & Cuetos, F. (2012). The effect of motion content in action naming by Parkinson's disease patients. *Cortex, 48*, 900–904.

Hume, D. (1985). *A treatise of human nature*. London: Penguin Classics.

Ioannidis, J. P. (2005). Why most published research findings are false. *PLoS Medicine, 2*, e124.

Kant, I. (1995). *Kritik der reinen vernunft*. Berlin: Suhrkamp Verlag.

Kaschak, M. P., Zwaan, R. A., Aveyard, M., & Yaxley, R. H. (2006). Perception of auditory motion affects language processing. *Cognitive Science, 30*, 733–744.

Kemmerer, D., Castillo, J. G., Talavage, T., Patterson, S., & Wiley, C. (2008). Neuroanatomical distribution of five semantic components of verbs: Evidence from fMRI. *Brain and Language, 107*, 16–43.

Kemmerer, D., Rudrauf, D., Manzel, K., & Tranel, D. (2010). Behavioral patterns and lesion sites associated with impaired processing of lexical and conceptual knowledge of actions. *Cortex, 48* (7), 826–848.

Kiefer, M., & Martens, U. (2010). Attentional sensitization of unconscious cognition: Task sets modulate subsequent masked semantic priming. *Journal of Experimental Psychology General, 139*, 464–489.

Kiefer, M., & Pulvermuller, F. (2011). Conceptual representations in mind and brain: Theoretical developments, current evidence and future directions. *Cortex, 48*(7), 805–825.

Kiefer, M., Sim, E. J., Herrnberger, B., Grothe, J., & Hoenig, K. (2008). The sound of concepts: Four markers for a link between auditory and conceptual brain systems. *The Journal of Neuroscience, 28*, 12224–12230.

Kiefer, M., Trumpp, N., Herrnberger, B., Sim, E. J., Hoenig, K., & Pulvermuller, F. (2012). Dissociating the representation of action-and sound-related concepts in middle temporal cortex. *Brain and Language, 122*, 120–125.

Kousta, S. T., Vigliocco, G., Vinson, D. P., Andrews, M., & Del Campo, E. (2011). The representation of abstract words: Why emotion matters. *Journal of Experimental Psychology General, 140*, 14–34.

Labruna, L., Fernandez-del-Olmo, M., Landau, A., Duque, J., & Ivry, R. B. (2011). Modulation of the motor system during visual and auditory language processing. *Experimental Brain Research, 211*, 243–250.

Lacey, S., Stilla, R., & Sathian, K. (2012). Metaphorically feeling: Comprehending textural metaphors activates somatosensory cortex. *Brain and Language, 120*, 416–421.

Lakoff, G. (1987). *Women, fire, and dangerous things: What categories reveal about the mind*. Chicago, IL: University of Chicago Press.

Mahon, B. Z., & Caramazza, A. (2008). A critical look at the embodied cognition hypothesis and a new proposal for grounding conceptual content. *Journal of Physiology, Paris, 102*, 59–70.

Mahon, B. Z., & Caramazza, A. (2009). Concepts and categories: A cognitive neuropsychological perspective. *Annual Review of Psychology, 60*, 27–51.

Martin, A. (2007). The representation of object concepts in the brain. *Annual Review of Psychology, 58*, 25–45.

Meteyard, L., Cuadrado, S. R., Bahrami, B., & Vigliocco, G. (2012). Coming of age: A review of embodiment and the neuroscience of semantics. *Cortex, 48*, 788–804.

Meteyard, L., Zokaei, N., Bahrami, B., & Vigliocco, G. (2008). Visual motion interferes with lexical decision on motion words. *Current Biology, 18*, R732–R733.

Meyer, K., & Damasio, A. (2009). Convergence and divergence in a neural architecture for recognition and memory. *Trends in Neurosciences, 32*, 376–382.

Mirabella, G., Iaconelli, S., Spadacenta, S., Federico, P., & Gallese, V. (2012). Processing of hand-related verbs specifically affects the planning and execution of arm reaching movements. *PLoS One, 7*, e35403.

Mollo, G., Pulvermuller, F., & Hauk, O. (2011). Language in action or action in language? An EMEG study on interactions between language and motor system. In *Annual Meeting of the Cognitive Neuroscience Society*. San Francisco, CA.

Moscoso del Prado Martin, F., Hauk, O., & Pulvermuller, F. (2006). Category specificity in the processing of color-related and form-related words: An ERP study. *NeuroImage, 29*, 29–37.

Moseley, R., Carota, F., Hauk, O., Mohr, B., & Pulvermuller, F. (2012). A role for the motor system in binding abstract emotional meaning. *Cerebral Cortex, 22*, 1634–1647.

Neininger, B., & Pulvermüller, F. (2003). Word-category specific deficits after lesions in the right hemisphere. *Neuropsychologia, 41*, 53–70.

Noppeney, U., & Price, C. J. (2003). Functional imaging of the semantic system: Retrieval of sensory-experienced and verbally learned knowledge. *Brain and Language, 84*, 120–133.

Patterson, K., Nestor, P. J., & Rogers, T. T. (2007). Where do you know what you know? The representation of semantic knowledge in the human brain. *Nature Reviews Neuroscience, 8*, 976–987.

Pecher, D., Boot, I., & Van Dantzig, S. (2011). Abstract concepts: Sensory-motor grounding, metaphors, and beyond. *Psychology of Learning and Motivation: Advances in Research and Theory, 54*(54), 217–248.

Poeppel, D., & Hickok, G. (2004). Towards a new functional anatomy of language. *Cognition, 92*, 1–12.

Postle, N., McMahon, K. L., Ashton, R., Meredith, M., & de Zubicaray, G. I. (2008). Action word meaning representations in cytoarchitectonically defined primary and premotor cortices. *NeuroImage, 43*(3), 634–644.

Price, C. J. (2000). The anatomy of language: Contributions from functional neuroimaging. *Journal of Anatomy, 197*(Pt 3), 335–359.

Pulvermuller, F. (1999). Words in the brain's language. *Behavioral and Brain Science, 22*, 253–279 [discussion 280–336].

Pulvermüller, F. (2005). Brain mechanisms linking language and action. *Nature Reviews Neuroscience, 6*, 576–582.

Pulvermuller, F. (2013a). How neurons make meaning: Brain mechanisms for embodied and abstract-symbolic semantics. *Trends in Cognitive Sciences, 17*(9), 458–470.

Pulvermuller, F. (2013b). Semantic embodiment, disembodiment or misembodiment? In search of meaning in modules and neuron circuits. *Brain and Language.*

Pulvermuller, F., Cooper-Pye, E., Dine, C., Hauk, O., Nestor, P. J., & Patterson, K. (2010). The word processing deficit in semantic dementia: All categories are equal, but some categories are more equal than others. *Journal of Cognitive Neuroscience, 22*, 2027–2041.

Pulvermüller, F., Härle, M., & Hummel, F. (2001). Walking or talking? Behavioral and neurophysiological correlates of action verb processing. *Brain and Language, 78*, 143–168.

Pulvermuller, F., & Hauk, O. (2006). Category-specific conceptual processing of color and form in left fronto-temporal cortex. *Cerebral Cortex, 16*, 1193–1201.

Pulvermuller, F., Hauk, O., Nikulin, V. V., & Ilmoniemi, R. J. (2005). Functional links between motor and language systems. *The European Journal of Neuroscience, 21*, 793–797.

Pulvermüller, F., Shtyrov, Y., & Ilmoniemi, R. (2005). Brain signatures of meaning access in action word recognition. *Journal of Cognitive Neuroscience, 17*, 884–892.

Raposo, A., Moss, H. E., Stamatakis, E. A., & Tyler, L. K. (2009). Modulation of motor and premotor cortices by actions, action words and action sentences. *Neuropsychologia, 47*, 388–396.

Rogers, T. T., Lambon Ralph, M. A., Garrard, P., Bozeat, S., McClelland, J. L., Hodges, J. R., et al. (2004). Structure and deterioration of semantic memory: A neuropsychological and computational investigation. *Psychological Review, 111*, 205–235.

Ruschemeyer, S. A., Brass, M., & Friederici, A. D. (2007). Comprehending prehending: Neural correlates of processing verbs with motor stems. *Journal of Cognitive Neuroscience, 19*, 855–865.

Santana, E., & de Vega, M. (2011). Metaphors are embodied, and so are their literal counterparts. *Frontiers in Psychology, 2*, 90.

Scorolli, C., Jacquet, P. O., Binkofski, F., Nicoletti, R., Tessari, A., & Borghi, A. M. (2012). Abstract and concrete phrases processing differentially modulates cortico-spinal excitability. *Brain Research, 1488*, 60–71.

Shallice, T., & Cooper, R. P. (2013). Is there a semantic system for abstract words? *Frontiers in Human Neuroscience, 7*, 175.

Shebani, Z., & Pulvermuller, F. (2013). Moving the hands and feet specifically impairs working memory for arm- and leg-related action words. *Cortex, 49*, 222–231.

Simmons, W. K., Hamann, S. B., Harenski, C. L., Hu, X. P., & Barsalou, L. W. (2008). fMRI evidence for word association and situated simulation in conceptual processing. *Journal of Physiology, Paris, 102*, 106–119.

Simmons, W. K., Ramjee, V., Beauchamp, M. S., McRae, K., Martin, A., & Barsalou, L. W. (2007). A common neural substrate for perceiving and knowing about color. *Neuropsychologia, 45*, 2802–2810.

Tettamanti, M., Manenti, R., Della Rosa, P. A., Falini, A., Perani, D., Cappa, S. F., et al. (2008). Negation in the brain: Modulating action representations. *NeuroImage, 43*, 358–367.

Tomasino, B., Fink, G. R., Sparing, R., Dafotakis, M., & Weiss, P. H. (2008). Action verbs and the primary motor cortex: A comparative TMS study of silent reading, frequency judgments, and motor imagery. *Neuropsychologia, 46*, 1915–1926.

Tomasino, B., Werner, C. J., Weiss, P. H., & Fink, G. R. (2007). Stimulus properties matter more than perspective: An fMRI study of mental imagery and silent reading of action phrases. *NeuroImage, 36*(Suppl. 2), T128–T141.

Trumpp, N. M., Kliese, D., Hoenig, K., Haarmeier, T., & Kiefer, M. (2012). Losing the sound of concepts: Damage to auditory association cortex impairs the processing of sound-related concepts. *Cortex, 49*(2), 474–486.

Tyler, L. K., Bright, P., Dick, E., Tavares, P., Pilgrim, L., Fletcher, P., et al. (2003). Do semantic categories activate distinct cortical regions? Evidence for a distributed neural semantic system. *Cognitive Neuropsychology, 20*, 541–559.

van Dam, W. O., van Dijk, M., Bekkering, H., & Rueschemeyer, S. A. (2012). Flexibility in embodied lexical-semantic representations. *Human Brain Mapping, 33*, 2322–2333.

van der Laan, L. N., de Ridder, D. T., Viergever, M. A., & Smeets, P. A. (2011). The first taste is always with the eyes: A meta-analysis on the neural correlates of processing visual food cues. *NeuroImage, 55*, 296–303.

van Elk, M., van Schie, H. T., Zwaan, R. A., & Bekkering, H. (2010). The functional role of motor activation in language processing: Motor cortical oscillations support lexical-semantic retrieval. *NeuroImage, 50*, 665–677.

Vul, E., Harris, C., Winkielman, P., & Pashler, H. (2009). Puzzlingly high correlations in fMRI studies of emotion, personality, and social cognition. *Perspectives on Psychological Science, 4*, 274–290.

Wagenmakers, E. J., Wetzels, R., Borsboom, D., van der Maas, H. L. J., & Kievit, R. A. (2012). An Agenda for purely confirmatory research. *Perspectives on Psychological Science, 7*, 632.

Warrington, E. K., & Shallice, T. (1984). Category specific semantic impairments. *Brain, 107*(Pt 3), 829–854.

Wheatley, T., Weisberg, J., Beauchamp, M. S., & Martin, A. (2005). Automatic priming of semantically related words reduces activity in the fusiform gyrus. *Journal of Cognitive Neuroscience, 17*, 1871–1885.

Willems, R. M., Hagoort, P., & Casasanto, D. (2010). Body-specific representations of action verbs: Neural evidence from right- and left-handers. *Psychological Science, 21*, 67–74.

Willems, R. M., Labruna, L., D'Esposito, M., Ivry, R., & Casasanto, D. (2011). A functional role for the motor system in language understanding: Evidence from theta-burst transcranial magnetic stimulation. *Psychological Science, 22*, 849–854.

Willems, R. M., Toni, I., Hagoort, P., & Casasanto, D. (2010). Neural dissociations between action verb understanding and motor imagery. *Journal of Cognitive Neuroscience, 22*, 2387–2400.

Wilson, M. (2002). Six views of embodied cognition. *Psychonomic Bulletin and Review, 9*, 625–636.

Wilson-Mendenhall, C. D., Barrett, L. F., Simmons, W. K., & Barsalou, L. W. (2011). Grounding emotion in situated conceptualization. *Neuropsychologia, 49*, 1105–1127.

Wilson-Mendenhall, C. D., Simmons, W. K., Martin, A., & Barsalou, L. W. (2013). Contextual processing of abstract concepts reveals neural representations of nonlinguistic semantic content. *Journal of Cognitive Neuroscience, 25*, 920–935.

Zwaan, R. A., van der Stoep, N., Guadalupe, T., & Bouwmeester, S. (2012). Language comprehension in the balance: The robustness of the action-compatibility effect (ACE). *PLoS One, 7*, e31204.

SECTION L

WRITTEN LANGUAGE

CHAPTER

63

Acquired Dyslexia

H. Branch Coslett[1] *and Peter Turkeltaub*[2,3]

[1]Department of Neurology, Perelman School of Medicine at the University of Pennsylvania, Philadelphia, PA, USA; [2]Department of Neurology, Georgetown University School of Medicine, Washington, DC, USA; [3]MedStar National Rehabilitation Hospital, Washington, DC, USA

63.1 INTRODUCTION

Investigations of acquired dyslexia began in the late 19th century with Dejerine's report of two patients with quite different patterns of reading impairment. Dejerine's first patient (Dejerine, 1891) developed dyslexia and dysgraphia, (i.e., deficits in reading and spelling/writing respectively) after an infarction involving the left parietal lobe. Dejerine termed this disorder "alexia with agraphia"[1] and attributed the disturbance to a disruption of the "optical image for words," which he thought to be supported by the left angular gyrus. Anticipating some contemporary psychological accounts, Dejerine concluded that reading and writing required the activation of these "optical images" and that the loss of the images resulted in the inability to recognize or write even familiar words.

Dejerine's second patient (Dejerine, 1892) was quite different. This patient demonstrated a remarkable dissociation between reading and writing; he was unable to read aloud or for comprehension, yet he could write remarkably well. Like some subsequently reported patients, he was unable to read what he had written. Invoking a prominent theoretical construct of the time, "alexia without agraphia" was attributed to a disconnection between visual input regarding the letter string and the "optical images" for words supported by the left angular gyrus, the brain region he thought to be critical for word recognition.

After decades of relative obscurity, the study of acquired dyslexia was revitalized by the elegant and insightful analyses by Marshall and Newcombe (1966, 1973). These seminal studies were significant for at least two reasons. First, they demonstrated that detailed assessment of multiple aspects of performance could provide crucial insights; they demonstrated, for example, that systematic analyses of the types of errors made by dyslexics were crucial in the assessment of performance. By scrutinizing the effects of variables (e.g., part of speech, imageability) that had been largely ignored in prior studies, Marshall and Newcombe contributed greatly to the emerging discipline of cognitive neuropsychology. Second, these investigators described novel reading disorders including "deep" and "surface" dyslexias.

On the basis of these data, Marshall and Newcombe (1973) concluded that the pronunciation of written words could be derived by two distinct procedures. The first was a direct procedure whereby familiar words activated the appropriate stored representation (or visual word form) that, in turn, activated meaning directly; reading in deep dyslexia was assumed to involve this procedure. The second procedure was assumed to be a phonologically based process in which "grapheme to phoneme" or "print-to-sound" correspondences were used to derive the appropriate phonology (i.e., "sound out" the word); the reading of surface dyslexics was assumed to be mediated by this nonlexical procedure.

The conceptual framework proposed by Marshall and Newcombe may be illustrated by the "box and arrow" model presented in Figure 63.1. Because this account provides the general framework with which the various acquired dyslexias will be discussed and motivated one of the leading computationally instantiated current models of reading (Coltheart, Rastle, Perry, Langdon, & Ziegler, 2001; Rastle, Tyler, & Marslen-Wilson, 2006),

[1]Acquired disorders of reading have been designated "dyslexia" and "alexia" by different authors; we use the terms interchangeably.

Neurobiology of Language. DOI: http://dx.doi.org/10.1016/B978-0-12-407794-2.00063-8

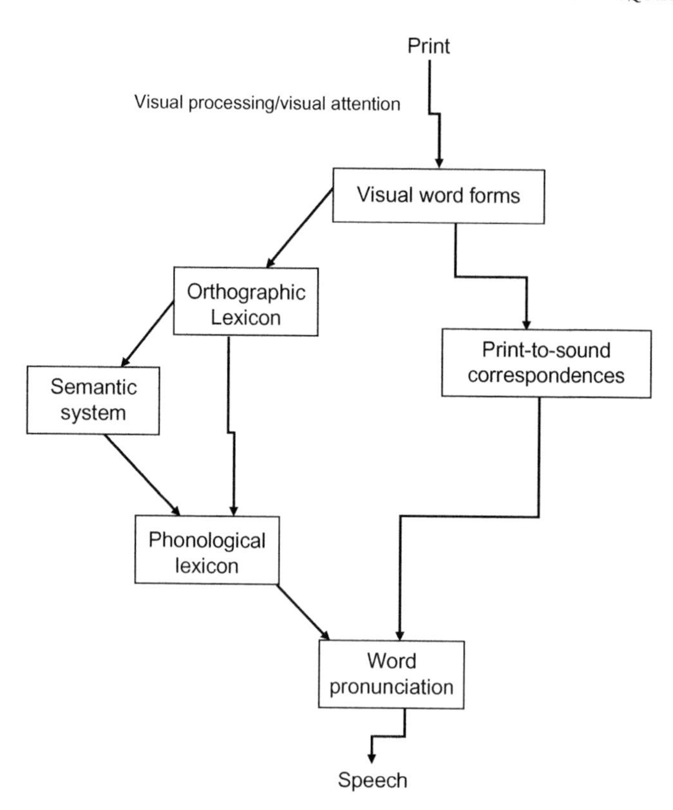

FIGURE 63.1 A simplified information-processing model of the procedures involved in reading provided for illustrative purposes.

we briefly describe it here. It should be emphasized that we use this model as a heuristic only; alternative accounts of reading such as the "triangle model" (Harm & Seidenberg, 2004; Plaut, McClelland, Seidenberg, & Patterson, 1996) that make very different assumptions about the mechanisms underlying reading are also discussed.

The information-processing model of reading depicted in Figure 63.1 provides three distinct procedures for oral reading. Two of these procedures correspond to those described by Marshall and Newcombe (1973). The first involves the activation of a stored entry in the visual word form system and the subsequent access to semantic information and, ultimately, activation of the stored sound of the word at the level of the phonologic output lexicon. The second involves the nonlexical grapheme-to-phoneme or print-to-sound translation process; this procedure does not entail access to any stored information about words, but rather is assumed to be mediated by access to a catalog of correspondences between letters or letter sequences and speech sounds. Some accounts of the language mechanisms subserving reading incorporate a third reading procedure. This mechanism is lexically based in that it is assumed to involve the activation of the visual word form system and the phonologic output lexicon. The procedure differs from the lexical procedure described here, however, in that there is no intervening activation of semantic information. This procedure has been called the "direct" reading mechanism or route. Support for the direct lexical mechanism comes from a number of sources including observations that some subjects read aloud words that cannot be readily "sounded out" that they do not appear to comprehend (Coslett, 1991; Schwartz, Marin, & Saffran, 1979). More recently, Law, Wong, Sung, and Hon (2006) provided evidence for a lexical but nonsemantic route in reading. They reported data from two Chinese patients who had suffered strokes and whose performance on tests of phonology and access to semantics for verbal materials argued for independent lexical mechanisms.

This chapter focuses on disorders of reading rather than language more generally; it should be noted, however, that disorders of reading are rarely observed in isolation. In the large majority of instances, for example, phonologic dyslexia is associated with deficits in phonology that can be demonstrated on a range of nonreading tasks (Woollams, Hoffman, Roberts, Lambon Ralph, & Patterson, 2014; but see Cipolotti & Warrington, 1996). Similarly, even "pure alexia" is not pure in the sense that most patients with this disorder also exhibit deficits in other aspects of visual processing (Behrmann & Plaut, 2014). The association of reading with more basic cognitive and perceptual deficits is not surprising given that the first evidence of written language dates to approximately 4,000 years ago, whereas spoken language and the other perceptual and motor faculties that underpin reading evolved over a far greater amount of time. Although most investigators would likely agree that many of the processes involved in reading are "domain general" in that they are crucial for oral language or other faculties, the degree to which reading invokes reading-specific faculties remains a matter of significant debate (Rastle et al., 2006; Woollams et al., 2014).

63.2 PERIPHERAL DYSLEXIAS

A useful starting point in the discussion of the dyslexias is the distinction offered by Shallice and Warrington (1980) between "peripheral" and "central" dyslexias. We take the former to be conditions characterized by a deficit in the processing of visual aspects of the stimulus that prevent the patient from reliably matching a familiar word to its stored representation or "visual word form" (Shallice & Warrington, 1980). In contrast, central dyslexias reflect impairment procedures by which visual word forms gain access to meaning or speech production mechanisms. We start by considering the major types of peripheral dyslexia.

63.2.1 Alexia Without Agraphia or Pure Alexia

The classical syndrome of alexia without agraphia or pure alexia is perhaps the prototypical peripheral dyslexia. The traditional account (Dejerine, 1892; Geschwind & Fusillo, 1966) of this disorder attributes the syndrome to a "disconnection" of visual information; as conceptualized by Dejerine (1892), this syndrome is attributed to the combination of two impairments. First, the left hemisphere is deprived of visual input by a left posterior lesion. Second, visual information from the right hemisphere is disconnected from the left hemisphere structures mediating word recognition (the left angular gyrus on Dejerine's account) by a lesion that involves the splenium of the corpus callosum or the extension of these fibers in the deep white matter of the occipital lobe (forceps major).

Although these patients do not appear to be able to read in the sense of fast, automatic word recognition, many are able to use a compensatory strategy that involves naming the letters of the word in serial fashion; they read, in effect, "letter-by-letter" (Patterson & Kay, 1982). Using the slow and inefficient letter-by-letter procedure, pure alexics typically exhibit significant effects of word length, requiring more time to read long words as compared to short words. Although patients with this disorder have demonstrated features of "surface" (Friedman & Hadley, 1992) or "deep" dyslexic phenomena (Buxbaum & Coslett, 1996), in most subjects performance is not influenced by factors such as part of speech (e.g., noun vs functor), the extent to which the referent of the word is concrete (e.g., "table") or abstract (e.g., "destiny"), or whether the word is orthographically regular (i.e., can be "sounded out"). It should be noted that pure alexia may be differentiated from hemianopic dyslexia (see Schuett, Heywood, Kentridge, & Zihl, 2008 for review). The latter syndrome is associated with significant right or left visual field defects and is manifested primarily in text reading. Based on evidence demonstrating that subjects with the disorder have impaired saccadic eye movements as well as the observation that the disorder is observed in subjects with lesions that involve more than calcarine cortex (i.e., V1), hemianopic dyslexia is thought to be caused by a combination of reduced visual fields and impaired systems controlling eye movements and visual attention.

A number of alternative accounts of the processing deficit in pure alexia have been proposed. Thus, some investigators have proposed that the impairment is attributable to a limitation in the transmission of letter identity information to the visual word system (Patterson & Kay, 1982) or an inability to directly encode visual letters as abstract orthographic types (Bub & Arguin, 1995). A number of investigators have proposed that pure alexia is attributable to a low-level visual impairment (Chialant & Caramazza, 1996; Farah & Wallace, 1991). Although many patients with the disorder have no clinically evident visual deficits, more careful assessments of vision have consistently demonstrated visual deficits. Mycroft, Behrmann, and Kay (2009), for example, they demonstrated that these subjects have difficulty processing letter-like stimuli as well as matching checkerboards. Additionally, Behrmann and Plaut (2014) demonstrated that these patients have difficulty with demanding facial recognition tasks. Finally, Roberts et al. (2013) demonstrated that pure alexics have difficulty processing complex visual stimuli and the degree of impairment roughly paralleled their word recognition deficits. It would seem that, at least in many instances, pure alexia is associated with a deficit in processing stimuli comprising easily confusable components that require high spatial frequency, the prototypical example of which is words.

Shallice and Warrington (1980) argued that the disorder is attributable to a disruption of the visual word form system. Subsequent work has supported their hypothesis. Numerous functional imaging studies have demonstrated that the posterior, inferior temporo-occipital cortex, particularly on the left, is activated in response to letter strings (Cohen & Dehaene, 2004; Cohen, Dehaene, Chochon, Lehéricy, & Naccache, 2000); it has been proposed that this region supports the visual word form area (VWFA). Several studies have suggested that damage to the VWFA itself can cause pure alexia (Gaillard et al., 2006; Tsapkini & Rapp, 2010; Turkeltaub et al., 2014). In most of these cases, the white matter pathways neighboring the VWFA may also have been damaged, although in our recent case the stroke was restricted to the cortex itself (Turkeltaub et al., 2014). This patient recovered quickly and was not available for extensive testing, limiting the value of this case. Based on these cases, a reasonable hypothesis is that different processing deficits associated with pure alexia may result from either disconnection of, or direct damage to, the text-sensitive areas of the left mid-fusiform cortex. For instance, damage to the VWFA itself may result in loss of orthographic representations, manifested by relatively greater impairment of spelling than is seen when an intact VWFA is disconnected from visual input (Rapcsak & Beeson, 2004).

The "disconnection" account originally offered by Dejerine has received support from an elegant neuroimaging study of a subject who was examined before and after a surgical procedure that resulted in dyslexia (Epelbaum et al., 2008). The surgical lesion was just posterior to the VWFA (localized using preoperative fMRI), sparing the cortex of the VWFA itself, but damaging inferior longitudinal fasciculus fibers connecting it to more posterior areas of occipital cortex. This analysis supported the long hypothesized role of the inferior longitudinal fasciculus in normal reading.

An important contribution to the understanding of the anatomic basis of pure alexia has recently come from the comparison of eye movements during reading of subjects with pure alexia and subjects with right hemianopia who did not exhibit the full syndrome of pure alexia (Pflugshaupt et al., 2009). Whereas all subjects exhibited low-amplitude rightward saccades during text reading, the substantial increases in fixation frequency and viewing time were specific to subjects with pure alexia. Furthermore, supporting claims about the role of damage to the visual word form system in pure alexia, they found that the region corresponding to the visual word form system in the left fusiform gyrus was disrupted in pure alexia but not hemianopic subjects without dyslexia.

Similarly, in a prior lesion overlap comparison, individuals with hemianopic alexia had damage restricted to medial occipital lesions, whereas those with pure alexia had ventral occipital lesions extending more laterally into the fusiform gyrus and the neighboring white matter (Leff, Spitsyna, Plant, & Wise, 2006). This again implicates lateral ventral occipital pathways in reading, but these studies do not clarify whether pure alexia results from direct disruption of text-sensitive areas of cortex, such as the VWFA or from disconnection of these cortical areas from visual input.

We note that the different accounts of pure alexia are not mutually exclusive. As with many syndromes defined on the basis of a single, cardinal deficit, there is reason to believe that pure alexia may be attributable to different processing impairments in different individuals (see Rosazza, Appollonio, Isella, & Shallice, 2007).

Although most reports of pure alexia have emphasized the profound nature of the reading deficit, often stating that patients were utterly incapable of reading without recourse to a letter-by-letter procedure (Geschwind & Fusillo, 1966; Patterson & Kay, 1982), we and other investigators have reported data demonstrating that pure alexic patients are able to comprehend words that they are unable to explicitly identify (Coslett & Saffran, 1989a; Landis, Regard, & Serrat, 1980; Shallice & Saffran, 1986). The interpretation of these data remains controversial. Behrmann, Plaut, and Nelson (1998) have suggested that the preserved access to meaning in pure alexia is mediated by the (lesioned) normal reading system. By this account, the imageability and part of speech effects as well as the implicit nature of the reading performance are assumed to be attributable to weak activation of stored word forms by the visual input.

Alternatively, we (Coslett & Saffran, 1989a, p. 19; Saffran & Coslett, 1998) and others (Landis et al., 1980) have argued that the implicit reading that emerges in pure alexia reflects the lexical processing of the right hemisphere. We reported data, for example, from four patients with pure alexia who performed well above chance on a number of lexical decision and semantic categorization tasks with briefly presented words that they could not explicitly identify (Coslett & Saffran, 1989b). Three of the patients who regained the ability to explicitly identify rapidly presented words exhibited a pattern of performance consistent with the right hemisphere reading hypothesis (Coltheart, 2000; Coltheart, Patterson, & Marshall, 1980; Saffran, Bogyo, Schwartz, & Marin, 1980) in that they read nouns better than functors and words of high imageability (e.g., *chair*) better than words of low imageability (e.g., *destiny*). Additionally, both patients for whom data are available demonstrated a deficit in the reading of suffixed (e.g., "flower") as compared to pseudo-suffixed (e.g., "flowed") words. We return to this issue of right hemisphere contributions to reading in a discussion of deep dyslexia.

Recent imaging studies provide a complex picture of the anatomic bases of reading in recovering pure alexics. In a case of an intracerebral hemorrhage in the area of the VWFA, brain activity during a text lexical decision task was observed in the right hemisphere homolog of the VWFA 7 days after stroke but shifted back to the left VWFA after 50 days (Ino et al., 2008). We recently reported a patient with chronic pure alexia in whom transcranial direct current stimulation was provided in combination with multiple oral rereading treatment. In this case, ventral occipito-temporal activity shifted from the right to the left hemisphere after only 5 days of treatment (Lacey et al., 2015). Other case reports have demonstrated posterior displacement of reading-related activity in the left fusiform cortex (Gaillard et al., 2006) and bilateral activity in posterior fusiform and anterior temporal areas after left focal left mid-fusiform resection (Tsapkini, Vindiola, & Rapp, 2011).

63.2.1.1 Neglect Dyslexia

Neglect dyslexia, first described by Brain in 1941, is characterized by a failure to explicitly identify the initial portion of a letter string. The disorder is most frequently encountered in subjects with the syndrome of left-sided spatial neglect. Interestingly, just as the recognition of objects may be influenced by "top-down" factors, the performance of patients with neglect dyslexia is often influenced by the lexical status of the letter string (Sieroff, Pollatsek, & Posner, 1988); thus, patients with this disorder may fail to report the "ti-" in nonwords such as "tiggle" but read the word "giggle" correctly. The fact that performance is affected by the lexical status of the stimulus suggests that neglect dyslexia is not attributable to a failure to register letter information, but reflects an attentional impairment at a higher level of representation (Behrmann, Moscovitch, Black, & Mozer, 1990; Làdavas, Shallice, & Zanella, 1997). Haywood and Coltheart (2000) suggested that neglect dyslexia can be caused by deficits at three types of spatial representations; alternatively, Mozer (2002) simulated in a connectionist model the error patterns reported in neglect dyslexia using an explicit attentional mechanism based on a single viewer-centered reference frame and implemented. It also should be noted that reading errors in neglect dyslexia cannot be taken as evidence that access to meaning is entirely abolished because Ladavas, Paladini, and Cubelli (1993, 1997) demonstrated that neglected words influenced performance (see also Lie & Coslett, 2006).

Finally, it should be noted that neglect dyslexia for the right side of letter strings is occasionally observed in subjects with left hemisphere lesions. As demonstrated by Petrich, Greenwald, and Berndt (2007), this deficit may be severe for written words but not evident for any other type of stimulus.

63.2.1.2 Attentional Dyslexia and Related Disorders

Perhaps the least studied of the acquired dyslexias, attentional dyslexia is characterized by the relative preservation of single-word reading in the context of a gross disruption of reading when words are presented as text or in the presence of other words or letters (Price & Humphreys, 1993; Saffran, 1996; Warrington, Cipolotti, & McNeil, 1993). Patients with this disorder may also exhibit difficulties identifying letters within words, even though the words themselves are read correctly (Saffran, 1996), and be impaired in identifying words flanked by extraneous letters (e.g., "lboat"). Finally, we (Warrington et al., 1993) reported a patient with attentional dyslexia secondary to autopsy-proven Alzheimer disease who produced frequent "blend" errors in which letters from one word of a two-word display intruded into the other word (e.g., "bake lime" read as "like"). A number of investigators have explored additional reading deficits in subjects with degenerative disorders such as the "visual variant" of Alzheimer disease (Price & Humphreys, 1995). Crutch and Warrington (2007) reported data from three subjects in whom the effects of visual "crowding" were systematically investigated. They found that the proximity (but not identity) of flanking letters significantly influenced the report of a target letter.

63.3 CENTRAL DYSLEXIAS

63.3.1 Phonological Dyslexia

First described in 1979 by Derouesne and Beauvois, phonologic dyslexia is attributable to a selective deficit in the procedure translating between print and sound (see Coltheart, 1996; Nickels, Biedermann, Coltheart, Saunders, & Tree, 2008). In the context of the reading model depicted in Figure 63.1, the account of this disorder is relatively straightforward. Good performance with real words suggests that the processes involved in normal "lexical" reading—that is, visual analysis, the visual word form system, semantics, and the phonological output lexicon—are at least relatively preserved. The impairment in nonword reading suggests that the print-to-sound translation procedure is disrupted.

As the procedure by which stored representations of words contact semantics or output phonology are preserved, the cardinal deficit in phonologic dyslexia is impaired reading aloud of unfamiliar letter strings or nonwords. Thus, patients with this disorder correctly read 85–95% of real words (Bub, Black, Howell, & Kertesz, 1987; Funnell, 1983). Some patients with this disorder read all different types of words with equal facility (Bub et al., 1987), whereas other patients are relatively impaired in the reading of functors (or "little words") (Glosser & Friedman, 1990). Because stored lexical representations for familiar words are available, the regularity of print-to-sound correspondences is not relevant to the performance of phonologic dyslexics; thus, these patients are as likely to correctly pronounce orthographically irregular words such as *colonel* as words with standard print-to-sound correspondences such as *administer*. Many errors in response to real words bear a visual similarity to the target word (e.g., *topple* read as "table"); errors with nonwords typically involve the incorrect application of print to sound correspondences (e.g., *stime* read as "stim" [to rhyme with "him"]) or the substitution of a visually similar real word (e.g., *flig* read as "flag").

The cause of the impairment in translating between print and sound remains controversial. Some investigators have emphasized the role of a general phonologic deficit in the genesis of phonological dyslexia (e.g., Farah, 1996; Harm & Seidenberg, 1999; Woollams et al., 2014). Consistent with this claim, Patterson and Marcel (1977) and Crisp and Lambon Ralph (2006) demonstrated that the severity of the reading deficit in phonologic dyslexia is correlated with the degree of impairment on nonreading phonologic tasks. By this account, the disorder is attributable to a general phonologic impairment (see also Rapcsak et al., 2009). With other accounts (e.g., Castles, Bates, & Coltheart, 2006; Coltheart et al., 2001), phonologic dyslexia is attributable to impairment of a reading-specific processing module that translates between letters and sounds. Evidence in support of this comes from several reports of patients with phonologic dyslexia who do not demonstrate deficits on nonreading tasks (Coltheart, 2006; Tree & Kay, 2006). We return to this issue later.

Phonologic dyslexia has been observed in association with lesions in a number of sites in the dominant peri-Sylvian cortex (Rapcsak et al., 2009) and, on occasion, with lesions of the right hemisphere (e.g., Patterson, 1982). Damage to the superior temporal lobe and angular and supramarginal gyri in particular is found in many patients with this disorder, including those with logopenic progressive aphasia, which is commonly associated with phonological alexia (Brambati, Ogar, Neuhaus, Miller, & Gorno-Tempini, 2009). Damage to inferior frontal cortex, including the frontal operculum, has also been associated with phonological dyslexia (Fiez, Tranel, Seager-Frerichs, & Damasio, 2006; Rapcsak et al., 2009), as has the insula (Ripamonti et al., 2014). Although quantitative data are lacking, the lesions associated with phonological dyslexia appear, on average, to be smaller than those associated with deep dyslexia.

A few studies have examined imaging changes associated with rehabilitation of phonological dyslexia. In one case, semantic mediation treatment resulted in recruitment of right perisylvian areas, but subsequent over-training of items was associated with a shift in activity back to left perilesional areas (Kurland et al., 2008). Another study found increased right posterior perisylvian and bilateral inferior frontal activity during pseudoword reading after an auditory discrimination treatment for phonological dyslexia (Adair et al., 2000).

63.3.2 Surface Dyslexia

Surface dyslexia, first described by Marshall and Newcombe (1973), is a disorder characterized by the inability to read words with "irregular" or exceptional print-to-sound correspondences. Languages differ with respect to the consistency with which letters are mapped to sounds; in some languages (e.g., Italian), letter-to-sound correspondences are predictable, whereas in other languages (e.g., English) rules govern the mapping between print and sound, but there are also numerous exceptions. Patients with surface dyslexia fail to read words such as *colonel, yacht, island, have,* and *borough,* for which the mapping between print and sound is not governed by rules. In contrast, these patients perform well with words containing regular correspondences (e.g., *state, hand, abdominal*) as well as nonwords (e.g., *blape*). A significant effect of word frequency is typically encountered such that regularization errors are far more common with low-frequency words. The great majority of patients with surface dyslexia also demonstrate semantic impairment on tasks that do not involve reading (Woollams, Ralph, Plaut, & Patterson, 2007; see Cipolotti & Warrington, 1996 for an exception).

The fact that the nature of the print-to-sound correspondences significantly influences performance in surface dyslexia demonstrates that the deficit in this syndrome is in the semantically mediated and "direct" reading mechanisms. Similarly, the preserved ability to read words and nonwords is taken as evidence that the procedures by which words are "sounded out" are at least relatively preserved.

In the context of the model depicted in Figure 63.1, surface dyslexia is attributed to disruption of the direct and semantically mediated mechanisms with a reliance on print-to-sound correspondences. In the context of dual-route models of reading, these disruptions may occur at different levels of representation within the lexical system. Thus, for example, surface dyslexia has been attributed to a disruption of semantics in conjunction with a deficit in the "direct" route (Schwartz et al., 1979; Shallice, Warrington, & McCarthy, 1983) or with a lesion involving the phonologic output lexicon (Howard & Franklin, 1987). On the "triangle" connectionist model, surface dyslexia is attributed to a disruption of semantics (Woollams et al., 2014).

Surface dyslexia is infrequently observed in patients with focal lesions. In a sample of 59 patients with dyslexia due to left hemisphere stroke, Ripamonti et al. (2014) identified five with surface dyslexia. These patients had lesions primarily involving the temporal lobe, including both posterior and anterior regions. Surface dyslexia is most often encountered in patients with progressive, degenerative dementias such as Alzheimer disease or fronto-temporal dementia. Surface dyslexia is characteristic of semantic dementia, a disorder associated with left temporal atrophy in which subjects slowly but inexorably lose knowledge of the world (Breedin, Saffran, & Coslett, 1994; Hodges, Patterson, Oxbury, & Funnell, 1992).

Few studies have explicitly examined the neuroanatomical basis of surface dyslexia in semantic dementia. One examined atrophy associated with exception word and pseudoword reading in a sample of patients with semantic dementia and the "logopenic" variant of primary progressive aphasia, and found that anterior temporal lobe atrophy was associated with exception word reading deficits (Brambati et al., 2009). Similarly, another recent study identified an area of left anterior middle temporal cortex that was active in control subjects during an exception versus pseudoword reading comparison, and it was also atrophied in a patient with surface dyslexia related to semantic dementia (Wilson et al., 2012). Another study examined the neural basis of over-regularization in surface dyslexia due to semantic dementia (Wilson et al., 2009). In control subjects, they identified an area of intraparietal sulcus that is active during pseudoword and low-frequency regular word reading, suggesting this area is involved in grapheme-to-phoneme conversions. In patients with surface dyslexia, this area was activated in association with regularization errors on exception words.

63.3.3 Deep Dyslexia

Deep dyslexia, initially described by Marshall and Newcombe (1973), is an intriguing but complex acquired dyslexia, the hallmark of which is the production of semantic errors in which a word related in meaning is substituted for the target word. Thus, shown the word *castle*, a deep dyslexic may respond "knight"; presented with *bird*, the patient may respond "canary." At least for some deep dyslexics, it is clear that these errors are not circumlocutions. Semantic errors may represent the most frequent error type in some deep dyslexics, whereas in other patients they comprise a small proportion of reading errors. Deep dyslexics also typically produce frequent "visual" errors (e.g., *skate* read as "scale") and morphological errors in which a prefix or suffix is added, deleted, or substituted (e.g., scolded read as "scolds"; governor read as "government").

In contrast to surface dyslexia and phonologic dyslexia that, on traditional accounts, appear to be largely attributable to disorders of print-to-sound correspondence and stored word representations, respectively, deep dyslexia has a number of deficits that, at least on some accounts, are attributed to multiple processing impairments. All deep dyslexics exhibit a severe impairment in phonologic processing. Reading of nonwords, for example, is typically profoundly impaired. Nonword letter strings such as *flig* or *churt* frequently elicit "lexicalization" errors (e.g., *flig* read as "flag"), perhaps reflecting a reliance on lexical reading in the absence of access to reliable print-to-sound correspondences. Performance is often poor on other phonologic tasks such as naming the sound associated with a letter or even rhyme judgments for auditory words.

Another consistent finding in deep dyslexia is that patients are typically far more successful reading words of high imageability as compared to low imageability. Thus, words such as *table*, *chair*, *ceiling*, and *buttercup*, the referent of which is concrete or imageable, are read more successfully than words such as *fate*, *destiny*, *wish*, and *universal*, which denote abstract concepts.

Another characteristic feature of deep dyslexia is a part of speech effect such that nouns are read more reliably than modifiers (adjectives and adverbs) that are, in turn, read more accurately than verbs. Deep dyslexics manifest particular difficulty in the reading of functors (a class of words that includes pronouns, prepositions, conjunctions, and interrogatives including *that*, *which*, *they*, *because*, *under*, etc.). The striking part of speech effect may be illustrated by the patient reported by Saffran and Marin (1977) who correctly read the word *chrysanthemum* but was unable to read the word *the*. Many errors to functors involve the substitution of a different functor (*that* read as "which") rather than the production of words of a different class such as nouns or verbs. As functors are, in general, less imageable than nouns, verbs, or adjectives, some investigators have claimed that the apparent effect of part of speech is in reality a manifestation of the pervasive imageability effect (Allport & Funnell, 1981). We (Coslett, 1991) have reported a patient, however, whose performance suggests that the part of speech effect is not simply a reflection of a more general deficit in the processing of low-imageability words.

Several alternative explanations have been proposed for deep dyslexia. Some investigators have argued that deep dyslexia is on a continuum with phonologic dyslexia (Crisp, Howard, & Lambon Ralph, 2011; Patterson & Lambon Ralph, 1999). With this account, the two disorders share as the primary disorder an impairment in phonologic processing, but the deficit is more severe in deep dyslexics. Friedman (1996) noted that as subjects with deep dyslexia recovered, they often stopped making semantic errors, thereby meeting criteria for phonologic dyslexia (see also Glosser & Friedman, 1990). On the basis of 12 subjects with acquired dyslexia, Crisp and Lambon Ralph (2006) argued that deep and phonologic dyslexia were distinguished by the severity of their semantic and phonologic impairments.

Other investigators believe that deep dyslexia reflects the effects of multiple deficits (Morton & Patterson, 1980; Shallice, 1988). First, the strikingly impaired performance in reading nonwords and other tasks assessing phonologic function suggest that the ability to translate between print and sound is severely disrupted. Second, the presence of semantic errors and the effects of imageability (a variable thought to influence processing at the level of semantics) is taken as evidence that these subjects also suffer from a semantic impairment. Finally, the production of visual errors suggests that these patients suffer from impairment in the visual word form system or in the processes mediating access of the stimulus to the visual word form system.

A third potential account is that deep dyslexia derives from a post-semantic impairment. Caramazza and Hillis (1990) reported two patients who made frequent semantic errors yet demonstrated intact comprehension of written and spoken words. They argued that semantic errors can arise after intact semantic information is contacted; they attribute the deficit to an output lexicon that specifies phonologic form. Buchanan and colleagues (Buchanan, McEwen, Westbury, & Libben, 2003; Caramazza & Hillis, 1990; Colangelo & Buchanan, 2007) have taken a similar position.

Finally, a number of investigators have argued that deep dyslexics' reading is mediated by a system not normally used in reading—the right hemisphere (Coltheart, 2000; Coltheart et al., 1980; Saffran et al., 1980). As articulated by Coltheart (2000), this account proposes that orthographic processing is mediated by the right hemisphere in deep dyslexics; semantic information is accessed in the right hemisphere as well. Speech production may depend on the left inferior frontal region or, in some individuals, the right inferior frontal region.

Although long considered to be "word blind" (Dejerine, 1892; Geschwind, 1965), several lines of evidence support the claim that the right hemisphere may possess some capacity to read (see Lindell, 2006 for review). One seemingly incontrovertible line of evidence comes from the performance of a patient who underwent a left hemispherectomy at age 15 for treatment of seizures caused by Rasmussen encephalitis (Patterson, Vargha-Khadem, & Polkey, 1989). After the hemispherectomy, the patient was able to read approximately 30% of single words and exhibited an effect of part of speech; she was also utterly unable to use a grapheme-to-phoneme conversion process. Thus, in many respects this patient's performance was similar to that of acquired deep dyslexia.

The performance of some patients with partial or complete hemispheric disconnection are also consistent with the claim that the right hemisphere is literate. These patients may, for example, be able to match printed words presented to the right hemisphere with an appropriate object (Zaidel, 1978; Zaidel & Peters, 1981). Interestingly, the patients are apparently unable to derive sound from the words presented to the right hemisphere; thus, they are unable to determine if a word presented to the right hemisphere rhymes with an auditorally presented word. Michel, Hénaff, and Intriligator (1996) also reported a patient with a lesion of the posterior portion of the corpus callosum whose reading of words presented to the right hemisphere was similar to that of deep dyslexics.

Finally, we reported data from an investigation with a patient with pure alexia whose recovered reading was disrupted by transcranial magnetic stimulation to the right, but not the left, hemisphere, suggesting that the right hemisphere mediated reading in this subject (Coslett & Monsul, 1994).

Although deep dyslexia has occasionally been associated with posterior lesions, the disorder is typically encountered in association with large peri-Sylvian lesions extending into the frontal lobe (Coltheart et al., 1980). As might be expected given the lesion data, deep dyslexia is usually associated with Global or Broca aphasia but may rarely be encountered in patients with fluent aphasia.

Several functional imaging studies have addressed the anatomic basis of deep dyslexia. Weekes, Carusi, and Zaidel (1997) reported measured regional blood flow with Xenon-133 in a deep dyslexic performing a variety of reading tasks. By subtracting blood flow related to viewing a false font from a condition involving lexical decision to visually presented words, they demonstrated that for the deep dyslexic subject, blood flow was greater in the right hemisphere as compared to the left, supporting the role of the right hemisphere in word recognition; additional comparisons suggested that word production in this subject was likely to be mediated by the left frontal lobe. Subsequently, Price et al. (1998) reported an fMRI investigation from which they concluded that reading in deep dyslexia was mediated by the left hemisphere. This interpretation was subsequently challenged by Coltheart (2000), who argued that their data, in fact, supported the right hemisphere account of deep dyslexia.

63.4 COMPUTATIONAL MODELS OF READING

For ease of exposition, the discussion to this point has been framed with reference to an information processing or "box and arrow" account of the processes involved in reading. In the past two decades, however, there have been substantial advances in the precision and specificity of theoretical accounts of reading; several computationally explicit models have been developed that not only account for normal reading but also can be "lesioned" in an effort to simulate patient data (e.g., Lambon Ralph, Patterson, & Plaut, 2011; Nickels et al., 2008; Patterson, Seidenberg, & McClelland, 1989). One model, originally developed by Seidenberg and McClelland (1989) and subsequently elaborated by a number of investigators (Harm & Seidenberg, 2004; Plaut et al., 1996; Woollams et al., 2014), belongs to the general class of parallel distributed processing or connectionist models. This model is often termed the "triangle model" because it incorporates three major types of representations—orthographic units, semantic units, and phonological units (see Figure 63.2). Crucially, this model does not incorporate word-specific representations; instead, "knowledge" is distributed across connections between units. In this account, there are two procedures by which subjects may learn to pronounce a word. First, by means of repeated exposure to letter strings, subjects learn to map letter sequences to phonology. The probabilistic mapping between letters and sounds is assumed to provide the means by which both familiar and unfamiliar words are pronounced. Second, the model incorporates a

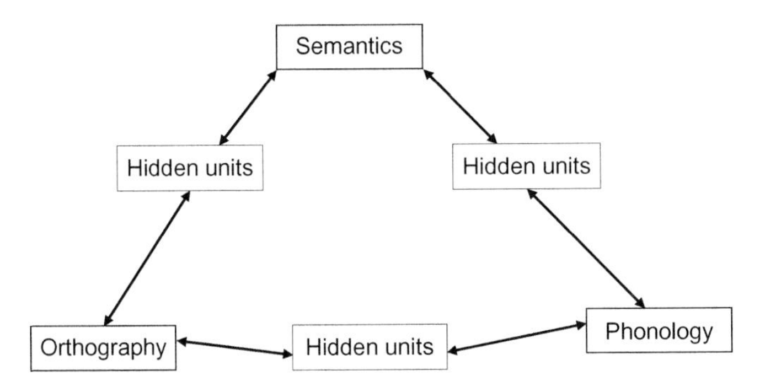

FIGURE 63.2 A simplified version of the "triangle model" of reading.

"semantic" reading procedure in which meaning is initially accessed; subsequently, patterns of activation across semantic and orthographic units are mapped onto phonology (Harm & Seidenberg, 2004). This model not only accommodates many of the classic findings in the literature on normal reading but also has been "lesioned" in an attempt to reproduce the patterns of reading impairment characteristic of different types of acquired dyslexia (Plaut & Shallice, 1993; Welbourne & Lambon Ralph, 2007; Woollams et al., 2007). Woollams et al. (2007) provide evidence that the triangle model accommodates many aspects of the performance of surface dyslexia exhibited by patients with semantic dementia. Welbourne and Lambon Ralph (2007) attempted to model phonologic dyslexia with the "triangle" model. Reasoning that the performance of subjects with acquired dyslexia reflects not only the effect of the lesion but also the effects of recovery, they re-trained a lesioned version of the Plaut et al. (1996; simulation 4) connectionist model; they argue that if the network was permitted to "recover," then a pattern of deficits characteristic of phonological dyslexia was observed. The adequacy of these simulations remains a topic of substantial debate. Coltheart (2006) has argued that "lesioned" versions of these models do not faithfully reproduce the pattern of deficits observed in phonological or surface dyslexia.

An alternative computational account of reading has been developed by Coltheart and colleagues (Castles et al., 2006; Coltheart & Rastle, 1994; Coltheart et al., 2001; Rastle & Coltheart, 1999). The "dual-route cascaded" or DRC model represents a computationally instantiated version of the dual-route theory presented in Figure 63.1. This "localist" account incorporates a "lexical" route that includes a listing of familiar orthographic and phonological word forms; in addition, the model incorporates a "nonlexical" route by which the pronunciation of letter strings is generated by the serial application of rules specifying position-specific grapheme-to-phoneme correspondences. Finally, like the triangle model, the DRC model incorporates a semantic system that may mediate between orthography and phonologic output representations; to this point, semantic representations have not been implemented in the model. The model also assumes that both procedures operate in parallel and that there is interaction between the lexical and nonlexical procedures.

The DRC model accommodates a wide range of findings from the literature on normal reading. For example, Coltheart et al. (2001) demonstrated that the DRC model generates accounts for phenomena seen in normal reading, including effects of frequency, regularity, position, position of regularity, neighborhood size, pseudohomophony, lexicality, length, and a number of interactions of these variables. They developed an algorithm that yields quantitative predictions about reading of orthographically regular words from performance of orthographically irregular words and nonwords; they demonstrated that this equation accurately predicts performance of young normal readers, children with developmental dyslexia, and children with brain tumors (Castles et al., 2006). Rapcsak, Henry, Teague, Carnahan, and Beeson (2007) extended this work to a heterogeneous but well-studied group of 33 adults with acquired dyslexia from stroke. They demonstrated that the performance was well-predicted by the equation derived from dual-route theory. Finally, Nickels et al. (2008) reported simulations of the data from three subjects with phonological dyslexia with the DRC model. Interestingly, no single lesion generated adequate simulations for all three subjects; consistent with the basic principle that different sites of pathology are likely to generate different processing deficits, different "lesions" to the model were required to reproduce the pattern of deficits exhibited by the different patients.

Finally, a computationally instantiated model of reading aloud based on Glushko's (1979) "reading by analogy" hypothesis has also been developed (Damper, Marchand, Adamson, & Gustafson, 1999). In this account, there is a single procedure for deriving the pronunciation of letter strings that is based on analogy to familiar words (Glushko, 1979; Kay & Marcel, 1981). This account proposes that when confronted with the nonword "paze," subjects generate a response that rhymes with gaze, reflecting the fact that "-aze" has a single phonological instantiation in English; when presented the letter string "tave," in contrast, subjects are both slower to respond and more variable as "-ave" has multiple phonological instantiations in English (e.g., have versus gave). Marchand and Friedman (2005) demonstrated that a computationally implemented version of the analogy account can be lesioned to account for the data of two dyslexic subjects with unusual patterns of performance.

Although the DRC and triangle models are based on fundamentally different perspectives regarding the brain basis of reading and are instantiated in very different computational architectures, both models accommodate a wide range of empirical data. Additional data will be required to adjudicate between the models and the theoretical perspectives that animate them.

63.5 ASSESSMENT OF READING

As previously noted, the specific types of dyslexia are distinguished on the basis of performance with different types of stimuli. For example, deep dyslexia is characterized by impaired performance on nonwords, an effect of part of speech such that nouns are read better than modifiers or functors and an effect of imageability such that words denoting more abstract objects or concepts are read less well than words of high imageability. The assessment of patients with dyslexia should include stimuli varying along the dimensions discussed here.

The effect of imageability or concreteness should be assessed by presenting words of high (e.g., *desk*, *frog*, *mountain*) and low (e.g., *fate*, *universal*, *ambiguous*) imageability. Part of speech should be assessed by presenting nouns (e.g., *table*, *meatloaf*), modifiers (e.g., *beautiful*, *early*), verbs (e.g., *ambulate*, *thrive*), and functors (e.g., *because*, *their*). The effect of orthographic regularity should be assessed by presenting regular words that can be "sounded out" (e.g., *flame*, *target*) and irregular words that cannot (e.g., *come*, *tomb*). The ability to "sound out" words is also assessed by presenting nonword letter strings that may sound like a real word (e.g., *phish*) or may not (e.g., *blape*).

Because word frequency is typically an important determinant of performance, a wide range of word frequencies should also be used. To obtain a reliable assessment of performance, testing should include at least 10 words of each of the stimulus types noted in this chapter. The compilation of the appropriate lists of stimuli may be time-consuming; consequently, many investigators use published word lists, some of which are commercially available (Kay, Lesser, & Coltheart, 1996). Clinical assessments of acquired dyslexia should also include sentence and paragraph reading, because real-life reading requires reading words in context rather than in isolation. In the research setting, additional tools such as eye-tracking, neuroimaging, or electrophysiology can complement behavioral performance.

Finally, it should be noted that there is an emerging literature on the treatment of acquired reading disorders. Whereas there is no widely recognized and effective treatment for any of the acquired dyslexias, there have been a number of reports of theoretically motivated interventions that appear to offer promise (see Leff & Behrmann, 2008 for a review).

References

Adair, J. C., Nadeau, S. E., Conway, T. W., Gonzalez-Rothi, L. J., Heilman, P. C., Green, I. A., et al. (2000). Alterations in the functional anatomy of reading induced by rehabilitation of an alexic patient. *Neuropsychiatry, Neuropsychology, and Behavioral Neurology, 13*(4), 303–311.

Allport, D., & Funnell, E. (1981). Components of the mental lexicon. *Philosophical Transactions of the Royal Society of London B: Biological Sciences, 295*(1077), 397–410.

Behrmann, M., Moscovitch, M., Black, S. E., & Mozer, M. (1990). Perceptual and conceptual mechanisms in neglect dyslexia. Two contrasting case studies. *Brain: A Journal of Neurology, 113*(Pt 4), 1163–1183.

Behrmann, M., & Plaut, D. C. (2014). Bilateral hemispheric processing of words and faces: Evidence from word impairments in prosopagnosia and face impairments in pure alexia. *Cerebral Cortex (New York, NY: 1991), 24*(4), 1102–1118. Available from: http://dx.doi.org/10.1093/cercor/bhs390.

Behrmann, M., Plaut, D. C., & Nelson, J. (1998). A literature review and new data supporting an interactive account of letter-by-letter reading. *Cognitive Neuropsychology, 15*(1–2), 7–51.

Brambati, S. M., Ogar, J., Neuhaus, J., Miller, B. L., & Gorno-Tempini, M. L. (2009). Reading disorders in primary progressive aphasia: A behavioral and neuroimaging study. *Neuropsychologia, 47*(8–9), 1893–1900. Available from: http://dx.doi.org/10.1016/j.neuropsychologia.2009.02.033.

Breedin, S. D., Saffran, E. M., & Coslett, H. B. (1994). Reversal of the concreteness effect in a patient with semantic dementia. *Cognitive Neuropsychology, 11*(6), 617–660.

Bub, D., Black, S., Howell, J., & Kertesz, A. (1987). Speech output processes and reading. In M. Coltheart, G. Sartori, & R. Job (Eds.), *The cognitive neuropsychology of language* (pp. 79–102). Hove, East Sussex: Lawrence Erlbaum Associates.

Bub, D. N., & Arguin, M. (1995). Visual word activation in pure alexia. *Brain and Language, 49*(1), 77–103. Available from: http://dx.doi.org/10.1006/brln.1995.1022.

Buchanan, L., McEwen, S., Westbury, C., & Libben, G. (2003). Semantics and semantic errors: Implicit access to semantic information from words and nonwords in deep dyslexia. *Brain and Language, 84*(1), 65–83.

Buxbaum, L. J., & Coslett, H. B. (1996). Deep dyslexic phenomena in a letter-by-letter reader. *Brain and Language, 54*(1), 136–167.

Caramazza, A., & Hillis, A. E. (1990). Where do semantic errors come from? *Cortex; a Journal Devoted to the Study of the Nervous System and Behavior, 26*(1), 95–122.

Castles, A., Bates, T., & Coltheart, M. (2006). John Marshall and the developmental dyslexias. *Aphasiology, 20*(9), 871–892. Available from: http://dx.doi.org/10.1080/02687030600738952.

Chialant, D., & Caramazza, A. (1996). Low-level perceptual factors in letter-by-letter reading. *Brain and Cognition, 30*(3), 332–334.

Cipolotti, L., & Warrington, E. K. (1996). Does recognizing orally spelled words depend on reading? An investigation into a case of better written than oral spelling. *Neuropsychologia, 34*(5), 427–440.

Cohen, L., & Dehaene, S. (2004). Specialization within the ventral stream: The case for the visual word form area. *NeuroImage, 22*(1), 466–476. Available from: http://dx.doi.org/10.1016/j.neuroimage.2003.12.049.

Cohen, L., Dehaene, S., Chochon, F., Lehéricy, S., & Naccache, L. (2000). Language and calculation within the parietal lobe: A combined cognitive, anatomical and fMRI study. *Neuropsychologia*, *38*(10), 1426–1440.

Colangelo, A., & Buchanan, L. (2007). Localizing damage in the functional architecture: The distinction between implicit and explicit processing in deep dyslexia. *Journal of Neurolinguistics*, *20*(2), 111–144. Available from: http://dx.doi.org/10.1016/j.jneuroling.2006.08.001.

Coltheart, M. (1996). Phonological dyslexia: Past and future issues. *Cognitive Neuropsychology*, *13*, 749–762.

Coltheart, M. (2000). Deep dyslexia is right-hemisphere reading. *Brain and Language*, *71*(2), 299–309. Available from: http://dx.doi.org/10.1006/brln.1999.2183.

Coltheart, M. (2006). Acquired dyslexias and the computational modelling of reading. *Cognitive Neuropsychology*, *23*(1), 96–109. Available from: http://dx.doi.org/10.1080/02643290500202649.

Coltheart, M., Patterson, K., & Marshall, J. C. (Eds.). (1980). Deep dyslexia: A right-hemisphere hypothesis. In *Deep Dyslexia*. London: Routledge and Kegan Paul.

Coltheart, M., & Rastle, K. (1994). Serial processing in reading aloud: Evidence for dual-route models of reading. *Journal of Experimental Psychology: Human Perception and Performance*, *20*(6), 1197.

Coltheart, M., Rastle, K., Perry, C., Langdon, R., & Ziegler, J. (2001). DRC: A dual route cascaded model of visual word recognition and reading aloud. *Psychological Review*, *108*(1), 204–256.

Coslett, H. B. (1991). Read but not write "idea": Evidence for a third reading mechanism. *Brain and Language*, *40*(4), 425–443.

Coslett, H. B., & Monsul, N. (1994). Reading with the right hemisphere: Evidence from transcranial magnetic stimulation. *Brain and Language*, *46*(2), 198–211. Available from: http://dx.doi.org/10.1006/brln.1994.1012.

Coslett, H. B., & Saffran, E. M. (1989a). Evidence for preserved reading in "pure alexia." *Brain*, *112*(2), 327–359.

Coslett, H. B., & Saffran, E. M. (1989b). Preserved object recognition and reading comprehension in optic aphasia. *Brain: A Journal of Neurology*, *112*(Pt 4), 1091–1110.

Crisp, J., Howard, D., & Lambon Ralph, M. A. (2011). More evidence for a continuum between phonological and deep dyslexia: Novel data from three measures of direct orthography-to-phonology translation. *Aphasiology*, *25*(5), 615–641. Available from: http://dx.doi.org/10.1080/02687038.2010.541470.

Crisp, J., & Lambon Ralph, M. A. (2006). Unlocking the nature of the phonological-deep dyslexia continuum: The keys to reading aloud are in phonology and semantics. *Journal of Cognitive Neuroscience*, *18*(3), 348–362. Available from: http://dx.doi.org/10.1162/089892906775990543.

Crutch, S. J., & Warrington, E. K. (2007). Foveal crowding in posterior cortical atrophy: A specific early-visual-processing deficit affecting word reading. *Cognitive Neuropsychology*, *24*(8), 843–866. Available from: http://dx.doi.org/10.1080/02643290701754240.

Damper, R. I., Marchand, Y., Adamson, M. J., & Gustafson, K. (1999). Evaluating the pronunciation component of text-to-speech systems for English: A performance comparison of different approaches. *Computer Speech & Language*, *13*(2), 155–176. Available from: http://dx.doi.org/10.1006/csla.1998.0117.

Dejerine, J. J. (1891). Sur en case de Cecite verbal avec agraphie, suivi d'autopsie. *Compte Rendu Des Seances de La Societe de Biologie*, *3*, 197–201.

Dejerine, J. J. (1892). Contribution à l'étude anatomo-pathologique et clinique des différentes variétés de cécité verbale. *Comptes rendus des séances de la Société de Biologie et de ses filiales*, *44*(9), 61–90.

Derouesne, J., & Beauvois, M. F. (1979). Phonological processing in reading: Data from alexia. *Journal of Neurology, Neurosurgery & Psychiatry*, *42*(12), 1125–1132.

Epelbaum, S., Pinel, P., Gaillard, R., Delmaire, C., Perrin, M., Dupont, S., et al. (2008). Pure alexia as a disconnection syndrome: New diffusion imaging evidence for an old concept. *Cortex*, *44*(8), 962–974. Available from: http://dx.doi.org/10.1016/j.cortex.2008.05.003.

Farah, M. J. (1996). Phonological dyslexia: Loss of a reading-specific component of the cognitive architecture? *Cognitive Neuropsychology*, *13*(6), 849–868.

Farah, M. J., & Wallace, M. A. (1991). Pure alexia as a visual impairment: A reconsideration. *Cognitive Neuropsychology*, *8*(3–4), 313–334.

Fiez, J. A., Tranel, D., Seager-Frerichs, D., & Damasio, H. (2006). Specific reading and phonological processing deficits are associated with damage to the left frontal operculum. *Cortex; a Journal Devoted to the Study of the Nervous System and Behavior*, *42*(4), 624–643.

Friedman, R. B. (1996). Recovery from deep alexia to phonological alexia: Points on a continuum. *Brain and Language*, *52*(1), 114–128. Available from: http://dx.doi.org/10.1006/brln.1996.0006.

Friedman, R. B., & Hadley, J. A. (1992). Letter-by-letter surface alexia. *Cognitive Neuropsychology*, *9*(3), 185–208. Available from: http://dx.doi.org/10.1080/02643299208252058.

Funnell, E. (1983). Phonological processes in reading: New evidence from acquired dyslexia. *British Journal of Psychology*, *74*(2), 159–180.

Gaillard, R., Naccache, L., Pinel, P., Clémenceau, S., Volle, E., Hasboun, D., et al. (2006). Direct intracranial, FMRI, and lesion evidence for the causal role of left inferotemporal cortex in reading. *Neuron*, *50*(2), 191–204. Available from: http://dx.doi.org/10.1016/j.neuron.2006.03.031.

Geschwind, N. (1965). Disconnexion syndromes in animals and man, Part II. *Brain*, *88*(3), 585–644.

Geschwind, N., & Fusillo, M. (1966). Color-naming defects in association with alexia. *Archives of Neurology*, *15*(2), 137–146.

Glosser, G., & Friedman, R. B. (1990). The continuum of deep/phonological alexia. *Cortex*, *26*(3), 343–359.

Glushko, R. J. (1979). The organization and activation of orthographic knowledge in reading aloud. *Journal of Experimental Psychology: Human Perception and Performance*, *5*(4), 674–691. Available from: http://dx.doi.org/10.1037/0096-1523.5.4.674.

Harm, M. W., & Seidenberg, M. S. (1999). Phonology, reading acquisition, and dyslexia: Insights from connectionist models. *Psychological Review*, *106*(3), 491–528.

Harm, M. W., & Seidenberg, M. S. (2004). Computing the meanings of words in reading: Cooperative division of labor between visual and phonological processes. *Psychological Review*, *111*(3), 662–720. Available from: http://dx.doi.org/10.1037/0033-295X.111.3.662.

Haywood, M., & Coltheart, M. (2000). Neglect dyslexia and the early stages of visual word recognition. *Neurocase*, *6*(1), 33–44.

Hodges, J. R., Patterson, K., Oxbury, S., & Funnell, E. (1992). Semantic dementia. Progressive fluent aphasia with temporal lobe atrophy. *Brain: A Journal of Neurology*, *115*(Pt 6), 1783–1806.

Howard, D., & Franklin, S. (1987). Three ways for understanding written words, and their use in two contrasting cases of surface dyslexia (together with an odd routine for making "orthographic" errors in oral word production). In A. Allport, D. Mackay, W. Prinz, & E. Scheerer (Eds.), *Language Perception and Production*. New York, NY: Academic Press.

Ino, T., Tokumoto, K., Usami, K., Kimura, T., Hashimoto, Y., & Fukuyama, H. (2008). Longitudinal fMRI study of reading in a patient with letter-by-letter reading. *Cortex; a Journal Devoted to the Study of the Nervous System and Behavior*, *44*(7), 773–781. Available from: http://dx.doi.org/10.1016/j.cortex.2007.03.002.

Kay, J., Lesser, R., & Coltheart, M. (1996). Psycholinguistic assessments of language processing in aphasia (PALPA): An introduction. *Aphasiology*, *10*(2), 159–180.

Kay, J., & Marcel, A. (1981). One process, not two, in reading aloud: Lexical analogies do the work of non-lexical rules. *The Quarterly Journal of Experimental Psychology Section A*, *33*(4), 397–413. Available from: http://dx.doi.org/10.1080/14640748108400800.

Kurland, J., Cortes, C. R., Wilke, M., Sperling, A. J., Lott, S. N., Tagamets, M. A., et al. (2008). Neural mechanisms underlying learning following semantic mediation treatment in a case of phonologic alexia. *Brain Imaging and Behavior*, *2*(3), 147. Available from: http://dx.doi.org/10.1007/s11682-008-9027-2.

Lacey, E. H., Jiang, X., Friedman, R. B., Snider, S. F., Parra, L. C., Huang, Y., et al. (2015). Transcranial direct current stimulation for pure alexia: Effects on brain and behavior. *Brain Stimulation*, *8*(2), 305–307. Available from: http://dx.doi.org/10.1016/j.brs.2014.10.019.

Ladavas, E., Paladini, R., & Cubelli, R. (1993). Implicit associative priming in a patient with left visual neglect. *Neuropsychologia*, *31*, 1307–1320.

Làdavas, E., Shallice, T., & Zanella, M. T. (1997). Preserved semantic access in neglect dyslexia. *Neuropsychologia*, *35*(3), 257–270.

Lambon Ralph, M. A., Patterson, K., & Plaut, D. C. (2011). Finite case series or infinite single-case studies? Comments on "Case series investigations in cognitive neuropsychology" by Schwartz and Dell (2010). *Cognitive Neuropsychology*, *28*(7), 466–474; discussion 515–520. Available from: http://dx.doi.org/10.1080/02643294.2012.671765

Landis, T., Regard, M., & Serrat, A. (1980). Iconic reading in a case of alexia without agraphia caused by a brain tumor: A tachistoscopic study. *Brain and Language*, *11*(1), 45–53.

Law, S.-P., Wong, W., Sung, F., & Hon, J. (2006). A study of semantic treatment of three Chinese anomic patients. *Neuropsychological Rehabilitation*, *16*(6), 601–629. Available from: http://dx.doi.org/10.1080/09602010543000046.

Leff, A. P., & Behrmann, M. (2008). Treatment of reading impairment after stroke. *Current Opinion in Neurology*, *21*(6), 644–648. Available from: http://dx.doi.org/10.1097/WCO.0b013e3283168dc7.

Leff, A. P., Spitsyna, G., Plant, G. T., & Wise, R. J. S. (2006). Structural anatomy of pure and hemianopic alexia. *Journal of Neurology, Neurosurgery, and Psychiatry*, *77*(9), 1004–1007. Available from: http://dx.doi.org/10.1136/jnnp.2005.086983.

Lie, E., & Coslett, H. B. (2006). The effect of gaze direction on sound localization in brain-injured and normal adults. *Experimental Brain Research*, *168*(3), 322–336. Available from: http://dx.doi.org/10.1007/s00221-005-0100-4.

Lindell, A. K. (2006). In your right mind: Right hemisphere contributions to language processing and production. *Neuropsychology Review*, *16*(3), 131–148. Available from: http://dx.doi.org/10.1007/s11065-006-9011-9.

Marchand, Y., & Friedman, R. B. (2005). Impaired oral reading in two atypical dyslexics: A comparison with a computational lexical-analogy model. *Brain and Language*, *93*(3), 255–266. Available from: http://dx.doi.org/10.1016/j.bandl.2004.10.013.

Marshall, J. C., & Newcombe, F. (1966). Syntactic and semantic errors in paralexia. *Neuropsychologia*, *4*(2), 169–176.

Marshall, J. C., & Newcombe, F. (1973). Patterns of paralexia: A psycholinguistic approach. *Journal of Psycholinguistic Research*, *2*(3), 175–199.

Michel, F., Hénaff, M. A., & Intriligator, J. (1996). Two different readers in the same brain after a posterior callosal lesion. *Neuroreport*, *7*(3), 786–788.

Morton, J., & Patterson, K. (1980). A new attempt at an interpretation, or, an attempt at a new interpretation. In M. Coltheart, K. Patterson, & J. C. Marshall (Eds.), *Deep dyslexia* (pp. 91–118). London: Routledge & Kegan Paul.

Mozer, M. C. (2002). Frames of reference in unilateral neglect and visual perception: A computational perspective. *Psychological Review*, *109*(1), 156–185.

Mycroft, R. H., Behrmann, M., & Kay, J. (2009). Visuoperceptual deficits in letter-by-letter reading? *Neuropsychologia*, *47*, 1733–1744.

Nickels, L., Biedermann, B., Coltheart, M., Saunders, S., & Tree, J. J. (2008). Computational modeling of phonological dyslexia: How does the DRC model fare? *Cognitive Neuropsychology*, *25*(2), 165–193. Available from: http://dx.doi.org/10.1080/02643290701514479.

Patterson, K. (1982). The relation between reading and phonological coding: Further neuropsychological observations. In: *Normality and pathology in cognitive functions*. London: Academic Press.

Patterson, K., & Kay, J. (1982). Letter-by-letter reading: Psychological descriptions of a neurological syndrome. *The Quarterly Journal of Experimental Psychology. A, Human Experimental Psychology*, *34*(Pt 3), 411–441.

Patterson, K., & Lambon Ralph, M. A. (1999). Selective disorders of reading? *Current Opinion in Neurobiology*, *9*(2), 235–239. Available from: doi:10.1016/S0959-4388(99)80033-6.

Patterson, K., Seidenberg, M. S., & McClelland, J. L. (1989). Connections and disconnections: Acquired dyslexia in a computational model of reading processes. In R. Morris (Ed.), *Parallel distributed processing: Implications for psychology and neurobiology*. Oxford: Oxford University Press.

Patterson, K., Vargha-Khadem, F., & Polkey, C. E. (1989). Reading with one hemisphere. *Brain*, *112*(1), 39–63.

Patterson, K. E., & Marcel, A. J. (1977). Aphasia, dyslexia and the phonological coding of written words. *The Quarterly Journal of Experimental Psychology*, *29*(2), 307–318. Available from: http://dx.doi.org/10.1080/14640747708400606.

Petrich, J. A. F., Greenwald, M. L., & Berndt, R. S. (2007). An investigation of attentional contributions to visual errors in right "neglect dyslexia." *Cortex; a Journal Devoted to the Study of the Nervous System and Behavior*, *43*(8), 1036–1046.

Pflugshaupt, T., Gutbrod, K., Wurtz, P., von Wartburg, R., Nyffeler, T., de Haan, B., et al. (2009). About the role of visual field defects in pure alexia. *Brain: A Journal of Neurology*, *132*(Pt 7), 1907–1917. Available from: http://dx.doi.org/10.1093/brain/awp141.

Plaut, D. C., McClelland, J. L., Seidenberg, M. S., & Patterson, K. (1996). Understanding normal and impaired word reading: Computational principles in quasi-regular domains. *Psychological Review*, *103*(1), 56–115.

Plaut, D. C., & Shallice, T. (1993). Deep dyslexia: A case study of connectionist neuropsychology. *Cognitive Neuropsychology*, *10*(5), 377–500.

Price, C. J., Howard, D., Patterson, K., Warburton, E. A., Friston, K. J., & Frackowiak, S. J. (1998). A functional neuroimaging description of two deep dyslexic patients. *Journal of Cognitive Neuroscience*, *10*(3), 303–315.

Price, C. J., & Humphreys, G. W. (1993). Attentional dyslexia: The effect of co-occurring deficits. *Cognitive Neuropsychology*, *10*(6), 569–592.

Price, C. J., & Humphreys, G. W. (1995). Contrasting effects of letter-spacing in alexia: Further evidence that different strategies generate word length effects in reading. *The Quarterly Journal of Experimental Psychology. A, Human Experimental Psychology*, *48*(3), 573–597.

Rapcsak, S. Z., & Beeson, P. M. (2004). The role of left posterior inferior temporal cortex in spelling. *Neurology*, *62*(12), 2221–2229.

Rapcsak, S. Z., Beeson, P. M., Henry, M. L., Leyden, A., Kim, E., Rising, K., et al. (2009). Phonological dyslexia and dysgraphia: Cognitive mechanisms and neural substrates. *Cortex; a Journal Devoted to the Study of the Nervous System and Behavior*, *45*(5), 575–591. Available from: http://dx.doi.org/10.1016/j.cortex.2008.04.006.

Rapcsak, S. Z., Henry, M., Teague, S., Carnahan, S. D., & Beeson, P. M. (2007). Do dual-route models accurately predict reading and spelling performance in individuals with acquired alexia and agraphia?. *Neuropsychologia*, *45*(11), 2519–2524. Available from: http://dx.doi.org/10.1016/j.neuropsychologia.2007.03.019.

Rastle, K., & Coltheart, M. (1999). Serial and strategic effects in reading aloud. *Journal of Experimental Psychology: Human Perception and Performance*, *25*(2), 482.

Rastle, K., Tyler, L. K., & Marslen-Wilson, W. (2006). New evidence for morphological errors in deep dyslexia. *Brain and Language*, *97*(2), 189–199. Available from: http://dx.doi.org/10.1016/j.bandl.2005.10.003.

Ripamonti, E., Aggujaro, S., Molteni, F., Zonca, G., Frustaci, M., & Luzzatti, C. (2014). The anatomical foundations of acquired reading disorders: A neuropsychological verification of the dual-route model of reading. *Brain and Language*, *134*, 44–67. Available from: http://dx.doi.org/10.1016/j.bandl.2014.04.001.

Roberts, D. J., Woollams, A. M., Kim, E., Beeson, P. M., Rapcsak, S. Z., & Lambon Ralph, M. A. (2013). Efficient visual object and word recognition relies on high spatial frequency coding in the left posterior fusiform gyrus: Evidence from a case-series of patients with ventral occipito-temporal cortex damage. *Cerebral Cortex (New York, NY: 1991)*, *23*(11), 2568–2580. Available from: http://dx.doi.org/10.1093/cercor/bhs224.

Rosazza, C., Appollonio, I., Isella, V., & Shallice, T. (2007). Qualitatively different forms of pure alexia. *Cognitive Neuropsychology*, *24*(4), 393–418. Available from: http://dx.doi.org/10.1080/02643290701377877.

Saffran, E., Bogyo, L. C., Schwartz, M. F., & Marin, O. S. (1980). *Does deep dyslexia reflect right-hemisphere reading*. In *Deep dyslexia* (pp. 381–406). London: Routledge and Kegan Paul.

Saffran, E. M. (1996). "Attentional Dyslexia" in Alzheimer's disease: A case study. *Cognitive Neuropsychology*, *13*(2), 205–228. Available from: http://dx.doi.org/10.1080/026432996382006.

Saffran, E. M., & Coslett, H. B. (1998). Implicit vs. letter-by-letter reading in pure alexia: A tale of two systems. *Cognitive Neuropsychology*, *15*(1–2), 141–165.

Saffran, E. M., & Marin, O. S. (1977). Reading without phonology: Evidence from aphasia. *The Quarterly Journal of Experimental Psychology*, *29*(3), 515–525. Available from: http://dx.doi.org/10.1080/14640747708400627.

Schuett, S., Heywood, C. A., Kentridge, R. W., & Zihl, J. (2008). The significance of visual information processing in reading: Insights from hemianopic dyslexia. *Neuropsychologia*, *46*(10), 2445–2462. Available from: http://dx.doi.org/10.1016/j.neuropsychologia.2008.04.016.

Schwartz, M. F., Marin, O. S. M., & Saffran, E. M. (1979). Dissociations of language function in dementia: A case study. *Brain and Language*, *7*(3), 277–306.

Seidenberg, M. S., & McClelland, J. L. (1989). A distributed, developmental model of word recognition and naming. *Psychological Review*, *96*(4), 523–568.

Shallice, T. (1988). *From neuropsychology to mental structure*. Cambridge: Cambridge University Press.

Shallice, T., & Saffran, E. (1986). Lexical processing in the absence of explicit word identification: Evidence from a letter-by-letter reader. *Cognitive Neuropsychology*, *3*(4), 429–458.

Shallice, T., Warrington, E., & McCarthy, R. (1983). Reading without semantics. *The Quarterly Journal of Experimental Psychology*, *35*(1), 111–138.

Shallice, T., & Warrington, E. K. (1980). Single and multiple component central dyslexic syndromes. In M. Coltheart, K. Patterson, & J. Marshall (Eds.), *Deep dyslexia*. London: Routledge and Kegan Paul.

Sieroff, E., Pollatsek, A., & Posner, M. I. (1988). Recognition of visual letter strings following injury to the posterior visual spatial attention system. *Cognitive Neuropsychology*, *5*(4), 427–449.

Tree, J. J., & Kay, J. (2006). Phonological dyslexia and phonological impairment: An exception to the rule? *Neuropsychologia*, *44*(14), 2861–2873. Available from: http://dx.doi.org/10.1016/j.neuropsychologia.2006.06.006.

Tsapkini, K., & Rapp, B. (2010). The orthography-specific functions of the left fusiform gyrus: Evidence of modality and category specificity. *Cortex; a Journal Devoted to the Study of the Nervous System and Behavior*, *46*(2), 185–205. Available from: http://dx.doi.org/10.1016/j.cortex.2009.02.025.

Tsapkini, K., Vindiola, M., & Rapp, B. (2011). Patterns of brain reorganization subsequent to left fusiform damage: fMRI evidence from visual processing of words and pseudowords, faces and objects. *NeuroImage*, *55*(3), 1357–1372. Available from: http://dx.doi.org/10.1016/j.neuroimage.2010.12.024.

Turkeltaub, P. E., Goldberg, E. M., Postman-Caucheteux, W. A., Palovcak, M., Quinn, C., Cantor, C., et al. (2014). Alexia due to ischemic stroke of the visual word form area. *Neurocase*, *20*(2), 230–235. Available from: http://dx.doi.org/10.1080/13554794.2013.770873.

Warrington, E. K., Cipolotti, L., & McNeil, J. (1993). Attentional dyslexia: A single case study. *Neuropsychologia*, *31*(9), 871–885.

Weekes, N. Y., Carusi, D., & Zaidel, E. (1997). Interhemispheric relations in hierarchical perception: A second look. *Neuropsychologia*, *35*(1), 37–44.

Welbourne, S. R., & Lambon Ralph, M. A. (2007). Using parallel distributed processing models to simulate phonological dyslexia: The key role of plasticity-related recovery. *Journal of Cognitive Neuroscience*, *19*(7), 1125–1139. Available from: http://dx.doi.org/10.1162/jocn.2007.19.7.1125.

Wilson, M. A., Joubert, S., Ferré, P., Belleville, S., Ansaldo, A. I., Joanette, Y., et al. (2012). The role of the left anterior temporal lobe in exception word reading: Reconciling patient and neuroimaging findings. *NeuroImage*, *60*(4), 2000–2007. Available from: http://dx.doi.org/10.1016/j.neuroimage.2012.02.009.

Wilson, S. M., Brambati, S. M., Henry, R. G., Handwerker, D. A., Agosta, F., Miller, B. L., et al. (2009). The neural basis of surface dyslexia in semantic dementia. *Brain: A Journal of Neurology*, *132* (Pt 1), 71–86. Available from: http://dx.doi.org/10.1093/brain/awn300.

Woollams, A. M., Hoffman, P., Roberts, D. J., Lambon Ralph, M. A., & Patterson, K. E. (2014). What lies beneath: A comparison of reading aloud in pure alexia and semantic dementia. *Cognitive Neuropsychology*, *31*(5–6), 461–481. Available from: http://dx.doi.org/10.1080/02643294.2014.882300.

Woollams, A. M., Ralph, M. A. L., Plaut, D. C., & Patterson, K. (2007). SD-squared: On the association between semantic dementia and surface dyslexia. *Psychological Review*, *114*(2), 316–339. Available from: http://dx.doi.org/10.1037/0033-295X.114.2.316.

Zaidel, E. (1978). Lexical organization in the right hemisphere. *Cerebral Correlates of Conscious Experience*, 177–197.

Zaidel, E., & Peters, A. M. (1981). Phonological encoding and ideographic reading by the disconnected right hemisphere: Two case studies. *Brain and Language*, *14*(2), 205–234.

CHAPTER

64

Imaging Brain Networks for Language: Methodology and Examples from the Neurobiology of Reading

Anjali Raja Beharelle[1] *and Steven L. Small*[2]

[1]Laboratory for Social and Neural Systems Research, University of Zurich, Zurich, Switzerland;
[2]Department of Neurology, University of California, Irvine, CA, USA

64.1 INTRODUCTION

Functional neuroimaging is a powerful tool for answering questions regarding how the brain implements language. Standard functional neuroimaging methods tend to treat the brain from a modular perspective, identifying sets of individual regions that are active during particular language tasks. The majority of studies examining the neurobiology of language have used standard functional neuroimaging methods that can help map out areas involved in reading and spoken language. Anatomically, the brain consists of dense and complex connections among these areas, suggesting that interregional communication plays a key role in cognitive function. More recent advances in neuroimaging methods have been developed that can be used to examine the interacting and overlapping networks of the brain regions that support reading. This permits the exploration of the functional properties of any single region with respect to the activity of the other regions within the network (neural context; McIntosh, 1998, 2000), which is essential for characterizing the integration of function among regions. In this chapter, we summarize some of the principal methods of assessing functional and effective brain connectivity and give examples of current studies using these approaches to investigate reading-related networks. We also discuss the advantage each method confers relative to the other methods.

Functional and effective connectivity can be assessed in the context of a particular behavior (task-dependent) or during rest (task-independent). Task-dependent approaches detect synchronization of activity in neural regions in response to extrinsic stimulation. Approaches involving the absence of a task examine intrinsic functional connections that are formed via spontaneous activity arising during the "resting state" (Biswal, Yetkin, Haughton, & Hyde, 1995). Resting state functional connectivity is thought to reflect a history of coactivation among regions across a wide range of tasks and time (Fox & Raichle, 2007) and can be used to characterize regions across the whole brain into groupings that show high correlations of spontaneous activity (these groups have been referred to as communities, modules, subnetworks, or clusters in network analyses; Power et al., 2011).

Connectivity analyses require the selection of a set of relevant nodes within the network to focus the investigation. This is a nontrivial undertaking because there is a lack of consensus regarding how best to define fundamental neural elements (Craddock et al., 2013) and connectivity results can vary based on choices that are made regarding node definition. The specific brain subunits encompassed by the nodes can range from patches of cortex as small as one voxel in size to macroscopic brain regions (such as the pars opercularis of the inferior frontal gyrus). When macroscopic brain regions are used, parcellation schemes have varied widely and can be anatomically based (e.g., automated anatomic labeling (AAL), Tzourio-Mazoyer et al., 2002; the Talairach Daemon, Lancaster et al., 2000; or the "Destrieux" atlas, Fischl et al., 2004) or functionally based (e.g., the CC200 and CC400 atlases that are derived from 200- and 400-unit

Neurobiology of Language. DOI: http://dx.doi.org/10.1016/B978-0-12-407794-2.00064-X

functional parcellations, respectively; Craddock, James, Holtzheimer, Hu, & Mayberg, 2012). Although these atlases share some similarities macroscopically, the specific details of the parcellations can vary considerably. Meta-analyses can also be conducted to define nodes based on previous task-based neuroimaging studies (e.g., Vogel et al., 2013).

64.2 FUNCTIONAL CONNECTIVITY ANALYSES: A SET OF EXPLORATORY TECHNIQUES

64.2.1 Overview

Functional connectivity can be defined as the synchronization between spatially remote neurophysiological events (Friston, Frith, Liddle, & Frackowiak, 1993). First introduced via early electroencephalography (EEG) and multiunit recording studies, functional connectivity analyses were applied to positron emission tomography and functional magnetic resonance imaging (fMRI) in 1993. Analyses of functional connectivity identify reliable patterns of covarying brain signals that index neural activity. These techniques are exploratory in nature because they are mostly data-driven and are not based on an explicit hypothesis or model about the relationships among neural regions or the effects of task conditions or subject groups. Therefore, a functional connection does not necessarily arise from direct communication between the two regions, because their covariance (or correlation) could be due to input from a third region (or a variety of other inputs), and thus causal inferences cannot be made about the association.

64.2.2 Independent Components Analysis

64.2.2.1 *ICA Method*

Independent components analysis (ICA) is used to take a large data set consisting of many variables and reduce it into smaller number dimensions that can be understood as self-organized functional networks (Beckmann & Smith, 2004). Unlike principal components analysis (PCA), which assumes that the components are uncorrelated in both spatial and temporal domains, ICA components are maximally statistically independent in only one domain. The rationale for ICA is that blood-oxygen–dependent (BOLD) signal measured within the voxels can be regarded as a linear combination of a smaller number of independent component sources. The independent components are identified to be maximally statistically independent, but they are not necessarily uncorrelated as principal components are (McIntosh & Mišić, 2013).

For neuroimaging analyses, independence among components can be imposed in either the spatial (spatial ICA) or the temporal (temporal ICA) domain. Spatial ICA is used more often for fMRI analyses because neural activity is assumed to be sparse among a large number of voxels. Therefore, the independent components isolate coherent networks that overlap as little as possible. However, this assumption of sparseness in the brain can be problematic because spatial ICA will push each noncontiguous activity cluster into separate components. Temporal ICA is more often used for event-related potential (ERP) data because scalp recordings have distinct time courses; therefore, the underlying components are assumed to be temporally independent but may have overlapping spatial topographies.

To compare components across participants, the ICA can be performed on all participants as a group. Here, the data from all subjects are concatenated so that each subject is treated as an observation of the same underlying system (Calhoun, Adali, Pearlson, & Pekar, 2001; Kovacevic & McIntosh, 2007). If data are concatenated along the spatial dimension, subjects will have unique spatial maps but common time courses. The converse is true if data are concatenated along the temporal dimension. The concatenated group data are then decomposed into independent components. This puts all subjects in the same space and allows them to be directly compared. Statistical inference on the independent components is then possible.

This method is useful for extracting functionally segregated networks supporting a cognitive function. The independence of the regions identified as active during task performance relative to the rest of the brain cannot be inferred from the other connectivity methods.

64.2.2.2 *ICA: Reading Network Example*

In an fMRI study, Ye, Doñamayor, and Münte (2014) used ICA to examine connectivity across the whole brain underlying semantic integration during a sentence reading task with either semantically congruent or incongruent endings. The authors extracted a functional network consisting of the supplementary motor area, left basal ganglia, left inferior frontal gyrus, left middle temporal gyrus, and left angular gyrus that was modulated by the semantic manipulation in the semantic reading task. The time courses of these regions were highly correlated and their activity was greater for incongruent versus congruent sentence endings.

64.2.3 Seed Partial Least Squares (PLS)

64.2.3.1 *PLS Method*

In general, seed-based functional connectivity techniques examine the correlation with an *a priori* region of interest (ROI) or "seed" region. In its most basic form, an averaged ROI time series is correlated with the time series of all other voxels in the brain or with the average time series of a set of ROIs. Determining the seed region can be done based on functional activity or anatomical parcellation.

A specific type of seed-based functional connectivity can be performed with partial least squares (PLS) (McIntosh, Bookstein, Haxby, & Grady, 1996). PLS is a multivariate analysis technique that can be used to identify a set of variables (called latent variables (LVs)) that optimally link spatiotemporal brain data to the task design or to behavioral measures, or, in the case of seed PLS, that link functional connectivity to other neural seed regions by extracting commonalities between them. PLS is similar to a PCA with several important differences: (i) PLS analysis is constrained to the part of the covariance matrix that is related to the time series of the given neural seed region that allows for interpretation of brain connectivity results as relating to each experimental condition; (ii) statistical inferences regarding the significance of the experimental manipulations are made using nonparametric permutation methods that allow one to select the LVs that significantly express task or behavioral effects on connectivity; and (iii) bootstrap resampling is used to retain only voxels that robustly express the task or the behavioral effects. Finally, PLS is specialized to handle larger data sets where the dependent measures are highly correlated; therefore, it is well-suited for the analysis of neuroimaging data (McIntosh & Lobaugh, 2004).

Brain activity data for PLS are organized into a 2D matrix where the rows contain scan data for each participant within each task condition and the columns consist of voxels × time. A second matrix consisting of time series of the seed neural region is similarly stacked by participant within condition within participant group. The data from the seed region are then correlated with the overall brain data of the participants and subjected to singular value decomposition (SVD). From the input matrix, SVD creates a set of orthogonal singular vectors (the LVs), which represent the entire covariance of the mean-centered matrix in decreasing order of magnitude and whose number is equal to the total number of task conditions times groups. Thus, the LVs can be thought of as being similar to the eigenvectors generated by PCA. Each LV consists of a pair of left- and right-singular vectors that relate brain connectivity to the experimental design.

The weights within the LV at each voxel–time point combination are referred to as voxel saliences. The voxel saliences identify a collection of voxels that, as a group, have connectivity to the seed region most related to the task design. The task saliences indicate the degree to which each task is related to the pattern BOLD connectivity differences. The saliences are similar to PCA eigenvector weights.

To determine how often the singular value matrix for an LV generated from the original analysis is larger than singular value matrices generated from random data, permutation testing is used. Permutation testing involves resampling *without* replacement, where data are shuffled to reassign the order of task conditions for each participant (Good, 2004). PLS is then re-run a certain number of times on each set of reordered data, and the number of times the permuted singular values exceed the original values is calculated and given a probability. The 95th percentile of the resulting probability distribution of singular values is used as the significance threshold. In this scenario, the assumption of a normal distribution is not required (McIntosh et al., 1996).

In a second step, one assesses the reliability of each voxel's contribution to the LV by estimating the standard error of the voxel saliences using bootstrapping. Bootstrapping consists of resampling *with* replacement where participant data are shuffled while the experiment conditions remain fixed. SVD is then performed on the resampled matrix consisting of participant subsets of the original data set, and the standard error of the voxels contributing to the task effects are calculated (Efron & Tibshirani, 1993). The ratio of the salience to the standard error of the voxels is used to threshold the data and can be thought of as similar to a z-score if the data are normally distributed.

Although this method is quite data-driven, the advantage of seed PLS relative to the other methods is that it permits the testing of a hypothesis focused on a particular ROI.

64.2.3.2 PLS: *Reading Network Example*

Reinke, Fernandes, Schwindt, O'Craven, and Grady (2008) examined how whole brain functional connectivity changed with the visual word form area (VWFA) based on whether participants viewed English words, meaningful symbols such as $ and %, digits, words in an unfamiliar language (Hebrew), and a control set of stimuli consisting of geometric shapes. They were able to show that while neural activity in the VWFA did not differ significantly for words and meaningful symbols, a specific functional network of regions including the left hippocampus, left lateral temporal, and left prefrontal cortices was specific to words. This study underscored the fact that the neural context of the VWFA, specifically the broader distributed brain activity that is correlated with the VWFA, is specific for visual word processing but not for activity in the focal brain region itself.

64.2.4 Synchronization of Neuronal Oscillations

64.2.4.1 SNO Method

Scalp recordings techniques such as magnetoencephalography (MEG) and EEG, which afford greater temporal resolution than fMRI, allow for real-time investigation of brain network dynamics during reading. Constituent neuronal populations involved in the same functional network are identifiable because of the fact that they fire in synchrony at a given frequency. The specificity of this frequency allows the neuronal population to participate in a variety of representations at different points in time. In addition, oscillatory synchrony can also serve to bind together information that is represented in the different neuronal populations (Gray, Konig, Engel, & Singer, 1989).

In its simplest form, one can investigate the linear interdependencies or correlations between the amplitudes of various EEG or MEG signals. This avoids the necessity to band-pass filter or extract instantaneous phase. One can also examine the cross-correlation, which includes further information about the systematic time shifts between the amplitudes of the two signals. However, studies often examine phase synchronization in frequency space. Cross-spectral density can be computed by multiplying the Fourier-transformed signals of the time series. Coherence is assessed by normalizing the cross-spectral density with the power spectral density of both time series. This value ranges from 0, if the signals have no similarity, to 1, if the signals are identical. It is critical for such a synchronization analysis to select the frequency bands of interest. Often, one can assess reliability via confidence levels by comparing to synthetic data, which are created by shuffling the time point of the original data while preserving spatial relationships (Gross et al., 2001).

The advantage of this method is that it allows the researcher to characterize dynamics among regions involved in reading on a faster temporal scale than the other methods.

64.2.4.2 SNO: Reading Network Example

Kujala et al. (2007) found the strongest synchronization among regions often implicated in reading (occipital temporal, medial, superior, and inferior temporal, prefrontal, and orbital cortices, face motor areas, insula, and cerebellum) at 8–13 Hz during a rapid reading task. Notably, regions such as the supramarginal gyrus or the posterior superior temporal cortex, which are thought to be involved in grapheme-to-phoneme conversion, were not a part of the network, potentially because the nature of the task required a more lexical-semantic than phonological reading strategy.

64.3 EFFECTIVE CONNECTIVITY ANALYSES: A SET OF CONFIRMATORY TECHNIQUES

64.3.1 Overview

Effective connectivity is defined as a directed causal influence of one region on another (Aertsen, Gerstein, Habib, & Palm, 1989). Analyses of effective connectivity involve confirmation of hypotheses. Unlike exploratory analyses, a confirmatory approach begins with the construction of an explicit model of interregional neural relationships. The model is then tested for goodness of fit with the observed data and/or whether it can fit the observed data better than an alternative model. Therefore, effective connectivity analyses test precise hypotheses that take into account external inputs and the neuroanatomical architecture rather than being data-driven like functional connectivity analyses (McIntosh & Mišić, 2013).

64.3.2 Psychophysiological Interactions (PPI)

64.3.2.1 PPI Method

If an experimental manipulation relates to significant changes in the correlation between a pair of brain regions, then this suggests an interaction between the psychological variable and the neural or physiological connectivity, termed a psychophysiological interaction (PPI) (Friston et al., 1997). In a typical functional connectivity analysis, regions may have some baseline correlations due to anatomical connections as in the case of resting state networks (Biswal et al., 1995), common sensory inputs, or neuromodulatory influences. Furthermore, if a change in correlation of activity in the seed is observed with another region, it could be caused by a change in another functional connection, a change in the level of observation noise, or a change in the amplitude of endogenous neuronal fluctuations (Friston, 2011). Therefore, a significant correlation or even a significant change in correlation cannot always be interpreted as a change in the underlying coupling between the two regions. PPI seeks to address this issue by moving beyond task-independent correlations and examining changes in correlations that occur as a result of an imposed task manipulation.

In PPI, the activity of the seed region is regressed onto the activity of another brain region across different experimental conditions and the change in slope is calculated. Just like seed PLS, the first step of PPI involves selecting a seed region and extracting its time course. The goal is to find regions with which the seed has a stronger relationship during a particular experimental condition than during the others (i.e., a task by seed region interaction). For this purpose, an interaction (PPI) regressor is created by taking the scalar product of the mean-centered task time course with the mean-corrected seed region time course. Voxels whose activity correlates only with the seed region or that show an effect of task will have some correlation to the predictor. Therefore, it is necessary to include the experimental task design and the physiological time courses from which the interaction term was created as covariates of no interest in the model. This will ensure that the variance explained by the interaction term goes significantly beyond what can already be explained by the main effects of task and correlation to the seed.

The advantage of this method is that, much like seed PLS, it allows the researcher to test a hypothesis focused on a particular ROI or voxel. However, it specifically examines how the task modulates the network based on that ROI.

64.3.2.2 *PPI: Reading Network Example*

Callan, Callan, and Masaki (2005) trained native Japanese speakers to learn the character-to-sound associations of an unknown orthography (either Thai or Korean phonograms, i.e., a grapheme that represents a phoneme or a combination of phonemes) and examined changes in brain connectivity pre- and post-training with fMRI. They found significant changes in activation post-training relative to pre-training in the left angular gyrus, and then used this region as a seed for a PPI analysis. The authors then identified a network showing greater integration of left angular gyrus activity with activity in the primary visual cortex and superior temporal gyrus for the trained phonograms after training. This finding underscored the importance of the left angular gyrus in grapheme-to-phoneme conversion.

64.3.3 Structural Equation Modeling (SEM)

64.3.3.1 *SEM Method*

The purpose of structural equation modeling (SEM) is to define a theoretical causal model consisting of a set of predicted covariances between variables and then test whether it is plausible when compared to the observed data (Jöreskog, 1970; Wright, 1934). In neuroimaging, these causal models consist of the brain activity signal of interest in a subset of ROIs and the pattern of directional influences among them (McIntosh & Gonzalez-Lima, 1991, 1994). The influences are constrained anatomically so that a direct connection between two regions is only possible if there is a known white matter pathway between them.

The first step in defining an SEM is to specify the brain regions, which are treated as variables, and the causal influences between them in terms of linear regression equations. There is always one equation for each dependent variable (activity in the ROI), and some variables can be included in more than one equation. This system of equations can be expressed in matrix notation as $\mathbf{Y} = \beta\mathbf{Y} + \psi$, where $\mathbf{Y}$ contains the variances of the regional activity for the ROIs, β is a matrix of connection strengths that defines the anatomical network model, and ψ contains residual effects, which can be thought of as either the external influences from other brain regions that cannot be stipulated in the model or the influence of the brain region on itself. Because the model is underspecified, having more unknown than known parameters, it is not possible to construct the model in a completely data-driven manner, and thus some constraints are needed on the model parameters. The most common approach is to arbitrarily restrict some elements of the residual matrix ψ to a constant, usually 35–80% of the variance for a given brain region, and to set the covariances between residuals to zero (McIntosh & Gonzalez-Lima, 1994). It is also common in neuroimaging to keep the path coefficients in both directions equal for regions that have mutually coupled paths.

The main idea of SEM is that the system of equations takes on a specific causal order, which can be used to generate an implied covariance matrix (McArdle & McDonald, 1984). Unlike in multiple regression models, where the regression coefficients are derived from the minimization of the sum of squared differences from the observed and predicted dependent variables, SEM minimizes the difference between the observed covariance structure and the one implied by the structural or path model. This is done by modifying the path coefficients and residual variances iteratively until there is no further improvement in fit. In most cases, a method such as maximum likelihood estimation or weighted least-squares is used to establish a fit criterion that needs to be maximized. The identified best-fitting path coefficient has a meaning similar to a semipartial correlation in that it reflects the influence of one region onto a second region with the influences from all other regions to the second region held constant. SEM can be conceptualized as a method that uses patterns of functional connectivity (covariances) to derive information about effective connectivity (path coefficients) (McIntosh & Mišić, 2013).

Model inference is done in SEM by comparing the goodness-of-fit between the model implied covariance matrix and the empirical covariance matrix using a χ^2 test. It is also possible to compare model fits using a χ^2 difference test, and this can be done to examine whether one or more causal influences change as the result of a task or group effect. To this end, models are combined in a single multigroup or stacked run.

The null hypothesis is that the effective connections do not differ between groups or task conditions and the null model is constructed so that path coefficients are set to be equal across groups or task conditions. The alternative hypothesis is that the effective connections are significantly different between groups or task conditions. Implied covariance matrices are generated for each group- or task-specific model. An alternative χ^2 that is significantly lower (better fitting) than the null χ^2 implies a significant group or task effect on the effective connections that were specified differently in the models. It is possible that the omnibus test can indicate a poor overall fit, but the difference test shows a significant change from one task to another. SEM has been shown to be robust in these cases and is able to detect changes in effective connectivity, even if the absolute fit of the model is insufficient (Protzner & McIntosh, 2006). Finally, it is possible to use an alternative approach to model selection, where nodes of the network are selected *a priori*, but the paths are connected in a data-driven manner (see Bullmore et al., 2000).

The advantage of SEM is that one can identify directionality in the influence of activity from one region to that of another. In addition, SEM allows the researcher to test the validity of a theoretical model regarding network interactions among regions supporting the task under investigation.

64.3.3.2 SEM: Reading Network Example

Levy et al. (2009) used SEM to test neuroanatomical predictions made by the dual-route cascade reading model (Coltheart, Rastle, Perry, Langdon, & Ziegler, 2001) on reading skill. Their effective connectivity models consisted of four left hemisphere ROIs: middle occipital gyrus (MOG); occipito-temporal junction (LOT); parietal cortex (LP); and inferior frontal gyrus (IFG). For reading words, MOG→LP, MOG→LOT, and LP→IFG pathways were significantly more involved than the LOT→LP path, suggesting that information traffics along both ventral and dorsal pathways during word reading. For pseudoword reading, MOG→LOT and LOT→LP were significantly more involved than MOG→LP, suggesting that information first flows to LOT before being transferred to the dorsal pathway. In addition, increased reliance on the "word pathway" (MOG→LP) positively correlated with reading skill and increased reliance on the "pseudoword pathway" (MOG→LOT) correlated with the pseudoword reading ability. Their findings are in agreement with the DRC model, suggesting that regular words can be read in two ways (via parallel dorsal and ventral stream processing); however, the dorsal pathway is selective for word stimuli and increased connectivity in this pathway is related to better word reading. Pseudowords, however, undergo letter/sublexical analysis in the posterior ventral pathway before being fed to the dorsal path.

64.3.4 Dynamic Causal Modeling (DCM)

64.3.4.1 DCM Method

The key concept behind Dynamic Causal Modeling (DCM) is that brain networks comprise an input-state-output system, where causal interactions are mediated by unobservable neuronal dynamics (Friston, Harrison, & Penny, 2003). Referred to as a causal model, these "hidden" interactions are specified by coupling parameters, which denote the degree of synaptic coupling and model effective connectivity. The local neuronal dynamics underlying these interactions are defined by a set of differential equations. This causal model is then combined with a forward, or observation, model that relates the mapping from the neuronal activity to the observed responses (Friston, Moran, & Seth, 2013). It is important to note that causality is inferred at the neuronal population level and not at the level of the observed responses.

In the causal model, each ROI consists of neuronal populations that are intrinsically coupled to each other and extrinsically coupled to neuronal populations of the other regions in the network. Stochastic or ordinary differential equations relate the present state of a neuronal population to the future state of the same neuronal population and to the states of the other populations. The coupling parameters can be thought of as rate constants, which determine how rapidly one population affects another (McIntosh & Mišić, 2013). Experimental effects are modeled as external perturbations to the system, which can cause either a change in coupling or a change in neuronal activity of a population. The underlying causal model is represented as a system of coupled differential equations, where the rate of changes of state x is a function of the states of the other populations (x), the external inputs, and the coupling parameters, which are unknown and need to be inferred similarly to the path coefficients in SEM. DCMs do not stipulate any particular biophysical model of neuronal activity and only require the model to be biologically plausible and sufficiently able to explain the external perturbations and the interactions between neuronal populations.

In the second step of DCM, the forward model is used to map neuronal states into the observed signal measurements while incorporating unknown parameters. The form of the forward model depends on the imaging modality. For ERPs, for example, the mapping function is the lead field matrix that models the propagation and volume conduction of electromagnetic fields through neural tissue, cerebrospinal fluid, skull, and skin. The unknown parameters that are introduced are the location and orientation of the source dipole (Kiebel, David, & Friston, 2006). If BOLD signal contrast is the observed measure, then the mapping function models how changes in neuronal activity engender changes in local blood flow, which result in an influx of oxygenated blood and a reduction in deoxygenated hemoglobin (Buxton, Wong, & Frank, 1998). Here, the unknown parameters can specify factors such as the rate constants of vasodilatory signal decay and capillary resting net oxygen extraction (Stephan, Weiskopf, Drysdale, Robinson, & Friston, 2007).

DCMs use a Bayesian approach (see Friston et al., 2002) for estimating the unknown parameters in the models. Essentially, a posterior distribution is estimated for each parameter using an optimization algorithm and taking into account prior beliefs about the value the parameter can realistically take on and the observed data. The observed data are used to update the model (i.e., estimate the parameters) to maximize model evidence in a procedure known as Bayesian model inversion. The model evidence accounts for the ability of the model to explain the data as accurately as possible and to have the fewest parameters (i.e., parsimonious models are rewarded). Models can be compared by taking the ratio of their respective model evidences or the difference in their respective log evidences, and model comparisons can be made to determine group or task effects on anatomical connections in a similar manner to SEMs; however, DCMs can also vary in the specification of their priors for different task or group treatments.

Model inversion is performed on each subject individually. Therefore, an experimenter must decide whether to keep the same model for all subjects in a between-subjects analysis. If experimenters choose this, then they can multiply the model evidences or add the log evidences across subjects to get the group model evidence, and this essentially represents a fixed-effects analysis. In this case, one approach to obtaining group-level estimates of the parameters is to compute a joint density for the subject-specific posterior distribution estimates. If the experimenters want to treat the subjects as heterogeneous, then they can take the ratio of the number of subjects who show positive model evidence for a given model to the number of subjects who show greater model evidence for another model (Stephan et al., 2007), and this is essentially a random-effects analysis. In this case, it is common to take a summary statistic of the subject-specific posterior distribution estimates (such as the median or mode of the distribution) as the parameter estimate and conduct traditional random-effects analyses comparing the group means of these summary statistics (e.g., t-test).

Like SEM, DCM can show the directional influences of regions on one another. Whereas SEM is used to test theoretical models of network interactions based on what is known about anatomical connections, DCM is meant to provide a more explanatory understanding of the relationships among regions identified to be active during task-based fMRI analyses.

64.3.4.2 DCM: Reading Network Example

The triangular part of the inferior frontal gyrus is thought to be particularly involved in semantic processing, and the opercular part is thought to have more of a role in phonological tasks (e.g., Poldrack et al., 1999). Mechelli et al. (2005) tested this theoretical framework using DCM. Effective connectivity from the anterior fusiform gyrus to the pars triangularis increased for exception words (e.g., PINT or STEAK), necessitating more lexical-semantic processing relative to pseudowords and between dorsal premotor cortex and posterior fusiform gyrus for pseudowords (e.g., RINT or MAVE) requiring more phonological mapping relative to exception words. This finding demonstrated distinct neuronal mechanisms for semantic and phonological processing and confirmed previous theories of dissociation of function in the inferior frontal gyrus.

64.4 TECHNIQUES SPANNING BOTH FUNCTIONAL AND EFFECTIVE DOMAINS

64.4.1 Granger Causality (GC)

64.4.1.1 GC Method

Granger causality (GC) does not fit easily into the functional or effective connectivity because it has both exploratory and confirmatory characteristics. The main idea behind GC is that B "Granger causes" A if B contains information that helps predict the future of A better than information in the past of A predicts or information in the past of other conditioning variables, C (Friston et al., 2013). The GC measure is based on the relative change in the model error when new time series are added to improve the prediction of the dependent signal (Granger, 1969). Essentially, GC is the ratio of the variance of the model before and after the addition of the new time series (time series "B") in this case:

$$Fy \rightarrow x = \ln \frac{\mathrm{Var}(ea||a)}{\mathrm{Var}(ea||ab)}$$

In the context of brain networks, multiple predictors comprising the past time series of all ROIs can be specified to account for the present time series of all ROIs. If one assumes the effects to be linear, then the relationships can be specified using multivariate linear regression (through what is termed multivariate vector autoregressive (MVAR) modeling; Goebel, Roebroeck, Kim, & Formisano, 2003). An MVAR model contains every possible connection in the network, and each connection is tested to determine which ones are nonzero. This allows subnetworks to be extracted without having to specify connectivity patterns *a priori*. For any given connection, the influence of all other nodes are partialled out, allowing one to obtain an estimate for whether the past time series of B helps predict the time series of A more than what is accounted for by all other variables combined, C. The coefficients in the MVAR can be estimated using ordinary least-squares (by minimizing the difference in the sum of squared errors between the predicted and observed values of the present times series).

The advantage of Granger causality is that it allows a researcher to pinpoint directional influences of regions on another without any *a priori* hypothesis regarding which regions are involved in particular subnetworks. This is the case because subnetworks are identified in a data-driven manner.

64.4.1.2 GC: Reading Network Example

In a MEG study, Frye, Liederman, McGraw Fisher, and Wu (2012) used GC to examine the connectivity of bilateral temporoparietal areas (TPAs) in dyslexic and typical readers during a nonword reading task. The important feature of this study is that GC allowed them to examine hierarchical network structure (i.e., which nodes show dominant influences of the other nodes) rather than connectivity alone. In the beta frequency band, those participants with greater connectivity from the left TPA to other regions (left TPA dominant) were more likely to show improved phonological decoding performance, and those with greater connectivity from other regions into the TPA were more likely to show poorer phonological decoding performance across both groups of participants. It may be the case that participants in which a hierarchical network topography is manifested (i.e., those with greater outward connectivity of TPA) also show more stable network processing, allowing them to optimally process stimuli, because neural networks with hierarchical structures have been shown to be more stable compared to neural networks with nonhierarchical topography. Second, greater relative outward connectivity of the right TPA to other brain areas was associated with worse performance in dyslexic readers. This finding extends previous studies of dyslexic readers that reported greater TPA activity by indicating that the direction of influence of the right TPA onto other brain regions may play a key role in dyslexia.

64.4.2 Graph Theory

64.4.2.1 Graph Theory Method

Graph theoretic measures can be applied to structural, functional, and effective connectivity matrices (Bullmore & Sporns, 2009). Unlike the metrics previously discussed, which are often limited by which seeds or subset of nodes are chosen *a priori*, graph theoretic measures provide information regarding changes in the topology of the whole brain as well as the role that the individual nodes play within that topology, thereby providing a comprehensive assessment of patterns of information flow. This approach allows for the investigation of functional integration and functional segregation within the global brain network architecture.

The first step of such an analysis is to construct a graph that consists of nodes (i.e., the neural elements of interest) and edges, representing the statistical dependencies or connections between the nodes. For structural connectivity graphs, the edges correspond to anatomical white matter pathways; for functional connectivity graphs, the edges are pairwise associations between the brain signals of the nodes; and for effective connectivity graphs, the edges consist of pairwise measures of the causal influence of one node onto another. Structural brain connectivity yields a sparse and directed (or asymmetric) graph. Functional and effective brain connectivity give rise to full undirected (or symmetric) or directed graphs, respectively, which can be further reduced by setting a threshold to control the degree of sparsity. All graphs may be weighted, with the weights representing connection densities or efficacies, or binary, indicating the presence or absence of a connection.

Once the graph is constructed, various graph theoretic measures can be computed (see Rubinov & Sporns, 2010). One can simply measure the connectedness of each node by counting its total number of connections (degree). To investigate integration, one can examine the least number of steps necessary to get from one node to another (shortest path length) as well as the characteristic path length for the entire network. The network's global efficiency is a related measure that is the average inverse shortest path length. To examine segregation, one can compute the clustering coefficient, which is the fraction of the node's neighbors that are also neighbors of each other. One can also partition the brain into subnetworks or communities within the whole brain and then examine their interactions by computing modularity, which is the ratio of the density of connections within specific subnetworks compared to the density of connections between subnetworks. Once the brain has been partitioned into subnetworks, it is possible to identify which nodes act as connector hubs (the terminology for these nodes varies, but here we mean nodes that play a strong role in linking subnetworks to each other) by computing their participation coefficient and which nodes act as modular hubs (nodes that have greater connections within their own subnetwork) by computing their within-module degree *z*-score. Finally, it is possible to assess the frequency with which certain combinations of nodes or edges occur by looking at motifs. Some metrics, such as modularity or characteristic path length, produce one value per subject and can be submitted to standard between-subject univariate tests. The node-specific measures can be used to generate topological maps for each subject, and some correction for multiple comparisons must be performed for inferential analyses about robust effects.

The advantage of graph theory, unlike most of the other methods that require confining analyses to *a priori* regions thought to be involved in a functional network, is that we have the capability of examining and quantifying interactions among regions on a whole-brain level, thus giving us a comprehensive view of functional network architecture while still being able to assign specific connectivity roles within the network to certain nodes.

64.4.2.2 Graph Theory: Reading Network Example

Vogel et al. (2013) examined the functional subnetwork structure during resting state in an attempt to identify a reading-dedicated community. Unlike previous studies that have examined this, the authors only included prominent reading-related regions derived from a meta-analysis (including left supramarginal gyrus, angular gyrus, middle temporal gyrus, and inferior frontal gyrus) as nodes in the analysis. Rather than cluster into their own community, reading-related regions were assigned to other communities whose primary function is more general than reading (e.g., the VWFA was assigned to a visual community, and the angular gyrus was assigned to a default mode community), suggesting the lack of an intrinsic reading community in the brain. This was the case with mature as well as developmental subjects, suggesting that the regions implicated in reading are broadly used across many tasks, including reading. Because this study was undertaken during resting state, it does not rule out that there may be special relationships among some of these regions during task-based reading.

64.5 CONCLUSIONS

Connectivity analyses identifying neural networks move beyond isolating a collection of individual regions that are associated with reading and provide an account of how neural regions interact with one another. Exploratory methods such as ICA and PLS aim to reduce the original data into interpretable functional networks. Effective connectivity analyses like SEM and DCM are useful, particularly when trying to test *a priori* anatomical or theoretical hypotheses about the neurobiology underlying reading (e.g., the DRC model). Graph theoretic measures take into account the global topology of neural networks while also specifying the role of individual regions within a network or subnetwork. These measures can identify the patterns of information flow among regions across the whole brain rather than examining interactions with a seed or within a subset of region, and they have been used to test whether reading-specific subnetworks exist in the brain.

References

Aertsen, A., Gerstein, G., Habib, M., & Palm, G. (1989). Dynamics of neuronal firing correlation: Modulation of "effective connectivity.". *Journal of Neurophysiology, 61*, 900–917.

Beckmann, C., & Smith, S. (2004). Probabilistic independent component analysis for functional magnetic resonance imaging. *IEEE Transactions on Medical Imaging, 23*, 137–152.

Biswal, B., Yetkin, F. Z., Haughton, V. M., & Hyde, J. S. (1995). Functional connectivity in the motor cortex of resting human brain using echo-planar MRI. *Magnetic Resonance in Medicine, 34*, 537–541.

Bullmore, E., Horwitz, B., Honey, G., Brammer, M., Williams, S., & Sharma, T. (2000). How good is good enough in path analysis of fMRI data? *NeuroImage, 11*, 289–301.

Bullmore, E., & Sporns, O. (2009). Complex brain networks: Graph theoretical analysis of structural and functional systems. *Nature Reviews Neuroscience, 10*, 186–198.

Buxton, R. B., Wong, E. C., & Frank, L. R. (1998). Dynamics of blood flow and oxygen metabolism during brain activation: The balloon model. *Magnetic Resonance in Medicine, 39*, 855–864.

Calhoun, V., Adali, T., Pearlson, G., & Pekar, J. (2001). A method for making group inferences from functional MRI data using independent component analysis. *Human Brain Mapping, 14*, 140–151.

Callan, A. M., Callan, D. E., & Masaki, S. (2005). When meaningless symbols become letters: Neural activity change in learning new phonograms. *NeuroImage, 28*(3), 553–562.

Coltheart, M., Rastle, K., Perry, C., Langdon, R., & Ziegler, J. (2001). DRC: A dual route cascaded model of visual word recognition and reading aloud. *Psychological Review, 108*, 204–256.

Craddock, R. C., James, G. A., Holtzheimer, P. E., Hu, X. P., & Mayberg, H. S. (2012). A whole brain fMRI atlas generated via spatially constrained spectral clustering. *Human Brain Mapping, 33*, 1914–1928.

Craddock, R. C., Jbabdi, S., Yan, C. G., Vogelstein, J. T., Castellanos, F. X., Di Martino, A., et al. (2013). Imaging human connectomes at the macroscale. *Nature Methods, 10*(6), 524–539.

Efron, B., & Tibshirani, R. (1993). *An introduction to the bootstrap*. New York, NY: Chapman & Hall.

Fischl, B., van der Kouwe, A., Destrieux, C., Halgren, E., Segonne, F., Salat, D. H., et al. (2004). Automatically parcellating the human cerebral cortex. *Cerebral Cortex, 14*, 11–22.

Fox, M. D., & Raichle, M. E. (2007). Spontaneous fluctuations in brain activity observed with functional magnetic resonance imaging. *Nature Reviews Neuroscience, 8*, 700–711.

Friston, K., Buechel, C., Fink, G., Morris, J., Rolls, E., & Dolan, R. (1997). Psychophysiological and modulatory interactions in neuroimaging. *NeuroImage, 6*, 218–229.

Friston, K., Harrison, L., & Penny, W. (2003). Dynamic causal modelling. *NeuroImage, 19*, 1273–1302.

Friston, K., Moran, R., & Seth, A. K. (2013). Analysing connectivity with Granger causality and dynamic causal modeling. *Current Opinions in Neurobiology, 23*(2), 172–178.

Friston, K. J. (2011). Functional and effective connectivity: A review. *Brain Connectivity, 1*(1), 13–36.

Friston, K. J., Frith, C. D., Liddle, P. F., & Frackowiak, R. S. (1993). Functional connectivity: The principal-component analysis of large (PET) data sets. *Journal of Cerebral Blood Flow and Metabolism, 13*, 5–14.

Friston, K. J., Penny, W., Phillips, C., Kiebel, S., Hinton, G., & Ashburner, J. (2002). Classical and Bayesian inference in neuroimaging: Theory. *NeuroImage, 16*(2), 465–483.

Frye, R. E., Liederman, J., McGraw Fisher, J., & Wu, M. H. (2012). Laterality of temporoparietal causal connectivity during the prestimulus period correlates with phonological decoding task performance in dyslexic and typical readers. *Cerebral Cortex, 22*(8), 1923–1934.

Goebel, R., Roebroeck, A., Kim, D., & Formisano, E. (2003). Investigating directed cortical interactions in timeresolved fMRI data using vector autoregressive modeling and Granger causality mapping. *Magnetic Resonance Imaging, 21*, 1251–1261.

Good, P. I. (2004). *Permutation, parametric, and bootstrap tests of hypotheses* (3rd ed.). New York, NY: Springer-Verlag.

Granger, C. W. J. (1969). Investigating causal relations by econometric models and cross-spectral methods. *Econometrica, 37*, 424–438.

Gray, C. M., Konig, P., Engel, A. K., & Singer, W. (1989). Oscillatory responses in cat visual cortex exhibit inter-columnar synchronization which reflects global stimulus properties. *Nature, 338*, 334–337.

Gross, J., Kujala, J., Hämäläinen, M., Timmermann, L., Schnitzler, A., & Salmelin, R. (2001). Dynamic imaging of coherent sources: Studying neural interactions in the human brain. *Proceedings of the National Academy of Sciences, 98*, 694–699.

Jöreskog, K. G. (1970). A general method for analysis of covariance structures. *Biometrika, 57*, 239–251.

Kiebel, S., David, O., & Friston, K. (2006). Dynamic causal modelling of evoked responses in EEG/MEG with lead field parameterization. *NeuroImage, 30*, 1273–1284.

Kovacevic, N., & McIntosh, A. (2007). Groupwise independent component decomposition of EEG data and partial least square analysis. *NeuroImage, 35*, 1103–1112.

Kujala, J., Pammer, K., Cornelissen, P., Roebroeck, A., Formisano, E., & Salmelin, R. (2007). Phase coupling in a cerebro-cerebellar network at 8–13 Hz during reading. *Cerebral Cortex, 17*(6), 1476–1485.

Lancaster, J. L., Woldorff, M. G., Parsons, L. M., Liotti, M., Freitas, C. S., Rainey, L., et al. (2000). Automated Talairach atlas labels for functional brain mapping. *Human Brain Mapping, 10*, 120–131.

Levy, J., Pernet, C., Treserras, S., Boulanouar, K., Aubry, F., Démonet, J. F., et al. (2009). Testing for the dual-route cascade reading model in the brain: An fMRI effective connectivity account of an efficient reading style. *PLoS One, 18*(4), e6675.

McArdle, J., & McDonald, R. (1984). Some algebraic properties of the reticular actionmodel for moment structures. *The British Journal of Mathematical and Statistical Psychology, 37*, 234–251.

McIntosh, A. (1998). Understanding neural interactions in learning and memory using functional neuroimaging. *Annals of the New York Academy of Sciences, 855*, 556–571.

McIntosh, A. (2000). Towards a network theory of cognition. *Neural Networks, 13*, 861–870.

McIntosh, A., & Gonzalez-Lima, F. (1991). Structural modeling of functional neural pathways mapped with 2-deoxyglucose: Effects of acoustic startle habituation on the auditory system. *Brain Research, 547*, 295–302.

McIntosh, A., & Gonzalez-Lima, F. (1994). Structural equation modeling and its application to network analysis in functional brain imaging. *Human Brain Mapping, 2*, 2–22.

McIntosh, A. R., Bookstein, F. L., Haxby, J. V., & Grady, C. L. (1996). Spatial pattern analysis of functional brain images using Partial Least Squares. *NeuroImage, 3*, 143–157.

McIntosh, A. R., & Lobaugh, N. J. (2004). Partial least squares analysis of neuroimaging data: Applications and advances. *NeuroImage, 23*, S250–S263.

McIntosh, A. R., & Mišić, B. (2013). Multivariate statistical analyses for neuroimaging data. *Annual Review of Psychology, 64*, 499–525.

Mechelli, A., Crinion, J. T., Long, S., Friston, K. J., Lambon Ralph, M. A., Patterson, K., et al. (2005). Dissociating reading processes on the basis of neuronal interactions. *Journal of Cognitive Neuroscience, 17*, 1753–1765.

Poldrack, R. A., Wagner, A. D., Prull, M. W., Desmond, J. E., Glover, G. H., & Gabrieli, J. D. (1999). Functional specialization for semantic knowledge and phonological processing in the left inferior prefrontal cortex. *NeuroImage, 10*, 15–35.

Power, J. D., Cohen, A. L., Nelson, S. M., Wig, G. S., Barnes, K. A. B., Church, J. A., et al. (2011). Functional network organization of the human brain. *Neuron, 72*(4), 665–678.

Protzner, A., & McIntosh, A. (2006). Testing effective connectivity changes with structural equation modeling: What does a bad model tell us? *Human Brain Mapping, 27*, 935–947.

Reinke, K., Fernandes, M., Schwindt, G., O'Craven, K., & Grady, C. L. (2008). Functional specificity of the visual word form area: General activation for words and symbols but specific network activation for words. *Brain and Language, 104*, 180–189.

Rubinov, M., & Sporns, O. (2010). Complex network measures of brain connectivity: Uses and interpretations. *NeuroImage, 52*, 1059–1069.

Stephan, K., Weiskopf, N., Drysdale, P., Robinson, P., & Friston, K. (2007). Comparing hemodynamic models with DCM. *NeuroImage, 38*, 387–401.

Tzourio-Mazoyer, N., Landeau, B., Papathanassiou, D., Crivello, F., Etard, O., Delcroix, N., et al. (2002). Automated anatomical labeling of activations in SPM using a macroscopic anatomical parcellation of the MNI MRI single-subject brain. *NeuroImage, 15*, 273–289.

Vogel, A. C., Church, J. A., Power, J. D., Miezin, F. M., Petersen, S. E., & Schlaggar, B. L. (2013). Functional network architecture of reading-related regions across development. *Brain and Language, 125*(2), 231–243.

Wright, S. (1934). The method of path coefficients. *The Annals of Mathematical Statistics, 5*, 161–215.

Ye, Z., Doñamayor, N., & Münte, T. F. (2014). Brain network of semantic integration in sentence reading: Insights from independent component analysis and graph theoretical analysis. *Human Brain Mapping*.

CHAPTER

65

Developmental Dyslexia

Guinevere F. Eden, Olumide A. Olulade, Tanya M. Evans, Anthony J. Krafnick and Diana R. Alkire

Center for the Study of Learning, Georgetown University, Washington, DC, USA

65.1 INTRODUCTION

65.1.1 What Is Developmental Dyslexia?

Unlike spoken language, available to humans for many thousands of years, written language is a recent (less than 6,000 years) cultural invention and has not been subjected to evolutionary pressure. As such, existing brain regions that subserve other functions, such as spoken language and object recognition, are utilized for reading acquisition, which occurs over a protracted period of time with formal schooling. However, for some children, learning to read is especially difficult.

> Developmental dyslexia is characterized by difficulties with accurate and/or fluent word recognition and by poor spelling and decoding abilities. These difficulties typically result from a deficit in the phonological component of language that is often unexpected in relation to other cognitive abilities and the provision of effective classroom instruction. Secondary consequences may include problems in reading comprehension and reduced reading experience that can impede growth of vocabulary and background knowledge (Lyon, Shaywitz, & Shaywitz, 2003).

This is one of several research definitions of dyslexia and was developed by the International Dyslexia Association and endorsed by the US-based National Institute of Child Health and Human Development. Like most definitions, it emphasizes problems with decoding, a critical piece in research and diagnostic evaluation of dyslexia that is attributed to poor phonological awareness (PA). Comprehension of text can be impacted as a secondary consequence of poor decoding (this distinguishes dyslexia from specific language impairment, wherein poor language comprehension can directly lead to reading problems; see Nation, Cocksey, Taylor, & Bishop, 2010).

Dyslexia is the most prevalent learning disability. It affects 5–12% of the English-speaking population, and the incidence is two to three times higher in males than in females (Rutter et al., 2004). These rates are slightly lower for other languages and other writing systems (Brunswick, McDougall, & Davies, 2010), with reading being espetrun cially challenging in English, where the orthography is deep (the mapping between sound and print is not one-to-one; see Richlan, 2014). Dyslexia is highly heritable, with an estimated 30–50% chance of being passed from parent to child (Fisher & DeFries, 2002). Linkage and association studies have identified candidate genes and established relationships between genotypical variance and the dyslexia endophenotype (for review, see Scerri & Schulte-Körne, 2010). Because reading provides the key to learning almost all subject materials, reading failure can be a limiting factor in almost all of a child's academic learning experiences.

65.1.2 Skills That Support Typical Reading Acquisition are Impaired in Dyslexia

There is a rich behavioral literature describing normal reading acquisition (Ehri, 1999) and the skills that promote learning to read (Schatschneider, Fletcher, Francis, Carlson, & Foorman, 2004; Wagner & Torgesen, 1987). These include broad language skills such as expressive and receptive language, vocabulary, morphology, and syntax (Scarborough, 2005). Further, specific skills such as PA (Bradley & Bryant, 1983; Schatschneider et al., 2004) and orthographic awareness (Badian, 1994) have been shown to be critical to reading acquisition. Understanding the phonological code allows grapheme–phoneme mapping of unfamiliar words, and visual word form recognition aids in mapping orthography of familiar

Neurobiology of Language. DOI: http://dx.doi.org/10.1016/B978-0-12-407794-2.00064-X

words to the mental lexicon, together providing access to semantic representations. Good orthographic awareness, whereby words are recognized without decoding, leads to better reading fluency and, in turn, reading comprehension. PA refers to a "broad class of skills that involve attending to, thinking about, and intentionally manipulating the phonological aspects of spoken language" (Scarborough & Brady, 2002). PA, typically measured by sound manipulation at the phoneme level (e.g., phoneme deletion), predicts future reading acquisition in normal readers with a high degree of confidence (Schatschneider et al., 2004; Wagner & Torgesen, 1987) and is especially impaired in dyslexic readers (Lyon et al., 2003; Peterson & Pennington, 2012; Vellutino, Fletcher, Snowling, & Scanlon, 2004). These studies have also shown that two other measures of phonological processing, speeded lexical retrieval (rapid naming of letters and numbers) and verbal short-term memory, often have a moderating role in reading outcome in addition to PA.

It has become widely accepted that weak PA is the core deficit in dyslexia, causing reading impairment by interfering with the grapheme–phoneme mapping required for decoding. A causal role of poor PA in dyslexia has been demonstrated by a combination of evidence: (i) young children's PA skills predict later reading outcome; (ii) poorly developed PA skills are found in children with dyslexia as early as kindergarten and often prevail into adulthood; (iii) reading level-match design studies demonstrate that children with dyslexia have weaker PA than younger children matched on reading level; and (iv) interventions addressing these weaknesses in PA are largely successful in bringing about gains in decoding in individuals with dyslexia. Together, these have led to the theory that a phonological core deficit best describes the condition of dyslexia (McCardle, Scarborough, & Catts, 2001; Stanovich & Siegel, 1994). Although this account is not likely to represent the entire explanation of this reading disability (Peterson & Pennington, 2012; Scarborough, 2005), and although alternative theories exist, it is most relevant to the neurobiology of language and thus provides the framework for this chapter. For discussion of theoretical frameworks such as auditory temporal, motor timing, automaticity-based cerebellar, and visual magnocellular deficits, we refer the reader to other in-depth reviews (Ramus, 2004; Vellutino et al., 2004).

65.2 FUNCTIONAL ANATOMY OF READING

As described in detail in the present volume (Cathy Price and Karalyn Patterson) and elsewhere (Price, 2012), reading is supported by a network of regions in the left hemisphere, including ventral (occipitotemporal), dorsal (temporoparietal), and inferior frontal cortices (Figure 65.1). The occipitotemporal cortex (OTC) holds the so-called visual word form system (VWFS), specifically the "visual word form area" (VWFA), which is responsible for visual identification of words. Both the temporoparietal (TPC) and inferior frontal (IFC) cortices play a role in phonological and semantic processing of words, with IFC also involved in articulatory processes. These areas have all been shown to be altered in dyslexia (for review see Gabrieli, 2009; Pugh et al., 2001; Sandak, Mencl, Frost, & Pugh, 2004).

65.3 NEUROANATOMICAL BASES OF DYSLEXIA

A neuroanatomical basis of dyslexia was first discovered during postmortem examinations of gross anatomy of brains of adults who had dyslexia during their lifetime. Most notably, the asymmetry typically seen in the planum temporale, favoring the left

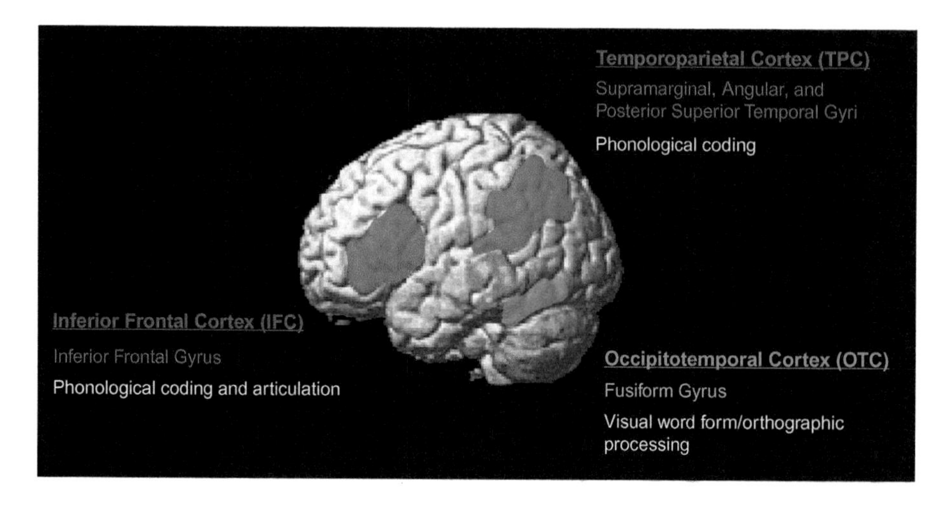

FIGURE 65.1 Schematic representation of brain regions involved in reading and reading-related processes.

hemisphere in size, was not found in the brains of these dyslexics (Galaburda & Kemper, 1979). The same investigators also discovered ectopias on the cortical surfaces of these brains, presumed to be a result of improper cortical migration (Galaburda, Sherman, Rosen, Aboitiz, & Geschwind, 1985), a finding that has received renewed interest in the context of dyslexia-associated genes involved in axonal guidance during development (Galaburda, LoTurco, Ramus, Fitch, & Rosen, 2006).

Manual tracing of magnetic resonance imaging (MRI) structural images later allowed for the replication of differences in the planum temporale *in vivo*, along with more general differences in TPC (Hynd, Semrud-Clikeman, Lorys, Novey, & Eliopulos, 1990; Larsen, Høien, Lundberg, & Ødegaard, 1990; Leonard et al., 2001). New observations were reported for the cerebellum (Eckert et al., 2003; Leonard et al., 2001) and inferior frontal gyrus (IFG) (Robichon, Levrier, Farnarier, & Habib, 2000). Not surprisingly, given the variability in measurement techniques, some reports conflicted. The advent of automated image processing techniques has enabled more quantitative and investigator-independent examinations of brain differences in dyslexia. Notably, voxel-based morphometry (VBM), using fully or semiautomated algorithms, has now been used in several studies of whole-brain gray matter volume (GMV) in adults and children (for recent review of details see Evans, Flowers, Napoliello, & Eden, 2014). These studies suggest that the most reliable differences in dyslexia are located in bilateral temporal lobe structures (inferior, middle, and superior gyri), inferior parietal lobes, and cerebellum. This has been confirmed by meta-analyses of existing VBM studies of GMV in dyslexia (Linkersdörfer, Lonnemann, Lindberg, Hasselhorn, & Fiebach, 2012; Richlan, Kronbichler, & Wimmer, 2013). For example, Richlan and colleagues (2013) found, as shown in Figure 65.2A, less GMV in right superior temporal gyrus (STG) and left superior temporal sulcus (STS).

VBM has also been used to examine white matter volume (WMV) anomalies in dyslexia, and less WMV has been found within left temporoparietal regions in dyslexic children (Eckert et al., 2005). Diffusion tensor imaging (DTI) studies have shown results of lesser fractional anisotropy (FA) in dyslexic individuals compared with controls, indicating differences in white matter integrity primarily in temporoparietal and frontal areas (Klingberg et al., 2000; for review see Vandermosten, Boets, Wouters, & Ghesquière, 2012). A meta-analysis of DTI studies in dyslexia localized the most common finding to a left temporoparietal region (Figure 65.2B) and used tractography to demonstrate that this region hosts the arcuate fasciculus (also referred to as superior longitudinal fasciculus) and the corona radiata (Vandermosten et al., 2012). Combined with related observations of correlations between FA values and reading skill in left temporoparietal tracts (Odegard, Farris, Ring, McColl, & Black, 2009), these findings suggest a loss of connections between temporoparietal and frontal areas and interruption of networks that subserve phonological processing (Boets, 2014).

65.4 NEUROFUNCTIONAL BASES OF DYSLEXIA

Early functional studies were limited to adults because they used xenon (Flowers, Wood, & Naylor, 1991) and positron emission tomography (PET) techniques (Gross-Glenn et al., 1991; Paulesu et al., 1996; Rumsey et al., 1992). Then, magnetoencelography (MEG) and functional magnetic resonance imaging (fMRI) became available in the 1990s and provided better temporal and spatial resolution, respectively, without the concern of radioactive tracers. Reading and its constituent components (phonology, orthography, and semantics) were examined using a range of overt and covert tasks, mostly using single word presentation. Despite variations in tasks and participants, a substantial corpus of publications from different countries achieved convergence in their findings of left hemisphere areas, including: (i) the inferior ventral visual stream, broadly referred to here as OTC; (ii) posterior dorsal TPC, including the posterior superior temporal gyrus (pSTG: within Wernicke's area), the angular gyrus (AG), supramarginal gyrus (SMG), and the inferior parietal lobule (IPL); and (iii) IFC, including IFG within Broca's area. For example, lower signals in dyslexic compared with control groups have been demonstrated in the left OTC (Cao, Bitan, Chou, Burman, & Booth, 2006; Georgiewa et al., 1999; Hu et al., 2010; Paulesu, 2001; Richlan et al., 2010; Shaywitz et al., 1998; Siok, Niu, Jin, Perfetti, & Tan, 2008; Van der Mark et al., 2009; Wimmer et al., 2010; see also Salmelin, Service, Kiesilä, Uutela, & Salonen, 1996) and the left TPC (Brunswick, McCrory, Price, Frith, & Frith, 1999; Cao et al., 2006; Eden et al., 2004; Georgiewa et al., 1999; Paulesu, 2001; Richlan et al., 2010; Shaywitz et al., 1998, 2003; Wimmer et al., 2010). OTC dysfunction has been interpreted in terms of problems with word form processing, whereas less TPC activity has been attributed to poor phonological processing (Pugh et al., 2001; Shaywitz et al., 1998). Findings in the IFG have conflicting results, with some studies reporting hypoactivation in dyslexics relative to controls (Eden et al., 2004; Siok et al., 2008) and others reporting hyperactivation (Shaywitz et al., 1998) or no differences (Paulesu, 2001; Paulesu et al., 1996).

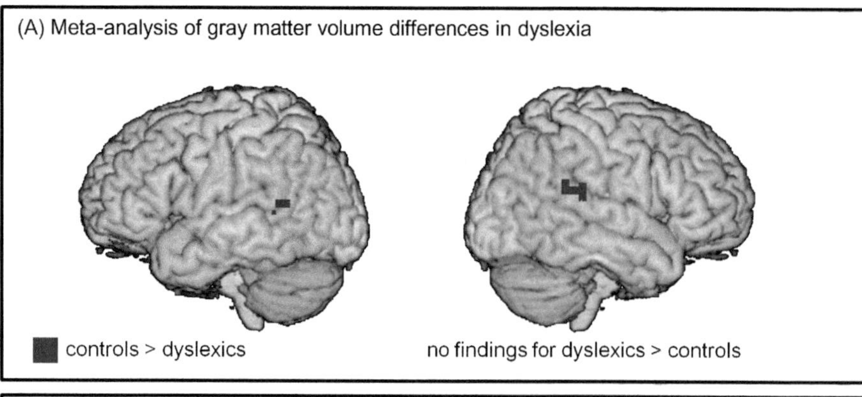

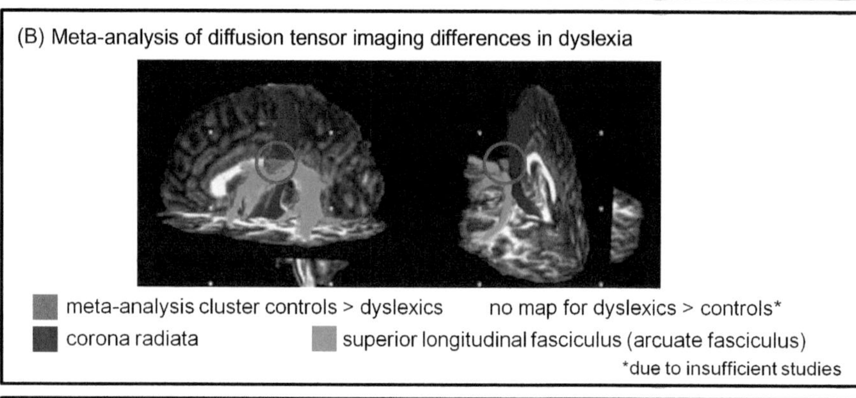

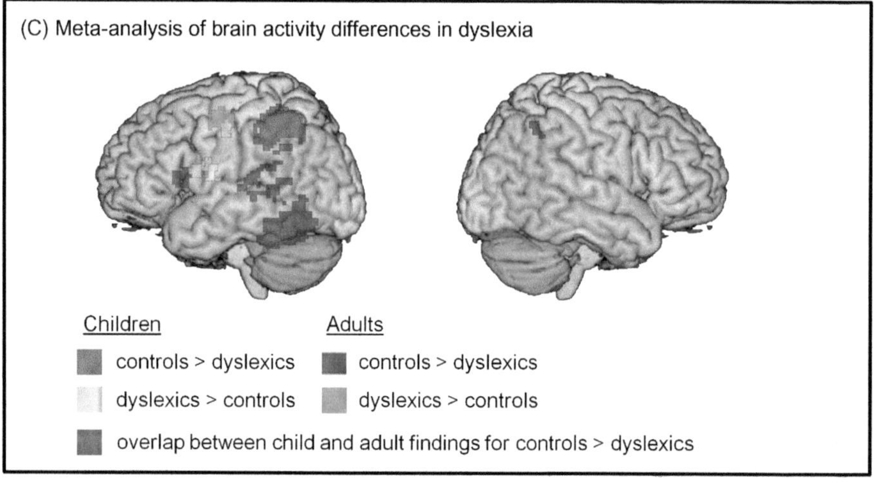

FIGURE 65.2 (A) Surface rendering of meta-analysis results for GMV differences in dyslexics versus controls (children and adults combined). Areas of less GMV in dyslexics are shown in blue; there were no findings for the reverse comparison. (B) Two angles of the same sagittal/horizontal view of an anatomical scan with superimposed three-dimensional fibretracking data (obtained from one representative adult control subject) through the left temporoparietal cluster identified in a meta-analysis (depicted in red: controls more than dyslexics; reverse comparison not conducted). The cluster contained fibers belonging to the corona radiata (blue) and to the superior longitudinal fasciculus/arcuate fasciculus (green). (C) Surface rendering of two meta-analyses results for brain activity differences in dyslexics versus controls (children and adults analyzed separately). Areas of less activity in children with dyslexia are shown in red, and areas of more activity are shown in yellow. Areas of less activity in adults with dyslexia are shown in blue, and areas of more activity are shown in green. The overlap between children and adults in areas of relatively less activity in dyslexics is shown in violet. *Adapted from (A) Richlan et al. (2013); (B) Vandermosten et al. (2012); (C) Richlan et al. (2011) with permission from the publisher.*

Meta-analyses have again captured the most salient observations (Linkersdörfer et al., 2012; Maisog, Einbinder, Flowers, Turkeltaub, & Eden, 2008; Richlan, Kronbichler, & Wimmer, 2011). Richlan and colleagues (2011) performed separate meta-analyses of dyslexic children and adults (Figure 65.2C). They found left IFG hypoactivation in adults but not in children with dyslexia (and hyperactivation in precentral/premotor regions for both children and adults) and TPC hypoactivation in pSTG in adults and in IPL in children, together suggesting a dynamic developmental course of dyslexia. Notably, hypoactivation of the left OTC was found in both children and adults, suggesting early alteration in this region.

Significant interest in the VWFS has influenced studies of dyslexia. This region is not only underactivated in dyslexia; specific patterns have also been uncovered using a region-of-interest (ROI) approach: whereas typically reading children and adults demonstrate a posterior-to-anterior gradient of increasing word selectivity in medial left OTC (i.e., relative signal increase for words compared with false font/symbol string stimuli along the posterior-to-anterior axis; Brem et al., 2006; Olulade, Flowers, Napoliello, & Eden, 2013; Van der Mark et al., 2009; Vinckier et al., 2007), this gradient is disrupted (Van der Mark et al., 2009) and fails to become tuned in the early phases of learning (Maurer et al., 2007) in children with dyslexia.

Functional connectivity has been used to better assess interregional correlations or cooperation between different brain areas in typical and dyslexic readers. Using PET and fMRI, respectively, two early investigations reported reduced connectivity between the AG and several sites, including the pSTG, ventral OTC, and early visual areas in dyslexics during phonological tasks (Horwitz, Rumsey, & Donohue, 1998; Pugh et al., 2000). Horwitz et al. (1998) also reported reduced AG connectivity with the IFG and cerebellum. There are several similar seed-based studies in adults, and disruption of connectivity has also been observed in children, with some (Cao, Bitan, & Booth, 2008) noting reduced connectivity between the ventral OTC and the IPL and others (Richards & Berninger, 2008) reporting abnormal connectivity between the left IFG and multiple bilateral brain regions. These studies and others (Van der Mark et al., 2011; Vourkas et al., 2011; see also Koyama et al., 2011, for resting-state connectivity) illustrate the complex nature of the reading process and the fact that dyslexia is associated with focal differences as well as disruptions of connections among regions.

The origin of these differences remains unknown. Several theories on the etiology and processes by which the phonological deficit might be operating have been put forward; these range from low-level perceptual problems to higher meta-cognitive dysfunction and are beyond the scope of this chapter (we refer the reader to specific studies and comprehensive reviews, e.g., Ahissar, 2007; Boets, 2014; Giraud & Ramus, 2013; Goswami, 2011, 2015; McArthur & Bishop, 2001; McNorgan, Randazzo-Wagner, & Booth, 2013). Multimodal brain imaging is proving a valuable tool for testing these theoretical frameworks. For example, using a multivoxel pattern analysis of fMRI data in adult dyslexics, Boets and colleagues (2013) found intact phonetic representations in bilateral auditory cortices but disrupted functional and anatomical connections between these regions and the left IFG, suggesting a problem of access. These findings echo earlier brain imaging reports of "disconnection" in dyslexia (Paulesu et al., 1996). As such, the search for the neurobiological basis for the phonological core deficit in dyslexia has expanded beyond the left TPC regions known to be engaged in phonological processing in typical readers to tracts that connect posterior to frontal brain regions.

Specifically, the developmental trajectory of the neural basis of dyslexia is still poorly understood. Given that reading is acquired over a protracted period of time and during development, it is likely that brain-based findings are going to be dynamic, sometimes presenting proximally and other times distally to the point of origin. For example, if normal reading acquisition is achieved by phonological assembly of novel words in left dorsal regions, with a shift to more automatic word recognition in OTC with increasing reading expertise, then dyslexia might represent failure to advance to the use of OTC due to disruption of dorsal TPC for phonological assembly (Pugh et al., 2001; Sandak et al., 2004). It could be that the OTC underactivity represents a primary deficit, as suggested by impaired OTC function in both children and adults (Richlan et al., 2011). However, the problem may not originate in cortex *per se*, but rather in abnormal connecting fibers preventing the normal development of skills subserved by the cortex that feeds into them (Boets, 2014). Hyperactivations (e.g., in left frontal as well as right hemisphere regions) have been interpreted as compensatory mechanisms to offset underactivity elsewhere (Pugh et al., 2001), but their role is not clear. Many of the findings depend on the age of participants, and together these factors indicate the need for longitudinal studies of dyslexia to shed light on the timing and etiology of these differences.

65.5 GENETIC AND PHYSIOLOGICAL MECHANISMS IN DYSLEXIA

Genetic linkage studies have found several loci that may be involved in dyslexia, and at some of these loci genetic variants associated with disease risk have been identified (Scerri & Schulte-Körne, 2010). Several genes in the *DYX2* (dyslexia susceptibility-2) locus on chromosome *6p22* have been associated with dyslexia (Eicher & Gruen, 2013), including *KIAA0319*, *TTRAP*, and *DCDC2*.

Some of these dyslexia-associated genes have been studied in knockout mice and have been found to be implicated in abnormal neuronal migration (Galaburda et al., 2006). A recent study suggests increased excitability and decreased temporal precision in action potential firing in neocortex of *DCDC2* knockout mice, implicating the *N*-methyl-D-aspartate (NMDA) receptor (Che, Girgenti, & Loturco, 2014).

In humans, magnetic resonance spectroscopy (MRS) has been used to identify which neurometabolites might be altered in dyslexia. *N*-acetyl-aspartate and choline were found to be abnormal in adults with dyslexia (Bruno, Lu, & Manis, 2013), and choline and glutamate were abnormal in children with dyslexia (Pugh et al., 2014). These reports highlight the connections between choline levels and abnormal white matter (dove-tailing with the described WMV and FA studies), as well as between increased glutamate and hyperexcitability at the level of the synapse (Pugh et al., 2014). More direct connections have been made between brain anatomy and dyslexia-associated genes. For example, Meda et al. (2008) investigated associations between *DCDC2* and GMV in typical readers and found that individuals heterozygous for the deletion, compared with those homozygous for no deletion, had significantly higher GMV in brain regions related to reading/language and symbol decoding. Looking at white matter integrity, Darki, Peyrard-Janvid, Matsson, Kere, and Klingberg (2012) found that *DYX1C1*, *DCDC2*, and *KIAA0319* were significantly associated with WMV in the left temporoparietal region and that WMV was positively correlated with reading ability. In functional studies, participants with the *KIAA0319/TTRAP/THEM2* variants showed a reduced left-hemispheric asymmetry of the STS (Pinel et al., 2012), and Cope et al. (2012) found activity in several brain areas to be influenced by variants in the *DYX2* locus. Together these studies are beginning to reveal the possible connections between molecular mechanisms and behavior in dyslexia, as well as the brain's mediating role.

65.6 NEUROBIOLOGY OF READING INTERVENTIONS

Understanding the brain-based changes underlying successful reading intervention demonstrated by behavioral studies (Alexander & Slinger-Constant, 2004; Bradley & Bryant, 1983) provides insights into treatment mechanisms of dyslexia and, potentially, its etiology. Functional neuroimaging studies in children and adults have reported several regions of increased activation after reading intervention. These have often been discussed in terms of "compensation" versus "normalization" processes and in the context of language function, with some discussion on skills outside of language that might support reading. As summarized by Barquero, Davis, and Cutting (2014), areas of postintervention increases include bilateral inferior frontal, superior temporal, middle temporal, middle frontal, superior frontal, and postcentral gyri, as well as bilateral occipital cortex, IPL, thalami, and insulae. A meta-analysis of fMRI studies (also reported by Barquero et al. (2014)) sheds light on the most salient postintervention increases, namely left thalamus, right insula/IFG, left IFG, left middle occipital gyri, and right posterior cingulate (Figure 65.3). There are also studies examining anatomical changes after reading intervention in dyslexia, demonstrating GMV increases in children in left anterior fusiform gyrus/hippocampus and precuneus and right hemisphere hippocampus and cerebellum (Krafnick, Flowers, Napoliello, & Eden, 2011), and increased FA in left anterior centrum semiovale in adults (Keller & Just, 2009). These anatomical changes do not directly colocalize to the functional changes described, suggesting that the neurobiological mechanisms of reading remediation are complex. For example, even if the intervention involves training of phonological skills, the brain mechanisms underlying successful reading gains may rely on memory and other learning mechanisms. Further, the roles of participant age and the method of treatment in the neural correlates of reading intervention have yet to be fully explored.

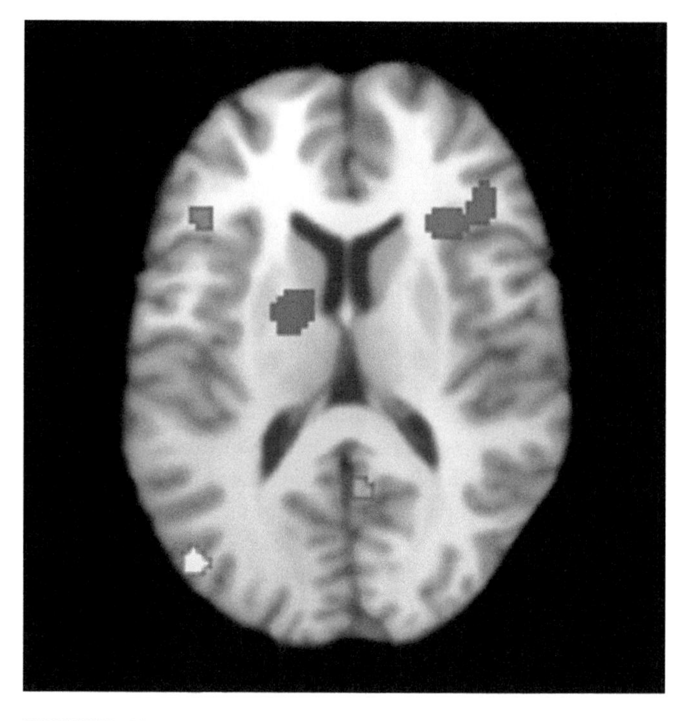

FIGURE 65.3 Axial view of meta-analysis results for brain activity increases in dyslexics after reading intervention: left middle occipital gyrus, thalamus, inferior frontal gyrus; right posterior cingulate and insula/inferior frontal gyrus. *From Barquero et al. (2014) with permission from the publisher.*

There is increasing interest in correlating success of reading acquisition with neurobiological measures. For example, adult compensated dyslexics followed from childhood were shown to have more activation in right superior frontal and middle temporal gyri and left anterior cingulate gyrus compared with persistently poor readers (Shaywitz et al., 2003). Some studies have compared postintervention brain activity in dyslexic children who showed a favorable behavioral response to intervention with activity in children who did not benefit from treatment (Davis et al., 2011; Odegard, Ring, Smith, Biggan, & Black, 2008). More activity in responders than in nonresponders was found by Odegard et al. (2008) in left IPL and in left STG by Davis et al. (2011).

Furthermore, neuroimaging data have been used to predict long-term outcomes of reading in typically reading (McNorgan, Alvarez, Bhullar, Gayda, & Booth, 2011) and dyslexic children (Hoeft et al., 2011). The latter study found that right IFG activation and right superior longitudinal fasciculus integrity at the beginning of the study predicted reading scores in dyslexics 2.5 years later. As such, brain-based predictive studies bear important information on the mechanisms of successful treatment of dyslexia and also hold potential for contributing to the decision for a specific intervention strategy.

65.7 CAUSE VERSUS CONSEQUENCE?

A pressing question for the neurobiological bases of dyslexia is in regard to whether the findings that distinguish dyslexics from nondyslexics represent a cause or consequence of the disorder. Literate and illiterate adults differ in brain function (Dehaene et al., 2010) and anatomy, including brain regions known to be involved in dyslexia, such as the arcuate fasciculus (Thiebaut de Schotten, Cohen, Amemiya, Braga, & Dehaene, 2014), suggesting that reading acquisition results in significant learning-induced plasticity. Adults who were illiterate but then learned to read exhibit greater GMV in TPC compared with illiterate adults (Carreiras et al., 2009). Longitudinal studies of typical readers reveal widespread decrease in cortical thickness with age (Giedd et al., 1999), yet language regions in perisylvian cortex thicken (Sowell et al., 2004) in correlation with improvements in phonological skills (Lu et al., 2007). As such, it is possible that typical readers, as a consequence of learning to read, experience changes in brain anatomy and function that are not realized in dyslexic readers (who read less), leading to a relative difference that is the consequence, and not the cause, of dyslexia.

Therefore, brain imaging studies have implemented experimental designs that disambiguate those characteristics causal to dyslexia from those that are secondary or, in fact, the consequence of experience (or lack of experience). This question can be addressed by: (i) longitudinal studies of typical readers and children at risk for dyslexia (based on a family history or early signs of weaknesses in skills that support reading) and (ii) studies using the reading level-match design, where children with dyslexia are compared with younger typical readers who are matched on reading level (Goswami & Bryant, 1989). The noninvasive nature of fMRI and event-related potential (ERP) techniques has allowed pediatric studies to flourish, and brain imaging studies on the causal nature of dyslexia have begun to emerge. For example, Raschle, Chang, and Gaab (2011) report reduced GMV in left OTC, bilateral TPC, left fusiform gyrus, and right lingual regions in prereading children with, compared with children without, a family history of dyslexia (see Hosseini and colleagues, 2013, for similar work measuring network differences in surface area). The same group of children also showed less brain activity in bilateral OTC and left TPC (Raschle, Zuk, & Gaab, 2012). Interestingly, maternal history of reading disability is associated with smaller bilateral prefrontal and TPC GMV of 5- to 6-year-olds, and because these replicate for the left IPL on a measure of surface area but not cortical thickness, they are likely due to prenatal influences (Black et al., 2012). In studies of white matter structure, it has been shown that the volume and FA of the arcuate fasciculus is positively correlated with PA in children before they learn how to read (Saygin et al., 2013). The reading level-match design has also been applied: Hoeft et al. (2007) found left IPL GMV differences in dyslexia for both control group comparisons (age-matched and reading level-matched) in a study that used fMRI data to determine the ROIs for GMV analyses. However, using a whole-brain approach (as well as a follow-up analysis using ROIs), Krafnick, Flowers, Luetje, Napoliello, and Eden (2014) found that few GMV and none of the WMV differences identified between dyslexics and age-matched controls emerged when dyslexics were compared with controls matched on reading abilities. Also, the older controls (age-matched to the dyslexics) had more GMV than the younger controls in several of the areas identified as variant in the dyslexics when compared with age-matched controls, raising the possibility that not all differences can be attributed to dyslexia *per se*. Interestingly, in a sample of children eventually diagnosed with dyslexia, the structural abnormalities that preceded learning to read were found in early visual and auditory cortical areas and not in regions associated with reading (Clark et al., 2014). As such, the question remains whether anomalies in neural migration or synaptic activity affect brain anatomy and function directly or via reading experience.

65.8 IMPORTANT VARIABLES IN STUDIES OF DYSLEXIA

Due to the heterogeneity in dyslexia, researchers have attempted to identify subgroups. For example, Jednoróg, Gawron, Marchewka, Heim, and Grabowska (2014) have shown that cognitive subtypes of dyslexia are characterized by distinct patterns of GMV. This concept is also being applied to functional neuroimaging studies of reading intervention based on subtypes (Heim, Pape-Neumann, van Ermingen-Marbach, Brinkhaus, & Grande, 2014). Although the phonological deficit theory also dominates such studies, it must be noted again that there are criticisms of and theoretical alternatives to the phonological deficit theory, and careful consideration must be given to dyslexia in languages other than English (Brunswick et al., 2010), particularly those in which grapheme–phoneme mapping is more direct (orthographically transparent). A behavioral study of dyslexia in six languages revealed that high orthographic complexity of a language exacerbates phoneme deletion and rapid naming problems in dyslexia (Landerl et al., 2013). In the case of German, where mapping between graphemes and phonemes is highly consistent, dyslexia is associated with poor reading speed and spelling, whereas phonological deficits have a lesser role (Wimmer & Schurz, 2010). It is also debated whether the demands of a logographic writing system such as Chinese result in brain-based differences that are specific to dyslexia in that orthography (Siok, Perfetti, Jin, & Tan, 2004) or not (Hu et al., 2010).

Because of higher prevalence of dyslexia in males, some studies are based on samples that are dominated by males, yet the results are generalized to both sexes with dyslexia. For example, less than 20% of the participants in the studies included in the meta-analysis by Linkersdörfer et al. (2012) were female. However, there is evidence to suggest sex-specific differences in brain anatomy in dyslexia, as illustrated by GMV differences in dyslexic females (children and adults) compared with controls that reside outside of the language regions typically reported in males with dyslexia (Evans et al., 2014). Differences in cortical thickness have been reported for girls but not boys in ventral OTC (Altarelli et al., 2013), and altered asymmetry (rightward) of the planum temporale surface area in dyslexic boys but not girls (Altarelli et al., 2014). Given the sexual dimorphism in the general population for brain anatomy (Good et al., 2001) and brain function underlying language, including phonological processing (Shaywitz et al., 1995), these findings may call for sex-specific models of dyslexia and, at least, highlight the need to include more females in dyslexia research.

Another understudied group includes those of low socioeconomic status (SES). This has recently been addressed in a study showing that the differences observed in both OTC and STG in dyslexia exist independently of SES (Monzalvo, Fluss, Billard, Dehaene, & Dehaene-Lambertz, 2012), even though SES is associated with differences in GMV (but not WMV) in areas relevant to language (Jednoróg et al., 2012). Together, these and other variables will continue to play an increasingly important role in studies of the brain bases of dyslexia.

65.9 CONCLUSION

Building on decades of behavioral work, brain imaging has provided a neurobiological basis for understanding reading and dyslexia. Consistent with the predominant phonological core deficit theory of dyslexia, anatomical and functional studies have revealed differences in left temporoparietal, inferior frontal, and ventral occipitotemporal cortices, together with disruptions of their connections. This work has been integrated with efforts in reading intervention, predictors of reading outcome, dyslexia-associated genes, and neurometabolites. Studies conducted at different ages suggest an age-dependent neurobiological profile of dyslexia. Although this is captured by current putative neurocognitive developmental models of dyslexia, longitudinal investigations in young children are ultimately needed to ascertain how these differences behave over time (e.g., a given anomaly may be a different manifestation of an earlier anomaly) and to determine which differences are attributed to dyslexia *per se* (rather than altered reading experience) using methods that capture the neurodevelopmental origin of these. Further, diversity of writing systems and orthographic depth is likely to play a modulating role, suggesting that beyond the universal aspects of dyslexia, orthography-specific aspects of reading disability will need to be integrated into neurobiological models (Richlan, 2014).

Acknowledgments

The authors are supported by the Eunice Kennedy Shriver National Institute of Child Health and Human Development (RO1HD37890, P50HD40095, RO1HD056107) and the National Science Foundation (grant SBE 0541953).

References

Ahissar, M. (2007). Dyslexia and the anchoring-deficit hypothesis. *Trends in Cognitive Sciences, 11*(11), 458–465. Available from: http://dx.doi.org/10.1016/j.tics.2007.08.015.

Alexander, A. W., & Slinger-Constant, A.-M. (2004). Current status of treatments for dyslexia: Critical review. *Journal of Child Neurology, 19*(10), 744–758. Retrieved from: <http://www.ncbi.nlm.nih.gov/pubmed/15559890>.

Altarelli, I., Leroy, F., Monzalvo, K., Fluss, J., Billard, C., Dehaene-Lambertz, G., et al. (2014). Planum temporale asymmetry in developmental dyslexia: Revisiting an old question. *Human Brain Mapping, 35*(12), 5715–5735. Available from: http://dx.doi.org/10.1002/hbm.22579.

Altarelli, I., Monzalvo, K., Iannuzzi, S., Fluss, J., Billard, C., Ramus, F., et al. (2013). A functionally guided approach to the morphometry of occipitotemporal regions in developmental dyslexia: Evidence for differential effects in boys and girls. *The Journal of Neuroscience: The Official Journal of the Society for Neuroscience, 33*(27), 11296–11301. Available from: http://dx.doi.org/10.1523/JNEUROSCI.5854-12.2013.

Badian, N. A. (1994). Preschool prediction: Orthographic and phonological skills, and reading. *Annals of Dyslexia, 44*(1), 1–25. Available from: http://dx.doi.org/10.1007/BF02648153.

Barquero, L. A., Davis, N., & Cutting, L. E. (2014). Neuroimaging of reading intervention: A systematic review and activation likelihood estimate meta-analysis. *PLoS ONE, 9*(1), e83668. Available from: http://dx.doi.org/10.1371/journal.pone.0083668.

Black, J. M., Tanaka, H., Stanley, L., Nagamine, M., Zakerani, N., Thurston, A., et al. (2012). Maternal history of reading difficulty is associated with reduced language-related gray matter in beginning readers. *NeuroImage, 59*(3), 3021–3032. Available from: http://dx.doi.org/10.1016/j.neuroimage.2011.10.024.

Boets, B. (2014). Dyslexia: Reconciling controversies within an integrative developmental perspective. *Trends in Cognitive Sciences, 18* (10), 501–503. Available from: http://dx.doi.org/10.1016/j.tics.2014.06.003.

Boets, B., Op de Beeck, H. P., Vandermosten, M., Scott, S. K., Gillebert, C. R., Mantini, D., et al. (2013). Intact but less accessible phonetic representations in adults with dyslexia. *Science (New York, NY), 342*(6163), 1251–1254. Available from: http://dx.doi.org/10.1126/science.1244333.

Bradley, L., & Bryant, P. E. (1983). Categorizing sounds and learning to read—a causal connection. *Nature, 301*(5899), 419–421. Available from: http://dx.doi.org/10.1038/301419a0.

Brem, S., Bucher, K., Halder, P., Summers, P., Dietrich, T., Martin, E., et al. (2006). Evidence for developmental changes in the visual word processing network beyond adolescence. *NeuroImage, 29*(3), 822–837. Available from: http://dx.doi.org/10.1016/j.neuroimage.2005.09.023.

Bruno, J. L., Lu, Z.-L., & Manis, F. R. (2013). Phonological processing is uniquely associated with neuro-metabolic concentration. *NeuroImage, 67*, 175–181. Available from: http://dx.doi.org/10.1016/j.neuroimage.2012.10.092.

Brunswick, N., McCrory, E., Price, C. J., Frith, C. D., & Frith, U. (1999). Explicit and implicit processing of words and pseudowords by adult developmental dyslexics: A search for Wernicke's Wortschatz? *Brain: A Journal of Neurology, 122*(Pt 10), 1901–1917.

Brunswick, N., McDougall, S., & Davies, P. de M. (2010). *Reading and dyslexia in different orthographies*. New York, NY: Psychology Press.

Cao, F., Bitan, T., & Booth, J. R. (2008). Effective brain connectivity in children with reading difficulties during phonological processing. *Brain and Language, 107*(2), 91–101. Available from: http://dx.doi.org/10.1016/j.bandl.2007.12.009.

Cao, F., Bitan, T., Chou, T.-L., Burman, D. D., & Booth, J. R. (2006). Deficient orthographic and phonological representations in children with dyslexia revealed by brain activation patterns. *Journal of Child Psychology and Psychiatry, and Allied Disciplines, 47*(10), 1041–1050. Available from: http://dx.doi.org/10.1111/j.1469-7610.2006.01684.x.

Carreiras, M., Seghier, M. L., Baquero, S., Estévez, A., Lozano, A., Devlin, J. T., et al. (2009). An anatomical signature for literacy. *Nature, 461*(7266), 983–986. Available from: http://dx.doi.org/10.1038/nature08461.

Che, A., Girgenti, M. J., & Loturco, J. (2014). The dyslexia-associated gene DCDC2 is required for spike-timing precision in mouse neocortex. *Biological Psychiatry, 76*(5), 387–396. Available from: http://dx.doi.org/10.1016/j.biopsych.2013.08.018.

Clark, K. A., Helland, T., Specht, K., Narr, K. L., Manis, F. R., Toga, A. W., et al. Neuroanatomical precursors of dyslexia identified from pre-reading through to age 11. *Brain: A Journal of Neurology*. Available from: http://dx.doi.org/10.1093/brain/awu229

Cope, N., Eicher, J. D., Meng, H., Gibson, C. J., Hager, K., Lacadie, C., et al. (2012). Variants in the DYX2 locus are associated with altered brain activation in reading-related brain regions in subjects with reading disability. *NeuroImage, 63*(1), 148–156. Available from: http://dx.doi.org/10.1016/j.neuroimage.2012.06.037.

Darki, F., Peyrard-Janvid, M., Matsson, H., Kere, J., & Klingberg, T. (2012). Three dyslexia susceptibility genes, DYX1C1, DCDC2, and KIAA0319, affect temporo-parietal white matter structure. *Biological Psychiatry, 72*(8), 671–676. Available from: http://dx.doi.org/10.1016/j.biopsych.2012.05.008.

Davis, N., Barquero, L., Compton, D. L., Fuchs, L. S., Fuchs, D., Gore, J. C., et al. (2011). Functional correlates of children's responsiveness to intervention. *Developmental Neuropsychology, 36*(3), 288–301. Available from: http://dx.doi.org/10.1080/87565641.2010.549875.

Dehaene, S., Pegado, F., Braga, L. W., Ventura, P., Nunes Filho, G., Jobert, A., et al. (2010). How learning to read changes the cortical networks for vision and language. *Science, 330*(6009), 1359–1364. Available from: http://dx.doi.org/10.1126/science.1194140.

Eckert, M. A., Leonard, C. M., Richards, T. L., Aylward, E. H., Thomson, J., & Berninger, V. W. (2003). Anatomical correlates of dyslexia: Frontal and cerebellar findings. *Brain, 126*(2), 482–494. Available from: http://dx.doi.org/10.1093/brain/awg026.

Eckert, M. A., Leonard, C. M., Wilke, M., Eckert, M., Richards, T., Richards, A., et al. (2005). Anatomical signatures of dyslexia in children: Unique information from manual and voxel based morphometry brain measures. *Cortex, 41*(3), 304–315. Available from: http://dx.doi.org/10.1016/S0010-9452(08)70268-5.

Eden, G. F., Jones, K. M., Cappell, K., Gareau, L., Wood, F. B., Zeffiro, T. A., et al. (2004). Neural changes following remediation in adult developmental dyslexia. *Neuron, 44*(3), 411–422.

Ehri, L. C. (1999). Phases of development in learning to read words. In J. Oakhill, & R. Beard (Eds.), *Reading development and the teaching of reading: A psychological perspective* (pp. 79–108). Oxford, England: Blackwell Science.

Eicher, J. D., & Gruen, J. R. (2013). Imaging-genetics in dyslexia: Connecting risk genetic variants to brain neuroimaging and ultimately to reading impairments. *Molecular Genetics and Metabolism, 110*(3), 201–212. Available from: http://dx.doi.org/10.1016/j.ymgme.2013.07.001.

Evans, T. M., Flowers, D. L., Napoliello, E. M., & Eden, G. F. (2014). Sex-specific gray matter volume differences in females with developmental dyslexia. *Brain Structure & Function, 219*(3), 1041–1054. Available from: http://dx.doi.org/10.1007/s00429-013-0552-4. PMCID: PMC3775969.

Fisher, S. E., & DeFries, J. C. (2002). Developmental dyslexia: Genetic dissection of a complex cognitive trait. *Nature Reviews. Neuroscience, 3*(10), 767–780. Available from: http://dx.doi.org/10.1038/nrn936.

Flowers, D. L., Wood, F. B., & Naylor, C. E. (1991). Regional cerebral blood flow correlates of language processes in reading disability. *Archives of Neurology, 48*(6), 637–643.

Gabrieli, J. D. E. (2009). Dyslexia: A new synergy between education and cognitive neuroscience. *Science (New York, NY), 325*(5938), 280–283. Available from: http://dx.doi.org/10.1126/science.1171999.

Galaburda, A. M., & Kemper, T. L. (1979). Cytoarchitectonic abnormalities in developmental dyslexia: A case study. *Annals of Neurology, 6*(2), 94–100. Available from: http://dx.doi.org/10.1002/ana.410060203.

Galaburda, A. M., LoTurco, J., Ramus, F., Fitch, R. H., & Rosen, G. D. (2006). From genes to behavior in developmental dyslexia. *Nature Neuroscience, 9*(10), 1213–1217. Available from: http://dx.doi.org/10.1038/nn1772.

Galaburda, A. M., Sherman, G. F., Rosen, G. D., Aboitiz, F., & Geschwind, N. (1985). Developmental dyslexia: Four consecutive patients with cortical anomalies. *Annals of Neurology, 18*(2), 222–233. Available from: http://dx.doi.org/10.1002/ana.410180210.

Georgiewa, P., Rzanny, R., Hopf, J. M., Knab, R., Glauche, V., Kaiser, W. A., et al. (1999). fMRI during word processing in dyslexic and normal reading children. *Neuroreport, 10*(16), 3459–3465.

Giedd, J. N., Blumenthal, J., Jeffries, N. O., Castellanos, F. X., Liu, H., Zijdenbos, A., et al. (1999). Brain development during childhood and adolescence: A longitudinal MRI study. *Nature Neuroscience, 2*(10), 861–863. Available from: http://dx.doi.org/10.1038/13158.

Giraud, A.-L., & Ramus, F. (2013). Neurogenetics and auditory processing in developmental dyslexia. *Current Opinion in Neurobiology, 23*(1), 37–42. Available from: http://dx.doi.org/10.1016/j.conb.2012.09.003.

Good, C. D., Johnsrude, I., Ashburner, J., Henson, R. N., Friston, K. J., & Frackowiak, R. S. (2001). Cerebral asymmetry and the effects of sex and handedness on brain structure: A voxel-based morphometric analysis of 465 normal adult human brains. *NeuroImage, 14*(3), 685–700. Available from: http://dx.doi.org/10.1006/nimg.2001.0857.

Goswami, U. (2011). A temporal sampling framework for developmental dyslexia. *Trends in Cognitive Sciences, 15*(1), 3–10. Available from: http://dx.doi.org/10.1016/j.tics.2010.10.001.

Goswami, U. (2015). Sensory theories of developmental dyslexia: Three challenges for research. *Nature Reviews Neuroscience, 16*(1), 43–54. Available from: http://dx.doi.org/10.1038/nrn3836.

Goswami, U., & Bryant, P. (1989). The interpretation of studies using the reading level design. *Journal of Literacy Research, 21*(4), 413–424. Available from: http://dx.doi.org/10.1080/10862968909547687.

Gross-Glenn, K., Duara, R., Barker, W. W., Loewenstein, D., Chang, J. Y., Yoshii, F., et al. (1991). Positron emission tomographic studies during serial word-reading by normal and dyslexic adults. *Journal of Clinical and Experimental Neuropsychology, 13*(4), 531–544. Available from: http://dx.doi.org/10.1080/01688639108401069.

Heim, S., Pape-Neumann, J., van Ermingen-Marbach, M., Brinkhaus, M., & Grande, M. (2014). Shared vs. specific brain activation changes in dyslexia after training of phonology, attention, or reading. *Brain Structure and Function*. Advance online publication. Available from: http://dx.doi.org/10.1007/s00429-014-0784-y.

Hoeft, F., McCandliss, B. D., Black, J. M., Gantman, A., Zakerani, N., Hulme, C., et al. (2011). Neural systems predicting long-term outcome in dyslexia. *Proceedings of the National Academy of Sciences of the United States of America, 108*(1), 361–366. Available from: http://dx.doi.org/10.1073/pnas.1008950108.

Hoeft, F., Meyler, A., Hernandez, A., Juel, C., Taylor-Hill, H., Martindale, J. L., et al. (2007). Functional and morphometric brain dissociation between dyslexia and reading ability. *Proceedings of the National Academy of Sciences of the United States of America, 104*(10), 4234–4239. Available from: http://dx.doi.org/10.1073/pnas.0609399104.

Horwitz, B., Rumsey, J. M., & Donohue, B. C. (1998). Functional connectivity of the angular gyrus in normal reading and dyslexia. *Proceedings of the National Academy of Sciences of the United States of America, 95*(15), 8939–8944. Available from: http://dx.doi.org/10.1073/pnas.95.15.8939.

Hosseini, S. M. H., Black, J. M., Soriano, T., Bugescu, N., Martinez, R., Raman, M. M., et al. (2013). Topological properties of large-scale structural brain networks in children with familial risk for reading difficulties. *NeuroImage, 71*, 260–274. Available from: http://dx.doi.org/10.1016/j.neuroimage.2013.01.013.

Hu, W., Lee, H. L., Zhang, Q., Liu, T., Geng, L. B., Seghier, M. L., et al. (2010). Developmental dyslexia in Chinese and English populations: Dissociating the effect of dyslexia from language differences. *Brain, 133*(6), 1694–1706. Available from: http://dx.doi.org/10.1093/brain/awq106.

Hynd, G. W., Semrud-Clikeman, M., Lorys, A. R., Novey, E. S., & Eliopulos, D. (1990). Brain morphology in developmental dyslexia and attention deficit disorder/hyperactivity. *Archives of Neurology, 47*(8), 919–926.

Jednoróg, K., Altarelli, I., Monzalvo, K., Fluss, J., Dubois, J., Billard, C., et al. (2012). The influence of socioeconomic status on children's brain structure. *PLoS ONE, 7*(8), e42486. Available from: http://dx.doi.org/10.1371/journal.pone.0042486.

Jednoróg, K., Gawron, N., Marchewka, A., Heim, S., & Grabowska, A. (2014). Cognitive subtypes of dyslexia are characterized by distinct patterns of grey matter volume. *Brain Structure and Function, 219*(5), 1697–1707. Available from: http://dx.doi.org/10.1007/s00429-013-0595-6.

Keller, T. A., & Just, M. A. (2009). Altering cortical connectivity: Remediation-induced changes in the white matter of poor readers. *Neuron, 64*(5), 624–631. Available from: http://dx.doi.org/10.1016/j.neuron.2009.10.018.

Klingberg, T., Hedehus, M., Temple, E., Salz, T., Gabrieli, J. D., Moseley, M. E., et al. (2000). Microstructure of temporo-parietal white matter as a basis for reading ability: Evidence from diffusion tensor magnetic resonance imaging. *Neuron, 25*(2), 493–500.

Koyama, M. S., Di Martino, A., Zuo, X.-N., Kelly, C., Mennes, M., Jutagir, D. R., et al. (2011). Resting-state functional connectivity indexes reading competence in children and adults. *The Journal of Neuroscience: The Official Journal of the Society for Neuroscience, 31*(23), 8617–8624. Available from: http://dx.doi.org/10.1523/JNEUROSCI.4865-10.2011.

Krafnick, A. J., Flowers, D. L., Luetje, M. M., Napoliello, E. M., & Eden, G. F. (2014). An investigation into the origin of anatomical differences in dyslexia. *The Journal of Neuroscience, 34*(3), 901–908. Available from: http://dx.doi.org/10.1523/JNEUROSCI.2092-13.2013. PMCID: PMC3891966.

Krafnick, A. J., Flowers, D. L., Napoliello, E. M., & Eden, G. F. (2011). Gray matter volume changes following reading intervention in dyslexic children. *NeuroImage, 57*(3), 733–741. Available from: http://dx.doi.org/10.1016/j.neuroimage.2010.10.062.

Landerl, K., Ramus, F., Moll, K., Lyytinen, H., Leppänen, P. H. T., Lohvansuu, K., et al. (2013). Predictors of developmental dyslexia in European orthographies with varying complexity. *Journal of Child Psychology and Psychiatry, and Allied Disciplines, 54*(6), 686–694. Available from: http://dx.doi.org/10.1111/jcpp.12029.

Larsen, J. P., Høien, T., Lundberg, I., & Ødegaard, H. (1990). MRI evaluation of the size and symmetry of the planum temporale in adolescents with developmental dyslexia. *Brain and Language, 39*(2), 289–301. Available from: http://dx.doi.org/10.1016/0093-934X(90)90015-9.

Leonard, C. M., Eckert, M. A., Lombardino, L. J., Oakland, T., Kranzler, J., Mohr, C. M., et al. (2001). Anatomical risk factors for phonological dyslexia. *Cerebral Cortex, 11*(2), 148–157. Available from: http://dx.doi.org/10.1093/cercor/11.2.148.

Linkersdörfer, J., Lonnemann, J., Lindberg, S., Hasselhorn, M., & Fiebach, C. J. (2012). Grey matter alterations co-localize with functional abnormalities in developmental dyslexia: An ALE meta-analysis. *Public Library of Science One, 7*(8), e43122. Available from: http://dx.doi.org/10.1371/journal.pone.0043122.

Lu, L. H., Leonard, C. M., Thompson, P. M., Kan, E., Jolley, J., Welcome, S. E., et al. (2007). Normal developmental changes in inferior frontal gray matter are associated with improvement in phonological processing: A longitudinal MRI analysis. *Cerebral Cortex, 17*(5), 1092–1099. Available from: http://dx.doi.org/10.1093/cercor/bhl019.

Lyon, G. R., Shaywitz, S. E., & Shaywitz, B. A. (2003). A definition of dyslexia. *Annals of Dyslexia, 53*, 1–14.

Maisog, J. M., Einbinder, E. R., Flowers, D. L., Turkeltaub, P. E., & Eden, G. F. (2008). A meta-analysis of functional neuroimaging studies of dyslexia. *Annals of the New York Academy of Sciences, 1145*, 237–259. Available from: http://dx.doi.org/10.1196/annals.1416.024.

Maurer, U., Brem, S., Bucher, K., Kranz, F., Benz, R., Steinhausen, H.-C., et al. (2007). Impaired tuning of a fast occipito-temporal response for print in dyslexic children learning to read. *Brain, 130*(12), 3200–3210. Available from: http://dx.doi.org/10.1093/brain/awm193.

McArthur, G. m., & Bishop, D. v. m. (2001). Auditory perceptual processing in people with reading and oral language impairments: Current issues and recommendations. *Dyslexia, 7*(3), 150–170. Available from: http://dx.doi.org/10.1002/dys.200.

McCardle, P., Scarborough, H. S., & Catts, H. W. (2001). Predicting, explaining, and preventing children's reading difficulties. *Learning Disabilities Research and Practice*, *16*(4), 230–239. Available from: http://dx.doi.org/10.1111/0938-8982.00023.

McNorgan, C., Alvarez, A., Bhullar, A., Gayda, J., & Booth, J. R. (2011). Prediction of reading skill several years later depends on age and brain region: Implications for developmental models of reading. *The Journal of Neuroscience: The Official Journal of the Society for Neuroscience*, *31*(26), 9641–9648. Available from: http://dx.doi.org/10.1523/JNEUROSCI.0334-11.2011.

McNorgan, C., Randazzo-Wagner, M., & Booth, J. R. (2013). Cross-modal integration in the brain is related to phonological awareness only in typical readers, not in those with reading difficulty. *Frontiers in Human Neuroscience*, *7*, 388. Available from: http://dx.doi.org/10.3389/fnhum.2013.00388.

Meda, S. A., Gelernter, J., Gruen, J. R., Calhoun, V. D., Meng, H., Cope, N. A., et al. (2008). Polymorphism of DCDC2 reveals differences in cortical morphology of healthy individuals—A preliminary voxel based morphometry study. *Brain Imaging and Behavior*, *2*(1), 21–26. Available from: http://dx.doi.org/10.1007/s11682-007-9012-1.

Monzalvo, K., Fluss, J., Billard, C., Dehaene, S., & Dehaene-Lambertz, G. (2012). Cortical networks for vision and language in dyslexic and normal children of variable socio-economic status. *NeuroImage*, *61*(1), 258–274. Available from: http://dx.doi.org/10.1016/j.neuroimage.2012.02.035.

Nation, K., Cocksey, J., Taylor, J. S. H., & Bishop, D. V. M. (2010). A longitudinal investigation of early reading and language skills in children with poor reading comprehension. *Journal of Child Psychology and Psychiatry, and Allied Disciplines*, *51*(9), 1031–1039. Available from: http://dx.doi.org/10.1111/j.1469-7610.2010.02254.x.

Odegard, T. N., Farris, E. A., Ring, J., McColl, R., & Black, J. (2009). Brain connectivity in non-reading impaired children and children diagnosed with developmental dyslexia. *Neuropsychologia*, *47* (8–9), 1972–1977. Available from: http://dx.doi.org/10.1016/j.neuropsychologia.2009.03.009.

Odegard, T. N., Ring, J., Smith, S., Biggan, J., & Black, J. (2008). Differentiating the neural response to intervention in children with developmental dyslexia. *Annals of Dyslexia*, *58*(1), 1–14. Available from: http://dx.doi.org/10.1007/s11881-008-0014-5.

Olulade, O. A., Flowers, D. L., Napoliello, E. M., & Eden, G. F. (2013). Developmental differences for word processing in the ventral stream. *Brain and Language*, *125*(2), 134–145. Available from: http://dx.doi.org/10.1016/j.bandl.2012.04.003. PMCID: PMC3426643.

Paulesu, E. (2001). Dyslexia: Cultural diversity and biological unity. *Science*, *291*(5511), 2165–2167. Available from: http://dx.doi.org/10.1126/science.1057179.

Paulesu, E., Frith, U., Snowling, M., Gallagher, A., Morton, J., Frackowiak, R. S., et al. (1996). Is developmental dyslexia a disconnection syndrome? Evidence from PET scanning. *Brain: A Journal of Neurology*, *119*(Pt 1), 143–157.

Peterson, R. L., & Pennington, B. F. (2012). Developmental dyslexia. *Lancet*, *379*(9830), 1997–2007. Available from: http://dx.doi.org/10.1016/S0140-6736(12)60198-6.

Pinel, P., Fauchereau, F., Moreno, A., Barbot, A., Lathrop, M., Zelenika, D., et al. (2012). Genetic variants of FOXP2 and KIAA0319/TTRAP/THEM2 locus are associated with altered brain activation in distinct language-related regions. *The Journal of Neuroscience: The Official Journal of the Society for Neuroscience*, *32* (3), 817–825. Available from: http://dx.doi.org/10.1523/JNEUROSCI.5996-10.2012.

Price, C. J. (2012). A review and synthesis of the first 20 years of PET and fMRI studies of heard speech, spoken language and reading. *NeuroImage*, *62*(2), 816–847. Available from: http://dx.doi.org/10.1016/j.neuroimage.2012.04.062.

Pugh, K. R., Frost, S. J., Rothman, D. L., Hoeft, F., Tufo, S. N. D., Mason, G. F., et al. (2014). Glutamate and choline levels predict individual differences in reading ability in emergent readers. *The Journal of Neuroscience*, *34*(11), 4082–4089. Available from: http://dx.doi.org/10.1523/JNEUROSCI.3907-13.2014.

Pugh, K. R., Mencl, W. E., Jenner, A. R., Katz, L., Frost, S. J., Lee, J. R., et al. (2001). Neurobiological studies of reading and reading disability. *Journal of Communication Disorders*, *34*(6), 479–492.

Pugh, K. R., Mencl, W. E., Shaywitz, B. A., Shaywitz, S. E., Fulbright, R. K., Constable, R. T., et al. (2000). The angular gyrus in developmental dyslexia: Task-specific differences in functional connectivity within posterior cortex. *Psychological Science*, *11*(1), 51–56. Available from: http://dx.doi.org/10.1111/1467-9280.00214.

Ramus, F. (2004). Neurobiology of dyslexia: A reinterpretation of the data. *Trends in Neurosciences*, *27*(12), 720–726. Available from: http://dx.doi.org/10.1016/j.tins.2004.10.004.

Raschle, N. M., Chang, M., & Gaab, N. (2011). Structural brain alterations associated with dyslexia predate reading onset. *NeuroImage*, *57*(3), 742–749. Retrieved from: <http://www.sciencedirect.com/science/article/pii/S1053811910012553>.

Raschle, N. M., Zuk, J., & Gaab, N. (2012). Functional characteristics of developmental dyslexia in left-hemispheric posterior brain regions predate reading onset. *Proceedings of the National Academy of Sciences of the United States of America*, *109*(6), 2156–2161. Available from: http://dx.doi.org/10.1073/pnas.1107721109.

Richards, T. L., & Berninger, V. W. (2008). Abnormal fMRI connectivity in children with dyslexia during a phoneme task: Before but not after treatment. *Journal of Neurolinguistics*, *21*(4), 294–304. Available from: http://dx.doi.org/10.1016/j.jneuroling.2007.07.002.

Richlan, F. (2014). Functional neuroanatomy of developmental dyslexia: The role of orthographic depth. *Frontiers in Human Neuroscience*, *8*, 347. Available from: http://dx.doi.org/10.3389/fnhum.2014.00347.

Richlan, F., Kronbichler, M., & Wimmer, H. (2011). Meta-analyzing brain dysfunctions in dyslexic children and adults. *NeuroImage*, *56*(3), 1735–1742. Available from: http://dx.doi.org/10.1016/j.neuroimage.2011.02.040.

Richlan, F., Kronbichler, M., & Wimmer, H. (2013). Structural abnormalities in the dyslexic brain: A meta-analysis of voxel-based morphometry studies. *Human Brain Mapping*, *34*(11), 3055–3065. Available from: http://dx.doi.org/10.1002/hbm.22127.

Richlan, F., Sturm, D., Schurz, M., Kronbichler, M., Ladurner, G., & Wimmer, H. (2010). A common left occipito-temporal dysfunction in developmental dyslexia and acquired letter-by-letter reading? *PLoS ONE*, *5*(8), e12073. Available from: http://dx.doi.org/10.1371/journal.pone.0012073.

Robichon, F., Levrier, O., Farnarier, P., & Habib, M. (2000). Developmental dyslexia: Atypical cortical asymmetries and functional significance. *European Journal of Neurology*, *7*(1), 35–46. Available from: http://dx.doi.org/10.1046/j.1468-1331.2000.00020.x.

Rumsey, J. M., Andreason, P., Zametkin, A. J., Aquino, T., King, A. C., Hamburger, S. D., et al. (1992). Failure to activate the left temporoparietal cortex in dyslexia. An oxygen 15 positron emission tomographic study. *Archives of Neurology*, *49*, 527–534.

Rutter, M., Caspi, A., Fergusson, D., Horwood, L. J., Goodman, R., Maughan, B., et al. (2004). Sex differences in developmental reading disability: New findings from 4 epidemiological studies. *JAMA: The Journal of the American Medical Association*, *291*(16), 2007–2012. Available from: http://dx.doi.org/10.1001/jama.291.16.2007.

Salmelin, R., Service, E., Kiesilä, P., Uutela, K., & Salonen, O. (1996). Impaired visual word processing in dyslexia revealed with magnetoencephalography. *Annals of Neurology, 40*(2), 157–162. Available from: http://dx.doi.org/10.1002/ana.410400206.

Sandak, R., Mencl, W. E., Frost, S. J., & Pugh, K. R. (2004). The neurobiological basis of skilled and impaired reading: Recent findings and new directions. *Scientific Studies of Reading, 8*(3), 273–292. Available from: http://dx.doi.org/10.1207/s1532799xssr0803_6.

Saygin, Z. M., Norton, E. S., Osher, D. E., Beach, S. D., Cyr, A. B., Ozernov-Palchik, O., et al. (2013). Tracking the roots of reading ability: White matter volume and integrity correlate with phonological awareness in prereading and early-reading kindergarten children. *The Journal of Neuroscience: The Official Journal of the Society for Neuroscience, 33*(33), 13251–13258. Available from: http://dx.doi.org/10.1523/JNEUROSCI.4383-12.2013.

Scarborough, H. S. (2005). Developmental relationships between language and reading: Reconciling a beautiful hypothesis with some ugly facts. In H. W. Catts, & A. G. Kamhi (Eds.), *The connections between language and reading disabilities* (pp. 3–24). Mahwah, NJ: Lawrence Erlbaum Associates.

Scarborough, H. S., & Brady, S. A. (2002). Toward a common terminology for talking about speech and reading: A glossary of the "phon" words and some related terms. *Journal of Literacy Research, 34*(3), 299–336. Available from: http://dx.doi.org/10.1207/s15548430jlr3403_3.

Scerri, T. S., & Schulte-Körne, G. (2010). Genetics of developmental dyslexia. *European Child and Adolescent Psychiatry, 19*(3), 179–197. Available from: http://dx.doi.org/10.1007/s00787-009-0081-0.

Schatschneider, C., Fletcher, J. M., Francis, D. J., Carlson, C. D., & Foorman, B. R. (2004). Kindergarten prediction of reading skills: A longitudinal comparative analysis. *Journal of Educational Psychology, 96*(2), 265. Retrieved from: <http://psycnet.apa.org/journals/edu/96/2/265/>.

Shaywitz, B. A., Shaywitz, S. E., Pugh, K. R., Constable, R. T., Skudlarski, P., Fulbright, R. K., et al. (1995). Sex differences in the functional organization of the brain for language. *Nature, 373*(6515), 607–609. Available from: http://dx.doi.org/10.1038/373607a0.

Shaywitz, S. E., Shaywitz, B. A., Fulbright, R. K., Skudlarski, P., Mencl, W. E., Constable, R. T., et al. (2003). Neural systems for compensation and persistence: Young adult outcome of childhood reading disability. *Biological Psychiatry, 54*, 25–33. Available from: http://dx.doi.org/10.1016/S0006-3223(03)01836-X.

Shaywitz, S. E., Shaywitz, B. A., Pugh, K. R., Fulbright, R. K., Constable, R. T., Mencl, W. E., et al. (1998). Functional disruption in the organization of the brain for reading in dyslexia. *Proceedings of the National Academy of Sciences of the United States of America, 95*(5), 2636–2641.

Siok, W. T., Niu, Z., Jin, Z., Perfetti, C. A., & Tan, L. H. (2008). A structural-functional basis for dyslexia in the cortex of Chinese readers. *Proceedings of the National Academy of Sciences of the United States of America, 105*(14), 5561–5566. Available from: http://dx.doi.org/10.1073/pnas.0801750105.

Siok, W. T., Perfetti, C. A., Jin, Z., & Tan, L. H. (2004). Biological abnormality of impaired reading is constrained by culture. *Nature, 431*(7004), 71–76. Available from: http://dx.doi.org/10.1038/nature02865.

Sowell, E. R., Thompson, P. M., Leonard, C. M., Welcome, S. E., Kan, E., & Toga, A. W. (2004). Longitudinal mapping of cortical thickness and brain growth in normal children. *The Journal of Neuroscience, 24*(38), 8223–8231. Available from: http://dx.doi.org/10.1523/JNEUROSCI.1798-04.2004.

Stanovich, K. E., & Siegel, L. S. (1994). Phenotypic performance profile of children with reading disabilities: A regression-based test of the phonological-core variable-difference model. *Journal of Educational Psychology, 86*(1), 24–53. Available from: http://dx.doi.org/10.1037/0022-0663.86.1.24.

Thiebaut de Schotten, M., Cohen, L., Amemiya, E., Braga, L. W., & Dehaene, S. (2014). Learning to read improves the structure of the arcuate fasciculus. *Cerebral Cortex (New York, NY: 1991), 24*(4), 989–995. Available from: http://dx.doi.org/10.1093/cercor/bhs383.

Van der Mark, S., Bucher, K., Maurer, U., Schulz, E., Brem, S., Buckelmüller, J., et al. (2009). Children with dyslexia lack multiple specializations along the visual word-form (VWF) system. *NeuroImage, 47*(4), 1940–1949. Available from: http://dx.doi.org/10.1016/j.neuroimage.2009.05.021.

Van der Mark, S., Klaver, P., Bucher, K., Maurer, U., Schulz, E., Brem, S., et al. (2011). The left occipitotemporal system in reading: Disruption of focal fMRI connectivity to left inferior frontal and inferior parietal language areas in children with dyslexia. *NeuroImage, 54*(3), 2426–2436. Available from: http://dx.doi.org/10.1016/j.neuroimage.2010.10.002.

Vandermosten, M., Boets, B., Wouters, J., & Ghesquière, P. (2012). A qualitative and quantitative review of diffusion tensor imaging studies in reading and dyslexia. *Neuroscience and Biobehavioral Reviews, 36*(6), 1532–1552. Available from: http://dx.doi.org/10.1016/j.neubiorev.2012.04.002.

Vellutino, F. R., Fletcher, J. M., Snowling, M. J., & Scanlon, D. M. (2004). Specific reading disability (dyslexia): What have we learned in the past four decades? *Journal of Child Psychology and Psychiatry, and Allied Disciplines, 45*(1), 2–40.

Vinckier, F., Dehaene, S., Jobert, A., Dubus, J. P., Sigman, M., & Cohen, L. (2007). Hierarchical coding of letter strings in the ventral stream: Dissecting the inner organization of the visual word-form system. *Neuron, 55*(1), 143–156. Available from: http://dx.doi.org/10.1016/j.neuron.2007.05.031.

Vourkas, M., Micheloyannis, S., Simos, P. G., Rezaie, R., Fletcher, J. M., Cirino, P. T., et al. (2011). Dynamic task-specific brain network connectivity in children with severe reading difficulties. *Neuroscience Letters, 488*(2), 123–128. Available from: http://dx.doi.org/10.1016/j.neulet.2010.11.013.

Wagner, R. K., & Torgesen, J. K. (1987). The nature of phonological processing and its causal role in the acquisition of reading skills. *Psychological Bulletin, 101*(2), 192–212. Available from: http://dx.doi.org/10.1037/0033-2909.101.2.192.

Wimmer, H., & Schurz, M. (2010). Dyslexia in regular orthographies: Manifestation and causation. *Dyslexia, 16*(4), 283–299. Available from: http://dx.doi.org/10.1002/dys.411.

Wimmer, H., Schurz, M., Sturm, D., Richlan, F., Klackl, J., Kronbichler, M., et al. (2010). A dual-route perspective on poor reading in a regular orthography: An fMRI study. *Cortex; A Journal Devoted to the Study of the Nervous System and Behavior, 46*(10), 1284–1298. Available from: http://dx.doi.org/10.1016/j.cortex.2010.06.004.

SECTION M

ANIMAL MODELS FOR LANGUAGE

CHAPTER

66

Rodent Models of Speech Sound Processing

Crystal T. Engineer, Tracy M. Centanni and Michael P. Kilgard

School of Behavioral and Brain Sciences, The University of Texas at Dallas, Richardson, TX, USA

66.1 RODENT MODELS ARE IMPORTANT FOR STUDYING NEURAL CORRELATES OF SPEECH PERCEPTION

Numerous studies have documented the ability of rodents to perform complex speech discrimination tasks as accurately as humans. The observation that rodents exhibit categorical perception of speech sounds indicates that the basic neural mechanisms used to distinguish speech sounds are present in rodents. Speech sounds evoke unique spatiotemporal response patterns in both human and rodent auditory cortex. Rodent models of speech sound processing provide the opportunity to precisely control the genetic and environmental factors that can impair speech sound processing in humans.

66.2 SPEECH SOUND DISCRIMINATION BY RODENTS

66.2.1 Consonants

Numerous studies of speech sound processing in rodents have shown that humans and rodents discriminate speech sounds similarly. Kuhl and Miller (1975) performed one of the most well-known studies of rodent speech perception, which went against the theories of the time that speech perception was unique to humans. Chinchillas were trained to categorize /d/ and /t/ syllables in multiple vowel and talker contexts. The acoustic cue voice onset time (VOT), which is the time between the onset of the consonant and the onset of voicing, can be used by humans and rodents to categorize stop consonants. The voiced consonants (/b/, /d/, /g/) have a short VOT, whereas the voiceless consonants (/p/, /t/, /k/) have a long VOT. Chinchillas were trained on an avoidance conditioning task during which they had to cross a barrier when presented with a target sound and refrain from crossing when presented with a nontarget sound. The chinchillas could accurately categorize voiced *versus* voiceless sounds, and they were even able to generalize when presented with synthetic syllables, syllables spoken by novel talkers, and syllables with novel vowels. The chinchillas were then tested using synthetic /d/ and /t/ syllables with VOTs ranging from 0 to 80 ms to determine their phonetic boundary, the VOT where they were unable to distinguish the voicing category. Chinchillas and humans had a nearly identical phonetic boundary (33.5 ms for chinchillas and 35.2 ms for humans). A follow-up study determined the phonetic boundary for chinchillas using a /b/-/p/ continuum and a /g/-/k/ continuum, and found that chinchillas and humans have equivalent boundaries for these sounds as well (Kuhl & Miller, 1978). These findings support the idea that speech categories make use of natural boundaries in the auditory system that are present in rodents and are not unique to humans.

Studies testing the phonetic boundaries of synthetic speech sound continuums in gerbils provided additional evidence that rodents and humans discriminate speech sounds similarly. Gerbils were trained to categorize vowel, liquid, and stop consonant speech sounds (Sinnott & Mosteller, 2001). The gerbils listened to a repeating background speech sound and had to jump off a platform when the repeating background sound changed into a target speech sound to receive a food pellet reward. The phonetic boundaries in gerbils were similar to the boundaries using the same sounds in humans, particularly for the stop consonant task (on a continuum of 8 stimuli: stimulus 4.7 boundary for gerbils and stimulus 4.5 boundary for humans).

In addition to stop consonants, speech sound continuums can also be made for affricate and fricative

Neurobiology of Language. DOI: http://dx.doi.org/10.1016/B978-0-12-407794-2.00066-3

speech sounds by altering the rise time and duration of the sounds. The affricate "ch" has a shorter duration and faster rise time compared with the fricative "sh." Rats were trained to press a left lever for one sound (e.g., "ch") and a right lever for the other sound ("sh"), and they were tested on their ability to generalize to sounds midway along the continuum (Reed, Howell, Sackin, Pizzimenti, & Rosen, 2003). The investigators concluded that rats, like humans, appear to use the rise time cue to accurately categorize these sounds.

As in humans, speech sound discrimination by rats can be affected by previous exposure to speech sounds. Infants who were exposed to a unimodal distribution of sounds at the midpoint of a /d/-/t/ continuum are less able to discriminate the endpoints of the continuum compared with infants who were exposed to a bimodal distribution of sounds near the edges of the /d/-/t/ continuum (Maye, Werker, & Gerken, 2002). After the same exposure, rats, like infants, were able to discriminate the sounds more accurately when exposed to a bimodal distribution compared with a unimodal distribution (Pons, 2006).

While operant speech discrimination training in rodents can be time-consuming, rats can rapidly learn to discriminate between consonant sounds using a prepulse inhibition (PPI) task (Floody & Kilgard, 2007). During the PPI task, a standard speech sound is played ("pa") in the background, and this sound rarely predicts a loud startle stimulus (a 102-dB burst of white noise). A second oddball speech sound ("ba") occurs rarely, but it is always followed by the startle stimulus. After just a few minutes of training, rats startled less to the startle stimulus when it was preceded by the oddball sound compared with trials in which the startle stimulus was preceded by the standard sound. These results suggest that rats could reliably detect the change from the standard speech sound to the oddball speech sound.

Rats have also been tested on consonant discrimination using pairs of sounds that differ in one articulatory feature (place of articulation, voicing, or manner of articulation). Studies of speech sound processing in rats often use speech sounds that have been shifted higher by one octave to better match the rat hearing range. Rats were trained to press the lever in response to a target consonant ("d"), and to refrain from pressing the lever in response to a nontarget sound ("t"). As seen in previous studies, rats could easily discriminate most consonant contrasts, such as "d" *versus* "t" or "sh" *versus* "ch" (Engineer et al., 2008). Rats were also able to accurately categorize the sounds when additional temporal or speaker variation was introduced (Engineer et al., 2013). Rats can discriminate consonant sounds in the presence of background noise or after spectral and temporal degradation (Ranasinghe, Vrana, Matney, & Kilgard, 2012; Shetake et al., 2011). However, rats were unable to discriminate consonant contrasts that are difficult for individuals learning English as a second language (Engineer et al., 2008), such as "r" *versus* "l" (Logan, Lively, & Pisoni, 1991), suggesting that rodents are a good animal model of speech perception.

66.2.2 Vowels

Rats are also able to discriminate between vowel sounds. Rats were trained to discriminate between two vowel categories, and were able to generalize to novel vowel sounds (Eriksson & Villa, 2006; Perez et al., 2013). Next, novel vowels were introduced where the formant frequency correlation to the fundamental frequency was reversed, altering the normal relationship between the formant and fundamental frequencies. The results showed that rats, like humans, use the fundamental frequency as a cue in vowel discrimination (Eriksson & Villa, 2006).

Chinchillas are also able to discriminate between vowel sounds with a high degree of accuracy. Similar to the Kuhl consonant discrimination study, chinchillas were trained to categorize /a/ and /i/ vowels (Burdick & Miller, 1975). There was no decrement in performance when /a/ and /i/ vowels in multiple pitch (low, natural, and high) and multiple talker contexts (two male, two female) were introduced. The chinchillas were able to generalize to vowels produced by novel talkers (12 male, 12 female), as well as synthetic vowels. This extensive set of experiments demonstrated that rodents, like humans, are able to accurately categorize vowels while ignoring large amounts of irrelevant variability in the sounds.

66.2.3 Complex Tasks

Rodents not only are able to discriminate between individual consonant and vowel sounds but also are able to perform complex speech sound discrimination tasks. Toro and colleagues found that rats were able to discriminate between sentences spoken in Dutch and sentences spoken in Japanese, but were unable to discriminate sentences in the two languages when the sentences were presented backwards (Toro, Trobalon, & Sebastián-Gallés, 2003, 2005). The same observation has also been reported in infants and monkeys (Ramus, Hauser, Miller, Morris, & Mehler, 2000).

Toro and colleagues also examined the ability of rats to segment a speech stream using statistical regularities (Toro & Trobalón, 2005). After being passively exposed to words from an artificial language (e.g., "tupiro" or "pigola"), rats were tested using words they had been exposed to, part-words that were composed of the last syllable of one word with the first two syllables of another word, or nonwords that were composed of syllables in a different order than the order the rats had previously heard. Rats pressed the lever more often during the test phase in response to a word that they had been exposed to compared with part-words or nonwords, suggesting that the rats were able to detect syllable regularities in the speech stream.

Rats are also able to learn rules involving repeating consonants and vowels, and they can generalize the rule to novel words (de la Mora & Toro, 2013). In the first experiment, rats were trained on CVCVCV words ("C," consonant; "V," vowel) and learned to press the lever for words that had repeating vowels (AAB structure, "bakasi"), but not for words with three different vowels (ABC structure, "bakisœ"). The rats were then able to generalize to new words composed of novel consonants and vowels (e.g., lever press for "felepu," but not for "felopu"). The second experiment focused on consonants instead of vowels, and rats learned to press the lever for words that had repeating consonants (AAB structure, "sasœki"), but not for words that did not have repeating consonants (ABC structure, "sikabœ"). Rats were then able to generalize the repeating consonant rule to new words composed of novel consonants and vowels (press for "fefulo" but not "fepulo"). The same set of experiments was performed in humans, and although humans could accurately generalize the repeating vowel rule, they were unable to generalize the repeating consonant rule, suggesting that humans process consonants and vowels differently because of their experience with language. This study replicates the findings of a previous study showing that rats can learn rules for sequences of auditory (3.2 or 9 kHz tone) or visual stimuli (light or darkness) and transfer the learned rule to novel sequences (Murphy, Mondragón, & Murphy, 2008).

66.3 SPEECH SOUND NEURAL CODING

The early stages of speech sound processing in rodents closely parallel early speech processing in humans (Johnson, Nicol, & Kraus, 2005). Responses to speech sounds in both humans and animals closely resemble the physical characteristics of the sound early in the auditory system (e.g., in the auditory nerve or inferior colliculus [IC]), and they grow more abstract to better represent the perceptual characteristics of the sound at higher levels of the auditory system (e.g., in secondary auditory cortex or prefrontal cortex).

Responses in the auditory nerve of chinchillas were recorded in response to synthetic speech sounds differing in VOT (Sinex & McDonald, 1988). As in previous studies, a /ga/–/ka/ or /da/–/ta/ continuum of speech sounds with VOTs ranging from 0 to 80 ms in 10 ms steps was presented. Neurons with a low characteristic frequency (CF) near the first formant of the speech sounds increased their firing rate in response to the onset of voicing, and the strength of this onset response can be used to accurately predict the onset of voicing when compared with baseline. Neurons with higher frequency tuning, however, did not reliably change their firing rate and were not good predictors of the onset of voicing. A second study of auditory nerve responses to sounds differing in VOT in chinchillas found differing results (Stevens & Wickesberg, 1999). This study used the naturally spoken syllables /da/ and /ta/ and found that both low- and high-frequency neurons accurately encode the onset of voicing. This finding that naturally spoken speech and synthetic speech produce different results in the same area of the same species suggests that auditory nerve responses are highly dependent on the acoustics of the sound.

Slightly higher along the auditory pathway, in the ventral cochlear nucleus, responses were recorded in anesthetized rats in response to six naturally spoken consonants differing in VOT (/bot/, /dot/, /got/, /pot/, /tot/, and /kot/) (Clarey, Paolini, Grayden, Burkitt, & Clark, 2004). Responses recorded in three ventral cochlear nucleus cell types (primary-like, chopper, and onset) were similar to auditory nerve responses. The consonant release evoked a peak of activity followed by a second peak of activity in response to voicing onset.

Responses in the IC to speech sounds differing in VOT also accurately encode the onset of voicing. However, IC neurons have less sustained responses through both the consonant and vowel portions of the sound compared with auditory nerve responses (Portfors & Sinex, 2005). A separate study recorded IC responses to 28 CVC (consonant–vowel–consonant) speech sounds differing in either the vowel or the initial consonant (Perez et al., 2013). IC response patterns predicted both consonant and vowel discrimination performance. Speech sounds that evoke similar IC response patterns ("rad" *versus* "lad" or "dad" *versus* "dead") are more difficult for rats to discriminate than speech sounds that evoke distinct IC response patterns ("dad" *versus* "bad" or "dad" *versus* "deed"). Consonant discrimination was well-predicted by IC response pattern similarity when spike timing information was preserved, whereas vowel discrimination was well-predicted by IC response pattern similarity when spike timing information was eliminated. The observation that consonants and vowels are processed differently parallels similar findings in humans (Caramazza, Chialant, Capasso, & Miceli, 2000; Poeppel, 2003).

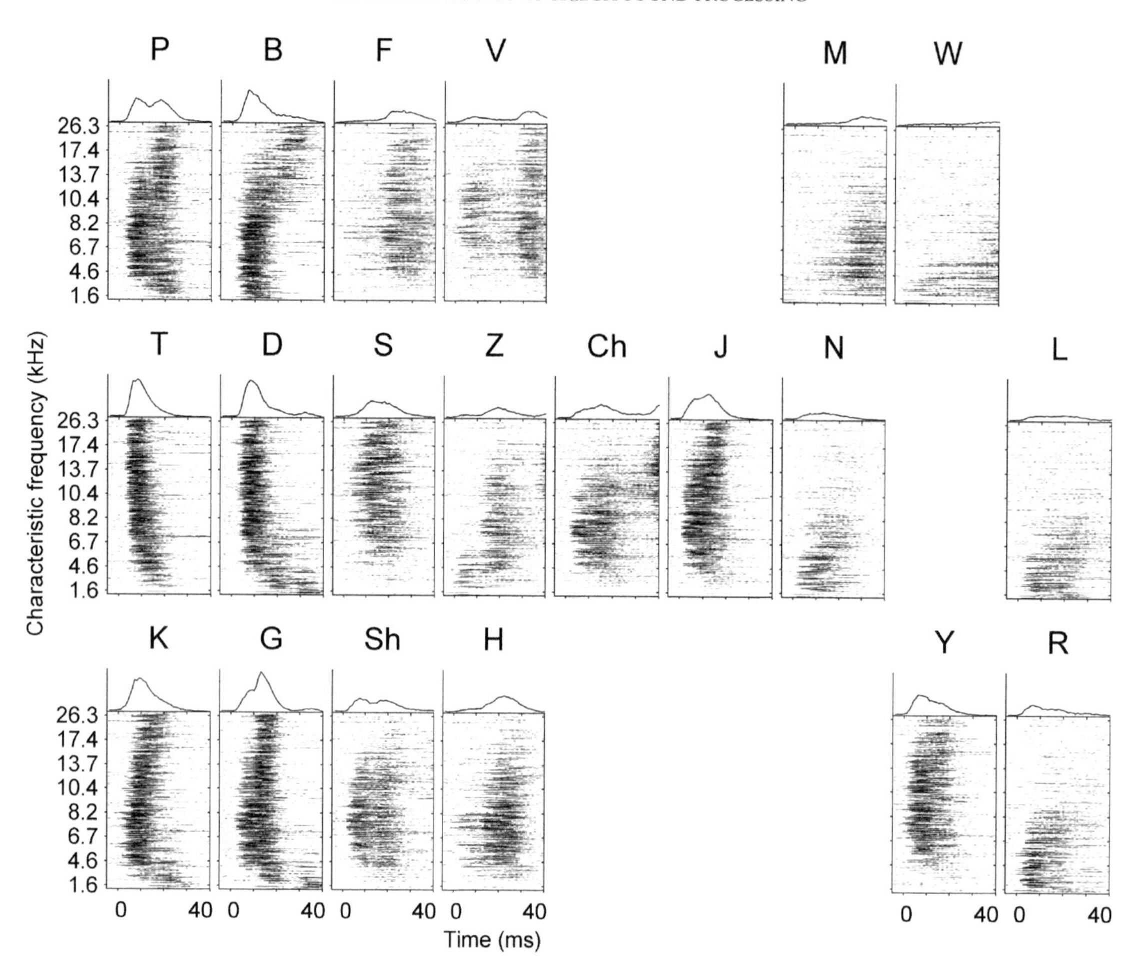

FIGURE 66.1 Primary auditory cortex responds uniquely to individual consonant sounds. The *y*-axis orders A1 sites by CF (kHz), and the *x*-axis is time (ms). The average population response is plotted above each neurogram. *From Engineer et al. (2008, Figure 2) with permission from the publisher.*

Although auditory nerve and IC neurons respond throughout a speech sound, primary auditory cortex (A1) neurons more strongly represent the onset and transition parts of speech sounds. The A1 response pattern evoked by each consonant and vowel sound is unique (Engineer et al., 2008; Perez et al., 2013). For example, the stop consonants /b/, /d/, and /g/ differ in their place of articulation, and A1 neurons respond distinctly to each of the consonants. For the consonant /b/, A1 low-frequency neurons respond first, followed by high-frequency neurons; for the consonant /d/, A1 high-frequency neurons respond first, followed by low-frequency neurons; for the consonant /g/, A1 mid-frequency neurons respond first, followed by high- and low-frequency neurons (Figure 66.1). As in IC, A1 neural response patterns accurately predict speech sound discrimination accuracy ($R^2 = 0.75$; Figure 66.2).

A recent study compared the amount of redundancy in the responses to speech sounds in IC neurons compared with A1 neurons (Ranasinghe, Vrana, Matney, & Kilgard, 2013). The firing rate evoked by each speech

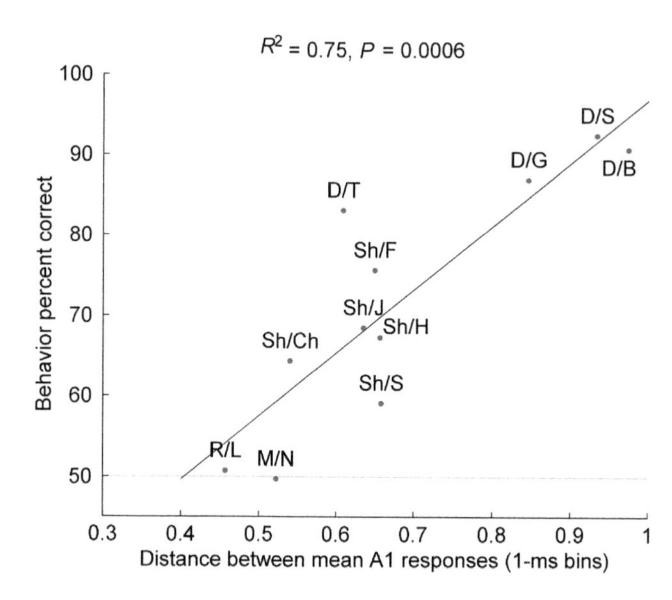

FIGURE 66.2 Neural similarity predicts consonant discrimination ability. The dotted line indicates chance performance (50%). *From Engineer et al. (2008, Figure 5a) with permission from the publisher.*

sound was compared in pairs of IC neurons that were tuned to the same CF. The response strength to a speech sound was nearly identical in pairs of IC neurons tuned to the same frequency. The same comparison was made using pairs of A1 neurons tuned to the same frequency, but the response strength to a speech sound in one A1 neuron could not be predicted by the response strength to a speech sound in a similarly tuned A1 neuron. This finding indicates that A1 neurons evoke response patterns that are more diverse than the patterns evoked by IC neurons. The result that IC neurons tuned to the same frequency evoke identical response patterns while A1 neurons tuned to the same frequency evoke distinct response patterns supports the finding that responses in A1 and higher auditory areas are more abstract and do not just represent the physical characteristics of the sound (Chechik et al., 2006).

The various auditory cortical fields in the rat brain respond differently to speech sounds (Centanni, Engineer, & Kilgard, 2013). The cortical response patterns in A1, anterior auditory field (AAF), ventral auditory field (VAF), and posterior auditory field (PAF) are correlated with behavioral discrimination ability. As shown previously in IC and A1, the firing rate evoked by each speech sound in pairs of neurons tuned to the same CF was compared across the four auditory cortex fields. A1 and AAF neurons tuned to the same frequency evoked more similar response patterns than neurons in PAF and VAF, suggesting that PAF and VAF are higher along the auditory hierarchy and represent speech sounds in a more abstract manner.

The similarity of the basic behavioral and neural responses to speech in rodents and humans can be explained by the conservation of basic auditory coding principles across mammals. There is evidence that humans process speech sounds differently than animals at later stages, but processing at earlier stages is remarkably well-conserved (Davis, Ford, Kherif, & Johnsrude, 2011; Obleser, Wise, Dresner, & Scott, 2007). This fact provides an opportunity to use animal models to better understand speech processing under difficult conditions such as background noise, hearing loss, central nervous system damage, and genetic disorders.

66.4 SPEECH SOUND PROCESSING PROBLEMS

66.4.1 Acoustic Degradation

Human and animal communication requires spectrotemporal precision in auditory cortex (Ranasinghe et al., 2013; Sinex & McDonald, 1988; Steinschneider, Fishman, & Arezzo, 2003; Steinschneider, Volkov, Noh, Garell, & Howard, 1999). Auditory-evoked responses in A1 of gerbils were recorded using gerbil vocalizations that were spectrally degraded to 16, 8, or 4 bands using a noise vocoder or temporally degraded by reversing the sound using a variety of bin sizes (5, 10, 20, 30, 50, 100, 150, 200, 250 ms, or the entire sound). Spike rate could not be used to identify stimulus condition, but a Euclidean distance measure comparing pairs of spike trains (Victor & Purpura, 1997) was able to identify the calls with a high degree of accuracy. This method was more accurate on the spectrally degraded calls than the temporally degraded calls (Ter-Mikaelian, Semple, & Sanes, 2013). These results demonstrate that the timing of evoked action potentials contains more information about gerbil call identity than the number of action potentials that occur over the same period of time, and they suggest that A1 spike timing could also be used to accurately identify spectrally and temporally degraded speech sounds.

As discussed, chinchillas are able to discriminate between human consonant sounds and could be a useful model to understand neural correlates of speech sound discrimination in humans. The initial consonant from the words "ball," "dirty," "persons," and "today" were degraded into one, two, three, or four spectral bands, and recordings were acquired from individual auditory nerve fibers. Responses to one- and two-band spectrally degraded stimuli were significantly different than responses to sounds with greater spectral information (Loebach & Wickesberg, 2006). Because rodents are highly accurate at discriminating human speech sounds (Ranasinghe et al., 2012), these results suggest that a rodent would be able to accurately discriminate human speech sounds with more than two spectral bands.

Rats are able to behaviorally discriminate human speech sounds after significant spectral or temporal degradation. Ranasinghe et al. (2012) demonstrated that rats discriminate human speech sounds even when only four bands of spectral information are preserved, as is seen in human participants (Figure 66.3). The Euclidean distance between pairs of auditory cortical responses to the degraded sounds was able to predict the accuracy of the rats when discriminating the same sounds (Ranasinghe et al., 2012).

Rats are also able to accurately discriminate speech sounds in high levels of white and speech-shaped background noise with thresholds that are comparable with those of humans (Shetake et al., 2011). Like humans, rats are more impaired both neurally and behaviorally at discriminating sounds in speech-shaped noise compared with white noise (Busch & Eldredge, 1967). Both humans and rats are able to discriminate some speech sound contrasts even when the background noise is 12 dB louder than the speech signal (Miller & Nicely, 1955; Shetake et al., 2011). The observation that rodents are able to discriminate human speech sounds in a variety of difficult listening

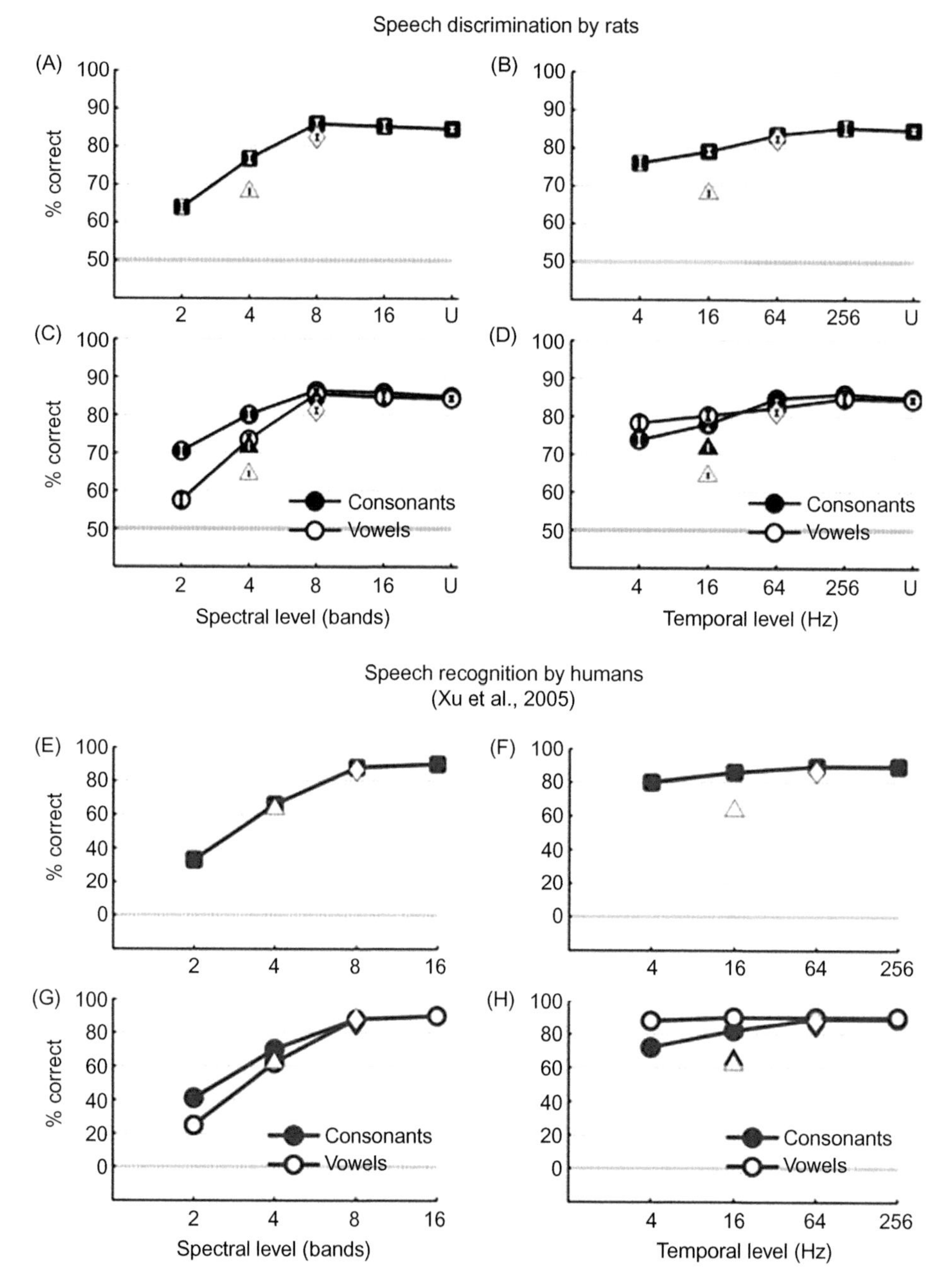

FIGURE 66.3 Rats and humans can discriminate consonant and vowel sounds even after significant stimulus degradation. Rats are able to discriminate a target sound from several distractors when spectral information is reduced to only two bands of spectral information (top left) or when the temporal information is reduced by low pass filtering of the speech envelope at 4 Hz (top right). The human pattern of speech discrimination performance after spectral degradation (bottom left) and temporal degradation (bottom right) was similar to rat performance, which suggests that rats are a good model for speech sound processing. *Data from Ranasinghe et al. (2012, Figure 3) and (Xu, Thompson, & Pfingst, 2005, Figures 1 and 2) with permission from the publisher.*

situations with thresholds that mimic those seen in humans suggests that rodents are good animal models of speech sound processing.

66.4.2 Cortical Lesions

Damage to auditory cortex has been a useful tool in rats to investigate the role of specific auditory areas. Removal or inactivation of auditory cortex causes deficits specific to the affected region. Aspiration lesions involve physically removing the gray matter section of cortex while leaving the white matter underneath intact. Bilateral removal of auditory cortex by this method does not impair rats' ability to reflexively discriminate consonant–vowel–consonant speech sounds (Floody, Ouda, Porter, & Kilgard, 2010), as measured with a gap-detection startle task. Gap detection is a task that relies on the rat's reflexive startle response rather than an operant go/no-go task. When tested on the operant task described, rats with auditory cortex lesions were unable to discriminate between consonant onsets when tested using an operant go/no-go task (Porter, Rosenthal, Ranasinghe, & Kilgard, 2011). These results suggest that auditory cortex is important for categorization of consonant sounds, but it is not required for reflexive responses to a change in consonant acoustics. Vowel discrimination ability is not impaired by the removal of A1 through electrolytic lesions, but it is impaired by removal of dorsal and rostral auditory areas (Kudoh, Nakayama, Hishida, & Shibuki, 2006). These results suggest that the different auditory areas in rodents may extract different information from speech, similar to humans (Obleser, Leaver, VanMeter, & Rauschecker, 2010).

66.4.3 Rodent Models of Speech Processing Impairments

Exposure to loud, repetitive stimuli during the early postnatal period is known to cause changes in the structure and function of auditory cortex. A1 in rats is organized from neurons tuned to low frequencies at the posterior end to neurons tuned to high frequencies at the anterior end (Centanni et al., 2013; Polley, Read, Storace, & Merzenich, 2007). When young rat pups were raised in an environment with constant exposure to pulsed noise, A1 developed with a significantly different tonotopic organization (Chang & Merzenich, 2003; Ranasinghe et al., 2012). However, despite significant degradation of tone response in A1, neural and behavioral discrimination of speech sounds remained highly accurate in these rats (Ranasinghe et al., 2012).

As discussed thus far, rodents have become a useful model for studying perception and discrimination of human speech sounds in a variety of contexts. Rodent models also have the advantage that environmental and genetic conditions can be strictly controlled to a level not possible in human participants. Many researchers have used such models to make progress in understanding the biology and function of many human speech sound processing impairments. In addition to the cortical and stimulus degradation conditions already discussed, rodents exposed to high doses of valproic acid *in utero* have been used to study autism and valproate syndrome (Engineer et al., 2014; Markram & Markram, 2010). A recent study used rodents to evaluate the neural and behavioral effects of temporary or permanent hearing loss on speech processing (Reed et al., 2014). Many other environmental insults, including perinatal anoxia (Strata et al., 2005), and prenatal exposure to antidepressants (Simpson et al., 2011) or pollutants (Kenet, Froemke, Schreiner, Pessah, & Merzenich, 2007) are known to impair auditory system function in rats, but the impact of these insults on speech processing have not been evaluated. Rodent models are likely to aid in understanding the neural mechanisms responsible for a variety of communication disorders.

66.4.4 Genetic Manipulation

Rodents may prove to be particularly useful for understanding communication disorders with a genetic etiology. Rodents have a homolog for the majority of the genes that cause communication disorders in humans (Kere, 2011; Polleux & Lauder, 2004; Shearer & Smith, 2012). Genetic disruption is now routinely performed in rodents. The genes *FMR1*, *MeCP2*, and *KIAA0319* have been the focus of intense study in rodent models in recent years, because disruption of these genes can cause Fragile X syndrome, Rett syndrome, and dyslexia, respectively. Rats with reduced expression of *Kiaa0319* (the rat homolog of *KIAA0319)* have cortical abnormalities similar to those seen in dyslexia, including microgyria and heterotopia (Szalkowski et al., 2013). Neural responses in the auditory brainstem to speech sounds in dyslexic individuals exhibit much more trial-by-trial variability (Figure 66.4A) than in control subjects (Hornickel & Kraus, 2013). It has been proposed that the inefficient mapping from phonemes (sounds) to graphemes (letters) is a major cause of reading problems. The recent demonstration that the dyslexia gene *Kiaa0319* can increase trial-by-trial variability in rats provides the first direct link between the genetics of dyslexia and a potential mechanistic explanation (Centanni et al., 2014). Reduced expression of *Kiaa0319* reduced the amplitude of the average neural response to speech and delayed the time to peak response, which is consistent with human studies. Use of the rodent model made it possible to demonstrate that these differences in the average response are caused by elevated variability and not by a reduction in the neural response (Figure 66.4B). The total number of cortical action potentials evoked in response to speech sounds in rats with reduced *Kiaa0319* expression was not significantly different from that of control rats. This important experiment confirms that rodent models of speech sound processing are well-suited for evaluating critical and long-standing hypotheses regarding the neural basis of speech sound processing disorders.

It was not initially expected that rodents would provide a valuable animal model of speech sound processing because rodents do not use language. However, the body of experimental evidence described suggests that rodents are an appropriate model of

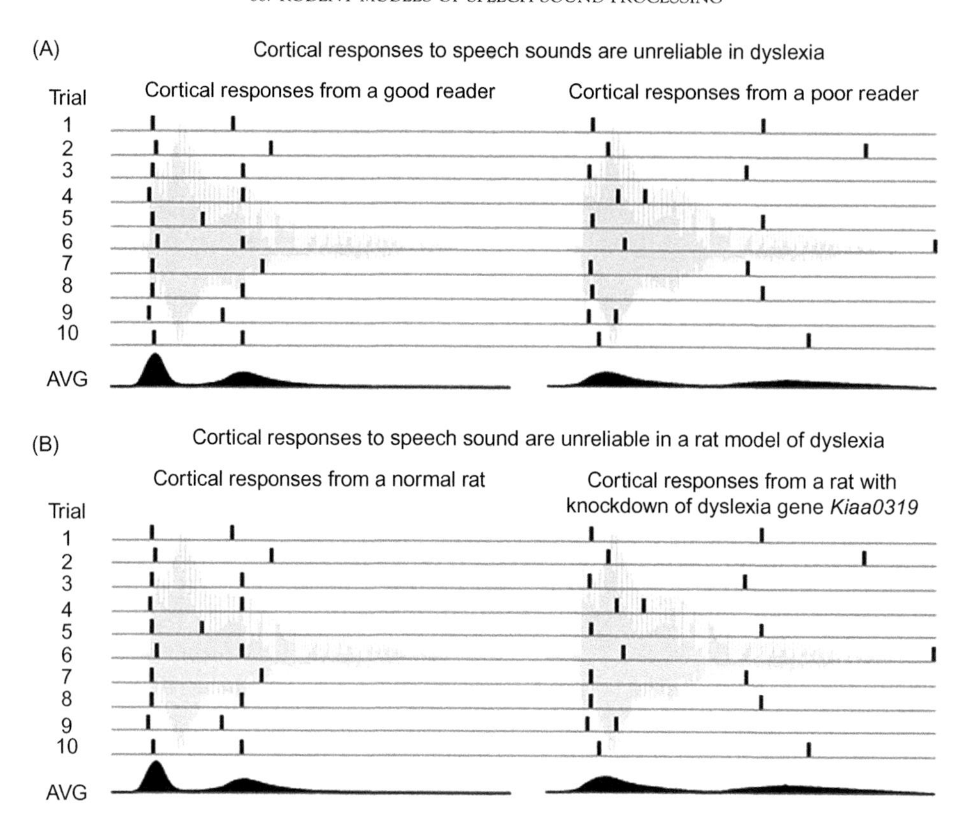

FIGURE 66.4 Schematic of increased neural variability in dyslexia and in a rat model of dyslexia. (A) In normal readers, neural responses to speech sounds (gray) are consistently timed across multiple presentations of the same stimulus. In individuals with dyslexia, the average neural response is significantly delayed and weaker due to greater trial-by-trial variability (Hornickel & Kraus, 2013). (B) In control rats, neural responses to speech sounds are consistently timed across multiple presentations of the same stimulus. In rats with reduced expression of the dyslexia gene *Kiaa0319* (caused by *in utero* RNA interference), neural responses to speech sounds were more variable, which caused the average neural response to be delayed and reduced in amplitude, even though the number of evoked action potentials (black tick marks) was the same as in control rats (Centanni et al., 2014).

human speech sound processing. The early stages of speech sound processing (e.g., the spatiotemporal response patterns evoked by human consonant and vowel sounds) in humans and rodents rely on conserved neural mechanisms that are shared by all mammals. Additionally, the shared behavioral discrimination thresholds between rats and humans allow researchers to investigate neural correlates of behavioral ability while maintaining strict control of genetic and environmental variables that cannot be controlled in human participants.

Future research should take advantage of this model, especially as candidate genes for a variety of disorders are identified. Genetic knockdown models, such as the one described, will be useful in disorders where a specific protein is produced in lower quantities than in typically developing individuals. Knockout models, such as the fragile X model described, will be useful in evaluating the neural and behavioral impairments of disorders in which the protein is not translated at all (Engineer et al., 2014). Both models will provide the opportunity to not only elucidate the neural mechanisms behind these disorders but also help evaluate the effectiveness of behavioral and pharmacological interventions. Future work in rodent models may also make valuable predictions about ways to predict response to treatment in humans with the same conditions.

References

Burdick, C. K., & Miller, J. D. (1975). Speech perception by the chinchilla: Discrimination of sustained /a/ and /i/. *The Journal of the Acoustical Society of America*, *58*, 415.

Busch, A. C., & Eldredge, D. (1967). The effect of differing noise spectra on the consistency of identification of consonants. *Language and Speech*, *10*(3), 194–202.

Caramazza, A., Chialant, D., Capasso, R., & Miceli, G. (2000). Separable processing of consonants and vowels. *Nature*, *403* (6768), 428–430.

Centanni, T., Booker, A., Sloan, A., Chen, F., Maher, B., Carraway, R., et al. (2014). Knockdown of the dyslexia-associated gene *Kiaa0319* impairs temporal responses to speech stimuli in rat primary auditory cortex. *Cerebral Cortex*, *24*, 1753–1766.

Centanni, T. M., Engineer, C. T., & Kilgard, M. P. (2013). Cortical speech-evoked response patterns in multiple auditory fields are correlated with behavioral discrimination ability. *Journal of Neurophysiology, 110*(1), 177–189.

Chang, E. F., & Merzenich, M. M. (2003). Environmental noise retards auditory cortical development. *Science, 300*(5618), 498–502.

Chechik, G., Anderson, M. J., Bar-Yosef, O., Young, E. D., Tishby, N., & Nelken, I. (2006). Reduction of information redundancy in the ascending auditory pathway. *Neuron, 51*(3), 359–368.

Clarey, J. C., Paolini, A. G., Grayden, D. B., Burkitt, A. N., & Clark, G. M. (2004). Ventral cochlear nucleus coding of voice onset time in naturally spoken syllables. *Hearing Research, 190*(1), 37–59.

Davis, M. H., Ford, M. A., Kherif, F., & Johnsrude, I. S. (2011). Does semantic context benefit speech understanding through "top-down" processes? Evidence from time-resolved sparse fMRI. *Journal of Cognitive Neuroscience, 23*(12), 3914–3932.

de la Mora, D. M., & Toro, J. M. (2013). Rule learning over consonants and vowels in a non-human animal. *Cognition, 126*(2), 307–312.

Engineer, C., Centanni, T., Im, K., Borland, M., Moreno, N., Carraway, R., et al. (2014). Degraded auditory processing in a rat model of autism limits the speech representation in non-primary auditory cortex. *Developmental Neurobiology, 74*, 972–986.

Engineer, C. T., Centanni, T. M., Im, K. W., Rahebi, K. C., Buell, E. P., & Kilgard, M. P. (2014). Degraded speech sound processing in a rat model of fragile X syndrome. *Brain Research, 1564*, 72–84.

Engineer, C. T., Perez, C. A., Carraway, R. S., Chang, K. Q., Roland, J. L., Sloan, A. M., et al. (2013). Similarity of cortical activity patterns predicts generalization behavior. *PLos One, 8*(10), e78607.

Engineer, C. T., Perez, C. A., Chen, Y. H., Carraway, R. S., Reed, A. C., Shetake, J. A., et al. (2008). Cortical activity patterns predict speech discrimination ability. *Nature Neuroscience, 11*(5), 603–608.

Eriksson, J. L., & Villa, A. E. (2006). Learning of auditory equivalence classes for vowels by rats. *Behavioural Processes, 73*(3), 348–359.

Floody, O. R., & Kilgard, M. P. (2007). Differential reductions in acoustic startle document the discrimination of speech sounds in rats. *The Journal of the Acoustical Society of America, 122*, 1884.

Floody, O. R., Ouda, L., Porter, B. A., & Kilgard, M. P. (2010). Effects of damage to auditory cortex on the discrimination of speech sounds by rats. *Physiology and Behavior, 101*(2), 260–268.

Hornickel, J., & Kraus, N. (2013). Unstable representation of sound: A biological marker of dyslexia. *The Journal of Neuroscience, 33*(8), 3500–3504.

Johnson, K. L., Nicol, T. G., & Kraus, N. (2005). Brain stem response to speech: A biological marker of auditory processing. *Ear and Hearing, 26*(5), 424–434.

Kenet, T., Froemke, R., Schreiner, C., Pessah, I., & Merzenich, M. (2007). Perinatal exposure to a noncoplanar polychlorinated biphenyl alters tonotopy, receptive fields, and plasticity in rat primary auditory cortex. *Proceedings of the National Academy of Sciences of the United States of America, 104*(18), 7646–7651.

Kere, J. (2011). Molecular genetics and molecular biology of dyslexia. *Wiley Interdisciplinary Reviews: Cognitive Science, 2*(4), 441–448.

Kudoh, M., Nakayama, Y., Hishida, R., & Shibuki, K. (2006). Requirement of the auditory association cortex for discrimination of vowel-like sounds in rats. *Neuroreport, 17*(17), 1761–1766.

Kuhl, P. K., & Miller, J. D. (1975). Speech perception by the chinchilla: Voiced-voiceless distinction in alveolar plosive consonants. *Science, 190*(4209), 69–72.

Kuhl, P. K., & Miller, J. D. (1978). Speech perception by the chinchilla: Identification functions for synthetic VOT stimuli. *The Journal of the Acoustical Society of America, 63*, 905.

Loebach, J. L., & Wickesberg, R. E. (2006). The representation of noise vocoded speech in the auditory nerve of the chinchilla: Physiological correlates of the perception of spectrally reduced speech. *Hearing Research, 213*(1), 130–144.

Logan, J. S., Lively, S. E., & Pisoni, D. B. (1991). Training Japanese listeners to identify english /r/and/l: A first report. *The Journal of the Acoustical Society of America, 89*(2), 874.

Markram, K., & Markram, H. (2010). The intense world theory—A unifying theory of the neurobiology of autism. *Frontiers in Human Neuroscience, 4*, 224. Available from: http://dx.doi.org/10.3389/fnhum.2010.00224.

Maye, J., Werker, J. F., & Gerken, L. (2002). Infant sensitivity to distributional information can affect phonetic discrimination. *Cognition, 82*(3), B101–B111.

Miller, G. A., & Nicely, P. E. (1955). An analysis of perceptual confusions among some English consonants. *The Journal of the Acoustical Society of America, 27*(2), 338–352.

Murphy, R. A., Mondragón, E., & Murphy, V. A. (2008). Rule learning by rats. *Science, 319*(5871), 1849–1851.

Obleser, J., Leaver, A. M., VanMeter, J., & Rauschecker, J. P. (2010). Segregation of vowels and consonants in human auditory cortex: Evidence for distributed hierarchical organization. *Frontiers in Psychology, 1*, 232. Available from: http://dx.doi.org/10.3389/fpsyg.2010.00232.

Obleser, J., Wise, R. J., Dresner, M. A., & Scott, S. K. (2007). Functional integration across brain regions improves speech perception under adverse listening conditions. *The Journal of Neuroscience, 27*(9), 2283–2289.

Perez, C. A., Engineer, C. T., Jakkamsetti, V., Carraway, R. S., Perry, M. S., & Kilgard, M. P. (2013). Different timescales for the neural coding of consonant and vowel sounds. *Cerebral Cortex, 23*(3), 670–683.

Poeppel, D. (2003). The analysis of speech in different temporal integration windows: Cerebral lateralization as "asymmetric sampling in time". *Speech Communication, 41*(1), 245–255.

Polleux, F., & Lauder, J. M. (2004). Toward a developmental neurobiology of autism. *Mental Retardation and Developmental Disabilities Research Reviews, 10*(4), 303–317.

Polley, D. B., Read, H. L., Storace, D. A., & Merzenich, M. M. (2007). Multiparametric auditory receptive field organization across five cortical fields in the albino rat. *Journal of Neurophysiology, 97*(5), 3621–3638.

Pons, F. (2006). The effects of distributional learning on rats' sensitivity to phonetic information. *Journal of Experimental Psychology: Animal Behavior Processes, 32*(1), 97.

Porter, B. A., Rosenthal, T. R., Ranasinghe, K. G., & Kilgard, M. P. (2011). Discrimination of brief speech sounds is impaired in rats with auditory cortex lesions. *Behavioural Brain Research, 219*(1), 68–74.

Portfors, C. V., & Sinex, D. G. (2005). Coding of communication sounds in the inferior colliculus. In J. A. Winer, & C. E. Schreiner (Eds.), *The inferior colliculus* (pp. 411–425). New York: Springer.

Ramus, F., Hauser, M. D., Miller, C., Morris, D., & Mehler, J. (2000). Language discrimination by human newborns and by cotton-top tamarin monkeys. *Science, 288*(5464), 349–351.

Ranasinghe, K. G., Carraway, R. S., Borland, M. S., Moreno, N. A., Hanacik, E. A., Miller, R. S., et al. (2012). Speech discrimination after early exposure to pulsed-noise or speech. *Hearing Research, 289*(1), 1–12.

Ranasinghe, K. G., Vrana, W. A., Matney, C. J., & Kilgard, M. P. (2012). Neural mechanisms supporting robust discrimination of spectrally and temporally degraded speech. *Journal of the Association for Research in Otolaryngology, 13*(4), 527–542.

Ranasinghe, K. G., Vrana, W. A., Matney, C. J., & Kilgard, M. P. (2013). Increasing diversity of neural responses to speech sounds across the central auditory pathway. *Neuroscience, 252*, 80–97.

Reed, A. C., Centanni, T. M., Borland, M. S., Matney, C. J., Engineer, C. T., & Kilgard, M. P. (2014). Behavioral and neural discrimination of speech sounds after moderate or intense noise exposure in rats. *Ear and Hearing*. Available from: http://dx.doi.org/10.1097/AUD.0000000000000062.

Reed, P., Howell, P., Sackin, S., Pizzimenti, L., & Rosen, S. (2003). Speech perception in rats: Use of duration and rise time cues in labeling of affricate/fricative sounds. *Journal of the Experimental Analysis of Behavior*, *80*(2), 205–215.

Shearer, A. E., & Smith, R. J. (2012). Genetics: Advances in genetic testing for deafness. *Current Opinion in Pediatrics*, *24*(6), 679–686.

Shetake, J. A., Wolf, J. T., Cheung, R. J., Engineer, C. T., Ram, S. K., & Kilgard, M. P. (2011). Cortical activity patterns predict robust speech discrimination ability in noise. *European Journal of Neuroscience*, *34*(11), 1823–1838.

Simpson, K. L., Weaver, K. J., de Villers-Sidani, E., Lu, J. Y., Cai, Z., Pang, Y., et al. (2011). Perinatal antidepressant exposure alters cortical network function in rodents. *Proceedings of the National Academy of Sciences of the United States of America*, *108*(45), 18465–18470.

Sinex, D. G., & McDonald, L. P. (1988). Average discharge rate representation of voice onset time in the chinchilla auditory nerve. *The Journal of the Acoustical Society of America*, *83*, 1817.

Sinnott, J. M., & Mosteller, K. W. (2001). A comparative assessment of speech sound discrimination in the mongolian gerbil. *The Journal of the Acoustical Society of America*, *110*, 1729.

Steinschneider, M., Fishman, Y. I., & Arezzo, J. C. (2003). Representation of the voice onset time (VOT) speech parameter in population responses within primary auditory cortex of the awake monkey. *The Journal of the Acoustical Society of America*, *114*, 307.

Steinschneider, M., Volkov, I. O., Noh, M. D., Garell, P. C., & Howard, M. A. (1999). Temporal encoding of the voice onset time phonetic parameter by field potentials recorded directly from human auditory cortex. *Journal of Neurophysiology*, *82*(5), 2346–2357.

Stevens, H. E., & Wickesberg, R. E. (1999). Ensemble responses of the auditory nerve to normal and whispered stop consonants. *Hearing Research*, *131*(1), 47–62.

Strata, F., Bonham, B., Chang, E., Liu, R., Nakahara, H., & Merzenich, M. (2005). Perinatal anoxia degrades auditory system function in rats. *Proceedings of the National Academy of Sciences of the United States of America*, *102*(52), 19156–19161.

Szalkowski, C. E., Booker, A. B., Truong, D. T., Threlkeld, S. W., Rosen, G. D., & Fitch, R. H. (2013). Knockdown of the candidate dyslexia susceptibility gene homolog Dyx1c1 in rodents: Effects on auditory processing, visual attention, and cortical and thalamic anatomy. *Developmental Neuroscience*, *35*(1), 50–68.

Ter-Mikaelian, M., Semple, M. N., & Sanes, D. H. (2013). Effects of spectral and temporal disruption on cortical encoding of gerbil vocalizations. *Journal of Neurophysiology*, *110*, 1190–1204.

Toro, J. M., & Trobalón, J. B. (2005). Statistical computations over a speech stream in a rodent. *Perception and Psychophysics*, *67*(5), 867–875.

Toro, J. M., Trobalon, J. B., & Sebastián-Gallés, N. (2003). The use of prosodic cues in language discrimination tasks by rats. *Animal Cognition*, *6*(2), 131–136.

Toro, J. M., Trobalon, J. B., & Sebastián-Gallés, N. (2005). Effects of backward speech and speaker variability in language discrimination by rats. *Journal of Experimental Psychology: Animal Behavior Processes*, *31*(1), 95.

Victor, J. D., & Purpura, K. P. (1997). Metric-space analysis of spike trains: Theory, algorithms and application. *Network: Computation in Neural Systems*, *8*(2), 127–164.

Xu, L., Thompson, C. S., & Pfingst, B. E. (2005). Relative contributions of spectral and temporal cues for phoneme recognition. *The Journal of the Acoustical Society of America*, *117*, 3255.

SECTION N

MEMORY FOR LANGUAGE

CHAPTER

67

Introduction to Memory

Shauna M. Stark and Craig E.L. Stark

Department of Neurobiology and Behavior, University of California, Irvine, CA, USA

67.1 INTRODUCTION: AMNESIA AND PATIENT H.M.

Amnesic patients gave us the first insight into the role of the hippocampus and surrounding medial temporal lobe cortex in the formation of long-term memory (Penfield & Milner, 1958; Ribot, 1882). Most notably, in the 1950s, the hippocampus and most of the surrounding medial temporal cortex were surgically removed in epileptic patient H.M. in an attempt to cure his seizures (Scoville & Milner, 1957). His resulting profound amnesia and inability to learn new things and recall them at a later time emphasized the importance of these brain regions for memory encoding and retrieval. The following decades of research in humans, nonhuman primates, and rodents have contributed to a model of the medial temporal lobe memory system, which includes the hippocampal region, perirhinal, entorhinal, and parahippocampal cortices (Cohen & Eichenbaum, 1993; Squire, 1992; Squire, Stark, & Clark, 2004).

After his surgery, H.M. suffered from both anterograde and retrograde amnesia. Anterograde amnesia refers to an inability to learn new information and retrieve it later as part of a long-term memory trace (Scoville & Milner, 1957). H.M. was unable to learn the names of the hospital staff, even years after his surgery and after repeated exposure to these individuals (Milner, Corkin, & Teuber, 1968). He was severely impaired regarding episodic memory, or memory for autobiographical events that are associated with a specific time, place, and emotion (Tulving, 1984), and semantic memory, consisting of general knowledge about the world (Tulving & Schacter, 1990). He was also unable to learn new vocabulary or apply new meanings to existing words (such as the computer-related meaning for Windows®) (Gabrieli, Cohen, & Corkin, 1988). In contrast, retrograde amnesia refers to memory loss extending back in time from the surgery or accident that caused the amnesia (Ribot, 1882; Squire, Clark, & Knowlton, 2001). For example, H.M. could recall early childhood memories, but not events that happened in the years immediately prior to his surgery.

Although H.M. was severely impaired on most tests of memory (including, but not limited to, verbal, spatial, and object-base stimuli presented in visual and auditory domains), he was able to learn some new information (Corkin, 2002). He could acquire tasks that tapped skill or habit learning, demonstrating decreased reaction times and increased accuracy on a mirror-tracing task over repeated sessions. Although he could not explicitly recall the task from one session to the next, his performance improved with practice (Milner, 1962). The dichotomy between these two types of memory prompted the hypothesis that there are multiple forms of memory supported by multiple memory systems in the brain, broadly divided into declarative memories (both episodic and semantic information), which rely on the medial temporal lobes, and nondeclarative or procedural memories, which rely on an array of other brain structures outside the medial temporal lobe system (Squire, 1992). In this chapter, we review these types of memory and the brain regions that support them. In addition, we discuss how these learning systems might be relevant for language learning, such as vocabulary acquisition, and the flexible and creative expression of language.

67.2 MEDIAL TEMPORAL LOBE MEMORY SYSTEM

Squire and Zola-Morgan (1991) proposed the "medial temporal lobe memory system" to describe the network of brain structures critically involved in

Neurobiology of Language. DOI: http://dx.doi.org/10.1016/B978-0-12-407794-2.00067-5

long-term memory formation. This neural system consists of the hippocampus and the adjacent entorhinal, perirhinal, and parahippocampal cortices (which together form the parahippocampal gyrus). These structures form a loose hierarchy such that hippocampus receives projections from entorhinal cortex, which in turn receives projections from the perirhinal and parahippocampal cortices. These connections are bidirectional, feeding information forward into the hippocampus and returning it back through the cortices. In addition, heavy projections run between the perirhinal and parahippocampal cortices that, in turn, receive widespread projections from unimodal and polymodal areas in the frontal, temporal, and parietal lobes (Amaral, 1999).

Although these medial temporal structures are all involved in declarative memory, some specificity has been assigned to individual structures; however, it remains a matter of debate how exclusive that specificity is for each region. The parahippocampal cortex receives input from the dorsal visual stream, often dubbed the "where" visual stream because it is involved in the spatial relationships of visual stimuli. Thus, this region (or a portion of this region) has been coined the parahippocampal place area (Kanwisher, McDermott, & Chun, 1997) and is often associated with memory tasks that involve a spatial component. In contrast, the perirhinal cortex receives input from the ventral visual stream, or the "what" visual stream, which has been associated with object-based representations. Lesion studies have identified a double dissociation between these two regions on spatial and object-based memory tasks (see Malkova & Mishkin, 2003 for a review). In addition, neuroimaging studies have reported activation in the parahippocampal cortex when participants view scenes (Davachi, Mitchell, & Wagner, 2003; Epstein, Harris, Stanley, & Kanwisher, 1999) or when scene representations are evoked (Buffalo, Bellgowan, & Martin, 2006) and there is preferential activity in the perirhinal cortex for objects (Awipi & Davachi, 2008). However, this dichotomy may be oversimplified and dependent on task demands. Some studies have reported comparable imaging-related activity for objects in both the parahippocampal and perirhinal cortices (Litman, Awipi, & Davachi, 2009), whereas others have reported parahippocampal activation for associations that are not restricted to spatial information (Aminoff, Gronau, & Bar, 2007). An alternative view is that the perirhinal cortices and parahippocampal cortices are preferentially involved in item-based and contextual information, respectively (Aggleton & Brown, 1999; Cohen & Eichenbaum, 1993; Ranganath, 2010; Wang, Yonelinas, & Ranganath, 2013).

The entorhinal cortex is the main interface between the hippocampus and rest of the neocortex. It consists of six layers that contain different types of cells and innervate the various subfields of the hippocampus. Most notably, the entorhinal cortex contains "grid cells." These cells code for spatial locations in an environment by responding to a grid of locations oriented at a specific angle and with specific spacing (Hafting, Fyhn, Molden, Moser, & Moser, 2005). The universal spatial representation of these grid cells may reflect a context-specific code within the hippocampal network that is crucial for successful storage of episodic memories (Fyhn, Molden, Witter, Moser, & Moser, 2004). The entorhinal cortex can also be divided into the lateral and medial portions, which may be involved in differential processing (Kerr, Agster, Furtak, & Burwell, 2007). The medial entorhinal cortex, the primary location of the grid cells (and "head direction cells"), has been heavily associated with spatial processing (Hafting et al., 2005; McNaughton, Battaglia, Jenson, Moser, & Moser, 2006). In contrast, the lateral entorhinal cortex exhibits little spatial modulation (Hargreaves, Rao, Lee, & Knierim, 2005; Yoganarasimha, Rao, & Knierim, 2010) and may be more involved in object-based processing (Hunsaker, Mooy, Swift, & Kesner, 2007).

The hippocampus sits at the end of the medial temporal lobe network, sharing reciprocal connections with the entorhinal cortex. It is shaped somewhat like a seahorse (*hippocampus* is the Greek term for seahorse) and comprises the subfields CA1-CA4, the dentate gyrus, and the subiculum (Cajal, 1911). These subfields interact with each other in two primary loops, often referred to as the direct and indirect pathways. In the direct pathway, projections from the entorhinal cortex feed forward to the CA1, onto the subiculum, and return back to the entorhinal cortex. In the indirect pathway, the entorhinal cortex feeds forward to the dentate gyrus and then onto the CA3. The CA3 has strong recurrent connections and also projects to CA1 (which then completes the cycle projecting to the subiculum, and then back to the entorhinal cortex). The various layers of the entorhinal cortex interact differentially with these subfields and the various subfields also share reciprocal connections with one another, but these two pathways represent the primary flow of information through the hippocampus (Amaral & Lavenex, 2007; Duvernoy et al., 2005).

Some of these subfields exhibit unique properties that contribute to memory formation and retrieval. Computational models have proposed pattern separation and pattern completion as primary mechanisms of the hippocampus (Marr, 1971; Treves & Rolls, 1994). Pattern separation refers to the process whereby similar representations are transformed into distinct, nonoverlapping representations. Pattern completion, in contrast, refers to transforming incomplete or degraded representations into previously stored representations by filling in the missing information. Both mechanisms are critical in forming new associative memories, storing memories independently of each other, retrieving memories from partial cues, and flexibly applying stored memories to novel situations (Yassa & Stark, 2011).

The dentate gyrus has been associated with pattern separation in rodents (Leutgeb, Leutgeb, Treves, Moser, & Moser, 2004; Vazdarjanova & Guzowski, 2004) and humans (Bakker, Kirwan, Miller, & Stark, 2008; Lacy, Yassa, Stark, Muftuler, & Stark, 2011). The dentate gyrus receives input from the entorhinal cortex via the perforant path (Witter, 1993). Activity in the dentate gyrus is markedly sparse, with very few neurons firing at any given time (Jung & McNaughton, 1993), leading to a strong reduction in potential overlap between patterns (i.e., pattern separation). It has sparse and powerful projections to the CA3 (Blackstad, Brink, Hem, & Jeune, 1970; Swanson, Wyss, & Cowan, 1978) that give it the potential to strongly drive representations in the CA3 (McNaughton & Morris, 1987). These properties make the dentate gyrus optimal for coding unique representations and sensitive to small changes in input. The CA3 contains an extensive recurrent collateral network of neurons postulated to act as an auto-associative pattern completion network (Marr, 1971; Treves & Rolls, 1992). These recurrent collaterals may be involved in matching the input from the dentate gyrus with other stored representations (Rolls, 2007). The winner of the competition between the information sent from the dentate gyrus and the stored representation from the CA3 is then fed onto the CA1, which does not appear to perform either pattern separation or completion. One theory suggests that the CA1 plays a role in matching sensory input with an existing memory trace, whereas the CA3 and dentate gyrus are involved in the detection of a mismatch with a stored representation (Duncan, Ketz, Inati, & Davachi, 2011; Hasselmo, 2005; Kumaran & Maguire, 2007).

These subfields may be specialized in other ways as well. The CA1 and CA3 subfields both exhibit place fields in rats, such that these neurons exhibit a higher rate of firing when an animal is in a specific location in the environment (O'Keefe & Dostrosky, 1971). These place cells are thought to help form a map of the environment (O'Keefe & Nadel, 1978). There is evidence that CA3 may also be involved in learning the sequence of events over time (Lisman, 1999), and possibly more involved in tasks involving spatial locations (Hunsaker, Rosenberg, & Kesner, 2008). CA1 may also be involved in sequence learning, although to a lesser extent (Farovik, Dupont, & Eichenbaum, 2010), and possibly more so for object-based learning (Hoge & Kesner, 2007; Manns, Howard, & Eichenbaum, 2007). There is conflicting evidence for these theories partly because of the interconnections between regions, which makes it difficult to isolate the functioning of any one region. Nevertheless, the subfields contain unique properties that support the optimal functioning of the hippocampus as a unit in memory encoding and retrieval.

67.3 EPISODIC MEMORY

Episodic memory (and the related notion of autobiographical memory) refers to memory for specific experiences, usually associated with a time, place, and emotion (Tulving, 1984). Tulving (2002) likened the capacity of remembering specific episodes to "mental time travel," as if the individual is able to reexperience individual events. This type of memory is commonly associated with the subjective mental experience that requires a sense of self and an awareness that this event happened in the past. For example, recalling a birthday party that you attended, who was there, where it occurred, and your own personal interactions there, reflects your episodic memory of that event. In this way, episodic memories represent one-time episodes (i.e., one-trial learning) and rely heavily on the hippocampus. Similarly, the first exposure to a new vocabulary word will be encoded as an episodic memory and be associated with the context in which it was learned.

Episodic memory impairment is the hallmark symptom of amnesic patients. For example, patient H. M. was unable to recall any events from his daily life after his hippocampal resection (Corkin, 2002). Although the formation and retention of new episodic memories is clearly impaired in amnesia, the preservation of older episodic memories is less clear. It is clear that even in cases of extensive medial temporal lobe damage including the hippocampus, old autobiographical memories are typically still present. For example, H.M. was able to produce well-formed autobiographical memories from ages 16 years and younger (Corkin, 1984). Other amnesic patients have also produced rich autobiographical memories from earlier years prior to their amnesic episodes (Bayley, Hopkins, & Squire, 2006; Kirwan, Bayley, Galvan, & Squire, 2008). However, there is still debate regarding whether these memories are truly episodic and contain the same recollective detail found in healthy individuals (Gilboa et al., 2006; Moscovitch, Nadel, Winocur, Gilboa, & Rosenbaum, 2006). Whether the hippocampus is required to truly reexperience an event retrieved from memory (as suggested by Multiple Trace Theory, see below), it is clear that damage to this system would have profound effects on the learning and memory processes used in language (which need not be truly episodic).

There is debate in the literature regarding whether animals also possess episodic memory. Because episodic memory is defined as having insight into the memory, to travel back in time and re-experience a memory, it is nearly impossible to test in animals. Therefore, based on this definition, Tulving (2002) argues that episodic memory is unique to humans and may be one of the factors that make humans unique from other animals. However, an alternative theory of episodic memory proposes that there is shared neural history between animals and humans, focusing on memory for events in context (Allen & Fortin, 2013). Episodic memories can instead be defined by the unique characteristics of their what, where, and when components. Utilizing this definition, Clayton and Dickinson (1998) demonstrated that scrub jays could remember what food they stored (worms or peanuts), as well as where (the location of the cage) and when (4 or 124 h ago) it was cached. This information can be updated and used flexibly (Clayton & Dickinson, 1999; Clayton, Yu, & Dickinson, 2003) and expressed spontaneously (Singer & Zentall, 2007), and it has been demonstrated in other species, such as rats (Eacott, Easton, & Zinkivskay, 2005; Ergorul & Eichenbaum, 2004), mice (Dere, Huston, & De Souza Silva, 2005), and nonhuman primates (Hoffman, Beran, & Washburn, 2009), much like the definition applied to episodic memories.

67.4 SEMANTIC MEMORY

The counterpart to episodic memory is semantic memory, which is defined as the recollection of facts and generalized knowledge about the world (Squire, 1992; Tulving, 1983). Semantic memories reflect the information that remains after the details of the learning experience has been lost. Like autobiographical memories, the acquisition of semantic memories is impaired in amnesia (Gabrieli et al., 1988; Hamann & Squire, 1995). Amnesic patients are impaired in learning new vocabulary words (Verfaellie, Croce, & Milberg, 1995), recognizing famous faces (Smith et al., 2014), and recalling current public events (Reed & Squire, 1998). Although this learning does not emphasize the source details of the learning event (the what-where-when context of the learning), the semantic knowledge is still not retained.

Most memories begin as relatively specific, episodic memories and gradually become more general or semantic over time (Cermak, 1984). The loss of source details may be the result of some form of averaging of learning experiences, where many details of the learning experience are irrelevant to the learned information. For example, most people know that George Washington was the first president of the United States, or the meaning of the word "retribution." However, most of us do not recall the details regarding when or where we first learned that information. As we encounter information over multiple learning episodes, in different contexts, those details are lost, perhaps during the process of systems consolidation. Systems consolidation refers to the process by which memories are transferred from the hippocampus to areas of the cortex as part of long-term retention (McClelland, McNaughton, & O'Reilly, 1995). Through this mechanism, most declarative memories are acquired through episodic learning but become semanticized over time.

Although this mechanism is generally accepted, there is some debate regarding the role of the hippocampus in retrieving semantic and episodic memories over time. The Standard Model of Systems Consolidation (Squire & Alvarez, 1995) asserts that the initial memory trace is critical in the early encoding of the memory; however, as that memory is transferred to the cortex, the hippocampus is no longer required for retrieval. In contrast, the Multiple Trace Theory (Nadel, Samsonovich, Ryan, & Moscovitch, 2000) proposes that the hippocampus is involved in the retrieval of all episodic memories, including remote ones. The hippocampal traces are presumed to be contextually rich in spatial and temporal details, critical for episodic memories, whereas the cortical traces are presumed to be semantic and largely context-free. In this framework, the hippocampus acts as an index to retrieve older episodic memories, along with their contextual details.

There is support for both these models in the amnesia literature (in many domains both make the same predictions) with no clear consensus regarding which is more accurate. The key test of these models relies on the nature of retrograde amnesia. If the hippocampus is required for the retrieval of all episodic memories, regardless of age, then there should not be evidence for a temporally graded retrograde amnesia for episodic memory. In other words, without a hippocampus, all memories should be semanticized, with no qualitative difference as a function of the age of the memory. Although early reports on H.M. indicated that he demonstrated temporally graded retrograde amnesia (Scoville & Milner, 1957), more refined testing later revealed that much of H.M.'s memories had become semanticized. For example, Corkin (2002) asserted that "H.M. was unable to supply an episodic memory of his mother or his father—he could not narrate even one event that occurred in a specific time and place." Studies of other amnesic patients, however, have revealed older episodic memories that appear to be qualitatively intact and demonstrate a temporal gradient such that older memories are intact, whereas more recent memories are impaired (Bayley et al., 2006; Kirwan et al., 2008). The discrepancies in these findings are sometimes attributed to variability in the etiology and location of the brain damage reported in various amnesic patients. Partial hippocampal damage is not uncommon, and there is considerable variability in the condition of the surrounding medial temporal cortex and lateral temporal cortex. However, even in lesions of the hippocampus in rodents, where the extent and location of the hippocampal damage can be better controlled and quantified, the evidence is mixed for temporally graded amnesia (Frankland & Bontempi, 2005; Sutherland, Sparks, & Lehmann, 2010). Recently, the Competitive Trace Theory has been proposed to account for these disparate data that combine aspects of both of them (Yassa & Reagh, 2013). Future studies are necessary to settle the debate regarding the role of the hippocampus in episodic memory retrieval.

Regardless of the differences between these models, it is clear that the hippocampus and surrounding medial temporal cortices are critical for the encoding of new information and then transferring this information into the cortex over time. These episodic experiences are thought to become semanticized over time through a process of "replay" in which the hippocampus trains the cortex directly (McClelland et al., 1995). There is also evidence, however, that the cortex can learn directly, without relying on the hippocampus to relay the information first. Although H.M. performed poorly on tests of current events or naming people who had become famous after his surgery, he was able to identify some of them, indicating some new learning (Corkin, 2002; Mayes et al., 1994). Furthermore, he was also able to draw an accurate and detailed map of his home, which he lived in only after his hippocampal resection (Corkin, 2002).

There are two main theories regarding how this new learning was accomplished. The first is that this learning was supported by residual medial temporal lobe structures that are still intact. In H.M., for example, the posterior parahippocampal gyrus was largely spared (Corkin, Amaral, Gonzales, Johnson, & Hyman, 1997; Salat et al., 2006). Another well-characterized amnesic patient, E.P., had a more complete medial temporal lesion (resulting from herpes simplex encephalitis) (Insausti, Annese, Amaral, & Squire, 2013; Stefanacci, Buffalo, Schmolck, & Squire, 2000), with very little residual functional tissue. Yet, he was able to identify some household items that had been acquired after the onset of his amnesia (Bayley, O'Reilly, Curran, & Squire, 2008). In this case, the medial temporal lobe is unlikely to have contributed to this new learning, leading to the second theory of how this learning was accomplished, which is by the neocortex directly.

Complimentary Learning Systems models of memory have proposed two learning systems in the brain: the hippocampal system (encompassing the hippocampus and surrounding medial temporal lobe regions) and the neocortical learning system (McClelland et al., 1995). As mentioned, the hippocampal system is responsible for episodic and semantic learning, rapidly acquiring information in a single episode and gradually transferring this information to the cortex over time for permanent storage. This process involves gradual, interleaved learning, possibly involving replay of events during wakefulness (Karlsson & Frank, 2009) and sleep (Frankland & Bontempi, 2005). In the absence of an intact hippocampal memory system, the cortex may be able to learn the statistical regularities or "similarity structure" in the environment through repeated exposure, mimicking the interleaved training normally provided by the hippocampus.

One could imagine that through repeated exposure via television and magazines, how a severely amnesic patient like H.M. could learn the names of presidents and other major events that occurred after his surgery by utilizing such a mechanism. A more rigorous demonstration of this learning comes from the training of vocabulary words to memory-impaired patients. Using a technique called "errorless learning" in which a new vocabulary word and its definition are presented repeatedly and without an opportunity for incorrect associations, amnesic patients have demonstrated new learning of vocabulary words (Glisky, Schacter, & Tulving, 1986). However, this learning is hyper-specific, such that performance plummets with any variation from the original training in the context in which the word is presented or any rewording of the definition (Baddeley & Wilson, 1994; Glisky et al., 1986). Nevertheless, these studies provide evidence that some new learning can be acquired outside of the medial temporal lobes and can be applied for rehabilitation (Clare, Wilson, Breen, & Hodges, 1999; Wilson, Baddeley, & Evans, 1994), particularly if specific training techniques are utilized to increase the generalization of the information (Stark & Gordon, 2008; Stark, Stark, & Gordon, 2005).

67.5 PROCEDURAL MEMORY

Learning mechanisms that operate outside the medial temporal lobe represent a diverse array of other types of learning, beyond the semantic learning noted. Procedural memory refers to several forms of learning that occurs during the performance of various tasks, which is typically expressed in enhanced or speeded performance (Squire, 2004). For example, learning a sequence of movements, such as swinging a golf club or riding a bike, is expressed as a gradual improvement in speed and accuracy. Patients with amnesia perform normally on these procedural learning tasks because they do not rely on the medial temporal lobes, but instead they involve a variety of other structures. For example, patient H.M. was able to improve his accuracy on a mirror-tracing task over the course of several trials and several days, despite having no memory for the learning events themselves (Milner, 1962).

Procedural memory can take many different forms, sometimes called "habit learning." Importantly, it operates outside of conscious awareness and typically requires several learning trials, in contrast to the hippocampal memory system, which requires only one exposure to acquire new information and involves conscious information processing. Procedural memory includes skills and habits, largely dependent on the striatum and basal ganglia, and refers to those tasks that involve a series of motor movements, such as riding a bike or playing the piano (Seger & Spiering, 2011). Priming and perceptual learning occur in the neocortex and are reflected in faster reaction times in responding to a repeated word or picture. The initial presentation "primes" the memory trace such that it is more accessible during a second presentation (Hamann & Squire, 1997; Levy, Stark, & Squire, 2004; Stark & Squire, 2000). Simple classical conditioning operates in the amygdala, such as fear conditioning in which an animal can learn to associate a painful shock with a particular context in one exposure and without the use of the hippocampus (Fanselow, 1994; Phillips & LeDoux, 1992), and the cerebellum, such as eye-blink conditioning (Steinmetz, Lavond, Ivkovich, Logan, & Thompson, 1992), Additionally, some tasks that require gradually acquiring the statistical regularities of a stimulus set can operate outside of the medial temporal lobe. For example, amnesic patients are able to learn new cognitive skills (Dienes, Baddeley, & Jansari, 2012; Squire & Frambach, 1990), artificial grammar (Knowlton & Squire, 1996), and categorical assignment (Hopkins, Myers, Shohamy, Grossman, & Gluck, 2004; Knowlton & Squire, 1993). This same network of brain regions also serves to support the rules associated with the sequential order of lexical items, the structure of phrases and sentences (syntax), and the morphology of complex words (Ullman, 2004).

67.6 MEMORY CONSOLIDATION AND SLEEP

As discussed, the process of memory consolidation at a systems level involves the transfer of memories from the hippocampus and medial temporal lobe structures to the cortex for long-term storage. There is considerable debate regarding the role of the hippocampus in the retrieval of old episodic memories, but it is widely accepted that the final storage site of much of declarative memory is in the cortex. However, it is unclear how long it takes for a memory to be fully transferred to the cortex. Studies of retrograde amnesia indicate that it may take 10 or 20 years for memories to be fully consolidated and stable, such that they are resilient in the face of damage to the hippocampus (Bayley, Hopkins, & Squire, 2003; Cipolotti et al., 2001; Manns, Hopkins, & Squire, 2003; Nadel et al., 2000). This consolidation process occurs without intent or awareness, and it does not refer to conscious or behavioral rehearsal.

Systems consolidation is only one form of memory consolidation, however. At a cellular level, learning initially leads to an "early" form of Long-Term Potentiation (LTP) of synapses that will decay within hours unless specific forms of protein synthesis occur that stabilize or consolidate the memory at a cellular level within the hippocampus (Nicoll & Roche, 2013). Retrieval and replay of a memory can also "consolidate" or stabilize a memory, perhaps explaining at least some of why memory consolidation is facilitated by sleep (Stickgold & Walker, 2007).

Memory consolidation serves to stabilize the memory in the initial phase and then to enhance and integrate the memory at later stages. Motor skill memories have been shown to be disrupted by training on an alternate task within the first hours after training on the initial task, indicating the importance of the stabilization phase in reducing interference (Brashers-Krug, Shadmehr, & Bizzi, 1996; Walker, Brakefield, Hobson, & Stickgold, 2003). Memory consolidation may also serve to enhance memories, improving behavioral performance independent of further practice (Walker, Liston, Hobson, & Stickgold, 2002). Additionally, the memories gradually become integrated into existing memory networks (Dumay & Gaskell, 2005; Dumay & Gaskell, 2007; Stickgold, 2002). There is also evidence that the mere act of recalling a memory can destabilize it, making it vulnerable to interference and degradation, in a process called reconsolidation (Nader, 2003).

The stabilization phase of memory consolidation appears to occur largely during wakefulness (Brashers-Krug et al., 1996; Walker et al., 2003). The enhancement stage appears to occur primarily during sleep, either restoring previously lost memories (Fenn, Nusbaum, & Margoliash, 2003) or producing additional learning (Walker et al., 2003, 2002), both without need for additional practice. Although many of these studies involve a motor-learning task, such as finger tapping, and assess the improvement in performance over time, similar benefits have been reported for the learning of word pairs after sleep (Gais, Molle, Helms, & Born, 2002). The "slow wave sleep" (SWS) phase of the rapid eye movement (REM) cycle of sleep seems to be particularly important (Gais & Born, 2004; Gais, Plihal, Wagner, & Born, 2000), with sleep deprivation resulting in learning impairments (Zimmerman, Stoyva, & Metcalf, 1970; Zimmerman, Stoyva, & Reite, 1978). Neuroimaging studies have revealed reactivation during sleep of regions active during learning of a task (Peigneux et al., 2003). This sleep reactivation, or replay, has also been observed in neural firing rates in rodent studies, particularly in the hippocampus, but also in other cortical regions (Euston, Tatsuno, & McNaughton, 2007; Ji & Wilson, 2007; Wilson & McNaughton, 1994). The process of sleep reactivation is thought to strengthen the cortical memory trace, thereby allowing memories to become independent of the hippocampus (Frankland & Bontempi, 2005). There is also evidence for replay during wakefulness that may play a similar role in consolidation (Karlsson & Frank, 2009).

67.7 NEUROGENESIS

Neurogenesis refers to the birth of new neurons in the brain and occurs primarily in the dentate gyrus of the hippocampus and the olfactory bulb (Altman, 1962, 1963). Recent estimates indicate that one-third of the neurons in the hippocampus are regularly renewed in the hippocampus throughout life, amounting to approximately 700 new neurons added to the hippocampus per day (Spalding et al., 2013). Adult-born hippocampal neurons have enhanced synaptic plasticity for a period of time (Ge, Yang, Hsu, Ming, & Song, 2007; Schmidt-Hieber, Jonas, & Bischofberger, 2004), indicating that they may provide an optimal mechanism for encoding new memories. The long-term storage capacity of the human brain is unknown, but the process of neurogenesis may provide an explanation for how a relatively small brain structure can continuously encode new experiences and integrate them into long-term storage.

There is some evidence for the functional significance of these newborn neurons for efficient pattern separation and the ability to distinguish and store similar experiences as distinct memories (Clelland et al., 2009; Sahay et al., 2011). These new neurons have been directly associated with spatial pattern separation (Clelland et al., 2009) and in the encoding of time in new memories (Aimone, Wiles, & Gage, 2006). In contrast, there is some evidence to suggest that older granule cells in the dentate gyrus are necessary for pattern completion, which serves to generalize similar memories with each other (Nakashiba et al., 2012). Computational models of neurogenesis in hippocampal learning also propose that this mechanism can increase memory capacity (Becker, 2005). Taken together, neurogenesis provides a mechanism for encoding new memories over the course of the lifespan.

Neurogenesis also provides a mechanism to explain memory deficits and a target for improving memory performance. A decrease in adult-born neurons has been observed in cases of severe depression and is associated with an increased stress response (Sahay & Hen, 2007; Schloesser, Lehmann, Martinowich, Manji, & Herkenham, 2010; Snyder, Soumier, Brewer, Pickel, & Cameron, 2011). A neurogenesis deficit could bias hippocampal encoding and retrieval toward a narrow, predominately negative representation of context, generalizing across all positive or negative experiences (Becker & Wojtowicz, 2006). In addition, a decrease in neurogenesis may be linked to age-related declines in hippocampal function and associated memory performance (Kuhn, Dickinson-Anson, & Gage, 1996; Spalding et al., 2013; Zitnik & Martin, 2002).

The potential for rescue of neurogenesis is an important target for improving memory performance. Antidepressants have been shown to increase neurogenesis and rescue memory deficits (Dranovsky & Hen, 2006; Surget et al., 2011). Interestingly, environmental enrichment and physical exercise have proven to be extremely beneficial in increasing the production of newborn neurons and in their long-term integration in the brain. Environmental enrichment in animals involving larger cages with exposure to balls, tunnels, and other novelties, have resulted in greater neurogenesis (Kempermann, Kuhn, & Gage, 1997; Leal-Galicia, Castaneda-Bueno, Quiroz-Baez, & Arias, 2008). Likewise, voluntary wheel-running in rodents has been shown to increase neurogenesis and enhance learning (van Praag, 2008; van Praag, Shubert, Zhao, & Gage, 2005). These findings have been extended to humans, such that physical exercise has been associated with improved memory performance in healthy adults (Hertzog, Kramer, Wilson, & Lindenberger, 2009) and has been shown to offset some of the memory declines associated with aging (Erickson et al., 2011; Kramer, Erickson, & Colcombe, 2006) and Alzheimer's disease (Cotman, Berchtold, & Christie, 2007; Lautenschlager et al., 2008).

67.8 AGING AND MEMORY

One of the key concerns of older adults is the experience of memory loss, both in the normal course of aging and as it is associated with Alzheimer's disease and other forms of dementia. Age-related memory impairment affects various types of memory, including source memory (i.e., the knowledge of where or when information was encoded) (Johnson, Hashtroudi, & Lindsay, 1993; Schacter, Koutstaal, Johnson, Gross, & Angell, 1997). Older adults are also more prone to false recollections (i.e., miscombining features of different events that are confidently held as true) (Koutstaal, Schacter, & Brenner, 2001; Lyle, Bloise, & Johnson, 2006). Changes in the hippocampus may be responsible for these errors in binding in older adults (Kessels, Hobbel, & Postma, 2007). It is worth noting that some domains of memory remain unchanged with age, such as procedural memory (Fleischman, Wilson, Gabrieli, Bienias, & Bennett, 2004) and working memory (Nilsson, 2003), whereas semantic knowledge (e.g., vocabulary) actually improves (Verhaeghen, 2003).

Although several brain regions undergo changes associated with aging, such as the frontal and temporal cortices (Fjell & Walhovd, 2005), changes in the hippocampus may be particularly relevant for memory-related declines. Hippocampal volume is reduced in older adults (Raz, Rodrigue, Head, Kennedy, & Acker, 2004), but there is no evidence that this volume loss is due to fewer neurons (Burke & Barnes, 2006). Instead, there may be synaptic loss, although only in selective regions. For example, the synapses in the CA3 subfield of the hippocampus that form the recurrent collaterals in the auto-associative network are not reduced in aging (Smith, Adams, Gallagher, Morrison, & Rapp, 2000). In contrast, the number of synapses from the entorhinal cortex into the dentate gyrus via the perforant path is reduced (Scheff, Price, Schmitt, & Mufson, 2006). This pattern was also observed in the integrity of the perforant path in humans using neuroimaging and was tied to memory performance, such that reduced perforant path integrity in older adults was correlated with worse memory performance (Yassa, Mattfeld, Stark, & Stark, 2011). In addition to this reduction in input, there is also reduced activity in the dentate gyrus, resulting in weak pattern separation processes (Wilson, Gallagher, Eichenbaum, & Tanila, 2006; Yassa et al., 2011; Yassa & Stark, 2011). In contrast, neuroimaging of the CA1 subfield of the hippocampus appears unaffected and is consistent with robust pattern completion performance (Stark, Yassa, Lacy, & Stark, 2013; Yassa et al., 2011).

In addition to age-related memory changes associated with normal aging, there are memory deficits and brain changes associated with Alzheimer's disease. The earliest brain changes in Alzheimer's disease appear to be targeted in the entorhinal cortex, demonstrating neuronal loss (Gomez-Isla et al., 1996) and an increase in neurofibrillary tangles associated with the disease (Braak, Alafuzoff, Arzberger, Kretzschmar, & Del Tredici, 2006). Whereas memory-related changes associated with aging appear to be somewhat confined (with preserved performance on most tests of recognition memory), Alzheimer's disease results in fairly general memory deficits (McKhann et al., 1984) that progress over the course of the disease and extend to other cognitive domains (Bowen et al., 1997). Discriminating between those individuals with healthy aging and those with early Alzheimer's disease remains a target of vigorous research (Bauer, Cabral, Greve, & Killiany, 2013; Colliot et al., 2008; Morris et al., 1991), particularly that aiming for earlier detection and, ultimately, earlier intervention and treatment.

67.9 LANGUAGE LEARNING AND THE MEDIAL TEMPORAL LOBE

The hippocampus and surrounding medial temporal cortices are involved in the acquisition of new facts and events, so it serves to reason that damage to these regions would impair the learning of language as well. Amnesic patient H.M. demonstrated the relative integrity of his expressive and comprehensive language capacity (Kensinger, Ullman, & Corkin, 2001). However, more detailed investigations revealed impairments in sentence ambiguity (MacKay, James, Taylor, & Marian, 2007; Schmolck, Stefanacci, & Squire, 2000) and increases in grammatical errors (MacKay et al., 2007; MacKay, James, Hadley, & Fogler, 2011). Studies of patients with damage limited to just the hippocampus perform much like healthy controls (Schmolck et al., 2000), indicating that other medial temporal structures or lateral temporal lobe structures, such as the anterolateral cortex (often also damaged in these patients), may be more involved (Schmolck, Kensinger, Corkin, & Squire, 2002). Consistent with this interpretation, patients with semantic dementia, a progressive neurodegenerative disorder characterized by loss of semantic memory in both verbal and nonverbal domains, have atrophy in the lateral temporal cortex (Bozeat, Lambon Ralph, Patterson, Garrard, & Hodges, 2000; Hodges, Patterson, Oxbury, & Funnell, 1992). These patients act as a counterpart to amnesia in that they have semantic learning impairments, whereas their episodic memory for recent events remains intact. In terms of consolidation, the anterolateral cortex may be critical as the long-term repository of perceptual and semantic categorical knowledge that is transferred from the hippocampus (Murre, Graham, & Hodges, 2001). Damage to the hippocampal memory system prevents the learning of new vocabulary, whereas damage to the anterolateral cortex impairs the semantic knowledge for previously acquired words and concepts.

The procedural memory system, composed of a large network of brain structures, including the basal ganglia, frontal cortex, parietal cortex, and the cerebellum, may subserve the rule-based structure underlying grammar. These regions may play a role in the maintenance in working memory for linguistic elements and in learning of grammatical rules underlying the regularities of complex structures (Ullman, 2004). This system is largely spared in amnesia, yet there has been mixed evidence regarding the integrity of grammar in these patients. Amnesic patient H.M. displayed normal syntactic processing and morphological production of irregular and plural words (Kensinger et al., 2001). Further investigation revealed that H.M. produced fewer grammatical sentences and described pictures less completely and accurately than controls, particularly for unfamiliar stimuli, challenging the conclusion that his language production is intact (MacKay, James, & Hadley, 2008). However, similar investigations of other amnesic patients have not replicated this deficit in grammar, identifying cerebellar (or other) degeneration in H.M. that may account for these impairments (Schmolck et al., 2002). The interaction between these two memory systems likely accounts for some of the deficits observed in H.M., particularly because his errors are more pronounced in reference to novel stimuli.

The hippocampus and surrounding medial temporal lobe may be critical for the acquisition of language and the ability to use it creatively and flexibly in novel ways. Certainly, the medial temporal lobe is necessary for the acquisition of new vocabulary words, including the meaning, lexical status, and pronunciation of the words (Gabrieli et al., 1988). In addition, the medial temporal lobe memory system may be critical for relational binding (Shimamura, 2010) and representational flexibility (Bunsey, 2002). Duff and Brown-Schmidt (2012) have made a strong argument for the necessity of the medial temporal lobe memory system in online language processing. They argue that language is a system of arbitrary associations arranged in a temporal pattern, consistent with the ample evidence that the hippocampus is critical for the learning of temporal sequences of events (Fortin, Agster, & Eichenbaum, 2002). Without a medial temporal lobe memory system, individuals with amnesia exhibit rigidity in their referential expressions, with a lack of communal knowledge during discourse (Duff et al., 2008). However, language is flexible and creative, allowing for a nearly unlimited combination of words in novel constructions and contexts. Combined, these data support the theory that the medial temporal lobe memory system contains unique properties, making it a key contributor to language learning and production.

References

Aggleton, J. P., & Brown, M. W. (1999). Episodic memory, amnesia, and the hippocampal-anterior thalamic axis. *Behavioral and Brain Sciences, 22*, 425–489.

Aimone, J. B., Wiles, J., & Gage, F. H. (2006). Potential role for adult neurogenesis in the encoding of time in new memories. *Nature Neuroscience, 9*(6), 723–727. Available from: http://dx.doi.org/10.1038/nn1707.

Allen, T. A., & Fortin, N. (2013). The evolution of episodic memory. *Proceedings of the National Academy of Sciences of the United States of America, 110*, 10379–10386.

Altman, J. (1962). Are new neurons formed in the brains of adult mammals? *Science, 135*(3509), 1127–1128.

Altman, J. (1963). Autoradiographic investigation of cell proliferation in the brains of rats and cats. *The Anatomical Record, 145*(4), 573–591.

Amaral, D., & Lavenex, P. (2007). Hippocampal neuroanatomy. In P. Andersen, R. Morris, D. Amaral, T. Bliss, & J. O'Keefe (Eds.), *The hippocampus book* (pp. 37–131). New York, NY: Oxford University Press.

Amaral, D. G. (1999). Introduction: What is where in the medial temporal lobe? *Hippocampus, 9*, 1–6.

Aminoff, E., Gronau, N., & Bar, M. (2007). The parahippocampal cortex mediates spatial and nonspatial associations. *Cerebral Cortex, 27*, 1493–1503.

Awipi, T., & Davachi, L. (2008). Content-specific source encoding in the human medial temporal lobe. *Journal of Experimental Psychology: Learning, Memory and Cognition, 34*(4), 769–779.

Baddeley, R. J., & Wilson, B. A. (1994). When implicit learning fails: Amnesia and the problem of error elimination. *Neuropsychologia, 32*, 53–68.

Bakker, A., Kirwan, C. B., Miller, M. I., & Stark, C. E. L. (2008). Pattern separation in the human hippocampal CA3 and dentate gyrus. *Science, 319*, 1640–1642.

Bauer, C. M., Cabral, H. J., Greve, D. N., & Killiany, R. J. (2013). Differentiating between normal aging, mild cognitive impairment, and Alzheimer's disease with FDG-PET: Effects of normalization region and partial volume correction model. *Journal of Alzheimer's Disease and Parkinsonism, 3*(1), 113. Available from: http://dx.doi.org/10.4172/2161-0460.1000113.

Bayley, P. J., Hopkins, R. O., & Squire, L. R. (2003). Successful recollection of remote autobiographical memories by amnesic patients with medial temporal lobe lesions. *Neuron, 38*, 135–144.

Bayley, P. J., Hopkins, R. O., & Squire, L. R. (2006). The fate of old memories after medial temporal lobe damage. *Journal of Neuroscience, 26* (51), 13311–13317.

Bayley, P. J., O'Reilly, R. C., Curran, T., & Squire, L. R. (2008). New semantic learning in patients with large medial temporal lobe lesions. *Hippocampus, 18*, 575–583.

Becker, S. (2005). A computational principle for hippocampal learning and neurogenesis. *Hippocampus, 15*, 722–738.

Becker, S., & Wojtowicz, J. M. (2006). A model of hippocampal neurogenesis in memory and mood disorders. *Trends in Cognitive Sciences, 11*(2), 70–76.

Blackstad, T. W., Brink, K., Hem, J., & Jeune, B. (1970). Distribution of hippocampal mossy fibers in the rat. An experimental study with silver impregnation methods. *The Journal of Comparative Neurology, 138*(4), 433–449. Available from: http://dx.doi.org/10.1002/cne.901380404.

Bowen, J., Teri, L., Kukull, W., McCormick, W., McCurry, S. M., & Larson, E. B. (1997). Progression to dementia in patients with isolated memory loss. *The Lancet, 349*(9054), 763–765.

Bozeat, S., Lambon Ralph, M. A., Patterson, K., Garrard, P., & Hodges, J. R. (2000). Non-verbal semantic impairment in semantic dementia. *Neuropsychologia, 38*(9), 1207–1215.

Braak, H., Alafuzoff, I., Arzberger, T., Kretzschmar, H., & Del Tredici, K. (2006). Staging of Alzheimer's disease-associated neurofibrillary pathology using paraffin sections and immunocytochemistry. *Acta Neuropathologica*, *112*, 389–404.

Brashers-Krug, T., Shadmehr, R., & Bizzi, E. (1996). Consolidation in human motor memory. *Nature*, *382*(6588), 252–255.

Buffalo, E. A., Bellgowan, P. S., & Martin, A. (2006). Distinct roles for medial temporal lobe structures in memory for objects and their locations. *Learning and Memory*, *13*, 638–643.

Bunsey, M. (2002). Conservation of a hippocampal role in representational flexibility. In S. B. Fountain, M. Bunsey, J. H. Danks, & M. K. McBeath (Eds.), *Animal cognition and sequential behavior: Behavioral, biological, and computational perspectives* (pp. 229–247). New York, NY: Springer.

Burke, S. N., & Barnes, C. A. (2006). Neural plasticity in the ageing brain. *Nature Reviews. Neuroscience*, *7*(1), 30–40.

Cajal, R. S. (1911). *Histologie du systeme nerveux de l'homme et des vertebres*. Paris: Maloine.

Cermak, L. S. (1984). The episodic-semantic distinction in amnesia. In L. R. Squire, & N. Butters (Eds.), *Neuropsychology of memory* (pp. 55–62). New York, NY: Guilford Press.

Cipolotti, L., Shallice, T., Chan, D., Fox, N., Scahill, R., Harrison, G., et al. (2001). Long-term retrograde amnesia. The crucial role of the hippocampus. *Neuropsychologia*, *39*, 151–172.

Clare, L., Wilson, B. A., Breen, K., & Hodges, J. R. (1999). Errorless learning of face-name association in early Alzheimer's disease. *Neurocase*, *5*, 37–46.

Clayton, N. S., & Dickinson, A. (1998). Episodic-like memory during cache recovery by scrub jays. *Nature*, *395*(6699), 272–274.

Clayton, N. S., & Dickinson, A. (1999). Memory for the content of caches by scrub jays (*Aphelocoma coerulescens*). *Journal of Experimental Psychology: Animal Behavior Processes*, *25*(1), 82–91.

Clayton, N. S., Yu, K. S., & Dickinson, A. (2003). Interacting cache memories: Evidence for flexible memory use by Western Scrub-Jays (*Aphelocoma californica*). *Journal of Experimental Psychology: Animal Behavior Processes*, *29*(1), 14–22.

Clelland, C. D., Choi, M., Romberg, C., Clemenson, G. D. J., Fragniere, A., Tyers, P., et al. (2009). A functional role for adult hippocampal neurogenesis in spatial pattern separation. *Science*, *325*(5937), 210–213. Available from: http://dx.doi.org/10.1126/science.1173215.

Cohen, N. J., & Eichenbaum, H. (1993). *Memory, amnesia, and the hippocampal system*. Cambridge, MA: MIT Press.

Colliot, O., Chetelat, G., Chupin, M., Desgranges, B., Magnin, B., Benali, H., et al. (2008). Discrimination between Alzheimer's disease, mild cognitive impairment, and normal aging by using automated segmentation of the hippocampus. *Radiology*, *248*, 194–201.

Corkin, S. (1984). Lasting consquences of bilateral medial temporal lobectomy: Clinical course and experimental findings in H.M. *Seminars in Neurology*, *4*(2), 249–259.

Corkin, S. (2002). What's new with the amnesic patient H.M.? *Nature Reviews Neuroscience*, *3*, 153–160.

Corkin, S., Amaral, D. G., Gonzales, R. G., Johnson, K. A., & Hyman, B. T. (1997). H.M.'s medial temporal lobe lesion: Findings from magnetic resonance imaging. *The Journal of Neuroscience*, *17*(10), 3964–3979.

Cotman, C. W., Berchtold, N. C., & Christie, L. A. (2007). Exercise builds brain health: Key roles of growth factor cascades and inflammation. *Trends in Neuroscience*, *30*(9), 464–472.

Davachi, L., Mitchell, J. P., & Wagner, A. D. (2003). Multiple routes to memory: Distinct medial temporal lobe processes build item and source memories. *Proceedings of the National Academy of Sciences of the United States of America*, *100*(4), 2157–2162.

Dere, E., Huston, J. P., & De Souza Silva, M. A. (2005). Integrated memory for objects, places, and temporal order: Evidence for epsodic-like memory in mice. *Neurobiology of Learning and Memory*, *84*(3), 214–221.

Dienes, Z., Baddeley, R. J., & Jansari, A. (2012). Rapidly measuing the speed of unconscious learning: Amnesics learn quickly and happy people slowly. *PLoS One*, *7*(3), 1–9. Available from: http://dx.doi.org/10.1371/journal.pone.0033400.

Dranovsky, A., & Hen, R. (2006). Hippocampal neurogenesis: Regulation by stress and antidepressants. *Biological Psychiatry*, *59* (12), 1136–1143.

Duff, M. C., & Brown-Schmidt, S. (2012). The hippocampus and the flexible use and processing of language. *Frontiers in Human Neuroscience*, *6*(69), 1–9. Available from: http://dx.doi.org/10.1371/journal.pone.0033400.

Duff, M. C., Hengst, J., Tengshe, C., Krema, A., Tranel, D., & Cohen, N. J. (2008). Hippocampal amnesia disrupts the flexible use of procedural discourse in social interaction. *Aphasiology*, *22*(7–8), 866–880.

Dumay, N., & Gaskell, M. G. (2005). Do words go to sleep? Exploring consolidation of spoken forms through direct and indirect measures. *Behavioral and Brain Sciences*, *28*, 69–70.

Dumay, N., & Gaskell, M. G. (2007). Sleep-associated changes in the mental representation of spoken words. *Psychological Science*, *18* (1), 35–39.

Duncan, K., Ketz, N., Inati, S. J., & Davachi, L. (2011). Evidence for area CA1 as a match/mismatch detector: A high-resolution fMRI study of the human hippocampus. *Hippocampus*, *22*. (3). Available from: http://dx.doi.org/10.1002/hipo.20933.

Duvernoy, H., Cattin, F., Naidich, T. P., Raybaud, C. R., Salvolini, U., Scarabino, U., et al. *The human hippocampus: Functional anatomy, vascularization and serial sections with MRI* (3rd ed.). New York, NY: Springer.

Eacott, M. J., Easton, A., & Zinkivskay, A. (2005). Recollection in an episodic-like memory task in the rate. *Learning and Memory*, *12*(3), 221–223.

Epstein, R., Harris, A., Stanley, D., & Kanwisher, N. (1999). The parahippocampal place area: Recognition, navigation, or encoding? *Neuron*, *23*, 115–125.

Ergorul, C., & Eichenbaum, H. (2004). The hippocampus and memory for "what", "where", and "when". *Learning and Memory*, *11* (4), 397–405.

Erickson, K. I., Voss, M. W., Prakash, R. S., Basak, C., Szabo, A., Chaddock, L., et al. (2011). Exercise training increases size of hippocampus and improves memory. *Proceedings of the National Academy of Sciences of the United States of America*, *108*, 3017–3022. Available from: http://dx.doi.org/10.1073/pnas.1015950108.

Euston, D. R., Tatsuno, M., & McNaughton, B. L. (2007). Fast-forward playback of recent memory sequences in prefrontal cortex during sleep. *Science*, *318*, 1147–1150.

Fanselow, M. S. (1994). Neural organization of the defensive behavior system responsible for fear. *Psychonomic Bulletin and Review*, *1*, 429–438.

Farovik, A., Dupont, L. M., & Eichenbaum, H. (2010). Distinct roles for dorsal CA3 and CA1 in memory for sequential nonspatial events. *Learning and Memory*, *17*, 12–17.

Fenn, K. M., Nusbaum, H. C., & Margoliash, D. (2003). Consolidation during sleep of perceptual learning of spoken language. *Nature*, *425*, 614–616.

Fjell, A. M., & Walhovd, K. B. (2005). Age-sensitivity of P3 in high-functioning adults. *Neurobiology of Aging*, *26*, 1297–1299.

Fleischman, D. A., Wilson, R. S., Gabrieli, J. D., Bienias, J. L., & Bennett, D. A. (2004). A longitudinal study of implicit and explicit memory in old persons. *Psychology and Aging*, *19*(4), 617–625.

Fortin, N., Agster, K. L., & Eichenbaum, H. (2002). Critical role of the hippocampus in memory for sequences of events. *Nature Neuroscience*, *5*, 458–462.

Frankland, P. W., & Bontempi, B. (2005). The organization of recent and remote memory. *Nature Reviews Neuroscience*, *6*, 119–130.

Fyhn, M., Molden, S., Witter, M. P., Moser, E. I., & Moser, M. (2004). Spatial representation in the entorhinal cortex. *Science*, *305*, 1258–1264.

Gabrieli, J. D., Cohen, J. D., & Corkin, S. (1988). The impaired learning of semantic knowledge following bilateral medial temporal-lobe resection. *Brain Cognition*, *7*, 157–177.

Gais, S., & Born, J. (2004). Low acetylcholine during slow-wave sleep is critical for declarative memory consolidation. *Proceedings of the National Academy of Sciences of the United States of America*, *101*(7), 2140–2144.

Gais, S., Molle, M., Helms, K., & Born, K. (2002). Learning-depending increases in sleep spindle density. *Journal of Neuroscience*, *22*, 6830–6834.

Gais, S., Plihal, W., Wagner, U., & Born, J. (2000). Early sleep triggers memory for early visual discrimination skills. *Nature Neuroscience*, *3*, 1335–1339.

Ge, S., Yang, C. H., Hsu, K. S., Ming, G. L., & Song, H. (2007). A critical period for enhanced synaptic plasticity in newly generated neurons of the adult brain. *Neuron*, *54*, 559–566.

Gilboa, A., Winocur, G., Rosenbaum, R. S., Poreh, A., Gao, F., Black, S. E., et al. (2006). Hippocampal contributions to recollection in retrograde and anterograde amnesia. *Hippocampus*, *16*(11), 966–980.

Glisky, E. L., Schacter, D. L., & Tulving, E. (1986). Computer learning by memory-impaired patients: Acquisition and retention of complex knowledge. *Neuropsychologia*, *24*(3), 313–328.

Gomez-Isla, T., Price, J. L., McKeel, D. W. J., Morris, J. C., Growdon, J. H., & Hyman, B. T. (1996). Profound loss of layer II entorhinal cortex neurons occurs in very mild Alzheimer's disease. *The Journal of Neuroscience*, *16*(14), 4491–4500.

Hafting, T., Fyhn, M., Molden, S., Moser, M. B., & Moser, E. I. (2005). Microstructure of a spatial map in the entorhinal cortex. *Nature*, *436*(7052), 801–806.

Hamann, B., & Squire, L. R. (1997). Intact perceptual memory in the absence of conscious memory. *Behavioral Neuroscience*, *111*(4), 850–854.

Hamann, S. B., & Squire, L. R. (1995). On the acquisition of new declarative knowledge in amnesia. *Behavioral Neuroscience*, *109*(6), 1027–1044.

Hargreaves, E. L., Rao, G., Lee, I., & Knierim, J. J. (2005). Major dissociation between medial and lateral entorhinal input to dorsal hippocampus. *Science*, *308*, 1792–1794.

Hasselmo, M. E. (2005). The role of hippocampal regions CA3 and CA1 in matching entorhinal input with retrieval of associations between objects and context: Theoretical comment on Lee et al. *Behavioral Neuroscience*, *119*(1), 342–345.

Hertzog, C., Kramer, A. F., Wilson, R. S., & Lindenberger, U. (2009). Enrichment effects on adult cognitive development: Can the functional capacity of older adults be preserved and enhanced? *Psychological Science*, *9*(1), 1–65.

Hodges, J. R., Patterson, K., Oxbury, S., & Funnell, E. (1992). Semantic dementia: Progressive fluent aphasia with temporal lobe atrophy. *Brain*, *115*(6), 1783–1806.

Hoffman, M. L., Beran, M. J., & Washburn, D. A. (2009). Memory for "what", "where", and "when" information in rhesus monkeys (*Macaca mulatta*). *Journal of Experimental Psychology: Animal Behavior Processes*, *35*(2), 143–152.

Hoge, J. A., & Kesner, R. P. (2007). Role of CA3 and CA1 subregions of the dorsal hippocampus on temporal processing of objects. *Neurobiology of Learning and Memory*, *88*, 225–231.

Hopkins, R. O., Myers, C. E., Shohamy, D., Grossman, S., & Gluck, M. A. (2004). Impaired probabilistic category learning in hypoxic subjects with hippocampal damage. *Neuropsychologia*, *42*, 524–535.

Hunsaker, M. R., Mooy, G. G., Swift, J. S., & Kesner, R. P. (2007). Dissociations of the medial and lateral perforant path projections into dorsal DG, CA3, and CA1 for spatial and nonspatial (visual object) information processing. *Behavioral Neuroscience*, *121*, 742–750.

Hunsaker, M. R., Rosenberg, J. S., & Kesner, R. P. (2008). The role of the dentate gyrus, CA3a,b, and CA3c for detecting spatial and environmental novelty. *Hippocampus*, *18*(10), 1064–1073. Available from: http://dx.doi.org/10.1002/hipo.20464.

Insausti, R., Annese, J., Amaral, D. G., & Squire, L. R. (2013). Human amnesia and the medial temporal lobe illuminated by neuropsychological and neurhistological findings for patient E.P. *Proceedings of the National Academy of Sciences of the United States of America*, *110*, E1953–E1962. Available from: http://dx.doi.org/10.1073/pnas.1306244110.

Ji, D., & Wilson, M. A. (2007). Coordinated memory replay in the visual cortex and hippocampus during sleep. *Nature Neuroscience*, *10*, 100–107.

Johnson, M. K., Hashtroudi, S., & Lindsay, D. S. (1993). Source monitoring. *Psychological Bulletin*, *114*, 3–28.

Jung, M. W., & McNaughton, B. L. (1993). Spatial selectivity of unit activity in the hippocampal granular layer. *Hippocampus*, *3*(2), 165–182.

Kanwisher, N., McDermott, J., & Chun, M. (1997). The fusiform face area: A module in human extrastriate cortex specialized for the perception of faces. *Journal of Neuroscience*, *17*, 4302–4311.

Karlsson, M. P., & Frank, L. M. (2009). Awake replay of remote experiences in the hippocampus. *Nature Neuroscience*, *12*, 913–918.

Kempermann, G., Kuhn, H. G., & Gage, F. H. (1997). More hippocampal neurons in adult mice living in an enriched environment. *Nature*, *386*, 493–495.

Kensinger, E. A., Ullman, M. T., & Corkin, S. (2001). Bilateral medial temporal lobe damage does not affect lexical or grammatical processing: Evidence from amnesic patient H.M. *Hippocampus*, *11*, 347–360.

Kerr, K. M., Agster, K. L., Furtak, S., & Burwell, R. D. (2007). Functional neuranatomy of the parahippocampal region: The lateral and medial entorhinal areas. *Hippocampus*, *17*, 697–708.

Kessels, R. P., Hobbel, D., & Postma, A. (2007). Aging, context memory and binding: A comparison of "what, where and when" in young and older adults. *International Journal of Neuroscience*, *117*(6), 795–810.

Kirwan, C. B., Bayley, P. J., Galvan, V. V., & Squire, L. R. (2008). Detailed recollection of remote autobiographical memory after damage to the medial temporal lobe. *Proceedings of the National Academy of Sciences of the United States of America*, *105*(7), 2676–2680.

Knowlton, B. J., & Squire, L. R. (1993). The learning of categories: Parallel brain systems for item memory and category knowledge. *Science*, *262*, 1747–1749.

Knowlton, B. J., & Squire, L. R. (1996). Artificial grammar learning depends on implicit acquisition of both abstract and exemplar-specific information. *Journal of Experimental Psychology: Learning, Memory, and Cognition*, *22*, 169–181.

Koutstaal, W., Schacter, D. L., & Brenner, C. (2001). Dual task demands and gist-based false recognition of pictures in younger and older adults. *Journal of Memory and Language*, *44*, 399–426.

Kramer, A. F., Erickson, K. I., & Colcombe, S. (2006). Exercise, cognition, and the aging brain. *Journal of Applied Physiology*, *101*, 1237–1242.

Kuhn, H. G., Dickinson-Anson, H., & Gage, F. H. (1996). Neurogenesis in the dentate gyrus of the adult rat: Age-related decrease of neuronal progenitor proliferation. *The Journal of Neuroscience, 16*(6), 2027–2033.

Kumaran, D., & Maguire, E. A. (2007). Match mismatch processes underlie human hippocampal responses to associative novelty. *The Journal of Neuroscience, 27*(32), 8517–8524. Available from: http://dx.doi.org/10.1523/JNEUROSCI.1677-07.2007.

Lacy, J. W., Yassa, M. A., Stark, S. M., Muftuler, L. T., & Stark, C. E. (2011). Distinct pattern separation related transfer functions in human CA3/dentate and CA1 revealed using high-resolution fMRI and variable mnemonic similarity. *Learning and Memory, 18*(1), 15–18. Available from: http://dx.doi.org/10.1101/lm.1971111.

Lautenschlager, N. T., Cox, K. L., Flicker, L., Foster, J. K., van Bockxmeer, F. M., Xiao, J., et al. (2008). Effect of physical activity on cognitive function in older adults at risk for Alzheimer's disease. *The Journal of the American Medical Association, 300*(9), 1027–1037.

Leal-Galicia, P., Castaneda-Bueno, M., Quiroz-Baez, C., & Arias, C. (2008). Long-term exposure to environmental enrichment since youth prevents recognition memory decline and increases synaptic plasticity markers in aging. *Neurobiology of Learning and Memory, 90*, 511–518.

Leutgeb, S., Leutgeb, J. K., Treves, A., Moser, M. B., & Moser, E. I. (2004). Distinct ensemble codes in hippocampal areas CA3 and CA1. *Science, 305*(5688), 1295–1298. Available from: http://dx.doi.org/10.1126/science.1100265.

Levy, D. A., Stark, C. E. L., & Squire, L. R. (2004). Intact conceptual priming in the absence of declarative memory. *Psychological Science, 15*, 680–686.

Lisman, J. E. (1999). Relating hippocampal circuitry to function: Recall of memory sequences by reciprocal dentate-CA3 interactions. *Neuron, 22*, 233–242.

Litman, L., Awipi, T., & Davachi, L. (2009). Category-specificity in the human medial temporal lobe. *Hippocampus, 19*(3), 308–319.

Lyle, K. B., Bloise, S. M., & Johnson, M. K. (2006). Age-related binding deficits and the content of false memories. *Psychology and Aging, 21*, 86–95.

MacKay, D. G., James, L. E., & Hadley, C. B. (2008). Amnesic H.M.'s performance on the language competence test: Parallel deficits in memory and sentence production. *Journal of Clinical and Experimental Neuropsychology, 30*(3), 280–300.

MacKay, D. G., James, L. E., Hadley, C. B., & Fogler, K. A. (2011). Speech errors of amnesic H.M.: Unlike everyday slips-of-the-tongue. *Cortex, 47*, 377–408.

MacKay, D. G., James, L. E., Taylor, J. K., & Marian, D. E. (2007). Amnesic H.M. exhibits parallel deficits and sparing in language and memory: Systems versus binding theory account. *Language and Cognitive Processes, 22*(3), 377–452.

Malkova, L., & Mishkin, M. (2003). One-trial memory for object-place associations after separate lesions of hippocampus and posterior parahippocampal region in the monkey. *Journal of Neuroscience, 23*, 1956–1965.

Manns, J., Hopkins, R. O., & Squire, L. R. (2003). Semantic memory and the human hippocampus. *Neuron, 38*(1), 127–133.

Manns, J. R., Howard, M. W., & Eichenbaum, H. (2007). Gradual changes in hippocampal activity support remembering the order of events. *Neuron, 56*, 530–540.

Marr, D. (1971). Simple memory: A theory for archicortex. *Philosophical Transactions of the Royal Society of London Series B, Biological Sciences, 262*(841), 23–81.

Mayes, A. R., Downes, J. J., McDonald, C., Poole, V., Rooke, S., Sagar, H. L., et al. (1994). Two tests for assessing remote public knowledge: A tool for assessing retrograde amnesia. *Memory, 2*(2), 183–210.

McClelland, J. L., McNaughton, B. L., & O'Reilly, R. C. (1995). Why there are complementary learning systems in the hippocampus and neocortex: Insights from the successes and failures of connectionist models of learning and memory. *Psychological Review, 102*(3), 419–457.

McKhann, G., Drachman, D., Folstein, M., Katsman, R., Price, D., & Stadian, E. (1984). Clinical diagnosis of Alzheimer's disease: Report of the NINCDS-ADRDA work group under the auspices of department of health and human services task force on Alzheimer's disease. *Neurology, 34*, 939–944.

McNaughton, B. L., Battaglia, F. P., Jenson, O., Moser, E. I., & Moser, M. B. (2006). Path integration and the neural basis of the "cognitive map". *Nature Reviews Neuroscience, 7*, 663–678.

McNaughton, B. L., & Morris, R. G. (1987). Hippocampal synaptic enhancement and information storage within a distributed memory system. *Trends in Neurosciences, 10*(10), 408–415.

Milner, B. (1962). Physiologie de l'hippocampe. In P. Passouant (Ed.), *Centre National de la Recherche Scientifique* (pp. 257–272). Paris.

Milner, B., Corkin, S., & Teuber, H. L. (1968). Further analysis of the hippocampal amnesic syndrome: 14-year follow-up study of H. M. *Neuropsychologia, 6*, 215–234.

Morris, J. C., McKeel, D. W., Storandt, M., Rubin, E. H., Price, J. L., Grant, E. A., et al. (1991). Very mild Alzheimer's disease: Informant-based clinical, psychometric, and pathologic distinction from normal aging. *Neurology, 41*(4), 469.

Moscovitch, M., Nadel, L., Winocur, G., Gilboa, A., & Rosenbaum, R. S. (2006). The cognitive neuroscience of remote episodic, semantic and spatial memory. *Current Opinion in Neurobiology, 16*(2), 179–190.

Murre, J. M. J., Graham, K. S., & Hodges, J. R. (2001). Semantic dementia: Relevance to connectionist models of long-term memory. *Brain, 124*, 647–675.

Nadel, L., Samsonovich, A., Ryan, L., & Moscovitch, M. (2000). Multiple trace theory of human memory: Computational, neuroimaging, and neuropsychological results. *Hippocampus, 10*(4), 352–368.

Nader, K. (2003). Memory traces unbound. *Trends in Neuroscience, 26*(2), 65–72.

Nakashiba, T., Cushman, J. D., Pelkey, K. A., Renaudineau, S., Buhl, D. L., McHugh, T. J., et al. (2012). Young dentate granule cells mediate pattern separation, whereas old granule cells facilitate pattern completion. *Cell, 149*(1), 188–201.

Nicoll, R. A., & Roche, K. W. (2013). Long-term potentiation: Peeling the onion. *Neuropharmacology, 74*, 18–22. Available from: http://dx.doi.org/10.1016/j.neuropharm.2013.02.010.

Nilsson, L. G. (2003). Memory function in normal aging. *Acta Neurologica Scandinavica, Supplementum, 179*, 7–13.

O'Keefe, J., & Dostrosky, J. (1971). The hippocampus as a spatial map: Preliminary data from unit activity in the freely-moving rat. *Brain Research, 34*(1), 171–175.

O'Keefe, J., & Nadel, L. (1978). *The hippocampus as a cognitive map*. Oxford University Press.

Peigneux, P., Laureys, S., Fuchs, S., Destrebecqz, A., Collette, F., & Delbeuck, C. (2003). Learned material content and acquisition level modulate cerebral reactivation during posttraining rapid-eye-movement sleep. *NeuroImage, 20*, 125–134.

Penfield, W., & Milner, B. (1958). Memory deficit produced by bilateral lesions in the hippocampal zone. *A.M.A. Archives of Neurology and Psychiatry, 79*(5), 475–497.

Phillips, R. G., & LeDoux, J. E. (1992). Differential contribution of amygdala and hippocampus to cued and contextual fear conditioning. *Behavioral Neuroscience, 106*(2), 274–285.

Ranganath, C. (2010). A unified framework for the functional organization of the medial temporal lobes and the phenomenology of episodic memory. *Hippocampus, 20*(11), 1263–1290.

Raz, N., Rodrigue, K. M., Head, D., Kennedy, K. M., & Acker, J. D. (2004). Differential aging of the medial temporal lobe: A study of a five-year change. *Neurology, 62*(3), 433–438.

Reed, J. M., & Squire, L. R. (1998). Retrograde amnesia for facts and events: Findings from four new cases. *Journal of Neuroscience, 18*, 3943–3954.

Ribot, T. A. (1882). *Disease of memory, an essay in the positive psychology*. New York, NY: Appleton.

Rolls, E. T. (2007). An attractor network in the hippocampus: Theory and neurophysiology. *Learning and Memory, 14*(11), 714–731.

Sahay, A., & Hen, R. (2007). Adult hippocampal neurogenesis in depression. *Nature Neuroscience, 10*(9), 1110–1115.

Sahay, A., Scobie, K. N., Hill, A. S., O'Carroll, C. M., Kheirbek, M. A., Burghardt, N. S., et al. (2011). Increasing adult hippocampal neurogenesis is sufficient to improve pattern separation. *Nature, 472*(7344), 466–470. Available from: http://dx.doi.org/10.1038/nature09817.

Salat, D. H., van der Kouwe, A. J., Tuch, D. S., Quinn, B. T., Fischl, B., Dale, A. M., et al. (2006). Neuroimaging of H.M.: A 10-year follow-up investigation. *Hippocampus, 16*(11), 936–945.

Schacter, D. L., Koutstaal, W., Johnson, M. K., Gross, M. S., & Angell, K. A. (1997). False recollection induced by photographs: A comparison of older and younger adults. *Psychology and Aging, 12*, 203–215.

Scheff, S. W., Price, D. A., Schmitt, F. A., & Mufson, E. J. (2006). Hippocampal synaptic loss in early Alzheimer's disease and mild cognitive impairment. *Neurobiology of Aging, 27*(10), 1372–1384.

Schloesser, R. J., Lehmann, M., Martinowich, K., Manji, H. K., & Herkenham, M. (2010). Environmental enrichment requires adult neurogenesis to facilitate the recovery from psychosocial stress. *Molecular Psychiatry, 15*(12), 1152–1163.

Schmidt-Hieber, C., Jonas, P., & Bischofberger, J. (2004). Enhanced synaptic plasticity in newly generated granule cells of the adult hippocampus. *Nature, 429*, 184–187.

Schmolck, H., Kensinger, E. A., Corkin, S., & Squire, L. R. (2002). Semantic knowledge in patient H.M. and other patients with bilateral medial and temporal lobe lesions. *Hippocampus, 15*, 520–533.

Schmolck, H., Stefanacci, L., & Squire, L. R. (2000). Detection and explanation of sentence ambiguity are unaffected by hippocampal lesions but are impaired by larger temporal lobe lesions. *Hippocampus, 10*, 759–770.

Scoville, W. B., & Milner, B. (1957). Loss of recent memory after bilateral hippocampal lesions. *Journal of Neurology Neurosurgery Psychiatry, 20*(1), 11–21.

Seger, C. A., & Spiering, B. J. (2011). A critical review of habit learning and the basal ganglia. *Frontiers in Systems Neuroscience, 5*(66), 1–9. Available from: http://dx.doi.org/10.3389/fnsys.2011.00066.

Shimamura, A. P. (2010). Hierarchical relational binding in the medial temporal lobe: The strong get stronger. *Hippocampus, 20* (11), 1206–1216.

Singer, R. A., & Zentall, T. R. (2007). Pigeons learn to answer the question "where did you just peck?" and can report peck location when unexpectedly asked. *Learning and Behavior, 35*(3), 184–189.

Smith, C. N., Jeneson, A., Frascino, J. C., Kirwan, C. B., Hopkins, R. O., & Squire, L. R. (2014). When recognition memory is independent of hippocampal function. *Proceedings of the National Academy of Sciences of the United States of America, 111*(27), 9935–9940.

Smith, T. D., Adams, M. M., Gallagher, M., Morrison, J. H., & Rapp, P. R. (2000). Circuit-specific alterations in hippocampal synaptophysin immunoreactivity predict spatial learning impairment in aged rats. *Journal of Neuroscience, 20*(17), 6587–6593.

Snyder, J. S., Soumier, A., Brewer, M., Pickel, J., & Cameron, H. (2011). Adult hippocampal neurogenesis buffers stress responses and depressive behaviour. *Nature, 476*, 458–461.

Spalding, K. L., Bergmann, O., Alkass, K., Bernard, S., Salehpour, M., Huttner, H. B., et al. (2013). Dynamics of hippocampal neurogenesis in adult humans. *Cell, 153*, 1219–1227.

Squire, L. R. (1992). Declarative and nondeclarative memory: Multiple brain systems supporting learning and memory. *Journal of Cognitive Neuroscience, 4*(3), 232–243.

Squire, L. R. (2004). Memory systems of the brain: A brief history and current perspective. *Neurobiology of Learning and Memory, 82*, 171–177.

Squire, L. R., & Alvarez, P. (1995). Retrograde amnesia and memory consolidation: A neurobiological perspective. *Current Opinion in Neurobiology, 5*, 169–177.

Squire, L. R., Clark, R. E., & Knowlton, B. J. (2001). Retrograde amnesia. *Hippocampus, 11*, 50–55.

Squire, L. R., & Frambach, M. (1990). Cognitive skill learning in amnesia. *Psychobiology, 18*, 109–117.

Squire, L. R., Stark, C. E. L., & Clark, R. E. (2004). The medial temporal lobe. *Annual Review of Neuroscience, 27*, 279–306.

Squire, L. R., & Zola-Morgan, S. (1991). The medial temporal lobe memory system. *Science, 253*, 1380–1386.

Stark, C. E. L., & Squire, L. R. (2000). Recognition memory and familiarity judgments in severe amnesia: No evidence for a contribution of repetition priming. *Behavioral Neuroscience, 114*, 459–467.

Stark, C. E. L., Stark, S., & Gordon, B. (2005). New semantic learning and generalization in an amnesic patient. *Neuropsychology, 19*, 139–151.

Stark, S., & Gordon, B. S. (2008). A case study of amnesia: Exploring a paradigm for new semantic learning and generalization. *Brain Injury, 22*(3), 283–292.

Stark, S. M., Yassa, M. A., Lacy, J. W., & Stark, C. E. L. (2013). A task to assess behavioral pattern separation (BPS) in humans: Data from healthy aging and mild cognitive impairment. *Neuropsychologia, 51*, 2442–2449. Available from: http://dx.doi.org/10.1016/j.neuropsychologia.2012.12.014.

Stefanacci, L., Buffalo, E. A., Schmolck, H., & Squire, L. R. (2000). Profound amnesia after damage to the medial temporal lobe: A neuroanatomical and neuropsychological profile of patient E. P. *Journal of Neuroscience, 20*, 7024–7036.

Steinmetz, J. E., Lavond, D. G., Ivkovich, D., Logan, C. G., & Thompson, R. F. (1992). Disruption of classical eyelid conditioning after cerebellar lesions: Damage to a memory trace system or a simple performance deficit? *Journal of Neuroscience, 12*(11), 4403–4426.

Stickgold, R. (2002). Emdra putative neurobiological mechanism of action. *Journal of Clinical Psychology, 58*, 61–75.

Stickgold, R., & Walker, M. P. (2007). Sleep-dependent memory consolidation and reconsolidation. *Sleep Medicine, 8*(1), 331–343.

Surget, A., Tanti, A., Leonardo, E. D., Laugeray, A., Ranier, Q., Touma, C., et al. (2011). Antidepressants recruit new neurons to improve stress response regulation. *Molecular Psychiatry, 16*(12), 1177–1188.

Sutherland, R. J., Sparks, F. T., & Lehmann, H. (2010). Hippocampus and retrograde amnesia in the rat model: A modest proposal for the situation of systems consolidation. *Neuropsychologia, 48*(8), 2357–2369.

Swanson, L. W., Wyss, J. M., & Cowan, W. M. (1978). An autoradiographic study of the organization of intrahippocampal association pathways in the rat. *The Journal of Comparative Neurology, 181*(4), 681–715. Available from: http://dx.doi.org/10.1002/cne.901810402.

Treves, A., & Rolls, E. T. (1992). Computational constraints suggest the need for two distinct input systems to the hippocampal CA3 network. *Hippocampus, 2*(2), 189–199. Available from: http://dx.doi.org/10.1002/hipo.450020209.

Treves, A., & Rolls, E. T. (1994). Computational analysis of the role of the hippocampus in memory. *Hippocampus, 4*(3), 374–391.

Tulving, E. (1983). *Elements of episodic memory*. Oxford: Clarendon Press.

Tulving, E. (1984). Relations among components and processes of memory. *Behavioral and Brain Sciences*, *7*, 257–268.

Tulving, E. (2002). Episodic memory: From mind to brain. *Annual Review of Psychology*, *53*, 1–25.

Tulving, E., & Schacter, D. L. (1990). Priming and human memory systems. *Science*, *247*(4940), 301–306.

Ullman, M. T. (2004). Contributions of memory circuits to language: The declarative/procedural model. *Cognition*, *92*(1–2), 231–270.

van Praag, H. (2008). Neurogenesis and exercise: Past and future directions. *Neuromolecular Medicine*, *10*(2), 128–140.

van Praag, H., Shubert, T., Zhao, C., & Gage, F. H. (2005). Exercise enhances learning and hippocampal neurogenesis in aged mice. *Journal of Neuroscience*, *25*, 8680–8685.

Vazdarjanova, A., & Guzowski, J. F. (2004). Differences in hippocampal neuronal population responses to modifications of an environmental context: Evidence for distinct, yet complementary, functions of CA3 and CA1 ensembles. *The Journal of Neuroscience*, *24*(29), 6489–6496. Available from: http://dx.doi.org/10.1523/JNEUROSCI.0350-04.2004.

Verfaellie, M., Croce, P., & Milberg, W. P. (1995). The role of episodic memory in semantic learning: An examination of vocabulary acquisition in a patient with amnesia due to encephalitis. *Neurocase*, *1*(4), 291–304.

Verhaeghen, P. (2003). Aging and vocabulary scores: A meta-analysis. *Psychology and Aging*, *18*(2), 332–339.

Walker, M. P., Brakefield, T., Hobson, H. A., & Stickgold, R. (2003). Dissociable stages of human memory consolidation and reconsolidation. *Nature*, *425*(6958), 616–620.

Walker, M. P., Brakefield, T., Seidman, J., Morgan, A., Hobson, J. A., & Stickgold, R. (2003). Sleep and the time course of motor skill learning. *Learning and Memory*, *10*, 275–284.

Walker, M. P., Liston, C., Hobson, J. A., & Stickgold, R. (2002). Cognitive flexibility across the sleep-wake cycle: REM-sleep enhancement of anagram problem solving. *Cognitive Brain Research*, *14*, 317–324.

Wang, W. C., Yonelinas, A. P., & Ranganath, C. (2013). Dissociable neural correlates of item and context retrieval in the medial temporal lobes. *Behavioural Brain Research*, *254*, 102–107.

Wilson, B. A., Baddeley, A., & Evans, J. (1994). Errorless learning in the rehabilitation of memory-impaired people. *Neuropsychological Rehabilitation*, *4*(3), 307–326.

Wilson, I. A., Gallagher, M., Eichenbaum, H., & Tanila, H. (2006). Neurocognitive aging: Prior memories hinder new hippocampal encoding. *Trends in Neurosciences*, *29*(12), 662–670.

Wilson, M. A., & McNaughton, B. L. (1994). Reactivation of hippocampal ensemble memories during sleep. *Science*, *265*, 676–679.

Witter, M. P. (1993). Organization of the entorhinal-hippocampal system: A review of current anatomical data. *Hippocampus*, *3*, 33–44.

Yassa, M. A., Lacy, J. W., Stark, S. M., Albert, M. S., Gallagher, M., & Stark, C. E. L. (2011). Pattern separation deficits associated with increased hippocampal CA3 and dentate gyrus activity in nondemented older adults. *Hippocampus*, *21*(9), 968–979.

Yassa, M. A., Mattfeld, A. T., Stark, S. M., & Stark, C. E. (2011). Age-related memory deficits linked to circuit-specific disruptions in the hippocampus. *Proceedings of the National Academy of Sciences of the United States of America*, *108*(21), 8873–8878. Available from: http://dx.doi.org/10.1073/pnas.1101567108.

Yassa, M. A., & Reagh, Z. M. (2013). Competitive trace theory: A role for the hippocampus in contextual interference during retrieval. *Frontiers in Behavioral Neuroscience*, *7*(107), 1–13.

Yassa, M. A., & Stark, C. E. (2011). Pattern separation in the hippocampus. *Trends in Neurosciences*, *34*(10), 515–525. Available from: http://dx.doi.org/10.1016/j.tins.2011.06.006.

Yoganarasimha, D., Rao, G., & Knierim, J. J. (2010). Lateral entorhinal neurons are not spatially selective in cue-rich environments. *Hippocampus*, *12*, 1363–1374. Available from: http://dx.doi.org/10.1002/hipo.20839.

Zimmerman, J. T., Stoyva, J. M., & Metcalf, D. (1970). Distorted visual feedback and augmented REM sleep. *Psychophysiology*, *7*, 298–303.

Zimmerman, J. T., Stoyva, J. M., & Reite, M. L. (1978). Spatial rearranged vision and REM sleep: A lack of effect. *Biological Psychiatry*, *13*, 301–316.

Zitnik, G., & Martin, G. (2002). Age-related decline in neurogenesis: Old cells or old environment? *Journal of Neuroscience Research*, *70*(3), 258–263.

C H A P T E R

68

Neural Basis of Phonological Short-Term Memory

Julie A. Fiez

Department of Psychology, Department of Neuroscience, Department of Communication Science and Disorders, Center for Learning Research and Development and Center for the Neural Basis of Cognition, University of Pittsburgh, PA, USA

The short-term maintenance of phonological (speech-related) information is crucial to the performance of many tasks. These include acts of communication, such as formulating and producing a spoken sentence (Acheson & McDonald, 2009), and acts of cognition, such as using inner speech to help mentally solve an arithmetic problem (Frank, Fedorenko, Lai, Saxe, & Gibson, 2012). Thus, understanding how phonological content is dynamically stored (i.e., understanding *short-term phonological memory*) is central to understanding how language is used for communication and verbally mediated cognition.

This chapter provides an overview of phonological memory that draws on cognitive psychological and cognitive neuroscience theories, methods, and results. Cognitive psychology, which emerged in the mid-1900s, marked a shift from behaviorist approaches that treated the mind as a "black box" to cognitive approaches that focused on internal representations and processes that underlie observable human behavior. The increase of cognitive neuroscience in the late 1900s provided the opportunity to look inside the "black box" through neuropsychological studies that examined the effects of brain damage on cognitive performance and neuroimaging studies that examined local changes in brain activity associated with the performance of cognitive tasks. As these two disciplines have continued to mature and interact, so has our understanding of phonological memory and its neural basis. The chapter begins by describing different theoretical perspectives on phonological memory. It then reviews how theoretical aspects of phonological memory have been mapped onto particular brain regions and considers how the functions of implicated brain regions can be further contextualized by looking outside of the verbal working memory literature.

68.1 THEORETICAL PERSPECTIVES ON PHONOLOGICAL SHORT-TERM MEMORY

68.1.1 Phonological Memory as a Passive Process

Early research on phonological memory tackled basic questions, including how much verbal information can be stored in short-term memory (i.e., its capacity), how long it can be stored before the content is forgotten (i.e., its duration), and whether the loss of information is due to the decay of a memory trace or interference between similar traces. These questions can be posed for situations in which short-term memory storage is passive and when it occurs as part of a deliberate and effortful strategy to maintain presented information (Crowder, 1982; Jones, Macken, & Nicholls, 2004). In this first section, the passive storage of auditory and phonological content is discussed.

The Brown–Petersen task provides a classic example of how passive short-term phonological memory can be probed. In this task, subjects view three letters in each trial while at the same time performing a demanding concurrent task (e.g., mental arithmetic). The secondary task is designed to prevent (or minimize) active efforts to remember the presented letter trigram. After a short delay, subjects respond to a probe letter by indicating whether it was or was not presented in

Neurobiology of Language. DOI: http://dx.doi.org/10.1016/B978-0-12-407794-2.00068-7

the previously viewed trigram. Findings from the Brown–Peterson task indicate that verbal content is rapidly forgotten unless active measures are taken to prevent forgetting, which can occur due to both trace decay and item interference effects (Crowder, 1982).

One important twist is that there are modality effects in the passive storage of verbal content. Specifically, recall is typically better when list items are spoken rather than written. The benefit of auditory presentation is particularly pronounced for the final one or two items on a list. This enhanced recency effect can be observed even when subjects are required to perform a concurrent task that reduces their overall memory performance (Penney, 1989). Moreover, the size of an observed recency effect depends on whether an auditory stimulus is perceived as speech, as opposed to a nonspeech sound (Neath, Surprenant, & Crowder, 1993). Together, these findings suggest that passive phonological memory mechanisms may differ across sensory systems, such that there may be an "echoic store" that specifically supports the passive, short-term storage of spoken input (Crowder & Morton, 1969). This idea would be consistent with computational models of speech perception that emphasize the importance of temporal integration in the linguistic perception of an evolving acoustic signal (McClelland & Elman, 1986).

68.1.2 Phonological Memory as a Dedicated Repository That Can Be "Refreshed" via Inner Speech

Short-term phonological memory can also be investigated under conditions that allow the use of volitional processes, such as executive control and inner speech, as a way to mitigate information loss. Baddeley and Hitch (1974) used the term *working memory* to distinguish this more active type of short-term memory. Their seminal work popularized the use of serial recall tasks as a tool for studying the short-term maintenance of verbal information. For the prototypical immediate serial recall (ISR) task, subjects are given a short list of verbal items (e.g., a list of nine or fewer letters, digits, or words) that are presented one at a time (usually 1 every 1–2 s). After presentation of the list, subjects are prompted to immediately recall the items in the order they were presented through spoken, written, or pointed responses.

Baddeley and Hitch proposed the Multicomponent Model of Working Memory as a way to account for a body of experimental results obtained using the ISR task (Baddeley, 1986; Baddeley & Hitch, 1974). With nearly 10,000 citations to date, this model continues to be a highly influential conceptualization of phonological memory. The original model contains a central executive component and two "slave" systems: a phonological loop for the maintenance of verbal content and a visuospatial sketchpad for the maintenance of nonverbal content. The phonological loop comprises a phonological store that is dedicated to working memory and that serves to temporarily hold verbal information, and an articulatory loop, through which inner speech is used to reactivate, or "refresh," the representations in the phonological store. The store can be directly accessed by spoken stimuli, whereas written stimuli must be recoded into the store using the articulatory loop. Items in the phonological store are subject to decay unless they are refreshed by the articulatory loop. As a consequence, the number of items that can be maintained in phonological short-term memory will reflect how well the items can gain access to the phonological store, how well items represented within the store can resist interference and decay, and how quickly and effectively they can be reactivated by articulatory rehearsal.

68.1.3 Short-Term Phonological Memory as Attention-Based Activation

The Baddeley and Hitch model treats phonological memory as a process distinct from language processing. However, other theoretical models have regarded it as an emergent product of phonological knowledge stored in long-term memory and general attention mechanisms. For example, in the Embedded Processes model developed by Cowan (1999), there is a main memory repository that can be approximately equated with the long-term memory system and a very brief sensory store. Information in the long-term memory system can be brought into working memory, which occurs when an "embedded" subset of information in the long-term store takes on a temporarily heightened state of activation. A further embedded subset of the activated information can be made particularly salient when it falls under the *focus of attention*. Working memory is assumed to comprise all information in a readily accessible state by virtue of its activation, including information within the focus of attention and also information in an activated state outside of attention.

68.1.4 Short-Term Phonological Memory via Speech Planning and Efferent Reactivation

A third perspective on phonological memory regards it as the natural by-product of an interactive neural architecture for speech perception and production (for review, see Acheson & McDonald, 2009). In this view, degradation within the speech architecture not only will affect linguistic performance but also will imperil the maintenance of phonological information as an emergent by-product of processing. This perspective is consistent with the fact that ISR and repetition deficits are pervasive in individuals with an acquired central disorder of language (e.g., some form of aphasia), and these deficits can be related to aspects of word and sentence-level performance (Martin, Saffran, & Dell, 1996; Martin, 2005).

The model proposed by Gupta and MacWhinney (1997) illustrates an integrated model of phonological memory and speech perception and production. In this model, phoneme-level representations are interconnected with lexical-phonology representations, which are interconnected with semantic representations. Speech input automatically activates phonemic and lexical representations of the spoken input, and then connections between the phoneme representations and an optional speech planning pathway allow the encoded phonological information to guide spoken output. Rehearsal is conceptualized as the volitional activation of lexical-phonological representations, which then activate phonemic representations and output planning. The engagement of output planning yields an efference copy of the planned production or, in other words, an internal representation of the planned articulation. The efference copy is processed in the same way that would occur if the planned output were presented as spoken input. Thus, verbal rehearsal yields automatic reactivation of the lexical-phonological and phoneme representations by using the internal outputs of speech planning to mimic the repeated presentation of the list items.

68.2 NEURAL PERSPECTIVES ON SHORT-TERM PHONOLOGICAL MEMORY

68.2.1 Passive Phonological Memory

The neural underpinnings of passive short-term phonological storage remain unclear, in part because they have received little direct investigation. One possibility is that the neurons involved in representing the experienced content (e.g., representing the phonological forms of spoken stimuli) exhibit persistent activity across a delay interval. There are neural mechanisms that could produce such persistent activity, such as top-down signals from prefrontal cortex or dynamically assembled oscillatory networks (Benchenane, Tiesinga, & Battaglia, 2011; Gazzaley & Nobre, 2012). However, such mechanisms are generally regarded as attention demanding, and therefore they would be difficult to sustain in combination with the types of effortful concurrent tasks that have been used to reveal passive short-term phonological memory. Another possibility is that neurons involved in representing phonological content can undergo a form of short-term synaptic plasticity, leaving them more easily reactivated in response to a subsequent probe stimulus or internal recall attempt. This type of mechanism would be consistent with cellular and molecular studies of memory formation, which have shown that changes in neural connection strengths involve a variety of mechanisms that occur across different time scales (milliseconds to days). Thus, passive phonological memory could reflect intrinsic cellular processes that occur after a period of active neuronal firing (Nee & Jonides, 2013).

The neural basis of echoic memory has also received relatively little attention. The neuropsychological literature provides examples of patients with brain injury who did not show modality and auditory recency effects, suggesting that their brain injury disrupted the functions of an echoic store (Vallar & Papagno, 1986). However, the lesion sites of such patients have not been systematically investigated and, moreover, the impairments of these patients have not generally been interpreted from an echoic memory perspective. Neuroimaging studies have examined the short-term retention of spoken information but have generally used active conditions (e.g., repetition tasks) that do not specifically probe the passive storage of speech content (Hickok & Poeppel, 2004). Studies involving the auditory short-term memory abilities of nonhuman primates are similarly limited, but intriguing: results from neurophysiological recordings of single auditory cortex neurons indicate that a prior sound can influence the neural processing of a subsequent sound for at least 5 s, irrespective of whether an auditory stimulus sequence is presented within an active or passive behavioral context (Werner-Reiss, Porter, Underhill, & Groh, 2006).

68.2.2 Active Short-Term Phonological Memory: Differing Perspectives

The differing theoretical perspectives on active phonological memory lead to different proposed structure–function mappings in the brain. Much of the early research on this topic used the Baddeley and Hitch model to motivate experimental designs and explain observed findings. For instance, early neurobiological support for the phonological loop subcomponent came from case studies of patients with brain injury, typically due to stroke. These studies provided evidence of dissociations in the behavioral profile of ISR performance across different patients, lending support to the idea that different components of verbal working memory could be dissociated from each other (Baddeley, 1986).

As neuroimaging became available in the 1980s and 1990s, it offered a new tool for investigating verbal working memory. Early work tested various aspects of the Baddeley and Hitch model, such as the assumption of a central executive processor that mediates the behavior of the subsidiary maintenance subsystems, the existence of dissociable verbal and visuospatial maintenance subsystems, and the dissociability of storage and rehearsal processes within a phonological loop. The results from such efforts generally placed the phonological storage component of the verbal maintenance subsystem into the left inferior parietal cortex, the speech-based rehearsal process into the inferior frontal gyrus (with possible additional contributions from the premotor, supplementary motor, and cerebellar areas), and the executive control system into the dorsolateral prefrontal cortex (D'Esposito et al., 1995; Paulesu, Frith, & Frackowiak, 1993; Smith & Jonides, 1998). Overall, this body of work provided convergent support for the general tenets of the Baddeley and Hitch model.

However, arguments have also been made against simple mappings of the Baddeley and Hitch model onto a corresponding set of brain regions. For instance, Chein and Fiez (2010) used neuroimaging and behavioral approaches to investigate the effects of three different manipulations on ISR performance and brain activation: irrelevant speech presentation, irrelevant nonspeech presentation, and concurrent articulation. These effects were chosen because different models of working memory posit unique patterns of association and dissociation between them. By gathering empirical evidence for how the effects patterned together, they tested which theory best predicted the observed data. The results indicated that the Embedded Process model (Cowan, 1999) provided the most straightforward fit to the data. Chein and Fiez combined key components of the Embedded Process model with their information about the neural locus of the different manipulations to assign functions to the brain regions that are active during verbal working memory tasks. They proposed that dorsolateral prefrontal cortex supports executive control, the dorsal sector of the inferior parietal cortex supports attentional scanning, the inferior frontal gyrus supports the activation of phonological information from long-term memory, and motor-related regions (e.g., premotor cortex, supplementary motor cortex, and the cerebellum) are specifically used when a volitional rehearsal-based strategy is used. One appealing aspect of this structure–function mapping is its closer alignment to the concept of working memory in the basic neuroscience literature, which has focused on the role of a frontoparietal attention system in maintaining the active neuronal firing that is necessary to keep information in conscious awareness (Goldman-Rakic, 1995).

Other investigators have approached the neuropsychological and neuroimaging literature on phonological memory from a psycholinguistic perspective (for reviews, see Buchsbaum & D'Esposito, 2008; Gupta & MacWhinney, 1997; Hickok & Poeppel, 2004; Jacquemot & Scott, 2006; Martin, 2005). All of these reviews present a cogent analysis of the overlap between brain regions involved in speech perception and production, leading them to argue against the idea of a dedicated phonological store. Instead, they posit that the speech perception and production system has an intrinsic ability to support the sustained activation of phonological information, and they share the view that the brain regions involved in phonological memory are highly interactive but at the same time tuned to different types of phonological representation. A psycholinguistic perspective on phonological memory motivates the following sections, which draw on results from outside the verbal working memory literature to further contextualize the neural basis of phonological memory.

68.2.3 A Core Phonological System

Although differing theoretical perspectives on phonological memory give rise to differing structure–function mappings, there is a high level of agreement on the set of involved brain regions (Figure 68.1). Anteriorly, the critical tissue can be broadly labeled as the left frontal operculum (L-FO) and considered to include the inferior frontal gyrus, the ventral portions of the precentral gyrus, and the anterior portions of the insula. Posteriorly, the critical tissue can be broadly labeled as the left temporoparietal junction (L-TPJ) and considered to include the posterior superior temporal gyrus and sulcus, the posterior sector of the insula, and the ventral portions of the supramarginal gyrus. Together, the L-FO and L-TPJ territories can be regarded as constituents of a core phonological system for language, which can be complemented by nonlinguistic areas that have been associated with inner speech (e.g., the supplementary motor area, the premotor cortex, and the cerebellum) and brain regions involved in attentional control (e.g., dorsolateral prefrontal cortex).

Turning first to the frontal portion of the core phonological system, the L-FO territory is generally thought to be involved in output-based phonological processing (i.e., processing that is weighted toward supporting speech production) (Benson, 1979; Gupta & MacWhinney, 1997; Hickok & Poeppel, 2007; Martin, 2005). A variety of evidence suggests that the L-FO supports phonological representation at the level of articulatory gestures and that it is particularly important for

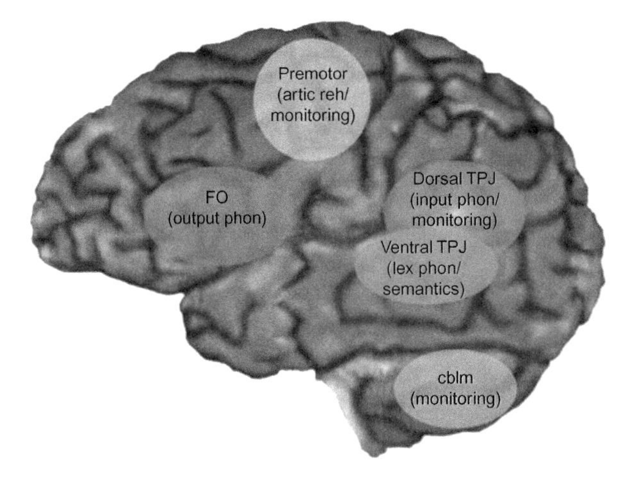

FIGURE 68.1 Neural regions associated with active short-term phonological memory. A variety of evidence indicates that tissue within the L-FO and the L-TPJ contributes to core phonological processing across a variety of language tasks, including verbal ISR tasks. It is thought that the dorsal L-TPJ territory is weighted toward representations and processes associated with speech perception (input-based phonological processing), the ventral L-TPJ territory is weighted toward representations and processes associated with lexical-level phonology and semantics, and the L-FO territory is weighted toward representations and processes associated with speech production. Phonological memory may also involve the use of motor-related regions (such as premotor cortex and the cerebellum) to take advantage of speech production mechanisms for reactivating input-based phonological representations and providing efferent-based signals that can be used to detect and correct errors in a rehearsed sequence. Not shown are regions thought to contribute to domain-general aspects of active phonological memory, such as parietal and dorsolateral prefrontal regions associated with attention and executive control. Abbreviations: artic reh, articulatory rehearsal; phon, phonology; lex, lexical; cbm, cerebellum; FO, frontal operculum; TPJ, temporoparietal junction.

tasks that require sublexical phonetic analysis (Hickok & Poeppel, 2007; Indefrey & Levelt, 2004). Although the L-FO is usually linked to rehearsal-related processing in verbal working (Paulesu et al., 1993; Smith, Jonides, Marshuetz, & Koeppe, 1998), there are reasons to believe that it may automatically support the output-based coding of visual and auditory stimuli in phonological memory even when rehearsal is not used. For instance, Chein and Fiez (2010) found that this region was affected differently by a concurrent articulation task than other motor speech areas in the frontal lobe.

There is less agreement about the role of posterior components of the core phonological system (Becker, MacAndrew, & Fiez, 1999; Buchsbaum & D'Esposito, 2008), most likely because specific areas within the L-TPJ territory make distinct contributions that require further delineation (Buchsbaum & D'Esposito, 2008; Jacquemot & Scott, 2006; Martin, 2005). Setting aside the issue of precise localization, the L-TPJ territory is commonly associated with input-based phonological processing (i.e., processing that is weighted toward supporting speech perception) (Gupta & MacWhinney, 1997; Hickok & Poeppel, 2007; Jacquemot & Scott, 2006; Martin, 2005). Neuropsychological and neuroimaging studies provide compelling, although not completely consistent, evidence that the L-TPJ territory is active during the maintenance of phonological information and that it is crucial for normal ISR task performance (for review, see Buchsbaum & D'Esposito, 2008). It has often been proposed as the site of a phonological store: either one that is dedicated to verbal working memory (Paulesu et al., 1993; Vigneau et al., 2006) or one that reflects demands for the buffering of phonological and order information associated not only with verbal working memory but also with speech planning and production (Buchsbaum & D'Esposito, 2008; Gupta & MacWhinney, 1997; Jacquemot & Scott, 2006; Martin, 2005).

Anatomically, the dorsal sector of the L-TPJ is interconnected with regions associated with auditory, motor, and speech processing (Guenther, 2006; Petrides & Pandya, 1988; Rauschecker, 2011; Simonyan & Jurgens, 2002; Yeterian & Pandya, 1998); furthermore, neuroimaging studies have demonstrated that both sensory and motor-related responses can be found within this region (for review, see Guenther, 2006; Hickok, Houde, & Rong, 2011; Rauschecker, 2011). These and other findings suggest that this dorsal territory functions as a sensorimotor bridge between frontal tissue associated with speech production and temporal tissue associated with speech recognition and semantic analysis (Acheson, Hamidi, Binder, & Postle, 2011; Buchsbaum & D'Esposito, 2008; Hickok & Poeppel, 2007; Rauschecker, 2011). In the context of an ISR task, the dorsal L-TPJ may be weighted toward phonological encoding and the maintenance of order information, whereas the ventral L-TPJ may be weighted toward lexical-level coding (Acheson et al., 2011; Buchsbaum & D'Esposito, 2008; Gupta & MacWhinney, 1997; Jacquemot & Scott, 2006; Martin, 2005). Engagement of the ventral L-TPJ could provide a pathway that accounts for lexical and semantic influences on ISR task performance and speech production (Corley, Brockehurst, & Moat, 2011; Martin, 2008; Roelofs, 2004).

68.2.4 Neural Mechanisms for Articulatory Rehearsal

Moving outside the core phonological network, areas implicated in motor processing may be used to strategically implement verbal rehearsal. The motoric coding that is associated with intentional rehearsal could provide an efferent-based mechanism by which phonological information can be reactivated (Guenther, 1995; Gupta & MacWhinney, 1997; Jacquemot & Scott, 2006; Rauschecker, 2011). This is a conceptually simple and compelling idea, but there is surprisingly little neural data to support it. One idea might be that premotor or motor association areas would be a source of excitatory input into one or more areas within the L-TPJ territory. Cortical surface electrode recordings from individuals undergoing presurgical planning for epilepsy treatment provide intriguing evidence for motor-related changes in L-TPJ tissue (Edwards et al., 2010). However, fMRI studies have not consistently revealed increased L-TPJ activation associated with verbal rehearsal and silent speech; if anything, there appears to be a suppression of activity within the superior temporal gyrus during speech production (Curio, Neuloh, Numminen, Jousmaki, & Hari, 2000; Paus, Marrett, Worsley, & Evans, 1996; Yinen et al., 2014). The disparate results may reflect the fact that speech production leads to highly focal regions of activation that can become obscured by surrounding areas of suppression and individual neuroanatomical variability (Buchsbaum, Olsen, Koch, & Berman, 2005; Flinker, Chang, Barbaro, Berger, & Knight, 2011).

A related idea is that efferent-based information may be used to monitor and correct for errors that occur during either overt speech production or internal speech. Theories of motor control and motor learning emphasize the need for neural mechanisms that allow motor representations to be derived from sensory goals (inverse models) and for sensory consequences to be predicted based on actual or planned actions (forward models); inverse and forward models are thought to play a critical role in detecting and correcting breakdowns in performance that occur when there are mismatches between sensory and motor representational spaces (Kawato & Wolpert, 1998; Wolpert & Ghahramani, 2000). A relatively large amount of data provide evidence that the L-TPJ territory plays a critical role in the monitoring of internal speech, most likely as recipient of information about the planned utterance (for review, see Guenther, 2006; Hickok et al., 2011; Rauschecker, 2011). The cerebellum is also likely to be important, given that this brain structure features prominently in many models of motor control and learning (Doya, 2000). Particularly relevant is the model of verbal rehearsal proposed by Desmond and collaborators (Desmond, Gabrieli, Wagner, Ginier, & Glover, 1997). This model proposes a cerebro-cerebellar loop involving frontal motor areas, which provides the superior cerebellum with information about the actual contents of the rehearsal loop, whereas a second loop involving the L-TPJ provides the inferior cerebellum with information about the intended contents. Circuitry within the cerebellum is used to compare the two pieces of information and to create an error correction signal (when there is a discrepancy) that is relayed to frontal motor areas to adaptively modify the contents of the rehearsal loop. Neuropsychological studies have shown that individuals with cerebellar damage exhibit modest decrements in ISR performance, as would be expected if core phonological processing was left intact but the capacity for error correction was impaired (Justus, Ravizza, Fiez, & Ivry, 2005; Ravizza et al., 2006).

68.3 SUMMARY

Language and verbally mediated cognition involve short-term phonological storage in the form of passive and active mechanisms for maintaining phonological content. Three general classes of models provide differing theoretical perspectives on how short-term phonological memories are actively stored and maintained, and how different functional aspects of phonological memory map onto specific brain regions. In general, however, phonological memory is thought to rely on a core phonological network that includes the tissue surrounding the L-TPJ and the L-FO, brain territories that are also associated with speech perception and production, respectively. In addition, brain regions associated with motor aspects of speech production (e.g., premotor cortex, the cerebellum) may contribute to phonological memory, most likely by providing the neural substrates for inner speech. The volitional use of inner speech may mitigate memory loss by reactivating phonological content and providing the information needed to monitor and correct degraded representations in phonological memory.

References

Acheson, D. J., Hamidi, M., Binder, J. R., & Postle, B. R. (2011). A common neural substrate for language production and verbal working memory. *Journal of Cognitive Neuroscience*, *23*, 1358–1367.

Acheson, D. J., & McDonald, M. C. (2009). Verbal working memory and language production: Common approaches to the serial ordering of verbal information. *Psychological Bulletin*, *135*, 50–68.

Baddeley, A. D. (1986). *Working memory*. New York, NY: Oxford University Press.

Baddeley, A. D., & Hitch, G. (1974). Working memory. In G. H. Bower (Ed.), *The psychology of learning and motivation* (Vol. 8, pp. 47–89). New York, NY: Academic Press.

Becker, J. T., MacAndrew, D. K., & Fiez, J. A. (1999). A comment on the functional localization of the phonological storage subsystem of working memory. *Brain and Cognition*, *41*, 27–38.

Benchenane, K., Tiesinga, P. H., & Battaglia, F. P. (2011). Oscillations in the prefrontal cortex: A gateway to memory and attention. *Current Opinion in Neurobiology*, *21*, 475–485.

Benson, D. F. (1979). *Aphasia, alexia, and agraphia*. New York, NY: Churchill Livingstone.

Buchsbaum, B., Olsen, R. K., Koch, P., & Berman, K. F. (2005). Human dorsal and ventral auditory streams subserve rehearsal-based and echoic processes during verbal working memory. *Neuron*, *48*, 687–697.

Buchsbaum, B. R., & D'Esposito, M. (2008). The search for the phonological store: From loop to convolution. *Journal of Cognitive Neuroscience, 20*, 762–778.

Chein, J. M., & Fiez, J. A. (2010). Evaluating models of working memory through the effects of concurrent irrelevant information. *Journal of Experimental Psychology: General, 139*, 117–137.

Corley, M., Brockehurst, P. H., & Moat, H. S. (2011). Error biases in inner and overt speech: Evidence from tongue twisters. *Journal of Experimental Psychology: Learning, Memory, and Cognition, 37*, 162–175.

Cowan, N. (1999). An embedded-processes model of working memory. In A. Miyake, & P. Shah (Eds.), *Models of working memory: Mechanisms of active maintenance and executive control* (pp. 62–101). New York, NY: Cambridge University Press.

Crowder, R. G. (1982). The demise of short-term memory. *Acta Psychologica, 50*, 291–323.

Crowder, R. G., & Morton, J. (1969). Precategorical acoustic storage. *Perception and Psychophysics, 5*, 365–373.

Curio, G., Neuloh, G., Numminen, J., Jousmaki, V., & Hari, R. (2000). Speaking modifies voice-evoked activity in the human auditory cortex. *Human Brain Mapping, 9*, 183–191.

Desmond, J. E., Gabrieli, J. D. E., Wagner, A. D., Ginier, B. L., & Glover, G. H. (1997). Lobular patterns of cerebellar activation in verbal working-memory and finger-tapping tasks as revealed by functional MRI. *Journal of Neuroscience, 17*, 9675–9685.

D'Esposito, M., Detre, J. A., Alsop, D. C., Shin, R. K., Atlas, S., & Grossman, M. (1995). The neural basis of the central executive system of working memory. *Nature, 378*, 279–281.

Doya, K. (2000). Complementary roles of basal ganglia and cerebellum in learning and motor control. *Current Opinion in Neurobiology, 10*, 732–739.

Edwards, E., Nagarajan, S. S., Dalal, S. S., Canolty, R. T., Kirsch, H. E., Barbaro, N. M., et al. (2010). Spatiotemporal imaging of cortical activation during verb generation and picture naming. *NeuroImage, 50*, 291–301.

Flinker, A., Chang, E. F., Barbaro, N. M., Berger, M. S., & Knight, R. T. (2011). Sub-centimeter language organization in the human temporal lobe. *Brain and Language, 117*, 103–109.

Frank, M. C., Fedorenko, E., Lai, P., Saxe, R., & Gibson, E. (2012). Verbal interference suppresses exact numerical representation. *Cognitive Psychology, 64*, 74–92.

Gazzaley, A., & Nobre, A. C. (2012). Top-down modulation: Bridging selective attention and working memory. *Trends in Cognitive Science, 16*, 129–135.

Goldman-Rakic, P. S. (1995). Architecture of the prefrontal cortex and the central executive. *Annals of the New York Academy of Sciences, 769*, 71–83.

Guenther, F. H. (1995). Speech sound acquisition, coarticulation, and rate effects in a neural network model of speech production. *Psychological Review, 102*, 594–621.

Guenther, F. H. (2006). Cortical interactions underlying the production of speech sounds. *Journal of Communication Disorders, 39*, 350–365.

Gupta, P., & MacWhinney, B. (1997). Vocabulary acquisition and verbal short-term memory: Computational and neural bases. *Brain and Language, 59*, 267–333.

Hickok, G., Houde, J., & Rong, F. (2011). Sensorimotor integration in speech processing: Computational basis and neural organization. *Neuron, 69*, 407–422.

Hickok, G., & Poeppel, D. (2004). Dorsal and ventral streams: A framework for understanding aspects of the functional anatomy of language. *Cognition, 92*, 67–99.

Hickok, G., & Poeppel, D. (2007). The cortical organization of speech processing. *Nature Reviews. Neuroscience, 8*, 393–402.

Indefrey, P., & Levelt, W. J. M. (2004). The spatial and temporal signatures of word production components. *Cognition, 92*, 101–104.

Jacquemot, C., & Scott, S. K. (2006). What is the relationship between phonological short-term memory and speech processing. *Trends in Cognitive Sciences, 10*, 480–486.

Jones, D. M., Macken, W. J., & Nicholls, A. P. (2004). The phonological store of working memory: Is it phonological and is it a store? *Journal of Experimental Psychology: Learning, Memory, and Cognition, 30*, 656–674.

Justus, T., Ravizza, S. M., Fiez, J. A., & Ivry, R. B. (2005). Reduced phonological similarity effects in patients with damage to the cerebellum. *Brain and Language, 95*, 304–318.

Kawato, M., & Wolpert, D. (1998). Internal models for motor control. *Novartis Foundation Symposium, 218*, 291–304.

Martin, N. (2008). The role of semantic processing in short-term memory and learning: Evidence from aphasia. In A. Thorn, & M. Page (Eds.), *Interactions between short-term and long-term memory in the verbal domain*. New York, NY: Psychology Press.

Martin, N., Saffran, E. M., & Dell, G. S. (1996). Recovery in deep dysphasia: evidence for a relation between auditory—Verbal STM capacity and lexical errors in repetition. *Brain and Language, 52*, 83–113.

Martin, R. C. (2005). Components of short-term memory and their relation to language processing: Evidence from neuropsychology and neuroimaging. *Current Directions in Psychological Science, 14*, 204–208.

McClelland, J. L., & Elman, J. L. (1986). The TRACE model of speech perception. *Cognitive Psychology, 18*, 1–86.

Neath, I., Surprenant, A. M., & Crowder, R. G. (1993). The context-dependent stimulus suffix effect. *Journal of Experimental Psychology: Learning, Memory, and Cognition, 19*, 698–703.

Nee, D. E., & Jonides, J. (2013). Trisecting representational states in short-term memory. *Frontiers in Human Neuroscience, 7*, 796.

Paulesu, E., Frith, C. D., & Frackowiak, R. S. (1993). The neural correlates of the verbal component of working memory. *Nature, 362*, 342–345.

Paus, T., Marrett, S., Worsley, K., & Evans, A. (1996). Imaging motor-to-sensory discharges in the human brain: An experimental tool for the assessment of functional connectivity. *NeuroImage, 4*, 78–86.

Penney, C. G. (1989). Modality effects and the structure of short-term verbal memory. *Memory and Cognition, 17*, 398–422.

Petrides, M., & Pandya, D. N. (1988). Association fiber pathways to the frontal cortex from the superior temporal region in the rhesus monkey. *Journal of Comparative Neurology, 273*, 52–66.

Rauschecker, J. P. (2011). An expanded role for the dorsal auditory pathway in sensorimotor control and integration. *Hearing Research, 271*, 16–25.

Ravizza, S. M., McCormick, C. A., Schlerf, J., Justus, T., Ivry, R. B., & Fiez, J. A. (2006). Cerebellar damage produces selective deficits in verbal working memory. *Brain, 129*, 306–320.

Roelofs, A. (2004). Error biases in spoken word planning and monitoring by aphasic and nonaphasic speakers: Comment on Rapp and Goldrick (2000). *Psychological Review, 111*, 561–572.

Simonyan, K., & Jurgens, U. (2002). Cortico-cortical projections of the motorcortical larynx area in the rhesus monkey. *Brain Research, 949*, 23–31.

Smith, E. E., & Jonides, J. (1998). Neuroimaging analyses of human working memory. *Proceedings of the National Academy of Sciences of the United States of America, 95*, 12061–12068.

Smith, E. E., Jonides, J., Marshuetz, C., & Koeppe, R. A. (1998). Components of verbal working memory: Evidence from neuroimaging. *Proceedings of the National Academy of Sciences of the United States of America, 95*, 876–882.

Vallar, G., & Papagno, C. (1986). Phonological short-term store and the nature of the recency effect: Evidence from neuropsychology. *Brain and Cognition, 5*, 428–442.

Vigneau, M., Beaucousin, V., Herve, P. Y., Duffau, H., Crivello, F., Houde, O., et al. (2006). Meta-analyzing left hemisphere language areas: Phonology, semantics, and sentence processing. *NeuroImage*, *30*, 1414–1432.

Werner-Reiss, U., Porter, K. K., Underhill, A. M., & Groh, J. M. (2006). Long lasting attenuation by prior sounds in auditory cortex of awake primates. *Experimental Brain Research*, *168*, 272–276.

Wolpert, D. M., & Ghahramani, Z. (2000). Computational principles of movement neuroscience. *Nature Neuroscience*, *3*, 1212–1217.

Yeterian, E. H., & Pandya, D. N. (1998). Corticostriatal connections of the superior temporal region in rhesus monkeys. *The Journal of Comparative Neurology*, *399*, 384–402.

Yinen, S., Nora, A., Leminen, A., Hakala, T., Huotilainen, M., Shtyrov, Y., et al. (2014). Two distinct auditory-motor circuits for monitoring speech production as revealed by content-specific suppression of auditory cortex. *Cerebral Cortex*. Available from: http://dx.doi.org/doi:10.1093/cercor/bht351.

CHAPTER

69

Working Memory and Language

Bradley R. Buchsbaum

Rotman Research Institute, Baycrest, University of Toronto, Toronto, ON, Canada

69.1 INTRODUCTION

In human verbal communication there is frequently some period of time that intervenes between the sensory perception of a speech message and an appropriate response to that message. Consider the plight of the Starbucks *barista*: a fastidious customer makes his way to the front of the line and orders a rather complex espresso concoction. The cashier rings up the order and repeats it aloud: "double decaf lightly foamed soy milk cappucino"—a verbal message containing all of 7 words and 13 syllables. The *barista*, meanwhile busy preparing a *peppermint caramel machiatto* for the previous customer, must mentally record and retain in memory the content of the new request. Although the spoken message describing an elaborate coffee drink has only a brief existence as a stream of acoustic vibrations in the physical world, it must nevertheless be transmitted and stably represented in the minds of three successive people. The customer *produces* the message, the cashier *repeats* the message, and the *barista* must *maintain* the message "in mind" until he or she has completed the order. When the drink arrives at the counter with a cheerful: "Sir, your double decaf lightly foamed soy milk capuccino is ready!" a bystander, uninitiated to this routine modern coffee-ordering spectacle, might be forgiven for considering the fulfillment of the customer's request to be something of a miracle. From the standpoint of the study of human cognition, the success of this commercial exchange depends critically not only on language competence (i.e., the ability to perceive and produce speech) but also on the ability to consciously hold onto a sequence of verbal information over a short period of time. In the parlance of modern cognitive psychology and neuroscience, we refer to this ability to retain information in an accessible state over short periods of time as "working memory," a cognitive faculty that enables humans and other complex animals to temporarily store, process, and manipulate important pieces of information that are no longer readily available in the sensory environment.

In this chapter, I first review how working memory emerged as a concept in cognitive psychology. I then describe the preeminent cognitive model of verbal working memory, the phonological loop, and the logic by which it explains critical laboratory phenomena associated with memory for verbal material. I then discuss the attempts and associated difficulties in mapping and otherwise situating the phonological loop in the brain. Finally, I cover the recent movement in the cognitive neuroscience of language to view verbal working memory as emerging from the language circuitry that underpins core language functions such as the perception and production on speech.

69.2 THE EMERGENCE OF THE CONCEPT OF SHORT-TERM MEMORY

The idea of memory as consisting of two main compartments, one for the current contents of consciousness and another for a permanent record of experience, has gone in and out of fashion in the past century. James (2011) coined the terms "primary memory" and "secondary memory" to refer to these two basic concepts, setting off a long-standing debate in the psychological sciences as to whether memory is best viewed a unitary or mechanistically divisible phenomenon. In the middle part of the 20th century, most theorists viewed memory as a unitary system governed by a single set of principles that were largely invariant over time (Melton, 1963; Underwood, 1957). However, in the 1960s, evidence from cognitive

Neurobiology of Language. DOI: http://dx.doi.org/10.1016/B978-0-12-407794-2.00069-9

psychology began to point to the existence of two memory systems, one for very recent events (short-term memory) and one for events that occurred in the more distant past (long-term memory).

A critical piece of evidence supporting the "dual-store" view of memory came from studies of free recall. It was shown that when subjects are presented a list of words and must recall them in any order (free recall), performance is best for the first few items (the *primacy* effect) and for the last few items (the *recency* effect). When accuracy is plotted as a function of input order, it reveals a characteristic U-shaped (Davelaar, Goshen-Gottstein, Ashkenazi, Haarmann, & Usher, 2005; Glanzer & Cunitz, 1966; Waugh & Norman, 1965) pattern, which is referred to as the serial position curve. However, if a short delay (e.g., 10 s) is placed between stimulus presentation and recall during which subjects are required to engage in some distracting activity, the shape of the serial positive curve changed. Performance on early items (primacy) is relatively unaffected, but the recency effect is abolished (Glanzer & Cunitz, 1966; Postman & Phillips, 1965). Recency effects are attributed to the readout of the last few items in a list from short-term memory (STM), and primacy effects are reflected in the long-term memory (LTM) advantage for the first few items in a list due to the greater rehearsal devoted to those items. Moreover, recall from the long-term store requires a more effortful and slow probabilistic form of retrieval that largely depends on associative, semantic, and contextual retrieval cues than is retrieval from the short-term store. It would be remiss not to mention that this interpretation of patterns of recency effects in immediate and delayed recall as reflected in the operation of two stores has long been disputed and is complicated by the demonstration of recency effects that can span across minutes or even days (Bjork & Whitten, 1974; Crowder, 1982), although it has yet to be shown that these "long-term recency effects" have the same underlying mechanism as standard recency effects.

In summary, short-term memory is essentially a limitation of the online capacity of an information processing system. Thus, short-term memory can be viewed as a cup into which sensory information flows. The capacity of the cup is fixed and is prone to overflowing. The precise capacity of the cup varies across individuals (Unsworth & Engle, 2007), although as Miller (1956) memorably pointed out, it tends to hover around a "magical number" of 7 plus or minus 2 (but see also Cowan, 2001). When incoming information exceeds the capacity of the cup, the spillover may still be recorded in a secondary container, that is, long-term memory.

69.3 NEUROLOGICAL EVIDENCE FOR A SEPARATION OF SHORT-TERM AND LONG-TERM MEMORY

One of the critical pieces of evidence supporting the existence of separable systems for short-term and long-term memory was the discovery of patients with brain damage that appeared to have selective deficits affecting only long-term or only short-term memory. The most famous example of such dissociation was the case of the patient H.M., whose medial temporal lobes were famously removed as a treatment of intractable epilepsy. The surgery resulted in nearly complete loss in the ability to form new long-term declarative memories (Corkin, 2002; Scoville & Milner, 1957). H.M. and other patients with bilateral medial temporal lobe lesions that have subsequently been described (Squire, Stark, & Clark, 2004) live in a kind of "permanent present tense" (Corkin, 2013), unable to consciously recall events that occurred even a few minutes ago. Notwithstanding this severe impairment in the ability to form new long-term declarative memories, such patients appear to have little or no deficit on tests of short-term memory such as repeating back short strings of digits (Baddeley & Warrington, 1970; Wickelgren, 1968), although deficits in short-term memory have been sometimes observed in these patients in tests using novel visual objects (Ranganath & Blumenfeld, 2005) or in tests of short-term associative memory (Olson, Page, Moore, Chatterjee, & Verfaellie, 2006; Ryan & Cohen, 2004; Yee, Hannula, Tranel, & Cohen, 2014).

Further bolstering support for the idea of a neurobiological distinction between short-term and long-term memory was the discovery of patients with severely impaired short-term memory for numbers and words together with a preserved ability to learn supra-span (e.g., more than 10 items) word lists with repeated study (Baddeley, Papagno, & Vallar, 1988; Basso, Spinnler, Vallar, & Zanobio, 1982; Shallice & Warrington, 1970; Warrington & Shallice, 1969). Thus, the behavioral pattern evidenced in these subjects contrasted with that of H.M. in that they could form new long-term (verbal) memories but had little or no verbal short-term memory. However, whereas H.M.'s memory deficit was a general memory impairment that applied to all forms of declarative information (e.g., verbal, visual, and spatial), these particular "short-term memory patients," as they would later be called, had deficits that were confined to the auditory-verbal modality. It is also important to emphasize that the short-term memory deficits by these patients, in the purest cases (Shallice & Butterworth, 1977; Shallice & Vallar, 1990; Shallice & Warrington, 1977; Takayama,

Kinomoto, & Nakamura, 2004; Vallar & Baddeley, 1984; Vallar, Di Betta, & Silveri, 1997), are not accompanied by severe deficits in language comprehension and production. Thus, patient J.B. (Shallice & Butterworth, 1977) was able to have conversations normally and to speak fluently without abnormal pauses, errors, or other aphasic symptoms. What this seemed to show was that verbal storage is not "built in" to the language processing system but is an independent entity in its own right, an informational buffer that exists not to support language *per se*, but rather as a passive "holding place" where recently encountered linguistic information can be temporarily stored. Thus, the discovery of "short-term memory patients," as they were to be called (Vallar, 2006), established a double dissociation both in brain localization (LTM, medial temporal lobe; verbal STM, temporoparietal cortex) and patterns of performance, between short-term and long-term memory systems. In addition, the short-term memory disorder could be distinguished from severe disorders of language production and comprehension such as Broca's aphasia, Wernicke's aphasia, and other classic neurological language impairments.

69.4 THE EMERGENCE OF THE CONCEPT OF WORKING MEMORY

The Working Memory model of Baddeley and colleagues (Baddeley, 1986, 2000; Baddeley & Hitch, 1974) was developed with the aim of explaining the relevant behavioral findings in the memory literature while also taking into account important neuropsychological case study reports such as those reviewed here. In addition, whereas prior models of short-term memory tended to emphasize storage buffers as the receptacles for information arriving from the senses, Baddeley and Hitch (1974) focused on rehearsal processes, that is, strategic mechanisms for the maintenance of items in memory. Baddeley and Hitch (1974) attempted to account for a system that could simultaneously manipulate the current contents of memory and update information in working memory in the service of task goals. Such a system is especially important when one needs to maintain information over short periods in many complex cognitive activities such as reading, mental calculation, spatial reasoning, and so forth.

In the canonical "real world" example, when one is trying to keep a telephone number in mind, one will often mentally rehearse the contents of the numeric sequence. Research has shown that in tests of serial recall, when subjects are prevented from engaging in subvocal rehearsal during a delay period that is inserted between stimulus presentation and recall, overall performance suffers (Baddeley, Thomson, & Buchanan, 1975). In the case of verbal material, then, this suggested that the ability to keep verbal sequences in working memory depends on covert articulatory processes. This insight was central to the development of the verbal component of Working Memory, the "phonological loop," and led to a broader conceptualization of short-term memory that sought to explain not only how and why information enters and exits awareness but also how resources are deployed in a strategic effort to capture and maintain the objects of memory in conscious awareness.

The key tenets of the Working Memory model are as follows: (i) it is a limited capacity system; at any moment in time, there is only a finite amount of information directly available for processing in memory; (ii) specialized subsystems devoted to the representation of information of a particular type, for instance, verbal or visuospatial, are structurally independent of one another (i.e., the integrity of information represented in one domain is protected from the interfering effects of information that may be arriving to another domain); and (iii) storage of information in memory is distinct from the processes that underlie sensory perception; instead, there is two-stage process whereby sensory information is first analyzed by perceptual modules and then transferred into specialized storage buffers that have no other role but to temporarily "hold" preprocessed units of information. Moreover, the pieces of information that reside in these specialized buffers are subject to passive, time-based decay as well as inter-item interference (e.g., similar sounding words like "man, mad, map, cap, mad" can lead to interference within a specialized phonological storage structure). Finally, the storage buffers have no built-in or internal mechanism for maintaining or otherwise refreshing their contents, rather, this must occur from without, through the process of rehearsal, which might be a motor or top-down control mechanism that can sequentially access and refresh the contents that remain active within the store.

The Working Memory model, first proposed by Baddeley and Hitch (1974) and later refined (Baddeley, 1986, 2000; Salame & Baddeley, 1982), argued for the existence of three functional components of working memory. The "central executive" was envisioned as a control system of limited attentional capacity responsible for coordinating and controlling two subsidiary slave systems, a *phonological loop* and a *visuospatial sketchpad*. The phonological loop was responsible for the storage and maintenance of information in a verbal form, and the visuospatial sketchpad was dedicated to the storage and maintenance of visuospatial information. In the past decade, a fourth component, the "episodic buffer," has been added to the model to capture a number of phenomena related to interactions between short-term and long-term memory that could not be readily explained within the original framework. In the next section, the phonological loop is described in more detail because it is the central component underlying working memory for verbal material.

69.5 THE PHONOLOGICAL LOOP

As noted, the Working Memory model entails a separation of domain-specific mechanisms of memory maintenance and domain-general mechanisms of executive control (Figure 69.1). Thus, the verbal component of working memory, or the phonological loop, is regarded as a "slave" system that is under the supervisory control of the central executive component. Within the phonological loop, two interacting components—the phonological store and the articulatory rehearsal process—enable verbal representations to be maintained in an active state.

The phonological store is a passive buffer wherein verbal information can be stored for brief (approximately 2 s) periods. The articulatory control process serves to refresh the contents of the store, thereby allowing the system to maintain sequences of verbal items in memory over some interval of time. This division of labor between two interlocking components, one an active *process* and the other a *passive* store, is crucial to the model's explanatory power. For instance, when the articulatory control process is interfered with through the method of articulatory suppression (e.g., by requiring subjects to say "hiya" over and over again), items in the store rapidly decay and recall performance suffers greatly. The phonological store, then, lacks a mechanism of reactivating its own contents but possesses memory capacity, whereas the articulatory rehearsal process lacks an intrinsic memory capacity of its own but can exert its effect indirectly by refreshing the contents of the store.

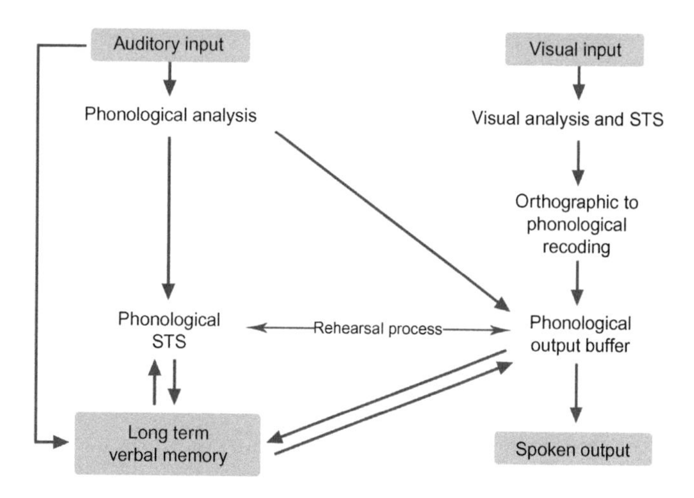

FIGURE 69.1 An anatomo-functional model of phonological short-term memory based on Vallar et al. (1997) and Baddeley, Gathercole, and Papagno (1998).

The phonological loop model of verbal working memory has stood the test of time, largely because it explains many of the behavioral phenomena associated with verbal memory performance in a simple and intuitive way. It is important to briefly note what these core behavioral phenomena are and how the phonological loop model accounts for them. The appeal of the model comes partly from its parsimony—with only a very minimal set of functional specifications, it is able to account for a large number of behavioral findings. It therefore provides a benchmark that any competing model, whether neural or purely cognitive, must be able to match. An overview of how the phonological loop explains certain well-established behavioral phenomena associated with verbal working memory, namely, the *phonological similarity effect*, the *word-length effect*, the effect of *articulatory suppression*, and the *irrelevant sound effect* (Repovs & Baddeley, 2006) is provided here. The phonological similarity effect refers to the finding that similar sounding sets of words are more difficult to retain in memory than sets of phonologically dissimilar words (Conrad & Hull, 1964). The locus of this effect is the phonological store, and it results from the increased amount of interference that occurs between memory traces that share overlapping representational (e.g., phonemic) features, relative to those that do not.

The word-length effect simply refers to the fact that lists of words that take more time to articulate—longer words—are more poorly remembered than words that take less time to articulate (Baddeley et al., 1975; Mueller, Seymour, Kieras, & Meyer, 2003). This occurs not only between sets of words that have different numbers of syllables but also in sets of words that are equated for number of syllables but are, nevertheless, unequal in absolute articulatory duration. The effect is explained by assuming that items in the phonological store suffer time-based decay that can only be reversed by way of articulation. Thus, as the articulatory loop cycles through a set of long words, the overall time elapsed between successive iterations will be greater and, therefore, the probability that one of the several items in the store may have (irretrievably) decayed will be consequently increased. This effect, then, is jointly determined by the properties of the rehearsal process (rate of articulation) and that of the phonological store (rate of decay).

The negative effect of articulatory suppression on recall performance is observed when subjects are prevented from using inner speech either during presentation or during a delay inserted before recall. Thus, as articulatory suppression interferes with the articulatory rehearsal process, the mechanisms that are ordinarily used to refresh the items in the phonological store and the system are therefore unable to counteract trace decay, thus leading to a decline in recall performance.

The irrelevant sound effect occurs when the to-be-remembered verbal stimuli are accompanied by a stream of unattended auditory information (Macken, Mosdell, & Jones, 1999; Salame & Baddeley, 1982; Tremblay, Nicholls, Alford, & Jones, 2000). These "irrelevant sounds" need not be in the speaker's native language or even phonemic to be disruptive, provided there is some degree of variation in the sound stream. For instance, a single tone or even white noise does not have an effect, although a changing sequence of tones does cause impairment (Jones, Madden, & Miles, 1992). The locus of the irrelevant sound effect is in the phonological store, where the incoming acoustic information interferes with the to-be-remembered items in the store. Because the presentation of irrelevant visual–verbal information does not have an effect on recall, it is assumed that auditory information has obligatory access to the store, whereas visual–verbal information does not. How, then, does visual–verbal information enter the phonological store? The answer, supported by several lines of evidence (Baddeley, Lewis, & Vallar, 1984; Levy, 1971), is that textual information must first be recoded phonologically before it can enter the store. This recoding process requires the involvement of the articulatory rehearsal process, because subvocalization is necessary to reroute visually derived verbal information into the phonological store. In support of this contention is the finding that articulatory suppression abolishes the phonological similarity effect for visual, but not auditory, presentation. Because auditory information has obligatory access to the store, articulatory suppression has no effect on its deposition within the store. For visual presentation, however, articulatory suppression ties up the rehearsal system, preventing phonological recoding of visual–verbal material and, consequently, blocking subvocally mediated access to the store. The main components of the phonological loop, as well as the manner in which its architecture and functional characteristics account for certain reliable effects observed in studies of verbal STM, have been briefly outlined. It should be made clear that it is not universally accepted that every detail of the phonological loop is perfectly supported by available evidence. For instance, there is a great deal of debate about whether the word-length effect is actually caused by an increase in the absolute spoken duration of the items, or whether it is better explained by, for instance, the phonological complexity of the items (Caplan, Rochon, & Waters, 1992; Mueller et al., 2003).

In summary, the Working Memory model of Baddeley and colleagues describes a system for the maintenance and manipulation of information that is stored in domain-specific memory buffers. Separate cognitive components are dedicated to the functions of storage, rehearsal, and executive control. Informational encapsulation and domain segregation dictates that auditory-verbal and visual information is kept in separate storage subsystems—the phonological loop and the visuospatial sketchpad, respectively. These storage subsystems themselves comprise specialized components for the passive storage of memory traces, which are subject to time and interference-based decay, and for the reactivation of these memory traces by way of simulation, or rehearsal. Thus, storage components represent memory traces but have no internal means of refreshing them, whereas rehearsal processes (e.g., articulatory) have no mnemonic capacity of their own but can reactivate the decaying traces held in temporary stores. How neuroscience has built on the cognitive foundation of the Working Memory model of Baddeley and colleagues to refine our understanding of how information is maintained and manipulated in the brain is discussed here. We see that, in some cases, neuroscientific evidence has bolstered and reinforced aspects of the Working Memory model, whereas in other cases neuroscience has compelled a departure from certain core principles of the Baddeleyan concept.

69.6 NEURAL BASIS OF VERBAL WORKING MEMORY

Research on the neural basis verbal working memory has some unique challenges and is in several ways more difficult to study than, for example, visual or spatial working memory. For example, whereas in visual working memory many of the most influential ideas and concepts have derived from work in nonhuman primates and other animals, verbal working memory is a uniquely human phenomenon; therefore, it has benefited from animal research only in terms of broad neuroscience principles.

Even research on the primary modality relevant to verbal working memory, that of audition, is surprisingly scarce in the monkey literature because of the difficulty in training nonhuman primates to perform delayed response tasks with auditory stimuli, which can take upwards of 15,000 learning trials (Fritz, Mishkin, & Saunders, 2005). Because of the lack of animal work on verbal working memory, neuroscience has often looked to cognitive psychological models, such as the Working Memory model of Baddeley colleagues, for a sensible framework for interpreting neuroscientific evidence about verbal working memory. Of course, many of the classic neurological studies of language were performed before "working memory" existed as a concept, let alone an object of avid neuroscience enquiry. Nevertheless, the core idea of Baddeley and colleagues phonological loop, namely, that verbal information can be reciprocally transferred between auditory and motor components, has a clear cognate in the Wernicke–Lichtehim–Geschwind model of language organization, the core of which originated in Carl Wernicke's 1874 monograph on the aphasias (Geschwind, 1965b; Wernicke, 1874). A fundamental challenge to the cognitive neuroscience of verbal working memory has been to integrate both empirical data and theoretical constructs that have emerged from different subfields—cognitive, neurological, neuropsychological—of enquiry in the brain and behavioral sciences.

69.7 NEUROLOGICAL STUDIES OF LANGUAGE AND VERBAL SHORT-TERM MEMORY

Early neurological investigations of patients with language disturbances, or aphasia, revealed that lesions to specific parts of the cerebral cortex could cause selective deficits in language abilities (Goodglass, 1993). Thus, lesions to the inferior frontal gyrus and surrounding cortex are associated with Broca's aphasia, a disorder that causes severe impairments in speech production. Broca's aphasia is not, however, a disorder of peripheral motor coordination, such as the ability to move and control the tongue and mouth, but rather is a disorder of the ability to plan, program, and access the motor codes required for the production of speech (Goodglass, 1993; Mohr et al., 1978). Lesions to the posterior superior temporal gyrus and surrounding cortex, however, are associated with Wernicke's aphasia, a complex syndrome that is characterized by fluent, but error-filled production, and poor comprehension and perception of speech. A third, less common, syndrome called conduction aphasia, typically caused by lesions in the auditory cortex and posterior Sylvian region (generally less extensive and relatively superior to lesions causing Wernicke's aphasia), is associated with relatively preserved speech perception and comprehension, occasional errors in otherwise fluent spontaneous speech (e.g., phoneme substitutions), and severe difficulties with verbatim repetition of words and sentences (Axer, Keyserlingk, Berks, & Keyserlingk, 2001; Baldo & Dronkers, 2006; Damasio & Damasio, 1980).

From the standpoint of verbal short-term memory, there are a number of important points to be drawn from these three classic aphasic syndromes. First, the neural structures that underlie the perception and production of speech are partly dissociable. Although it is tempting to postulate that posterior temporal lesions primarily affect receptive language functions and anterior lesions affect productive language functions, this is not quite true. Both Wernicke's aphasia and conduction aphasia are caused by posterior lesions, yet only the former is associated with a receptive language disturbance, whereas both syndromes involve a deficit in speech production. Moreover, lesions in and around the inferior frontal gyrus (Broca's area), although normally associated with deficits in speech production (labored, nonfluent speech), can also lead to deficits in the comprehension of grammatically complex sentences and subtle deficits in speech perception (Berndt, Mitchum, & Wayland, 1997; Caplan, Baker, & Dehaut, 1985). Second, the aforementioned disorders affect basic aspects of language processing, such as the comprehension, production, and perception of speech. Third, the classical Wernicke–Lichteim–Geschwind (Geschwind, 1965a) model of language explains each of these three syndromes as disruptions to components of a neuroanatomical network of areas in the inferior frontal and superior temporal cortices that subserve language function. Finally, it should not be surprising that aphasic syndromes associated with poor language performance have also been shown to affect verbal working memory (Burgio & Basso, 1997; De Renzi & Nichelli, 1975). This cannot be taken to necessarily imply that verbal working memory is a function of the core language network. For example, as noted, input to the storage component of the phonological loop theoretically depends on a functioning language system while not participating in core language processes such as speech perception and speech production. Thus, the association between language disturbances and verbal working memory impairment does not prove that the two systems are functionally equivalent.

This is especially true in light of evidence from neuropsychology showing that verbal short-term memory impairment need not accompany a basic impairment to the language faculty. An interpretation consistent with the phonological loop model of verbal working memory is that a selective deficit to verbal short-term memory is caused by lesions that have damaged the phonological store while leaving language perception and production centers intact. Moreover, these patients offer credence to the conceptualization of memory, exemplified by the phonological loop, as a distinct entity in its own right, whose functional purpose is to store and maintain information in mind, rather than to analyze and process incoming sensory input. Although the evidence for such a functional dissociation on the basis of these "short-term memory patients" is intriguing and compelling, the anatomical localization of these lesions presents cognitive neuroscience with a difficult puzzle.

With respect to the "short-term memory patient," the critical question from the standpoint of cognitive neuroscience is how a lesion in the middle of the perisylvian speech center encompassing the temporoparietal area could produce such a pure short-term memory deficit without any collateral impairment in basic online language functioning. One possibility is that the precise location of the brain injury is determinative, so that a particularly focal and well-placed lesion in temporoparietal cortex might spare cortex critical for speech perception and speech production while damaging a region dedicated to the temporary storage of auditory-verbal information. The number of patients described with a selective impairment to auditory-verbal short-term memory is small, however, and the lesion locations that have been reported are not clearly distinguishable from those that might, in another patient, have led to conduction or Wernicke's aphasia (Baldo & Dronkers, 2006; Damasio, 1992; Goodglass, 1993). This would seem to be a question particularly well-suited for high-resolution functional neuroimaging.

69.8 FUNCTIONAL NEUROIMAGING INVESTIGATIONS OF VERBAL WORKING MEMORY

The first study that attempted to localize the components of phonological loop in the brain was that of Paulesu and colleagues (1993). In one task, English letters were visually presented on a monitor and subjects were asked to remember them. In a second task, letters were presented and rhyming judgments were made about them (press a button if letter rhymes with "B"). In a baseline condition, Korean letters were visually presented and subjects were asked to remember them using a visual code. According to the authors' logic, the first task would require the contribution of all the components of the phonological loop—subvocal rehearsal, phonological storage, and executive processes—whereas the second (rhyming) task would only require subvocal rehearsal and executive processes. This reasoning was based on previous research showing that when letters are presented visually (Vallar & Baddeley, 1984), rhyming decisions engage the subvocal rehearsal system, but not the phonological store. Thus, a subtraction of the rhyming condition from the letter-rehearsal condition should isolate the neural locus of the phonological store. First, results were presented for the two tasks requiring phonological processing with the baseline tasks (viewing Korean letters) that did not. Several areas were shown to be significantly more active in the "phonological" tasks, including (in all cases, bilaterally) Broca's area (BA 44/45), the supplementary motor cortex (SMA), the insula, the cerebellum, Brodmann area 22/42, and Brodmann area 40. Subtracting the rhyming condition from the phonological short-term memory condition left a single brain area, Brodmann area 40, which is the neural correlate of the phonological store.

The articulatory rehearsal process recruited a distributed neural circuit that included the inferior frontal gyrus, cerebellum, supplementarty motor area, and premotor cortex. Activation of multiple brain regions during articulatory rehearsal is not surprising given the complexity of the process and the variety of lesion sites associated with a speech production deficit. However, the localization of the phonological store in a single brain region, BA 40 (or the supramarginal gyrus of the parietal lobe), fit well with the idea of a "receptacle" where phonological information is temporarily stored. A number of follow-up positron emission tomography (PET) studies using various tasks and design logic generally replicated the basic finding of the Paulesu et al. study, namely, a fronto-insular-cerebellar network associated with rehearsal processes and a parietal locus for the phonological store (Awh et al., 1996; Jonides et al., 1998; Salmon et al., 1996; Schumacher et al., 1996; Smith & Jonides, 1999). Early PET studies of verbal working memory were interpreted as broadly supporting the architecture of the phonological loop, with storage being subserved by a left parietal lobe structure outside the core language network and articulatory rehearsal associated with regions known to be involved in motor speech planning.

In a trenchant review of these initial PET investigations of verbal working memory work, however, Becker, MacAndrew, and Fiez (1999) questioned whether the localization of the phonological store in the left parietal cortex could be reconciled with the logical architecture of the phonological loop. For instance, as reviewed, a key element of the phonological loop model is that auditory information (whether it be speech, tones, music, or white noise), but not visual information, has *obligatory access* to the phonological store. The reason for this difference is to account for dissociations in memory performance that depend on the modality in which information is presented. For instance, the presentation of distracting auditory information while subjects attempt to retain a list of verbal items in memory impairs performance on tests of recall. In contrast, the presentation of distracting visual information during verbal memory retention has no impact on verbal recall.

This phenomenon—the irrelevant sound effect—is explained by assuming that auditory information always enters the phonological store, but that visual–verbal information only enters the store when it is explicitly subvocalized. Becker et al., however, argued that if auditory information has automatic access to the phonological store, then its "neural correlate" should be active even during *passive auditory perception*. Functional neuroimaging studies of passive auditory listening (e.g., with no memory component), however, do not show activity in the region of the parietal that had been associated in previous studies with phonological storage, but rather show activation that is largely confined to the superior temporal lobe (Binder et al., 2000).

A second difficulty with a parietal locus of the phonological store is that efforts to show verbal mnemonic specificity to maintenance-related activity in the parietal lobe have not been successful (Chein, Ravizza, & Fiez, 2003). Instead, it has been shown that working memory for words, visual objects, and spatial locations all activate the area (Badre, Poldrack, Pare-Blagoev, Insler, & Wagner, 2005; Niendam et al., 2012; Nystrom et al., 2000; Zurowski et al., 2002). Thus, if there were a perfect "neural correlate" of the phonological store, then it must reside within the confines of the auditory cortical zone of the superior temporal cortex.

69.9 EVENT-RELATED fMRI STUDIES OF VERBAL AND AUDITORY WORKING MEMORY

Studies using event-related functional magnetic resonance imaging (fMRI), with its ability to isolate delay-period activity during working memory, have greatly improved our understanding of the neural circuitry associated with verbal working memory maintenance. Postle, Berger, and D'Esposito (1999) showed, with visual–verbal presentation of letter stimuli, that delay-period activity in single subjects was often localized in the posterior superior temporal cortex rather than the parietal lobe that was typically identified in the early PET studies reviewed here. Buchsbaum, Hickok, and Humphries (2001) also used an event-related fMRI paradigm in which, during each trial, subjects were presented with acoustic speech information that they then rehearsed subvocally for 27 s, followed by a rest period. Analysis focused on identifying regions that were responsive *both* during the perceptual phase and the rehearsal phase of the trial. Activation occurred in two regions in the posterior superior temporal cortex, one in the posterior superior temporal sulcus (STS) bilaterally and one along the dorsal surface of the left posterior planum temporale, that is, in the Sylvian fissure at the parietal-temporal boundary (area "Spt"). Notably, whereas the parietal lobe did show delay-period activity, it was unresponsive during auditory stimulus presentation. In a follow-up study, Hickok, Buchsbaum, Humphries, and Muftuler (2003) showed that the same superior temporal regions (posterior STS and Spt) were active both during the perception and delay-period maintenance of short (5 s) musical melodies, suggesting that these posterior temporal storage sites are not restricted to speech-based, or "phonological," information (Figure 69.2).

Several subsequent studies have confirmed the role of Spt in internal rehearsal of musical and speech sequences (Hashimoto, Lee, Preus, McCarley, & Wible, 2010; Hickok, Okada, & Serences, 2009; Koelsch et al., 2009). Acheson, Hamidi, Binder, and Postle (2011) used fMRI to identify posterior temporal regions activated during verbal working memory maintenance, and then used repetitive transcranial magnetic stimulation (TMS) to these sites while subjects performed a rapid paced reading task that involved language production but no memory load. TMS applied to the posterior temporal area significantly interfered with paced reading, arguing for common neural substrate for language production and verbal working memory.

Stevens (2004) and Rama et al. (2004) have shown that memory for voice identity, independent of phonological content (i.e., matching speaker identity as opposed to word identity), selectively activates the mid-STS and the anterior STG of the superior temporal region, but not the more posterior and dorsally situated Spt region. Buchsbaum, Olsen, Koch, and Berman (2005) have further shown that the mid-STS is more active when subjects recall verbal information that is acoustically presented than when the information is visually presented, whereas area Spt shows equally strong delay-period activity for both auditory and visual forms of input (Hashimoto et al., 2010). This finding is supported by regional analyses of structural magnetic resonance imaging (MRI) in large groups of patients with brain lesions that have showed that damage to the STG is most predictive of auditory short-term memory impairment (Koenigs et al., 2011; Leff et al., 2009). Leung and Alain (2011) have also shown dissociation between auditory object and spatial working memory, with the former activating more ventral stream auditory areas and the latter activating the dorsal parietal lobe. Thus, it appears that different regions in the temporoparietal area are attuned to different qualities or features of a verbal stimulus, such as voice information, input modality, phonological content, spatial location, and lexical status (Martin & Freedman, 2001); all of these codes may play a role in the short-term maintenance of verbal information.

Additional support for a feature-based topography of auditory association cortex comes from neuroanatomical tract-tracing studies in both monkeys and

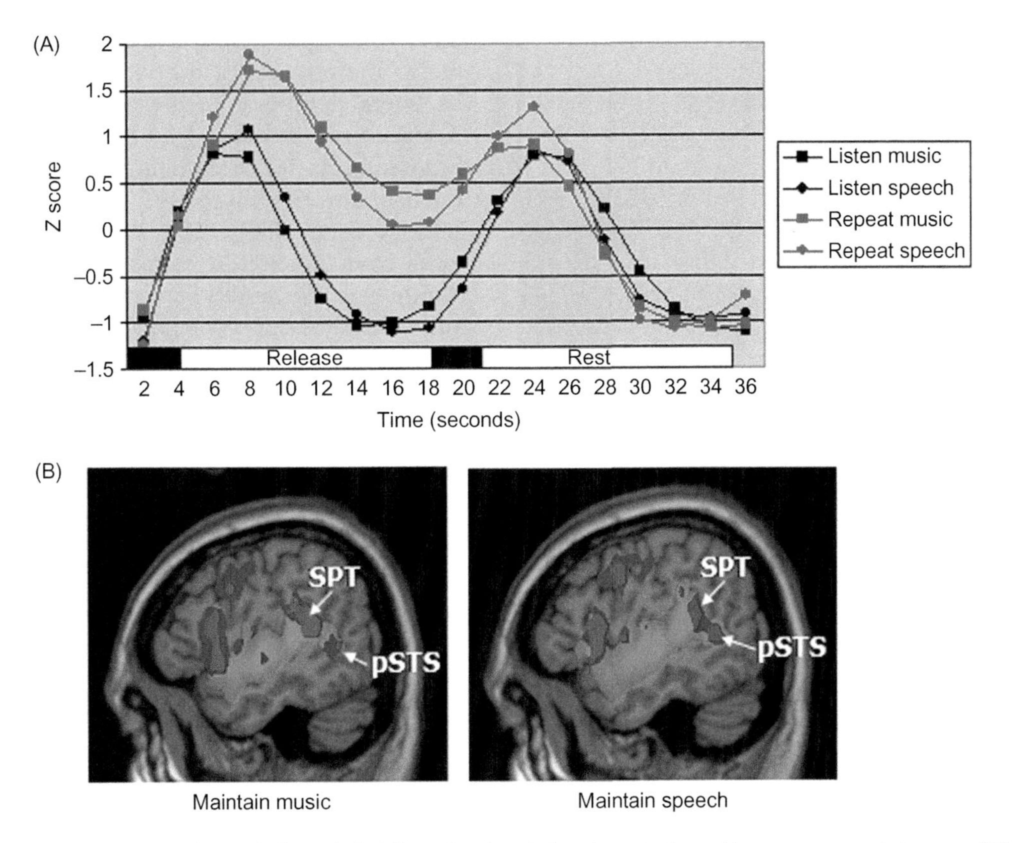

FIGURE 69.2 Main results from the Hickok et al. (2003) study of verbal and musical working memory maintenance. (A) Averaged time-course of activation over the course of a trial in area Spt for speech and music conditions. Timeline at the bottom shows the structure of each trial; black bars indicate auditory stimulus presentation. Red traces indicate activation during rehearsal trials, and black traces indicate activity during listen-only trials in which subjects did not rehearse stimuli at all. (B) Activation maps of the left hemisphere (sagittal slices) showing three response patterns for both music rehearsal (left) and speech rehearsal trials (right): auditory-only responses shown in green; delay-period responses shown in blue; and auditory + rehearsal responses shown in red. Arrows indicate the location of area Spt.

humans that have revealed separate temporo-prefrontal pathways arising along the anterior-posterior axis of the superior temporal region (Bavelier et al., 1998; Rauschecker, 2011; Romanski, 2004; Romanski et al., 1999; Saur et al., 2008). The posterior part of the STG projects to dorsolateral PFC (BA 46, 8), whereas neurons in the anterior STG are more strongly connected to the ventral PFC, including BA 12 and 47. Several authors have suggested, similar to the visual system, a dichotomy between ventral-going auditory-object and a dorsal-going auditory-spatial processing streams (Arnott, Binns, Grady, & Alain, 2004; Rauschecker & Tian, 2000; Tian, Reser, Durham, Kustov, & Rauschecker, 2001). Thus, studies have shown that the neurons in the rostral STG show more selective responses to classes of complex sounds, such as vocalizations, whereas more caudally located regions show more spatial selectivity (Chevillet, Riesenhuber, & Rauschecker, 2011; Rauschecker & Tian, 2000; Tian et al., 2001).

Hickok and Poeppel (2007) have proposed that human speech processing also proceeds along diverging auditory dorsal and ventral streams, although they emphasize the distinction between perception for action, or auditory-motor integration, in the dorsal stream and perception for comprehension in the ventral stream. Buchsbaum et al. (2005) has shown with fMRI time series data that, consistent with the monkey connectivity patterns, the most posterior and dorsal part of the superior temporal cortex, area Spt, shows the strongest functional connectivity with dorsolateral and posterior (premotor) parts of the PFC, whereas the mid portion of the STS is most tightly coupled with BA 12 and BA 47 of the ventrolateral PFC (Figure 69.3). Moreover, gross distinctions between anterior (BA 47) and posterior (BA 44/6), parts of the PFC have been associated with conceptual-semantic and phonological-articulatory aspects of verbal processing (Poldrack et al., 1999; Wagner, Pare-Blagoev, Clark, & Poldrack, 2001). The fMRI studies have also shown that maintenance of verbal-semantic information relies to a greater extent on the anteroventral aspects of the temproal lobe than does mainteance of phonological information (e.g., nonword sequences; Fiebach, Rissman, & D'Esposito, 2006; Shivde & Thompson-Schill, 2004).

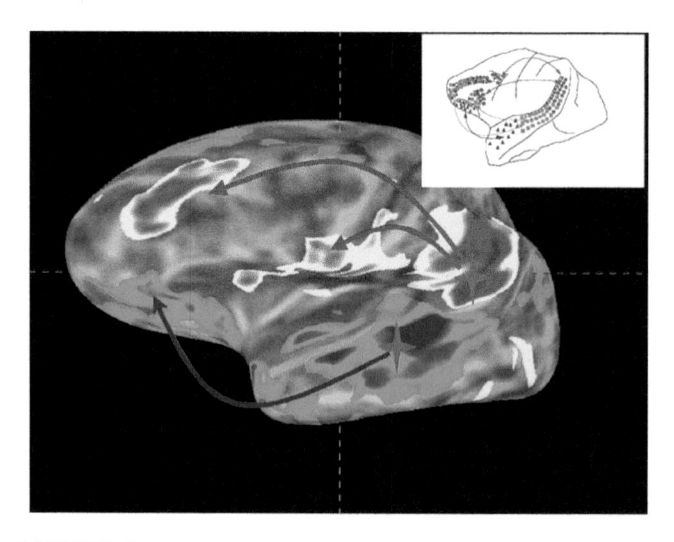

FIGURE 69.3 Map of functional connectivity delay period maintenance of verbal stimuli from Buchsbaum et al. (2005). Seed regions for correlation analysis are denoted by stars located in area Spt and the middle part of the STS. Warm (red, orange, and yellow) colors show areas more strongly correlated with Spt than with STS; cold (blue and green) colors show areas more strongly correlated with STS than Spt. Inset shows temporal-prefrontal connectivity in the monkey (Romanski et al., 1999).

Taken together, findings from functional neuroimaging have shown that the maintenance of verbal information in working memory relies of a distributed network of primarily frontal, temporal, and parietal brain regions. The particular topography of activation depends on the content matter of the to-be-remembered stimuli and/or current task goals. Moreover, activation patterns during memory for musical or tonal sequences overlap considerably with that for phonological sequences. There does not appear to be a single brain region where verbal information is passively stored, as would be predicted by the phonological loop model. Rather, it seems that short-term mnemonic storage is embedded in the very neural structures that support the auditory-verbal perception and production, and that these areas comprise a distributed fronto-temporo-parietal network.

69.10 RECONCILING NEUROPSYCHOLOGICAL AND FUNCTIONAL NEUROIMAGING DATA

The question regarding how a lesion in posterior Sylvian cortex, an area of known importance for online language processing, could occasionally produce an impairment restricted to phonological short-term memory has been posed previously. One solution to this puzzle is that subjects with selective verbal short-term memory deficits from temporoparietal lesions retain their perceptual and comprehension abilities due to the sparing of the ventral stream pathways in the lateral temporal cortex, whereas the preservation of speech production is due to an unusual capacity in these subjects for right-hemisphere control of speech (Buchsbaum & D'Esposito, 2008; Hickok & Poeppel, 2004; Nadeau, 2001). The short-term memory deficit arises from a selective deficit in auditory-motor integration—or the ability to translate between acoustic and articulatory speech codes—a function that is especially taxed during tests of repetition and short-term memory (Buchsbaum & D'Esposito, 2008). Conduction aphasia, the aphasic syndrome most often associated with a deficit in auditory repetition and verbal short-term memory in the absence of any difficulty with speech perception, may reflect a disorder of auditory-motor integration. It has recently been shown that the lesion site most often implicated in conduction aphasia circumscribes area Spt in the posteriormost portion of the superior temporal lobe, and is a link between a disorder of verbal repetition and a region in the brain often implicated in tasks of verbal working memory (Buchsbaum et al., 2011; see Figure 69.4). Thus, impairment in the ability to temporarily store verbal information, as occurs in conduction aphasia, may result from damage to a system, area Spt, which is critical for the interfacing of auditory and motor representations of sound.

69.11 SUMMARY AND CONCLUSION

Elucidation of the cognitive and neural architectures underlying verbal working memory has been an important focus of neuroscience research for much of the past two decades. The emergence of the concept of working memory, with its emphasis on the utilization of the information stored in memory in the service of behavioral goals, has enlarged our understanding and broadened the scope of neuroscience research of short-term memory. Data from numerous studies have been reviewed and have demonstrated that a network of brain regions, principally in the temporal and frontal lobes, is critical for the active maintenance of internal verbal representations. It is clear from these investigations that verbal memory cannot be localized to a single brain region, but rather it is an emergent property of the functional interactions between the frontal and posterior neocortical regions. Numerous questions remain about the neural basis of this complex cognitive system, but studies such as those reviewed in this chapter should continue to provide converging evidence that may provide answers to the many residual questions.

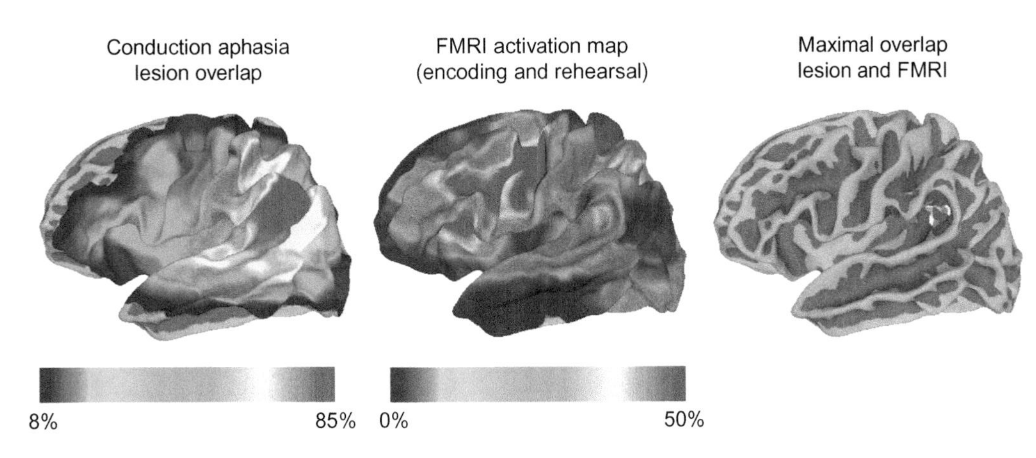

FIGURE 69.4 A comparison of conduction aphasia, phonological working memory in fMRI, and their overlap. The leftmost panel surface shows the regional distribution lesion overlap in patients with conduction aphasia (max is 12/14 or 85% overlap). Middle panel shows the percentage of subjects showing maintenance-related activity in a phonological working memory task. The right panel shows the area of maximal overlap between the lesion and fMRI surfaces (lesion >85% overlap and significant fMRI activity for conjunction of encoding and rehearsal).

References

Acheson, D. J., Hamidi, M., Binder, J. R., & Postle, B. R. (2011). A common neural substrate for language production and verbal working memory. *Journal of Cognitive Neuroscience, 23*(6), 1358–1367.

Arnott, S. R., Binns, M. A., Grady, C. L., & Alain, C. (2004). Assessing the auditory dual-pathway model in humans. *NeuroImage, 22*(1), 401–408.

Awh, E., Jonides, J., Smith, E. E., Schumacher, E. H., Koeppe, R. A., & Katz, S. (1996). Dissociation of storage and rehearsal in working memory: PET evidence. *Psychological Science, 7*, 25–31.

Axer, H., Keyserlingk, A. G. V., Berks, G., & Keyserlingk, D. G. V. (2001). Supra-and infrasylvian conduction aphasia. *Brain and Language, 76*(3), 317–331.

Baddeley, A. (2000). The episodic buffer: A new component of working memory? *Trends in Cognitive Sciences, 4*(11), 417–423. Available from: http://dx.doi.org/10.1016/S1364-6613(00)01538-2.

Baddeley, A., Gathercole, S., & Papagno, C. (1998). The phonological loop as a language learning device. *Psychological Review, 105*(1), 158.

Baddeley, A., Lewis, V., & Vallar, G. (1984). Exploring the articulatory loop. *The Quarterly Journal of Experimental Psychology, 36*(2), 233–252.

Baddeley, A., Papagno, C., & Vallar, G. (1988). When long-term learning depends on short-term storage. *Journal of Memory and Language, 27*(5), 586–595.

Baddeley, A. D. (1986). *Working memory*. Oxford, Oxfordshire/New York, NY: Clarendon Press; Oxford University Press.

Baddeley, A. D., & Hitch, G. J. (1974). Working memory. In G. Bower (Ed.), *The psychology of learning and motivation* (Vol. 7, pp. 47–90). New York, NY: Academic Press.

Baddeley, A. D., Thomson, N., & Buchanan, M. (1975). Word length and the structure of short-term memory. *Journal of Verbal Learning and Verbal Behavior, 14*(6), 575–589.

Baddeley, A. D., & Warrington, E. K. (1970). Amnesia and the distinction between long- and short-term memory. *Journal of Verbal Learning and Verbal Behavior, 9*(2), 176–189.

Badre, D., Poldrack, R. A., Pare-Blagoev, E. J., Insler, R. Z., & Wagner, A. D. (2005). Dissociable controlled retrieval and generalized selection mechanisms in ventrolateral prefrontal cortex. *Neuron, 47*(6), 907–918.

Baldo, J. V., & Dronkers, N. F. (2006). The role of inferior parietal and inferior frontal cortex in working memory. *Neuropsychology, 20*(5), 529–538. 2006-10978-003 [pii] Available from: http://dx.doi.org/10.1037/0894-4105.20.5.529.

Basso, A., Spinnler, H., Vallar, G., & Zanobio, M. (1982). Left hemisphere damage and selective impairment of auditory verbal short-term memory. A case study. *Neuropsychologia, 20*(3), 263–274.

Bavelier, D., Corina, D., Jezzard, P., Clark, V., Karni, A., Lalwani, A., et al. (1998). Hemispheric specialization for English and ASL: Left invariance-right variability. *Neuroreport, 9*(7), 1537–1542.

Becker, J. T., MacAndrew, D. K., & Fiez, J. A. (1999). A comment on the functional localization of the phonological storage subsystem of working memory. *Brain and Cognition, 41*(1), 27–38.

Berndt, R. S., Mitchum, C. C., & Wayland, S. (1997). Patterns of sentence comprehension in aphasia: A consideration of three hypotheses. *Brain and Language, 60*(2), 197–221.

Binder, J. R., Frost, J. A., Hammeke, T. A., Bellgowan, P. S., Springer, J. A., Kaufman, J. N., et al. (2000). Human temporal lobe activation by speech and nonspeech sounds. *Cerebral Cortex, 10*(5), 512–528.

Bjork, R. A., & Whitten, W. B. (1974). Recency-sensitive retrieval processes in long-term free recall. *Cognitive Psychology, 6*(2), 173–189.

Buchsbaum, B. R., Baldo, J., Okada, K., Berman, K. F., Dronkers, N., D'Esposito, M., et al. (2011). Conduction aphasia, sensory-motor integration, and phonological short-term memory—an aggregate analysis of lesion and fMRI data. *Brain and Language, 119*(3), 119–128.

Buchsbaum, B. R., Hickok, G., & Humphries, C. (2001). Role of left posterior superior temporal gyrus in phonological processing for speech perception and production. *Cognitive Science, 25*(5), 663–678.

Buchsbaum, B. R., Olsen, R. K., Koch, P., & Berman, K. F. (2005). Human dorsal and ventral auditory streams subserve rehearsal-based and echoic processes during verbal working memory. *Neuron, 48*(4), 687–697.

Buchsbaum, B. R., & D'Esposito, M. (2008). The search for the phonological store: From loop to convolution. *Journal of Cognitive Neuroscience, 20*(5), 762–778. Available from: http://dx.doi.org/10.1162/jocn.2008.20501.

Burgio, F., & Basso, A. (1997). Memory and aphasia. *Neuropsychologia, 35*(6), 759–766.

Caplan, D., Baker, C., & Dehaut, F. (1985). Syntactic determinants of sentence comprehension in aphasia. *Cognition, 21*(2), 117–175.

Caplan, D., Rochon, E., & Waters, G. S. (1992). Articulatory and phonological determinants of word length effects in span tasks. *The Quarterly Journal of Experimental Psychology A, 45*(2), 177–192.

Chein, J. M., Ravizza, S. M., & Fiez, J. A. (2003). Using neuroimaging to evaluate models of working memory and their implications for language processing. *Journal of Neurolinguistics, 16*, 315–339.

Chevillet, M., Riesenhuber, M., & Rauschecker, J. P. (2011). Functional correlates of the anterolateral processing hierarchy in human auditory cortex. *The Journal of Neuroscience, 31*(25), 9345–9352. 31/25/9345 [pii] Available from: http://dx.doi.org/10.1523/JNEUROSCI.1448-11.2011.

Conrad, R., & Hull, A. J. (1964). Information, acoustic confusion and memory span. *British Journal of Psychology, 55*, 429–432.

Corkin, S. (2002). What's new with the amnesic patient H.M.? *Nature Reviews Neuroscience, 3*(2), 153–160. Available from: http://dx.doi.org/10.1038/nrn726nrn726.

Corkin, S. (2013). *Permanent present tense: The unforgettable life of the amnesic patient* (Vol. 1000). New York, NY: Basic Books.

Cowan, N. (2001). The magical number 4 in short-term memory: A reconsideration of mental storage capacity. *The Behavioral and Brain Sciences, 24*(1), 87–114. discussion 114–185.

Crowder, R. G. (1982). The demise of short-term memory. *Acta Psychologica, 50*(3), 291–323.

Damasio, A. R. (1992). Aphasia. *The New England Journal of Medicine, 326*(8), 531–539.

Damasio, H., & Damasio, A. R. (1980). The anatomical basis of conduction aphasia. *Brain, 103*(2), 337–350.

Davelaar, E. J., Goshen-Gottstein, Y., Ashkenazi, A., Haarmann, H. J., & Usher, M. (2005). The demise of short-term memory revisited: Empirical and computational investigations of recency effects. *Psychological Review, 112*(1), 3–42. 2004-22409-001 [pii] Available from: http://dx.doi.org/10.1037/0033-295X.112.1.3.

De Renzi, E., & Nichelli, P. (1975). Verbal and non-verbal short-term memory impairment following hemispheric damage. *Cortex, 11*(4), 341–354.

Fiebach, C. J., Rissman, J., & D'Esposito, M. (2006). Modulation of inferotemporal cortex activation during verbal working memory maintenance. *Neuron, 51*(2), 251–261. S0896-6273(06)00460-0[pii] Available from: http://dx.doi.org/10.1016/j.neuron.2006.06.007.

Fritz, J., Mishkin, M., & Saunders, R. C. (2005). In search of an auditory engram. *Proceedings of the National Academy of Sciences of the United States of America, 102*(26), 9359–9364.

Geschwind, N. (1965a). Disconnexion syndromes in animals and man. *Brain, 88*, 237–294.

Geschwind, N. (1965b). Disconnexion syndromes in animals and man. *Brain, 88*, 585–644.

Glanzer, M., & Cunitz, A.-R. (1966). Two storage mechanisms in free recall. *Journal of Verbal Learning and Verbal Behavior, 5*, 351–360.

Goodglass, H. (1993). *Understanding aphasia.* San Diego, CA: Academic Press.

Hashimoto, R., Lee, K., Preus, A., McCarley, R. W., & Wible, C. G. (2010). An fMRI study of functional abnormalities in the verbal working memory system and the relationship to clinical symptoms in chronic schizophrenia. *Cerebral Cortex, 20*(1), 46–60. bhp079 [pii] Available from: http://dx.doi.org/10.1093/cercor/bhp079.

Hickok, G., Buchsbaum, B., Humphries, C., & Muftuler, T. (2003). Auditory–motor interaction revealed by fMRI: Speech, music, and working memory in area Spt. *Journal of Cognitive Neuroscience, 15*(5), 673–682.

Hickok, G., Okada, K., & Serences, J. T. (2009). Area Spt in the human planum temporale supports sensory-motor integration for speech processing. *Journal of Neurophysiology, 101*(5), 2725–2732.

Hickok, G., & Poeppel, D. (2004). Dorsal and ventral streams: A framework for understanding aspects of the functional anatomy of language. *Cognition, 92*(1–2), 67–99.

Hickok, G., & Poeppel, D. (2007). The cortical organization of speech processing. *Nature Reviews Neuroscience, 8*(5), 393–402.

James, W. (2011). *The Principles of Psychology (Volume 2 of 2).* Digireads. com Publishing.

Jones, D., Madden, C., & Miles, C. (1992). Privileged access by irrelevant speech to short-term memory: The role of changing state. *The Quarterly Journal of Experimental Psychology A, 44*(4), 645–669.

Jonides, J., Schumacher, E. H., Smith, E. E., Koeppe, R. A., Awh, E., Reuter-Lorenz, P. A., et al. (1998). The role of parietal cortex in verbal working memory. *Journal of Neuroscience, 18*(13), 5026–5034.

Koelsch, S., Schulze, K., Sammler, D., Fritz, T., Muller, K., & Gruber, O. (2009). Functional architecture of verbal and tonal working memory: An FMRI study. *Human Brain Mapping, 30*(3), 859–873. Available from: http://dx.doi.org/10.1002/hbm.20550.

Koenigs, M., Acheson, D. J., Barbey, A. K., Solomon, J., Postle, B. R., & Grafman, J. (2011). Areas of left perisylvian cortex mediate auditory–verbal short-term memory. *Neuropsychologia, 49*(13), 3612–3619.

Leff, A. P., Schofield, T. M., Crinion, J. T., Seghier, M. L., Grogan, A., Green, D. W., et al. (2009). The left superior temporal gyrus is a shared substrate for auditory short-term memory and speech comprehension: Evidence from 210 patients with stroke. *Brain, 132*(Pt 12), 3401–3410. awp273 [pii] Available from: http://dx.doi.org/10.1093/brain/awp273.

Leung, A. W. S., & Alain, C. (2011). Working memory load modulates the auditory "What" and "Where" neural networks. *NeuroImage, 55*(3), 1260–1269.

Levy, B. A. (1971). Role of articulation in auditory and visual short-term memory. *Journal of Verbal Learning and Verbal Behavior, 10*(2), 123–132.

Macken, W. J., Mosdell, N., & Jones, D. M. (1999). Explaining the irrelevant-sound effect: Temporal distinctiveness or changing state? *Journal of Experimental Psychology: Learning, Memory, and Cognition, 25*(3), 810.

Martin, R. C., & Freedman, M. L. (2001). Short-term retention of lexical-semantic representations: Implications for speech production. *Memory, 9*(4), 261–280.

Melton, A. W. (1963). Memory. *Science, 140*(356), 82.

Miller, G. A. (1956). The magical number seven plus or minus two: Some limits on our capacity for processing information. *Psychological Review, 63*(2), 81–97.

Mohr, J. P., Pessin, M. S., Finkelstein, S., Funkenstein, H. H., Duncan, G. W., & Davis, K. R. (1978). Broca aphasia pathologic and clinical. *Neurology, 28*(4), 311–324.

Mueller, S. T., Seymour, T. L., Kieras, D. E., & Meyer, D. E. (2003). Theoretical implications of articulatory duration, phonological similarity, and phonological complexity in verbal working memory. *Journal of Experimental Psychology Learning, Memory, and Cognition, 29*(6), 1353–1380.

Nadeau, S. E. (2001). Phonology: A review and proposals from a connectionist perspective. *Brain and Language, 79*(3), 511–579.

Niendam, T. A., Laird, A. R., Ray, K. L., Dean, Y. M., Glahn, D. C., & Carter, C. S. (2012). Meta-analytic evidence for a superordinate cognitive control network subserving diverse executive functions. *Cognitive, Affective, and Behavioral Neuroscience, 12*(2) 241–268.

Nystrom, L. E., Braver, T. S., Sabb, F. W., Delgado, M. R., Noll, D. C., & Cohen, J. D. (2000). Working memory for letters, shapes, and locations: fMRI evidence against stimulus-based regional organization in human prefrontal cortex. *NeuroImage, 11*(5 Pt 1), 424–446.

Olson, I. R., Page, K., Moore, K. S., Chatterjee, A., & Verfaellie, M. (2006). Working memory for conjunctions relies on the medial temporal lobe. *The Journal of Neuroscience, 26*(17), 4596–4601.

Paulesu, E., Frith, C. D., & Frackowiak, R. S. (1993). The neural correlates of the verbal component of working memory. *Nature, 362* (6418), 342–345.

Poldrack, R. A., Wagner, A. D., Prull, M. W., Desmond, J. E., Glover, G. H., & Gabrieli, J. D. (1999). Functional specialization for semantic and phonological processing in the left inferior prefrontal cortex. *NeuroImage, 10*(1), 15–35.

Postle, B. R., Berger, J. S., & D'Esposito, M. (1999). Functional neuroanatomical double dissociation of mnemonic and executive control processes contributing to working memory performance. *Proceedings of the National Academy of Sciences, 96*(22), 12959–12964.

Postman, L., & Phillips, L.-W. (1965). Short-term temporal changes in free recall. *Quarterly Journal of Experiemental Psychology, 17*, 132–138.

Rama, P., Poremba, A., Sala, J. B., Yee, L., Malloy, M., Mishkin, M., et al. (2004). Dissociable functional cortical topographies for working memory maintenance of voice identity and location. *Cerebral Cortex, 14*(7), 768–780.

Ranganath, C., & Blumenfeld, R. S. (2005). Doubts about double dissociations between short- and long-term memory. *Trends in Cognitive Sciences, 9*(8), 374–380.

Rauschecker, J. P. (2011). An expanded role for the dorsal auditory pathway in sensorimotor control and integration. *Hearing Research, 271*(1), 16–25.

Rauschecker, J. P., & Tian, B. (2000). Mechanisms and streams for processing of "what" and "where" in auditory cortex. *Proceedings of the National Academy of Sciences of the United States of America, 97* (22), 11800–11806.

Repovs, G., & Baddeley, A. (2006). The multi-component model of working memory: Explorations in experimental cognitive psychology. *Neuroscience, 139*(1), 5–21.

Romanski, L. M. (2004). Domain specificity in the primate prefrontal cortex. *Cognitive, Affective and Behavioral Neuroscience, 4*(4), 421–429.

Romanski, L. M., Tian, B., Fritz, J., Mishkin, M., Goldman-Rakic, P. S., & Rauschecker, J. P. (1999). Dual streams of auditory afferents target multiple domains in the primate prefrontal cortex. *Nature Neuroscience, 2*(12), 1131–1136.

Ryan, J. D., & Cohen, N. J. (2004). The nature of change detection and online representations of scenes. *Journal of Experimental Psychology: Human Perception and Performance, 30*(5), 988.

Salame, P., & Baddeley, A. D. (1982). Disruption of short-term memory by unattended speech: Implications for the structure of working memory. *Joruual of Verbal Learning and Verbal Behavior, 21*, 150–164.

Salmon, E., Van der Linden, M., Collette, F., Delfiore, G., Maquet, P., Degueldre, C., et al. (1996). Regional brain activity during working memory tasks. *Brain, 119*(Pt 5), 1617–1625.

Saur, D., Kreher, B. W., Schnell, S., Kümmerer, D., Kellmeyer, P., Vry, M. S., et al. (2008). Ventral and dorsal pathways for language. *Proceedings of the National Academy of Sciences, 105*(46), 18035–18040.

Schumacher, E. H., Lauber, E., Awh, E., Jonides, J., Smith, E. E., & Koeppe, R. A. (1996). PET evidence for an amodal verbal working memory system. *NeuroImage, 3*(2), 79–88.

Scoville, W. B., & Milner, B. (1957). Loss of recent memory after bilateral hippocampal lesions. *Journal of Neurology, Neurosurgery, and Psychiatry, 20*(1), 11–21.

Shallice, T., & Butterworth, B. (1977). Short-term-memory impairment and spontaneous speech. *Neuropsychologia, 15*(6), 729–735.

Shallice, T., & Vallar, G. (1990). The impairment of auditory-verbal short-term storage. In G. Vallar, & T. Shallice (Eds.), *Neuropsychological impairments of short-term memory* (pp. 11–53). Cambridge: Cambridge University Press.

Shallice, T., & Warrington, E. K. (1970). Independent functioning of verbal memory stores: A neuropsychological study. *The Quarterly Journal of Experimental Psychology, 22*(2), 261–273.

Shallice, T., & Warrington, E. K. (1977). Auditory-verbal short-term-memory impairment and conduction aphasia. *Brain and Language, 4*(4), 479–491.

Shivde, G., & Thompson-Schill, S. L. (2004). Dissociating semantic and phonological maintenance using fMRI. *Cognitive, Affective, and Behavioral Neuroscience, 4*(1), 10–19.

Smith, E. E., & Jonides, J. (1999). Storage and executive processes in the frontal lobes. *Science, 283*(5408), 1657–1661.

Squire, L. R., Stark, C. E., & Clark, R. E. (2004). The medial temporal lobe. *Annual Review of Neuroscience, 27*, 279–306. Available from: http://dx.doi.org/10.1146/annurev.neuro.27.070203.144130.

Stevens, A. A. (2004). Dissociating the cortical basis of memory for voices, words and tones. *Brain Research Cognitive Brain Research, 18*(2), 162–171.

Takayama, Y., Kinomoto, K., & Nakamura, K. (2004). Selective impairment of the auditory-verbal short-term memory due to a lesion of the superior temporal gyrus. *European Neurology, 51*(2), 115–117.

Tian, B., Reser, D., Durham, A., Kustov, A., & Rauschecker, J. P. (2001). Functional specialization in rhesus monkey auditory cortex. *Science, 292*(5515), 290–293.

Tremblay, S., Nicholls, A. P., Alford, D., & Jones, D. M. (2000). The irrelevant sound effect: Does speech play a special role? *Journal of Experimental Psychology: Learning, Memory, and Cognition, 26*(6), 1750.

Underwood, B. J. (1957). Interference and forgetting. *Psychological Review, 64*(1), 49–60.

Unsworth, N., & Engle, R. W. (2007). The nature of individual differences in working memory capacity: Active maintenance in primary memory and controlled search from secondary memory. *Psychological Review, 114*(1), 104–132. Available from: http://dx.doi.org/10.1037/0033-295X.114.1.104.

Vallar, G. (2006). Mind, brain, and functional neuroimaging. *Cortex, 42*(3), 402–405; discussion 422-407.

Vallar, G., & Baddeley, A. (1984). Fractionation of working memory: Neuropsychological evidence for a phonological short-term store. *Journal of Verbal Learning and Verbal Behavior, 23*, 151–161.

Vallar, G., Di Betta, A. M., & Silveri, M. C. (1997). The phonological short-term store-rehearsal system: Patterns of impairment and neural correlates. *Neuropsychologia, 35*(6), 795–812.

Wagner, A. D., Pare-Blagoev, E. J., Clark, J., & Poldrack, R. A. (2001). Recovering meaning: Left prefrontal cortex guides controlled semantic retrieval. *Neuron, 31*(2), 329–338. S0896-6273(01)00359-2 [pii].

Warrington, E. K., & Shallice, T. (1969). The selective impairment of auditory verbal short-term memory. *Brain, 92*(4), 885–896.

Waugh, N. C., & Norman, D. A. (1965). Primary memory. *Psychological Review, 72*, 89–104.

Wernicke, C. (1874). The symptom complex of aphasia: A psychological study on an anatomical basis. In R. S. Cohen, & M. W. Wartofsky (Eds.), *Boston studies in the philosophy of science*. Dordecht: D. Reidel Publishing Company.

Wickelgren, W. A. (1968). Sparing of short-term memory in an amnesic patient: Implications for strength theory of memory. *Neuropsychologia, 6*(3), 235–244.

Yee, L. T. S., Hannula, D. E., Tranel, D., & Cohen, N. J. (2014). Short-term retention of relational memory in amnesia revisited: Accurate performance depends on hippocampal integrity. *Frontiers in Human Neuroscience, 8*.

Zurowski, B., Gostomzyk, J., Gron, G., Weller, R., Schirrmeister, H., Neumeier, B., et al. (2002). Dissociating a common working memory network from different neural substrates of phonological and spatial stimulus processing. *NeuroImage, 15*(1), 45–57.

SECTION O

LANGUAGE BREAKDOWN

CHAPTER

70

Language Development in Autism

Morton Ann Gernsbacher[1], *Emily M. Morson*[2] *and Elizabeth J. Grace*[3]

[1]Psychology, University of Wisconsin-Madison, Madison, WI, USA; [2]Psychology and Neuroscience, Indiana University, Bloomington, IN, USA; [3]Special Education, National Louis University, Chicago, IL, USA

The diagnostic criteria for autism in the American Psychiatric Association's (APA) most recent *Diagnostic and Statistical Manual* (APA, 2013) are mute regarding language development. Instead, the most recent diagnostic criteria mention only aspects of sociocultural communication, such as eye contact, facial expressions, or hand gestures. Language itself—the perception and production of speech or writing—is not referenced, and neither is language development.

In contrast, more than 30 years ago, when autism first appeared in the *Diagnostic and Statistical Manual* (APA, 1980), "gross deficits in language development" were considered so dispositive of the autistic phenotype that this diagnostic criterion not only was required but also was one of only a few criteria (Gernsbacher, Dawson, & Goldsmith, 2005; Gernsbacher, Geye, & Ellis Weismer, 2005). Seven years later, the focus moved away from how language developed to how language was produced. "Marked abnormalities in the production of speech, including volume, pitch, stress, rate, rhythm, and intonation" and "marked abnormalities in the form or content of speech, including stereotyped and repetitive use of speech" appeared as diagnostic criteria. However, these two criteria were among more than a dozen other criteria, and they need not be met to warrant a diagnosis (APA, 1987).

Another 7 years passed, and language development recurred in the diagnostic criteria ("delay in, or total lack of, the development of spoken language"), but again as only one of numerous criteria for which only a subset needed to be met (APA, 1994, 2000). Thus, abnormal language development at one time defined autism and then became an optional means for making a diagnosis. Now, it no longer figures into contemporary diagnostic criteria.

In this chapter, we review recent empirical research on language development in autism. To paint a contemporary picture, we restrict our review to studies published in the twenty-first century. We conclude that language development in autism is often delayed, but not deviant; that a delay in language development is not unique to autism; and that language development in autism is remarkably heterogeneous and variable.

70.1 DELAY IN AUTISTIC LANGUAGE DEVELOPMENT

Many studies that have measured language abilities at one discrete point in time have suggested that autistic language development is delayed compared with typical language development. The most consistently reported delays are in producing and expressing language, what is often referred to as productive or expressive language. For example, autistic children have been reported to be delayed in speaking their first words (Charman, Drew, Baird, & Baird, 2003; Matson, Mahan, Kozlowski, & Shoemaker, 2010), speaking their first phrases (e.g., *blue car*, Grandgeorge et al., 2009; Kenworthy et al., 2012; Pry, Peterson, & Baghdadli, 2011), and speaking their first grammatical utterances (e.g., *go bye-bye*) or sentences (Anderson et al., 2007; Wodka, Mathy, & Kalb, 2013).

Therefore, studies that have measured the size of young, autistic children's expressive vocabularies at specific points during development have often reported that young, autistic children have smaller expressive vocabularies than typically developing children of the same age (Charman et al., 2003; Fulton & D'Entremont, 2013; Kover, McDuffie, Hagerman, & Abbeduto, 2013; Luyster, Kadlec, Carter, & Tager-Flusberg, 2008; Luyster, Lopez, & Lord, 2007; Miniscalco, Fränberg, Schachinger-Lorentzon, & Gillberg, 2012; Sandercock, 2013; Stone & Yoder, 2001).

Neurobiology of Language. DOI: http://dx.doi.org/10.1016/B978-0-12-407794-2.00070-5

Other studies that have examined other expressive language skills, for instance, expressing relations such as big and little, correctly producing grammatical morphemes for plurals and verb tenses, and using rising intonation when asking questions, have also reported that autistic children are less skilled than typically developing children (Fulton & D'Entremont, 2013; Hudry et al., 2010; Sigman & McGovern, 2005; Sutera et al., 2007; Vanvuchelen, Roeyers, & DeWeerdt, 2011; Walton & Ingersoll, 2013).

That young, autistic children are often characterized by smaller expressive vocabularies should be of little surprise given that delays in the number of words and phrases children are saying are some of the most notable 'red flags' for autism (Baird, Cass, & Slonims, 2003; Filipek et al., 1999). Delays in early expressive language are also the primary concern that motivates parents to seek diagnostic evaluation of their children (Agin, 2004).

Regarding receptive language (the ability to understand language rather than produce it), reliable measurements are more difficult to obtain, particularly for very young children. Consequently, valid conclusions are more difficult to draw. One problem is that measuring receptive language in young children often relies heavily on parent report measures, such as the MacArthur-Bates Communicative Development Inventory (Fenson et al., 1993, 2007), which asks parents to check-off which of a list of words or phrases the parents believe their children can understand.

There are multiple reasons why parents might misestimate the number of words that their children understand (Feldman et al., 2000); for children who respond atypically, such misestimates are more likely (Akhtar & Gernsbacher, 2007, 2008; Bruckner, Yoder, Stone, & Saylor, 2007). Nonetheless, many studies using the MacArthur-Bates Communicative Development Inventory have reported that young, autistic children are delayed in their receptive language development (Charman et al., 2003; Fulton & D'Entremont, 2013; Luyster et al., 2008, 2007; Maljaars, Noens, Scholte, & Van Berckelaer-Onnes, 2012; Miniscalco et al., 2012; Paul, Chawarska, Cicchetti, & Volkmar, 2008; Paul, Chawarska, Fowler, Cicchetti, & Volkmar, 2007; Vanvuchelen et al., 2011).

Using more objective measures, other studies have also reported that autistic children are delayed in their receptive language development. These studies have used standardized assessments, such as the Reynell Language Development scale (RLDS: Miniscalco et al., 2012; Reynell & Gruber, 1990; Vanvuchelen et al., 2011), the Peabody Picture Vocabulary Test or British Picture Vocabulary Scale (PPVT: Dunn & Dunn, 1997; BPVS: Dunn, Dunn, Whetton, & Burley, 1997; Grigorenko et al., 2002; Howlin, 2003; Kover et al., 2013), the Clinical Evaluations of Language Fundamentals (Aman et al., 2004; CELF: Semel, Wiig, & Secord, 1992, 1995, 2000, 2006; Sigman & McGovern, 2005; Wisdom, Dyck, Piek, Hay, & Hallmayer, 2007), the Mullen Scale of Early Learning (Luyster et al., 2008; MSEL: Mullen, 1995; Sutera et al., 2007; Swensen, Kelley, Fein, & Naigles, 2007), the Preschool Language Scale (Hudry et al., 2010; Jasmin et al., 2009; Walton & Ingersoll, 2013; PLS: Zimmerman, Steiner, & Pond, 1992, 2002), or the Psychoeducational Profile-3 (Fulton & D'Entremont, 2013; PEP-3: Schopler, Lansing, Reichler, & Marcus, 2005).

However, other studies that have also measured language abilities at one discrete point in time have not shown that autistic language development differed from typical language development. For example, autistic toddlers were not reported to differ from typically developing toddlers in the number of words that they produced (Goodwin, Fein, & Naigles, 2012); autistic teenagers did not differ from typically developing teenagers in the number of words that they understood (Åsberg, 2010; Henderson, Clarke, & Snowling, 2011; Paul, Augustyn, Klin, & Volkmar, 2005); and autistic children, teens, and adults did not differ from typically developing participants in the quality or quantity of their written language production, be it number of words, length of words, length of sentences, or complexity of sentences (Troyb, 2011).

To summarize, several contemporary studies have suggested that autistic language development is delayed compared with typical language development. These studies have suggested delays in both expressive (producing) language and receptive (understanding) language. However, other studies have not shown that autistic language development is delayed compared with typical language development. Thus, delayed language development is a common, but not a universal, characteristic of autism. In fact, there is good evidence that language develops independently from autistic traits. For example, in a recent large-scale study involving 3,000 pairs of twins, amount of language development was both phenotypically and genetically unrelated to degree of autistic traits (so-called severity; Taylor et al., 2014). The empirically demonstrated independence between language and autism underlies not only the variability in research findings but also the variability in autistic language development.

70.2 HETEROGENEITY AND VARIABILITY IN AUTISTIC LANGUAGE DEVELOPMENT

Several studies have reported observing language delays in some subgroups of autistic participants but not in others (Kjelgaard & Tager-Flusberg, 2001).

For example, the majority of a sample of autistic children achieved normative size expressive vocabularies, but 15% of that sample were delayed between one and two standard deviations below normal (Jones & Schwartz, 2009). Most of a sample of autistic teenagers did not differ from a sample of typically developing teenagers in their ability to read, but one-third of the sample did (Åsberg & Dahlgren Sandberg, 2012). Half of a sample of autistic children and teenagers produced scores on the BPVS in the average range, but one-quarter of the sample performed one to two standard deviations below average, and another one-quarter of the sample performed more than two standard deviations above average (McCann, Peppé, Gibbon, O'Hare, & Rutherford, 2005).

Autistic language development often demonstrates extreme variability. For example, in a large sample of autistic toddlers with a wide range of reported IQ scores, more than three-fourths of the sample had spoken their first words before 18 months, which is within the range of typical development. However, a bit more than 5% of the sample had still not spoken their first words at 6 years of age (Wilson et al., 2003), which is far beyond the range of typical development. In two large samples of autistic preschool-age children, some children scored 2 years below age level on measures of expressive and receptive language, whereas other children scored nearly 2 years above age level (Fulton & D'Entremont, 2013; Hudry et al., 2010).

In a sample of autistic children and teenagers whose receptive vocabulary was, on average, in the normal range, some autistic children scored as low as four standard deviations below normal, whereas other autistic children scored as high as two standard deviations above normal (Nation, Clarke, Wright, & Williams, 2006). Similarly, in a sample of autistic preschool-age children, their receptive vocabulary ranged from four standard deviations below normal to two standard deviations above normal (Jasmin et al., 2009). In a sample of autistic school-age children, their expressive vocabulary also ranged from four standard deviations below normal to two standard deviations above normal (Joseph, McGrath, & Tager-Flusberg, 2005). In a sample of autistic teens, their reading vocabulary ranged from three standard deviations below normal to one standard deviation above normal (Ricketts, Jones, Happé, & Charman, 2013).

To summarize, several studies have reported observing language delays in some subgroups of autistic participants but not others; autistic language development often demonstrates extreme variability. In large samples of autistic participants, it is not unusual to find scores on various language measures that range from as low as two standard deviations below the norm to as high as two standard deviations above the norm.

70.3 TRAJECTORIES OF LANGUAGE DEVELOPMENT

Within the first few years of life, autistic children's language development trajectories have been reported to be flatter than that of typically developing children (or children with other developmental disabilities; Landa & Garrett-Mayer, 2006). However, examining development over a longer period of time sometimes shows that autistic language development trajectories can subsequently become steeper. After an initial delay, there was accelerated growth.

For example, autistic boys of grade-school age began with lower receptive, expressive, and overall language skills than typically developing boys. However, over a 9-year period, the autistic boys' language skills improved, on average, by 10% per year, whereas the typically developing boys' language skills improved by only 1.6% per year (Cariello et al., 2011). Autistic children's language development continued on an upward trajectory at age 9, whereas nonautistic children, who were characterized by other types of atypical development, began to plateau (Anderson et al., 2007).

Vocabulary development continued to improve through adulthood for a sample of autistic adults first studied during grade school and followed-up in their early 20s; however, for a comparable sample of adults with language impairment, vocabulary development stagnated (Mawhood, Howlin, & Rutter, 2000). A steeper trajectory of language development is, of course, expected if the starting point is low—but only if the final measurement point also indicates improvement.

Figure 70.1 (modified from data reported by Dockrell, Ricketts, Palikara, Charman, & Lindsay, 2010) illustrates the growth of expressive language (Figure 70.1A) and receptive language (Figure 70.1B) during a 4-year period for a sample of more than 100 autistic children of grade-school age. Illustrated along with the autistic grade-school children is a matched sample of more than 200 grade-school children who were not autistic but had language disabilities. The data presented are the two groups' performances on expressive and receptive measures from the Clinical Evaluation of Language Fundamentals (Semel et al., 2006), presented in z-scores, based on a norms-based mean of zero and a norms-based standard deviation of one.

As Figure 70.1 illustrates, the older the autistic grade-school children, the better their expressive and receptive language skills. In contrast, for the language-disabled children who were not autistic, neither their expressive nor their receptive language skills improved with age. Although these data are cross-sectional, they represent a steep trajectory of increasing language development for the autistic children, such

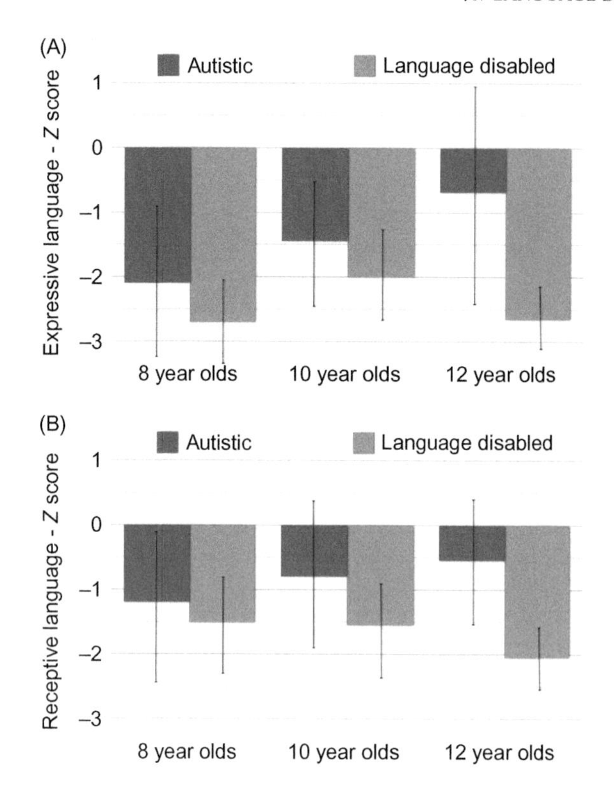

FIGURE 70.1 Growth of expressive language (A) and receptive language (B) during a 4-year period for a sample of more than 100 autistic children of grade-school age. *Modified from data reported by Dockrell et al. (2010) with permission from the publisher.*

that by the end point of measurement the autistic children's language skills did not differ from normal.

However, trajectories of language development in autism, similar to static measures of language ability in autism, show great individual variability. Figure 70.2 (modified from data reported by Smith, Mirenda, & Zaidman-Zait, 2007) illustrates individual trajectories in expressive vocabulary development for 35 autistic preschool-age children. Each child was assessed (by the MacArthur-Bates Communicative Development Inventory, Fenson et al., 1993) at four time points over a 24-month period. Although children were enrolled into the study only if their expressive vocabulary was less than 60 words, the children showed a wide range of vocabulary growth over the subsequent 2-year period, as Figure 70.2 demonstrates.

Some children, indicated by blue lines in Figure 70.2, showed a steep rate of expressive vocabulary development, with expressive vocabularies of nearly 700 words at the last time point. Other children, indicated by green lines in Figure 70.2, showed a steady increase in vocabulary development, with expressive vocabularies of 400 to a bit more than 600 words at the last time point. Still other children, indicated by gold lines in Figure 70.2, showed a slow increase in vocabulary development. The remaining children,

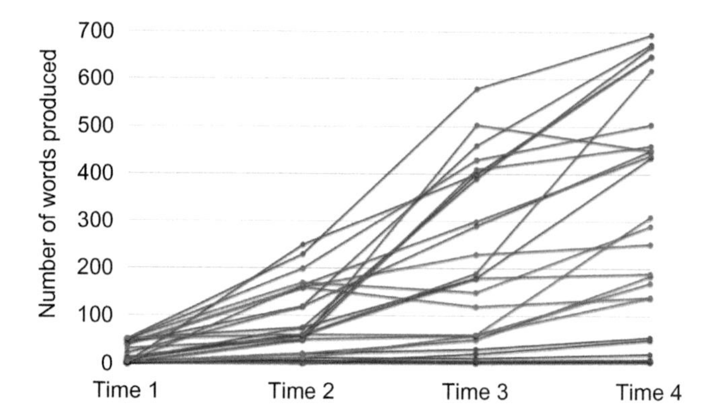

FIGURE 70.2 Individual trajectories in expressive vocabulary development for 35 autistic preschool-age children. *Modified from data reported by Smith et al. (2007) with permission from the publisher.*

indicated by red lines, showed a flat rate of vocabulary development, with little change in the number of words they produced over the 2-year period.

The four color-coded clusters illustrated in Figure 70.2 were produced via statistical cluster analysis (Smith et al., 2007) based on the trajectory of the individual children's vocabulary development. However, the average age of the children within each of the four clusters did not differ; neither did their level of cognitive development nor their degree of autistic traits. Most markedly, all children were undergoing the same therapy. Therefore, these data show just how variable the course of language development can be for autistic children, even when the children are all receiving 15–20 h per week of early behavioral intervention, including speech-language therapy.

Very few studies of autistic language development have followed participants through adulthood; indeed, very few studies have followed autistic participants into adulthood at all regarding any type of development (Dawson, Mottron, & Gernsbacher, 2008). However, one of the few longitudinal studies that has extended through adulthood followed 29 autistic adults who had marked language delays in childhood and 35 autistic adults who did not. In adulthood, the two groups did not differ in either their expressive or their receptive vocabulary (Howlin, 2003).

To summarize, within the first few years of life, autistic children's language development trajectories have been reported to be flatter than that of typically developing children (or children with other developmental disabilities); however, examining development over a longer period of time sometimes shows that autistic language development trajectories can also be steeper. After an initial delay, there is accelerated growth. However, trajectories of language development in autism, similar to static measures of language ability in autism, show great individual variability.

70.4 LANGUAGE DELAY VERSUS LANGUAGE DEVIANCE

Evidence suggests that language development can be delayed in autism, but is it deviant? Does it proceed in the same general sequence? Are there qualitative differences? Most studies that have investigated specific nuances of language development, rather than gross measures on standardized tests, have suggested that autistic children's language development proceeds in the same order and is qualitatively similar in its developmental course to the language development of nonautistic children at the same stages of development.

For example, although autistic toddlers and preschool-age children might understand fewer words than age-matched typically developing children, when compared with younger, typically developing children with the same size vocabulary, autistic children's receptive vocabularies contain the same relative proportion of words from different grammatical categories (e.g., nouns, verbs, pronouns, and the like) and the same relative proportion of words from various semantic categories (e.g., people, games and routines, body parts, sound effects, and the like; Charman et al., 2003).

Similarly, although autistic toddlers and preschool-age children might produce fewer words than age-matched typically developing children, when compared with younger, typically developing children with the same vocabulary size, there is substantial overlap in autistic and typically developing children's most frequently spoken words (Wicklund, 2012). When autistic toddlers are compared with nonautistic toddlers with delays in their language development ("late talkers"), the two groups are just as likely to produce words from a range of grammatical and semantic categories, including emotion terms, and the autistic toddlers do not differ from the nonautistic late talkers in the complexity of their grammatical utterances (Ellis Weismer et al., 2011).

When autistic preschool-age children are compared with nonautistic preschool-age children who have other developmental disabilities, the autistic children do not differ from the nonautistic children in their correct production of noun phrases, sentence structures, plurals, singulars, or past-tense inflections (Park, Yelland, Taffe, & Gray, 2012).

Norbury and colleagues have demonstrated in several studies that when compared with nonautistic children of grade-school age, autistic children do not differ from nonautistic children in their sequential achievement of a variety of language comprehension processes. For instance, autistic children of grade-school age acquire the understanding of idioms (e.g., "it's raining cats and dogs") at the same time as language ability–matched nonautistic children do (Norbury, 2004; see also Gernsbacher & Pripas-Kapit, 2012).

Autistic children acquire the understanding of metaphors (e.g., "[Because] John spent too long in the swimming pool, he was a prune") at the same time as language ability–matched nonautistic children (Norbury, 2005b). Autistic children acquire the ability to draw inferences from stories (Norbury & Bishop, 2002; see also Young, Diehl, Morris, Hyman, & Bennetto, 2005), negotiate ambiguities in language (Norbury, 2005a), and structure the stories that they tell (Norbury & Bishop, 2003) at the same time as language-ability matched nonautistic children.

Norbury's studies also illustrate the independence between language development and autistic traits. In these studies, Norbury and colleagues typically assemble four participant groups: autistic children with language impairment; autistic children without language impairment; nonautistic children with language impairment; and nonautistic children without language impairment. By definition, the autistic participants differ from the nonautistic participants in their degree of autistic traits. For instance, in Norbury's (2005a) study, the difference in degree of autistic traits between the autistic and nonautistic participants is more than two standard deviations (which is beyond the effect size of the difference between average height of men versus women). Thus, the autistic children in these studies clearly have more autistic traits than the nonautistic children and most likely are "autistic enough."

However, in each of Norbury and colleagues' studies, it is the participants' degree of language impairment, not their degree of autistic traits, that predicts their ability to understand idioms (Norbury, 2004), understand metaphors (Norbury, 2005b), draw inferences from stories (Norbury & Bishop, 2002), negotiate ambiguities in language (Norbury, 2005a), and structure stories that they tell (Norbury & Bishop, 2003). Thus, autistic children do not differ from language-matched nonautistic children in their sequential achievement of important language comprehension processes.

Historically it was assumed that unusual, perhaps aberrant, features of language development characterized autism, such as echolalia (repeating words and expressions verbatim) and pronoun reversal (using *you* when *I* is intended). More recently, some researchers have suggested another aberrant feature might characterize autistic language development. Expressive language might proceed abnormally ahead of receptive language (i.e., autistic children say more than they understand). However, none of these putatively aberrant characteristics are empirically reliable or universal among autistic children, as Gernsbacher, Morson, and Grace (in press) recently argued.

Another characteristic of autistic language development that is sometimes assumed to be specific to autism is regression, or loss, of language skills. Unfortunately, codifying language loss is complicated (is it a consistent loss or a fluctuating loss?), and measuring language loss is difficult, particularly if measurements are not taken prospectively (Lord, Shulman, & DiLavore, 2004). Despite classification and measurement difficulties, two patterns appear to be consistent in the data on autistic language loss.

First, autistic children who seemed to lose language are those whose early language was developing on time or with little delay (Baird et al., 2008; Pickles et al., 2009). For example, autistic toddlers who were coded as having lost language had produced their first words at approximately 1 year of age, whereas autistic toddlers who were not coded as having lost language had not produced their first words until 2 years of age (Pickles et al., 2009).

Second, autistic children who seemed to lose language do not progress as quickly in subsequent language development as autistic children who do not seem to lose language. For example, at age 6, many of the autistic children who seemed to lose language earlier were still using single words, whereas other autistic children had advanced to producing sentences (Bernabei, Cerquiglini, Cortesi, & D'Ardia, 2007).

To summarize, most studies that have investigated specific nuances of language development, rather than gross measures on standardized tests, have suggested that autistic children's language development proceeds in the same order and is qualitatively similar in its developmental course to the language development of nonautistic children at the same stages of development. Although it has been assumed that aberrant features of language development characterize autism, such as echolalia and pronoun reversal, neither these features nor language loss is unique to autism.

References

Agin, M. C. (2004). The "late talker"—When silence isn't golden. *Contemporary Pediatrics, 21*, 22–32.

Akhtar, N., & Gernsbacher, M. A. (2007). Joint attention and vocabulary learning: A critical look. *Linguistics and Language Compass, 1* (3), 195–207.

Akhtar, N., & Gernsbacher, M. A. (2008). On privileging the role of gaze in infant social cognition. *Child Development Perspectives, 2*, 60–66.

Aman, M. G., Novotny, S., Samango-Sprouse, C., Lecavalier, L., Leonard, E., Gadow, K. D., et al. (2004). Outcome measures for clinical drug trials in autism. *CNS Spectrums, 9*, 36–47.

American Psychiatric Association (1980). *Diagnostic and statistical manual of mental disorders* (3rd ed.). Washington, DC: American Psychiatric Association.

American Psychiatric Association (1987). *Diagnostic and statistical manual of mental disorders* (3rd ed., Revised). Washington, DC: American Psychiatric Association.

American Psychiatric Association (1994). *Diagnostic and statistical manual of mental disorders* (4th ed.). Washington, DC: American Psychiatric Association.

American Psychiatric Association (2000). *Diagnostic and statistical manual of mental disorders* (4th ed., Text Revision ed.). Washington, DC: American Psychiatric Association.

American Psychiatric Association (2013). *Diagnostic and statistical manual of mental disorders* (5th ed.). Arlington, VA: American Psychiatric Publishing.

Anderson, K., Lord, C., Risi, S., DiLavore, P. S., Shulman, C., Thurm, A., et al. (2007). Patterns of growth in verbal abilities among children with autism spectrum disorder. *Journal of Counseling and Clinical Psychology, 75*, 594–604.

Åsberg, J. (2010). Patterns of language and discourse comprehension skills in school-aged children with autism spectrum disorders. *Scandinavian Journal of Psychology, 51*, 534–539.

Åsberg, J., & Dahlgren Sandberg, A. (2012). Dyslexic, delayed, precocious, or just normal? Word reading skills of children with autism spectrum disorders. *Journal of Research in Reading, 35*, 20–31.

Baird, G., Cass, H., & Slonims, V. (2003). Clinical review: Diagnosis of autism. *British Medical Journal, 327*, 488–493.

Baird, G., Charman, T., Pickles, A., Chandler, S., Loucas, T., Meldrum, D., et al. (2008). Regression, developmental trajectory and associated problems in disorders in the autism spectrum: The SNAP study. *Journal of Autism and Developmental Disorders, 38*, 1827–1836.

Bernabei, P., Cerquiglini, A., Cortesi, F., & D'Ardia, C. (2007). Regression versus no regression in the autistic disorder: Developmental trajectories. *Journal of Autism and Developmental Disorders, 37*, 580–588.

Bruckner, C., Yoder, P., Stone, W., & Saylor, M. (2007). Construct validity of the MCDI-I receptive vocabulary scale can be improved: Differential item functioning between toddlers with autism spectrum disorders and typically developing infants. *Journal of Speech, Language, and Hearing Research, 50*, 1631–1638.

Cariello, A., Bigler, E. D., Tolley, S. E., Prigge, M. D., Neeley, E. S., Lange, N, et al. (2011, May). *A longitudinal look at expressive, receptive, and total language development in individuals with autism spectrum disorders.* Paper presented at the International Meeting for Autism Research, San Sebastian, Spain.

Charman, T., Drew, A., Baird, C., & Baird, G. (2003). Measuring early language development in preschool children with autism spectrum disorder using the MacArthur Communicative Development Inventory (Infant Form). *Journal of Child Language, 30*, 213–236.

Dawson, M., Mottron, L., & Gernsbacher, M. A. (2008). Learning in autism. In H. L. Roediger, III (Ed.), Cognitive psychology of memory *(pp. 759–772, Vol. 2, Learning and memory: A comprehensive reference, J. Byrne Editor).* Oxford: Elsevier.

Dockrell, J., Ricketts, J., Palikara, O., Charman, T., & Lindsay, G. (2010). *Profiles of need and provision for children with language impairments and autism spectrum disorders in mainstream schools: A prospective study (Department for Education No. DFE-RR247-BCRP9).* London: HMSO.

Dunn, L. M., & Dunn, L. M. (1997). *Peabody picture vocabulary test* (3rd ed.). Circle Pines, MN: American Guidance Service.

Dunn, L. M., Dunn, L. M., Whetton, C., & Burley, J. (1997). *The British picture vocabulary scale-II.* Windsor, UK: NFER-Nelson.

Ellis Weismer, S., Gernsbacher, M. A., Stronach, S., Karasinski, K., Eernisse, E. R., Venker, C. E., et al. (2011). Lexical and grammatical skills in toddlers on the autism spectrum compared to late talking toddlers. *Journal of Autism and Developmental Disorders, 41*, 1065–1075.

Feldman, H. M., Dollaghan, C. A., Campbell, T. F., Kurs-Lasky, M., Janosky, J. E., & Paradise, J. L. (2000). Measurement properties of the MacArthur communicative development inventories at ages one and two years. *Child Development, 71*, 310–322.

Fenson, L., Dale, P. S., Reznick, J. S., Thal, D., Bates, E., Hartung, J. P, et al. *The MacArthur communicative development inventories: User's guide and technical manual*. Baltimore, MD: Paul H. Brookes.

Fenson, L., Marchman, V. A., Thal, D. J., Dale, P. S., Reznick, J. S., & Bates, E. (2007). *The MacArthur–Bates communicative development inventories: User's guide and technical manual* (2nd ed.). Baltimore, MD: Paul Brookes Publishing.

Filipek, P. A., Accardo, P. J., Baranek, G. T., Cook, E. H., Jr., Dawson, G., Gordon, B., et al. (1999). The screening and diagnosis of autism spectrum disorders. *Journal of Autism and Developmental Disorders, 29*, 439–484.

Fulton, M. L., & D'Entremont, B. (2013). Utility of the psychoeducational profile-3 for assessing cognitive and language skills of children with autism spectrum disorder. *Journal of Autism and Developmental Disorders, 43*, 2460–2471.

Gernsbacher, M. A., Dawson, M., & Goldsmith, H. H. (2005). Three reasons not to believe in an autism epidemic. *Current Directions in Psychological Science, 14*, 55–58.

Gernsbacher, M. A., Geye, H. M., & Ellis Weismer, S. (2005). The role of language and communication impairments within autism. In P. Fletcher, & J. C. Miller (Eds.), *Language disorders and developmental theory* (pp. 73–93). Philadelphia, PA: John Benjamins.

Gernsbacher, M.A., Morson, E. M., & Grace, E. J. (in press). Clinical linguistics: Speech and language. *Annual Review of Linguistics, 1*.

Gernsbacher, M. A., & Pripas-Kapit, S. (2012). Who's missing the point? A commentary on claims that autistic persons have a specific deficit in figurative language comprehension. *Metaphor and Symbol, 27*, 93–105.

Goodwin, A., Fein, D., & Naigles, L. R. (2012). Comprehension of wh-questions precedes their production in typical development and autism spectrum disorders. *Autism Research, 5*, 109–123.

Grandgeorge, M., Hausberger, M., Tordjman, S., Deleau, M., Lazartigues, A., & Lemonnier, E. (2009). Environmental factors influence language development in children with autism spectrum disorders. *PLoS ONE, 4*.

Grigorenko, E. L., Klin, A., Pauls, D. L., Senft, R., Hooper, C., & Volkmar, F. (2002). A descriptive study of hyperlexia in a clinically referred sample of children with developmental delays. *Journal of Autism and Developmental Disorders, 32*, 3–12.

Henderson, L. M., Clarke, P. J., & Snowling, M. J. (2011). Accessing and selecting word meaning in autism spectrum disorder. *Journal of Child Psychology and Psychiatry, 52*, 964–973.

Howlin, P. (2003). Outcome in high-functioning adults with autism with and without early language delays: Implications for the differentiation between autism and asperger syndrome. *Journal of Autism and Developmental Disorders, 33*, 3–13.

Hudry, K., Leadbitter, K., Temple, K., Slonims, V., McConachie, H., Aldred, C., et al. (2010). Preschoolers with autism show greater impairment in receptive compared with expressive language abilities. *International Journal of Communication Disorders, 45*, 681–690.

Jasmin, E., Couture, M., McKinely, P., Reid, G., Fombonne, E., & Gisel, E. (2009). Sensori-motor and daily living skills of preschool children with autism spectrum disorders. *Journal of Autism and Developmental Disorders, 39*, 231–241.

Jones, C. D., & Schwartz, I. S. (2009). When asking questions is not enough: An observational study of social communication differences in high functioning children with autism. *Journal of Autism and Developmental Disorders, 39*, 432–443.

Joseph, R. M., McGrath, L. M., & Tager-Flusberg, H. (2005). Executive dysfunction and its relation to language ability in verbal school-age children with autism. *Developmental Neuropsychology, 27*, 361–378.

Kenworthy, L., Wallace, G. L., Powell, K., Anselmo, C., Martin, A., & Black, D. O. (2012). Early language milestones predict later language, but not autism symptoms in higher functioning children with autism spectrum disorders. *Research in Autism Spectrum Disorders, 6*, 1194–1202.

Kjelgaard, M. M., & Tager-Flusberg, H. (2001). An investigation of language impairment in autism: Implications for genetic subgroups. *Language and Cognitive Processes, 16*, 287–308.

Kover, S. T., McDuffie, A. S., Hagerman, R. J., & Abbeduto, L. (2013). Receptive vocabulary in boys with autism spectrum disorder: Cross-sectional developmental trajectories. *Journal of Autism and Developmental Disorders, 43*, 2696–2709.

Landa, R., & Garrett-Mayer, E. (2006). Development in infants with autism spectrum disorders: A prospective study. *Journal of Child Psychology and Psychiatry, 47*, 629–638.

Lord, C., Shulman, C., & DiLavore, P. (2004). Regression and word loss in autistic spectrum disorders. *Journal of Child Psychology and Psychiatry, 45*, 936–955.

Luyster, R., Lopez, K., & Lord, C. (2007). Characterizing communicative development in children referred for autism spectrum disorder using the Mac-Arthur Bates Communicative Development Inventory (CDI). *Journal of Child Language, 34*, 623–654.

Luyster, R. J., Kadlec, M. B., Carter, A., & Tager-Flusberg, H. (2008). Language assessment and development in toddlers with autism spectrum disorders. *Journal of Autism and Developmental Disorders, 38*, 1426–1438.

Maljaars, J., Noens, I., Scholte, E., & Van Berckelaer-Onnes, I. (2012). Language in low-functioning children with autistic disorder: Differences between receptive and expressive skills and concurrent predictors of language. *Journal of Autism and Developmental Disorders, 42*, 2181–2191.

Matson, J. L., Mahan, S., Kozlowski, A. M., & Shoemaker, M. (2010). Developmental milestones in toddlers with autistic disorder, pervasive developmental disorder-not otherwise specified, and atypical development. *Developmental Neurorehabilitation, 13*, 239–247.

Mawhood, L., Howlin, P., & Rutter, M. (2000). Autism and developmental receptive language disorder—A comparative follow-up in early adult life. I: Cognitive and language outcomes. *Journal of Child Psychology and Psychiatry, 41*, 547–559.

McCann, J., Peppé, S., Gibbon, F. E., O'Hare, A., & Rutherford, M. (2005). *Prosody and its relationship to language in school-aged children with high-functioning autism* (working paper WP-3). Queen Margaret University College Speech Science Research Center.

Miniscalco, C., Fränberg, J., Schachinger-Lorentzon, U., & Gillberg, C. (2012). Meaning what you say? Comprehension and word production in young children with autism. *Research in Autism Spectrum Disorders, 6*, 204–211.

Mullen, E. (1995). *Mullen scales of early learning*. Circle Pines, MN: American Guidance Service, Inc.

Nation, K., Clarke, P., Wright, B., & Williams, C. (2006). Patterns of reading ability in children with autism spectrum disorder. *Journal of Autism and Developmental Disorders, 36*, 911–919.

Norbury, C. F. (2004). Factors supporting idiom comprehension in children with communication disorders. *Journal of Speech, Language, and Hearing Research, 47*, 1179–1193.

Norbury, C. F. (2005a). Barking up the wrong tree? Lexical ambiguity resolution in children with language impairments and autistic spectrum disorders. *Journal of Experimental Child Psychology, 90*, 142–171.

Norbury, C. F. (2005b). The relationship between theory of mind and metaphor: Evidence from children with language impairment and autistic spectrum disorder. *British Journal of Developmental Psychology*, *23*, 383–399.

Norbury, C. F., & Bishop, D. V. M. (2002). Inferential processing and story recall in children with communication problems: A comparison of specific language impairment, pragmatic language impairment and high-functioning autism. *International Journal of Language Communication Disorders*, *37*, 227–251.

Norbury, C. F., & Bishop, D. V. M. (2003). Narrative skills of children with communication impairments. *International Journal of Language Communication Disorders*, *38*, 287–313.

Park, C. J., Yelland, G. W., Taffe, J. R., & Gray, K. M. (2012). Morphological and syntactic skills in language samples of preschool aged children with autism: Atypical development? *International Journal of Speech-Language Pathology*, *14*, 95–108.

Paul, R., Augustyn, A., Klin, A., & Volkmar, F. R. (2005). Perception and production of prosody by speakers with autism spectrum disorders. *Journal of Autism and Developmental Disabilities*, *35*, 205–220.

Paul, R., Chawarska, K., Cicchetti, D., & Volkmar, F. (2008). Language outcomes of toddlers with autism spectrum disorders: A two year follow-up. *Autism Research*, *1*, 97–107.

Paul, R., Chawarska, K., Fowler, C., Cicchetti, D., & Volkmar, F. (2007). "Listen my children and you shall hear": Auditory preferences in toddlers with autism spectrum disorders. *Journal of Speech, Language, and Hearing Research*, *50*, 1350–1364.

Pickles, A., Simonoff, E., Conti-Ramsden, G., Falcaro, M., Simkin, Z., Charman, T., et al. (2009). Loss of language in early development of autism and specific language impairment. *Journal of Child Psychology and Psychiatry*, *50*, 843–852.

Pry, R., Peterson, A. F., & Baghdadli, A. M. (2011). On general and specific markers of lexical development in children with autism from 5 to 8 years of age. *Research in Autism Spectrum Disorders*, *5*, 1243–1252.

Reynell, J. K., & Gruber, C. P. (1990). *Reynell developmental language scales (U.S. edition)*. Los Angeles, CA: Western Psychological Services.

Ricketts, J., Jones, C. R. G., Happé, F., & Charman, T. (2013). Reading comprehension in autism spectrum disorders: The role of oral language and social functioning. *Journal of Autism and Developmental Disorders*, *43*, 807–816.

Sandercock, R. K. (2013). *Gesture as a predictor of language development in infants at high risk for autism spectrum disorders* (unpublished doctoral dissertation). Pittsburgh, PA: University of Pittsburgh.

Schopler, E., Lansing, M. D., Reichler, R. J., & Marcus, L. M. (2005). *Examiner's manual of the psychoeducational profile* (Vol. 3). Austin, TX: Pro-Ed Inc.

Semel, E., Wiig, E. H., & Secord, W. (2000). *Clinical evaluation of language fundamentals* (3rd ed.). London: Psychological Corporation.

Semel, E., Wiig, E. H., & Secord, W. (2006). *Clinical evaluation of language fundamentals* (4th ed.). London: Pearson Assessment.

Semel, E., Wiig, E. H., & Secord, W. A. (1995). *Clinical evaluation of language fundamentals* (3rd ed.). San Antonio, TX: Psychological Corporation.

Semel, E. M., Wiig, E. H., & Secord, W. (1992). *Clinical evaluation of language fundamentals—Preschool*. San Antonio, TX: The Psychological Corporation.

Sigman, M., & McGovern, C. W. (2005). Improvement in cognitive and language skills from preschool to adolescence in autism. *Journal of Autism and Developmental Disorders*, *35*, 15–23.

Smith, V., Mirenda, P., & Zaidman-Zait, A. (2007). Predictors of expressive vocabulary growth in children with autism. *Journal of Speech, Language, and Hearing Research*, *50*, 149–160.

Stone, W. L., & Yoder, P. J. (2001). Predicting spoken language level in children with autism spectrum disorders. *Autism*, *5*, 341–361.

Sutera, S., Pandey, J., Esser, E. L., Rosenthal, M. A., Wilson, L. B., Barton, M., et al. (2007). Predictors of optimal outcome in toddlers diagnosed with autism spectrum disorders. *Journal of Autism and Developmental Disorders*, *37*, 98–107.

Swensen, L. D., Kelley, E., Fein, D., & Naigles, L. R. (2007). Processes of language acquisition in children with autism: Evidence from preferential looking. *Child Development*, *78*, 542–557.

Taylor, M. J., Charman, T., Robinson, E. B., Hayiou-Thomas, M. E., Happé, F., Dale, P. S., et al. (2014). Language and traits of autism spectrum conditions: Evidence of limited phenotypic and etiological overlap. *American Journal of Medical Genetics*, *65B*(7), 587–595. Available from: http://dx.doi.org/10.1002/ajmg.b.32262.

Troyb, E. (2011). *Academic abilities in children and adolescents with a history of autism spectrum disorders who have achieved optimal outcomes* (unpublished master's thesis, paper 189). Storrs, CT: University of Connecticut.

Vanvuchelen, M., Roeyers, R., & DeWeerdt, W. (2011). Imitation assessment and its utility to the diagnosis of autism: Evidence from consecutive clinical preschool referrals for suspected autism. *Journal of Autism and Developmental Disorders*, *41*, 484–496.

Walton, K. M., & Ingersoll, B. R. (2013). Expressive and receptive fast-mapping in children with autism spectrum disorders and typical development: The influence of orienting cues. *Research in Autism Spectrum Disorders*, *7*, 687–698.

Wicklund, M. D. (2012). *Use of referring expressions by autistic children in spontaneous conversations: Does impaired metarepresentational ability affect reference production?* (unpublished doctoral dissertation). Minneapolis, MN: University of Minnesota.

Wilson, S., Djukic, A., Shinnar, S., Dharmani, C., & Rapin, I. (2003). Clinical characteristics of language regression in children. *Developmental Medicine & Child Neurology*, *45*(8), 508–514.

Wisdom, S. N., Dyck, M. J., Piek, J. P., Hay, D., & Hallmayer, J. (2007). Can autism, language, and coordination disorders be differentiated based on ability profiles? *European Child and Adolescent Psychiatry*, *16*, 178–186.

Wodka, E. L., Mathy, P., & Kalb, L. (2013). Predictors of phrase and fluent speech in children with autism and severe language delay. *Pediatrics*, *131*, e1128–e1134.

Young, E. C., Diehl, J. J., Morris, D., Hyman, S. L., & Bennetto, L. (2005). The use of two language tests to identify pragmatic language problems in children with autism spectrum disorders. *Language, Speech, and Hearing Services in Schools*, *36*, 62–72.

Zimmerman, I., Steiner, V., & Pond, R. (1992). *PLS-3: Preschool language scale-3*. San Antonio, TX: The Psychological Corporation.

Zimmerman, I. L., Steiner, V. G., & Pond, R. E. (2002). *Preschool language scale* (4th ed.). San Antonio, TX: The Psychological Corporation.

CHAPTER

71

Symptoms and Neurobiological Models of Language in Schizophrenia

Arne Nagels and Tilo Kircher

Department of Psychiatry and Psychotherapy, Philipps University Marburg, Marburg, Germany

71.1 INTRODUCTION

Schizophrenia is a common, severe, and complex mental disorder that affects approximately 1% of the population worldwide (Sawa & Snyder, 2002), irrespective of cultural background. The course of disease differs among patients; however, symptoms usually first occur in young adulthood (20–30 years). Before the acute onset of the disorder, patients experience a subclinical prodromal phase, lasting from weeks to years, comprising cognitive impairments, depression, subclinical psychotic symptoms, and psychosocial impairments. The diagnosis is made from interview-based clinical observations—the way in which patients communicate thoughts and emotions via language—often supplemented by third party reports. Overt organic causes, determined by physical examination, CT/MRI, blood and cerebrospinal fluid tests, or drug intoxication have to be absent. The course of illness varies widely between individuals, ranging from one single episode lasting a couple of weeks to episodic (e.g., a few episodes in a lifetime with symptoms present for several weeks with complete or partial remission) and severe, primary chronic, detrimental courses (patient lives in supported housing and work, dependent on social welfare). Treatment consists of antipsychotic medication, psychoeducation, cognitive behavioral psychotherapy, and social support, tailored to the individual needs of the patient.

Schizophrenia psychopathology can be subdivided into positive (e.g., delusions, hallucinations), negative (e.g., affective blunting, social withdrawal, anergia, laconic speech) and disorganized symptoms. The latter are reflected in disorganized behavior and/or incoherent, disorganized, or dysfluent speech patterns (e.g., thought interference, derailment), referred to as (positive) formal thought disorder (FTD). Other language-related symptoms are auditory hallucinations—usually hearing the voices of other people—that are not covered in this chapter. Not all symptoms are present in all patients and the expression of a specific symptom in an individual also varies over time. FTD represents a *hallmark syndrome* playing a crucial role in the diagnosis of schizophrenia (DSM V, ICD 10). It is therefore not surprising that researchers with different backgrounds such as linguists, speech scientists, (neuro-)psychologists, psychiatrists, neuroscientists, and others are interested in the description and pathophysiology of the wide variety of language and speech impairments in these patients.

After more than a century of schizophrenia research, it has become clear that the interplay between genetic vulnerability and environmental risk factors contributes to the etiology of the disorder. The concordance rate of monozygotic twins, for instance, is approximately 50–80%, indicating a considerable influence of genetic predisposition. A number of risk genes have been identified, such as CACNA1C, NRG1, or MHC regions (each with a low odds ratio [OR] of approximately 1.2–1.6). Environmental risk factors further include high paternal age, pregnancy and birth complications, childhood maltreatment, urban upbringing, migration, and cannabis abuse. These risk factors lead—among others—to synaptic rarefication and dysconnectivity, as well as to a neurotransmitter imbalance (particularly in the dopaminergic and glutamatergic system), giving rise to a vulnerable brain system. MRI brain volumetric investigations found reduced gray matter, most prominently in the frontal operculae, anterior cingulate cortex (ACC), and middle/superior lateral as well as medial temporal lobes. In addition, ventricular enlargements (Huber, 1955, 1961) as well as a dysconnection between brain areas—already

Neurobiology of Language. DOI: http://dx.doi.org/10.1016/B978-0-12-407794-2.00071-7

hypothetically postulated by Carl Wernicke (Wernicke, 1906)—are observed (Friston & Frith, 1995). These points are addressed in detail in the following sections.

Language-related dysfunctions in patients with schizophrenia can be observed throughout the different, multi-faceted domains of speech production and perception, ranging from basic auditory perception mechanisms (such as mismatch negativity [MMN]) to complex pragmatic information processing deficits (such as metaphors, jokes, irony, or Theory of Mind). Thus, relating language deficits to neural dysfunctions in a heterogeneous disorder is challenging and requires both reduction and abstraction. We therefore restrict our framework to the following premises. If available, we will focus on patients with (versus without) FTD. This approach controls for confounding factors such as general aspects of disease pathology and medication. Due to the complexity of the neural architecture of the language system, we will restrict ourselves primarily to "traditional" language-related brain regions (e.g., DLPFC, VLPFC, MTG, STG) and their connections, being aware that the neural network underlying human speech comprises multiple and widespread regions throughout the brain (e.g., corpus callosum, IPL, hippocampus, cerebellum, motor areas, subcortical loops). We further conceptualize schizophrenia—besides localized (focal) structural alterations—as a brain disconnection syndrome from synaptic to gross white matter levels.

Regarding specific linguistic domains, special consideration is given to neurobiological correlates of semantics and pragmatics while only briefly touching on the other domains such as acoustic perception, phonological processing, and syntax. First, we focus on the descriptive phenomenology of FTDs as well as on their structural and functional correlates (see Sections 71.2–71.4). Thereafter, results of a representative sample of studies on semantic impairments are discussed (see Section 71.5), whereupon we move to the superordinate level of pragmatic deficits (see Section 71.6). Dysfunctions on the phonological level as well as on the syntactic level are briefly referred to in Sections 71.7 and 71.8. Subsequently, we review neurobiological and molecular models of schizophrenia, relating them to language deficits and touching briefly on imaging genetics and pharmacological imaging approaches (Sections 71.9 and 71.10). Inversed language lateralization and functional asymmetries are summarized separately (Section 71.11).

71.2 PHENOMENOLOGY, ASSESSMENT, AND COURSE OF FORMAL THOUGHT AND LANGUAGE DISORDER

In patients with schizophrenia, impairments in the *production* of language and subjective alterations in the thinking process are clinically referred to as FTD—a disorder in the *form* of thought and not the *content* (i.e., delusions). On a phenomenological level, severe positive FTD in schizophrenia is very similar to the speech of neurologically impaired patients with Wernicke's aphasia (Faber et al., 1983) (e.g., *vagueness, looseness of associations, neologisms,* etc.) (Gerson, Benson, & Frazier, 1977).

One can distinguish positive from negative FTD. The latter mostly reflects a quantitative deficit in speech production (*poverty of speech, laconic speech, slowed thinking*) or a lack of ideas. Negative FTD, particularly *alogia,* are often seen in patients with chronic schizophrenia (McKenna & Oh, 2005). In contrast, positive FTD usually refers to a larger amount of produced speech (*logorrhea, pressure of speech*), loosening of associations, the use of new words (*neologisms*), or stilted speech (*manneristic speech*) phenomena. The total breakdown of language, referred to as *schizophasia* or *word salad* typically occurs in acute states of the disease. In extremely rare cases, patients even use their own terminology, invented words, and novel morpho-syntactic constructions (new, private language), resulting in a largely or even entirely unintelligible speech.

FTD symptoms may vary markedly between patients and also during the individual course of illness, yet moderate consistency of FTD over time was also reported (Docherty, Cohen, Nienow, Dinzeo, & Dangelmaier, 2003). The presence of negative FTD may be predictive of a poor outcome (Andreasen & Grove, 1986).

Along with the introduction of operationalized diagnostic systems (the DSM III and ICD 10), the development of validated clinical rating scales for FTD took hold (Andreasen, 1986; Liddle et al., 2002; Parnas et al., 2005). An example is the Scale for the Assessment of Thought, Language, and Communication (Andreasen, 1986), translated and validated for German (Nagels et al., 2013), French (Bazin, Lefrere, Passerieux, Sarfati, & Hardy-Bayle, 2002), and Greek (Andreou et al., 2008). The existing rating instruments, however, did not consider subjectively reported FTD in patients that is the self-reported dysfunction in the production and perception of language and speech. For this purpose, a new comprehensive and nosologically open clinical rating scale for the comprehensive assessment of formal Thought and Language Dysfunctions (TALD), available in English and German, has been developed and validated (Kircher et al., 2014). The 30-item TALD scale, based on a clinical interview, has good psychometric properties and a four-dimensional structure of FTD symptoms, reflecting a positive and a negative factor as well as an objective and subjective dimension.

71.3 STRUCTURAL BRAIN CHANGES AND FTD

Regional cortical thinning in the frontal operculum and lateral temporal (language-related) cortices in patients with schizophrenia versus healthy subjects are consistent findings from structural MRI meta- and mega-analyses (Fusar-Poli, Radua, McGuire, & Borgwardt, 2012; Shepherd, Laurens, Matheson, Carr, & Green, 2012). Shenton et al. (1992) reported for the first time a relation between FTD and a reduction in grey matter volume of the posterior superior temporal region in schizophrenia patients with FTD. This finding has been replicated in other independent samples (Horn et al., 2010; Horn et al., 2009; Shenton, Dickey, Frumin, & McCarley, 2001). For example, Sans-Sansa et al. (2013) found correlations between volume reductions in Broca's and Wernicke's areas and *fluent disorganization* phenomena, whereas *poverty of content of speech* was related to reductions in bilateral medial frontal and orbito-frontal cortical regions, respectively. Altogether, these results show that volumetric structural aberrations in language related areas (superior temporal, lateral prefrontal) are the most consistent structural findings in schizophrenia and are correlated with the presence of FTD.

Besides cortical grey matter thinning in the superior temporal gyrus (STG), the integrity of the left white matter fiber bundle (fractional anisotropy measured with MRI Diffusion Tensor Imaging) of the left middle longitudinal fascicle—a long association fiber connecting the STG and temporal pole with the angular gyrus through the white matter of the STG—demonstrated a negative association with disorganized thoughts (Asami et al., 2013). Thus, the connection within the left STG and to the angular gyrus is correlated with the amount of FTD in patients, providing further evidence for the disconnection hypothesis previously mentioned.

71.4 NEURAL CORRELATES OF FTD (SYMPTOM CATCHING)

To date, very few studies have investigated the neural substrates of naturally produced speech in FTD patients, and brain activation using H_3O-PET or fMRI is measured at the same time (symptom catching approach). In these rare but revealing studies, patients were asked to speak about pictures in the scanner and their verbal output was recorded online and transcribed subsequently so that psychopathological phenomena could be assessed, evaluated, and precisely time-locked with the brain signal changes. Accordingly, in one study the amount of FTD as

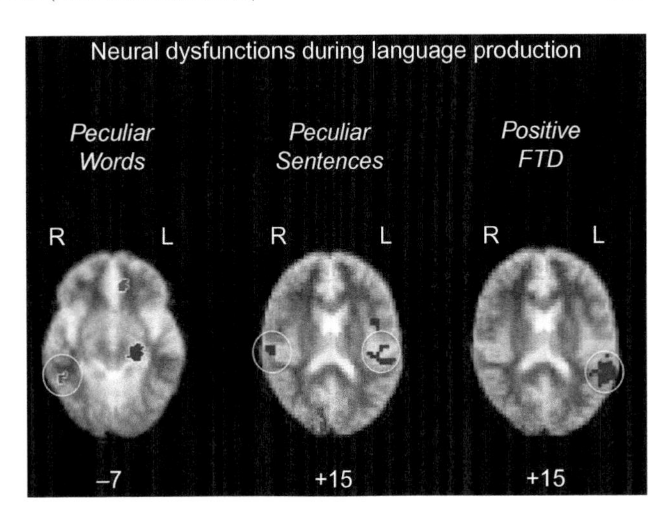

FIGURE 71.1 The "symptom catching" approach in schizophrenia patients during natural speech production. Brain signal changes were correlated with the production of "peculiar words" (left), "peculiar sentences" (middle), and "positive FTD" (right). Red voxels indicate positive correlations; blue voxels refer to negative correlations. Note that according to the neuroradiological convention, the left hemisphere (L) is presented on the right side (R) of the brain image. *Based on Kircher (2003) and Kircher et al. (2001).*

measured with a clinical rating scale (Thought and Language Index) (Liddle et al., 2002) was correlated with brain activation during 20-s epochs. The expression of positive FTD was found to be negatively correlated with activation in the left STG, part of Wernicke's area (Figure 71.1) (Kircher et al., 2001). Thus, during the production of positive FTD, brain activation in the STG fluctuated and was reduced during those short time intervals when language dysfunction (i.e., positive FTD) was maximal (Kircher et al., 2001; McGuire et al., 1998).

Using an fMRI event-related approach, the occurrence of particular FTD symptoms, recorded in continuously speaking patients, has been correlated with brain activation (see Figure 71.1). Here, a correlation was found between semantic paraphasias ("peculiar words") and BOLD enhancements in the left anterior cingulate (ACC) and the right middle temporal gyrus (MTG) (Kircher, 2003). The ACC is associated with decision processes or error monitoring, as well as with the detection of paraphasias in this study. The retrieval of semantic information in the RH, in particular in the right MTG, was found to be related to the processing of wide associative semantic fields, giving rise to the selection of "unintended" but semantically related peculiar notions (Beeman, 1993). During the articulation of peculiar sentences (Liddle et al., 2002) as opposed to morpho-syntactically correct lexical sentences, the left and right STG were deactivated (Kircher, 2003) (for further details, see Section 71.8).

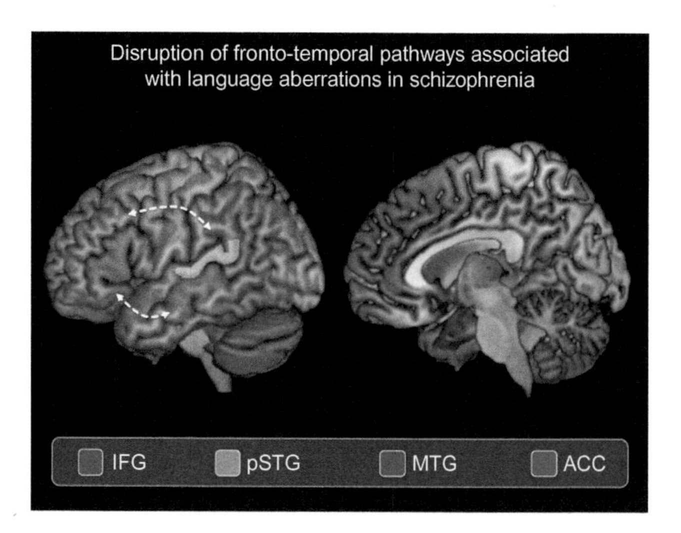

FIGURE 71.2 Brain regions (orange, inferior frontal gyrus [IFG]; yellow, superior temporal gyrus [STG]/inferior parietal lobe [IPL]; green, middle temporal gyrus [MTG]; blue, anterior cingulate cortex [ACC]) being associated with functional and anatomical aberrations in schizophrenia patients related to language impairments. Arrows depict fronto-temporal dysconnections (arcuate fasciculus, uncinate fasciculus, superior longitudinal fasciculus), particularly between the IFG and the pSTG/MTG. Note that brain regions are schematically depicted.

These areas are generally involved in both language production and perception, however deactivated, during aberrant, nonlexical speech production in patients.

The same approach has also been applied for negative FTD (poverty of speech), detecting positive correlations in the right inferior parietal lobe, middle frontal gyrus, cuneus, and the left posterior cingulate. Inverse correlations were found in the left hippocampal and fusiform region. Results were interpreted in terms of rich memories and associations being experienced subjectively by the patient, but not verbalized (Kircher, Liddle, Brammer, Murray, & McGuire, 2003). A recent investigation examined the neural activity associated with pauses that occurred between clauses and with pauses that were filled (Matsumoto et al., 2013). An attenuated involvement of the STG during between-clause pauses indicates defective speech planning and monitoring mechanisms being related to FTD symptomatology (Matsumoto et al., 2013). In summary, structural and functional imaging investigations point to an involvement of IFG and, in particular, STG structures in the pathophysiology of FTD (Figure 71.2).

71.5 SEMANTICS

Dysfunctions in semantic associations—measured behaviorally (Ketteler, Theodoridou, Ketteler, & Jager, 2012)—and semantic memory (Rossell & David, 2006) are frequently reported and account for some of the language impairments. One way to access the functionality of the associative semantic network connections is realized by behavioral semantic priming tasks. Here, the participants decide whether a presented letter string is a word or a nonword and indicate their decision via button press. Beforehand, a prime word is briefly presented (e.g., 150 ms), which is either semantically related to the target word or not. The dependent variable is the reaction time that is significantly shorter when the priming word is semantically related to the target. Patients with positive FTD (versus non-TD patients) show a shorter reaction time during short Stimulus Onset Asynchrony (SOA) (<250 ms automatic processing versus more controlled processing with SOA >500 ms) for semantically indirectly (versus directly) related words. Indirectly related words are, for instance, *anvil–nail*, directly related words are *picture–frame*. Based on the assumptions of the spreading semantic activation model (Collins & Loftus, 1975), it has been shown in these semantic priming studies that altered associative processes within the semantic network or, in other words, decreases in inhibition in the spreading and associational activation are related to positive FTD in schizophrenia patients (Manschreck et al., 1988; Spitzer, Braun, Maier, Hermle, & Maher, 1993). It is important to note that this hyperpriming only occurs under the conditions that patients display positive FTD, there is automatic processing, and the semantic relation of words is indirect. In FTD patients, the automatic spread of activation happens faster and more widely in the semantic network (most likely related to the right lateral temporal lobe, see parts 4 and 11).

Brain imaging studies investigated the neural correlates of semantic association (priming) tasks in moderately thought-disordered patients with schizophrenia and healthy controls (Sass et al., 2014). Here, an effect of semantic relation was found within the right angular gyrus and precuneus. Moreover, semantic distance (direct as opposed to indirect semantic association of the prime target relation) revealed distinct patterns of activations encompassing the left middle temporal and right precuneus, among others. Here, direct relations induced higher activations in controls. Results were interpreted in terms of aberrant priming-related brain responses in schizophrenia patients. Both delayed and enhanced spreading activations within the semantic framework may influence semantic processing impairments. Kuperberg and colleagues took a similar approach using a semantic priming paradigm with directly related, indirectly related, and unrelated word pairs in schizophrenia patients (Kuperberg, Deckersbach, Holt, Goff, & West, 2007). They revealed hemodynamic dysactivation in inferior prefrontal and temporal regions in connection with direct and indirect (relative to unrelated) word pairs in chronic schizophrenia.

Speech does not consist of single, isolated semantic units, but rather of coherent discourse (i.e., several words in context). Concerning language comprehension on the sentence level, a recent review of fMRI studies revealed activation changes in the IFG as well as in the STG (left "fronto-temporal network") in FTD versus non-FTD patients with schizophrenia (Rapp & Steinhauser, 2013). Weinstein and colleagues investigated a receptive language processing task, asking patients as well as healthy controls to listen to 30-s speech samples. Only FTD was reported to correlate positively with BOLD response in the left pSTG, indicating that compensatory mechanisms are involved to allow for normal performance (Weinstein, Werker, Vouloumanos, Woodward & Ngan, 2006). An association between fronto-temporal network dysfunctions and the presence of FTD was also found in the context of semantic decision tasks (Arcuri et al., 2012), so that a fronto-temporal dysconnection in the semantic network is suggested to be involved in FTD (Horn et al., 2012).

On a speech production level, lexico-semantic deficits are often investigated in highly controlled word generation tasks. A frequently used approach for assessing the functionality of word retrieval according to given stimuli (e.g., semantic categories such as "animals" = semantic verbal fluency; words to a given initial letter such as "P" = lexical verbal fluency) and rules (e.g., no word stem repetitions) within a predetermined time window is referred to as verbal fluency task (Nagels, Kircher, et al., 2012). In schizophrenia patients, deficits in semantic verbal fluency performance represent the most consistent neuropsychological finding (Henry & Crawford, 2005; Szoke et al., 2008). Larger deficits for semantic relative to lexical fluency performance suggest that, in addition to generalized retrieval difficulties of verbal information, schizophrenia is associated with particular impairments in the semantic store (Henry & Crawford, 2005). Neural evidence for impaired relational memory processes was derived from impaired semantic verbal fluency performance (Kircher, Whitney, Krings, Huber, & Weis, 2008). Here, patients revealed attenuated hippocampal activity indicating dysfunctions in the retrieval of semantic verbal knowledge. In general, neural dysfunctions—hypoactivations as well as hyperactivations, particularly in left prefrontal regions—during verbal fluency performance were reported (Broome et al., 2009; Curtis et al., 1998, 1999, 2001; Fu, Suckling, et al., 2005; Ragland et al., 2008), depending on the investigated task (e.g., semantic versus lexical fluency, continuous versus single word production, silent versus overt) and the specific cognitive demands (Curtis et al., 1999). In one study, the severity of psychotic symptoms was associated with the recruitment of the anterior cingulate (Fu, Suckling, et al., 2005). Dysfunctions in the latter region as well as in the right prefrontal cortex for patients as compared with controls were interpreted in terms of compensatory mechanisms (Schaufelberger et al., 2005). According to this model, additional neural resources are recruited to countervail cognitive deficits. The inverted U-shaped model—originally introduced to explain task-related hypoactivations and hyperactivations in working memory paradigms (Manoach, 2003)—suggests that with increasing cognitive demand, frontal neural responses increase. In the case of schizophrenia patients, however, the U-shaped curve is shifted to the left, resulting in temporarily earlier activations as observed in healthy controls.

On the whole, behavioral study results point to impairments in retrieval and inhibition of contextually irrelevant semantic information in schizophrenia, particularly in patients with positive FTD. Neural aberrations were mainly reported in left frontal as well as in temporal brain regions. However, different experimental approaches (indirect versus direct priming, semantic decision, verbal fluency), varying semantic task demand (single word production versus continuous generation), production versus perception tasks, and FTD symptom severity all contribute to different neural responses and must be meticulously taken into consideration when interpreting the results.

71.6 PRAGMATICS

Apart from dysfunctions on the semantic level, contextual information processing in schizophrenia patients is impaired as well. In everyday communication, pragmatic aspects of communication such as interpretations in situations and conversational implicatures play an important role. Here, the intended meaning is often conveyed in an indirect way inasmuch as information exchange may comprise indirect meanings ("Could you open the window?"), abstract figurative expressions (proverbs), and nonliteral meanings (irony, sarcasm) asking the interlocutor to read the intended message between the lines (Rapp et al., 2013). In schizophrenia patients, the ability to transfer the abstract semantic content of a metaphoric expression ("Life is a journey.") or a proverb ("You can lead a horse to water, but you can't make him drink.") to the intended concrete meaning is impaired. As a result, affected patients tend to interpret the figurative meaning literally, which is clinically referred to as *concretism*, a commonly observed phenomenon in schizophrenia.

Nevertheless, imaging studies on the neural basis of metaphor processing dysfunctions in schizophrenia are still rare, because many potentially confounding factors need to be controlled for (word frequency, syntactic complexity, familiarity, etc.). The few studies available report the left IFG to be dysactivated during metaphor processing in patients (Kircher, Leube, Erb, Grodd, & Rapp, 2007; Mashal, Vishne, Laor, & Titone, 2013) (Figure 71.3).

Further evidence for neural dysfunctions in metaphor processing in schizophrenia was found in an fMRI study on abstract gesture processing (e.g., hearing "The talk was on a high level" while watching the actor lift his hand to indicate the "high level") (Straube, Green, Sass, Kirner-Veselinovic, & Kircher, 2013). Patients and a healthy control group were presented with either concrete iconic (form-descriptive) gestures or abstract metaphoric gesture material, as described. Imaging results indicate that the neural integration of gesture and speech is not impaired *per se*, because the processing of concrete iconic co-speech gesture was intact. However, patients failed to recruit the fronto-temporal neural network in the metaphoric gesture context, which leads to the assumption that the multi-modal integration of audio-visual language material is specifically impaired with respect to the abstract figurative domain. Evidence for a functional fronto-temporal dysconnectivity in patients was found in a subsequent investigation of metaphoric gesture processing (Straube, Green, Sass, & Kircher, 2014) (Figure 71.4). Here, the left superior temporal sulcus was misconnected to the IFG, supporting the view of a dysfunctional integration of abstract language material.

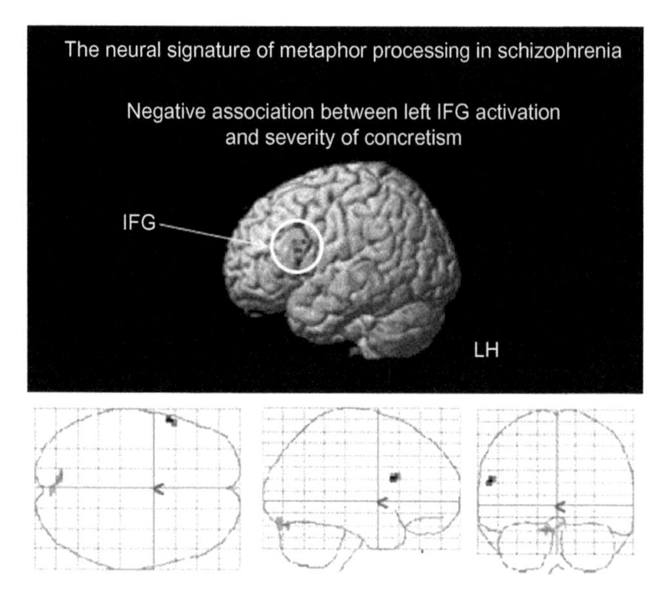

FIGURE 71.3 A negative association between left IFG recruitment and the severity of concretistic thinking in schizophrenia patients was found in the left hemisphere (LH). *Based on Kircher et al. (2007).*

71.7 AUDITORY SENSORY, PHONOLOGICAL, AND PROSODIC PROCESSING

Patients with schizophrenia exhibit aberrations in the most basic, early stages of auditory signal processing such as mere tone perception, distinction, and

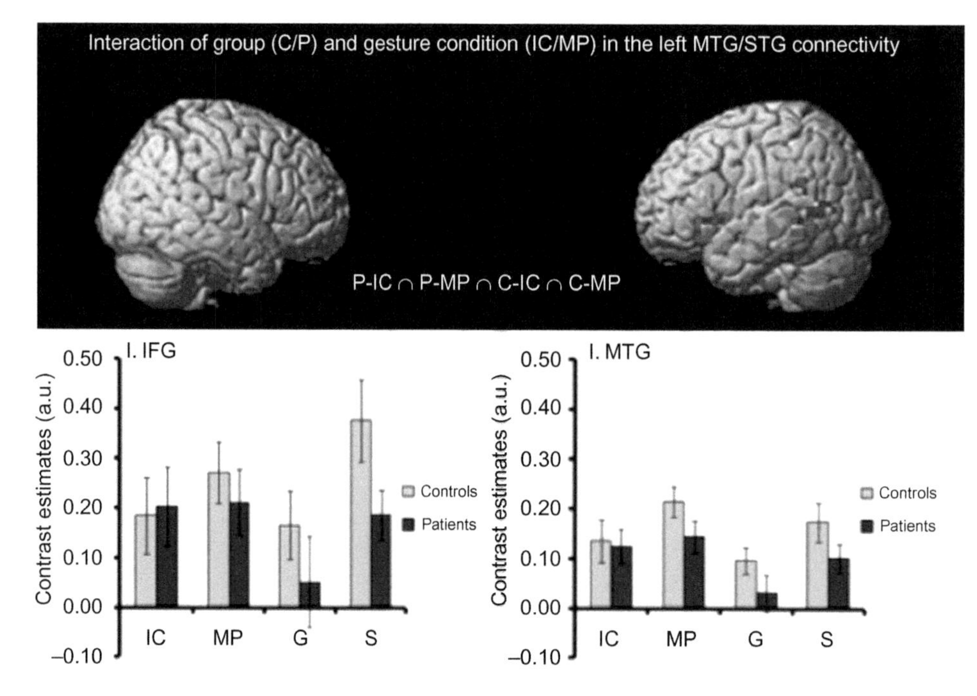

FIGURE 71.4 Dysconnectivity between the left IFG and the left MTG/STG region during metaphoric co-speech gesture processing (MP) in patients with schizophrenia (P) as compared with healthy controls (C). Apart from the metaphoric (MP) co-speech gesture video clips, an iconic (IC) condition was also presented. *Based on Straube et al. (2014).*

discrimination. This has been widely demonstrated using event-related potentials. For instance, "MMN" deficits represent an often replicated finding in schizophrenia patients (Kircher et al., 2004; Umbricht & Krljes, 2005). MMN is a correlate of auditory (sensory) memory and has been linked to coordinated neural mass synchronization (several thousands of neurons firing coordinated when a deviant tone occurs). MMN is generated in the posterior portion of the STG, as demonstrated in a combined MEG and fMRI study (Kircher et al., 2004) in healthy subjects, with a dysregulation in patients in this area. Further, the glutamate system, particularly the *N*-methyl-D-aspartate (NMDA) receptor (see Section 71.9), is fundamentally involved in the generation of MMN (Javitt, 2009). We therefore propose NMDA receptor dysfunction and synaptic rarefication in the STG including the primary (sensory) auditory circuits as the origin of altered tone processing, giving rise to difficulties in acoustic discrimination (e.g., phones) and phonological working memory capacities. Note that this is the key area of dysfunction in auditory hallucinations and positive FTD (see Sections 71.3 and 71.4). Further auditory deficits such as prosody discrimination and detection impairments (Leitman et al., 2007) together with a reduced sensitivity to alterations in pitch processing (Kantrowitz, Hoptman, Leitman, Silipo, & Javitt, 2013), correlated with dysactivations in the STG in patients, also indicate impairments in the primary auditory cortex (Leitman et al., 2007). Taken together, it can be assumed that dysactivations in the left STG particularly contribute to phonological aspects of aberrant speech processing.

71.8 SYNTAX

More complex syntactic processing is impaired in some patients; in others, reduced syntactic complexity of spoken language is observed (Morice & McNicol, 1985). The articulation of grammatically more simple speech patterns in patients versus healthy control subjects was found to be associated with an absence of activation in the right posterior temporal as well as in the left superior frontal cortex in the patients only (Kircher, Oh, Brammer, & McGuire, 2005).

Neural dysfunctions underlying impaired syntactic processing may potentially be associated with impaired verbal working memory capacities as well as with aberrations in the online processing of morphosyntactic relationships between words. Apart from these impairments, aberrations in the complex interaction between semantic memory and the build-up of sentence and discourse (Kuperberg, 2008) may play an influential role leading to dysfunctions on the syntactic level within a sentence ("Who does what to whom") (Kuperberg, 2008). These different cognitive, hierarchically organized, and serially linked processes cannot be easily disentangled and controlled for in an experimental setting. Thus, imaging studies investigating this particular field are still rare, so that the validity of obtained results is still limited to date.

71.9 NEUROTRANSMITTER DYSFUNCTION

So far, patient studies have been discussed focusing on different levels of language impairments in connection with the neural dysfunctions. To further probe particular neurotransmitter systems, pharmacological model psychoses have been tested with psychoactive substances administered to healthy volunteers. Schizophrenia-like symptoms such as hallucinations or FTD can be elicited in healthy subjects in a controlled experimental context by administering substances such as LSD, PCP, psylocybin, or ketamine (for review see Gouzoulis-Mayfrank, Hermle, Thelen, & Sass, 1998; Vollenweider & Kometer, 2010). The main neurotransmitter systems implicated in schizophrenia are dopamine, glutamate, and acetylcholine, which of course interact in a complex way.

Ketamine, a glutamatergic NMDA receptor antagonist, was used to test the hypothesis that an imbalance in the glutamatergic system is associated with the presence of psychotic symptoms. Thus, a number of PET (Vernaleken et al., 2012; Vollenweider, Leenders, Oye, Hell, & Angst, 1997; Vollenweider, Leenders, Scharfetter, et al., 1997; Vollenweider, Vontobel, Oye, Hell, & Leenders, 2000) and fMRI studies (Abel, Allin, Kucharska-Pietura, Andrew, et al., 2003; Abel, Allin, Kucharska-Pietura, David, et al., 2003; Fu, Abel, et al., 2005; Musso et al., 2011) investigated the effects of an experimentally controlled NMDA receptor blockade on the psychopathological, behavioral, and neural level. Ketamine administration particularly elicited FTD symptoms, largely resembling those observed in schizophrenia patients (Nagels, Kirner-Veselinovic, et al., 2012). Language-related fMRI ketamine studies used overt verbal fluency tasks (Fu, Abel, et al., 2005; Nagels, Kirner-Veselinovic, Krach, & Kircher, 2011), revealing quantitative impairments in verbal fluency performance on the behavioral performance level (Krystal et al., 1994; Nagels et al., 2011). In addition, positive correlations were found between left STG activations and ketamine-induced FTD symptoms during an overt verbal fluency task. Altogether, effects on the neural and behavioral level for visual field (VF) were comparable with those observed in patients with schizophrenia (e.g., fronto-temporal dysfunctions, impaired VF performance), strengthening the hypothesis of a glutamatergic

NMDA receptor dysfunction being involved in the pathophysiology of schizophrenia. However, more pharmaco-imaging studies are needed to elucidate the connection between NMDA receptor dysfunction, symptom formation, and language impairments, particularly using multi-modal imaging approaches such as combined EEG-fMRI (Musso et al., 2011).

71.10 GENETIC INFLUENCE ON SPEECH AND LANGUAGE DYSFUNCTIONS IN SCHIZOPHRENIA

Alterations in language use are the only symptoms in schizophrenia that are genetically inherited (Arboleda & Holzman, 1985; Cardno, Sham, Murray, & McGuffin, 2001; Kinney et al., 1997). Research has recently focused on genetic variations of candidate genes for schizophrenia and their effects on brain structure and function. The majority of these common risk variants are found in the general population (>10%), and each conveys only very little increase in risk (OR 1.2–1.6). For many risk variants, their effects on the brain are unknown; therefore, imaging genetics studies comprise healthy subjects carrying a genetic risk single nucleotide polymorphism tagging marker for schizophrenia. This approach allows for an assessment of behavioral and imaging data in the absence of potentially confounding effects due to medication, duration of illness, or psychopathological differences between subjects.

Functional imaging tasks often investigate paradigms that are known to be sensitive for cognitive impairments in schizophrenia, such as semantic verbal fluency tasks. Thus, in the context of a genetic risk variant in CACNA1c, Krug et al. (2010) reported dysfunctions in the left IFG. Another susceptibility gene, NRG1, was also found to modulate VF performance as well as neural activations (Kircher et al., 2009). The authors report decreased activation encompassing the left IFG, right middle temporal gyri, and the anterior cingulate being correlated with the number of risk alleles in an overt semantic verbal fluency task. In this study, an association was found for both neural activation patterns and VF impairments similar to schizophrenia patients—but to a lesser degree—leading to the assumption that NRG1 has an impact on semantic language capacities.

Further imaging genetics investigations are needed to confirm the obtained results. Moreover, future studies should investigate follow-up, longitudinal designs to further elucidate the influence of candidate risk variants, as well as environmental effects on functional, structural, and behavioral aberrations over time.

71.11 LATERALIZATION ASYMMETRY IN SCHIZOPHRENIA

A well-replicated finding is the functional asymmetry or decreased language lateralization in schizophrenia (Bleich-Cohen, Hendler, Kotler, & Strous, 2009; Bleich-Cohen et al., 2012) that is already present in patients with a first episode. In healthy subjects, the left STG as opposed to its contralateral homologue in the right hemisphere is comparatively thicker. However, in patients with schizophrenia this distribution is symmetrical or even reversed (Ratnanather et al., 2013). These structural asymmetry deficits as well as the cortical gray matter volume deficits in the STG region may contribute to FTD symptom formation (Horn et al., 2010; Sans-Sansa et al., 2013; Subotnik, Bartzokis, Green, & Nuechterlein, 2003). Diminished functional trans-hemispheric connectivity, due to structural and functional abnormalities in the corpus callosum, for example, between the VLPFC or lateral temporal lobes (Curcic-Blake et al., 2013), are other potential reasons for language impairments, particularly with respect to higher-order contextual discourse processing. Likely, the interaction deficit between hemispheres results in these language impairments (Strelnikov, 2010).

71.12 CONCLUSIONS AND FUTURE PERSPECTIVES

Language-related dysfunctions in patients with schizophrenia can be observed throughout all the multi-layered domains of speech production and perception. Structural, functional, and pharmacological imaging investigations mostly found aberrations in fronto-temporal neural circuits, particularly encompassing the left IFG and portions of the left STG, together with their interhemispheric and transhemispheric connections. Other areas, such as the hippocampi, precuneus, anterior cingulate, cerebellar, and motor areas, as well as subcortical structures are further involved in this language pathology.

With the exception of a few imaging studies, the majority of studies used highly controlled experimental perception/processing (rather than production) paradigms eliminating the "natural" complex context (Andric & Small, 2012). Imaging genetics approaches may play a role in relating genetic with neurophysiological and anatomical findings. Follow-up study designs are needed to illuminate the longitudinal progression of neural changes as well as its influencing factors.

Considering that the complex and multi-facetted human speech capacity and its neurobiological foundations are not yet fully understood in healthy participants, it is not surprising that there is no single unified explanatory model for speech and language deficits in schizophrenia patients available yet. Further multi-modal and translational approaches—combining results from different scientific fields—are needed to explore the nature of the underlying pathophysiological mechanisms resulting in language aberrations in schizophrenia.

References

Abel, K. M., Allin, M. P., Kucharska-Pietura, K., Andrew, C., Williams, S., David, A. S., et al. (2003). Ketamine and fMRI BOLD signal: distinguishing between effects mediated by change in blood flow versus change in cognitive state. *Human Brain Mapping, 18*(2), 135–145.

Abel, K. M., Allin, M. P., Kucharska-Pietura, K., David, A., Andrew, C., Williams, S., et al. (2003). Ketamine alters neural processing of facial emotion recognition in healthy men: an fMRI study. *Neuroreport, 14*(3), 387–391.

Andreasen, N., & Grove, W. M. (1986). Thought, language, and communication in schizophrenia: diagnosis and prognosis. *Schizophrenia Bulletin, 12*(3), 348–359.

Andreasen, N. C. (1986). Scale for the assessment of thought, language, and communication (TLC). *Schizophrenia Bulletin, 12*(3), 473–482.

Andreou, C., Bozikas, V. P., Papouliakos, I., Kosmidis, M. H., Garyfallos, G., Karavatos, A., et al. (2008). Factor structure of the Greek translation of the scale for the assessment of thought, language and communication. *The Australian and New Zealand Journal of Psychiatry, 42*, 636–642.

Andric, M., & Small, S. L. (2012). Gesture's neural language. *Frontiers in Psychology, 3*, 99.

Arboleda, C., & Holzman, P. S. (1985). Thought disorder in children at risk for psychosis. *Archives of General Psychiatry, 42*(10), 1004–1013.

Arcuri, S. M., Broome, M. R., Giampietro, V., Amaro, E., Jr., Kircher, T. T., Williams, S. C., et al. (2012). Faulty suppression of irrelevant material in patients with thought disorder linked to attenuated frontotemporal activation. *Schizophrenia Research and Treatment, 2012*, 176290.

Asami, T., Saito, Y., Whitford, T. J., Makris, N., Niznikiewicz, M., McCarley, R. W., et al. (2013). Abnormalities of middle longitudinal fascicle and disorganization in patients with schizophrenia. *Schizophrenia Research, 143*(2–3), 253–259.

Bazin, N., Lefrere, F., Passerieux, C., Sarfati, Y., & Hardy-Bayle, M. C. (2002). [Formal thought disorders: French translation of the thought, language and communication assessment scale]. *Encephale, 28*(2), 109–119.

Beeman, M. (1993). Semantic processing in the right hemisphere may contribute to drawing inferences from discourse. *Brain and Language, 44*(1), 80–120.

Bleich-Cohen, M., Hendler, T., Kotler, M., & Strous, R. D. (2009). Reduced language lateralization in first-episode schizophrenia: an fMRI index of functional asymmetry. *Psychiatry Research, 171*(2), 82–93.

Bleich-Cohen, M., Sharon, H., Weizman, R., Poyurovsky, M., Faragian, S., & Hendler, T. (2012). Diminished language lateralization in schizophrenia corresponds to impaired interhemispheric functional connectivity. *Schizophrenia Research, 134*(2–3), 131–136.

Broome, M. R., Matthiasson, P., Fusar-Poli, P., Woolley, J. B., Johns, L. C., Tabraham, P., et al. (2009). Neural correlates of executive function and working memory in the 'at-risk mental state'. *The British Journal of Psychiatry, 194*(1), 25–33.

Cardno, A. G., Sham, P. C., Murray, R. M., & McGuffin, P. (2001). Twin study of symptom dimensions in psychoses. *The British Journal of Psychiatry, 179*, 39–45.

Collins, A., & Loftus, E. (1975). A spreading-activation theory of semantic processing. *Psychological Review, 82*(6), 407–428.

Curcic-Blake, B., Liemburg, E., Vercammen, A., Swart, M., Knegtering, H., Bruggeman, R., et al. (2013). When Broca goes uninformed: Reduced information flow to Broca's area in schizophrenia patients with auditory hallucinations. *Schizophrenia Bulletin, 39*(5), 1087–1095.

Curtis, V. A., Bullmore, E. T., Brammer, M. J., Wright, I. C., Williams, S. C., Morris, R. G., et al. (1998). Attenuated frontal activation during a verbal fluency task in patients with schizophrenia. *The American Journal of Psychiatry, 155*(8), 1056–1063.

Curtis, V. A., Bullmore, E. T., Morris, R. G., Brammer, M. J., Williams, S. C., Simmons, A., et al. (1999). Attenuated frontal activation in schizophrenia may be task dependent. *Schizophrenia Research, 37*(1), 35–44.

Curtis, V. A., Dixon, T. A., Morris, R. G., Bullmore, E. T., Brammer, M. J., Williams, S. C., et al. (2001). Differential frontal activation in schizophrenia and bipolar illness during verbal fluency. *Journal of Affective Disorders, 66*(2–3), 111–121.

Docherty, N. M., Cohen, A. S., Nienow, T. M., Dinzeo, T. J., & Dangelmaier, R. E. (2003). Stability of formal thought disorder and referential communication disturbances in schizophrenia. *Journal of Abnormal Psychology, 112*(3), 469–475.

Faber, R., Abrams, R., Taylor, M. A., Kasprison, A., Morris, C., & Weisz, R. (1983). Comparison of schizophrenic patients with formal thought disorder and neurologically impaired patients with aphasia. *The American Journal of Psychiatry, 140*(10), 1348–1351.

Friston, K. J., & Frith, C. D. (1995). Schizophrenia: a disconnection syndrome? *Clinical Neuroscience, 3*(2), 89–97.

Fu, C. H., Abel, K. M., Allin, M. P., Gasston, D., Costafreda, S. G., Suckling, J., et al. (2005). Effects of ketamine on prefrontal and striatal regions in an overt verbal fluency task: a functional magnetic resonance imaging study. *Psychopharmacology (Berl), 183*(1), 92–102.

Fu, C. H., Suckling, J., Williams, S. C., Andrew, C. M., Vythelingum, G. N., & McGuire, P. K. (2005). Effects of psychotic state and task demand on prefrontal function in schizophrenia: an fMRI study of overt verbal fluency. *The American Journal of Psychiatry, 162*(3), 485–494.

Fusar-Poli, P., Radua, J., McGuire, P., & Borgwardt, S. (2012). Neuroanatomical maps of psychosis onset: voxel-wise meta-analysis of antipsychotic-naive VBM studies. *Schizophrenia Bulletin, 38*(6), 1297–1307.

Gerson, S. N., Benson, F., & Frazier, S. H. (1977). Diagnosis: schizophrenia versus posterior aphasia. *The American Journal of Psychiatry, 134*(9), 966–969.

Gouzoulis-Mayfrank, E., Hermle, L., Thelen, B., & Sass, H. (1998). History, rationale and potential of human experimental hallucinogenic drug research in psychiatry. *Pharmacopsychiatry, 31*(Suppl. 2), 63–68.

Henry, J. D., & Crawford, J. R. (2005). A meta-analytic review of verbal fluency deficits in schizophrenia relative to other neurocognitive deficits. *Cognitive Neuropsychiatry, 10*(1), 1–33.

Horn, H., Federspiel, A., Wirth, M., Muller, T. J., Wiest, R., Walther, S., et al. (2010). Gray matter volume differences specific to formal thought disorder in schizophrenia. *Psychiatry Research, 182*(2), 183–186.

Horn, H., Federspiel, A., Wirth, M., Muller, T. J., Wiest, R., Wang, J. J., et al. (2009). Structural and metabolic changes in language areas linked to formal thought disorder. *The British Journal of Psychiatry, 194*(2), 130–138.

Horn, H., Jann, K., Federspiel, A., Walther, S., Wiest, R., Muller, T., et al. (2012). Semantic network disconnection in formal thought disorder. *Neuropsychobiology, 66*(1), 14–23.

Huber, G. (1955). [The pneumoencephalogram at the onset of schizophrenic disease]. *Archiv fur Psychiatrie und Nervenkrankheiten, vereinigt mit Zeitschrift fur die gesamte Neurologie und Psychiatrie, 193* (4), 406–426.

Huber, G. (1961). [Clinical and neuroradiological research on chronic schizophrenics]. *Nervenarzt, 32*, 7–15.

Javitt, D. C. (2009). Sensory processing in schizophrenia: neither simple nor intact. *Schizophrenia Bulletin, 35*(6), 1059–1064.

Kantrowitz, J. T., Hoptman, M. J., Leitman, D. I., Silipo, G., & Javitt, D. C. (2013). The 5% difference: early sensory processing predicts sarcasm perception in schizophrenia and schizo-affective disorder. *Psychological Medicine*, 1–12.

Ketteler, D., Theodoridou, A., Ketteler, S., & Jager, M. (2012). High order linguistic features such as ambiguity processing as relevant diagnostic markers for schizophrenia. *Schizophrenia Research and Treatment, 2012*, 825050.

Kinney, D. K., Holzman, P. S., Jacobsen, B., Jansson, L., Faber, B., Hildebrand, W., et al. (1997). Thought disorder in schizophrenic and control adoptees and their relatives. *Archives of General Psychiatry, 54*(5), 475–479.

Kircher, T. (2003). *Neuronale Korrelate psychopathologischer Syndrome Denk- und Sprachprozesse bei Gesunden und Patienten mit Schizophrenie*. Darmstadt: Steinkopff Verlag.

Kircher, T., Krug, A., Markov, V., Whitney, C., Krach, S., Zerres, K., et al. (2009). Genetic variation in the schizophrenia-risk gene neuregulin 1 correlates with brain activation and impaired speech production in a verbal fluency task in healthy individuals. *Human Brain Mapping, 30*(10), 3406–3416.

Kircher, T., Krug, A., Stratmann, M., Ghazi, S., Schales, C., Frauenheim, M., et al. (2014). A rating scale for the assessment of objective and subjective formal Thought and Language Disorder (TALD). *Schizophrenia Research, 160*(1–3), 216–221.

Kircher, T., Liddle, P., Brammer, M., Murray, R., & McGuire, P. (2003). [Neural correlates of "negative" formal thought disorder]. *Nervenarzt, 74*(9), 748–754.

Kircher, T., Whitney, C., Krings, T., Huber, W., & Weis, S. (2008). Hippocampal dysfunction during free word association in male patients with schizophrenia. *Schizophrenia Research, 101*(1–3), 242–255.

Kircher, T. T., Leube, D. T., Erb, M., Grodd, W., & Rapp, A. M. (2007). Neural correlates of metaphor processing in schizophrenia. *Neuroimage, 34*(1), 281–289.

Kircher, T. T., Liddle, P. F., Brammer, M. J., Williams, S. C., Murray, R. M., & McGuire, P. K. (2001). Neural correlates of formal thought disorder in schizophrenia: preliminary findings from a functional magnetic resonance imaging study. *Archives of General Psychiatry, 58*(8), 769–774.

Kircher, T. T., Oh, T. M., Brammer, M. J., & McGuire, P. K. (2005). Neural correlates of syntax production in schizophrenia. *The British Journal of Psychiatry, 186*, 209–214.

Kircher, T. T., Rapp, A., Grodd, W., Buchkremer, G., Weiskopf, N., Lutzenberger, W., et al. (2004). Mismatch negativity responses in schizophrenia: a combined fMRI and whole-head MEG study. *The American Journal of Psychiatry, 161*(2), 294–304.

Krug, A., Nieratschker, V., Markov, V., Krach, S., Jansen, A., Zerres, K., et al. (2010). Effect of CACNA1C rs1006737 on neural correlates of verbal fluency in healthy individuals. *Neuroimage, 49*(2), 1831–1836.

Krystal, J. H., Karper, L. P., Seibyl, J. P., Freeman, G. K., Delaney, R., Bremner, J. D., et al. (1994). Subanesthetic effects of the noncompetitive NMDA antagonist, ketamine, in humans. Psychotomimetic, perceptual, cognitive, and neuroendocrine responses. *Archives of General Psychiatry, 51*(3), 199–214.

Kuperberg, G. R. (2008). Building meaning in schizophrenia. *Clinical EEG and Neuroscience, 39*(2), 99–102.

Kuperberg, G. R., Deckersbach, T., Holt, D. J., Goff, D., & West, W. C. (2007). Increased temporal and prefrontal activity in response to semantic associations in schizophrenia. *Archives of General Psychiatry, 64*(2), 138–151.

Leitman, D. I., Hoptman, M. J., Foxe, J. J., Saccente, E., Wylie, G. R., Nierenberg, J., et al. (2007). The neural substrates of impaired prosodic detection in schizophrenia and its sensorial antecedents. *The American Journal of Psychiatry, 164*(3), 474–482.

Liddle, P. F., Ngan, E. T., Caissie, S. L., Anderson, C. M., Bates, A. T., Quested, D. J., et al. (2002). Thought and language index: an instrument for assessing thought and language in schizophrenia. *The British Journal of Psychiatry, 181*, 326–330.

Manoach, D. S. (2003). Prefrontal cortex dysfunction during working memory performance in schizophrenia: Reconciling discrepant findings. *Schizophrenia Research, 60*(2–3), 285–298.

Manschreck, T. C., Maher, B. A., Milavetz, J. J., Ames, D., Weisstein, C. C., & Schneyer, M. L. (1988). Semantic priming in thought disordered schizophrenic patients. *Schizophrenia Research, 1*(1), 61–66.

Mashal, N., Vishne, T., Laor, N., & Titone, D. (2013). Enhanced left frontal involvement during novel metaphor comprehension in schizophrenia: evidence from functional neuroimaging. *Brain and Language, 124*(1), 66–74.

Matsumoto, K., Kircher, T. T., Stokes, P. R., Brammer, M. J., Liddle, P. F., & McGuire, P. K. (2013). Frequency and neural correlates of pauses in patients with formal thought disorder. *Frontiers in Psychiatry, 4*, 127.

McGuire, P. K., Quested, D. J., Spence, S. A., Murray, R. M., Frith, C. D., & Liddle, P. F. (1998). Pathophysiology of 'positive' thought disorder in schizophrenia. *The British Journal of Psychiatry, 173*, 231–235.

McKenna, P., & Oh, T. (2005). *Schizophrenic speech* (Vol. 1). Cambridge: Cambridge University Press.

Morice, R., & McNicol, D. (1985). The comprehension and production of complex syntax in schizophrenia. *Cortex, 21*(4), 567–580.

Musso, F., Brinkmeyer, J., Ecker, D., London, M. K., Thieme, G., Warbrick, T., et al. (2011). Ketamine effects on brain function—simultaneous fMRI/EEG during a visual oddball task. *Neuroimage, 58*(2), 508–525.

Nagels, A., Kircher, T., Dietsche, B., Backes, H., Marquetand, J., & Krug, A. (2012). Neural processing of overt word generation in healthy individuals: the effect of age and word knowledge. *Neuroimage, 61*(4), 832–840.

Nagels, A., Kirner-Veselinovic, A., Krach, S., & Kircher, T. (2011). Neural correlates of *S*-ketamine induced psychosis during overt continuous verbal fluency. *Neuroimage, 54*(2), 1307–1314.

Nagels, A., Kirner-Veselinovic, A., Wiese, R., Paulus, F. M., Kircher, T., & Krach, S. (2012). Effects of ketamine-induced psychopathological symptoms on continuous overt rhyme fluency. *European Archives of Psychiatry and Clinical Neuroscience, 262*(5), 403–414.

Nagels, A., Stratmann, M., Ghazi, S., Schales, C., Frauenheim, M., Turner, L., et al. (2013). The German translation and validation of the scale for the assessment of thought, language and communication: A factor analytic study. *Psychopathology, 46*(6), 390–395.

Parnas, J., Moller, P., Kircher, T., Thalbitzer, J., Jansson, L., Handest, P., et al. (2005). EASE: examination of anomalous self-experience. *Psychopathology, 38*(5), 236–258.

Ragland, J. D., Moelter, S. T., Bhati, M. T., Valdez, J. N., Kohler, C. G., Siegel, S. J., et al. (2008). Effect of retrieval effort and switching demand on fMRI activation during semantic word generation in schizophrenia. *Schizophrenia Research, 99*(1–3), 312–323. [Epub 2007 Dec 2026].

Rapp, A. M., Langohr, K., Mutschler, D. E., Klingberg, S., Wild, B., & Erb, M. (2013). Isn't it ironic? Neural correlates of irony comprehension in schizophrenia. *PLoS One*, *8*(9), e74224.

Rapp, A. M., & Steinhauser, A. E. (2013). Functional MRI of sentence-level language comprehension in schizophrenia: a coordinate-based analysis. *Schizophrenia Research*, *150*(1), 107–113.

Ratnanather, J. T., Poynton, C. B., Pisano, D. V., Crocker, B., Postell, E., Cebron, S., et al. (2013). Morphometry of superior temporal gyrus and planum temporale in schizophrenia and psychotic bipolar disorder. *Schizophrenia Research*, *150*(2–3), 476–483.

Rossell, S. L., & David, A. S. (2006). Are semantic deficits in schizophrenia due to problems with access or storage? *Schizophrenia Research*, *82*(2–3), 121–134.

Sans-Sansa, B., McKenna, P. J., Canales-Rodriguez, E. J., Ortiz-Gil, J., Lopez-Araquistain, L., Sarro, S., et al. (2013). Association of formal thought disorder in schizophrenia with structural brain abnormalities in language-related cortical regions. *Schizophrenia Research*, *146*(1–3), 308–313.

Sass, K., Heim, S., Sachs, O., Straube, B., Schneider, F., Habel, U., et al. (2014). Neural correlates of semantic associations in patients with schizophrenia. *European Archives of Psychiatry and Clinical Neuroscience*, *264*(2), 143–154.

Sawa, A., & Snyder, S. H. (2002). Schizophrenia: diverse approaches to a complex disease. *Science*, *296*(5568), 692–695.

Schaufelberger, M., Senhorini, M. C., Barreiros, M. A., Amaro, E., Jr., Menezes, P. R., Scazufca, M., et al. (2005). Frontal and anterior cingulate activation during overt verbal fluency in patients with first episode psychosis. *Revista Brasileira de Psiquiatria*, *27*(3), 228–232.

Shenton, M. E., Dickey, C. C., Frumin, M., & McCarley, R. W. (2001). A review of MRI findings in schizophrenia. *Schizophrenia Research*, *49*(1–2), 1–52.

Shenton, M. E., Kikinis, R., Jolesz, F. A., Pollak, S. D., LeMay, M., Wible, C. G., et al. (1992). Abnormalities of the left temporal lobe and thought disorder in schizophrenia. A quantitative magnetic resonance imaging study. *The New England Journal of Medicine*, *327*(9), 604–612.

Shepherd, A. M., Laurens, K. R., Matheson, S. L., Carr, V. J., & Green, M. J. (2012). Systematic meta-review and quality assessment of the structural brain alterations in schizophrenia. *Neuroscience and Biobehavioral Reviews*, *36*(4), 1342–1356.

Spitzer, M., Braun, U., Maier, S., Hermle, L., & Maher, B. A. (1993). Indirect semantic priming in schizophrenic patients. *Schizophrenia Research*, *11*(1), 71–80.

Straube, B., Green, A., Sass, K., & Kircher, T. (2014). Superior temporal sulcus disconnectivity during processing of metaphoric gestures in schizophrenia. *Schizophrenia Bulletin*, *40*(4), 936–944.

Straube, B., Green, A., Sass, K., Kirner-Veselinovic, A., & Kircher, T. (2013). Neural integration of speech and gesture in schizophrenia: Evidence for differential processing of metaphoric gestures. *Human Brain Mapping*, *34*(7), 1696–1712.

Strelnikov, K. (2010). Schizophrenia and language--shall we look for a deficit of deviance detection? *Psychiatry Research*, *178*(2), 225–229.

Subotnik, K. L., Bartzokis, G., Green, M. F., & Nuechterlein, K. H. (2003). Neuroanatomical correlates of formal thought disorder in schizophrenia. *Cognitive Neuropsychiatry*, *8*(2), 81–88.

Szoke, A., Trandafir, A., Dupont, M. E., Meary, A., Schurhoff, F., & Leboyer, M. (2008). Longitudinal studies of cognition in schizophrenia: meta-analysis. *The British Journal of Psychiatry*, *192*(4), 248–257.

Umbricht, D., & Krljes, S. (2005). Mismatch negativity in schizophrenia: a meta-analysis. *Schizophrenia Research*, *76*(1), 1–23.

Vernaleken, I., Klomp, M., Moeller, O., Raptis, M., Nagels, A., Rosch, F., et al. (2012). Vulnerability to psychotogenic effects of ketamine is associated with elevated D2/3-receptor availability. *The International Journal of Neuropsychopharmacology*, 1–10.

Vollenweider, F. X., & Kometer, M. (2010). The neurobiology of psychedelic drugs: implications for the treatment of mood disorders. *Nature Reviews Neuroscience*, *11*(9), 642–651.

Vollenweider, F. X., Leenders, K. L., Oye, I., Hell, D., & Angst, J. (1997). Differential psychopathology and patterns of cerebral glucose utilisation produced by (S)- and (R)-ketamine in healthy volunteers using Positron Emission Tomography (PET). *European Neuropsychopharmacology*, *7*(1), 25–38.

Vollenweider, F. X., Leenders, K. L., Scharfetter, C., Antonini, A., Maguire, P., Missimer, J., et al. (1997). Metabolic hyperfrontality and psychopathology in the ketamine model of psychosis using Positron Emission Tomography (PET) and [^{18}F]fluorodeoxyglucose (FDG). *European Neuropsychopharmacology*, *7*(1), 9–24.

Vollenweider, F. X., Vontobel, P., Oye, I., Hell, D., & Leenders, K. L. (2000). Effects of (S)-ketamine on striatal dopamine: a [11C]raclopride PET study of a model psychosis in humans. *Journal of Psychiatric Research*, *34*(1), 35–43.

Weinstein, S., Werker, J. F., Vouloumanos, A., Woodward, T. S., & Ngan, E. T. (2006). Do you hear what I hear? Neural correlates of thought disorder during listening to speech in schizophrenia. *Schizophrenia Research*, *86*(1–3), 130–137.

Wernicke, C. (1906). *Grundriss der Psychiatrie in klinischen Vorlesungen*. Leipzig: Thieme.

CHAPTER

72

Specific Language Impairment

Julia L. Evans[1,2] *and Timothy T. Brown*[3]

[1]School of Behavioral and Brain Sciences, The University of Texas at Dallas, Richardson, TX, USA; [2]Center for Research in Language, University of California, San Diego, La Jolla, CA, USA; [3]Department of Neurosciences, University of California, San Diego, School of Medicine, La Jolla, CA, USA

72.1 INTRODUCTION

Specific language impairment (SLI) is a developmental language disorder characterized by the inability to master spoken and written language expression and comprehension, despite normal nonverbal intelligence, hearing acuity, and speech motor skills, and no overt physical disability, recognized syndrome, or other mitigating medical factors known to cause language disorders in children (Tager-Flusberg & Cooper, 1999). Although numbers vary slightly across countries, in the United States SLI is estimated to occur in ~7% of English-speaking 5-year-old children (Tomblin, Smith, & Zhang, 1997). It persists, fully or partially, into adulthood, placing individuals with SLI at risk for poor academic performance, difficulty developing and maintaining friendships and significant relationships, difficulty in the work environment, and reduced earning potential and standard of living. The stress that accompanies coping with the disorder also places both individuals with SLI and their families at risk for secondary stress-related physical, social, and emotional problems (Catts, Bridges, Little, & Tomblin, 2008; Conti-Ramsden, 2013; Durkin & Conti-Ramsden, 2007; Tomblin, Freese, & Records, 1992).

SLI has been described in the literature for more than a century; however, despite decades of study, the cause of the disorder is still unknown (Leonard, 2014). Historically, hypothesized accounts of SLI have been based predominantly on experimental data and behavioral observations. Taking advantage of advances in noninvasive brain imaging techniques that allow for increased spatial and temporal resolution, a growing number of SLI researchers have begun to focus on brain structure and brain function in this population. From these studies of the neurobiology of SLI, a richer understanding of the nature of brain–language relationships is beginning to develop.

Some of the techniques that have been used in these investigations of the neurobiology of SLI include structural magnetic resonance imaging (sMRI), functional magnetic resonance imaging (fMRI), diffusion tensor imaging (DTI), functional transcranial Doppler ultrasound (fTCD), single-photon emission computed tomography (SPECT), event-related potentials (ERPs), magnetoencephalography (MEG), and anatomically constrained magnetoencephalography (aMEG). Although each of these techniques has limitations, when taken together a profile emerges for SLI that is characterized not by a single pattern of gross abnormality but also by patterns of atypical brain morphology and tissue properties for some brain regions but not others, coupled with atypical patterns of neural activity mediating some aspects of language but not others.

72.2 NEUROPSYCHOLOGICAL PROFILE

There is no universally agreed on set of diagnostic criteria or terminology for SLI; however, there is general agreement among clinicians and researchers that these are children who have notable language disabilities that are below age-level expectations and are significantly out of line with other aspects of the child's development (for detailed discussion see Bishop, 2014; Reilly, Bishop, & Tomblin, 2014).[1] As with any clinical disorder,

[1]In this chapter, we use the term "specific" to denote "ideiopathic" or "functional" language impairment—of unknown cause—as recently proposed by Bishop (2014).

Neurobiology of Language. DOI: http://dx.doi.org/10.1016/B978-0-12-407794-2.00072-9

there are individual differences in severity and symptomatology of SLI, yet despite this heterogeneity, there are some common deficit profiles. The most common profile is delayed onset and slower acquisition of the lexical, syntax, and grammatical morphological aspects of spoken and written language coupled with nonlinguistic cognitive processing, learning, and memory impairments (Leonard, 2000). These nonlinguistic deficits typically include slower speed of processing, poor phonological and verbal working memory, poor auditory processing, and slow and inefficient sequential procedural learning and memory (Leonard & Weber-Fox, 2012).

72.3 STRUCTURAL IMAGING OF SLI

Although the anatomical correlates of acquired language disorders in other clinical populations have received considerable attention over the past century (Geschwind, 1970, 1979), studies of brain morphology in SLI began in earnest in the early 1990s. In unimpaired populations, asymmetries in brain morphology within left hemisphere perisylvian areas are the norm, most notably expressed as relatively greater volumetric measures as compared with homologous right hemisphere regions, and they have been viewed as support for a left hemisphere dominance model of language (Dorsaint-Pierre et al., 2006; Szaflarski et al., 2012). Guided by Orton's proposal that a lack of left hemisphere dominance for language could result in the failure to develop language (Molfese & Segalowitz, 1988; Orton, 1937; Vallortigara, Rogers, & Bisazza, 1999), these initial studies focused on reversed or a lack of hemispheric asymmetry in SLI. Early studies used MRI and manual slice-wise tracing to investigate left–right hemisphere (a)symmetry in those regions specifically hypothesized to support language (Clark & Plante, 1998; Jernigan, Hesselink, Sowell, & Tallal, 1991; Plante, Swisher, Vance, & Rapcsak, 1991). These studies found a higher prevalence of "atypical" cerebral asymmetry (i.e., $R > L$) in the language-impaired participants as compared with normal language controls. Follow-up studies continue to report lacking or rightward asymmetry, as well as abnormal patterning, shape, and volume of brain morphology for SLI, most notably in anterior and posterior perisylvian areas such as pars triangularis, planum temporale, and the posterior ascending ramus in individuals with SLI (De Fossé et al., 2004; Gauger, Lombardino, & Leonard, 1997; Soriano-Mas et al., 2009).

With advances in imaging technologies, automated postprocessing, and statistical and modeling approaches to the analysis of brain imaging data, researchers are now able to quantify anatomical features of the brain in more nuanced units. Recent automated techniques can reliably separate cerebral and cerebellar gray and white matter into cortical and subcortical compartments, measure the volumes of deep nuclei and cerebrospinal fluid (CSF) compartments, accurately reconstruct the cortex and subdivide its volume into measures of surface area and thickness at every region, and map the tissue properties and volumes of white matter connections across different cortical and subcortical regions. These morphological approaches all begin with noninvasive neuroimaging data, typically from MRI data—with one of the most common methods being voxel-based morphometry (VBM).

Using largely VBM-based processing and analysis techniques, data from studies of brain morphology in SLI consistently show atypical gray and white matter volumes in some brain regions but not in all brain regions (Badcock, Bishop, Hardiman, Barry, & Watkins, 2012; Girbau-Massana, Garcia-Marti, Marti-Bonmati, & Schwartz, 2014; Lee, Nopoulos, & Tomblin, 2013; Mayes, Reilly, & Morgan, 2015; Soriano-Mas et al., 2009). For instance, Badcock et al. (2012) observed that although total gray matter volume was no different for SLI and normal controls, gray matter volume was significantly different in some regions for SLI as compared with normal controls. Specifically, Badcock and colleagues observed that their SLI group had significantly *more* gray matter than controls in the left inferior gyrus (IFG), right insula, and left intraparietal sulcus, but significantly *less* gray matter in medial frontal pole, medial superior parietal cortex, posterior superior temporal sulcus (STS), superior temporal gyrus (STG), and subcortical regions such as the caudate nucleus and in the midbrain at the level of the substantia nigra.

One advantage of the VBM approach over previously common manual tracing methods is that researchers can more readily attempt to control for individual differences in intracranial and/or intrahemisphere volume by norm-referencing the morphology of specific regions of interest to individual intracranial and/or intrahemisphere volumes. For instance, Lee et al. (2013) controlled for differences in intracranial volume (ICV) in their SLI and normal control groups. They observed absolute volume of the bilateral caudate nucleus, left globus pallidus, bilateral thalamus, and cerebral lobes to be *less* for SLI as compared with normal controls. However, ICV was also significantly *less* for the SLI as compared with normal controls. To address this potential confound of differences in ICV between the two groups, Lee and colleagues adjusted region-wise measures based on each participant's total ICV. Using regional measures as a proportion of total volume, Lee and colleagues instead found that the putamen, right globus pallidus, nucleus accumbens, and hippocampus adjusted volumes were significantly *greater* for the SLI group as compared with normal controls, whereas adjusted volume of the cerebral lobes were now no longer different for SLI and normal controls.

A small number of researchers are beginning to examine the CSF volume in SLI; however, findings from these studies are somewhat inconsistent (Soriano-Mas et al., 2009). Comparing absolute values, Soriano-Mas et al. (2009) observed CSF volumes to be normal for children with SLI. However, using VBM and controlling for individual differences in ICV, Girbau-Massana et al. (2014) found a different pattern. They examined CSF volumes in a group of children with SLI, a group with SLI and reading disabilities, and a group of normal controls and found *greater* CSF volume for both SLI groups as compared with normal controls. Although the possible cognitive implications of greater CSF volume are not clear, similar effects in age-matched comparisons have been associated with a variety of clinical groups and may reflect relatively decreased total gray matter.

In addition to investigating cortical and subcortical structural volumes in SLI, researchers have also examined white matter in SLI using DTI measures. Based on anisotropic water movement within fibrous tissue, DTI measures provide indices of the overall *magnitude* of diffusion within a given white matter tract regardless of direction and indices of the *directionality* of diffusion (Basser, Mattiello, & LeBihan, 1994; Lim, Han, Uhlhaas, & Kaiser, 2013; Beaulieu, 2002). Measures of diffusion magnitude within white matter, such as mean diffusivity (MD), tend to show age-related decreases across childhood development, and measures of diffusion directionality, such as fractional anisotropy (FA), tend to show developmental increases (Brown et al., 2012). These two age-related effects are believed to be related in part to the increasing degree of myelination that occurs around the axons of neurons, and their specific time courses vary by brain region and tract. Using DTI tractography, maps of fiber tracts in the brain can be virtually reconstructed, revealing streamlines of white matter that purportedly connect one brain region to another. Although commonly used tractography methods require manual guidance to reveal white matter tract regions of interest (Mori & Van Zijl, 2002; Mori & Zhang, 2006), recent methods provide fully automated tractography, thus avoiding the subjective biases inherent in choosing interregional streamlines by hand (Hagler et al., 2009).

One of the first studies to use DTI to examine white matter pathways in SLI was conducted by Kim et al. (2006). They found decreased anisotropy in the genu of the corpus callosum for SLI despite the appearance of "normal" brain characteristics based on MRI scans, indicating that individuals with SLI may have grossly normal brain morphology but still have that abnormalities in the brain at the microstructural level that are not evident on conventional MRI scans. For example, abnormal corpus callosum white matter microstructure might contribute to abnormal integration of information between the left and right hemispheres.

Similar to studies of brain morphology, data from DTI studies of SLI also have found diffusion properties within white matter tracts that are outside typical ranges in some, but not all, regions of the brain (Kim et al., 2006; Lee et al., 2013; Verhoeven et al., 2012). For instance, comparing FA values in individuals with autism spectrum disorder, those with SLI, and normal controls, Verhoeven and colleagues (Verhoeven et al., 2012) observed no differences in FA values in the superior longitudinal fascicle (SLF) between the normal control and autism spectrum groups, but they found that SLF FA values were significantly reduced in the SLI group as compared with normal controls. Similarly, Lee and colleagues (2013) found FA values in the hippocampus, caudate nucleus, putamen, and nucleus accumbens to be similar for those with SLI and normal controls, but FA values for thalamus, globus pallidus, and superior longitudinal fasciculus were found to be significantly *lower* in the SLI group as compared with normal controls.

Taken together, imaging studies of brain morphology in SLI suggest a notable absence of gross brain abnormalities. Instead, these studies suggest a pattern of abnormal left/right hemisphere asymmetry in some brain regions for SLI, reduced cortical and subcortical volumetric values for some brain regions but not others, and reduced measures of diffusion directionality (i.e., FA) for SLI as compared with controls, again in some regions but not in others. Data from more recent studies indicate that comparison of wide-ranging or global values and/or absolute values alone may not capture the region-specific nature of many differences and suggest that norm-referencing volume in regions of interest to individual interhemisphere/intrahemisphere/cranial volumes may be an important methodological consideration in future research.

Another important issue for future studies of SLI brain morphology and tissue properties will be to use more consistent brain segmentation techniques, including cortical surface-based reconstruction and analysis, which can accurately separate cortical volume into independent measures of surface area and thickness, which vary regionally and have very distinct normal developmental trajectories (Brown et al., 2012). Manual tracing and VBM-based methods in some regions can lead to inaccurate volume and asymmetry measurements because of partial voluming of gray matter, white matter, and CSF (or all three) at the edges of structures of interest. Further, manual tracking methods rely on subjective visual determination of specific cortical structures, and this assumes that gross structural features are clearly visible relative to the boundaries of cortical regions to be defined. Even issues such as head motion and position can introduce noise into the data. For instance, even when researchers use standardized head positioning, the slightest head rotation will result in qualitative differences in the appearance of brain structures along the plane being imaged and this will lead to spurious "within plane" asymmetries of apparently comparable structures in the two hemispheres (Jernigan, Hesselink, et al., 1991). The methods used to measure cortical volume, area, and thickness and how these measures are related and/or distinguished from one another can also influence researcher's interpretations of the data. New techniques now make it possible for researchers to obtain topologically accurate measurements of cortical brain morphology and relate this information to subcortical anatomy.

To gain a more accurate picture of cortical morphology in SLI, future research should use three-dimensional surface-based reconstructions to examine cortical thickness and surface area separately, instead of simply measuring cortical volume. Cortical surface area and thickness have distinct cellular mechanisms and genetic etiologies (Chen et al., 2013, 2012, 2011). For typical individuals, total cortical surface area, average cortical thickness, and total intracranial volume (TIV) (global measures and head size) are each highly heritable (Brans et al., 2010; Panizzon et al., 2009), with genetic influences accounting for 89% of the variance in cortical surface area, 81% of the variance for cortical thickness, and 78% of the variance for TIV.

A complete picture of the topological organization of brain morphology for SLI will require a developmental approach in which both age-related and gender-related aspects of regionally varying brain morphology and tissue properties are factored into any study of SLI brain structure (Brown et al., 2012; Khundrakpam et al., 2013; Squeglia, Jacobus, Sorg, Jernigan, & Tapert, 2013; Tamnes et al., 2013). Before child brain imaging became commonplace, it was assumed that the majority of changes in brain tissue occurred during the immediate postnatal brain growth spurt; however, studies comparing gray and white matter morphology and tissue properties in children and adults now show that considerable changes in brain maturation continue to take place throughout adolescence and well into adulthood, and that these changes vary considerably in girls and boys (Brown & Jernigan, 2012; Jernigan, Archibald, et al., 1991; Jernigan & Tallal, 2010).

72.4 FUNCTIONAL IMAGING OF SLI

72.4.1 Localization of Brain Activity in SLI

Degree of hemispheric specialization, so-called cerebral dominance, for language functioning in SLI has been examined using techniques such as fTCD and SPECT. fTCD compares blood flow velocity within cerebral arteries related to regional increases in associated neural activity linked to the performance of cognitive tasks. Often in language studies, the left and right middle cerebral arteries are measured during fTCD, accounting for more than 70% of the blood supply to the respective hemispheres. As with other hemodynamic and perfusion-sensitive neuroimaging techniques, fTCD works under the premise that regional increases in neural activity are associated with greater glucose and oxygen consumption that must be replenished via enhanced blood flow to the area (Müller, Neumann, Lohmann, Mildner, & Cramon, 2005).

SPECT relies on an injection of a radioactive tracer that travels through the brain's vasculature and provides a reading of regional blood flow concentrations detected using a ring of radioactivity sensors. The development of three-dimensional techniques used in SPECT and positron emission tomography (PET) allow for visualization of blood flow in any region of the body, including cortical, subcortical, and cerebellar brain structures. The decay of the radiotracer depends on which radionuclide is used, but typically a series of scans is completed within several minutes, providing measures of regional cerebral blood flow (rCBF) at the time of injection. Because change in the cerebral blood flow is linearly related to glucose consumption, changes in cerebral blood flow are believed to be an indication of local neuronal activity (Saper et al., 2000).

In unimpaired populations, language processing is most often characterized by relatively greater activity in the left as compared with the right cerebral hemisphere (Brown et al., 2005; Knecht et al., 1998). SLI studies using fTCD more commonly show abnormal patterns of cerebral blood flow, characterized either by relatively greater right hemisphere lateralization or by bilateral cerebral blood flow. For example, one fTCD study of 11 individuals with SLI found that two of the subjects (18.2%) had left lateralization, six subjects (54.5%) showed right lateralization, and three subjects with SLI (27.3%) showed bilateral or mixed dominance (Whitehouse & Bishop, 2008). SLI studies using SPECT also show an abnormal pattern of cerebral blood flow, characterized by reduced asymmetry and/or relative hypoperfusion of the left hemisphere—an atypical pattern as compared with unimpaired populations (Chiron et al., 1999; Denays et al., 1989; Lou, Henriksen, & Bruhn, 1990; Ors et al., 2005; Tzourio, Heim, Zilbovicius, Gerard, & Mazoyer, 1994).

A small number of SLI studies have used fMRI to investigate the cerebral functional organization of language and cognitive processing (Badcock et al., 2012; Ellis Weismer, Plante, Jones, & Tomblin, 2005; Hugdahl et al., 2004; Schmithorst, Yuan, & Plante, 2007). fMRI is a functional neuroimaging procedure using MRI technology that measures regional brain activity by detecting associated changes in the relative concentrations of oxygenated and deoxygenated hemoglobin within blood vessels. This technique, similar to other hemodynamic methods such as PET and SPECT, relies on the fact that deoxygenated hemoglobin is paramagnetic and that increases and decreases in its concentrations are regionally coupled with changes in neural activity. The primary form of fMRI uses the blood oxygen level–dependent (BOLD) contrast. Because of the high spatial resolution of MRI, fMRI provides for a very precise localization of brain activity. However, because fMRI must rely on a sluggish hemodynamic proxy of neural activity, it provides indirect information about brain activation that is limited to much poorer temporal resolution (i.e., smeared over several seconds) than neural activity itself (i.e., in milliseconds).

Similar to the fTCD and SPECT studies, data from fMRI studies of SLI indicate abnormal patterns of brain activation during language and cognitive processing tasks (Badcock et al., 2012; de Guibert et al., 2011; Ellis Weismer et al., 2005; Hugdahl et al., 2004). These studies report reduced left lateralization in those areas viewed as core to language functions, normal lateralization, and significantly reduced activation in the left hemisphere (relative to the right) in frontal, parietal, or temporal areas (Ellis Weismer et al., 2005; Hugdahl et al., 2004). For instance, Hugdahl et al. (2004) observed reduced brain activation in left temporal and frontal lobes, primarily in the middle temporal gyrus bordering the STS in SLI as compared with normal controls during lexical processing and phonological awareness tasks. Badcock et al. (2012) also observed reduced left hemisphere as compared with right hemisphere activation in SLI as compared with normal controls, as well as reduced activation for SLI as compared with normal controls, particularly in the left IFG (pars orbitalis), right putamen, and the STS/G bilaterally. Studies of children with a documented delay in onset of spoken language (e.g., late talkers) also demonstrate significantly reduced activation in the left superior temporal gyrus as compared with normal controls (Preston et al., 2010).

What is notable about the Badcock et al. (2012) study is that the researchers also directly examined the relationship between brain structure and brain function in children with SLI, their siblings, and normal language controls. As discussed, the SLI group had significantly more gray matter as compared with normal controls in the left inferior frontal gyrus (IFG) and significantly less gray matter in the right caudate nucleus and the STS bilaterally. The SLI group also had atypical brain activation during language processing in the left IFG and posterior temporal cortex bilaterally as well. The pattern of linkage between atypical structure and function was not evident in Broca's area, however. Specifically, in regions where the children with SLI had increased gray matter as compared with controls, they showed decreased brain activation during language processing. Thus, despite the coincident atypical structure and function in SLI, gray matter volume and brain activity may be more closely related in some brain regions than in others.

Taken together, brain activation patterns for SLI appear to be characterized by atypically reduced levels of activity within the left hemisphere as compared with the right, as well as significantly reduced overall amplitude of brain activation when compared with normal controls not only in the left hemisphere but in contralateral cortical and subcortical areas as well. One question is whether the functional brain organization is atypical for all aspects of language in SLI or only for some subcomponents or processing operations of language tasks. Studies examining the relationships between brain organization and language performance now indicate that this relationship is a dynamic one involving complex, distributed brain systems, and that simplistic concepts such as hemispheric asymmetry or cerebral dominance are not likely to capture the subtleties of these brain-behavior relationships (Dorsaint-Pierre et al., 2006; Keller et al., 2011). In particular, studies of the neurobiology of language in typical individuals suggest that while the left hemisphere is clearly dominant for many aspects of language, the functional organization of language is more complex than a simple left–right dichotomy and comprises multiple brain systems that also vary meaningfully according to the anterior and posterior dimensions of both hemispheres (Brown et al., 2014; Federmeier, Wlotko, & Meyer, 2008; Friederici, 2005; Friederici, Wang, Herrmann, Maess, & Oertel, 2000; Helenius, Parviainen, Paetau, & Salmelin, 2009; Kuhl & Rivera-Gaxiola, 2008; Poeppel, Emmorey, Hickok, & Pylkkänen, 2012; Pylkkänen & Marantz, 2003).

72.4.2 Time Course of Cortical Activation in SLI

Although the high spatial resolution of SPECT and fMRI make them well-suited to investigate the location of brain activation in SLI, the temporal resolution of SPECT and fMRI is on the order of seconds to minutes, making these techniques poorly suited to investigate the time course of brain activity. In contrast, EEG and MEG, which measure electrical and magnetic fluctuations, respectively, have a temporal resolution on the order of milliseconds, making them better suited to examine the time course of brain activity (Hansen, Kringelbach, & Salmelin, 2010; Näätänen, Ilmoniemi, & Alho, 1994). The time-locked changes in EEG and MEG activity that are triggered by an external stimulus or event can be classified in various ways according to their amplitude, timing relative to stimulus onset, polarity, and anatomical site of generation. These "event-related" potentials (ERPs) are commonly referred to by their polarity (negative/positive) and latency in milliseconds. For instance, the large, negative-going EEG-evoked potential measured peaks in adults between 80 and 120 milliseconds after the onset of a stimulus that are referred to as the N100 or N1 (its equivalent in MEG is referred to as M100).

ERP and MEG components can also be classified in terms of the "functional" processes that are believed to be associated with each of them (Luck & Kappenman, 2012). From a functional perspective, differences in the time course of a component are often viewed as a reflection of the time course of the cognitive processes that are being engaged across different experimental conditions or experimental stimuli. Similarly, differences in the degree of amplitude modulation of a component are often viewed as a measure of cognitive processing "effort" inherent in the stimuli and/or experimental task, or expectation (Frisch, Kotz, Yves von Cramon, & Friederici, 2003; Kutas & Federmeier, 2011; Polich & Kok, 1995).

The ERP components that have been examined in SLI include, but are not limited to, mismatch negativity (MMN), T-complex, late discriminative negativity (LDN), early left anterior negativity (ELAN), N1, P2, N1-P2, P3, N400, and P600. In these studies, brain activity has been investigated within the context of different hypothesized deficits in SLI. For instance, researchers have examined the T-complex and MMN components in investigations of central auditory processing deficits in SLI (Bishop, 2007; Bishop, Hardiman, & Barry, 2012), the N400 in studies of lexical-semantic knowledge in SLI (Neville, Coffey, Holcomb, & Tallal, 1993; Weber-Fox, Leonard, Wray, & Tomblin, 2010), the P3 and N2 in studies of cognitive and working memory deficits in SLI (Evans, Selinger, & Pollak, 2011; Epstein, Shafer, Melara, & Schwartz, 2014), and the P600 and left-anterior negativity (LAN) in studies of syntactic knowledge in SLI (Epstein, Hestvik, Shafer, & Schwartz, 2013; Friederici, 2006; Fonteneau & van der Lely, 2008; Purdy, Leonard, Weber-Fox, & Kaganovich, 2014; Sabisch, Hahne, Glass, Suchodoletz von, & Friederici, 2009; Weber-Fox et al., 2010).

Much of the electrophysiological data indicate abnormal cerebral information processing for SLI, characterized by patterns of both atypical timing and peak amplitudes for the early occurring components such as MMN, N1-P2 (Kaganovich, Schumaker, Leonard, Gustafson, & Macias, 2014; Kaganovich, Schumaker, Macias, & Gustafson, 2014; Schwartz & Shafer, 2012) and N2 (Epstein et al., 2014). For the later occurring components (e.g., P3, N400, P600), the pattern is somewhat different, characterized broadly as one of more normal peak latency but of abnormal amplitude modulation and lateralization of cortical activation for SLI as compared with normal controls. There are inconsistencies in these findings, however. For instance, atypical attenuation and latency of MMN have been observed for SLI in some, but not all, studies (Bishop, 2007). Similarly, the latency and amplitude modulation of the N1-P2 has been reported to be normal for SLI in some studies, but not others (Adams, Courchesne, Elmasian, & Lincoln, 1987; Çeponiené, Cummings, Wulfect, Ballantyne, & Townsend, 2009; McArthur, Atkinson, & Ellis, 2009; McArthur & Bishop, 2005; Neville et al., 1993; Tonnquist-Uhlen, 1996).

P3 studies consistently report similar peak latencies for SLI and normal controls (Courchesne & Plante, 1996; Evans et al., 2011; Jirsa & Clontz, 1990; Lincoln Courchesne, Harms, & Allen, 1993), but some studies report normal modulation of the P300 for SLI (Lincoln et al., 1993), whereas others report abnormal modulation of the P3 (Evans et al., 2011). Similarly, for the N400, left-anterior negativity (LAN), and P600, some studies report both latency and amplitude modulation of N400, left-anterior negativity (LAN), and P600 to be abnormal for SLI as compared with normal controls, but others find that only the peak amplitudes of scalp potentials, but not the latencies, are abnormal for SLI (Epstein et al., 2013; Cummings & Çeponiené, 2010; Neville et al., 1993; Plante, Petten, & Śenkfor, 2000; Polse, Sizemore, Burns, & Evans, 2011; Sabisch et al., 2009; Sizemore, Polse, Burns, & Evans, 2011; Weber-Fox et al., 2010).

Although differences in stimuli and participant selection criteria coupled with low statistical power may be contributing to the inconsistencies in ERP studies, there are methodological issues that are unique to EEG that may also be contributing to the lack of a clear pattern of evoked activity for SLI. For instance, one might find it striking, given that the wealth of behavioral data consistently show slower speed of processing in SLI (Leonard et al., 2007), that there seems to be a notable absence of a pattern of slower scalp-recorded activity in EEG studies of SLI. Although one of the advantages of EEG is that the temporal resolution is on the order of milliseconds, analysis of latency measures of the EEG components are particularly unreliable because a sharp peak often cannot be defined in the waveform (Luck, 2005; Luck & Kappenman, 2012).

This problem of identification of "peak amplitude" is compounded further in studies where individuals differ in the time course of spoken word identification. For instance, lexical-phonological processing in children with SLI is characterized not only by atypical activation of lexical-phonological competitors as compared with normal controls but also by the need for up to more than 200 ms more of the acoustic signal as compared with normal controls before spoken word are recognized (Mainela-Arnold & Evans, 2013; Mainela-Arnold, Evans, & Coady, 2008). If one assumes that semantic and/or syntactic processing of spoken words begins at the point when the child is able to pick out the word from the acoustic signal, and the point at which children with SLI and typical children can uniquely identify a word differs on the order of 200 ms, then the time course of ERP components should be differentially affected for children with SLI as compared with normal controls, and comparison of amplitude modulation will be out of synchronization.

In addition to the timing features of ERP waveforms not necessarily reflecting the actual point in time when the brain first distinguishes the conditions, differences in the amplitude of an ERP component also may not correspond to differences in component size. For instance, it is possible that neural activity differed before a given time point but that the ERP is not sensitive to this difference. Further issues arise in comparing the time courses of ERP components across groups. A primary question of interest is differences in latency, and the measure most often used is comparing peak latencies (Jemel, Oades, Oknina, Achenbach, & Röpcke, 2003; Luck, 2005); however, there are significant issues inherent in using peak latency to infer differences in the time course of the ERP components across groups or conditions (Bishop & Hardiman, 2010; Luck, 2005).

The degree of modulation of an ERP component is measured by subtracting the average response between two conditions (i.e., standard/deviant; semantically congruent/semantically incongruent; grammatical/agrammatical) and comparing the amplitude of this difference wave in a given time window. Differences in modulation of a component for SLI as compared with normal controls may also be confounded by learning and memory in SLI, which will manifest in differences in repetition effects over the course of the trials for SLI as compared with normal controls. Finally, the topography of activity for SLI as compared with normal controls may be swamped by factors such as atypical laterality for SLI as compared with normal controls.

To address some of these methodological problems, researchers have begun to explore alternative research designs and data analysis techniques. For instance, Bishop and Hardiman (2010) recently presented a novel analysis of MMN data using an independent component analysis to reduce artifacts and remove those components having high trial-by-trial variance. Using this analysis, Bishop and Hardmin were able to show that the sensitivity and specificity of MMN could be used to identify individuals with auditory processing deficits. In an analysis of experimental data, as well as of a dummy dataset, their criterion for MMN identified 2 of 17 (12%) false-positive results and showed that some individuals evidenced atypical MMN despite showing good ability to discriminate the stimuli on behavioral tests (Bishop & Hardiman, 2010).

One limitation of ERP studies is their inherently poor spatial resolution, which significantly constrains the degree to which one can make inferences about specific locations of brain activity. MEG is a neurophysiological technique similar to EEG in that it measures neural activity with potentially sub-millisecond temporal resolution, limited only by the digitization rate. It differs from EEG in that it measures magnetic fields from electrical currents that are produced by the brain during sensory, motor, or cognitive tasks. Because magnetic fields pass through biological tissues with essentially no perturbation, the spatial relationship between active cortical sources and the sensor layout is quite correspondent, providing greater precision in the localization of brain activity as compared with EEG (Huang et al., 2007).[2] As a result, MEG provides a means of observing not only the temporal patterns in cortical activity underlying language processing but also, simultaneously, information regarding where in the brain this activity is occurring (Marinkovic et al., 2003).

To date, only a few studies have used MEG to investigate language processing in SLI (Brown et al., 2014; Helenius et al., 2009, 2014; Larson & Lee, 2014). For instance, Helenius et al. (2014) used MEG to investigate the spatiotemporal course of cortical activation in SLI. Peak latency and amplitude of the N400 m response in the SLI children did not differ from normal controls; however, detailed analysis revealed that functionally, the N400 m response was qualitatively different for SLI as compared with the normal controls. Specifically, although the SLI group showed activity similar to that of normal controls for the N400 m effect on words and nonwords at 600–900 ms after word onset in the left and right hemispheres, there was no repetition effect evident in the left hemisphere for the SLI group during the initial phase of the N400 m response (300–600 ms). This nearly nonexistent repetition priming in the left hemisphere in the SLI children is consistent with other studies of brain function in SLI, suggesting abnormal or reduced involvement of the left hemisphere while processing words. The findings from this study differ from the study of Helenius and colleagues involving adults with a history of SLI (Helenius et al., 2009). In this earlier study, they observed that although repetition effects were less robust for adults with a history of SLI as compared with normal controls, there was nonetheless a repetition suppression effect, albeit abnormally small but evident bilaterally, in contrast to the lack of normal lateralization and uniquely right hemisphere suppression for the children with SLI.

[2]Unlike PET and fMRI, MEG activity maps are still dependent on the modeling assumptions about the number and complexity of the activity sources being examined.

A technology known as aMEG (Dale & Halgren, 2001) has also recently been used to investigate individual differences in language in SLI (Brown et al., 2014). aMEG is a technique that uses anatomical MRI information about an individual subject's cortical structure to constrain the space of potential activity sources for MEG or EEG signals (Dale et al., 2000; Dale & Sereno, 1993) to generate a dynamic statistical parametric map (dSPM) similar to maps generated for fMRI. These map movies provide millisecond-wise temporal resolution with much better spatial resolution than standard EEG of real-time language and cognitive processing, revealing the dynamic unfolding of task-locked brain activity for an individual subject. Noise-normalized, anatomically constrained statistical parametric maps of MEG-derived brain activity have been shown to have strong spatial correspondence with recordings from direct intracranial EEG for a variety of stimulus types and sensory and cognitive components (Halgren et al., 1994).

In addition to its sub-millisecond temporal resolution, aMEG provides excellent signal-to-noise properties and enhanced localization of brain activity through the use of noise-normalized source estimates constrained to the cortical reconstruction of each individual subject and aligned using sulcal and gyral surface-based registration (Dale & Halgren, 2001; Dale et al., 2000). Unlike single-dipole fitting MEG methods, the aMEG technique assumes multiple, distributed, and simultaneous cortical generators, which multiple functional neuroimaging and recording studies overwhelmingly show is an appropriate assumption for complex cognitive and language tasks. Many researchers now using MEG use some model constraints that attempt to inform activity localization with information from the same subject's MRI scan.

Recently, Brown et al. (2014) used a novel technique that allowed them to investigate the dynamic functional brain organization for semantic processing for individual subjects with SLI using aMEG. In using this approach, they were able to compare the dynamic time course and cortical activity linked to semantic processing of a single participant with SLI to his sibling with normal language abilities as well as to a normal control group. Brown and colleagues found that the dynamic pattern of brain activity for the child with SLI was subjectively and statistically different from his sibling and from the average of the group of unrelated normal controls. Specifically, during the semantic processing of pictures of concrete objects, the SLI participant showed a spatiotemporal pattern of activity that was most notable for being strongly right-lateralized and delayed as compared with the normal control group and his sibling. When performing the same semantic task with word stimuli, cortical activity for the child with SLI was characterized by an even greater degree of right hemisphere lateralization and was similarly delayed in time as compared with his siblings as well as the normal control group.

In a direct comparison with the distribution of neural activity at all cortical locations and time points within the typically developing group using *z*-scores, Brown and colleagues observed that the child with SLI showed marked differences from the typical dynamic functional organization that agreed with qualitative comparisons of the dSPMs. Consistent with previous studies, Brown and colleagues observed the child with SLI to evidence statistically greater right hemisphere cortical activation and statistically significant under-recruitment of left hemisphere cortical regions including perisylvian, anterior temporal, opercular, and lateral and superior frontal cortex as compared with normal controls. However, this study was novel in its characterization of the dynamic and localized patterns of cortical activity over time, using the combined spatial and temporal sensitivities of aMEG to make inferences about specific components of language that might be affected in this individual with SLI.

72.5 CONCLUSION

Structurally, a pattern that emerges across studies of brain structure in SLI is one of atypical cortical and subcortical morphology and tissue properties for some, but not all, areas of the brain. Abnormal findings in SLI have been demonstrated in measures of cortical and subcortical gray matter volumes and in the microstructural characteristics of white matter as revealed by diffusion imaging and tractography. Recent technological advances that now enable researchers to differentiate cortical and subcortical surface area and thickness, and genetic influences on these aspects of cortical and subcortical structure, suggest an important future direction in the study of SLI. Two important questions that need to be addressed are the following: (i) are the characteristics of cortical and subcortical neuronal structure in SLI reflective of "immature" or delayed cortical development, or do they reflect qualitatively different brain development and (ii) to what extent is atypical brain structure and functional organization in SLI a proximal "cause" of the language impairments, or are there some aspects of the structural and functional brain abnormalities that are more accurately thought of as merely the expression or product of the behavioral deficits of the child with SLI?

Overall, brain activity for SLI appears to be characterized by reduced activation levels for some, but not all, cortical and subcortical regions based on studies using hemodynamic measures, atypical modulation of cerebral peak latencies and amplitudes when measured by EEG, and differences in the lateralization and timing of cortical activation patterns when assessed via MEG. Electrophysiological studies suggest that the latency of brain activity in SLI appears to be atypical during early stages, but not at later points in the time course of speech and language processing. If future research shows that this lack of difference in the time course of brain activity for SLI and typical peers at later time points (i.e., P3, N400, P600) is not an artifact of data analysis methods, then a model of SLI needs to be developed that accounts for atypical brain activation, the time course of early and late stages of cortical activity, and the substantial body of behavioral data consistently showing a pattern of generalized slower speed of processing for children with SLI.

Findings from recent studies using MEG and aMEG that suggest the language systems of individuals with SLI may be organized quantitatively and qualitatively in a different manner from those of their typically developing peers, and that, at least for some individuals, this may involve atypical right hemisphere specialization for specific subcomponents of language such as semantic representations for word meanings and object concepts. So, future research should incorporate the use of these multimodal imaging techniques that allow for a dynamic approach to examine the multiple language systems in the working brain. This may be advantageous in understanding the cortical and subcortical organization of both normal and abnormal language systems.

72.6 TOWARDS A NEUROBIOLOGY OF SLI

Historically, the focus of studies of brain structure and function in SLI has been on characterizing the underlying neurobiology in these children with the hope that a more detailed characterization of the brain structure and function might ultimately lead to the discovery of the cause of SLI. There is a debate among researchers regarding whether SLI comprises a distinct diagnostic clinical category, or if children with SLI merely represent those children whose language abilities fall at the lower end of the normal distribution (for detailed discussion see Bishop, 2014; Dollaghan, 2011). This debate rests largely on the use of behavioral measures to classify SLI. From the rapidly growing body of work investigating brain structure and function in SLI, a picture is emerging that suggests that, although there are individual differences in the manifestation of the behavioral symptoms of SLI, the term SLI does not merely represent children on the low end of the normal distribution but is instead a neurodevelopmental language disorder, characterized by a disturbance in the dynamics of brain development that collectively affect cortical and subcortical morphology in selected regions of the brain and the tissue properties of white matter connections, and that this pattern of neurobiological aberrations leads to a characteristic combination of delayed onset of language acquisition, uneven language acquisition profile, and learning and memory deficits.

Advances in noninvasive, safe, functional brain imaging technologies coupled with a richer integration between neuroscience, cognitive psychology, and linguistics have led to an explosion of new research in the neuroscience of language and in our understanding of the nature of language and cognition in the brain (Devor et al., 2013; Eicher et al., 2013; Erus et al., 2014; Fjell et al., 2012, 2013; Lieberman, 2002; Poeppel et al., 2012). With the advent of better imaging technologies and a growing body of research in SLI, neurobiological models of the language impairments in SLI are being reconceptualized and are beginning to rest on richer theoretical foundations. Instead of looking for a direct link between abnormal brain morphology and brain function and language impairments in SLI, these new models are shifting away from a "single-cause" model of SLI and are beginning to take advantage of these technologies to unravel the genetic and neurobiological characteristics of SLI (Badcock et al., 2012; Bishop, 2013; Bishop, 2009; Li & Bartlett, 2012; Paracchini, 2011; Poeppel, 2011; Rice, 2013).

Focusing on general purpose learning systems that appear to support language acquisition in normal populations, this new work suggests that abnormal brain development and, specifically, atypical cortical structural organization may result in subsequent abnormal cerebral functional organization. In line with this new view, studies directly comparing structural and functional brain organization in SLI show both anatomical and physiological abnormalities in some regions, but in other regions abnormalities in brain structure were not correlated with brain function during language processing, raising the question regarding whether and in which direction abnormal brain morphology and atypical functional lateralization in SLI are causally linked (Badcock et al., 2012; Bishop, Holt, Whitehouse, & Groen, 2014). This work also suggests that the language impairments seen for some children with SLI may be the result of a domain-general learning and memory deficit—mediated by the corticostriatal system—where abnormal functioning of this system manifests as atypical cortical organization of language (Brown et al., 2014; Lum, Conti-Ramsden, Page, & Ullman, 2012; Lum, Ullman, & Conti-Ramsden, 2015; van der Lely & Pinker, 2014).

Importantly, similar to a host of neurodegenerative disorders, multiple genetic and environmental factors will likely be shown to contribute to disturbances in the dynamics of brain development in children with

SLI. Thus, the dynamic between these factors and atypical brain development in SLI is a complex one. However, moving away from both a "single-cause" deficit account of SLI and the classic left hemisphere dominant model of language and moving toward characterizing this dynamic neurodevelopmental phenomenon as a complex system may ultimately be the key to discovering the cause of SLI.

References

Adams, J., Courchesne, E., Elmasian, R., & Lincoln, A. (1987). Increased amplitude of the auditory P2 and P3b components in adolescents with developmental dysphasia. *Electroencephalography and Clinical Neurophysiology Supplement, 40*, 577–583.

Badcock, N. A., Bishop, D. V. M., Hardiman, M. J., Barry, J. G., & Watkins, K. E. (2012). Co-localisation of abnormal brain structure and function in specific language impairment. *Brain and Language, 120*(3), 310–320. Available from: http://dx.doi.org/10.1016/j.bandl.2011.10.006.

Basser, P. J., Mattiello, J., & LeBihan, D. (1994). MR diffusion tensor spectroscopy and imaging. *Biophysical Journal, 66*, 259–267.

Beaulieu, C. (2002). The basis of anisotropic water diffusion in the nervous system: A technical review. *NMR in Biomedicine, 15*(7), 435–455.

Bishop, D. V. M. (2007). Using mismatch negativity to study central auditory processing in developmental language and literacy impairments: Where are we, and where should we be going? *Psychological Bulletin, 133*(4), 651–672.

Bishop, D. V. M. (2009). Genes, cognition, and communication: Insights from neurodevelopmental disorders. *Annals of the New York Academy of Sciences, 1156*, 1–18. Available from: http://dx.doi.org/10.1111/j.1749-6632.2009.04419.x.

Bishop, D. V. M. (2013). Cerebral asymmetry and language development: Cause, correlate, or consequence? *Science, New Series, 340* (6138), 1230531–1230538. Available from: http://dx.doi.org/10.1126/science.1230531.

Bishop, D. V. M. (2014). Ten questions about terminology for children with unexplained language problems. *International Journal of Language and Communication Disorders, 49*(4), 381–415. Available from: http://dx.doi.org/10.1111/1460-6984.12101.

Bishop, D. V. M., Hardiman, M. J., & Barry, J. G. (2012). Auditory deficit as a consequence rather than endophenotype of specific language impairment: Electrophysiological evidence. *PLoS ONE, 7*(5), e35851. Available from: http://dx.doi.org/10.1371/journal.pone.0035851.

Bishop, D. V. M., & Hardiman, M. J. (2010). Measurement of mismatch negativity in individuals: A study using single-trial analysis. *Psychophysiology, 47*(4), 697–705. Available from: http://dx.doi.org/10.1111/j.1469-8986.2009.00970.x.

Bishop, D. V. M., Holt, G., Whitehouse, A. J. O., & Groen, M. (2014). No population bias to left-hemisphere language in 4-year-olds with language impairment. *PeerJ, 2*(3), e507. Available from: http://dx.doi.org/10.7717/peerj.507/supp-1.

Brans, R. G. H., Kahn, R. S., Schnack, H. G., van Baal, G. C. M., Posthuma, D., van Haren, N. E. M., et al. (2010). Brain plasticity and intellectual ability are influenced by shared genes. *Journal of Neuroscience, 30*(16), 5519–5524. Available from: http://dx.doi.org/10.1523/JNEUROSCI.5841-09.2010.

Brown, T. T., Lugar, H. M., Coalson, R. S., Miezin, F. M., Petersen, S. E., & Schlaggar, B. L. (2005). Developmental changes in human cerebral functional organization for word generation. *Cerebral Cortex, 15*, 275–290.

Brown, T. T., Erhart, M., Avesar, D., Dale, A. M., Halgren, E., & Evans, J. L. (2014). Atyipcial right hemisphere specialization for object representations in an adolescent with specific language impairment. *Frontiers in Human Neuroscience, 82*(8), 1–48. Available from: http://dx.doi.org/10.3389/fnhum.2014.00082.

Brown, T. T., & Jernigan, T. L. (2012). Brain development during the preschool years. *Neuropsychology Review, 22*(4), 313–333. Available from: http://dx.doi.org/10.1007/s11065-012-9214-1.

Brown, T. T., Kuperman, J. M., Chung, Y., Erhart, M., McCabe, C., Hagler, D. J., et al. (2012). Neuroanatomical assessment of biological maturity. *Current Biology, 22*(18), 1693–1698. Available from: http://dx.doi.org/10.1016/j.cub.2012.07.002.

Catts, H. W., Bridges, M. S., Little, T. D., & Tomblin, J. B. (2008). Reading achievement growth in children with language impairments. *Journal of Speech, Language, and Hearing Research, 51*(6), 1569–1579. Available from: http://dx.doi.org/10.1044/1092-4388 (2008/07-0259).

Çeponiené, R., Cummings, A., Wulfect, B., Ballantyne, A., & Townsend, J. (2009). Spectral vs. temporal auditory processing in specific language impairment: A developmental ERP study. *Brain and Language, 110*(3), 107–120.

Chen, C.-H., Fiecas, M. J. A., Gutierrez, E. D., Panizzon, M. S., Vouksimaa, E., Thompson, W. K., et al. (2013). Genetic topography of brain morphology. *Proceedings of the National Academy of Sciences of the United States of America, 110*(42), 17089–17094. Available from: http://dx.doi.org/10.1073/pnas.1308091110/-/DCSupplemental.

Chen, C. H., Gutierrez, E. D., Thompson, W., Panizzon, M. S., Jernigan, T. L., Eyler, L. T., et al. (2012). Hierarchical genetic organization of human cortical surface area. *Science, 335*(6076), 1634–1636. Available from: http://dx.doi.org/10.1126/science.1215330.

Chen, C.-H., Panizzon, M. S., Eyler, L. T., Jernigan, T. L., Thompson, W., Fennema-Notestine, C., et al. (2011). Genetic Influences on cortical regionalization in the human brain. *Neuron, 72*(4), 537–544. Available from: http://dx.doi.org/10.1016/j.neuron.2011.08.021.

Chiron, C., Pinton, F., Masure, M. C., Duvelleroy-Hommet, C., Leon, F., & Billard, C. (1999). Hemispheric specialization using SPECT and stimulation tasks in children with dysphasia and dystrophia. *Developmental Medicine and Child Neurology, 41*(8), 512–520.

Clark, M. M., & Plante, E. (1998). Morphology of the inferior frontal gyrus in developmentally language-disordered adults. *Brain and Language, 61*(2), 288–303. Available from: http://dx.doi.org/10.1006/brln.1997.1864.

Conti-Ramsden, G. (2013). Commentary: Increased risk of later emotional and behavioural problems in children with SLI—reflections on Yew and O'Kearney (2013). *Journal of Child Psychology and Psychiatry, and Allied Disciplines, 54*(5), 525–526. Available from: http://dx.doi.org/10.1111/jcpp.12027.

Courchesne, E., & Plante, E. (1996). Measurement and analysis issues in neurodevelopmental magnetic resonance imaging. *Developmental Neuroimaging: Mapping the Development of Brain and Behavior*, 43–65.

Cummings, A., & Çeponiené, R. (2010). Verbal and nonverbal semantic processing in children with developmental language impairment. *Neuropsychologia, 48*(1), 77–85. Available from: http://dx.doi.org/10.1016/j.neuropsychologia.2009.08.012.

Dale, A. M., & Halgren, E. (2001). Spatiotemporal mapping of brain activity by integration of multiple imaging modalities. *Current opinion in neurobiology, 11*, 202–208. Available from: http://dx.doi.org/10.1016/S0959-4388(00)00197-5.

Dale, A. M., Liu, A. K., Fischl, B. R., Buckner, R. L., Belliveau, J. W., Lewine, J. D., et al. (2000). Dynamic statistical parametric mapping: Combining fMRI and MEG for high-resolution imaging of cortical activity. *Neuron, 26*(1), 55–67.

Dale, A. M., & Sereno, M. I. (1993). Improved localization of cortical activity by combining EEG and MEG with MRI cortical surface reconstruction: A linear approach. *Journal of Cognitive Neuroscience*, *5*(2), 162–176.

De Fossé, L., Hodge, S. M., Makris, N., Kennedy, D. N., Caviness, V. S., McGrath, L., et al. (2004). Language-association cortex asymmetry in autism and specific language impairment. *Annals of Neurology*, *56*(6), 757–766. Available from: http://dx.doi.org/10.1002/ana.20275.

de Guibert, C., Maumet, C., Jannin, P., Ferre, J. C., Treguier, C., Barillot, C., et al. (2011). Abnormal functional lateralization and activity of language brain areas in typical specific language impairment (developmental dysphasia). *Brain*, *134*(10), 3044–3058. Available from: http://dx.doi.org/10.1093/brain/awr141.

Denays, R., Tondeur, M., Foulon, M., Verstraeten, F., Ham, H., Piepsz, A., et al. (1989). Regional brain blood flow in congenital dysphasia: Studies with technetium-99m HM-PAO SPECT. *Journal of Nuclear Medicine*, *30*(11), 1825–1829.

Devor, A., Bandettini, P. A., Boas, D. A., Bower, J. M., Buxton, R. B., Cohen, L. B., et al. (2013). The challenge of connecting the dots in the B.R.A.I.N. *Neuron*, *80*(2), 270–274. Available from: http://dx.doi.org/10.1016/j.neuron.2013.09.008.

Dollaghan, C. (2011). Taxometric analysis of specific language impairment. *Journal of Speech Language and Hearing Research*, *54*, 1361–1371.

Dorsaint-Pierre, R., Penhune, V. B., Watkins, K. E., Neelin, P., Lerch, J. P., Bouffard, M., et al. (2006). Asymmetries of the planum temporale and Heschl's gyrus: Relationship to language lateralization. *Brain*, *129*(5), 1164–1176. Available from: http://dx.doi.org/10.1093/brain/awl055.

Durkin, K., & Conti-Ramsden, G. (2007). Language, social behavior, and the quality of friendships in adolescents with and without a history of specific language impairment. *Child Development*, *78*(5), 1441–1457. Available from: http://dx.doi.org/10.1111/j.1467-8624.2007.01076.x.

Eicher, J. D., Powers, N. R., Miller, L. L., Akshoomoff, N., Amaral, D. G., Bloss, C. S., et al. (2013). Genome-wide association study of shared components of reading disability and language impairment. *Genes, Brain and Behavior*, *12*(8), 792–801. Available from: http://dx.doi.org/10.1111/gbb.12085.

Ellis Weismer, S., Plante, E., Jones, M., & Tomblin, J. B. (2005). A functional magnetic resonance imaging investigation of verbal working memory in adolescents with specific language impairment. *Journal of Speech, Language, and Hearing Research*, *48*(2), 405–425. Available from: http://dx.doi.org/10.1044/1092-4388(2005/028).

Epstein, B., Hestvik, A., Shafer, V. L., & Schwartz, R. G. (2013). ERPs reveal atypical processing of subject versus object Wh-questions in children with specific language impairment. *International Journal of Language & Communication Disorders*, *48*(4), 351–365. Available from: http://dx.doi.org/10.1111/1460-6984.12009.

Epstein, B., Shafer, V. L., Melara, R. D., & Schwartz, R. G. (2014). Can children with SLI detect cognitive conflict? Behavioral and electrophysiological evidence. *Journal of Speech Language and Hearing Research*, *57*(4), 1453. Available from: http://dx.doi.org/10.1044/2014_JSLHR-L-13-0234.

Erus, G., Battapady, H., Satterthwaite, T. D., Hakonarson, H., Gur, R. E., Davatzikos, C., et al. (2014). Imaging patterns of brain development and their relationship to cognition. *Cerebral Cortex*. Available from: http://dx.doi.org/10.1093/cercor/bht425.

Evans, J. L., Selinger, C., & Pollak, S. D. (2011). P300 as a measure of processing capacity in auditory and visual domains in specific language impairment. *Brain Research*, *1389*, 93–102. Available from: http://dx.doi.org/10.1016/j.brainres.2011.02.010.

Federmeier, K. D., Wlotko, E. W., & Meyer, A. M. (2008). What's "right" in language comprehension: Event-related potentials reveal right-hemisphere language capabilities. *Language and Linguistics Compass*, *2*, 1–17.

Fjell, A. M., Walhovd, K. B., Brown, T. T., Kuperman, J. M., Chung, Y., Hagler, D. J., et al. (2012). Multimodal imaging of the self-regulating developing brain. *Proceedings of the National Academy of Sciences of the United States of America*, *109*(48), 19620–19625. Available from: http://dx.doi.org/10.1073/pnas.1208243109.

Fjell, A. M., Westlye, L. T., Grydeland, H., Amlien, I., Espeseth, T., Reinvang, I., et al. (2013). Critical ages in the life course of the adult brain: Nonlinear subcortical aging. *Neurobiology of Aging*, *34*(10), 2239–2247. Available from: http://dx.doi.org/10.1016/j.neurobiolaging.2013.04.006.

Fonteneau, E, & van der Lely, H. K. J. (2008). Electrical brain responses in language-impaired children reveal grammar-specific deficits. *PLoS One*, *3*(3), e1832. Available from: http://dx.doi.org/10.1371/journal.pone.0001832.s012.

Friederici, A. D. (2005). Neurophysiological markers of early language acquisition: From syllables to sentences. *Trends in Cognitive Sciences*, *9*(10), 481–488. Available from: http://dx.doi.org/10.1016/j.tics.2005.08.008.

Friederici, A. D. (2006). The neural basis of language development and its impairment. *Neuron*, *52*(6), 941–952. Available from: http://dx.doi.org/10.1016/j.neuron.2006.12.002.

Friederici, A. D., Wang, Y., Herrmann, C. S., Maess, B., & Oertel, U. (2000). Localization of early syntactic processes in frontal and temporal cortical areas: A magnetoencephalographic study. *Human Brain Mapping*, *11*(1), 1–11. Available from: http://dx.doi.org/10.1002/1097-0193(200009)11:1<1::AID-HBM10>3.0.CO;2-B.

Frisch, S., Kotz, S. A., Yves von Cramon, D., & Friederici, A. D. (2003). Why the P600 is not just a P300: The role of the basal ganglia. *Clinical Neurophysiology*, *114*(2), 336–340.

Gauger, L. M., Lombardino, L. J., & Leonard, C. M. (1997). Brain morphology in children with specific language impairment. *Journal of Speech, Language, and Hearing Research*, *40*(6), 1272–1284. Available from: http://dx.doi.org/10.1044/jslhr.4006.1272.

Geschwind, N. (1970). The organization of language and the brain. *Science*, *170*(3961), 940–944.

Geschwind, N. (1979). Specializations of the human brain. *Scientific American*, *241*(3), 180–199. Available from: http://dx.doi.org/10.1038/scientificamerican0979-180.

Girbau-Massana, D., Garcia-Marti, G., Marti-Bonmati, L., & Schwartz, R. G. (2014). Gray-white matter and cerebrospinal fluid volume differences in children with Specific Language Imapairment and/or Reading Disability. *Neuropsychologia*, *56*(C), 90–100. Available from: http://dx.doi.org/10.1016/j.neuropsychologia.2014.01.004.

Hagler, D. J., Halgren, E., Martinez, A., Huang, M., Hillyard, S. A., & Dale, A. M. (2009). Source estimates for MEG/EEG visual evoked responses constrained by multiple, retinotopically-mapped stimulus locations. *Human Brain Mapping*, *30*(4), 1290–1309.

Halgren, E., Baudena, P., Heit, G., Clarke, J. M., Marinkovic, K., & Chauvel, P. (1994). Spatio-temporal stages in face and word processing 2 Depth-recorded potentials in the human frontal and rolandic cortices. *Journal of Physiology*, *88*, 51–80.

Hansen, P. C., Kringelbach, M. L., & Salmelin, R. (2010). *MEG: An introduction to methods*. New York: Oxford University Press Inc.

Helenius, P., Parviainen, T., Paetau, R., & Salmelin, R. (2009). Neural processing of spoken words in specific language impairment and dyslexia. *Brain*, *132*(7), 1918–1927. Available from: http://dx.doi.org/10.1093/brain/awp134.

Helenius, P., Sivonen, P., Parviainen, T., Isoaho, P., Hannus, S., Kauppila, T., et al. (2014). Abnormal functioning of the left temporal lobe in language-impaired children. *Brain and Language, 130*, 11–18. Available from: http://dx.doi.org/10.1016/j.bandl.2014.01.005.

Huang, M. X., Song, T., Hagler, D. J., Jr., Podgorny, I., Jousmaki, V., Cui, L., et al. (2007). A novel integrated MEG and EEG analysis method for dipolar sources. *NeuroImage, 37*, 731–748.

Hugdahl, K., Gundersen, H., Brekke, C., Thomsen, T., Rimol, L. M., Ersland, L., et al. (2004). FMRI brain activation in a finnish family with specific language impairment compared with a normal control group. *Journal of Speech, Language, and Hearing Research, 47*(1), 162–172. Available from: http://dx.doi.org/10.1044/1092-4388(2004/014).

Jemel, B., Oades, R. D., Oknina, L., Achenbach, C., & Röpcke, B. (2003). Frontal and temporal Lobe sources for a marker of controlled auditory attention: The negative difference (Nd) event-related potential. *Brain Topography, 15*(4), 249–262.

Jernigan, T. L., Archibald, S. L., Berhow, M. T., Sowell, E. R., Foster, D. S., & Hesselink, J. R. (1991). Cerebral structure on MRI, Part I: Localization of age-related changes. *Biological Psychiatry, 29*(1), 55–67. Available from: http://dx.doi.org/10.1016/0006-3223(91)90210-D.

Jernigan, T. L., Hesselink, J. R., Sowell, E., & Tallal, P. (1991). Cerebral structure on magnetic resonance imaging in language-and learning-impaired children. *Archives of Neurology, 48*(5), 539–545. Available from: http://dx.doi.org/10.1001/archneur.1991.00530170103028.

Jernigan, T. L., & Tallal, P. (2010). Late childhood changes in brain morphology observable with MRI. *Developmental Medicine and Child Neurology, 32*(5), 379–385. Available from: http://dx.doi.org/10.1111/j.1469-8749.1990.tb16956.x.

Jirsa, R. E., & Clontz, K. B. (1990). Long latency auditory event-related potentials from children with auditory processing disorders. *Ear and Hearing, 11*(3), 222–223.

Kaganovich, N., & Schumaker, J. (2014). Audiovisual integration for speech during mid-childhood: Electrophysiological evidence. *Brain and Language, 139C*, 36–48. Available from: http://dx.doi.org/10.1016/j.bandl.2014.09.011.

Kaganovich, N., Schumaker, J., Leonard, L. B., Gustafson, D., & Macias, D. (2014). Children with a history of SLI show reduced sensitivity to audiovisual temporal asynchrony: an ERP study. *Journal of Speech Language and Hearing Research, 57*(4), 1480–1502. Available from: http://dx.doi.org/10.1044/2014_JSLHR-L-13-0192.

Kaganovich, N., Schumaker, J., Macias, D., & Gustafson, D. (2014). Processing of audiovisually congruent and incongruent speech in school-age children with a history of specific language impairment: a behavioral and event-related potentials study. *Developmental Science*. Available from: http://dx.doi.org/10.1111/desc.12263.

Keller, S. S., Roberts, N., García-Fiñana, M., Mohammadi, S., Ringelstein, E. B., Knecht, S., et al. (2011). Can the language-dominant hemisphere be predicted by brain anatomy? *Journal of Cognitive Neuroscience, 23*(8), 2013–2029. Available from: http://dx.doi.org/10.1162/jocn.2010.21563.

Khundrakpam, B. S., Reid, A., Brauer, J., Carbonell, F., Lewis, J., Ameis, S., et al. (2013). Developmental changes in organization of structural brain networks. *Cerebral Cortex, 23*(9), 2072–2085. Available from: http://dx.doi.org/10.1093/cercor/bhs187.

Kim, J., Kim, Y. W., Park, C., Park, E. S., Kim, H., Lee, S., et al. (2006). Diffusion-tensor magnetic resonance imaging in children with language impairment. *Neuroreport, 17*(12), 1279–1282.

Knecht, S., Deppe, M., Ebner, A., Henningsen, H., Huber, T., Jokeit, H., et al. (1998). Non-invasive determination of language lateralisation by functional transcranial Doppler sonography: A comparison with the Wada test. *Stroke, 29*(1), 82–86.

Kuhl, P., & Rivera-Gaxiola, M. (2008). Neural substrates of language acquisition. *Annual Review of Neuroscience, 31*(1), 511–534. Available from: http://dx.doi.org/10.1146/annurev.neuro.30.051606.094321.

Kutas, M., & Federmeier, K. D. (2011). Thirty years and counting: Finding meaning in the N400 component of the event-related brain potential (ERP). *Annual Review of Psychology, 62*(1), 621–647. Available from: http://dx.doi.org/10.1146/annurev.psych.093008.131123.

Larson, E., & Lee, A. K. (2014). Potential use of MEG to understand abnormalities in auditory function in clinical populations. *Frontiers in Human Neuroscience, 8*(151), 1–15. Available from: http://dx.doi.org/10.3389/fnhum.2014.00151.

Lee, J. C., Nopoulos, P. C., & Tomblin, J. B. (2013). Abnormal subcortical components of the corticostriatal system in young adults with DLI: A combined structural MRI and DTI study. *Neuropsychologia, 51*(11), 2154–2161. Available from: http://dx.doi.org/10.1016/j.neuropsychologia.2013.07.011.

Leonard, L. B. (2014). *Children with specific language impairment* (2nd ed.). Cambridge, MA: MIT Press.

Leonard, L. B., Ellis Weismer, S., Miller, C. A., Francis, D. J., Tomblin, J. B., & Kail, R. V. (2007). Speed of processing, working memory, and language impairment in children. *Journal of Speech, Language, and Hearing Research, 50*(2), 408. Available from: http://dx.doi.org/10.1044/1092-4388(2007/029).

Leonard, L. B., & Weber-Fox, C. (2012). Specific language impairment: Processing deficits in linguistic, cognitive, and sensory domains. In M. Faust (Ed.), *The handbook of the neuropsychology of language* (Vol. 1&2, pp. 826–846). Oxford, UK: Blackwell Plublishing Ltd.

Li, N., & Bartlett, C. W. (2012). Defining the genetic architecture of human developmental language impairment. *Life Sciences, 90*(13–14), 469–475. Available from: http://dx.doi.org/10.1016/j.lfs.2012.01.016.

Lieberman, P. (2002). On the nature and evolution of the neural bases of human language. *American Journal of Physical Anthropology, 119*(S35), 36–62. Available from: http://dx.doi.org/10.1002/ajpa.10171.

Lim, S., Han, C. E., Uhlhaas, P. J., & Kaiser, M. (2013). Preferential detachment during human brain development: Age- and sex-specific structural connectivity in diffusion tensor imaging (DTI) data. *Cerebral Cortex*. Available from: http://dx.doi.org/10.1093/cercor/bht333 (first published online December 15, 2013).

Lincoln, A. J., Courchesne, E., Harms, L., & Allen, M. (1993). Context probability evaluation in autistic, receptive developmental language disorder, and control children: Event-related brain potential evidence. *Journal of Autism and Developmental Disorders, 23*(1), 37–58.

Lou, H. C., Henriksen, L., & Bruhn, P. (1990). Focal cerebral dysfunction in developmental learning disabilities. *Lancet, 335*(8680), 8–11.

Luck, S. J. (2005). *An introduction to the event-related potential technique*. Cambridge, MA: The MIT Press.

Luck, S. J., & Kappenman, E. S. (Eds.), (2012). *The Oxford handbook of event-related potential components* Oxford University Press p. 664. ISBN 9780195374148.

Lum, J. A. G., Conti-Ramsden, G., Page, D., & Ullman, M. T. (2012). Working, declarative and procedural memory in specific language impairment. *Cortex, 48*(9), 1138–1154. Available from: http://dx.doi.org/10.1016/j.cortex.2011.06.001.

Lum, J. A. G., Ullman, M. T., & Conti-Ramsden, G. (2015). Verbal declarative memory impairments in specific language impairment are related to working memory deficits. *Brain and Language, 142*(1), 76–85. Available from: http://dx.doi.org/10.1016/j.bandl.2015.01.008.

Mayes, A. K., Reilly, S., & Morgan, A. T. (2015). Neural correlates of childhood language disorder: a systematic review. *Developmental Medicine & Child Neurology, n/a–n/a*. Available from: http://dx.doi.org/10.1111/dmcn.12714.

Mainela-Arnold, E., & Evans, J. L. (2013). Do statistical segmentation abilities predict lexical-phonological and lexical-semantic abilities in children with and without SLI? *Journal of Child Language, 41* (02), 327–351. Available from: http://dx.doi.org/10.1017/S0305000912000736.

Mainela-Arnold, E., Evans, J. L., & Coady, J. A. (2008). Lexical representations in children with SLI: Evidence from a frequency-manipulated gating task. *Journal of Speech Language and Hearing Research, 51*(2), 381–393. Available from: http://dx.doi.org/10.1044/1092-4388(2008/028).

Marinkovic, K., Dhond, R. P., Dale, A. M., Glessner, M., Carr, V., & Halgren, E. (2003). Spatiotemporal dynamics of modality-specific and supramodal word processing. *Neuron, 38*(3), 487–497.

McArthur, G., & Bishop, D. V. M. (2005). Speech and non-speech processing in people with specific language impairment. *Brain and Language, 94*, 260–273.

McArthur, G., Atkinson, C., & Ellis, D. (2009). Atypical brain responses to sounds in children with specific language impairment and reading impairments. *Developmental Science, 12* (5), 768–783.

Molfese, D. L., & Segalowitz, S. J. (1988). *Brain lateralization in children: Developmental implications*. New York, NY: Guilford Press.

Mori, S., & Van Zijl, P. C. (2002). Fiber tracking: Principles and strategies – a techinical review. *NMR in Biomedicine, 15*, 468–480. Available from: http://dx.doi.org/10.1002/nmb.781.

Mori, S., & Zhang, J. (2006). Principles of Diffusion Tensor Imaging and its application to basic neuroscience research. *Neuron, 51*(5), 527–539.

Müller, K., Neumann, J., Lohmann, G., Mildner, T., & Yves von Cramon, D. (2005). The correlation between blood oxygenation level-dependent signal strength and latency. *Journal of Magnetic Resonance Imaging, 21*(4), 489–494.

Näätänen, R., Ilmoniemi, R. J., & Alho, K. (1994). Magnetoencephalography in studies of human cognitive brain function. *Trends in Neuroscience, 17*(9), 389–395.

Neville, H. J., Coffey, S. A., Holcomb, P. J., & Tallal, P. (1993). The neurobiology of sensory and language processing in language-impaired children. *Journal of Cognitive Neuroscience, 5*(2), 235–253. Available from: http://dx.doi.org/10.1016/0013-4694(91)90136-R.

Ors, M., Ryding, E., Lindgren, M., Gustafsson, P., Blennow, G., & Rosén, I. (2005). Spect findings in children with specific language impairment. *Cortex, 41*(3), 316–326. Available from: http://dx.doi.org/10.1016/S0010-9452(08)70269-7.

Orton, S. T. (1937). *Reading, writing and speech problems in children*. New York, NY: W. W. Norton & Co.

Panizzon, M. S., Fennema-Notestine, C., Eyler, L. T., Jernigan, T. L., Prom-Wormley, E., Neale, M., et al. (2009). Distinct genetic influences on cortical surface area and cortical thickness. *Cerebral Cortex, 19*(11), 2728–2735. Available from: http://dx.doi.org/10.1093/cercor/bhp026.

Paracchini, S. (2011). Dissection of genetic associations with language-related traits in population-based cohorts. *Journal of Neurodevelopmental Disorders, 3*(4), 365–373. Available from: http://dx.doi.org/10.1007/s11689-011-9091-6.

Plante, E., Petten, C. V., & Senkfor, A. J. (2000). Electrophysiological dissociation between verbal and nonverbal semantic processing in learning disabled adults. *Neuropsychologia, 38*(13), 1669–1684. Available from: http://dx.doi.org/10.1016/S0028-3932(00)00083-X.

Plante, E., Swisher, L., Vance, R., & Rapcsak, S. (1991). MRI findings in boys with specific language impairment. *Brain and Language, 41*, 52–66.

Poeppel, D. (2011). Genetics and language: A neurobiological perspective on the missing link (-ing hypotheses). *Journal of Neurodevelopmental Disorders, 3*(4), 381–387. Available from: http://dx.doi.org/10.1007/s11689-011-9097-0.

Poeppel, D., Emmorey, K., Hickok, G., & Pylkkänen, L. (2012). Towards a new neurobiology of Language. *Journal of Neuroscience, 32*(41), 14125–14131. Available from: http://dx.doi.org/10.1523/JNEUROSCI.3244-12.2012.

Polich, J., & Kok, A. (1995). Cognitive and biological determinants of P300: An integrative review. *Biological Psychology, 41*(2), 103–146.

Polse, L., Sizemore, M., Burns, E. L., & Evans, J. L. (2011). ERP evidence for both similar and distinct cortical networks underlying semantic integration in adolescents with Specific Language Impairment. Poster presented at the neurobiology of language conference in Baltimore, MD.

Purdy, J. D., Leonard, L. B., Weber-Fox, C., & Kaganovich, N. (2014). Decreased sensitivity to long-distance dependencies in children with a history of specific language impairment: Electrophysiological evidence. *Journal of Speech Language and Hearing Research*1040–1059. Available from: http://dx.doi.org/10.1044/2014_JSLHR-L-13-0176.

Preston, J. L., Frost, S. J., Mencl, W. E., Fulbright, R. K., Landi, N., Grigorenko, E., et al. (2010). Early and late talkers: School-age language, literacy and neurolinguistic differences. *Brain, 133*(8), 2185–2195. Available from: http://dx.doi.org/10.1093/brain/awq163.

Pylkkänen, L., & Marantz, A. (2003). Tracking the time course of word recognition with MEG. *Trends in Cognitive Sciences, 7*(5), 187–189. Available from: http://dx.doi.org/10.1016/S1364-6613(03)00092-5.

Reilly, S., Bishop, D. V. M., & Tomblin, B. (2014). Terminological debate over language impairment in children: Forward movement and sticking points. *International Journal of Language and Communication Disorders, 49*(4), 452–462. Available from: http://dx.doi.org/10.1111/1460-6984.12111.

Rice, M. L. (2012). Toward epigenetic and gene regulation models of specific language impairment: looking for links among growth, genes, and impairments. *Journal of Neurodevelopmental Disorders, 4* (1), 1–14. Available from: http://dx.doi.org/10.1186/1866-1955-4-27.

Rice, M. L. (2013). Invited paper: Language growth and genetics of specific language impairment. *International Journal of Speech-Language Pathology, Early Online*, 1–11. Available from: http://dx.doi.org/10.3109/17549507.2013.783113.

Sabisch, B., Hahne, C. A., Glass, E., Suchodoletz von, W., & Friederici, A. D. (2009). Children with specific language impairment: The role of prosodic processes in explaining difficulties in processing syntactic information. *Brain Research, 1261*(C), 37–44. Available from: http://dx.doi.org/10.1016/j.brainres.2009.01.012.

Schmithorst, V. J., Yuan, W., & Plante, E. (2007). Functional MRI of language lateralization during development in children. *International Journal of Audiology, 46*(9), 533–551.

Schwartz, R. G., & Shafer, V. L. (2012). The neurobiology of specific language impairment. In M. Faust (Ed.), *The handbook of the neuropsychology of language* (1st ed., pp. 847–867). Oxford, UK: Blackwell Publishing, Ltd.

Sizemore, M., Polse, L., Burns, E. L., & Evans, J. L. (2011). Frequency and imageability effects on N400 amplitudes in adolescents with SLI. Poster presented at the Neurobiology of Language Conference in Baltimore, MD.

Soriano-Mas, C., Pujol, J., Ortiz, H., Deus, J., López-Sala, A., & Sans, A. (2009). Age-related brain structural alterations in children with specific language impairment. *Human Brain Mapping,*

30(5), 1626–1636. Available from: http://dx.doi.org/10.1002/hbm.20620.

Squeglia, L. M., Jacobus, J., Sorg, S. F., Jernigan, T. L., & Tapert, S. F. (2013). Early adolescent cortical thinning is related to better neuropsychological performance. *Journal of the International Neuropsychological Society*, *19*(9), 1–9. Available from: http://dx.doi.org/10.1017/S1355617713000878.

Szaflarski, J. P., Rajagopal, A., Altaye, M., Byars, A. W., Jacola, L., Schmithorst, V. J., et al. (2012). Left-handedness and language lateralization in children. *Brain Research*, *1433*, 85–97. Available from: http://dx.doi.org/10.1016/j.brainres.2011.11.026.

Tager-Flusberg, H., & Cooper, J. (1999). Present and future possibilities for defining a phenotype for specific language impairment. *Journal of Speech, Language, and Hearing Research*, *42*, 1275–1278. Available from: http://dx.doi.org/10.1044/jslhr.4205.1275.

Tamnes, C. K., Walhovd, K. B., Grydeland, H., Holland, D., Ostby, Y., Dale, A. M., et al. (2013). Longitudinal working memory development is related to structural maturation of frontal and parietal cortices. *Journal of Cognitive Neuroscience*, *25*(10), 1611–1623. Available from: http://dx.doi.org/10.1162/jocn_a_00434.

Tomblin, J. B., Freese, P. R., & Records, N. L. (1992). Diagnosing specific language impairment in adults for the purpose of pedigree analysis. *Journal of Speech, Language, and Hearing Research*, *35*(4), 832–843. Available from: http://dx.doi.org/10.1044/jshr.3504.832.

Tomblin, J. B., Smith, E., & Zhang, X. (1997). Epidemiology of specific language impairment: Prenatal and perinatal risk factors. *Journal of Communication Disorders*, *30*(4), 325–344. Available from: http://dx.doi.org/10.1016/S0021-9924(97)00015-4.

Tonnquist-Uhlen, I. (1996). Topography of auditory evoked long-latency potentials in children with severe language impairment: The P2 and N2 components. *Ear & Hearing*, *17*(4), 314–326.

Tzourio, N., Heim, A., Zilbovicius, M., Gerard, C., & Mazoyer, B. M. (1994). Abnormal regional CBF response in left hemisphere of dysphasic children during a language task. *Pediatric Neurology*, *10*(1), 20–26.

Vallortigara, G., Rogers, L. J., & Bisazza, A. (1999). Possible evolutionary origins of cognitive brain lateralization. *Brain Research Reviews*, *30*(2), 164–175. Available from: http://dx.doi.org/10.1016/S0165-0173(99)00012-0.

van der Lely, H., & Pinker, S. (2014). The biological basis of language: Insights from developmental grammatical impairments. *Trends in Cognitive Sciences*, *18*(11), 586–595.

Verhoeven, J. S., Rommel, N., Prodi, E., Leemans, A., Zink, I., Vandewalle, E., et al. (2012). Is there a common neuroanatomical substrate of language deficit between autism spectrum disorder and specific language impairment? *Cerebral Cortex*, *22*(10), 2263–2271. Available from: http://dx.doi.org/10.1093/cercor/bhr292.

Weber-Fox, C., Leonard, L. B., Wray, A. H., & Tomblin, J. B. (2010). Electrophysiological correlates of rapid auditory and linguistic processing in adolescents with specific language impairment. *Brain and Language*, *115*(3), 162–181. Available from: http://dx.doi.org/10.1016/j.bandl.2010.09.001.

Whitehouse, A. J. O., & Bishop, D. V. M. (2008). Cerebral dominance for language function in adults with specific language impairment or autism. *Brain*, *131*(12), 3193–3200. Available from: http://dx.doi.org/10.1093/brain/awn266.

CHAPTER

73

Vascular Aphasia Syndromes

Donna C. Tippett[1,2,3] *and Argye E. Hillis*[2,3,4]

[1]Department of Otolaryngology—Head and Neck Surgery, Johns Hopkins University School of Medicine, Baltimore, MD, USA; [2]Department of Physical Medicine and Rehabilitation, Johns Hopkins University School of Medicine, Baltimore, MD, USA; [3]Department of Neurology, Johns Hopkins University School of Medicine, Baltimore, MD, USA; [4]Department of Cognitive Science, Johns Hopkins University, Baltimore, MD, USA

73.1 INTRODUCTION

The study of aphasia and its associated lesions in the 19th century by Dax (1936), Broca (1861, 1865), Wernicke (1874, 1881) and others revealed new insights about the neural organization of language. Perhaps the most reliable finding was that patients with language impairment typically had damage to the left hemisphere. Damage to the more anterior parts of the brain, particularly the left posterior inferior frontal gyrus (IFG), was often found in those whose spoken output was limited or poorly articulated (Broca, 1865); damage to the more posterior regions in the left temporal lobe (in the absence of damage to frontal regions) was found in those whose spoken output was well-articulated but meaningless (Wernicke, 1881). These early observations established that language functions are localized in the left cerebral hemisphere and provided the groundwork for Lichtheim's model of aphasia, later adapted by Geschwind (1965). Geschwind put together early observations, along with subsequent reports of behavior and associated lesions (based largely on autopsy), to characterize the collection of frequently co-occurring language characteristics that result from damage to particular areas of the brain. With the advent of computerized tomography (CT) technology in the early 1970s, it was confirmed that classic aphasic syndromes correlated with particular vascular territories (Damasio & Geschwind, 1984).

In this chapter, the vascular aphasia syndromes are described. It is emphasized that the various characteristics frequently co-occur because they all depend on areas of the brain supplied by the same artery (or branch of an artery). If only part of the territory is damaged, then only a subset of the characteristics is present. That is, the characteristics are dissociable. Many of the vascular aphasia syndromes are associated with damage to nonlanguage functions that also depend on brain tissue supplied by the same arterial branch. For example, Broca's aphasia is associated with left arm weakness or spasticity, because left arm function (like the speech and language functions affected in Broca's aphasia) depends on territory supplied by the superior division of the left middle cerebral artery (MCA). Therefore, one can see from the outset that these syndromes are not theoretically coherent syndromes in the sense that they have a single underlying functional basis; rather, they have a shared anatomical basis. After describing the vascular syndromes, we discuss how they can be understood within more current concepts of language and aphasia, and whether they have any usefulness in aphasia research, management of stroke, or rehabilitation of aphasia.

73.2 CLASSIC APHASIA CATEGORIZATION: VASCULAR SYNDROMES

73.2.1 Broca's Aphasia

Broca's aphasia is characterized by nonfluent, telegraphic, poorly articulated verbal output in which meaning is conveyed by content or information-carrying words, such as nouns and verbs; however, nouns are named more accurately than verbs (Berndt, Mitchum, Haendiges, & Sandson, 1997; Miceli, Silveri, Villa, & Caramazza, 1984). Morphological inflections (e.g., past tense, plurals) and function words (e.g., articles, conjunctions, prepositions, auxiliary verbs,

Neurobiology of Language. DOI: http://dx.doi.org/10.1016/B978-0-12-407794-2.00073-0

pronouns) are often omitted or incorrectly produced, and word order may be incorrect. Thus, speech production is characterized as agrammatic. Although it is widely agreed that individuals with Broca's aphasia are nonfluent, fluency is a multidimensional construct. "Nonfluency" may be associated with diminished number of words per minute, reduced phrase length, impaired melody, disrupted articulatory agility, and/or agrammatic sentence production (Goodglass & Kaplan, 1983). The factors that affect speech fluency may vary from one individual to the next. Also, individuals with aphasia may have preserved islands of fluency, particularly for rote and overlearned speech, in otherwise nonfluent novel speech output. Because speech fluency can be defined by a complex set of features, it may be difficult to judge. Experienced speech-language pathologists analyzed the content and fluency ratings on the Western Aphasia Battery for 20 individuals with aphasia. Inter-rater reliability ratings for fluency were poor using published criteria (Trupe, 1984).

It is important to note that the severity of language impairment can vary across individuals with Broca's aphasia. In its mildest form, Broca's aphasia presents with reduced phrase length with agrammatic sentence production and relatively retained language comprehension, although there may be deficits in comprehension of multistep commands or sentences with complex syntax (e.g., The flowers that Bob gave to Susan were roses). In more severe presentations, there is reduction of all speech output, with speech being limited to one or more recurrent utterances and automatic sequences (Goodglass, Kaplan, & Barresi, 2001). Repetition is usually halting and dysfluent. Reading comprehension may parallel auditory comprehension. The mechanics of written expression are compromised because patients must write with the left, nondominant hand, because of frequently co-occurring right hemiparesis. Spelling is often impaired, and written output is agrammatic.

Broca's aphasia, as originally described by Paul Broca, was attributed to lesions of the posterior half of the left third frontal convolution (IFG; Brodmann area 44). This area became known as Broca's area and, along with surrounding areas in the posterior frontal lobe, is supplied by the superior division of the left MCA. Subsequently, neuroanatomists and aphasiologists challenged the causative association between this circumscribed lesion site and the nature of the aphasia reported by Broca. Mohr et al. (1978) reported that a more widespread region must be damaged to result in the constellation of impairments seen in the syndrome of Broca's aphasia as originally described. Damage to Broca's area alone, according to Mohr, results in only a subset of the symptoms, most notably impaired motor speech (Keller, Crow, Foundas, Amunts, & Roberts, 2009). Furthermore, advances in neuroimaging, permitting studies of neural activity in normal adults during cognitive/language tasks, have led away from classic structure/function relationships to the role of brain networks and regions in the performance of cognitive/language tasks. Increasingly, Broca's area is tied to a variety of language functions (Davis et al., 2008; Grodzinsky & Amunts, 2006), functions that might be critical to language such as verbal working memory (Rogalsky & Hickok, 2011), and even nonlinguistic functions, such as visual searching and visual spatial cognition (Grodzinsky & Amunts, 2006). Many studies distinguish between functions of pars opercularis and pars triangularis (which are distinct cytoarchitectural fields that comprise Broca's area but have a great deal of anatomical overlap across individuals; Grodzinsky & Amunts, 2006). In individuals with stroke involving Broca's area, there is often concomitant right hemiparesis or monoparesis of the arm (and right facial weakness) due to ischemia of the nearby precentral gyrus.

73.2.2 Wernicke's Aphasia

Wernicke's aphasia is characterized by fluent, effortless, but relatively meaningless, spontaneous speech and repetition, and impaired comprehension at the word, sentence, and discourse levels. Spoken language may be limited to jargon comprising either real words or neologisms (nonwords such as "klimorata"), or a combination of the two. In milder forms, paraphasic errors are present as well as intermittent coherent "social" phrases, such as, "yes, that's right." Melodic contour of spoken language is often preserved, initially giving listeners the impression that output is intact. Comprehension may be severely compromised and impaired phonological analysis is thought to underlie the comprehension impairment in Wernicke's aphasia (Robson, Keidel, Lambon Ralph, & Sage, 2012). Repetition is generally similar to spontaneous speech—fluent jargon. These deficits have been attributed to impaired inhibition of lexical activation, so that the person cannot select the appropriate word, sound, or meaning from competing linguistic units that are also activated (Blumstein & Milberg, 2000). Although such an underlying impairment would account for many of the observed language deficits, it could not easily account for all cases. For example, there have been some reported cases of Wernicke's aphasia with relatively preserved or relatively impaired categories of words, such as animals or tools (Hillis & Caramazza, 1991a), or selectively impaired nouns relative to verbs (Zingeser & Berndt, 1990). Selective impairment or preservation of particular semantic categories of words indicates a deficit at the level of accessing lexical-semantics (Hillis & Caramazza, 1991a). Reading comprehension and written expression are typically impaired in a very similar pattern as the impairment in auditory comprehension and verbal expression, respectively, although some individuals may be able to read aloud and spell to dictation regular words that they fail to understand (Hillis & Caramazza, 1991a). In contrast to those with Broca's aphasia, individuals with Wernicke's aphasia are typically unaware of their errors; they may have only a shallow awareness that they have some kind of difficulty communicating.

This collection of deficits is usually caused by neural dysfunction in regions supplied by the inferior division of the left MCA, including Wernicke's area (most of Brodmann area 22, in the posterior, superior temporal gyrus), and often inferior parietal cortex (angular and supramarginal gyri) and inferior and middle temporal gyri. Because the Meyer's loop of the optic radiations runs through the temporal cortex, there is often a concomitant right homonymous superior quadrantanopsia. The fact that there is individual variability in the cerebral vasculature and the areas supplied by particular arteries can account for occasional dissociations between the typical deficits in Wernicke's aphasia (i.e., only some of the usual territory is affected by stroke in some cases, such that only some of the symptoms occur) as well as occasional anomalous lesion sites in patients with Wernicke's aphasia.

73.2.3 Global Aphasia

Global aphasia refers to a profound impairment of all modalities of receptive and expressive language. Individuals with global aphasia typically present with marked impairments of comprehension of single words, sentences, and conversations, as well as severely limited spoken output. Spontaneous verbal output may be restricted to single words, nonwords, or undifferentiated phonation and some individuals' speech only consists of perseverative utterances (e.g., "no, no"). In addition, reading and writing are typically profoundly compromised. Because of the severity of language impairment, communication partners sometimes need to anticipate the communicative intentions of individuals with global aphasia or rely primarily on gestures or drawing. In most cases, both Broca's area and Wernicke's area are damaged (Mazzocchi & Vignolo, 1979) or functionally compromised (Hillis et al., 2004) because of occlusion or stenosis of the proximal MCA (affecting both the inferior and superior divisions) or the internal carotid artery (ICA).

73.2.4 Conduction Aphasia

Conduction aphasia is usually defined as a language impairment characterized by relatively fluent, although paraphasic, spontaneous speech, intact auditory comprehension, and disproportionately impaired speech repetition. Secondary features include reading impairments, variable writing difficulties, and ideomotor apraxia (Benson et al., 1973). Individuals with conduction aphasia display well-articulated responses that are phonemically similar to target words and repetitive self-corrections resulting in increasingly closer approximations to targets. This phenomenon is termed "conduit d'approache" (Goodglass, 1992). Traditionally, conduction aphasia is thought to be caused by a lesion in the arcuate fasciculus, a white matter tract that runs between Broca's and Wernicke's areas, and thus is considered a disconnection syndrome because a lesion in the arcuate fasciculus is assumed to interrupt communication between the sensory and motor modules of the classically defined speech-language system (Geschwind, 1965). This hypothesis has been challenged. More recent evidence shows that conduction aphasia is not the result of damage only to the arcuate fasciculus. Although damage to the arcuate fasciculus is reported be present in the setting of conduction aphasia in some contemporary accounts (Geldmacher, Quigg, & Elias, 2007; Yamada et al., 2007; Zhang et al., 2010), it is not reported in others. In fact, cortical lesions alone may produce conduction aphasia (Anderson et al., 1999; Quigg, Geldmacher, & Elias, 2006), indicating that damage to the arcuate fasciculus is not a prerequisite condition for conduction aphasia. Furthermore, lesions of the arcuate fasciculus do not always cause conduction aphasia (Epstein-Peterson, Vasconcellos-Faria, Mori, Hillis, & Tsapkini, 2012; Selnes, van Zijl, Barker, Hillis, & Mori, 2002). Substantial anatomical evidence suggests that conduction aphasia is caused by damage to the left superior temporal gyrus and/or the left supramarginal gyrus (Axer, von Keyserlingk, Berks, & von Keyserlingk, 2001; Baldo & Dronkers, 2006) due to occlusion of a branch of the inferior division of the left MCA.

73.2.5 Anomic Aphasia

Impairment of word retrieval is the primary feature of anomic aphasia with relatively well-preserved language function in other realms. Individuals with anomic aphasia may use circumlocutions for targets and display protracted pauses in verbal output and use of fillers (e.g., "What is that called?"; "You know what I mean"), resulting in empty or low-content verbal output. Anomia is present in other aphasia syndromes and may be the residual language symptom as individuals recover language abilities. Consequently, anomic aphasia is the least reliably localized aphasia in chronic stroke. However, acute, isolated anomia is most often associated with small areas of ischemia in left inferior temporal cortex or thalamus (Hillis et al., 2006).

73.2.6 Transcortical Aphasias

Three transcortical aphasias have been characterized: transcortical motor (TCM); transcortical sensory (TCS); and mixed transcortical (MTC). The transcortical aphasias are distinguished by intact repetition ability. TCM aphasia shares many characteristics of Broca's aphasia but has the distinctive feature of fluent, grammatical repetition. This vascular syndrome is caused by lesions just anterior or superior to (surrounding) Broca's area (Freedman, Alexander, & Naeser, 1984), often caused by occlusion of the anterior cerebral artery (ACA) (Masdeu, Schoene, & Funkenstein, 1979; Rubens, 1976) or "watershed" areas between the ACA and the MCA. TCS aphasia is similar to Wernicke's aphasia, except for the presence of accurate repetition. TCS aphasia is usually attributed to lesions in areas surrounding Wernicke's area, in the watershed territories between the MCA and posterior cerebral artery (PCA), or the PCA territory (Alexander, Hiltbrunner, & Fischer, 1989). Induction of transient TCS aphasia during cortical mapping of individuals with seizure disorders is associated with multiple sites along the posterior superior and middle temporal gyri in classical Wernicke's area. Electrical interference mapping studies indicate that sparing of Wernicke's area is not a necessary condition for TCS aphasia (Boatman et al., 2000). MTC aphasia is analogous to global aphasia, with reduced or absent spontaneous speech, severely impaired language comprehension, and preserved repetition with consequent echolalia (Albert, Goodglass, Helm, Rubens, & Alexander, 1981). Lesion sites include both anterior and posterior left hemisphere cortical association areas, sparing the perisylvian language core and producing an "isolation of the speech area" (Rapcsak, Krupp, Rubens, & Reim, 1990).

73.2.7 Subcortical Aphasias

Subcortical aphasias can be classified into three groups: striato-capsular aphasia; thalamic aphasia; and aphasia associated with white matter paraventricular lesions (Kuljic-Obradovic, 2003). Preservation of repetition is common to all three subtypes, although there are features specific to each subtype. Striato-capsular aphasia and aphasia associated with paraventricular lesions are characterized by impairment of fluency, semantic paraphasias, and generally preserved comprehension. Thalamic aphasia is characterized by impaired comprehension and naming with fluent output containing predominantly semantic paraphasias (Benson & Ardila, 1995; Demonet, 1997). The role of subcortical structures in language remains controversial and different mechanisms have been offered to explain how such subcortical lesions can lead to aphasia. Some ascribe a direct role to subcortical areas in language function (Cappa, Cavallotti, Guidotti, Papagno, & Vignolo, 1983; Damasio, Damasio, Rizzo, Varney, & Gersh, 1982). Others contend that subcortical lesions lead to aphasia through diaschisis—dysfunction of a remote area of cortex caused by impaired input from the subcortical region (Perani, Vallar, Cappa, Messa, & Fazio, 1987; von Monakow, 1914). For example, unilateral lesions of the ventromedial thalamic nucleus produce a major ipsilateral metabolic depression in the cortex and adjacent thalamic nuclei, bilateral metabolic reduction in the basal ganglia, and relatively minor effects in the contralateral cortex (Girault, Savaki, Desban, Glowinski, & Besson, 1985). Language impairments in nonthalamic subcortical infarcts have been attributed to hypoperfusion of cortical structures caused by stenosis or occlusion of a large cerebral vessel responsible at the same time as the subcortical infarct (Hillis et al., 2002).

73.2.8 Variability of Vascular Syndromes

These classic aphasia classifications have also been reviewed in detail by Damasio (1992), Goodglass (1993), and Hillis (2007). The early accounts of vascular syndromes used the lesion method in which abnormal behavior is documented in the context of brain pathology, and localization of normal function is extrapolated by assuming that the lesioned area was responsible for whatever function was impaired at the time of assessment (Benson, 1994). It is important to note that individual variability in the shape of the brain as well as the patterns of sulci and gyri renders only approximate localization of cytoarchitectural fields, so even lesions that appear to be in the identical area may not affect identical functions (Amunts et al., 1999; Rademacher, Caviness, Steinmetz, & Galaburda, 1993). Furthermore, human vasculature varies considerably, more so in disease states, so that blockage of a particular branch does not reliably affect the identical areas (Caplan & Van Gijn, 2012). Not surprisingly, studies have variably confirmed the relationship between vascular territories and the vascular syndromes (Gavrilescu & Kase, 1995; Kumral, Bayulkem, Evyapan, & Yunten, 2002). Furthermore, recovery from aphasia varies substantially among patients (Hochstenback, den Otter, & Mulder, 2003; Lazar, Speizer, Festa, Krakauer, & Marshall, 2008). Depending on the time of evaluation, a person with damage to an entire vascular territory might present with the entire vascular syndrome, a partial syndrome, or none of it (if the person recovered completely). One study showed that the vascular syndromes correspond to vascular territories more reliably in the acute than in the chronic stages of stroke (Ochfeld et al., 2010). In acute stroke, language impairments reflect the entire dysfunctional tissue (both infarcted and hypoperfused tissue) not always visible on structural imaging.

Beginning in the 1980s, developments in functional neuroimaging, including PET, functional MRI (fMRI), and magnetoencephalography, expanded understanding of the functional neuroanatomy of language. Safe, noninvasive imaging of the brain reveals that areas in both hemispheres of the brain are activated specifically during language tasks, although the left hemisphere shows more activation in the majority of neurologically normal adults (Binder, 1997; Crinion, Lambon Ralph, Warburton, Howard, & Wise, 2003; Fridriksson & Morrow, 2005; Petersen, Fox, Posner, Mintun, & Raichle, 1988), and that more distant areas of the cortex, such as inferior and anterior temporal cortex (Wise, 2003) and the basal ganglia and thalamus (Kraut et al., 2002), are also activated during language tasks.

73.3 VASCULAR SYNDROMES AND CONTEMPORARY PARADIGMS

How do the vascular syndromes fit with more contemporary frameworks of the functional neuroanatomy of language that postulate a dual stream of language processing (Hickok & Poeppel, 2000, 2004, 2007)? The brain is assumed to compute a transformation between thought and an acoustic signal (Schwartz, Faseyitan, Kim, & Coslett, 2012), and it executes parallel processing to synthesize input via interconnected neural networks (Ueno, Saito, Rogers, & Lambon Ralph, 2011). In the neuroanatomical model of speech processing, proposed by Hickok and Poeppel (2007), speech perception involves auditory-responsive areas in the superior temporal gyrus bilaterally, left more so than right. The processing system then diverges into two streams. The ventral stream is a sound-meaning interface responsible for processing speech signals for comprehension. In the dorsal stream, acoustic speech signals are translated into articulatory representations essential for speech development and production involving auditory-motor integration. These streams are also thought to be bidirectional; the ventral stream mediates the relationship between sound and meaning for perception and production, and the dorsal system can also map motor speech representations onto auditory speech representations (Hickok & Poeppel, 2004, 2007).

When a map of the vascular territories is superimposed onto this neuroanatomical model of speech processing (Figure 73.1), one can see that the dorsal stream is supplied by the superior division of the left MCA, and the ventral stream is supplied largely by the inferior division of the left MCA. Therefore, it is not surprising that individuals with the vascular syndrome of Broca's aphasia typically have deficits that can be attributed to disruption of the dorsal stream: the articulatory network or sensorimotor interface. Likewise, it is not surprising that individuals with the vascular syndrome of Wernicke's aphasia have deficits that can be attributed to the lexical interface and/or combinatorial network to map sound onto meaning. The repetition deficit in conduction aphasia is attributed to damage in the Sylvian parietal–temporal (Spt) area, which is located within the Sylvian fissure at the parietal–temporal boundary, reflecting disruption of the dorsal stream route (Hickok & Poeppel, 2004). This hypothesis is supported by fMRI data that show that the region of maximal overlap in lesion distribution in 14 conduction aphasics includes area Spt (Buchsbaum et al., 2011). This is consistent with recent evidence that suggests that conduction aphasia is caused by damage to the left superior temporal gyrus and/or the left supramarginal gyrus (Axer et al., 2001; Baldo & Dronkers, 2006), both of which are anatomical regions associated with the dorsal stream (Schwartz et al., 2012). Further support is found in studies using computational models to synthesize aphasia. Conduction aphasia is produced after damage to the dorsal pathway in a neuroanatomically constrained computational dual dorsal-ventral pathway computational model (Ueno et al., 2011). Other relevant work includes voxelwise lesion behavior mapping to investigate the association of language impairments and dorsal-ventral streams. Although these studies do not typically associate lesions in particular voxels with specific aphasia syndromes, findings are that deficits in naming and repetition are associated with the dorsal stream (Hanley, Kay, & Edwards, 2002; Ueno et al., 2011) and deficits in comprehension are associated with the ventral pathway (Hillis & Caramazza, 1991b). Finally, it should not be surprising that unilateral vascular syndromes do not typically disrupt the conceptual system, which is bilaterally and widely represented in this model. Rather, conceptual meaning is disrupted in neurodegenerative disease, such as semantic variant primary progressive aphasia.

73.4 COGNITIVE PROCESSES UNDERLYING APHASIA

What is lost by characterizing an individual as an exemplar of a particular vascular syndrome? As noted, each vascular syndrome is a collection of frequently

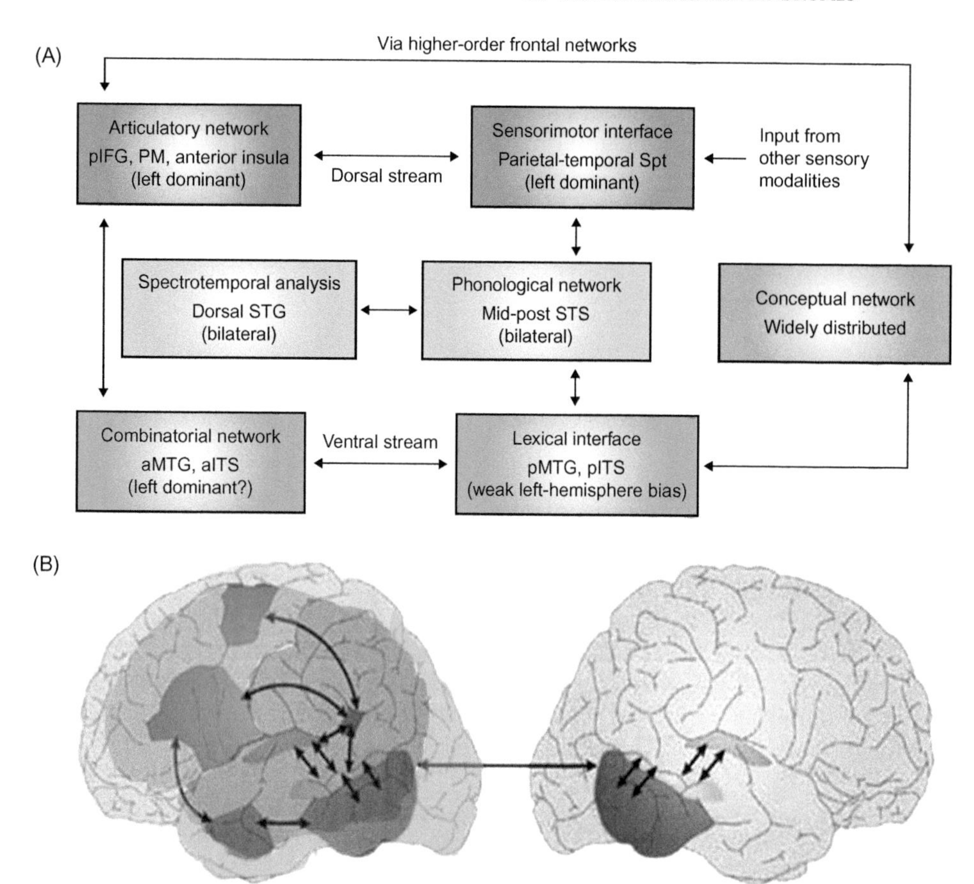

FIGURE 73.1 A schematic representation of the cortical organization of speech processing proposed by Hickok and Poeppel (2007), on which we have superimposed a map of the vascular territories on the left hemisphere only. The left ACA is shown in transparent yellow; the left superior division MCA is shown in transparent blue; the left inferior division MCA is shown in transparent pink; and the left PCA is shown in green.

co-occurring impairments to functions that depend on an area of brain supplied by a particular blood vessel. In the past, these deficits were characterized as impairments to particular tasks (e.g., sentence repetition). However, it is now understood that each language task comprises a number of cognitive representations and processes that might depend on different areas of the brain. For example, some of the controversy regarding the anatomical correlates of conduction aphasia may have arisen by characterizing patients as having impaired sentence repetition, rather than identifying what cognitive processes underlying sentence repetition were impaired in individual patients. Cognitive models of language processing include both a semantic and nonlexical phonological mechanism for word and sentence repetition (Hanley et al., 2002). Evidence for this model is found in case reports of performance on language tasks in individuals with neurologic impairments (Hanley, Dell, Kay, & Baron, 2004; Hillis & Caramazza, 1991b). More recently, in a study of sentence recall in healthy adults and an individual with aphasia, phonological information was judged to contribute to sentence recall performance as a complement to semantic and conceptual information (Schweppe, Rummer, Bormann, & Martin, 2011). Some patients may have a disrupted phonological short-term storage system, whereas others may have a disrupted semantic working memory system or disrupted "central executive" component of working memory as the cause of impaired sentence repetition. Each of these impairments might be associated with different lesion sites. Error rates on syllable and pseudoword repetition tasks were increased with inhibitory transcranial magnetic stimulation (TMS) to the posterior part of the superior temporal sulcus and temporoparietal junction in normal volunteers, implicating the role of the dorsal pathway. Differences in error rates were not seen during sentence repetition tasks, suggesting that the ventral pathway may have been recruited for sentence repetition (Murakami, Kell, Restile, Ugawa, & Siemann, 2013).

73.5 POTENTIAL USEFULNESS OF VASCULAR SYNDROMES

73.5.1 Aphasia Research

Grouping patients for research assumes that individuals in the group are homogeneous with respect to something of theoretical interest. The vascular syndromes are frequently co-occurring symptoms.

Individuals with Broca's aphasia often have both apraxia of speech and more difficulty naming verbs than nouns (Miceli et al., 1984); they also often have weakness of the right arm. These symptoms may all stem from damage to the posterior frontal lobe, supplied by the left superior division MCA. But they do not have a common underlying cause, and they dissociate in individuals. It has been argued that because only a subset of deficits in a vascular syndrome is likely to be present in an individual (depending on the portion of the vascular territory affected by stroke, the degree of recovery, individual variation in anatomy), the syndrome approach has limited usefulness for aphasia research. For example, patients in the group "Broca's aphasia" might each have different disruptions to cognitive processes underlying articulation and/or sentence production (Caramazza & Badecker, 1989), all due to lesions in areas supplied by the superior division of the left MCA. Alternatives include single subject designs (Caramazza & Coltheart, 2006) and case series analysis (Schwartz & Dell, 2010), or grouping by cognitive impairment (DeLeon et al., 2007). However, classification by vascular syndrome remains the standard basis for describing groups of participants in aphasia research.

73.5.2 Treatment of Stroke

Furthermore, grouping patients by affected vascular territory may be quite useful for a stroke neurologist. For example, grouping by vascular aphasia syndromes allowed the discovery that inferior division MCA strokes (associated with Wernicke's aphasia) were more likely due to cardioembolism (Bogousslavsky, Van Melle, & Regli, 1989), and superior division MCA strokes (associated with Broca's aphasia) were more likely due to carotid dissection and carotid occlusion (Trupe et al., 2013). Because it is often essential to initiate treatment before the lesion is seen on imaging, a good idea of the volume and location of the affected territory is useful in planning intervention. Furthermore, acute presentations of vascular aphasia syndromes provide the stroke neurologist with information about the likely areas of brain that are dysfunctional (either infarcted or hypoperfused). When there is mismatch between the vascular aphasia syndrome and the structural lesion seen on diffusion-weighted imaging (DWI; which is very sensitive to ischemic tissue even early in stroke), the stroke neurologist knows there is brain tissue that is hypoperfused and at risk for further ischemia. This "diffusion-clinical mismatch" is frequently a prompt for intervention (Reineck, Agarwal, & Hillis, 2005). To illustrate, the patient whose scans are shown in Figure 73.2 had Wernicke's aphasia at onset but only a tiny lesion on DWI in the insula that could not account for his deficits. His "diffusion-clinical mismatch" prompted further imaging with perfusion-weighted imaging (PWI) the same day and intervention to reperfuse Wernicke's area. His Wernicke's aphasia resolved by day 3 with restored blood flow to Wernicke's area.

73.5.3 Aphasia Treatment

Is it useful to group patients by vascular aphasia syndrome for treatment of their aphasia? Certainly, there have been treatments described for Broca's aphasia, Wernicke's aphasia, and so on (Carlomagno, Zulian, Razzano, DeMercurio, & Marini, 2013; Conklyn, Novak, Boissy, Bethoux, & Chemali, 2012; Helm-Estabrooks, Fitzpatrick, & Barresi, 1982; Helm-Estabrooks & Ramsberger, 1986). Treatment on the basis of syndrome classification alone would obviously neglect important individual differences known to exist in each of the aphasia syndromes as well as

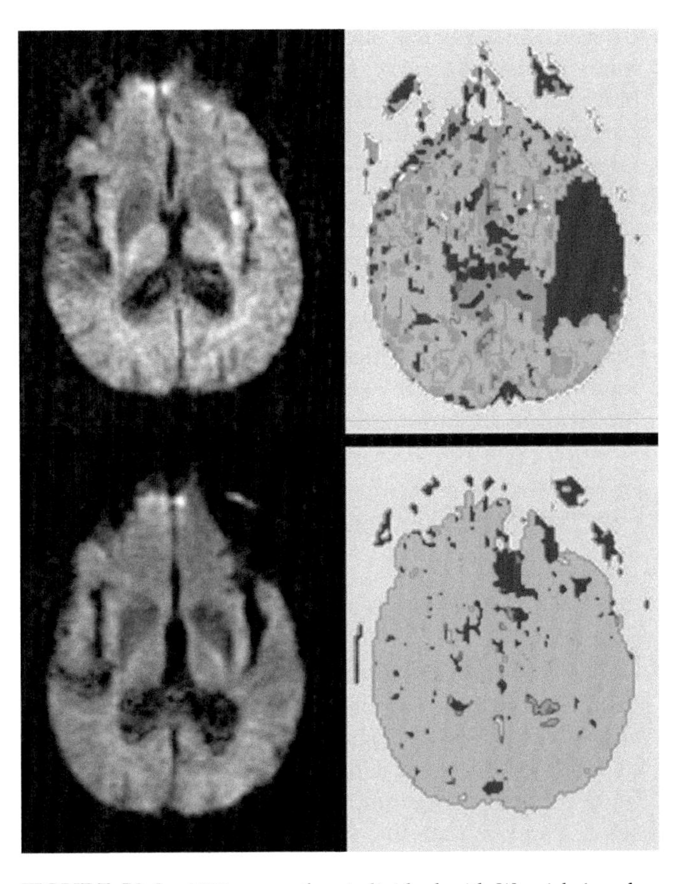

FIGURE 73.2 MRI scans of an individual with Wernicke's aphasia at day 1 (top panel). DWI (left) shows a tiny area of ischemia in the left insula. PWI the same day shows severe hypoperfusion of Wernicke's area and the surrounding left superior temporal cortex. Repeat imaging at day 3 (lower panel) shows restored blood flow to the entire area in association with resolution of Wernicke's aphasia.

personal preferences of individual patients, vital aspects of therapy consistent with evidence-based practice. In fact, most of these treatments are therapies for particular deficits within each syndrome. For example, in a case series study including individuals with Broca's, Wernicke's, and anomic aphasia, a cueing hierarchy was developed to treat anomia due to impaired access to word meaning or word form in all participants (Best et al., 2013). It would be at least as rational in most cases to describe the intervention as a therapy for "constructing a sentence planning frame" (rather than Broca's aphasia) or "mapping lexical representations to meaning" (rather than Wernicke's aphasia). Therapy for specific aphasia syndromes may be optimized by considering the respective roles of the ventral and dorsal streams. For example, the ventral pathway could be leveraged by encouraging patients to process the meaning of a target word during a repetition task in the treatment of conduction aphasia (Ueno & Lambon Ralph, 2013).

However, characterizing patients as exemplars of a vascular aphasia syndrome might be useful for neuromodulatory treatments that are, by hypothesis, directed to particular sites of the brain. Aphasia treatment is just beginning to take advantage of targeted brain-based interventions, such as transcranial direct current stimulation (tDCS) and TMS designed to facilitate synaptic plasticity (Mottaghy, Sparing, & Topper, 2006; Schlaug, Renga, & Nair, 2008). Most studies have selected patients for treatment on the basis of vascular syndrome (Vines, Norton, & Schlaug, 2011) and have carefully targeted particular areas of the brain based on fMRI (Dmochowski et al., 2013; Fridriksson, Richardson, Baker, & Rorden, 2011) or predetermined regions. However, it is yet to be shown whether the localization of these treatments or the site of lesion/vascular syndrome of patients receiving these treatments (in conjunction with behavioral therapies) is critical to their effectiveness.

73.6 CONCLUSION

The value of classifying aphasias as vascular syndromes continues to be deliberated as the field of aphasiology evolves. The usefulness of classification depends on the goal. Characterizing someone as having a particular vascular syndrome communicates: (i) a set of likely symptoms; (ii) the area of the brain most likely to be affected (at least if the individual has not shown substantial change in either the lesion or the language symptoms since onset); and (iii) the arterial branch most likely to be involved if the cause of aphasia is ischemic stroke. Advances in neuroimaging have allowed refinements in characterizing lesions, so we can now identify areas of the brain associated with impairments in specific cognitive functions underlying language. We are no longer limited to describing the symptoms associated with a vascular territory. New theories of language processing and production complement long-established classification systems, and these novel and emerging perspectives will certainly translate to changes in the conducting of research and treatment of stroke.

Acknowledgments

This publication was made possible by NIH grants R01 DC 05375 and R01 DC 03681 from NIDCD. We gratefully acknowledge this support.

References

Albert, M. L., Goodglass, H., Helm, N. A., Rubens, A. B., & Alexander, M. P. (1981). *Clinical aspects of dysphasia*. Wien: Springer-Verlag.

Alexander, M. P., Hiltbrunner, B., & Fischer, R. S. (1989). Distributed anatomy of transcortical sensory aphasia. *Archives of Neurology, 46*, 885–892.

Amunts, K., Schleicher, A., Burgel, U., Mohlberg, H., Uylings, H. B. M., & Zilles, K. (1999). Broca's region revisited: Cytoarchitecture and intersubject variability. *Journal of Comparative Neurology, 412*, 319–341.

Anderson, J. M., Gilmore, R., Roper, S., Crosson, B., Bauer, R. M., Nadeau, S., et al. (1999). Conduction aphasia and the arcuate fasciculus: A reexamination of the Wernicke–Geschwind model. *Brain and Language, 70*, 1–12.

Axer, H., von Keyserlingk, A. G., Berks, G., & von Keyserlingk, D. G. (2001). Supra- and infrasylvian conduction aphasia. *Brain and Language, 76*, 317–331.

Baldo, J. V., & Dronkers, N. F. (2006). The role of inferior parietal and inferior frontal cortex in working memory. *Neuropsychology, 20*, 529–538.

Benson, D. F. (1994). *The neurology of thinking*. New York, NY: Oxford University Press.

Benson, D. F., & Ardila, A. (1995). *Aphasia: A clinical perspective*. New York, NY: Oxford University Press.

Benson, D. F., Sheremata, W. A., Bouchard, R., Segarra, J. M., Price, D., & Geschwind, N. (1973). Conduction aphasia. A clinicopathological study. *Archives of Neurology, 28*, 339–346.

Berndt, R. S., Mitchum, C. C., Haendiges, A. N., & Sandson, J. (1997). Verb retrieval in aphasia. *Brain and Language, 56*, 68–106.

Best, W., Greenwood, A., Grassly, J., Herbert, R., Hickin, J., & Howard, D. (2013). Aphasia rehabilitation: Does generalisation from anomia therapy occur and is it predictable? A case series study. *Cortex, 49*, 2345–2357. Available online February 4, 2012 at www.sciencedirect.com. Available from: http://dx.doi.org/10.1016/j.cortex.2013.01.005.

Binder, J. R. (1997). Neuroanatomy of language processing studied with functional MRI. *Clinical Neuroscience, 4*, 87–94.

Blumstein, S. E., & Milberg, W. P. (2000). Language deficits in Broca's and Wernicke's aphasia: A singular impairment. In Y. Grodzinsky, L. P. Shapiro, & D. Swinney (Eds.), *Language and the Brain* (pp. 167–183). New York, NY: Academic Press.

Boatman, D., Gordon, B., Hart, J., Selnes, O., Miglioretti, D., & Lenz, F. (2000). Transcortical sensory aphasia: Revisited and revised. *Brain, 123*, 634–1642.

Bogousslavsky, J., Van Melle, G., & Regli, F. (1989). Middle cerebral artery pial territory infarcts: A study of the Lausanne Stroke Registry. *Annals of Neurology, 25*, 555–560.

Broca, P. (1861). Perte de la parole: Ramollissement chronique et destruction partielle du lobe anterieur gauche du cerveau. *Bulletins de la Sociéte d'anthropologie, 2*, 235–238.

Broca, P. (1865). Sur le siège de la faculté du langage articulé. *Bulletins de la Sociéte d'anthropologie, 6*, 337–393.

Buchsbaum, B. R., Baldo, J. V., Okada, K., Berman, K. F., Dronkers, N., D'Esposito, M., et al. (2011). Conduction aphasia, sensory-motor integration and phonological short-term memory—An aggregate analysis of lesion and fMRI data. *Brain and Language, 119*, 119–128.

Caplan, L., & Van Gijn, J. (2012). *Stroke syndromes*. Cambridge, UK: Cambridge University Press.

Cappa, S. F., Cavallotti, G., Guidotti, M., Papagno, C., & Vignolo, L. A. (1983). Subcortical aphasia: Two clinical-CT scan correlation studies. *Cortex, 19*, 227–241.

Caramazza, A., & Badecker, W. (1989). Patient classification in neuropsychological research. *Brain and Cognition, 10*, 256–295.

Caramazza, A., & Coltheart, M. (2006). Cognitive neuropsychology twenty years on. *Cognitive Neuropsychology, 23*, 3–12.

Carlomagno, S., Zulian, A. N., Razzano, C., DeMercurio, I. D., & Marini, A. (2013). Coverbal gestures in the recovery from severe fluent aphasia: A pilot study. *Journal of Communication Disorders, 46*, 84–99.

Conklyn, D., Novak, E., Boissy, A., Bethoux, F., & Chemali, K. (2012). The effects of modified melodic intonation therapy on nonfluent aphasia: A pilot study. *Journal of Speech, Language and Hearing Research, 55*, 1463–1471.

Crinion, J. T., Lambon Ralph, M. A., Warburton, E. A., Howard, D., & Wise, R. J. (2003). Temporal lobe regions engaged during normal speech comprehension. *Brain, 126*(Pt 5), 1193–1201.

Damasio, A. R. (1992). Aphasia. *New England Journal of Medicine, 326*, 531–539.

Damasio, A. R., Damasio, H., Rizzo, M., Varney, N., & Gersh, F. (1982). Aphasia with nonhemorrhagic lesions in the basal ganglia and internal capsule. *Archives of Neurology, 39*, 15–24.

Damasio, A. R., & Geschwind, N. (1984). The neural basis of language. *Annual Review of Neuroscience, 7*, 127–147.

Davis, C., Kleinman, J. T., Newhart, M., Gingis, L., Pawlak, M., & Hillis, A. E. (2008). Speech and language functions that require a functioning Broca's area. *Brain and Language, 105*, 50–58.

Dax, M. (1936). Lésions de la moitié gauche de l'encêphale coincedent avec l'oublie des signes de la pensée. Montpelier. *Gazette Hebdomadaire de Medecine et de Chirurgie, 17*, 259–260.

DeLeon, J., Gottesman, R. F., Kleinman, J. T., Newhart, M., Davis, C., Heidler-Gary, J., et al. (2007). Neural regions essential for distinct cognitive processes underlying picture naming. *Brain, 130*, 1408–1422.

Demonet, J. F. (1997). Subcortical aphasia(s): A controversial and promising topic. *Brain and Language, 58*, 410–417.

Dmochowski, J. P., Datta, A., Huang, Y., Richardson, J. D., Bikson, M., Fridriksson, J., et al. (2013). Targeted transcranial direct current stimulation for rehabilitation after stroke. *NeuroImage, 75*, 12–19.

Epstein-Peterson, Z., Vasconcellos-Faria, A., Mori, S., Hillis, A. E., & Tsapkini, K. (2012). Relatively normal repetition performance despite severe disruption of the left arcuate fasciculus. *Neurocase, 18*, 521–526.

Freedman, M., Alexander, M. P., & Naeser, M. A. (1984). Anatomic basis of transcortical motor aphasia. *Neurology, 34*, 409–417.

Fridriksson, J., & Morrow, L. (2005). Cortical activation associated with language task difficulty in aphasia. *Aphasiology, 19*, 239–250.

Fridriksson, J., Richardson, J. D., Baker, J. M., & Rorden, C. (2011). Transcranial direct current stimulation improves naming reaction time in fluent aphasia: A double-blind, sham-controlled study. *Stroke, 42*, 819–821.

Gavrilescu, T., & Kase, C. S. (1995). Clinical stroke syndromes: Clinical anatomical correlations. *Cerebrovascular and Brain Metabolism Reviews, 7*, 218–239.

Geldmacher, D. S., Quigg, M., & Elias, W. J. (2007). MR tractography depicting damage to the arcuate fasciculus in a patient with conduction aphasia. *Neurology, 69*, 321–322.

Geschwind, N. (1965). Disconnexion syndromes in animals and man. *Brain, 88*, 237–294 585–644.

Girault, J. A., Savaki, H. E., Desban, M., Glowinski, J., & Besson, M. J. (1985). Bilateral cerebral metabolic alterations following lesion of the ventromedial thalamic nucleus: Mapping by the 14C-deoxyglucose method in conscious rats. *Journal of Comparative Neurology, 231*, 137–149.

Goodglass, H. (1992). Diagnosis of conduction aphasia. In S. E. Kohn (Ed.), *Conduction aphasia* (pp. 39–49). Hillsdale, NJ: Lawrence Erlbaum Associates.

Goodglass, H. (1993). *Understanding aphasia*. San Diego, CA: Academic Press.

Goodglass, H., & Kaplan, E. (1983). *The assessment of aphasia and related disorders* (2nd ed.). Philadelphia, PA: Lea & Febiger.

Goodglass, H., Kaplan, E., & Barresi, B. (2001). *Boston diagnostic aphasia examination* (3rd ed.). Baltimore, MD: Lippincott Williams & Wilkins.

Grodzinsky, Y., & Amunts, K. (2006). *Broca's region*. New York, NY: Oxford University Press.

Hanley, J. R., Dell, G. S., Kay, J., & Baron, R. (2004). Evidence for the involvement of a nonlexical route in the repetition of familiar words: A comparison of single and dual route models of auditory repetition. *Cognitive Neuropsychology, 21*, 147–158.

Hanley, J. R., Kay, J., & Edwards, M. (2002). Imageability effects, phonological errors, and the relationship between auditory repetition and picture naming: Implications for models of auditory repetition. *Cognitive Neuropsychology, 19*, 193–206.

Helm-Estabrooks, N., Fitzpatrick, P., & Barresi, B. (1982). Visual action therapy for global aphasia. *Journal of Speech and Hearing Disorders, 47*, 385–398.

Helm-Estabrooks, N., & Ramsberger, G. (1986). Treatment of agrammatism in long-term Broca's aphasia. *British Journal of Disorders of Communication, 21*, 39–45.

Hickok, G., & Poeppel, D. (2000). Towards a functional neuroanatomy of speech perception. *Trends in Cognitive Sciences, 4*, 131–138.

Hickok, G., & Poeppel, D. (2004). Dorsal and ventral streams: A framework for understanding aspects of the functional anatomy of language. *Cognition, 92*, 67–99.

Hickok, G., & Poeppel, D. (2007). The cortical organization of speech processing. *Nature Reviews Neuroscience, 8*, 393–402.

Hillis, A. E. (2007). Aphasia: Progress in the last quarter of a century. *Neurology, 69*, 200–213.

Hillis, A. E., Barker, P. B., Wityk, R. J., Aldrich, E. M., Restrepo, L., Breese, E. L., et al. (2004). Variability in subcortical aphasia is due to variable sites of cortical hypoperfusion. *Brain and Language, 89*, 524–530.

Hillis, A. E., & Caramazza, A. (1991a). Category-specific naming and comprehension impairment: A double dissociation. *Brain, 114*, 2081–2094.

Hillis, A. E., & Caramazza, A. (1991b). Mechanisms for accessing lexical representations for output: Evidence from a category-specific semantic deficit. *Brain and Language, 40*, 106–144.

Hillis, A. E., Kleinman, K. T., Newhart, M., Heidler-Gary, J., Gottesman, R., Barker, P. B., et al. (2006). Restoring cerebral blood flow reveals neural regions critical for naming. *Journal of Neuroscience, 26*, 8069–8073.

Hillis, A. E., Wityk, R. J., Barker, P. B., Beauchamp, N. J., Gailloud, P., Murphy, K., et al. (2002). Subcortical aphasia and neglect in acute stroke: The role of cortical hypoperfusion. *Brain, 125*, 1094–1104.

Hochstenbach, J. B., den Otter, R., & Mulder, T. W. (2003). Cognitive recovery after stroke: A 2-year follow up. *Archives of Physical Medical Rehabilitation, 84*, 1499–1504.

Keller, S. S., Crow, T., Foundas, A., Amunts, K., & Roberts, N. (2009). Broca's area: Nomenclature, anatomy, typology and asymmetry. *Brain and Language, 109*, 29–48.

Kraut, M. A., Kremen, S., Mon, L. R., Segal, J. B., Calhoun, V., & Hartt, J. J. (2002). Object activation in semantic memory from visual multimodal feature input. *Journal of Cognitive Neuroscience, 14*, 37–47.

Kuljic-Obradovic, D. C. (2003). Subcortical aphasia: Three different language disorder syndromes? *European Journal of Neurology, 10*, 445–448.

Kumral, E., Bayulkem, G., Evyapan, D., & Yunten, N. (2002). Spectrum of anterior cerebral artery territory infarction: Clinical and MRI findings. *European Journal of Neurology, 9*, 615–624.

Lazar, R. M., Speizer, A. E., Festa, J. R., Krakauer, J. W., & Marshall, R. S. (2008). Variability in language recovery after first-time stroke. *Journal of Neurology, Neurosurgery and Psychiatry, 79*, 530–534.

Masdeu, J. C., Schoene, W. C., & Funkenstein, H. H. (1979). Aphasia following infarction of the left supplementary motor area. *Neurology, 15*, 627–653.

Mazzocchi, F., & Vignolo, L. A. (1979). Localisation of lesions in aphasia: Clinical CT scan correlations in stroke patients. *Cortex, 15*, 627–653.

Miceli, G., Silveri, M. C., Villa, G., & Caramazza, A. (1984). On the basis for the agrammatic's difficulty in producing main verbs. *Cortex, 20*, 207–220.

Mohr, J. P., Pessin, M. S., Finkelstein, S., Funkenstein, H. H., Duncan, G. W., & Davis, K. R. (1978). Broca aphasia: Pathologic and clinical. *Neurology, 28*(4), 311–324.

Mottaghy, F. M., Sparing, R., & Topper, R. (2006). Enhancing picture naming with transcranial magnetic stimulation. *Behavioral Neurology, 17*, 177–189.

Murakami, T., Kell, C., Restile, J., Ugawa, Y., & Siemann, U. (2013). Functional causality of the dorsal stream in sensorimotor integration of speech repetition. *Clinical Neurophysiology, 124*, e82–e83.

Ochfeld, E., Newhart, M., Molitoris, J., Leigh, R., Cloutman, L., Davis, C., et al. (2010). Ischemia in Broca area is associated with Broca aphasia more reliably in acute than in chronic stroke. *Stroke, 41*, 325–3300.

Perani, D., Vallar, G., Cappa, S., Messa, C., & Fazio, F. (1987). Aphasia and neglect after subcortical stroke. A clinical/cerebral perfusion correlation study. *Brain, 110*, 1211–1229.

Petersen, S. E., Fox, P. T., Posner, M. I., Mintun, M., & Raichle, M. E. (1988). Positron emission tomography studies of the cortical anatomy of single word processing. *Nature, 331*, 585–589.

Quigg, M., Geldmacher, D. S., & Elias, W. J. (2006). Conduction aphasia as a function of the dominant posterior perisylvian cortex: Report of two cases. *Journal of Neurosurgery, 104*, 845–848.

Rademacher, J., Caviness, V. S., Steinmetz, H., & Galaburda, A. M. (1993). Topographical variation in the human primary cortices: Implications for neuroimaging, brain mapping, and neurobiology. *Cerebral Cortex, 3*, 313–329.

Rapcsak, S. Z., Krupp, L. B., Rubens, A. B., & Reim, J. (1990). Mixed transcortical aphasia without anatomic isolation of the speech area. *Stroke, 21*, 953–956.

Reineck, L. A., Agarwal, S., & Hillis, A. E. (2005). "Diffusion-clinical mismatch" is associated with potential for early recovery of aphasia. *Neurology, 64*, 828–833.

Robson, H., Keidel, J. L., Lambon Ralph, M. A., & Sage, K. (2012). Revealing and quantifying the impaired phonological analysis underpinning impaired comprehension in Wernicke's aphasia. *Neuropsychologia, 50*, 276–288.

Rogalsky, C., & Hickok, G. (2011). The role of Broca's area in sentence comprehension. *Journal of Cognitive Neuroscience, 23*, 1664–1680.

Rubens, A. B. (1976). Transcortical motor aphasia. In H. Whitaker, & H. A. Whitaker (Eds.), *Studies in neurolinguistics* (pp. 293–306). New York, NY: Academic Press.

Schlaug, G., Renga, V., & Nair, D. (2008). Transcranial direct current stimulation in stroke recovery. *Archives of Neurology, 65*, 1571–1576.

Schwartz, M. F., & Dell, G. S. (2010). Case series investigations in cognitive neuropsychology. *Cognitive Neuropsychology, 27*, 477–494.

Schwartz, M. F., Faseyitan, O., Kim, J., & Coslett, H. B. (2012). The dorsal stream contribution to phonological retrieval in object naming. *Brain, 135*, 3799–3814.

Schweppe, J., Rummer, R., Bormann, T., & Martin, R. C. (2011). Semantic and phonological information in sentence recall: Converging psycholinguistic and neuropsychological evidence. *Cognitive Neuropsychology, 28*, 521–545.

Selnes, O. A., van Zijl, P. C. M., Barker, P. B., Hillis, A. E., & Mori, S. (2002). MR diffusion tensor imaging documented arcuate fasciculus lesion in a patient with normal repetition performance. *Aphasiology, 16*, 897–902.

Trupe, E. H. (1984). Reliability of rating spontaneous speech in the western aphasia battery: Implications for classification. In R. Brookshire (Ed.), *Clinical aphasiology: Proceedings of the conference* (pp. 55–69). Minneapolis, MN: BRK Publishers.

Trupe, L. A., Varma, D. D., Gomez, Y., Race, D., Leigh, R., Hillis, A. E., et al. (2013). Chronic apraxia of speech and Broca's area. *Stroke, 44*, 740–744.

Ueno, T., & Lambon Ralph, M. A. (2013). The roles of the "ventral" semantic and "dorsal" pathways in conduit d'approche: A neuroanatomically-constrained computational modeling investigation. *Frontiers in Human Neuroscience, 7*, 422. Available from: http://dx.doi.org/10.3389/fnhum.2013.00422.

Ueno, T., Saito, S., Rogers, T. T., & Lambon Ralph, M. A. (2011). Lichtheim 2: Synthesizing aphasia and the neural basis of language in a neurocomputational model of the dual dorsal-ventral language pathways. *Neuron, 72*, 385–396.

Vines, B. W., Norton, A. C., & Schlaug, G. (2011). Non-invasive brain stimulation enhances the effects of melodic intonation therapy. *Frontiers in Psychology, 26*, 230. Available from: http://dx.doi.org/10.3389/fpsyg.2011.00230.

von Monakow, C. (1914). *Die Lokalisation im Grosshirn und der Abbau der Funktion durch Kortikale Herde.* Wiesbaden: Bergmann.

Wernicke, C. (1874). *Der Aphasische Symptomenkomplex.* Breslau (Poland): Cohen and Weigert.

Wernicke, C. (1881). *Lehrbuch der Gehirnkrankheiten.* Berlin: Theodor Fisher.

Wise, R. J. S. (2003). Language systems in normal and aphasic human subjects: Functional imaging studies and inferences from animal studies. *British Medical Bulletin, 65*, 95–119.

Yamada, K., Nagakane, Y., Mizuno, T., Hosomi, A., Nakagawa, M., & Nishimura, T. (2007). MR tractography depicting damage to the arcuate fasciculus in a patient with conduction aphasia. *Neurology, 68*, 789–790.

Zhang, Y., Wang, C., Zhao, X., Chen, H., Han, Z., & Wang, Y. (2010). Diffusion tensor imaging depicting damage to the arcuate fasciculus in patients with conduction aphasia: A study of the Wernicke–Geschwind model. *Neurological Research, 32*, 775–778.

Zingeser, L. B., & Berndt, R. S. (1990). Retrieval of nouns and verbs in agrammatism and anomia. *Brain and Language, 39*, 14–32.

CHAPTER

74

Psycholinguistic Approaches to the Study of Syndromes and Symptoms of Aphasia

Sheila E. Blumstein

Department of Cognitive Linguistic and Psychological Sciences, Brown University, Brown Institute for Brain Sciences, Providence, RI, USA

74.1 INTRODUCTION

One of the fascinating aspects of aphasia is that language breakdown is not unitary. Not all aspects of language are affected in the same way and, crucially, depending on the site and extent of the lesion, brain injury to the left hemisphere results in a clinically diverse set of impairments and abilities. Persons with aphasia may show a constellation of impaired and spared abilities in auditory comprehension, speech output, repetition, naming, reading, and writing that together comprise a particular syndrome.

There is a long history of the characterization of the aphasia syndromes (Geschwind, 1965; Benson, 1979 for review), and although there are differences in nomenclature, there is general consensus on the clinical features that comprise a particular syndrome. However, syndromes provide a description of clinical behaviors. They do not provide an explanation of the language deficits that give rise to these behaviors. For example, a failure to understand language could be due to any number of factors—among them, impairment in processing the sound structure of language, impairment in mapping sounds to word meaning, and/or failure to process the syntactic structure or the semantic structure of words. To understand the nature of language deficits in aphasia requires taking a different approach, one that takes into consideration the potential language processes and mechanisms underlying these deficits. Roman Jakobson (1956) was probably the first linguist to take such a psycholinguistic approach to aphasia. Since then, there has been a history and tradition of using the theoretical framework and experimental tools provided by psycholinguistics to examine the nature of deficits among the different types of aphasia (Goodglass, 1993). As discussed, such a study has provided the framework for much of the recent work using functional neuroimaging and electrophysiological approaches to the neurobiology of language.

This chapter reviews the contributions that this approach has made to our understanding of the neurobiology of language. We begin by describing those classical aphasia syndromes and their associated lesion profiles that have served as the focus of much of the research on language impairments in aphasia. The psycholinguistic studies that were conducted were designed to understand the basis of the underlying deficits giving rise to particular clinical features of these syndromes and the functional role of the brain areas involved. The findings have also been used to provide insight into the functional architecture of language, that is, how the system fractionates provides a window into its structural properties and the mechanisms and processes involved in normal language use.

74.2 THE APHASIA SYNDROMES

The aphasia syndromes typically result from damage to perisylvian areas of the left hemisphere. Although there have been many syndromes described in the literature (see Geschwind, 1965 for a review), the three that have probably been studied in the most detail and have served as the foundation for psycholinguistic studies of aphasia are Broca's, conduction, and Wernicke's aphasia. These syndromes provide a rich tapestry of impaired and spared language abilities. Of importance, they are defined in terms of the *relative* performance of patients among a set of language functions, including speaking, auditory comprehension, repetition, naming,

Neurobiology of Language. DOI: http://dx.doi.org/10.1016/B978-0-12-407794-2.00074-2

and comprehension and production in reading and writing. Thus, an absolute score on a particular language function cannot be used to classify a patient. Rather, it is performance on a particular language function in relation to the other clinical characteristics that defines a patient's syndrome.

Focusing solely on a particular language function will result in grouping patients together who have different syndromes and potentially different underlying deficits. For example, there are a number of aphasia syndromes that are characterized by poor auditory comprehension, including Wernicke's aphasia, transcortical sensory aphasia, and global aphasia (Goodglass & Kaplan, 1972). Similarly, there are a number of syndromes that are characterized by fluent speech output, including Wernicke's aphasia, conduction aphasia, transcortical sensory aphasia, and anomic aphasia. It is the relationship between these various language functions that differentiate between and among the aphasia types.

Broca's aphasia is characterized by impairment in language expression in the face of relatively good auditory comprehension. Speech output of these patients is typically nonfluent: production is slow, labored, with many speech errors, and is often dysarthric, (i.e., characterized by motor weakness that affects speech articulation). Additionally, these patients often display agrammatism in their speech output, with a tendency to omit freestanding function words such as *the* and *is* and to either delete or substitute grammatical endings. Repetition is usually similar to or a little better than spontaneous speech output. Lesions associated with Broca's aphasia typically involve the frontal operculum (Broca's area, i.e., BA44 and 45) and premotor and motor regions posterior and superior to the frontal operculum, and extend to the white matter structures including the basal ganglia and insula (Damasio, 1998).

Wernicke's aphasics show a very different clinical picture. They have fluent well-articulated speech in the context of impairment in auditory language comprehension. These patients often produce paraphasias, which are errors in their output that are either phonologically based (phonemic paraphasias, e.g., *top* → *dop*) or semantically based (verbal paraphasia, e.g., *wife* → *sister*). Some Wernicke's aphasics produce jargon or neologisms, which are productions that are phonologically possible but are not words in the language (e.g., *tufbei*). Although speech output is fluent, containing grammatical words and endings, sentences are often described as paragrammatic, characterized by the inappropriate juxtaposition of words often rendering the sentence ungrammatical. Additionally, the content of the discourse is typically empty semantically, partly because of the overuse of semantically empty high-frequency words such as *thing, is, this*. Wernicke's patients also have repetition impairment as well as a naming deficit. The lesions associated with Wernicke's aphasia include the posterior superior temporal gyrus (BA22), often extending to the middle temporal, supramarginal, and angular gyri (Damasio, 1998; Dronkers, Redfern, & Ludy, 1995; Dronkers, Wilkins, Van Valin, Redfern, & Jaeger, 2004).

In conduction aphasia, repetition is the presenting deficit in the context of fluent, well-articulated speech and relatively good auditory language comprehension. Speech output contains phonemic paraphasias and relatively few semantic paraphasias. The patient appears to be aware of these errors because they often attempt to correct them, producing *conduite d'approche* or successive approximations to the target word. The lesions associated with conduction aphasia include the supramarginal gyrus and the white matter structures deep to it (the arcuate fasciculus). Importantly, the posterior portion of the superior temporal gyrus (Wernicke's area) is typically spared (Damasio, 1998; but see Hickok et al., 2000).

These clinical characteristics and associated lesion loci have raised a series of questions about the functional and neural architecture of language. With respect to the functional architecture of language, it is generally assumed that both expressive and receptive language functions comprise different linguistic domains involving multiple stages of processing. These domains include phonological/phonetic, lexical, syntactic, and conceptual/semantic. It is also generally assumed that information flows from one stage of processing to the other and that this information flow is interactive, with activation at one stage of processing influencing activation at other stages of processing both upstream and downstream from it (Dell, 1986; Marslen-Wilson & Warren, 1994). For example, phonetic/phonological information affects lexical processing (bottom-up processing), and lexical processing, in turn, can affect phonetic/phonological processing (top-down processing).

The clinical characteristics of Broca's, conduction, and Wernicke's aphasia suggest potential deficits in the representations and processes involved in the reception and/or expression of speech, words, and syntax. And it is these domains that we review here. At the same time, they also provide potential insight into the neural systems underlying these domains, allowing for an examination of whether the neural areas are functionally autonomous.

74.3 SOME CAVEATS AND CHALLENGES

The use of the aphasia syndromes as the framework to investigate the neurobiology of language has been met with challenges in the literature that are worthwhile to briefly review and consider. In particular, there have been challenges to the assumption that clinical syndromes can be used to reliably classify and study participants with aphasia (Caramazza, 1984), and there have been challenges to the assumption that there is one-to-one mapping between clinical syndrome and underlying neuropathology (Willmes & Poeck, 1993). This has led to two other approaches over the past 30 years in the investigation of language deficits in aphasia. The first has focused on detailed case studies of patients as a means of informing current theories of language processing (Caramazza, 1986; see Rapp & Goldrick, 2006 for a review). The single case methodology has been used throughout the history of neuropsychology and has provided many insights into not only how language may fractionate but also what neural areas give rise to the deficit (Geschwind, 1965 discussion of Dejerine). However, the more recent case study approach has largely been agnostic with respect to the underlying neuropathology of the patient (Rapp & Goldrick, 2006). Thus, a downside of this approach is that it cannot provide insight into the neural architecture underlying the patient's deficit, nor can it provide *predictions* of patterns of deficits to new patients. In the second approach, a psycholinguistic question is raised and then studied in a group of patients unselected for either syndrome or lesion (see Schuell & Jenkins, 1959). The assumption here is that the underlying deficit is the same irrespective of lesion localization. Under these assumptions, it is impossible to determine whether there are distinct deficits that arise as a function of a particular area of damage and whether the basis of the deficit differs as a function of lesion site.

In the end, it is lesions that produce language impairments in aphasia. And the rationale for focusing on syndromes is that they not only present with a constellation of impairments but also are the result of lesions to particular neural areas. With advances in neuroimaging techniques over the past 20 years, we know that lesions of patients are rarely focal and typically include both cortical and subcortical structures. No individual with aphasia has exactly the same lesion profile, and there are differences not only in the extent of the lesion but also in the degree of damage to a particular area. With regard to behavioral effects of lesions, we know that within a syndrome there are differences in severity and that not all patients can be classified into a particular syndrome, in both cases presumably because of the extent and location of the lesion. We also have learned that damage in one area can result in hypometabolism in areas distant from it, even in the absence of overt structural damage (Metter, Hanson, Jackson, & Kempler, 1990). Such findings suggest that deficits may reflect neural systems, rather than solely being due to local pathology.

Taken together, such observations identify the complexities of this research endeavor. That the picture is more complex than what the classical aphasiologists from the 1920s to the 1960s had proposed in terms of clinical diagnosis and lesion locus is not surprising. However, these facts do not obviate the existence of the aphasia syndromes or that lesions are localizable and ultimately produce different patterns of language impairment broadly in line with the classical aphasia syndromes (Kreisler et al., 2000; Richardson, Fillmore, Rorden, LaPointe, & Fridriksson, 2012). The results of this approach converge with recent neuroimaging findings focusing on the neural substrates of language.

74.4 LANGUAGE DEFICITS UNDERLYING APHASIA SYNDROMES

It is beyond the scope of this article to review the extensive literature on psycholinguistic investigations of aphasia. Rather, we examine two main classes of findings that have been shown in each of the domains of phonetics/phonology, the lexicon, and syntax. The first class of findings has shown similarities in patterns of deficits across patients presenting with different aphasia syndromes. Such results provide insight into the functional architecture of language, typically showing integrity of the structural properties of language. They also suggest that such processing recruits a broadly distributed neural system. The second class of findings has shown deficits in patients presenting with different aphasia syndromes; however, of interest and importance, the patterns of deficits differ as a function of clinical syndrome. Such findings suggest that the nature of the deficit giving rise to the pathological performance is due to a different functional impairment presumably reflecting the neural locus of the lesion.

74.4.1 Phonetic/Phonological Impairments

74.4.1.1 Speech Production

Although all aphasic patients make speech production errors across a number of language tasks, based on the clinical picture of the patients, it was generally assumed that the source of those errors differed. In particular, given that the lesions of Broca's aphasics involved frontal structures typically including motor areas, they were considered to have phonetic impairments reflecting articulatory planning and articulatory implementation deficits. In contrast, given that Wernicke's and conduction aphasics' lesions involved posterior areas, these patients were considered to have phonological impairments reflecting selection deficits. Experimental results showed that these characterizations were both correct and wrong. In particular, acoustic analysis of the patterns of production of a number of parameters of speech showed clear-cut phonetic impairments for Broca's aphasics that were not present in either Wernicke's or conduction aphasics. Broca's aphasics showed deficits in the timing relations required for voicing in stop consonants (voice-onset time) (Blumstein, Cooper, Goodglass, Statlender, & Gottlieb, 1980; Gandour & Dardarananda, 1984; Kent & Rosenbek, 1983), timing of voicing onset and amplitude properties required for voicing in fricative consonants (Kurowski, Hazen, & Blumstein, 2003), and duration and amplitude measures required for the production manner of articulation for nasal consonants (Kurowski, Blumstein, Palumbo, Waldstein, & Burton, 2007). Although the two posterior aphasic groups showed normal articulatory implementation, they did show more variability in their productions than normal individuals. Taken together, these findings support the view that Broca's aphasics have articulatory planning and implementation deficits, and that frontal brain structures including Broca's area, premotor and motor regions posterior and superior to the frontal operculum, and white matter structures deep to them are functionally involved in these stages of production. That the posterior patients only showed variability in their productions in the context of normal articulatory implementation suggests that speech production not only recruits frontal structures but also invokes posterior structures, presumably as part of a feedback mechanism for error detection, monitoring, and sensorimotor integration from auditory (temporal cortex) and somatosensory (inferior parietal cortex) areas (Guenther, Ghosh, & Tourville, 2006; Hickok, 2012; Hickok, Houde, & Rong, 2011; Houde & Nagarajan, 2011; see Blumstein & Baum for discussion of neuroimaging studies supporting this view, Chapter 55, this volume).

In contrast to phonetic deficits, studies of phonological analysis of the speech of Broca's, conduction, and Wernicke's aphasia failed to show different patterns of errors. Analyses of speech output errors showed that patients from all three groups made similar phonological errors, including phoneme substitutions (phonemic paraphasias), addition or deletions of sounds, transposition of sounds either within a word or between words, and contextual errors based on the phonological context of the word (Blumstein, 1973). For all patients, errors reflected structural principles of the language. In particular, sound substitutions were more likely to occur between sounds that were distinguished by a single phonetic feature, and both addition and deletion errors were more likely to result in the canonical syllable structure CV (e.g., segments were typically added if a word began with a vowel, e.g., "elevator" → /kɛləvetr/, sound segments were added to produce a word with a CV onset, e.g., "cloudy" → /kəlawdi/, and consonant clusters were simplified, "French" → /fɛnč/). The similar pattern of errors irrespective of clinical syndrome suggests that whether the error occurred at selection, articulatory planning, or implementation stages of production, the basic structure of words and the phonological principles underlying them are preserved. It is not surprising that when errors occur, they are more likely to be manifest among phonologically similar sounds and to result in "simpler" phonological structures.

74.4.1.2 Speech Perception

As described, one of the distinguishing clinical features for Wernicke's aphasics is poor auditory comprehension. The question is, what is the basis or bases of this impairment? Because the primary auditory areas surface in the temporal lobe and the superior temporal gyrus has been implicated in speech perception (Binder, 2000; Hickok & Poeppel, 2007), one likely candidate deficit that has been studied in detail is an impairment in the processing of the sounds of speech. Luria (1966) proposed that Wernicke's aphasics have a deficit in phonemic hearing. In this view, an inability to correctly perceive the phonological properties of speech sounds leads to poor auditory comprehension and results in what appears to be semantic impairments. For example, it is not uncommon for such patients to select the incorrect picture of a word such as "pea" if it is presented in an array of phonologically similar items such as "bee," "T," or "key."

A series of studies was conducted investigating discrimination and identification of both naturally produced and synthetically constructed stimuli (Csepe, Osman-Sagi, Molnar, & Gosy, 2001; Leeper, Shewan, & Booth, 1986). Results showed that Wernicke's aphasics displayed severe speech perception deficits (Basso, Casati, & Vignolo, 1977; Blumstein, Baker, & Goodglass, 1977; Caplan, Gow, & Makris, 1995; Gow & Caplan, 1996; Robson, Keidel, Lambon Ralph, & Sage, 2012; see Hickok, 2009 for an alternative view). In addition, however, Broca's and conduction aphasics also showed deficits, although they were milder (see Hickok, Costanzo, Capasso, & Miceli, 2011 for an alternative view). Of interest, the predictive relationship between performance on these tasks and comprehension ability was inconsistent across studies; some failed to show a relationship (Basso et al., 1977; Blumstein et al., 1977), whereas others did show a relationship (Miceli, Gainotti, Caltagirone, & Masullo, 1980; Robson et al., 2012). These findings suggest that Wernicke's aphasics do have a speech perception deficit, but that other aspects of language, particularly semantic processing, may also be impaired (Baker, Blumstein, &

Goodglass, 1981; Basso et al., 1977; Robson, Sage, & Lambon Ralph, 2012; but see Walker et al., 2011). The possibility that Wernicke's aphasics have not only speech perception impairment but also a deficit in semantic processing is also supported by the neuroimaging literature. As described, it is not uncommon for the lesion profile of Wernicke's aphasics to extend into the middle temporal gyrus, an area that is involved in semantic processing and appears to be recruited in accessing stored semantic representations (see Binder, Desai, Graves, & Conant, 2009 for a review).

The findings that speech perception impairments emerge not only in Wernicke's aphasics but also in milder forms in Broca's and conduction aphasics are consistent with neuroimaging results showing that multiple neural areas are recruited in the processing of speech. Studies examining phonological contrasts (Burton, Small, & Blumstein, 2000) and acoustic phonetic properties of speech (Blumstein, Myers, & Rissman, 2005; Joanisse, Zevin, & McCandliss, 2007; Liebenthal et al., 2010) have shown activation in a neural network, including temporal (superior temporal gyrus), parietal (supramarginal gyrus), and frontal (inferior frontal gyrus) areas. It has been argued that the functional role of these areas differs with superior temporal areas (potentially bilaterally) recruited in earlier stages of speech processing, temporoparietal areas (posterior superior temporal and supramarginal gyri) involved in phonological processing (Buchsbaum, Hickok, & Humphries, 2001; Hickok et al., 2008; Hickok, 2009), and frontal areas engaged in executive processes related to phonetic category decisions (Burton et al., 2000; Myers, Blumstein, Walsh, & Eliassen, 2009; see also Venezia, Saberi, Chubb, & Hickok, 2012).

Nonetheless, of interest, despite differences in overall performance of the patients in speech perception tasks, their patterns of impairment were similar. Namely, all patients were more likely to make discrimination errors on stimulus pairs that were distinguished by a single phonetic feature than several features, and they were more likely to make errors discriminating stimulus pairs contrasting in the features corresponding to place of articulation than for the feature voicing (Blumstein et al., 1977). Finally, all patients displayed more deficits in perceiving the sound structure of nonsense syllables compared with real words. That Wernicke's aphasics showed a systematic pattern of impairment similar to other aphasic patients indicates that their behavior is not random and does not reflect a loss of sensitivity to the phonetic/phonological properties of speech.

The patterns of performance displayed by the patients reflect the integrity of the structural properties of the sound structure of language. Sounds distinguished by a single phonetic feature not only share more phonological features but also are more similar acoustically than are sounds distinguished by multiple features. Thus, it is not surprising that they are more difficult to discriminate. Nonetheless, despite the difficulty that all patients have in either discriminating or categorizing acoustic cues associated with either voicing or place of articulation, the locus and shape of the phonetic boundary are similar to those of normal individuals. The superiority of performance for real words compared with nonsense syllables is consistent with current models of the functional architecture of language (Dell, 1986; McClelland & Elman, 1986). Here, information flow is interactive; namely, information flows from phonetic/phonological analysis stages to activate potential lexical candidates. These candidates, in turn, boost the activation of phonological units downstream from them. Because nonwords do not match any words in the lexicon, they may only weakly activate phonologically similar words (Milberg, Blumstein, & Dworetzky, 1988). In such a case, they do not have the same degree of support of the lexical-semantic network; hence, they are more vulnerable in tasks that focus on phonological/phonetic properties.

74.5 LEXICAL IMPAIRMENTS

Models of the cognitive architecture of language have proposed that the words of a language (the mental lexicon) are organized in terms of a network-like architecture of shared or partially overlapping sound structure or semantic properties (Gaskell & Marslen-Wilson, 1999; Plaut, 1995). In this view, a word not only activates its phonological and semantic representations but also partially activates words that share sound structure and semantic properties with it. As a consequence, accessing a word for either spoken production or auditory word recognition requires selecting the target word from this set of activated competitors. Thus, both the production and recognition of words require a multistage process including access to the mental lexicon, activation of a network of potential word candidates, and, ultimately, the selection of the target word from the set of semantically related and phonologically related competitors.

One of the most common and least localizing clinical features in aphasia is a word retrieval deficit. This may be shown either in spoken word production or in auditory word recognition. In spoken word production, patients may fail to come up with a word either in spontaneous speech output or when presented with a picture or verbal description of a word (naming) (Goodglass & Kaplan, 1972). Typically, naming errors include phonemic paraphasias, where the patient makes a sound error on the target word, or verbal (semantic) paraphasias, where the patient produces an incorrect word, often semantically or associatively related to the target word. Clinically, Wernicke's aphasics tend to make more semantic paraphasias, and conduction aphasics make more phonemic paraphasias. In auditory word recognition, Broca's, conduction, and Wernicke's aphasics may fail to select a picture of a word from an array of pictures whose names are either phonemically or semantically related to the target (Baker et al., 1981).

Evidence suggests that although Broca's, conduction, and Wernicke's aphasics display word processing deficits, it is not because they have lost either the "concept" of a word or its phonological representation. Rather, they appear to have difficulty accessing or retrieving the word. Improved naming occurs for these patients with contextual support provided by either a phonological cue (e.g., producing [bə] for the word "bear") or a semantic cue ("Smoky, the _.").

The relative preservation of semantic/conceptual representations has been shown in studies examining semantic priming. Both Broca's and Wernicke's aphasics show semantic priming in a lexical decision task (Blumstein, Milberg, & Shrier, 1982; de Salles, Holderbaum, Parente, Mansur, & Ansaldo, 2012; Hagoort, 1997; Milberg & Blumstein, 1981). That is, they display shorter reaction time latencies to target words preceded by semantically related ("dog–cat") compared with semantically unrelated ("ring-cat") words. These findings support the integrity of the lexical-semantic network in these patients. In contrast to the classical aphasias, however, recent research suggests that degradation of semantic structure does occur, but in patients with semantic dementia or aphasic patients with lesions extending to the anterior temporal lobe (an area not included in the lesion profile of Wernicke's aphasics) (Jefferies & Lambon Ralph, 2006; Walker et al., 2011).

The relative preservation of phonological representations has been shown in studies examining the tip-of-the-tongue state in aphasic patients. The tip-of-the-tongue state is a phenomenon in which subjects are unable to come up with a word but "feel" that they know what the word is and, in fact, that the word is on the "tip of their tongue." Brown and McNeil (1966) showed that normal individuals retain the sound structure properties of words that they failed to retrieve; they can identify its first letter, the number of syllables it has, and words that are semantically related to it. Although aphasics show a similar sensitivity to the sound structure of words they cannot name (Barton, 1971), Goodglass, Kaplan, Weintraub, and Ackerman (1976) showed that conduction aphasics were better able to recall the sound properties of words compared with both Broca's and Wernicke's aphasics. Thus, they suggested that the differences in the pattern of performance reflected different stages in word retrieval.

Despite the relative preservation of the lexical-semantic and phonological structure of words in Broca's, conduction, and Wernicke's aphasics, these patients do show deficits in the various stages involved in lexical access. In particular, a series of studies have shown that Wernicke's aphasics are able to activate lexical candidates; however, the lexical competitors remain active longer (or fail to get inhibited). In contrast, Broca's aphasics also activate lexical candidates; however, they are unable to resolve competition in selecting the target word from among competing lexical candidates. These findings have been shown using a variety of paradigms, including lexical decision (Janse, 2006; Milberg, Blumstein, & Dworetzky, 1987) and eyetracking (Yee, Blumstein, & Sedivy, 2008). Additional support for selection deficits comes from both verb generation (Thompson-Schill et al., 1998) and lexical decision studies (Bedny, Hulbert, & Thompson-Schill, 2007) that focused on lesion location, in this case a portion of the inferior frontal gyrus (BA45), and not clinical diagnosis of aphasia.

The lexical processing deficits of Broca's and Wernicke's aphasics emerge whether the source of the competition is semantic or phonological. For example, in the semantic case, the patient must select a word from competing meanings of ambiguous words presented in congruent and incongruent contexts (e.g., the subject is required to make a lexical decision on the third word of a triplet such as "coin-bank-money" versus "river-bank-money") (Bedny et al., 2007; Milberg et al., 1987). Another study required the patient to select words that have high versus low selection demands (e.g., the subject is asked to generate a verb for the word "scissors" versus "ball") (Thompson-Schill et al., 1998).

Several paradigms have been used to investigate the effects of resolving phonological competitors. One set of experiments used eyetracking and examined the potential effects of onset competitors in selecting a target word (Yee et al., 2008). Here, the subject was asked to point to a picture given the auditory presentation of a word from an array that included the picture of the target word, a word with an onset competitor, and two semantically and phonologically unrelated foils (e.g., the target word is "hammock" and the pictures include "hammock," "hammer," "monkey," and "chocolate"). Another series of experiments examined the effects of acoustically degraded prime stimuli on the magnitude of semantic priming (Misiurski, Blumstein, Rissman, & Berman, 2005; Utman, Blumstein, & Sullivan, 2001). Stimulus pairs included semantically related stimuli with and without voicing competitors, "time-clock" (with a voiced competitor "dime") and "cat–dog" (with no voiced competitor, "gat" is not a word). Similar to normal individuals, Broca's aphasics showed semantic priming for phonologically clear, semantically related pairs, and reduced semantic priming for degraded prime stimuli without a voiced competitor. In contrast to normal individuals, Broca's aphasics lost priming only when the degraded prime had a voiced competitor.

Taken together, these findings indicate that aphasic patients have lexical processing impairments. In particular, aphasics retain the underlying semantic and phonological representations of words, but they show impairments in *accessing* them. Moreover, different patterns of performance emerge between Broca's and Wernicke's aphasics, suggesting that the basis of their deficit differs (see Blumstein, 2009, for review; Janse, 2006). For Wernicke's aphasics, word candidates stay active longer either due to an inability to inhibit word competitors or due to their overactivation. In contrast, Broca's aphasics show a deficit in selection processes and an inability to select among competing semantic and phonological competitors.

Neuroimaging findings support the view that both spoken word production and word recognition processes engage a temporoparietal and frontal network. Semantic processing recruits both temporal and frontal structures (see Binder et al., 2009 for a review); selection among competing semantic alternatives recruits the inferior frontal gyrus (Thompson-Schill, D'Esposito, Aguirre, & Farah, 1997); and selection among competing phonological alternatives also recruits the inferior frontal gyrus as well as the posterior superior temporal and supramarginal gyri (Righi, Blumstein, Mertus, & Worden, 2010).

74.6 SYNTACTIC IMPAIRMENTS

As described, one of the clinical characteristics of some Broca's aphasics is agrammatism in speech output in the context of generally good auditory comprehension. There is a long and controversial history of the potential basis of the grammatical deficit in these patients. Early hypotheses (Kolk & Heeschen, 1990; Kolk & Van Grunsven, 1985) suggested that the deficit reflected a compensatory mechanism of the patient to provide the most semantic content with the least amount of speech. In this view, the output disorder reflects an "economy of effort" and thus is secondary to nonfluent output and difficulty in producing and articulating speech. Evidence in support of this view came from analyses of the error patterns in production. Results showed that there was not only a tendency to omit function words but also a tendency to simplify morphological structures, particularly in contexts where the morphological ending was redundant (e.g., "two books" → "two book") (Dick, Bates, Wulfeck, Utman, & Gernsbacher, 2001). As shown by analyses of inflected languages, morphological errors produced by agrammatic aphasics were in fact substitutions of one morphological ending for another, not a "loss" of endings (Grodzinksy, 1990; Menn & Obler, 1990). Analyses showed that there was a tendency to produce a linguistically less marked structure such as a verb in the present tense or in infinitival form rather than a verb with a past tense or future tense inflection.

A series of seminal studies by Zurif and colleagues (Caramazza & Zurif, 1976; Goodenough, Zurif, & Weintraub, 1977; Zurif, Caramazza, & Myerson, 1972), however, suggested that the agrammatic deficit of Broca's aphasics was not limited to speech production, but rather was a "central" impairment affecting not only speech production but also comprehension. This was originally shown using a hierarchical clustering paradigm in which subjects were presented the written form of a sentence such as "the dog chased a cat" (Zurif et al., 1972). With the sentence always in display, subjects were given a random selection of three cards, each containing one of the words in the sentence. They were asked to put "the two words that went best together." Results showed that Broca's aphasics did not know where/how to cluster the function words. They were as likely to cluster "the" and "a" with each other than within their associated noun phrases. Thus, Broca's aphasics showed impairment in their linguistic "intuitions" about the syntactic structure of sentences.

From there, a plethora of studies examined sentence comprehension in aphasia focusing on syntactic structures. Results have shown that Broca's aphasics displayed impairments in comprehending sentences when the only cue to comprehension was syntax (e.g., "the lion chased the tiger" versus "the boy ate the hamburger") (Caramazza & Zurif, 1976). They had difficulty in understanding noncanonical syntactic structures such as passive sentences compared with active sentences ("the girl is liked by the boy" versus "the boy likes the girl"), syntactically complex compared with simple sentences ("the boy who sees the man likes the girl" versus "the boy likes the girl") object-embedded compared with subject-embedded sentences ("the boy the girl likes reads a book" versus "the girl likes the boy who reads the book") and sentences that did and did not contain traces (Caplan, Baker, & Dehaut, 1985).

These findings gave rise to a large number of hypotheses to characterize the underlying impairment, the details of which are beyond the scope of this article. The proposals are far-reaching, invoking either representational (Grodzinsky, 1986, 2000; Mauner, Fromkin, & Cornell, 1993) or processing deficits (Grodzinsky & Friederici, 2006; Love, Swinney, Walenski, & Zurif, 2008) involving potential impairments in syntactic structures governing movement and/or binding (Choy & Thompson, 2010; Swinney & Zurif, 1995), thematic role assignment (Saffran, Schwartz, & Linebarger, 1998), working memory or resource limitations (Caplan & Waters, 1995; Carpenter, Miyake, & Just, 1995), and time-course delay of processing (Ferrill, Love, Walenski, & Shapiro, 2012; Love, Swinney, & Zurif, 2001). See Chapter 47 (Rogalsky) of this volume for further discussion.

Beyond the debate concerning the underlying deficit in Broca's aphasics, there is a more critical issue—namely, is it truly the case that only Broca's aphasics display syntactic comprehension impairments? Unfortunately, much of the literature examining the basis of syntactic impairments has tested only Broca's aphasics. Thus, it is not clear whether other types of patients also show impairments. Those studies that have looked at other patient groups or patients with different lesion sites show similar patterns of impairment as those of Broca's aphasics (Caplan et al., 1995; Caplan, Hildebrandt, & Makris, 1996; Dick et al., 2001; Zurif & Caramazza, 1976). It is not surprising to find that structurally complex sentences are more difficult to understand not only for aphasic patients but also for neurologically intact subjects tested under adverse listening conditions (Dick et al., 2001; cf. also Obleser, Meyer, & Friederici, 2011).

What is not clear from these studies is whether the *basis* of the impairment is different across aphasia syndromes. Although some hypotheses have been proposed (Friederici, 2011), no studies have yet been conducted that distinguish behavioral performance of patients based on some operational measure of the purported functional deficit. One challenge inherent in this research is assessing syntactic comprehension independent of meaning.

The neuroimaging literature has shown similar conflicting findings. Some studies have shown selective activation of the inferior frontal gyrus in auditory processing of syntactic structure (Moro et al., 2001; Stromswold, Caplan, Alpert, & Rauch, 1996), and others have shown a broad fronto-temporo-parietal network (Fedorenko, Nieto-Castañon, & Kanwisher, 2012; Friederici, Meyer, & von Cramon, 2000; Just, Carpenter, Keller, Eddy, & Thulborn, 1996; see Kaan & Swaab, 2002 for a review).

Although it remains unclear whether there are functional distinctions in the auditory processing of syntactic structure as a function of clinical syndrome and/or lesion site, the original observation that only Broca's aphasics display agrammatism in production remains. This leaves open the possibility that these patients do have a selective syntactic impairment, but it is restricted to spoken language production. It is for future research to determine whether this is the case and what the underlying basis of this impairment may be.

74.7 CONCLUSION

Psycholinguistic studies of the clinical syndromes of aphasia have provided a unique window into the neurobiology of language. Such studies offer insights that behavioral and neuroimaging studies alone cannot. Behavioral studies do not provide evidence of the neural systems underlying a particular deficit. Neuroimaging studies are unable to determine whether activation of a neural area indicates that it is *necessary* for a particular linguistic function. Coupled with these approaches, technological advances now available for detailed mapping of lesion profiles coupled with careful clinical examination and classification of patients hold the promise of not only gaining a deeper understanding of the functional and neural architecture of language but also providing critical insights into the bases of language deficits that can be used in developing rehabilitation programs for patients with aphasia.

Acknowledgments

This research was supported in part by grants NIH R01 DC 006220 and R21DC013100, and grants from the Dana Foundation to Brown University. The content is solely the responsibility of the authors and does not necessarily represent the official views of the National Institute on Deafness and Other Communication Disorders or the National Institutes of Health.

References

Baker, E., Blumstein, S. E., & Goodglass, H. (1981). Interaction between phonological and semantic factors in auditory comprehension. *Neuropsychologia*, *19*, 1–15.

Barton, M. (1971). Recall of generic properties of words in aphasic patients. *Cortex*, *7*, 73–82.

Basso, A., Casati, G., & Vignolo, L. A. (1977). Phonemic identification defect in aphasia. *Cortex*, *13*, 85–95.

Bedny, M., Hulbert, J. C., & Thompson-Schill, S. L. (2007). Understanding words in context: The role of Broca's area in word comprehension. *Brain Research*, *18*, 101–114.

Benson, D. F. (1979). Aphasia. In K. M. Heilman, & E. Valenstein (Eds.), *Clinical neuropsychology* (pp. 22–58). New York, NY: Oxford.

Binder, J. (2000). The new neuroanatomy of speech perception. *Brain*, *123*, 2371–2372.

Binder, J. R., Desai, R. H., Graves, W. W., & Conant, L. L. (2009). Where is the semantic system? A critical review and meta-analysis of 120 functional neuroimaging studies. *Cerebral Cortex*, *19*, 2767–2796.

Blumstein, S. E. (1973). *A phonological investigation of aphasic speech.* The Hague: Mouton.

Blumstein, S. E. (2009). Auditory word recognition: Evidence from aphasia and functional neuroimaging. *Language and Linguistics Compass, 3*, 824–838.

Blumstein, S. E., Baker, E., & Goodglass, H. (1977). Phonological factors in auditory comprehension in aphasia. *Neuropsychologia, 15*, 19–30.

Blumstein, S. E., Cooper, W., Goodglass, H., Statlender, S., & Gottlieb, J. (1980). Production deficits in aphasia: A voice onset time analysis. *Brain and Language, 9*, 153–170.

Blumstein, S. E., Milberg, W., & Shrier, R. (1982). Semantic processing in aphasia: Evidence from an auditory lexical decision task. *Brain and Language, 17*, 301–315.

Blumstein, S. E., Myers, E. B., & Rissman, J. (2005). The perception of voice onset time: An fMRI investigation of phonetic category structure. *Journal of Cognitive Neuroscience, 17*, 1353–1366.

Brown, R., & McNeil, D. (1966). The "tip of the tongue" phenomenon. *Journal of Verbal Learning and Verbal Behavior, 5*, 325–337.

Buchsbaum, B. R., Hickok, G., & Humphries, C. (2001). Role of left posterior superior temporal gyrus in phonological processing for speech perception and production. *Cognitive Science, 25*, 663–678.

Burton, M. W., Small, S. L., & Blumstein, S. E. (2000). The role of segmentation in phonological processing: An fMRI investigation. *Journal of Cognitive Neuroscience, 12*, 679–690.

Caplan, D., Baker, C., & Dehaut, F. (1985). Syntactic determinants of sentence comprehension in aphasia. *Cognition, 21*, 117–175.

Caplan, D., Gow, D., & Makris, N. (1995). Analysis of lesions by MRI in stroke patients with acoustic-phonetic processing deficits. *Neurology, 45*, 293–298.

Caplan, D., Hildebrandt, N., & Makris, N. (1996). Location of lesions in stroke patients with deficits in syntactic processing in sentence comprehension. *Brain, 119*, 933–949.

Caplan, D., & Waters, G. S. (1995). Aphasic disorders of syntactic comprehension and working memory capacity. *Cognitive Neuropsychology, 12*, 637–649.

Caramazza, A. (1984). The logic of neuropsychological research and the problem of patient classifications in aphasia. *Brain and Language, 21*, 9–20.

Caramazza, A. (1986). On drawing inferences about the structure of normal cognitive systems from the analysis of impaired performance: The case for single patient studies. *Brain and Cognition, 5*, 41–66.

Caramazza, A., & Zurif, E. B. (1976). Dissociation of algorithmic and heuristics processes in sentence comprehension: Evidence from aphasia. *Brain and Language, 3*, 572–582.

Carpenter, P. A., Miyake, A., & Just, M. A. (1995). Language comprehension: Sentence and discourse processing. *Annual Review of Psychology, 46*, 91–120.

Choy, J. J., & Thompson, C. K. (2010). Binding in agrammatic aphasia: Processing to comprehension. *Aphasiology, 24*, 551–579.

Csepe, V., Osman-Sagi, J., Molnar, M., & Gosy, M. (2001). Impaired speech perception in aphasic patients: Event-related potential and neuropsychological assessment. *Neuropsychologia, 39*, 1194–1208.

Damasio, H. (1998). Neuronanatomical correlates of the aphasias. In M. T. Sarno (Ed.), *Acquired aphasia* (pp. 43–70). New York, NY: Academic Press.

Dell, G. S. (1986). A spreading activation theory of retrieval in sentence production. *Psychological Review, 93*, 283–321.

de Salles, J. F., Holderbaum, C. F., Parente, M. A., Mansur, L. L., & Ansaldo, A. I. (2012). Lexical-semantic processing in the semantic priming paradigm in aphasic patients. *Arquivos de Neuro-Psiquiatria, 70*, 718–726.

Dick, F., Bates, E., Wulfeck, B., Utman, J., & Gernsbacher, M. A. (2001). Evidence for a distributed model of language breakdown in aphasic patients and neurologically intact individuals. *Psychological Review, 108*, 759–788.

Dronkers, N. F., Redfern, B. B., & Ludy, C. A. (1995). Lesion localization in chronic Wernicke's aphasia. *Brain and Language, 51*, 62–65.

Dronkers, N. F., Wilkins, D. P., Van Valin, R. D., Jr., Redfern, B. B., & Jaeger, J. J. (2004). Lesion analysis of the brain areas involved in language comprehension. *Cognition, 92*, 145–177.

Fedorenko, E., Nieto-Castañon, A., & Kanwisher, N. (2012). Lexical and syntactic representations in the brain: An fMRI investigation with multi-voxel pattern analyses. *Neuropsychologia, 50*, 499–513.

Ferrill, M., Love, T., Walenski, M., & Shapiro, L. P. (2012). The time-course of lexical activation during sentence comprehension in people with aphasia. *American Journal of Speech and Language Pathology, 21*, S179–S189.

Friederici, A. D. (2011). The brain basis of language processing: From structure to function. *Physiological Review, 91*, 1357–1392.

Friederici, A. D., Meyer, M., & von Cramon, D. Y. (2000). Auditory language comprehension: An event-related fMRI study on the processing of syntactic and lexical information. *Brain and Language, 74*, 289–300.

Gandour, J., & Dardarananda, R. (1984). Voice-onset time in aphasia: Thai, II: Production. *Brain and Language, 18*, 389–410.

Gaskell, M. G., & Marslen-Wilson, W. D. (1999). Ambiguity, competition, and blending in spoken word recognition. *Cognitive Science, 23*, 439–462.

Geschwind, N. (1965). Disconnexion syndromes in animals and man. *Brain, 88*, 237–294, 585–944.

Goodenough, C., Zurif, E., & Weintraub, S. (1977). Aphasics' attention to grammatical morphemes. *Language and Speech, 20*, 11–19.

Goodglass, H. (1993). *Understanding aphasia.* New York, NY: Academic Press.

Goodglass, H., & Kaplan, E. (1972). *The assessment of aphasia and related disorders.* Philadelphia, PA: Lea & Febiger.

Goodglass, H., Kaplan, E., Weintraub, S., & Ackerman, N. (1976). The "tip of the tongue" phenomenon in aphasia. *Cortex, 12*, 145–153.

Gow, D. W., & Caplan, D. (1996). An examination of impaired acoustic-phonetic processing in aphasia. *Brain and Language, 52*, 386–407.

Grodzinksy, Y. (1990). The formal description of agrammatism. In Y. Grodzinsky (Ed.), *Theoretical perspectives on language deficits* (pp. 37–108). Cambridge, MA: MIT Press.

Grodzinsky, Y. (1986). Language deficits and the theory of syntax. *Brain and Language, 27*, 135–159.

Grodzinsky, Y. (2000). The neurology of syntax: Language use without Broca's area. *Behavioral and Brain Sciences, 23*, 1–71.

Grodzinsky, Y., & Friederici, A. D. (2006). Neuroimaging of syntax and syntactic processing. *Current Opinions in Neurobiology, 15*, 240–246.

Guenther, F., Ghosh, S., & Tourville, J. (2006). Neural modeling and imaging of the cortical interactions underlying syllable production. *Brain and Language, 96*, 280–301.

Hagoort, P. (1997). Semantic priming in Broca's aphasics at a short SOA: No support for an automatic access deficit. *Brain and Language, 56*, 287–300.

Hickok, G. (2009). The functional neuroanatomy of language. *Physics of Life Reviews, 6*, 121–143.

Hickok, G. (2012). Computational neuroanatomy of speech production. *Nature Reviews Neuroscience, 13*(2), 135–145.

Hickok, G., Costanzo, M., Capasso, R., & Miceli, G. (2011). The role of Broca's area in speech perception: Evidence from aphasia revisited. *Brain and Language, 119*, 214–220.

Hickok, G., Erhard, P., Kassubek, J., Helms-Tilleryd, K., Naeve-Velguth, S., Strupp, J. P., et al. (2000). A functional magnetic resonance imaging study of the role of left posterior superior temporal gyrus in speech production: Implications for the explanation of conduction aphasia. *Neuroscience Letters, 287*, 156–160.

Hickok, G., Houde, J., & Rong, F. (2011). Sensorimotor integration in speech processing: Computational basis and neural organization. *Neuron, 69*, 407–422.

Hickok, G., Okada, K., Barr, W., Rogalsky, C., Donnelly, K., Barde, L., et al. (2008). Bilateral capacity for speech sound processing in auditory comprehension: Evidence from Wada procedures. *Brain and Language, 107*, 179–184.

Hickok, G., & Poeppel, D. (2007). The cortical organization of speech processing. *Nature Reviews Neuroscience, 8*, 393–402.

Houde, J. F., & Nagarajan, S. S. (2011). Speech production as state feedback control. *Frontiers in Human Neuroscience, 5*, 82.

Jakobson, J. (1956). Two aspects of language and two types of aphasic disturbances. In R. Jakobson, & M. Halle (Eds.), *Fundamentals of language* (pp. 55–82). The Hague: Mouton.

Janse, E. (2006). Lexical competition effects in aphasia: Deactivation of lexical candidates in spoken word processing. *Brain and Language, 97*, 1–11.

Jefferies, E., & Lambon Ralph, M. A. (2006). Semantic impairment in stroke aphasia versus semantic dementia: A case-series comparison. *Brain, 129*, 2132–2147.

Joanisse, M. F., Zevin, J. D., & McCandliss, B. D. (2007). Brain mechanisms implicated in the preattentive categorization of speech sounds revealed using fMRI and a short-interval habituation trial paradigm. *Cerebral Cortex, 17*, 2084–2093.

Just, M. A., Carpenter, P. A., Keller, T. A., Eddy, W. F., & Thulborn, K. R. (1996). Brain activation modulated by sentence comprehension. *Science, 274*, 114–116.

Kaan, E., & Swaab, T. Y. (2002). The brain circuitry of syntactic comprehension. *Trends in Cognitive Science, 6*, 350–356.

Kent, R., & Rosenbek, J. (1983). Acoustic patterns of apraxia of speech. *Journal of Speech and Hearing Research, 26*, 231–248.

Kolk, H., & Heeschen, O. (1990). Adaptation symptoms and impairment symptoms in Broca's aphasia. *Aphasiology, 4*, 221–231.

Kolk, H., & Van Grunsven, M. J. F. (1985). Agrammatism as a variable phenomenon. *Cognitive Neuropsychology, 2*, 347–384.

Kreisler, A., Godefroy, O., Delmaire, C., Debachy, B., Leclerq, M., Pruvo, J. P., et al. (2000). The anatomy of aphasia revisited. *Neurology, 54*, 1117–1123.

Kurowski, K., Blumstein, S. E., Palumbo, C. L., Waldstein, R., & Burton, M. W. (2007). Nasal production in anterior and posterior aphasics: Speech deficits and neuroanatomical correlates. *Brain and Language, 100*, 262–275.

Kurowski, K., Hazen, E., & Blumstein, S. E. (2003). The nature of speech production impairments in anterior aphasics: An acoustic analysis of voicing in fricative consonants. *Brain and Language, 84*, 353–371.

Leeper, H. A., Shewan, C. M., & Booth, J. C. (1986). Altered acoustic cue discrimination in Broca's and conduction aphasia. *Journal of Communication Disorders, 19*, 83–103.

Liebenthal, E., Desai, R., Ellingson, M. M., Ramachandran, B., Desai, A., & Binder, J. R. (2010). Specialization along the left superior temporal sulcus for auditory categorization. *Cerebral Cortex, 20*, 2958–2970.

Love, T., Swinney, D., Walenski, & Zurif, E. (2008). How left inferior frontal cortex participates in syntactic processing: Evidence from aphasia. *Brain and Language, 107*, 203–219.

Love, T., Swinney, D., & Zurif, E. (2001). Aphasia and the time-course of processing long distance dependencies. *Brain and Language, 79*, 169–171.

Luria, A. R. (1966). *Higher cortical functions in man*. New York, NY: Basic Books (Chapter 2).

Marslen-Wilson, W., & Warren, P. (1994). Levels of representation and process in lexical access. *Psychological Review, 101*, 653–675.

Mauner, G., Fromkin, V. A., & Cornell, T. L. (1993). Comprehension and acceptability judgments in agrammatism: Disruptions in the syntax of referential dependency. *Brain and Language, 45*, 340–370.

McClelland, J., & Elman, J. (1986). The TRACE model of speech perception. *Cognitive Psychology, 18*, 1–86.

Menn, L., & Obler, L. (1990). Cross-language data and theories of agrammatism. In L. Menn, & L. Obler (Eds.), *Agrammatic aphasia: A cross-language narrative source book* (Vol. 2, pp. 1369–1389). Philadelphia, PA: John Benjamin.

Metter, E. J., Hanson, W. R., Jackson, C. A., & Kempler, D. (1990). Temporoparietal cortex in aphasia: Evidence from positron emission tomography. *Archives of Neurology, 47*, 1235–1238.

Miceli, G., Gainotti, G., Caltagirone, C., & Masullo, C. (1980). Some aspects of phonological impairment in aphasia. *Brain and Language, 11*, 159–169.

Milberg, W., & Blumstein, S. E. (1981). Lexical decision and aphasia: Evidence for semantic processing. *Brain and Language, 14*, 371–385.

Milberg, W., Blumstein, S. E., & Dworetzky, B. (1987). Processing of lexical ambiguities in aphasia. *Brain and Language, 31*, 138–150.

Milberg, W., Blumstein, S. E., & Dworetzky, B. (1988). Phonological processing and lexical access in aphasia. *Brain and Language, 34*, 279–293.

Misiurski, C., Blumstein, S. E., Rissman, J., & Berman, D. (2005). The role of lexical competition and acoustic-phonetic structure in lexical processing: Evidence from normal subjects and aphasic patients. *Brain and Language, 93*, 64–78.

Moro, A., Tettamanti, M., Perani, D., Donati, C., Cappa, S. F., & Fazio, F. (2001). Syntax and the brain: Disentangling grammar by selective anomalies. *NeuroImage, 13*, 110–118.

Myers, E. B., Blumstein, S. E., Walsh, E., & Eliassen, J. (2009). Inferior frontal regions underlie the perception of phonetic category invariance. *Psychological Science, 20*, 895–903.

Obleser, J., Meyer, L., & Friederici, A. D. (2011). Dynamic assignment of neural resources in auditory comprehension of complex sentences. *NeuroImage, 56*, 2310–2320.

Plaut, D. (1995). Semantic and associative priming in a distributed attractor network. *Proceedings of the 17th annual conference of the cognitive science society* (pp. 37–42). Pittsburgh, Pennsylvania.

Rapp, B., & Goldrick, M. (2006). Speaking words: Contributions of cognitive neuropsychological research. *Cognitive Neuropsychology, 23*, 39–73.

Richardson, J. D., Fillmore, P., Rorden, C., LaPointe, L. L., & Fridriksson, J. (2012). Re-establishing Broca's initial findings. *Brain and Language, 123*, 125–130.

Righi, G., Blumstein, S. E., Mertus, J., & Worden, M. S. (2010). Neural systems underlying lexical competition: An eye tracking and fMRI study. *Journal of Cognitive Neuroscience, 22*, 213–224.

Robson, H., Keidel, J., Lambon Ralph, M. A., & Sage, K. (2012). Revealing and quantifying the impaired phonological analysis underpinning impaired comprehension in Wernicke's aphasia. *Neuropsychologia, 50*, 276–288.

Robson, H., Sage, K., & Lambon Ralph, M. A. (2012). Wernicke's aphasia reflects a combination of acoustic-phonological and semantic control deficits: A case-series comparison of Wernicke's aphasia, semantic dementia and semantic aphasia. *Neuropsychologia, 50*, 266–275.

Saffran, E. M., Schwartz, M. F., & Linebarger, M. C. (1998). Semantic influences on thematic role assignment: Evidence from normals and aphasics. *Brain and Language, 62*, 255–297.

Schuell, H., & Jenkins, J. J. (1959). The nature of language deficit in aphasia. *Psychological Review, 66*, 45–67.

Stromswold, K., Caplan, D., Alpert, N., & Rauch, S. (1996). Localization of syntactic comprehension by positron emission tomography. *Brain and Language, 52*, 452–473.

Swinney, D., & Zurif, E. (1995). Syntactic processing in aphasia. *Brain and Language, 50*, 225–239.

Thompson-Schill, S. L., D'Esposito, M., Aguirre, G. K., & Farah, M. J. (1997). Role of left inferior prefrontal cortex in retrieval of semantic knowledge: A reevaluation. *Proceedings of the National Academy of Sciences of the United States of America, 94*, 14792–14797.

Thompson-Schill, S. L., Swick, D., Farah, M., D'Esposito, M., Kan, I. P., & Knight, R. T. (1998). Verb generation in patients with focal frontal lesions: A neuropsychological test of neuroimaging findings. *Proceedings of the National Academy of Sciences of the United States of America, 95*, 15855–15860.

Utman, J. A., Blumstein, S. E., & Sullivan, K. (2001). Mapping from sound to meaning: Reduced lexical activation in Broca's aphasics. *Brain and Language, 79*, 444–472.

Venezia, J. H., Saberi, K., Chubb, C., & Hickok, G. (2012). Response bias modulates the speech motor system during syllable discrimination. *Frontier in Psychology, 3*, 157.

Walker, G. M., Schwartz, M. F., Kimberg, D. Y., Faseyitan, O., Brecher, A., Dell, G. S., et al. (2011). Support for anterior temporal involvement in semantic error production in aphasia: New evidence from VLSM. *Brain and Language, 117*, 110–122.

Willmes, K., & Poeck, K. (1993). To what extent can aphasic syndromes be localized? *Brain, 11*, 1527–1540.

Yee, E., Blumstein, S. E., & Sedivy, J. C. (2008). Lexical-semantic activation in Broca's and Wernicke's aphasia: Evidence from eye movements. *Journal of Cognitive Neuroscience, 20*, 592–612.

Zurif, E. B., & Caramazza, A. (1976). Dissociation of algorithmic and heuristic processes in language comprehension: Evidence from aphasia. *Brain and Language, 3*, 572–582.

Zurif, E. B., Caramazza, A., & Myerson, R. (1972). Grammatical judgments of agrammatic aphasics. *Neuropsychologia, 10*, 405–417.

CHAPTER

75

Introduction to Primary Progressive Aphasia

Maria Luisa Gorno-Tempini[1] *and Peter Pressman*[2]

[1]UCSF Memory and Aging Center, Sandler Neurosciences Center, University of California, San Francisco, CA, USA;
[2]Memory and Aging Center, Department of Neurology, University of California, San Francisco

75.1 INTRODUCTION AND HISTORY OF PRIMARY PROGRESSIVE APHASIA

The sudden appearance of aphasia due to stroke or brain trauma is prominent in neurological history. Such acute injuries permitted seminal insights into the neural underpinnings of language. The left inferior frontal gyrus injury described by Paul Broca (Broca, 1861; Dronkers, Plaisant, Iba-Zizen, & Cabanis, 2007) and the left temporoparietal lesions described by Karl Wernicke (Wernicke, 1874) both led to distinctive and now-eponymous aphasias, and introduced the anatomy of the first left perisylvian language circuit.

Injury to the language brain system, however, does not always occur abruptly. Neurodegenerative diseases, such as Alzheimer disease (AD) and frontotemporal degeneration (FTD) spectrum disorders, more widely recognized to cause insidious changes in memory or personality, have been associated with selective decline of language functions since the earliest descriptions. In 1892, the Czech psychiatrist Arnold Pick described the first case of progressive aphasia, which he called "semantic aphasia" (Pick, 1892).

Despite Pick's early discovery, for several decades the prevailing view in medicine was that neurodegenerative dementia was virtually synonymous with AD, which primarily impacted memory. While aphasia was a recognized complication of dementia, a more serious amnesia was believed to precede and accompany any language deficit. This paradigm was contested in 1982 when Dr. M. Marcel Mesulam and colleagues published a case series describing slowly progressive aphasia associated with degeneration of the perisylvian region of the left hemisphere (Mesulam, 1982). Rather than being accompanied by signs of a more generalized dementia, aphasia remained the predominant symptom for many years after disease onset. Such patients were otherwise functional in everyday life and had relatively preserved nonverbal neuropsychological scores.

Reports of similar cases were found from around the world (Heath, Kennedy, & Kapur, 1983; Mesulam, 1987, 2007). The term "primary progressive aphasia" (PPA) was subsequently created, and more refined diagnostic criteria for the syndrome were established (Mesulam, 1987, 2007). The PPA diagnosis can be made in a patient with a disorder of language (aphasia) due to a neurodegenerative disorder (progressive) and in whom the aphasia is the most prominent symptom for about 2 years (primary) (Mesulam, 2001, 2003) (Table 75.1). Mesulam's original description required that the most frequent initial symptom be anomia.

Initially, PPA was a unified diagnosis, but as more cases became documented the concept of "fluent and nonfluent" cases was introduced. However, much confusion arose over the years as it became clear that the stroke aphasia classification scheme was not applicable to PPA. PPA was initially distributed into two main variants: fluent and nonfluent. Such a schema preliminarily attempted to correlate PPA symptoms with those caused by vascular lesions. Almost from the outset, the existence of a third variant was suspected, and in 2004 Gorno-Tempini and colleagues published a study that compared cognitive and neuroimaging features of a large group of PPA patients, thereby describing the features of what was dubbed the logopenic variant of PPA (Gorno-Tempini, Dronkers, et al., 2004). What had previously been known as the nonfluent variant was refined into the nonfluent/agrammatic variant, sometimes also called the PPA-grammatical subtype or progressive nonfluent aphasia (nfvPPA, PPA-G, PNFA). The more fluent variant was reclassified as semantic variant, sometimes called PPA-semantic subtype or semantic

Neurobiology of Language. DOI: http://dx.doi.org/10.1016/B978-0-12-407794-2.00075-4

TABLE 75.1 Inclusion and Exclusion Criteria for the Diagnosis of PPA

Inclusion: criteria 1–3 must be answered positively
1. Most prominent clinical feature is difficulty with language
2. These deficits are the principal cause of impaired daily living activities
3. Aphasia should be the most prominent deficit at symptom onset and for the initial phases of the disease
Exclusion: criteria 1–4 must be answered negatively for a PPA diagnosis
1. Pattern of deficits is better accounted for by other nondegenerative nervous system or medical disorders
2. Cognitive disturbance is better accounted for by a psychiatric diagnosis
3. Prominent initial episodic memory, visual memory, and visuoperceptual impairments
4. Prominent, initial behavioral disturbance

Updated diagnostic criteria for PPA from Gorno-Tempini et al. (2011).
Source: *Modified from Mesulam (2003).*

TABLE 75.2 Characteristics of PPA Variants

PPA subtype	Supporting clinical features	Relevant neuropsychological tests	Anatomical correlate	Common histopathology
lvPPA	Poor repetition of prolonged phrases, intermediate fluency, phonemic paraphasias, spared motor speech and semantic knowledge	Boston Naming, repetition tasks, digit span	Dominant temporo-parietal junction	AD
svPPA	Poor object knowledge and confrontation naming, limited single-word comprehension, surface dyslexia, spared motor speech and grammar	Pyramids and Palm Trees, Boston Naming, Famous Faces testing	Dominant anterior temporal lobe	TDP-43 type C
nfvPPA	Agrammatism, effortful speech, impaired comprehension of syntactically complex sentences, spared single word comprehension and object knowledge	Grammatically complex sentences, Northwestern Anagram Test, Boston Naming	Dominant posterior fronto-insular region	Tau (CBD, PSP, Picks), TDP-43-A

A table depicting some major characteristics of the variants of PPA. CBD, corticobasal degeneration; lvPPA, logopenic variant primary progressive aphasia; nfvPPA, nonfluent variant primary progressive aphasia; PSP, progressive supranuclear palsy; svPPA, semantic variant PPA; TDP-43, trans-activator regulatory deoxyribonucleic acid (TAR-DNA) binding protein 43.

dementia (svPPA, PPA-S, SD). The third variant is the logopenic variant (lvPPA, PPA-L), many members of which had originally been incorporated into the other two subtypes (Table 75.2).

The combination of speech and language symptoms in PPA depends on patterns of neuroanatomical degeneration. Although regions of neurodegeneration in PPA share some anatomical boundaries with acute aphasias due to ischemic stroke, the symptoms produced in PPA are not identical to a vascular counterpart. In addition to neural susceptibility to neurodegeneration differing from vascular anatomy, this symptomatic dissociation is due to the brain's ability to adapt with time to a slowly evolving injury. Neuronal loss in PPA is gradual and partial, allowing the synaptic connections of the brain to reorganize as the disease progresses. As such compensation is incomplete, however, discrete language deficits emerge.

Language consists of many subdomains. Comprehension entails a cascade of processes including correctly identifying relevant sounds and words, retrieving conceptual information conveyed by these symbols, and furthermore identifying the meaning behind different syntactical and grammatical word arrangements. Language production depends on an intricate process that relays and arranges semantic, lexical, and phonological information, and furthermore coordinates the muscles of the pharynx, larynx, and face required for speech, all the while relying on sensory feedback to guarantee the movements are correctly performed. Each of these tasks relies on different, highly interacting neural subnetworks within the greater left hemispheric language network. Early loci of degeneration in gray and white matter within this network lead to distinctive patterns of language symptoms, which can reveal the underlying neuroanatomy to the trained diagnostician. The regions of degeneration in PPA are predominantly left hemispheric, reflecting high hemispheric specialization of language networks in the general population, although interaction with right hemisphere structures assures proper use of language and speech in social context. As the disease progresses, atrophy becomes more generalized, extending through the language network and ultimately involving the opposite hemisphere as well. Among the approximately 40% of left-handed individuals in whom language dominance localizes to the right hemisphere, PPA is characterized by atrophy of that part of the brain (Mesulam, Weintraub, Parrish, & Gitelman, 2005).

In addition to each of these clinical syndromic subtypes resulting from atrophy of a distinct neuroanatomical location, each subtype is associated with different probabilities of associated pathological findings on autopsy (Gorno-Tempini et al., 2011; Mesulam, Wieneke, Thompson, Rogalski, & Weintraub, 2012).

Overall, PPA patients are found to most often have frontotemporal lobar degeneration (FTLD)-type pathology, with abnormal precipitates of either tau or of transactive response DNA binding protein (FTLD-TDP-43). However, 30% to 40% of PPA patients have been found to have either Alzheimer pathology or biomarkers indicative of brain amyloid deposition (Mesulam, 2013). As discussed later, many of the PPA cases with AD pathology have the clinical-anatomical presentation of the logopenic variant.

Although PPA has taught us much about the language network, many debates are still being held regarding needed adjustments to diagnostic schematics and neuroanatomy (Chare et al., 2014). The hope is that accurate diagnoses will guide counseling regarding disease progression and may have implications for treatments now and as targeted therapies become available in the future.

75.2 THE NONFLUENT/AGRAMMATIC VARIANT

Damage to the left inferior frontal lobe underlies some of the earliest noted vascular and neurodegenerative aphasias. Also called Broca's area, this region was also atrophied in at least two patients described in Dr. Mesulam's seminal 1982 description of what is now called PPA (Mesulam, 1982).

75.2.1 Demographics

The incidence and prevalence of nonfluent/agrammatic variant (nfvPPA) can be estimated from those of FTLD, which are approximately 2.2 to 3.5 per 100,000 and 2.7 to 15 per 100,000, respectively. Approximately 45% of FTLD cases have PPA, with a little less than half of these having nfvPPA (Grossman, 2012). These statistics permit an estimated nfvPPA prevalence of 0.65 to 3.9 per 100,000, with an incidence of 0.5 to 0.9 per 100,000, although admittedly the estimate is rough.

Patients with nfvPPA may be more likely to be female (Johnson et al., 2005), although this is contested—other studies have found no gender bias (Grossman, 2012). Some have also suggested that the age of onset may be later than other forms of FTLD (Johnson et al., 2005; Rosso et al., 2003). One study described that nfvPPA had a mean age of onset of 63 years compared with 57.5 in bvFTD and 59.3 in svPPA (Johnson et al., 2005). The range is quite broad, however, ranging from the 30s to the early 80s. Survival is quite variable but is thought to average around 7 years (Grossman, 2012).

75.2.2 Clinical Characteristics

One of the first descriptions of the nonfluent variant was in 1996 by Murray Grossman and colleagues, who noted agrammatism and nonfluent speech, but also mentioned paraphasias and repetition deficits as prominent features (Grossman et al., 1996). Neary and colleagues included descriptions of the diagnostic criteria for "progressive aphasia," which was described as nonfluent but that likely included what we now call nonfluent/agrammatic and logopenic variants (Neary et al., 1998). The consensus classification introduced by Gorno-Tempini and a group of international colleagues in 2011 de-emphasize any problems with paraphasias and repetition in nfvPPA, proposing that effort speech (as in apraxia of speech [AOS]) and/or agrammatism should be the core features of this variant and further specify that single word comprehension and object knowledge are typically unimpaired (Gorno-Tempini et al., 2011) (Table 75.3).

In most cases, the most immediately notable symptom among those with nfvPPA is the effortful and halting quality of their speech. The average rate of speech produced by patients with nfvPPA is about 45 words per minute compared with 140 words per minute in healthy controls (Grossman et al., 2013). Patients mainly experience difficulty articulating words that they "have in their head." A combination of motor speech hesitations and grammatical processing problems are the main contributors to diminished fluency, although in mild cases patients could still be considered fluent by classic aphasiology batteries because symptoms are very mild (Gunawardena et al., 2010).

Coordinating the movements needed for speech is frequently difficult among those with nfvPPA. AOS refers to this incoordination of subtle movements required for production of distinct speech, even in the absence of any primary muscular dysfunction to cause dysarthria (Duffy, Peach, & Strand, 2007). Patients with AOS may describe a sensation of knowing what they want to say, but being unable to form the desired

TABLE 75.3 Diagnostic Features for the Nonfluent Variant of PPA

I. Clinical diagnosis of nonfluent/agrammatic variant PPA
At least one of the following core features must be present:
1. Agrammatism in language production
2. Effortful, halting speech with inconsistent speech sound errors (AOS)
At least two of three of the following other features must be present:
1. Impaired comprehension of syntactically complex sentences
2. Spared single-word comprehension
3. Spared object knowledge
II. Imaging-supported nonfluent/agrammatic variant diagnosis
Both of the following criteria must be present:
1. Clinical diagnosis of nonfluent/agrammatic variant PPA
2. Imaging must show one or more of the following results:
a. Predominant left posterior fronto-insular atrophy on MRI or
b. Predominant left posterior fronto-insular hypoperfusion or hypometabolism on SPECT or PET
III. Nonfluent/agrammatic variant PPA with definite pathology
Clinical diagnosis (criterion 1 below) and either criterion 2 or 3 must be present:
1. Clinical diagnosis of nonfluent/agrammatic variant PPA
2. Histopathologic evidence of a specific neurodegenerative pathology (e.g., FTLD-tau, FTLD-TDP, AD, other)
3. Presence of a known pathogenic mutation

AD, Alzheimer's disease; FTLD, frontotemporal lobar degeneration; MRI, magnetic resonance imaging; PET, positron emission tomography; SPECT, single proton emission computed tomography.
Source: *From Gorno-Tempini et al. (2011).*

words with their lips and tongue. AOS may occur in spontaneous speech, but may be most pronounced with particularly difficult articulatory phrases, such as "first British field artillery," or by repeating multisyllabic words multiple times (Ogar, Slama, Dronkers, Amici, & Gorno-Tempini, 2005). Some authors propose a separate classification when patients have mainly progressive motor speech deficits without agrammatism (Josephs, Duffy, et al., 2013).

Agrammatism is usually present, although it may be mild, and has been called a core feature nfvPPA, to the point that Mesulam and colleagues refers to this variant as the agrammatic subtype (PPA-G) (Mesulam, 2013). Abnormal syntax, inappropriate use of pronouns, misused pronouns, decreased verb use, and a dearth of article and prepositions correlate with nonfluency in nfvPPA. Limited verb production also correlates with decreased speech rate, perhaps due to the crucial role of verbs in sentence structure. Difficulty constructing complex sentences and words may only be apparent in writing in early stages, but they will ultimately involve both written and spoken language (Grossman, 2012). Patients may attempt to avoid these deficits by using shortened sentences, resulting in an abbreviated mean utterance length as patients attempt to simplify their speech.

Comprehension of single words and simple sentences is generally spared in nfvPPA. Comprehension deficits only become apparent with complex syntactic constructions such as passive or relative sentences (e.g., "The lion was not eaten by the tiger") (Weintraub et al., 2009). Such errors are among the most distinctive equalities of nfvPPA used to distinguish the subtype from other PPA variants (Mesulam et al., 2009; Wilson, Henry, et al., 2010). In addition to spontaneous speech, such comprehensive agrammatism may be present in reading and writing as well. Overall, however, there is little difficulty reading written words aloud. Spelling is also largely preserved in nfvPPA, although nonword spelling may be more problematic (Shim, Hurley, Rogalski, & Mesulam, 2012).

As time passes, speech becomes increasingly effortful (Sapolsky et al., 2010). Working memory and executive control also become involved. Specifically, working memory, planning, and dual tasking become impaired. Letter-guided naming fluency seems to decline over time as well. Episodic memory seems to be relatively preserved, as is visuospatial functioning—although some exceptions occur in the context of the corticobasal syndrome (CBS) (Murray et al., 2007). Early symptoms of PPA may erroneously be attributed to depression, anxiety, or stress. Such symptoms are not uncommon in nfvPPA, although these may represent a natural response to the deep awareness that these patients often demonstrate for their language deficit. As the disease progresses, damage to frontal circuits may lead to more evident depression and frustration (Neary et al., 1998). Apathy, diminished empathy, and lack of insight can also emerge (Rohrer & Warren, 2010).

On the neurological examination, mild motor symptoms affecting the right side of the body are not uncommon. Mild rigidity, motor slowing, and reduced dexterity may be present. Limb apraxia may also be present, as it often is when the left parietofrontal network is disrupted (Zadikoff & Lang, 2005). Although AOS is not to be confused with dysarthria, dysarthria may co-occur with nfvPPA. Features of CBS or progressive supranuclear palsy syndrome (PSPS) often occur during disease progression, causing extrapyramidal findings such as rigidity and bradykinesia. Frontotemporal dementias such as nfvPPA have been associated with symptoms of motor neuron diseases such as amyotrophic lateral sclerosis (ALS), and symptoms of this disease may also appear.

75.2.3 Neuroanatomy and Imaging

The peak atrophy site in nfvPPA is the posterior inferior frontal gyrus, also known as Broca's area. Such regions of atrophy are usually studied using volumetric magnetic resonance imaging (MRI) studies. Clinically, neuroimaging also helps exclude other nondegenerative potential causes of aphasia such as tumor or vascular disease.

The area of degeneration in nfvPPA typically extends from the inferior frontal gyrus to the adjacent frontal operculum, premotor cortex, and anterior insula, and sometimes extends into prefrontal regions and into the superior aspect to the left anterior temporal lobe (Rogalski, Cobia, Harrison, Wieneke, Weintraub, et al., 2011) (Figure 75.1). These regions of atrophy are closely related to reduced speech fluency (Ash et al., 2009; Rogalski, Cobia, Harrison, Wieneke, Thompson, et al., 2011; Wilson, Henry, et al., 2010). Premotor area and supplementary motor area (SMA) and basal ganglia are frequently also impacted when motor speech difficulties are especially pronounced (Josephs et al., 2006). While the left hemisphere is usually more damaged than the right in nfvPPA, bilateral damage is common, impacting up to 31% of cases (Westbury & Bub, 1997).

Although grey matter losses have been emphasized in the MRI studies described, the white matter is not spared in nfvPPA. Diffusion tensor imaging (DTI) demonstrates fractional anisotropy and mean diffusivity changes in many projections related to the inferior frontal lobe (Galantucci et al., 2011). Intrafrontal connections between inferior frontal, SMA, and basal ganglia are related to speech difficulties in nfvPPA (Mandelli et al., 2014). A dorsal stream in the peri-Sylvian language network seems instead related to major grammatical category and syntactic features of language, respectively (Agosta et al., 2012; Borroni et al., 2007; Galantucci et al., 2011; Grossman, 2012; Mahoney et al., 2013; Schwindt et al., 2013; Whitwell et al., 2010) (Figure 75.2).

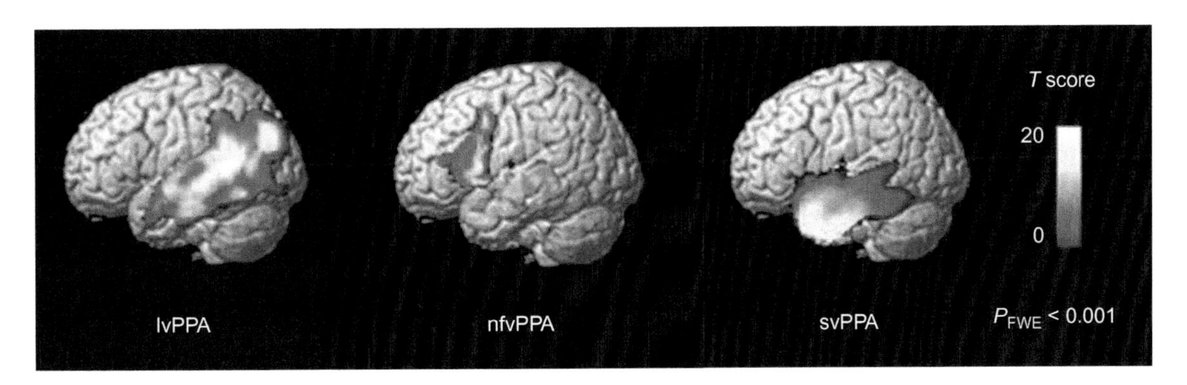

FIGURE 75.1 Atrophy patterns in patients with PPA variants versus controls. The statistical parametrical mapping depicts patterns of grey matter atrophy in patients with logopenic variant, nonfluent variant, and semantic variant PPA compared with healthy controls matched for age, gender, scan, and sample size, thresholded at a family-wise error (FWE) rate of $P < 0.001$. *From Miller, Mandelli, et al. (2013) with permission from the publisher.*

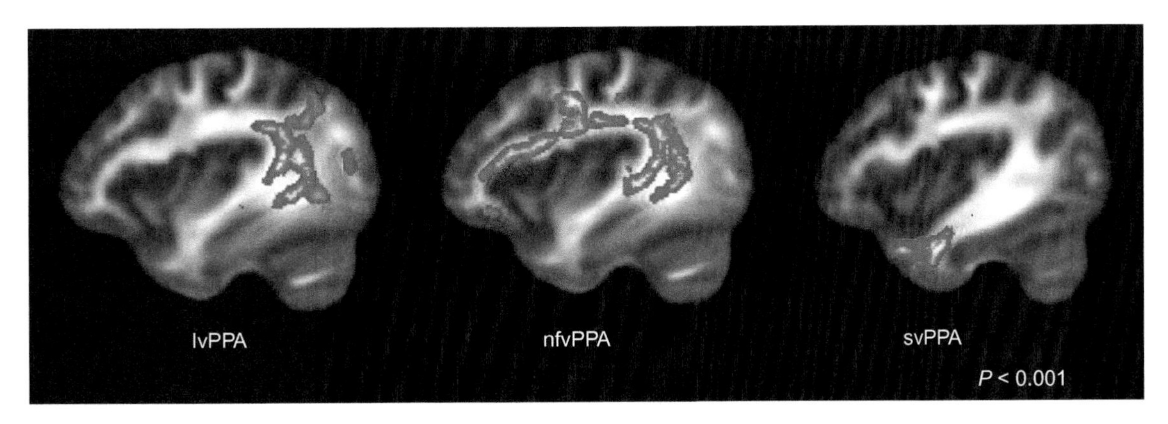

FIGURE 75.2 Changes in DTI measurements in three variants of PPA compared with controls. From left to right, there is a significant increase in mean diffusivity in patients with logopenic variant ($P < 0.05$) and a significant decrease in fractional anisotropy in both ($P < 0.001$) semantic variant and nonfluent variant of PPA, with distinctive anatomical distributions for each variant. *Image courtesy of Nico Papinutto.*

In addition to the structural imaging modalities described, nuclear medicine techniques have been used to investigate how brain activity is altered by nfvPPA. Positron emission tomography, for example, may demonstrate decreased metabolism in the left frontal operculum and anterior insula even before atrophy becomes apparent on structural neuroimaging. PiB binding can also examine whether amyloid is present, suggestive of an underlying Alzheimer etiology, although dual pathology is also possible (Rabinovici et al., 2008). SPECT can also demonstrate blood flow changes in the left inferior frontal lobe in nfvPPA (Nestor et al., 2003).

Although fMRI studies have not yet found a place in the clinical diagnosis of nfvPPA, studies have shown networks in the ventral aspect of the frontal lobe to be inactive during a task of syntactically complex sentence comprehension among patients with nfvPPA compared with controls (Cooke et al., 2003). Resting state functional connectivity MRI has also been used to outline in healthy controls the left frontal network impacted in nfvPPA (Seeley, Crawford, Zhou, Miller, & Greicius, 2009).

75.2.4 Histopathology and Genetics

The nonfluent/agrammatic variant is considered a subtype of frontotemporal dementia. There are three forms of frontotemporal dementia: the behavioral variant (bvFTD), semantic variant of PPA (svPPA), and nfvPPA. Whereas both svPPA and AD are fairly histopathologically homogenous, both bvFTD and nfvPPA can be caused by a number of different pathological changes. The nonfluent variant most often heralds neurodegenerative disease such as FTLD-4R tauopathies, such as PSP or corticobasal degeneration (CBD). Pathological changes of FTLD-TDP-A are also found in different frequencies depending on the research group. Less frequently, AD may be involved (Caso et al., 2013).

Pathologically, the nonfluent variant of PPA is most commonly associated with deposits of tau or transactive-response DNA-binding protein of approximately 43 kD (TDP-43) types A or B. Causative tauopathies include PSP, CBD, and less commonly Pick's disease (Figure 75.3). Among those cases of nfvPPA associated with TDP-43, type A is most

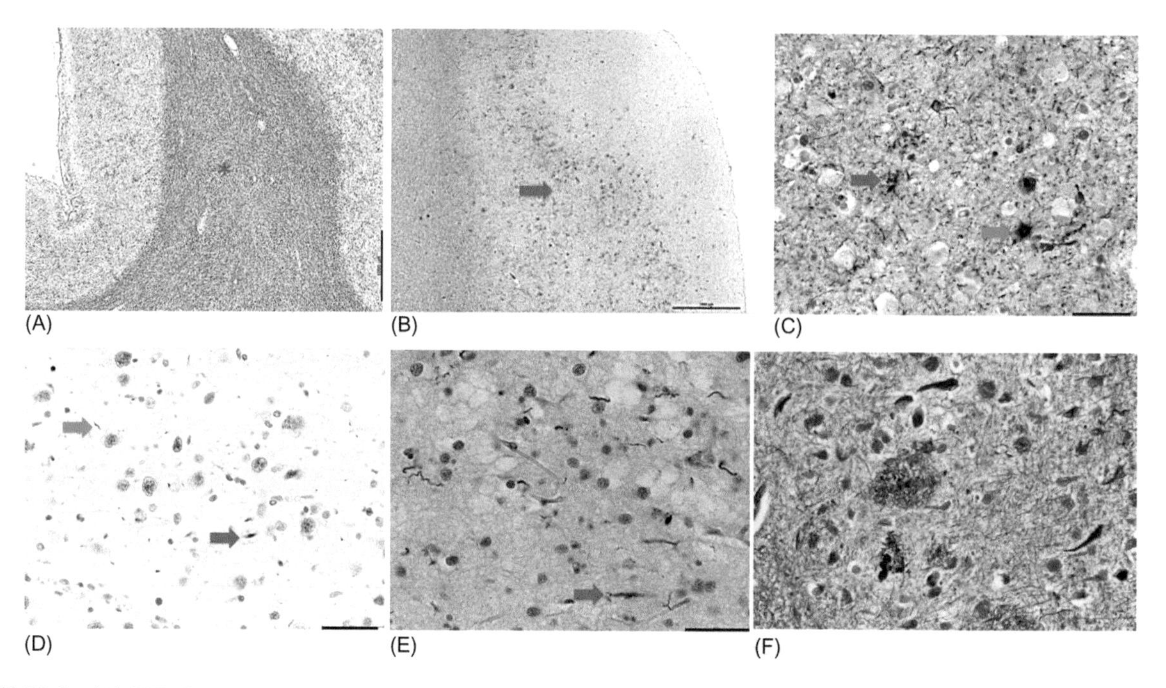

FIGURE 75.3 (A) CBD. Immunostaining for phospho-tau. The * demonstrates white matter, where tau density is higher than in the gray matter of the outer layer. Scale bar: 1,000 μm. (B) PSP. Immunostaining for phospho-tau. The red arrow indicates a higher number of inclusion bodies in gray matter than in white matter. Scale bar: 1,000 μm. (C) Pick disease. Immunostaining for phospho-tau. The red arrow indicates an astrocytic inclusion, and the green arrow indicates a Pick body. Scale bar: 50 μm. (D) Transactive response DNA-binding protein of approximately 43 kD type A (TDP-A). Immunostaining for TDP-43. The red arrow indicates a typical "cat-eye" intranuclear inclusion, and the green arrow indicates short neurites. (E) Transactive response DNA-binding protein of approximately 43 kD type C (TDP-C). Immunostaining for TDP-43. The red arrow denotes granular neuronal cytoplasmic inclusions. (F) A typical Alzheimer disease plaque is depicted left of center, revealed by Bielschowsky silver staining. Several neurofibrillary tangles are also present. *Slides (A)–(E) courtesy of Dr. Lea T. Grinberg, UCSF Memory and Aging Center. From Pressman and Miller (2014) with permission from the publisher. (F) From Pressman and Rabinovici (2014).*

common (Josephs, Stroh, Dugger, & Dickson, 2009) (Figure 75.3). Although both of these histological patterns are more commonly associated with syndromes of the same name, it is possible that no symptoms of those syndromes may appear in nfvPPA. Beyond typical FTLD patterns of tau and TDP-43, reports show that many cases diagnosed with nfvPPA are ultimately found to have AD pathology, although some of these reports were retrospective and may not have investigated specifically for the logopenic pattern of impairment (Chare et al., 2014). In addition, while rare, other pathologies such as dementia with Lewy bodies have been associated with nfvPPA (Grossman, 2010).

Distinguishing *in vivo* among these pathological causes of nfvPPA is difficult. One small study comparing nfvPPA due to tau or TDP pathology indicated that the presence of extrapyramidal signs and agrammatism plus imaging signs of white matter damage might help in early differential diagnosis of nfvPPA caused by a 4R-tauopathy (Caso et al., 2013). The near-availability of tau radioligands will soon permit better distinction between TDP and tau pathologies *in vivo*.

Despite family history being strongly positive in up to 40% of FTD syndromes (Goldman et al., 2011; Rohrer et al., 2009), there is little to suggest that nfvPPA is usually inherited in an autosomal dominant manner. *MAPT* mutations have occasionally been associated with nfvPPA. The H1/H1 haplotypes on chromosome 17 may be associated with PPA (Sobrido et al., 2003). *MAPT* mutations do not seem to result in disproportionately high rates of nfvPPA, however, and phenotypes may be highly variable.

75.3 THE SEMANTIC VARIANT

The semantic variant of PPA (svPPA) has gone under various identities over the past century, perhaps including Pick's original descriptions, which included pronounced temporal lobe atrophy associated with semantic aphasia. Several similar reports followed (Compston, 2008; Rosenfeld, 1909; Stertz, 1926). In 1943, the Japanese researcher Tsuneo Imura further described an aphasia in which knowledge of the kanji characters depicting relatively complete concepts was lost while the more phonemic hiragana were preserved in Japanese patients. Originally named "Gogi" aphasia, this is now thought to represent the semantic variant of PPA in the Japanese population (Ralph & Howard, 2000; Sasanuma & Monoi, 1975). In 1975, Elizabeth Warrington proposed that such patients' word loss, associated agnosia, and impaired comprehension resulted from a fundamental loss of semantic knowledge (Warrington, 1975). Such knowledge refers to our ability to place an object in a network of associations with words, meanings, and concepts.

Until the 1980s, the language network had been widely perceived as consisting of Wernicke's and Broca's areas, connected by the arcuate fasciculus. The role of the anterior temporal lobe was not appreciated. This region rarely suffers isolated damage due to strokes, and when included in vascular injuries larger swaths of cortex are typically also damaged, confounding previous clinicopathological correlations.

In recent decades, however, the anterior temporal lobe has been recognized as being involved with several aspects of semantic and conceptual processing. Soon after Mesulam's 1982 description of what he would ultimately term PPA, it became recognized that those with fluent aphasias frequently had predominant degeneration of the left anterior temporal lobe, although mixed cases were also described (Adlam et al., 2006; Hodges, Patterson, Oxbury, & Funnell, 1992).

The term "semantic dementia" was proposed in 1989 by Julie Snowden to describe those with fluent aphasia and a loss of comprehension (Snowden, Goulding, & Neary, 1989). The diagnosis of semantic dementia was later included in the 1998 Neary criteria as one of the subtypes of frontotemporal dementia, in addition to being considered a subtype of PPA (Neary et al., 1998). The criteria were updated again in 2011 to further describe the distinctive language deficits in these patients, solidifying the term "svPPA" (Gorno-Tempini et al., 2011). Although aphasic symptoms, such as anomia, are a key component of this syndrome when it initiates in the left temporal lobe, both temporal lobes ultimately become involved in the disease, and the disease may also start in the right lobe instead of the left (Seeley et al., 2005). The right temporal variant of this disorder may demonstrate how semantic loss can extend beyond verbal abilities, including diminished social pragmatics, empathy, and complex behavioral symptoms (Seeley et al., 2005).

75.3.1 Demographics

The age of onset for svPPA has been thought to be generally earlier than that for nfvPPA or lvPPA, but later than bvFTD (Johnson et al., 2005). Studies suggest that the mean onset may be about 60 years of age (Hodges et al., 2010; Rohrer et al., 2011). The rate of progression is thought to be relatively slow in svPPA, with an average time from symptom onset to death of about 10 to 14 years, with much individual variation (Davies et al., 2005; Hodges et al., 2010; Nunnemann et al., 2011). There may be a male predominance, with approximately 60% of svPPA patients being men.

As with nfvPPA, the incidence and prevalence of svPPA can be estimated from those of FTLD because approximately 8.6% to 20% of all FTLD comprise the semantic variant (Forman et al., 2006; Johnson et al., 2005; Rohrer et al., 2011; Shi et al., 2005; Snowden et al., 2011). This allows for an estimated population incidence of 0.19 to 0.7 per 100,000 and an estimated prevalence of 0.23 to 3.0 per 100,000.

75.3.2 Clinical Characteristics

Although some people presenting with svPPA may initially report "loss of memory," it soon becomes apparent that their problem is actually "loss of words" (Hodges et al., 2010). Severe anomia is the clearest deficit in neuropsychological testing. On further examination, orientation and visuospatial skills are well-preserved in svPPA, as is episodic memory. Knowledge of facts that are detached from a spatio-temporal and emotional context, that is, concepts or semantics, is diminished. In addition to words, patients have difficulty identifying objects and famous people. By doing specific tests that avoid linguistic content, a broader semantic deficit can be demonstrated. For example, the Pyramid and Palm Trees test shows a picture of an object and then requires the participant to choose which of two additional objects best goes with the initial stimulus (Howard & Patterson, 1992). Loss of conceptual knowledge for objects and words is highly influenced by familiarity of the object or concept. Early on, only unfamiliar, atypical concepts are lost, whereas knowledge of high-frequency or prototypical words and objects can be spared. Because classic stroke aphasia test batteries mainly contain high-frequency items, they might miss the comprehension and semantic deficits in early cases.

Linguistically, svPPA features impaired object naming and single-word comprehension, with preservation of fluency, phonology, and grammar (Gorno-Tempini et al., 2011) (Table 75.4). The discrepancy between conversational speech and object naming can be quite striking, especially early in the disease. The subject may be able to fluently discuss his or her life events, but then react to a commonplace word such as "watermelon" as if it were part of a foreign language. Spontaneous speech is often filled by nonspecific or "passe-partout" words, such as "this," "that," "the place where I used to go," or "thing."

In a study of language and semantic loss in left temporal atrophy, object-naming deficits were the most severe and consistent deficit (Mesulam et al., 2013). Evidence has suggested that this results predominantly from a loss of semantic knowledge rather than pure word-retrieval deficits. Patients with svPPA are less likely to think of the word later than those with other forms of PPA, and they are consistent in the names they cannot recall (Ralph & Howard, 2000). Cuing with the initial sounds of a word is less beneficial, and providing multiple choice options provides little help compared with other types of aphasia (Graham, Patterson, & Hodges, 1995).

However, in at least one study addressing object naming deficits in very early left temporal atrophy, many errors were pure retrieval failures as judged by the denoting word still being understood, the object recognized, and object–word matching remaining intact. The authors suggest that this may be due to a more bilateral or right-hemispheric localization of object knowledge (Mesulam et al., 2013). An alternative explanation could be that the information embedding for more general semantic loss differs from that between word and concept. A concept is connected to

TABLE 75.4 Diagnostic Features of the Semantic Variant of PPA

I. Clinical diagnosis of semantic variant PPA
Both of the following core features must be present:
1. Impaired confrontation naming
2. Impaired single-word comprehension
At least three of the following other diagnostic features must be present:
1. Impaired object knowledge, particularly for low-frequency or low-familiarity items
2. Surface dyslexia or dysgraphia
3. Spared repetition
4. Spared speech production (grammar and motor speech)
II. Imaging-supported semantic variant PPA diagnosis
Both of the following criteria must be present:
1. Clinical diagnosis of semantic variant PPA
2. Imaging must show one or more of the following results:
a. Predominant anterior temporal lobe atrophy
b. Predominant anterior temporal hypoperfusion or hypometabolism on SPECT or PET
III. Semantic variant PPA with definite pathology
Clinical diagnosis (criteria 1 below) and either criterion 2 or 3 must be present:
1. Clinical diagnosis of semantic variant PPA
2. Histopathologic evidence of a specific neurodegenerative pathology (e.g., FTLD-tau, FTLD-TDP, AD, other)
3. Presence of a known pathogenic mutation

AD, Alzheimer's disease; FTLD, frontotemporal lobar degeneration; MRI, magnetic resonance imaging; PET, positron emission tomography; SPECT, single proton emission computed tomography.
Source: *From Gorno-Tempini et al. (2011).*

several other concepts, of which a word is only one. If a connection between the concept of an apple and the word "apple" is lost, the concept is still connected to a number of other neuronal representations for red, fruit, sweet, and so on. Compared to these sensory associations, the association with the word "apple" may be relatively arbitrary and weak, and relatively susceptible to loss in the face of a disease. When conceptual knowledge erodes, the more arbitrary and isolated connection vanishes first (Hodges, Davies, & Patterson, 2009). The appearance of a predominant naming deficit in svPPA, then, may result less from neuroanatomic localization than relative robustness of connections within the neural network.

In addition to loss of single-word comprehension, loss of object knowledge, and object naming deficits, the 2011 criteria specifically mention surface dyslexia, which is a tendency to read irregularly spelled words phonetically. Patients with surface dyslexia regularize words with an atypical spelling pattern. The word "yacht," for example, may be pronounced to rhyme with "latch (lætʃ)." This phenomenon stems from an inability to remember atypical word spellings as a type of semantic information, and so needing to read the word phonologically. Normally, words not spelled phonologically must be identified as an object in and of themselves, which svPPA patients are unable to endow with meaning. In addition to reading errors, when writing such patients may struggle with spelling exception words and make phonetically plausible spelling errors (e.g., "fruit" and "froot") (Shim et al., 2012). In languages that use a more pictographic system of writing, those symbols that encapsulate a more semantic meaning, such as Japanese kanji, lose meaning before those symbols that are phonological, such as Japanese hiragana and katakana (Sasanuma & Monoi, 1975). Similarly, the inability to recall irregular past tense forms of verbs, such as "knew," has also been considered as a symptom of semantic loss. In this instance patients tend to apply general morphological rules, thereby producing verbs such as "knowed."

Phonemic paraphasias are relatively rare in svPPA compared to other PPA subtypes (Mesulam et al., 2013). Semantic paraphasias, however, in which one word is switched for another rather than one part of a word for another, are more common in svPPA. Specific and low-frequency words are replaced by the prototypical exemplar of that category of its prototypical exemplar or supraordinate, so all "cow" might become "dog" or "animal." As svPPA progresses, semantic paraphasias tend to become more vague. Accordingly, conversation becomes less specific, more superficial, and difficult to understand, although speech might still be fluent.

In addition to conceptual knowledge for objects and words, the anterior temporal lobe is involved with identification of known people from names, voices, and faces (Evans, Heggs, Antoun, & Hodges, 1995; Imaizumi et al., 1997), high-level visual and auditory processing (Horel, Keating, & Misantone, 1975; Lambon Ralph, 2014), socioaffective concepts such as empathy and theory of mind (Duval et al., 2012; Gorno-Tempini, Rankin, et al., 2004; Irish, Hodges, & Piguet, 2014), and the regulation of eating and sexual behavior. Dysfunction in these arenas become more readily apparent in the right temporal variant of svPPA but also occur in left svPPA as disease progresses and starts involving the right anterior temporal lobe and ventromedial frontal regions. As might be predicted, language is initially not as impaired in the right temporal variant, and it can be initially diagnosed as bvFTD. However, considering that identification of people and social concepts can be considered as part of semantic knowledge (one does not know when they learned to identify facial expressions), the right temporal form of svPPA should still be considered a semantic variant as opposed to bvFTD. In general, patients with svPPA tend to underestimate the extent of their behavioral and semantic changes, often only reporting word-finding deficits (Eslinger et al., 2005; Zamboni, Grafman, Krueger, Knutson, & Huey, 2010). Although they may also overestimate their empathic ability, data here are mixed (Duval et al., 2012).

Despite their deficits, it is important to recognize that many aspects of cognition remain intact in patients with svPPA. For example, they may remember life events well, find their way around without difficulty, and engage in complex hobbies with retention of many practical, visuo-spatial and creative skills. As stated in the 2011 criteria, repetition and grammar are relatively spared.

75.3.3 Anatomy and Imaging

MRI most often shows left greater than right anterior temporal lobe atrophy. Volume loss is greatest in the polar, parahippocampal middle, and inferior temporal regions, including the anterior fusiform gyrus (Pereira et al., 2009). This is typically more on the left than the right (Figure 75.1). Some have postulated that this apparent predilection toward the left lobe reflects a referral bias for a plausibly more obvious clinical presentation, that is, aphasia, rather than the comparatively subtle behavioral changes inherent in a right-temporal variant. Others, however, have pointed out that there is no significant difference between right and left predominant cases in the mean length of time from symptom onset to diagnosis (Hodges et al., 2010). Nevertheless, as awareness of these diseases increases, more cases of right greater than left anterior temporal atrophy are starting to be referred to neurology clinics. Even at this greater rate or recognition, however, the proportion of right to left temporal variants in our UCSF cohort is about 1:4.

Although the appearance of the temporal lobes is asymmetric, atrophy is often bilateral in most cases, even early in the disease course (Chan et al., 2001). As time passes, those who begin with language impairment increasingly develop a social and emotional behavioral syndrome increasingly develop language problems. This corresponds with a spread of atrophy from the predominant lobe toward the opposite lobe, with subsequent spread to posterior temporal regions and orbitofrontal cortex as well (Brambati et al., 2009).

Initial reports described relative sparing of the hippocampus in svPPA, although volumetric analyses, in fact, demonstrate some atrophy of the hippocampus, which may be as severe if not more so in svPPA than AD when patients are matched for disease duration (Davies, Graham, Xuereb, Williams, & Hodges, 2004). Such hippocampal loss averages about 20%, compared with 50%, loss suffered by the temporopolar and perirhinal cortices, giving the illusion or relative preservation. The amygdala is also universally involved with svPPA. This may contribute to some of the behavioral symptoms and lack of social cues such as reading of facial emotions (Rosen et al., 2002).

The anterior temporal lobe is described as a hub that binds together modality-specific information to create a unitary concept (Patterson, Nestor, & Rogers, 2007). The existence of an anterior temporal semantic hub does not exclude the possibility of some differential hemispheric specialization, depending on the connectivity of role of neighboring regions (Gainotti, 2006). In this framework, nonverbal and socio/emotional semantic information might be more right dominant or bilateral (Gainotti, 2006; Mesulam et al., 2013). A resting-state functional connectivity MRI study performed by Guo and colleagues suggests that the anterior temporal hub is a conglomerate or "transmodal" hub but with multimodal upstream contributing regions. For example, the superior temporal lobe was demonstrated to be more involved with auditory processing, and the inferior was demonstrated to be more involved with visual processing (Guo et al., 2013). DTI studies have shown that the inferior longitudinal fasciculus, the uncinate fasciculus, and the temporal portion of the arcuate bundle are particularly impacted in svPPA. The fronto-parietal portions of the superior longitudinal fasciculus (SLF) and the intrafrontal speech network are instead relatively spared (Galantucci et al., 2011) (Figure 75.2).

75.3.4 Histopathology and Genetics

SvPPA has one of the tightest associations between clinical syndrome and pathology. TDP-43 type C is the most common cause of svPPA and is found in about 90% of affected patients (Rohrer, Geser, et al., 2010) (Figure 75.3). Although this pathology most commonly localizes to the temporal lobes, cases have been described in which TDP-43 type C pathology is associated with corticospinal tract degeneration (Josephs, Whitwell, et al., 2013). Pick's disease can also present as semantic variant, although atrophy is usually more fronto-temporal. AD can also rarely affect the anterior temporal lobe and cause selective semantic loss.

While up to 40% of FTD cases are said to have a family history of a neurodegenerative disorder, of which 10% to 20% show a clear positive family history of FTD, there is a much lower familial rate in svPPA compared with other subtypes (Goldman et al., 2005). This suggests a strong role of neurodevelopmental, environmental exposure, and the possibility of an autoimmune process. Interestingly, at 18%, the rate of certain autoimmune diseases in svPPA is higher than that found in healthy controls or AD (Miller, Rankin, et al., 2013). As autoimmune diseases frequently cluster together, there is a possible role for immunotherapy in svPPA in the future. Furthermore, there is a disproportionate number of left-handed individuals in those with svPPA, at nearly twice the incidence of the general population (Miller, Mandelli, et al., 2013).

75.4 THE LOGOPENIC VARIANT (lvPPA)

When PPA was initially divided into the nonfluent and fluent subtypes, non-Alzheimer pathology was considered the likeliest cause, although Alzheimer presenting with aphasia was also discussed (Kertesz, Davidson, McCabe, Takagi, & Munoz, 2003; Weintraub, Rubin, & Mesulam, 1990). Following the clearer descriptions of the semantic and nonfluent subtypes, Gorno-Tempini and colleagues published a comparison of cognitive and neuroimaging features of a large group of PPA patients in 2004, therein describing a third variant called logopenic PPA (lvPPA or PPA-L) (Gorno-Tempini et al., 2004). Patients with lvPPA have specific cognitive and neuroimaging characteristics that distinguish them from the other variants. In retrospect, many cases of lvPPA had previously been included among nonfluent or fluent aphasia cases.

In the decade that followed, the cognitive mechanism and resulting symptoms of lvPPA have been further elucidated. The role of phonological processes and phonological short-term memory was elucidated (Foxe, Irish, Hodges, & Piguet, 2013; Gorno-Tempini et al., 2008; Leyton, Piguet, Savage, Burrell, & Hodges, 2012), and in contrast to the FTLD underlying svPPA and nfvPPA, the most common pathological process in lvPPA has been found to be AD (Kirshner, 2012; Mesulam et al., 2008) (Figure 75.3). Possible links between PPA, especially lvPPA, and neurodevelopmental disorders have been suggested to predispose to this focal presentation of AD (Miller, Mandelli, et al., 2013; Rogalski, Johnson, Weintraub, & Mesulam, 2008; Rogalski, Weintraub, & Mesulam, 2013). Despite this progress, many questions remain regarding the neuroanatomical, linguistic, and histopathological characteristics of lvPPA.

75.4.1 Demographics

At our center, the mean age of patients with lvPPA is 68 years, with a standard deviation of 8.6 years; 77.3% of our lvPPA patients have been right-handed, and about 47% have been male. The incidence and prevalence of lvPPA are uncertain.

75.4.2 Clinical Features

The term logopenia is derived from Greek, meaning "lack of words." The neologism was coined to describe those PPA patients who were fluent when engaging in small talk and when permitted to use circumlocutions, but who had word-finding hesitations when precise word selection was required (Mesulam, 1982, 2007). The result is a decrease in verbal output, with decreased rate of speech production, although fluency is usually intermediate between svPPA and nfvPPA, that appears similar to the vascular conduction aphasia syndrome (Gorno-Tempini et al., 2004).

As a symptom of phonological impairment, deficits in sentence and nonword repetition are key features of lvPPA, although this deficit may appear mild unless rigorously tested, which has led some to question this criterion's inclusion in the current classification scheme (Mesulam et al., 2009; Mesulam, 2013; Mesulam et al., 2012). Phrase repetition is especially detectible, however, if phrases are long or have improbable word combinations. For example, patients with lvPPA will struggle more to repeat the phrase "the pastry cook was elated" compared to "the telephone is ringing" (Gorno-Tempini et al., 2008). They may also engage in circumlocution, returning a phrase that uses more common words, such as "something about a happy baker." Although other forms of PPA, such as nfvPPA, may involve difficulty with repetition, the nature of the errors differs. Those with nfvPPA tend to err even in short sentences if they are sufficiently grammatically complex, typically dropping articles or pronouns (Wilson, Dronkers, et al., 2010).

The observation of the pronounced repetition deficits in those with lvPPA guided the hypothesis that auditory phonological short-term phonological memory deficits are the key mechanism behind the disorder. Short-term phonological memory (or the phonological loop) refers to our ability to briefly retain a certain amount of verbal auditory information, usually between five and nine components in length, until we can manipulate that information or store it for a longer period. lvPPA involves a disruption in the short-term memory circuit dedicated to manipulation of phonemes, which are small contrastive units in the sound of a word, such as "fo" and "neem." This process is particularly necessary to arrange sounds into long words, to accurately reproduce sequences of words and digits (as in digit span tasks), and to compose sentences (Baddeley, 1988; Vallar, Di Betta, & Silveri, 1997). Patients with lvPPA cannot retain this information long enough to repeat the exact phonemic sequence of a long pseudoword or the sequence of words into a phrase, although they can usually comprehend the general meaning if the sentence is not too long. Such patients may be unable to retain verbal information long enough to integrate the end of a long sentence with the sentence's beginning, and so shorter and simpler sentences are preferentially used conversationally by lvPPA patients. Comprehension of sentences in lvPPA is improved by providing the stimuli in written form, thus decreasing the load on the phonological loop. The neural basis of the phonological loop is likely sustained to the left temporo-parietal junction to inferior frontal circuit.

In lvPPA, lexical retrieval deficits and phonological errors in speech production likely result from greater impairment in phonological processing, likely related to posterior temporal atrophy. Word retrieval can be assisted by phonological cues (e.g., "pa" to cue "palette"). Similarly, single word comprehension deficits in lvPPA are initially subtle, but they become more pronounced as the disease progresses involving more anterior and ventral temporal regions. Unlike svPPA, nonverbal semantic association and single-word comprehension are initially spared, although exceptions may exist in cases with progranulin mutations who show logopenic symptoms often accompanied by other impairments (Rohrer, Crutch, Warrington, & Warren, 2010). Although lvPPA patients may seem to have difficulty comprehending syntax, the cognitive mechanism differs from the agrammatism of nfvPPA. Sentence comprehension deficits in lvPPA more likely result from difficulty integrating long strings of words in short-term memory, rather than a true grammatical problem (Gorno-Tempini et al., 2008; Wilson, Dronkers, et al., 2010).

Phonological processing deficits also distinguish the paraphasias sprinkled through the speech of those with lvPPA. Unlike the production errors in svPPA, in which an entire word is initially replaced with something taxonomically similar, those with lvPPA stumble over part of the word. The intended word is thereby replaced with a similar sounding substitute, such as "cantaloupe" for "antelope" (Wilson, Henry, et al., 2010). That said, phonological paraphasias are typical of, but not specific to, the lvPPA phenotype, as mentioned, likely in relation to the spreading of anatomical damage to posterior temporal cortex (Croot, Ballard, Leyton, & Hodges, 2012; Mesulam et al., 2012). lvPPA patients with such severe phonological issues may have had a relatively weak phonological system in the form of lifelong neurodevelopmental disorders such as dyslexia, which are further weakened by neurodegeneration.

In lvPPA, phonological processing deficits likely underlie reading and spelling errors that take the form of phonological dyslexia, characterized by greater difficulty in pseudo-words (meaningless words that still follow general rules of English) that require the use of grapheme to phoneme mechanisms and cannot rely on whole word semantic processes (Brambati, Ogar, Neuhaus, Miller, & Gorno-Tempini, 2009; Sepelyak et al., 2011; Shim et al., 2012).

Additional nonverbal symptoms may accompany language problems in lvPPA due to anatomical proximity to the left temporoparietal region most atrophied in the disease. Functionality of nearby cortical regions may result in dyscalculia and limb apraxia, for example (Rohrer et al., 2008). Degrees of impairment in cognitive domains such as visuospatial ability, divided attention, and cognitive flexibility have also been reported (Galantucci et al., 2011; Gorno-Tempini et al., 2004; Machulda et al., 2013; Rohrer & Warren, 2010; Wicklund, Johnson, Rademaker, Weitner, & Weintraub, 2007; Wilson, Henry, et al., 2010). Possibly reflecting the underlying Alzheimer pathology, those with lvPPA also tend to have greater general memory impairment than other PPA variants (Mesulam et al., 2008). Mood changes are not uncommon, especially anxiety, agitation, and irritability (Rohrer & Warren, 2010).

The international classification scheme requires the presence of both word-finding difficulties and repetition deficits for a diagnosis of lvPPA (Table 75.5). The utility of including both these criteria has been recently challenged (Wicklund et al., 2014), as sentence repetition deficits appear to be not as common as initially thought (Mesulam & Weintraub, 2014). More evidence from prospectively collected cohorts is needed to determine the frequency of sentence repetition deficits in lvPPA because particular sentence types are necessary to uncover the problem. It is likely that, as is often the case for operationalized diagnostic criteria of complex cognitive syndromes, sensitivity might come to the detriment of specificity.

TABLE 75.5 Features of the Logopenic Variant of PPA

I. Clinical diagnosis of logopenic variant PPA
Both of the following core features must be present:
1. Impaired single-word retrieval in spontaneous speech and naming
2. Impaired repetition of sentences and phrases
At least three of the following other features must be present:
1. Speech (phonologic) errors in spontaneous speech and naming
2. Spared single-word comprehension and object knowledge
3. Spared motor speech
4. Absence of frank agrammatism
II. Imaging-supported logopenic variant diagnosis
Both criteria must be present:
1. Clinical diagnosis of logopenic variant PPA
2. Imaging must show at least one of the following results:
a. Predominant left posterior perisylvian or parietal atrophy on MRI
b. Predominant left posterior perisylvian or parietal hypoperfusion or hypometabolism on SPECT or PET
III. Logopenic variant PPA with definite pathology
Clinical diagnosis (criterion 1 below) and either criterion 2 or 3 must be present:
1. Clinical diagnosis of logopenic variant PPA
2. Histopathologic evidence of a specific neurodegenerative pathology (e.g., AD, FTLD-tau, FTLD-TDP, other)
3. Presence of a known pathogenic mutation

AD, Alzheimer's disease; FTLD, frontotemporal lobar degeneration; MRI, magnetic resonance imaging; PET, positron emission tomography; SPECT, single proton emission computed tomography.
Source: *From Gorno-Tempini et al. (2011).*

75.4.3 Anatomy and Imaging

MRI studies in lvPPA demonstrate volume loss in the left hemisphere, specifically in the posterior temporal and inferior parietal regions (Figure 75.1). Detailed voxel-based morphometry (VBM) methods demonstrate grey matter loss in the angular gyrus, supramarginal gyrus posterior third of the middle temporal gyrus, and superior temporal gyrus, as well as some involvement of the left anterior hippocampus and precuneus. More generally, the atrophy can be said to predominantly involve the left temporoparietal junction (Gorno-Tempini et al., 2008; Rohrer et al., 2013). As suggested by the term "frontotemporal degeneration," the pattern of atrophy of lvPPA is decidedly unlike the other two forms of PPA, which are also considered subtypes of FTD.

The rate of whole brain atrophy has been shown to be greater in those with lvPPA compared with controls

(2.0% per year compared with 0.3% in controls). The rate of volume loss is significantly greater in the left hemisphere compared with the right (2.3% compared with 1.6%). Time leads to involvement of more anterior regions, including caudate, hippocampus, and medial parietal lobes (Rohrer et al., 2013). DTI suggests white matter damage in the left temporo-parietal component of the arcuate fasciculus (Galantucci et al., 2011) (Figure 75.2). These white matter deficits may be less severe than in other forms of PPA, however, again in relation to the fact that lvPPA is most often caused by AD. However, other studies have found more widespread white matter changes, including the whole arcuate and the inferior longitudinal fasciculus (Agosta et al., 2012).

FDG-PET studies have confirmed hypometabolism in the left temporoparietal lobe. Similar patterns have been described in early-onset AD. Consistent with those findings, binding with the nuclear ligand tracer 11C-labeled Pittsburgh Compound-B is more likely to show evidence of cortical amyloid in lvPPA compared with controls (REFS). There is little correlation between location of amyloid, however, and the patient's symptoms. Although lvPPA symptoms do not correlate with location of amyloid plaques, as evidenced on PiB-PET imaging and pathology, the temporoparietal atrophy in lvPPA correlates with increased neurofibrillary tangles in that region.

75.4.4 Histopathology and Genetics

Even in some of the earliest detailed descriptions of the logopenic subgroup, the potential for causative Alzheimer pathology was apparent. As mentioned, Gorno-Tempini and colleagues discovered a posterior pattern of volume loss (Gorno-Tempini et al., 2004). Moreover, they found a 67% frequency of the APOE4 haplotype, commonly associated with Alzheimer pathology, in the logopenic group, compared with 0% in the semantic group and only 20% of svPPA (Gorno-Tempini et al., 2004). Although this connection between ApoE4 and lvPPA is not always found in different research groups, the association between Alzheimer pathology and lvPPA was subsequently demonstrated by several groups by pathological examination (Gefen et al., 2012; Josephs, Dickson, et al., 2013; Mesulam et al., 2008). In addition, cases of lvPPA have been associated with argyrophilic thorny astrocyte clusters on microscopic examination, a finding that may be associated with asymmetric and focal presentations of degenerative disease (Munoz, Woulfe, & Kertesz, 2007).

Retrospective studies of patients with PPA who were later found to have AD found significant logopenic symptoms (Rohrer, Rossor, & Warren, 2012). Other supporting evidence includes cortical amyloid binding in nuclear imaging studies and cerebrospinal fluid studies consistent with a diagnosis of Alzheimer disease in logopenic patients (Gil-Navarro et al., 2013; Gorno-Tempini et al., 2004; Rohrer, Rossor, & Warren, 2010). Thus, lvPPA is more correctly classified as an atypical presentation of early-age-of-onset AD than a variant of the FTD-spectrum diseases (Migliaccio et al., 2012; Rabinovici et al., 2008).

The relationship between lvPPA and Alzheimer pathology is not absolute, however. Other pathological processes may also lead to syndromes that at least resemble lvPPA, including Creutzfeldt-Jakob disease and Lewy body dementia (Martory et al., 2012; Teichmann, Migliaccio, Kas, & Dubois, 2013). VBM studies suggest that there is more posterior-superior temporal atrophy in lvPPA, whereas non-AD etiologies of lvPPA have more severe perisylvian atrophy (Hu et al., 2010).

Furthermore, in some cases of lvPPA, atrophy may extend into areas commonly associated with other PPA variants, leading to a blurring of clinical syndromes that can start to resemble a global aphasia (Rogalski, Cobia, Harrison, Wieneke, Thompson, et al., 2011; Rogalski, Cobia, Harrison, Wieneke, Weintraub, et al., 2011). Rohrer and colleagues discovered that 2 out of 9 patients with lvPPA had a mutation in the progranulin gene (*PGN*). Progranulin has been associated with frontotemporal dementia and is thought to predispose toward asymmetric cortical atrophy such as that in PPA. Although those who had *PGN* mutations had symptoms that best fit lvPPA, they also had severe single word comprehension deficits, severe anomia, and irregular word reading that was more suggestive of svPPA. VBM analysis revealed more anterior temporal atrophy in these patients than other forms of lvPPA, explaining the semantic deficits and highlighting that the syndrome is really logopenic "plus" (Rohrer, Crutch, et al., 2010).

75.5 FUTURE DIRECTIONS IN PPA

The classification scheme for PPA devised in 2011 has given clarity to diagnosing these disorders, but intentionally leaves room for clinical judgment. Objective classification based on test cutoffs, for example, is difficult and perhaps inappropriate in such complex cognitive syndromes in which deficient performance can be caused by different underlying mechanisms. The criteria are best applied to patients at early disease stages and for sporadic (nongenetic) cases of PPA.

The current classification scheme is not intended to limit the variety of presentations of PPA. One report by Wicklund and colleagues suggests that while almost 70% of patients are assigned to an existing category, 31% did not meet the criteria of any of the categories (Wicklund et al., 2014). Some patients appear to be unclassifiable, and others fulfill criteria for more than one subtype (Harris et al., 2013; Mesulam & Weintraub, 2014; Sajjadi, Patterson, Arnold, Watson, & Nestor, 2012). In fact, each patient will present slightly differently, and it may be possible to further subtype these clinical presentations that relate to the patterns of underlying atrophy. For example, some have suggested that there is a motor speech disorder that impacts speech more than underlying language processes, and a primary progressive AOS diagnosis is more appropriate (Josephs et al., 2010; Josephs, Duffy, et al., 2013). Some have suggested that PGN mutations or TDP-C with corticospinal tract involvement might involve distinct diagnoses.

Although the current criteria allow some prediction of different probabilities of underlying pathology, different histopathologies sometimes still appear clinically indistinguishable within the same subgroup of PPA. Biomarkers assist *in vivo* molecular diagnosis and will increase our accuracy in predicting pathology. Dedicated PET scans will soon be available that permit tau imaging in a manner currently performed with amyloid, which will allow further distinction between TDP-43 and tau pathology (Maruyama et al., 2013).

Accurate diagnoses are of more than academic interest—the hope is that they will permit for better treatments. Unfortunately, treatments are still elusive. Patient care in PPA is currently symptomatic, meaning that no treatment currently addresses the underlying neurodegeneration. A knowledgeable speech therapist can guide the patient toward useful communication techniques and devices, and small studies indicate that early language rehabilitation has proven useful when different strategies are applied for each variant (Beeson et al., 2011; Henry et al., 2013). For those with word-retrieval deficits, focusing on personally relevant words may provide sufficient relief to merit the time and effort involved with such therapy (Henry et al., 2013). In cases of Alzheimer disease, a cholinesterase inhibitor such as galantamine or donepezil may be of some benefit (Kertesz et al., 2008). Preliminary benefits of techniques such as transcranial magnetic stimulation have been described, at least for nfvPPA (Cotelli et al., 2014).

PPA is still relatively rare, and it can be difficult to connect patients with appropriate resources. Web sites such as the international PPA connection (ppaconnection.org) may offer opportunities for patients and families to access information about the disorder and any clinical trials that may become available.

References

Adlam, A. L., Patterson, K., Rogers, T. T., Nestor, P. J., Salmond, C. H., Acosta-Cabronero, J., et al. (2006). Semantic dementia and fluent primary progressive aphasia: Two sides of the same coin? *Brain: A Journal of Neurology*, *129*(Pt. 11), 3066–3080.

Agosta, F., Scola, E., Canu, E., Marcone, A., Magnani, G., Sarro, L., et al. (2012). White matter damage in frontotemporal lobar degeneration spectrum. *Cerebral Cortex*, *22*(12), 2705–2714.

Ash, S., Moore, P., Vesely, L., Gunawardena, D., McMillan, C., Anderson, C., et al. (2009). Non-fluent speech in frontotemporal lobar degeneration. *Journal of Neurolinguistics*, *22*(4), 370–383.

Baddeley, A. (1988). Cognitive psychology and human memory. *Trends in Neurosciences*, *11*(4), 176–181.

Beeson, P. M., King, R. M., Bonakdarpour, B., Henry, M. L., Cho, H., & Rapcsak, S. Z. (2011). Positive effects of language treatment for the logopenic variant of primary progressive aphasia. *Journal of Molecular Neuroscience*, *45*(3), 724–736.

Borroni, B., Brambati, S. M., Agosti, C., Gipponi, S., Bellelli, G., Gasparotti, R., et al. (2007). Evidence of white matter changes on diffusion tensor imaging in frontotemporal dementia. *Archives of Neurology*, *64*(2), 246–251.

Brambati, S. M., Ogar, J., Neuhaus, J., Miller, B. L., & Gorno-Tempini, M. L. (2009). Reading disorders in primary progressive aphasia: A behavioral and neuroimaging study. *Neuropsychologia*, *47*(8–9), 1893–1900.

Brambati, S. M., Rankin, K. P., Narvid, J., Seeley, W. W., Dean, D., Rosen, H. J., et al. (2009). Atrophy progression in semantic dementia with asymmetric temporal involvement: A tensor-based morphometry study. *Neurobiology of Aging*, *30*(1), 103–111.

Broca, P. (1861). Perte de la parole: Ramollissement chronique et destruction partielle du lobe anterieur gauche du cerveau. *Bulletins de la Societe d'anthropologie*, *2*, 235–238.

Caso, F., Gesierich, B., Henry, M., Sidhu, M., LaMarre, A., Babiak, M., et al. (2013). Nonfluent/agrammatic PPA with *in-vivo* cortical amyloidosis and Pick's disease pathology. *Behavioural Neurology*, *26*(1–2), 95–106.

Chan, D., Fox, N. C., Scahill, R. I., Crum, W. R., Whitwell, J. L., Leschziner, G., et al. (2001). Patterns of temporal lobe atrophy in semantic dementia and Alzheimer's disease. *Annals of Neurology*, *49*(4), 433–442.

Chare, L., Hodges, J. R., Leyton, C. E., McGinley, C., Tan, R. H., Kril, J. J., et al. (2014). New criteria for frontotemporal dementia syndromes: Clinical and pathological diagnostic implications. *Journal of Neurology, Neurosurgery, and Psychiatry*, *85*(8), 865–870.

Compston, A. (2008). On aphasia due to atrophy of the cerebral convolutions. By Dr G. Mingazzini. (1914), Professor of Neuropathology at the Royal University of Rome. Brain 36, 493–524. *Brain: A Journal of Neurology*, *131*(3), 600–603.

Cooke, A., DeVita, C., Gee, J., Alsop, D., Detre, J., Chen, W., et al. (2003). Neural basis for sentence comprehension deficits in frontotemporal dementia. *Brain and Language*, *85*(2), 211–221.

Cotelli, M., Manenti, R., Petesi, M., Brambilla, M., Cosseddu, M., Zanetti, O., et al. (2014). Treatment of primary progressive aphasias by transcranial direct current stimulation combined with language training. *Journal of Alzheimer's Disease*, *39*(4), 799–808.

Croot, K., Ballard, K., Leyton, C. E., & Hodges, J. R. (2012). Apraxia of speech and phonological errors in the diagnosis of nonfluent/agrammatic and logopenic variants of primary progressive aphasia. *Journal of Speech, Language, and Hearing Research*, *55*(5), S1562–S1572.

Davies, R. R., Graham, K. S., Xuereb, J. H., Williams, G. B., & Hodges, J. R. (2004). The human perirhinal cortex and semantic memory. *The European Journal of Neuroscience*, *20*(9), 2441–2446.

Davies, R. R., Hodges, J. R., Kril, J. J., Patterson, K., Halliday, G. M., & Xuereb, J. H. (2005). The pathological basis of semantic dementia. *Brain: A Journal of Neurology*, *128*(Pt. 9), 1984–1995.

Dronkers, N. F., Plaisant, O., Iba-Zizen, M. T., & Cabanis, E. A. (2007). Paul Broca's historic cases: High resolution MR imaging of the brains of Leborgne and Lelong. *Brain: A Journal of Neurology, 130*(Pt. 5), 1432–1441.

Duffy, J. R., Peach, R. K., & Strand, E. A. (2007). Progressive apraxia of speech as a sign of motor neuron disease. *American Journal of Speech-Language Pathology/American Speech-Language-Hearing Association, 16* (3), 198–208.

Duval, C., Bejanin, A., Piolino, P., Laisney, M., de La Sayette, V., Belliard, S., et al. (2012). Theory of mind impairments in patients with semantic dementia. *Brain: A Journal of Neurology, 135*(Pt. 1), 228–241.

Eslinger, P. J., Dennis, K., Moore, P., Antani, S., Hauck, R., & Grossman, M. (2005). Metacognitive deficits in frontotemporal dementia. *Journal of Neurology, Neurosurgery, and Psychiatry, 76* (12), 1630–1635.

Evans, J. J., Heggs, A. J., Antoun, N., & Hodges, J. R. (1995). Progressive prosopagnosia associated with selective right temporal lobe atrophy. A new syndrome? *Brain: A Journal of Neurology, 118*(Pt. 1), 1–13.

Forman, M. S., Farmer, J., Johnson, J. K., Clark, C. M., Arnold, S. E., Coslett, H. B., et al. (2006). Frontotemporal dementia: Clinicopathological correlations. *Annals of Neurology, 59*(6), 952–962.

Foxe, D. G., Irish, M., Hodges, J. R., & Piguet, O. (2013). Verbal and visuospatial span in logopenic progressive aphasia and Alzheimer's disease. *Journal of the International Neuropsychological Society, 19*(3), 247–253.

Gainotti, G. (2006). Anatomical functional and cognitive determinants of semantic memory disorders. *Neuroscience and Biobehavioral Reviews, 30*(5), 577–594.

Galantucci, S., Tartaglia, M. C., Wilson, S. M., Henry, M. L., Filippi, M., Agosta, F., et al. (2011). White matter damage in primary progressive aphasias: A diffusion tensor tractography study. *Brain: A Journal of Neurology, 134*(Pt. 10), 3011–3029.

Gefen, T., Gasho, K., Rademaker, A., Lalehzari, M., Weintraub, S., Rogalski, E., et al. (2012). Clinically concordant variations of Alzheimer pathology in aphasic versus amnestic dementia. *Brain: A Journal of Neurology, 135*(Pt. 5), 1554–1565.

Gil-Navarro, S., Llado, A., Rami, L., Castellvi, M., Bosch, B., Bargallo, N., et al. (2013). Neuroimaging and biochemical markers in the three variants of primary progressive aphasia. *Dementia and Geriatric Cognitive Disorders, 35*(1–2), 106–117.

Goldman, J. S., Farmer, J. M., Wood, E. M., Johnson, J. K., Boxer, A., Neuhaus, J., et al. (2005). Comparison of family histories in FTLD subtypes and related tauopathies. *Neurology, 65*(11), 1817–1819.

Goldman, J. S., Rademakers, R., Huey, E. D., Boxer, A. L., Mayeux, R., Miller, B. L., et al. (2011). An algorithm for genetic testing of frontotemporal lobar degeneration. *Neurology, 76*(5), 475–483.

Gorno-Tempini, M. L., Brambati, S. M., Ginex, V., Ogar, J., Dronkers, N. F., Marcone, A., et al. (2008). The logopenic/phonological variant of primary progressive aphasia. *Neurology, 71*(16), 1227–1234.

Gorno-Tempini, M. L., Dronkers, N. F., Rankin, K. P., Ogar, J. M., Phengrasamy, L., Rosen, H. J., et al. (2004). Cognition and anatomy in three variants of primary progressive aphasia. *Annals of Neurology, 55*(3), 335–346.

Gorno-Tempini, M. L., Hillis, A. E., Weintraub, S., Kertesz, A., Mendez, M., Cappa, S. F., et al. (2011). Classification of primary progressive aphasia and its variants. *Neurology, 76*(11), 1006–10014.

Gorno-Tempini, M. L., Rankin, K. P., Woolley, J. D., Rosen, H. J., Phengrasamy, L., & Miller, B. L. (2004). Cognitive and behavioral profile in a case of right anterior temporal lobe neurodegeneration. *Cortex: A Journal Devoted to the Study of the Nervous System and Behavior, 40*(4–5), 631–644.

Graham, K., Patterson, K., & Hodges, J. (1995). Progressive pure anomia: Insufficient activation of phonology by meaning. *Neurocase, 1*, 25–38.

Grossman, M. (2010). Primary progressive aphasia: Clinicopathological correlations. *Nature Reviews Neurology, 6*(2), 88–97.

Grossman, M. (2012). The non-fluent/agrammatic variant of primary progressive aphasia. *Lancet Neurology, 11*(6), 545–555.

Grossman, M., Mickanin, J., Onishi, K., Hughes, E., D'Esposito, M., Ding, X. S., et al. (1996). Progressive nonfluent aphasia: Language, cognitive, and PET measures contrasted with probable Alzheimer's disease. *Journal of Cognitive Neuroscience, 8*(2), 135–154.

Grossman, M., Powers, J., Ash, S., McMillan, C., Burkholder, L., Irwin, D., et al. (2013). Disruption of large-scale neural networks in non-fluent/agrammatic variant primary progressive aphasia associated with frontotemporal degeneration pathology. *Brain and Language, 127*(2), 106–120.

Gunawardena, D., Ash, S., McMillan, C., Avants, B., Gee, J., & Grossman, M. (2010). Why are patients with progressive nonfluent aphasia nonfluent? *Neurology, 75*(7), 588–594.

Guo, C. C., Gorno-Tempini, M. L., Gesierich, B., Henry, M., Trujillo, A., Shany-Ur, T., et al. (2013). Anterior temporal lobe degeneration produces widespread network-driven dysfunction. *Brain: A Journal of Neurology, 136*(Pt. 10), 2979–2991.

Harris, J. M., Gall, C., Thompson, J. C., Richardson, A. M., Neary, D., du Plessis, D., et al. (2013). Classification and pathology of primary progressive aphasia. *Neurology, 81*(21), 1832–1839.

Heath, P. D., Kennedy, P., & Kapur, N. (1983). Slowly progressive aphasia without generalized dementia. *Annals of Neurology, 13*(6), 687–688.

Henry, M. L., Rising, K., Demarco, A. T., Miller, B. L., Gorno-Tempini, M. L., & Beeson, P. M. (2013). Examining the value of lexical retrieval treatment in primary progressive aphasia: Two positive cases. *Brain and Language, 127*(2), 145–156.

Hodges, J. R., Davies, R. R., & Patterson, K. (2009). Semantic dementia. In B. L. Miller, & B. F. Boeve (Eds.), *The behaivoral neurology of dementia* (pp. 264–278). New York: Cambridge University Press.

Hodges, J. R., Mitchell, J., Dawson, K., Spillantini, M. G., Xuereb, J. H., McMonagle, P., et al. (2010). Semantic dementia: Demography, familial factors and survival in a consecutive series of 100 cases. *Brain: A Journal of Neurology, 133*(Pt. 1), 300–306.

Hodges, J. R., Patterson, K., Oxbury, S., & Funnell, E. (1992). Semantic dementia. Progressive fluent aphasia with temporal lobe atrophy. *Brain: A Journal of Neurology, 115*(Pt. 6), 1783–1806.

Horel, J., Keating, E., & Misantone, L. (1975). Partial Kluver-Bucy syndrome produced by destroying temporal neocortex or amygdala. *Brain Research, 94*, 347–359.

Howard, D., & Patterson, K. (1992). *Pyramids and palm trees: A test of semantic access from pictures and words*. Bury St Edmunds, UK: Thames Valley Test Company.

Hu, W. T., McMillan, C., Libon, D., Leight, S., Forman, M., Lee, V. M., et al. (2010). Multimodal predictors for Alzheimer disease in nonfluent primary progressive aphasia. *Neurology, 75*(7), 595–602.

Imaizumi, S., Mori, K., Kiritani, S., Kawashima, R., Sugiura, M., Fukuda, H., et al. (1997). Vocal identification of speaker and emotion activates different brain regions. *Neuroreport, 8*(12), 2809–2812.

Irish, M., Hodges, J. R., & Piguet, O. (2014). Right anterior temporal lobe dysfunction underlies theory of mind impairments in semantic dementia. *Brain: A Journal of Neurology, 137*(Pt. 4), 1241–1253.

Johnson, J. K., Diehl, J., Mendez, M. F., Neuhaus, J., Shapira, J. S., Forman, M., et al. (2005). Frontotemporal lobar degeneration: Demographic characteristics of 353 patients. *Archives of Neurology, 62*(6), 925–930.

Josephs, K. A., Dickson, D. W., Murray, M. E., Senjem, M. L., Parisi, J. E., Petersen, R. C., et al. (2013). Quantitative neurofibrillary tangle density and brain volumetric MRI analyses in Alzheimer's disease presenting as logopenic progressive aphasia. *Brain and Language, 127*(2), 127–134.

Josephs, K. A., Duffy, J. R., Fossett, T. R., Strand, E. A., Claassen, D. O., Whitwell, J. L., et al. (2010). Fluorodeoxyglucose F18 positron emission tomography in progressive apraxia of speech and primary progressive aphasia variants. *Archives of Neurology*, *67*(5), 596–605.

Josephs, K. A., Duffy, J. R., Strand, E. A., Machulda, M. M., Senjem, M. L., Lowe, V. J., et al. (2013). Syndromes dominated by apraxia of speech show distinct characteristics from agrammatic PPA. *Neurology*, *81*(4), 337–345.

Josephs, K. A., Duffy, J. R., Strand, E. A., Whitwell, J. L., Layton, K. F., Parisi, J. E., et al. (2006). Clinicopathological and imaging correlates of progressive aphasia and apraxia of speech. *Brain: A Journal of Neurology*, *129*(Pt. 6), 1385–1398.

Josephs, K. A., Stroh, A., Dugger, B., & Dickson, D. W. (2009). Evaluation of subcortical pathology and clinical correlations in FTLD-U subtypes. *Acta Neuropathologica*, *118*(3), 349–358.

Josephs, K. A., Whitwell, J. L., Murray, M. E., Parisi, J. E., Graff-Radford, N. R., Knopman, D. S., et al. (2013). Corticospinal tract degeneration associated with TDP-43 type C pathology and semantic dementia. *Brain: A Journal of Neurology*, *136*(Pt. 2), 455–470.

Kertesz, A., Davidson, W., McCabe, P., Takagi, K., & Munoz, D. (2003). Primary progressive aphasia: Diagnosis, varieties, evolution. *Journal of International Neuropsychological Society*, *9*(5), 710–719.

Kertesz, A., Morlog, D., Light, M., Blair, M., Davidson, W., Jesso, S., et al. (2008). Galantamine in frontotemporal dementia and primary progressive aphasia. *Dementia and Geriatric Cognitive Disorders*, *25*(2), 178–185.

Kirshner, H. S. (2012). Primary progressive aphasia and Alzheimer's disease: Brief history, recent evidence. *Current Neurology and Neuroscience Reports*, *12*(6), 709–714.

Lambon Ralph, M. A. (2014). Neurocognitive insights on conceptual knowledge and its breakdown. *Philosophical Transactions of the Royal Society of London Series B, Biological Sciences*, *369*(1634), 20120392.

Leyton, C. E., Piguet, O., Savage, S., Burrell, J., & Hodges, J. R. (2012). The neural basis of logopenic progressive aphasia. *Journal of Alzheimer's Disease*, *32*(4), 1051–1059.

Machulda, M. M., Whitwell, J. L., Duffy, J. R., Strand, E. A., Dean, P. M., Senjem, M. L., et al. (2013). Identification of an atypical variant of logopenic progressive aphasia. *Brain and Language*, *127*(2), 139–144.

Mahoney, C. J., Malone, I. B., Ridgway, G. R., Buckley, A. H., Downey, L. E., Golden, H. L., et al. (2013). White matter tract signatures of the progressive aphasias. *Neurobiology of Aging*, *34*(6), 1687–1699.

Mandelli, M. L., Caverzasi, E., Binney, R. J., Henry, M. L., Lobach, I., Block, N., et al. (2014). Frontal white matter tracts sustaining speech production in primary progressive aphasia. *Journal of Neuroscience*, *34*(29), 9754–9767.

Martory, M. D., Roth, S., Lovblad, K. O., Neumann, M., Lobrinus, J. A., & Assal, F. (2012). Creutzfeldt-Jakob disease revealed by a logopenic variant of primary progressive aphasia. *European Neurology*, *67*(6), 360–362.

Maruyama, M., Shimada, H., Suhara, T., Shinotoh, H., Ji, B., Maeda, J., et al. (2013). Imaging of tau pathology in a tauopathy mouse model and in Alzheimer patients compared to normal controls. *Neuron*, *79*(6), 1094–1108.

Mesulam, M., Weintraub, S., Parrish, T., & Gitelman, D. (2005). Primary progressive aphasia: Reversed asymmetry of atrophy and right hemisphere language dominance. *Neurology*, *64*(3), 556–557.

Mesulam, M., Wicklund, A., Johnson, N., Rogalski, E., Leger, G. C., Rademaker, A., et al. (2008). Alzheimer and frontotemporal pathology in subsets of primary progressive aphasia. *Annals of Neurology*, *63*(6), 709–719.

Mesulam, M., Wieneke, C., Rogalski, E., Cobia, D., Thompson, C., & Weintraub, S. (2009). Quantitative template for subtyping primary progressive aphasia. *Archives of Neurology*, *66*(12), 1545–1551.

Mesulam, M.-M. (1982). Slowly progressive aphasia without generalized dementia. *Annals of Neurology*, *11*(6), 592–598.

Mesulam, M. M. (1987). Primary progressive aphasia—differentiation from Alzheimer's disease. *Annals of Neurology*, *22*(4), 533–534.

Mesulam, M. M. (2001). Primary progressive aphasia. *Annals of Neurology*, *49*(4), 425–432.

Mesulam, M. M. (2003). Primary progressive aphasia—a language-based dementia. *The New England Journal of Medicine*, *349*(16), 1535–1542.

Mesulam, M. M. (2007). Primary progressive aphasia: A 25-year retrospective. *Alzheimer Disease & Associated Disorders*, *21*(4), S8–S11.

Mesulam, M. M. (2013). Primary progressive aphasia and the language network: The 2013 H. Houston Merritt Lecture. *Neurology*, *81*(5), 456–462.

Mesulam, M. M., & Weintraub, S. (2014). Is it time to revisit the classification guidelines for primary progressive aphasia? *Neurology*, *82*(13), 1108–1109.

Mesulam, M. M., Wieneke, C., Hurley, R., Rademaker, A., Thompson, C. K., Weintraub, S., et al. (2013). Words and objects at the tip of the left temporal lobe in primary progressive aphasia. *Brain: A Journal of Neurology*, *136*(Pt. 2), 601–618.

Mesulam, M. M., Wieneke, C., Thompson, C., Rogalski, E., & Weintraub, S. (2012). Quantitative classification of primary progressive aphasia at early and mild impairment stages. *Brain: A Journal of Neurology*, *135*(Pt. 5), 1537–1553.

Migliaccio, R., Agosta, F., Possin, K. L., Rabinovici, G. D., Miller, B. L., & Gorno-Tempini, M. L. (2012). White matter atrophy in Alzheimer's disease variants. Alzheimer's & dementia. *The Journal of the Alzheimer's Association*, *8*(5 Suppl.), S78–S87, e1-2.

Miller, Z. A., Rankin, K. P., Graff-Radford, N. R., Takada, L. T., Sturm, V. E., Cleveland, C. M., et al. (2013). TDP-43 frontotemporal lobar degeneration and autoimmune disease. *J Neurol Neurosurg Psychiatry*, *84*(9), 956–962. Available from: http://dx.doi.org/10.10.1136/jnnp-2012-304644.

Miller, Z. A., Mandelli, M. L., Rankin, K. P., Henry, M. L., Babiak, M. C., Frazier, D. T., et al. (2013). Handedness and language learning disability differentially distribute in progressive aphasia variants. *Brain: A Journal of Neurology*, *136*(Pt. 11), 3461–3473.

Munoz, D. G., Woulfe, J., & Kertesz, A. (2007). Argyrophilic thorny astrocyte clusters in association with Alzheimer's disease pathology in possible primary progressive aphasia. *Acta Neuropathologica*, *114*(4), 347–357.

Murray, R., Neumann, M., Forman, M. S., Farmer, J., Massimo, L., Rice, A., et al. (2007). Cognitive and motor assessment in autopsy-proven corticobasal degeneration. *Neurology*, *68*(16), 1274–1283.

Neary, D., Snowden, J., Gustafson, L., Passant, U., Stuss, D., Black, S., et al. (1998). Frontotemporal lobar degeneration: A consensus on clinical diagnostic criteria. *Neurology*, *51*, 1546–1554.

Nestor, P. J., Graham, N. L., Fryer, T. D., Williams, G. B., Patterson, K., & Hodges, J. R. (2003). Progressive non-fluent aphasia is associated with hypometabolism centred on the left anterior insula. *Brain: A Journal of Neurology*, *126*(Pt. 11), 2406–2418.

Nunnemann, S., Last, D., Schuster, T., Forstl, H., Kurz, A., & Diehl-Schmid, J. (2011). Survival in a German population with frontotemporal lobar degeneration. *Neuroepidemiology*, *37*(3–4), 160–165.

Ogar, J., Slama, H., Dronkers, N., Amici, S., & Gorno-Tempini, M. L. (2005). Apraxia of speech: An overview. *Neurocase*, *11*(6), 427–432.

Patterson, K., Nestor, P. J., & Rogers, T. T. (2007). Where do you know what you know? The representation of semantic knowledge in the human brain. *Nature Reviews Neuroscience*, *8*(12), 976–987.

Pereira, J. M., Williams, G. B., Acosta-Cabronero, J., Pengas, G., Spillantini, M. G., Xuereb, J. H., et al. (2009). Atrophy patterns in histologic vs clinical groupings of frontotemporal lobar degeneration. *Neurology, 72*(19), 1653–1660.

Pick, A. (1892). Uber die beziehungen der senilen hirnatropie zur aphasie. *Pragen Medizinischen Wochenschrift, 17*, 165–167.

Pressman, P., & Miller, B. (2014). Diagnosis and management of behavioral variant frontotemporal dementia. *Biological Psychiatry, 75*(7), 574–581.

Pressman, P., & Rabinovici, G. (2014). Alzheimer's Disease, In R. B. Baroff & M. J. Aminoff (Eds.), *Encyclopedia of the Neurological Sciences* (Vol. 1, pp. 122–127). New York: Elsevier.

Rabinovici, G. D., Jagust, W. J., Furst, A. J., Ogar, J. M., Racine, C. A., Mormino, E. C., et al. (2008). Abeta amyloid and glucose metabolism in three variants of primary progressive aphasia. *Annals of Neurology, 64*(4), 388–401.

Ralph, M. A., & Howard, D. (2000). Gogi aphasia or semantic dementia? Simulating and assessing poor verbal comprehension in a case of progressive fluent aphasia. *Cognitive Neuropsychology, 17*(5), 437–465.

Rogalski, E., Cobia, D., Harrison, T. M., Wieneke, C., Thompson, C. K., Weintraub, S., et al. (2011). Anatomy of language impairments in primary progressive aphasia. *The Journal of Neuroscience: The Official Journal of the Society for Neuroscience, 31*(9), 3344–3350.

Rogalski, E., Cobia, D., Harrison, T. M., Wieneke, C., Weintraub, S., & Mesulam, M. M. (2011). Progression of language decline and cortical atrophy in subtypes of primary progressive aphasia. *Neurology, 76*(21), 1804–1810.

Rogalski, E., Johnson, N., Weintraub, S., & Mesulam, M. (2008). Increased frequency of learning disability in patients with primary progressive aphasia and their first-degree relatives. *Archives of Neurology, 65*(2), 244–248.

Rogalski, E., Weintraub, S., & Mesulam, M. M. (2013). Are there susceptibility factors for primary progressive aphasia? *Brain and Language, 127*(2), 135–138

Rohrer, J. D., Caso, F., Mahoney, C., Henry, M., Rosen, H. J., Rabinovici, G., et al. (2013). Patterns of longitudinal brain atrophy in the logopenic variant of primary progressive aphasia. *Brain and Language, 127*(2), 121–126.

Rohrer, J. D., Crutch, S. J., Warrington, E. K., & Warren, J. D. (2010). Progranulin-associated primary progressive aphasia: A distinct phenotype? *Neuropsychologia, 48*(1), 288–297.

Rohrer, J. D., Geser, F., Zhou, J., Gennatas, E. D., Sidhu, M., Trojanowski, J. Q., et al. (2010). TDP-43 subtypes are associated with distinct atrophy patterns in frontotemporal dementia. *Neurology, 75*(24), 2204–2211.

Rohrer, J. D., Guerreiro, R., Vandrovcova, J., Uphill, J., Reiman, D., Beck, J., et al. (2009). The heritability and genetics of frontotemporal lobar degeneration. *Neurology, 73*(18), 1451–1456.

Rohrer, J. D., Knight, W. D., Warren, J. E., Fox, N. C., Rossor, M. N., & Warren, J. D. (2008). Word-finding difficulty: A clinical analysis of the progressive aphasias. *Brain: A Journal of Neurology, 131*(Pt. 1), 8–38.

Rohrer, J. D., Lashley, T., Schott, J. M., Warren, J. E., Mead, S., Isaacs, A. M., et al. (2011). Clinical and neuroanatomical signatures of tissue pathology in frontotemporal lobar degeneration. *Brain: A Journal of Neurology, 134*(Pt. 9), 2565–2581.

Rohrer, J. D., Rossor, M. N., & Warren, J. D. (2010). Syndromes of nonfluent primary progressive aphasia: A clinical and neurolinguistic analysis. *Neurology, 75*(7), 603–610.

Rohrer, J. D., Rossor, M. N., & Warren, J. D. (2012). Alzheimer's pathology in primary progressive aphasia. *Neurobiology of Aging, 33*(4), 744–752.

Rohrer, J. D., & Warren, J. D. (2010). Phenomenology and anatomy of abnormal behaviours in primary progressive aphasia. *Journal of the Neurological Sciences, 293*(1–2), 35–38.

Rosen, H. J., Perry, R. J., Murphy, J., Kramer, J. H., Mychack, P., Schuff, N., et al. (2002). Emotion comprehension in the temporal variant of frontotemporal dementia. *Brain: A Journal of Neurology, 125*(Pt. 10), 2286–2295.

Rosenfeld, M. (1909). Die partielle gorsshirnatrophie. *Journal fur Psychologie und Neurologie, 14*, 115–130.

Rosso, S. M., Donker Kaat, L., Baks, T., Joosse, M., de Koning, I., Pijnenburg, Y., et al. (2003). Frontotemporal dementia in The Netherlands: Patient characteristics and prevalence estimates from a population-based study. *Brain: A Journal of Neurology, 126* (Pt. 9), 2016–2022.

Sajjadi, S. A., Patterson, K., Arnold, R. J., Watson, P. C., & Nestor, P. J. (2012). Primary progressive aphasia: A tale of two syndromes and the rest. *Neurology, 78*(21), 1670–1677.

Sapolsky, D., Bakkour, A., Negreira, A., Nalipinski, P., Weintraub, S., Mesulam, M. M., et al. (2010). Cortical neuroanatomic correlates of symptom severity in primary progressive aphasia. *Neurology, 75*(4), 358–366.

Sasanuma, S., & Monoi, H. (1975). The syndrome of Gogi (word meaning) aphasia. Selective impairment of kanji processing. *Neurology, 25*(7), 627–632.

Schwindt, G. C., Graham, N. L., Rochon, E., Tang-Wai, D. F., Lobaugh, N. J., Chow, T. W., et al. (2013). Whole-brain white matter disruption in semantic and nonfluent variants of primary progressive aphasia. *Human Brain Mapping, 34*(4), 973–984.

Seeley, W. W., Bauer, A. M., Miller, B. L., Gorno-Tempini, M. L., Kramer, J. H., Weiner, M., et al. (2005). The natural history of temporal variant frontotemporal dementia. *Neurology, 64*(8), 1384–1390.

Seeley, W. W., Crawford, R. K., Zhou, J., Miller, B. L., & Greicius, M. D. (2009). Neurodegenerative diseases target large-scale human brain networks. *Neuron, 62*(1), 42–52.

Sepelyak, K., Crinion, J., Molitoris, J., Epstein-Peterson, Z., Bann, M., Davis, C., et al. (2011). Patterns of breakdown in spelling in primary progressive aphasia. *Cortex: A Journal Devoted to the Study of the Nervous System and Behavior, 47*(3), 342–352.

Shi, J., Shaw, C. L., Du Plessis, D., Richardson, A. M., Bailey, K. L., Julien, C., et al. (2005). Histopathological changes underlying frontotemporal lobar degeneration with clinicopathological correlation. *Acta Neuropathologica, 110*(5), 501–512.

Shim, H., Hurley, R. S., Rogalski, E., & Mesulam, M. M. (2012). Anatomic, clinical, and neuropsychological correlates of spelling errors in primary progressive aphasia. *Neuropsychologia, 50*(8), 1929–1935.

Snowden, J., Goulding, P., & Neary, D. (1989). Semantic dementia: A form of circumscribed cerebral atrophy. *Behavioural Neurology, 2* (3), 167–182.

Snowden, J. S., Thompson, J. C., Stopford, C. L., Richardson, A. M. T., Gerhard, A., Neary, D., et al. (2011). The clinical diagnosis of early-onset dementias: Diagnostic accuracy and clinicopathological relationships. *Brain: A Journal of Neurology, 134*(Pt. 9), 2478–2492.

Sobrido, M. J., Abu-Khalil, A., Weintraub, S., Johnson, N., Quinn, B., Cummings, J. L., et al. (2003). Possible association of the tau H1/H1 genotype with primary progressive aphasia. *Neurology, 60*(5), 862–864.

Stertz, G. (1926). Uber die Picksche atrophie. *Aeitschrift fur die Gesamte Neurologie und Psychiatrie, 101*, 729–747.

Teichmann, M., Migliaccio, R., Kas, A., & Dubois, B. (2013). Logopenic progressive aphasia beyond Alzheimer's—an evolution towards dementia with Lewy bodies. *Journal of Neurology, Neurosurgery, and Psychiatry, 84*(1), 113–114.

Vallar, G., Di Betta, A. M., & Silveri, M. C. (1997). The phonological short-term store-rehearsal system: Patterns of impairment and neural correlates. *Neuropsychologia, 35*(6), 795–812.

Warrington, E. K. (1975). The selective impairment of semantic memory. *The Quarterly Journal of Experimental Psychology, 27*(4), 635–657.

Weintraub, S., Mesulam, M. M., Wieneke, C., Rademaker, A., Rogalski, E. J., & Thompson, C. K. (2009). The northwestern anagram test: Measuring sentence production in primary progressive aphasia. *American Journal of Alzheimer's Disease and Other Dementias, 24*(5), 408–416.

Weintraub, S., Rubin, N. P., & Mesulam, M. M. (1990). Primary progressive aphasia. Longitudinal course, neuropsychological profile, and language features. *Archives of Neurology, 47*(12), 1329–1335.

Wernicke, C. (1874). *Der aphasische symptomencomplex*. Breslau: Kohn and Weigert.

Westbury, C., & Bub, D. (1997). Primary progressive aphasia: A review of 112 cases. *Brain and Language, 60*(3), 381–406.

Whitwell, J. L., Avula, R., Senjem, M. L., Kantarci, K., Weigand, S. D., Samikoglu, A., et al. (2010). Gray and white matter water diffusion in the syndromic variants of frontotemporal dementia. *Neurology, 74*(16), 1279–1287.

Wicklund, A. H., Johnson, N., Rademaker, A., Weitner, B. B., & Weintraub, S. (2007). Profiles of decline in activities of daily living in non-Alzheimer dementia. *Alzheimer Disease and Associated Disorders, 21*(1), 8–13.

Wicklund, M. R., Duffy, J. R., Strand, E. A., Machulda, M. M., Whitwell, J. L., & Josephs, K. A. (2014). Quantitative application of the primary progressive aphasia consensus criteria. *Neurology, 82*(13), 1119–1126.

Wilson, S. M., Dronkers, N. F., Ogar, J. M., Jang, J., Growdon, M. E., Agosta, F., et al. (2010). Neural correlates of syntactic processing in the nonfluent variant of primary progressive aphasia. *Journal of Neuroscience, 30*(50), 16845–16854.

Wilson, S. M., Henry, M. L., Besbris, M., Ogar, J. M., Dronkers, N. F., Jarrold, W., et al. (2010). Connected speech production in three variants of primary progressive aphasia. *Brain: A Journal of Neurology, 133*(Pt. 7), 2069–2088.

Zadikoff, C., & Lang, A. E. (2005). Apraxia in movement disorders. *Brain: A Journal of Neurology, 128*(Pt. 7), 1480–1497.

Zamboni, G., Grafman, J., Krueger, F., Knutson, K. M., & Huey, E. D. (2010). Anosognosia for behavioral disturbances in frontotemporal dementia and corticobasal syndrome: A voxel-based morphometry study. *Dementia & Geriatric Cognitive Disorders, 29*(1), 88–96.

CHAPTER

76

The Declarative/Procedural Model: A Neurobiological Model of Language Learning, Knowledge, and Use

Michael T. Ullman

Brain and Language Laboratory, Department of Neuroscience, Georgetown University, Washington, DC, USA

Not surprisingly, most research on the neurobiology of language has focused on language itself. Yet even after decades of investigation, we know much less about the neurobiological correlates of language than we do about the neurobiology of many other domains, such as vision, motor function, and memory. This relative lack of progress is likely due not only to the complexity of language and to a lack of animal models, but also to an overly narrow "isolationist" approach to language research. A complementary approach of examining links between language and other domains and their neurobiological substrates could significantly advance our understanding of the neurobiology of language, particularly if the neurobiology of these other domains is well-understood. Such an approach would likely be very powerful in that it could generate novel well-motivated predictions about language based on our independent knowledge of these other domains.

There is, in fact, no clear evidence that the neurobiological bases of language underlie language alone, that is, that they are domain-specific (Ullman, Lum, & Conti-Ramsden, 2014). On the contrary, we should expect language to depend heavily on neurobiological substrates that also subserve(d) other domains. In evolution, mechanisms and structures are constantly being reused for new purposes. For example, fins evolved into limbs, limbs into hands and wings, and scales into feathers (Woltering, Noordermeer, Leleu, & Duboule, 2014). Such co-optation of a given substrate for new functions takes place not only phylogenetically (evolutionarily) but also ontogenetically (developmentally). For example, reading likely depends on brain circuitry that is co-opted for this function during learning and development. A given structure can even be co-opted for new functions without any further changes in the underlying mechanism (this has been referred to as exaptation) (Gould & Vrba, 1982). For example, certain species of heron spread their wings to provide shade so they can better see their prey.

Therefore, language should depend importantly on previously existing neurobiological systems, whether or not these systems have subsequently become further specialized for language, either phylogenetically or ontogenetically. I will call this the *co-optation hypothesis of language*. Various neurobiological systems may be good candidates for such co-optation, including working memory (Caplan & Waters, 1999; Just & Carpenter, 1992), and dorsal and ventral stream processing (Bornkessel-Schlesewsky & Schlesewsky, 2013; Hickok & Poeppel, 2007; Petrides & Pandya, 2009).

Two learning and memory systems in the brain—declarative memory and procedural memory—are also excellent candidates. Most of language must be learned, whether or not there are innately specified aspects of this domain. Moreover, declarative and procedural memory seem to be the two most important learning and memory systems in the mammalian brain, including in humans, in terms of the range of domains, tasks, and functions that they underlie (Eichenbaum, 2012; Mishkin, Malamut, & Bachevalier, 1984; Squire & Wixted, 2011). The declarative/procedural (DP) model simply posits that these systems should therefore play wide-ranging roles in language learning, knowledge, and use. That is, the DP model posits that these two memory systems have been co-opted for language—whether or not they have become further specialized phylogenetically or ontogenetically for this domain.

Neurobiology of Language. DOI: http://dx.doi.org/10.1016/B978-0-12-407794-2.00076-6

Importantly, these memory systems have been well-studied both in humans and in nonhuman animals, and are thus quite well-understood at many levels—including their behavioral, computational, neuroanatomical, electrophysiological, cellular, biochemical, and genetic correlates. Many of these levels are far better understood for the two systems than for language. Because the posited co-optation of the memory systems for language leads to the expectation that the systems play similar roles in language as in other domains (although not necessarily identical roles, in part because the systems may have become further specialized for language), our understanding of the memory systems should generate a wide range of specific predictions for language. Crucially, these predictions are generated independently from the study of language itself, and are well-motivated from this independent knowledge. Moreover, many of them are likely to be novel because there would often be no reason to make such predictions based on the more limited study of language alone. Finally, linking language to the memory systems should not only generate new predictions, but may also help to account for already-observed language phenomena for which no good explanation independent of language currently exists.

Here, I first present an overview of the two memory systems, whose neurobiological and other correlates constitute the foundation of our predictions. Next, I present key predictions regarding the roles of these systems in language, with a focus on functional neuroanatomy. Then, I summarize a range of evidence testing these predictions. Converging evidence from multiple methodological approaches provides support for many of the predictions, thus supporting the DP model as well as the co-optation hypothesis more generally, and advancing our understanding of language and its underlying neurobiology. Note that the purpose of this chapter is to present the theory and its predictions; therefore, I focus on the motivating background (the memory systems) and the predictions, and more briefly summarize the evidence.

76.1 THE MEMORY SYSTEMS

Here, I provide an overview of the declarative and procedural memory systems, and discuss interactions between the two systems as well as with other neurocognitive systems. Note that the declarative and procedural memory systems refer here to the entire neurocognitive systems involved in the learning, representation, and use of the relevant knowledge, not just to those portions underlying learning and consolidating new knowledge, which is how some researchers refer to the systems. For additional information on the memory systems, see Stark and Stark chapter in this book (Chapter 67).

76.1.1 The Declarative Memory System

The declarative memory brain system is quite well-understood (Cabeza & Moscovitch, 2013; Eichenbaum, 2012; Eichenbaum, Sauvage, Fortin, Komorowski, & Lipton, 2012; Henke, 2010; Mishkin et al., 1984; Squire & Wixted, 2011; Stark and Stark chapter in this book; Ullman, 2004). Briefly, the hippocampus and other medial temporal lobe (MTL) structures are critical for learning and consolidating new knowledge that depends on this system, although ultimately the storage of this knowledge depends largely on neocortical regions, particularly in the temporal lobes.

Within the MTL, evidence from human and animal studies suggests that the hippocampus underlies the rapid linking (binding, associating) of different bits of knowledge or experience across multiple domains and modalities, including what may be characterized as knowledge of "what" (e.g., facts, meanings), "where" (e.g., landmarks), and "when" (when an event occurred) (Henke, 2010; Squire & Wixted, 2011). Other MTL structures closely connected with the hippocampus are also important, including the entorhinal, perirhinal, and parahippocampal cortices (Stark and Stark chapter in this book). Perirhinal cortex may underlie the familiarity of newly learned information, whereas the hippocampus subserves its explicit recollection (Brown, Warburton, & Aggleton, 2010; but see Wixted & Squire, 2011). Perirhinal cortex may support memories of single items (at least in the visual modality), whereas the hippocampus underlies more complex relational associations. MTL structures also appear to play a role in detecting and learning novel stimuli, perhaps perirhinal cortex for novel individual items and the hippocampus for novel relational information (Yonelinas, Aly, Wang, & Koen, 2010). Different nonhippocampal MTL regions may preferentially underlie memories in different domains: for example, perirhinal cortex for object recognition and parahippocampal cortex for spatial recognition as well as temporal information (Eichenbaum & Lipton, 2008; Eichenbaum et al., 2012; Squire & Wixted, 2011). Auditory information may particularly rely on parahippocampal cortex (Munoz-Lopez, Mohedano-Moriano, & Insausti, 2010), perhaps due to the temporal nature of this information. In contrast, as mentioned, the hippocampus binds information from a wide variety of domains and modalities, including time (which may explain its importance in episodic memory). More complex high-level concepts (e.g., about Jennifer Aniston) may also be represented in the hippocampus (Quiroga, Reddy, Kreiman, Koch, & Fried, 2005). More generally, the hippocampus and other MTL structures may be not only involved but also required for learning arbitrary bits of information and binding them, as evidenced by the virtual lack of such information acquired by densely amnesic individuals with extensive MTL lesions such as patient H.M. (Henke, 2010; Squire & Wixted, 2011; although some such information may still be learnable; see Stark and Stark chapter in this book). Finally, other structures connected to the MTL also play a role in declarative memory, including the fornix and various diencephalic structures and tracts (especially the mammillary bodies, the mammilothalamic tract, and the medial dorsal and anterior thalamic nuclei) (Squire & Wixted, 2011).

The MTL, including the hippocampus, is not just involved in learning and memory. As we have seen, the MTL plays a role in novelty detection, perirhinal cortex may subserve object recognition, and parahippocampal cortex may underlie spatial recognition and temporal information. It has been suggested that perirhinal cortex plays both memory and perceptual roles that link the declarative memory system with the ventral stream ("what pathway") in the visual modality (Bussey & Saksida, 2007). In contrast, parahippocampal cortex is closely linked to the dorsal stream ("where") pathway, particularly in the representation of landmarks (Kravitz, Saleem, Baker, & Mishkin, 2011). Additionally, the hippocampus may underlie other functions not strictly related to long-term memory, including short-term memory and aspects of imagination and prediction (Eichenbaum & Fortin, 2009; Kumaran & Maguire, 2009). Thus, MTL structures may play a number of (very possibly interrelated) roles in learning, memory, and other functions.

As mentioned above regarding the long-term memory functions of this brain system, knowledge that critically depends on the MTL during learning and consolidation eventually relies largely on neocortex. The MTL may rapidly bind neocortical representations together, which, over time and/or experience, eventually develop cortical-cortical links, thereby no longer requiring the MTL (McClelland, McNaughton, & O'Reilly, 1995). However, the MTL continues to play a role in this knowledge. As we have seen, it seems to be involved in object recognition (perirhinal cortex), spatial recognition (parahippocampal cortex), and recognition of higher-level concepts (hippocampus). Moreover, it has been suggested that the MTL continues to underlie long-term memories, particularly for autobiographical (episodic) knowledge (Winocur & Moscovitch, 2011), although this claim has been disputed (Squire & Wixted, 2011).

Within neocortex, different regions appear to underlie different types of knowledge (Martin, 2007; Squire & Wixted, 2011). For example, knowledge for living and nonliving things seems to involve different neocortical regions. An important organizational principle appears to be that neocortex proximate to particular sensory cortices underlies knowledge closely linked to those sensory modalities. Thus, knowledge with strong auditory associations depends on superior temporal regions (near primary auditory cortex), whereas knowledge with visual associations involves temporal and other regions downstream from visual cortex. Higher-level knowledge may also be organized neuroanatomically. For example, knowledge of faces involves the "fusiform face area" and knowledge of written words involves the "visual word form area." More generally, higher-level concepts may rely on more anterior temporal lobe areas (Barense et al., 2012). Information may flow from posterior to anterior regions of the temporal lobe, such that, at least in vision (and possibly in audition, Rauschecker & Scott, 2009), features are represented hierarchically in increasingly complex conjunctions, with low-level features represented more posteriorly and higher-level features represented more anteriorly—perhaps with the most complex conjunctions (e.g., higher-level concepts) in MTL regions, such as in the hippocampus.

Neocortical regions outside of temporal cortex also play roles in declarative memory—not only in the representation of long-term knowledge but also in particular processes associated with declarative memory. A region in inferior frontal neocortex corresponding largely to Brodmann's areas (BAs) 45 and 47 (within and near classical Broca's area) seems to underlie the encoding as well as the selection or recall of declarative memories (Ullman, 2004). Portions of the basal ganglia, perhaps with connections to these areas, may play analogous roles (Ullman, 2006). And a posterior parietal region seems to underlie aspects of encoding or retrieval (Uncapher & Wagner, 2009; Wagner, Shannon, Kahn, & Buckner, 2005).

The behavioral correlates of this network of brain structures rooted in the MTL are reasonably well-characterized (Eichenbaum, 2012; Henke, 2010; Squire & Wixted, 2011; Ullman, 2004). The system may be specialized for learning arbitrary bits of information and associating them; it may even be necessary for learning this information. It underlies the learning, representation, and use of knowledge about both facts (semantic memory) and events (episodic memory), such as the fact that Catalan is derived from Latin, or the event of you having a bowl of delicious pho yesterday for lunch. More broadly, the system can learn a wide range of information across sensory modalities and cognitive domains, and may also support various non-long-term memory functions. Unlike other types of long-term memory such as procedural memory or fear conditioning, multiple types of knowledge can be learned very rapidly in this system with as little as a single exposure to the stimulus, although additional exposures strengthen memories. The acquired knowledge is at least partly, although not completely (Chun, 2000; Henke, 2010; Schendan, Searl, Melrose, & Stern, 2003), explicit—that is, available to conscious awareness. Indeed, this appears to be the only long-term memory system that underlies explicit knowledge; thus, any knowledge that is explicit should have been learned in declarative memory. Once learned, information in declarative memory can be generalized and used flexibly across different contexts.

The molecular bases of declarative memory have also been reasonably well-studied in humans and animals (Green & Dunbar, 2012; Pezawas et al., 2004; Ullman, 2004). Various genes, including those for brain-derived neurotrophic factor (BDNF) and apolipoprotein E (APOE), play roles in declarative memory and hippocampal function, as do the neurotransmitter acetylcholine and the hormone estrogen (higher levels are associated with better declarative memory in humans and rats). BDNF may play a role in consolidation (as may estrogen). Estrogen may modulate declarative memory functionality via BDNF and/or acetylcholine.

Finally, various subject-level factors appear to modulate learning and retention in declarative memory, including not only genetic variability but also age (declarative memory improves during childhood, plateaus in adolescence/early adulthood, and then declines), sex (females seem to have an advantage at declarative memory over males), handedness (left-handedness may be associated with a declarative memory advantage), sleep (memory consolidation seems to improve during sleep), and exercise (which can enhance declarative memory) (Erickson et al., 2011; Marshall & Born, 2007; Ullman, 2005; Ullman et al., under revision; Ullman, Miranda, & Travers, 2008).

76.1.2 The Procedural Memory System

Although procedural memory is still not as well-characterized as declarative memory, its neurobiological and behavioral correlates are becoming clearer (Ashby, Turner, & Horvitz, 2010; Doyon et al., 2009; Ullman, 2004). Procedural memory involves a network of interconnected brain structures rooted in frontal/basal-ganglia circuits, including frontal premotor and related regions, particularly BA 6 and BA 44. (Note that we use the term procedural memory to refer to a particular brain system and its characteristics, rather than implicit memory more generally, which is how some researchers use the term.) The basal ganglia play a critical role in the learning and consolidation of motor and cognitive skills, whereas neocortical regions may be more important for processing skills after they have been automatized. Within the basal ganglia, the caudate nucleus (and the anterior putamen) may be especially important for skill acquisition.

This circuitry underlies the implicit (nonconscious) learning and processing of a wide range of perceptual-motor and cognitive skills, tasks, and functions (Eichenbaum & Cohen, 2001; Henke, 2010; Ullman, 2004), including navigation, sequences, rules, and categories. It may be specialized for learning to predict (perhaps especially probabilistic outcomes), for example, the next item in a sequence or the output of a rule. Learning in the system requires practice, and thus is slower than learning in declarative memory—though what is eventually learned seems to be processed more rapidly and automatically than knowledge in declarative memory. Although the system is rooted in the basal ganglia, the cerebellum may also play a role; however, exactly how and in what ways it interacts with the basal ganglia remain unclear.

Some aspects of the molecular bases of procedural memory are also beginning to emerge. The neurotransmitter dopamine plays an important role, particularly in learning and consolidation (Ashby et al., 2010). Certain genes involved in procedural memory have been identified, including *FOXP2*, *PPP1R1B* (for the protein DARPP-32), and *DRD2* (Meyer-Lindenberg et al., 2007; Ullman & Pierpont, 2005; Wong, Ettlinger, & Zheng, 2013). A recent study found that humanized *Foxp2* in mice (i.e., with human-specific amino acid substitutions) sped up learning, in particular by promoting the transition from declarative to procedural memory (Schreiweis et al., 2014). Finally, other factors may also affect procedural memory, including age. Unlike declarative memory, procedural memory functionality seems to be well-established early in life, after which learning or consolidation in this system may attenuate (Adi-Japha, Badir, Dorfberger, & Karni, 2014; Ullman, 2005). In contrast, sleep and exercise, among other factors, appear to show similar effects in the two memory systems, improving learning in both.

76.1.3 Interactions Between the Memory Systems

The declarative and procedural memory systems interact in a number of ways. First, evidence suggests that the two systems can, to some extent, acquire the same or analogous knowledge or skills (Poldrack & Packard, 2003; Ullman, 2004). According to the *redundancy hypothesis*, they therefore play at least partly redundant roles. Such redundancy can be found for multiple tasks and functions, including navigation, sequences, rules, and categories. Perhaps not surprisingly, the type and form of knowledge learned in the two systems are often quite different, even while this knowledge underlies the same or similar outcomes. For example, evidence from rodents suggests that navigation can be learned in procedural memory, such that animals learn to turn at a particular point (response strategy), or in declarative memory, by using landmarks (place strategies). Similarly, humans can learn sequences, rules, and categories implicitly in procedural memory or explicitly (and perhaps also implicitly) in declarative memory.

Various factors appear to modulate which of the two systems is used for a given task or function that can be learned or processed by either system. The declarative memory system often acquires knowledge initially, thanks to its fast acquisition abilities, whereas the procedural system gradually learns analogous knowledge that is eventually processed rapidly and automatically. The knowledge in declarative memory seems to remain intact even when procedural memory takes over; for example, lesions of the basal ganglia can lead to a reversion of dependence on knowledge that was initially learned in declarative memory (Packard, 2008). The learning context can also affect which system is relied on more. Explicit instruction (e.g., of sequences), or even just paying attention to the stimuli and underlying rules or patterns, can increase learning in declarative memory. Conversely, a lack of explicit instruction, as well as manipulations that reduce attention to the stimuli (e.g., in dual task paradigms), or a high level of complexity of rules or patterns (thus decreasing the subject's ability to explicitly detect patterns) may all shift learning toward procedural memory.

Many other factors likely also play roles affecting which system is relied on more. Any factor that enhances learning, retention, or retrieval preferentially in one of the memory systems should lead to an increased dependence on that system. Thus, the relative functionality of the two systems can affect which one is relied on more. For example, likely due in part to a female advantage at declarative memory (perhaps thanks to higher estrogen levels), females may rely more on this system, while males correspondingly rely more on procedural memory, for tasks that can be performed by either system (Ullman et al., 2008). And disorders that affect one system can lead to a compensatory role for the other. For example, individuals with Specific Language Impairment (SLI), dyslexia, autism, or obsessive-compulsive disorder, all of which have been associated with abnormalities of procedural memory but relatively intact declarative memory, appear to rely more on this intact system (Ullman & Pullman, 2015). Thus, multiple within- and between-subject factors may modulate the relative dependence on the two systems.

Second, animal and human studies suggest that declarative and procedural memory also interact *competitively*, resulting in a "seesaw effect" (Ullman, 2004). The dysfunction of one system may lead not only to an increased dependence on the other system for those tasks and functions that can depend on either one, but *also* to the enhanced functioning of that system. Evidence for such a seesaw effect comes from both animal and human studies (Packard, 2008; Poldrack & Packard, 2003; Ullman, 2004). Additionally, estrogen may not only enhance declarative memory but also inhibit procedural memory. Note that the seesaw effect, and compensation due to redundancy, are distinct notions: if only one system is dysfunctional, then the other can compensate whether or not its functionality is enhanced—although, of course, any enhancement from the seesaw effect would bolster any such compensation.

Third, the learning and/or retrieval of knowledge in declarative memory may block (inhibit) the learning and/or retrieval of analogous knowledge in procedural memory (Ullman, 2004). The converse may hold as well. For example, even when a task is learned initially in declarative memory, it can be overridden by procedural memory when it is subsequently learned in that system (Packard, 2008).

The two memory systems are also linked to and interact with *other* neurocognitive brain systems. First, working memory seems to be closely related to (at least) declarative memory. For example, frontal brain structures involved in working memory also underlie declarative memory in both encoding and recall (Ullman, 2004). And deficits of working memory are associated with deficits of declarative memory (Lum, Ullman, & Conti-Ramsden, 2015; Ullman & Pullman, 2015). More generally, we suggest that working memory may constitute input and output mechanisms for at least explicit knowledge in declarative memory. Second, priming depends importantly on knowledge initially learned in the memory systems. For example, the priming of concepts and word forms seems to rely on representations learned in declarative memory, as suggested by the inability of dense amnesics to learn new information of this sort. Finally, as indicated above, there are links between the memory systems and the ventral and dorsal streams (Ullman, 2004). The ventral stream seems to be linked strongly to the declarative memory system. The dorsal stream may also interact with this system, with projections to parahippocampal cortex, which plays a role in representing landmarks. However, the dorsal stream may be particularly tied to procedural memory. While "what" knowledge seems to depend on a combination of the ventral stream (for processing) and declarative memory (for learning), "how" knowledge may depend on a combination of the dorsal stream (for processing) and procedural memory (for learning). Indeed, learning supported by the basal ganglia may link inputs from parietal structures to motor regions (Ashby et al., 2010; Doyon et al., 2009). This posited interdependence between learning in the declarative and procedural memory systems and respective processing of this knowledge in the ventral and dorsal streams may be referred to as the *memory-processing interdependence hypothesis*.

76.2 PREDICTIONS FOR LANGUAGE

Here, I summarize some key predictions of the DP model—that is, predictions for language that are derived from our understanding of the two memory systems. For each memory system I first present predictions regarding *what* types of language-related knowledge and functions the system should underlie, and then *how* exactly these should be subserved by the system, with a focus on functional neuroanatomy.

Note that the DP model does *not* predict that the memory systems underlie language and other domains in identical ways—both because language is not identical to other domains and because portions of the systems may have become specialized for language (either evolutionarily or developmentally). Rather, because the co-optation hypothesis leads to the expectation that the systems play at least similar if not identical roles across domains, our substantial independent knowledge of the systems' roles in other domains is an excellent starting point for making predictions about language. Empirical studies will reveal exactly where and how the systems' roles in language might diverge from those in other domains.

76.2.1 Declarative Memory: Predictions for Language

76.2.1.1 What Should Declarative Memory Underlie in Language?

First, because declarative memory seems to be critical for learning, storing, and accessing arbitrary bits of information, as well as arbitrary associations among them, aspects of language that involve such bits or associations should critically depend on this memory system. Thus, declarative memory should be crucial for all learned idiosyncratic linguistic knowledge at the word or multi-word level (though presumably not for any such knowledge that may be purely innately specified, if such knowledge exists). Simple (i.e., not rule-governed and fully derivable) content words (e.g., *cat*, *devour*), including their phonological forms, meanings, (sub)categorization knowledge (e.g., *devour* requires a complement), and mappings between them (e.g., sound-meaning mappings), should be learned in this system. Knowledge about irregular morphological forms, both inflectional and derivational (e.g., *dig-dug*, *solemn-solemnity*), should be stored in declarative memory, as should knowledge about idioms, proverbs, and so on. In principle, such stored knowledge could be represented in a variety of ways, even in parallel for the same forms in the same individuals, such as structured or unstructured, as wholes or as collocations with probabilistic associations between their parts, or even as stored links to rules as suggested by Distributed Morphology (Halle & Marantz, 1993). Storing representations in declarative memory does not preclude the additional involvement of rule-governed aspects of procedural memory, for example, for inflecting forms within larger stored sequences (e.g., *jumps the gun*). Finally, in addition to these types of idiosyncratic language knowledge, which correspond broadly to traditional notions of semantic memory (world knowledge not bound to a particular personal experience), declarative memory should also underlie aspects of episodic knowledge in language—for example, memories regarding whether and in which context one has recently encountered or learned a particular word.

Second, due to its ability to learn a wide range of information, declarative memory should be able to acquire much more than idiosyncratic knowledge. Indeed, there may be few constraints on the types of linguistic knowledge that this system can learn. At the least, all the types of information that it can learn about idiosyncratic aspects of language should also be learnable for nonidiosyncratic, rule-governed aspects. Thus, just like simple and irregular words, one should be able to store fully rule-governed complex forms (e.g., "*walked*," "*the cat*," and even longer sequences). More abstract representations could also be stored, such as portions of linguistic hierarchies (e.g., Noun Phrase), as has been posited by linguistic theories such as Tree-Adjoining Grammar (Joshi & Rambow, 2003). More productive aspects of grammar may also be achieved by relying on declarative memory, for example, by generalizing across already-stored representations to new representations (e.g., analogic generalization across similar forms) (Hartshorne & Ullman, 2006) or by "shallow" parsing of sentences (Clahsen & Felser, 2006), which relies on lexical-semantic knowledge (which is learned in declarative memory). Grammatical rules and constraints themselves should also be learnable by declarative memory, either explicitly or implicitly, even though these are generally acquired by procedural memory. Other types of linguistic knowledge or functions normally learned in procedural memory should also be learnable in declarative memory, such as word segmentation. However, just because such grammatical and other knowledge *can* be learned by declarative memory does not mean it is, or if it is learned, that it is consistently used; rather, this will depend on factors that modulate the relative dependence on declarative and procedural memory and the interactions between them (e.g., blocking).

As we have seen, the brain structures underlying declarative memory, including the MTL, underlie not only learning and long-term memory but also other cognitive processes, including object recognition, spatial recognition, novelty detection, short-term memory, and prediction. Analogous roles for such functions may thus be found in language. Additionally, given the rapidity with which new associations can be learned in this memory system, we might expect it to play an active role in online processes involving integration and binding in language.

76.2.1.2 How Should Declarative Memory Underlie Language?

Based on our independent knowledge of declarative memory, we can make numerous specific neurobiological and other predictions about those aspects of language that should depend on declarative memory. Here, I summarize some of them. First, the functional neuroanatomy of those aspects of language that are learned, stored, and processed by declarative memory should reflect the functional neuroanatomy of this system. Thus, linguistic knowledge learned in this system should depend on the hippocampus and/or other MTL structures, at least during learning and consolidation. The hippocampus itself may be heavily engaged in language-related learning, given the multiple types of information and modalities involved in this domain and the importance of linking and integrating this information. Individual items such as words, however, might rely particularly on perirhinal cortex. Novel linguistic items or relations should involve the MTL, perhaps especially the hippocampus (novel associations) and perirhinal cortex (novel items). These structures should be active mainly during learning, but perhaps also for storage and use. MTL structures may also play roles in language-related short-term memory and prediction. Other structures linked to MTL, such as the fornix and mammillary bodies, may also play learning roles in language.

With time and/or experience, the MTL should play a decreasingly important role for linguistic knowledge learned in declarative memory, with an increasing role for neocortical structures. Linguistic knowledge should show similar neuroanatomical patterns as nonlinguistic knowledge of the same concepts. For example, words, phrases, or sentences referring to living or nonliving things, or with strong visual attributes of particular sorts (e.g., color, form), should involve neocortical regions independently linked to these features. More complex, higher-level linguistic representations may depend on more anterior temporal lobe regions, and perhaps also on MTL structures, particularly the hippocampus. For example, abstract linguistic categories such as Noun or Verb might depend on anterior temporal or even more upstream areas. Inferior frontal cortex, especially BA 45/47, should underlie the encoding of new linguistic information being learned in declarative memory, as well as the recall of that knowledge once it is learned. Posterior parietal cortex may also play a role in the encoding and retrieval of this information.

Linguistic knowledge should be rapidly learnable in declarative memory, even from a single presentation of the information, although repeated exposures should improve learning and retention. Thus, we expect that words can be quickly acquired. Linguistic knowledge in declarative memory could be either explicit (e.g., verbalizable word or rules) or implicit. Conversely, however, if linguistic knowledge is explicit, then it must be stored in declarative memory, because this is the only long-term memory system to underlie explicit knowledge. Once learned in this system, linguistic knowledge can be used flexibly across different contexts.

Finally, molecular and other factors that modulate declarative memory should play analogous roles in language. For example, polymorphisms of BNDF and APOE should modulate declarative memory–dependent aspects of language in a similar manner as for nonlinguistic knowledge, and likewise for estrogen and acetylcholine. For example, higher levels of estrogen may improve language learning in this system. Based on the available evidence, it seems likely that language learning in declarative memory should ameliorate during childhood, plateau in adolescence/early adulthood, and then decline. Thus, word learning should follow this pattern. Females may show advantages at language learning in this system as compared to males. Sleep, exercise, and other factors should improve language learning and retention in declarative memory.

76.2.2 Procedural Memory: Predictions for Language

It is important to emphasize that we know less about this system than about declarative memory; therefore, our predictions are more tentative and less specific.

76.2.2.1 What Should Procedural Memory Underlie in Language?

Procedural memory should underlie the learning and processing of sequences and rules in language, perhaps especially those that are probabilistic rather than deterministic. The system may be particularly important in "learning to predict" in language, such as predicting the next item in a sequence or the output of a linguistic rule. Only rules or sequences that are implicit should be learned by procedural memory (I emphasize that implicit procedural knowledge of a rule or sequence does not preclud either explicit or implicit analogous knowledge in declarative memory). Given that grammar involves largely implicit rules, both probabilistic and deterministic, particularly ones that involve (hierarchical) sequencing, procedural memory should play a major role in this aspect of language. This should hold across linguistic subdomains, including syntax, morphology, and phonology. Exactly what computational roles procedural memory should play in linguistic sequencing and rules cannot be clearly predicted at this point because of our lack of understanding of these issues regarding procedural memory itself. However, also based on our understanding of grammar, it seems likely that procedural memory somehow underlies the learning of rules and the implicit rule-governed composition of both hierarchical and nonhierarchical sequences.

Other aspects of language may also be learned in procedural memory, including categories and other knowledge, especially if the knowledge is implicit and involves probabilistic patterns, sequences, and learning to predict. For example, the implicit learning of word boundaries in a speech stream (Saffran, Aslin, & Newport, 1996) should depend on procedural memory.

76.2.2.2 How Should Procedural Memory Underlie Language?

As with declarative memory, predictions for language follow from what we know about procedural memory from animal and human studies. First, linguistic skills and knowledge that are learned, stored, and processed by procedural memory should reflect this system's underlying functional neuroanatomy. Thus, these should involve frontal and basal ganglia structures, and perhaps the cerebellum. Learning and consolidation should engage the basal ganglia, especially the caudate nucleus and the anterior putamen. (Note that this learning role of the basal ganglia does not preclude other language roles for this set of subcortical structures, including grammar, because the structures subserve other functions as well, including working memory and attention.) Once automatized, knowledge and skills should rely especially on neocortical regions, particularly BA 6 and BA 44. Only implicit, not explicit, linguistic knowledge should rely on procedural memory (although of course not all implicit knowledge should depend on this system). Because procedural memory learns only with repeated exposure, this knowledge should be learned and automatized gradually.

Given its role in procedural memory, dopamine should play a role in grammar and other aspects of language, especially in learning and consolidation. Certain genes, such as *FOXP2*, *PPP1R1B*, and *DRD2*, should also be involved. Because procedural memory learns gradually and declarative memory learns rapidly, grammar rules should generally be acquired more slowly than words. Given the developmental trajectory of procedural memory, linguistic learning and consolidation in this system should be better in childhood than adolescence or adulthood, and thus proceduralization of grammar should be slower and more incomplete in later (e.g., second language) learners. Sleep, exercise, and other factors should improve language learning and retention in procedural memory.

76.2.3 Interactions Between the Memory Systems: Predictions for Language

Our understanding of interactions between the two memory systems, and between these and other systems, also leads to various predictions for language.

First, to some extent, we expect the two memory systems to acquire the same or analogous knowledge or skills, that is, to play at least partly redundant roles. According to the redundancy hypothesis, in language as in nonlinguistic domains, such redundancy may be found for any tasks or functions that could be subserved by either system. Given the learning power of declarative memory, and the fact that it can underlie implicit as well as explicit knowledge, it may be able to at least partly support most if not all aspects of language subserved by procedural memory, including grammar and word segmentation.

Various factors should modulate which memory system is relied on more for linguistic knowledge that can be learned by both systems. Such knowledge should often be learned first by declarative memory, but eventually by procedural memory, at which point it should be more automatized. Thus, both first and second language learners should generally depend initially on declarative memory for grammatical functions (e.g., by chunking or explicit rules, with the exact nature of this dependence perhaps differing between first and second language learners), but both should gradually learn grammar in procedural memory.

(For more on the DP model and second language, see Ullman, 2005; Ullman, 2015). After such proceduralization, the declarative knowledge may still remain intact and could become accessible again in certain circumstances, such as subsequent procedural memory dysfunction. Explicit instruction or attention to the input may increase learning in declarative memory, whereas a lack of such instruction or attention, or greater complexity of rules or patterns (e.g., more complex grammatical rules or constraints), may lead to a greater dependence on procedural memory. Estrogen may promote linguistic learning in declarative memory at the expense of procedural memory. Females may rely more on declarative memory than males for aspects of language (e.g., grammar) that can rely on either system, and they may show superior learning of idiosyncratic linguistic knowledge (which must be learned in declarative memory). The developmental trajectories of the two systems suggest that young children should more easily proceduralize their grammar (in first or second language) than adults. And a relative dysfunction of one system should lead to a greater (compensatory) dependence of language on the other.

Second, we might expect a seesaw effect in language. Estrogen might not only promote language learning or use in declarative memory but may also inhibit it in procedural memory. Similarly, a dysfunction of one system might lead not only to language compensation by the other but also to its enhanced functioning.

Third, learning or processing in one system may block or inhibit the other. For example, grammar learning in declarative memory may inhibit grammar learning in procedural memory. And successful retrieval of irregular forms (e.g., *dug*), or chunked rule-governed forms (*walked*), should block the rule-based computation of corresponding forms in procedural memory (*dig + -ed*, *walk + -ed*). Conversely, a highly automatized rule in procedural memory should tend to override the use of analogous declarative knowledge, especially if this declarative knowledge is not well-learned. Thus, over the course of language learning, grammar should depend increasingly on procedural memory and correspondingly less on declarative memory.

Finally, the two memory systems should interact with other neurocognitive brain systems in the learning and processing of language. First, working memory in language processing should be closely related to declarative memory. For example, some of the same frontal structures should play roles in verbal working memory and in the encoding and recall of language knowledge learned in declarative memory. Second, linguistic knowledge that can be primed (e.g., lexical or grammatical knowledge) should, in most cases, have been learned by one or the other memory system. Finally, there should be close links between language learning in the two memory systems and language processing in the dorsal and ventral streams. For example, language knowledge processed in the ventral stream should be learned mainly in declarative memory, which in turn should facilitate further processing of this knowledge in the ventral stream, while a similar relation may hold for procedural memory and the dorsal stream.

76.3 EVIDENCE

Parallel to the predictions presented above, for each memory system I summarize first, evidence regarding *what* types of language knowledge and functions depend on it, and second, *how* the system subserves these, with a focus on functional neuroanatomy.

76.3.1 Declarative Memory in Language: Evidence

76.3.1.1 What Does Declarative Memory Underlie in Language?

Evidence from various methodological approaches, including behavioral, neurological, neuroimaging, and electrophysiological studies, has implicated declarative memory in the learning, storage, and retrieval of idiosyncratic aspects of language. First, behavioral evidence indicates that words can be learned very quickly (fast mapping) and are generally acquired earlier than grammatical rules (in both first and second language) (Bloom, 2000; Marcus et al., 1992; Ullman, 2015). More direct behavioral evidence comes from studies of children that reveal correlations, across participants, between vocabulary abilities and learning abilities in declarative memory (but not procedural memory), whereas grammar shows the opposite pattern (Kidd, 2012a; Lum, Conti-Ramsden, Page, & Ullman, 2012). Behavioral studies using other techniques, such as the examination of frequency and imageability effects, have also revealed links between lexical and declarative memory (Babcock, Stowe, Maloof, Brovetto, & Ullman, 2012; Prado & Ullman, 2009). Second, neurological evidence shows that dense amnesia from substantial MTL lesions seems to preclude word learning, as demonstrated in H.M. and other patients (Davis & Gaskell, 2009; Postle & Corkin, 1998). In contrast, H.M. retained idiosyncratic word forms that were common in English prior to his surgery, as expected if the MTL does not remain crucial for knowledge well after it has been learned (Kensinger, Ullman, Locascio, & Corkin, 1999). Third, functional neuroimaging evidence strongly implicates the MTL in word learning

(Breitenstein et al., 2005; Davis & Gaskell, 2009; Raboyeau, Marcotte, Adrover-Roig, & Ansaldo, 2010). Fourth, the presentation of words, including novel words (pseudowords), reliably leads to N400 event-related potentials (ERPs) (Kutas & Federmeier, 2011), which have independently been tied to nonlinguistic idiosyncratic knowledge (e.g., faces and objects) and the MTL (Kutas & Federmeier, 2011; McCarthy, Nobre, Bentin, & Spencer, 1995; Meyer et al., 2005).

Declarative memory has also been tied to nonidiosyncratic aspects of language. First, behavioral evidence suggests that regular morphological forms can be stored (e.g., as chunks), generalized across similar forms, or computed from explicit rules, all indicating a reliance on declarative memory. For example, like irregulars, regular inflected forms can show frequency, imageability, and phonological neighborhood effects (Alegre & Gordon, 1999; Babcock et al., 2012; Dye, Walenski, Prado, Mostofsky, & Ullman, 2013; Hartshorne & Ullman, 2006; Prado & Ullman, 2009; Ullman et al., under revision). However, whereas these effects are found reliably for irregulars, consistent with their obligatory storage in declarative memory, regulars show them inconsistently, and mainly for those conditions where a dependence on declarative memory is expected (e.g., higher frequency forms, females, left-handers, second language learners). Evidence suggests that even more complex rule-governed forms, that is, surface syntactic structures, can also be learned in declarative memory (Hamrick, 2014). And a recent study found that learning an "analogic grammar" (posited to involve generalization over stored forms) correlated with abilities at declarative but not procedural memory, whereas learning a rule-governed concatenative grammar showed the opposite pattern (Wong et al., 2013). However, as would be expected given faster learning by declarative than procedural memory, syntactic processing at early stages of learning a rule-governed artificial language correlated with declarative (and not procedural) memory, whereas the reverse pattern was found at later stages (Morgan-Short, Faretta-Stutenberg, Brill-Schuetz, Carpenter, & Wong, 2014). Second, neurological evidence from SLI, dyslexia, autism, and agrammatic aphasia suggests that individuals with these disorders compensate for grammatical impairments by relying on declarative memory, via chunking, explicit rules, or other means (Ullman & Pullman, 2015). Third, neuroimaging studies of artificial grammar learning have found MTL activation (Lieberman, Chang, Chiao, Bookheimer, & Knowlton, 2004; Yang & Li, 2012). And neuroimaging evidence has implicated the MTL in online sentence integration and processing, including of syntax (Duff & Brown-Schmidt, 2012; Meyer et al., 2005). Fourth, electrophysiological evidence from ERPs has found that (morpho)syntactic processing can elicit N400s, primarily in those conditions where a dependence on declarative memory is expected (e.g., females, second language learners, and individuals with SLI, dyslexia, or agrammatic aphasia) (Ullman, 2015; Ullman et al., 2008; Ullman & Pullman, 2015).

76.3.1.2 How Does Declarative Memory Underlie Language?

Thus, significant language knowledge is learned and even processed in the MTL. But which portions of the MTL are involved and in which circumstances? First, the hippocampus is heavily implicated. Multiple neuroimaging studies of word learning report hippocampal activation (Breitenstein et al., 2005; Davis & Gaskell, 2009), as has a study of chunking in artificial grammar learning (Lieberman et al., 2004). Similarly, the hippocampus has been implicated in the integration of linguistic knowledge (Duff & Brown-Schmidt, 2012; Meyer et al., 2005).

However, other MTL structures also play roles in language. Although lesions restricted to the hippocampus can impair word learning, at least some such patients can still learn words (Davis & Gaskell, 2009; Vargha-Khadem, Gadian, & Mishkin, 2001). It may be that MTL lesions that extend beyond the hippocampus, such as H.M.'s, are required to eliminate word learning altogether. Some evidence suggests that rhinal cortex, in particular perirhinal cortex, may be important for words, perhaps especially for word learning. One study found that remembering the context in which a word was recently presented engaged the hippocampus, whereas the word itself activated perirhinal cortex (Davachi, Mitchell, & Wagner, 2003). In the MTL, N400s have been tied mainly to perirhinal cortex, particularly for novel words (Fernandez, Klaver, Fell, Grunwald, & Elger, 2002). In one study, lexical/semantic violations that typically elicit N400s were tied to rhinal cortex, whereas syntactic violations were linked to the hippocampus, likely due to P600 effects that involve controlled (conscious) syntactic integration (Meyer et al., 2005).

Other structures in the declarative memory system also play roles in language. One study found deficits in word learning after lesions to the mammillary bodies (Martins, Guillery-Girard, Jambaque, Dulac, & Eustache, 2006). Linguistic as well as nonlinguistic stimuli associated with particular concepts engage neocortical regions linked to those concepts (e.g., living things) (Martin, 2007). BA 45/47 is involved in word encoding and recall. For example, word encoding and recall tasks activate this region, and lesions to this area are associated with word recall deficits (Buckner,

Wheeler, & Sheridan, 2001; Wagner et al., 1998). Posterior parietal cortex has been found to correlate with vocabulary size in monolinguals (Lee et al., 2007) and to be larger in bilinguals than in monolinguals (possibly due to the larger total vocabulary of the former) (Mechelli et al., 2004), and it has been implicated in lexical/semantics in both first and second language (Abutalebi et al., 2012; Chee, Hon, Lee, & Soon, 2001) as well as in grammar in second (but not first) language (Wartenburger et al., 2003). For a functional neuroanatomical meta-analysis of first and second language revealing such a pattern, see Tagarelli, Turkeltaub, Grey, and Ullman (in preparation).

There has also been some work on the genetic and molecular bases of linguistic aspects of declarative memory. Quite a few studies have implicated BDNF and APOE, as well as estrogen and acetelycholine, in episodic memory tasks involving words (Ullman, 2007; Ullman et al., 2008). For example, performance at such tasks improves with cholinesteraste inhibitors (Freo, Pizzolato, Dam, Ori, & Battistin, 2002). And estrogen has been found to improve retrieval of irregular (but not regular) inflected forms (Estabrooke, Mordecai, Maki, & Ullman, 2002). More research is needed to examine links between language and the genetic and molecular bases of declarative memory.

Finally, evidence suggests that females may be better than males at learning words (Kaushanskaya, Marian, & Yoo, 2011; Ullman et al., 2008), and likewise left-handers as compared with right-handers (Ullman et al., under revision). And, consistent with the developmental trajectory of declarative memory, word learning improves during childhood into adolescence (Bloom, 2000).

76.3.2 Procedural Memory in Language: Evidence

Consistent with our more impoverished understanding of procedural than declarative memory, there is less empirical evidence thus far regarding the role of this system in language.

76.3.2.1 What Does Procedural Memory Underlie in Language?

Evidence from various methodologies suggests links between grammar and procedural memory. First, behavioral evidence has revealed correlations between grammar measures (e.g., syntactic priming and processing) and learning in procedural memory, but not with declarative memory in the same individuals (Kidd, 2012b; Lum et al., 2012). Procedural (but not declarative) memory has also been found to correlate with learning a rule-governed concatenative grammar (Wong et al., 2013). In another artificial language study, syntactic processing correlated with procedural (and not declarative) memory, but only at later stages of learning (Morgan-Short et al., 2014). In an interference study, syntactic (but not word) processing interfered with procedural memory (Nemeth et al., 2011). And, as mentioned above, regular morphological forms generally do not show signs of reliance on declarative memory; rather they show evidence for composition, consistent with a primary dependence on procedural memory (Walenski, Prado, Ozawa, Steinhauer, & Ullman, under revision).

Second, neurological evidence has tied grammar to procedural memory. (Here, I focus on SLI; for other disorders, including agrammatic aphasia and Parkinson's and Huntington's diseases, see Ullman, 2004, 2013). Children with SLI, who typically have grammatical deficits (of syntax, morphology, and phonology) but less consistent lexical impairments, show reliable procedural memory impairments (Lum, Conti-Ramsden, Morgan, & Ullman, 2014; Ullman & Pierpont, 2005) and consistent brain abnormalities only in frontal cortex and the basal ganglia, as revealed by a recent neuroanatomical meta-analysis (Ullman et al., under review). Moreover, the grammar difficulties in SLI have been directly linked to procedural memory deficits (Hedenius et al., 2011).

Third, neuroimaging evidence has tied artificial grammar learning to the basal ganglia (Lieberman et al., 2004; Petersson, Folia, & Hagoort, 2012; Yang & Li, 2012). Basal ganglia activation is also reliably elicited by grammatical (but not lexical) processing in second language learners, but not in native speakers, as revealed by our recent neuroanatomical meta-analysis of first and second language (Tagarelli et al., in preparation). This is consistent with the prediction that the basal ganglia play a particularly important role in grammar acquisition.

Finally, some research has begun to examine links between procedural memory and nongrammatical implicit aspects of language. For example, a recent fMRI study implicated the basal ganglia in the implicit learning of word boundaries in a speech stream (Karuza et al., 2013).

76.3.2.2 How Does Procedural Memory Underlie Language?

Although evidence is still limited, some specific neurobiological patterns appear to be emerging regarding the role of procedural memory in language. First, within the basal ganglia, the caudate nucleus, and perhaps the caudate head in particular, may play a particularly important role in grammar learning. These structures have been implicated in neuroimaging studies of artificial grammar learning and of grammar in second (but not first) language (Tagarelli et al., in preparation). And our neuroanatomical meta-analysis of brain abnormalities in SLI revealed that the consistent structural abnormalities in the basal ganglia occur only in the caudate nucleus, with particular involvement of the caudate head (Ullman et al., under review).

Once learned, grammar depends heavily on BA 44, as well as BA 6 (especially the frontal operculum), particularly on the left side. Activation in these regions is strongly associated with syntactic processing in both first and second language (Friederici, 2006; Tagarelli et al., in preparation). And a recent functional neuroimaging meta-analysis of regular and irregular morphology strongly implicates BA 44 in the former but not the latter (Ullman, Campbell, McQuaid, Tagarelli, & Turkeltaub, in preparation).

There has been far less work to date examining links between nongrammatical aspects of language and procedural memory. Nevertheless, as mentioned above, one recent study of the implicit learning of word segmentation reported basal ganglia activation (Karuza et al., 2013)—indeed, mainly in the caudate head and anterior putamen.

Genetic evidence has also tied language to procedural memory. The *FOXP2* gene has been linked to grammar, including syntax, morphology, and phonology, as well as to procedural memory (Ullman & Gopnik, 1999; Ullman & Pierpont, 2005). Moreover, the recent finding that humanized *Foxp2* speeds up learning by promoting the transition from declarative to procedural memory (Schreiweis et al., 2014) suggests that evolutionary changes to procedural memory may be critical for the evolution of language, particularly of grammar. This underscores the utility of the co-optation hypothesis by showing that investigating preexisting systems, and these memory systems in particular, can reveal not only how they work similarly in language as in nonlanguage functions, but also how they might become further specialized for language itself. In other words, the systems constitute *targets* for studying the potential evolutionary changes that may facilitate language (whether or not those changes were due to adaptation for language alone). Finally, *DRD2* (for the dopamine receptor D_2) has also been linked to grammar learning, as well as to basal ganglia activation during the learning of a rule-governed concatenative grammar (Wong et al., 2013).

Some research has also examined the effect on language of subject-level factors that modulate procedural memory. For example, as would be expected if learning and/or consolidating in procedural memory becomes more difficult between early childhood and adulthood, adult second language learners have particular difficulty with grammar (Ullman, 2005).

76.3.3 Interactions Between the Memory Systems in Language: Evidence

First, evidence from multiple methodologies suggests that declarative and procedural memory play redundant roles for grammar, but not for lexical/semantics, which seems to require declarative memory. Much of the evidence for such redundancy has been discussed above. In brief, rule-governed compositional forms can be not only learned and computed by procedural memory but also stored and processed by declarative memory, via chunking, analogical generalization in associative memory, composition by explicit rules, and other processes. This dependence on declarative memory seems to be modulated by various factors. It occurs more for higher than lower frequency or imageability items; more for second than for first language learners; more for early versus later stages of learning (at least in second language; it remains unclear whether this predicted pattern is also found in first language); more for females than for males; more for left-handers than for right-handers; and more in disorders associated with a procedural memory system dysfunction but relatively intact declarative memory (e.g., SLI, dyslexia, autism, agrammatic aphasia).

Second, some evidence, though still limited, supports the predicted seesaw effect in language. Thus far, the only evidence we are aware of, which is somewhat indirect, is findings from neurodevelopmental disorders. Children with dyslexia or autism, both of which are linked to grammatical and procedural memory abnormalities (Lum, Ullman, & Conti-Ramsden, 2013; Walenski, Tager-Flusberg, & Ullman, 2006), may also show enhanced lexical or declarative memory abilities (Hedenius, Ullman, Alm, Jennische, & Persson, 2013; Ullman & Pullman, 2015; Walenski, Mostofsky, Gidley-Larson, & Ullman, 2008). Similarly, children with SLI may show not only grammatical and procedural memory deficits but also enhanced declarative memory, in particular at consolidation (Lukacs, Kemeny, Lum, & Ullman, in preparation; Lum, Hedenius, Tomblin, & Ullman, in preparation).

Third, some evidence suggests linguistic inhibition by one or the other system. Perhaps the best studied phenomenon is blocking. For example, the retrieval of a stored irregular form blocks the computation of its corresponding overregularization (e.g., retrieving *dug* blocks *digged*), whereas retrieval difficulties due to lexical/declarative memory deficits yield overregularizations (Ullman, 2004, 2013). Similarly, retrieval of a chunked regular may at least partially block the rule-based computation of the same form (Prado & Ullman, 2009). Inhibition between the systems has also been observed in learning. In a recent study of second language learning, explicit training (which should encourage learning in declarative memory) delayed the development of automatic syntactic processing that has been associated with procedural memory (Morgan-Short, Finger, Grey, & Ullman, 2012; Morgan-Short, Steinhauer, Sanz, & Ullman, 2012). Conversely, even when an N400 is found for grammar at early stages of second language learning, the later emergence of automatic syntactic processing overrides this lexical/semantic process (Morgan-Short, Finger, et al., 2012; Morgan-Short, Steinhauer, et al., 2012).

Finally, some evidence exists regarding language-related interactions between the two memory systems and other neurocognitive systems. For example, BA 45/47 seems to be involved not only in word encoding and word recall but also in verbal working memory. And words that entered the language after H.M.'s resection do not show priming, whereas amnesic patients do show priming both for previously learned words and for syntax (Ferreira, Bock, Wilson, & Cohen, 2008; Postle & Corkin, 1998).

76.4 SUMMARY AND CONCLUSION

The DP model is premised on three principles of biology and language. First, new biological functions commonly recruit pre-existing biological mechanisms, whether or not those mechanisms then become further specialized—either evolutionarily or developmentally—for the functions. That is, biological mechanisms are often co-opted for new purposes. Second, most of language must be learned, whether or not aspects of this capacity are innately specified. Third, declarative memory and procedural memory are perhaps the two most important learning and memory systems in the brain, in terms of the range of domains and functions that they subserve. Based on these principles, the DP model simply posits that these two memory systems are highly likely to play important roles in language learning, knowledge, and use.

If language depends heavily on the two memory systems, then what we know independently about these memory systems should tend to apply to language as well. That is, the memory systems should play similar roles in language as in nonlanguage functions, and thus our independent knowledge of how these memory systems work should generate predictions for language. Because the memory systems are quite well-understood at many levels, from work with both humans and animal models, the theory can generate a wide range of well-motivated specific predictions, many of which there would be no reason to make based on the more limited study of language. For example, we can make predictions about MTL, or perirhinal cortex, or certain genes, or estrogen, or sex differences, or age effects in first versus second language, that there might be no *independent* reason to make based on the study of language alone. Thus, this is a very powerful theoretical approach.

As we have seen, converging evidence from multiple methodologies provides support for many of the general as well as specific predictions of the theory. The findings suggest that language does indeed depend on the two memory systems. More generally, the results yield insights that advance our understanding of language and its underlying neurobiology. Importantly, unlike language-specific accounts, the theory can predict and explain aspects of linguistic and neurolinguistic phenomena in the broader context of our understanding of the brain and mind. Thus, the theory has substantial explanatory power.

Finally, many of the predictions I have summarized in this chapter have not yet been tested. In fact, additional predictions have not even been discussed. For example, the theory predicts that behavioral or pharmacological interventions that have been shown to enhance learning or processing in the memory systems are likely to also enhance the learning or processing of language. This prediction may have important educational and translational/clinical outcomes, in particular for second language learning and language disorders. Thus, the theoretical approach presented here promises to continue to be fruitful and to lead to advances in multiple aspects of language and its underlying neurobiology.

Acknowledgments

I thank Goldie Ann McQuaid, Jarrett Lovelett, Kaitlyn Tagarelli, Scott Miles, and Başak Karatas for input on this chapter.

References

Abutalebi, J., Della Rosa, P. A., Green, D. W., Hernandez, M., Scifo, P., Keim, R., et al. (2012). Bilingualism tunes the anterior cingulate cortex for conflict monitoring. *Cerebral Cortex*, *22*(9), 2076–2086.

Adi-Japha, E., Badir, R., Dorfberger, S., & Karni, A. (2014). A matter of time: Rapid motor memory stabilization in childhood. *Developmental Science*, *17*(3), 424–433.

Alegre, M., & Gordon, P. (1999). Frequency effects and the representational status of regular inflections. *Journal of Memory and Language*, *40*(1), 41–61.

Ashby, F. G., Turner, B. O., & Horvitz, J. C. (2010). Cortical and basal ganglia contributions to habit learning and automaticity. *Trends in Cognitive Sciences*, *14*(5), 208–215.

Babcock, L., Stowe, J. C., Maloof, C. J., Brovetto, C., & Ullman, M. T. (2012). The storage and composition of inflected forms in adult-learned second language: A study of the influence of length of residence, age of arrival, sex, and other factors. *Bilingualism: Language and Cognition*, *15*(4), 820–840.

Barense, M. D., Groen, I. I. A., Lee, A. C. H., Yeung, L. K., Brady, S. M., Gregori, M., et al. (2012). Intact memory for irrelevant information impairs perception in amnesia. *Neuron*, *75*(1), 157–167. Available from: http://dx.doi.org/10.1016/j.neuron.2012.05.014.

Bloom, P. (2000). *How children learn the meanings of words*. Cambridge, MA: The MIT Press.

Bornkessel-Schlesewsky, I., & Schlesewsky, M. (2013). Reconciling time, space and function: A new dorsal-ventral stream model of sentence comprehension. *Brain and Language, 125*(1), 60–76.

Breitenstein, C., Jansen, A., Deppe, M., Foerster, A.-F., Sommer, J., Wolbers, T., et al. (2005). Hippocampus activity differentiates good from poor learners of a novel lexicon. *NeuroImage, 25*(3), 958–968.

Brown, M. W., Warburton, E. C., & Aggleton, J. P. (2010). Recognition memory: Material, processes, and substrates. *Hippocampus, 20*(11), 1228–1244.

Buckner, R. L., Wheeler, M. E., & Sheridan, M. A. (2001). Encoding processes during retrieval tasks. *Journal of Cognitive Neuroscience, 13*(3), 406–415.

Bussey, T. J., & Saksida, L. M. (2007). Memory, perception, and the ventral visual-perirhinal-hippocampal stream: Thinking outside of the boxes. *Hippocampus, 17*(9), 898–908. Available from: http://dx.doi.org/10.1002/Hipo.20320.

Cabeza, R., & Moscovitch, M. (2013). Memory systems, processing modes, and components: Functional neuroimaging evidence. *Perspectives on Psychological Science, 8*(1), 49–55. Available from: http://dx.doi.org/10.1177/1745691612469033.

Caplan, D., & Waters, G. S. (1999). Verbal working memory and sentence comprehension. *Behavioral and Brain Sciences, 22*(1), 77–126.

Chee, M. W., Hon, N., Lee, H. L., & Soon, C. S. (2001). Relative language proficiency modulates BOLD signal change when bilinguals perform semantic judgments. *NeuroImage, 13*(6), 1155–1163.

Chun, M. M. (2000). Contextual cueing of visual attention. *Trends in Cognitive Sciences, 4*(5), 170–178.

Clahsen, H., & Felser, C. (2006). Grammatical processing in language learners. *Applied Psycholinguistics, 27*(1), 3–42.

Davachi, L., Mitchell, J. P., & Wagner, A. D. (2003). Multiple routes to memory: Distinct medial temporal lobe processes build item and source memories. *Proceedings of the National Academy of Sciences of the United States of America, 100*(4), 2157–2162.

Davis, M. H., & Gaskell, M. G. (2009). A complementary systems account of word learning: Neural and behavioural evidence. *Philosophical Transactions of the Royal Society of London, 364*(1536), 3773–3800.

Doyon, J., Bellec, P., Amsel, R., Penhune, V. B., Monchi, O., Carrier, J., et al. (2009). Contributions of the basal ganglia and functionally related brain structures to motor learning. *Behavioural Brain Research, 199*, 61–75.

Duff, M. C., & Brown-Schmidt, S. (2012). The hippocampus and the flexible use and processing of language. *Frontiers in Human Neuroscience, 6*, 69.

Dye, C. D., Walenski, M., Prado, E. L., Mostofsky, S. H., & Ullman, M. T. (2013). Children's computation of complex linguistic forms: A study of frequency and imageability effects. *Public Library of Science ONE, 8*, e74683.

Eichenbaum, H. (2012). *The cognitive neuroscience of memory: An introduction* (2nd ed.). Oxford: Oxford University Press.

Eichenbaum, H., & Cohen, N. J. (2001). *From conditioning to conscious recollection: Memory systems of the brain*. New York, NY: Oxford University Press.

Eichenbaum, H., & Fortin, N. J. (2009). The neurobiology of memory based predictions. *Philosophical Transactions of the Royal Society of London, 364*(1521), 1183–1191.

Eichenbaum, H., & Lipton, P. A. (2008). Towards a functional organization of the medial temporal lobe memory system: Role of the parahippocampal and medial entorhinal cortical areas. *Hippocampus, 18*(12), 1314–1324.

Eichenbaum, H., Sauvage, M., Fortin, N., Komorowski, R., & Lipton, P. (2012). Towards a functional organization of episodic memory in the medial temporal lobe. *Neuroscience and Biobehavioral Reviews, 36* (7), 1597–1608. Available from: http://dx.doi.org/10.1016/j.neubiorev.2011.07.006.

Erickson, K. I., Voss, M. W., Prakash, R. S., Basak, C., Szabo, A., Chaddock, L., et al. (2011). Exercise training increases size of hippocampus and improves memory. *Proceedings of the National Academy of Sciences of the United States of America, 108*(7), 3017–3022.

Estabrooke, I. V., Mordecai, K., Maki, P. M., & Ullman, M. (2002). The effect of sex hormones on language processing. *Brain and Language, 83*, 143–146.

Fernandez, G., Klaver, P., Fell, J., Grunwald, T., & Elger, C. E. (2002). Human declarative memory formation: Segregating rhinal and hippocampal contributions. *Hippocampus, 12*(4), 514–519. Available from: http://dx.doi.org/10.1002/Hipo.10050.

Ferreira, V. S., Bock, K., Wilson, M. P., & Cohen, N. J. (2008). Memory for syntax despite amnesia. *Psychological Science, 19*(9), 940–946.

Freo, U., Pizzolato, G., Dam, M., Ori, C., & Battistin, L. (2002). A short review of cognitive and functional neuroimaging studies of cholinergic drugs: Implications for therapeutic potentials. *Journal of Neural Transmission, 109*(5–6), 857–870.

Friederici, A. D. (2006). Broca's area and the ventral premotor cortex in language: Functional differentiation and specificity. *Cortex, 42* (4), 472–475.

Gould, S. J., & Vrba, E. S. (1982). Exaptation—A missing term in the science of form. *Paleobiology, 8*(1), 4–15.

Green, A. E., & Dunbar, K. N. (2012). Mental function as genetic expression: Emerging insights from cognitive neurogenetics. In K. J. Holyoak, & R. G. Morrison (Eds.), *The Oxford handbook of thinking and reasoning* (pp. 90–111). Oxford: Oxford University Press.

Halle, M., & Marantz, A. (1993). Distributed morphology and the pieces of inflection. In K. Hale, & S. J. Keyser (Eds.), *The view from building 20*. Cambridge, MA: MIT Press.

Hamrick, P. (2014). Recognition memory for novel syntactic structures. *Canadian Journal of Experimental Psychology, 68*(1), 2–7.

Hartshorne, J. K., & Ullman, M. T. (2006). Why girls say "holded" more than boys. *Developmental Science, 9*(1), 21–32.

Hedenius, M., Persson, J., Tremblay, A., Adi-Japha, E., Veríssimo, J., Dye, C. D., et al. (2011). Grammar predicts procedural learning and consolidation deficits in children with specific language impairment. *Research in Developmental Disabilities, 32*(6), 2362–2375.

Hedenius, M., Ullman, M. T., Alm, P., Jennische, M., & Persson, J. (2013). Enhanced recognition memory after incidental encoding in children with developmental dyslexia. *Public Library of Science ONE, 8*(5), e63998.

Henke, K. (2010). A model for memory systems based on processing modes rather than consciousness. *Nature Reviews Neuroscience, 11*, 523–532.

Hickok, G., & Poeppel, D. (2007). The cortical organization of speech processing. *Nature Reviews Neuroscience, 8*(5), 393–402.

Joshi, A., & Rambow, O. (2003). A formalism for dependency grammar based on tree adjoining grammar. Meaning Text Theory Conference, Paris.

Just, M. A., & Carpenter, P. A. (1992). A capacity theory of comprehension: Individual differences in working memory. *Psychological Review, 99*(1), 122–149.

Karuza, E. A., Newport, E. L., Aslin, R. N., Starling, S. J., Tivarus, M. E., & Bavelier, D. (2013). The neural correlates of statistical learning in a word segmentation task: An fMRI study. *Brain and Language, 127*(1), 46–54.

Kaushanskaya, M., Marian, V., & Yoo, J. (2011). Gender differences in adult word learning. *Acta Psychologica, 137*(1), 24–35.

Kensinger, E. A., Ullman, M. T., Locascio, J. J., & Corkin, S. (1999). What is the relation between medial temporal lobe structures and lexical memory? Evidence from amnesic patient H.M. *Society for Neuroscience Abstracts*, *25*, 357.

Kidd, E. (2012a). Implicit statistical learning is directly associated with the acquisition of syntax. *Developmental Psychology*, *48*(1), 171–184.

Kidd, E. (2012b). Individual differences in syntactic priming in language acquisition. *Applied Psycholinguistics*, *33*, 393–418.

Kravitz, D. J., Saleem, K. S., Baker, C. I., & Mishkin, M. (2011). A new neural framework for visuospatial processing. *Nature Reviews Neuroscience*, *12*(4), 217–230.

Kumaran, D., & Maguire, E. A. (2009). Novelty signals: A window into hippocampal information processing. *Trends in Cognitive Sciences*, *13*(2), 47–54.

Kutas, M., & Federmeier, K. D. (2011). Thirty years and counting: Finding meaning in the N400 component of the event-related brain potential (ERP). *Annual Review of Psychology*, *62*, 621–647.

Lee, H., Devlin, J. T., Shakeshaft, C., Stewart, L. H., Brennan, A., Glensman, J., et al. (2007). Anatomical traces of vocabulary acquisition in the adolescent brain. *The Journal of Neuroscience*, *27*(5), 1184–1189.

Lieberman, M. D., Chang, G. Y., Chiao, J., Bookheimer, S. Y., & Knowlton, B. J. (2004). An event-related fMRI study of artificial grammar learning in a balanced chunk strength design. *Journal of Cognitive Neuroscience*, *16*(3), 427–438.

Lukacs, A., Kemeny, F., Lum, J. A. G., & Ullman, M. T. (in preparation). Effective declarative memory consolidation in children with SLI.

Lum, J. A. G., Hedenius, M., Tomblin, J. B., & Ullman, M. T. (in preparation). Recognition memory in children with SLI.

Lum, J. A. G., Conti-Ramsden, G., Morgan, A. T., & Ullman, M. T. (2014). Procedural learning deficits in Specific Language Impairment (SLI): A meta-analysis of serial reaction time task performance. *Cortex*, *51*, 1–10. Available from: http://dx.doi.org/10.1016/j.cortex.2013.10.011.

Lum, J. A. G., Conti-Ramsden, G., Page, D., & Ullman, M. T. (2012). Working, declarative and procedural memory in specific language impairment. *Cortex*, *48*(9), 1138–1154.

Lum, J. A. G., Ullman, M. T., & Conti-Ramsden, G. (2013). Procedural learning is impaired in dyslexia: Evidence from a meta-analysis of serial reaction time studies. *Research in Developmental Disabilities*, *34*(10), 3460–3476. Available from: http://dx.doi.org/10.1016/j.ridd.2013.07.017.

Lum, J. A. G., Ullman, M. T., & Conti-Ramsden, G. (2015). Verbal declarative memory impairments in specific language impairment are related to working memory deficits. *Brain and Language*, *142*, 76–85.

Marcus, G. F., Pinker, S., Ullman, M. T., Hollander, M., Rosen, T. J., & Xu, F. (1992). Overregularization in language acquisition. *Monographs of the Society for Research in Child Development*, *57*(4), 1–165, Serial No. 228.

Marshall, L., & Born, J. (2007). The contribution of sleep to hippocampus-dependent memory consolidation. *Trends in Cognitive Sciences*, *11*(10), 442–450.

Martin, A. (2007). The representation of object concepts in the brain. *Annual Review of Psychology*, *58*, 25–45.

Martins, S., Guillery-Girard, B., Jambaque, I., Dulac, O., & Eustache, F. (2006). How children suffering severe amnesic syndrome acquire new concepts? *Neuropsychologia*, *44*(14), 2792–2805. Available from: http://dx.doi.org/10.1016/j.neuropsychologia.2006.05.022.

McCarthy, G., Nobre, A. C., Bentin, S., & Spencer, D. D. (1995). Language-related field potentials in the anterior-medial temporal lobe: I. Intracranial distribution and neural generators. *Journal of Neuroscience*, *15*(2), 1080–1089.

McClelland, J. L., McNaughton, B. L., & O'Reilly, R. C. (1995). Why there are complementary learning systems in the hippocampus and neocortex: Insights from the successes and failures of connectionist models of learning and memory. *Psychological Review*, *102*(3), 419–457.

Mechelli, A., Crinion, J. T., Noppeney, U., O'Doherty, J., Ashburner, J., Frackowiak, R. S., et al. (2004). Neurolinguistics: Structural plasticity in the bilingual brain. *Nature*, *431*(7010), 757.

Meyer-Lindenberg, A. E. S. R., Lipska, B. K., Verchinski, B. A., Goldberg, T., Callicott, J. H., Weinberger, D., et al. (2007). Genetic evidence implicating DARPP-32 in human frontostriatal structure, function, and cognition. *The Journal of Clinical Investigation*, *117*(3), 672–682.

Meyer, P., Mecklinger, A., Grunwald, T., Fell, J., Elger, C. E., & Friederici, A. D. (2005). Language processing within the human medial temporal lobe. *Hippocampus*, *15*, 451–459.

Mishkin, M., Malamut, B., & Bachevalier, J. (1984). Memories and habits: Two neural systems. In G. Lynch, J. L. McGaugh, & N. M. Weinburger (Eds.), *Neurobiology of learning and memory* (pp. 65–77). New York, NY: The Guilford Press.

Morgan-Short, K., Faretta-Stutenberg, M., Brill-Schuetz, K. A., Carpenter, H., & Wong, P. C. M. (2014). Declarative and procedural memory as individual differences in second language acquisition. *Bilingualism: Language and Cognition*, *17*(1), 56–72.

Morgan-Short, K., Finger, I., Grey, S., & Ullman, M. T. (2012). Second language processing shows increased native-like neural responses after months of no exposure. *Public Library of Science ONE*, *7*(3), e32974.

Morgan-Short, K., Steinhauer, K., Sanz, C., & Ullman, M. T. (2012). Explicit and implicit second language training differentially affect the achievement of native-like brain activation patterns. *Journal of Cognitive Neuroscience*, *24*(4), 933–947.

Munoz-Lopez, M. M., Mohedano-Moriano, A., & Insausti, R. (2010). Anatomical pathways for auditory memory in primates. *Frontiers in Neuroanatomy*, *4*(129).

Nemeth, D., Janacsek, K., Csifcsak, G., Szvoboda, G., Howard, J., James, H., et al. (2011). Interference between sentence processing and probabilistic implicit sequence learning. *Public Library of Science ONE*, *6*(3), e17577.

Packard, M. G. (2008). Neurobiology of procedural learning in animals. In J. H. Byrne (Ed.), *Concise learning and memory: The editor's selection* (pp. 341–356). London: Elsevier Science & Technology Books.

Petersson, K. M., Folia, V., & Hagoort, P. (2012). What artificial grammar learning reveals about the neurobiology of syntax. *Brain and Language*, *120*(2), 83–95.

Petrides, M., & Pandya, D. N. (2009). Distinct parietal and temporal pathways to the homologues of Broca's area in the monkey. *Public Library of Science Biology*, *7*(8), e1000170.

Pezawas, L., Verchinski, B. A., Mattay, V. S., Callicott, J. H., Kolachana, B. S., Straub, R. E., et al. (2004). The brain-derived neurotrophic factor Val66met polymorphism and variation in human cortical morphology. *Journal of Neuroscience*, *24*(45), 10099–10102.

Poldrack, R. A., & Packard, M. G. (2003). Competition among multiple memory systems: Converging evidence from animal and human brain studies. *Neuropsychologia*, *41*(3), 245–251.

Postle, B. R., & Corkin, S. (1998). Impaired word-stem completion priming but intact perceptual identification priming with novel words: Evidence from the amnesic patient H.M. *Neuropsychologia*, *15*, 421–440.

Prado, E., & Ullman, M. T. (2009). Can imageability help us draw the line between storage and composition? *Journal of Experimental Psychology: Language, Memory, and Cognition*, *35*(4), 849–866.

Quiroga, R. Q., Reddy, L., Kreiman, G., Koch, C., & Fried, I. (2005). Invariant visual representation by single neurons in the human brain. *Nature*, *435*(7045), 1102–1107. Available from: http://dx.doi.org/10.1038/nature03687.

Raboyeau, G., Marcotte, K., Adrover-Roig, D., & Ansaldo, A. I. (2010). Brain activation and lexical learning: The impact of learning phase and word type. *NeuroImage, 49*(3), 2850–2861.

Rauschecker, J. P., & Scott, S. K. (2009). Maps and streams in the auditory cortex: Nonhuman primates illuminate human speech processing. *Nature Neuroscience, 12*, 718–724.

Saffran, J. R., Aslin, R. N., & Newport, E. L. (1996). Statistical learning by 8-month-old infants. *Science, 274*, 1926–1928.

Schendan, H., Searl, M., Melrose, R., & Stern, C. (2003). An fMRI study of the role of the medial temporal lobe in implicit and explicit sequence learning. *Neuron, 37*(6), 1013–1025.

Schreiweis, C., Bornschein, U., Burguiere, E., Kerimoglu, C., Schreiter, S., Dannemann, M., et al. (2014). Humanized Foxp2 accelerates learning by enhancing transitions from declarative to procedural performance. *Proceedings of the National Academy of Sciences of the United States of America*. Available from: http://dx.doi.org/10.1073/pnas.1414542111.

Squire, L. R., & Wixted, J. T. (2011). The cognitive neuroscience of human memory since H.M. *Annual Review of Neuroscience, 34*, 259–288.

Tagarelli, K., Turkeltaub, P. E., Grey, S., & Ullman, M. T. (in preparation). The functional neuroanatomy of adult second language: An activation likelihood estimation meta-analysis.

Ullman, M. T. (2004). Contributions of memory circuits to language: The declarative/procedural model. *Cognition, 92*(1–2), 231–270.

Ullman, M. T. (2005). A cognitive neuroscience perspective on second language acquisition: The declarative/procedural model. In C. Sanz (Ed.), *Mind and context in adult second language acquisition: Methods, theory and practice* (pp. 141–178). Washington, DC: Georgetown University Press.

Ullman, M. T. (2006). Is Broca's area part of a basal ganglia thalamocortical circuit? *Cortex, 42*(4), 480–485.

Ullman, M. T. (2007). The biocognition of the mental lexicon. In M. G. Gaskell (Ed.), *The Oxford handbook of psycholinguistics* (pp. 267–286). Oxford, UK: Oxford University Press.

Ullman, M. T. (2013). The role of declarative and procedural memory in disorders of language. *Linguistic Variation, 13*(2), 133–154.

Ullman, M. T. (2015). The declarative/procedural model: A neurobiologically-motivated theory of first and second language. In B. VanPatten & J. Williams (Eds.), *Theories in second language acquisition* (2nd ed., pp. 135–158). New York, London: Routledge.

Ullman, M. T., Campbell, R., McQuaid, G. A., Tagarelli, K. M., & Turkeltaub, P. E. (in preparation). The functional neuroanatomy of regular and irregular morphology.

Ullman, M. T., Lovelett, J. T., Gelfand, M. P., Litcofsky, K. A., Pullman, M. Y., Moffa, M., et al. (under revision). The influence of handedness on language: Storage versus composition differences in left- and right-handers.

Ullman, M. T., & Gopnik, M. (1999). Inflectional morphology in a family with inherited specific language impairment. *Applied Psycholinguistics, 20*(1), 51–117.

Ullman, M. T., Lum, J. A. G., & Conti-Ramsden, G. (2014). Domain specificity in language development. In P. Brooks, & V. Kempe (Eds.), *Encyclopedia of language development* (Vol. 1, 1st ed.). Los Angeles, CA: Sage Publications.

Ullman, M. T., Miranda, R. A., & Travers, M. L. (2008). Sex differences in the neurocognition of language. In J. B. Becker, K. J. Berkley, N. Geary, E. Hampson, J. Herman, & E. Young (Eds.), *Sex on the brain: From genes to behavior* (pp. 291–309). New York, NY: Oxford University Press.

Ullman, M. T., & Pierpont, E. I. (2005). Specific language impairment is not specific to language: The procedural deficit hypothesis. *Cortex, 41*(3), 399–433.

Ullman, M. T., Pullman, M., Lovelett, J. T., McQuaid, G. A., Pierpont, E. I., & Turkeltaub, P. E. (under review). Brain abnormalities in specific language impairment are localized to frontal regions and the caudate nucleus. *Neuroscience and Biobehavioral Reviews.*

Ullman, M. T., & Pullman, M. Y. (2015). A compensatory role for declarative memory in neurodevelopmental disorders. *Neuroscience and Biobehavioral Reviews, 51*, 205–222.

Uncapher, M. R., & Wagner, A. D. (2009). Posterior parietal cortex and episodic encoding: Insights from fMRI subsequent memory effects and dual-attention theory. *Neurobiology of Learning and Memory, 91*(2), 139–154. Available from: http://dx.doi.org/10.1016/j.nlm.2008.10.011.

Vargha-Khadem, F., Gadian, D. G., & Mishkin, M. (2001). Dissociations in cognitive memory: The syndrome of developmental amnesia. *Philosophical Transactions of the Royal Society of London. Series B, Biological Sciences, 356*(1413), 1435–1440.

Wagner, A. D., Schacter, D. L., Rotte, M., Koutstaal, W., Maril, A., Dale, A. M., et al. (1998). Building memories: Remembering and forgetting of verbal experiences as predicted by brain activity. *Science, 281*(5380), 1188–1191.

Wagner, A. D., Shannon, B. J., Kahn, I., & Buckner, R. L. (2005). Parietal lobe contributions to episodic memory retrieval. *Trends in Cognitive Sciences, 9*(9), 445–453. Available from: http://dx.doi.org/10.1016/j.tics.2005.07.001.

Walenski, M., Mostofsky, S. H., Gidley-Larson, J. C., & Ullman, M. T. (2008). Brief report: Enhanced picture naming in autism. *Journal of Autism and Developmental Disorders, 38*(7), 1395–1399.

Walenski, M., Prado, E. L., Ozawa, K., Steinhauer, K., & Ullman, M. T. (under revision). The compositionality and storage of inflected forms: Evidence from working memory effects. *Cognition.*

Walenski, M., Tager-Flusberg, H., & Ullman, M. T. (2006). Language in autism. In S. O. Moldin, & J. L. R. Rubenstein (Eds.), *Understanding autism: From basic neuroscience to treatment* (pp. 175–203). Boca Raton, FL: Taylor and Francis Books.

Wartenburger, I., Heekeren, H. R., Abutalebi, J., Cappa, S. F., Villringer, A., & Perani, D. (2003). Early setting of grammatical processing in the bilingual brain. *Neuron, 37*, 159–170.

Winocur, G., & Moscovitch, M. (2011). Memory transformation and systems consolidation. *Journal of the International Neuropsychological Society, 17*(5), 766–780. Available from: http://dx.doi.org/10.1017/S1355617711000683.

Wixted, J. T., & Squire, L. R. (2011). The medial temporal lobe and the attributes of memory. *Trends in Cognitive Sciences, 15*(5), 210–217.

Woltering, J. M., Noordermeer, D., Leleu, M., & Duboule, D. (2014). Conservation and divergence of regulatory strategies at Hox Loci and the origin of tetrapod digits. *Public Library of Science Biology, 12*(1), e1001773. Available from: http://dx.doi.org/10.1371/journal.pbio.1001773.

Wong, P. C. M., Ettlinger, M., & Zheng, J. (2013). Linguistic grammar learning and DRD2-TAQ-IA polymorphism. *Public Library of Science ONE, 8*(5), e64983.

Yang, J., & Li, P. (2012). Brain networks of explicit and implicit learning. *Public Library of Science ONE, 7*(8), e42993.

Yonelinas, A. P., Aly, M., Wang, W.-C., & Koen, J. D. (2010). Recollection and familiarity: Examining controversial assumptions and new directions. *Hippocampus, 20*(11), 1178–1194.

CHAPTER

77

Perinatal Focal Brain Injury: Scope and Limits of Plasticity for Language Functions

Susan C. Levine[1], *Anjali Raja Beharelle*[2,5], *Özlem Ece Demir*[3,6] *and Steven L. Small*[4]

[1]Department of Psychology, Department of Comparative Human Development, and Committee on Education, University of Chicago, Chicago, IL, USA; [2]Department of Economics, University of Zürich, Zürich, Switzerland; [3]Department of Psychology, University of Chicago, Chicago, IL, USA; [4]Department of Neurology, The University of California, Irvine, CA, USA; [5]Laboratory for Social and Neural Systems Research, University of Zurich, Zurich, Switzerland; [6]Department of Communication Sciences and Disorders, Northwestern University, Evanston, IL, USA

77.1 PERINATAL FOCAL BRAIN INJURY: LANGUAGE DEVELOPMENT AND NEURAL PLASTICITY

The study of children with perinatal focal brain injury provides a unique window into the plasticity of language development. A large body of research reveals normal or near-normal language development in these children even when lesions are large and encompass classic left hemisphere perisylvian language networks. Strikingly, children with unilateral perinatal lesions (PLs) do not exhibit the marked aphasias that are common when anatomically comparable lesions are incurred during adulthood (Basser, 1962; Bates et al., 2001; Bishop, 1993; Lenneberg, 1967; Reed & Reitan, 1969; Rowe, Levine, Fisher, & Goldin-Meadow, 2009). Plasticity for language functions after early brain injury presents a challenge to theories that posit an immutable brain basis for human language by demonstrating that alternative neural networks can support language, at least when unilateral injuries occur at certain developmental stages. Thus, although the neural networks that are typically involved in language may be optimal, they do not appear to be necessary for the development of normal or near-normal levels of language function.

The normal or near-normal language skills of children with PL are consistent with the view of a dynamically developing, plastic brain—a developing brain capable of responding to internal biological signals, including those associated with the lesion, and to information provided by the environment (Demir, Fisher, Levine, & Goldin-Meadow, 2013; Demir, Rowe, Heller, Levine, & Goldin-Meadow, 2015; Elman et al., 1996; Huttenlocher, 2002; Rowe et al., 2009; Stiles, Reilly, Levine, Trauner, & Nass, 2012; Stiles, Reilly, Paul, & Moses, 2005). This adaptive view of brain development has dramatically expanded the focus of research on children with PL from one that considers plasticity as a lesion-induced phenomenon (Teuber, 1974; Witelson, 1985) to one that considers plasticity after early lesions in the context of biological and environmental factors that affect brain and behavior over the course of development in typically developing (TD) children (Greenough, Black, & Wallace, 1987; Rowe et al., 2009; Stiles et al., 2012, 2005; Thomas & Karmiloff-Smith, 2002; Witte, 1998). Thomas and Karmiloff-Smith (2002) argue that the processes of development cannot be ignored in constructing models of developmental disorders. Within this broader view, our understanding of brain and behavioral development in children with early brain injury can inform and be informed by our understanding of normative brain development.

Language functioning in TD children is known to depend on the development of widespread neural networks, and development of these networks is influenced by variations in the language experiences of individual children (Hart & Risley, 1992; Hoff, 2006; Huttenlocher, Haight, Bryk, Seltzer, & Lyons, 1991). When a lesion is superimposed on the developing brain, the lesion itself becomes a factor in this dynamic development. Outcomes are likely to depend on all of the factors that affect language development in TD children

Neurobiology of Language. DOI: http://dx.doi.org/10.1016/B978-0-12-407794-2.00077-8

as well as many factors introduced by the lesion, including the timing of the lesion within the developmental period and the location and extent of the lesion.

It is important to point out that the study of children with PL remains an important scientific enterprise even in the context of powerful brain imaging tools that can be used to study brain–behavior relationships in TD children. In particular, whereas brain imaging studies of children developing language with an intact brain can tell us about the brain networks that are typically involved in language functioning at particular ages, they cannot tell us about the alternative networks that are able to support language functioning after an early brain injury, or which regions are necessary or sufficient for development of language. In addition, although animal studies have played a major role in increasing our knowledge about the plasticity of the developing nervous system with respect to biological perturbations and with respect to variations in experience (Greenough et al., 1987; Kolb & Gibb, 2010), these studies do not provide an ideal model for specific questions about the plasticity of language networks because humans are the only species with language.

77.2 FOUR CENTRAL QUESTIONS

Studies of language development in children with PL have largely focused on examining the scope and limits of plasticity for language functions in this population (Bates et al., 1997; Bates, Vicari, & Trauner, 1999; Reilly, Levine, Nass, & Stiles, 2008; Stiles et al., 2012; Trauner, Eshagh, Ballantyne, & Bates, 2013). To the extent that language is disrupted in similar ways after early and later lesions, we gain evidence for the importance of particular brain regions for language, beginning early in life, and for the limits of early functional plasticity. In contrast, to the extent that the process of language development proceeds normally in children with PL, regardless of lesion location, we gain evidence for the robustness of language development and for the ability of the young brain to adapt to injury.

We organize this chapter around four key questions that have been addressed by researchers studying language development in children with perinatal stroke, reviewing the literature relevant to each.

These questions are:

1. How do focal PLs affect language development?
2. How do biological characteristics of early focal lesions relate to language development?
3. What is the role of language input on the language development of children with PLs?
4. What are potential mechanisms of language plasticity after early lesions?

77.3 HOW DO FOCAL PERINATAL LESIONS AFFECT LANGUAGE DEVELOPMENT?

During the 19th century, Broca (1865) and others noted that aphasias observed in adults with brain injury were not present when lesions were acquired early in life. Jules Cotard, a colleague of Broca's, wrote: "Intelligence may be normal when a hemisphere is destroyed during infancy…in these cases one never encounters aphasia" (Ritti, 1894, see Stiles et al., 2012). These observations have largely stood the test of time. That is, the language deficits of children with early focal lesions are subtle and do not resemble the aphasias observed when lesions are acquired during adulthood.

Bates et al. (1997) and Bates, Thal, Finlay, and Clancy (2003) pointed out a fundamental but often ignored difference between the situations faced by the individual with an early versus later brain injury. When an injury occurs during adulthood, it is superimposed on a developed brain that has already acquired a language. In contrast, when a lesion occurs early in life, it is superimposed on a developing brain that has yet to acquire language. Thus, the adult with an acquired lesion is faced with the task of relearning language with a brain that is already mature and has become specialized over time. The neuroanatomical changes that occur over the course of childhood and adolescence have been well-documented and many changes in both gray and white matter structures support the observed specialization (Giedd et al., 1999; Huttenlocher, 2002; Huttenlocher & Dabholkar, 1997; Sowell, Trauner, Gamst, & Jernigan, 2002). In contrast, the child with PL is faced with the task of learning language for the first time with an injured brain, which, although compromised by the injury, is still a highly flexible, adaptable neural substrate. The fundamental differences between the learning task as well as the developmental state of the brain at the time of injury undoubtedly affect the different outcomes that are observed in the postlesion language functioning of these groups.

77.3.1 Severity of Language Deficits in Children with perinatal lesions

Children with PL have delays when beginning language and show delays in babbling, gesturing, and productive and expressive vocabulary and syntax (Bates et al., 1997; Feldman, 1994; Feldman, Holland, Kemp, & Janowsky, 1992; Marchman, Miller, & Bates, 1991; Trauner et al., 2013). These initial delays are followed by a period of "catch-up" by the start of elementary school, when children's vocabulary and syntactic skills generally fall in the low-average to average range (Bates & Roe, 2001; Bates et al., 1997,

1999; Demir, Levine, & Goldin-Meadow, 2010; Levine, Huttenlocher, Banich, & Duda, 1987).

The catch-up that has been found for some linguistic tasks does not extend to all language tasks. At the group level, children with PL experience difficulty on a variety of complex linguistic tasks during the elementary school period and adolescence, although there is wide individual variability. For example, on a demanding morpho-syntactic task, by generating tag endings to questions (e.g., "John wants to go the 6 o'clock movie, *doesn't he*?"), children with PL perform significantly worse than TD children at ages 12–16 years, but not between 4 and 7 years of age or between 8 and 11 years of age, suggesting that linguistic deficits may emerge or become more apparent over the course of development (Weckerly, Wulfeck, & Reilly, 2004).

Children with PL also have difficulty on narrative tasks. In one study, the narrative production skills of young elementary school children with PL were compared with those of TD control children (Demir et al., 2010). Children were asked to generate a story in response to a story stem such as, "Once there was a big gray fox who lived in a cave near a forest...." (Stein & Albro, 1997). The children with PL produced significantly shorter narratives than the TD children and included fewer unique words and clauses. Additionally, they were less likely to tell goal-based stories than their TD peers, even though the groups did not differ in receptive vocabulary or in their syntactic skills. Similar results have been reported on other tasks tapping narrative skill such as retelling a story from a wordless picture book, from a wordless cartoon, or from a story told by a storyteller (Chapman, Max, McGlothlin, Gamino, & Cliff, 2003; Demir, et al., 2013; Reilly, Bates, & Marchman, 1998; Reilly, Losh, Bellugi, & Wulfeck, 2004).

Producing a well-structured narrative may be a challenging task because it puts heavy demands on planning and organizing events in a hierarchical manner. It also involves producing an extended monologue that is removed from the here and now. Consistent with the view that narratives are computationally demanding, narrative tasks engage a wide neural network (Nichelli et al., 1995), and early brain injury may compromise such networks (Feldman, 2005; Feldman, MacWhinney, & Sacco, 2002; Levine, Kraus, Alexander, Suriyakham, & Huttenlocher, 2005).

Children with PL also have difficulty on other complex linguistic tasks during elementary school and beyond. For example, a study by MacWhinney, Feldman, Sacco, and Valdes-Perez (2000) revealed that 5- to 11-year-old children with PL scored significantly lower than TD children in "language planning" skills, including difficulty composing sentences that include specific words and using the linguistic information in a sentence to perform a set of actions. These language planning skills go beyond the basic lexical and syntactic skills that are tapped by vocabulary and syntax tasks on which children with PL perform relatively well. Interestingly, producing a coherent narrative also puts demands on language planning skills.

A few studies have examined online language processing skills of children with perinatal brain injury and TD children, allowing them to examine not only errors in processing but also speed of processing of various kinds of linguistic structures. Generally, children with PL are less accurate and slower at processing complex sentences (Dick, Wulfeck, Krupa-Kwiatkowski, & Bates, 2004; Feldman et al., 2002; MacWhinney et al., 2000). For example, Dick et al. (2004) presented children with PL (7–18 years of age; mean, 10.8 years) and TD children (5–17 years of age; mean, 10–8 years) with a task involving pointing out the agent of sentences. The sentences had subjects and objects that were either in a canonical order (active: *The dog is biting the cat* or subject cleft: *It is the dog that is biting the cat*) or in noncanonical order (passive: *The cat is bitten by the dog* or object cleft: *It is the cat that the dog is biting*). Findings showed no difference in performance between the groups on the canonical sentences, but showed significantly more errors and slower reaction times among children with PL on the noncanonical sentences. The performance of the group with PLs was found to be similar to that of the younger TD children, suggesting that their language processing skills follow a normal course, but at a slower rate.

Although very few studies have examined written language skills of children with PL, there is some indication that they experience difficulty with reading comprehension and spelling (Aram & Ekelman, 1988; Demir, Carlson, Levine, & Goldin-Meadow, 2013; Woods & Carey, 1979). This is clearly an area in need of further research.

It is possible that the language processing problems of children with PL, which are apparent on complex language tasks, are related to subtle deficits in executive functioning, notably verbal working memory after PL (Nichols et al., 2004; Westmacott, Askalan, MacGregor, Anderson, & Devebar, 2010). However, there has been very little work probing this relationship, and more is needed before concluding that this is the case.

77.3.2 Relation of Gesture and Language

Examining the role of gesture in language learning provides another way to compare the process of language learning in children with PL with that of TD children. Several studies show that the production of gestures not only precedes but also predicts the production of words in TD children (Acredolo & Goodwyn, 1989; Bates, Benigni, Bretherton, Camaioni, & Volterra, 1979). Further, in TD children, gesture–speech combinations in which the gesture supplements the information conveyed in speech precede and predict early sentences expressed in speech alone (Goldin-Meadow & Butcher, 2003; Iverson & Goldin-Meadow, 2005; Özcaliskan & Goldin-Meadow, 2005). If these relations hold for children with PLs, this would suggest that similar processes govern language acquisition in the two groups of children.

With respect to vocabulary development, Sauer, Levine, and Goldin-Meadow (2010) examined the relation of early gesture and later vocabulary development in children with perinatal brain injury. Those whose gesture production was above the 25th percentile of TD children at 14 months of age had productive vocabularies in the normal range when tested again at 22 and 26 months of age and receptive vocabularies in the normal range when tested at 30 months. In contrast, those whose gesture production was below the 25th percentile of TD children at 14 months had productive and receptive vocabularies below the normal range at these same ages. Importantly, these later productive and receptive vocabulary difficulties were not predicted by children's productive vocabulary at 14 months. Thus, as for TD children, early gesture predicts later vocabulary development at an age when early language does not.

With respect to syntactic development, Özcaliskan, Levine, and Goldin-Meadow (2013) found that children with PL, like TD children, produce simple sentences across gesture and speech (point at box + "open") several months before producing them entirely in speech ("open box"). However, unlike TD children, children with PL produced complex sentences (sentences with two predicates) in speech alone before producing them in gesture and speech, albeit approximately 8 months later than their TD peers.

Özcaliskan et al. propose that children with PL may have difficulty producing the complex gestures that occur in predicate–predicate utterances because of their hemiparesis. Studies that examined children with congenital hemiplegia (which is primarily due to periventricular, cortical, and subcortical lesions that can be unilateral or have asymmetric bilaterality) have shown that, to varying degrees, these children experience hand weakness and decreased dexterity in the hand contralateral to the lesion, difficulties that affect both unimanual and bimanual coordination (Gordon & Steenbergen, 2008; Sakzewski, Ziviani, & Boyd, 2010; Stiles et al., 2012). Some studies also report impaired strength and coordination in the hand ipsilateral to the lesion (the "unimpaired" hand) (Brown et al., 1989; Duque et al., 2003; Kuhtz-Buschbeck, Sundholm, Eliasson, & Forssberg, 2000).

These hand-motor difficulties may account for the paucity of predicate–predicate utterances in gesture–speech produced by children with PL, which in turn may account for their delay in producing complex predicate–predicate utterance in speech alone. This scenario is consistent with the theory that gesture milestones not only predict language milestones but also promote these milestones. An interesting possibility raised by these findings is that an intervention that involves encouraging young children with PL to gesture may support their language development (Goldin-Meadow et al., 2014).

77.3.3 Summary

Considered together, the findings that have emerged from the study of children with PL clearly show that their language difficulties are much more subtle than those seen in adults with similar lesions. However, early brain injuries do appear to have some cost to language development—that is, plasticity is not complete. The degree to which one observes plasticity for language appears to depend on the demands of the tasks administered and the age of the child when they are administered (Banich, Levine, Kim, & Huttenlocher, 1990; Feldman, 2005; Levine et al., 2005). For example, when one assesses a complex linguistic skill prior to the age at which TD children have much success with it, deficits may not be apparent. However, when one assesses this skill later, children with PL may exhibit delays or deficits. At this later time point, the neural networks that underlie efficient processing may be sufficiently mature to support strong performance in TD children but remain compromised in the children with PL (Goldman, 1974; Levine et al., 2005). As we have seen, some of the difficulties that children with PL experience appear to be transient (e.g., the delay in beginning language). Other difficulties emerge later, such as difficulty with tag questions, narrative production, and language planning, but longitudinal data are needed to determine whether these difficulties represent an iterative pattern of delay and catch-up or more persistent delay (Stiles et al., 2005).

77.4 HOW DO BIOLOGICAL CHARACTERISTICS OF EARLY FOCAL LESIONS RELATE TO LANGUAGE FUNCTIONING?

77.4.1 Lesion Location

There are two prevalent views concerning the association of lesion laterality and language difficulties in children versus adults with focal injury (Bates et al., 1997;

Bates, Vicari, & Trauner, 1999; Rowe et al., 2009). According to one view, early left hemisphere lesions result in more marked language deficits than early right hemisphere lesions (Annett, 1973; Aram & Ekelman, 1986; Aram, Ekelman, Rose, & Whitaker, 1985; Dennis & Whitaker, 1976; Levine et al., 1987; Rankin, Aram, & Horwitz, 1981; Riva & Cazzaniga, 1986; Vargha-Khadem, O'Gorman, & Watter, 1985; Woods & Teuber, 1978). For example, Chilosi, Cipriani, Bertuccelli, Pfanner, and Cioni (2001) found that 2-year-old and 4-year-old children with left hemisphere injury generally scored lower than those with right hemisphere injury on tasks assessing lexical and syntactic skills. The other view is that language delays and deficits are not associated with the laterality of early lesions (Ballantyne, Spilkin & Trauner, 2007; Bates et al., 2001; Dall'Oglio, Bates, Volterra, De Capua & Pezzini 1994; Dick et al., 2004; Feldman et al., 1992; Marchman et al., 1991; Reilly et al., 1998; Vargha-Khadem, Isaacs, & Muter, 1994). Although there is inconsistency across studies with respect to these views, the strong relation between language deficits and left hemisphere lesions that emerges from studies of adult stroke patients is clearly much weaker after PL.

Two prospective studies have examined whether language profiles of children with PL correspond to the site-specific profiles that have been described in adults with focal brain injury. Thal et al. (2001) examined the receptive and productive vocabularies of children with PL based on parents' reports. They found stronger and more consistent delays in productive than receptive vocabulary when children were 12–35 months old. They also found that children with right hemisphere damage were particularly delayed in receptive vocabulary, a surprising finding in view of the adult literature indicating that left posterior lesions are associated with comprehension deficits. Another surprising finding was that expressive vocabulary was more delayed in 17- to 35-month-olds with left posterior lesions (retrorolandic, with or without prerolandic involvement) than in those with left anterior lesions. This is the opposite of what one expects from the adult aphasia literature. However, some caveats are that these results were based on a relatively small sample and the group of children without left posterior damage was extremely heterogeneous, including children with lesions restricted to anterior left hemisphere and children with unilateral right hemisphere lesions.

Bates et al. (1997) reexamined these site-specific relations with a larger sample that included some children from the Thal et al. (1991) study. They again found that between 10 and 17 months of age, children with right hemisphere lesions, regardless of location, had lower receptive vocabularies than those with left hemisphere lesions, according to parental report. In addition, children with right hemisphere lesions were particularly delayed in the production of communicative gestures. With respect to expressive vocabulary in this early period, children with PL were delayed regardless of lesion location. These delays were also found later, between 19 and 31 months age, which encompasses the typical vocabulary burst. However, during this later period, having a lesion that involved the left temporal cortex increased the magnitude and the likelihood of delay. Further, children with lesions of the frontal cortex of either hemisphere showed marked delays in expressive vocabulary. Parallel results were found for a measure of syntax production based on parental report, indicating that lexical and syntactic skills are associated in children with PL, as is the case in TD children (Fenson et al., 1994).

Bates et al. (1997) also examined the mean length of utterance (MLU) of children with PL from free speech samples between 20 and 44 months. Results were consistent with prior findings showing productive delays in children with PL as a group, but with an added risk when lesions involved the left temporal lobe. As noted by the investigators, it is possible that the lesions that involved the left temporal lobe were larger than those that did not, simply by requiring that this particular area be damaged in one group but not in the comparison group. A subsequent study showed that children with left temporal lesions continued to have a disadvantage in grammar up to 6 years of age, but that after age 6 years there were no site-specific deficits (Reilly et al., 1998).

77.4.2 Lesion Size

Not surprisingly, most studies report that the size of perinatal focal lesions is associated with the severity of deficits in gesture, language, reading, and other cognitive functions (Booth et al., 2000; Cioni et al., 1999; Levine et al., 2005; Rowe et al., 2009; Sauer et al., 2010), although there are exceptions (Ballantyne, Spilkin, Hesselink, & Trauner, 2008). One small study reports a curvilinear rather than a linear relationship between lesion size and language function, such that medium lesions are associated with more marked language deficits than either small or large lesions (Thal et al., 1991). There is some support for the curvilinear hypothesis from studies of monkeys with brain lesions (Irle, 1990). However, a later study that expanded the size of the sample in the Thal et al. (1991) study did not replicate the finding of better language function in children with small and large PL than in children with medium lesions (Bates et al., 1999). Lesion size and lesion type are correlated, which may explain some of the inconsistencies in findings.

77.4.3 Lesion Type

Early vascular lesions (stroke) result from hemorrhage (bleeding in the brain) or ischemic infarction (death of brain tissue from lack of blood delivery). The hemorrhages typically occur in highly cellular regions of the developing brain adjacent to fluid-filled spaces (ventricles) and primarily affect white matter tracts, and have been called *periventricular*. The ischemic infarctions typically involve disruptions of blood flow (or *ischemia*) to parts of the brain supplied by the middle cerebral artery and primarily affect gray matter structures including the frontal, temporal, and parietal lobes as well as the basal ganglia, and have been called *vascular* lesions (Goldin-Meadow et al., 2014). In addition to these differences in location, these lesions also tend to differ in size, with vascular (ischemic) lesions tending to be larger (Rowe et al., 2009). These two different lesion types also differ with respect to the time of lesion onset: periventricular lesions tend to occur mainly during the early third trimester of pregnancy, whereas ischemic infarctions tend to occur mainly during the late third trimester of pregnancy or perinatally (Krägeloh-Mann, 2004; Staudt et al., 2004). Rowe et al. (2009) examined the growth of productive vocabulary and MLU in children with PL and in TD children between 14 and 46 months of age. They found that children with vascular lesions produced significantly fewer word types (unique words) and shorter utterances than TD children and children with periventricular lesions who did not differ from the control children. Additionally, Sauer et al. (2010) found that children with vascular lesions were more likely to be in the low gesture group at 14 months of age than children with periventricular lesions. Because of the correlations noted, it is difficult to disentangle whether the better language functioning of the children with periventricular lesions is due to lesion location, lesion size, or lesion timing (Staudt et al., 2004). Further, although Sauer et al. did not include children born before 36 weeks of gestation, periventricular lesions are often associated with prematurity, which introduces an additional factor that needs consideration when comparing the effects of these different lesion types.

77.4.4 Seizure History

There is some inconsistency in reports concerning the association of seizures with language difficulties in children with PL. Bates et al. (1999) stress that the effects of early unilateral brain injury are found in children who have never experienced a seizure and are therefore not solely the result of seizures. Some studies found no significant effects of seizures on language function or IQ in this population (Bates et al., 1997; Levine et al., 2005). Other studies, in contrast, found that seizures negatively affect functioning in children with early brain injury (Ballantyne et al., 2008; Ballantyne et al., 2007; Chilosi et al., 2001; Cioni et al., 1999; Cohen & Duffner, 1981; Dall'Oglio et al., 1994; Vargha-Khadem, Isaacs, Van Der Werf, Robb, & Wilson, 1992). Notably, Vargha-Khadem et al. (1992) found that among children with early unilateral lesions, those with seizures had lower verbal and performance IQ than that of children with lesions who never experienced seizures and lower than that of TD children. Although the seizure-free group had lower performance IQ than TD children, this was not the case for their verbal IQ.

These findings are difficult to interpret because of the heterogeneity of the seizure groups, which consisted of children with a wide range of severity of seizure disorders. A large study of the relation of seizures with IQ suggests that it may specifically be the children with recurrent seizures who experience intellectual deficits (Huttenlocher & Hapke, 1990). It is possible that epilepsy may interfere with plasticity by transforming a focal lesion into a more global pathology (Stiles et al., 2012). Consistent with this possibility, Ballantyne et al. (2008) found that the full-scale IQ of children with PL and seizures declined over time, whereas this was not the case for children who did not have seizures. The potential role of seizure medications also needs to be considered (Levine et al., 2005).

77.4.5 Motor and Language Functions

A number of studies report a negative relation between degree of motor impairment and verbal and performance IQ (Levine et al., 1987; Riva & Cazzaniga, 1986), but other studies do not report such a relationship (Carlsson et al., 1994) in children with PL. Levine et al. (1987) found that degree of hemiparesis, lesion size, and Electroencephalogram (EEG) abnormality were all negatively correlated with IQ scores. However, hemiparesis was the strongest predictor of IQ, and neither lesion size nor EEG abnormality accounted for added variance above this variable. It is possible that degree of hemiparesis is an index of underlying neural impairment. It is also possible that motor difficulties impede the development of language and cognitive skills, a theory that would align with theories of embodiment, which posit a close relation between sensorimotor representations and more abstract concepts (Barsalou, 1999; Glenberg, 1997; Zwaan, 2008). What is clear from studies of children with PL and TD children is that motor and language functions can dissociate. In TD children, there is not a significant correlation between motor and language milestones between 9 and

13 months of age (Bates et al., 1979). Further, in children with PL, hemiparesis commonly persists even though language functions develop quite normally.

77.4.6 Lesion Timing

Although most studies report greater plasticity for language functions after PL than after lesions later in life, few studies have compared the effects of variations in lesion timing during the perinatal period, when we know that there are rapid changes in brain development (Bates et al., 2003; Stiles et al., 2012). The answer to the question of whether there is greater plasticity after lesions early in life than later in life may depend on several factors. One of these factors is the developmental status of brain structures that are compromised by the lesion and the dependence of undamaged regions on connectivity to these damaged regions.

Studies involving animal models and studies of humans show that the "Kennard Principle"—the general rule that early lesions result in less severe deficits than later lesions—is an oversimplification (Kennard, 1936). For example, Goldman and colleagues showed that neonatal lesions of the dorsolateral prefrontal cortex of monkeys have minimal effects on spatial working memory tasks compared with later lesions (when the monkeys were tested between 12 and 18 months of age). However, when testing was performed later, the monkeys with early lesions showed clear deficits, although they were still less severe than those exhibited by the monkeys with later lesions. The emerging deficits of the monkeys with early lesions may reflect the increasing maturation of the dorsolateral frontal lobes in the control monkeys, allowing them to outperform the monkeys with early lesions to this region. Goldman found a very different pattern when a subcortical structure, the caudate nucleus, was lesioned in the early compared with the later period. That is, caudate lesions resulted in deficits that were just as devastating after early and later lesions (Goldman, 1974), perhaps due to the greater maturity of the caudate than the prefrontal cortex at the time of injury.

Other findings highlight the interaction of lesion timing with the extent of the lesion (Kolb, Holmes, & Wishaw, 1987; Kolb & Tomie, 1988). In these studies, rats received lesions involving hemidecortication or bilateral lesions of the frontal or parietal cortex at different postnatal ages. Early hemidecortication resulted in less severe spatial memory deficits on the Morris water maze (Morris, 1984) than later hemidecortication, consistent with greater plasticity early in life. In contrast, for bilateral lesions of frontal or parietal cortex, early lesions resulted in greater deficit than later lesions. Thus, the reorganizational capacity of the young brain after early lesions may depend on lesions being unilateral. It may also depend on maturity of the structures that are damaged at the time of the injury as well as at the time of testing.

A few studies have probed the effects of timing of lesions during the perinatal period in humans. In a landmark study, Staudt et al. (2004) studied hand-motor function in children whose lesions date from different periods during prenatal development—congenital malformations, which date from the first and second trimesters of pregnancy, periventricular lesions, which date mainly from the early third trimester, and middle cerebral artery infarctions, which date mainly from the late third trimester or perinatally. They found better function of the hand contralateral to the lesion after periventricular lesions, largely based on the fact that with these earlier lesions, ipsilateral corticospinal pathways are able to contribute to the functioning of the contra-lesional hand. By late in the third trimester, ipsilateral corticospinal connections may be sufficiently reduced, based on normal neuronal competition, that they are no longer sufficient to enable good hand-motor control.

Another recent study examined the IQ scores of children who sustained a lesion at different ages. The investigators compared the IQs of children who had lesions dating from before versus after age 3 years (Anderson et al., 2014). Results showed that the late lesion group had higher full-scale, verbal, and performance IQs (controlling for lesion location, volume, and seizure history) than the early lesion group. The researchers interpret these findings as supporting an early vulnerability hypothesis rather than an early plasticity hypothesis. The etiologies of lesions at various ages differed markedly, from brain trauma in the older lesion groups to dysplasia in the early lesion groups. Additionally, the extent of the lesions was larger in the early lesion groups. It is unclear whether similar results would be obtained if different lesion age categories were chosen, for example, categories that are more likely to include children with particular lesion etiologies, as was performed by Staudt et al. (2004), who examining the effects of age at lesion on hand-motor functioning. At the very least, it seems important to use finer age categories in view of the rapid changes in brain development that are occurring during the time spanning the prenatal period to age 3 years (Bates et al., 2003).

Considered together, these studies suggest that the answer to the question of whether there is greater plasticity for language functions after early than after later focal lesions is a resounding yes when one compares children with PL with patients with adult-onset lesions. However, the answer to the question of when during the early developmental period lesions are least disruptive to the development of language functions is not as firmly established, and it appears to depend on the developmental status of the areas of the brain that are injured, their connectivity with other regions, and the etiology, location, and extent of the injury

77.5 WHAT IS THE ROLE OF LANGUAGE INPUT ON THE LANGUAGE DEVELOPMENT OF CHILDREN WITH PERINATAL LESIONS?

Animal models of brain injury provide strong evidence that input can mitigate the effects of brain injury, and that the effects of experience vary depending on the age of the animal at the time of injury (Kleim & Jones, 2008; Kolb & Gibb, 2010; Kolb, Gibb, & Gorny, 2003). One would think that such findings would lead to research on how the effects of early variations in language input can result in different outcomes after early PL. However, even though variation in language input is known to be an important predictor of the language growth trajectories of TD children (Hart & Risley, 1992; Hoff, 2006; Huttenlocher et al., 1991), studies of children with PL have focused almost exclusively on biological characteristics of children's lesions in attempting to explain variations in language development among these children. Recently, this has started to change as researchers have begun to consider how variations in the language input that children experience may affect the development of language functions in children with early unilateral lesions, and how input variations and the biological characteristics of lesions may interact to predict language development trajectories.

By comparing the effects of language experiences on language development in children with perinatal brain injury and TD children, one can examine whether the effects of environmental factors differ depending on the biological characteristics of the learner. One possibility is that environmental input plays the same role in children with and without early brain injury. Such a finding would suggest that the environment plays a role in language learning in children with early lesions that is similar to that of TD children, supporting the robustness of language development in the face of early injury. A second possibility is that environmental input plays a less important role than it does in TD children, possibly because the lesion limits the child's ability to profit from the input provided. Finally, a third possibility is that environmental input may play a more important role in supporting language development after early brain injury, suggesting that input can help compensate for the deleterious effects of brain injury. Existing studies present a complex picture such that language input appears to play a similar role in the development of some aspects of language in TD children and children with early brain injury. For other aspects of language function, input appears to play a more important role for children with early brain injury, particularly for the subset of children who show initial delays.

The small number of studies that have examined the effect of environmental factors on cognitive outcomes in children with early brain injury have largely focused on global indices of input (e.g., socioeconomic status, stability of the home environment, parental attitudes, patterns of management) and global indices of child outcomes (e.g., IQ, behavioral and psychiatric problems) (Seidel, Chadwick, & Rutter, 1975; Thomas & Chess, 1975). Recent studies have explored specific relations between parental input and children's later language outcomes. For example, Rowe et al. (2009) examined the impact of lesion characteristics and parent–child-directed talk on the growth of productive vocabulary and syntax between 14 and 46 months of age. Both lesion characteristics and input variations predicted vocabulary growth and syntactic growth in children with PL. Controlling for parental socioeconomic status and characteristics of children's lesions, the diversity of parent vocabulary did not differentially predict growth in the vocabulary of children who were in the brain lesion and control groups. In contrast, the syntactic complexity of parent input played a larger role in predicting later syntax in children with PL than in TD children. Importantly, the speech of children with PL was lower in syntactic complexity than that of TD children, but it did not differ in terms of vocabulary diversity.

Building on this research, another recent study explored the effect of a complex type of parental talk—decontextualized talk during naturalistic parent–child interactions—on subsequent child vocabulary, syntax, and narratives, again comparing input effects for children with PLs and TD children. Decontextualized language refers to talk that parents produce regarding abstract and invisible entities, is typically seen in parents' conversations about the past and future, pretend play, and explanations, and is a strong predictor of later language outcomes in TD control children (Rowe, 2012; Snow, 1991). In this study, Demir et al. (2015) showed that for both groups of children, decontextualized talk by a parent to a 30-month-old child was a significant predictor of child language performance in kindergarten, controlling for parental contextualized talk, demographic factors, and the preschool language skill of the child. Decontextualized talk played a greater role in predicting kindergarten narrative outcomes for children with PL than for control children. This difference stemmed primarily from the fact that the children with PL had lower narrative (but not vocabulary or syntax) scores than the control children. When the two groups were matched in terms of narrative skill at kindergarten, the impact that decontextualized talk had on their narrative skills did not differ for children with PL and for TD children.

These two studies suggest that differences in the effects of input on the language skills of children with PL and control children may be due to the relatively low level of language skills on certain tasks in some children with PL. This hypothesis was supported in another study that experimentally manipulated the input children receive. Demir, Fisher, et al. (2013) found that co-speech gesture of an experimenter provided in a narrative retelling task was particularly helpful in supporting the narratives of children with larger lesions (associated with vascular [ischemic] infarcts) who had the most difficulty telling a well-structured narrative.

Only a small number of studies have examined later language-related academic achievement in children with PL. In general, these studies report lower academic achievement scores in the PL group, with some reporting specific deficits associated with left hemisphere lesions (Frith & Vargha-Khadem, 2001; Woods & Carey, 1979) and others reporting no laterality effects (Ballantyne et al., 2008). A recent study examined the effect of school input on development of reading skills by charting growth of decoding (grapheme–phoneme conversion; "phonics") and reading comprehension skills in TD children and children with PL between kindergarten and second grade (Demir, Carlson, et al., 2013). Children with PL performed lower than control children on reading comprehension, but not on decoding. Moreover, the difference between school and summer growth rates in reading comprehension was greater for the children with PL. This finding suggests that the structured learning environment of school may play a more crucial role in the development of reading comprehension for children with PL than for TD children.

Overall, a small but increasing number of studies on the role of language input in the language development of children with early brain injury show that variations in language development in children with PL are influenced not only by the biological characteristics of their lesions but also by the environmental input they receive. These studies also show that the effect of the environmental input might vary depending on what language skill is examined; input might play a greater role for children with PL than for TD children in language tasks that are challenging for them.

77.6 WHAT IS THE MECHANISM OF LANGUAGE PLASTICITY AFTER EARLY LESIONS?

The lack of association of left hemisphere lesions with persistent language difficulties is one of the most striking findings that has emerged from the study of children with early brain injury. Various theories have been proposed to explain the lack of association of left hemisphere lesions with language difficulties after lesions early in life. Until recently, these theories focused mainly on the degree to which language functions were lateralized to the left hemisphere during early development as a potential mechanism for age-related differences in functional plasticity (see Bates et al., 1999; Bishop, 1993 and Stiles et al., 2012 for reviews).

77.6.1 Equipotentiality

One theory, equipotentiality, primarily associated with Lenneberg (1967), posited that left hemisphere specialization for language functions emerged over the course of development as a consequence of language learning and was achieved by puberty. At this point, the period of plasticity is diminished, making it difficult to withstand an injury to the left hemisphere without showing marked language deficits. This critical period theory provides an explanation for why language deficits would not be more common after left than after right hemisphere lesions in children, as well as for why the language deficits observed after early lesions would be more subtle when lesions occur early in life.

77.6.2 Left Hemisphere Specialization from the Start

Lenneberg's theory has been challenged by a variety of evidence supporting a more nativist position—that the left hemisphere may be specialized for language from the start (see Feldman, 2005 for review). Witelson and Pallie (1973), for example, reported that the left planum temporale, a region of the left hemisphere associated with language functioning, is larger than the corresponding region in the right hemisphere in infants, just as it is in adults (Geschwind & Levitsky, 1968). Further, differential left hemisphere sensitivity to speech sounds has been supported by behavioral and EEG measures (Entus, 1977; Molfese & Molfese, 1980). More recently, neuroimaging studies reveal that infants' speech processing is supported by the same left frontotemporal brain networks as that in adults (Dehaene-Lambertz, Dehaene, & Hertz-Pannier, 2002; Dehaene-Lambertz et al., 2006). Thus, the strong form of Lenneberg's hypothesis—that the two hemispheres initially provide equally attractive neural real estate for language—does not appear to be correct. Further, as pointed out by Bates et al. (1999), the equipotentiality hypothesis fails to provide an explanation for the emergence of left hemisphere specialization for language. Such a predominant pattern is much more consistent with the presence of a bias from the beginning.

77.6.3 Emergent Specialization of Language Networks

Recent studies provide some support for a middle ground. That is, they show that the left hemisphere is specialized for some language functions early, but that there is also developmental change characterized by emergent left hemisphere specialization (Szaflarski, Holland, Schmithorst, & Byars, 2006). Consistent with this view, diffusion tensor imaging (DTI) shows that some of the fiber tracts that connect language-relevant brain areas in adults are in place in infants whereas others are not (Perani et al., 2011). Further, an Event-related potential (ERP) study showed a shift from more bilateral to more left hemisphere patterns of activation for processing words between 13 and 20 months of age, with the timing of the shift dependent on the level of the child's vocabulary development (Mills, Coffey-Corina, & Neville, 1997). Similarly, several studies report a shift from more bilateral to more left hemisphere activation between ages 5 years and adolescence on a verb generation task (Brown et al., 2005; Holland et al., 2001). Thus, certain aspects of language are lateralized early in life, whereas for others lateralization emerges much more slowly. Moreover, the timing of this emergent specialization may vary with the individual's level of language function, which, in turn, is influenced by his or her language learning experiences (Hart & Risley, 1992; Huttenlocher et al., 1991).

77.6.4 Neural Underpinnings of Language in Children with perinatal lesions

Functional magnetic resonance imaging (fMRI) studies provide a way to examine the functional neural organization supporting language after early brain injury to perisylvian regions of the left hemisphere. Many of these studies report bilateral activity or increased activity in homotopic right hemisphere areas (Booth et al., 1999; Duncan et al., 1997; Fair, Brown, Petersen, & Schlaggar, 2006; Jacola et al., 2006; Müller et al., 1999; Papanicolaou et al., 2001; Staudt et al., 2001, 2002), suggesting that right hemisphere language-related activity is associated with better language outcomes after early focal injury to the left hemisphere.

Lidzba and Staudt (2008) suggest that early left hemisphere injury may disrupt emergent left hemisphere specialization, and the maintained bilateral involvement in language processing may serve as a potential mechanism underlying functional plasticity. However, a few studies have shown that injury to the left hemisphere does not necessitate recruitment of right hemisphere homologues (Hertz-Pannier et al., 2002; Liégeois et al., 2004; Rasmussen & Milner, 1977). For example, Liégeois et al. found that damage to Broca's area in the left hemisphere does not always lead to right lateralization of activity during a verb generation task. In fact, patients with lesions within this classical language area showed greater perilesional activation in the left hemisphere, whereas patients with lesions remote from the classical language areas demonstrated interhemispheric language reorganization. Overall, the outcome patterns of neural organization for language after early focal brain injury are highly variable (Fair et al., 2010; Raja Beharelle et al., 2010), and analyses investigating brain organization postinjury and relationships with behavior need to focus on interindividual differences rather than group averages.

Although there is heterogeneity of language-related organization after early focal insult, few studies have tested whether certain patterns of organization are more optimal for language functioning than others. Staudt and colleagues (2002) found that greater whole brain activation predicted poorer verbal and full-scale IQ; however, there were no relationships between lateralized activity and outcome measures, and the sample size was likely too small (n = 5) to examine these associations. In a larger study of individuals with PL, recruitment of left inferior frontal regions and bilateral inferior parietal and posterior temporal areas during a category fluency task predicted better language outcomes across a range of measures (Raja Beharelle et al., 2010). These relationships held regardless of lesion size or extent of damage to classical language areas, suggesting that maintaining a more typical neural organization by recruiting classical left hemisphere language areas may better support language function than recruiting right hemisphere areas. However, a recent study with a relatively small number of participants combined patients with perinatal and childhood stroke and found that increased left lateralization was related to lower expressive language scores (Ilves et al., 2014). Thus, it is not clear whether certain patterns of neural organization are optimal for language after early focal brain injury, or how a variety of factors (e.g., age at lesion, age at testing, lesion aetiology, lesion location, lesion size, and seizure history) might influence the answer to this question.

Current neuroimaging studies have largely focused on left versus right hemisphere differences in brain activity after early focal brain injury. However, research shows that early injury causes both local and distributed effects in the brain, leading to global changes of the neural architecture. Structurally, early injury can result in changes in anatomic connections to areas remote from the site of injury (Pascual-Leone & Amedi, 2005), and repair processes involve regeneration around the lesion site as well as extensively throughout the brain (Anderson, Spencer-Smith, & Wood, 2011). Thus, to more comprehensively understand the changes in language organization occurring after early injury, it will be necessary to move beyond the narrow lens of laterality of activation and more broadly investigate postinjury patterns of organization and their relation to functioning.

77.7 SUMMARY AND FUTURE DIRECTIONS

Our review indicates that considerable progress has been made in understanding how focal lesions of perinatal onset impact language functioning. Some general patterns emerge from the existing literature. In particular, on a background of remarkable plasticity, children with PL do experience some language difficulties, and these tend to be more marked during the period when language is beginning and for later developing, more complex language skills. This perhaps represents a cycle of difficulties that are apparent when one examines the aspects of language that are particularly challenging at different developmental time points. That is, once a particular aspect of language is acquired, a later, more challenging aspect of development may show a subsequent delay (Stiles et al., 2005).

In terms of lesion characteristics that are associated with greater delays and deficits, lesion size appears to be a more important predictor of language functioning in this population than lesion laterality. Further, early language experiences predict language outcomes and, for some aspects of language, high-quality input appears to be even more important for children with PL than for TD children, perhaps related to the lower levels of functioning of some children with PL (Wilcox, Hadley, & Ashland, 1996). An important next step will be to examine how variations in language experiences map onto variations in the neural networks that are recruited for language processing after a PL.

The tools provided by sensitive behavioral measures used in combination with brain imaging are allowing for new ways to advance our understanding of the mechanisms of plasticity for language functions after PL. It is clear that these advances will depend on increasing our understanding of normal neurodevelopmental processes and how these processes combine with processes set in motion by an early brain injury to impact plasticity and language learning trajectories. It will also require increasing our understanding of how experiences at particular times during development can exacerbate or ameliorate the effects of lesions.

Exciting new findings are emerging from animal research concerning the mechanisms of plasticity (Hensch, 2005), including studies showing how periods of plasticity can be delayed and extended. Notably, these studies have shown that gamma-aminobutyric acid (GABA), the primary inhibitory neurotransmitter in the brain, plays a key role in the opening and closing of periods of plasticity. Although we are not yet at the point of applying these findings to humans, these findings hold promise for boosting plasticity in populations such as children with PL.

In the meantime, we do have evidence that rich language input has the potential to boost plasticity, and we can test whether this is the case using rigorous experimental designs. Further, by examining the neural signatures of language improvements after interventions, we can gain insights into the mechanisms of change. Because language difficulties in children with PL may only become apparent when the child confronts demanding linguistic tasks, it may be best to intervene proactively, before children experience difficulties. In this way, we may be able to close developmental gaps before they emerge and help children attain optimal outcomes.

References

Acredolo, L. P., & Goodwyn, S. W. (1989). Symbolic gesturing in normal infants. *Child Development, 59*, 450–466.

Anderson, V., Spencer-Smith, M., & Wood, A. (2011). Do children really recover better? Neurobehavioural plasticity after early brain insult. *Brain, 134*, 2197–2221.

Anderson, V. A., Spencer-Smith, M. M., Coleman, L, Anderson, P. J., Greenham, M., Jacobs, R., et al. (2014). Predicting neurocognitive and behavioural outcome after early brain insult. *Developmental Medicine and Child Neurology*.

Annett, M. (1973). Laterality of childhood hemiplegia and the growth of speech an intelligence. *Cortex, 9*, 4–33.

Aram, D. M., & Ekelman, B. (1986). Cognitive profiles of children with early onset of unilateral lesions. *Developmental Neuropsychology, 2*, 155–172.

Aram, D. M., & Ekelman, B. (1988). Scholastic aptitude and achievement among children with unilateral brain lesions. *Neuropsychologia, 26*, 903–916.

Aram, D. M., Ekelman, B., Rose, D., & Whitaker, H. (1985). Verbal and cognitive sequelae following unilateral lesions acquired in early childhood. *Journal of Clinical and Experimental Neuropsychology, 7*, 55–78.

Ballantyne, A, Spilkin, A, & Trauner, D. (2007). Language outcome after perinatal stroke: Does side matter? *Child Neuropsychology, 13*, 494–509.

Ballantyne, A. O., Spilkin, A. M., Hesselink, J., & Trauner, D. A. (2008). Plasticity in the developing brain: Intellectual, language, and academic functions in children with ischemic perintal stroke. *Brain, 131*, 2975–2985.

Banich, M. T., Levine, S. C., Kim, H., & Huttenlocher, P. (1990). The effects of developmental factors on IQ in hemiplegic children. *Neuropsychologia, 28*, 35–48.

Barsalou, L. W. (1999). Perceptual symbol systems. *Behavioral and Brain Sciences, 22*, 577–660.

Basser, L. (1962). Hemiplegia of early onset and the faculty of speech with special reference to the effects of hemispherectomy. *Brain, 85*, 427–460.

Bates, E., Benigni, L., Bretherton, I., Camaioni, L., & Volterra, V. (1979). *The emergence of symbols: Cognition and communication in infancy.* New York, NY: Academic Press.

Bates, E., Reilly, J., Wulfeck, B., Dronkers, N., Opie, M., Fenson, J., et al. (2001). Differential effects of unilateral lesions on language production in children and adults. *Brain and language, 79*(2), 223–265.

Bates, E., & Roe, K. (2001). Language development in children with unilateral brain injury. In C. A. Nelson, & M. Luciana (Eds.), *Handbook of developmental cognitive neuroscience.* Cambridge, MA: MIT Press.

Bates, E., Thal, D., Trauner, D., Fenson, J., Aram, D., Eisle, J., et al. (1997). From first words to grammar in children with focal brain injury. *Developmental Neuropsychology, 13*, 447–476.

Bates, E., Thal, T., Finlay, B., & Clancy, B. (2003). Early language development and its neural correlates. In Rapin, & S. Segalowitz (Eds.), *Handbook of neuropsychology, child neurology* (Vol. 6, 2nd ed.). Amsterdam: Elsevier.

Bates, E., Vicari, S., & Trauner, D. (1999). Neural mediation of language development. Perspectives from lesion studies of infants and children. In H. Tager-Flusberg (Ed.), *Neurodevelopmental disorders* (pp. 533–581). Cambridge, MA: MIT Press.

Bishop, D. (1993). Language development after focal brain damage. In D. Bishop, & K. Mogford (Eds.), *Language development in exceptional circumstances.* Hove, UK: Lawrence Erlbaum Associates.

Booth, J. R., MacWhinney, B., Thulborn, K., Sacco, K., Voyvodic, J., & Feldman, H. M. (1999). Functional organization of activation patterns in children: Whole brain fMRI imaging during three different cognitive tasks. *Progress in Neuropsychopharmacology and Biological Psychiatry, 23*, 669–682.

Booth, J. R., MacWhinney, B., Thulborn, K. R., Sacco, K., Voyvodic, J. T., & Feldman, H. M. (2000). Developmental and lesion effects in brain activation during sentence comprehension and mental rotation. *Developmental Neuropsychology, 18*, 139–169.

Broca, P. (1865). Sur le siège de la faculté du langage articulé. *Bulletins de la Société d'Anthropologie de Paris, 6*(1), 377–393.

Brown, J. V., Schumacher, U., Rohlmann, A., Ettlinger, G., Schmidt, R. C., & Skreczek, W. (1989). Aimed movements to visual targets in hemiplegic and normal children: Is the "good" hand of children with infantile hemiplegia also normal? *Neuropsychologia, 27* (3), 283–302.

Brown, T. T., Lugar, H. M., Coalson, R. S., Miezin, F. M., Petersen, S. E., & Schlaggar, B. L. (2005). Developmental changes in human cerebral functional organization for word generation. *Cerebral Cortex, 15*, 275–290.

Carlsson, G., Uvebrant, P., Hugdahl, K., Arvidsson, J., Wiklund, L. M., & Wendt, L. (1994). Verbal and non-verbal function of children with right-versus left-hemiplegic cerebral palsy of pre-and perinatal origin. *Developmental Medicine & Child Neurology, 36*(6), 503–512.

Chapman, S. B., Max, J. E., McGlothlin, J. H., Gamino, J. F., & Cliff, S. (2003). Discourse plasticity in children after stroke: Age at injury and lesion effects. *Pediatric Neurology, 29*, 34–41.

Chilosi, A. M., Cipriani, P., Bertuccelli, B., Pfanner, P. L., & Cioni, G. (2001). Early cognitive and communication development in children with focal brain lesions. *Journal of Child Neurology, 16*, 309–316.

Cioni, G., Sales, B., Paolicelli, P. B., Petacchi, E., Scusa, M. F., & Canapicchi, R. (1999). MRI and clinical characteristics of children with hemiplegic cerebral palsy. *Neuropediatrics, 30*, 249–255.

Cohen, M. E., & Duffner, P. K. (1981). Prognostic indicators in hemiparetic cerebral palsy. *Annals of Neurology, 9*, 353–357.

Dall'Oglio, A. M., Bates, E., Volterra, V., De Capua, M., & Pezzini, G. (1994). Early cognition, communication, and language in children with focal brain injury. *Developmental Medicine and Child Neurology, 35*, 1076–1098.

Dehaene-Lambertz, G., Dehaene, S., & Hertz-Pannier, L. (2002). Functional neuroimaging of speech perception in infants. *Science, 298*, 2013–2015.

Dehaene-Lambertz, G., Hertz-Pannier, L., Dubois, J., Meriaux, S., Roche, A., Sigman, M., et al. (2006). Functional organization of perisylvian activation during presentation of sentences in preverbal infants. *Proceedings of the National Academy of Sciences of the United States of America, 103*, 14240–14245.

Demir, E., Levine, S. C., & Goldin-Meadow, S. (2010). Narrative skill in children with early unilateral brain injury: A possible limit to functional plasticity. *Developmental Science, 13*(4), 636–647.

Demir, Ö. E., Carlson, M., Levine, S. C., & Goldin-Meadow, S. (2013). *Summer setback: The impact of time off from school on reading in children with brain injury.* Poster presented at the biennial meeting of the Society for Research in Child Development, Seattle, WA.

Demir, Ö. E., Fisher, J., Levine, S., & Goldin-Meadow, S. (2013). Narrative processing in typically-developing children and children with early unilateral brain injury: Seeing gesture matters. *Developmental Psychology, 50*(3), 815–828.

Demir, Ö. E., Rowe, M. L., Heller, G., Goldin-Meadow, S., & Levine, S. C. (2015). Vocabulary, syntax, and narrative development in typically developing children and children with early unilateral brain injury: Early parental talk about the "there-and-then" matters. *Developmental psychology, 51*(2), 161.

Dennis, M., & Whitaker, H. A. (1976). Language acquisition following hemidecortication: Linguistic superiority of the left over the right hemisphere. *Brain and Language, 3*, 404–433.

Dick, F., Wulfeck, B., Krupa-Kwiatkowski, M., & Bates, E. (2004). The development of complex sentence interpretation in typically developing children compared with children with specific language impairments or early unilateral focal lesions. *Developmental Science, 7*(3), 360–377.

Duncan, J. D., Moss, S. D., Bandy, D. J., Manwaring, K., Kaplan, A. M., Reinian, E. M., et al. (1997). Use of positron emission tomography for presurgical localization of eloquent brain areas in children with seizures. *Pediatric Neurosurgery, 26*(3), 144–156.

Duque, J., Thonnard, J. L., Vandermeeren, Y., Sébire, G., Cosnard, G., & Olivier, E. (2003). Correlation between impaired dexterity and corticospinal tract dysgenesis in congenital hemiplegia. *Brain, 126*(3), 732–747.

Elman, J. L., Bates, E., Johnson, M. H., Karmiloff-Smith, A., Parisi, D., & Plunkett, K. (1996). *Rethinking innateness: A connectionist perspective on development.* Cambridge, MA: MIT Press.

Entus, A. (1977). Hemispheric asymmetry in processing of dichotically presented speech and nonspeech stimuli by infants. In S. Segalowitz, & F. Gruber (Eds.), *Language development and neurological theory* (pp. 63–73). New York, NY: Academic Press.

Fair, D. A., Brown, T. T., Petersen, S. E., & Schlaggar, B. L. (2006). FMRI reveals novel functional neuroanatomy in a child with perinatal stroke. *Neurology, 67*, 2246–2249.

Fair, D. A., Choi, A. H., Yannic, B. L., Coalson, R. S., Meizin, F. M., Petersen, S. E., et al. (2010). The functional organization of trial-related activity in lexical processing after early left hemispheric brain lesions: An event-related fMRI study. *Brain and Language, 114*, 135–146.

Feldman, H. (2005). Language learning with an injured brain. *Language Learning and Development, 1*, 265–288.

Feldman, H. M. (1994). Language development after early brain injury: A replication study. In H. Tager-Flusberg (Ed.), *Constraints on language acquisition: Studies of atypical children* (pp. 75–90). Hillsdale, NJ: Lawrence Erlbaum Associates, Inc.

Feldman, H. M., Holland, A. L., Kemp, S. S., & Janowsky, J. E. (1992). Early language and communicative abilities of children with periventricular leukomalacia. *American Journal of Mental Retardation, 97*, 222–234.

Feldman, H. M., MacWhinney, B., & Sacco, K. (2002). Sentence processing in children with early unilateral brain injury. *Brain and Language, 42*, 89–102.

Fenson, L., Dale, P. A., Reznick, J. S., Bates, E., Thal, D., & Pethick, S. J. (1994). Variability in early communicative development. *Monographs of the Society for Research in Child Development, 59*(5), 1–173.

Frith, U., & Vargha-Khadem, F. (2001). Are there sex differences in the brain basis of literacy related skills? Evidence from reading and spelling impairments after early unilateral brain damage. *Neuropsychologia, 39*, 1485–1488.

Geschwind, N., & Levitsky, W. (1968). Human brain: Left-right asymmetries in temporal speech region. *Science, 161*, 186–187.

Giedd, J. N., Blumenthal, J., Jeffries, N. O., Castellanos, F. X., Liu, H., Zijdenbos, A., et al. (1999). Brain development during childhood and adolescence: A longitudinal MRI study. *Nature Neuroscience, 2*, 861–863.

Glenberg, A. M. (1997). What memory is for. *Behavioral and Brain Sciences, 20*(1), 1–55.

Goldin-Meadow, S., & Butcher, C. (2003). Pointing toward two-word speech in young children. In S. Kita (Ed.), *Pointing: Where language, culture, and cognition meet* (pp. 85–107). Mahwah, NJ: Lawrence Erlbaum Associates.

Goldin-Meadow, S., Levine, S. C., Hedges, L. V., Huttenlocher, J., Raudenbush, S. W., & Small, S. L. (2014). New evidence about language and cognitive development based on a longitudinal study: Hypotheses for intervention. *American Psychologist, 69*(6), 588.

Goldman, P. (1974). An alternative to developmental plasticity: Heterology of CNS structures in infants and adults. In N. Butters (Ed.), *Plasticity and recovery of function in the central nervous system* (pp. 149–174). New York, NY: Academic Press.

Gordon, A. M., & Steenbergen, B. (2008). Bimanual coordination in children with cerebral palsy. In A. C. Eliasson, & P. Burtner (Eds.), *Improving hand function in children with cerebral palsy: Theory, evidence & intervention. Clinics in developmental medicine* (pp. 160–175). London: MacKeith Press.

Greenough, W. T., Black, J. E., & Wallace, C. S. (1987). Experience and brain development. *Child Development, 58*, 539–559.

Hart, B., & Risley, T. R. (1992). American parenting of language-learning children: Persisting difference in family-child interactions observed in natural home environments. *Developmental Psychology, 28*, 1096–1105.

Hensch, T. K. (2005). Critical period plasticity in local cortical circuits. *Nature Reviews Neuroscience, 6*, 877–888.

Hertz-Pannier, L., Chiron, C., Jamabaque, I., Renaux-Kieffer, V., Van de Moortele, P., Delalande, O., et al. (2002). Late plasticity for language in a child's non-dominant hemisphere: A pre- and post-surgery fMRI study. *Brain, 125*, 361–372.

Hoff, E. (2006). How social contexts support and shape language development. *Developmental Review, 26*, 55–88.

Holland, S. K., Plante, E., Byars, A. W., Strawburg, R. H., Schmithorst, V. J., & Ball, W. S., Jr. (2001). Normal fMRI brain activation patterns in children performing a verb generation task. *NeuroImage, 14*, 837–843.

Huttenlocher, J., Haight, W., Bryk, A., Seltzer, M., & Lyons, T. (1991). Early vocabulary growth: Relation to language input and gender. *Developmental Psychology, 27*, 236–248.

Huttenlocher, P. R. (2002). Neural plasticity: The effects of the environment on the development of cerebral cortex. Cambridge, MA: Harvard University Press.

Huttenlocher, P. R., & Dabholkar, A. S. (1997). Regional difference in synaptogenesis in human cerebral cortex. *Journal of Comparative Neurology, 387*, 167–178.

Huttenlocher, P. R., & Hapke, R. J. (1990). A follow-up study of intractable seizures in chldhood. *Annals of Neurology, 28*, 699–705.

Ilves, P., Tomberg, T., Kepler, J., Laugesaar, R., Kaldoja, M. L., Kepler, K., et al. (2014). Different plasticity patterns of language function in children with perinatal and childhood stroke. *Journal of Child Neurology, 29*, 756–764.

Irle, E. (1990). An analysis of the correlation of lesion size, localization and behavioral effects in 283 published studies of cortical and subcortical lesions in old-world monkeys. *Brain Research Review, 15*, 181–213.

Iverson, J. M., & Goldin-Meadow, S. (2005). Gesture paves the way for language development. *Psychological Science, 16*, 368–371.

Jacola, L. M., Schapiro, M. B., Schmithorst, V. J., Byars, A. W., Strawsburg, R. H., Szaflarski, J. P., et al. (2006). Functional magnetic resonance imaging reveals atypical language organization in children following perinatal left middle cerebral artery stroke. *Neuropediatrics, 37*, 46–52.

Kennard, M. A. (1936). Age and other factors in motor recovery from precentral lesions in monkeys. *American Journal of Physiology, 115*, 138–146.

Kleim, J. A., & Jones, T. A. (2008). Principles of experience-dependent neural plasticity: Implications for rehabilitation after brain damage. *Journal of Speech, Language, and Hearing Research, 51* (1), S225–S239. Available from: http://dx.doi.org/10.1044/1092-4388(2008/018).

Kolb, B., & Gibb, R. (2010). Tactile stimulation after frontal or parietal cortical injury in infant rats facilitates functional recovery and produces synaptic changes in adjacent cortex. *Behavioral Brain Research, 14*, 115–120.

Kolb, B., Gibb, R., & Gorny (2003). Experience-dependent changes in dendritic arbor and spine density in neorcortex vary qualitatively with age and sex. *Neurobiology of Learning and Memory, 79*, 1–10.

Kolb, B., Holmes, C., & Wishaw, IQ. (1987). Recovery from early cortical lesions in rats. III. Neonatal removal of posterior parietal cortex has greater behavioral and anatomical effects than similar removals in adulthood. *Behavioral Brain Research, 26*, 119–137.

Kolb, B., & Tomie, J. A. (1988). Recovery from early cortical damage in rats. IV. Effects of hemidecorication at 1, 5, or 10 days of age on cerebral anatomy and behavior. *Behavioral Brain Research, 28*, 259–274.

Krägeloh-Mann, I. (2004). Imaging of early brain injury and cortical plasticity. *Experimental Neurology, 190*(Suppl. 1), S84–S90.

Kuhtz-Buschbeck, J. P., Sundholm, L. K., Eliasson, A. C., & Forssberg, H. (2000). Quantitative assessment of mirror movements in children and adolescents with hemiplegic cerebral palsy. *Developmental Medicine and Child Neurology, 42*(11), 728–736.

Lenneberg, E. H. (1967). *Biological foundations of language*. New York, NY: Wiley.

Levine, S. C., Huttenlocher, P. R., Banich, M. T., & Duda, E. (1987). Factors affecting cognitive functioning of hemiplegic children. *Developmental Medicine and Child Neurology, 27*, 27–35.

Levine, S. C., Kraus, R., Alexander, E., Suriyakham, L., & Huttenlocher, P. R. (2005). IQ decline following early unilateral brain injury: A longitudinal study. *Brain and Cognition, 59*, 114–123.

Lidzba, K., & Staudt, M. (2008). Development and (re)organization of language after early brain lesions: Capacities and limitation of early brain plasticity. *Brain and Language, 106*, 167–176.

Liégeois, F., Connelly, A., Cross, J. H., Boyd, S. G., Gadian, D. G., Vargha-Khadem, F., et al. (2004). Language reorganization in children with early-onset lesions of the left hemisphere: An fMRI study. *Brain, 127*, 1229–1236.

MacWhinney, B., Feldman, H., Sacco, K., & Valdes-Perez, R. (2000). Online measures of basic language skills in children with early focal brain lesions. *Brain and Language, 71*, 400–431.

Marchman, V. A., Miller, R., & Bates, E. (1991). Babble and first words words in children with focal brain injury. *Applied Psycholinguistics, 12*, 1–22.

Mills, D. L., Coffey-Corina, S. A., & Neville, H. J. (1997). Language comprehension and cerebral specialization from 13 to 20 months. *Developmental Neuropsychology, 13*, 397–445.

Molfese, D., & Molfese, J. (1980). Cortical responses of preterm infants to phonetic and nonphonetic speech stimuli. *Developmental Psychology, 16*, 574–581.

Morris, R. (1984). Developments of a water-maze procedure for studying spatial learning in the rat. *Journal of neuroscience methods, 11*(1), 47–60.

Müller, R.-A., Behen, M. E., Rothermel, R. D., Muzik, O., Chakraborty, P. K., & Chugani, H. T. (1999). Brain organization for language in children, adolescents, and adults with left hemisphere lesions: A PET study. *Progress in Neuropsychopharmacology and Biological Psychiatry, 23*, 657–668.

Nichelli, P., Grafman, J., Petrini, P., Clark, K., Lee, K. Y., & Miletich, R. (1995). Where the brain appreciates the moral of a story. *Neuroreport, 6*, 2309–2313.

Nichols, S., Jones, W., Roman, M., Wulfeck, B., Delis, D. C., Reilly, J., et al. (2004). Mechanisms of verbal memory impairment in four neurodevelopmental disorders. *Brain and Language, 88*, 180–189.

Özcaliskan, S., & Goldin-Meadow, S. (2005). Gesture is at the cutting edge of early language development. *Cognition, 96*, B101–B113.

Özcaliskan, S., Levine, S. C., & Goldin-Meadow, S. (2013). Gesturing with an injured brain: How gesture helps children with early brain injury learn linguistic constructions. *Journal of Child Language, 40*, 69–105.

Papanicolaou, A. C., Simos, P. G., Breier, J. I., Wheless, J. W., Mancias, P., Baumgartner, J. E., et al. (2001). Brain plasticity for sensory and linguistic functions: A functional imaging study using magnetoencephalography with children and young adults. *Journal of Child Neurology, 16*, 241–252.

Pascual-Leone, A., & Amedi, B. (2005). The plastic human brain cortex. *Annual Reviews Neuroscience, 28*, 377–401.

Perani, D., Saccuman, M. C., Scifo, P, Anwander, A., Spada, D., Baldoli, C., et al. (2011). Neural language networks at birth. *Proceedings of the National Academy of Sciences of the United States of America, 108*, 16056–16061.

Raja Beharelle, A., Dick, A. S., Josse, G., Solodkin, A., Huttenlocher, P. R., Levine, S. C., et al. (2010). Left hemisphere regions are critical for language in the face of early left focal brain injury. *Brain, 133*, 1707–1716.

Rankin, J. M., Aram, D. M., & Horwitz, S. J. (1981). Language ability in right and left hemiplegic children. *Brain and Language, 14*, 292–306.

Rasmussen, T., & Milner, B. (1977). The role of early left-brain injury in determining lateralization of cerebral speech functions. *Annals of the New York Accademy of Sciences, 299*, 355–369.

Reed, J. C., & Reitan, R. M. (1969). Verbal and performance differences among brain-injured children with lateralized motor deficits. *Perceptual and Motor Skills, 29*, 747–752.

Reilly, J., Bates, E., & Marchman, V. (1998). Narrative discourse in children with early focal brain injury. *Brain and Language, 61*, 335–375.

Reilly, J., Levine, S. C., Nass, R., & Stiles, J. (2008). Brain plasticity: Evidence from children with prenatal brain injury. In J. Reed, & J. Warner (Eds.), *Child neuropsychology*. Oxford: Blackwell Publishing.

Reilly, J., Losh, M., Bellugi, U., & Wulfeck, B. (2004). "Frog, where are you?" Narratives in children with specific language impairment, early focal brain injury, and Williams syndrome. *Brain and Language, 88*, 229–247.

Ritti, A. (1894). Eloge du Docteur Jules Cotard (From a paper read at the Annual Public Lecture of the Societe Medico Psychologique on 30 April 1894). Copy held in Rare Manuscripts Section, Library of School of Medicine, Paris 35 pages, catalog no. 56613/6. Imprimerie de la Cour d'Appel, Paris.

Riva, D., & Cazzaniga, I. (1986). Late effects of unilateral brain lesions sustained before and after age one. *Neuropsychologia, 24*, 423–428.

Rowe, M. L. (2012). A Longitudinal Investigation of the role of quantity and quality of child directed speech in vocabulary development. *Child Development, 83*(5), 1762–1774.

Rowe, M. L., Levine, S. C., Fisher, J. A., & Goldin-Meadow, S. (2009). Does linguistic input play the same role in language learning for children with and without early brain injury? *Developmental Psychology, 45*(1), 90–102.

Sakzewski, L., Ziviani, J., & Boyd, R. (2010). The relationship between unimanual capacity and bimanual performance in children with congenital hemiplegia. *Developmental Medicine and Child Neurology, 58*, 811–816.

Sauer, E., Levine, S. C., & Goldin-Meadow, S. (2010). Early gesture predicts language delay in children with pre- or perinatal brain lesions. *Child Development, 81*(2), 528–539.

Seidel, U. P., Chadwick, O. F. D., & Rutter, M. (1975). Psychological disorders in crippled children. A comparative study of children with and without brain damages. *Developmental Medicine and Child Neurology, e17*(5), 563–573.

Snow, C. E. (1991). The theoretical basis for relationships between language and literacy in development. *Journal of Research in Childhood Education, 6*, 5–10.

Sowell, E. R., Trauner, D. A., Gamst, A., & Jernigan, T. L. (2002). Development of cortical and subcortical brain structures in childhood and adolescence: A structural MRI study. *Developmental Medicine and Child Neurology, 44*, 4–16.

Staudt, M., Gerloff, C., Grodd, W., Holthausen, H., Niemann, G., & Krägeloh-Mann, I. (2004). Reorganization in cognenital hemiparesis acquired at different gestational ages. *Annals of Neurology, 56*, 854–863.

Staudt, M., Grodd, W., Niemann, G., Wildgruber, D., Erb, M., & Krägeloh-Mann, I. (2001). Early left periventricular brain lesions induce right hemispheric organization of speech. *Neurology, 57*, 122–125.

Staudt, M., Lidzba, K., Grodd, W., Wildgruber, D., Erb, M., & Krägeloh-Mann, I. (2002). Right-hemispheric organization of language following early left-sided brain lesions: Functional MRI topography. *NeuroImage, 16*, 954–967.

Stein, N. L., & Albro, E. R. (1997). Building complexity and coherence: Children's use of goal-structured knowledge in telling stories. In M. Bamburg (Ed.), *Narrative development: Six approaches* (pp. 5–44). Mahwah, NH: Lawrence Erlbaum.

Stiles, J., Reilly, J. S., Levine, S. C., Trauner, D. A., & Nass, R. (2012). *Neural plasticity and cognitie development: Insights from children with perinatal brain injury*. New York, NY: Oxford University Press.

Stiles, J., Reilly, J. S., Paul, B., & Moses, P. (2005). Cognitive development following early brain injury: Evidence for neural adaptation. *Trends in Cognitive Sciences, 9*, 136–143.

Szaflarski, J. P., Holland, S. K., Schmithorst, V. J., & Byars, A. W. (2006). FMRI study of language lateralization in children and adults. *Human Brain Mapping, 27*, 202–212.

Teuber, H.-L. (1974). Functional recovery after lesions of the nervous system. II. Recovery of function after lesions of the central nervous system: History and prospects. *Neurosciences Research Program Bulletin, 12*, 197–211.

Thal, D., Marchman, V. A., Stiles, J., Aram, D., Trauner, D., Nass, R., et al. (1991). Early lexical development in children with focal brain injury. *Brain and Language, 40*, 491–527.

Thomas, A., & Chess, S. (1975). A longitudinal study of three brain damaged children: Infancy to adolescence. *Archives of General Psychiatry, r32*(4), 457.

Thomas, M., & Karmiloff-Smith, A. (2002). Are developmental disorders like cases of adult brain damage? Implications from connectionist modeling. *Behavioral and Brain Sciences, 25*, 727–788.

Trauner, D. A., Eshagh, K., Ballantyne, A. O., & Bates, E. (2013). Early language development after peri-natal stroke. *Brain and language, 127*(3), 399–403.

Vargha-Khadem, F., Isaacs, E., & Muter, F. (1994). A review of cognitive outcome after unilateral lesions sustained during childhood. *Journal of Child Neurology, 9*(Suppl.), 2S67-2S73.

Vargha-Khadem, F., Isaacs, E., Van Der Werf, S., Robb, S., & Wilson, J. (1992). Development of intelligence and memory in children with hemiplegic cerebral palsy: The deleterious consequences of early seizures. *Brain, 115*, 315–329.

Vargha-Khadem, F., O'Gorman, A., & Watter, G. (1985). Aphasia and handedness in relation to hemispheric side, age at injury, and severity of cerebral lesion during childhood. *Brain, 108*, 667–696.

Weckerly, J., Wulfeck, B., & Reilly, J. (2004). The development of morphosyntactic ability in atypical populations: The acquisition of tag questions in children with early focal lesions and children with specific-language impairment. *Brain and Language, 88*, 190–201.

Westmacott, R., Askalan, R., MacGregor, D., Anderson, P., & Devebar, G. (2010). Cogntiive outcome following unilateral arterial ischaemic stroke in childhood: Effects of age at stroke and lesion location. *Developmental Medicine and Child Neurology, 52*, 386–393.

Wilcox, M., Hadley, P., & Ashland, J. (1996). Communication and language development in infants and toddlers. In M. Hanson (Ed.), *Atypical infant development*. Austin, TX: Pro-Ed.

Witelson, S. F. (1985). On hemisphere specialization and cerebral plasticity from birth: Mark II. In C. T. Best (Ed.), *Hemispheric function and collaboration in the child* (pp. 33–77). Orlando, FL: Academic Press.

Witelson, S. F., & Pallie, W. (1973). Left-hemisphere specialisation for language in the new-born: Neuroanatomical evidence of asymmetry. *Brain, 88*, 653–662.

Witte, O. W. (1998). Lesion-induced plasticity as a potential mechanism for recovery and rehabilitative training. *Current Opinion in Neurology, 11*, 655–662.

Woods, B. T., & Carey, S. (1979). Language deficits after apparent clinical recovery from childhood aphasia. *Annals of Neurology, 6*, 405–409.

Woods, B. T., & Teuber, H.-L. (1978). Changing patterns of childhood aphasia. *Annals of Neurology, 3*, 273–280.

Zwaan, R. A. (2008). Experiential traces and mental simulations in language comprehension. In M. DeVega, A. M. Glenberg, & A. C. Graesser (Eds.), *Symbols, emobidment and meaning* (pp. 165–180). Oxford: Oxford University Press.

CHAPTER

78

Motor Speech Impairments

Wolfram Ziegler and Anja Staiger

Clinical Neuropsychology Research Group, Clinic for Neuropsychology, City Hospital, Munich, Germany

78.1 INTRODUCTION

According to conventional clinical taxonomies for neurological speech and language disorders, the term *motor speech impairment* comprises the different *dysarthria* syndromes as well as the syndrome of *apraxia of speech* (AOS). Although the dysarthria syndromes are considered to result from pathologies afflicting the control and execution of speech movements, apraxia of speech is usually ascribed to a dysfunction of speech motor planning or programming functions (Duffy, 2013).

Historically, many neurolinguistic theories have explicitly or implicitly embraced a fundamental divide between the biological foundations of *language* and those of *motor speech* (and *auditory processing*). Aphasiologists have willingly adopted the *langue–parole* distinction made by de Saussure and its continuance in structuralist and early generative phonology, in which the motor aspects of speaking (and the auditory aspects of understanding) are almost completely excluded from the arena of language biology. From this standard perspective, the dysarthrias and AOS have been neglected because they are viewed as disorders of a physical organ whose relationship to language is incidental or external rather than structurally or functionally linked.

In this chapter, we regard motor speech disorders from three vantage points. First, we describe them as syndromes resulting from the recognized neuropathologies of body movement disorders, more or less following the standard view of speech as a motor function that is sealed from its overarching linguistic framework. Second, we discuss how the speech motor system is specialized to serve its linguistic-communicative goal. Third, we expand on the sensorimotor aspects of speech motor impairments with the aim of illuminating the different neural stages during which auditory, somatosensory, and motor information is integrated.

78.2 MOTOR IMPAIRMENTS WITHIN A NEUROLOGICAL FRAMEWORK

Since Darley's seminal work, the taxonomy of neurogenic motor speech impairments largely mirrors the taxonomy of (body) movement disorders, with the dysarthric syndromes corresponding to the paretic (flaccid, spastic), ataxic, akinetic, and dyskinetic motor syndromes. AOS, in this terminology, is considered to correspond with the syndrome of limb apraxia (Darley, Aronson, & Brown, 1975).

In this section we describe the major CNS neuropathologies leading to speech impairment, focusing on the question of how a pathomechanism described for disorders of body movements may translate into speech motor mechanisms (for a fully comprehensive survey and detailed descriptions of clinical symptoms, see Duffy, 2013).

78.2.1 Spastic Paresis

Lesions on areas representing the speech muscles in the ventral part of the Rolandic motor cortex and on the corticobulbar motor pathways cause a syndrome characterized by a paresis of the musculature involved in speaking. This dysarthria type, termed *spastic dysarthria*, may arise from lesions either at the motor cortex level or along the descending motor neuron fiber tracts to the pontine and medullary motor nuclei. Such lesions can be caused, for example, by infarctions or by disseminated MS plaques, by traumatic brain injuries, by congenital or very early brain damage, as in cerebral palsy (CP), or by progressive disorders affecting the upper motor neuron pathways, such as progressive supranuclear palsy (PSP) or motor neuron disease (MND). Because all bulbar motor nuclei, except the facial nucleus, receive considerable bilateral motor cortical input, lesions restricted to one hemisphere often

Neurobiology of Language. DOI: http://dx.doi.org/10.1016/B978-0-12-407794-2.00078-X

lead to only mild and transient speech impairment (Muellbacher, Artner, & Mamoli, 1999), whereas lesions affecting the upper motor neuron system of both hemispheres, such as in the Foix–Chavany–Marie syndrome (with infarctions in the left and right motor cortices) or after bilateral brainstem stroke or traumatic brain injury, may lead to a persisting syndrome including, among others, severe dysarthria and dysphagia (Duffy, 2013).

Although the presence of spasticity, as defined in the limb muscles, cannot easily be verified in the lingual, pharyngeal, or laryngeal musculature, the general understanding is that lesions to the corticobulbar motor system cause a spastic speech syndrome characterized by muscle weakness and loss of fine motor skill (as a consequence of lesions to the monosynaptic fibers of the upper motor neuron system) in combination with excessive muscle tone (as a consequence of lesions to indirect fibers targeting the motor nuclei via multiple "extrapyramidal" synapses; Duffy, 2013). The combination of upper motor neuron weakness and spasticity is considered to cause reduction of respiratory support, slowing of articulator movements, imprecise consonant articulation, or, when the velopharynx is affected, hypernasality. Increased muscle tone can be visible in the lower face muscles or, via endoscopic inspection, in the larynx. Hypertonicity may lead to a strained or strangled voice because of glottal hyperadduction or increased tension in the hypopharynx.

78.2.2 Ataxia

Dysarthria may also arise when cerebellar contributions to speech motor control are compromised due to lesions of either the cerebellum itself or its efferent or afferent projections. Such lesions may result, for example, from cerebellar infarctions, multiple sclerosis, or hereditary ataxic disorders (e.g., Friedreich's ataxia, spino-cerebellar ataxias). Conflicting theories exist regarding the lateralization and particular parts of the cerebellum implicated in ataxic dysarthria (Mariën's contribution in Manto et al., 2012). Recently, two functional subsystems have been hypothesized on the basis of functional imaging data. They include a *superior cerebellar circuit* (encompassing connections of superior parts of the cerebellar hemispheres with the inferior frontal gyrus, anterior insular cortex, and the supplementary area), mainly involved in preparatory and motor planning aspects of speech production, and an *inferior cerebellar circuit* (encompassing inferior-cerebellar thalamo-cortical connections), mainly involved in the motor execution aspects of speaking (Ackermann, 2008). On the basis of clinical considerations, a similar distinction between two functional levels of cerebellar contributions to motor speech has been proposed that distinguishes between a motor planning/programming circuit (mainly involving connections of the right cerebellar hemisphere with left inferior-frontal speech planning centers in the cerebral cortex) and a motor execution circuit involving superior parts of both cerebellar hemispheres (Spencer & Slocomb, 2007).

The clinical pattern of ataxic dysarthria may vary considerably across patients. Among the ataxic pathomechanisms, impaired motor timing, sequencing, and movement coordination have been considered preeminent explanations for speech characteristics such as irregular articulatory inaccuracy, slow articulatory rates, prolonged phonemes, inappropriate pitch and loudness variation, voice tremor, or temporally disorganized or paradoxical respiratory movement patterns during speech breathing (Brendel et al., 2013; Duffy, 2013). Other symptoms may result from compensatory mechanisms serving to suppress tremor or dysmetria, such as a strained or strangled voice quality or a regularly paced, scanning speech rhythm. Overall, the pattern of cerebellar speech impairment cannot be easily divided into a group of features reflecting a planning deficit and others reflecting impaired motor execution, as would be predicted by the two-level model of cerebellar speech motor functions mentioned. Cerebellar speech signs fit into the symptom pattern traditionally described as *ataxia*, with dysmetria, incoordination, and deficient timing as its preeminent pathomechanisms; they differ substantially from the *apraxic* signs observed after lesions to the anterior perisylvian and subsylvian cortex of the left cerebral hemisphere (Ziegler's contribution in Mariën et al., 2013; Section 78.2.5).

78.2.3 Akinesia

The pathomechanism of akinesia in movement disorders has been ascribed to dysfunction at the level of the striato-thalamo-cortical motor circuit, with idiopathic Parkinson's disease as the prototypic clinical model of the akinetic condition (Jankovic, 2008). Akinesia is considered to encompass a *hypokinetic* component, mainly characterized by a reduction in the range of rhythmical movements (e.g., gait, breathing) as a result of excessive inhibition of central pattern generators in the brainstem, and a *bradykinetic* component, characterized by slowness of movements as a result of reduced striato thalamic "energisation" of appropriately selected motor commands at the motor cortical level, through a defect of the striatal motor circuit that facilitates recruitment of cortical motoneurons for an intended movement. Both mechanisms, hypokinesia and bradykinesia, are presumably the result of a progressive loss of dopaminergic striatal innervation (Rodriguez-Oroz et al., 2009). It is not clear whether parkinsonian speech impairment originates from exactly the same mechanisms, especially because dysarthria in Parkinson's disease is much less (if at all) responsive to dopaminergic drugs and deep brain stimulation treatment than akinetic motor impairment of the limbs and trunk (Rodriguez-Oroz et al., 2005). Moreover, symptoms such as a slowing of the serial control of sequential movements involving different body parts, which are considered characteristic of Parkinson's disease (Rodriguez-Oroz et al., 2009), are not typically present in hypokinetic dysarthria. On the contrary, Parkinson's patients often demonstrate a normal or even accelerated speaking rate and chains of tightly conjoined and rapidly produced syllables ("short rushes of speech"; Ackermann & Ziegler, 1992). Nonetheless, the overall pattern of speech motor signs in Parkinson's disease, with its visible undershooting of labial and mandibular movements, its loss of sufficiently distinct consonant and vowel articulations, its hypophonic voice, and its monotonous intonation, is largely compatible with the predicted consequences of an akinetic condition. In addition, Rodriguez-Oroz et al. (2009) suggest that increased responsiveness of the subthalamic nucleus (STN) and internal segment of the globus pallidus (GPi) in Parkinsonism may contribute to a progressive attenuation of sequential movements as in handwriting (micrographia), which might also serve as an explanation of progressive hypophonia and articulatory undershoot and progressive acceleration phenomena occurring across stretches of speech.

Reduced spontaneous speech and hypophonia may also be observed in patients with lesions on the medial premotor areas (supplementary motor area [SMA]) and the anterior cingulate gyrus (ACG), for instance, during recovery from akinetic mutism (Krainik et al., 2003). These data fit within a model of a limbic striatal circuit encompassing the ACG and SMA, which may serve as a starting mechanism for speech production and as a gateway for motivational and affective modulation of vocal and articulatory processes (Ackermann, Hage, & Ziegler, 2014; see Section 78.3.1).

78.2.4 Dyskinesia

Although akinesia is attributed to excessive striatal inhibition of thalamo-cortical and brainstem mechanisms, opposite mechanisms may lead to a loss of inhibition or an imbalance of thalamo-cortical motor activations as a consequence of reduced STN-GPi activity (Hallett, 2011). These pathophysiological conditions result in hyperkinetic or dystonic motor impairments characterized by abnormal postures, uncontrollable movements or muscle spasms, and a loss of selectivity of muscular activation. They are classified into rhythmic (*tremor*) and arrhythmic variants, with the latter including the *dystonias* as a form of sustained motor abnormalities and *chorea*, *myoclonus*, and *tics* as rapid forms (Albanese & Jankovic, 2012). Dyskinetic conditions may result from a variety of neuropathologies, including genetic, drug-induced, toxic or metabolic, traumatic, or vascular etiologies. In many cases, especially in the dystonias, the etiology is unknown.

Speech can be afflicted by all types of dyskinetic syndromes. Leaving aside the different variants of vocal tremor (e.g., in Parkinson's disease, cerebellar ataxia, essential tremor), the dyskinetic dysarthria syndromes that have received particular consideration are the choreatic and athetotic forms occurring in Huntington's disease (Duffy, 2013) and in a subtype of CP, and the focal dystonias of the laryngeal and the oromandibular muscles (*spasmodic dysphonia*, *oromandibular lingual dystonia*; Duffy, 2013). Choreatic hyperkinesias of the speech muscles may lead to intermittent disruptions of respiratory activity, uncontrolled vocalizations, intermittent noise productions with the tongue or lips, excessive pitch and loudness variations, overshooting and undershooting of articulatory movements, intermittent hyponasality and hypernasality, and irregular pauses or sound prolongations (Duffy, 2013). Spasmodic dysphonia may affect the adductor and/or abductor muscles of the larynx and lead to a strained–strangled or breathy or aphonic voice quality, respectively (Simonyan, Berman, Herscovitch, & Hallett, 2013). Oromandibular lingual dystonia may interfere with speaking secondary to, for example, excessive jaw opening or closing and involuntary tongue protrusion (Ushe & Perlmutter, 2012).

78.2.5 Apraxia of Speech

The concept of AOS dates back to Liepmann (1900), who considered the articulation impairment of a patient with Broca's aphasia (then termed "motor aphasia") an "apraxia of the language muscles" ("*Apraxie der Sprachmuskeln*"; Liepmann, 1900, p. 129). Much later, in their neurologically based classification of motor speech impairments, Darley et al. (1975) resurrected the term to describe a speech impairment occurring after infarction of the anterior branch of the left middle cerebral artery whose characteristics were not compatible with any of the neuromotor pathomechanisms of the dysarthrias. Similar to limb apraxia, the original definition of AOS consisted of largely negative designations; the speech symptoms cannot be explained by "slowness, weakness, incoordination, or change of tone" of the involved musculature (Darley et al., 1975, p. 251). The problem was interpreted as a consequence of impaired motor programming (Darley et al., 1975, p. 255).

A specific feature of limb apraxia is that lesions of the left hemisphere may lead to motor impairment of not only the contralesional (right) but also the ipsilesional (left) limb. In speech this corresponds, in a way, with the observation that AOS is a left hemisphere syndrome and that, unlike (hemi-) paretic dysarthria, it cannot be compensated for by innervations from intact right hemisphere homologues. However, modern theories of apraxia, with their strong focus on failures of motor functions of the upper extremities, such as pantomime or tool use (Goldenberg, 2013), confuse the issue of drawing analogies between speech and limb apraxia. Without any specific reference to the latter, AOS may best be characterized as a loss of the acquired implicit knowledge of how the muscular aerodynamic apparatus of the speech organs is manipulated for the generation of syllables, words, and phrases (Ziegler, Aichert, & Staiger, 2012).

Regarding the sites of the lesions responsible for AOS, there remains controversy about the roles of left inferior frontal gyrus (area 44) and left anterior insular cortex (Richardson, Fillmore, Rorden, Lapointe, & Fridriksson, 2012). Greater agreement exists about the implication of left ventral premotor and motor regions in the origin of AOS (Graff-Radford et al., 2014). In the majority of cases, AOS results from left middle cerebral artery stroke, but other etiologies have also been reported. Furthermore, speech abnormalities consistent with AOS have been observed as a primary progressive condition (Josephs et al., 2006). However, the patient groups with alleged primary progressive AOS described so far are probably rather heterogeneous. In particular, patients showing substance loss in the region of the SMA, as described, for instance, by Josephs et al. (2006), may suffer from dysfluencies due to mesiofrontal speech initiation problems (Ziegler, Kilian, & Deger, 1997) rather than from a frontolateral speech motor planning impairment.

In modern psycholinguistic terminology, AOS is allocated to the phonetic planning stage of speech production (Ziegler, 2008). Speech errors in AOS differ from dysarthria in that they can be inconsistent and often are only evident at a segmental level rather than spreading over larger parts of an utterance. For instance, excess nasality may occur selectively and locally on single phonemes in AOS, whereas in dysarthric speakers hypernasality, if present, occurs more as a global feature that extends, as a consequence of velar weakness or slowness, almost uniformly across the segments of a word or phrase. This makes apraxic speech errors less predictable than dysarthric distortions (Staiger, Finger-Berg, Aichert, & Ziegler, 2012). Nonetheless, unlike the phoneme errors observed in many aphasic patients without AOS, the distortions observed in AOS appear motoric because they often lack the quality of well-articulated phonemes through, for example, excess plosive aspirations, audible phoneme transitions, or nasal releases in stop consonants. AOS patients usually are fully aware of their speech problems and, unlike individuals with dysarthria, tend to grope for the correct articulation before they start speaking. They often self-correct their false starts and speech errors, which renders their speech dysfluent and halting.

78.3 MOTOR IMPAIRMENTS FOR SPOKEN LANGUAGE PRODUCTION

Speaking differs from other motor functions in several ways. It is more skillful than many other activities within the human motor repertoire. Among the more advanced motor skills, such as playing a musical instrument, it is the only one that every healthy child acquires without specific instruction. Speaking is acquired over more than the first decade of life, is usually exercised daily, and is continuously adapted to the gradual and sometimes abrupt anatomical changes that occur from childhood to old age. Speech movements are tuned to the generation of (speech) sounds through aerodynamic mechanisms and, hence, are conducted within an acoustic rather than a spatial reference frame (Perkell, 2012). Finally, the evolution of the speech motor system is intrinsically tied to the requirements of spoken communication and is entrenched with the linguistic framework of human language and of a speaker's native language. For these reasons, speaking should be considered a highly specific motor activity that, through mechanisms of practice-related plasticity, shapes a neural basis dedicated to its linguistic and communicative goals (Ziegler & Ackermann, 2013). In the following subsections we discuss evidence from speech disorders that illuminate the particularities of speaking across the different motor activities of the respiratory, laryngeal, and vocal tract muscles.

78.3.1 Speech and Emotional Expression

The laryngeal and facial muscles are used not only in speech but also in emotional expressions such as laughter and crying. According to a dual pathway model of acoustic communication developed by Jürgens and Ploog (Jürgens, 2002), the motor activities of these muscles during emotional expression versus propositional speech are tied to different brain networks. The motor pathways engaged in emotional and intrinsic vocalizations, as understood from primate vocalization studies, have their origin in mesiofrontal cortex (ACG) and travel through the midbrain periaqueductal gray and adjacent tegmentum to the reticular formation and brainstem motor nuclei (Ackermann et al., 2014). The motor pathway involved in voluntary motor activities of the vocal tract, especially in speaking, takes a separate route encompassing corticobulbar, striatal, and cerebellar systems (Section 78.2). Several clinical observations corroborate the distinct courses of these two motor systems. A striking example in this regard relates to the Foix–Chavany–Marie syndrome, which results from bilateral lesions to the upper motor neuron system (Section 78.2.1). Patients with this syndrome are severely dysarthric or anarthric because of bilateral upper motor neuron paralysis of the bulbar speech muscles. At the same time, these patients are able to completely adduct their vocal folds and activate their facial musculature during emotionally driven laughter or crying (Mao, Coull, Golper, & Rau, 1989). This so-called automatic–voluntary movement dissociation is taken as evidence for the separate courses of emotional and volitional motor pathways for the speech muscles. A less dramatic but equally convincing dissociation may occur after unilateral cerebral lesions that may cause contralateral lower facial paresis during speaking and volitional mouth spreading ("show your teeth") but symmetric spreading during spontaneous smiles ("volitional facial paresis") or, conversely, asymmetric smiling but symmetric lip spreading in speech and the "show your teeth" task ("emotional facial paresis"; Hopf, Müller-Forell, & Hopf, 1992).

Although these dissociations relate to distinct actions with either a volitional/linguistic or an emotional content, natural speaking is usually linked to motivational and emotional states. The impact of a speaker's attitudes, motivations, and emotions on his/her speech movements is reflected in the prosodic modulation of spoken utterances. This interaction implies the existence of a neural interface through which the emotional/intrinsic ("limbic") vocalization system modulates speech motor pathway activity. In a recent extension of Jürgens' (2002) dual pathway model, Ackermann et al. (2014) proposed that the basal ganglia provide a platform for the integration of limbic mechanisms of acoustic communication with articulate speech. According to this theory, a cascade of striato-nigro-striatal circuits extending from ventromedial (limbic) to dorsolateral (motor) components of the striatum is the substrate for the limbic-motor integration process. These circuits interconnect two parallel cortico–basal ganglia–thalamo–cortical loops, one conveying motivation-related information via ventromedial–dorsolateral pathways and the other conveying speech motor information via the corticostriatal motor loop (Ackermann et al., 2014). The hypokinetic dysarthria associated with Parkinson's disease may illustrate how this integration can fail. That is, the depletion of striatal dopamine leads to a diminished impact of attitudinal, motivational, and emotional states on speech motor control and thereby results in the flattened, monotonous prosody and the hypophonic voice that is characteristic of many people with Parkinson's disease. This illustrates how evidence from speech motor impairments may contribute to a deeper understanding of the neural organization and the interaction of laryngeal and oral motor activity for speech and for emotional expression.

78.3.2 Speech Versus Volitional Nonspeech Vocal Tract Movements

In the automatic–voluntary motor dissociations described in the preceding section, speech was subsumed among a broader class of willed motor actions involving the vocal tract, including movements such as volitional lip spreading, tongue protrusion, and the like. If we assume, as outlined, that the speech motor system co-evolves within the structural framework of linguistic communication and that the domain-specific properties of speech motor control are represented at the neural level, then we would expect that brain lesions may selectively affect or preserve speech relative to nonspeech volitional motor activities.

There is substantial evidence that speech and nonspeech motor impairments can be dissociated. Patients with brain lesions may show impairments of the oral (voluntary) phase of swallowing or may have problems imitating tongue protrusion or other labial or lingual displays, but at the same time have normal or almost normal speech. Conversely, some patients may present with marked dysarthric or speech apraxic symptoms but have a relatively preserved ability to perform nonspeech vocal tract movements. Dissociations have been found for a number of nonspeech motor tasks, including chewing and swallowing, movement imitation, strength and endurance tasks, rapid syllable production, or visuomotor tracking (Ziegler, 2003). One of the reasons for these findings is that nonspeech oromotor activities used in research and assessment differ from speech along a number of dimensions, such as timing, strength, and airflow requirements, the degree of interaction between subsystems, the rhythmical entrainment of movements, and the role of acoustic output as a reference frame. The specific requirements of producing intelligible and naturally sounding speech by manipulating the respiratory, laryngeal, and supralaryngeal muscles in a particular way, and the fact that the highly adaptive interplay of these muscles in speaking is acquired over time during childhood, call for the engagement of a specialized neural network in adult motor speech. This conclusion receives strong theoretical support from investigations into experience-dependent neuroplasticity (Ostry, Darainy, Mattar, Wong, & Gribble, 2010; Zatorre, Fields, & Johansen-Berg, 2012). For example, neuroimaging studies comparing speech with oral nonspeech movements converge on the observation that motor speech is more lateralized to the left hemisphere and is associated with less neural activation than nonspeech oral motor tasks (Moser et al., 2009; for a more extensive review of these arguments see Ziegler and Ackermann, 2013).

In summary, the selective and domain-specific nature of motor speech impairments is consistent with highly specialized neural organization of the motor aspects of speaking. This contributes to evidence that the speech motor system constitutes an integral part of the biology of human language.

78.3.3 Language-Specific Phonological Structure Interacts with Speech Motor Impairment

From early childhood, speech acquisition is shaped by the phonological structures the child encounters in his/her native language. The maturation of vocal tract, laryngeal, and respiratory functions for speaking separates rather early from nonspeech metabolic motor patterns (Moore, Caulfield, & Green, 2001) and takes a course toward mastering the specific motor requirements of the language's phoneme repertoire and the phonotactic and prosodic patterns of the child's ambient language (Astruc, Payne, Post, Vanrell, & Prieto, 2013). As a consequence, the speech motor system in the adult brain is shaped by the properties of the speaker's native language. When the system breaks down after a brain lesion, observed motor speech failures reflect universal and language-particular aspects of phonological structure (Ziegler & Ackermann, 2013).

This is quite obviously the case in AOS. The sound level errors and phonetic distortions of speakers with AOS are sensitive to syllable structure, respect syllable boundaries, and are influenced by language-particular frequency-related properties of syllables and words (Aichert & Ziegler, 2004; Romani & Galluzzi, 2005; Schoor, Aichert, & Ziegler, 2012; Staiger & Ziegler, 2008). A further potential source of influence on apraxic errors is the metrical pattern of the speaker's language. For example, in German speakers with AOS, more errors are observed on disyllabic nouns with stress on the second (iambic) as compared with the first syllable (trochaic), which conforms to expectations because trochees are by far more frequent than iambs in German (Aichert, Büchner, & Ziegler, 2011).

We examined the influences of phonological structure on speech errors in German speakers with AOS using a nonlinear probabilistic model based on the hierarchical architecture of words, extending from the level of articulatory gestures to the level of metrical feet (Ziegler, 2009). The model was built to estimate the likelihood that a word with a particular syllabic and metrical structure would be produced accurately by an apraxic speaker. It was fitted to a large sample of words for which accuracy data from 120 apraxic productions were available. The findings revealed a motor planning hierarchy that was consistent with phonological models of syllable constituency and metrical structure of German with, for instance, relatively strong bonds between nucleus and coda gestures compared with onset–rime gesture combinations or between syllable pairs forming a trochaic foot relative to nontrochaic combinations.

Analyses of apraxic speech errors demonstrate that the architecture of speech motor programs conforms to the phonological architecture of words, implying that speech motor control and linguistic structure are mutually interconnected. These findings are at odds with theories postulating a strict dualism of linguistic versus motor functions.

78.4 SENSORY-MOTOR ASPECTS OF SPEECH SOUND PRODUCTION IMPAIRMENT

The interaction between sensory and motor processing mechanisms has been a core issue in the understanding of the biology of language ever since the development of the Wernicke–Lichtheim model (Lichtheim, 1885). In that model, language production relies on the activation of information from sensory centers in which the *sound images* (German *Klangbilder*) of words are stored (Lichtheim, 1885, p. 211). This theory was derived from Wernicke's observation that lesions to auditory association centers in the left posterior superior temporal lobe not only caused auditory comprehension problems but also was associated with paraphasic language production. Later, Liepmann (1900) developed his influential theory that motor action (of the upper extremities) relies on a posterior-to-anterior stream of information located in the left hemisphere that conveys an *ideatory blueprint* (German: *ideatorischer Entwurf*) based primarily on sensory information (Liepmann, 1913, pp. 488–490). In modern accounts of speech production, these ideas are vested in computational models based on *auditory goals*—not of words but rather of phonemes or syllables—that provide an acoustic reference frame for speaking (Guenther, Hampson, & Johnson, 1998). More recently, new techniques enabling online perturbations of somatosensory information during articulation have led to an extension of sensory-motor theories of speech production by including a proprioceptive feedback processing route (Tremblay, Shiller, & Ostry, 2003). Speech production models based on experimental data relating to auditory and somatosensory feedback processing have invoked the concept of *internal models* that guide motor action (Hickok, 2014; Tourville & Guenther, 2011). Parallel to these developments, a new research focus based on fiber-tracking methods has refined the connectional anatomy related to Liepmann's model of anterior-posterior information processing streams in language production (Dick & Tremblay, 2012; Rauschecker, 2012). Clinical data from patients with speech sound impairments have not yet had much influence on this research.

78.4.1 Auditory and Somatosensory Feedback and Speech Impairment

The role of *auditory* feedback in the genesis of speech impairment has been addressed in research focused on stuttering (Cai et al., 2012) and the effects on speech in those with hearing loss and cochlear implants (Perkell, 2012). Although this research is clearly relevant to the field, it is not considered further here.

Regarding the role of *somatosensory* afferent information in speech, only scarce clinical data exist because lesions causing a complete extinction of trigeminal somatosensory input are rare and the degree to which such information is still available cannot be assessed reliably enough by clinical methods. Duffy (2013) conjectures the existence of a "sensory dysarthria" syndrome resulting from impaired oral somatosensory processing that may lead to imprecise articulation that could reflect compensations for reduced sensory input through increased range of articulatory movement (e.g., exaggerated jaw movements). Hoole (1987) described a patient who, after a closed head trauma and whiplash injury, had experienced substantial sensory deficits in the oral-facial region as assessed by standard clinical methods. After initial severe dysarthria the patient's speech recovered quickly, although his sensory deficits remained unresolved. Experimental investigations of isolated vowel articulation after motor speech recovery revealed reduced ability of this patient to compensate for proprioceptive perturbations with a biteblock, especially when auditory feedback was blocked by noise-masking. This result corroborates assumptions that somatosensory processing supports adaptive mechanisms in speaking and that auditory and somatosensory processing can partially complement each other in this role (Perkell, 2012). Yet, clinical cases of this kind will always leave the possibility that residual sensory information that evades clinical physiological detection may still provide sufficient afferent information to support normal speech, at least when the lesion is acquired in adulthood.

78.4.2 Cerebellar Sensorimotor Integration Mechanisms in Speech Impairment

The cerebellum is classically considered as a site in the brain where sensorimotor integration takes place (Bhanpuri, Okamura, & Bastian, 2013). Ataxic dysarthria after cerebellar lesions might therefore be considered as a clinical model of impaired integration of proprioceptive and motor information in speech. However, most current theories conceptualize cerebellar dysarthria to result from impaired *feedforward* processing mechanisms (Spencer & Slocomb, 2007; Mariën, in Manto et al., 2012). Yet, for instance, the fact that Friedreich's ataxia is primarily viewed as an afferent ataxic syndrome suggests that the dysarthric impairment observed in these patients may at least partly reflect an impairment of proprioceptive feedback processing in speech (Pandolfo, 2009).

The disturbance of sensory feedback mechanisms in ataxic patients becomes more apparent in paraspeech or nonspeech oral motor tasks that involve strong feedback integration capacities, such as sustained vowel production, visuomotor tracking, or rapid syllable repetition. Maintenance of a stable pitch and loudness level in sustained vowels is often disproportionately impaired in ataxic patients who often demonstrate fluctuating pitch or voice tremor during sustained phonation over several seconds, but not necessarily during speaking (Ackermann & Ziegler, 1991; Brendel et al., 2013). This may reflect a failure of correction mechanisms based on reafferent laryngeal proprioceptive or auditory information. Severely impaired adaptive sensorimotor mechanisms of patients with hereditary ataxias were also observed in a visuomotor tracking task requiring control of airflow velocity during expiration to track a ramp signal (Deger, Ziegler, & Wessel, 1999), but there was no correlation between the tracking and the speech impairment. Finally, in several studies patients with cerebellar pathology have had difficulty adapting to the specific demands of a task requiring repetition of a syllable (e.g., puh, tuh, kuh) at maximum speed (Brendel et al., 2013; Ziegler & Wessel, 1996), a task considered to rely strongly on sensory mechanisms for the selection of a jaw angle that supports maximally rapid labial or lingual opening and closing movements. The patients had dramatically reduced acceleration ratios of syllable repetition relative to speaking rates (Ziegler, in Mariën et al., 2013). Taken together, these results point to vulnerability of patients with cerebellar lesions to motor demands requiring a strong reliance on sensorimotor adaptation mechanisms.

78.4.3 Striatal Mechanisms of Sensorimotor Integration in Speech Impairment

The motor functions of the basal ganglia are considered to rely, at least partly, on striatal and pallidal sensory processing mechanisms involved in kinaesthesia or somatosensory discrimination (Maschke, Gomez, Tuite, & Konczak, 2003). Altered sensory processing is thought to contribute to the motor deficits of patients with Parkinson's and Huntington's disease (Boecker et al., 1999), and a somatosensory disinhibition mechanism has been proposed to underlie dystonic motor impairment (Frasson et al., 2001). According to a hypothesis advanced by Yin (2014), the basal ganglia control movement velocity through kinaesthetic reafferent input.

It is not known if similar sensory mechanisms can also serve as an explanation for the speech motor patterns of patients with basal ganglia dysfunction. Patients with Parkinson's disease demonstrate consistent proprioceptive deficits in the oral region, but the relationship of these abnormalities to speech impairment is unclear (Schneider, Diamond, & Markham, 1986). In a recent report of two patients with oromandibular dystonia (Møller et al., 2013), evidence was found supporting abnormal sensorimotor integration or somatosensory dysfunction for afferent input from the oral region as an explanation for the motor impairment.

Other sensory processing functions of the basal ganglia relevant for speech relate to the self-perception of speech loudness. Ho, Bradshaw, and Iansek (2000) examined the hypothesis that reduced loudness (hypophonia) in Parkinson's disease is due to impaired motor scaling mechanisms based on patients' misperception of their own speech loudness. In an "immediate self-perception rating," Parkinson's patients overestimated their loudness relative to normal subjects, which the authors interpreted as an exaggeration of self-perceived effort during speaking. Likewise, the patients also overestimated their speech volume in a playback condition, even though they had spoken more quietly than controls. Ho et al. (2000) suggested that their findings provide support for the presence of sensory anomalies in Parkinson's disease, which may cause inappropriate scaling of loudness.

78.4.4 Sensorimotor Connectivity at the Cortical Level

As already discussed, recent functional neuroanatomic accounts of the dorsal stream system connecting posterior superior-temporal cortex with inferior-parietal and posterior-inferior frontal areas constitute a modern version of Liepmann's idea of a posterior-to-anterior stream of information governing motor actions through sensory-based goals (Saur et al., 2008). The cortical target areas of this system are interconnected by a massive fiber bundle, the superior longitudinal fascicle, which constitutes a circuit that includes auditory, somatosensory, and motor association areas. The dorsal pathway of the left (dominant) hemisphere is considered to be involved in higher sensorimotor integration by mapping acoustic speech sounds, and eventually also somatosensory representations, onto their corresponding articulatory actions. The prototype task targeting this system in functional imaging studies is the word repetition task (Saur et al., 2008), but more complex tasks involving phonological transformations of words have also been used (Kellmeyer et al., 2013).

Although a simplification, the anterior target area of the left dorsal stream approximately coincides with the lesion site reported for AOS. This could support an inference that the pathomechanism of AOS predominantly reflects damage to feedforward processing components of speech motor control while leaving sensory feedback mechanisms intact. This is compatible with clinical experience and some experimental evidence that patients with AOS have intact monitoring of their speech errors and preserved auditory speech processing. Jacks (2008) performed a biteblock experiment with apraxic speakers to test this feedforward hypothesis and found that AOS participants compensated for the biteblock perturbation in a manner similar to normal speakers. This normal adaptation to proprioceptive perturbation was taken as evidence of intact somatosensory feedback.

Historically, and in modern research, the dorsal pathway of the left hemisphere has been associated with aphasic phonological impairment. A longstanding hypothesis is that the phonemic paraphasias of patients with conduction aphasia result from a disconnection of the auditory from the motor representation areas of words (Geschwind, 1965). In conventional aphasiology, a strict boundary is drawn between aphasic phonological impairment as purely *abstract-symbolic*, and AOS as purely *motor* by nature. However, one may question whether such a clear-cut dichotomy is tenable, taking the aforementioned dorsal pathway hypothesis into consideration. As said, the dorsal pathway *interfaces* motor with sensory information (Hickok, 2014; Rauschecker, 2012). This interface can be assumed to act on a representational level, which is sufficiently abstract to make any additional assumptions about symbolic representations dispensable.

78.5 CONCLUSION

In this chapter we have characterized the nature of the speech motor system and its integration with sensory processes as a part of human linguistic behavior. Future theories and clinical models of motor speech disorders should consider, to a greater extent, the evidence consistent with such an account.

Acknowledgment

We are grateful to Joe Duffy for his valuable comments and suggestions regarding a former version of this manuscript.

References

Ackermann, H. (2008). Cerebellar contributions to speech production and speech perception: Psycholinguistic and neurobiological perspectives. *Trends in Neurosciences, 31*, 265–272.

Ackermann, H., Hage, S. R., & Ziegler, W. (2014). Brain mechanisms of acoustic communication in humans and nonhuman primates: An evolutionary perspective. *Behavioral and Brain Sciences, 37*, 529–604.

Ackermann, H., & Ziegler, W. (1991). Cerebellar voice tremor: An acoustic analysis. *Journal of Neurology, Neurosurgery, and Psychiatry, 54*, 74–76.

Ackermann, H., & Ziegler, W. (1992). Articulatory deficits in parkinsonian dysarthria: An acoustic analysis. *Journal of Neurology, Neurosurgery, and Psychiatry, 54*, 1093–1098.

Aichert, I., Büchner, M., & Ziegler, W. (2011). Why is ['ju:do] easier than [ju've:l]? Perceptual and acoustic analyses of word stress in patients with apraxia of speech. *Stem-, Spraak- en Taalpathologie, 17*, 15.

Aichert, I., & Ziegler, W. (2004). Syllable frequency and syllable structure in apraxia of speech. *Brain and Language, 88*, 148–159.

Albanese, A., & Jankovic, J. (2012). *Hyperkinetic movement disorders: Differential diagnosis and treatment*. Hoboken, NJ: John Wiley & Sons.

Astruc, L., Payne, E., Post, B., Vanrell, M., & Prieto, P. (2013). Tonal targets in early child English, Spanish, and Catalan. *Language and Speech, 56*, 229–253.

Bhanpuri, N. H., Okamura, A. M., & Bastian, A. J. (2013). Predictive modeling by the cerebellum improves proprioception. *The Journal of Neuroscience, 33*, 14301–14306.

Boecker, H., Ceballos-Baumann, A., Bartenstein, P., Weindl, A., Siebner, H. R., Fassbender, T., et al. (1999). Sensory processing in Parkinson's and Huntington's disease investigations with 3D H215O-PET. *Brain, 122*, 1651–1665.

Brendel, B., Ackermann, H., Berg, D., Lindig, T., Scholderle, T., Schols, L., et al. (2013). Friedreich ataxia: Dysarthria profile and clinical data. *Cerebellum, 12*, 475–484.

Cai, S., Beal, D. S., Ghosh, S. S., Tiede, M. K., Guenther, F. H., & Perkell, J. S. (2012). Weak responses to auditory feedback perturbation during articulation in persons who stutter: Evidence for abnormal auditory-motor transformation. *PLoS ONE, 7*(7), e41830. Available from: http://dx.doi.org/10.1371/journal.pone.0041830.

Darley, F. L., Aronson, A. E., & Brown, J. R. (1975). *Motor speech disorders*. Philadelphia, PA: W.B. Saunders.

Deger, K., Ziegler, W., & Wessel, K. (1999). Airflow tracking in patients with ataxic disorders. *Clinical Linguistics and Phonetics, 13*, 433–447.

Dick, A. S., & Tremblay, P. (2012). Beyond the arcuate fasciculus: Consensus and controversy in the connectional anatomy of language. *Brain, 135*, 3529–3550.

Duffy, J. R. (2013). *Motor speech disorders: Substrates, differential diagnosis, and management* (3rd ed.). St. Louis, MO: Elsevier Mosby.

Frasson, E., Priori, A., Bertolasi, L., Mauguiere, F., Fiaschi, A., & Tinazzi, M. (2001). Somatosensory disinhibition in dystonia. *Movement Disorders, 16*, 674–682.

Geschwind, N. (1965). Disconnexion syndromes in animal and man. Part I. *Brain, 88*, 237–294.

Graff-Radford, J., Jones, D. T., Strand, E. A., Rabinstein, A. A., Duffy, J. R., & Josephs, K. A. (2014). The neuroanatomy of pure apraxia of speech in stroke. *Brain and Language, 129*, 43–46.

Goldenberg, G. (2013). *Apraxia. The cognitive side of motor control. Oxford: Oxford University Press, 79*.

Guenther, F. H., Hampson, M., & Johnson, D. (1998). A theoretical investigation of reference frames for the planning of speech movements. *Psychological Review, 105*, 611–633.

Hallett, M. (2011). Neurophysiology of dystonia: The role of inhibition. *Neurobiology of Disease, 42*, 177–184.

Hickok, G. (2014). Towards an integrated psycholinguistic, neurolinguistic, sensorimotor framework for speech production. *Language, Cognition and Neuroscience, 29*, 52–59.

Ho, A. K., Bradshaw, J. L., & Iansek, R. (2000). Volume perception in parkinsonian speech. *Movement Disorders, 15*, 1125–1131.

Hoole, P. (1987). Bite-block speech in the absence of oral sensibility. In *Proceedings of the 11th International Congress of Phonetic Sciences* (Vol. 4, pp. 16–19). Tallinn: Academy of Sciences of the Estonian SSR.

Hopf, H. C., Müller-Forell, W., & Hopf, N. J. (1992). Localization of emotional and volitional facial paresis. *Neurology, 42*, 1918–1923.

Jacks, A. (2008). Bite block vowel production in apraxia of speech. *Journal of Speech Language and Hearing Research, 51*, 898–913.

Jankovic, J. (2008). Parkinson's disease: Clinical features and diagnosis. *Journal of Neurology, Neurosurgery and Psychiatry, 79*, 368–376.

Josephs, K. A., Duffy, J. R., Strand, E. A., Whitwell, J. L., Layton, K. F., Parisi, J. E., et al. (2006). Clinicopathological and imaging correlates of progressive aphasia and apraxia of speech. *Brain, 129*, 1385–1398.

Jürgens, U. (2002). Neural pathways underlying vocal control. *Neuroscience and Biobehavioral Reviews, 26*, 235–258.

Kellmeyer, P., Ziegler, W., Peschke, C., Juliane, E., Schnell, S., Baumgaertner, A., et al. (2013). Fronto-parietal dorsal and ventral pathways in the context of different linguistic manipulations. *Brain and Language, 127*, 241–250.

Krainik, A., Lehericy, S., Duffau, H., Capelle, L., Chainay, H., Cornu, P., et al. (2003). Postoperative speech disorder after medial frontal surgery role of the supplementary motor area. *Neurology, 60*, 587–594.

Lichtheim, L. (1885). Ueber Aphasie. Aus der medicinischen Klinik in Bern. *Deutsches Archiv für klinische Medicin, 36*, 204–268.

Liepmann, H. (1900). Das Krankheitsbild der Apraxie ("motorischen Asymbolie") auf Grund eines Falles von einseitiger Apraxie (II). *Monatsschrift für Psychiatrie und Neurologie, VIII*, 102–132.

Liepmann, H. (1913). Motorische Aphasie und Apraxie. *Monatsschrift für Psychiatrie und Neurologie, XXXIV*, 485–494.

Manto, M., Bower, J. M., Conforto, A. B., Delgado-Garcia, J. M., Farias da Guarda, S. N., Gerwig, M., et al. (2012). Consensus paper: Roles of the cerebellum in motor control—The diversity of ideas on cerebellar involvement in movement. *Cerebellum, 11*, 457–487.

Mao, C.-C., Coull, B. M., Golper, L. A. C., & Rau, M. T. (1989). Anterior operculum syndrome. *Neurology, 39*, 1169–1172.

Mariën, P., Ackermann, H., Adamaszek, M., Barwood, C. H., Beaton, A., Desmond, J., et al. (2013). Consensus paper: Language and the cerebellum: An ongoing enigma. *The Cerebellum*, 1–25.

Maschke, M., Gomez, C. M., Tuite, P. J., & Konczak, J. (2003). Dysfunction of the basal ganglia, but not the cerebellum, impairs kinaesthesia. *Brain, 126*, 2312–2322.

Moore, C. A., Caulfield, T. J., & Green, J. R. (2001). Relative kinematics of the rib cage and abdomen during speech and nonspeech behaviors of 15-month-old children. *Journal of Speech, Language, and Hearing Research, 44*, 80–94.

Moser, D., Fridriksson, J., Bonilha, L., Healy, E. W., Baylis, G., Baker, J. M., et al. (2009). Neural recruitment for the production of native and novel speech sounds. *NeuroImage, 46*, 549–557.

Muellbacher, W., Artner, C., & Mamoli, B. (1999). The role of the intact hemisphere in recovery of midline muscles after recent monohemispheric stroke. *Journal of Neurology, 246*, 250–256.

Møller, E., Bakke, M., Dalager, T., Werdelin, L. M., Lonsdale, M. N., Højgaard, L., et al. (2013). Somatosensory input and oromandibular dystonia. *Clinical Neurology and Neurosurgery, 115*, 1141–1143.

Ostry, D. J., Darainy, M., Mattar, A. A., Wong, J., & Gribble, P. L. (2010). Somatosensory plasticity and motor learning. *The Journal of Neuroscience, 30*, 5384–5393.

Pandolfo, M. (2009). Friedreich ataxia: The clinical picture. *Journal of Neurology, 256*(Suppl. 1), 3–8.

Perkell, J. S. (2012). Movement goals and feedback and feedforward control mechanisms in speech production. *Journal of Neurolinguistics, 25*, 382–407.

Rauschecker, J. P. (2012). Ventral and dorsal streams in the evolution of speech and language. *Frontiers in Evolutionary Neuroscience, 4*, 7.

Richardson, J. D., Fillmore, P., Rorden, C., Lapointe, L. L., & Fridriksson, J. (2012). Re-establishing Broca's initial findings. *Brain Language, 123*, 125–130.

Rodriguez-Oroz, M. C., Jahanshahi, M., Krack, P., Litvan, I., Macias, R., Bezard, E., et al. (2009). Initial clinical manifestations of Parkinson's disease: Features and pathophysiological mechanisms. *Lancet Neurology, 8*, 1128–1139.

Rodriguez-Oroz, M. C., Obeso, J. A., Lang, A. E., Houeto, J. L., Pollak, P., Rehncrona, S., et al. (2005). Bilateral deep brain stimulation in Parkinson's disease: A multicentre study with 4 years follow-up. *Brain, 128*, 2240–2249.

Romani, C., & Galluzzi, C. (2005). Effects of syllabic complexity in predicting accuracy of repetition and direction of errors in patients with articulatory and phonological difficulties. *Cognitive Neuropsychology, 22*, 817–850.

Saur, D., Kreher, B. W., Schnell, S., Kummerer, D., Kellmeyer, P., Vry, M. S., et al. (2008). Ventral and dorsal pathways for language. *Proceedings of the National Academy of Sciences of the United States of America, 105*, 18035–18040.

Schneider, J. S., Diamond, S. G., & Markham, C. H. (1986). Deficits in orofacial sensorimotor function in Parkinson's disease. *Annals of Neurology, 19*, 275–282.

Schoor, A., Aichert, I., & Ziegler, W. (2012). A motor learning perspective on phonetic syllable kinships: How training effects transfer from learned to new syllables in severe apraxia of speech. *Aphasiology, 26*, 880–894.

Simonyan, K., Berman, B. D., Herscovitch, P., & Hallett, M. (2013). Abnormal striatal dopaminergic neurotransmission during rest and task production in spasmodic dysphonia. *The Journal of Neuroscience, 33*, 14705–14714.

Spencer, K. A., & Slocomb, D. L. (2007). The neural basis of ataxic dysarthria. *The Cerebellum, 6*, 58–65.

Staiger, A., Finger-Berg, W., Aichert, I., & Ziegler, W. (2012). Error variability in apraxia of speech: A matter of controversy. *Journal of Speech, Language, and Hearing Research, 55*, S1544–S1561.

Staiger, A., & Ziegler, W. (2008). Syllable frequency and syllable structure in the spontaneous speech production of patients with apraxia of speech. *Aphasiology, 22*, 1201–1215.

Tourville, J. A., & Guenther, F. H. (2011). The DIVA model: A neural theory of speech acquisition and production. *Language and Cognitive Processes, 26*, 952–981.

Tremblay, S., Shiller, D. M., & Ostry, D. J. (2003). Somatosensory basis of speech production. *Nature, 423*, 866–869.

Ushe, M., & Perlmutter, J. S. (2012). Oromandibular and lingual dystonia associated with spinocerebellar ataxia type 8. *Movement Disorders, 27*, 1741–1742.

Yin, H. H. (2014). Action, time and the basal ganglia. *Philosophical Transactions of the Royal Society B: Biological Sciences, 369*, 20120473.

Zatorre, R. J., Fields, R. D., & Johansen-Berg, H. (2012). Plasticity in gray and white: Neuroimaging changes in brain structure during learning. *Nature Neuroscience, 15*, 528–536.

Ziegler, W. (2003). Speech motor control is task-specific. Evidence from dysarthria and apraxia of speech. *Aphasiology, 17*, 3–36.

Ziegler, W. (2008). Apraxia of speech. In G. Goldenberg, & B. Miller (Eds.), *Handbook of clinical neurology* (pp. 269–285). London: Elsevier.

Ziegler, W. (2009). Modelling the architecture of phonetic plans: Evidence from apraxia of speech. *Language and Cognitive Processes, 24*, 631–661.

Ziegler, W., & Ackermann, H. (2013). Neuromotor speech impairment: It's all in the talking. *Folia Phoniatrica et Logopaedica, 65*, 55–67.

Ziegler, W., Aichert, I., & Staiger, A. (2012). Apraxia of speech: Concepts and controversies. *Journal of Speech, Language, and Hearing Research, 55*, S1485–S1501.

Ziegler, W., Kilian, B., & Deger, K. (1997). The role of the left mesial frontal cortex in fluent speech: Evidence from a case of left supplementary motor area hemorrhage. *Neuropsychologia, 35*, 1197–1208.

Ziegler, W., & Wessel, K. (1996). Speech timing in ataxic disorders: Sentence production and rapid repetitive articulation. *Neurology, 47*, 208–214.

CHAPTER

79

The Neurobiology of Developmental Stuttering

Kate E. Watkins, Jennifer Chesters and Emily L. Connally

Department of Experimental Psychology, University of Oxford, Oxford, UK

79.1 INTRODUCTION

Dysfluency occurs when the normal flow and smooth delivery of speech are disrupted. Often, normal speech dysfluencies, such as silent pauses and nonlexical vocalizations (e.g., "uh" or "um"), can usefully add emphasis or draw attention to the content of upcoming utterances. In some people, however, speech dysfluencies are pathological and interfere with speech communication to such an extent that a fluency disorder is diagnosed. The most commonly diagnosed fluency disorder is *developmental stuttering*, which is distinguished from *acquired or neurogenic* stuttering that is associated with brain disease or injury. In this chapter, we describe the current state of our understanding of the neural basis of developmental stuttering. It is rare to be able to study a developmental disruption to a brain function such as speech that does not have consequences for other aspects of cognition. However, such specificity is the case for developmental stuttering. A better understanding of the neurobiological bases of stuttering will assist the development of novel and effective therapies. Furthermore, it will increase our understanding of the neural basis of normal speech production. Here, we first describe the key features of developmental stuttering. We then outline the evidence for the neurobiological differences between people who stutter and fluent speakers, and consider these differences in the context of a number of theories that attempt to explain stuttering.

79.2 DEVELOPMENTAL STUTTERING

Stuttering is a long-documented speech disorder that appears to exist in all languages and cultures. Records of this type of halted or repetitious speech date back to the Old Testament; the speech of Moses and the Roman Emperor Claudius have been described as showing features of stuttering. A current definition describes stuttering as:

> *Speech that is characterized by frequent repetition or prolongation of sounds or syllables or words, or by frequent hesitations or pauses that disrupt the rhythmic flow of speech. It should be classified as a disorder only if its severity is such as to markedly disturb the fluency of speech.* ***The International Classification of Diseases, Version 10 (World Health Organisation, 2010; http://www.who.int/classifications/icd/en/).***

People who stutter produce utterances that are characterized by periods of normal fluency interspersed with occasions when speech is interrupted temporarily. On such occasions, the person knows what he/she wants to say but has difficulty moving forward in the speech sequence.

The moments of dysfluency experienced by a person who stutters are not randomly distributed in the speech stream. Dysfluencies tend to occur at the beginnings of words or sentences. An example of a sentence containing three types of dysfluency is shown in Figure 79.1. As can be seen from the spectrogram, the first word in the sentence is preceded by a tense pause indicating that speech is blocked. There is a low-amplitude band of acoustic energy visible in the spectrogram that would be heard as a "creaking" voice, but often these pauses are inaudible. The "l" sound before the successful production of the word "like" is prolonged, resulting in a longer band of higher-amplitude energy. Finally, the first sounds of the words "didn't" and "like" are repeated and shown as brief bursts of energy.

It should be noted that the occasional dysfluencies of fluent speakers also typically occur in sentence or

Neurobiology of Language. DOI: http://dx.doi.org/10.1016/B978-0-12-407794-2.00079-1

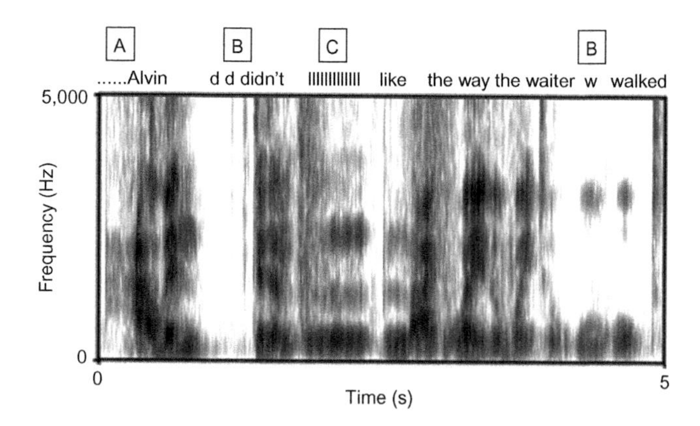

FIGURE 79.1 Sound spectrogram of a sentence spoken by a male who stutters. The sentence is transcribed above the spectrogram. Dysfluencies are indicated above the spectrogram by the letters (A–C). (A) Tense pause or "block". (B) Repetition. (C) Prolongation. This visual representation of the sentence shows the variation in frequencies (y-axis) in the acoustic signal across time (x-axis), with the amplitude (loudness) of the signal indicated by the intensity of the grayscale.

word initial positions. The pattern of dysfluency in developmental stuttering suggests difficulty in the initiation of speech segments or in transition between segments. The characteristic repetition of speech sounds in stuttered speech may reflect repeated attempts to successfully transition to the next sound in the speech sequence. There are also linguistic and affective influences on stuttered speech. Children's stuttering is increased if the planned speech utterance is syntactically complex or long (Zackheim & Conture, 2003). Speech fluency is also negatively affected by stress and fatigue in people who stutter, as it is for fluent speakers.

In addition to these core characteristics of disrupted speech in development stuttering, secondary behaviors are often evident. These may include increased or abnormal muscular tension in the face and neck and extraneous movements of the head or body when speaking. It is also common for people who stutter to avoid words or sounds that they find difficult to produce, or situations in which they are more likely to stutter (Bloodstein & Ratner, 2008). Anxiety is more commonly reported in people who stutter than in the general population, to the extent that some early theories suggested that "nervousness" or anxiety caused stuttering. However, arguments that a shy or nervous disposition may be a precursor to development of stuttering have now been debunked by prospective studies (Yairi & Ambrose, 2013). Furthermore, the strongest evidence to date is that social anxiety specifically, not anxiety in general, is related to stuttering (Iverach, Menzies, O'Brian, Packman, & Onslow, 2011). Because speech is fundamental to normal everyday functioning, it is not surprising that living with a speech disorder can cause anxiety. As such, anxiety in social contexts would be a likely consequence of continual negative experience of speaking situations, as is often the case for someone who stutters.

Developmental stuttering typically starts early in childhood and affects approximately 1 in 20 children. It develops between the ages of 2 and 4 years, although onset later in childhood has often been described. The early age at which stuttering starts might reflect vulnerability in the speech motor system at approximately the time when language skills rapidly expand. The majority of children recover from stuttering during early childhood, with another half of the remainder recovering before puberty (Yairi & Ambrose, 2013). The disorder persists to adulthood in approximately 1%. A major challenge for researchers and clinicians is to determine factors that predict persistence or recovery in stuttering. One clear predictor of persistence in developmental stuttering is male gender. Approximately twice as many boys as girls stutter during early childhood. In adulthood, the ratio increases to approximately four or five males to one female who stutters. Genetic differences between males and females might confer greater or lesser mechanisms of plasticity or reorganizational capacity, which in turn would affect recovery and persistence.

79.3 ENHANCING FLUENCY IN PEOPLE WHO STUTTER

There is no "cure" for stuttering, but many therapeutic interventions have been designed to increase fluency. Early historic approaches included holding pebbles in the mouth, as practiced by Demosthenes, an orator of Ancient Greece. The more extreme practice of excision of parts of the tongue was popular in 18th century and 19th century France. Contemporary fluency therapies that directly target speech focus primarily on changing speech motor patterns. These may include slowing speech, producing gentle onsets to syllables, or producing continual voicing during speech.

It is well-known that the characteristic symptoms of stuttering are absent during singing. Fluency can be enhanced in people who stutter through a variety of other modes that involve changing the way speech is produced, including adopting a different accent or using an external stimulus such as a metronome or another speaker. Using a metronome to pace speaking (known as "syllable-timed speech") was a popular method in stuttering therapy in the late 20th century. Choral speech (speaking in time with another person) is also very effective, perhaps because it also provides external timing cues, although hearing another speaker's voice may also serve to mask that of the person who stutters. It is likely that these different methods work to enhance fluency because they involve changing speech rhythm or prosody or both. People who stutter report that they are fluent when speaking to babies or to animals. This might be because these are contexts in which prosody is usually exaggerated. Altering the rhythm and stress patterns of speech involves moving from a habitual overlearned motor speech pattern to a novel one. This may engage distinct neural systems for the timing and sequencing of speech sounds.

Another way to increase fluency in people who stutter is to alter the sensory feedback of their own speech production. Simple masking to prevent auditory feedback of a stuttering speaker's voice is effective. Recording speech production and feeding it back with either a short (50 ms) delay or a shift in frequency can achieve complete or nearly complete fluency and also results in natural-sounding speech (Lincoln, Packman, & Onslow, 2006). The mechanism by which altered auditory feedback is effective in inducing fluency is not understood. Delayed auditory feedback can reduce speech rate and even induce dysfluency in otherwise fluent speakers (Lee, 1951). Frequency-shifted feedback, however, does not alter the rate of speech production but could "trick" the brain into thinking another person is speaking, producing an effect similar to choral speech. Devices to alter speech feedback in these ways have become commercially available and smartphone applications used to aid fluency are increasing in popularity. However, it can be cognitively demanding to use these devices, and there is evidence that the fluency enhancing effects may "wear off" over time. The effectiveness of altered auditory feedback in enhancing fluency suggests that the core deficit in developmental stuttering may be one of integrating sensory and motor information. Consistent with this hypothesis are findings that providing or changing visual feedback through a mirror and somatosensory feedback through vibro-tactile devices can also reduce stuttering (Snyder, Blanchet, Waddell, & Ivy, 2009; Snyder, Hough, Blanchet, Ivy, & Waddell, 2009).

Pharmacological interventions have been investigated to treat developmental stuttering. Medications that modulate dopamine can be effective, although there is considerable population heterogeneity with regard to the direction of dopamine regulation. A recent study trialed a partial GABA-A agonist (pagoclone) as a potential therapeutic agent in stuttering, but the results were not compelling (Maguire et al., 2010). Pharmacological approaches might benefit from genetic studies that could yield insights into the molecular basis of developmental stuttering. Drugs could also be effective if paired with therapy. New methods to noninvasively stimulate the brain, such as transcranial direct current stimulation, have been identified as promising adjuncts for treating neurological conditions. Noninvasive brain stimulation has been applied to other communication disorders (predominantly aphasia due to stroke) along with speech therapy. It is an interesting possibility that this combination could also be used to enhance fluency in people who stutter.

79.4 GENETIC STUDIES OF DEVELOPMENTAL STUTTERING

There is strong evidence that developmental stuttering can be inherited. The risk of stuttering is considerably higher if a first-degree relative stutters (Kidd, Heimbuch, & Records, 1981). Twin studies show high heritability, with concordance rates in identical (monozygotic) twins being six-times greater than in fraternal (dizygotic) twins (Andrews, Morris-Yates, Howie, & Martin, 1991). Genetic models of the twin data estimate that 70% of the variance in liability to stuttering can be attributed to genetic effects, and 30% can be attributed to the nonshared environment (Andrews et al., 1991; Felsenfeld, 2002). Despite this evidence, the search for genes that may cause stuttering has been hampered by complex patterns of inheritance, early recovery, and the imbalance in the numbers of males and females affected.

Linkage analysis of large pedigrees with a high density of stuttering individuals has been used to identify several chromosomal regions where genes for stuttering might be located. Unfortunately, the outcomes of these studies rarely produced similar or overlapping loci. One study examined a very large pedigree from Cameroon containing 71 individuals, 33 of whom stuttered (Raza et al., 2013). Genome-wide linkage failed to identify a single locus; rather, analysis of smaller subfamilies indicated susceptibility for stuttering on previously identified loci on chromosomes 3q and 15q, as well as novel loci. This pattern of linkage to several loci and the high density of stuttering in this pedigree can be explained by nonrandom mate selection, which may have introduced variants of stuttering to the family.

Another approach for linkage analysis involves the study of inbred extended families and has yielded identification of the first causative genes for stuttering (Kang et al., 2010). Initially, significant linkage to chromosome 12q was identified in 44 families in Pakistan. Further analysis of this chromosomal region identified a missense mutation in the *GNPTAB* gene in 28 family members who stuttered. The same mutation was found in three members of other inbred families in Pakistan, and another nine unrelated individuals of

Pakistani or Indian origin. Subsequently, other mutations of the *GNPTAB* gene and related genes *GNPTG* or *NAGPA* were identified in approximately 10% of unrelated Pakistani and North American individuals who stutter. Mutations in these genes cause lysosomal storage disorders resulting in serious and severe disease. None of the affected individuals who stuttered had any disease normally related to mutations in this gene, and how their speech-specific impairment is explained by these findings is still unknown (Kang & Drayna, 2012).

Another potentially fruitful approach to identifying genes for stuttering is to look for *de novo* mutations in individuals with no family history of stuttering. The theory is that such people will have new, highly penetrant mutations that are causative. To our knowledge, this approach has not yet been implemented, but it is attractive because it does not require large sample sizes and thus may be more suitable in combination with further behavioral and imaging investigations.

79.5 THE NEURAL BASIS OF DEVELOPMENTAL STUTTERING

The advent of neuroimaging offered unprecedented research opportunities and insights into the neurobiology of developmental stuttering. There are no reports of *post mortem* examinations of the brains of people with persistent developmental stuttering that predate the imaging reports. Regardless of the lack of physical evidence for a neurological explanation of stuttering, hypotheses were widespread during the 19th and early 20th century. The link between left-handedness and stuttering and abnormal cerebral dominance was first proposed at this time. The relationship between diseases affecting the striatum, such as the *encephalitis lethargica* (which was epidemic in the 1920s), and speech disturbances, including stuttering, may have been the origin of theories implicating basal ganglia abnormalities as causes of developmental stuttering.

The earliest functional imaging studies of developmental stuttering used positron emission tomography (PET) to measure regional cerebral blood flow. More recent studies have used functional MRI to scan people who stutter. MRI has the advantage of being widely available, but it should be noted that whereas PET scans are silent, MRI is very noisy and therefore could affect feedback mechanisms and even enhance fluency. The functional imaging literature on developmental stuttering was summarized in 2005 by a meta-analysis of eight studies, six of which used PET and two used functional MRI (Brown, Ingham, Ingham, Laird, & Fox, 2005). The review identified three "neural signatures" of stuttering: (i) overactivation of the right frontal operculum or anterior insula or both; (ii) overactivation of the cerebellar vermis; and (iii) an "absence" of activity in auditory cortex (Figure 79.2). The first of these neural signatures of stuttering—overactivation of the right hemisphere homologue of Broca's area—is consistent with the proposal that stuttering is due to incomplete cerebral dominance for speech processing (Travis, 1978). The second and third neural signatures—abnormal cerebellar and auditory cortex activation—might reflect abnormalities in the brain's ability to integrate the sensory consequences of the motor acts that produce speech. This deficit could be due to poor internal models, noisy feedback, or both (Max, Guenther, Gracco, Ghosh, & Wallace, 2004). The cerebellum is thought to play a critical role in integrating sensory afference and motor efference copy to build forward models that predict the consequences of motor acts (Wolpert, Miall, & Kawato, 1998). Under normal circumstances, the prediction signal (corollary discharge) produced by the internal model should match the sensory consequences (sensory reafference) of speech and should be utilized for purposes such as attenuating the activity in sensory cortex to self-generated signals. A fourth candidate neural signature of developmental stuttering, which was absent from the imaging meta-analysis, is abnormal basal ganglia activity (Figure 79.2). Basal ganglia dysfunction has been suspected to be related to stuttering for several decades due, in large part, to studies using dopamine blockers, such as haloperidol, to treat stuttering. The early imaging studies of dopamine, oxygen, and glucose metabolism in developmental stuttering supported this hypothesis (Wu et al., 1997). We discuss each of these theories of developmental stuttering in more detail and evaluate the degree to which they are supported by findings from brain imaging studies.

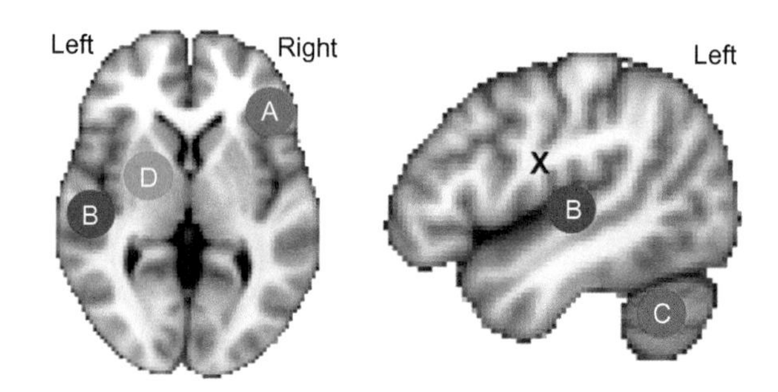

FIGURE 79.2 Brain abnormalities in people who stutter. (A) Overactive right inferior frontal cortex. (B) Underactive auditory cortex. (C) Overactive cerebellum. (D) Dysfunctional basal ganglia. X indicates reduced integrity of white matter tracts underlying left sensorimotor cortex.

79.5.1 Incomplete Cerebral Lateralization in Developmental Stuttering

The hypothesis that stuttering is caused by incomplete cerebral lateralization dates back to Samuel Orton, who made a similar proposal to explain reading disability. The proposal appears to be based on observations of stuttering emerging after enforced use of the right hand for writing in children who earlier showed a tendency for left-handedness. The idea was that the enforced use of the nondominant hand weakened the (presumably reversed) pattern of cerebral lateralization, producing conflict between the two hemispheres. Stuttering results, therefore, from incomplete lateralization of function between the two cerebral hemispheres. This theory fell out of favor with stuttering researchers until it received something of a revival when it found support from the results of early imaging studies (Pool, Devous, Freeman, Watson, & Finitzo, 1991; Wood, Stump, McKeehan, Sheldon, & Proctor, 1980). For example, one study of two young adults indicated right-lateralized blood flow in the homologue of Broca's area during stuttered speech, which reversed to become left-lateralized when fluency was induced using haloperidol medication (Wood et al., 1980). Subsequent imaging work used PET to measure regional cerebral blood flow during stuttered speech and while fluency was induced by chorus reading (Fox et al., 1996). Right-greater-than-left asymmetry of motor and auditory areas was observed during stuttered speech but was normalized when fluency was induced. In recent functional MRI studies, overactive right frontal opercular cortex extending to the anterior insula and the orbito-frontal surface is commonly described in adults who stutter (Kell et al., 2009; Watkins, Smith, Davis, & Howell, 2008). This rightward overactivity can be reduced if fluency is improved by therapy, however, resulting in more typical patterns of left-lateralized activity (De Nil, Kroll, Lafaille, & Houle, 2003; Kell et al., 2009; Preibisch et al., 2003).

Perhaps the best recent evidence supporting altered cerebral lateralization in stuttering comes from a study using near-infrared spectroscopy (Sato et al., 2011). Asymmetry in auditory processing was determined for speech that contained either a phonetic or a prosodic contrast. Adults and children who do not stutter show left-hemisphere lateralization for the phonemic contrast and right-hemisphere lateralization for the prosodic one. None of the adults and children who stutter showed the expected pattern of leftward lateralization, however.

It is unclear, however, whether the consistently observed right hemisphere overactivity and the reduced pattern of lateralization of function is a cause of stuttering, a consequence, or merely a correlate. Activation of the right hemisphere homologue of Broca's area in some patients with nonfluent aphasia is known to be maladaptive. In stuttering, the right hemisphere activity could also be compensatory, reflecting reorganization of function in response to a left-hemisphere structural abnormality (Preibisch et al., 2003).

A reliable finding of a brain structural abnormality associated with developmental stuttering is a reduction in the integrity of the white matter underlying the sensorimotor cortex close to the Sylvian fissure in the left hemisphere (Figure 79.2). Fractional anisotropy, a measure of white matter microstructure, was first described as reduced in this area by Sommer and colleagues. They proposed it reflected a disconnection of the white matter tracts connecting the frontal and temporal cortical areas in the left hemisphere that are typically involved in speech (Buchel & Sommer, 2004; Sommer, Koch, Paulus, Weiller, & Buchel, 2002). Subsequent diffusion imaging studies in children who stutter, including a group who had recovered, and in adolescents and adults who stutter found a disruption to white matter integrity in the same region (Chang, Erickson, Ambrose, Hasegawa-Johnson, & Ludlow, 2008; Connally, Ward, Howell, & Watkins, 2013; Cykowski, Fox, Ingham, Ingham, & Robin, 2010; Watkins et al., 2008).

Another robust finding in structural imaging studies of white matter tracts in developmental stuttering is an abnormality of the corpus callosum (Choo, Chang, Zengin-Bolatkale, Ambrose, & Loucks, 2012; Choo et al., 2011; Connally et al., 2013; Cykowski et al., 2010). The integrity of interhemispheric connections is thought to be critical for integrating information across the two hemispheres and for functional lateralization and changes in cortical asymmetry.

Structural imaging studies in people who stutter have also reported reduced leftward asymmetry in the planum temporale, which is more rightward in those with severe stuttering (Foundas et al., 2004). However, it is possible that the inclusion of left-handed individuals and females influenced the findings of altered planum temporale asymmetry in stuttering. Two studies using only right-handed males found the typical leftward asymmetry of the planum temporale in adults and children who stutter compared with fluent controls (Chang et al., 2008; Cykowski et al., 2008).

Further structural analyses of cortical areas involved in speech and language have not noted any additional alteration in underlying asymmetries, for example, in Broca's area. However, they have described unusual sulcal patterns in the brains of people who stutter (Foundas, Bollich, Corey, Hurley, & Heilman, 2001).

An extra diagonal sulcus (a shallow indentation on the surface of the pars opercularis) and extra gyri along the upper bank of the Sylvian fissure have been described in both hemispheres of adults who stutter (Cykowski et al., 2008; Foundas et al., 2001).

In summary, although initially the reports of overactivity in the right hemisphere in developmental stuttering revealed by brain imaging seemed consistent with Orton's proposal of incomplete cerebral dominance, further studies are needed to elaborate on the precise nature of this activity. Brain stimulation techniques could be used to address questions relating to compensatory or maladaptive reorganization to the right hemisphere in response to left hemisphere damage. Furthermore, longitudinal analyses starting early in development are needed to chart the course of changes in lateralized brain function and brain structure in people who stutter.

79.5.2 Abnormal Sensorimotor Integration in Developmental Stuttering

Another of the neural signatures revealed by the meta-analysis of functional imaging studies in developmental stuttering was reduced or absent activation of the auditory cortex (Brown et al., 2005). The authors of the meta-analysis attributed this attenuation to increased inhibitory efference copy input from overactive motor areas. Such a mechanism is used to explain how we distinguish between sensory inputs that result from our own actions and external ones. During speech production, the sensitivity of a speaker's auditory cortex is altered in expectation of the sensory consequences of speech, which are predicted from motor commands sent to the articulators. Magnetoencephalography (MEG) can be used to measure changes in the sensitivity of auditory cortex to speech-evoked signals. These signals are suppressed when the participant speaks relative to when they listen to recordings of the same speech sounds (Houde, Nagarajan, Sekihara, & Merzenich, 2002). This speech-induced suppression has been measured in children and adults who stutter to test the hypothesis that stuttering is related to impairment in the efference copy mechanism. The size of the effect was not different in people who stutter compared with fluent-speaking controls, although there were abnormalities in the timing of the effect in the stuttering groups (Beal et al., 2010, 2011).

In addition to reduced activity in the auditory cortex, overactivity of the cerebellar vermis was noted as a neural signature of stuttering according to the meta-analysis described. The cerebellum is thought to be involved in using the efference copy of motor commands and sensory information about the current state of the periphery to predict the sensory consequences of movement (Wolpert et al., 1998). Differences between the predicted outcomes and the intended ones are then used to adjust ongoing movements without the need for long latency feedback. Mismatches between the actual sensory consequences and the expected ones generated by the internal model are used to train and update the model. Such models have been developed to explain sensorimotor control in general and have focused primarily on visuo-motor control of the limbs. Models based on the same principles have been developed for speech motor control (Guenther & Vladusich, 2012; Hickok, Houde, & Rong, 2011). The overactivation of the cerebellum in stuttering is consistent with impairment in the internal model (Max et al., 2004). This could lead to excessive error detection when comparing the model's predictions with the actual sensory consequences. When therapy is successful in treating stuttering, the overactivity in the cerebellum, evident even at rest, normalizes to the levels seen in fluent speakers (Lu et al., 2012).

The cerebellum receives copies of motor commands conveyed directly from the primary motor cortex via the pons and the middle cerebellar peduncles and sensory inputs from the periphery via the inferior cerebellar peduncles. Signals from the cerebellum return to the cerebral cortex from the cerebellar nuclei via the superior cerebellar peduncles and the thalamus. In people who stutter, diffusion tensor imaging revealed abnormal white matter microstructure in each of the three pairs of cerebellar peduncles (Connally et al., 2013). Reduced integrity of these important tracts could affect the quality or timing of the cerebellar inputs and outputs in developmental stuttering.

It is worth considering, in the context of deficient internal models being a potential cause of stuttering, how the techniques known to temporarily enhance fluency might be effective. As noted in the introduction, these techniques change either the sensory feedback of speech or the way speech is produced, typically encouraging slower speech rate. Producing slower movements could be helpful if sensory inputs are noisy or delayed by allowing the relevant information to accumulate during speech production and online adjustments to be made. Slowing speech could also compensate for a weak or unstable model by allowing time for the actual sensory feedback to be used to inform the system of the consequences of speech rather than relying on the model's predictions. Altering auditory feedback or masking it with noise tends to activate the auditory cortex, which, as noted, is typically underactivated in people who stutter. Sensory input that is very obviously different from that expected could result in elimination of the feedback stage of internal modeling and minimize the error signals. This is presumably what happens when noise is used to mask feedback and, as a result, fluency is enhanced in people who stutter. It is necessary to use feedback to build the internal model and maintain it, however. Delayed auditory feedback was shown to increase the degree to which the time courses of activity in the auditory and motor cortex (at the level of the representation of the face) were synchronized in people who stutter. MRI signals from these areas in people who stutter were less correlated during speech production with normal auditory feedback than they were during silent rest. When speaking with delayed auditory feedback, the correlation between the time courses of activity in these areas increased to the same levels as seen in control speakers (Watkins, 2011).

The idea that stuttering is caused by an impairment in internal modeling of speech movements and its sensory consequences has obvious appeal. So far, however, tests of this theory have not found measureable differences in the response of the auditory cortex to self-generated speech. Structural abnormalities in the auditory cortex and in the pathways that link auditory cortex and the motor speech areas in the frontal cortex (i.e., the arcuate fasciculus) have been described in people who stutter. White matter tracts conveying sensory and motor signals to and from the cerebellum, a key structure in integrating this information, also show reduced integrity in developmental stuttering (Connally et al., 2013). Whether these white matter abnormalities play a causal role in the disorder or are an effect of it remains unknown, however.

79.5.3 Abnormal Basal Ganglia Function in Developmental Stuttering

The evidence that developmental stuttering is caused by impairment in basal ganglia function or dopamine activity, or both, has been reviewed by Alm (2004). He proposed that the core impairment in stuttering related to a deficit in the basal ganglia output to the supplementary motor cortex, which disrupted the production of timing cues used to initiate the next speech sound in a sequence. Consistent with the idea that this circuit is functionally abnormal, imaging studies report overactivation of the supplementary motor area and several nuclei in the basal ganglia circuitry in people who stutter (Fox et al., 1996; Giraud et al., 2008; Watkins et al., 2008). Activity in these areas normalizes when fluency is induced temporarily in people who stutter or is improved by successful therapy (Fox et al., 1996; Giraud et al., 2008; Kell et al., 2009). Furthermore, a small group of males who stutter showed a three-fold increase in dopaminergic activity in the medial frontal cortex and caudate nucleus relative to controls (Wu et al., 1997).

Reports that drugs that block dopamine improve fluency in people who stutter are consistent with the notion that there is excessive dopaminergic activity in the disorder. Specifically, drugs that block D2 receptors can alleviate the symptoms of stuttering (Maguire, Yu, Franklin, & Riley, 2004). These receptors are found primarily on striatal neurons in the indirect pathway, which is involved in the inhibition of competing motor programs. Unfortunately, D2 antagonists also produce problematic side effects, meaning they are not well-tolerated as a treatment for stuttering. In some people who stutter, drugs that stimulate dopamine, either directly or indirectly, can also improve fluency, suggesting the possibility of different subtypes of the disorder (Alm, 2004). Similarly, reports of patients with Parkinson's disease who develop stuttering with disease onset (i.e., during dopamine depletion) or who worsen on dopamine replacement therapy indicate that the relationship between speech fluency and dopaminergic activity is not straightforward. As with other cognitive functions, dopamine levels appear to show an inverted U-shape relationship with speech fluency.

Interestingly, patients with Parkinson's disease also benefit from external cues to initiate and perform sequences of movements fluently. For example, gait can markedly improve with either auditory rhythmic (e.g., music with a strong beat) or visual structured (e.g., stripes on the floor) cues (Glickstein & Stein, 1991). This effect bears striking similarity to the known fluency enhancing effects of external cues such as a metronome in developmental stuttering. Alm (2004) proposes that external cues for speech initiation are effective in enhancing fluency in people who stutter because they engage a circuit comprising the cerebellum and lateral premotor cortex. The premotor-basal ganglia circuit is important for producing speech patterns that are overlearned motor sequences. Dysfunction in the basal ganglia loops could be compensated for by engaging the lateral premotor-cerebellar circuit to produce speech motor sequences in a novel way, such as with an accent or melody, or in time with an external cue.

Stuttering can be aggravated by deep brain stimulation of the subthalamic nucleus in patients with Parkinson's disease (Toft & Dietrichs, 2011) or by stimulating the globus pallidus or thalamus to relieve dystonia (Allert, Kelm, Blahak, Capelle, & Krauss, 2010; Nebel, Reese, Deuschl, Mehdorn, & Volkmann, 2009). The symptoms are reversible, however, and are relieved by turning off the stimulator. As with the pharmacological treatment studies, the opposite effects have also been obtained; stuttering can also be relieved by deep brain stimulation of the subthalamic nucleus in Parkinson's disease (Walker et al., 2009) and stimulation of the ventral intermediate nucleus of the thalamus (the main cerebellar thalamic relay) in the case of essential tremor (Maguire et al., 2012).

Neurogenic stuttering can be acquired due to brain damage or disease and results in symptoms that are sometimes indistinguishable from the symptoms of developmental stuttering. Cases of neurogenic stuttering have been reported to be primarily due to left hemisphere cortical and subcortical areas. In a recent lesion-symptom mapping study, patients who acquired stuttering-like dysfluencies due to stroke were found to have damage to nine left-hemisphere areas. These areas overlapped a cortico-striatal-thalamo-cortical loop involving frontal and temporal lobe cortex, the basal ganglia, and the white matter tracts connecting them (Theys, De Nil, Thijs, van Wieringen, & Sunaert, 2013).

There is evidence from brain imaging, pharmacological, and lesion studies implicating abnormal function of the basal ganglia circuitry in stuttering. Previously described structural abnormalities such as the white matter abnormalities underlying sensorimotor cortex could affect cortico-striatal projections in developmental stuttering, and damage to these tracts can cause neurogenic stuttering. A recent neuro-computational model of stuttering proposed that either impaired cortico-striatal projections or dopaminergic hyperactivity could result in dysfluent syllable sequencing (Civier, Bullock, Max, & Guenther, 2013). Both of these abnormalities would affect function in cortico-striatal-thalamo-cortical loops and cause stuttering either independently or in concert. The evidence appears to suggest that stuttering can occur due to impairment in many different locations that each contribute to the function of basal ganglia circuitry. This might explain why imaging studies that rely on averaging data across subjects do not reliably produce evidence of functional abnormality in these nuclei. The possible relationship between different basal ganglia circuit abnormalities and stuttering subtypes remains an important challenge for future studies.

79.6 CONCLUSION

Neuroscientific investigation of developmental stuttering is providing useful information regarding possible causes. The first genes have been identified. Imaging studies reveal consistent patterns of structural and functional correlates of the communication disorder. These findings lend support to several different models of the neurobiological causes of developmental stuttering. Future studies will explore whether these different models can be reconciled or, in fact, reflect heterogeneity in the disorder due to different subtypes. Because stuttering is a developmental disorder, there is also a considerable opportunity for compensatory and maladaptive reorganization that could further add to individual differences in the underlying brain abnormalities. Longitudinal investigations during development and therapy are needed to address outstanding questions relating to cause and effect.

References

Allert, N., Kelm, D., Blahak, C., Capelle, H. H., & Krauss, J. K. (2010). Stuttering induced by thalamic deep brain stimulation for dystonia. *Journal of Neural Transmission, 117*(5), 617–620. Available from: http://dx.doi.org/10.1007/s00702-010-0380-0.

Alm, P. A. (2004). Stuttering and the basal ganglia circuits: A critical review of possible relations. *Journal of Communication Disorders, 37*(4), 325–369.

Andrews, G., Morris-Yates, A., Howie, P., & Martin, N. G. (1991). Genetic factors in stuttering confirmed. *Archives of General Psychiatry, 48*(11), 1034–1035.

Beal, D. S., Cheyne, D. O., Gracco, V. L., Quraan, M. A., Taylor, M. J., & De Nil, L. F. (2010). Auditory evoked fields to vocalization during passive listening and active generation in adults who stutter. *NeuroImage, 52*(4), 1645–1653. Available from: http://dx.doi.org/10.1016/j.neuroimage.2010.04.277.

Beal, D. S., Quraan, M. A., Cheyne, D. O., Taylor, M. J., Gracco, V. L., & De Nil, L. F. (2011). Speech-induced suppression of evoked auditory fields in children who stutter. *NeuroImage, 54*(4), 2994–3003. Available from: http://dx.doi.org/10.1016/j.neuroimage.2010.11.026.

Bloodstein, O., & Ratner, N. B. (2008). *A handbook on stuttering*. Clifton Park, NY: Thomson Delmar Publishing.

Brown, S., Ingham, R. J., Ingham, J. C., Laird, A. R., & Fox, P. T. (2005). Stuttered and fluent speech production: An ALE meta-analysis of functional neuroimaging studies. *Human Brain Mapping, 25*, 105–117.

Buchel, C., & Sommer, M. (2004). What causes stuttering? *PLoS Biology, 2*(2), 0159–0163.

Chang, S. E., Erickson, K. I., Ambrose, N. G., Hasegawa-Johnson, M. A., & Ludlow, C. L. (2008). Brain anatomy differences in childhood stuttering. *NeuroImage, 39*(3), 1333–1344.

Choo, A. L., Chang, S. E., Zengin-Bolatkale, H., Ambrose, N. G., & Loucks, T. M. (2012). Corpus callosum morphology in children who stutter. *Journal of Communication Disorders, 45*(4), 279–289. Available from: http://dx.doi.org/10.1016/j.jcomdis.2012.03.004.

Choo, A. L., Kraft, S. J., Olivero, W., Ambrose, N. G., Sharma, H., Chang, S. E., et al. (2011). Corpus callosum differences associated with persistent stuttering in adults. *Journal of Communication Disorders, 44*(4), 470–477. Available from: http://dx.doi.org/10.1016/j.jcomdis.2011.03.001.

Civier, O., Bullock, D., Max, L., & Guenther, F. H. (2013). Computational modeling of stuttering caused by impairments in a basal ganglia thalamo-cortical circuit involved in syllable selection and initiation. *Brain and Language, 126*(3), 263–278. Available from: http://dx.doi.org/10.1016/j.bandl.2013.05.016.

Connally, E. L., Ward, D., Howell, P., & Watkins, K. E. (2013). Disrupted white matter in language and motor tracts in developmental stuttering. *Brain and Language, 131*, 25–35. Available from: http://dx.doi.org/10.1016/j.bandl.2013.05.013.

Cykowski, M. D., Fox, P. T., Ingham, R. J., Ingham, J. C., & Robin, D. A. (2010). A study of the reproducibility and etiology of diffusion anisotropy differences in developmental stuttering: A potential role for impaired myelination. *NeuroImage, 52*(4), 1495–1504. Available from: http://dx.doi.org/10.1016/j.neuroimage.2010.05.011.

Cykowski, M. D., Kochunov, P. V., Ingham, R. J., Ingham, J. C., Mangin, J. F., Riviere, D., et al. (2008). Perisylvian sulcal morphology and cerebral asymmetry patterns in adults who stutter. *Cerebral Cortex*, *18*(3), 571–583 bhm093 [pii]. Available from: http://dx.doi.org/10.1093/cercor/bhm093.

De Nil, L. F., Kroll, R. M., Lafaille, S. J., & Houle, S. (2003). A positron emission tomography study of short- and long-term treatment effects on functional brain activation in adults who stutter. *Journal of Fluency Disorders*, *28*(4), 357–379 (quiz 379–380).

Felsenfeld, S. (2002). Finding susceptibility genes for developmental disorders of speech: The long and winding road. *Journal of Communication Disorders*, *35*, 329–345.

Foundas, A. L., Bollich, A. M., Corey, D. M., Hurley, M., & Heilman, K. M. (2001). Anomalous anatomy of speech-language areas in adults with persistent developmental stuttering. *Neurology*, *57*(2), 207–215.

Foundas, A. L., Bollich, A. M., Feldman, J., Corey, D. M., Hurley, M., Lemen, L. C., et al. (2004). Aberrant auditory processing and atypical planum temporale in developmental stuttering. *Neurology*, *63*(9), 1640–1646.

Fox, P. T., Ingham, R. J., Ingham, J. C., Hirsch, T. B., Downs, J. H., Martin, C., et al. (1996). A PET study of the neural systems of stuttering. *Nature*, *382*(6587), 158–161.

Giraud, A. L., Neumann, K., Bachoud-Levi, A. C., von Gudenberg, A. W., Euler, H. A., Lanfermann, H., et al. (2008). Severity of dysfluency correlates with basal ganglia activity in persistent developmental stuttering. *Brain and Language*, *104*(2), 190–199.

Glickstein, M., & Stein, J. (1991). Paradoxical movement in Parkinson's disease. *Trends in Neurosciences*, *14*(11), 480–482.

Guenther, F. H., & Vladusich, T. (2012). A neural theory of speech acquisition and production. *Journal of Neurolinguistics*, *25*(5), 408–422. Available from: http://dx.doi.org/10.1016/j.jneuroling.2009.08.006.

Hickok, G., Houde, J., & Rong, F. (2011). Sensorimotor integration in speech processing: Computational basis and neural organization. *Neuron*, *69*(3), 407–422. Available from: http://dx.doi.org/10.1016/j.neuron.2011.01.019.

Houde, J. F., Nagarajan, S. S., Sekihara, K., & Merzenich, M. M. (2002). Modulation of auditory cortex during speech: An MEG study. *Journal of Cognitive Neuroscience*, *14*(8), 1125–1138.

Iverach, L., Menzies, R. G., O'Brian, S., Packman, A., & Onslow, M. (2011). Anxiety and stuttering: Continuing to explore a complex relationship. *American Journal of Speech-Language Pathology*, *20*(3), 221–232. Available from: http://dx.doi.org/10.1044/1058-0360(2011/10-0091).

Kang, C., & Drayna, D. (2012). A role for inherited metabolic deficits in persistent developmental stuttering. *Molecular Genetics and Metabolism*, *107*(3), 276–280. Available from: http://dx.doi.org/10.1016/j.ymgme.2012.07.020.

Kang, C., Riazuddin, S., Mundorff, J., Krasnewich, D., Friedman, P., Mullikin, J. C., et al. (2010). Mutations in the lysosomal enzyme-targeting pathway and persistent stuttering. *The New England Journal of Medicine*, *362*(8), 677–685. Available from: http://dx.doi.org/10.1056/NEJMoa0902630.

Kell, C. A., Neumann, K., von Kriegstein, K., Posenenske, C., von Gudenberg, A. W., Euler, H., et al. (2009). How the brain repairs stuttering. *Brain*, *132*(Pt 10), 2747–2760 awp185 [pii]. Available from: http://dx.doi.org/10.1093/brain/awp185.

Kidd, K. K., Heimbuch, R. C., & Records, M. A. (1981). Vertical transmission of susceptibility to stuttering with sex-modified expression. *Proceedings of the National Academy of Sciences of the United States of America*, *78*(1), 606–610.

Lee, B. S. (1951). Artificial stutter. *Journal of Speech and Hearing Disorders*, *15*, 53–55.

Lincoln, M., Packman, A., & Onslow, M. (2006). Altered auditory feedback and the treatment of stuttering: A review. *Journal of Fluency Disorders*, *31*(2), 71–89 S0094-730X(06)00030-1 [pii]. Available from: http://dx.doi.org/10.1016/j.jfludis.2006.04.001.

Lu, C., Chen, C., Peng, D., You, W., Zhang, X., Ding, G., et al. (2012). Neural anomaly and reorganization in speakers who stutter: A short-term intervention study. *Neurology*, *79*(7), 625–632. Available from: http://dx.doi.org/10.1212/WNL.0b013e31826356d2.

Maguire, G. A., Franklin, D., Vatakis, N. G., Morgenshtern, E., Denko, T., Yaruss, J. S., et al. (2010). Exploratory randomized clinical study of pagoclone in persistent developmental stuttering: The EXamining pagoclone for peRsistent dEvelopmental stuttering study. *Journal of Clinical Psychopharmacology*, *30*(1), 48–56. Available from: http://dx.doi.org/10.1097/JCP.0b013e3181caebbe.

Maguire, G. A., Ngo, J., Fonsworth Iii, P. K., Doan, J., Birch, J. A., & Fineman, I. (2012). Alleviation of developmental stuttering following deep brain stimulation of the ventral intermediate nucleus of the thalamus. *The American Journal of Psychiatry*, *169*(7), 759–760. Available from: http://dx.doi.org/10.1176/appi.ajp.2012.12010016.

Maguire, G. A., Yu, B. P., Franklin, D. L., & Riley, G. D. (2004). Alleviating stuttering with pharmacological interventions. *Expert Opinion on Pharmacotherapy*, *5*(7), 1565–1571. Available from: http://dx.doi.org/10.1517/14656566.5.7.1565.

Max, L., Guenther, F., Gracco, V. L., Ghosh, S. S., & Wallace, M. E. (2004). Unstable or insufficiently activated internal models and feedback-biased motor control as sources of dysfluency: A theoretical model of stuttering. *Contemporary Issues in Communicative Sciences and Disorders*, *31*, 105–122.

Nebel, A., Reese, R., Deuschl, G., Mehdorn, H. M., & Volkmann, J. (2009). Acquired stuttering after pallidal deep brain stimulation for dystonia. *Journal of Neural Transmission*, *116*(2), 167–169. Available from: http://dx.doi.org/10.1007/s00702-008-0173-x.

Pool, K. D., Devous, M. D., Sr., Freeman, F. J., Watson, B. C., & Finitzo, T. (1991). Regional cerebral blood flow in developmental stutterers. *Archives of Neurology*, *48*(5), 509–512.

Preibisch, C., Neumann, K., Raab, P., Euler, H. A., von Gudenberg, A. W., Lanfermann, H., et al. (2003). Evidence for compensation for stuttering by the right frontal operculum. *NeuroImage*, *20*(2), 1356–1364.

Raza, M. H., Gertz, E. M., Mundorff, J., Lukong, J., Kuster, J., Schaffer, A. A., et al. (2013). Linkage analysis of a large African family segregating stuttering suggests polygenic inheritance. *Human Genetics*, *132*(4), 385–396. Available from: http://dx.doi.org/10.1007/s00439-012-1252-5.

Sato, Y., Mori, K., Koizumi, T., Minagawa-Kawai, Y., Tanaka, A., Ozawa, E., et al. (2011). Functional lateralization of speech processing in adults and children who stutter. *Frontiers in Psychology*, *2*, 70. Available from: http://dx.doi.org/10.3389/fpsyg.2011.00070.

Snyder, G. J., Blanchet, P., Waddell, D., & Ivy, L. J. (2009). Effects of digital vibrotactile speech feedback on overt stuttering frequency. *Perceptual and Motor Skills*, *108*(1), 271–280.

Snyder, G. J., Hough, M. S., Blanchet, P., Ivy, L. J., & Waddell, D. (2009). The effects of self-generated synchronous and asynchronous visual speech feedback on overt stuttering frequency. *Journal of Communication Disorders*, *42*(3), 235–244. Available from: http://dx.doi.org/10.1016/j.jcomdis.2009.02.002.

Sommer, M., Koch, M. A., Paulus, W., Weiller, C., & Buchel, C. (2002). Disconnection of speech-relevant brain areas in persistent developmental stuttering. *Lancet*, *360*(9330), 380–383.

Theys, C., De Nil, L., Thijs, V., van Wieringen, A., & Sunaert, S. (2013). A crucial role for the cortico-striato-cortical loop in the pathogenesis of stroke-related neurogenic stuttering. *Human Brain Mapping*, *34*(9), 2103–2112. Available from: http://dx.doi.org/10.1002/hbm.22052.

Toft, M., & Dietrichs, E. (2011). Aggravated stuttering following subthalamic deep brain stimulation in Parkinson's disease—Two cases. *BMC Neurology, 11*, 44. Available from: http://dx.doi.org/10.1186/1471-2377-11-44.

Travis, L. E. (1978). The cerebral dominance theory of stuttering: 1931–1978. *The Journal of Speech and Hearing Disorders, 43*(3), 278–281.

Walker, H. C., Phillips, D. E., Boswell, D. B., Guthrie, B. L., Guthrie, S. L., Nicholas, A. P., et al. (2009). Relief of acquired stuttering associated with Parkinson's disease by unilateral left subthalamic brain stimulation. *Journal of Speech, Language, and Hearing Research, 52*(6), 1652–1657. Available from: http://dx.doi.org/10.1044/1092-4388(2009/08-0089).

Watkins, K. E. (2011). Developmental disorders of speech and language: From genes to brain structure and function. *Progress in Brain Research, 189*, 225–238. Available from: http://dx.doi.org/10.1016/B978-0-444-53884-0.00027-0.

Watkins, K. E., Smith, S. M., Davis, S., & Howell, P. (2008). Structural and functional abnormalities of the motor system in developmental stuttering. *Brain, 131*(Pt 1), 50–59 awm241 [pii]. Available from: http://dx.doi.org/10.1093/brain/awm241.

Wolpert, D. M., Miall, R. C., & Kawato, M. (1998). Internal models in the cerebellum. *Trends in Cognitive Sciences, 2*(9), 338–347.

Wood, F., Stump, D., McKeehan, A., Sheldon, S., & Proctor, J. (1980). Patterns of regional cerebral blood flow during attempted reading aloud by stutterers both on and off haloperidol medication: Evidence for inadequate left frontal activation during stuttering. *Brain and Language, 9*(1), 141–144.

Wu, J. C., Maguire, G., Riley, G., Lee, A., Keator, D., Tang, C., et al. (1997). Increased dopamine activity associated with stuttering. *Neuroreport, 8*(3), 767–770.

Yairi, E., & Ambrose, N. (2013). Epidemiology of stuttering: 21st century advances. *Journal of Fluency Disorders, 38*(2), 66–87. Available from: http://dx.doi.org/10.1016/j.jfludis.2012.11.002.

Zackheim, C. T., & Conture, E. G. (2003). Childhood stuttering and speech disfluencies in relation to children's mean length of utterance: A preliminary study. *Journal of Fluency Disorders, 28*(2), 115–141 quiz 141–142.

SECTION P

LANGUAGE TREATMENT

CHAPTER

80

Neuroplasticity Associated with Treated Aphasia Recovery

Julius Fridriksson and Kimberly Smith

The Aphasia Lab, Department of Communication Sciences and Disorders, University of South Carolina, Columbia, SC, USA

Aphasia therapy typically refers to clinician administered behavioral approaches that use language-based drills or the learning of compensatory strategies to ameliorate the effects of impairment. Aphasia treatment is most often administered through didactic interactions between a patient and a clinician—a speech-language pathologist—who provides targeted language stimulation or task instructions focusing on a specific process or communication skill. Other increasingly common forms of treatment administration include group treatment or computerized language therapy. Although its effectiveness has been debated, recent meta-analyses suggest that aphasia treatment is generally beneficial, although much work is needed to better understand which treatment approach might work best for a specific patient (Brady, Kelly, Godwin, & Enderby, 2012; Kelly, Brady, & Enderby, 2010; Robey, 1998). That is, aphasia therapy promotes improved language function even though only a few therapy techniques have been established to tackle the different aspects of aphasic language impairment.

Regardless of treatment mode or maximum efficacy, it is an implicit assumption that treatment-assisted recovery from aphasia is supported by plastic brain changes, either functional or structural. Whether these changes are transient or permanent may determine the long-term effectiveness of aphasia treatment. The brain mechanisms that support aphasia recovery are not clear, and research in this area has yielded seemingly conflicting results (Saur & Hartwigsen, 2012; Saur et al., 2006; Szaflarski, Allendorfer, Banks, Vannest, & Holland, 2013). In this chapter, we address treated aphasia recovery in the context of contemporary understanding of neuroplasticity.

80.1 NEUROPLASTICITY

A paramount characteristic of the human brain, among other species, is its ability to continuously adapt its structure and function based on internal and external environmental changes. Typically, this ability is known as neuroplasticity. A generally accepted assumption of neuroplasticity states that to change behavior, the brain must also change. Considerable evidence suggests that structural changes to existing neural circuits or the generation of new circuits at the neuronal level underlie the behavioral changes associated with neuroplasticity (Cramer, 2008; Kolb & Whishaw, 1998). Dendritic morphology (i.e., dendritic form and structure) has been stressed as a critical contributor to these neuronal changes, whereas additional neuronal changes are likely contributory, such as number of synapses, synapse size, and metabolic activity. Neuroplasticity is made possible by coordinating changes of neuronal morphology, glia, and vascular and metabolic processes. These modifications are stimulated by several factors, including sensory-motor experiences, task learning, gonadal and stress hormones, psychoactive drugs, neurotrophic factors, cortical stimulation, aging, and diet (Cramer, 2008; Johansson, 2000; Kolb & Whishaw, 1998; Kleim & Jones, 2008; Rijntjes & Weiller, 2002). In broad terms, synaptic and dendritic changes allow the brain to be structurally and functionally plastic as a response to experiences during development, learning, recovery from injury, and aging (Hebb, 1949).

Typically, neuroplasticity has been considered in the context of behavioral stimulation and response to the external environment. Although it is straightforward to see why this is the case—much of what we know about

Neurobiology of Language. DOI: http://dx.doi.org/10.1016/B978-0-12-407794-2.00080-8

neural plasticity is derived from highly controlled animal studies with optimized experimental manipulations and carefully controlled environments—it is important to consider aphasia recovery in the context of potentially adaptive and maladaptive sequelae of recovery. Most cases of aphasia are caused by ischemic stroke in which the vascular supply to the cortex has been interrupted. Aphasia is often viewed as the consequence of frank brain damage to the cortical language network. However, John Hughlings Jackson, the eminent English physician who studied many forms of neurological disorders in the late 19th century, suggested that aphasic language impairment should be considered not only the direct behavioral result of the brain damage but also a reflection of the function of remaining brain tissue in the absence of cortical areas damaged by the stroke (Jackson, 1874).

The cortical reorganization and plastic changes after stroke probably occur to optimize function; however, the location and extent of injury dictate what kinds of recovery processes can be engaged and may influence whether cortical maladaptation occurs. In the case of maladaptation, it could be that plastic changes in response to cortical damage actually hinder recovery or promote suboptimal reorganization in the early phases after stroke. Moreover, it is possible that early training (e.g., aphasia therapy) may actually negatively affect long-term recovery. Adverse consequences of acute rehabilitation have been demonstrated in a mouse model after motor stroke (Allred & Jones, 2008) in which early training of the spared forelimb had a negative effect on long-term improvement of the affected limb. Whether the same principle could be attributed to early aphasia treatment is questionable, yet this study demonstrates that we know very little about the effects—positive or negative—of plastic brain changes on recovery, and we also do not know the optimal timing for initiation of aphasia treatment. Several studies have suggested that neuroplasticity is greatest soon after brain damage, with lesser changes seen over time (Hartman, 1981; Kim, Ko, Parrish, & Kim, 2002; Robey, 1998). This could mean that targeted treatment seeking to capitalize on these changes should be dispensed as early as possible after aphasia onset. Clearly, there is an urgent need for a better understanding of the capacity for adaptive neuroplasticity as a consequence of initial brain damage and subsequent aphasia treatment, as well as the risk of maladaptive neuroplasticity that could stifle spontaneous and treated recovery.

80.2 ACUTE AND CHRONIC CONSIDERATIONS

In some cases of acute stroke, the recovery from aphasia can be very substantial in the first few hours and days after onset. As demonstrated by Hillis et al. (2001), Hillis and Heidler (2002), and Hillis et al. (2004), some of this recovery is supported by vascular changes leading to reperfusion of the ischemic penumbra and critical language areas. For patients who make it to the hospital within 4.5 hours of ischemic stroke onset and who are candidates for thrombolysis with tissue plasminogen activator (tPA), the effects of stroke may be partially ameliorated by dissolving the blood clot that caused the stroke. For some of these patients, the return of language function can be quite dramatic and, in some cases, complete (Felberg et al., 2002; Maas et al., 2012; Saur et al., 2006). However, thrombolysis is administered in approximately 10% of cases (Bray et al., 2013; Minnerup et al., 2011; Monks, Pitt, Stein, & James, 2012), which means that many stroke patients do not benefit from early intervention and may ultimately face a life-altering injury to which they must adapt.

Immediate efforts to dissolve or remove the blood clot in acute ischemic stroke are crucial for the preservation of as much neural tissue as possible and to prevent subsequent negative effects on behavior. However, neuroplasticity that occurs after cell death is completed probably takes place over weeks, months, and even years (Cramer, 2008; Johansson, 2000; Kleim & Jones, 2008). Although some of these plastic changes may occur in response to inherent homeostatic factors, it is likely that neuroplasticity supporting aphasia recovery primarily occurs in subacute (e.g., during the first 90 days after onset) and chronic phases of stroke recovery. That is, very early recovery from aphasia may be driven in part by neurovascular changes, whereas later recovery appears to rely more on actual neuroplasticity and, possibly, angiogenesis (the formation of new blood vessels from preexisting ones) (Carmichael, 2008; Cramer, 2008; Wei, Erinjeri, Rovainen, & Woolsey, 2001) as well as neurogenesis (the formation of new neurons) (Carmichael, 2008; Johansson, 2000). At this time, far more work is needed to determine whether and to what extent angiogenesis and neurogenesis play roles in aphasia recovery.

In chronic aphasia, recovery is mediated by relearning lost information, retraining specific processes that were impaired as a result of brain damage, learning compensatory strategies that aid in communication, and psychosocial adaptations such as those afforded by environmental enrichments and increased access to appropriate social environments (Turkstra, Holland, & Bays, 2003). Regardless of the kinds of behavior modifications that take place, neuroplasticity is at the crux of mediating and supporting such changes. Specific patterns of neuroplasticity that support aphasia recovery are being explored and considerable research has been devoted to this issue (Crosson et al., 2005, 2009;

Fridriksson, 2010; Fridriksson et al., 2012; Fridriksson, Morrow-Odom, Moser, Fridriksson, & Baylis, 2006; Heiss, Kessler, Thiel, Ghaemi, & Karbe, 1999; Heiss & Thiel, 2006; Heiss, Thiel, Kessler, & Herholz, 2003; Hillis et al., 2001, 2004; Hillis & Heidler, 2002; Kim et al., 2002; Marcotte et al., 2012; Meinzer & Breitenstein, 2008; Meinzer et al., 2004; Meinzer, Streiftau, & Rockstroh, 2007; Menke et al., 2009; Musso et al., 1999; Postman-Caucheteux et al., 2010; Saur & Hartwigsen, 2012; Saur et al., 2006; Schlaug, Marchina, & Norton, 2009; Szaflarski et al., 2013; Thompson, 2000a, 2000b; Thompson, den Ouden, Bonakdarpour, Garibaldi, & Parrish, 2010).

80.3 STRUCTURAL BRAIN CHANGES AND APHASIA RECOVERY

A few studies have examined structural brain changes associated with aphasia recovery. Probably the first, and perhaps best, known study of this kind was conducted by Schlaug and colleagues (2009), who treated six aphasic patients using Melodic Intonation Therapy (MIT) (Albert, Sparks, & Helm, 1973; Helm-Estabrooks, Nicholas, & Morgan, 1989), a treatment approach that targets nonfluent speech production and emphasizes the patient imitating intoned speech modeled by a clinician. The main idea behind MIT is that it targets activation of the right hemisphere, where melody is thought to be processed (Albert et al., 1973; Conklyn, Novak, Boissy, Bethoux, & Chemali, 2012; Helm-Estabrooks et al., 1989; Norton, Zipse, Marchina, & Schlaug, 2009; van der Meulen, van de Sandt-Koenderman, & Ribbers, 2012; Schlaug et al., 2009). Diffusion tensor imaging (DTI) was used to assess changes in white matter density in the right hemisphere before and after a therapy program consisting of 75 MIT sessions. In summary, this study revealed an increase in volume and number of fibers in the right arcuate fasciculus, suggesting that structural connectivity was increased as a result of aphasia treatment. A case study of a 12-year-old patient with a large left hemisphere lesion and severely nonfluent speech yielded similar results (Zipse, Norton, Marchina, & Schlaug, 2012). That patient underwent 120 hours of treatment using MIT, and DTI was used to assess white matter changes in the right hemisphere. Increased volume was found at the mid-point of the treatment phase and at 1 year after treatment in the right arcuate fasciculus and the uncinate fasciculus, a white matter tract that, in the left hemisphere, is commonly associated with semantic processing (Catani et al., 2013; Harvey, Wei, Ellmore, Cris Hamilton, & Schnur, 2013). Intriguingly, these studies suggest that the type and location of neuroplastic changes associated with treated aphasia recovery might be dependent on the type of treatment. That is, treatments that involve processes that are supported by the right hemisphere (e.g., intonation) are more likely to yield plastic changes in the right hemisphere, whereas approaches that focus more on language processes (which primarily tax the left hemisphere in neurologically intact subjects) may be more likely to recruit preserved areas of the injured left hemisphere.

In another study that specifically examined structural brain changes with DTI, Allendorfer et al. (2012) used intermittent theta burst transcranial magnetic stimulation (iTBS) to target preserved anterior left hemisphere regions in eight patients with different types of chronic aphasia. Although the specific mechanism is unknown, iTBS has been shown to enhance motor-evoked potentials, suggesting that it has excitatory effects on neural tissue. All patients underwent 10 iTBS sessions (without behavioral language treatment), after which increased fractional anisotrophy (FA) was found in the targeted regions, including left inferior and superior frontal gyri, as well as in the right midbrain and several bilateral regions such as the temporal and parietal cortices. With regard to language changes, improvements were found on a semantic fluency test but not on the Boston Naming Test (Kaplan, Goodglass, & Weintraub, 2001) or the Peabody Picture Vocabulary Test (Dunn & Dunn, 2007). Importantly, no relationship was revealed between the extent of FA changes and language improvement. Nevertheless, studies of this kind suggest that treatment of aphasic patients, using either behavioral language treatment or transcranial cortical stimulation, can change the structure of the brain. However, it is far too early to postulate what might be the specific patterns of structural brain changes that support aphasia recovery. More research including a larger number of patients, detailed descriptions of language ability, and consistent methods across studies is needed to better understand structural neuroplasticity in aphasia recovery.

80.4 FUNCTIONAL BRAIN CHANGES AND APHASIA RECOVERY

Compared with studies of structural neuroplasticity, far more research has focused on functional brain changes associated with spontaneous and treated recovery from aphasia. One of the challenges of understanding how functional brain changes support aphasia recovery is that many different approaches, methods, and behavioral tasks have been used to relate cortical modulation to behavioral changes. This can be expected because of the many types and varying severity levels of aphasia. Future investigations cannot ignore these differences because each may contribute uniquely to plasticity.

It is also a concern that most studies of functional brain changes associated with aphasia treatment have relied on single case studies or very small sample sizes. This is because of the extensive variability that characterizes both the language behaviors and the associated patterns of brain tissue either spared or destroyed by stroke and other causes of aphasia. A case study by Epstein-Peterson, Vasconcellos Faria, Mori, Hillis, and Tsapkini (2012) demonstrates this point well. Their patient had a large left hemisphere stroke affecting most of the structures supplied by the middle cerebral artery, including areas that typically are thought to be crucial for normal language processing (Broca's area and most of the middle and superior temporal lobe). Despite this severe injury to the left hemisphere, the patient had a relatively mild language impairment characterized mostly by somewhat restricted speech fluency, relatively spared auditory comprehension of canonical sentences, and only very limited anomia. Although the patient was diagnosed with global aphasia (or severe impairment in all language modalities) immediately after stroke, testing at the time of study inclusion (3 years after stroke) indicated fairly mild aphasia. This case is unusual in that most patients with similar damage continue to present with more severe aphasia during the chronic phase, most typically consistent with Broca's or even global aphasia (Fridriksson, Fillmore, Guo, & Rorden, 2014; Fridriksson et al., 2012). However, such rarities clearly do exist and remain unexplained. Such mild language impairment after extensive left hemisphere damage likely demonstrates premorbid differences in language organization and unique behavioral characteristics. The presence of such patients also demonstrates extensive variability in the process(es) by which recovery from severe brain damage can be achieved. Finally, and not insignificantly, it must be remembered that environmental manipulations, in many guises ranging from the quality, extent, and intensity of treatment to the amount and type of external personal support, also play a role. At the very least, this case and others like it (Berthier, 2001; Heilman, Rothi, McFarling, & Rottmann, 1981; Pulvermüller & Schönle, 1993) demonstrate the current difficulty and uncertainty of ascribing language recovery to a single process or to specific changes in functional or structural neuroplasticity.

As stated, many different methods have been used to investigate changes in functional activation associated with aphasia treatment. These include functional magnetic resonance imaging (fMRI), magnetoencephalography (MEG), positron emission tomography (PET), and transcranial cortical stimulation. Two patterns of functional brain activation have been described to support language recovery in aphasia: (i) functional reactivation, which suggests greater reliance on preserved language areas, including cortex immediately adjacent to the lesion (Cappa, 2000; Heiss et al., 2003); and (ii) functional reorganization, which refers to activation of nontraditional language areas, either residual left hemisphere structures or right hemisphere homologues (Marcotte et al., 2012). Several studies have provided evidence that supports a more favorable outcome when left hemisphere perilesional areas are recruited (Fridriksson, 2010; Fridriksson et al., 2009; Fridriksson, Richardson, Fillmore, & Cai, 2012; Postman-Caucheteux et al., 2010; Warburton, Price, Swinburn, & Wise, 1999) and, in fact, show detrimental outcomes when right hemisphere regions are recruited (Naeser et al., 2005), whereas others report better outcomes associated with right hemisphere modulation (Crosson et al., 2005; Leff et al., 2002; Musso et al., 1999; Peck et al., 2004). There is also evidence that suggests aphasia recovery is supported by bilateral hemispheric recruitment (Belin et al., 1996; Fridriksson et al., 2006, 2007; Thulborn, Gindin, Davis, & Erb, 1999; Weiller et al., 1995). Saur and colleagues (2006) discussed a plausible explanation for these seemingly incompatible findings in detail, suggesting three phases of language recovery where neural recruitment transitions: weak activation of intact left hemisphere regions in the early acute phase; strong activation of the entire language network—but especially right hemisphere regions—in the subacute phase; and, finally, normalization of activation as peak cortical recruitment returns to the left hemisphere during the chronic phase of recovery. In this study, language recovery systematically improved at each phase of recovery. Thus, there seem to be implications of different patterns of recovery that are potentially adaptive at different times.

Based on the evidence discussed, there is support for functional brain changes associated with both spontaneous and treatment-induced recovery of language in patients with aphasia. Many studies over the past several years have documented recovery in both scenarios, although fewer have looked exclusively at spontaneous recovery during the first few weeks after stroke. Despite the limited research, the studies seem to support a hierarchical model of language recovery during the acute and subacute stages, because dramatic improvements in language function are demonstrated within the first 2 weeks up to a few months after onset (Fernandez et al., 2004; Heiss et al., 1999; Karbe, Herholz, Halber, & Heiss, 1998; Pedersen, Stig Jørgensen, Nakayama, Raaschou, & Olsen, 1995; Saur et al., 2006). These improvements may be accounted for by diaschisis resolution and strong bilateral activation (Saur & Hartwigsen, 2012). Most patients recruited for these studies received speech-language therapy as part of standard poststroke care, although the details of the implemented therapies were not reported; therefore, it is nearly impossible to identify the functional outcomes that occur purely as a result of spontaneous recovery as opposed to being driven by aphasia treatment.

In contrast, an increasing number of studies have assessed treatment-induced neural plasticity. Treatment approaches associated with neural plastic changes have focused on anomia training (Crosson et al., 2007, 2005, 2009; Fridriksson, 2010; Leger et al., 2002; Menke et al., 2009), semantic feature analysis (Marcotte et al., 2012), auditory comprehension (Musso et al., 1999), and sentence processing (Thompson et al., 2010), and changing activation patterns are most commonly assessed using fMRI. Meinzer and Breitenstein (2008) completed a review of 13 fMRI studies that implemented intervention paradigms to investigate language recovery in patients with chronic aphasia and reported that most studies found treatment-induced neural changes in both hemispheres. Because these findings were predominantly based on word retrieval interventions and not other aspects of impaired language, and because there were only three studies with more than 10 patients in the sample, we must interpret these findings with caution.

MEG, EEG, and PET studies have also documented treatment-based functional brain changes. MEG was used to assess treatment-induced plasticity after constraint-induced language treatment (CILT) (Breier et al., 2009; Meinzer et al., 2004). Results suggest that although the right hemisphere may support language recovery immediately after treatment, recruitment of perilesional regions is fundamental for prolonged, stable effects. This assumption holds true for the majority of studies described, and it supports the premise that treatment-induced changes occurring during the chronic stage of recovery may be attributed to the application of model-based therapies compared with generalized stimulation techniques applied during the acute stage of recovery (Saur & Hartwigsen, 2012).

Further, to optimize neural plasticity post-injury, new approaches to enhance treatment-induced recovery have been explored. Brain stimulation, a class of techniques that modifies cortical excitability, has been coupled with traditional speech-language therapy with the goal of facilitating increased learning and treatment outcomes, either by suppressing or by exciting targeted cortical regions. For example, Baker, Rorden, and Fridriksson (2010) found improved naming outcomes when anomia training was paired with anodal transcranial direct cortical stimulation (atDCS). Similarly, Meinzer and colleagues (2007) paired repetitive transcranial magnetic stimulation (rTMS) with CILT in two patients with aphasia and reported that both patients demonstrated improved outcomes when CILT was paired with rTMS compared with receiving rTMS in isolation.

In summary, the current evidence is encouraging yet nascent. Given the heterogeneity and small sample size, and given the variability in treatment approaches, more controlled group studies are required to confirm the present findings for both spontaneous and treatment-induced recovery. Despite variability in the current literature, as a whole, the implication is that the neural reorganization that occurs during spontaneous recovery is the same as therapy-induced reorganization, which takes place in the bilateral temporo-frontal network (Saur & Hartwigsen, 2012). Future work is necessary to further inform our knowledge of functional language recovery and to contribute to improved treatment efficacy for individuals with aphasia.

References

Albert, M. L., Sparks, R. W., & Helm, N. A. (1973). Melodic intonation therapy for aphasia. *Archives of Neurology, 29*(2), 130.

Allendorfer, J. B., Lindsell, C. J., Siegel, M., Banks, C. L., Vannest, J., Holland, S. K., et al. (2012). Females and males are highly similar in language performance and cortical activation patterns during verb generation. *Cortex, 48*(9), 1218–1233.

Allred, R. P., & Jones, T. A. (2008). Maladaptive effects of learning with the less-affected forelimb after focal cortical infarcts in rats. *Experimental Neurology, 210*(1), 172–181.

Baker, J. M., Rorden, C., & Fridriksson, J. (2010). Using transcranial direct-current stimulation to treat stroke patients with aphasia. *Stroke, 41*(6), 1229–1236.

Belin, P., Zilbovicius, M., Remy, P., Francois, C., Guillaume, S., Chain, F., et al. (1996). Recovery from nonfluent aphasia after melodic intonation therapy A PET study. *Neurology, 47*(6), 1504–1511.

Berthier, M. L. (2001). Unexpected brain-language relationships in aphasia: Evidence from transcortical sensory aphasia associated with frontal lobe lesions. *Aphasiology, 15*(2), 99–130.

Brady, M. C., Kelly, H., Godwin, J., & Enderby, P. (2012). Speech and language therapy for aphasia following stroke. *Cochrane Database of Systematic Reviews, 5*, CD000425.

Bray, B. D., Campbell, J., Geoffrey, C. C., Hoffman, A., Tyrrell, P. J., Wolfe, C. D., et al. (2013). Bigger, faster? Associations between hospital thrombolysis volume and speed of thrombolysis administration in acute ischemic stroke. *Stroke, 44*, 3129–3135.

Breier, J. I., Juranek, J., Maher, L. M., Schmadeke, S., Men, D., & Papanicolaou, A. C. (2009). Behavioral and neurophysiologic response to therapy for chronic aphasia. *Archives of Physical Medicine and Rehabilitation, 90*(12), 2026–2033.

Cappa, S. F. (2000). Neuroimaging of recovery from aphasia. *Neuropsychological Rehabilitation, 10*(3), 365–376.

Carmichael, S. T. (2008). Themes and strategies for studying the biology of stroke recovery in the poststroke epoch. *Stroke, 39*(4), 1380–1388.

Catani, M., Mesulam, M. M., Jakobsen, E., Malik, F., Martersteck, A., Wieneke, C., et al. (2013). A novel frontal pathway underlies verbal fluency in primary progressive aphasia. *Brain, 136*(8), 2619–2628.

Conklyn, D., Novak, E., Boissy, A., Bethoux, F., & Chemali, K. (2012). The effects of modified melodic intonation therapy on nonfluent aphasia: A pilot study. *Journal of Speech, Language and Hearing Research*, *55*(5), 1463.

Cramer, S. C. (2008). Repairing the human brain after stroke: I. Mechanisms of spontaneous recovery. *Annals of Neurology*, *63*(3), 272–287.

Crosson, B., McGregor, K., Gopinath, K. S., Conway, T. W., Benjamin, M., Chang, Y. L., et al. (2007). Functional MRI of language in aphasia: A review of the literature and the methodological challenges. *Neuropsychology Review*, *17*(2), 157–177.

Crosson, B., Moore, A. B., Gopinath, K., White, K. D., Wierenga, C. E., Gaiefsky, M. E., et al. (2005). Role of the right and left hemispheres in recovery of function during treatment of intention in aphasia. *Journal of Cognitive Neuroscience*, *17*(3), 392–406.

Crosson, B., Moore, A. B., McGregor, K. M., Chang, Y. L., Benjamin, M., Gopinath, K., et al. (2009). Regional changes in word-production laterality after a naming treatment designed to produce a rightward shift in frontal activity. *Brain and Language*, *111* (2), 73–85.

Dunn, L. M., & Dunn, D. M. (2007). *Peabody picture vocabulary test, (PPVT-4)*. Minneapolis, MN: Pearson Assessments.

Epstein-Peterson, Z., Vasconcellos Faria, A., Mori, S., Hillis, A. E., & Tsapkini, K. (2012). Relatively normal repetition performance despite severe disruption of the left arcuate fasciculus. *Neurocase*, *18*(6), 521–526.

Felberg, R. A., Okon, N. J., El-Mitwalli, A., Burgin, W. S., Grotta, J. C., & Alexandrov, A. V. (2002). Early dramatic recovery during intravenous tissue plasminogen activator infusion clinical pattern and outcome in acute middle cerebral artery stroke. *Stroke*, *33*(5), 1301–1307.

Fernandez, B., Cardebat, D., Demonet, J. F., Joseph, P. A., Mazaux, J. M., Barat, M., et al. (2004). Functional MRI follow-up study of language processes in healthy subjects and during recovery in a case of aphasia. *Stroke*, *35*(9), 2171–2176.

Fridriksson, J. (2010). Preservation and modulation of specific left hemisphere regions is vital for treated recovery from anomia in stroke. *The Journal of Neuroscience*, *30*(35), 11558–11564.

Fridriksson, J., Baker, J. M., Whiteside, J., Eoute, D., Moser, D., Vesselinov, R., et al. (2009). Treating visual speech perception to improve speech production in nonfluent aphasia. *Stroke*, *40*(3), 853–858.

Fridriksson, J., Fillmore, P., Guo, D., & Rorden, C. (2014). Chronic Broca's aphasia is caused by damage to Broca's and Wernicke's areas. *Cortex*Jul 11. pii: bhu152. [Epub ahead of print].

Fridriksson, J., Hubbard, H. I., Hudspeth, S. G., Holland, A. L., Bonilha, L., Fromm, D., et al. (2012). Speech entrainment enables patients with Broca's aphasia to produce fluent speech. *Brain*, *135* (12), 3815–3829.

Fridriksson, J., Morrow-Odom, L., Moser, D., Fridriksson, A., & Baylis, G. (2006). Neural recruitment associated with anomia treatment in aphasia. *NeuroImage*, *32*(3), 1403–1412.

Fridriksson, J., Moser, D., Bonilha, L., Morrow-Odom, K. L., Shaw, H., Fridriksson, A., et al. (2007). Neural correlates of phonological and semantic-based anomia treatment in aphasia. *Neuropsychologia*, *45* (8), 1812–1822.

Fridriksson, J., Richardson, J. D., Fillmore, P., & Cai, B. (2012). Left hemisphere plasticity and aphasia recovery. *NeuroImage*, *60*(2), 854–863.

Hartman, J. (1981). Measurement of early spontaneous recovery from aphasia with stroke. *Annals of Neurology*, *9*(1), 89–91.

Harvey, D. Y., Wei, T., Ellmore, T. M., Cris Hamilton, A., & Schnur, T. T. (2013). Neuropsychological evidence for the functional role of the uncinate fasciculus in semantic control. *Neuropsychologia*, *51*, 789–801.

Hebb, D. O. (1949). *The organization of behavior*. New York, NY: Wiley.

Heilman, K. M., Rothi, L., McFarling, D., & Rottmann, A. L. (1981). Transcortical sensory aphasia with relatively spared spontaneous speech and naming. *Archives of Neurology*, *38*(4), 236.

Heiss, W. D., Kessler, J., Thiel, A., Ghaemi, M., & Karbe, H. (1999). Differential capacity of left and right hemispheric areas for compensation of poststroke aphasia. *Annals of Neurology*, *45*(4), 430–438.

Heiss, W. D., & Thiel, A. (2006). A proposed regional hierarchy in recovery of post-stroke aphasia. *Brain and Language*, *98*(1), 118–123.

Heiss, W. D., Thiel, A., Kessler, J., & Herholz, K. (2003). Disturbance and recovery of language function: Correlates in PET activation studies. *NeuroImage*, *20*, S42–S49.

Helm-Estabrooks, N, Nicholas, M, & Morgan, A. (1989). *MIT, melodic intonation therapy manual*. San Antonio, TX: Special Press.

Hillis, A. E., Barker, P. B., Beauchamp, N. J., Winters, B. D., Mirski, M., & Wityk, R. J. (2001). Restoring blood pressure reperfused Wernicke's area and improved language. *Neurology*, *56*(5), 670–672.

Hillis, A. E., Barker, P. B., Wityk, R. J., Aldrich, E. M., Restrepo, L., Breese, E. L., et al. (2004). Variability in subcortical aphasia is due to variable sites of cortical hypoperfusion. *Brain and Language*, *89* (3), 524–530.

Hillis, A. E., & Heidler, J. (2002). Mechanisms of early aphasia recovery. *Aphasiology*, *16*(9), 885–895.

Jackson, J. H. (1874). On the nature of the duality of the brain. *Med Press Circular* 1–63. Reprinted in The Classics of Neurology and Neurosurgery Library, 1984; Birmingham, AL, 1984.

Johansson, B. B. (2000). Brain plasticity and stroke rehabilitation. The Willis lecture. *Stroke*, *31*(1), 223–230.

Kaplan, E., Goodglass, H., & Weintraub, S. (2001). *Boston naming test, 2nd ed Pro-Ed. Peabody Picture Vocabulary Test*. Austin, TX: Pro-Ed.

Karbe, H., Herholz, K., Halber, M., & Heiss, W. D. (1998). Collateral inhibition of transcallosal activity facilitates functional brain asymmetry. *Journal of Cerebral Blood Flow and Metabolism*, *18*(10), 1157–1161.

Kelly, H., Brady, M. C., & Enderby, P. (2010). Speech and language therapy for aphasia following stroke. *Cochrane Database of Systematic Reviews*, *12*(5), CD000425.

Kim, Y. H., Ko, M. H., Parrish, T. B., & Kim, H. G. (2002). Reorganization of cortical language areas in patients with aphasia: A functional MRI study. *Yonsei Medical Journal*, *43*(4), 441–445.

Kleim, J. A., & Jones, T. A. (2008). Principles of experience-dependent neural plasticity: Implications for rehabilitation after brain damage. *Journal of Speech, Language and Hearing Research*, *51*(1), S225.

Kolb, B., & Whishaw, I. Q. (1998). Brain plasticity and behavior. *Annual Review of Psychology*, *49*(1), 43–64.

Leff, A., Crinion, J., Scott, S., Turkheimer, F., Howard, D., & Wise, R. (2002). A physiological change in the homotopic cortex following left posterior temporal lobe infarction. *Annals of Neurology*, *51*(5), 553–558.

Leger, A., Demonet, J. F., Ruff, S., Aithamon, B., Touyeras, B., Puel, M., et al. (2002). Neural substrates of spoken language rehabilitation in an aphasic patient: An fMRI study. *Neuroimage*, *17*(1), 174–183.

Maas, M. B., Lev, M. H., Ay, H., Singhal, A. B., Greer, D. M., Smith, W. S., et al. (2012). The prognosis for aphasia in stroke. *Journal of Stroke and Cerebrovascular Diseases*, *21*(5), 350–357.

Marcotte, K., Adrover-Roig, D., Damien, B., de Preaumont, M., Genereux, S., Hubert, M., et al. (2012). Therapy-induced neuroplasticity in chronic aphasia. *Neuropsychologia*, *50*(8) 1776–1786.

Meinzer, M., & Breitenstein, C. (2008). Functional imaging studies of treatment induced recovery in chronic aphasia. *Aphasiology, 22* (12), 1251–1268.

Meinzer, M., Elbert, T., Wienbruch, C., Djundja, D., Barthel, G., & Rockstroh, B. (2004). Intensive language training enhances brain plasticity in chronic aphasia. *BMC Biology, 2*(1), 20.

Meinzer, M., Streiftau, S., & Rockstroh, B. (2007). Intensive language training in the rehabilitation of chronic aphasia: Efficient training by laypersons. *Journal of the International Neuropsychological Society, 13*(05), 846–853.

Menke, R., Meinzer, M., Kugel, H., Deppe, M., Baumgartner, A., Schiffbauer, H., et al. (2009). Imaging short-and long-term training success in chronic aphasia. *BMC Neuroscience, 10*(1), 118.

Minnerup, J., Wersching, H., Ringelstein, E. B., Schilling, M., Schäbitz, W. R., Wellmann, J., et al. (2011). Impact of the extended thrombolysis time window on the proportion of recombinant tissue-type plasminogen activator-treated stroke patients and on door-to-needle time. *Stroke, 42*(10), 2838–2843.

Monks, T., Pitt, M., Stein, K., & James, M. (2012). Maximizing the population benefit from thrombolysis in acute ischemic stroke-a modeling study of in-hospital delays. *Stroke, 43*(10), 2706–2711.

Musso, M., Weiller, C., Kiebel, S., Müller, S. P., Bülau, P., & Rijntjes, M. (1999). Training-induced brain plasticity in aphasia. *Brain, 122* (9), 1781–1790.

Naeser, M. A., Martin, P. I., Nicholas, M., Baker, E. H., Seekins, H., Kobayashi, M., et al. (2005). Improved picture naming in chronic aphasia after TMS to part of right Broca's area: An open-protocol study. *Brain and Language, 93*(1), 95–105.

Norton, A., Zipse, L., Marchina, S., & Schlaug, G. (2009). Melodic intonation therapy. *Annals of the New York Academy of Sciences, 1169*(1), 431–436.

Peck, K. K., Moore, A. B., Crosson, B. A., Gaiefsky, M., Gopinath, K. S., White, K., et al. (2004). Functional magnetic resonance imaging before and after aphasia therapy shifts in hemodynamic time to peak during an overt language task. *Stroke, 35*(2), 554–559.

Pedersen, P. M., Stig Jørgensen, H., Nakayama, H., Raaschou, H. O., & Olsen, T. S. (1995). Aphasia in acute stroke: Incidence, determinants, and recovery. *Annals of Neurology, 38*(4), 659–666.

Postman-Caucheteux, W. A., Birn, R. M., Pursley, R. H., Butman, J. A., Solomon, J. M., Picchioni, D., et al. (2010). Single-trial fMRI shows contralesional activity linked to overt naming errors in chronic aphasic patients. *Journal of Cognitive Neuroscience, 22*(6), 1299–1318.

Pulvermüller, F., & Schönle, P. W. (1993). Behavioral and neoronal changes during treatment of mixed transcortical aphasia: A case study. *Cognition, 48*(2), 139–161.

Rijntjes, M., & Weiller, C. (2002). Recovery of motor and language abilities after stroke: The contribution of functional imaging. *Progress in Neurobiology, 66*(2), 109–122.

Robey, R. R. (1998). A meta-analysis of clinical outcomes in the treatment of aphasia. *Journal of Speech, Language and Hearing Research, 41*(1), 172.

Saur, D., & Hartwigsen, G. (2012). Neurobiology of language recovery after stroke: Lessons from neuroimaging studies. *Archives of Physical Medicine and Rehabilitation, 93*(1), S15–S25.

Saur, D., Lange, R., Baumgaertner, A., Schraknepper, V., Willmes, K., Rijntjes, M., et al. (2006). Dynamics of language reorganization after stroke. *Brain, 129*(6), 1371–1384.

Schlaug, G., Marchina, S., & Norton, A. (2009). Evidence for plasticity in white matter tracts of patients with chronic Broca's aphasia undergoing intense intonation based speech therapy. *Annals of the New York Academy of Sciences, 1169*(1), 385–394.

Szaflarski, J. P., Allendorfer, J. B., Banks, C., Vannest, J., & Holland, S. K. (2013). Recovered vs. not-recovered from post-stroke aphasia: The contributions from the dominant and non-dominant hemispheres. *Restorative Neurology and Neuroscience, 31*(4), 347–360.

Thompson, C. K. (2000a). Neuroplasticity: Evidence from aphasia. *Journal of Communication Disorders, 33*(4), 357.

Thompson, C. K. (2000b). The neurobiology of language recovery in aphasia. *Brain and Language, 71*(1), 245.

Thompson, C. K., den Ouden, D. B., Bonakdarpour, B., Garibaldi, K., & Parrish, T. B. (2010). Neural plasticity and treatment-induced recovery of sentence processing in agrammatism. *Neuropsychologia, 48*(11), 3211–3227.

Thulborn, K. R., Gindin, T. S., Davis, D., & Erb, P. (1999). Comprehensive MR imaging protocol for stroke management: tissue sodium concentration as a measure of tissue viability in nonhuman primate studies and in clinical studies 1. *Radiology, 213*(1), 156–166.

Turkstra, L. S., Holland, A. L., & Bays, G. A. (2003). The neuroscience of recovery and rehabilitation: What have we learned from animal research? *Archives of Physical Medicine and Rehabilitation, 84*(4), 604–612.

van der Meulen, I., van de Sandt-Koenderman, M. E., & Ribbers, G. M. (2012). Melodic intonation therapy: Present controversies and future opportunities. *Archives of Physical Medicine and Rehabilitation, 93*(1), S46–S52.

Warburton, E., Price, C. J., Swinburn, K., & Wise, R. J. (1999). Mechanisms of recovery from aphasia: Evidence from positron emission tomography studies. *Journal of Neurology, Neurosurgery and Psychiatry, 66*(2), 155–161.

Wei, L., Erinjeri, J. P., Rovainen, C. M., & Woolsey, T. A. (2001). Collateral growth and angiogenesis around cortical stroke. *Stroke, 32*(9), 2179–2184.

Weiller, C., Isensee, Ch., Rijntjes, M., Huber, W., Müller, S. P., Bier, D., et al. (1995). Recovery from Wernicke's aphasia: a positron emission tomographic study. *Annals of Neurology, 37*(6), 723–732.

Zipse, L., Norton, A., Marchina, S., & Schlaug, G. (2012). When right is all that is left: Plasticity of right hemisphere tracts in a young aphasic patient. *Annals of the New York Academy of Sciences, 1252* (1), 237–245.

CHAPTER

81

Melodic Intonation Therapy

Gottfried Schlaug

Department of Neurology, Neuroimaging, and Stroke Recovery Laboratories, Beth Israel Deaconess Medical Center and Harvard Medical School, Boston, MA, USA

81.1 THE IMPACT OF NONFLUENT APHASIA

Of the estimated 750,000–800,000 new strokes occurring in the United States each year, approximately 25–50% present with some form of aphasia, as estimated based on studies performed in other countries (Engelter, Gostynski, Papa, Frei, & Born, 2006; Pedersen, Jorgensen, Nakayama, Raaschou, & Olsen, 1995; Pedersen, Vinter, & Olsen, 2004). Approximately 40% of these acute patients were available for follow-up at 1 year (attrition was due to death or other inability to participate in trial). Approximately two-thirds of these 40% of patients from the original cohort showed abnormal scores on aphasia testing, with approximately one-quarter of them being in the severe, nonfluent category (Kertesz & McCabe, 1977; Pedersen et al., 1995, 2004; Wade, Hewer, David, & Enderby, 1986). In right-handed individuals, nonfluent aphasia generally results from lesions in the left frontal lobe, including the portion of the left frontal lobe known as Broca's region. Named after Paul Broca (1961), who first linked this area of the brain with nonfluent aphasia, this region is thought to consist of the posterior inferior frontal gyrus (IFG) encompassing Brodmann's areas 44 and 45. However, subsequent reports have shown that different lesion locations and larger lesion volumes involving not only the inferior and middle frontal gyrus but also the inferior peri-rolandic regions and subcortical brain structures can have a clinical presentation similar to that of Broca's aphasia (Kertesz, Lesk, & McCabe, 1977). Recent research has shown that the location of the lesions in relation to the arcuate fasciculus (AF), a white matter tract connecting frontal and temporal brain regions, is a strong predictor of fluency impairment (Fridriksson, Guo, Fillmore, Holland, & Rorden, 2013; Marchina et al., 2011; Wang, Marchina, Norton, Wan, & Schlaug, 2013). The AF lesion load (AF-LL) might serve as a biomarker for fluency impairment in chronic aphasic patients (Figure 81.1). The AF-LL can also be used to predict fluency outcomes in patients with acute aphasia and to stratify patients (by AF-LL) for experimental studies.

Patients with large left hemispheric lesions or a high AF-LL resulting in severe nonfluent aphasia typically do not show good natural recovery from such an insult, nor do they appear to be as responsive to nonintonation-based speech therapy as patients with smaller lesions or other types of aphasia. Although there is no generally agreed on definition of nonfluent aphasia, for the purpose of this chapter nonfluency is defined as having less than 10 correct information units (CIUs) per minute in propositional speech assessments.

There are also no universally accepted methods or "gold standards" for the treatment or treatment outcomes of severe nonfluent aphasia against which new or existing interventions could be compared, nor have any criteria been established for measuring meaningful treatment efficacy. Nevertheless, most therapists, clinicians, and researchers in the field of aphasiology would probably agree that treatment could be considered effective if patients show improvements in speech output that generalizes to untrained language structures and/or contexts (Thompson & Shapiro, 2007) and not just an improvement in items that were practiced with the patient. Although a meta-analysis by Robey (1998) determined that an array of treatment methods for all kinds of aphasic syndromes is, on average, beneficial, treatment effect sizes have varied widely (Moss & Nicholas, 2006; Robey, 1998). Furthermore, a meta-analysis by Bhogal, Teasell, and Speechley (2003) concluded that aphasia treatments are more likely to achieve positive results if the total

Neurobiology of Language. DOI: http://dx.doi.org/10.1016/B978-0-12-407794-2.00081-X

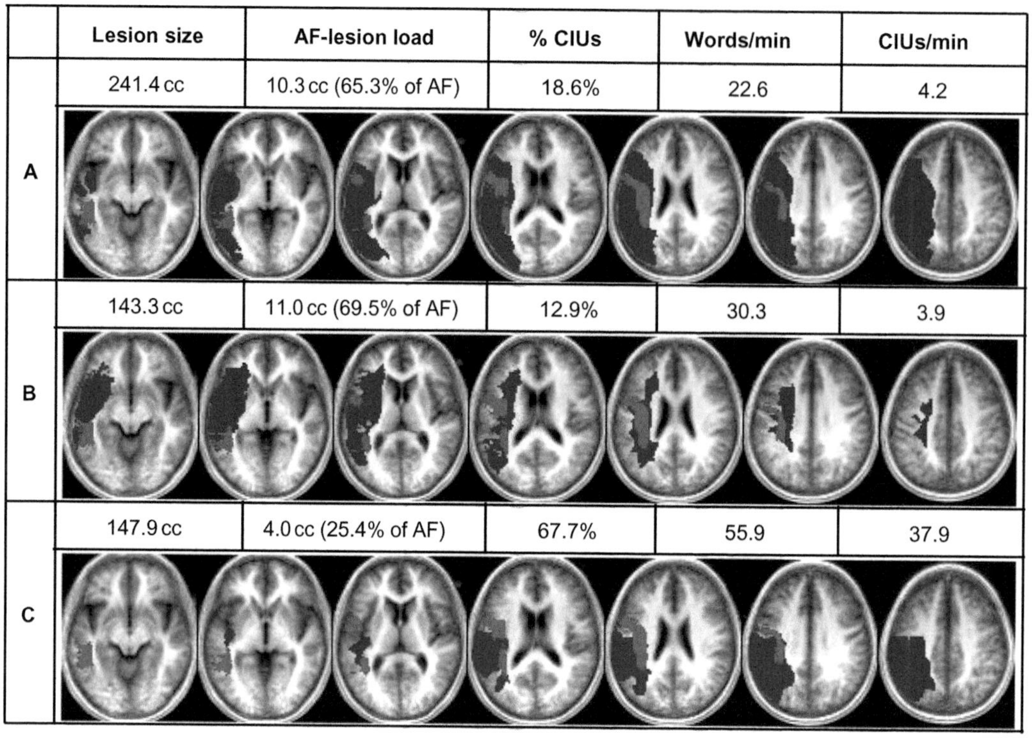

	Lesion size	AF-lesion load	% CIUs	Words/min	CIUs/min
A	241.4 cc	10.3 cc (65.3% of AF)	18.6%	22.6	4.2
B	143.3 cc	11.0 cc (69.5% of AF)	12.9%	30.3	3.9
C	147.9 cc	4.0 cc (25.4% of AF)	67.7%	55.9	37.9

FIGURE 81.1 AF-LL and fluency. Examples (A–C) of behavioral scores, lesion sizes, and AF-LLs of three patients, as well as their individual lesion maps (depicted in blue) overlaid onto the probabilistic AF map (depicted in red). Overlap between lesion and AF is displayed in purple. The axial slices depicted correspond to $z = -10$, -2, 8, 18, 26, 34, and 42 in Talairach space. Comparison of patients A and B shows how two patients can display comparable AF-LLs and behavioral scores despite drastically different overall lesion volumes. Similarly, comparison of patients B and C shows how a similar lesion size can produce two markedly different AF-LLs and, accordingly, result in very different levels of impairment.

amount of therapy exceeds 55 h, emphasizing the importance of intensity and duration of the treatment applied.

81.2 THE BASIS AND COMPONENTS OF INTONATION-BASED SPEECH THERAPY FOR PATIENTS WITH NONFLUENT APHASIA

Melodic Intonation Therapy (MIT) is an intonation-based treatment method for nonfluent or dysfluent patients with aphasia that was developed in response to the observation that severely aphasic patients can often produce well-articulated, linguistically accurate words while singing but not during speech (Gerstman, 1964; Geschwind, 1971; Hebert, Racette, Gagnon, & Peretz, 2003; Keith & Aronson, 1975; Kinsella, Prior, & Murray, 1988; Yamadori, Osumi, Masuhara, & Okubo, 1977). MIT is a hierarchically structured treatment that uses intoned (sung) patterns that exaggerate the normal melodic content of speech across three levels of increasing difficulty. These different levels (Elementary Level, Intermediate Level, Advanced Level) are characterized by an increase in the length of the phrase, an increase in the speed of production, and a decrease in the level of support provided by the therapist. The intonation works by translating prosodic speech patterns (spoken phrases) into melodically intoned patterns using just two pitches. The higher pitch represents the syllables that would naturally be stressed (accented) during speech. At the simplest level, patients learn to intone (sing) a series of two-syllable words/phrases (e.g., "Water," "Ice cream," "Bathroom") or simple two- or three-syllable social phrases (e.g., "Thank you," "I love you"). As each level is mastered, patients move to the next, and phrases gradually increase in length (e.g., "I am thirsty," "A cup of coffee, please"). Beyond the increased phrase length, the primary differences between the three levels of MIT lie in the way the treatment is administered and the level of support that is provided by the therapist. More details regarding the intervention and the differences between the three levels can be found elsewhere (Norton, Zipse, Marchina, and Schlaug, 2009).

Compared with nonintonation-based speech therapies, MIT contains two unique components: (i) the melodic intonation (singing) with its inherent continuous voicing and (ii) the rhythmic tapping of each syllable (using the patient's left hand) while phrases are intoned and repeated. The relative contribution of either intonation (pitch) or rhythm in a therapeutic context has not been examined in any study, mostly because it would be difficult to power such a longitudinal study considering the similarities between the melodic intonation, which is typically done at one syllable per second using stressed and unstressed syllables, and the rhythmic tapping with the unaffected hand, which is also done at one tap per second. Furthermore, melodic intonation and rhythmic tapping might have an additive effect and, ultimately, rhythm-based or melody-/pitch-based intervention would have to be compared with a combined intervention. Although the melodic intonation and the rhythmic tapping have been recognized as the unique features of MIT, there are other aspects of the MIT method that might be of equal importance. MIT has been introduced as an intensive treatment program that requires a commitment of up to 1.5 h/day, 5 days/week over a period of several months (as recommended by the developers of MIT) (Helm-Estabrooks, Nicholas, & Morgan, 1989), resulting in at least 75 treatment sessions until the patient has mastered all three levels of MIT. In addition to the intensity of the therapy, the slow rate of vocalization (one syllable per second) and an administration protocol that includes one-on-one sessions with a therapist who introduces and practices words/phrases using picture cues while giving continuous feedback to the patient make MIT stand out among particular intervention programs that have been described with such elaborate details within the speech therapy literature. All of these features mentioned (some of them are shared between MIT and other speech therapies) must be carefully considered when the efficacy of MIT is tested against a control intervention (Schlaug, Marchina, & Norton, 2008).

81.3 EXPERIENCES WITH THE APPLICATION OF MIT

Because the initial account of its successful use with three chronic, nonfluent (Broca's) aphasic patients (Albert, Sparks, & Helm, 1973), reports have outlined a comprehensive program of MIT (Helm-Estabrooks & Albert, 1991; Helm-Estabrooks et al., 1989; Norton et al., 2009; Sparks & Holland, 1976) including strict patient selection criteria (Helm-Estabrooks et al., 1989) and data that showed significant improvement on the Boston Diagnostic Aphasia Examination (BDAE) (Goodglass & Kaplan, 1983) after treatment (Bonakdarpour, Eftekharzadeh, & Ashayeri, 2000; Sparks, Helm, & Albert, 1974). In a case study comparing MIT to nonintonation-based control therapy (Wilson, Parsons, & Reutens, 2006), the authors found that MIT had a general facilitating effect on articulation and a longer-term effect on phrase production that they attributed specifically to its melodic component. However, the outcomes of that study were measured by the patient's ability to produce practiced phrases prompted by the therapist rather than by the transfer of language skills to untrained structures and/or contexts. In one of our own studies (Schlaug et al., 2008), we compared two patients with similar speech output impairments and similar lesion size and location that completed MIT or a control intervention also geared toward enhancing verbal output termed Speech-Repetition-Therapy (SRT). After treatment, both interventions yielded significant improvements in propositional speech that generalized to nonpracticed words and phrases; importantly, the MIT-treated patient's gains surpassed those of the control-treated patient. Another case series of six patients treated with MIT (Schlaug, Marchina, & Norton, 2009) showed a more than 200% improvement in propositional speech, measured as CIUs. In a comparison between after therapy and before therapy, the behavioral improvement was highly significant compared with baseline variations in repeated test assessments. The change in CIUs showed a strong trend for a correlation ($P = 0.08$) with the change in the size of the right AF, a fiber bundle connecting temporal and frontal brain regions (Schlaug, Marchina, et al., 2009).

van der Meulen, van de Sandt-Koenderman, Heijenbrok-Kal, Visch-Brink, and Ribbers (2014) recently conducted the first randomized controlled trial (RCT) examining the efficacy of MIT with regard to language production in subacute patients with nonfluent aphasia contrasting MIT against a wait-list control in which patients were focused on writing, language comprehension, and nonverbal communication strategies, but not on spoken output. Considering that this RCT might have been underpowered, the observed findings are still very encouraging in that the MIT-treated group did significantly better in repetition tests of trained items than the control group. Strong nonsignificant trends in the group comparisons were found in a test measuring verbal communication in daily life.

The main differences in these single cases and case series (Bonakdarpour et al., 2000; Schlaug et al., 2008; Sparks et al., 1974; Wan, Zheng, Marchina, Norton, & Schlaug, 2014; Wilson et al., 2006) and randomized trials (van der Meulen et al., 2014) are whether the studies adhere to the original description of MIT. Variations in the MIT technique can include differences in the intonation pattern and rhythmic tapping with the unimpaired hand, the intensity and total amount of treatment sessions as well as the type of control intervention with which MIT is compared, and the outcome variables that assess whether there is not only an improvement in treated words/phrases but also transfer to untreated items or improvements in propositional speech.

MIT has also been examined in combination with other recovery-enhancing interventions. In a cross-over design, Vines, Norton, and Schlaug (2011) combined MIT with simultaneous noninvasive anodal transcranial direct current stimulation (tDCS) in six patients with severe nonfluent aphasia to increase excitability in the right posterior frontal region. Significant effects in fluency were seen after only three sessions compared with MIT simultaneously applied with sham stimulation. Al-Janabi et al. (2014) used three treatment sessions of intermittent theta-burst stimulation with transcranial magnetic stimulation (TMS), which increases excitability in the underlying right posterior frontal gyrus, followed by 40 min of MIT. These three sessions were compared with sham TMS also followed by 40 min of MIT. The two patients treated in this study showed somewhat different effects. Improvements in verbal fluency and repetition were only seen in one of the two patients. Functional imaging changes were seen in both patients using overt speech tasks, although only the patient with the behavioral improvement showed right hemispheric functional changes. Overall, the combination of MIT with noninvasive brain stimulation might be a fruitful development; however, target regions that drive the therapeutic effect of MIT should be clearly identified and more patients should be examined to control for variability in lesion location, lesion size, and baseline impairment level.

81.4 EXAMINING ASPECTS OF RHYTHM AND MELODY IN CROSS-SECTIONAL STUDIES

There are several cross-sectional studies that have examined particular aspects of intonation (melody) or rhythm to better understand mechanisms that underlie the therapeutic effects of an intonation-based therapy such as MIT. Previous studies suggested that patients with severe nonfluent aphasia were able to produce words in familiar songs, but they had reduced speech output without the use of songs. However, when this was tested in several cross-sectional studies, no clear advantage of the singing condition over normal speech condition for word production was found (Hebert et al., 2003; Stahl, Kotz, Henseler, Turner, & Geyer, 2011). These authors report that the automatic status of lyrics in over-learned songs could account for the earlier clinical descriptions of better verbal production in familiar songs compared with spontaneous speech. This was taken one step further by Racette, Bard, and Peretz (2006) in a study examining the effects on speech output of "singing along" with other singers versus "singing alone." Singing along with other singers produced more speech output than singing alone. Racette and colleagues concluded that sung lyrics are more regular in rhythm than spoken words and greater temporal regularity would allow for better synchronization. Others have used this to suggest that rhythmic aspects of MIT might be a more important factor than melodic aspects and that intoned speech facilitation could depend on the rhythmic properties of the underlying spoken language (Stahl et al., 2011). One should not forget that one important aspect of MIT is that it slows the verbal output to less than one syllable per second, which is quite slow compared with natural speech. This slowing of articulation, together with an underlying melodic contour, leads to more bihemispheric involvement in vocal-motor tasks (Ozdemir, Norton, & Schlaug, 2006) and could even lead to a right hemisphere dominance by making use of the increased sensitivity of the right hemisphere to slow temporal features in acoustic signals (Abrams, Nicol, Zecker, & Kraus, 2008; Poeppel, 2003; Zatorre & Gandour, 2008).

81.5 NEURAL CORRELATES OF MIT: NEUROIMAGING FINDINGS

It was originally thought that the success of MIT lies in its ability to engage areas for articulation and speech output in the right hemisphere, and that these regions on the right hemisphere could be engaged, because they are involved in music processing (Albert et al., 1973; Sparks et al., 1974). An alternative interpretation is that MIT might exert its effect either by unmasking existing music/language connections in both hemispheres or by engaging preserved language-capable regions in either or both hemispheres due to particular aspects of music. Because MIT incorporates both the melodic and rhythmic aspects of music (Albert et al., 1973; Boucher, Garcia, Fleurant, & Paradis, 2001; Cohen & Masse, 1993; Helm-Estabrooks et al., 1989; Norton et al., 2009; Sparks et al., 1974; Sparks & Holland, 1976), it may be unique in its potential for engaging not just the right but both hemispheres. Belin, Van Eeckhout, Zilbovicius, Remy, and Francois (1996) suggested that MIT-facilitated recovery is associated with the reactivation of left hemisphere language-related regions, most notably the left prefrontal cortex, just anterior to Broca's region. Although Belin et al. (1996) were the first to examine patients

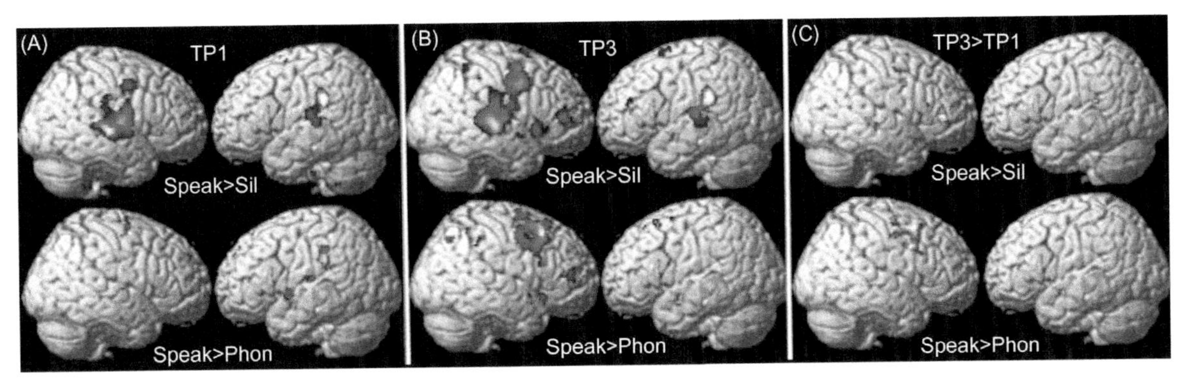

FIGURE 81.2 Activation maps of overt speaking tasks. fMRI activation maps (superimposed onto the surface projections of a spatially standardized normal brain) of the contrast between overt speaking versus silence (control condition) in the top row and overt speaking versus vowel production in the bottom row ($P<0.05$; Family-Wise Error (FWE) corrected) before therapy (A) and after therapy (B), as well as a voxel-by-voxel comparison of the two time points (C). The color codes represent different magnitudes of activation: yellow indicates stronger activation than red. All comparisons are thresholded at $P<0.05$ (FWE-corrected). See Ozdemir et al. (2006) for more details regarding the functional MRI tasks and fMRI data analysis. TP1 = before therapy, TP3 = after therapy. TP3 > TP1 = voxel-by-voxel comparison of the two fMRI acquisitions.

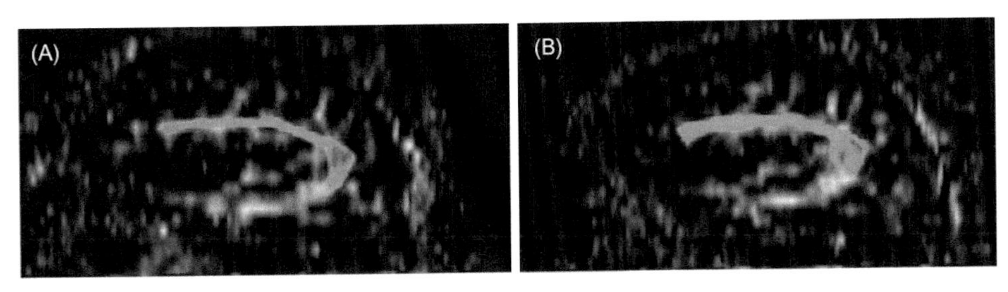

FIGURE 81.3 AF before and after therapy. AF was reconstructed from diffusion tensor imaging scans of a patient before and after an intense course of MIT. There is a visible increase in the size (number of fibers, volume of tract, and length of fibers) of the right AF when the acquisition before therapy (A) is compared with those acquired after treatment (B).

treated with an MIT-like intervention using functional neuroimaging, their findings were surprising and somewhat contrary to the hypotheses that had been put forth by the original developers of MIT and those that have been observed using MIT in a clinical research setting (Albert et al., 1973; Bonakdarpour et al., 2000; Schlaug, Marchina, et al., 2009; Sparks et al., 1974). It is interesting to note that although the primary finding Belin and colleagues was an activation of left prefrontal regions when participants were asked to repeat intoned words, there is an important aspect of their study that is not often reported. In their analysis comparing the repetition of spoken words with the hearing of those words, Belin and colleagues found blood flow changes that occurred predominantly in the right hemisphere (including the right temporal lobe and the right central operculum), which is consistent with some of our findings (Schlaug et al., 2008; Schlaug, Marchina, et al., 2009). Figure 81.2 shows fMRI activation maps (superimposed onto the surface projections of a spatially standardized normal brain) of one of our other patients treated with MIT. Results of two imaging contrasts are shown: overt speaking versus silence (control condition) and overt speaking versus vowel production ($P<0.05$ FWE) before (Figure 81.2A) and after (Figure 81.2B) therapy. Furthermore, a direct voxel-by-voxel comparison of the two acquisitions is shown in Figure 81.2C. The color codes represent different magnitudes of activation: yellow indicates stronger activation than red. The pronounced differences in activation seen in the comparison between after therapy and before therapy show that there is more activation in the right temporal, premotor, and posterior inferior frontal regions after therapy. Details of our functional imaging tasks using overt speech repetition tasks as well as the analysis of sparse temporally acquired fMRI data are outlined in Ozdemir et al. (2006).

Figure 81.3 shows a diffusion tensor imaging study of one of our patients before and after 75 sessions of

MIT. The AF is typically not as strongly developed in the right hemisphere as it is in the left hemisphere in right-handed individuals (Glasser & Rilling, 2008). The treatment-induced increase is evident when the pretherapy (Figure 81.3A) and posttherapy (Figure 81.3B) images are compared. An even stronger effect of structural plasticity was seen in a teenager with a very large left hemisphere lesion who completed our intense MIT protocol. The therapeutic effects were quite remarkable and the structural changes seen in the AF could be clearly attributed to the treatment and observed improvements in verbal output, because normal developmental changes over the treatment period could be controlled by imaging the unimpaired monozygotic twin of this patient (Zipse, Norton, Marchina, & Schlaug, 2012).

The AF is a structure that plays an important role in auditory-motor mapping and, therefore, in language development and in the feedforward and feedback control of any vocal output. Furthermore, it may well be the remodeling of this fiber tract as a result of the intense therapy that supports long-term behavioral effects in severely nonfluent patients (Schlaug, Forgeard, Zhu, Norton, & Winner, 2009; Schlaug, Marchina, et al., 2009). The most likely explanation for these imaging results are changes in myelination (Alexander, Lee, Lazar, & Field, 2007; Jito et al., 2008; Vorisek & Sykova, 1997); however, changes in axon diameter due to changes in myelin or axon density due to axonal sprouting are also possible (Carmichael, 2006; Dancause, Barbay, Frost, Plautz, & Chen, 2005; Fields, 2008). In patients with large left hemispheric lesions, the right AF may play a crucial role in facilitating the mapping of sounds to motor actions and its feedback control.

81.6 POSSIBLE MECHANISMS EXPLAINING THE EFFECTS OF AN INTONATION-BASED SPEECH THERAPY

The traditional explanation for the dissociation between speaking and singing in aphasic patients is the presence of two routes for word articulation: one for spoken words through the brain's left hemisphere and a separate route for sung words that uses either the right or both hemispheres. The small amount of empirical data available support a bihemispheric role in the execution and sensorimotor control of vocal production for both speaking and singing (Bohland & Guenther, 2006; Brown, Martinez, Hodges, Fox, & Parsons, 2004; Guenther, Hampson, & Johnson, 1998; Jeffries, Fritz, & Braun, 2003; Ozdemir et al., 2006), with a tendency for greater left-lateralization for speaking under normal physiological conditions (i.e., faster rates of production during speaking than singing). The representation of the sensory elements of music and language may be either separate or in different locations with smaller degrees of overlap (for more details on this see also Koelsch, Fritz, Schulze, Alsop, & Schlaug, 2005; Koelsch et al., 2002; Patel, 2003; Peretz & Coltheart, 2003). Nevertheless, if there is a bihemispheric representation for speech production, then the question of why an intervention that uses singing, or a form of singing such as MIT, has the potential to facilitate syllable and word production still remains. In theory, there are four possible mechanisms by which the therapeutic effect of MIT could be achieved. The first is *reduction of speed*. In singing, words are articulated at a slower rate than in speaking, thereby reducing dependence on the left hemisphere. The rate of one syllable per second is the rate suggested by the developers of MIT (Helm-Estabrooks et al., 1989). Although we have made some adjustments to the original MIT protocol (Norton et al., 2009), we did find the slow rate of vocalization particularly useful as a starting rate for our patients. In fact, many of our patients are so severely impaired at baseline that one syllable per second might even be too fast for them initially. Although rate is not used as an outcome measure in the daily sessions, we do train our therapists to adhere to the one syllable per second rate. When patients reach the "advanced level" of MIT, the rate is gradually increased to approximately two syllables per second as they are transitioned from singing back to speaking. The second is *syllable lengthening*. This provides the opportunity to distinguish the individual phonemes that together form words and phrases while the continuous vocalization inherent in singing "strings" the sound together and thereby encourages fluency. This connected segmentation (i.e., overemphasizing the individual phonemes but still connecting them into meaningful words and phrases), coupled with the reduction of speed in singing, may help nonfluent aphasic patients practice auditory-motor mapping under feedback control and increase fluency and may receive greater support from right hemisphere structures. The third is *syllable "chunking."* Prosodic features such as intonation, change in pitch, and syllabic stress may help patients group syllables into words and words into phrases, and this "chunking" may also enlist more right hemisphere support. The fourth is *left hand tapping (one tap per syllable, one syllable per second)*. It is likely that the tapping component engages a right-hemispheric sensorimotor network that may, in turn, provide an impulse for verbal production in much the same way that a metronome has been shown to serve as a "pacemaker" in other motor activities (rhythmic anticipation, rhythmic entrainment) (Thaut & Abiru, 2010; Thaut, Kenyon,

Schauer, & McIntosh, 1999). In addition, there may be a set of shared neural correlates that control both hand movements and articulatory movements (Gentilucci, Benuzzi, Bertolani, Daprati, & Gangitano, 2000; Meister et al., 2003; Tokimura, Tokimura, Oliviero, Asakura, & Rothwell, 1996; Uozumi, Tamagawa, Hashimoto, & Tsuji, 2004). Further, the sound produced by the tapping may encourage auditory-motor coupling (Lahav, Saltzman, & Schlaug, 2007). In theory, reduction of speed, syllable lengthening, and syllable chunking can be applied to other nonintonation-based speech techniques. However, these components are often not featured in the traditional therapeutic context.

How might MIT facilitate recovery and what do its unique elements contribute to the process? Functional imaging tasks targeting the perception of musical components that require a more global than local processing strategy (e.g., melodic contour, musical phrasing, and/or meter) tend to elicit greater activity in right than in left hemispheric brain regions. It has also been shown that tasks emphasizing spectral information over temporal information have shown more right than left hemispheric activation (Meyer, Alter, Friederici, Lohmann, & von Cramon, 2002; Zatorre & Belin, 2001). Further, patients with right hemisphere lesions have greater difficulty with global processing (e.g., melody and contour processing) than those with left hemisphere lesions (Peretz, 1990; Schuppert, Munte, Wieringa, & Altenmuller, 2000). Last, but not least, the right hemisphere might have increased sensitivity to slow temporal features in acoustic signals (Abrams et al., 2008; Zatorre & Gandour, 2008). Thus, it is possible that the melodic element of MIT engages the right hemisphere, particularly the right temporal lobe, more than therapies that do not incorporate use of pitch or melody.

The effects of the left hand tapping should be considered in the same context. Once the right temporal lobe is specifically engaged by the melodic intonation and contour, it is conceivable that the role of the left hand tapping could be the activation and priming of a right hemispheric sensorimotor network for articulation. Because concurrent speech and hand use occurs in daily life, and because gestures are frequently used during speech, hand movements, possibly in synchrony with articulatory movements, may have a facilitating effect on speech production, but the precise role of this facilitation is unknown (Benjamin, Towler, Garcia, Park, & Sudhyadhom, 2014; Meister, Buelte, Staedtgen, Boroojerdi, & Sparing, 2009; Rose, 2013). We hypothesize that tapping the left hand may engage a right hemispheric sensorimotor network that coordinates not only hand movements but also orofacial and articulatory movements, and may facilitate speech production through rhythmic anticipation, rhythmic entrainment, or auditory-motor coupling (Benjamin et al., 2014; Lahav et al., 2007; Schlaug, Altenmuller, & Thaut, 2010; Thaut & Abiru, 2010; Thaut et al., 1999; Wan, Rüber, Hohmann, & Schlaug, 2010).

81.7 CONCLUSION

The clinical observation that patients with nonfluent aphasia are better at singing lyrics than they are at speaking the same words inspired the development of MIT. Despite several small case series, the efficacy of MIT, particularly with regard to untrained items or with regard to improvements in propositional speech, has not been substantiated. Neural correlates of the therapy have emerged in several studies, but differences between studies remain. Because of its potential to engage or unmask speech-motor regions in the unaffected right hemisphere, MIT appears to be well-suited for patients with large left hemisphere lesions whose only chance to recover is through recruitment of the right hemisphere. The observed brain changes after treatment indicate that MIT's engagement of predominantly right hemispheric brain regions (including the superior temporal region, primary sensorimotor and premotor cortices, as well as the IFG) and the connections between these regions (mainly through the AF) accounts for its therapeutic effect.

Acknowledgment

The author gratefully acknowledges support from NIH (1RO1 DC008796, 3R01DC008796-02S1, R01 DC009823-01), the Grammy Foundation, the Suzanne and Tom McManmon Family Foundation, the Rosalyn and Richard Slifka Family Foundation, and the Matina R. Proctor Foundation.

References

Abrams, D. A., Nicol, T., Zecker, S., & Kraus, N. (2008). Right-hemisphere auditory cortex is dominant for coding syllable patterns in speech. *The Journal of Neuroscience: The Official Journal of the Society for Neuroscience*, *28*, 3958–3965.

Albert, M. L., Sparks, R. W., & Helm, N. A. (1973). Melodic intonation therapy for aphasia. *Archives of Neurology*, *29*, 130–131.

Alexander, A. L., Lee, J. E., Lazar, M., & Field, A. S. (2007). Diffusion tensor imaging of the brain. *Neurotherapeutics*, *4*, 316–329.

Al-Janabi, S., Nickels, L. A., Sowman, P. F., Burianova, H., Merrett, D. L., & Thompson, W. F. (2014). Augmenting melodic intonation therapy with non-invasive brain stimulation to treat impaired left-hemisphere function: Two case studies. *Frontiers in Psychology*, *5*, 37.

Belin, P., Van Eeckhout, P., Zilbovicius, M., Remy, P., Francois, C., et al. (1996). Recovery from nonfluent aphasia after melodic intonation therapy: A PET study. *Neurology*, *47*, 1504–1511.

Benjamin, M. L., Towler, S., Garcia, A., Park, H., Sudhyadhom, A., et al. (2014). A behavioral manipulation engages right frontal cortex during aphasia therapy. *Neurorehabilitation and Neural Repair*, *28*, 545–553.

Bhogal, S. K., Teasell, R., & Speechley, M. (2003). Intensity of aphasia therapy, impact on recovery. *Stroke: A Journal of Cerebral Circulation, 34*, 987–993.

Bohland, J. W., & Guenther, F. H. (2006). An fMRI investigation of syllable sequence production. *NeuroImage, 32*, 821–841.

Bonakdarpour, B., Eftekharzadeh, A., & Ashayeri, H. (2000). Preliminary report on the effects of melodic intonation therapy in the rehabilitation of Persion aphasic patients. *Iranian Journal of Medical Sciences, 25*, 156–160.

Boucher, V., Garcia, L. J., Fleurant, J., & Paradis, J. (2001). Variable efficacy of rhythm and tone in melody-based interventions: Implications for the assumption of a right-hemisphere facilitation in nonfluent aphasia. *Aphasiology, 15*, 131–149.

Broca, P. (1961). Remarks on the seat of the faculty of articulated language, following an observation of aphasia (loss of speech). *Bulletin de la Societe Anatomique, 6*, 330–357. (Translation by Christopher D. Green).

Brown, S., Martinez, M. J., Hodges, D. A., Fox, P. T., & Parsons, L. M. (2004). The song system of the human brain. *Brain Research. Cognitive Brain Research, 20*, 363–375.

Carmichael, S. T. (2006). Cellular and molecular mechanisms of neural repair after stroke: Making waves. *Annals of Neurology, 59*, 735–742.

Cohen, N. S., & Masse, R. (1993). The application of singing and rhythmic instruction as a therapeutic intervention for persons with neurogenic communication disorders. *Journal of Music Therapy, 30*, 81–99.

Dancause, N., Barbay, S., Frost, S. B., Plautz, E. J., Chen, D., et al. (2005). Extensive cortical rewiring after brain injury. *The Journal of Neuroscience: The Official Journal of the Society for Neuroscience, 25*, 10167–10179.

Engelter, S. T., Gostynski, M., Papa, S., Frei, M., Born, C., et al. (2006). Epidemiology of aphasia attributable to first ischemic stroke: Incidence, severity, fluency, etiology, and thrombolysis. *Stroke: A Journal of Cerebral Circulation, 37*, 1379–1384.

Fields, R. D. (2008). White matter in learning, cognition and psychiatric disorders. *Trends in Neurosciences, 31*, 361–370.

Fridriksson, J., Guo, D., Fillmore, P., Holland, A., & Rorden, C. (2013). Damage to the anterior arcuate fasciculus predicts nonfluent speech production in aphasia. *Brain: A Journal of Neurology, 136*, 3451–3460.

Gentilucci, M., Benuzzi, F., Bertolani, L., Daprati, E., & Gangitano, M. (2000). Language and motor control. *Experimental Brain Research, 133*, 468–490.

Gerstman, H. L. (1964). A case of aphasia. *The Journal of Speech and Hearing Disorders, 29*, 89–91.

Geschwind, N. (1971). Current concepts: Aphasia. *The New England Journal of Medicine, 284*, 654–656.

Glasser, M. F., & Rilling, J. K. (2008). DTI tractography of the human brain's language pathways. *Cerebral Cortex, 18*, 2471–2482.

Goodglass, H., & Kaplan, E. (1983). *Boston diagnostic aphasia examination* (2nd ed.). Philadelphia, PA: Lea & Febiger.

Guenther, F. H., Hampson, M., & Johnson, D. (1998). A theoretical investigation of reference frames for the planning of speech movements. *Psychological Review, 105*, 611–633.

Hebert, S., Racette, A., Gagnon, L., & Peretz, I. (2003). Revisiting the dissociation between singing and speaking in expressive aphasia. *Brain: A Journal of Neurology, 126*, 1838–1850.

Helm-Estabrooks, N., & Albert, M. L. (1991). *Manual of aphasia therapy*. Austin, TX: Pro-Ed.

Helm-Estabrooks, N., Nicholas, M., & Morgan, A. (1989). *Melodic intonation therapy*. Austin, TX: Pro-Ed.

Jeffries, K. J., Fritz, J. B., & Braun, A. R. (2003). Words in melody: An H(2)15O PET study of brain activation during singing and speaking. *Neuroreport, 14*, 749–754.

Jito, J., Nakasu, S., Ito, R., Fukami, T., Morikawa, S., & Inubushi, T. (2008). Maturational changes in diffusion anisotropy in the rat corpus callosum: Comparison with quantitative histological evaluation. *Journal of Magnetic Resonance Imaging, 28*, 847–854.

Keith, R. L., & Aronson, A. E. (1975). Singing as therapy for apraxia of speech and aphasia: Report of a case. *Brain and Language, 2*, 483–488.

Kertesz, A., Lesk, D., & McCabe, P. (1977). Isotope localization of infarcts in aphasia. *Archives of Neurology, 34*, 590–601.

Kertesz, A., & McCabe, P. (1977). Recovery patterns and prognosis in aphasia. *Brain: A Journal of Neurology, 100*(Pt 1), 1–18.

Kinsella, G., Prior, M. R., & Murray, G. (1988). Singing ability after right and left sided brain damage. A research note. *Cortex; A Journal Devoted to the Study of the Nervous System and Behavior, 24*, 165–169.

Koelsch, S., Fritz, T., Schulze, K., Alsop, D., & Schlaug, G. (2005). Adults and children processing music: An fMRI study. *NeuroImage, 25*, 1068–1076.

Koelsch, S., Gunter, T. C., v Cramon, D. Y., Zysset, S., Lohmann, G., & Friederici, A. D. (2002). Bach speaks: A cortical "language-network" serves the processing of music. *NeuroImage, 17*, 956–966.

Lahav, A., Saltzman, E., & Schlaug, G. (2007). Action representation of sound: Audiomotor recognition network while listening to newly acquired actions. *The Journal of Neuroscience: The Official Journal of the Society for Neuroscience, 27*, 308–314.

Marchina, S., Zhu, L. L., Norton, A., Zipse, L., Wan, C. Y., & Schlaug, G. (2011). Impairment of speech production predicted by lesion load of the left arcuate fasciculus. *Stroke: A Journal of Cerebral Circulation, 42*, 2251–2256.

Meister, I. G., Boroojerdi, B., Foltys, H., Sparing, R., Huber, W., & Topper, R. (2003). Motor cortex hand area and speech: Implications for the development of language. *Neuropsychologia, 41*, 401–406.

Meister, I. G., Buelte, D., Staedtgen, M., Boroojerdi, B., & Sparing, R. (2009). The dorsal premotor cortex orchestrates concurrent speech and fingertapping movements. *The European Journal of Neuroscience, 29*, 2074–2082.

Meyer, M., Alter, K., Friederici, A. D., Lohmann, G., & von Cramon, D. Y. (2002). FMRI reveals brain regions mediating slow prosodic modulations in spoken sentences. *Human Brain Mapping, 17*, 73–88.

Moss, A., & Nicholas, M. (2006). Language rehabilitation in chronic aphasia and time postonset: A review of single-subject data. *Stroke: A Journal of Cerebral Circulation, 37*, 3043–3051.

Norton, A., Zipse, L., Marchina, S., & Schlaug, G. (2009). Melodic intonation therapy: Shared insights on how it is done and why it might help. *Annals of the New York Academy of Sciences, 1169*, 431–436.

Ozdemir, E., Norton, A., & Schlaug, G. (2006). Shared and distinct neural correlates of singing and speaking. *NeuroImage, 33*, 628–635.

Patel, A. D. (2003). Language, music, syntax and the brain. *Nature Neuroscience, 6*, 674–681.

Pedersen, P. M., Jorgensen, H. S., Nakayama, H., Raaschou, H. O., & Olsen, T. S. (1995). Aphasia in acute stroke: Incidence, determinants, and recovery. *Annals of Neurology, 38*, 659–666.

Pedersen, P. M., Vinter, K., & Olsen, T. S. (2004). Aphasia after stroke: Type, severity and prognosis. The Copenhagen aphasia study. *Cerebrovascular Diseases, 17*, 35–43.

Peretz, I. (1990). Processing of local and global musical information by unilateral brain-damaged patients. *Brain: A Journal of Neurology, 113*(Pt 4), 1185–1205.

Peretz, I., & Coltheart, M. (2003). Modularity of music processing. *Nature Neuroscience, 6*, 688–691.

Poeppel, D. (2003). The analysis of speech in different temporal integration windows: Cerebral lateralization as "asymmetric sampling in time". *Speech Communication, 41*(1), 245–255.

Racette, A., Bard, C., & Peretz, I. (2006). Making non-fluent aphasics speak: Sing along!. *Brain: A Journal of Neurology, 129*, 2571–2584.
Robey, R. R. (1998). A meta-analysis of clinical outcomes in the treatment of aphasia. *Journal of Speech, Language, and Hearing Research, 41*, 172–187.
Rose, M. L. (2013). Releasing the constraints on aphasia therapy: The positive impact of gesture and multimodality treatments. *American Journal of Speech-Language Pathology/American Speech-Language-Hearing Association, 22*, S227–S239.
Schlaug, G., Altenmuller, E., & Thaut, M. (2010). Music listening and music making in the treatment of neurological disorders and impairments. *Music Perception, 27*, 249–250.
Schlaug, G., Forgeard, M., Zhu, L., Norton, A., & Winner, E. (2009). Training-induced neuroplasticity in young children. *Annals of the New York Academy of Sciences, 1169*, 205–208.
Schlaug, G., Marchina, S., & Norton, A. (2008). From singing to speaking: Why patients with Broca's aphasia can sing and how that may lead to recovery of expressive language functions. *Music Perception, 25*, 315–323.
Schlaug, G., Marchina, S., & Norton, A. (2009). Evidence for plasticity in white-matter tracts of patients with chronic Broca's aphasia undergoing intense intonation-based speech therapy. *Annals of the New York Academy of Sciences, 1169*, 385–394.
Schuppert, M., Munte, T. F., Wieringa, B. M., & Altenmuller, E. (2000). Receptive amusia: Evidence for cross-hemispheric neural networks underlying music processing strategies. *Brain: A Journal of Neurology, 123*(Pt 3), 546–559.
Sparks, R., Helm, N., & Albert, M. (1974). Aphasia rehabilitation resulting from melodic intonation therapy. *Cortex: A Journal Devoted to the Study of the Nervous System and Behavior, 10*, 303–316.
Sparks, R. W., & Holland, A. L. (1976). Method: Melodic intonation therapy for aphasia. *The Journal of Speech and Hearing Disorders, 41*, 287–297.
Stahl, B., Kotz, S. A., Henseler, I., Turner, R., & Geyer, S. (2011). Rhythm in disguise: Why singing may not hold the key to recovery from aphasia. *Brain: A Journal of Neurology, 134*, 3083–3093.
Thaut, M. H., & Abiru, M. (2010). Rhythmic auditory stimulation in rehabilitation of movement disoders: A review of current research. *Music Perception, 27*, 263–269.
Thaut, M. H., Kenyon, G. P., Schauer, M. L., & McIntosh, G. C. (1999). The connection between rhythmicity and brain function. *IEEE Engineering in Medicine and Biology Magazine, 18*, 101–108.
Thompson, C. K., & Shapiro, L. P. (2007). Complexity in treatment of syntactic deficits. *American Journal of Speech-Language Pathology/American Speech-Language-Hearing Association, 16*, 30–42.
Tokimura, H., Tokimura, Y., Oliviero, A., Asakura, T., & Rothwell, J. C. (1996). Speech-induced changes in corticospinal excitability. *Annals of Neurology, 40*, 628–634.
Uozumi, T., Tamagawa, A., Hashimoto, T., & Tsuji, S. (2004). Motor hand representation in cortical area 44. *Neurology, 62*, 757–761.
van der Meulen, I., van de Sandt-Koenderman, W. M., Heijenbrok-Kal, M. H., Visch-Brink, E. G., & Ribbers, G. M. (2014). The efficacy and timing of melodic intonation therapy in subacute aphasia. *Neurorehabilitation and Neural Repair, 28*, 536–544.
Vines, B. W., Norton, A. C., & Schlaug, G. (2011). Non-invasive brain stimulation enhances the effects of melodic intonation therapy. *Frontiers in Auditory Cognitive Neuroscience, 2*, 230.
Vorisek, I., & Sykova, E. (1997). Evolution of anisotropic diffusion in the developing rat corpus callosum. *Journal of Neurophysiology, 78*, 912–919.
Wade, D. T., Hewer, R. L., David, R. M., & Enderby, P. M. (1986). Aphasia after stroke: Natural history and associated deficits. *Journal of Neurology, Neurosurgery, and Psychiatry, 49*, 11–16.
Wan, C. Y., Rüber, T., Hohmann, A., & Schlaug, G. (2010). The therapeutic effects of singing in neurological disorders. *Music Perception, 27*, 287–295.
Wan, C. Y., Zheng, X., Marchina, S., Norton, A., & Schlaug, G. (2014). Intensive therapy induces contralateral white matter changes in chronic stroke patients with Broca's aphasia. *Brain and Language, 136*, 1–7.
Wang, J., Marchina, S., Norton, A. C., Wan, C. Y., & Schlaug, G. (2013). Predicting speech fluency and naming abilities in aphasic patients. *Frontiers in Human Neuroscience, 7*, 831.
Wilson, S. J., Parsons, K., & Reutens, D. C. (2006). Preserved singing in aphasia: A case study of the efficacy of the Melodic intonation therapy. *Music Perception, 24*, 23–36.
Yamadori, A., Osumi, Y., Masuhara, S., & Okubo, M. (1977). Preservation of singing in Broca's aphasia. *Journal of Neurology, Neurosurgery, and Psychiatry, 40*, 221–224.
Zatorre, R. J., & Belin, P. (2001). Spectral and temporal processing in human auditory cortex. *Cerebral Cortex, 11*, 946–953.
Zatorre, R. J., & Gandour, J. T. (2008). Neural specializations for speech and pitch: Moving beyond the dichotomies. *Philosophical Transactions of the Royal Society of London. Series B, Biological Sciences, 363*, 1087–1104.
Zipse, L., Norton, A., Marchina, S., & Schlaug, G. (2012). When right is all that is left: Plasticity of right-hemisphere tracts in a young aphasic patient. *Annals of the New York Academy of Sciences, 1252*, 237–245.

C H A P T E R

82

Constraint-Induced Aphasia Therapy: A Neuroscience-Centered Translational Method

Friedemann Pulvermüller[1,2], *Bettina Mohr*[3] *and Edward Taub*[4]

[1]Brain Language Laboratory, Department of Philosophy and Humanities, Freie Universität Berlin, Berlin, Germany; [2]Graduate School of Mind and Brain, Humboldt Universität zu Berlin, Berlin, Germany; [3]Department of Psychiatry, Campus Benjamin Franklin, Charité Universitätsmedizin, Berlin, Germany; [4]Department of Psychology, University of Alabama at Birmingham, Birmingham, AL, USA

82.1 APHASIA THERAPY: RELEVANCE AND CLASSIC PARADIGMS

Aphasia is the most common language deficit attributable to neurological disease. It is most commonly caused by stroke. Approximately 1 out of 2,000 individuals suffer from poststroke aphasia (PSA), with a much higher prevalence in the elderly population (Dickey et al., 2010; Law et al., 2009). During the first year after stroke, during the acute and subacute stages, there is still a good chance that PSA will "spontaneously" improve. In the chronic phase (>1 year after onset), only minimal, if any, spontaneous improvement occurs (Kertesz, 1984). The traditional view was that language therapy delivered during the chronic stage does not lead to significant improvements, or that the evidence for such improvement is weak (Basso, Capitani, & Vignolo, 1979; Greener, Enderby, Whurr, & Grant, 1998; Holland & Wertz, 1986; Lincoln et al., 1984; Wertz et al., 1986). However, in 2001, a controlled clinical trial showed that this belief was incorrect and that patients with chronic PSA can improve their language performance over a short therapy interval of only 2 weeks (Pulvermüller et al., 2001). The main focus of this therapeutic approach was practicing behaviorally relevant language in an intensive way and administering language therapy tailored to the patients' preserved abilities and communication needs. The new method, called Constraint-Induced Aphasia Therapy (CIAT), had been developed on the basis of theory and data in neuroscience (Taub, 1977, 1980) and in the neuroscience of language (Pulvermüller & Berthier, 2008). The theories and neuroscience data behind this new therapeutic approach are introduced, the basic principles and methods of CIAT are addressed, and a summary of data concerning its efficacy is presented. Finally, two extensive methodological revisions of the original CIAT protocol, namely CIAT II (Johnson et al., 2014) and intensive language action therapy (ILAT) (DiFrancesco, Pulvermüller, & Mohr, 2012; Pulvermüller & Berthier, 2008) are described and their supporting preliminary data are discussed.

82.2 NEUROSCIENCE AND LANGUAGE EVIDENCE

82.2.1 Language Structure and Function in Aphasia Rehabilitation

CIAT was developed on the basis of two related methods of neurological rehabilitation. One is Constraint-Induced Movement Therapy (CIMT) (Taub et al., 1993; Taub, Uswatte, & Elbert, 2002; Taub, Uswatte, Mark, & Morris, 2006) and the other is pragmatic-linguistic or communicative aphasia therapy (CAT) (Aten, Caligiuri, & Holland, 1982; Davis & Wilcox, 1985; Pulvermüller & Berthier, 2008; Pulvermüller & Roth, 1991). These two therapeutic approaches target different neurological syndromes; CIMT is for limb paresis and CAT for language deficits. However, the two approaches have significant features in common. Most importantly, they put emphasis on the *behavioral relevance* of the actions performed during therapy. In CIMT, motor action relevant in

Neurobiology of Language. DOI: http://dx.doi.org/10.1016/B978-0-12-407794-2.00082-1

everyday life, such as pouring water from a pitcher into a glass, are targeted, although elementary finger movements may be practiced in preparation for the behaviorally relevant action (Taub, Uswatte, Bowman, et al., 2013). In CAT, linguistic utterances are used in action contexts where they carry similar functions as in everyday life, for example, requesting water from a communication partner; however, word and sentence repetition may be used as a springboard for such communicative actions (DiFrancesco et al., 2012; Pulvermüller & Schönle, 1993). This pragmatic aphasia therapy contrasts with other approaches to language therapy, where the emphasis is on the production or comprehension of linguistic forms (sounds, words, sentences, texts) rather than on the communicative functions these utterances carry in dialogues. The reasons why behavioral relevance and, in the specific case of language, communicative function play an important role is addressed. General findings in basic neuroscience research along with recent data from a large number of neuroimaging studies contributed to the further development and conceptual grounding of CIAT. In recent years, studies involving healthy participants as well as brain-damaged individuals demonstrate that the human brain's language and action systems are functionally linked and interdependent (Pulvermüller, 2005; Pulvermüller & Fadiga, 2010) and therefore cannot be viewed as having encapsulated functions, as once postulated by modularistic approaches (Newmeyer, 1988). For example, lesions in motor or sensory systems may lead to impairment in the semantic domain, for example, in processing words with specific action-related meaning (Bak, 2013; Kemmerer, Rudrauf, Manzel, & Tranel, 2012). It could even be demonstrated that stimulating the motor system facilitates specific language functions in healthy people (D'Ausilio et al., 2009; Pulvermüller, Hauk, Nikulin, & Ilmoniemi, 2005). These results not only show that the brain's language and action systems are functionally interwoven, as depicted in Figure 82.1, but also suggest that lesions to one of these systems may, in part, be compensated by supportive activity from the functionally linked system. In addition, brain imaging studies with healthy participants confirm that when the same utterences are used for *naming* exercises and for communicative speech acts such as *requesting*, the cortical motor system is activated by behaviorally relevant requesting, but not by naming (Egorova, Pulvermüller, & Shtyrov, 2014; Egorova, Shtyrov, & Pulvermüller, 2013). These studies show that everyday speech acts such as requesting engage both language and motor systems and, therefore, their putative synergistic between-systems links. From these and similar data, it seems evident that practicing *behaviorally relevant*, action-embedded language in social and communicative contexts in everyday life is important in aphasia therapy.

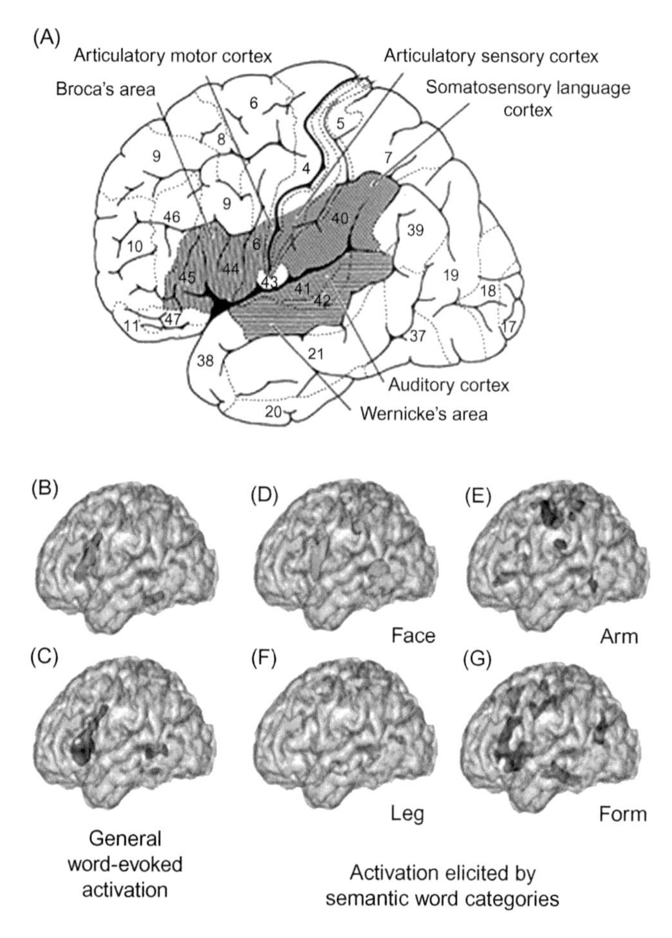

FIGURE 82.1 Brain areas involved in language processing. (A) The perisylvian language cortex, defined as the first convolution surrounding the Sylvian fissure, is indicated on a lateral view of the brain (shaded area). Broca's area and Wernicke's area are indicated along with the somatosensory language area. Lesions in left perisylvian areas typically cause aphasia. (B)–(G) Language processing involves a wide range of brain areas in addition to the left perisylvian areas. (B) and (C) Silently reading words activates the left inferior frontal and superior temporal areas and the left inferior temporal and fusiform gyri. Reading specific semantic word categories activates additional specific bilateral brain areas. (D)–(F) Action words related to the face, arm, and leg, such as "lick," "pick," and "kick," activate the motor system in a somatotopic manner, together with specific foci in temporal cortex. (G) Some semantic word categories, for example, form words such as "square," excite a wide range of fronto-parieto-temporal foci. Brain activity related to the processing of semantic information has been reported in both cortical hemispheres. *Adapted with permission from John Wiley and Sons, Inc. © Pulvermüller, Kherif, Hauk, Mohr, & Nimmo-Smith (2009).*

82.2.2 Guidance by Constraints

Originally, the CIMT model was thought to be applicable to PSA because of an apparent similarity in characteristics of the expressive deficit in aphasia and the motor deficit of the more affected upper extremity after stroke. CIMT is said to be efficacious for improving the motor deficit of the extremities, partly because it overcomes a learned inhibition of limb use, termed "learned nonuse," that develops in the acute period after substantial neurological injury takes place. When coordination of the movement of a limb is very poor, as is typical soon after injury, attempts to use that limb fail and, in behavioral terms, are thereby punished, reducing the frequency of future attempts to use that limb. Strategies to accomplish tasks that are successful, such as using the limb on the other side of the body or asking a family member to complete tasks for the patient, are rewarded and become more frequent. This learning process results, some time after injury, in greatly reduced real-world use of an extremity compared with the motor ability that the extremity can be shown to have in the laboratory by a test during which patients are requested to make their best possible effort. For example, a patient with hemiparesis primarily affecting the right arm would typically use the left arm for most activities as a compensatory strategy. Research involving animals and humans has shown that avoidance of use of the extremity affected by neurological disease can markedly reduce functional motor ability after central nervous system damage (Taub, Ellman, & Berman, 1966; Taub, Perrella, & Barro, 1973; Taub, Perrella, Miller, & Barro, 1975); (Taub, 1977, 1980; Taub et al., 2006). Based on these data, one of the key features of CIMT is the guidance of patients by constraints that are administered to overcome *learned nonuse*. One way to use constraints to target the negative effects of *learned nonuse* is to constrain the patient's unaffected arm by a sling or other restraining device so that the patient is "forced" to use the impaired arm in everyday activities as much as possible (Taub et al., 1993). Another way that is even more important for adult humans is intensive training of the more affected arm, especially by the behavioral technique termed shaping. Persistent use of the affected limb encouraged by these two techniques markedly reduced poststroke motor deficits. In fact, in the chronic stages after a stroke, dramatic improvements in purposive movements have been documented after intensive (several hours per day) CIMT (Taub et al., 1993; Taub, Uswatte, King, et al., 2006). These motor improvements have been shown to be accompanied by functional (Liepert et al., 2000) and structural changes in the brain (Gauthier et al., 2008; Sterling et al., 2013), especially in the sensorimotor cortices.

In a similar way, *learned nonuse* would appear to also occur when there is language impairment after neurological injury. Because of halting and slow verbal production and incomplete understanding, speech becomes effortful and often unsuccessful. The person compensates by remaining silent, by using gestures as a nonverbal means of communication, or by restricting himself/herself to words and (a)grammatical constructions easily available to them (Kolk & Heeschen, 1990). Moreover, caregivers often "speak for" the person with aphasia to facilitate communication, thereby further suppressing the remaining verbal abilities of the patients (Croteau & Le Dorze, 2006; Croteau, Vychytil, Larfeuil, & Le Dorze, 2004). Given the likely similarity of the mechanism underlying part of the deficit of both the extremities and language, it seemed reasonable to attempt to improve speech after damage to the central nervous system (CNS) by developing a treatment for language parallel in as many respects as possible to CIMT. This was the conceptual basis of CIAT and its modifications (CIAT II and ILAT), which are described in more detail in this chapter.

In conventional aphasia therapy, it is standard practice to choose materials appropriate for the patient undergoing treatment. For example, in the naming context, pictures and language materials can be tailored to the needs of the patients. These constrain the patients' use of language in a drill-like exercise far from the context of behaviorally relevant communicative interactions. As already noted, the constraints in CIMT are given in contexts that are behaviorally relevant. The methods of CIAT, which are explained here, allow for constraining and guiding patients' language actions, exploiting their invididual abilities and addressing their specific needs in social-interactive communication while at the same time providing language therapy in social-communicative contexts similar to everyday language use. To this end, so-called *language games* are used (Pulvermüller, 1990; Pulvermüller & Roth, 1991). Language games are schemas for communicative interaction defined by common goals using linguistic and nonlinguistic materials, including materials depicting action sequences. The concept of language games is derived from the work of the philosopher Ludwig Wittgenstein, who used them to illustrate the intrinsic relationship of language and action and, in the philosophic context, for semantic and conceptual analysis (Wittgenstein, 1953). In the context of language games, language carries its normal function as a tool in interaction—for example, when a builder asks a partner to bring him building blocks as part of an interactive building process. In CIAT, forms of interaction similar to Wittgenstein's language games are used to practice language use and communication with PSA patients. In this case, the communicative constraints—or guidance tools—include the game context and goals (language is used to obtain objects or object pictures from partners), the objects, and object pictures themselves (further features of the interaction setting are explained in this chapter). Another important feature of language games is that linguistic actions can be practiced repeatedly. Use of language games in the aphasia therapy context was adapted from CAT (Pulvermüller et al., 2001; Pulvermüller & Roth, 1991; Pulvermüller & Schönle, 1993). Note that language games serve as models of behaviorally relevant everyday language interactions in which language is embedded into action contexts—both overt body actions (such as handing over something) and linguistic actions or speech acts (such as requesting or rejecting a request).

In essence, CIAT seeks to *constrain and guide* patients' language use by communicative settings that make it possible to tailor therapy to the individual's level of ability and communicative needs. For CIAT, it is essential that such guidance be made available in interactions that resemble communicative interactions in daily living. Moreover, this guidance is done by shaping, that is, by successive approximations of the desired behavior or communicative interaction. Even small improvements of language functions are continously positively reinforced. In this way, language (re-)learning follows a behavioral scheme similar to the one used in CIMT.

82.2.3 Therapy Intensity and Frequency

One of the key features of CIAT is its high intensity or frequency. It has been demonstrated that the same amount of aphasia rehabilitation delivered in a short time period results in better outcomes than the same amount of therapy stretched over a longer period (Pulvermüller et al., 2001). In CIMT and CIAT, the standard is to provide therapy at least 3 h/day for at least 2 weeks or more. In the language therapy context, it has also been argued that "more helps more," that is, independent of type of therapy, the frequency of the therapeutic exercises determines its success (Bhogal, Teasell, & Speechley, 2003). The principles of *maximal amount (in hours overall) and especially that of maximal frequency of therapy (in hours per day)* can be justified on the basis of well-known Hebbian learning principles (Kempter, Gerstner, & Van Hemmen, 1999; Tsumoto, 1992). Assuming that the controlled environment of language therapy can support correlations between linguistic utterances and their appropriate contexts, the Hebbian principle "what fires together wires together" predicts that the crucial neuronal connections between linguistic and context representations strengthen. If therapy is applied with high frequency, then the time between consecutive therapy sessions is minimized. Between therapy, language contexts are not controlled and patients may therefore fail more frequently, so that the correlation between linguistic form and context is low. The anti-Hebb rule that "neurons out of sync delink" implies synaptic weakening of crucial language-context connections if linguistic signs appear in inappropriate contexts. Therefore, it is advantageous to provide the controlled environment of language therapy for as long as possible and to minimize the intervening time, thus resulting in high therapy frequency. The principles of *massed practice and high therapy frequency* are therefore important key features of CIAT.

82.3 CIAT: METHODS AND EFFICACY

82.3.1 CIAT Methods

CIAT uses interactive language game targeting requests. Requests or, more generally, directive speech acts represent one of the most common types of language use (Searle, 1979). Evidently, asking for assistance, asking for objects, or asking for actions by others are even more important for neurological patients than for healthy individuals. In natural communication, people make requests, for example, by asking for a beverage in a coffee shop, for apples in a grocery store, or for a specific dish to be served to them in a restaurant. Requests call for one out of a set of specific response actions by communication partners, for example, the delivery of the requested object, the rejection of the request ("We are out of tea, sorry"), or a request for clarification ("What did you say?") (Fritz & Gloning, 1992).

These different options to act are built into the therapeutic language game (Figure 82.2). A therapist and a

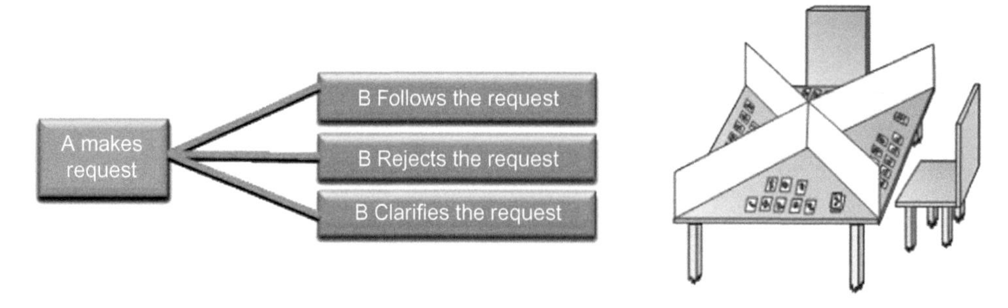

FIGURE 82.2 CIAT for chronic PSA. (Left) Action sequence structure of therapeutic language game used for CIAT. Participants are trained to make verbal requests and respond to requests by passing object cards or by rejecting or clarifying the request. (Right) Typical setting for the CIAT game. Barriers prevent the participants from seeing each other's cards, prompting them to communicate verbally. *Adapted with permission from Taylor & Francis Group © Pulvermüller and Berthier (2008).*

small group of patients sit around a table, with each having a set of picture cards depicting objects. Players are only permitted to see their own items; to this end, barriers are erected between players. Two copies of each item—object or picture card—are in the game, so that players can assume that any card in front of them has a copy that belongs to one of the other game participants. The aim of the game is to obtain items from the other participants by making verbal requests. A round of the game starts with one of the players (requester) directing a request for the twin of one of the cards in his/her hand to any of the other players (responders). If the selected target item is the picture of an apple, then the requester's utterance can be "apple," or "please give me an apple," or "might you possibly have a big red apple for me please," or perhaps "red round fruit," or even "a thing—well—sweet, red." The other player (responder 1) can then respond by handing over the requested object/picture. This requires responder 1 to have understood the request and to have the requested card/item. If this is not the case, then responder 1 would reject the request ("Sorry, no apple") or initiate a repair sequence ("I did not understand"). The requester can then address his/her request to another player (responder 2). The task of responder 2 is to help the requester by asking, "Do you mean yes or no," or by asking alternative questions referring to his/her own card set ("Would you like a pear?" "an apple?" "a knife?"). The round of the game is completed when one of the responders picks up the specific card in response to an appropriate request from the requester. Matching cards are then compared and shown to all players in the game to give feedback about communicative success. If the two cards are mismatched, then the round fails and the requester and responder take back their cards. In any case, it is the last responder's turn to make a new request (for more details, see Difrancesco et al., 2012).

Typically, the game is played with four players, one therapist and three aphasic patients, but it is also possible to have one additional co-therapist present who is not a player but who assists patients when necessary (DiFrancesco et al., 2012; Pulvermüller et al., 2001). If severely aphasic patients participate in therapy, then it might be better to decrease group size—or to work on a one-to-one basis (Pulvermüller & Schönle, 1993). However, it has been reported that including patients with global aphasia in CIAT groups with mixed performance levels can work well (Sickert, Anders, Münte, & Sailer, 2014). The therapist who is an active participant in the game acts as a model for the aphasic participants, shapes the actions of the patients, defines specific rules for each player, positively reinforces patients for making successsful linguistic actions, and selects and hands out card sets. The last three tasks can also be undertaken by a nonparticipating co-therapist.

Constraints are imposed by the game, for example, players cannot see the cards of the other players, nor can they see or easily make any manual gestures that communicate relevant information. In this way, verbal communication becomes the only method for achieving success. The therapist also successively shapes the patients' actions toward using spoken language by explicitly encouraging speech and discouraging nonverbal gestures. Still, manual gestures are only allowed as a means of communication when they are used to accompany appropriate verbal output (DiFrancesco et al., 2012). Further constraints or guidance are achieved by the design of the materials. A progression of difficulty levels in the game materials may increase the phonological, syntactic, lexical-semantic, or pragmatic complexity of the request that must be made. For example, we normally start using items that can be named by high-frequency words and then proceed to items with low word frequency, which are more difficult to process. Moreover, items from dissimilar semantic categories (e.g., animals and fruits) are used initially. With improved language use as the therapy progresses, items from the same semantic domain are used, which are more difficult to differentiate verbally (e.g., different mammals). Pragmatically, the game can first be a simple two-step process (request followed by handing over), progressing toward more complex interaction schemas. The rules and materials of the game are adjusted to the level of deficit of the aphasic players. If performance levels markedly differ between participants, then different rules can be established for different players within the same game. For example, a severely affected player may have few constraints placed on their performance, whereas better performers may be required to make requests using politeness formulas or whole sentences.

Problems in playing the linguistic card game may be encountered by including patients with severe attention or memory deficits, apraxia, visual agnosia, or other severe cognitive deficits. These problems therefore may be used as exclusion criteria. However, we recommend having prospective patients engage in a short practice game of low complexity to assess their ability to participate (for further details of the CIAT methods, please see Berthier, Green, Juárez, Lara, & Pulvermüller, 2014; DiFrancesco et al., 2012; Neininger, 2002; Pulvermüller & Berthier, 2008).

Part of the card set used in the first CIAT trial is available on the Internet (http://www.ub.uni-konstanz.de/kops/volltexte/2002/879/), and a Spanish version is now published in book form, together with a complete set of pictures (Berthier et al., 2014).

82.3.2 Evidence: Is CIAT Effective?

Randomized controlled trials (RCTs) are the strongest form of evidence. The first RCT of CIAT was performed in 2001 (Pulvermüller et al., 2001). CIAT was delivered for 2 weeks for 3 h/week to patients with mild to moderate PSA who had a poststroke onset mean of 6 years. A control group received the same amount of conventional therapy distributed over several weeks. The CIAT group demonstrated significant improvements on clinical language tests and a communication questionnaire that elicited information about the use of language in the life situation termed the *communicative activity log* (CAL) (Pulvermüller et al., 2001); the control group did not show these improvements. The difference between the two groups was also significant. Meinzer, Djundja, Barthel, Elbert, and Rockstroh (2005) performed a further RCT comparing two versions of CIAT, the original and a version including training of written language. A similar improvement was found in both groups. Moreover, the improvement persisted over 6 months, indicating long-term benefits of CIAT. Berthier et al. (2009) performed a placebo-controlled RCT comparing CIAT therapy alone with CIAT plus *memantine*, a competitive dopamine inhibitor. The significant improvement after CIAT was found to be further amplified by the administration of memantine (Figure 82.3). Further studies involving single cases and small groups have provided additional support for the efficacy of CIAT, and some of these have suggested that it might be more efficacious than other methods of speech-language therapy (Breier et al., 2009; Breier, Maher, Schmadeke, Hasan, & Papanicolaou, 2007; DiFrancesco et al., 2012; Kirmess & Maher, 2010; Kurland, Pulvermüller, Silva, Burke, & Andrianopoulos, 2012; Maher et al., 2006; Meinzer, Streiftau, & Rockstroh, 2007; Richter, Miltner, & Straube, 2008; Szaflarski et al., 2008).

Cochrane reviews have concluded that aphasia therapy is effective during the chronic stage (Brady, Kelly, Godwin, & Enderby, 2012; Kelly, Brady, & Enderby, 2010). Important evidence on which this conclusion was based is derived from the aforementioned RCTs on CIAT. However, CIAT is frequently described *only* as an intensive method of providing therapy. This ignores two of its other distinctive features, *behavioral relevance* and *guidance/constraint*. Whether the *high-frequency* principle alone can be sufficient for therapy success has been asked in a recent study in which single patients were treated with a traditional utterance-based approach in an intensive fashion comparable with CIAT (Barthel, Meinzer, Djundja, & Rockstroh, 2008). This study reports improvements comparable with those reported after CIAT, although the generality of the conclusions is limited because of the lack of a control group and the resultant possibility of a bias in patient selection. Furthermore, training was between a therapist and one patient only, so that therapy intensity per patient was greater in this study than in a typical CIAT group therapy. A recent RCT with early subacute PSA patients compared group therapy with CIAT delivered with somewhat reduced intensity (2 h/day) with an utterance-centered approach and found similar improvements with both methods (Sickert et al., 2014). Although the authors suggest that therapy intensity is an important factor and that other methodological differences might not be relevant, it is not possible to determine whether the improvements seen in their

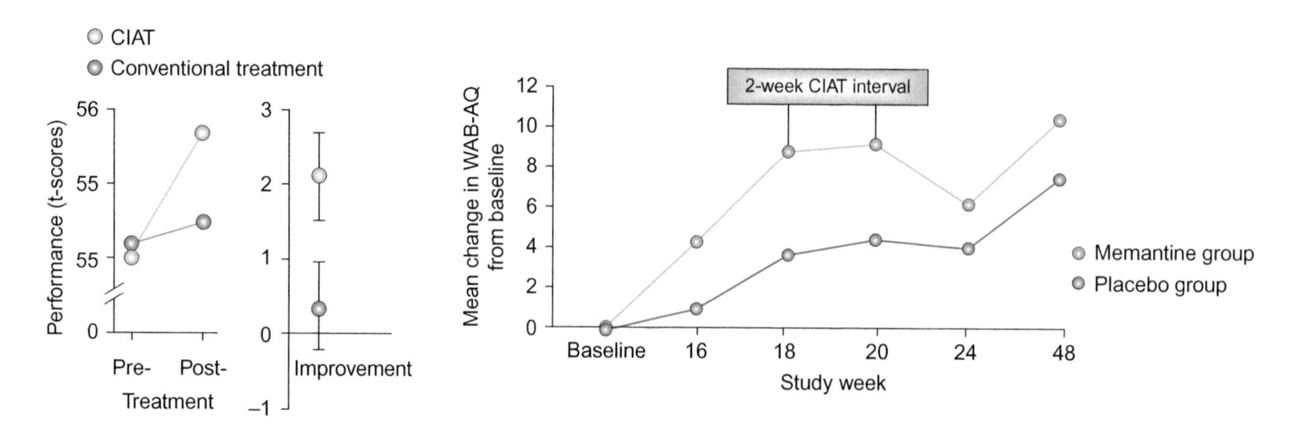

FIGURE 82.3 Efficiency of CIAT. (Left) Language performance before and after 2 weeks of CIAT (30 h) compared with same amount of conventional PSA therapy. (Right) Effects of memantine, CIAT, and concomitant memantine and CIAT. Drug therapy is indicated by line color (blue, on drug; red, placebo; gray, off drug). The results show stronger improvement with CIAT than with conventional treatment, synergistic effects of drug treatment and CIAT, and long-term stability of CIAT, but no long-term stability for drug treatment. CIAT, constraint-induced aphasia therapy; WAB-AQ, Western Aphasia Battery Aphasia Quotient. *Adapted with permission from American Heart Association © Pulvermüller et al. (2001) and from John Wiley and Sons, Inc. © Berthier et al. (2009).*

study were a result of therapy *per se*, because the patients were in the subacute phase; therefore, improvements could have been attributable to spontaneous recovery independent of the treatment. Previous studies of the effect of pragmatic and CAT suggest that behavioral relevance alone can lead to a significant improvement in language performance in chronic PSA (Pulvermüller & Roth, 1991; Pulvermüller & Schönle, 1993), although no RCT is available to strongly support this suggestion.

82.4 VARIANTS AND EXTENSIONS OF ORIGINAL CIAT METHODS

The basic CIAT design has been modified in a number of studies. For example, the application of the standard CIAT methods is sometimes difficult in clinical settings for organizational reasons. Therefore, CIAT methods were applied with slightly reduced frequency, and the CIAT methods were also taught to communication partners of patients (Sickert et al., 2014). The unexpected outcome was that lay-person–led training yielded improvements that were similar to those of CIAT administered by professional speech therapists (Meinzer et al., 2007). There have also been attempts to include a wider spectrum of therapy methods in the basic CIAT protocol. In one study, training of written language use was included (Meinzer et al., 2005). Finally, the supplementary use of additional methods that might have a beneficial effect on language recovery was studied, for example, the application of pharmacological or magnetic and electric stimulation of the brain during the CIAT period. As mentioned, use of drugs can produce synergistic effects enhancing the improvements brought about by CIAT (Berthier et al., 2014; Berthier & Pulvermüller, 2011; Berthier, Pulvermuller, Davila, Casares, & Gutierrez, 2011).

82.4.1 ILAT: Widening the Scope of Communicative Actions and Linguistic Materials

ILAT is not just a new term introduced to highlight two (intensity and action-embedding) of the three key principles of CIAT (as opposed to only one highlighted by CIAT, i.e., constraint). The new label ILAT also indexes a move toward a broader spectrum of social-communicative actions, interaction schemes, and linguistic materials targeted (Berthier & Pulvermüller, 2011; Pulvermüller & Berthier, 2008). In the original form of CIAT, participants always ask for objects or picture cards and, although the requests practiced are of eminent importance in everyday life, there is a wide range of social-communciative speech acts that are also relevant but outside the scope of the original method. Likewise, linguistically, CIAT's request game focuses on the use of nouns for objects, with other volcabulary types being somewhat underrepresented. New materials have been developed to also allow systematic coverage of adjectives (Berthier et al., 2014), but verbs and function words are typically difficult to integrate. ILAT overcomes these limitations by offering additional language games that now focus on further important linguistic actions such as planning, bargaining, making a proposal, rejecting or supporting a proposal, and arguing the pros and cons of a given possiblity to act. For example, in the "joint planning" game (DiFrancesco et al., 2012; Pulvermüller, 1990), participants use action pictures and the aim is to propose and agree on an action to be performed collectively by the participants. This new method makes it possible to specifically target action verbs and function words relevant for argumentation (because, therefore, thus, since...) and therefore helps constrain therapy toward word types specifically impaired in some subtypes of PSA (Kemmerer et al., 2012; Miceli, Silveri, Villa, & Caramazza, 1984). Concordant with the broadening of communicative methods, ILAT materials have been extended by newly developed card materials covering a variety of individual or group actions ranging from leisure to work-related and household-related activities (Berthier et al., 2014; DiFrancesco et al., 2012). In a recent clinical trial, ILAT, including both request and planning games, was administered for 10 consecutive weekdays and led to significant improvements of language functions in a group of 14 patients with chronic mildly to moderately impaired PSA (DiFrancesco et al., 2012).

82.4.2 CIAT II: A Revision of the Original CIAT Protocol Bringing It Closer to the CIMT Model

The original CIAT protocol was a partial translation of the methods applied in CIMT (Taub et al., 1994; Taub, Uswatte, King, et al., 2006). Although CIAT produces significantly larger real-world improvements of language functions than conventional speech and language therapies, its outcome seems smaller compared with the results reported in CIMT studies. Therefore, as a further development of CIAT, the initial protocol was modified to more closely resemble the methods used in CIMT. This new protocol is termed CIAT II (Johnson et al., 2014). Modifications involved addition of new exercises, including a final exercise in which everyday verbal interactions were simulated and modeled. There was increased emphasis on the shaping of responses and the primary caregiver was trained as an alternate therapist, with training beginning in the laboratory but focused largely on the at-home practice of verbal behavior. In addition, a "transfer package" (TP) parallel to that used in CIMT was introduced (Taub, Uswatte, King, et al., 2006; Taub, Uswatte, Mark, et al., 2013). The latter was probably the most important change in the CIAT protocol. The TP is a set of techniques designed to facilitate transfer of therapeutic gains from the treatment setting to life situations. The TP techniques consisted of behavioral contracts with the patient and caregiver to perform specified activities using only speech and not gestures or other means of communication, monitoring the amount of out-of-laboratory speech by daily maintenance of a verbal behavioral diary, daily administration of a structured questionnaire (Verbal Activity Log [VAL]) derived from the original CAL, problem-solving to circumvent apparent barriers to participation in speech in the life situation, home practice exercises, periodic phone contact after the end of treatment, and involvement of a caregiver in all phases of the treatment both in the laboratory and during home practice during formal treatment and after its end. These behavioral techniques, singly and in combination, have been used extensively for the treatment of a number of clinical conditions, including, for example, the outpatient treatment of cocaine dependence (Higgins, Budney, & Bickel, 1997), autism, and adherence to behavior modification training for parents of children with behavior problems (Eyberg & Johnson, 1974), and the control of obesity, smoking, and alcoholism.

To date, six patients have been treated with CIAT II. All patients showed significant improvements in the use of speech during life situations (measured by VAL) (Johnson et al., 2014). VAL data recorded an approximate 300% increase in spontaneous speech in the life situation compared with 30% reported by previous CIAT studies using the Communicative Aphasia Log. Data from the first study with CIAT II are presented in Figure 82.4. Interestingly, further improvements of verbal activity scores were found during the 6 months after the completion of treatment. This increase would appear to be attributable to the continuation of training by the caregivers in the real-world environment, a procedure that is part of CIAT II.

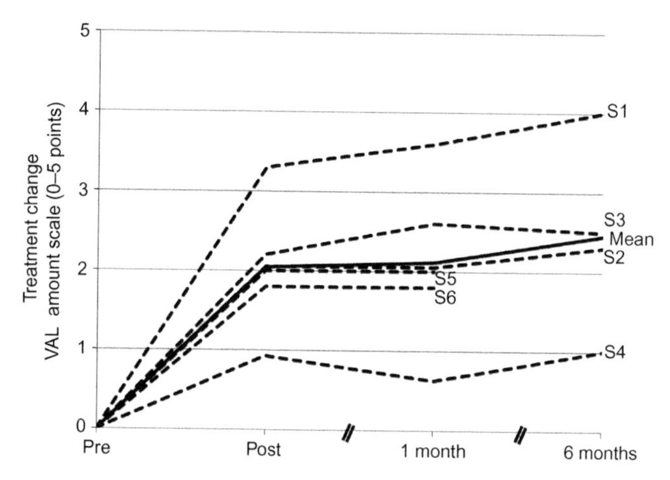

FIGURE 82.4 Changes in amount of participation in speech in life situations after CIAT II. Each line indexes results from one subject. The mean gain from pretreatment on the VAL Amount scale was significant after treatment. After treatment, there was an additional significant performance gain. Data in the graph are ipsitized (i.e., pretreatment scores are set to 0 for each participant and subsequent scores are reported as changes from pretreatment). *Modified from Johnson et al. (2014).*

82.5 SUMMARY AND OUTLOOK

The development of CIAT and the consistently positive outcome in a large number of studies published during the past decade demonstrate that this therapy approach is succssful in the treatment of PSA. Studies have shown that chronic and possibly also subacute patients greatly benefit from this therapy method, and that patients with a broad spectrum of symptoms can participate in the treatment and show significant improvements of language functions. The short duration and treatment intensity of CIAT, combined with its strong behavioral relevance in everyday communication, seem to be the factors that lead to the beneficial effects. However, it is not yet entirely clear which of these factors contribute most because they have been used in combination in CIAT. New developments of the original CIAT protocol include increased practice of behaviorally relevant language in everyday life, as applied in CIAT II, which is focused on translation of therapeutic gains achieved in the clinic to performance of spontaneous speech in everyday life activities. In ILAT, a greater focus on language–action relationships and a broadening of the repertoire of everyday communicative speech acts and vocabulary types has been implemented. As a next step, the positive preliminary findings of CIAT II and ILAT need to be confirmed in randomized controlled clinical trials with a larger number of aphasia patients. Moreover, it seems most promising to combine the newly developed techniques of CIAT II and ILAT to maximize treatment effects and transfer of newly acquired language skills in relevant daily communicative contexts.

References

Aten, J. L., Caligiuri, M. P., & Holland, A. L. (1982). The efficacy of functional communication therapy for chronic aphasic patients. *Journal of Speech and Hearing Disorders*, *47*(1), 93–96.

Bak, T. H. (2013). The neuroscience of action semantics in neurodegenerative brain diseases. *Current Opinion in Neurology*, *26*, 671–677.

Barthel, G., Meinzer, M., Djundja, D., & Rockstroh, B. (2008). Intensive language therapy in chronic aphasia: Which aspects contribute most? *Aphasiology*, *22*(4), 408–421.

Basso, A., Capitani, E., & Vignolo, L. A. (1979). Influence of rehabilitation on language skills in aphasic patients: A controlled study. *Archives of Neurology, 36*, 190–196.

Berthier, M. L., García-Casares, N., Walsh, S. F., Nabrozidis, A., Ruíz de Mier, R. J., Green, C., et al. (2014). Recovery from post-stroke aphasia: Lessons from brain imaging and implications for rehabilitation and biological treatments. *Discovery Medicine, 12*(65), 275–289.

Berthier, M. L., Green, C., Juárez, R., Lara, J. P., & Pulvermüller, F. (2014). *REGIA: Rehabilitación grupal intensiva de la afasia.* Madrid: TEA Ediciones, S.A.

Berthier, M. L., Green, C., Lara, J. P., Higueras, C., Barbancho, M. A., Davila, G., et al. (2009). Memantine and constraint-induced aphasia therapy in chronic poststroke aphasia. *Annals of neurology, 65* (5), 577–585.

Berthier, M. L., & Pulvermüller, F. (2011). Neuroscience insights improve neurorehabilitation of post-stroke aphasia. *Nature Reviews Neurology, 7*(2), 86–97.

Berthier, M. L., Pulvermuller, F., Davila, G., Casares, N. G., & Gutierrez, A. (2011). Drug therapy of post-stroke aphasia: A review of current evidence. *Neuropsychology Review, 21*(3), 302–317. Available from: http://dx.doi.org/10.1007/s11065-011-9177-7.

Bhogal, S. K., Teasell, R., & Speechley, M. (2003). Intensity of aphasia therapy, impact on recovery. *Stroke, 34*(4), 987–993.

Brady, M. C., Kelly, H., Godwin, J., & Enderby, P. (2012). Speech and language therapy for aphasia following stroke. *Cochrane Database of Systematic Reviews (Online), 5*, CD000425. Available from: http://dx.doi.org/10.1002/14651858.CD000425.pub3.

Breier, J. I., Juranek, J., Maher, L. M., Schmadeke, S., Men, D., & Papanicolaou, A. C. (2009). Behavioral and neurophysiologic response to therapy for chronic aphasia. *Archives of Physical Medicine and Rehabilitation, 90*(12), 2026–2033.

Breier, J. I., Maher, L. M., Schmadeke, S., Hasan, K. M., & Papanicolaou, A. C. (2007). Changes in language-specific brain activation after therapy for aphasia using magnetoencephalography: A case study. *Neurocase, 13*(3), 169–177.

Croteau, C., & Le Dorze, G. (2006). Overprotection, "speaking for," and conversational participation: A study of couples with aphasia. *Aphasiology, 20*, 327–336.

Croteau, C., Vychytil, A., Larfeuil, C., & Le Dorze, G. (2004). "Speaking for" behaviours in spouses of people with aphasia: A descriptive study of six couples in an interview situation. *Aphasiology, 18*, 291–312.

D'Ausilio, A., Pulvermüller, F., Salmas, P., Bufalari, I., Begliomini, C., & Fadiga, L. (2009). The motor somatotopy of speech perception. *Current Biology, 19*(5), 381–385.

Davis, G. A., & Wilcox, M. J. (1985). *Adult aphasia rehabilitation: Applied pragmatics.* San Diego, CA: College-Hill Press.

Dickey, L., Kagan, A., Lindsay, M. P., Fang, J., Rowland, A., & Black, S. (2010). Incidence and profile of inpatient stroke-induced aphasia in Ontario, Canada. *Archives of Physical Medicine and Rehabilitation, 91* (2), 196–202.

DiFrancesco, S., Pulvermüller, F., & Mohr, B. (2012). Intensive language action therapy: The methods. *Aphasiology, 26*(11), 1317–1351.

Egorova, N., Pulvermüller, F., & Shtyrov, Y. (2014). Neural dynamics of speech act comprehension: An MEG study of naming and requesting. *Brain Topography, 27*, 375–392.

Egorova, N., Shtyrov, Y., & Pulvermüller, F. (2013). Early and parallel processing of pragmatic and semantic information in speech acts: Neurophysiological evidence. *Frontiers in Human Neuroscience, 7*(86), 1–13. Available from: http://dx.doi.org/10.3389/fnhum.2013.00086.

Eyberg, S. M., & Johnson, S. M. (1974). Multiple assessment of behavior modification with families: Effects of contingency contracting and order of treated problems. *Journal of Consulting and Clinical Psychology, 42*, 594–606.

Fritz, G., & Gloning, T. (1992). Principles of linguistic communication analysis. In S. Stati, & E. Weigand (Eds.), *Methodologie der Dialoganalyse* (pp. 41–56). Tübingen: Max Niemeyer Verlag.

Gauthier, L., Taub, E., Perkins, C., Ortmann, M., Mark, V., & Uswatte, G. (2008). Remodeling the brain: Plastic structural brain changes produced by different motor therapies after stroke. *Stroke, 39*, 1520–1525. Available from: http://dx.doi.org/10.1161/STROKEAHA.107.502229.

Greener, J., Enderby, P., Whurr, R., & Grant, A. (1998). Treatment for aphasia following stroke: Evidence for effectiveness. *International Journal of Language and Communication Disorders, 33*(Suppl.), 158–161.

Higgins, S. T., Budney, A. J., & Bickel, W. K. (1997). Outpatient behavioral treatment for cocaine dependence: One-year outcome. In G. A. Marlatt, & G. R. VandenBos (Eds.), *Addictive behaviors: Readings on etiology, prevention, and treatment* (pp. 629–645). Washington, DC: American Psychological Association.

Holland, A. L., & Wertz, R. T. (1986). Measuring aphasia treatment effects—Case-studies and group studies. *Archives of Neurology, 43* (11), 1207.

Johnson, M., Taub, E., Harper, L. H., Wade, J. T., Bowman, M. H., Bishop-McKay, S., et al. (2014). An enhanced protocol for constraint-induced aphasia therapy II: A case series. *American Journal of Speech-Language Pathology, 23*, 60–72. Available from: http://dx.doi.org/10.1044/1058-0360(2013/12-0168).

Kelly, H., Brady, M. C., & Enderby, P. (2010). Speech and language therapy for aphasia following stroke. *Cochrane Database of Systematic Reviews (Online), 5*, CD000425.

Kemmerer, D., Rudrauf, D., Manzel, K., & Tranel, D. (2012). Behavioural patterns and lesion sites associated with impaired processing of lexical and conceptual knowledge of action. *Cortex, 48*(7), 826–848. Available from: http://dx.doi.org/10.1016/j.cortex.2010.11.001.

Kempter, R., Gerstner, W., & Van Hemmen, J. L. (1999). Hebbian learning and spiking neurons. *Physical Review E, 59*(4), 4498.

Kertesz, A. (1984). Recovery from aphasia. *Advances in Neurology, 42*, 23–39.

Kirmess, M., & Maher, L. M. (2010). Constraint induced language therapy in early aphasia rehabilitation. *Aphasiology, 24*(6–8), 725–736.

Kolk, H. H. J., & Heeschen, C. (1990). Adaptation symptoms and impairment symptoms in Broca's aphasia. *Aphasiology, 4*, 221–231.

Kurland, J., Pulvermüller, F., Silva, N., Burke, K., & Andrianopoulos, M. (2012). Constrained vs. unconstrained intensive language therapy in two individuals with chronic, moderate-to-severe aphasia and apraxia of speech: Behavioral and fMRI outcomes. *American Journal of Speech-Language Pathology, 21*(S), 65–87. Available from: http://dx.doi.org/10.1044/1058-0360(2012/11-0113).

Law, J., Huby, G., Irving, A. M., Pringle, A. M., Conochie, D., Haworth, C., et al. (2009). Reconciling the perspective of practitioner and service user: Findings from the aphasia in Scotland study. *International Journal of Language and Communication Disorders, 45*, 551–560.

Liepert, J., Bauder, H., Wolfgang, H. R., Miltner, W. H., Taub, E., & Weiller, C. (2000). Treatment-induced cortical reorganization after stroke in humans. *Stroke, 31*(6), 1210–1216.

Lincoln, N. B., McGuirk, E., Muller, G. P., Lendrem, W., Jones, A. C., & Mitchell, J. R. A. (1984). Effectiveness of speech therapy for aphasic stroke patients: A randomized controlled trial. *Lancet, 1*, 1197–1200.

Maher, L. M., Kendall, D., Swearengin, J. A., Rodriguez, A., Leon, S. A., Pingel, K., et al. (2006). A pilot study of use-dependent learning in the context of constraint induced language therapy. *Journal of the International Neuropsychological Society, 12*(6), 843–852.

Meinzer, M., Djundja, D., Barthel, G., Elbert, T., & Rockstroh, B. (2005). Long-term stability of improved language functions in chronic aphasia after constraint-induced aphasia therapy. *Stroke, 36*(7), 1462–1466.

Meinzer, M., Streiftau, S., & Rockstroh, B. (2007). Intensive language training in the rehabilitation of chronic aphasia: Efficient training by laypersons. *Journal of the International Neuropsychological Society, 13*(5), 846–853.

Miceli, G., Silveri, M., Villa, G., & Caramazza, A. (1984). On the basis of agrammatics' difficulty in producing main verbs. *Cortex, 20*, 207–220.

Neininger, B. (2002). *Sprachverarbeitung außerhalb der klassischen Sprachzentren [Speech processing outside the traditional core language areas]* (PhD Dr. rer.nat.). Konstanz: University of Konstanz. Retrieved from: http://www.ub.uni-konstanz.de/kops/volltexte/2002/879/.

Newmeyer, F. (Ed.), (1988). *Linguistics: The Cambridge Survey* Cambridge: Cambridge University Press.

Pulvermüller, F. (1990). *Aphasische Kommunikation. Grundfragen ihrer Analyse und Therapie.* Tübingen: Gunter Narr Verlag.

Pulvermüller, F. (2005). Brain mechanisms linking language and action. *Nature Reviews Neuroscience, 6*(7), 576–582.

Pulvermüller, F., Kherif, F., Hauk, O., Mohr, B., & Nimmo-Smith, I. (2009). Cortical cell assemblies for lexical and category-specific semantic processing as revealed by fMRI cluster analysis. *Human Brain Mapping, 30*(12), 3837–3850.

Pulvermüller, F., & Berthier, M. L. (2008). Aphasia therapy on a neuroscience basis. *Aphasiology, 22*(6), 563–599.

Pulvermüller, F., & Fadiga, L. (2010). Active perception: Sensorimotor circuits as a cortical basis for language. *Nature Reviews Neuroscience, 11*(5), 351–360.

Pulvermüller, F., Hauk, O., Nikulin, V., & Ilmoniemi, R. J. (2005). Functional interaction of language and action: a TMS study. *European Journal of Neuroscience, 21*(3), 793–797.

Pulvermüller, F., Neininger, B., Elbert, T., Mohr, B., Rockstroh, B., Koebbel, P., et al. (2001). Constraint-induced therapy of chronic aphasia following stroke. *Stroke, 32*(7), 1621–1626.

Pulvermüller, F., & Roth, V. M. (1991). Communicative aphasia treatment as a further development of PACE therapy. *Aphasiology, 5*, 39–50.

Pulvermüller, F., & Schönle, P. W. (1993). Behavioral and neuronal changes during treatment of mixed-transcortical aphasia: A case study. *Cognition, 48*, 139–161.

Richter, M., Miltner, W. H., & Straube, T. (2008). Association between therapy outcome and right-hemispheric activation in chronic aphasia. *Brain, 131*(Pt 5), 1391–1401.

Searle, J. R. (1979). *Expression and meaning*. Cambidge: Cambridge University Press.

Sickert, A., Anders, L.-C., Münte, T. F., & Sailer, M. (2014). Constraint-induced aphasia therapy following sub-acute stroke: A single-blind, randomised clinical trial of a modified therapy schedule. *Journal of Neurology, Neurosurgery and Psychiatry, 85*(1), 51–55.

Sterling, C., Taub, E., Davis, D., Rickards, T., Gauthier, L., Griffin, A., et al. (2013). Structural neuroplastic change following constraint-induced movement therapy in children with cerebral palsy. *Pediatrics, 131*, e1664–e1669.

Szaflarski, J. P., Ball, A., Grether, S., Al-Fwaress, F., Griffith, N. M., Neils-Strunjas, J., et al. (2008). Constraint-induced aphasia therapy stimulates language recovery in patients with chronic aphasia after ischemic stroke. *Medical Science Monitor: International Medical Journal of Experimental and Clinical Research, 14*(5), CR243–CR250.

Taub, E. (1977). Movement in nonhuman primates deprived of somatosensory feedback. *Exercise and Sports Science Reviews, 4*, 335–374. Available from: http://dx.doi.org/10.1249/00003677-197600040-00012.

Taub, E. (1980). Somatosensory deafferentation research with monkeys: Implications for rehabilitation medicine. In L. P. Ince (Ed.), *Behavioral psychology in rehabilitation medicine: Clinical applications* (pp. 371–401). New York, NY: Williams & Wilkins.

Taub, E., Crago, J., Burgio, L., Groomes, T., Cook, E. W., DeLuca, S., et al. (1994). An operant approach to overcoming learned nonuse after CNS damage in monkeys and man: The role of shaping. *Journal of the Experimental Analysis of Behavior, 61*, 281–293. Available from: http://dx.doi.org/10.1901/jeab.1994.61-281.

Taub, E., Ellman, S. J., & Berman, A. J. (1966). Deafferentation in monkeys: Effect on conditioned grasp response. *Science, 151*(710), 593–594.

Taub, E., Miller, N. E., Novack, T. A., Cook, E. W., III., Fleming, W. C., Nepomuceno, C. S., et al. (1993). Technique to improve chronic motor deficit after stroke. *Archives of Physical Medicine and Rehabilitation, 74*, 347–354.

Taub, E., Perrella, P. N., & Barro, G. (1973). Behavioral development following forelimb deafferentation on day of birth in monkeys with and without blinding. *Science, 181*(103), 959–960.

Taub, E., Perrella, P. N., Miller, E. A., & Barro, G. (1975). Diminution of early environmental control through perinatal and prenatal somatosensory deafferentation. *Biological Psychiatry, 10*(6), 609–626.

Taub, E., Uswatte, G., Bowman, M., Mark, V., Delgado, A., Bryson, C., et al. (2013). Constraint-induced movement therapy combined with conventional neurorehabilitation techniques in chronic stroke patients with plegic hands: A case series. *Archives of Physical Medicine and Rehabilitation, 94*(1), 86–94. Available from: http://dx.doi.org/10.1016/j.apmr.2012.07.029.

Taub, E., Uswatte, G., & Elbert, T. (2002). New treatments in neurorehabilitation founded on basic research. *Nature Reviews Neuroscience, 3*(3), 228–236.

Taub, E., Uswatte, G., King, D., Morris, D., Crago, J., & Chatterjee, A. (2006). A placebo-controlled trial of constraint-induced movement therapy for upper extremity after stroke. *Stroke, 37*(4) 1045–1049.

Taub, E., Uswatte, G., Mark, V., Morris, D., Barman, J., Bowman, M., et al. (2013). Method for enhancing real-world use of a more affected arm in chronic stroke: Transfer package of constraint-induced movement therapy. *Stroke, 44*, 1383–1388.

Taub, E., Uswatte, G., Mark, V. W., & Morris, D. (2006). The learned nonuse phenomenon: Implications for rehabilitation. *Europa Medicophysica, 42*(2), 241–255.

Tsumoto, T. (1992). Long-term potentiation and long-term depression in the neocortex. *Progress in Neurobiology, 39*, 209–228.

Wertz, R., Weiss, D. G., Aten, J. L., Brookshire, R. H., Garcia-Bunuel, L., Holland, A. L., et al. (1986). Comparison of clinic, home, and deferred language treatment for aphasia: A veterans administration cooperative study. *Archives of Neurology, 43*, 653–658.

Wittgenstein, L. (1953). *Philosophical investigations.* Oxford: Blackwell Publishers.

CHAPTER

83

Noninvasive Brain Stimulation in Aphasia Therapy: Lessons from TMS and tDCS

H. Branch Coslett

Department of Neurology, Perelman School of Medicine at the University of Pennsylvania, Philadelphia, PA, USA

83.1 INTRODUCTION

Disorders of language are quite common in the context of brain dysfunction. It is estimated that approximately 33% of individuals who experience a left hemisphere stroke develop aphasia (e.g., Dickey et al., 2010). Impairments of language are also observed in traumatic brain injury as well as degenerative brain disorders, some of which are characterized by profound language deficits in the absence of generalized dementia.

Although speech-language therapy (SLT) is of proven benefit (Kelly, Brady, & Enderby, 2010; Robey & Dalebout, 1998), remediation of aphasia has proven to be remarkably difficult. There are many potential reasons for this. First, variability in the severity and nature of an individual's language deficits may have important implications for the type of therapy that will prove most beneficial; at present, however, there is little information regarding the most effective mapping between type of SLT and aphasic performance. Second, there is considerable intersubject variability in factors such as the site and size of the lesion as well as other more difficult to quantify factors such as degree of atrophy, white matter burden, and number of lesions that are likely to be important determinants of recovery and response to therapy. Third, individual differences in genetic and developmental factors that underlie differences in the degree of lateralization of language and the capacity for reorganization after brain injury (one aspect of brain plasticity) are likely to significantly influence language recovery. Finally, access to SLT is limited for many people with aphasia. As conventionally delivered, SLT is resource-intensive; it has been estimated that maximal benefit for SLT requires 100 hours of therapy, whereas the average patient in the United States receives 30 hours of therapy.

In light of the urgent need for more effective forms of therapy, there has been substantial interest over the past decade in the use of noninvasive brain stimulation as a therapeutic intervention, often in combination with conventional SLT. This effort parallels the use of noninvasive brain stimulation as therapy for a wide range of disorders of brain function, including hemiparesis (Ayache et al., 2012; Khedr, Ahmed, Fathy, & Rothwell, 2005; Khedr et al., 2013), pain (Pérez-Borrego et al., 2014), and neglect (Koch et al., 2012; Mylius et al., 2012; Sunwoo et al., 2013; see Coslett & Hamilton, 2011). In this chapter we review recent work exploiting two types of noninvasive brain stimulation—transcranial magnetic stimulation (TMS) and transcranial direct current stimulation (tDCS)—that have shown promise in the treatment of aphasia. TMS and tDCS have been used in normal subjects to address questions regarding the functional architecture. This literature is outside the scope of this chapter; interested readers are referred to recent reviews of TMS (Torres, Drebing, & Hamilton, 2013) and tDCS (e.g., Flöel et al., 2012) studies of normal language.

83.2 TMS AS TREATMENT FOR APHASIA

83.2.1 Basics of TMS

TMS is a procedure in which a brief surge of current through a coil held near the skull induces a magnetic field that penetrates the skull and soft tissue of the head and ultimately generates a small electrical current in the brain (see Jahanshahi & Rothwell, 2000; Miniussi

Neurobiology of Language. DOI: http://dx.doi.org/10.1016/B978-0-12-407794-2.00091-2

et al., 2008). The electrical current in the brain causes neurons in a circumscribed region of the brain to depolarize, thereby generating action potentials.

The effects of TMS are dependent on a number of parameters including stimulus intensity, the frequency with which the pulses are applied, the configuration of the coil, and the distance between the coil and the cortex. Stimulus intensity is most frequently expressed as a percentage of maximal machine output that is required to generate a measurable motor response from an intrinsic hand muscle (e.g., the first dorsal interosseus); this value is called the motor threshold (MT). Depending on the stimulus intensity, individual characteristics of the subject (e.g., skull thickness), and coil design, a TMS pulse is assumed to depolarize approximately 1 cm^3 of brain tissue. It is important to note that although TMS directly influences only a small region of cortex, depolarization of neurons in one brain region may alter, by virtue of direct axonal connections to distant brain regions, function in multiple distal sites in the brain (George, Wassermann, & Post, 1996).

As the strength of the magnetic field induced by the current drops off sharply as a function of distance from the coil, TMS is most useful for stimulating cortex that is close to the skull. TMS may be administered as a single pulse or as a series of pulses, a paradigm termed repetitive TMS (rTMS). Whereas a single pulse produces only a very brief effect (e.g., 100 ms), rTMS produces effects that persist beyond the period of stimulation. With rTMS the frequency of the pulses determines the effect of the train of pulses. Low-frequency rTMS, usually administered at 1 Hz, causes a suppression of the activity of the underlying brain activity that has been likened to a "virtual lesion." Stimulation at higher frequencies (typically >5 Hz) is associated with enhancement of brain function. Finally, theta-burst TMS (TBS) (Huang & Rothwell, 2004) is a pulse sequence in which brief trains of 50 Hz (theta frequency) pulses are delivered with variable interpulse intervals; "continuous" TBS is thought, like 1 Hz TMS, to be "inhibitory," whereas "intermittent" TBS, like high-frequency rTMS, is thought to be "excitatory."

The effects of rTMS are transient (typically minutes to an hour in the motor system) when administered in a single session; crucial for the present discussion, with appropriate stimulus parameters, a series of TMS sessions may induce alterations in brain function that far outlast the administration of the stimulation (Pascual-Leone et al., 1998). The use of TMS as a therapy for aphasia builds on this crucial observation.

83.2.2 TMS as Therapy for Aphasia

To date there have been 22 reports including more than 200 subjects describing the results of TMS as a therapy for aphasia from stroke. Additionally, there is some literature describing beneficial effects of TMS for treatment of aphasia in the context of degenerative disorders of the brain such as Alzheimer's disease (Cotelli et al., 2011; Cotelli, Manenti, Cappa, Zanetti, & Miniussi, 2008). This chapter focuses on the use of TMS in the treatment of aphasia resulting from stroke.

The first reports of therapeutic effects of TMS in aphasic subjects were provided by Naeser and colleagues (Martin et al., 2004, 2009; Naeser et al., 2005a, 2005b); because these investigations not only provided the initial evidence that TMS can be beneficial in this context but also have served as the model for most subsequent studies, the basic approach of Naeser and colleagues is described in some detail (see also Naeser et al., 2012).

The rationale for Naeser et al.'s approach and for much of the work with tDCS comes from accounts from the motor domain that attribute aphasia, at least in part, to transcallosal inhibition of the lesioned hemisphere by the contralesional hemisphere (Heiss & Thiel, 2006; Van der Knaap & van der Ham, 2011). Naeser et al. (2005b, 2012) reasoned that if chronic aphasia is characterized by an inhibition of left hemisphere language structures by an overactive right hemisphere, suppression of the right hemisphere would permit the left hemisphere language structures to function optimally, thereby improving language performance. To that end, Naeser and colleagues administered 1 Hz repetitive TMS, a procedure thought to inhibit underlying cortex, to the right frontal lobe. As they had no strong *apriori* hypothesis regarding the optimal site to stimulate, they used a "site finding" procedure (see also Hamilton et al., 2010; Medina et al., 2012) in which multiple sites in the RH including the mouth region of M1, pars triangularis, pars opercularis, and pars orbitalis were stimulated, in turn, to identify a location at which TMS would improve picture naming. Those subjects for whom a site at which 10 minutes of 1 Hz TMS lead to significant improvement in naming could be identified were subsequently treated with 20 minutes of 1 Hz TMS at 90% of motor threshold for 10 sessions over 2 weeks.

In a series of studies involving a small number of subjects, Naeser and colleagues (Martin et al., 2004; Klein et al., 2007; Naeser et al., 2004, 2005a, 2005b) demonstrated consistent benefit in naming from 10 sessions of 1 Hz TMS to the right pars triangularis. Additionally, we and others have demonstrated significant improvement in multiple measures of language function (e.g., Hamilton et al., 2010) in controlled studies using a partial cross-over design with sham stimulation (Medina et al., 2013) or vertex stimulation (Barwood et al., 2013). These positive findings provided the impetus for a number of additional studies of the role of TMS in aphasia therapy that have used a wide range of experimental designs. In the following sections, we briefly review the more recent studies involving TMS as a therapy for aphasia.

83.2.3 Blinded, Controlled Studies in Subacute and Chronic Aphasia

As is typically the case when new therapies are under development, initial studies of TMS were "proof of principle" investigations in which small groups of people with chronic aphasia were enrolled and control conditions were not used (see Table 83.1 for details). There is clear evidence from these studies that TMS may benefit subjects with chronic, nonfluent aphasia. To determine if therapy is of more general benefit, however, blinded studies with adequate controls are needed. Since 2011 there have been nine such studies, six of which included 10 or more subjects. Six of the studies used sham stimulation (Cotelli et al., 2011; Khedr et al., 2014; Kindler et al., 2012; Medina et al., 2012; Seniów et al., 2013; Waldowski, Seniów, Leśniak, Iwański, & Członkowska, 2012), in some instances using a cross-over design so that all subjects eventually received active TMS; in three studies (Abo et al., 2012; Barwood et al., 2013; Thiel et al., 2013), subjects were randomized to RIFG or vertex stimulation.

Four studies enrolled people with chronic aphasia; three of these (Barwood et al., 2012; Medina et al., 2012; Thiel et al., 2013) involving a total of 46 subjects used a paradigm modeled on that of Naeser and colleagues: 1-Hz stimulation at 90% MT to the right IFG for several weeks. The three studies demonstrated a consistent and significant beneficial effect across a variety of tasks, including naming, picture description and a variety of other measures included in aphasia batteries such as the Aachen Aphasia Test. Although there is variability with respect to the magnitude of the benefit, many of the significant improvements were in the range of 20–30% relative to the pretest baseline. As in previous reports, benefit was observed at the final testing period, varying between 6 and 10 months after treatment (Barwood et al., 2011; Medina et al., 2012). As is discussed, in many instances the improvement noted after therapy increased over the course of the follow-up period.

There have been four studies of subacute stroke (Khedr et al., 2014; Seniów et al., 2013; Waldowski et al., 2012; Weiduschat et al., 2011) involving a total of 78 subjects.[1] Two of the larger studies, Seniów et al. (2013) with 38 subjects and Waldowski et al. (2012) with 26 subjects, demonstrated no main effect of group; significant benefit was observed by Weiduschat et al. (2011) and Khedr et al. (2014). Subgroup analyses reported by Seniów et al. (2013) demonstrated that nonfluent and more severe cases derived the greatest benefit.

83.2.4 Variables Relevant to the Response to TMS

Although a number of studies have used the approach pioneered by Naeser and colleagues, multiple other approaches have been tried. An overall assessment of the efficacy of TMS therapy for aphasia is complicated by substantial variability between investigations with respect to a wide range of factors including aphasia type, aphasia chronicity, site of stimulation, TMS stimulation parameters, and the use of speech therapy in conjunction with TMS. We review the effects of these variables before attempting to synthesize the lessons from studies to date.

83.2.4.1 Effect of Aphasia Subtype, Severity, and Lesion Location

Most studies have not offered precise characterizations of the type of aphasia exhibited by participants. Most commonly, participants have been described as "fluent" or "nonfluent," designations that offer little insight into the nature of the language processing deficits exhibited by the participants. Many studies included nonfluent subjects exclusively, but several more recent studies have included subjects with different types of aphasia (e.g., Abo et al., 2012; Khedr et al., 2014; Kindler et al., 2012; Seniów et al., 2013; Szaflarski et al., 2011; Thiel et al., 2013; Waldowski et al., 2012; Weiduschat et al., 2011). Only one small study of two subjects (Kakuda, Abo, Uruma, Kaito, & Watanabe, 2010) has been restricted to people with chronic "sensory" aphasia.

Studies including people with nonfluent aphasia have, with one exception (Kakuda, Abo, Kaito, Watanabe, & Senoo, 2010), demonstrated improvement in language function. The exception was a small study of four subjects that differed from most other studies not only with respect to subject characteristics but also in that the stimulated site was selected on the basis of fMRI findings; therefore, it differed across subjects. Furthermore, Waldowski et al. (2012) and Seniów et al. (2013) demonstrated greater improvement in people with nonfluent aphasia as compared to people with fluent aphasia. Thus, there is considerable evidence that TMS benefits subjects with nonfluent aphasia, whereas evidence for benefit in fluent subjects is less

[1]Note that the subjects reported by Waldowski et al. are included in the later Seniow et al. manuscript.

TABLE 83.1 Initial Proof of Priniciple Studies for TMS Treatment of Aphasia.

Author	Subjects	Time from stroke	Stimulated areas	Stimulus parameters	Number of sessions	Study design	Task	Speech therapy	Time of evaluation	Language effect
Martin et al., 2004	6 nonfluent	Chronic	R IFG	1 Hz, 20 min, 90% MT	10	No control group/ condition	Picture naming	No	2 months	Significant improvement
Naeser et al., 2005	1 nonfluent	Chronic	R IFG	1 Hz, 20 min, 90% MT	10	No control group/ condition	Picture naming	No	2 months	Significant improvement
Naeser et al., 2005	4 nonfluent	Chronic	R IFG	1 Hz, 20 min, 90% MT	10	No control group/ condition	Picture naming	No	8 months	Significant improvement
Martin et al., 2009	2 nonfluent	Chronic	R IFG	1 Hz, 20 min, 90% MT	10	No control group/ condition	Picture naming	No	2 months	Significant improvement
Naeser et al., 2011	1 nonfluent	Chronic	R IFG	1 Hz, 20 min, 90% MT	10	No control group/ condition	Picture naming, BDAE	No	6 months	Significant improvement in naming, auditory comp.
Martin et al., 2009	1 nonfluent	Chronic	R IFG	1 Hz, 20 min, 90% MT	10	No control group/ condition	Picture naming	Yes—CILT	No	Significant improvement
Kakuda, Abo, Kaito, et al., 2010	4 nonfluent	Chronic	Contralateral to area of max fMRI activation	1 Hz, 20 min, 90% MT	10 (6 days)	No control group/ condition	Picture naming	No	N0	No benefit
Kakuda, Abo, Uruma, et al., 2010	2 fluent	Chronic	Post-STG	1 Hz, 20 min, 90% MT	10 (6 days), then weekly for 3 months	No control group/ condition	Token test	Yes		"Modest" improvement
Hamilton et al., 2010	1 nonfluent	Chronic	R IFG	1 Hz, 20 min, 90% MT	10	No control group/ condition	Picture description	No	10 months	Significant improvement
Kakuda et al., 2011	4 nonfluent	Chronic	R IFG	600 pulses 6 Hz, then 1 Hz, 20 min, 90%MT	18 (2 × /day for 9 days)	No control group/ condition	SLTA, WAB	Yes		Modest improvement, no statistics, moderate better than mild aphasia
Weiduschat et al., 2011	10 multiple types	Subacute	R IFG, vertex	1 Hz, 20 min, 90% MT	10	Randomized: IFG or vertex	Aachen aphasia battery	Yes	4 months	Significantly greater improvement in IFG group than vertex (19.8 versus 8.5)
Cotelli et al., 2011	3 nonfluent	Chronic	L DLPFC	20 Hz trains, 2000 pulses, 90% MT	20 (50% sham, two subjects)	2/3 with sham TMS in cross-over design	Picture naming	Yes	Up to 48 weeks	Significantly better naming persisting to 48 weeks for trained and untrained but observed before TMS in two subjects; no delay between sham and real TMS

(Continued)

TABLE 83.1 (Continued)

Author	Subjects	Time from stroke	Stimulated areas	Stimulus parameters	Number of sessions	Study design	Task	Speech therapy	Time of evaluation	Language effect
Szaflarski et al., 2011	8 multiple types	Chronic	LH frontal defined by fMRI	Excitatory theta-burst	10	No control group/ condition	Multiple measures	No	2 weeks	Significant improvement in semantic fluency
Barwood et al., 2012	7 nonfluent (6 from 2011)	Chronic	R IFG	1 Hz, 20 min, 90% MT	10	No control group/ condition	BDAE, BNT	No	8 months	Significant benefit across number of measures; global greatest increase
Waldowski et al., 2012	26 multiple types	Subacute	Right pars opercularis & triangularis	1 Hz, 30 min, 90% MT	10	Randomized, double blind, sham controlled	Naming, other	Yes	15 weeks	No main effect group; improvement in anterior lesions
Abo et al., 2012	24 multiple types (13 hemorrhagic)	Chronic	Contralateral to max fMRI; fluent at post-STG, nonfluent at IFG	40 min, 90% MT	10	No control group/ condition	SLTA battery	Yes	4 weeks	Significant improvement in naming, comprehension, spontaneous speech; no major difference between fluent/ nonfluent
Kindler et al., 2012	18 multiple types	Variable (0.5–57 months)	R IFG	Continuous thetaburst TMS	1	Randomized, sham-controlled, cross-over	Timed picture naming	No	Immediate	Modest effects but <6 months after better response
Medina et al., 2012	10 nonfluent	Chronic	R IFG	1 Hz, 20 min, 90% MT	10	Randomized, cross-over, sham controlled	Picture description	No	2 months	Significant benefit from TMS
Thiel et al., 2013	24 multiple types	Chronic	R IFG, vertex	1 Hz , 20 min, 90% MT	10	Randomized, vertex active site	Aachen aphasia test	Yes	2 weeks	Significant improvement in AAT, including subtests
Seniów et al., 2013	38 multiple types	Subacute	R IFG, vertex	1 Hz, 30 min, 90% MT	15	Randomized, double-blinded, sham control		Yes		No main effect group but improved in severe and nonfluent aphasics
Barwood et al., 2013 (see also 2011)	12 nonfluent	Chronic	R IFG	1 Hz, 20 min, 90% MT	10	Randomized, blinded; vertex control	Language battery	No	12 months	Treated group improved in naming, comprehension, repetition, picture description
Khedr et al., 2014	30, multiple types	Subacute (mean, 5.3 weeks)	R IFG (500 at pars triangularis and opercularis), then L IFG (pars triangularis and opercularis	1-Hz, 16.6 min, 110% MT (1000 pulses), then 10 trains 20-Hz for 5 seconds (1000 pulses)	10	Randomized, sham controlled	Language battery	Yes	2 months	Improvement in comprehension, naming, repetition and fluency noted in first post-stimulation session and sustained.

convincing. However, this is clearly an area in which much remains to be learned; future studies should address the role of linguistic factors in predicting response to therapy.

Because many "nonfluent" subjects are likely to have lesions that involve the frontal lobe, it is not surprising that there is some evidence that subjects with frontal lesions respond better to TMS. Weiduschat et al. (2011), for example, reported that subjects with lesions involving the IFG responded better than subjects whose lesion did not involve the IFG. Similarly, although no formal data are provided, Waldowski et al. (2012) reported that subjects with anterior lesions had a better response.

There has been little discussion in the literature about subjects who fail to respond. The single exception to this comes from a report by Martin et al. (2009), who reported detailed analyses of the lesion characteristics of two individual subjects, one of whom benefited from TMS whereas the other did not. Unfortunately, multiple differences in the extent of the lesion in brain regions thought to be crucial for naming such as the posterior superior frontal gyrus and middle temporal gyrus make the interpretation of the differences in response problematic.

Although systematic analyses are lacking, there is suggestive evidence that TMS may be most beneficial in subjects with severe but not profound deficits. Seniów et al. (2013) and Barwood et al. (2012) reported greater benefit in subjects with more severe deficits; it may also be relevant in this context that the only study that failed to find benefit in people with nonfluent aphasia (Kakuda, Abo, Kaito, et al., 2010) included subjects with very mild deficits. Although the evidence is limited, some data suggest that subjects with truly profound deficits (e.g., one-word phrase length) (Martin et al., 2009) fail to benefit from TMS.

83.2.4.2 *Site of Stimulation*

As noted, most studies have targeted the right IFG. However, even in those studies stimulating this region, a potentially important methodologic difference may be identified. Naeser and colleagues (2005a, 2005b, 2011) used a "site-finding" approach in which picture naming is assessed before and after rTMS in an effort to find the optimal site for stimulation as defined by a neuro-navigation system. We believe that identifying the optimal structure on an individual basis may be important; for example, we have reported that subjects differ substantially with respect to the specific location at which the best effects are elicited (see Figure 83.1). Whereas most of our subjects demonstrated best performance with sites in the pars triangularis, we often observed different effects from stimulation of nearby sites and even from different sites in the pars triangularis. We also identified sites in the right IFG at which stimulation adversely affected performance (Hamilton et al., 2010; Medina et al., 2012). Random selection of a stimulation site in the IFG entails the risk that an inactive or even potentially deleterious site could be targeted.

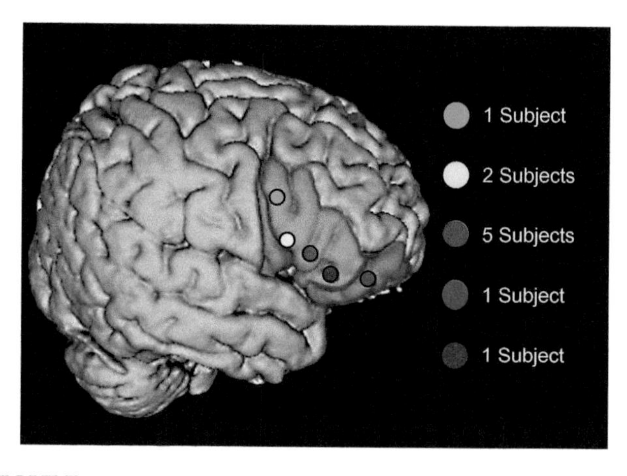

FIGURE 83.1 A lateral view of the right hemisphere indicating the sites at which the best response in object naming was found after 1-Hz TMS in 10 subjects (see Medina et al., 2012).

Other investigators have taken a different approach to right IFG stimulation. For example, citing TMS evidence that the pars triangularis and opercularis were relevant to different aspects of language function (e.g., Devlin, Matthews, & Rushworth, 2003; Gough, Nobre, & Devlin, 2005), Waldowski et al. (2012) stimulated both of these regions in all subjects. Furthermore, as many investigators targeted the right IFG using external coordinate systems (e.g., the International 10/20 system) rather than on the basis of the individual's anatomy, the precise site at which stimulation is delivered cannot be stated with certainty.

Still another approach to selecting the site to stimulate is to use fMRI activations to target TMS. Abo et al. (2012) stimulated the hemisphere contralateral to the hemisphere demonstrating the greatest fMRI activation on a language task; subjects with "fluent" aphasia were then stimulated in the posterior STG, whereas subjects with "nonfluent" aphasia were stimulated over the IFG of the hemisphere contralateral to the site of greatest BOLD activation. Kakuda, Abo, Kaito, et al. (2010) also identified the region of maximal fMRI activation on a naming task and stimulated over the homologous region of the contralateral hemisphere. Finally, Szaflarski et al. (2011) stimulated over the left frontal region, at which fMRI activation was greatest during a language task. There is mixed evidence regarding the efficacy of selecting the activation site on the basis of fMRI activation. Szaflarski et al. (2011) and Abo et al. (2012) reported improvement, but Kakuda, Abo, Kaito, et al. (2010) was the only group to fail to find benefit from TMS in people with chronic nonfluent aphasia.

Finally, Khedr et al. (2014) reported a randomized, sham controlled study in which they stimulated the right pars opercularis (500 pulses) and pars triangularis (500 pulses) of inhibitory 1 Hz stimulation followed by excitatory stimulation of the same regions on the left. They reported benefit across a number of language parameters in patients with subacute aphasia.

In an attempt to address the issue concerning the relative benefit of the stimulation site, we calculated effect sizes for those studies for which data regarding picture naming were available. There were eight published studies (Abo et al., 2012; Barwood et al., 2013; Heiss et al., 2013; Kakuda, Abo, Momosaki, & Morooka, 2011; Khedr et al., 2014; Kindler et al., 2012; Seniów et al., 2013; Szaflarski et al., 2011) and our unpublished picture naming data (see Medina et al., 2012 for picture description data from these subjects). Effect sizes ranged from small (0.181) to moderately large (0.889), with an average of 0.379 (0.22, 0.54). Restricting the analysis to those studies using inhibitory right IFG stimulation (see Table 83.1) yielded a mean Cohen's D of 0.584.

83.2.4.3 Stimulus Parameters

Most studies have used 20 minutes of 1-Hz stimulation at 90% of motor threshold for 10 sessions. Kakuda et al. (2011) used a strategy of "pretreating" the right IFG with 600 pulses at 6 Hz prior to 20 minutes of 1-Hz stimulation. Reflecting the widely held perspective that the goal of TMS is to increase the efficiency and involvement of the remaining left hemisphere tissue, two groups have used theta-burst TMS to alter the balance of activity in the right and left hemispheres. Szaflarski et al. (2011) delivered excitatory (intermittent) theta-burst TMS to the left frontal lobe with beneficial effects, whereas Kindler et al. (2012) administered inhibitory (continuous) theta-burst TMS to the right hemisphere with little benefit. Finally, Cotelli et al. (2011) took a different approach entirely, delivering "excitatory" 20 Hz rTMS to the left dorsolateral prefrontal cortex. Because differences in stimulus parameters were often confounded by other dissimilarities in experimental design, the data do not demonstrate a clear best practice.

83.2.4.4 Does Concurrent Speech Therapy Make a Difference?

Ten studies, including the five largest studies, administered speech language therapy during the same period during which TMS was administered. The potential role of SLT in conjunction with TMS was not systematically investigated in any of these studies, however. The evaluation of the potential contribution of concurrent speech therapy is complicated further by the fact that studies with and without concurrent SLT differed in a wide range of parameters. Thus, although there is a strong rationale for combining TMS with speech therapy, there is little convincing evidence that the presence of speech therapy is a significant determinant of response to therapy. This is an issue that warrants additional study.

83.2.4.5 Time Course of Benefit

A major weakness of many types of speech-language therapy is that the initial beneficial effect erodes over weeks to months. The beneficial effects of TMS in chronic aphasia stand in stark contrast to this general rule. In fact, one striking aspect of the data from TMS investigations is the finding that the benefit from TMS is typically not only sustained through the period of follow-up but also, at least in some instances, increases over time in the absence of ongoing therapy. For example, Martin et al. (2009) reported a subject with chronic aphasia whose performance at 16 months after therapy was significantly better than at 3 months. We reported a similar finding in a subject followed for 10 months (Hamilton et al., 2010; see also Naeser et al., 2005a and 2005b). Similarly, in an open-label case series involving 7 people with chronic aphasia tested at 1 week, 2 months, and 8 months after therapy, Barwood et al. (2012) found that subjects exhibited continued improvement; performance was better on 7 of 7 measures at 8 months as compared to 2 months after treatment.

Although the explanation for this finding remains unclear, we speculate that in the aftermath of a left hemisphere stroke residual components of the language circuits in the left and right hemispheres develop new patterns of interaction over weeks to months (see also Devlin and Watkins, 2007; Hamilton et al., 2010, 2011). Improvement in performance during this interval may reflect the optimization of performance of an inefficient architecture. Based on a meta-analysis of functional imaging studies of aphasic subjects, Turkeltaub et al. (2011; see also Hamilton et al., 2011) identified a region in the right anterior pars triangularis that, unlike most regions of activation in the right hemisphere, was not homotopic with sites of activation in the left hemisphere; one potential account, then, is that this brain region is recruited during language recovery but that activation at this site is deleterious. If this is true, then inhibiting this right IFG site would be expected to improve language function. Thus, we speculate that the subsequent slow and sustained improvement in language function after TMS is attributable to suppression of this site that serves to "reboot" the language network, permitting new and more adaptive patterns of connectivity to emerge. It is interesting to note in this context that optimal sites identified in our site finding studies of 10 aphasic subjects (Medina et al., 2012) were, in many cases, close to the site identified in the meta-analysis.

83.2.5 Summary of TMS Effects

In summary, multiple proof-of-principle studies as well as larger studies involving control conditions (e.g., sham or vertex stimulation) support the claim that TMS may improve picture naming and other indices of language function in people with aphasia. Although substantial differences in virtually all aspects of study design make it impossible to identify optimal procedures, there is compelling evidence that inhibitory stimulation to the right IFG produces reliable benefits in people with aphasia. Additional investigations will be necessary to determine whether aphasia type, chronicity, and severity are important predictors of response.

83.3 tDCS AS THERAPY FOR APHASIA

tDCS is a procedure in which a small current (e.g., 1–2 mA) is delivered to (usually) relatively large electrodes (e.g., 5 × 7 cm) on the scalp. Although much of the current is shunted from the anode to the cathode through skin, cerebrospinal fluid, and other tissues, the brain is subjected to tonic stimulation with, at least under some circumstances, polarity-dependent effects. Although the basic effects of small electrical currents applied to the cortex were first elucidated in the 1960s, electrical stimulation of the human brain with direct currents was largely ignored until approximately 2000, when groundbreaking work by Nitsche and Paulus (2000, 2001, Nitsche et al., 2003) demonstrated that the technique is capable of producing significant effects on behavior.

83.3.1 Basics of tDCS

Work in the 1960s with animals demonstrated that small positive and negative currents delivered to the cortex generated polarity-specific effects on resting membrane potentials; in particular, neurons near an anode exhibited lowered resting membrane potentials, thereby making neurons more likely to discharge; in contrast, the resting membrane potentials of neurons near the cathode were hyperpolarized, making neurons less likely to discharge (Purpura & McMurtry, 1965).

In a series of seminal reports, Nitzsche and Paulus (2000) demonstrated polarity-dependent effects on brain function from small direct currents applied to the scalp. They demonstrated that direct current of 1 mA with the anode over the motor cortex and the cathode over the contralateral supraorbital region for 12 minutes caused an increase in TMS-induced motor-evoked potentials, whereas reversing the electrode placement such that the cathode was over the motor strip caused a reduction in the motor-evoked potential. The effects of tDCS outlasted the stimulation by minutes to hours depending on the stimulation parameters.

Although much remains to be learned about the physiologic basis of tDCS effects, the persisting effects are thought to be caused by LTP and LTD-like processes, probably mediated by NMDA receptors and altered GABAergic activity (e.g., Nitsche et al., 2003; Stagg et al., 2009). The duration of tDCS effects is prolonged by combining stimulation with a task; this effect appears to be mediated, at least in part, by BDNF-dependent synaptic plasticity (Fritsch et al., 2010). There is substantial evidence that repeated administrations of tDCS produce effects that persist for weeks (e.g., Reis et al., 2009).

It is important to note that tDCS does not cause neurons to discharge but, by virtue of its effect on resting membrane potentials, it influences the rates at which neurons discharge. Because changes in firing rates may produce long-term effects by LTP and LTD-like mechanisms, the technique may have a disproportionate impact on circuits that are active during tDCS administration. For example, if neural circuits mediating language are active while tDCS is being administered, then anodal tDCS, which lowers resting membrane potentials, may increase the firing rates of the language-related neurons, thereby strengthening synaptic connections within the language circuits. For this reason, most of the investigations of tDCS as a therapy for aphasia have incorporated concurrent language therapy as part of the protocol.

There are a number of variables that determine the effect of tDCS. One crucial factor is electrode placement. In those instances in which there is reason to believe that enhancing neuronal activity is likely to be beneficial, the anode is typically placed near the brain region at which increased neuronal firing rate is desired and the cathode is often located at a site thought to be not relevant to task (e.g., supraorbital region) or off the head entirely (e.g., the shoulder). In those instances in which the goal is to decrease the firing rate of neurons in a specific brain region, the cathode is placed near the relevant tissue.

A second important variable is current intensity. Much of the pioneering work with tDCS in humans involved 1-mA currents; in recent years, many investigators have used larger currents (1.5–2 mA). Although it has been assumed that the effects of current intensity would be linear across intensities, recent data suggest that this is not true. Batsikadze et al. (2013) recently reported data from motor cortex stimulation with

2-mA currents for 20 minutes. They found that with the anode over the motor cortex, 2-mA stimulation produced the same enhancement of motor-evoked potential produced by 1-mA currents; surprisingly, however, reversing the placement of the electrodes so that the cathode was over the motor cortex generated effects that were quite different from those generated by 1-mA stimulation. Thus, with 2-mA currents, cathodal and anodal stimulation were indistinguishable in their effects on the motor cortex. Although the generalizability of this effect to different brain regions is not known, the findings of Batsikadze et al. (2012) complicate the interpretation of studies involving 2-mA stimulation. We return to this point later.

Finally, many investigators use sham stimulation as a control for nonspecific effects of the testing environment as well as subject expectations. Sham stimulation permits subjects and, in some instances, investigators to be blinded with respect to the type of stimulation (Kessler, Turkeltaub, Benson, & Hamilton, 2012). In most instances, sham stimulation involves a ramp-up of the current for 30–45 seconds to provide the tingling and local sensation associated with tDCS before extinguishing the current.

83.4 tDCS AS A TREATMENT FOR APHASIA

tDCS has proven to be beneficial in a number of clinical disorders such as epilepsy (Fregni et al., 2006), chronic pain (Bolognini, Olgiati, Maravita, Ferraro, & Fregni, 2013), major depression (Brunoni et al., 2011), and hemiparesis (Madhavan, Weber, & Stinear, 2011). In this section we review a total of 13 studies in which the effects of tDCS on aphasia have been reported. Studies have ranged in size from 3 to 37 subjects, with a total of approximately 140 subjects. As was the case with TMS investigations, there is substantial variability with respect to study design; indeed, no two studies used the same design. Nonetheless, certain commonalities in approach may be identified. We review the major variables in tDCS design and how they differ across studies before discussing the effectiveness of the intervention (see also de Aguiar, Paolazzi, & Miceli, 2015).

83.4.1 Electrode Placement

The most fundamental decision in the design of a noninvasive brain stimulation study is the electrode placement because this will determine the pattern of current flow and, ultimately, the brain regions that are stimulated. There has been striking variable with respect to electrode placement in the 13 studies reviewed here. Nine tDCS studies were motivated by the same perspective that animated TMS studies—the conviction that excitation of the left hemisphere or inhibition of the right hemisphere will improve performance in chronic aphasia. However, the manner in which this goal was achieved differed. Many studies (see Table 83.2) attempted to enhance left hemisphere function with anodal stimulation of the left hemisphere and a "neutral" cathodal placement (e.g., supraorbital region, shoulder). Kang, Kim, Sohn, Cohen, and Paik (2011) took a different approach to altering the balance between right and left hemisphere function, using right hemisphere cathode stimulation to "inhibit" the right hemisphere. You, Kim, Chun, Jung, and Park (2011) used a left anode and right cathode (both with supraorbital reference) in separate conditions. Finally, Marangolo et al. (2014) used a simultaneous left anode and right cathode to the bilateral IFG to both potentiate the left and inhibit the right hemisphere (see Lindenberg et al., 2010).

Other investigators systematically investigated the role of the left and right hemispheres by applying anodal and cathodal stimulation to the same structure in different conditions. Thus, Monti et al. (2008) administered both anodal and cathodal stimulation to the left frontal region in separate conditions while Flöel et al. (2011) administered both anodal and cathodal stimulation to the right temporal/parietal region in separate conditions.

Not all studies were motivated by the desire to alter the balance between the hemispheres in favor of greater activation in the left hemisphere. In the largest study reported to date (37 subjects), Jung, Lim, Kang, Sohn, and Paik (2011) used cathodal stimulation to left frontal regions to down-regulate neural activity in the left hemisphere. Additionally, based on the assumption that Melodic Intonation Therapy relies on right hemisphere structures, Vines, Norton, and Schlaug (2011) administered anodal stimulation to the right IFG to enhance the value of this concurrently administered therapy.

Although most studies reported significant benefit, no definitive conclusion regarding the optimal electrode placement can be drawn from studies that used only one active montage. Five studies that contrasted the effects of different montages do, however, provide some important insights. Marangolo, Fiori, Calpagnano, et al. (2013) and Marangolo, Fiori, Di Paola, et al. (2013) contrasted the effects of left Broca and left Wernicke anodal stimulation (both with right supraorbital cathode) on verb naming (2013) and a variety of measures of speech production (2013); in both studies, they found significantly greater benefit from Broca area anodal stimulation, whereas Wernicke area anodal stimulation did not differ from sham.

Two studies speak to the hypothesis that activating the left hemisphere or inhibiting the right hemisphere

TABLE 83.2 Studies for the Use of tDCS in the Treatment of Aphasia.

Author	Subjects	Anode*	Cathode*	Design	Intensity and Duration	Task	Concurrent Speech Therapy?	Time at Testing	Effects
Monti et al., 2008	8 Chronic aphasics:	Condition 1: L frontal	Condition 1: R shoulder	Single session, 1 wk between sessions	2 mA/ 10 min	Picture naming	No	Immediately after stimulation	Cathode 33% better accuracy
	4 Broca	Condition 2: R shoulder	Condition 2: L frontal	6 Subjects assigned to both condition 1 and 2					
	4 Wernicke	Control: sham	Control: sham						
Baker et al., 2010	10 Chronic aphasics (> 12 months)	Condition 1: L perilesional (25 cm^2)	Condition 2: R shoulder (25 cm^2)	20 min of 1 mA tDCS; 5 days in each condition	1 mA/ 20 min	Spoken word-picture matching	Yes	Immediately after treatment and 1 week later	Improved naming with anodal tDCS for treated
		Control: sham	Control: sham						Marginal effect for untreated words
You et al., 2011	21 Subacute global aphasics	Condition 1: L superior temporal gyrus (CP5)	Condition 1: R supraorbital	Subjects randomized to 1 of 3 conditions; 10 daily sessions each condition	2 mA/ 30 min	Naming	Yes		Significant benefit of right cathodal stimulation versus anode and sham for aud comp.; no effect on other measures
		Control: sham							
Flöel et al., 2011	12 Chronic aphasics	Condition 1: R temp/par	Condition 1: L supraorbital (100 cm^2)	Randomized, double-blind cross-over with 3 conditions; 2 sessions/day for 3 days in each condition	1 mA/ 20 min	Picture naming	Yes	2 weeks after treatment	All conditions show >80% improvement, right anodal greatest and significantly better than sham
		Condition 2: L supraorbital (100 cm^2)	Condition 2: R temp/par						
		Control: sham	Control: sham						
Fiori et al., 2011	3 Chronic aphasics	Condition 1: L Wernicke's (CP5)	Condition 1: R supraorbital	Cross-over design	1 mA/ 20 min	Picture naming	Yes	1 and 3 weeks after therapy	Anodal 20% better than sham
		Control: sham	Control: sham	2 Conditions separated by 1 week					
Fridriksson et al., 2011	8 Chronic fluent aphasics with postcortical. or subcortical lesions	Condition 1: posterior margin of lesion at max fMRI activation	Condition 1: R supraorbital (100 cm^2)	Double-blind, cross-over with 5 sessions	1 mA/20 min	Spoken word-picture matching	Yes	3 weeks after	Anodal reduces RT for trained nouns relative to sham; no generalization
		Control: sham	Control: sham						

(Continued)

TABLE 83.2 (Continued)

Author	Subjects	Anode*	Cathode*	Design	Intensity and Duration	Task	Concurrent Speech Therapy?	Time at Testing	Effects
Kang et al., 2011	10 Chronic aphasics	Condition 1: L supraorbital	Condition 1: R "Broca's" (F8)	Double-blind, randomized, cross-over design; 20 min daily for 5 days in each condition	2 mA	Picture naming	Yes	Immediately after treatment	Significant improvement in naming after tDCS
		Control: sham	Control: sham						
Vines et al., 2011	6 Nonfluent chronic aphasics (> 12 months)	Condition 1: R IFG (16.2 cm^2)	Condition 1: L supraorbital (30 cm^2)	Double-blind, cross-over design; 3 sessions each condition	1.2 mA	Speech production time	Yes, melodic intonation therapy	Immediately after treatment	Anodal reduced production time
		Control: sham	Control: sham						No effect of sham
Jung et al., 2011	37 Aphasics, average: 27 <90 days after and 26 nonfluent	Condition 1: R supraorbital	Condition 1: L Broca's area	10 Sessions of 20 min over 2–3 weeks	1 mA	Korean WAB	Yes	Immediately after treatment	Significant improvement of 14.9% or 65.2% of baseline
		Control: none	Control: none						
Santos et al., 2013	19 Chronic aphasics	L supraorbital	R motor strip	10 Sessions of 20 minutes over 2 weeks	2 mA	Multiple measures	No	Immediately after treatment	Significant improvement in 3 of 18 measures, trend in another 4 measures
		Control: none	Control: none						
Volpato et al., 2013	8 Chronic mild or moderate aphasics	Condition 1: L Broca's area	Condition 1: R supraorbital	Cross-over design; 10 daily sessions over 2 weeks in 2 conditions	2 mA/ 20 min	Object and action naming	No	Immediately after treatment	No group effects on action or object naming
		Control: sham	Control: sham						
Polanowska et al., 2013	24 Subacute (2–24 weeks) nonfluent aphasics	L "Broca's area"	R supraorbital	Randomized, sham controlled. 15 sessions	1 mA/10 min	Picture naming	Yes	Immediately after treatment and 3 months later	No significant difference between groups
Marangolo, Fiori, Calpagnano, et al., 2013	7 Chronic "nonfluent" aphasics	Condition 1: L "Broca" area (F5)	Condition 1: R supraorbital	5 Daily sessions in each of three conditions for all subjects	1 mA/20 min	Verb naming	Yes	Immediately after treatment, 1 and 4 weeks later	Broca's stimulation significantly better than sham and Wernicke stimulation (these two did not differ)
		Condition 2: L "Wernicke" area (CP5)	Condition 2: R supraorbital						

(Continued)

TABLE 83.2 (Continued)

Author	Subjects	Anode*	Cathode*	Design	Intensity and Duration	Task	Concurrent Speech Therapy?	Time at Testing	Effects
		Control: sham	Control: sham						
Marangolo, Fiori, Di Paola, et al., 2013	12 Chronic "nonfluent" aphasics	Condition 1: L "Broca's" area (F5)	Condition 1: R supraorbital	10 Daily sessions in each condition	1 mA/20 min	Content units in disclosure	Yes	Immediately after treatment	C-units, verbs, and sentences: Broca's area stimulation greater than Wernicke's and sham, which did not differ
		Condition 2: L "Wernicke's" area	Condition 2: R supraorbital					2 weeks later	Benefit maintained at follow-up
		Control: sham	Control: sham						Effect seen in trained and untrained materials
Marangolo et al., 2014	7 Chronic "nonfluent" aphasics	Condition 1: left IFG	Condition 1: right IFG	10 Daily sessions	2 mA/20 min	Description of videotaped vignettes	Yes: "pragmatic therapy"	Immediately after therapy and 1 week later	Improved noun, verb, description relative to sham; average increase across 3 measures 16.3% versus 5.3%
									Some generalization
		Control: sham	Control: sham						
Campana, Caltagirone, and Marangolo, 2015	20 Chronic nonfluent aphasics	L lateral frontal (F5)	R supraorbital	10 Daily sessions	2 mA/20 min.	Picture description, verb and noun naming	Yes	Before and after 10 day sessions	Significant improvement in all three measures

will enhance performance. Thus, Monti et al. (2008) reported greater benefit from cathodal as compared to anode placement over the left frontal region (both conditions paired with right shoulder electrode). Similarly, Flöel et al. (2011) contrasted the effects of placement of the anode and cathode over the posterior right hemisphere (both paired with left supraorbital) and found that anodal stimulation of the right hemisphere provided significantly better performance.

Finally, Hamilton and colleagues (in press) adopted a "site finding" approach similar to that pioneered by Naeser et al. (2005) in TMS investigations. Reasoning that extant data did not provide a clear "best" approach, these investigators assessed naming under a total of five conditions. On separate days, all subjects were tested before and after 20 minutes of 2 mA tDCS in the following conditions: (1) anode over the left IFG and cathode over the right IFG; (2) anode over the right IFG and cathode over the left IFG; (3) anode over the left posterior STG and cathode over the right posterior STG; (4) anode over the right STG and cathode over the left STG; and (5) sham stimulation using the configuration described in (1). Subjects who demonstrated significant benefit in one of the conditions received therapy consisting of daily sessions of 20 minutes at 2 mA using the montage that provided the greatest benefit.

One additional point regarding electrode placement concerns the limitation in specificity of stimulation resulting from the large electrode size and patterns of current flow. As most studies have used large (5×7 cm) electrodes that cover brain regions larger than those typically hypothesized to be relevant to language function, claims regarding the site of stimulation should be interpreted with caution. Furthermore, efforts to model current distribution across the brain (e.g., Datta et al., 2009; Miranda, Lomarev, & Hallett, 2006) suggest that effects are not likely to be restricted to the site of the electrode and may, in fact, be maximal at sites adjacent to rather than under the electrode. Although presently unsupported by empirical data, these simulations suggest that large swaths of regions between electrodes are subjected to current. Although the recent advent of "high-definition" tDCS using small and relatively closely space electrodes may afford substantially greater precision than has been available to date (Datta, Bikson, & Fregni, 2010; Suh, Lee, & Kim, 2012), it is not possible to make strong statements regarding the role of specific brain regions in the genesis of the reported effects.

83.4.2 Control Conditions and Follow-up

A strength of tDCS studies to date has been the inclusion of multiple experimental conditions. As previously noted, tDCS affords the possibility of a sham stimulation in which current is delivered at the start of the session, thereby inducing cutaneous symptoms of itching and/or burning before being extinguished. Several studies have demonstrated that, at least with current intensities of 1 mA, subjects cannot distinguish between real and sham conditions (Kessler et al., 2012). Eleven studies have used a sham condition. Four studies used two active conditions in addition to a sham condition to address hypotheses regarding the relative benefit of anterior versus posterior electrode placement (Marangolo, Fiori, Calpagnano, et al., 2013) or contrast the benefit of anodal or cathodal stimulation of the same (Flöel et al., 2011; Monti et al., 2008) structure or homologous structures of the right and left hemispheres (You et al., 2011).

A shortcoming of the tDCS literature to date is the lack of long-term follow-up. Whereas many TMS studies report data from at least 4 months after stimulation, a significant number of tDCS studies report data only from testing immediately after the end of treatment. Only four studies reported data from 3–4 weeks after treatment, rendering it impossible to determine the long-term effects of the intervention.

83.4.3 Concurrent Treatment

As previously noted, tDCS is assumed to modulate neural activity. Therefore, it is not surprising that 14 of 16 tDCS investigations have paired stimulation with speech therapy. Although the small numbers of studies argue for caution in drawing conclusions, it is noteworthy that Volpato et al. (2013) demonstrated no benefit relative to a sham condition from stimulation with the anode over the left Broca's area and the cathode over the right supraorbital region. Santos et al. (2013) reported modest effects, finding significant improvement in only 3 of 18 language measures. Thus, we believe that evidence to date argues for the inclusion of language therapy in future tDCS studies.

83.4.4 Outcomes

What can these studies tell us about who will benefit and the magnitude of that benefit? As with TMS studies, although striking differences between stimulation parameters and study designs confound attempts to draw generalizations from the investigations published to date, tentative conclusions may be drawn.

A closer look at the studies demonstrates that 2 of the 16 studies do not speak directly to the potential of tDCS as a therapy for aphasia. Subjects reported by Monti et al. (2008) received only one stimulation in each of three conditions and the only assessment was immediately after therapy. Additionally, Jung et al.

(2011) reported benefit in the largest series published to date (n = 37), but the absence of a control group in a population that included 73% subacute (less than 90 days poststroke) cases in subjects who would be expected to improve under any circumstances makes this finding interpretable. The fact that the subjects were receiving concurrent speech-language therapy further confounds efforts to discern effects of tDCS in this group.

Thirteen of the remaining 14 studies reported benefit on at least some measures. The only study that reported a negative result (Volpato et al., 2013) included many subjects with mild aphasia in whom beneficial effects could have been obscured by ceiling effects. There appears to be substantial variability with respect to the magnitude of the benefit, however. In some investigations, the effects are relatively modest. You et al. (2011), for example, reported a benefit in auditory comprehension in subjects treated with the cathode over the left hemisphere but no benefit in other measures; other investigators report improvement in naming accuracy on the order of 10–15% or less. In two studies (Fridriksson, Richardson, Baker, & Rorden, 2011; Vines et al., 2011), the beneficial effect was manifested as a modest reduction in speech production times.

Are the positive effects of tDCS clinically relevant? A definitive answer would require additional information regarding magnitude and duration of the benefit and the range of language processes that are influenced by the intervention. Although effect sizes are typically not reported, several studies (Baker, Rorden, & Fridriksson, 2010; Jung et al., 2011; Kang et al., 2011; Polanowska, Kesniak, Seniow, Czepiel, & Czlonkowska, 2013; Santos et al., 2013; Volpato et al., 2013; You et al., 2011) provide information regarding changes in accuracy and such a measure of effect size (Cohen's D) may be calculated. The effect sizes range from 0.175 (−0.76–1.11) (Volpato et al., 2013) to 1.064 (−0.31–2.44) (Baker et al., 2010), with a mean of approximately 0.489. As was the case with TMS studies, the effect size data must be considered preliminary because of differences in methods, small sample sizes, and other factors. Nonetheless, findings to date suggest that, at least in some instances, tDCS generates clear benefit.

Promising data regarding the range of language faculties influenced by tDCS are starting to emerge. Most early studies used only naming tasks to assess the effects of the intervention. You et al. (2011) reported data from a more comprehensive language assessment, the Korean version of the Western Aphasia Battery. The former demonstrated a benefit from auditory comprehension but, more generally, not the WAB Aphasia Quotient. More recently, Marangolo and colleagues have reported studies that have extended the range of phenomena that are positively influenced by tDCS. For example, Marangolo and colleagues reported benefit from stimulation with the anode over Broca's area on verb naming (Marangolo, Fiori, Calpagnano, et al., 2013) and "content units" (Marangolo, Fiori, Di Paola, et al., 2013) in discourse. Most recently these investigators also reported data from seven "nonfluent" subjects with left IFG anode and right IFG cathode demonstrating improvements in speech elicited in attempts to describe videotaped vignettes (Marangolo et al., 2014). In both of the latter two studies there was some evidence of generalization to unpracticed clips. Interestingly, "traditional" measures of language function did not change with the interventions in two of these studies (Marangolo, Fiori, Di Paola, et al., 2013, Fiori, Cipollari, Caltagirone, and Marangolo, 2014; Marangolo et al., 2014). Thus, recent data extend the range of phenomena shown to be effected by tDCS to include language capacities that are of clear ecological significance. Furthermore, the evidence of generalization found by these investigators, while not robust, is promising.

A second issue pertaining to the clinical relevance of the intervention is that only short-term outcome data are available. Most studies have assessed tDCS effects either immediately or several weeks after stimulation. Only one study reports findings persisting for up to 1 month (Marangolo, Fiori, Calpagnano, et al., 2013). The lack of information about duration of effects limits the assessment of the clinical utility of the benefit of tDCS and should be addressed in future studies.

There are a number of other fundamental questions regarding subject selection and study design that have yet to be addressed. For example, does the duration of aphasia influence the response to tDCS? Most studies have involved patients with chronic aphasia; however, people with subacute aphasia have been included in several studies. In a study of 21 subacute patients, You et al. (2011) reported no main effect of tDCS and a relatively modest benefit on only one measure, auditory comprehension. As previously noted, Jung et al. (2011) reported data from 37 subjects, 73% of whom were subacute; although the subacute subjects improved more than the people with chronic aphasia, the absence of a control group clouds the interpretation of this finding. Finally, Polanowska et al. (2013) demonstrated no significant differences between real and sham-treated groups. Thus, limited evidence to date does not demonstrate the utility of tDCS in people with subacute aphasia. There is no *a priori* reason to believe that the technique should not be beneficial in this setting and further study seems warranted.

Another important issue about which relevant information is lacking is the importance of subject characteristics such as type of aphasia. Many studies are heterogeneous with respect to aphasia subtype, whereas other studies report subjects only as "fluent" or "nonfluent," a dichotomy that fails to capture a range of potentially relevant dimensions of language function. Fridriksson et al. (2011) reported benefit on a spoken word to picture matching task in people with "fluent" aphasia and multiple studies involving people with "nonfluent" aphasia have demonstrated benefit in that (heterogeneous) group. Therefore, at present, there is little information regarding aphasia subtype and the basis on which to select subjects for tDCS therapy.

The optimal electrode montage has not been established. One factor that seriously undermines efforts to identify optimal sites of stimulation is the inherent imprecision of tDCS when administered with the large electrodes used in the studies reviewed here. Tentative conclusions may be drawn regarding electrode placement, however. First, the data do not unambiguously support the hypothesis that stimulation of the left or inhibition of the right hemisphere will be of benefit. Although a number of left hemisphere stimulation and right hemisphere inhibition (Kang et al., 2011) studies have demonstrated benefit, studies involving right hemisphere stimulation (Flöel et al., 2011; Vines et al., 2011) have also produced benefit. Furthermore, in a direct comparison of left frontal anodal and cathodal stimulation, Monti et al. (2008) reported that a single session of cathodal stimulation of the left frontal region produced significantly better performance on a naming task. However, because they were using 2-mA current, the interpretation of this finding is currently unclear.

Second, there is limited evidence regarding the effects of stimulation of the anterior as opposed to the posterior language cortices. Of greatest relevance in this context are the studies by Marangolo, Fiori, Calpagnano, et al. (2013) and Marangolo, Fiori, Di Paola, et al. (2013), in which the effects of anodal placement in the anterior and posterior regions and cathode over the contralateral forehead were directly contrasted. In both studies, anodal placement over the frontal lobe was associated with significant benefit, whereas performance with placement over the posterior regions did not differ from the sham condition.

83.5 GENERAL DISCUSSION

Data from more than 25 studies, many with appropriate control conditions, demonstrate that noninvasive brain stimulation may be of benefit in the treatment of aphasia. There is clear evidence that rTMS is of persistent benefit for people with subacute and chronic nonfluent aphasia with respect to a wide range of language tasks and what, at least in some instances, is estimated to be a large effect size. However, there is also evidence of benefit from tDCS in well-controlled, randomized studies at present regarding the duration and range of the positive effects.

83.5.1 TMS and tDCS: Advantages and Disadvantages (Wasserman, 1998)

As there is evidence for the effectiveness of both TMS and tDCS as treatments for aphasia, which one is preferable? Although there is no clear answer to this question, there are a number of factors that must be considered.

One crucial issue is safety. At this point, both techniques appear to be safe and well-tolerated. In particular, tDCS has an exemplary safety record. The technique has been investigated in several studies in normal subjects (e.g., Kessler et al., 2012; Lang et al. 2005) with no major adverse effects noted and there has been no evidence of significant issues in the growing literature documenting its use in hundreds of subjects with brain pathology. It is not surprising, then, that no adverse effects have been demonstrated in any study involving aphasic subjects. Similarly, there have been no reports of adverse effects in the more than 200 subjects participating in TMS studies to date. We are also unaware of reports of subjects dropping out of TMS studies because of adverse effects. Although one might be more concerned about seizures in subjects with a neurologic condition such as stroke, which is known to be associated with seizures, it is noteworthy that no significant adverse effects have been reported in any TMS involving people with aphasia. Furthermore, TMS has been administered to the lesioned hemisphere using more aggressive stimulation parameters than those in most studies of people with aphasia in a variety of other rehabilitation settings without adverse effects (e.g., Khedr et al., 2005, 2014). At present, therefore, both tDCS and TMS appear to be safe in this setting.

A second parameter relevant to the decision to use tDCS or TMS is cost. tDCS is less resource-intensive in several respects. The apparatus is less costly and easier to administer. Whereas TMS often requires administration by two experimenters, tDCS can be administered by one person.

A third factor that may be relevant to a choice between TMS and tDCS is the ease with which the technique may be paired with speech-language therapy. The effects of TMS are substantially more intrusive and distracting than tDCS, making the latter much easier to combine with therapy. The import of this factor is not clear. For theoretical reasons discussed herein, concurrent therapy may be important for tDCS but not for TMS, for which therapy immediately after stimulation may be adequate.

Finally, along with patient safety, the most important consideration regarding the decision to use tDCS or TMS is efficacy. Although there are no definitive data at this time, informal assessment of reported benefits suggest that TMS may be somewhat more effective, at least as judged by picture naming. Furthermore, there is evidence of sustained benefit from TMS, but the issue has not been addressed for tDCS.

83.5.2 NIBS and Mechanisms of Recovery

There has been controversy regarding the neural basis of the recovery of language function after stroke, at least since the initial report by Barlow (1877) of a patient who, after recovering from stroke-induced aphasia, became aphasic once again after a new right hemisphere stroke. In recent years there has been substantial debate regarding the anatomic basis of recovery, with some evidence suggesting that recovery is mediated by remaining left hemisphere tissue and other studies suggesting that the right hemisphere is crucial for recovery (Hamilton et al., 2010, 2011; Thiel et al., 2006; Turkeltaub et al., 2011; Winhuisen et al., 2007). The approach adopted by Naeser and many subsequent investigators using both TMS and tDCS has been motivated by the hypothesis that the right hemisphere adversely influences language performance, because of inhibition of remaining left hemisphere language structures, an increasing reliance on right hemisphere regions that are not optimized for language, or both. TMS studies involving "inhibitory" 1-Hz stimulation of the right hemisphere as well as "excitatory" stimulation of the left hemisphere are consistent with the predictions of this account. Furthermore, imaging data are in accord with this prediction. For example, Allendorfer et al. (2012) reported that excitatory theta-burst TMS to the left frontal cortex was associated with an increase in white matter connectivity as assessed by fractional anisotropy (FA) in regions near the stimulation site. However, as acknowledged by the authors, the interpretation of these data is complicated by the fact that other distant regions (e.g., right midbrain) also demonstrated increases or decreases in FA. Weiduschat et al. (2011) performed PET scans to measure blood flow during a verb generation task in 10 subacute stroke patients before and after TMS. They calculated a Laterality Index reflecting the difference between blood flow in language-related areas of the right and left hemispheres. They found that TMS to the left hemisphere was associated with a change in the LI reflecting greater blood flow in the left IFG (or preserved tissue nearby), whereas no such change was observed in the sham group. Similar findings were reported by Naeser et al. (2009) and Thiel et al. (2013). However, not all data support the hypothesis that recovery after TMS is attributable to greater left hemisphere involvement. We (Turkeltaub et al., 2011) reported a patient who improved after receiving TMS to the right IFG but became acutely and permanently worse after a subsequent right hemisphere stroke. These data strongly suggest that, at least for some subjects, right hemisphere structures are crucial components of the distributed language system.

Data from tDCS studies are also not definitive. As noted, improvement in language function after tDCS has been observed with electrode montages that, on traditional assumptions regarding the role of anodal and cathodal stimulation, would be expected to stimulate or inhibit both the right and left hemispheres. Therefore, at this juncture, the data from noninvasive brain stimulation studies do not unambiguously address the relative contributions of the right and left hemispheres to language recovery. Rather, the data reinforce the view that multiple factors such as infarct size, location, and premorbid functional anatomy determine the composition of language networks mediating recovery from aphasia.

In conclusion, although much remains to be learned regarding the patient selection and optimal procedures, there is compelling evidence that both TMS and tDCS improve language function in aphasic subjects. Ultimately, a randomized, controlled clinical trial involving subjects with multiple aphasia types with comprehensive language assessments extending for at least 6 months will be needed to assess the relative efficacy of the techniques.

References

Abo, M., Kakuda, W., Watanabe, M., Morooka, A., Kawakami, K., & Senoo, A. (2012). Effectiveness of low-frequency rTMS and intensive speech therapy in poststroke patients with aphasia: a pilot study based on evaluation by fMRI in relation to type of aphasia. *European Neurology*, *68*(4), 199–208. Available from: http://dx.doi.org/10.1159/000338773.

Allendorfer, J. B., Storrs, J. M., & Szaflarski, J. P. (2012). Changes in white matter integrity follow excitatory rTMS treatment of post-stroke aphasia. *Restorative Neurology and Neuroscience*, *30*(2), 103–113. Available from: http://doi.org/10.3233/RNN-2011-0627.

Ayache, S. S., Farhat, W. H., Zouari, H. G., Hosseini, H., Mylius, V., & Lefaucheur, J.-P. (2012). Stroke rehabilitation using noninvasive cortical stimulation: motor deficit. *Expert Review of Neurotherapeutics*, *12*(8), 949–972. Available from: http://dx.doi.org/10.1586/ern.12.83.

Baker, J. M., Rorden, C., & Fridriksson, J. (2010). Using transcranial direct-current stimulation to treat stroke patients with aphasia. *Stroke; a Journal of Cerebral Circulation*, *41*(6), 1229–1236. Available from: http://dx.doi.org/10.1161/STROKEAHA.109.576785.

Barlow, T. (1877). On a case of double hemiplesig, with cerebral symmetrical lesions. *The British Medical Journal, 2*, 103–104.

Barwood, C. H. S., Murdoch, B. E., Riek, S., O'Sullivan, J. D., Wong, A., Lloyd, D., et al. (2013). Long term language recovery subsequent to low frequency rTMS in chronic non-fluent aphasia. *NeuroRehabilitation, 32*(4), 915–928. Available from: http://dx.doi.org/10.3233/NRE-130915.

Barwood, C. H. S., Murdoch, B. E., Whelan, B. M., Lloyd, D., Riek, S., O'Sullivan, J. D., et al. (2011). The effects of low frequency Repetitive Transcranial Magnetic Stimulation (rTMS) and sham condition rTMS on behavioural language in chronic non-fluent aphasia: Short term outcomes. *NeuroRehabilitation, 28*(2), 113–128. Available from: http://dx.doi.org/10.3233/NRE-2011-0640.

Barwood, C. H. S., Murdoch, B. E., Whelan, B. M., Lloyd, D., Riek, S., O'Sullivan, J. D., et al. (2012). Improved receptive and expressive language abilities in nonfluent aphasic stroke patients after application of rTMS: An open protocol case series. *Brain Stimulation, 5*(3), 274–286. Available from: http://dx.doi.org/10.1016/j.brs.2011.03.005.

Batsikadze, G., Moliadze, V., Paulus, W., Kuo, M. F., & Nitsche, M. A. (2013). Partially non-linear stimulation intensity-dependent effects of direct current stimulation on motor cortex excitability in humans. *Journal of Physiology-London, 591*(7), 1987–2000. Available from: http://dx.doi.org/10.1113/jphysiol.2012.249730.

Bolognini, N., Olgiati, E., Maravita, A., Ferraro, F., & Fregni, F. (2013). Motor and parietal cortex stimulation for phantom limb pain and sensations. *Pain, 154*(8), 1274–1280. Available from: http://dx.doi.org/10.1016/j.pain.2013.03.040.

Brunoni, A. R., Valiengo, L., Baccaro, A., Zanao, T. A., de Oliveira, J. F., Vieira, G. P., et al. (2011). Sertraline vs. ELectrical Current Therapy for Treating Depression Clinical Trial—SELECT TDCS: design, rationale and objectives. *Contemporary Clinical Trials, 32*(1), 90–98. Available from: http://dx.doi.org/10.1016/j.cct.2010.09.007.

Campana, S., Caltagirone, C., & Marangolo, P. (2015). Combining voxel-based lesion symptoms mapping (VLSM) with A-tDCS language treatment: Predicting outcome of recovery in nonfluent chronic aphasia. *Brain Stimulation*, Jan 30. pii: S1935-861X(15)00877-3. http://dx.doi.org/10.1016/j.brs.2015.01.413. [Epub ahead of print].

Coslett, H. B., & Hamilton, R. (2011). Non-invasive brain current stimulation in neurorehabilitation. *Restorative Neurology and Neuroscience, 29*(6), 361–363. Available from: http://dx.doi.org/10.3233/RNN-2011-0626.

Cotelli, M., Calabria, M., Manenti, R., Rosini, S., Zanetti, O., Cappa, S. F., et al. (2011). Improved language performance in Alzheimer disease following brain stimulation. *Journal of Neurology, Neurosurgery, and Psychiatry, 82*(7), 794–797. Available from: http://dx.doi.org/10.1136/jnnp.2009.197848.

Cotelli, M., Manenti, R., Cappa, S. F., Zanetti, O., & Miniussi, C. (2008). Transcranial magnetic stimulation improves naming in Alzheimer disease patients at different stages of cognitive decline. *European Journal of Neurology, 15*(12), 1286–1292. Available from: http://dx.doi.org/10.1111/j.1468-1331.2008.02202.x.

Datta, A., Bansal, V., Diaz, J., Patel, J., Reato, D., & Bikson, M. (2009). Gyri-precise head model of transcranial direct current stimulation: improved spatial focality using a ring electrode versus conventional rectangular pad. *Brain Stimulation, 2*(4), 201–207, 207.e1. http://dx.doi.org/10.1016/j.brs.2009.03.005

Datta, A., Bikson, M., & Fregni, F. (2010). Transcranial direct current stimulation in patients with skull defects and skull plates: High-resolution computational FEM study of factors altering cortical current flow. *NeuroImage, 52*(4), 1268–1278. Available from: http://dx.doi.org/10.1016/j.neuroimage.2010.04.252.

de Aguiar, V., Paolazzi, C. L., & Miceli, G. (2015). tDCS in post-storke aphasia: The role of stimulation parameters, behavioral treatment and patient characteristics. *Cortex, 63*, 296–316.

Devlin, J. T., Matthews, P. M., & Rushworth, M. F. S. (2003). Semantic processing in the left inferior prefrontal cortex: A combined functional magnetic resonance imaging and transcranial magnetic stimulation study. *Journal of Cognitive Neuroscience, 15*(1), 71–84. Available from: http://dx.doi.org/10.1162/089892903321107837.

Devlin, J. T., & Watkins, K. E. (2007). Stimulating language: insights from TMS. *Brain: A Journal of Neurology, 130*(Pt 3), 610–622. Available from: http://doi.org/10.1093/brain/awl331.

Dickey, L., Kagan, A., Lindsay, M. P., Fang, J., Rowland, A., & Black, S. (2010). Incidence and profile of inpatient stroke-induced aphasia in Ontario, Canada. *Archives of Physical Medicine and Rehabilitation, 91*(2), 196–202. Available from: http://dx.doi.org/10.1016/j.apmr.2009.09.020.

Fiori, V., Coccia, M., Marinelli, C. V., Vecchi, V., Bonifazi, S., Ceravolo, M. G., et al. (2011). Transcranial direct current stimulation improves word retrieval in healthy and nonfluent aphasic subjects. *Journal of Cognitive Neuroscience, 23*(9), 2309–2323. Available from: http://dx.doi.org/10.1162/jocn.2010.21579.

Flöel, A., Meinzer, M., Kirstein, R., Nijhof, S., Deppe, M., Knecht, S., et al. (2011). Short-term anomia training and electrical brain stimulation. *Stroke, 42*(7), 2065–2067. Available from: http://dx.doi.org/10.1161/STROKEAHA.110.609032.

Flöel, A., Suttorp, W., Kohl, O., Kürten, J., Lohmann, H., Breitenstein, C., et al. (2012). Non-invasive brain stimulation improves object-location learning in the elderly. *Neurobiology of Aging, 33*(8), 1682–1689. Available from: http://dx.doi.org/10.1016/j.neurobiolaging.2011.05.007.

Fregni, F., Gimenes, R., Valle, A. C., Ferreira, M. J. L., Rocha, R. R., Natalle, L., et al. (2006). A randomized, sham-controlled, proof of principle study of transcranial direct current stimulation for the treatment of pain in fibromyalgia. *Arthritis and Rheumatism, 54*(12), 3988–3998. Available from: http://dx.doi.org/10.1002/art.22195.

Fridriksson, J., Richardson, J. D., Baker, J. M., & Rorden, C. (2011). Transcranial direct current stimulation improves naming reaction time in fluent aphasia: A double-blind, sham-controlled study. *Stroke, 42*(3), 819–821. Available from: http://dx.doi.org/10.1161/STROKEAHA.110.600288.

Fritsch, B., Reis, J., Martinowich, K., Schambra, H. M., Ji, Y., Cohen, L. G., et al. (2010). Direct current stimulation promotes BDNF-dependent synaptic plasticity: Potential implications for motor learning. *Neuron, 66*(2), 198–204. Available from: http://dx.doi.org/10.1016/j.neuron.2010.03.035.

George, M. S., Wassermann, E. M., & Post, R. M. (1996). Transcranial magnetic stimulation: A neuropsychiatric tool for the 21st century. *The Journal of Neuropsychiatry and Clinical Neurosciences, 8*(4), 373–382, PMID: 9116472.

Gough, P. M., Nobre, A. C., & Devlin, J. T. (2005). Dissociating linguistic processes in the left inferior frontal cortex with transcranial magnetic stimulation. *Journal of Neuroscience, 25*(35), 8010–8016. Available from: http://dx.doi.org/10.1523/JNEUROSCI.2307-05.2005.

Hamilton, R. H., Chrysikou, E. G., & Coslett, B. (2011). Mechanisms of aphasia recovery after stroke and the role of noninvasive brain stimulation. *Brain and Language, 118*(1–2), 40–50. Available from: http://dx.doi.org/10.1016/j.bandl.2011.02.005.

Hamilton, R. H., Sanders, L., Benson, J., Faseyitan, O., Norise, C., Naeser, M., et al. (2010). Stimulating conversation: Enhancement of elicited propositional speech in a patient with chronic non-fluent aphasia following transcranial magnetic stimulation. *Brain and Language, 113*(1), 45–50. Available from: http://dx.doi.org/10.1016/j.bandl.2010.01.001.

Heiss, W.-D., & Thiel, A. (2006). A proposed regional hierarchy in recovery of post-stroke aphasia. *Brain and Language, 98*(1), 118–123. Available from: http://dx.doi.org/10.1016/j.bandl.2006.02.002.

Heiss, W.-D., Hartmann, A., Rubi-Fessen, I., Anglade, C., Kracht, L., Kessler, J., et al. (2013). Noninvasive brain stimulation for treatment of right- and left-handed poststroke aphasics. *Cerebrovascular Diseases* (Basel, Switzerland), *36*(5-6), 363–372. Available from: http://doi.org/10.1159/000355499.

Huang, Y.-Z., & Rothwell, J. C. (2004). The effect of short-duration bursts of high-frequency, low-intensity transcranial magnetic stimulation on the human motor cortex. *Clinical Neurophysiology, 115*(5), 1069–1075. Available from: http://dx.doi.org/10.1016/j.clinph.2003.12.026.

Jahanshahi, M., & Rothwell, J. (2000). Transcranial magnetic stimulation studies of cognition: An emerging field. *Experimental Brain Research, 131*(1), 1–9. Available from: http://dx.doi.org/10.1007/s002219900224.

Jung, I.-Y., Lim, J. Y., Kang, E. K., Sohn, H. M., & Paik, N.-J. (2011). The factors associated with good responses to speech therapy combined with transcranial direct current stimulation in post-stroke aphasic patients. *Annals of Rehabilitation Medicine, 35*(4), 460–469. Available from: http://dx.doi.org/10.5535/arm.2011.35.4.460.

Kakuda, W., Abo, M., Kaito, N., Watanabe, M., & Senoo, A. (2010). Functional MRI-based therapeutic rTMS strategy for aphasic stroke patients: A case series pilot study. *The International Journal of Neuroscience, 120*(1), 60–66. Available from: http://dx.doi.org/10.3109/00207450903445628.

Kakuda, W., Abo, M., Momosaki, R., & Morooka, A. (2011). Therapeutic application of 6-Hz-primed low-frequency rTMS combined with intensive speech therapy for post-stroke aphasia. *Brain Injury: [BI], 25*(12), 1242–1248. Available from: http://dx.doi.org/10.3109/02699052.2011.608212.

Kakuda, W., Abo, M., Uruma, G., Kaito, N., & Watanabe, M. (2010). Low-frequency rTMS with language therapy over a 3-month period for sensory-dominant aphasia: Case series of two post-stroke Japanese patients. *Brain Injury: [BI], 24*(9), 1113–1117. Available from: http://dx.doi.org/10.3109/02699052.2010.494587.

Kang, E. K., Kim, Y. K., Sohn, H. M., Cohen, L. G., & Paik, N.-J. (2011). Improved picture naming in aphasia patients treated with cathodal tDCS to inhibit the right Broca's homologue area. *Restorative Neurology and Neuroscience, 29*(3), 141–152. Available from: http://dx.doi.org/10.3233/RNN-2011-0587.

Kelly, H., Brady, M. C., & Enderby, P. (2010). Speech and language therapy for aphasia following stroke. *The Cochrane Database of Systematic Reviews*(5), CD000425. Available from: http://dx.doi.org/10.1002/14651858.CD000425.pub2.

Kessler, S. K., Turkeltaub, P. E., Benson, J. G., & Hamilton, R. H. (2012). Differences in the experience of active and sham transcranial direct current stimulation. *Brain Stimulation, 5*(2), 155–162. Available from: http://dx.doi.org/10.1016/j.brs.2011.02.007.

Khedr, E. M., Ahmed, M. A., Fathy, N., & Rothwell, J. C. (2005). Therapeutic trial of repetitive transcranial magnetic stimulation after acute ischemic stroke. *Neurology, 65*(3), 466–468. Available from: http://dx.doi.org/10.1212/01.wnl.0000173067.84247.36.

Khedr, E. M., El-Fetoh, N. A., Ali, A. M., El-Hammady, D. H., Khalifa, H., Atta, H., et al. (2014). Dual-Hemisphere Repetitive Transcranial Magnetic Stimulation for Rehabilitation of Post-stroke Aphasia: A randomied, double-blind clinical trial. *Neurorehabilitation and Neural Repair, 28*, 740–750.

Khedr, E. M., Shawky, O. A., El-Hammady, D. H., Rothwell, J. C., Darwish, E. S., Mostafa, O. M., et al. (2013). Effect of anodal versus cathodal transcranial direct current stimulation on stroke rehabilitation: A pilot randomized controlled trial. *Neurorehabilitation and Neural Repair, 27*(7), 592–601. Available from: http://dx.doi.org/10.1177/1545968313484808.

Kindler, J., Schumacher, R., Cazzoli, D., Gutbrod, K., Koenig, M., Nyffeler, T., et al. (2012). Theta burst stimulation over the right Broca's homologue induces improvement of naming in aphasic patients. *Stroke, 43*(8), 2175–2179. Available from: http://dx.doi.org/10.1161/STROKEAHA.111.647503.

Koch, G., Bonnì, S., Giacobbe, V., Bucchi, G., Basile, B., Lupo, F., et al. (2012). θ-burst stimulation of the left hemisphere accelerates recovery of hemispatial neglect. *Neurology, 78*(1), 24–30. Available from: http://dx.doi.org/10.1212/WNL.0b013e31823ed08f.

Lindenberg, R., Renga, V., Zhu, L. L., Betzler, F., Alsop, D., & Schlaug, G. (2010). Structural integrity of corticospinal motor fibers predicts motor impairment in chronic stroke. *Neurology, 74*(4), 280–287. Available from: http://dx.doi.org/10.1212/WNL.0b013e3181ccc6d9.

Madhavan, S., Weber, K. A., & Stinear, J. W. (2011). Non-invasive brain stimulation enhances fine motor control of the hemiparetic ankle: Implications for rehabilitation. *Experimental Brain Research, 209*(1), 9–17. Available from: http://dx.doi.org/10.1007/s00221-010-2511-0.

Marangolo, P., Fiori, V., Calpagnano, M. A., Campana, S., Razzano, C., Caltagirone, C., et al. (2013). tDCS over the left inferior frontal cortex improves speech production in aphasia. *Frontiers in Human Neuroscience, 7*, 539. Available from: http://dx.doi.org/10.3389/fnhum.2013.00539.

Marangolo, P., Fiori, V., Di Paola, M., Cipollari, S., Razzano, C., Oliveri, M., et al. (2013). Differential involvement of the left frontal and temporal regions in verb naming: A tDCS treatment study. *Restorative Neurology and Neuroscience, 31*(1), 63–72. Available from: http://dx.doi.org/10.3233/RNN-120268.

Marangolo, P., Fiori, V., Camapana, S., Calpagnano, M. A., Razzano, C., Caltagirone, C., et al. (2014). Something to talk about: Enhancement of linguistic cohesion through tDCS in chronic non-fluent aphasia. *Neuropsychologia, 53*, 246–256.

Martin, P. I., Naeser, M. A., Ho, M., Doron, K. W., Kurland, J., Kaplan, J., et al. (2009). Overt naming fMRI pre- and post-TMS: Two nonfluent aphasia patients, with and without improved naming post-TMS. *Brain and Language, 111*(1), 20–35. Available from: http://dx.doi.org/10.1016/j.bandl.2009.07.007.

Martin, P. I., Naeser, M. A., Theoret, H., Tormos, J. M., Nicholas, M., Kurland, J., et al. (2004). Transcranial magnetic stimulation as a complementary treatment for aphasia. *Seminars in Speech and Language, 25*(2), 181–191. Available from: http://dx.doi.org/10.1055/s-2004-825654.

Medina, J., Beauvais, J., Datta, A., Bikson, M., Coslett, H. B., & Hamilton, R. H. (2013). Transcranial direct current stimulation accelerates allocentric target detection. *Brain Stimulation, 6*(3), 433–439. Available from: http://dx.doi.org/10.1016/j.brs.2012.05.008.

Medina, J., Norise, C., Faseyitan, O., Coslett, H. B., Turkeltaub, P. E., & Hamilton, R. H. (2012). Finding the right words: Transcranial magnetic stimulation improves discourse productivity in non-fluent aphasia after stroke. *Aphasiology, 26*(9), 1153–1168. Available from: http://dx.doi.org/10.1080/02687038.2012.710316.

Miniussi, C., Cappa, S. F., Cohen, L. G., Floel, A., Fregni, F., Nitsche, M. A., et al. (2008). Efficacy of repetitive transcranial magnetic stimulation/transcranial direct current stimulation in cognitive neurorehabilitation. *Brain Stimulation, 1*(4), 326–336. Available from: http://dx.doi.org/10.1016/j.brs.2008.07.002.

Miranda, P. C., Lomarev, M., & Hallett, M. (2006). Modeling the current distribution during transcranial direct current stimulation. *Clinical Neurophysiology: Official Journal of the International Federation of Clinical Neurophysiology, 117*(7), 1623–1629. Available from: http://dx.doi.org/10.1016/j.clinph.2006.04.009.

Monti, A., Cogiamanian, F., Marceglia, S., Ferrucci, R., Mameli, F., Mrakic-Sposta, S., et al. (2008). Improved naming after transcranial direct current stimulation in aphasia. *Journal of Neurology, Neurosurgery, and Psychiatry*, *79*(4), 451–453. Available from: http://dx.doi.org/10.1136/jnnp.2007.135277.

Mylius, V., Ayache, S. S., Zouari, H. G., Aoun-Sebaïti, M., Farhat, W. H., & Lefaucheur, J.-P. (2012). Stroke rehabilitation using noninvasive cortical stimulation: Hemispatial neglect. *Expert Review of Neurotherapeutics*, *12*(8), 983–991. Available from: http://dx.doi.org/10.1586/ern.12.78.

Naeser, M. A., Martin, P. I., Ho, M., Treglia, E., Kaplan, E., Bashir, S., et al. (2012). Transcranial magnetic stimulation and aphasia rehabilitation. *Archives of Physical Medicine and Rehabilitation*, *93*(1 Suppl.), S26–34. Available from: http://dx.doi.org/10.1016/j.apmr.2011.04.026.

Naeser, M. A., Martin, P. I., Nicholas, M., Baker, E. H., Seekins, H., Helm-Estabrooks, N., et al. (2005a). Improved naming after TMS treatments in a chronic, global aphasia patient—case report. *Neurocase*, *11*(3), 182–193. Available from: http://dx.doi.org/10.1080/13554790590944663.

Naeser, M. A., Martin, P. I., Nicholas, M., Baker, E. H., Seekins, H., Kobayashi, M., et al. (2005b). Improved picture naming in chronic aphasia after TMS to part of right Broca's area: An open-protocol study. *Brain and Language*, *93*(1), 95–105. Available from: http://dx.doi.org/10.1016/j.bandl.2004.08.004.

Naeser, M. A., Martin, P. I., Theoret, H., Kobayashi, M., Fregni, F., Nicholas, M., et al. (2011). TMS suppression of right pars triangularis, but not pars opercularis, improves naming in aphasia. *Brain and Language*, *119*(3), 206–213. Available from: http://dx.doi.org/10.1016/j.bandl.2011.07.005.

Nitsche, M. A., & Paulus, W. (2000). Excitability changes induced in the human motor cortex by weak transcranial direct current stimulation. *Journal of Physiology-London*, *527*(3), 633–639. Available from: http://dx.doi.org/10.1111/j.1469-7793.2000.t01-1-00633.x.

Nitsche, M. A., & Paulus, W. (2001). Sustained excitability elevations induced by transcranial DC motor cortex stimulation in humans. *Neurology*, *57*(10), 1899–1901.

Nitsche, M. A., Liebetanz, D., Antal, A., Lang, N., Tergau, F., & Paulus, W. (2003). Modulation of cortical excitability by weak direct current stimulation—technical, safety and functional aspects. *Supplements to Clinical Neurophysiology*, *56*, 255–276, PMID: 14677403.

Pascual-Leone, A., Tormos, J. M., Keenan, J., Tarazona, F., Cañete, C., & Catalá, M. D. (1998). Study and modulation of human cortical excitability with transcranial magnetic stimulation. *Journal of Clinical Neurophysiology*, *15*(4), 333–343, PMID: 9736467.

Pérez-Borrego, Y. A., Campolo, M., Soto-León, V., Rodriguez-Matas, M. J., Ortega, E., & Oliviero, A. (2014). Pain treatment using tDCS in a single patient: Tele-medicine approach in non-invasive brain simulation. *Brain Stimulation*, *7*(2), 334–335. Available from: http://dx.doi.org/10.1016/j.brs.2013.11.008.

Polanowska, K. E., Kesniak, M. M., Seniow, J. B., Czepiel, W., & Czlonkowska, A. (2013). Anodal transcranial direct current stimulation in early rehabilitation of patients with post-stroke non-fluent aphasia: A randomized, double-blind, sham-controlled pilot study. *Restorative Neurology and Neuroscience*, *31*, 761–771.

Purpura, D. P., & McMurtry, J. G. (1965). Intracellular activities and evoked potential changes during polarization of motor cortex. *Journal of Neurophysiology*, *28*, 166–185, PMID:14244793.

Reis, J., Schambra, H. M., Cohen, L. G., Buch, E. R., Fritsch, B., Zarahn, E., et al. (2009). Noninvasive cortical stimulation enhances motor skill acquisition over multiple days through an effect on consolidation. *Proceedings of the National Academy of Sciences of the United States of America*, *106*(5), 1590–1595. Available from: http://dx.doi.org/10.1073/pnas.0805413106.

Robey, R. R., & Dalebout, S. D. (1998). A tutorial on conducting meta-analyses of clinical outcome research. *Journal of Speech, Language, and Hearing Research: JSLHR*, *41*(6), 1227–1241, PMID: 9859880.

Santos, M. D., Gagliardi, R. J., Mac-Kay, A. P. M. G., Boggio, P. S., Lianza, R., & Fregni, F. (2013). Transcranial direct-current stimulation induced in stroke patients with aphasia: a prospective experimental cohort study. *Sao Paulo Medical Journal = Revista Paulista de Medicina*, *131*(6), 422–426. Available from: http://doi.org/10.1590/1516-3180.2013.1316595.

Seniów, J., Waldowski, K., Leśniak, M., Iwański, S., Czepiel, W., & Członkowska, A. (2013). Transcranial magnetic stimulation combined with speech and language training in early aphasia rehabilitation: A randomized double-blind controlled pilot study. *Topics in Stroke Rehabilitation*, *20*(3), 250–261. Available from: http://dx.doi.org/10.1310/tsr2003-250.

Stagg, C. J., Best, J. G., Stephenson, M. C., O'Shea, J., Wylezinska, M., Kincses, Z. T., et al. (2009). Polarity-sensitive modulation of cortical neurotransmitters by transcranial stimulation. *Journal of Neuroscience*, *29*(16), 5202–5206. Available from: http://dx.doi.org/10.1523/JNEUROSCI.4432-08.2009.

Suh, H. S., Lee, W. H., & Kim, T.-S. (2012). Influence of anisotropic conductivity in the skull and white matter on transcranial direct current stimulation via an anatomically realistic finite element head model. *Physics in Medicine and Biology*, *57*(21), 6961–6980. Available from: http://dx.doi.org/10.1088/0031-9155/57/21/6961.

Sunwoo, H., Kim, Y.-H., Chang, W. H., Noh, S., Kim, E.-J., & Ko, M.-H. (2013). Effects of dual transcranial direct current stimulation on post-stroke unilateral visuospatial neglect. *Neuroscience Letters*. Available from: http://dx.doi.org/10.1016/j.neulet.2013.08.064.

Szaflarski, J. P., Eaton, K., Ball, A. L., Banks, C., Vannest, J., Allendorfer, J. B., et al. (2011). Poststroke aphasia recovery assessed with functional magnetic resonance imaging and a picture identification task. *Journal of Stroke and Cerebrovascular Diseases*, *20*(4), 336–345. Available from: http://dx.doi.org/10.1016/j.jstrokecerebrovasdis.2010.02.003.

Thiel, A., Habedank, B., Herholz, K., Kessler, J., Winhuisen, L., Haupt, W. F., et al. (2006). From the left to the right: How the brain compensates progressive loss of language function. *Brain and Language*, *98*(1), 57–65. Available from: http://doi.org/10.1016/j.bandl.2006.01.007.

Thiel, A., Hartmann, A., Rubi-Fessen, I., Anglade, C., Kracht, L., Weiduschat, N., et al. (2013). Effects of noninvasive brain stimulation on language networks and recovery in early poststroke aphasia. *Stroke*, *44*(8), 2240–2246. Available from: http://dx.doi.org/10.1161/STROKEAHA.111.000574.

Torres, J., Drebing, D., & Hamilton, R. (2013). TMS and tDCS in post-stroke aphasia: Integrating novel treatment approaches with mechanisms of plasticity. *Restorative Neurology and Neuroscience*, *31*(4), 501–515. Available from: http://dx.doi.org/10.3233/RNN-130314.

Turkeltaub, P. E., Messing, S., Norise, C., & Hamilton, R. H. (2011). Are networks for residual language function and recovery consistent across aphasic patients? *Neurology*, *76*(20), 1726–1734. Available from: http://doi.org/10.1212/WNL.0b013e31821a44c1.

Van der Knaap, L. J., & van der Ham, I. J. M. (2011). How does the corpus callosum mediate interhemispheric transfer? A review. *Behavioural Brain Research*, *223*(1), 211–221. Available from: http://dx.doi.org/10.1016/j.bbr.2011.04.018.

Vines, B. W., Norton, A. C., & Schlaug, G. (2011). Non-invasive brain stimulation enhances the effects of melodic intonation therapy. *Frontiers in Psychology*, *2*, 230. Available from: http://dx.doi.org/10.3389/fpsyg.2011.00230.

Volpato, C., Cavinato, M., Piccione, F., Garzon, M., Meneghello, F., & Birbaumer, N. (2013). Transcranial direct current stimulation (tDCS) of Broca's area in chronic aphasia: A controlled outcome study. *Behavioural Brain Research*, *247*. Available from: http://dx.doi.org/10.1016/j.bbr.2013.03.029.

Waldowski, K., Seniów, J., Leśniak, M., Iwański, S., & Członkowska, A. (2012). Effect of low-frequency repetitive transcranial magnetic stimulation on naming abilities in early-stroke aphasic patients: A prospective, randomized, double-blind sham-controlled study. *TheScientificWorldJournal*, *2012*, 518568. Available from: http://dx.doi.org/10.1100/2012/518568.

Wassermann, E. M. (1998). Risk and safety of repetitive transcranial magnetic stimulation: report and suggested guidelines from the International Workshop on the Safety of Repetitive Transcranial Magnetic Stimulation, June 5-7, 1996. *Electroencephalography and Clinical Neurophysiology*, *108*(1), 1–16, PMID: 9474057.

Weiduschat, N., Thiel, A., Rubi-Fessen, I., Hartmann, A., Kessler, J., Merl, P., et al. (2011). Effects of repetitive transcranial magnetic stimulation in aphasic stroke: A randomized controlled pilot study. *Stroke*, *42*(2), 409–415. Available from: http://dx.doi.org/10.1161/STROKEAHA.110.597864.

Winhuisen, L., Thiel, A., Schumacher, B., Kessler, J., Rudolf, J., Haupt, W. F., & Heiss, W. D. (2007). The right inferior frontal gyrus and poststroke aphasia: a follow-up investigation. *Stroke; a Journal of Cerebral Circulation*, *38*(4), 1286–1292. Available from: http://dx.doi.org/10.1161/01.STR.0000259632.04324.6c.

You, D. S., Kim, D.-Y., Chun, M. H., Jung, S. E., & Park, S. J. (2011). Cathodal transcranial direct current stimulation of the right Wernicke's area improves comprehension in subacute stroke patients. *Brain and Language*, *119*(1), 1–5. Available from: http://dx.doi.org/10.1016/j.bandl.2011.05.002.

CHAPTER

84

Imitation-Based Aphasia Therapy

E. Susan Duncan[1] *and Steven L. Small*[2]

[1]Solodkin/Small Brain Circuits Laboratory, Department of Neurology, University of California, Irvine, Irvine, CA, USA; [2]Department of Neurology, University of California, Irvine, Irvine, CA, USA

84.1 INTRODUCTION: REPETITION AND IMITATION IN APHASIA

In 1683, the German physician Peter Rommel was the first to write about repetition deficits in a patient with nonfluent aphasia (Benton & Joynt, 1963). Imitation has since been a key diagnostic and treatment tool for such acquired language disorders. All popular standardized instruments for the assessment of aphasia, such as the Western Aphasia Battery-Revised (WAB-R) (Kertesz, 2006), Boston Diagnostic Aphasia Examination (Goodglass & Kaplan, 1983), and Aachen Aphasia Test (Huber, Poeck, & Willmes, 1984), include repetition ability in their classification scheme.

Imitation was fundamental to the nascent field of aphasia therapy at the turn of the 20th century, and it remains so today (Duffy, 1995). This chapter describes the neurobiological rationales for, and current implementations of, imitation in aphasia therapy. "Imitation" and "repetition" are used interchangeably, and modes of stimulus presentation are clarified as needed. Additionally, it should be noted that acquired apraxia of speech, a motor planning deficit frequently accompanying nonfluent aphasia (Duffy, 1995), is not specifically addressed in this chapter.

84.2 NEUROBIOLOGICAL APPROACHES TO LANGUAGE AND APHASIA

The earliest approaches were based on behavioral and educational principles, and this philosophy dominates aphasia therapy today (Small, 2004). Aphasia is a neurological impairment resulting from brain damage, typically stroke (Ellis, Dismuke, & Edwards, 2010), yet treatment programs are rarely biologically motivated. Since the end of the 20th century, studies in aphasia have been relying less on applied psychology and linguistic models, instead seeking to link observed deficits to impairments in the underlying neural systems (Blumstein, 1997). Rehabilitation of the behavioral deficits of aphasia must target the plasticity and repair of affected biological systems. Two main biological models characterizing these systems are considered here, the human mirror system and the dual-stream hypothesis for speech.

84.3 MIRROR NEURON SYSTEM

84.3.1 Macaque

Mirror neurons were discovered serendipitously during single-cell recordings of hand motor representations in the macaque. Rizzolatti and colleagues found neurons firing in premotor cortex (area F5) in a motionless monkey during observation of the experimenter (Di Pellegrino, Fadiga, Fogassi, Gallese, & Rizzolatti, 1992). Individual neurons were active during observation and execution for hand and mouth movements (Ferrari, Gallese, Rizzolatti, & Fogassi, 2003; Gallese, Fadiga, Fogassi, & Rizzolatti, 1996). Additional mirror neurons possessing visuomotor properties were subsequently identified in the inferior parietal region of the macaque (Fogassi, Gallese, Fadiga, & Rizzolatti, 1998), primarily in subcomponents PF and PFG (Rozzi, Ferrari, Bonini, Rizzolatti, & Fogassi, 2008), which have strong anatomical projections to the ventral premotor cortex (F5). These findings led to the suggestion of a functional "mirror" network (Rozzi et al., 2006).

The existence of mirror neurons immediately prompted hypotheses about their role in action recognition (Rizzolatti, Fadiga, Gallese, & Fogassi, 1996). Further support for this has been provided by the discovery that some mirror neurons in macaque F5 have auditory as well as visual and motor properties

Neurobiology of Language. DOI: http://dx.doi.org/10.1016/B978-0-12-407794-2.00091-2

(Kohler et al., 2002), firing in response to observation and execution of actions, and for sounds associated with those actions. This multimodal integration at the level of a single cell may form the basis for action understanding and motor learning (Jeannerod, 1994).

84.3.2 Human

Ethical considerations prohibit systematic human studies investigating individual mirror neurons. However, support for the existence of a human parieto-frontal mirror neuron system is converging from behavioral, neurophysiological, and brain imaging studies (Small, Buccino, & Solodkin, 2012). The "direct matching hypothesis" postulates that imitation is subserved by simple neural mechanisms mapping observed actions onto internal motor representations of the same action by neurons with mirror properties, which are more strongly activated for actions elicited by preceding observations (Iacoboni et al., 1999).

84.3.2.1 *Behavioral*

Behavioral studies demonstrate motor facilitation when action execution immediately follows observation, supporting the existence of a mirror system in humans. Finger movements are faster if the stimulus cue is a modeled finger movement compared to an unrelated symbol (Brass, Bekkering, Wohlschläger, & Prinz, 2000). Response speed further increases as the modeled movement more closely resembles the target, even when the stimulus image is flipped upside-down (Brass, Bekkering, & Prinz, 2001). Grasping response speed increases when subjects are shown a picture of a hand with optimal orientation for their own final hand position (Craighero, Bello, Fadiga, & Rizzolatti, 2002). On language tasks, response times for plausibility judgments are faster when the action response required is similar to the action described in the stimulus sentence (Glenberg & Kaschak, 2002).

84.3.2.2 *Neurophysiology*

Studies using electroencephalography (EEG) demonstrate a central mu rhythm in the alpha frequency range (8–13 Hz) present when the subject is at rest. This is suppressed during action observation, as was first described in 1954 (Cohen-Seat, Gastaut, Faure, & Heuyer, 1954). These findings have since been replicated for observation, imitation, and execution of actions with EEG (Altschuler et al., 2000; Cochin, Barthelemy, Roux, & Martineau, 1999). Responses measured via implanted subdural electrodes show a reduction in absolute power in the alpha band over primary motor cortex and Broca's area for both observation and execution of finger movements (Tremblay et al., 2004).

Mu suppression is stronger for grasping than for movements that are not goal-oriented in adults (Muthukumaraswamy & Johnson, 2004) and in children (Lepage & Théoret, 2007). A precursor to the mu rhythm, with overlapping reduction for observed and executed grasping movements, is found in infants in the frequency range of 6–9 Hz (Marshall, Young, & Meltzoff, 2011). It is proposed that mirror neurons underlie early childhood imitation, language acquisition, and the development of other social and cognitive functions (Williams, Whiten, Suddendorf, & Perrett, 2001).

Magnetoencephalography (MEG) reveals a band of activity of 15–25 Hz in the precentral motor cortex during rest. Suppressing this activity via upper extremity median nerve stimulation allows study of its rebound in varying contexts. This rebound is extinguished during object manipulation after stimulation and is significantly reduced during passive observation of the same task (Hari et al., 1998).

Transcranial magnetic stimulation (TMS) manipulates cortical responses by either inducing or inhibiting action potentials. TMS-induced motor-evoked potentials (MEPs) demonstrate increased excitability when observing grasping actions and arm movements using the same muscles (Fadiga, Fogassi, Pavesi, & Rizzolatti, 1995). These results have been replicated using observation of handwriting and arm movements compared to rest (Strafella & Paus, 2000).

These neurophysiological findings suggest the influence of mirror properties on the human motor system extends to primary motor areas, in addition to the postulated premotor homologues of the frontal regions where macaque mirror neurons have been identified. Greater extension still has been proposed. In single-cell recordings from subjects with medically intractable epilepsy, a significant number of neurons in the supplementary motor area (SMA) and the medial temporal lobe (MTL) respond to the observation and execution of a single action (Mukamel, Ekstrom, Kaplan, Iacoboni, & Fried, 2010). These regions, with clinical rather than theoretical determination of electrode placement, have not previously shown mirror properties. Additionally, some neurons responded with increased excitation for execution but suppressed firing rate for observation, unlike macaque studies. The authors propose that these findings may provide evidence of multiple mirroring systems in the brain, with reduced activity of some neurons during observation playing a role in suppressing socially inappropriate imitation.

84.3.2.3 *Brain Imaging*

Brain imaging studies permit greater spatial localization of mirror properties in humans. Early evidence came from a positron emission tomography (PET) study contrasting object observation with action observation (Rizzolatti, Fadiga, Matelli, et al., 1996). Activation in response to action observation was found in Broca's area and left hemisphere temporal regions (middle temporal gyrus, superior temporal sulcus). Mirror properties have since been demonstrated via functional magnetic resonance imaging (fMRI) in two regions of the human brain active during both passive observation and imitation of finger movements, the left frontal operculum of Broca's area and the right anterior parietal region (Iacoboni et al., 1999).

Broca's area, the putative frontal oral-motor and speech area in localist language models, plays a role in hand motor representation (Binkofski et al., 1999) and is typically identified as the human homologue of macaque F5 in which mirror neurons have been recorded (Rizzolatti, Fadiga, Gallese, et al., 1996). However, the consistent finding of activation in response to observation of hand and arm actions (Decety et al., 1997; Grafton, Arbib, Fadiga, & Rizzolatti, 1996) in Broca's region, in combination with its long history in the neuroscience of language, has raised the question of whether this increased neural response is perhaps an epiphenomenal artifact of internal speech during these tasks.

FMRI investigation of responses to actions performed by the hand, foot, or mouth reveals a somatotopic organization of premotor cortex similar to that found in the primary sensory and motor cortices with ventral mouth movements and dorsal foot movements (Buccino et al., 2001). Similar organization is found in the posterior parietal lobe for object-related actions. These findings ground single neuron measures from macaque in a broader network of motor circuitry underlying both action observation and execution in humans.

Macaque mirror neurons fire for observation of grasping only in the presence of a graspable object, even if it is not visible (Umiltá et al., 2001). It has been suggested that this system is not encoding simple movements, but goal-oriented motor acts (Gallese et al., 1996). In human PET scans, left frontal and temporal regions are activated for meaningful, but not meaningless, actions (Decety et al., 1997). With fMRI, actions embedded in contexts show increased activation in the posterior inferior frontal gyrus (IFG) and ventral premotor cortex (vPM) compared to viewing either the action or the context alone (Iacoboni et al., 2005). This is especially pertinent to the discussion of language, in which we use our motor systems to transmit meaningful messages, with mirror neurons bridging the gap between "doing" and "communicating" (Rizzolatti & Arbib, 1998).

84.4 MIRROR NEURON SYSTEM AND LANGUAGE

Language, as a uniquely human property, lacks an ideal animal model, and biological theories of language cannot be directly tested. However, indirect evidence supports a role for the mirror neuron system in human language ability, both phylogenetically through evolutionary selection processes (Rizzolatti & Arbib, 1998) and ontogenetically in facilitating child language acquisition (Kuhl & Rivera-Gaxiola, 2008). It is suggested that mirror deficits may underlie developmental disorders of language and social interaction, notably autism (Williams et al., 2001), although such issues are controversial and beyond the scope of this text.

84.4.1 Perception and Production of Articulated Speech

FMRI reveals somatotopy in human frontal and parietal regions having mirror properties with ventral mouth activation compared to hands or feet, as in primary motor and somatosensory cortices (Buccino et al., 2001). The observation–execution or direct matching hypothesis suggests that action perception, including speech, depends on previous experience producing those actions or sounds (Iacoboni et al., 1999). One model of this is the "inverse-forward model pairs" (IFMPs) (Skipper, Nusbaum, & Small, 2006). These are mechanistic components of the mirror system, in which speech sounds, heard or observed, are transformed into corresponding articulatory gestures (inverse) and motor predictions (forward) with resultant sensory consequences affecting perception. These IFMPs operate in the multisensory contexts in which we experience language, consisting of acoustic signals and also the visual cues of oral, facial, manual, and body gestures, particularly when auditory information is distorted or ambiguous.

84.4.1.1 Neurophysiology

Listening to the lingual trill /r/ results in significantly increased amplitude in tongue muscle MEPs in neurologically intact participants compared to the non-lingual labiodental phoneme /f/ or thumb muscle MEPs (Fadiga, Craighero, Buccino, & Rizzolatti, 2002). This response is more pronounced for real words compared to pseudowords. Similarly, increased MEPs in oral muscles, but not finger muscles, are found during listening to connected speech and viewing silent video of speech-related lip movements (Watkins, Strafella, & Paus, 2003). This contrast is not found for nonspeech control conditions, including nonverbal sounds and

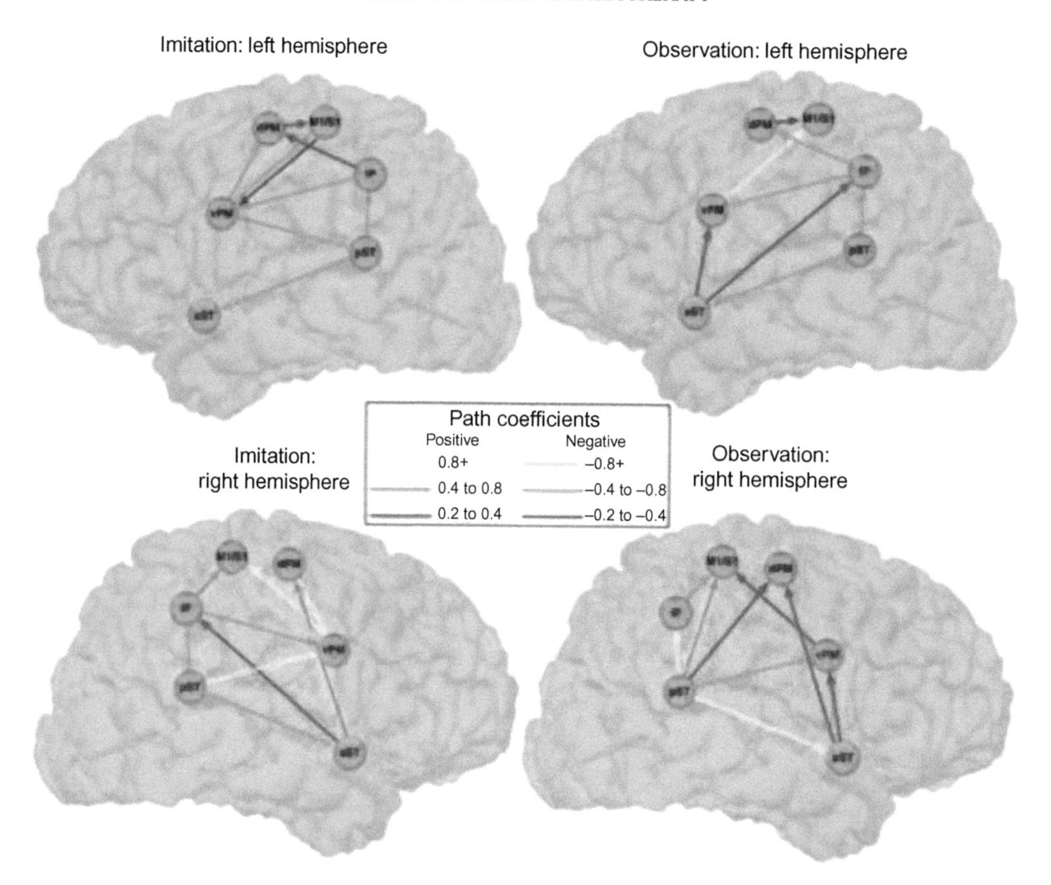

FIGURE 84.1 Weighted connections obtained from SEM of fMRI during observation (right) and imitation (left) of audiovisual syllables. Connections are shown for the left (top) and right (bottom) hemispheres. Both models share connections between pST, aST, IP, vPM, dPM, and M1S1. Abbreviations are as follows: IP, inferior parietal lobule; M1S1, primary motor/somatosensory cortex; pST, posterior superior temporal gyrus and sulcus; aST, anterior superior temporal gyrus and sulcus; vPM, ventral premotor cortex; dPM, dorsal premotor cortex; M1/S1, primary motor/somatosensory cortex. *Figure as originally published in Mashal et al. (2012).*

observation of eye movements. Consistent with widely accepted lateralization theories of speech and language, this follows stimulation of the left, but not the right, hemisphere.

84.4.1.2 *Brain Imaging*

84.4.1.2.1 MOTOR REGIONS ENGAGED DURING SPEECH PERCEPTION

Brain imaging studies provide indirect evidence of a relationship between speech observation and execution. Speech motor regions are engaged in response to audiovisual (Skipper, van Wassenhove, Nusbaum, & Small, 2007; Watkins et al., 2003), visual (Nishitani & Hari, 2002), and auditory (Fadiga et al., 2002; Tettamanti et al., 2005) speech perception. Figure 84.1 shows regions active during both syllable production and passive observation of audiovisual, visual, or auditory speech. Bilateral brain activation is present in premotor regions and Broca's area during silent lip-reading (Buccino et al., 2004), indicating that frontal motor cortices are activated in response to multimodal aspects of speech perception.

Structural equation modeling (SEM) shows common functional connections shared by observation and imitation of audiovisual syllables during fMRI, differing in connection strength but sharing the same essential structure (Mashal, Solodkin, Dick, Chen, & Small, 2012). Like imitation, speech observation engages dorsal and ventral premotor cortices and primary motor cortex.

84.4.1.2.2 TEMPORAL AND AUDITORY REGIONS ENGAGED DURING SPEECH PRODUCTION

Regions in posterior auditory cortex are active during speech production, including covert speech, as well as speech perception (Okada & Hickok, 2006; Papathanassiou et al., 2000). Further, posterior lesions of the left temporal cortex, as in Wernicke's aphasia, are associated with verbal expression and comprehension errors (Damasio & Geschwind, 1984). Temporal regions have classically been excluded from the putative mirror neuron system, because there has been no finding of motor activation in the temporal lobe in macaque studies (Keysers & Perrett, 2004). Still, the existence of individual mirror neurons remains poorly defined in humans due to limitations of appropriately noninvasive methods. It is possible that their cortical distribution is more extensive than that in our primate cousins (Mukamel et al., 2010).

84.4.2 Comprehension of Action Language

The human mirror system operates in tandem with low-level sensorimotor aspects of speech and also higher-level language comprehension (Barsalou, 2008). Listening to sentences describing motor activity activates a broad left-lateralized network of frontal, temporal, and parietal regions, as do action observation and execution, which do not occur for sentences not encoding action (Tettamanti et al., 2005). Listening to or reading action-related language evokes somatotopic motor cortex activation consistent with the described effector (Aziz-Zadeh, Wilson, Rizzolatti, & Iacoboni, 2006; Hauk, Johnsrude, & Pulvermüller, 2004).

Theories proposing that cognition is grounded in (superimposed on) basic sensory and motor processes also apply to language. Priming effects are found for objects sharing affordances, such as a piano and a typewriter, even when the task does not address the object's use (Myung, Blumstein, & Sedivy, 2006). Subjects receiving verbal or visual cues to assume certain handshapes are faster to respond to the plausibility of action–object pairings congruous with the simulated grasp (Klatzky, Pellegrino, McCloskey, & Doherty, 1989). This difference disappears when a verbal response replaces the motor response, indicating that the interplay between language processing and the motor system confers the advantage, rather than the semantic relationship.

84.5 DUAL STREAMS FOR SPEECH

The dual-stream theory of vision has long been dominant (Mishkin, Ungerleider, & Macko, 1983), with a ventral "what" pathway for object identification and a dorsal "where" or "how" pathway for visuomotor integration (Goodale, 1993). More recently, two stream models have been identified in the study of audition (Rauschecker & Tian, 2000), speech perception (Hickok & Poeppel, 2004), speech production (Hickok & Poeppel, 2007), and sentence comprehension (Friederici, 2009). There is considerable debate among varying theories regarding specific functions, regions, connections, and the role of feedback. Broadly, however, the dorsal stream progresses from temporoparietal regions to frontal premotor areas, whereas the ventral stream progresses through temporal lobes to prefrontal cortex.

The present discussion only superficially describes the putative roles of the two streams to elucidate their role in imitation. The interested reader is referred to the original sources, including those described here, and the Large-Scale Models section of this book. The streams are typically discussed separately, as here, but it should be understood that this division is an artificial one for the sake of simplicity rather than accuracy. In actuality, the streams must be integrated for successful functioning, operating through "cooperative computation" (Fagg & Arbib, 1998). The strongest neurobiological models underpinning language processing in the brain presently comprise the dual-stream model and the mirror neuron system.

84.5.1 The Ventral Stream

The ventral pathway is conceived in terms of semantics, extracting meaning from the communicative signal (Saur et al., 2008). It is considered to be bilateral in some models (Hickok & Poeppel, 2007), whereas others identify ventral auditory language pathways only in the left hemisphere (Parker et al., 2005). In the temporal lobe, it includes anterior portions of the superior temporal gyrus and sulcus, the middle and inferior temporal gyri, and the temporal poles. In the macaque, these areas connect to frontal regions including orbitofrontal cortex and pars orbitalis via the uncinate fasciculus, and pars opercularis and pars triangularis via the extreme capsule (Petrides & Pandya, 2009).

Imitation of familiar and meaningful actions is positively correlated with ventral activity in the inferior temporal cortex (Decety et al., 1997). This is consistent with the object identification, or "what," role of the visual ventral stream, which may share connectivity analogous to that of the auditory ventral stream (Seltzer & Pandya, 1978). In contrast, dorsal parieto-occipital activation occurs with imitation of novel and meaningless actions (Rumiati et al., 2005).

Words and sentences, which can be conceived of as meaningful "gestures" or speech "objects," also represent a domain of the ventral stream. Consistent with this, temporal lobe atrophy is associated with semantic dementia, a variant of frontotemporal dementia characterized by progressive deficits in confrontation naming and single word comprehension, and loss of the concepts associated with the language (Mummery et al., 2000), whereas speech repetition remains intact (Gorno-Tempini et al., 2011). Although the role of the ventral stream in imitation may seem minimal, the stimuli to be repeated might engage regions and enhance connectivity patterns differently depending on semantic meaning and social relevance (Kilner, Marchant, & Frith, 2006), such as personal significance, ecological validity, and familiarity.

84.5.2 The Dorsal Stream and Parietal Cortical Connectivity

In contrast with the semantic role of the ventral stream, analogous to the "where" or "how" role of the visual dorsal stream, the dorsal stream for speech is proposed as a sensorimotor network mapping sounds onto motor plans for production (Hickok & Poeppel, 2007). Repetition, especially of meaningless pseudowords, is the prototypical task of dorsal stream function (Saur et al., 2008).

The dorsal stream projects from primary auditory regions to posterior superior temporal and inferior parietal regions, and then to more posterior regions of the frontal lobe (including pars opercularis of the IFG and premotor and motor cortices) compared with ventral stream projections (Skipper et al., 2006). At a gross anatomical level, temporal and parietal regions of the dorsal stream associated with speech and language functions are shared by the dorsal stream for vision. All of these regions show strong activation on fMRI during action observation, especially in the left hemisphere (Decety & Grèzes, 1999). These regions, specifically pars opercularis and inferior parietal cortex, are considered the human homologues of macaque F5 and PF/PFG in which mirror neurons have been identified (Rizzolatti, Fogassi, & Gallese, 2001; Rozzi et al., 2006). Some models consider the dorsal stream for speech to be strongly left-dominant (Hickok & Poeppel, 2007).

In conduction aphasia, the ability to repeat is disproportionately impaired. This is classically attributed to damage to the left arcuate fasciculus (Geschwind, 1965), traditionally thought to serve as the primary dorsal pathway (Anderson et al., 1999); however, this is debated due to recent anatomical work implicating the superior longitudinal fasciculus (Schmahmann & Pandya, 2006). Voxelwise lesion symptom mapping with perfusion and diffusion-weighted imaging shows damage involving the left supramarginal gyrus or underlying white matter is most strongly related to repetition deficits (Fridriksson et al., 2010). The temporoparietal region has been identified by other investigators (Buchsbaum et al., 2011) to be implicated in repetition impairment, supporting the critical role of the dorsal stream in repetition and the presence of mirror neurons within this functional network. This also suggests that the arcuate fasciculus may not serve the crucial role once suggested for the interconnection of Broca's and Wernicke's areas.

84.6 APHASIA THERAPY: SPEECH IMITATION AS THERAPEUTIC TOOL

Imitation has a long history in therapy for communication disorders, including aphasia, using visual input to complement other sensory modalities and enhance the patient's ability to produce accurate speech output (Duffy, 1995). Early approaches to aphasia therapy were developed prior to the fundamental work of the 20th century in learning theory and the unfortunate consequences of World War II, which produced many young veterans with head injuries, and often relied solely on repetition and drilling (Basso, 2003). Later researchers continued to use repetition in aphasia rehabilitation, but within a better-defined theoretical framework. For example, in the Helm Elicited Language Program for Syntax Stimulation (Helm-Estabrooks, 1981) designed to treat agrammatism, increasingly more complex syntactic forms are introduced by imitation at level A before the same forms are elicited in context at level B.

Love and Webb (1977) found imitation to elicit the most accurate picture naming in patients with severe Broca's aphasia. Repetition is often the simplest level of a cueing hierarchy (Linebaugh, Shisler, & Lehner, 2005). Although imitation is sometimes promoted as a technique to be used only when no other prompts cue correct responses, its ability to facilitate speech output make it inherently error-reducing. Thus, it is a useful tool and desirable starting point in errorless learning designs, in which every response, regardless of accuracy, is viewed as self-reinforcing, and the therapy environment is structured to produce the greatest possible successes (Sigurðardóttir & Sighvatsson, 2006). However, the benefit of errorless learning remains debated in aphasia rehabilitation (Fillingham, Sage, & Lambon Ralph, 2005).

Many aphasia therapies used in research do not cite imitation as a rationale for their use or theorized effectiveness, yet they still rely heavily on imitation or choral reading (online imitation) in their implementation. Semantic Feature Analysis (SFA) seeks to improve word retrieval by targeting conceptual connections of individually trained words, using modeling and repetition of the target word and its semantic associations when these are not produced independently (Boyle & Coelho, 1995). Melodic Intonation Therapy (MIT), recommended for patients with nonfluent aphasia and poor repetition, uses melody and rhythm to increase speech output, relying on choral productions of intoned targets before progressing to imitation and more naturalistic contexts (Helm-Estabrooks, Morgan, & Nicholas, 1989). Conversational script training introduces scripts to be learned via online and delayed imitation (Youmans, Holland, Munoz, & Bourgeois, 2005). Choral reading and imitation are also paired with written stimuli in some therapy programs, such as Oral Reading Treatment (Orjada & Beeson, 2005) and Oral Reading for Language in Aphasia (Cherney, 2004).

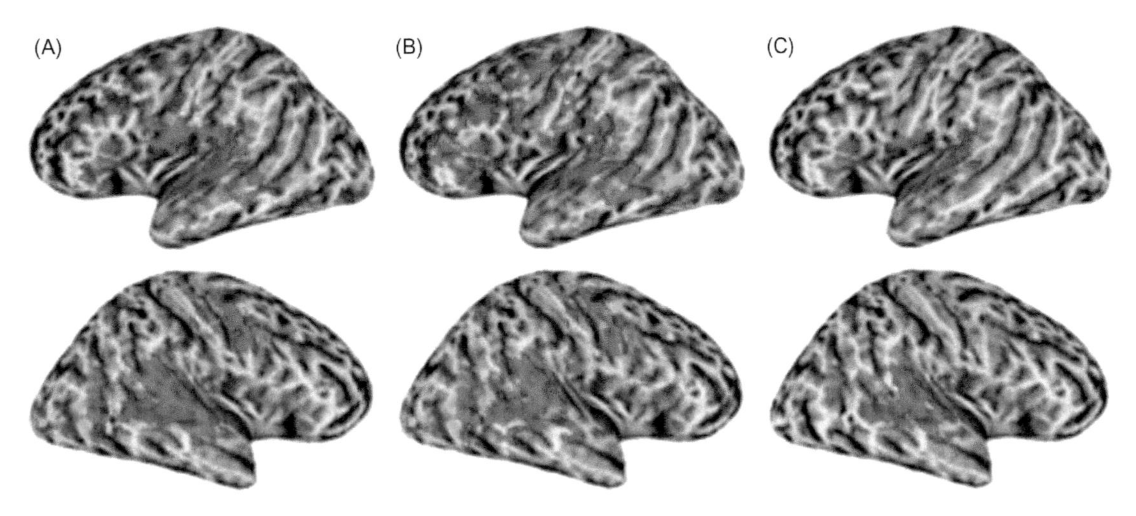

FIGURE 84.2 Logical conjunction analyses from fMRI of production and perception of the same syllables. Orange indicates regions of activation overlap between production and perception (thresholded at $p < 0.05$). Blue indicates regions active during passive perception but not during production (thresholded at $p < 0.05$). Stimuli for speech perception were audiovisual (A), visual only (B), or auditory only (C). *Reproduced with permission from Oxford University Press from Skipper et al. (2007).*

84.7 MIRROR NEURON SYSTEM AND REHABILITATION

Given evidence for motor system activation during action observation (Buccino et al., 2001), and given identification of neural circuits active during observation–execution of oral movements (Ferrari et al., 2003), there is a sound biological basis for speech imitation as an aphasia rehabilitation technique. Connections between inferior parietal and ventral premotor regions are active during observation and imitation of syllables, as seen in Figure 84.2, and may represent a human mirror neuron network for speech (Mashal et al., 2012). Although the most straightforward implication of engaging this system may be for the direct motor act of speech production, the role of this network in speech perception (Möttönen & Watkins, 2012) and comprehension of action language (Tettamanti et al., 2005) could result in a broader impact on more general aspects of language rehabilitation (Small et al., 2012).

A similar approach undertaken in hand motor rehabilitation following stroke comprises viewing videos of daily actions followed by therapist-assisted performance of observed actions with the impaired upper extremity (Ertelt et al., 2007). Patients demonstrate significant improvement following therapy compared to baseline performance or controls, with maintenance of at least 8 weeks after intervention. Increased activation during object manipulation was found with fMRI in contralateral supramarginal gyrus and bilateral ventral premotor cortices, SMA, and superior temporal gyri, consistent with human correlates of the macaque mirror neuron system (Small et al., 2012).

84.8 APHASIA THERAPY: SPEECH IMITATION AS THERAPEUTIC THEORY

84.8.1 IMITATE

IMITATE (Intensive Mouth Imitation and Talking for Aphasia Therapeutic Effect) is a novel computer-based aphasia therapy program to improve communication skills in aphasia by repetition of audiovisual words and phrases, motivated by neurophysiological findings in human and nonhuman primates (Lee, Fowler, Rodney, Cherney, & Small, 2010). The stimuli are presented by video featuring a view of the speaker's head and shoulders. The therapy is intense and uses ecologically valid stimuli presented by a variety of human talkers, and difficulty increases are graded overall yet are variable within a level.

The control therapy, REPEAT, uses similar principles but audio-only stimuli with a still image of the talker. This therapy also varies the stimulus presentation, such that subjects hear a single presentation by a single talker before each cued repetition, in contrast with the IMITATE group, which hears six consecutive talkers present each stimulus before repeating the target word or phrase several times. Each group hears the same overall number of stimuli and the same number of presentations.

Nineteen subjects completed a 6-week course of therapy (9 h weekly). The IMITATE group made significant gains in the Aphasia Quotient composite score of the WAB-R but the REPEAT group did not, although there were no significant differences between groups (Duncan, Schmah, & Small, in prep).

In a sleep study, high-density electroencephalographic (EEG) recordings were taken for 13 subjects with aphasia on two consecutive nights, before and after participating in a single, highly intensive 3.5-h session of IMITATE (Sarasso et al., 2014). Findings indicate a significant increase in slow wave activity (SWA), associated with synaptic plasticity (Huber, Ghilardi, Massimini, & Tononi, 2004), in regions active during observation–execution of speech in healthy controls (Mashal et al., 2012) in the right (intact) hemisphere. A positive correlation was found between increased SWA over the left ventral premotor cortex and improvement on the Repetition subtest of the WAB-R. This finding is of interest due to premotor cortex involvement in imitation and the inclusion of this region in the lesion extent of most of the participants.

84.8.2 Speech Entrainment

Citing previously mentioned findings of activation in left frontal speech-motor areas when visual observation accompanies auditory speech, Fridriksson et al. (2009) hypothesized that better performance would be elicited when a computer-based naming treatment for patients with nonfluent aphasia included audiovisual compared to auditory-only stimuli. Findings indicated statistically significant gains for audiovisual treatment only, including trained and untrained items.

Fridriksson et al. (2012) coined the term "speech entrainment" to describe the ability of some subjects with nonfluent aphasia to produce more fluent speech with an audiovisual model compared to spontaneous speech. Subjects performed online imitation of scripts, which are heard while viewing the speaker on an iPod screen. Only the speaker's mouth is visible to emphasize visual perception of the speech act. This therapy resulted in production of twice as many words during entrainment for 13 patients with Broca's aphasia. Significant increases in word variety were maintained for 1 week after treatment for production of practiced scripts during entrainment and spontaneous speech. Generalization to entrainment of untrained scripts remained significantly improved for 6 weeks.

Using fMRI to explore the neural mechanisms underlying behavioral findings, Fridriksson et al. (2012) found greater activation in left BA 37 and bilateral anterior insula/BA 47 for the speech entrainment condition compared to spontaneous speech. Imitation of speech may facilitate word retrieval (BA 37) and visceral speech support (anterior insula/BA 47) for rapid, online lexical processing and/or airflow modification or for lexical prediction and anticipation of respiratory demands. Broca's area may be an internal temporal gating device, which, although injured, can be compensated for by external temporal gating offered by real-time imitation of an observed speaker, entraining the requisite regions to again function as part of a coordinated network.

84.9 APHASIA THERAPY: NONSPEECH MOTOR OBSERVATION AND IMITATION

Speech is a motor activity and gestures are a rich aspect of human communication, whether for independent information transmission or to supplement spoken language (Goldin-Meadow, 1999). Verbal communication among humans may have evolved on top of existing gestural communication systems relying on the observation–execution matching system (Rizzolatti & Arbib, 1998). Gesture has thus been targeted as a means of treatment for aphasia.

Visual Action Therapy is a nonverbal therapeutic intervention for global aphasia, using real and drawn objects in a hierarchy (Helm-Estabrooks, Fitzpatrick, & Barresi, 1982). Manipulation of real objects or associated pantomimed gestures is imitated by the subject, with the ultimate goal of producing a pantomimed action to represent an unseen object. The rationale is that gestures, requiring unilateral gross motor control compared to speech, may be used symbolically (Helm-Estabrooks et al., 1982). Although Visual Action Therapy is not a contemporary subject of research, similar gesture-based therapies continue to be investigated in aphasia rehabilitation.

Patients with nonfluent aphasia improve in verb retrieval abilities after training of gesture labeling when observing or imitating the target gesture, but not when they observe a gesture and produce a meaningless movement (Marangolo et al., 2010). There is no significant difference between therapies using observation or meaningful imitation, with improvement maintained for 2 months. These findings may support a bilateral distribution of frontoparietal connections engaged in action observation and execution given the damage to regions associated with the human mirror neuron network of the left hemisphere (Aziz-Zadeh, Koski, Zaidel, Mazziotta, & Iacoboni, 2006). Action observation alone engages this network, but addition of a meaningless gesture interferes with this process, eliminating the beneficial result of observation and the resultant therapeutic gains in verb retrieval.

Observation therapy only improves verb production for actions within the human motor repertoire, such as dancing compared to printing (Marangolo, Cipollari, Fiori, Razzano, & Caltagirone, 2012). FMRI findings also demonstrate differences between activation resulting from observation of actions within the realm of human behavior, even when performed by nonhumans (e.g., a dog biting), compared to those that are not (Buccino et al., 2004). These findings further reinforce the role of a mirror neuron network implicated in action observation and execution that subserves language production.

84.10 CONCLUSION

Although aphasia is a biological disorder resulting from neurological damage, aphasia rehabilitation has traditionally neglected biological approaches to treatment in favor of behavioral and educational models. However, increasing understanding of the neurobiology underlying language is shifting the discourse toward biological mechanisms.

The two main biologically based models with empirical support at this time are the human mirror neuron system and the dual-stream hypothesis. These models support imitation as a powerful tool to rehabilitate the speech and language deficits of aphasia. Action observation engages the mirror properties of the same neural networks that are activated during execution, which is as true for speech and oral motor actions as for the grasping behaviors for which they were initially discovered. Higher-level language skills are also grounded in motor systems. Both observation and imitation of speech engage a similar network including components of the dorsal and ventral pathways for language.

Imitation has long been, and continues to be, used in many aphasia interventions. More recently, several researchers have developed neurophysiologically motivated aphasia therapy programs targeting online or delayed imitation as a strategy to improve speech output and language function. Some aphasia therapies have also used nonspeech imitation of actions to enhance gestural communication and production of action labels.

References

Altschuler, E. L., Vankov, A., Hubbard, E. M., Roberts, E., Ramachandran, V. S., & Pineda, J. A. (2000). Mu wave blocking by observation of movement and its possible use as a tool to study theory of other minds. *Society for Neuroscience, 68*.

Anderson, J. M., Gilmore, R., Roper, S., Crosson, B., Bauer, R. M., Nadeau, S., et al. (1999). Conduction aphasia and the arcuate fasciculus: A reexamination of the Wernicke–Geschwind model. *Brain and Language, 70*(1), 1–12.

Aziz-Zadeh, L., Koski, L., Zaidel, E., Mazziotta, J., & Iacoboni, M. (2006). Lateralization of the human mirror neuron system. *The Journal of Neuroscience, 26*(11), 2964–2970.

Aziz-Zadeh, L., Wilson, S. M., Rizzolatti, G., & Iacoboni, M. (2006). Congruent embodied representations for visually presented actions and linguistic phrases describing actions. *Current Biology, 16*(18), 1818–1823.

Barsalou, L. W. (2008). Grounded cognition. *Annual Review of Psychology, 59*, 617–645.

Basso, A. (2003). *Aphasia and its therapy*. New York, NY: Oxford University Press.

Benton, A. L., & Joynt, R. J. (1963). Three pioneers in the study of aphasia. *Journal of the History of Medicine and Allied Sciences, 18*(4), 381–383.

Binkofski, F., Buccino, G., Posse, S., Seitz, R. J., Rizzolatti, G., & Freund, H. J. (1999). A fronto-parietal circuit for object manipulation in man: Evidence from an fMRI-study. *European Journal of Neuroscience, 11*(9), 3276–3286.

Blumstein, S. E. (1997). A perspective on the neurobiology of language. *Brain and Language, 60*(3), 335–346.

Boyle, M., & Coelho, C. A. (1995). Application of semantic feature analysis as a treatment for aphasic dysnomia. *American Journal of Speech-Language Pathology, 4*(4), 94.

Brass, M., Bekkering, H., & Prinz, W. (2001). Movement observation affects movement execution in a simple response task. *Acta Psychologica, 106*(1), 3–22.

Brass, M., Bekkering, H., Wohlschläger, A., & Prinz, W. (2000). Compatibility between observed and executed finger movements: Comparing symbolic, spatial, and imitative cues. *Brain and Cognition, 44*(2), 124–143.

Buccino, G., Binkofski, F., Fink, G. R., Fadiga, L., Fogassi, L., Gallese, V., et al. (2001). Action observation activates premotor and parietal areas in a somatotopic manner: An fMRI study. *European Journal of Neuroscience, 13*(2), 400–404.

Buccino, G., Lui, F., Canessa, N., Patteri, I., Lagravinese, G., Benuzzi, F., et al. (2004). Neural circuits involved in the recognition of actions performed by nonconspecifics: An fMRI study. *Journal of Cognitive Neuroscience, 16*(1), 114–126.

Buchsbaum, B. R., Baldo, J., Okada, K., Berman, K. F., Dronkers, N., D'Esposito, M., et al. (2011). Conduction aphasia, sensory-motor integration, and phonological short-term memory–an aggregate analysis of lesion and fMRI data. *Brain and language, 119*(3), 119–128.

Cherney, L. R. (2004). Aphasia, alexia, and oral reading. *Topics in Stroke Rehabilitation, 11*(1), 22–36.

Cochin, S., Barthelemy, C., Roux, S., & Martineau, J. (1999). Observation and execution of movement: Similarities demonstrated by quantified electroencephalography. *European Journal of Neuroscience, 11*(5), 1839–1842.

Cohen-Seat, G., Gastaut, H., Faure, J., & Heuyer, G. (1954). Etudes expérimentales de l'activité nerveuse pendant la projection cinématographique. *Revue International de Filmologie, 5*, 7–64.

Craighero, L., Bello, A., Fadiga, L., & Rizzolatti, G. (2002). Hand action preparation influences the responses to hand pictures. *Neuropsychologia, 40*(5), 492–502.

Damasio, A. R., & Geschwind, N. (1984). The neural basis of language. *Annual Review of Neuroscience, 7*(1), 127–147.

Decety, J., & Grèzes, J. (1999). Neural mechanisms subserving the perception of human actions. *Trends in Cognitive Sciences, 3*(5), 172–178.

Decety, J., Grèzes, J., Costes, N., Perani, D., Jeannerod, M., Procyk, E., et al. (1997). Brain activity during observation of actions. Influence of action content and subject's strategy. *Brain, 120*(10), 1763–1777.

Di Pellegrino, G., Fadiga, L., Fogassi, L., Gallese, V., & Rizzolatti, G. (1992). Understanding motor events: A neurophysiological study. *Experimental Brain Research, 91*(1), 176–180.

Duffy, J. R. (1995). *Motor speech disorders: Substrates, differential diagnosis, and management* (1st ed.). Philadelphia, PA: Mosby.

Duncan, E. S., Schmah, T., & Small, S. L. (in prep). Performance variability as a predictor of response to aphasia treatment.

Ellis, C., Dismuke, C., & Edwards, K. K. (2010). Longitudinal trends in aphasia in the United States. *NeuroRehabilitation, 27*(4), 327–333.

Ertelt, D., Small, S., Solodkin, A., Dettmers, C., McNamara, A., Binkofski, F., et al. (2007). Action observation has a positive impact on rehabilitation of motor deficits after stroke. *NeuroImage, 36*, T164–T173.

Fadiga, L., Craighero, L., Buccino, G., & Rizzolatti, G. (2002). Speech listening specifically modulates the excitability of tongue muscles: A TMS study. *European Journal of Neuroscience, 15*(2), 399–402.

Fadiga, L., Fogassi, L., Pavesi, G., & Rizzolatti, G. (1995). Motor facilitation during action observation: A magnetic stimulation study. *Journal of Neurophysiology, 73*(6), 2608–2611.

Fagg, A. H., & Arbib, M. A. (1998). Modeling parietal–premotor interactions in primate control of grasping. *Neural Networks, 11*(7), 1277–1303.

Ferrari, P. F., Gallese, V., Rizzolatti, G., & Fogassi, L. (2003). Mirror neurons responding to the observation of ingestive and communicative mouth actions in the monkey ventral premotor cortex. *European Journal of Neuroscience, 17*(8), 1703–1714.

Fillingham, J., Sage, K., & Lambon Ralph, M. (2005). Further explorations and an overview of errorless and errorful therapy for aphasic word-finding difficulties: The number of naming attempts during therapy affects outcome. *Aphasiology, 19*(7), 597–614.

Fogassi, L., Gallese, V., Fadiga, L., & Rizzolatti, G. (1998). Neurons responding to the sight of goal-directed hand/arm actions in the parietal area PF (7b) of the macaque monkey. *Society of Neuroscience Abstracts, 24*, 257.5.

Fridriksson, J., Baker, J. M., Whiteside, J., Eoute, D., Moser, D., Vesselinov, R., et al. (2009). Treating visual speech perception to improve speech production in nonfluent aphasia. *Stroke, 40*(3), 853–858.

Fridriksson, J., Hubbard, H. I., Hudspeth, S. G., Holland, A. L., Bonilha, L., Fromm, D., et al. (2012). Speech entrainment enables patients with Broca's aphasia to produce fluent speech. *Brain, 135*(12), 3815–3829.

Fridriksson, J., Kjartansson, O., Morgan, P. S., Hjaltason, H., Magnusdottir, S., Bonilha, L., et al. (2010). Impaired speech repetition and left parietal lobe damage. *The Journal of Neuroscience, 30*(33), 11057–11061.

Friederici, A. D. (2009). Pathways to language: Fiber tracts in the human brain. *Trends in Cognitive Sciences, 13*(4), 175–181.

Gallese, V., Fadiga, L., Fogassi, L., & Rizzolatti, G. (1996). Action recognition in the premotor cortex. *Brain, 119*(2), 593–609.

Geschwind, N. (1965). Disconnexion syndromes in man and animals. *Brain, 88*, 237–294.

Glenberg, A. M., & Kaschak, M. P. (2002). Grounding language in action. *Psychonomic Bulletin and Review, 9*(3), 558–565.

Goldin-Meadow, S. (1999). The role of gesture in communication and thinking. *Trends in Cognitive Sciences, 3*(11), 419–429.

Goodale, M. A. (1993). Visual pathways supporting perception and action in the primate cerebral cortex. *Current Opinion in Neurobiology, 3*(4), 578–585.

Goodglass, H., & Kaplan, E. (1983). *Boston diagnostic aphasia examination booklet*. Philadelphia, PA: Lea & Febiger.

Gorno-Tempini, M. L., Hillis, A. E., Weintraub, S., Kertesz, A., Mendez, M., Cappa, S. F., et al. (2011). Classification of primary progressive aphasia and its variants. *Neurology, 76*(11), 1006–1014.

Grafton, S. T., Arbib, M. A., Fadiga, L., & Rizzolatti, G. (1996). Localization of grasp representations in humans by positron emission tomography. *Experimental Brain Research, 112*(1), 103–111.

Hari, R., Forss, N., Avikainen, S., Kirveskari, E., Salenius, S., & Rizzolatti, G. (1998). Activation of human primary motor cortex during action observation: A neuromagnetic study. *Proceedings of the National Academy of Sciences of the United States of America, 95*(25), 15061–15065.

Hauk, O., Johnsrude, I., & Pulvermüller, F. (2004). Somatotopic representation of action words in human motor and premotor cortex. *Neuron, 41*(2), 301–307.

Helm-Estabrooks, N. (1981). *Helm elicited language program for syntax stimulation*. Austin, TX: Exceptional Resources.

Helm-Estabrooks, N., Fitzpatrick, P. M., & Barresi, B. (1982). Visual action therapy for global aphasia. *Journal of Speech and Hearing Disorders, 47*(4), 385.

Helm-Estabrooks, N., Morgan, A. R., & Nicholas, M. (1989). *Melodic intonation therapy*. Rolling Meadows, IL: Riverside Publishing Company.

Hickok, G., & Poeppel, D. (2004). Dorsal and ventral streams: A framework for understanding aspects of the functional anatomy of language. *Cognition, 92*(1), 67–99.

Hickok, G., & Poeppel, D. (2007). The cortical organization of speech processing. *Nature Reviews Neuroscience, 8*(5), 393–402.

Huber, R., Ghilardi, M. F., Massimini, M., & Tononi, G. (2004). Local sleep and learning. *Nature, 430*(6995), 78–81.

Huber, W., Poeck, K., & Willmes, K. (1984). The Aachen aphasia test. *Advances in Neurology, 42*, 291.

Iacoboni, M., Molnar-Szakacs, I., Gallese, V., Buccino, G., Mazziotta, J. C., & Rizzolatti, G. (2005). Grasping the intentions of others with one's own mirror neuron system. *PLoS Biology, 3*(3), e79.

Iacoboni, M., Woods, R. P., Brass, M., Bekkering, H., Mazziotta, J. C., & Rizzolatti, G. (1999). Cortical mechanisms of human imitation. *Science, 286*(5449), 2526–2528.

Jeannerod, M. (1994). The representing brain: Neural correlates of motor intention and imagery. *Behavioral and Brain Sciences, 17*(2), 187–201.

Kertesz, A. (2006). *Western Aphasia Battery-Revised (WAB-R)*. San Antonio, TX: Pearson.

Keysers, C., & Perrett, D. I. (2004). Demystifying social cognition: A Hebbian perspective. *Trends in Cognitive Sciences, 8*(11), 501–507.

Kilner, J. M., Marchant, J. L., & Frith, C. D. (2006). Modulation of the mirror system by social relevance. *Social Cognitive and Affective Neuroscience, 1*(2), 143–148.

Klatzky, R. L., Pellegrino, J. W., McCloskey, B. P., & Doherty, S. (1989). Can you squeeze a tomato? The role of motor representations in semantic sensibility judgments. *Journal of Memory and Language, 28*(1), 56–77.

Kohler, E., Keysers, C., Umilta, M. A., Fogassi, L., Gallese, V., & Rizzolatti, G. (2002). Hearing sounds, understanding actions: Action representation in mirror neurons. *Science, 297*(5582), 846–848.

Kuhl, P., & Rivera-Gaxiola, M. (2008). Neural substrates of language acquisition. *Annual Review of Neuroscience, 31*, 511–534.

Lee, J., Fowler, R., Rodney, D., Cherney, L., & Small, S. L. (2010). IMITATE: An intensive computer-based treatment for aphasia based on action observation and imitation. *Aphasiology, 24*(4), 449–465.

Lepage, J. F., & Théoret, H. (2007). The mirror neuron system: Grasping others' actions from birth? *Developmental Science, 10*(5), 513–523.

Linebaugh, C. W., Shisler, R. J., & Lehner, L. H. (2005). CAC classics: Cueing hierarchies and word retrieval: A therapy program. *Aphasiology, 19*(1), 77–92.

Love, R. J., & Webb, W. G. (1977). The efficacy of cueing techniques in Broca's aphasia. *Journal of Speech and Hearing Disorders, 42*(2), 170.

Marangolo, P., Bonifazi, S., Tomaiuolo, F., Craighero, L., Coccia, M., Altoè, G., et al. (2010). Improving language without words: First evidence from aphasia. *Neuropsychologia, 48*(13), 3824–3833.

Marangolo, P., Cipollari, S., Fiori, V., Razzano, C., & Caltagirone, C. (2012). Walking but not barking improves verb recovery: Implications for action observation treatment in aphasia rehabilitation. *PLoS One*, *7*(6), e38610.

Marshall, P. J., Young, T., & Meltzoff, A. N. (2011). Neural correlates of action observation and execution in 14-month-old infants: An event-related EEG desynchronization study. *Developmental Science*, *14*(3), 474–480.

Mashal, N., Solodkin, A., Dick, A. S., Chen, E. E., & Small, S. L. (2012). A network model of observation and imitation of speech. *Frontiers in Psychology*, *3*, 84. Available from: http://dx.doi.org/10.3389/fpsyg.2012.00084.

Mishkin, M., Ungerleider, L. G., & Macko, K. A. (1983). Object vision and spatial vision: Two cortical pathways. *Trends in Neurosciences*, *6*, 414–417.

Möttönen, R., & Watkins, K. E. (2012). Using TMS to study the role of the articulatory motor system in speech perception. *Aphasiology*, *26*(9), 1103–1118.

Mukamel, R., Ekstrom, A. D., Kaplan, J., Iacoboni, M., & Fried, I. (2010). Single-neuron responses in humans during execution and observation of actions. *Current Biology*, *20*(8), 750–756.

Mummery, C. J., Patterson, K., Price, C. J., Ashburner, J., Frackowiak, R. S. J., & Hodges, J. R. (2000). A voxel-based morphometry study of semantic dementia: Relationship between temporal lobe atrophy and semantic memory. *Annals of Neurology*, *47*(1), 36–45.

Muthukumaraswamy, S. D., & Johnson, B. W. (2004). Changes in rolandic mu rhythm during observation of a precision grip. *Psychophysiology*, *41*(1), 152–156.

Myung, J. Y., Blumstein, S. E., & Sedivy, J. C. (2006). Playing on the typewriter, typing on the piano: Manipulation knowledge of objects. *Cognition*, *98*(3), 223–243.

Nishitani, N., & Hari, R. (2002). Viewing lip forms: Cortical dynamics. *Neuron*, *36*(6), 1211–1220.

Okada, K., & Hickok, G. (2006). Left posterior auditory-related cortices participate both in speech perception and speech production: Neural overlap revealed by fMRI. *Brain and Language*, *98*(1), 112–117.

Orjada, S., & Beeson, P. (2005). Concurrent treatment for reading and spelling in aphasia. *Aphasiology*, *19*(3–5), 341–351.

Papathanassiou, D., Etard, O., Mellet, E., Zago, L., Mazoyer, B., & Tzourio-Mazoyer, N. (2000). A common language network for comprehension and production: A contribution to the definition of language epicenters with PET. *NeuroImage*, *11*(4), 347–357.

Parker, G. J., Luzzi, S., Alexander, D. C., Wheeler-Kingshott, C. A., Ciccarelli, O., & Lambon Ralph, M. A. (2005). Lateralization of ventral and dorsal auditory-language pathways in the human brain. *NeuroImage*, *24*(3), 656–666.

Petrides, M., & Pandya, D. N. (2009). Distinct parietal and temporal pathways to the homologues of Broca's area in the monkey. *PLoS Biology*, *7*(8), e1000170.

Rauschecker, J. P., & Tian, B. (2000). Mechanisms and streams for processing of "what" and "where" in auditory cortex. *Proceedings of the National Academy of Sciences of the United States of America*, *97*(22), 11800–11806.

Rizzolatti, G., & Arbib, M. A. (1998). Language within our grasp. *Trends in Neurosciences*, *21*(5), 188–194.

Rizzolatti, G., Fadiga, L., Gallese, V., & Fogassi, L. (1996). Premotor cortex and the recognition of motor actions. *Cognitive Brain Research*, *3*(2), 131–141.

Rizzolatti, G., Fadiga, L., Matelli, M., Bettinardi, V., Paulesu, E., Perani, D., et al. (1996). Localization of grasp representations in humans by PET: 1. Observation versus execution. *Experimental Brain Research*, *111*(2), 246–252.

Rizzolatti, G., Fogassi, L., & Gallese, V. (2001). Neurophysiological mechanisms underlying the understanding and imitation of action. *Nature Reviews Neuroscience*, *2*(9), 661–670.

Rozzi, S., Calzavara, R., Belmalih, A., Borra, E., Gregoriou, G. G., Matelli, M., et al. (2006). Cortical connections of the inferior parietal cortical convexity of the macaque monkey. *Cerebral Cortex*, *16*(10), 1389–1417.

Rozzi, S., Ferrari, P. F., Bonini, L., Rizzolatti, G., & Fogassi, L. (2008). Functional organization of inferior parietal lobule convexity in the macaque monkey: Electrophysiological characterization of motor, sensory and mirror responses and their correlation with cytoarchitectonic areas. *European Journal of Neuroscience*, *28*(8), 1569–1588.

Rumiati, R. I., Weiss, P. H., Tessari, A., Assmus, A., Zilles, K., Herzog, H., et al. (2005). Common and differential neural mechanisms supporting imitation of meaningful and meaningless actions. *Journal of Cognitive Neuroscience*, *17*(9), 1420–1431.

Sarasso, S., Määtä, S., Ferrarelli, F., Poryazova, R., Tononi, G., & Small, S. L. (2014). Plastic changes following Imitation-Based speech and language therapy for aphasia a High-Density sleep EEG study. *Neurorehabilitation and Neural Repair*, *28*(2), 129–138.

Saur, D., Kreher, B. W., Schnell, S., Kümmerer, D., Kellmeyer, P., Vry, M. S., et al. (2008). Ventral and dorsal pathways for language. *Proceedings of the National Academy of Sciences of the United States of America*, *105*(46), 18035–18040.

Schmahmann, J. D., & Pandya, D. (2006). *Fiber pathways of the brain*. Oxford, UK: Oxford University Press.

Seltzer, B., & Pandya, D. N. (1978). Afferent cortical connections and architectonics of the superior temporal sulcus and surrounding cortex in the rhesus monkey. *Brain Research*, *149*(1), 1–24.

Sigurðardóttir, Z. G., & Sighvatsson, M. B. (2006). Operant conditioning and errorless learning procedures in the treatment of chronic aphasia. *International Journal of Psychology*, *41*(6), 527–540.

Skipper, J. I., Nusbaum, H. C., & Small, S. L. (2006). Lending a helping hand to hearing: Another motor theory of speech perception. *Action to Language Via the Mirror Neuron System*250–285.

Skipper, J. I., van Wassenhove, V., Nusbaum, H. C., & Small, S. L. (2007). Hearing lips and seeing voices: How cortical areas supporting speech production mediate audiovisual speech perception. *Cerebral Cortex*, *17*(10), 2387–2399. Available from: http://dx.doi.org/10.1093/cercor/bhl147.

Small, S. (2004). A biological model of aphasia rehabilitation: Pharmacological perspectives. *Aphasiology*, *18*(5–7), 473–492.

Small, S. L., Buccino, G., & Solodkin, A. (2012). The mirror neuron system and treatment of stroke. *Developmental Psychobiology*, *54*(3), 293–310.

Strafella, A. P., & Paus, T. (2000). Modulation of cortical excitability during action observation: A transcranial magnetic stimulation study. *Neuroreport*, *11*(10), 2289–2292.

Tettamanti, M., Buccino, G., Saccuman, M. C., Gallese, V., Danna, M., Scifo, P., et al. (2005). Listening to action-related sentences activates fronto-parietal motor circuits. *Journal of Cognitive Neuroscience*, *17*(2), 273–281.

Tremblay, C., Robert, M., Pascual-Leone, A., Lepore, F., Nguyen, D. K., Carmant, L., et al. (2004). Action observation and execution intracranial recordings in a human subject. *Neurology*, *63*(5), 937–938.

Umiltá, M. A., Kohler, E., Gallese, V., Fogassi, L., Fadiga, L., Keysers, C., et al. (2001). I know what you are doing: A neurophysiological study. *Neuron*, *31*(1), 155–165.

Watkins, K. E., Strafella, A. P., & Paus, T. (2003). Seeing and hearing speech excites the motor system involved in speech production. *Neuropsychologia*, *41*(8), 989–994.

Williams, J. H., Whiten, A., Suddendorf, T., & Perrett, D. I. (2001). Imitation, mirror neurons and autism. *Neuroscience and Biobehavioral Reviews*, *25*(4), 287–295.

Youmans, G., Holland, A., Munoz, M., & Bourgeois, M. (2005). Script training and automaticity in two individuals with aphasia. *Aphasiology*, *19*(3–5), 435–450.

CHAPTER

85

Pharmacotherapy for Aphasia

Daniel A. Llano[1] *and Steven L. Small*[2]

[1]Department of Molecular and Integrative Physiology, University of Illinois at Urbana-Champaign, Champaign, IL, USA; [2]Department of Neurology, University of California at Irvine, Irvine, CA, USA

85.1 INTRODUCTION

Aphasia is one of the most common consequences of stroke, and it is also one of the most feared. Approximately 10–30% of stroke victims suffer from a language deficit after stroke (Bersano, Burgio, Gattinoni, Candelise, & P.S.G., 2009; Pedersen, Stig Jørgensen, Nakayama, Raaschou, & Olsen, 1995; Wade, Hewer, David, & Enderby, 1986), and recovery from a stroke-related language deficit is highly variable and dependent on a range of factors. For example, lesion-related variables, such as the size and location of the stroke, and patient-related variables, such as the patient's age and history of strokes, are likely major determinants in recovery from stroke (Caplan, Hildebrandt, & Makris, 1996; Heiss & Thiel, 2006; Kertesz, Harlock, & Coates, 1979; Lazar & Antoniello, 2008) and are essentially unmodifiable in the postacute setting. Other aspects important in the long-term language recovery of patients, such as prevention of future strokes, design and implementation of a language therapy plan, and pharmacologic therapy, can all be points of successful intervention by the clinician.

In this chapter, we review the literature on pharmacologic approaches to aphasia recovery. Specifically, we attempt to integrate a body of literature on motor stroke recovery in animal models with clinical data examining the effects of drug interventions on aphasia recovery. Herein, we argue that a primary focus for the rehabilitation of patients with aphasia should be pharmacological enhancement of reorganization of neural circuits rather than simply the development of compensatory strategies. As such, we adopt a medical model to the treatment of stroke whereby the therapeutic aim is to guide plasticity with behavioral therapy and to facilitate this plasticity with pharmacological intervention.

85.2 MAJOR CHALLENGES

It is worth noting early in this chapter that there are two significant challenges to making progress in the area of aphasia recovery: limited clinical trial data and lack of animal models for language. Compared with other neurological diseases for which the underlying disease processes may be more homogeneous across subjects and the outcome measures have been validated to be sensitive to pharmacologic intervention, aphasia recovery is in a somewhat disadvantageous position. Strokes are highly variable in location, size, and even etiology. In addition, the term "aphasia" is quite broad and can include difficulties ranging from dysfluency to patients who are highly (or overly) fluent but with difficulties in the understanding and formulation of language. Such heterogeneity in language deficits likely reflects heterogeneity in lesion location, size, and etiology. The latter point is important because the term "stroke" may encompass a range of underlying pathophysiological mechanisms, such as large-vessel atherosclerotic disease, cardioembolic disease, small vessel disease, venous thrombosis, and intracranial hemorrhage. It is likely that stroke-related deficits caused by different underlying pathophysiologic mechanisms will have different time courses of recovery and different sensitivities to pharmacological therapy. Similarly, although there are many different rating scales available to evaluate the severity of aphasia (Shewan & Kertesz, 1980; Strauss, 2006), the underlying heterogeneity in aphasia types and etiologies and the lack of validation of such scales for sensitivity to drug manipulations create challenges for the use of a single instrument to evaluate the efficacy of drug therapies for aphasia. These factors have contributed to the relatively small number of clinical trials performed to examine the

Neurobiology of Language. DOI: http://dx.doi.org/10.1016/B978-0-12-407794-2.00085-7

potential for drugs to treat aphasia. For example, a recent (July 2013) search on www.clinicaltrials.gov revealed four studies actively recruiting subjects for drug interventions for aphasia. When evaluating the clinical literature on aphasia drug therapies, these challenges make it difficult to apply the same criteria as those for diseases that are more amenable to study using standard clinical trial approaches.

Another factor affecting progress in this area is the lack of adequate animal models. Animal models serve as a proving ground for putative new therapies and have been shown to be indispensable for the development of drug therapies for epilepsy, multiple sclerosis, and Parkinson disease (Dauer & Przedborski, 2003; Gold, Linington, & Lassmann, 2006; McNamara, 1985), and have led to great insights to the underlying pathophysiology of other diseases such as Alzheimer disease (Price & Sisodia, 1998). Currently, however, there are no established animal models for language—at least language as humans experience it. Although this is a topic that has been hotly debated in the literature (Hauser, Chomsky, & Fitch, 2002), and although the details of this are well beyond the scope of this chapter, it is safe to say that there are no established models with a high enough throughput to allow the vetting of potential drug therapies. Instead, the field has relied on animal models of motor stroke. Here, the literature is quite mature, many approaches to inducing stroke exist (middle cerebral artery ligation, photothrombosis, etc.) (Carmichael, 2005), and the outcome measures of motor recovery are highly quantifiable. Therefore, for the current discussion we assume that at least some of the mechanisms of repair and reorganization of neural structures involved in recovery from motor stroke are similar to those involved with language recovery, although more work is greatly needed in this area to validate this assumption.

Despite the challenges described, it is worth noting that now, more than ever, we are strategically positioned for great advances in the development of drug therapy for aphasia. Human and animal imaging technologies have advanced to the point that the impact of drug and/or behavioral therapy on neural networks can be studied in humans and animals using very similar imaging technologies, facilitating comparisons of the pharmacodynamics of putative therapies across species (Bifone, Gozzi, & Schwarz, 2010; Rumple et al., 2013). Therefore, the impact of drug and behavioral interventions, both on task-related activation patterns and resting states, as well as structural imaging of gray and white matter (Jang, 2011; Schwarz et al., 2011) can be assessed and compared. It is hoped that this chapter will serve as a starting point to initiate future mechanistic studies in this area.

85.3 MECHANISMS OF RECOVERY AND PHARMACOTHERAPY

There is a substantial body of work documenting the natural history and the impact of pharmacological modulation of stroke recovery in animal models. Most of this research has involved lesions of the motor cortex and has used motor output as its focus for efficacy. Once a small area of cortex is lost, a host of short-range and long-range mechanisms are engaged to facilitate reorganization of cortical function (Hermann & Chopp, 2012; Murphy & Corbett, 2009). These mechanisms are activity-dependent and include axonal sprouting (Dancause et al., 2005; Overman et al., 2012), which can extend distances >1 cm across the adult primate cortex, elaboration of dendritic spines (Brown & Murphy, 2008; Ueno et al., 2012), migration of subventricular stem cells to the infarction zone (Danilov, Kokaia, & Lindvall, 2012; Kahle & Bix, 2013; Lichtenwalner & Parent, 2005), and modulation of the strength or excitability of existing synapses (Di Filippo et al., 2008; Jaenisch, Witte, & Frahm, 2010; Yao et al., 2005). Because such mechanisms are likely to be differentially sensitive to pharmacological manipulation, one might postulate that different etiological mechanisms of infarction would require different forms of intervention. It is also possible that some forms of modulation of synaptic strength may be maladaptive (Costigan, Scholz, & Woolf, 2009; Di Filippo et al., 2008) and therapeutic modalities, appropriately targeted and timed, may be used to interfere with such forms of pathological plasticity.

Several concepts have arisen from the animal literature on stroke recovery (and lesion recovery more generally) that may inform our review of the literature on drug therapy for aphasia. First, it is clear that maximal benefit is derived not from drugs or rehabilitation approaches on their own, but with *combination* therapy. There are many studies supporting the idea that synaptic plasticity, which is presumed to be the dominant mechanism underlying synaptic rewiring responsible for stroke recovery, is greatly facilitated when neurotransmitter manipulations are accompanied by behavioral training (Bao, Chan, & Merzenich, 2001; Kilgard & Merzenich, 1998; Schultz, 2002). Further, many of the classes of drugs discussed within this review have been explicitly modeled to help support and strengthen neural networks being modified by behavioral training (Korchounov & Ziemann, 2011), such that their efficacy in the absence of behavioral training may be greatly reduced. Taken further, one might even speculate that drugs that support synaptic plasticity, when used in the absence of targeted behavioral therapy, may actually reinforce maladaptive patterns of neuronal activity. In addition, it is of high clinical relevance that the cerebral cortex undergoes plastic changes for at least *months* after stroke, and that these adaptive changes occur not only in the tissue immediately surrounding the lesion but also in areas remote from the site of injury (Jenkins & Merzenich, 1987; Nudo & Friel, 1999; Nudo, Wise, SiFuentes, & Milliken, 1996; Xerri, Merzenich, Peterson, & Jenkins, 1998). Finally, the animal studies suggest that certain classes of drugs commonly used in clinical practice, particularly sedating drugs such as benzodiazepines, can *diminish* the reorganization of brain networks after injury (Goldstein, 2000; Larson & Zollman, 2010). Given the emerging recognition of poststroke depression and anxiety (Aben et al., 2003; Barker-Collo, 2007) and the relatively high incidence of seizures after stroke (Lossius, Rønning, Mowinckel, & Gjerstad, 2002; Olsen, 2001), both of which may be treated with benzodiazepines, these data indicate that nonsedating alternatives may be more appropriate to avoid interference with poststroke recovery mechanisms.

85.3.1 Animal Studies: Catecholamine-Based Therapy

There is a rich history of the use of catecholamine augmentation as adjunctive therapy to enhance stroke recovery. Catecholamines are a natural target for stroke therapeutics because it has long been known that there are significant alterations in brainstem and cortical catecholamine levels in the acute and subacute periods after stroke (Brown, Carlson, Ljunggren, Siesjö, & Snider, 1974; Cohen, Waltz, & Jacobson, 1975; Robinson, Shoemaker, & Schlumpf, 1980; Robinson, Shoemaker, Schlumpf, Valk, & Bloom, 1975). In addition, lesions to the noradrenergic afferents from the locus ceruleus have been shown to impair recovery of animals with contralateral sensory-motor cortical injuries (Goldstein & Bullman, 1997), and pharmacological antagonism of alpha adrenergic receptors with phenoxybenzamine delays spontaneous recovery from cortical damage (Feeney & Westerberg, 1990).

A number of neurophysiological mechanisms are postulated to underlie the beneficial effects of catecholamine enhancement of stroke recovery. Norepinephrine and dopamine modulate multiple forms of synaptic plasticity, including long-term potentiation and long-term depression, and spike timing-dependent plasticity (Carey & Regehr, 2009; Clem & Huganir, 2013; Dommett, Henderson, Westwell, & Greenfield, 2008; Edelmann & Lessmann, 2013; Ghanbarian & Motamedi, 2013; Gu, 2002; Wolf, Mangiavacchi, & Sun, 2003). These data would suggest that the impact of catecholamine modulation on recovering neural circuits would be highly activity-dependent, which is consistent with the data presented. Other studies have shown that catecholamines may enhance neural regeneration (Hiramoto, Ihara, & Watanabe, 2006; Lloyd, Balest, Corotto, & Smeyne, 2010; Spiegel et al., 2007) or axonal sprouting (Papadopoulos et al., 2009). Overall, these data suggest that catecholamine augmentation may function to alter neuronal plasticity on several different temporal and spatial scales.

The paradigm for the study of catecholamine-based therapy was established by early work demonstrating that administration of a single dose of dextroamphetamine (D-amphetamine, which causes the release of stored norepinephrine, dopamine, and, to a lesser degree, serotonin) can facilitate recovery of beam-walking behavior after lesion of motor cortex (Feeney, Gonzalez, & Law, 1982). Importantly, drug effects on recovery were only seen when drug administration was coupled with the promotion of physical activity. This basic finding of the dependency between active training and catecholamine augmentation for maximum motor recovery has been reproduced in animals several times in several stroke models (Barbay et al., 2006; Beltran, Papadopoulos, Tsai, Kartje, & Wolf, 2010; Papadopoulos et al., 2009; Ramic et al., 2006; Rasmussen, Overgaard, Hildebrandt-Eriksen, & Boysen, 2006). For example, in the visual system, bilateral visual cortex lesions in cats leading to deficits in depth perception can be ameliorated with a combination of visual experience and dextroamphetamine administration (Feeney & Hovda, 1985). No benefit was seen when either the drug or behavioral training were given independently of each other.

The receptor pharmacology of the beneficial effects D-amphetamine is not yet entirely clear. The effect of D-amphetamine on motor recovery can be blocked by the haloperidol (Feeney et al., 1982), a D2 antagonist, which also has weaker antagonism at the alpha1 adrenoreceptor. In addition, as described, alpha1 receptor blockade impaired spontaneous recovery (Feeney & Westerberg, 1990) and treatment with intraventricular norepinephrine, but not dopamine, and reproduced the beneficial effect of D-amphetamine (Boyeson & Feeney, 1984). Further, findings in animals that have recovered from brain trauma that show that alpha adrenergic antagonists, such as phenoxybenzamine, can cause the recrudescence of lesion-related deficits long after functional recovery has occurred (Feeney, De Smet, & Rai, 2004) and suggest that a long-standing and tonic increase in norepinephrine is needed to restore function in certain neural networks. Most recently, motor recovery in a rat stroke model was facilitated by atipamezole and alpha-2 blocker, which elevates synaptic norepinephrine levels, with little effect on dopamine (Beltran et al., 2010; Gobert, Billiras, Cistarelli, & Millan, 2004). These data suggest that norepinephrine, acting on alpha adrenergic receptors, is the primary driver of functional recovery in these models. However, it has been suggested, primarily based on data from levodopa-enhanced word-list learning in normal humans and the low rates of conversion of levodopa to norepinephrine (~5%), that dopamine may also play a significant role (Breitenstein et al., 2006). The mixed clinical picture surrounding drugs like amphetamine and levodopa, which activate multiple receptor types, suggests that there is a need to more precisely define the mechanism of action of these drugs to target the efficacious mechanisms.

85.3.2 Animal Studies: Cholinergic Mechanisms

Luria postulated that enhancement of acetylcholine levels via the natural cholinesterase inhibitor galanthamine (precursor to modern Alzheimer's drug, galantamine), may promote functional language recovery after stroke (Luria, Naydyn, Tsvetkova, & Vinarskaya, 1969). In general, these agents have been particularly relevant to the treatment of Alzheimer disease, for which there is a cholinergic model that relates atrophy in the nucleus basalis of Meynert (the source of all cerebral cortical acetylcholine) to the pathophysiology of Alzheimer disease (Whitehouse, Price, Clark, Coyle, & DeLong, 1981). Acetylcholine is thought to be involved in a number of aspects of cognition, including sensory perception, selective attention, associative learning, and memory, as well as experience-dependent plasticity (Baskerville, Schweitzer, & Herron, 1997; Yu & Dayan, 2002). Very suggestive evidence for the potential for acetylcholine augmentation to be used as a therapeutic modality to enhance plasticity is derived from findings in the auditory cortex, where experience-dependent alterations of sensory maps were greatly enhanced when sensory stimulation was coupled with stimulation of cholinergic fibers from the basal forebrain (Kilgard & Merzenich, 1998). Notably, map reorganization parameters were directly related to the specific training stimuli used, emphasizing the importance of understanding the interactions between specific speech and language therapies with pharmacologic intervention. One view is that cholinergic neurons from the basal forebrain serve a modulatory function by *marking* behaviorally salient stimuli (Kilgard & Merzenich, 1998). This experience-dependent and acetylcholine-dependent map reorganization learning is probably mediated by muscarinic cholinergic receptors, rather than nicotinic receptors, because this learning can be blocked by scopolamine, a muscarinic antagonist (Thiel, Friston, & Dolan, 2002).

Despite the well-recognized role for acetylcholine in activity-dependent plasticity, there are comparatively little data supporting enhancement of cholinergic activity for motor stroke recovery in animal models. A study using galantamine in a motor stroke model did not show efficacy (Zhao, Puurunen, Schallert, Sivenius, & Jolkkonen, 2005), whereas the data for nicotine administration are mixed (Gonzalez, Gharbawie, & Kolb, 2006; Lim, Alaverdashvili, & Whishaw, 2009). It is interesting to note that the muscarinic blocker scopolamine has been shown to have some *beneficial* effects in early recovery in animal models of traumatic injury (Lyeth et al., 1992), possibly by reducing cholinergic neuronal activation (Saija et al., 1988), but the window for such an advantage is very short (15 min) (Hamm, O'Dell, Pike, & Lyeth, 1993). This is similar to what has been seen with several "neuroprotective" agents, such as NMDA(N-methyl-D-aspartate)-receptor antagonists (Villmann & Becker, 2007) and gamma-amino butyric acid (GABA)-potentiators (Green, Hainsworth, & Jackson, 2000), reinforcing the idea that the *timing* of intervention will play a critical role in the development of aphasia treatment strategies.

85.3.3 Animal Studies: Serotonin and Brain-Derived Neurotrophic Factor

Serotonin regulates some forms of cortical map reorganization (Gu & Singer, 1995; Jitsuki et al., 2011; Vetencourt, Tiraboschi, Spolidoro, Castrén, & Maffei, 2011) and adult neurogenesis in the hippocampus (Daszuta, 2011; Li et al., 2009), although the evidence supporting a role for serotonin in promoting neural reorganization in the chronic stroke setting is relatively sparse. More attention has been given to the ability of serotonin to enhance the expression of the ubiquitously expressed Brain-Derived Neurotrophic Factor (BDNF). Rodent studies have shown that BDNF increases both locally and remotely after the induction of stroke (Béjot et al., 2011), and that inhibition of BDNF using antisense oligonucleotides impair stroke functional recovery from motor cortex stroke (Ploughman et al., 2009). This study also revealed a synergistic interaction between physical therapy and BDNF expression. Other rodent work has demonstrated that exogenous BDNF administration facilitates motor recovery from acute stroke (Schabitz et al., 2007). Despite the suggestive findings regarding BDNF and stroke, there are major hurdles to the translation of this therapy to the clinic. Delivery of such a large molecule to the central nervous system (CNS) poses a problem, particularly in the chronic phase of stroke when the blood–brain barrier has reconstituted. This provides an opportunity to drive BDNF expression indirectly via serotonin augmentation (e.g., through the use of selective serotonin reuptake inhibitors [SSRIs]). It is not clear, however, whether serotonin augmentation through SSRIs is sufficient to drive BDNF expression to sufficient levels to promote recovery because SSRI studies in chronic stroke have had a mixed track record (Boyeson, Harmon, & Jones, 1994; Windle & Corbett, 2005). It may be that extrasynaptic serotonin is necessary to promote BDNF expression, or that only specific serotonin receptor subtypes are responsible for BDNF expression (there are at least seven subtypes of the serotonin receptor) and should be targeted more specifically. There is another unanswered question regarding the ability of BDNF to promote recovery in the chronic phase of stroke. BDNF has neuronal protectant properties (Marini et al., 2004), and most studies to date have been performed in the acute phase.

85.3.4 Animal Studies: GABAergic Mechanisms

Another agent that has received attention in animal studies is GABA. GABA is particularly interesting because its action on neurons is primarily inhibitory (Hendry, Schwark, Jones, & Yan, 1987). Augmentation of GABAergic neurotransmission, either via benzodiazepines or via barbiturates, has been a popular drug development strategy to promote neuronal protection during the acute phase of stroke, presumably by limiting neuronal metabolic demand during periods of high metabolic stress (Green et al., 2000). In the chronic phase, there is little evidence that GABA potentiation promotes reorganization after stroke. For example, intracortical infusion of GABA exacerbates the hemiparesis produced by a small motor cortex lesion in rats (Schallert et al., 1992). In addition, the short-term administration of diazepam (a benzodiazepine and indirect GABA agonist) permanently impedes sensory cortical recovery from neocortical injury (Schallert, Hernandez, & Barth, 1986). Furthermore, the GABA potentiator phenobarbital also interferes with recovery from brain injury (Hernandez & Holling, 1994; Montanez, Kline, Gasser, & Hernandez, 2000). Another approach involves removal of GABAergic inhibition, which may promote cortical map reorganization, for which there is some evidence (Jacobs & Donoghue, 1991). However, such an approach would pose a challenging translational strategy because blockade of GABA significantly lowers seizure threshold (Seger, 2004) and chronic stroke patients have a risk of seizure that is higher than baseline (Asconapé, 1991; Lossius et al., 2002; Olsen, 2001).

85.3.5 Animal Studies: Extracellular Matrix-Based Mechanisms

Stroke induces the expression of a number of extracellular molecules that are potentially hostile to axonal outgrowth. These include myelin-associated extracellular molecules, such as Nogo-A, myelin-associated glycoprotein, and oligodendrocyte-myelin glycoprotein, and extracellular matrix proteins, such as tenascin and chondroitin sulfate proteoglycans (Carmichael, 2006). Among these, the Nogo-A system has been well-characterized, and inhibitors of this system have moved forward into clinical trials of spinal cord injury and are under consideration for stroke. In animal models of subacute motor cortical stroke, inhibition of Nogo-A activity via passive immunization resulted in enhanced recovery of motor function in several studies (Brenneman et al., 2008; Papadopoulos et al., 2009; Tsai et al., 2007). Of potential relevance for the therapy of aphasia are the findings that intraventricular administration of anti-Nogo antibodies promoted recovery in a rodent model of hemispatial neglect (Brenneman et al., 2008) and poststroke spatial memory deficit (Gillani et al., 2010), suggesting that Nogo inhibition may ameliorate cognitive dysfunction. No studies of anti-Nogo therapy, to our knowledge, have been performed during the late chronic phase of stroke. However, because the studies referenced have not demonstrated any alteration in infarct size after anti-Nogo therapy, anti-Nogo therapy may promote neural reorganization rather than neuronal protection. This supposition suggests that anti-Nogo therapy represents at least a feasible path forward for studies of chronic stroke patients.

One potential concern with the body of rat literature documenting the benefits of anti-Nogo therapy is that in both the motor stroke studies (Papadopoulos et al., 2009; Tsai et al., 2007) and at least one of the "cognitive" stroke studies (Brenneman et al., 2008), recovery was in large part mediated by activation of contralesional structures. This suggests that anti-Nogo therapy may promote sprouting of long-range axons of contralesional origin rather than remaining ipsilateral axons. Because recovery in human motor stroke (Fridman et al., 2004), aphasia (Heiss et al., 1997; Karbe et al., 1998; Szaflarski, Allendorfer, Banks, Vannest, & Holland, 2013), and neglect (Corbetta, Kincade, Lewis, Snyder, & Sapir, 2005) is likely optimally mediated by ipsilesional structures, promotion of sprouting of contralesional axons may not be an optimal recovery strategy. It is not known if the preference for the contralesional structures is specific to anti-Nogo therapy or to rodent stroke recovery in general.

85.3.6 Animal Studies: Combining Drug and Behavioral Therapy

A feature common to most pharmacotherapy for chronic stroke is the requirement that physical therapy must be used to materialize the benefits of drug administration. By extension, speech and language therapy (SLT) may have a similar role in the pharmacotherapy for aphasia. Because certain forms of neuronal plasticity can be maladaptive (Costigan et al., 2009; Di Filippo et al., 2008), it is crucial that alterations of synaptic strength or number are not haphazard. In addition, because most drugs uniformly bathe the brain without the ability to specifically modify damaged circuitry, SLT is necessary to target particular areas for synaptic modification. As such, drug therapy likely plays a permissive role in stroke recovery. There may be exceptions to this. For example, stem cell therapy may be able to target specific areas of the brain that overexpress molecular factors in damaged areas (see Chapter 86 on cell-based therapies). It is unknown whether these molecular targets are sufficiently specific to allow neural repair without the guidance of behavioral training. It will be important to clarify this issue in animal models because it has therapeutic implications. For example, for patients unable to participate in rehabilitation, it may very well be counterproductive to offer permissive drug therapy, such as catecholamine augmentation, because these drugs may reinforce inappropriate circuitry. However, such patients may be candidates for therapies that can target damaged circuitry, such as stem cell therapy. Continued exploration of the combination of drug therapy and physical therapy in animal models will assist in guiding these decisions.

85.4 HUMAN STUDIES: PHARMACOTHERAPY FOR APHASIA

There is a long history of attempts at aphasia pharmacotherapy (Bergman & Green, 1951; Linn, 1947; Sarno, Sarno, & Diller, 1972; West & Stockel, 1965) (for a review of these studies, see Small, 1994 or Small, 2001). The outcomes of these studies were generally negative, likely related to inadequate scientific rationales or poor study design (e.g., underpowered studies, poor dose selection, inappropriate outcome measures). The animal data would suggest that the most promising targets for pharmacotherapy for aphasia would be accentuation of norepinephrine and dopamine levels coupled with behavioral therapy. Fortunately, this hypothesis lends itself to direct clinical translation because of the existence of several approved drugs that increase brain catecholamine levels (D-amphetamine, levodopa, etc.). Regarding study design, in general, the optimum study design to establish that a pharmacological agent promotes brain reorganization to enhance language processing would be a double-blind, placebo-controlled, adequately powered, parallel-group study that contains at least one outcome measure that is assessed after drug washout to ensure that any benefit observed is due to plastic changes in the brain. Unfortunately, very few studies have had this type of design. In addition, much of the literature utilizes heterogeneous outcome measures and many studies provide few details regarding the results, making meta-analyses very challenging. This section reviews the existing literature on drug therapy of aphasia. To our knowledge, there have been a total of 18 prospective, double-blind studies involving patients with subchronic or chronic stroke that used a language metric as a primary outcome measure, and we review these studies here. The attention here is restricted to small-molecule therapy in the subacute and chronic phases of aphasia due to stroke.

85.4.1 Human Studies: Noradrenergic Agents

The animal data tell a compelling story that enhancement of catecholamine levels, coupled with physical therapy, can promote neural reorganization and functional recovery even during the chronic phase of motor stroke. The two most highly studied catecholamines for stroke recovery are dopamine and norepinephrine. Dopamine and norepinephrine poorly cross the blood–brain barrier and are therefore rarely administered in these studies. Instead, other drugs such as D-amphetamine, which indirectly elevates synaptic catecholamine levels, or related compounds are administered. Other approaches have been to block the reuptake of norepinephrine and dopamine with methylphenidate. None of these drugs permit the isolation of noradrenergic from dopaminergic effects, although it has been argued that dopamine may play a dominant role in language recovery.

We are aware of four prospective, placebo-controlled, double-blind studies that examine the effects of D-amphetamine on language function in aphasic patients. The largest and most frequently cited study is that by Walker-Batson et al. (2001). In this study, subjects with subacute stroke were treated with D-amphetamine or placebo, and treatment was coupled with traditional speech therapy over a 5-week period.

A greater percentage of active subjects demonstrated improvement on the Porch Index of Communicative Ability (PICA) scale at the 6-week time point than their counterparts who received placebo (83% vs. 22%); improvement was assessed 1 week after the last dose of drug. There was a nonsignificant trend for a persistent benefit at 6 months after dosing. This study was confounded by differences in age (D-amphetamine patients were 9.5 years younger) and amount of therapy received (D-amphetamine patients received 21% more therapy time). The authors found that significant differences were maintained after adjusting for baseline age; unfortunately, there was no adjustment performed for therapy time. A more recent study by Whiting, Chenery, Chalk, and Copland (2007) using a double-blind, placebo-controlled, crossover design in two patients with chronic aphasia from stroke also demonstrated improvements in language function associated with D-amphetamine administration. In this study, naming improved in both patients (but reached statistical significance in only one patient) during periods when they received D-amphetamine plus language therapy compared with periods when they received placebo plus language therapy. This benefit persisted 1 month after cessation of drug and language therapy. Clearly, a crossover study with a behavioral outcome such as this study is potentially compromised by a period effect because of learning by the patient, the investigator, or both, over the course of the trial. These two small studies, each with design inadequacies, are at least suggestive of signals of efficacy for D-amphetamine when coupled with SLT for chronic aphasia.

Two earlier placebo-controlled studies did not show a benefit of D-amphetamine therapy. McNeil et al. (1997) conducted a crossover, multiple-baseline study that examined the effects of D-amphetamine or selegiline, with and without lexical-semantic activation inhibition therapy (L-SAIT), in two patients with chronic aphasia from stroke. Both patients responded well to L-SAIT, but there was not consistent improvement across several language measures when L-SAIT was combined with pharmacotherapy. Darley, Keith, and Sasanuma (1977) examined the effects of 3 days of 20 mg of daily methylphenidate administration on performance on PICA in 14 subchronic or chronic aphasia patients. The trial was a crossover design with no washout period, and no SLT was given. The subjects also underwent 3 days of daily 20-mg chlordiazepoxide administration. The absence of a washout period and the inclusion of traumatic brain injury patients in this study make this study particularly difficult to interpret.

In addition to the data on aphasic subjects, D-amphetamine coupled with SLT has been shown to improve language performance and alter the activation of language-related networks in the brain of healthy nonaphasic subjects. Breitenstein et al. (Breitenstein et al., 2004) taught 40 healthy subjects an artificial vocabulary of 50 words via word–picture matching and coupled this training with D-amphetamine or placebo administration. They found that D-amphetamine enhanced learning of the artificial words in the active group compared with placebo, and that this difference persisted 1 month after drugs, suggesting that enhanced performance was not simply a nonspecific effect of arousal. In a similar study, Whiting, Chenery, Chalk, Darnell, and Copland (2007) found that amphetamines enhanced new word learning in healthy subjects, and that there was no correlation between word learning and attention, mood, or cardiovascular arousal that would suggest a specific effect of D-amphetamine on the trained networks. Supporting the idea that D-amphetamine specifically influences the activity of behaviorally activated networks are two pharmacological MRI studies. Uftring et al. (2001) demonstrated that D-amphetamine specifically increased activation in auditory cortical regions during tone discrimination tasks and enhanced activation of motor cortical areas during motor tasks. Similarly, Sommer et al. (2006) found that D-amphetamine administration during verb generation and semantic decision task (using a conjoint analysis) increased overall left hemispheric activation and increased activation of Broca's areas, the right homolog of Broca's area, and left supramarginal gyrus, but it did not significantly increase activation of multiple other volumes of interest. These data suggest that D-amphetamine can act to potentiate activity and plasticity of behaviorally activated networks rather than nonspecifically promoting arousal.

Of importance, and in concordance with the animal model studies (Feeney et al., 1982; Feeney & Hovda, 1985), the studies showing beneficial effects of sympathomimetics (on motor or language) (Crisostomo, Duncan, Propst, Dawson, & Davis, 1988; Grade, Redford, Chrostowski, Toussaint, & Blackwell, 1998; Walker-Batson, Devous, Curtis, Unwin, & Greenlee, 1991; Walker-Batson, Smith, Curtis, Unwin, & Greenlee, 1995) share the common feature of evaluating D-amphetamine as an adjunct to behavioral or physical therapy rather than alone. This is consistent with the view espoused at the outset regarding the role of behavior in modifying neural circuits and the notion that, without training, even extensively reorganized or remodeled neural circuits are not likely to improve performance.

Although not reviewed here in detail, it is worth noting that after early promising trials, D-amphetamine has failed to show improvement in several more recent well-designed, appropriately powered trials of recovery of motor function after stroke in humans

(Gladstone et al., 2006; Platz et al., 2005; Sonde & Lökk, 2007; Sprigg et al., 2007). It is difficult to predict whether D-amphetamine for aphasia would meet a similar fate if tested in larger trials. One factor that differs between motor and language studies, and across language studies, is the type and magnitude of behavioral therapy given to patients. Might some forms of behavioral therapy be more amenable to drug facilitation than others? The constraint-induced movement therapy typically used in modern stroke motor recovery trials may promote plasticity near or around damaged neural networks, whereas other forms of therapy may be more likely to promote reorganization of different compensatory neural networks. Given the paucity of efficacious drug therapy for chronic aphasia, the good safety profile of D-amphetamine and the promising data from Walker-Batson et al. and Whiting et al., larger trials are merited. It may be interesting to independently vary the type of SLT offered to D-amphetamine patients to probe the drug–behavior interactions mentioned.

Although most of the basic and clinical literature has been directed toward the investigation of the benefits of sympathomimetic compounds as described, an additional report has been published documenting the effect of propranolol, a beta1/beta2 adrenergic antagonist, on language function (Beversdorf et al., 2007). In this double-blind crossover study, four chronic Broca's aphasic patients were administered single 40-mg doses of propranolol or placebo and had language assessed via performance on the Boston Naming Test (BNT) across three separate drug trials. The authors found consistent small increases in naming performance (average BNT before drug = 26.3, after drug = 29.0). This report is consistent with a previous abstract demonstrating benefit of propranolol on language performance (Porch, Wyckes, & Feeney, 1985). The authors speculate that the benefits of beta antagonism may be related to suppression of background activity (as discussed by Hasselmo, Linster, Patil, Ma, & Cekic, 1997) or, less likely, in their view, to the anxiolytic properties of propranolol.

85.4.2 Human Studies: Dopamine Agonists and Levodopa

Several studies have examined the role of dopamine in aphasia recovery by either providing bromocriptine, a D2 agonist, or levodopa, the precursor to dopamine, with a peripheral decarboxylase inhibitor to prevent peripheral conversion of levodopa to dopamine. These drugs have a long track record with a good safety profile and less abuse liability than D-amphetamine. We have found six prospective, double-blind, placebo-controlled studies of these drugs in chronic aphasia.

Bragoni et al. (2000) studied the effects of high-dose bromocriptine (up to 30 mg three times per day) on 11 chronic nonfluent aphasics. They utilized a single cohort study, and each subject was compared with his/her own baseline. They found that bromocriptine plus SLT for 60 days improved performance on several language metrics over SLT alone. The benefit was sustained after a 60-day washout of the drug for several metrics but was only statistically significant for reading comprehension. As suggested by the Whiting et al. (2007) study of D-amphetamine, a single cohort study such as this is potentially compromised by a period effect. A recent study by Seniów, Litwin, Litwin, Lesniak, and Czlonkowska (2009) utilized a parallel design of 39 patients with subacute stroke randomized to receive either 100 mg levodopa or placebo. Drug therapy was timed to precede five-times-weekly SLT by 30 min and was continued for 3 weeks. The Boston Diagnostic Aphasia Examination (BDAE) was used as the primary outcome measure. They found improvement on all metrics, but this only reached statistical significance for verbal fluency, repetition of phrases and sentences, and repetition of words. Washout performance was not assessed. Levodopa patients were younger than placebo subjects by 6.3 years, potentially confounding the analysis, but the investigators found that age was not associated with BDAE outcome.

Several other studies of dopamine-based therapy did not show efficacy. Ashtary, Janghorbani, Chitsaz, Reisi, and Bahrami (2006) examined the impact of bromocriptine, 10 mg daily, started during the acute phase and continued for 4 months; 19 active patients and 19 placebo patients were enrolled. SLT was not required during this study, and it is not clear if any of the subjects received SLT. The investigators did not find any benefit of bromocriptine administration on a standardized Persian language test. Sabe, Salvarezza, García Cuerva, Leiguarda, and Starkstein (1995) reported the results of a crossover study of seven subjects with chronic nonfluent aphasia who received up to 60 mg daily of bromocriptine for 6 weeks. There was no requirement for SLT. The authors found similar performance during both the bromocriptine periods and the placebo periods. Unfortunately, their randomization scheme placed all subjects into the drug arm first, raising the possibility that a practice effect benefited the placebo period. In a similarly designed study, Gupta, Mlcoch, Scolaro, and Moritz (1995) studied the effect of up to 45 mg of bromocriptine over 10 weeks in patients with chronic aphasia. There was no requirement for SLT. This study had a more balanced randomization scheme than that of Sabe et al., and the authors found no influence of bromocriptine over a wide range of language metrics. Most recently, Leemann et al. observed no benefit of 2 weeks of levodopa plus benserazide versus placebo when coupled with intensive computer therapy in the subacute period after stroke (Leemann, Laganaro, Chetelat-Mabillard, & Schnider, 2011).

It is notable that the two studies that demonstrated efficacy for dopamine therapy explicitly coupled dopamine therapy with SLT, whereas three out of four studies that did not show efficacy had no requirement for SLT. This is consistent with the animal literature described previously. These data are also consistent with much of the data from the human motor recovery literature, where levodopa, when coupled with physical therapy, has improved motor outcomes (Rösser et al., 2008; Scheidtmann, Fries, Müller, & Koenig, 2001). As recently noted (Gill & Leff, 2014), a relatively narrow range of dopaminergic agents and doses has been explored in these studies. The totality of the data, although incomplete, suggests that bromocriptine or levodopa therapy, coupled with SLT, may hold promise for aphasia treatment.

85.4.3 Human Studies: Cholinergics and Anticholinergics

Acetylcholine-augmentation as a therapeutic approach is supported by findings that there might be a relative lateralization of cholinergic projections (Bracco, Tiezzi, Ginanneschi, Campanella, & Amaducci, 1984) and that methylscopolamine, an anticholinergic drug, can impair phonological and lexical processing in normal adults (Aarsland, Larsen, Reinvang, & Aasland, 1994). Despite these findings and the body of neurophysiological data on the role of cholinergic projections in modulating neural plasticity and the neuropsychological data documenting its role in learning and memory described, there have been relatively few attempts to modulate this system for the treatment of aphasia. Inhibitors of acetylcholinesterase have widespread use for Alzheimer disease and, in general, have good safety and tolerability profiles. There have been several open-label pilot studies that showed signals of efficacy (Berthier, Hinojosa, Martin Mdel, & Fernandez, 2003; Pashek & Bachman, 2003; Tanaka, Miyazaki, & Albert, 1997; Tsz-Ming & Kaufer, 2001). One single case study involved a patient with a small subcortical lacunar infarction. Dopamine agonist therapy was ineffective, but donepezil led to a significant improvement in fluency (Hughes, Jacobs, & Heilman, 2000).

We are aware of one randomized controlled trial that examined the utility of cholinesterase therapy for stroke-related aphasia. Berthier et al. (2006) studied the effect of 16 weeks of donepezil (up to 10 mg daily) in patients with chronic aphasia. This was a parallel study design with 13 individuals in each group, and all patients received standard SLT. The subjects were well-matched at baseline. The investigators found that the donepezil group improved significantly on the Western Aphasia Battery (WAB), the Communicative Activity Log, and the picture-naming subtest of the Psycholinguistic Assessment of Language Processing in Aphasia test. There were trends for improvement regarding spoken word–picture matching and spoken sentence–picture matching. The improvements noted at week 16 were not present at week 20, suggesting that the benefits of donepezil are not related to neural reorganization. This lack of persistent benefit is similar to what has been seen with the effects of donepezil on the improvement of patients with Alzheimer disease (Rogers, Farlow, Doody, Mohs, & Friedhoff, 1998). Given these short-term improvements, it would be interesting to determine if the improvements on language tests are related to an overall improvement across multiple domains of cognition, which has been seen with donepezil, or whether they are specific to language. If so, then that would imply that the benefits of donepezil may not be dependent on SLT and therefore may become a therapeutic option for those patients who may not have access to SLT. These data are consistent with a more recent open-label study demonstrating efficacy of another cholinesterase inhibitor, galantamine, relative to control subjects, for the treatment of chronic poststroke aphasia (Hong, Shin, Lim, Lee, & Huh, 2012).

85.4.4 Human Studies: Piracetam

Piracetam is a derivative of GABA and has a range of effects on the CNS. Piracetam facilitates cholinergic and excitatory amine neurotransmission (Giurgea, Greindl, & Preat, 1983; Vernon & Sorkin, 1991), increases regional cerebral blood flow (Jordaan, Oliver, Dormehl, & Hugo, 1996), and alters neuronal membrane properties (Müller, Eckert, & Eckert, 1999). It has been claimed that this agent improves learning and memory, but it is not clear which of its biological effects (e.g., neuroprotective, circulatory, or others) are responsible for the purported cognitive benefit (Malykh & Sadaie, 2010). Piracetam is currently available as a nutritional supplement in the United States and is approved for the treatment of myoclonus in Europe.

The data on piracetam for aphasia are mixed. One large multicenter trial (n = 927) aimed to treat all stroke patients within 12 h and used a variety of outcome measures, including assessment of aphasia. This study showed no effect on the primary outcome measure of neurological status (Barthel Index and

Orgogozo scale) at 4 weeks (De Deyn, Reuck, Deberdt, Vlietinck, & Orgogozo, 1997). A post hoc analysis of an "early treatment subgroup" (defined prospectively as within 6 h, but retrospectively as within 7 h) showed some benefit of piracetam. This was particularly true in the moderate to severe subgroup (De Deyn et al., 1997; Orgogozo, 1999). Of these patients, approximately one-third (n = 373) were aphasic, and aphasia recovery at 12 weeks was better in the piracetam group than in the control group, particularly for the early treatment subgroup (Huber, 1999; Orgogozo, 1999).

In postacute and chronic aphasia, several randomized controlled trials have been performed. Enderby et al. observed significant improvements on a multivariate analysis of Aachen Aphasia subtest scores relative to baseline in favor of piracetam ($P = 0.02$) at 12 weeks. This effect was no longer present at 24 weeks (Enderby, Broeckx, Hospers, Schildermans, & Deberdt, 1994). A later double-blind, placebo-controlled study involving chronic aphasia showed trends for improvements across all subsets of the Aachen Aphasia Test, which only reached statistical significance written language (Huber, Willmes, Poeck, Van Vleymen, & Deberdt, 1997). Integrating functional imaging measures into a treatment trial, another study showed an increase in task-related blood flow in several left hemisphere regions generally associated with language over the course of the treatment period; more increase in blood flow in the treatment group than the placebo group. The piracetam group improved on six language measures and the placebo group improved on three (Kessler, Thiel, Karbe, & Heiss, 2000). Most recently, Güngör, Terzi, and Onar (2011), in a single-blind design, examined the impact of 6 months of piracetam given after ischemic stroke causing aphasia and found no benefit across any of the primary language or disability outcome measures, although subjects receiving piracetam showed improvement in auditory comprehension ($P = 0.023$). Thus, there were signals for efficacy in three out of four trials, three of which were relatively small (less than 25 active subjects each) and were thus likely underpowered to see small differences between groups.

85.4.5 Human Studies: Memantine

Memantine is a noncompetitive NMDA-receptor antagonist currently approved for the treatment of Alzheimer disease. A single trial has been performed examining the efficacy of memantine to treat chronic aphasia due to stroke. Berthier et al. (2009) found that 20 mg daily of memantine for 16 weeks, in the absence of SLT, produced enhanced performance on the WAB. Incorporation of constraint-induced aphasia therapy (CIAT) for 2 weeks produced further separation of the memantine group from the placebo group. After a 4-week washout, the memantine group's WAB performance declined substantially, but it was slightly better than that of the placebo group ($P = 0.041$). This study is suggestive of an effect of memantine in the absence of SLT, although evidence for a synergistic relationship between CIAT and memantine is weakened by the differences in WAB scores at the onset of CIAT. Given the good efficacy and tolerability profile of combination use of donepezil and memantine for Alzheimer disease (Tariot et al., 2004), and given the positive studies for both drugs and aphasia described here, it would be interesting to examine the effects of combination therapy on aphasia recovery.

85.4.6 Human Studies: Zolpidem

Zolpidem is a short-acting nonbenzodiazepine hypnotic that potentiates GABA. Although its hypnotic effects are similar to those of the benzodiazepines, it is classified as an imidazopyridine and is molecularly distinct from the classical benzodiazepine molecule. It binds to the alpha1 subunit of the type A GABA receptor, which is preferentially localized to the middle cortical layers of the neocortex when compared with other GABA A-receptor subtypes (Akbarian et al., 1995). There have been case reports documenting substantial clinical improvements in patients with deficits in arousal or awareness after receiving zolpidem (Clauss, Güldenpfennig, Nel, Sathekge, & Venkannagari, 2000; Thomas, Rascle, Mastain, Maron, & Vaiva, 1997). One recent double-blind, placebo-controlled report of one patient with akinetic mutism documented enhanced motor and language performance after the patient received 20 mg of zolpidem. This patient improved from having no speech output with placebo to having limited naming and repetition ability during the zolpidem period. The rCBF study of this patient using $H_2{}^{15}O$ positron emission tomography demonstrated increased blood flow to the anterior cingulate and orbital frontal cortices during the naming tasks (Brefel-Courbon et al., 2007). Another single, open-label case study reported the improvement in language function in an aphasic patient with a single dose of zolpidem (Cohen, Chaaban, & Habert, 2004). In this case report, an individual with nonfluent aphasia and a lesion in the left insula, putamen, and superior temporal gyrus had mild insomnia and was prescribed zolpidem, which led to sudden and unexpected improvement in her speech and naming ability. This remitted when the zolpidem wore off and was reproducible. An electroencephalography (EEG) failed to show any changes after zolpidem administration. Technetium-99 SPECT scanning demonstrated an increase in blood flow to Broca's area, left middle frontal gyrus, left supramarginal gyrus, and the bilateral orbitofrontal and mesial frontal cortices. The commonality of increases in blood flow to the orbitofrontal cortices in both patients that had improved language ability under zolpidem administration raises the possibility that zolpidem influenced the planning or motivational aspects of language behavior in these patients. It is worth noting that these reports are reminiscent of the very early literature on aphasia pharmacotherapy, in which many single cases were reported, mostly involving sedative hypnotics (e.g., amobarbital) (Linn, 1947), suggesting cures of aphasia (see Small, 1994). As yet, no large-scale studies of zolpidem have been reported.

85.4.7 Human Studies: Vasopressin

Vasopressin is produced in the hypothalamus, is released by the posterior pituitary, and is thought to be important in mediating social behavior (Donaldson & Young, 2008) as well as behavior in multiple cognitive domains (Born, Pietrowsky, & Fehm, 1998; Kovacs & De Wied, 1994). Tsikunov and Belokoskova (2007) examined the effects of intranasal desmopressin (a V2-receptor agonist) administration in 26 patients with chronic stroke-related aphasia. This was a single cohort crossover design, and comparisons were made between active treatment periods and placebo periods that always preceded the active periods. The authors observed "good" responses (improvements on at least 3 out of 10 language tests) for 13 out of 26 subjects. SLT was not incorporated into this trial. Because placebo periods always preceded drug periods, this study is subject to a period effect. Review of their Figure 1 shows no improvement between the baseline and placebo periods, but improvement during the drug periods, suggesting minimal period effect. These data are promising and should be studied in a confirmatory trial. One intriguing factor here is the fact that V2 receptors (the probable target for desmopressin) are sparse in the adult human brain. Therefore, it is not clear if the beneficial effects seen here are due to desmopressin interacting with V1 receptors (which are expressed in the brain) or if a peripheral effect of desmopressin (e.g., hyponatremia) is responsible for the benefits. If desmopressin is, in fact, acting in the brain, and given the speculation of the role of vasopressin in social behavior, then it would be interesting to determine the degree to which the improvements in language function correlate with indices of social functioning in these patients. If vasopressin-based recovery has a different mechanism of action than "traditional" neurotransmitter-based therapy, this approach holds promise in being a nonredundant form of pharmacotherapy for subjects receiving one of these classes of drugs.

85.4.8 Drugs to Avoid

If certain pharmacological manipulations have the potential to improve language outcomes in aphasic patients, then it is likely that other manipulations may worsen or delay recovery. Early studies of the inadvertent pharmacological interference with aphasia recovery suggested that haloperidol and hydrochlorothiazide diuretics were associated with worse language outcomes after aphasic stroke (Porch & Feeney, 1986; Porch et al., 1985). Another early report noted that several drugs that impair recovery in experimental stroke (e.g., drugs that affect catecholamine or GABA systems) are commonly given to stroke patients for coincident medical problems (Goldstein, 1993). This led to a formal retrospective (chart review) study of patients using these specific drugs at the time of their strokes (Goldstein, 1995). A total of 96 patient records were reviewed and patients were grouped regarding whether they were using one or more of the following drugs: clonidine, prazosin, any dopamine receptor antagonist (e.g., neuroleptics), benzodiazepines, phenytoin, or phenobarbital. Statistical analysis revealed that although patient demographics and stroke severity were similar between groups, motor recovery time was significantly shorter in the group not using one of these drugs.

Several other drugs have been associated with language difficulties of one type or another and in one context or another. Certainly, the effects of anticholinergic medications on memory function (Koller et al., 2003; Sherman, Atri, Hasselmo, Stern, & Howard, 2003; Taffe, Weed, & Gold, 1999) would suggest avoiding agents with these effects. Anticonvulsant medications, notably vigabatrin and topiramate, have potentially serious cognitive effects, including causing or exacerbating aphasia (Gil & Neau, 1995; Jambaqué, Chiron, Kaminska, Plouin, & Dulac, 1998; Wong & Lhatoo, 2000). Topiramate appears to occupy a unique niche in its ability to produce cognitive disturbances, often manifesting as aphasia. Mula, Trimble, Thompson, and Sander (2003) found word-finding difficulty in 7.2% of more than 400 patients with epilepsy while using topiramate, and a case study demonstrated reversible focal left frontal hypoperfusion and motor aphasia in a seizure patient using topiramate (Cappa, Ortelli, Garibotto, & Zamboni, 2007). Language disturbances have been observed in migraine patients using topiramate as well. Coppola et al. (2008) reported that 26.7% of migraine patients in their clinic who were treated with topiramate had some form of language disturbance compared with 0% for patients using other prophylactic therapy (and matched for headache syndrome severity). These studies do not specifically address the ability for topiramate or other anticonvulsants to alter the recovery from a lesion in the language network. Given the importance of SLT on language recovery, it is likely that topiramate would interfere with the ability of patients to optimally participate in SLT, and therefore it should be avoided.

85.5 CONCLUSION

Existing studies on pharmacological approaches to the treatment of aphasia do not yet paint a clear picture to guide current therapy. Nonetheless, there are increasingly reliable data suggesting a potential beneficial effect potentiation of catecholaminergic transmission on animal motor recovery and aphasia rehabilitation. The data are also promising for drugs that potentiate acetylcholine, as well as for compounds in which the scientific rationale and mechanisms are less clear, such as memantine, and vasopressin. Importantly, cholinesterase inhibitors and memantine have a long track record of safe use in elderly populations with cardiac comorbidities, suggesting few safety concerns about the use of these drugs. It is important to note that despite some of the encouraging data in the aphasia trials, the effect sizes are generally small, and there are few data demonstrating an impact of drug therapy on quality of life or functional outcome measures. Further, despite the common practice in most chronic diseases to use multidrug therapy to attack different pathophysiological pathways, there have been no randomized trials of combination therapy for aphasic stroke therapy. In addition, different etiological forms of aphasia probably have different natural histories and, therefore, different sensitivities to pharmacotherapy, and this has not yet been explored in any depth.

In most cases, drug efficacy has only been seen when coupled with active SLT, which was strongly predicted by the animal literature. SLT is likely the "behavioral engine" that drives pharmacological responses (Nadeau & Wu, 2006). Therefore, pharmacotherapy should not be used as a substitute for speech therapy, and any biological intervention should be used only in concert with individually tailored behavioral therapy, preferably carefully designed adaptive learning approaches. In addition, it is likely that different forms of drug therapy are likely to interact optimally with different forms of SLT, and this should be explored. Further, aphasic patients have a host of comorbidities, such as hypertension, depression, seizures, behavioral disturbances, and others, and they are often using other medications for these. To ensure maximal recovery of language function, it will be important to balance the need for each of these medications with their potential to worsen or delay language recovery and to seek alternatives when they exist. Finally, it should be noted that advances in imaging technology in both animals and humans will make it easier to directly translate findings from rodent studies into human study design, permitting development of novel therapeutics in the future.

Acknowledgments

The authors thank Dr. Christopher Grindrod for comments on a previous version of this chapter.

References

Aarsland, D., Larsen, J. P., Reinvang, I., & Aasland, A. M. (1994). Effects of cholinergic blockade on language in healthy young women. Implications for the cholinergic hypothesis in dementia of the Alzheimer type. *Brain, 117*(Pt 6), 1377–1384.

Aben, I., Verhey, F., Strik, J., Lousberg, R., Lodder, J., & Honig, A. (2003). A comparative study into the one year cumulative incidence of depression after stroke and myocardial infarction. *Journal of Neurology, Neurosurgery and Psychiatry, 74*(5), 581–585. Available from: http://dx.doi.org/10.1136/jnnp.74.5.581.

Akbarian, S., Huntsman, M. M., Kim, J. J., Tafazzoli, A., Potkin, S. G., Bunney, W. E., et al. (1995). GABAA receptor subunit gene expression in human prefrontal cortex: Comparison of schizophrenics and controls. *Cerebral Cortex, 5*(6), 550–560. Available from: http://dx.doi.org/10.1093/cercor/5.6.550.

Asconapé, J. J., & Penry, J. K. (1991). Poststroke seizures in the elderly. *Clinics in Geriatric Medicine, 7*(3), 483–492.

Ashtary, F., Janghorbani, M., Chitsaz, A., Reisi, M., & Bahrami, A. (2006). A randomized, doubleblind trial of bromocriptine efficacy in nonfluent aphasia after stroke. *Neurology, 66*(6), 914–916. Available from: http://dx.doi.org/10.1212/01.wnl.0000203119.91762.0c.

Bao, S., Chan, V. T., & Merzenich, M. M. (2001). Cortical remodelling induced by activity of ventral tegmental dopamine neurons. *Nature, 412*(6842), 79–83.

Barbay, S., Zoubina, E. V., Dancause, N., Frost, S. B., Eisner-Janowicz, I., Stowe, A. M., et al. (2006). A single injection of D-amphetamine facilitates improvements in motor training following a focal cortical infarct in squirrel monkeys. *Neurorehabilitation and Neural Repair, 20*(4), 455–458. Available from: http://dx.doi.org/10.1177/1545968306290773.

Barker-Collo, S. L. (2007). Depression and anxiety 3 months post stroke: Prevalence and correlates. *Archives of Clinical Neuropsychology, 22*(4), 519–531. Available from: http://dx.doi.org/10.1016/j.acn.2007.03.002.

Baskerville, K. A., Schweitzer, J. B., & Herron, P. (1997). Effects of cholinergic depletion on experience-dependent plasticity in the cortex of the rat. *Neuroscience, 80*(4), 1159–1169. Available from: http://dx.doi.org/10.1016/S0306-4522(97)00064-X.

Béjot, Y., Prigent-Tessier, A., Cachia, C., Giroud, M., Mossiat, C., Bertrand, N., et al. (2011). Time-dependent contribution of non neuronal cells to BDNF production after ischemic stroke in rats. *Neurochemistry International, 58*(1), 102–111. Available from: http://dx.doi.org/10.1016/j.neuint.2010.10.019.

Beltran, E. J., Papadopoulos, C. M., Tsai, S.-Y., Kartje, G. L., & Wolf, W. A. (2010). Long-term motor improvement after stroke is enhanced by short-term treatment with the alpha-2 antagonist, atipamezole. *Brain Research, 1346*(0), 174–182. Available from: http://dx.doi.org/10.1016/j.brainres.2010.05.063.

Bergman, P. S., & Green, M. (1951). Aphasia: Effect of intravenous sodium amytal. *Neurology, 1*, 471–475.

Bersano, A., Burgio, F., Gattinoni, M., Candelise, L., & P.S.G. (2009). Aphasia burden to hospitalised acute stroke patients: Need for an early rehabilitation programme. *International Journal of Stroke, 4*(6), 443–447. Available from: http://dx.doi.org/10.1111/j.1747-4949.2009.00349.x.

Berthier, M., Green, C., Lara, J., Higueras, C., Barbancho, M., Dávila, G., et al. (2009). Memantine and constraint-induced aphasia therapy in chronic poststroke aphasia. *Annals of Neurology, 65*(5), 577–585.

Berthier, M., Hinojosa, J., Martin Mdel, C., & Fernandez, I. (2003). Open-label study of donepezil in chronic poststroke aphasia. *Neurology, 60*(7), 1218–1219.

Berthier, M. L., Green, C., Higueras, C., Fernandez, I., Hinojosa, J., & Martin, M. C. (2006). A randomized, placebo-controlled study of donepezil in poststroke aphasia. *Neurology, 67*(9), 1687–1689. Available from: http://dx.doi.org/10.1212/01.wnl.0000242626.69666.e2.

Beversdorf, D. Q., Sharma, U. K., Phillips, N. N., Notestine, M. A., Slivka, A. P., Friedman, N. M., et al. (2007). Effect of propranolol on naming in chronic Broca's aphasia with anomia. *Neurocase: The Neural Basis of Cognition, 13*(4), 256–259.

Bifone, A., Gozzi, A., & Schwarz, A. J. (2010). Functional connectivity in the rat brain: A complex network approach. *Magnetic Resonance Imaging, 28*(8), 1200–1209. Available from: http://dx.doi.org/10.1016/j.mri.2010.07.001.

Born, J., Pietrowsky, R., & Fehm, H. (1998). Neuropsychological effects of vasopressin in healthy humans. *Progress in Brain Research, 119*, 619–643.

Boyeson, M., Harmon, R., & Jones, J. (1994). Comparative effects of fluoxetine, amitriptyline and serotonin on functional motor recovery after sensorimotor cortex injury. *American Journal of Physical Medicine and Rehabilitation/Association of Academic Physiatrists, 73*(2), 76–83.

Boyeson, M. G., & Feeney, D. M. (1984). The role of norepinephrine in recovery from brain injury (abstract). *Paper presented at the Annual Meeting of the Society for Neuroscience, 10*(68), 3.

Bracco, L., Tiezzi, A., Ginanneschi, A., Campanella, C., & Amaducci, L. (1984). Lateralization of choline acetyltransferase (ChAT) activity in fetus and adult human brain. *Neuroscience Letters, 50*(1–3), 301–305.

Bragoni, M., Altieri, M., Di Piero, V., Padovani, A., Mostardini, C., & Lenzi, G. (2000). Bromocriptine and speech therapy in non-fluent chronic aphasia after stroke. *Neurological Sciences: Official Journal of the Italian Neurological Society and of the Italian Society of Clinical Neurophysiology, 21*(1), 19–22.

Brefel-Courbon, C., Payoux, P., Ory, F., Sommet, A., Slaoui, T., Raboyeau, G., et al. (2007). Clinical and imaging evidence of zolpidem effect in hypoxic encephalopathy. *Annals of Neurology, 62*(1), 102–105.

Breitenstein, C., Flöel, A., Korsukewitz, C., Wailke, S., Bushuven, S., & Knecht, S. (2006). A shift of paradigm: From noradrenergic to dopaminergic modulation of learning? *Journal of the Neurological Sciences, 248*(1–2), 42–47.

Breitenstein, C., Wailke, S., Bushuven, S., Kamping, S., Zwitserlood, P., Ringelstein, E. B., et al. (2004). D-Amphetamine boosts language learning independent of its cardiovascular and motor arousing effects. *Neuropsychopharmacology, 29*(9), 1704–1714.

Brenneman, M. M., Wagner, S. J., Cheatwood, J. L., Heldt, S. A., Corwin, J. V., Reep, R. L., et al. (2008). Nogo-A inhibition induces recovery from neglect in rats. *Behavioural Brain Research, 187*(2), 262–272.

Brown, C. E., & Murphy, T. H. (2008). Livin' on the edge: Imaging dendritic spine turnover in the peri-infarct zone during ischemic stroke and recovery. *Neuroscientist, 14*(2), 139–146. Available from: http://dx.doi.org/10.1177/1073858407309854.

Brown, R. M., Carlson, A., Ljunggren, B., Siesjö, B. K., & Snider, S. R. (1974). Effect of ischemia on monoamine metabolism in the brain. *Acta Physiologica Scandinavica, 90*(4), 789–791.

Caplan, D., Hildebrandt, N., & Makris, N. (1996). Location of lesions in stroke patients with deficits in syntactic processing in sentence comprehension. *Brain, 119*(3), 933–949. Available from: http://dx.doi.org/10.1093/brain/119.3.933.

Cappa, S. F., Ortelli, P., Garibotto, V., & Zamboni, M. (2007). Reversible nonfluent aphasia and left frontal hypoperfusion during topiramate treatment. *Epilepsy and Behavior, 10*(1), 192–194.

Carey, M. R., & Regehr, W. G. (2009). Noradrenergic control of associative synaptic plasticity by selective modulation of instructive signals. *Neuron, 62*(1), 112–122.

Carmichael, S. (2006). Cellular and molecular mechanisms of neural repair after stroke: Making waves. *Annals of Neurology, 59*(5), 735–742.

Carmichael, S. T. (2005). Rodent models of focal stroke: Size, mechanism, and purpose. *NeuroRx, 2*(3), 396–409.

Clauss, R., Güldenpfennig, W., Nel, H., Sathekge, M., & Venkannagari, R. (2000). Extraordinary arousal from semi-comatose state on zolpidem. A case report. *South African Medical Journal, 90*(1), 68–72.

Clem, R. L., & Huganir, R. L. (2013). Norepinephrine enhances a discrete form of long-term depression during fear memory storage. *The Journal of Neuroscience, 33*(29), 11825–11832. Available from: http://dx.doi.org/10.1523/jneurosci.3317-12.2013.

Cohen, H. P., Waltz, A. G., & Jacobson, R. L. (1975). Catecholamine content of cerebral tissue after occlusion or manipulation of middle cerebral artery in cats. *Journal of Neurosurgery, 43*(1), 32–36.

Cohen, L., Chaaban, B., & Habert, M.-O. (2004). Transient improvement of aphasia with zolpidem. *The New England Journal of Medicine, 350*(9), 949–950. Available from: http://dx.doi.org/10.1056/nejm200402263500922.

Coppola, F., Rossi, C., Mancini, M., Corbelli, I., Nardi, K., Sarchielli, P., et al. (2008). Language disturbances as a side effect of prophylactic treatment of migraine. *Headache: The Journal of Head and Face Pain, 48*(1), 86–94.

Corbetta, M., Kincade, M., Lewis, C., Snyder, A., & Sapir, A. (2005). Neural basis and recovery of spatial attention deficits in spatial neglect. *Nature Neuroscience, 8*(11), 1603–1610.

Costigan, M., Scholz, J., & Woolf, C. J. (2009). Neuropathic pain: A maladaptive response of the nervous system to damage. *Annual Review of Neuroscience, 32*, 1–32. Available from: http://dx.doi.org/10.1146/annurev.neuro.051508.135531.

Crisostomo, E. A., Duncan, P. W., Propst, M., Dawson, D. V., & Davis, J. N. (1988). Evidence that amphetamine with physical therapy promotes recovery of motor function in stroke patients. *Annals of Neurology, 23*, 94–97.

Dancause, N., Barbay, S., Frost, S. B., Plautz, E. J., Chen, D., Zoubina, E. V., et al. (2005). Extensive cortical rewiring after brain injury. *The Journal of Neuroscience, 25*(44), 10167–10179. Available from: http://dx.doi.org/10.1523/jneurosci.3256-05.2005.

Danilov, A. I., Kokaia, Z., & Lindvall, O. (2012). Ectopic ependymal cells in striatum accompany neurogenesis in a rat model of stroke. *Neuroscience, 214*(0), 159–170. Available from: http://dx.doi.org/10.1016/j.neuroscience.2012.03.062.

Darley, F., Keith, R., & Sasanuma, S. (1977). The effect of alerting and tranquilizing drugs upon the performance of aphasic patients. *Clinical Aphasiology, 7*, 91–96.

Daszuta, A. (2011). *Serotonergic control of adult neurogenesis: Focus on 5-HT2C receptors 5-HT2C receptors in the pathophysiology of CNS disease*. New York: Human Press.

Dauer, W., & Przedborski, S. (2003). Parkinson's disease: Mechanisms and models. *Neuron, 39*(6), 889–909.

De Deyn, P. P., Reuck, J. D., Deberdt, W., Vlietinck, R., & Orgogozo, J. M. (1997). Treatment of acute ischemic stroke

with piracetam. Members of the Piracetam in Acute Stroke Study (PASS) group. *Stroke, 28*(12), 2347–2352.
Di Filippo, M., Tozzi, A., Costa, C., Belcastro, V., Tantucci, M., Picconi, B., et al. (2008). Plasticity and repair in the post-ischemic brain. *Neuropharmacology, 55*(3), 353–362.
Dommett, E. J., Henderson, E. L., Westwell, M. S., & Greenfield, S. A. (2008). Methylphenidate amplifies long-term plasticity in the hippocampus via noradrenergic mechanisms. *Learning and Memory, 15*(8), 580–586. Available from: http://dx.doi.org/10.1101/lm.1092608.
Donaldson, Z., & Young, L. (2008). Oxytocin, vasopressin, and the neurogenetics of sociality. *Science, 322*(5903), 900–904.
Edelmann, E., & Lessmann, V. (2013). Dopamine regulates intrinsic excitability thereby gating successful induction of spike timing-dependent plasticity in CA1 of the hippocampus. *Frontiers in Neuroscience, 7*, 1–11.
Enderby, P., Broeckx, J., Hospers, W., Schildermans, F., & Deberdt, W. (1994). Effect of piracetam on recovery and rehabilitation after stroke: A double-blind, placebo-controlled study. *Clinical Neuropharmacology, 17*(4), 320–331.
Feeney, D., De Smet, A., & Rai, S. (2004). Noradrenergic modulation of hemiplegia: Facilitation and maintenance of recovery. *Restorative Neurology and Neuroscience, 22*, 175–190.
Feeney, D., & Westerberg, V. (1990). Norepinephrine and brain damage: Alpha noradrenergic pharmacology alters functional recovery after cortical trauma. *Canadian Journal of Psychology, 44*(2), 233–252.
Feeney, D. M., Gonzalez, A., & Law, W. A. (1982). Amphetamine, haloperidol, and experience interact to affect rate of recovery after motor cortex injury. *Science, 217*(4562), 855–857.
Feeney, D. M., & Hovda, D. A. (1985). Reinstatement of binocular depth perception by amphetamine and visual experience after visual cortex ablation. *Brain Research, 342*(2), 352–356.
Fridman, E. A., Hanakawa, T., Chung, M., Hummel, F., Leiguarda, R. C., & Cohen, L. G. (2004). Reorganization of the human ipsilesional premotor cortex after stroke. *Brain, 127*(4), 747–758. Available from: http://dx.doi.org/10.1093/brain/awh082.
Ghanbarian, E., & Motamedi, F. (2013). Ventral tegmental area inactivation suppresses the expression of CA1 long term potentiation in anesthetized rat. *PLoS One, 8*(3), e58844.
Gil, R., & Neau, J. (1995). Rapid aggravation of aphasia by vigabatrin. *Journal of Neurology, 242*(4), 251–252.
Gill, S. K., & Leff, A. P. (2014). Dopaminergic therapy in aphasia. *Aphasiology, 28*, 155–170.
Gillani, R. L., Tsai, S.-Y., Wallace, D. G., O'Brien, T. E., Arhebamen, E., Tole, M., et al. (2010). Cognitive recovery in the aged rat after stroke and anti-Nogo-A immunotherapy. *Behavioural Brain Research, 208*(2), 415–424. Available from: http://dx.doi.org/10.1016/j.bbr.2009.12.015.
Giurgea, C. E., Greindl, M. G., & Preat, S. (1983). Nootropic drugs and aging. *Acta Psychiatrica Belgica, 83*(4), 349–358.
Gladstone, D. J., Danells, C. J., Armesto, A., McIlroy, W. E., Staines, W. R., Graham, S. J., et al. (2006). Physiotherapy coupled with dextroamphetamine for rehabilitation after hemiparetic stroke: A randomized, double-blind, placebo-controlled trial. *Stroke, 37*(1), 179–185. Available from: http://dx.doi.org/10.1161/01.str.0000195169.42447.78.
Gobert, A., Billiras, R., Cistarelli, L., & Millan, M. J. (2004). Quantification and pharmacological characterization of dialysate levels of noradrenaline in the striatum of freely-moving rats: Release from adrenergic terminals and modulation by α2-autoreceptors. *Journal of Neuroscience Methods, 140*(1–2), 141–152. Available from: http://dx.doi.org/10.1016/j.jneumeth.2004.04.040.
Gold, R., Linington, C., & Lassmann, H. (2006). Understanding pathogenesis and therapy of multiple sclerosis via animal models: 70 years of merits and culprits in experimental autoimmune encephalomyelitis research. *Brain, 129*(8), 1953–1971.
Goldstein, L. B. (1993). Basic and clinical studies of pharmacologic effects on recovery from brain injury. *Journal of Neural Transplantation and Plasticity, 4*(3), 175–192.
Goldstein, L. B. (1995). Common drugs may influence motor recovery after stroke. The sygen in acute stroke study investigators. *Neurology, 45*(5), 865–871.
Goldstein, L. B. (2000). Effects of amphetamines and small related molecules on recovery after stroke in animals and man. *Neuropharmacology, 39*(5), 852–859. Available from: http://dx.doi.org/10.1016/S0028-3908(99)00249-X.
Goldstein, L. B., & Bullman, S. (1997). Effects of dorsal noradrenergic bundle lesions on recovery after sensorimotor cortex injury. *Pharmacology, Biochemistry, and Behavior, 58*(4), 1151–1157.
Gonzalez, C. L., Gharbawie, O. A., & Kolb, B. (2006). Chronic low-dose administration of nicotine facilitates recovery and synaptic change after focal ischemia in rats. *Neuropharmacology, 50*(7), 777–787.
Grade, C., Redford, B., Chrostowski, J., Toussaint, L., & Blackwell, B. (1998). Methylphenidate in early poststroke recovery: A double-blind, placebo-controlled study. *Archives of Physical Medicine and Rehabilitation, 79*(9), 1047–1050.
Green, A. R., Hainsworth, A. H., & Jackson, D. M. (2000). GABA potentiation: A logical pharmacological approach for the treatment of acute ischaemic stroke. *Neuropharmacology, 39*(9), 1483–1494.
Gu, Q. (2002). Neuromodulatory transmitter systems in the cortex and their role in cortical plasticity. *Neuroscience, 111*(4), 815–835.
Gu, Q., & Singer, W. (1995). Involvement of serotonin in developmental plasticity of kitten visual cortex. *European Journal of Neuroscience, 7*, 1146–1153.
Güngör, L., Terzi, M., & Onar, M. K. (2011). Does long term use of piracetam improve speech disturbances due to ischemic cerebrovascular diseases? *Brain and Language, 117*(1), 23–27. Available from: http://dx.doi.org/10.1016/j.bandl.2010.11.003.
Gupta, S., Mlcoch, A., Scolaro, C., & Moritz, T. (1995). Bromocriptine treatment of nonfluent aphasia. *Neurology, 45*(12), 2170–2173.
Hamm, R. J., O'Dell, D. M., Pike, B. R., & Lyeth, B. G. (1993). Cognitive impairment following traumatic brain injury: The effect of pre- and post-injury administration of scopolamine and MK-801. *Brain Research Cognitive Brain Research, 1*(4), 223–226.
Hasselmo, M. E., Linster, C., Patil, M., Ma, D., & Cekic, M. (1997). Noradrenergic suppression of synaptic transmission may influence cortical signal-to-noise ratio. *Journal of Neurophysiology, 77*(6), 3326–3339.
Hauser, M. D., Chomsky, N., & Fitch, W. T. (2002). The faculty of language: What is it, who has it, and how did it evolve? *Science, 298*(5598), 1569–1579.
Heiss, W.-D., Karbe, H., Weber-Luxenburger, G., Herholz, K., Kessler, J., Pietrzyk, U., et al. (1997). Speech-induced cerebral metabolic activation reflects recovery from aphasia. *Journal of the Neurological Sciences, 145*(2), 213–217.
Heiss, W.-D., & Thiel, A. (2006). A proposed regional hierarchy in recovery of post-stroke aphasia. *Brain and Language, 98*(1), 118–123.
Hendry, S. H., Schwark, H. D., Jones, E. G., & Yan, J. (1987). Numbers and proportions of GABA-immunoreactive neurons in different areas of monkey cerebral cortex. *The Journal of Neuroscience, 7*(5), 1503–1519.
Hermann, D. M., & Chopp, M. (2012). Promoting brain remodelling and plasticity for stroke recovery: Therapeutic promise and

potential pitfalls of clinical translation. *The Lancet Neurology, 11*(4), 369–380.

Hernandez, T. D., & Holling, L. C. (1994). Disruption of behavioral recovery by the anti-convulsant phenobarbital. *Brain Research, 635*(1–2), 300–306.

Hiramoto, T., Ihara, Y., & Watanabe, Y. (2006). [alpha]-1 Adrenergic receptors stimulation induces the proliferation of neural progenitor cells *in vitro*. *Neuroscience Letters, 408*(1), 25–28.

Hong, J. M., Shin, D. H., Lim, T. S., Lee, J. S., & Huh, K. (2012). Galantamine administration in chronic post-stroke aphasia. *Journal of Neurology, Neurosurgery and Psychiatry, 83*(7), 675–680.

Huber, W. (1999). The role of piracetam in the treatment of acute and chronic aphasia. *Pharmacopsychiatry, 32*(Suppl. 1), 38–43.

Huber, W., Willmes, K., Poeck, K., Van Vleymen, B., & Deberdt, W. (1997). Piracetam as an adjuvant to language therapy for aphasia: A randomized double-blind placebo-controlled pilot study. *Archives of Physical Medicine and Rehabilitation, 78*(3), 245–250.

Hughes, J. D., Jacobs, D. H., & Heilman, K. M. (2000). Neuropharmacology and linguistic neuroplasticity. *Brain and Language, 71*(1), 96–101.

Jacobs, K. M., & Donoghue, J. P. (1991). Reshaping the cortical motor map by unmasking latent intracortical connections. *Science, 251*(4996), 944–947.

Jaenisch, N., Witte, O. W., & Frahm, C. (2010). Downregulation of potassium chloride cotransporter KCC2 after transient focal cerebral ischemia. *Stroke, 41*(3), e151–e159.

Jambaqué, I., Chiron, C., Kaminska, A., Plouin, P., & Dulac, O. (1998). Transient motor aphasia and recurrent partial seizures in a child: Language recovery upon seizure control. *Journal of Child Neurology, 13*(6), 296–300.

Jang, S. (2011). A review of diffusion tensor imaging studies on motor recovery mechanisms in stroke patients. *NeuroRehabilitation, 28*(4), 345–352.

Jenkins, W. M., & Merzenich, M. M. (1987). Reorganization of neocortical representations after brain injury: A neurophysiological model of the bases of recovery from stroke. *Progress in Brain Research, 71*, 241–266.

Jitsuki, S., Takemoto, K., Kawasaki, T., Tada, H., Takahashi, A., Becamel, C., et al. (2011). Serotonin mediates cross-modal reorganization of cortical circuits. *Neuron, 69*(4), 780–792.

Jordaan, B., Oliver, D., Dormehl, I., & Hugo, N. (1996). Cerebral blood flow effects of piracetam, pentifylline, and nicotinic acid in the baboon model compared with the known effect of acetazolamide. *Arzneimittelforschung, 46*(9), 844–847.

Kahle, M. P., & Bix, G. J. (2013). Neuronal restoration following ischemic stroke influences, barriers, and therapeutic potential. *Neurorehabilitation and Neural Repair, 27*(5), 469–478.

Karbe, H., Thiel, A., Weber-Luxenburger, G., Herholz, K., Kessler, J., & Heiss, W.-D. (1998). Brain plasticity in poststroke aphasia: What is the contribution of the right hemisphere? *Brain and Language, 64*(2), 215–230.

Kertesz, A., Harlock, W., & Coates, R. (1979). Computer tomographic localization, lesion size, and prognosis in aphasia and nonverbal impairment. *Brain and Language, 8*(1), 34–50. Available from: http://dx.doi.org/10.1016/0093-934X(79)90038-5.

Kessler, J., Thiel, A., Karbe, H., & Heiss, W. D. (2000). Piracetam improves activated blood flow and facilitates rehabilitation of poststroke aphasic patients. *Stroke, 31*(9), 2112–2116.

Kilgard, M. P., & Merzenich, M. M. (1998). Cortical map reorganization enabled by nucleus basalis activity. *Science, 279*(5357), 1714–1718.

Koller, G., Satzger, W., Adam, M., Wagner, M., Kathmann, N., Soyka, M., et al. (2003). Effects of scopolamine on matching to sample paradigm and related tests in human subjects. *Neuropsychobiology, 48*(2), 87–94.

Korchounov, A., & Ziemann, U. (2011). Neuromodulatory neurotransmitters influence LTP-like plasticity in human cortex: A pharmaco-TMS study. *Neuropsychopharmacology, 36*(9), 1894–1902.

Kovacs, G. L., & De Wied, D. (1994). Peptidergic modulation of learning and memory processes. *Pharmacological Reviews, 46*(3), 269–291.

Larson, E. B., & Zollman, F. S. (2010). The effect of sleep medications on cognitive recovery from traumatic brain injury. *The Journal of Head Trauma Rehabilitation, 25*(1), 61–67.

Lazar, R. M., & Antoniello, D. (2008). Variability in recovery from aphasia. *Current Neurology and Neuroscience Reports, 8*(6), 497–502.

Leemann, B., Laganaro, M., Chetelat-Mabillard, D., & Schnider, A. (2011). Crossover trial of subacute computerized aphasia therapy for anomia with the addition of either levodopa or placebo. *Neurorehabilitation and Neural Repair, 25*(1), 43–47.

Li, W.-L., Cai, H.-H., Wang, B., Chen, L., Zhou, Q.-G., Luo, C.-X., et al. (2009). Chronic fluoxetine treatment improves ischemia-induced spatial cognitive deficits through increasing hippocampal neurogenesis after stroke. *Journal of Neuroscience Research, 87*(1), 112–122.

Lichtenwalner, R. J., & Parent, J. M. (2005). Adult neurogenesis and the ischemic forebrain. *Journal of Cerebral Blood Flow and Metabolism, 26*(1), 1–20.

Lim, D. H., Alaverdashvili, M., & Whishaw, I. Q. (2009). Nicotine does not improve recovery from learned nonuse nor enhance constraint-induced therapy after motor cortex stroke in the rat. *Behavioural Brain Research, 198*(2), 411–419. Available from: http://dx.doi.org/10.1016/j.bbr.2008.11.038.

Linn, L. (1947). Sodium amytal in treatment of aphasia. *Archives of Neurology and Psychiatry, 58*, 357–358.

Lloyd, S. A., Balest, Z. R., Corotto, F. S., & Smeyne, R. J. (2010). Cocaine selectively increases proliferation in the adult murine hippocampus. *Neuroscience Letters, 485*(2), 112–116.

Lossius, M., Rønning, O., Mowinckel, P., & Gjerstad, L. (2002). Incidence and predictors for post-stroke epilepsy. A prospective controlled trial. The Akershus stroke study. *European Journal of Neurology, 9*(4), 365–368.

Luria, A., Naydyn, V., Tsvetkova, L., & Vinarskaya, E. (1969). *Restoration of higher cortical function following local brain damage*. Amsterdam: North-Holland Publishing Company.

Lyeth, B. G., Ray, M., Hamm, R. J., Schnabel, J., Saady, J. J., Poklis, A., et al. (1992). Postinjury scopolamine administration in experimental traumatic brain injury. *Brain research, 569*(2), 281–286.

Malykh, A., & Sadaie, M. R. (2010). Piracetam and piracetam-like drugs. *Drugs, 70*(3), 287–312. Available from: http://dx.doi.org/10.2165/11319230-000000000-00000.

Marini, A., Jiang, X., Wu, X., Tian, F., Zhu, D., Okagaki, P., et al. (2004). Role of brain-derived neurotrophic factor and NF-kappaB in neuronal plasticity and survival: From genes to phenotype. *Restorative Neurology and Neuroscience, 22*(2), 121–130.

McNamara, J. (1985). Kindling model of epilepsy. *Advances in Neurology, 44*, 303–318.

McNeil, M. R., Doyle, P. J., Spencer, K. A., Jackson, G. A., Flores, D., & Small, S. L. (1997). A double-blind, placebo-controlled study of pharmacological and behavioral treatment of lexical-semantic deficits in aphasia. *Aphasiology, 11*(4/5), 358–400.

Montanez, S., Kline, A. E., Gasser, T. A., & Hernandez, T. D. (2000). Phenobarbital administration directed against kindled seizures delays functional recovery following brain insult. *Brain Research, 860*(1–2), 29–40.

Mula, M., Trimble, M. R., Thompson, P., & Sander, J. W. A. S. (2003). Topiramate and word-finding difficulties in patients with epilepsy. *Neurology, 60*(7), 1104–1107.

Müller, W., Eckert, G., & Eckert, A. (1999). Piracetam: Novelty in a unique mode of action. *Pharmacopsychiatry, 32*, 2–9.

Murphy, T. H., & Corbett, D. (2009). Plasticity during stroke recovery: From synapse to behaviour. *Nature Reviews Neuroscience*, *10*(12), 861–872.

Nadeau, S., & Wu, S. (2006). CIMT as a behavioral engine in research on physiological adjuvants to neurorehabilitation: The challenge of merging animal and human research. *Neurorehabilitation*, *21*(2), 107–130.

Nudo, R. J., & Friel, K. M. (1999). Cortical plasticity after stroke: Implications for rehabilitation. *Revue Neurologique*, *155*(9), 713–717.

Nudo, R. J., Wise, B. M., SiFuentes, F., & Milliken, G. W. (1996). Neural substrates for the effects of rehabilitative training on motor recovery after ischemic infarct. *Science*, *272*(5269), 1791–1794.

Olsen, T. S. (2001). Post-stroke epilepsy. *Current Atherosclerosis Reports*, *3*, 340–344.

Orgogozo, J. M. (1999). Piracetam in the treatment of acute stroke. *Pharmacopsychiatry*, *32*(Suppl. 1), 25–32.

Overman, J. J., Clarkson, A. N., Wanner, I. B., Overman, W. T., Eckstein, I., Maguire, J. L., et al. (2012). A role for ephrin-A5 in axonal sprouting, recovery, and activity-dependent plasticity after stroke. *Proceedings of the National Academy of Sciences of the United States of America*, *109*(33), E2230–E2239.

Papadopoulos, C. M., Tsai, S.-Y., Guillen, V., Ortega, J., Kartje, G. L., & Wolf, W. A. (2009). Motor recovery and axonal plasticity with short-term amphetamine after stroke. *Stroke*, *40*(1), 294–302. Available from: http://dx.doi.org/10.1161/strokeaha.108.519769.

Pashek, G. V., & Bachman, D. L. (2003). Cognitive, linguistic and motor speech effects of donepezil hydrochloride in a patient with stroke-related aphasia and apraxia of speech. *Brain and Language*, *87*(1), 179–180.

Pedersen, P. M., Stig Jørgensen, H., Nakayama, H., Raaschou, H. O., & Olsen, T. S. (1995). Aphasia in acute stroke: Incidence, determinants, and recovery. *Annals of Neurology*, *38*(4), 659–666. Available from: http://dx.doi.org/10.1002/ana.410380416.

Platz, T., Kim, I., Engel, U., Pinkowski, C., Eickhof, C., & Kutzner, M. (2005). Amphetamine fails to facilitate motor performance and to enhance motor recovery among stroke patients with mild arm paresis: Interim analysis and termination of a double blind, randomised, placebo-controlled trial. *Restorative Neurology and Neuroscience*, *23*, 271–280.

Ploughman, M., Windle, V., MacLellan, C. L., White, N., Dore, J. J., & Corbett, D. (2009). Brain-derived neurotrophic factor contributes to recovery of skilled reaching after focal ischemia in rats. *Stroke*, *40*(4), 1490–1495. Available from: http://dx.doi.org/10.1161/strokeaha.108.531806.

Porch, B., Wyckes, J., & Feeney, D. (1985). Haloperidol, thiazides and some antihypertensives slow recovery from aphasia. *Paper presented at the Society for Neuroscience Abstracts*, *11*(1), 52.

Porch, B. E, & Feeney, D. M. (1986). Effects of antihypertensive drugs on recovery from aphasia. *Clinical Aphasiology*, *16*, 309–314.

Price, D. L., & Sisodia, S. S. (1998). Mutant genes in familial Alzheimer's disease and transgenic models. *Annual Review of Neuroscience*, *21*(1), 479–505.

Ramic, M., Emerick, A. J., Bollnow, M. R., O'Brien, T. E., Tsai, S.-Y., & Kartje, G. L. (2006). Axonal plasticity is associated with motor recovery following amphetamine treatment combined with rehabilitation after brain injury in the adult rat. *Brain Research*, *1111*(1), 176–186.

Rasmussen, R., Overgaard, K., Hildebrandt-Eriksen, E., & Boysen, G. (2006). D-Amphetamine improves cognitive deficits and physical therapy promotes fine motor rehabilitation in a rat embolic stroke model. *Acta Neurologica Scandinavica*, *113*(3), 189–198.

Robinson, R. G., Shoemaker, W. J., & Schlumpf, M. (1980). Time course of changes in catecholamines following right hemispheric cerebral infarction in the rat. *Brain Research*, *181*(1), 202–208.

Robinson, R. G., Shoemaker, W. J., Schlumpf, M., Valk, T., & Bloom, F. E. (1975). Effect of experimental cerebral infarction in rat brain on catecholamines and behaviour. *Nature*, *255*, 332–334.

Rogers, S., Farlow, M., Doody, R., Mohs, R., & Friedhoff, L. (1998). A 24-week, double-blind, placebo-controlled trial of donepezil in patients with Alzheimer's disease. Donepezil study group. *Neurology*, *50*(1), 136–145.

Rösser, N., Heuschmann, P., Wersching, H., Breitenstein, C., Knecht, S., & Flöel, A. (2008). Levodopa improves procedural motor learning in chronic stroke patients. *Archives of Physical Medicine and Rehabilitation*, *89*(9), 1633–1641.

Rumple, A., McMurray, M., Johns, J., Lauder, J., Makam, P., Radcliffe, M., et al. (2013). 3-Dimensional diffusion tensor imaging (DTI) atlas of the rat brain. *PLoS One*, *8*(7), e67334.

Sabe, L., Salvarezza, F., García Cuerva, A., Leiguarda, R., & Starkstein, S. (1995). A randomized, double-blind, placebo-controlled study of bromocriptine in nonfluent aphasia. *Neurology*, *45*, 2272–2274.

Saija, A., Robinson, S. E., Lyeth, B. G., Dixon, C. E., Yamamoto, T., Clifton, G. L., et al. (1988). The effects of scopolamine and traumatic brain injury on central cholinergic neurons. *Journal of Neurotrauma*, *5*(2), 161–170.

Sarno, M. T., Sarno, J. E., & Diller, L. (1972). The effect of hyperbaric oxygen on communication function in adults with aphasia secondary to stroke. *Journal of Speech and Hearing Research*, *15*, 42–48.

Schabitz, W.-R., Steigleder, T., Cooper-Kuhn, C. M., Schwab, S., Sommer, C., Schneider, A., et al. (2007). Intravenous brain-derived neurotrophic factor enhances poststroke sensorimotor recovery and stimulates neurogenesis. *Stroke*, *38*(7), 2165–2172. Available from: http://dx.doi.org/10.1161/strokeaha.106.477331.

Schallert, T., Hernandez, T. D., & Barth, T. M. (1986). Recovery of function after brain damage: Severe and chronic disruption by diazepam. *Brain Research*, *379*(1), 104–111. Available from: http://dx.doi.org/10.1016/0006-8993(86)90261-1.

Schallert, T., Jones, T., Weaver, M., Shapiro, L., Crippens, D., & Fulton, R. (1992). Pharmacologic and anatomic considerations in recovery of function. *Physical Medicine and Rehabilitation*, *6*, 375–393.

Scheidtmann, K., Fries, W., Müller, F., & Koenig, E. (2001). Effect of levodopa in combination with physiotherapy on functional motor recovery after stroke: A prospective, randomised, double-blind study. *The Lancet*, *358*(9284), 787–790.

Schultz, W. (2002). Getting formal with dopamine and reward. *Neuron*, *36*(2), 241–263.

Schwarz, A. J., Becerra, L., Upadhyay, J., Anderson, J., Baumgartner, R., Coimbra, A., et al. (2011). A procedural framework for good imaging practice in pharmacological fMRI studies applied to drug development #2: Protocol optimization and best practices. *Drug Discovery Today*, *16*(15), 671–682.

Seger, D. (2004). Flumazenil—Treatment or toxin. *Journal of Toxicology and Clinical Toxicology*, *42*(2), 209–216.

Seniów, J., Litwin, M., Litwin, T., Lesniak, M., & Czlonkowska, A. (2009). New approach to the rehabilitation of post-stroke focal cognitive syndrome: Effect of levodopa combined with speech and language therapy on functional recovery from aphasia. *Journal of the Neurological Sciences*, *283*, 214–218.

Sherman, S. J., Atri, A., Hasselmo, M. E., Stern, C. E., & Howard, M. W. (2003). Scopolamine impairs human recognition memory: Data and modeling. *Behavioral Neuroscience*, *117*(3), 526–539.

Shewan, C. M., & Kertesz, A. (1980). Reliability and validity characteristics of the Western Aphasia Battery (WAB). *Journal of Speech and Hearing Disorders*, *45*(3), 308.

Small, S. L. (1994). Pharmacotherapy of aphasia: A critical review. *Stroke*, *25*(6), 1282–1289.

Small, S. L. (2001). Biological approaches to the treatment of aphasia. In A. Hillis (Ed.), *Handbook on adult language disorders: Integrating cognitive neuropsychology, neurology, and rehabilitation* (pp. 397–411). Philadelphia, PA: Psychology Press.

Sommer, I., Oranje, B., Ramsey, N., Klerk, F., Mandl, R., Westenberg, H., et al. (2006). The influence of amphetamine on language activation: An fMRI study. *Psychopharmacology, 183*, 387–393.

Sonde, L., & Lökk, J. (2007). Effects of amphetamine and/or l-dopa and physiotherapy after stroke; A blinded randomized study. *Acta Neurologica Scandinavica, 115*(1), 55–59.

Spiegel, A., Shivtiel, S., Kalinkovich, A., Ludin, A., Netzer, N., Goichberg, P., et al. (2007). Catecholaminergic neurotransmitters regulate migration and repopulation of immature human CD34+ cells through Wnt signaling. *Nature Immunology, 8*(10), 1123–1131.

Sprigg, N., Willmot, M. R., Gray, L. J., Sunderland, A., Pomeroy, V., Walker, M., et al. (2007). Amphetamine increases blood pressure and heart rate but has no effect on motor recovery or cerebral haemodynamics in ischaemic stroke: A randomized controlled trial (ISRCTN 36285333). *Journal of Human Hypertension, 21*(8), 616–624.

Strauss, E. H. (2006). *A compendium of neuropsychological tests: Administration, norms, and commentary*. Oxford, UK: Oxford University Press.

Szaflarski, J. P., Allendorfer, J. B., Banks, C., Vannest, J., & Holland, S. K. (2013). Recovered vs. not-recovered from post-stroke aphasia: The contributions from the dominant and non-dominant hemispheres. *Restorative Neurology and Neuroscience, 31*(4), 347–360.

Taffe, M. A., Weed, M. R., & Gold, L. H. (1999). Scopolamine alters rhesus monkey performance on a novel neuropsychological test battery. *Brain Research Cognitive Brain Research, 8*(3), 203–212.

Tanaka, Y., Miyazaki, M., & Albert, M. L. (1997). Effects of increased cholinergic activity on naming in aphasia. *Lancet, 350*(9071), 116–117.

Tariot, P. N., Farlow, M. R., Grossberg, G. T., Graham, S. M., McDonald, S., & Gergel, I. (2004). Memantine treatment in patients with moderate to severe Alzheimer disease already receiving donepezil: A randomized controlled trial. *Jama, 291*(3), 317–324. Available from: http://dx.doi.org/10.1001/jama.291.3.317.

Thiel, C. M., Friston, K. J., & Dolan, R. J. (2002). Cholinergic modulation of experience-dependent plasticity in human auditory cortex. *Neuron, 35*(3), 567–574.

Thomas, P., Rascle, C., Mastain, B., Maron, M., & Vaiva, G. (1997). Test for catatonia with zolpidem. *The Lancet, 349*(9053), 702.

Tsai, S., Markus, T., Andrews, E., Cheatwood, J., Emerick, A., Mir, A., et al. (2007). Intrathecal treatment with anti-Nogo-A antibody improves functional recovery in adult rats after stroke. *Experimental Brain Research, 182*, 261–266.

Tsikunov, S. G., & Belokoskova, S. G. (2007). Psychophysiological analysis of the influence of vasopressin on speech in patients with post-stroke aphasias. *The Spanish Journal of Psychology, 10*(1), 178–188.

Tsz-Ming, C., & Kaufer, D. (2001). Effects of donepezil on aphasia, agnosia, and apraxia in patients with cerebrovascular lesions abstract. *The Journal of Neuropsychiatry and Clinical Neurosciences, 13*, 140.

Ueno, Y., Chopp, M., Zhang, L., Buller, B., Liu, Z., Lehman, N. L., et al. (2012). Axonal outgrowth and dendritic plasticity in the cortical peri-infarct area after experimental stroke. *Stroke, 43*(8), 2221–2228.

Uftring, S., Wachtel, S., Chu, D., McCandless, C., Levin, D., & de Wit, H. (2001). An fMRI study of the effect of amphetamine on brain activity. *Neuropsychopharmacology, 25*(6), 925–935.

Vernon, M. W., & Sorkin, E. M. (1991). Piracetam. An overview of its pharmacological properties and a review of its therapeutic use in senile cognitive disorders. *Drugs and Aging, 1*(1), 17–35.

Vetencourt, J. F. M., Tiraboschi, E., Spolidoro, M., Castrén, E., & Maffei, L. (2011). Serotonin triggers a transient epigenetic mechanism that reinstates adult visual cortex plasticity in rats. *European Journal of Neuroscience, 33*(1), 49–57.

Villmann, C., & Becker, C.-M. (2007). On the hypes and falls in neuroprotection: Targeting the NMDA receptor. *Neuroscientist, 13*(6), 594–615. Available from: http://dx.doi.org/10.1177/1073858406296259.

Wade, D. T., Hewer, R. L., David, R. M., & Enderby, P. M. (1986). Aphasia after stroke: Natural history and associated deficits. *Journal of Neurology, Neurosurgery and Psychiatry, 49*(1), 11–16. Available from: http://dx.doi.org/10.1136/jnnp.49.1.11.

Walker-Batson, D., Curtis, S., Natarajan, R., Ford, J., Dronkers, N., Salmeron, E., et al. (2001). A double-blind, placebo-controlled study of the use of amphetamine in the treatment of aphasia editorial comment. *Stroke, 32*(9), 2093–2098. Available from: http://dx.doi.org/10.1161/hs0901.095720.

Walker-Batson, D., Devous, M. D., Curtis, S., Unwin, D. H., & Greenlee, R. G. (1991). Response to amphetamine to facilitate recovery from aphasia subsequent to stroke. *Clinical Aphasiology, 21*, 137–143.

Walker-Batson, D., Smith, P., Curtis, S., Unwin, H., & Greenlee, R. (1995). Amphetamine paired with physical therapy accelerates motor recovery after stroke. Further evidence. *Stroke, 26*(12), 2254–2259.

West, R., & Stockel, S. (1965). The effect of meprobamate on recovery from aphasia. *Journal of Speech and Hearing Research, 8*, 57–62.

Whitehouse, P. J., Price, D. L., Clark, A. W., Coyle, J. T., & DeLong, M. R. (1981). Alzheimer disease: Evidence for selective loss of cholinergic neurons in the nucleus basalis. *Annals of Neurology, 10*(2), 122–126.

Whiting, E., Chenery, H., Chalk, J., Darnell, R., & Copland, D. (2007). Dexamphetamine enhances explicit new word learning for novel objects. *The International Journal of Neuropsychopharmacology, 10*(6), 805–816. Available from: http://dx.doi.org/10.1017/S1461145706007516.

Whiting, E., Chenery, H. J., Chalk, J., & Copland, D. A. (2007). Dexamphetamine boosts naming treatment effects in chronic aphasia. *Journal of the International Neuropsychological Society, 13*(6), 972–979. Available from: http://dx.doi.org/10.1017/S1355617707071317.

Windle, V., & Corbett, D. (2005). Fluoxetine and recovery of motor function after focal ischemia in rats. *Brain Research, 1044*(1), 25–32.

Wolf, M., Mangiavacchi, S., & Sun, X. (2003). Mechanisms by which dopamine receptors may influence synaptic plasticity. *Annals of the New York Academy of Sciences, 1003*, 241–249. (Glutamate and Disorders of Cognition and Motivation).

Wong, I. C., & Lhatoo, S. D. (2000). Adverse reactions to new anticonvulsant drugs. *Drug Safety, 23*(1), 35–56.

Xerri, C., Merzenich, M. M., Peterson, B. E., & Jenkins, W. (1998). Plasticity of primary somatosensory cortex paralleling sensorimotor skill recovery from stroke in adult monkeys. *Journal of Neurophysiology, 79*(4), 2119–2148.

Yao, C., Williams, A. J., Hartings, J. A., Lu, X. C. M., Tortella, F. C., & Dave, J. R. (2005). Down-regulation of the sodium channel Nav1.1 [alpha]-subunit following focal ischemic brain injury in rats: *In situ* hybridization and immunohistochemical analysis. *Life Sciences, 77*(10), 1116–1129.

Yu, A. J., & Dayan, P. (2002). Acetylcholine in cortical inference. *Neural Networks: The Official Journal of the International Neural Network Society, 15*(4–6), 719–730.

Zhao, C.-s., Puurunen, K., Schallert, T., Sivenius, J., & Jolkkonen, J. (2005). Effect of cholinergic medication, before and after focal photothrombotic ischemic cortical injury, on histological and functional outcome in aged and young adult rats. *Behavioural Brain Research, 156*(1), 85–94. Available from: http://dx.doi.org/10.1016/j.bbr.2004.05.011.

CHAPTER

86

Cell-Based Therapies for the Treatment of Aphasia

Hal X. Nguyen[1] *and Steven C. Cramer*[2]

[1]Mind Research Unit, Sue and Bill Gross Stem Cell Research Center, University of California, Irvine, CA, USA;
[2]Departments of Neurology, Anatomy and Neurobiology, and PM&R, Sue and Bill Gross Stem Cell Research Center, University of California, Irvine, CA, USA

86.1 INTRODUCTION

Aphasia is a consequence of injury to brain language networks. Many different forms of pathology can produce aphasia, including ischemia, degenerative disease, infection, and trauma, with the most common cause being ischemic stroke; approximately 1 in 4 patients with stroke have some degree of aphasia on acute presentation (Rathore, Hinn, Cooper, & Tyroler, 2002). The natural course of aphasia varies with the pathogenesis. When the cause is an ischemic stroke, some degree of spontaneous behavioral recovery is often seen. However, this is generally incomplete, and aphasia remains an important source of human disability.

A number of training and compensation strategies are available to reduce aphasia-related deficits, as reviewed elsewhere in this book. However, although some data suggest the potential for pharmacotherapy to improve outcomes (Allen, Mehta, McClure, & Teasell, 2012; and reviewed elsewhere in this book), there are no convincing data that any drugs improve speech, language, and communication; therefore, no drugs have been approved for this indication (Dobkin & Dorsch, 2013; Klein & Albert, 2004; Small, 1994). There are a number of possible reasons for this. For example, specific challenges that arise in studying therapies that target aphasia include the large number of languages and patterns of communication across our species, the paucity of directly relevant preclinical models of aphasia, current limitations in use of drugs to modulate restorative brain events in a targeted way, and the enormous heterogeneity by which aphasia presents clinically.

Increasing evidence suggests the potential for stem cell therapies to improve outcome after neural injury such as stroke. However, there has been very little research published examining stem cell therapies specifically in relation to aphasia. A review of potential stem cell therapies and general consideration of issues surrounding translation of stem cell therapy therapies in humans are presented.

86.2 STEM CELL THERAPIES: INTRODUCTION

Stem cells are found in all multicellular organisms across animal and plant kingdoms and can be defined on the basis of two key features. The first is self-renewal; cell division produces an additional stem cell as well as a differentiated cell. The second is potency, meaning that stem cells can become many different cell types. The extent of potency varies between different types of stem cells. Stem cells are normally present across the lifespan, from embryo to adult, and in healthy subjects are found within many different tissues, including the brain (Gonzalez-Perez, 2012).

These stem cells may be isolated and grown in culture for use as a therapeutic agent. Many variations are undergoing study, such as stem cell therapies that include gene modification, transformation *in vitro* such as by exposure to selected chemicals, or addition of a bioscaffold. Stem cell therapies have been examined in relation to a vast number of different diseases. This includes numerous neurological disorders and specific neurological deficits. The stage of development for

Neurobiology of Language. DOI: http://dx.doi.org/10.1016/B978-0-12-407794-2.00086-9

stem cell therapies ranges from preclinical to phase III trials, and at least some cell-based therapies have been approved in selected western countries (De Feo, Merlini, Laterza, & Martino, 2012; Eckert et al., 2013; Mankikar, 2010; Savitz, 2013). Numerous routes of stem cell administration have been advocated, from intravenous to highly invasive. Many different mechanisms of action have been described across the many formulations of stem cell therapy. In the brain, stem cell therapies generally promote repair through local or sometimes systemic pathways. In some non-neurological conditions, stem cell therapy can also be used to replace disease-affected tissue; however, to date, actual replacement of injured or dead neural cells by stem cell therapy has been suggested in only a very restricted set of conditions (De Feo et al., 2012). The current review focuses on classes of stem cells that have received a relatively greater amount of study in relation to neurological diseases such as stroke, and thus are the stem cells that may be of greatest relevance to treatment of aphasia.

86.3 HUMAN NEURAL STEM CELLS

Various types of human neural stem cells (hNSCs) have been proposed for cell replacement therapy in a number of neurological conditions, including traumatic injuries to the brain or spinal cord injury, stroke, multiple sclerosis, Parkinson's disease, Huntington's disease, amyotrophic lateral sclerosis, and Alzheimer's disease (Cummings et al., 2005; Gold et al., 2013; Karussis, Petrou, & Kassis, 2013; Keirstead et al., 2005; Pluchino et al., 2003). These hNSCs or neural derivatives are isolated from the donor human organism at different developmental stages and are classified according to their specific origin or the potential of what they can become. Many different types of end-point have been studied for hNSC and their neural derivatives, ranging from a goal as simple as synthesis of a vital protein to the complex goal of establishing synaptic connections and integrating with host circuitry.

86.4 ADULT/FETAL hNSCs

hNSCs can differentiate into neurons and glia and are multipotent cells, meaning that they can differentiate into multiple different types of cells, but not all types of cells found in the body. An important issue for therapeutic applications is that hNSCs obtained later in development, such as those harvested from the central nervous system (CNS) of a human fetus or adult, are likely to have cell fate restrictions that could limit what they can become. In some cases, cell fate restriction is more desirable and may reflect a lower tendency to form tumors or non-neural cell types. This inherent characteristic of fetal and adult hNSCs may be appealing to regulatory agencies and might have contributed to these cells having been approved for clinical trials for several neurological conditions. Ironically, the potential benefits of such hNSC populations can be limiting, mainly because their inherent fate restriction does not permit the realization of their full potential. Critically, fetal or adult hNSCs harvested at different developmental periods or from specific CNS regions are likely to yield specific neural progenitors or cell types that have therapeutic utility for a certain condition or CNS region. Accordingly, hNSCs that have been harvested from the late-stage fetal period or from the adult CNS mainly differentiate into astrocytes, whereas those harvested from an earlier stage of fetal development mainly differentiate into neurons (Nguyen et al., 2014). Furthermore, the fate restriction observed for fetal/adult hNSCs often coincides with a decrease in cell expansion potential, making it more difficult to scale-up production for human patients. In some instances, advances have been made to increase the expansion potential of fetal/adult hNSCs via the selection of the "stemness" marker CD133 (Wang, O'Bara, Pol, & Sim, 2013). However, cells selected in this manner are limited regarding the number of times that they can divide under culture conditions, and thus repeated hNSC cell division can, in the long-term, affect their differentiation potential as well as their chromosomal stability, issues common to many forms of stem cell therapy.

86.5 HUMAN EMBRYONIC STEM CELLS AND NEURAL DERIVATIVES

Human embryonic stem cells (hESCs) are harvested from the inner cell mass of a blastocyst and retain several inherent characteristics that permit long-term expansion. These cells are pluripotent, meaning that they have the capacity to become any of the specialized cell types that comprise the human body (Nichols & Smith, 2012). hESCs may be a limitless source for generating hNSCs and hNSC derivatives such as neurons or glia. This could represent a major potential advantage for human stem cell therapy applications (Brunt, Weisel, & Li, 2012). However, the tendency for hESCs and their derivatives to form tumors in the CNS has been a major concern for cell replacement. hESC-derived hNSCs and derivatives are heterogeneous and are often thought to consist of undifferentiated cells with retained tumor tendency. In an effort to prevent tumor formation, a number of neuralization protocols have been established whereby hESCs are differentiated into more neural-restricted derivatives. These neural derivatives have the potential to differentiate into astrocytes and neurons (Denham & Dottori, 2009; Zhang & Zhang, 2010), as well as specific neural progenitors that can give rise to motoneurons (Chambers et al., 2009; Ebert et al., 2009; Erceg et al., 2008; Li et al., 2005), dopaminergic neurons (Schwartz et al., 2012; Swistowski & Zeng, 2012; Zhang & Zhang, 2010), or oligodendrocytes (Erceg et al., 2010; Hatch, Nistor, & Keirstead, 2009; Okamura et al., 2007); note that neural derivatives generated from these protocols can have variable neural cell fate and expansion potential. Moreover, the resulting neural precursors are typically exposed to undefined factors from mouse feeder cells and bovine serum. In fact, recent advances for cell differentiation into hNSCs and specific neural progenitors or cell types have been made predominantly from cells maintained under conditions containing undefined animal-derived products, which can increase the risk of host immune responses, graft rejection, or infection by nonhuman pathogens (Cobo et al., 2005; Martin, Muotri, Gage, & Varki, 2005; Skottman & Hovatta, 2006). Furthermore, animal-derived products may have added unknown effects on cell characteristics (e.g., proliferation and differentiation), whether the exposure is *in vitro* or *in vivo*. Hence, establishing xeno-free (XF), or more defined, cell culture conditions and protocols for neuralization represents an important step to further the development of therapeutically relevant cell lines.

Several factors have been defined that have established value for differentiating hESCs into neural derivatives (e.g., motoneurons, dopaminergic neurons, or oligoprogenitors), including retinoic acid, sonic hedgehog, and the bone morphogenetic protein and/or signaling inhibitors such as noggin (Chiba, Lee, Zhou, & Freed, 2008; Li et al., 2005; Sundberg et al., 2011; Wada et al., 2009). One attractive feature of using such factors to establish neural derivatives is that they appear to reduce risks of transplant rejection and infection by nonhuman pathogens. However, the practicality of these factors to generate neural derivatives or hNSCs for the clinical human population may depend on their ability to induce neural differentiation while maintaining chromosomal stability of cell lines. A recent study has reported increased incidents of aneuploidy in retinoic acid–treated hESCs (Sartore et al., 2011); hence, the practical use of some of these factors may be limited and more research is needed to better assess the effect of long-term exposures of these factors. Alternatively, growth factors such as those normally found in humans (e.g., epidermal growth factor, basic fibroblast growth factor, and leukemia inhibitory factor) can induce efficient neural differentiation and therefore have been used successfully to generate neuralized spheres and adherent hNSCs under defined or XF conditions (Ebert et al., 2013). These neural derivatives are highly proliferative, retained normal karyotype and have the potential to differentiate in neurons, astrocytes, and/or oligoprogenitors. Overall, recent advances have demonstrated the feasibility of neural differentiation under defined or XF conditions that are attractive to cell replacement therapy for human patients. However, more work is needed to further characterize and test the safety and translational feasibility of these defined or XF neural derivatives, particularly of how they behave in an animal system and how they respond to the cues and toxicity of the brain or spinal cord after trauma or disease.

86.6 HUMAN-INDUCED PLURIPOTENT STEM CELLS AND NEURAL DERIVATIVES

Reprogramming technology has elevated excitement for cell replacement therapy because its advances have allowed somatic cells of the adult human body, such as fibroblasts derived from skin, to be modified *in vitro* into hESC-like stem cells referred to as human-induced pluripotent stem cells (hiPSCs). Like hESCs, hiPSCs can, in principle, become any cell type of the body, providing unprecedented access to *in vitro* human models and thus new opportunities (e.g., to study human genetic diseases or response of patient-specific cells to selected pharmacological compounds) (Bellin, Marchetto, Gage, & Mummery, 2012; Yamanaka, 2012). The hiPSCs also have the potential for autologous transplantation (i.e., use of cells derived from a patient's own body). In contrast with allogeneic transplantation (i.e., use of cells derived from a different organism but of the same species), autologous transplantation generally avoids the need for immunosuppression and therefore avoids the many complications associated with immunosuppression (Morizane et al., 2013).

Therapies using hiPSCs have many potential advantages such as ease of availability and of individualizing therapy. However, the properties of these cells have been incompletely characterized; therefore, concerns remain in relation to translation of hiPSCs for cell replacement therapy. Critics of hiPSC note that some lines are not fully reprogrammed, which might potentially cause mutations. Accordingly, some hiPSC lines previously generated have shown high variation in gene expression, DNA methylation, immunogenicity, and pluripotent potential (Yamanaka, 2012). In some cases, hiPSCs and their neural derivatives have shown an increased tendency to form tumors after cell transplantation (Miura et al., 2009), a finding that has been attributed to selected features of cell reprogramming and culture conditions. Recent advances include generating integration-free hiPSCs through induction methods that include plasmids, small molecules, and synthesized RNAs and proteins. An advantage of such approaches is that they do not involve vector integration into the host genome.

The potential benefits of using hiPSCs as a means to provide a patient-specific therapy remain attractive and contribute, in part, to the continued high level of research activity focused on hiPSCs for regenerative medicine and disease-specific cell model systems.

86.7 MESENCHYMAL STEM CELLs

Mesenchymal stromal cells (MSCs), also known as marrow stromal cells and marrow stem cells, are a form of cellular therapy that have received considerable attention, in part because they are found in many different body tissues (e.g., bone marrow or adipose) and are relatively easy to isolate. MSCs are also immunoprivileged; therefore, concomitant immunosuppression is not required. MSCs are multipotent (i.e., can differentiate into a limited number of cell types, for example, osteoblasts, adipocytes, and chondroblasts) (Dominici et al., 2006). Exogenously administered MSCs are selectively attracted to sites of injury, where they transform and provide benefit through multiple parallel processes, including paracrine delivery of growth factors, immunomodulation, reduced apoptosis, and promotion of cellular remodeling.

86.8 ISSUES RELATED TO CLINICAL APPLICATION OF STEM CELL THERAPIES

A number of issues arise when evaluating stem cell therapies in the treatment of neurological diseases such as those causing aphasia. Some are important across many classes of restorative therapy, whereas others are more specific to cell-based therapies.

Whatever efficacy is ultimately found for stem cells, this will be compared with any additional therapies also proposed for the same clinical indication. There are some unique challenges to providing stem cell therapy, and these are included in any cost-benefit analysis. For example, manufacture and storage of stem cells require very specific laboratory conditions that can be expensive to maintain (Chase et al., 2012; Ilic et al., 2011; Murdoch et al., 2012). For many stem cell therapies, the protocol calls for invasive methods of cell introduction; therefore, the risk of the stem cell therapy is determined by the combined risk of the cells themselves as well as the invasive procedure (Savitz, Cramer, & Wechsler, 2014). Similarly, for any stem cell therapy that requires concomitant immunosuppression, the risk must be measured as that of the stem cells combined with that of the immunosuppressive regimen, a consideration that increases in significance when the target disease is associated with an immunosuppressed state, as has been described early after stroke (Vogelgesang & Dressel, 2011). Many forms of stem cells persist in the body for months or longer, in contrast to most pharmacological compounds that are largely excreted within hours or days. As a result, the period of time needed to reasonably assess the true safety profile of a cell-based therapy may be measured in years. Some types of stem cells have an increased risk for tumor formation, adding to the potential cost and complexity of such a therapeutic approach (Hess, 2009).

A number of issues exist that are of broad importance across clinical investigation of many types of restorative therapies targeting neurological disease. Clinical trials must be carefully designed to enroll a population in which any treatment effect, if present, can be identified. This might be informed by preclinical studies or by careful mechanistic human studies. Studies enrolling patients who lack the substrate necessary for the effects of cell therapy are unlikely to identify a treatment effect. Dose-response studies are critical to translation of restorative therapies and are no less critical when the intervention is cell-based (Benowitz & Carmichael, 2010; Cramer, 2008; Cramer et al., 2011; Savitz et al., 2014). Stroke, the most common cause of aphasia, is a very heterogeneous condition, and methods for selecting or stratifying patients are critical to detecting treatment effects (Cramer, 2010). Efforts to reduce the many sources of potential inter-subject variance are critical to insure adequate study power (Barak & Duncan, 2006; Bath et al., 2012). A related issue is the need for clinical trials of stem cells to use as outcome measure tests that have established validity and reliability (Bath et al., 2012).

The outlook for cell-based therapies in the treatment of aphasia is uncertain at this time. Several factors suggest that stem cells may be found useful for improving language function. For example, several types of cell therapy appear promising in preclinical studies of diseases that affect language, such as stroke, and they are being evaluated in early phase human translational studies. A large body of evidence documents the many forms of behaviorally useful neural plasticity that occur in the human language system in response to neurological disease (Heiss & Thiel, 2006; Lubrano, Draper, & Roux, 2010; Naeser et al., 1998; Rosen et al., 2000; Vandenbulcke, Peeters, Van Hecke, & Vandenberghe, 2005), suggesting the capacity of the language network to undergo remodeling in support of improved language function, at least in some patients. Furthermore, other classes of restorative therapy might also have the potential to improve language function in selected cases and, when combined with a cell-based therapy, might provide further options (Bhogal, Teasell, & Speechley, 2003; Liepert, 2008; Naeser et al., 2005; Shah, Szaflarski, Allendorfer, & Hamilton, 2013; Small & Llano, 2009; Thiel et al., 2013). Other points temper enthusiasm at this time. There are no approved drugs to improve language function, increasing the challenges of identifying an effective cell-based therapy. Many language functions rely on eloquent cortex, and when such brain regions are critically devastated by disease, plasticity can be of limited value to behavioral status. Although the region of the brain that can contribute to language function in humans is broad and variable (Ojemann, Ojemann, Lettich, & Berger, 1989), the key areas needed to be functional in language might actually subtend a rather restricted distribution (Fridriksson, 2010; Hillis et al., 2006). Many of the more advanced stem cell therapeutic applications are based on the cells performing a specific action, such as immune system modulation or paracrine release of neurotrophins. If aphasia recovery requires regrowth of a new area of six-layered cortex with proper white matter connections, then the goal may be elusive in the near future.

Neural repair occurs on the basis of experience-dependent brain plasticity. As a result, it is critical to ensure that behavioral reinforcement is provided concomitant with a restorative therapy such as stem cell therapy (Kleim & Jones, 2008). As an extension of this principle, the amount of therapy provided outside of therapy (e.g., as part of standard of care) must be carefully measured to understand behavioral effects of stem cell therapy (Cramer, 2011).

References

Allen, L., Mehta, S., McClure, J. A., & Teasell, R. (2012). Therapeutic interventions for aphasia initiated more than six months post stroke: A review of the evidence. *Topics in Stroke Rehabilitation, 19*, 523–535.

Barak, S., & Duncan, P. W. (2006). Issues in selecting outcome measures to assess functional recovery after stroke. *NeuroRx, 3*, 505–524.

Bath, P. M., Lees, K. R., Schellinger, P. D., Altman, H., Bland, M., Hogg, C., et al. (2012). Statistical analysis of the primary outcome in acute stroke trials. *Stroke; a Journal of Cerebral Circulation, 43*, 1171–1178.

Bellin, M., Marchetto, M. C., Gage, F. H., & Mummery, C. L. (2012). Induced pluripotent stem cells: The new patient? *Nature Reviews Molecular Cell Biology, 13*, 713–726

Benowitz, L. I., & Carmichael, S. T. (2010). Promoting axonal rewiring to improve outcome after stroke. *Neurobiology of Disease, 37*, 259–266.

Bhogal, S., Teasell, R., & Speechley, M. (2003). Intensity of aphasia therapy, impact on recovery. *Stroke; a Journal of Cerebral Circulation, 34*, 987–993.

Brunt, K. R., Weisel, R. D., & Li, R. K. (2012). Stem cells and regenerative medicine – future perspectives. *Canadian Journal of Physiology and Pharmacology, 90*, 327–335.

Chambers, S. M., Fasano, C. A., Papapetrou, E. P., Tomishima, M., Sadelain, M., & Studer, L. (2009). Highly efficient neural conversion of human es and ips cells by dual inhibition of smad signaling. *Nature Biotechnology, 27*, 275–280.

Chase, L. G., Yang, S., Zachar, V., Yang, Z., Lakshmipathy, U., Bradford, J., et al. (2012). Development and characterization of a clinically compliant xeno-free culture medium in good manufacturing practice for human multipotent mesenchymal stem cells. *Stem Cells Translational Medicine, 1*, 750–758.

Chiba, S., Lee, Y. M., Zhou, W., & Freed, C. R. (2008). Noggin enhances dopamine neuron production from human embryonic stem cells and improves behavioral outcome after transplantation into parkinsonian rats. *Stem Cells, 26*, 2810–2820.

Cobo, F., Stacey, G. N., Hunt, C., Cabrera, C., Nieto, A., Montes, R., et al. (2005). Microbiological control in stem cell banks: Approaches to standardisation. *Applied Microbiology and Biotechnology, 68*, 456–466.

Cramer, S. C. (2008). Repairing the human brain after stroke. Ii. Restorative therapies. *Annals of Neurology, 63*, 549–560.

Cramer, S. C. (2010). Stratifying patients with stroke in trials that target brain repair. *Stroke; a Journal of Cerebral Circulation, 41*, S114–S116.

Cramer, S. C. (2011). An overview of therapies to promote repair of the brain after stroke. *Head and Neck, 33*(Suppl. 1)), S5–S7.

Cramer, S. C., Sur, M., Dobkin, B. H., O'Brien, C., Sanger, T. D., Trojanowski, J. Q., et al. (2011). Harnessing neuroplasticity for clinical applications. *Brain: A Journal of Neurology, 134*, 1591–1609.

Cummings, B. J., Uchida, N., Tamaki, S. J., Salazar, D. L., Hooshmand, M., Summers, R., et al. (2005). Human neural stem cells differentiate and promote locomotor recovery in spinal cord-injured mice. *Proceedings of the National Academy of Sciences of the United States of America, 102*, 14069–14074.

De Feo, D., Merlini, A., Laterza, C., & Martino, G. (2012). Neural stem cell transplantation in central nervous system disorders: From cell replacement to neuroprotection. *Current Opinion in Neurology, 25*, 322–333.

Denham, M., & Dottori, M. (2009). Signals involved in neural differentiation of human embryonic stem cells. *Neurosignals, 17*, 234–241.

Dobkin, B. H., & Dorsch, A. (2013). New evidence for therapies in stroke rehabilitation. *Current Atherosclerosis Reports, 15*, 331.

Dominici, M., Le Blanc, K., Mueller, I., Slaper-Cortenbach, I., Marini, F., Krause, D., et al. (2006). Minimal criteria for defining multipotent mesenchymal stromal cells. The international society for cellular therapy position statement. *Cytotherapy, 8*, 315–317.

Ebert, A. D., Shelley, B. C., Hurley, A. M., Onorati, M., Castiglioni, V., Patitucci, T. N., et al. (2013). Ez spheres: A stable and expandable culture system for the generation of pre-rosette multipotent stem cells from human ESCs and iPSCs. *Stem Cell Research, 10*, 417–427.

Ebert, A. D., Yu, J., Rose, F. F., Jr., Mattis, V. B., Lorson, C. L., Thomson, J. A., et al. (2009). Induced pluripotent stem cells from a spinal muscular atrophy patient. *Nature, 457*, 277–280.

Eckert, M. A., Vu, Q., Xie, K., Yu, J., Liao, W., Cramer, S. C., et al. (2013). Evidence for high translational potential of mesenchymal stromal cell therapy to improve recovery from ischemic stroke. *Journal of Cerebral Blood Flow and Metabolism: Official Journal of the International Society of Cerebral Blood Flow and Metabolism, 33*, 1322–1334.

Erceg, S., Lainez, S., Ronaghi, M., Stojkovic, P., Perez-Arago, M. A., Moreno-Manzano, V., et al. (2008). Differentiation of human embryonic stem cells to regional specific neural precursors in chemically defined medium conditions. *PLoS One*, *3*, e2122.

Erceg, S., Ronaghi, M., Oria, M., Rosello, M. G., Arago, M. A., Lopez, M. G., et al. (2010). Transplanted oligodendrocytes and motoneuron progenitors generated from human embryonic stem cells promote locomotor recovery after spinal cord transection. *Stem Cells*, *28*, 1541–1549.

Fridriksson, J. (2010). Preservation and modulation of specific left hemisphere regions is vital for treated recovery from anomia in stroke. *The Journal of Neuroscience*, *30*, 11558–11564.

Gold, E. M., Su, D., Lopez-Velazquez, L., Haus, D. L., Perez, H., Lacuesta, G. A., et al. (2013). Functional assessment of long-term deficits in rodent models of traumatic brain injury. *Regenerative Medicine*, *8*, 483–516.

Gonzalez-Perez, O. (2012). Neural stem cells in the adult human brain. *Biological and Biomedical Reports*, *2*, 59–69.

Hatch, M. N., Nistor, G., & Keirstead, H. S. (2009). Derivation of high-purity oligodendroglial progenitors. *Methods in Molecular Biology*, *549*, 59–75.

Heiss, W. D., & Thiel, A. (2006). A proposed regional hierarchy in recovery of post-stroke aphasia. *Brain and Language*, *98*, 118–123.

Hess, P. G. (2009). Risk of tumorigenesis in first-in-human trials of embryonic stem cell neural derivatives: Ethics in the face of long-term uncertainty. *Accountability in Research*, *16*, 175–198.

Hillis, A. E., Kleinman, J. T., Newhart, M., Heidler-Gary, J., Gottesman, R., Barker, P. B., et al. (2006). Restoring cerebral blood flow reveals neural regions critical for naming. *The Journal of Neuroscience: The Official Journal of the Society for Neuroscience*, *26*, 8069–8073.

Ilic, N., Brooke, G., Murray, P., Barlow, S., Rossetti, T., Pelekanos, R., et al. (2011). Manufacture of clinical grade human placenta-derived multipotent mesenchymal stromal cells. *Methods in Molecular Biology*, *698*, 89–106.

Karussis, D., Petrou, P., & Kassis, I. (2013). Clinical experience with stem cells and other cell therapies in neurological diseases. *Journal of the Neurological Sciences*, *324*, 1–9.

Keirstead, H. S., Nistor, G., Bernal, G., Totoiu, M., Cloutier, F., Sharp, K., et al. (2005). Human embryonic stem cell-derived oligodendrocyte progenitor cell transplants remyelinate and restore locomotion after spinal cord injury. *The Journal of Neuroscience*, *25*, 4694–4705.

Kleim, J. A., & Jones, T. A. (2008). Principles of experience-dependent neural plasticity: Implications for rehabilitation after brain damage. *Journal of Speech, Language, and Hearing Research*, *51*, S225–S239.

Klein, R. B., & Albert, M. L. (2004). Can drug therapies improve language functions of individuals with aphasia? A review of the evidence. *Seminars in Speech and Language*, *25*, 193–204.

Li, X. J., Du, Z. W., Zarnowska, E. D., Pankratz, M., Hansen, L. O., Pearce, R. A., et al. (2005). Specification of motoneurons from human embryonic stem cells. *Nature Biotechnology*, *23*, 215–221.

Liepert, J. (2008). Pharmacotherapy in restorative neurology. *Current Opinion in Neurology*, *21*, 639–643.

Lubrano, V., Draper, L., & Roux, F. E. (2010). What makes surgical tumor resection feasible in Broca's area? Insights into intra-operative brain mapping. *Neurosurgery*, *66*, 868–875 (discussion 875).

Mankikar, S. D. (2010). Stem cells: A new paradigm in medical therapeutics. *Journal of Long-Term Effects of Medical Implants*, *20*, 219–250.

Martin, M. J., Muotri, A., Gage, F., & Varki, A. (2005). Human embryonic stem cells express an immunogenic nonhuman sialic acid. *Nature Medicine*, *11*, 228–232.

Miura, K., Okada, Y., Aoi, T., Okada, A., Takahashi, K., Okita, K., et al. (2009). Variation in the safety of induced pluripotent stem cell lines. *Nature Biotechnology*, *27*, 743–745.

Morizane, A., Doi, D., Kikuchi, T., Okita, K., Hotta, A., Kawasaki, T., et al. (2013). Direct comparison of autologous and allogeneic transplantation of iPSC-derived neural cells in the brain of a non-human primate. *Stem Cell Reports*, *1*, 283–292.

Murdoch, A., Braude, P., Courtney, A., Brison, D., Hunt, C., Lawford-Davies, J., et al. (2012). The procurement of cells for the derivation of human embryonic stem cell lines for therapeutic use: Recommendations for good practice. *Stem Cell Reviews*, *8*, 91–99.

Naeser, M. A., Martin, P. I., Nicholas, M., Baker, E. H., Seekins, H., Kobayashi, M., et al. (2005). Improved picture naming in chronic aphasia after TMS to part of right Broca's area: An open-protocol study. *Brain and Language*, *93*, 95–105.

Naeser, M. A., Palumbo, C. L., Prete, M. N., Fitzpatrick, P. M., Mimura, M., Samaraweera, R., et al. (1998). Visible changes in lesion borders on CT scan after five years poststroke, and long-term recovery in aphasia. *Brain and Language*, *62*, 1–28.

Nguyen, H. X., Nekanti, U., Haus, D. L., Funes, G., Moreno, D., Kamei, N., et al. (2014). Induction of early neural precursors and derivation of tripotent neural stem cells from human pluripotent stem cells under xeno-free conditions. *The Journal of Comparative Neurology*, *522*, 2767–2783.

Nichols, J., & Smith, A. (2012). Pluripotency in the embryo and in culture. *Cold Spring Harbor Perspectives in Biology*, *4*, a008128.

Ojemann, G., Ojemann, J., Lettich, E., & Berger, M. (1989). Cortical language localization in left, dominant hemisphere. An electrical stimulation mapping investigation in 117 patients. *Journal of Neurosurgery*, *71*, 316–326.

Okamura, R. M., Lebkowski, J., Au, M., Priest, C. A., Denham, J., & Majumdar, A. S. (2007). Immunological properties of human embryonic stem cell-derived oligodendrocyte progenitor cells. *Journal of Neuroimmunology*, *192*, 134–144.

Pluchino, S., Quattrini, A., Brambilla, E., Gritti, A., Salani, G., Dina, G., et al. (2003). Injection of adult neurospheres induces recovery in a chronic model of multiple sclerosis. *Nature*, *422*, 688–694.

Rathore, S., Hinn, A., Cooper, L., Tyroler, H., & Rosamond, W. (2002). Characterization of incident stroke signs and symptoms: Findings from the atherosclerosis risk in communities study. *Stroke; a Journal of Cerebral Circulation*, *33*, 2718–2721.

Rosen, H., Petersen, S., Linenweber, M., Snyder, A., White, D., Chapman, L., et al. (2000). Neural correlates of recovery from aphasia after damage to left inferior frontal cortex. *Neurology*, *55*, 1883–1894.

Sartore, R. C., Campos, P. B., Trujillo, C. A., Ramalho, B. L., Negraes, P. D., Paulsen, B. S., et al. (2011). Retinoic acid-treated pluripotent stem cells undergoing neurogenesis present increased aneuploidy and micronuclei formation. *PLoS One*, *6*, e20667.

Savitz, S. I. (2013). Cell therapies: Careful translation from animals to patients. *Stroke; a Journal of Cerebral Circulation*, *44*, S107–S109.

Savitz, S. I., Cramer, S. C., & Wechsler, L. (2014). Stem cells as an emerging paradigm in stroke 3: Enhancing the development of clinical trials. *Stroke; a Journal of Cerebral Circulation*, *45*, 634–639.

Schwartz, C. M., Tavakoli, T., Jamias, C., Park, S. S., Maudsley, S., Martin, B., et al. (2012). Stromal factors sdf1alpha, sfrp1, and vegfd induce dopaminergic neuron differentiation of human pluripotent stem cells. *Journal of Neuroscience Research*, *90*, 1367–1381.

Shah, P. P., Szaflarski, J. P., Allendorfer, J., & Hamilton, R. H. (2013). Induction of neuroplasticity and recovery in post-stroke aphasia by non-invasive brain stimulation. *Frontiers in Human Neuroscience*, *7*, 888.

Skottman, H., & Hovatta, O. (2006). Culture conditions for human embryonic stem cells. *Reproduction*, *132*, 691–698.

Small, S. (1994). Pharmacotherapy of aphasia. A critical review. *Stroke; a Journal of Cerebral Circulation*, *25*, 1282–1289.

Small, S. L., & Llano, D. A. (2009). Biological approaches to aphasia treatment. *Current Neurology and Neuroscience Reports*, *9*, 443–450.

Sundberg, M., Hyysalo, A., Skottman, H., Shin, S., Vemuri, M., Suuronen, R., et al. (2011). A xeno-free culturing protocol for pluripotent stem cell-derived oligodendrocyte precursor cell production. *Regenerative Medicine*, *6*, 449–460.

Swistowski, A., & Zeng, X. (2012). Scalable production of transplantable dopaminergic neurons from hESCs and iPSCs in xeno-free defined conditions. *Current Protocols in Stem Cell Biology* (Chapter 2: Unit2D. 12).

Thiel, A., Hartmann, A., Rubi-Fessen, I., Anglade, C., Kracht, L., Weiduschat, N., et al. (2013). Effects of noninvasive brain stimulation on language networks and recovery in early poststroke aphasia. *Stroke; a Journal of Cerebral Circulation*, *44*, 2240–2246.

Vandenbulcke, M., Peeters, R., Van Hecke, P., & Vandenberghe, R. (2005). Anterior temporal laterality in primary progressive aphasia shifts to the right. *Annals of Neurology*, *58*, 362–370.

Vogelgesang, A., & Dressel, A. (2011). Immunological consequences of ischemic stroke: Immunosuppression and autoimmunity. *Journal of Neuroimmunology*, *231*, 105–110.

Wada, T., Honda, M., Minami, I., Tooi, N., Amagai, Y., Nakatsuji, N., et al. (2009). Highly efficient differentiation and enrichment of spinal motor neurons derived from human and monkey embryonic stem cells. *PLoS One*, *4*, e6722.

Wang, J., O'Bara, M. A., Pol, S. U., & Sim, F. J. (2013). Cd133/cd140a-based isolation of distinct human multipotent neural progenitor cells and oligodendrocyte progenitor cells. *Stem Cells and Development*, *22*, 2121–2131.

Yamanaka, S. (2012). Induced pluripotent stem cells: Past, present, and future. *Cell Stem Cell*, *10*, 678–684.

Zhang, X. Q., & Zhang, S. C. (2010). Differentiation of neural precursors and dopaminergic neurons from human embryonic stem cells. *Methods in Molecular Biology*, *584*, 355–366.

SECTION Q

PROSODY, TONE, AND MUSIC

CHAPTER

87

Processing Tone Languages

Jackson T. Gandour and Ananthanarayan Krishnan

Department of Speech Language Hearing Sciences, Purdue University, West Lafayette, IN, USA

87.1 INTRODUCTION

Speech perception is important because it provides multiple windows along the auditory pathway into the cerebral cortex regarding how continuous, acoustic signals are transformed into representations on which computations are based at different levels of the brain. Pitch is one of the most important information-bearing components of speech. Tone languages offer advantages for investigating neural mechanisms underlying pitch at different levels of processing because of their phonemic status at the word level (Yip, 2002).

With respect to tonal processing in the brain, almost all experiments performed since 2000 have focused on speech perception or recognition using techniques of functional brain imaging (positron emission tomography [PET]; functional magnetic resonance imaging [fMRI]) and neurophysiology (electroencephalography [EEG]; magnetoencephalography [MEG]). This review focuses primarily on the articles published within that time frame that address four topics related to lexical tone. We evaluate the effects of the phonological status of pitch information on pitch processing in the brain. These experiments tease apart sublexical tonal processing from other cognitive processes involved in speech perception, especially lexical semantic processing. Experimental findings reveal patterns of cortical activation that may vary as a function of acoustic features associated with types of phonological units (i.e., tone versus subsyllabic and segmental units, e.g., consonants and vowels). Because pitch is multidimensional, it is important that we evaluate the effects of pitch features in addition to tonal categories. Those experiments using methods with high temporal resolution reveal the role played by pitch features at early, preattentive stages of processing. There are other suprasegmental units besides tone. The question arises whether common or distinct neural substrates underlie the processing of different suprasegmental units (e.g., intonation, rhythm). Experimental findings to date support the view that speech prosody perception involves a dynamic interplay among widely distributed regions not only within a single hemisphere but also between the two hemispheres, and even different levels of the brain (e.g., midbrain). Moreover, it becomes clear that the time window is pivotal for revealing how hemispheric laterality patterns may reflect higher-level and lower-level stages of auditory processing.

This review on tonal processing in the brain extends previous surveys that have covered dichotic listening (Gandour, 2007; Wang, Behne, Jongman, & Sereno, 2004; Wang, Jongman, & Sereno, 2001), tonal breakdown in production and perception after brain damage (Gandour, 1987, 1994, 1998a, 1998b; Wong, 2002), brain mapping of speech prosody (Gandour, 2006a, 2006b; Zatorre & Gandour, 2008), meta-analysis of lesion literature of linguistic and emotional prosody perception (Witteman, van Ijzendoorn, van de Velde, van Heuven, & Schiller, 2011), and communication disorders in speakers of tone languages (Wong, Perrachione, Gunasekera, & Chandrasekaran, 2009).

87.2 TONE LANGUAGES OF EAST AND SOUTHEAST ASIA

This review focuses exclusively on *lexical tone* languages, that is, those in which the pitch of a word can change the meaning of a word. They are distinguished from *pitch accent* languages (e.g., Japanese), which have a smaller number of contrasting tones, narrower word distribution, and co-occurring syllable structure constraints (Yip, 2002, pp. 1–4). Mandarin Chinese (Beijing), hereafter referred to as Mandarin, has four contrastive tones: *ma*[1] "mother," *ma*[2] "hemp," *ma*[3] "horse," and *ma*[4] "scold." Tones 1 to 4 are high-level (M1), high-rising (M2),

Neurobiology of Language. DOI: http://dx.doi.org/10.1016/B978-0-12-407794-2.00087-0

low-falling–rising (M3), and high-falling (M4) (Xu, 1997). Cantonese (Hong Kong) has six contrastive tones: *ji*[1] "cure," *ji*[2] "chair," *ji*[3] "opinion," *ji*[4] "son," *ji*[5] "ear," and *ji*[6] "two." Tones 1 to 6 are high-level (C1), high-rising (C2), mid-level (C3), low-falling (C4), low-rising (C5), and low-level (C6) (Zee, 1999). Thai (Bangkok) has five contrastive tones: *khaa*[M] "stuck," *khaa*[L] "galangal," *khaa*[F] "kill," *khaa*[H] "trade," and *khaa*[R] "leg". Tones 1 to 5 are mid-level (T1), low-falling (T2), high-falling (T3), high-rising (T4), and low-rising (T5) (Tingsabadh & Abramson, 1999). Voice fundamental frequency (f_0) contours provide the dominant cue for tone recognition (Gandour, 1994).

87.3 LEXICAL VERSUS SUBLEXICAL UNITS

With technological advances in functional brain imaging and auditory neurophysiology at the turn of the century, the aim of the research agenda was to establish that the processing of pitch information in the brain could vary depending on its functional status (linguistic versus nonlinguistic). At that time, it was already well-known that nonlinguistic pitch perception was mediated by neural mechanisms in the right hemisphere (RH) (Zatorre, Belin, & Penhune, 2002, review).

Almost all functional imaging studies of lexical tones have been performed on Thai and Mandarin. Subjects were required to make active judgments (same–different, word recognition) involving later stages of cognitive processing (working memory, decision-making). In discrimination judgments of Thai tones embedded in real words, Thai natives activated the left inferior frontal gyrus (IFG), but Mandarin-speaking Chinese and English did not (1: Gandour et al., 2000; Gandour, Wong, & Hutchins, 1998).[1] This leftward asymmetry in the Thai group is not restricted to a phonemic contrast in tone. Vowel length (/*bat*^{low}/ "card," /*baat*^{low}/ "monetary unit") is phonemic in Thai. When asked to discriminate pitch and timing patterns in Thai pseudowords and nonlinguistic hums (4: Gandour et al., 2002), Thai natives, but not Chinese, similarly activated the left IFG. Chinese natives, however, activated the left IFG when presented with Mandarin tones (3: Klein, Zatorre, Milner, & Zhao, 2001); in contrast, homologous regions in the RH were activated by English (2: Hsieh, Gandour, Wong, & Hutchins, 2001). To isolate processing of lexical tone, Chinese and English listeners were presented with Mandarin tones embedded in actual Mandarin words and in English pseudowords (5: Wong, Parsons, Martinez, & Diehl, 2004). When Chinese listeners were asked to discriminate Mandarin tones embedded in Mandarin words, the left anterior insula was the most active; when embedded in English pseudowords, the right anterior insula was the most active. English listeners activated the right insula and IFG regardless of whether the pitch patterns were embedded in Mandarin or English words. This finding is strengthened by an experiment in which English-speaking adults were trained to use Mandarin tones (M1, M2, M4) to signal lexical meaning on English pseudosyllables (7: Wong, Perrachione, & Parrish, 2007). Good English learners of Mandarin tones showed increased activity in the left posterior superior temporal gyrus (STG); poor learners showed increased activity in the right STG and IFG. Thus, pitch processing engages the left hemisphere (LH) when the pitch patterns signal lexical meaning; otherwise, they are lateralized to the RH.

To isolate sublexical tonal processing, hybrid stimuli were created by superimposing Thai tones onto Mandarin syllables (*tonal chimeras*) and Mandarin tones onto the same syllables (Mandarin words) (6: Xu, Gandour, Talavage, et al., 2006). The tonal chimeras were nonwords in both Mandarin and Thai. In a comparison of native versus nonnative tones, overlapping activity between Mandarin and Thai listeners was identified in the left planum temporale (PT). In this area, a double dissociation between language experience and neural representation of pitch occurred such that stronger activity was elicited in response to native as compared with non-native tones. This neural activity arguably reflects sublexical, phonological processing and is consistent with the view that neural responses to acoustic stimuli can be modulated by their linguistic function (Griffiths & Warren, 2002). Converging evidence that the left PT plays a role in tonal processing comes from an fMRI study in which Chinese listened *attentively* to normal and pitch-flattened sentences (8: Xu, Zhang, Shu, Wang, & Li, 2013). Pitch-flattened sentences elicited greater activation in the left PT compared with normal sentences. Moreover, this activation began to increase and reach its peak earlier than activations in other areas responsible for lexical semantic processing (right PT activation for passive listening to pitch-flattened German sentences; Meyer, Alter, Friederici, Lohmann, & von Cramon, 2002; Meyer, Steinhauer, Alter, Friederici, & von Cramon, 2004). The time course of activation suggests that access to lexical meaning in pitch-flattened sentences is accomplished by the recovery of long-term tonal representations.

[1]The *number:* notation preceding a citation indicates its location in Table 87.1.

TABLE 87.1 Selected References on the Neurobiology of Tonal Processing (2000−2013)

Study	Year	Stimuli	Tasks	Methods	Conclusions
LEXICAL VS. SUBLEXICAL UNITS					
1: Gandour	2000	THA: word, hum; onset consonant, tone	Same−different	PET	Activity in left IFG (frontal operculum) varies depending on linguistic status of pitch
2: Hsieh	2001	MAN: word sequences, hum; onset, rime, tone	Same−different	PET	Leftward asymmetry in inferior frontal cortex for pitch processing depends on its language functions
3: Klein	2001	MAN: word, silence; tone	Same−different	PET	Hemispheric specialization for pitch varies as a function of its linguistic relevance
4: Gandour	2002	THA: pseudowords, hum; tone, vowel length	Same−different	fMRI	Thai group shows more activity in left inferior frontal cortex than Mandarin for processing tone and vowel length in nonlexical contexts
5: Wong	2004	ENG: pseudosyllable; MAN: tones m1, m2, m4	Same−different	PET	Activity in anterior insula indexes whether stimulus is a word (LH) or nonword (RH) in the Mandarin group, but not English group
6: Xu	2006	MAN: word; tonal chimera: MAN syllable + Thai tones	Same−different	fMRI	Double dissociation in the left PT reflects stronger activity to native (Mandarin or Thai) than nonnative tones
7: Wong	2007	ENG, MAN: pseudowords with m1, m2, m4	Same−different	fMRI	Good English learners of Mandarin tones show increased activity in the left posterior STG; poor learners, in the right STG and IFG
8: Xu	2013	MAN: normal, pitch-flattened sentences	Active listening	fMRI	Pitch-flattened sentences elicit greater activity than normal in the left PT, reflecting its role in automatic tonal decoding
TONAL VS. SEGMENTAL UNITS					
9: Gandour	2003	MAN: word	Same−different	fMRI	Activity is greater for rimes versus onsets and tones in left posterior middle frontal gyrus, tones versus onsets and rimes in posterior IFG bilaterally
10: Li	2003	MAN: word	Auditory probe	fMRI	Tone extraction relative to the syllable elicits activity in dorsal frontoparietal areas of the LH
11: Liang	2004	MAN: word	Identification tone, vowel	Aphasia	Differential breakdown of vowels (spared) and tones (impaired) in spoken word production of Chinese aphasic supports a dissociation of tonal and segmental processing
12: Schirmer	2005	CAN: word	Passive oddball	N400	Tonal and segmental information play comparable roles for word processing in Cantonese
13: Luo	2006	MAN: word	Passive oddball	MMN	Opposite laterality for onsets (LH) and tones (RH) indicates acoustic basis for hemispheric dominance at early stage of processing
14: Liu	2006	MAN: word	Naming	fMRI	Tones elicit more activity than vowels in the right IFG in spoken word production
15: Li	2010	MAN: word	Auditory probe	fMRI	RH asymmetry in frontoparietal areas for tones versus onsets or rimes supports role of RH in speech prosody processing
16: Zhao	2011	MAN: word	Passive oddball	N400	Rimes, tones, and syllables equally modulate the amplitude and time course of N400
17: Malins	2012	MAN: word	Match picture	N400	Tonal and phonemic (onsets, rimes) information, not syllabic, constrain spoken word recognition
18: Hu	2012	MAN: word	Semantic congruity	N400	N400 and LPC support functional dissociation of vowel and tone processing in spoken word recognition

(*Continued*)

TABLE 87.1 (Continued)

Study	Year	Stimuli	Tasks	Methods	Conclusions
TONAL FEATURES					
19: Chandrasekaran	2007	MAN: m1/m3, m2/m3	Passive oddball	MMN	Language experience (M > E) influences early, preattentive cortical processing of pitch
20: Chandrasekaran	2007	MAN: m1/2, m1/3	Passive oddball	MMN	Language-dependent weighting of specific, perceptual features of tone may influence its early cortical processing
21: Chandrasekaran	2007	MAN: m1/2, m1/3, m2/3	Passive oddball	MMN, MDS	Effects of language experience vary depending on specific pitch dimensions (height, contour)
22: Kaan	2007	THA: word; t1/t2, t1/t4	Passive oddball	MMN	English and Mandarin listeners, respectively, are sensitive to pitch height and pitch contour of Thai tones
23: Tsang	2011	CAN: c6/1, c6/3, c1/2, c6/2	Passive oddball	MMN, P3a	Change in pitch contour (P3a) and height (MMN) indicate that both tonal attributes are important to tonal processing
TONAL PROCESSING IN THE BRAINSTEM					
24: Swaminathan	2008	MAN: nonspeech; m1, m2, m3, m4	Passive listening	FFR	Pitch representation in the brainstem is sensitive to specific features across speech/nonspeech contexts
25: Krishnan	2009	MAN: nonspeech; m2, m2 inverted, m2 linear, m2 trilinear	Passive listening	FFR	Brainstem pitch encoding is sensitive to time-varying perceptually salient features of pitch patterns
26: Krishnan	2009	MAN: nonspeech; m1, m2, m3, m4	Passive listening	FFR	Degree of acceleration is a critical variable that influences pitch extraction in the brainstem
27: Krishnan	2010	MAN: click trains; m2	Passive listening	FFR	Mandarin listeners' pitch encoding advantage extends to higher acceleration rates beyond the speech domain
28: Krishnan	2011	MAN: m2, m2i; [œ]	Passive listening	FFR	Functional ear (a)symmetries in the brainstem vary depending upon the linguistic status of pitch contours
CATEGORICAL PERCEPTION OF TONE					
29: Xi	2010	MAN: m2, m4	Passive oddball	MMN	Acoustic and phonological information is processed in parallel within the MMN time window
30: Zhang	2011	MAN: m2, m4	Passive oddball	fMRI	Across-category deviants elicit stronger activity in the mid portion of the left middle temporal gyrus; within-category deviants in the right Heshchl's gyrus and STG
31: Zheng	2012	MAN,CAN: m1, m2 and c1, c2	Active detection of deviants	P300	Cantonese (not Mandarin) show strong categorical perception effect in P300 amp that may reflect differences in tonal inventories between the two tone languages
TONE VS. OTHER SUPRASEGMENTAL UNITS					
32: Gandour	2004	MAN: pseudosentences; tone, sentence meaning (statement, question)	Same–different	fMRI	Speech prosody perception is mediated primarily by the RH, but is left-lateralized to task-dependent regions when language processing is required beyond the auditory analysis of the complex sound
33: Tong	2005	MAN: sentence focus (initial, final); intonation (statement, question)	Same–different	fMRI	Speech prosody perception involves a dynamic interplay among widely distributed regions not only within a single hemisphere but also between the two hemispheres

(Continued)

TABLE 87.1 (Continued)

Study	Year	Stimuli	Tasks	Methods	Conclusions
34: Gandour	2007	MAN, English: sentence focus (initial, final); intonation (statement, question)	Same–different	fMRI	Phonetic discrimination of functionally equivalent prosodic contrasts in Mandarin and English by unequal Chinese/English bilinguals reveals essentially a unitary neural system that can adapt to stimulus-specific and task-specific demands for processing a lower-proficiency second language
35: Fournier	2010	RD: rd1, rd2; statement, question	Active listening	MMNm	Lateralization of pitch processing is condition-dependent (tone, LH; intonation, RH) in the Roermond Dutch tone dialect group only, suggesting that language experience determines how processes should be distributed between hemispheres according to the functions available in the grammar
36: Zhang	2010	French: CV sequences; intonation, rhythm	Passive oddball	fMRI	Both rhythm and intonation activated a common area in the right mid portion of the STG for Mandarin listeners, whereas intonation elicited additional activation in the right anterior STS

Key to Table 87.1
List of abbreviations:

BRAIN	
MMN	Mismatch negativity, index of automatic auditory change detection
MMNm	Magnetic mismatch negativity
N400	ERP component associated with later-going lexical semantic processing
P300	ERP component indexes ease of updating memory of stimulus context in response to changes in stimulus attributes
P3a	ERP component associated with automatic switching of attention induced by unexpected change in stimulus event
METHODS	
ERP	Event-related potentials
FFR	Frequency following response generated from the auditory brainstem
fMRI	Functional magnetic resonance imaging
MDS	Multidimensional scaling
MEG	Magnetoencephalography
PET	Positron emission tomography
LANGUAGES	
CAN	Cantonese
ENG	English
MAN	Mandarin
RD	Roermond Dutch
THA	Thai
LEXICAL TONES	
c1	Cantonese Tone 1, high level
c2	Cantonese Tone 2, high rising
c3	Cantonese Tone 3, mid level
c4	Cantonese Tone 4, low falling
c5	Cantonese Tone 5, low rising
c6	Cantonese Tone 6, low level

(Continued)

TABLE 87.1 (Continued)

Key to Table 87.1
List of abbreviations:

m1	Mandarin Tone 1, high level
m2	Mandarin Tone 2, high rising, curvilinear
m2i	Mandarin Tone 2 inverted
m2l	Mandarin Tone 2 linear
m2tl	Mandarin Tone 2 trilinear
m2up	Mandarin Tone 2 transposed up 2 semitones
m3	Mandarin Tone 3, low falling–rising
m4	Mandarin Tone 4, high falling
rd1	Roermond Dutch Accent 1, falling
rd2	Roermond Dutch Accent 2, falling–rising
t1	Thai Tone 1, mid level
t2	Thai Tone 2, low falling
t3	Thai Tone 3, high falling
t4	Thai Tone 4, high rising
t5	Thai Tone 5, low rising

87.4 TONAL VERSUS SEGMENTAL UNITS

Linguistic theory informs us that the onset and rime of a syllable contain segmental units. They differ in their duration and the order in which their information unfolds in time over the duration of a syllable. Rimes and tones, however, overlap substantially in the order in which their information unfolds in time. Tones are suprasegmental; they are mapped onto (morpho) syllables.

Depending on task demands, tones elicit effects that differ from those of segments. The time course and amplitude of N400 (a negative component associated with lexical semantic processing that peaks approximately 400 ms after the auditory stimulus) were the same for consonant, rime, and tone violations in Cantonese (12: Schirmer, Tang, Penney, Gunter, & Chen, 2005). Their findings were replicated in Mandarin, but syllable violations elicited an earlier and stronger N400 than tone (17: Malins & Joanisse, 2012; cf. 16: Zhao, Guo, Zhou, & Shu, 2011). This separation of tone from its carrier syllable was also reported in an auditory verbal recognition paradigm in which subjects selectively attended to either the syllable or the tone (10: Li et al., 2003). In a spoken word recognition paradigm, tones elicited larger late positive event-related potential (ERP) component than vowels (19: Hu, Gao, Ma, & Yao, 2012). In a left brain-damaged Chinese aphasic, vowels were spared and tones were severely impaired (11: Liang & van Heuven, 2004). These findings together support a functional dissociation of tonal and segmental information.

It is well-known that hemispheric specialization may be driven by differences in acoustic features associated with segments. The question is whether hemispheric specialization for tone can be dissociated from segments. Tones induce greater activation in the right posterior middle frontal gyrus (MFG) for English speakers when compared with consonants or rimes (9: Gandour et al., 2003). This area has been implicated in pitch perception (Zatorre et al., 2002). Their increased activation is presumably due to their lack of experience with Chinese tones. Using a tone identification task, the right IFG was found to be activated in English learners of Mandarin tone only *after* training (Wang, Sereno, Jongman, & Hirsch, 2003). This finding demonstrates early cortical effects of learning a second language that involve recruitment of cortical regions implicated in tonal processing. Focusing on hemispheric specialization for tone production (14: Liu et al., 2006), Mandarin tones elicited more activity in the right IFG than vowels. This rightward preference for tonal processing converges more broadly with the role of the RH in mediating speech prosody (Friederici & Alter, 2004; Glasser & Rilling, 2008; Wildgruber, Ackermann, Kreifelts, & Ethofer, 2006).

As measured by the mismatch negativity (MMN), a fronto-centrally distributed cortical ERP that indexes a change in auditory detection, it is well-known that language experience may influence the automatic, involuntary processing of consonants and vowels (Naatanen, 2001, review). Therefore, one would expect language experience to modulate the automatic cortical processing of lexical tones. Tones evoked stronger MMN in the RH relative to the LH, whereas consonants produced the opposite pattern (13: Luo et al., 2006). An fMRI study showed that Mandarin tones, relative to consonants or rimes, elicited increased activation in right frontoparietal areas (15: Li et al., 2010). Taken together, these data suggest the balance of hemispheric specialization may be modulated by distinct acoustic features associated with tonal as compared with segmental units.

87.5 TONAL FEATURES

The notion that a phonetic segment can be decomposed into a set of features is universally accepted among linguists. Tone, a suprasegmental unit, has also been characterized as being made up of features (Wang, 1967; Yip, 2002, pp. 39–64). Their ontological status in tone perception is well-established (Gandour, 1983; Gandour & Harshman, 1978) and confirmed in more recent studies of tone perception (Huang & Johnson, 2011; Khouw & Ciocca, 2007) and tone learning (Chandrasekaran, Sampath, & Wong, 2010; Francis, Ciocca, Ma, & Fenn, 2008). The brain, however, is a neurophysiological apparatus. Features, however, are not to be confused with neural mechanisms.

How they are implemented in the brain depends on the anatomical level to which they are being applied and their functional status in a particular language. Using nonspeech homologues of Mandarin tones (19: Chandrasekaran, Krishnan, & Gandour, 2007a), native Chinese exhibited larger MMN responses than English in response to a deviant representing a natural *curvilinear* rising pitch contour representative of M2, but not in response to a *linear* rising ramp that is a crude approximation of M2 that does not occur in natural speech. This finding demonstrates that experience-dependent plasticity is sensitive to the shape of pitch contours. To further probe the stimulus attributes that trigger these language-dependent effects (20: Chandrasekaran, Krishnan, & Gandour, 2007b), two passive oddball conditions were presented to Mandarin and English listeners. One contained two tones that are acoustically dissimilar to one another (M1/M3); the other contained two tones that are acoustically similar (M2/M3). MMN responses of Chinese listeners were larger than those of English for the high dissimilarity condition only (M1/M3). An explanation based on tonal categories does not tell us why MMN amplitude is reduced for one condition but not the other. All three stimuli exhibited pitch contours exemplary of their tonal category. Language group differences may be attributed to the relative saliency of perceptual features. To test the hypothesis of separate neural processing of pitch dimensions, another oddball condition (M1/M2) was added. A multidimensional scaling analysis of pairwise dissimilarities of MMN responses to Mandarin tones revealed that Chinese listeners, relative to English, are more sensitive to pitch contour than pitch height (21: Chandrasekaran, Gandour, & Krishnan, 2007). Thus, MMN may serve as a neural index of the relative saliency of underlying features of pitch that are differentially weighted by language experience.

In Cantonese (23: Tsang, Jia, Huang, & Chen, 2011), MMN and P3a (an automatic attention shift induced by the detection of deviant features in the passive oddball paradigm) were elicited from two Cantonese tonal pairs: one differing in pitch height (height-large, C6/C1; height-small, C6/C3) and the other differing in pitch contour (contour-early, C1/C2; contour late, C6/C2). The size and latency of MMN were sensitive to the size of pitch level change, whereas the latency of P3a captured the presence of pitch contour change. Their findings confirm that pitch contour and pitch height are important tonal features in early lexical tone processing. Most importantly, MMN and P3a are revealed to be independent neural components that are differentially sensitive to pitch height and contour, respectively. In another study (22: Kaan, Wayland, Bao, & Barkley, 2007), two oddball conditions were presented to Mandarin and English listeners to assess the effects of perceptual training of Thai tones as a function of language background. One condition contained two tones that are acoustically dissimilar (T1/T4), mid versus high-rising; the other contained two tones that are acoustically similar (T1/T2), mid versus low-falling. After training, the high-rising deviant (T4) elicited a larger MMN amplitude for English listeners in contrast to a later MMN latency for Mandarin listeners. Their findings suggest that English listeners are more sensitive to early differences in pitch height, whereas Mandarin Chinese are more sensitive to later rapid changes in pitch contour.

87.6 TONAL PROCESSING AT THE LEVEL OF THE AUDITORY BRAINSTEM

Pitch processing may also be subject to experience-dependent effects at the level of the brainstem before the auditory signal reaches the cerebral cortex. Electrophysiological responses to tonal features may emerge no later than 5 to 8 ms from the time the auditory signal enters the ear (Krishnan & Gandour, 2009, review). The frequency following response (FFR) reflects sustained phase-locked activity in a population of neural units within the brainstem and is characterized by a periodic waveform that follows the individual cycles of the stimulus waveform (Chandrasekaran & Kraus, 2010; Krishnan, 2007, tutorials). Experience-dependent pitch encoding mechanisms in the brainstem are especially sensitive to the *curvilinear* shape of pitch contours that occur in speech and nonspeech contexts (24: Swaminathan, Krishnan, & Gandour, 2008). Linear approximations of Mandarin tones (M2, M4) fail to elicit a language-dependent effect (25: Krishnan, Gandour, Bidelman, & Swaminathan, 2009; Xu, Krishnan, & Gandour, 2006).

Neural mechanisms in the brainstem show enhanced language-dependent pitch encoding in response to particular time-varying acoustic properties within tonal *subsections*. Using nonspeech homologues of Mandarin tones (26: Krishnan, Swaminathan, & Gandour, 2009), pitch strength (magnitude of the normalized autocorrelation peak) of 40-ms subsections revealed that Chinese listeners, relative to English, exhibit more robust pitch representation of those subsections containing rapid changes in pitch. This heightened sensitivity to rapid changes in pitch by Chinese listeners was maintained even in severely degraded stimuli (Krishnan, Gandour, & Bidelman, 2010a). This experience-dependent enhancement of pitch encoding may transfer to other tone languages. Pitch strength of tonal subsections containing moderate rises in pitch were most important in distinguishing tonal (Mandarin, Thai) from nontonal language (English) groups (Krishnan, Gandour, & Bidelman, 2010b). Neuroplasticity for pitch processing in the brainstem is not necessarily limited to the domain in which the pitch contours are perceptually relevant. Mandarin listeners had an advantage over English not only in response to a click-train homologue of M2 but also in response to scaled variants with increasingly higher acceleration rates that fall proximal to or outside the boundary of natural speech (27: Krishnan, Gandour, Smalt, & Bidelman, 2010). Moreover, changes to the acoustic periodicity of a stimulus directly influence brainstem encoding and its corresponding perceptual responses to pitch (Krishnan, Bidelman, & Gandour, 2010). Neural pitch strength in the brainstem and perceptual pitch salience, as reflected by f_0 difference limen estimates, improved systematically with increasing temporal regularity of the M2 stimulus. This strong correlation between neural and behavioral measures supports the view that pitch encoding at a subcortical sensory level of processing plays an important role in shaping tone perception.

Hemispheric asymmetries in the cerebral cortex are predictable based on low-level, spectrotemporal features of stimuli, but they can also be modulated by their linguistic function (Meyer, 2008; Poeppel, Idsardi, & van Wassenhove, 2008; Wildgruber et al., 2006; Zatorre & Gandour, 2008, reviews). It is also well-known that there are fixed, structural asymmetries in the auditory pathway. Whether *ear asymmetries* at the level of the brainstem can be modulated by functional changes in pitch is an open question. Using two synthetic speech stimuli (native M2; nonnative flipped variant of M2), magnitude of the f_0 component in the FFR (amplitude of the spectral component at f_0) was obtained from a perceptually salient portion of M2 that exhibits rapidly changing pitch (28: Krishnan, Gandour, Ananthakrishnan, Bidelman, & Smalt, 2011). The native tone (M2) evoked a comparatively larger degree of rightward ear asymmetry in pitch encoding than the non-native pitch pattern. In response to left-ear and right-ear stimulation, the FFR evoked by M2 was larger than its flipped variant with right ear stimulation only. On an absolute scale, asymmetry favoring left ear stimulation was evoked by the non-native pitch contour. These differences in ear asymmetry may reflect an emerging functional separation of periodicity and spectral representations at the midbrain level.

87.7 CATEGORICAL PERCEPTION OF TONE

Categorical perception is believed to reflect fundamental aspects of the processing of speech sounds (Harnad, 1987, review). It refers to the phenomenon whereby a specific step along a continuous sensory dimension may signal the boundary between separate categories. The bulk of research has focused on consonants and vowels (tones; Francis, Ciocca, & Ng, 2003; Peng et al., 2010; Xu, Gandour, & Francis, 2006). An ERP study of the categorical perception of Mandarin tones provides a window to the interplay between phonetic and phonological processing (29: Xi, Zhang, Shu, Zhang, & Li, 2010). Xi et al. created an 11-step f_0 continuum with M2 (high-rising) and M4 (high-falling) as the endpoint stimuli in both speech and nonspeech conditions. Using a passive oddball paradigm, both within-category and across-category deviants elicited larger MMNs in the RH sites; however, at the same time, larger MMNs were elicited by across-category than by within-category deviants in the LH. Given their low spatial resolution and methodological constraints that limit unambiguous interpretation of hemispheric dominance of ERPs based on scalp topographical maps, it was necessary to use a method with high spatial resolution to clearly identify cortical activation in different brain regions. In a companion fMRI study (30: Zhang et al., 2011), brain areas activated by acoustic variation within tonal categories were located in the dorsal and posterior-lateral STG bilaterally, especially the right middle STG. In contrast, brain areas activated by phonological variation across tonal categories, as compared with within-category acoustic variation, were located in the left middle temporal gyrus (MTG). These findings are consistent with the view that the dorsal STG and lateral mSTS/MTG are responsible for acoustic analysis and phonological processing, respectively. Superior regions of the temporal lobe are known to be responsible for initial stages of auditory analysis, whereas the superior temporal sulcus (STS) and MTG have been implicated in higher-level phonological processing (Hickok & Poeppel, 2004, 2007; Liebenthal, Binder, Spitzer, Possing, & Medler, 2005; Liebenthal et al., 2010). A cross-language ERP study focused on the influence of language experience (Mandarin, Cantonese) on categorical perception of a three-step f_0 continuum consisting of rising pitch contours common to both tone languages (31: Zheng, Minett, Peng, & Wang, 2012). Deviant responses were measured by the P300 amplitude, a voluntary attention-switching response elicited by an *active* oddball paradigm (Polich, 2007). As reflected by P300 amplitude, Cantonese listeners discriminated the tonal stimuli better than Mandarin. Zheng et al. speculate that Cantonese listeners make finer distinctions in f_0 height and slope because the Cantonese tonal space (six contrasts) is more dense that that of Mandarin (four contrasts).

87.8 TONE VERSUS OTHER SUPRASEGMENTAL UNITS

There are other suprasegmental units of speech besides tone (Lehiste, 1996). In comparison with tone, we are especially interested in those units that may also be signaled by variations in pitch (e.g., stress, intonation, sentence focus). In tone languages, pitch variations can be used to signal differences in the meaning of sentences as well as words. In an fMRI study of Mandarin tone and intonation (32: Gandour et al., 2004), Chinese listeners exhibited greater activity than English in the left ventral aspects of the inferior parietal lobule regardless of the level of prosodic representation. Both language groups, however, showed activity within the right STS and MFG (Ren, Yang, & Li, 2009). This right-sided preference may reflect shared mechanisms underlying early processing of complex pitch patterns irrespective of language experience (Zatorre, Mondor, & Evans, 1999). The LH activity in the inferior parietal lobule is likely to reflect higher-level, language-dependent phonological processing (Jacquemot, Pallier, LeBihan, Dehaene, & Dupoux, 2003).

In tone languages, pitch variations can be used to signal differences in the meaning of sentences as well as words. The MEG study of tone and intonation in Roermond Dutch provides an account of a tonal dialect that unambiguously encodes both contrasts phonologically (35: Fournier, Gussenhoven, Jensen, & Hagoort, 2010). That is, Roermond has two lexical tones that are phonetically distinct in statements and questions. When asked to listen *attentively* to oddball sequences, native Roermond listeners showed a stronger MMNm (150–250 ms) over the left temporal cortex for tone and a predominantly RH response for intonation. Non-native listeners showed a stronger response over the left temporal cortex irrespective of prosodic unit. Using a *passive* oddball paradigm, the MMN (120–240 ms) yielded RH dominance in Mandarin for both word-level (tone) and sentence-level (intonation) prosodic functions (Ren et al., 2009). These conflicting findings between Roermond Dutch and Mandarin are likely due to task demands (attentive versus preattentive) rather than the lack of a prosodic function in their phonological system.

Mandarin and English differ structurally in their use of prosody at the word level. However, both languages exploit prosody at the sentence level to distinguish focus and discourse meaning. A cross-language fMRI study of the perception of Mandarin sentence focus and intonation demonstrated that Mandarin listeners exhibited greater activity in the left supramarginal gyrus and posterior MTG than English across conditions (33: Tong et al., 2005). This leftward specialization is consistent with the notion of a dorsal processing stream that emanates from auditory cortex, projects to the inferior parietal lobule, and ultimately projects to frontal lobe regions, and with a ventral processing stream that projects to the posterior MTG (Hickok & Poeppel, 2007). Rightward preferences were observed in the middle portion of the MFG for both language groups, implicating more general attention and working memory processes associated with pitch perception. Because both sentence-level phenomena occur in Mandarin and English, it is also possible to compare the processing of the same prosodic contrasts in late-onset Chinese/English bilinguals' first (Mandarin) and second languages (English). Any differences in neural activity associated with auditory processing of the same prosodic contrast in the bilinguals' native language and second language may serve as an index of whether the neural substrates are shared or segregated for the two languages. Chinese/English bilinguals displayed overlapping activation between Mandarin and English stimuli in frontal, parietal, and temporal areas regardless of language (34: Gandour et al., 2007). The sentence focus task, however, elicited greater activation for English stimuli than Mandarin in the bilateral anterior insula and MFG. This is presumably attributable to differences in the way sentence focus is signaled phonetically in the two languages (Xu, 2006). Increased computational demands for the lower-proficiency language lead to greater activation in frontal areas implicated in attention (Shaywitz et al., 2001) and working memory (Smith & Jonides, 1999).

Another suprasegmental feature of speech is rhythm (pattern of timing variations over phrases). Both rhythm and intonation (pattern of pitch variations over phrases) span a number of segments over a relatively long time interval and, therefore, are expected to be preferentially processed in the RH (Poeppel, 2003). The question is whether overlapping or distinct regions of the RH are involved in the processing of rhythm and intonation. Using a passive listening task, a common area in the right middle portion of the STG was activated by Mandarin listeners for both rhythm and intonation conditions (36: Zhang, Shu, Zhou, Wang, & Li, 2010). Compared with rhythm, intonation elicited additional activation in the right anterior STS. This isolation of a particular brain region in the processing of intonation suggests that it is responsive to specific acoustic features associated with dynamic variations in pitch (Humphries, Love, Swinney, & Hickok, 2005; Lattner, Meyer, & Friederici, 2005).

87.9 CONCLUSION

Language experience shapes processing of pitch information at both cortical and subcortical levels. Tones play a role comparable with that of segments in word processing. Whereas both engage the LH in attention-modulated, task-dependent processing, tones show a distinctive rightward asymmetry relative to segments, especially at early stages of processing. Pitch is a multidimensional perceptual attribute that affords us an opportunity to investigate pitch features. Tonal processing reveals experience-dependent sensitivity to specific features that are linguistically relevant at the level of the cerebral cortex and the brainstem. Specific cortical regions may index whether variation in pitch contour falls within or between tonal category boundaries. Neural representations of pitch information are already extracted by early preattentive sensory level processing in both the brainstem and auditory cortex. Neural substrates of tone and other units of speech prosody that are manifested by variations in pitch share widely distributed cortical regions in common. However, when compared directly with a unit of speech prosody based primarily on timing variations (e.g., rhythm), segregated brain regions appear that are responsive to acoustic features associated with dynamic variations in pitch.

The importance of pitch features in gaining a fuller understanding of tonal processing in the brain cannot be overemphasized. We argue that it is necessary to develop a neural response specific to pitch features and, moreover, one that is capable of indexing dynamic variations in pitch that are ecologically representative of those that occur in natural speech (Krishnan, Bidelman, Smalt, Ananthakrishnan, & Gandour, 2012; Krishnan, Gandour, Ananthakrishnan, & Vijayaraghavan, 2014). With respect to speech perception, each pitch feature is defined by an auditory pattern that triggers its detection (Poeppel et al., 2008, pp. 1082–1082, review). Their precise definition, however, varies depending on the level of brain structure, time window, and functional representation in speech perception.

We hasten to acknowledge that the fundamental elements of linguistic theory are not easily reduced to or matched up with the fundamental biological units identified by neuroscience (Poeppel & Embick, 2006). The challenge is to formulate hypotheses about linguistic computations that underlie *real-time* tonal processing at different levels of biological structure in the brain.

Acknowledgment

Research was supported by the National Institutes of Health 5R01DC008549 (A.K.).

References

Chandrasekaran, B., Gandour, J. T., & Krishnan, A. (2007). Neuroplasticity in the processing of pitch dimensions: A multidimensional scaling analysis of the mismatch negativity. *Restorative Neurology and Neuroscience*, *25*(3–4), 195–210.

Chandrasekaran, B., & Kraus, N. (2010). The scalp-recorded brainstem response to speech: Neural origins and plasticity. *Psychophysiology*, *47*(2), 236–246. Available from: http://dx.doi.org./10.1111/j.1469-8986.2009.00928.x.

Chandrasekaran, B., Krishnan, A., & Gandour, J. T. (2007a). Experience-dependent neural plasticity is sensitive to shape of pitch contours. *Neuroreport*, *18*(18), 1963–1967. Available from: http://dx.doi.org/10.1097/WNR.0b013e3282f213c5.

Chandrasekaran, B., Krishnan, A., & Gandour, J. T. (2007b). Mismatch negativity to pitch contours is influenced by language experience. *Brain Research*, *1128*(1), 148–156. Available from: http://dx.doi.org/10.1016/j.brainres.2006.10.064.

Chandrasekaran, B., Sampath, P. D., & Wong, P. C. (2010). Individual variability in cue-weighting and lexical tone learning. *Journal of the Acoustical Society of America*, *128*(1), 456–465. Available from: http://dx.doi.org/10.1121/1.3445785.

Fournier, R., Gussenhoven, C., Jensen, O., & Hagoort, P. (2010). Lateralization of tonal and intonational pitch processing: An MEG study. *Brain Research*, *1328*, 79–88. Available from: http://dx.doi.org/10.1016/j.brainres.2010.02.053.

Francis, A. L., Ciocca, V., Ma, L., & Fenn, K. (2008). Perceptual learning of Cantonese lexical tones by tone and non-tone language speakers. *Journal of Phonetics*, *36*(2), 268–294. Available from: http://dx.doi.org/10.1016/j.wocn.2007.06.005.

Francis, A. L., Ciocca, V., & Ng, B. K. (2003). On the (non)categorical perception of lexical tones. *Perception and Psychophysics, 65*(7), 1029–1044.

Friederici, A. D., & Alter, K. (2004). Lateralization of auditory language functions: A dynamic dual pathway model. *Brain and Language, 89*(2), 267–276. Available from: http://dx.doi.org/10.1016/S0093-934X(03)00351-1.

Gandour, J. T. (1983). Tone perception in far Eastern languages. *Journal of Phonetics, 11*, 149–175.

Gandour, J. T. (1987). Tone production in aphasia. In J. Ryalls (Ed.), *Phonetic approaches to speech production in aphasia and related disorders* (pp. 45–57). Boston, MA: College-Hill.

Gandour, J. T. (1994). Phonetics of tone. In R. Asher, & J. Simpson (Eds.), *The encyclopedia of language & linguistics* (Vol. 6, pp. 3116–3123). New York, NY: Pergamon Press.

Gandour, J. T. (1998a). Aphasia in tone languages. In P. Coppens, A. Basso, & Y. Lebrun (Eds.), *Aphasia in atypical populations* (pp. 117–141). Hillsdale, NJ: Lawrence Erlbaum.

Gandour, J. T. (1998b). Phonetics and phonology. In B. Stemmer, & H. A. Whitaker (Eds.), *Handbook of neurolinguistics* (pp. 207–219). San Diego, CA: Academic Press.

Gandour, J. T. (2006a). Brain mapping of Chinese speech prosody. In P. Li, L. H. Tan, E. Bates, & O. J. L. Tzeng (Eds.), *Handbook of East Asian psycholinguistics* (Vol. 1, pp. 308–319). Cambridge, UK: Cambridge University Press (Chinese).

Gandour, J. T. (2006b). Tone: Neurophonetics. In K. Brown (Ed.), *Encyclopedia of language and linguistics* (Vol. 12, 2nd ed.). Oxford, UK: Elsevier.

Gandour, J. T. (2007). Neural circuitry underlying the perception of linguistic prosody. In C. Gussenhoven, & T. Raid (Eds.), *Tones and tunes: Experimental studies in word and sentence prosody* (Vol. 2, pp. 3–25). Berlin: Mouton de Gruyter.

Gandour, J. T., & Harshman, R. A. (1978). Crosslanguage differences in tone perception: A multidimensional scaling investigation. *Language and Speech, 21*(1), 1–33.

Gandour, J. T., Tong, Y., Talavage, T., Wong, D., Dzemidzic, M., Xu, Y., et al. (2007). Neural basis of first and second language processing of sentence-level linguistic prosody. *Human Brain Mapping, 28*(2), 94–108. Available from: http://dx.doi.org/10.1002/hbm.20255.

Gandour, J. T., Tong, Y., Wong, D., Talavage, T., Dzemidzic, M., Xu, Y., et al. (2004). Hemispheric roles in the perception of speech prosody. *NeuroImage, 23*(1), 344–357. Available from: http://dx.doi.org/10.1016/j.neuroimage.2004.06.004.

Gandour, J. T., Wong, D., Hsieh, L., Weinzapfel, B., Van Lancker, D., & Hutchins, G. D. (2000). A crosslinguistic PET study of tone perception. *Journal of Cognitive Neuroscience, 12*(1), 207–222.

Gandour, J. T., Wong, D., & Hutchins, G. (1998). Pitch processing in the human brain is influenced by language experience. *Neuroreport, 9*(9), 2115–2119.

Gandour, J. T., Wong, D., Lowe, M., Dzemidzic, M., Satthamnuwong, N., Tong, Y., et al. (2002). A cross-linguistic fMRI study of spectral and temporal cues underlying phonological processing. *Journal of Cognitive Neuroscience, 14*(7), 1076–1087. Available from: http://dx.doi.org/10.1162/089892902320474526.

Gandour, J. T., Xu, Y., Wong, D., Dzemidzic, M., Lowe, M., Li, X., et al. (2003). Neural correlates of segmental and tonal information in speech perception. *Human Brain Mapping, 20*(4), 185–200. Available from: http://dx.doi.org/10.1002/hbm.10137.

Glasser, M. F., & Rilling, J. K. (2008). DTI tractography of the human brain's language pathways. *Cerebral Cortex, 18*(11), 2471–2482. Available from: http://dx.doi.org/10.1093/cercor/bhn011.

Griffiths, T. D., & Warren, J. D. (2002). The planum temporale as a computational hub. *Trends in Neurosciences, 25*(7), 348–353.

Harnad, S. R. (Ed.), (1987). *Categorical perception: The groundwork of cognition* New York, NY: Cambridge University Press.

Hickok, G., & Poeppel, D. (2004). Dorsal and ventral streams: A framework for understanding aspects of the functional anatomy of language. *Cognition, 92*(1–2), 67–99.

Hickok, G., & Poeppel, D. (2007). The cortical organization of speech processing. *Nature Reviews Neuroscience, 8*(5), 393–402. Available from: http://dx.doi.org/10.1038/nrn2113.

Hsieh, L., Gandour, J. T., Wong, D., & Hutchins, G. D. (2001). Functional heterogeneity of inferior frontal gyrus is shaped by linguistic experience. *Brain and Language, 76*(3), 227–252. Available from: http://dx.doi.org/10.1006/brln.2000.2382.

Hu, J. H., Gao, S., Ma, W. Y., & Yao, D. Z. (2012). Dissociation of tone and vowel processing in Mandarin idioms. *Psychophysiology, 49*(9), 1179–1190. Available from: http://dx.doi.org/10.1111/j.1469-8986.2012.01406.x.

Huang, T., & Johnson, K. (2011). Language specificity in speech perception: Perception of Mandarin tones by native and nonnative listeners. *Phonetica, 67*, 243–267.

Humphries, C., Love, T., Swinney, D., & Hickok, G. (2005). Response of anterior temporal cortex to syntactic and prosodic manipulations during sentence processing. *Human Brain Mapping, 26*(2), 128–138. Available from: http://dx.doi.org/10.1002/hbm.20148.

Jacquemot, C., Pallier, C., LeBihan, D., Dehaene, S., & Dupoux, E. (2003). Phonological grammar shapes the auditory cortex: A functional magnetic resonance imaging study. *Journal of Neuroscience, 23*(29), 9541–9546.

Kaan, E., Wayland, R., Bao, M., & Barkley, C. M. (2007). Effects of native language and training on lexical tone perception: An event-related potential study. *Brain Research, 1148*, 113–122. Available from: http://dx.doi.org/10.1016/j.brainres.2007.02.019.

Khouw, E., & Ciocca, V. (2007). Perceptual correlates of Cantonese tones. *Journal of Phonetics, 35*(1), 104–117. Available from: http://dx.doi.org/10.1016/j.wocn.2005.10.003.

Klein, D., Zatorre, R. J., Milner, B., & Zhao, V. (2001). A cross-linguistic PET study of tone perception in Mandarin Chinese and English speakers. *NeuroImage, 13*(4), 646–653. Available from: http://dx.doi.org/10.1006/nimg.2000.0738.

Krishnan, A. (2007). Human frequency following response. In R. F. Burkard, M. Don, & J. J. Eggermont (Eds.), *Auditory evoked potentials: Basic principles and clinical application* (pp. 313–335). Baltimore, MD: Lippincott Williams & Wilkins.

Krishnan, A., Bidelman, G. M., & Gandour, J. T. (2010). Neural representation of pitch salience in the human brainstem revealed by psychophysical and electrophysiological indices. *Hearing Research, 268*(1–2), 60–66. Available from: http://dx.doi.org/10.1016/j.heares.2010.04.016.

Krishnan, A., Bidelman, G. M., Smalt, C. J., Ananthakrishnan, S., & Gandour, J. T. (2012). Relationship between brainstem, cortical and behavioral measures relevant to pitch salience in humans. *Neuropsychologia, 50*(12), 2849–2859. Available from: http://dx.doi.org/10.1016/j.neuropsychologia.2012.08.013.

Krishnan, A., & Gandour, J. T. (2009). The role of the auditory brainstem in processing linguistically-relevant pitch patterns. *Brain and Language, 110*(3), 135–148. Available from: http://dx.doi.org/10.1016/j.bandl.2009.03.005.

Krishnan, A., Gandour, J. T., Ananthakrishnan, S., Bidelman, G. M., & Smalt, C. J. (2011). Functional ear (a)symmetry in brainstem neural activity relevant to encoding of voice pitch: A precursor for hemispheric specialization? *Brain and Language, 119*(3), 226–231. Available from: http://dx.doi.org/10.1016/j.bandl.2011.05.001.

Krishnan, A., Gandour, J. T., Ananthakrishnan, S., & Vijayaraghavan, V. (2014). Cortical pitch response components index stimulus onset/offset and dynamic features of pitch contours. *Neuropsychologia, 59C*, 1–12. Available from: http://dx.doi.org/10.1016/j.neuropsychologia.2014.04.006.

Krishnan, A., Gandour, J. T., & Bidelman, G. M. (2010a). Brainstem pitch representation in native speakers of Mandarin is less susceptible to degradation of stimulus temporal regularity. *Brain Research, 1313*, 124–133. Available from: http://dx.doi.org/10.1016/j.brainres.2009.11.061.

Krishnan, A., Gandour, J. T., & Bidelman, G. M. (2010b). The effects of tone language experience on pitch processing in the brainstem. *Journal of Neurolinguistics, 23*(1), 81–95. Available from: http://dx.doi.org/10.1016/j.jneuroling.2009.09.001.

Krishnan, A., Gandour, J. T., Bidelman, G. M., & Swaminathan, J. (2009). Experience-dependent neural representation of dynamic pitch in the brainstem. *Neuroreport, 20*(4), 408–413. Available from: http://dx.doi.org/10.1097/WNR.0b013e3283263000.

Krishnan, A., Gandour, J. T., Smalt, C. J., & Bidelman, G. M. (2010). Language-dependent pitch encoding advantage in the brainstem is not limited to acceleration rates that occur in natural speech. *Brain and Language, 114*(3), 193–198. Available from: http://dx.doi.org/10.1016/j.bandl.2010.05.004.

Krishnan, A., Swaminathan, J., & Gandour, J. T. (2009). Experience-dependent enhancement of linguistic pitch representation in the brainstem is not specific to a speech context. *Journal of Cognitive Neuroscience, 21*(6), 1092–1105. Available from: http://dx.doi.org/10.1162/jocn.2009.21077.

Lattner, S., Meyer, M. E., & Friederici, A. D. (2005). Voice perception: Sex, pitch, and the right hemisphere. *Human Brain Mapping, 24*(1), 11–20. Available from: http://dx.doi.org/10.1002/hbm.20065.

Lehiste, I. (1996). Suprasegmental features of speech. In N. J. Lass (Ed.), *Principles of experimental phonetics* (pp. 226–244). St. Louis, MO: Mosby.

Li, X., Gandour, J., Talavage, T., Wong, D., Dzemidzic, M., Lowe, M., et al. (2003). Selective attention to lexical tones recruits left dorsal frontoparietal network. *Neuroreport, 14*(17), 2263–2266. Available from: http://dx.doi.org/10.1097/01.wnr.0000097045.56589.31.

Li, X., Gandour, J. T., Talavage, T., Wong, D., Hoffa, A., Lowe, M., et al. (2010). Hemispheric asymmetries in phonological processing of tones versus segmental units. *Neuroreport, 21*(10), 690–694. Available from: http://dx.doi.org/10.1097/WNR.0b013e32833b0a10.

Liang, J., & van Heuven, V. J. (2004). Evidence for separate tonal and segmental tiers in the lexical specification of words: A case study of a brain-damaged Chinese speaker. *Brain and Language, 91*(3), 282–293. Available from: http://dx.doi.org/10.1016/j.bandl.2004.03.006.

Liebenthal, E., Binder, J. R., Spitzer, S. M., Possing, E. T., & Medler, D. A. (2005). Neural substrates of phonemic perception. *Cerebral Cortex, 15*(10), 1621–1631. Available from: http://dx.doi.org/10.1093/cercor/bhi040.

Liebenthal, E., Desai, R., Ellingson, M. M., Ramachandran, B., Desai, A., & Binder, J. R. (2010). Specialization along the left superior temporal sulcus for auditory categorization. *Cerebral Cortex, 20*(12), 2958–2970. Available from: http://dx.doi.org/10.1093/cercor/bhq045.

Liu, L., Peng, D., Ding, G., Jin, Z., Zhang, L., Li, K., et al. (2006). Dissociation in the neural basis underlying Chinese tone and vowel production. *NeuroImage, 29*(2), 515–523.

Luo, H., Ni, J. T., Li, Z. H., Li, X. O., Zhang, D. R., Zeng, F. G., et al. (2006). Opposite patterns of hemisphere dominance for early auditory processing of lexical tones and consonants. *Proceedings of the National Academy of Sciences of the United States of America, 103*(51), 19558–19563.

Malins, J. G., & Joanisse, M. F. (2012). Setting the tone: An ERP investigation of the influences of phonological similarity on spoken word recognition in Mandarin Chinese. *Neuropsychologia, 50*(8), 2032–2043. Available from: http://dx.doi.org/10.1016/j.neuropsychologia.2012.05.002.

Meyer, M. (2008). Functions of the left and right posterior temporal lobes during segmental and suprasegmental speech perception. *Zeitshcrift fur Neuropsycholgie, 19*(2), 101–115.

Meyer, M., Alter, K., Friederici, A. D., Lohmann, G., & von Cramon, D. Y. (2002). FMRI reveals brain regions mediating slow prosodic modulations in spoken sentences. *Human Brain Mapping, 17*(2), 73–88.

Meyer, M., Steinhauer, K., Alter, K., Friederici, A. D., & von Cramon, D. Y. (2004). Brain activity varies with modulation of dynamic pitch variance in sentence melody. *Brain and Language, 89*(2), 277–289. Available from: http://dx.doi.org/10.1016/S0093-934X(03)00350-X.

Naatanen, R. (2001). The perception of speech sounds by the human brain as reflected by the mismatch negativity (MMN) and its magnetic equivalent (MMNm). *Psychophysiology, 38*(1), 1–21.

Peng, G., Zheng, H.-Y., Gong, T., Yang, R.-X., Kong, J.-P., & Wang, W. S.-Y. (2010). The influence of language experience on categorical perception of pitch contours. *Journal of Phonetics, 38*, 616–624.

Poeppel, D. (2003). The analysis of speech in different temporal integration windows: Cerebral lateralization as "asymmetric sampling in time". *Speech Communication, 41*(1), 245–255.

Poeppel, D., & Embick, D. (2006). Defining the relation between linguistics and neuroscience. In A. Cutler (Ed.), *Twenty-first century psycholinguistics. Four cornerstones* (pp. 103–118). Mahwah, NJ: Lawrence Erlbaum.

Poeppel, D., Idsardi, W. J., & van Wassenhove, V. (2008). Speech perception at the interface of neurobiology and linguistics. *Philosophical Transactions of the Royal Society of London. Series B: Biological Sciences, 363*(1493), 1071–1086. Available from: http://dx.doi.org/10.1098/rstb.2007.2160.

Polich, J. (2007). Updating P300: An integrative theory of P3a and P3b. *Clinical Neurophysiology, 118*(10), 2128–2148. Available from: http://dx.doi.org/10.1016/j.clinph.2007.04.019.

Ren, G. Q., Yang, Y., & Li, X. (2009). Early cortical processing of linguistic pitch patterns as revealed by the mismatch negativity. *Neuroscience, 162*(1), 87–95. Available from: http://dx.doi.org/10.1016/j.neuroscience.2009.04.021.

Schirmer, A., Tang, S. L., Penney, T. B., Gunter, T. C., & Chen, H. C. (2005). Brain responses to segmentally and tonally induced semantic violations in Cantonese. *Journal of Cognitive Neuroscience, 17*(1), 1–12.

Shaywitz, B. A., Shaywitz, S. E., Pugh, K. R., Fulbright, R. K., Skudlarski, P., Mencl, W. E., et al. (2001). The functional neural architecture of components of attention in language-processing tasks. *NeuroImage, 13*(4), 601–612.

Smith, E. E., & Jonides, J. (1999). Storage and executive processes in the frontal lobes. *Science, 283*(5408), 1657–1661.

Swaminathan, J., Krishnan, A., & Gandour, J. T. (2008). Pitch encoding in speech and nonspeech contexts in the human auditory brainstem. *Neuroreport, 19*(11), 1163–1167. Available from: http://dx.doi.org/10.1097/WNR.0b013e3283088d31.

Tingsabadh, K., & Abramson, A. S. (1999). Thai. In I.P.A. (Ed.), *Handbook of the international phonetic association* (pp. 147–149). Cambridge, UK: Cambridge University Press.

Tong, Y., Gandour, J. T., Talavage, T., Wong, D., Dzemidzic, M., Xu, Y., et al. (2005). Neural circuitry underlying sentence-level linguistic prosody. *NeuroImage, 28*(2), 417–428. Available from: http://dx.doi.org/10.1016/j.neuroimage.2005.06.002.

Tsang, Y. K., Jia, S., Huang, J., & Chen, H. C. (2011). ERP correlates of pre-attentive processing of Cantonese lexical tones: The effects of pitch contour and pitch height. *Neuroscience Letters, 487*(3), 268–272. Available from: http://dx.doi.org/10.1016/j.neulet.2010.10.035.

Wang, W. S.-Y. (1967). Phonological features of tone. *International Journal of American Linguistics, 33*(2), 93–105.

Wang, Y., Behne, D. M., Jongman, A., & Sereno, J. A. (2004). The role of linguistic experience in the hemispheric processing of lexical tone. *Applied Psycholinguistics, 25*, 449–466.

Wang, Y., Jongman, A., & Sereno, J. A. (2001). Dichotic perception of Mandarin tones by Chinese and American listeners. *Brain and Language, 78*(3), 332–348.

Wang, Y., Sereno, J. A., Jongman, A., & Hirsch, J. (2003). fMRI evidence for cortical modification during learning of Mandarin lexical tone. *Journal of Cognitive Neuroscience, 15*(7), 1019–1027.

Wildgruber, D., Ackermann, H., Kreifelts, B., & Ethofer, T. (2006). Cerebral processing of linguistic and emotional prosody: fMRI studies. *Progress in Brain Research, 156*, 249–268. Available from: http://dx.doi.org/10.1016/S0079-6123(06)56013-3.

Witteman, J., van Ijzendoorn, M. H., van de Velde, D., van Heuven, V. J., & Schiller, N. O. (2011). The nature of hemispheric specialization for linguistic and emotional prosodic perception: A meta-analysis of the lesion literature. *Neuropsychologia, 49*(13), 3722–3738. Available from: http://dx.doi.org/10.1016/j.neuropsychologia.2011.09.028.

Wong, P. C. (2002). Hemispheric specialization of linguistic pitch patterns. *Brain Research Bulletin, 59*(2), 83–95.

Wong, P. C., Parsons, L. M., Martinez, M., & Diehl, R. L. (2004). The role of the insular cortex in pitch pattern perception: The effect of linguistic contexts. *Journal of Neuroscience, 24*(41), 9153–9160.

Wong, P. C., Perrachione, T. K., Gunasekera, G., & Chandrasekaran, B. (2009). Communication disorders in speakers of tone languages: Etiological bases and clinical considerations. *Seminars in Speech and Language, 30*(3), 162–173. Available from: http://dx.doi.org/10.1055/s-0029-1225953.

Wong, P. C., Perrachione, T. K., & Parrish, T. B. (2007). Neural characteristics of successful and less successful speech and word learning in adults. *Human Brain Mapping, 28*(10), 995–1006. Available from: http://dx.doi.org/10.1002/hbm.20330.

Xi, J., Zhang, L., Shu, H., Zhang, Y., & Li, P. (2010). Categorical perception of lexical tones in Chinese revealed by mismatch negativity. *Neuroscience, 170*(1), 223–231. Available from: http:dx.doi.org/10.1016/j.neuroscience.2010.06.077.

Xu, G., Zhang, L., Shu, H., Wang, X., & Li, P. (2013). Access to lexical meaning in pitch-flattened Chinese sentences: An fMRI study. *Neuropsychologia, 51*(3), 550–556. Available from: http://dx.doi.org/10.1016/j.neuropsychologia.2012.12.006.

Xu, Y. (1997). Contextual tonal variations in Mandarin. *Journal of Phonetics, 25*, 61–83.

Xu, Y. (2006). Tone in connected discourse. In (2nd ed.K. Brown (Ed.), *Encyclopedia of language and linguistics* (Vol. 12Oxford, UK: Elsevier.

Xu, Y., Gandour, J., Talavage, T., Wong, D., Dzemidzic, M., Tong, Y., et al. (2006). Activation of the left planum temporale in pitch processing is shaped by language experience. *Human Brain Mapping, 27*(2), 173–183.

Xu, Y., Gandour, J. T., & Francis, A. L. (2006). Effects of language experience and stimulus complexity on the categorical perception of pitch direction. *Journal of the Acoustical Society of America, 120*(2), 1063–1074.

Xu, Y., Krishnan, A., & Gandour, J. T. (2006). Specificity of experience-dependent pitch representation in the brainstem. *Neuroreport, 17*(15), 1601–1605. Available from: http://dx.doi.org/10.1097/01.wnr.0000236865.31705.3a.

Yip, M. (2002). *Tone*. New York, NY: Cambridge University Press.

Zatorre, R. J., Belin, P., & Penhune, V. B. (2002). Structure and function of auditory cortex: Music and speech. *Trends in Cognitive Sciences, 6*(1), 37–46.

Zatorre, R. J., & Gandour, J. T. (2008). Neural specializations for speech and pitch: Moving beyond the dichotomies. *Philosophical Transactions of the Royal Society of London. Series B: Biological Sciences, 363*(1493), 1087–1104. Available from: http://dx.doi.org/10.1098/rstb.2007.2161.

Zatorre, R. J., Mondor, T. A., & Evans, A. C. (1999). Auditory attention to space and frequency activates similar cerebral systems. *NeuroImage, 10*(5), 544–554.

Zee, E. (1999). *Handbook of the international phonetic association*. Cambridge, UK: Cambridge University Press.

Zhang, L., Shu, H., Zhou, F., Wang, X., & Li, P. (2010). Common and distinct neural substrates for the perception of speech rhythm and intonation. *Human Brain Mapping, 31*(7), 1106–1116. Available from: http://dx.doi.org/10.1002/hbm.20922.

Zhang, L., Xi, J., Xu, G., Shu, H., Wang, X., & Li, P. (2011). Cortical dynamics of acoustic and phonological processing in speech perception. *Public Library of Science One, 6*(6), e20963 Retrieved from: http://www.ncbi.nlm.nih.gov/pubmed/21695133. Available from: doi:10.1371/journal.pone.0020963.

Zhao, J., Guo, J., Zhou, F., & Shu, H. (2011). Time course of Chinese monosyllabic spoken word recognition: Evidence from ERP analyses. *Neuropsychologia, 49*(7), 1761–1770. Available from: http://dx.doi.org/10.1016/j.neuropsychologia.2011.02.054.

Zheng, H.-Y., Minett, J. W., Peng, G., & Wang, W. S.-Y. (2012). The impact of tone systems on the categorical perception of lexical tones: An event-related potentials study. *Language and Cognitive Processes, 27*(2), 184–209. Available from: http://dx.doi.org/10.1080/01690965.2010.520493.

CHAPTER

88

The Neurocognition of Prosody

Silke Paulmann

Department of Psychology, Centre for Brain Science, University of Essex, Colchester, UK

88.1 INTRODUCTION

Verbal communication is often a question of "tone." Modulating parameters such as vocal pitch (high/low), loudness (loud/silent), tempo (fast/slow), or voice quality (clear/harsh) allows us to give the correct meaning to what we are saying and help the listener to interpret a message correctly. This is true not only when expressing how we feel (e.g., angry, nervous, happy) but also when conveying nonemotional information. For instance, by raising (or not raising) our voice when articulating a string of words, we can alter the interpretation of an utterance (e.g., changing from a statement such as *You finished writing the chapter.* to a question *You finished writing the chapter?*). Suprasegmental parameters of speech (prosody) can also be used to convey lexical meanings (e.g., *hot dog* vs. *hot dog*) or discourse information (e.g., new information is often accented while old information is de-accented; prosodic phrasing guides syntactic sentence interpretation) while expressing emotions and attitudes (e.g., often a raised voice is associated with an angry speaker, whereas a lowered voice might indicate that the speaker feels sad). Thus, prosody serves several linguistic and nonlinguistic (emotional) functions; however, it is often an undervalued component of spoken language and brain-based models that take into account the role prosody plays in dynamic speech comprehension are still rare (but see Friederici & Alter, 2004) and controversially discussed. In fact, much of the controversy around the neural basis of prosody is probably attributable to the fact that it fulfils several communicative functions (often at once); thus, the question that has been driving past research is whether the different prosodic functions are independent or interdependent. This chapter reviews past research on each function and outlines our current understanding of the neurocognitive architecture underlying prosody.

88.2 BRAIN MAPPING OF PROSODY

Historically, investigations into the cerebral representation of prosody aimed to specify whether one hemisphere dominates control over linguistic and/or emotional prosody processing by examining lesion data. These early *simple hemispheric models* (Sidtis & Van Lancker Sidtis, 2003) resulted in three main hypotheses:

1. Prosody, irrespective of communicative function (e.g., linguistic, emotional), is lateralized to the right cerebral hemisphere. Interactions with other linguistic information such as syntactic or semantic information are mediated through the corpus callosum (Friederici & Alter, 2004; Klouda, Robin, Graff-Radford, & Cooper, 1988).
2. Emotional prosody is processed in the right hemisphere (RH) (Blonder, Bowers, & Heilman, 1991; Ross, 1981).
3. Linguistic prosody is processed in the left hemisphere (LH), whereas emotional or affective prosody processing can be linked predominantly to the RH (known as Functional Lateralization Hypothesis) (Van Lancker, 1980).

Although it has since been shown that a simple cortical hemispheric distinction is not substantiated by the available data (for reviews see Baum & Pell, 1999; Sidtis & Van Lancker Sidtis, 2003), the effect of these proposals can still be felt when scanning through the more recent literature. Many researchers base their hypotheses about the brain network underlying prosody processing on the premise that linguistic and emotional prosody are fully distinct

Neurobiology of Language. DOI: http://dx.doi.org/10.1016/B978-0-12-407794-2.00088-2

entities in the language system and that these processes are (each) fully lateralized to one cerebral hemisphere (but see Seddoh, 2002, arguing why emotional and linguistic prosody should not be considered distinct categories). However, an alternative to simple hemispheric models was put forward by Van Lancker and Sidtis (1992): hemispheric lateralization is based on physical features. Pitch is preferably processed by right hemispheric brain structures, whereas duration and intensity are primarily processed by left hemispheric structures (known as the Cue or Physical Feature–Dependent Hypothesis).

An additional problem with the early simple hemispheric models is that they neglect the role that subcortical brain regions play during prosody processing. However, accumulating evidence from lesion patients showed that impaired prosody processing is often associated with damage to structures such as the caudate nucleus, putamen, and/or globus pallidus (i.e., the basal ganglia [BG]) (Breitenstein, Daum, & Ackermann, 1998; Cancelliere & Kertesz, 1990; Paulmann, Pell, & Kotz, 2008, 2009). Hence, a fifth hypothesis posits that prosody processing is heavily mediated by subcortical brain regions without a strong hemispheric lateralization.

In an attempt to consolidate the different hypotheses, Sidtis and Van Lancker Sidtis (2003) proposed a Neurobehavioral Approach to Dysprosody. This framework suggests that prosody processing is not driven by one single mechanism, but instead relies on a complex conglomerate of motor, perceptual, and more cognitively based functions. Thus, a widespread, bilateral brain network might be implicated in prosody processing and discrepancies across lesion studies are probably due to the fact that dysprosody can materialize after disruption to any of the involved mechanisms linked to different brain regions.

More recently, similar working models, that is, frameworks that suggest a highly differentiated brain network underlying *emotional* prosody processing, have been put forward by researchers who based their hypotheses primarily on evidence obtained from neuroimaging studies instead of data from the lesion approach (Brück, Kreifelts, & Wildgruber, 2011; Kotz & Paulmann, 2011; Schirmer & Kotz, 2006; Wildgruber, Ethofer, Grandjean, & Kreifelts, 2009). Moreover, some of these models also hypothesized the temporal dynamics of emotional prosody processing. For instance, Kotz and Paulmann (2011) suggest that an initial extraction of acoustic cues (e.g., fundamental frequency, loudness, voice quality) takes place within 100 ms of stimulus onset. This process is argued to be mediated by primary and secondary auditory cortices (bilaterally). Once acoustic properties have been accessed, derivation of emotional salience/meaning (established through integration of emotionally relevant acoustic cues) occurs within 200 ms after stimulus onset. This process has been linked to the right anterior superior temporal sulcus/superior temporal gyrus. Finally, more elaborate processes (e.g., integration of information from prosody with semantics or broad context) could start approximately 400 ms after stimulus onset. These higher cognitive processes are presumed to be mediated by inferior frontal and orbitofrontal cortex (bilaterally). Thus, researchers moved away from the assumption that prosodic processing is one single mechanism and instead suggest that different emotional prosody processing stages are subserved by different brain areas. Ideally, future models on the neural circuitry regulating emotional and linguistic prosody will be able to integrate findings from both lesion and imaging fields of the literature.

The following review aims to show how results from clinical and empirical neuroscience studies have helped shape our understanding of prosody processing. To provide an integrative view of key findings, evidence from important past and more recent studies are discussed, followed by a summary of how available evidence supports prosody processing models.

88.3 THE NEURAL BASIS OF LINGUISTIC PROSODY PROCESSING

88.3.1 Clinical Evidence

Initially, scientists predominantly relied on the lesion approach to specify which brain regions might underlie prosody processing. For instance, one of the earlier studies to explore the contribution of the RH to linguistic prosody processing was conducted by Weintraub, Mesulam, & Kramer (1981). They tested RH patients and healthy controls (HCs) and assessed their ability to discriminate between phonemic (e.g., DARK room versus dark ROOM) and sentential stress (e.g., STEVE drives the car versus Steve drives the CAR). Patients were found to be outperformed by controls on these tasks, prompting the authors to suggest that the RH plays a strong role in linguistic prosody perception. At approximately the same time, Baum, Daniloff, Daniloff, and Lewis (1982) investigated phonemic and sentential stress comprehension in LH patients and HCs. Their results showed that LH patients *also* suffer from difficulties on these tasks, challenging the view that the RH alone is involved in linguistic prosody perception. Heilman, Bowers, Speedie, and Coslett (1984) thus tested performance of *both* LH and RH patients on linguistic prosody processing in one study. The authors reported difficulties in identifying the modality (question, statement, command) of filtered sentences by both LH and RH patients when compared with HCs. Again, these results challenged the view that the RH is solely responsible for linguistic prosodic processing and instead point to a possible additional involvement of the LH during linguistic prosody perception. In fact, evidence from a subsequent study comparing LH and RH patient performance on phonemic stress identification suggests that LH patients can be even more strongly impaired than RH patients (Emmorey, 1987). However, only 2 years later, conflicting evidence emerged when Bryan (1989) reported results from a range of tasks (e.g., sentential and phonemic stress identification, discrimination, and identification of sentence modalities) that showed that RH patients were outperformed by LH patients on most of the tasks administered. Although generally better than RH patients, LH patients performed poorly on the majority (but not all) of the tasks when compared with HCs, suggesting that both RH and LH patients show impairments for linguistic prosody processing. The importance of task effects was further examined years later. Pell and Baum (1997) asked RH and LH patients as well as HCs to either identify or discriminate between different sentence modalities (interrogative, declarative, imperative). Both patient groups performed comparable with the HC group when discriminating between different prosodic patterns. However, when examining the identification task, patients performed significantly worse than HCs. These data point to the possibility that linguistic prosody is processed bilaterally in the brain, and they also show that severity of impairment might depend on task instructions.

In the years that followed, researchers continued to obtain conflicting results when exploring the influence of the LH and RH on linguistic prosody processing. Borod and colleagues (1992) reported findings obtained from LH and RH patients and HCs. Nonsense syllable strings (e.g., pa-da-ka) that were intoned in three different ways (declarative, interrogative, emphatic) had to be discriminated by participants using multiple-choice response cards. Results showed that RH patients made significantly more errors than LH patients and HCs who did not differ from each other. In contrast, Breitenstein et al. (1998) report data from LH and RH patients who performed comparably with HCs when discriminating between sentence modality pairs. Yet again, Pell (1998) showed that LH patients are worse at identifying emphatic stress patterns than RH patients or controls. Similarly, Walker, Daigle, and Buzzard (2002) showed that LH patients performed significantly worse than RH patients and HCs when identifying lexical or sentential stress. In addition, LH patients also suffered from difficulties in prosodic phrasing. As outlined, task differences (e.g., instructions, complexity) between studies are likely to affect results; however, discrepancies between findings from clinical research can also likely be linked to differences in patients' lesion locations and size as well as to differences in their speech and language abilities (see Baum & Pell, 1999; Kotz, Meyer, & Paulmann, 2006, for similar observations). Thus, Amebu Seddoh (2006) subdivided his LH patient population into three subgroups: Wernicke's aphasia patients, Broca's aphasia patients, and global aphasia patients. Their ability to identify the sentence modality (question versus statement) was assessed. Results confirmed that Wernicke's aphasia patients had no difficulties with the task, whereas Broca's aphasia patients suffered from difficulties in identifying questions and global aphasics suffered from difficulties in identifying statements and questions. Results suggest once more that the LH can be critically tied to linguistic prosody processing, but they also support the view that patients' lesion locations need to be controlled for better than by LH/RH distinctions. The latter conclusion is also underpinned by recent data by Rymarczyk and Grabowska (2007), who subdivided RH patients into three groups (patients with lesions to frontal, temporo-parietal, or subcortical brain structures). They looked at the performance of identifying as well as discriminating between sentence modalities and discriminating between empathic stress patterns. Although all patient groups perform significantly worse than HCs on all three tasks, results confirm the importance of controlling for lesion location because patients with lesions to temporo-parietal sites performed worse than the two other groups. The idea that we need to examine patients whose lesion delineation is comparable was followed-up by Kho et al. (2008). They looked at RH and LH patients who underwent anterior temporal cortex resection. The authors reported no differences between RH and LH patients when detecting word or contrastive stress, or during sentential discrimination.

Taken together, the evidence elaborated here clearly shows that clinical evidence provides very little convergent evidence that linguistic prosody is processed solely by one hemisphere. Moreover, the evidence from lesion patients that suggests that not only cortical but also subcortical brain structures play a critical role during prosody processing (Brådvik et al., 1991; Cancelliere & Kertesz, 1990; Ross & Mesulam, 1979; Rymarczyk & Grabowska, 2007; Starkstein, Federoff, Price, Leiguarda, & Robinson, 1994) also challenges all simple hemispheric models, although the idea that prosody is primarily mediated by subcortical structures is also clearly not substantiated. In short, although telling, clinical evidence alone has not helped to provide support for either of the original hypotheses. This conclusion is supported by a recent activation likelihood estimation (ALE) meta-analysis by Witteman, van Ijzendoorn, van de Velde, van Heuven, and Schiller (2011), who reported that lesions to the LH or RH have similar detrimental effects on linguistic prosody perception.

88.3.2 Brain Imaging Evidence

Over the past two decades, fMRI and PET techniques have become popular tools for observing normal brain function. Motivated by the heterogeneous results from clinical studies, several investigations have been conducted to further delineate the brain network underlying prosody processing. However, in contrast to emotional prosody, imaging studies on linguistic prosody processing have been rare. It is likely that the limited number of imaging studies on linguistic prosody are linked to the problem that linguistic prosody is an umbrella term that refers to studies exploring a variety of processes linked to suprasegmental changes including (but not limited to) sentence type, phrase boundary, word stress, and pitch contour processing. Some of the existing research is summarized to show that despite using more fine-grained methodologies, the brain network underlying linguistic prosodic processing is still not fully specified.

In an early PET study, Gandour et al. (2000) investigated pitch perception of one-syllable Thai words in speakers of tone and nontone languages (Thai, Chinese, English). Their results suggest that neural mechanisms underlying pitch perception differ depending on linguistic relevance of stimuli. Specifically, left frontal operculum activity was found for Thai speakers when discriminating between pitch patterns, but the same activity was not found for English speakers. Also, Thai speakers failed to show the same left-lateralized activation when discriminating pitch patterns of nonspeech stimuli, suggesting that "linguistic relevance" might modulate lateralized activation patterns. This hemispheric laterality effect was confirmed in an fMRI study a few years later (Gandour et al., 2004) when brain activity of Chinese and English speakers was measured. Participants had to discriminate between one- and three-syllable-long utterances and four different Chinese tones. Chinese (i.e., the tone language group), but not English, participants showed left-lateralized activation in inferior parietal and posterior superior temporal, anterior temporal, and frontopolar brain regions. Both English and Chinese participants showed right-lateralized activation hot spots in parts of the superior temporal sulcus as well as the middle frontal gyrus. Similar to the previous study (Gandour et al., 2000), the authors interpreted their results to suggest that linguistic relevance/knowledge can influence brain activation patterns for prosody. In particular, the tone language group is argued to have an implicit understanding about the relationship between acoustic cues and internal representations of suprasegmental sentence information. In contrast, listeners from nontone language backgrounds cannot possess the same higher-order prosodic representations. Thus, it was argued that LH lateralization is linked to higher-order prosodic processing, whereas RH lateralization might represent lower-order acoustic feature processing (Gandour et al., 2004).

Imaging studies looking at sentences rather than syllables or words often report RH dominance for linguistic prosodic processing. For instance, Meyer, Alter, Friederici, Lohmann, and von Cramon (2002) investigated the neuroanatomical correlates of slow prosodic modulations. They presented participants with delexicalized (i.e., stimuli that contain no lexical-semantic information) and normal speech. When comparing stimuli from both conditions, the authors found increased activation in the right superior temporal brain region as well as in the fronto-opercular cortex for filtered speech, suggesting a strong RH involvement during linguistic prosodic speech processing (and slow pitch movements in particular). Similar RH dominance in response to prosodic stimuli was reported by Plante, Creusere, and Sabin (2002), who also explored the neural correlates of sentential prosody. Participants were again presented with low-pass filtered (i.e., delexicalized speech) and normal speech and had to perform tasks high (remember and recognize words) or low (no task) in memory load. When comparing hemodynamic responses for prosodic speech with responses to unfiltered speech in the "no task" condition, the authors found a stronger bilateral activation within the superior temporal gyrus for prosodic speech. Moreover, when looking at frontal lobe activation patterns, results showed that the tasks high in memory load affected processing of filtered and unfiltered speech differently: although both speech stimuli resulted in bilateral activation of the frontal lobes, processing filtered stimuli resulted in stronger RH activation than processing unfiltered stimuli.

Influence of task instructions on prosodic processes was also reported by other authors who looked at sentence material (Geiser, Zaehle, Jancke, & Meyer, 2008; Wildgruber et al., 2004), albeit they failed to confirm a strongly right-lateralized network for linguistic prosodic processes. Instead, they showed that participants who focus on linguistic prosodic aspects of stimuli exhibit greater activation in the LH. Specifically, in their fMRI study, Wildgruber et al. (2004) showed right-lateralized activation of the dorsolateral frontal cortex and bilateral activation of thalamic and temporal regions in contrast to a rest condition for stimuli that were synthetically manipulated to exhibit different sentence foci (second word versus final word focus). However, when participants were asked to focus on linguistic characteristics of stimuli, activation of the left inferior frontal gyrus was reported (as opposed to activation of bilateral orbitofrontal areas when the task focus was emotional). Similarly, Tracy et al. (2011) reported that processing pitch information from lexical stimuli predominantly recruits left-lateralized structures, including the cingulated gyrus and middle temporal and superior temporal gyri, whereas pitch processing from tone sequences relies on right frontal and temporal cortices.

Taken together, data from neuroimaging studies clearly suggest that a complex neural network spanning both hemispheres underlies linguistic prosodic processing. Findings further imply that various factors can impact lateralization of effects. Specifically, it has been shown that lateralization of prosody can depend on task focus (linguistic/nonlinguistic, active/passive; Wildgruber et al., 2004), task demands (high/low; Plante et al., 2002), language background/experience (tone/nontone language; Gandour et al., 2000), acoustic cue (pitch/duration; Van Lancker & Sidtis, 1992), and stimulus type (syllable/word/sentence; Gandour et al., 2004; Meyer et al., 2002), as well as on methodological factors, including design (event-related/blocked; Kotz et al., 2006) and contrast/comparison conditions (rest/alternative prosodic function).

88.3.3 ERP Findings

In addition to exploring the brain structures involved, research has also tried to specify the time course of linguistic prosody using event-related brain potentials (ERPs). The temporal online dynamics of prosody are of particular interest because information might consolidate the different proposals on how prosody is represented in the brain considering that research suggests that *emotional* prosodic processing is a multistage process (Kotz & Paulmann, 2011; Schirmer & Kotz, 2006; Wildgruber, Ackermann, Kreifelts, & Ethofer, 2006; Wildgruber et al., 2009) with each stage linked to different brain areas. It is likely that a similar multilayered mechanism applies to linguistic prosody processing.

Several ERP components with different onset latencies have been described. For instance, Steinhauer, Alter, and Friederici (1999) explored how prosody helps listeners to establish a syntactic structure during language comprehension. They were the first to report that the so-called closure positive shift (CPS) is elicited quickly after prosodic phrase boundaries (which usually coincide with syntactic boundaries). This marker of prosodic boundary processing is also found in delexicalized speech (Steinhauer & Friederici, 2001), as well as during implicit prosodic processing situations (e.g., reading; Hwang & Steinhauer, 2011). Generally speaking, the CPS is elicited rapidly (between −100 and 0 ms) after the offset of a preboundary word and is thus argued to be triggered by preboundary syllable lengthening (Pauker, Itzhak, Baum, & Steinhauer, 2011). A slightly later prosody-related component was recently reported by Li et al. (2011), who observed a fronto-centrally distributed negative ERP between 270 and 510 ms in response to prosodic prominence manipulations. Responses to prosodic boundary violations resulted in a longer negative ERP effect lasting from 270 to 660 ms. For a similar time window, Böcker, Bastiaansen, Vroomen, Brunia, and Gelder (1999) found that extraction of metrical stress from bi-syllabic words can be linked to the N325, which is elicited under active (discrimination) and passive (listening) tasks.

In addition, processing of prosodic contour expectancy violations has been studied. For instance, Paulmann, Jessen, and Kotz (2012) recently violated linguistic prosodic expectancy by merging the beginning of a declarative sentence with the end of a question. Results revealed a frontally distributed Prosodic Expectancy Positivity (PEP) 620 ms after the onset of prosodic violations irrespective of task focus (linguistic/emotional), suggesting that listeners can detect an abrupt change in prosodic contour and that discrepancies in contours are quickly reanalyzed (Paulmann et al., 2012). Similar late positivities were found by Astésano, Besson, and Alter (2004), who reported a P800 for similar prosodic expectancy violations, and by Eckstein and Friederici (2006), who reported a P600 response to prosodic incongruity of the final word of a sentence.

These findings lend support to the assumption that participants not only use prosody to build information about the sentence structure and modality but also realize quickly if the expectation is not fulfilled. However, these prosodic processes can sometimes be influenced by task focus, because the P800 in the study by Astésano et al. was only observed when participants focused on the prosody, whereas the positivity reported by Paulmann et al., (2012) was found even without participants' explicit focus on prosody. In sum, ERP results confirm that prosody interfaces with other language functions such as semantics (Paulmann et al., 2012) and syntax (Eckstein & Friederici, 2006; Steinhauer et al., 1999) during online language comprehension. Given the different distributions of prosody-related ERP effects and their differing temporal dynamics, it seems reasonable to assume that linguistic prosody processing not only is multifaceted but also at least partly hinges on differing neural mechanisms depending on the precise function that could recruit brain structures from both hemispheres.

88.4 THE NEURAL BASIS OF EMOTIONAL PROSODY PROCESSING

88.4.1 Clinical Evidence

Emotional prosody has long been of special interest to researchers exploring hemispheric specialization of brain processes, probably because this type of prosodic function cannot easily be associated with pure emotion (historically linked to the RH; Borod, Cicero, et al., 1998; Borod, Bloom, & Santschi-Haywood, 1998) or pure language (historically linked to the LH; Friederici, 2002) processing. Instead, emotional prosody is at the intersection of both domains. Early research on how emotional prosody is anchored in the brain was performed by Heilman, Scholes, and Watson (1975). They asked patients with right and left temporo-parietal lesions to identify the emotional tone a speaker used when intoning semantically neutral sentences. Results revealed that RH patients performed significantly worse on the task than LH patients, suggesting that emotional prosody is primarily processed in the RH, although the lack of a HC group posits a problem to this conclusion. Thus, a few years later, the same group (Heilman et al., 1984) investigated the comprehension of emotional prosody in RH and LH patients and this time compared their performance with that of HCs. Results confirmed a deficit for RH patients when compared with LH patients and HCs. Similarly, Bowers, Coslett, Bauer, Speedie, and Heilman (1987) reported that RH patients performed significantly worse than LH patients and HCs when discriminating between emotional categories of prosodically and semantically emotional congruent and incongruent sentences as well as low-pass filtered sentences. This once more indicated an important role for the RH in emotional prosody perception. Results from Blonder et al. (1991) also revealed that RH patients have difficulties discriminating among different emotional categories for emotionally intoned neutral sentences when compared with LH and HCs. Although subsequent evidence has often confirmed the RH involvement in emotional prosody perception (Blonder et al., 1991; Borod, Cicero, et al., 1998; Borod, Bloom, et al., 1998; Lalande, Braun, Charlebois, & Whitaker, 1992; Ross & Monnot, 2008; Rymarczyk & Grabowska, 2007), there is some clinical data that question the unique role of the RH (Breitenstein et al., 1998; Cancelliere & Kertesz, 1990; Schlanger, Schlanger, & Gerstman, 1976; Starkstein et al., 1994).

For instance, Schlanger et al. (1976) described no differences between LH and RH aphasic patients for identifying the emotional tone of a speaker, suggesting that LH brain structures can also play a role during emotional prosody perception. Van Lancker and Sidtis (1992) asked LH and RH patients to identify emotional prosodic speech samples by matching them to facial expressions and emotional labels. They also fail to report differences between groups with respect to emotional prosody perception, indicating that emotional prosody perception is mediated through a bilateral network of brain structures. Support for this hypothesis comes from recent work by Kho et al. (2008), who showed that both RH and LH temporal lobe epilepsy patients were impaired on emotional prosody recognition when compared with HCs. Similarly, Starkstein et al. (1994) report that both LH and RH patients displayed emotional prosody comprehension difficulties; however, it should be noted that more detailed analyses also showed that RH patients with lesions in the BG and tempo-parietal cortex were most severely affected. Corroborating findings that subcortical brain structures are implied in an emotional prosodic network, Cancelliere and Kertesz (1990) report emotional prosody processing difficulties in patients with LH and RH lesions involving the BG. Paulmann et al. (2008) also report emotional prosody recognition deficits in patients with LH lesions in the BG (for a detailed review on the role of the BG in emotional prosody processing, see Kotz, Hasting, & Paulmann, 2013), rendering it unlikely that emotional prosody perception is uniquely mediated through RH cortical regions. In fact, in an attempt to illuminate how the LH and RH might contribute differently to emotional prosody recognition, Van Lancker and Sidtis (1992) explored whether patients with lesions in either hemisphere made similar errors in an emotional prosody identification task. Their findings suggest that patients used acoustic cues differently when judging the emotional tone of the speaker (Van Lancker & Sidtis, 1992). Specifically, a discrimination analysis on error patterns revealed that patients with lesions in the LH relied on pitch information to infer emotionality of stimuli, whereas patients with lesions in the RH seemed to predominantly use durational cues to infer emotionality. These data were argued to support the physical feature–dependent hypothesis; however, following the same methodology, Pell and Baum (1997) failed to find differences between LH and RH patients in their misclassifications of stimuli.

In short, results from clinical studies do not support simple hemispheric models, but instead suggest that emotional prosody processing recruits a broad network of cortical and subcortical brain regions possibly slightly more right-lateralized than left-lateralized. The modestly bigger RH involvement is corroborated by a recent meta-analysis (Witteman et al., 2011). When comparing results from studies testing RH and LH patients directly, findings implied that damage to the RH results in more severe problems for emotional prosody perception than damage to the LH. Finally, task focus, stimuli differences, differences in experimental paradigms, and, most critically, differences in patients' lesion size and location can heavily impact findings and therefore might contribute to equivocal findings.

88.4.2 Brain Imaging Evidence

In line with the studies reviewed, neuroimaging findings also tend to reveal that emotional prosody perception is mediated by a complex bilateral network involving both cortical and subcortical brain structures, with some studies suggesting a slightly more right-lateralized network. For instance, an early PET study explored emotional prosody categorization of sentences (George et al., 1996). Results revealed significant blood flow changes in the right prefrontal cortex. Similarly, Buchanan et al. (2000) asked participants to listen to emotionally intoned words while either detecting the emotionality of the speaker (emotion task) or detecting a probe word (nonemotion task). They report right-lateralized frontal lobe and right-lateralized anterior auditory cortex activation for emotionally intoned words when comparing the emotion task with the nonemotion task; however, a bilateral activation is reported for emotionally intoned words when comparing the emotion task to a rest baseline. Also, Mitchell, Elliott, Barry, Cruttenden, and Woodruff (2003) reported right-lateralized activation of superior and middle temporal gyri when participants listened to emotional sentences spoken in different emotions. A right-lateralized network for emotional prosody perception was also confirmed by Beaucousin et al. (2007), who reported greater activation in the right temporal lobe for emotional prosody processing when comparing emotional speech with text-to-text speech lacking emotional attributes. Similarly, Wildgruber et al. (2005) outline that recognition of emotionally intoned and semantically neutral sentences resulted in right-lateralized activation of the posterior superior temporal sulcus, as well as dorsolateral and orbitobasal frontal areas when participants engaged in an emotional identification task. However, when comparing activation patterns for the emotional recognition task with a rest baseline, frontal, temporal, and parietal brain areas were activated bilaterally, again demonstrating that task effects (and/or condition comparisons) play a crucial role when looking at brain activation patterns for emotional prosody comprehension.

In fact, as is expected based on neuropsychological findings, bilateral brain activation is also reported frequently. For instance, Kotz et al. (2003) compared sentences spoken in an emotional tone of voice with filtered sentences that contained only prosodic information. Comparisons revealed a bilateral frontal and subcortical (BG) activation pattern for the emotional prosodic condition. In addition, Sander et al. (2005) compared angry and neutral prosody processing in a dichotic listening paradigm. Irrespective of whether participants attended to or ignored the meaningless angry speech stimuli, right amygdala and bilateral superior temporal sulcus activation was reported. However, the same stimuli elicited greater bilateral activation in the orbitofrontal cortex and the cuneus (in the medial occipital cortex) when participants paid attention to presented stimuli as opposed to when they ignored them. Moreover, greater bilateral activation of mid-superior temporal sulcus has been reported for angry in contrast to neutral prosody (Grandjean et al., 2005). Also, Wiethoff et al. (2008) investigated activation for emotionally arousing prosody during passive listening. Specifically, they looked at happy, erotic, fearful, and angry prosody, and thus included stimuli of both positive and negative valence. The authors report greater activation for arousing prosody in the right primary auditory cortex, the mid-superior temporal gyrus, and the left temporal pole, as well as in the hypothalamus. Leitman et al. (2010) tried to compare stimuli that were either rich or low in emotional acoustic cue saliency (i.e., stimuli that are easily recognizable by a single acoustic parameter such as fundamental frequency or intensity). They reported greater activation for emotional prosodic stimuli that are rich in cue saliency for the planum temporale, posterior superior temporal, and middle gyri (i.e., superior temporal cortex), as well as the amygdala, whereas participants engaged in an emotional sentence identification task. In contrast, greater activation was found in inferior and temporofrontal areas when participants processed stimuli with less salient or dominant acoustic cues. Interestingly, their results indicate that lateralization can depend on the specific emotion investigated, because stronger left-lateralized activation patterns were found for angry prosody in contrast to more right-lateralized hot spots for fearful and happy prosody.

In addition to exploring the influence of individual acoustic cues (or their saliency) on activation patterns, more recent research has also tried to illuminate the role of task effects in neural responses linked to emotional prosody processing. For instance, Bach et al. (2008) compared activation patterns for emotionally intoned pseudowords while participants had to decide either which gender the speaker voice was or which emotion the speaker was trying to convey. They found bilateral activation of the amygdala, left superior temporal sulcus, and right parietal areas when participants judged the gender of the speaker, whereas activation for left inferior frontal gyrus, bilateral parietal, anterior cingulate, and supplemental motor cortex was found when participants focused on the emotionality of the stimuli. Moreover, when comparing emotional with neutral prosody, neural activity was found in right superior temporal gyrus and left inferior frontal gyrus, as well as in the anterior cingulate (bilateral), the insula, and the putamen (bilateral). Interestingly, subcortical brain activation was particularly strong when participants focused on the emotionality of the stimuli, once again leading to the impression that task effects can critically impact activation patterns. Finally, Ethofer, Van De Ville, Scherer, and Vuilleumier (2009) presented words spoken in an angry or neutral prosody to participants while they engaged in a valence discrimination or a word classification task. Activation in response to angry prosody was found not only in voice-sensitive areas of temporal cortices but also in the amygdala, insula, and mediodorsal thalami, irrespective of the task in which participants engaged. However, when comparing the valence discrimination with the word classification task, Ethofer et al. (2009) found stronger activation in the right middle temporal gyrus as well as in the orbitofrontal cortex (bilateral), suggesting that evaluation of emotional aspects activates these areas in particular.

In sum, neuroimaging data generally confirm that emotional prosody processing involves a bilateral temporo-frontal brain network, with some studies describing activation of subcortical structures. Activation hot spots between studies seem to differ depending on task focus and stimuli quality on activation patterns. Latter acoustic cue influence is confirmed in a meta-analysis by Witteman, Van Heuven, and Schiller (2012), who advocated that higher activation likelihood of the RH might stem from lateralized activation of primary and secondary auditory cortices (i.e., sensory cue processing).

88.4.3 ERP Findings

The time course underlying emotional prosody processing has been explored in recent electrophysiological studies. Several ERP components have been of special interest. The N100 is generally assumed to reflect processing of frequency (e.g., pitch) and loudness information (i.e., it is linked to the extraction of acoustic cues). This early component is followed by the P200, a fronto-centrally distributed component peaking 200 ms after stimulus onset (i.e., far before a sentence is completed), and that has been shown to be responsive to emotional prosodic (Paulmann et al., 2008; Schirmer, Chen, Ching, Tan, & Hong, 2013) and arousal (Paulmann, Bleichner, & Kotz, 2013) attributes of stimuli. Specifically, it has been outlined that different emotional prosodies can be distinguished from one another (Paulmann et al., 2013) and from neutral (Paulman & Kotz, 2008; Schirmer et al., 2013) within 200 ms of stimuli onsets. Although the sensitivity of the P200 to pitch (Pantev, Elbert, Ross, Eulitz, & Terhardt, 1996) and loudness (Picton, Woods, Baribeau-Braun, & Healey, 1977) variations has been demonstrated, research on emotional prosody implies that listeners rely on more than just one acoustic parameter when detecting emotional salience from auditory stimuli. However, which specific acoustic cues (configurations) are needed to detect the valence or even one particular emotion from speech still needs further clarification. Although the P200 is elicited under attentive processing conditions (albeit irrespective of implicit or explicit emotional tasks), the mismatch negativity has been linked to emotional category change detection under preattentive processing conditions (Schirmer, Striano, & Friederici, 2005). Both components have repeatedly been linked to early emotional salience detection (Kotz & Paulmann, 2011; Paulmann & Kotz, 2008; Schirmer & Kotz, 2006).

This early emotional evaluation or appraisal of vocal expressions is followed by more elaborated stimulus evaluations. In particular, meaning evaluation or access to (emotional) memory representations have been linked to later ERP components such as the P300 (Wambacq & Jerger, 2004), N300 (Bostanov & Kotchoubey, 2004), N400 (Schirmer & Kotz, 2003; Schirmer, Kotz, & Friederici, 2002; Schirmer et al., 2005; Paulmann & Pell, 2010), and the late positive complex (LPC) (Schirmer et al., 2013; Paulmann, Bleichner, & Kotz, 2013). For instance, Bostanov and Kotchoubey (2004) assessed how emotional meaning is extracted from exclamations such as "Wow" and "Oooh" by presenting participants with emotionally congruent or incongruent stimuli. They reported an enhanced N300 in response to incongruous exclamations, suggesting that emotional prosodic meaning is extracted approximately 300 ms after stimulus onset. Similarly, larger N400 amplitudes were found for prosodically/semantically incongruent emotional words (e.g., "happy" spoken in an angry voice; Schirmer & Kotz, 2003). To explore how *much* prosodic information is needed to infer emotional meaning, Paulmann and Pell (2010) presented participants with emotionally intoned sentence fragments that were either 200 or 400 ms long. Sentence fragments served as primes and were followed by emotionally matching or mismatching facial expressions. N400-like priming effects were found for faces that were preceded by emotionally mismatching sentence fragments, although priming from shorter fragments led to a reversed effect. The findings nicely showed that listeners can extract emotions from both short and somewhat longer sentence fragments, supporting the view that 200 ms is sufficient to build emotional context (see P200 results) and that emotionally relevant cues are extracted rapidly. Moreover, the N400-like priming effects are in line with studies proposing that emotional meaning is processed at approximately 300 to 400 ms after stimulus onset. Finally, two recent studies report differently modulated LPCs in response to sentences differing in emotional tones, suggesting continued exhaustive processing of emotional prosodic information at late processing stages (Paulmann et al., 2013; Schirmer et al., 2013). Moreover, Schirmer et al. (2013) further showed that P200 modulations can predict modulation of subsequently elicited LPCs, suggesting that successful early emotional salience detection goes hand-in-hand with later more in-depth processing of emotional prosody. Arguably, the latter step is necessary to ensure appropriate social behavior.

Finally, to investigate how far specific brain areas can be linked to individual emotional prosody processing steps, ERP lesion studies have been performed (Paulmann et al., 2008, 2009; Paulmann, Seifert, & Kotz, 2010). In a nutshell, these studies revealed that early emotional prosodic appraisal as reflected in the P200 component does not seem to be critically tied to the BG or orbitofrontal cortex. In contrast, later meaning-related processes (as reflected in the PEP and N400-like components) seem to be affected by lesions to the BG (Paulmann et al., 2008, 2009) and later emotional prosody recognition processes (as reflected in behavioral responses such as emotion recognition rates) are affected by lesions to the BG (Paulmann et al., 2008, 2009) and the orbitofrontal cortex (Paulmann et al., 2010). These findings highlight the potential of ERP lesion studies because they allow exploration of the function of specific brain areas with methodologies that have excellent temporal resolution, thereby helping to illuminate which processing stages might be modulated via specific neural structures. This methodology can thus help to consolidate conflicting findings from lesion and neuroimaging findings because the latter two lack the high temporal resolution of ERPs.

The studies reviewed in this section confirm that emotional prosodic processing includes processes such as rapid early emotional appraisal as well as comprehensive emotional meaning processing, that these processes occur irrespective of directed attention of listeners, and that different functions can be linked to different underlying brain areas. Moreover, comparing findings described in this section (emotional prosody) and the previous one (linguistic prosody), it also seems as if the two processes generally elicit different ERPs (with different onset latencies). When exploring the comparative nature of the time course linked to the two functions, results from a recent ERP study (Paulmann et al., 2012) revealed that emotional prosodic expectancy violations are detected approximately 150 ms earlier than expectancy violation of linguistic prosody and that both violations resulted in different PEP distributions. Linguistic PEPs were elicited predominantly at anterior electrode sites, whereas emotional PEPs were most dominant at posterior electrode sites, suggesting that at least partly different neural mechanisms are at play during emotional and linguistic prosody processing. Taken together, and similar to the imaging literature reviewed, differences in ERP studies (e.g., ERP polarity and latency) seem to be influenced by stimuli (e.g., words/sentences, normal/filtered/pseudo-speech), tasks (e.g., implicit/explicit emotional evaluation), and designs (e.g., blocked/randomized). However, despite methodological differences across studies, electrophysiological research has helped delineate the time course underlying emotional prosodic processing, clearly supporting the idea that emotional prosody processing comprises different subprocesses.

88.5 SUMMARY

Years of research have shown that initially advocated simple hemispheric models fail to adequately describe brain mechanisms underlying prosody processing. Research has revealed that neural mechanisms of prosody are vulnerable to external influences, such as task demands, stimulus quality, and experimental design, thereby explaining some of the discrepant literature reports. Given the functional complexity of linguistic prosody processing, its neural specifications seem to be less clearly delineated than neural structures for emotional prosody. However, clinical and imaging results seem to suggest that both functions of prosody cover different subprocesses that are each anchored in different parts of the brain. Future development of brain-based language models requires that the field continues to move away from understanding prosody as a holistic concept and focuses instead on portraying each function as multilayered.

Acknowledgments

I thank Rick Hanley for helpful comments on a previous version of this chapter.

References

Amebu Seddoh, S. (2006). Basis of intonation disturbance in aphasia: Perception. *Journal of Neurolinguistics*, *19*(4), 270–290.

Astésano, C., Besson, M., & Alter, K. (2004). Brain potentials during semantic and prosodic processing in French. *Cognitive Brain Research*, *18*, 172–184.

Bach, D. R., Grandjean, D., Sander, D., Herdener, M., Strik, W. K., & Seifritz, E. (2008). The effect of appraisal level on processing of emotional prosody in meaningless speech. *NeuroImage*, *42*(2), 919–927.

Baum, S. R., Daniloff, J. K., Daniloff, R., & Lewis, J. (1982). Sentence comprehension by Broca's aphasics: Effects of some suprasegmental variables. *Brain and Language*, *17*(2), 261–271.

Baum, S. R., & Pell, M. D. (1999). The neural bases of prosody: Insights from lesion studies and neuroimaging. *Aphasiology*, *13*(8), 581–608.

Beaucousin, V., Lacheret, A., Turbelin, M. R., Morel, M., Mazoyer, B., & Tzourio-Mazoyer, N. (2007). FMRI study of emotional speech comprehension. *Cerebral Cortex*, *17*(2), 339–352.

Blonder, L. X., Bowers, D., & Heilman, K. M. (1991). The role of the right hemisphere in emotional communication. *Brain*, *114*(3), 1115–1127.

Böcker, K. B., Bastiaansen, M., Vroomen, J., Brunia, C. H., & Gelder, B. (1999). An ERP correlate of metrical stress in spoken word recognition. *Psychophysiology*, *36*(6), 706–720.

Borod, J. C., Bloom, R. L., & Santschi-Haywood, C. (1998). Verbal aspects of emotional communication. In M. Beeman, & C. Chiarello (Eds.), *Right hemisphere language comprehension: Perspectives from cognitive neuroscience* (pp. 285–307). Hillsdale, NJ: Lawrence Erlbaum Associates.

Borod, J. C., Andelman, F., Obler, L. K., Tweedy, J. R., & Welkowitz, J. (1992). Right hemisphere specializations for the identification of emotional words and sentences: Evidence from stroke patients. *Neuropsychologia*, *30*, 827–844.

Borod, J. C., Cicero, B. A., Obler, L. K., Welkowitz, J., Erhan, H. M., Santschi, C., et al. (1998). Right hemisphere emotional perception: Evidence across multiple channels. *Neuropsychology*, *12*(3), 446.

Bostanov, V., & Kotchoubey, B. (2004). Recognition of affective prosody: Continuous wavelet measures of event-related brain potentials to emotional exclamations. *Psychophysiology*, *41*(2), 259–268.

Bowers, D., Coslett, H., Bauer, R. M., Speedie, L. J., & Heilman, K. M. (1987). Comprehension of emotional prosody following unilateral hemispheric lesions: Processing defect versus distraction defect. *Neuropsychologia*, *25*(2), 317–328.

Brådvik, B., Dravins, C., Holtås, S., Rosen, I., Ryding, E., & Ingvar, D. H. (1991). Disturbances of speech prosody following right hemisphere infarcts. *Acta Neurologica Scandinavica*, *84*(2), 114–126.

Breitenstein, C., Daum, I., & Ackermann, H. (1998). Emotional processing following cortical and subcortical brain damage: Contribution of the fronto-striatal circuitry. *Behavioural Neurology*, *11*(1), 29–42.

Brück, C., Kreifelts, B., & Wildgruber, D. (2011). Emotional voices in context: A neurobiological model of multimodal affective information processing. *Physics of Life Reviews*, *8*(4), 383–403.

Bryan, K. L. (1989). Language prosody and the right hemisphere. *Aphasiology*, *3*(4), 285–299.

Buchanan, T. W., Lutz, K., Mirzazade, S., Specht, K., Shah, N. J., Zilles, K., et al. (2000). Recognition of emotional prosody and verbal components of spoken language: An fMRI study. *Cognitive Brain Research*, *9*(3), 227–238.

Cancelliere, A. E., & Kertesz, A. (1990). Lesion localization in acquired deficits of emotional expression and comprehension. *Brain and Cognition*, *13*(2), 133–147.

Eckstein, K., & Friederici, A. D. (2006). It's early: Event-related potential evidence for initial interaction of syntax and prosody in speech comprehension. *Journal of Cognitive Neuroscience*, *18*(10), 1696–1711.

Emmorey, K. D. (1987). The neurological substrates for prosodic aspects of speech. *Brain and Language*, *30*(2), 305–320.

Ethofer, T., Van De Ville, D., Scherer, K., & Vuilleumier, P. (2009). Decoding of emotional information in voice-sensitive cortices. *Current Biology*, *19*(12), 1028–1033.

Friederici, A. D. (2002). Towards a neural basis of auditory sentence processing. *Trends in Cognitive Sciences*, *6*(2), 78–84.

Friederici, A. D., & Alter, K. (2004). Lateralization of auditory language functions: A dynamic dual pathway model. *Brain and Language*, *89*(2), 267–276.

Gandour, J., Tong, Y., Wong, D., Talavage, T., Dzemidzic, M., Xu, Y., et al. (2004). Hemispheric roles in the perception of speech prosody. *NeuroImage*, *23*(1), 344–357.

Gandour, J., Wong, D., Hsieh, L., Weinzapfel, B., Van Lancker, D., & Hutchins, G. D. (2000). A crosslinguistic PET study of tone perception. *Journal of Cognitive Neuroscience*, *12*(1), 207–222.

Geiser, E., Zaehle, T., Jancke, L., & Meyer, M. (2008). The neural correlate of speech rhythm as evidenced by metrical speech processing. *Journal of Cognitive Neuroscience*, *20*(3), 541–552.

George, M. S., Parekh, P. I., Rosinsky, N., Ketter, T. A., Kimbrell, T. A., Heilman, K. M., et al. (1996). Understanding emotional prosody activates right hemisphere regions. *Archives of Neurology*, *53*(7), 665.

Grandjean, D., Sander, D., Pourtois, G., Schwartz, S., Seghier, M. L., Scherer, K. R., et al. (2005). The voices of wrath: Brain responses to angry prosody in meaningless speech. *Nature Neuroscience*, *8*(2), 145–146.

Heilman, K. M., Bowers, D., Speedie, L., & Coslett, H. B. (1984). Comprehension of affective and nonaffective prosody. *Neurology*, *34*(7), 917–921.

Heilman, K. M., Scholes, R., & Watson, R. T. (1975). Auditory affective agnosia. Disturbed comprehension of affective speech. *Journal of Neurology, Neurosurgery and Psychiatry*, *38*(1), 69–72.

Hwang, H., & Steinhauer, K. (2011). Phrase length matters: The interplay between implicit prosody and syntax in Korean "Garden Path" sentences. *Journal of Cognitive Neuroscience*, *23*(11), 3555–3575.

Kho, K. H., Indefrey, P., Hagoort, P., Van Veelen, C. W. M., Van Rijen, P. C., & Ramsey, N. F. (2008). Unimpaired sentence comprehension after anterior temporal cortex resection. *Neuropsychologia*, *46*(4), 1170–1178.

Klouda, G. V., Robin, D. A., Graff-Radford, N. R., & Cooper, W. E. (1988). The role of callosal connections in speech prosody. *Brain and Language*, *35*(1), 154–171.

Kotz, S. A., Hasting, A. S., & Paulmann, S. (2013). *On the orbito-striatal interface in (acoustic) emotional processing. The evolution of emotional communication: From sounds in nonhuman mammals to speech and music in man*. New York: Oxford University Press.

Kotz, S. A., Meyer, M., Alter, K., Besson, M., von Cramon, D. Y., & Friederici, A. D. (2003). On the lateralization of emotional prosody: An event-related functional MR investigation. *Brain and Language*, *86*(3), 366–376.

Kotz, S. A., Meyer, M., & Paulmann, S. (2006). Lateralization of emotional prosody in the brain: An overview and synopsis on the impact of study design. *Progress in Brain Research*, *156*, 285–294.

Kotz, S. A., & Paulmann, S. (2011). Emotion, language, and the brain. *Language and Linguistics Compass*, *5*(3), 108–125.

Lalande, S., Braun, C. M. J., Charlebois, N., & Whitaker, H. A. (1992). Effects of right and left hemisphere cerebrovascular lesions on discrimination of prosodic and semantic aspects of affect in sentences. *Brain and Language*, *42*(2), 165–186.

Leitman, D. I., Wolf, D. H., Ragland, J. D., Laukka, P., Loughead, J., Valdez, J. N., et al. (2010). "It's Not What You Say, But How You Say It": A reciprocal temporo-frontal network for affective prosody. *Frontiers in Human Neuroscience*, *4*.

Li, X., Chen, Y., & Yang, Y. (2011). Immediate integration of different types of prosodic information during on-line spoken language comprehension: An ERP study. *Brain Research, 1386*, 139–152.

Meyer, M., Alter, K., Friederici, A. D., Lohmann, G., & von Cramon, D. Y. (2002). FMRI reveals brain regions mediating slow prosodic modulations in spoken sentences. *Human Brain Mapping, 17*(2), 73–88.

Mitchell, R. L., Elliott, R., Barry, M., Cruttenden, A., & Woodruff, P. W. (2003). The neural response to emotional prosody, as revealed by functional magnetic resonance imaging. *Neuropsychologia, 41*(10), 1410–1421.

Pantev, C., Elbert, T., Ross, B., Eulitz, C., & Terhardt, E. (1996). Binaural fusion and the representation of virtual pitch in the human auditory cortex. *Hearing Research, 100*, 164–170.

Pauker, E., Itzhak, I., Baum, S. R., & Steinhauer, K. (2011). Effects of cooperating and conflicting prosody in spoken English garden path sentences: ERP evidence for the boundary deletion hypothesis. *Journal of Cognitive Neuroscience, 23*(10), 2731–2751.

Paulmann, S., Bleichner, M., & Kotz, S. A. (2013). Valence, arousal, and task effects in emotional prosody processing. *Frontiers in Psychology, 4*, 345.

Paulmann, S., Jessen, S., & Kotz, S. A. (2012). It's special the way you say it: An ERP investigation on the temporal dynamics of two types of prosody. *Neuropsychologia, 50*(7), 1609–1620.

Paulmann, S., & Kotz, S. A. (2008). Early emotional prosody perception based on different speaker voices. *Neuroreport, 19*(2), 209–213.

Paulmann, S., & Pell, M. D. (2010). Contextual influences of emotional speech prosody on face processing: How much is enough? *Cognitive, Affective, and Behavioral Neuroscience, 10*(2), 230–242.

Paulmann, S., Pell, M. D., & Kotz, S. A. (2008). Functional contributions of the basal ganglia to emotional prosody: Evidence from ERPs. *Brain Research, 1217*, 171–178.

Paulmann, S., Pell, M. D., & Kotz, S. A. (2009). Comparative processing of emotional prosody and semantics following basal ganglia infarcts: ERP evidence of selective impairments for disgust and fear. *Brain Research, 1295*, 159–169.

Paulmann, S., Seifert, S., & Kotz, S. A. (2010). Orbito-frontal lesions cause impairment during late but not early emotional prosodic processing. *Social Neuroscience, 5*(1), 59–75.

Pell, M. D. (1998). Recognition of prosody following unilateral brain lesion: Influence of functional and structural attributes of prosodic contours. *Neuropsychologia, 36*(8), 701–715.

Pell, M. D., & Baum, S. R. (1997). The ability to perceive and comprehend intonation in linguistic and affective contexts by brain-damaged adults. *Brain and Language, 57*(1), 80–99.

Picton, T. W., Woods, D. L., Baribeau-Braun, J., & Healey, T. M. G. (1977). Evoked potential audiometry. *Journal of Otolaryngology, 6*(2), 90–119.

Plante, E., Creusere, M., & Sabin, C. (2002). Dissociating sentential prosody from sentence processing: Activation interacts with task demands. *NeuroImage, 17*(1), 401–410.

Ross, E. D. (1981). The aprosodias. Functional-anatomic organization of the affective components of language in the right hemisphere. *Archives of Neurology, 38*(9), 561–569.

Ross, E. D., & Mesulam, M. M. (1979). Dominant language functions of the right hemisphere? Prosody and emotional gesturing. *Archives of Neurology, 36*(3), 144.

Ross, E. D., & Monnot, M. (2008). Neurology of affective prosody and its functional–anatomic organization in right hemisphere. *Brain and Language, 104*(1), 51–74.

Rymarczyk, K., & Grabowska, A. (2007). Sex differences in brain control of prosody. *Neuropsychologia, 45*(5), 921–930.

Sander, D., Grandjean, D., Pourtois, G., Schwartz, S., Seghier, M. L., Scherer, K. R., et al. (2005). Emotion and attention interactions in social cognition: Brain regions involved in processing anger prosody. *NeuroImage, 28*(4), 848–858.

Schirmer, A., Chen, C. B., Ching, A., Tan, L., & Hong, R. Y. (2013). Vocal emotions influence verbal memory: Neural correlates and interindividual differences. *Cognitive, Affective, and Behavioral Neuroscience, 13*(1), 80–93.

Schirmer, A., & Kotz, S. A. (2003). ERP evidence for a sex-specific Stroop effect in emotional speech. *Journal of Cognitive Neuroscience, 15*(8), 1135–1148.

Schirmer, A., & Kotz, S. A. (2006). Beyond the right hemisphere: Brain mechanisms mediating vocal emotional processing. *Trends in Cognitive Sciences, 10*(1), 24–30.

Schirmer, A., Kotz, S. A., & Friederici, A. D. (2002). Sex differentiates the role of emotional prosody during word processing. *Cognitive Brain Research, 14*(2), 228–233.

Schirmer, A., Striano, T., & Friederici, A. D. (2005). Sex differences in the preattentive processing of vocal emotional expressions. *Neuroreport, 16*(6), 635–639.

Schlanger, B. B., Schlanger, P., & Gerstman, L. J. (1976). The perception of emotionally toned sentences by right hemisphere-damaged and aphasic subjects. *Brain and Language, 3*(3), 396–403.

Seddoh, S. A. (2002). How discrete or independent are "affective prosody" and "linguistic prosody"? *Aphasiology, 16*(7), 683–692.

Sidtis, J. J., & Van Lancker Sidtis, D. (2003). *A neurobehavioral approach to dysprosody, Seminars in speech and language* (Vol. 24, No. 2, pp. 93–106). New York: Theime Medical Publishers Inc.

Starkstein, S. E., Federoff, J. P., Price, T. R., Leiguarda, R. C., & Robinson, R. G. (1994). Neuropsychological and neuroradiologic correlates of emotional prosody comprehension. *Neurology, 44*(3 Part 1), 515–522.

Steinhauer, K., Alter, K., & Friederici, A. D. (1999). Brain potentials indicate immediate use of prosodic cues in natural speech processing. *Nature Neuroscience, 2*(2), 191–196.

Steinhauer, K., & Friederici, A. D. (2001). Prosodic boundaries, comma rules, and brain responses: The closure positive shift in ERPs as a universal marker for prosodic phrasing in listeners and readers. *Journal of Psycholinguistic Research, 30*(3), 267–295.

Tracy, D. K., Ho, D. K., O'Daly, O., Michalopoulou, P., Lloyd, L. C., Dimond, E., et al. (2011). It's not what you say but the way that you say it: An fMRI study of differential lexical and non-lexical prosodic pitch processing. *BMC Neuroscience, 12*(1), 128.

Van Lancker, D. (1980). Cerebral lateralization of pitch cues in the linguistic signal. *Research on Language and Social Interaction, 13*(2), 201–277.

Van Lancker, D., & Sidtis, J. J. (1992). The identification of affective-prosodic stimuli by left-and right-hemisphere-damaged subjects: All errors are not created equal. *Journal of Speech, Language and Hearing Research, 35*(5), 963.

Walker, J. P., Daigle, T., & Buzzard, M. (2002). Hemispheric specialisation in processing prosodic structures: Revisited. *Aphasiology, 16*(12), 1155–1172.

Wambacq, I. J., & Jerger, J. F. (2004). Processing of affective prosody and lexical-semantics in spoken utterances as differentiated by event-related potentials. *Cognitive Brain Research, 20*(3), 427–437.

Weintraub, S., Mesulam, M., & Kramer, L. (1981). Disturbances in prosody: A right-hemisphere contribution to language. *Archives of Neurology, 38*(12), 742.

Wiethoff, S., Wildgruber, D., Kreifelts, B., Becker, H., Herbert, C., Grodd, W., et al. (2008). Cerebral processing of emotional prosody—Influence of acoustic parameters and arousal. *NeuroImage, 39*(2), 885–893.

Wildgruber, D., Ackermann, H., Kreifelts, B., & Ethofer, T. (2006). Cerebral processing of linguistic and emotional prosody: fMRI studies. *Progress in Brain Research, 156*, 249–268.

Wildgruber, D., Ethofer, T., Grandjean, D., & Kreifelts, B. (2009). A cerebral network model of speech prosody comprehension. *International Journal of Speech-Language Pathology, 11*(4), 277–281.

Wildgruber, D., Hertrich, I., Riecker, A., Erb, M., Anders, S., Grodd, W., et al. (2004). Distinct frontal regions subserve evaluation of linguistic and emotional aspects of speech intonation. *Cerebral Cortex, 14*(12), 1384–1389.

Wildgruber, D., Riecker, A., Hertrich, I., Erb, M., Grodd, W., Ethofer, T., et al. (2005). Identification of emotional intonation evaluated by fMRI. *NeuroImage, 24*(4), 1233–1241.

Witteman, J., Van Heuven, V. J., & Schiller, N. O. (2012). Hearing feelings: A quantitative meta-analysis on the neuroimaging literature of emotional prosody perception. *Neuropsychologia, 50*(12), 2752–2763.

Witteman, J., van Ijzendoorn, M. H., van de Velde, D., van Heuven, V. J., & Schiller, N. O. (2011). The nature of hemispheric specialization for linguistic and emotional prosodic perception: A meta-analysis of the lesion literature. *Neuropsychologia, 49*(13), 3722–3738.

C H A P T E R

89

Environmental Sounds

Frederic Dick[1,2], *Saloni Krishnan*[1,2], *Robert Leech*[3] *and Ayşe Pinar Saygin*[4]

[1]Birkbeck/UCL Centre for NeuroImaging (BUCNI), London, UK; [2]Centre for Brain and Cognitive Development (CBCD), Department of Psychological Sciences, Birkbeck College, University of London, London, UK; [3]Computational, Cognitive and Clinical Neuroimaging Laboratory (C3NL), Imperial College London, London, UK; [4]Department of Cognitive Science, University of California—San Diego, La Jolla, CA, USA

89.1 WHAT ARE ENVIRONMENTAL SOUNDS?

An environmental sound is historically defined by what it is not. An environmental sound is not speech. Nor is it a decontextualized sine tone or white noise burst often used in psychoacoustics experiments. But any other sound is fair game.[1] As Ballas and Howard (1987) commented, such a definition by exclusion is not very satisfying. This is particularly true given that perceiving environmental sounds is presumably the primary job of the vertebrate auditory system. Thus, by understanding more about environmental sounds, we should arrive at a better understanding of the auditory system that evolved to perceive them—a system parasitized in our own species by spoken language.

What, then, are some general characteristics of environmental sounds? Until recently, the acoustical character of environmental sounds was determined by interactions between solids, liquids, and gases (Gaver, 1993a), thereby intrinsically binding the sounds to their physical sources. Thanks to modern computers and sound synthesizers, this no longer needs to be the case. Nonetheless, because of the tight correspondence between a sound and its generating substance, environmental sounds can be thought of as a naturally occurring "auditory icon" for that object (Gaver, 1993b). Alternatively, an environmental sound can be conceived of as an auditory pointer to a physical referent, one that might convey meaning to the hearer. However, it is important to keep in mind that with the exception of some nonhuman vocalizations, human nonspeech sounds, or alert sounds, environmental sounds are not intentional communicative signals. Moreover, spoken language has a fairly rigid temporal frame in which to convey action concepts, whereas environmental sounds vary hugely, with semantic identity unambiguously conveyed extremely quickly (a cork pop) to quite slowly (buttering toast).

As with visual objects and arrays, a useful means of characterizing environmental sounds is to partition them along acoustical, perceptual, and conceptual dimensions. Such an exercise can provide a rough guide of what type of information can and cannot be derived from environmental sounds, what the "grain" of that information is, and how it is distributed across sounds.

The acoustical and semantic space that environmental sounds inhabit has been mapped out by various groups over the past few decades, primarily by asking participants to perform similarity or overt category ratings on a set of recorded sounds and also by measuring a sizeable number of acoustic characteristics of the rated sounds. These studies tend to focus on one main class of environmental sounds often called *"auditory objects"*—a term that remains amorphous and controversial in equal measures (Giordano, McAdams, Zatorre, Kriegeskorte, & Belin, 2012; Griffiths & Warren, 2004; Lewis, Talkington, Tallaksen, & Frum, 2012; Wightman & Jenison, 1995). These

[1]Although music and nonhuman vocalizations are often used as environmental sounds (Leaver & Rauschecker, 2010; Lewis, Talkington, Puce, Engel, & Frum, 2011; Talkington, Rapuano, Hitt, Frum, & Lewis, 2012), both are described in depth in other parts of this volume and therefore will not be discussed here.

Neurobiology of Language. DOI: http://dx.doi.org/10.1016/B978-0-12-407794-2.00089-4

auditory objects are often naturally occurring,[2] short *auditory events* and can have definable and reasonably obvious beginning and ending points. Moreover, as stated by Lewis et al. (2012), "auditory pattern analyses should allow for perceptual categorization and auditory objects should be separable by perceptual boundaries." For example, the sound of a lead crystal wine glass smashing on a marble floor might begin with the main impact of the base of the glass and end when all of the glass shards have stopped bouncing. If asked to identify or name the sound, the typical English-speaking listener will name the source or material from which that the sound emanates, often along with a verb indicating what is happening with that source (e.g., "cow mooing," "glass breaking").

However, many, if not most, auditory objects have a more ambiguous event structure. Take the seemingly simple case of the sound of someone typing on a computer keyboard. A single keystroke might constitute an auditory object or event. Unless the sound recording is of the most hesitant of typists, keystrokes usually play out in clusters, and even in long sequences of bursts if the keyboardist is inspired (or under a tight deadline). Somewhat akin to speech, the sound of each keystroke will often blend with the subsequent one; with reasonably speedy typists, the sound of the release of one key will overlap in time with the sound of another being depressed. Thus, the "auditory object" becomes less straightforward to identify—is it the keystroke, a cluster of keystrokes, or the entire sequence of keystrokes that begins and ends with a prolonged silence?[3] Even in the seemingly more straightforward case of the wine glass shattering, the event itself can be considered (and artificially reconstructed) as multiple overlapping repetitions of glass shards hitting the floor (Warren & Verbrugge, 1984).

This iterative repetition of a basic event (or waveform)—but over an indefinitely long period of time—is the acoustic signature of some very familiar auditory objects, such as a helicopter hovering or a diesel truck engine idling. These sounds could be considered boundary cases between auditory objects and the other major division of environmental sounds, *auditory scenes*. Some of these scenes can be acoustically quite simple and repetitive ambient human-created environments such as the distant sound of highway traffic, the buzz of an electrical transformer, or the aforementioned idling diesel engine. They also include complex natural *textures* or *"soundscapes"* (Gygi & Shafiro, 2011; Turner, 2010) such as waves crashing on the shore, wind blowing through leaves, a crackling fire, or gentle rain falling on tarmac. Remarkably, many of these natural soundscapes can be generated by combining "atomic elements" of sounds in different combinations (McDermott, Wrobleski, & Oxenham, 2011; Turner, 2010). These simple scenes or soundscapes are also distinguished by their *lack* of event structure in that nothing particularly noteworthy happens or changes over a long period of time. Instead, soundscapes provide a particular auditory *context*—something we return to later.[4]

More complex auditory scenes or *backgrounds*—such the ambient sounds of a busy park, a quiet office, or tropical rainforest—may contain several of these more basic soundscape features along with repeating, shorter auditory objects embedded in the scene.[5] In natural scenes, the probability that a given *category* of auditory object will be heard is partly contingent on the type of soundscape—a contextual property to which listeners are sensitive (Gygi & Shafiro, 2011; Krishnan, Leech, Aydelott, & Dick, 2013; Leech, Gygi, Aydelott, & Dick, 2009; discussed in more depth later). However, these scenes typically have no real event structure in that there is a reasonably uniform temporal probability that auditory objects will appear within the (indefinitely long) soundscape.

These relatively static auditory backgrounds contrast with *dynamic auditory scenes* in which the probability of an auditory event occurring at a particular time is *contingent* on what has been heard previously. A classic example of such a scene is a car crash sequence in an action movie (Humphries, Willard, Buchsbaum, & Hickok, 2001), where the sounds of a car racing is followed by squealing brakes and tires, crinkling metal and glass, and—depending on the director's penchant for drama—the whoof of igniting gasoline that precedes a death-dealing explosion. Needless to say, the temporal contingencies in such dynamic scenes are entirely dependent on the "real-world" event structure. Therefore, they are a physical

[2]An increasing number of auditory objects and events are completely artificial (with the most obvious being the novel alerts and "auditory icons" that accompany daily computing and telephone use. In addition, many people now have their only experience with natural sounds via recordings or synthesized examples, rather than with the animal or environment itself.

[3]A typical label from an English speaker for *any* of these keyboard sounds might be "typing."

[4]Lewis et al. (2012) have noted that mechanical sounds classified by listeners as auditory objects (a clock or fax machine) tend to be size-delimited, whereas auditory scenes depict things that are large relative to the size of the observer (wind, rain, etc).

[5]These scenes are ubiquitous in radio plays, TV shows, and the movies, with vast numbers of prerecorded scenes available from the BBC and other studios.

reflection of the changing state of the world as opposed to language's more abstract representation of it, one whose surface form might change ("The car hit the post" versus "The post was hit by the car") but whose semantic content remains very similar. These cases highlight a clear difference between what is conveyed by environmental sounds and spoken language, namely the directness and invariability of the mapping between the information-carrying acoustic signal and the semantic content and/or event structure that is conveyed by that signal.

89.2 PERCEPTUAL, COGNITIVE, AND NEURAL PROCESSING OF ENVIRONMENTAL SOUNDS

Because the academic field studying spoken language is so much larger and more established, it has been natural for the study of environmental sound to mirror the study of language, and for the general subtopics (e.g., perceptual, cognitive, and neural factors) implicated in processing environmental sounds to be similar to those for language research. One consequence of this homologous organization is that it enables comparisons between language and environmental sounds. Therefore, what follows is structured around theoretical and methodological themes that roughly parallel ones in the psycholinguistics literature. We first summarize results from a number of environmental sound norming studies (somewhat similar to picture naming; Bates, D'Amico, et al., 2003) that give an idea of the perceptual and semantic characteristics and categories that seem to underlie environmental sounds recognition. In studies that roughly parallel some in the spoken word recognition literature, we examine how environmental sounds recognition and identification develops and changes over the lifespan. We then turn to the literature on perceptual and semantic priming with environmental sounds using unimodal, cross-modal, and cross-domain priming paradigms. We finally examine context effects—both acoustic and "real-world knowledge"-based—on environmental sound recognition.

In the second section on the neural bases of environmental sound comprehension, we begin with a brief overview of the historical literature comparing hemispheric asymmetries for environmental sounds and language processing, and then proceed to a more extensive review on auditory agnosias and the effect of brain damage on environmental sound comprehension, and how these effects compare with those on language processing. We finish with an in-depth look at the neuroimaging literature on environmental sounds, with particular emphasis on comparisons with language and on regional neural biases for particular acoustical or semantic characteristics.

89.3 SECTION ONE: PERCEPTUAL AND COGNITIVE FACTORS IN PROCESSING ENVIRONMENTAL SOUNDS

89.3.1 Identification and Categorization of Environmental Sounds

As has been carried out with corpora of meaningful visual objects such as pictures (Bates, Burani, D'Amico, & Barca, 2001; Bates, D'Amico, et al., 2003; Cycowicz, Friedman, Rothstein, & Snodgrass, 1997; Szekely et al., 2004) as well as words (Balota et al., 2007), several groups have undertaken norming studies of environmental sound recordings to establish baseline characteristics such as discriminability, identifiability, nameability, and imageability (Ballas, 1993; Fabiani, Kazmerski, Cycowicz, & Friedman, 1996; Giordano, Mcdonnell, & McAdams, 2010; Gygi, Kidd, & Watson, 2004; Hocking, Dzafic, Kazovsky, & Copland, 2013; Marcell, Borella, Greene, Kerr, & Rogers, 2000; Saygin, Dick, & Bates, 2005; Schneider, Engel, & Debener, 2008; Shafiro, 2008; Shafiro & Gygi, 2004). These norming studies tend to use similar experimental paradigms in which participants listen to one of X sounds, press a key as soon as they think they know its identity, and then type or choose a verbal label. In some cases, participants are also asked to give alternative labels and perform a number of rating tasks on the sound. Among other results (comprehensively laid out in Hocking et al., 2013; Marcell et al., 2000), these ratings have shown: (i) imageable sounds are more identifiable; (ii) the more imageable a sound is, the faster participants will name it (and the more confidence they will have in their judgments about the sound); and (iii) familiar sounds are more identifiable and are also more pleasurable.

One useful metric derived in several of these studies is the "causal uncertainty" or Hcu[6] (Ballas, 1993) associated with a sound. Although its definition varies slightly across studies, the Hcu is essentially a measure of the number of potential sound sources associated with a given sound, as indexed by the number of semantically distinct labels for a sound that are

[6]As defined by Ballas, the Hcu or causal entropy of a given sound is calculated as follows: for each label provided for the sound, multiply the probability that the label was used by the binary log of that probability and then sum these products, e.g., Hcu = p_{ij} $\log_2 p_{ij}$. A similar metric has been used in picture-naming studies (Snodgrass & Vanderwart, 1980).

provided by participants. For instance, highly distinct and identifiable sounds like a doorbell would have a low Hcu, whereas the sound of a light switch—which in the study of Ballas (1993) was confused with a stapler or ballpoint pen—would have a higher Hcu. Perhaps unsurprisingly, a sound's Hcu is strongly correlated with the time it takes to identify and name it.

A major lacuna in all environmental sounds norming studies (and of environmental sounds studies more generally) is an estimate of a sound or sound category's frequency of occurrence—in psycholinguistics, word frequency is a fundamental variable of interest (Balota et al., 2007; Bates, D'Amico, et al., 2003; Roland, Dick, & Elman, 2007). To our knowledge, the only published frequency-of-occurrence data on environmental sounds are in the work by Ballas (1993).[7] Here, undergraduate volunteers were prompted (via timers and pagers) to write down what sound(s) they were hearing at different times over the course of a normal day. Ballas (1993) used these survey data to predict behavior in a subset of the sounds used in the norming studies and showed that more frequently occurring sounds were more identifiable (and identified more quickly) and had lower causal uncertainty. The exceptions to this trend were alert sounds and signals, which had shorter reaction times (RTs) and were more identifiable than would be expected from their low frequency of occurrence.

Another challenge and conundrum in environmental sounds studies is how to account for sound length—again, word or phrase length is a very important psycholinguistic property. As Marcell et al. (2000) write regarding their own stimuli, "[we edited] each sound to a duration that we believed allowed the 'sound event' or 'auditory object' [...] to unfold naturally [...] this was clearly more of an artistic than empirical endeavor." This ambiguity is reflected in the variability in sound duration across (and even within) norming studies, with durations varying by more than two orders of magnitude, ranging from consistently short or medium duration (Ballas, 1993, all sounds <625 ms; Fabiani et al., 1996, average duration ~330 ms; Hocking et al., 2013, all sounds set to 1,000 ms) to highly variable (Saygin et al., 2005, ~500–4,500 ms; Marcell et al., 2000, ~500–6,000 ms). This is the equivalent in duration of comparing single short spoken words to sentences with 20 or more words (a considerable potential difference in informativeness!)—yet duration does not appear to be correlated with recorded sounds' identifiability (either in accuracy or in response latency), familiarity, or confidence with which listeners identify the sound (Marcell et al., 2000).

A primary goal of many of these studies is to uncover the acoustical, perceptual, and cognitive category structure of environmental sounds. Explicit categorization tasks (Gygi, Kidd, & Watson, 2007; Hocking et al., 2013; Houix, Lemaitre, Misdariis, Susini, & Urdapilleta, 2012; Marcell et al., 2000) have revealed fairly consistent category groupings, for instance, Hocking et al. (2013) found that participants could reliably classify sounds as: (i) animal; (ii) human; (iii) nature; (iv) household/tool/accessory; (v) recreational; (vi) transport; (vii) weapon; (viii) alarm/signal; or (ix) musical instruments.[8] A somewhat more detailed category structure was captured in a free classification task by Marcell et al. (2000) (see Table 89.1).

When this category space is compared with that associated with visual objects (Huth, Nishimoto, Vu, & Gallant, 2012), it is striking how much more limited, sparse, and clumpy the semantic space is that is covered by environmental sounds (at least recognizable recorded ones). This may be due to a fundamental property of sound itself. As Wightman and Jenison (1995) noted, "In the case of vision the physical features of environmental objects map directly to patterns of stimulation on the retina [...] In contrast, hearing offers no direct peripheral representation of environmental

TABLE 89.1 Environmental Sounds Categories (in Alphabetical Order) Provided by Participants in Marcell et al. (2000)

Accident	Ground transportation	Pet	Household
Air transportation	Human	Reptile/amphibian	Machine
Bathroom	Hygiene	Sickness	Paper
Bird	Insect	Signal	Weapon
Farm animal	Kitchen	Sleep	Weather
Four-legged animal	Musical instrument	Tool	
Game/recreation	Nature	Water/liquid	

[7]Cummings et al. (2009) conducted a survey of infants' and toddlers' parents to try to establish frequency-of-occurrence norms for everyday sounds; however, there was little consistency in parents' reports, which may reflect a problem in the survey instrument or true variability in children's exposure to environmental sounds.

[8]In their fMRI studies of "action sounds," Lewis et al. (2012) have suggested a division into human, animal, mechanical, and environmental sounds, based on acoustical, perceptual, and cortical organization grounds.

objects." In other words, visible light will reflect off almost any surface and therefore create a "signal" that tells us about almost any surface or object. In contrast, for us to receive an acoustic signal from an object, it must *actively* emit vibrations that can travel in the air far enough to be detected, or it must interact with air movement and turbulence in a distinctive way (i.e., the vibrations induced by wind flowing through dry leaves versus chimes). This means a more restricted number of objects/situations create sounds, severely limiting the category space of the auditory objects and scenes relative to vision.

The perceptual and categorical similarity space of environmental sounds has been extensively characterized in a series of studies by Gygi et al. (2004, 2007). By using participants' pairwise similarity ratings for 50 sounds along with a large number of acoustic measures, they found that sounds tended to fall into three major acoustical categories—harmonic, discrete impact, and continuous sounds—and that the different underlying sound sources tended to cluster together in the multidimensional scaling (MDS) space. They also found that ordering of the sounds along different dimensions of the MDS space was associated with linear combinations of acoustic variables, such as harmonicity, amount of silence, and modulation depth.[9]

89.3.2 Environmental Sounds Comprehension Over the Lifespan

In a comprehensive study of children ages 5 to 16, young adults, older adults, and patients with Alzheimer's dementia, Fabiani et al. (1996) showed that with few exceptions (such as video game sounds), people across the lifespan generally categorize environmental sounds very similarly. However, older participants and those with Alzheimer's tended to be less accurate and consistent in their identification of sounds and often used superordinate categories to name sounds. This was true even in the case of animal vocalizations, which tend to be quite specific and easy to identify for younger adults. It is important to note that while there is general consistency in environmental sounds identification, there can be very significant regional and cultural differences in exposure to different sounds, which can have a real impact on causal uncertainty. For example, when norming our own set of environmental sounds in different cities (Dick et al., 2007), there were some sounds (such as cutting crusty bread and lighting a gas stove) that were easily recognizable by residents of Rome, but that flummoxed San Diegans. As Giordano et al. (2010) (p. 9) note, it is important to account for these effects because "uncontrolled differences in identification performance between environmental subcategories might be sufficient to produce patterns of neural selectivity."

In Childhood, a series of studies using behavioral picture–sound matching paradigms has shown that environmental sounds and spoken language comprehension generally go hand-in-hand over development. Cummings, Saygin, Bates, and Dick (2009) tested 60 infants (ages 15–25 months) listening to environmental sounds or spoken phrases in a preferential looking paradigm, where infants were presented with two photographs side by side and then heard an environmental sound or spoken phrase corresponding to one of the objects; their looking time to the correct picture was the dependent measure. Infants' looking accuracy for environmental sounds was correlated with that for spoken language after taking into account the effects of age. Regression analyses showed that environmental sounds comprehension improvements were associated with chronological age, whereas spoken phrase comprehension was associated with productive vocabulary as measured by the MacArthur-Bates Communicative Development Inventory. Using a match-to-sample sound–picture paradigm (Saygin et al., 2005) with children 6 to 18 years of age, Dick, Borovsky, Cummings, Trauner, and Saygin (unpublished data) found that RTs to environmental sounds and spoken phrases were tightly correlated, even after accounting for age-related decreases. Using the same behavioral paradigm, Saygin et al. (2005) found that RTs for the two domains were only marginally correlated in early and later adulthood. Nonetheless, studies involving adults with cochlear implants (or who were hearing simulated cochlear implant input) have shown that training with environmental sounds can generalize to language and speech comprehension to a surprising degree (Loebach & Pisoni, 2008; Shafiro, Sheft, Gygi, & Ho, 2012), suggesting that there are shared processing resources underlying the two domains.

89.3.3 Semantic and Conceptual Priming with Environmental Sounds

As shown by these recognition and identification studies, even very young listeners can extract consistent meaningful information from environmental sounds. A number of behavioral and electroencephalography (EEG)/ERP studies have asked how this semantic, conceptual, or "real-world" information conveyed by environmental sounds is processed and integrated with subsequently encountered information.

[9]Lemaitre, Houix, Misdariis, and Susini (2010) reported that "sound experts" (musicians, sound engineers, and so forth) tended to weight such acoustic properties in their sound categorization ratings somewhat more heavily than everyday listeners.

Many of these studies have also asked how similar or distinct semantic and conceptual processing of environmental sounds might be to that observed in spoken language comprehension. Typically in these studies a standard semantic priming paradigm is used in which a target word (written or spoken) is preceded by a prime stimulus that varies in terms of its semantic relationship with the target. A half-century of psycholinguistics research has shown that targets that are semantically related to the prime should show faster RTs and higher accuracy than unrelated (or less-related) targets. In the ERP literature, the N400, a negative wave peaking at approximately 400 ms post-stimulus onset, is commonly used as an indicator of semantic integration of the incoming word with the foregoing content: the more explicit the expectation for the next word, the larger the N400 amplitude for words violating the expectation (Kutas & Federmeier, 2000). The N400 can also be elicited by mismatching meaningful stimulus pairs: two words, a sentence and word, two pictures, or a picture and a word.

In general, priming and particularly ERP studies have shown very similar semantic effects with environmental sounds, alone and when directly compared with language stimuli. RT measures have shown that environmental sounds can prime semantically related words (Chiu & Schacter, 1995; Frey, Aramaki, & Besson, 2014; Orgs, Lange, Dombrowski, & Heil, 2006; van Petten & Rheinfelder, 1995) and pictures (Chen & Spence, 2011; Schneider et al., 2008), and may also prime other semantically related sounds (Stuart & Jones, 1995, but see Chiu & Schacter, 1995; Frey et al., 2014; Friedman, Cycowicz, & Dziobek, 2003). Written words (Orgs et al., 2006; Pizzamiglio et al., 2005) and pictures (Schneider et al., 2008) have also been reported to prime related environmental sounds (but see Chiu & Schacter, 1995; Schön, Ystad, Kronland-Martinet, & Besson, 2010). Using a masked visual word prime and environmental sound target, Galati et al., (2008) showed that cross-domain priming between words and environmental sounds can be highly semantically specific in that RTs to an identity prime (the masked word "laugh" followed by the sound of laughing) are faster than both within-category pairs ("whistle" followed by laughing) and across-category pairs ("boiling" followed by laughing).

Among ERP studies that use environmental sounds, there are quite consistent effects of semantic or conceptual relatedness on the N400. An "N400 effect" (more negative-going N400 waveform for a target that is conceptually unrelated to a context or prime compared with a related target) has been observed in several combinations. N400 semantic relatedness effects have been observed for short environmental sound targets preceded by pictures (Cummings, Ceponiene, Dick, Saygin, & Townsend, 2008; Cummings et al., 2006; Plante, van Petten, & Senkfor, 2000), longer environmental sounds (Schirmer, Soh, Penney, & Wyse, 2011), visually presented words (Orgs et al., 2006; Orgs, Lange, Dombrowski, & Heil, 2007; Schön et al., 2010; van Petten & Rheinfelder, 1995; Wu, Athanassiou, Dorjee, Roberts, & Thierry, 2012), and spoken words (van Petten & Rheinfelder, 1995, but see Frey et al., 2014). N400 relatedness effects with environmental sounds as primes have also been observed for target spoken words (Frey et al., 2014; Plante et al., 2000; Van Petten & Rheinfelder, 1995) and visually presented words (Orgs et al., 2006).[10]

Direct comparisons between environmental sound-evoked and language-evoked N400s have tended to show very similar overall N400 profiles (see Figure 89.1, adapted from Cummings et al., 2006) but with a somewhat earlier onset and more anterior distribution of the N400 effect for semantically unrelated environmental sounds targets compared with words (Cummings et al., 2006; Orgs et al., 2006; but Schön et al., 2010; van Petten

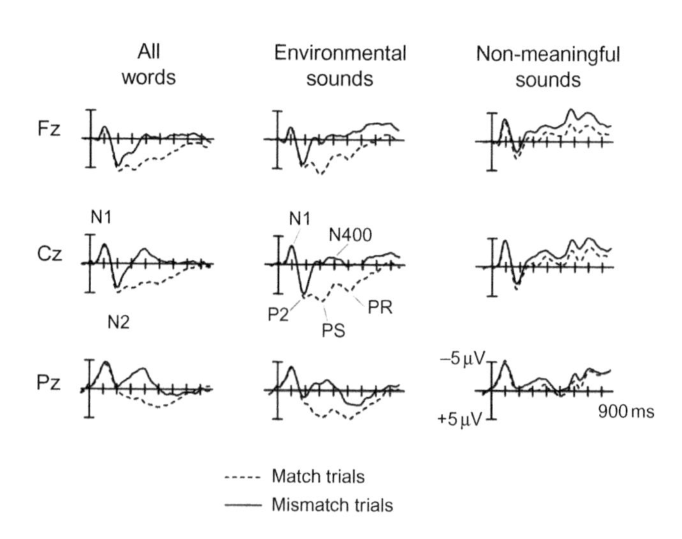

FIGURE 89.1 Matching and mismatching ERP responses to words, environmental sounds, and nonmeaningful sounds. *Redrawn from Cummings et al. (2006) with permission from the publisher.*

[10]Other ERP components have also been investigated with environmental sounds—for results and review, see Cummings et al. (2006), Schirmer et al. (2011), and Frey et al. (2014), among other studies. In addition, environmental sounds have been shown to evoke repetition priming effects in EEG and fMRI studies (Bergerbest, Ghahremani, & Gabrieli, 2004; Murray, Camen, Spierer, & Clarke, 2008). EEG with environmental sounds has also shown quite early discrimination of sounds derived from living versus human-made objects (Murray, Camen, Gonzalez Andino, Bovet, & Clarke, 2006) and differential localization of mouth-related and hand-related sounds (Pizzamiglio et al., 2005).

& Rheinfelder, 1995).[11] Some studies have found differential lateralization of N400 responses to environmental sounds and words (Plante et al., 2000; van Petten & Rheinfelder, 1995), but others have not (Cummings et al., 2006; Orgs et al., 2006; Schirmer et al., 2011). N400 effects with both linguistic and environmental sound stimuli are also similarly modulated by task demands (Frey et al., 2014; Wu et al., 2012).

There has been some question about whether such cross-domain similarities might be driven by covert naming of environmental sounds. However, this explanation does not have much empirical support. As mentioned, participants show longer RTs when asked to covertly name sounds during a sound–picture matching task, suggesting that this is not a default strategy for most participants (Dick, Bussiere, & Saygin, 2002). In addition, synthesized nonlinguistic sounds that are not easily associated with verbal labels evoke very similar N400 effects (Schön et al., 2010). Even novel cross-modal associations between "jagged" and "smooth" sounds and shapes can evoke a small N400-like effect (Cummings et al., 2006).

In contrast to results with typically developing individuals, some ERP studies with language-impaired (LI) and autistic children have shown cross-domain differences in N400 effects on spoken language and environmental sounds. Cummings & Ceponiene (2010) found that children with LI showed typical N400 effects of semantic incongruency for environmental sound/picture mismatches, but showed a delayed N400 latency to spoken phrase/picture mismatches when compared with typically developing age-matched children. McCleery et al., 2010 used a very similar paradigm with young highly functioning autistic children (ages 4–7) and found an even more striking difference: autistic children showed a typical N400 effect with environmental sound/picture mismatches but no discernible N400 effect with spoken word/ picture mismatches.[12] Thus, there is some indication that in certain developmental disorders, language and environmental sounds processing can become detoggled—a result that contrasts with much work with acquired language disorders such as aphasia.

89.3.4 Context Effects in Environmental Sounds Processing

Linguistic context affects performance in phoneme monitoring, lexical decision, and naming tasks (reviewed in Aydelott, Leech, & Crinion, 2010). In the speech perception literature, the effects of context are more obvious when ambiguity is present or when the target stimuli are impoverished. Ballas and Mullins (1991) investigated the effects of "semantic" contexts on the identification of ambiguous environmental sounds. They conducted a series of experiments where listeners were played nearly homonymous sounds (for instance, a fuse burning and food frying) and found that these sounds were more accurately identified when presented in isolation or embedded in consistent contexts (food frying in the context of kitchen sounds) relative to when context was biased toward the other alternative (food frying in a "fireworks" context).

The effects of competing backgrounds have also been extensively studied in speech perception and are often referred to as the cocktail party effect. Environmental sound identification also occurs in the presence of similar competing sounds and backgrounds. In everyday listening situations, listeners do not identify environmental sound targets in isolation, but from within a background of multiple competing sound sources. As demonstrated by Ballas and Mullins (1991), listeners are aware of the likelihood of certain sounds occurring in certain contexts because of their everyday listening experience, for instance, cow moos are likely to occur in the context of a barnyard than in an office.

When these more naturalistic listening conditions are simulated, the detection and identification of familiar unambiguous environmental sounds are enhanced in the presence of contextually *incongruent* environmental scenes in adults (Gygi & Shafiro, 2011; Leech et al., 2009) and school-age children (Krishnan et al., 2013). Gygi and Shafiro (2011) showed that the advantage for contextually incongruent sounds was level-dependent and interacted with the listener's familiarity with the sounds and the background. Congruent backgrounds might produce greater interference by increasing the uncertainty about the separation of target from background in a manner not dissimilar from speech. For example, listeners are adversely affected by competing speech in their native language (Van Engen & Bradlow, 2007).

In vision, some studies suggest that a contextual pop-out effect also occurs when viewing natural scenes, that is, inconsistent objects pop-out in these scenes (Underwood & Foulsham, 2006) However, other researchers have suggested that these differences can be accounted for by low-level visual perception (Võ & Henderson, 2011). Leech et al. (2009) investigated the low-level acoustic and perceptual features that may play a role in identifying environmental sounds within

[11]Interestingly, a developmental ERP study by Cummings et al. (2008) showed that this earlier latency for environmental sound–related N400 effects slowly emerges with increasing age, with latencies decreasing from young school-age children through adolescence to early adulthood.

[12]This is in contrast to results from behavioral results from the study by van Lancker et al. (1988) involving autistic children who showed similar performance on the two domains in a sound-to-picture matching task.

a context. In addition to experimenter-defined factors of congruence or incongruence, they found that spectral measures, such as the mean, standard deviation, and range of the sound's pitch, as well as temporal autocorrelation measures were significant predictors of the advantage for incongruent over congruent sounds. The results of these studies indicate that the identification of environmental sounds in context involves the integration of perceptual, attentional, and real-world knowledge.

89.4 SECTION TWO: NEURAL FACTORS IN PROCESSING ENVIRONMENTAL SOUNDS

89.4.1 Behavioral Studies on Hemispheric Asymmetries in Environmental Sounds Comprehension

Before exploring the literature on brain damage and environmental sound comprehension, we briefly review early studies on dichotic listening, ones that have been instrumental in establishing our notions regarding hemispheric differences for processing different aspects of meaningful sounds, both linguistic and nonlinguistic. The logic of these experiments is based on preferential projections from the cochlea to the contralateral hemisphere. In an important early study, Curry (1967) compared recall of words and environmental sounds in left-handed and right-handed English speakers. He presented triplets of spoken word or nonword pairs, with each word in a pair presented to only one ear; after listening to the word pairs, subjects had to list all the words that they recalled. They also listened to dichotically presented pairs of short sounds and wrote down the names of the sounds they heard. Curry found that right-handed individuals showed a significant right ear advantage (REA) for words (and an even more sizeable one for nonwords), with a significant left ear advantage (LEA) for identifying environmental sounds. This suggested that meaningful sounds were preferentially processed in the left hemisphere if they were linguistic and preferentially processed in the right if nonlinguistic. Complicating the story slightly, he also showed that left-handed individuals showed a weak but significant REA for words and no significant ear advantages for either nonwords or environmental sounds.[13]

When might these differential ear advantage effects for spoken language and environmental sounds emerge over development? Knox and Kimura (1970) conducted a comprehensive set of dichotic listening studies with large samples of children ages 5 to 8 years. Here, children listened to dichotically presented digit pairs, environmental sound pairs, and animal sounds pairs, as well as performed two word–picture match-to-sample tasks with minimal pairs, with one word presented to each ear. For all language tasks, children showed a considerable advantage in recalling or perceiving words presented to the right versus the left ear. By contrast, environmental sound identification showed a weak but significant LEA, with no real lateralization when listening to animal sounds. These data suggest that a hemispheric division of labor emerges for processing meaningful sounds by the beginning of the school years. However, as seen in neuropsychological and neuroimaging studies, hemispheric differences are less cut-and-dried than one might be led to believe from these dichotic listening studies.

89.4.1.1 Effects of Brain Damage on Environmental Sounds Processing

A specific deficit in recognizing environmental sounds is often termed *nonverbal auditory agnosia*. It (like word deafness) is a very rare phenomenon usually associated with bilateral (Albert, Sparks, Stockert, & Sax, 1972; Kazui, Naritomi, Sawada, Inoue, & Okuda, 1990; Spreen, Benton, & Fincham, 1965) and, rarely, with unilateral left (Saygin, Leech, & Dick, 2010) or right hemisphere (Fujii et al., 1990) lesions. Experimental studies with larger groups of patients (Saygin, Dick, Wilson, Dronkers, & Bates, 2003; Schnider, Benson, Alexander, & Schnider-Klaus, 1994; Varney & Damasio, 1986) have shown that dissociations between verbal and nonverbal domains are the exception rather than the norm.

Two forms of nonverbal auditory agnosia have been proposed: (i) *perceptual-discriminative*, with patients failing to identify whether two consecutive sounds are identical and (ii) *associative-semantic*, with patients being impaired at audiovisual matching or naming. Early work did not reveal clear lesion correlates of these agnosia types. Bilateral lesions have been implicated in severe discriminative disorders (Albert et al., 1972; Kazui et al., 1990; Lechevalier et al., 1984; Mendez & Geehan, 1988; Motomura, Yamadori, Mori, & Tamaru, 1986; Rosati et al., 1982; Taniwaki, Tagawa, Sato, & Lino, 2000; Vignolo, 1982). Unilateral right hemisphere lesions can lead to normal association with impaired discrimination (Eustache, Lechevalier, Viader, & Lambert, 1990; Vignolo, 1982), deficient association with normal discrimination (Spreen et al., 1965), or deficient association and deficient discrimination (Fujii et al., 1990). Unilateral left hemisphere

[13]The right ear advantage for listening to linguistic sounds may not hold cross-linguistically, at least in some circumstances (Hatta & Dimond, 1981).

lesions have been linked to deficient association (Saygin et al., 2003) and normal discrimination (Vignolo, 1982), although in most studies with this sample discrimination was not tested on left hemisphere–injured patients because of a focus on language processing (Saygin et al., 2003; Varney, 1980; Varney & Damasio, 1986). Vignolo, Spinnler, and Faglioni reported disturbances of environmental sound recognition after unilateral hemispheric damage in a group of patients (Faglioni, Spinnler, & Vignolo, 1969; Spinnler & Vignolo, 1966). They noted that right hemisphere–damaged (RHD) patients tended to perform significantly worse than controls on perceptual tests involving environmental sounds and that left hemisphere–damaged (LHD) patients performed significantly worse on associative or semantic tests.

Several studies have directly compared environmental sounds and spoken language comprehension after stroke. In 1980, Varney used environmental sounds to examine both verbal and nonverbal comprehension deficits in the same aphasic patients and found that impairments in environmental sound recognition were seen only in subjects with impaired verbal comprehension, and that aphasic patients with intact verbal comprehension also performed well on sound recognition. There were some patients who were impaired in verbal comprehension, but not in sound recognition. More recently, Schnider et al. (1994) observed that both LHD and RHD patients performed significantly worse than a group of normal controls on an environmental sound recognition test. They found no significant differences in the performances of the two patient groups; however, the pattern of errors appeared to differ across groups. LHD patients made more semantically based errors, whereas RHD patients and control subjects made almost exclusively acoustic errors. For all patients, accuracy in recognizing environmental sounds correlated with language comprehension as measured by the Western Aphasia Battery (Kertesz, 1979). Lesion behavior correlations showed that LHD patients with impaired environmental sound recognition tended to have damage to the posterior superior temporal gyrus (pSTG) and the inferior parietal lobe.

Although these studies support a link between aphasia and nonverbal auditory agnosia, none of these were precise comparisons between verbal and nonverbal auditory processing in the same patients and did not control for factors such as stimulus frequency and identifiability or the relationship between the auditory and visual stimuli. In a large neuropsychological and lesion-mapping study, our group addressed these gaps in knowledge and assessed the relationship between verbal and nonverbal comprehension of complex, meaningful information in the auditory modality by examining aphasic patients' abilities to match environmental sounds and corresponding linguistic phrases to associated pictures (Saygin et al., 2005; Saygin et al., 2003). Task demands, stimulus characteristics, and semantic features were all carefully controlled to reduce confounds and to focus on the relationship between verbal and nonverbal domains. Results from 30 LHD and 5 RHD patients (along with 21 neurologically intact age-matched control subjects) showed that RHD patients were only mildly impaired in the task, performing similarly to mild (anomic) aphasic LHD patients but significantly worse than controls. Patients with more severe aphasia (LHD) were impaired to the same extent in comprehending speech and environmental sounds, as measured by both accuracy and RT measures. Performance within the task between the two domains went hand-in-hand, with a strong correlation between accuracy and RT for speech and environmental sounds, suggesting that the two domains utilize some shared perceptual and neural mechanisms. Both lesion overlays (Saygin et al., 2003) and Voxel-Based Lesion-Symptom Mapping (Bates, Wilson, et al., 2003) demonstrated that damage to posterior regions in the left middle and superior temporal gyri and damage to the inferior parietal lobe were predictors of deficits for both speech and environmental sounds. Brodmann's area 22 and the surrounding middle temporal and inferior parietal regions (encompassing Wernicke's area) were also implicated in environmental sound processing. In fact, Wernicke's area itself, in the pSTG, was more strongly associated with performance in the nonverbal domain than in the verbal domain (Saygin et al., 2003).

It is not just within the stroke literature that impairments in spoken language and environmental sounds (ES) have been shown to be comorbid. Recent studies with neurodegenerative disorders (e.g., primary progressive aphasia and semantic dementia) have also revealed concomitant deficits in verbal and nonverbal meaningful sound processing (Bozeat, Lambon Ralph, Patterson, Garrard, & Hodges, 2000; Goll et al., 2010; Hsieh, Hornberger, Piguet, & Hodges, 2011).

Although an association is the norm between speech and nonverbal sounds, dissociations in neuropsychology can also be informative, even when rare (Bates, Appelbaum, Salcedo, Saygin, & Pizzamiglio, 2003; Bates, Saygin, Moineau, Marangolo, & Pizzamiglio, 2005). Varney (1980) had reported deficits in nonverbal comprehension only in patients who also exhibited deficits in verbal comprehension, but they did not find dissociations in the opposite direction. Clarke, Bellmann, De Ribaupierre, and Assal (1996) reported a patient who was deficient in the nonverbal auditory domain but had no diagnosed verbal comprehension deficits (although formal testing was not reported). When both the task and stimuli were closely matched across domains, as we have done, deficits in the two domains were largely associated with each other.

We assessed dissociations between environmental sounds and verbal comprehension both quantitatively and statistically (Bates, Appelbaum, et al., 2003; Bates et al., 2005) in the large patient sample of Saygin et al. (2003). Doing so, we found three outliers, of which only one, Patient M, showed dissociation between the two domains. M, a right-handed individual who suffered a left hemisphere stroke, had persistent difficulty with environmental sound comprehension but had recovered language function to within normal levels (Saygin et al., 2010). Detailed behavioral assessment verified that the patient exhibited persistent and severe auditory agnosia for nonverbal sounds in the absence of verbal comprehension deficits or peripheral hearing problems. Acoustical analyses suggested that his residual processing of a minority of environmental sounds might rely on his speech processing abilities. In the patient's brain, contralateral (right) temporal cortex as well as perilesional (left) anterior temporal cortex were strongly responsive to verbal, but not to nonverbal, sounds, a pattern that stood in marked contrast to the data of the controls, suggesting a selective reorganization of auditory processing for speech but not environmental sounds processing.

In summary, performance in verbal and nonverbal domains is, in general, highly correlated after brain injury. However, not only is it possible to identify patients who perform worse in the verbal domain (i.e., the expected result based on an aphasic sample; Varney, 1980) but also we can reliably identify patients who perform worse in the nonverbal domain, an unexpected and rarely reported outcome (Saygin et al., 2010). It is possible that these dissociations are due to variation between individuals' premorbid brain organization for these functions, as well as nonuniform poststroke recovery patterns across patients and across domains.[14]

89.4.2 Functional Neuroimaging of Environmental Sounds

Neuropsychological and lesion symptom mapping approaches are extremely useful for understanding what brain regions are necessary for processing environmental sounds, as well as for characterizing how plastic the brain's organization for understanding meaningful sounds may be after injury and subsequent learning and retraining. Complementary functional neuroimaging studies can tell us which brain networks are *involved* in environmental sounds (and shared with spoken word) comprehension. Equally, neuroimaging provides much finer-grained information about how these networks change in response to task demands and stimulus properties, allowing us to investigate how the brain decomposes and represents acoustical, perceptual, and cognitive dimensions of environmental sounds.

89.4.2.1 "Passive" Listening Neuroimaging Studies

In the simplest neuroimaging studies, participants were asked to listen to a series of single unordered environmental sounds without making any behavioral responses or covert judgments, with activation compared with rest. An early positron emission tomography (PET) regional cerebral blood flow (rCBF) study by Engelien et al. (1995) revealed bilateral but slightly right-lateralized perisylvian activation in the STG, superior temporal sulcus (STS), anterior insula/frontal operculum, and inferior parietal cortex. Using event-related fMRI, Specht and Reul (2003) also showed bilateral but more right-lateralized activation in primary and secondary auditory cortices, the superior temporal sulcus, and the lingual gyrus, and bilaterally in the supplementary motor area and precentral gyri. In a block-design study, Leech and Saygin (2011) showed bilateral but somewhat *left-lateralized* activation all along the temporal plane. Humphries et al. (2001) had participants passively listen to longer, artificially created auditory "events" (like the sound of a gun followed by the sound of fading footsteps) and found extensive bilateral activation in the middle and posterior portions of the STG/STS, along with the inferior frontal gyri. Engelien et al. (2005) compared PET rCBF for passive listening to environmental sounds that were presented either intact or temporally "scrambled" but with their overall spectrum and amplitude envelope intact, thereby rendering the sounds meaningless and unfamiliar. Temporally scrambled sounds evoked almost entirely right STG/STS/inferior frontal gyrus (IFG) activation, whereas the comparison of intact (meaningful/identifiable) minus scrambled sounds showed more left anterior STG/STS, IFG, and anterior parahippocampal gyrus activation, along with right anterior STG and mid-orbito-frontal gyrus activation.

Several of these passive listening paradigms also compared environmental sounds with spoken language activation, showing largely overlapping systems

[14]Dissociations have also been observed for different classes of environmental sound. Pazzaglia, Pizzamiglio, Pes, and Aglioti (2008) found that patients with buccofacial or limb apraxia showed deficits in environmental sounds comprehension related to their apraxia subtype, with buccofacial apraxics having somewhat greater difficulties with mouth-related sounds than limb-related sounds, and limb apraxics showing the converse pattern of deficits.

but with regional relative increases for either ES or spoken language. Specht and Reul (2003) compared the environmental sounds activation described with that of one- to two-syllable words spoken by a single talker (as well as with single tones). This revealed different graded activation profiles with: (i) the left orbital IFG and posterior STS and bilateral mid-STS/middle temporal gyrus (MTG) showing activation for words greater than environmental sounds and (ii) the left and right transverse gyri and planum temporale showing the opposite profile, with environmental sounds greater than spoken words. Humphries et al. (2001) compared their environmental sound sequences to sentential descriptions of the sequences and found more activation for sentences *bilaterally* in the anterior and mid-STS, as well as in the left inferior frontal and prefrontal gyri. Environmental sounds only showed more activation than sentences in the left inferior central sulcus and the right IFG. Leech and Saygin (2011) compared their environmental sound blocks to matched spoken phrases and also showed greater speech than environmental sound activation in bilateral STG/STS/MTG, as well as left anterior orbital IFG. In contrast to Humphries et al., they found more activation for environmental sounds than spoken phrases in medial auditory cortex. Leech and Saygin (2011) compared this simple subtraction-based analysis to a multivariate pattern analysis, which showed that finer-grained patterns of activation and deactivation within much of the left and right perisylvian cortex can disambiguate which class of sounds the listener is hearing, even if there is no overall activation advantage for one class within the region. Furthermore, these distributed patterns of activation to environmental sounds and language tend to be quite different across individuals. A promising model for understanding such differences—and what they might tell us about the functional organization cortex—might have been the approach of Charest, Kievit, Schmitz, Deca, and Kriegeskorte (2014), who showed that individual differences in visual representational structure in ventral inferior temporal regions was related to participants' individual interactions and associations with the represented objects.

89.4.2.2 "Active"/Task-Based Listening Neuroimaging

To our knowledge, all other PET and fMRI studies of environmental sounds use "active" tasks, whereby participants make a decision or judgment during scanning. Patterns of environmental sounds activation are strongly influenced by task and cognitive demands, even quite subtly different ones, although the trade-off is that this makes it more difficult to disambiguate "pure" sound processing from a range of metacognitive and attentional factors involved in performing the task. An early demonstration of this was in the Engelien et al. (1995) PET study, where a comparison of the passive listening task with one where participants performed covert categorization of the sounds showed strongly left-lateralized increases during categorization in left prefrontal and frontal regions, along with categorization-related left-lateralized activation in inferior parietal and middle temporal regions. Lewis et al. (2004) compared activation for forward and reverse environmental sounds while participants pressed buttons to indicate whether they had recognized the sound (without any visual stimulation). They showed extensive perisylvian activation for environmental sounds and considerably greater activation for forward than for backward sounds in highly left-lateralized IFG, posterior STS, and anterior fusiform regions along with right posterior STS and right orbital IFG (see also Lebrun et al., 2001). Lewis et al. (2004) compellingly demonstrated that the greater activation for forward versus backward sounds are related to perceptual rather than acoustic properties because comparing recognized to unrecognized *forward sounds* showed the same pattern of activation differences as forward versus backward sounds. Moreover, there were no significant differences between *unrecognized* forward and backward sounds in these regions. Using overlays of cognitive and visual activation from other studies, Lewis et al. (2004) also showed that environmental sound recognition involves multiple inferior and prefrontal, inferior parietal, and posterior middle/superior temporal regions that are involved in high-level spoken word semantics (Figure 89.2). Hocking, McMahon, and De Zubicaray (2011) showed that environmental sound activity in these regions is differentially modulated by quite specific types of category judgments. For instance, the anterior fusiform was selectively activated by making more visually based judgments (again without any systematic visual stimulation), and activation in the angular gyrus and retrosplenial cortex was modulated by making semantic judgments (i.e., whether the sound was generated by animals who live in Australia).

89.4.2.3 "Active" Tasks Comparing Environmental Sounds and Language

A number of studies using active tasks without a visual component have directly compared environmental sound and language-related activation. An early PET study by O'Leary et al. (1996) compared healthy participants' rCBF with consonant–vowel–consonant (CVC) words, CVC nonwords, and short environmental sounds. They showed no quantitative differences between activations in these conditions, but they did demonstrate that attending to input from one ear

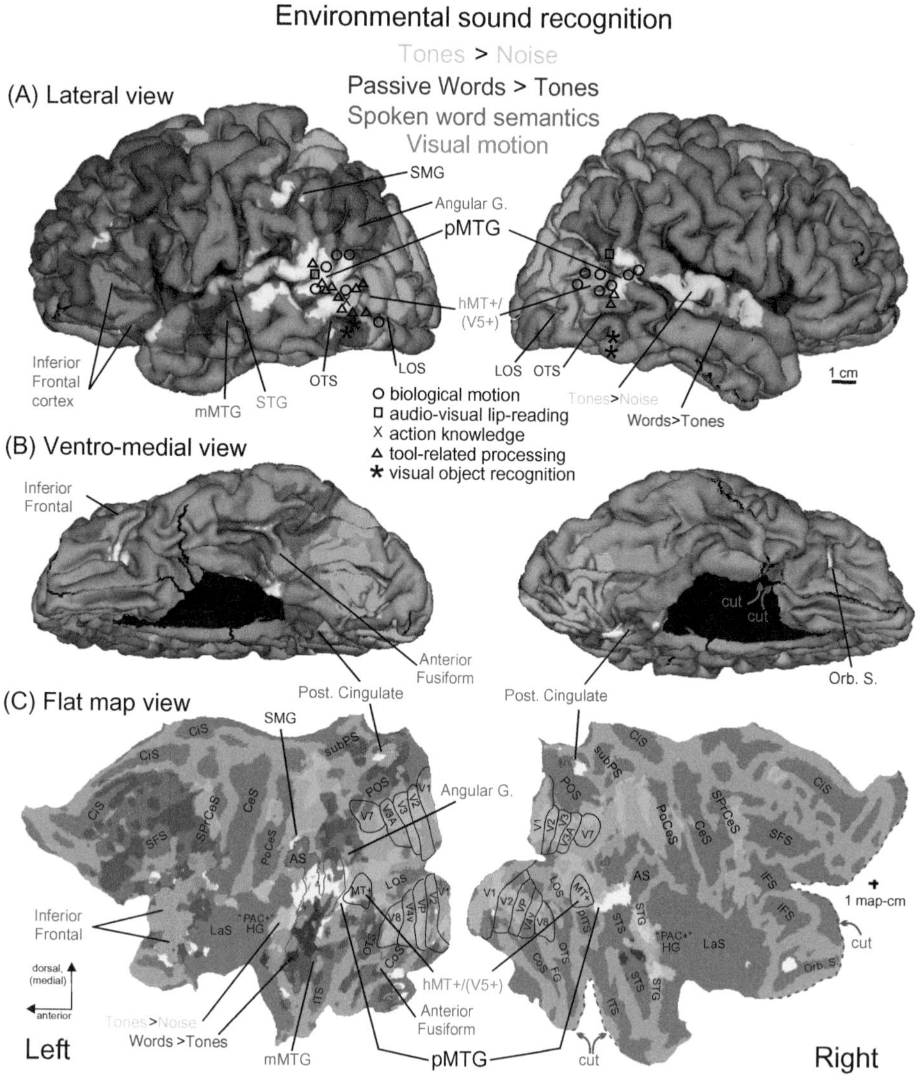

FIGURE 89.2 A comparison of environmental sound activation (yellow) with other tasks reported in the literature. Blue hues indicate early acoustic processing, including passive listening to tones versus white noise (light blue) and spoken words versus tones (dark blue). Red indicates semantic processing of spoken words (animal names). Green indicates visual motion processing of coherent versus random dot displays. Regions of overlap are indicated by intermediate colors. Symbols indicate previously reported activation centroids in Talairach coordinates (projected to the outer cortical surface for visibility). *Reprinted with permission from Lewis et al. (2004).*

substantially affects activation lateralization across domains. For instance, attending to the left ear when either environmental sounds or nonword CVCs are presented evoked very similar right-lateralized activation in the anterior STG/STS. Overlapping activation for environmental sounds and language activation was also reported by Visser and Lambon Ralph (2011). This fMRI study asked participants to perform a living/nonliving categorization task on diverse environmental sound and language stimuli (compared with low-level control conditions) and showed remarkably similar activity in both domains in the left and right anterior STG/STS, similar to the results of Engelien et al. (1995), but more activation for language than environmental sounds in a left anterior ventral temporal area known to be affected in semantic dementia.

In a PET study using a spoken recognition or verification task with environmental sounds, words, and meaningless syllables, Giraud and Price (2001) showed generally overlapping perisylvian networks, but they found that words and syllables evoked more activation than environmental sounds in the left aSTG and pSTG and very posterior STS. Building on these results, Thierry, Giraud, and Price (2003) used similar stimuli as Humphries et al. (2001), whereby participants listened to constructed sequences of environmental sounds or phrases that would or would not make sense conceptually depending on their order of presentation (e.g., the sequence of events in a car crash). Participants performed two different tasks. In the "easy" task they indicated if there was an animal in the sequence; in the "hard" task, they indicated whether the entire sequence made sense. As with previous studies, Thierry et al. found activation overlap for verbal phrases and sounds in the left IFG, cerebellum, and all along the left STG, right STS, and right cerebellum. Comparison across domains showed more activation for language than environmental sound in the left STS/STG, with the opposite pattern in right posterior STG. The authors performed a direct comparison with the Giraud and Price (2001) data and showed that the right hemisphere environmental sound greater than language activation was much more prominent in the sequence-judgment paradigm than in the single sounds recognition paradigm of Giraud and Price. Thierry et al. (2003)'s environmental sounds greater than language pattern was in the same right pSTG/STS region where Giraud and Price (2001) showed more activation for words than environmental sounds.

Cross-modal paradigms have also been used to compare environmental sound and language-related activation. Dick et al. (2007) used a picture–sound matching paradigm very similar to that of Saygin et al. (2003), whereby participants saw two semantically unrelated pictures and heard either an environmental sound or a short Italian phrase corresponding to one of the pictures; subjects pressed a button to indicate which picture matched the sound or phrase. Relative to the control condition (a simple match-to-sample task with nonsense shapes and tones), both environmental sounds and language stimuli evoked significant bilateral activation in the inferior frontal gyri, superior temporal gyri (anterior/transverse/posterior), and posterior middle and inferior temporal and fusiform gyri. Here, language stimuli evoked more activation than environmental sounds in the middle left MTG/STS/STG, the anterior left *and* right STG, and *right*-lateralized lateral fusiform gyrus. Environmental sounds evoked more activation than did language stimuli in patches along the right planum temporale, anterior, and superior-most extent of the right supramarginal gyrus, and the right IFG.

Dick et al. also analyzed relative lateralization for both domains across a number of language-associated perisylvian regions. Language-evoked activation was significantly L > R in all region of interests (ROIs) except the angular and supramarginal gyri, the transverse gyri, and the opercular part of the IFG. Unlike language, environmental sounds showed significant L > R activation only in the inferior temporal gyrus (ITG). However, environmental sounds did *not* show any significant lateralization effects in the opposite (R > L) pattern. Finally, using data derived from the lesion-mapping study of Saygin et al. (2003), Dick et al. (2007) also found that the more consistent the mapping between environmental sounds behavioral deficit and lesion was in a given left hemisphere region, the more environmental sounds activation they observed in typical participants. Notably, this held true not only *within* but *across* domains—that is, lesion maps associated with environmental sound behavioral deficits predicted degree of fMRI activation for matched verbal phrases sounds in healthy controls.

89.4.2.4 fMRI Studies of Cross-Modal Priming with Environmental Sounds

In an fMRI cross-modal and cross-domain priming paradigm that paralleled the behavioral and ERP/EEG studies reviewed, Noppeney, Josephs, Hocking, Price, and Friston (2008) presented participants with written words or pictures, and then with sounds or spoken words that were semantically congruent or incongruent with the visual display. To engage semantic processes, subjects made weight judgments about the sound source. Both spoken words and environmental sounds showed increased activation in the left inferior frontal sulcus and pre-supplementary motor area (pre-SMA) when semantically incongruent with the preceding written word or picture. Domain-specific effects were found in the left STS, where semantic incongruency effects were greater for spoken words than for environmental sounds, whereas the opposite effect was found in left posterior parietal cortex (environmental sounds >> words). These results contrast somewhat with those of Galati et al. (2008), who found action sound–specific increases for *congruent* versus incongruent written-word/environmental sound pairs in the left inferior frontal sulcus (IFS) as well as in the left ventral inferior frontal gyrus. It is possible that these results may be due to the particular stimuli used by Galati et al. (2008), who used highly somatotopically specific classes of environmental sounds. In addition to their semantic congruency effects, they found that "hand-related" sounds like clapping showed more activation than mouth-related vocalization-type sounds in the left inferior frontal sulcus, whereas the converse held true in bilateral STG/STS, where mouth-related sounds were greater than hand-related sounds. Gazzola, Aziz-Zadeh, and Keysers (2006) also demonstrated that mouth motor and somatosensory regions showed fMRI BOLD activation for mouth-related sounds than hand-related sounds, whereas hand movement–related somatomotor areas show activation for hand-produced sounds over mouth-produced sounds.

89.4.2.5 Categorical Representations of Sound Categories

The representation of sound category—particularly with reference to the body part involved in generating the sound—has been of particular interest in environmental sound research, given the relevance to "mirror neurons" (Rizzolatti, Luppino, & Matelli, 1998) and theories of embodiment in cognition (Glenberg & Robertson, 2000). In an influential fMRI study, Lewis, Brefczynski, Phinney, Janik, and DeYoe (2005) asked healthy participants to listen to a series of animal-related and tool-related sounds with their eyes closed and to respond silently in their heads regarding whether the likely sound source was a tool or an animal. They found that animal sounds (primarily vocalizations) evoked significantly stronger activity than tool sounds along middle portions of the STG bilaterally (similar to the results of Galati et al., 2008), whereas tool sounds preferentially activated multiple left motor and somatosensory regions (mIFS, vPMC, iPS, AIP) along with the right pMTG/STS.[15] By comparing the sound activation results with those of a "virtual tool use" functional localizer with a subset of the same participants, Lewis and colleagues (Lewis et al., 2005; Lewis, Phinney, Brefczynski-Lewis, & DeYoe, 2006; Lewis et al., 2011) showed that these tool sound activations overlapped extensively with regions engaged with virtual tool manipulation, that the lateralization of the tool sound activations corresponded closely to the handedness of the individual, and that these activation patterns can be tuned by training. Moreover, they showed that animal sounds *misperceived* as tools showed "tool-like" activation, suggesting that it is the *percept* of tool use and not the acoustics driving this effect. In contrast, activation patterns of miscategorized animal vocalizations were very similar to those for correctly perceived animal vocalizations (with more activation in mSTG); therefore, this animal sound response preference is likely to be due to acoustic rather than perceptual processing (a question more thoroughly analyzed in Lewis et al., 2012).

Direct analyses of the effect of different environmental sound semantic and acoustic properties on activation patterns has been investigated by Giordano et al. (2012) and Lewis et al. (2012), who showed that acoustics of sound objects appear to be coded in different temporal regions. Giordano et al.'s multivariate pattern classification results with different acoustical properties of auditory objects suggested that lateral Heschl's and mid-STG regions tend to code for cross-object similarity in median pitch, with perceived loudness also preferentially encoded in left lateral Heschl's STG and with right anterior STG particularly sensitive to structure in the flux of an auditory object's spectral centroid. Giordano et al. (2012) also suggest that some categorical information about auditory objects—for instance, whether they are generated by living or nonliving objects, or by human/nonhuman agents—is represented in lateral pSTG. The combination of acoustical and informational analyses toggled with high-resolution fMRI and multivariate pattern analysis (MVPA) of environmental sounds have also been very useful in unveiling fundamental coding principles of human auditory cortex (Moerel, De Martino, & Formisano, 2012; Moerel et al., 2013). One promising avenue for future research is to use parametric variation of different perceptual dimensions inherent in environmental sounds to characterize the informational divisions that might underlie auditory functional organization (for an example of this approach, see Lewis et al., 2011).

In summary, the functional neuropsychological and neuroimaging findings converge on several key results. Both environmental sound and spoken language processing largely share neural resources centered on classical language and auditory processing regions. Functional neuroimaging suggests a bilateral temporal-parietal auditory system with a slight bias toward more right-lateralized processing for environmental sounds. However, there are circumstances when the two dissociate, which is accentuated by the specific auditory or cognitive task required. The neural organization of environmental sound processing is particularly dependent on cross-modal processes, such as somatosensory and visual information, and is dependent on actions associated with the sounds. More generally, the neural organization of environmental sounds processing in auditory and auditory association regions, such as those in the lateral STG, reflects basic acoustical properties of the sound.

89.5 CONCLUSION

Many of the processes involved in processing environmental sounds are highly similar to those for spoken language. Behaviorally, we see that, across the lifespan, we are experts at identifying environmental sounds, and that individual variability in performance on environmental sounds can resemble that observed in word comprehension. Further, factors such as frequency, length, and imageability affect ES comprehension as they do word comprehension. Equally, semantic and contextual factors are major determinants in ES comprehension. From a neural perspective, although much has been made of differences between spoken language processing and environmental sound processing (e.g., a right-lateralized bias for ES), the similarities are more striking. Well-controlled studies of patient groups suggest that, in the majority of cases, language and environmental sound processing breakdown together. Most neuroimaging studies reveal a shared bilateral network of regions for spoken language and environmental sounds, with within-environmental sound variability in activation often greater than the

[15]Using an fMRI adaptation paradigm, Doehrmann, Naumer, Volz, Kaiser, and Altmann (2008) showed a somewhat different pattern of tool and animal response preference, where adaptation to animal sounds was shown in bilateral anterolateral STG and transverse gyri, whereas adaptation to tools was most prominent in several patches in the left insula.

difference between environmental sound and spoken language networks.

These similarities in the neural "territory" for both environmental sounds and language are interesting given the considerable differences between environmental sounds—which are highly iconic, source-bound, and acoustically variable—and spoken language—with its complex syntactic organization, almost obligatory communicative content, and more abstract sound-to-meaning mapping. It is an exciting challenge for the neurobiology of language to understand how highly overlapping brain networks are able to learn and represent such disparate domains.

Acknowledgments

Thanks to James Lewis for useful comments and for providing permission to reprint Figure 89.2.

References

Albert, M. L., Sparks, R., Stockert, T. V., & Sax, D. (1972). A case study of auditory agnosia: Linguistic and non-linguistic processing. *Cortex*, *8*, 427–443.

Aydelott, J., Leech, R., & Crinion, J. (2010). Normal adult aging and the contextual influences affecting speech and meaningful sound perception. *Trends in Amplification*, *14*(4), 218–232. Available from: http://dx.doi.org/10.1177/1084713810393751.

Ballas, J. A. (1993). Common factors in the identification of an assortment of brief everyday sounds. *Journal of Experimenta Psychology: Human Perception and Performance*, *29*(2), 250–267.

Ballas, J. A., & Howard, J. H. (1987). Interpreting the language of environmental sounds. *Environment and Behavior*, *19*(1), 91–114.

Ballas, J. A., & Mullins, T. (1991). Effects of context on the identification everyday sounds. *Human Performance*, *4*(3), 199–219.

Balota, D. A., Yap, M. J., Cortese, M. J., Hutchison, K. A., Kessler, B., Loftis, B., et al. (2007). The English Lexicon project. *Behavior Research Methods*, *39*(3), 445–459.

Bates, E., Appelbaum, M., Salcedo, J., Saygin, A. P., & Pizzamiglio, L. (2003). Quantifying dissociations in neuropsychological research. *Journal of Clinical and Experimental Neuropsychology*, *25*(8), 1128–1153. Available from: http://dx.doi.org/10.1076/jcen.25.8.1128.16724.

Bates, E., Burani, C., D'Amico, S., & Barca, L. (2001). Word reading and picture naming in Italian. *Memory and Cognition*, *29*(7), 986–999.

Bates, E., D'Amico, S., Jacobsen, T., Szekely, A., Andonova, E., Devescovi, A., et al. (2003). Timed picture naming in seven languages. *Psychonomic Bulletin and Review*, *10*(2), 344–380.

Bates, E., Saygin, A. P., Moineau, S., Marangolo, P., & Pizzamiglio, L. (2005). Analyzing aphasia data in a multidimensional symptom space. *Brain and Language*, *92*(2), 106–116. Available from: http://dx.doi.org/10.1016/j.bandl.2004.06.108.

Bates, E., Wilson, S. M., Saygin, A. P., Dick, F., Sereno, M. I., Knight, R. T., et al. (2003). Voxel-based lesion-symptom mapping. *Nature Neuroscience*, *6*(5), 448–450. Available from: http://dx.doi.org/10.1038/nn1050.

Bergerbest, D., Ghahremani, D. G., & Gabrieli, J. D. E. (2004). Neural correlates of auditory repetition priming: Reduced fMRI activation in the auditory cortex. *Journal of Cognitive Neuroscience*, *16*(6), 966–977. Available from: http://dx.doi.org/10.1162/0898929041502760.

Bozeat, S., Lambon Ralph, M. A., Patterson, K., Garrard, P., & Hodges, J. R. (2000). Non-verbal semantic impairment in semantic dementia. *Neuropsychologia*, *38*(9), 1207–1215. Available from: http://dx.doi.org/10.1016/S0028-3932(00)00034-8.

Charest, I., Kievit, R. A., Schmitz, T. W., Deca, D., & Kriegeskorte, N. (2014). Unique semantic space in the brain of each beholder predicts perceived similarity. *Proceedings of the National Academy of Sciences of the United States of America*, *111*(40), 14565–14570. Available from: http://dx.doi.org/10.1073/pnas.1402594111.

Chen, Y.-C., & Spence, C. (2011). Crossmodal semantic priming by naturalistic sounds and spoken words enhances visual sensitivity. *Journal of Experimental Psychology Human Perception and Performance*, *37*(5), 1554–1568. Available from: http://dx.doi.org/10.1037/a0024329.

Chiu, C. Y. P., & Schacter, D. L. (1995). Auditory priming for nonverbal information: Implicit and explicit memory for environmental sounds. *Consciousness and Cognition*, *4*(4), 440–458. Available from: http://dx.doi.org/10.1006/ccog.1995.1050.

Clarke, S., Bellmann, A., De Ribaupierre, F., & Assal, G. (1996). Nonverbal auditory recognition in normal subjects and brain-damaged patients: Evidence for parallel processing. *Neuropsychologia*, *34*(6), 587–603.

Cummings, A., & Ceponiene, R. (2010). Verbal and nonverbal semantic processing in children with developmental language impairment. *Neuropsychologia*, *48*(1), 77–85. Available from: http://dx.doi.org/10.1016/j.neuropsychologia.2009.08.012.

Cummings, A., Ceponiene, R., Dick, F., Saygin, A. P., & Townsend, J. (2008). A developmental ERP study of verbal and non-verbal semantic processing. *Brain Research*, *1208*, 137–149. Available from: http://dx.doi.org/10.1016/j.brainres.2008.02.015.

Cummings, A., Ceponiene, R., Koyama, A., Saygin, A. P., Townsend, J., & Dick, F. (2006). Auditory semantic networks for words and natural sounds. *Brain Research*, *1115*(1), 92–107. Available from: http://dx.doi.org/10.1016/j.brainres.2006.07.050.

Cummings, A., Saygin, A. P., Bates, E., & Dick, F. (2009). Infants' recognition of meaningful verbal and nonverbal sounds. *Language learning and development: The official journal of the society for language development*, *5*(3), 172–190. Available from: http://dx.doi.org/10.1080/15475440902754086.

Curry, F. K. W. (1967). A comparison of left-handed and right-handed subjects on verbal and non-verbal dichotic listening tasks. *Cortex; A Journal Devoted to the Study of the Nervous System and Behavior*, *3*(3), 343–352.

Cycowicz, Y. M., Friedman, D., Rothstein, M., & Snodgrass, J. G. (1997). Picture naming by young children: Norms for name agreement, familiarity, and visual complexity. *Journal of Experimental Child Psychology*, *65*(2), 171–237. Available from: http://dx.doi.org/10.1006/jecp.1996.2356.

Dick, F., Bussiere, J., & Saygin, A. P. (2002). The effects of linguistic mediation on the identification of environmental sounds. *Center for Research in Language Newsletter*, *14*(3), 3–9.

Dick, F., Saygin, A. P., Galati, G., Pitzalis, S., Bentrovato, S., D'Amico, S., et al. (2007). What is involved and what is necessary for complex linguistic and nonlinguistic auditory processing: Evidence from functional magnetic resonance imaging and lesion data. *Journal of Cognitive Neuroscience*, *19*(5), 799–816. Available from: http://dx.doi.org/10.1162/jocn.2007.19.5.799.

Doehrmann, O., Naumer, M. J., Volz, S., Kaiser, J., & Altmann, C. F. (2008). Probing category selectivity for environmental sounds in the human auditory brain. *Neuropsychologia*, *46*(11), 2776–2786. Available from: http://dx.doi.org/10.1016/j.neuropsychologia.2008.05.011.

Engelien, A., Sibersweig, D., Stern, E., Huber, W., Döring, W., Frith, C., et al. (1995). The functional anatomy of recovery from auditory agnosia. *Brain: A Journal of Neurology, 118*(6), 1395–1409. Available from: http://dx.doi.org/10.1093/brain/118.6.1395.

Engelien, A., Tüscher, O., Hermans, W., Isenberg, N., Eidelberg, D., Frith, C., et al. (2005). Functional neuroanatomy of non-verbal semantic sound processing in humans. *Journal of Neural Transmission, 113*(5), 599–608. Available from: http://dx.doi.org/10.1007/s00702-005-0342-0.

Eustache, F., Lechevalier, B., Viader, F., & Lambert, J. (1990). Identification and discrimination disorders in auditory perception: A report on 2 cases. *Neuropsychologia, 28*, 257–270.

Fabiani, M., Kazmerski, V. A., Cycowicz, Y. M., & Friedman, D. (1996). Naming norms for brief environmental sounds: Effects of age and dementia. *Psychophysiology, 33*(4), 462–475. Available from: http://dx.doi.org/10.1111/j.1469-8986.1996.tb01072.x.

Faglioni, P., Spinnler, H., & Vignolo, L. A. (1969). Contrasting behavior of right and left hemisphere-damaged patients on a discriminative and a semantic task of auditory recognition. *Cortex, 5*, 366–389.

Frey, A., Aramaki, M., & Besson, M. (2014). Conceptual priming for realistic auditory scenes and for auditory words. *Brain and Cognition, 84*(1), 141–152. Available from: http://dx.doi.org/10.1016/j.bandc.2013.11.013.

Friedman, D., Cycowicz, Y. M., & Dziobek, I. (2003). Cross-form conceptual relations between sounds and words: Effects on the novelty P3. *Brain Research Cognitive Brain Research, 18*(1), 58–64.

Fujii, T., Fukatsu, R., Watabe, S., Ohnuma, A., Teramura, K., Kimura, I., et al. (1990). Auditory sound agnosia without aphasia following a right temporal lobe lesion. *Cortex, 26*, 263–268.

Galati, G., Committeri, G., Spitoni, G., Aprile, T., Di Russo, F., Pitzalis, S., et al. (2008). A selective representation of the meaning of actions in the auditory mirror system. *NeuroImage, 40*(3), 1274–1286. Available from: http://dx.doi.org/10.1016/j.neuroimage.2007.12.044.

Gaver, W. W. (1993a). How do we hear in the world? Explorations in ecological acoustics. *Ecological Psychology, 5*(4), 285–313.

Gaver, W.W. (1993b). Synthesizing auditory icons. In *Proceedings of the INTERACT'93 and CHI'93 conference on human factors in computing systems* (pp. 228–235). Association for Computer Machinery.

Gazzola, V., Aziz-Zadeh, L., & Keysers, C. (2006). Empathy and the somatotopic auditory mirror system in humans. *Current Biology, 16*(18), 1824–1829. Available from: http://dx.doi.org/10.1016/j.cub.2006.07.072.

Giordano, B. L., McAdams, S., Zatorre, R. J., Kriegeskorte, N., & Belin, P. (2012). Abstract encoding of auditory objects in cortical activity patterns. *Cerebral Cortex, 23*(9), 2025–2037. Available from: http://dx.doi.org/10.1093/cercor/bhs162.

Giordano, B. L., Mcdonnell, J., & McAdams, S. (2010). Hearing living symbols and nonliving icons: Category specificities in the cognitive processing of environmental sounds. *Brain and Cognition, 73*(1), 7–19. Available from: http://dx.doi.org/10.1016/j.bandc.2010.01.005.

Giraud, A. L., & Price, C. J. (2001). The constraints functional neuroimaging places on classical models of auditory word processing. *Journal of Cognitive Neuroscience, 13*(6), 754–765. Available from: http://dx.doi.org/10.1093/cercor/6.1.21.

Glenberg, A. M., & Robertson, D. A. (2000). Symbol grounding and meaning: A comparison of high-dimensional and embodied theories of meaning. *Journal of Memory and Language, 43*, 379–401.

Goll, J. C., Crutch, S. J., Loo, J. H. Y., Rohrer, J. D., Frost, C., Bamiou, D.-E., et al. (2010). Non-verbal sound processing in the primary progressive aphasias. *Brain: A Journal of Neurology, 133*(Pt 1), 272–285. Available from: http://dx.doi.org/10.1093/brain/awp235.

Griffiths, T. D., & Warren, J. D. (2004). What is an auditory object? *Nature Reviews Neuroscience, 5*(11), 887–892. Available from: http://dx.doi.org/10.1038/nrn1538.

Gygi, B., Kidd, G. R., & Watson, C. S. (2004). Spectral-temporal factors in the identification of environmental sounds. *The Journal of the Acoustical Society of America, 115*(3), 1252–1265.

Gygi, B., Kidd, G. R., & Watson, C. S. (2007). Similarity and categorization of environmental sounds. *Perception and Psychophysics, 69*(6), 839–855. Available from: http://dx.doi.org/10.3758/BF03193921.

Gygi, B., & Shafiro, V. (2011). The incongruency advantage for environmental sounds presented in natural auditory scenes. *Journal of Experimental Psychology Human Perception and Performance, 37*(2), 551–565. Available from: http://dx.doi.org/10.1037/a0020671.

Hatta, T., & Dimond, S. J. (1981). The inferential interference effects of environmental sounds on spoken speech in Japanese and British people. *Brain and Language, 13*(2), 241–249. Available from: http://dx.doi.org/10.1016/0093-934X(81)90093-6.

Hocking, J., Dzafic, I., Kazovsky, M., & Copland, D. A. (2013). NESSTI: Norms for environmental sound stimuli. *PLoS ONE, 8*(9), e73382. Available from: http://dx.doi.org/10.1371/journal.pone.0073382.

Hocking, J., McMahon, K. L., & De Zubicaray, G. I. (2011). Cortical organization of environmental sounds by attribute. *Human Brain Mapping, 32*(5), 688–698. Available from: http://dx.doi.org/10.1002/hbm.21040.

Houix, O., Lemaitre, G., Misdariis, N., Susini, P., & Urdapilleta, I. (2012). A lexical analysis of environmental sound categories. *Journal of Experimental Psychology Applied, 18*(1), 52–80. Available from: http://dx.doi.org/10.1037/a0026240.

Hsieh, S., Hornberger, M., Piguet, O., & Hodges, J. R. (2011). Neural basis of music knowledge: Evidence from the dementias. *Brain: A Journal of Neurology, 134*(Pt 9), 2523–2534. Available from: http://dx.doi.org/10.1093/brain/awr190.

Humphries, C., Willard, K., Buchsbaum, B., & Hickok, G. (2001). Role of anterior temporal cortex in auditory sentence comprehension: An fMRI study. *NeuroReport, 12*(8), 1749–1752.

Huth, A. G., Nishimoto, S., Vu, A. T., & Gallant, J. L. (2012). A continuous semantic space describes the representation of thousands of object and action categories across the human brain. *Neuron, 76*(6), 1210–1224. Available from: http://dx.doi.org/10.1016/j.neuron.2012.10.014.

Kazui, S., Naritomi, H., Sawada, T., Inoue, N., & Okuda, J. (1990). Subcortical auditory agnosia. *Brain and Language, 38*, 476–487.

Kertesz, A. (1979). *Aphasia and associated disorders: Taxonomy, localization, and recovery.* New York, NY: Grune & Stratton.

Knox, C., & Kimura, D. (1970). Cerebral processing of nonverbal sounds in boys and girls. *Neuropsychologia, 8*(2), 227–237.

Krishnan, S., Leech, R., Aydelott, J., & Dick, F. (2013). School-age children's environmental object identification in natural auditory scenes: Effects of masking and contextual congruence. *Hearing Research, 300*, 46–55. Available from: http://dx.doi.org/10.1016/j.heares.2013.03.003.

Kutas, M., & Federmeier, K. (2000). Electrophysiology reveals semantic memory use in language comprehension. *Trends in Cognitive Sciences, 4*(12), 463–470.

Leaver, A. M., & Rauschecker, J. P. (2010). Cortical representation of natural complex sounds: Effects of acoustic features and auditory object category. *The Journal of Neuroscience: The Official Journal of the Society for Neuroscience, 30*(22), 7604–7612. Available from: http://dx.doi.org/10.1523/JNEUROSCI.0296-10.2010.

Lebrun, N., Clochon, P., Etévenon, P., Lambert, J., Baron, J.-C., & Eustache, F. (2001). An ERD mapping study of the neurocognitive processes involved in the perceptual and semantic analysis of environmental sounds and words. *Cognitive Brain Research, 11*(2), 235–248. Available from: http://dx.doi.org/10.1016/S0926-6410(00)00078-1.

Lechevalier, B., Rossa, Y., Eustache, F., Schupp, C., Boner, L., & Bazin, C. (1984). Cortical deafness with partial sparing of music: A case report. *Revue Neurologique, 140*, 190–201.

Leech, R., Gygi, B., Aydelott, J., & Dick, F. (2009). Informational factors in identifying environmental sounds in natural auditory scenes. *The Journal of the Acoustical Society of America*, *126*(6), 3147–3155. Available from: http://dx.doi.org/10.1121/1.3238160.

Leech, R., & Saygin, A. P. (2011). Distributed processing and cortical specialization for speech and environmental sounds in human temporal cortex. *Brain and Language*, *116*(2), 83–90. Available from: http://dx.doi.org/10.1016/j.bandl.2010.11.001.

Lemaitre, G., Houix, O., Misdariis, N., & Susini, P. (2010). Listener expertise and sound identification influence the categorization of environmental sounds. *Journal of Experimental Psychology Applied*, *16*(1), 16–32. Available from: http://dx.doi.org/10.1037/a0018762.

Lewis, J. W., Brefczynski, J. A., Phinney, R. E., Janik, J. J., & DeYoe, E. A. (2005). Distinct cortical pathways for processing tool versus animal sounds. *Journal of Neuroscience*, *25*(21), 5148–5158. Available from: http://dx.doi.org/10.1523/JNEUROSCI.0419-05.2005.

Lewis, J. W., Phinney, R. E., Brefczynski-Lewis, J. A., & DeYoe, E. A. (2006). Lefties get it "right" when hearing tool sounds. *Journal of Cognitive Neuroscience*, *18*(8), 1314–1330.

Lewis, J. W., Talkington, W. J., Puce, A., Engel, L. R., & Frum, C. (2011). Cortical networks representing object categories and high-level attributes of familiar real-world action sounds. *Journal of Cognitive Neuroscience*, *23*(8), 2079–2101. Available from: http://dx.doi.org/10.1162/jocn.2010.21570.

Lewis, J. W., Talkington, W. J., Tallaksen, K. C., & Frum, C. A. (2012). Auditory object salience: Human cortical processing of non-biological action sounds and their acoustic signal attributes. *Frontiers in Systems Neuroscience*, *6*. Available from: http://dx.doi.org/10.3389/fnsys.2012.00027.

Lewis, J. W., Wightman, F. L., Brefczynski, J. A., Phinney, R. E., Binder, J. R., & DeYoe, E. A. (2004). Human brain regions involved in recognizing environmental sounds. *Cerebral Cortex*, *14*(9), 1008–1021. Available from: http://dx.doi.org/10.1093/cercor/bhh061.

Loebach, J. L., & Pisoni, D. B. (2008). Perceptual learning of spectrally degraded speech and environmental sounds. *The Journal of the Acoustical Society of America*, *123*(2), 1126–1139. Available from: http://dx.doi.org/10.1121/1.2823453.

Marcell, M. M., Borella, D., Greene, M., Kerr, E., & Rogers, S. (2000). Confrontation naming of environmental sounds. *Journal of Clinical and Experimental Neuropsychology*, *22*(6), 830–864. Available from: http://dx.doi.org/10.1076/jcen.22.6.830.949.

McCleery, J. P., Ceponiene, R., Burner, K. M., Townsend, J., Kinnear, M., & Schreibman, L. (2010). Neural correlates of verbal and non-verbal semantic integration in children with autism spectrum disorders. *Journal of Child Psychology and Psychiatry*, *51*(3), 277–286. Available from: http://dx.doi.org/10.1111/j.1469-7610.2009.02157.x.

McDermott, J. H., Wrobleski, D., & Oxenham, A. J. (2011). Recovering sound sources from embedded repetition. *Proceedings of the National Academy of Sciences of the United States of America*, *108*(3), 1188–1193. Available from: http://dx.doi.org/10.1073/pnas.1004765108.

Mendez, M. F., & Geehan, G. R., Jr. (1988). Cortical auditory disorders: Clinical and psychoacoustic features. *Journal of Neurology, Neurosurgery, and Psychiatry*, *51*, 1–9.

Moerel, M., De Martino, F., & Formisano, E. (2012). Processing of natural sounds in human auditory cortex: Tonotopy, spectral tuning, and relation to voice sensitivity. *Journal of Neuroscience*, *32*(41), 14205–14216. Available from: http://dx.doi.org/10.1523/JNEUROSCI.1388-12.2012.

Moerel, M., De Martino, F., Santoro, R., Ugurbil, K., Goebel, R., Yacoub, E., et al. (2013). Processing of natural sounds: Characterization of multipeak spectral tuning in human auditory cortex. *Journal of Neuroscience*, *33*(29), 11888–11898. Available from: http://dx.doi.org/10.1523/JNEUROSCI.5306-12.2013.

Motomura, N., Yamadori, A., Mori, E., & Tamaru, F. (1986). Auditory agnosia: Analysis of a case with bilateral subcortical lesions. *Brain*, *109*, 379–391.

Murray, M. M., Camen, C., Gonzalez Andino, S. L., Bovet, P., & Clarke, S. (2006). Rapid brain discrimination of sounds of objects. *Journal of Neuroscience*, *26*(4), 1293–1302. Available from: http://dx.doi.org/10.1523/JNEUROSCI.4511-05.2006.

Murray, M. M., Camen, C., Spierer, L., & Clarke, S. (2008). Plasticity in representations of environmental sounds revealed by electrical neuroimaging. *NeuroImage*, *39*(2), 847–856. Available from: http://dx.doi.org/10.1016/j.neuroimage.2007.09.002.

Noppeney, U., Josephs, O., Hocking, J., Price, C. J., & Friston, K. J. (2008). The effect of prior visual information on recognition of speech and sounds. *Cerebral Cortex*, *18*(3), 598–609. Available from: http://dx.doi.org/10.1093/cercor/bhm091.

O'Leary, D. S., Andreasen, N. C., Hurtig, R. R., Hichwa, R. D., Watkins, G. L., Boles Ponto, L. L., et al. (1996). A positron emission tomography study of binaurally and dichotically presented stimuli: Effects of level of language and directed attention. *Brain and Language*, *53*(1), 20–39. Available from: http://dx.doi.org/10.1006/brln.1996.0034.

Orgs, G., Lange, K., Dombrowski, J., & Heil, M. (2007). Is conceptual priming for environmental sounds obligatory? *International Journal of Psychophysiology*, *65*(2), 162–166. Available from: http://dx.doi.org/10.1016/j.ijpsycho.2007.03.003.

Orgs, G., Lange, K., Dombrowski, J.-H., & Heil, M. (2006). Conceptual priming for environmental sounds and words: An ERP study. *Brain and Cognition*, *62*(3), 267–272. Available from: http://dx.doi.org/10.1016/j.bandc.2006.05.003.

Pazzaglia, M., Pizzamiglio, L., Pes, E., & Aglioti, S. M. (2008). The sound of actions in apraxia. *Current Biology*, *18*(22), 1766–1772. Available from: http://dx.doi.org/10.1016/j.cub.2008.09.061.

Pizzamiglio, L., Aprile, T., Spitoni, G., Pitzalis, S., Bates, E., D'Amico, S., et al. (2005). Separate neural systems for processing action- or non-action-related sounds. *NeuroImage*, *24*(3), 852–861. Available from: http://dx.doi.org/10.1016/j.neuroimage.2004.09.025.

Plante, E., van Petten, C., & Senkfor, A. (2000). Electrophysiological dissociation between verbal and nonverbal semantic processing in learning disabled adults. *Neuropsychologia*, *38*(13), 1669–1684.

Rizzolatti, G., Luppino, G., & Matelli, M. (1998). The organization of the cortical motor system: New concepts. *Electroencephalography and Clinical Neurophysiology*, *106*(4), 283–296.

Roland, D., Dick, F., & Elman, J. L. (2007). Frequency of basic English grammatical structures: A corpus analysis. *Journal of Memory and Language*, *57*(3), 348–379. Available from: http://dx.doi.org/10.1016/j.jml.2007.03.002.

Rosati, G., Debastiani, P., Paolino, E., Prosser, S., Arslan, E., & Artioli, M. (1982). Clinical and audiological findings in a case of auditory agnosia. *Journal of Neurology*, *227*, 21–27.

Saygin, A. P., Dick, F., & Bates, E. (2005). An on-line task for contrasting auditory processing in the verbal and nonverbal domains and norms for younger and older adults. *Behavior Research Methods*, *37*(1), 99–110.

Saygin, A. P., Dick, F., Wilson, S. W., Dronkers, N. F., & Bates, E. (2003). Neural resources for processing language and environmental sounds: Evidence from aphasia. *Brain: A Journal of Neurology*, *126*(Pt 4), 928–945.

Saygin, A. P., Leech, R., & Dick, F. (2010). Nonverbal auditory agnosia with lesion to Wernicke's area. *Neuropsychologia*, *48*(1), 107–113. Available from: http://dx.doi.org/10.1016/j.neuropsychologia.2009.08.015.

Schirmer, A., Soh, Y. H., Penney, T. B., & Wyse, L. (2011). Perceptual and conceptual priming of environmental sounds. *Journal of Cognitive Neuroscience*, *23*(11), 3241–3253. Available from: http://dx.doi.org/10.1162/jocn.2011.21623.

Schneider, T. R., Engel, A. K., & Debener, S. (2008). Multisensory identification of natural objects in a two-way crossmodal priming paradigm. *Experimental Psychology (Formerly "Zeitschrift Für Experimentelle Psychologie")*, *55*(2), 121–132. Available from: http://dx.doi.org/10.1027/1618-3169.55.2.121.

Schnider, A., Benson, F., Alexander, D. N., & Schnider-Klaus, A. (1994). Nonverbal environmental sound recognition after unilateral hemispheric stroke. *Brain*, *117*, 281–287.

Schön, D., Ystad, S., Kronland-Martinet, R., & Besson, M. (2010). The evocative power of sounds: Conceptual priming between words and nonverbal sounds. *Journal of Cognitive Neuroscience*, *22*(5), 1026–1035. Available from: http://dx.doi.org/10.1162/jocn.2009.21302.

Shafiro, V. (2008). Development of a large-item environmental sound test and the effects of short-term training with spectrally-degraded stimuli. *Ear and Hearing*, *29*(5), 775–790. Available from: http://dx.doi.org/10.1097/AUD.0b013e31817e08ea.

Shafiro, V., & Gygi, B. (2004). How to select stimuli for environmental sound research and where to find them. *Behavior Research Methods*, *36*(4), 590–598.

Shafiro, V., Sheft, S., Gygi, B., & Ho, K. T. N. (2012). The influence of environmental sound training on the perception of spectrally degraded speech and environmental sounds. *Trends in Amplification*, *16*(2), 83–101. Available from: http://dx.doi.org/10.1177/1084713812454225.

Snodgrass, J. G., & Vanderwart, M. (1980). A standardized set of 260 pictures: Norms for name agreement, image agreement, familiarity, and visual complexity. *Journal of Experimental Psychology: Human Learning and Memory*, *6*(2), 174–215.

Specht, K., & Reul, J. (2003). Functional segregation of the temporal lobes into highly differentiated subsystems for auditory perception: An auditory rapid event-related fMRI-task. *NeuroImage*, *20*(4), 1944–1954.

Spinnler, H., & Vignolo, L. A. (1966). Impaired recognition of meaningful sounds in aphasia. *Cortex*, *2*, 337–348.

Spreen, O., Benton, A. L., & Fincham, R. W. (1965). Auditory agnosia without aphasia. *Archives of Neurology*, *13*, 84–92.

Stuart, G. P., & Jones, D. M. (1995). Priming the identification of environmental sounds. *The Quarterly Journal of Experimental Psychology Section A*, *48*(3), 741–761. Available from: http://dx.doi.org/10.1080/14640749508401413.

Szekely, A., Jacobsen, T., D'Amico, S., Devescovi, A., Andonova, E., Herron, D., et al. (2004). A new on-line resource for psycholinguistic studies. *Journal of Memory and Language*, *51*(2), 247–250.

Talkington, W. J., Rapuano, K. M., Hitt, L. A., Frum, C. A., & Lewis, J. W. (2012). Humans mimicking animals: A cortical hierarchy for human vocal communication sounds. *Journal of Neuroscience*, *32*(23), 8084–8093. Available from: http://dx.doi.org/10.1523/JNEUROSCI.1118-12.2012.

Taniwaki, T., Tagawa, K., Sato, F., & Lino, K. (2000). Auditory agnosia restricted to environmental sounds following cortical deafness and generalized auditory agnosia. *Clinical Neurology and Neurosurgery*, *102*, 156–162.

Thierry, G., Giraud, A. L., & Price, C. (2003). Hemispheric dissociation in access to the human semantic system. *Neuron*, *38*(3), 499–506.

Turner, R.E. (2010). *Statistical models for natural sounds*. Ph.D. thesis. London: UCL, 1–216.

Underwood, G., & Foulsham, T. (2006). Visual saliency and semantic incongruency influence eye movements when inspecting pictures. *The Quarterly Journal of Experimental Psychology*, *59*(11), 1931–1949. Available from: http://dx.doi.org/10.1080/17470210500416342.

Van Engen, K. J., & Bradlow, A. R. (2007). Sentence recognition in native- and foreign-language multi-talker background noise. *The Journal of the Acoustical Society of America*, *121*(1), 519–526.

van Lancker, D., Cornelius, C., Kreiman, J., Tonick, I., Tanguay, P., & Schulman, M. L. (1988). Recognition of environmental sounds in autistic children. *Journal of the American Academy of Child and Adolescent Psychiatry*, *27*(4), 423–427.

van Petten, C., & Rheinfelder, H. (1995). Conceptual relationships between spoken words and environmental sounds: Event-related brain potential measures. *Neuropsychologia*, *33*(4), 485–508.

Varney, N. R. (1980). Sound recognition in relation to aural language comprehension in aphasic patients. *Journal of Neurology, Neurosurgery & Psychiatry*, *43*(1), 71–75.

Varney, N. R., & Damasio, H. (1986). CT scan correlates of sound recognition defect in aphasia. *Cortex*, *22*, 483–486.

Vignolo, L. A. (1982). Auditory agnosia. *Philosophical transactions of the Royal Society of London Series B, Biological sciences*, *298*, 49–57.

Visser, M., & Lambon Ralph, M. A. (2011). Differential contributions of bilateral ventral anterior temporal lobe and left anterior superior temporal gyrus to semantic processes. *Journal of Cognitive Neuroscience*, *23*(10), 3121–3131. Available from: http://dx.doi.org/10.1162/jocn_a_00007.

Võ, M. L. H., & Henderson, J. M. (2011). Object–scene inconsistencies do not capture gaze: Evidence from the flash-preview moving-window paradigm. *Attention, Perception, and Psychophysics*, *73*(6), 1742–1753. Available from: http://dx.doi.org/10.3758/s13414-011-0150-6.

Warren, W. H., & Verbrugge, R. R. (1984). Auditory perception of breaking and bouncing events: A case study in ecological acoustics. *Journal of Experimental Psychology Human Perception and Performance*, *10*(5), 704–712.

Wightman, F. L., & Jenison, R. L. (1995). Auditory spatial layout. In W. Epstein, & S. Rogers (Eds.), *Perception of space and motion* (pp. 365–400). San Diego, CA: Academic Press.

Wu, Y. J., Athanassiou, S., Dorjee, D., Roberts, M., & Thierry, G. (2012). Brain potentials dissociate emotional and conceptual cross-modal priming of environmental sounds. *Cerebral Cortex*, *22*(3), 577–583. Available from: http://dx.doi.org/10.1093/cercor/bhr128.

Index

Note: Page numbers followed by "*b*," "*f*," and "*t*" refer to boxes, figures, and tables, respectively.

B

C

D

G

M

N

O

P

Q

R

S